2014 IEEE International Solid-State Circuits Conference

AA002392

(ISSCC 2014)

San Francisco, California, USA
9-13 February 2014

IEEE Catalog Number: CFP14ISS-POD
ISBN: 978-1-4799-0917-9

Copyright © 2014 by the Institute of Electrical and Electronic Engineers, Inc
All Rights Reserved

Copyright and Reprint Permissions: Abstracting is permitted with credit to the source. Libraries are permitted to photocopy beyond the limit of U.S. copyright law for private use of patrons those articles in this volume that carry a code at the bottom of the first page, provided the per-copy fee indicated in the code is paid through Copyright Clearance Center, 222 Rosewood Drive, Danvers, MA 01923.

For other copying, reprint or republication permission, write to IEEE Copyrights Manager, IEEE Service Center, 445 Hoes Lane, Piscataway, NJ 08854. All rights reserved.

******This publication is a representation of what appears in the IEEE Digital Libraries. Some format issues inherent in the e-media version may also appear in this print version.***

IEEE Catalog Number: CFP14ISS-POD
ISBN 13: 978-1-4799-0917-9
ISSN: 0193-6530

Additional Copies of This Publication Are Available From:

Curran Associates, Inc
57 Morehouse Lane
Red Hook, NY 12571 USA
Phone: (845) 758-0400
Fax: (845) 758-2633
E-mail: curran@proceedings.com
Web: www.proceedings.com

2014 IEEE INTERNATIONAL SOLID-STATE CIRCUITS CONFERENCE

DIGEST OF TECHNICAL PAPERS

First Edition

February 2014

978-1-4799-0917-9/14 $31.00 © 2014 IEEE

TABLE OF CONTENTS

PAPER SESSIONS

TUTORIALS

FORUMS

EVENING SESSIONS

SHORT COURSE

978-1-4799-0917-9/14 $31.00 © 2014 IEEE

Reflections

What you see before you this year, is the result of many years of continuous iterative refinement of the submission process and information processing. This year, however, as the economic situation persists, we continue to provide a reduced-featured Digest, in which the continuation pages (typically including a micrograph and occasionally summary data) have been eliminated from the print version (only), but are available in the Digest download and in IEEE Xplore. Both to reduce cost and to be more-green, we continue to use partially recycled paper along with a just-in-time printing process.

Again, this year, we continue with a technical editorial group (listed below) under the direction of a managing editor (Laura Chizuko Fujino). However, to reduce costs, the amount of technical and language editing has been dramatically reduced.

In recognition of the large amount of work leading to the Digest open before you, I wish to acknowledge a great many individuals: Trudy Stetzler, Hoi-Jun Yoo, Anantha Chandrakasan, members of the ITPC, and all of the authors, for their individual contributions; Brad Philips, Alija Husic, and Mira Digital Publishing, for Web-based and other preparatory support, including continuing improvement and facilitation of the paper-review and pre-voting process, as well as the Digest printing and Digest DVD; Steve Bonney, and S3 Digital Publishing, for author and Session-Chair interaction, for figure layout, for paper formatting, for pre-press preparation, and for general assistance; Melissa Widerkehr, and Widerkehr and Associates for general interfacing, problem-solving, and coordination; Jason Anderson (University of Toronto), Vincent Gaudet (University of Waterloo), James Haslett (University of Calgary), Kostas Pagiamtzis (Semtech), our technical editors, with Dustin Dunwell (Kapik Integration), Glenn Gulak (University of Toronto), Shahriar Mirabbasi (University of British Columbia), and K.C. Smith (University of Toronto), our technical editors-at-large, for heroic effort on a very tight schedule.

My sincere thanks to you all!

Laura Chizuko Fujino
ISSCC Director of Publications and Presentations

February 2014

Foreword: Silicon Systems Bridging the Cloud

It is my pleasure to welcome you to the 61st International Solid-State Circuits Conference. The Conference continues its outstanding tradition of presenting the most-advanced and innovative work, both from industry and academe, worldwide, in the area of integrated circuits and systems. This year, the geographical distribution of the accepted technical papers illustrates the truly international character of the Conference: 41% of the accepted papers are from North America, 36% from the Far East, and 23% from Europe. Of all of these, 54% are from academe, 39% are from industry, and 7% from institutions/labs.

The Conference theme for 2014 is "**Silicon Systems Bridging the Cloud**". Traditional computing solutions tend to be very vertically oriented. To this, the cloud brings the promise of many benefits, such as flexible computing power, shared software applications, and centralized databases. Correspondingly, we need to unite these approaches – providing solutions for "bridging the gaps" in the evolving cloud, leaping from the technologies and systems we have today. Thus, new approaches to sharing networks, infrastructure, and data storage, will be required, as will new approaches to securing these shared resources. Such solutions will challenge systems designers to consider new system architectures, and will also require advances in circuits and technology to deal with increasing computational demands.

As part of the regular program, there are four plenary speakers exploring topics from the lowest level of the atom to the heights of the cloud: Mark Horowitz from Stanford University will present his vision for solving computing's energy problem; Ming-Kai Tsai from MediaTek will discuss where we are heading with Cloud 2.0; Erik Heijne from CERN will explore how ICs are enabling research into the attoworld; and, finally, Susie Wee from Cisco will discuss the network and technology requirements needed to support the Internet-of-Things (IoT). Moreover on Monday evening, there will be a special plenary panel discussion with eminent leaders from across the industry on the challenges that next-generation systems create for system and circuit designers.

Overall, the ISSCC 2014 Technical Program consists of 206 outstanding papers, distributed over 30 thematic sessions. In addition to regular paper presentations, the Conference continues to offer a wide variety of high-quality educational events, adding to the already significant value of attending ISSCC. This year, there are 10 Tutorials covering a wide range of topics, including filtering in RF transceivers, constraint and optimization in VLSI circuit design, 3D Integration and near-field coupling, power-optimized processor design, peripheral circuits for analog-to-digital converters, analog front-end design for wireline receivers, self-adapting design techniques, interference-robust receiver techniques, DC-DC converter design, and physical-to-digital converters. These Tutorials, which are given by top experts in the field, offer an opportunity for participants to experience an introduction and overview of important developments in each of these topics. Besides these 90-minute tutorials on Sunday, on Thursday, there is an all-day short course on "Biomedical and Sensor Interface Circuits".

The Conference also offers 6 Forums (2 on Sunday and 4 on Thursday) whose intent is to cover more-advanced topics for experts in the field. This year, Forums deal with digitally assisted analog and analog-assisted digital design, 3D stacking technologies, adaptive design for energy efficiency, mm-waves for active safety and communications systems, low-power radios for sensor networks, and energy-efficient I/O design.

In addition, there are two types of evening sessions available to all attendees: Panel discussions and Special-Topic Sessions, which provide participants with the opportunity to learn about a timely topic in a more-relaxed setting. As introduced earlier, the first panel on Monday will be held in a special plenary format. The topic is a systems-focused event entitled, "Next-Generation Networked Systems: Challenges for Silicon", where 6 experts from across the electronics industry will share their experiences and vision on the development of innovative systems based on semiconductor technology. The second panel on Tuesday entitled, "Anatomy of Innovation: Bug or Feature?" will provide an entertaining debate of whether innovation is more effective as a result of accidental discovery or intent. On Tuesday, the third panel deals with "Perspectives on the Future of Semiconductor Innovation": An ensemble of visionaries, experts (CEOs and VCs) will discuss the opportunities and challenges for innovation in our industry, particularly featuring the availability of funding for new semiconductor ventures, and from where innovation will spring. As well, two Special-Topic Evening Sessions will offer plenty of opportunities for attendees to be informed on topics of increased importance in the field of solid-state circuits. These topics have been carefully selected by the Program Committee for their timeliness and relevance. One (on Sunday) focuses on the data centers required to support tomorrow's cloud, and the other (on Tuesday) on wearable wellness devices.

Moreover, on Monday a Demonstration Session will provide further information on selected papers, during the Social Hour. Also, a special program for students has been organized to introduce them to ISSCC, preparing the way for our future presenters and participants. Held on Sunday evening, the Student Research Preview allows graduate students from around the world, who are beginning to work in the field, to interact with each other and with experts. The students are given the opportunity to present their work both orally and in a Poster Session, in which they benefit from immediate feedback from session attendees.

The quality and high standards that we associate with ISSCC is by-and-large due to the diligent work of the International Technical-Program Committee (ITPC). This year, ITPC has 161 members divided into ten technical subcommittees. The members come from industry and academe from around the world. Each member has made a tremendous contribution to the ITPC by reading and reviewing large numbers of submitted papers, planning and organizing evening sessions and educational events, preparing the Advance Program, Press-Kit, and Digest material, in a timely manner, as well as performing session chair/organizer duties. In particular, I would like to acknowledge the leadership and guidance of the Technical Subcommittee Chairs: Axel Thomsen (Analog), Boris Murmann (Data Converters), Stephen Kosonocky (Energy-Efficient Digital), Stefan Rusu (High-Performance Digital), Roland Thewes (Imagers/MEMS/Medical/Displays), Kevin Zhang (Memory), Andreia Cathelin (RF), Eugenio Cantatore (Technology Directions), Aarno Pärssinen (Wireless), and Daniel Friedman (Wireline). Also, I would like to thank the members of the Regional Committees: Alison Burdett and Maurits Ortmanns from the European Region; and Kazutami Arimoto, Jae-Youl Lee, and Tsung-Hsien Lin from the Far-East Region. Their help and support have been essential for the smooth operation of the Conference.

Many other individuals have played an essential role in making the Conference possible: Specifically, I would like to thank Bram Nauta, the ISSCC 2013 Program Chair, and Hoi-Jun Yoo, the ISSCC 2014 Program Vice-Chair for their help; Melissa Widerkehr and Widerkehr Associates for their invaluable assistance with Conference operations and arrangements. Thanks are also due to representatives of MIRA Digital Publishing for their assistance with the electronic manuscript submission, pre-voting, and in putting together and formatting the Advance Program as well as the Digest, and to S3 iPublishing for page layout and facilities coordination. I must also thank the Technical Editors: Jason Anderson, Dustin Dunwell, Vincent Gaudet, Glenn Gulak, James Haslett, Shahriar Mirabbasi, and Kostas Pagiamtzis, as well as the multi-media coordinator Dave Halupka. Also special thanks go to Laura Fujino and Kenneth Smith for their unrelenting help with many aspects of the Conference, including the paper submission process, the Advance Program, the Press Kit, the Digest and associated editing, and DVDs, and to Siva Narendra for his pursuit of the Press Kit and its preparation process. Also thanks to Ali Sheikholeslami for his coordination of the Tutorials and Short Course; to Willy Sansen for organizing the Short Course; to Bram Nauta for his guidance in putting the Forums together; to Jan van der Spiegel for organizing the Student Research Preview; to Bill Bowhill and Uming Ko for organizing the Demonstration Session; to Bill Bowhill for managing the Conference website; to John Trnka for his excellent work with planning and coordinating the audio-visual services for the Conference; to Bryant Griffin for his financial oversight.

We are indebted as well to an unusual group of volunteer graduate students from the University of Toronto, who through their individual technical expertise ensure the orderly conduct of the presentations in each session, as well as countless other behind the scenes activities.

Finally, an individual who deserves special recognition is Anantha Chandrakasan, the ISSCC Conference Chair, for his enthusiastic support that he has given me, and his visionary leadership that will ensure ISSCC continues as the premier Conference in solid-state circuits.

Enjoy ISSCC 2014!

Trudy Stetzler

Trudy Stetzler, ISSCC 2014 International Technical-Program Chair.

The Visuals Supplement and The Saratoga Group – A 25-Year History

Laura Chizuko Fujino
University of Toronto,
Toronto, Canada

Preamble:

ISSCC 2014 marks the 25th year of the Saratoga Group, a group annually dedicated to the in-conference real-time creation of the conference record (named the Slide Supplement, and now the Visuals Supplement) that first appeared for ISSCC 1990.

The idea of providing attendees with all presentation material arose as a result of a variety of comments made in an attendee survey conducted for ISSCC 1989. The survey crystallized ideas which had been expressed for many years earlier concerning a heartfelt need to reconcile two problems that had long bothered the attendees: The first was dissatisfaction with those who snapped photos of the slides during presentations; The second was the often-heard lament that the audience was shown marvelous things that were not recorded or depicted in the Digest.

Since the data reduction of the survey was assigned to K.C. Smith, Laura Fujino did all of the work, and was graciously allowed to present the findings at the August Executive Committee meeting in 1989. Amid a lot of discussion of various topics raised by the Survey, a decision was made by the Executive Committee that something needed to be done, and that a solution lay in some mechanism to provide copies of all the presentation slides to all attendees, after the Conference. Subsequent to the meeting in which Laura Fujino was present only for the Survey presentation, David Pricer as the Executive-Committee Chair was assigned to seek some mechanism for the resolution of this problem. Shortly thereafter, he contacted Laura Fujino with an invitation to join the Executive Committee with the role of addressing this challenge! Within a few weeks, a concept emerged on how to proceed: The existence of a photocopier with a 35mm slide projector attachment was identified from which medium-size paper copies of the slides could be made. As well, this photocopier had a paper to paper enlarging/reduction facility, which though of limited range could be used in multiple passes to acquire a paper image of the desired size.

A production process was conceived as follows: Soon after the completion of the talk, the author's slide carousel was used to produce a set of paper images from their 35mm slides; later these images were sized using the photocopier to provide paper images that were combined on an eight-per-page presentation, by a manual cut-and-paste process; this was combined with the title, author data, and abstract, in a nominal two-page format with extensions at the back of the compiled book, called the "*Slide Supplement to the Digest of Technical Papers*".

Correspondingly, during the 1990 Conference held at the San Francisco Hilton, a group of volunteers and one part-time casual employee met in the "Saratoga Room" to implement this process. The group included: John Eggert (Digest and Supplement printer) and his wife Shirely, John Wuorinen (Digest Editor) and his wife Susan, Nancy Pricer, wife of Dave, Henry Osborne (casual convention services employee), K.C. Smith, under the direction of Laura Fujino. Thus, thereafter, this group was called the "Saratoga Group". Since the process was machine limited it took very long days to maintain a reasonable schedule with two operators (Henry and Laura). John Eggert led the manual final layout process, while Nancy Pricer collected the slides and sized paper copy, and the others were involved primarily with image sizing. (It was pleasing to note that no blood was spilled during the cut-and-paste process!) The final product, the ISSCC 1990 Slide Supplement (193 pages) was printed, bound, and mailed to all attendees, about one month after the Conference.

Later Developments:

The acceptance of the first edition of the Slide Supplement was sufficiently strong that we were encouraged to continue the logical development of the original concept. For the next two years, ISSCC 1991 and 1992, the process continued with some modification in the "Saratoga Room" at the San Francisco Hilton. For ISSCC 1993, operations were moved to the San Francisco Marriott, where the Conference has remained to this day.

For ISSCC 1991 and 1992, the major change was replacement of some of the initial volunteers and casual employee in the "Saratoga Group" by graduate-student volunteers from the University of Toronto.

For ISSCC 1994, following a change of the printer, Pat Duplessis joined the team in charge of manual page layout for which he introduced the user of a waxer for image adhesion. This process continued with one major change, until the introduction of electronic projection in 2001: The major change, introduced at ISSCC 1993, was to use author provided hard copy of their 35mm slides to simplify image production but requiring a slide-to-paper checking process, which in turn required more volunteer students, but was otherwise faster and did not rely as much on unreliable equipment. For ISSCC 1997, Steve Bonney joined Pat Duplessis for page layout. For ISSCC 1998, 1999, and 2000, we replaced much of the photo reduction by scanning the author's hard copy, and thereafter, sizing and page layout by computer.

For ISSCC 2000, an experiment was conducted with electronic projection (replacing 35 mm slides by computer-driven projectors) in three sessions, necessitating some additional image processing for Supplement production. For ISSCC 2001, electronic projection for day sessions became the norm. This included the use of multiple computers for back up and reliability, with the need for more student volunteers to operate them, and provide set up and checking. At the same time, the name "Slide Supplement" was changed to "Visuals Supplement", and the task of page layout was assigned solely to professionals.

The Graduate-Student Volunteer Group:

Today, the graduate-student volunteers, the major part of the current Saratoga Group is involved in a diverse set of tasks: Before the Conference they set up and check the laptops of which three are used for each regular session to provide projection, projection back-up, and audio recording, as required; organize Speaker and Committee Registration material; handle corresponding Speaker and Committee Registration; set-up equipment for Speaker Rehearsal; assist with Speaker Rehearsals; assist with Plenary Speaker Rehearsal; unpack and check Award plaques, organize and transport to the ballroom; man the Press desk; interact with unionized A/V staff; check slides with Speakers prior to their presentation; operate the computer projection system for Tutorials, Forums, Plenary Session, Regular Sessions, Evening Sessions, Short Course; assist speakers during the question and answer period; operate the recording system for the Plenary Session, Tutorials, and Short Course; act as videographers for the Demonstration Sessions; take photos of on-going events; complete data reduction for Tutorials, Forums, Short Course, Regular Sessions; perform highlight data reduction for Regular Sessions for JSSC paper selection; help with crisis intervention.

Over the past 25 years, the number of graduate-student volunteers annually has ranged from 3 for ISSCC 1991 through a peak of 22 for ISSCC 2008 (to support the audio recording of the entire Conference), to 17 since ISSCC 2011.

An interesting vignette concerning the performance of the volunteer students occurred early on in the context of their handling of electronic-projection systems. Virtually at the moment that student operated projection began, there were many reports from attendees concerning a magical process that they had observed during the question and answer periods of sessions that they had attended: Their common comment was that miraculously, during the formation of the question by a questioner that a highly relevant slide illustrating the subject of the question and often its answer would appear suddenly on the screen. Their subsequent question to me was what marvelous artificial intelligence software had we acquired that allowed such instant insight and response! My retort was that it was not magic, but simply the consequence of care in selecting and assigning of the graduate-student volunteers: Each of them was highly-selected from amongst senior members of a large graduate-student body, typically in the PhD program, or at the end of a Masters degree from a range of circuit-related specializations; further they were assigned to sessions that were in their area of research interests. Thus, they were amongst the most astute listeners in the audience!

It is interesting to note that a large fraction of these Saratoga-Group graduate-student volunteers have gone on to very successful careers in the solid-state area, both in industry and academia. Increasingly, one has seen them presenting a paper at ISSCC!

Conclusion:

Upon reflection of the past 25 years, one is surprised that such a process evolved through technological change, personnel change, and vagaries of attendance variation. Beyond the annual appearance of increasingly higher quality of slide material, including videos, animation, and so on, the process has influenced an enormous number of young lives, hundreds of individuals whose outlook on life is different because of their week or so of frenetic, yet focused activity at ISSCC.

On the facing page, you will find a list of members of the Saratoga Group over this quarter century of hyper activity, with the years of their membership. This list is dominated by student-volunteer members, but includes others some of whom are volunteers, while others are assigned professionals.

978-1-4799-0917-9/14 $31.00 © 2014 IEEE

A Quarter Century of Participation

The Saratoga Group Membership from 1990 to 2014

2008	Mohammed Abdalla
2002	Sherif Abdalla
2009-2011	Karim Abdelhalim
2005	Mohamed Abdulla
2004-2008	Imran Ahmed
2007	Mehdi Alimadadi
2008, 2009	David Alldred
2013, 2014	Guy Alter
2014	Nadeesha Amarasinghe
1997	Jason Anderson
2006-2010	Ricardo Aroca
2003	Igor Arsovski
2003-2006	Navid Azizi
1995, 1996	Srinivasa Banna
2007-2011	Kevin Banovic
2007-2012	Mike Bichan
1996-1998	Jason Bickford
1994	Adi Bonen
1997-2014	Steve Bonney
1993	Bertil Brandin
2007-2014	Nancy Bush
2005-2009	Trevor Caldwell
1999, 2000	Anthony Carusone
2001	David Cassan
2010, 2011	Pearl Cao
2005-2007	Theodoros Chalvatzis
2003	Trevis Chandler
2012-2014	Jingxuan Chen
1995	Peter Chen
2008	Horace Cheng
1996, 2000	Ruth Cherian
2000	Sarah Cherian
2003	Dickson Cheung
1992, 1994	Raymond Chik
2014	Stephen Alexander Chin
2004	Jeffrey Chow
1992	Terry Choy
1994	Jeremy Cooperstock
2002, 2004-2007	Ahmad Darabiha
1997	Sandy Decker
2005, 2006	Tod Dickson
1991	Ralph Duncan
2007-2013	Dustin Dunwell
1994-2000	Pat Duplessis
1990, 1991, 1993	John Eggert
1990, 1991	Shirley Eggert
2002, 2005	Yadollah Eslami
2001	Tooraj Esmailian
2002	Kamran Farzan
2009	William Feng
2001	Ted Fill
2009, 2010	Armin Fomani
2013, 2014	Ying Ying Fu
1990-2014	Laura Fujino
2013, 2014	Michal Fulmyk
2006, 2007	Sean Garcia
2008, 2009	Ruslana Gelman
1997-2000, 2003	Vincent Gaudet
2004-2006	Dan Gerken
2011-2013	Chris Gooch
1999-2003	Warren Gross
2005, 2007	Afshin Haftbaradaran
2004-2012	David Halupka
2002	Anas Hamoui
2008	Adam Hart
2013, 2014	Robert Hesse
2007-2009	Derek Ho
1999, 2001	Steve Hranilovic
2010	Safeen Huda
2009, 2014	Alija Husic
2001-2006	Marcus van Ierssel
2004	Joel Ironstone
1998	Nadine Jackson
2009, 2011	Hamed Jafari
1991	Steve Jantzi
2014	Ge Jin
2008, 2009	Tony Kao
2004-2006	Rafal Karakiewicz
2008	Mehdi Khanpour
2012	Neno Kovacevic
1992	Karen Kozma
2012	Ivan Krotnev
2006	Fumie Kunimatsu
2006	Ian Kuon
2007	Mohamed Kwokgy
2005-2009	Katya Laskin
2001	Agustin Lebron
2003	Dora Lee
1992, 1993	Edward Lee
2012	Junmin Lee
2005-2014	Bertram Leesti
1998	Oliver Leung
1993	Kam-Wing Li
2013	Joshua Liang
1995	Tracy Liao
2013, 2014	Sevil Zeynep Lulec
1997-1999	Dave McCausland
2007-2009	Scott McLeod
2011-2014	Mario Milicevic
2001-2003	Shahriar Mirabbasi
1999	Elizabeth Morelli
2006-2008	Akram Nafee
1995	Cuong Nguyen
2006	Sean Nicholson
1998	Javad Omid
1990	Henry Osborne
2001-2006	Kostas Pagiamtzis
2006, 2007	Samir Parikh
2009	Dimpesh Patel
2006, 2007	Jennifer Pham
2007-2014	Brad Philips
1997	Jason Podaima
2008	Gail Prestidge
1990-1996	Nancy Pricer
2001-2008	Jennifer Rodrigues
2000	Jonathan Rogers
2007, 2008	Bill Romer
2011	Alain Rousson
2009	Martin Rozee
1996	Ricardo Saad
2001, 2002	Saman Sadr
1992	Reza Safaee-Rad
1995-1997	Mazen Saghir
2004-2008	Steve Samson
1995	Angela Schultz
1993-1995	Kenneth Schultz
2007	Mahdi Shabani
2007-2009	Shahriar Shahramian
2010	Shayan Shahramian
2012	Alireza Sharif-Bakhtiar
1996, 1998	Ali Sheikholeslami
2012	Ravi Shivnaraine
2012	Stefan Shopov
2011-2014	Andrew Shorten
2000	Jim Small
1990-2014	Kenneth Smith
2010	Vadim Smolyakov
2013, 2014	Dawei Song
2009	William Song
2009	Soroush Tabatabaei
2008, 2010	Tina Tahmoureszadeh
2011-2013	Clifford Ting
2012	Anita Tino
2009	Alex Tomkins
2005-2007	Olivier Trescases
1991	Patty Trnka
2012, 2013	Colin Tse
2005-2012	Aleksey Tyshchenko
2013, 2014	Jasmina Vaslijevic
2012-2014	Aynaz Vatankhahghadim
2012	Jeffrey Wang
2010, 2011	Jing Wang
2012	Luke Wang
2004	Robert Wang
2007	Guowen Wei
2011, 2013, 2014	Yue (Victor) Wen
2003	Joyce Wong
2001	Laurie Wood
2002, 2003	Sarah Wood
1991	Steve Wood
1990	John Wuorinen
1990	Susan Wuorinen
2004, 2008	Navid Yaghini
2007-2011	Kentaro Yamamoto
2010	Hemesh Yasotharan
2007, 2008	Kenneth Yau
2004	Ricky Yuen
2014	Mohammad Meysam Zargham
2010, 2011	Guangzhao (Andy) Zhang

978-1-4799-0917-9/14 $31.00 © 2014 IEEE

Session 1 Overview

Plenary Session

Chair: **Anantha Chandrakasan,** *Massachusetts Institute of Technology, Cambridge, MA*
ISSCC Conference Chair

Co-Chair: **Trudy Stetzler,** *Halliburton, Houston, TX,*
ISSCC Technical Program Chair

The Plenary Session begins with opening remarks and a welcome to attendees from the Conference Chair, Anantha Chandrakasan. Then, the Program Chair, Trudy Stetzler, introduces the first two of four plenary speakers. Following the second plenary speaker, the Session pauses for an Awards Presentation moderated by the Conference Chair. Following a short break, the third and fourth plenary speakers are introduced by the Program Chair, and their presentations will conclude the Session.

The Conference theme for 2014 is "**Silicon Systems Bridging the Cloud**". The cloud brings the promise of many benefits such as flexible computing power, shared software applications, and centralized databases. Solutions for bridging the current gaps in the cloud will need to evolve from the technologies and systems we have today. New approaches to networks, computing, and data storage, will be required, as will new approaches to securing these shared resources. The Internet-of-Things (IoT) will drive new applications, increasing computational needs while dealing with limited resources. These solutions will challenge systems designers to consider new system architectures, and will also require advances in circuits and technology to meet the demands of the next-generation Cloud.

In the first plenary presentation [1.1], Mark Horowitz of Stanford University observes that the technology-driven growth of electronic systems to its current state of societal dominance has now reached a barrier, that of energy limitation. Further, he describes why conventional technology change cannot rescue this situation, and that something more must be done. He identifies that the solution lies in a new dimension, that of system design, where application specialists ensure that electronic-product components are tuned to energy-efficient execution of the desired task. For this to occur, the solid-state industry must provide tools to bridge the gap between applications-intensive knowledge and electronics know how!

The second plenary presentation [1.2] by Ming-Kai Tsai of MediaTek deals with the emerging Cloud 2.0 that will be driven by wireless WAN and personal devices that together enable us to be always connected. After a review of the current technologies and systems, the presentation focuses on the energy, reliability, and security challenges for new Cloud 2.0 devices given a small form factor and increased computational requirements. Several solutions for addressing these challenges are presented.

Erik H.M. Heijne of CERN provides the third plenary presentation [1.3] on the topic of how the introduction of highly-segmented low capacitance silicon sensors and CMOS integrated circuits have been key contributors to the recent discovery of the Higgs boson at the Large Hadron Collider (LHC) at CERN. While providing an overview of the operation of the LHC, and the various sensors and custom CMOS circuits at the heart of the experiments, the presentation will highlight how new technology has created opportunities for scientific research.

The fourth plenary presentation [1.4] is delivered by Susie Wee of Cisco who will explore the next generation of networked experiences. This presentation will examine the network architectural challenges that must be addressed to meet the ever growing end users' expectations for the Internet-of-Things. The technology innovations required to support this new networked world will be examined for overall needs from underlying infrastructure to client devices.

Overall, the Plenary Session will provide an excellent overview of future directions in the entire technical field of integrated circuits and systems. With talks ranging from the advancement of basic research in the fundamentals of nature to the next generation networked experiences, this year's plenary session highlights the diversity of IC technology. We thank our distinguished plenary speakers for their eager willingness to lead off this year's exciting program!

978-1-4799-0917-9/14 $31.00 © 2014 IEEE

1.1 Computing's Energy Problem (and what we can do about it) 8:45AM

Mark Horowitz, *Yahoo! Founder's Chair, Stanford University, Stanford, CA, USA*

Almost all electronic systems are energy-limited, from the computers in your Bluetooth headset, to the ones that answer the questions that you type into Google. Performance optimization drove us to this energy limit, even before the energy scaling of gates slowed down. Now that gate energy is scaling slowly, the problem is even worse. While previous energy limits were overcome by changing technology (tubes, transistors, CMOS) I will explain why we can not count on technology saving the day this time.

We next look at making a modern web-server farm more efficient. As CPUs have improved, memory and I/O-system energy dominate. While reducing this energy is possible, this fact raises the question of how ASICs can be 1000× more efficient than processors, if memory energy dominates. Their secret lies in the ASIC's choice of applications: those that have many short-integer operations and extremely local storage. The rest of the presentation will explain how we have leveraged this insight, both to create a programmable unit with ASIC efficiencies, and to bound the amount of compute specialization needed to create energy-efficient hardware for other classes of applications.

1.2: Cloud 2.0 Clients and Connectivity – Technology and Challenges 9:20AM

Ming-Kai Tsai, *Chairman and CEO, MediaTek, Hsinchu, Taiwan*

Two key enablers for distributed web services, or Cloud 1.0, are clients such as smartphones and tablets, and connectivity. To satisfy growing computing requirements, mobile CPU clock frequencies have exceeded GHz. However, the technology will soon hit a frequency wall beyond which cost becomes prohibitively high. Thus, mobile clients are rapidly moving to multi-core CPU and GPU structures to enable consumer applications and mobile human-interface devices (HIDs), with system-adaptive power management, thermal throttling, and heterogeneous multi-processing, for optimal performance and energy efficiency within thermal limits. On the other hand, while wired communication, LAN, and WAN, all drove Cloud 1.0, it is wireless WAN technology that enabled users to stay connected anywhere at anytime. Soon, Cloud 1.0 will morph into Cloud 2.0, heralding the ubiquitous era where the central cloud will be distributed into personal (e.g., Smartphone, Tablet), home (e.g., Smart-TV), and local clouds. The insatiable computation need, coupled with the explosion of Internet-of-Things (IoT) devices and distributed clouds, demands orders-of-magnitude higher bandwidth for better user experience. But, with the constraints of portable-device form factor and limited battery-technology improvement, the energy and thermal gaps present major technical challenges. To overcome these challenges, many innovations are desperately needed to enable the ubiquitous Cloud 2.0 ecosystem which promises to provide ample possibilities to enhance and enrich everyone's life.

ISSCC, SSCS, IEEE AWARD PRESENTATIONS 9:55AM

BREAK 10:25AM

1.3 How Chips Pave the Road to the Higgs Particle and the Attoworld Beyond 10:40AM

Erik H.M. Heijne, *Instrumentation Physicist, CERN PH Department, Geneva, Switzerland, also with IEAP Czech Technical University Prague & Nikhef Amsterdam*

Fundamental research on elementary constituents of matter needs ever-higher energies, and sophisticated instrumentation. Segmented low capacitance silicon PIN diode sensors and CMOS circuits have proved essential to the recent discovery of the Higgs boson at the CERN Large Hadron Collider (LHC). Custom ICs provided a paradigm shift in experimental techniques, permitting event imaging and selection at unprecedented rates. Four experiments at the Collider act as giant cameras, each taking 40 million 3D pictures per second. In the largest experiment six concentric subsystems with ~500 million sensor elements allow precise localization of thousands of particles per collision. The readout chips need adequate timing, noise, power and matching performance and must be radiation tolerant. The Worldwide LHC Computing Grid (WLCG) enabled quick analysis of the massive data volume. This presentation describes how custom VLSI chips combined with silicon detectors contribute to our understanding of Nature.

1.4: The Next Generation of Networked Experiences 11:15AM

Susie Wee, *Vice President & Chief Technology Officer of Networked Experiences, Cisco, San Jose, CA*

The evolution of network technology has enabled impressive networked experiences, ranging from connected mobile context-aware experiences to those provided by large-screen interactive immersive video dislays. The rapid pace of innovation and the trends in consumerization have made these experiences affordable and widely available, greatly increasing end users' expectations. In the years ahead, networks will undergo the greatest architectural transition seen in the past two decades, with the advent of software-defined networks, big data, and the internet of things. By taking an experience-driven approach to this architectural shift, we can understand the network and technology requirements for the underlying compute and network infrastructure, client devices, and sensors and the interaction of applications with the network. In this presentation, I will discuss the next generation of networked experiences and the technology innovations that will be needed in the years to come.

PRESENTATION TO PLENARY SPEAKERS 11:50AM

CONCLUSION 11:55AM

978-1-4799-0917-9/14 $31.00 © 2014 IEEE

1.1 Computing's Energy Problem (and what we can do about it)

Mark Horowitz

Departments of Electrical Engineering and Computer Science,
Stanford University, Stanford, CA

1. Introduction

Technology scaling has decreased the cost of computing to the point where it can be included in almost anything. As a result, we now live in a world surrounded by computing devices. They power our searches on Google, connect to our friends on Facebook, answer our questions to Siri, and serve us our entertainment on Youtube; they are in our homes everywhere, in all our appliances (I recently had to reboot my refrigerator), cars, workplaces, and even in the cards we send to each other. We have become so accustomed to computing becoming faster, cheaper, and lower power, we simply assume it will continue. Already, smartphone capabilities are being embedded in eye glasses [1] and smart watches [2].

While scaling computing performance has never been easy, a number of factors have made scaling increasingly difficult this past decade, and have caused power to become the principal constraint on performance. Section 2 quickly reviews how computing became power limited, even before Dennard constant-field scaling [3] broke down, and explains the difficulties of using a technology change to fix our problems. The rest of the paper explores different approaches to addressing this computing-energy-consumption challenge, and shows that it will take more than parallelism to get the results we need. The new key to scaling computing performance is to create applications and hardware which are better matched to the task and each other. Accomplishing this will require tools that allow application experts to create these new efficient systems. While creating these tools is challenging, they will enable a renaissance in application-optimized computing!

2. Processor Scaling

Performance data for commercial microprocessors is now accessible on the web [4], and Figure 1.1.1 uses this data to show how gate speed and processor performance have scaled over time. The gate speed is an approximation of the FO4 (the delay of an inverter in that technology driving a load that is 4× its input capacitance) and the processor performance is normalized to that of a 386, using SPEC scores as a criterion. More details about how these numbers are derived are given in [5]. Note that the gate speed has improved by 100× since CMOS processors were introduced in the mid 80s, while application-level uniprocessor performance has increased by over 3000. During this same period, the number of transistors has followed a nice Moore's law exponential growth, and the rate of feature size scaling has also been remarkably consistent (see Figure 1.1.2). But, what has not scaled according to plan, is the power density of the processors (see Figure 1.1.3). The underlying cause of the observed exponential power growth can be traced to two factors: the fact that we did not scale power supply voltages at the constant field rate, which Dennard himself reported [6], and the fact that in our quest for performance, we scaled clock frequencies faster than dictated by constant-field scaling (see Figure 1.1.4). In the 1990s, voltage scaled down, but slower than technology (at about the square root of feature-size scaling), and frequency scaled up, but faster (at about the *square* of feature-size scaling). Because power is CV^2F, and C scales with technology, power should have become an issue much sooner than it did. However, during this same time frame, designers added many effective power saving techniques which extended this period of rapid performance gains.

But, this rapid scaling of clock frequency stopped in the early 2000s, as a result of two factors: First, processors hit the power wall for air cooling (using a low-cost heatsink, and air flow at a noise level acceptable for an office) which is around 100W. Worse still was the fact that voltage scaling also slowed down, since it was no longer possible to scale the threshold voltage due to rising leakage currents. To keep power in check, processor frequencies were reduced, and multiple processors were added to each die. However, since processor "performance" was changed to measure throughput, it continued to scale, as more cores were added each generation.

Like most chips today, processors used to run at a fixed supply voltage, and this voltage depended on the fabrication technology that was used. But, as processors became power constrained and leakage current grew, it became apparent that one could dramatically reduce the power dissipation, and improve the performance yield of a processor if each processor chip could specify the supply voltage that was required for it to operate at the desired performance. This would allow a chip fabricated with high-leakage, lower-average-V_{th} transistors, to run at a lower supply voltage, reducing both the dynamic and leakage power, for overall power optimization. Correspondingly, processors with higher V_{th} transistors, and lower leakage could run at a higher supply voltage while still operating within the total power budget, enabling these transistors to operate at the desired speed. While this has been good for processor specification, it has made it much more difficult to track how the average supply voltages have been scaling over the past decade. Thus, the numbers in the voltage plot in Figure 1.1.4 are the peak allowable supply voltages, and do not represent the average voltages used. From limited data, the actual operating supply voltages seem to remain in the 0.9 to 1.1 volt range for peak performance. But, the recent move to 3-D channel structures with reduced leakage currents, has enabled about a 100 to 200mV decrease in operating voltage.

3. Technology to the Rescue?

Given the current limitations of CMOS scaling, it is natural to look toward other technologies to enable computing performance to continue to scale. After all, computing started with mechanical devices, moved to electro-mechanical (relays), then to electronic tubes, transistors, bipolar ICs, nMOS, and finally to the CMOS technology we are using today. In fact, our CMOS technology has changed dramatically over the past 30 years: moving from single level Al connection to over 10 levels of Cu, away from SiO_2 gate oxide and back to metal gates, adding Ge for strain, and now to 3-D topologies. While CMOS will continue to evolve, and everyone should hope that the research in off-roadmap technologies is successful, there is a real possibility that, computing at least, will stay with CMOS-like technologies. The problem is not simply the potential abandonment of the manufacturing capability that the enormous investment has created: CMOS VLSI also has shaped our design abstractions that have allowed us to build functional artifacts of enormous complexity.

To move off the CMOS roadmap (for example, to quantum computing), would require a very different hardware platform. (Note, that we have already accepted that technology advances such as finFET are an extension of CMOS rather than a separate manufacturing process,) Such a new platform would disrupt the entire design abstraction hierarchy. Thus, the insertion cost for such a disruptive technology would include the creation of both a state-of-the-art manufacturing facility, and the tools and training in all the design abstraction layers that require change, which is not going to "come cheap". And, that is the problem. The fundamental challenge for any new startup (new idea) is insertion – how to minimize the investment needed to compete with the status quo. A radical idea must demonstrate its utility with only a modest investment since there is significant chance it will fail. The larger the investment, the lower the risk the investors are willing to tolerate, which is why large industries change incrementally. Thus, changing technology to fix the power problem is a perfect catch-22.

In some ways, the key problem is the capabilities of modern CMOS technologies: With CMOS, we can create chips with millions of transistors at nearly no cost, have all the transistors work, and run at GHz frequencies. While I am sure there are better technologies out there, I am not sure how we can afford the investment to find and develop them to the point where they are competitive. While there are many interesting new technologies, for any to develop will require a niche market different from computing, where they can be successful and earn the resources they need to grow. It is important to remember that most successful radical ideas create **new** markets. Only when these markets become large, do they challenge the players of the status quo, often indirectly, by changing the underlying rules of that market. So while it is possible that a new technology will eventually compete with CMOS for computing, it is not likely it will be from a frontal assault, and it will take time for this new technology to grow large enough to compete. Given this dynamic, CMOS-like technology will continue to dominate computing for the foreseeable future, and we will need to figure out ways to make our computing systems more energy efficient by means other than technology scaling.

978-1-4799-0917-9/14 $31.00 © 2014 IEEE

4. The Limits of Parallelism

When uniprocessors ran into the power wall, to continue to scale performance, manufacturers began to include more processor cores on each die. It had been known for decades [7] that if the application was parallel, a parallel machine would be more power-efficient. The reason can be easily seen from the data in Figure 1.1.5: Here, we have taken data on early processors, and plotted the energy/operation vs. the peak performance that the processor could achieve, and approximately normalize out the technology's effect on the energy and the performance. The curve clearly shows that achieving more operations/second means that each operation consumes more energy, which means that power increases super-linearly with performance, since power is energy/operation (ops/sec). This is quite contrary to what one needs if the system is power limited: In the power limited world in which we now live, increasing performance means we need to decrease the energy/operation to keep the total power constant. Thus, twice the performance requires each operation consume half the energy.

The move to parallel processing avoided the performance-energy correlation. It allowed each core to be more energy efficient by having a lower peak performance, and added multiple cores on the die to increase the overall performance. In addition, the move to multicore allowed processors to use the energy/delay scaling of changing supply voltages to their advantage. Remember that we need to reduce the energy/operation to enable us to keep the power in check when running at peak performance with all cores operating. To achieve this goal, the processor is run at a supply voltage that is below the peak safe operating voltage for this technology, at a level that is tuned to the individual die. However, now sequential applications run slower than before, since each uses only a single slower core. To improve the performance of sequential application, the supply voltage is increased, and that one core gets overclocked, while the other cores are parked, (that is put in a low-power state). While running at a higher voltage decreases its energy efficiency, since the other core(s) are idle, the overall chip remains within its power window.

Such turbo modes make the accounting of supply voltages even more difficult, since now there are at least three voltages to track, one each for turbo mode, normal mode, and low power/battery mode. For mobile processors, the power-supply voltage is even more complex. Since these systems are concerned about total energy usage (a.k.a. battery life), they need to do computation as efficiently as possible, that is at as low a voltage as possible, while still meeting performance targets. As a result, these systems support a large number of frequencies and supply voltages at which they can run, and have a number of control loops which set the supply voltages and frequencies. This technique is referred to as DVFS (dynamic voltage and frequency scaling). In systems which dissipate static power, the optimal frequency selection becomes a little more complex, since slower operation increases the energy/operation when static power is present,

Parallelism alone is not going to allow computing performance to scale; this is apparent from the graph in Figure 1.1.5 which shows diminishing returns at both extremes. As we try to lower energy, we soon need to forgo large performance factors for small energy savings. Said differently, once we have moved off the "bleeding edge", the energy gain for going even slower is modest. The same is true for supply voltage. This means that as we scale technology, reducing device switching capacitances, we still get an energy saving since everything is smaller. Thus, a 2× linear shrink will reduce energy by 2×, which should allow the number of cores to increase by a similar amount, if we run each core at the same frequency as they run today. If the intrinsic gate speed improves, small additional performance gain might be possible (by lowering supply to save a little more energy). Note that this scenario is growing the number of cores at half the rate possible from device scaling. This is much less performance scaling than we are used to, since a 2× shrink no longer enables 4× the number of cores on the same die area.

Yet even this slow performance scaling might be somewhat optimistic since it ignores the energy cost of the memory system that is attached to the processor. As shown below in Section 5, this memory-system energy can dwarf the energy of an efficient processor.

5. Don't Forget the Memory Energy

The data in Figure 1.1.5 is not completely accurate, since it assumes that the processor performance would scale linearly with the clock rate as technology scaled. This is not true unless the time the processor spends waiting on memory also scales with the clock rate. Since DRAM access times have scaled slowly, to scale memory stall time we need to decrease the miss rate of the processor's caches by adding a large last-level cache. Today this cache is on the order of 8MB, and needs to be added to the processor's energy budget. The leakage of the large last level cache is very dependent on technology and the circuit tricks used to reduce the leakage when the cache is idle (since most of the cache is idle most of the time). Thus, it is hard to predict how cache leakage will scale. We estimate the leakage to be around 100mW/MB, which matches data we had for 45 to 32nm parts. Correspondingly, we have added this leakage power to the energy of each processor presented in Figure 1.1.6. When this energy is added, the energy efficiency advantage of the older processors goes away, and makes the processor energy change with performance even flatter. A comparison of Figures 1.1.5 and 1.1.6 leads to a startling conclusion: the leakage power of a modern last-level cache is larger than the power of a simple core running full out. This means that a slower energy efficient processor is really not efficient when the entire memory system is considered, unless future technology can greatly reduce this leakage power. Without this development, it is better to use multiple processors sharing the cache such that its leakage cost can be amortized.

The importance of memory energy is shown in Figure 1.1.7 which gives the power breakdown of a recent 40nm, 8-core superscalar processor with an 8MB last-level cache. Over 50% of the processor die energy is dissipated in the caches and register files in this machine. While the cache hierarchy consumes a significant amount of energy, it reduces the overall system energy by eliminating energy intensive memory accesses. Given that the energy/operation is now critical, it is worth revisiting many of the cache design strategies with a view to minimizing the average memory access energy (AMAE) [8]. In this optimization, it is important to consider adding a small level-0 cache to reduce the energy cost of loads, and to consider leakage to determine the optimal cache size, since bigger is no longer always better!

AMAE minimization must also take into account the DRAM energy of the system. DRAM power is generally left out of most processor power analysis, since it is off-die, but must be included in any computing-system analysis. The energy cost of a DRAM access (1 to 2nJ) is a couple of orders-of-magnitude higher than the cost of an internal cache access or functional operation (10pJ). Part of this high cost comes from the very energy-inefficient I/O that DRAM systems use [8], which not only takes over 20pJ/bit, but also require static power to keep the I/O active. This static power makes the effective cost/bit even higher, especially for efficient computation, which minimizes the number of memory accesses. One hopes that this interface power problem can be resolved soon, since very efficient I/O has been demonstrated by many different companies [9], but changing standards can be difficult. Yet even when the I/O is improved, the energy cost of a DRAM access will still be large (10pJ/bit, 0.6nJ/8B) compared to a core processor operation, since requests and data still must travel a large distance on the processor and memory chips to reach this efficient I/O. Thus, much work still needs to be done in this area to find ways to minimize this energy, or at least minimize it for specific kinds of access patterns.

6. Gaining Efficiency through Specialization

Given the energy costs of the memory system, and the constraints on both parallelism and technology scaling, it might seem like there is not much room for energy improvement. Yet there are numerous examples of specialized hardware which is 2 to 3 orders-of-magnitude more efficient than a processor-based solution as shown in Figure 1.1.8 [10]. The key to understanding how much energy saving is available is to look at the energy costs for the fundamental operations of the application. It is critical to consider both the computation and the communication (memory accesses) that are required. The data for various operations in a 45nm technology are shown in Figure 1.1.9, which also gives the energy breakdown of a simple in-order processor. These numbers make two things very clear: First, the programmable nature of a processor has high energy overhead, 70pJ/instruction, vs. a few pJ for an operation. Fetching the instruction, and clocking the state registers that keep the data organized, add a significant cost. Second, if high energy efficiency is required, the application must have very good data locality, since a cache fetch is 20pJ. Fetches from the first level cache have an energy cost that is a significant fraction of that of an instruction, so if data needs to be fetched often, the energy improvement will be modest.

While the energy cost of programmability is high, the dollar cost of creating custom solutions is also high. This combination of forces has instigated a change from bare processor dies to SoCs which integrate a number of application accelerators and processors for energy efficiency. Most processor dies today contain the CPU, GPU, and image processing accelerator, and a number of audio, video, and wireless codecs. This growth in application accelerators has led to research into whether there is enough commonality between a set of applications to create a user programmable hardware engine for accelerating that entire application class. We have seen this sort of evolution before in graphics, where we went from units with limited programmability to a highly-programmable engine for a class of data-parallel floating-point (FP) applications. In fact, since FP arithmetic takes around 1/10 the energy of a simple instruction, forming a SIMD engine which performs the same operation on about 10 data lanes, easily makes the machine's instruction energy dominated by the FP operation, minimizing the cost of the programmability. This is the approach that GPUs have taken, so it is likely that this class of machine might be able to handle all data-parallel FP computation efficiently. Current machines cannot yet do this, since they were architected to maximize performance, and they do not yet leverage locality as much as they could. However, both hardware and software are migrating in this direction, since they need to exploit locality as well, to reduce their energy usage.

Getting the highest energy efficiency requires a very specific combination: very low energy operations, and extreme locality. 1000 MOPS/mW = 1pJ/operation. Such efficiency levels are possible only if the application works on short integer data (8 to16bits), and tens of data operations are completed for each local memory fetch, and roughly a thousand operations are completed for each DRAM fetch. While this degree of reuse seems unlikely, it actually is quite common, and is very similar to what is needed to implement convolution. A problem with extremely-efficient computation is that each computation does not consume much energy. Thus, to amortize the cost of each memory access, it is critical that the outputs of one operation are forwarded directly (or through a local buffer) to another intense computation. While this convolution-like data flow is a very restricted form of computation, most of the accelerators being proposed today fall into this form, including the image, video, and modem processors being integrated on SoCs.

As a result, my research group has been investigating different approaches to creating accelerators for convolution-like applications. We have created an abstract machine model for this class of stencil computations, and are working on a hardware generator and a programmable engine that can support it. Since generating software and drivers is always difficult for custom accelerators, we have created a domain specific language (DSL) for stencil computations that will enable application developers to take advantage of the special hardware in the final product.

7. Tools for Enabling Design

A convolution engine solves only one class of applications, and that is the problem with specialization. Any special solution does not work for all applications, so there are always more problems that need to be addressed. Generating solutions requires a deep understanding of the application, and the corresponding energy costs of computation. Rarely will an unmodified application or algorithm run, let alone reach the desired efficiency on an accelerator. It takes people thinking and working on their problem to find a truly efficient hardware implementation. Thus, if specialization is going to be one of the key strategies for continuing to scale computing performance, our principal task will be to enable a larger group of application experts to participate in creating the efficient hardware/software systems that they require. While the breadth of the application areas makes it unlikely that we can automatically generate highly-efficient hardware given just the algorithm, codifying the principles of efficient hardware implementations seems feasible. We are starting to see the precursors of these tools now, from the current generation of high-level synthesis systems, to hardware generators such as Chisel [11] and Genesis 2 [12] to more domain-specific systems like SPIRAL [13]. Nearly thirty years after the adoption of logic synthesis and place-and-route tools that opened the IC industry to creating application-specific integrated circuits (ASICs), we have come full circle, to the point where we need a new set of tools to enable application experts to access our technology.

As we work on creating these tools, we should remember that not all problems require the "bleeding edge" of performance or energy efficiency. For many applications, current processors/microcontrollers are more than adequate to meet the required system performance. Unfortunately, the people who know that these parts exist and how to use them are likely a distinct group from the people who have applications that they want to implement. Like the high-performance space, we again need tools that allow application experts access to technology, but now at the level of building board-level systems from existing parts. Imagine what would happen if creating a new hardware widget was as easy as writing an iPhone or Android application. This environment would drive a new wave of innovative uses of computing.

8. Conclusion

In summary, our challenge is clear: The drive for performance and the end of voltage scaling have made power, and not the number of transistors, the principal factor limiting further improvements in computing performance. Continuing to scale compute performance will require the creation and effective use of new specialized compute engines, and will require the participation of application experts to be successful. If we play our cards right, and develop the tools that allow our customers to become part of the design process, we will create a new wave of innovative and efficient computing devices.

Acknowledgements:
This work was partially funded by DARPA under the SEEC project, the STARnet CFAR center, and Stanford's Center for Pervasive Parallelism. The author would like to acknowledge the help from Dejan Markovic and Steven Richardson and all his current and former students which made this research and paper possible, especially Ofer Shacham, John Brunhaver, Andrew Danowitz, Sameh Galal, Steven Bell, Krishna Malladi, Wajahat Qadeer, and Rehan Hameed.

References:

[1] http://www.google.com/glass/start/
[2] https://getpebble.com
[3] R. H. Dennard, F. H. Gaensslen, H. N. Yu, V. L. Rideout, E. Bassous, and A. R. LeBlanc, "Design of ion-implanted MOSFETs with very small physical dimensions", IEEE J. Solid-State Circuits, vol. SC-9, pp.256 -268, 1974
[4] http://cpudb.stanford.edu/
[5] Andrew Danowitz, Kyle Kelley, James Mao, John P. Stevenson, and Mark Horowitz, "CPU DB: Recording Microprocessor History", Communications of the ACM 55, Issue 4, pp. 55-63, April 2012.
[6] Robert H. Dennard, Jin Cai, Arvind Kumar, "A Perspective on Today's Scaling Challenges and Possible Future Directions", Solid-State Electronics, Volume 51, Issue 4, pp. 518-525, April 2007.
[7] A. Chandrakasan, Low Power Digital CMOS Design, Ph.D. Thesis, UC Berkeley, 1994.
[8] Krishna T. Malladi, Benjamin C. Lee, Frank A. Nothaft, Christos Kozyrakis, Karthika Periyathambi, and Mark Horowitz, "Towards Energy-Proportional Datacenter Memory with Mobile DRAM", ISCA '12, Proc. of International [9] Symposium of Computer Architecture, pp. 37-48, June 2012.
[9] J. W. Poulton, W. J. Dally, Xi Chen, J. G. Eyles, T. H. Greer, S. G. Tell, C. T. Gray, "A 0.54pJ/b 20Gb/s Ground-Referenced Single-Ended Short-Haul Serial Link in 28nm CMOS for Advanced Packaging Applications," *ISSCC Dig. Tech. Papers*, pp. 404,405, Feb. 2013.
[10] Dejan Markovic, EE292E Lecture Notes, Lecture 5, Stanford University, 2013
[11] https://chisel.eecs.berkeley.edu/
[12] http://genesis2.stanford.edu/
[13]http://www.spiral.net/

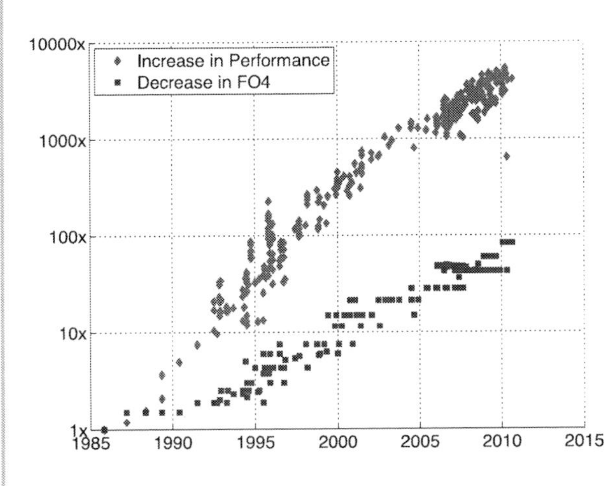

Figure 1.1.1: Improvement in microprocessor and gate performance vs. year.

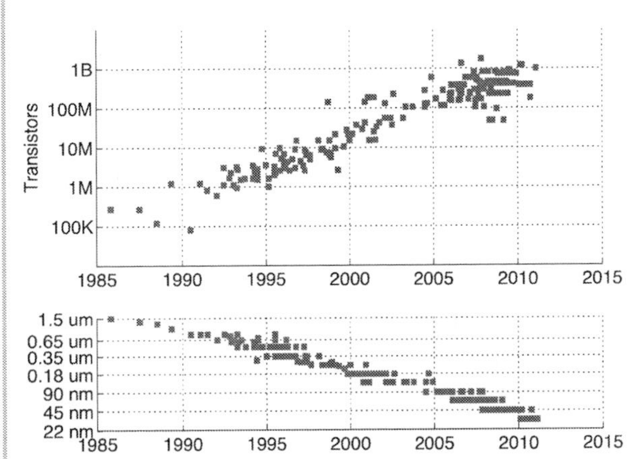

Figure 1.1.2: Number of transistors and feature size vs. year.

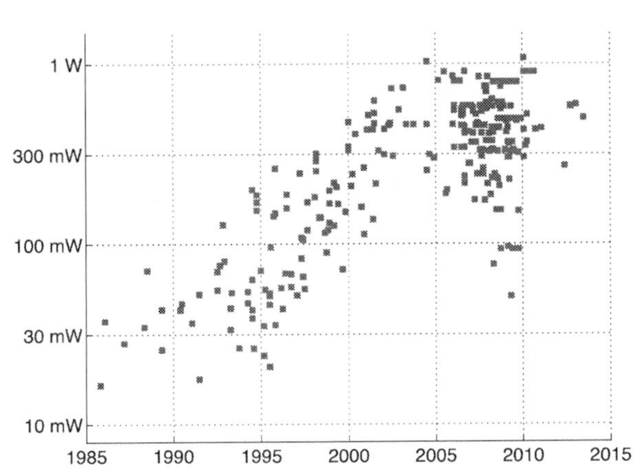

Figure 1.1.3: Power density in mW/mm² vs. year.

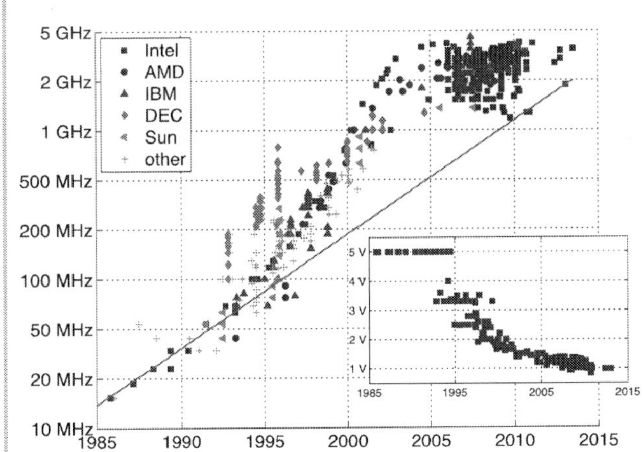

Figure 1.1.4: Clock frequency vs. year. The red line indicates frequency increase due to gate speed. The insert plot is Vdd vs. year.

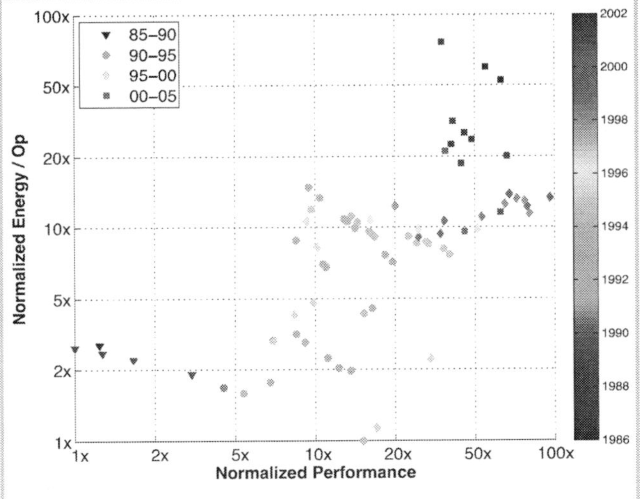

Figure 1.1.5: Instruction energy vs. peak performance (normalized).

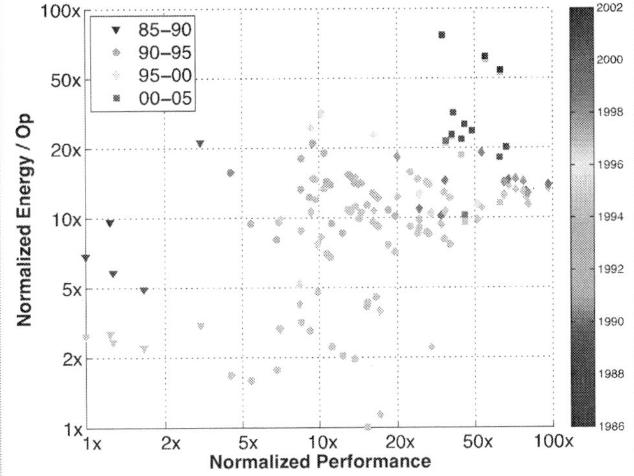

Figure 1.1.6: Instruction energy vs performance, with LLcache leakage added, with original points shown in grey for comparison.

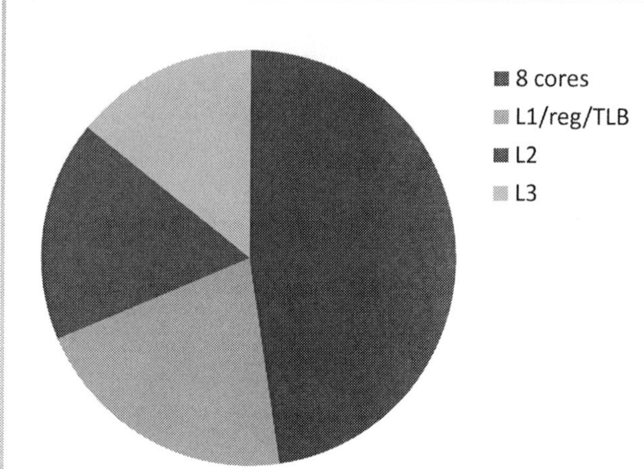

Figure 1.1.7: Power breakdown of an 8 core server chip.

Chip	Year	Paper	Description
1	2009	3.8	Dunnington
2	2010	5.7	MSG-Passing
3	2010	5.5	Wire-speed
4	2011	4.4	Godson-3B
5	2013	3.5	Godson-3B1500
6	2011	15.1	Sandy Bridge
7	2012	3.1	Ivy Bridge
8	2011	15.4	Zacate
9	2013	9.4	ARM-v7A

Chip	Year	Paper	Description
10	2012	10.6	3D Proc.
11	2013	9.3	H.264
12	2012	28.8	Razor SIMD
13	2011	7.1	3DTV
14	2011	7.3	Multimedia
15	2011	19.1	ECG/EEG
16	2010	18.4	Obj. Recog.
17	2012	12.4	Obj. Recog.
18	2013	9.8	Obj. Recog.
19	2011	7.4	Neural Network
20	2013	28.2	Visual. Recog.

Chip type:
Microprocessor
Microprocessor + GPU
General purpose DSP
Dedicated design

Energy Efficiency (MOPS/mW)
10000
1000
100
10
1
0.1
CPUs
CPUs+GPUs
GP DSPs
Dedicated
1000x
1 2 3 4 5 6 7 8 9 10 11 12 13 14 15 16 17 18 19 20

Figure 1.1.8: Energy efficiency of specialized processing, from [10].

Integer	
Add	
8 bit	0.03pJ
32 bit	0.1pJ
Mult	
8 bit	0.2pJ
32 bit	3.1pJ

FP	
FAdd	
16 bit	0.4pJ
32 bit	0.9pJ
FMult	
16 bit	1.1pJ
32 bit	3.7pJ

Memory	
Cache	(64bit)
8KB	10pJ
32KB	20pJ
1MB	100pJ
DRAM	1.3-2.6nJ

Instruction Energy Breakdown

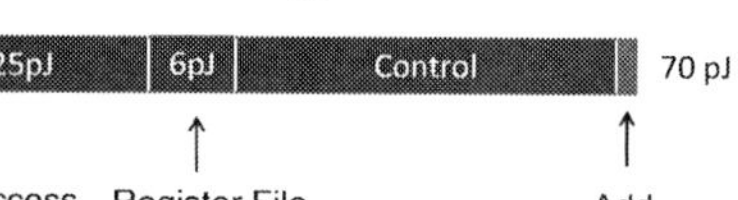

Figure 1.1.9: Rough energy costs for various operations in 45nm 0.9V.

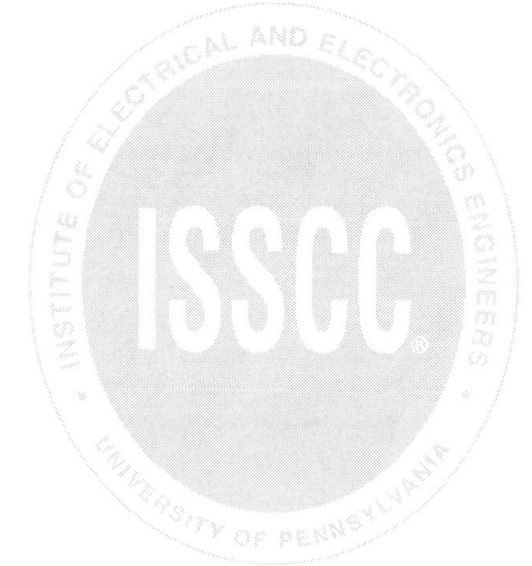

1.2 Cloud 2.0 Clients and Connectivity – Technology and Challenges

Ming-Kai Tsai

Chairman and CEO, MediaTek, Hsinchu, Taiwan

1.0 Introduction

With the dramatically increasing use of mobile and portable devices, the need for computation has intensified, motivating the transformation of traditional static services (Web and storage) to evolve toward distributed Web services, forming Cloud 1.0; in this process, the evolution into the Smart Device Era involved many changes: stationary computing devices are going mobile, standalone devices are becoming connected, and peer-to-peer communication (email) extending to many-to-many (social networking). Two of the biggest enablers for Cloud 1.0 have been clients such as Smartphones and tablets, connected through wired and wireless networks. Embedded within each of these clients are the CPU and GPU processors needed to enable consumer applications and mobile human-interface devices (HIDs). To satisfy the ever-growing computational requirements, mobile CPU clock frequencies have extended into the GHz region. To avoid this barrier, mobile clients are driving the downscaling of process technology while motivating the rapid rise of multi-core CPUs and GPUs. In this process, new architectures involving asymmetric-CPU and octa-cores are emerging. As well, investment is pouring into the hardware/software (HW/SW) infrastructure to provide adaptive power management, thermal throttling, and efficient heterogeneous multi-processing, all to enable maximum core usage and energy efficiency within the tight thermal limits of the Smartphone and tablet domains.

Another key technology enabling end-user connection to the Cloud is reliable high-data-rate communication. While wired communication historically offers more throughput compared to current wireless LAN and WAN, it lacks mobility. Correspondingly, as Smartphone penetration rate increased, mobile data communication growth outpaced PCs, and wireless gradually took the place of wireline for clients connecting to the Cloud. Although wireless LAN throughput is higher than that for cellular communications, it is not available everywhere. However, the latest cellular communication technology (e.g. LTE) finally enables users to stay connected to the Cloud anywhere and anytime with reasonable throughput.

Now, in a few short years, one can expect that Cloud 1.0 will morph into Cloud 2.0; in this new (ubiquitous) era, the central Cloud will be augmented, making greater use of our own Smartphones (personal cloud), tablets (local cloud), smart homes (home cloud), and so forth. But, these Cloud 2.0 elements will require orders-of-magnitude improvement in computational throughput to better enhance the user experience, demanding innovations ranging from system architectures to circuits. As well, the increased number of distributed clouds demands orders-of-magnitude-higher data bandwidth, motivating next-generation wireless LAN, WAN, and BAN technologies. Furthermore, these improved technologies must collaborate with wired network innovations to enable seamless connectivity and workload optimization. However, because of the form factor constraints on portable client devices, and linear (at best) battery technology improvement, the energy gap and thermal wall cannot simply be scaled through process improvement, alone. These are the major technical challenges for which we need major innovations at all abstraction levels required to enable the ubiquitous Cloud 2.0 ecosystem (including next generation Internet-of-Things (IoT), Life Technology (LF), Machine-to-Machine (M2M) solutions), providing ample possibilities to enhance and enrich everyone's life.

2.0 Cloud 1.0: The Evolutionary Transition to the Smart-Device Era

Scaling of semiconductor process technology (Figure 1.2.1) has been one of the key enablers of the evolution of the computational model — from mainframe computers in the 1960s to mini-computers in the 1980s, then desktop personal computers (PC) in the 1990s, and laptops in the 2000s [1]. As a result of the combination of lowering cost of ownership (Figure 1.2.2) and improving functionality (such as higher performance, lower power, and new applications and services), each generation of the computational cycle stated above, has sparked an order of magnitude increase in users and devices. At every turn, new high-end applications tend to push the performance limit of existing hardware systems, which in turn motivates the next generation of process technology. This "positive technology-market expansion loop" has been the fundamental driver of semiconductor growth over the past 50 years.

Since the 1990s, when desktops became mainstream at home, the demand for network connection for personal emails has fueled the development of the commercial server infrastructure. As users began to proliferate, their natural demand for more computation and storage capacity, motivated data-center creation and development. In the mid-1990s, internet-based applications (such as Amazon, Yahoo!, and Google) further spread like wildfire around the world, and eventually social networking (Facebook in 2003) began to prevail (Figure 1.2.3). About the same time, standalone mobile devices (such as the MP3 players, game consoles, and feature phones) became the new symbol of status and fashion, all with easy access. While the Smartphone first appeared in 2002 (with Blackberry email), a major new direction was set by the appearance of the Apple iPhone in 2007, (with 3G internet access), marking a new chapter for mobile computing. In retrospect, the computational model evolved into the smart device era (Cloud 1.0); from stationary desktop to mobile laptop, from standalone to connected devices, and from peer-to-peer email to many-to-many social networking. Cloud 1.0 incorporates three key components — infrastructure, client devices, and connectivity. The latter two elements are elaborated:

2.1 Cloud 1.0: Client Devices

One of the key clients within Cloud 1.0 has been the Smartphone. Connectivity made it possible for Smartphones to access the Cloud, leading the way to many new markets such as online shopping, online banking, online gaming, and social networking. With each new application, came new challenges in performance and power: One of the largest has been to satisfy the wide range of performance requirements. In the power arena, the challenge is more complex: During the typically one-day life of a single charge of a Smartphone battery, most of the time is spent in standby or idle mode, while for the average user the aggregate active use is only 1 to 2 hours. Thus, the inactive-mode power is just as important as active-mode power, and in fact even more. Figure 1.2.4a illustrates typical usage during a single day: 22 hours in standby, and 2 hours in active mode distributed amongst various activities (such as talk, text...). The battery energy consumption Pareto chart (Figure 1.2.4b) shows that all use-cases are equally important including standby-mode [2].

As the number of desired applications grew, the need for processor performance increased exponentially, whether to enhance user-experience or to resolve multimedia demands. However, at the same time battery capacity and technology stayed flat; thus, the challenge was to provide more features but within the same thermal limits and battery-charge life. In the development of the general purpose CPU, there has been a wall at 3.0GHz, where the cost of power and heat removal is prohibitive (Figure 1.2.5) [3][4]. However, CPU performance has continued to improve through architectural innovation such as multi-core techniques. The same trend can be observed in the Smartphone market beginning in 2012 when multi-core emerged. To balance power and performance for all use-cases, new asymmetric architectures were created for the Smartphone, using multiple heterogeneous cores, where a high performance CPU operates when needed, and a power efficient CPU is used when performance demands are low. As well, software has evolved to support Heterogeneous Multi-Processors (HMP): This software intelligently schedules threads to the high-performance and power-efficient cores to optimize both performance and power cost functions. A representative example appears in ISSCC 2014 (10.3 [5]), where a 28nm application multi-processor is described which optimizes Power, Thermal, and Performance (PTP) using circuit and architectural innovations, as well as the first use of the HMP in a mobile SoC.

2.2 Cloud 1.0: Wireless Connectivity

In addition to the client devices described above, another key technology that enables end users to access the computation and storage capability of the cloud is reliable high-data-rate communication, either through wires or wirelessly. Wired communication offers greater throughput compared to wireless. But, it has been wireless that gradually gained momentum in the Cloud 1.0 Era, allowing users to have the desired mobility.

978-1-4799-0917-9/14 $31.00 © 2014 IEEE

In wireless communication, wireless LAN offers higher throughput, but its coverage is, in general, more limited than cellular (wireless WAN). This exchange between communication range and throughput has been identified earlier in the ISSCC 2013 trend tracking report [6][7], as shown in Figure 1.2.6. Before the Cloud 1.0 era, the data rate improved linearly. But since Cloud 1.0, the data rate of WLAN and cellular connectivity have both steadily increased at a rate of approximately ten times every five years. Although the exact triggers of this data-rate increase are not easily identifiable, some possible contributors may be. First, there is the effect of the form factor of smart devices (including phones, laptops, and tablets) as they finally became truly portable. Multi-disciplinary technological advancement, including process technology, circuit innovation, display evolution, packaging innovation, thermal engineering, and more have enabled form-factor reduction. The second possible contributor has been the improvement in wireless infrastructure. Independent of form-factor reduction, the users before Cloud 1.0 had only limited ability to leverage the power of the cloud without satisfactory access to the Internet. Within the Cloud 1.0 Era, the deficiency of Internet access was reduced through the introduction of WLAN hot spots, the increased coverage of cellular infrastructure, and the evolution of wireless standard. Note that it is the cellular infrastructure that allows users to stay always connected (although the user experience may not be satisfactory). Third, on the client side, the increasing penetration rate of WLAN in laptops and tablets, and the penetration rate of mobile data usage in phones, have been critical. Once the minimum connectivity throughput requirement have been fulfilled, it is likely that users began to rely more on mobile data rather than on fixed Internet data, and gradually shifted their usage pattern. This gradual change of usage pattern could also be observed from the statistics and forecast for mobile data IP traffic. It had been predicted that mobile data IP traffic would grow at a compound annual growth rate (CAGR) of 66% from 2012 to 2017 [8], as shown in Figure 1.2.7. By contrast, the fixed-Internet data-traffic CAGR was predicted to be 21% during the same period of time. Other shifts in data traffic patterns worthy of note are that, in 2017: consumers are projected to provide more traffic (80%) than that provided by business (20%); and cloud servers will be responsible for more data traffic (69%) than will traditional servers (31%) [8]. Overall, the compound effect of form-factor reduction, improved wireless infrastructure and penetration rate, and increased mobile data usage, has been to trigger the demand for reliable wireless connectivity at higher data rates, and continues to push the envelope of connectivity technology. The evolution of connectivity technology is part of the "positive-market-expansion-loop" story.

While the form factor has been drastically improving, and the throughput has been increasing at an exponential rate, battery-technology improvement for client devices has been relatively slow. Correspondingly, an important question can be raised: "although a smart device is portable, and a wireless link can be established with satisfactory speed, how long can such a link last". Thus, high energy efficiency must be considered as another key enabling factor for Cloud 1.0 connectivity. Energy efficiency (for connectivity, only) of client devices is improving primarily through the following factors: First, channel bandwidth is increasing, and more complex modulation schemes, are appearing using new communication standards. By this means, not only does throughput is improved, but energy per bit is also reduced. Thus, for cellular standards, the evolution from 2G to LTE has increased the bandwidth from 200kHz to 20MHz (or even more, 40MHz using carrier aggregation), with carrier modulation upgrading from GMSK to 64QAM. Similarly, for the 802.11 standards, the same trends were on-going — The bandwidth increased from 20MHz for 802.11b to 2.16GHz for 802.11ad, and the modulation and coding scheme changed from CCK for 802.11b to 256QAM for 802.11ac. Secondly, process, radio architecture, and circuit-technique improvements also led to an increase in energy efficiency. A final observation at this time is that offloading mobile data traffic from cellular networks to Wi-Fi improves throughput, and energy efficiency.

At this point, it is convenient to use an example through which the impact to users of increased energy efficiency and throughput can be understood. Here, we judiciously compare the case of a 50GB data download (approximately equivalent to a dual layer blue ray DVD), where maximum theoretical throughput is assumed, and only the energy of connectivity devices is counted. This allows for easier evaluation of the severity of battery drain contributed by connectivity devices. Such a download using an 802.11ad 60GHz link, occupies several minutes, using a few percent of a 2000mAh battery. However, in contrast, if we had used the GPRS cellular data network to perform the same task, the time required would have been several months, consuming the energy of tens of fully charged batteries! Thus, many applications that are commonplace today, were impossible in the past.

3.0 Cloud 2.0: Evolution to the Ubiquitous Era

In Cloud 2.0, we will see the introduction of a new type of client — the IoT device armed with pervasive sensing technologies. These devices will face various energy limitations, and secure reliable connectivity challenges, as the number of such devices grows to tens to hundreds per person in the 2020s.

Cloud 1.0 client, such as, the Smartphone and Tablet will continue to evolve as they take on the role of the IoT hub as well as support more and more new applications, such as Natural user interfaces and evolving multimedia. These clients will depend on advanced communication protocols for more bandwidth, and will support high-computation signal-processing algorithms with co-processors to enable these new features for Cloud 2.0.

3.1 Cloud 2.0: Energy Limitations of IoT devices

IoT devices will be even more energy-challenged than current Cloud 1.0 clients such as Smartphones and Tablets. One reason for this is that their form factor will constrain these devices to batteries no larger than a coin cell, which has 10× lower capacity than that provided in traditional Smart devices. Also, with thousands of IoT radios per person, it is not reasonable to change batteries daily. Thus, the ideal way to power these devices using energy harvesting; however, both the reliability of such power sources and their current capacity is so low. Thus, to do anything meaningful, from sensing to data processing along with wireless transmission, using coin cells or energy harvesting, these devices require orders-of-magnitude-more energy-efficient protocols, architecture, and circuits.

Wireless charging has found early success in the phone market due, in part to the novelty of power without wires, but also because of a push by operators to mitigate the high power consumption of LTE phones, with the availability 'convenient' charging. Wireless chargers for phones today are based on the use of tightly coupled magnetic fields, known as inductive charging. Inductive charging is inherently one-to-one, meaning one transmitter coil is required for each receiver coil. But, for multiple devices, this becomes a cumbersome and costly approach to charging. In addition, the coils must be placed in very close proximity, concentrically aligned and of similar size in order to achieve reasonable efficiency of power transfer.

Correspondingly, forward-looking wireless charging systems are based on loosely coupled technology, using high-Q resonance, forming inherently forming a one-to-many system, where a single transmitter coil supports multiple receiver coils concurrently. Such coils do not need to be of similar size, nor be aligned concentrically, and can be some distance apart while providing reasonable efficiency. But, the challenge in designing a high-Q resonant system is to meet the requirement of maintaining high Q under all conditions.

Other challenges for wireless charging systems include efficiency and associated heat loss. Moreover, consumer devices will have strict limits on the allowable temperature within the case and battery. The majority of power loss and therefore heat, is in the coil and the power electronics (rectifier and regulator), both typically placed on the back inside cover of the phone. For Smartphones, the maximum power transfer is limited to about 3.5W due to these thermal effects. In the future, charging solutions may place the power electronics on the main PCB where they will dissipate heat more efficiently, providing to 5-to-7W power-transfer solutions. The transition of wireless chargers to the mass market with high shipment volumes may also require the adoption of a single global standard. Thus far, there are several competing standards existing for high-Q resonant wireless charging.

Beyond wireless charging, another energy challenge lies in the IoT devices themselves. Thus, we foresee, that by 2025, there will be not only humans but also massive machines connected to the cloud through wireless networks anytime and anywhere, for many different purposes (general sensing, health monitoring, automobile control, etc.). If the use of existing wireless networks (designed to support basic human communication such as voice call, email, etc.) continues, then the system may be highly inefficient. For example, a

978-1-4799-0917-9/14 $31.00 © 2014 IEEE

typical voice call last for minutes, but a machine call may only need to send or receive a few bytes. Thus, reducing the signaling overhead for setting up a connection, can greatly improve the network efficiency, especially when these types of machine communication become popular. In addition, some of these machine-communication devices such as sensors may need a very long battery life without re-charging for weeks or even months. Therefore, a "light weight" communication protocol, along with advanced low power technologies is needed to accomplish this goal.

3.2 Cloud 2.0: IoT Reliability

Beyond the issue of low energy consumption, the large number of devices (IoT) surrounding the smart hubs, forming local/personal clouds, implies reliability challenges. Large numbers of devices in close proximity are, in general, more prone to interference. Furthermore, not all such devices are likely to be using the same protocol to communicate with each other. Thus, avoiding conflict in spectrum or time becomes more difficult when multiple protocols are taken into account. As well, as noted above, these devices are likely to have much smaller batteries (such as a coin battery with capacity about 1/10 of that in a typical Smartphones), and would be expected to operate for months to years. Unfortunately, typical solutions to mitigate interference such as radio-linearity enhancement and interference-cancellation algorithm require energy.

Another difficulty is that these devices are expected to operate almost free of human intervention: imagine a situation in which one device goes wrong, forcing the user to reset his thousands of radios to re-establish communication, which would normally lack computation capability for complex network routing. Therefore, the smart hub of the associated network needs to be able to correct the network autonomously (for example, in the case of data traffic congestion). The hub also needs to sense and manage information flow, for all load conditions, and allocate resources for critical (such as life-related) information, in a way that devices and protocols are transparent.

3.3 Cloud 2.0: Security and Privacy

Concern for security is not new for any device that is connected to the Internet. However, in Cloud 2.0, the impact of insecure devices escalates. This is because more sensitive and even life-critical data (such as, medical-related information, payment data, and more), are now available, wirelessly. Moreover, Cloud 2.0 devices are used not only for sensory purposes, but more critically, for control. Potentially, any device in close proximity could receive the same information as the targeted device if they follow the same protocol being used (such as, BT-Smart). Now, to obtain such data, breaching infrastructure is not required. Furthermore, when such devices are used for control-related applications (home-security, industrial or infrastructure control), the security issue is not limited to just virtual data or privacy. Physical harm may arise due to insecure connection. In general, in the Cloud 1.0 Era, this issue was not as serious. The lack of energy, computation capability, and storage on most of these sensory devices put a limit on the amount of protection they could provide. In addition, that everyday objects become connected directly to Internet makes them potential security risks. But, Internet connection distributes these risks far more widely and easily. This situation calls for the use of energy-efficient encryption, and data-protection technologies at the wireless physical layer. Absence of human interface for most of the devices and the large number of devices makes authentication-key management unlikely; thus, automated key management is essential upon device deployment.

3.4 Cloud 2.0: Bandwidth

Bandwidth is one of the most-effective factors in improving throughput and energy efficiency. However, licensed spectrum is scarce and not universally available. It is therefore unclear if the throughput and energy efficiency of both the wireless infrastructure and client devices can continue improving at a rate similar to that forecast for mobile-data-traffic CAGR of 66% [8]. For instance, in 2012, it is estimated that global cellular infrastructure already consumes the same amount of energy as that used by the United Kingdom annually for all purposes [9][10]. On the client-device side, smart devices are expected to handle more computation and communication, while battery technology does not improve correspondingly. Clearly, new standards, protocols, and connectivity technologies, are required to continue the push for energy efficiency.

The cellular industry is currently investigating the next generation (5th, or 5G) of wireless-communication technology. Until now, 5G is a vague concept, but, what is generally expected is that 5G will be deployed around 2020 to support explosive mobile-data growth and massively growing numbers of connected devices. Intelligent spectrum reuse and denser cellular deployment have been discussed as the means to address the challenges of bandwidth. Previous study shows that significant portions of today's spectrum, (such as those used by governments), are used only at certain locations during specific times. Thus, mobile network capacity can be improved by accessing this under-utilized spectrum intelligently without sacrificing their primary usage. New software-defined networking technologies are required for the network to effectively detect the unused spectrum when needed and release the spectrum when done. How to manage interference will be the key to maintaining link quality while exploiting the capacity gain. Densely deployed cells bring base stations closer to the users. The spectrum could be reused, capacity can be improved, and device power reduced, if interference can be controlled. As discussed in Section 3.1, a good interference-cancellation receiver will be highly desirable here, too.

3.5 Cloud 2.0: Elastic Computing

The decision to operate an application on the Cloud or in the Client is based on two criteria: the magnitude of the computation, and the availability of communication bandwidth. If the application requires real-time response and low latency, the application is best run directly on the Client. If the latency to go to the Cloud is acceptable, and the data bandwidth is supportable, then it is preferable to go to the Cloud.

The issue can be illustrated by considering a specific example: the contrast between speech recognition and gesture recognition. Most speech data processing is done in the "Cloud", while all gesture data processing is carried out locally. For gesturing, game machines require real-time response and very-short latency, while the volume of video/depth data is too large to upload to the "Cloud" for processing. Hence, local processing for gesture recognition is a logical choice. On the other hand, speech recognition needs a large centralized database and considerable computational power to implement sophisticated speech modeling, machine learning, and natural-language processing/responding, leading to its cloud-based solution. Other new applications will continue to work their way, first onto the Cloud, and then selectively onto the Client. Typical applications include augmented reality, machine learning, and computer vision.

As applications continue to be developed for the global and centralized Cloud, the energy to support many data centers will increase exponentially over time. To reduce worldwide energy consumption, low-power local servers can un-load the global Cloud, assigning more of the centralized data to the Client. Whereas datacenters can consume 5000× more power than the Client, a low power server can operate with only 50× more power than the client, reducing overall system power dissipation.

In the future, both communication data rate and computational ability will continue to improve, and the upload of a larger volume of raw data to the "personal cloud" for processing becomes more feasible. Thus, a "thin" client is not a good direction for applications that require low latency or fast real-time performance. However, the "Cloud" is always more resourceful than a Client (in terms of computational power and storage capacity), and can easily support centralized databases or knowledge bases, especially those requiring adaptation or learning. Thus, different applications require different global, local, and personal Cloud/Client trade-offs.

Fingerprint authentication has been widely used in the past to identify individuals or for verifying personal identity. Correspondingly, there are several successful applications using fingerprint authentication for access control of building entrance or personal computer. As well, fingerprint authentication technology in Smartphone applications is emerging, such as for phone unlocking, remote entrance control, and for some cloud-based applications such as mobile payment or network banking.

The challenges of embedding finger print authentication within Smartphones and mobile devices include the following fundamental factors: sensor area, power consumption, and user experience coming from failure/pass rates for register/non-register users. Some driving requirements are becoming obvious, and are leading to a potentially wider adoption of fingerprint authentication within mobile devices in the near future: the fast growing demand for more-secure cloud applications, require higher complexity of user passwords, and more-advanced fingerprint-sensor technologies — moving toward smaller form factor, lower power consumption (<<1mA), and easier and friendlier user interfaces (area vs. swipe sensing).

As well, new multimedia co-processors will emerge to support new applications. One classic example from Cloud 1.0 is the Graphics Processing Unit (GPU). The GPU has been a vital accelerator, bringing advanced signal processing solutions to the Client in a power-efficient manner. Initially, the first GPUs were single-chip processors embodying many integrated engines — transform, lighting, triangle setup/clipping, and rendering — all capable of processing a minimum of 10 million polygons per second. As time evolved, GPUs added programmable shading to their capabilities. Each pixel could now be processed by a short program, called a "pixel shader" that could include additional textures as inputs, and each geometric vertex could likewise be processed by a short program, called a "vertex shader" before it was projected onto the screen. Current GPUs are highly programmable devices that can handle general-purpose programs in addition to shaders. Such a computing paradigm is referred to as GP-GPU, standing for General Purpose computing on GPU. It came at the right time when serial computing on a CPU hit the power wall.

As more general-purpose computing becoming available, then came the introduction of OpenCL (Open Computing Language). An OpenCL program written for a specific GPU can be ported to a different GPU, or even CPU. Furthermore, such programs would run faster, not just with higher clock rate, but using more cores and more parallelism. In 2012, HSA (Heterogeneous System Architecture) was founded to complement OpenCL, and extend it to the next level of efficiency and flexibility. Software that can efficiently schedule between completely different multi-media co-processors means that someday the bulk of the computing workload can be done on the GPUs at low power, while CPUs handle critical tasks such as load balancing and required-low-latency use-cases. Similarly, new multimedia co-processors with some limited programming capability will be developed to support the new applications demanded by Cloud 2.0

4 Summary

In summary, as we begin to evolve into Cloud 2.0, the number of radios will increase rapidly due to the explosion of IoT devices, which act like an army of tiny robots making our life easier, as a result of the positive technology-market expansion loop driven by lower cost of ownership and improved functionality. It can be expected that, by 2030, one thousand IoT devices/radios per person will be within reach, connected through various protocols to various cloud hubs (personal cloud, home cloud, work cloud, local cloud (u-server), and cloud center). Insatiable demand for computation and mobile data bandwidth will continue to push the next wave of technologies and innovations, for orders-of-magnitude improvement in performance and energy-efficiency of clients and cloud hubs. Furthermore, autonomous elastic computing between clients and cloud hubs will be critical to optimize the combination of workload, latency and capacity of computation, storage, and energy, all for analytics and for big data applications, ranging from daily-life enhancement (fitness and healthcare), computational photography, augmented reality, machine-to-machine exchange, five-sense multimedia, to machine learning. A reliable secure and energy-efficient connectivity technology covering all aspects of distance and data rates will be more important in Cloud 2.0, than ever. Overall, it is estimated that there is a U.S $100 Billion semiconductor opportunity employing 8 Trillion smart radios connected to various cloud hubs by 2030. For the first time ever, most people on the planet will be Internet-connected and have equal access to Rich media. Shortly, we will find ourselves not only at the beginning of the world's greatest revolution, we are unleashing unprecedented new opportunities — socially, culturally and in all fields of research, commerce, and education!

References

[1]"Technology-Semiconductors: The Internet of Things is now", Morgan Stanley Research Global, September 2013

[2] eMarketer, "Digital Set to Surpass TV in Time Spent with US Media", August 2013;

[3] "The Free Lunch Is Over: A Fundamental Turn Toward Concurrency in Software", Herb Sutter, Dr. Dobb's Journal, 30 (3), March 2005

[4] A. Danowitz et. al, "CPU DB: Recording Microprocessor History" , Communications of the ACM, Vol. 55, Pages 55-63.

[5] A. Wang et. al, "Heterogeneous Multi-Processing Quad-core CPU and Dual-GPU design for optimal Performance, Power and Thermal tradeoffs in a 28nm Mobile Application Processor", *ISSCC Dig. Tech. Papers*, , Paper 10.3, February 2014.

[6] K.C. Smith, A. Wang, L.C. Fujino, "Through the Looking Glass II: Part 1 of 2, Trend Tracking for ISSCC 2013," *IEEE Solid-State Circuits Magazine*, Vol. 5, Issue 1, pp. 71-89, March 2013.

[7] K.C. Smith, A. Wang, L.C. Fujino, "Through the Looking Glass: Part 2 of 2, Trend Tracking for ISSCC 2013," *IEEE Solid-State Circuits Magazine*, Vol. 5, Issue 2, pp. 33-43, June 2013.

[8]Cisco Visual Networking Index: Global Mobile Data Traffic Forecast Update, Feb. 2013.

[9] "Key World Energy Statistics", International Energy Agency, 2012.

[10] "Overview of ICT energy consumption", Network of Excellence in Internet Science, 2012.

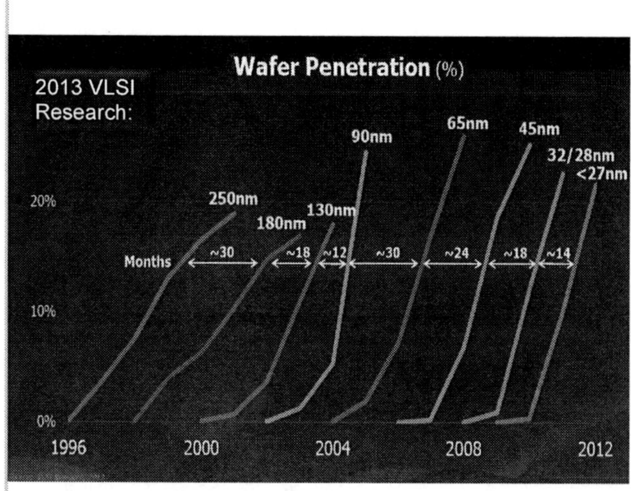

Figure 1.2.1: Wafer Penetration vs. Time.

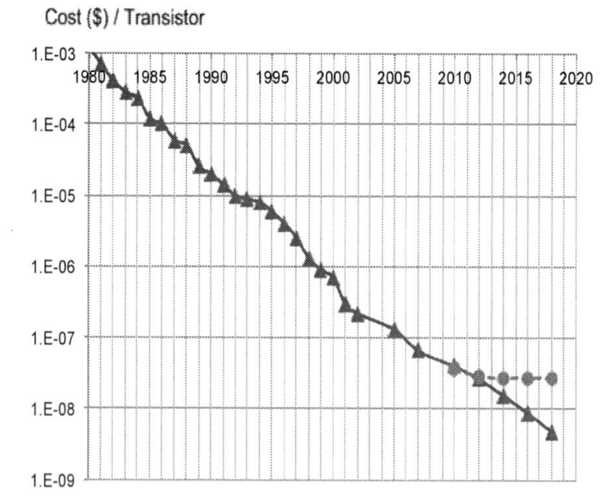

Figure 1.2.2: Transistor Cost vs. Time.

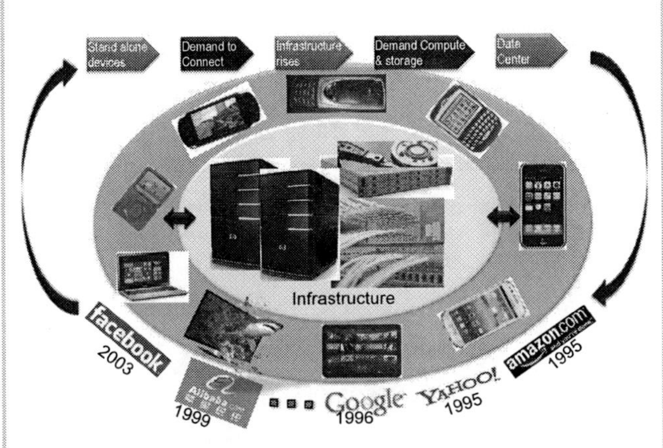

Figure 1.2.3: Cloud 1.0: Evolution.

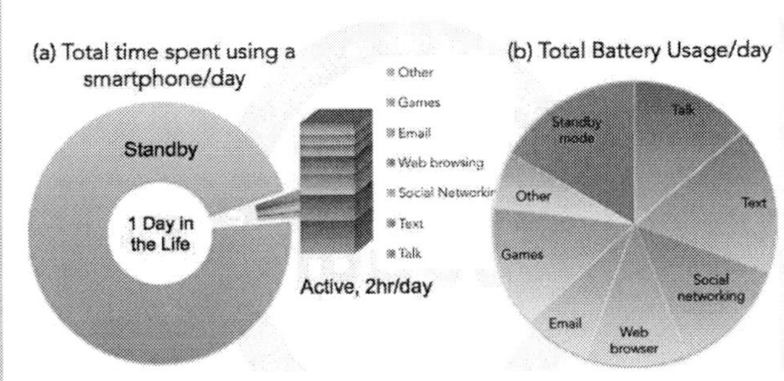

Figure 1.2.4: Smartphone Use Scenario and Energy Consumption.

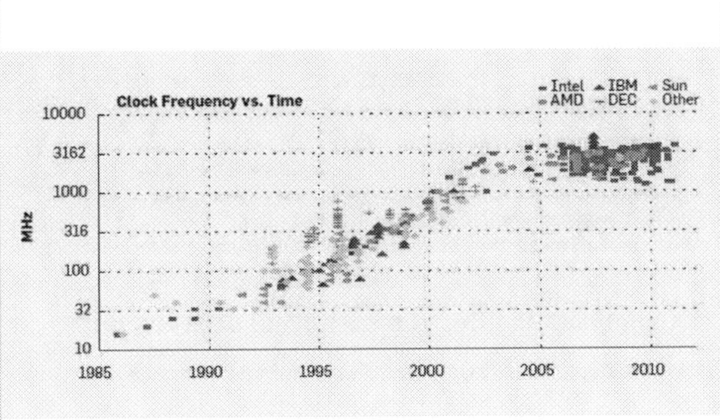

Figure 1.2.5: CPU Clock Frequency vs. Time.

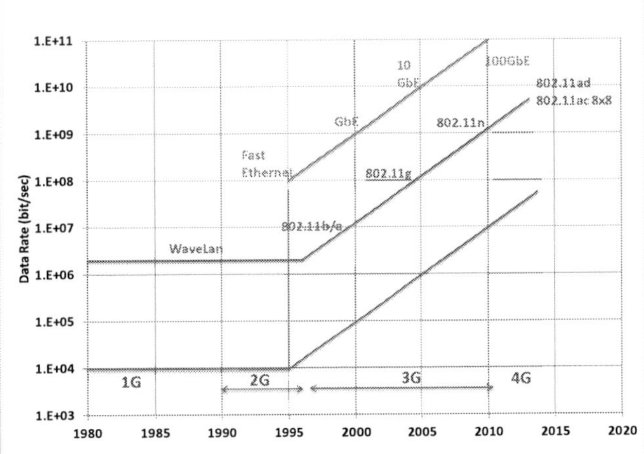

Figure 1.2.6: Communication Data Rate vs. Time.

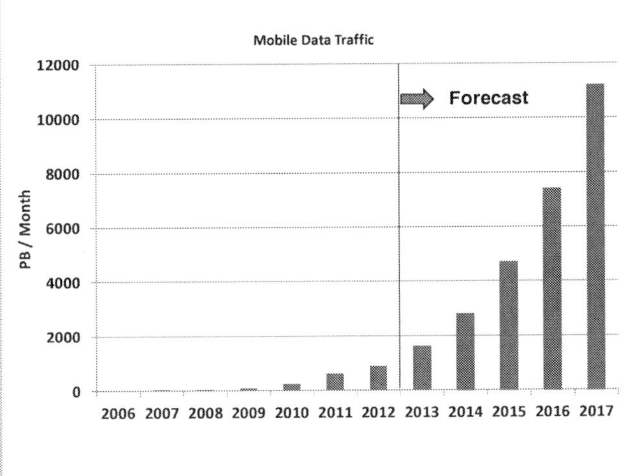

Figure 1.2.7: Mobile-Data-Traffic Growth.

978-1-4799-0917-9/14 $31.00 © 2014 IEEE

ISSCC AWARDS

2013 Lewis Winner Award for Outstanding Paper

"A 0.54pJ/b 20Gb/s Ground-Referenced Single-Ended Short-Haul Serial Link in 28nm CMOS for Advanced Packaging Applications"

John W. Poulton[1], William J. Dally[2], Xi Chen[2], John G. Eyles[1], Thomas H. Greer III[1], Stephen G. Tell[1], C. Thomas Gray[1]

[1]Nvidia, Durham, NC, [2]Nvidia, Santa Clara, CA

2013 Lewis Winner Award for Outstanding Paper

"A Fully Differential Charge-Balanced Accelerometer for Electronic Stability Control"

Vladimir P Petkov[1], Ganesh K Balachandran[1], Jochen Beintner[2]

[1]Robert Bosch, Palo Alto, CA
[2]Robert Bosch, Reutlingen, Germany

2013 Distinguished-Technical-Paper Award

"A Fully Integrated 8-Channel Closed-Loop Neural-Prosthetic SoC for Real-Time Epileptic Seizure Control"

Wei-Ming Chen[1], Herming Chiueh[1], Tsan-Jieh Chen[1], Chia-Lun Ho[1], Chi Jeng[1], Shun-Ting Chang[1], Ming-Dou Ker[1], Chun-Yu Lin[1], Ya-Chun Huang[1], Chia-Wei Chou[1], Tsun-Yuan Fan[1], Ming-Seng Cheng[1],Sheng-Fu Liang[2], Tzu-Chieh Chien[2], Sih-Yen Wu[2], Yu-Lin Wang[2],Fu-Zen Shaw[2], Yu-Hsing Huang[2], Chia-Hsiang Yang[1], Jin-Chern Chiou[1],Chih-Wei Chang[1], Lei-Chun Chou[1], Chung-Yu Wu[1]

[1]National Chiao Tung University, Hsinchu, Taiwan
[2]National Cheng Kung University, Tainan, Taiwan

2013 Jan Van Vessem Award for Outstanding European Paper

"A 20b Clockless DAC with Sub-ppm-Linearity 7.5nV/√Hz-Noise and 0.05ppm/°C-Stability"

Roddy C. McLachlan[1], Alan Gillespie[1], Michael C. W. Coln[2], Douglas Chisholm[1], Denise T. Lee[1]

[1]Analog Devices, Edinburgh, United Kingdom
[2]Analog Devices, Wilmington, MA

2013 Jack Raper Award for Outstanding Technology-Directions Paper

"Experimental Demonstration of a Fully Digital Capacitive Sensor Interface Built Entirely Using Carbon-Nanotube FETs"

Max Shulaker[1], Jelle Van Rethy[2], Gage Hills[1], Hong-Yu Chen[1], Georges Gielen[2], H.-S. Philip Wong[1], Subhasish Mitra[1]

[1]Stanford University, Stanford, CA
[2]KU Leuven, Heverlee, Belgium

2013 Jack Kilby Award for Outstanding Student Paper

"A Digitally Modulated mm-Wave Cartesian Beamforming Transmitter with Quadrature Spatial Combining"

Jiashu Chen[1], Lu Ye[1], Diane Titz[2], Fred Gianesello[3], Romain Pilard[3], Andreia Cathelin[3], Fabien Ferrero[2], Cyril Luxey[2], Ali M Niknejad[1]

[1]University of California, Berkeley, CA
[2]University of Nice, Nice, France
[3]STMicroelectronics, Crolles, France

2013 ISSCC Award for Outstanding Forum Presenter

"The BSIM6 MOSFET Compact Model and Its Use for Analog and RF Design"

Christian Enz, CSEM and EPFL, Neuchâtel, Switzerland

2013 Evening Session Award

"You're Hired! The Top 25 Interview Questions for Circuit Designers"

Organizers/Moderators: **Michael P. Flynn,** *University of Michigan, Ann Arbor, MI*
John Khoury, *Silicon Labs, Austin, TX*

Panelists:

Ali Hajimiri, *California Institute of Technology, Pasadena, CA*
Beomsup Kim, *Qualcomm, Santa Clara, CA*
Marcel Pelgrom, *Consultant, Helmond, The Netherlands*
Behzad Razavi, *University of California, Los Angeles, CA*
Eric Swanson, *Cirrus Logic, Austin, TX*
Sanroku Tsukamoto, *Fujitsu Laboratories, Kawasaki, Japan*

2013 Demonstration Session Certificate of Recognition

"A Fully Integrated 8-Channel Closed-Loop Neural-Prosthetic SoC for Real-Time Epileptic Seizure Control"

Wei-Ming Chen[1], Herming Chiueh[1], Tsan-Jieh Chen[1], Chia-Lun Ho[1], Chi Jeng[1], Shun-Ting Chang[1], Ming-Dou Ker[1], Chun-Yu Lin[1], Ya-Chun Huang[1], Chia-Wei Chou[1], Tsun-Yuan Fan[1], Ming-Seng Cheng[1], Sheng-Fu Liang[2], Tzu-Chieh Chien[2], Sih-Yen Wu[2], Yu-Lin Wang[2], Fu-Zen Shaw[2], Yu-Hsing Huang[2], Chia-Hsiang Yang[1], Jin-Chern Chiou[1], Chih-Wei Chang[1], Lei-Chun Chou[1], Chung-Yu Wu[1]

[1]National Chiao Tung University, Hsinchu, Taiwan
[2]National Cheng Kung University, Tainan, Taiwan

2013 Demonstration Session Certificate of Recognition

"5.5GHz System z Microprocessor and Multichip Module"

James Warnock[1], Yuen H Chan[2], Hubert Harrer[3], David Rude[2],Ruchir Puri[4], Sean Carey[2], Gerard Salem[5], Guenter Mayer[3], Yiu-Hing Chan[2], Mark Mayo[2], Adam Jatkowski[2], Gerald Strevig[6],Leon Sigal[4], Ayan Datta[7], Anne Gattiker[8], Aditya Bansal[4],Douglas Malone[2], Thomas Strach[3], Huajun Wen[6], Pak-Kin Mak[2],Chung-Lung Shum[2], Donald Plass[2], Charles Webb[2]

[1]IBM Systems and Technology Group, Yorktown Heights, NY
[2]IBM Systems and Technology Group, Poughkeepsie, NY
[3]IBM Systems and Technology Group, Boeblingen, Germany
[4]IBM Research, Yorktown Heights, NY
[5]IBM Systems and Technology Group, Williston, VT
[6]IBM Systems and Technology Group, Austin, TX
[7]IBM Systems and Technology Group, Bangalore, India
[8]IBM Research, Austin, TX

978-1-4799-0917-9/14 $31.00 © 2014 IEEE

2014 Silkroad Award

"A 2GHz 130mW Direct Digital Frequency Synthesizer with a Nonlinear DAC in 55nm CMOS"

Taegeun Yoo, Chung-Ang University, Seoul, Korea

2013 Student-Research Preview (SRP) Award

"A Fully Self-Powered Hybrid Sheet Based on CMOS ICs and Large-Area Electronics for Large-Scale Strain Sensing"

Yingzhe Hu, Princeton University

2013 Student-Research Preview (SRP) Award (Honorable Mention)

"A 12b 50MS/S 2.1mW SAR ADC with Redundancy and Digital Calibration"

Albert Chang, Massachusetts Institute of Technology

"Design and Demonstration of Scaled MEM Relay Multipliers"

Hossein Fariborzi, Massachusetts Institute of Technology

IEEE SOLID-STATE CIRCUITS SOCIETY AWARDS

2012 Journal of Solid-State Circuits Best Paper Award

"A Blocker-Tolerant, Noise-Cancelling Receiver Suitable for Wideband Wireless Applications"

David Murphy, Hooman Darabi, Asad Abidi, Amr A. Hafez, Ahmad Mirzaei, Mohyee Mikhemar, Mau-Chung Frank Chang

2014 IEEE Donald O. Pederson Award in Solid-State Circuits

Robert G. Meyer, University of California, Berekely, CA

"For pioneering contributions to the design and modeling of analog and radio-frequency circuits."

2014 IEEE Frederik Philips Award

Henry T. Nicholas, III, Newport, CA

"For exemplary leadership and entrepreneurial vision in the commercialization of communications semiconductors that enable ubiquitous broadband connectivity."

2014 IEEE Fellows

Krste Asanovic
University of California-Berkeley, Berkeley, CA

"For contributions to computer architecture"

Andrea Baschirotto
University of Milano-Bicocca, Milano, Italy

"For contributions to analog filters"

Jan Craninckx
Interuniversity Microelectronics Center (IMEC)

"For contributions to the design of CMOS RF transceivers"

Hooman Darabi
Broadcom, Irvine, CA

"For contributions to radio frequency integrated circuits and systems"

Ichiro Fujimori
Broadcom, Irvine, CA

"For contributions to oversampled data converters and gigabit wireline transceivers"

Xicheng Jiang
Broadcom, Irvine, CA

"For development of communication systems-on-chip products"

Tanay Karnik
Intel, Hillsboro, OR

"For contributions to error-tolerant circuits and near-load voltage regulators"

Howard Luong
Hong Kong University of Science and Technology, Hong Kong, China

"For contributions to CMOS radio-frequency transceiver design"

William McFarland
Qualcomm Atheros, San Jose, CA

"For leadership in single-chip wifi radio systems-on-a-chip development"

Philip Mok
Hong Kong University of Science & Technology, Hong Kong, China

"For contributions to the design of analog power-management integrated circuits"

Daniel Radack
Institute of Defense Analyses, Kensington, MD

"For leadership in microwave and millimeter-wave integrated circuit technologies and packaging techniques"

William Redman-White
University of Southampton, Southampton, UK

"For contributions to chip design aspects of telecommunications systems and RFIC design"

Toru Shimizu
Renesas Electronics, Tokyo, Japan

"For development of integrated multi-core microprocessors with large memories"

978-1-4799-0917-9/14 $31.00 © 2014 IEEE

1.3 How Chips Pave the Road to the Higgs Particle and the Attoworld Beyond

Erik H. M. Heijne

Instrumentation Physicist, CERN PH Department, Geneva, Switzerland
also with IEAP Czech Technical University Prague & Nikhef Amsterdam

1.0 Introduction

Scientific knowledge is the basis for new technology, but in return, new technology enables progress in science. One example has been the introduction of semiconductor imagers in astronomy. Telescopes now can go into space, taking stunning pictures and opening up new wavelength windows. Another example will be described in this paper. In the field of elementary particle physics, the experiments at the Large Hadron Collider (LHC) at CERN use custom-designed CMOS chips as key components for the analog signal processing and digital information extraction, and these chips enabled significant discoveries to be made, that of the Higgs particle in particular, after only a few years of operation.

The LHC accelerates counter-circulating beams of protons to 7 TeV/c^2 energy-mass within a 27 km long, circular tunnel, ~90m underground. By way of reference, the heaviest known elementary constituent prior to LHC operation was the top quark with a rest mass of 0.17 TeV/c^2. In proton collisions at such energies new phenomena occur, but at extremely low probabilities compared to interaction processes already known. After a collision, the reaction products that emerge from the original interaction (which are particles that often penetrate through meters of material) have to be characterized, and must be contained as much as possible within the sensitive detector volume. As a consequence, the physical volume required in an experiment becomes gigantic. Such an experimental setup is composed of large sub-systems arranged in cylindrical layers around the beam crossing point, where each layer is specialized to address a different type of secondary particle. The complete volume is instrumented with small sensing elements, to determine the trajectories, the momenta, the energies and from these the identities of these particles, in order to reconstruct the details of the original interaction process. A magnetic field encompasses this detector volume so that the charge and the momentum of each particle can be determined. The deployment of integrated circuits has enabled a dense packing of sensor layers, but most importantly, it has provided the on-detector processing power necessary to select interesting events while coping with the tremendous increase in repetition rate that is needed for the extremely rare Higgs boson to be found.

The aim of the LHC experiments is to study matter at the scale of the attometer, with the hope of reaching beyond the current level of understanding, embodied in the "Standard Model". Looking backward, it is apparent that without our knowledge of quantum mechanics, we would not be living with all of the benefits of present day electronics. Thus, is it thinkable that one day this science of Higgs, supersymmetry and dark matter will in turn lead to new technology?

2.0 The Accelerator and the Experiments

Ever bigger particle accelerators concentrate energy in ever smaller volumes, probing smaller dimensions; with this energy they can materialize ever heavier particles. A particle has a mass, which at the same time is a well-defined quantum of energy following $m=E/c^2$. By its dual nature, a quantum also is a localized wave with a frequency and wavelength. TeV energy corresponds to a dimension of order 10^{-18}m (attometer). At the LHC two beams of protons are injected in opposite directions using the older complex of lower energy accelerators. Each beam actually consists of a train of 2208 proton packets travelling around the ring, usually called "bunches", which are ~30cm long and contain 10^{11} protons. The beams are kept in their orbit by 1232 superconducting 8.3Tesla dipole bending magnets, each 14.3m long. In an RF accelerating cavity, a 400MHz synchronized longitudinal wave imparts energy to the bunches, each time they pass through this cavity. Once acceleration is complete at 7TeV, a steady state is achieved, and collisions are initiated; the experiments can take data for ~10 hours, until the beam intensity becomes too low due to the losses from the collisions. The beams are squeezed to ~10μm in diameter and steered to intersect at the four points where the detectors are located, as illustrated in Figure 1.3.1. This figure also shows the sequencing of incoming and outgoing bunches, which follow each other at 25ns time intervals, equivalent to a distance of 7.5m at the speed of light. Three consecutive incoming and outgoing bunches traverse a detector simultaneously. A few tens of protons collide (actually, it is the quarks or gluons inside the proton which collide, one from each side) and break up, and the energy in a collision is materialized into a spray of secondary particles, with an extremely small probability to create new elementary constituents, including the elusive Higgs boson. Because of this low probability, interactions, (or "events") are generated at a high rate of 40MHz, but the majority of events are not interesting enough to be selected for further analysis. The voluminous detectors, with diameters up to 25m and lengths of 44m, have to contain and measure the properties of all the secondary products from each event separately, in order to determine the precise amounts of energy that were involved in a given collision. Typically, six concentric detector sub-systems hermetically surround the interaction region, and serve to identify these secondary ionizing particles, neutrons, photons, etc. Spherical "waves" of secondary particles spread out from the interaction point successively through the detector. The resulting signals must be correctly assigned to the original bunch crossing, and therefore a timestamp from the 40MHz clock is associated with each signal. This clock is derived from the 400MHz timing of the accelerator, requiring a complex timing distribution network to be installed. The timing in the different areas of these systems must be adjusted so that all measurements that belong to the same event are tagged appropriately, taking into account the speed of the particles and the propagation time of signals in the cables. Altogether, the detector sub-systems act as a giant camera that images the processes going on inside its volume at a rate of 40 million pictures per second. Every 25ns, approximately 500 million sensor elements are used to record the trajectories and other parameters of a few thousands of particles, producing a massive volume of data. One of the tasks of the electronics is to make a first reduction of the data volume by a factor ~500. The overall data-reduction process is illustrated in Figure 1.3.2. The first step, the so-called Level 1 (L1) trigger, relies on local memory placed directly on the analog signal readout chips. After this first selection, 300 Gb of data per second are transmitted from the detector. In the last step of real-time analysis, a further narrowed set of promising frames is selected for permanent storage at ~1Gb/s. A cloud computing network has been developed, the Worlwide LHC Computing Grid (WLCG), permitting physicists across the globe to access and process the data.

3.0 Silicon Now at the Heart of the Experiments

Electronics has traditionally had a profound impact on nuclear and particle physics instrumentation, but dedicated integrated circuit chips have been employed only fairly recently. An early project in 1973 to equip a gas-filled multi-wire proportional chamber (MWPC) with chips made in a 10μm PMOS process [2] had no follow up. More than a decade later, the main driver for the use of CMOS integrated signal processing chips was the introduction of segmented silicon diode arrays: the silicon microstrip detector in 1980 [3] followed by the 2D silicon pixel detector in 1991 [4]. These sensor-matrix arrays contained hundreds or even thousands of cells, each of which needs an individual signal processing chain. Initiatives in custom chip design were developed in Stanford [5] in 1986 for use in an electron-positron-collider experiment, and in 1988 by our own team at CERN [6] in collaboration with the ESAT and IMEC Institutes in Leuven, Belgium. Adopting the Mead-Conway approach, a 3μm CMOS chip named "AMPLEX" (with 16 amplifier-shaper circuits and multiplexed readout) was designed in-house at CERN and fabricated by IMECs foundry service. In the UA2 proton-antiproton collider experiment ~200 units were in operation by Fall 1988. This demonstrated the feasibility of developing custom ICs for physics experiments, encouraging several physics teams to initiate IC design efforts. Since ~1990, custom CMOS chips have been slowly introduced into particle physics experiments worldwide. One special aspect of this application is that thousands of chips must operate in parallel, and uniform characteristics are needed. Improvements in signal/noise performance, compactness, lower power dissipation and even a reduction in system cost were progressively achieved. Once the full design efforts for the LHC experiments were underway, by the end of the 1990s, it was widely accepted that dedicated CMOS circuits would

978-1-4799-0917-9/14 $31.00 © 2014 IEEE

be optimal for the signal-processing functions in the various detector sub-systems. In all of these systems, it is essential to time-tag every event to a given bunch cross-over. In addition, radiation tolerance becomes a major issue for the sub-systems which are close to the interaction region, in particular the tracking detectors. Another constraint is that power consumption needs to be minimized as all power dissipated in the hermetically closed detector volume must be evacuated by cooling systems which bring excess material to the installation, spoiling the spatial precision of the tracker. For the same reason, the use of massive current conductors must be avoided. In some places aluminum conductors are used, instead of copper, because the disturbances of particle trajectories increase with the value of atomic number Z. Reducing disturbance imposes an effort through overall reduction of passive materials in the detector volume. Very light, but stiff carbon-based composites have been used for the mechanical supports. Integrating many functions in the miniaturized silicon chips saves material, and has been a decisive factor in the adoption of CMOS in place of the previously used printed-circuit boards with discrete components.

The detectors at the four beam crossing points in the LHC all incorporate large silicon sensor arrays in their inner region. Using these Si arrays, the magnetically-bent particle trajectories can be measured with a precision down to ~10μm. Together with the thousands of associated CMOS readout chips, the heart of the detector, up to a radius of ~1m, now consists mainly of layers of silicon devices, where in earlier generations of particle physics experiments gas- or liquid-filled devices had been employed.

3.1 Segmented Silicon Sensors

In semiconductors, even a small amount of deposited energy is sufficient to liberate electrons from their bound states. Since the 1950s, high-resistivity silicon diodes have been exploited as sensors for energy spectrometry of nuclear and elementary particles such as X-rays or protons. Now, the silicon detectors with 1D or 2D segmentation have also become the workhorses of particle tracking. These detectors typically consist of a large slab of lightly doped n-type bulk Si, with a single n^+ ohmic contact on one side, and a contiguous matrix of p^+ diodes on the other. A common bias voltage is applied to all diodes in parallel, allowing the entire volume of the sensor to be depleted of free charge carriers. Usually, the connections between the diodes and the readout electronics maintain the potential of the p^+ diodes around ground. The low doping and the low defect density of the Si result in minority carrier lifetimes of milliseconds so that electrical charge signals are collected from single particles that generate electron-hole pairs along their trajectory through the bulk of the chip. One electron-hole pair can be generated by a photon provided its energy exceeds the 1.12eV bandgap, which corresponds to wavelengths shorter than 1.1μm, in the near infrared. Particles and photons (quanta) with keV or MeV energies have been found to lose (on average) 3.64eV per liberated electron-hole pair, because a large fraction of their deposited energy is dissipated as heat.

Figure 1.3.3 depicts typical sensors for particle tracking in an LHC experiment. A 20 cm^2 silicon microstrip device can have as many as 1024 parallel linear diodes, isolated by GΩ surface oxides. In the case of pixel detectors, a single detector "ladder" (bump bonded to multiple readout chips) may contain ~10^5 diodes. A cross-section of one pixel cell is shown at the bottom of Figure 1.3.3. The full chip thickness (typically ~300μm) is depleted by applying a reverse bias voltage <100V over the silicon bulk of ~10kΩcm resistivity. The passage of a swift ionizing particle through this sensitive layer then results in a charge cloud of ~$2x10^4$ excess electrons and holes. These free carriers induce a signal current in the nearby diode contacts. Once all signal charge is collected (after ~15ns for this 300μm thickness) the bulk is again practically depleted of carriers, and ready for another signal. In the quiescent state, there remains only a small (~$1nA/cm^2$) dark current. The segmentation of the sensor into microscopic cells and the incorporation of individual analog signal processors, has had four positive consequences: (i) The reduction of the sensor capacitance leads to a reduction in noise, while the signal, generated very locally, remains the same; (ii) The smaller capacitance also allows faster signal collection; (iii) More particles can be detected simultaneously; and (iv) The particle positions can inherently be determined with micrometer accuracy, benefitting from the intrinsically precise MOS lithography used for the sensor manufacturing. Some silicon detector assemblies are illustrated in Figure 1.3.4. These silicon detectors for particle tracking present geometrical stability over long periods, and a wide operating temperature range. In LHC cooling below -0°C is often applied in order to reduce degradation under irradiation.

For future generations of experiments, new concepts for the sensor matrix can be developed by exploiting the benefits offered by reduced geometry silicon technology. One possibility is to shape the diodes as pillars, made by 3D etching through the wafer [7], with the diode fields in lateral direction over a distance that can be shorter than the still fully depleted wafer thickness. The resulting signal risetime will be faster and the overall collection time can be reduced to <1ns, enabling higher rates of particle detection. Moreover, trapping of charge on radiation-induced defects can be reduced. Employing more advanced, deep submicron planar CMOS processes would also further reduce input capacitance, thus reducing noise in a monolithic detector matrix. In order to maintain fully parallel operation, the associated amplifiers would have to be optimized for amplification of fA currents or even lower.

3.2 Associated CMOS ICs

A generic schematic diagram for a detector-segment readout channel is illustrated in the upper part of Figure 1.3.3. A tiny analog signal from an incident particle (usually between 5000 and 10^5 electron charges, corresponding to 1 and 20fC) is injected at a random moment in time to the input of a transimpedance or charge-sensitive single-ended amplifier. The signal, together with the timestamp of the associated bunch-crossing time, must be stored for a few μs in a local memory, awaiting a "trigger" decision about its possible usefulness. At the 40MHz bunch-crossing frequency, typically 120 analog memory elements are needed for each sensor cell to retain the full history of all event data during the ~3μs time delay needed to reach the L1 trigger decision level. This trigger process will be described briefly in section 4.2. On receiving a positive trigger, the selected signal is digitized, encoded and transmitted outside the detector for the next level of decision. A simplified local memory scheme, with only one or two local storage cells is possible for the silicon pixel detector matrices, employed as the innermost layers of a LHC detector. Because of the small pixel dimensions (~50μm), the probability of being hit more than twice in 3μs is sufficiently low, thus only the amplitudes and timestamps of one or two particles in each pixel must be retained until the time of the L1 decision. A quite different situation occurs in the so-called calorimeter systems also deployed within the LHC experiments, whose sensor cells are of cm-scale dimensions. Either charge or light is produced in proportion to the energy deposited, and up to 18-bit digitization is needed to cover the full dynamic range. Specific details of this wide variety of customized chips cannot be discussed in the framework of this presentation, but a few aspects are outlined below.

3.3 Radiation Hardness

In the LHC, a complex radiation environment is created by the circulating beams and the interactions at the cross-overs. Simulation programs were used to determine in advance the expected radiation intensity distribution, and now during operation all around the ring instruments continuously monitor the actual radiation accumulation. In the tunnel, the main radiation component is a relatively moderate neutron flux which may cause single event upsets in logic circuits or breakdown in protection diodes. The highest radiation levels (up to hundreds of kGy per year) are generated at the interaction regions. The intensity of the ionizing radiation falls off with the distance from the collision point, and the Si-based particle-tracking systems in the center are the most exposed. The sensors deteriorate mostly through radiation-induced crystal defects in the silicon, which act as generation centers for reverse bias leakage current. Cooling helps to keep these currents at acceptable levels, below a few μA per cm^2. The effects from ionizing radiation in the CMOS chips have been minimized by using thin gate oxides, in which hardly any threshold shift occurs, and by using enclosed layout for the transistors, NMOS in particular, so that there is no leakage path under the thick isolation oxides between devices. It has been found that modern trench isolation is more radiation hard than the very sensitive LOCOS. Single-event effects are caused by nuclear interactions of neutrons or by highly ionizing particles such as alpha particles or nuclear fragments. The large amount of dense, locally-generated charge can flip a logic state, or cause a temporary conducting path in a circuit. Mitigation of these effects is obtained by various counter measures in system design that can be integrated on-chip, including redundancy, guard banding and majority voting. Further, an extensive program of chip and system qualification has been implemented, and to date no major operational problems have been caused by radiation effects in the custom-designed electronics. The electronics systems have been designed for a projected lifetime of at least 10 years, with accumulated radiation levels that may reach 10MGy.

4.0 Components of a Typical Readout Scheme for a Silicon Microstrip Detector

4.1 Front-End

The detailed implementation varies from one LHC experiment to another, but here we present one example which illustrates the main components of a readout chip for a Si microstrip detector unit. The chip named "ABCD" was designed at CERN for the ATLAS Silicon Tracker (SCT). Figure 1.3.5 shows a schematic block diagram of this 128-channel readout chip. Each sensor element is connected to a dedicated signal processing chain beginning with a front-end amplifier that optimizes the charge-transfer efficiency for the random analog current signals from the sensor, while exhibiting minimum noise at the lowest possible power. The signal peaking time for LHC applications must be <25ns, in order to avoid pile-up of signals in successive 40MHz bunch crossings. Many details of the development of such amplifiers as well as a discussion of the merits of this transimpedance approach versus a charge-sensitive configuration can be found in [8]. The ABCD chip consumes 3.1mW per channel, most of which is in the front-end transistor. Following pre-amplification and noise shaping, the output signal of the front-end is compared with a globally defined threshold. If the signal passes the threshold and the channel is not masked as being noisy (as defined by a mask register) a "1" is saved in the L1 pipeline buffer. This pipeline stores the binary information for all channels for up to 132 bunch cross-overs. If an event is considered interesting, the entire 128-bit word is passed to another buffer, and may be sent off-chip, awaiting a higher level "trigger". If this event is ultimately not triggered, the buffer will be overwritten after 133 clock cycles. The total power consumption for the 128-channel IC is ~400mW. The ATLAS experiment has $5x10^4$ of these ICs installed which process signals from $61m^2$ of Si sensors in their SCT.

4.2 Trigger Selection and Data Links

As previously mentioned, there is only a small probability of interesting events occurring in any given bunch-cross-over. Correspondingly, a trigger-selection process is therefore designed to identify potentially interesting frames. This process operates in steps, as illustrated in Figure 1.3.2 for the CMS experiment. In CMS, tracker data are stored on-chip for 2.5µs, which is sufficient for the CMS L1 trigger latency time. The trigger decision is made on processors located in an underground counting room, next to the detector cave itself, using a subset of promptly available signals. The main ingredients for this process are signals from sensors that report energetic muons seen at the outer layers of the detector, and signals of large energy deposits in the calorimeter layers. Cable transmission time via fast links accounts for ~1µs, and the L1 processing itself takes also ~1µs. A positive trigger signal is sent back to all readout systems on the detector and the selected frame is earmarked or sometimes transferred into a separate on-chip memory. Depending on the experiment, the rate of L1 accepted frames varies between 75kHz to 3MHz. These frames are further filtered, typically taking ~40ms, with surviving frames being transferred, usually via optical links to the "Event Builder" and the "Event Filter" which are located in a surface building. The links and their driver chips have been also custom-designed because close to the detector they need to be radiation hard. This last "Event Building" step, taking ~4s per event, determines in parallel streams which frames must be stored off-line. Each second, between 300 and 3000 complete frames are written to permanent storage. Besides the ASICs and FPGAs, the trigger and data acquisition processes necessitate thousands of PCs.

4.3 Digitization

The choice of the signal digitization scheme depends on the characteristics of the detector, and on the requirements for the reconstruction of the particles and energies in the interaction. For some of the finely-segmented silicon sensors, such as the ATLAS SCT, it is sufficient to record the passage of a particle with 1-bit encoding using a comparator. However, even simple binary encoding results in many silicon design challenges. Random-noise hits must remain $<10^{-5}$ even for the millions of channels in the tracker, which necessitates elaborate on-chip test facilities and DACs for threshold setting. Precise parameter matching is required for the devices that make up these amplifiers and comparators, both within the 128-channel chip and between chips. The first version of the ABCD chip suffered performance limitations due to mismatch effects and additional calibration circuits, DACs and monitoring software had to be added to the design. The advantage of encoding data as binary information is a reduction of the data volume before transmission takes place.

Precise analog information is required if one wants to make a weighted interpolation of signals in adjacent sensor cells, or if the pulse heights represent a definite energy content, such as in the calorimetry systems. Depending on the requirements, the A/D conversion is often on-chip, and can be prior to pipelining, or only after the trigger selection. As an alternative example, in the ATLAS liquid argon calorimeter system, the preamplifier output signal with a large dynamic range, is copied into three parallel shapers with overlapping gains of 1,10 and 100, which feed into separate 144-cell switched-capacitor arrays that store the complementary signals during the L1 latency period. Following trigger selection, the three signals are digitized by 12-bit ADCs, and in combination they cover a range of 18 bits. Yet another approach is used for the CMS silicon tracker, where analog drivers and fiber links are employed, and conversions are made relatively far away in the counting room.

5.0 The Pixel Detector

The development of densely-integrated hybrid pixel detectors [9] has been essential in enabling excellent pattern recognition, even with thousands of particle tracks in successive 25ns exposures. In Figure 1.3.3 a cross-section of a single cell is given; in Figure 1.3.4, a half-barrel of this subsystem is illustrated, and Figure 1.3.6 shows the hybridization of the sensor and CMOS readout chips, as well as a typical schematic diagram for the circuitry contained in each single pixel. Using a programmable clock and a so-called Time-over-Threshold (ToT) circuit, both the arrival time of a particle, and the amplitude of the charge signal from the sensor can be digitized on-chip, within the pixel cell itself. Using this low-capacitance pixel cell (typically a few tens of fF) an equivalent rms noise <100 e^- can be obtained with signals of ~10 ke^-. At this SNR of 100, a comparator threshold of $1000e^-$ can be used, such that hardly any noise hits are recorded, even with $>10^8$ pixel cells. Three of the LHC experiments have such pixel detector sub-systems installed in the most central region of the apparatus, just around the vacuum beam pipe, as can be seen in Figure 1.3.7. In this central location, the highest radiation levels are encountered, and precautions have been taken to mitigate the various radiation effects. Numerous single-event upsets have indeed been observed and corrected, but until now no extensive damage has occurred in these parts of the detectors.

Due to the negligible noise and the good track resolution, the pixel systems provide a reliable starting point for precise event reconstruction, as illustrated in Figure 1.3.8, even with tens of interactions happening in a single beam crossing. Over the first years of operation, the accelerator has performed better than expected in terms of beam quality and intensity. In the next phase of experimentation further improvements in beam intensity might result in hundreds of interactions in each crossing, during the ~1.1ns exposure time. How can one deal with this in the experiments? Tracks might be associated with their original interaction by using time-tags, also within the 1.1ns period, if a timing resolution of <100ps can be achieved. More advanced CMOS processes will help to increase speed, but it is not yet clear if the silicon sensors will be able to deliver adequately-fast signals. To resolve the increasing ambiguities, an alternative possibility may be to measure a few directional vectors, using sets of close, parallel planes, as illustrated in Figure 1.3.6, instead of the usual single spacepoints along the track. Pattern recognition may be easier with such vectors, and achievable within ~3µs so that tracking information can be available for the elaboration of the L1 trigger decision. Such pixel detector layers would have to be mounted in a dense 3D stack, and thus power consumption and cooling will become critical issues.

The pixel detectors developed for the LHC have also begun to find other applications in science and industry. These include imaging and timing measurements of various types of radiation, such as X-ray diffraction patterns in materials analysis equipment, neutron time-of-flight imaging, or space radiation dosimetry.

6.0 Installation of Chips in an LHC Detector

The manufactured chips were assembled with sensors, power and cooling on low-mass mechanical supports, as shown in the photos of Figure 1.3.4. In the case of the tracking detectors, most chips have not been packaged, but are installed chip-on-board with wire bonding connections. This approach has been chosen in order to reduce the overall amount of material inside the detector volume, and also helps in lowering the input capacitance at the amplifier inputs. The basic units were built into larger system components, often in parallel in workshops at several universities. This involved a lot of logistics worldwide but allowed for redundancy and backup. The final construction of the detector subsystems took place at CERN, before they were placed in their final position underground in the caverns on the beam crossing points as shown in Figure 1.3.1. The actual introduction of the CMS pixel detector into the center of the largely complete CMS detector is illustrated in Figure 1.3.7. A number of workers manipulating the pixel detector on the platform can just be discerned by their white helmets. The purpose of this photograph is to provide an impression of the size of the equipment.

Several hundreds of thousands of chips have been installed in each experiment, covering a wide range of subsystems and functions. In the Table shown in Figure 1.3.9 a partial overview of installed chips with their code-names is given for the two large experiments, ATLAS and CMS. The chips are divided between the signal processing devices directly connected to the sensors in the different sub-systems, and the smaller number of control and monitoring chips for each of these sub-systems.

7.0 Finding Needles in a Haystack: Use the Grid

A heavy new particle such as W, Z or H (Higgs) cannot be observed directly, because its lifetime may be as short as 10^{-25} to 10^{-22} seconds before it decays into daughter quanta. Its range during such a short time is far below what any detector can measure. The existence of the particle is revealed statistically by the repeated occurrence of a fixed amount of energy of the combined quanta coming from the original mass value. Several competing processes, all well-understood, create a continuous background and the decays of the new particle appear as an added component superimposed on this continuum. Billions of interactions must be studied in order to have sufficient statistics to confirm that an observed excess value truly represents a definite particle mass. The reconstruction of an interaction of two colliding protons uses all the signals from the detector subsystems from the relevant 25ns timeframe. From the tracking devices, in particular the pixel detector, the particles can be traced back to their common primary vertex point. Such a reconstructed event is shown in Figure 1.3.8. Outside the field of this picture, the calorimeter subsystems capture the outgoing particles, and their energies are measured from the signal amplitudes. The outermost layers of wirechambers record the muons which penetrate tens of meters of material, and whose energy can be determined from the curvature in the magnetic field.

As described previously, typically ~300 event frames (but in some experiments up to 3000) are selected for permanent storage. Subsequently the process of event analysis causes the amount of data to again increase. Particle physicists are accustomed to dealing with large volumes of data, and computer scientists at CERN have long been developing strategies to process "big data" in a distributed way. It was in this context that in 1990, Tim Berners Lee, working with Robert Cailliau, came up with protocols for the World Wide Web. From 2010 to 2012 the descendant of the WWW, the Worldwide LHC Computing Grid (WLCG) has processed 70 Petabytes(PB) of data. This Grid is organized in three Tiers (Figure 1.3.10) with hundreds of processing sites at universities and national institutes around the world. The worldwide storage capacity in this Grid is currently 180PB; when work on this Grid was started in~2000, this seemed to be a big cloud in comparison with the Web. The use of the WLCG allowed effective international collaboration amongst more than 10,000 scientists, enabling the announcement in July 2012 of the discovery of a Higgs-like boson, while the first 3-year LHC run was still ongoing. This particle appears as an excess of decay events with an energy around 125GeV, and an example of one of the decay modes is illustrated in Figure 1.3.11, taken from the ATLAS publication [10].

8.0 Working with Single Quanta

The electron was discovered in 1897, and after Ernest Rutherford showed that most of the mass is in an atomic nucleus, only two elementary particles seemed to be needed: proton and electron. Surprisingly, more elementary particles were discovered, but eventually it was realized that these are built with just the up and down quarks. At higher energies it turned out that there are two additional families of quarks. Each family has a corresponding lepton particle: electron, muon and the abnormally heavy tau (with rest mass 1.7GeV, decay time 10^{-15}s). All this is described in the "Standard Model" which leads to a new "periodic table" of elementary constituents, as illustrated in Figure 1.3.12. In our experiments, we process signals from such single particles in the detector systems. Knowledge of the particles themselves has no clear application yet; however, the instrumental expertise can be exploited in a variety of situations. One example is in X-ray imaging, where pixel detectors are used to record individual photons, and pictures can show the photon energies, realizing color X-ray imaging. Using pixel detectors, single hot electrons, with energy above a few keV become easily observable. Such an energy level can be achieved in a simple vacuum tube, and in fact, energetic electrons are everywhere, hardly noticed, all around us. Instruments now allow us to work with single electrons, with single molecules, and single atoms, objects that populate a world with dimensions below 10^{-15}m. Atoms do not scale, but there is plenty of structure inside. Could we imagine that one day we will be able to exploit properties and energy levels inside these particles, and achieve further device miniaturization? In daily life, we are not accustomed to sub-nuclear phenomena and quantum effects; however, many commercially used technologies already exploit these phenomena (such as superconductivity, lasers and entangled photon transmission), and much more might come. At the current stage, one can only speculate; applications and technological advances often emerge decades later.

From human dimensions to nano-technology involves a dimensional step of 10^{-9}. From there to the attoworld involves another step of 10^{-9}. Will this step bring some useful technology? Can one deal with quantum behavior and particle-wave duality? Watch out: TeV interactions on an attometer scale involve a high energy density and appear to be violent, analogous to astrophysics phenomena (such as supernovae or quasars) at the cosmic scale. But, even beyond this attoscale, physicists like to theorize about strings, whose dimensions would be in the realm of the Planck scale, 10^{-34}m. Thus, there remains a long path of continued miniaturization ahead, and we may wonder if we will ever perceive the ultimate building blocks of existence!

9.0 Conclusions and Outlook

CMOS technology is now used extensively in the inner region of particle physics experiments, with finely-segmented silicon-diode array sensors for the measurement of trajectories and momenta of particles in successive frames of 25ns. The tiny signals generated by these sensors are processed by tens of thousands of low-power CMOS chips, with such chips also used with other types of sensors in the outer layers of the detector. Signal amplification, data storage, selection, and digitization functions, are integrated in these chips. Thanks to this integration approach, the improvement in rate capability of the experiments has been enormous when we consider that the large bubble chamber experiments filled with liquid hydrogen, operating in the 70s and early 80s, could typically take only a single photographic 3D exposure per second.

For the processing of the massive volume of data generated by the LHC detectors, large numbers of computers and fast data links must be employed; the WLGC allows the data to be promptly distributed for analysis by various remote physics institutes. Guided by the experience obtained over the first years of LHC operation, improvements are now underway, particularly in the areas of silicon sensors and integrated electronics. Next generation CMOS technologies will allow further increases in data rates, more on-chip intelligence, and a reduced power budget. Nanometer technologies achieve such low power that information could be fully treated in the digital domain, immediately after the front-end amplifier. Also, there is growing interest in picosecond time-stamping of signals, so that tighter coincidences can be exploited.

Despite the phenomenal size and complexity of the current instrumentation built into LHC experiments, the functionality delivered by the first generation of equipment remains fundamentally basic, in the sense that most processing and feature extraction of events until now is done offline, only after all data for one "image" are collected in a computer memory. To be able to quickly recognize specific features of collisions, such as particles with unusually high energies, or patterns with special features such as "jets" of particles, it has been a long dream of physicists to embed more local intelligence into the detectors. As an analogy, modern point-and-shoot cameras are capable of quick face recognition in order to help the focusing mechanisms of the camera. It is now conceivable to incorporate sufficient low-power digital processing in particle detectors so as to recognize specific local patterns, for example high momenta, characteristic of rare types of particles. Such a new tracking detector would make use of chips that are capable of mapping correlations between patterns in close planes of sensors. Particles that are bent only very slightly by the magnetic field have very high momenta, and could be immediately identified. Unlike the point-and-shoot camera example, such future detectors will have thousands of feature extraction processors embedded in the detectors themselves, all working in parallel. For this processing to be possible within the affordable power envelope, at the required frame rate of 40MHz, and with >1.2 billion channels present in a new detector, extremely careful architectural and circuit design is needed. The availability of advanced submicron technologies makes it possible to design chips where literally each microwatt of power is critically accounted for.

Microelectronics will allow continuous improvements in the architecture, design and performance of very complex scientific instruments, thus extending the useful lifetime of the huge investment in the LHC accelerator and further enhancing the potential for new discoveries!

Acknowledgements:
The scientific endeavor of LHC was made possible by the dedication, often day and night, of thousands of people worldwide. The introduction of microelectronics was the job of hundreds, and far too little space is available here for the references that need to do them justice. Many individuals from industry and institutes of microelectronics have assisted physics teams, to become acquainted with advanced CMOS. Especially instrumental have been Willy Sansen and Gilbert Declerck from Leuven, and Eric Vittoz from CSEM and EPFL. Exchanges with scientists at the Naval Research Lab in Washington, Sarnoff-RCA in Princeton, with Sandia in Albuquerque and with LETI in Grenoble, France, have led to "normal" CMOS chips that work well also under intense radiation. The development of innovative silicon sensor matrices has been made possible by long collaborations with companies such as Philips-NL, Enertec-Schlumberger-France, Micron Semiconductor Ltd-UK and Hamamatsu Photonics-Japan. Josef Kemmer, Paul Burger, Colin Wilburn and Koei Yamamoto were critical actors in this industrial interaction. The early support by the Italian-funded LAA project led by A. Zichichi, has been one of the key factors for starting the chip design in the early phase of LHC preparation. The stimulating environment at CERN and particularly in the microelectronics group has been an essential factor in all the work, preparing for the LHC, as well as for this presentation.

References:
[1] IOP *on-line* Journal of Instrumentation JINST 3, Special Issue S08001 to S08005, 2008.
[2] J. Borel, G. Merckel, P. Meunier, "A PMOS Eight-Channel Monolithic Instrumentation Amplifier and Signal Processing Circuit", *ISSCC Dig.Tech. Papers*, pp. 124-125, Feb.1974.
[3] J. Kemmer , P. Burger, R. Henck, E. Heijne, "Performance and Applications of Passivated Ion-Implanted Silicon Detectors", IEEE Trans. Nucl. Sci. NS-29, pp.733-737, 1982.
[4] F. Anghinolfi et al., "A 1006 Element Hybrid Silicon Pixel Detector with Strobed Binary Output", IEEE Trans. Nucl. Sci. NS-39, pp.654-661, 1992.
[5] G. Anzivino et al., "First Results from a Silicon-Strip Detector with VLSI Readout", Nucl. Instr. Meth. A243 pp.153-158,1986
[6] E. Heijne, P. Jarron, "A Low Noise CMOS Integrated Signal Processor for Multi-Element Particle Detectors", ESSCIRC'88, pp.66-69, 1988
[7] S. Parker, C.J. Kenney, J. Segal, "3D - A Proposed New Architecture for Solid-State Radiation Detectors", Nucl. Instr. Meth. A395 pp.328-343,1997.
[8] J. Kaplon, P. Jarron, *Nyquist AD converters, Sensor Interfaces, and Robustness*, Chapter 10, pp. 175-199, Springer, ISBN 978-1-4614-4586-9, 2013.
[9] Erik H.M. Heijne, "Semiconductor Micropattern Pixel Detectors: A Review of the Beginnings, Nucl. Instr. Meth. A465, pp. 1-26, 2001.
[10] ATLAS Collaboration, "Observation of a New Particle in the Search for the Standard Model Higgs Boson with the ATLAS Detector at the LHC", Phys. Lett. B 716, pp. 1–29, 2012.

978-1-4799-0917-9/14 $31.00 © 2014 IEEE

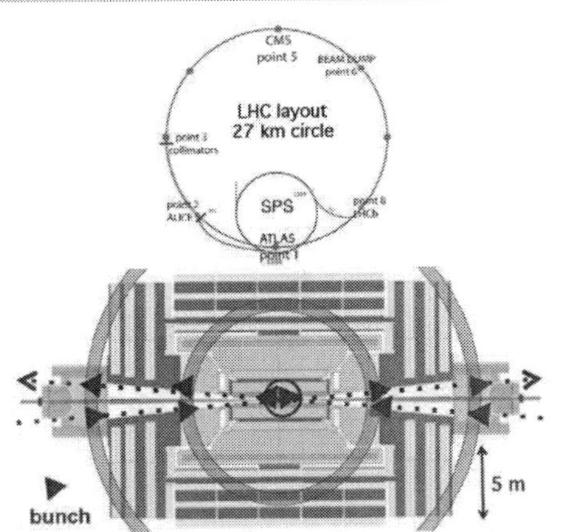

Figure 1.3.1: Schematic overview of the underground LHC installations: a cross-section of an experiment with proton bunches, cross-over-point and the successive outgoing spherical waves of secondary particles. CMS Experiment © 2013 CERN

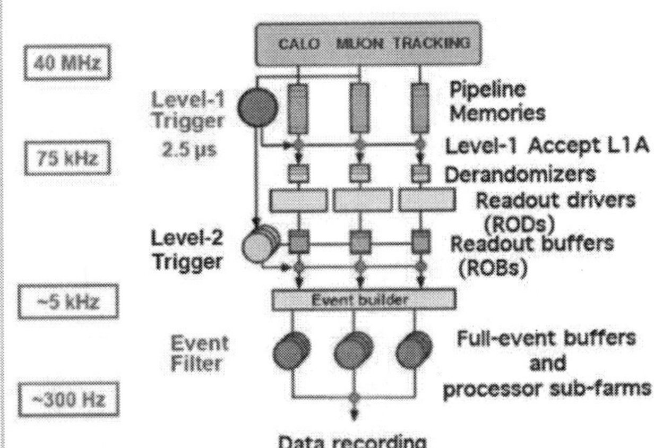

Figure 1.3.2: Successive trigger levels used to select promising exposure frames on the basis of programmable criteria in a subset of data. CMS Experiment © 2013 CERN

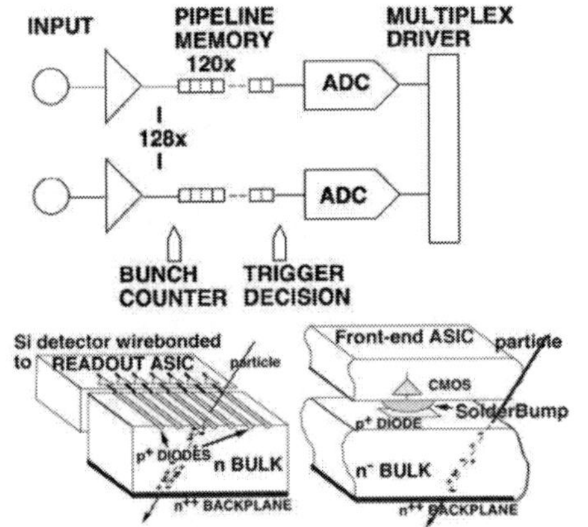

Figure 1.3.3: Bottom left: Passage of a swift particle through a Si microstrip detector with segmented p⁺-n Si diodes and fully depleted n-type bulk,with schematic amplifier connections. Bottom right: Cross-section of one pixel of a 2D matrix detector, with connection to its amplifier by a solder bump. Top: Block diagram of 128 generic readout channels on a single ASIC.

Figure 1.3.4: Photographs with partial view of a silicon detector module and the assembly of a pixel array. Left: Detector unit, showing sensor, flex connections and chips. Right: Assembly work on the outer half-barrel of the ATLAS pixel detector. The white caps towards the front are the connections for cooling pipes. © 2013 CERN

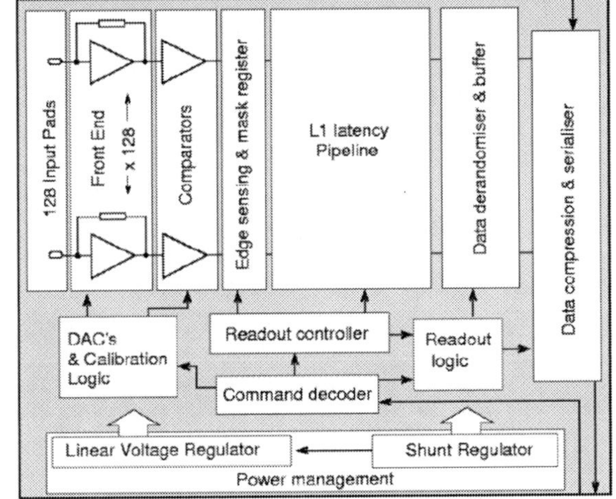

Figure 1.3.5: Schematic diagram of the readout chip "ABCD'"used for the ATLAS siicon microstrip tracker detector.

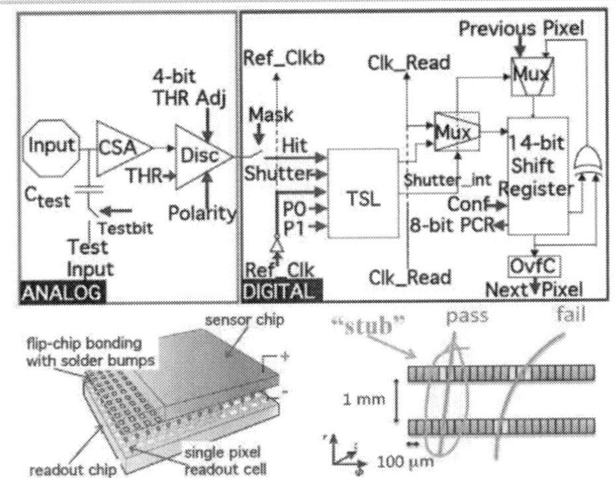

Figure 1.3.6 Schematic diagram of the circuit in a single 55µmx55µm pixel of the Timepix chip, with 256x256 cells. Stacked hybrid pixel detectors are intended to measure space vectors of the tracks, instead of points.

978-1-4799-0917-9/14 $31.00 © 2014 IEEE

Figure 1.3.7: Insertion of the pixel system into the center of the CMS experiment. A few people surrounding the device on the platform can be discerned because of their white helmets. CMS Experiment © 2013 CERN

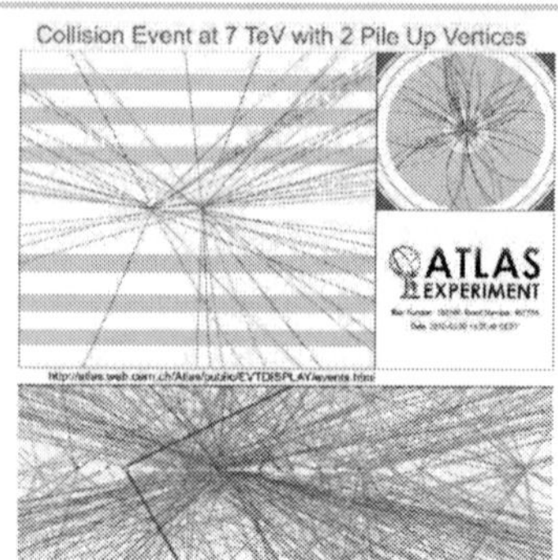

Figure 1.3.8: Reconstructed vertex points for collisions in ATLAS, with 2 and 25 different interactions. The six horizontal light bands in the top left represent the positions of the three concentric pixel detector layers at 5.0, 8.8, and 12.2 cm radial distance. The overall height is ~25cm, and only over the larger diameter shown at top right, can one see the bending of the particlesin the magnetic field. The height of the bottom reconstruction region is only ~10 cm, most of this region is inside the vacuum beamtube. ATLAS Experiment © 2013 CERN
Source http://www.atlas.ch/photos/events-collision-proton.html

TABLE 1 Installed chips in the large LHC experiments: ATLAS and CMS

Detector Subsystem	ATLAS	ATLAS	CMS	CMS
	Chip-ID	#	Chip-ID	#
Si Pixel Detector Tracker	FEI	28 000	PS146	16 800
Control & Monitoring	DORIC	2 700	TBM05	4 690
Si Microstrip Detector Tracker	ABCD	50 000	APV25	110 000
Control & Monitoring	DORIC	12 300		52 000
Gas-filled Tracker	ASDBLR	38 000		
Control & Monitoring	DTMROC	19 000		
Calorimeters (different types)		77 300	QIE8	220 400
Control & Monitoring		37 000		48 000
Muon Tracker	ASD	148 000	MAD BTI	181 034
Control & Monitoring	AMT TDC	30 000	RPC	857
TOTAL		**442 300**		**633 781**

Figure 1.3.9: Installed chips in the large LHC experiments (ATLAS and CMS)

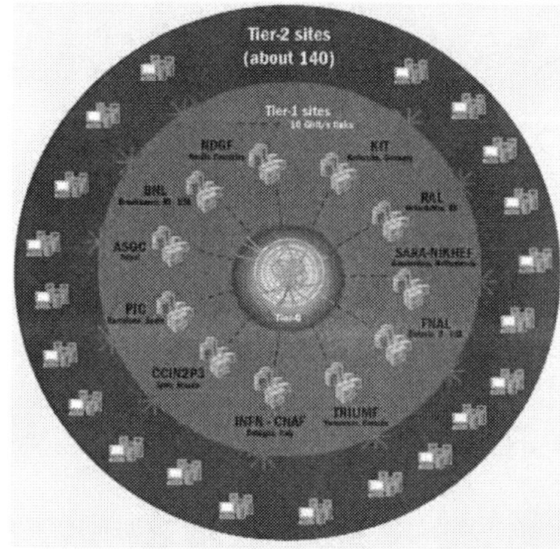

Figure 1.3.10: The Worldwide LHC Computing Grid WLCG is organized in central Tier-0 (CERN), Tier-1 (major national institutes) and Tier-2 (all participating institutes and universities worldwide).

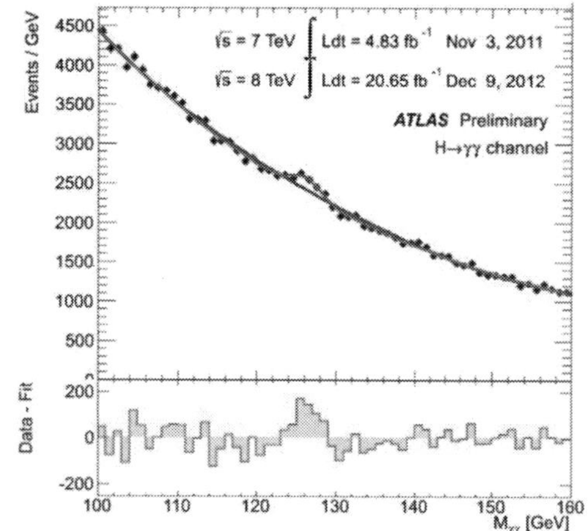

Figure 1.3.11: Histogram of energy distribution for interactions resulting in two gamma rays, showing excess due to the decay of the Higgs boson. ATLAS experiment, ref [10].

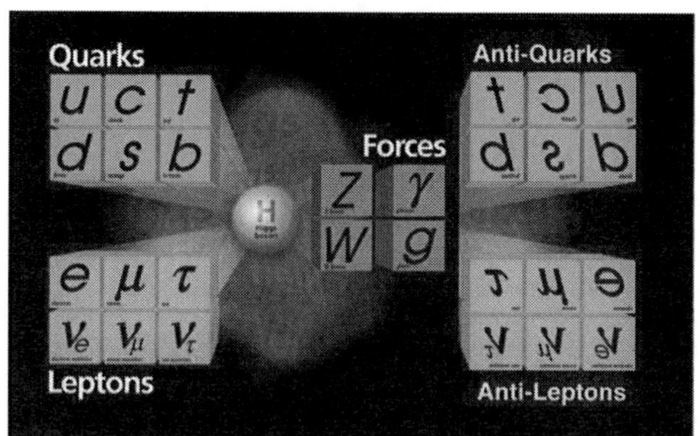

Figure: 1.3.12 Simplified periodic table of currently known elementary particles and forces.The red-blue-yellow color properties of the quarks and the gluon are not taken into account. The antiparicles are mirrored to the right, and an anti-gluon with colors is not represented separately . A graviton has not been found as yet. Source Fermilab, Batavia, IL, USA.

1.4 The Next Generation of Networked Experiences

Susie Wee, VP and CTO of Networked Experiences

Cisco Systems, San Jose, CA

Abstract

The evolution of networking technology has enabled impressive networked experiences, ranging from connected mobile context-aware experiences to immersive experiences provided by large-screen interactive displays. The rapid pace of innovation and trends in consumerization have made these experiences affordable and widely available, greatly increasing end users' expectations. In the years ahead, networks will undergo the greatest architectural transition seen in the past two decades, with the advent of software-defined networking (SDN) and the Internet of things (IoT). By taking an experience-driven approach to this architectural shift, we can understand the requirements for the underlying compute and network infrastructure, client devices, and sensors and the interaction of applications with the network. This paper discusses the next generation of networked experiences.

Introduction

Network connectivity has had a profound impact on how people communicate and access information. Networks have increasing bandwidth and reach, devices have become more capable and IP-enabled, and the cloud makes it easy to access people, content, and services from any device. Mobile devices have become a rich platform of innovation for mobile-application developers, who have application-programming-interface (API) access to device hardware, and API access to cloud services that, together, allow them to develop very compelling software applications. This rich mobile-application ecosystem has made mobile devices critical to everyone's everyday lives.

The computing industry has made a similar shift to open platforms, service-oriented architectures, and programmability. Compute virtualization has allowed elasticity and scale in the compute infrastructure, and cloud architectures have emerged for private-cloud, public-cloud, and hybrid-cloud implementations and offerings. Open-source cloud platforms such as OpenStack have emerged to allow a broader ecosystem of developers to contribute to the advancement of cloud technologies.

With the emergence of software-defined networking (SDN) programmability will become a central characteristic of network architectures, much like it has in the mobile and cloud-computing worlds. This provides the opportunity for networks to become a platform for innovation, allowing an ecosystem of software developers to develop network services and network applications that can be deployed in the network infrastructure. Network programmability provides a tremendous opportunity to improve the operational experience of networks, to increase the agility and speed of deploying new services on networks, and to tie applications to the network for improved performance and improved capabilities.

We define networked experiences as 1) the end user experiences people have with devices, applications, and services connected to the network; 2) the administrative and operational experiences associated with operating the network; and 3) the software developer's experience associated with a programmable network infrastructure, as shown in Figure 1.4.1. We describe each of these, next.

End-user applications are becoming more compelling as networks not only transport data but also provide applications with network information that can improve application performance. Today, many applications such as voice and video infer the network state, and adapt their network transmission accordingly in an over-the-top manner. SDN-based architectures will allow the network to provide this information to applications directly. End-user applications will also become richer as the network provides applications with contextual information about users, devices, and locations. For example, users often have to authenticate their devices to get access to wireless networks. This information about the user, their device, and their location, can be passed to applications that use that information to provide context-aware experiences.

Networked experiences also include the administrative and operational experience associated with configuring, managing, and operating networks. Today's networks are managed by configuring individual network devices through command-line interfaces (CLIs) that were developed over two decades ago. Since then, a number of standard and proprietary network protocols have been developed to provide information about networks and configure networks. The network programmability of SDN will advance network operations to an entirely new level. The multi-layer architecture of SDN provides various levels of abstraction of the network, ranging from network device-level abstractions to network wide-level abstractions to policy- and intent-level abstractions. Through these abstractions, SDN has the promise of simplifying the network operator's experience, but this will only be realized if simplification and ease of use are explicit design goals of SDN.

Finally, networked experiences also include the software developer's experience with creating end-user applications, management applications, and network services. The core of SDN lies in network programmability in a multi-layer API-based architecture. The APIs and target developer segments are different for each layer of the SDN architecture, as a developer who programs individual network devices can be quite different from a developer who is programming business-level policies. Enabling a developer ecosystem on a service-centric and application-centric network architecture for mission-critical networks is not a trivial design goal, as factors such as security, scalability, reliability, and performance, must be considered. However, if architected correctly, the network can become a rich platform for innovation with a vibrant developer ecosystem.

2.0 End User Experience

Perhaps the most rapid pace of innovation of the past five years has been in the end-user experiences enabled by new technology. The end-user experience has evolved along a number of dimensions as shown in Figure 1.4.2. By examining these dimensions, we can understand the opportunities for additional technology and experience innovations.

Experiences have become more mobile, more immersive, and more pervasive, as devices and applications become affordable and accessible by the masses. Advances in device technology have given the masses very capable mobile devices and immersive devices with high-quality cameras and displays, microphones and speakers, multi-touch and gestural interfaces, and impressive amounts of compute, storage, and networking capabilities. Smartphones, tablets, and even televisions, have rich application ecosystems connected to cloud services that give people pervasive access to their content and applications through these devices. Web technologies have allowed services to be easily accessible not only from your own devices but from any web browser, greatly expanding the pervasiveness of such experiences.

User experiences have become more social as new services and applications involve people from various locations. These social experiences can include asynchronous communication for messaging-based applications, such as instant messaging and social networking, and synchronous communication for voice and video conferencing. Social-networking services have become so popular that they have pushed the limits of technology in the compute infrastructure, and caused technology advancements to address their unique workloads.

Advances in networking and device technologies have introduced more ambient information through the development of sensors. This is resulting in interesting applications: these include, for fitness, wristbands with sensors to detect a person's movement and state of health; home-automation services that sense temperature, and learn people's activity patterns in the house, adapting thermostat settings accordingly. As connected sensors become more prevalent, a whole new breed of applications becomes possible – this is the area known as the Internet-of-Things (IoT). IoT creates a fundamental shift in the demands placed on the network, and the requirements for network programmability.

We will now dive deeper into two specific end-user experience and technology areas – video communication and augmented collaboration.

2.1 Use Case: Video-Communication Experience

Device technologies and networking technologies have made real-time IP-based video communications a reality. These systems let people connect with video when they are mobile through smartphones and tablets, when they are at their desk with PC software clients and desktop video-conference units, and when they are in conferences rooms using single-screen and multi-screen video-conferencing systems, as depicted in Figure 1.4.3. These video-communication systems offer a range of bandwidths and resolutions ranging from lower-data rate video used in mobile devices over wireless connections to higher-data-rate HDTV resolutions used in multi-screen room-based video-conferencing systems. Another technology advancement underway is Real-Time Communication for the Web (WebRTC), which will bring video conferencing to web browsers, furthering the pervasiveness of video conferencing.

It is now common to have a single video-conferencing session that involves participants using many of these various hardware and software clients; this enables video participants from multiple sites to be connected to the session with a wide range of device capabilities and networking bandwidths. Let us now consider the video technologies needed to suport this experience.

In sessions with large numbers of participants, rather than sending all the video streams to all the clients, it may be useful to have an MCU (multipoint conferencing unit) that detects the most recent active speakers, and sends only those video streams to the video clients. A three-screen system may show high-

978-1-4799-0917-9/14 $31.00 © 2014 IEEE

resolution video for the three most recent speakers and small thumbnails for the rest, while a mobile device may only receive one lower-resolution video of the active speaker.

Another technology that is needed is adaptive video streaming and multi-resolution video coding. Adaptive video streaming allows the bitrate of the video stream to be adapted based on the client capabilities and available network bandwidth. Various coding technologies may be used for multi-resolution video coding, ranging from simulcast video which transmits multiple-resolution streams at once to transcoding technologies which adapt higher-resolution streams to lower-resolution streams in the network to scalable video-coding technologies which code the video into layers that can be combined to create both lower- and higher-resolution video. The choice of video technology depends on system requirements, such as the compute capabilities available at the end-hosts and in the middle of the network. This provides much opportunity for technology improvements in the years ahead.

2.2 Use case: Augmented-Collaboration Experience

Collaboration is going beyond video conferencing and web conferencing to include interactive touch experiences. We are developing a next-generation collaboration experience that we call augmented collaboration, as illustrated in Figure 1.4.4. In augmented collaboration, we combine video conferencing and interactive-content collaboration into a single seamless experience. We use multiple displays tiled to provide a head-to-toe video experience. We use touch technology to allow touch interactivity. Users can share and annotate presentations, and have interactive whiteboard sessions across geographies.

The next generation of user interfaces is shifting as experiences become more social and more interactive, and this changes how we must think about I/O. Traditionally, a user interface was designed for one person using one device. But, now, networked experiences have brought on applications that allow multiple people to interact with one device, and applications that allow multiple people to participate from multiple devices.

There are many technologies and experiences contained in augmented collaboration, but in this paper we will focus only on the touch technology needed to enable this experience. Touch is an area where performance has a direct impact on the user experience, as even small lags can make applications and devices seem slow, unresponsive, and undesirable. For this reason, it is well worth the effort to optimize touch performance. To date, touch technology has primarily focused on detecting single- and multi-touch inputs on a single device. The sweet spot of touch technologies has been spearheaded in mobile devices, where capacitive touch displays provide high degrees of sensitivity and responsiveness. However, the price becomes prohibitive for large displays. At larger sizes, one may not need the same level of touch sensitivity as on a smaller personal device.

Touch sensors on mobile phones and tablets are designed for one user. In this case, the touch sensor is on the device, which has a driver that detects touch inputs and feeds them to applications. While raw touch data can be given to the application, this can provide much complexity to the software developer. It makes more sense to create a touch library that conveys touch gestures such as single tap, double tap, press and hold, two-finger pinch, two-finger swipe, and five-finger grab. Signal processing must be done to translate the raw touch data to the touch gestures that application developers can use. This processing can be done in hardware, embedded software, or application software.

Now consider the challenges that arise as we move to a large-screen touch experience with a single user. A person can use both hands and up to ten fingers for touch input. This enables a larger set of touch gestures, so additional signal processing must be done to detect these additional gestures.

As displays and touch sensors get even larger, it becomes possible to have multiple users interacting with the device. At this stage, another level of signal processing must be done to not only detect touch gestures, but also to determine which inputs belong to which users. This added level of processing is very difficult to do with today's technology in a responsive way. This creates an opportunity for technology advancements in detecting multi-user touch gestures.

Finally, let us consider collaborative touch systems that involve users collaborating in a shared session through devices in different locations. It is natural for such systems to have applications that allow collaborative whiteboarding, and collaborative annotation of documents from various locations. Users should be able to create, edit, manipulate, and erase annotations and objects on the screen. They should be able to manipulate not only the annotations that they created, but also those that others have created. This requires network coordination of touch gestures.

It is interesting to consider the touch inputs and gestures seen by the application in this scenario. The users are spread amongst various sites, but participate in a shared session. Their touch inputs must be brought together in the shared session. As mentioned previously, responsiveness is a key factor in the user experience of touch systems. However, now the touch system not only requires optimization of the touch inputs on a single device, but also on devices connected across a network. The architecture must determine what processing is done locally on the device (in hardware, embedded software, or application software) or, remotely, on another device in another location, or in a server in the cloud, and the network performance must be taken into account. The good thing is that there is higher tolerance for delay when receiving touch inputs from a remote attendee. This provides some leeway in performance.

As such, the progression of needs for touch technology has advanced from single touch to multi-touch to multi-person multi-touch to networked multi-person multi-touch experiences. These technology advances will allow touch-collaboration experiences to become pervasive across a range of devices much like video conferencing is today.

3.0 Network Administration and Operator Experience

IP networking provides end-to-end connectivity within and across enterprises, service providers, and data centers. It is useful to discuss today's IP networks to understand the various network architectures, and business and technology requirements and challenges, that arise in each of these domains, so we can understand the benefits that SDN can provide.

Many large enterprise networks and service-provider networks use the three-tier architecture shown in Figure 1.4.5. In a three-tier network, the core layer provides high-speed low-latency connectivity between several distribution/aggregation switches that may be spread amongst various locations. The core also provides connectivity to several external networks such as the Internet and wide-area networks (WANs) that may connect to other sites. The distribution/aggregation layer interfaces between the core and access layers. It provides intelligent switching and routing of packets, and connects network domains in adherence to network policy. The access layer provides network access to end-hosts such as computers, printers, tablets, and servers. It connects and provides isolation to groups of users, applications, and other endpoints according to policy. One key attribute of these networks is that the network traffic is primarily north-south traffic, carrying traffic between end-hosts up through the access, distribution, and core layers of the network. Today, an access switch may have 48 1 Gigabit Ethernet connections down to the hosts, and 2 10 Gigabit Ethernet connections up to the distribution/aggregation layer switches and routers.

Data centers often use the spine-leaf network architecture shown in Figure 1.4.6. Data centers are full of racks of servers that run applications and store data in virtualized environments. The leaf nodes are the network switches on the top of each rack connecting the servers below, and the spine nodes connect the leaf node switches to provide high-speed low-latency connectivity directly between the servers. The nature of these applications and their data access patterns causes the majority of network traffic to run east-west across the data center, as applications access other applications and data stored across the data center. Spine-leaf networks provide high degrees of connectivity across the data center. Because of the east-west traffic in data centers, leaf switches have more symmetric uplink and downlink bandwidths, for example 12 40GbE connections up to the spine switches, and 48 10GbE connections down to the servers in the rack.

Enterprise, service provider, and data center networks, can have hundreds, thousands, or tens of thousands of network devices that need to be managed and operated to deliver network traffic with reliability, performance, and scale. The means of configuring, managing, and operating today's networks are based on decades old technology, for example, the primary interface to many network devices is the command line interface (CLI) which is prone to human error – there is much room for improvement. In addition, networks need greater agility to be able to rapidly deliver new services at the speeds that businesses demand. In order to meet today's and tomorrow's business requirements, we need a fundamental shift in network architecture. The emergence of software-defined networking addresses this need.

4.0 The Emergence of Software-Defined Networking (SDN)

The emergence of software-defined networking (SDN) is bringing on the biggest architectural shift in networking of the past two decades. The definition of SDN is evolving rapidly, especially as real-world implementations of SDN are coming into play. Overall, SDN is centered on the principle of network programmability which in turn provides advantages in automating and orchestrating networks. The key principles of SDN are as follows:

- Network programmability enables software control of the network and provides automation of network configuration tasks.
- Networks are programmable at an individual network-device level, and at a network-wide level.
- Network functions are performed in software or hardware, but are controllable by software.

- Network decisions are made locally, or globally, based on performance requirements and network-device capabilities.
- Network collectors and controllers co-exist and act as interfaces to the network elements and the physical and virtual network infrastructure, through a set of southbound APIs.
- Network services that perform analytics and orchestration are written on top of the controller. Network applications can use these network services through programmatic northbound APIs.
- Cross-domain control and orchestration is done through a well-defined set of northbound APIs.
- The software-controlled infrastructure enables rapid deployment of new applications and services in the network.

Let us look at the SDN architecture in more detail:

4.1 SDN Architecture

Today's networks are made up of a physical infrastructure of network devices that perform network functions. Each network device can be accessed and configured through a command line interface (CLI) that provides detailed information about the network, and allows configuration and control of the network device. The network can be managed through a management application that interfaces with the various network devices using the CLI, and various standard and vendor-specific protocols, such as SNMP and Netflow. This is shown in the left of Figure 1.4.7.

SDN introduces a new network architecture shown on the right of Figure 1.4.7. Network devices are in the base layer of the network infrastructure. Above this is a network-interface layer that collects information from and controls the network devices within the infrastructure. This in turns provides a network-wide view and network-wide control of the network infrastructure. Above this is a network-services layer that provides basic and advanced network infrastructure, management, and orchestration services. At the top is the set of applications used to administer and operate the network. Let us look into these layers more closely.

4.1.1 Network Infrastructure

The network infrastructure embodies the physical and virtual network devices. In practice, these network devices are managed, monitored, and configured, using standardized protocols such as SNMP and NetFlow, industry-specific protocols such as CDP (Cisco Discovery Protocol), and command line interfaces (CLIs) accessed by logging into each network device. Network processors are used to add processing power to network devices. Task-specific network devices, such as firewalls are used to implement specific network tasks.

4.1.2 Network Controller

The network controller layer collects information from and controls the network devices in the network infrastructure. This information is collected and stored in a distributed database that can be accessed by the higher layer network services and network applications. This data collection and access is a critical part of the SDN architecture, as it allows the network devices to be viewed and controlled at a network-wide level.

One factor that has hampered the advancement of network operators to date is the heterogeneity of network devices, and the lack of a standard interface to these devices. One role of the network controller is to create a network abstraction that hides these differences by interfacing with the various devices, extracting network information from these devices with the appropriate protocols and interfaces, and storing the network information. This network information is accessible by APIs.

A network-information model is at the heart of the network collector/controller. The network-information model allows network information to be collected from heterogeneous devices, and stored in a way that is accessible by network services and network applications. When data in the database, such as network device configuration, is changed the controller applies the change to the network devices.

A number of network models exist, but one that is gaining momentum is Netconf/Yang. This provides a standard way to represent network information, such as topology and network-configuration settings. This defines not only the information model, but also the methods for writing and reading the information to and from the network devices.

One major advantage of the network controller is that it allows network-wide control of the network through device-by-device control. This enables a high degree of automation, which simplifies network operation, and greatly reduces network errors that are often caused by manually configuring the network.

4.1.3 Network Services

The true promise of SDN lies in the network-services layer of the SDN architecture shown in Figure 1.4.8. While the network controller performs elementary infrastructure functions to collect information from the network devices and program them, network services use the collected information and control to perform various functions (infrastructure, orchestration, and management), as shown in Figure 1.4.8. The infrastructure service functions can automate network-configuration tasks that are often done manually. The orchestration functions can perform analytics to optimize performance of the network. The management functions can represent and enforce policies in the network.

As an example, elementary infrastructure functions can perform device discovery to detect the network devices or end-hosts that are in the network; infrastructure services can generate network topology and calculate routing paths; orchestration functions can perform traffic engineering to optimize traffic flows; and management functions can represent policies for access control of network traffic, applications, users, and devices. More examples are given in Section 6.0.

In today's networks, the networking devices themselves make the majority of networking decisions. But, in the SDN architecture, while some network decisions continue to be made by the devices themselves, other decisions are made centrally by network services. Early definitions of SDN centralized all decision making of the network, however, real-world implementations are showing that this is not a scalable approach for all network decisions. Thus, the definition of SDN is evolving to include a hybrid of central and distributed decision-making based on the needs of the particular function and the capabilities of the network devices.

4.1.4 Network Applications

Network applications are built on the network services using the northbound APIs of the SDN network. Some applications are intended for use by network operators to show information about the network, and to allow actions to be performed on the network. Others are system applications that perform the data gathering, analysis, and control in a more automated manner, providing alerts when needed. The network applications can benefit from the network services available in the network. The services remove the need for every application to independently gather information about the network, interface with network elements, calculate topologies and routing tables, and perform analytics. Rather, it allows re-use of the functions provided as network services. The northbound APIs are often RESTful APIs passing information in a JSON format. The network applications contain the actual UI that a network operator would use.

5.0 Developer Experience: API-Centric Infrastructure

SDN provides network programmability through an API-centric infrastructure. There are different layers in the SDN architecture and each layer provides a different level of abstraction to the network. While highly-flexible software-development environments allow software developers to contribute to all layers, it is important to remember that networks are large-scale infrastructures that must perform well with high levels of reliability, performance, and scale. Thus, what is needed is a well-designed software platform that opens the correct level of abstraction and interfaces to developers. Similar to how products are designed for targeted customer segments, software platforms must be designed for targeted developer segments.

Figure 1.4.9 shows the developer segments for various layers of the SDN architecture. Network-application developers and IT and network administrators and operators may expose the lowest device-level APIs and the network-wide APIs provided by the controller. A high level of networking expertise is needed to ensure that the network is programmed properly, as mistakes at this level can take down the entire network. The controller APIs can provide much benefit to these developer segments as they absorb the heterogeneity of the interfaces to the individual network elements, and allow the network devices to be viewed and programmed as a whole, rather than in a device-by-device manner.

SDN classifies its APIs as "northbound" and "southbound". These are best understood by combining the controller and network services layers as shown in Figure 1.4.8. The southbound APIs are those that reach down into the network infrastructure, so that the controller can interact with the network devices. The northbound APIs are for controller and network services that are accessible by other network services, and by the higher-layer network applications.

We view that the majority of SDN-application developers will use the northbound interfaces above the network-services layer. This provides developers with a higher level of abstraction, using network services to translate the developers' desired actions into the network actions in a high-performance manner. This allows a broader range of developers to write applications on the network using higher-level constructs, such as users, devices, applications, and policies. An area under active development lies in creating languages that allow a developer to specify policy and intent with higher degrees of abstraction. Some examples will be given in the following section:

5.1 Open Daylight: An Open Platform for SDN
The SDN multi-layer architecture makes the network into an extensible software platform upon which network services and network applications can be built. Open Daylight is an open source SDN project spearheaded by a number of companies who recognized that the success of SDN requires interoperability in a multi-vendor environment, and an open architecture that allows it to be easily extended.

Open Daylight is an SDN controller framework that: 1) includes a set of northbound APIs that interfaces with applications; 2) supports a common set of southbound protocols to interface with physical and virtual devices in the network; 3) provides a framework for services and applications to be used with the controller; and 4) provides a service abstraction layer (SAL) that maps service requests to the southbound protocols that interact with the network infrastructure. It has an extensible, modular architecture that allows support for different network devices, protocols, and services to be added over time

6.0 SDN in Practice
The problems that must be solved for enterprise campus networks, service provider wide-area networks, and data-center networks are different because of their unique technology and business environments. While each has different business and technology requirements, SDN can provide value in all of these network domains.

6.1 Enterprise-Network Operators
The goal of an enterprise-network operator is to provide reliable networking services for the enterprise. It is important to understand that many enterprise networks are "brownfield" deployments with network devices that have been around for many years; these networks are expanded and upgraded as the enterprise's networking needs grow or change, for example, with a merger or acquisition. These networks have a diversity of network devices that have different features and capabilities, and even the same feature, such as QoS, may be implemented differently on each device.

Thus, a major pain point of enterprise-network operators lies in configuring, managing, and troubleshooting the network. Network operators often do this manually by using a command line interface to the network devices themselves, or by using custom software scripts that use the CLI in a more automated way. Some management tools exist, but many large enterprises have developed their own homegrown network-management tools.

As another example, it is important to note that while many interesting distributed algorithms exist for optimizing the delivery of network traffic, in reality it is very difficult to deploy them in real world networks, because of the heterogeneity in network devices and the lack of an infrastructure to deploy them. Even if advanced network features are already available on some networking devices, in brownfield deployments it can be difficult to turn them on if they are not available at the appropriate places in the network.

As networks move towards an SDN-based architecture, network programmability enables automation of networking tasks. As mentioned previously, an enterprise-network infrastructure contains a large number of routers and switches arranged in a three-tier architecture including the core, distribution/aggregation, and access networks. An SDN architecture brings programmability and ease of use in operating these three-tier networks.

At Cisco, we are building an enterprise-SDN system, as shown in Figure 1.4.10. The network controller interfaces with the network devices, collecting information from the device, forming a network-wide view, using scripted CLI as well as networking and management protocols such as NetFlow and SNMP. The network is programmed using scripted CLI and other protocols such as OpenFlow and network device-specific programmatic interfaces, for example, onePK developed for Cisco routers and switches. Elementary infrastructure functions use these interfaces to perform device discovery to collect an inventory of what devices are in the network, and loads their network information into a database. Infrastructure service functions use this information to calculate the topology of the network and routing paths in the network. We have three automation services available on this controller:

1) One of the network services is QoS. The management function stores the policy associated with applications and devices on the network so that functions such as QoS can be programmed onto the appropriate network elements based on this policy. As mentioned previously, there are many different implementations of QoS, so it is important for the controller to know the capabilities of each network element to ensure that the appropriate implementation of QoS is instantiated at each node to ensure the desired end-to-end QoS.

2) Another network service is access control. A key element of networking lies in the access-control lists (ACLs) found on every network element. ACLs are lists of entries that have a source and destination IP address and port number, traffic type, and a permit or deny designation. Network routers and switches have high-speed switching fabrics that read the ACLs to determine whether a packet can be forwarded or should be denied. Managing ACLs is one of the biggest pain points that network operators face today. The SDN ACL service allows easy viewing of ACLs across the devices in the network. It detects conflicts and shadows, and alerts them to the network operator. It also traces traffic flows between endpoints while checking the ACLs on the network devices along the path to see whether a flow will be permitted or denied. The ACL manager ensures that the ACLs are configured on the devices according to policy specifications.

3) Another network service is application visibility and control. There has long been a vision of tying applications to the network to improve the performance of the application and of the network. However, it has been difficult to realize these benefits because it is difficult to use these features in practice. The network programmability of SDN makes application-network optimizations possible. Cisco routers have a feature called application visibility and control (AVC) which in hardware identifies and classifies applications, monitors statistics of application flows, sets QoS priorities based on the applications, and dynamically chooses network paths based on performance. With SDN, we can leverage this information to optimize network and application performance.

This platform shows an example where hardware functions are leveraged in a software-controlled environment, using a desirable combination of hardware and software to achieve flexibility and performance.

6.2 Service-Provider-Network Operators
Network-service providers provide the network as a service to their customers. For example, they often provide WAN connectivity to enterprises to connect their campus, branches, and data centers, and provide broader connectivity to external networks and the public Internet. The service provider must reliably transport its customers' data according to business SLAs, as this directly impacts their revenue. Since the network is their product and the cost of the network is their main expense, managing and operating the network efficiently is of top concern.

At Cisco, we are building a service provider SDN that can be used for WAN orchestration. The system is shown in Figure 1.4.11. Service provider networks use protocols such as BGP and MPLS, and, correspondingly, use different interfaces to the network from an enterprise campus network. In our system, data collection is done using CLI, NetFlow, and SNMP, as in the enterprise campus controller, but it also uses BGP-LS to collect information about the topology and link-state information about the nodes and links. The network is controlled with protocols such as OpenFlow and onePK; in addition I2RS and PCEP are used to control the routes and paths of flows in the network.

Service providers must constantly optimize their network configuration and routing policies according to their customers' changing traffic usage and traffic demands. For example, a customer may need to run a large-scale video streaming service, or need to perform a backup operation across data centers connected by the service provider's WAN. In this scenario, the customer requests a bandwidth demand from the service provider, and the service provider performs the traffic engineering needed to deliver it successfully given the state and utilization of the network and other customers' network traffic. There is much room for traffic engineering to improve network utilization, providing a direct business benefit to the service provider.

A number of infrastructure functions are used to generate the topology and to provision and compute paths. The demand-admission service determines whether or not to admit the demand request based on the state of the network links. If the default path exceeds the capacity of one of the links, a traffic-engineering (TE) optimization service is used to identify a better route that may exist. This optimization creates traffic-engineering tunnels, and considers the resulting primary and secondary paths as options in its optimization calculations.

Another management service is bandwidth calendaring which is used to schedule regular events such as nightly data-center backups. Bandwidth-calendaring information can be used when performing demand admission, or TE optimization. This is useful as it provides optimization services with information about scheduled network traffic that may be started during the course of the requested demand.

6.3 Data-Center-Network Operators
In data centers, the goal is to provide scalable compute and storage service to customers. The data center may serve an enterprise's private cloud needs, and be owned and operated by the enterprise itself. Alternatively, a cloud service provider may offer compute and storage services to its customers as a cloud offering. A key factor of data centers is virtualization. Applications are run on virtual machines that can flexibly ramp up compute and storage resources as needed by the application.

Multi-tenant data centers virtualize resources for multiple customers on the same infrastructure. The amount of compute and storage resources allocated to a customer scales up and down according to their needs. When considering the east-west traffic patterns shown in Figure 1.4.12, it becomes obvious that the heaviest traffic will be between server racks that are allocated to the same customer. SDN provides the benefit of flexibly changing the connectivity between servers through software configuration changes, eliminating the need to physically change connections. In other words, the physical connectivity is fixed in the underlay network, but a virtual network overlay is configured through software configuration changes to adapt to the needs of the changing workloads. Network overlays provide a big benefit to data-center-network operators.

One advantage of data-center networks is that it is a more contained and homogeneous environment. As new data centers are being built, SDN can be implemented in a "greenfield" deployment where a homogeneous set of networking devices are chosen according to the needs. This greatly simplifies the controller, and allows more optimizations that can lead to improved performance.

Many optimization algorithms are used in data centers to place workloads according to the compute and storage requirements of the application. As more bandwidth intensive applications are moved to the data center, it is also important to consider networking requirements of the services and applications that are placed within the data center. This is an active area of investigation that will provide many opportunities for improved performance. For example, as mentioned previously, social-networking services have become so popular that they have pushed the limits of technology in the compute infrastructure, and caused technology advancements to address their unique workloads. Social-networking feeds provide personalized views that contain an aggregation of each person's friends' data. This significantly changes data-access patterns, and brings new requirements to the compute, storage, and networking infrastructure in data centers.

6.4 Hardware, Software, and Network-Function Virtualization

Virtualization has had a major impact in the computing world, and is now impacting the networking world, as well. The emerging area of Network-Function Virtualization (NFV) is allowing many network functions to be performed in a virtualized environment.

One highly-debated topic in SDN is the role of hardware and software. One school of thought is that SDN commoditizes network hardware as it places network intelligence in the software controller. Another view is that SDN provides software control of network functions that are performed by network services above the controller, or by the network devices themselves.

Cisco's SDN implementations emphasize the latter perspective. While virtualization can mean performing network functions in software, we view NFV as a means of abstracting and accessing network functions implemented in software or hardware. SDN provides a means of service discovery and service composition through a software-control layer. Specifically, a number of network functions are implemented in software as network services above the controller. In addition, network functions can be performed on the network elements themselves where hardware acceleration can be leveraged for performance-intensive operations. In essence, network functions can be virtualized and controlled by software, but implemented flexibly in software or hardware, leveraging all the capabilities that are available.

6.5 The Interaction between Applications and the Network

There has been a promise of application-centric networking that allows network optimizations for network traffic that takes the application knowledge into consideration. However, this advantage has been difficult to realize to date because of its deployment difficulties. Currently, applications are largely separated from the network, riding over the top of IP networks. But, we can see how the programmable multi-layer SDN architecture allows the network to be coupled to applications. The higher levels of network abstraction provided by SDN create a more natural coupling with applications. For example, consider the task of deploying a new video-conferencing system in an enterprise. SDN-management services can be used to configure dedicated ports for voice and video. Policies can be set to ensure that these applications are transmitted with a desired level of QoS. Furthermore, in service-provider networks bandwidth-orchestration services can be used to adaptively create tunnels and steer traffic for higher-bandwidth video-streaming sessions or bulk data-center operations.

7.0 The Internet of Things (IoT)

The next big fundamental shift in networking and applications will come from the emerging area referred to as the Internet-of-Things (IoT). In IoT, large numbers of sensors are brought to IP networks to collect and aggregate data that can then be used by various applications. These sensors are programmable and can be controlled by applications. Their data is analyzed to provide interesting intelligent services to end users for consumer and business applications.

For example, consider vertical areas of industry, such as manufacturing, and oil and gas. These are areas where sensors can collect information about what is happening on the plant floor and in the field, and this information can be used to monitor operations, forecast and report issues, and provide information for troubleshooting. In retail, video sensors can collect and analyze shoppers' traffic patterns to help create better store layouts of merchandise and highlight areas where a customer needs service. Transportation is another industry that will have significant impact, as traffic sensors and parking meters can detect vehicles, and redirect traffic accordingly.

Today, many of these areas are being addressed in an application-specific manner, where one company is providing the sensors, application, and service. Going forward, the sensors will become a platform in a programmable network infrastructure, where the sensors provide sensor data that can be used by multiple services. As IoT gains widespread deployment, it will fundamentally change network traffic patterns, as large amounts of data are being captured by large numbers of sensors, and this data needs to be sampled and analyzed according to each application's needs. This brings a new set of requirements to the network, including the need to have compute capabilities closer to the sensors at the edge of the network to analyze the data as it is being captured, rather than being transmitted across the network for central processing.

8.0 Final Remarks

In this paper, we have discussed the evolution of networked experiences. We have taken the approach of coupling user experience and technology, highlighting how increased expectations in user experience drive technology requirements, and how technology developments have greatly advanced user experiences. We have discussed this in terms of end-user experiences, using video communication and touch collaboration, as examples. We have also demonstrated the experience-technology relationship embodied in those network-operator experiences which focus on ease of operation, that can be achieved through the higher levels of abstraction provided by the emerging multi-layer SDN architecture.

The shift to SDN has a number of key principles, a few being: network programmability; network-wide control, rather than device-level configuration; a multi-layer SDN architecture that provides various levels of network abstraction; and a platform-based architecture that allows new network services to be flexibly deployed on the network, and new applications to be written using those services. This shift to software control provides greater agility in deploying new network services, making the network a true platform for innovation.

Acknowledgments

I would like to thank David Ward, Jeff Reed, Blue Lang, Kiran Yedavalli, Carl Solder, Jan Medved, Chris Metz, John Apostolopoulos, Qibin Sun, Zhishou Zhang, You Yi, Edwin Zhang, Zhongping Zhu, and Anantha Chandrakasan for their contributions to this work.

References

[1] "The Road to Immersive Communication", J. Apostolopoulos, P. Chou, B. Culbertson, T. Kalker, M. Trott, S. Wee, Proceedings of the IEEE, April 2012.

[2] "Software-Defined Networking: The New Norm for Networks". White paper. Open Networking Foundation. Apr 13, 2012. Retrieved Aug 22, 2013.

[3] SDN Software Defined Networks, T. D. Nadeau and K. Gray, O'Reilly, August 2013.

[4] Improving Network Management with Software Defined Networking, H. Kim and N. Feamster, IEEE Communications Magazine, Feb 2013.

[5] Network Configuration Protocol (NETCONF), R. Enns, M. Bjorklund, J. Schoenwaelder, A. Bierman, IETF RFC 6241, http://www.ietf.org/rfc/rfc6241.txt, Jun 2011.

[6] YANG – A Data Modeling Language for the Network Configuration Protocol, M. Bjorklund, IETF RFC 6020, http://www.ietf.org/rfc/rfc6020.txt, Oct 2010.

[7] Open Daylight project, http://www.opendaylight.org

[8] Frenetic: A Network Programming Language, N. Foster, R. Harrison, M. Freedman, C. Monsanto, J. Rexford, A. Story, D. Walker, ACM SIGPLAN ICFP, Sept 2011.

[9] Composing Software-Defined Networks, C. Monsanto, J. Reich, N. Foster, J. Rexford, D. Walker, USENIX conference on Networked Systems Design and Implementation, Apr 2013.

[10] NetFlow Services Export Version 9, B. Claise, IETF RFC 3954, http://www.ietf.org/rfc/rfc3954.txt, Oct 2004.

[11] NetFlow Services Solution Guide, Cisco NetFlow Collection Engine, http://www.cisco.com/en/US/partner/docs/ios/solutions_docs/netflow/nfwhite.html.

[12] North-Bound Distribution of Link-State and TE Information using BGP, H. Gredler, J. Medved, S. Previdi, A. Farrel, S. Ray, IETF draft, http://tools.ietf.org/id/draft-ietf-idr-ls-distribution-04.txt, Nov 2013.

[13] Path Computation Element (PCE) Communication Protocol (PCEP), J.P. Vasseur, J.L. Le Roux, IETF RFC 5440, http://www.ietf.org/rfc/rfc5440.txt, Mar 2009.

[14] A Demonstration of Virtual Machine Mobility in an OpenFlow network, D. Erickson, G. Gibb, B. Heller, D. Underhill, J. Naous, G. Appenzeller, G. Parulkar, N. McKeown, M. Rosenblum, M. Lam, S. Kumar, V. Alaria, P. Monclus, F. Bonomi, J. Tourrilhes, P. Yalagandula, S. Banerjee, C. Clark, R. McGeer, ACM SIGCOMM 2008.

978-1-4799-0917-9/14 $31.00 © 2014 IEEE

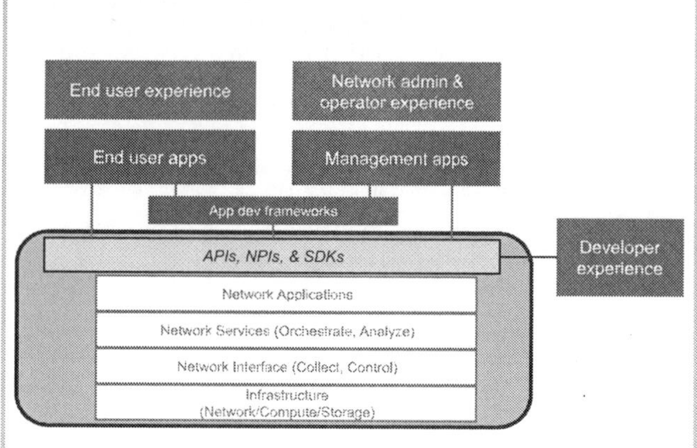

Figure 1.4.1: Networked experiences with a programmable network include: the end-user experience, network administrator and operator experience, and software developer experience.

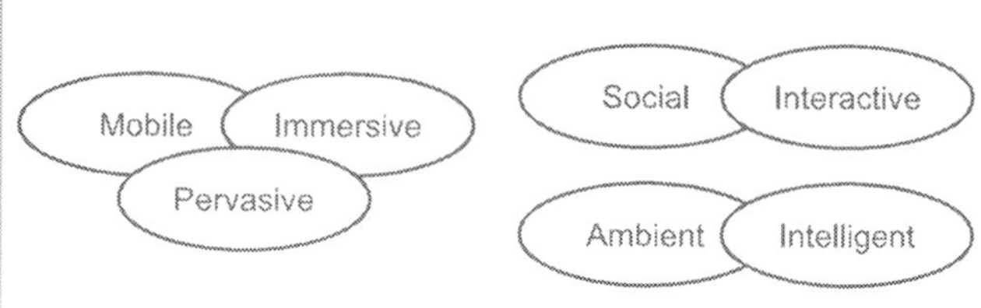

Figure 1.4.2: Technology progress has led advances in user applications and experiences in a number of dimensions.

Figure 1.4.3: Video-communication experiences have become mobile, immersive, and pervasive, through a range of software-, hardware-, and web-based clients.

Figure 1.4.4: "Augmented-Collaboration Experience" seamlessly blends video and content collaboration wtih head-to-toe video and shoulder-to-shoulder collaboration.

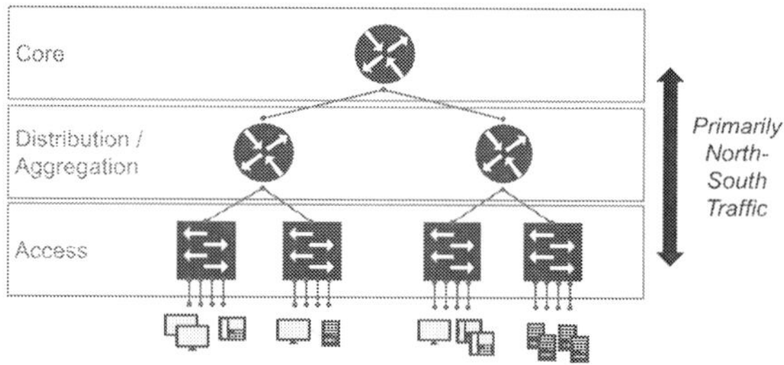

Figure 1.4.5: Enterprise and service-provider networks often use a three-tier architecture with primarily north-south traffic.

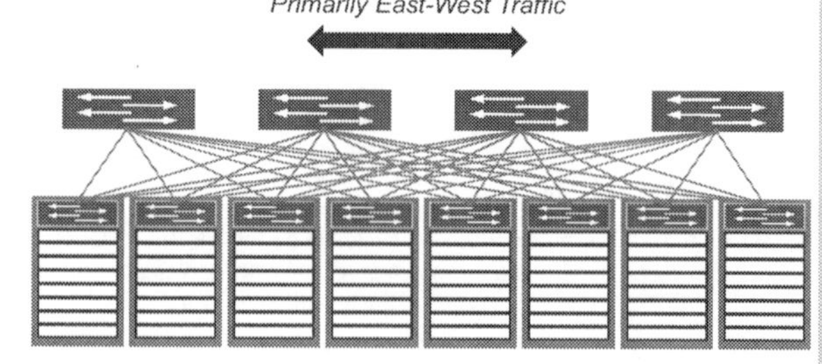

Figure 1.4.6: Data-center networks often use a spine-leaf architecture with primarily east-west traffic that travels between the data center-servers.

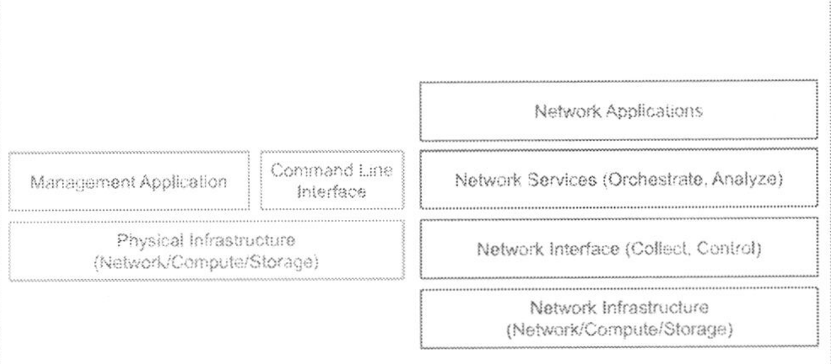

Figure 1.4.7: Today, many networks (at the left) have a physical infrastructure of networked devices coordinated by management applications that interact with the devices through a command-line interface (CLI). The SDN architecture (at the right) embodies a physical and virtual network infrastructure, a network collector, network services, and network applications.

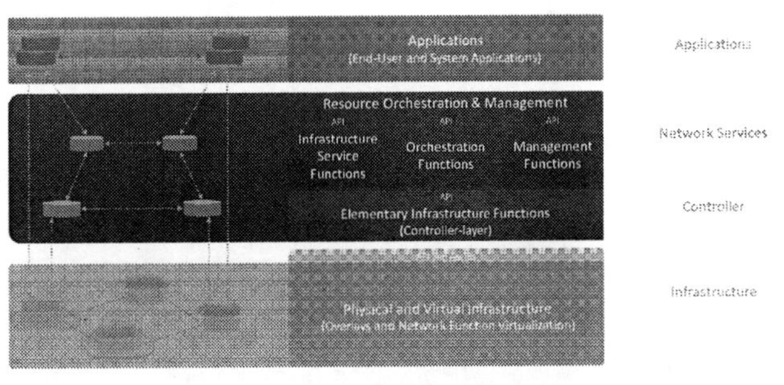

Figure 1.4.8: The network controller and network-services layers are often combined. The northbound interfaces are those that are exposed to network applications and other network services. The southbound interfaces are used to interface with the network infrastructure.

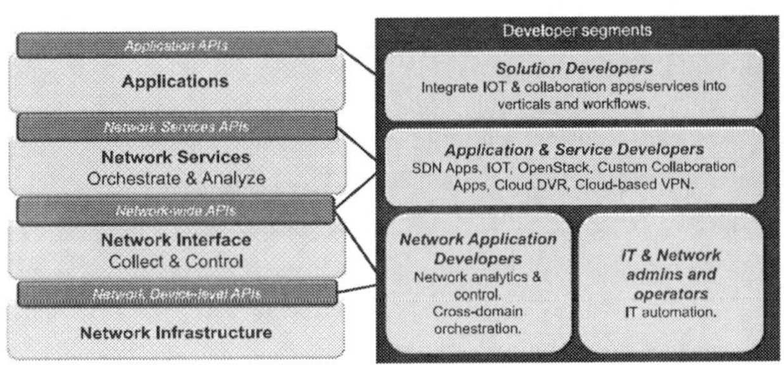

Figure 1.4.9: Various layers of the SDN architecture provide different levels of abstraction of the network. Each layer can be used by particular target developer segments depending on the needs of their application.

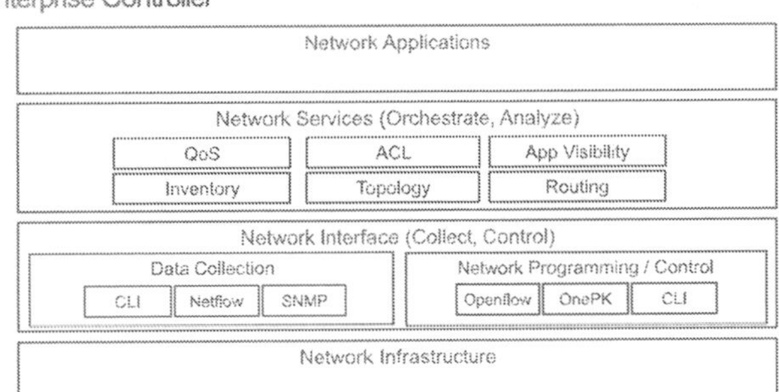

Figure 1.4.10: Cisco's enterprise SDN platform targets an enterprise operator's needs for automation. This includes: infrastructure services such as inventory, topology, and routing; management services such as QoS and ACLs; and orchestration services based on application visbility and control.

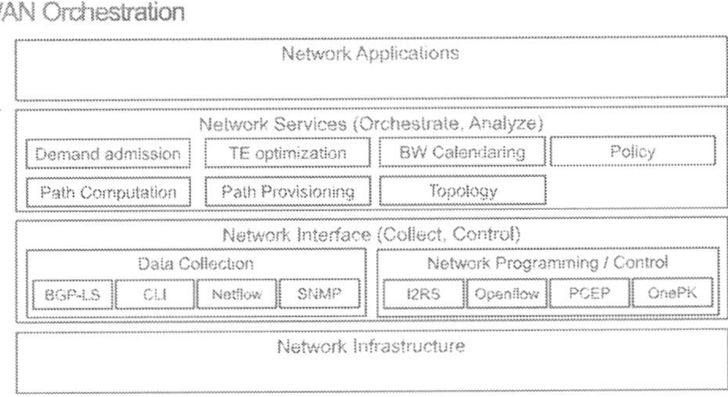

Figure 1.4.11: Cisco's service-provider WAN-orchestration SDN platform addresses the service provider's needs for bandwidth orchestration. This includes: infrastructure services for topology and path computation and provisioning; orchestration functions for demand admission, traffic optimization, and bandwidth calendaring; and a policy engine to ensure customer SLAs are being satisfied.

Session 2 Overview:
Ultra-High-Speed Transceivers and Techniques
WIRELINE SUBCOMMITTEE

Session Chair: *Ken Chang*
Xilinx, San Jose, CA

Session Co-Chair: *Koichi Yamaguchi*
Renesas Electronics, Kawasaki, Japan

The rise of cloud computing and mobile communications has driven an explosion in the need for data communication bandwidth, making efficient wireline transceivers that work at the limits of the process technology increasingly critical. Techniques to reduce the power consumed by the sophisticated equalization and clocking circuits necessary to operate over lossy electrical channels are therefore required. This session includes 9 papers in this area, addressing topics of 28Gb/s transceivers, a 60Gb/s transmitter, sub-250fJ/b equalizers at 16 to 25Gb/s, and robust TX equalization and clock generation/recovery designs.

2.1 28Gb/s 560mW Multi-Standard SerDes with Single-Stage Analog Front-End and 14-Tap Decision-Feedback Equalizer in 28nm CMOS **1:30 PM**

H. Kimura, LSI, San Jose, CA

In Paper 2.1, LSI presents a multi-standard 28Gb/s transceiver implemented in 28nm CMOS and consuming 560mW per channel. The design's AFE provides 15dB of boost, while the DFE uses a half-rate 1-tap unrolled architecture. Over a 34dB test channel, the horizontal eye margin is 0.49UI and the vertical margin is 99mV.

2.2 A 780mW 4×28Gb/s Transceiver for 100GbE Gearbox PHY in 40nm CMOS **2:00 PM**

U. Singh, Broadcom, Irvine, CA

In Paper 2.2, Broadcom describes a 4×28Gb/s transceiver supporting 100GbE/40GbE standards realized in 40nm CMOS. The TX incorporates a 3-tap FIR with output-phase adjustment, and the RX uses a half-rate CDR with a dedicated eye-monitor channel. The transceiver dissipates 780mW from a 0.9V supply and achieves a BER $<10^{-15}$ over a 20dB-loss channel at Nyquist.

2.3 60Gb/s NRZ and PAM4 Transmitters for 400GbE in 65nm CMOS **2:30 PM**

P-C. Chiang, National Taiwan University, Taipei, Taiwan
and Atilia Technology, Taipei, Taiwan

In Paper 2.3, National Taiwan University, introduces fully-integrated 60Gb/s NRZ and PAM4 transmitters in 65nm CMOS. The NRZ transmitter consumes 450mW of power and the PAM4 TX consumes 290mW, both from a 1.2V supply.

978-1-4799-0917-9/14 $31.00 © 2014 IEEE

2.4 A 25Gb/s 5.8mW CMOS Equalizer 3:15 PM

J. W. Jung, University of California, Los Angeles, CA

In Paper 2.4, UCLA presents a 25Gb/s receiver that includes a one-stage CTLE and a half-rate/quarter-rate two-tap DFE. The design is fabricated in 45nm CMOS and uses charge-steering techniques to equalize for a Nyquist channel loss of 24dB. The circuit delivers an eye opening of 0.4UI for BER $<10^{-12}$ and performs 4× demultiplexing while drawing 5.8mW from a 1V supply.

2.5 A 0.25pJ/b 0.7V 16Gb/s 3-Tap Decision-Feedback Equalizer in 65nm CMOS 3:45 PM

R. Bai, Oregon State University, Corvallis, OR

In Paper 2.5, Oregon State University describes a 16Gb/s 3-tap DFE implemented in 65nm GP CMOS that uses a charge-based sampling circuit and is capable of operating down to a 0.7V supply. The DFE achieves 0.46UI margin at BER $<10^{-12}$ over 18dB of channel loss with 0.25pJ/b energy efficiency.

2.6 A 5.67mW 9Gb/s DLL-Based Reference-less CDR with Pattern-Dependent Clock-Embedded Signaling for Intra-Panel Interface 4:15 PM

D. H. Baek, Pohang University of Science and Technology, Pohang, Korea and Samsung Electronics, Yongin, Korea

In Paper 2.6, POSTECH presents a DLL-based referenceless CDR for intra-panel interfaces. Using a pattern-dependent clock-embedded signaling scheme, clock information is transmitted with an overhead of only one bit for each data packet. The CDR is implemented in 65nm CMOS and achieves a power efficiency of 0.63mW/Gb/s at 9Gb/s.

2.7 A Coefficient-Error-Robust FFE TX with 230% Eye-Variation Improvement Without Calibration in 65nm CMOS Technology 4:30 PM

S. Han, Pohang University of Science and Technology, Pohang, KoreaIn Paper 2.7, POSTECH introduces a coefficient-error-robust transmitter architecture that uses channel loss to suppress the impact of coefficient errors. An implementation fabricated in 65nm CMOS operates at 8Gb/s over a 17.5dB-loss channel and improves the eye variation by more than 2× compared with a conventional FFE transmitter with only 27% area overhead.

2.8 A Pulse-Position-Modulation Phase-Noise-Reduction Technique for a 2-to-16GHz Injection-Locked Ring Oscillator in 20nm CMOS 4:45 PM

J-C. Chien, University of California, Berkeley, CA

In Paper 2.8, UC Berkeley/Xilinx describe a 16GHz injection-locked 20nm ring oscillator employing pulse-position modulation that reduces flicker noise through feedback control. Compared to conventional injection-locking, jitter reduces from 434 to $268fs_{rms}$. The circuit consumes 46.2mW with an area of $0.044mm^2$.

2.9 A Background Calibration Technique to Control Bandwidth in Digital PLLs 5:00 PM

G. Marzin, Politecnico di Milano, Milan, Italy

In Paper 2.9, Politecnico di Milano presents a circuit to control the bandwidth and frequency response of a digital PLL. The technique ensures that the loop bandwidth remains independent of analog parameters and does not require injection of extra signals into the loop. When embedded in a 65nm CMOS bang-bang 2.9-to-4.0-GHz PLL, it sets the bandwidth in the 100-to-2MHz range with 4% accuracy independent of input noise level.

978-1-4799-0917-9/14 $31.00 © 2014 IEEE

2.1 28Gb/s 560mW Multi-Standard SerDes with Single-Stage Analog Front-End and 14-Tap Decision-Feedback Equalizer in 28nm CMOS

Hiroshi Kimura, Pervez Aziz, Tai Jing, Ashutosh Sinha, Ram Narayan, Hairong Gao, Ping Jing, Gary Hom, Anshi Liang, Eric Zhang, Aniket Kadkol, Ruchi Kothari, Gordon Chan, Yehui Sun, Benjamin Ge, Jason Zeng, Kathy Ling, Michael Wang, Amaresh Malipatil, Shiva Kotagiri, Lijun Li, Chris Abel, Freeman Zhong

LSI, San Jose, CA

A high-speed SerDes must meet multiple challenges including high-speed operation, intensive equalization technique, low power consumption, small area and robustness. In order to meet new standards, such a OIF CEI-25G-LR, CEI-28G-MR/SR/VSR, IEEE802.3bj and 32G-FC, data-rates are increased to 25 to 28Gb/s, which is more than 75% higher than the previous generation of SerDes. For SerDes applications with several hundreds of lanes integrated in single chip, power consumption is very important factor while maintaining high performance. There are several previous works at 28Gb/s or higher data-rate [1-2]. They use an unrolled DFE to meet the critical timing margin, but the unrolled DFE structure increases the number of DFE slicers, increasing the overall power and die area. In order to tackle these challenges, we introduce several circuits and architectural techniques. The analog front-end (AFE) uses a single-stage architecture and a compact on-chip passive inductor in the transimpedance amplifier (TIA), providing 15dB boost. The boost is adaptive and its adaptation loop is decoupled from the decision-feedback equalizer (DFE) adaptation loop by the use of a group-delay adaptation (GDA) algorithm. DFE has a half-rate 1-tap unrolled structure with 2 total error latches for power and area reduction. A two-stage sense-amplifier-based slicer achieves a sensitivity of 15mV and DFE timing closure. We also develop a high-speed clock buffer that uses a new active-inductor circuit. This active-inductor circuit has the capability to control output-common-mode voltage to optimize circuit operating points.

Figure 2.1.1 shows the entire SerDes block diagram. For the receiver (RX) input stage, the use of a bump inductor improves the return-loss performance and saves die area since it is placed under the bump. Also, on-chip capacitance decouples the dc bias voltage from transmitter (TX) side and the AFE input-common-mode voltage can be set to the optimum voltage. After ac-coupling, the differential signal of long-run-length data can droop. A baseline-wander-correction (BLWC) circuit compensates the low-frequency drift. The AFE is a single-stage circuit and incorporates both the variable-gain adjustment (VGA) and linear equalizer (LEQ) functions. The buffer drives the DFE input stages. The offset-cancellation loop monitors the buffer output offset and feeds back to the AFE. The DFE uses a half-rate 1st-tap unrolled architecture with sign-sign least mean square (LMS) algorithm for tap adaptation. For a 1-tap unrolled architecture based on even/odd data path slices, 8 error latches are required, 4 per even/odd slice. However, in order to minimize power dissipation, this architecture employs one error latch per even/odd slice (total of 2 vs. 8) and allows the choice of error-latch threshold placement via programmable controls. Moreover, the error-latch threshold choices can be dynamically rotated to minimize sensitivity to pattern variations. For clock-and-data recovery (CDR), a mostly digital decimated CDR is used. We have two PLLs for different frequency ranges. Each PLL has two LC tank VCOs to generate the low-jitter clock. The PLL block generates a 2T IQ clock and sends it to RX and TX lanes. On the RX side, a phase interpolator (PI) block receives the 2T IQ clock and generates the sampling clock for DFE. There are several buffer stages before and after the PI to maintain clock quality. These clock buffers employ active-inductor circuits with a common-mode feedback (CMFB) circuit. An IQ calibration circuit is employed to maintain the phase relationship before the clocks reach the DFE latches. The DFE latches are calibrated according to their DC value with forced value onto the 1st-tap unrolling muxes. The TX block consists of a serializer and source-series-terminated (SST) driver. The clock path for TX uses a duty-cycle-distortion (DCD) correction circuit similar to the one described in [3].

The AFE is an important block not only for equalization but also in terms of robustness and power consumption. In order to address these aspects, a simple and minimum-stage construction is preferable. Hence, we use a single-stage TIA-based LEQ. The AFE needs to drive the DFE input stages where there are several data and error DFE slicer circuits connected. The TIA decouples the high load capacitance, improving the high-frequency response. Also, we develop a compact inductor and make it possible to implement in small silicon area. With the TIA structure plus compact inductor, we achieve 15dB boost at a 28Gb/s data-rate. The single-stage AFE circuit is shown in Fig. 2.1.2. It consists of NMOS differential pair (Mn1) with degeneration resistor (R_{gain}) and capacitor (C_{leq}). The gain is controlled by changing R_{gain} and the high-frequency boost is controlled by C_{leq}. This differential pair converts the input signal to a current and feeds it to a TIA stage. The TIA amplifier uses differential single-stage inverter-based circuits with tail current, i_2, to maintain the bias current. The TIA amplifier has a feedback path via the compact inductor (L_{fdbk}) and resistor (R_{fdbk}). The compact inductor uses two top metal layers to implement 800pH in 25×25μm². Also it has common-mode-feedback circuits and feedback to PMOS (Mp1). The buffer stage also has a shunt-peaking inductor and the total peaking from both AFE and buffer is 15dB. Figure 2.1.3 shows the silicon measurement of the AFE response. This includes about 2 to 3dB of package and PCB trace loss. For this measurement, we sweep the input sinusoidal signal frequency and run the adaptation without DFE. By monitoring the mean target amplitude derived from the error-latch value, an AFE response is calculated and plotted. The adaptation algorithm for the AFE peaking parameter is a simplified form of an LMS-type algorithm. Although this algorithm generally works well, the use of the error information at the data-sampling phase and past decision terms, which are also used by the DFE adaptation, can in certain situations lead to coupling of the adaptation loops when they are jointly adapted. To alleviate this potential coupling a GDA that uses error information at the transition sample is also available.

A half-rate DFE with H1-tap unrolled architecture is implemented to eliminate H1 timing from the critical path. The H2-tap is fed back to the DFE buffer and the H3 to H14 taps are fed back to the DFE summer. The slicer is the most challenging part of the DFE implementation as link performance directly depends on the slicer sensitivity. Figure 2.1.4 shows the slicer circuit diagram and clock sampling phases. The slicer has three differential inputs for the data, the DFE H3 to H14 feedback and the offset. The CK1 and CK2 generate the 3 phases: sampling phase, re-generation phase and preset phase. The mux also has the same circuit topology as the slicer circuits and this architecture achieves 15mV sensitivity. The high-speed clock buffer is a key building block and it is used in many blocks in this SerDes. The performance and power is an important contributor to overall SerDes power consumption. Figure 2.1.5 shows the active-inductor clock-buffer circuits with common-mode feedback. Conventionally, NMOS differential pair amplifiers use NMOS active inductors as loads. But this does not work with a low power-supply voltage. So we use PMOS active inductors as a load. Also, we use a common-mode control with this active inductor to implement a low-power-supply high-bandwidth clock buffer. This structure uses PMOS device (Mp) to create an active inductor with resistor, r_1, and capacitor, c_1. By adding the current source i_1, the output-common-mode voltage can be controlled by IR drop, i_1r_1. The output-common-mode voltage can be expressed by $V_{DD} - V_{GS} + i_1r_1$. The current source i_1 is controlled by the common-mode feedback circuit to maintain the optimum output voltage.

The measured eye diagram at 28Gb/s is shown in Fig. 2.1.6. The insertion loss of the test channel is 34dB at Nyquist frequency, the horizontal eye margin is 0.6UI, and the vertical margin is 110mV. Figure 2.1.6 also shows the sinusoidal jitter (SJ) tolerance data for 100G-KR standard. The transceiver is fabricated in a 28nm CMOS process and the worst power consumption is 560mW at 28Gb/s operation (Fig. 2.1.7). The die micrograph is shown in Fig. 2.1.7. The total area with 4 RX/TX lanes and 2 PLLs is 3.34mm².

References:

[1] J. Bulzacchelli, et al., "A 28-Gb/s 4-Tap FFE/15-Tap DFE Serial Link Transceiver in 32-nm SOI CMOS Technology," *ISSCC Dig. Tech. Papers*, pp. 324-325, Feb., 2012.

[2] S. Parikh, et al., "A 32Gb/s Wireline Receiver with a Low-Frequency Equalizer, CTLE and 2-Tap DFE in 28nm CMOS," *ISSCC Dig. Tech. Papers*, pp. 28-29, Feb., 2013.

[3] F. Zhong, et at., "A 1.0625-to-14.025Gb/s multimedia transceiver with full-rate source-series-terminated transmit driver and floating-tap decision-feedback equalizer in 40nm CMOS," *ISSCC Dig. Tech. Papers*, pp. 348-349, Feb., 2011.

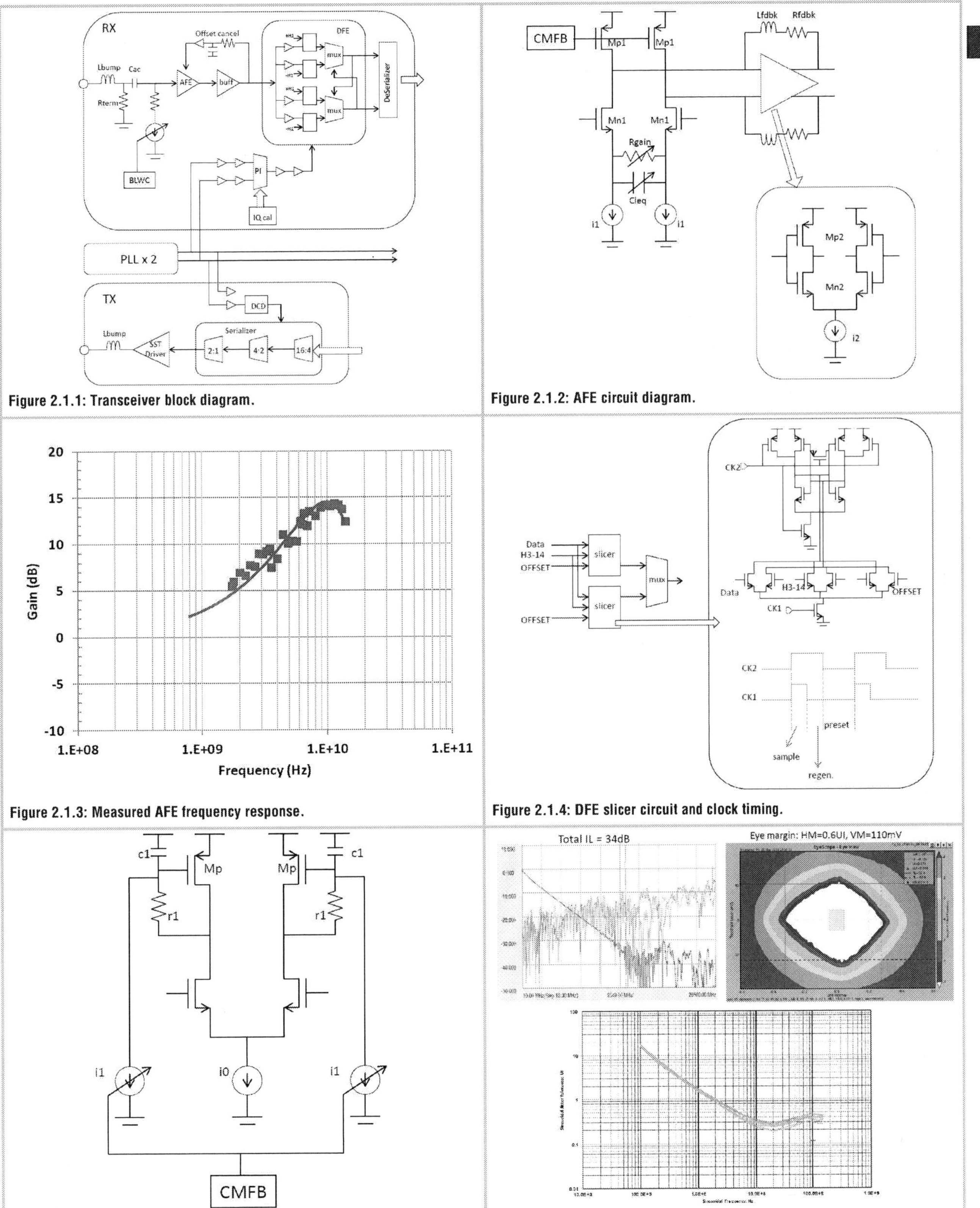

Figure 2.1.1: Transceiver block diagram.

Figure 2.1.2: AFE circuit diagram.

Figure 2.1.3: Measured AFE frequency response.

Figure 2.1.4: DFE slicer circuit and clock timing.

Figure 2.1.5: Active inductor clock buffer circuit with CMFB.

Figure 2.1.6: Measured test-channel response, eye diagram and SJ tolerance.

Technology	28nm CMOS
Power Supply	1.5V / 1.05V / 0.85V
Area (4 channel+2 pll)	3.34 mm2
Data Rate Range	1.25 ~ 28.5 Gbps
Channel Loss	30dB @ 14GHz
Measured Worst Power Consumption	560 mW (3 sigma ff corner, temp=125)

Figure 2.1.7: Transceiver summary table and chip micrograph.

2.2 A 780mW 4×28Gb/s Transceiver for 100GbE Gearbox PHY in 40nm CMOS

Ullas Singh, Adesh Garg, Bharath Raghavan, Nick Huang, Heng Zhang, Zhi Huang, Afshin Momtaz, Jun Cao

Broadcom, Irvine, CA

Network traffic speeds are increasing to meet the demands of data centers and network operators to support data-rich services like video streaming and social media. This has accelerated the adoption of 100Gb/s connectivity from the present 10Gb/s and 40Gb/s rates. One challenge that remains is the high power consumption of 100Gb/s systems. As mentioned in [1], power dissipation of the 100GbE gearbox transceiver is a significant portion of the optical module power. This paper demonstrates a low-power quad-lane 20-to-28Gb/s transceiver targeting 100GbE/40GbE (IEEE 802.3ba) standard. The transceiver features a low-jitter TX, half-rate calibrated RX slicer with folded active inductor and a wide-range PLL (20 to 28GHz) with low-power half-rate clock driver using programmable distributed inductors. It operates from a standard 0.9V supply and the power consumption for line-side transceiver is 780mW for 28Gb/s. Additionally the chipset integrates a system interface that is CAUI-compliant, composed of a 10-lane data bus operating at 9.95 to 11.2Gb/s. In default mode it converts 100GbE (10×10 Gb/s) signal to a 4×25Gb/s line signal and vice versa. The line-side interface can also be reconfigured as 40GbE, with both line- and system-side operating at 4×11.2Gb/s.

The block diagram of the 28Gb/s interface transmitter is shown in Fig. 2.2.1. It is composed of 64:4 CMOS MUX, 4:1 CML MUX and a driver. The driver is CML-based for high immunity to supply noise, eliminating the need for an LDO [2], which would require a higher supply voltage. To reduce power used by the driver, the design must reduce swing variation due to process. The driver uses a calibrated resistor to fix the load and a feedback loop to minimize current variations. The feedback loop adjusts the bias circuit to compensate process-related current-mirroring errors in the driver. As a result, the driver swing variation is limited to 10% over process corners. The output-driver stage provides multiple swing settings and a 3-tap FIR with up to 9dB post-cursor and 3dB pre-cursor boost. The final 2:1 MUX and FIR block is usually the most power hungry circuit in the TX. The main path needs to drive a large load compared to pre- and post-data path. Inductive peaking is used to increase the bandwidth and lower the power of latches and MUX. The inductor area is optimized to fit in the input-device pitch and does not increase interconnect parasitics between stages. For better ISI performance, the MAINMUX is realized using clock-up MUX as described in [3] and to reduce clock loading and power, PREMUX and POSTMUX are realized using conventional CML clock-down MUX.

The load of 14GHz clock in TX is very large due to the FIR loading. A single tuned-clock driver is used to supply the clock to the final latches and MUX. The tuned structure minimizes the power and provides large clock amplitude, which improves the ISI performance. Capacitor banks are connected to the tuned inductor to shift the resonant frequency adaptively to track the operating frequency and process variation by using VCO calibration results. Each TX also has a phase interpolator (PI), allowing for independent phase adjustment of each lane.

In most previously reported implementations [1-3], quarter-rate receiver architectures are implemented. This requires a significant amount of power to generate and distribute eight phases of clocks while maintaining precise 45° phase offset. It is more area- and power-efficient to distribute 4 phases of a half-rate clock through a resonant clock tree. As shown in Fig. 2.2.2, a half-rate CDR is used in this transceiver. Compared to a quarter-rate implementation, this reduces the number of slicers by 50%.

The RX front-end has a CTLE and a 5-stage limiting amplifier. The CTLE is implemented using programmable resistor and capacitor degeneration, providing up to 13dB boost at Nyquist. Channel equalization for the received data is performed adaptively by changing the CTLE settings. The limiting amplifier drives the slicer array of data, edge and eye-monitor channels. A CML-based slicer is used to meet the bandwidth requirements. The slicer schematic is shown in Fig. 2.2.2. It uses a folded active inductor to improve the slicer sensitivity. The gate resistance, R_{ind} can be adjusted to give peaking at high frequencies. The size of the peaking device is 15% of the main device, this causes negligible increase in output capacitance but the sensitivity is improved by almost 50%. Slicer sizes have been reduced to meet the minimum limiting-amplifier bandwidth of 20GHz. This can result in large input-referred offset. An offset-calibration scheme is implemented to automatically calibrate the offset by injecting a differential current at the output through a current DAC. The calibration occurs once upon startup.

The data stream is sampled using the 0°, 90°, 180°, and 270° clocks generated by phase interpolators. Each PI has a total of 128 steps with a step-size accuracy of 2.8°. The PI is implemented using a CML circuit similar to [3]. It is inductively peaked to drive the clocks to slicers. The peaking is optimized to give maximum clock amplitude at the frequency of operation. The PLL uses 2 LC-VCOs to generate a low-jitter clock ranging from 20 to 28GHz. Each VCO covers half the total frequency range, with only one operating at any time. Due to Q reduction at lower frequencies, a wide-range LC-VCO is usually less energy efficient than an LC-VCO with a narrow tuning range. Lower power and better phase noise can be achieved by using 2 VCOs in the PLL. The output of the 2 VCOs are multiplexed and delivered to the clock driver. A big challenge in the design is distributing the wide-range high-frequency 10-to-14GHz clock across 4 lanes of TX or RX. The traditional approach is to use a source-tuned circuit to drive the long interconnect. The major drawback is that amplitude is heavily impacted by the interconnect parasitic inductance. The clock amplitude varies along the interconnect and also drops significantly at lower operating frequencies. In this transceiver, we use a tuned circuit with distributed programmable inductors, as shown in Fig. 2.2.3. Multiple inductors are distributed along the clock path to absorb the parasitic inductance. This results in a stable amplitude at each lane. To adjust the operating frequency, the distributed inductors can be switched to a lower or higher value. This improves the clock amplitude and decrease the power consumption of the clock driver, especially at lower frequencies, compared to using switched capacitors. The power consumption of the global clock channel is 18mW, which is a 70% saving compared to previous work [1].

To achieve 4×11.2Gb/s operation on the line side, the transmitter, receiver, and clock driver need to be reconfigured. For the TX, the clock is running at 22.4GHz and duplicate data streams are sent from digital. For the RX, it operates in 2× oversampling mode, so the CDR function can be implemented using just the data-slicer path. The global-clock-driver inductor switch is turned off, thereby increasing the inductance value and lowering tuning frequency. This flexibility allows a single core to operate at 112Gb/s for 100GbE and 44.8Gb/s for 40GbE applications.

The transceiver is fabricated in a 40nm CMOS technology and is packaged in standard plastic BGA. A single power supply of 0.9V is used. The measured PLL phase noise is shown in Fig. 2.2.4. Integrating the phase noise from 10kHz to 100MHz yields a jitter of $0.16ps_{rms}$. The PLL operating range is 19 to 29GHz. The measured TX driver eye diagram, including the package and board losses, is shown in Fig. 2.2.4. The TX output achieves RJ of $0.2ps_{rms}$, $1.87ps_{pp}$ DJ, and $600mV_{pp\text{-}diff}$ amplitude with an 11ps rise/fall time at 28Gb/s. The out-of-band jitter tolerance is $0.46UI_{pp}$ at 80MHz as shown in Fig. 2.2.5. Measured RX input sensitivity is $<27mV_{pp\text{-}diff}$ at 28Gb/s. Measurements show <1.2ps of INL and 0.4ps of DNL for the phase interpolator. Using a fully adaptive equalizer, the transceiver achieves BER $<10^{-15}$ with 28Gb/s 2^{31}-1 PRBS input for a channel with 20dB insertion loss at Nyquist. This is 7dB higher than that required by the 100GbE specification. The entire 4×28Gb/s transceiver consumes 780mW. A performance comparison with prior publications is summarized in Fig. 2.2.6. The measured RJ and DJ in TX, input sensitivity, jitter tolerance and loss equalization capability are better than previous state-of-the-art implementations [1,2,4]. This performance is achieved with the lowest power consumption. The die micrograph in Fig. 2.2.7 shows RX and TX separately.

References:

G. Ono, et al., "A 10:4 MUX and 4:10 DEMUX Gearbox LSI for 100-Gigabit Ethernet Link", *ISSCC Dig. Tech. Papers*, pp. 148-150, Feb. 2011.

M. Harwood, et al., "A 225mW 28Gb/s Serdes in 40nm CMOS with 13dB of Analog Equalization for 10GBASE-LR4 and optical transport lane 4.4 applications", *ISSCC Dig. Tech. Papers*, pp. 326-327, Feb. 2012.

B. Raghavan, et al., "A Sub-2W 39.8-to-44.6Gb/s Transmitter and Receiver Chipset with SFI-5.2 Interface in 40nm CMOS", *ISSCC Dig. Tech. Papers*, pp. 32-33, Feb. 2013.

Jhih-Yu Jiang, et al., "100Gb/s Ethernet Chipsets in 65nm CMOS Technology", *ISSCC Dig. Tech. Papers*, pp. 120-121, Feb. 2013.

978-1-4799-0917-9/14 $31.00 © 2014 IEEE

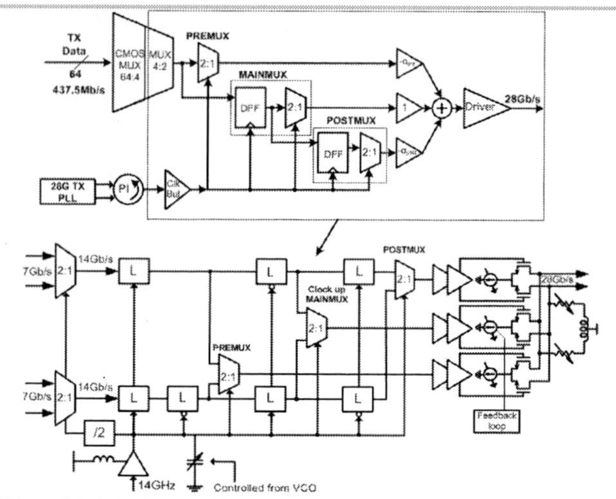

Figure 2.2.1: 28Gb/s CMOS transmitter block diagram and 2:1 MUX/driver schematic.

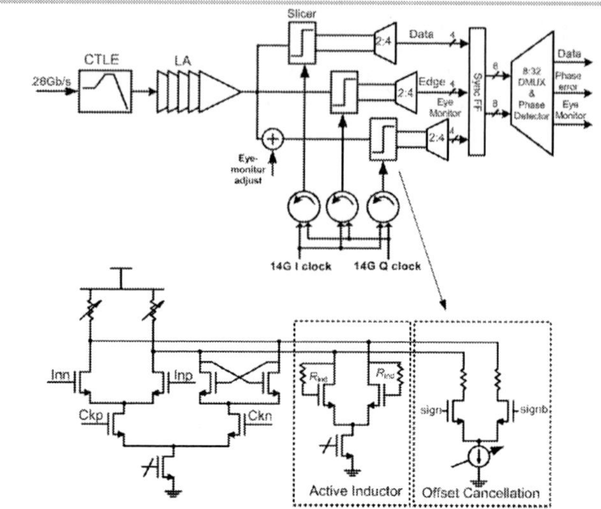

Figure 2.2.2: 28Gb/s CMOS receiver block diagram and slicer schematic.

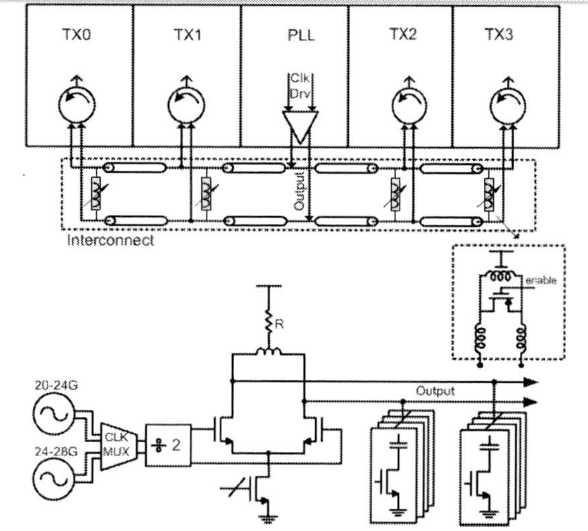

Figure 2.2.3: Distributed programmable tuned clock distribution scheme.

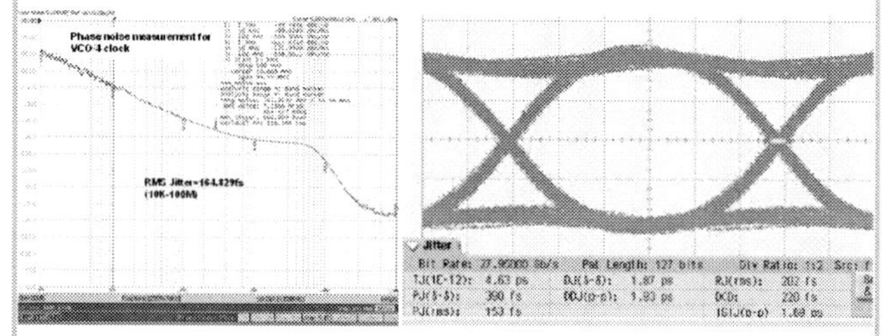

Figure 2.2.4: Measured PLL clock spectrum and transmitter eye diagram at 28Gb/s.

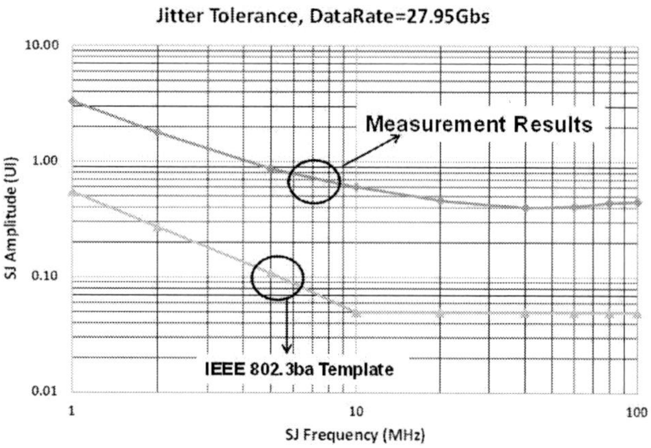

Jitter Tolerance, DataRate=27.95Gbs

Figure 2.2.5: Measured receiver jitter tolerance at 28Gb/s.

		[1]	[2]	[4]	This work
Process Technology		65nm CMOS	40nm CMOS	65nm CMOS	40nm CMOS
PLL RJ_{rms} (ps)		429fs* (10kHz to 100MHz)	350fs (100kHz to 1GHz)	187fs* (100Hz to 1GHz)	165fs (10kHz to 100MHz)
TX	DJ_{pp}(ps)	3.3*	NA	NA	1.87
	TJ_{pp}(ps) BER= 10^{-12}	NA	NA	6.67*	4.63
RX Jitter Tolerance (UI_{pp}) @ 80MHz		NA	0.4	0.2*	0.46
RX Input Sensitivity ($mV_{pp-diff}$)		34.4*	NA	NA	27
Channel loss compensated (dB) @ Nyquist (BER = 10^{-15})		NA	-13	NA	-20
Power of 28G Interface (W)		1.4*	0.9	1.84*	0.78
Supported Data Rate (Gb/s)		25	25 to 28	25	20 to 28

* Measured at 25.78Gb/s.

Figure 2.2.6: Performance comparison of 28Gb/s transceiver.

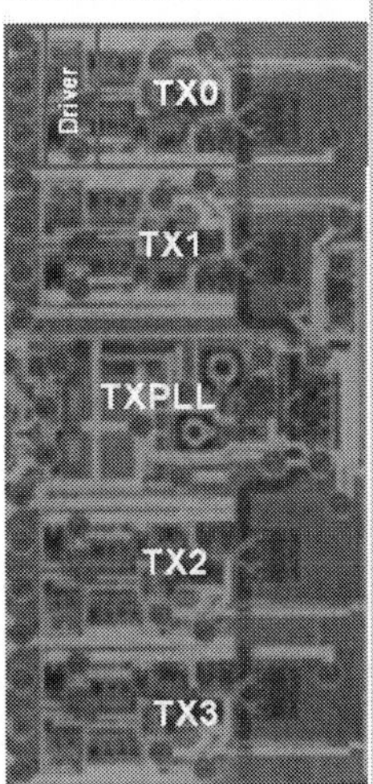

Figure 2.2.7: Chip micrograph of 28Gb/s interface.

2.3 60Gb/s NRZ and PAM4 Transmitters for 400GbE in 65nm CMOS

Ping-Chuan Chiang[1,2], Hao-Wei Hung[1], Hsiang-Yun Chu[1], Guan-Sing Chen[1], Jri Lee[1,2]

[1]National Taiwan University, Taipei, Taiwan,
[2]Atilia Technology, Taipei, Taiwan

Recent research indicates that data-link transceivers running at or below 40Gb/s are practical to implement in CMOS technology [1]. However, next-generation datacom and telecom systems require transceivers to operate at even higher data rates. For example, a 400Gb/s Ethernet system may need 8×50Gb/s PAM2 (NRZ) or PAM4 channels [2]. This paper introduces fully integrated solutions for NRZ and PAM4 transmitters. The 60Gb/s operating speed demonstrates sufficient bandwidth even for standards with coding overhead.

Figure 2.3.1 illustrates the NRZ transmitter architecture. It consists of a 4:1 multiplexer in a tree structure, a 60GHz PLL with adaptive phase aligner to optimize the clock phase in the last stage, and a built-in quarter-rate PRBS generator to facilitate testing. There are two independent modes to select the incoming data and clock by means of switches Sel_1 and Sel_2. In normal operation, the 4 input data ports are fed by 4 external independent data sequences (15Gb/s each). In self-testing mode, on the other hand, the inputs come from the built-in quarter-rate 2^7-1 PRBS generator. Similarly, the synchronizing clock can be selected from either from the internal PLL or from the external clock source. At 60Gb/s, the phase relationship between clock and data is critical to a functioning system. In the first multiplexing stage, delays ΔT_1 and ΔT_2 are inserted to balance the sample timing. These delays are designed to match the internal skews over a wide temperature range. At 60Gb/s, the phase-alignment issue becomes so severe that a static delay does not work. For instance, the acceptable sampling window in the last stage (60Gb/s output) is about 8 to 10ps, but the phase drifting caused by PVT variations can be as large as 15 to 20ps. To accommodate the random phase relationship, we put a phase aligner in front of the second multiplexing stage to dynamically track the optimal clock and data phases. The phase tracking operates as follows. First, the synchronization clock (wherever it comes from) is divided by 2 to generate quadrature clocks at 30GHz. The data transition is examined by using a roughly 16.5ps delay ΔT_3 with a mixer (M1) to detect the arrival of the internal 30Gb/s data. With the help of the 30GHz phase interpolator (PI) and the second mixer (M2), we arrive at a feedback loop that forces the PI to produce the clock phase that aligns with the data transition. As a result, the 60Gb/s multiplexer can properly sample and serialize the data, even under extreme PVT variations.

The final 2:1 selector stage needs to provide wide bandwidth and reasonable gain. As shown in Fig. 2.3.2, it is made of a 3-stage distributed amplifier with CML switching pair in the bottom. Two data inputs and one switching clock are applied into transmission lines, travelling along the 3 stages until the end terminations. The output ports are also connected through transmission lines. Here, one end is terminated while the other is open as an output port. The final D_{out60} can be ac coupled to external loading with 50Ω termination. Simulation shows that the output matching S_{22} is kept below −10dB from dc up to 70GHz. The data (L_G) and clock (L_C) paths are designed to have characteristic impedance of 50Ω as well, and the two paths have identical group velocity. Gate capacitances are absorbed into the transmission lines. It is essential to properly choose the number of stages (N) to achieve good performance. Taking into consideration the transmission-line loss and active device g_m, we determine that a 3-stage structure provides the best performance. As illustrated in Fig. 2.3.3(a), the total gain starts to roll off as N becomes larger than 3, whereas the overall power dissipation continues to increase. To avoid long routing, gate transmission lines are realized as lumped inductors in congested areas.

The PAM4 transmitter design is depicted in Fig. 2.3.4. It includes a built-in PLL for clock generation, and a two-path half-rate FFE with 3 taps and ×2 weighting factor. The original data is split into two sequences (D_{inA} and D_{inB}) of 28Gb/s, which are pre-emphasized (with the same coefficients) before combination. To ensure signal integrity, all high-speed paths are realized as transmission lines or equivalent peaking circuits. A key component that significantly affects the performance is the combiner (i.e., output driver). At tens of GHz, large-area elements such as inductors can no longer be considered lumped components, but rather distributed devices. In that sense, the peaking and signal-travelling circuits must be combined as a distributed network so as to minimize skews, reflection, and other non-idealities. Figure 2.3.3(b) reveals the combiner design. Here, peaking inductors L_D and L_G are inserted between taps to (1) absorb the gate and drain capacitance, (2) balance the travelling time. Back to the transmitter architecture in Fig. 2.3.4, we have the master clock designed in a way that it can be either provided externally or generated by the PLL (by switch Sel). The 28GHz nominal frequency is for some standard requirements defined in [3]. Again with the help of a SSB mixer-based PFD [4] and a sub-harmonic injection locking technique [5], the PLL provides a pure clock from 26.9 to 28.5GHz with jitter as low as $508fs_{rms}$. Based on the design of the inside matching network, the transmitter supports a wider operating range via the external clock. Simulation indicates the internal peaking and transmission lines behave well from dc to 105GHz, and the transmitter is verified by measurement to provide a flat data response from 1Gb/s to 60Gb/s. The tail currents in combiner II are twice as much as those in combiner I to realize PAM4 waveforms.

Both TX circuits are designed and fabricated in 65nm CMOS technology. The NRZ transmitter consumes 450mW of power and the PAM4 TX 290mW, both from a 1.2V supply. Figure 2.3.5 shows the measurement results for the NRZ transmitter. The 30Gb/s and 60Gb/s outputs are shown in Fig. 2.3.5(a) and (b), respectively, presenting output magnitude of 100mV with open eyes. The rms data jitter of 30Gb/s output measured from oscilloscope is equal to 1.08ps, and its peak-to-peak data jitter is measured as 5.33ps. The 30GHz clock output from the built-in PLL is also recorded as shown in Fig. 2.3.5(c). It presents rms jitter of 461fs (integrated from 1kHz to 20MHz offset), and phase noise of −100dBc/Hz at 1MHz offset. The PAM4 transmitter is also tested thoroughly. Figure 2.3.6(a) depicts the output waveform and phase noise plot of the 28GHz built-in PLL. It shows an integrated rms jitter (from 100Hz to 1GHz offset) of the divided-by-2 clock (i.e., 14GHz) of 508fs, and −98.5dBc/Hz phase noise at 1MHz offset. Using an external clock, we confirm the output waveform at different data rates. The PAM4 TX is verified to operate from less than 1Gb/s to 62Gb/s. Figure 2.3.6(b) and (c) reveal the PAM4 waveform at 32 and 60Gb/s, implying rising/falling time (20-to-80%) of 12.8ps and minimum eye opening of 50mV. The sharp transition and clean eyes ensure proper data delivery. Figure 2.3.7 shows the die micrograph of the two transmitter chips, which occupy 2.1×1.0mm² and 1.2×0.95mm², respectively. A table summarizing the performance of this work and that of other state-of-the-art transmitters is shown in Fig. 2.3.7 as well.

Acknowledgment:
The authors thank TSMC university shuttle program and National Chip Implementation Center (CIC) for chip fabrication.

References:

[1] J. Jiang et al., "100Gb/s Ethernet Chipsets in 65nm CMOS Technology," *ISSCC Dig. Tech. Papers*, pp. 120-121, Feb. 2013.
[2] S. Zhai et al., "The Requirement Analysis of 400GE FEC for Gen1 PMDs," *IEEE 400Gb/s Ethernet Study Group*, July 2013. [Online]. Available: http://www.ieee802.org/3/400GSG/public/13_07/zhai_400_01_0713.pdf
[3] *40 Gb/s and 100 Gb/s Ethernet Task Force.* [Online]. Available: http://www.ieee802.org/3/ba/index.html
[4] Jri Lee et al., "A 75-GHz Phase-Locked Loop in 90-nm CMOS Technique," *IEEE J. Solid-State Circuits*, vol. 43, pp. 1414-1426, Jun. 2008.
[5] Jri Lee et al., "Study of Subharmonically Injection-Locked PLLs," *IEEE J. Solid-State Circuits*, vol. 44, pp. 1539-1553, May 2009.
[6] D. Yamazaki et al., "A 25GHz Clock Buffer and a 50Gb/s 2:1 Selector in 90nm CMOS," *ISSCC Dig. Tech. Papers*, pp. 240-241, Feb. 2004.
[7] K. Kanda et al., "A Single-40Gb/s Dual-20Gb/s Serializer IC with SFI-5.2 Interface in 65nm CMOS," *ISSCC Dig. Tech. Papers*, pp. 360-361, Feb. 2009.
[8] C. Menolfi et al., "A 25Gb/s PAM4 Transmitter in 90nm CMOS SOI," *ISSCC Dig. Tech. Papers*, pp. 72-73, Feb. 2005.

978-1-4799-0917-9/14 $31.00 © 2014 IEEE

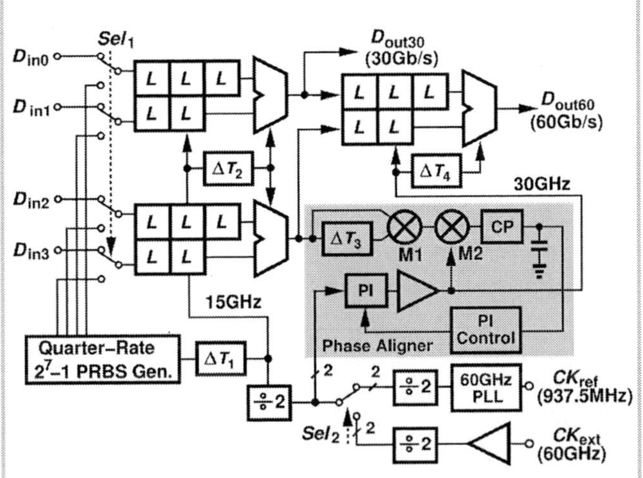

Figure 2.3.1: NRZ TX architecture.

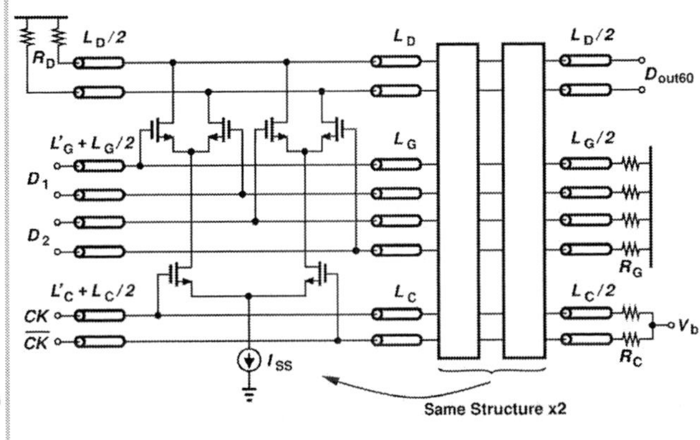

Figure 2.3.2: 60Gb/s 2:1 MUX.

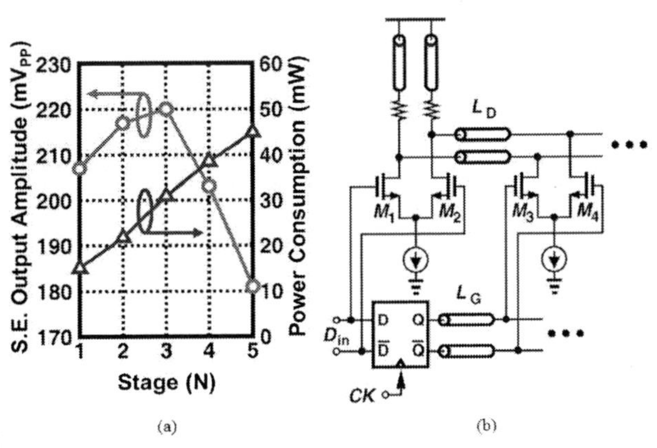

Figure 2.3.3: (a) Output amplitude and power consumption of distributed amplifier, (b) combiner.

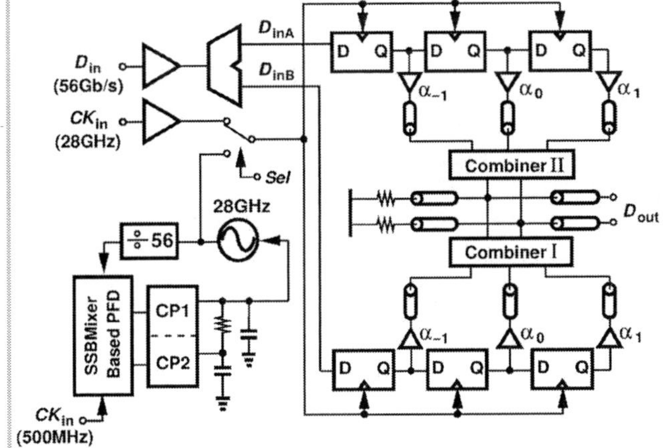

Figure 2.3.4: PAM4 TX architecture.

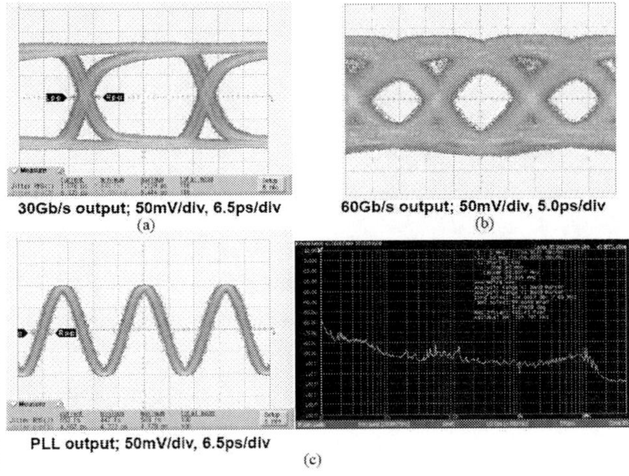

Figure 2.3.5: NRZ TX measurement results: (a) 30Gb/s output, (b) 60Gb/s output, (c) 30GHz clock waveform and its phase noise.

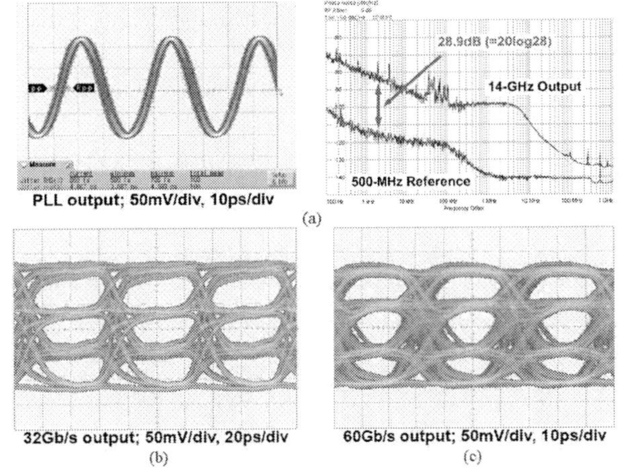

Figure 2.3.6: PAM4 TX measurement results: (a) 28GHz clock waveform and its phase noise, (b) 32Gb/s output, (c) 60Gb/s output.

978-1-4799-0917-9/14 $31.00 © 2014 IEEE

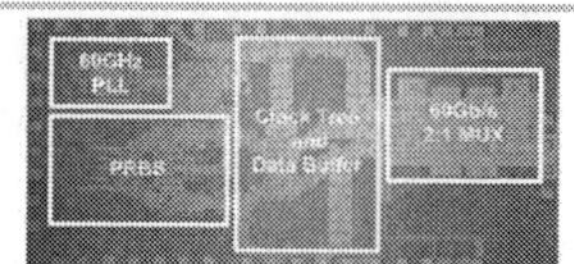

NRZ TX			
	[6]	[7]	This Work
Data Rate	50Gb/s	40Gb/s	60Gb/s
Function	2:1 MUX Only	SFI 5.2 Receiver + 4:1 MUX (40Gb/s) + 4:2 MUX (20Gb/s) + 20GHz PLL	4:1 MUX + 60GHz PLL + Built-in PRBS
TX Clock PNoise @ 1MHz	N/A	N/A	-100dBc/Hz
TX Clock RMS Jitter	N/A	N/A	461fs (1kHz~20MHz)
Data Input Range (S.E.)	$1V_{PP}$	N/A	$50\sim300mV_{PP}$
Clock Input Range (S.E.)	$1V_{PP}$	N/A	$100\sim300mV_{PP}$
Data Swing (S.E.) Full-Rate Half-Rate	$70mV_{PP}$ (50Gb/s) N/A	$325mV_{PP}$ (40Gb/s) $400mV_{PP}$ (20Gb/s)	$250mV_{PP}$ (60Gb/s) $250mV_{PP}$ (30Gb/s)
Data Jitter	N/A	783fs,rms (40Gb/s)	1.08ps,rms (30Gb/s)*
20-80% Rise/Fall Time Full Rate Half-Rate	10.4ps (50Gb/s) N/A	10.22ps (40Gb/s) 14.67ps (20Gb/s)	8.0ps (60Gb/s) 7.5ps (30Gb/s)
Power Consumption	43mW	1.8W (40Gb/s mode)	450mW (MUX: 31mW)
Chip Area	1.8 x 1mm²	4.2 x 4.2mm²	2.1 x 1mm²
Technology	90nm CMOS (Shrunk to 48nm)	65nm Digital CMOS	65nm Digital CMOS

* For half-rate D_{OUT}. Full-rate data jitter is not measurable due to the limited bandwidth of oscilloscope.

PAM4 TX		
	[8]	This Work
Data Rate	25Gb/s	60Gb/s
Function	Combiner + FFE	Combiner + FFE + Built-in PLL
TX Clock PNoise @ 1MHz	N/A	-98.5dBc/Hz
TX Clock RMS Jitter	N/A	508fs (100Hz~1GHz)
Data Input Range (S.E.)	N/A	$50\sim300mV_{PP}$
Clock Input Range (S.E.)	N/A	$100\sim300mV_{PP}$
D_{OUT} Full Mag. (4 Levels)	$430mV_{PP}$	$250mV_{PP}$
Min. Eye Opening	$90mV_{PP}$	$50mV_{PP}$
Horizontal Eye Opening	>0.4UI @ 25Gb/s	>0.6UI @ 60Gb/s
20-80% Rise/Fall Time	29ps @ 25Gb/s	12.8ps @ 60Gb/s
Power Consumption	101.8mW	290mW
Chip Area	1 x 0.5mm²	1.2 x 0.95mm²
Technology	90nm SOI CMOS	65nm Digital CMOS

Figure 2.3.7: Die micrograph and performance summary.

2.4 A 25Gb/s 5.8mW CMOS Equalizer

Jun Won Jung, Behzad Razavi

University of California, Los Angeles, CA

The power consumption of broadband receivers becomes particularly critical in multi-lane applications such as the 100 Gigabit Ethernet. However, the power-speed trade-off tends to intensify at higher rates, making it a greater challenge to reach the generally-accepted efficiency of 1mW/Gb/s. Prominent among the power-hungry receiver building blocks are the clock-and-data-recovery circuit, the deserializer, and the front-end equalizer. The use of charge-steering techniques has shown promise for the low-power implementation of the first two functions [1]. This paper introduces a half-rate 25Gb/s equalizer employing charge steering and achieving an efficiency of 0.232mW/Gb/s.

In addition to dealing with the generic delay bounds in direct or unrolled decision-feedback equalizers (DFEs), our architecture must also accommodate the return-to-zero (RZ) format inherent in certain charge-steering topologies [1]. Shown in Fig. 2.4.1, the overall system consists of a continuous-time linear equalizer (CTLE), a 1-to-2 demultiplexer ($DMUX_1$), and two half-rate/quarter-rate (HRQR) paths. Each path includes a summer, another level of demultiplexing (by means of charge-steering latches L_1-L_2 or L_3-L_4), and one more set of latches (L_5-L_6 or L_7-L_8). Operating with complementary clocks at 6.25GHz, L_1 and L_2 alternately apply their RZ outputs to the summer in the other path, thus realizing the first tap. This summer internally multiplexes the two data streams received from L_1 and L_2 and combines the result with the incoming data. This DMUX/MUX sequence ensures that the feedback information reaching the summing junction is correct and complete even though the RZ outputs of L_1-L_2 (or L_3-L_4) are reset for half a cycle. The second tap operates in a similar manner: charge-steering latches L_5-L_6 (or L_7-L_8) sample the demultiplexed data using the Q output of the divider and apply the results to the summer.

The architecture of Fig. 2.4.1 merits three remarks. First, while demultiplexing before the DFE is attractive [2], such a DMUX must maintain some linearity so as not to irreversibly corrupt the received dispersed data. For example, the designs in [2,3] employ simple passive samplers for this purpose. Second, this architecture merges the feedback MUX with the tap differential pairs within the summers, relaxing the loop timing. Third, to achieve low power consumption while generating quadrature phases, the divide-by-two circuit is based on the topology described in [1].

Figure 2.4.2 shows the implementation of the front-end. The one-stage CTLE incorporates degeneration to create a maximum high-frequency boost of 8dB as well as inductive peaking to drive the DMUX with sufficient bandwidth. This stage also realizes offset cancellation by imbalancing the tail currents and without adding devices in the signal path.

The DMUX employs passive switching but also boosts the sampled signal level by 6dB through the use of a regenerative charge-steering pair. With a 1dB-compression point of $180mV_{pp}$, this pair exhibits enough linearity for the odd and even DFEs to equalize the dispersed signal. Note that $DMUX_1$ delivers NRZ outputs because the cross-coupled charge-steering latches merge the reset and sampling phases [1].

Figure 2.4.3 presents the implementation of one half-rate/quarter-rate path (excluding tap 2 and RZ/NRZ conversion). The summing junction is driven by the input stage (running at 12.5Gb/s) and differential pairs comprising tap 1 and tap 2 (not shown), all of which steer charge and produce a single-ended output swing of about $150mV_{pp}$. The output is applied to the charge-steering DMUX consisting of L_1 and L_2.

We note several attributes of the circuit in Fig. 2.4.3. First, the charge-steering stages, and in particular the input pair, briefly draw a packet of charge and remain off for the rest of the time, dissipating low power and allowing operation across a wide frequency range. By contrast, integrating or dynamic summers [3,4] pull a continuous current from the output nodes for half a cycle, potentially consuming high power and making it difficult to run at different rates. Second, the degeneration network in the input pair also provides some linear equalization. Third, the cross-coupled PMOS pair tied to X and Y in Fig. 2.4.3 prevents collapse of these nodes when both tap 1 and tap 2 branches draw charge. Applied to all of the stages, this technique also increases the output swing by restoring the high level to V_{DD}. Fourth, the coefficients are adjusted by varying the tail capacitances in 25 discrete steps in tap 1 (and 10 in tap 2). Fifth, the multiplexing of the feedback components is accomplished through gating the tails in Fig. 2.4.3 by the 6.25GHz clock.

To ensure sufficient hold time throughout the cascade L_1-L_8, the quadrature phases of the 6.25GHz clock alternately sample the signals. The RZ/NRZ conversion circuit incorporates clocked comparators and RS latches similar to that in [1].

The equalizer is fabricated in TSMC's 45nm digital CMOS technology. Figure 2.4.7 shows the die core, which measures 100×100μm². The circuit is tested with a channel having a loss of 24dB at 12.5GHz. Figure 4 shows the received and output eye diagrams. The bit-error rate (BER) in this case is below 10^{-12}. Figure 2.4.5 plots the BER as a function of the external clock phase, revealing an eye opening of approximately 0.44UI. Since the input PRBS generator has a peak-to-peak jitter of about 7ps, an opening of about 0.18UI is lost.

Figure 2.4.6 summarizes the measured performance of the equalizer and compares it with that of prior art. The circuit consumes 5.8mW, of which 2.44mW is drawn by the CTLE, 1.25mW by the divide-by-2 circuit, and 2.11mW by the two HRQR paths. We note that [6] compensates for 10dB of loss and achieves an eye opening of 0.11UI for BER = 10^{-9}.

Acknowledgments:

This research was supported by Texas Instruments and Realtek Semiconductor. The authors are grateful to the TSMC University Shuttle Program for chip fabrication.

References:

[1] J. W. Jung and B. Razavi, "A 25-Gb/s 5-mW CDR/Deserializer," IEEE *J. Solid-State Circuits*, vol. 48, pp. 684-697, Mar., 2013.

[2] K. J. Wong et al., "A 5-mW 6-Gb/s Quarter-Rate Sampling Receiver With a 2-Tap DFE Using Soft Decisions," IEEE *J. Solid-State Circuits*, vol. 42, pp. 881-888 Apr., 2007.

[3] A. Agrawal et al., "A 19Gb/s Serial Link Receiver with Both 4-Tap FFE and 5-Tap DFE Functions in 45nm SOI CMOS," IEEE *ISSCC Dig. Tech. Papers*, Feb 2012, pp. 134-135.

[4] J. Bulzacchelli et al., "A 28 Gb/s 4-tap FFE/15-tap DFE serial link transceiver in 32 nm SOI CMOS technology," IEEE *ISSCC Dig. Tech. Papers*, Feb 2012, pp. 324-325.

[5] K. Jung et al., "A 0.94mW/Gb/s 22Gb/s 2-Tap Partial-Response DFE Receiver in 40nm LP CMOS," IEEE *ISSCC Dig. Tech. Papers*, Feb 2013, pp. 42-43.

[6] K. Kaviani et al., "A 27 Gb/s 0.41-mW/Gb/s 1-Tap Predictive Decision Feedback Equalizer in 40-nm Low-Power CMOS," IEEE *CICC*, Sep 2012.

[7] J. E. Proesel and T. O. Dickson, "A 20-Gb/s, 0.66-pJ/bit Serial Receiver with 2-Stage Continuous-Time Linear Equalizer and 1-Tap Decision Feedback Equalizer in 45nm SOI CMOS," IEEE *Symp. VLSI Circuits*, Jun 2011, pp. 206-207.

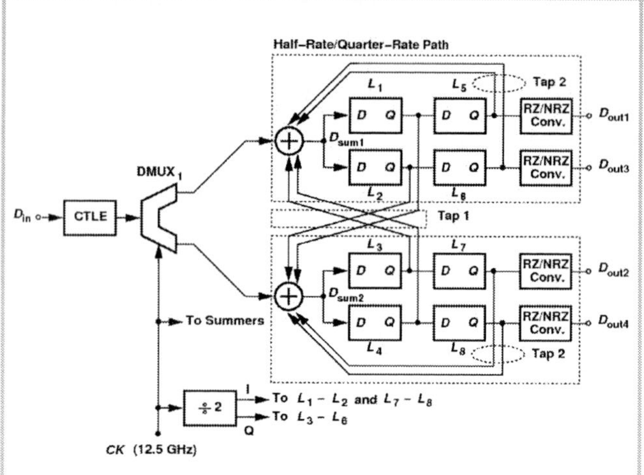

Figure 2.4.1: Equalizer architecture.

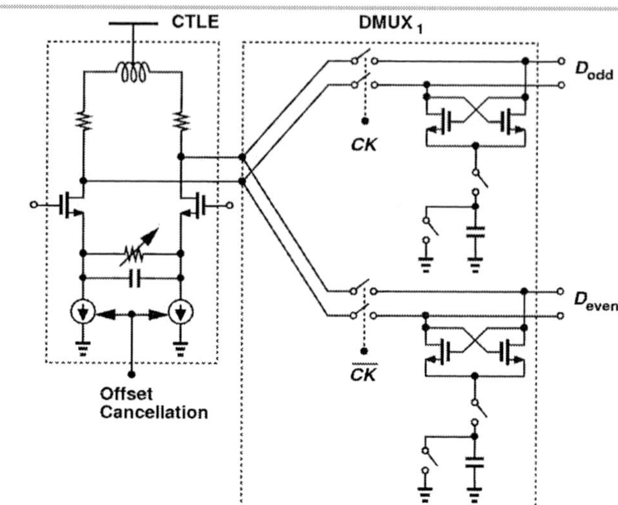

Figure 2.4.2: Implementation of front-end.

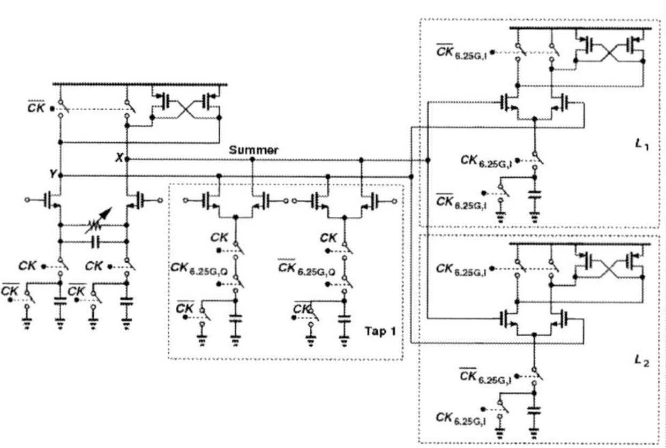

Figure 2.4.3: Implementation of one half-rate/quarter-rate path.

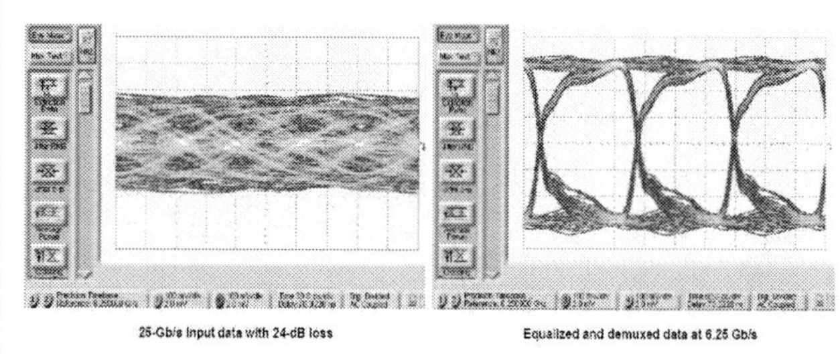

25-Gb/s input data with 24-dB loss
Equalized and demuxed data at 6.25 Gb/s

Figure 2.4.4: Measured eye diagrams of input and output data.

Figure 2.4.5: Measured bathtub curve at 25Gb/s with 24dB loss in channel.

Reference	[3]	[4]	[5]	[6]	[7]	This Work
Data Rate	19 Gb/s	28 Gb/s	22 Gb/s	27 Gb/s	20 Gb/s	25 Gb/s
Architecture	4-tap FFE + 5-tap DFE	CTLE + 15-tap DFE	CTLE + 2-tap DFE	1-tap DFE	CTLE + 1-tap DFE	CTLE + 2-tap DFE
DFE Clocking	Quarter Rate	Half Rate	Quarter Rate	Quarter Rate	Half Rate	Half Rate
Channel Loss @ Nyquist	25 dB	35 dB	16 dB	>10 dB	26.3 dB	24 dB
BER / horizontal eye opening	$< 10^{-9}$ / 36% UI	$< 10^{-9}$ / 35.6% UI	$< 10^{-12}$ / 26% UI	$< 10^{-9}$ / 11% UI	$< 10^{-8}$ / 26% UI	$< 10^{-12}$ / 44% UI
Supply (V)	1.1	1.05	1.15	1.1	1.2	1
Power (mW)	118	80*	20.6	11.1	13.2	5.8
Area (mm^2)	0.07	0.81**	0.016	0.015	0.012	0.01
Technology	45-nm SOI CMOS	32-nm SOI CMOS	40-nm CMOS	40-nm CMOS	45-nm SOI CMOS	45-nm CMOS

*Only for odd and even DFEs. Excludes CTLE, etc.

**Includes TX+RX+PLL/4

Figure 2.4.6: Performance summary and comparison with prior art.

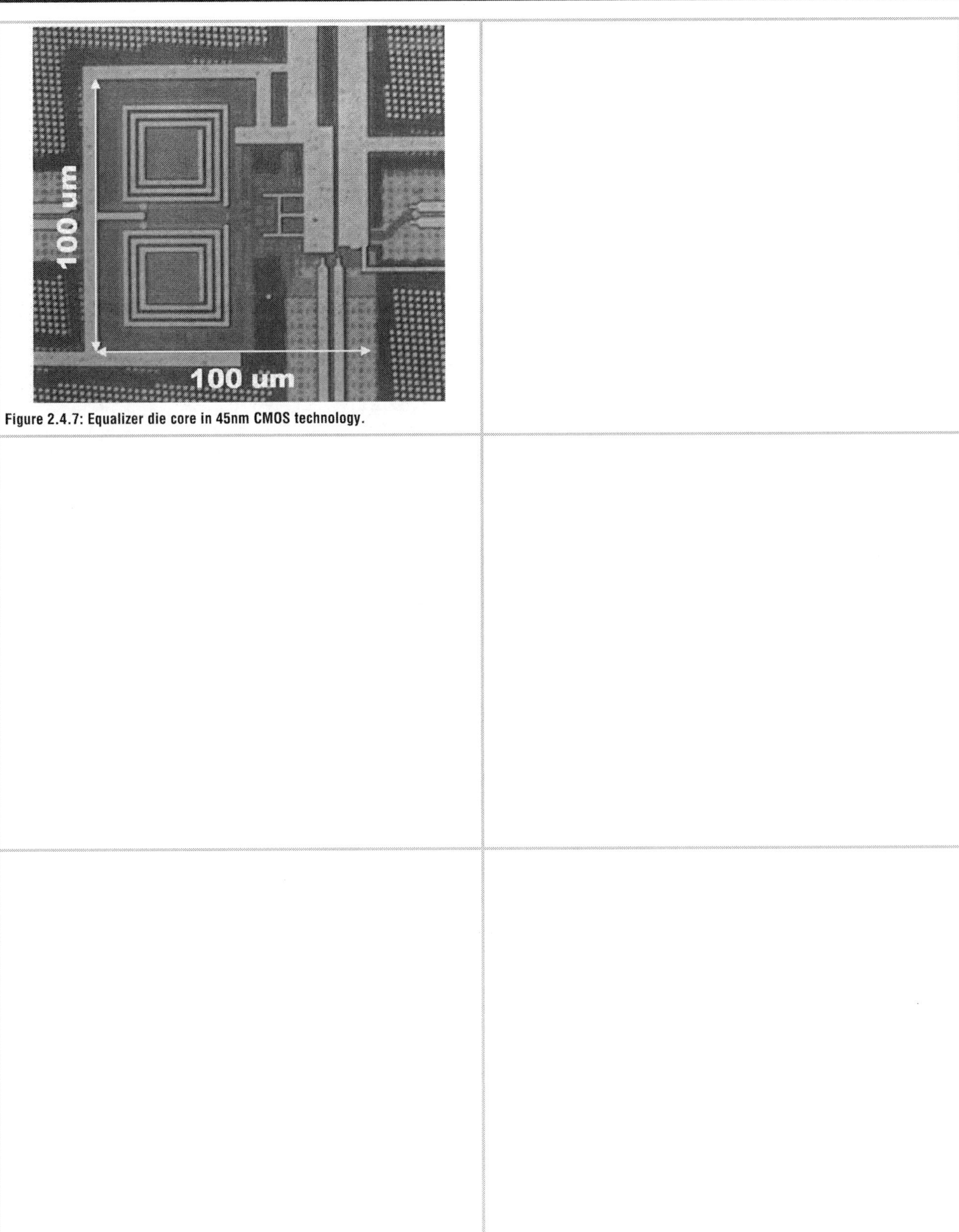

Figure 2.4.7: Equalizer die core in 45nm CMOS technology.

2.5 A 0.25pJ/b 0.7V 16Gb/s 3-Tap Decision-Feedback Equalizer in 65nm CMOS

Rui Bai[1], Samuel Palermo[2], Patrick Yin Chiang[1,3]

[1]Oregon State University, Corvallis, OR,
[2]Texas A&M University, College Station, TX,
[3]Fudan University, Shanghai, China

Supply-voltage scaling has become one of the most effective methods to improve the energy efficiency of power-constrained systems, motivating its application towards high-performance I/O links [1]. Due to the accelerating need for more off-chip I/O bandwidth, it is desirable to provide both higher data rate and low-V_{DD} operation to achieve optimal energy efficiency. Unfortunately, efficient implementations of equalization circuits are one of the major challenges faced in >10Gb/s serial-link systems that attempt to incorporate reduced-supply operation. For example, a continuous-time linear equalizer (CTLE), due to its analog nature, exhibits a rapid degradation in gain/bandwidth and only linear power scaling when operating at low V_{DD}. Decision-feedback equalizers (DFEs) also have to make significant compromise of speed from longer delay in the critical feedback path. Consequently, previous >10Gb/s equalizers have not pursued extensive supply voltage scaling [2-5].

In this work, a DFE is presented that is designed specifically to operate at low V_{DD} and scale well in energy-efficiency. To achieve this goal, the following innovations are introduced: 1) fast and energy-efficient charge-based latch and sample-and-hold (S/H) topologies; 2) a CMOS-clocked quarter-rate DFE architecture with summer gain and power optimization; 3) an integrating summer with a compact common-mode restoration circuit. Leveraging these techniques, the DFE is capable of operating at or below 0.7V, with an energy efficiency of or better than 0.25pJ/bit.

Charge-based circuits have been previously shown to be effective in achieving high-speed operation and low-power consumption [6], but not at low supply voltages. Our low-V_{DD} DFE leverages a charge-based approach for the critical quantization latches and input sample-and-hold circuits, as shown in Fig. 2.5.1. First, consider the operation of the two-stage charge-based quantization latch. Both the 1st-stage output VXN/VXP and 2nd-stage output VON/VOP are reset to V_{DD} when the clock CK is low. When CK goes high, both stages start discharging toward GND. However, if the 1st-stage is designed with a faster discharge rate, the 2nd-stage discharging will stop and a differential output voltage is maintained that is proportional to the differential input. One potential problem for low-V_{DD} operation is that a large differential gain results in a significant common-mode voltage drop at the latch output. To mitigate this, MOS caps are added to dump positive charge into the output nodes to elevate the output common-mode voltage level. Fast settling and small aperture time are achieved due to the charge-based latch's tail nodes rapidly being pulled to GND, resulting in an effective one-stack circuit. The latch draws almost no static current and exhibits quadratic power savings with supply, resulting in significant power savings relative to conventional CML latches whose power scales linearly with supply. Furthermore, the latch is able to achieve a gain of more than 2, relaxing the gain and hence power requirements of the integrating summer. A similar topology is used by charge-based S/H circuits at the DFE input, with additional cascode devices added to the 2nd-stage that allow for near unity gain, preserving DFE input linearity. Compared to conventional transmission-gate based S/Hs, the designed S/H does not suffer from severe bandwidth reduction at 0.7V, and, if desired, has the added benefit of providing larger than unity gain.

A quarter-rate DFE architecture (Fig. 2.5.2) is employed to allow for CMOS clocking with a 0.7V supply. As shown in the timing diagram, because the sampling delay of the S/H and the CK-Q delay of the latch are nearly identical, timing margin is saved by aligning these two phases. Each summer output is followed by a successive chain of three charge-based latches, which then provide feedback to the three other summers. Note that return-to-zero (RZ) operation of these latches is not a problem because the feedback values only have to be valid during the first half of the summation period, i.e., one quarter of the cycle.

Because four integrating summers are used, it is critical to optimize the summer static current consumption by achieving only the minimum gain required to make a correct decision. Since the decision amplitude does not need to be full swing but just large enough to slew the differential pairs of the feedback taps [5], the over-drive voltage of the summer feedback differential pairs are designed to be less than 200mV. Thus, assuming a 100mV input cursor amplitude, only a total gain of 2 is needed in the S/H-summer-latch path. Since the gain of the latch itself is more than 2, the gain and therefore power of the summer can be minimized.

The integrating summer consists of a linearized input tap, three feedback taps, and one offset-cancellation tap, with 7b control for both the input and 1st-tap, and 6b control for the remaining taps and offset cancellation. To avoid gain and linearity degradation at 0.7V operation, we connect a pair of common-mode restoration circuits to the summation nodes. During the summer reset phase, the capacitor is charged to V_{DD} by the M1 PMOS transistor. When the summer integration starts, the initial voltage on the capacitor top plate is bootstrapped to ~2V_{DD}, thereby keeping M2 in saturation in order to pump current into the summation node. By maintaining a high impedance for the current sources, linearity and differential gain are preserved while introducing more than 200mV boost in the common-mode voltage. Using high-density MOS varactors, each pair of restoration circuits adds 34μm^2 of area overhead.

Figure 2.5.7 shows a die micrograph of the DFE, fabricated in a 65nm CMOS process. The compact size of the 60×60μm^2 DFE core area is critical, as the speed, gain, and power consumption of charge-based circuits are all heavily affected by parasitic capacitance. During BER testing, a 16Gb/s 2^7-1 PRBS pattern is applied to the DFE inputs through an RF probe to evaluate the DFE performance. One of the four integrating summers is on-chip buffered and driven off-chip through another RF probe to produce the eye diagrams seen in Fig. 2.5.4. The integrating summer's output eye is completely closed when all the equalization taps are disabled, and gradually opens as more feedback taps are enabled. Note that the buffer gain is less than 0.4 to reduce loading to the summer. Figure 2.5.5 shows the measured BER bathtub curves for two test channels with 13dB and 18dB loss at 8GHz. With the 13dB-loss channel, the DFE operates at 0.65V with 3.3mW power consumption and achieves a 0.53UI timing margin. As more equalization and signal gain is required for the 18dB loss channel, a 0.7V supply and 4mW of power are required to obtain a 0.46UI timing margin. Fig. 2.5.6 shows the measured DFE power breakdown for 0.7V operation with the 18dB loss channel. The reported power includes the DFE core, DACs, DC biasing, and clock buffers, excluding only the CML clock divider that down-converts an 8GHz differential input clock for quadrature phase generation. A comparison with recently published low-power DFE designs shows that this work achieves a 2× improvement in energy-efficiency for a 0.7V supply, and is further improved when the supply is reduced to 0.65V.

Acknowledgements:
This work was funded, in part, by a Department of Energy Early CAREER grant, Intel Labs Wireline Signaling Program, and the Semiconductor Research Corporation (SRC) grant 1836.060 through the Texas Analog Center of Excellence (TxACE). The authors would like to thank E. Alon and Y. Lu of UC Berkeley; J. Cheng and M. Brown of OSU; K. Hu of Broadcom; and J. Calvin of Tektronix for generous loaning of high-speed BERT equipment.

References:
[1] M. Mansuri et al., "A Scalable 0.128-to-1Tb/s 0.8-to-2.6pJ/b 64-Lane Parallel I/O in 32nm CMOS," *ISSCC Dig. Tech. Papers*, pp. 402-403, Feb 2013.
[2] M. Nazari and A. Emami-Neyestanak, "A 15Gb/s 0.5mW/Gb/s 2-Tap DFE Receiver with Far-End Crosstalk Cancellation," *ISSCC Dig. Tech. Papers*, pp. 446-447, Feb 2011.
[3] J. Proesel and T. Dickson, "A 20-Gb/s, 0.66-pJ/bit Serial Receiver with 2-Stage Continuous-Time Linear Equalizer and 1-Tap Decision Feedback Equalizer in 45nm SOI CMOS," *Symposium on VLSI Circuits Dig. Of Tech. Papers*, pp. 206-207, June 2011.
[4] K. Kaviani et al., "A 0.4-mW/Gb/s Near-Ground Receiver Front-End With Replica Transconductance Termination Calibration for a 16-Gb/s Source-Series Terminated Transceiver," *Solid-State Circuits, IEEE Journal of*, vol.48, no.3, pp.636-648, March 2013.
[5] Y. Lu and E. Alon, "A 66Gb/s 46mW 3-Tap Decision-Feedback Equalizer in 65nm CMOS," *ISSCC Dig. Tech. Papers*, pp. 30-31, Feb 2013.
[6] J.W. Jung and B. Razavi, "A 25-Gb/s 5-mW CMOS CDR/Deserializer," *Solid-State Circuits, IEEE Journal of*, vol.48, no.3, pp.684-697, March 2013

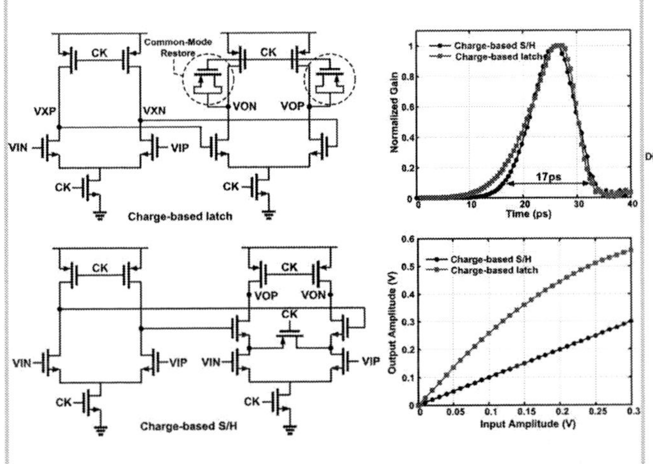

Figure 2.5.1: Charge-based latch and S/H with simulated Impulse Sensitivity Function (ISF) and gain at VDD=0.7V.

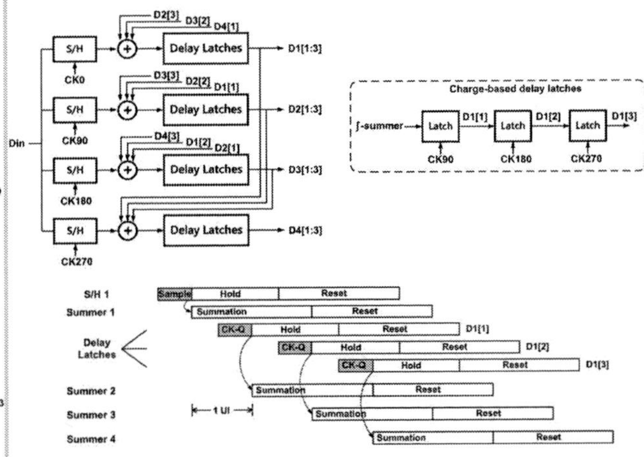

Figure 2.5.2: Block diagram and timing diagram of proposed DFE.

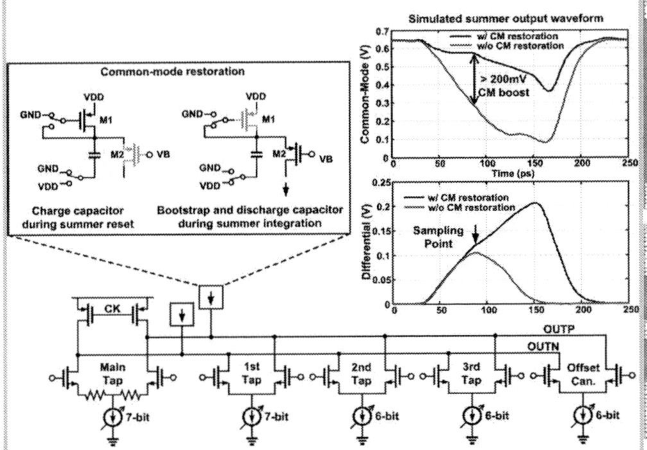

Figure 2.5.3: Integrating summer with common-mode restoration.

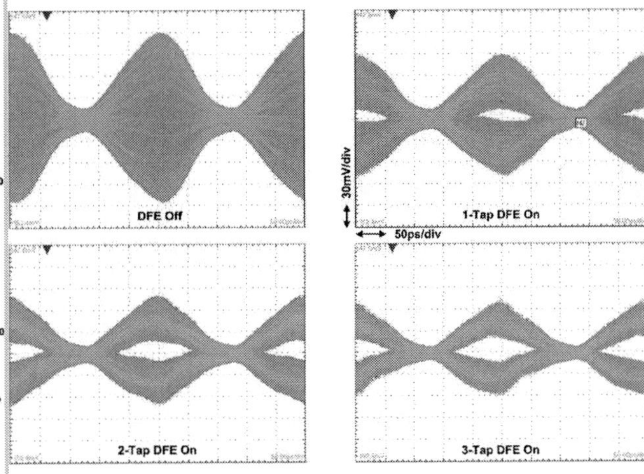

Figure 2.5.4: Internal eye diagrams of the integrating summer at the 4Gb/s quarter rate, measured from on-chip buffer with a high-speed probe.

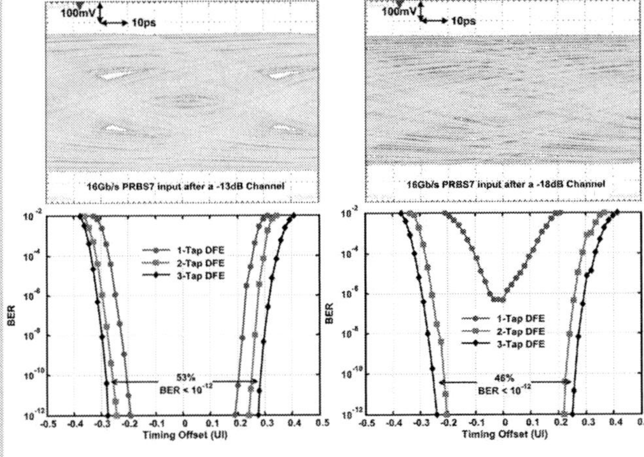

Figure 2.5.5: DFE input and BER bathtub curve with a 13dB loss channel, 0.65V supply and a 18dB loss channel, 0.7V supply.

Component	Power (mW)
Summers and DACs	1.4
Latches and biasing	1.1
Clocking	1.5
Total	4

References	[2]	[3]	[4]	[5]	This Work	
Data Rate (Gb/s)	15	20	16	66	16	
Process	45nm SOI	45nm SOI	40nm GP	65nm GP	65nm GP	
Equalization	2-tap DFE	CTLE + 1-tap DFE	Passive LE + 1-tap DFE	3-tap DFE	3-tap DFE	
Clocking	Half Rate	Half Rate	Half Rate	Half Rate	Quarter Rate	
Supply (V)	1.2	1.2	1.0	1.2	0.65	0.7
Channel Loss (dB)	14.5	26.3	15	N/A*	13	18
Timing Margin	34% BER < 10^{-8}	26% BER < 10^{-12}	>25% BER < 10^{-12}	60% BER < 10^{-12}	53% BER < 10^{-12}	46% BER < 10^{-12}
Power (mW) (Including Clocking)	7.5	13.2	9.25	55.1**	3.3	4
Energy Efficiency (pJ/b)	0.50	0.66	0.59	0.86**	0.21	0.25

* ISI emulated by a 3-tap transmitter FIR filter with a total post cursor 1.65x of main cursor.

** 9.1mW input from external clock source used for clocking power calculation

Figure 2.5.6: Power breakdown, performance summary and comparison.

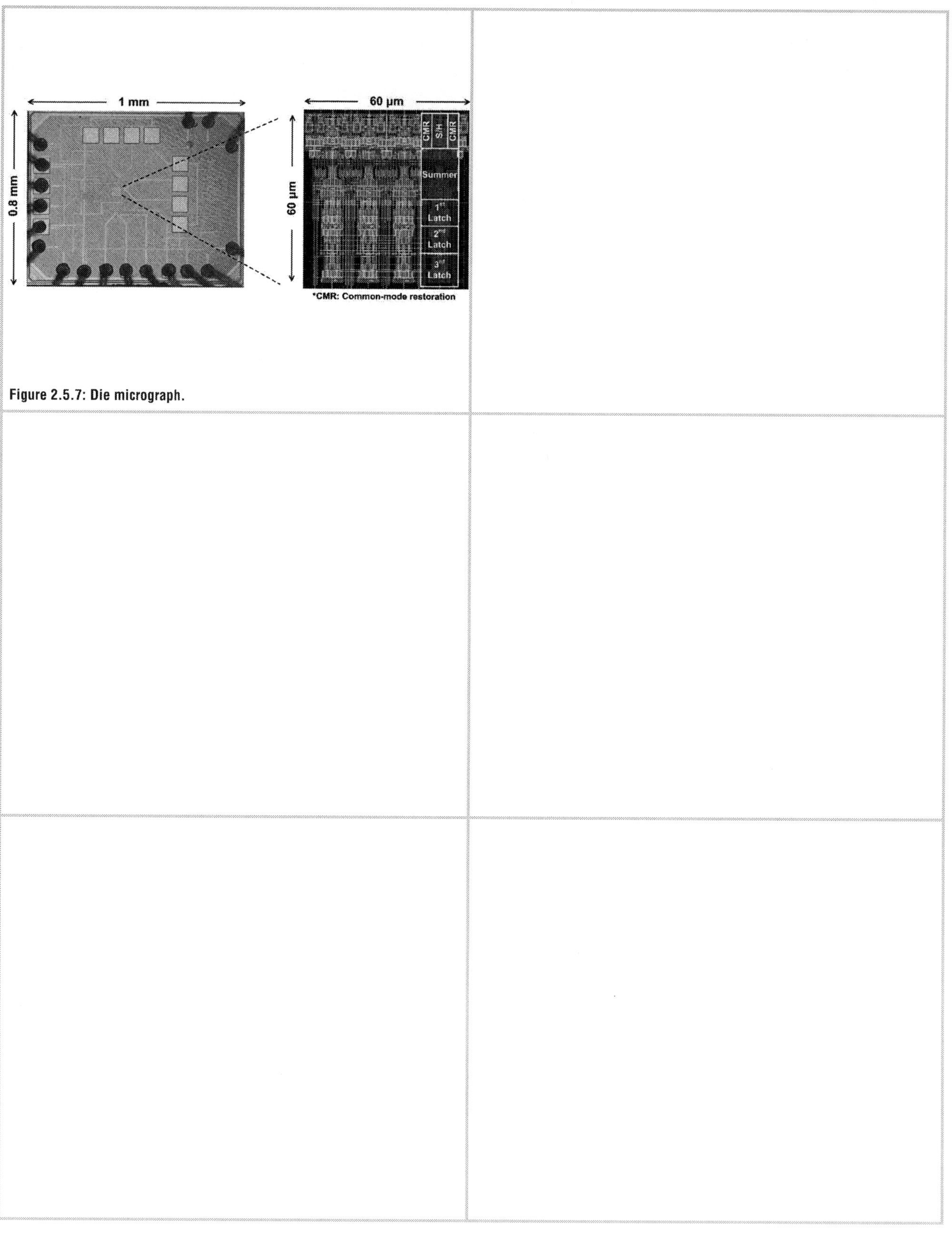

Figure 2.5.7: Die micrograph.

2.6 A 5.67mW 9Gb/s DLL-Based Reference-less CDR with Pattern-Dependent Clock-Embedded Signaling for Intra-Panel Interface

Dong Hoon Baek[1,2], Byungsub Kim[1], Hong-June Park[1], Jae-Yoon Sim[1]

[1]Pohang University of Science and Technology, Pohang, Korea,
[2]Samsung Electronics, Yongin, Korea

Point-to-point data transmission with clock-embedded signaling (CES) has been generally adopted in intra-panel interfaces, which need to support fine resolution, high frame rate, and large display size. Since CES embeds the clock-transition information in the data stream, it enables wide-range clock acquisition with PLL-based [1,2] or DLL-based [3-5] clock-and-data recovery (CDR) schemes. It also offers additional benefits of reduced EMI and low cost by eliminating the need for an additional clock channel or reference signal. For clock recovery, however, CES transmits a significant number of extra bits attached to each data packet to carry clock transition information. The number of extra bits is at least three [4] or four [3], sufficient to reduce the effect of inter-symbol interference (ISI) on clock transitions from adjacent random data patterns, providing cleaner reference to the clock recovery circuit. The repeated transitions in every data packet also intensify the spectral energy at the clock frequency as the data-rate increases, seriously aggravating EMI problem. This paper presents a DLL-based CDR with a new CES scheme that carries clock transitions with only one bit overhead but effectively sees the same ISI as a three- or four-bit overhead. By introducing a pattern-dependent clock embedding, our CES assimilates with random data transitions and almost eliminates the EMI issue. The CDR, implemented in 65nm CMOS, shows a lock range of 6.5 to 9Gb/s and a power efficiency of 0.63mW/Gb/s at 9Gb/s.

Figure 2.6.1 compares the conventional and proposed CES signaling schemes. The conventional CES is composed of N-bit data and extra M additional bits to embed the clock transition. The number of extra bits is chosen considering the effect of ISI on clock transition from random data pattern. This example shows a general CES with a 4b overhead for clock transition by allocating the first half to low and the next half to high [3]. This transition provides the reference (RCLK) to a DLL or a PLL, which generates multiphase clocks for data recovery. The multiphase should be out-phased by 0.5UI from the reference transition to place the generated phases at the center of data eyes. As an alternative, a 3b overhead is also considered by allocating 1.5 bits to both low and high [4]. This work develops a pattern-dependent CES. There are two different types of packet depending on MSB (D0<N>). Each packet is composed of N+1 bits. Therefore, effectively 1 bit of overhead is imposed on each N bits of data. If D0<N> is high (TYPE1), D1b<0> is attached after D0<N-1> held for 2b period, ensuring a transition before D1<1>. If D0<N> is low (TYPE2), D0<N-1> holds for 2b period followed by a toggled transition. Therefore, there is a guaranteed transition after the N+1th bit or the Nth bit, for TYPE1 and TYPE2, respectively. Both rising and falling transitions are meaningful, and a run length of more than 2 bits is always guaranteed before these transitions to reduce the effect of ISI. Since the selection of TYPE is determined by D<N>, which is random, the transmitted data stream shows almost the same EMI characteristics as random data. The MSB can be easily recovered in CDR by checking if there is a transition between two data sampled with ϕ_N and ϕ_{N+1}. N is taken to be 8 in this work.

Figure 2.6.2 shows the circuit diagram of the CDR, which consists of a clock-recovery block to extract RCLK from the input stream, an all-digital DLL (ADDLL) to generate 9 phases (ϕ1 to 9), a bias generator for 0.5UI delay, and samplers for 8b pixel data (RDATA<8:1>) recovery. Lock process begins with a training step for the initial frequency acquisition. In this step, the clock recovery operation is disabled and a training clock with a period of 9UI is directly received through RCLK_T1 path. ADDLL runs and stores the counter code for initial lock. The counter code is initially set to the maximum so that VCDL has the minimum delay, and it decreases as lock process goes on [6]. At the same time, the bias generator also finds out the code for generation of 0.5UI delay. Lock detector (LD) monitors the counter code and checks if the code stops decreasing which is the condition for the end of the initial lock. After the completion of training, the clock-recovery block is enabled and normal CDR operation starts. The type detector in the clock recovery block checks the type of incoming data packet and controls MUX gating so that the transition edge in the packet selects one of the two delays (1.5UI or 0.5UI), extracting RCLK from data packet to provide reference transitions to ADDLL. The extraction is performed using two window signals (WD1 and WD2). WD1 is a short pulse whose width is between ϕ7 and ϕ8 and used to detect TYPE1 packet by checking if there is a transition during pulse is high. WD2 is a pulse between ϕ6 and ϕ7 and used to detect TYPE2 packet. If the generated 9 sampling phases are synchronized with data transition edges, an extra delay of 0.5UI is added to all the sampling phases to place them at the center of the data eyes. To implement the 0.5UI delay, 2× delay cells are conventionally used. However, the use of a large number of delay cells in VCDL limits the maximum data-rate of CDR operation and increases power consumption. In this work, a 0.5UI delay is added to RCLK instead of adding to the sampling phases. Then, the sampling phases generated with 9 delay cells are automatically placed at the center of the 9 data eyes when locked. This 0.5UI delay in RCLK is achieved by adding one of 1.5UI or 0.5UI, instead of 1UI or 0UI, respectively. IN_D is a delayed IN by a replica delay of the type detector and MUX.

Figure 2.6.3 shows the circuit diagram of the type detector. When LOCK is low, IN (= training clock) always passes RCLK_T1 path for the initial locking of ADDLL. When LOCK is high, the extracted transition information passes RCLK_T2 or RCLK_T1, depending on TYPE of data packet. The CLK_FALL is a short pulse generated by ϕ5 to reset RCLK_T2 and RCLK_T1 to prepare for receiving the next transition information.

To compare frequency spectra with different CES schemes, test transmitter circuits for the conventional CES with 4b overhead [3] and our CES are designed and simulated. For reference, a PRBS transmitter without embedded clock is also simulated. A 2^{12}-1 PRBS pattern was taken as data to be sent. The proposed CES reduces the peak by 9.6dB compared with the conventional CES and shows almost the same EMI characteristics of pure PRBS pattern (Fig. 2.6.4). The designed CDR is fabricated with 65nm CMOS. An on-chip BER test block with a 2^{12} - 1 PRBS generator is also implemented. To generate a CES pattern, a built-in self-test (BIST) block is also included with a CES encoder, serializer, and another identical PRBS generator. The CDR shows a BER of less than 10^{-12} at data-rates from 6.5 to 9 Gb/s. The power efficiency is 0.63mW/Gb/s at 9Gb/s. The measured jitter of the recovered clock at 9Gb/s is 2.82ps_{rms} and 19.33ps_{pp} (Fig. 2.6.5). The systematically different two traces in jitter histogram are due to slightly mismatched reference transitions extracted from TYPE1 and TYPE2 packets. Figure 2.6.6 compares performance with previously reported CES-based CDRs for intra-panel interface. Active area of the CDR is 0.057mm^2 (Fig. 2.6.7).

Acknowledgement:
This work was supported in part by NRF of Korea under Grant 2011-0010685, Grant 2008-0062617 and a scholarship from Samsung Electronics.

References:
[1] K. Yamaguchi, Y. Hori, K. Nakajima, et al., "A 2.0Gb/s Clock-Embedded Interface for Full-HD 10b 120Hz LCD Drivers with 1/5-Rate Noise-Tolerant Phase and Frequency Recovery," *ISSCC Dig. Tech. Papers*, pp. 192-193, Feb., 2009.
[2] I. Jung, D. Shin, T. Kim, and C. Kim, "A 140-Mb/s to 1.82-Gb/s Continuous-Rate Embedded Clock Receiver for Flat-Panel Displays," *IEEE Trans. on Circuits and Systems II*, vol. 56, no. 10, pp. 773-777, Oct., 2009.
[3] H.-K. Jeon, Y. H. Moon, et al., "An Intra-Panel Interface With Clock-Embedded Differential Signaling for TFT-LCD Systems," *IEEE J. Display Technology.*, vol. 7, no. 10, pp. 562-571, Oct. 2011.
[4] S. Jang, H. Song, S. Ye, D.-K. Jung "A 13.8mW 3.0Gb/s Clock-Embedded Video Interface with DLL-Based Data-Recovery Circuit," *ISSCC Dig. Tech. Papers*, pp. 450-452, Feb., 2011.
[5] J.-W. Kwon, X. Jin, G.-C. Hwang, et al., "A 3.0Gb/s Clock Data Recovery Circuits Based on Digital DLL for Clock-Embedded Display Interface," *in Proc. ESSCIRC*, pp.454-457, Sept. 2012.
[6] Y.-S. Kim, S.-K. Lee, H.-J. Park, J.-Y Sim, "A 110MHz to 1.4 GHz Locking 40-Phase All-Digital DLL," *IEEE J. Solid-State Circuits*, vol. 46, no. 2, pp. 435-444, Feb., 2011.

978-1-4799-0917-9/14 $31.00 © 2014 IEEE

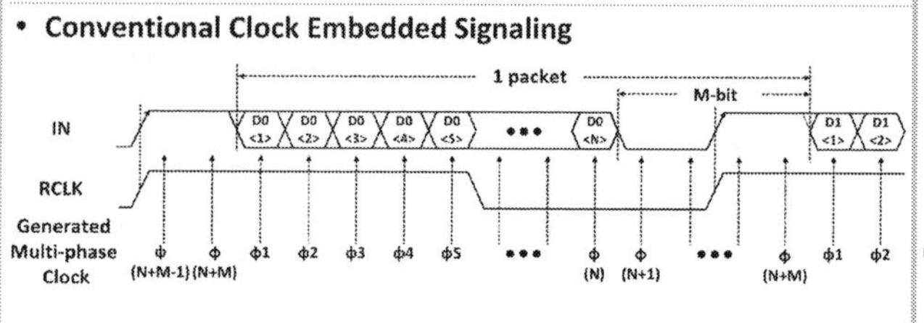

• Proposed Clock Embedded Signaling

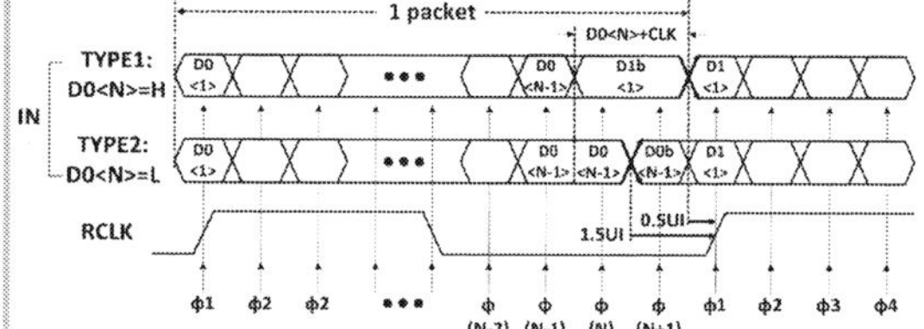

Figure 2.6.1: Conventional and proposed CES.

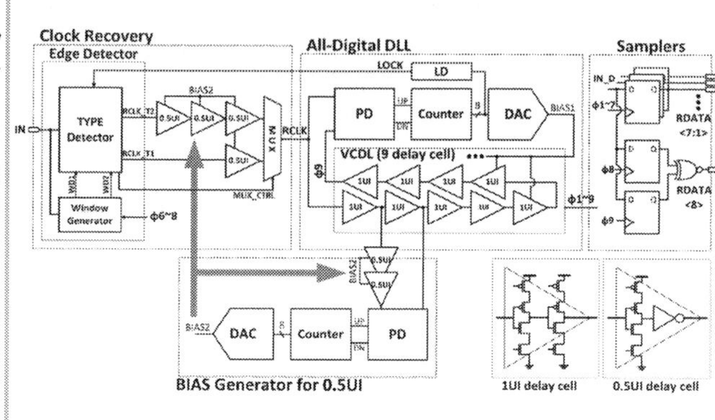

Figure 2.6.2: Circuit diagram of the CDR.

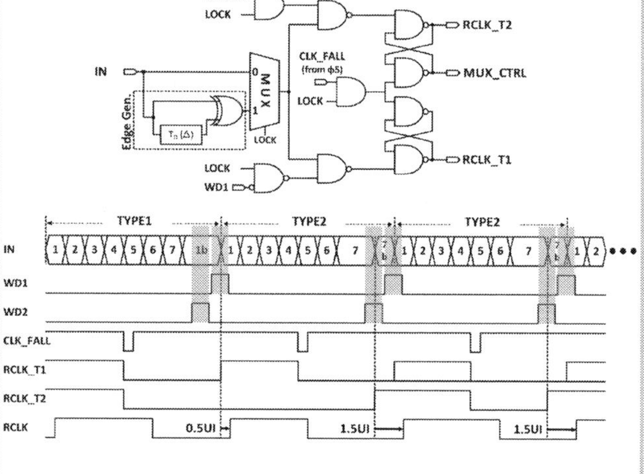

Figure 2.6.3: Type detector.

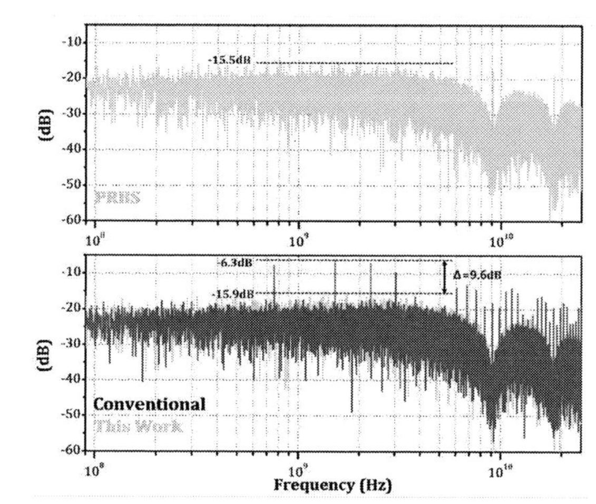

Figure 2.6.4: Simulated spectra at 9Gb/s.

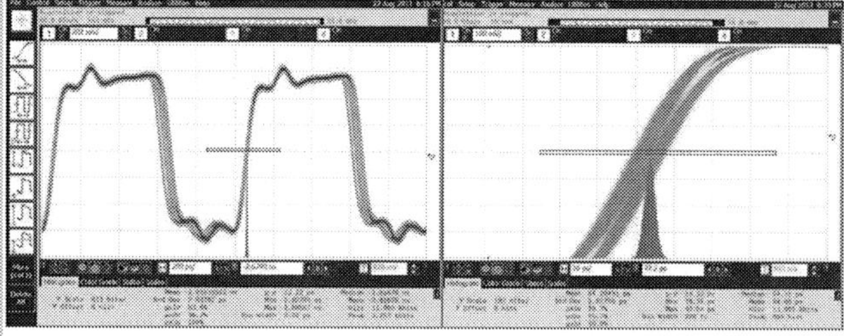

Figure 2.6.5: Measured recovered clock and jitter histogram at 9Gb/s.

	ISSCC '09 [1]	TCAS-II 09 [2]	ISSCC '11 [4]	ESSCIRC '12 [5]	This work
Technology	0.25 μm	0.25 μm	0.13 μm	0.13 μm	65 nm
Supply Voltage	3.0 V	2.5 V	1.2 V	1.2 V	0.9 V
CDR Type	PLL	PLL	Analog DLL	Digital DLL	Digital DLL
Overhead in packet	4B5B coding	4bit	3bit	2bit	1bit
Data Rate (Gb/s)	1.25 ~ 3.0	0.14 ~ 1.82	1.36 ~ 3.0	~ 3.0	6.5 ~ 9.0
RMS jitter (ps)	11 @ 2Gb/s	14.96 @ 1.82Gb/s	5.85 @ 3Gb/s	4.8 @ 3Gb/s	2.82 @ 9Gb/s
Power efficiency (mW/Gb/s)	46.5	75.27*	1.93	2.24	0.63
Area (mm^2)	0.45	2*	0.064	0.076	0.057

* : IN/OUT buffers included

Figure 2.6.6: Performance comparison table.

978-1-4799-0917-9/14 $31.00 © 2014 IEEE

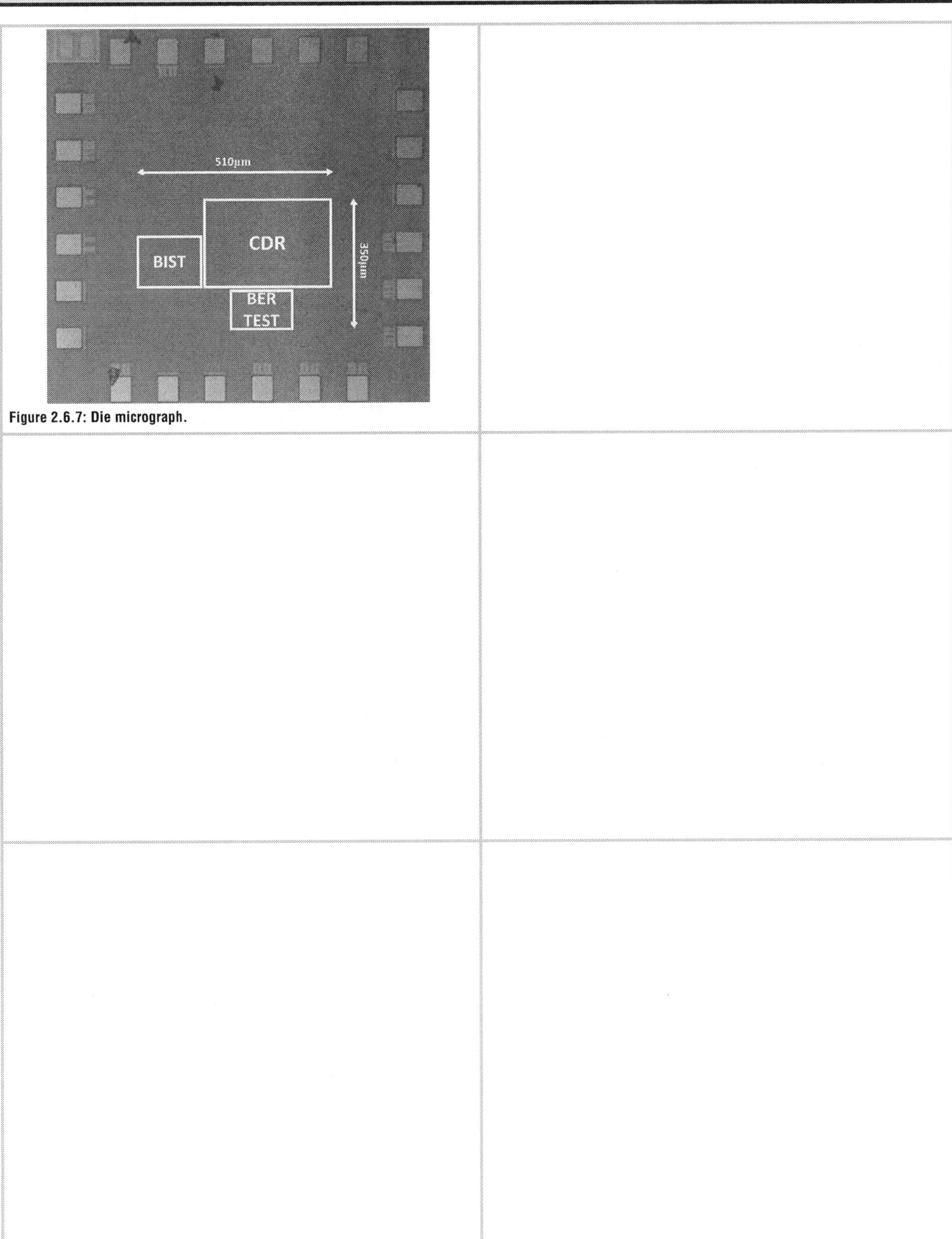

Figure 2.6.7: Die micrograph.

2.7 A Coefficient-Error-Robust FFE TX with 230% Eye-Variation Improvement Without Calibration in 65nm CMOS Technology

Seungho Han, Sooeun Lee, Minsoo Choi, Jae-Yoon Sim, Hong-June Park, Byungsub Kim

Pohang University of Science and Technology, Pohang, Korea

This paper presents a 4-tap coefficient-error-robust feed-forward equalization (FFE) transmitter (TX) for massively parallel links. Recently, massively parallel links such as on-chip links [1-3], silicon interposers [4,5], or wide I/Os [6] are gaining popularity to meet increasing demand for data transmission with a limited power budget. However, calibration overhead for thousands I/Os to compensate coefficient errors due to nano-scale variation has a high hardware cost. To reduce this overhead, we develop a coefficient-error-robust FFE (B-FFE) TX architecture that uses the channel loss to suppress eye perturbation due to coefficient errors while behaving identically to a conventional FFE.

Without coefficient errors, a B-FFE TX can be designed identical to any FFE TX. Figure 2.7.1 depicts the architectures of the B-FFE and a conventional FFE. A B-FFE TX contains a digital transition-detection (TD) filter that detects transitions of incoming data and generates transition signal: '1' for '-1'→'1' data transition; '-1' for '1'→'-1' data transition; and '0' for no transition. The transition signal is delayed by a chain of 1UI delay units. Each delayed transition signal is weighted by a coefficient (a_k, $k \neq 0$) and added to the incoming data weighted by a_0, to generate the output voltage. If designers configure the FFE coefficients (w_k) and the B-FFE coefficients (a_k) as $a_0 = \sum_{i=0}^{N\text{-}1} w_i$ and $a_{k\neq 0} = -2\sum_{i=k}^{N\text{-}1} w_i$ then the output voltages of both FFE and B-FFE TXs are identical.

Although the nominal behaviors of FFE and B-FFE are identical, the coefficient-error-tolerance of B-FFE is superior to FFEs. Figure 2.7.2 explains how B-FFE suppresses signal perturbation caused by a coefficient error. Typical coefficient errors due to mismatch, process-temperature variation, or supply-voltage drop can be modeled as an additive constant (Δw_k and Δa_k) to the nominal coefficient (w_k and a_k) as Fig. 2.7.1 shows. Since the FFE and B-FFE TXs are linear time-invariant systems, a coefficient error perturbs the pulse response of the TX by adding an error pulse, which is the pulse at the tap-position multiplied by the coefficient error as shown in Fig. 2.7.2. For the FFE TX, the error pulse at TX is square-shaped having large low-frequency portion, and thus is not significantly attenuated by the low pass filter (LPF) channel (-25dB loss at Nyquist frequency 4GHz). However, the error pulse of the B-FFE TX except the first tap (Δa_0) has large high-frequency portion after the TD modulation. Therefore, the LPF channel significantly attenuates the B-FFE error pulse, and thus, at the receiver, the impact of the B-FFE coefficient error is significantly attenuated, improving tolerance to the coefficient errors. The error caused by the first tap (Δa_0) of a B-FFE is typically insignificant since the first tap size, which determines the DC level, is typically small for a lossy channel.

To comparatively analyze the concept, we design a test-chip containing a 4-tap B-FFE TX and a 4-tap FFE TX. Figure 2.7.3 depicts the simplified block diagrams of both TXs. To demonstrate that B-FFE TXs can replace the industry-standard FFE TXs, the most popular current-mode logic (CML) drivers are used for both TXs. For speed and area-power efficiency, a latch-based half-rate architecture is used for both TXs. For the B-FFE TX, the TD block is designed with standard digital logic, which require only small hardware cost as shown in Fig. 2.7.3. The transition signal is represented by two digital bits: '10' for level '1', '00' for level '0', and '01' for level '-1.' Therefore, analog variation error is prevented in the TD signal just as the digital data signal in the FFE. To test and fairly compare the impact of mismatch, the tap coefficients of both TXs are configured by the same four reference currents selected from one of 128-by-4 current mirror sets. Therefore, we can acquire 128 mismatch data.

The test-chip is fabricated in 65nm CMOS technology, and operates at 8Gb/s with a 1.3V supply over a 96cm printed-circuit-board (PCB) trace. The measured channel response of the PCB trace is shown in Figure 2.7.2. The measured channel losses at Nyquist frequency (4GHz) are -17.5dB/-25dB with and without bonding wires and PADs. Figure 2.7.4 compares the measured nominal eye diagrams of both TXs. To acquire the nominal eye diagram, we optimize tap coefficients for each TX with the first current-mirror setting. To eliminate the tap-size scaling effect and to fairly compare the two TXs, we use the normalized eye sizes as a metric. The achieved vertical eye openings are 47% and 46% of the eye amplitude for the FFE TX and the B-FFE TX, respectively. Since both TXs are mathematically identical at nominal, the nominal eye openings are similar in size. To quantify how sensitive the TXs are to coefficient errors, we measure eye sensitivities [1]: the percentage of eye-size reduction divided by the percentage of coefficient reduction. The measured worst eye sensitivities are 1.35 and 0.56 for the FFE and the B-FFE TXs from the 3rd and the 2nd taps, respectively, implying that the B-FFE TX is 2.35× more tolerant to coefficient variation than the FFE TX.

Figure 2.7.5 shows the histograms of eye openings measured with 128 current mirror sets to acquire mismatch data. To see the effects of mismatch, the eye sizes are measured for 128 different current-mirror sets whose average currents are configured for the optimal coefficients to mimic global coefficient optimization in a parallel I/O bundle. Eight external digital inputs can switch the connection of 128 current mirror sets during the test. To prevent the effect of scaling up tap-coefficients during experimental setting, we use a vertical-eye percentage with respect to the eye amplitude as a metric for fair comparison. The means of both TXs are about the same because their nominal behaviors are theoretically identical: 43.7% for FFE and 43.8% for B-FFE. To achieve 99.73% yield (3σ yield) for a chip, each I/O must achieve about 99.9997% yield, which corresponds to 4.5σ, if a chip has 1000 I/Os. Therefore, from the perspective of yield, the 4.5σ eye deviations of both TXs are compared. The measured eye deviations (4.5σ) are 8.6% and 3.7% for the FFE TX and the B-FFE TX, respectively, showing that B-FFE TX improves by 230% the eye variation without calibration compared with the conventional FFE TX. Therefore, the worst expected 4.5σ-yield eyes of the FFE and the B-FFE TXs are 35.1% and 40.1%, respectively. Although 4.5σ eye is improved by only 14%, this improvement is significant considering that an eye size below 40% is vulnerable to noise.

Figure 2.7.6 compares the experimental results of the FFE and B-FFE TXs. The chip areas of the B-FFE and FFE TXs are 59×46μm² and 56×38μm², respectively, showing that the B-FFE TX is about 27% larger than the FFE TX while it improves the eye variation by 230%. Even with the larger chip area and the correspondingly larger parasitic capacitance, the B-FFE TX dissipates slightly less power than the FFE TX since the 2nd to 4th taps of the B-FFE are turned on only when a data transition occurs, similar to those of charge-injection TX [1,2] whereas FFE taps are always turned on. The measured supply currents of the FFE and the B-FFE TXs operating at 8Gb/s with 1.3V supply voltage are 17.3mA and 16.7mA, respectively. Figure 2.7.7 shows the die micrograph of the test chip.

Acknowledgement:
The authors acknowledge the support by National Research Foundation (NRF) of Korea grant (No. 2012R1A2A2A02010432) funded by the Korean Ministry of Education, Science and Technology (MEST), and EDA tool support by IDEC.

References:

[1] B. Kim et al., "An Energy-efficient Equalized Transceiver for RC-dominant Channels," *IEEE JSSC*, vol. 45, no. 6, pp. 1186-1197, June 2010.
[2] B. Kim et al., "A 4Gb/s/ch 356fJ/b 10mm Equalized On-chip Interconnect with Nonlinear Charge-Injecting Transmitter Filter and Transimpedance Receiver in 90nm CMOS Technology," *IEEE ISSCC Dig. Tech. Papers*, Feb 2009. pp. 66-67, 978.
[3] S. Lee et al., "A 95fJ/b Current-Mode Transceiver for 10mm On-Chip Interconnect," *IEEE ISSCC Dig. Tech. Papers*, Feb 2013, pp. 262-263.
[4] B. Kim et al., "A 10-Gb/s Compact Low-Power Serial I/O with DFE-IIR Equalization in 65-nm CMOS," *IEEE JSSC*, vol. 44, no. 12, pp. 3526-3538, Dec. 2011.
[5] Y. Liu et al., "A 0.1pJ/b 5-10Gb/s Charge-Recycling Stacked Low-Power I/O for On-Chip Signaling in 45nm CMOS SOI," *IEEE ISSCC Dig. Tech. Papers*, Feb 2013, pp. 400, 401.
[6] J. Kim et al., "A 1.2V 12.8GB/s 2Gb mobile Wide-I/O DRAM with 4x128 I/Os using TSV based stacking," *IEEE ISSCC Dig. Tech. Papers*, Feb 2011, pp. 495-498.

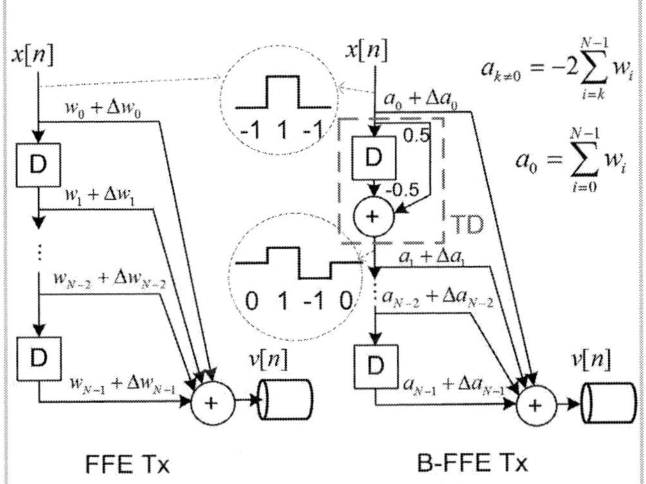

Figure 2.7.1: Block diagrams of an FFE TX and the B-FFE TX.

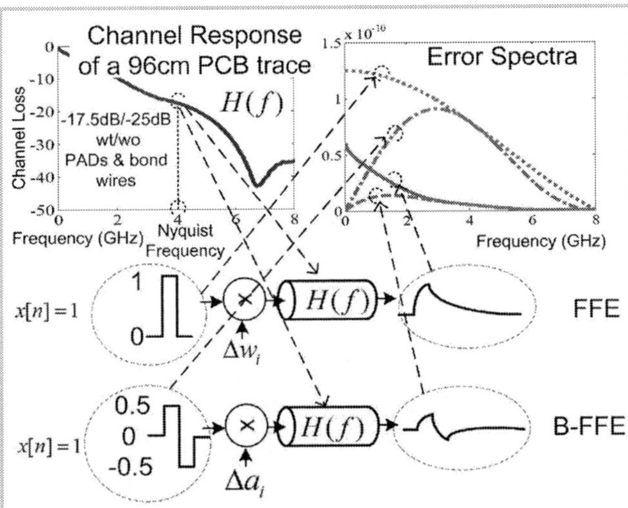

Figure 2.7.2: Error spectra of FFE and B-FFE for x[n]=1.

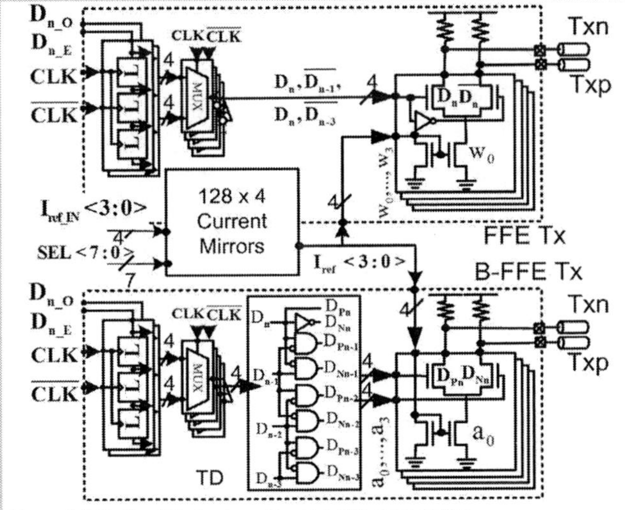

Figure 2.7.3: Simplified schematics of a FFE and a B-FFE.

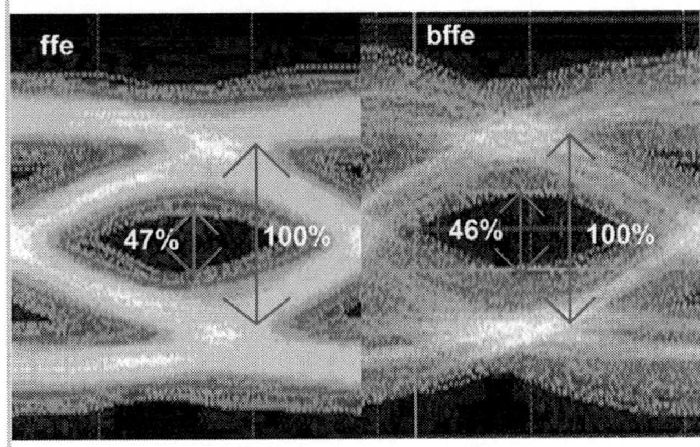

Figure 2.7.4: Eye diagrams of FFE and B-FFE.

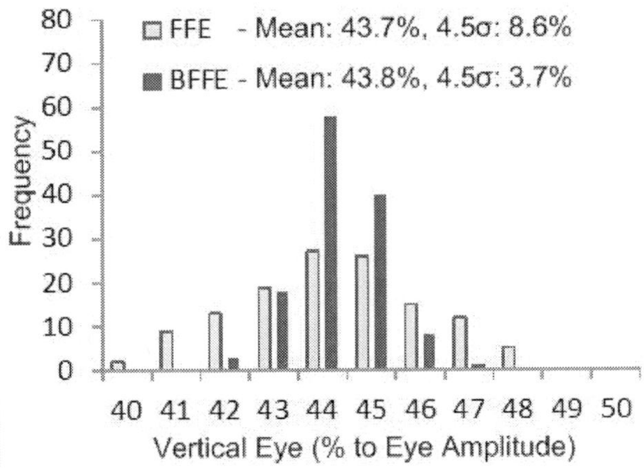

FFE - Mean: 43.7%, 4.5σ: 8.6%
BFFE - Mean: 43.8%, 4.5σ: 3.7%

Frequency

Vertical Eye (% to Eye Amplitude)

Figure 2.7.5: Eye size histograms from 128 samples.

	FFE	BFFE
Technology	65nm	65nm
Supply voltage	1.3V	1.3V
Data rate	8Gb/s	8Gb/s
Channel loss	25dB	25dB
Channel length	95.812cm	95.812cm
Vertical eye (average)	43.7%	43.8%
Worst eye sensitivity	1.35	0.56
4.5-σ eye deviations	8.6%	3.7%
Supply current	17.311mA	16.732mA
Area	2128um^2	2714um^2

Figure 2.7.6: Comparison of FFE TX and B-FFE TX.

Figure 2.7.7: Die micrograph (1×1.5mm²)

2.8 A Pulse-Position-Modulation Phase-Noise-Reduction Technique for a 2-to-16GHz Injection-Locked Ring Oscillator in 20nm CMOS

Jun-Chau Chien[1], Parag Upadhyaya[2], Howard Jung[2], Stanley Chen[2], Wayne Fang[2], Ali M. Niknejad[1], Jafar Savoj[2], Ken Chang[2]

[1]University of California, Berkeley, CA,
[2]Xilinx, San Jose, CA

High-speed transceivers embedded inside FPGAs require software-programmable clocking circuits to cover a wide range of data rates across different channels [1]. These transceivers use high-frequency PLLs with LC oscillators to satisfy stringent jitter requirements at increasing data rates. However, the large area of these oscillators limits the number of independent LC-based clocking sources and reduces the flexibility offered by the FPGA. A ring-based PLL occupies smaller area but produces higher jitter. With injection-locking (IL) techniques [2-3], ring-based oscillators achieve comparable performance with their LC counterparts [4-5] at frequencies below 10GHz. Moreover, addition of a PLL to an injection-locked VCO (IL-PLL) provides injection-timing calibration and frequency tracking against PVT [3,5]. Nevertheless, applying injection-locking techniques to high-speed ring oscillators in deep submicron CMOS processes, with high flicker-noise corner frequencies at tens of MHz, poses a design challenge for low-jitter operation. Shown in Fig. 2.8.1, injection locking can be modeled as a single-pole feedback system that achieves 20dB/dec of in-band noise shaping against intrinsic VCO phase noise over a wide bandwidth [6]. As a consequence, this technique suppresses the $1/f^2$ noise of the VCO but not its $1/f^3$ noise. Note that the conventional IL-PLL is capable of shaping the VCO in-band noise at 40dB/dec [6]; however, its noise shaping is limited by the narrow PLL bandwidth due to significant attenuation of the loop gain by injection locking. To achieve wideband 2nd-order noise shaping in 20nm ring oscillators, we present a circuit technique that applies pulse-position-modulated (PPM) injection through feedback control.

In such a ring oscillator, a voltage-controlled delay line (VCDL) preceding the pulse generator is controlled by a delay-locked loop (DLL) (Fig. 2.8.1). The DLL performs pulse-position modulation to correct residual phase error at a much higher level compared to injection with fixed position. The operation can be explained with two consecutive injection events. With fixed pulse position, the phase error between the VCO (CK_{vco}) and the reference (CK_{ref}) is gradually reduced from the instant when the injection takes place. As the DLL is enabled, the residual phase error between CK_{vco} and CK_{ref} is detected and stored to adjust the position of the following pulse. In this way, any noise with memory effect can be significantly suppressed upon the arrival of the next injection signal. Note that the same is achieved in a conventional 2nd-order PLL as well as any IL-PLL where the correction is performed through the control line of the VCO. Nevertheless, the PPM technique shows wider bandwidth for a given stability limit and is very effective in systems using oscillators with high intrinsic phase noise. Note that the introduction of the DLL along the injection path is similar to the injection-time calibration technique [5]. However, the noise-shaping characteristics of this architecture have not been previously explored.

Figure 2.8.2 shows the discrete-time model of the PPM injection-locked oscillator. Here $H_1(z)$ and $H_2(z)$ represent the phase-noise transfer function for the VCO and the reference noise under injection, respectively, but *without* PPM [6]. $H_3(z)$ models the loop gain of the DLL, which includes integration by the loop filter capacitor. Intuitively, the VCO-noise-transfer function can be considered as a cascade of two 1st-order high-pass filters. Therefore both filters can be designed with maximum bandwidth to reduce VCO noise without any stability issue. The complete VCO and reference transfer functions including discrete-to-continuous time conversion are derived and a comparison between different architectures is shown in Fig. 2.8.2. It is clear that the PPM architecture achieves the widest 2nd-order VCO noise shaping. The system indeed suffers from jitter peaking similar to a conventional DLL and systematic optimization for minimum integrated jitter is necessary. The wide bandwidth is achieved at a loop gain backed-off from the stability limit by 4.8×.

Figure 2.8.3 shows the circuit block diagram of the PPM injection-locked oscillator. A supply-regulated 4-stage pseudo-differential ring oscillator with injection transistor (M_1) forms the core of I/Q clock generation circuit. Through the noise-shaping property of the PPM technique, the VCO noise requirement can be relaxed and therefore power and area can be significantly reduced for a given oscillation frequency. The low-noise DLL is implemented with a split-tuned architecture to relax the trade-off between regulator power-supply-noise rejection (PSNR) and loop bandwidth [7]. An inverter-based VCDL with a level restorer preceding the CMOS pulse generator provides coarse delay control while a switched-capacitor bank with analog control is placed in the high-bandwidth path. A sub-sampling phase detector (SSPD) with 15fF of sampling capacitance reduces charge-pump noise significantly due to the fast transition edges offered by the advanced process.

To minimize VCDL delay upon locking, the DLL coarse control voltage is released from the maximum voltage (1.0V) after injection locking. Due to the periodicity in the transfer function of the SSPD, it is possible that the DLL drives the VCDL in a direction of reducing its delay. Figure 2.8.3 shows the timing waveform for each possible delay adjustment. Also, a pulsed injection-locked ring oscillator can be locked at either the rising or the falling edge of the VCO signal and the buffer delay can significantly vary across PVT. To keep the regulator within the linear range, a detection circuit is implemented by comparing the fine control voltage with its initial release after 10 reference cycles. If the VCDL delay is reduced, the loop is reset and the charge pump polarity is inverted prior to the next acquisition. The ring oscillator must remain injection-locked during loop reset to guarantee a consistent timing relationship.

To validate the effectiveness of PPM technique, the VCO control voltage is provided through an external supply. Figure 2.8.4 shows the measured spectrum at 3.76GHz under fundamental injection (N = 1) at different injection-strength settings. Excessive noise is intentionally introduced in the measurement setup and is observed as the ring oscillator is injection-locked with fixed pulse position. This noise is significantly reduced as PPM is enabled. As injection strength decreases, significant jitter peaking appears since the injection-locking pole is now at a frequency much closer to the DLL bandwidth. This proves the system to be of 2nd-order. Figure 2.8.5 shows the measured phase noise at 15GHz with N = 8. Integrated jitter (100kHz to 1GHz) is reduced from 434fs to 268fs with the PPM technique, measured using an Agilent E4433B signal generator. Compared with the reference source, jitter is lowered by 26.7% (366fs). Spur levels are less than -48dBc. The test-chip is implemented in 20nm CMOS and consumes 46.2mW from 1.1 and 1.25V supplies excluding drivers and occupies an area of $0.044mm^2$. Figure 2.8.6 shows the performance summary in comparison with the state-of-the-art PLLs. Fig. 2.8.7 shows the die micrograph.

Acknowledgements:
The authors would like to thank Kenny Hsieh, Fu-Tai An, Jalil Kamali, Daniel Wu, Jayesh Patil, Ying Shih, Gamal Said, Amy Chen, Kang-Wei Lai, and Yue Lu for their valuable suggestions and support.

References:
[1] J. Savoj, et al., "Design of high-speed wireline transceivers for backplane communications in 28nm CMOS," *Proc. CICC*, pp. 1-4, Sep. 2012.
[2] A. Elshazly, et al., "A 1.5GHz 890μW digital MDLL with 400fs_{rms} integrated jitter, -55.6dBc reference spur and 20fs/mV supply-noise sensitivity using 1b TDC," *ISSCC Dig. Tech. Papers*, pp. 242–243, Feb. 2012.
[3] W. Deng, et al., "A $0.022mm^2$ 970μW dual-loop injection-locked PLL with -243dB FOM using synthesizable all-digital PVT calibration circuits," *ISSCC Dig. Tech. Papers*, pp. 248–249, Feb. 2013.
[4] J. Lee, et al., "Study of subharmonically injection-locked PLLs," *IEEE J. Solid-State Circuits*, vol. 44, no. 5, pp. 1539–1553, May, 2009.
[5] Y.C. Huang, et al., "A 2.4GHz sub-harmonically injection-locked PLL with self-calibrated injection timing," *ISSCC Dig. Tech. Papers*, pp. 338-340, Feb. 2012.
[6] S. Ye, et al., "A multiple-crystal interface PLL with VCO realignment to reduce phase noise," *ISSCC Dig. Tech. Papers*, pp. 78–79, Feb. 2002.
[7] A. Arakali, et al., "Low-power supply-regulation techniques for ring oscillators in phase-locked loops using a split-tuned architecture," *IEEE J. Solid-State Circuits*, vol. 44, no. 8, pp. 2169–2181, Aug., 2009.

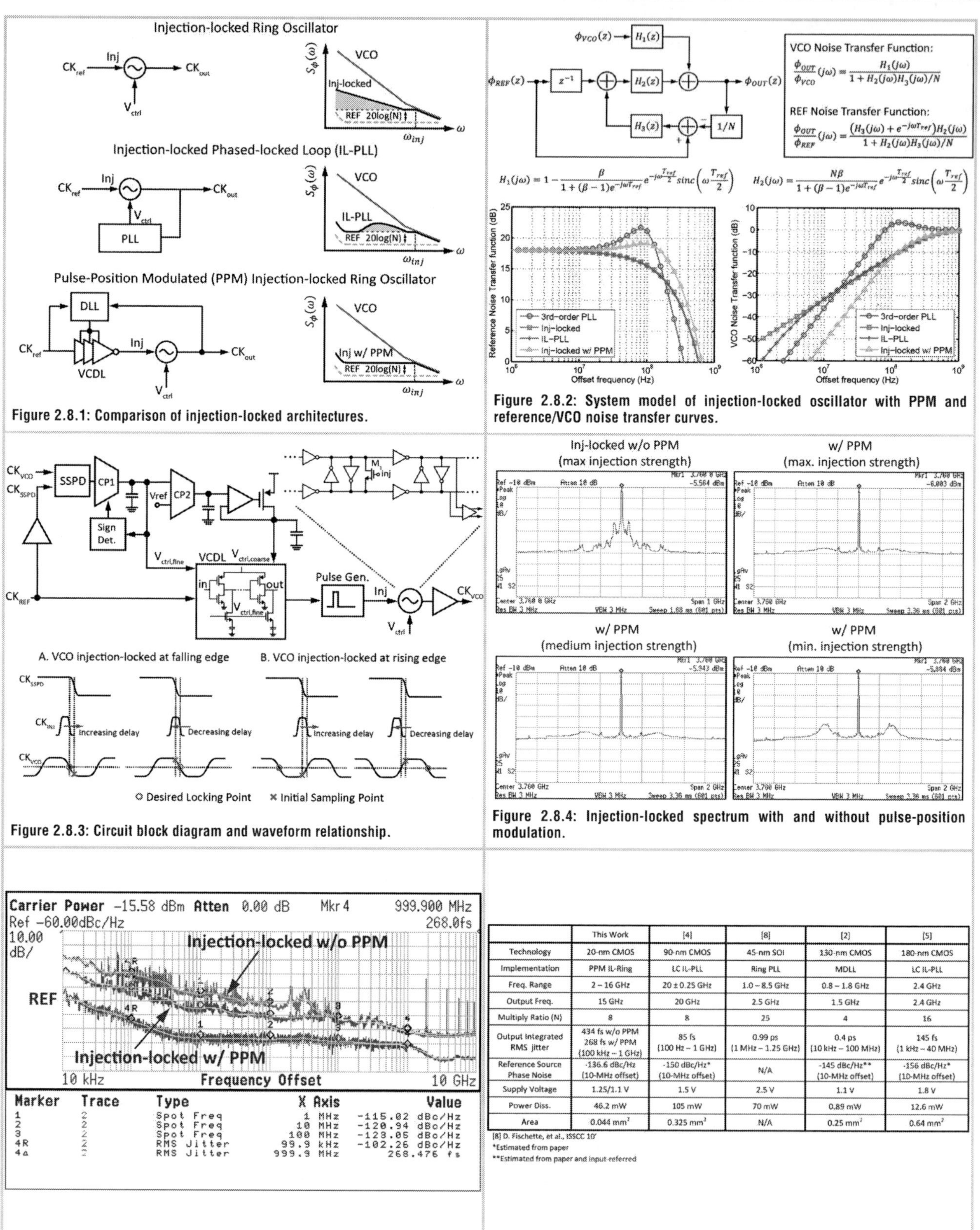

Figure 2.8.1: Comparison of injection-locked architectures.

Figure 2.8.2: System model of injection-locked oscillator with PPM and reference/VCO noise transfer curves.

Figure 2.8.3: Circuit block diagram and waveform relationship.

Figure 2.8.4: Injection-locked spectrum with and without pulse-position modulation.

Figure 2.8.5: Measured phase noise at 15GHz output.

	This Work	[4]	[8]	[2]	[5]
Technology	20-nm CMOS	90-nm CMOS	45-nm SOI	130-nm CMOS	180-nm CMOS
Implementation	PPM IL-Ring	LC IL-PLL	Ring PLL	MDLL	LC IL-PLL
Freq. Range	2 ~ 16 GHz	20 ± 0.25 GHz	1.0 ~ 8.5 GHz	0.8 ~ 1.8 GHz	2.4 GHz
Output Freq.	15 GHz	20 GHz	2.5 GHz	1.5 GHz	2.4 GHz
Multiply Ratio (N)	8	8	25	4	16
Output Integrated RMS jitter	434 fs w/o PPM 268 fs w/ PPM (100 kHz ~ 1 GHz)	85 fs (100 Hz ~ 1 GHz)	0.99 ps (1 MHz ~ 1.25 GHz)	0.4 ps (10 kHz ~ 100 MHz)	145 fs (1 kHz ~ 40 MHz)
Reference Source Phase Noise	-136.6 dBc/Hz (10-MHz offset)	-150 dBc/Hz* (10-MHz offset)	N/A	-145 dBc/Hz** (10-MHz offset)	-156 dBc/Hz* (10-MHz offset)
Supply Voltage	1.25/1.1 V	1.5 V	2.5 V	1.1 V	1.8 V
Power Diss.	46.2 mW	105 mW	70 mW	0.89 mW	12.6 mW
Area	0.044 mm^2	0.325 mm^2	N/A	0.25 mm^2	0.64 mm^2

[8] D. Fischette, et al., ISSCC 10'

*Estimated from paper

**Estimated from paper and input-referred

Figure 2.8.6: Comparison table.

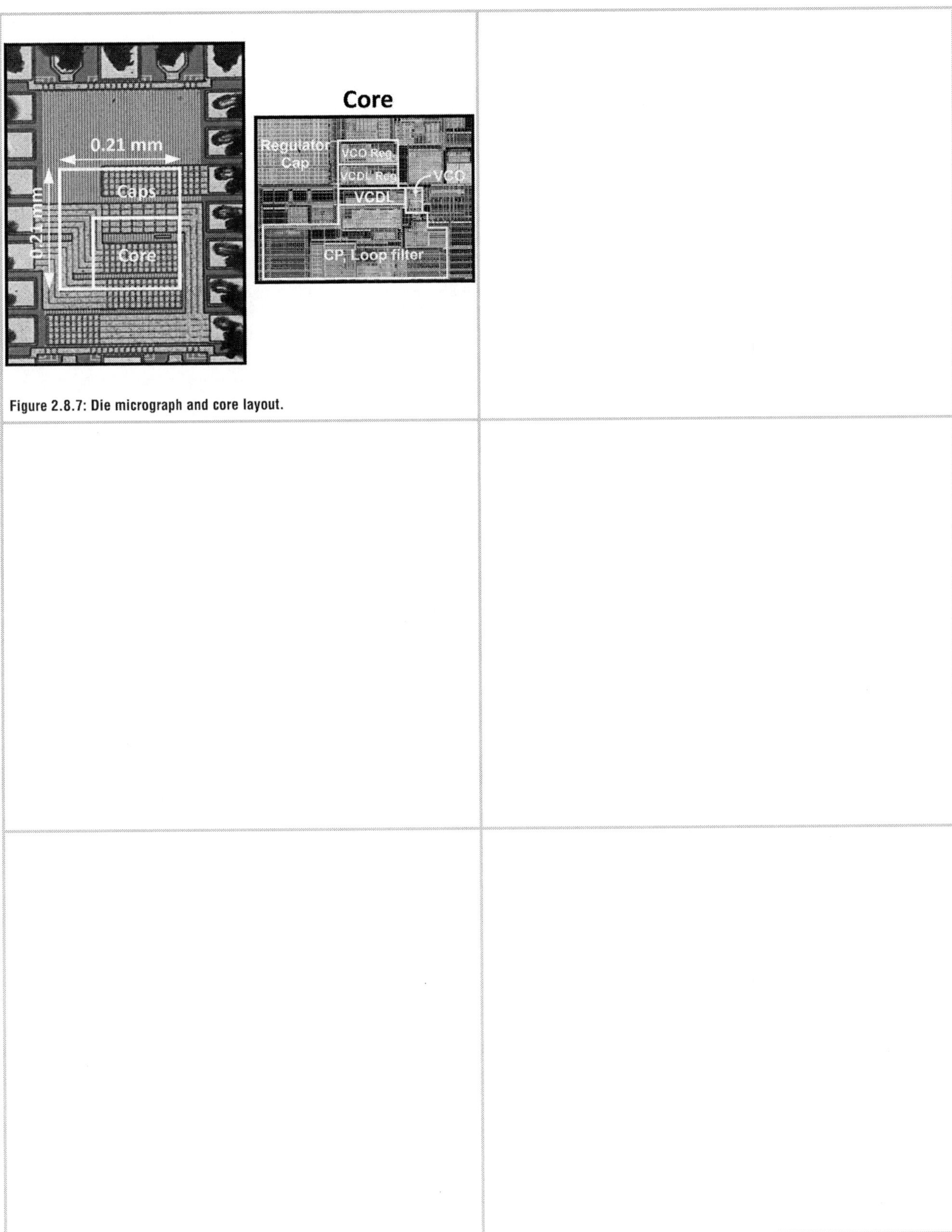

Figure 2.8.7: Die micrograph and core layout.

2.9 A Background Calibration Technique to Control Bandwidth in Digital PLLs

Giovanni Marzin, Salvatore Levantino, Carlo Samori, Andrea L. Lacaita

Politecnico di Milano, Milan, Italy

The bandwidth of a phased-locked loop (PLL) is dependent on several analog parameters that are subject to process, temperature and voltage spreads, as well as to variations along the frequency-tuning range. Even in digital PLLs, which rely on a digital loop filter, the bandwidth still depends on the gains of two mixed-signal building blocks, namely the time/digital converter (TDC) and the digitally-controlled oscillator (DCO), that have conversion characteristics that are not well-controlled. The situation is even more cumbersome employing a single-bit TDC, often referred to as bang-bang phase detector (BBPD), where the linearized gain is inversely proportional to the input jitter [1]. An accurate and repeatable value of the PLL bandwidth, and in the general of the frequency response, is essential to meet several specifications, such as stability margin, settling time, jitter and spur level. When the PLL is operated as a direct frequency modulator with pre-emphasis of the modulation signal, the accuracy requirement of the frequency response is even more demanding [2]. Previously disclosed methods to control PLL bandwidth require a modulation signal to be injected into the loop [2], compensate the gain variations of just a single block (e.g., VCO [3] or BBPD [4]), or operate in the foreground [5]. This paper presents a digital PLL employing a digital background normalization of loop gain, which makes it independent of any analog variable (except for the reference frequency, which often is available from an accurate source). This method requires no injection of additional test signals and operates at a low rate, achieving low-noise and low-power operation, and also is suitable even for bang-bang PLLs.

Figure 2.9.1 shows the developed automatic bandwidth control (ABWC) embedded into a typical digital PLL scheme. The main function of the ABWC is to estimate the gain of the DCO-divider-TDC cascade, and to divide the TDC output e[k] by this gain. Doing so, the gain of the PLL loop becomes independent of any analog variable and dependent only on the transfer function of the digital loop filter, which is perfectly predictable and repeatable. The gain estimation is based on a least-mean square (LMS) algorithm. However, instead of injecting a training sequence into the loop, potentially worsening PLL noise performance, we use the quantization error sequence introduced by the $\Delta\Sigma$ modulator. The latter, which is typically employed in digital PLLs to improve DCO frequency resolution, adds quantization noise -q[k] to its input signal s[k]. The sequence -q[k] circulates through the loop and appears at TDC output e[k]. Thus, correlating e[k] and q[k], we are able to estimate the loop gain.

To explain in more details the operating principle of the calibration technique, we refer to Fig. 2.9.2. The DCO operates as a zero-order hold (ZOH) block and an integrator, where K_{dco} is the DCO tuning sensitivity in units of rad/s/b. The divider/TDC ensemble operates as the cascade of a gain stage 1/N (where N is the division factor of the divider), a sampler at the reference rate (T_{ref} is the reference period) and a linear gain K_{tdc} (in b/rad). Thus, the whole DCO/divider/TDC cascade acts as a sampled-time integrator at reference rate and it can represented in the z-domain as an integrator with gain G, whose expression is given in Fig. 2.9.2. The purpose of the ABWC block is to estimate the gain G. To do so, the $\Delta\Sigma$ quantization error q[k] is integrated in the digital domain, similar to the integration of the $\Delta\Sigma$ modulator output tw[k] = s[k]-q[k] operated in the analog domain by the DCO/divider/TDC cascade. After the integration, the output a[k] is multiplied by a sequence g[k], once again similar to the multiplication by G performed by the analog blocks. The sum of e[k] and g[k]a[k], which contains the component (g[k]-G)a[k], is multiplied by a[k] and the result is fed to a digital integrator that produces g[k]. In this way, the difference (g[k]-G) is forced to zero at steady state and, multiplying e[k] by the inverse of g[k], the PLL loop gain is made independent of analog parameters. In the above analysis, the presence of the PLL control signal s[k] is neglected, since s[k] is uncorrelated with q[k], and the effect of the PLL closed loop, which alters the transfer function from tw[k] to e[k] at low frequencies, is not taken into account, because the spectrum of q[k] is high-pass shaped.

The practical implementation of the digital PLL and the bandwidth-calibration technique is illustrated in Fig. 2.9.3. The TDC is implemented as a bang-bang phase detector and the frequency divider is realized as a digital/time-converter-based fractional-N divider, adopting the same architecture disclosed in [1]. The digital loop filter is a programmable proportional-integral filter, followed by an IIR block. DCO coarse tuning is realized by a multi-bank capacitor topology, which covers the required dynamic range while maintaining low-area occupation. Fine tuning is realized by a MOS varactor, with sensitivity of about 2MHz/V, driven by a 5b DAC and a single-pole 12MHz RC filter. The latter is used to filter out the quantization noise of the $\Delta\Sigma$ modulator, which operates at the reference rate. The RC filter alters the transfer function of DCO/Divider/TDC cascade, which can be approximated, in this case, by a two-tap FIR filter followed by the discrete integrator with gain G. Based on this model, the ABWC block is built to estimate the gains $g_0[k]$ and $g_1[k]$ of the two filter taps. The sum of these two gains converges to the integrator gain G, as desired. The ABWC block is fully implemented in the digital domain and operates in the background of the PLL standard operation. The required operators are limited to 3 adders, 2 integrators, 4 multipliers and one inversion, and they work at the reference rate, thus allowing for low power consumption and easy realization.

The frequency synthesizer in Fig. 2.9.3 is integrated in a 65nm CMOS process. Starting from a 40MHz reference, it synthesizes frequencies from 2.9 to 4.0GHz with fractional resolution of about 70Hz. Figure 2.9.7 shows the die micrograph of the test-chip. The digital section has a placement density of about 40% so the core area of the overall synthesizer is estimated to be about 0.2mm^2. The ABWC block occupies about 5% of the total core area. The power consumption of the synthesizer is 4.5mW (excluding the pad driver), while the contribution of the ABWC is estimated to be 0.04mW. The ABWC block allows the PLL bandwidth to be digitally programmed in a predictable way from about 100kHz to 2MHz, by adjusting the parameters of the digital loop filter. Figure 2.9.4 shows a comparison between the measured phase-noise spectra and the traces expected from calculations (at different bandwidth settings). The good match between measured and theoretical spectra demonstrate the high accuracy of the calibration technique. The worst-case accuracy, which has been verified over all bandwidth settings and over five samples, is better than 4%. Since this PLL is based on bang-bang phase detection, in the absence of the ABWC correction, its loop gain and bandwidth depend on input jitter [1]. To assess the effectiveness of bandwidth calibration in the presence of jitter variations, we measure PLL phase noise at different levels of reference noise with the ABWC block both on and off. Figure 2.9.5 shows the resulting spectra. Using the 1MHz bandwidth setting, when the AWBC is disabled and the reference noise level is increased, the bandwidth shrinks to 200kHz, showing its dependence on input jitter. In contrast, when the ABWC is enabled, the bandwidth is always equal to 1MHz, independent of the level of reference noise. The achieved performance is summarized in Fig. 2.9.6. In the 300kHz bandwidth setting, jitter is 400fs$_{rms}$ and settling time for a 40MHz frequency step is 120µs.

Acknowledgments:
The authors wish to thank M. Zanuso for useful discussions and STMicroelectronics for the silicon donation.

References:

[1] D. Tasca et al., "A 2.9-to-4.0GHz fractional-N digital PLL with Bang-Bang phase detector and 560fsrms integrated jitter at 4.5mW power," *IEEE ISSCC Dig. Tech. Papers*, pp. 88-90, Feb. 2011.

[2] D. McMahill and C. G. Sodini, "A 2.5-Mb/s GFSK 5.0-Mb/s 4-FSK automatically calibrated $\Delta\Sigma$ frequency synthesizer," *IEEE J. Solid-State Circuits*, vol. 37, pp. 18–26, Jan. 2002.

[3] J. Shin and H. Shin, "A 1.9-3.8 GHz $\Delta\Sigma$ fractional-N PLL frequency synthesizer with fast auto-calibration of loop bandwidth and VCO frequency," *IEEE J. Solid-State Circuits*, vol. 47, pp. 665–675, Mar. 2012.

[4] D.-S. Kim et al., "A 0.3-1.4 GHz all-digital fractional-N PLL with adaptive loop gain controller," *IEEE J. Solid-State Circuits*, vol. 45, pp. 2300-2310, Nov. 2010.

[5] M. Ferriss et al., "An Integer Path Self-Calibration Scheme for a Dual-Loop PLL," *IEEE J. Solid-State Circuits*, vol. 48, pp. 996–1008, April 2013.

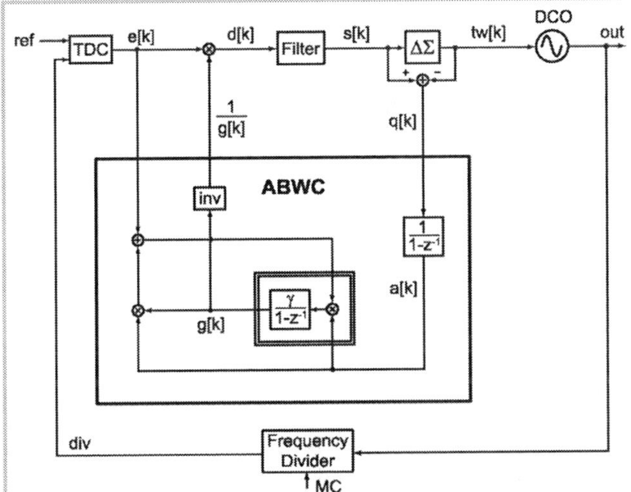

Figure 2.9.1: Operating principle of automatic bandwidth control.

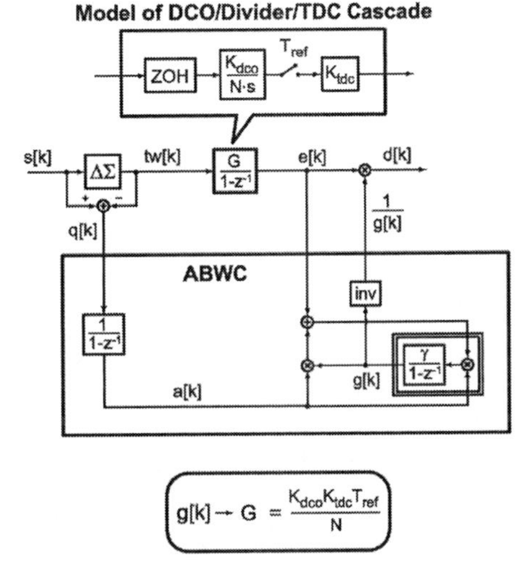

Figure 2.9.2: Equivalent model.

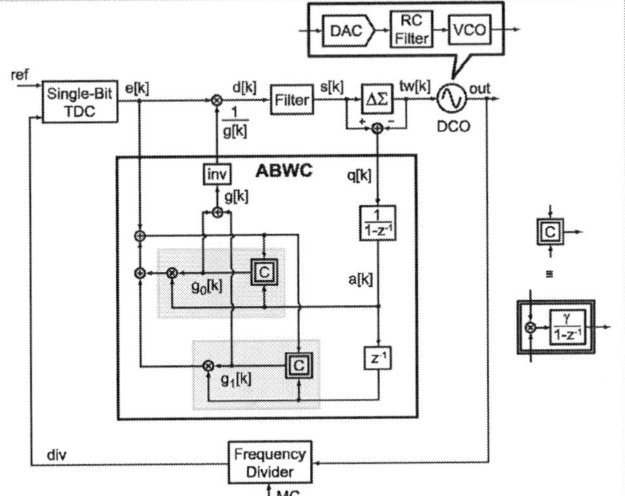

Figure 2.9.3: Implemented digital PLL.

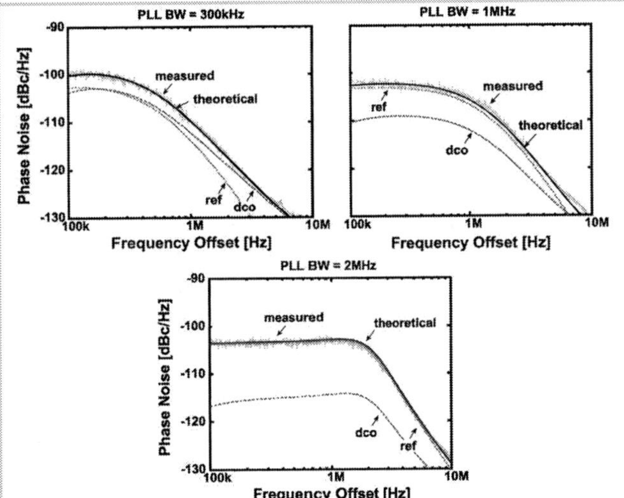

Figure 2.9.4: Measured and theoretical phase noise.

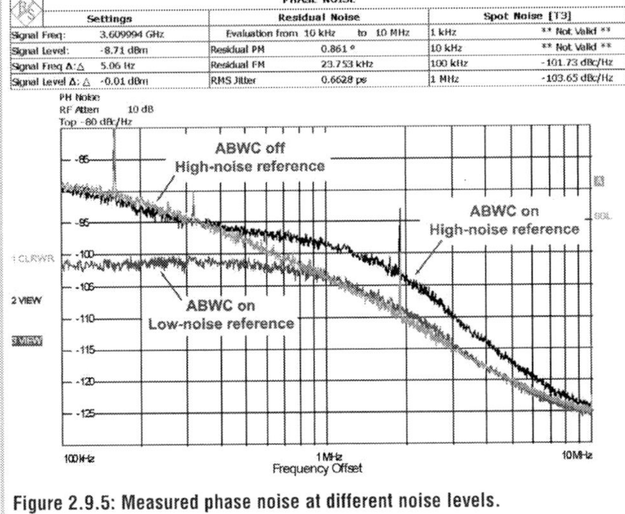

Figure 2.9.5: Measured phase noise at different noise levels.

Synthesizer architecture	Fractional-N Digital PLL
Bandwidth calibration technique	Background
Parameters detected	Loop Gain
Reference clock frequency	40 MHz
Synthesizer output frequency	2.9 - 4.0 GHz
Total power consumption	4.5 mW
ABWC power consumption	0.04 mW
Total area occupation	0.22 mm^2
PLL bandwidth range	100 kHz - 2 MHz
Bandwidth regulation accuracy	< 4%
Settling time (at 300-kHz PLL BW)	120 μs
RMS jitter (at 300-kHz PLL BW)	400 fs
Process	CMOS 65 nm

Figure 2.9.6: Performance table.

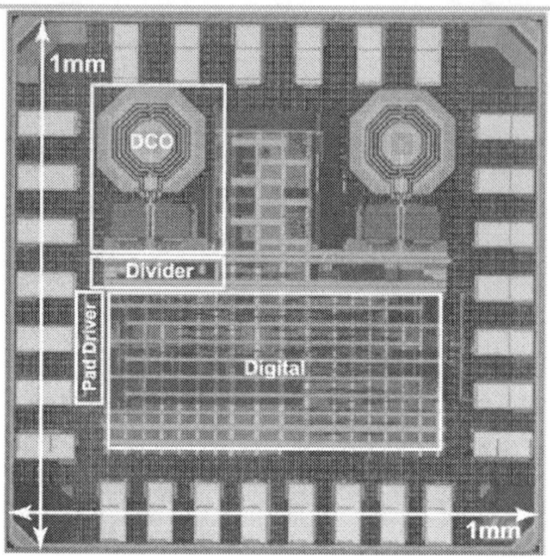

Figure 2.9.7: Die micrograph.

Session 3 Overview: *RF Techniques*

RF SUBCOMMITTEE

Session Chair: *Masoud Zargari*
Qualcomm-Atheros, Irvine, CA

Session Co-Chair: *Tae Wook Kim*
Yonsei University, Seoul, Korea

In recent years the use of CMOS technology has enabled the integration of low-cost, low power and multi-standard RF transceivers. While architectural innovation has allowed most of the transceiver blocks to be integrated in a single chip, full compliance to wireless standards still demands that many off-chip components be added to achieve the overall solution. The papers presented in this session focus on increasing the level of integration of several key building blocks of wireless transceivers. The first part of the session is focused on the transmitter side and in particular on several closed-loop techniques to maximize the output power delivered to the antenna. The second part of the session gives an overview of the most recent techniques applied to the receiver path. Three wideband solutions are presented, followed by a state-of-the-art tuner and a novel beam-forming receiver.

3.1 Polar Antenna Impedance Detection and Tuning for Efficiency Improvement in a 3G/4G CMOS Power Amplifier — **1:30 PM**

S. Kousai, Toshiba, Kawasaki, Japan

In Paper 3.1, Toshiba proposes a new on-chip antenna impedance detection technique, which improves PAE by 1.5x in a 0.13µm 3G/4G CMOS Power Amplifier.

3.2 A 1.95GHz Fully Integrated Envelope Elimination and Restoration CMOS Power Amplifier with Envelope/Phase Generator and Timing Aligner for WCDMA and LTE — **2:00 PM**

K. Oishi, Fujitsu Laboratories, Kawasaki, Japan

In Paper 3.2, Fujitsu Laboratories presents a fully integrated EER CMOS PA in 90nm for WCDMA and LTE. Power efficiencies of 34.1% and 32.2% were achieved for LTE10MHz and LTE20MHz respectively.

3.3 A Transformer-Coupled True-RMS Power Detector in 40nm CMOS — **2:30 PM**

B. Francois, KU Leuven, Leuven, Belgium

In Paper 3.3, KU Leuven presents a novel on-chip true-rms power detector in 40nm CMOS. The power detector allows measuring the power under varying antenna impedances. The measured input range for a linearity error within ±0.5dB is 32.5dB at 5GHz.

978-1-4799-0917-9/14 $31.00 © 2014 IEEE

3.4 A Dual-Mode Transformer-Based Doherty LTE Power Amplifier in 40nm CMOS **2:45 PM**

E. Kaymaksut, KU Leuven, Leuven, Belgium

In Paper 3.4, KU Leuven presents a dual-mode transformer-based Doherty LTE PA in 40nm CMOS that is 2x more efficient at power back-off compared to previous CMOS LTE PAs.

3.5 A 1.0-to-2.5GHz Beamforming Receiver with Constant-G_m Vector Modulator Consuming < 9mW per Antenna Element in 65nm CMOS **3:15 PM**

M. C. Soer, University of Twente, Enschede, The Netherlands

In Paper 3.5, The University of Twente and TNO present a 4-element beamforming receiver with new constant-G_m vector-modulator topology in 65nm CMOS. This enables a low power consumption of 6.5-to-9mW per element in a 1.0-to-2.5GHz RF band with a 73dB spurious-free dynamic range

3.6 A Noise-Cancelling Receiver with Enhanced Resilience to Harmonic Blockers **3:30 PM**

D. Murphy, Broadcom, Irvine, CA

In Paper 3.6, Broadcom presents a sub-2dB noise-cancelling receiver in 28nm CMOS that boosts large-signal resilience to blockers located at integer multiples of LO harmonics by more than 20dB.

3.7 A Fully Integrated TV Tuner Front-End with 3.1dB NF, >+31dBm OIP3, >83dB HRR3/5 and >68dB HRR7 **3:45 PM**

I-Y. Lee, KAIST, Daejeon, Korea

In Paper 3.7 KAIST, UT Dallas and PHYCHIPS present a TV tuner front-end in 0.13μm CMOS that shows an exceptional linearity performance over the whole front-end gain range with state-of-the-art harmonic-rejection ratio (>83dB HRR3, >89dB HRR5, and >68dB HRR7).The LNA shows >+27dBm OIP3 at 22dB of voltage gain. Contrary to the previous works the 2-stage harmonic rejection is done in baseband, which makes it more resilient to gain mismatch.

3.8 A Fully Integrated Highly Reconfigurable Discrete-Time Super-Heterodyne Receiver **4:15 PM**

M. Tohidian, Delft University of Technology, Delft, The Netherlands

In Paper 3.8 TU Delft presents the first discrete-time superheterodyne receiver with complex BPF in 65nm CMOS. This highly reconfigurable RX covers the input frequency range of 1.8 to 2.5GHz with 0.2-to-20MHz BW and achieves uncalibrated IIP2 of > +85dBm. The power dissipation of the entire chain (up to the ADC) is 55 to 65mW.

3.9 An RF-to-BB Current-Reuse Wideband Receiver with Parallel N-Path Active/Passive Mixers and a Single-MOS Pole-Zero LPF **4:45 PM**

P-I. Mak, University of Macau, Macao, China

In Paper 3.9 The University of Macau, UMTEC, and Instituto Superior Tecnico present an RF to BB current-re-use receiver with parallel N-path mixers to alleviate the tradeoff between NF and power. The 65nm CMOS receiver achieves 4.6±0.9dB NF, +61/+17.4dBm out-of-band IIP2/IIP3 and >51dB HRR2-6, without external parts or calibration.

3.1 Polar Antenna Impedance Detection and Tuning for Efficiency Improvement in a 3G/4G CMOS Power Amplifier

Shouhei Kousai, Kohei Onizuka, Takashi Yamaguchi, Yasuhiko Kuriyama, Masami Nagaoka

Toshiba, Kawasaki, Japan

One of the ultimate goals in power amplifier design is to enhance the effective efficiency and achieve a long battery life. Therefore, both the peak efficiency and the efficiency loss due to antenna impedance mismatch or power back-off are highly critical design issues. In particular, the challenge of the antenna impedance mismatch is becoming more severe, due to the increased frequency band and smaller antenna size. Moreover, the antenna mismatch also changes with time due to the user proximity effect [1].

Considering the aforementioned challenges, antenna impedance tuning has become a key design issue. Conventional techniques detect scalar magnitude or phase with costly and bulky components such as bidirectional couplers [2,3]. Therefore, it is difficult to tune the antenna impedance in a vector fashion to its optimal value precisely and promptly to track the time-varying mismatch. To overcome this issue, we propose an antenna impedance-tuning loop with detection circuits, which is implemented on a CMOS-PA chip, to track the antenna impedance. A low-loss impedance tuner is realized with an SOI switch, benefiting from the recent advance in SOI technology [4]. Thus, the entire tuning loop can be realized on a cost-effective Si platform. For reduced output power, the loop controls the tuner impedance automatically to the optimal value, and improves the back-off efficiency. With the proposed antenna tuning loop, the peak and linear efficiencies, including loss of the impedance tuner, are improved from 28% to 42% and from 16% to 31% at 1.95GHz, respectively, with a VSWR of 2.5. Also, back-off drain efficiency is improved from 30% to 40%, at a power back-off of 6dB.

Figure 3.1.1 describes the concept of the proposed impedance detection and tuning. The idea is to measure the impedance seen from the PA (Z_{PA}) with respect to the reference impedance (Z_{REF}) by detecting the phase and amplitude difference between voltages of X and Y. The result does not depend on the input power, and the PVT variations can be largely minimized due to the nature of the relative measurements. The attenuation ratio and phase offset are pre-determined such that the outputs of the amplitude comparator (V_{Z0}) and phase comparator (V_{θ}) are both at zero if Z_{PA} is optimal (usually 50 Ω). This mitigates the issues of delay, gain, and impedance mismatches between the PA and reference amplifier. When Z_{PA} deviates from the optimal value, the comparator outputs of V_{Z0} and V_{θ} represent the magnitude (Z_0) and phase (θ) of Z_{PA}, respectively, as shown in the Smith chart plots. Then the tuner can be orthogonally controlled so that the amplitude and phase detector outputs are both set back to zero, resulting in an optimal Z_{PA}. This scheme can be expanded to different frequency bands by applying different attenuation and phase offsets for different frequencies. Moreover, this scheme can be expanded to track complex optimum impedance for general applications. In this paper, the scheme is also utilized to improve back-off efficiency, as Z_{PA} is controlled to optimal values for reduced output power. In detail, when the effective gate width of the PA is reduced by deactivating certain numbers of PA units, the output current (i_{PA}) of the PA will be reduced linearly. The control loop thus automatically increases the impedance observed from the PA (Z_{PA}), such that the voltage swing at the PA output (Vx) is constant. Consequently, the tuned impedance is inversely proportional to the PA current ($Vx = i_{PA} \times Z_{PA}$), satisfying load-line theory and resulting in optimum impedance for reduced gate width [5]. For an ideal case the peak drain efficiency is independent of the gate width and is kept to the maximum value.

Figure 3.1.2 shows the implemented block diagram. The PA, driver amplifier (DA), and reference amplifier are differential amplifiers with cascode thick-gate-oxide devices. The comparisons of phase and amplitude are realized by an envelope OTA and phase detector [6]. The outputs of the envelope OTA and the phase detector (PD) are quantized in three levels, which indicate whether the magnitude (Z_0) and phase (θ) of Z_{PA} are too large, too small, or on the target. The reference levels for the quantizers are provided considering the sensitivity for impedance detection. The impedance tuner is realized as a separate board with SP5T antenna SOI SWs. The controller is implemented in an FPGA, which estimates the Z_{ANT} by 0.1 steps of the normalized real and imaginary reflection coefficients (Γ_{ANT}) based on the quantized result by using a look-up table, and updates the tuner setting by +0.1 / 0 / -0.1 at a time.

In this implementation, the above-mentioned tuning sequence is triggered by a burst detector. One quantization is executed with each burst, to track the VSWR drift. The trigger is generated at a defined RF swing, for which the PA is not distorted. This is effective to suppress the nonlinear effect of the power amplifier, as detected amplitude and phase drifts with a large output swing. In order to prevent false triggering by modulated signals, the burst is detected after output swing exceeds two trigger levels within a defined period of around 8μsec. Note that the accurate impedance measurement is possible even when the input power is being ramped up, since both amplitude and phase measurements are relative. The bandwidth for the impedance measurement is about 3MHz. After the burst detection, the reference amplifier, envelope OTA, and PD are also used as a part of the linearization loop of the PA-closed loop [6].

Since the envelope OTA and PD do not reject harmonics, voltage at the antenna side, instead of the PA side, is detected as shown in Fig. 3.1.2. Also differential detection is essential for minimizing the effect of large substrate noise. Figure 3.1.3 explains more details about the detector. Voltage is detected through the capacitive coupling between the detector coil and the antenna coil. Thanks to the strong magnetic coupling within the output transformer, current detection due to the magnetic coupling between the detector coil and transformer is negligible. Therefore the voltages at nodes A and A' are mostly linearly related to the antenna impedance (Z_{ANT}). The differential-mode signal between A and A' is enhanced by utilizing the series inductance ($L_{d1,DIFF}$) and parallel inductance ($L_{d2,DIFF}$), while the common-mode signal is reduced according to the ratio of the capacitors, as the common-mode inductance for $L_{d1,CM}$ is negligibly small. Simulated voltage gain is about -13dB at 1.95GHz, and the CMRR is 35dB. The 3rd-order harmonic is rejected due to the filtering effect of the output transformer.

Figure 3.1.4 shows the measurement results of the impedance detection. The quantized result is plotted on Smith charts for impedance seen from the PA (Z_{PA}). The magnitude of 50Ω and phase of 0 degrees, which are "Line-1" and "Line-2" on the figure, respectively, are accurately measured. The measured accuracy is less than VSWR of 1.2. Figure 3.1.5 shows the efficiency measurement result at 1.95GHz. The linear efficiency represents the PAE where the output spectrum satisfies the ACLR of 40dBc with a WCDMA signal of 3.5dB PAPR. The loss of the tuner is included for the results with tuning, and the PA-closed loop is activated with the impedance tuning. When the VSWR is about 2.5, the peak PAE is kept higher than 40% with tuning, whereas it varies from 28% to 45% without tuning. The linear efficiency without the tuning is 15% as the worst case, while more than 30% of PAE is maintained when the tuning and PA-closed loop are enabled. The VSWR dependencies are also shown for different two angles, where both the peak and linear efficiencies are drastically improved. Figure 3.1.6 shows the result of the back-off efficiency improvement. When the PA unit is decreased from 10 to 5, the drain efficiency is improved from 30% to 40%, at an output power of 24dBm. In this case, the tuned impedance is around 100Ω, aligned with the load-line theory. Theoretically, the backed-off peak drain efficiency is as high as full-power peak efficiency. The increased loss of the on-chip transformer is the major reason for the discrepancy. Figure 3.1.7 shows the micrograph of the die and a photo of the impedance tuner board. Performance summary and comparison are also shown.

Acknowledgement:
Authors would like to thank Y. Aizawa and M. Sugiura from Toshiba and H. Wang from Georgia Tech. for their support.

References:
[1] K. Boyle, et al., "Analysis of Mobile Phone Antenna Impedance Variations with User Proximity", *IEEE Trans. Antennas and Propagation,* pp. 364-372, Feb. 2007.
[2] H. Song, et al., "A CMOS Adaptive Antenna-Impedance-Tuning IC Operating in the 850MHz-to-2GHz band", *ISSCC Dig. Tech Papers,* pp. 384-385, Feb. 2009.
[3] A. Bezooijen, et al., "A GSM/EDGE/WCDMA Adaptive Series-LC Matching Network Using RF-MEMS Switches", *IEEE J. Solid-State Circuits,* pp. 2259-2268, Oct 2008.
[4] http://www.semicon.toshiba.co.jp/eng/product/rf/rf_sw/index.html
[5] S. Cripps, "RF Power Amplifiers for Wireless Communications", 2nd edition, Artech House, 2006.
[6] S. Kousai, "A 28.3mW PA-Closed Loop for Linearity and Efficiency Improvement Integrated in a 27.1dBm WCDMA CMOS PA", *ISSCC Dig. Tech. Papers,* pp. 84-86, Feb. 2012.

978-1-4799-0917-9/14 $31.00 © 2014 IEEE

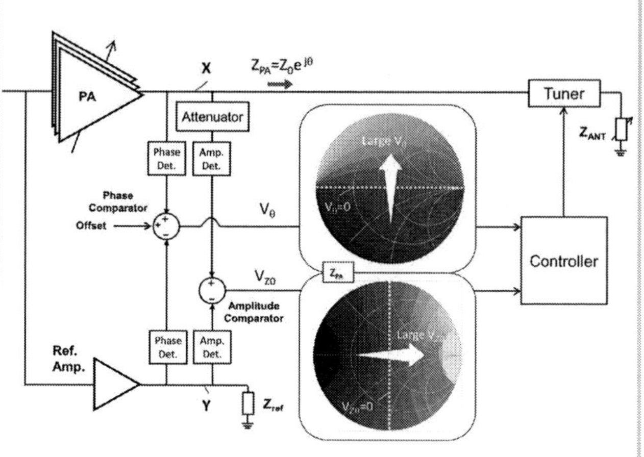

Figure 3.1.1: Conceptual block diagram for proposed impedance detection and tuning.

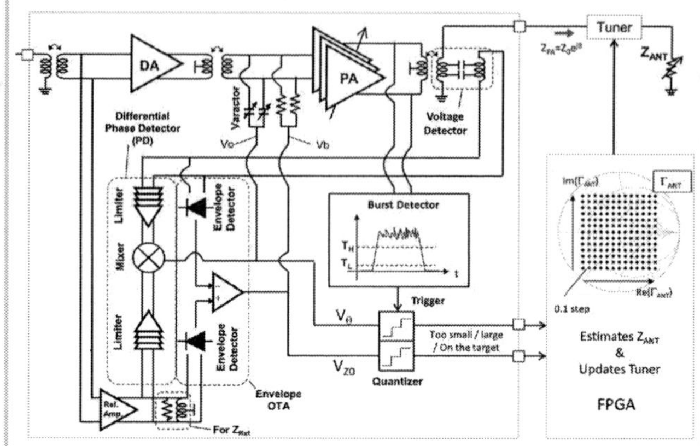

Figure 3.1.2: Detailed block diagram of the implemented circuit.

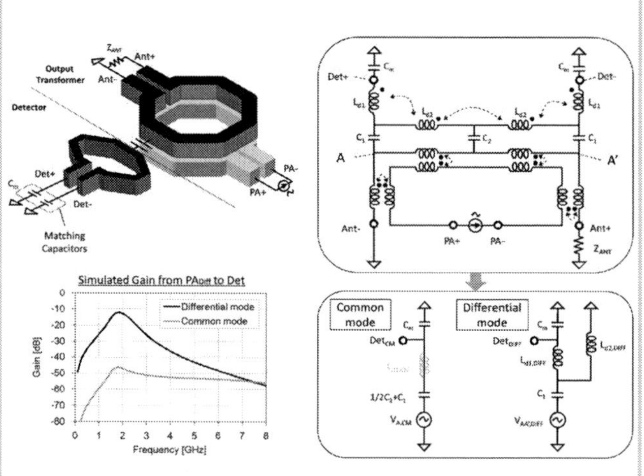

Figure 3.1.3: Differential voltage detector.

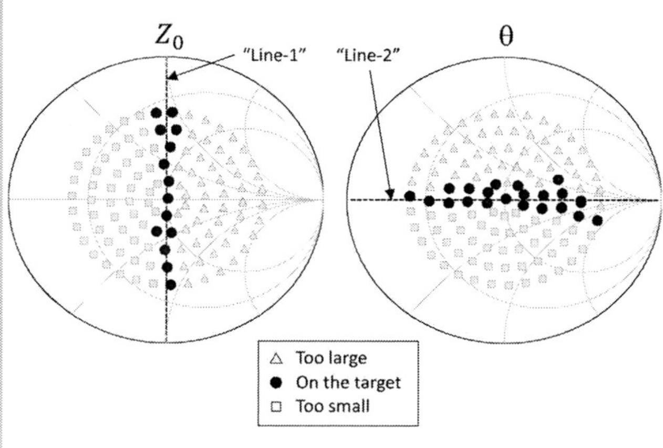

Figure 3.1.4: Impedance measurement results plotted on the Smith chart for Z_{PA}.

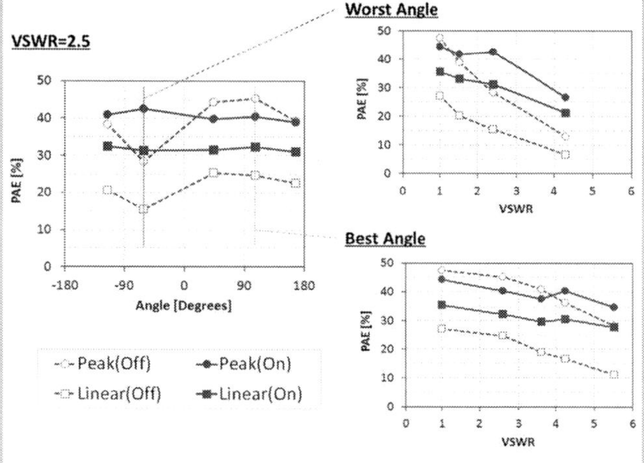

Figure 3.1.5: Efficiency dependence on VSWR with tuning/PA-closed loop on and off.

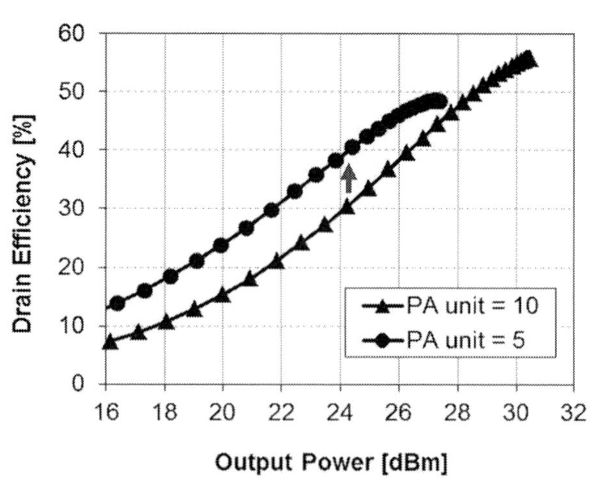

Figure 3.1.6: Drain efficiency for reduced PA unit.

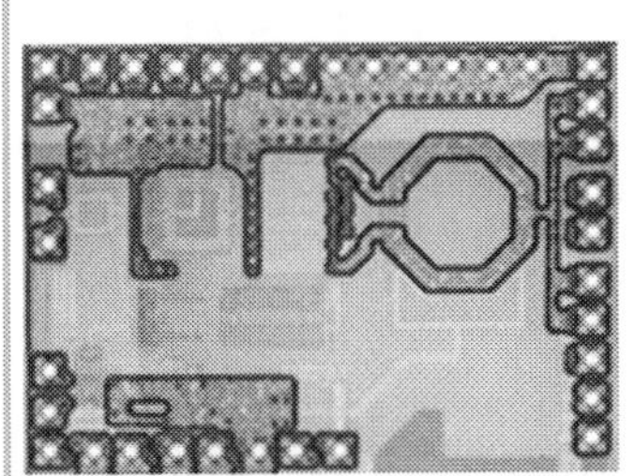

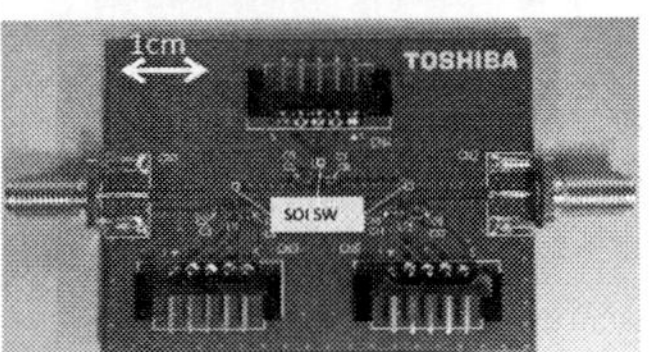

Performance Summary (Measured)

PAchip	
Technology	0.13 μm CMOS
Area	1.6 mm x 2.2 mm
Frequency	1.95 GHz
Peak Output Power	30.8 dBm
Peak PAE	47%
WCDMA*1 Linear Output Power *2	27.9 dBm
WCDMA PAE @Linear	38 %
LTE *3 Linear Output Power *4	25.5 dBm
LTE PAE @Linear	30 %
Tuning	
Technology for SP5T SOI SW	0.18 μm
Tuning range	VSWR of <6
Impedance measurement accuracy	VSWR of 1.2
Tuner loss	0.3-1.6 dB

*1 PAPR is 3.5dB. *2 ACLR is 40dB.

*3 Bandwidth is 5MHz and 1% PAPR is 5.5dB. *4 ACLR is 35dB.

Performance comparison for impedance detection

	ISSCC2009 [2]	This work
Detection target	Magnitude only	Polar
Optimization method	Exhaustive search	Successive approximation (Thermometer)
Required calibration steps	$O(N)$	$O(N^{0.5})$
Calibration time per step	1 μsec	8 μsec *5
Non-50Ω tuning	No	Yes
Time-varying input	No	Yes
Power consumption	27.5 mW	30 mW

*5 Limited by the burst detector

Figure 3.1.7: Photographs of the PA chip and tuner board with performance summary and comparison.

3.2 A 1.95GHz Fully Integrated Envelope Elimination and Restoration CMOS Power Amplifier with Envelope/Phase Generator and Timing Aligner for WCDMA and LTE

Kazuaki Oishi[1], Eiji Yoshida[1], Yasufumi Sakai[1], Hideki Takauchi[1], Yoichi Kawano[1], Noriaki Shirai[2], Hideki Kano[2], Masahiro Kudo[2], Tomotoshi Murakami[2], Tetsuro Tamura[2], Shigeaki Kawai[2], Shinji Yamaura[2], Kazuo Suto[2], Hiroshi Yamazaki[1], Toshihiko Mori[1]

[1]Fujitsu Laboratories, Kawasaki, Japan,
[2]Fujitsu Semiconductor, Yokohama, Japan

In recent years, the demand for low cost and system-on-a-chip for mobile terminals has led to the development of a highly-integrated, low-distortion, and high-power-efficiency CMOS power amplifier (PA). To improve the power efficiency of the conventional linear PA [1-4], an envelope tracking (ET) technique, which modulates supply voltage of a linear PA, has attracted attention. However, the published power efficiency, gain and output power are not sufficient for LTE applications [5], and its typical implementation requires an external supply modulator that is a high-speed power supply circuit [6]. Envelope elimination and restoration (EER) is an alternative supply modulation technique that can further improve the power efficiency over ET by replacing the linear PA with a switching PA driven by a phase signal [7]. However, to meet the specified low distortion, especially for LTE with a wide bandwidth baseband signal, an EER PA generally has difficulty achieving a wide bandwidth for the phase signal path, and requires a high-speed supply modulator, and highly accurate timing between envelope and phase signals. To overcome these problems, this paper introduces an envelope / phase generator based on a mixer and a timing aligner based on a delay-locked loop. Additionally, they were integrated with a switching PA and a supply modulator on the same die.

The fully integrated EER CMOS PA depicted in Fig. 3.2.1 consists of a switching PA, a supply modulator, an envelope / phase generator and a timing aligner. The envelope / phase generator divides an input-modulated RF signal into envelope and phase signals. The switching PA is a two-stage amplifier with a driver and an output stage. The driver is composed of a differential source grounded amplifier with an inductive load using 90nm MOS transistors. A supply voltage of 0.9V is applied for the center tap of the inductive load and becomes an input bias voltage for the output stage. The layout distance between the driver and the output stage was minimized by the small number of elements between them, which results in high-speed switching. The output stage is a differential source grounded amplifier using high breakdown voltage transistors. Similar to an inverse Class-F amplifier, the differential third harmonic is suppressed and the impedance of the common second harmonic is set relatively high by an L-C resonator between the differential drain terminals to improve efficiency. The differential outputs of the output stage are combined into a single-ended signal by an on-chip output transformer. The supply modulator output is applied to the center tap of the transformer to control instant output power in accordance with the envelope signal. In the supply modulator, an internal switching regulator with an external 1μH inductor sources or sinks current in accordance with the current sensing result at an internal linear regulator output. The timing aligner compensates for timing mismatch between the envelope and phase signals. An external impedance tuner was used for obtaining the optimal load impedance for the switching PA.

Figure 3.2.2 shows a simplified schematic of the envelope / phase generator and simulated waveforms. The phase signal is generated by a limiter based on CMOS inverters with resistive feedback. The limiter and the driver can handle the phase signal with a bandwidth that is almost 5 to 10 times wider than that of the baseband signal, because these are high-speed circuits with a simulated bandwidth of over 300MHz. This can suppress degradation of the phase signal. The envelope signal is generated by a full-wave rectifier, which multiplies a modulated RF signal by the limiter output. It consists of a passive mixer, which is an adequately biased NMOS and PMOS combination, with low distortion and small delay. High-frequency components at the rectifier output are eliminated by finite frequency response of the envelope path. The DC offset voltage generated by self-mixing at the rectifier is cancelled at the supply modulator by subtracting a replica rectifier's DC offset voltage from the main rectifier's DC offset voltage.

Another major problem of EER, which deteriorates linearity, is timing mismatch between envelope and phase signals. According to simulated results of an ideal model, timing mismatch needs to be less than around 1ns for 20MHz-BW LTE compared with around 4ns mismatch for WCDMA. Simulated delay of the phase path was less than 0.3ns and estimated delay of the supply modulator is 4 to 16ns. Thus, timing mismatch is mostly generated by the envelope path. The timing aligner depicted in Fig. 3.2.3 compensates for delay in the envelope path by a delay-locked loop with a variable high-pass filter (HPF) before the supply modulator following the envelope signal. It uses negative delay of the HPF, as opposed to positive delay of a low-pass filter (LPF). The rectifier output with small delay and the supply modulator output with large delay are both converted into full-swing signals by limiters. Then they are compared by a phase detector and are converted into a cut-off frequency (fc) control signal for the variable HPF through a charge pump and a loop filter. The cut-off frequency fc of the variable HPF is automatically adjusted by feedback control so the delay of the envelope path becomes sufficiently small. After convergence, the timing aligner not only compensates for delay but also expands bandwidth of the envelope path. Therefore, bandwidth of the supply modulator, which generally needs to be a large bandwidth, needs only around the main bandwidth of the envelope signal. The bandwidths of the supply modulator were designed at around 10MHz, 20MHz and 40MHz for WCDMA, 10MHz-BW LTE, and 20MHz-BW LTE respectively. Moreover, unexpected delays caused by parasitic elements on an evaluation board were minimized by single-chip integration.

A fully integrated EER PA chip was fabricated in a 90nm CMOS process with a 3μm-thick metal option. The chip was assembled on an evaluation board and measured at 1.95GHz. Figure 3.2.4 shows the measurement results for the WCDMA uplink. Power-added efficiency (PAE) and adjacent-channel leakage ratio (ACLR) were 39.2% and -41.1dBc, which are higher PAE and lower ACLR than previously reported values [1-3]. Both 1.2V and 3.7V supply voltages were applied to the envelope / phase generator and the timing aligner, and supply voltage for the supply modulator was 3.7V. PAE was obtained from power consumption of all supplies including the 0.9V supply for the driver. Figure 3.2.5 shows the measurement results for a 20MHz-BW LTE uplink. Sufficient ACLR under -33dBc was achieved by the envelope / phase generator and the timing aligner, although baseband bandwidth was wide. PAE was 32.2% at 25.6dBm output power with sufficient gain of 28.1dB, which is more than 14% higher than the PAE reported in [5]. Figure 3.2.6 summarizes the performances including 10MHz-BW LTE results. PAE for 10MHz-BW LTE was higher than previously reported values [4,6] although they do not include power consumption of the driver and use a relatively low-loss off-chip output transformer. Figure 3.2.7 shows a micrograph of the test chip.

Acknowledgements:
The authors would like to thank Hiroyuki Nakamoto and Shoichi Masui for their fruitful discussions and advice.

References:

[1] Bonhoon Koo, et al., "A Fully Integrated Dual-Mode CMOS Power Amplifier for WCDMA Applications," *ISSCC Dig. Tech. Papers*, pp. 82-83, Feb. 2012.
[2] Kouichi Kanda, et al., "A Fully Integrated Triple-Band CMOS Power Amplifier for WCDMA Mobile Handsets," *ISSCC Dig. Tech. Papers*, pp. 86-87, Feb. 2012.
[3] Shouhei Kousai, et al., "A 28.3mW PA-Closed Loop for Linearity and Efficiency Improvement Integrated in a +27.1dBm WCDMA CMOS Power Amplifier," *ISSCC Dig. Tech. Papers*, pp. 84-85, Feb. 2012.
[4] Byungjoon Park, et al., "A 31.5 %, 26 dBm LTE CMOS Power Amplifier with Harmonic Control," *European Microwave Integrated Circuits Conference*, pp.341-344, Oct. 2012.
[5] Kohei Onizuka, et al., "A 1.8GHz Linear CMOS Power Amplifier with Supply-Path Switching Scheme for WCDMA/LTE Applications," *ISSCC Dig. Tech. Papers*, pp. 90-91, Feb. 2013.
[6] Daehyun Kang, et al., "A 34% PAE, 26-dBm Output Power Envelope-Tracking CMOS Power Amplifier for 10-MHz BW LTE applications," *IEEE International Microwave Symposium Dig.* 2012.
[7] Vincent Pinon, et al., "A Single-Chip WCDMA Envelope Reconstruction LDMOS PA with 130MHz Switched-Mode Power Supply," *ISSCC Dig. Tech. Papers*, pp. 564-565, Feb. 2008.

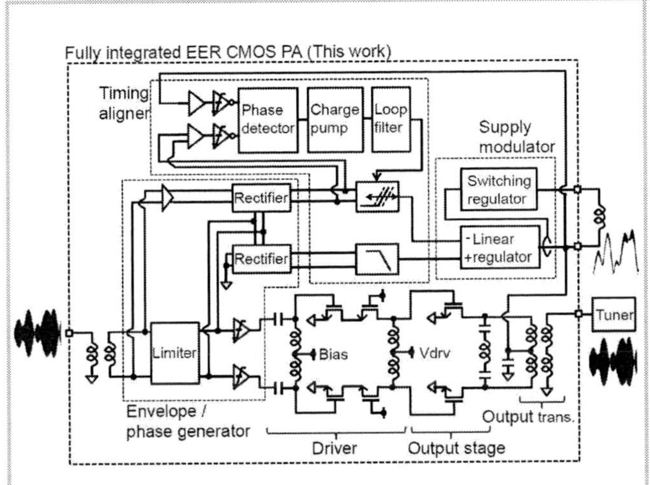

Figure 3.2.1: Block diagram of single-chip EER PA.

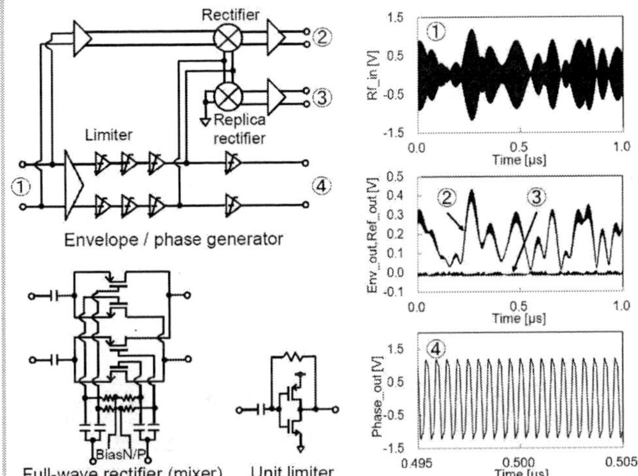

Figure 3.2.2: Schematic of envelope / phase generator.

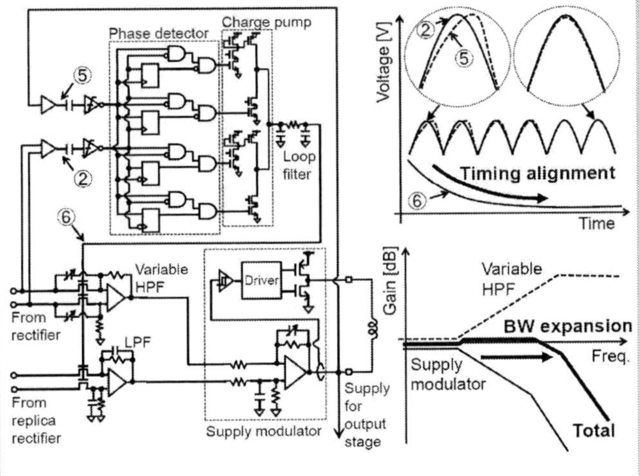

Figure 3.2.3: Schematic of timing aligner.

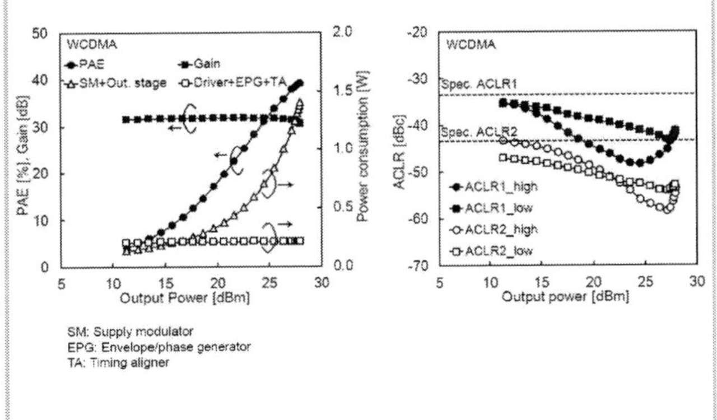

Figure 3.2.4: Measured PAE and ACLR (WCDMA uplink).

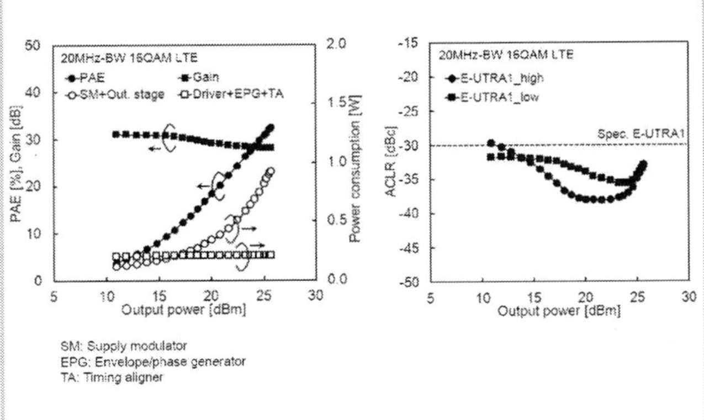

Figure 3.2.5: Measured PAE and ACLR (20MHz-BW LTE uplink).

WCDMA CMOS PAs

	Ref. [1]	Ref. [2]	Ref. [3]	This work
Frequency [GHz]	1.95	2	1.88	1.95
Output Power [dBm]	28	27.4	27.1	28
Gain [dB]	23.7	27.3	28.3	30.5
PAE [%]	36.4	28.5	28	39.2
ACLR1 [dBc]	-35	-34	-40	-41.1
ACLR2 [dBc]	-	-	-52.1	-52.8
Output transformer	on-chip	on-chip	on-chip	on-chip
Technology	0.18μm CMOS	90nm CMOS	0.13μm CMOS	90nm CMOS
Architecture	Linear	Linear	Linear	EER

LTE CMOS PAs

	Ref.[4]	Ref. [5]	Ref. [6]	This work
Frequency [GHz]	1.85	1.8	1.8	1.95
Modulation	10M 16QAM	20M 64QAM	10M 16QAM	10M/ 20M 16QAM
Output Power [dBm]	26	21.3	26	26/ 25.6
Gain [dB]	~10 (w/o driver)	~13 (w/o driver)	~10 (w/o driver)	29/ 28.1 (w/ driver)
PAE [%] (w/o driver)	31.5	18	34	37.2/ 35.3*
PAE [%] (w/ driver)	-	-	-	34.1/ 32.2
E-UTRA ACLR1 [dBc]	-32	-	-32.5	-33.2/ -33
Output transformer	off-chip	on-chip	off-chip	on-chip
Number of chips	1 chip	1 chip	2 chips PA+Supply	1 chip
Technology	0.18μm CMOS	65nm CMOS	0.18μm CMOS	90nm CMOS
Architecture	Linear	Supply path switching (ET mode)	ET	EER

* Calculated without power consumption of the driver.

Figure 3.2.6: Performance summary.

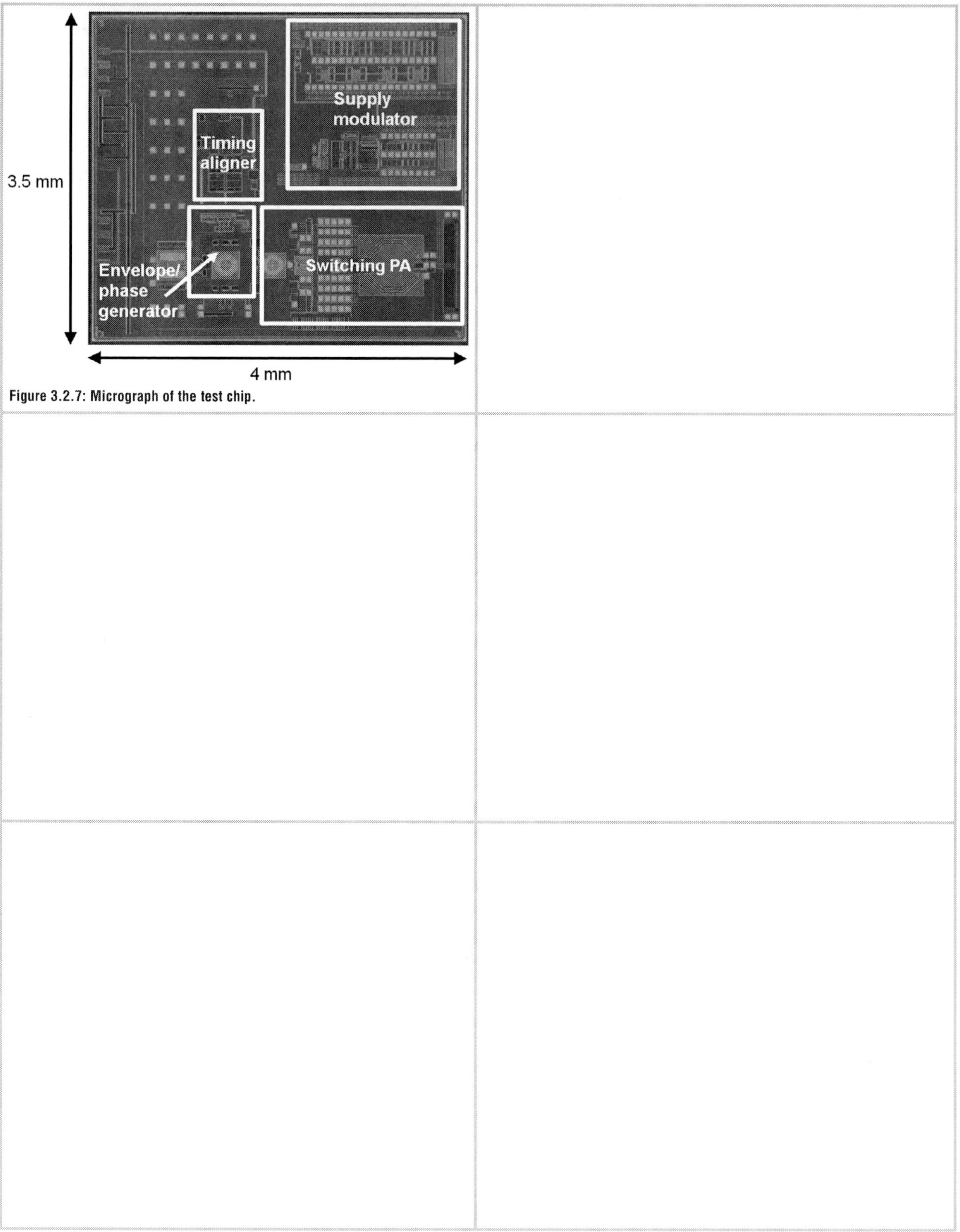

Figure 3.2.7: Micrograph of the test chip.

3.3 A Transformer-Coupled True-RMS Power Detector in 40nm CMOS

Brecht Francois, Patrick Reynaert

KU Leuven, Leuven, Belgium

To optimize the power consumption and system performance of battery-supplied devices, it is required to monitor and adjust the transmitted RF power accurately and continuously. This is typically done by an external power detector (PD), which increases area and cost. On the other hand, fully integrated power detectors are typically voltage-based [1-5] and only give the correct RF output power for a fixed load impedance. But in practice, antenna impedance variations will occur, causing VSWR mismatches that introduce an error in these voltage-based RF output power measurements.

This paper presents a 5GHz WLAN PA with an on-chip true-RMS Power Detector, without any additional power loss or area overhead. The power detector is based on a magnetically coupled sense winding and takes advantage of transformer-based power combining and impedance transformation that has become common practice in nanometer CMOS RF PAs. The proposed power detector performs both an RF voltage and RF current measurement at the PA output and is therefore capable of performing a True power measurement, even under VSWR mismatches or load variations. This proposed power detector is implemented in 40nm standard CMOS and unlike earlier reported power detectors [1-4], it is integrated together with a 5GHz RF PA targeting the WLAN (IEEE 802.11a) communication standard.

To perform the RF current measurement, a sense winding is introduced inside the RF PA output transformer to measure the RF output current. The RF voltage measurement is done through a capacitive division at the transformer output. The current signal and voltage signal are subsequently mixed on chip, as shown in Fig. 3.3.1. This results in a DC output voltage that represents the true RMS output power of the RF PA, without requiring any rectification. The RF PA design is based on transformers and thick oxide cascode transistors. This allows the RF PA to deliver up to 24dBm with a 38.8% PAE. The average output power when IEEE 802.11a WLAN-signals are applied, is 17.2dBm while achieving 18.5% PAE for an EVM of -25dB and obeying IEEE 802.11a WLAN spectral requirements. The impedance transformation at the output is realized with a 2:1 planar output transformer. All transformers and windings are implemented in the 3.5μm-thick top copper layer to achieve highest quality factor.

Figure 3.3.2 shows how the third sense winding is introduced inside the output transformer. The sense winding picks up the magnetic field and thus measures the output current. Similar to the output transformer, the sense winding is also designed in the thick top metal. In order to minimize the loss in the RF path, the sense winding is designed much smaller than the secondary winding, as shown in Fig. 3.3.2. This way, the additional insertion loss is below 0.1dB. The coupling factor (k_{out}) of the RF path from primary to secondary reaches up to 0.84 and is not affected by the third winding, while the coupling to the sense winding is only 0.04 (k_{sense}).

The RF output voltage of the secondary winding is capacitively divided and coupled to the mixer, where it is multiplied with the RF output current from the sense winding. This results in a DC output voltage representing the true RMS output power of the RF PA. The power detector incorporates a mixer and an operational amplifier (op-amp), where the output buffer is only required for performing off-chip measurements. The mixer and power detector operate from a 0.9V supply, whereas the output buffer uses the 2.5V supply. The designed mixer is a pseudo-differential double balanced mixer. Since the down-converted flicker noise of the cross-coupled quad transistors is proportional to their bias current and their parasitic capacitances at the common source node, a cross-coupled pair was introduced in the double-balanced mixer to minimize this flicker noise. The OTA is a two-stage Miller-OTA that amplifies the small signal. The dynamic range of the detector is limited by the linearity and noise of the mixer and OTA. Therefore, biasing is chosen carefully to guarantee linear operation and to maximize the dynamic range of the power detector. The value of the Miller capacitor, C_c, in the OTA is selected larger than for optimal GBW, to minimize the noise integration which maximizes the detection range and accuracy. The bandwidth of the detector is 1MHz, which is more than sufficient for wireless applications.

The magnetic coupled power detector accurately detects the power of the RF PA from -10.5dBm to 22dBm with a maximum inaccuracy of ±0.5 dB. Thus, the measured input range for a linearity error within ±0.5dB is 32.5dB at 5GHz. The dynamic range of the power detection is >25dB when 802.11a WLAN signals are applied. The power detector consumes only 349μA from a 0.9V supply, which is at least 1000 times smaller than the peak power consumption of the RF PA, and it occupies an active area of only 45 x 20μm^2, directly located under the output connections of the output transformer, which is generally unused area anyhow. The benefits of the true-RMS power detector are further verified by load-pull measurements for different VSWR mismatches. Figure 3.3.5 shows the measured detected power versus the real RF output power for different antenna impedances. For antenna impedances close to 50Ω, less than ±0.1dB error is observed. With a much larger impedance variation (VSWR 2.5:1), the maximum inaccuracy is still limited to ±0.6dB. This proves that the presented power detector can accurately sense the transmitted power even under varying load conditions.

Figure 3.3.6 compares the measured performance with recently reported power detectors. Unlike other reported power detectors, the proposed on-chip power detector was fabricated in a 40nm GP CMOS technology and integrated together with an RF power amplifier. Furthermore, the power detection principle is not solely based on the detection of the RF output voltage but on both voltage and current sensing. Moreover, the proposed power detector is measured with both fixed and varying antenna impedances and regardless of the antenna impedance, the linearity or accuracy of the detector still remains, without sacrificing the PA efficiency. To conclude, this design uses both current and voltage sensing and realizes a true-RMS power detection by multiplying both signals. In addition, to reduce cost, the proposed power detector can be completely incorporated inside the PA output transformer resulting in zero area overhead.

References:

[1] Y. Zhou and M. Y -W Chia, "A Low-Power Ultra-Wideband CMOS True RMS Power Detector," *IEEE Trans. on Microwave Theory and Techniques*, vol. 56, no. 5, pp. 1052-1058, May 2008.

[2] K. A. Townsend and J. W. Haslett, "A Wideband Power Detection System Optimized for the UWB Spectrum," *IEEE J. Solid-State Circuits*, vol. 44, pp. 371-381, Jan. 2009.

[3] J. Gorisse, et al., "A 60GHz 65nm CMOS RMS Power Detector for Antenna Impedance Mismatch Detection," *Proc. of ESSCIRC 2009*, pp.172-175, Sept. 2009.

[4] Li Chaojiang, et al., "A Low-Power Ultrawideband CMOS Power Detector With an Embedded Amplifier," *IEEE Trans. on Instrumentation and Measurement*, vol. 59, no. 12, pp. 3270-3278, Dec. 2010.

[5] H. Nakamoto, et al., "A Real-Time Temperature-Compensated CMOS RF On-Chip Power Detector with High Linearity for Wireless Applications," *Proc. of ESSCIRC 2012*, pp. 349-352, Sept. 2012.

978-1-4799-0917-9/14 $31.00 © 2014 IEEE

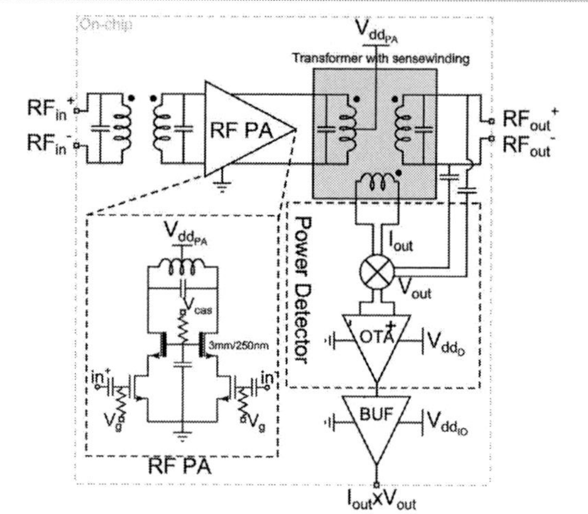

Figure 3.3.1: System diagram of the true-RMS power detector based on a transformer-coupled sense winding with the 5GHz-WLAN RF PA.

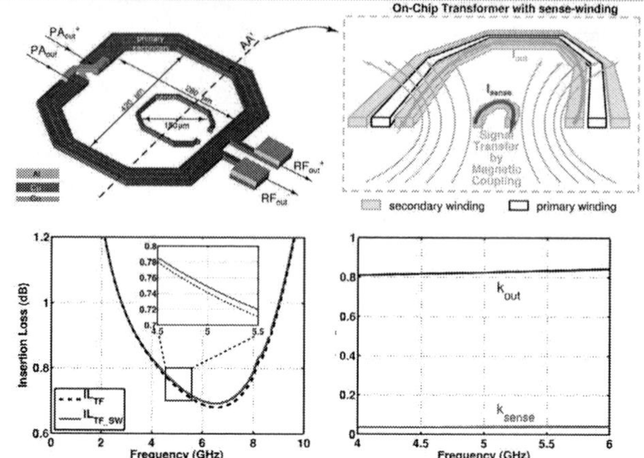

Figure 3.3.2: Layout of the sense winding inside the output transformer and illustrating the magnetic coupling; insertion loss with, IL_{TF}, and without, IL_{TF_SW}, sense winding; coupling factor k_{out} (from primary to secondary winding) and k_{sense} (from secondary to sense winding).

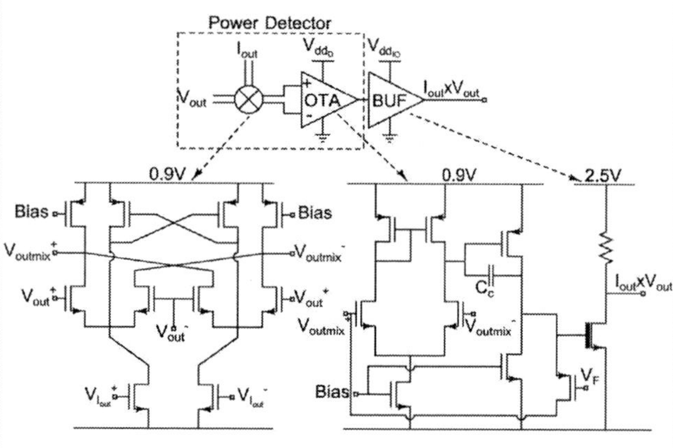

Figure 3.3.3: Schematic of the proposed true RMS power detector.

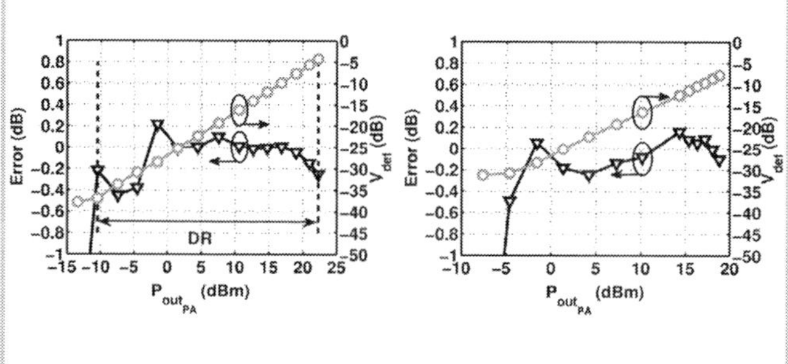

Figure 3.3.4: Measured detector output voltages (dB) and error (dB) versus output power of the RF PA (dBm) for CW (lhs) and 20MHz (IEEE 802.11a) (rhs) WLAN modulated signals at 5GHz.

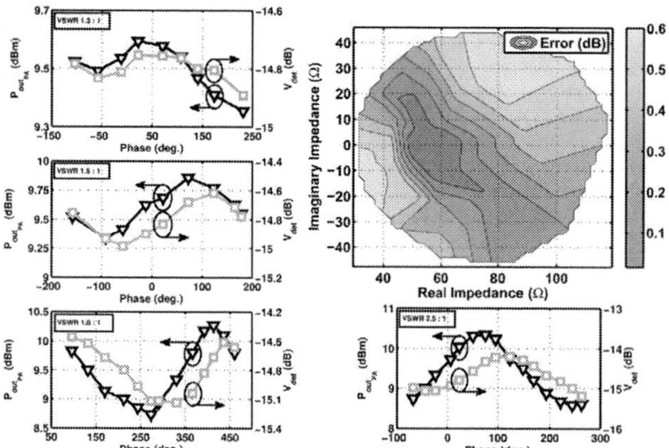

Figure 3.3.5: Measured detector output voltages V_{det} (dB) versus output power of the RF PA, P_{outPA}, for CW at 5GHz for different antenna mismatches: VSWR 1.3:1, VSWR 1.5:1, VSWR 1.8:1, VSWR 2.5:1; the error on the detector output voltage for different antenna impedances.

	[1] MTT'08	[2] JSSC'09	[3] ESSCIRC'10	[4] TIM'09	[5] ESSCIRC'12	This work
CMOS Technology	130 nm	180 nm	65 nm	130 nm	90 nm	**40 nm GP**
PA integrated	No	No	No	No	Yes	**Yes**
Coupling topology	Capacitive	Capacitive	Capacitive	Capacitive	Capacitive	**Capacitive & Inductive**
Detector Principle	Voltage	Voltage	Voltage	Voltage	Voltage	**Voltage & Current**
Power consumption [mW]	0.18	3.8	0.06	0.12	0.3-0.63	**0.31**
Dynamic Range [dB]	20	20	25	21	27	**32.5**
Linearity Error [dB]	±0.5	±2.4	-	±1%	±0.5	**±0.5**
Error under VSWR mismatch [dB]	-	-	-	-	-	**±0.6**
Active Area [μm^2]	12600	360000	6400	85000	39000	**900**
Area Overhead	Yes	Yes	Yes	Yes	Yes	**No**

Figure 3.3.6: Comparison Table.

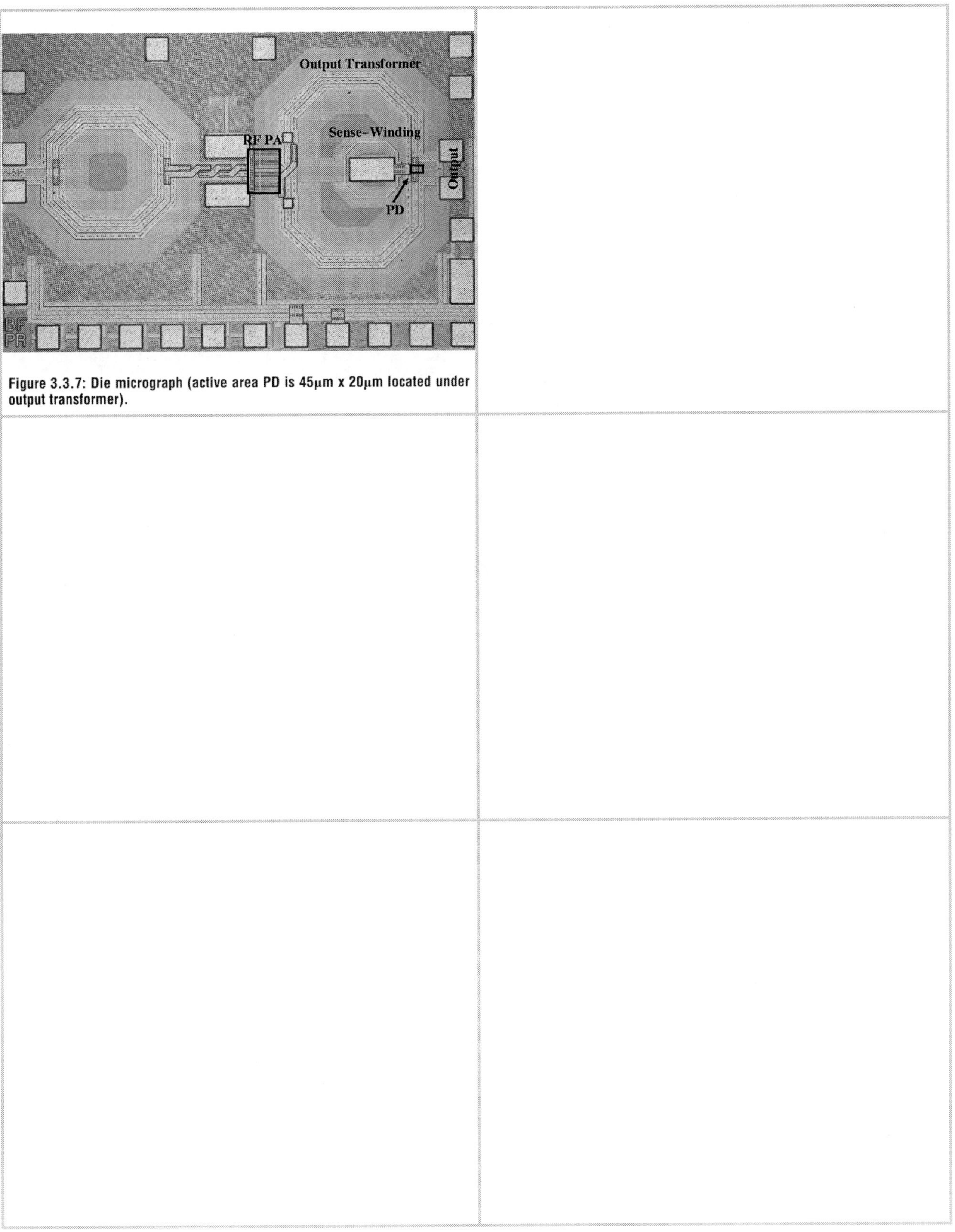

Figure 3.3.7: Die micrograph (active area PD is 45μm x 20μm located under output transformer).

3.4 A Dual-Mode Transformer-Based Doherty LTE Power Amplifier in 40nm CMOS

Ercan Kaymaksut, Patrick Reynaert

KU Leuven, Leuven, Belgium

Modern high-data-rate communication systems such as LTE use spectrally efficient modulation schemes with a high peak-to-average power ratio (PAPR), placing stringent linearity demands on the RF power amplifiers (PA). The main challenge for LTE power amplifiers is therefore to achieve high efficiency and high linearity for a wide power range. In addition, delivering Watt-level output power is another challenge for CMOS RF power amplifiers due to the low breakdown voltage of the transistors. Series-combining transformers (SCT) enable high output power levels by summing up the output voltages of low-voltage CMOS power amplifiers [1-4].

This paper presents a 40nm CMOS LTE RF PA with a hybrid series-combining transformer that consists of two Distributed Active Transformers (DAT) with Doherty operation as shown in Fig. 3.4.1. The Doherty combiner demonstrates high efficiency even when only one of four amplifiers is operational. The large (main) DAT is composed of two main PAs that operate in Class-AB bias. The small (auxiliary) DAT combines two auxiliary amplifiers that are biased in Class-C mode. The PA is implemented in 40nm CMOS technology and can operate in 2 modes: High-Power (HP) mode and Low-Power (LP) mode. The Doherty operation is implemented for both modes. As such, compared to previously published CMOS power amplifiers, this work combines two techniques that improve back-off efficiency and linearity, resulting in a higher efficiency with modulated signals over a wide power range. Furthermore, the proposed Doherty architecture does not suffer from baseband bandwidth expansion that occurs in other RF PA architectures (outphasing, envelope tracking, polar).

The full-wave EM simulation results of the proposed transformer show that the combining efficiency of the proposed transformer is still as high as 74% in the low-power mode when only one amplifier (the main PA1) is operational, including the losses of the non-operational amplifiers and the switches as shown in Fig. 3.4.1. A conventional figure-8-shaped 4-way combiner is also simulated for comparison purposes. Although the figure-8-shaped combiner demonstrates similar peak efficiency as the proposed transformer; the back-off efficiency of the figure-8-shaped transformer is only 58% when only 1 of 4 PAs is operational.

In HP mode, the transformer combines all four amplifiers: two main and two auxiliary amplifiers. In LP mode, only one main and one auxiliary amplifier are operational. The transition between the HP and LP mode is realized by using the switches S1-4. Switches S1 and S3 are used to adjust the tuning capacitance of the main PA1 and aux PA1 respectively. Therefore, they are shorted in HP mode and open in LP mode. The main PA2 and the aux PA2 do not require a switched capacitor since they only operate in the HP mode. The switches S2 and S4 are used to short the output of the main PA2 and the aux PA2 in the LP mode since these amplifiers are not operational in LP mode.

In the HP mode, the two Class-AB main PAs start to saturate and demonstrate high efficiency at 6dB back-off from the peak output power level. At this power level, the two Class-C auxiliary PAs start to deliver power, which reduces the load impedance seen by the main PAs due to active load modulation. Therefore, the Doherty PA achieves high efficiency in the upper 6dB power range as shown in Fig. 3.4.1. The Doherty operation in the LP mode is similar to the HP mode except that only one main PA and only one auxiliary PA are operational. In LP mode, the Doherty PA delivers around 6dB lower power compared to the HP mode. Thus, the high efficiency is implemented over a 12dB range.

Each amplifier and their drivers are implemented by using cascode PA units as shown in Fig. 3.4.2. The power amplifier uses 40 nm NMOS transistors for both the common-source and common-gate amplifiers to achieve high efficiency and high gain. The Doherty amplifier uses a supply voltage of 1.5V for reliable operation. The high-frequency performance of the switches S1-4 is less critical since they toggle only during the transition between the LP and HP modes. The switches are implemented by using thick-oxide transistors (W/L=4mm/250nm).

The Class-C auxiliary amplifier is the key for achieving back-off efficiency enhancement in Doherty operation. It should be biased below the threshold voltage so it does not consume DC power until the main PA saturates. However, Class-C-biased amplifiers cause more distortion and deliver less output power at high power levels compared to their Class-AB counterparts. Therefore, an adaptive bias circuit is implemented in the same chip as shown in Fig. 3.4.2 to keep the auxiliary amplifier's bias voltage below the threshold voltage until the main PA saturates and then increase the bias voltage of the Class-C auxiliary PA with increasing input power.

The CW measurement results of the Doherty PA at 1.9GHz are given in Fig. 3.4.3, showing a peak output power of 28(24.4) dBm with a PAE of 34(30.4)% in the HP(LP) mode. The PAE of the Doherty PA at 6dB back-off is still as high as 25.5(21.9)% in HP(LP) mode. In addition, in the LP mode the PAE at 12dB back-off from the HP peak power level is as high as 19.7%. At 6(12)dB back-off, these results are 1.5(2.3) times better than an ideal Class-B back-off curve with the same peak performance. The CW frequency response of the PA is presented in Fig. 3.4.4. The PAE at 6dB back-off (HP mode) is higher than 20.3% and the PAE at 12dB back-off (LP mode) is above 15.4% from 1.7GHz to 2.1GHz as shown in Fig. 3.4.4. This wideband characteristic is a result of the highly coupled overlying transformer layout shown in Fig. 3.4.1.

The PA is tested with 16QAM, 20MHz BW LTE signals centered at 1.9GHz. The PA satisfies the E-ULTRA specification of ACLR1 lower than -30dBc up to 23.4 (19.8)dBm average output power levels with a maximum PAE of 23.3 (21.1)% in the HP (LP) mode as shown in Fig. 3.4.5. In addition, in the whole measured power range, the measured EVM is 5dB lower than the -18dB EVM specification and the measured ACLR2 is 9dB lower than -36dB ACLR2 specification. The measured efficiency with an LTE signal is still as high as 18.4% at 17.4dBm (6dB back-off) average output power level and 11.1% at 11.4 dBm (12dB back-off) average output power level. This high efficiency with a high PAPR (8.36dB) LTE signal proves the linearity and back-off efficiency enhancement behavior of the proposed Doherty amplifier. Furthermore, the combination of dual-mode operation together with the Doherty technique minimizes the average DC power consumption under the tight power control scheme of LTE systems. In all these measurements, no predistortion was applied.

Figure 3.4.6 compares this work with other CMOS PAs and LTE PAs. Among the CMOS PAs, this work achieves the highest efficiency at 6dB back-off. When compared with state-of-the-art LTE power amplifiers, this PA achieves the highest efficiency among the CMOS LTE PAs. The PAE at 12dB back-off with LTE signals is two times higher than previous LTE CMOS PAs.

Acknowledgments:
This work was supported by the European Union's Seventh Framework (FP7/2007-2013) under grant agreement 248277.

References:

[1] D.Chowdhury et al., "A Fully Integrated Dual-Mode Highly Linear 2.4 GHz CMOS Power Amplifier for 4G WiMax Applications," *IEEE J. Solid-State Circuits*, pp. 3393-3402, Dec. 2009.
[2] W. Tai et al., "A Transformer-Combined 31.5 dBm Outphasing Power Amplifier in 45nm LP CMOS with Dynamic Power Control for Back-off Power Efficiency Enhancement", *IEEE J. Solid-State Circuits*, vol. 47, no.7, July 2012.
[3] E. Kaymaksut, et al., "Transformer-Based Uneven Doherty Power Amplifier in 90 nm CMOS for WLAN Applications", *IEEE J. Solid-State Circuits*, vol. 47, no. 7, pp.1659-1671, July 2012.
[4] K. Onizuka, et al., "A 1.8GHz Linear CMOS Power Amplifier with Supply-Path Switching Scheme for WCDMA/LTE Applications", *ISSCC Dig. Tech. Papers*, pp. 90-91, Feb. 2013.
[5] B. Koo, et al., "A Fully Integrated Dual-Mode CMOS Power Amplifier for WCDMA Applications," *ISSCC Dig. Tech. Papers*, pp. 81-82, Feb. 2012.
[6] C. Yunsung, et. al, "A Dual Power-Mode Multi-Band Power Amplifier With Envelope Tracking for Handset Applications," *IEEE Trans on Microwave Theory and Techniques*, vol. 61, no. 4, pp.1608-1619, April 2013.
[7] R. Wu,; Liu, Y.-T.; Lopez, J.; Schecht, C.; Li, Y.; Lie, D.Y.C., "High-Efficiency Silicon-Based Envelope-Tracking Power Amplifier Design With Envelope Shaping for Broadband Wireless Applications," *IEEE J. Solid-State Circuits*, pp. 2030-2040, Sept. 2013.
[8] L. Yan Li et al. "A SiGe Envelope-Tracking Power Amplifier with an Integrated CMOS Envelope Modulator for Mobile WiMAX/3GPP LTE Transmitters", *IEEE Trans. on Microwave Theory and Techniques*, vol. 59, no.10, Oct. 2011.

978-1-4799-0917-9/14 $31.00 © 2014 IEEE

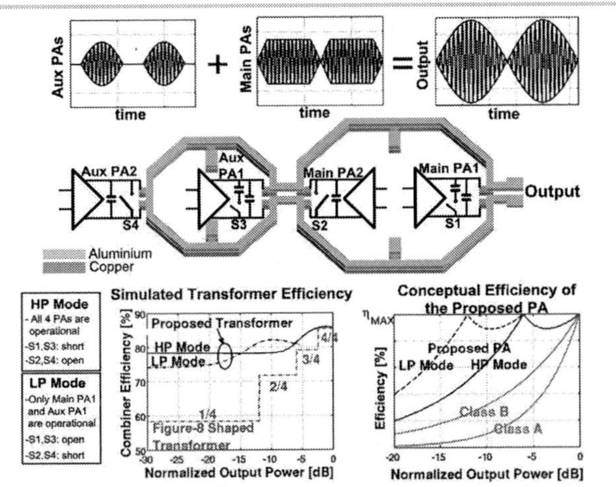

Figure 3.4.1: The proposed 4-way transformer-combiner for Dual-Mode Doherty Operation and the EM simulation results of the proposed transformer in comparison with a conventional 4-way figure-8-shaped transformer.

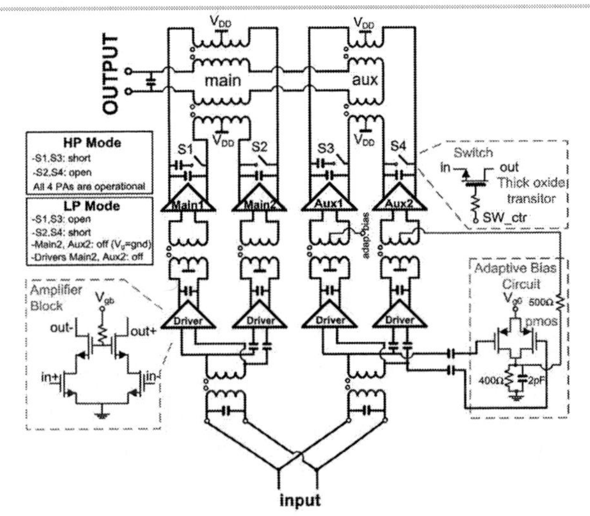

Figure 3.4.2: Schematic of the proposed Dual-Mode Doherty PA.

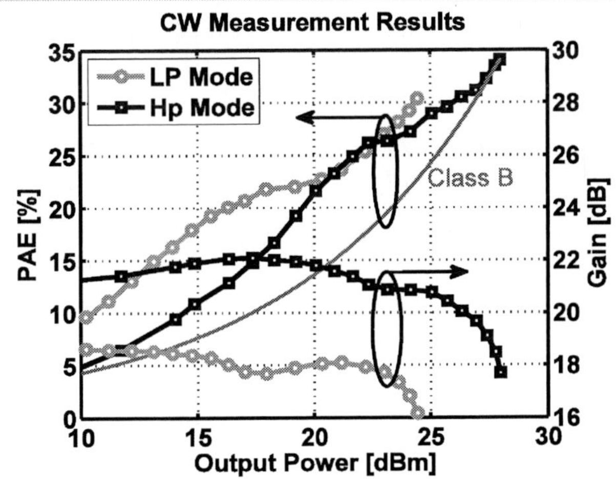

Figure 3.4.3: CW measurement results at 1.9GHz and Class-B efficiency curve with same peak performance.

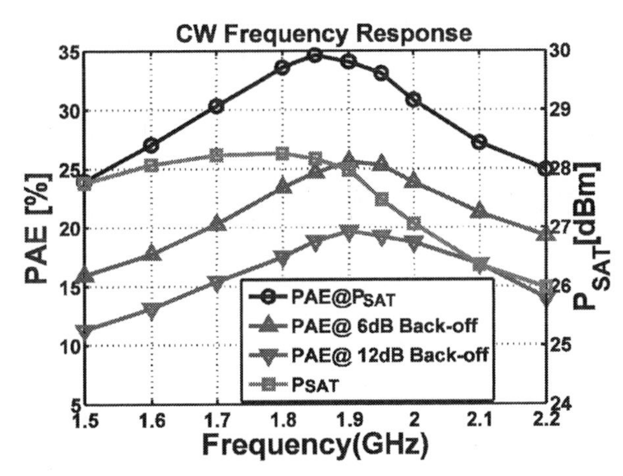

Figure 3.4.4: Measured CW frequency response of the PA.

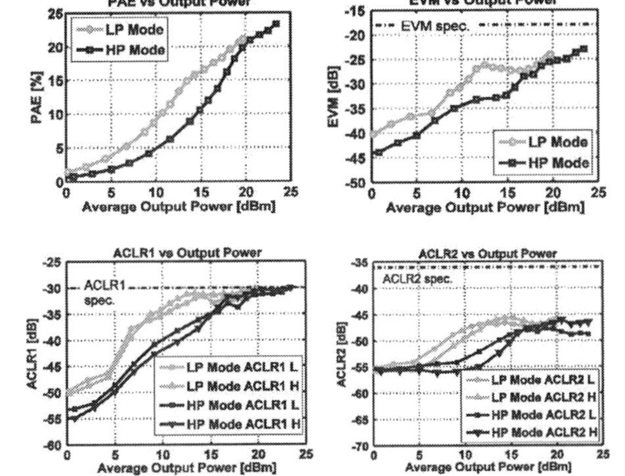

Figure 3.4.5: Measured performance with (16-QAM, 20MHz) LTE signal at 1.9GHz.

CMOS PAs with Back-Off Efficiency Enhancement (CW measurements)

	Freq [GHz]	P_{SAT} [dBm]	Peak PAE[%]	PAE@6dB back-off	PAE@12dB back-off	Area [mm²]	Technology, Supply [V]
[2]	2.4	31.5	27	20	12	3.12	45 nm, 2.4
[3]	2.4	26.3	33	25.1	13	1.85	90 nm, 2
[4]	1.8	27.2	30	22*	9*	5.2	65 nm, 3.3
[5]	1.95	30.5	42.1	24	33*	2.7	0.18um, 3.4
This work	**1.9**	**28.0**	**34.0**	**25.5**	**19.7***	**3**	**40 nm, 1.5**

*Low Power (LP) mode.

LTE PAs (Measured with LTE signal)

	Gain [dB]	P_{MAX}# [dBm]	PAE@ P_{MAX} [%]	PAE@ P_{MAX}-6 dB [%]	PAE@ P_{MAX}-12 dB [%]	Fully Integrated?	Technology, Supply [V]	Technique
[6]	24 /11*	27	37	22	21*	No	180nm CMOS + InGaP/GaAs,3.4V	dual mode envelope tracking (dual chip)
[7]	17	27.9	40.2	20	10	No	0.35μm BCD + 0.35μm SiGe, 5V	envelope tracking (dual chip)
[8]	17	24.3	42	16	6	No	0.35μm SiGe BiCMOS,4.2V	envelope tracking
[4]	12	21.3	18	13*	5*	Yes	65nm CMOS, 3.3V	envelope tracking
This work	**21.3 /18.2***	**23.4**	**23.3**	**18.4***	**11.1***	**Yes**	**40 nm CMOS,1.5V**	**Doherty + Mode Switching**

P_{MAX} is defined as the measured average output power for which the ACLR1, ACLR2 and EVM specs are met
*Low Power (LP) Mode

Figure 3.4.6: Performance comparison with CMOS PAs with back-off efficiency enhancement (CW performance) and the state-of-the-art LTE power amplifiers (LTE performance).

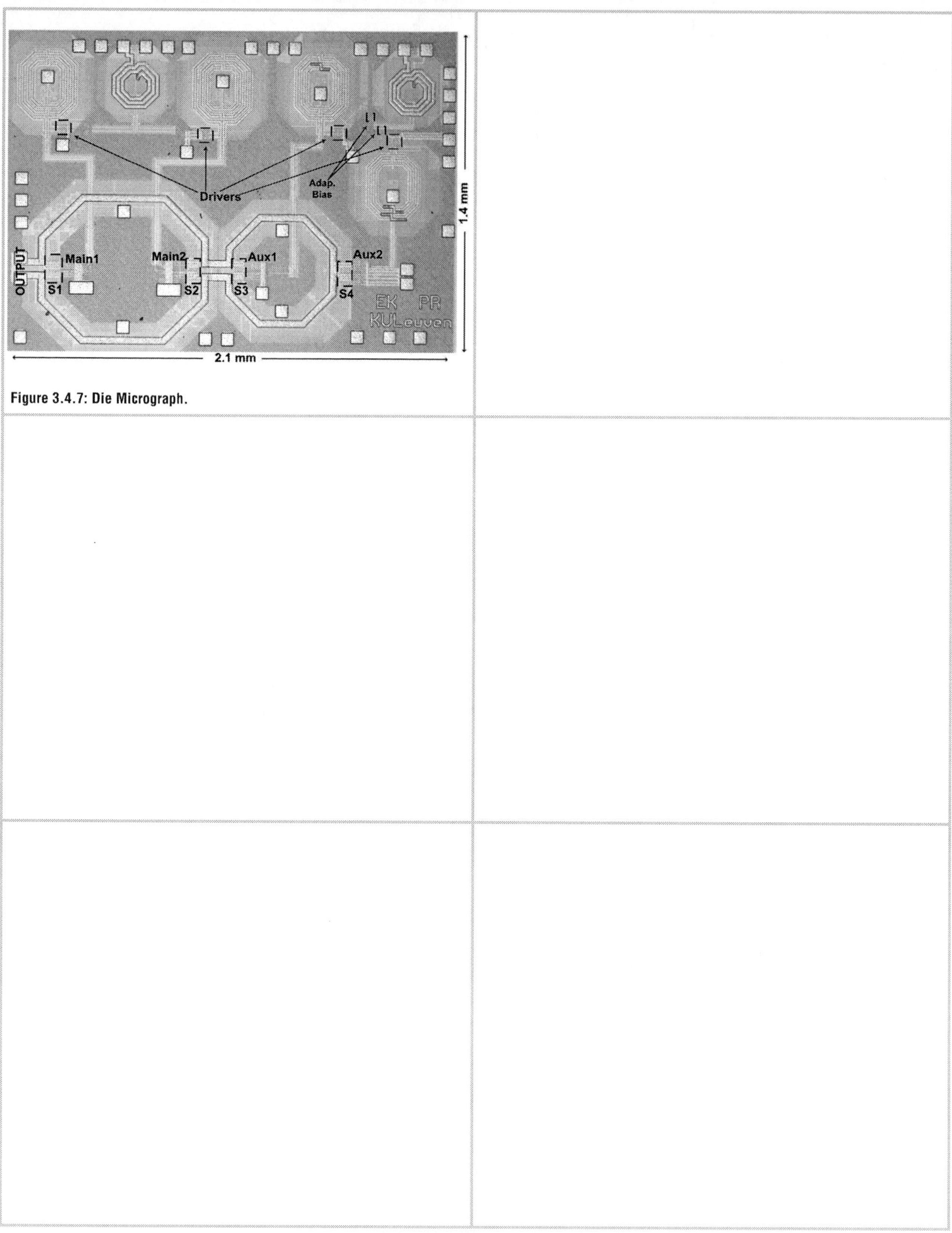

Figure 3.4.7: Die Micrograph.

3.5 A 1.0-to-2.5GHz Beamforming Receiver with Constant-G_m Vector Modulator Consuming < 9mW per Antenna Element in 65nm CMOS

Michiel C. M. Soer[1], Eric A. M. Klumperink[1], Bram Nauta[1], Frank E. van Vliet[1,2]

[1]University of Twente, Enschede, The Netherlands,
[2]TNO Science and Industry, The Hague, The Netherlands

Beamforming phased-array receivers aim to increase receiver sensitivity and reject interferers in the spatial domain [1,2]. A receiver with programmable phase shift and high linearity is crucial to cope with interference. Switched-capacitor vector modulators can provide adequate phase shift and linearity [3,4], but so far, at the cost of a high power consumption. As power consumption increases linearly with the number of antenna elements, it is one of the bottlenecks hindering commercialization of beamforming. In this paper, we demonstrate several design techniques on architectural and circuit levels, to reduce the power consumption per element, while still achieving competitive Spurious Free Dynamic Range (SFDR).

A vector modulator creates a weighted sum of two 90 degrees out-of-phase signals, I and Q, to generate an output with adjustable phase and gain. In the top part of Fig. 3.5.1, the classical Cartesian vector modulator is depicted, with *separate*, digitally controllable transconductors at baseband for I and Q, and current summing. Each constellation point in the phasor diagram is a vector summation of the weighted I and Q vectors. By allowing smaller steps in the tunable G_m, a finer resolution in the phasor diagram can be reached, while the polarity switch in I and Q allows for reaching all four quadrants. The phasors for the actual phase shifter are picked as those lying close to a circle. In this architecture, the total G_m varies with the applied phase shift, so the output impedance will vary with the phase-shifter setting. We propose a constant-G_m vector modulator with transconductances that are *shared between* I and Q, as is shown on the bottom of Fig. 3.5.1. The transconductances are binary weighted and the output is a summation of binary-weighted vectors. The reconfiguration switches control the orientation of each of these vectors, along the I+/I-/Q+ or Q- axis. The successive phasor diagrams show how the number of phase points grows as a function of the number of transconductances. More resolution is achieved by adding more, increasingly smaller, transconductances. If N is the number of binary weighted transconductors, a 2^N by 2^N rectangular grid of points in the phasor diagram can be made. Similar to the traditional Cartesian scheme, a rectangular constellation is generated (a Cartesian modulator has extra points on the axis, corresponding to disabled G_m's), but rotated by 45 degrees. This rotation of the constellation has no impact on the beamforming, since phase *differences* between antennas are required. Since now the total G_m is constant, the impedance on the output is also constant. We will show that this scheme is better suited for our architecture.

With the concept of a constant-G_m vector modulator in mind, we propose the Zero-IF beamforming receiver architecture in Fig. 3.5.2. Compared to Fig. 3.5.1, the transconductance has now been placed at RF before the I/Q generation done by the mixer [1]. The reconfiguration switches are positioned at baseband, so that their parasitic capacitance can be easily absorbed in the baseband capacitors. This N=4 design (16x16 constellation points) uses thermometer-coded unit slices to improve matching, in groups of 1,2,4 and 8, for a total of 15 slices. A slice is worked out in detail in the circuit diagram in Fig. 3.5.3. A passive 25% duty cycle current-mixer converts the RF transconductor current to zero-IF, while also providing single-to-differential conversion to I+, I-, Q+ and Q- (allowing for image rejection). Note that good 1/f noise can be expected as the transconductor operates at RF and passive mixers are used. As the inverter transconductor amplifies the RF signal first, the impedance level seen by the passive mixer can be much higher than the 50Ω antenna impedance. Therefore, compared to a mixer-first design [3,4], the mixer switches can be smaller, resulting in a large reduction in dynamic power consumption in the clocks. It is important to note that the current of the transconductor is always steered to only one of the outputs and the baseband capacitors retain their voltage when the current is routed to another path. If the transconductance were to be implemented in baseband instead of RF, each of the four outputs would need its own, resulting in four times the total G_m.

The inverter transconductor in Fig. 3.5.3 is self-biased, and consists of standard-V_t transistors, but is supplied with a low 1.0V supply voltage. This biases the transistors in moderate inversion for a better transconductance/current ratio at the cost of some increased capacitive parasitics. Impedance matching is realized by the bias resistor without extra static power consumption. The passive mixer upconverts the low-pass filtering in baseband, to a band-pass filtering on the inverter output node. This lowers the out-of-band RF voltage swing on the inverter output node, improving out-of-band linearity and compression.

The RF input of the mixer is AC coupled and biased at 0V. Full-swing inverters directly drive the mixer gates without any AC coupling or bootstrap circuits that would otherwise increase dynamic power consumption. Moreover, the AC coupling at RF helps in filtering IM2 products of closely spaced interferers which could otherwise potentially feed through the mixer to baseband. Conveniently, the element summing required for beamforming is automatically achieved by connecting the element outputs to shared capacitors [3]. The baseband bandwidth is defined by the output resistance of the transconductor (Rout), the baseband capacitance and the mixer duty cycle [5]. Due to the constant-G_m principle, the output resistance is constant, resulting in a well-defined bandwidth. Moreover, the voltage swing at the outputs of the inverters is also constant, resulting in a constant input impedance of the IC. This prevents phase-shifter-dependent gain errors caused by a change in input matching, so that no calibration is required (in contrast to [1]) and the resolution of the vector modulator is optimally used.

The 4-element beamforming receiver was implemented in 65nm CMOS (see Fig. 3.5.7). The measured phase-shifter constellation is plotted in Fig. 3.5.4. Also, phase and gain errors respective to uniform, constant amplitude phase steps are shown. These errors are mostly due to the mapping of the ideal phase shifts to a quantized constellation with a finite number of points. The measured gain, Fig. 3.5.5 top, of 12dB falls off by 1dB in an RF bandwidth from 1.0 to 2.5GHz. Over this frequency range, the S11 matching is better than -10dB. The NF of 6dB fits to simulation and is constant over the phase-shifter settings. The in/out-of-band input-referred IP3 and 1dB compression point are plotted in Fig. 3.5.5. The baseband capacitors are digitally tunable and can change the baseband bandwidth from 30 to 300MHz. In this figure, the bandwidth was set to 30MHz and the increase in out-of-band linearity is clearly visible in the measurement. In/out-of-band IIP2 was measured to be >52dBm. Figure 3.5.6 shows a comparison table with high-SFDR low-GHz beamforming receivers. Only [3] achieves higher SFDR (but for 13dB lower gain), while its phase shifter resolution is low. In comparison to [2,4], the presented receiver achieves similar or better SFDR, at a significantly lower power consumption of 6.5 to 9mW per element for 1.0 to 2.5GHz.

Acknowledgements:
This research was conducted as part of the Sensor Technology Applied in Reconfigurable systems for sustainable Security (STARS) project funded by the Dutch Government. We thank STMicroelectronics for silicon donation and CMP for assistance. Also, thanks go to Gerard Wienk and Henk de Vries for CAD support and lab assistance, and to Jean-Francois Paillotin for packaging.

References:

[1] J. Paramesh, et al., "A 1.4V 5GHz Four-Antenna Cartesian-Combining Receiver in 90nm CMOS for Beamforming and Spatial Diversity Applications", *ISSCC Dig. Tech. Papers*, pp. 210-211, Feb. 2005.
[2] R. Tseng, et al., "A Four-Channel Beamforming Down-Converter in 90nm CMOS Utilizing Phase-Oversampling", *IEEE J. Solid State Circuits,* vol. 45, no. 11, pp. 2262 - 2272, Nov. 2010.
[3] A. Ghaffari, et al., "Simultaneous Spatial and Frequency Domain Filtering at the Antenna Inputs Achieving up to +10dBm Out-of-Band/Beam P1dB", *ISSCC Dig. Tech. Papers*, pp. 84-85, Feb. 2013.
[4] M.C.M. Soer, et al., "A 1.0-4.0GHz 65nm CMOS Four-Element Beamforming Receiver Using a Switched-Capacitor Vector Modulator with Approximate Sine Weighting via Charge Redistribution", *ISSCC Dig. Tech. Papers*, pp. 64-65, Feb. 2011.
[5] A. Ghaffari, et al., "Tunable High-Q N-Path Band-Pass Filters: Modeling and Verification", *IEEE J. Solid State Circuits,* vol. 46, no. 5, pp. 998 - 1010, May. 2011.

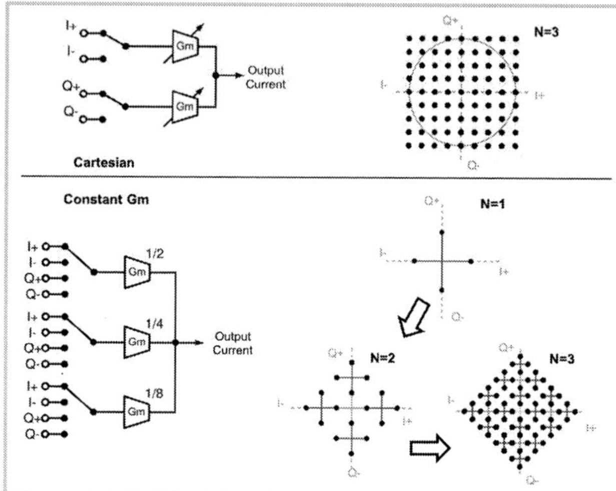

Figure 3.5.1: Traditional Cartesian (top) and proposed Constant-G_m (bottom) vector modulator, example for N=3.

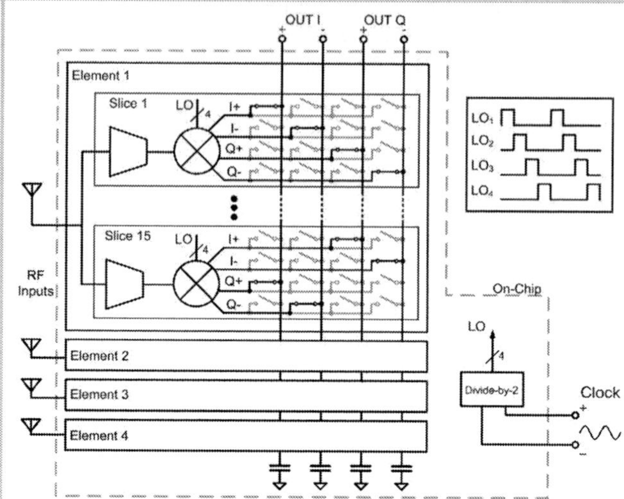

Figure 3.5.2: Architecture of the 4-element beamforming receiver.

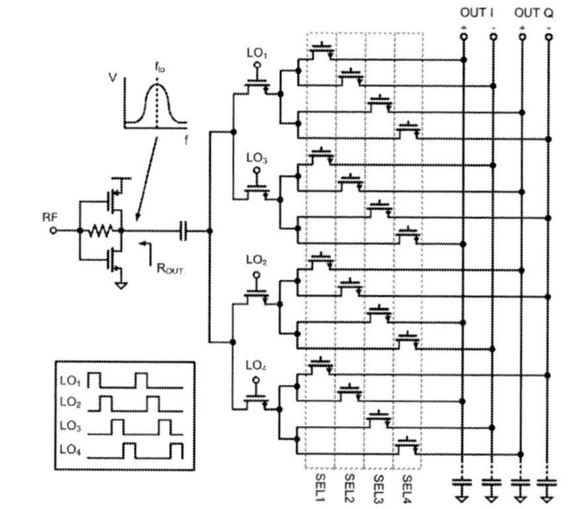

Figure 3.5.3: Circuit diagram of a vector-modulator slice.

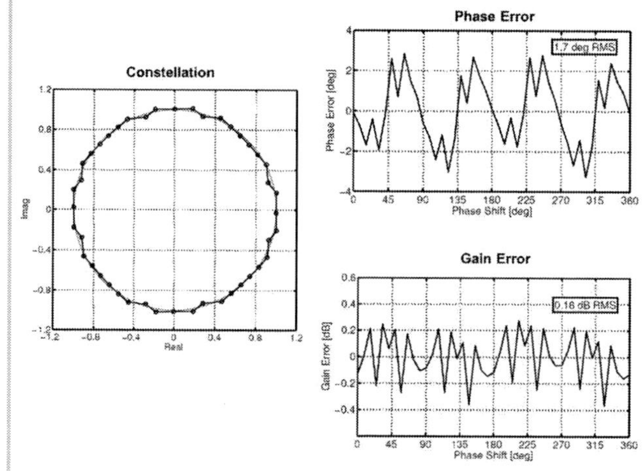

Figure 3.5.4: Measured phase-shifter constellation and phase/gain errors.

Figure 3.5.5: Measured single-element gain, NF (top), IIP3, compression point (bottom) and chip power consumption (top).

		[1]	[2]	[3]	[4]	This Work	
Technology		90nm	90nm	65nm	65nm	65nm	
Active area		1.3 [a]	1.4 [a]	0.97	0.18	0.20	mm²
Supply voltage		1.4	1.2	1.2	1.0 - 1.2	1.0	V
RF frequency band		5.0	4.0	0.6 - 3.6	1.5 – 5.0	1.0 – 2.5	GHz
Power consumption		140	166	68 – 195	65 – 168	26 - 36	mW
Array elements		4	4	4	4	4	
Single Element Performance @			4.0	2.0	3.0	1.5	GHz
Gain		N/A	15	-1	-6	12	dB
Noise figure		N/A	13	4 [c]	18	6	dB
Compression point	in-band	N/A	N/A	-5	2	-9	dBm
	out-of-band			10	N/A	-3	
IIP3	in-band	N/A	2	6	13	1	dBm
	out-of-band			N/A	N/A	5	
In-band SFDR in 1MHz BW		N/A	69	77	73	73	dB
# Phase shifter steps		N/A	32	8	32	44	
Phase error (RMS)		N/A	N/A	N/A	2.0	1.7	deg
Amplitude error (RMS)		N/A	N/A	N/A	0.2	0.18	dB
Vector modulator EVM		N/A	2 [b]	N/A	4	4	%

[a] Estimated from chip photo. [b] After calibration of the LO phases.
[c] Degraded by 2 dB for 4 elements, due to correlated noise.

Figure 3.5.6: Comparison table.

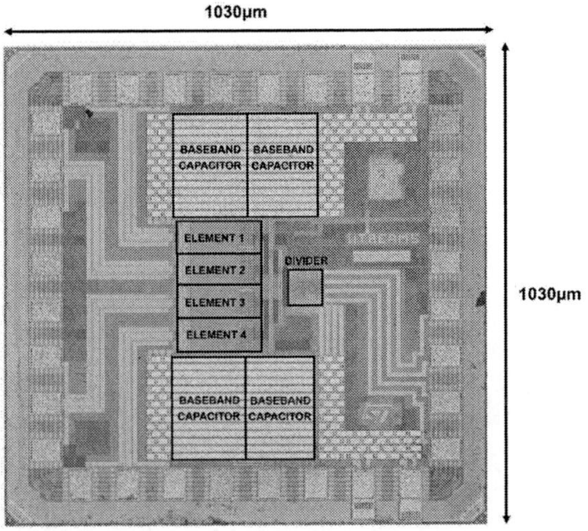

1030µm

BASEBAND CAPACITOR
BASEBAND CAPACITOR
ELEMENT 1
ELEMENT 2
ELEMENT 3
ELEMENT 4
DIVIDER
BASEBAND CAPACITOR
BASEBAND CAPACITOR

1030µm

Figure 3.5.7: Chip micrograph.

3.6 A Noise-Cancelling Receiver with Enhanced Resilience to Harmonic Blockers

David Murphy, Hooman Darabi, Hao Xu

Broadcom, Irvine, CA

By employing two passive-mixer-based downconversion paths, the frequency-translational noise-cancelling receiver (FTNC-RX) achieves a low noise figure and can tolerate most out-of-band blockers up to 0dBm with little performance degradation [1]. However, like most wideband passive-mixer-based designs, the architecture is far less tolerant of harmonic blockers, that is *blockers located at or around precise integer multiples of the LO frequency*. In a typical *M*-phase passive mixer, shown in Fig. 3.6.1a, most out-of-band blockers are heavily attenuated by large shunt capacitors at the inputs of the baseband TIAs. Harmonic blockers are an exception and do not experience this attenuation since they are downconverted inside the TIA bandwidth, are amplified along with the wanted signal, and are only rejected by the subsequent harmonic-rejection circuitry. Since TIA gain is generally large in order to maintain a low noise figure, moderate harmonic blockers will saturate the TIAs and consequently the receiver.

In order to overcome this limitation, this work presents an enhanced architecture that employs two separate techniques. The first technique, shown in Fig. 3.6.1b, is suitable for all *voltage-driven* passive-mixer-based designs, including mixer-first receivers [2]. Instead of each TIA being excited with a single output of the passive-mixer, the proposed harmonic-rejection TIAs (HR-TIAs) utilize a G_M-stage whose output is the weighted combination of all *M* output terminals of an *M*-phase passive mixer. If the G_M-stage weightings are appropriately chosen (Fig. 3.6.1c), the gain to output of the HR-TIAs is large in response to the wanted signal, but zero in response to harmonic blockers up to $(M-1)*F_{LO}$. This is analogous to a differential amplifier that reacts to differential signals and ignores common-mode excitations, but instead of using two terminals to reject 1-mode, *M* terminals are used to reject *(M-2)* modes. To further enhance this technique, an active resistor that is a function of all *M*-outputs of the TIAs is placed at the output of each unit G_M-stage. This active resistor, using a similar scheme to the weighted-G_M approach, ensures that the output impedance of each G_M-stage is large for wanted signal excitations and close to zero for harmonic blockers. Indeed, the active resistor can be designed to cancel the parasitic output impedance of each G_M-stage.

Such a scheme is suitable for the main-path of the FTNC-RX, but it is not sufficient for the auxiliary path, which operates as a *current-driven* passive mixer. The problem with using only these HR-TIAs is that when the gain of the G_M cells is zero, as in response to a harmonic blocker, the input impedance of the HR-TIAs becomes large. This is not particularly troubling in a voltage-driven passive-mixer (i.e. the main path), but in a current-driven design it can result in large voltage swings at the input of the HR-TIAs. Therefore, in the auxiliary path of the FTNC-RX, a weighted set of RF-G_M cells is used to cancel harmonic blocker currents before they reach the TIAs (Fig. 3.6.2). The technique is adapted from [3] where it was used to boost small-signal harmonic rejection (its beneficial large-signal rejection properties were not emphasized in [3]). Importantly, unlike [3], the RF-G_M unit cells do not need to provide impedance matching in this adaptation and can be optimized for low noise operation. Figure 3.6.2 also shows how the technique is extended to a single-ended design, so that even-order harmonic blockers are also cancelled. As simple (low noise and linear) positive transconductors are not available, an effective DC offset is introduced when selecting the weighting constants of the RF-G_M cells for a single-ended design.

The schematic of a fully-differential prototype, fabricated in 28nm, is shown in Fig. 3.6.3. In a fully differential design, even-order harmonic blockers are naturally rejected by the balun and differential passive-mixers before they reach the baseband TIAs and, so, the proposed circuit enhancements need only to be configured to reject odd harmonics. The RF-G_M cells are realized using a Class-AB design and use non-minimum length transistors in order to boost the cell's output impedance, which is critical to suppressing the noise of the auxiliary path mixers and TIAs. Cross-coupled feedback capacitors are used to cancel the Miller effect. A prototype with a single-ended RF input was also fabricated, which employed 6 auxiliary down-conversion paths in order to suppress even-order harmonics. To maintain the same RF bandwidth as the differential design, minimum length devices are used in the RF-G_M cells along with a negative resistance at the output at the cost of a small increase in power.

When the enhanced FTNC-RX is correctly configured the only significant noise contributors are the auxiliary path RF-G_M cells. The optimized noise-factor is given by:

$$F = \left(1 + \frac{\kappa\gamma}{G_M R_A}\right) \left|\frac{1}{\mathrm{sinc}(\pi/M)}\right|^2$$

where G_M is the linear sum of all the RF transconductances, R_A is the antenna resistance, γ is the noise coefficient of the G_M cell, and κ is a variable that defines a specific configuration. In the standard FTNC-RX [1], where only one RF-G_M cell is used, $\kappa = 1$. In a modified FTNC-RX, that is designed to rejected odd-order harmonic blockers (for use in a fully differential design) $\kappa = (3+2\sqrt{2}) \approx 1.46$, while in a modified FTNC-RX, that is designed to reject both even and odd harmonic blockers (for use in an RX with a single-ended input) $\kappa = 4$. Consequently, the large-signal harmonic rejection does come at the cost of either an increase in noise figure or RF-G_M current consumption. For this reason, the single-ended prototype can be configured to operate in either a low-noise or harmonic-blocker-tolerant mode. The differential prototype did not employ this re-configurability because of the more modest power/noise trade-off.

Focusing on the differential prototype, Fig. 3.6.4 shows the measured 1dB small-signal compression of a 500.2MHz wanted signal in the presence of $3F_{LO}$ and $5F_{LO}$ harmonic blockers (500.2MHz is chosen to ensure all the harmonic blockers fall in-band). The tolerance to these harmonic blockers is estimated to improve by more than 20dB when compared to the original FTNC-RX. The measured blocker noise figure for a -5dBm $3F_{LO}$ blocker was 9dB. Note that these measurements are in response to *precise harmonic blockers only*; the receiver can tolerate standard 0dBm non-harmonic blockers with far less performance degradation. For example, at an 80MHz offset from a 2GHz carrier, a blocker P_{1dB} of -2.5dBm and a 0dBm blocker NF of 5dB were measured. While the focus of this work was on large-signal harmonic rejection, the uncalibrated small-signal rejection also improves from around 40dB to 60dB. An S_{11} of better than -10dB was measured across the entire receive band without the use of any off-chip matching components.

The prototype also employs a technique to optimize noise cancellation, which relies on the fact that when the noise figure is optimized, the noise contribution of the main path TIAs is ideally nulled [1]. To model the TIA noise, small phase-shifted current sources are introduced at the output of the main path TIAs (Fig. 3.6.3). The relative gain and phase of the two paths are then swept until the resultant voltage at the receiver's output is minimized. As shown in Fig. 3.6.5, the measured NF after the calibration technique is applied is within 0.2dB of the lab-optimized NF measurement.

Figure 3.6.6 contrasts this work with other recent highly linear GHz designs. Compared to the existing solutions, the architecture retains the superior small-signal and blocker noise figure of the standard FTNC-RX, but is also tolerant of large-signal harmonic blockers without any RF filtering. The measured performance metrics of the single-ended prototype are also listed. The larger chip area in both designs is attributable to the increased baseband filtering where *0dBm blockers as close as 8MHz* to carrier are tolerated (Fig. 3.6.4).

References:

[1] D. Murphy, A. Hafez, A. Mirzaei, M. Mikhemar, H. Darabi, M.F. Chang, and A. Abidi, "A Blocker-Tolerant Wideband Noise-Cancelling Receiver with a 2dB Noise Figure," *ISSCC Dig. Tech. Papers*, pp. 74-76, Feb. 2012.

[2] C. Andrews, and A.C. Molnar, "A Passive-Mixer-First Receiver with Baseband-Controlled RF Impedance Matching, < 6dB NF, and > 27dBm Wideband IIP3," *ISSCC Dig. Tech. Papers*, pp.46-47, Feb. 2010.

[3] Z. Ru, E. Klumperink, G. Wienk, and B. Nauta, "A Software-Defined Radio Receiver Architecture Robust to Out-of-Band Interference," *ISSCC Dig. Tech. Papers*. pp. 230–231, 231a, Feb. 2009.

[4] J. Borremans, B. van Liempd, E. Martens, S. Cha, and J. Craninckx, "A 0.9V Low-Power 0.4–6GHz Linear SDR Receiver in 28nm CMOS," *Symposium on VLSI Circuits (VLSIC)*, pp. C146-C147, 12-14 June 2013.

978-1-4799-0917-9/14 $31.00 © 2014 IEEE

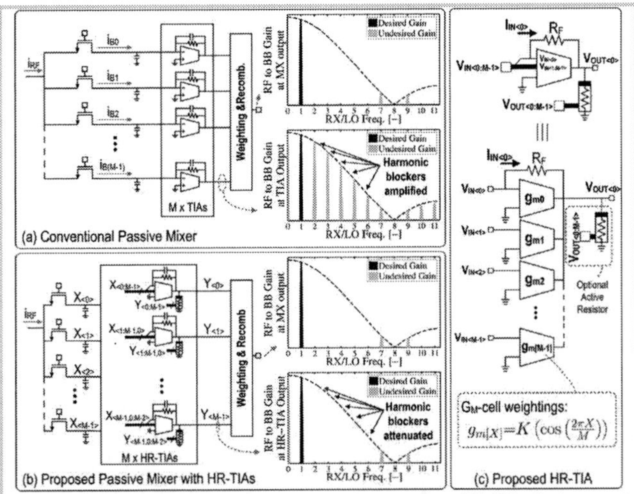

Figure 3.6.1: Proposed M-phase passive mixer employs HR-TIAs that prevent amplification of harmonic blockers at the TIA outputs.

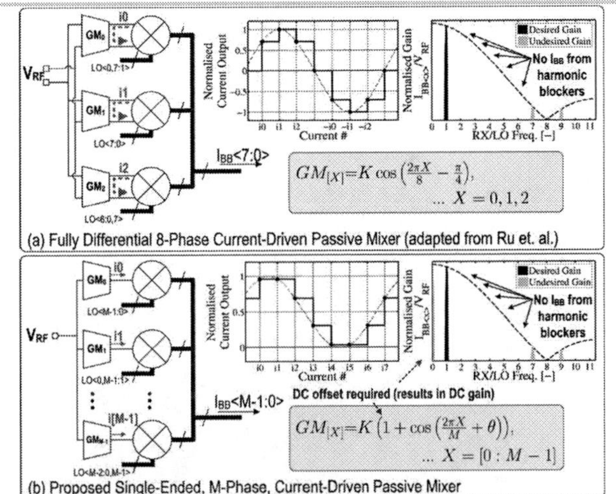

Figure 3.6.2: The auxiliary path utilizes an adapted version of the differential current-driven passive mixer proposed in [3].

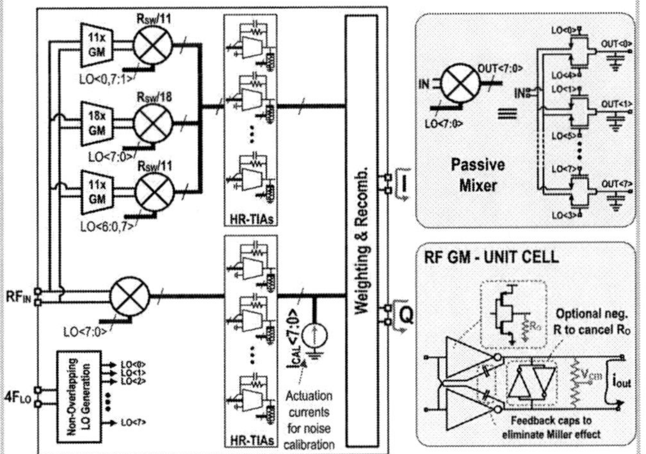

Figure 3.6.3: Fabricated fully differential prototype with enhanced resilience to large-signal harmonic blockers.

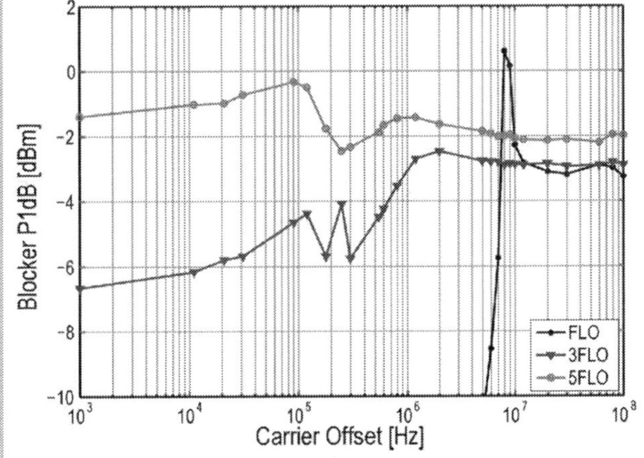

Figure 3.6.4: Measured blocker 1dB compression versus carrier offset for various harmonic blockers. Measurement setup assumes maximum available baseband filtering at an IF of 200kHz.

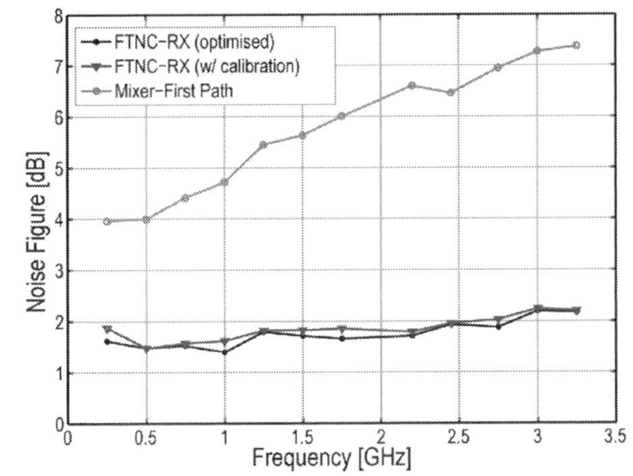

Figure 3.6.5: Measured noise figure versus frequency. Calibration algorithm results in a noise figure that is within 0.2dB of the lab-optimized noise figure.

	Andrews et. al. ISSCC'10 [2]	Borremans et. al. VLSI'13 [4]	Murphy et. al. ISSCC'12 [1]	This Work	
Topology Description	Mixer-First	Resistive Feedback	FTNC-RX	Modified FTNC-RX	
CMOS Technology	65nm	28nm	40nm	28nm	
RF Input	Single-Ended	Differential	Single-Ended	Differential	Single-Ended
RX Frequency [MHz]	100-2400	400-3000	80-2700	100-3300	600-3000
NF @ 2GHz [dB]	7	2.3-2.9**	1.9	1.7	1.8 (low NF mode) 3 (HR mode)
OB-P1dB [dBm]	4	N/A	-2	-2.5	-6
0dBm OB-Blocker NF [dB]	N/A	15	4.1	5	9 (low NF mode) 13 (HR mode)
Harmonic Blocker P1dB [dBm]	N/A	N/A	N/A	-6.5 ($3F_{LO}$) -3 ($5F_{LO}$)	-10 ($2F_{LO}$) -8 ($3F_{LO}$)
Harmonic Blocker NF [dB]	N/A	N/A	N/A	9@-5dBm ($3F_{LO}$)	7@-7.5dBm ($2F_{LO}$) 9@-7.5dBm ($3F_{LO}$)
RF Power [mW] BB Power [mW]	0 30	N/A	12 20	18 18	22 12-24
LO Power [mW/GHz]	13.8*	N/A	17.3	8	8
Total Power [mW]	37-70	40	35.1-78	36.8-62.4	38.8-70
Supply Voltage [V]	1.2/2.5	0.9	1.3	1.0	1.0
OB-IIP3 [dBm]	+25	+3	+13.5	+11.5	+10
OB-IIP2 [dBm]	+58	85¥	+54	+55	+49.5
Harmonic Rejection [dB] $3F_{LO}/5F_{LO}$	35.4/42.6	70/55¥	42/45	≈60/60	≈52/54 ($3F_{LO}/5F_{LO}$) ≈60/60 ($2F_{LO}/4F_{LO}$)
Active Area [mm²]	2	0.6	1.2	5.2§	5.0§

* Estimated and/or interpreted from plots, figures and/or reported numbers. ** In high linearity mode. ¥ With calibration.
§ Large size is due to increased baseband filtering. If only 80MHz blocker tolerance, area would be comparable to [1].

Figure 3.6.6: Comparison with other blocker-tolerant GHz receivers.

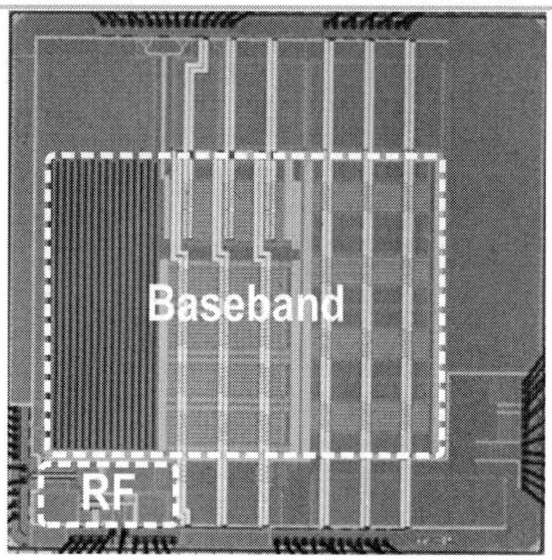

Baseband

RF

Figure 3.6.7: Die micrograph.

3.7 A Fully Integrated TV Tuner Front-End with 3.1dB NF, >+31dBm OIP3, >83dB HRR3/5 and >68dB HRR7

In-Young Lee[1], Sang-Sung Lee[1], Donggu Im[2], Seungjin Kim[1], Jeongki Choi[3], Sang-Gug Lee[1], Jinho Ko[3]

[1]KAIST, Daejeon, Korea,
[2]University of Texas, Dallas, Richardson, TX,
[3]PHYCHIPS, Daejeon, Korea

In TV tuner systems, the RF front-end design has been a challenging issue since it must simultaneously satisfy over 65dB of harmonic rejection (HR), and have high linearity for high-power input and low noise over wide bandwidth (48-to-870MHz). In terms of harmonic rejection, even though the state-of-the-art work reports over 60dB rejections on the 3rd- and 5th- order harmonics with a single mixer [1], higher-than-5th-order harmonic rejections are still required for the low-band channels in TV tuners and thereby RF filters are indispensable at the RF front-end. However, due to the difficulties of integrating RF filters satisfying low noise and high linearity over wide bandwidth, the previous works inevitably had to use external inductors [2-4]. Although a recent work successfully integrates an RF filter satisfying all the stringent specifications by current-domain signal flow from the LNA output to the baseband stage [5], the transconductance stage at the filter input is not linear enough to drive the high-power input and thus the input signal needs to be attenuated at the RF front-end, which eventually degrades system SNR.

This paper presents a TV tuner front-end that shows exceptional linearity and harmonic-rejection performances over the whole frequency range while sustaining low noise. Figure 3.7.1 shows the block diagram of the TV tuner front-end. As shown in Fig. 3.7.1, the RF front-end consists of a shunt feedback LNA with post-nonlinearity correction scheme, a highly linear 6-b tunable RF filter and a 2-stage baseband HR mixer. In the front-end, a new-topology RF filter shows a cutting-edge linearity performance with low noise while the LNA also shows high linearity performance even in the high-gain mode. In addition, two-stage harmonic rejection in the baseband allows itself more robust performance to the gain mismatch compared to the previous two-stage harmonic rejection in [1].

Figure 3.7.2 shows the schematic of the two-stage resistive-feedback LNA. In Fig. 3.7.2, the LNA is composed of two gain stages (A_1 and A_2) with a resistive feedback (R_F) loop, where A_1 provides sufficiently high loop gain for low noise figure, and A_2 ensures near-ideal feedback operation by removing the loading effects. Basically, each gain stage adopts a common-source amplifier (M_{1N} or M_{3N}) cross-stacked with a source follower (M_{2N} or M_{4N}) considering its stable loop gain over PVT variations (the gain is determined by the transconductance (g_m) ratio of the M_{1N} to M_{2N}) and smaller amount of odd-order harmonic distortion by the post-linearization process [3]. In the first gain stage (A_1) as shown in Fig. 3.7.2, the combination of M_{1N} and M_{1P} configures an inverter-type amplifier, which inherently helps to cancel out the even-order harmonics, thus further improving the post-linearization process. Additionally, the cross-coupling current-bleeding transistors (M_{1P}) helps to boost the loop gain of the LNA by reducing the amount of current in M_{2N} (I_{2N} in Fig. 3.7.2), thus maximizing g_{m1N}/g_{m2N}, which also leads to the additional linearity improvement of the LNA. The LNA is designed to have 23dB of maximum gain, >+27dBm of OIP3 over the whole gain range with 1.4dB of noise figure at maximum gain while consuming 36mA from a 1.5V supply.

Figure 3.7.3 shows the schematic of the 4th-order active-RC low-pass filter. In Fig. 3.7.3, the filter biquad is composed of an RC ladder with a unity-gain buffer feedback. Contrary to the typical Sallen-Key (SK) filter, where the active circuits are placed in the feed-through path, the presented filter places it within the feedback path and thus not only eliminates the inherent stop-band limitation of the SK filter but also architecturally filters out the noise and nonlinear components of the active circuits. Moreover, since C_1 presents an open circuit at frequencies near DC, the filter is also free from the DC-offset problem in the unity-gain buffer. In order to sustain higher filter Q and better linearity, a sub-1-ohm-output-impedance source follower is used as the unity-gain buffer. As can be seen in Fig. 3.7.3, the source follower boosts the loop-gain and thus minimizes its output impedance by adopting common-gate active feedback (M_5, M_6), input cross-coupling through C_x, and source degeneration (R_s) to M_7, M_8. Since the low output impedance of the source follower allows sufficient filter Q at high frequencies even with small value of R in Fig. 3.7.3, the filter noise figure can be substantially reduced. Besides, the boosted loop gain reduces the harmonic distortion in the source follower, thus resulting in additional linearity improvement in the filter. The filter is designed to have <15dB of noise figure and >+25dBm of IIP3 for -9dBm of two-tone inputs with 6b bandwidth control over 40-to-660MHz of frequency range while the sub-1-ohm source follower consumes 8mA from a 1.5V supply and sustains its half-circuit output impedance below 1 ohm up to 900MHz.

The harmonic rejection mixer in Fig. 3.7.1 adopts the current-driven passive mixer topology for its low flicker noise and good linearity performance where the two-stage harmonic-rejection scheme is implemented in the baseband. In the harmonic rejection stage, $1:\sqrt{2}:1$ gain control and 45° phase shift are implemented by the $\sqrt{2}:1:\sqrt{2}$ weighted resistors and divider-by-4 block with phase-trimming circuits. Compared to the transconductance ratio control at high frequency as in [1], the resistor ratio control in baseband is more accurate and less susceptible to the PVT variations, especially the threshold voltage variation. The mixer is designed to have 15dB of gain, 12dB of noise figure, >+33dBm of OIP3, and >60dB of 3rd- and 5th-order harmonic rejections in simulation while consuming 62mA for I/Q generation from a 1.5V supply.

Figure 3.7.4 shows the measured frequency response of the filter and P_{1dB} comparison up to the front-end gain. The bandwidth of the filter ranges from 40MHz to 520MHz sustaining steep roll-off characteristics. The measured output P_{1dB} in the LNA maximum-gain mode sustains the same value of +11dBm as the minimum-gain mode. This indicates that the linearity of the LNA is high enough not to degrade the linearity of the front-end. As seen in Fig. 3.7.4, in the low-gain mode, the front-end can endure up to -4dBm of input power without gain compression. As shown in Fig. 3.7.5, the IMD_3 of the front-end at 300MHz filter cut-off frequency is -68dBc with -12dBm two-tone inputs which corresponds to +37.6dBm of OIP3. Also in Fig. 3.7.4, the front-end OIP3 is >+31dBm over the whole gain and frequency range. The measured harmonic-rejection ratio of the RF front-end shows the 3rd- and 5th-order harmonic rejections of >83dB and >89dB respectively while the 7th and higher order is rejected by >68dB over 40-to-300MHz bandwidth. Figure 3.7.6 shows the summary of the front-end performances in comparison with state-of-the-art works. As can be seen in Fig. 3.7.6, the TV tuner front-end shows robust linearity performance regardless of its gain change. Even in case of 15dB of RF stage gain, the front-end can receive up to -4dBm of input signal. Figure 3.7.7 shows the chip micrograph where the tuner front-end occupies 2.7mm^2.

Acknowledgements:
This work was supported by a National Research Foundation of Korea (NSF) grant funded by the Korean Government [Ministry of Education, Science and Technology (MEST)] (Grant No. 2010-0018899).

References:

[1] Z. Ru et al., "A Software-Defined Radio Receiver Architecture Robust to Out-of-Band Interference," *ISSCC Dig. Tech. Papers*, pp. 230-232, Feb. 2009.
[2] T. Choke et al., "A Multiband Mobile Analog TV Tuner SoC With 78-dB Harmonic Rejection and GSM Blocker Detection in 65-nm CMOS," *IEEE J. Solid-State Circuits*, vol. 48, no. 5, pp. 1174-1187, May 2013.
[3] D. Im et al., "A Broadband CMOS RF Front-End for Universal Tuners Supporting Multi-Standard Terrestrial and Cable Broadcasts," *IEEE J. Solid-State Circuits*, vol. 47, no. 2, pp. 392-406, Feb. 2012.
[4] H. Cha et al., "A CMOS Wideband RF Front-End With Mismatch Calibrated Harmonic Rejection Mixer for Terrestrial Digital TV Tuner Applications," *IEEE Trans. on Microwave Theory and Techniques*, vol. 58, no. 8, pp. 2143-2151, Aug. 2010.
[5] J. Greenberg et al., "A 40MHz-to-1GHz Fully Integrated Multistandard Silicon Tuner in 80nm CMOS," *ISSCC Dig. Tech. Papers*, pp. 162-164, Feb. 2012.

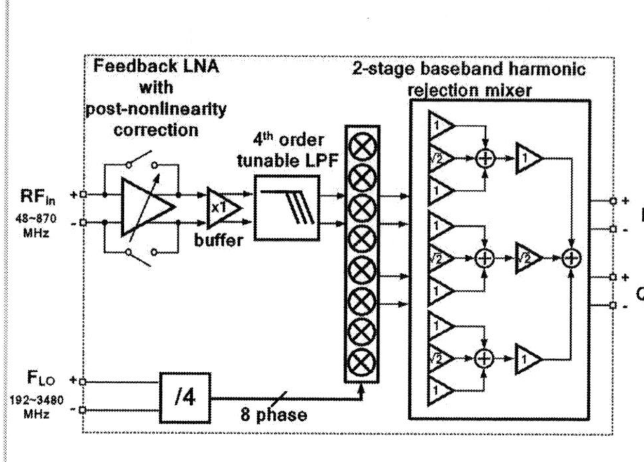

Figure 3.7.1: Block diagram of the TV tuner front-end.

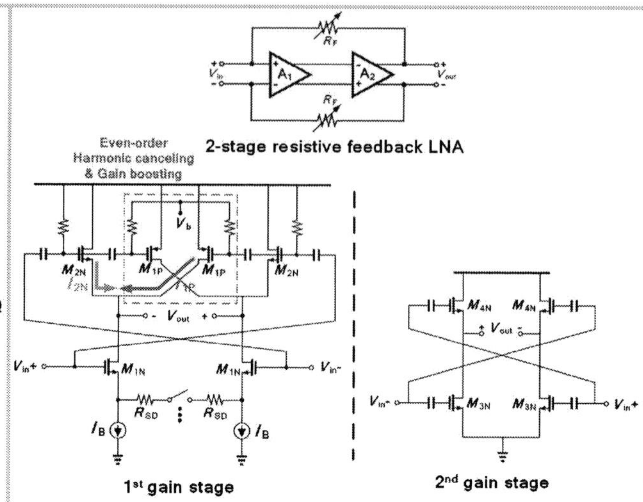

Figure 3.7.2: Schematic of the LNA and gain stages.

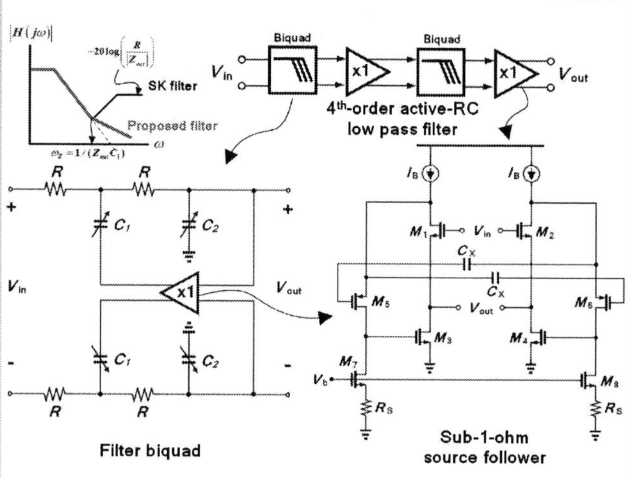

Figure 3.7.3: Schematic of the 4th-order low-pass filter.

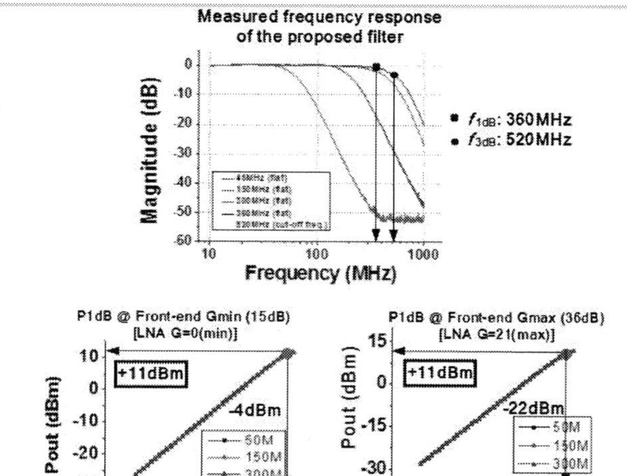

Figure 3.7.4: The measured filter shape and P_{1dB} of the front-end.

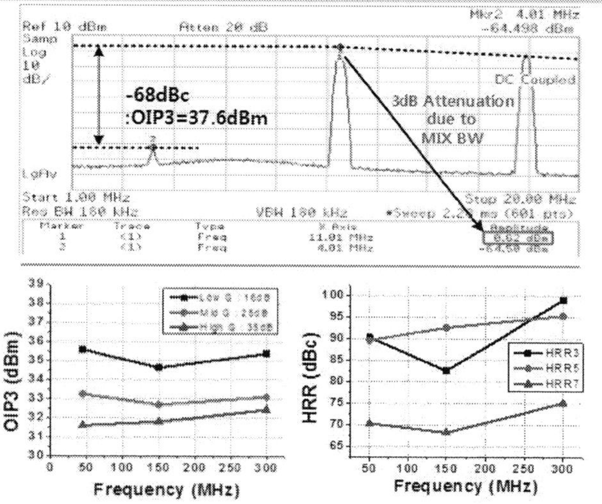

Figure 3.7.5: The measured OIP3 and HRR of the front-end.

Parameter	This work	[5] ISSCC'12	[2] JSSC'13	[3] JSSC'12	[4] T-MTT'10
Front-End Gain max/min [dB]	+36/+15	+36/-19	+42/-36	+42/-16	+40/-22
NF @ Max Gain [dB]	3.1	3	5	6	5.5
OIP3 min/max [dBm]	+31 / +34	+21 / +30	+32.5 / N/A	+29 / N/A	+30 / N/A
OIP2 min/max [dBm]	+59 / +69	+56 / +66	N/A / +62	N/A / +51	N/A
Ext. inductor in filter (#)	N	N	Y(3)	Y(1)	Y(1)
3rd HRR [dB]	>83	>65	>78	>70	>70
5th HRR [dB]	>89	>65	>84	>70	>70
7th HRR [dB]	>68	>65	N/A	>65	>60
Power [mW]	183 @1.5 V	[2]340 @1.8 V	[1]55.5 @1.5 V	[1]115 @1.8 V	[1]140 @1.8 V
Technology	0.13μm CMOS	80nm CMOS	65nm CMOS	0.18μm CMOS	0.18μm CMOS

[1]: the power consumption excluding RF filters.
[2]: the power consumption of all analog parts.

Figure 3.7.6: Performance table and comparison with state-of-the-art.

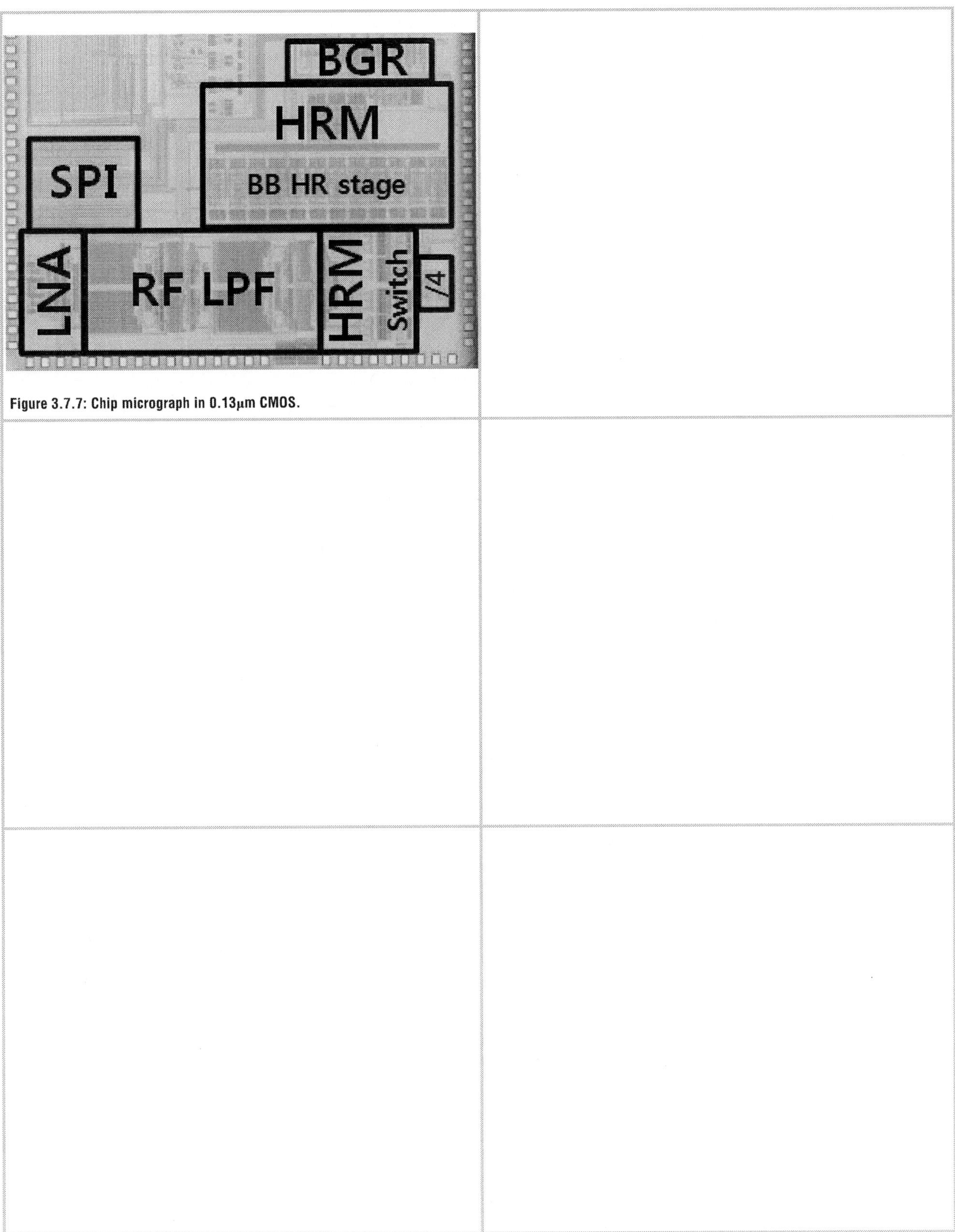

Figure 3.7.7: Chip micrograph in 0.13µm CMOS.

3.8 A Fully Integrated Highly Reconfigurable Discrete-Time Super-Heterodyne Receiver

Massoud Tohidian, Iman Madadi, Robert Bogdan Staszewski

Delft University of Technology, Delft, The Netherlands

Since the invention of radio, superheterodyne has been the architecture of choice for receivers (RX). Thanks to its high intermediate-frequency (IF), the problems related to flicker noise, time-varying dc offsets, in-band LO leakage and sensitivity to 2nd-order intermodulation are simply avoided. Unfortunately, the high IF requires high-quality-factor (Q) band-pass filters for image rejection, which cannot be easily integrated in CMOS. This forced the CMOS receivers to migrate to zero (or low) IF and suffer from the abovementioned problems. Recently, there have been attempts to revisit the high IF operation by exploiting N-path filtering [1] and a combination of a discrete-time (DT) band-pass charge-sharing filtering with feedback filtering [2]. Here, we propose a superheterodyne RX architecture with full DT operation using only g_m stages, switches and capacitors. The transfer function is accurate and controlled by the clock frequency and precise capacitor ratios.

In DT zero-IF receivers, the signal is commonly sampled in quadrature at Nyquist rate ($2\times f_{RF}$), but then immediately decimated [3,4]. While this decimation might be acceptable in narrowband ZIF RX, it creates images without enough attenuation in a high-IF receiver. In the proposed receiver structure, the input signal is sampled at $4\times f_{LO}$ directly at RF, giving an OSR of more than 2. This ensures no signal aliasing up to more than $3\times f_{RF}$.This very high sampling rate is kept throughout the IF stage to create enough attenuation of unwanted frequencies before any decimation. The passive sampling mixer shown in Fig. 3.8.1 does both sampling of the signal and DT quadrature mixing. On each of the phases φ1-4, the signal is sampled and then mixed with [1 0 -1 0] and [0 1 0 -1] in I and Q branches, respectively. The point of keeping the 4x sampling rate is that all the data at the output of each mixer, including 0's data, should be read out and processed by the next stage. This avoids any early decimation. Output data of this mixer are DT charge packets stored on capacitors of subsequent stages.

After the mixer, the signal is then fed to a DT complex BPF, which attenuates blockers and image of the signal. The BPF also attenuates frequencies that will be folded by a later decimation. The structure of the charge sharing BPF (CS-BPF) is shown in Fig. 3.8.2. On each of the phases φ1-4, part of the charge stored on C_H is transferred by C_R from I to Q branch with positive sign, and from Q to I branch with negative sign. As drawn in Fig. 3.8.2, this filter has a band-pass transfer function centered at $f_c=f_{IF}$, i.e., $f_{LO}/16$ in this design. This transfer function does not have any replica and folding image frequencies within –fs/2 to +fs/2, in contrast with a frequency-translated BPF used in the superheterodyne receiver in [1]. Thanks to the fully passive implementation of this filter, it works up to 10GS/s, in contrast with the 200MS/s opamp-based structure in [5]. The center frequency of this filter can be accurately set by the C_R/C_H ratio, making it insensitive to PVT.

Four of such CS-BPFs are cascaded by three g_m-cells to form a total 4th-order complex BPF. This band-pass filtering at IF improves out-of-band linearity of the rest of the chain. Thanks to the high IF used in this structure, the g_m-cells provide flicker noise free gain at IF, reducing input-referred noise of subsequent stages. An inverter-based structure is used for these g_m-cells to make them process-scalable in addition to providing a good linearity.

To operate at a lower frequency, the signal is decimated by 16 after the last IF filter (Fig. 3.8.1). This decimation is simply done by reducing clock frequency of the following stages, and integrating 16 samples. This temporal decimation makes a *sinc*-type antialiasing filter before downsampling the signal. This decimation creates unwanted folding images located at multiples of $4f_{IF}$ from the wanted signal, which have been attenuated more than 56dB by the IF BPF and antialiasing filter in addition to the input preselect filter. Later, a second DT quadrature mixer downconverts the signal from IF to baseband. Since this mixer is clocked at IF frequency, it should have a higher IIP2 than that of a ZIF receiver, whose mixer is clocked at RF. Also, IIP2 of this mixer is improved by the preceding band-pass filters.

Baseband (BB) signal processing of the receiver consists of the total 12th-order DT filtering, gain stages and decimations. After the IF mixer, a 6th-order passive charge-rotating filter, [6], selects a desired channel, and filters the rest of the frequencies heavily (Fig. 3.8.3). Thanks to the passive implementation, this filter has a very high linearity. In high-BB-rate mode, this filter works up to 667MS/s. For narrow-bandwidth signals, such as GSM, a decimation by 4 is placed inside this stage (low BB rate). In this mode, $\varphi 1\text{-}8_H$ are disabled and $\varphi 1{,}5_L$ are enabled. After each 4 phases, four C_S are shorted together, making a spatial decimation. Then one of them continues connecting to other $C_{H2\text{-}6}$ for the rest of the filtering. This also reduces the amount of required C_H to have a narrow bandwidth. A high order of filtering at this stage relaxes linearity of the following stages. Sampling rate is further reduced by using a lower clock frequency for the rest of the chain. It makes another temporal decimation, 4 and 16 times for high and low BB rates, respectively. Each of LPF2 and LPF3 is a 3rd-order filter, also similar to [6]. The output signal after LPF3 goes to an ADC (off-chip). Note that the total of the 12th-order passive filtering in this receiver consumes less than 1.5mW, while relaxing the number of required bits for the ADC considerably. Since only real poles are realized, BB filtering requires a simple digital equalizer after the ADC to flatten the passband. Passband loss caused by the high-order passive filtering is compensated using a high gain in the receiver.

The implemented RX is able to receive RF signals from 1.8GHz to 2.5GHz, i.e., up to 10GS/s, with 200kHz–to-20MHz bandwidth. Capacitors of the BPFs and LPFs, different gains at RF, IF and BB, decimation factors and waveform generator circuit are programmable via a 128-b UART port. Although we use a fixed ratio IF frequency, i.e., $f_{LO}/16$, it could be programmable. In this way, the receiver is able to switch from one IF to another in the presence of a large blocker, to prevent desensitization of the receiver.

The LNTA circuit comprises a differential common-gate LNA with cross-coupled capacitor, followed by an inverter-based g_m-cell. The LNA and g_m-cell use 2V and 1.2V power supplies, respectively. To reduce area, this wideband LNTA does not use any tuned LC circuits. Simulated NF of the LNTA is 2 to 2.5dB.

The measured wideband transfer function (TF) of the receiver is plotted in Fig. 3.8.4 (top). There is only a discrete number of frequency points that can fold into the received band. They are located at multiples of $4\times f_{IF}$ away from f_{RF}. The reminder of the f_{IF} multiples are much smaller, except for the one (37dB rejection) that is due to the uncalibrated I/Q clock mismatch. With a preselect antenna filtering the total image rejection could be better than 70dB.

Measured close-in TF of the receiver in low/high BB rate is shown in Fig. 3.8.4 (mid). This TF can be digitally equalized afterwards. As shown in Fig. 3.8.4 (bottom), measured in-band IIP3 of the receiver is -7dBm. Uncalibrated in-band and out-of-band IIP2 of the receiver are measured as +45dBm and over +85dBm respectively.

The receiver implemented in TSMC 1P7M 65nm CMOS occupies an active area of $1.1mm^2$. The receiver consists mostly of MOS switches, capacitors and inverter-based g_m-cells, making it process scalable and friendly to digital nanoscale CMOS. Most of the capacitors are of metal-oxide-metal (MOM) type implemented differentially. Figure 3.8.5 summarizes measured performance of the RX and compares it with other state-of-the-art. The analog part of the receiver consumes 34mW for the highest gain setting. The clock waveform generator consumes 21 to 30mW, linearly changing with f_{LO}. All the required clocks are generated from an external clock at $2\times f_{LO}$, using programmable multiphase dividers. Figure 3.8.6 depicts the RX chip micrograph.

Acknowledgment:
We thank the RF Dept. of HiSilicon for technical and financial support.

References:

[1] A. Mirzaei, H. Darabi, and D. Murphy, "A Low-Power Process-Scalable Superheterodyne Receiver with Integrated High-Q Filters," *ISSCC Dig. Tech. Papers*, pp. 60-62, Feb. 2011.

[2] I. Madadi, M. Tohidian, R.B. Staszewski, "A 65nm CMOS High-IF Superheterodyne Receiver with a High-Q Complex BPF," *IEEE RFIC Symposium*, pp. 323-326, June 2013.

[3] K. Muhammad, D. Leipold, R.B. Staszewski, et. al., "A Discrete-Time Bluetooth Receiver in a 0.13μm Digital CMOS Process," *ISSCC Dig. Tech. Papers*, pp. 268, 527, Feb. 2004.

[4] A. Geis, et. al., "A 0.5 mm^2 Power-Scalable 0.5–3.8-GHz CMOS DT-SDR Receiver with Second-Order RF Band-Pass Sampler," *IEEE J. Solid-State Circuits*, vol.45, no.11, pp. 2375-2387, Nov. 2010.

[5] S. Karvonen, et. al., "A Quadrature Charge-Domain Sampler with Embedded FIR and IIR Filtering Functions," *IEEE J. Solid-State Circuits*, vol. 41, no. 2, pp. 507-515, Feb. 2006.

[6] M. Tohidian, I. Madadi, R.B. Staszewski, "A 2mW 800MS/s 7th-Order Discrete-Time IIR Filter with 400kHz-to-30MHz BW and 100dB Stop-Band Rejection in 65nm CMOS," *ISSCC Dig. Tech. Papers*, pp. 174-175, Feb. 2013.

978-1-4799-0917-9/14 $31.00 © 2014 IEEE

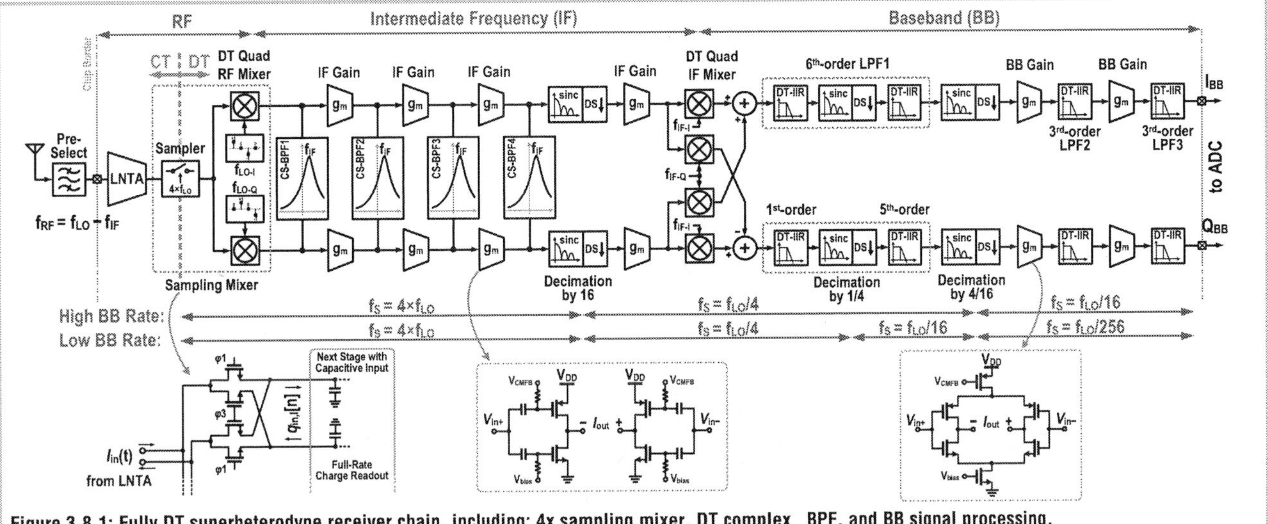

Figure 3.8.1: Fully DT superheterodyne receiver chain, including: 4x sampling mixer, DT complex BPF, and BB signal processing.

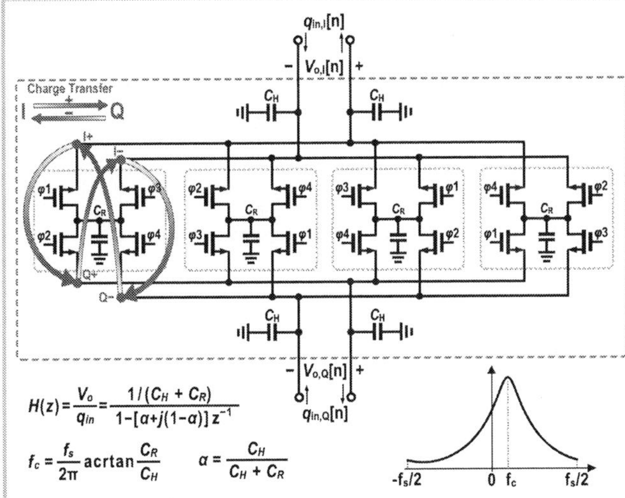

Figure 3.8.2: DT Complex BPF using I/Q charge sharing.

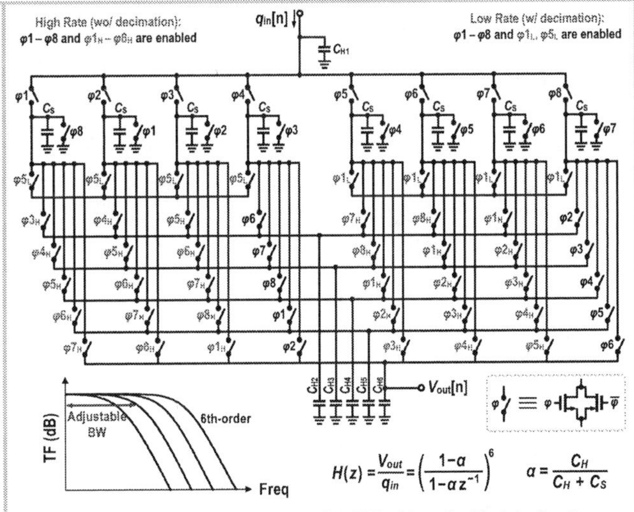

Figure 3.8.3: DT Charge-rotating 6th-order LPF with embedded decimation.

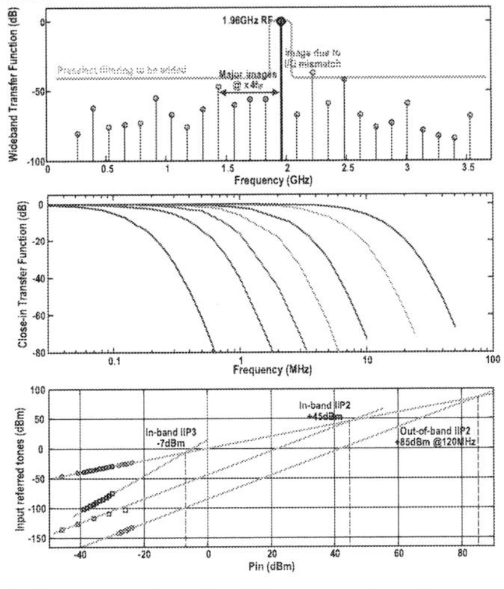

Figure 3.8.4: Measurement results of the proposed receiver.

	This Work	[2]	[1]	[4]
Technology	65nm	65nm	65nm	90nm
Architecture	Superheterodyne (High-IF)	Superheterodyne (High-IF)	Superheterodyne (High-IF)	Homodyne (Zero-IF)
Analog Baseband	Yes	No	No	Yes
RF Frequency (GHz)	1.8 – 2.5	0.5 – 1.2	1.8 – 2.2	0.5 – 3.8
Supply Voltage (V)	1.2 / 2	1.2	1.2 / 2.5	1.2
Power Consumption (mW)	55 – 65	24.5	34	67 – 115
NF (dB)	3.2 – 4.5	7.5	2.8	5.3 – 6.0
Max Gain (dB)	82	35	55	58 / 64
In-band IIP3 (dBm)	-7	+10	-8.5	+1 / +2.5
In-band IIP2 (dBm)	+45 §	-- *	-- *	-- *
Out-of-band IIP2 (dBm)	+85 §	-- *	-- *	+38 / +52
Channel BW (MHz)	0.2 – 20	4.5	4	0.2 – 20
S11 (dB)	< -10	< -10	< -10	< -10
Area (mm^2)	1.1	0.45	0.76	0.5

* Not reported § Without calibration

Figure 3.8.5: Performance summary and comparison with state-of-the-art.

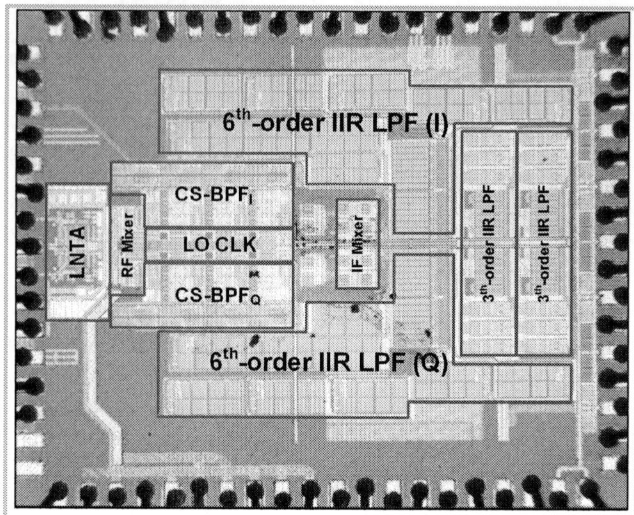

Figure 3.8.6: Chip micrograph.

3.9 An RF-to-BB Current-Reuse Wideband Receiver with Parallel N-Path Active/Passive Mixers and a Single-MOS Pole-Zero LPF

Fujian Lin[1], Pui-In Mak[1,2], Rui Martins[1,2,3]

[1]University of Macau, Macao, China,
[2]UMTEC, Macao, China,
[3]Instituto Superior Tecnico, Lisbon, Portugal

The latest passive-mixer-first wideband receiver (RX) [1] has managed to squeeze the power (10 to 12mW) via resonant multi-phase LO and current-reuse harmonic rejection BB, but the removals of *RF gain* and *virtual ground* severely penalize its NF (10.5±2.5dB), while devaluing its original IIP3 benefits (+10dBm). The described wideband RX exploits an RF-to-BB current-reuse topology, with parallel N-path active/passive mixers, to leverage such power-performance tradeoffs. Specifically, the RX features: 1) a *current-reuse RF front-end* with an N-path active mixer to realize most RF-to-BB functions in the current domain, resulting in better power efficiency and linearity; 2) a *feedforward N-path passive mixer* to enable LO-defined input matching with zero external components, while offering frequency-translated band-pass filtering and noise cancelling; 3) a *single-MOS pole-zero LPF* to perform current-mode BB filtering while alleviating the tradeoff between the in-/out-of-band linearity, and 4) a *BB-only two-stage harmonic-recombination (HR) amplifier* to boost the 3rd and 5th harmonic rejection ratios ($HRR_{3,5}$) with low hardware intricacy. Targeting the TV-band (0.15 to 0.85GHz) cognitive radios for IEEE 802.22/802.11af, the RX manifests favorable NF (4.6±0.9dB) and out-of-band IIP2/IIP3 (+61/+17.4dBm) under low power dissipation (10.6 to 16.2mW).

Figure 3.9.1 depicts the RF front-end that unifies most functions. A single-ended RF input (V_{in}) avoids the balun and its insertion loss. The $-g_{m,CS}$ stage (M_{CS}) serves as the LNA, which is stacked by an 8-path active mixer (M_{A1-8}) for down-conversion, and an 8-path current-mode LPF for channel selection before BB I-to-V conversion at R_L (V_{BB0}, V_{BB45}...V_{BB315}). M_{A1-8} driven by an 8-phase 12.5%-duty-cycle LO (V_{LO0}, V_{LO45}...V_{LO315}) allow HR at BB to enhance the critical HRR_{2-6}. A feedforward 8-path passive mixer (M_{P1-8}) driven by the same set of LOs, but anti-phased with M_{A1-8}, is added for three intents:

Input Matching and Out-of-Band Filtering: Owing to the bidirectional transparency of passive mixers [1-4], M_{P1-8} can frequency-translate the low-pass impedance ($Z_{in,LPF}$) of V_x to a band-pass one ($Z_{in,RF}$) at V_{in}, allowing LO-defined input matching and RF filtering. Moreover, the BB signal at V_x is upconverted to V_{in} and downconverted back to V_x by the $-g_{m,CS}$ stage and M_{A1-8}. The resultant positive loop gain is $G_{loop}=g_{m,CS}R_{in,LPF}/8$, where the factor 8 comes from the number of mixer paths. To ensure stability (i.e., $Z_{in,RF}>0$), G_{loop} can be set as 0.55 (i.e., $g_{m,CS}$=20mS, $R_{in,LPF}$=220Ω), resulting in a 2.2x increment of $Z_{in,RF}$. This act permits a smaller R_{SW} (6Ω) to enhance the ultimate stopband rejection at V_{in}, which is theoretically 13.3dB [$2R_{sw}/(R_{sw}+R_S)$]. To vacate more voltage headroom and enhance the linearity, M_{A1-8} were biased in the *triode* region. This approach brings down the swing at V_y (drain node of M_{CS}) and frequency-translates the low-pass $Z_{in,LPF}$ from V_x, to a band-pass one at V_y, aiding the out-of-band rejection.

Input Biasing: unlike the TV-band RXs in [5-6] that entail a bulky external inductor for bias and wideband impedance matching, here the gate of M_{CS} is biased via the passive mixers copying the dc voltage from V_x to V_{in}, avoiding any external parts while giving adequate overdrive voltage (V_{DS}=420mV) on M_{CS} for better linearity. Moreover, owing to no AC-coupling capacitors at V_{in}, the RF bandwidth easily covers the low-frequency range.

Noise Cancelling: the anti-LO-phased active and passive mixers allow *concurrent* noise cancellation of R_{sw} and LPF under $g_{m,CS}R_s=1$ (Fig. 3.9.2). For the former, R_{SW} induces a noise current to R_S, and is sensed by the $-g_{m,CS}$ stage to produce an anti-phased output nullifying the noise inherently. For the latter, when the RX is operated differentially, the LPF's noise current on R_S will be copied to another path with the same phase, being a cancellable common-mode noise. Hence, the RX NF is dominated by the thermal noise of M_{CS} (i.e., Noise Factor=$1+\gamma$, where γ is the channel noise factor), which can be upsized (W/L:120/0.18) to reduce the 1/f noise. M_{A1-8} contribute insignificant noise and are small in size (W/L:12/0.06) to save the LO power.

The current-mode Biquads in [5,6] only can synthesize two complex poles, while the proposed single-MOS LPF (Fig. 3.9.3) offers stronger pole-zero filtering, being more cost-effective than its real-pole-only counterparts [1-4]. The LPF's transfer function relies on R_B, C_B, and an intentionally large transistor M_{LPF} (W/L: 768/0.5). The latter brings in large parasitics C_{gd} (~0.3pF) and C_{ds} (~0.3pF) under bulk-source connection, introducing two stopband zeros. By tuning the zeros to 150MHz, stopband rejection at 100-to-200MHz offset can be enhanced, suitable for filtering the GSM850/900 bands when the RX is to operate up to 710MHz (IEEE 802.11af). The simulated bandwidth has a mean value of 14.6MHz (σ=0.48MHz). Another useful property of the LPF is the peaking of $Z_{in,LPF}$ around the cutoff, which avoids the fast roll-off shape when it is translated to RF. The grounded C_B essentially suppresses the out-of-band interferers before they see the active device (M_{LPF}). Without affecting most in-band metrics, C_B can be upsized to concurrently narrow the RF and BB bandwidths at the expense of die area. Elegantly, the stopband profile of this LPF is highly insensitive to R_L, easing the tradeoff between the in-/out-of-band linearity. V_{bias} from a replica RF front-end handily aids the biasing.

Single-stage HR shows an uncalibrated $HRR_{3,5}$ of 34 to 45dB [1,3]. The proposed BB-only two-stage HR amplifier (Fig. 3.9.4) boosts the $HRR_{3,5}$ without the gain scaling at RF [4], resulting in simpler layout and lower parasitics. Owing to the embedded BB channel selectivity at the RF front-end, the linearity of such HR amplifiers is highly relaxed, resulting in power savings. The latter also leads to limited BB bandwidth assisting the stopband rejection. The gain weighting is based on a PMOS amplifier {2:3:2} followed by an NMOS amplifier {5:7:5} to approximate the gain ratio {1:√2:1} with <0.1% error [4]. Thus, the total relative gain error becomes insignificant due to the multiplication: $(\varepsilon_o+\varepsilon_{1,HR})\varepsilon_{2,HR}/4$, where ε_o is the relative gain error of the RF front-end, and $\varepsilon_{1,HR}/\varepsilon_{2,HR}$ is the relative gain error of the 1st/2nd stage HR amplifier. The simulated worst $HRR_{3,5}$ is >53dB (mean=62dB). Thus, the $HRR_{3,5}$ is dominated by the LO phase error.

The RX fabricated in 65nm CMOS occupies a small die area of 0.55mm² that is dominated by the 8-path LPFs with C_B=24pF. From 0.15 to 0.85GHz, all LO-defined S_{11} are <–12.5dB. The RF-to-IF gain is 51±1dB and NF is 4.6±0.9dB. The power rises with the frequency from 10.6 to 16.2mW, in which 7.5mW due to the static power (RF+BB). At 0.7GHz RF and maximum gain, the IIP2/IIP3 raises from +15/–12dBm (in-band) to +61/+17.4dBm (out-of-band). The BB bandwidth is ~9MHz with 86.3dB stopband rejection at 150MHz offset, thanks to the dual stopband zeros. The out-of-band P_{1dB} is –2.5dBm at 50MHz offset, enhanced by the RF filtering.

The chip summary is given in Fig. 3.9.6. Benchmarking with the passive-mixer-based RXs [1-4], this work succeeds in saving the power without sacrificing the NF, out-of-band linearity and HRR. No external component is entailed and stronger BB filtering is achieved in a small die size. Figure 3.9.7 shows the RX die micrograph.

Acknowledgements:
This work was funded by the Macao FDCT and UM - MYRG114-FST13-MPI.

References:
[1] C. Andrews, et al., "A Wideband Receiver with Resonant Multi-Phase LO and Current Reuse Harmonic Rejection Baseband," *IEEE J. Solid-State Circuits*, vol. 48, pp. 1188-1198, May 2013.
[2] J. Borremans et al., "A 0.9V Low-Power 0.4-6GHz Linear SDR Receiver in 28nm CMOS," *Symp. on VLSI Circuits, Dig. Tech. Papers*, pp. 146-147, June 2013.
[3] D. Murphy et al., "A Blocker-Tolerant Wideband Noise-Cancelling Receiver with a 2dB Noise Figure," *ISSCC Dig. Tech. Papers*, pp. 74-75, Feb. 2012.
[4] Z. Ru et al., "A Software-Defined Radio Receiver Architecture Robust to Out-of-Band Interference," *ISSCC Dig. Tech. Papers*, pp. 230-231, Feb. 2009.
[5] P.-I. Mak et al., "A 0.46mm² 4-dB NF Unified Receiver Front-End for Full-Band Mobile TV in 65nm CMOS," *ISSCC Dig. Tech. Papers*, pp. 172-173, Feb. 2011.
[6] J. Greenberg et al., "A 40MHz-to-1GHz Fully Integrated Multistandard Silicon Tuner in 80nm CMOS," *ISSCC Dig. Tech. Papers*, pp. 162-163, Feb. 2012.

978-1-4799-0917-9/14 $31.00 © 2014 IEEE

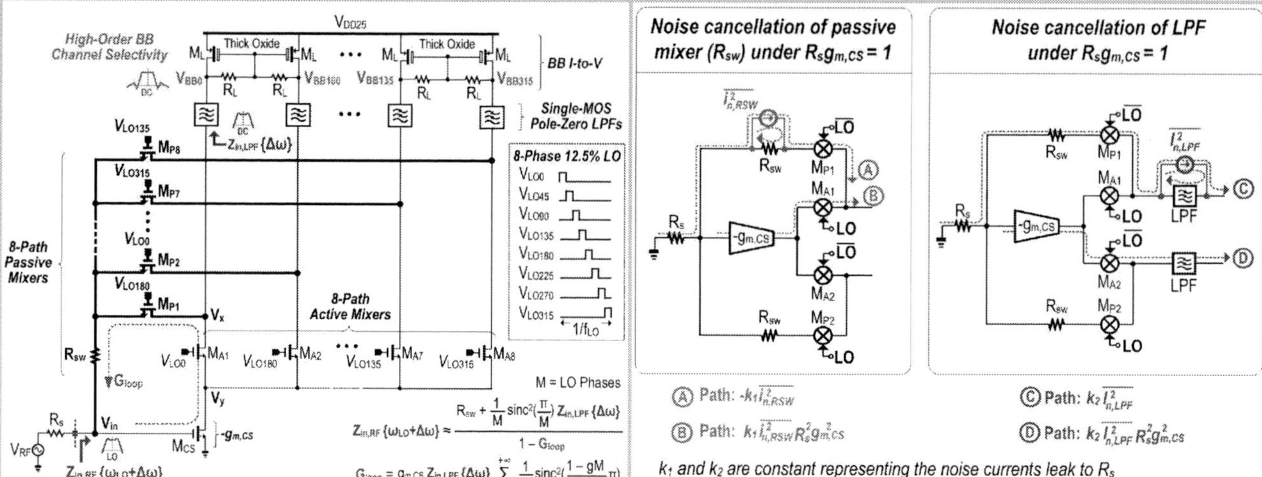

Figure 3.9.1: A current-reuse RF front-end with a single-ended RF input.

Figure 3.9.2: Simplified two-phase noise equivalent circuits of the RF front-end showing the noise cancellation of R_{sw} (left) and LPF (right).

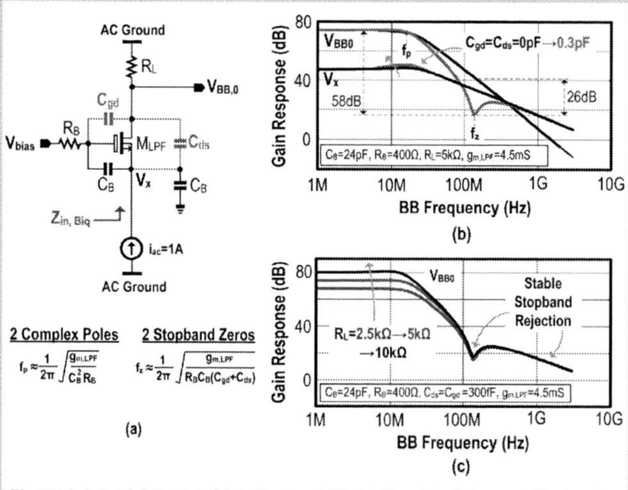

Figure 3.9.3: a) Single-MOS pole-zero LPF. b) Simulated V_{BB0} and V_x showing the rejection added by the stopband zeros. c) Sizing R_L for In-band gain.

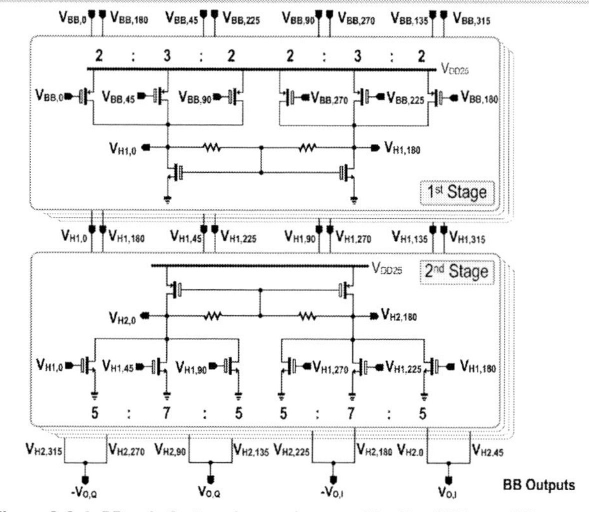

Figure 3.9.4: BB-only 2-stage harmonic-recombination (HR) amplifier.

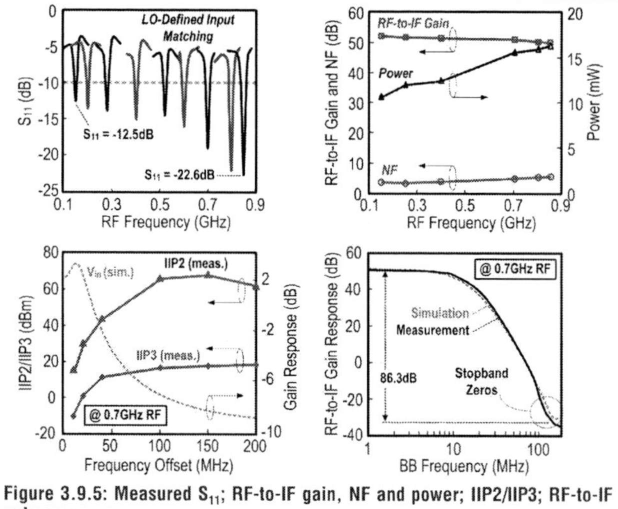

Figure 3.9.5: Measured S_{11}; RF-to-IF gain, NF and power; IIP2/IIP3; RF-to-IF gain response.

	This Work	JSSC'13 [1]	VLSI'13 [2]	ISSCC'12 [3]	ISSCC'09 [4]
RX Architecture	Current-Reuse RF Front-End + Feedforward Passive Mixer	Passive Mixer + BB LNA	RF LNA + Passive Mixer + G_m-C + Op-Amp	2-Path Noise-Canceling+ Passive-Mixer + OpAmp	RF LNA + Passive Mixer + Op-Amp
Downconversion	Active // Passive	Passive	Passive	Passive	Passive
RF Input Style	Single-Ended	Single-Ended	Differential	Single-Ended	Differential
RF Range (GHz)	0.15 to 0.85	0.7 to 1.6 (8-phase path)	0.4 to 3 (8-phase path)	0.08 to 2.7	0.4 to 0.9
Power (mW) @ RF	10.6 @ 0.15GHz 16.2 @ 0.85GHz	10~12 @ 0.7GHz 10~12 @ 1.6GHz	20 @ 0.4GHz 40 @ 3GHz	37 @ 0.08GHz 70 @ 2.7GHz	49 @ 0.4GHz 60 @ 0.9GHz
DSB NF (dB)	4.6±0.9	10.5±2.5	1.8 to 2.4	1.9±0.4	4±0.5
Ultimate Out-of-Band IIP3 (dBm)	+17.4	+10	+3	+13.5	+16
Ultimate Out-of-Band IIP2 (dBm)	+61	+26.6	+65 (calibrated)	+54	+56
External Parts	Zero	Zero	Transformer	Zero	2 inductors and 1 Transformer
Active Area (mm²)	0.55	2.9 (inc. VCOs)	~0.5 (from Fig.)	1.2	1
BB Filtering Style	2 Complex Poles + 2 Stopband Zeros (Current-Mode)	1 Real Pole (Passive-RC)	2 Real Poles (Active/ Passive-RC)	2 Real Poles (Active/Passive-RC)	2 Real Poles (Active-RC)
$HRR_{3,5}$ (dB)	>53, >51	34, 34	70, 55 (calibrated)	42, 45	60, 64
BB Bandwidth (MHz)	9	20	0.5 to 50	2	12
RF-to-IF Gain (dB)	51±1	37	36	72	34.4±0.2
Supply (V)	1.2, 2.5	1.3	0.9	1.3	1.2
CMOS Technology	65 nm	65 nm	28 nm	40 nm	65 nm

Figure 3.9.6: Chip summary and benchmark with recent passive-mixer-based RXs [1-4].

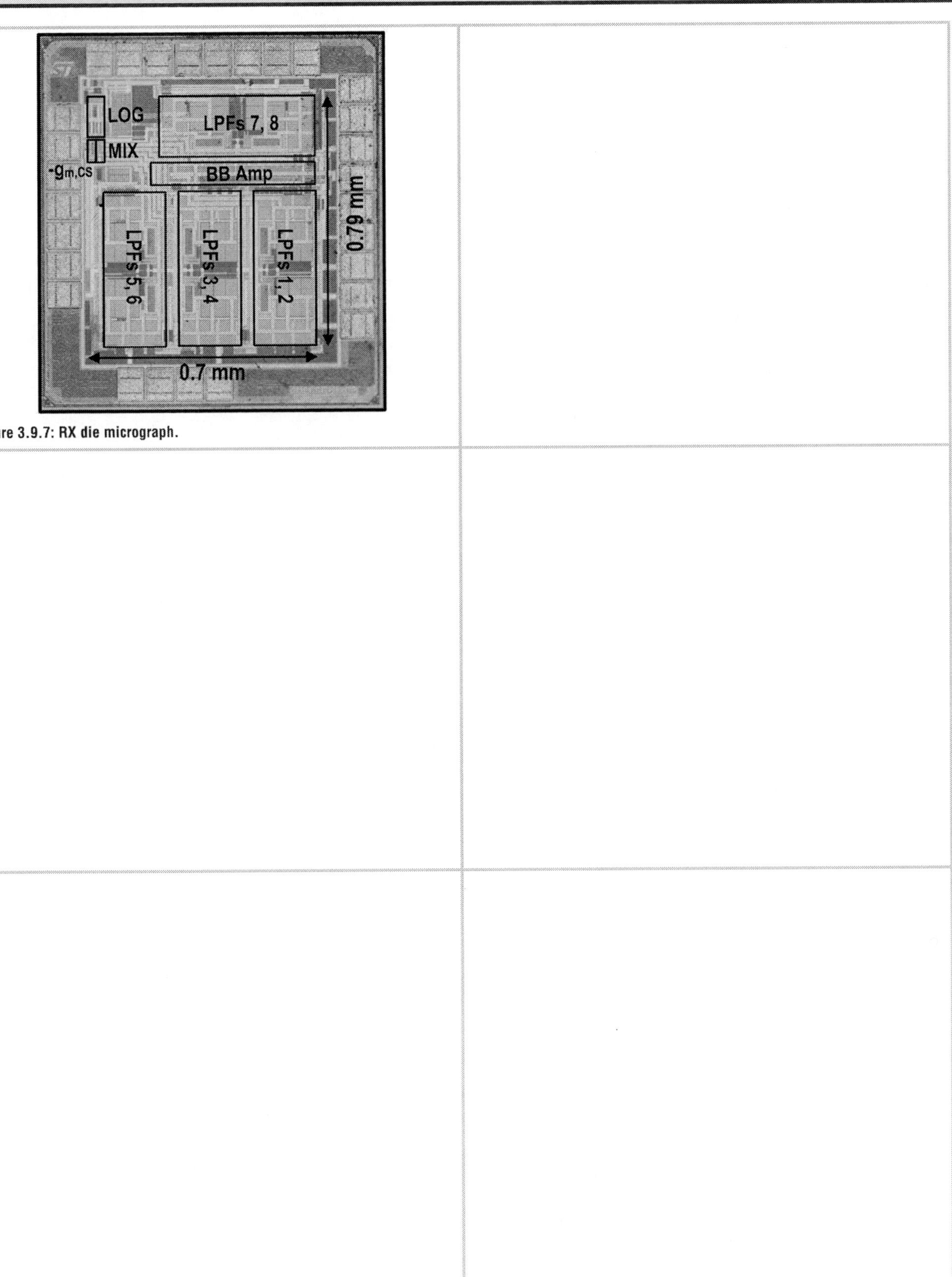

Figure 3.9.7: RX die micrograph.

Session 4 Overview: *DC-DC Converters*

ANALOG SUBCOMMITTEE

Session Chair: *Wing-Hung Ki*
HKUST, Hong Kong, China

Session Co-Chair: *Christoph Sandner*
Infineon, Villach, Austria

To meet the challenges of dynamic power requirements of diverse electronic applications, both high performance switched-mode and switched-capacitor DC-DC converters are indispensable. In the first part of the session, switched-mode power converters catering to multi-core SoCs (system-on-chips) are presented. They have to switch at frequencies into the 10MHz regime for small form factor, to have multi-phase for ripple reduction, fast control in response to large and fast load current changes, and fast reference tracking for dynamic voltage scaling.

The second part of the session is dedicated to switched-capacitor (SC) DC-DC converters. With technical advances allowing power levels of several watts, they find their way into applications where inductor-based DC-DC converters have dominated over the last decades. Several techniques are shown to increase both efficiency and power density. This is achieved by making use of multi-phase and reconfigurable architectures.

4.1 A 3-Phase Digitally Controlled DC-DC Converter with 88% Ripple Reduced 1-Cycle Phase Adding/Dropping Scheme and 28% Power Saving CT/DT Hybrid Current Control **1:30 PM**

C. K. Teh, Toshiba, Kawasaki, Japan

In Paper 4.1, by Toshiba, a 3-phase digitally controlled DC-DC converter with 84%-to-90% efficiency over 0.4A-to-9A load range is presented. A 1-cycle fast phase adding/dropping scheme is proposed, offering less than 10mV ripple, 88% reduction from the optimal PID control.

4.2 A 6A 40MHz Four-Phase ZDS Hysteretic DC-DC Converter with 118mV Droop and 230ns Response Time for a 5A/5ns Load Transient **2:00 PM**

M. K. Song, University of Texas, Dallas, Richardson, TX

In Paper 4.2, by the University of Dallas, Texas, a 40-MHz 4-phase ZDS hysteretic converter achieves 1% settling time of 230ns for 5A/5ns load step with C_{OUT}=940nF and 9.8% droop at 1.2V. The ZDS control enables 6x lower switching frequency without affecting the transient response.

4.3 An 87%-Peak-Efficiency DVS-Capable Single-Inductor 4-Output DC-DC Buck Converter with Ripple-Based Adaptive Off-Time Control **2:30 PM**

D. Lu, Fudan University, Shanghai, China

In Paper 4.3, by Fudan University, a ripple-based adaptive on-time control method is proposed and proven in a DVS-capable single-inductor 4-output DC-DC buck converter to overcome the cross-regulation problem. The converter achieves ripple below 30mV and 87% peak efficiency.

978-1-4799-0917-9/14 $31.00 © 2014 IEEE

4.4 A 10/30MHz Wide-Duty-Cycle-Range Buck Converter with DDA-Based Type-III Compensator and Fast Reference-Tracking Responses for DVS Applications 3:15 PM

L. Cheng, Hong Kong University of Science and Technology, Hong Kong, China

In Paper 4.4, by HKUST, a 10/30MHz buck converter that features a high-accuracy delay-compensated ramp generator and an area-efficient Type-III compensator built around a differential difference amplifier (DDA) that also facilitates reference-tracking is presented. The measured maximum output power is 3.6W.

4.5 A 2-Phase Resonant Switched-Capacitor Converter Delivering 4.3W at 0.6W/mm² with 85% Efficiency 3:45 PM

K. Kesarwani, Dartmouth College, Hanover, NH

In Paper 4.5, by Dartmouth College, a chip-scale resonant switched-capacitor (ReSC) converter is presented that can deliver over 4 watts at 0.6W/mm² with 85% efficiency. The 2-phase, nominally 2:1 converter supports input voltages from 3.6-6V, and provides output voltage regulation and high efficiency in light load through dynamic off-time modulation. The converter is implemented in 0.18µm HVCMOS and uses MIM capacitors with die-attached air-core solenoid inductors.

4.6 An 85%-Efficiency Fully Integrated 15-Ratio Recursive Switched-Capacitor DC-DC Converter with 0.1-to-2.2V Output Voltage Range 4:15 PM

L. G. Salem, University of California, San Diego, La Jolla, CA

In Paper 4.6, by the University of California, San Diego, a recursive switched-capacitor topology is implemented that provides 15 conversion ratios for high efficiency over a wide output voltage range. The topology opportunistically connects individual 2:1 SC cells while maximizing capacitance utilization.

4.7 A Sub-ns Response On-Chip Switched-Capacitor DC-DC Voltage Regulator Delivering 3.7W/mm² at 90% Efficiency Using Deep-Trench Capacitors in 32nm SOI CMOS 4:45 PM

T. M. Andersen, ETH Zurich and IBM Research, Rüschlikon, Switzerland

In Paper 4.7, by ETH Zurich, an on-chip switched-capacitor DC-DC voltage regulator in 32nm SOI CMOS for microprocessor power delivery is presented. A reconfigurable power stage regulates the output voltage between 0.7 to 1.1V from a 1.8V input supply. The power density is up to 3.1 W/mm².

4.8 3-Phase 6/1 Switched-Capacitor DC-DC Boost Converter Providing 16V at 7mA and 70.3% Efficiency in 1.1mm³ 5:00 PM

R. Karadi, NXP Semiconductors, Eindhoven, The Netherlands

In Paper 4.8, by NXP Semiconductors, a 3-phase switched-capacitor boost converter using only 2 external floating capacitors and providing 16V output from a 3.3V input is presented. It reaches an efficiency of 70.3% at 7mA load current. The volume is 15x smaller compared to competitive inductive converters.

978-1-4799-0917-9/14 $31.00 © 2014 IEEE

4.1 A 3-Phase Digitally Controlled DC-DC Converter with 88% Ripple Reduced 1-Cycle Phase Adding/Dropping Scheme and 28% Power Saving CT/DT Hybrid Current Control

Chen Kong Teh, Atsushi Suzuki, Manabu Yamada, Mototsugu Hamada, Yasuo Unekawa

Toshiba, Kawasaki, Japan

Multiphase DC-DC converters are essential to provide good efficiency over a wide range of load current, especially for today's multicore SoCs, which usually have a wide range of current profile. Active-phase-count (APC) control, as proposed in [1-3], is the key technique that offers the wide load range, and dynamically adjusts the number of phases according to load conditions. However, an APC transition induces a voltage disturbance due to current redistribution among phases. This makes the APC control only suitable for a voltage regulator with large output capacitors beyond 1000μF, or with high switching frequency beyond 10MHz [4] that hardly delivers more than 2A current with a good efficiency. To mitigate the disturbance impact, [2] has proposed adding or dropping a phase slowly, to allow the PID control to gradually redistribute the phase currents. However, the load current may change direction spontaneously before the completion of the transition, leading to deterioration in the transient response. This paper presents a fast APC scheme which performs the transition within 1 switching cycle, and only requires 66μF capacitors to limit the voltage ripple within 10mV, an 88% ripple reduction with respect to the optimal PID-only control. The proposed APC scheme utilizes digital phase current data, which is also needed to determine the optimal phase count [1-3]. However, the A/D conversions consume 3mW, which is 28% of the total power consumption. A continuous-time/discrete-time (CT/DT) hybrid current control architecture is introduced to eliminate the 3mW penalty while producing A/D converted data by time sharing of the analog circuitry in the control loop. Moreover, a fast transient response configuration is designed, offering less than 40mV fluctuation during an 8A load transition and less than 5mV fluctuation during a 2V line transition.

Figure 4.1.1 shows the overall block diagram of the 3-phase DC-DC converter that delivers load current up to 9A. Digital control architecture is chosen for simplifying the implementation of the APC scheme, especially in realizing seamless overwrite of the PID internal state during an APC transition. The converter is based on the current-programmed-mode (CPM) control, which ensures current sharing among phases. The high-side FETs are turned on by their respective switching clocks clk_sw_i, and are turned off by the CPM control when the respective phase currents $isense_i$ are greater than a peak control value determined by the PID. The active phase FSM controls the activity of each phase, and overwrites the internal state of the PID upon changing of the phase count, with a calculated new control value as close as to that of the new steady-state value after such change. The CT/DT hybrid current control provides a continuous-time and a discrete-time CPM control, interleaved within a switching cycle. The former is used to provide a CT resolution of PWM duty cycle for a limit-cycle free regulation, and the latter is used for SAR-ADC operation, controlled by the time-sharing FSM. The clock management unit has a phase adjuster actively adjusting the phase intervals among clk_sw_i according to the active phase count.

Figure 4.1.2 shows the proposed APC scheme. The key concept is to ensure the charge balance of the converter as accurate as possible during an APC transition. However, to reduce the integral calculation cost while producing a nearly optimal result, the calculation is approximated to a simplified equation such that the total of the average phase currents after the transition is equal to that of before the transition. The scheme offers less than 10mV ripple with only 66μF capacitors, which is quite challenging for the conventional PID control to provide even by using big capacitors beyond 1000μF. The average phase currents $iavg_{before}$ before the transition are calculated by using the digital peak and valley values of $isense_i$, and the digital value corresponding to 0A (*ioffset*) measured during inactive state, respectively. The current difference ΔI after the transition is then calculated based on the simplified equation. Finally, the PID internal state is overwritten with a new ΔI different from the old one. Concurrently, the APC information is fed to the clock management unit, where the phase intervals among clk_sw_i are reconfigured within 1 cycle. The timing diagram of the counter-based clk_sw_i generator is shown in the figure, where the internal states of the counters are overwritten respectively according to the active phase count, for shifting the rising edges of clk_sw_i.

Figure 4.1.3 shows the architecture of the CT/DT hybrid current control (HCC). The 11b R-2R DAC, comparator and S/H unit are time shared, where CT peak searching is performed by the comparator when $isense_i$ is close to its peak value, and DT peak searching as well as 8MS/s 11b SAR-ADC operation is performed when $isense_i$ is far from its peak value, determined by the control value $ictrl\_ramp_i$. A mode decision test is performed in every 12 cycles of the 100MHz clock, at which the DAC is fed by a judging value $ictrl\_th_i$ constantly smaller than $ictrl\_ramp_i$, for comparing with $isense_i$, and the result will determine the mode to proceed. The distance of $ictrl\_th_i$ from $ictrl\_ramp_i$ is self adjusted according to the steepness of the rising slope of the digital $isense\_d_i$. In CT peak searching mode, the S/H is temporarily switched to the conduction state, and CT comparison is performed between $ictrl\_ramp_i$ and $isense_i$. In SAR searching mode, the DAC is successively fed by 11 different binary search values for A/D conversion to produce $isense\_d_i$. In DT peak searching mode, comparison is performed between $ictrl\_ramp_i$ and the linear interpolated value of $isense\_d_i$ calculated in 100MHz. The HCC architecture eliminates 3mW that amounts to 28% of the total power consumption of the converter, offering 0.4% efficiency improvement at 0.5A.

Figure 4.1.5 shows the measured results of the APC scheme. The efficiency results indicate that the converter can offer an efficiency from 84% to 90% over a wide load range from 0.4 to 9A, in 5-to-1.5V step-down regulation. The figure also illustrates the transient waveforms during the APC transitions, where in all combinations the ripple is observed to stay within 10mV. The transition is completed in 1 switching cycle, and good current sharing among phases is observed in the steady state.

Figure 4.1.5 shows the configuration for boosting the transient response as well as the noise immunity. A 4b 20MS/s delay-line ADC is designed to produce the digital *error* from *Vout*, which is then lowpass filtered and is down-sampled to 500kHz times the active phase count by the decimator. This lowers the loop delay of the HCC control and raises the SNR of the feedback signal. The decimated signal is then fed to the digital PID to produce *ictrl* for all phases, which is then ramped down to produce $ictrl\_ramp_i$. In order to maximize immunity to current-sensing noise, the falling slope of $ictrl\_ramp_i$ is gradually adjusted to be as close as that of $isense_i$ by using the smoothed slope data of $isense\_d_i$. The measured transient responses indicate less than 40mV fluctuations during load transient between 1A and 9A, and less than 5mV fluctuations during line transient between 3.5V and 5V.

Figure 4.1.6 shows the performance summary of the proposed converter for a tablet PC application, fabricated in 2.5V/5V CMOS technology. We design the pre-driver and the current-sensing amplifier by using 5V transistors, and the rest by using 2.5V transistors. The converter consumes 7.6mW, and the line and load regulations are within 1mV. The peak efficiency is 90.2%, with more than 84% efficiency over 0.4-to-9A load range. The 500kHz DC-DC converter with only 66μF capacitors demonstrates 1-cycle fast-APC transition with less than 10mV ripple, the best performance ever published.

Acknowledgments:
The authors would like to thank T. Morishige, T. Namekawa, T. Takayama, T. Inoue, Y. Satoh for their support.

References:
[1] W. Qiu, C. Cheung, S. Xiao, and G. Miller, "Power Loss Analyses for Dynamic Phase Number Control in Multiphase Voltage Regulators," *Applied Power Electronics Conf.*, pp. 102-108, Feb. 2009.
[2] P. Zumel, C. Fernnndez, A. de Castro, and O. Garcia, "Efficiency Improvement in Multiphase Converter by Changing Dynamically the Number of Phases," *Power Electronics Specialists Conf.*, pp. 2845-2850, Jun. 2006.
[3] W.Y. Wang, H.H.C. Iu, W. Du, and V. Sreeram, "Multiphase DC-DC Converter with High Dynamic Performance and High Efficiency," *IET Power Electronics*, vol. 4, no. 1, pp. 101-110, Jan. 2011.
[4] C. Huang and P.K.T. Mok, "An 82.4% Efficiency Package-Bondwire-Based Four-Phase Fully Integrated Buck Converter with Flying Capacitor for Area Reduction," *ISSCC Dig. Tech. Papers*, pp. 362-363, Feb. 2013.

978-1-4799-0917-9/14 $31.00 © 2014 IEEE

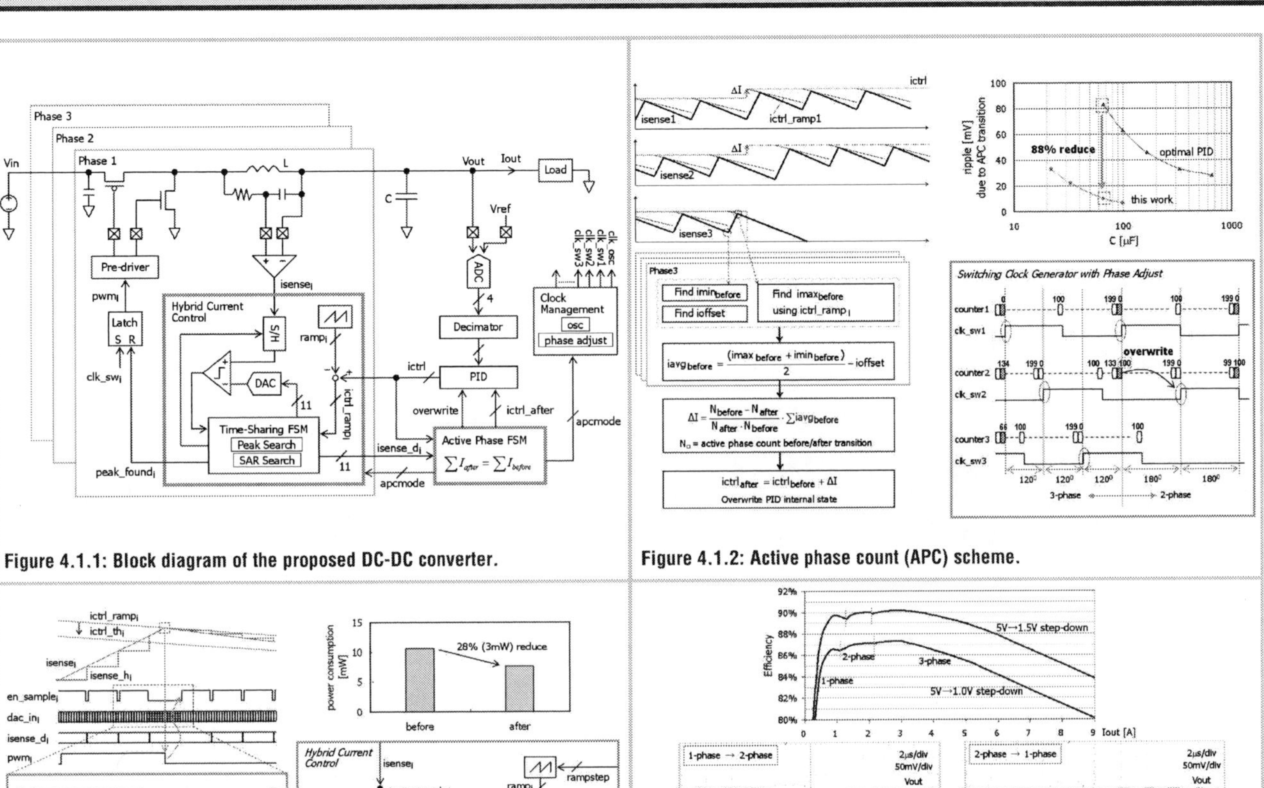

Figure 4.1.1: Block diagram of the proposed DC-DC converter.

Figure 4.1.2: Active phase count (APC) scheme.

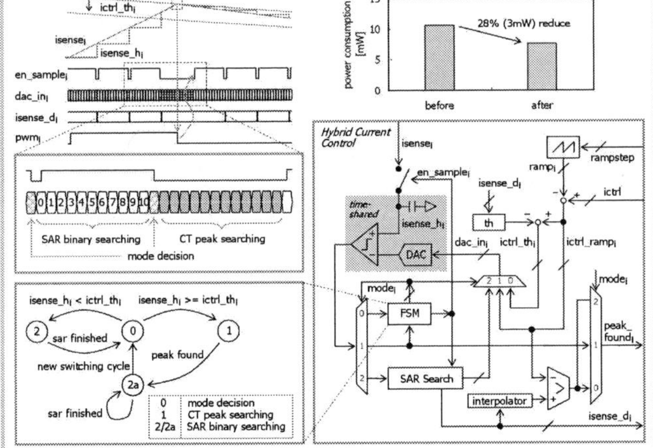

Figure 4.1.3: Architecture of CT/DT hybrid current control.

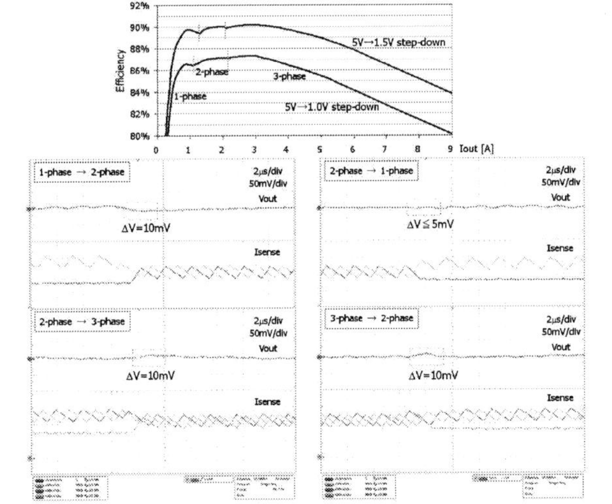

Figure 4.1.4: Measured efficiency and APC transient waveforms.

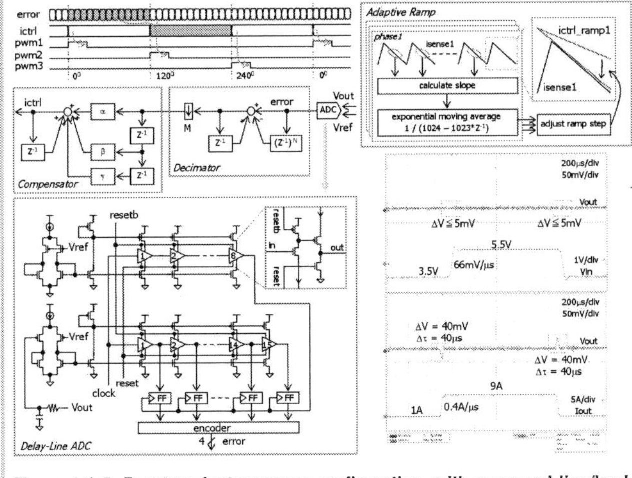

Figure 4.1.5: Fast transient response configuration, with measured line/load responses.

Features	
Technology	0.25µm 2.5V/5V CMOS
Number of Phases	3
Max. Iout	9A
Switching Frequency	500kHz
Core Frequency	100MHz
Vin	2.8V to 5.5V
Vout	0.2V to 3.3V
L (DCR)	2.2µH (13mΩ)
C	66uF
Logic Gate Count	24kgate
Area	1.3mm^2
Delay-Line ADC	4bit, 20Msps
SAR-ADC	11bit, 8Msps

Performance	
Peak Efficiency (η)	90.2% @ 5V→1.5V step-down 87.3% @ 5V→1.0V step-down
Load Range @ good η	0.4A to 9A @ η>84%, 5V→1.5V 0.3A to 9A @ η>80%, 5V→1.0V
Load Transient Response (ΔV, $\Delta\tau_{settling}$)	40mV, 40µs @ 8A step, 0.4A/µs
Line Transient Response	≦5mV @ 2V step, 66mV/µs
Vout Ripple	≦5mV @ 1-phase
Ripple @ APC Change	≦10mV
Line Regulation	1mV @ 2.8V to 5.5V
Load Regulation	1mV @ 0A to 9A
Power Consumption	7.6mW

Figure 4.1.6: Features and performance summary.

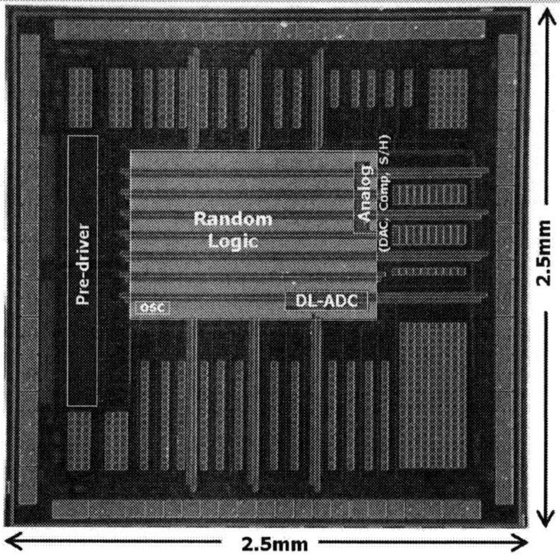

Figure 4.1.7: Die Micrograph.

4.2 A 6A 40MHz Four-Phase ZDS Hysteretic DC-DC Converter with 118mV Droop and 230ns Response Time for a 5A/5ns Load Transient

Min Kyu Song, Joseph Sankman, Dongsheng Ma

University of Texas, Dallas, Richardson, TX

In recent years, the clock frequency, the number of cores, and the power dissipation of application processors (APs) for portable electronics have dramatically increased. As a result, peak processor currents have reached several amperes with slew rates on the order of 1A/ns. These fast large steps incur large output voltage (V_{OUT}) droops, which induce failed paths and cause processor black-outs. Present voltage regulators (VRs) combat these challenges by using bulky output capacitor (C_{OUT}) arrays that could add up to over 100μF. However, this practice is untenable for next-generation APs, which have severely limited PCB area and require fast dynamic voltage scaling. These challenges have led to great demand for ultra-fast VRs. PWM control requires a bandwidth 5 to 10× less than the switching frequency, f_{SW}, which results in a slow response [1]. Hysteretic control has been proposed to achieve faster response [2, 3], however, it still suffers from an inherent delay (t_{delay}) up to the discharge period, (1−D)T, due to realistic hysteretic window size and inductor current (I_L) slew limit. Consider, for example, a fast hysteretic VR achieving 10% voltage droop for an instant load step of 5A. For an inductor (L) chosen to have I_L ripple < 200mA, and C_{OUT} under a few μF, t_{delay} cannot exceed a few ns. This requires a f_{SW} from 0.5 to 1GHz, which in turn causes a large switching loss and restricts the feasible power level of the converter. This is against the power demand trend of APs. An interleaved multiphase topology can be the most effective way to improve both the system response and the equivalent I_L slew rate by changing the number of phases; however, clock and phase synchronization and current sharing for conventional hysteretic control are challenging.

A single-phase block implementation of the current-mode zero-delay synchronized (ZDS) hysteretic control is shown in Fig. 4.2.1. This approach mitigates the challenges associated with extremely fast response and clock synchronization in hysteretic control. The operation principle can be explained using the waveforms that are shown in Fig. 4.2.1. Rather than adopting a fixed hysteretic window formed by V_H and V_L, the ZDS scheme adaptively adjusts its hysteretic window size, ΔV_{HYS}, as wide as necessary for clock synchronization during the charge period, DT, and narrows the window to near zero for delay-less response during the discharge period, (1−D)T. An I_L down-slope tracking signal, V_{SL}, is subtracted from the upper boundary of the physical hysteretic limit, V_H, and creates I_L-slope tracking reference, V_{HYS}, which is compared to the sensed inductor current, V_{SENS}. In the steady state, V_{SENS} rises from V_L during DT (V_{SW} high) until it hits V_{HYS}. The moment when V_{SENS} and V_{HYS} intersect determines the start of (1−D)T (V_{SW} low), and V_{SENS} falls from V_H. During (1−D)T, V_{HYS} down tracks with the same slope of V_{SENS}, which is proportional to $R_I \times V_{OUT}/L$, creating a near-zero hysteretic window for the entire (1−D)T. Thus, when a load step-up occurs, V_{SENS} is able to exit the hysteretic boundary without t_{delay} of the conventional hysteretic control, that results in quicker system response. The charge loss, ΔQ_{extra}, caused by t_{delay} is eliminated, alleviating the need for 0.5 to 1GHz f_{SW} by up to an order of magnitude.

During the process of synchronization (Fig. 4.2.1), V_{HYS} slides down, tracking V_{SENS}, until the next V_{CLK} pulse. When the pulse occurs, V_{HYS} is reset to V_H, triggering V_{SW} to high. In this way, the ZDS control scheme ensures the leading edge of each cycle is synchronized to V_{CLK}. This synchronization scheme accommodates the immediate hysteretic response to a load change that causes a break from V_{CLK} synchronization, and is capable of synchronization recovery. Right after the I_{OUT} step-up, the injected fourth V_{CLK} pulse causes V_{HYS} to reset to V_H so that the leading edge is re-synchronized with V_{CLK}, rather than waiting for V_{SENS} to fall to the fixed V_L causing de-synchronization similar to the conventional hysteretic control. This process defines V_{L_SYNC}, which is the clock-injected lower hysteretic limit. After a few cycles, the trajectory of V_{L_SYNC} settles to V_L and the converter returns to its steady state. Although maintaining clock synchronization, ZDS control is still hysteretic based, and achieves stability over the entire duty ratio range in contrast to peak/valley current-mode PWM control.

To ensure proper operation at very high f_{SW}, an accurate high-speed I_L sensor is required. A widely used technique for sensing I_L is to utilize the voltage drop on the inductor direct-current resistance (DCR). As depicted in Fig. 4.2.2, by having a matching $R_F C_F$=L/DCR, the voltage across C_F emulates V_{DCR}. To achieve sufficient current sense gain ($R_I = V_{SENS}/I_L$) for system robustness and stability, the DCR value must be large, which immensely adds to conduction loss. Alternatively, an additional amplifier can be implemented, however, the power dissipation vastly grows as f_{SW} increases [4]. To overcome these challenges, an emulated AC+DC I_L sensor is proposed in Fig. 4.2.2. The idea is to split I_L into an average (DC) component and an AC ripple, amplify each independently, and then recombine the two. The average I_L, I_{L_AVG}, cannot change quickly due to physical I_L slew limit, therefore, the sensor only implements a simple low-power current conveyor to amplify I_{L_AVG}. The amplifier regulates the voltage across R_{DC} to the average voltage of $V_{CF} = V_{DCR}$, resulting in current output I_{DC}, which achieves DC gain of $R_\Sigma/R_{DC} = A_{SENS\_DC}$. Meanwhile, the AC ripple, V_{AC}, can be amplified by another passive $R_{AC}C_F$ network across L with a smaller time constant to yield an AC gain of $R_F/R_{AC} = A_{SENS\_AC}$. By setting $A_{SENS\_DC} = A_{SENS\_AC}$, the output V_{SENS} properly emulates I_L with sense gain, R_I=DCR×A_{SENS} which enables the use of ultra-low DCR inductors for high efficiency.

The complete 4-phase implementation of this work is detailed in Fig. 4.2.3. The converter contains 4 ZDS single-phase sub-converters, which are synchronized by a 4-phase CLK synchronizer. With the inherent clock synchronization ability of each ZDS sub-converter, 4-phase synchronization of the converter is easily facilitated. Current-mode hysteretic operation of each sub-converter enables cycle-by-cycle current sharing among all 4 sub-converters (unlike [1, 3]) and allows using ultra-low equivalent series resistance (ESR) for C_{OUT} that mitigates V_{OUT} droop. An error compensator is added to compensate for both system response change, caused by different R_I value and systematic offsets. An active feedback dead-time (t_{DT}) control gate driver with dual slew-rates (SRs) is implemented to maintain minimum body diode conduction time across a wide load range. I_L-sensed burst-mode control with DCM operation is implemented for light load efficiency, and is made possible by high-speed sensing of the AC+DC I_L sensor. In addition, adaptive transistor sizing (ATS) is employed for a wide range of efficiency, which is facilitated by monitoring I_{L_AVG}.

The converter is fabricated in 0.18μm CMOS. It employs four 78nH inductors with ultra-low 42mΩ DCR and two 470nF C_{OUT} capacitors with 5mΩ ESR. As shown in Fig. 4.2.4, under a 5A/5ns load step (V_{OUT}=1.2V), the droop is 118mV with a 1% settling time (t_{settle}) of 230ns. The node V_{LX} shows instant duty saturation that lasts until I_L rises by 5A (~41ns). Under the opposite load step, the overshoot is 155mV with a t_{settle} of 278ns. Forced CCM (which exclusively occurs at the moment of load step-down) from the I_L-sensed burst-mode control allows a fast C_{OUT} discharge. In the steady state, the converter displays proper 4-phase synchronization with each phase operating at 40MHz with 90° phase shift. The measured efficiency is shown in Fig. 4.2.5, demonstrating 86.1% peak efficiency and increased efficiency in light load due to the burst-mode control. Furthermore, 71% efficiency (V_{OUT}=1.6V) at maximum load current (I_{MAX}) of 6A is achieved. V_{LX} is measured to verify proper transition from burst mode to CCM, as shown in Fig. 4.2.5. A comparison with the prior art is given in Fig. 4.2.6. The ZDS hysteretic control enables 6× lower f_{SW} for commensurable transient response with [2]. Lower f_{SW} facilitates higher efficiency at 2× lower conversion ratio and accommodates much larger I_{MAX} of 6A. Figure 4.2.7 shows the chip micrograph with a die size of 2.5mm × 3.1mm (including test circuits).

Acknowledgements:
This work is jointly supported by the U.S. National Science Foundation under the research contracts CCF-0844557 and DGE-1147385.

References:

[1] C. Huang and P. K. T. Mok, "An 82.4% Efficiency Package-Bondwire-Based Four-Phase Fully Integrated Buck Converter with Flying Capacitor for Area Reduction," *ISSCC Dig. Tech. Papers*, pp. 362-364, Feb. 2013.

[2] P. Hazucha, G. Schrom, J. Hahn, et al., "A 233-MHz 80%-87% Efficient Four-Phase DC-DC Converter Utilizing Air-Core Inductors on Package," *IEEE J. Solid-State Circuits*, vol. 40, no. 4, pp. 838-845, Apr. 2005.

[3] P. Li, L. Xue, P. Hazucha, et al., "A Delay-Locked Loop Synchronization Scheme for High-Frequency Multiphase Hysteretic DC-DC Converters," *IEEE J. Solid-State Circuits*, vol. 44, no. 11, pp. 3131-3145, Nov. 2009.

[4] S.-W. Wang, G.-H. Cho, and G.-H. Cho, "A High-Stability Emulated Absolute Current Hysteretic Control Single-Inductor 5-Output Switching DC-DC Converter with Energy Sharing and Balancing," *ISSCC Dig. Tech. Papers*, pp. 276-277, Feb. 2012.

978-1-4799-0917-9/14 $31.00 © 2014 IEEE

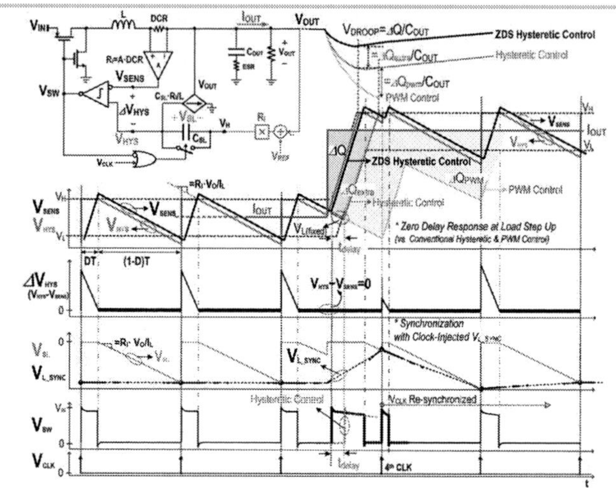

Figure 4.2.1: Zero-delay synchronized (ZDS) hysteretic control block diagram and waveforms.

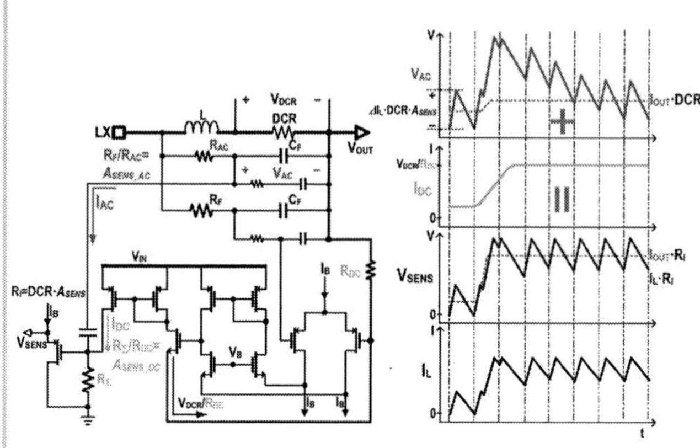

Figure 4.2.2: Emulated AC+DC (average) current-sensor circuit with key waveforms.

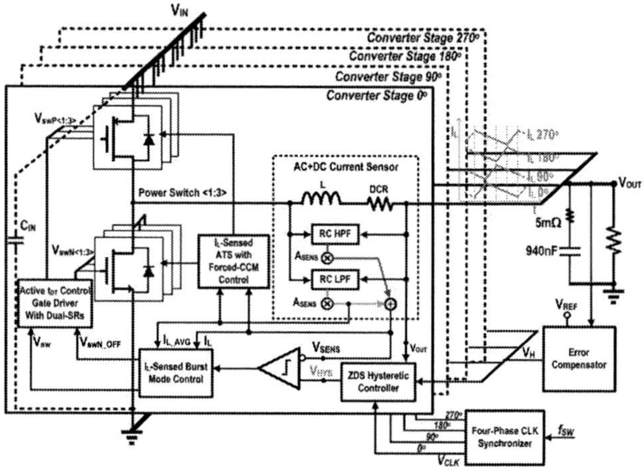

Figure 4.2.3: Complete 4-phase implementation of the converter with I_L-sensed burst-mode control and adaptive transistor sizing.

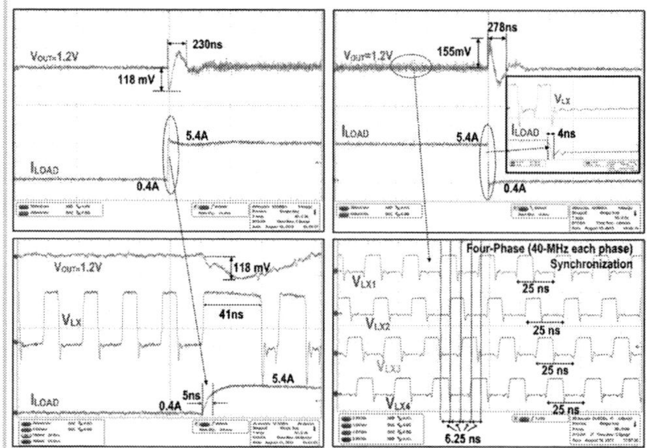

Figure 4.2.4: Measurement of V_{OUT} response to 5A/5ns load step transients and steady-state 4-phase synchronization.

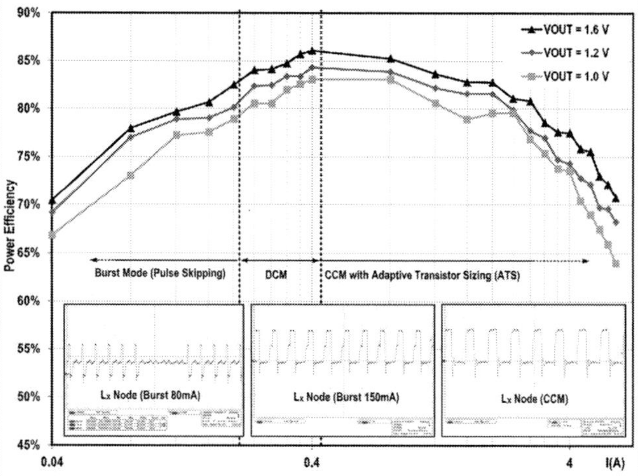

Power Efficiency: 90%, 85%, 80%, 75%, 70%, 65%, 60%, 55%, 50%, 45%

VOUT = 1.6 V
VOUT = 1.2 V
VOUT = 1.0 V

Burst Mode (Pulse Skipping) | DCM | CCM with Adaptive Transistor Sizing (ATS)

Lx Node (Burst 80mA) | Lx Node (Burst 150mA) | Lx Node (CCM)

0.04 | 0.4 | 4 | I(A)

Figure 4.2.5: Efficiency plot and V_{LX} measurements in each operation mode.

	ISSCC '13 [1]	JSSC '05 [2]	JSSC '09 [3]	This Work
Control	PWM	Hysteretic	Hysteretic	ZDS Hysteretic
Current Sharing	Master-Slave	Cycle-by-Cycle	None	Cycle-by-Cycle
$V_{IN\,(MAX)}$ (V)	1.2	1.2	4.9	3.3
V_{OUT} (V)	0.6-1.05	0.9	0.86-3.93	0.7-2.5
f_{SW} (MHz) (phases)	100 (×4)	233 (×4)	32-35 (×4)	40 (×4)
L (nH)	8	6.8	110	78
C_{OUT} (μF)	0.00187	0.0025	0.2	0.94
I_{MAX} (A)	1.2	0.3	1	6
Load Step (mA/ns)	180 / 800	150 / 0.1	300 / 30	5000 / 5
1% t_{settle} (ns)	~2000	~30	~350	230
V_{OUT} Droop (%)	6.7% (V_{OUT}=0.9V)	10% (V_{OUT}=0.9V)	10% (V_{OUT}=1.8V)	9.8% (V_{OUT}=1.2V)
Peak Efficiency (%)	82.4	83.2	80	86.1

Figure 4.2.6: Performance summary and comparison.

978-1-4799-0917-9/14 $31.00 © 2014 IEEE

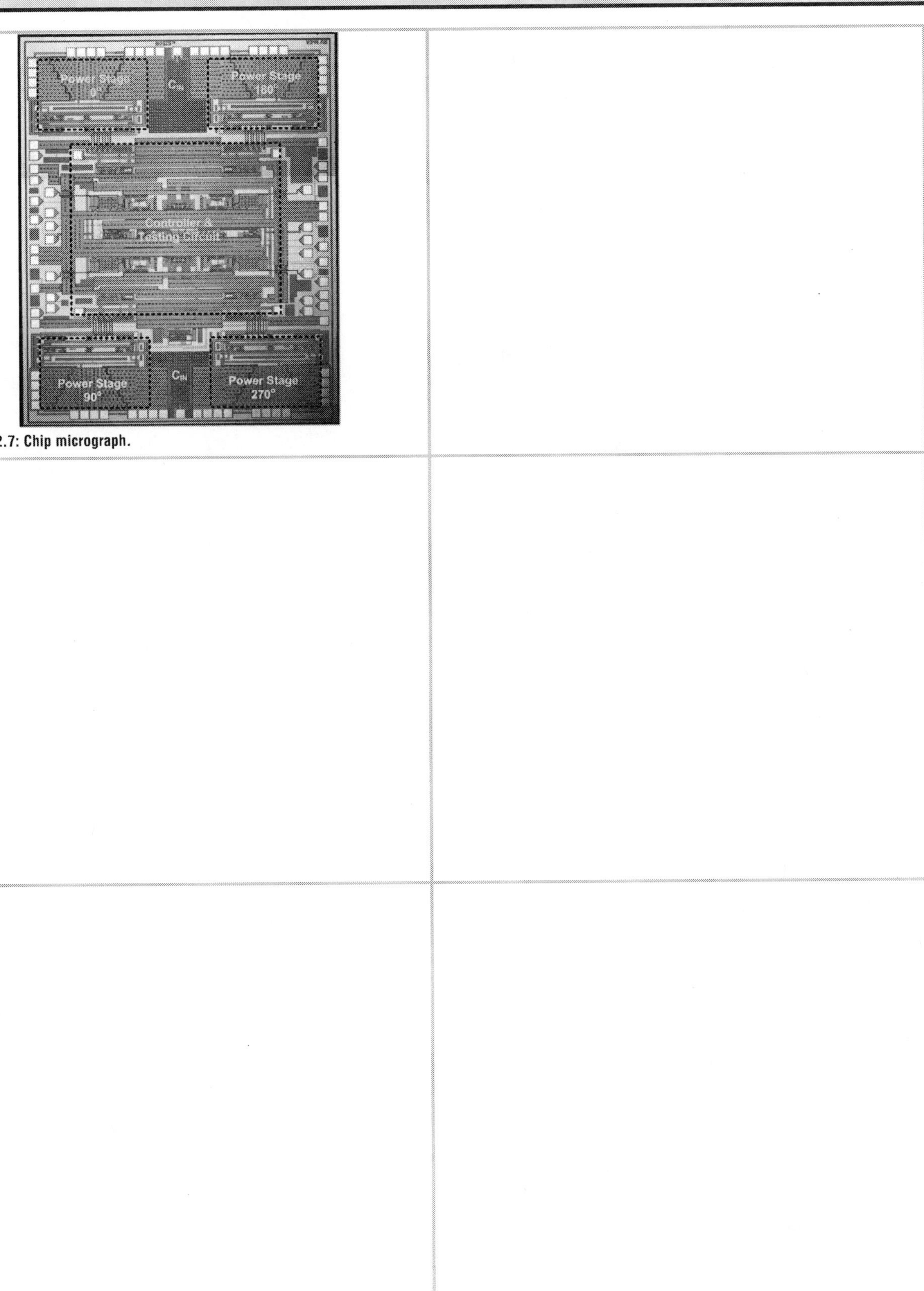

Figure 4.2.7: Chip micrograph.

4.3 An 87%-Peak-Efficiency DVS-Capable Single-Inductor 4-Output DC-DC Buck Converter with Ripple-Based Adaptive Off-Time Control

Danzhu Lu, Yao Qian, Zhiliang Hong

Fudan University, Shanghai, China

Improving battery longevity in portable devices usually requires the use of different voltage levels with a wide range of load capability for various functional blocks. Since a single-inductor-multiple-output (SIMO) converter can support multiple output voltages while using only one inductor, it is an excellent candidate to minimize the component count and thus the production cost. However, the cross-regulation and power consumption are two main issues of the previously reported SIMO converters [1-5]. Although pseudo-continuous conduction-mode (PCCM) control with a freewheel period [1] tries to augment power density and eliminate cross-regulation, associated power dissipation of freewheel switch exacerbates its overall efficiency. The charge-control technique with energy recovery presented [2] decouples the output channels between each switching cycle, at the expense of additional switching loss and slow response. The comparator-based controlled SIMO converters are investigated in [3, 4] and the cross-regulation in most channels is improved due to the fast response of the comparator. However, since the channel that is last connected to inductor is inevitably regulated by the accumulative error of all channels to balance the overall inductor current, every load transition at other outputs will introduce serious cross-regulation [3] and load-regulation problem [4] in the channel. In addition, cross-regulation and slow response also limit the application of dynamic voltage scaling (DVS) technique, which is widely used in single-output converters to improve the system power efficiency by providing variable voltage with fast reference tracking.

In this paper, a ripple-based adaptive off-time (RBAOT) control method is realized to overcome the cross-regulation problem in each sub-channel without any efficiency degeneration. The technique is suitable for buck or boost converters and can realize fast load and reference transient responses. Figure 4.3.1 illustrates the overall architecture of the 4-output SIMO DC-DC buck converter. Unlike conventional comparator-based control, all the output voltages V_{o1} to V_{o4} are directly regulated by comparators to time-share the magnetic energy stored in the single inductor with four energy distribution switches (M_{s1} to M_{s4}). The duty signal of the energy generation switches (M_n and M_p) is determined by the off-time control unit and adjusts the charging and discharging periods of the inductor current. The operating frequency is locked to the reference clock with the adaptive off-time generated by the PLL unit. Since only the comparison result of output ripple is utilized and no error amplifier or PWM compensation is needed, RBAOT control circuit is simple to implement and has a wide bandwidth control loop.

The operating principle of RBAOT control can be introduced using the timing waveforms in Fig. 4.3.2. As the inductor current ramps up and down, all energy distribution switches (M_{s1} to M_{s4}) are turned on one by one in ascending order according to the states of the corresponding comparator outputs. In energy generation part, the main power switch M_p is turned off when M_{s1} is turned on to charge the output capacitor of V_{o1}.The setting of off-time is created by the PLL unit and when it expires, M_p is turned on and M_n is turned off to increase the inductor current I_L until V_{o4} is higher than the reference voltage and V_{o1} is being charged again. The working frequency of the energy-generation and energy-distribution switches are the same and they can both be fixed in different loads and I/O conditions by adjusting the off-time values.

In RBAOT control, the operating frequency is not fixed during the transient response to decouple the output channels, however, it is locked in the steady-state response to avoid unpredictable switching noise spectrum, as shown in Fig. 4.3.2. Since the bandwidth of the PLL is relatively narrow (usually less than one-tenth of the operating frequency), the off-time (T_{off}) can be considered as a constant value in the beginning of the load and reference transition. Considering a load rising at channel V_{o2}, the first conversion period is lengthened as the required charging time of V_{o2} increases, which leads to a longer on-time of switch M_p and the inductor current I_L goes up immediately. Therefore, V_{o2} recovers quickly from the change in the load and the average inductor current increases from I_{avg1} to I_{avg2}, which means that the SIMO converter is temporarily stable at a new period (Ts'). Then, after several cycles of adjustment, the switching period is automatically modified by the PLL unit and goes back to the original value (Ts) for the adaptive off-time. For DVS up-tracking, the output capacitor of V_{out2} is directly charged by I_L to follow the reference and the long charging time results in I_L to increase, as illustrated in Fig. 4.3.2. Then I_L goes down due to the reduction of the charging time when V_{o2} reaches V_{ref2} and the switching frequency can finally return to Ts in a similar way. Since the charge error cancellation of all the outputs during the load and reference transient response is achieved by switching frequency adjustment (and no freewheel switch [1, 2] is used which in turn avoids the additional power loss of such a switch,) RBAOT control can achieve a low cross-regulation and a high efficiency simultaneously.

Figure 4.3.3 shows the schematic of the PLL control loop in RBAOT control. The DUTY signal is triggered by the gate control signal of M_{s1} (S_{o1} in Fig. 4.3.1) to start the off-time. The phase-frequency detector (PFD) measures the phase difference between the DUTY and CLK signals. Filtered by the charge-pump low-pass filter (CPLPF), the phase error is converted into V_{TH} and then the off-time is regulated through a voltage-controlled delay cell (VCDC) until the switching frequency is consistent with the reference clock. The operating process of the PLL control loop are shown in Fig. 4.3.3, while its bandwidth should be designed to be much lower than the switching frequency to attenuate the voltage hopping on V_{TH}.

The SIMO buck converter is fabricated in a 2P4M 0.35μm CMOS process. The external inductor is 4.7μH and the capacitors are 10μF each. Figure 4.3.4 shows the measured steady-state waveforms of the output ripples, and VLX1 and VLX2 node voltages. With RBAOT control, stable output ripples below 30mV are observed including the spikes and the switching frequency is locked to the target of 1MHz in different I/O conditions. Figure 4.3.5 shows the measured dynamic characteristics including load and reference transient waveforms. A load transient measurement (I_{o1}=100 to 350mA) shows only 40mV overshoot/undershoot at a 250mA load step and ultra-fast response which is less than 7 switching cycles. The operating frequency adjustment of the RBAOT control can be clearly observed in the waveforms and the worst cross regulation of 0.04mV/mA is measured at V_{out3}. For DVS reference tracking, the up-tracking speed is as fast as 60mV/μs and a 5% (50mV/V) maximum cross regulation is achieved. The maximum output current is 400mA for each channel and the peak efficiency reaches 87% with regulated voltages of 0.9V, 1.2V, 1.5V, and 1.8V and regulated currents of 100mA, 85mA, 100mA and 90mA, respectively. The complete performance comparison to the state-of-the-art SIMO buck converters is shown in Fig. 4.3.6. The chip micrograph is depicted in Fig. 4.3.7. The chip, including all the pads, occupies 1.8×3mm².

Acknowledgements:
This project is supported by Analog Devices and the State Key Laboratory of ASIC & System, Fudan University. The authors would like to thank these two organizations for their help and support.

References:
[1] D. Ma, W.-H. Ki, and C.-Y. Tsui, "A Pseudo-CCM/DCM SIMO Switching Converter with Freewheel Switching," *ISSCC Dig. Tech. Papers*, pp. 390-391, February 2002.
[2] C.-W. Kuan and H.-C. Lin, "Near-Independently Regulated 5-Output Single-Inductor DC-DC Buck Converter Delivering 1.2W/mm² in 65nm CMOS," *ISSCC Dig. Tech. Papers*, pp.274-275, February 2012.
[3] H.-P. Le, C.-S. Chae, K.-C. Lee, et al., "A Single-Inductor Switching DC-DC Converter with 5 Outputs and Ordered Power-Distributive Control," *ISSCC Dig. Tech. Papers*, pp. 534-535, February 2007.
[4] K.-C. Lee, C.-S. Chae, G.-H. Cho, and G.-H. Cho, "A PLL-Based High-Stability Single-Inductor 6-Channel Output DC-DC Buck Converter," *ISSCC Dig. Tech. Papers*, pp. 200-201, February 2010.
[5] M. Belloni, E. Bonizzoni, E. Kiseliovas, et al., "A 4-Output Single-Inductor DCDC Buck Converter with Self-Boosted Switch Drivers and 1.2A Total Output Current," *ISSCC Dig. Tech. Papers*, pp. 444-445, February 2008.

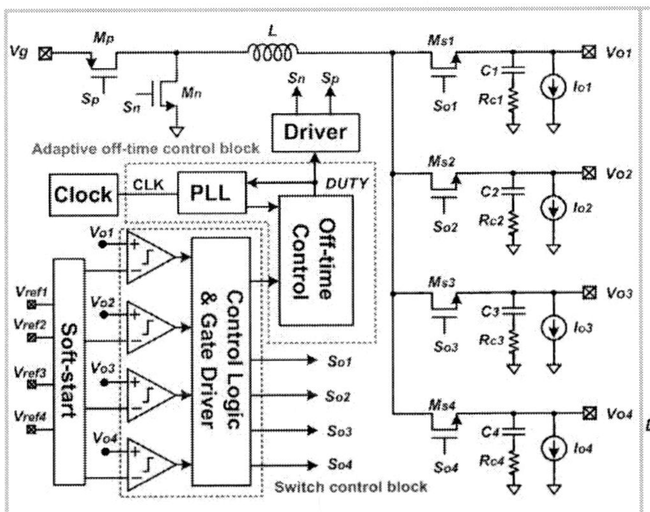

Figure 4.3.1: Architecture of the 4-output SIMO DC-DC buck converter.

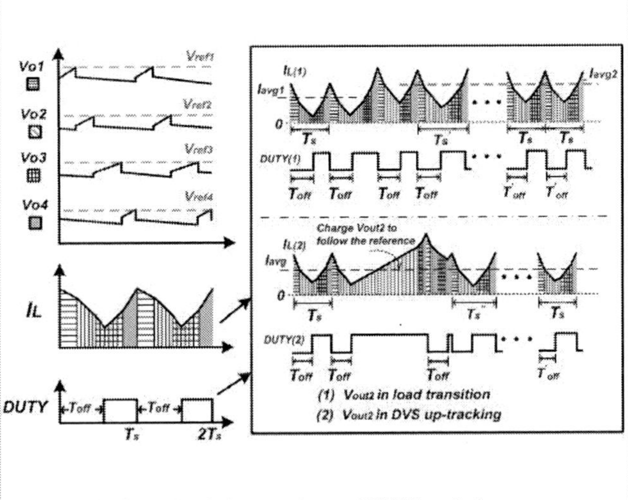

Figure 4.3.2: Operation timing waveforms of RBAOT control.

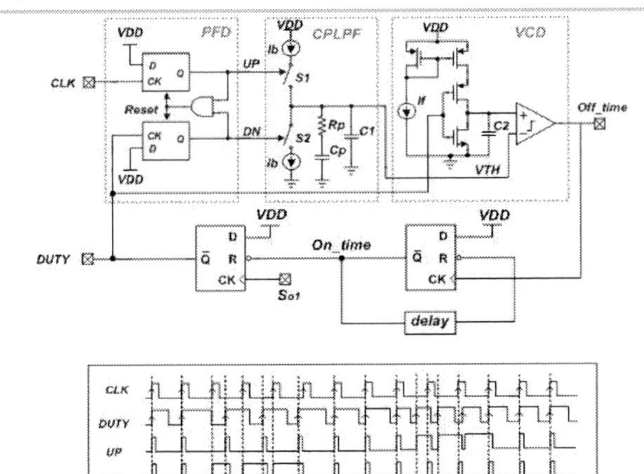

Figure 4.3.3: Operation principle and schematic of PLL control loop in RBAOT control.

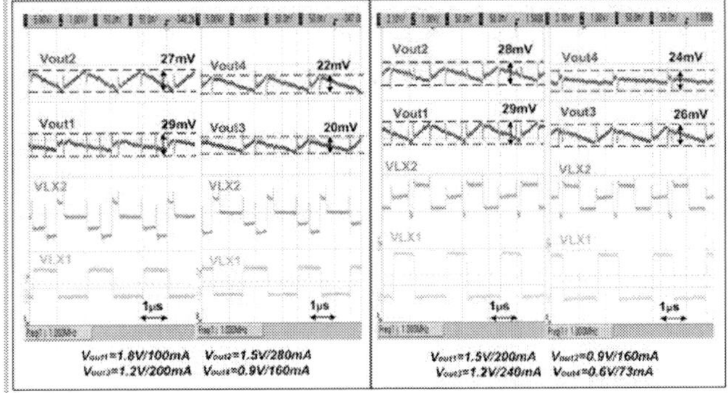

Figure 4.3.4: Measured steady-state waveforms and node voltages in different I/O condition.

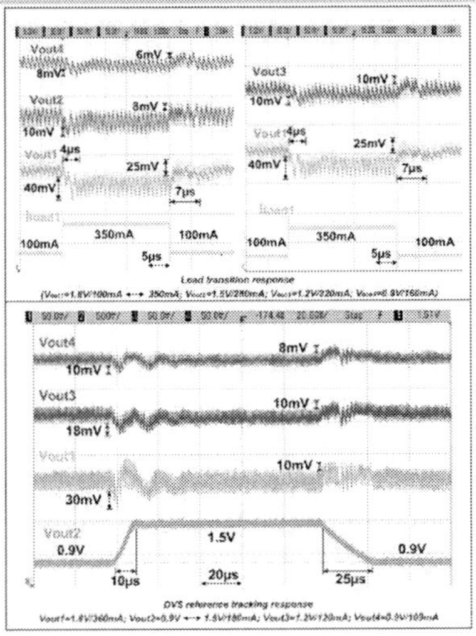

Figure 4.3.5: Measured waveforms of dynamic characteristics.

	[2]	[4]	[5]	This work
Process	65nm	0.35μm	0.5μm	**0.35μm**
Control method	Charge control	Comparator -basd control	PWM loop control	**RBAOT control**
Topology	5 buck outputs	6 buck outputs	4 buck outputs	**4 buck outputs**
Supply voltage (V)	3.4~4.3	5.0	2.3~3.6	**2.7~5.0**
Frequency (MHz)	1.2	2	3	**1**
Inductor & Capacitor	2.2μH & 4.7μF	4.7μH & 10μF	4.7μH& 10μF	**4.7μH & 10μF**
Output ripple (mV)	<40	<25*	<80	**<30**
Load transient (mV/mA)	0.6	0.93	0.82	**0.16**
Cross regulation (mV/mA)	0.067	0.31	0.41	**0.04**
DVS tracking Speed (mV/μs)	no	no	no	**60**
Max. Efficiency (%)	83.1	N/A	82	**87**
Max. Output power(W)	2.232	1.2	1.88	**2.16**

*ripple without spike

Figure 4.3.6: Performance comparison.

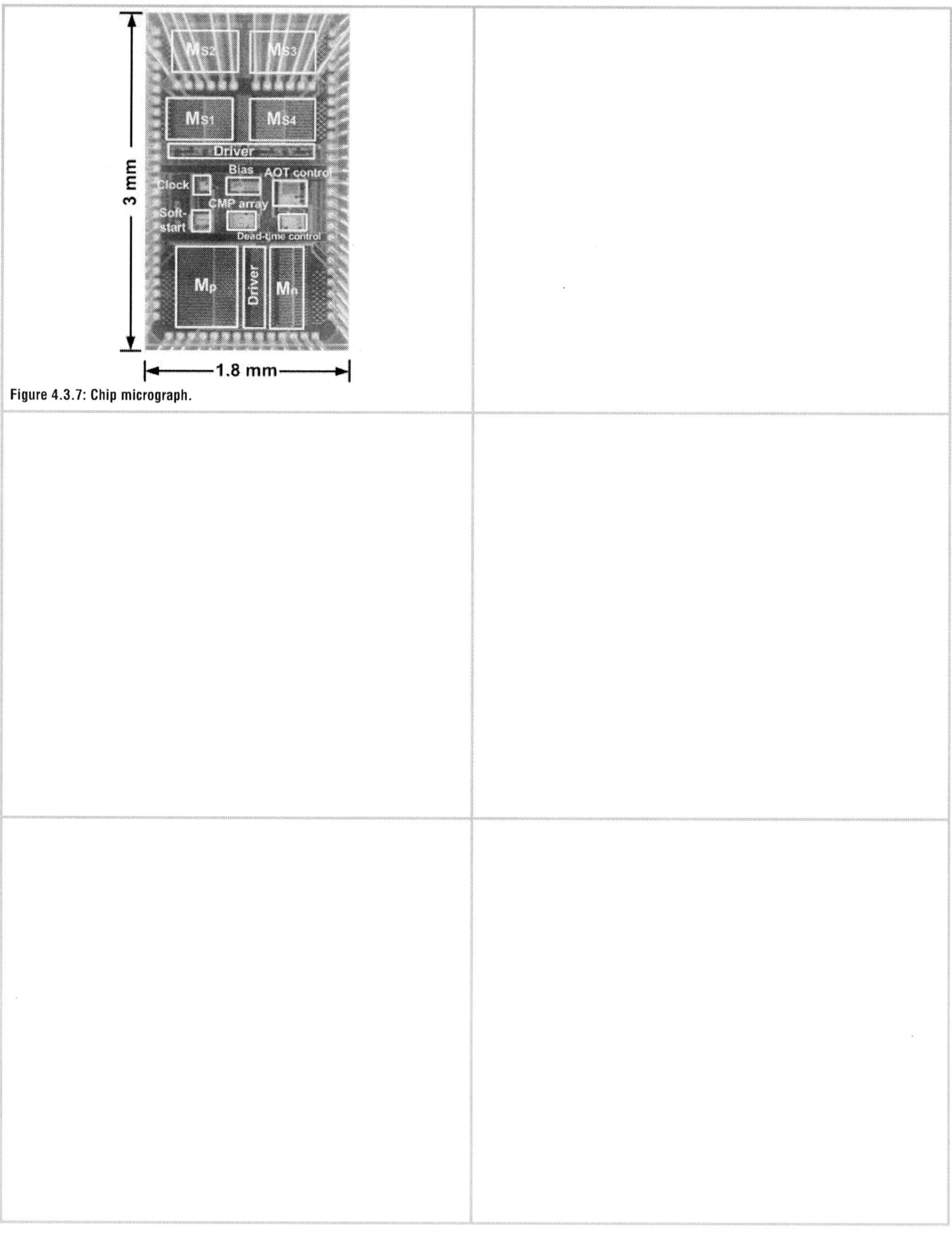

Figure 4.3.7: Chip micrograph.

4.4 A 10/30MHz Wide-Duty-Cycle-Range Buck Converter with DDA-Based Type-III Compensator and Fast Reference-Tracking Responses for DVS Applications

Lin Cheng, Yonggen Liu, Wing-Hung Ki

Hong Kong University of Science and Technology, Hong Kong, China

Dynamic voltage scaling (DVS) is an effective strategy in reducing the power consumption of a processor through adjusting its supply voltage at runtime. The power converter that drives the processor should have a wide output voltage range and fast reference-tracking response while maintaining a high efficiency. For such applications, inductive switching converters are suitable candidates [1-4]. To enhance the transient response and to reduce the size of off-chip components, DC-DC converters switching in the 10MHz range are studied recently. When switching at a high frequency, efficiency is compromised as the switching loss becomes significant. Moreover, comparator delay of a few ns causes serious over-charging and over-discharging of the ramp capacitor and the delay also limits the maximum duty cycle of the converter. A current-mode control needs an on-chip current sensor that would consume too much power when running in the 10MHz range. In [5], a 5MHz converter employs auto-selectable peak- and valley-current control, but the duty cycle range is limited to 0.6. A higher switching frequency (f_s) would further reduce this range and would not be applicable for DVS. Therefore, a voltage-mode control with Type-III compensation for loop bandwidth extension becomes an attractive solution. However, a conventional Type-III compensator needs large on-chip compensation capacitors and resistors. In [6], a pseudo Type-III compensator reduces the value and thus size of those components, however, it requires adding lowpass and bandpass functions that complicates the design, and furthermore it does not use any special technique for fast reference tracking.

In this paper, a 10/30MHz buck converter with an area-efficient Type-III compensator and a fast reference-tracking scheme is proposed. An accurate delay-compensated ramp generator allows the converter to switch at 30MHz. Figure 4.4.1 shows the proposed DC-DC converter with the Type-III compensator. The compensator consists of a G_m amplifier and a differential difference amplifier (DDA) [7] with compensation resistors and capacitors. The negative feedback dictates that $(v_{1+}-v_{1-})=-(v_{2+}-v_{2-})$ in the steady state. The transfer function of the compensator $A(s)=V_{ea}/V_{fb}$ is also shown in Fig. 4.4.1. The low-frequency pole $-1/C_{mos}r_o$ and zero $-G_m/C_{mos}$ are realized by the G_m amplifier, and the high-frequency pole $-1/C_2R_1$ and zero $-1/[(C_1+C_2)R_1]$ are realized by the C_1, C_2, and R_1 network. There is a third pole at a higher frequency due to the parasitic components of the DDA. Area reduction is achieved by using the G_m amplifier to realize the low-frequency pole and zero, as the transconductance G_m and the MOS capacitor C_{mos} are more area-efficient than a poly resistor and a MIM capacitor, respectively. When compared to the case where A(s) is realized using a conventional Type-III compensator, the area is reduced by 60% ($0.019mm^2$ versus $0.048mm^2$).

Based on the proposed DDA Type-III compensator, a fast reference-tracking scheme is further developed. From Fig. 4.4.1, V_{fb} ($=bV_o$) is forced to be equal to V_{ref} by the G_m amplifier, and the DC gain from V_{com} to V_{ea} is 1. The DDA output voltage V_{ea} is determined by the duty ratio $D=V_o/V_g=V_{ref}/bV_g=(V_{ea}-V_L)/V_M$, where the parameters are defined in Fig. 4.4.1 and Fig. 4.4.2. If V_{com} is made equal to $V_{ea}=(V_MV_{ref}/bV_g+V_L)$ through the tracking circuit shown in Fig. 4.4.2, then $v_{1+}=v_{1-}$, forcing $v_{2+}=v_{2-}$ and hence, V_{fb} is equal to V_{Gm}. This means that V_{ref}, V_{fb} and V_{Gm} are all equal in the steady state. In addition, V_M is designed to vary with V_g such that V_M/V_g is a constant k, and $V_{com}=V_{ea}=(kV_{ref}/b+V_L)$. The accuracy of V_M and V_L is guaranteed by the accurate ramp generator, as will be explained shortly. Reference tracking is realized as follows: when V_{ref} is changed to adjust V_o for DVS, V_{ea} is predicted by adjusting V_{com}; meanwhile, V_{Gm} is connected to V_{ref} through a buffer that predicts the final value quickly and accurately, without relying on the loop response of the converter. Due to this prediction, the tracking speed is greatly improved and is only limited by the LC filter of the power stage.

Figure 4.4.3 shows the conventional and the proposed ramp generator. For the conventional design, the capacitor voltage should be bounded by V_H and V_L, but the comparator delays result in over-charging and over-discharging. The comparator delays will change due to PVT variations, and f_s and V_M will change accordingly. For the proposed design, two feedback loops are added. Consider the lower branch as an example. In the steady state, C_{ls} and C_{lh} sample the valley value of the ramp signal V_{vy}. For the voltage stored by C_{lh} (V_{vy}) to be equal to V_L, the comparator CMP_L should trip a little earlier, when C_r reaches $V_{L'}$ (instead of V_L). The feedback amplifier G_m_L compares V_{vy} with V_L, and drives the output accordingly to be $V_{L'}$. Therefore, while the tripping voltages are $V_{H'}$ and $V_{L'}$, the peak and valley voltages of C_r are V_H and V_L, respectively, as designed. Both comparator delays and logic delays are accounted for, and low-power comparators can be used. Moreover, the feedback loops do not need to be fast, and the amplifiers can be biased with small currents.

The converter is fabricated in a 0.13µm CMOS process using 3.3V devices to handle the Li-Ion battery voltage range. Figure 4.4.4 shows the measured ramp signal through an on-chip buffer. Over-charging and over-discharging are both significantly reduced. The ramp generator is designed to run at 10 to 30MHz, and measurement results confirm that it can work up to 70MHz. Figure 4.4.4 also shows the measured steady-state waveforms and load transient responses of the converter. For V_g=3.3V and I_o=300mA, V_o can be regulated from 0.37V to 2.85V when f_s is 10MHz, and the controllable range of D is 0.86–0.11=0.75; and from 0.45V to 2.4V when f_s is 30MHz, and the range of D is 0.73–0.14=0.59, a great improvement when compared to [5]. The measured maximum output power is 3.6W (V_o=2.4V and I_o=1.5A) and the measured peak efficiency for f_s=10MHz and f_s=30MHz are 91.8% and 86.6%, respectively. The load transient responses are measured at V_o=1.8V with a load current step of 340mA. The DDA Type-III compensator works as designed, with a small amount of overshoot/undershoot and a fast recovery time.

Figure 4.4.5 shows the measurement results of reference-tracking at I_o=500mA, and V_{ref} is switched between 0.8V to 1.4V with rising and falling times of 20ns. If the conventional Type-III compensator was used, large overshoot and undershoot of around 0.3V would be observed, and the recovery time would be long. By using the DDA Type-III compensator, V_o settled to its final value quickly with negligible amount of overshoot and undershoot. The up-tracking speeds for f_s=10MHz and f_s=30MHz are 1.67µs/V and 0.67µs/V, respectively, which are more than 12 times faster than those of the converters that use the conventional Type-III compensator. The down-tracking speeds for f_s=10MHz and f_s=30MHz are 4.44µs/V and 1.56µs/V, respectively. The measured inductor current waveforms during the tracking period show that a large positive peak inductor current is generated to fast-charge the output capacitor during up-tracking, but it is limited by the reverse current protection circuit of the power supply during down-tracking. Figure 4.4.6 compares the performance of the proposed converter with that of the state-of-the-art designs. The tracking speeds of the proposed converter are faster than those of the fully integrated converter of [2] that switches at f_s=300MHz. Although [4] reported tracking speeds of 0.02µs/V, the results are measured in open-loop, and no direct comparison can be made. Figure 4.4.7 shows the die micrograph of the converter.

Acknowledgement:
This work is supported by Hong Kong RGC under GRF 613810.

References:

[1] C. Zheng and D. Ma, "A 10-MHz Green-Mode Automatic Reconfigurable Switching Converter for DVS-Enabled VLSI Systems," *IEEE J. Solid-State Circuits*, vol. 46, no. 6, pp. 1464-1477, Jun. 2011.

[2] S. S. Kudva and R. Harjani, "Fully-Integrated On-Chip DC-DC Converter With a 450X Output Range," *IEEE J. Solid-State Circuits*, vol. 46, no. 8, pp. 1940-1951, Aug. 2011.

[3] Y.-H. Lee, S.-C. Huang, S.-W. Wang, et al., "Power-Tracking Embedded Buck-Boost Converter With Fast Dynamic Voltage Scaling for the SoC System," *IEEE Trans. Power Electronics*, vol. 27, no. 3, pp. 1271-1282, Mar. 2012.

[4] W. Kim, D. Brooks, and G.-Y. Wei, "A Fully-Integrated 3-Level DC-DC Converter for Nanosecond-Scale DVFS," *IEEE J. Solid-State Circuits*, vol. 47, no. 1, pp. 206-219, Jan. 2012.

[5] M. Du, H. Lee, and J. Liu, "A 5-MHz 91% Peak-Power-Efficiency Buck Regulator With Auto-Selectable Peak- and Valley-Current Control," *IEEE J. Solid-State Circuits*, vol. 46, no. 8, pp. 1928-1939, Aug. 2011.

[6] P.Y. Wu, S.Y.S. Tsui, and P.K.T. Mok, "Area- and Power-Efficient Monolithic Buck Converters With Pseudo-Type III Compensation," *IEEE J. Solid-State Circuits*, vol. 45, no. 8, pp. 1446-1455, Aug. 2010.

[7] E. Sackinger and W. Guggenbuhl, "A Versatile Building Block: The CMOS Differential Difference Amplifier," *IEEE J. Solid-State Circuits*, vol. 22, no. 2, pp. 287-294, Apr. 1987.

978-1-4799-0917-9/14 $31.00 © 2014 IEEE

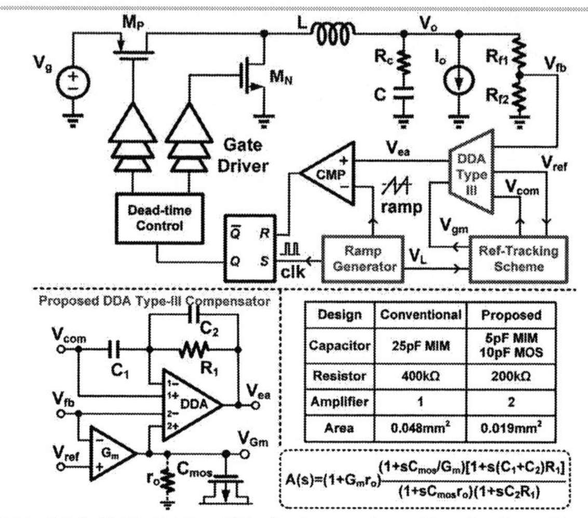

Figure 4.4.1: Block diagram of the buck converter and the structure of the proposed DDA Type-III compensator.

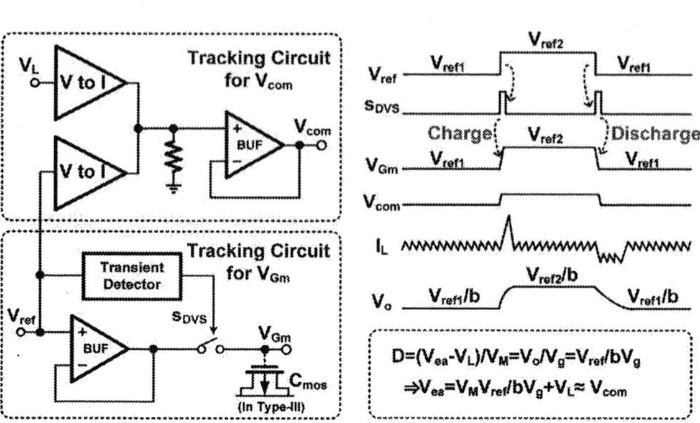

Figure 4.4.2: Proposed reference-tracking scheme.

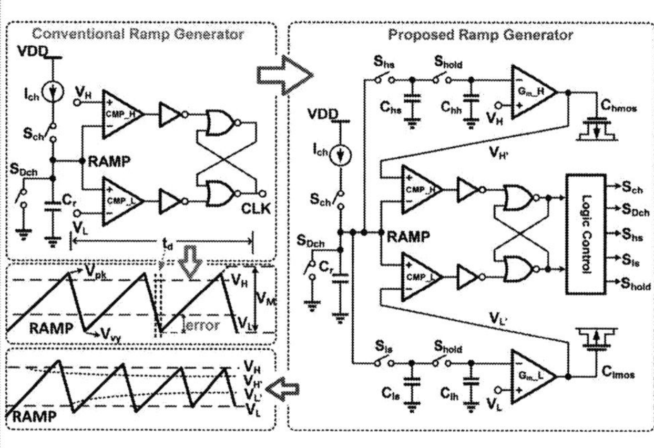

Figure 4.4.3: Schematics of the conventional and the proposed ramp generator.

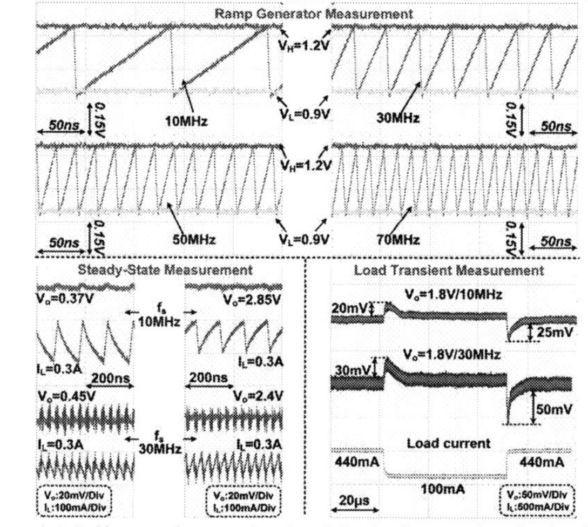

Figure 4.4.4: Measured ramp signal and the buck converter steady-state as well as load transient response.

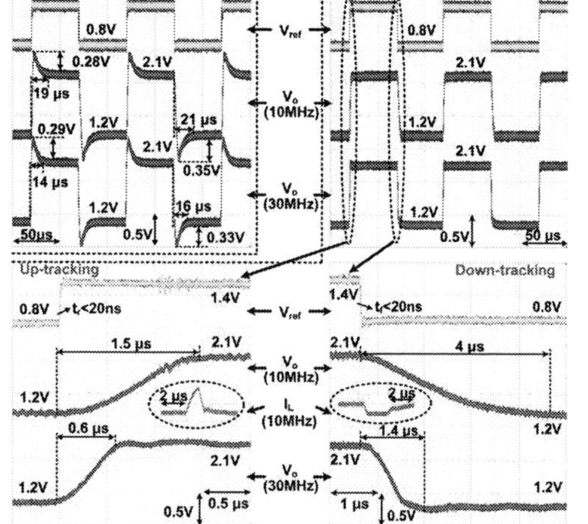

Figure 4.4.5: Measured reference tracking response of both the conventional and the proposed Type-III compensator.

Publication	[1] 2011	[2] 2011	[3] 2012	This Work	
CMOS Technology	0.13μm	0.13μm	0.25μm	**0.13μm**	
Topology	Buck-Boost	Buck	Buck-Boost	**Buck**	
Control Scheme	Hysteresis	Voltage Mode	Current Mode	**Voltage Mode**	
Maximum Output Power	0.4W	0.27W	1.4W	**3.6W**	
Switching Frequency	10MHz	300MHz	5MHz	**10MHz**	**30MHz**
Inductor	1μH	2nH	1μH	**0.33μH**	**0.33μH**
Capacitor	1μF	5nF	0.88μF	**3.3μF**	**1μF**
Nominal Input Voltage	1.5V	1.2V	2.5~4.5V	**3.3V**	**3.3V**
Output Voltage Range	0.9~2.2V	0.3~0.88V	3V	**0.37~2.85V**	**0.45~2.4V**
Peak Efficiency	92.1%	74.5%	91%	**91.8%**	**86.6%**
Up-tracking	93.3μs/V	2.6μs/V	20μs/V	**1.67μs/V**	**0.67μs/V**
Down-tracking	26.7μs/V	3.6μs/V	15μs/V	**4.44μs/V**	**1.56μs/V**

Figure 4.4.6: Performance summary and comparison.

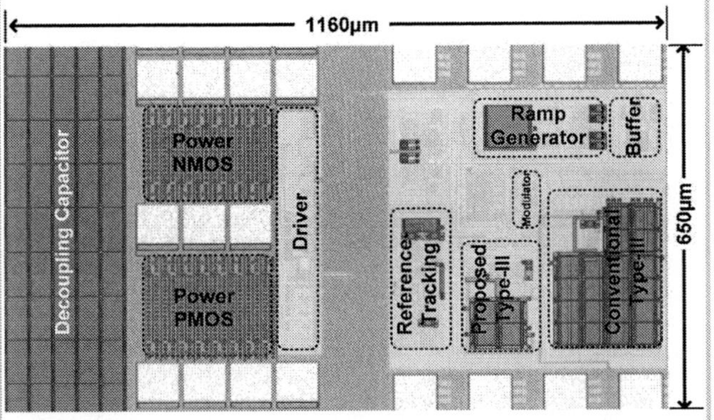

Figure 4.4.7: Die micrograph.

4.5 A 2-Phase Resonant Switched-Capacitor Converter Delivering 4.3W at 0.6W/mm² with 85% Efficiency

Kapil Kesarwani, Rahul Sangwan, Jason T. Stauth

Dartmouth College, Hanover, NH

There is an increasing need for fully integrated power converters to reduce form factor in mobile devices and provide high-density point-of-load power delivery in performance computing applications [1]. While the number of microprocessor cores on a single die continues to increase, power delivery subsystems have not scaled as favorably, and there remain significant challenges to providing complex multi-core regulation with high efficiency, fast dynamic response, and at low cost.

Switched-capacitor (SC) converters have emerged as a prime candidate for high-power-density and high-conversion-ratio power delivery due to favorable tradeoffs between device utilization and conversion ratio, and the inherently higher energy density of capacitors as compared to inductors in many voltage and current ranges [1-3]. However, silicon-integrated magnetic components continue to improve in terms of power density, efficiency, and operating frequency due to the use of nano-composite materials and new processing techniques [3]. Recently, resonant switched-capacitor (ReSC) converters have shown significant promise for fully integrated power management, as they can leverage advantages of both high-density capacitors and integrated magnetics [5-6]. Compared to SC converters, the ReSC approach can better utilize the available energy density of integrated passives, can significantly reduce bottom-plate switching loss, and enable 'zero-current switching' which can reduce device stress and power loss. Importantly, ReSC converters require only a small amount (<<10nH) of high-Q inductance to achieve substantial performance benefits [6]. ReSC topologies have the same advantages of SC topologies in terms of device utilization: they can be built using similar hierarchical or cascaded architectures where submicron devices are exposed only to a fraction of the supply, providing scalability across process technologies and voltage ranges. However, lower frequency operation also improves performance for longer-channel-length, higher voltage, and lower f_t devices, and increases the viability of moderate-bottom-plate technologies such as bulk MOS capacitors.

Figure 4.5.1 shows a comparison of the ReSC to SC architecture in a 2:1 configuration where converters operate with the same flying capacitance and series resistance, R_{ESR}. A key FoM, effective output resistance (R_{EFF}) captures the resistive loadline and resulting converter conduction loss. In Fig. 4.5.1, R_{EFF} is normalized to R_{ESR} and frequency is normalized to the SC '*fast switching limit*' (FSL) boundary. The ReSC approach outperforms comparable SC converters in the frequency-R_{EFF} space by operating at fundamental and subharmonic resonant minima. By tuning out the reactive impedance of the flying capacitor(s), the ReSC approach achieves the best-case limiting conduction loss of an SC converter, but with much lower switching loss.

In this work, we present a 2-phase ReSC converter that operates with supply voltages from 3.6 to 6V, providing compatibility for a range of applications including Li-Ion battery supplies. Figure 4.5.2 shows the power train of the 2-phase ReSC converter. The architecture is similar to the 2:1 SC converters in [2-3], but uses inductance, L_X, to resonate with the on-chip flying capacitor, C_X. On-chip bypass capacitance, C_{bp}, is used to filter the output voltage and complete the resonant loop in the energy transfer process. The timing of key signals in converter operation is shown in Fig. 4.5.3. In normal operation at the fundamental resonant frequency, $\omega_o=(L_XC_X)^{-1/2}$, resonant impedance Z_X is configured in parallel with *Vin-Vout* in ϕ_1; in ϕ_2, Z_X is configured in parallel with *Vout*. If there is a voltage difference between *Vin-Vout* and *Vout-GND*, voltage V_X appears as a square wave at the resonant frequency. In ϕ_1, a positive half wave current flows into Z_X, drawing energy from Vin; in ϕ_2, a negative half wave current flows out of Z_X, supplying energy to the load. Similar to the SC topology, this process can be modeled as an effective resistance, R_{EFF}, the details of which are discussed in [6]. Operation at the fundamental mode provides the lowest achievable R_{EFF}, which is approximately $R_{ESR}\cdot\pi^2/8$ for the 2:1 configuration, near the minimum achievable R_{EFF} for a comparable SC converter.

The converter uses modulation of R_{EFF} to regulate the output voltage. In addition to the use of subharmonic modes, we also demonstrate the use of '*dynamic off-time modulation*' (DOTM). This technique enables both variable high-resolution voltage regulation as well as higher efficiency in light load operation. In DOTM, the deadtime where both N- and P-stages are in high impedance mode is extended. This process achieves the maximum energy transfer per switching cycle, but the switching frequency can be reduced arbitrarily. In this mode, R_{EFF} follows the linear trend shown in Fig. 4.5.1, but switching losses are reduced in proportion to switching frequency. This enables high-resolution voltage regulation and reduction of power loss in light-load.

Figure 4.5.4 shows the system diagram for the ReSC converter. An off-chip MCU is used to control the chip and measure the output voltage for regulation purposes. The MCU controls the on-chip timing and synchronization circuit that generated the clocks for each of the two phases that were operated with 180° interleaving. Also, on-chip, 4 level-shift circuits are designed to control each set of power devices. For example Clk_{1a} drives the N-channel power devices in ϕ_1; Clk_{1a} and $Clk_{1a'}$ are in-phase, but operate between the different voltage domains required for the high-side and low-side devices. Clk_{1a} and Clk_{2a} operate with a variable deadtime that is controlled by the off-time modulation block. ϕ_2 operates similarly, but is 180° out of phase to minimize output voltage ripple and maximally utilize on-chip bypass capacitance. Therefore, the resonant energy transfer process is controlled by *Ton,n* and *Ton,p*, but R_{EFF} and switching loss can be scaled by the off-time, which can have a range from 1 ns to over 10 µs.

The converter is implemented in a 0.18µm HVCMOS process using triple MIM capacitors with ~6.6 fF/µm². All capacitors, C_X(~9nF per phase) and C_{bp} (~11nF), are implemented on-chip, as was the CMOS power train, level shift, and timing control blocks. Inductors, L_X, are implemented in a 3D die stack using a bump process on the die surface. We used air-core solenoid inductors with 1.9nH to 5.5nH per phase and area from ~0.9-1.75mm². Measured AC resistance was around 38mΩ in the operating frequency range (20-40 MHz), providing quality factor (Q) of around 10-20. Output voltage ripple was measured to be less than 10% peak-peak at power densities up to 0.6W/mm².

Figure 4.5.5 Shows measured efficiency versus power density at input voltage of 6V, using an on-chip four terminal connection. At 0.6W/mm², measured efficiency was 85.0% for Lx = 5.5nH and 82.5% for Lx = 1.9nH. In light load, subharmonic operation was able to maintain efficiency over 82% for power density down to 0.11W/mm². Using the voltage regulation loop and off-time modulation, the system achieves regulation within 2% of its nominal conditions for load currents between 50mA and 1.2A. Shown in Fig. 4.5.5, efficiency remains high in DOTM because switching losses scale down as off-time is increased.

Figure 4.5.6 shows a comparison table of this work with previously published and comparable designs. This work shows higher absolute power level compared to the published works. Also, it achieves efficiency-power-density performance comparable to designs in deep submicron SOI technologies, but is implemented in a moderate-channel bulk CMOS. Compared to other designs using MIM capacitors, it achieves higher power density and absolute power, and can operate with higher input voltages. As integrated magnetic components continue to improve, the ReSC topology appears to be a viable means to achieve fully integrated power management and will complement both deep-submicron MOS-capacitor designs, and deep-trench technology.

References:

[1] G. Villar-Pique, H. J. Bergveld, and E. Alarcon, "Survey and Benchmark of Fully Integrated Switching Power Converters: Switched-Capacitor versus Inductive Approach," IEEE Trans. Power Electronics, vol. 28, no. 9, pp. 4156-4167, Sept. 2013.

[2] H. P. Le, S. R. Sanders, and E. Alon, "Design Techniques for Fully Integrated Switched-Capacitor DC-DC converters," *IEEE J. Solid-State Circuits*, vol. 46, no. 9, pp. 2120-2131, Sep. 2011.

[3] H. Meyvaert, T. V. Breussegem, and M. Steyaert, "A Monolithic 0.77 W/mm² Power Dense Capacitive DC-DC Step-Down Converter in 90nm Bulk CMOS," *Proc. ESSCIRC*, pp. 483-486, Sept. 2011

[4] C. R. Sullivan, D. V. Harburg, J. Qiu, C. G. Levey, and D. Yao, "Integrating Magnetics for on-Chip Power: A Persepctive," *IEEE Trans. Power Electronics*, vol. 28, no. 9, pp. 4342-4353, Sep. 2013.

[5] J. T. Stauth, M. D. Seeman, K. Kesarwani, "A Resonant Switched-Capacitor IC and Embedded System for Sub-Module Photovoltaic Power Management," *IEEE J. Solid-State Circuits*, vol. 47, no. 12, pp. 3043-3054, Dec. 2012.

[6] K. Kesarwani, R. Sangwan, and J.T. Stauth, "Resonant Switched-Capacitor Converters for Chip-Scale Power Delivery: Modeling and Design," *IEEE Workshop on Control and Modeling for Power Electronics*, June 2013.

978-1-4799-0917-9/14 $31.00 © 2014 IEEE

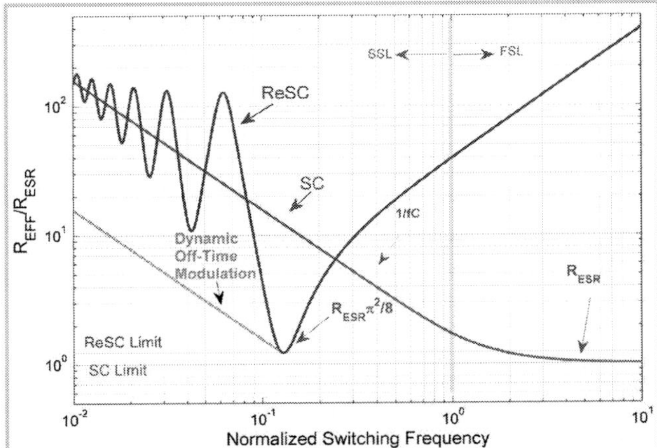

Figure 4.5.1: Comparison of SC and ReSC: effective output resistance (REFF) versus frequency. REFF is normalized to loop resistance (RESR) (includes switch, interconnect, and passive component resistance); frequency is normalized to the SSL-FSL boundary.

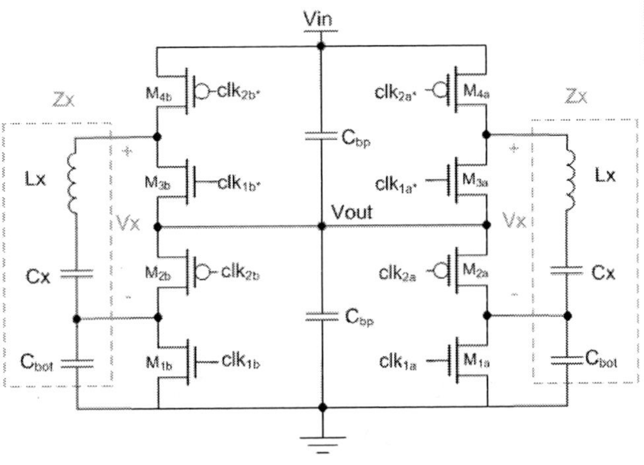

Figure 4.5.2: 2-phase resonant switched-capacitor (ReSC) power train for 2:1 configuration. Devices have a maximum rating of 3.6V (Vds and Vgs).

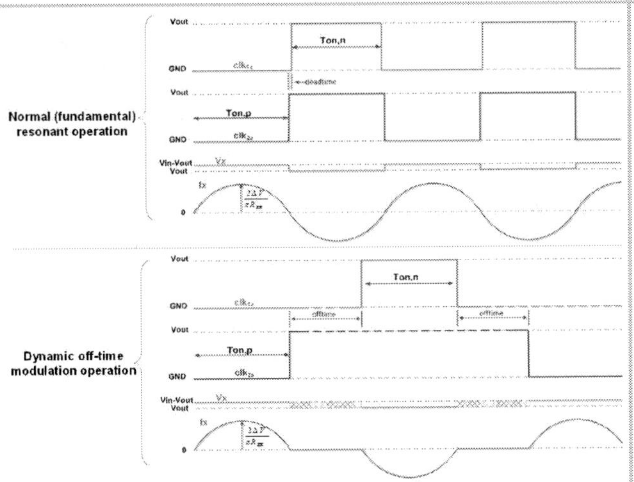

Figure 4.5.3: Timing waveforms for normal (fundamental) mode operation and operation with dynamic off-time modulation.

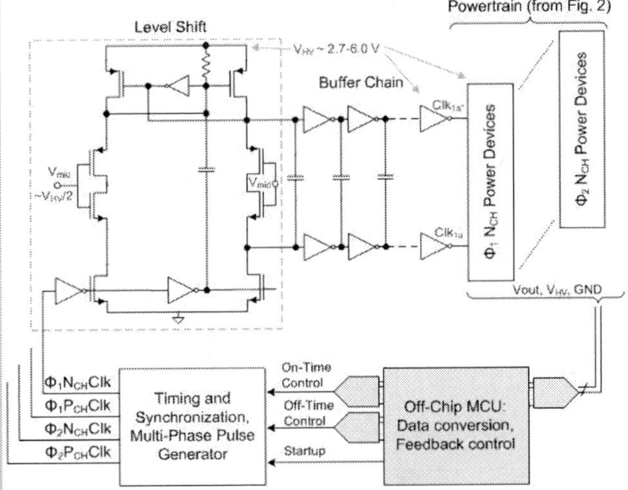

Figure 4.5.4: System diagram and level-shift schematic.

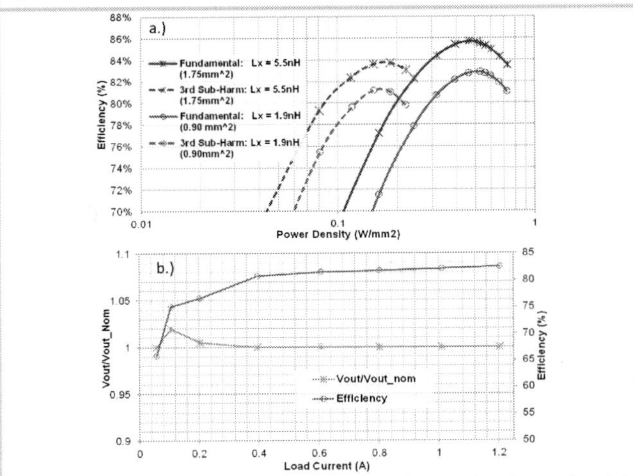

Figure 4.5.5: Measured circuit performance in various modes of operation: (a) power density–efficiency curves, (b) output voltage regulation across load current range and efficiency.

Work	[2]	[3]	Van Breussegem[1]	Pique[2]	Chang[3]	This Work
Topology	2:1, 3:2, 3:1	2:1	2:1	2:1, 3:2	2:1	2:1
Input (V)/Output (V)	3V/1V 2V/1V	2V/1V	3.6V/1.5V	1.6V/0.7V 1.2V/0.7V	2V/1V	6V/3V 3.6V/1.8V
Capacitor Technology	MOS	MOS	MIM	MOS	Deep Trench	MIM
Process Technology	32nm SOI	90nm bulk CMOS	90nm bulk CMOS	90nm bulk CMOS	45nm SOI	0.18µm bulk HVCMOS
Interleaved Phases	32	21	1	41	1	2
Power Density (PD)	0.86 W/mm²	0.77 W/mm²	0.046 W/mm²	0.039 W/mm²	2.185 W/mm²	0.60 W/mm²
Efficiency @ PD	79.76%	69%	74%	81% (3:2 mode)	90%	85%
Maximum Output Power	0.325 W	1.65 W	0.150 W	0.010 W	2.62 W	4.3 W
Notes	85% Eff @ 0.28 W/mm²	-	-	75% Peak Eff in 2:1 mode	External Decoupling	Resonant Topology

[1] Breussegem, et al., ESSCIRC, 2010.
[2] Villar-Piqué, G., ISSCC, 2012.
[3] Chang, R.K., VLSI, 2012

Figure 4.5.6: Comparison with recently published fully integrated SC converters.

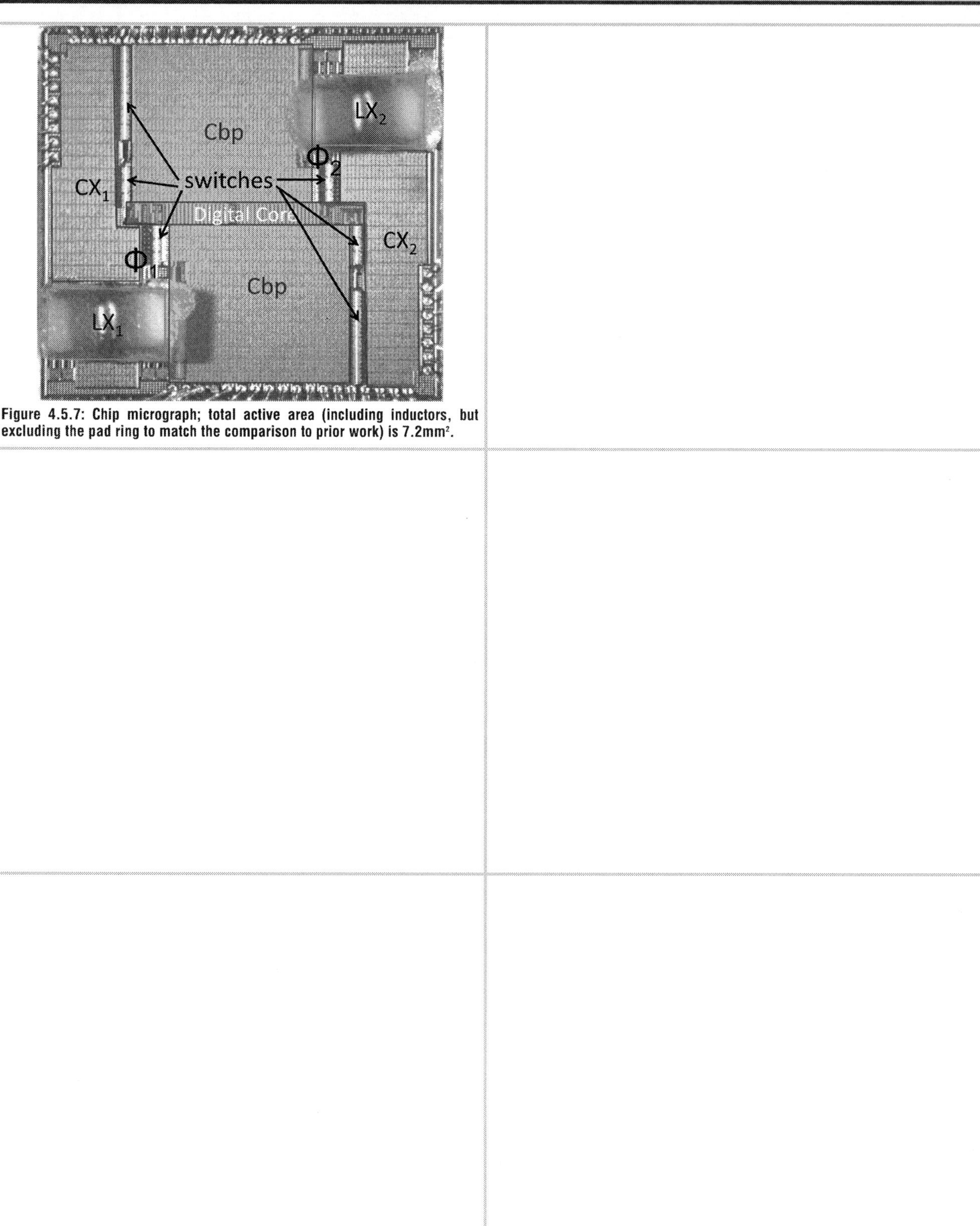

Figure 4.5.7: Chip micrograph; total active area (including inductors, but excluding the pad ring to match the comparison to prior work) is 7.2mm².

4.6 An 85%-Efficiency Fully Integrated 15-Ratio Recursive Switched-Capacitor DC-DC Converter with 0.1-to-2.2V Output Voltage Range

Loai G. Salem, Patick P. Mercier

University of California, San Diego, La Jolla, CA

The growing demand for both performance and battery life in portable consumer electronics requires SoCs and power management circuits to be small, efficient, and dynamically powerful. Dynamic voltage scaling (DVS) can help achieve these goals in load circuits, though generally at the expense of increased DC-DC converter size (through use of external inductors) or loss (through linear regulation). While switched-capacitor (SC) DC-DC converters can offer conversion in small fully integrated form factors [1-5], their efficiencies are only high at discrete ratios between the input and output voltages. To increase an SC converter efficiency across its output voltage range, multiple conversion ratios can be utilized to realize a finer output voltage resolution. For instance, many converters employ a small handful of conversion ratios [1-4]. However, more conversion ratios are generally necessary to achieve high efficiency across the wide output range necessary for DVS, as converter efficiencies can otherwise fall by more than 20% between unloaded ratios [1-4]. Unfortunately, increasing the number of ratios beyond a small handful using standard topologies can significantly increase the number of components, escalating converter complexity and adding losses in the additional switching elements. To overcome this, a successive approximation (SAR) SC topology was proposed in [6] which cascades several 2:1 SC stages to provide a large number of conversion ratios with minimal hardware overhead. However, the linear cascading of stages introduces cascaded losses, limiting overall efficiency. For example, the minimum R_{out} is more than 30X R_{out} of a similar ratio Series-Parallel topology using the same silicon area. Additionally, current density is limited to that of a single stage, and capacitance utilization can be low for many conversion ratios.

In this paper, we propose a *recursive switched capacitor* (RSC) DC-DC converter topology that achieves high efficiency across a wide output voltage range by providing 2^N-1 conversion ratios using N 2:1 SC cells with minimal hardware overhead. High efficiency is primarily achieved by: 1) opportunistically connecting individual 2:1 SC converters either in series or parallel, where the number of connections to V_{in} and ground is maximized in order to minimize the total charge transferred through the flying capacitors thereby minimizing cascaded losses; 2) utilizing 100% of the flying capacitance in charge transfer across all 2^N-1 conversion ratios; and 3) ensuring optimal relative sizing between flying capacitor sizes and constituent switches such that 2:1 SC stages with the highest current are dynamically allocated the majority of the capacitance and switch resources.

The proposed RSC topology is illustrated in Fig. 4.6.1. Each iteration through the RSC topology pseudo-code generator instantiates a new 2:1 SC cell, where the output of each cell, MID, produces the average voltage of its input port IN_{top} and "ground" port IN_{bottom}. The first instantiated 2:1 SC cell is connected between V_{in} and circuit ground. The IN_{top} ports of all subsequent 2:1 cells are either connected to V_{in} or another stage's MID port, while the IN_{bottom} ports are either connected to circuit ground or another stage's MID port. Through these connections, the amount of charge through the flying capacitors is minimized, maximizing the capacitance utilization factor and minimizing cascaded losses. The number of iterations (i.e., recursion depth N) defines the resolution of the output voltage as $V_{in}/2^N$, where V_{out} is obtained through the MID port of the final conversion stage. Figure 4.6.1 illustrates simplified examples of 1/2, 3/4, 3/8, and 11/16 ratios.

The 8-cell block in Fig. 4.6.1 illustrates the implemented recursive SC that can achieve 15 ratios. Although four 2:1 SC converters are only technically necessary to realize 15 ratios, eight cells are employed in this work to permit optimization of relative capacitor and switches sizes, ensuring the latter stages that handle increased current have larger capacitors and switches, which increases the efficiency by > 5%. Two types of cells are used in this converter: a) boundary cells that have port V_{int} for connection to the MID port of the prior stage; and b) transfer cells that have an extra port V_{int2} for connections from non-adjacent cells. All modes are opportunistically configured for a maximum number of V_{in} and 0 connections. For example, in the 1/2 ratio, all 8 cells are connected in parallel to V_{out} for maximum efficiency. In $n_{odd}/4$ ratios (i.e., 1/4 and 3/4), cells 1, 2 are connected in cascade as well as the cells 3_1, 3_2; 4_1, 4_2; and 4_3, 4_4, while in total four 2-cell cascades are connected in parallel. In $n_{odd}/8$ ratios, cells 1-3 are connected in cascade while cell 4 is configured as a paralleled configuration to cells 1-3. Finally, in $n_{odd}/16$ modes, cells 1-4 are connected in cascade (though sub-cells within cells 3 and 4 are connected in parallel).

Figure 4.6.2 illustrates the implementation of the boundary and transfer cells. Each cell uses two out-of-phase 2:1 SC converters to provide charge balance between stages. Transmission gates are used for MID switches to maintain constant on-resistance among various ratios. Two non-overlap phases $\phi1$, $\phi2$ are used to drive the switches. To realize the various states of the boundary and transfer cells, digital selection logic is used to semi-permanently enable/disable a switch, or select the $\phi1$, $\phi2$ signals for a transmission gate; hence there is no need for extra series reconfiguration switches. A decoder is used to provide the selection signals.

The overall architecture of the implemented chip is shown in Fig. 4.6.3. An all-digital binary search controller is used to dynamically select between the 15 modes. After a mode is selected, the comparator fine tunes the output through R_{out} frequency control. The coarse controller uses the SC power stage itself to produce the different comparison levels, which simplifies the implementation and provides an accurate control that minimizes R_{out} and provides robustness against process variation. Once STROBE is activated, EN is set and the SC switches at the highest frequency where R_{out} is minimum for a given mode. The coarse controller takes 3 decision cycles if the target mode is $n_{odd}/16$. A fourth correction cycle may result when the desired mode is $n_{even}/16$, where the back-off logic defines the amount of shift-left to the latest ($V_{mode} > V_{ref}$) occurrence. The illustrated mode-code simplifies the controller implementation where the code registers the consecutive comparison decisions.

The recursive switched capacitor converter is fully integrated in a 0.25μm bulk CMOS with MIM capacitor densities of 0.9fF/μm^2 and a maximum input voltage of 2.5V. Measurement results in Fig. 4.6.4 show the efficiency of the converter versus the output voltage (at 2mA) and the output current (at 1.15V), illustrating the principal advantages of the proposed architecture: greater than 70% efficiency is achieved over an output range from 0.9 to 2.2V at a load current of 2mA, while achieving greater than 80% efficiency for currents ranging from 30μA to 1mA. The converter achieves a peak efficiency of 85%. Measured results match those from a theoretical model within 1%. The RSC topology achieves 4.5% and 28% efficiency improvements over similarly-modeled 3-ratio Series-Parallel and SAR topologies for the same silicon area in 0.25μm CMOS, respectively, while achieving an output operating range that is 38% larger than the 3-ratio topology. Figure 4.6.5 shows the transient response of the converter for a variable step control voltage; the response time of the controller is 8μs. A table of comparisons is shown in Fig. 4.6.6. A die photo of the chip utilizing 4.645mm^2 of area for 3nF of on-chip capacitance is shown in Fig. 4.6.7.

References:

[1] D. El-Damak, S. Bandyopadhyay, and A.P. Chandrakasan, "A 93% Efficiency Reconfigurable Switched-Capacitor DC-DC Converter using On-Chip Ferroelectric Capacitors," *ISSCC Dig. Tech. Papers*, pp. 374-375, Feb. 2013.

[2] H.-P. Le, J. Crossley, S.R. Sanders, and E. Alon, "A Sub-ns Response Fully Integrated Battery-Connected Switched-Capacitor Voltage Regulator Delivering 0.19W/mm^2 at 73% Efficiency," *ISSCC Dig. Tech. Papers*, pp. 372-373, Feb. 2013.

[3] Y.K. Ramadass, A. Fayed, B. Haroun, and A. Chandrakasan, "A 0.16mm^2 Completely On-Chip Switched-Capacitor DC-DC Converter Using Digital Capacitance Modulation for LDO Replacement in 45nm CMOS," *ISSCC Dig. Tech. Papers*, pp. 208-209, Feb. 2010.

[4] Y.K. Ramadass and A.P. Chandrakasan, "Voltage Scalable Switched Capacitor DC-DC Converter for Ultra-Low-Power On-Chip Applications," *IEEE Power Electronics Specialists Conference*, pp. 2353-2359, Jun. 2007.

[5] V. Ng and S. Sanders, "A 92%-Efficiency Wide-Input-Voltage-Range Switched-Capacitor DC-DC Converter," *ISSCC Dig. Tech. Papers*, pp. 282-283, Feb. 2012.

[6] S. Bang, A. Wang, B. Giridhar, D. Blaauw, and D. Sylvester, "A Fully-Integrated Successive-Approximation Switched-Capacitor DC-DC Converter with 31mV Output Voltage Resolution," *ISSCC Dig. Tech. Papers*, pp. 370-371, Feb. 2013.

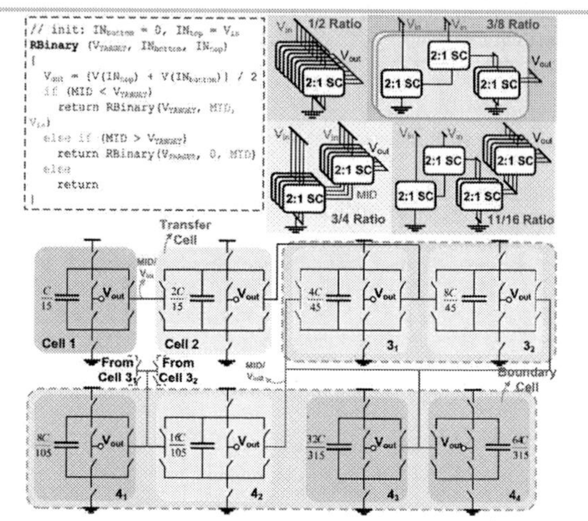

Figure 4.6.1: Recursive topology pseudo-code generator (top left). An example of 4-bit RSC operation across four different ratios (top right). Implemented 4-bit RSC block diagram (bottom).

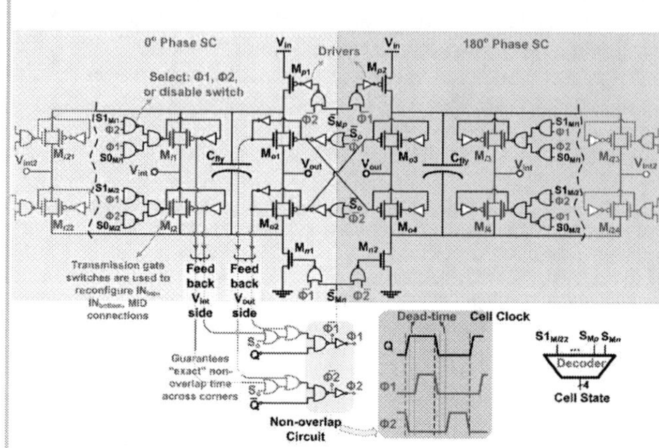

Figure 4.6.2: Implementation of the standard boundary and transfer cells, non-overlap circuit and state decoder.

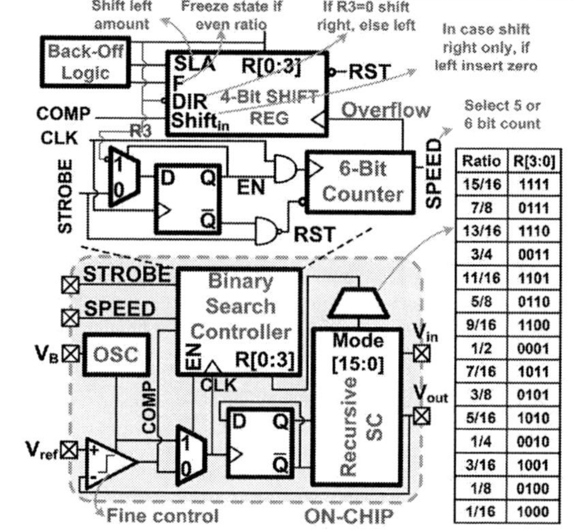

Figure 4.6.3: The overall architecture of the implemented converter chip.

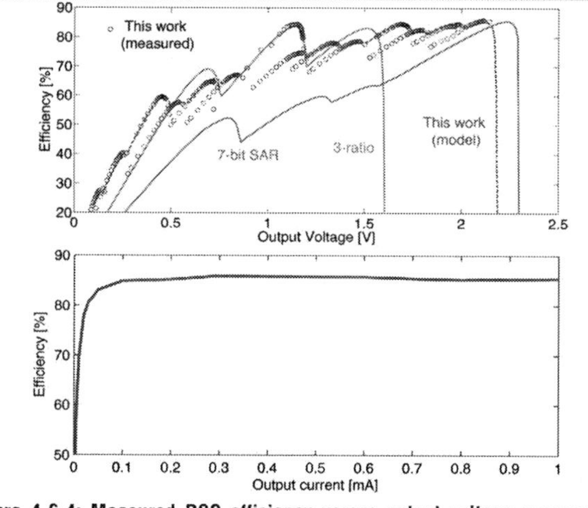

Figure 4.6.4: Measured RSC efficiency versus output voltage compared to model of RSC and previous state-of-art SC topologies (top). Measured efficiency versus load current (bottom).

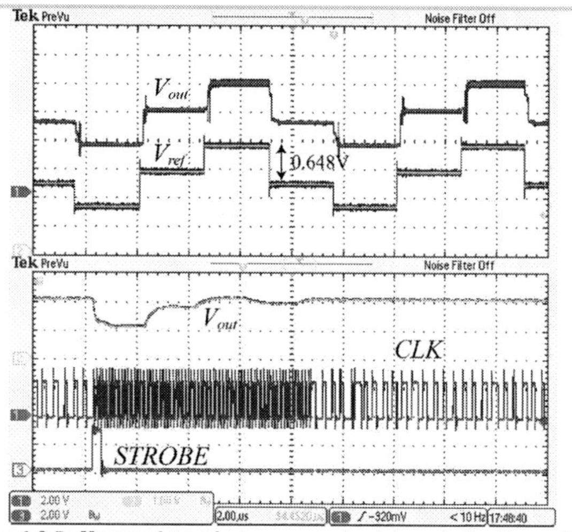

Figure 4.6.5: Measured transient response to stair control voltages (top). Measured 4-cycle, 8μs converter step response (bottom).

Design	[1]	[2]	[5]	This Work
Technology	130nm CMOS	65nm CMOS	180nm CMOS	0.25um CMOS
Cap Type	Ferroelectric	Bulk PMOS	On-Chip	MIM
Chip Area (mm^2)	0.366	0.64	1.69	4.645
Total Capacitance	8 nF	3.88 nF	2.24 nF	3 nF
Topology	1,2/3,1/2,1/3 Series-Parallel	1/3, 2/5 Series-Parallel	7-bit SAR	4-bit Recursive
V_{in}	1.5 V	3 - 4 V	3.4 – 4.3 V	2.5 V
V_{out}	0.4 - 1.1 V	1 V	0.9 – 1.5 V	0.1 - 2.18 V
Quoted Efficiency (η)	93%	74%	72%	85%
Load Current @ η	1mA	32mA	10μA	2mA

Figure 4.6.6: Comparison with prior art.

Figure 4.6.7: Micrograph of the recursive binary switched-capacitor DC-DC converter.

4.7 A Sub-ns Response On-Chip Switched-Capacitor DC-DC Voltage Regulator Delivering 3.7W/mm^2 at 90% Efficiency Using Deep-Trench Capacitors in 32nm SOI CMOS

Toke Meyer Andersen[1,2], Florian Krismer[1], Johann Walter Kolar[1], Thomas Toifl[2], Christian Menolfi[2], Lukas Kull[2], Thomas Morf[2], Marcel Kossel[2], Matthias Brändli[2], Peter Buchmann[2], Pier Andrea Francese[2]

[1]ETH, Zurich, Switzerland, [2]IBM Research, Rüschlikon, Switzerland

For an on-chip or fully integrated microprocessor power-delivery system, the on-chip power converter must 1) be designed using the same technology as the microprocessor, 2) deliver high power density to supply a microprocessor core with small area overhead, 3) achieve high efficiency, and 4) perform fast regulation over a wide voltage range for dynamic voltage and frequency scaling (DVFS). On-chip switched-capacitor (SC) converters have gained increasing popularity for this application due to their ease of integration using only transistors and capacitors readily available in the chosen technologies [1-6].

Historically, on-chip SC converters have been perceived as low power converters with output powers below 150mW [1-5]. However, the scalability of output power with chip area does not limit SC converters to being low power; the 1.65W maximum output power in [6] and the 840mW maximum output power presented in this paper exemplify the feasibility of high power on-chip SC converters. SC designs [1-4] utilize reconfigurable power stages for increased output and/or input voltage ranges as well as interleaving techniques to minimize the output voltage ripple, e.g., in [1], where 3.8mV$_{pp}$ output ripple is reported for a 41-phase interleaved SC converter. Using bulk CMOS, designs are limited in efficiency to 81% in [1] and in power density to 0.19W/mm^2 in [2]. Regarding on-chip capacitor technologies, MIM capacitors are used in a 22nm tri-gate technology in [3], and 93% efficiency is reported using ferroelectric capacitors in [4], but both designs achieve low power densities (<0.1W/mm^2). Employing deep trench capacitors has shown superior efficiency and power density performances, e.g., 4.6W/mm^2 at 86% efficiency for a single phase unregulated on-chip SC converter [5]. Furthermore, multi-GHz sampling frequencies are used in hysteretic control loops to achieve fast response times to transient events, e.g., 3-to-5ns response time in [3] and <1ns response time in [2]. In this paper, we utilize the deep trench capacitor and thin-oxide transistors available in 32nm SOI CMOS to design a high power (840mW) and fast response (<1ns) 16-phase interleaved reconfigurable on-chip SC converter that achieves 86.4% maximum efficiency at 2.2W/mm^2 in the 2:1 configuration and 90.0% maximum efficiency at 3.7W/mm^2 in the 3:2 configuration.

The overall system diagram of the implemented SC converter is depicted in Fig. 4.7.1. Two capacitors can be configured to provide either a 2:1 or a 3:2 ideal voltage conversion ratio by toggling between a charging and a discharging state at 50% duty cycle. A 16-phase interleaving technique is employed to reduce the input current and output voltage ripples, thereby omitting the need for a dedicated output decoupling capacitor. The on-chip load consists of a programmable resistor array, which can be externally programmed by the digital configuration interface. Also the gear signal, which sets the power stage in the 2:1 or 3:2 configuration, is externally controlled. The clocked comparator compares V_{out} with V_{hys} and produces a clock signal clk_{trig} for the digital clock interleaver, which generates the clock signals $clk_{0\text{-}15}$ for each SC converter unit. The 250ps sampling period (4GHz) of the comparator ensures <1ns response time to transient events.

The power stage implementation shown in Fig. 4.7.2 consists of 2 capacitors and 12 switches, and it can be reconfigured between the 2:1 and 3:2 ideal voltage conversion ratios [2]. A gate driver generates the gate signals $v_{g1\text{-}9(s)}$ for the charging and discharging states for each transistor as shown in the table in Fig. 4.7.2. In recent works, the generation of $v_{g1\text{-}9}$ depends on several internal nodes and/or additional external voltage supplies [1,2]. However, this design proposes a simplified gate driver implementation that only depends on V_{in}, V_{out}, and gnd. The level-shifted non-overlapping gate signals $v_{g,(n/p)(H/L)}$ are generated as in [5], and multiplexers controlled by the gear configuration signal are used to change the clock feeds for $v_{g(4,5,7,7s)}$. All transistors $M_{1\text{-}9(s)}$ are thin-oxide devices (V_{max}≈1.2V) for low on-state resistance and fast transition times. Since transistor M_5 should always be off in the 2:1 configuration, M_{5s} is implemented to protect the gate-source of M_5 in the charging state against overvoltage (gnd–Vin). With a nominal input voltage of V_{in}=1.8V and an output voltage range of 0.7 to 1.1V, node V_X approximately equals $V_{out}/2$ in the discharging state of the 3:2 configuration, thereby exposing the drain-source of M_6 to overvoltage ($V_{in}-V_{out,min}/2$=1.45V). The stacking transistor M_{6s} effectively protects M_6 against this overvoltage situation. Due to symmetrical transistors, M_7 undesirably turns on in the discharging state of the 3:2 configuration due to a positive gate-drain voltage ($V_{out}/2$); M_{7s} ensures that M_7 remains turned off.

The 16-phase digital clock interleaver, of which a 4-phase example implementation is shown in Fig. 4.7.3, produces 16 time-interleaved clock phases. Every second output of the shift register is inverted to average out dissimilar currents delivered in the charging and discharging states of the 3:2 configuration, thereby keeping the output voltage ripple V_{ripple} low. The comparator clock frequency f_{trig} is determined by the number of interleaved stages N and a specified maximum switching frequency $f_{sw,max}$ of each converter unit following the equation in Fig. 4.7.3. Furthermore, f_{trig} is upper limited by the total loop latency t_{lat} (combined latency of the comparator, the digital controller, and the gate driver) to ensure that the SC converter has time to react on a trigger signal before the next comparison event. Having 16 phases in this design, f_{trig}=4GHz results in $f_{sw,max}$=125MHz; furthermore, the loop latency is minimized to t_{lat}≈200ps.

The on-chip programmable load, which is implemented as an array of 31 switchable resistors (resulting in 32 different load values including the no load), can provide a load step between any two load levels within 50ps. Such fast load steps are used to evaluate the <1ns response under worst-case conditions. In Fig. 4.7.4, the measured transient responses when stepping between 0.1x and 1x nominal load, corresponding to 30mA and 365mA output current, are shown. As observed, the output voltage is maintained several nanoseconds after the transient event, verifying the <1ns response of the regulation loop. The output voltage droop is caused by the large input voltage droop, which is due to 1) the slower response of the external input power supply and 2) the parasitic capacitances and inductances of the power distribution network connecting to the chip.

The measured efficiencies and power densities are shown in Fig. 4.7.5 for three different load levels, where resistances and voltages are measured using Kelvin contacts. For V_{in}=1.8V, the efficiency is >70% over the specified output voltage range of 0.7 to 1.1V with maximum efficiencies of 86.4% and 90.0% in the 2:1 and 3:2 configurations, respectively. For high loads, the maximum power density, and thereby the maximum output voltage $V_{out,max}$, is limited by the 4GHz comparator clock frequency, whereas for $V_{out}<V_{out,max}$, the maximum power density is limited by the maximum on-chip load.

Figure 4.7.6 shows how this design compares to prior art, revealing >9× improvement in power density while providing <1ns response time and 90% efficiency. An output ripple of 30mV$_{pp}$ is achieved without using a dedicated decoupling capacitor, and the 840mW maximum output power verifies the feasibility of high power on-chip SC converters.

The on-chip SC converter micrograph is shown in Fig. 4.7.7. The total converter area including gate drivers and the digital controller is 0.15mm^2.

References

[1] G.V. Piqué, "A 41-Phase Switched-Capacitor Power Converter with 3.8mV Output Ripple and 81% Efficiency in Baseline 90nm CMOS," *IEEE ISSCC Dig. Tech. Papers*, pp. 98-100, Feb. 2012.

[2] H.-P. Le, J. Crossley, S.R. Sanders, and E. Alon, "A Sub-ns Response Fully Integrated Battery-Connected Switched-Capacitor Voltage Regulator Delivering 0.19W/mm² at 73% Efficiency," *IEEE ISSCC Dig. Tech. Papers*, pp. 372-373, Feb. 2013.

[3] R. Jain, B. Geuskens, M. Khellah, et al., "A 0.45-1V Fully Integrated Reconfigurable Switched Capacitor Step-Down DC-DC Converter with High Density MIM Capacitor in 22nm Tri-Gate CMOS," *IEEE Symp. VLSI Circuits*, pp. 174-175, Jun. 2013.

[4] D. El-Damak, S. Bandyopadhyay, and A.P. Chandrakasan, "A 93% Efficiency Reconfigurable Switched-Capacitor DC-DC Converter Using On-Chip Ferroelectric Capacitors," *IEEE ISSCC Dig. Tech. Papers*, pp. 374-375, Feb. 2013.

[5] T. M. Andersen, F. Krismer, J.W. Kolar, et al., "A 4.6W/mm² Power Density 86% Efficiency On-Chip Switched Capacitor DC-DC Converter in 32 nm SOI CMOS," *IEEE Applied Power Electronics Conf. and Exposition (APEC)*, pp. 692-699, Mar. 2013.

[6] H. Meyvaert, T. Van Breussegem, and M. Steyaert, "A 1.65W Fully Integrated 90nm Bulk CMOS Intrinsic Charge Recycling Capacitive DC-DC Converter: Design & Techniques for High Power Density," *IEEE Energy Conversion Congress and Exposition (ECCE)*, pp. 3234-3241, Sep. 2011.

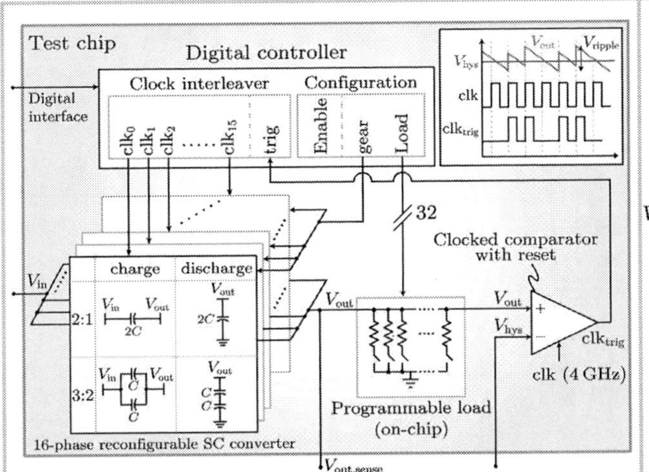

Figure 4.7.1: Overall system diagram of the on-chip SC voltage regulator showing the reconfigurable SC power stage with digital controller.

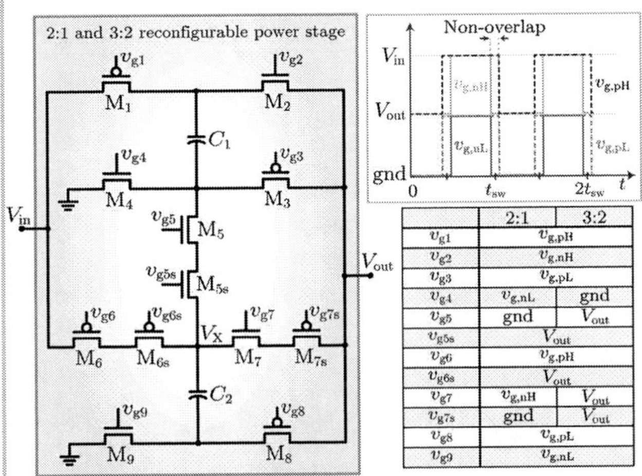

Figure 4.7.2: Reconfigurable SC converter power stage with gate signals shown for the 2:1 and 3:2 configurations.

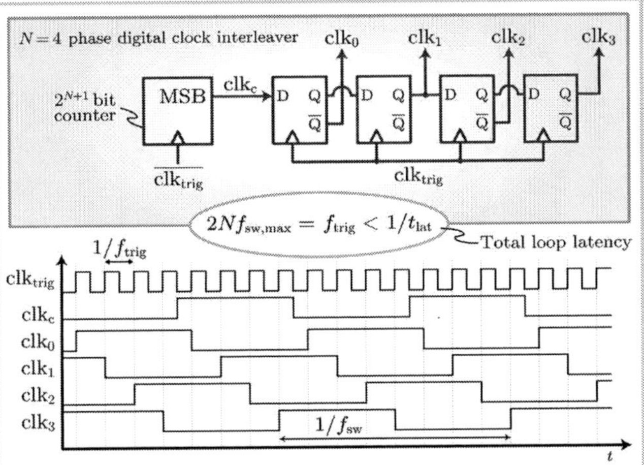

Figure 4.7.3: Example 4-phase digital clock interleaver and timing diagram. The presented converter implements a 16-phase clock interleaver following the same principles.

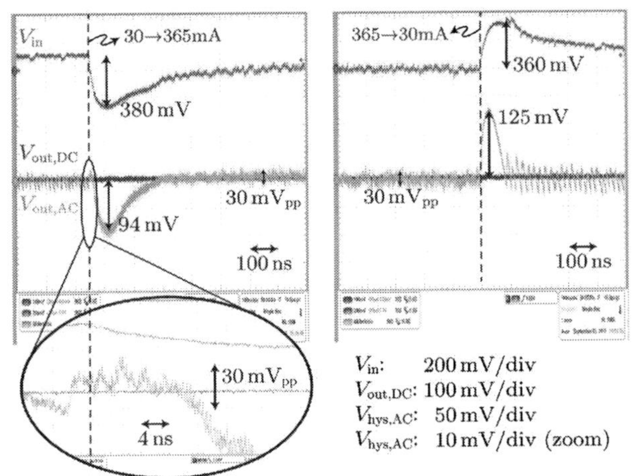

Figure 4.7.4: Transient responses for V_{in}=1.8V and V_{out}=840mV.

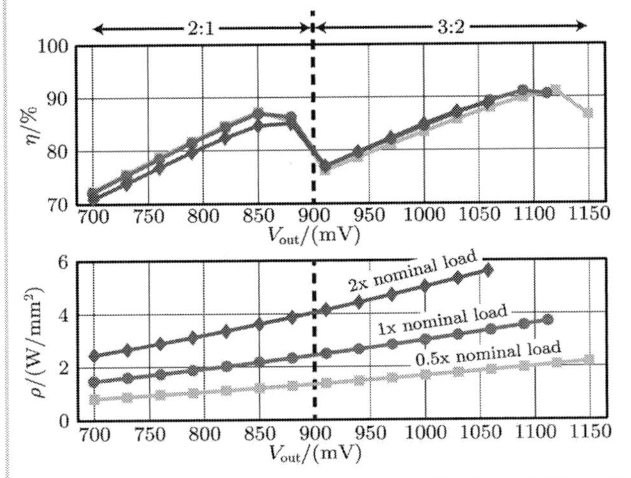

Figure 4.7.5: Measured efficiencies and power densities for V_{in}=1.8V over the full output voltage range.

Design	Piqué [1] ISSCC 2012	Le [2] ISSCC 2013	El-Damak [3] ISSCC 2013	Jain [4] VLSI 2013	Meyvaert [6] ECCE 2011	This Work
Technology	90nm bulk	65nm bulk	32nm SOI	22nm tri-gate	90nm bulk	**32nm SOI**
Conversion ratios (M)	2:1, 3:2	5:2, 3:1	3:1, 2:1, 3:2, 1:1	2:1, 3:2, 5:4, 1:1	2:1	**2:1, 3:2**
Capacitor type	MOS + fringe metal	MOS	Ferroelectric	MIM	MOS	**Deep Trench**
Interleaving	41	18	4	8	21	**16**
C_{fly} / C_{out}	14pF / 85pF	3.88nF / 0	1nF / 10nF	- / 100pF	0.57nF / 0	**1nF / 0**
V_{in}	1.1 to 2V	3 to 4V	1.5V	1.23V	2.35 to 2.6V	**1.8V**
V_{out}	0.7V	1V	0.4 to 1.1V	0.45 to 1V	1.03	**0.7 to 1.1V**
$P_{out,max}$	9.5mW	121mW	1.1mW	88mW	1.65W	**840mW**
$t_{response}$	-	<1ns	<1ms	3 to 5ns	≈15μs	**<1ns**
V_{droop}	-	76mV	-	<25mV	95mV	**94mV (due to V_{in} droop)**
$V_{ripple,pp}$ @ nom. load	3.8mV	-	-	43mV	-	**30mV**
η_{max} @ M	75%, 81%	71.5%, 73%	90%, 91%, 93%, 80%	82%, 71%, 73%, 68%	69%	**86.4%, 90.0%**
ρ (W/mm²) @ η_{max}	0.038, 0.025	0.19, 0.19	0.0006, 0.0010, 0.0013, 0.0016	0.062, 0.100, 0.126, 0.243	0.42	**2.17, 3.71**

Figure 4.7.6: Performance summary and comparison with recently published on-chip SC voltage regulators.

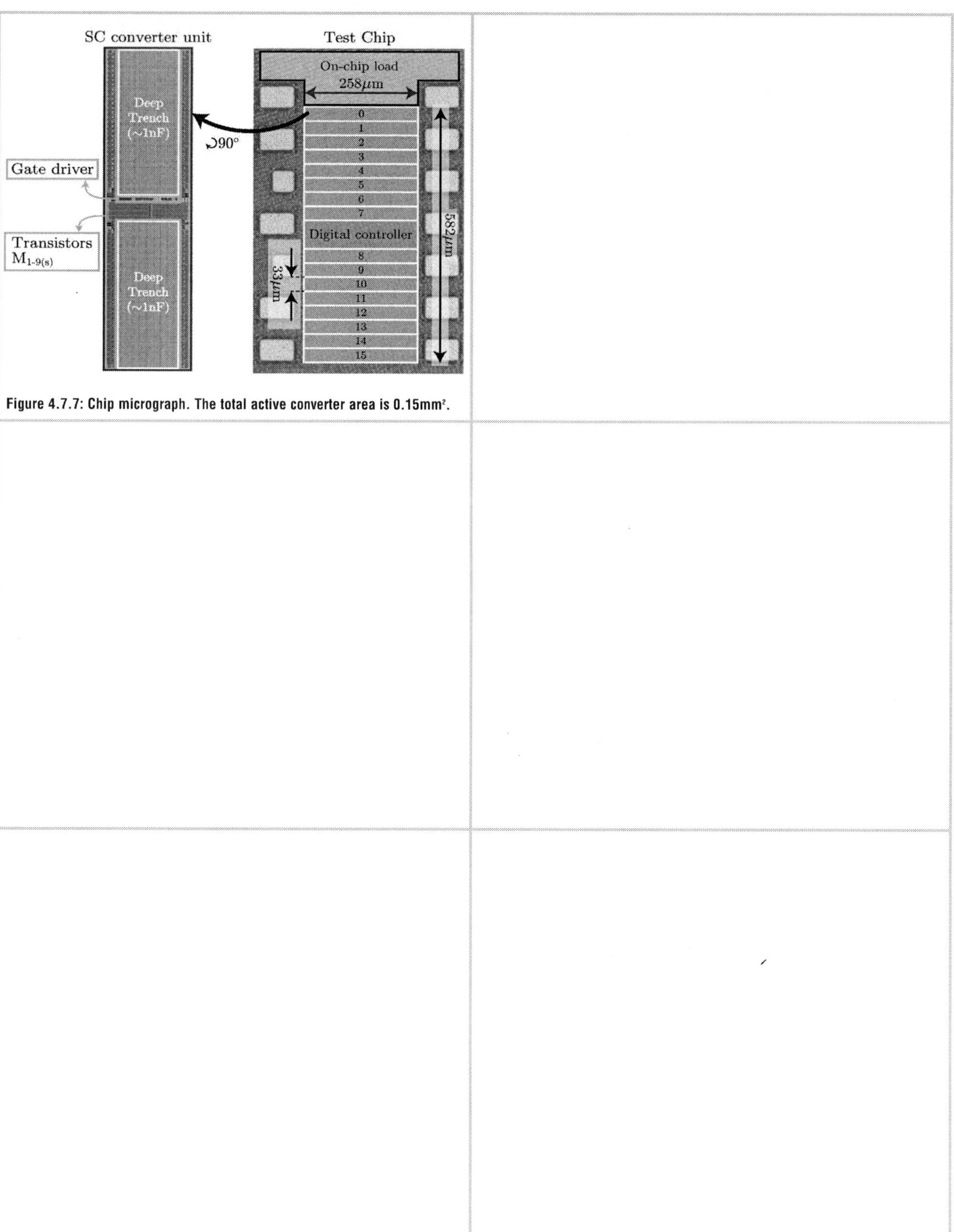

Figure 4.7.7: Chip micrograph. The total active converter area is 0.15mm².

4.8 3-Phase 6/1 Switched-Capacitor DC-DC Boost Converter Providing 16V at 7mA and 70.3% Efficiency in 1.1mm³

Ravi Karadi, Gerard Villar Pique

NXP Semiconductors, Eindhoven, The Netherlands

In this paper, a 3-phase switched-capacitor (SC) boost converter that uses 2 external floating capacitors to provide 16V output from a 3.3V input is presented. It achieves an efficiency of 70.3% at 7mA load current while the output ripple is 1% of the output voltage. To minimize the size of the capacitors the converter operates at a switching frequency of 6.67MHz. The chip is implemented in a 0.16μm CMOS technology and occupies 1.65mm². The total volume of the PCB components consisting of the chip in wafer-level chip-scale package (WLSCP) and 4 external SMD capacitors (1 input, 1 output and 2 floating) is only 1.1mm³ making it attractive for space-constrained mobile applications. The volume is more than 15× smaller compared to competitive inductive converters [1-3]. The maximum thickness of the solution is 550μm.

Mobile applications require a 16V supply rail providing power in the range of 100mW for LCD bias supply and OLED display drivers etc. Currently, this is realized using inductive boost converters [1-3], however, such converters have the disadvantage of low power density (output power per total component volume) due to poor scalability of inductive converters at lower power levels [4]. In contrast, SC power converters (SCPC) can achieve high power density at these lower power levels. However, conventional 2-phase SCPCs need a large number of floating capacitors for high conversion ratios (> 3/1) [4, 6]. Present work benefits from a 3-phase topology shown in Fig. 4.8.1 that uses only 3 floating capacitors instead of a minimum of 4 floating capacitors required in a 2-phase topology to achieve a conversion ratio of 6/1.

Figure 4.8.1 shows the operation of the converter: in phase 1, capacitors C_1 and C_2 are charged to input V_{in}. In phase 2 capacitors C_1 and C_2 are connected in series with V_{in} to charge capacitor C_3 to $3V_{in}$. In the last phase, C_3 is also connected in series with C_1 and C_2 to generate $6V_{in}$ and charge is supplied to the output, V_o. The resulting conversion ratio is $M=V_o/V_{in}=6/1$.

The converter is implemented in a 0.16μm CMOS technology that has 1.8V, 3.3V, and 5V (gate-oxide) transistors. Among the 3 floating capacitors, C_1 is chosen for integration in the chip to reduce the number of external components and pins. This choice minimizes the bottom plate losses, since the bottom plate swing of C_1 is only V_{in} while C_2, C_3 each have a bottom-plate swing of $2V_{in}$, and $3V_{in}$ respectively. C_1=2.3nF is integrated in the chip as a 3.3V MOS capacitor and occupies 0.45mm². C_2=1μF and C_3=220nF are external SMD components in compact 0201 packages. The use of large external capacitors enables the reduction of C_1 to save silicon area and still provide required output power.

The converter requires 10 switches as shown in the power-plant diagram of Fig. 4.8.1. The switches are implemented with single 3.3V or 5V transistors (shown with thick gate), or as cascode connection of 2 transistors. Choice of the switch type depends on its location in the power-plant and the required blocking voltage. The switch sizes are optimized to minimize switching losses while attaining the required output power capability. Floating NMOS transistors are placed in individual deep nwells for isolation.

The drivers of the switching transistors in the power-plant are of diverse complexity as indicated in Fig. 4.8.1. The drivers of S1, S3, S6 and S9_2 are simple tapered buffers supplied from V_{in}. The driver of S2 is also a tapered buffer, but is supplied from C_1. The gate of the cascode transistor S9_1 is tied to V_{in}. The driver of the switch S4 is supplied from capacitor C_2 itself and is controlled by a floating level-shifter (LS). Other drivers require auxiliary rails of 5.5V (V6) and 10.5V (V11). The gate of the cascode transistor S10_1 is connected to auxiliary rail V11. The driver of transistor S10_2 operates in the (V_o-V11) domain and a level-shifter is used to shift the control signal from V_{in} domain to (V_o-V11) domain. The drivers of the switches S7 and S8 are particularly complex since both transistors of the cascode connection in each switch need changing voltages to avoid exceeding their V_{gs} and V_{ds} limitations. The drivers of S7_1 and S7_2 are shown in Figure 4.8.2. S7_2 is controlled by the cascode inverter structure M1-M2 and M3-M4. The gates of M2 and M3 are simply tied to auxiliary rail V6. M1 and M4 are driven by tapered buffers in (V11-V6) and (V6-Vin) domain respectively, and 2 level-shifters controlled by the same signal D1 are used to shift the control signal from V_{in} domain. S7_1 is controlled by M7 and cascode transistors of M5-M6. The gate of M6 is tied to V6, while M5 and M7 both have drivers in (V11-V6) domain. The AND gate combines the two control signals D2 and D3 to avoid over-stressing M7. Two level-shifters are used to shift D2 and D3 from V_{in} domain to (V11-V6) domain. The drivers of S8_1 and S8_2 are similar.

Figure 4.8.3 shows the block diagram of the converter with power-plant, controller and converters for auxiliary rails of V6 and V11. The controller implements pulse skipping control chosen for its robustness, fast transient response and its ability to maintain high efficiency over a wide range of load currents. The controller consists of a resistive divider, a band-gap reference, a clocked comparator, a state-machine and a signal generator. To attain fast response, the controller proceeds through phase 1 and phase 2, charging the floating capacitors C_1 to C_3 , and keeps waiting while $V_{fb}>V_{ref}$. When V_{fb} drops below V_{ref} , it immediately goes to phase 3 and supplies charge to the output. Clock signal of 20MHz ($3 \times f_s$) is supplied from outside to the comparator and to the state-machine that generates 3 signals corresponding to the 3 different phases. The signal generator produces the required switching signals in the V_{in} domain ensuring non-overlapping while avoiding stressing of the power switches during phase transitions.

Non-overlap in the power-plant is obtained by inserting delays in the signal generation block to obtain proper timing of the switching signals, instead of using feedback. This requires considering all the possible process and temperature corners. Feedback is not used due to the high frequency operation (3-phase operation at f_s=6.67MHz) and the presence of different voltage domains, requiring level-shifters that would add delays and reduce the effective duration of each phase.

The additional voltage rails V6 of 5.5V and V11 of 10.5V required by the drivers are generated with internal 2-phase SC converters. The switching activity of these converters is synchronized to the main converter (charging in phase 2, discharging in phase 3), so that their output power capability scales along with the current required by the drivers. This simplifies the design, since they do not need separate controllers, and keeps their power consumption proportional to the output current of the main converter.

To validate the concept a test chip is designed and fabricated in bulk 0.16μm CMOS technology with 3μm-thick top interconnect metal. The chip uses a WLSCP package of 400μm pitch with 12 bump pads (including pins for test purposes). The efficiency results with V_{in}=3.3V and load current varied from 0 to 10mA as well as the output voltage are shown in Fig. 4.8.4. The efficiency is above 70% with load current from 1mA to 7mA, and a peak efficiency of 70.8% is attained for a load current of 4mA. The low load efficiency at 0.1mA current is 58.5%. The output ripple at I_o=7mA is 1% of V_o, even with a small output capacitor of 100nF. The control loop provides fast transient response. Figure 4.8.5 shows 2 different load steps: 0→1mA shows no voltage drop; and 0→7mA shows 1.5% voltage drop due to output power limitations of the converter at DC level (as observed in the results of Fig. 4.8.4). In high-to-low load steps there are no overshoots. Figure 4.8.6 shows the performance summary of the present work as well as comparison with inductive converters with similar specifications. Figure 4.8.7 shows the micrograph of the test chip.

Acknowledgements:
The authors would like to acknowledge the technical support of Jaume Tornila Oliver, Brad Gunter, Arnoud van der Wel and Stan Meeuwsen.

References:

[1] Linear Technology, LT3494/LT3494A Data sheet, "Micropower Low Noise Boost Converters with Output Disconnect," Nov. 2006, Accessed Dec. 2013,<http://cds.linear.com/docs/en/datasheet/3494fb.pdf>.

[2] Texas Instruments, TPS61045 Data sheet Rev. B, "Digitally Adjustable Boost Converter," Mar. 2009, Accessed Dec. 2013, <http://www.ti.com/lit/ds/symlink/tps61045.pdf>.

[3] Linear Technology, LT3460 Data sheet, "1.3MHz/650kHz Step-Up DC-DC Converter in SC70, ThinSOT and DFN," 2007, Accessed Dec. 2013, <http://cds.linear.com/docs/en/datasheet/34601fb.pdf>.

[4] Gerard Villar Pique and Ravi Karadi, "Potential Benefits of Integrated Switching Power Converters: Inductive vs. Switched-Capacitor," *International Workshop on Power Supply on Chip, Power SoC 12*, Nov. 2012.

[5] M.S. Makowski and D. Maksimovic, "Performance Limits of Switched-Capacitor DC-DC Converters," *IEEE Trans. Power Electronics*, vol.2, pp.1215-1221, Mar. 1995.

[6] V. Ng and S. Sanders, "A 92%-Efficient Wide-Input-Voltage-Range Switched-Capacitor DC-DC Converter," *ISSCC Dig. Tech. Papers*, pp.282-283, Feb. 2012.

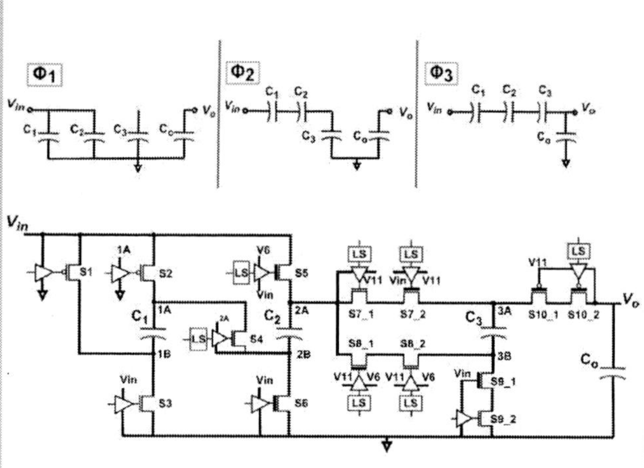

Figure 4.8.1: 3-phase switched-capacitor boost converter and its transistor-level implementation.

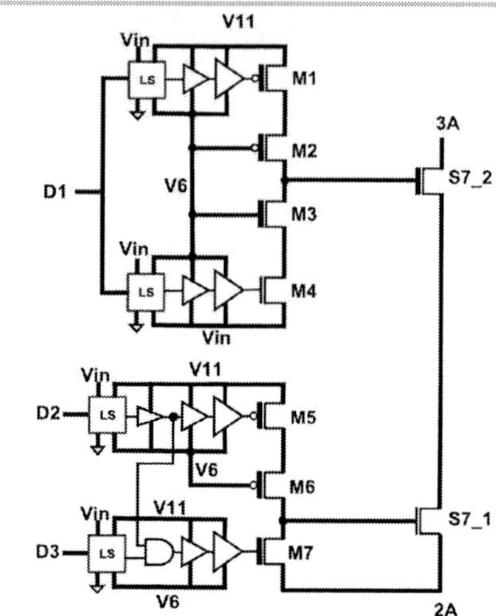

Figure 4.8.2: Drivers of the cascode-connected transistors S7_1 and S7_2.

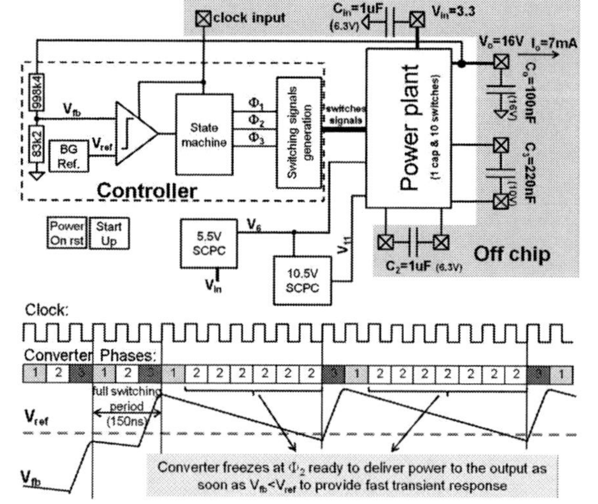

Figure 4.8.3: Converter block diagram.

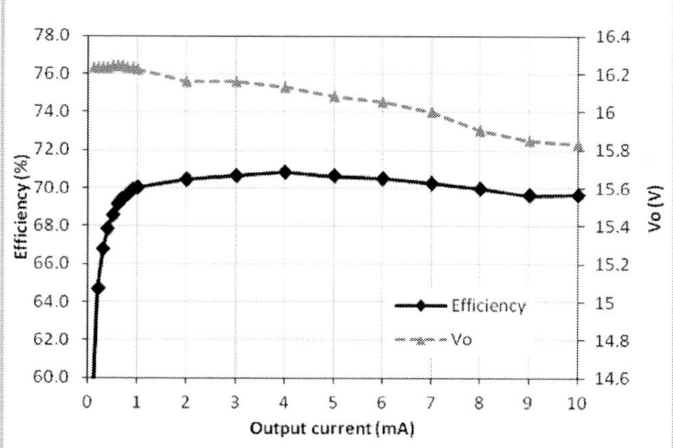

Figure 4.8.4: Measurement results showing efficiency and output voltage versus load current.

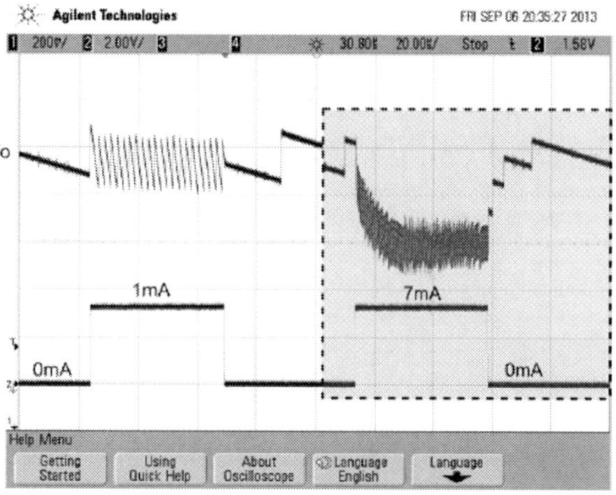

Figure 4.8.5: Composition with the same axes of 2 oscilloscope screenshots corresponding to 2 different load step response: 0 to 1mA (left); and 0 to 7mA (right, darker square).

Converter specifications	LT3494 [1]	TI TPS61045 [2]	LT3460 [3]	This work
Total Component volume (mm³)*	32	37.5	18.3	1.1
Maximum thickness (mm)	1.8	1.6	1.8	0.55
Efficiency at 16V@7mA	75.8%	84%	79%	70.3%
Efficiency at 16V@0.1mA	45.6%	--	13.2%	58.5%
Output ripple at 7mA (relative to V_o=16V)	0.06% (C_o=2.2µF)	0.28% (C_o=1µF)	0.3% (C_o=1µF)	1% (C_o=100nF)
Components*	1XDFN 1x1210 2x1206 1x0603	1xOFN-8 1x1210 1x0805 1x1206 1xSOD123	1xDFN-6 1x1206 2x0603 1xSOD-323	1xWLCSP 4x0201

*Considering the components recommended in the datasheets

Figure 4.8.6: Performance summary and comparison.

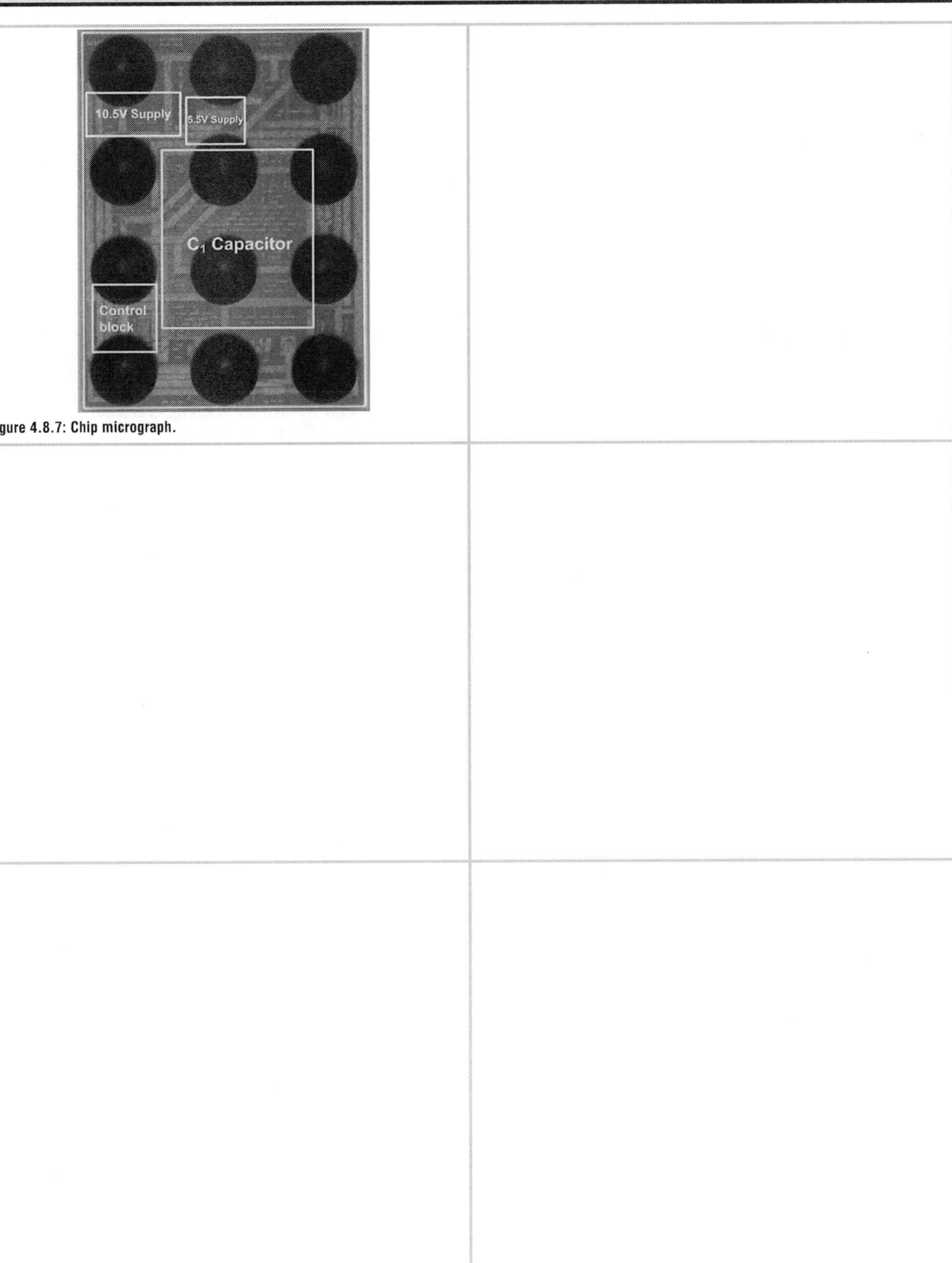

Figure 4.8.7: Chip micrograph.

Session 5 Overview: *Processors*

HIGH-PERFORMANCE DIGITAL SUBCOMMITTEE

Session Chair: *Atsuki Inoue*
Fujitsu, Kawasaki, Japan

Session Co-Chair: *Christopher Gonzalez*
IBM, Yorktown, NY

As compute power is increasingly migrated to large data centers and the cloud, microprocessors face progressively more stringent design constraints. This year's processor session introduces 5 new processors providing increased performance and power efficiency. In addition to growing core counts, cache size, and thread count, the historical theme of integration continues as voltage regulators are now being implemented on chip. Other papers in this session demonstrate creative self-monitoring and adaptive techniques to meet power and performance design goals.

5.1 POWER8™: A 12-Core Server-Class Processor in 22nm SOI with 7.6Tb/s Off-Chip Bandwidth **1:30 PM**

E. J. Fluhr, IBM STG, Austin, TX

In Paper 5.1, IBM presents the POWER8™ microprocessor featuring 12 cores and 96MB of on-chip L3 cache. The chip is implemented in 22nm SOI eDRAM technology with 15 levels of metal. The processor features 7.6Tb/s off-chip bandwidth, integrated voltage regulation and resonant clocking.

5.2 Distributed System of Digitally Controlled Microregulators Enabling Per-Core DVFS for the POWER8™ Microprocessor **2:00 PM**

Z. Toprak-Deniz, IBM, Yorktown Heights, NY

In Paper 5.2, IBM demonstrates a distributed system of micro-regulators implemented in 22nm SOI technology for power management of the POWER8™ microprocessor. With a 11.9A load, the micro-regulators achieve a power efficiency of 90.5% and power density of 34.5W/mm^2.

5.3 Wide-Frequency-Range Resonant Clock with On-the-Fly Mode Changing for the POWER8™ Microprocessor **2:15 PM**

P. Restle, IBM Research, Yorktown Heights, NY

In Paper 5.3, IBM describes the resonant clocking architecture implemented in the POWER8™ microprocessor. The multimode resonant design can oscillate from 2.5GHz to greater than 5GHz, reduce clock grid power by 33% and can dynamically switch between high and low resonant frequency modes without idle cycles.

978-1-4799-0917-9/14 $31.00 © 2014 IEEE

5.4 Ivytown: A 22nm 15-Core Enterprise Xeon® Processor Family **2:30 PM**

S. Rusu, Intel, Santa Clara, CA

In Paper 5.4, Intel introduces the Ivytown Xeon® processor containing 15 cores, and 37.5MB shared L3 cache. The processor has 4.31B transistors in a high-κ metal-gate tri-gate 22nm CMOS technology with 9 metal layers. It features an enhanced ring bus topology, a 1 PLL per column clocking structure and a multimode memory interface.

5.5 Steamroller: An x86-64 Core Implemented in 28nm Bulk CMOS **3:15 PM**

K. Gillespie, AMD, Boxborough, MA

In Paper 5.5, AMD presents their latest x86 core, Steamroller. It is implemented in 28nm occupying 29.47mm^2, with more than 236 million transistors. A shared 96KB 3-way instruction cache and 10KB L2 branch target buffer improve single and multi-threaded performance as compared to a previous 32nm design. Power reduction is achieved by resonant clocking, power supply monitors and power gating techniques.

5.6 Adaptive Clocking System for Improved Power Efficiency in a 28nm x86-64 Microprocessor **3:45 PM**

A. Grenat, AMD, Austin, TX

In Paper 5.6, AMD demonstrates a real-time voltage-droop detector which triggers the clock to adapt to falling voltage, allowing frequency to be maintained at lower voltage and achieving power efficiency improvements of 7 to 15%.

5.7 A Graphics Execution Core in 22nm CMOS Featuring Adaptive Clocking, Selective Boosting and State-Retentive Sleep **4:00 PM**

C. Tokunaga, Intel, Hillsboro, OR

In Paper 5.7, Intel presents a graphics execution core implemented in 22nm tri-gate CMOS incorporating an adaptive clocking technique for fast voltage-droop mitigation, and an all-digital retention voltage clamp for reduced leakage during state-retentive sleep mode. The core offers 12 to 27% power savings, wide voltage operation down to 0.38V, and 40% improvement in peak GFLOPS/W.

5.8 A 3GHz 64b ARM v8 Processor in 40nm Bulk CMOS Technology **4:15 PM**

H. Partovi, Applied Micro, Sunnyvale, CA

In Paper 5.8, Applied Micro reveals the first generation 64b v8 ARM 8-core processor. Featuring a 4-wide out-of-order superscalar micro-architecture, the processor is fabricated in 40nm bulk CMOS technology, and operates at 3GHz with a 0.9V supply, consuming 4.5W.

5.9 Haswell: A Family of IA 22nm Processors **4:45 PM**

N. Kurd, Intel, Hillsboro, OR

In Paper 5.9, Intel describes the 4th Generation Intel® Core™ processor Haswell implemented in a 22nm tri-gate process. It fully integrates voltage regulators improving battery life by ~50%, improves graphics by extending the cache hierarchy with eDRAM providing 102GB/s bandwidth at 1.22pJ/b, adds lower power states reducing standby power by 95%, has optimized I/O interfaces, and doubles floating point operation capability.

978-1-4799-0917-9/14 $31.00 © 2014 IEEE

5.1 POWER8™: A 12-Core Server-Class Processor in 22nm SOI with 7.6Tb/s Off-Chip Bandwidth

Eric J. Fluhr[1], Joshua Friedrich[1], Daniel Dreps[1], Victor Zyuban[2], Gregory Still[3], Christopher Gonzalez[2], Allen Hall[1], David Hogenmiller[1], Frank Malgioglio[4], Ryan Nett[1], Jose Paredes[1], Juergen Pille[5], Donald Plass[4], Ruchir Puri[2], Phillip Restle[2], David Shan[1], Kevin Stawiasz[2], Zeynep Toprak Deniz[2], Dieter Wendel[5], Matt Ziegler[2]

[1]IBM STG, Austin, TX, [2]IBM T. J. Watson, Yorktown Heights, NY,
[3]IBM STG, Raleigh, NC, [4]IBM STG, Poughkeepsie, NY,
[5]IBM STG, Boeblingen, Germany

The 12-core 649mm^2 POWER8™ leverages IBM's 22nm eDRAM SOI technology [1], and microarchitectural enhancements to deliver up to 2.5× the socket performance [2] of its 32nm predecessor, POWER7+™ [3]. POWER8 contains 4.2B transistors and 31.5μF of deep-trench decoupling capacitance. Three thin-oxide transistor V_ts are used for power/performance tuning, and thick-oxide transistors enable high-voltage I/O and analog designs. The 15-layer BEOL contains 5-80nm, 2-144nm, 3-288nm, and 3-640nm pitch layers for low-latency communication as well as 2-2400nm ultra-thick-metal (UTM) pitch layers for low-resistance distribution of power and clocks.

The POWER8 C4 array contains 15823 total pads: 5982 power, 7742 ground, and 2099 signal. As shown in Fig. 5.1.1, POWER8 has 7 input voltages: 1) core / cache logic (V_{DD}), 2) core / cache arrays (V_{CS}), 3) nest & I/O (V_{IO}), 4) PCI (V_{PCI}), 5) stand-by logic (V_{SB}), 6) high-voltage analog (AV_{DD}) and 7) a high-precision reference (V_{ref}). The core+L2 and L3 voltage are partitioned into independent power-gating regions using 5 header columns. A charge pump raises the header gate voltage enabling >1000× leakage reduction over non-gated mode. Multi-stage analog muxing allows measurement of all internal regulated-voltage domains.

Four specialized RAM cells are used: 1) 0.160μm^2 6T with split-wordline for banked-read high-performance SRAMs, 2) 0.144μm^2 6T for density-optimized SRAMs, 3) 0.192μm^2 8T for two-port SRAMs, and 4) 0.026μm^2 eDRAM. Ground-rule clean CAM and register file cells enable efficient architecture support, such as the 8-read, 6-write general-purpose register (GPR) cell.

In addition to increasing core count, POWER8 enhances the core microarchitecture to provide a 1.6× thread and 2× max-SMT per-core performance boost. The core supports 8-way simultaneous multithreading and has a 32KB instruction cache and a 64KB data cache that can support up to 4 loads/cycle. Each core is supported by a private 512KB SRAM L2 and a 96MB shared eDRAM L3. The 22nm eDRAM implements a 2-level bitline hierarchy with 66 cells per bitline to provide a 30% subarray area reduction versus the prior 3-level, 34-cell-per-bitline design. Leveraging 22nm density, POWER8 also integrates in-core hardware transactional memory and on-chip accelerators for cryptography, memory compression, virtual memory management, and data movement, as well as a coherent interface (CAPI) for external hardware acceleration.

POWER8 maintains a balanced system by scaling the nest architecture and IO bandwidths with the increased compute capacity. Each of the 12 cores connects to the constant-frequency fabric via a 192GB/s asynchronous data interface. Eight differential memory interfaces supply 230GB/s sustained memory data bandwidth to the chip. The on-node and off-node SMP buses enable 494 GB/s of chip-to-chip communication that includes data, command, control, error correction, and sparing. Integrated PCIe Gen-3 provides high-bandwidth, low-latency IO.

The clock topology, shown in Fig. 5.1.2, contains 29 domains. A redundant system reference clock drives 12 chiplets × 2 local meshes (1:1 resonant and 2:1 non-resonant), the nest mesh, fast (resonant) and slow meshes for the socket-to-socket SMP links, and a mesh for node-to-node SMP links. To minimize transfer latency, the socket-to-socket SMP mesh can operate synchronously with the nest using a control loop with voltage-regulated programmable delay to align the nest clock to the IO clock within 5ps across PVT without adding delay (and therefore jitter) to the IO clock. A redundant PCI reference clock is used for the PCI mesh which operates asynchronous to the nest clock.

To deliver its 951GB/s (7.6Tb/s) of raw off chip bandwidth (> 160% POWER7™), while reducing total I/O power, POWER8 uses proprietary, pseudo source-synchronous interfaces with a form of clean-up exhibiting a very high tracking bandwidth of at least 150 to 200MHz. The source synchronous clocking is run as double data rate (DDR) or quad data rate (QDR). The memory interface is 5pJ/b in functional mode with a 20dB reach at Nyquist. It invokes all equalization in the receiver [4] and DFE-1 (decision feedback equalization) as indicated in Fig. 5.1.3 and runs 1.5× faster than POWER7+, while consuming 50% less power. All proprietary interfaces have the feature of dynamically providing constant bandwidth even under lane fault conditions. All differential interfaces invoke T-coils for return loss management, as in Fig. 5.1.3. Noise control is achieved by use of deep trench (DT) capacitors, and step down regulators are only used by the PLL's. The write memory interface invokes a low-power 4.8GHz CMOS resonant clock distribution that can drive up to 72 lanes running at 9.6Gb/s. The socket-to-socket interface is single-ended running 4.8Gb/s (1.5× faster than POWER7+) with synchronous data transfer for very low latency.

To maintain the power dissipation of POWER7+ in spite of its large increase in performance and bandwidth, POWER8 invested significantly in power-management innovations. A new on-chip controller (OCC) utilizing an embedded PowerPC™ core with 512KB of SRAM runs real-time control firmware to respond to workload variations by adjusting the per-core frequency and voltage based on activity, thermal, voltage, and current sensors. The on-die nature of the OCC allows for ~100× speed up in response to workload changes over POWER7+, enabling reaction under the timescale of a typical OS timeslice and allowing for multi-socket, scalable systems to be supported.

POWER8 also includes an internal voltage regulation capability (Fig. 5.1.4) that enables each core to run at unique voltage. Optimizing both voltage and frequency for workload variation enables ~50% increase in power saving at ½ frequency versus optimizing frequency only. The distributed regulator includes a central voltage controller (VREGC), which receives an output sense line from the grid, a code representing the target voltage, and an external high-precision reference voltage. To minimize the error between target and output voltages, VREGC transmits a 7b up/down correction code (UPDN) to 64 charge-pump-based microregulators (UREGs), which are distributed throughout the 5 header columns in Fig. 5.1.1 and control the headers using a mix of fast-switching and RC filtered control lines.

Figure 5.1.5 highlights the >18% chip power reduction achieved by key power reduction efforts: cache architecture, V_t and device size tuning, and PHY optimization. Low and high-frequency resonant clocking modes enabled resonating the core mesh between 2.5 to 5+GHz and saved an additional 4% of chip power. Each clock sector includes programmable-strength clock buffers, two identical inductors built from the 3mm-thick UTM, two mode-switches, and a common large decoupling capacitor node. Automatic changes between non-resonant, low-frequency resonant, and high-frequency resonant modes during functional operation are supported with no impact to performance through a gradual 15-step process to turn-on the switches connecting the LC tank to the mesh.

To manage higher integration and technology complexity, POWER8 improved its synthesis, sign-off, and timing methodologies. The new synthesis approach, "large block structured synthesis (LBSS)", incorporated algorithms to structure dataflow, congestion-mitigation techniques, and embedded hard-IP blocks such as arrays. Enhanced gate-level sign-off techniques for timing, power, and noise provided 3-to-10+× speedup over traditional transistor-based sign-off without measurable accuracy loss. In turn, these capabilities reduced the number of custom, transistor-level, digital blocks by 3× relative to POWER7 and allowed dramatic growth in macro size. Entire units such as the instruction fetch unit (IFU) and vector scalar unit (VSU) were designed as flat synthesizable objects, and the overall number of unique blocks on the chip was reduced by 30% relative to POWER7. Statistical timing methods ensured design functionality in the face of both the systematic variations caused by the wide voltage operating range and the high random variability inherent in 22nm technology.

Results from 1st pass POWER8 silicon at lab conditions are shown in Fig. 5.1.6. Hardware is fully functional over the target voltages across a wide test frequency range, well in excess of 4GHz.

References:

[1] S. Narasimha, "22nm High-Performance SOI Technology Featuring Dual-Embedded Stressors, Epi-Plate High-K Deep-Trench Embedded DRAM and Self-Aligned Via 15LM BEOL," *IEDM Dig. Tech. Papers*, pp. 52-55, 2012.

[2] J. Stuecheli, *et al.*, "Next Generation POWER microprocessor," *Hot Chips 23*, 2013.

[3] S. Taylor, "POWER7+™: IBM's next generation POWER microprocessor," *Hot Chips 24*, 2012

[4] T. Toifl, "A 2.6mW/Gbps 12.5 Gbps RX With 8-Tap Switched-Capacitor DFE in 32nm CMOS," *IEEE J. Solid-State Circuits*, vol. 47, no. 4, pp. 897-910, 2012.

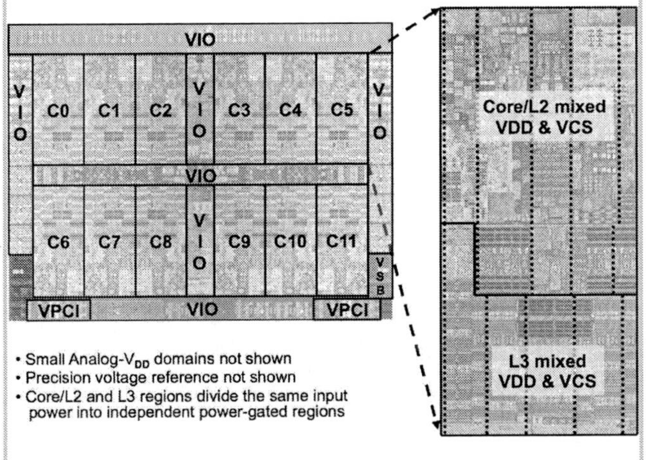

Figure 5.1.1: Voltage regions and chiplet power header columns.

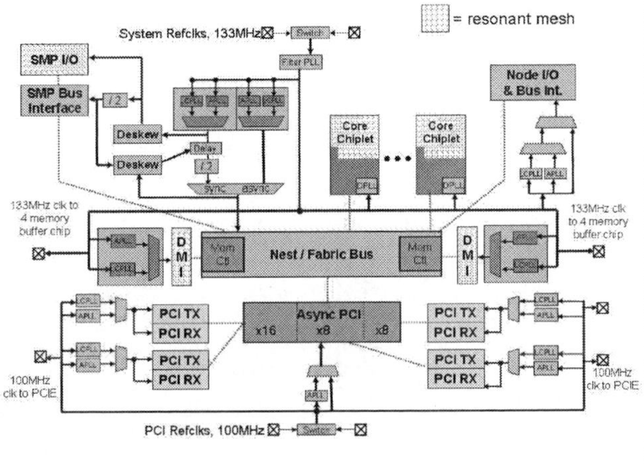

Figure 5.1.2: Clocking topology.

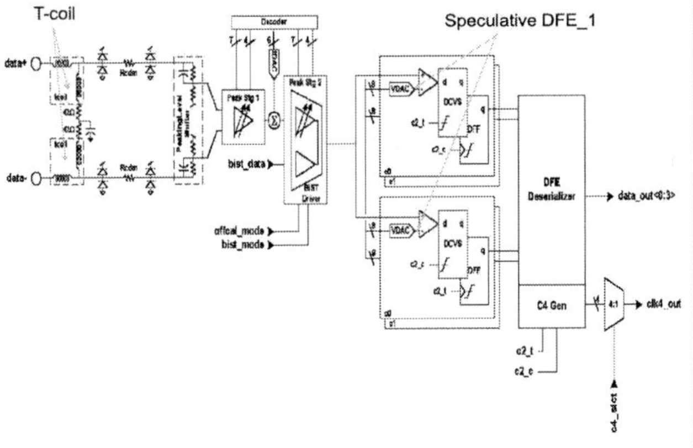

Figure 5.1.3: 9.6GB/s memory interface receiver.

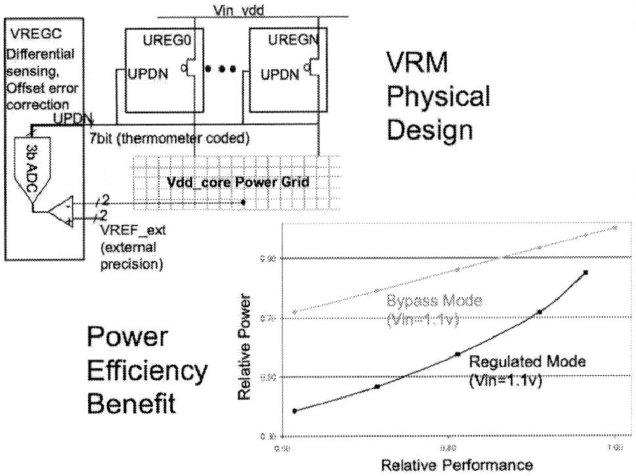

Figure 5.1.4: On-die per-core voltage regulation design and benefit.

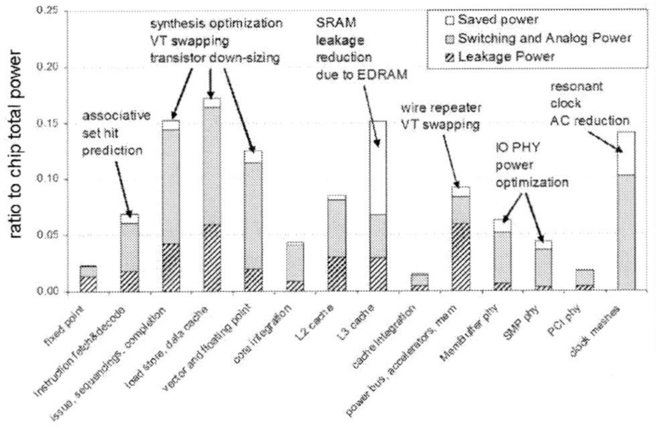

Figure 5.1.5: Power breakdown.

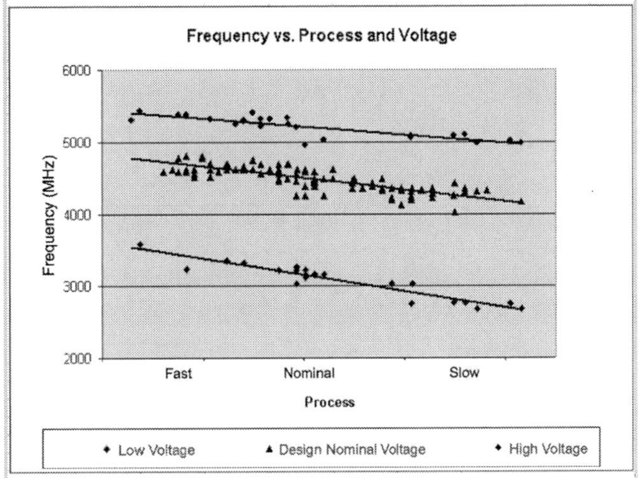

Figure 5.1.6: Frequency versus process and voltage.

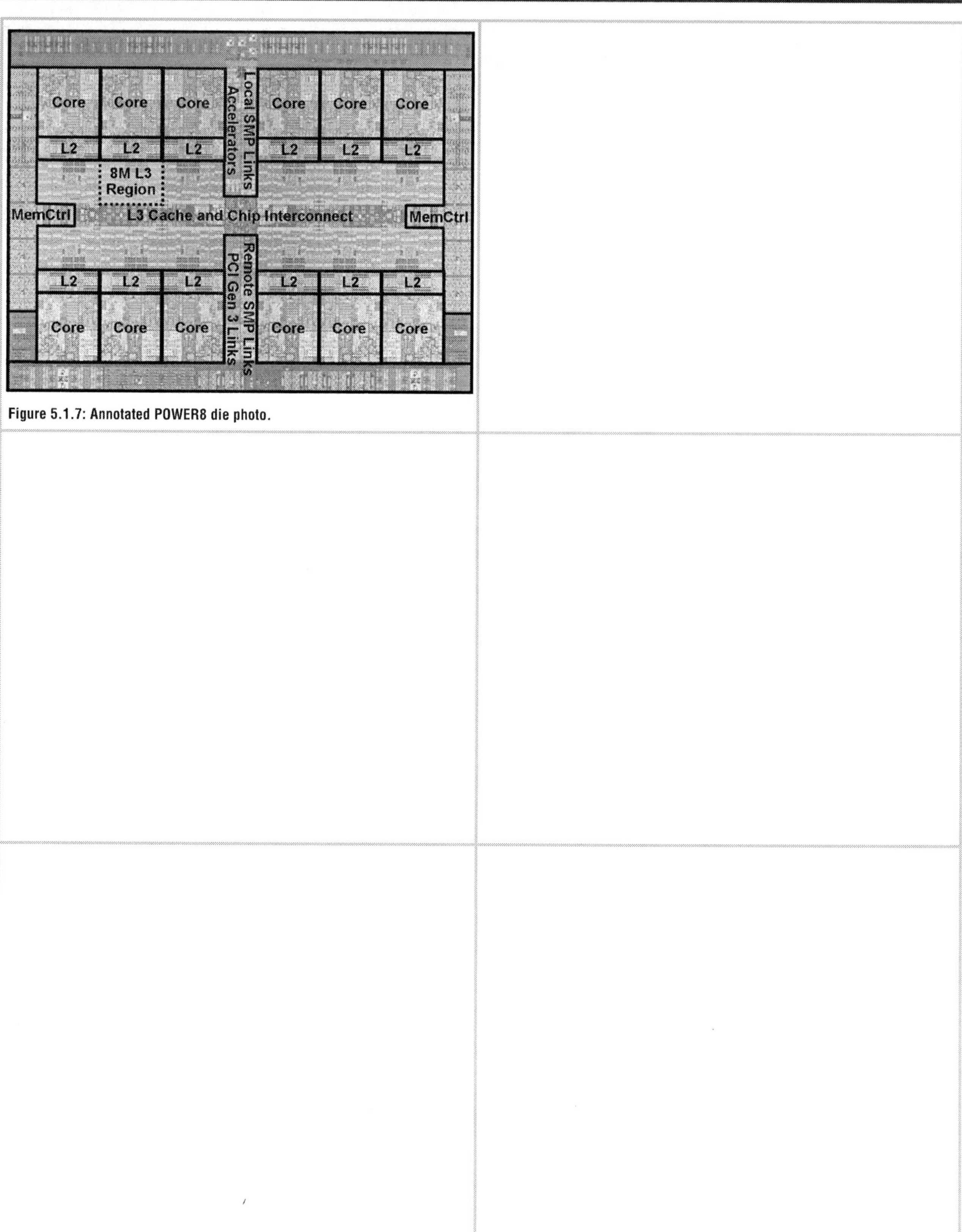

Figure 5.1.7: Annotated POWER8 die photo.

5.2 Distributed System of Digitally Controlled Microregulators Enabling Per-Core DVFS for the POWER8™ Microprocessor

Zeynep Toprak-Deniz[1], Michael Sperling[2], John Bulzacchelli[1], Gregory Still[3], Ryan Kruse[4], Seongwon Kim[1], David Boerstler[4], Tilman Gloekler[5], Raphael Robertazzi[1], Kevin Stawiasz[1], Timothy Diemoz[2], George English[2], David Hui[2], Paul Muench[2], Joshua Friedrich[4]

[1]IBM, Yorktown Heights, NY, [2]IBM, Poughkeepsie, NY, [3]IBM, Raleigh, NC, [4]IBM, Austin, TX, [5]IBM, Boeblingen, Germany

Integrated voltage regulator modules (iVRMs) [1] provide a cost-effective path to realizing per-core dynamic voltage and frequency scaling (DVFS), which can be used to optimize the performance of a power-constrained multi-core processor. This paper presents an iVRM system developed for the POWER8™ microprocessor, which functions as a very fast, accurate low-dropout regulator (LDO), with 90.5% peak power efficiency (only 3.1% worse than an ideal LDO). At low output voltages, efficiency is reduced but still sufficient to realize beneficial energy savings with DVFS. Each iVRM features a bypass mode so that some of the cores can be operated at maximum performance with no regulator loss. With the iVRM area including the input decoupling capacitance (DCAP) (but not the output DCAP inherent to the cores), the iVRMs achieve a power density of 34.5W/mm², which exceeds that of inductor-based or SC converters by at least 3.4× [2].

The POWER8™ microprocessor comprises 12 chiplets. Within each chiplet, the power grids for the logic supply (Vdd) and SRAM supply (Vcs) are divided into two regions – one for the main core (Vdd_core, Vcs_core) and one for the L3 cache (Vdd_cache, Vcs_cache). This allows the L3 cache to remain on (for data retention) while the main core is power gated. The 48 regulated domains (4 per chiplet) are powered from 2 external supplies: Vdd_in for the Vdd_core and Vdd_cache domains, and Vcs_in for the Vcs_core and Vcs_cache domains. The power manager (PM), which programs the iVRMs to the desired voltage levels for DVFS, also controls the voltage levels of the external VRMs to maximize iVRM efficiency. Fig. 5.2.1 shows the iVRM systems for the Vdd_core and Vdd_cache domains of one chiplet. The iVRM of each domain is implemented as a distributed system with a single voltage-regulator controller (VREGC) governing the operation of multiple microregulators (UREGs). The input voltage grids, UREGs, and power headers (PFETs) are placed in 5 columns. The UREGs do not receive an accurate DC reference voltage; instead, they receive digital up/down correction signals from VREGC that affect the trip point of each UREG. VREGC compares the regulated voltage (e.g., Vdd_core) at a sense point (VS_{Vdd_core}) on the grid to a programmable voltage derived from a high-precision external reference (V_{REF_ext}) and feeds back a digital code ($UPDN_{Vdd\_core}$) to all the UREGs. This digital distribution of up/down codes is more suitable for noisy processor environments than the analog distribution of up/down currents used in [3]. To optimize iVRM performance over a wide range of operating conditions, PMOS strength (PS) calibration is used to adjust the active width of the regulator passgate. An FSM employing look-up tables predictively calculates the optimum passgate width as a function of core frequency (f_{core}) and input and output voltages – an approach that is inherently faster than the analog calibration loop of [3].

Figure 5.2.2 shows the block diagram of VREGC. To avoid errors due to ground drops, the voltage at VS_{Vdd_core} is sampled differentially and converted to a single-ended signal V_{SAMP} (referenced to local ground) with a S/H. An RC filter in front of the S/H ensures that high-frequency ripple on Vdd_core is not aliased to a lower frequency inside the regulator control loop bandwidth. A similar S/H converts V_{REF_ext} to a single-ended signal, from which is generated a programmable reference level (V_{REFPRG}) set by a 7b code (VID_{Vdd_core}) from the PM. Vdd_core can be programmed with 6.25mV nominal resolution. A preamplifier senses the error between V_{SAMP} and V_{REFPRG}, and its output is converted to a 7b thermometer code $UPDN_{Vdd\_core}$ with a 3b flash ADC. With the preamplifier, a ±4mV error on Vdd_core drives the ADC to full scale. The auto-zeroed (AZ) preamplifier employs a ping-pong architecture in which one amplifier (e.g., AZ1) is in use, while the other (e.g., AZ2) is being offset compensated. Similar circuitry (not shown in figure) is used for offset compensation of amplifiers AZ3/AZ4.

The UREG (Fig. 5.2.3) features a comparator with sub-ns response time [3], which turns a PMOS passgate M0 on and off in a bang-bang fashion. The comparator trip point is tuned for high DC accuracy with a local charge pump (CP), whose output (V_{CP}) serves as a reference voltage for an error amplifier (common-gate stage M1). A current-steering IDAC converts the UPDN code from VREGC to IUP and IDN currents for the CP. If D is the duty cycle of M0 conduction, CP balance is achieved when IUP/IDN=D/(1-D). Since every UREG receives the same UPDN code, the UREG CP voltages are automatically adjusted to ensure equal duty cycles (balanced load sharing) even in the face of comparator offsets. The M1 stage output is amplified to rail-to-rail levels and then level-shifted (LS) to the Vdd_in domain. Driving the M0 gate capacitance with CMOS inverters and gates is power-efficient in modern processes [4]. For a UREG of this power level (≈40× greater than that in [3]), the power overhead of the sensing stages is negligible, which greatly increases current efficiency. Supplementing the fast switching passgate M0 with another passgate M0SL, whose gate is not fully modulated, improves the tradeoff between self-generated ripple and current handling. The slower M0SL gate signal is generated locally within each UREG using a 2nd-order RC filter instead of being globally distributed as in [3]. PMOS strength calibration by the FSM further reduces self-generated ripple by adjusting the active widths of M0 and M0SL to handle the maximum load current without oversizing at strong corners. A binary-weighted code PS_{Vdd_core}<4:0> and a thermometer code PSL_{Vdd_core}<3:0> set the active widths of M0 and M0SL.

The iVRMs were integrated into the POWER8™ chiplets (Fig. 5.2.7) and fabricated in a 22nm SOI CMOS process. The highest-current iVRM (Vdd_core) uses 64 UREGs and 90nF of deep-trench (DT) input DCAP (shared with the Vdd_cache domain). These components and the Vdd_core VREGC occupy about 1% of the chiplet area. The output DCAP (also DT) for this domain is 750nF. Fig. 5.2.4 shows DC measurements of Vdd_core as a function of VID_{Vdd_core} with different loading conditions and values of Vdd_in. High loading is achieved both with custom test code intended to stress the current capacity of the iVRM and by raising f_{core} above its rated operating range. Low loading (>4× reduction in current) is achieved by gating off the clocks of the core. With Vdd_in=1.1V and 0.61V≤Vdd_core≤1.05V, load-regulation error is less than 3mV. With adequate headroom (Vdd_in-Vdd_core>50mV), absolute voltage error (Fig. 5.2.4(b)) is below 9mV, and the variation with Vdd_in is less than 5mV. Fig. 5.2.5 shows the measured power efficiency as a function of Vdd_core (with high load). With Vdd_in=1.1V, the iVRM achieves peak power efficiency of 90.5% supplying 11.9A at Vdd_core=1.03V, at a power density of 34.5W/mm².

Dynamic tracking between Vdd_core and Vdd_cache is important to avoid the overhead of level shifters between domains. Since the output slew rate of each iVRM depends on loading, the reference voltages of the domains are moved in small steps slowly to ensure tracking. Fig. 5.2.6 shows measurements of Vdd_core being moved up and down in 12.5mV steps. When Vdd_core is lowered, the PM decreases f_{core} with a DPLL before updating VID_{Vdd_core}. Because the PM must wait for the DPLL to respond, the downward movement is slower than the upward one. Finally, measurements of maximum core operating frequency (F_{max}) show virtually identical results in bypass and regulated modes for the same values of Vdd_core, indicating that iVRM dynamic performance meets application requirements.

Acknowledgments:
The authors thank L. Acevedo and A. Wu for verification work and the IBM processor design and manufacturing teams for project support.

References:
[1] W. Kim, D. M. Brooks, G.-Y. Wei, "A Fully-Integrated 3-Level DC/DC Converter for Nanosecond-Scale DVS with Fast Shunt Regulation," *ISSCC Dig. Tech. Papers*, pp. 268-269, 2011.
[2] S. R. Sanders, *et al.*, "The Road to Fully Integrated DC-DC Conversion via the Switched-Capacitor Approach," *IEEE Trans. Power Electron.*, vol. 28, pp. 4146-4155, 2013.
[3] J. F. Bulzacchelli, *et al.*, "Dual-Loop System of Distributed Microregulators with High DC Accuracy, Load Response Time Below 500 ps, and 85-mV Dropout Voltage," *IEEE J. Solid-State Circuits*, vol. 47, pp. 863-874, 2012.
[4] P. Hazucha, *et al.*, "A Linear Regulator with Fast Digital Control for Biasing Integrated DC-DC Converters," *ISSCC Dig. Tech. Papers*, pp. 536-537, 2006.

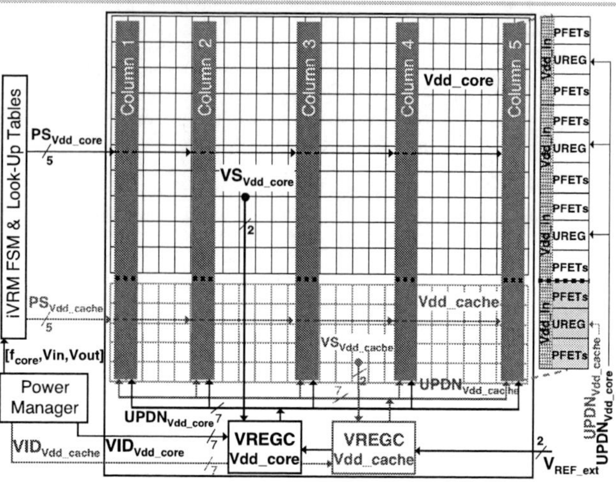

Figure 5.2.1: Distributed iVRMs for the Vdd_core and Vdd_cache domains of a single chiplet.

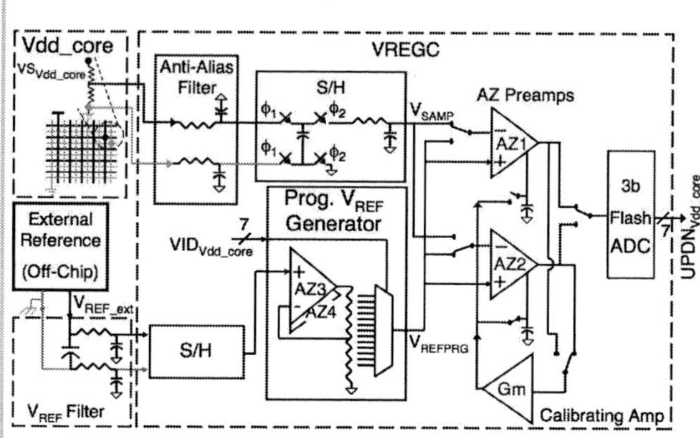

Figure 5.2.2: Block diagram of VREGC for the Vdd_core domain.

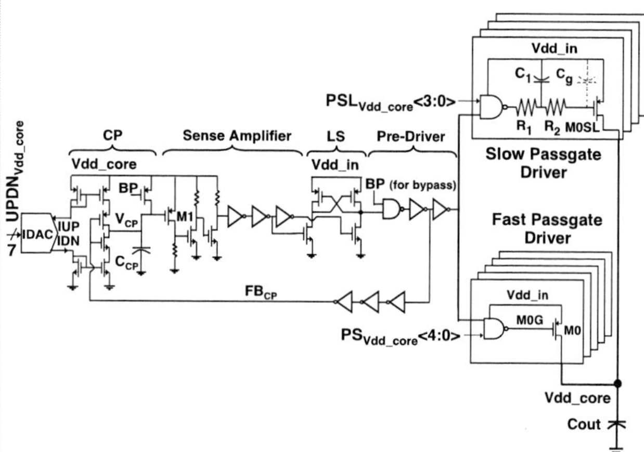

Figure 5.2.3: Simplified schematic of UREG.

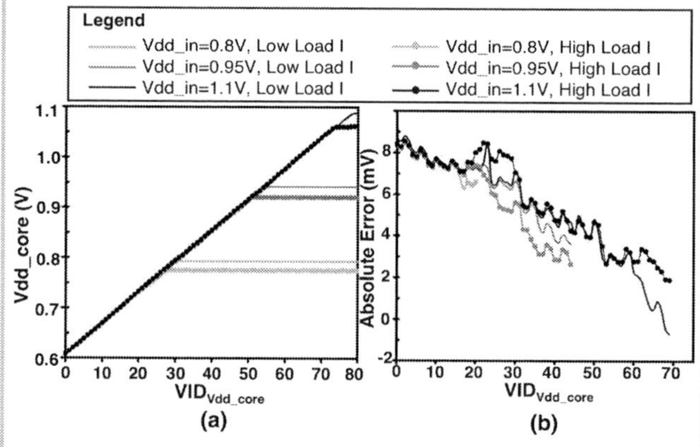

Figure 5.2.4: Measured (a) Vdd_core voltage and (b) deviation from its nominal value as function of VID_{Vdd_core}. Deviation from nominal value is plotted only for cases with Vdd_in-Vdd_core>50mV.

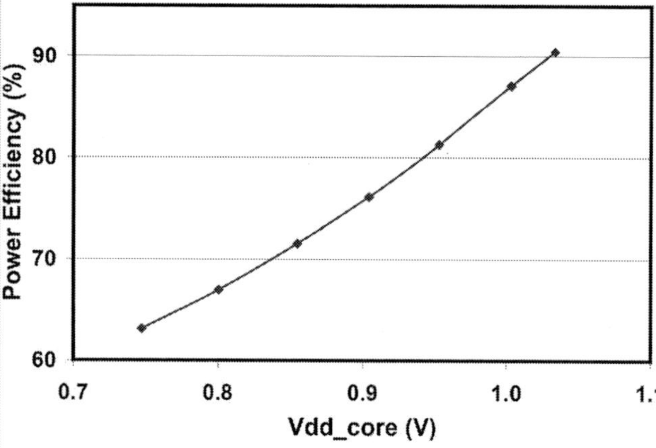

Figure 5.2.5: Measured power efficiency as function of regulated output voltage under high load conditions with Vdd_in=1.1V.

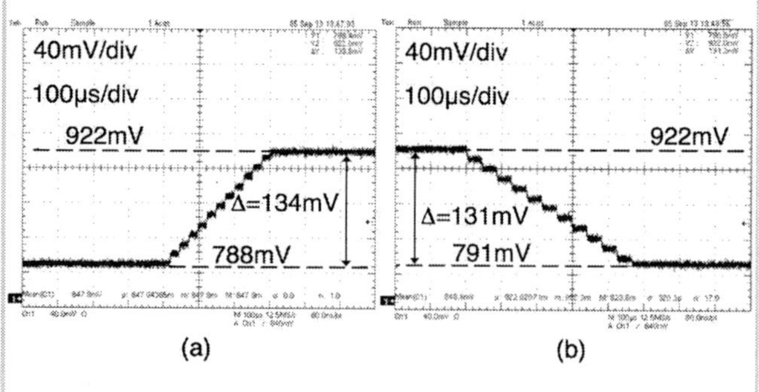

Figure 5.2.6: Measured Vdd_core voltage showing 12.5mV steps in (a) upward and (b) downward directions. At each step, PS_{Vdd_core} is updated by the FSM.

978-1-4799-0917-9/14 $31.00 © 2014 IEEE

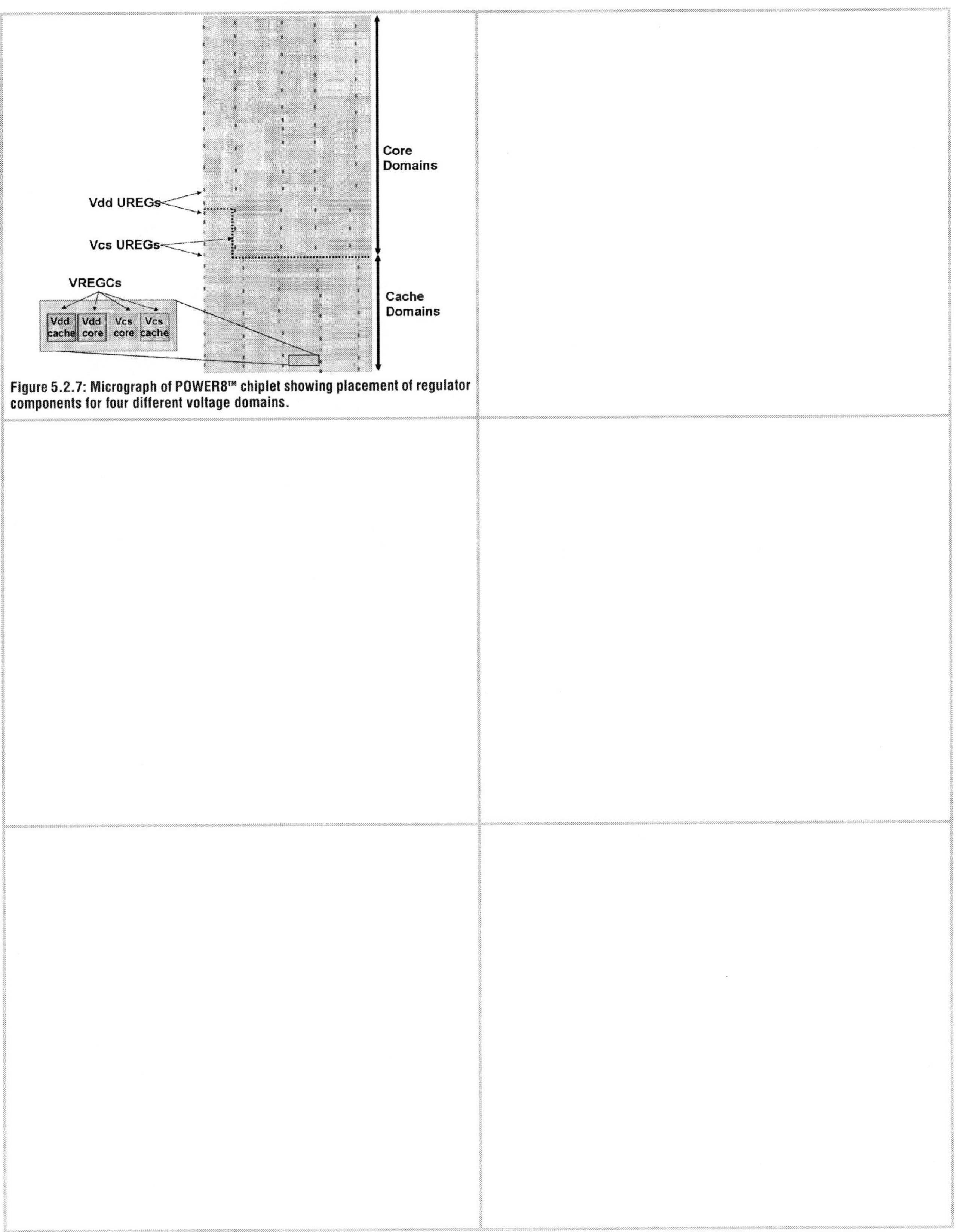

Figure 5.2.7: Micrograph of POWER8™ chiplet showing placement of regulator components for four different voltage domains.

5.3 Wide-Frequency-Range Resonant Clock with On-the-Fly Mode Changing for the POWER8™ Microprocessor

Phillip Restle[1], David Shan[2], David Hogenmiller[2], Yong Kim[2], Alan Drake[3], Jason Hibbeler[4], Thomas Bucelot[1], Gregory Still[5], Keith Jenkins[1], Joshua Friedrich[2]

[1]IBM Research, Yorktown Heights, NY, [2]IBM STG, Austin, TX, [3]IBM Research, Austin, TX, [4]IBM STG, Williston, VT, [5]IBM STG, Raleigh, NC

A resonant-clock design for the IBM POWER8 processor core was implemented with 2 resonant modes (and a non-resonant mode), saving clock power over a wide frequency range from 2.5GHz to more than 5GHz. The POWER8 microprocessor is composed of 12 chiplets, each containing a single resonant clock grid for one core and its L2 cache, and a half-frequency, non-resonant clock grid for the L3 cache. The clock grids drive the local clock buffers (LCBs) that in turn drive the latches. The LCBs are gated off to measure the global clock power from the PLL to the LCBs. The resonant core communicates synchronously with the L3, requiring low skew between the domains. The chip was designed in a 22nm SOI process, including two ultra-thick-metal (UTM) layers (3 microns thick) for power distribution, I/O, all long global clock wires, and the resonant clock inductors. The UTM technology reduces wire resistance and simplifies inductor design, but requires accurate transmission line modeling and special routing.

Resonant clocking has been shown to reduce processor global clock power by using on-chip inductors to recycle some of the energy required to switch the global clock capacitance [1-2]. However, to make resonant clocking attractive for POWER8, resonant clocking must coexist with rapid voltage-frequency scaling over a wide range of supply voltages and more than a 2× range in clock frequency. To support dynamic voltage-frequency scaling, the POWER8 design allows on-the-fly mode changing managed by an on-chip controller. The controller automatically switches between the non-resonant mode (NRclk), the low-frequency resonant mode (LFclk) and the high-frequency resonant mode (HFclk), with no idle cycles or performance degradation. A circuit topology, which merges the mode switches with the resonant capacitor, reduces power in the HFclk mode. Changing from the NRclk to HFclk mode reduces the measured global clock power by 37%, corresponding to 28% power savings compared to the simulated power of a non-resonant baseline design with no inductors. Considering only the final-stage power of the sector buffers (SBs) driving the grid, the peak power savings is 33% over the baseline.

The resonant core and L2 clock grid was designed as 57 sectors, each tuned to drive the clock load distribution within that sector. Fig. 5.3.1 shows the global clock circuits contained in each sector. All sectors are shorted together by abutment of the UTM grid wires in each sector.

A single resonant mode using a single inductor optimized for high frequency could not save power over the required core clock frequency range from 2.5GHz to >5.0GHz. Therefore, two identical inductors and two mode-switches are used in each sector, sharing a decoupling capacitor. In HFclk mode, both switches are closed. In LFclk mode, the resonant frequency is reduced by opening the smaller HF switch. Both switches are opened for NRclk mode.

The top horizontal UTM layer was largely consumed by power distribution, so the inductors use only the vertical UTM layer. Fig. 5.3.2 shows the inductor and sector layouts. The inductors were placed near the mode switches to reduce wire resistance, but could be placed over any circuit, and required only minor modification of the regular power grid to reduce eddy currents. Based on detailed clock load distribution, up to 4 driving locations were chosen in each sector, each driven by a SB. To drive the non-uniform loads in each sector, 9 SB sizes were designed with a 5× range in final device sizes. Each SB has 16 programmable-strength settings to optimize power savings and enable on-the-fly mode changing. The SBs used a non-pulsed design similar to that used on 32nm IBM processors [3] (but with finer strength control), allowing robust resonant operation over a wide frequency range with no local delay circuits. Careful tuning of the wire lengths from the SB to the grid reduced skew and increased resonant power savings by decoupling the SB driver from the more sinusoidal resonant grid waveforms. All global clock UTM wires were shielded and length-tuned using a special-purpose router to control transmission line effects [4].

Supporting on-the-fly mode changing with no impact on F_{max} required less than 1% cycle compression for any cycle during a mode change. This was accomplished through the detailed design of the switches, SBs, and their controls. A common switch and resonant capacitor (C_{decap}) design was used in all sectors, containing HF and LF mode switches, each with 16 resistance settings using 16 transmission gates with relative sizes ranging from 1 to 64 (see Fig. 5.3.1). To perform a mode change from NRclk mode to HFclk mode, the sectors pass through the 16 states shown in Fig. 5.3.3. In the first step, the smallest part of each switch in each sector is closed. This allows some current to flow through the inductors to C_{decap}, causing power and latency reductions, and a waveform change. In the next mode-changing step, the next smallest transmission gate is also closed, and so on, until all 16 transmission gates are closed. In about half of these steps, the sector buffer-strength settings are also reduced, maintaining the clock transition time and amplitude in each state, and reducing power. On-chip diagnostic circuits were used to measure the latency reduction for each step with 0.5ps accuracy. Some of the steps do produce a latency reduction of more than 1% of the cycle. However, the control signals were timed such that adjacent sectors made each step in different cycles. This smoothed the cycle compression across 2 or 3 cycles, reducing cycle compressions to less than 1%. To verify that mode changing has no effect on F_{max}, cores were run through 100,000 mode changes while running a high-coverage exerciser, within 1% of F_{max}, with no faults observed. Doubling and quadrupling the cycle compression by skipping ½ or ¾ of the 16 states also produced no observable effect on F_{max}.

Changing modes from NRclk to HFclk reduces the clock latency up to 18ps total, as shown in Fig. 5.3.3, affecting skew between the resonant core grid and the non-resonant L3 grid. By adjusting the latency to the L3 clock grid to the average of the NRclk and HFclk latency, this became ±9ps latency change from the mean, consuming only a small fraction of the total grid-to-grid clock skew margin.

The choice of circuit topology (merging the switches with the resonant capacitor) was important to optimize the HFclk mode. Unlike previous work [2], the mode switches leave the inductors connected to the clock grid in non-resonant mode. The direct UTM connection between the inductor and the clock grid eliminates two sets of via stacks that would be needed if the switches were between the grid and the inductors. In HFclk mode, all wiring and device capacitances associated with the switches become part of C_{decap} and are not parasitic. The parasitic capacitances of unutilized inductors increase global clock power by 6% in NRclk mode and 3% in LFclk mode. However, because UTM allowed small, low-capacitance inductor designs, power savings in the low-frequency LFclk band is still significant. Due to the wide frequency range of the two resonant bands, NRclk mode will rarely be used. The low resistance of UTM also enables larger sectors, with more load capacitance per inductor, reducing both the number and size of the tuned inductors. The mode switches increase the power by 10% in the NRclk mode, and 3.3% in LFclk.

Figure 5.3.4 shows the measured 37% power reduction from NRclk to HFclk mode at 60°C, corresponding to a 28% power savings compared to a simulated non-resonant baseline design or 33% final-stage savings. Fig. 5.3.5 shows the total chiplet AC power reduction including voltage-frequency scaling for both resonant modes. The chiplet AC power reduction reaches a maximum of 7.7% from 3.8 to 4.3GHz, with 5.8% power reduction measured in LFclk mode down to 2.4GHz.

Monte Carlo simulations of the LCBs showed that global clock transition time affects LCB delay variability. Fig. 5.3.6 shows the hardware-measured global clock power vs. the simulated transition time for all modes at the 16 sector buffer strength settings. For the chips tested in detail, no functionality or F_{max} degradation was observed at transition times well above the original design target of 20 to 25ps.

The 12-core chip resonant clock power reduction in HFclk mode approaches the AC power of one core.

References:

[1] S. C. Chan, P. Restle, *et al.*, "A resonant global clock distribution for the cell broadband engine processor," *IEEE J. Solid-State Circuits*, vol. 44, no. 1, pp. 64–72, 2009.

[2] V. S. Sathe, S. Arekapudi, *et al.*, "Resonant-Clock Design for a Power-Efficient, High-Volume x86-64 Microprocessor," *IEEE J. Solid-State Circuits*, vol. 48, no. 1, pp. 140-149, 2013.

[3] James Warnock, *et al.*, "Circuit and Physical Design of the zEnterprise™ EC12 Microprocessor Chips and Multi-Chip Module," *IEEE J. Solid-State Circuits*, vol. 49, no. 1, 2014.

[4] H. Qian , P. Restle, *et al.*, "Subtractive Router for Tree-Driven-Grid Clocks," *IEEE Trans. on CAD*, vol. 31, no. 6, pp. 868-877, 2012.

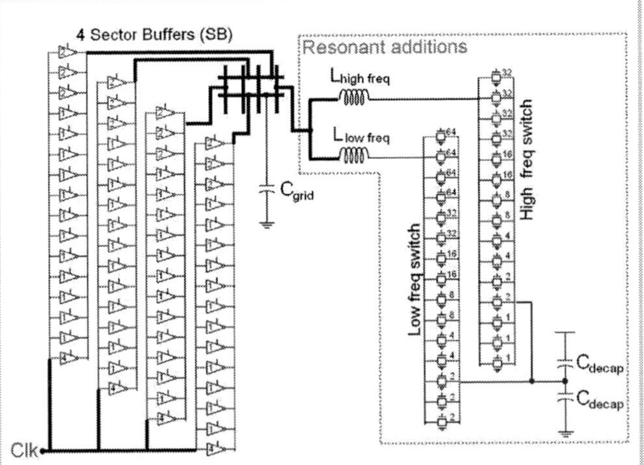

Figure 5.3.1: Resonant circuits in each sector, with relative devices sizes annotated. UTM connections are drawn with wide lines.

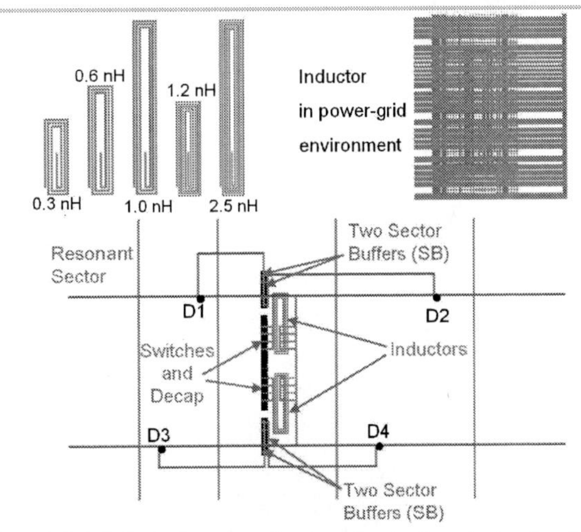

Figure 5.3.2: Inductor layouts and resonant-sector layout, showing 4 driving locations (D).

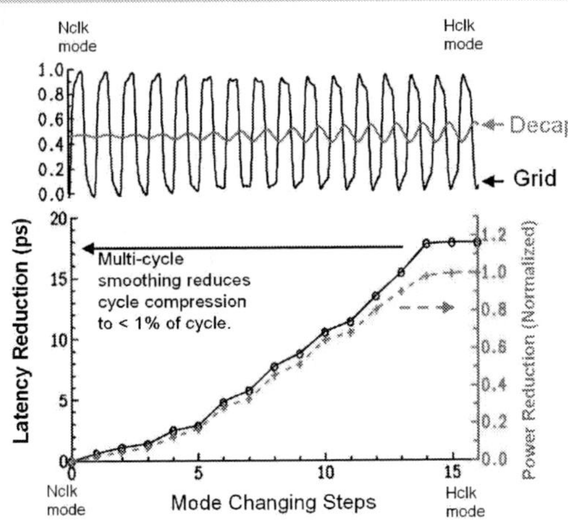

Figure 5.3.3: Simulation waveforms, latency reduction, and power reduction for a mode change from NR_{clk} to HR_{clk} mode.

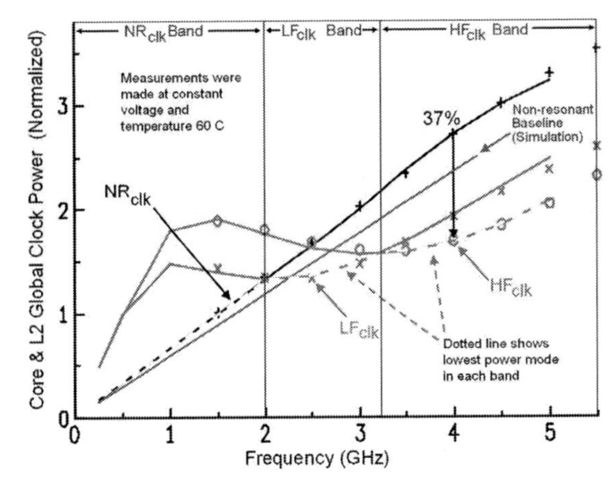

Figure 5.3.4: Measured (points) and simulated (lines) power in all modes, with non-resonant baseline, 60°C.

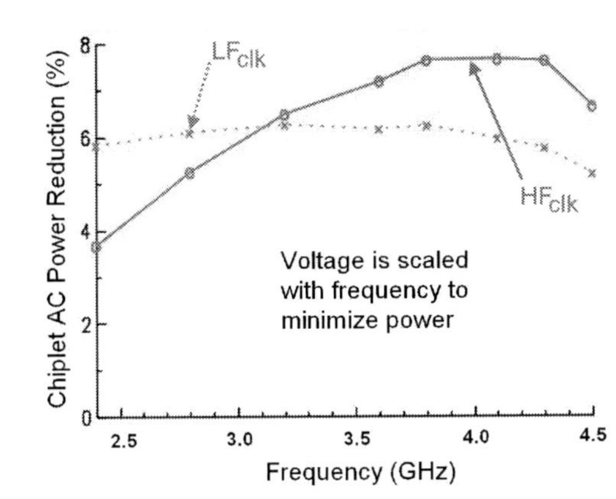

Figure 5.3.5: Chiplet AC power reduction from resonant modes.

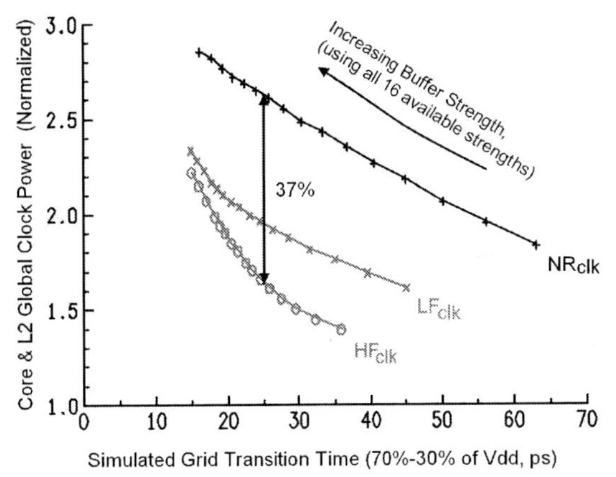

Figure 5.3.6: Measured power vs. simulated grid transition time.

978-1-4799-0917-9/14 $31.00 © 2014 IEEE

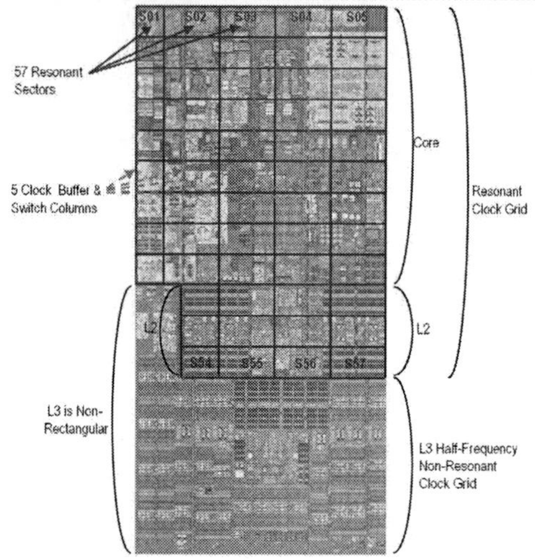

Figure 5.3.7: Die photo of POWER8 chiplet.

5.4 Ivytown: A 22nm 15-Core Enterprise Xeon® Processor Family

Stefan Rusu, Harry Muljono, David Ayers, Simon Tam, Wei Chen, Aaron Martin, Shenggao Li, Sujal Vora, Raj Varada, Eddie Wang

Intel, Santa Clara, CA

The next-generation enterprise Xeon® server processor has 15 dual-threaded 64b Ivybridge cores [1] and 37.5MB shared L3 cache. The system interface includes two on-chip memory controllers, each with two memory channels and supports multiple system topologies. The processor has 4.31B transistors in a high-κ metal-gate tri-gate 22nm CMOS technology with 9 metal layers [2]. The design supports a wide array of product offerings with thermal design power ranging from 40 to 150W and frequencies ranging from 1.4 to 3.8GHz. Fig. 5.4.1(a) shows the processor block diagram. The floorplan (Fig. 5.4.1(b)) is driven by the ring bus routability and latency, as well as the chop requirements to smaller core counts. The cores and associated L3 cache are organized in columns of five, with the ring bus segment embedded. The fully populated die has 15-cores in three columns. The 10-core chop removes the rightmost 3rd column and its dedicated top and bottom IOs. CMOS muxes embedded in the ring bus are programmably operable in a 2-or-3-columns configuration. The 6-core chop removes the 2nd and 4th rows from the 10-core die.

The L3 cache is built from 15 slices, each having 80 data arrays, 2048 sets and 20 ways (Fig. 5.4.2(a)). Each cache line is 64 bytes constructed into 2 chunks. Data arrays use 0.108μm² cells and in-line double-error-correction and triple-error-detection (DECTED) with variable latency. Tag arrays use 0.130μm² cells and in-line single-error-correction and double-error detection (SECDED) with fixed latency. The core-valid least-recently used (CVLRU) arrays use 0.170μm² cells due to their phase read/write operation and in-line SECDED with fixed latency. The data array has both column and row redundancy, while the tag and CVLRU arrays have only row redundancy. Redundant resources are used for both hard defects and low voltage fixes. Programmable transient-voltage-collapse write-assist and wordline under-drive read-assist circuits (RWA) are used [3]. Due to die-to-die variations and the large cache size, one static RWA fuse setting is not sufficient to get the best V_{min} across all parts. Therefore, we use per-die post-silicon RWA fusing, which together with the per-die redundancy flow has improved L3 cache V_{min} by more than 150mV. P-type sleep transistors are used in SRAM cells and word-line drivers for all arrays. Local bitlines in the data array are floated when the array is not accessed to further reduce leakage power. They are pre-charged and equalized one cycle before the bitlines start developing signals for an array access. Fig. 5.4.2(b) summarizes the leakage reduction features. To minimize the dynamic power associated with waking up the arrays, way hit and set signals are used to qualify SRAM, local bitline float and word-line wake up. Only 1.25% (1/80) of SRAM and 2.5% (1/40) of wordline drivers and bitlines are brought out of sleep one cycle before each array access. The shut-off mode minimizes the leakage power in disabled cache slices by lowering the SRAM voltage as low as 0.056V, which saves approximately 87% of leakage power.

The processor clocking architecture includes 13 PLLs as shown in Fig. 5.4.3. There are three clock entry points shorted together in the package. Each column of cores has a single PLL in the center core to save power and minimize the clock-crossing deskew points. The post-layout simulated uncore skew across each core column is under 15ps, without any clock compensation enabled. Fig. 5.4.4(a) presents the clock domains and the crossings located in the north and south channels. A zone-to-zone skew budgeting methodology was employed in the uncore timing verification flow. Fig. 5.4.4(b) shows the five power supplies in this processor: one for the cores, L3 caches, ring structure, and core-ring connections; a second supply for the system agent IO support logic; a third for the Intel Quick Path Interconnect® (QPI) and lower-voltage DDR IO logic; a fourth for the DDR IO buffers; and finally, a supply for the on-chip regulated PLL voltages. Level shifters are placed between voltage domains to ensure robust signaling and enable per-part and real-time adjustment of the voltage levels on the major supplies. The design uses lower-leakage transistors in non-timing-critical paths, achieving 63% usage in the cores and over 90% in the non-core area. Overall, leakage accounts for about 22% of the total power at the typical process corner. MIM capacitors [2] in the upper metal stack provide over 20nF/mm² to control the power noise at the natural response frequencies of the die/package system which are <100MHz. The high-speed serial IO's consist of 40 lanes of PCIe (2.5/5.0/8.0Gbps), 4 lanes of direct media interface (DMI) (2.5/5.0Gbps), and 60 lanes of QPI (6.4/7.2/8.0Gbps), with an architecture based on [4]. The measured rms jitter of the clock generator (LC-VCO based) is 0.224ps (integrated from 10MHz to 1GHz). The serial I/O active power is 11pJ/b.

The 4-channel memory interface includes multi-mode support for both DDR3 1.5V and 1.35V operation running at 800 to 1867MT/s, as well as a voltage-mode single-ended (VMSE) interface to a memory extension buffer at frequencies up to 2667MT/s on the same pins. To achieve these data rates at minimal power, we implemented Tx and Rx equalization, bit deskew per rank for read (Rd) and write (Wr) training, offset adjustment per rank for Rd training, closed-loop replica-biased receivers and wide-range delay-locked loops with programmable delay cells. The receiver architecture (Fig. 5.4.5(a)) allows for equalization tuning via a dual-differential-pair input stage and offset tuning combined into a common folded-load output. Receiver measurements indicate 12mV deterministic and 2.8mV random input voltage uncertainty with 1100 input pattern at 2667MT/s. The DDR transmitter is a voltage-mode design with stacked output transistors for EOS protection and a shared-resistor final-stage architecture (Fig. 5.4.5(b)) that reduces output capacitance and area but requires care in minimizing shoot-through current. Transmit linear equalization is implemented by re-tasking some segments of the driver to switch in opposite polarity to the main driver, depending on the data pattern. Jitter specifications are met by placing CMOS clock trees and all receive path circuits on a separate "quiet" supply, filtered from external noise by an on-package inductor and on-die capacitance. VMSE command bus training uses a multiplicative scrambler/descrambler architecture with real-time feedback to the host, enabling VMSE command pins to run at the same rate as the data pins with adequate margin.

The IO pad capacitance (C_{pad}) is a key performance limiter for both serial and parallel links and needs to be measured accurately to ensure compliance with electrical specs. The Ivytown processor employs a C_{pad} metering circuit for both serial and memory links. The measurement is taken at wafer sort to avoid biasing from package traces and enables speed binning and product segment differentiation. A replica IO buffer uses a current source I_{source} to pull up the IO pad node to a voltage level (V_{out}) for a programmable duration of T_{pulse} time (Fig. 5.4.6). The V_{out} is compared to a series of reference voltages V_{ref} until the final V_{out} level is found. The pad capacitance value is calculated as $C_{pad}=I_{source}*T_{pulse}/V_{out}$. Measured C_{pad} values using this circuit are accurate to within 10% of the conventional TDR measurements.

The die has over 30k C4 bumps placed in a pattern synchronized with the die upper-metal layers and capacitance layer. The C4 bump pitch of each IO type is optimized to ensure the IO arrays are not the critical die dimension. The DDR bump pitch is 184μm and the QPI/PCIe bump pitch is 162μm. The processor supports two 2011-land, 40-mil pitch organic flip-chip LGA package options: a 6-8-6 stack, 45.0x52.5mm package supporting four VMSE memory channels, and a 5-8-5 stack, 52.5x51.0mm package enabling four DDR3 channels. The packages include inductive structures to provide filtered power supplies to the IO clock networks. Each package includes an opening in the land array which allows for placement of decoupling capacitors directly opposite the circuits to minimize impedance and enhance effectiveness. In addition to core and cache recovery that are used to improve yields [5], this processor introduces IO yield recovery: IO ports that are not used in a particular package version are clock gated and shorted to ground in the package to reduce power consumption and enhance yields. Design-for-test and debug features include scan, scan-out observability registers, IO loopback and an IO test generator, on-die clock shrink, within-die process monitors, multiple TAP controllers and voltage droop detectors and inducers. A programmable on-die engine tests all caches and a secondary path dumps all cache content out of the processor. Fig. 5.4.7 shows the die photo.

Acknowledgement:
The authors gratefully acknowledge the work of the talented and dedicated Intel team that implemented this processor.

References:

[1] S. Damaraju, *et al.*, "A 22nm IA Multi-CPU and GPU System-on-Chip," *ISSCC Dig. Tech. Papers*, pp. 56-57, 2012
[2] C. Auth, *et al.*, "A 22nm High Performance and Low Power CMOS Technology Featuring Fully-Depleted Tri-Gate Transistors, Self-Aligned Contacts and High Density MIM Capacitors," *IEEE Symp. VLSI Tech.*, pp. 131-132, 2012
[3] W. Chen, *et al.*, "A 22nm 2.5MB Slice On-Die L3 Cache for the Next Generation Xeon® Processor," *IEEE Symp. VLSI Circuits*, pp. C132-C133, 2013.
[4] F. Spagna, *et al.*, "A 78mW 11.8Gb/s serial link transceiver with adaptive RX equalization and baud-rate CDR in 32nm CMOS," *ISSCC Dig. Tech. Papers*, pp. 366-367, 2010
[5] S. Rusu, *et al.*, "A 45nm 8-core enterprise Xeon® processor," *ISSCC Dig. Tech Papers*, pp. 56-57, 2009.

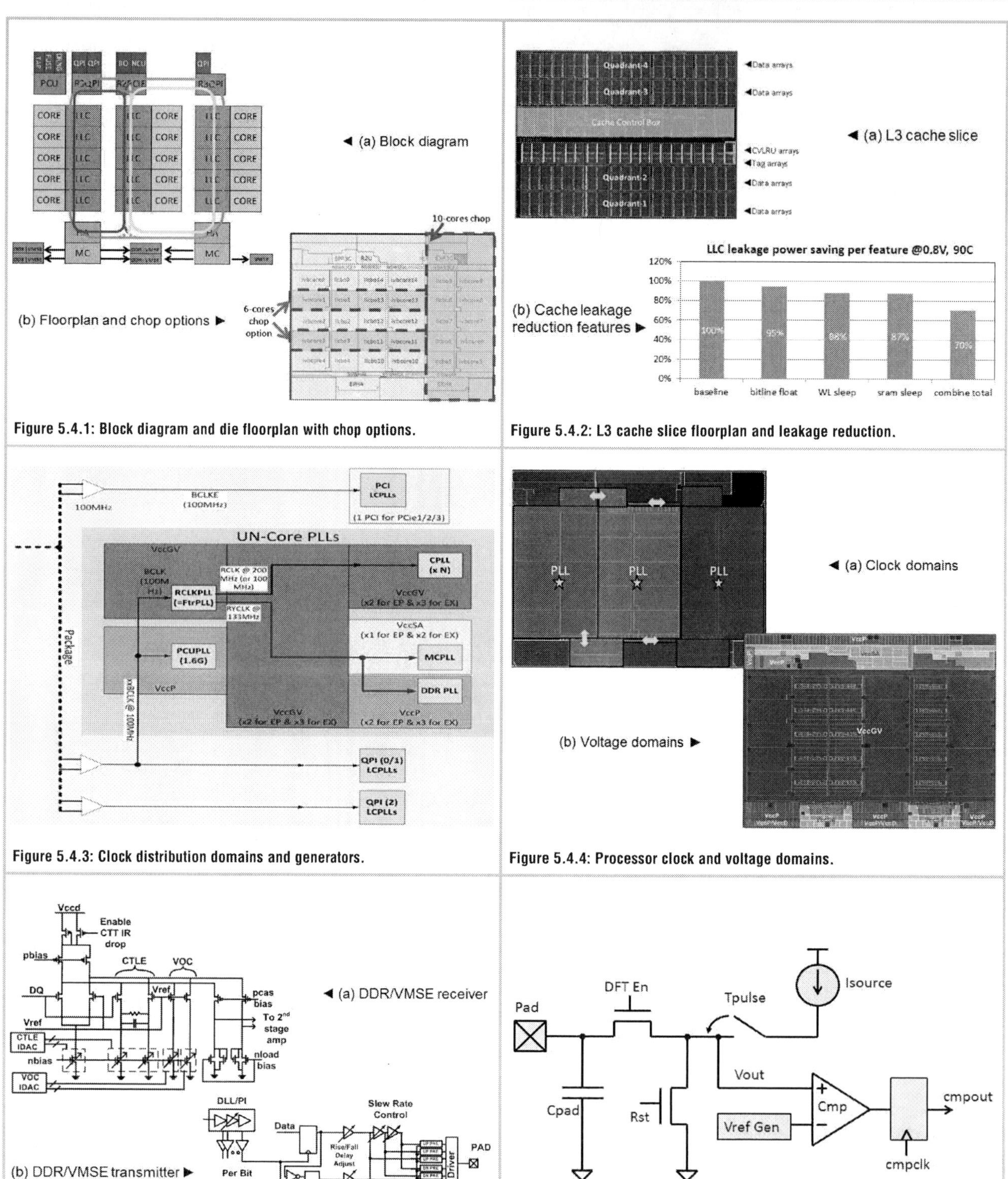

Figure 5.4.1: Block diagram and die floorplan with chop options.

Figure 5.4.2: L3 cache slice floorplan and leakage reduction.

Figure 5.4.3: Clock distribution domains and generators.

Figure 5.4.4: Processor clock and voltage domains.

Figure 5.4.5: DDR/VMSE IO circuit architecture.

Figure 5.4.6: C_{pad} measuring circuit.

Figure 5.4.7: Ivytown die photo.

5.5 Steamroller: An x86-64 Core Implemented in 28nm Bulk CMOS

Kevin Gillespie[1], Harry R. Fair III[1], Carson Henrion[2], Ravi Jotwani[3], Stephen Kosonocky[2], Robert S. Orefice[1], Donald A. Priore[1], Jonathan White[1], Kathryn Wilcox[1]

[1]AMD, Boxborough, MA, [2]AMD, Fort Collins, CO, [3]AMD, Austin, TX

The AMD two-core x86-64 CPU module, codenamed "Steamroller", contains 236 million transistors implemented in 28nm high-κ metal gate (HKMG) bulk CMOS using 12 levels of metal. It is designed to operate from 0.8 to 1.45V. The CPU module occupies 29.47 mm², which includes two independent integer cores, two instruction decode units and shared instruction fetch, floating-point, and 2MB 16-way L2 cache units (Fig. 5.5.7). Along with the second instruction decode unit, this design includes a larger shared 96KB 3-way instruction cache and a 10KB L2 branch target buffer for improved single-threaded performance and multi-threaded throughput compared to a previous 32nm AMD x86-64 CPU codenamed "Bulldozer" [1].

The module design contains 63 unique custom or compiled macros and 436,770 scan-able flip-flops. The design includes a new error-tolerant (ET) flop family to address the higher soft error rate (SER) in bulk CMOS than in SOI CMOS. A distributed power-gating technique is used throughout the module, including the gating of 4-way (512KB) sections within the L2 cache. Another addition to the design is the power-supply monitoring (PSM) circuits.

Design challenges included a new metallization compared to the 32nm process. The 28nm metal stack included more 1× metal layers than the previous design, which was closer to optimal for the GPU design on the APU SoC (Fig. 5.5.1). The Steamroller design used two threshold voltage devices (LV_t, RV_t) for the majority of the design, with the addition of an increased channel length on the RV_t device. In addition, HV_t devices were used specifically for the distribution of the power-header enable signals. The percentage of different device types for the 32nm design is similar to the 28nm Steamroller design (Fig. 5.5.2).

The flops used in this design were based on flops used in the previous generation [2]. In addition to the dual-clock soft-edge flop, a single-clock soft-edge flop was used as well. The advantage of the single-clock soft-edge flop was a reduction in routing overhead for the clock signals from the gater to the flops. As already mentioned, ET flops (Fig. 5.5.3) were also added to the design due to the move from SOI to bulk. ET flops comprise about 3% of the total core flops. This design showed a 5× reduction in error rate (as measured with neutron beam experiments) in 28nm bulk compared to the default flop. The additional redundancy is needed only on the slave latch due to the fact that approximately 90% of the flops are clock-gated. Additionally, a fault insertion flow was created to identify flops sensitive to SER. The sensitive flops were replaced with a footprint-compatible flop that had an improved failure-in-time (FIT) value on the slave node of about 20% compared to the default flops with little timing overhead. These improved SER flops comprise ~70% of the total flops in the core.

An improved resonant clocking design [3] was incorporated into Steamroller. In addition to the core clock, the L2 clock was also designed to resonate. In the frequency range of 3 to 4GHz, measured Cac savings (defined as $Cac=P_{dynamic}/V^2f$) were 150pF.

Similar to other processor designs, more synthesis and fewer custom macros were used in this design than in previous generations. Based on a total gate count, the percentage of synthesized gates has increased by about 4% from generation to generation on the past three implementations of the core.

Reduction of static and dynamic power consumption is critical in supporting thermally power-constrained products, enabling voltage boost states, as well as enhancing battery life for mobile products. To this end, the Steamroller core module focused significant design effort in these areas. In addition to the Steamroller core improvements, the core is instantiated in an SoC that makes use of an adaptive clocking system to improve power efficiency [4].

Rather than use a power-gating ring as used on previous designs [5], an integrated power-gating solution is implemented that reduces core leakage by 90% when gated at a 5% area penalty. A programmable state machine controls in-rush current during initial power-up of the gated supply by sequencing various amounts of header devices (Fig. 5.5.4). To ease the test burden, a serial status chain validates the header-enable signal is void of stuck-at faults.

Initial estimates projected the second-level cache to contribute as much as 27% of the total core leakage power. Implementation of a traditional retention mode bitcell sleep solution [6] is challenging due to the latency sensitivity of L2 accesses and the lack of advance notice to wake the bitcells. Additionally, bitcell sleep does not reduce active power, which is a substantial component of total power in an L2 with high access rates. A way-based power-gating scheme (Fig. 5.5.5) was implemented to address the power challenges specific to a large and active L2. The 2MB L2 cache is constructed from eight 256KB banks. Each bank is 16-way set associative. Within each bank, groups of 4 ways called "way domains" (WD) are individually power-gated. To reduce both leakage and active power, only the desired numbers of WD are dynamically enabled. This feature is especially valuable for periods of frequent C6 usage in which the cache is not fully utilized. This results in lower core energy per operation for applications such as Blu-ray playback that do not require maximum performance. Power gating by index would require an L2 flush to increase the cache size; however, power gating by way requires only that the tags of the newly powered ways be initialized.

In-rush current must be carefully managed when dynamically increasing the cache size so state is retained in the enabled ways (especially during operation near the bitcell V_{min}). To limit in-rush current, each WD is controlled by four wake signals and one run signal tied to progressively larger headers. The L2 has its own programmable state machine similar to the core solution. Additionally, each WD is divided into two power regions that have independent wake signals. This solution allows way-based power gating to function in down-cache scenarios in which banks are disabled. Each data macro contains two power domains of 4 ways each and a central power domain that is independent of the way gating.

Each tag macro contains 4 ways of tag information in one domain and a central power domain. Simulations show a reduction in total core power of 5% leakage and 0.5% active per gated WD.

Each Steamroller core contains ten PSM circuits distributed within each core pair to allow high-speed monitoring and digitization of the power supply grid during a running workload. The PSM block consists of a 32-stage differential ring oscillator (RO) that is reset at the start of every core clock (cclk) cycle (Fig. 5.5.6). At the end of each clock cycle, the state of the RO is captured and encoded with a priority encoder to generate the least-significant 6b of its digital output. The upper 8b are generated by dual binary counters offset by 180 degrees, which count the number of RO oscillations. On the next cycle, the RO state information is used to select automatically the binary counter that is fully settled to avoid propagating invalid data to the downstream logic. PSM configuration registers are accessed by a test and functional serial interface (JTAG/SPMI), while real-time streaming supply data can be transmitted onto a debug bus and stored in a dedicated L2 cache bank for debug and timing analysis.

Instantaneous measurement of the power supply for the two Steamroller core pairs for a virus workload has been measured (Fig. 5.5.6). The PSM also contains a clock divider for down-sampling and additional functions for calculating the minimum, maximum, average approximation, and a voltage-crossing alarm during a specified polling period.

Acknowledgements:
The authors thank David Akeson, Michael Bates, Steven Bakke, Tom Burd, Rob Dupcak, Ross McCoy, Richard McGowen, Spence Oliver, Jeshuah Sniderman, and all the members of the Steamroller physical design team for their contributions.

References:
[1] T. Fischer, *et al.*, "Design Solutions for the Bulldozer 32nm SOI 2-Core processor module in an 8-Core CPU," *ISSCC Dig. Tech. Papers*, pp. 78-80, 2011.
[2] S. Dillen, *et al.*, "Design and Implementation of Soft-Edge Flip-Flops for x86-64 AMD Microprocessor Modules," *IEEE Custom Integrated Circuits Conf.*, pp. 1-4, 2012.
[3] V. Sathe, *et al.*, "Resonant Clock Design for a Power-Efficient High-Volume x86-64 Microprocessor," *ISSCC Dig. Tech. Papers,* pp. 68-70, 2012.
[4] A. Grenat, *et al.*, "Adaptive Clocking System for Improved Power Efficiency in a 28nm x86-64 Microprocessor," to appear in *ISSCC Dig. Tech Papers*, 2014.
[5] R. Jotwani, *et al.*, "An x86-64 Core Implemented in 32nm SOI CMOS," *ISSCC Dig. Tech. Papers*, pp. 106-107, 2010.
[6] Y. Wang, *et al.*, "A 4.0 GHz 291Mb Voltage-Scalable SRAM Design in a 32nm High-κ + Metal-Gate CMOS Technology with Integrated Power Management," *ISSCC Dig. Tech. Papers*, pp. 456-457, 2009.

978-1-4799-0917-9/14 $31.00 © 2014 IEEE

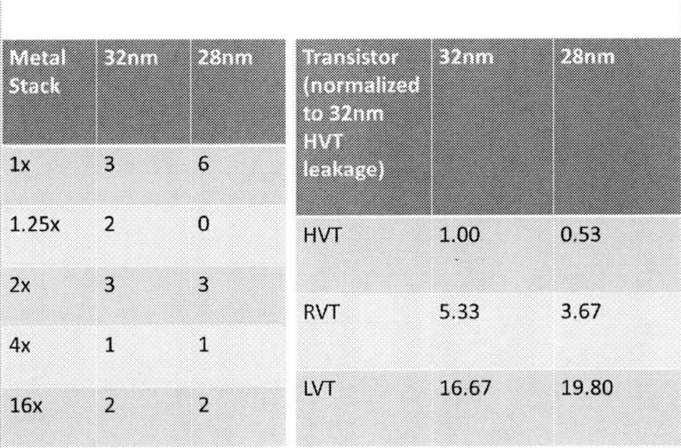

Metal Stack	32nm	28nm
1x	3	6
1.25x	2	0
2x	3	3
4x	1	1
16x	2	2

Transistor (normalized to 32nm HVT leakage)	32nm	28nm
HVT	1.00	0.53
RVT	5.33	3.67
LVT	16.67	19.80

Figure 5.5.1: Metal stack and transistor comparison from 32nm to 28nm.

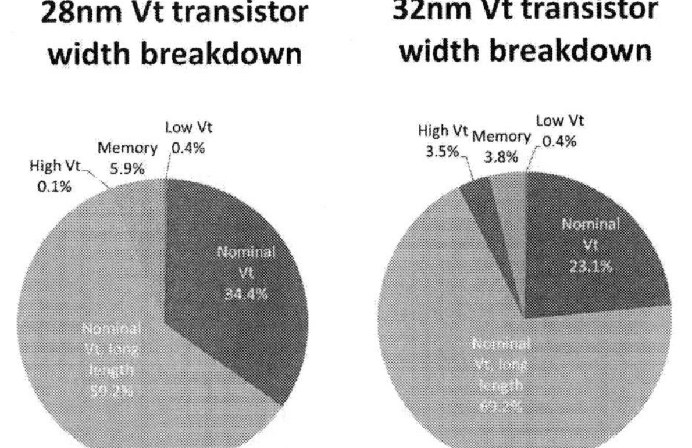

Figure 5.5.2: Transistor type breakdown.

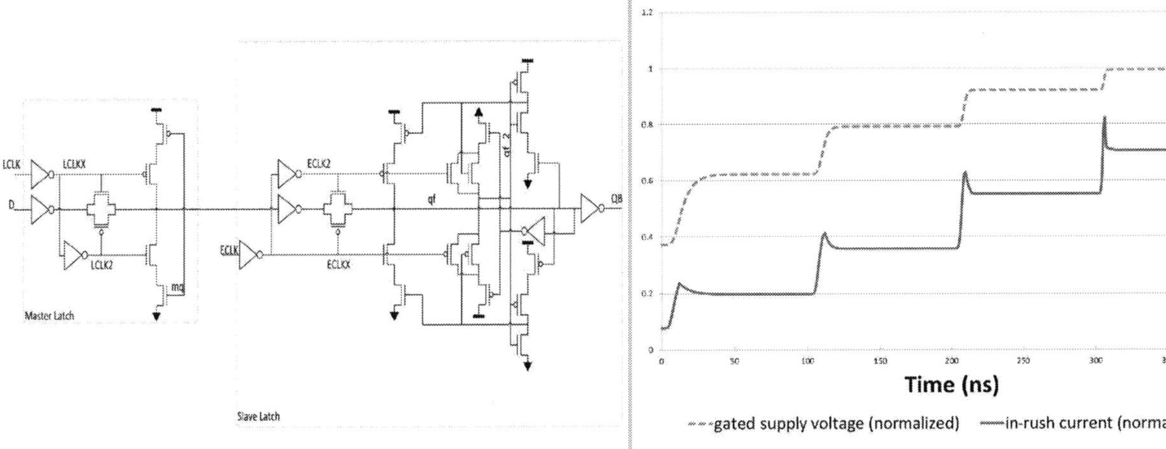

Figure 5.5.3: Error-tolerant flop.

Figure 5.5.4: In-rush example based on one possible power-header state machine configuration.

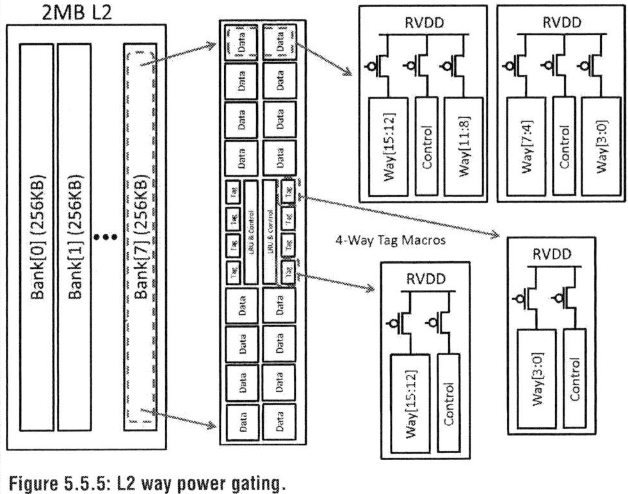

Figure 5.5.5: L2 way power gating.

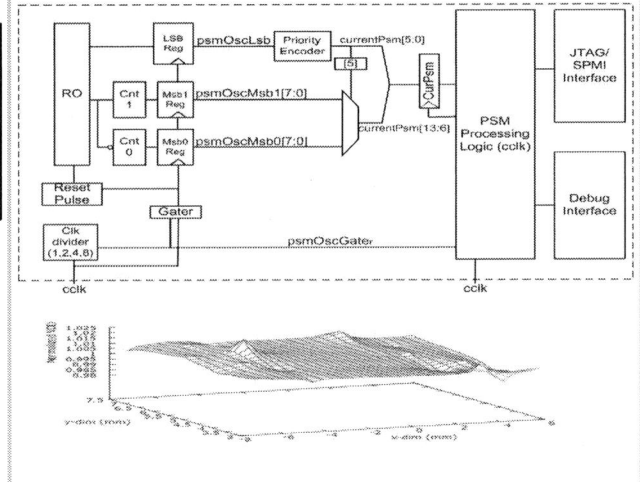

Figure 5.5.6: PSM logic and instantaneous measurement.

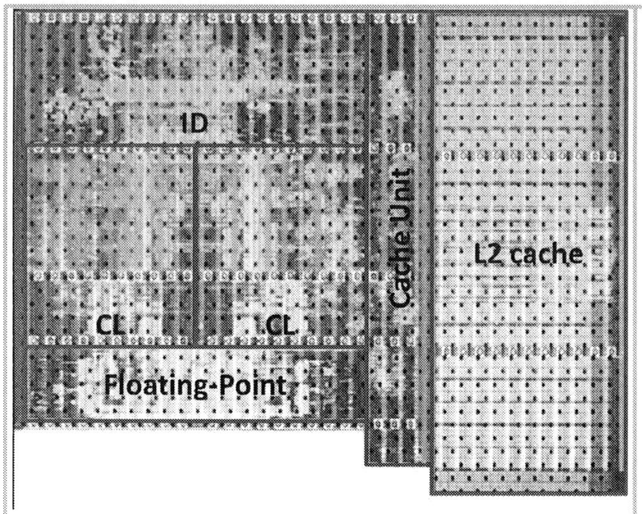

Figure 5.5.7: Die plot.

5.6 Adaptive Clocking System for Improved Power Efficiency in a 28nm x86-64 Microprocessor

Aaron Grenat[1], Sanjay Pant[1], Ravinder Rachala[1], Samuel Naffziger[2]

[1]AMD, Austin, TX, [2]AMD, Fort Collins, CO

In high-performance microprocessor cores, the on-die supply voltage seen by the transistors is non-ideal and exhibits significant fluctuations. These supply fluctuations are caused by sudden changes in the current consumed by the microprocessor in response to variations in workloads. This non-ideal supply can cause performance degradation or functional failures. Therefore, a significant amount of margin (10-15%) needs to be added to the ideal voltage (if there were no AC voltage variations) to ensure that the processor always executes correctly at the committed voltage-frequency points. This excess voltage wastes power proportional to the square of the voltage increase.

Several techniques have been proposed and implemented either to mitigate or compensate for supply noise. The techniques include the following: (1) Improving the supply network impedance by addition of decoupling capacitance or better packaging. (2) Static voltage margining: set the VRM output to be higher by some amount (guard band); this method costs significant additional power. (3) VR load-lining: the VRM supply is increased during periods of low processor activity to provide voltage headroom for potential supply droops. This approach can recover in the range of one-quarter to one-third of the F_{max} margin with the downside of increased power during low-load conditions. This method may be susceptible to workloads with both high average power and large transients [7]. (4) Architectural techniques for increasing the ramp-up or ramp-down time of current surges/drops of the processor by throttling the instructions issued [3, 5, 6]. (5) Adaptive clocking: clock period adjustment in response to supply variation [1, 2]; Intel's PLL clock adjustment in Nehalem appears to fall into this category of supply-droop compensation. These methods adaptively tune the processor clock cycle based on the supply droop. In Intel's solution, the supply voltage of the processor core is mixed into the supply of the PLL VCO, thus increasing the clock period during supply droops and decreasing the clock period during supply overshoots. These five approaches address only part of the droop spectrum, are limited in their efficacy by recovering only a small part of the frequency loss (5), or have significant costs such as packaging costs (1), IPC loss (4) or power increases (2 and 3).

The adaptive clocking technique implemented in this AMD x86-64 microprocessor core addresses both 1st and 2nd droops and can recover a larger percentage of the frequency loss at virtually no cost. The design consists of two circuits (Fig. 5.6.1). A configurable droop detector (e.g., 2.5% or 5% of supply voltage) which detects that the supply is drooping, and a digital frequency synthesizer which increases the clock period by a configurable amount (e.g., 5% or 10%). When a voltage droop is detected, the clock runs at a lower frequency (generated by the DFS) until the supply voltage comes back above the threshold (Fig. 5.6.2). This avoids timing failures due to the decreased voltage, which translates to higher overall frequency. The interface between the two circuits is asynchronous because the droop-threshold crossing can occur at any time relative to the clock edge. Fig. 5.6.7 shows the placement on die.

The droop detector circuit uses a DLL to compare the locked output of a delay line with the phase of the PLL output clock. It has 3 configuration bits available for programming a "droop threshold" below which "stretch" events are triggered, which is used to optimize circuit performance post silicon. To detect a drop in the supply, a programmable phase (determined by the droop threshold setting) of the DLL is compared to phase-0 using a simple phase detector. Whenever the core supply is below the set droop threshold, a stretch signal is asserted that requests the clock-phase picker to pick appropriate phases in the phase generator to generate a stretched clock. This droop-detected signal is asynchronous; it is synchronized using a high-t synchronizer with a configurable number of latches into the picker clock domain to minimize latency in the reaction time to generate a slowed clock. The reaction time to begin stretching the clock is one to three cycles depending on the configuration.

To create an instantly reduced clock frequency, a phase generator with a DLL is used to generate 20 phases of the PLL output clock. The DLL operates at a regulated voltage. To support finer stretch amounts, the 20 clock phases are converted to 40 phases using an interpolator chain. In the event of a droop, a glitch-less clock picker circuit takes the 40 DLL phases and generates a stretched clock by selecting appropriate phases. The clock picker always performs a complete loop through all the phases before selecting the 0th phase going back to the default clock frequency. The additional logic in the clock path does lead to additional jitter measured in silicon as 0.5-1% of cycle time; however, this jitter is encountered only when the clock is stretched and the core already is operating at a lower frequency. Both the phase-generator and droop-detector DLLs were verified to be stable under intentional dynamic voltage-frequency transitions as well as power-supply noise transients and input-clock jitter.

An on-die power supply monitor (PSM) circuit is used to monitor the voltage rail during the stretch event. The PSM is a high-speed time-to-digital converter that counts the number of gate transitions in a clock period [4]. Once calibrated, the PSM result can be translated directly to voltage at the transistor. Fig. 5.6.3 shows voltage at the transistor plotted against time, indicating stretch being triggered as the voltage is falling in a droop event.

The adaptive clocking system provides a benefit of up to 9% V_{min} reduction. Fig. 5.6.4 shows the power benefit across the operating range with the adaptive clocking system enabled when running production speed-binning workloads. At lower voltages and frequencies, the amount of di/dt-induced high-speed transients is less than at higher voltage. This data is taken on a high-power desktop platform. Increases in series inductance in lower-power platforms not designed for such high currents may increase the adaptive clocking system's performance at lower voltage and frequency.

The optimal setting for droop threshold presents a balance between stretching the clock aggressively for even small droop events (thereby reducing the average frequency) and using a higher threshold so stretch events are rare (average frequency is not affected) but less margin is recovered. For typical applications, we determined a droop threshold value of 2.5% to be most beneficial to overall core performance for most workloads. Fig. 5.6.5 displays the amount of stretching incurred with different droop thresholds across workloads. At 1.25%, many workloads spent longer periods in the slowed-clock state, leading to additional performance loss. It also shows average core frequency while running standard application suites. Fig. 5.6.6 shows the system benefit (reduced V_{min}) with programmable "droop threshold" and "stretch amount" parameters.

In summary, this adaptive clocking system enables the reduction of voltage at a given frequency by 3% to 6%, resulting in core power efficiency increasing by 7% to 15% (Fig. 5.6.4).

Acknowledgments:
Visvesh Sathe, Stephen Kosonocky, and Frank Huang

References:

[1] T. Fischer, *et al.*, "A 90nm Variable-Frequency Clock System for a Power-Managed Itanium®-Family Processor," *ISSCC Dig. Tech. Papers*, pp. 294-295, 2005.

[2] N. Kurd, *et al.*, "Next Generation Intel Core Micro-Architecture (Nehalem) Clocking," *IEEE J. Solid-State Circuits*, vol. 44, no. 4, pp. 1121-1129, 2009.

[3] M. Floyd et al., "Introducing the Adaptive Energy Management Features of the Power7 Chip," *IEEE Micro*, vol. 31, no. 2, pp.60-75, March-April 2011

[4] K. Gillespie, *et al.*, "Steamroller, an x86-64 Core Implemented in 28nm Bulk CMOS," to appear in *ISSCC Dig. Tech.* Papers, 2014.

[5] M. Gupta, *et al.*, "An Event-Guided Approach to Reducing Voltage Noise in Processors," *IEEE Design, Automation and Test Conf.*, pp. 160-165, 2009.

[6] V.J. Reddi, *et al.*, "Predicting Voltage Droops Using Recurring Program and Microarchitectural Event Activity," *IEEE Micro*, vol. 30, no.1, pp.110, 2010

[7] X. Zhang, *et al.*, "A Novel VRM Control with Direct Load Current Feedback," *IEEE Applied Power Electronics Conf.*, pp. 267-271, 2004.

[8] J. Tschanz, *et al.*, "Adaptive Frequency and Biasing Techniques for Tolerance to Dynamic Temperature-Voltage Variations and Aging," *ISSCC Dig. Tech. Papers*, pp. 292-293, 2007.

978-1-4799-0917-9/14 $31.00 © 2014 IEEE

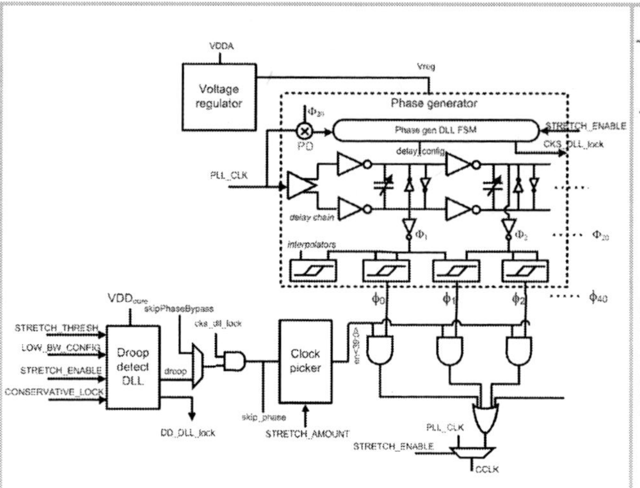

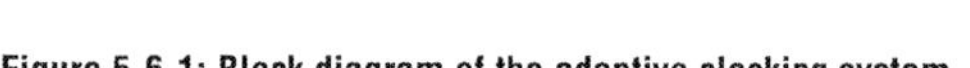

Figure 5.6.1: Block diagram of the adaptive clocking system.

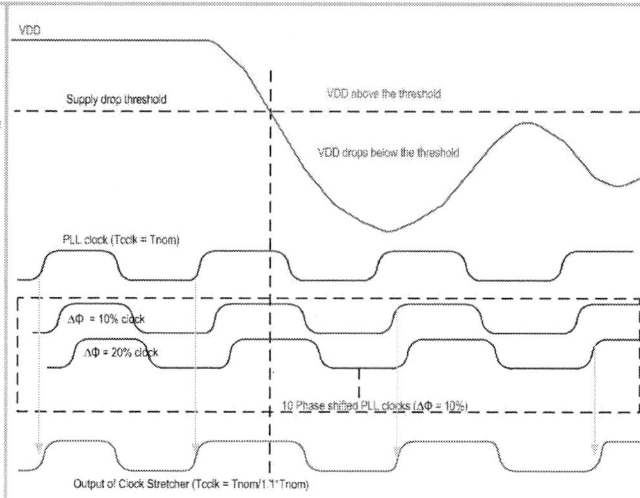

Figure 5.6.2: Principle of adaptive clocking system operation.

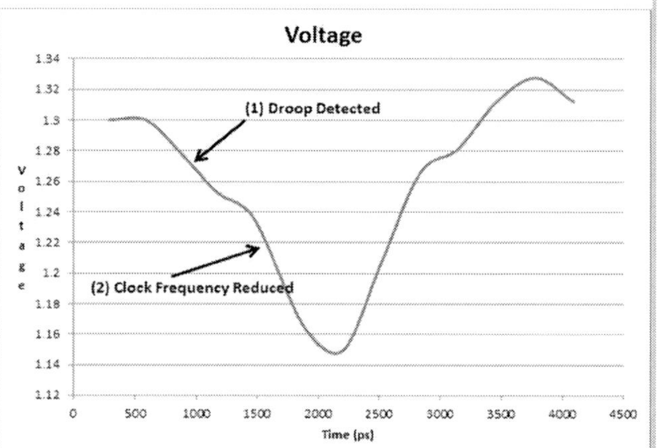

Figure 5.6.3: PSM measurements of the voltage at the transistor.

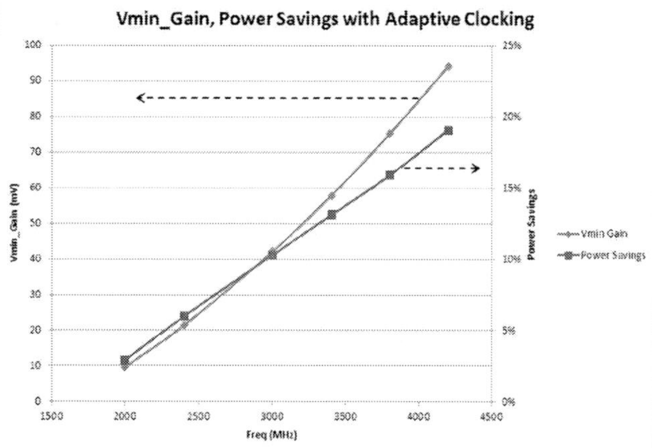

Vmin_Gain, Power Savings with Adaptive Clocking

Figure 5.6.4: V_{min} and corresponding power savings with 7% stretch and 2.5% droop threshold.

DroopThreshold	1.25%	2.50%	3.75%	5.00%
Stretch%	0.50%	0.22%	0.02%	0.0002430%

	CU0-avg-freq	CU1-avg-freq
3DMark	3398	3396
CineBench	3397	3394
iTunes_aac	3396	3393
iTunes_mp3	3396	3394
POVRay	3394	3391
WinRAR	3400	3400

Figure 5.6.5: (Top) Number of cycles stretched while running standard workloads; (Bottom) Average frequency observed on standard apps with adaptive clocking enabled (input frequency = 3400 MHz).

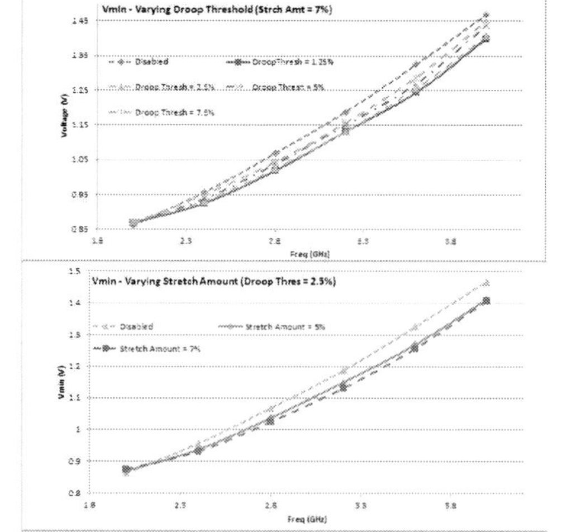

Figure 5.6.6: V_{min} benefit with varying "droop threshold" and "stretch amount" settings.

Figure 5.6.7: Die plot showing adaptive clock IP placement and corresponding cores.

5.7 A Graphics Execution Core in 22nm CMOS Featuring Adaptive Clocking, Selective Boosting and State-Retentive Sleep

Carlos Tokunaga, Joseph F. Ryan, Charles Augustine, Jaydeep P. Kulkarni, Yi-Chun Shih, Stephen T. Kim, Rinkle Jain, Keith Bowman, Arijit Raychowdhury, Muhammad M. Khellah, James W. Tschanz, Vivek De

Intel, Hillsboro, OR

The demand for high-performance graphics capability even in extremely power-constrained platforms such as smartphones and tablets requires circuit techniques that scale from efficient operation at low voltage to high performance when needed. It is well known that energy efficiency improves as supply voltage is scaled down, reaching a maximum near the device threshold voltage where switching energy savings from voltage reduction is balanced by increased leakage energy from frequency loss. Achieving this voltage reduction, however, requires techniques that address intrinsic V_{MIN} limitations in arrays (SRAM, register file arrays, ROMs), voltage droop guardband reduction in logic, as well as techniques for reducing leakage energy, which can dominate at low voltage. It is important that these techniques, while providing energy-efficient operation at low voltage, do not impact the high-performance mode, which is also critical for graphics workloads.

In this paper, we present a low-power graphics processing core that achieves a 40% improvement in peak energy efficiency using dual-V_{CC} arrays, adaptive clocking for voltage droop mitigation, and state retention capability with an integrated retention clamping circuit for low-power sleep mode. The 22nm testchip (Fig. 5.7.1) includes a graphics execution core [1] connected to an SRAM array and test controller used for storage and delivery of at-speed test vectors. Correct execution of the tests is validated through a multiple-input signature register (MISR), which accumulates key signals in the core and generates a 32b signature at test completion.

To mitigate the impact of high-frequency voltage droops, an adaptive clocking technique is implemented to proactively gate or divide the core clock when a droop is detected. This clocking technique (Fig. 5.7.1) consists of an adaptive clock distribution (ACD) block [2], a tunable replica circuit (TRC) at the root of the clock distribution for timing margin detection, and a clock gate/divider to locally gate or divide the clock frequency by two when a droop is detected by the TRC. The ACD extends the clock-data delay compensation effect [3] that naturally occurs during a droop by inserting additional, programmable delay in the clock distribution. The clock-data compensation effect, in which the critical-path slowdown is compensated by clock-period stretching that occurs during the onset of the droop, allows core paths to continue to operate correctly for several cycles while the adaptive clock circuits detect the droop and initiate the clock response. In this work we enhanced the clock response by implementing the clock division mode, which – compared to gating the clock – reduces throughput loss when a droop occurs and also reduces transients on the V_{CC} grid when the clock is ungated. The adaptive clock control maintains the clock response for a programmable time period before returning to full-frequency mode.

Embedded inside the execution core are a 32KB graphics register file (GRF) and a 6KB ROM array used for extended math operations. These arrays typically limit the minimum operating voltage (V_{MIN}) and hence the energy efficiency. As an alternative to sizing up the bitcells in the GRF and ROM to reduce impact of variations on intrinsic V_{MIN}, the dual-V_{CC} approach [4] instead selectively supplies key V_{MIN}-limiting nodes in the arrays with a higher voltage (V_{BOOST}), available as a second supply in the core and routed as a sparse grid for minimal impact on the power and signal routing (Fig. 5.7.2). The power overhead of this selective boosting technique is minimal because the second V_{BOOST} supply does not power the entire array. In read mode, the GRF RWL is boosted to compensate for the impact of V_t variations in the stacked NMOS read port and/or the PMOS keeper on the local Read BL. In write mode, the WWL is boosted to mitigate contention between the bitcell NMOS pass and PMOS pull-up devices. The ROM is also implemented using the dual-V_{CC} approach, where both the selected RWL and column mux input are boosted using embedded dynamic, level-shifting drivers. Active leakage can be mitigated through the use of fast power gating; however this requires local storage for key state nodes to eliminate long latencies for saving and restoring context to always-on storage. For retention of GRF contents during sleep mode, a V_{CC} mux – implemented per local GRF column – allows all GRF bitcells to be connected to V_{BOOST} which acts as an always-on supply (Fig. 5.7.2). The bitcells are also disconnected from the WBLs. For storage of critical state distributed in the execution core, state-retention sequentials (Fig. 5.7.3) isolate the slave storage node during sleep and connect to an always-on "V_{RET}" grid. Power during sleep is further reduced through an all-digital, fully-synthesized active retention clamp design (Fig. 5.7.3) which gradually transitions the voltage of the GRF bitcells and state-retention sequentials to a pre-set retention voltage that guarantees correct retention for the worst-case state element on the die. A hysteretic control maintains the V_{RET} between two reference inputs VrefLow/High, and implements a low-power voltage-to-time converter using a ring oscillator, which is time-multiplexed to reduce leakage power and eliminate variation-induced offset. The clamp is designed to support an extremely wide range of output current on the retention grid, covering more than three orders of magnitude to guarantee operation across process skew, voltage, and temperature.

The 3.38mm^2 testchip (Fig. 5.7.7) is fabricated in a 22nm, tri-gate SoC technology [5] and is validated with test sequences ported from pre-silicon validation. The ability to save context locally in retention flops and in the GRF allows fast power gating for active leakage reduction, showing 8× power reduction compared to clock gating only (Fig. 5.7.4). Further reduction of sleep power is obtained by enabling the retention clamp, improving sleep power savings to 10×, including the overhead power for the clamp operation. Operation of the retention clamp across multiple skewed parts and temperatures demonstrates leakage savings from 4× to 20× while guaranteeing correct flip-flop retention.

Dual-V_{CC} capability allows the graphics core to operate across a wide voltage range (Fig. 5.7.5) by optimizing V_{BOOST} such that arrays do not limit V_{MIN} of the core logic. Here a slight word-line under-drive is used for the baseline case to model the higher V_{MIN} that would be observed with a larger sample size and down-sized bitcell. V_{MIN} improves up to 270mV when boost is employed allowing the core voltage to scale below 0.4V for a test dominated by the GRF. For a test that uses the ROM and GRF, the V_{MIN} is reduced up to 350mV. Failure rate data indicate that small amounts of boost can provide significant V_{MIN} reduction as the tail of the bitcell distribution is compensated with the higher boost voltage.

Adaptive clocking effectively compensates the frequency loss due to fast voltage droops as long as the ACD length is sufficient to provide clock-data compensation for the required response time. While both clock-gating and frequency-division modes achieve this goal, frequency-division demonstrates the best performance and is able to recover 90% of the frequency loss incurred by a 10% voltage droop (Fig. 5.7.6). By nearly eliminating this droop guardband, adaptive clocking improves power at high voltage by 12.4%. Dual-V_{CC} arrays extend the efficient operating range down to 0.38V, where 54% power savings are achieved at 100MHz. The combination of dual-V_{CC} and adaptive clocking improves energy efficiency up to 2.7× at low voltage, with peak energy efficiency gain of 40% GFLOPS/W.

Acknowledgements:

The authors thank K. Ikeda, L. Peake, Jijin T, A. Sandra, T.-H. Foo, L. Avery, C. Parsons, I. Mirza, D. Jenkins, and D. Finan for implementation, B. Matush, J. McCoskey, and W.C. Chee for graphics validation, T. Nguyen and P. Aseron for lab assistance, and R. Forand for encouragement and support. This research was, in part, funded by the U.S. Government (DARPA). The views and conclusions contained in this document are those of the authors and should not be interpreted as representing the official policies, either expressed or implied, of the U.S. Government.

References:

[1] S. Damaraju, *et al.*, "A 22nm IA Multi-CPU and GPU System-on-Chip," *ISSCC Dig. Tech. Papers*, pp. 56-57, 2012.

[2] K. Bowman, *et al.*, "A 22nm Dynamically Adaptive Clock Distribution for Voltage Droop Tolerance," *IEEE Symp. VLSI Circuits*, pp. 94-95, 2012.

[3] D. Jiao, *et al.*, "A Programmable Adaptive Phase-Shifting PLL for Clock Data Compensation Under Resonant Supply Noise," *ISSCC Dig. Tech. Papers*, pp. 272-274, 2011.

[4] J. Kulkarni, *et al.*, "Dual-Vcc 8T bitcell SRAM Array in 22nm Tri-Gate CMOS for Energy-Efficient Operation Across Wide Dynamic Range," *IEEE Symp. VLSI Circuits*, pp. C126-C127, 2013.

[5] C.-H. Jan, *et al.*, "A 22nm SoC Platform Technology Featuring 3-D Tri-Gate and High-k/Metal Gate, Optimized for Ultra Low Power, High Performance and High Density SoC Applications," *IEDM Dig. Tech. Papers*, pp. 4-7, 2012.

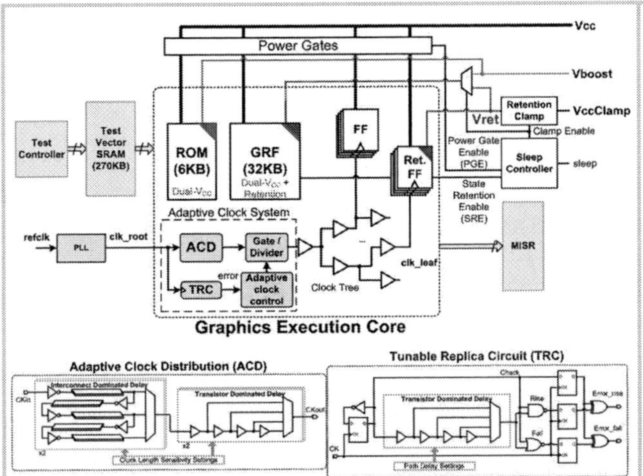

Figure 5.7.1: Block diagram of the low-voltage graphics execution core, and details of the adaptive clock distribution and tunable replica circuit.

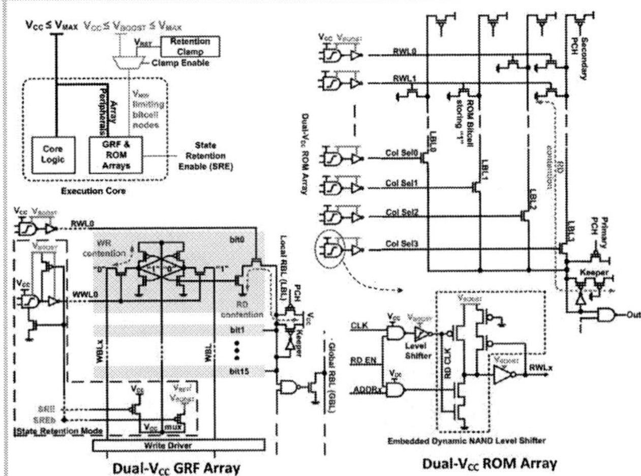

Figure 5.7.2: Dual-V_{CC} arrays design overview, dual-V_{CC} GRF with retention mode, and dual-V_{CC} ROM.

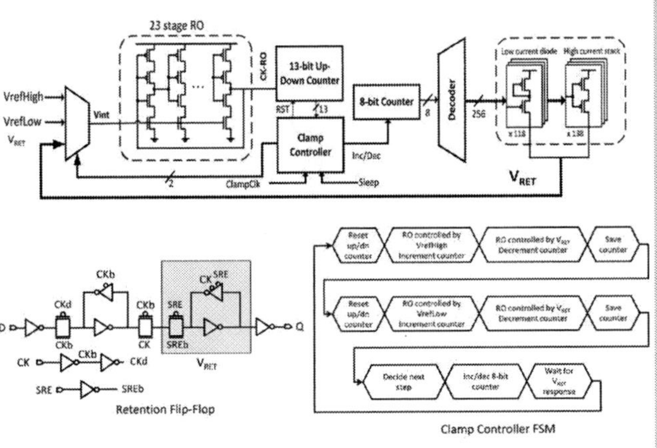

Figure 5.7.3: Overview of the retention clamp design, timing sequence for the clamp controller, and retention flip-flop circuit.

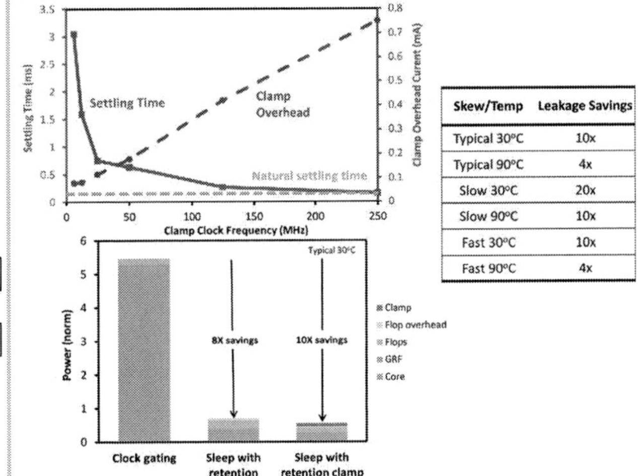

Skew/Temp	Leakage Savings
Typical 30ºC	10x
Typical 90ºC	4x
Slow 30ºC	20x
Slow 90ºC	10x
Fast 30ºC	10x
Fast 90ºC	4x

Figure 5.7.4: Measured standby power savings with retention FFs, GRF and retention clamp. Measured clamp settling time and overhead.

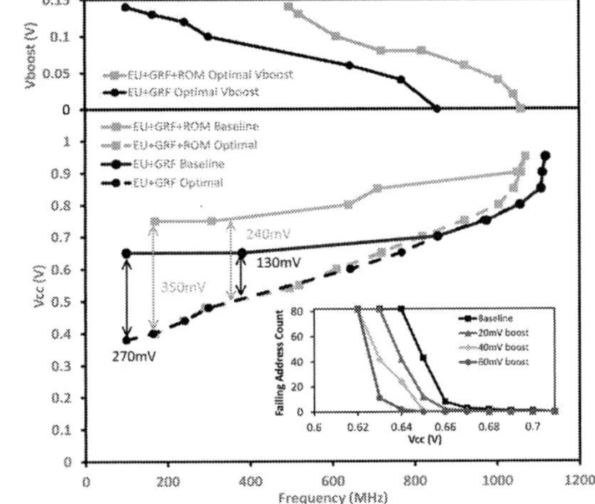

Figure 5.7.5: Impact of selective dual-V_{CC} boost of GRF and ROM on VMin. Read failure rate, obtained with BIST at 600MHz.

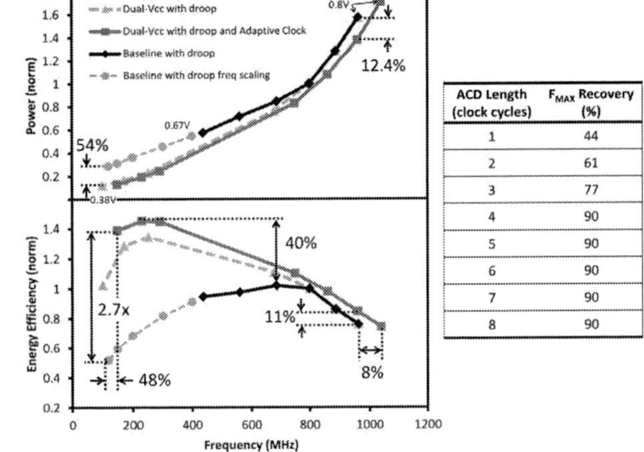

ACD Length (clock cycles)	F_{MAX} Recovery (%)
1	44
2	61
3	77
4	90
5	90
6	90
7	90
8	90

Figure 5.7.6: Measured power and energy efficiency for baseline design, and dual-V_{CC} design with adaptive clocking. F_{MAX} recovery for different adaptive clock distribution lengths.

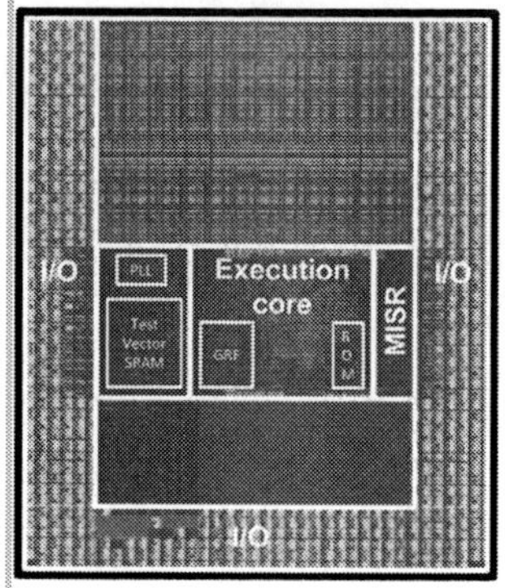

Technology	22nm, 9-metal layer tri-gate high-K/MG CMOS
Area: testchip die	4.0 x 5.8 mm²
Area: core + test	2.6 x 1.3 mm²
Core transistor count	22.8M
Target voltage, frequency	0.7V, 800MHz
Retention sequential count	14,411
Package	FCBGA13 951

Figure 5.7.7: Testchip die micrograph and design details.

5.8 A 3GHz 64b ARM v8 Processor in 40nm Bulk CMOS Technology

Alfred Yeung, Hamid Partovi, Qawi Harvard, Luca Ravezzi, John Ngai, Russ Homer, Matthew Ashcraft, Greg Favor

Applied Micro, Sunnyvale, CA

Potenza is a first generation 64b ARM v8 processor and memory sub-system of the X-Gene™ server platform [1]. The Potenza processor module (PMD) is an integrated design unit, comprising two identical cores sharing a 256KB L2 cache, and is designed to be scalable for different server configurations. Each PMD contains 84 million transistors, occupying over 14.8mm², and averages 4.5W under representative workloads. The initial platform is configured with 4 PMDs, a shared 8MB L3 cache, and 4 DRAM channels arranged around a central switch. Potenza can operate up to 3GHz at 0.9V supply and is fabricated in a 40nm bulk CMOS technology using 10 metal layers (Fig. 5.8.7).

Each core of the PMD features a four-wide, out-of-order superscalar micro-architecture. Execution units are crafted with pipeline designs for concurrent handling of one load, one store, two integer, as well as multiple ASIMD/floating-point operations. Micro-architectural elements include branch predication, separate L1 instruction and data-caches, L1 and L2 data pre-fetch, and hardware table walk. A full set of power management features incorporate techniques ranging from fine-grained clock gating, to software-hinted power states, to full DVFS.

In order to minimize the development cycle for this clean-slate design, a portfolio of circuit components was optimized to be highly reusable. Efficient reuse dictated the inclusion of integrated features to address numerous unique applications, yet effecting negligible power and area overhead.

As device variability poses a major problem in low-voltage sub-micron technologies, use of self-timing techniques and pulse-mode circuits was disallowed [2]. Based on sensitivity, the design was partitioned into different classes for tier-based statistical analysis. The most sensitive circuits, tier-one, were optimized based on Monte-Carlo (MC) analysis, including supply noise. Traditional MC analysis was used to guide tier-two circuits. For typically robust circuits in the lowest tier, convergence was first achieved with PVT-based analysis, after which, designs with less than 15% cycle-time slack were highlighted for MC analysis and hardened to tolerate device variation.

The PMD's principal memory building block is a 2KB RAM (R2K) and incorporates a 0.374μm² 6T SRAM cell. R2K is instantiated over 200 times in a PMD, with numerous applications varying from 32KB L1 instruction cache, 32KB L1 data cache, 256KB L2 cache, and conditional branch predictors. It features built-in column redundancy, byte-wise write mask and bypass, power-reduction clock gating, and native BIST/scan support. R2K is designed to accommodate late-arriving address signals without negatively affecting the L1 cache latency. The late addresses, comprising an 8b one-hot bus, drive the R2K column selects. By crafting these signals to fall monotonically, they can arrive up to half a cycle later than other addresses without incurring speed penalty.

Another memory structure is the 1/2 KB tag ram (TgR). While similar to R2K in its feature set, TgR also includes 8 sets of 9b comparators with per-bit compare masks. Two different implementations of the comparator logic support specific requirements between one-hot signal transitions and speed. The former case addresses the critical path in set-associative memories where the monotonic hit outputs of a TgR are used as column-select signals of the R2K, as previously described (Fig. 5.8.1). The latter case uses a fast dynamic miss implementation for cache hit prediction; in this application, comparator outputs incorporate a logic-integrated flip-flop to complete the prediction logic [3].

In addition to the R2K and TgR, a high-performance phase-pipelined translation lookaside buffer (TLB) (Fig. 5.8.2) is designed to operate in conjunction with the instruction and data caches, as well as the shared L2 cache. The TLB has 20 entries, containing 38b tags and 46b data, and concurrently supports 1 write, 1 read, and 2 content searches. Tag array entries incorporate valid and tagsize bits, while data array entries permit tag-to-data bypass based on the tagsize and write bypass for read/look-up operations. An address translation completes in a single clock cycle, spanning across two arrays while maintaining phase-alignment. In the tag array, reads and content searches occur in the clock-high phase, while invalidates and writes happen in the clock-low phase. These operations are phase-delayed in the data array to prevent hazards during a translation. The tag array uses a two-stage dynamic miss circuit with delayed keepers to facilitate the content search, and adopts a static cross-entry hit detection for optimum speed. Together with R2K and TgR, this optimized TLB provides the low cache latency necessary in high performance processors.

The execution of integer and floating point instructions comprise many PMD critical paths and require access to an efficient, high-speed register file (RF). A single RF design with a superset of features that satisfies both applications is realized in a non-pipelined, 5-read 3-write 80-entry, 68b general purpose memory (Fig. 5.8.3). Key features include full cross-port, full/half-field controllable write bypass, per port clock gating, address conflict detection, and native BIST/scan test support. Cell-writes use a single-ended structure to reduce area and track congestion. Cell-reads use a low-swing differential topology to provide excellent noise rejection, less sensitivity to device variation, and the required low latency. Read effectively operates at 4GHz to support data distribution external to the RF.

A large set of library cells were designed to provide improved performance at negligible area and power overhead. A family of flip-flops (FF) capable of integrating complex logic was implemented by utilizing an efficient master latch and built-in dynamic stage [4]. The resulting design integrates logic into the FF while dissolving the associated latency, as illustrated by a FF which incorporates an 8-way mux function (Fig. 5.8.4). While logic-integrated FFs are especially beneficial in feedback paths, where time-borrowing is unfeasible, they are also incorporated into critical timing paths which results in 20% overall frequency improvement. Other library cells include high performance compressor and comparator structures that are critical in arithmetic and wake logic, respectively. Common to these circuits is a low latency XOR design that concurrently generates the XNOR signal at a minimal area overhead, resulting in 50% performance improvements when cascaded.

The high frequency target calls for a low skew/jitter clock design. The prohibition of self-timed circuits necessitates dependence to the clock falling edge, imposing strict requirements on duty-cycle control. A combination of star, H-tree, and mesh clock distribution of common-mode differential (CML) and single-ended CMOS signaling was employed (Fig. 5.8.5). The CML signaling technique is used to deliver the clock to each PMD, which is locally converted into CMOS clocks. This architecture delivers the clock over a distance of 18mm from the PLL, to synchronous clock domains over 15mm², while producing less than 5% skew and minimal period jitter over each PMD. To address duty-cycle distortion due to process variability, operating conditions and aging, a duty-cycle correction and programming circuit (DCC) was added to each processor core. The DCC detects distortion accumulated from preceding elements and dynamically corrects the falling clock-edge while maintaining rising-edge synchronicity of the full PMD. This circuit also has a silicon-programming feature that can deliberately distort the 50% duty-cycle by up to ±6% to favor the critical clock-phase. By effecting the desired distortion, this programming feature has been instrumental to silicon debug of phase-critical timing paths as well as increasing frequency by 150MHz.

The processor uses a mesh-based on-chip power delivery model throughout the stack-up to normalize metal variability. While the top two metal layers provide over 95% of the planar distribution from supply bumps, lower layers balance between vertical resistance to transistors and signal route-ability. Analyses in both frequency and transient domains are used to address inductive voltage drop issues (Fig. 5.8.6). Band-reject filters are placed off-chip to control system impedance up to sub-gigahertz range, while on-chip decoupling and power-grid dictate the impedance in the remaining spectrum. Functional-mode changes are refined in both hardware and software to avoid stimulation of resonant frequencies.

Acknowledgements:
The authors would like to thank the entire Potenza team for their dedication to the success of the project.

References:
[1] G. Singh, "X-Gene™: Applied Micro's 64bit ARM CPU and Soc", *Hot Chips Symposium*, 2012

[2] D. Krueger, *et al.*, "Circuit Design for Voltage Scaling and SER Immunity on a Quad-Core Itanium Processor," *ISSCC Dig. Tech. Papers*, pp. 94-95, 2008.

[3] A.R. Pelella, *et al.*, "Dynamic Hit Logic with Embedded 8Kb SRAM in 45nm SOI for the zEnterprise Processor", *ISSCC Dig. Tech. Papers*, pp. 72-73, 2011.

[4] H. Partovi, *et al.*, "A 3-stage Pseudo Single-phase Flip-flop Family," *IEEE Symp. VLSI Circuits*, pp. 172-173, 2012.

978-1-4799-0917-9/14 $31.00 © 2014 IEEE

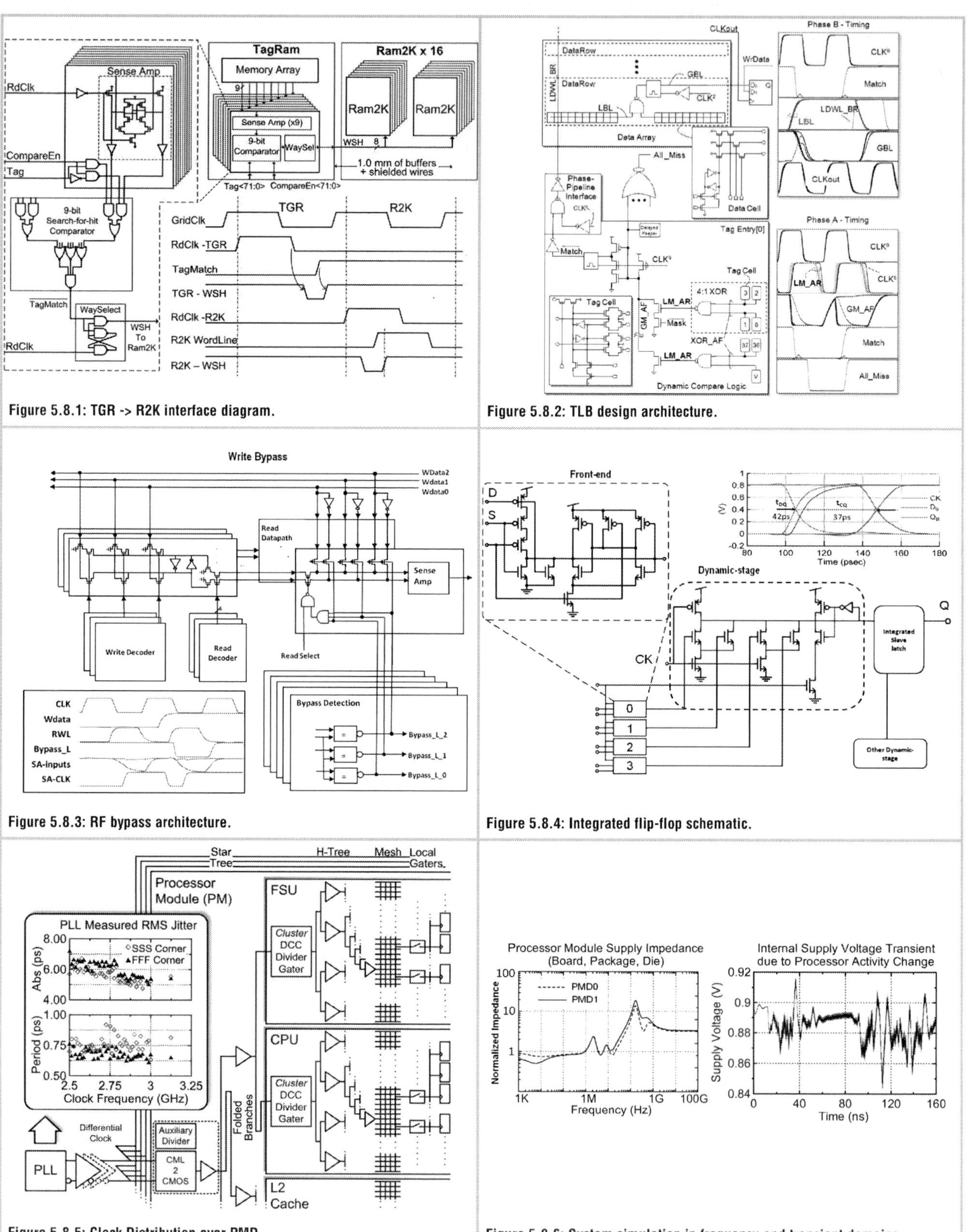

Figure 5.8.1: TGR -> R2K interface diagram.

Figure 5.8.2: TLB design architecture.

Figure 5.8.3: RF bypass architecture.

Figure 5.8.4: Integrated flip-flop schematic.

Figure 5.8.5: Clock Distribution over PMD.

Figure 5.8.6: System simulation in frequency and transient domains.

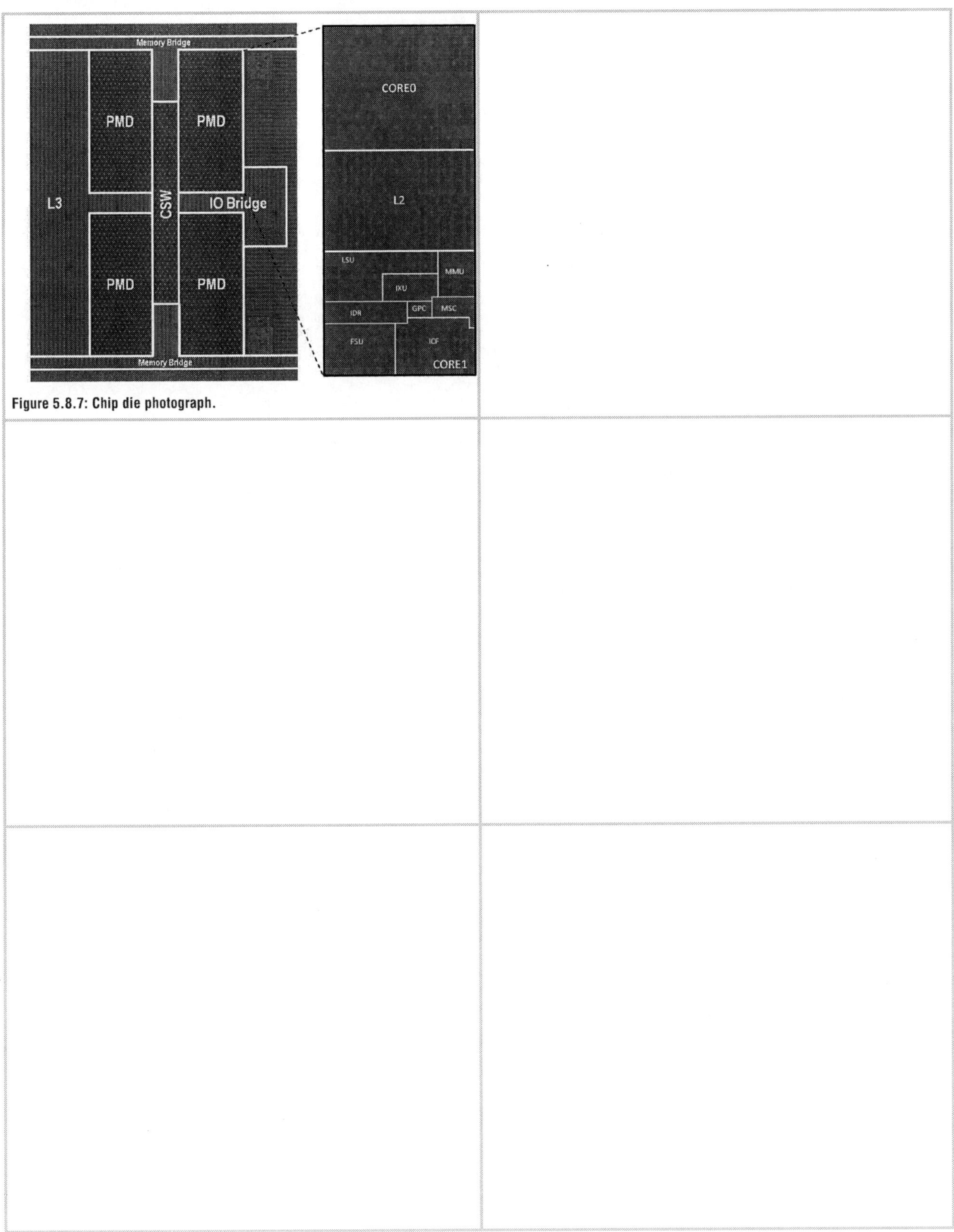

Figure 5.8.7: Chip die photograph.

5.9 Haswell: A Family of IA 22nm Processors

Nasser Kurd, Muntaquim Chowdhury, Edward Burton, Thomas P. Thomas, Christopher Mozak, Brent Boswell, Manoj Lal, Anant Deval, Jonathan Douglas, Mahmoud Elassal, Ankireddy Nalamalpu, Timothy M. Wilson, Matthew Merten, Srinivas Chennupaty, Wilfred Gomes, Rajesh Kumar

Intel, Hillsboro, OR

The 4th Generation Intel® Core™ processor, codenamed Haswell, is a family of products implemented on Intel 22nm Tri-gate process technology [1]. The primary goals for the Haswell program are platform integration and low power to enable smaller form factors. Haswell incorporates several building blocks, including: platform controller hubs (PCHs), memory, CPU, graphics and media processing engines, thus creating a portfolio of product segments from fan-less Ultrabooks™ to high-performance desktop, as shown in Fig. 5.9.1. It also integrates a number of new technologies: a fully integrated voltage regulator (VR) consolidating 5 platform VRs down to 1, on-die eDRAM cache for improved graphics performance, lower-power states, optimized IO interfaces, an Intel AVX2 instruction set that supports floating-point multiply-add (FMA), and 256b SIMD integer achieving 2× the number of floating-point and integer operations over its predecessor. The 22nm process is optimized for Haswell and includes 11 metal layers (2 additional metal layers vs. Ivy Bridge [2]), high-density metal-insulator-metal (MIM) capacitors, and is tuned for different leakage/speed targets based on the market segment. For example, in some low-power products, the process is optimized to reduce leakage by 75% at V_{min}, while paying only 12% intrinsic device degradation at the high-voltage corner.

The CPU incorporates architectural features not present in the prior generation: FMA, transactional synchronization extensions (TSX), hardware lock elision (HLE), increased throughput via 2 dispatch ports, a larger L2 TLB, and other targeted performance enhancements. For similar configurations compared to the previous generation, Haswell improves integer performance by more than 13% (specint*). The graphics engine improvement comes from architectural enhancements and an increase in the number of execution units, achieving over 40% performance improvement vs. the previous generation (specviewperf*). A new category of graphics halo products incorporates a 128MB eDRAM cache die in-package, boosting performance to that of entry/mid-level discrete graphics cards. For Ultrabooks, the PCH is modified for low power and paired with the CPU over a low-power bus in a multichip package (MCP), reducing PCH active power by 33% and standby by ~94%. This, in addition to deeper low-power and low-exit-latency power states (i.e. C9 250µs and C10 500µs-5ms), enables Haswell Ultrabooks to achieve 95% standby power reduction.

Haswell improves the floating-point performance by doubling the maximum number of floating-point operations per clock cycle, and reducing the latency of dependent multiply-add chains by 38%. This is made possible by replacing a 5-cycle FP multiply functional unit with one capable of executing a 5-cycle FP FMA or a 5-cycle multiply, and replacing the 3-cycle FP add with an FMA capable of executing a 5-cycle FMA or a 3-cycle add. This allows two 256b FMAs to begin each clock cycle. Memory bandwidth from the L0 data cache to the execution hardware has also doubled, allowing two 256b loads, and one 256b store per clock cycle. Compared to Ivy Bridge, the addition of the FMA and wider memory-to-execution datapath gives Haswell a >10% performance improvement (specfp*).

To meet the halo graphics segment bandwidth/capacity demands, Haswell extends its cache hierarchy using a high-density eDRAM L4 cache to provide 102GB/s peak bandwidth [3]. Intel eDRAM process technology provides both high-density memory and high-speed logic for IOs. The CPU and eDRAM are in an MCP connected with a full-duplex on-package IO (OPIO), as shown in Fig. 5.9.2. The 128MB array is constructed from eight instances of 16MB macros, which also includes charge pumps and their regulators. The array supports simultaneous read and write accesses. Four data macros (16MB per instance) are accessed in parallel to deliver a cache line.

The OPIO technology provides 4GB/s bandwidth between the CPU and PCH at 1pJ/b, and provides 102GB/s bandwidth between CPU and eDRAM at 1.22pJ/b. By keeping CPU and PCH or eDRAM close together in the MCP (1.5mm), it is possible to simplify IO while providing high bandwidth. OPIO with high-speed single-ended signaling reduces silicon area by 7× compared to DDR3. The OPIO uses no receiver termination but the driver is impedance-matched to the channel. Since OPIO high-speed traces are not exposed outside the package, ESD protection and pad capacitance are reduced by 3×. The transmitter is a CMOS push-pull single-ended driver transmitting at 6.4GT/s using both edges of a 3.2GHz clock. The OPIO data receiver is responsible for sampling, retiming and deserialization of data from the transmitter. Data is received by an inverter with an adjustable P:N drive strength and sampled by a latch driven by differential clocks generated from a DLL. Four clusters, each containing 19 instances of transmitters and receivers for 16b of data, include a valid signal and a forwarded clock per cluster. A request cluster and a low-speed sideband provide commands, address and control for eDRAM.

Haswell products have up to 13 fully integrated voltage regulators (FIVRs): synchronous bucks switching at 140MHz (Fig. 5.9.3). Higher frequency enables LC output filters small enough to be implemented on the CPU using package trace inductors and, in most cases, on-die MIM capacitors. The non-magnetic trace inductors are built using standard package technology. The FIVR's high current density (32A/mm^2) requires a modest silicon area, making it very affordable. The FIVR's have a programmable type-III controller, delivering up to 80MHz unity gain bandwidth (UGB). FIVRs powering CPU cores include a non-linear transient control, which responds to current surges too fast for the control loop. The 1.7-to-1.8V input rail supplying power to Haswell's FIVRs replaced several of the prior platform's rails. The resulting reductions in platform area, cost, and thickness enable thinner platforms with more features and larger batteries. Since the input voltage rail is used solely to power the FIVR, it can have generous noise and ripple specifications, without the normal power-performance burden. These relaxed voltage specifications lead to an order-of-magnitude reduction in decoupling, which allows the input rail to quickly ramp the input voltage off to save power and quickly ramp back on to resume operation. The ability to quickly and efficiently cycle the input power rail is a key enabler for Haswell's roughly 50% battery life gains. Fig. 5.9.4 shows the efficiency as phases are shed from 16 down to 2. As seen, the composite efficiency stays at about 90% over the full load range. It is important to note the voltage regulator integration allows much larger CPU and graphics architectures to be shipped in smaller platforms. This is enabled by the FIVR roughly doubling the input voltage (halving the current), and delivering the low-output impedance required to run such architectures efficiently.

The Haswell memory controller supports 2 channels of DDR3(L) or LPDDR3, using a 1.2-to-1.5V supply. The transmitter, implemented with thin-gate transistors, is a voltage-mode cascode driver as shown in Fig. 5.9.5. The PMOS predriver is a pulsed level shifter that uses a locally generated V_{ss}Hi supply to set the V_{OL} level, improving performance, power and area compared to prior level shifter designs [4]. The V_{ss}Hi voltage is generated using an on-die digital linear regulator with 97% peak power efficiency. DDR PHY can be configured to support DDR3 or LPDDR3, as well as different platform configurations like DIMMs for traditional form factors or memory down in side-by-side configuration for Ultrabooks. As in prior implementations, DDR uses an analog DLL with phase interpolators to precisely control IO timing. During idle periods, the clock, phase detector and charge pump are turned off but the bias voltages are preserved on the loop-filter capacitors. This allows the DLL to periodically wake up and refresh the bias voltage in a few ns, while reducing idle power consumption by 90% (typical). On-die power gates are implemented to reduce leakage in lower power states. As shown in Fig. 5.9.6, the high-voltage power gate uses stacked transistors to both handle the high voltage and achieve < 1mW leakage. The supply clamps are on the gated supply (V_{DD}QG) near the IO buffers for optimal ESD performance; the clamp is modified to keep the RC timer on the ungated supply, allowing the gated supply to wake up in under 100ns, while maintaining an RC time constant of several µs for a human body model (HBM). Fig. 5.9.7 shows Haswell quad and dual-die micrographs.

References:

[1] C. Auth, *et al.*, "A 22nm High Performance and Low-Power CMOS Technology Featuring Fully-Depleted Tri-Gate Transistors, Self-Aligned Contacts and High Density MIM Capacitors", *IEEE Symp. VLSI Tech.*, pp. 131-132, 2012.
[2] S. Damaraju, *et al.*, "A 22nm IA Multi-CPU and GPU System-on-Chip", *ISSCC Dig. Tech. Papers,* pp. 56-57, 2012.
[3] F. Hamzaoglu, *et al.*, "1Gb 2GHz Embedded DRAM in 22nm Tri-Gate CMOS Technology", *ISSCC Dig. Tech. Papers,* 2014.
[4] N. Kurd, *et al.*, "Westmere: A Family of 32nm IA Processors", *ISSCC Dig. Tech. Papers*, pp. 96-76, 2010.

978-1-4799-0917-9/14 $31.00 © 2014 IEEE

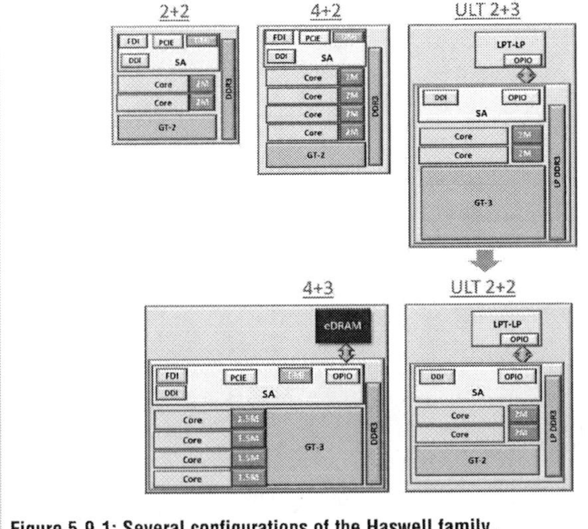

Figure 5.9.1: Several configurations of the Haswell family.

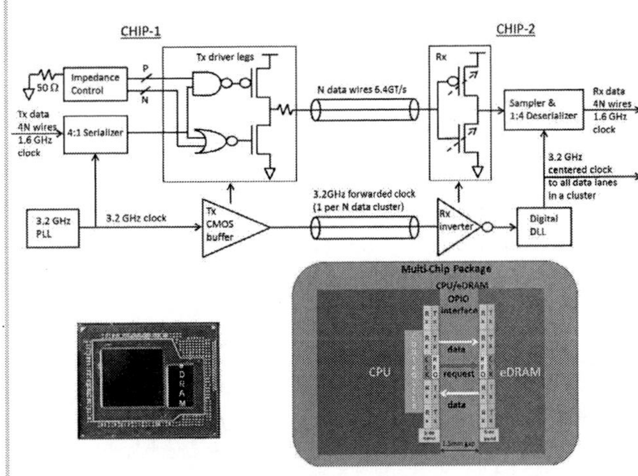

Figure 5.9.2: Haswell CPU-eDRAM MCP package with OPIO.

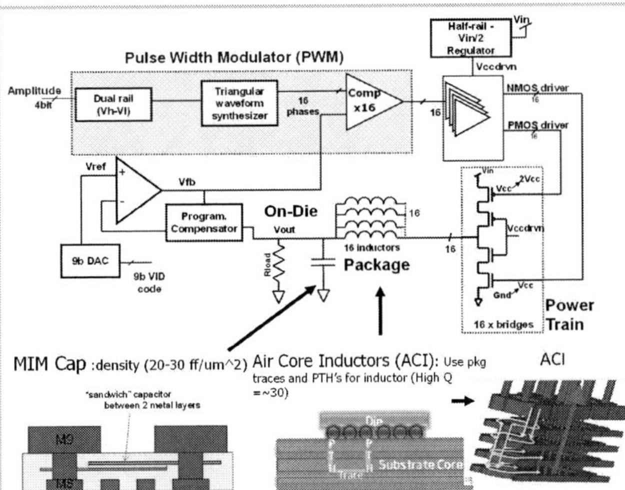

Figure 5.9.3: FIVR controller architecture.

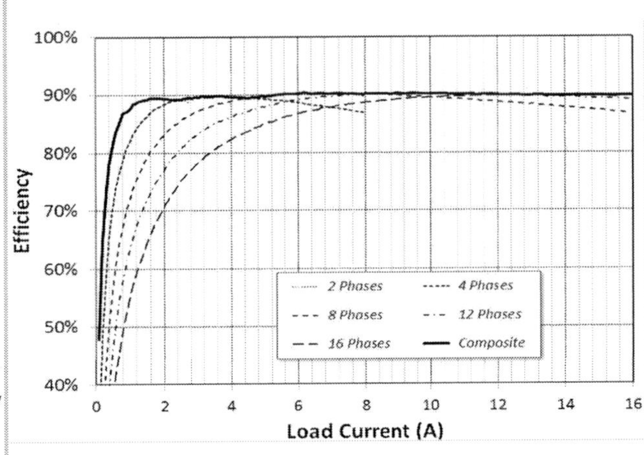

Figure 5.9.4: Measured efficiency for two, four, and eight phases active on a core voltage domain (V_{CCIN} = 1.7V, V_{out} = 1.05V).

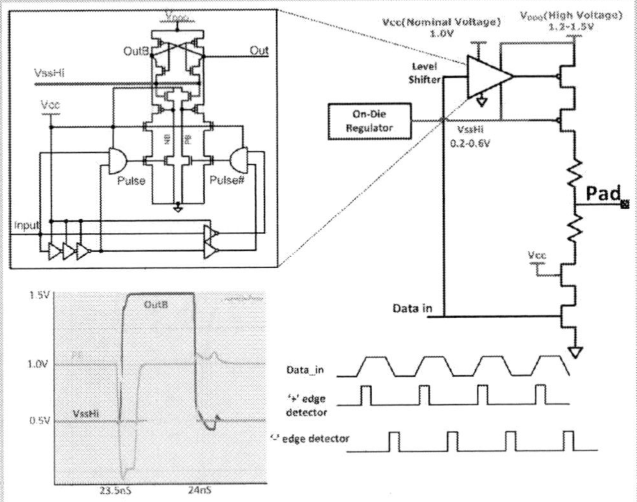

Figure 5.9.5: DDR cascode output driver and pre-driver level shifter.

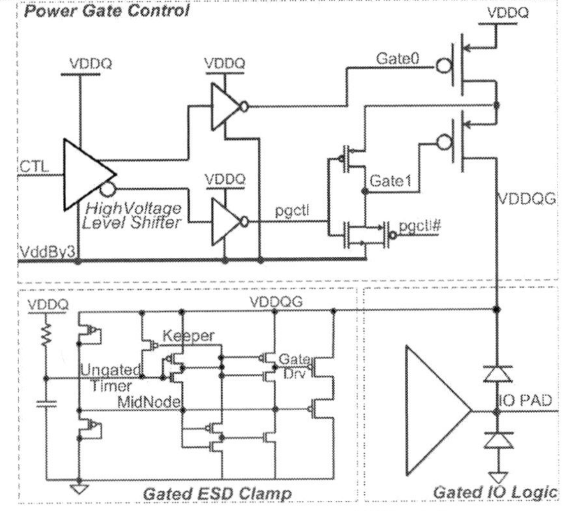

Figure 5.9.6: DDR high-voltage power gate and ESD solution.

978-1-4799-0917-9/14 $31.00 © 2014 IEEE

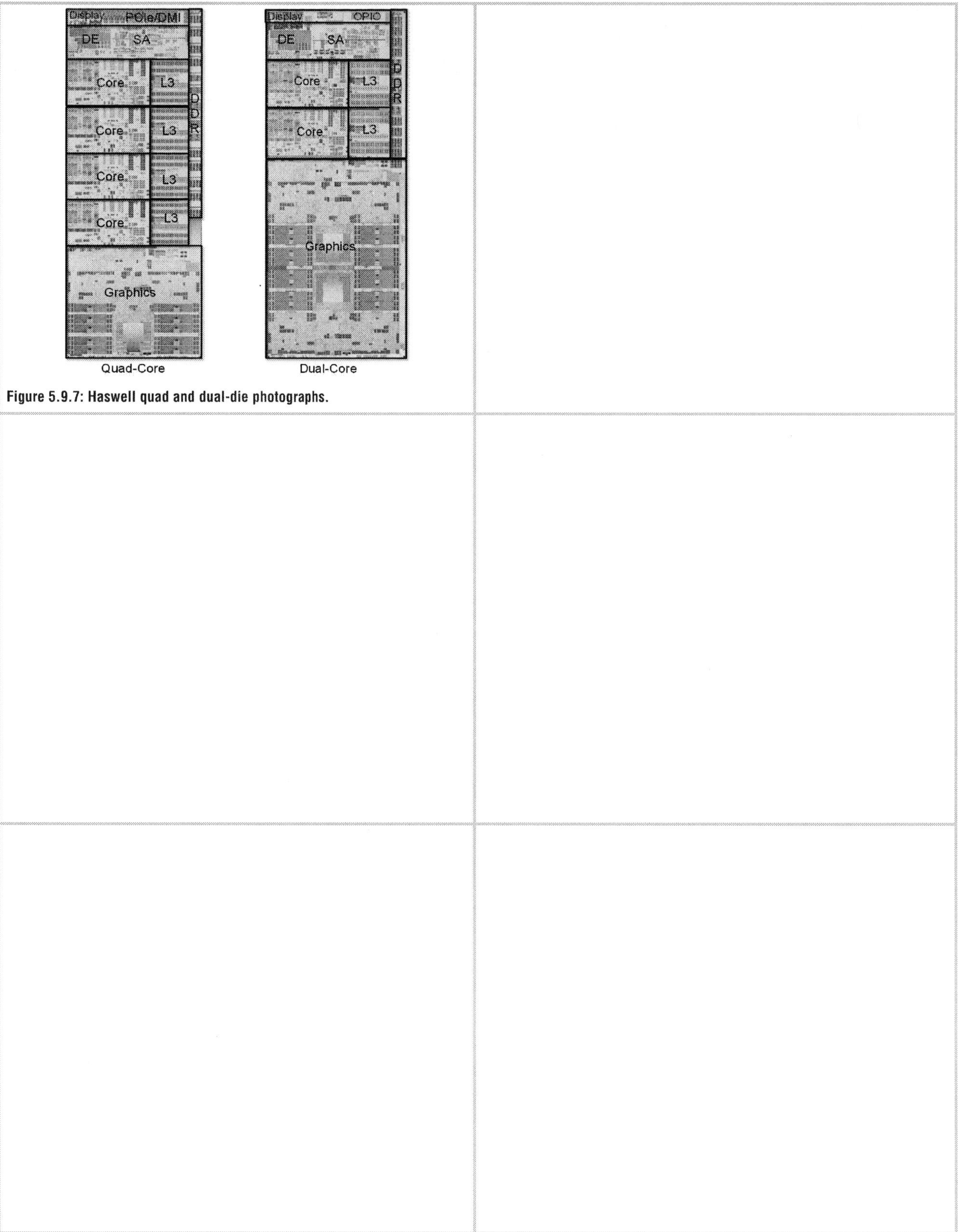

Figure 5.9.7: Haswell quad and dual-die photographs.

Session 6 Overview: *Technologies for High-Speed Data Networks*

TECHNOLOGY DIRECTIONS SUBCOMMITTEE

Session Chair: *Pirooz Parvarandeh*
Maxim Integrated, San Jose, CA

Session Co-Chair: *Chris Nicol*
Wave Semiconductor, Sunnyvale, CA

Increasing data rates require the use of a number of technologies to achieve high data throughput while minimizing power consumption. The papers in this session showcase industry-led innovations that underpin future high-speed data networks. The first paper maximizes inter-chip data rates while lowering power consumption for memory architectures in high-speed switches. The second paper utilizes advanced DSP techniques to enable high-capacity optical networking systems. The third paper demonstrates the integration of high-performance data converters with 28nm FPGAs. The first and third papers also demonstrate the benefits of 3D integration.

978-1-4799-0917-9/14 $31.00 © 2014 IEEE

6.1 Memory and System Architecture for 400Gb/s Networking and Beyond **1:30 PM**

D. Maheshwari, Cypress Semiconductor, San Jose, CA

In Paper 6.1, *Cypress Semiconductor*, a 3D IC for 400Gb/s Ethernet improves bandwidth between a packet processor and memory by using an interposer, thus lowering power in router line cards. This paper provides a perspective on the challenges of implementing high-speed data networking systems.

6.2 High-Capacity Scalable Optical Communication for Future Optical Transport Network **2:00 PM**

Y. Miyamoto, NTT, Yokosuka, Japan

In Paper 6.2, *NTT Network Innovation Laboratories,* a digital signal processor ASIC for 100Gb/s coherent optical communications is reported. It incorporates Low Density Parity Check (LDPC) Forward Error Correction (FEC) and Multiple Input Multiple Output (MIMO) detection, and enables high-capacity optical transport networks achieving more than 8Tb/s/fiber. This paper additionally discusses the advancements that are required to achieve a 1Pb/s/fiber capacity.

6.3 A Heterogeneous 3D-IC Consisting of Two 28nm FPGA Die and 32 Reconfigurable High-Performance Data Converters **2:30 PM**

C. Erdmann, Xilinx, Dublin, Ireland

In Paper 6.3, *Xilinx,* ultra-high integration of two 28nm FPGA dice and two 65nm mixed-signal dice on a 65nm interposer realizing a 3D solution in a BGA package is demonstrated. The digital-to-analog converters achieve SFDR > 63.8dBc to 400MHz at 1.6GS/s. The ADC attains SNDR > 61.6dBFS to Nyquist at 500MS/s with an interface power of 0.3mW/Gb/s and analog isolation > 92dB.

6.1 Memory and System Architecture for 400Gb/s Networking and Beyond

Dinesh Maheshwari

Cypress Semiconductor, San Jose, CA

Networking relies on fast line card packet rates that are directly proportional to and limited by the Random Transaction Rate (RTR) of the memory system. Networking line cards to date are ≤200Gb/s and were able to use memories optimized for latency (SRAM) and bandwidth (SDRAM) designed for computing systems. Next generation line cards are ≥400Gb/s and the memory system for these line cards need to be explicitly architected and designed for delivering the required high RTR.

Networking line card packet rate approximately doubles from one generation to next while scaling the functional features proportionally. This requires that not only the RTR of the memory system doubles but also the density doubles from one generation to next. RTR is doubling at a faster pace than the corresponding compute metric (latency). Similarly, networking memory density is doubling at a faster pace than compute cache density.

Networking routers and switches (Fig. 6.1.1) deal with a sequence of random packets that need to be routed to destinations dictated by their header fields after appropriate security checks and classification. While the functions to be performed on each packet are predetermined and can be pipelined, the data that is accessed, such as from the forwarding information base (FIB) for routing and access tables for classifications, is dependent on the packet header field itself; and the packet headers are different from one packet to the next at the ingress of the switch. Each packet requires 15-to-25 memory transactions to random locations before it is routed from the ingress to an egress of the switch.

A 100Gb/s line card is required to support a continuous stream of 64B min size Ethernet packets with an inter-frame gap of 20B – this results in 148.8 million packets per second. The total # of random memory transactions for a continuous stream of min-size packet traffic in a 100Gb/s line card is ~2250-~3750 million transactions per second (MT/s); if the line card cannot support this magnitude of random memory transactions then it cannot support min-size packet rate of 148.8MP/s and would not qualify as 100Gb/s line card. Unlike the computing world where the saleable compute performance is a continuous range, the line card performance is measured in quantum steps (10Gb/s, 100Gb/s, etc.) – there is no 99.99Gb/s line card, for example. The main requirement placed on the memory sub-system in a networking dataplane linecard is # of memory transactions to random locations per second (random transaction rate, RTR) and not access latency. As RTR increases for networking systems, Bandwidth (BW) also increases (BW=RTR*Transaction_Width). For computing systems, BW is high due to wider Transaction_Width (64B) while the RTR is lower.

Memory RTR is one of the major limiting factors for the continued increase in networking switching rates. High performance networking ASIC/ASSPs (≥ 400Gb/s) will require segregating logic and memory, while re-integrating the overall system by 2.5D stacking of heterogeneous KGDs (Fig. 6.1.3) on an interposer-based SIP (System in Package).

Networking functions that stress memory sub-systems are all in the networking data plane line card (Fig. 6.1.3) and are listed as (1) Table Lookups (LU), (2) Statistics/State (SS) Measurements, (3) Scheduling, and (4) Packet Buffering (PB). The table in Fig. 6.1.2 captures random transaction rates required for various functions for 100Gb/s to 400Gb/s – it illustrates that the memory RTR and density needs to at least double with every line card generation. It shows that 400Gb/s needs an RTR of 9600 MT/s to support WAN/MAN statistics.

In addition to high random transaction rate, packet buffer memory sub-system has a secondary requirement in that it requires large density for large Round Trip Time (RTT) and high packet rates. There exists a head-tail buffering algorithm (Fig. 6.1.4) that uses SRAM for the head-tail buffer to serve the RTR requirement and SDRAM to serve the density requirement. As shown in the table in Fig. 6.1.4, a combination of SRAM and SDRAM is the most cost effective solution for line cards.

QDR-II+ synchronous SRAM [1] was architected to optimize for the traditional SRAM metric, latency, established from the compute days and designed in 65nm process technology enabling RTR of 666MT/s (Fig. 6.1.7). QDR-IV [2] was re-architected to optimize RTR to 2133MT/S and designed on the same process technology node of 65nm improving the networking relevant performance metric of RTR by over 3x at the same Mb/mm^2 memory density (Fig. 6.1.7). SRAM RTR is inversely proportional to the duration between local word line activation and sense amp output. For a given # of cells/bitline, RTR can be further improved by reducing capacitance on the bit line, improving cell read current and sense amp speed.

Only SRAM technology is capable of meeting the RTR quanta required for networking line cards for ≥ 200Gb/s (Fig. 6.1.5) and has the improvement rate required from one generation to the next to keep up with a minimum of doubling in the RTR requirement. While memories that provide RTR less than the packet rate (R) can still be used to perform random read operations, this requires replication of data in different banks/memory chips to serve a single memory access per packet at the packet rate. Fig. 6.1.6 shows that performing a single read per packet @ 100Gb/s using SDRAM DDR3-1600 would require 8X-to-12X replication of the database using 4 SDRAM DDR3-1600 x16 chips. Algorithmic FIB lookup for IPv4 requires 4-to-8 accesses/packet and to support this using DDR3-1600 would require 16-to-32 chips. Replication eliminates the cost advantage of SDRAM as compared with SRAMs ($/Effective_Mb), using multiple DDR3 chips increases the board space, ASIC/ASSP I/O requirements and memory subsystem power consumption.

For networking functions like statistics/states and scheduler, the memory subsystem needs to support a read+write operation to the same data structure. For 200Gb/s, a single read+write operation would require a RTR of 600MT/s. Only SRAMs can serve the read+write RTR requirements of 600MT/s.

ASICs/ASSPs for 200Gb/s networking line cards (and some for 100Gb/s) have therefore integrated up to 512Mb of SRAM on die to serve the required RTR and density. These ASICs/ASSPs are ~500mm^2 in 28nm node with ~66% of the area occupied by SRAM. ASICs/ASSPs for 400Gb/s line cards will need to double SRAM RTR and the density to ~1Gb.

While logic density in advanced process technology nodes continues to double with every node, memory density increase is less than 2X. This would cause the networking ICs with integrated memory to continue to increase in size resulting in yield and manufacturing issues. An effective solution is to separate big memory instances out into a separate die. High RTR requirements on these memory instances require a large number (≥8) of logical channels to these die which when mapped onto a parallel interface result in large # of signal pins (≥1500). Interfacing such large # of pins using traditional parallel interfaces to packaged die is impractical. PCB routing complexity, power (~5mW/Gb/s), and silicon area on both sides for large drivers (~15mm^2) are limitations. The # of signals for interconnection can be reduced by mapping the channels to serial interfaces on packaged parts – while serial interfaces can reduce the package pins and PCB routing complexity somewhat, they exacerbate the power (>5mW/Gb/s) and the silicon area (~25mm^2) required further over external parallel interfaces.

An effective solution is to adopt 2.5D packaging wherein the big memory instances are separated out into a separate die that is packaged along with the ASIC/ASSP on a cost effective 2.5D interposer-based SIP as shown in Fig. 6.1.3. The interposer supports fine pitch routing of large # of signals between closely placed die thus reducing the loading on the drivers. The required drivers are very similar to the repeater buffers on the die itself and therefore require a small diffusion footprint on die (~3mm^2) and consume very low power (≤ 1mW/Gb/s) – High Bandwidth Memory (HBM) interface, as being defined by JEDEC, is one such interface. A low silicon footprint interface for interconnection on 2.5D interposer allows for the support of multiple HBM interfaces – thereby allowing for higher memory density and creating multiple SKUs (Stock Keeping Unit) from the same base ASIC/ASSP die and multiple memory die supporting multiple configurations of all SRAM die or a mix of SRAM and SDRAM die.

Heterogeneous integration of ASIC/ASSP on 2.5D interposer with KGD SRAM supporting standardized wide interfaces is capable of delivering the requisite RTR of 9600MT/s for the ≥ 400Gb/s line cards while keeping the individual die sizes manageable, reducing PCB routing complexity and reducing NRE for multiple SKUs.

Acknowledgement:
The author thanks Dr. Ali Keshavarzi for valuable technical discussions.

References:

[1] Cypress Semiconductor QDR-II+ datasheet, Updated 11/05/2012.
[2] Cypress Semiconductor QDR-IV datasheet, Updated 11/15/2013.
[3] J. D'Ambrosia, "Ethernet's Next Step: 400 Gigabit Ethernet", Ethernet Technology Summit 2013.
[4] D. Yu, "Innovative Wafer-based Interconnect Enabling System Integration and Semiconductor Paradigm Shifts", *IEEE International Interconnect Technology Conference* 2013, pp. 1-3
[5] J. Sun, "System Scaling and Collaborative Open Innovation", *Symposium on VLSI Technology,* 2013, pages T2-T7.
[6] C. Hou, "3DIC Integration: A Foundry Perspective", *International Conference on Electronics Packaging* 2012. Keynote Speech.

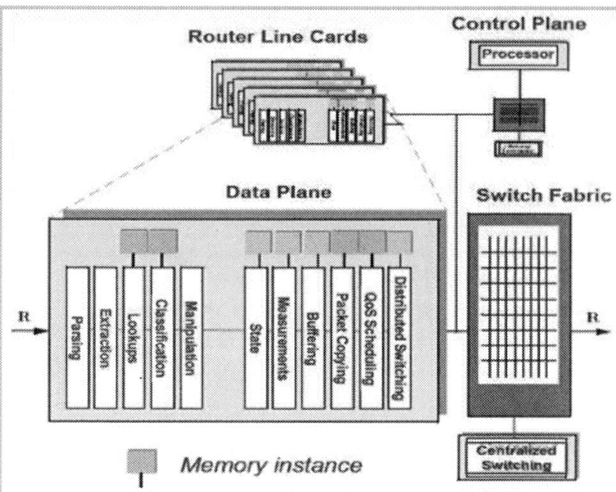

Figure 6.1.1: Networking System consisting of data plane, control plane and switch fabric.

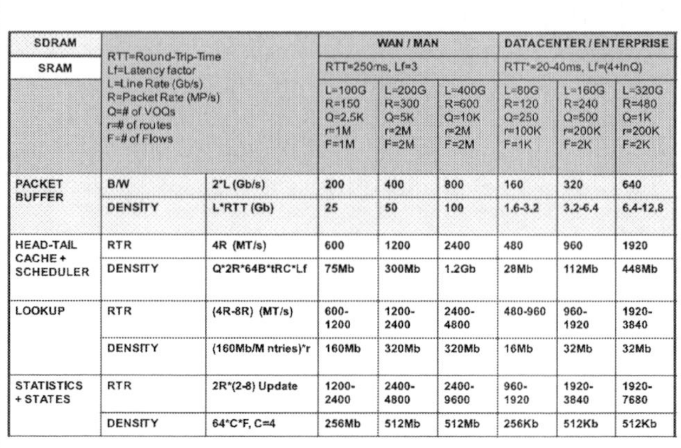

SDRAM / SRAM	RTT=Round-Trip-Time Lf=Latency factor L=Line Rate (Gb/s) R=Packet Rate (MP/s) Q=# of VOQs r=# of routes F=# of Flows		WAN / MAN RTT=250ms, Lf=3 L=100G R=150 Q=2.5K r=1M F=1M	 L=200G R=300 Q=5K r=2M F=2M	 L=400G R=600 Q=10K r=2M F=2M	DATACENTER / ENTERPRISE RTT*=20-40ms, Lf=(4+lnQ) L=80G R=120 Q=250 r=100K F=1K	 L=160G R=240 Q=500 r=200K F=2K	 L=320G R=480 Q=1K r=200K F=2K
PACKET BUFFER	B/W	2*L (Gb/s)	200	400	800	160	320	640
	DENSITY	L*RTT (Gb)	25	50	100	1.6-3.2	3.2-6.4	6.4-12.8
HEAD-TAIL CACHE + SCHEDULER	RTR	4R (MT/s)	600	1200	2400	480	960	1920
	DENSITY	Q*2R*64B*tRC*Lf	75Mb	300Mb	1.2Gb	28Mb	112Mb	448Mb
LOOKUP	RTR	(4R-8R) (MT/s)	600-1200	1200-2400	2400-4800	480-960	960-1920	1920-3840
	DENSITY	(160Mb/M ntries)*r	160Mb	320Mb	320Mb	16Mb	32Mb	32Mb
STATISTICS + STATES	RTR	2R*(2-8) Update	1200-2400	2400-4800	2400-9600	960-1920	1920-3840	1920-7680
	DENSITY	64*C*F, C=4	256Mb	512Mb	512Mb	256Kb	512Kb	512Kb

Figure 6.1.2: RTR and density requirements for networking system line card.

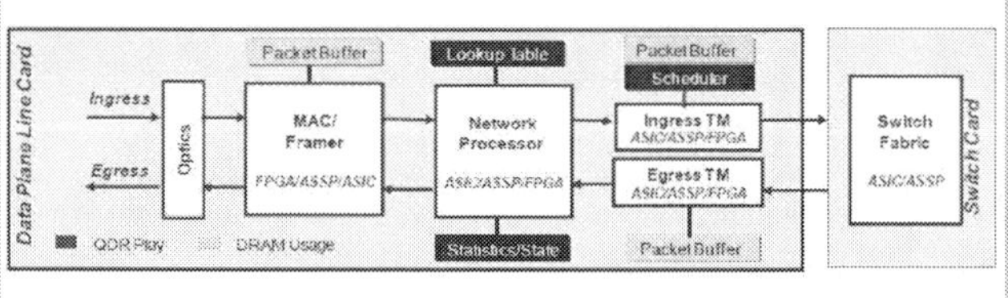

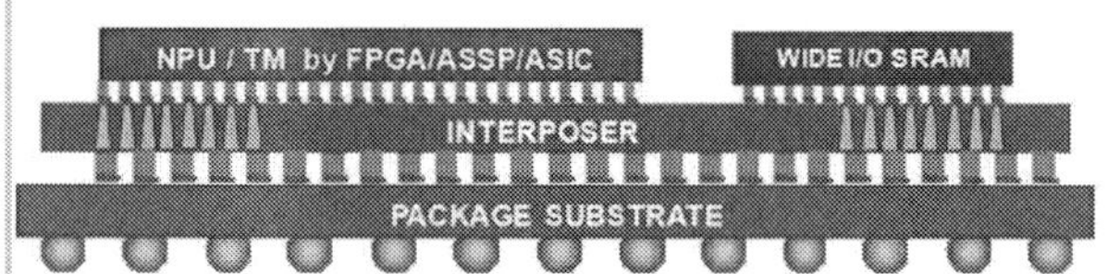

Figure 6.1.3: Future Networking Line Card Memory System in 2.5D Stacking Topology utilizing advanced Silicon technology integrated on 2.5D interposer-based SIP (System in Package) solution.

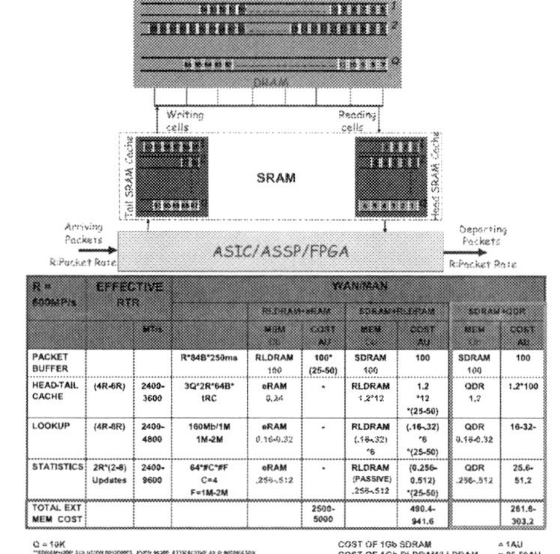

R = 600MP/s	EFFECTIVE RTR	MT/s	WAN/MAN	RLDRAM+eRAM MEM	COST AU	SDRAM+RLDRAM MEM	COST AU	SDRAM+QDR MEM	COST AU
PACKET BUFFER			R*84B*250ms	RLDRAM 100	100* (25-50)	SDRAM 100	100	SDRAM 100	100
HEAD-TAIL CACHE	(4R-6R)	2400-3600	3Q*2R*64B* tRC	eRAM 0.24	-	RLDRAM 1.2*12	1.2 *12 *(25-50)	QDR 1.2	1.2*100
LOOKUP	(4R-8R)	2400-4800	160Mb/1M 1M-2M	eRAM 0.16-0.32	-	RLDRAM (.16-.32) *6	(.16-.32) *6 *(25-50)	QDR 0.16-0.32	16-32-
STATISTICS	2R*(2-8) Updates	2400-9600	64*#C*#F C=4 F=1M-2M	eRAM .256-.512	-	RLDRAM (PASSIVE) .256-.512	(0.256-0.512) *(25-50)	QDR .256-.512	25.6-51.2
TOTAL EXT MEM COST					2500-5000		490.4-941.6		261.6-303.2

Q = 10K

COST OF 1Gb SDRAM	= 1AU
COST OF 1Gb RLDRAM/LLDRAM	= 25-50AU
COST OF 1Gb QDR	= 100AU

Figure 6.1.4: Packet Buffer Memory Sub-System.

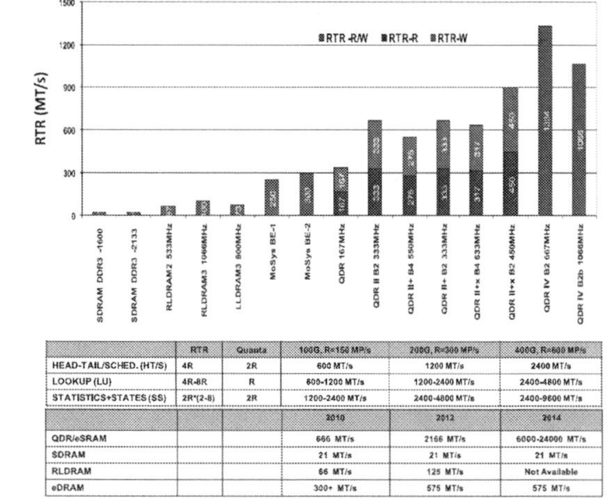

	RTR	Quanta	100G, R=150 MP/s	200G, R=300 MP/s	400G, R=600 MP/s
HEAD-TAIL/SCHED. (HT/S)	4R	2R	600 MT/s	1200 MT/s	2400 MT/s
LOOKUP (LU)	4R-8R	R	600-1200 MT/s	1200-2400 MT/s	2400-4800 MT/s
STATISTICS+STATES (SS)	2R*(2-8)	2R	1200-2400 MT/s	2400-4800 MT/s	2400-9600 MT/s
			2010	2012	2014
QDR/eSRAM			666 MT/s	2166 MT/s	6000-24000 MT/s
SDRAM			21 MT/s	21 MT/s	21 MT/s
RLDRAM			66 MT/s	125 MT/s	Not Available
eDRAM			300+ MT/s	575 MT/s	575 MT/s

Figure 6.1.5: RTR of different various commercially available memory products and future requirements.

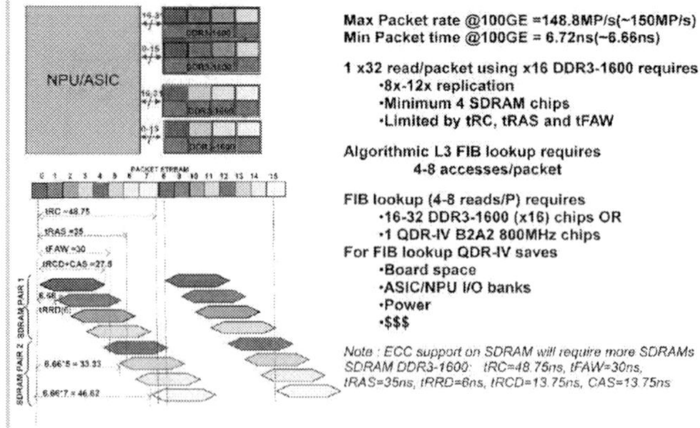

Figure 6.1.6: Lookup for 100GE using SDRAM DDR3-1600.

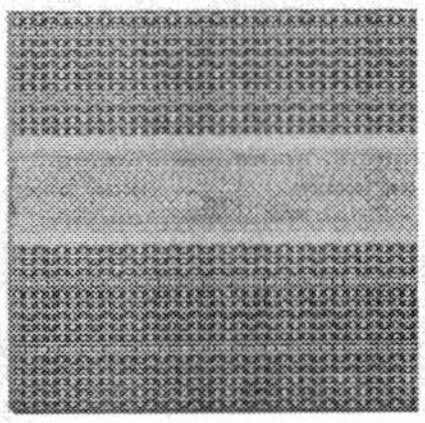

KEY INFORMATION	
TECHNOLOGY	65nm
CORE / IO VDD	1.2V / 1.2V
CELL TYPE	0.525um^2 SRAM

QDR-II+ QDR-IV

Summary Table

Products	Process Tech	RTR (MT/s)	Optimization	Status
QDR-II+	65nm	666		Shipping
QDR-IV	65nm	2133	Memory array supporting high RTR: Micro-architecture change	Sampling

Figure 6.1.7: QDR-II+ and QDR-IV die micrographs in 65nm technology including key parameters summary tables.

6.2 High-Capacity Scalable Optical Communication for Future Optical Transport Network

Yutaka Miyamoto, Masahito Tomizawa

NTT, Yokosuka, Japan

The future penetration of long-term-evolution mobile phone services and various data cloud services will continuously accelerate the present traffic evolution. Figure 6.2.1 shows the commercial system capacity evolution of NTT's network over the last 30 years. The transmission capacity of today's Optical Transport Network (OTN) exceeds 1Tbit/s based on the conventional single-core single-mode fiber (SMF) at the growth rate about 1.4-to-1.5 times a year. In 10 years however, we will encounter the fundamental capacity limit of conventional SMF at around 100 Tb/s due to optical fiber nonlinearity and the limitation of allowable fiber launched power into the SMF.

In this paper, the impact and future scaling of digital signal processing (DSP) on high-capacity OTN are investigated. A high-speed optical communication system with coherent detection aided by DSP ASIC, that is a digital coherent system, has great potential to change the transmission system design [1,2]. This approach can enhance total commercial system capacity to more than 8 Tbit/s. Digital coherent systems will evolve to both high-speed applications and ultralow-power consumption applications in the near future. Further future enhancements to the key DSP are discussed enabling 1Pb/s/fiber capacity in combination with space division multiplexing

Figure 6.2.2 shows the block diagram of the developed 100Gb/s DSP ASIC, which enables coherent detection of PDM-QPSK, 20%-redundant FEC concatenation with a 7% framer FEC, and fast recovery time [2]. The 100G digital coherent system employed the polarization division multiplexing and quadrature phase-shift keying (PDM-QPSK) modulation format. In the developed DSP ASIC, the encoder block receives optical channel transport unit 4 (OTU4) client data from the OTN framer, inserts FEC parity bits for a LDPC code with 13% redundancy and a 0.7% dispersion estimator overhead. The encoded data are mapped into a 4-logical/4-physical parallel lane distribution format of optical channel transport lane (OTL) for transmission over physical 4 lanes, which correspond to sets of in-phase and quadrature-phase lanes for each of two polarizations of PDM-QPSK signal format. On the decoder side, the 4 lanes of the input analog signals are digitized by a 4-channel ADC at a over-sampling rate of 2. The frequency domain equalizer (FDE) in the fixed equalizer block compensates for chromatic dispersion (CD). The CD range of ± 40,000ps/nm (equivalent to the CD of 2000km SMF) can be compensated. The total CD of the received signal is estimated within 5ms, by detecting the arrival time delay calculated from the frequency spectrum of the dispersion estimator overhead. The estimation time does not depend on the amount of dispersion. An adaptive equalizer fulfills the roles of polarization de-multiplexing, PMD compensation, frequency offset compensation between signal LD and local LD, carrier phase recovery, and sampling clock recovery. Target specifications are a PMD compensation range of 100ps, and polarization tracking speed of 50kHz, frequency offset of ± 5GHz. Then, the SD-LDPC decoder dramatically reduces the number of bit errors even in a high bit error rate (BER) range such as 10^{-2}, and the remaining errors can be completely corrected by the framer FEC. The NCG of 10.8dB was achieved at the BER of 10^{-15}, which results in the input Q-limit of 6.4dB. The target recovery time is less than 50ms, and the frame synchronization for OTL is established within 50ms, even when the transmission route of the optical channel is switched. The above functions are implemented based on LSI using a 40nm CMOS process. Figure 6.2.7 shows the summary of the ASIC, the chip die photograph and the packaged ASIC overview.

Figure 6.2.3 shows high-speed optical channel recovery capability in the case of optical route switch due to planned network operation or disaster recovery. The received signal is switched from the 140km route (route 1) to the 420km route (route 2) using an optical switch. The cumulated amount of chromatic dispersion changes from 350ps/nm (route 1) to 1,097ps/nm (route 2) after optical signal loss. The OTL4.4 lane alignment alarm disappears after the recovery of the optical input signal with the time delay of 12 ms, as shown in lower part of the Fig. 6.2.4. This includes re-convergence of adaptive filters, frame synchronization, and frame alignment among the OTL4.4 multilanes in addition to the chromatic dispersion estimation and FDE block renewal. In this experiment, total outage time was approximately 15ms, including the period of signal loss of 3ms.

Figure 6.2.4 shows how to enhance the channel capacity for future high spectral efficient OTN. Since today's OTN relies on SMF, the use of much higher order quadrature amplitude modulation (QAM), and parallel bulk data transmission scheme using several optical carriers (multi-carriers) will be very beneficial considering the usable 40Gbaud symbol rate of the ADC and DAC of the CMOS ASIC. In ten years, however, the OTN capacity severely limited by the ultimate capacity limit of SMF due to optical fiber nonlinearity and the allowable launched power limit. Recently the introduction of space division multiplexing (SDM) in optical fiber transmission is proposed to offset these limitations of SMF [3]. Two kinds of SDM enhancements have been proposed. One increases the number of cores (*M*) in a single strand of fiber (that is multi-core-fiber (MCF)), and the other enhances the number of propagation modes (*N*) in a multimode core.

Figure 6.2.5 shows recent R&D challenges of high-capacity OTN based on SDM. Pb/s-class capacity capability has been recently demonstrated. With regard to the MCF approach shown by the filled circles and diamonds in Fig. 6.2.5, the system typically requires MCF, fan-in fan-out devices, multicore optical amplifiers and optical nodes. Since the SDM signals can be multiplexed and demultiplexed without digital signal processing, each tributary signal to be SDM-multiplexed can be treated as an individual optical channel. Therefore, it has great potential to enhance the number of optical paths by the factor of *M*, which can exceed 10. Recently, 1.01Pb/s transmission was first reported over 52.4km 12-core fiber [6]. The 12-core fiber was designed to minimize the core-to-core crosstalk to be low enough to employ the 32 quadrature amplitude modulation (32QAM) multicarrier digital coherent format (offline DSP processing) achieving high spectral efficiency of 7.6bit/s/Hz/core. 84Tb/s WDM transmission was achieved in each core with 222 channels of 456Gb/s multi-carrier digital coherent signals. In terms of the multi-mode approach shown by the triangles in Fig. 6.2.5, the system requires multimode fiber, mode couplers/splitters, and transceivers with multi-input multi-output (MIMO) processing. Since MIMO processing realizes the demultiplexing of each mode individually in a receiver, the SDM-multiplexed tributary signals naturally have the same destination. The scalability of mode number *N* is mainly determined by the scaling of MIMO processing DSP circuits integrated in a CMOS ASIC chip that can manage the differential mode group delay (DMGD) [4] and its dynamic property [5].

Figure 6.2.6 shows the possible future evolution of DSP for high-capacity OTN. In the today's coherent DSP ASIC, two types of equalizers (namely adaptive equalizer and fixed equalizer) are implemented. It should be noted that the adaptive equalizer using 2x2 MIMO processing with more than 10 taps realizes the reliable operation of polarization division de-multiplexing in the receiver for dynamic polarization properties in terms of polarization state and polarization mode dispersion. The fixed equalizer is mainly utilized to equalize static large-signal distortion by chromatic dispersion. In DSP for future OTN based on SDM, especially SDM using multi-modes with the mode number N, the adaptive equalizer of 2N x 2N MIMO has to operate dynamically with large tap number or FFT points that is 10 times larger than that of a fixed equalizer found in today's digital coherent transport system, since today's multimode fiber has a large DMGD of 1-to-10ns/km with the dynamic change as fast as the polarization issue in today's digital coherent system [5]. The excellent design of DMGD of multimode fiber and MIMO processing algorithms are crucial for a cost-effective realization of future OTN based on SDM.

Acknowledgements:
This work is partly supported by the MIC of Japan and by the NICT.

References:

[1] H. Sun, et al.," Real-time measurements of a 40 Gb/s coherent system," *Optics Express*, vol. 16, no. 2, pp. 873-879, 2008.
[2] E. Yamazaki, et al.," Fast optical channel recovery in field demonstration of 100-Gbit/s Ethernet over OTN using real-time DSP," *Optics Express*, vol. 19, no. 14, pp. 13179-13184, 2011.
[3] T. Morioka, "New Generation optical infrastructure technologies: EXAT initiative towards 2020 and beyond," *OECC2009*, FT4, 2009.
[4] V. Sleiffer, et al., "73.7 Tb/s (96X3x256-Gb/s) mode-division-multiplexed DP-16QAM transmission with inline MM-EDFA," *ECOC2012*, Th.3.C.4. 2012.
[5] Xi Chen, et al., "Characterization of Dynamic Evolution of Channel Matrix" *OFC/NFOEC2013*, OM2C.3, 2013.
[6] H. Takara, et al., "1.01-Pb/s (12 SDM/222 WDM/456 Gb/s) Crosstalk-managed Transmission with 91.4-b/s/Hz Aggregate Spectral Efficiency," *ECOC2012*, Th.3.C.1, 2012.

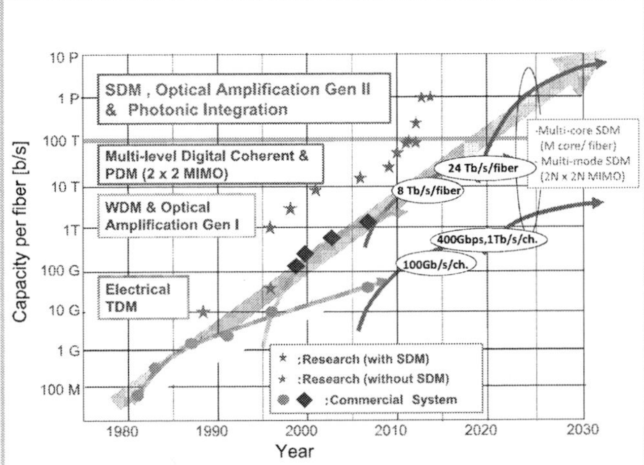

Figure 6.2.1: Commercial system capacity evolution of Optical Transport Network.

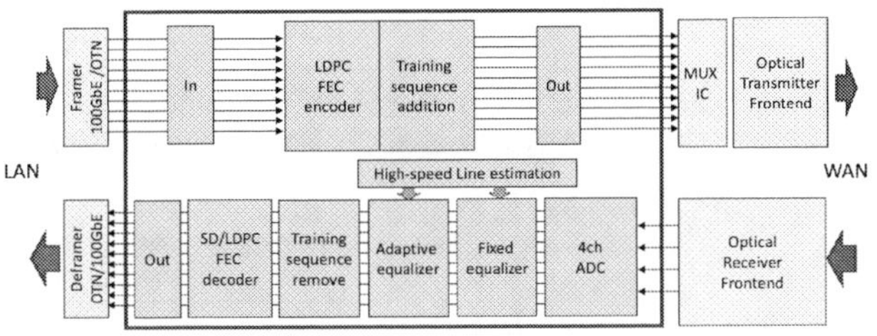

Figure 6.2.2: Functional block diagram of the developed 100Gbit/s DSP ASIC for long-haul transport.

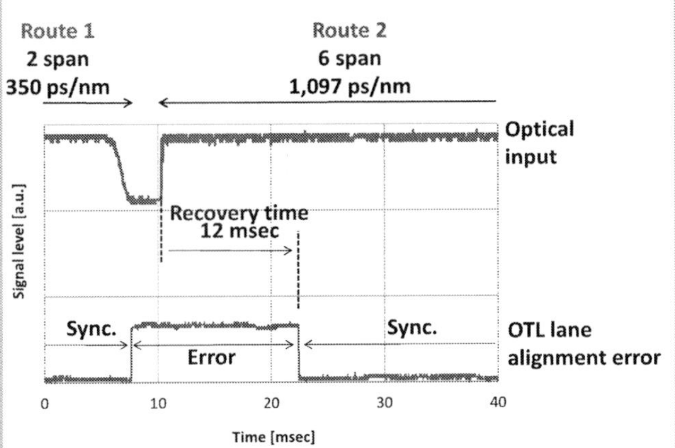

Figure 6.2.3: Fast Optical Channel Recovery by 100Gbps realtime DSP AISC.

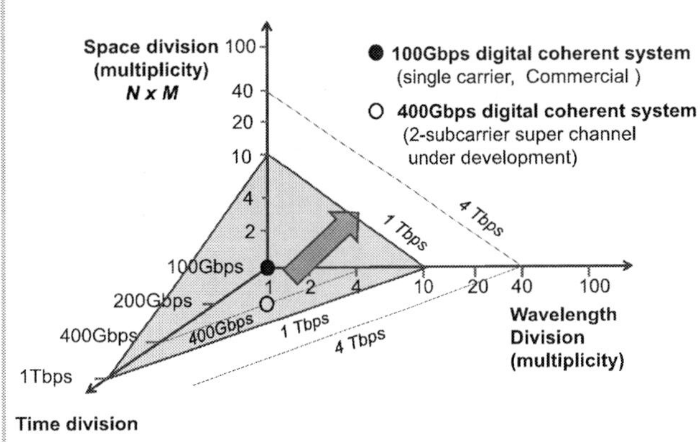

Figure 6.2.4: The evolution of high-capacity optical channel using SDM.

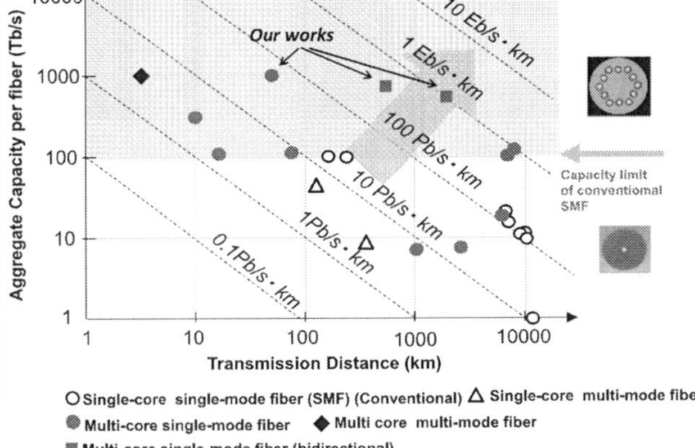

Figure 6.2.5: The recent experimental challenges of high-capacity transmission based on SDM.

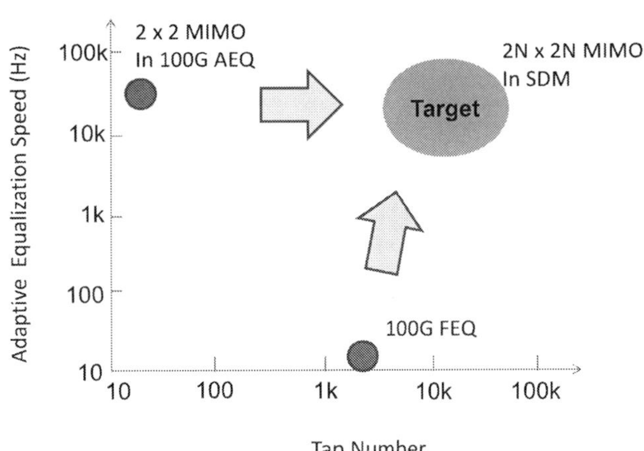

Figure 6.2.6: The requirement of DSP for future capacity evolution by SDM based on MIMO processing.

Item	Specifications
Modulation format	PDM-QPSK
Line side signal bit rate	127.156 Gb/s
Oversampling ratio	2.0
CD compensation	±40,000 ps/nm
Differential group delay (Polarization mode dispersion)	100 ps
Polarization tracking speed	50 kHz
Frequency offset compensation	± 5 GHz
FEC-NCG	10.8 dB @10^{-15} SD-LDPC + Enhanced-FEC
Recovery time	< 50 ms
Process	40nm CMOS

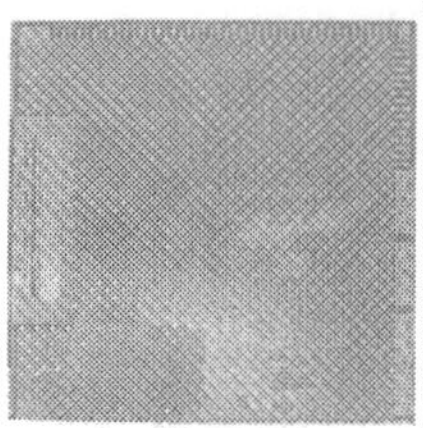

Figure 6.2.7: The summaries of the concept proof of 100Gb/s DSP ASIC, the chip die photograph and the packged ASIC overview.

6.3 A Heterogeneous 3D-IC Consisting of Two 28nm FPGA Die and 32 Reconfigurable High-Performance Data Converters

Christophe Erdmann[1], Donnacha Lowney[1], Adrian Lynam[1], Aidan Keady[1], John McGrath[1], Edward Cullen[1], Daire Breathnach[1], Denis Keane[1], Patrick Lynch[1], Marites De La Torre[1], Ronnie De La Torre[1], Peng Lim[1], Anthony Collins[1], Brendan Farley[1], Liam Madden[2]

[1]Xilinx, Dublin, Ireland, [2]Xilinx, San Jose, CA

Data converters are required to interface digital processing engines, for example FPGAs, to the real world. Data conversion is typically accomplished using discrete devices that are interfaced to the FPGA using various IO standards. However, exponential growth in bandwidth as a result of increasing channel count and higher sample rate means this IO interface is becoming a limiting factor in the system budget with respect to interconnect complexity and associated power. The integration of flexible data converters with FPGA eliminates this IO cost and also offers a dynamically scalable, power efficient platform solution that addresses diverse application needs.

In this paper, we demonstrate an aggregate bandwidth in excess of 400Gb/s using sixteen 16b DAC instances running at 1.6GS/s with an FPGA-to-die interface power of 0.3mW/Gb/s. We introduce a reconfigurable receive system that allows channel count to trade with system sample rate. Specifically, we demonstrate a 500MS/s ADC by interleaving four 125MS/s units.

An overview of the system is depicted in Fig. 6.3.1. There are two separate 65nm die, one containing sixteen 125MS/s 13b ADCs and the other containing sixteen 1.6GS/s 16b DACs. A 65nm CMOS process was chosen for the mixed signal die as it offers a favourable compromise between digital circuit speed/area and the analog requirements of gain/matching. For the digital logic, two 350T FPGA dies with a total logic cell count greater than 580k were fabricated with a 28nm CMOS process optimized for high performance and low power operation mode. Six IO banks, distributed over 300 package balls, support an array of single-ended and differential interface standards. Eight SerDes transceivers support individual lane rates of 11.182 Gb/s, which can be used for backplane communication, PCIe and Gigabit Ethernet. On-die sensors are used to monitor temperature and supply voltages, making use of separate dual 12b 1MS/s ADCs. The four die were micro-bumped to allow assembly using Stacked Silicon Interconnect (SSI) technology [1,2] on a 65nm passive interposer within a 35mmx35mm CS-BGA package. Performance degradation due to micro-bump strain was mitigated using strategic bump placement such that no performance impact could be determined at 16b level in the DAC output transfer function. Power integrity was optimized throughout the 3D-IC chip based on simulated current profiles. This determined the choice of chip-scale decoupling capacitors at package level, MOM-cap decoupling at the interposer level, and various decoupling strategies at the die level. Electrical interconnect between dies on the interposer was made using single-ended copper Super Long Lines (SLLs) with nominal capacitance less than 150fF/mm. Per 16b DAC, 128 SLLs were required with an 8:1 data gearing scheme to transfer the data for 1.6GS/s DAC sampling speed. Only 13 SLLs per ADC were used to send data across the interposer at 125MS/s sample rate. Noise isolation between mixed-signal and FPGA die was improved in the interposer using a grounded shield array of redundant Through Silicon Vias (TSVs) in addition to separate power and ground domains. This strategy was replicated through the package substrate out to the PCB.

Systems using discrete high-speed high-performance data converters traditionally use power intensive LVDS or JESD204B interfaces and require careful PCB routing. Multi-chip module (MCM) approaches are limited by package interconnect density to approximately 30 signals/mm. In contrast, SSI increases interconnect density to greater than 1200 signals/mm for an equivalent data rate. This solution completely integrates the FPGA-data converter interface with a measured power of only 0.3mW/Gb/s, which is about 2 orders of magnitude improvement compared to discrete data converter interfaces. Heterogeneous integration also has a much smaller PCB footprint, and allows for an optimal choice of technology for each major system function.

Crosstalk is the main challenge with co-integration of digital VLSI with sensitive mixed-signal functions. In the presented chip, FPGA clock and data related interference spurs of less than -92dBc were measured at the DAC output for 12W of FPGA switching power using a toggling data aggressor scheme. Fig. 6.3.2 shows FPGA-to-analog isolation, as defined by the ratio of DAC output amplitude to FPGA interfering tone, while the DAC is synthesizing a 70MHz full-scale output tone. Crosstalk to the ADC was lower than the ADC noise floor and thus not observable.

Each mixed-signal die were divided into two sets of quads. On-die termination is provided for the full speed clock receivers. Each die has its own set of bandgap derived current and voltage references, which ensures de-correlation of reference related noise for specific applications. DACs and ADCs are arranged in quad pairs and share common resources as shown in Fig. 6.3.1.

The DAC architecture [3, 4] is composed of a synthesised 6:10 segmentation decoder, a current mode logic bit-slice and intrinsically sized current source array which demonstrates DNL and INL better than 4 LSB on a 16b level. Spectral performance was measured for 20mA full-scale current into 50 Ohm load with a 2:1 impedance transformer. Fig. 6.3.3 illustrates SFDR, IM3 and NSD vs. output signal frequency at 1.6GS/s over 400MHz output signal frequency range. For low output frequencies the SFDR > 80dBc and exceeds 64dBc at 400MHz. Two tone (+/-200kHz) IM3 tests demonstrate better than -75dBc; whilst NSD varies from -162dBm/Hz to -157dBm/Hz over the spectral range.

Each ADC operates up to 125MS/s and consists of a 6-stage pipeline: 5 stages resolving 2.5b with a final 3b flash stage yielding 13b resolution [5]. Signal swing of 1Vpp-diff was applied on die to a buffered or un-buffered input depending on user selection. At 125MS/s, the measured INL is +/-0.86LSBs, while the DNL is +/-0.46LSBs. Spectral performance at 125MS/s is shown in Fig. 6.3.4 with THD < -70dBc and SNDR > 66dBFS over the Nyquist frequency range. For both DACs and ADCs, cross-talk measurements showed all spurious components were smaller than the dominant intrinsic converter spur.

ADC channel count can easily be traded against conversion speed, power or signal-to-noise ratio. ADCs were combined on a time-interleaved clocking scheme to facilitate 250MS/s and 500MS/s operation. Amplitude, offset and time skew errors were background calibrated on the FPGA. Comparable spectral performance metrics were achieved for a 500MS/s ADC system over its Nyquist range to those of a 125MS/s unit. Fig. 6.3.5 illustrates a typical FFT amplitude spectrum of a 220MHz input tone sampled at 500MS/s achieving SNDR > 61.6dBFS and THD < -72dBc.

The ADC uses a switched-capacitor bias generator [6] and allows linear power scaling with sample rate, varying from 49mW at 40MS/s to 134mW at 125MS/s. Owing to the use of CML circuits in the DAC bit-slice, power was largely independent of sample rate and optimization traded swing amplitude of current steering switch driver with performance metric. For a performance optimized setting at 1.6GS/s, 465mW and Fout=100MHz, IM3=-85dBc was achieved while for the power optimized setting of 330mW, IM3=-70dBc was measured. Note the termination power of 66mW is included in these figures.

In conclusion, we report the co-integration of 32 high-performance data converters with high-density digital circuitry in a reconfigurable processing system achieving performance comparable to best-in-class converters. We provide a summary of specifications and comparison to prior art in Fig. 6.3.6; noting state of the art interface power of only 0.3mW/Gb/s and analog-to-digital die isolation better than 92dB. The heterogeneous system architecture allows new partitioning at the system level. The interposer stratum facilitates high density interconnect between subsystems thereby addressing two key interconnect challenges; namely power reduction and bandwidth extension. The reconfigurable nature of the digital system provides for a solution that is inherently scalable and can be optimized against user-defined performance metrics: channel count, power and speed.

References:

[1] L. Madden, et al., "Advancing High Performance Heterogeneous Integration Through Die Stacking", *Proc. ESSCIRC*, pp. 18-24, Sept. 2012.

[2] M. Santarini, "Stacked & Loaded", *XCell Journal*, pp. 8-13, Q1-2011.

[3] C-H. Lin, et al., "A 12b 2.9GS/s DAC with IM3 <-60dBc Beyond 1GHz in 65nm CMOS", *ISSCC Dig. Tech. Papers*, pp. 74-75, Feb. 2009.

[4] K. Doris, et al., "A 12b 500MS/s DAC with >70dB SFDR up to 120MHz in 0.18μm CMOS", *ISSCC Dig. Tech. Papers*, pp. 116-117, Feb. 2005.

[5] J-W. Nam, et al., "A 12-bit 100-MS/s pipelined ADC in 45nm CMOS", *SoC Design Conference (ISOCC)*, pp. 405-407, 2011.

[6] T-N Andersen, et al., "A cost-efficient high-speed 12-bit pipeline ADC in 0.18μm digital CMOS", *IEEE J. Solid-State Circuits*, vol. 40, no.7, pp. 1506-1513, 2005.

978-1-4799-0917-9/14 $31.00 © 2014 IEEE

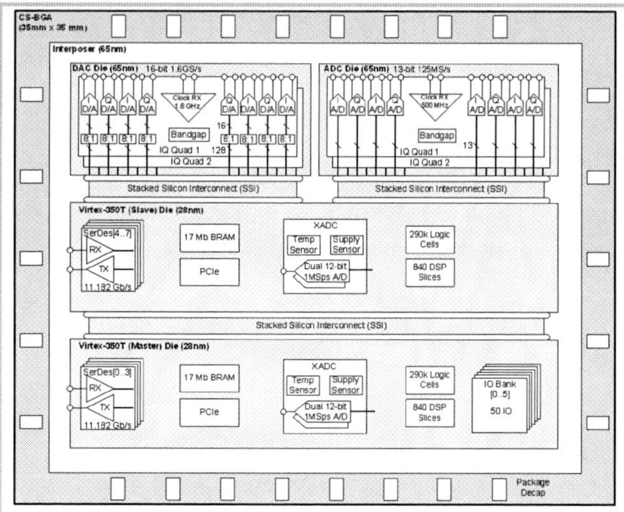

Figure 6.3.1: Overview of the chip.

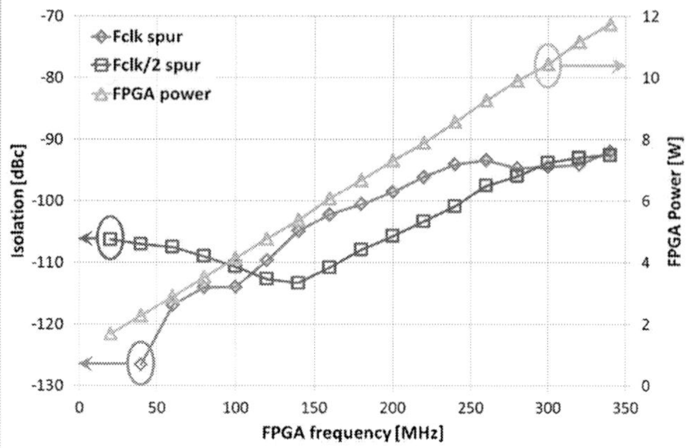

Figure 6.3.2: Measured FPGA-to-analog isolation while the DAC is synthesizing a 70MHz full scale output tone. 100k DFF within the 28nm logic die and 2k SLL interconnect were toggling at the FPGA clock rate.

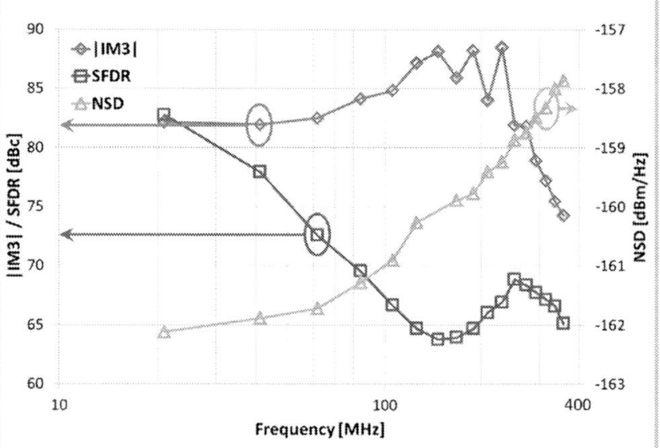

Figure 6.3.3: Measured DAC dynamic performance (SFDR, IM3 and NSD) versus output signal frequency sampled at 1.6GS/s. Note negative complement of IM3 is shown for clarity.

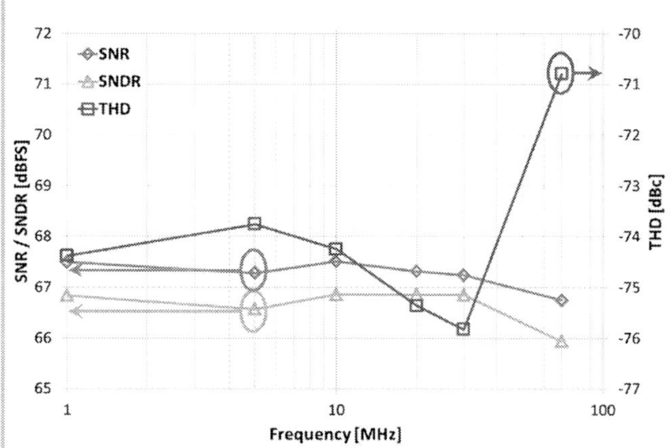

Figure 6.3.4: Measured ADC dynamic performance (SNR, SNDR and THD) versus input signal frequency sampled at 125MS/s.

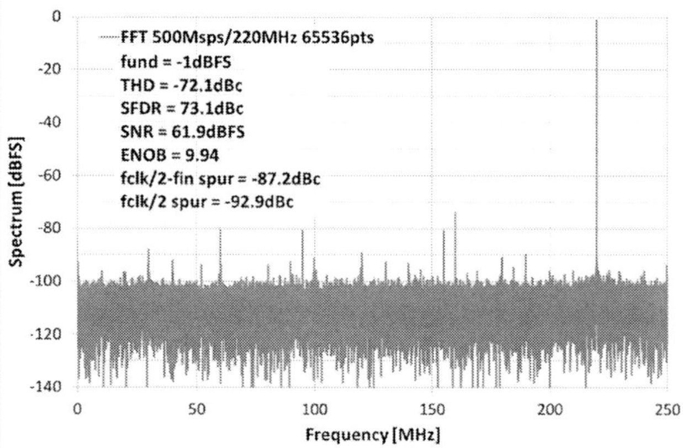

Figure 6.3.5: Typical FFT spectrum of 4x interleaved ADC for 220MHz input signal frequency sampled at 500MS/s.

DAC Parameter	This work	[3]	ADC Parameter	This work	[5]	FPGA Parameter	This work
Resolution [bits]	16	12	Resolution [bits]	13	12	Logic cells	580480
INL referred to 16b	4	8	Fclk [MS/s]	125	100	CLB slices	90700
DNL referred to 16b	4	4.8	SNDR [dBFS]	66.6 (64.4)	59.0	BRAM [kb]	33840
Fclk [GS/s]	1.6	2.9	SNR [dBFS]	67.3 (65.2)	-	DSP slices	1680
SFDR [dBc] @ 350 MHz	65.1 (59.9)	63	THD [dBc]	-73.8 (-68.8)	-	11 Gb/s SerDes	8
IM3 [dBc] @ 300 MHz	-74.3 (-70.2)	-75	SFDR [dBc]	76.6 (69)	63.2	IO pins	300
NSD [dBm/Hz]	-157.9 (-157.5)	N/A	Unit Power [mW]	134	30.4	Interface power [mW/Gb/s]	0.3
Unit Power [mW]	399	188	FOM [pJ/conv]	0.61	0.41	FPGA to analog isolation [dB]	>92

Figure 6.3.6: Summary of system specifications, measured performance and comparison to prior art. Worst values over process, voltage and temperature noted in brackets.

Figure 6.3.7: Device micrograph.

Session 7 Overview: *Image Sensors*

IMMD SUBCOMMITTEE

Session Chair: *Makoto Ikeda*
University of Tokyo, Tokyo, Japan

Session Co-Chair: *David Stoppa*
Fondazione Bruno Kessler, Trento, Italy

This session presents recent advancements in the field of image sensors such as: process improvements for back-side-illuminated CMOS sensors, low-power sensors for wireless applications, high-speed image processing for feature extraction and recognition, as well as recent advancements in time-of-flight sensors.

7.1 A 1/4-inch 8Mpixel CMOS Image Sensor with 3D Backside-Illuminated 1.12μm Pixel with Front-Side Deep-Trench Isolation and Vertical Transfer Gate **3:15 PM**

J. Ahn, Samsung Electronics, Yongin, Korea

In Paper 7.1, Samsung Electronics presents a 1/4-inch 8Mpixel CMOS image sensor with back-side-illuminated 1.12μm pixels. Each pixel is fully surrounded by front-side deep-trench isolation and includes a vertical transfer gate. This technology leads to a reduction of the optical crosstalk from 19% to 12.5%, and to an increase in linear full-well capacity from 5,000 to 6,200 electrons.

7.2 243.3pJ/pixel Bio-Inspired Time-Stamp-Based 2D Optic Flow Sensor for Artificial Compound Eyes **3:45 PM**

S. Park, University of Michigan, Ann Arbor, MI

In Paper 7.2, the University of Michigan reports a bio-inspired analog/digital mixed-mode 2D optical flow sensor for artificial compound eyes. The sensor estimates 2D optical flows from the integrated mixed-mode algorithm core, providing 2D 16b optical flows or compressed frames up to 120fps. The 2D flow estimation core consumes 243.3pJ/pixel.

978-1-4799-0917-9/14 $31.00 © 2014 IEEE

7.3 A 1000fps Vision Chip Based on a Dynamically Reconfigurable Hybrid Architecture Comprising a PE Array and Self-Organizing Map Neural Network **4:00 PM**

C. Shi, Chinese Academy of Sciences, Beijing, China and Tsinghua University, Beijing, China

In Paper 7.3, the Chinese Academy of Sciences (with Tsinghua University) presents a vision chip in a 0.18μm process, which integrates an image sensor consisting of 256×256 pixels together with three von Neumann-type parallel processors and a self-organizing map (SOM) neural network. The SOM network can be reconfigured into an array processor in 3 clock cycles. The SOM network enables face recognition at 1340fps with a recognition accuracy of 86%.

7.4 A 413×240-Pixel Sub-Centimeter Resolution Time-of-Flight CMOS Image Sensor with In-Pixel Background Canceling Using Lateral-Electric-Field Charge Modulators **4:15 PM**

S-M. Han, Shizuoka University, Hamamatsu, Japan

In Paper 7.4, Shizuoka University (with Brookman Technology) presents a 413×240-pixel CMOS Time-of-Flight (ToF) range imager with pinned photodiode high-speed lock-in pixels. The lock-in pixel structure uses lateral electric field control to implement a multiple-tap charge modulator. A range resolution of 4-to-6.4mm, with a light pulsewidth of 13ns, is measured for the range of 0.8 to 1.8m. Range data can be calculated for every frame.

7.5 A 0.3mm-Resolution Time-of-Flight CMOS Range Imager with Column-Gating Clock-Skew Calibration **4:30 PM**

K. Yasutomi, Shizuoka University, Hamamatsu, Japan

In Paper 7.5, Shizuoka University presents a 132×120-pixel CMOS ToF range imager with 0.3mm resolution at 32mm range. The ToF imager employs an indirect ToF measurement technique using pulse photocurrent response and draining-only modulation pixels. Furthermore, a column gating-clock skew calibration scheme allows a timing skew of 8ps rms.

7.6 A 512×424 CMOS 3D Time-of-Flight Image Sensor with Multi-Frequency Photo-Demodulation up to 130MHz and 2GS/s ADC **4:45 PM**

A. Payne, Microsoft, Mountain View, CA

In Paper 7.6, Microsoft describes pixel and signal path of a 512×424-pixel ToF image sensor. Consecutive frames with different modulation frequencies are combined, which enables a range error of less than 1% over ranges from 0.8 to 4.2m. Modulation frequencies up to 130MHz are supported, and the modulation contrast at 50MHz is 67%.

7.1 A 1/4-inch 8Mpixel CMOS Image Sensor with 3D Backside-Illuminated 1.12μm Pixel with Front-Side Deep-Trench Isolation and Vertical Transfer Gate

JungChak Ahn, Kyungho Lee, Yitae Kim, Heegeun Jeong, Bumsuk Kim, Hongki Kim, Jongeun Park, Taesub Jung, Wonje Park, Taeheon Lee, Eunkyung Park, Sangjun Choi, Gyehun Choi, Haeyong Park, Yujung Choi, Seungwook Lee, Yunkyung Kim, Y. Jay Jung, Donghyuk Park, Seungjoo Nah, Youngsun Oh, Mihye Kim, Yooseung Lee, Youngwoo Chung, Ihara Hisanori, Joonhyuk Im, Daniel-KJ Lee, Byunghyun Yim, GiDoo Lee, Heesang Kown, Sungho Choi, Jeonsook Lee, Dongyoung Jang, Youngchan Kim, Tae Chan Kim, Goto Hiroshige, Chi-Young Choi, Duckhyung Lee, GabSoo Han

Samsung Electronics, Yongin, Korea

According to the trend towards high-resolution CMOS image sensors, pixel sizes are continuously shrinking, towards and below 1.0μm, and sizes are now reaching a technological limit to meet required SNR performance [1-2]. SNR at low-light conditions, which is a key performance metric, is determined by the sensitivity and crosstalk in pixels. To improve sensitivity, pixel technology has migrated from frontside illumination (FSI) to backside illumiation (BSI) as pixel size shrinks down. In BSI technology, it is very difficult to further increase the sensitivity in a pixel of near-1.0μm size because there are no structural obstacles for incident light from micro-lens to photodiode. Therefore the only way to improve low-light SNR is to reduce crosstalk, which makes the non-diagonal elements of the color-correction matrix (CCM) close to zero and thus reduces color noise [3]. The best way to improve crosstalk is to introduce a complete physical isolation between neighboring pixels, e.g., using deep-trench isolation (DTI). So far, a few attempts using DTI have been made to suppress silicon crosstalk. A backside DTI in as small as 1.12μm-pixel, which is formed in the BSI process, is reported in [4], but it is just an intermediate step in the DTI-related technology because it cannot completely prevent silicon crosstalk, especially for long wavelengths of light. On the other hand, front-side DTIs for FSI pixels [5] and BSI pixels [6] are reported. In [5], however, DTI is present not only along the periphery of each pixel, but also invades into the pixel so that it is inefficient in terms of gathering incident light and providing sufficient amount of photodiode area. In [6], the pixel size is as large as 2.0μm and it is hard to scale down with this technology for near 1.0μm pitch because DTI width imposes a critical limit on the sufficient amount of photodiode area for full-well capacity. Thus, a new technological advance is necessary to realize the ideal front DTI in a small size pixel near 1.0μm.

In our work, a small pixel with fully surrounding and full-depth DTI is demonstrated. As shown in Fig. 7.1.1, in the conventional 2-dimensional (2D) pixel structure, if DTI is placed along the periphery of each pixel, the effective photodiode area is reduced by the amount of DTI width in addition to the fixed pixel transistor area. In a pixel size near 1.0μm, there is little area remaining for a photodiode and full-well capacity. In this paper, to overcome the small photodiode fill factor by the presence of DTI, a vertical transfer gate (VTG) and buried photodiode are combined with front DTI technology, forming a 3-dimensional (3D) pixel, which is realized in a single wafer, contrary to the previous silicon stack structure [7]. In this 3D pixel, transistors and photodiode are separated. Transistors are present in the silicon surface plane, as in conventional 2D pixels, but the photodiode is placed and buried beneath the transistor plane. A VTG connects both planes (photodiode and transistors) and thus accumulated charges in the buried photodiode can be transferred vertically into the floating-diffusion node in the transistor plane. Moreover, photodiode fill factor is increased to over 70%, compared with 50% for conventional pixels, due to the separation of photodiode and transistor planes. This vertical structure is the key to realize front DTI in a small pixel size of around 1.0μm.

Figure 7.1.2 illustrates the front DTI pixel structure, comparing it to that of a conventional pixel. The pixel has a DTI-introduced physical isolation barrier to prevent optical and electrical crosstalk, as well as blooming between photodiodes. Prevention of optical crosstalk by relying on the total internal reflection of DTI is illustrated. Moreover, there is no blooming due to inherently perfect electrostatic isolation of DTI, so that there is no need for a blooming path in the photodiode design, and thus the potential height of the photodiode full well is increased, even with the maximum potential of the photodiode remaining as low as around 1.0V, lower than that of a conventional pixel (1.7V). This accounts for the low-voltage operation of the VTG. A cross-sectional view of the fabricated 1.12μm 3D DTI pixel is shown in Fig. 7.1.2. For this 3D pixel, 2 different process modules are added into the conventional BSI pixel process. First, a narrow DTI is formed by etching silicon along the periphery of each pixel and filling it with oxide and polysilicon for optical and electrical isolation, and total internal reflection of light. Second, a vertical cylindrical hole in the middle of the pixel is etched away from the silicon surface, deep into the buried photodiode. After that, conventional pixel processes are followed and then in the backside grinding process step, backside silicon is grinded away to the bottom of front DTI so that each pixel is isolated in all directions.

Figure 7.1.3 shows the normalized QE spectrum of the fabricated 1.12μm 3D DTI pixel compared with that of a conventional BSI pixel. A crosstalk of 12.5%, in the presence of front DTI, is measured, while that of conventional BSI is 19.0%. With this reduced crosstalk, the YSNR10 is greatly improved, compared with a conventional BSI. YSNR10 is measured by applying white balance and CCM to capture raw images from an 18% gray patch under 3200K light source and F/2.8 lens [8]. Also, the evaluation of angular responses shows that the front DTI pixel has better crosstalk and relative illumination characteristics as incidence angle increases than the conventional pixel, which allows use of a low F-number lens for higher sensitivity as well as lower module height.

Figure 7.1.4 presents the measured electrical and dark characteristics in a front DTI pixel compared with those of a conventional pixel. First, full-well capacity of $6{,}200e^-$ in a DTI pixel is measured as 24% larger than for a conventional one, from the photon-transfer-curve. Second, blooming is not observed even at highly negative TG bias due to the DTI potential barrier. Third, an almost zero lag is measured while reducing TG high voltage as low as 2.5V. In general, the deep-trench etching process might cause a degradation in dark characteristics. To avoid such degradation, careful cleaning and passivation processes are applied along the silicon and DTI interface. A dark current of $6e^-/s$ is achieved and similar white spot counts to those of a conventional BSI pixel are observed.

The sensor performance is summarized in Fig. 7.1.5. YSNR10 is improved down to 105 lux and full-well capacity is increased up to $6{,}200e^-$, which are 30% and 24% improvements, respectively, while maintaining overall dark characteristics as low as the conventional pixel. Figure 7.1.6 is a reproduced image at 1/4-inch 8Mpixel CIS with front DTI 3D pixel. The measured SNR of the front DTI 3D pixel is 2.0dB higher than that of the conventional one at both high- and low-illumination conditions.

In summary, a small 3D backside-illuminated pixel having fully surrounding front-side deep trench isolation and a vertical transfer gate is developed. YSNR10 of as low as 105lux at F/2.8 lens is achieved due to inherently zero silicon crosstalk using deep trench isolation, and full-well capacity of as high as $6{,}200e^-$ is reached using a vertical transfer gate and buried photodiode. This pixel is successfully demonstrated in a 1/4-inch 8Mpixel CMOS image sensor.

References:

[1] J.C. Ahn, *et al.*, "Advanced Image Sensor Technology for Pixel Scaling Down Toward 1.0μm (Invited)," *Proc. IEDM*, pp. 1-4, 2008.

[2] H. Wakabayashi, *et al.*, "A 1/2.3-inch 10.3Mpixel 50 frame/s Back-Illuminated CMOS Image Sensor," *ISSCC Dig. Tech. Papers*, pp. 410-411, Feb. 2010.

[3] J.C. Ahn, *et al.*, "SNR Metric and Crosstalk in Color Image Sensor of Small Size Pixel," *Proc. VLSI-TSA*, pp. 1-2, 2013.

[4] Y. Kitamura, *et al.*, "Suppression of Crosstalk by Using Backside Deep Trench Isolation for 1.12μm Backside Illuminated CMOS Image Sensor," *Proc. IEDM*, 2012.

[5] A. Tournier, *et al.*, "Pixel-to-Pixel Isolation by Deep Trench Technology: Application to CMOS Image Sensor," *Proc. IISW*, pp. 12-15, 2011.

[6] R. Fontaine, "Innovative Technology Elements for Large and Small Pixel CIS Devices," *Proc. IISW*, p. 1-4, 2013.

[7] V. Suntharalingam, *et al.*, "A 4-side Tileable Back Illuminated 3D-Integrated Mpixel CMOS Image Sensor," *ISSCC Dig. Tech. Papers*, pp. 38-39, Feb. 2009.

[8] J. Alakarhu, "Image Sensors and Image Quality in Mobile Phones," *Proc. IISW*, pp. 1-4, 2007.

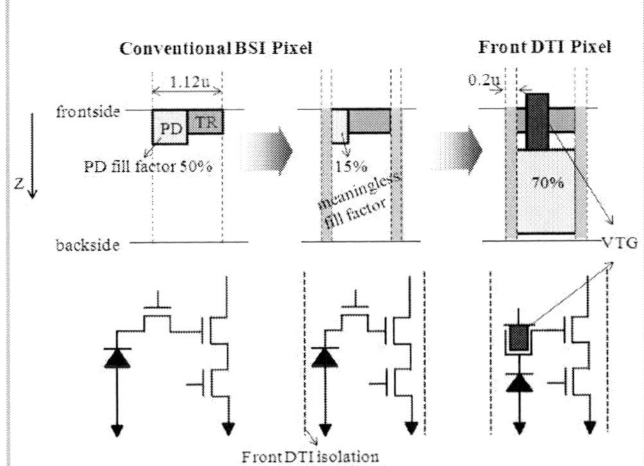

Figure 7.1.1: 3D pixel with DTI, VTG, and buried PD.

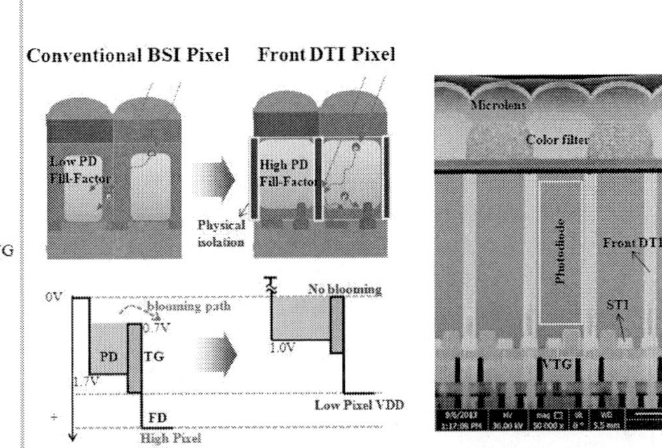

Figure 7.1.2: Concept, operation, and structure.

7

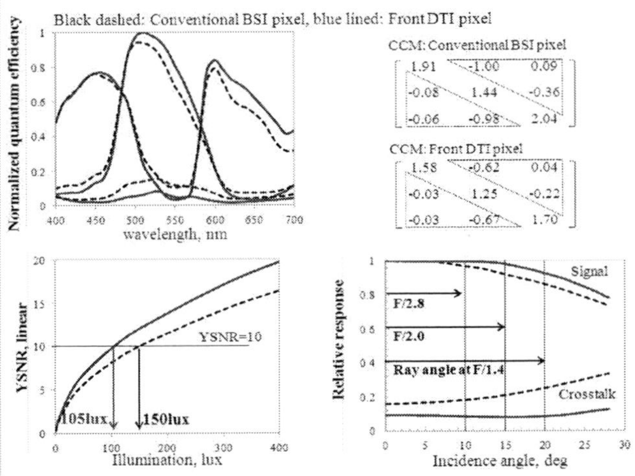

Figure 7.1.3: Optical characteristics.

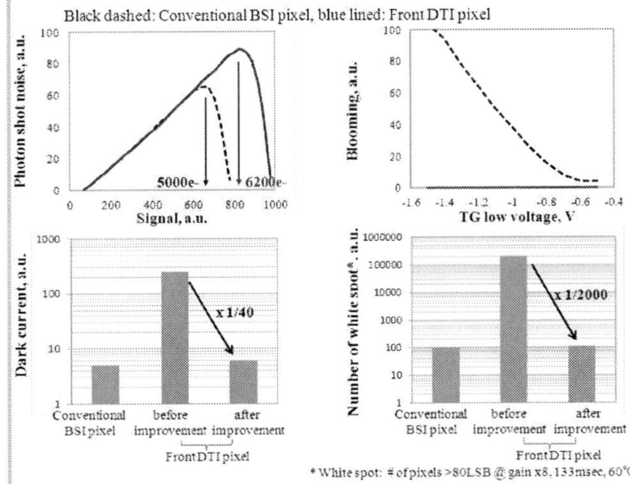

Figure 7.1.4: Electrical and dark characteristics.

	unit	1.12u Conventional BSI	1.12u Front DTI
YSNR10*	lux	150	105
G-Sensitivity @D65-light	e-/lux.sec	3960	4080
Crosstalk	%	19.0%	12.5%
Linear full well	e-	5,000	6,200
Dark temporal noise	e-	1.7	1.7
Dark fixed pattern noise	e-	1.0	1.0
Dynamic range	dB	68.1	69.9
Dark current @60°C	e-/s	5	6
White spot**	ea/Mp	100	110

* YSNR10: illumination for Y-SNR=10 with AWB, CCM, F/2.8, 3200K-light, 18% reflectance gray patch. 80% lens transmittance, IR filter of 96% max. transmittance and 650nm cut-off, color accuracy(△E2000)=2.5

** White spot: # of pixels >80LSB @ gain x8, 133msec, 60°C, dark

Figure 7.1.5: Sensor performance table.

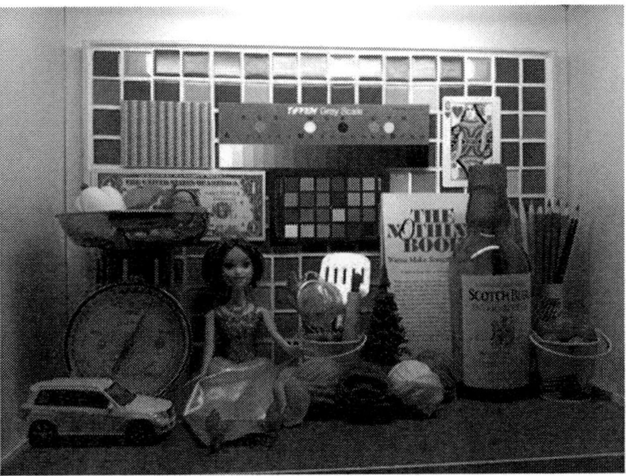

Figure 7.1.6: Reproduced image.

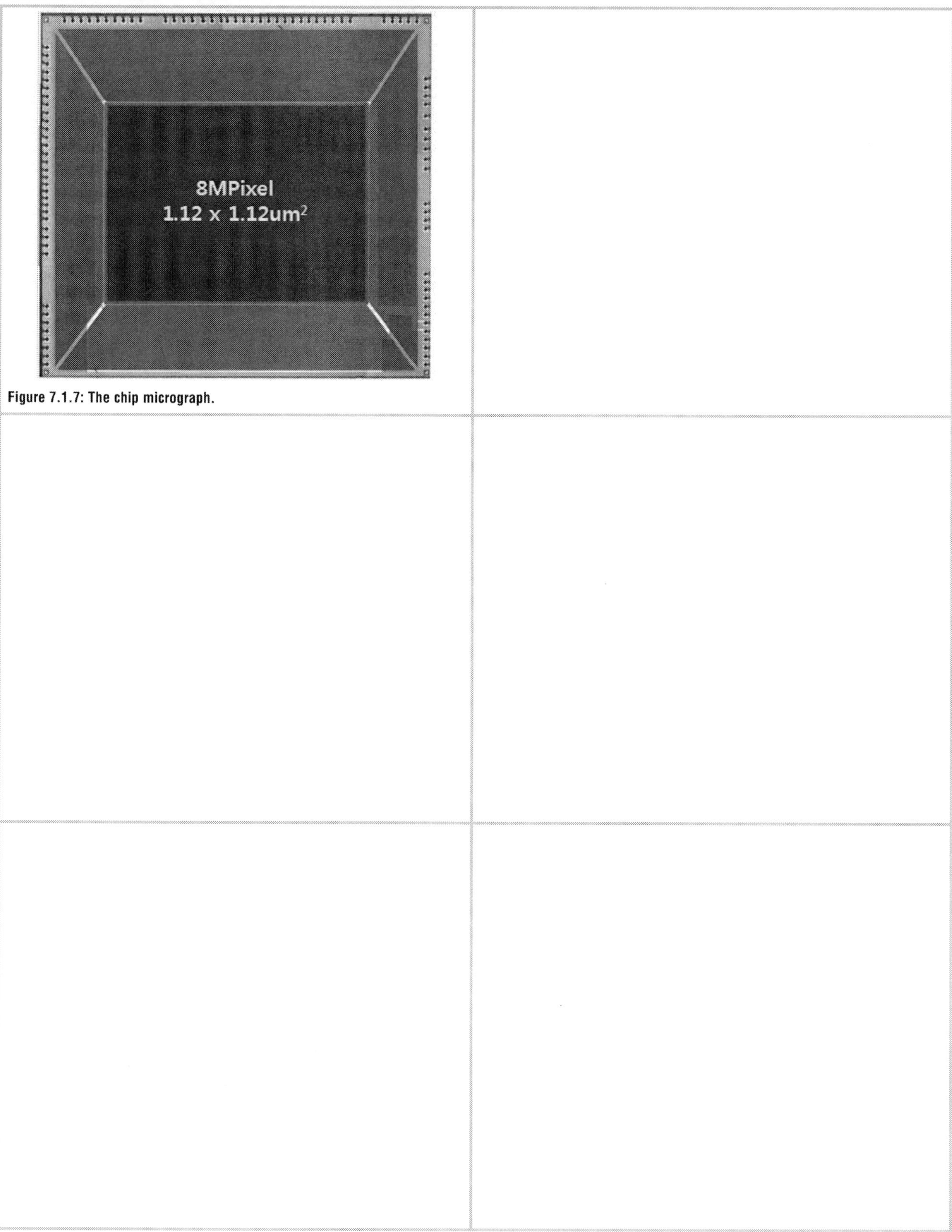

Figure 7.1.7: The chip micrograph.

7.2 243.3pJ/pixel Bio-Inspired Time-Stamp-Based 2D Optic Flow Sensor for Artificial Compound Eyes

Seokjun Park, Jihyun Cho, Kyuseok Lee, Euisik Yoon

University of Michigan, Ann Arbor, MI

Miniaturized low-power artificial compound eyes in a small form factor and a low payload can be a promising approach to provide wide-field information for micro-air-vehicle (MAV) applications. Recently, research efforts have been made to realize bio-inspired artificial compound eyes to mimic the wide field of view (FoV) of insect visual organs by implementing photoreceptors to independently face different angles [1-2]. However, these approaches have drawbacks. They use complicated fabrication processes to form a hemispherical lens configuration and secure an independent optical path to each photoreceptor. We take a simple and practical approach to realize wide-field optic flow sensing in a pseudo-hemispherical configuration by mounting a number of 2D array optic flow sensors on a flexible PCB module as shown in Figure 7.2.1. In this scheme, the 2D optic flow sensor should meet the requirements of MAV applications: extremely low power consumption while maintaining robust optic flow generation. Conventional optic flow algorithms, such as Lucas-and-Kanade, require huge amounts of numerical calculations; therefore, they require substantial digital hardware (CPU and/or FPGA), resulting in large power consumption [3-4]. As an alternative approach for low-power implementation, bio-inspired elementary motion detector (EMD) based algorithms (or neuromorphic algorithms) have been studied and implemented in analog VLSI circuits for autonomous navigation [5-6]. However, pure analog signal processing is easily susceptible to temperature and process variations and it is difficult to scale the pixel size or apply low-power design techniques because extensive analog processing is implemented in pixel-level circuits. In this work, we have devised and implemented a time-stamp-based optic flow algorithm, which is modified from the conventional EMD algorithm to give an optimum partitioning of hardware blocks in analog and digital domains as well as assign adequate allocation of pixel-level, column-parallel, and chip-level processing. Temporal filtering, which may require huge hardware resources if implemented in the digital domain, remains in a pixel-level analog processing unit. Feature detection is implemented using digital circuits that are column parallel. The embedded digital core decodes the 2D time-stamp information into velocity using chip-level processing. Finally, the estimated 16b optic flow data are compressed and transmitted to the host through a 4-wired Serial Peripheral Interface (SPI) bus.

Figure 7.2.1 shows the conceptual view of the semi-hemispherical artificial compound eyes and its system architecture. The system is designed to provide wide-field optic flow sensing, which covers 180° FoV per each module like a compound eye in flying insects. To realize the wide FoV, the system mounts multiple optic flow sensors on a flat flexible PCB; then, the PCB is bent to form the desired inter-sensor angle by origami packaging. The host controller, which may be located at the bottom of the PCB package, communicates with all the sensors by 4-wired SPI. The 3MB/s SPI, currently the fastest bus available in MAV systems, can only send full resolution (8KB) raw optic flow data at 120fps from 3 sensors. Thus, the sensor integrates the lossless data compression algorithm to reduce the data rate down to 12.0% of the total data on average, so that at least 25 sensors can be connected in the same bus.

Figure 7.2.2 shows the pixel architecture and system block diagram of the bio-inspired time-stamp-based optic flow senor. In normal image mode, the sensor generates 8b digital images from the embedded single slope (SS) ADCs. In optic flow mode, the sensor operates as a bio-inspired vision chip emulating insect vision by estimating optic flows from the time-of-travel of a moving feature that is detected by temporal contrast changes. First, the temporal contrast change is monitored from two consecutive frames in the pixel array. The pixel includes a sampling capacitor (C_1) and the gain capacitor (C_2) for setting a PGA gain. A frame difference is acquired by sampling the pixel voltage ($V_p(t_1)$) of the previous frame from C1 first, and then applying the current pixel voltage ($V_p(t_2)$) to the PGA. Two supply voltages are used in the pixel array: 3.3V for photodiodes and source followers, and 1.8V for PGAs. To reduce static power consumption, the PGAs are enabled only during the signal transfer period. The column-level digital circuits use a 0.9V supply. A 1b comparator detects the moving feature when there is a significant change in the temporal contrast beyond a threshold that can be dynamically adjusted. The column-parallel 1b feature information enables the embedded 2D optic flow digital core to update recent time-stamp information when a moving feature appears in each pixel. The 1.8V supply digital core estimates 16b raw optic flows, representing 8b for each x and y direction. The generated raw optic flows are compressed and sent to the host.

Figure 7.2.3 shows the block diagram of the embedded digital processing unit in detail. The 2D optic flow computation core first stores in SRAM the time-stamp information of the recent occurrence of a moving feature in each pixel. The updated time-stamp information is aligned to those from the previous ones in a raster scan order to estimate 2D optic flows by a 3×3 masking operation. Two computation examples are illustrated at different times at t=12 and 14. At time t=12, a vertically moving feature enters to the center with a speed of 1 pixel/frame; the time-stamp value reflects this change. As a result, only the y-direction time-of-travel is measured at the updated location, and the final estimated optic flow is $(V_x, V_y) = (0, 1)$ [pixel/frame]. In the same manner, at time t=14, the optic flow of $(V_x, V_y) = (0.5, 0)$ [pixel/frame] is estimated. The calculated 16b raw optic flows can be compressed in the lossless data compression block. We employ a compression algorithm that utilizes sparsity in signals, which is the nature of typical optic flows. As shown in an example presented in the histogram, the 94.6% of pixels in one frame are at zero (i.e., no optic flows). The compression algorithm sends only a 1b code for zero optic flows and sends a 2b header plus a 16b raw optic flow value (total 18b) for non-zero optic flows. As a result, the algorithm allocates 1.92b on average instead of a uniform allocation of 16b for each pixel, achieving 88% reduction in bandwidth. Because the compressor output is either 1b or 18b depending on optic flow values, the packetizing block piles up the compressed data and outputs 64b packets to be stored in the output buffer SRAM.

Figure 7.2.4 shows the measured performance of the fabricated 2D optic flow sensor. The linearity curve was characterized by averaging the output optic flows while applying the fixed velocity input patterns. (We applied the moving bar patterns for characterization.) The circles show the measured data from the whole sensor signal path. The triangles show only for the digital core by directly loading the input patterns to the core using the test ports. The mismatch between these two results may come from the imperfection of optics and/or non-uniformity in input patterns projected from LCD monitors. We also tested simple computer-generated patterns: translating, diagonally moving, and rotating patterns at 30fps (Fig. 7.2.4). We also successfully captured optic flows in 3 test cases: a fan rotating at 160 rpm, a bouncing ball, letters written by tracking a laser pointer source, demonstrating the performance and feasibility for deploying the fabricated sensor in actual MAV platforms (Fig. 7.2.5).

A prototype chip was fabricated using a 0.18μm 1P4M process. The performance of the sensor is summarized in Fig. 7.2.6. We achieved a figure of merit of 243.3pJ/pixel to estimate 16b 2D optic flows. A chip micrograph is shown in Fig. 7.2.7.

References:

[1] D. Floreano, *et al.*, "Miniature Curved Artificial Compound Eyes," *Proceedings of the National Academy of Sciences of the United States of America*, vol. 110, no. 23, pp. 9267–9272, June 2013.

[2] Y. M. Song, *et al.*, "Digital Cameras with Designs Inspired by the Arthropod Eye," *Nature*, vol. 497, no. 7447, pp. 95–99, May 2013.

[3] J. Conroy, G. Gremillion, B. Ranganathan, and J. S. Humbert, "Implementation of Wide-Field Integration of Optic Flow for Autonomous Quadrotor Navigation," *Autonomous Robots*, vol. 27, no. 3, pp. 189–198, Aug. 2009.

[4] V. Mahalingam, K. Bhattacharya, N. Ranganathan, H. Chakravarthula, R. R. Murphy, and K. S. Pratt, "A VLSI Architecture and Algorithm for Lucas–Kanade-Based Optical Flow Computation," *IEEE Trans. VLSI*, vol. 18, no. 1, pp. 29–38, 2010.

[5] J. Krammer and C. Koch, "Pulse-Based Analog VLSI Velocity Sensors," *IEEE Trans. Circuits and Systems II: Analog and Digital Signal Processing*, vol. 44, no. 2, pp. 86–101, 1997.

[6] R. R. Harrison, "A Biologically Inspired Analog IC for Visual Collision Detection," *IEEE Trans. Circuits and Systems I: Regular Papers*, vol. 52, no. 11, pp. 2308–2318, Nov. 2005.

978-1-4799-0917-9/14 $31.00 © 2014 IEEE

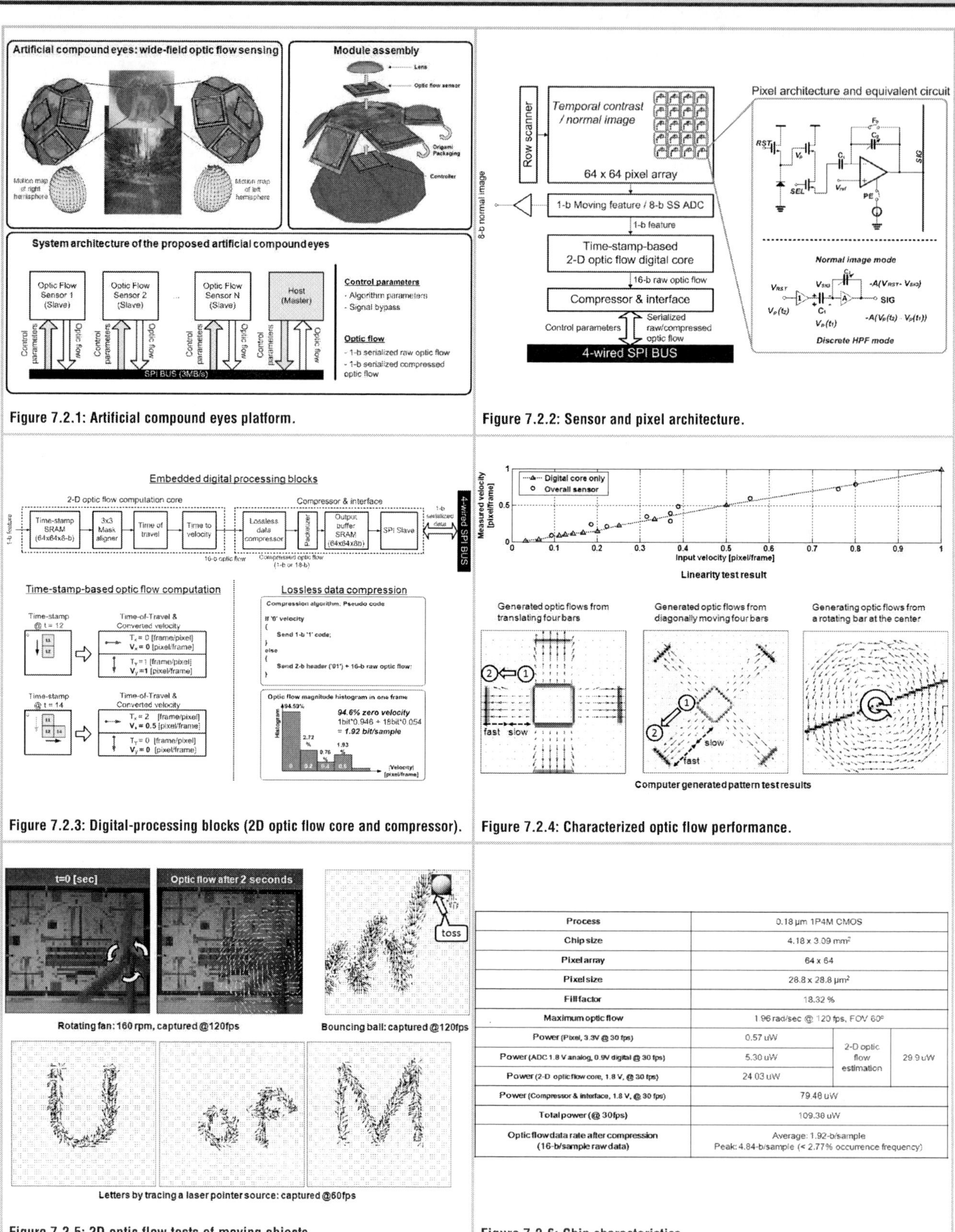

Figure 7.2.1: Artificial compound eyes platform.

Figure 7.2.2: Sensor and pixel architecture.

Figure 7.2.3: Digital-processing blocks (2D optic flow core and compressor).

Figure 7.2.4: Characterized optic flow performance.

Figure 7.2.5: 2D optic flow tests of moving objects.

Process	0.18 µm 1P4M CMOS		
Chip size	4.18 x 3.09 mm²		
Pixel array	64 x 64		
Pixel size	28.8 x 28.8 µm²		
Fill factor	18.32 %		
Maximum optic flow	1.96 rad/sec @ 120 fps, FOV 60°		
Power (Pixel, 3.3V @ 30 fps)	0.57 uW	2-D optic flow estimation	29.9 uW
Power (ADC 1.8 V analog, 0.9V digital @ 30 fps)	5.30 uW		
Power (2-D optic flow core, 1.8 V, @ 30 fps)	24.03 uW		
Power (Compressor & interface, 1.8 V, @ 30 fps)	79.48 uW		
Total power (@ 30fps)	109.38 uW		
Optic flow data rate after compression (16-b/sample raw data)	Average: 1.92-b/sample Peak: 4.84-b/sample (< 2.77% occurrence frequency)		

Figure 7.2.6: Chip characteristics.

978-1-4799-0917-9/14 $31.00 © 2014 IEEE

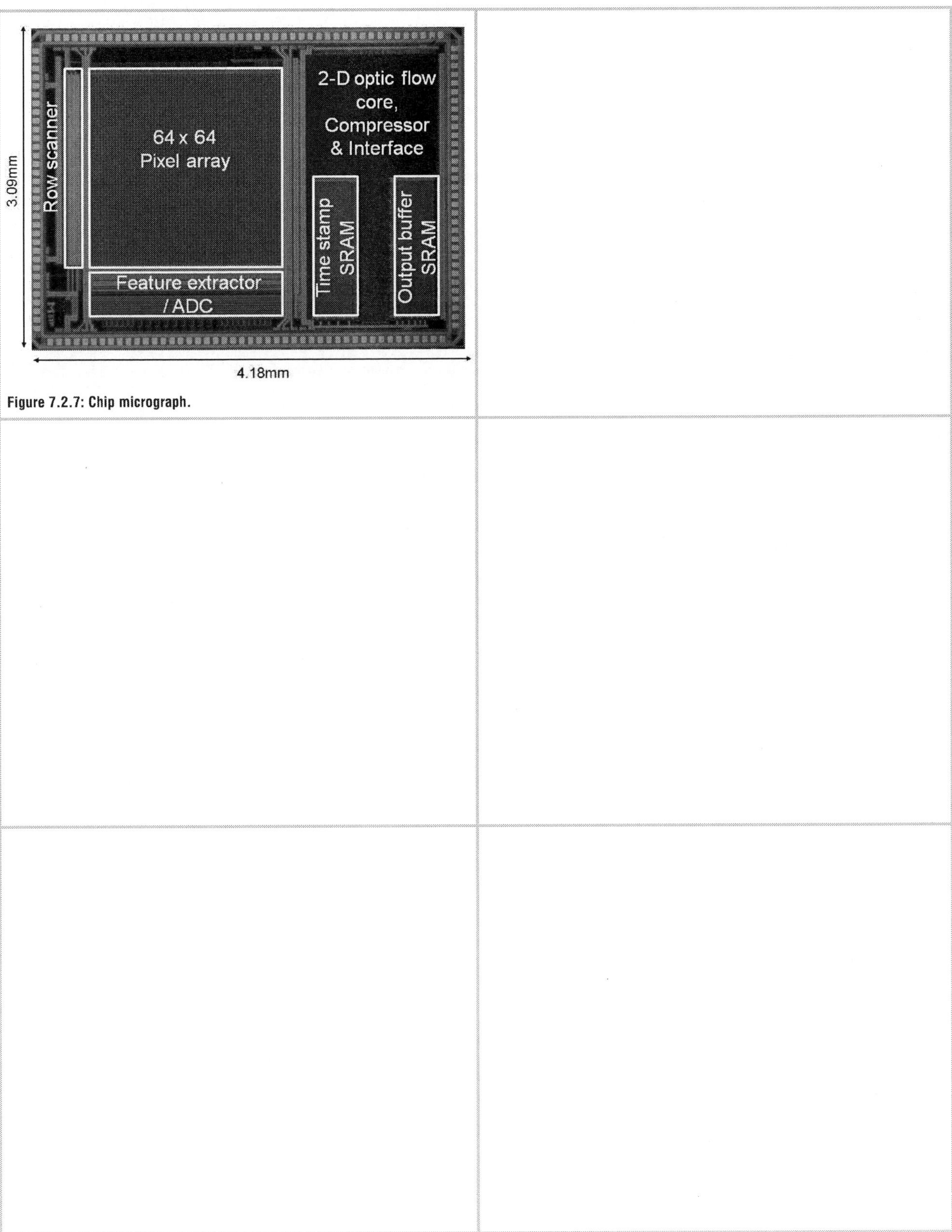

Figure 7.2.7: Chip micrograph.

7.3 A 1000fps Vision Chip Based on a Dynamically Reconfigurable Hybrid Architecture Comprising a PE Array and Self-Organizing Map Neural Network

Cong Shi[1,2], Jie Yang[1], Ye Han[1], Zhongxiang Cao[1], Qi Qin[1], Liyuan Liu[1], Nan-Jian Wu[1], Zhihua Wang[2]

[1]Chinese Academy of Sciences, Beijing, China,
[2]Tsinghua University, Beijing, China

A vision chip is a high-speed and compact vision system that integrates an image sensor and parallel image processors on a single silicon die. Nowadays, high-speed vision chips with powerful recognition capabilities are greatly demanded in applications such as: industrial automation, security, entertainment, robotic vision, and human-machine interaction. Some 100-to-1,000fps vision chips have been reported [1-4]. These chips integrate pixel-parallel and row-parallel SIMD array processors to speed up low- and mid-level image processing [1,2]. Recently, microprocessors (MPU) have been embedded to carry out high-level image processing [3,4]. Although excellent in low- and mid-level processing, these systems are poor in high-level feature vector (FV) recognition tasks due to the von Neumann bottleneck of the MPU. As a consequence, these chips can no longer achieve 1,000fps system-level performance, from image acquisition to high-level feature-recognition processing.

This paper reports on a 1,000fps vision chip based on a dynamically reconfigurable hybrid architecture. It integrates a high-speed image sensor, von Neumann-type pixel-parallel and row-parallel array processors, and a non-von Neumann-type Self-Organizing Map (SOM) neural network. The SOM network can significantly speed up the high-level image processing in a vector-parallel fashion [5]. Furthermore, the SOM network can be dynamically reconfigured from the pixel-parallel array processor at negligible costs of ~2% total chip area and 3 clock cycle latency. The reconfiguration technique extends the capability of conventional pixel-parallel array processors from low-level to high-level processing, and avoids significant chip area budget for a standalone SOM network. The chip can achieve 1,000fps system-level performance even when complicated high-level recognition tasks are involved.

Figure 7.3.1 shows the system architecture of the vision chip. It mainly consists of a 256×256 4T-APS rolling-shutter pixel array, a 256 10b cyclic-ADC array with CDS and PGA circuits, a 64×64 pixel-parallel 1b processing element (PE) array processor, a 64 row-parallel 8b row processor (RP) array, a thread-parallel dual-core 32b RISC MPU, and a 16×16 vector-parallel SOM neural network. The pixel-parallel PE array processor can be dynamically reconfigured as the SOM network for high-level image processing, and then back as the PE array for low-level image processing, as demand requires. This reconfiguration method is enabled by the similar mesh grid topology between the two components. As the PE array is smaller than the pixel array in size, dynamically mapping relationships with different pixel sampling intervals can be established between the two arrays (Fig. 7.3.1, bottom). The PE and RP array processors perform low- and mid-level processing, such as image filtering, image segmentation, mathematical morphology, and FV extraction. The SOM network recognizes FVs with a speedup of 16×16, avoiding the serious bottleneck in the high-level recognition processing. The dual-core MPU performs other simpler non-recognition high-level processing tasks, as well as the overall chip management.

The PE array processor can be divided into 16×16 sub-arrays. Each sub-array contains 4×4 PEs and constitutes one neuron along with a small-size condition generator (CG). Figure 7.3.2 illustrates the dynamic reconfiguration scheme from the PE array processor to the SOM network, and vice versa. If R-signal is logic low, the solid paths among the PEs are activated to form a 2D-meshed 64×64 pixel-parallel PE array processor. Otherwise, the dashed paths among the PEs are activated to reconfigure each sub-array with CG into one SOM neuron. Thus the 64×64 PE array processor is reconfigured into the SOM network with 16×16 neurons. The 16 1b PEs in one neuron are chained in a snaky style to form a 16b processing engine. Each PE has a unique bit-position (bp) in the neuron. The CG is dedicated for conditional operations involved in the SOM network training and recognition procedures. The dynamic reconfiguration can be completed in 3 clock cycles by switching the topological connections among the PEs and the CG.

Figure 7.3.3 shows the PE and RP circuit schematics. The reconfiguration multiplexers (gray colored) in the PE circuit can switch the topological connections between neighboring PEs for different modes. For PE array processor mode, the PE circuit can not only accomplish single-bit operations but also multiple-bit operations by bit-serial processing. For SOM network mode, the 1b ALUs and 1b-wide memories of the 16 PEs in one neuron equivalently constitute a 16b ALU and a 16b-wide memory. The inter-PE connections of *LL*, *LH, AH* support I/O operations and multiplier-free multiplication/division operations based on a bit-shifting method for the SOM neuron. The RP processor contains an 8b ALU and can support some nonlinear and *skip* chain operations.

The SOM network is trained by an online LVQ training procedure [5]. First, the random reference vectors (RV) are inputted to all of the neurons. After a sample FV with known class label is broadcasted to the SOM network, the Manhattan distances between the FV and RVs are computed simultaneously in all the neurons. Then the RP array processor determines the neuron with the minimum distance as the winner, which represents the recognized class of the FV. Finally, RVs of the neurons within the winner neighborhood are updated based on the consistency of the recognized class with the known class. After the network is trained, it can be used to recognize the FV of the real-time sensor image by the above procedure without the RV updating stage.

The vision chip is fabricated in a 0.18μm 1P5M CIS process. Figure 7.3.4 shows the experimental results of >1,000fps hand gesture recognition and face recognition based on 16D and 32D PPED [6] FVs. The high processing speed and ~86% recognition accuracy for hand gesture and face recognition can facilitate many natural human-machine interactions. Figure 7.3.5 compares the image-recognition speed of the SOM network with that of the 32b dual-core MPU at 50MHz on the fabricated chip. The SOM network reduces the processing time by >98% for the hand gesture and face recognitions. The SOM network achieves 45.4 and 31.6 kilo recognitions per second (KRPS) for 16D and 32D vectors, respectively. While the speed of dual-core MPU is only 0.84 and 0.46 KRPS, respectively. The performance improvement of the SOM network over the dual-core MPU for recognition tasks becomes more impressive with the increase of the FV dimension.

Figure 7.3.6 compares the vision chip with the state of the art. Our chip avoids the high-level image-processing bottleneck in previous chips and makes the high-level processing speed well matched with that of the low- and mid-level processing. It exhibits the best system-level performance of >1000fps from image acquisition to high-level feature-recognition processing. Figure 7.3.7 shows the chip micrograph and summarizes the chip specifications.

Acknowledgement:
The authors would like to thank Quanliang Li, Zhe Chen and Yongxing Yang for their technical suggestions. This work was supported by the National Natural Science Foundation of China (Grant No. 61234003), and Special Funds for Major State Basic Research Project of China (No. 2011CB932902).

References:
[1] W. Miao, *et al.*, "A Programmable SIMD Vision Chip for Real-Time Vision Applications", *IEEE J. Solid-State Circuits*, vol. 43, no. 6, pp. 1470-1479, June 2008.
[2] W. Jendernalik, *et al.*, "An Analog Sub-Miliwatt CMOS Image Sensor with Pixel-Level Convolution Processing", *IEEE Trans. Circuits and Systems—I: Regular Papers*, vol. 60, no. 2, pp. 279-289, Feb. 2013.
[3] C. Cheng, *et al.*, "iVisual: An Intelligent Visual Sensor SoC with 2790fps CMOS Image Sensor and 205GOPS/W Vision Processor", *ISSCC Dig. Tech. Papers*, pp. 306-307, Feb. 2008.
[4] W. Zhang, *et al.*, "A Programmable Vision Chip Based on Multiple Levels of Parallel Processors", *IEEE J. Solid-State Circuits*, vol. 46, no. 9, pp. 2132-2147, Sept. 2011.
[5] D. C. Hendry, *et al.*, "IP Core Implementation of a Self-Organizing Neural Network", *IEEE Trans. Neural Networks*, vol. 14, no. 5, pp. 1085-1096, Sept. 2003.
[6] H. Yamasaki, *et al.*, "A Real-Time Image-Feature-Extraction and Vector-Generation VLSI Employing Arrayed-Shift-Register Architecture", *IEEE J. Solid-State Circuits*, vol. 42, no. 9, pp. 2046-2053, Sept. 2007.

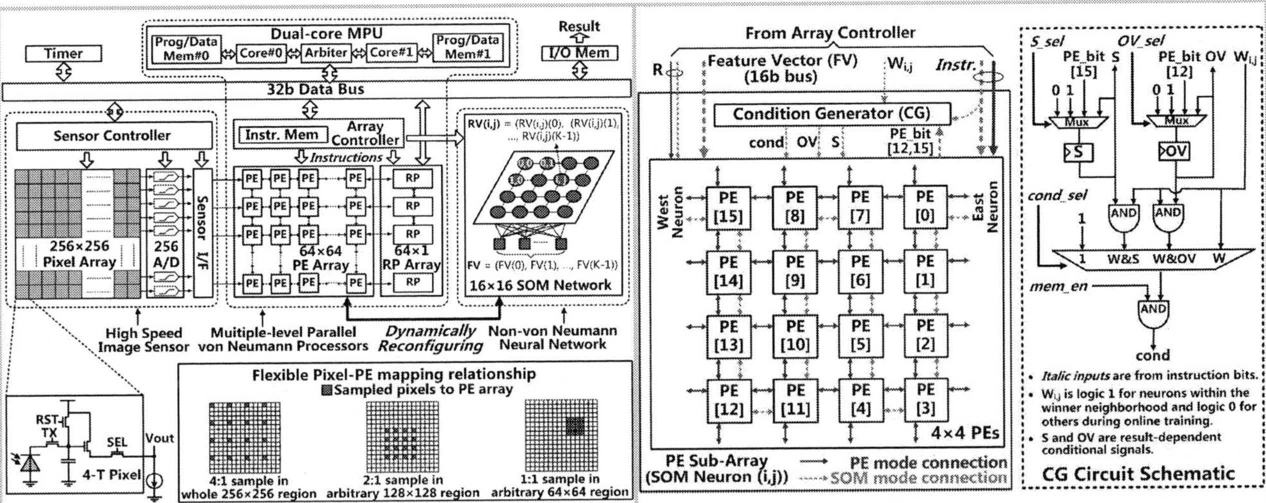

Figure 7.3.1: Vision chip architecture.

Figure 7.3.2: The reconfiguration between PE array and SOM network.

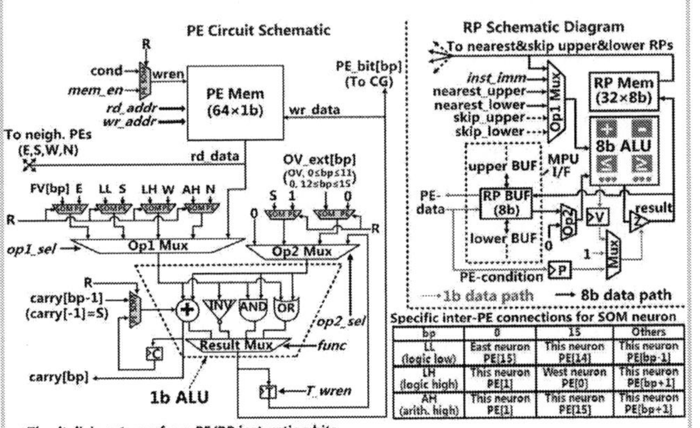

Figure 7.3.3: The PE and RP circuit schematics.

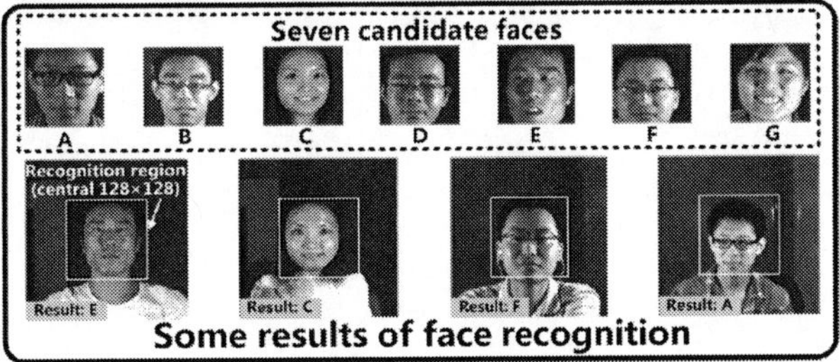

Figure 7.3.4: Experimental results of the fabricated vision chip.

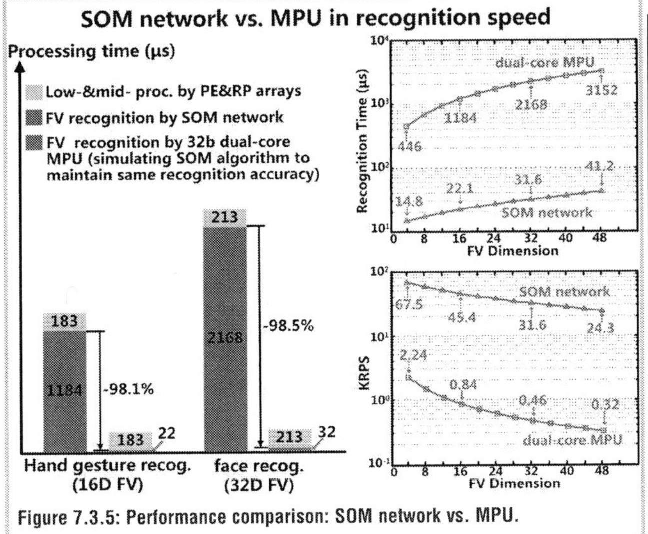

Figure 7.3.5: Performance comparison: SOM network vs. MPU.

		This work	[1]	[2]	[3]	[4]
Technology		0.18μm 1P5M	0.18μm 1P6M	0.35μm 2P4M	0.18μm 2P4M	0.18μm 1P6M
Chip area		82.3mm²	2.3mm²	9.8mm²	70.5mm²	13.5mm²
Power consumption		630mW	8.72mW	0.28mW	455mW	460mW
Clock frequency		50MHz	20MHz	N/A	50MHz	100MHz
FOM[1] (with 16D FV)		0.1834	~0	~0	0.0380	0.0273
FOM (with 32D FV)		0.1675	~0	~0	0.0192	0.0137
High speed image sensor	Sensor resolution	256×256	16×16	64×64	128×128	128×128
	Pixel fill factor	60%	3%	23%	N/A	58%
	Sensitivity	8.1V/Lux·s	N/A	1.4V/Lux·s	N/A	N/A
	Dynamic Range	48dB	N/A	58dB	N/A	N/A
	ADC resolution	10b	1b	N/A	8b	8b
Parallel image processors	Parallelism	Pixel-, row-, thread-, vector-	Pixel-, row-	Pixel-	Row-	Pixel-, row-
	Processor reconfigurability	Dynamically between PE array processor and SOM network	No	No	No	Statically among different-grained PE array processors
	Neural network	16×16 SOM network	No	No	No	No
	MPU	32b dual-core	No	No	32b single-core	8b single-core
	Low-level proc.	Fast	No	Moderate	Moderate	Fast
	Mid-level proc.	Fast	Fast	No	Fast	Moderate
	High-level proc.	Fast	No	No	Slow	Slow
	Performance in low-,mid-levels	12GOPS	0.2GOPS	0.1ms, 3x3conv. (~0.45GOPS)	76.8GOPS	44GOPS
	Performance in high-level recog.	45.4KRPS[2]@16D 31.6KRPS@32D	0	0	1.24KRPS@16D 0.60KRPS@32D	0.16KRPS@16D 0.08KRPS@32D
	system-level frame rate	1340fps@face recognition	N/A	N/A	360fps@posture recognition	76fps@face recognition

[1]FOM = $(GOPS^{-1}+KRPS^{-1})^{-1}$/Area(mm²)·Power(W).

[2]KRPS = kilo recognitions per second.

Figure 7.3.6: Comparison with the state of the art.

978-1-4799-0917-9/14 $31.00 © 2014 IEEE

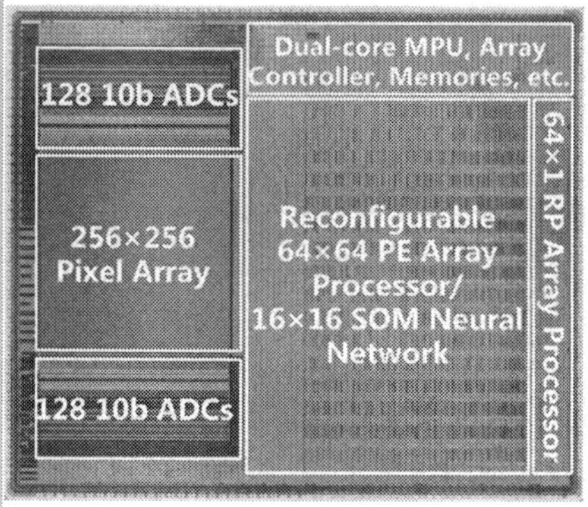

Technology	0.18μm 1P5M CIS
Chip area (pad incl.)	9.8mm×8.4mm
Package	BGA 312pin
clock frequency	50MHz
Supply voltages	1.8V for digital and ADC 3.3V for I/O and pixel
Power consumption	630mW
Sensor resolution	256×256
Max. sensor frame rate	1500fps@256×256 3000fps@128×128 6000fps@64×64
Pixel size	10μm×10μm
Pixel fill factor	60%
Pixel Sensitivity	8.13V/Lux.s
Pixel Dynamic Range	48dB
ADC resolution	10b
PE array size	64×64
Memory per PE	64×1b
RP array size	64×1
SOM network size	16×16 neurons
Count of MPU cores	2
Processor performance	PE array processor: 12GOPS @8b gray img 109GOPS @1b binary img RP array processor: 1.6GOPS @8b data Dual core MPU ~0.1GOPS @32b data SOM network: 45.4KRPS* @16D vector 31.2KRPS* @32D vector

Figure 7.3.7: Chip micrograph and specifications.

7.4 A 413×240-Pixel Sub-Centimeter Resolution Time-of-Flight CMOS Image Sensor with In-Pixel Background Canceling Using Lateral-Electric-Field Charge Modulators

Sang-Man Han[1], Taishi Takasawa[1], Tomoyuki Akahori[2], Keita Yasutomi[1], Keiichiro Kagawa[1], Shoji Kawahito[1,2]

[1]Shizuoka University, Hamamatsu, Japan,
[2]Brookman Technology, Hamamatsu, Japan

Time-of-Flight (ToF) range imagers have a wide range of applications, such as 3D mice, gesture-based remote controllers, amusement, robots, security systems, and automobiles. Numerous ToF range imager developments have been reported [1-4]. Recent developments are often based on CMOS image sensor technology with pinned photodiode options [5-7], which are suitable for cost-effective mass production. Reported CMOS ToF range imagers use single-tap or two-tap lock-in pixels; to cancel the influence of background light, two or four sub-frames are used to produce a background-canceled range image. These architectures, however, have difficulty with precise range measurements of moving objects, because background light cancelation is not guaranteed for moving objects. Lock-in pixels without any charge-draining gate suffer from background light during the readout time of the operation. Another important issue with CMOS ToF range imagers for high range resolution is the speed of lock-in pixels, which must be improved to use high-modulation-frequency light or short-duration light pulses.

To address these problems and requirements, this paper presents a CMOS ToF range imager using pinned-photodiode high-speed lock-in pixels with background light-canceling capability. The lock-in pixel structure, which uses lateral electric field control, is suitable for implementing a multiple-tap charge modulator while achieving high-speed charge transfer for high time resolution.

Figure 7.4.1 shows the charge modulator using lateral electric field control with three-tap outputs and a drain. In this lateral electric field modulator (LEFM), 4 sets of gates (G_1, G_2, G_3 and G_D) are used for applying a lateral electric field in the channel region created in a pinned diode. The gates are not used for transferring photo charge under the gates, but for controlling the electric field in the Y-Y` and $X_1$-$X_1$` directions in Fig. 7.4.1. To do this, a small positive voltage (HIGH=1.8V) and negative voltage (Low=-0.8V) are used for the operation. As shown in the cross-section and potential profile of X_2-X_2` in Fig. 7.4.1, the depleted potential of the pinned diode can be modulated by applying a negative or small positive voltage to the gates while maintaining the potential barrier to the gate region [8]. For example, when G_1 is HIGH and the others are LOW, the potential of the LEFM (see profiles of Y-Y` and $X_1$-$X_1$`) attracts photo electrons generated in the aperture region to be transferred to a floating diffusion of the terminal T_1. Similarly, photo electrons can be transferred to the terminal T_2, terminal T_3 or the drain by applying a HIGH level to G_2, G_3, G_D, respectively, and LOW levels to the others. During signal readout, a HIGH level is applied to G_D, and LOW levels to the other gates for preventing the influence of background light. To realize a large potential modulation in the pinned diode, the doping concentration of the p-type epitaxial layer and surface p+ layer for hole pinning are optimized.

Figure 7.4.2 shows the timing diagram with a small duty ratio light pulse. The gate pulse width of G_1, G_2, and G_3 is given by $R_D T_C$, where T_C is the cycle time and R_D is duty ratio of the gate pulse to the cycle time. The gate pulse width of G_D is given by (1-3R_D) T_C. The signal light pulse width and the time of flight of the received light are denoted by T_0 and T_d, respectively. The difference of the amount of charges between the two nodes (T_2, T_3) reflects the time of flight of the light pulse. The range is calculated by the delay-dependent charges. The background light is assumed to be constant with time. The background light charge may disturb the accuracy of the range calculation. In order to cancel the background light, the T_1 node is used for taking background light charges only. By subtracting the output of T_1 from T_2, T_3, the background light can be canceled. The equation for estimating the range in each pixel is given by:

$$L = \frac{cT_0}{2} \cdot \frac{S_2 - S_1}{S_2 + S_3 - 2S_1}$$

where c is the speed of light, and S_1, S_2 and S_3 are outputs of T_1, T_2, and T_3, respectively.

The sensor architecture and the pixel schematic are shown in Fig. 7.4.3. The ToF imager consists of a 413×240 pixel array, vertical and horizontal shift register, analog-to-digital converter (ADC), and a LEF charge modulator driver. Each pixel has three floating-diffusion (FD) node outputs (T_1, T_2, T_3), which are shared by 11 charge modulators (CMs) and are connected to inputs of source followers and reset transistors. In order to increase full well capacity, these FD nodes are also connected to MOS capacitors. The full-well capacities can be changed by switching the MOS switches. Each of the three source followers' outputs from each pixel are connected to each of three column ADCs and are converted to digital codes in parallel using high-resolution folding integration/cyclic ADCs [9].

A prototype sensor is fabricated using a 0.11μm 1P 4M standard CMOS image sensor process. A relatively thick lightly doped p-type epitaxial layer is used for enhancing the response of photo-carriers generated by the near infrared light.

Figure 7.4.4 shows range linearity and range resolution as a function of distance with background light of 4000lx (right) and without background light (left). Measurements are done using 10×10 pixels. The influence of background light is cancelled by the operation explained above. The non-linearity errors with and without background light are 1.5cm at most for the distance range of 0.8 to 1.8m. A good range resolution of 4 to 6.4mm is obtained if there is no background light. Under background illumination of 4,000lx, range resolution is from 0.8 to 2.7cm depending on the distance to be measured.

Figure 7.4.5 shows range images of a moving hand under the background illumination of 2000lx using in-pixel background cancelling (upper) and background cancelling with two consecutive frames (lower). The in-pixel background canceling nicely works so that the range image is captured without visible errors, whereas the background canceling with two consecutive frames causes large motion artifact due to the time difference between the two frames. Figure 7.4.6 shows a performance comparison of the present ToF range imager chip with recently reported works. Because of the short light pulse (13ns) modulation with a small duty cycle, the average illumination power is relatively small while achieving a high range resolution of less than 10mm. The microphotograph of the ToF range imager chip is shown in Fig. 7.4.7.

Acknowledgment:
This work was partly supported by the Grant-in-Aid for Scientific Research (S), No. 25220905.

References:

[1] P. Seitz, *et al.*, "Demodulation Pixels in CCD and CMOS technologies for Time-of-Flight Ranging", *Proc. SPIE*, vol. 3965, pp. 177-188, 2000.
[2] S. Kawahito, *et al.*,"A CMOS Time-of-Flight Range Image Sensor with Gates-on-Field-Oxide Structure," *IEEE Sensors J.*, vol. 7, no. 12, pp. 1578-1586, Dec. 2007.
[3] D. Stoppa, *et al.*, "A Range Image Sensor Based on 10μm Lock-In Pixels in 0.18μm CMOS Imaging Technology," *IEEE J. Solid-State Circuits*, vol. 46, no. 1, pp. 248-258, Jan. 2011.
[4] O. Shcherbakova, *et al.*, "3D Camera Based on Linear-Mode Gain-Modulated Avalanche Photodiodes," *ISSCC Dig. Tech. Papers*, pp. 490-491, Feb. 2013.
[5] S. J. Kim, *et al.*, "A CMOS Image Sensor Based on Unified Pixel Architecture with Time-Division Multiplexing Scheme for Color and Depth Image Acquisition," *IEEE J. Solid-State Circuits*, vol. 47, no. 11, pp. 2834-2845, Nov. 2012.
[6] S.J. Kim, *et al.*, "A 1920×1080 3.65μm-Pixel 2D/3D Image Sensor with Split and Binning Pixel Structure in 0.11μm Standard CMOS," *ISSCC Dig. Tech. Papers*, pp. 396-397, Feb. 2012.
[7] W. Kim, *et al.*, "A 1.5Mpixel RGBZ CMOS Image Sensor for Simultaneous Color and Range Image Capture," *ISSCC Dig. Tech. Papers*, pp. 393-394, Feb. 2012.
[8] S. Kawahito, *et al.*, "CMOS Lock-in Pixel Image Sensors with Lateral Eelectric Field Cotrol for Time-Resolved Imaging, " *Proc. 2013 Int. Image Sensor Workshop*, pp. 361-364, June 2013.
[9] M.-W. Seo, *et al.*, "A Low-Noise High Intrascene Dynamic Range CMOS Image Sensor With a 13 to 19b Variable-Resolution Column-Parallel Folding-Integration/Cyclic ADC," *IEEE J. Solid-State Circuits*, vol. 47, no. 1, pp. 272-283, Jan. 2012.

978-1-4799-0917-9/14 $31.00 © 2014 IEEE

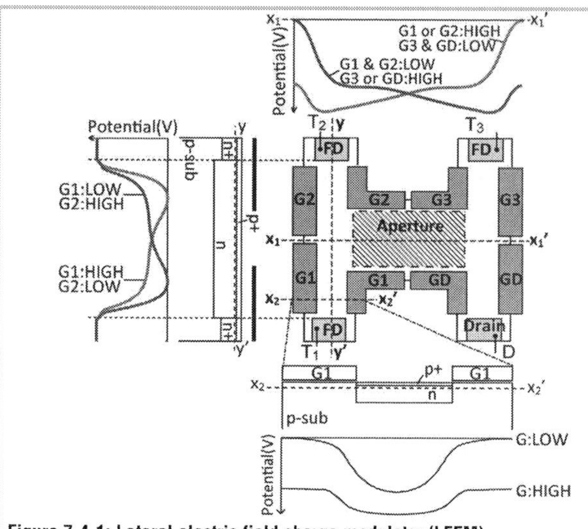

Figure 7.4.1: Lateral electric field charge modulator (LEFM).

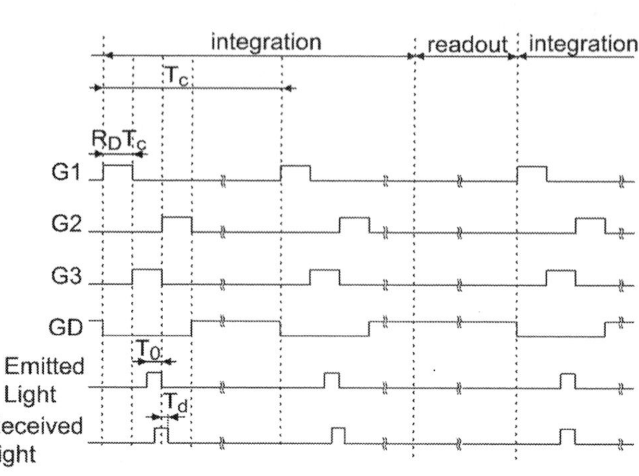

Figure 7.4.2: Timing diagram for the lock-in pixel operation.

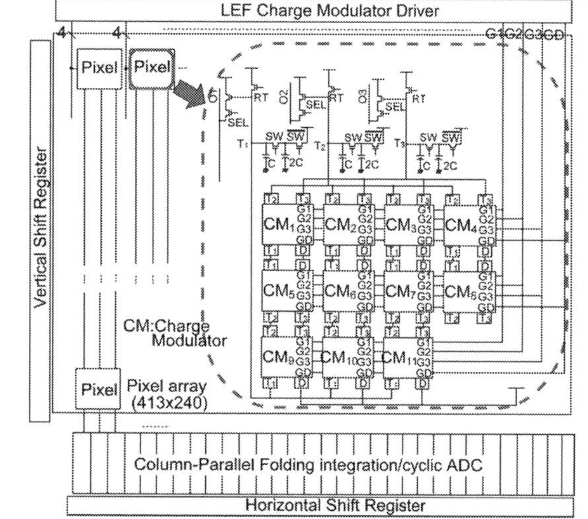

Figure 7.4.3: Pixel and sensor schematic diagram.

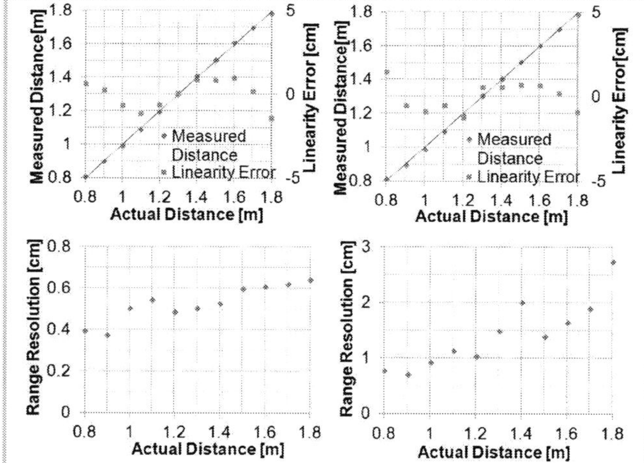

Figure 7.4.4: Range linearity and range resolution as a function of actual distance without background light (left) and with background light of 4,000lx (right).

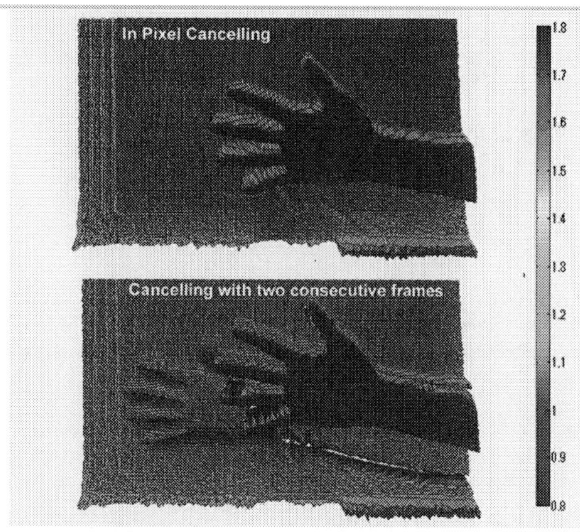

Figure 7.4.5: Range images of a moving hand under the background illumination of 2,000lx.

Parameter	This work	Stoppa et al.[4]	S. Kim et al.[6]	W. Kim et al.[6]
Resolution	413 X 240	64 X 64	500 X 274	480 X 270
Pixel size (um²)	16.8 X 16.8	30 X 30	14.6 X 14.6	2.25 X 9
process	0.11 um 1P4M	0.35 um 1P4M	0.11 um 1P4M	0.13 um
Fill factor (%)	50.5 (with microlens)	25.75	38.5	48
Lens f-number	1.4	1.2	1.6	1.2
Illumination wavelength (nm)	870	850	850	850
Illumination power (mW)	250	21 W/m² (peak power @ 1m)	700	450
Integration time (ms)	50	-	20/40/80	10
Frame rate (fps)	15	50/200	11 (Int. time 40ms)	-
Modulation frequency(MHz), Light Pulse width(ns)	13 ns (Light width) 30 ns (Gate width)	25	20	20
Measured Range (m)	0.8 ~ 1.8	2 ~ 4.75 @ 50fps	0.75 ~ 4.5	1~7
Non-linearity (%)	1.03	1.7	0.93	2.25
Range resolution (mm)	4.0 ~ 6.4	19 ~ 57 @ 50fps	10 ~ 38 (with different int. time)	5~160
In pixel Background Cancelling	Yes	No	No	No
Frames to make Range image	1 frame	4 frame	2 frame	4 frame

Figure 7.4.6: Performance comparison with recently reported works.

978-1-4799-0917-9/14 $31.00 © 2014 IEEE

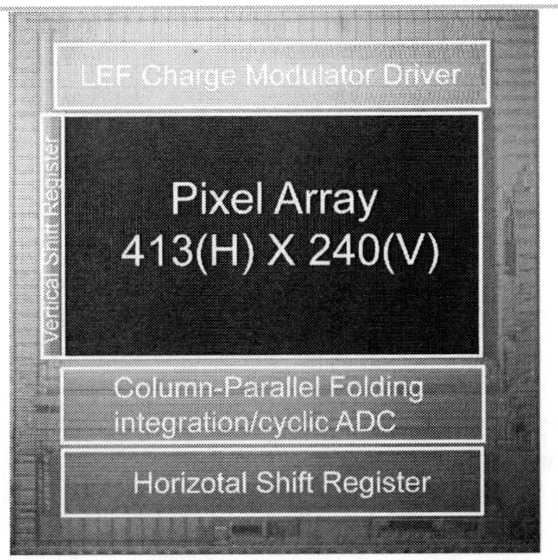

Figure 7.4.7: Chip microphotograph.

7.5 A 0.3mm-Resolution Time-of-Flight CMOS Range Imager with Column-Gating Clock-Skew Calibration

Keita Yasutomi, Takahiro Usui, Sang-Man Han, Taishi Takasawa, Keiichiro Kagawa, Shoji Kawahito

Shizuoka University, Hamamatsu, Japan

Recently, 3D scanning systems have attracted rapidly rising attention in combination with 3D printers. One of the common technologies in contactless 3D scanners is the light-section method, which has advantages in term of accuracy. The method, however, requires a long base line between a camera and light source to achieve high resolution and a mechanical scanning system. A high range resolution Time-of-Flight (ToF) imager provides new possibilities of implementing a miniature head, which allows flexible scanning of an object with a complicated structure. The range resolution of reported CMOS ToF imagers [1-3] is limited to a few centimeters. For higher resolution, higher modulation frequency is required. However, the modulation frequency used for CMOS ToF imagers is limited to several tens of MHz.

This paper presents a ToF imager with 0.3mm range resolution, which corresponds to 2ps time resolution. To achieve this high resolution, the imager uses a ToF measurement technique based on an impulse photocurrent response [4] and draining-only modulation (DOM) pixels [5]. To realize a range imager with 2D pixel array, column-wise gating-clock skew calibration is implemented to demonstrate simultaneous sub-mm ToF measurements for the whole pixel array.

Figure 7.5.1 shows an indirect ToF measurement technique, phase diagram and the DOM detector. In the technique, a short pulse laser is used as a light source, which can be regarded as an impulse input. Since the range calculation is determined only by the photocurrent response, distortion of the light source is negligible. When the photocurrent response is assumed to be linear, a time of flight (t_{TOF}) is given by:

$$t_{TOF} = T_{offset} - T_0 \cdot \sqrt{\frac{2(N_2 - N_3)}{N_1 - N_3}} \quad (1)$$

where N_1, N_2 and N_3 are signal electrons whose accumulation is controlled by the gating clock signals, TD(1), TD(2) and TD(3), respectively. The 3 signals are accumulated and read out sequentially as shown in Fig. 7.5.1(b). T_{offset} determines measureable range. T_0 corresponds to the photocurrent response of the DOM devices.

Since T_0 corresponds to the photocurrent response itself, it should be reduced to attain a higher resolution. For this reason, the modulator employs the DOM structure as shown in Fig. 7.5.1(c). In the DOM detector, when a draining gate (TD) is opened, all of the generated electrons in the photodiode are drained out. While the TD is closed, generated electrons are transferred into a floating diffusion (FD). The charge transfer and draining are controlled by a lateral electric field. Since the DOM detector does not have any transfer gate in its signal path, high-speed charge modulation and loss-less repetitive accumulation are achieved.

Figure 7.5.2 shows the sensor architecture with skew calibration and pixel circuitry. Unlike in [5], signal charge is stored in the FD, which has large capacitance (>100fF) by adding a MOS capacitor, in order to enhance the handling charge capacity. To satisfy both high-speed charge transfer and high sensitivity, unit size of the DOM detector is designed to be 2.8×2.8μm². Therefore, 48 detectors are arrayed and connected in parallel in the pixel size of 22.4×22.4μm² Since the response of TD clocks is also involved in the total photocurrent response, an in-pixel buffer is implemented in each pixel. The rest of pixel area is filled by MOS capacitors for power-supply stabilization and in-pixel circuits.

As shown in Fig. 7.5.2, the gating clock of the DOM detector, TD, is provided through an inverter tree, skew-calibration circuit, and clock drivers, causing a different delay from pixel to pixel. In particular, the skew due to device mismatch and voltage drop of the power supply line in the modulation clock driver are inevitable. In the sensor, the available range is a few tens of millimeters, which corresponds to a few hundred picoseconds. If the skew is comparable or larger than the measurable range, each pixel has a different measureable range; thereby, simultaneous capturing for all pixels cannot be achieved. To calibrate the skew column by column, 2-stage voltage-controlled delay lines (VCDLs) with 7b current-steering DACs are implemented in every column. The 1st and 2nd stages of VCDL are used as coarse and fine calibration, respectively.

The skew-calibration procedures are as follows; first, the skew is measured when all registers are set to one, which is the fastest condition. The skew is extracted from modulation characteristics shown in [4] by using an external digital delay generator. The register setting of the 1st-stage VCDL (Coarse) is adjusted, in which the delay meets the slowest column. Then, the skew with the coarse calibration is measured in the same way as the first skew measurement. The register of the 2nd-stage VCDL (Fine) is also adjusted to the slowest column.

The ToF imager is fabricated in 0.11μm CIS technology. The pixel array is 310(H)×120(V), in which main pixel count is 132(H)×120(V). Figure 7.5.3 shows a measured skew distribution with and without the skew calibration. As expected, a large skew of $850ps_{p\text{-}p}$ or $173ps_{rms}$ is observed before calibration. The skew includes both random and systematic components, which are presumed to be due to device mismatches and a voltage drop of the power supply line, respectively. After the calibration, the skews are reduced to $81ps_{p\text{-}p}$ or $8ps_{rms}$, which are within the measurable range.

Figure 7.5.5 shows a measured distance and resolution as a function of the distance to a mirror target. In this measurement, data of all the main pixels, i.e., 132×120 pixels, are taken into account. The integration time is set to 33ms at approximately 12.5μW/cm² on the focal plane. From the measurement results, the non-linearity is below 3%FS at 32mm range. The resolution within the measurable range is measured to be 0.3mm on average. Figure 7.5.6 shows a sample image taken by the test imager. The target is composed of three gauges with different thickness of 1mm, 2mm, and 3mm.

The sensor performance is summarized in Fig. 7.5.6. The frame rate is determined by the total pixel count, i.e., 310×120 pixels. Although pixels other than the main ones are unused, their signals are read out and their calibration circuit registers are set to a low value to reduce power consumption. The wavelength and pulse width of the laser are 443nm and 71.6ps, respectively. A die micrograph is shown in Fig. 7.5.7.

Acknowledgments:
The authors thank to M. Fukuda for his support in the measurements and T. Akahori and Y. Kaneko, Brookman Technology Inc. for their help in the chip design. This work was partly supported by the Grant-in-Aid for Scientific Research (S), No. 25220905 of the Ministry of Education, Science, Sports and Culture (MEXT) and supported by VLSI Design and Education Center (VDEC), the University of Tokyo in collaboration with Cadence Design Systems, Inc. and Mentor Graphics, Inc.

References:
[1] D. Stoppa, *et al.*, "A Range Image Sensor Based on 10-μm Lock-In Pixels in0.18-μm CMOS Imaging Technology," *IEEE J. Solid-State Circuits*, vol. 46, no. 1, Jan. 2011.
[2] S-J. Kim, *et. al.*, "A CMOS Image Sensor Based on Unified Pixel Architecture With Time-Division Multiplexing Scheme for Color and Depth Image Acquisition," *IEEE J. Solid-State Circuits*, vol. 47, no.11, pp. 2834-2845, Nov. 2012.
[3] S. Kawahito, *et al.*, "A CMOS Time-of-Flight Range Image Sensor with Gates-on-Field-Oxide Structure," *IEEE Sensors J.*, vol. 7, no.12, pp. 1578-1586, Dec. 2007.
[4] K. Yasutomi, *et al.*, "A Time-of-Flight Image Sensor with Sub-mm Resolution Using Draining Only Modulation Pixels," *Proc. 2013 Int. Image Sensor Workshop*, pp. 357-360, June 2013.
[5] Z. Li, S. Kawahito, *et al.*, "A Time-Resolved CMOS Image Sensor with Draining-only Modulation Pixels for Fluorescence Lifetime Imaging," *IEEE Trans. Electron Devices*, vol. 59, no. 10, pp. 2715-2722, Oct. 2012.

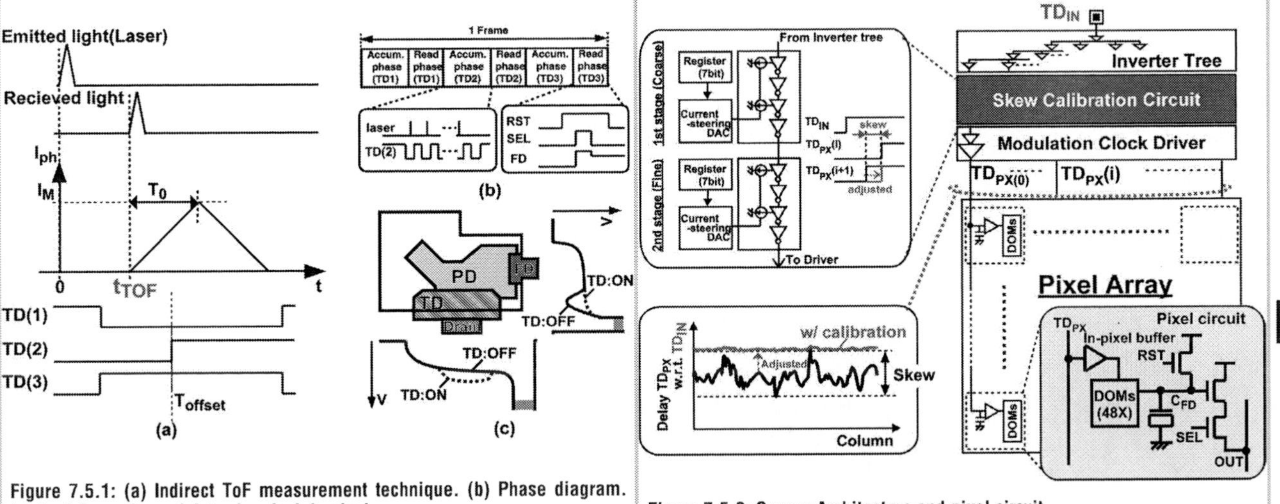

Figure 7.5.1: (a) Indirect ToF measurement technique. (b) Phase diagram. (c) Draining-only modulator in a lock-in pixel.

Figure 7.5.2: Sensor Architecture and pixel circuit.

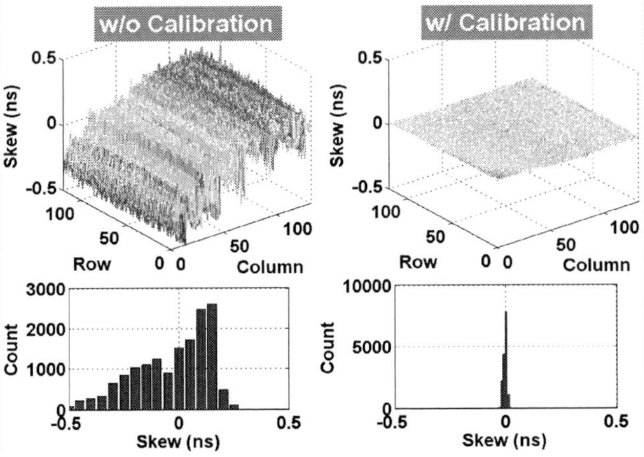

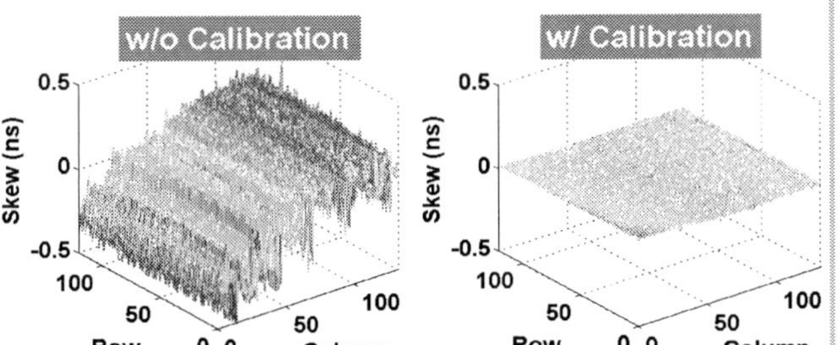

Figure 7.5.3: Measured skew with and without calibration.

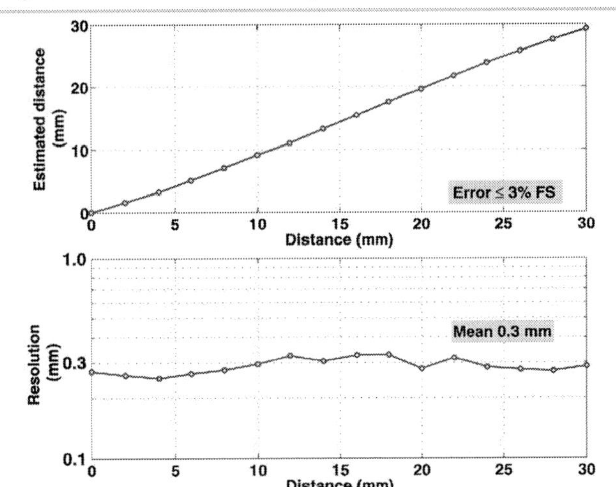

Figure 7.5.4: Measured distance and resolution as a function of the distance to a mirror target.

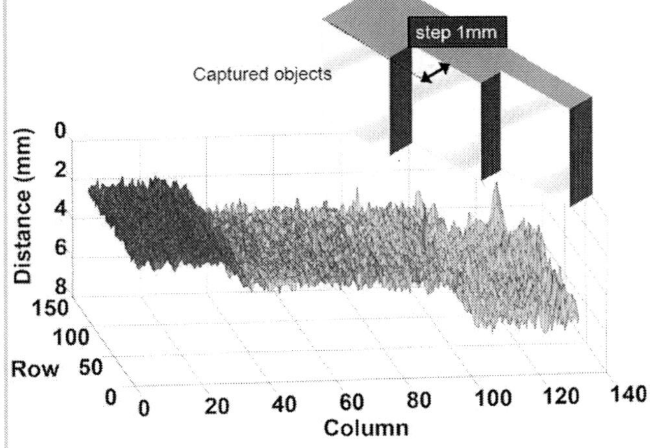

Figure 7.5.5: Sample 3D image in 500-frame average.

Technology		0.11-µm CIS
Total pixels		310(H) × 120(V)
Effective pixels		132(H) × 120(V)
Pixel size		22.4 x 22.4 µm²
Frame rate (Integration time 13.3ms/33ms)		22 fps / 9.6 fps
Repetition frequency		7.5 MHz
Fill factor (without micro-lens)		24 %
Emitter	Wavelength	443 nm
	Pulse width	71.6 ps
	Power	12.5 µW/cm² at focal plane
Measurable Range		32 mm at error < 3% Full scale
Range Resolution		0.3 mm (mean in measurable range)

Figure 7.5.6: Performance summary.

978-1-4799-0917-9/14 $31.00 © 2014 IEEE

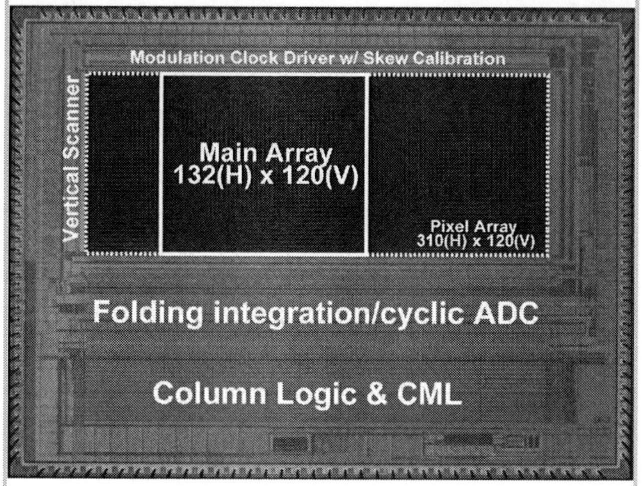

Figure 7.5.7: Chip micrograph.

7.6 A 512×424 CMOS 3D Time-of-Flight Image Sensor with Multi-Frequency Photo-Demodulation up to 130MHz and 2GS/s ADC

Andrew Payne, Andy Daniel, Anik Mehta, Barry Thompson, Cyrus S. Bamji, Dane Snow, Hideaki Oshima, Larry Prather, Mike Fenton, Lou Kordus, Pat O'Connor, Rich McCauley, Sheethal Nayak, Sunil Acharya, Swati Mehta, Tamer Elkhatib, Thomas Meyer, Tod O'Dwyer, Travis Perry, Vei-Han Chan, Vincent Wong, Vishali Mogallapu, William Qian, Zhanping Xu

Microsoft, Mountain View, CA

Interest in 3D depth cameras has been piqued by the release of the Kinect motion sensor for the Xbox 360 gaming console [1,2,3]. This paper presents the pixel and 2GS/s signal paths in a state-of-the-art Time-of-Flight (ToF) sensor suitable for use in the latest Kinect sensor for Xbox One. ToF cameras determine the distance to objects by measuring the round trip travel time of an amplitude-modulated light from the source to the target and back to the camera at each pixel. ToF technology provides an accurate high pixel resolution, low motion blur, wide field of view (FoV), high dynamic range depth image as well as an ambient light invariant brightness image (active IR) that meets the highest quality requirements for 3D motion detection.

Depth and active IR images are produced by combining multiple images that are captured at different phase relationships of the clocks provided to the light source and pixel array. The captures are taken in rapid temporal succession to avoid motion blur. In addition, high differential dynamic range is necessary to simultaneously render high-reflectivity objects near the camera and low-reflectivity objects far from the camera. High dynamic range is realized by allowing each pixel to independently select the best shutter time (*multi-shutter*) and the best amplifier gain setting (*multi-gain*) at each capture.

Due to the multiple captures that need to be taken in rapid succession and the high dynamic range requirements, ADC conversion must be performed many times per capture and due to noise considerations cannot happen simultaneously with integration. Therefore a high-bandwidth 2GS/s 10b, column-parallel ADC is employed. Noise and mismatches are cancelled by using a completely differential design from pixel through ADC.

The ToF chip includes a 512×424 pixel array with 10μm pixel pitch fabricated in a standard TSMC 0.13μm CMOS LP 1P5M process. The 60% fill-factor (effective with μLens) pixel achieves a modulation contrast (MC) of 67% (measured at 50MHz) and a responsivity of 0.14A/W at 860nm. The chip can operate at high modulation frequencies of up to 130MHz to extract maximum depth quality while minimizing system light-source power. The schematic of the fully differential pixel design with a simplified detector plan is shown in Fig. 7.6.1. Capacitors $CInt_A$ and $CInt_B$, are MIM caps and MSF_A & MSF_B are native source followers. Special care was taken in the pixel layout to maximize symmetry.

The pixel-timing diagram for an exemplary capture is also shown in Fig. 7.6.1. Correlated double sampling (CDS) is used to cancel the differential reset kT/C and fixed-pattern noises. During integration of a capture, ClkA & ClkB are modulated 180° out of phase for time t^{int1} at the chosen modulation frequency and relative clock phase between the light source and pixel array. Exemplary integrated signals D_A & D_B are also shown in Fig. 7.6.1. At the end of integration, ClkA & ClkB are turned OFF and Read is turned ON to take an integration sample ($BitlineA^{int1}$ – $BitlineB^{int1}$). If multi-shutter is used, one or more integrations with different exposure times (e.g., t^{int2}, t^{int3}) may follow before the pixel is Reset again for the next light phase/frequency. To further avoid pixel Common-Mode saturation in the presence of large amounts of ambient light, *Common-Mode-Reset* (CMR), which cancels common mode while preserving differential mode may be performed intermittently between integration cycles (as shown in Fig. 7.6.1) [4].

Figure 7.6.2 shows a cut view of the pixel detector when modulation clock ClkA (connected to poly-gates A shaped like fingers normal to the figure) is at ground and ClkB (connected to poly-gates B) is at a higher positive bias. Under these bias conditions, negative photo charges are collected under the gate oxide of the poly-gates. Upon collection under a poly gate, charges diffuse (in a direction normal to the plane of the figure) to a floating diffusion (FD) n+ collection node not in the plane of the figure. Potential barriers created by p+ doped areas between gates ensure that charges collected by one gate are never transferred to an adjacent gate even if it is at a higher potential.

Electric field lines from gates A (respectively B) shown in black (respectively white) define two distinct non-overlapping zones: a large ZoneA where the field lines terminate under gates A and a much smaller ZoneB for B. Notice that the electric field lines run tangent to the boundary between ZoneA and ZoneB (dotted line in Fig. 7.6.2). Photo charges generated in ZoneA are collected under gates A and similarly under gates B for ZoneB. Because ZoneA is much larger than ZoneB, most photo charges created by light arriving while ClkA is high are collected by gates A (and similarly for B). This assigns photo charges to either A or B depending on their arrival time with respect to ClkA and ClkB. The ratio (ZoneA-ZoneB)/(ZoneA+ZoneB) is approximately the pixel modulation contrast (MC) at low frequencies.

Since charges are never transferred between A and B, charge assignment to A or B occurs concurrently with charge collection under the gates and does not need an additional step of shifting charges between gates [3]. This makes our method (called: Quantum Efficiency Modulation) suitable for high modulation frequency. Ultimately, charges are collected at an FD node, but the timing is decoupled from charge assignment and can thus be performed *leisurely*.

The chip signal path starts with the 10μm-pitch differential column amplifiers shown in Fig. 7.6.3. First, the amplifier offset and pixel array column output voltages are sampled simultaneously onto 320fF input capacitors. Next, the input and feedback circuits are configured for high gain. Finally, the high gain output is compared with the ADC reference. If it would result in a saturated digital value at the output of the ADC, the amplifier is switched to a lower gain. This overflow event is reported to the system along with the ADC data.

The offset-cancellation switched-capacitor amplifier architecture allows programmable gain and provides better than 1% gain matching. It has a rail-to-rail input common-mode range and an output-referred offset that is nearly independent of gain. The amplifier also shifts the operating voltage from 3.3V to 1.5V to allow use of more efficient core transistors in the ADC.

The outputs from each group of 4 column amplifiers drive a 10b 8MS/s space-efficient 0.027mm^2 [5] successive-approximation ADC. A bank of 6 sampling capacitors capture the 4 column amplifier outputs on a round-robin basis. The complete 4:1 multiplexed converter is drawn on a 40μm pitch with an LSB capacitance of 4fF. 10b conversions with 8b ENOB complete every 10 clock cycles at 80MHz. A total of 256 ADCs on the chip produce over 2GS/s thus converting a full chip image capture in approximately 100μs.

The digitized output from the ADCs flows into the Shutter Engine, which choreographs CDS, multi-shutter, and multi-gain. The captured image is accumulated in an on-chip buffer and transferred via 8-lane 1Gb/s MIPI DPHY. The image capture rate is doubled by placing readout circuits at both the top and bottom of the pixel array.

Accuracy error of the system is generally better than 1% as presented in Fig. 7.6.4, which shows the mean measured distance vs. actual distance up to a range of 3.5m. This data is an average of 100 frames collected at 30fps where the reflectivity of the target was 10%. The standard deviation, also shown in Fig. 7.6.4, is 1cm at 3.5m with indoor fluorescent lighting and 1.5cm with 2.2μW/nm/cm^2. The high resolution, low motion blur, and wide FoV features of the sensor appear clearly in the captured depth and active IR images shown in Fig. 7.6.5. This ToF camera is capable of detecting up to 6 different persons in the FoV at a distance of about 3m. Figure 7.6.5 also shows a depth map generated at closer range that demonstrates the fine granularity of the depth data, which can detect not only individual fingers but wrinkles in clothing as well. A summary of important key metrics of this ToF sensor are listed in Fig. 7.6.6. Finally, a chip micrograph is presented in Fig. 7.6.7.

References:

[1] C. Niclass, *et al.*, "A 0.18μm CMOS SoC for a 100m-Range 10fps 200×96-Pixel Time-of-Flight Depth Sensor", *ISSCC Deg. Tech. Papers*, pp. 488-489, Feb. 2013.

[2] L. Pancheri, *et al.*, "A QVGA-Range Image Sensor Based on Buried-Channel Demodulator Pixels in 0.18μm CMOS with Extended Dynamic Range", *ISSCC Deg. Tech. Papers*, pp. 394-395, Feb. 2012.

[3] W. Kim, *et al.*, "A 1.5Mpixel RGBZ CMOS Image Sensor for Simultaneous Color and Range Image Capture," *ISSCC Dig. Tech. Papers*, pp. 392-393, Feb. 2012.

[4] C. Bamji, *et al.*, "Method and System to Differentially Enhance Sensor Dynamic Range" US. Patent 6,919,549 B2, July 2005.

[5] Y. Suh, *et al.*, "A 10-bit 25-MS/s 1.25-mW Pipelined ADC With a Semidigital Gm-Based Amplifier," *IEEE Trans Circuits and Systems II*: Express Briefs, Vol. 60, No. 3, pp. 142-146, Mar. 2013.

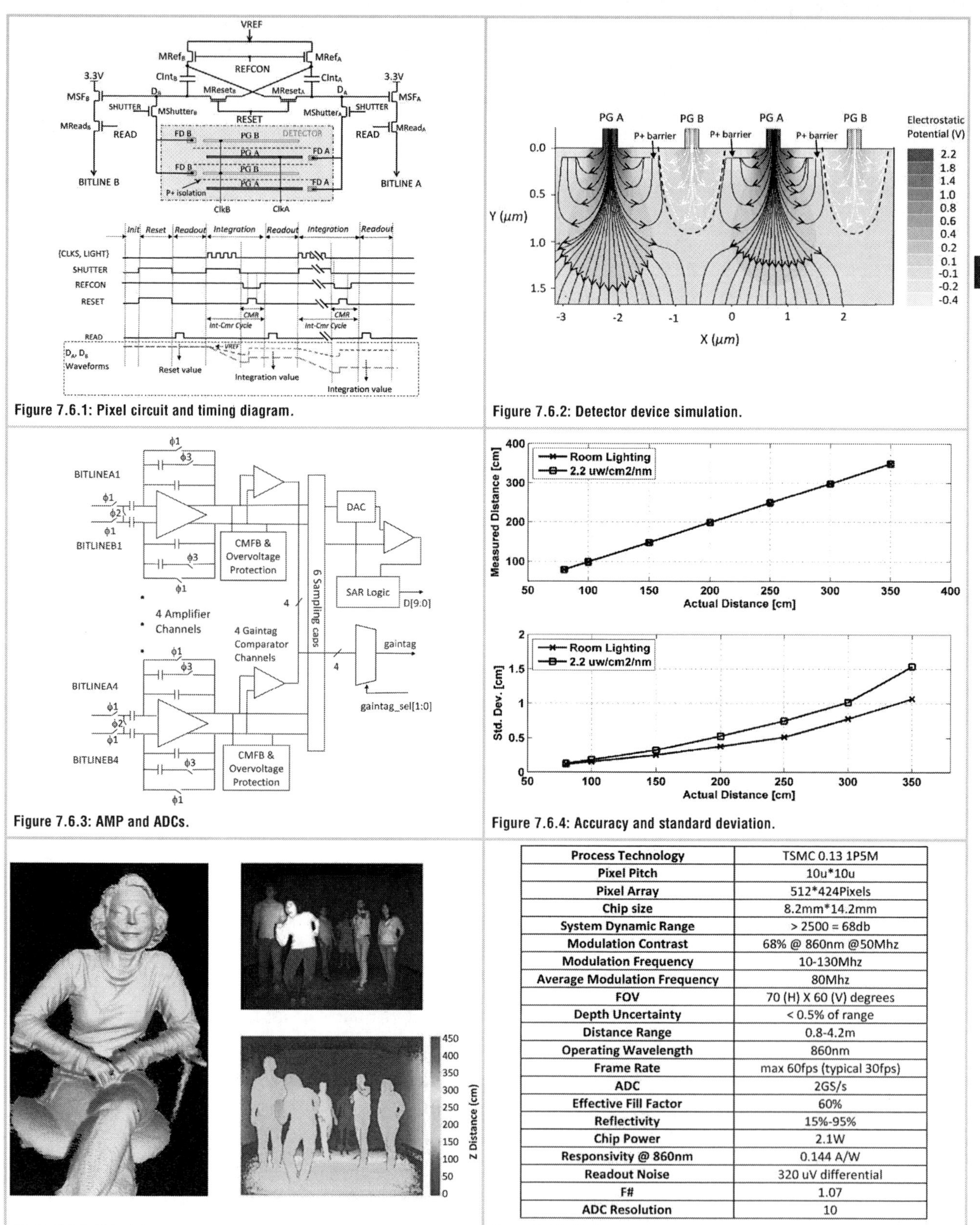

Figure 7.6.1: Pixel circuit and timing diagram.

Figure 7.6.2: Detector device simulation.

Figure 7.6.3: AMP and ADCs.

Figure 7.6.4: Accuracy and standard deviation.

Figure 7.6.5: Chip data.

Process Technology	TSMC 0.13 1P5M
Pixel Pitch	10u*10u
Pixel Array	512*424Pixels
Chip size	8.2mm*14.2mm
System Dynamic Range	> 2500 = 68db
Modulation Contrast	68% @ 860nm @50Mhz
Modulation Frequency	10-130Mhz
Average Modulation Frequency	80Mhz
FOV	70 (H) X 60 (V) degrees
Depth Uncertainty	< 0.5% of range
Distance Range	0.8-4.2m
Operating Wavelength	860nm
Frame Rate	max 60fps (typical 30fps)
ADC	2GS/s
Effective Fill Factor	60%
Reflectivity	15%-95%
Chip Power	2.1W
Responsivity @ 860nm	0.144 A/W
Readout Noise	320 uV differential
F#	1.07
ADC Resolution	10

Figure 7.6.6: Performance parameters.

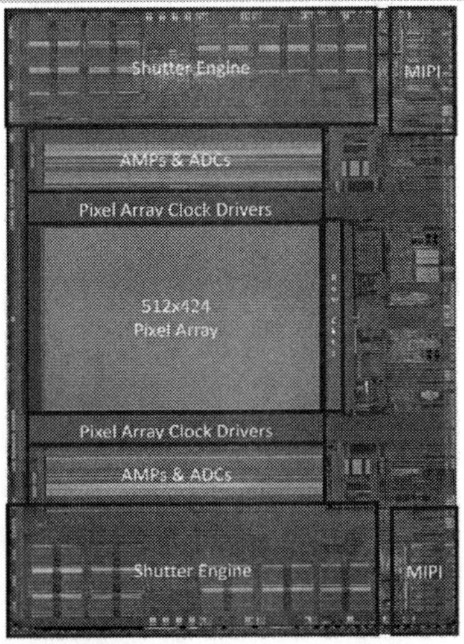

Shutter Engine

MIPI

AMPs & ADCs

Pixel Array Clock Drivers

512x424
Pixel Array

Pixel Array Clock Drivers

AMPs & ADCs

Shutter Engine

MIPI

Figure 7.6.7: Chip micrograph.

Session 8 Overview: *Optical Links and Copper PHYs*

WIRELINE SUBCOMMITTEE

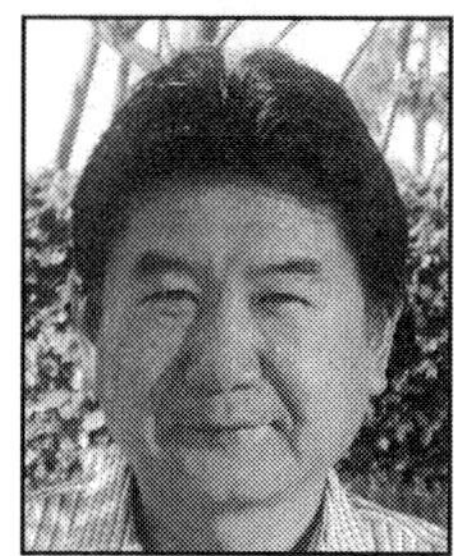

Session Chair: *Ichiro Fujimori*
Broadcom, Irvine, CA

Session Co-Chair: *Hideyuki Nosaka*
NTT, Atsugi, Japan

Cloud computing requires interconnect solutions for distances ranging from tens of kilometers to less than one meter, driving the demand for advances in both optical links and copper-based Ethernet PHYs. The power consumption, cost, data-rate, and density of these designs must all be improved simultaneously, driving innovation in both the circuit and system architectures. This session includes 9 papers, including demonstrations of a highly parallel short-reach 60×10Gb/s optical interconnect, linear as well as non-linear techniques for electronic dispersion compensation of optical-component non-idealities at 6 to 25Gb/s, and a driver enabling a VCSEL rated for 25Gb/s to operate at 40Gb/s. Similarly, a new 10GBASE-T analog front-end and a GPHY driver architecture with rail-to-rail full-duplex operation demonstrate power savings for Ethernet PHY applications. Finally, two new CDR architectures enable robust frequency acquisition and low-jitter clock recovery in reference-less systems.

8.1 A 6Gb/s Transceiver with a Nonlinear Electronic Dispersion Compensator for Directly Modulated Distributed-Feedback Lasers **8:30 AM**
K. Kwon, KAIST, Daejeon, Korea
In Paper 8.1, KAIST presents an electronic-dispersion-compensation circuit (EDC) for directly modulated distributed feedback lasers in 90nm CMOS. At 6Gb/s, the EDC consumes 68mW and achieves OSNR gains of 11.7dB and 14.7dB for 50km and 75km length optical cables, respectively.

8.2 A 12×5 Two-Dimensional Optical I/O Array for 600Gb/s Chip-to-Chip Interconnect in 65nm CMOS **9:00 AM**
H. Morita, Sony, Tokyo, Japan
In Paper 8.2, Sony demonstrates a 600Gb/s optical interface using a 12×5 two-dimensional arrays of VCSELs, photodiodes, and opto-electronic drive/receive circuits. Using 60 links operating at 10Gb/s fabricated in 65nm CMOS, the input sensitivity is 13.3μA_{pp} at a BER of 10^{-12}.

8.3 A Power-Scalable 7-Tap FIR Equalizer with Tunable Active Delay Line for 10-to-25Gb/s Multi-Mode Fiber EDC in 28nm LP-CMOS **9:30 AM**
E. Mammei, University of Pavia, Pavia, Italy
In Paper 8.3, the University of Pavia describes a power-scalable continuous delay-based equalizer for electronic dispersion compensation in multi-mode fibers that operates over a wide range of data-rates. Fabricated in 28nm LP CMOS, the equalizer power scales from 55 to 90mW as the data-rate scales from 10 to 25Gb/s.

978-1-4799-0917-9/14 $31.00 © 2014 IEEE

8.4 A 28Gb/s 1pJ/b Shared-Inductor Optical Receiver with 56% Chip-Area Reduction in 28nm CMOS **9:45 AM**

T-C. Huang, TSMC Design Technology, San Jose, CA

In Paper 8.4, TSMC presents an area- and power-efficient optical receiver. Using a shared inductor, the 28nm CMOS prototype achieves a power efficiency of 1pJ/b and an area reduction of 56% compared to an implementation without inductor sharing. At a BER of 10^{-12}, the vertical eye opening is >0.25UI at 28Gb/s.

8.5 A Sub-1.75W Full-Duplex 10GBASE-T Transceiver in 40nm CMOS **10:15 AM**

J. R. Westra, Broadcom, Bunnik, The Netherlands

In Paper 8.5, Broadcom describes an analog front-end (AFE) for a 4-port 10GBASE-T transceiver chip in 40nm CMOS including transmitter, receiver, and hybrid. For a cable length of 100m supporting full 10Gb/s traffic, the AFE dissipates <1.75W, the lowest power for a 10GBASE-T AFE reported to date.

8.6 A Full-Duplex Line Driver for Gigabit Ethernet with Rail-to-Rail Class-AB Output Stage in 28nm CMOS **10:45 AM**

H. Pan, Broadcom, Irvine, CA

In Paper 8.6, Broadcom presents a driver architecture that enables rail-to-rail full-duplex operation to improve driver efficiency. The architecture is demonstrated in a 2.5V GPHY driver in 28nm CMOS, with the reduced supply resulting in a 24% power reduction as compared to the mainstream 3.3V drivers.

8.7 A 4-to-10.5Gb/s 2.2mW/Gb/s Continuous-Rate Digital CDR with Automatic Frequency Acquisition in 65nm CMOS **11:15 AM**

G. Shu, University of Illinois, Urbana, IL

In Paper 8.7, University of Illinois at Urbana-Champaign demonstrates a BBPD-based automatic frequency acquisition scheme that has unlimited acquisition range and is immune to transition-density variation. The digital CDR, implemented in 65nm CMOS, has a range of 4 to 10.5Gb/s, consumes 22.5mW, and achieves 0.2MHz jitter-transfer bandwidth and 9MHz jitter tolerance.

8.8 An 8.2-to-10.3Gb/s Full-Rate Linear Reference-less CDR Without Frequency Detector in 0.18μm CMOS **11:45 AM**

S. Huang, University of California, Irvine, CA

In Paper 8.8, the UC Irvine describes an 8.2-to-10.3Gb/s full-rate reference-less CDR in 0.18μm CMOS that uses an asymmetric phase-detector transfer curve for frequency detection. The 10.3Gb/s CDR achieves an out-of-band jitter tolerance of $0.58UI_{pp}$ with 2^{31}-1 PRBS.

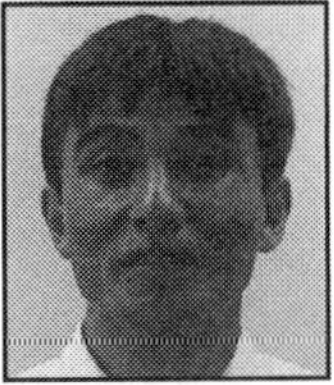

8.9 A 40Gb/s VCSEL Over-Driving IC with Group-Delay-Tunable Pre-Emphasis for Optical Interconnection **12:00 PM**

Y. Tsunoda, Fujitsu Laboratories, Atsugi, Japan

In Paper 8.9, Fujitsu presents a 40Gb/s driver IC over-driving a 25Gb/s VCSEL using a group-delay-compensation circuit for the pre-emphasis. The VCSEL driver achieves 40Gb/s operation with optical modulation amplitude of 2.3dBm and power consumption of 312mW.

978-1-4799-0917-9/14 $31.00 © 2014 IEEE

8.1 A 6Gb/s Transceiver with a Nonlinear Electronic Dispersion Compensator for Directly Modulated Distributed-Feedback Lasers

Kyeongha Kwon, Jonghyeok Yoon, Soon-Won Kwon, Jaehyeok Yang, Joon-Yeong Lee, Hyosup Won, Hyeon-Min Bae

KAIST, Daejeon, Korea

The directly modulated distributed-feedback laser (DML) is widely employed in medium-reach optical links due to its cost effectiveness. However, DMLs are not appropriate for use in fiber links longer than 20km at 6Gb/s or equivalent, because the SNR penalty increases abruptly due to excessive chromatic dispersion caused by frequency chirp. Therefore externally modulated lasers (EMLs), which are more costly, have been a natural choice for applications requiring extended reach. In this paper, a clock and data recovery (CDR) IC that compensates for chromatic dispersion caused by the frequency chirp of the DML is presented. The CDR with EDC is fabricated in a 90nm CMOS process, and the test-chip consumes 226mW at 6Gb/s.

In a DML, the optical wavelength is strongly affected by the emission power of the optical signal, due to a phenomenon called adiabatic chirp. A frequency-chirped optical waveform is dispersed through an optical fiber because waves with different wavelengths propagate at different velocities. Chirp-induced dispersion can be categorized as linear dispersion, called *rabbit-ear*, and nonlinear dispersion, called *tilting*, as shown in Fig. 8.1.1. The rabbit-ear typically appears in every rising edge in widely deployed SMF-28 optical fiber at a wavelength of 1590nm. The difference in power levels between "0" and "1" induces a deviation in the emission frequency, causing "1" to propagate faster than "0," and overlapping between two pulses is eventually created. The tilting effect, with asymmetrical rising and falling times, is caused by power-dependent propagation velocity. Rise time shortens as a signal with larger amplitude propagates faster than the one with the smaller amplitude. Fall time increases because the propagation speed of subsequent pieces of a falling edge decreases gradually [1]. Tilting degrades sampling-point SNR causing a BER penalty because the maximum eye openings for "1"s and "0"s are not aligned.

Figure 8.1.2 shows the overall architecture of the EDC-based CDR. The receiving EDC consists of a linear filter for the elimination of the rabbit-ear and a nonlinear block for the mitigation of the tilting problem. The dispersion-compensated data is sampled and demultiplexed by a factor of 4 by using multiphase clock signals and then 4:1 serialized for transmission. A phase-rotator-based all-digital quadrature CDR is used for robustness and power efficiency. CDR logic and a digital loop filter (DLF) are synthesized and operate at 375MHz. In the transmitter, pattern-dependent pre-emphasis and pulse-widening are performed to mitigate the rabbit-ear effect and duty-cycle distortion, respectively. The rabbit-ear effect increases the BER of the receiver by reducing the sampling-point SNR and impeding proper clock recovery via multiple-pattern-dependent zero crossings (see Fig. 8.1.1). The rabbit-ear effect can be modeled by $r(t) = d_0(t)+d_1(t+\Delta t_{0\leftrightarrow 1})$, where $d_0(t)$ and $d_1(t)$ denote the pulses for "0" and "1", respectively. The model indicates that the rabbit-ear appears during $\Delta t_{0\leftrightarrow 1}$ when two pulses overlap. Therefore, the rabbit-ear effect can be reduced by pre-emphasizing $d_0(t)$ during $\Delta t_{0\leftrightarrow 1}$ before every rising edge as shown in Fig. 8.1.2. In our design, the duration of the pre-emphasis is set at 0.5UI. The rising pattern detector detects the rising edge by comparing the half-UI delayed main-tap (A) and the pre-tap data (B) and creates an adjustment signal (C). Then, the adjustment signal is weighted and subtracted from the main-tap data in the analog domain.

As the traveling optical wave becomes dispersed through the optical fiber, severe duty-cycle distortion occurs and the zero-crossing points of the received data shifts towards zero. This phenomenon impairs the edge-detection process of the CDR. Thus, a 4:1 multiplexer (MUX) featuring a pattern-dependent pulse-width control scheme is used as shown in Fig. 8.1.3. The 4:1 MUX operates using quadrature phase clock signals. Because a proper input path is selected through the clock overlaps among the quadrature-phase clock signals, a pattern-dependent pulse-widening scheme can be implemented efficiently via dynamic phase control of the overlapping clock signals. Once falling edges are detected, a phase-shifted clock signal (CLK + W in Fig. 8.1.3), instead of the quadrature clock signal, is chosen in the 2:1 MUX for the multiplexing to increase the pulse-width. The width of the output pulse is programmable via digitally controlled phase interpolators.

The transmitted NRZ signal can be decomposed as sequences of '"0"s and "1"s as shown in Fig. 8.1.4. Because the streams of "0"s and "1"s are complementary, the sequence of "0"s can be represented by an inverted and power-scaled sequence of "1"s. Then the rabbit-ear equation depends on only $d(t)$ replacing $d_1(t)$ as $r(t) = [ER \cdot d(t+\Delta t_{0\leftrightarrow 1}) - d(t)]/(ER-1)$, where ER is the extinction ratio. By using backward substitutions, the desired signal $d(t)$ can be represented by the received signal $r(t)$ as

$$d(t) = (1-1/ER)[r(t) + (1/ER) \cdot r(t-\Delta t_{0\leftrightarrow 1})].$$

This indicates that the weighted sum of current and delayed data eliminates the rabbit-ear. Considering the implementation complexity, the above equation is converted to the frequency domain by using a bilinear transformation,

$$H_{lin} = (1-2/ER)[\{s + 2(ER+1)/(ER-1)/\Delta t_{0\leftrightarrow 1}\} / (s + 2/\Delta t_{0\leftrightarrow 1})].$$

For a large extinction ratio, the response of H_{lin} approaches that of an all-pass filter. However, the locations of pole and zero split as the extinction ratio decreases. Typical extinction ratio of a distributed-feedback laser is 6dB.

The tilting problem can be modeled by Burger's equation [2],

$$dP/dt + [\, c+\alpha(P-L)](dP/dx)=0$$

where $P(x,t)$ denotes the power of the particle at position x at time t, c is the speed of light, L denotes the emitted power for "0"s and $c+\alpha(P-L)$ is the velocity in the $+x$ direction. Burger's equation is identical to a 1st-order wave equation except that the propagation constant is replaced with the power of the propagating signal, which implies that the propagation speed is in proportion to the power of the waveform. To mitigate the tilting problem, we create a virtual channel in which the velocity of a particle is inversely proportional to its power, which is given by $dP/dt + \beta(H+L-P)(dP/dx)=0$, where H denotes the emitted power for "1"s. The appropriate value of β varies depending on the amount of frequency chirp in a laser and the fiber characteristics as shown in Fig. 8.1.4. In this virtual channel, a high-power signal propagates slower than one with low power, and the signal tilts in the opposite direction as it does in a fiber. Then the virtual channel is spatially discretized using an Euler formula as given by

$$dP_o/dt + \beta(H+L-P_i)(P_o-P_i)/\Delta=0,$$

where P_i and P_o denote the respective powers of the input and output particles of the quantized channel, and Δ denotes the spatial quantization parameter, which varies depending on the fiber length. The spatially quantized nonlinear wave equation is converted into second order polynomials using a Volterra series expansion as given by

$$P_o(t) = [p_c/(s+p_c)]\circ P_i(t)-[p_c \cdot s/(s+p_c)^2/(H+L)]\circ P_i^2(t)$$

where $\circ$ denotes the Volterra operator and the pole location is represented by $p_c = c(H+L)/(zHLDC_{chirp}\lambda_c^2)$ where λ_c is the center emission wavelength, D is the fiber dispersion coefficient, z is the fiber length and C_{chirp} is the adiabatic chirp coefficient. It follows that the tilting compensator can be implemented using a multiplier, adder, low-pass filter, and zero-pole filter. An intuitive time-domain waveform reconstruction procedure is shown in Fig. 8.1.4. Finally, a conventional continuous-time linear equalizer (CTLE) is placed subsequent to the nonlinear equalizer to compensate for the chromatic dispersion.

Figure 8.1.5 shows the experimental setup and the measured eye diagrams before and after the EDC-based CDR after 75km SMF transmission of 6Gb/s 2^{31}-1 PRBS when the launch power of the DML is 5dBm. The SFP module LIGHTRON HTR6G-LE200-SV3 is used as a DML source. With the EDC, the rabbit-ear and tilting are completely compensated. Figure 8.1.6 shows the measured BER versus optical SNR (OSNR) with and without the EDC. At a BER of 10^{-6}, the EDC achieves 11.7dB and 14.69dB OSNR gains at 50km and 75km transmission, respectively. The BER-OSNR curve at 0km shows that the DML laser has slight back-to-back OSNR penalty. A micrograph of the chip, fabricated in a 90nm CMOS process is shown in Fig. 8.1.7. The chip consumes 226mW of which the EDC consumes 68mW and the global clock generator consumes 102mW.

References:

[1] P. Krehlik, "Characterization of semiconductor laser frequency chirp based on signal distortion in dispersive optical fiber," *Opto-Electron. Rev.*, vol. 14, no. 2, pp.123-128, June. 2006.

[2] R.C. McOwen, *Partial Differential Equations: Methods and Applications*, Prentice Hall, 1996.

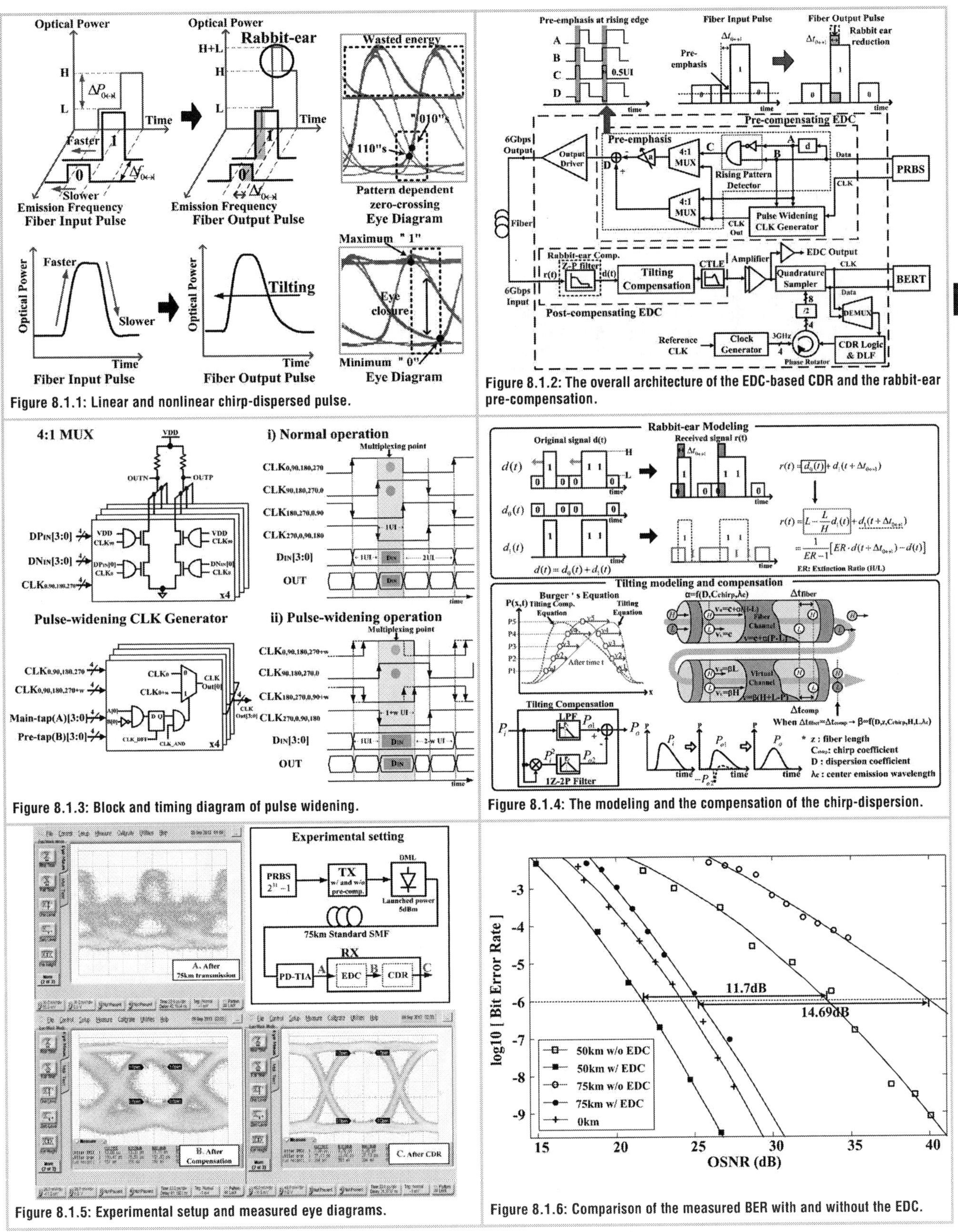

Figure 8.1.1: Linear and nonlinear chirp-dispersed pulse.

Figure 8.1.2: The overall architecture of the EDC-based CDR and the rabbit-ear pre-compensation.

Figure 8.1.3: Block and timing diagram of pulse widening.

Figure 8.1.4: The modeling and the compensation of the chirp-dispersion.

Figure 8.1.5: Experimental setup and measured eye diagrams.

Figure 8.1.6: Comparison of the measured BER with and without the EDC.

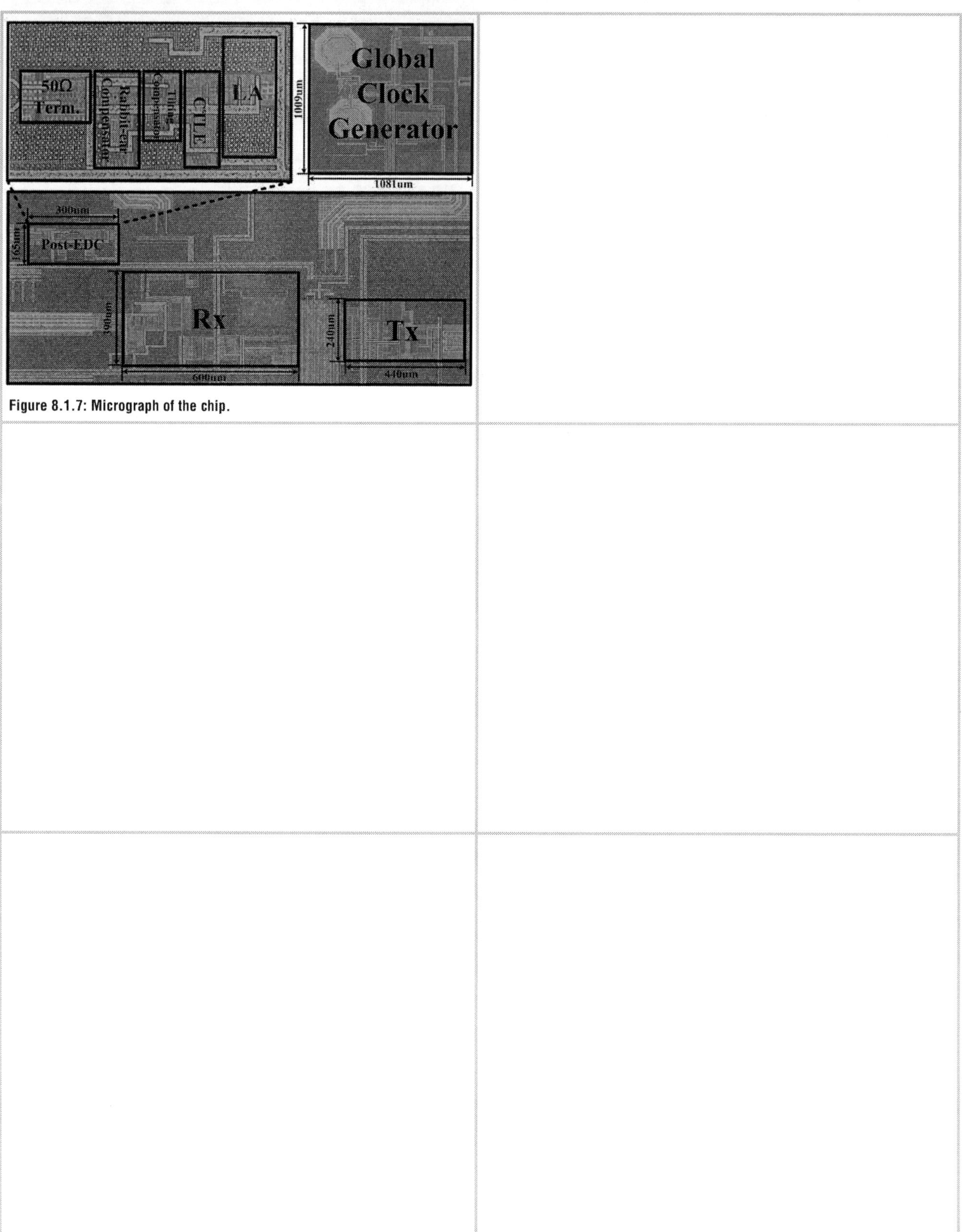

Figure 8.1.7: Micrograph of the chip.

8.2 A 12×5 Two-Dimensional Optical I/O Array for 600Gb/s Chip-to-Chip Interconnect in 65nm CMOS

Hiroshi Morita, Koki Uchino, Eiji Otani, Hiizu Ohtorii, Takeshi Ogura, Kazunao Oniki, Shuichi Oka, Shusaku Yanagawa, Hideyuki Suzuki

Sony, Tokyo, Japan

High-performance systems require high-bandwidth interconnections. The aggregate bandwidth required between two processors, for example, is expected to extend into the terabit-per-second range or higher [1]. Bandwidth is typically the bottleneck in such situations. Optical interconnect technologies have the potential to overcome bandwidth limitations for such chip-to-chip or board-to-board communication [2] through increased channel speed and/or multiple channels. Channel speeds have reached 25Gb/s and higher [3,4], in addition, a 24-channel transmitter and 24-channel receiver is disclosed [5] that employs optical vias in silicon to couple the lens array. Two possible structures to implement a multichannel system are shown in Fig. 8.2.1. A conventional multi-channel architecture places the laser diode drivers (LDD) and VCSELs on the same side of the interposer [1]. This paper describes a 12x5 two-dimensional optical I/O array for 600Gb/s, utilizing 60 channels, each with an operating speed of 10Gb/s. The physical limitation in the number of channels is relaxed by connecting the LDDs through vias to the VCSELs placed on the opposite side of the interposer. The arrangement of the RX, in relation to the two-dimensional photo detector (PD) and TIA array, is the same as the TX. Key elements of each channel are the LDD consuming 2.17mW/Gb/s and the TIA that consumes 0.96mW/Gb/s while achieving an input-referred noise of 0.95μA_{rms}. The low power of the LDD and TIA improve the package reliability while the high sensitivity of the TIA enables the transmission via a long optical waveguide.

Figure 8.2.2 shows the block diagram of the chip fabricated in a standard 65nm logic CMOS process. The chip size is 18×10mm^2 as shown in Fig. 8.2.7. The transmitter and the receiver each have 60 channels organized in a 12x5 matrix. The pitch of LDDs and TIAs is 250μm in the vertical direction and 1500μm in the horizontal direction. The optical devices also have the same pitch. The transmitter channel consists of an input buffer (IB), a pre-driver and an LDD. The equalizer in IB compensates for high-frequency loss with a range of 0 to 9dB at 6GHz. The IB also includes a polarity selector and a signal duty-cycle corrector. The IB amplifies the signal to 1.2V_{pp}, the pre-driver drives the input capacitance of the LDD and the LDD drives the 850nm GaAs VCSEL. A laser-diode supervisor in each channel detects an open or short to GND at the output node as abnormalities and, under these conditions, the LDD is disabled. The automatic power control (APC) maintains constant optical output power. The receiver channel consists of a TIA, a limiting amplifier (LA), and an output buffer (OB). The TIA receives the PD output current, which is in the order of 10μA_{pp}. The LA is CML-type, has a level shifter as the final stage, and amplifies the signal to a full-swing level of 1.2V_{pp}, with the OB driving the 50Ω transmission line. The polarity selector is implemented in the LA. The DC component of the current from the PD is eliminated by the DC canceller, Amp_{DC} in Fig. 8.2.4, at the TIA input, and the A/D in the received-signal-strength Indicator (RSSI) block converts the current to digital.

The LDD, shown in Fig. 8.2.3, consists of a high-frequency driver (HFD) and a low-frequency driver (LFD). The HFD outputs the current I_H, which is the high-frequency component of the modulation current (I_m). The LFD outputs I_L, which consists of bias current I_b and the low frequency component of the I_m. A flat gain versus frequency response is obtained by summing I_H and I_L. This arrangement realizes low power consumption by supplying the power to HFD from 1.2V and separating the HFD from the LFD by an on-chip AC-coupling capacitor C_{HFD}. Due to the high cut-off frequency of LFD (f_{c1}) and HFD (f_{c2}), the value C_{HFD} need not be large. During the initialization process in the power-up sequence, the cut-off frequency f_{c1} is automatically adjusted to the optimum value corresponding to f_{c2} by controlling the value of C_{LFD}. The gain of HFD is also matched to the gain of LFD by controlling the LDO output voltage that determines the amplitude of I_H.

Figure 8.2.4 shows the schematic of the TIA, which consists of a pre-amplifier, a post-amplifier and a single-ended-to-differential converter. All of the buffers are composed of CMOS inverters. Low input-referred noise is achieved by the high gain of CMOS and the low thermal noise of PMOS. In order to further reduce input-referred noise, the pre-amplifier size becomes large, which also has the effect of increasing capacitance and causing bandwidth to decrease. High-frequency boost is therefore applied via the post-amplifier with positive feedback, Amp_1 to compensate for the bandwidth degradation. In the initialization process, Amp_1 is automatically adjusted to the optimum value to cancel the effect of MOS process. To maintain a constant bandwidth through the TIA, a PMOS current source (I_{CORE}) proportional to temperature is added between V_{DD} and V_{CORE}, thus absorbing the effect of process, voltage and temperature variations. Conventional single-ended-to-differential converters rely on a large-area reference-voltage-generation block, for example an RC filter or a replica of the pre-amplifier. To reduce the area cost, the implemented reference-less single-ended-to-differential converter generates the differential signal by the combination of a wideband amplifier, Amp_2, with a gain of -1, and two self-biased amplifiers, Amp_3 and Amp_4, with feedback resistors R_1 and R_2, which keep the four nodes, V_0, V_1, V_{o+} and V_{o-}, at the same bias voltage.

Figure 8.2.5 shows the measured eye diagrams and receiver sensitivity at 10Gb/s, PRBS7 with one channel enabled. The TX optical output jitter is 22.4ps_{pp}, the output average power is 5.9dBm, extinction ratio (ER) is 5.6dB and optical modulation amplitude (OMA) is 6.4dBm, with a VCSEL bias current of 7mA and a modulation current of 8mA. The range of modulation and bias current of the LDD is 2 to 10mA and 2.5 to 12mA, respectively. The power consumption of the LDD is 21.7mW/ch not including the power dissipation of VCSEL at 2V. The pre-driver and LDD occupy an area of 510×250μm^2. An electrical 10Gb/s eye diagram of the RX is measured with a load of 50Ω and the loss of the transmission line de-embedded. The TIA power consumption is 9.6mW/ch. The transimpedance of the TIA is 63.2dBΩ. The area of the TIA is 170×25μm^2 without DC canceller, Amp_{DC}. The maximum input current signal is 2.2mA_{pp}. The input sensitivity reached 13.3μA_{pp} (BER = 10^{-12}), which corresponds to an input-referred noise of 0.95μA_{rms}. The PD capacitance is 270fF and its responsivity 0.55A/W. The power consumption is 69.5mW/ch for transmitter including VCSEL and 68.2mW/ch for receiver.

Figure 8.2.6 shows the measured eye diagrams at the TX and RX outputs with all LDDs enabled at the TX BGA package, and all TIAs enabled at the RX BGA package simultaneously. The BGA package has 862 pins and measures 4.0×3.8cm^2. The 60ch optical waveguide, which is 6mm in width and 400mm in length with 60ch optical connectors at either end, is connected between the TX and RX. The TX and RX motherboards used for measurement have an aperture of 16×12mm^2 and lens coupling is employed to obtain large tolerance between the VCSEL and the waveguide. Transmission at 600Gb/s is confirmed with BER < 10^{-12} for all 60 channels at 10Gb/s operation with PRBS7.

References:

[1] I. Young, et al., "Optical I/O Technology for Tera-Scale Computing," *J. Solid-State Circuits*, vol. 45, no. 1, pp. 235-248, Jan. 2010.

[2] C. Schow, et al., "A 24-Channel, 300 Gb/s, 8.2 pJ/bit, Full-Duplex Fiber-Coupled Optical Transceiver Module Based on a Single "Holey" CMOS IC," *J. Lightwave Technology*, vol. 29, no. 4, pp. 542-553, Feb. 2011.

[3] T. Takemoto, et al., "A 4×25-to-28Gb/s 4.9mW/Gb/s -9.7dBm High-Sensitivity Optical Receiver Based on 65nm CMOS for Board-to-Board Interconnects," *ISSCC Dig. Tech. Papers*, pp. 118-119, Feb. 2013.

[4] J. Proesel, et al., "25Gb/s 3.6pJ/b and 15Gb/s 1.37pJ/b VCSEL-based optical links in 90nm CMOS," *ISSCC Dig. Tech. Papers*, pp. 418-419, Feb. 2012.

[5] F. Doany, et al., "Terabit/Sec VCSEL-Based 48-Channel Optical Module Based on Holey CMOS Transceiver IC," *J. Lightwave Technology*, vol. 31, no. 4, pp. 672-680, Feb. 2013.

978-1-4799-0917-9/14 $31.00 © 2014 IEEE

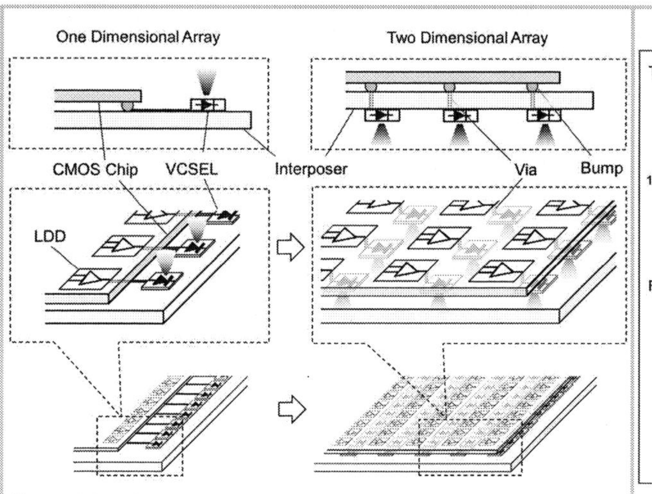

Figure 8.2.1: Conceptual structure of two-dimensional optical transmitter array.

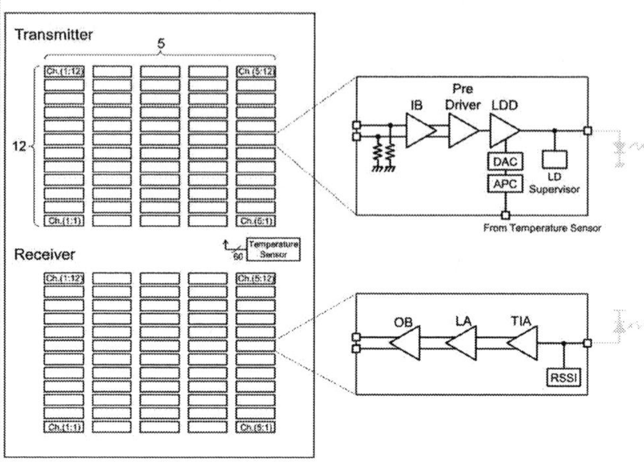

Figure 8.2.2: Block diagram of 60-channel optical I/O chip.

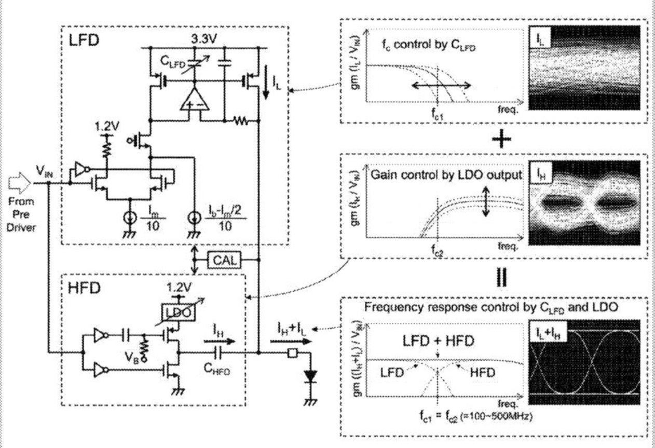

Figure 8.2.3: Schematic of low-power 10Gb/s laser-diode driver in the transmitter.

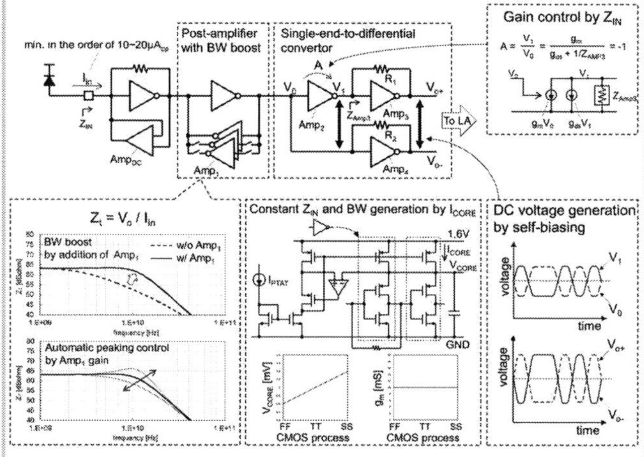

Figure 8.2.4: Schematic of low-power low-noise 10Gb/s TIA in the receiver.

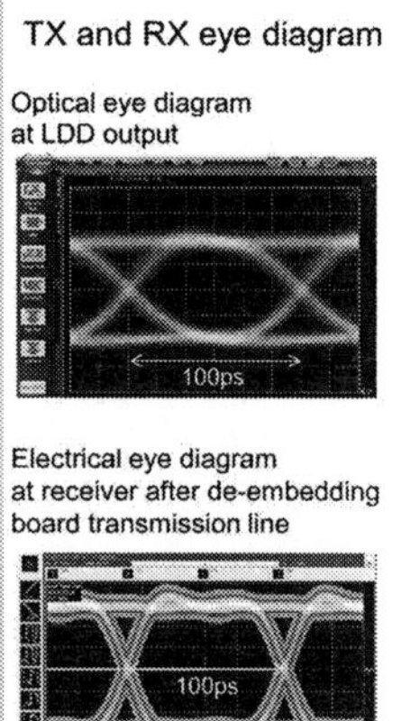

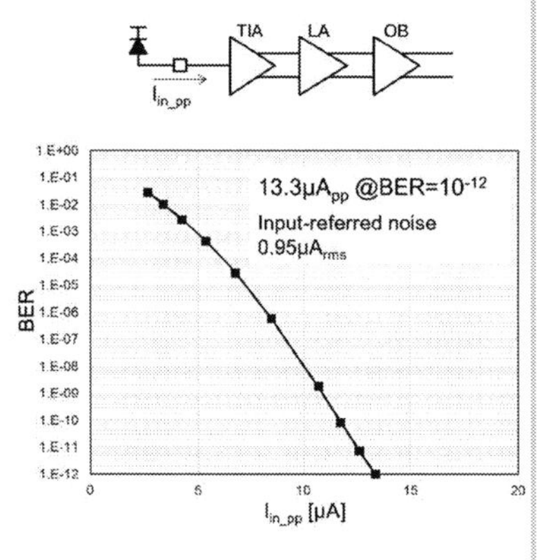

Figure 8.2.5: Measured results of LDD and TIA with one channel operating at 10Gb/s.

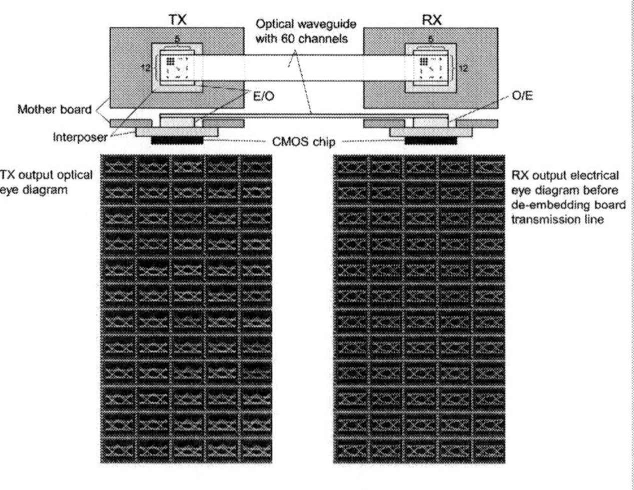

Figure 8.2.6: Eye diagrams at 600Gb/s operation.

978-1-4799-0917-9/14 $31.00 © 2014 IEEE

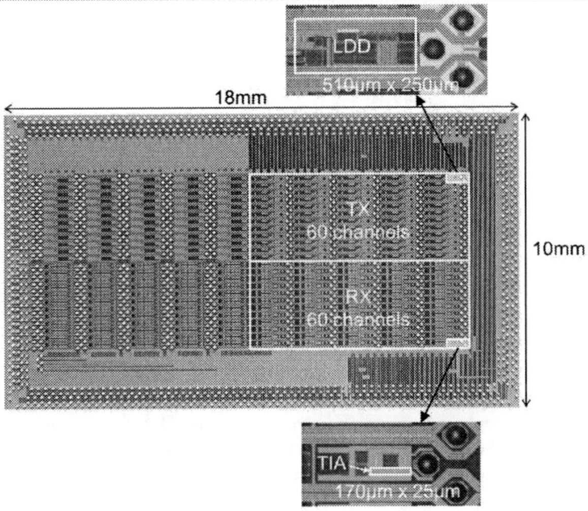

Figure 8.2.7: Die micrograph.

8.3 A Power-Scalable 7-Tap FIR Equalizer with Tunable Active Delay Line for 10-to-25Gb/s Multi-Mode Fiber EDC in 28nm LP-CMOS

Enrico Mammei[1], Fabrizio Loi[1], Francesco Radice[2], Angelo Dati[2], Melchiorre Bruccoleri[2], Matteo Bassi[1], Andrea Mazzanti[1]

[1]University of Pavia, Pavia, Italy,
[2]STMicroelectronics, Cornaredo, Italy

Multi-mode fiber (MMF) is the most cost-effective fiber for high-speed LANs. Modal dispersion leads to optical-energy spreading over several symbol periods, drastically limiting distance and data-rate. Compared with copper channels, equalization is challenging because the channel response varies enormously from fiber to fiber and also over time [1]. These aspects, paired with the practical difficulty of implementing TX pulse shaping, increase the equalization burden at the receiver. To date, electronic dispersion compensation (EDC) consisting of an FIR filter cascaded with a nonlinear equalizer, such as DFE, enables 10Gb/s up to 300m according to the 10GBASE-LRM standard. To satisfy the demand for greater network capacity, solutions to reach 25Gb/s on a single fiber, and up to 400Gb/s aggregated throughput with space-division multiplexing on 16 fibers are being investigated [2]. At this data-rate, robust DSP-based EDCs still need high power, indicating an analog approach to signal processing to reduce power. To have market impact and economic feasibility, the interface must be flexible, accommodating a variable data-rates for compatibility with legacy channels and different standards [2]. In addition, achieving high energy efficiency at each standard (i.e., data rate) is fundamental.

The most critical block in the EDC of Fig. 8.3.1 is the FIR equalizer. Realizations with delay lines based on lumped-LC networks [3-7] need high-quality passive components and large area, making them impractical if multiple I/Os are stepped on the same chip to reach 400Gb/s. Furthermore, since delay is proportional to $\sqrt{LC}$, a wide tuning range requires a large change in component values, trading with quality factor, impedance magnitude and power dissipation. A 4-tap sampled-time FIR equalizer has been recently proposed for high-speed backplane communications [8] but increasing the number of taps, as required by MMF EDCs, makes clock generation and distribution challenging. In this work a 7-tap FIR equalizer based on a tunable active delay line is presented. The solution allows wide tuning range and small area. The equalizer is realized in 28nm low-power (LP) CMOS and proves successful operation from 10 to 25Gb/s. Power dissipation ranges from 55 to 90mW at 1V supply and core silicon area is 0.085mm^2.

The unit element of the delay line is shown in Fig. 8.3.2. It is realized with a one-pole low-pass filter with time constant τ, and a unity-gain path. Ideally, the transfer function provides an in-band group delay of 2τ with flat magnitude. The differential circuit implementation is simple and amenable to high-frequency operation. g_{m1} and the RC load form the low pass filter, while g_{m2}-g_{m3} are used to subtract input signals. Peaking inductors extend the bandwidth, limited by the capacitive loading of the subsequent stage. Compared with a straightforward realization where the low-pass filter outputs are directly connected to g_{m2}, crossing the wires limits the voltage swing at the input of g_{m2} from $2V_{in}$ to $V_{in}/2$. In this case g_{m1} sustains the maximum swing, equal to V_{in}, and 1dB compression point is enhanced by ~6dB. From system-level simulations, a tap-to-tap delay T_d between 2/3T and 3/4T (where T is the symbol period), provides the best performance. By changing the RC time constant of the low-pass filter, T_d can be programmed from 75 down to 30ps, covering the 10-to-25Gb/s data-rates. The roll-off of the group delay close to Nyquist frequency has negligible impact, as confirmed by simulations. By changing biasing currents and load resistors, bandwidth and power dissipation are scaled according to the data-rate from 14GHz with 5.5mW to 25GHz with 9mW. Input voltage at 1dB compression point is 220mV while the integrated equivalent input noise is lower than 700μV_{rms}.

The tap amplifiers shown in Fig. 8.3.1 are 6b digitally programmable transconductors implemented with several thermometer-sized Gilbert cells switched ON and OFF. Summing of the output currents is challenging. The transconductors and their relatively long interconnections result in ~250fF parasitic capacitance on node A (Fig. 8.3.1) introducing a severe gain-bandwidth trade-off with a simple resistive load. To keep high gain with sufficient bandwidth, the feedback transimpedance amplifier (TIA) of Fig. 8.3.3 is designed. M_{n1}-M_{n2} form the core amplifier while R_T is the feedback resistor. M_{p3}-M_{p4} provide the biasing current to the taps. The high output conductance of minimum-length transistors used in the design (g_m/g_{ds}=4-6) impair the TIA performance. Instead of cascoding, which would introduce parasitic poles and limit signal swing, negative resistors (M_{n3}-M_{n6}) are added to partially cancel the device output conductances. Gain, bandwidth and dissipation of the TIA can be configured by programming R_T and switching ON and OFF unit elements of M_{n1}-M_{n2}. The transimpedance gain is programmable from 33 to 46dBΩ while bandwidth scales from 9 to 17GHz. Power dissipation ranges from 10 to 25mW. For measurement purposes, a buffer follows the TIA.

Test-chips (micrograph shown in Fig. 8.3.7) are encapsulated in plastic flip-chip BGA packages and mounted on PCBs. For measurements, a first FIR filter driven by a PRBS generator mimics the channel dispersion of a MMF link. Examples of pulse responses, spread over 4 to 5 symbol periods, are shown in the insets of Fig. 8.3.4 (fully symmetric split pulses are disregarded, being DFE essential for good eye opening [6]). A second chip performs equalization. The output is connected to a sampling scope while a PC running an MMSE adaptation algorithm acquires the scope waveforms and controls the tap gains. Measured eye diagrams before and after equalization at 25Gb/s with T_d set to 30ps are shown in Fig. 8.3.4. Eyes amplitude is ~100mV with peak-to-peak vertical and horizontal openings better than 43% and 57% respectively. From simulations, integrated output noise is less than 4mV_{rms}.

Figure 8.3.5 shows the importance of a wide-tuning-rage delay line to maintain optimal performance at different data-rates. Equalization of two different channel responses at 10Gb/s is compared with T_d = 30ps and T_d = 75ps. With the "Postcursor" channel, performance is quite similar. But in "Split-A" case, raising T_d to the optimal value yields improvement. The peak-to-peak horizontal eye opening increases from 48% to 69%, improving significantly the timing margin. The reason can be qualitatively explained by looking at the channel FFTs, shown on the right of Fig. 8.3.5. Despite the large attenuation at Nyquist, the "Postcursor" channel has a regular low-pass shape. This channel can be equalized with a simple high-pass response, easily implemented also without adjusting T_d. The "Split-A" channel has an in-band notch. Frequencies at which each tap of the FIR introduces boost or attenuation are inversely proportional to T_d, and setting a larger T_d shifts the full FIR equalizing capability to lower frequency allowing better inversion of channels with significant in-band distortion.

Measured results are summarized and compared with published FIR equalizers in the table of Fig. 8.3.6. Thanks to the tunable active delay line, the presented equalizer is flexible, proving successful operation over a large variation of the input data-rate with scalable dissipation. These features are particularly effective to accommodate the demand of variable transfer rates, back-compatibility and energy efficiency of emerging 400Gb/s standards.

References:

[1] R. E. Freund et al., "High-Speed Transmission in Multimode Fibers," *IEEE J. of Lightwave Technology*, pp.569-586, vol. 28, no. 4, Feb 2010.
[2] http://www.ieee802.org/3/400GSG
[3] H. Wu et al., "Integrated transversal equalizers in high-speed fiber-optic systems," *IEEE J. of Solid State Circuits*, pp. 2131-2137, vol. 38, no. 12, Dec. 2003.
[4] S. Reynolds et al., "A 7-Tap Transverse Analog-FIR Filter in 0.13μm CMOS for Equalization of 10Gb/s Fiber-Optic Data Systems," *ISSCC Dig. Tech. Papers*, pp. 330-331, Feb. 2005.
[5] J. Sewter and A. C. Carusone, "A CMOS Finite Impulse Response Filter With a Crossover Traveling Wave Topology for Equalization up to 30 Gb/s," *IEEE J. of Solid State Circuits*, pp. 909-917, vol. 41, no. 4, April 2006.
[6] J. Sewter and A. C. Carusone, "A 3-Tap FIR Filter With Cascaded Distributed Tap Amplifiers for Equalization Up to 40Gb/s in 0.18μm CMOS," *IEEE J. of Solid State Circuits*, pp. 1919-1929, vol. 41, no. 8, August 2006.
[7] A. Momtaz and M. M. Green, "An 80 mW 40 Gb/s 7-Tap T/2-Spaced Feed-Forward Equalizer in 65 nm CMOS," *IEEE J. of Solid State Circuits*, pp. 629-639, vol. 45, no. 3, March 2010.
[8] A. Agrawal et al., "A 19-Gb/s Serial Link Receiver With Both 4-Tap FFE and 5-Tap DFE Functions in 45-nm SOI CMOS", *IEEE J. of Solid State Circuits*, pp. 3220-3231, vol. 47, no. 12, December 2012.

978-1-4799-0917-9/14 $31.00 © 2014 IEEE

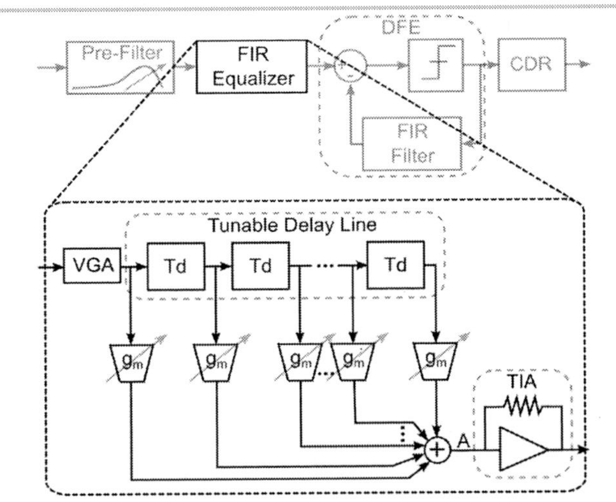

Figure 8.3.1: Block diagram of a typical EDC interface and the designed FIR equalizer.

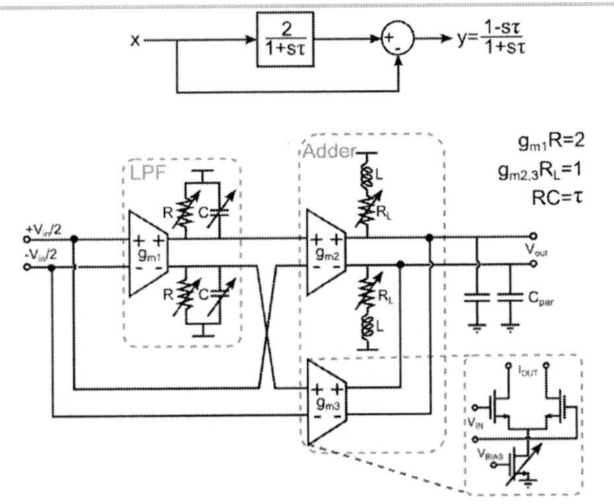

Figure 8.3.2: Block diagram and schematic of the unit element of the delay line.

8

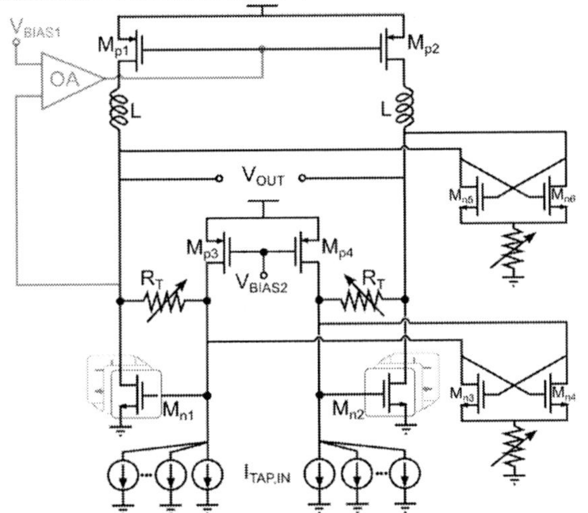

Figure 8.3.3: Schematic of the transimpedance amplifier.

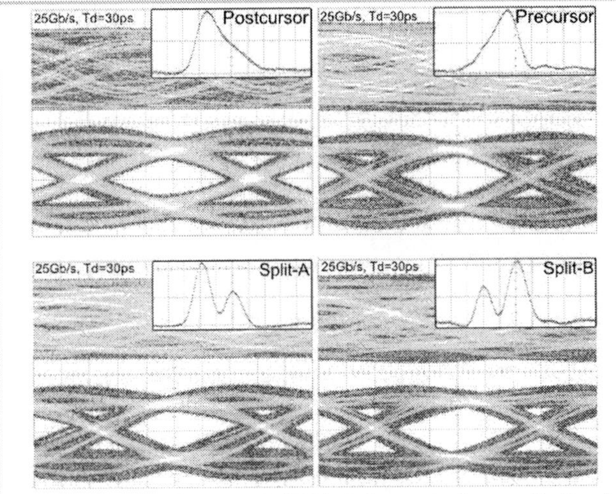

Figure 8.3.4: Eye diagrams at 25Gb/s before and after equalization for different channel responses (Vert. scale 40mV/div).

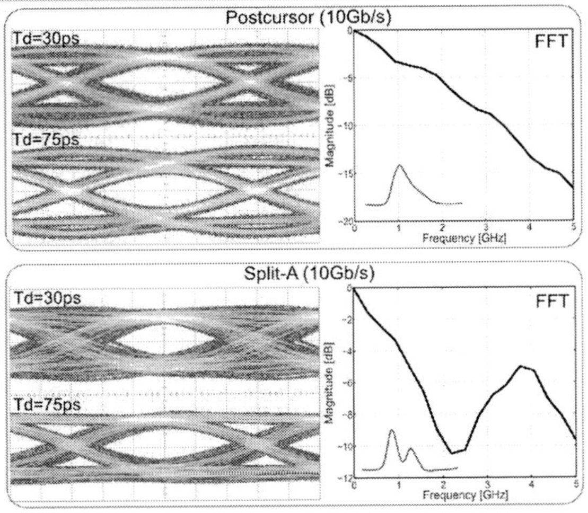

Figure 8.3.5: Comparison of equalization performance at 10Gb/s with and without adjusting T_d (Vert. scale 40mV/div).

Ref.	Tech.	Data Rate [Gb/s]	# Taps	Total Delay [ps]	Power [mW]	Power / (DataRate•TotalDelay) [mW]	Core Area* [mm²]
[3]	180n SiGe	10	7	300	40	13.3	1.9
[4]	130n CMOS	10	7	450	325	72.2	3.8
[5]	90n CMOS	24 - 30	3	70	25	14.8 - 11.9	0.3
[6]	180n CMOS	30 - 40	3	50	70	46.6 - 35	0.45
[7]	65n CMOS	40	7	75	65	21.6	0.75
This Work	**28n CMOS**	**10 - 25**	**7**	**450 - 180**	**55 - 90**	**12.2 - 20**	**0.085**

* Estimated from chip micrograph

Figure 8.3.6: Performance summary and comparison with high-speed FIR equalizers.

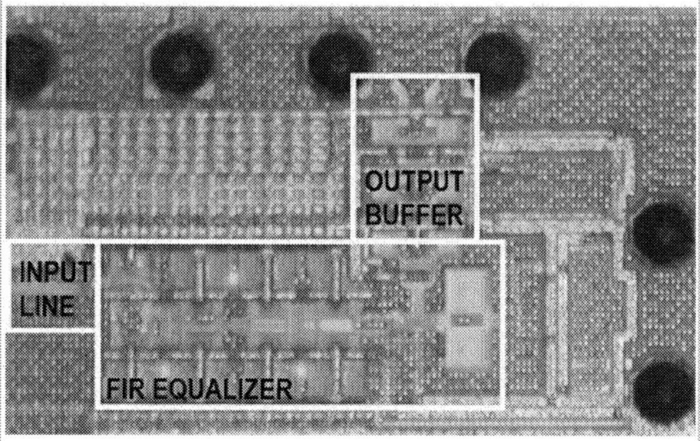

Figure 8.3.7: Chip micrograph.

8.4 A 28Gb/s 1pJ/b Shared-Inductor Optical Receiver with 56% Chip-Area Reduction in 28nm CMOS

Tsung-Ching Huang[1], Tao-Wen Chung[2], Chan-Hong Chern[1], Ming-Chieh Huang[1], Chih-Chang Lin[1], Fu-Lung Hsueh[3]

[1]TSMC Design Technology, San Jose, CA, [2]nVidia, San Jose, CA, [3]TSMC, Hsinchu, Taiwan

Next-generation high-performance computing systems require high-bandwidth serial links to transport high-speed data streams among computational blocks. Optical links have recently attracted attention due to their low channel loss at high frequencies, requiring simpler equalization circuits than electrical links. The energy-efficiency of optical links can thus be significantly improved [1-5]. Broadband techniques such as inductive peaking are commonly used in high-speed optical transceivers for bandwidth enhancement at the expense of the chip area. Inductor-less receivers have been proposed [4,6] to reduce chip area but they usually consume more power or have lower data rates at given technology nodes.

In this paper, we present two optical receivers that each consists of a pseudo-differential CMOS push-pull transimpedance amplifier (TIA), a DC offset-cancellation circuit, a limiting amplifier (LA) with interleaving active-feedback [6], and a T-Coil f_T-doubler output buffer. The block diagram and experimental setup are shown in Fig. 8.4.1. The capacitance of the off-chip GaAs PIN photo-detector (PD), which is wire-bonded to the CMOS receiver, is 100fF with 0.4A/W responsivity. The two optical receivers have identical designs except for the LA, in which two different inductive peaking techniques, conventional and shared-inductor, are designed and fabricated on the same die in 28nm CMOS technology.

Figure 8.4.2 shows the circuit schematics of the pseudo-differential CMOS push-pull TIA with series-peaking inductors. The CMOS push-pull TIA has good signal gain and input-referred noise at low supply voltage because of current re-use of NMOS and PMOS [4]. Compared with CMOS inverter TIA in [4], the presented TIA employs a current tail to make the g_m of *M1~M4* refer to the bias current instead of the supply voltage for better supply noise rejection. In order to provide better single-ended to differential conversion for LA input, we include the cross-coupled pair M_7 and M_8, which act as common-source amplifiers to provide negative voltage gain through the feedback resistor R_F. The pseudo-differential configuration of the TIA provides better supply-noise rejection as well as jitter performance than the singled-ended TIA. The simulated gain of TIA is 46dBΩ and the input-referred noise is $2.5\mu A_{rms}$ with 20GHz BW. The DC-offset cancellation circuit in Fig. 8.4.2 receives pseudo-differential outputs from the TIA and adjusts the output DC levels for the LA, based on the offset voltage provided by the LPF as shown in Fig. 8.4.1.

Circuit schematics of LAs using both conventional and shared-inductor peaking are shown in Fig. 8.4.2. For the LA using conventional inductive peaking, two inductors L_1 are required for each stage. On the other hand, the LA using shared-inductor peaking requires only two inductors L_2 for every two adjacent stages, wherein the inductance value of L_2 is half of L_1. The reason that L_2 can be only half of L_1 is because by sharing the inductor L_2 between two adjacent stages, the in-phase current I_L through the inductor L_2 is doubled. Particularly when the voltage swing in LA is sufficiently large and the switching current I_L through the inductor is close to the bias current I_B, the switching current flowing through the shared-inductor becomes $2I_B$ because of sharing two bias currents I_B. Since the voltage swing V_L across the inductor is proportional to the time rate of change of the current I_L through the inductor L_2, doubling the in-phase current I_L will simply need only half of the inductance to obtain the same voltage swing V_L across the inductor. The chip area of the shared-inductor LA can therefore be reduced by 56% if including the output buffer and by 60% if not including the buffer, when compared with conventional inductive peaking. Main pole locations of the transfer functions for two different types of LA are illustrated in Fig. 8.4.3. Here L_2 is half of L_1 and the output capacitance C_L and drain resistance R_D are identical for both cases. Identical pole locations for both cases imply that the bandwidth enhancement of shared-inductor peaking is not degraded, compared to conventional inductive peaking, but the chip area is significantly reduced. To further improve the signal bandwidth of LA, both LA circuits apply interleaving active-feedback as shown in Fig. 8.4.3 [6]. The simulated BW and gain for LA with conventional peaking are 19.4GHz and 30dB, while those for LA with shared-inductor peaking are 19.1GHz and 30.5dB.

Measured eye diagrams for both optical receivers are shown in Fig. 8.4.4. Due to optical test capability limitation, the optical test is conducted up to 10Gb/s but the electrical test is conducted up to 28Gb/s. PRBS-7 NRZ input pattern is provided by a pattern generator and used to directly drive VCSEL module for generating the optical input with measured jitter $2.5ps_{rms}$. The optical input level is adjusted with an optical attenuator for investigating input sensitivity. Measured rise and fall times as well as the jitter performance of both receivers are close to each other, implying comparable signal bandwidth of both receivers. When supplying $400\mu A_{PP}$ electrical input, the maximum error-free (BER $<10^{-12}$) data-rate of the receiver using conventional peaking is 25Gb/s with 38.8mW power consumption for TIA/LA. On the other hand, for the receiver using shared-inductor peaking, 28Gb/s data-rate can be achieved with 28.8mW power consumption for TIA/LA. The T-coil output buffer consumed 18mW from a 1V supply and provided larger than 150mV voltage swing on a 50Ω load. Lower data rate and higher power consumption for the conventional inductive peaking receiver can be attributed to larger capacitive wire loading because of longer signal routing paths needed between peaking inductors.

Measured bathtub and sensitivity curves are shown in Fig. 8.4.5. Optical as well as electrical input levels are swept to obtain the sensitivity curves; bathtub curves are taken with optical input power of -6dBm and electrical input of $400\mu A_{PP}$. For optical input at 10Gb/s, the eye opening at BER $<10^{-12}$ is around 0.5UI, while for electrical input at 25 to 28Gb/s, the opening reduces to around 0.2UI due to ISI. To compare the input sensitivity of electrical and optical inputs for the shared-inductor receiver, the optical power is converted to equivalent input current by means of 0.4A/W responsivity of the PIN-PD, wherein the best sensitivity is -20dBm at 10Gb/s for optical input and -8dBm for electrical input at 28Gb/s. Optimized energy-efficiency to achieve BER $<10^{-12}$ as well as output swing $>100mV_{pp}$ is 1.55pJ/bit at 25Gb/s for conventional peaking RX and 1.03pJ/bit at 28Gb/s for shared-inductor peaking RX. The plot of optimized energy efficiency versus data rate in Fig. 8.4.5 reveals a power penalty at 10Gb/s because of the high-bandwidth LA design, necessitating over ten LA stages, but at higher data rates, the energy-efficiency is improved to <2pJ/b for >16Gb/s data rates.

In this paper, we present two optical receivers with different inductive-peaking techniques in 28nm CMOS technology. We develop the shared-inductor peaking technique that can reduce the chip area by 56% while maintaining similar bandwidth enhancement compared to conventional inductive peaking. Figure 8.4.6 shows the technical summary of 25-28Gb/s optical receivers in recent published literature. The die photo shown in Fig. 8.4.7 includes a 4-channel GaAs PIN-PD that is wire-bonded to CMOS RX. The inductor details are obscured for propriety information protection. We believe that the presented shared-inductor peaking technique offers high potential for next-generation high-speed low-power optical link transceivers.

Acknowledgment:
The authors thank Y.-H. Chen, Y.-H. Kuo, S.-H. Huang, and W.-Z. Chen for chip testing support.

References:

[1] Jhih-Yu Jiang, et al., "100Gb/s Ethernet Chipset in 65nm CMOS Technology," *ISSCC Dig. Tech Papers*, pp. 120-121, 2013.
[2] G. Kalogerakis, et al., "A Quad 25Gb/s 270mW TIA in 0.13µm BiCMOS with <0.15dB Crosstalk Penalty," *ISSCC Dig. Tech Papers*, pp. 116-117, 2013.
[3] T. Takemoto, et al., "A 4× 25-to-28Gb/s 4.9mW/Gb/s -9.7dBm High-Sensitivity Optical Receiver Based on 65nm CMOS for Board-to-Board Interconnects," *ISSCC Dig. Tech Papers*, pp. 118-119, 2013.
[4] J. Proesel, et al., "25Gb/s 3.6pJ/b and 15Gb/s 1.37pJ/b VCSEL-based optical links in 90nm CMOS," *ISSCC Dig. Tech Papers*, pp. 418-419, 2012.
[5] T. Takemoto,et al., "A 25-Gb/s 2.2-W Optical Transceiver Using an Analog FE Tolerant to Power Supply Noise and Redundant Data Format Conversion in 65-nm CMOS," *IEEE Symposium of VLSI Circuits*, pp. 106-107, 2012
[6] Huei-Yang Huang, et al., "A 10-Gb/s Inductorless CMOS Limiting Amplifier With Third-Order Interleaving Active Feedback," *IEEE J. Solid-State Circuits*, vol. 42, no.5, pp. 1111-1120, May 2007.

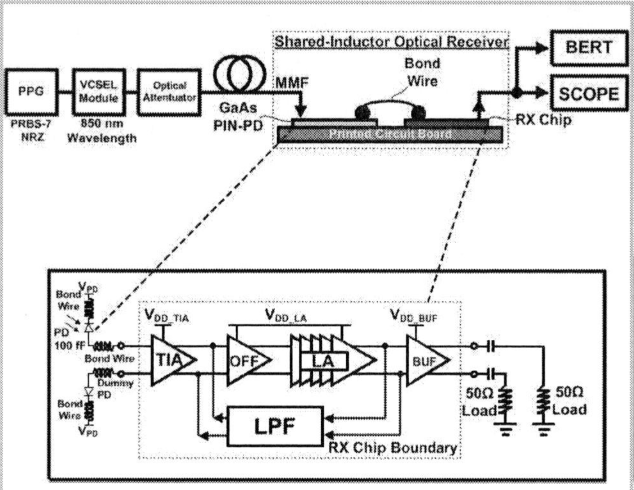

Figure 8.4.1: Block diagram and test setup of receiver.

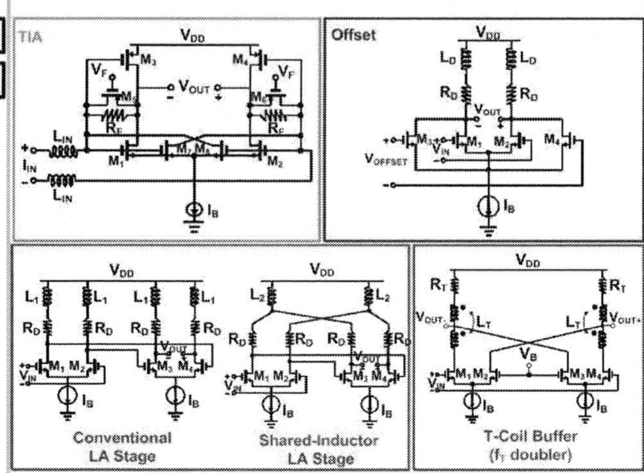

Figure 8.4.2: Schematics of TIA, LA, and output buffer.

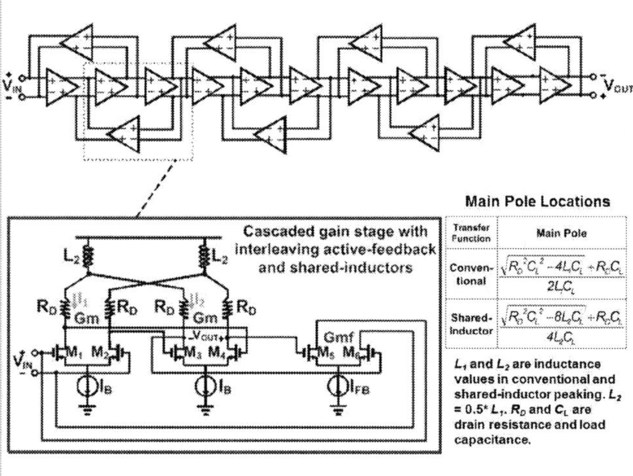

Figure 8.4.3: Schematics and transfer functions of LA.

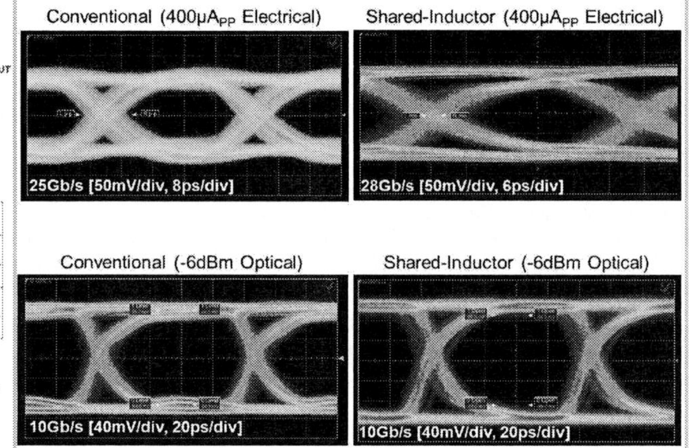

Figure 8.4.4: Measured optical and electrical eye diagrams.

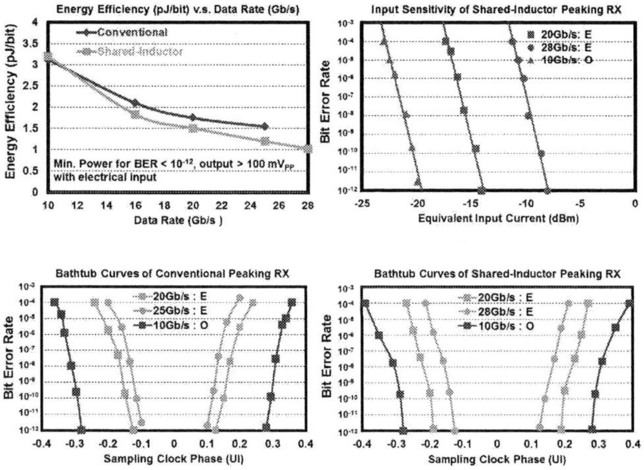

Figure 8.4.5: Measured bathtub and sensitivity curves.

Reference	Data Rate (Gb/s)	Power (mW/ channel)	Energy Efficiency (pJ/bit)	Sensitivity @ 10^{-12} BER (dBm)	PD Cap (fF)	Chip Area (mm^2)	Tech-nology
This work (w/ shared-inductor)	28	28.8*	1.03*	-6 @ 10Gb/s (O) -7 @ 28Gb/s (E)	100	0.6 x 0.53	28nm CMOS
This work (w/o shared-inductor)	25	38.8*	1.55*	-6 @ 10Gb/s (O) -7 @ 25Gb/s (E)	100	0.89 x 0.81	28nm CMOS
[1] ISSCC'13	25	69*	2.76*	-6.8	50	1.6 x 0.65***	65nm CMOS
[2] ISSCC'13	25	67.5 (TIA)	2.7 (TIA)	-12	65	3.3 x 1.5***	130nm SiGe
[3] ISSCC'13	28	137.5**	4.9**	-9.7 @ 25Gb/s	—	2.66 x 2.28***	65nm CMOS
[4] ISSCC'12	25	33.6*/ 44.4**	1.34*/ 1.78**	-4 @ 22Gb/s	80	0.25 x 0.39	90nm CMOS
[5] VLSI'12	25	59*	2.36*	—	—	3.6 x 5.3***	65nm CMOS

Figure 8.4.6: Technical summary of 25Gb/s+ optical RX.

Figure 8.4.7: Die micrograph of CMOS RX and off-chip PIN-PD.

8.5 A Sub-1.75W Full-Duplex 10GBASE-T Transceiver in 40nm CMOS

Jan R. Westra, Jan Mulder, Yi Ke, Davide Vecchi, Xiaodong Liu, Erol Arslan, Jiansong Wan, Qiongna Zhang, Sijia Wang, Frank M.L. van der Goes, Klaas Bult

Broadcom, Bunnik, The Netherlands

The IEEE802.3an 10GBASE-T standard describes full-duplex 10Gb/s Ethernet transmission over four pairs of up to 100m UTP cable. For the implementation of high-density 10GBASE-T network switches, highly integrated transceivers are required that have both a small form factor and high power efficiency. This paper describes an analog front-end (AFE) that is used in a quad-port 10GBASE-T transceiver chip. The small form factor of the AFE allows for the use of a 23×23mm² BGA package, enabling implementation of 48-port switches with all transceivers in a single row on the PCB pitch-matched to the RJ45 connector arrays. The design achieves >62dBc transmitter SFDR, >62dBc echo cancellation (EC) SFDR, and >60dBc receiver SFDR up to 400MHz. It occupies an area of 15.1mm² per port in a 40nm CMOS process. At 100m full 10Gb/s traffic, the AFE dissipates less than 1.75W.

The architecture of the transceiver is depicted in Fig. 8.5.1, detailing one of the four channels. The transmitter (TX) consists of a 12b main DAC directly driving the UTP cable. The 5th-order elliptic low-pass filter, which is implemented on the BGA substrate, prevents high-frequency disturbers like cell phones from coupling into the receiver (RX). Due to the cable attenuation, the power received from the far-end TX can be more than 45dB below the locally transmitted power, especially at high frequencies. To increase the dynamic range of the local RX, the TX signal is cancelled by two hybrid DACs creating the opposite of the TX signal across hybrid resistors R_h. Any linear TX signal remaining after this analog cancellation is subsequently cancelled by an adaptive filter in the RX DSP engine. For power efficiency, the hybrid DACs are scaled 8× with respect to the main DAC.

The IEEE 802.3an specification requires that the TX signal meet an SFDR of only 54dB at low frequency, degrading to 38dB close to 400MHz. Though sufficient for the far-end RX, these distortion levels are too high for the local RX in this full-duplex system. For this reason, the TX linearity must be significantly better than these specifications. Alternatively, one can rely on distortion cancellation that can be achieved from either matched DACs [1,2] or matched output drivers [3,4], such that the resulting SFDR after the analog echo cancellation (EC SFDR) is increased [1,3-4]. Cancellation of nonlinear signal components, however, depends to a large extent on the matching of impedances. In a practical application, the finite return loss of chip package and PCB traces, transformers, connectors, and patch cables makes this impedance matching very hard to achieve. For that reason, our AFE uses an inherently high-linear TX/hybrid combination.

Figure 8.5.2 illustrates the design and layout of the merged main/hybrid DACs. The 12b DACs are clocked at 1.6GS/s and implemented as a 6+6 segmented current-steering architecture, similar to the one described in [5]. Each unit element consists of one main DAC section and two hybrid DAC sections. In the layout, the main/hybrid unit elements are interdigitated to achieve good matching, reducing the effect of layout gradients on the achievable hybrid cancellation. To minimize the effect of layout gradients on the main DAC output signal, a spatial scrambling is used in the layout. Each DAC cell is driven from a cascade of CMOS latches. This minimizes the data dependency of the transition time for the signals driving the switches and further improves the linearity of the DAC output signal.

Clocking the DACs at 1.6GS/s enables 2× oversampling, which significantly reduces the high-frequency echo energy and common-mode glitches due to the switching of the DAC current cells. Together with the low-pass filter, this improves the EMI performance of the AFE. By directly driving the cable from the main DAC, cancelling the TX signal with scaled hybrid DACs, and implementing all clocking and latches in CMOS, the TX reaches high power efficiency. At 200mW per channel, including all bias circuits, the power dissipation approaches the theoretical limit for a Class-A circuit.

The receiver chain, illustrated in Fig. 8.5.3, starts with a programmable-gain amplifier (PGA) that optimizes the loading of the subsequent ADC. In addition, it provides filtering of the RX input signal. At the PGA input, capacitor C_{in}, in combination with R_h, creates a 50MHz 1st-order high-pass. This high-pass improves the required dynamic range of the RX. The feedback impedance, comprising R_{fb} and C_{fb}, creates a 400MHz 1st-order low-pass characteristic, reducing noise above the Nyquist frequency. To adjust the gain of the PGA, capacitor C_{in} is split into many segments. By switching some segments to ground, instead of the amplifier virtual ground, the PGA gain can be adjusted without changing the input impedance or the transfer function of the RX.

The PGA drives a 4× time-interleaved sample-and-hold (SH) circuit, each slice running at 200MS/s. The sampling circuit uses bootstrapped input switches to improve its linearity. The sampled voltage is buffered by a 1× gain hold-buffer to drive the ADC. The ADC is an optimized, 6-stage version of the 4× time-interleaved, 800MS/s dual-residue architecture described in [6]. Its MDAC stages do not require full settling, leading to a power-efficient design. For further power efficiency, the PGA, the SH, and the hold-buffer are implemented using thin-oxide (1.0V) transistors to optimize gain, bandwidth, and noise at low bias current levels, while using the higher 2.5V I/O supply voltage for increased output swing. The reliability of all core transistors is guaranteed by ensuring that the maximum voltage difference between any two transistor terminals is always below 1.0V, both during normal operation and during start-up, power-down, and Ethernet speed transitions.

Figure 8.5.4 depicts a simplified schematic of the two-stage PGA. Being the first circuit in the RX chain, it contributes significantly to the RX noise floor and therefore potentially exhibits high power consumption. Besides employing thin-oxide transistors and using the high I/O supply, further power savings are obtained by using complementary amplifier stages. The input stage consists of both an NMOS (M1-M2) and a PMOS (M3-M4) differential pair, effectively reusing the bias current in two pairs. The folded-cascode stage (M5-M6) couples the input stage to the output stage, which uses a complementary design as well. Besides the PMOS differential pair (M7-M8), the NMOS transistors (M9-M10) are also signal-driven through capacitors (C1-C2). Frequency stability is ensured using an Ahuja compensation scheme, which has an inherently better distortion behavior when compared to standard Miller compensation. The implemented RX consumes a total of 200mW per channel.

The measured two-tone output spectrum of the actual TX DAC is shown in Fig. 8.5.5 at the maximum output voltage swing of $2.1V_{pp\text{-}diff}$. At 90MHz, the TX SFDR is above 73dBc. It gradually decreases to 62dBc for a 350MHz two-tone signal. Figure 8.5.6 illustrates the measured TX SFDR over the full frequency range between 50 and 360MHz. It shows that the SFDR of the TX is above 62dBc over the whole frequency range of interest, and well above the IEEE 802.3an TX distortion limit. The hybrid achieves a linear cancellation of over 20dB up to 200MHz, staying above 10dB over the entire Nyquist band. The EC SFDR measured after the hybrid cancellation is higher than the TX SFDR, demonstrating that the nonlinearity of the TX DAC is further suppressed by the hybrid. The SFDR of the RX is measured to be 75dBc at low frequencies, staying above 60dBc over the complete Nyquist band. The achieved BER is better than 10^{-12} over PVT corners, under alien crosstalk. Figure 8.5.7 shows a micrograph of one complete port of the 10GBASE-T transceiver, which is implemented in a standard 40nm CMOS process. Including PLL and bias circuit, it occupies an area of 15.1mm² and consumes 1.75W per port.

Acknowledgements:
The authors thank Hakan Ilgaz and the Broadcom Longmont, San Jose and Irvine teams for their support.

References:

[1] G. Chandra, M. Malkin, "A Full-Duplex 10GBase-T Transmitter Hybrid with SFDR >65dBc Over 1 to 400MHz in 40nm CMOS," *ISSCC Dig. Tech Papers*, pp. 144-145, Feb. 2011.

[2] S. Gupta et al., "A 10Gb/s IEEE 802.3an-Compliant Ethernet Transceiver for 100m UTP Cable in 0.13μm CMOS," *ISSCC Dig. Tech Papers*, pp.106-107, Feb. 2008.

[3] T. Gupta et al., "A Sub-2W 10GBASE-T Analog Front End in 40nm CMOS Process," *ISSCC Dig. Tech Papers*, pp.410-411, Feb. 2012.

[4] F. Gerfers et al., "A 16-Port FCC-Compliant 10GBASE-T Transmitter and Hybrid with 76dBc SFDR up to 400MHz Scalable to 48 Ports," *ISSCC Dig. Tech Papers*, pp. 412-414, Feb. 2012.

[5] C-H Lin et al., "A 12b 2.9GS/s DAC with IM3<<-60dBc beyond 1 GHz in 65nm CMOS," *ISSCC Dig. Tech Papers*, pp. 74-75, Feb. 2009.

[6] J. Mulder et al., "An 800MS/s Dual-Residue Pipeline ADC in 40nm CMOS," *ISSCC Dig. Tech Papers*, pp.184-185, Feb. 2011.

978-1-4799-0917-9/14 $31.00 © 2014 IEEE

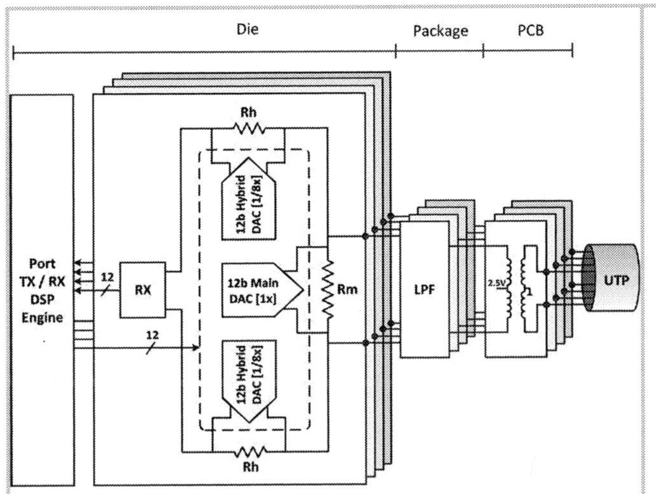

Figure 8.5.1: Overall architecture of the presented 10Gb/s Ethernet transceiver.

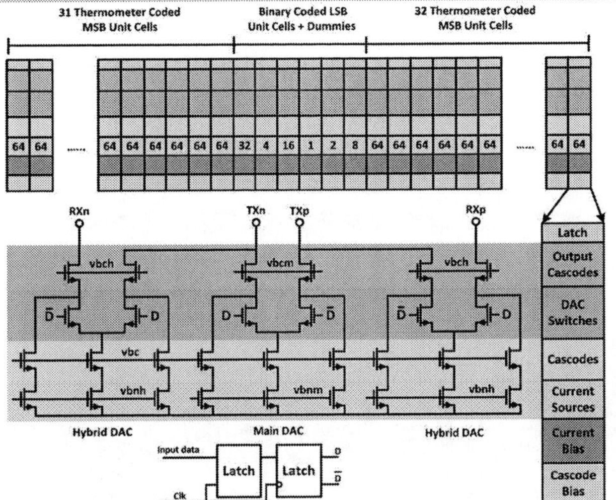

Figure 8.5.2: Design and layout of the main/hybrid DAC combination.

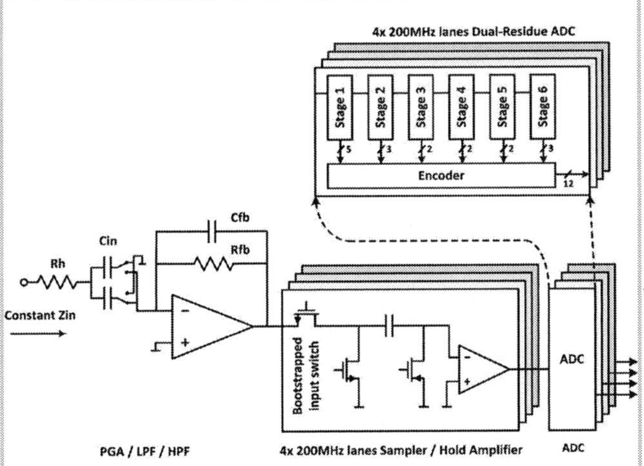

Figure 8.5.3: Architecture of the receiver with PGA and 4× interleaved samplers and dual-residue ADCs.

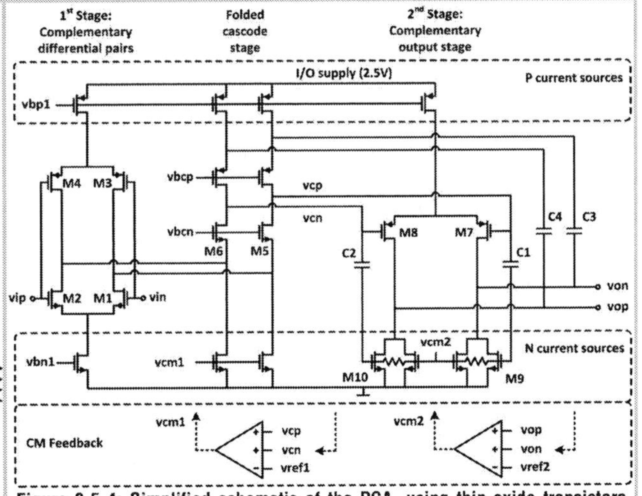

Figure 8.5.4: Simplified schematic of the PGA, using thin-oxide transistors from the high I/O supply.

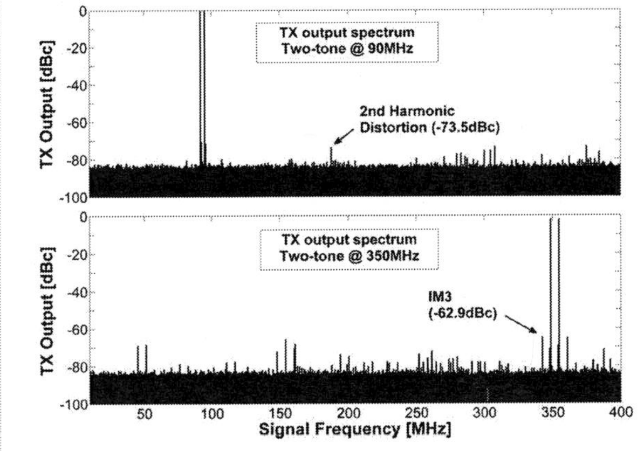

Figure 8.5.5: Output spectrum of the TX measured at low (100MHz) and high (350MHz) input frequency.

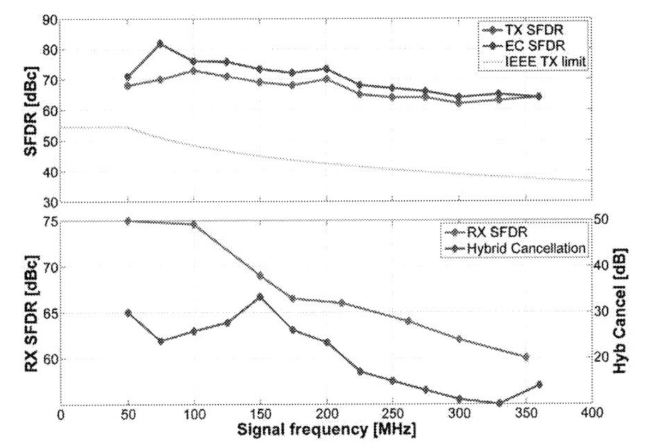

Figure 8.5.6: Measured spurious-free dynamic range of TX, hybrid, RX and hybrid cancellation.

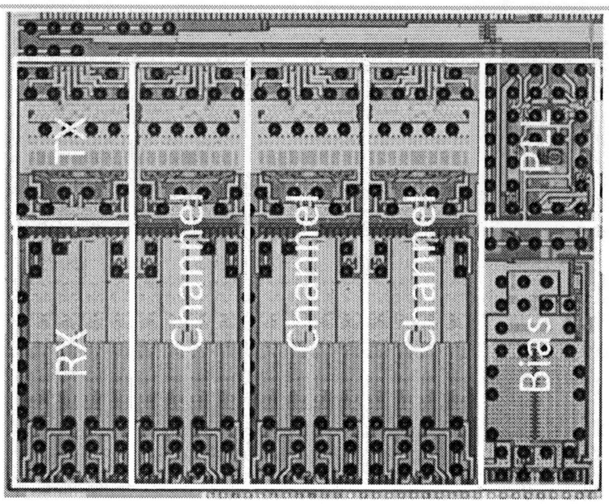

Figure 8.5.7: Chip micrograph of the implemented 10GBASE-T port in 40nm CMOS. RX and TX are shown in one of the four channels.

8.6 A Full-Duplex Line Driver for Gigabit Ethernet with Rail-to-Rail Class-AB Output Stage in 28nm CMOS

Hui Pan, Yuan Yao, Mostafa Hammad, Junhua Tan, Karim Abdelhalim, Evelyn Wang, Rick Hsu, Jenny Yu, Joseph Aziz, Derek Tam, Ichiro Fujimori

Broadcom, Irvine, CA

Gigabit Ethernet PHY (GPHY) transceivers find wide use in SoCs and standalone PHY chips with hundreds of millions of ports shipped every year. Transceiver design has recently focused on power reduction driven by the need for higher port density and throughput with minimum energy and thermal cost. The line drivers that deliver power from a high voltage supply to remote 100Ω differential loads dominate the GPHY power consumption. The supply voltage determined by the transmit amplitude specs (e.g., $2V_{ppdiff}$ for 1000BASE-T/100BASE-TX Ethernet) does not scale with technology. This paper presents an architecture that enables rail-to-rail full-duplex operation for high voltage efficiency resulting in a 2.5V GPHY driver in 28nm CMOS that saves 24% power from the mainstream 3.3V drivers.

High voltage efficiency is achieved when a variable supply is reactively synthesized to track the signal voltage. However, variable supply is costly, if feasible, for wideband applications such as GPHY. Given a steady supply, the highest voltage efficiency occurs in rail-to-rail operation, which is often impossible because of extra voltage headroom needed at maximum voltage swing to source simultaneous maximum current with specified linearity. In full-duplex transmissions such as 1000BASE-T Ethernet, the driver output current i_s (or i_{s1}) and voltage v_s (or v_{s1}) could peak simultaneously as is shown in Fig. 8.6.1. However, simultaneous maxima do not happen between the line current i_{line} and the line voltage v_{line}. This is generally true for transmission lines that have

$$i_{line} = i_{tx} - i_{rx} = (v_{tx} - v_{rx})/Z_0 \quad (1)$$

across any intersecting plane and $v_{line} = v_{tx} + v_{rx} = (i_{tx} + i_{rx})Z_0$ at the intersecting point, where Z_0 is the line characteristic impedance, v_{tx} (i_{tx}) and v_{rx} (i_{rx}) are the voltages (currents) including reflections propagating from the driver and the remote link partner, respectively, to the intersecting plane. When Z_0 = source R_S = load R_L without line attenuation, v_{tx} and v_{rx} have the same amplitude in full-duplex transmission. According to eq. (1), $i_{line} = (v_{tx} - v_{rx})/R_L = 0$ at peak v_{line} when $v_{tx} = v_{rx}$ = the peak voltage, which is +1V or -1V for 1000BASE-T. Imagine a driver with output current and voltage equal to i_{line} and v_{line}, respectively: output rail-to-rail operation becomes possible with zero line current sourced at maximum line voltage swing, and self-termination is inherent due to equivalence to drivers with passive source or shunt termination.

Generation of i_{line} seems impossible because the i_{rx} term in $i_{line} = i_{tx} - i_{rx}$ is presumably unknown from the remote partner. However, due to finite signal bandwidth (BW), i_{rx} can be approximated from the received voltage v_{rx} available from the full-duplex transceiver hybrid: $i_{rx} \sim v_{rx}/R_L$. The hybrid extracts v_{rx} from v_{line} with output $v_h = v_{rx} + v_{leak}$, where the hybrid leakage $v_{leak} = 0$ when v_{tx} is completely rejected. To take advantage of duplex current cancellation between i_{tx} and i_{rx}, i_{line} is generated as a whole from $v_{rx} - v_{tx}$ using a closed-loop G_m circuit (Fig. 8.6.2). According to eq. (1), $G_m = -1/R_L$ with $Z_0 = R_L$. The G_m input voltage $v_c = v_{rx} - v_{tx}$ is synthesized by superposition of v_{line} and $(-2v_{tx})$. The former is sensed by the $2R$ ($>>R_L$) feedback resistance; the latter is developed by the current DAC (IDAC) output i'_{tx} ($<<i_{tx}$) flowing through $2R$.

The resulting driver is *full duplex* as it extracts v_{rx} at the feedback midpoint (hybrid output) while transmitting v_{tx}. At the midpoint v_{tx} is rejected by resistive interpolation between G_m output v_{tx} and G_m input $v_{tx}/A \sim (-v_{tx})$. When the transmit voltage gain A is normalized: $A \sim G_mR_L = -1$, the hybrid leakage of v_{tx} is nulled: $v_{leak} = 0$, and the impedances are matched: $R_S = -1/G_m = R_L$, all simultaneously. Those inherently merged full-duplex functionalities enable rail-to-rail operation and distinguish this driver from other types [1].

For maximum current efficiency, the G_m circuit does not use the popular differential current-steering topology, which wastes common-mode (CM) current in the same amount as the differential-mode (DM) and usually needs class-A current biasing [2-3] to minimize CM emission. Class-AB push-pull topology delivers current from supply to load at almost 100% efficiency, but it usually operates in closed-loop voltage mode [4-5]. Figure 8.6.3 shows a conversion to the desired current mode. A line-current replica $i'_{line} = i_{line}/m$ ($m >> 1$) is generated by the class-AB output stages of a voltage (v_c) buffer driving a differential load replica $2R'_L = mR_L$. The m factor is maximized with technology scaling to minimize the replica overhead. The push-pull MOSFETs M1-M4 mirror i'_{line} m times to drive the differential load $2R_L$ in an H-bridge topology. The output CM is forced to $V_{DD}/2$ by the g_{cm} block injecting CM error current to the center tap of the load replica.

The difference between $v'_c = v'_{cp} - v'_{cn}$ and $v_{line} = v_{linep} - v_{linen}$ limits the accuracy of i'_{line}-to-i_{line} mirroring especially at rail-to-rail output swing. To meet the IEEE 802.3ab 10mV transmit distortion spec, an auxiliary op-amp loop is added to equalize the master-side drain voltage of each current mirror to the slave side as is shown in Figure 8.6.4. The op-amp senses the drain voltage difference and drives the cascode on the master side with little overhead. Voltage headroom is not an issue for the master cascodes M'c1–M'c4 with v'_c swing attenuated by finite R_0 (Figure 8.6.3). The output cascodes Mc1–Mc4 enter deep triode region at rail-to-rail output swing, but they limit the drain voltage swing of the current mirrors and provide over-voltage (OV) protection.

The driver consists of many loops such as the *outer* G_m loops, the *inner* v_c buffer loops, and the output CM loop formed with g_{cm} in Fig. 8.6.3. It is a challenge to mitigate loop interactions while achieving the required loop BWs on a tight power budget. The outer loop DM BW is 200MHz, high enough for the DM termination to pass the GPHY return-loss mask of -16dB from 1 to 40MHz but low enough with respect to the inner loop BW to preserve the phase margin. The output CM loop targets a CM termination of 25Ω with a BW similar to the DM termination to ensure outer-loop CM stability and an EMI performance like voltage-mode drivers. This precludes application of conventional center-tap (AC) grounding to the load replica for CM stability of the pseudo-differential v_c buffer driving it. Figure 8.6.5 shows the inner loop CM is compensated with the Miller caps (C_{cm}) terminated to an output CM node of the 2-stage op-amps. Input differential pairs driven by input and output CM cancel part of the CM g_m for better stability. This costs no power by current reuse and dynamic biasing. The resulting CM Miller caps present negligible loading to the inner loop DM and the output CM loop.

Implemented in standard 28nm CMOS with a 1.8V thick-oxide option, the full-duplex driver passes the IEEE 802.3 specs for 1000BASE-T/100BASE-TX Ethernet with 5.6mA quiescent current from a 2.5V supply. The closest reported is 8mA from 3.3V [5]. Figure 8.6.6 shows the transmit pulses pass the template test with <10mV distortion. The hybrid leakage $(v_{leak})_{pp}/(v_h)_{pp}$ is measured below -31dB at 31.25MHz for $2R_L$ = 100Ω. With 3.3V supply the driver is backward compatible to 10BASE-T Ethernet that specifies $5V_{ppdiff}$ transmit amplitude. Voltage-mode drivers [4-5] with 100Ω differential termination can neither provide the 10BASE-T voltage swing from a 3.3V supply nor support 1000BASE-T at 2.5V. Figure 8.6.7 shows a die micrograph of the drivers integrated in a GPHY quad-port AFE that achieves 150m reach over standard CAT-5/5e cables across corners with less than 10.8mA supply current per transmitter (TX). Each TX channel occupies 160×430μm².

Acknowledgements:
The authors thank Yue Yu and Lan Tran for the layout, Justin Nguyen, Phuong-Quynh Phan, and Ning Wang for the testing.

References:

[1] B. Nauta, et al., "Analog line driver with adaptive impedance matching," *IEEE J. Solid-State Circuits*, vol. 33, no. 12, pp. 1992–1998, Dec. 1998.

[2] J. Aziz, et al., "A 65nm CMOS Self-Terminated Open-Drain IDAC Line Driver Suitable for Fast Ethernet Applications," *CICC* 2011.

[3] R. Mahadeva and D. A. Johns, "A differential 160-MHz Self-Terminating Adaptive CMOS Line Driver," *IEEE J. Solid-State Circuits*, vol. 35, no. 12, pp. 1889–1894, Dec. 2000.

[4] J. N. Babanezhad, "A 100-MHz, 50-Ω, -45-dB Distortion, 3.3-V CMOS Line Driver for Ethernet and Fast Ethernet Networking Applications," *IEEE J. Solid-State Circuits*, vol. 34, no. 8, pp. 1044–1050, Aug. 1999.

[5] D. Stiurca, "A fully differential line driver with on-chip calibrated source termination for gigabit and fast Ethernet in a standard 0.13μm CMOS process," *ISCAS* 2005, pp. 2176 – 2179.

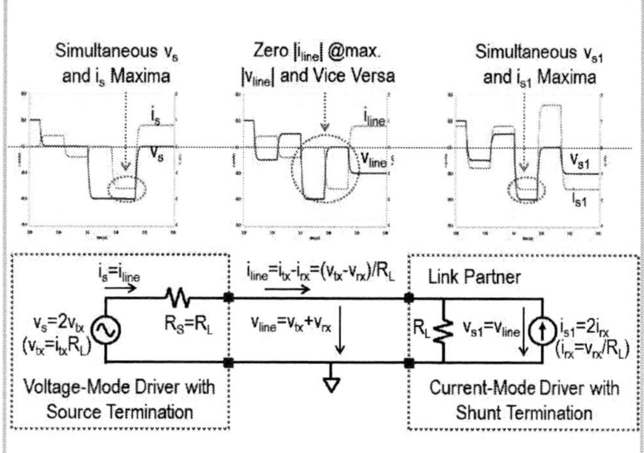

Figure 8.6.1: Zero line current at maximum line voltage in full-duplex transmission.

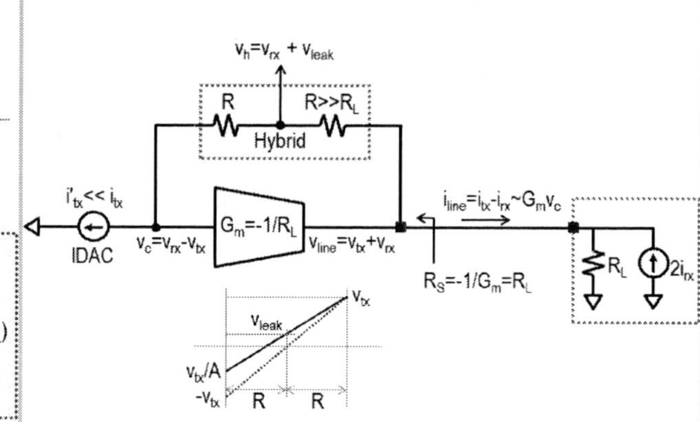

Figure 8.6.2: Full-duplex line driver with inherent hybrid and self-termination.

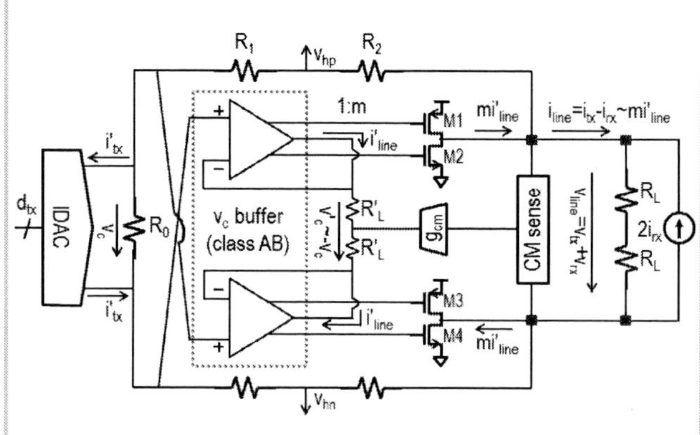

Figure 8.6.3: Schematic of the differential class-AB push-pull current-mode driver.

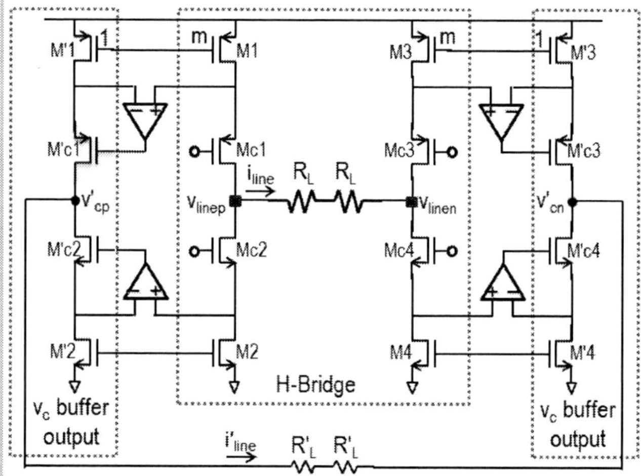

Figure 8.6.4: Schematic of the differential rail-to-rail output stage.

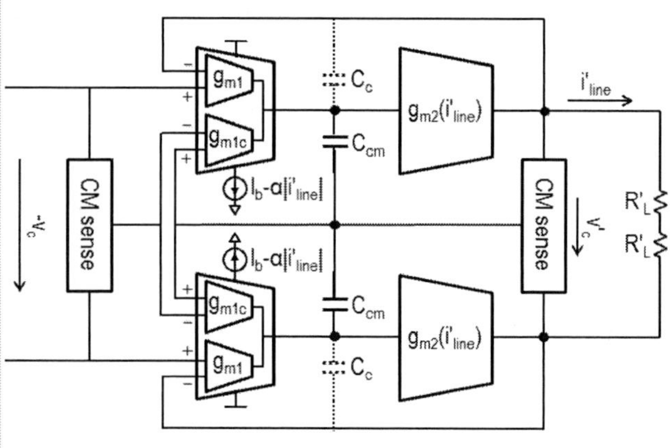

Figure 8.6.5: CM compensation of the pseudo-differential v_c buffer.

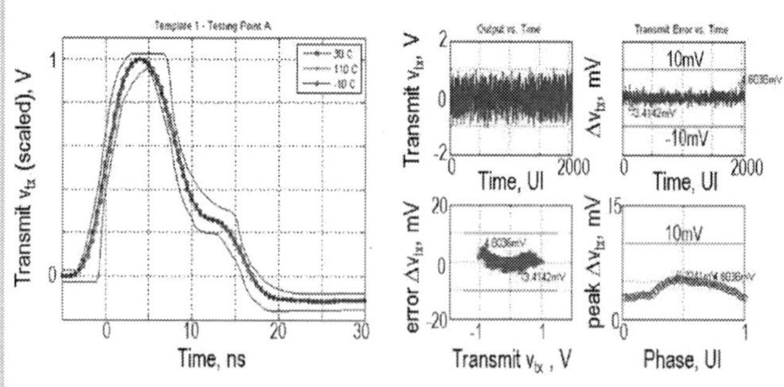

1000BASE-T transmit pulses vs. template with v_{rx} = 1.4Vppd @31.25MHz, Vdd = 2.5V, T = -10°C, 30°C, 110°C

1000BASE-T transmit distortion with v_{rx} = 2.7Vppd @20.83MHz, Vdd = 2.5V, T = 110°C

Figure 8.6.6: GPHY driver template test and distortion measurement.

978-1-4799-0917-9/14 $31.00 © 2014 IEEE

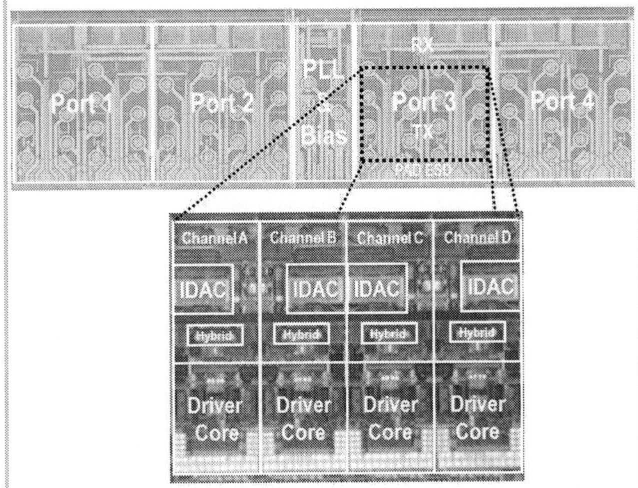

Figure 8.6.7: Die micrograph of the quad-port GPHY analog front-end (AFE).

8.7 A 4-to-10.5Gb/s 2.2mW/Gb/s Continuous-Rate Digital CDR with Automatic Frequency Acquisition in 65nm CMOS

Guanghua Shu[1], Woo-Seok Choi[1], Saurabh Saxena[1], Tejasvi Anand[1], Amr Elshazly[2], Pavan Kumar Hanumolu[1]

[1]University of Illinois, Urbana, IL, [2]Intel, Hillsboro, OR

Continuous-rate clock-and-data recovery (CDR) circuits with automatic frequency acquisition offer flexibility in both optical and electrical communication networks, and minimize cost with a single-chip multi-standard solution. The two major challenges in the design of such a CDR are: (a) extracting the bit-rate from the incoming random data stream, and (b) designing a wide-tuning-range low-noise oscillator. Among all available frequency detectors (FDs), the stochastic divider-based approach has the widest frequency acquisition range and is well suited for sub-rate CDRs [1]. However, its accuracy strongly depends on input transition density ($0 \le \rho \le 1$), with any deviation of ρ from 0.5 (50% transition density) causing $2\times(\rho\text{-}0.5)\times10^6$ppm of frequency error. In this paper, we present an automatic frequency-acquisition scheme that has unlimited range and is immune to variations in transition density. Implemented using a conventional bang-bang phase detector (BBPD), it requires minimum additional hardware and is applicable to sub-rate CDRs as well. Instead of using multiple LC oscillators that are carefully designed to cover a wide frequency range [2,3], a ring-oscillator-based fractional-N PLL is used as a digitally controlled oscillator (DCO) to achieve both wide range and low noise, and to decouple the tradeoff between jitter transfer (JTRAN) bandwidth and ring-oscillator-noise suppression.

Figure 8.7.1 depicts the developed digital CDR architecture. It is composed of a frequency-locked loop (FLL), and a delay- and phase-locked loop (D/PLL). Both the FLL and D/PLL are updated using early/late (E/L) signals provided by the BBPD. In the FLL, frequency-detection logic block (FDL) operates on E/L signals and drives the DCO to within the pull-in range of the D/PLL through accumulator ACC_F. Both loss-of-lock detection (LOLD) and lock detection (LD) needed to ensure seamless switching between data-rates are also implemented in the FDL. LOLD triggers a new frequency acquisition when the error ($\Delta F = F_{DCO} - F_{DIN}$) between DCO frequency and data rate exceeds 1000ppm. Lock is declared when ΔF is smaller than 500ppm.

The D/PLL is composed of a digital DLL and a digital PLL, and can be viewed as the digital equivalent of the architecture reported in [2]. Similar to its analog counterpart, the digital D/PLL features a decoupled JTRAN bandwidth and jitter tolerance (JTOL) corner frequency, and exhibits well-controlled JTRAN bandwidth even in the presence of BBPD gain variations caused by input jitter [3]. Unlike [2], our D/PLL does not need large on-chip capacitors and the DCO is implemented using a fractional-N PLL employing a single ring oscillator instead of multiple LC oscillators. Additionally, to maximize JTOL, the digitally controlled delay line (DCDL) is biased at its mid-delay point in steady state by the path containing gain block K_0 and accumulator ACC_0. The path containing divide-by-H and accumulator ACC_H is used to prevent false locking as discussed later.

The principle behind the BBPD-based frequency detector (FD) is illustrated in Fig. 8.7.2. Consider the transfer function (TF) of a conventional BBPD in which the output changes sign at $\Delta\Phi = n\pi$ for all integer values of n. Due to this, BBPD output is typically considered to be valid only if $\Delta\Phi$ lies between $-\pi$ and π. This condition is violated in the presence of frequency error since the phase error accumulates indefinitely, causing BBPD to periodically switch between consecutive E and L signals as indicated in Fig. 8.7.2. However, the number of consecutive E (or L) is dictated by the magnitude of ΔF, with a larger ΔF resulting in a fewer number of consecutive E (or L) signals and vice versa. Based on this behavior, we seek to estimate the magnitude of ΔF by measuring the number of consecutive E (or L) signals. Inside FD, an accumulator, $ACC_{E/L}$, integrates the BBPD output and is reset whenever the BBPD output changes sign. The peak value of $ACC_{E/L}$ can be calculated to be $N_P = \rho F_{DIN}/2\Delta F$ and therefore, $\Delta F = \rho F_{DIN}/2N_P$. During frequency acquisition, ΔF is reduced to be within the pull-in range (ΔF_P) of the D/PLL by increasing the DCO frequency until N_P exceeds the desired threshold $N_{TH} = \rho F_{DIN}/2\Delta F_P$. This is implemented by incrementing the DCO frequency control accumulator, ACC_F, if $ACC_{E/L}$ is less than N_{TH} when the BBPD output changes sign (see Fig. 8.7.2). Lock is declared as soon as $ACC_{E/L}$ exceeds N_{TH}. Note that this frequency detection scheme does not provide the sign of ΔF. The DCO is reset to its lowest frequency at the start of acquisition process, so that ΔF is guaranteed to be always negative. This also prevents harmonic locking.

The accuracy of the frequency detection scheme depends on ρ and data/clock jitter (Φ_j), as quantified by the tabulated frequency error in Fig. 8.7.2. However, setting N_{TH} corresponding to $\rho = 1$ (i.e., $N_{TH} = F_{DIN}/2\Delta F_P$) ensures that residual frequency error will always be smaller than ΔF_P for any ρ. For example, $N_{TH} = 500$ ensures that the DCO is always locked within 1000ppm to target data rate. Interestingly, Φ_j improves accuracy as it is equivalent to setting a larger N_{TH} with $\Phi_j = 0$. Very large Φ_j can cause false updates of the DCO frequency, which can be prevented by not incrementing ACC_F when the peak value of $ACC_{E/L}$ is smaller than its previous peak. Potential false locking caused by some degenerate input patterns manifests as reduced $ACC_{E/L}$ count. Therefore, separately counting the number of transitions using divider H and ACC_H, and comparing to $ACC_{E/L}$ can detect false locking. Under this condition, incrementing the DCO frequency will pull the CDR away from false locking.

The DCO is implemented using a fractional-N PLL as shown in Fig. 8.7.3. Frequency control word (FCW), provided by the CDR logic, tunes the fractional-N PLL output frequency by varying its feedback divider from 4 to 15. When operated with a 500MHz reference clock, this translates to a wide DCO tuning range of 5.5GHz (2 to 7.5GHz). Since more than 2× frequency range is achieved, lower data rates can be easily accommodated using a divider chain [2]. Further, using a high reference clock extends the PLL bandwidth to adequately suppress ring-oscillator phase noise while maintaining the same quantization error of the fractional divider [4], and provides the freedom to use small JTRAN to filter input noise without degrading performance due to DCO phase noise. An on-chip digital multiplying DLL (MDLL) generates the 500MHz reference clock from a 50MHz crystal. It is important to note the crystal oscillator does not aid frequency acquisition, as its frequency has no relation to the input data rate. Fractional-N PLL helps suppress oscillator phase noise and can be eliminated if the ring oscillator meets JGEN specification (FCW drives ring oscillator directly in that case). Compared to using multiple LC oscillators [2,3], this approach covers a wide range with only one ring oscillator and has a linear relationship between FCW and data rate. A second-order $\Delta\Sigma$ modulator is used to truncate FCW and drive the feedback divider. A 2nd-order loop filter along with the 3rd pole located at the drain of current-source transistor, M_1, is used to suppress $\Delta\Sigma$ truncation error.

The prototype CDR is implemented in a 65nm CMOS process, occupies an active area of 1.63mm^2, and is packaged in QFN88 package. At 10Gb/s, the CDR consumes 22.5mW and achieves a BER < 10^{-12}. Measured residual frequency error versus locking threshold N_{TH} (Fig. 8.7.4) shows that the FLL is immune to transition density. With N_{TH} > 600, the frequency error is always less than 500ppm. Jitter transfer curves measured with different input jitter amplitudes illustrate that JTRAN bandwidth is independent of jitter amplitude even when using a BBPD (Fig. 8.7.5). The measured JTOL plot in Fig. 8.7.5 indicates a corner frequency of about 9MHz, which is much larger than JTRAN bandwidth of 0.2MHz. From 1.1 to 2.5MHz, JTOL is limited by DCDL range, and low frequency JTOL is restricted to $2UI_{pp}$ due to instrument limitation. Figure 8.7.6 tabulates the performance summary and the comparison. Compared to the results cited in the table, this work achieves the highest power efficiency, and lowest jitter while using ring-based oscillators. The die micrograph is shown in Fig. 8.7.7.

Acknowledgment:

Intel Labs University Research Office, Kawasaki Microelectronics America, Inc., and NSF under CAREER Award EECS-0954969 supported this work. Berkeley Design Automation provided Analog Fast Spice (AFS) simulator. Twisted Traces Inc. and Seong-Joong Kim provided testing assistance.

References:

[1] R. Inti, et al., "A 0.5-to-2.5-Gb/s reference-less half-rate digital CDR with unlimited frequency acquisition range and improved input duty-cycle error tolerance," in *IEEE ISSCC Dig. Tech. Papers,* Feb. 2011, pp. 438-439.

[2] D. Dalton, et al., "A 12.5-Mb/s to 2.7-Gb/s continuous-rate CDR with automatic frequency acquisition and data-rate read back," in *IEEE ISSCC Dig. Tech. Papers*, Feb. 2005, pp. 230-231.

[3] J. Kenney, et al., "A 9.95-11.1-Gb/s XFP transceiver in 0.13-μm CMOS," in *IEEE ISSCC Dig. Tech. Papers*, Feb. 2006, pp. 232-233.

[4] D. Park, S. Cho, "A 14.2mW 2.55-to-3GHz cascaded PLL with reference injection, 800MHz delta-sigma modulator and 255fs$_{rms}$ integrated jitter in 0.13μm CMOS," in *IEEE ISSCC Dig. Tech. Papers*, Feb. 2012, pp. 344-346.

978-1-4799-0917-9/14 $31.00 © 2014 IEEE

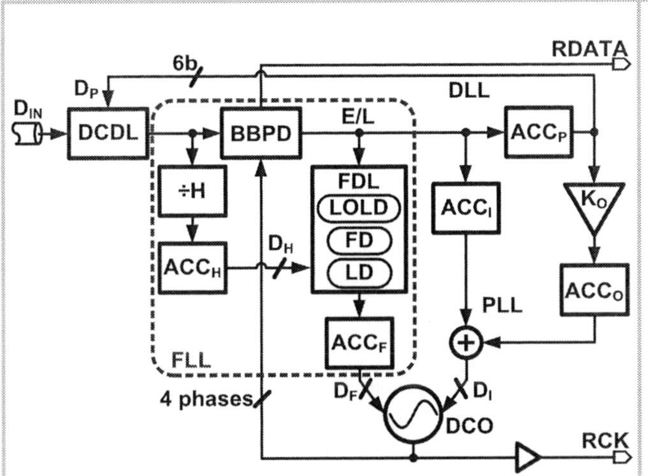

Figure 8.7.1: Block diagram of the continuous-rate digital CDR architecture.

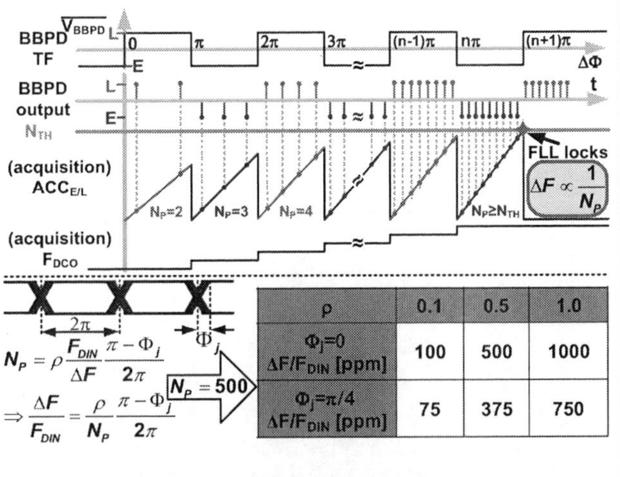

ρ	0.1	0.5	1.0
$\Phi_j=0$ $\Delta F/F_{DIN}$ [ppm]	100	500	1000
$\Phi_j=\pi/4$ $\Delta F/F_{DIN}$ [ppm]	75	375	750

$$N_P = \rho \frac{F_{DIN}}{\Delta F} \frac{\pi - \Phi_j}{2\pi} \Rightarrow \frac{\Delta F}{F_{DIN}} = \frac{\rho}{N_P} \frac{\pi - \Phi_j}{2\pi}$$

Figure 8.7.2: Principle of the automatic frequency acquisition using BBPD outputs and the sensitivity to transition density and jitter.

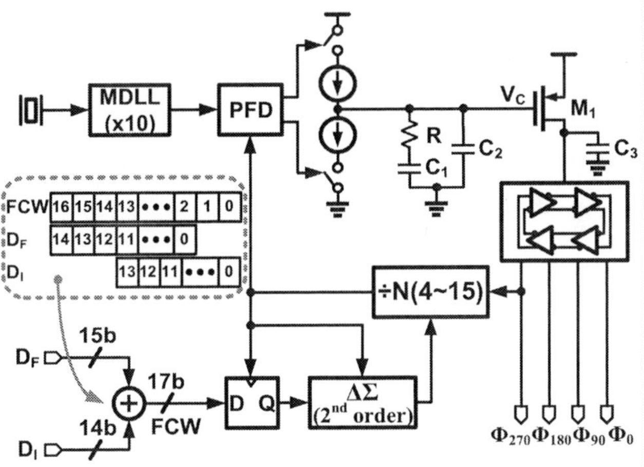

Figure 8.7.3: Schematic of fractional-N PLL-based wide-range DCO.

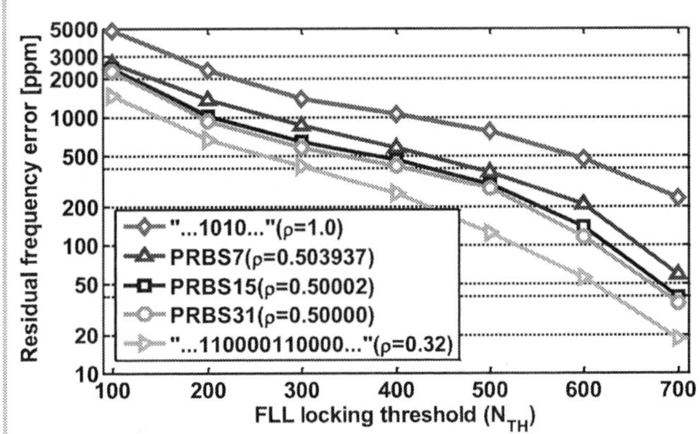

Figure 8.7.4: Measured residual frequency error versus locking threshold N_{TH} at different transition densities.

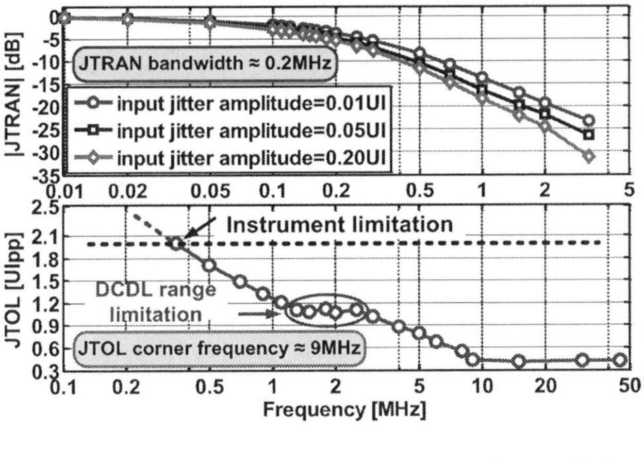

Figure 8.7.5: Measured jitter transfer with different input jitter amplitudes, and jitter tolerance with PRBS7 data (BER threshold of 10^{-9}).

	ISSCC'05 [2]	ISSCC'06 [3]	ISSCC'11 [1]	ISSCC'09 S.-K. Lee	This work
Technology	0.35μm	0.13μm	0.13μm	65nm	65nm
Supply [V]	3.3	3.3/1.8	1.2/0.8	1.2	1.2/1.0
FD type	RFD	Counter	Divider	DLL	BBPD
Architecture	Full-rate	Half-rate	Half-rate	Full-rate	Half-rate
Oscillator	LC	LC	Ring	Ring	Ring
JTRAN [MHz]	0.5	1.2	N/A	N/A	0.2
Jitter [ps_{rms}/ps_{pp}]	0.4/8.0	0.5/4.5	5.4/44.0	9.7/53.3	2.2/24.0
Data rate [Gb/s]	0.0125-2.7	9.95-11.3	0.5-2.5	0.65-8	4-10.5
Power [mW]	775@2.5Gb/s	800@11.4Gb/s	6.1@2.1Gb/s	88.6@8Gb/s	22.5@10Gb/s
FoM [mW/Gb/s]	310.2	70.2	3.05	11.08	2.25
Area [mm²]	9.0	8.0	0.39	0.11	1.63

Figure 8.7.6: CDR performance summary and comparison.

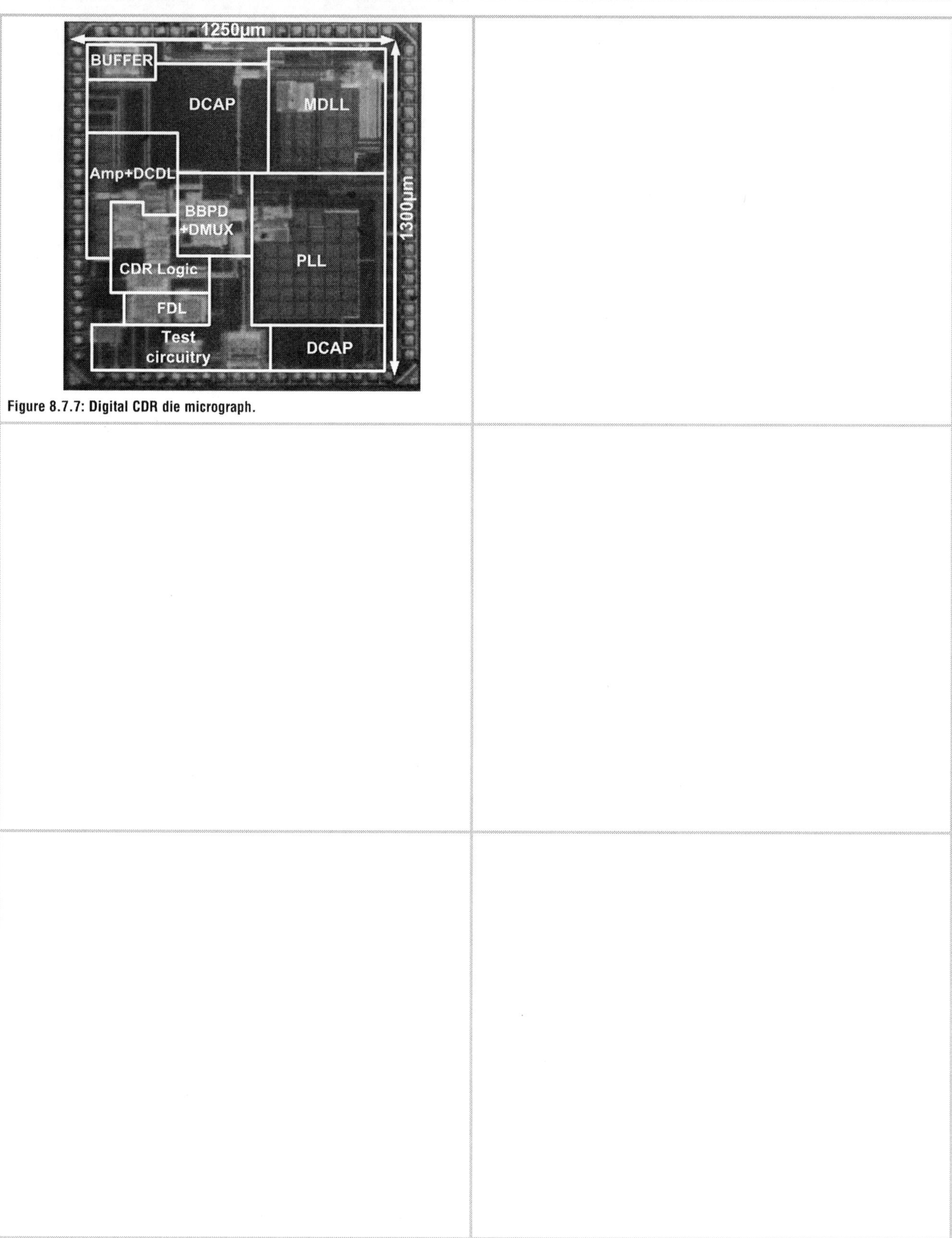

Figure 8.7.7: Digital CDR die micrograph.

8.8 An 8.2-to-10.3Gb/s Full-Rate Linear Reference-less CDR Without Frequency Detector in 0.18μm CMOS

Sui Huang[1], Jun Cao[2], Michael M. Green[1]

[1]University of California, Irvine, CA,
[2]Broadcom, Irvine, CA

As an alternative to the conventional dual-loop architecture, reference-less CDR architectures have become more popular in industry because of their simplicity and flexibility [1-5]. However, the robustness of the transition between frequency acquisition and phase locking is always a concern, particularly for the linear CDR, which has an extremely limited capture range. Many works, based mainly on the Pottbacker frequency detector (FD) [1], have been reported. In [3] the capture range of the FD is only ±2.4% at 20Gb/s with no capacitor bank in the VCO; in [4] the capture range of the FD is about ±6.4% at 2.75Gb/s, with an 8b resolution of the capacitor bank in the VCO; in [5] the capture range is ±15% at 10Gb/s, with an 11b resolution of the capacitor bank. Thus the Pottbacker FD inherently suffers from a limited capture range, requiring a dedicated FD and a stringent tradeoff between the CDR capture range and the number of VCO bands. In the presence of input jitter and phase-detector (PD) non-idealities, it is difficult to design an architecture where the resolution of the capacitor bank and the turnoff mechanism can guarantee that the VCO frequency will eventually fall within the pull-in range of the CDR.

This paper introduces a full-rate reference-less CDR architecture with neither an FD nor a lock detector. Its operation is instead based on the theory that if an offset (or "strobe point") is deliberately introduced into the PD characteristic, the pull-in range will be enhanced as long as the initial frequency offset is the appropriate polarity [6]. For example, if the strobe point (SP) is negative as illustrated in curve (b) of Fig. 8.8.1 and the initial VCO frequency is higher than the input data bit rate, then the VCO frequency will naturally decrease toward the correct frequency since there will be a net discharge of the loop filter during each cycle slip compared with the ideal case as shown in curve (a) of Fig. 8.8.1. Therefore, the linear PD itself can function as an FD with a very high capture range if the polarity of the SP is set appropriately, consistent with the initial VCO frequency.

The CDR architecture used to implement this concept is shown in Fig. 8.8.2. Other than those in the digital control circuit (DCC), all signals are differential. The SP of the PD is controlled by voltage V_{SP}, which is generated by the DCC. A frequency-acquisition algorithm is used to set the polarity of the SP and search the correct band in the capacitor bank while in the frequency-acquisition mode (FAM). The resolution of the capacitor bank is only 5 bits, since the "single-sided" pull-in range is sufficiently wide and the requirement for the tuning range in each band does not need to be very stringent.

The combined phase detector/strobe point detector (SPD) circuit is shown in Fig. 8.8.3. DFF1, DFF2, BUF1, XOR1, and XOR2 compose a standard Hogge PD. By inserting the two tunable buffers BUF2 and BUF3, the value of the SP can be adjusted overall range of ±15ps by changing the difference between their delays via $V_{SP}+$ and $V_{SP}-$. These buffers are realized by CML, but using triode-biased PMOS transistors in place of resistors. By changing $V_{SP}+$ and $V_{SP}-$, the output RC time constants and thus the delay times of the buffer cells vary. Once the frequency acquisition has completed, the clock signal is then nearly frequency-locked to the data, but with a strobe point that might be far from zero. Thus it is necessary to have a phase adjustment mode (PAM) after locking. The SPD circuit, which consists of DFF1, DFF2, DFF3, BUF4, XOR2, and XOR3, has a bang-bang characteristic with a nearly zero strobe point and is much less sensitive to delay mismatches than the linear PD. The average value of (*UP2 – DN*) provides a voltage that has the same sign as the SP. As shown in Fig. 8.8.2, this voltage is converted to a current I_{SPD} and then integrated onto either C1 or C2 to feed back to the PD in order to bring the SP close to zero in the PAM.

The frequency-acquisition mode functions as follows. If $V_{CTRL} > V_{REF}+$ ($V_{REF}+$ is the maximum VCO control voltage allowed in a band), then Comp+ is high and the counter is incremented to shift the VCO to a lower frequency band while at the same time switches S1, S2, and S6 are closed. Since C1 has been precharged, this sets V_{SP} to $V_{EXT}+$, which sets the PD strobe point to approximately -15ps, and the voltage on C2 is set to $V_{EXT}-$, which will be used for frequency acquisition in next band. If the VCO has been set to the correct band, the CDR will lock, since the VCO frequency will be higher than the input bit rate before locking, and the negative SP can guarantee that the capture range in this case is larger than the frequency range of this band; otherwise V_{CTRL} will continue decreasing until it goes below $V_{REF}-$ ($V_{REF}-$ is the minimum VCO control voltage allowed in a band), at which time the counter is incremented again, changing the VCO to a new, lower frequency band. At this moment, switches S3, S4, and S5 are closed, while V_{SP} is set to $V_{EXT}-$, which sets the SP to be a large positive value to pull up the VCO frequency. This process continues until the appropriate band has been reached, the frequency settles to the correct value with V_{CTRL} close to its final value, and a large SP is no longer needed. At this time the PAM takes over, and V_{SP} adjusts itself to set the PD strobe point to be very close to zero. The measured waveforms of $V_{CTRL}+$ and $V_{SP}+$, illustrating the FAM and PAM processes, are shown in Fig. 8.8.4. Initially the CDR is locked at 9.7Gb/s, which corresponds to band 6 (where band 1 is the highest and band 32 is the lowest frequency band of the VCO). At t_0, the input bit-rate is switched to 8.4Gb/s, and the loop starts to lose lock. At t_1, the FAM is activated. The correct band (band 24) is found at t_2, and then the PAM takes over. The system is phase locked at t_3.

The chip is fabricated in the Jazz Semiconductor SBC18 BiCMOS technology using only 0.18μm CMOS transistors and tested with a 1.8V supply. The chip consumes 174mW at 10.3Gb/s, not including the output buffer. The CDR capture range is from 8.2 to 10.3Gb/s covering the entire VCO range. The eye diagram of the recovered data is shown in Fig. 8.8.5 in a response to a 2^{31}-1 PRBS. The measured random jitter and pattern-dependent deterministic jitter at 10.3Gb/s are $0.336ps_{rms}$ and $7.7ps_{pp}$, respectively.

Figure 8.8.6 shows the jitter-tolerance comparison for operation both with and without PAM. In the latter case, the differential-mode component of V_{SP} is set to 0. Setting a BER of 10^{-12}, the out-of-band jitter tolerance is 0.58UI with the SPD activated, improved from 0.25UI with the SPD deactivated. The active area of the chip, whose die micrograph is shown in Fig. 8.8.7, is $0.9{\times}0.6mm^2$. The measurement results prove that with the phase adjustment, this architecture has a better jitter tolerance than simply using the initial strobe point of the linear PD.

Acknowledgements:
The authors would like to thank Broadcom Corp. for measurement facilities and the TowerJazz Shuttle Program for providing chip fabrication.

References:

[1] A. Pottbacker, et al., "A Si Bipolar Phase and Frequency Detector IC for Clock Extraction up to 8 Gb/s," *IEEE J. Solid-State Circuits*, vol. 27, pp. 1747–1751, Dec. 1992.

[2] R. Inti, et al., "A 0.5-to-2.5Gb/s Reference-Less Half-Rate Digital CDR With Unlimited Frequency Acquisition Range and Improved Input Duty-Cycle Error Tolerance," *ISSCC Dig. Tech. Papers*, pp. 438–439, Feb. 2011.

[3] J. Lee and K. C. Wu, "A 20 G/s Full-Rate Linear Clock and Data Recovery Circuit With Automatic Frequency Acquisition," *IEEE J. Solid-State Circuits*, vol. 44, pp. 3590–3602, Dec. 2009.

[4] S. B. Anond, and B. Razavi, "A 2.75 Gb/s CMOS Clock Recovery Circuit With Broadband Capture Range," *ISSCC Dig. Tech. Papers*, pp. 214-215, Feb. 2001.

[5] N. Kocaman, et al., "An 8.5–11.5-Gbps SONET Transceiver With Referenceless Frequency Acquisition," *IEEE J. Solid-State Circuits*, vol. 48, pp. 1875-1884, Aug. 2013.

[6] J. Cao, et al., "Non-Idealities in Linear CDR Phase Detectors," *Int. J. of Circuit Theory and Design*, vol. 41, pp. 331-346, Apr. 2013.

978-1-4799-0917-9/14 $31.00 © 2014 IEEE

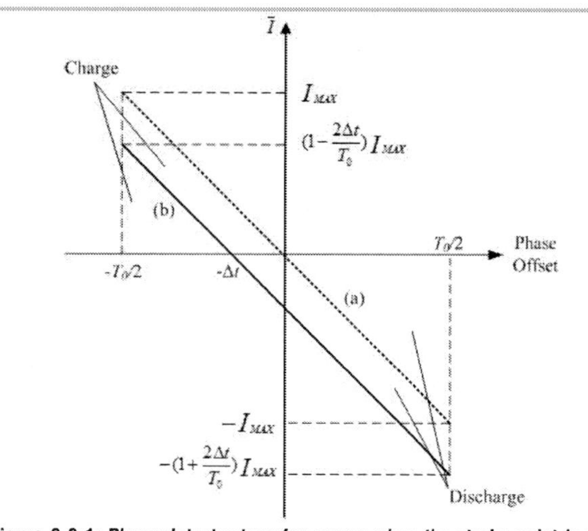

Figure 8.8.1: Phase-detector transfer curves when the strobe point is (a) 0, (b) -Δt.

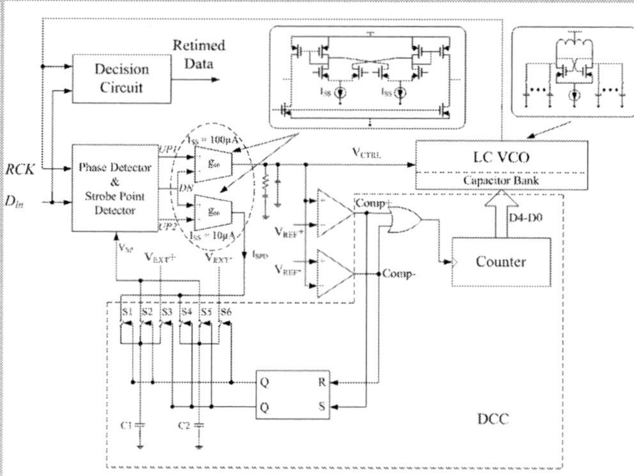

Figure 8.8.2: Block diagram of the designed reference-less CDR architecture with strobe-point control and phase-adjustment circuit.

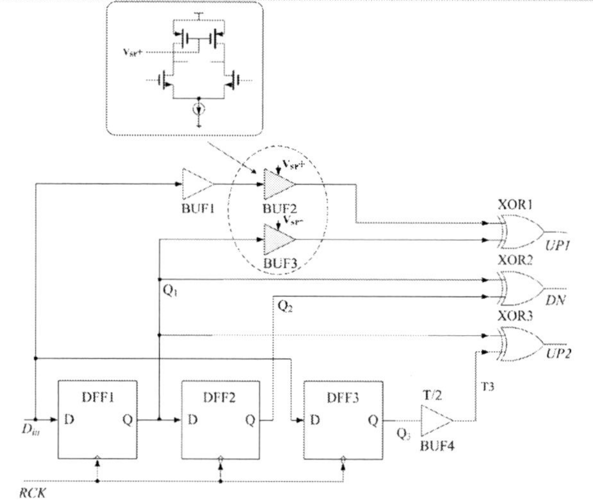

Figure 8.8.3: Hogge phase detector combined with the SPD circuit.

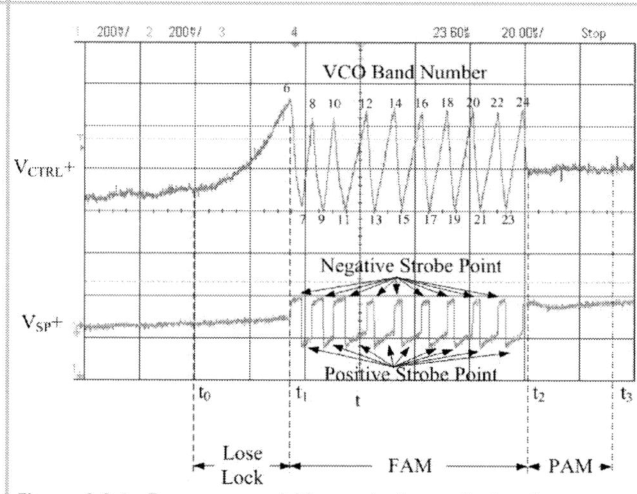

Figure 8.8.4: Frequency-acquisition and phase-adjustment processes (horizontal scale: 20µs/div, vertical scale: 200mV/div).

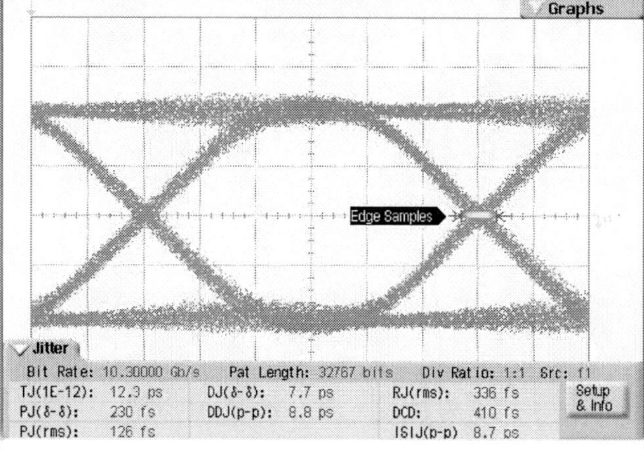

Figure 8.8.5: 10.3Gb/s recovered data output (horizontal scale: 16.5ps/div, vertical scale: 100mV/div).

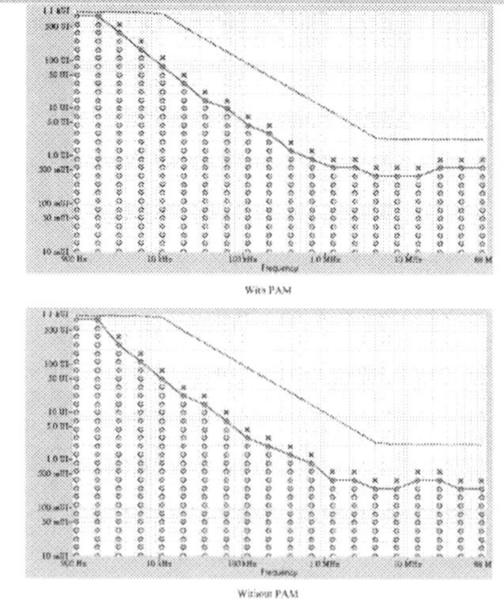

Figure 8.8.6: Jitter tolerance (a) with SPD activated, and (b) with SPD deactivated.

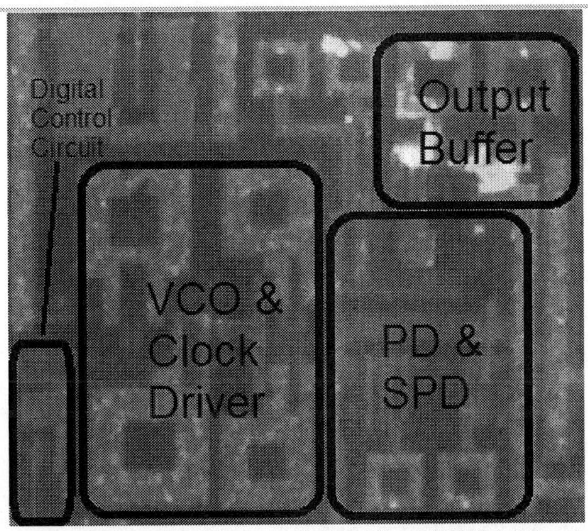

Figure 8.8.7: Micrograph of the test-chip in 0.18μm BiCMOS (all the circuits use only CMOS transistors).

8.9 A 40Gb/s VCSEL Over-Driving IC with Group-Delay-Tunable Pre-Emphasis for Optical Interconnection

Yukito Tsunoda, Mariko Sugawara, Hideki Oku,
Satoshi Ide, Kazuhiro Tanaka

Fujitsu Laboratories, Atsugi, Japan

High-speed and high-density interconnections between racks and modules in the high-performance computing systems and data centers are currently being developed. The transmission range of conventional electrical interconnections is limited due to the bandwidth of electrical channels. VCSEL-based optical interconnection technologies are a promising solution for overcoming bandwidth bottlenecks in large scale computing systems [1-3]. Although it is anticipated that the next challenge for optical interconnections is to move to a serial data-rate of 40Gb/s, there are few 40Gb/s class VCSELs at present. Over-driving is a method that boosts high-frequency response to overcome the VCSEL speed limit [4,5]. To develop high-density optical interconnections, a low-power over-driving IC is a key technology. In addition, the optical modulation amplitude (OMA) must be increased to enable long-distance transmission in large scale computing systems as a data center. To achieve this large modulation amplitude, we must overcome the jitter issue caused by the intrinsic group delay of VCSELs. In this paper, we present a 40Gb/s driver IC for over-driving a 25Gb/s VCSEL using a new 2-tap pre-emphasis circuit with tunable group-delay compensation. This circuit compensates for the complex group delay of VCSELs. With this circuit, we achieve 40Gb/s low-jitter operation with 2.3dBm OMA and reduce the power consumption to as low as 312mW/ch.

Figure 8.9.1(a) shows the block diagram of the driver IC. The driver IC consists of a pre-amplifier stage, pre-emphasis stage, and output stage. To drive 25Gb/s VCSELs at 40Gb/s, the pre-emphasis stage is implemented to boost high-frequency response. Here, we explain the VCSEL output power dependence on the VCSEL current as show in Fig. 8.9.1(b). Because the saturation of the output power of VCSELs occurs over 10mA current, VCSELs must be used under low-bias condition such as 5mA to realize the large modulation amplitude of 2.3dBm. The frequency response of a VCSEL is largely dependent on the VCSEL bias current. Peaking in the group delay becomes significant when the bias current is low. This group-delay peak increases the jitter, which is a serious issue at high data rates, such as 40Gb/s. Therefore, the compensation of this group delay is necessary to realize a large modulation amplitude. Although the 3-tap pre-emphasis [6] can compensate for this group-delay peak and reduces jitter, it results in an increase in the power consumption. We develop a 2-tap FFE architecture with a tunable group-delay compensation circuit. This circuit increased the flexibility of 2-tap pre-emphasis and lets us adjust the complex group delay of VCSELs.

Figure 8.9.2(a) shows the group-delay compensation circuit. It consists of an emitter peaking circuit, a positive-feedback circuit, and a limiting amplifier. The emitter peaking circuit decreases the group delay at low frequency range depending on the amount of emitter peaking. The positive-feedback circuit increases the group delay at low frequency depending on the amount of positive feedback. By adjusting the amount of emitter peaking and positive feedback, the group delay can be tuned from negative to positive value. These circuits change not only the frequency dependence of group delay, but also the rising and falling edge property of the pulse signal. The limiting amplifier enables elimination of the peaking and degradation of the rising and falling edge, and realizes group delay tuning with constant pulse amplitude. The simulated result of this circuit's tunable group delay compensation property is shown in Fig. 8.9.2(b). The circuit simulation result indicates that the group delay can be adjusted.

Figure 8.9.3 shows the simulated frequency response of the 25Gb/s VCSEL model and the pre-emphasis circuit model including the VCSEL model. The 25Gb/s VCSEL model is created from measurement results. This VCSEL has a group-delay peak, for example, $9ps_{pp}$ at high bias (9mA) and $25ps_{pp}$ at low bias (5mA) in the 0 to 20GHz range. Using the pre-emphasis technique, the bandwidth of the VCSEL output can be increased by about 6GHz. However, the group-delay peak of VCSELs under low bias condition is too high and is not improved by the pre-emphasis. This large group-delay peak results in high jitter in the VCSEL eye diagram. The phase-compensation circuit creates a reverse group delay for the VCSEL to improve this jitter. Owing to this compensation circuit, our pre-emphasis decreases the group-delay peak to below $5ps_{pp}$.

We design a new 40Gb/s output stage circuit for VCSEL anode driving. The circuit is shown in Fig. 8.9.4(a). The anode-driving architecture helps to reduce power consumption because this circuit allows VCSEL operation with a low supply voltage. We develop an anode-driving output-stage circuit with bridge-type back terminations [7]. There are additional issues with the lack of bandwidth due to the parasitic capacitance of the PMOS bias circuit when designing the 40Gb/s over-drive IC because it requires very large bias transistors to supply current to the delayed stage as well as the main stage and the VCSEL. This improved anode-driving architecture, in which the main bias transistor is connected to the middle of the back termination resistors, enables us to neglect the parasitic capacitance because the middle of the back-termination resistors acts as a virtual AC ground. The peaking Inductors are optimized to operate at 40Gb/s. The electrical characteristics of output-stage circuit are shown in Fig. 8.9.4(b). The input data are 40Gb/s 2^7-1 pseudo random bit sequences (PRBS). A clear eye opening at 40Gb/s is confirmed.

The above-described 40Gb/s VCSEL driver IC is fabricated in 0.13μm SiGe BiCMOS technology. The power supply for the pre-amplifier and pre-emphasis stage is 2.5V and that for output stage is 3.3V. We evaluate the 40Gb/s electrical and optical property with 25Gb/s VCSELs using pre-emphasis signals. We use a GaAs VCSEL with an 850nm wavelength that has a 3dB bandwidth of 16GHz. The electrical and optical eye diagrams without group delay compensation are shown in Fig. 8.9.5(a). The rising and the falling edge of the electrical signal are emphasized, and a clear electrical pre-emphasis eye opening is observed. However, though the rising and the falling edge of the optical signal are boosted, significant deterministic jitter is observed. This jitter is caused by the VCSEL group-delay peak, so we apply our group-delay compensation, as shown in Fig. 8.9.5(b). Owing to this group-delay compensation, jitter in the optical eye is improved even with this low bias and large OMA. The average output power is 3.0dBm and the OMA is 2.3dBm. These results indicate that the driver IC not only boosts the VCSEL speed, but also compensates for the group-delay peak of VCSELs under low bias current. The pre-amplifier and pre-emphasis circuits consume 166mW, and the output stage consumes 146mW. We also achieve power consumption as low as 312mW/ch with the pre-emphasis driver due to the low-power output stage with anode-driving architecture.

We present a 40Gb/s VCSEL over-driving IC with group-delay tunable pre-emphasis for high-speed optical interconnections. The performance and comparison with other results are summarized in Fig. 8.9.6. Due to our new pre-emphasis driver, we achieve 40Gb/s operation of 25Gb/s VCSELs with a 40% reduction in power consumption and double OMA compared to conventional technology. Figure 8.9.7 shows a chip micrograph of the driver IC. The chip size is 2.0×2.0mm^2, and the core area of each channel is 0.250×2.0mm^2. This driver IC is promising for high-speed and high-density optical interconnection.

References:

[1] I. Young, et al., "Optical I/O Technology for Tera-Scale Computing," *IEEE J. Solid-State Circuit,* vol. 45, no. 1, pp. 235-248, Jan. 2010.

[2] J. Proesel, C. Schow, A. Rylyakov, "25Gb/s 3.6pJ/b and 15Gb/s 1.37pJ/b VCSEL-Based Optical Links in 90nm CMOS," *ISSCC Dig. Tech. Papers,* pp. 418-420, Feb. 2012.

[3] J. Jiang, et al., "100Gb/s Ethernet Chipsets in 65nm CMOS Technology" *ISSCC Dig. Tech. Papers,* pp. 120-122, Feb. 2013.

[4] A. Rylyakov, et al., "A 40-Gb/s, 850-nm, VCSEL-Based Full Optical Link," *Proc. OFC/NFOFC 2012,* OTh1E1, Mar. 2012.

[5] D. Kuchta, et al., "A 56.1Gb/s NRZ Modulated 850nm VCSEL-Based Optical Link," *Proc. OFC/NFOFC 2013,* OW1B5, Mar. 2013.

[6] Y. Tsunoda, et al., "25-Gb/s Transmitter for Optical Interconnection with "10-Gb/s VCSEL Using Dual Peak-Tunable Pre-Emphasis," *Proc. OFC/NFOFC 2011,* OThZ2, Mar. 2011.

[7] M. Sugawara, et al., "Novel VCSEL driving technique with virtual back termination for high-speed optical interconnection," *Proc. of SPIE Photonics West 2012,* vol. 8267 826713-1, Feb. 2012.

978-1-4799-0917-9/14 $31.00 © 2014 IEEE

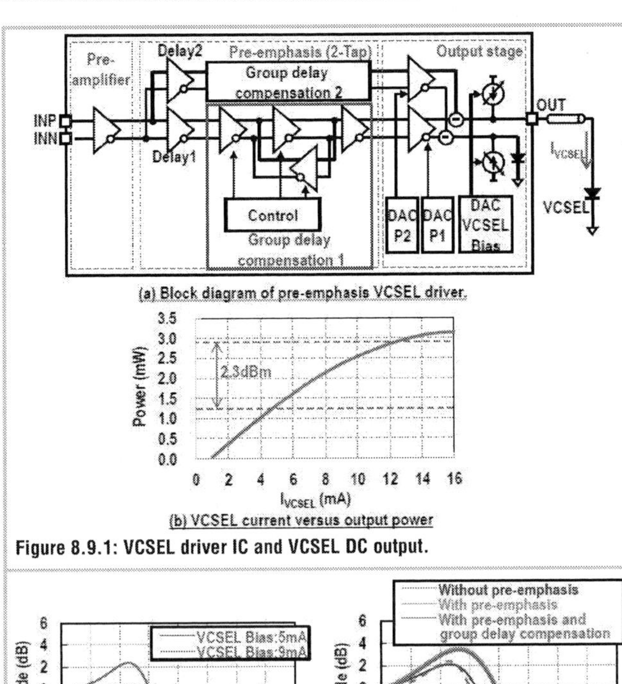

(a) Block diagram of pre-emphasis VCSEL driver.

(b) VCSEL current versus output power

Figure 8.9.1: VCSEL driver IC and VCSEL DC output.

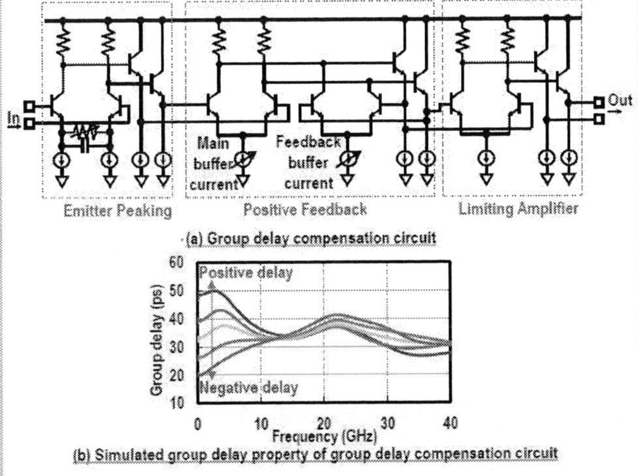

(a) Group delay compensation circuit

(b) Simulated group delay property of group delay compensation circuit

Figure 8.9.2: Group-delay compensation circuit.

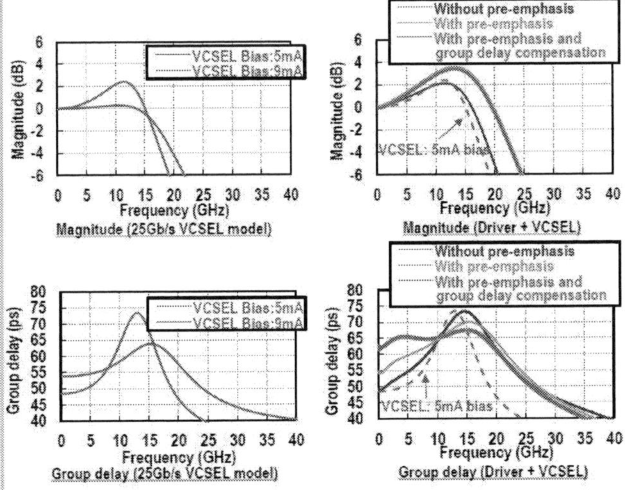

Figure 8.9.3: Simulated frequency response.

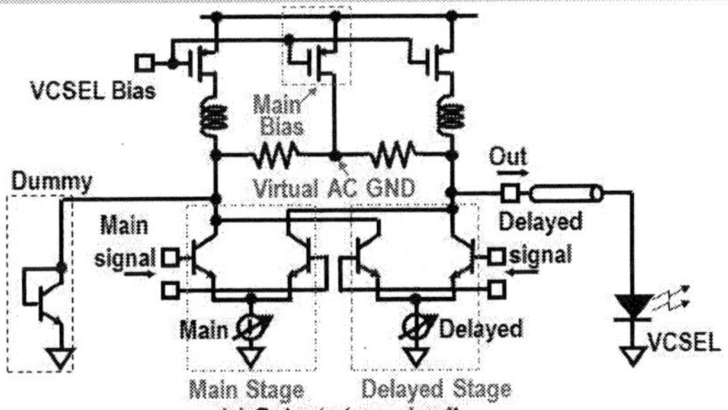

(a) Output stage circuit

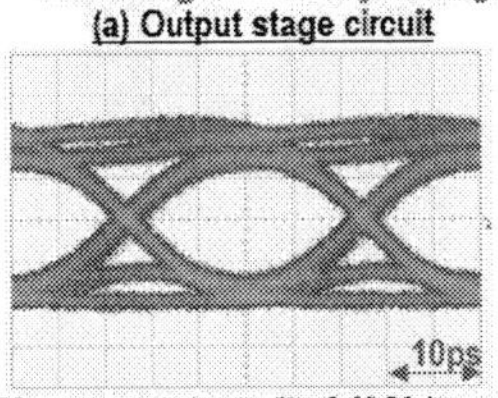

(b) Measurement result of 40Gb/s output eye

Figure 8.9.4: Output-stage circuit and eye diagram.

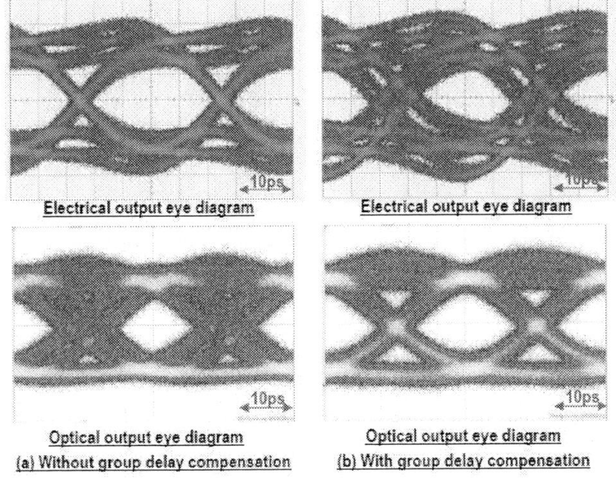

Optical output eye diagram
(a) Without group delay compensation

Optical output eye diagram
(b) With group delay compensation

Figure 8.9.5: Electrical and optical eye at 40Gb/s.

	[3]	[4]	[5]	This work
Technology	CMOS 65nm	SiGe 0.13µm	SiGe 0.13µm	SiGe 0.13µm
Data Rate (Gb/s)	25	40	56.1	40
Supply Voltage (V)	1.2/3.6	4.0/5.8	No data	2.5/3.3
Power Consumption (mW/ch)	99	530	682	312
Used VCSEL Bandwidth (GHz)	No data	16	24	16
Optical Modulation Amplitude (dBm)	0.8	-1	No data	2.3
VCSEL Driver Type	Cathode Drive	Cathode Drive	Cathode Drive	Anode Drive

Figure 8.9.6: Performance summary.

978-1-4799-0917-9/14 $31.00 © 2014 IEEE

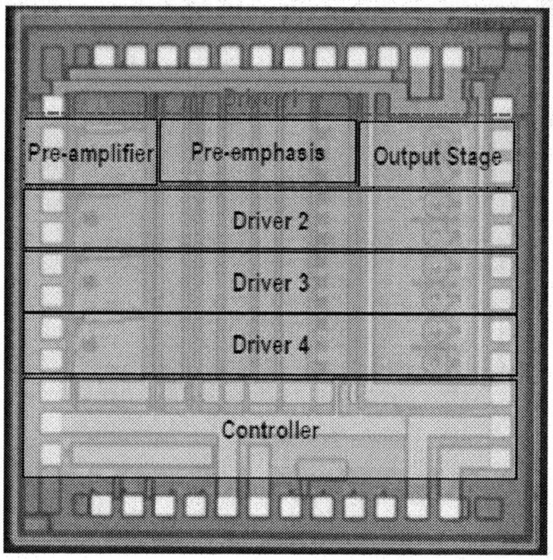

Figure 8.9.7: Chip micrograph of our driver IC.

Session 9 Overview: *Low-Power Wireless*

WIRELESS SUBCOMMITTEE

Session Chair: *Jan Crols*
AnSem, Heverlee, Belgium

Session Co-Chair: *Alyosha Molnar*
Cornell University, Ithaca NY

Low-power transceivers for short-range wireless communication are emerging in a wide range of frequencies and bandwidths. This session starts with a full SOC for near-field communication (NFC). It then covers two low-power ultrawideband radios. After that, five radios for ISM band operation at 433MHz, 900MHz and 2.4GHz are presented, each demonstrating different architecture and circuit choices to achieve the lowest possible power consumption without compromising bit rate and output power (TX) or sensitivity level (RX).

9.1 A Self-Calibrating NFC SoC with a Triple-Mode Reconfigurable PLL and a Single-Path PICC-PCD Receiver in 0.11μm CMOS **8:30 AM**
W. L. Lien, MediaTek, Singapore, Singapore
In Paper 9.1, Mediatek presents a near-field communication (NFC) transceiver SOC supporting card emulation mode, reader mode and peer-to-peer mode. The radio supports cm-range communication at 13.6MHz and fills 1.1mm² in 0.11μm CMOS.

9.2 A 13.3mW 500Mb/s IR-UWB Transceiver with Link-Margin Enhancement Technique for Meter-Range Communications **9:00 AM**
D. Liu, Tsinghua University, Beijing, China
In Paper 9.2, Tsinghua University presents an ultra-wideband transceiver in 65nm CMOS that supports 500Mb/s at 1m range while consuming only 13.3mW. The system meets FCC spectral mask requirements through a combination of high-bit-rate OOK modulation and frequency hopping.

9.3 A 1mW 1Mb/s 7.75-to-8.25GHz Chirp-UWB Transceiver with Low Peak Power Transmission and Fast Synchronization Capability **9:30 AM**
F. Chen, Tsinghua University, Beijing, China
In Paper 9.3, Tsinghua University and Samsung present an ultra-wideband transceiver in 65nm CMOS that provides 1Mb/s communication while consuming only 1mW. The system enables low-power synchronization through a low duty cycle, pulsed-chirp spread-spectrum approach.

978-1-4799-0917-9/14 $31.00 © 2014 IEEE

9.4 A 0.5V 1.15mW 0.2mm² Sub-GHz ZigBee Receiver Supporting 433/860/915/960MHz ISM Bands with Zero External Components **10:15 AM**

Z. Lin, University of Macau, Macao, China

In Paper 9.4, The University of Macau and Instituto Superior Technico present a ZigBee receiver in 65nm CMOS for the 433-to-960MHz bands, achieving an 8.1dB Noise Figure while consuming 1.15mW. The receiver demonstrates new techniques for current reuse, and employs N-path feedback and downconversion to reuse the input LNA for both RF and baseband signals.

9.5 A 1.2nJ/b 2.4GHz Receiver with a Sliding-IF Phase-to-Digital Converter for Wireless Personal/Body-Area Networks **10:45 AM**

Y-H. Liu, Holst Centre/imec, Eindhoven, The Netherlands

In Paper 9.5, imec presents a 2.4GHz receiver in 65nm CMOS employing a digital phase-tracking loop to perform direct phase-to-digital conversion. The receiver achieves -92dBm sensitivity for 2Mb/s signals in the IEEE 802.15.4 standard while consuming 2.4mW.

9.6 A 1.3mW 0.6V WBAN-Compatible Sub-Sampling PSK Receiver in 65nm CMOS **11:15 AM**

J. Cheng, Oregon State University, Corvallis, OR

In Paper 9.6, Oregon State University and Fudan University present a 2.4-to-2.7GHz WBAN receiver in 65nm CMOS, able to demodulate 971kb/s signals with -90dBm sensitivity while consuming 1.05mW. The receiver combines a Q-enhanced LNA with sub-sampling to provide low power downconversion with minimal noise folding.

9.7 A 0.33nJ/b IEEE802.15.6/Proprietary-MICS/ISM-Band Transceiver with Scalable Data-Rate from 11kb/s to 4.5Mb/s for Medical Applications **11:45 AM**

M. Vidojkovic, Holst Centre/imec, Eindhoven, The Netherlands

In Paper 9.7, imec, Eindhoven University of Technology and Fujitsu present a 400-to-450MHz IEEE 802.15.6 transceiver in 40nm CMOS for medical applications, employing digitally reconfigurable circuits to support modulation schemes and data rates from 11.7kb/s to 4.5Mb/s. The receiver achieves a sensitivity as low as -112dBm while consuming 2.2mW.

9.8 An 860μW 2.1-to-2.7GHz All-Digital PLL-Based Frequency Modulator with a DTC-Assisted Snapshot TDC for WPAN (Bluetooth Smart and ZigBee) Applications **12:00 PM**

V. K. Chillara, Holst Centre/imec, Eindhoven, The Netherlands and Delft University of Technology, Delft, The Netherlands, now at Analog Devices, Limerick, Ireland

In Paper 9.8, imec and Delft University of Technology present a 2.1-to-2.7GHz Frequency Modulator in 40nm CMOS for ZigBee and Bluetooth Smart, employing an all-digital fractional-N PLL with two-point modulation. The PLL employs a snapshot TDC to provide <1.7ps jitter while consuming 0.86mW.

9.1 A Self-Calibrating NFC SoC with a Triple-Mode Reconfigurable PLL and a Single-Path PICC-PCD Receiver in 0.11μm CMOS

Wee Liang Lien[1], Tieng Ying Choke[1], Ying Chow Tan[1], Ming Kong[1], Eng Chuan Low[1], Dan Ping Li[1], Liming Jin[1], Huajiang Zhang[1], Chin Heng Leow[1], Soong Lin Chew[1], Uday Dasgupta[1], Chee Hong Yong[1], Tian Bao Gao[1], Geok Teng Ong[1], Wee Guan Tan[1], Weimin Shu[1], Chee Lee Heng[1], Osama Shana'A[1,2]

[1]MediaTek, Singapore, Singapore,
[2]MediaTek, San Jose, CA

The popularity of the Near-Field Communication (NFC) system stems from being able to establish communication by merely being in the vicinity of another NFC device, an operation known as "tap and go". An NFC device is quite complex: it has to support both ASK/BPSK modulation, variable data rates from 106kb/s to 848kb/s, different ASK modulation indices (8% to 100%), different card types (NFC-A/B/F), and various coding. It also has different operating modes such as Proximity-Inductively-Coupled Card (PICC) or card-emulation mode, Proximity-Coupled-Device (PCD) or reader mode and Peer-to-Peer (P2P) mode. Furthermore, some PCDs transmit NFC-A/B/F ASK data in a polling loop manner. Therefore, PICC receivers must support joint data-type detection for successful communication with such PCDs. The device shown in Fig. 9.1.1 supports all these operating modes and complies with ISO-14443, ISO-18092 and NFC Forum standards. The SoC has an ultra-low current receiver, a digital transmitter with a 250mA-maximum-current-drive Class-D PA, a Single-Wire Protocol (SWP) supporting two external UICC SIM cards and one micro-SD chip, an energy-harvesting rectifier unit, an agile synthesizer, and a digital modem. Traditionally, two separate receivers with respective synthesizers and clock recovery/generation circuits are adopted for PICC, PCD and P2P modes [1,2]. In addition, three parallel analog demodulators are needed (one each for NFC-A, B and F) to support joint data-type detection, thus increasing die area significantly. The focus of this paper is the adoption of a single receiver with one reconfigurable PLL to support all NFC modes for compact die area.

The RX taps the antenna matching network via a digitally programmable impedance divider. Both PICC and PCD modes use the same direct- conversion RX path shown in Fig. 9.1.2, which uses an I/Q passive sample&dump mixer to downconvert the 13.56MHz NFC carrier to baseband via a 13.56MHz LO. The resulting large DC offset is blocked by a high-pass RC stage followed by an op-amp biquad low-pass filter, with a digitally programmable gain. The output is fed to an op-amp PGA stage with 24dB maximum voltage gain and 24dB gain control range in 3dB steps and has an analog DC-offset-removal servo loop followed by a 6b SAR ADC driven by a 13.56MHz clock. The I/Q RX path consumes ~840μA from a 1.5V supply. In PCD RX mode, the AGC is performed based on load modulation amplitude at 848kHz offset subcarrier by changing the low-pass filter and PGA gains. In PICC RX mode, the AGC is disabled because the high-pass filters distort the received ASK signal, generating signal peaks/dips whose amplitude is a function of the modulation index, slew rate and ASK envelope. These factors are mitigated through several methods. Since the ASK signal amplitude varies, depending on H-field strength and antenna-matching quality factor, an automatic impedance control (AZC) scheme is used to maintain a relatively constant RX amplitude by detecting the peak amplitude and adjusting the programmable impedance divider. Due to the large difference in modulation index between NFC-A and NFC-B/F, it is difficult to use a single receiver gain setting for all data types. This is resolved by using the pause detector in the reconfigurable PLL as an analog demodulator only when the modulation index is large (100% ASK for NFC-A). When the modulation index is small (30%~10% ASK for NFC-B/F), the RX voltage gain is fixed at 12dB in PICC mode to ensure sufficient amplitude at the ADC input. In this way, the digital demodulator can simultaneously detect the received data type (NFC-A/B/F).

A triple-mode reconfigurable ring-oscillator-based PLL with a dual-path active loop filter is implemented to support all NFC modes as shown in Fig. 9.1.1. In PCD mode, the PLL is configured as a fractional-N with VCO oscillating at 162.72MHz, locked to a 13MHz-to-52MHz reference clock. In PICC mode, the PLL is configured as integer-N clock recovery locked to the 13.56MHz H-Field. In P2P active initiator mode, the PLL is configured initially as fractional-N for TX data transmission but when the TX is off, the PLL loop is opened and the charge pump is set to tri-state to hold the VCO tune voltage to maintain the oscillating frequency at 162.72MHz. Upon detection of the H-field from the responding P2P active target, the PLL loop is closed and is synchronized quickly to the H-field for data reception within 15μs, which fulfills the frame-delay-time requirement of less than 188μs [3] with sufficient margin. Since the VCO frequency is maintained at 162.72MHz throughout the entire P2P active communications, the digital baseband clock is continuous and synchronous. To satisfy a $1mV_{rms}$ sensitivity for the PCD receiver and 12dB SNR needed to fulfill a communication distance requirement of 5cm, the PLL LO phase-noise specification at 848kHz offset is calculated to be <-124dBc/Hz, which is met using a three-stage differential ring-VCO topology, shown in Fig. 9.1.3. The basic delay cell is composed of a CMOS differential inverter with PMOS latches. The 162.72MHz ring VCO is tuned by adjusting its bias current via a linear V-to-I bias circuit, which reduces the VCO gain, supply pushing and phase noise. The PLL loop gain and bandwidth are configured according to NFC operating modes for optimum loop dynamics and phase noise. To overcome the challenge of pause or no clock during NFC-A communication with 100% modulation index, an adaptive pause detector is used to ensure a continuous PLL clock to both ADC and digital baseband. The clock extractor with auto-threshold pause-detection circuit shown in Fig. 9.1.4 generates accurate start- and end-of-pause indicators to the PLL. When the output of the pause detector is "high", the PLL loop is closed, and when it is "low", the PLL loop is opened. Based on the detected RX input level through the level detector in Fig. 9.1.4, the reference voltage ΔV of comparator A2 is adjusted automatically. This coarse adjustment is followed by a fine adjustment in the delay circuit to set the optimum threshold of the pause detector so that the end-of-pause is accurately detected within 295ns. With such accuracy, the end-of-pause detection scheme can also be used as the start of trigger for FDT to fulfill the stringent FDT variance requirement of 400ns [4], and also as an analog demodulator for NFC-A.

The SoC incorporates many automatic calibrations. For example, the TX modulation index is automatically calibrated through a loop-back TX-to-RX scheme. Under this condition, the PLL is configured to produce an RX LO signal whose frequency is different from the TX 13.56MHz LO so as to operate the receiver in low-IF mode. Based on the received signal level, the TX level is accurately adjusted to meet the desired modulation index. This SoC also incorporates a low-power polling mode in which it constantly detects the presence of a card in the vicinity in a sub-millisecond operation before it performs a full-card communication when a card is detected, an operation that takes a few tens of milliseconds to complete. This reduces the battery drain current to a mere 70μA during this mode, which is realized through the same loop-back scheme used for TX modulation-index calibration.

This NFC SoC is fabricated in a 0.11μm CMOS process and is housed in a 4mm×4mm 32-pin QFN package. The performance is characterized using the Reference Poller-0 and Listener-1 specified in [4] using a 50mm x 46mm antenna. In PCD mode, the communication distances for NFC-A and NFC-B are 5.0cm and 10.3cm, respectively (Reference Listener-1 connected with 820ohm load). In the PICC mode, the communication distances for NFC-A/ NFC-B and NFC-F212/424 are 6.2cm and 5.5cm, respectively (Reference Poller-0 calibrated to provide nominal power). The SoC has been subjected to an extensive interoperability test involving more than a hundred commercial readers and cards with no communication holes/issues. The measured LO phase noise in PCD mode of -125dBc/Hz at 848kHz offset is shown in Fig. 9.1.5. The performance summary is shown in Fig. 9.1.6 with comparison to state of the art showing a comparable or better performance but at a much smaller die size, even using a relatively old process node thanks to the single RX and reconfigurable PLL architecture. The SoC die size is $4.7mm^2$ of which only $1.1mm^2$ is occupied by RF/analog/PMU/SWP circuits. The die micrograph is shown in Fig. 9.1.7, showing no on-chip inductors.

References:

[1] Yogesh Darwhekar, *et al.* "A 45nm CMOS Near Field Communication Radio with 0.15A/m RX sensitivity and 4mA current consumption in card emulation mode," *ISSCC Dig. Tech. Papers*, pp. 422-423, Feb. 2013.

[2] S. Morris and A. Lefley, "A 90nm CMOS 13.56MHz NFC Transceiver," *IEEE ASSCC*, Nov. 2009, pp. 25-28.

[3] Peer-to-Peer Specification: ISO-IEC 18092-NFCIP-1(ECMA-340), 2004.

[4] Test Cases for NFC RF Analogue Specification, NFC FORUM ANALOGUE_TC 1.0.00, 2012-07-0.

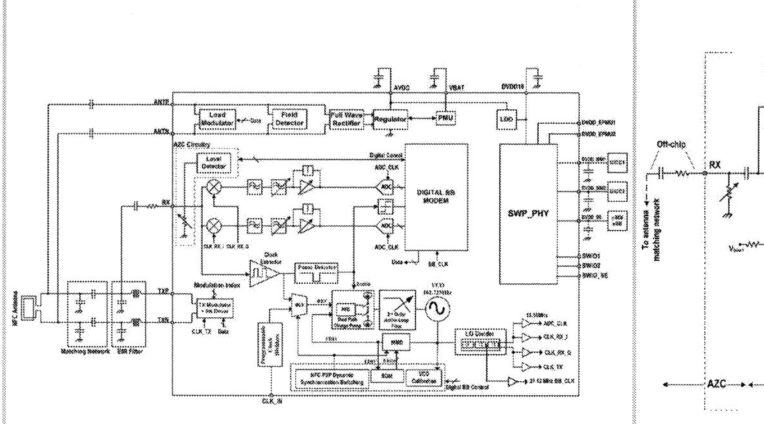

Figure 9.1.1: The NFC SoC-architecture block diagram.

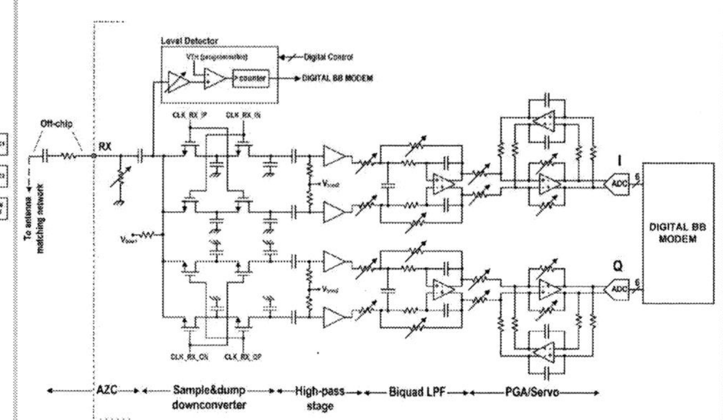

Figure 9.1.2: PCD/PICC Mixer-First Receiver schematic.

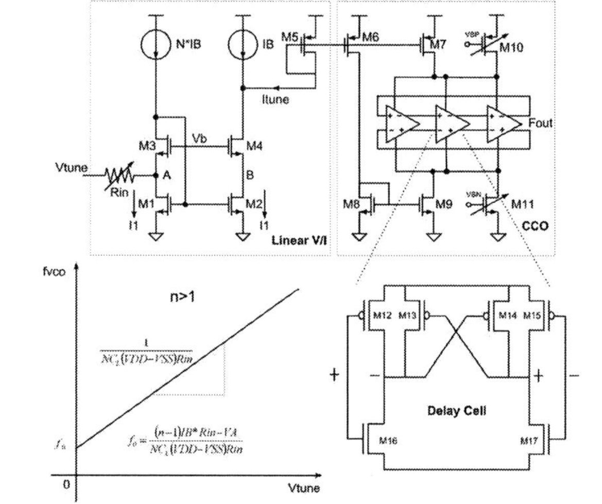

Figure 9.1.3: Ring VCO with V-to-I tuning circuit.

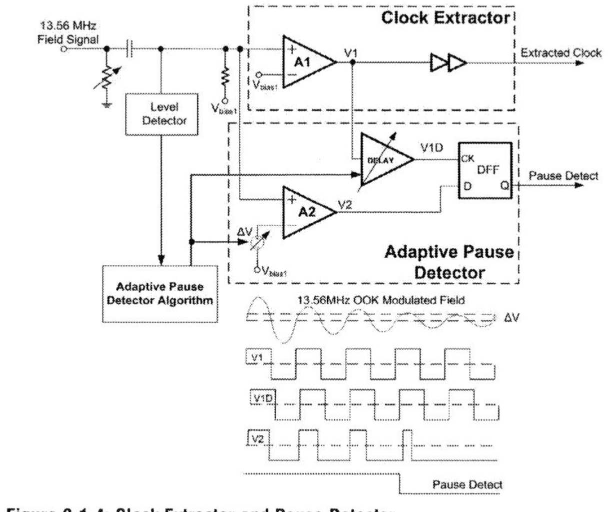

Figure 9.1.4: Clock Extractor and Pause Detector.

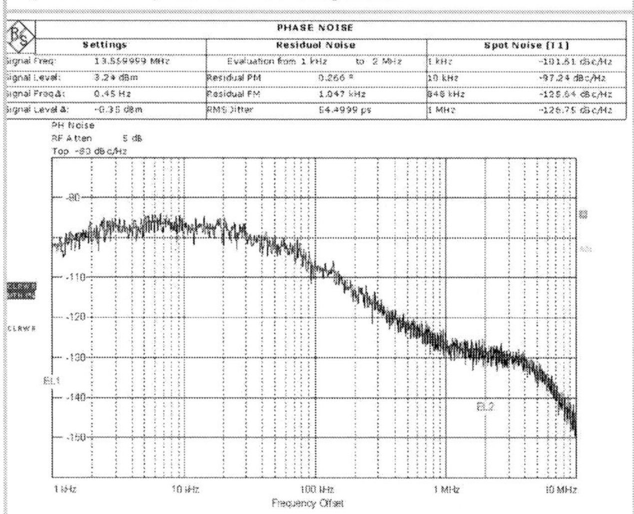

PHASE NOISE

Settings		Residual Noise		Spot Noise [T1]	
Signal Freq:	13.559999 MHz	Evaluation from 1 kHz	to 2 MHz	1 kHz	-101.61 dBc/Hz
Signal Level:	3.24 dBm	Residual PM	0.266 °	10 kHz	-97.24 dBc/Hz
Signal Freq Δ:	0.45 Hz	Residual FM	1.047 kHz	848 kHz	-125.64 dBc/Hz
Signal Level Δ:	-0.35 dBm	RMS Jitter	54.4999 ps	1 MHz	-126.75 dBc/Hz

PH Noise
RF Atten 5 dB
Top -80 dBc/Hz

Frequency Offset

Figure 9.1.5: Measured PCD 13.56MHz LO phase noise.

	This Work	ISSCC13 [1]
Technology	110nm CMOS	45nm CMOS
RF + Analog Area	1.1 mm²	3.4 mm²
Supported modulation depth range	8%~100%	8%~100%
Supported standards	ISO-14443 ISO-18092 ISO-15693 NFC forum	ISO-14443 ISO-18092 ISO-15693 NFC forum
PLL VCO topology	ring	LC
Reader PA VDD	3.3V	NA
Digital Supply voltage	1.2V	1.0V
Analog Supply voltage	1.8/1.5 V	1.8
PICC mode SoC current	3.5mA	4mA
Max. reader drive current	250mA	200mA
Card detection loop average supply current	70μA	N/A
Communication distance (P2P, PCD & PICC)	≤5cm	N/A
ESD	+/-3kV HBM, +/-300V MM, +/-500V CDM	+/-2kV HBM, +/-500V CDM

Figure 9.1.6: Measured performance & comparison summary.

Digital

RF

Figure 9.1.7: NFC SoC die micrograph.

9.2 A 13.3mW 500Mb/s IR-UWB Transceiver with Link-Margin Enhancement Technique for Meter-Range Communications

Shuli Geng, Dang Liu, Yanfeng Li, Huiying Zhuo, Woogeun Rhee, Zhihua Wang

Tsinghua University, Beijing, China

Unlike consumer electronics, transfusing the advanced wireless technology into medical equipment has not been rapidly developed. For wireless medical applications, lossless connection and noninvasive transmission are important factors. Moreover, in medical imaging applications such as 4D ultrasound imaging, high-information-rate transmission with a real-time display are demanded. To support the high-definition video format, a data-rate as high as 500Mb/s is desirable for raw data transmission. The mm-Wave transceiver [1] or the MB-OFDM transceiver [2] can provide high data-rate, but high power consumption and high transmission power are shortcomings for medical applications.

On the other hand, high-band IR-UWB technology features low transmission power, robustness against multipath fading, and resilience to other narrowband wireless standards. Especially, the low-power transmission is less harmful than narrowband transmission and enables frequency reuse in other medical rooms with the same frequency channel. In addition, the data-rate of a few hundred Mb/s can be designed with low power consumption [3-5]. However, the spectrum regulation limits the link margin of high-data-rate UWB transceivers and extending the communication range has not been well addressed in the literature. In contrast to low-data-rate transmitters, high-data-rate transmitters have more difficulty in increasing the transmitting power since the average power can easily violate the spectrum mask. At the same time, the sensitivity of the receivers cannot be further improved because of wide signal bandwidth, thus limiting the communication distance of high-data-rate IR-UWB transceivers to several decimeters. A 3m-range wireless transmission can be found in a commercial 7.8GHz UWB product [6] but it is equipped with multiple antennas and a VGA display. In this paper, we present a low-power 500Mb/s UWB transceiver with a spectrum-efficient frequency-hopping (FH) technique to significantly extend the communication range with a single antenna.

Figure 9.2.1 illustrates the proposed overlapped FH scheme in both time and frequency domains. The center frequency of the IR-UWB signal hops successively from one to another at a rate of 50MHz, having the duration of the FH burst with a timing period of 20ns. Eight overlapped sub-bands are used from 7.5GHz to 9.25GHz with a signal bandwidth of 500MHz and an overlapped frequency band of 250MHz as depicted in Fig. 9.2.1. By employing the FH technique, the proposed transmitter achievers higher output power than the conventional one by 8 times without spectrum violation. For the reception of the transmitted signal, a noncoherent-energy-detection receiver that does not require power-hungry LO generation is designed. The RF signal in each sub-band can be uniformly converted to the baseband if the frequency response of the RF front-end is flat with enough bandwidth. The bandwidth of the LPF in the baseband is designed as 250MHz to get the best SNR. Compared with the conventional single-band UWB transceiver, the output power of the transmitter is 8 times higher, while the SNR in the receiver remains the same, resulting in an improved link margin by 9dB in theory.

Figure 9.2.2 shows the block diagram of the proposed transceiver. Since frequency inaccuracy is tolerable in the noncoherent receiver, the 8 sub-bands are generated by an LC DCO with a 9b control word *FCW [8:0]* and a frequency step of 5MHz. The FH modulator will generate the *FCW* as well as an 8b amplitude control word *ACW[7:0]* to control the center frequency and the output power for each sub-band. The BPSK modulation is used to smoothen the spectrum. The polarity control bit *SCR* determines the polarity of the pulse and the gating control bit *TXDAT* controls the OOK modulation. The receiver consists of a wideband LNA, a fully differential squarer, a programmable gain amplifier (PGA), an integrator, and a comparator. The PGA has 6-level gain control with a bandwidth of 250MHz. The duration and phase of the timing window for the integrator can be configured by the integration window controller (IWC) with a time step of 100ps. The IWC consists of an edge combiner and other logic circuits to combine multiphase signals into a short pulse with configurable pulse width and phase. The multiphase signals are generated by a 500MHz PLL with a ring VCO.

Figure 9.2.3 shows the schematics of the transmitter and the receiver front-ends. The resonant tank in the DCO consists of an on-chip inductor and a MIM-capacitor-based binary-coded capacitor array. The frequency is monotonically changed with the 9b *FCW*. The PA provides BPSK/OOK modulation, pulse shaping and power control with 8 different levels. The baseband pulse modulates the differential LO output with a triangular envelope. In the receiver, a current-reused 4-stage LNA is designed for low power. The first stage (STG1) is a balun-combined LNA with lower Q factor and larger bandwidth than those of the following stages (STG2-STG4). The center frequencies of the STG1, the STG2, and the STG3 are tuned at the center band, the low band, and the high band respectively. The center frequency of the STG4 is tuned at the center band, so that it works as a buffer to drive the squarer with a low Q factor. A fully differential squarer is designed with cross-coupled subthreshold transistors to achieve high conversion gain.

A prototype 7.5-to-9.5GHz IR-UWB transceiver was implemented in 65nm CMOS. With OOK modulation, the transceiver achieves a wide-range data-rate of 125-to-500Mb/s with good energy efficiency. Figure 9.2.4 shows the measured results of the transmitter with a data-rate of 500Mb/s. When the FH function is disabled, the measured peak-to-peak amplitude of the transmitted pulses is larger than 400mV with a 50Ω load. However, the power spectral density will exceed the FCC spectrum mask by 9dB. When the FH function is enabled, the UWB spectrum is expanded to 2GHz and meets the FCC mask. By adjusting the amplitude control bit *ACW*, a flat in-band spectrum is achieved with the overlapped FH method. Figure 9.2.5 shows the measured performance of the receiver. The measured sensitivity variation of each sub-band is less than 1dB while worse case from 7.75GHz to 9.75GHz is lower than –60dB. The measured communication distance is about 40cm without the FH function. The output power has to be lower than –41.3dBm/MHz to comply with the FCC mask. The measured communication distance with the FH technique is more than 1.2m, which is three times larger than single-band transmission.

The transceiver including the baseband PLL consumes 13.3mW from a 1V supply at a data-rate of 500Mb/s, achieving an energy efficiency of 26.6pJ/bit. Figure 9.2.6 shows the performance summary and comparison with existing high-data-rate UWB transceivers. A chip micrograph is shown in Fig. 9.2.7. The transceiver occupies an area of 2.25mm^2. By employing the overlapped FH method, the transmission distance is extended by three times, making it possible for the high-data-rate UWB transceivers to achieve meter-range wireless communication.

Acknowledgments:
This work was partly supported by the Global Research Outreach (GRO) Program of the Samsung Advanced Institute of Technology (SAIT), Suwon, Korea.

Reference:
[1] T. Tsukizawa *et al.*, "A Fully Integrated 60GHz CMOS Transceiver Chipset Based on WiGig/IEEE802.11ad with Built-in Self Calibration for Mobile Applications," *ISSCC Dig. Tech. Papers*, pp. 230-231, Feb. 2013.
[2] D. Leenaerts *et al.*, "A 65nm CMOS Inductorless Triple-Band-Group WiMedia UWB PHY," *ISSCC Dig. Tech. Papers*, pp. 410-411, Feb. 2009.
[3] M. Tamura *et al.*, "A 1V 357Mb/s-Throughput TransferJet SoC with Embedded Transceiver and Digital Baseband in 90nm CMOS," *ISSCC Dig. Tech. Papers*, pp. 440-441, Feb. 2012.
[4] T. Abe *et al.*, "A 2Gb/s 150mW UWB Direct-Conversion Coherent Transceiver with IQ-Switching Carrier Recovery Scheme," *ISSCC Dig. Tech. Papers*, pp. 442-443, Feb. 2012.
[5] C. Hu *et al.*, "A 90 nm-CMOS, 500 Mb/s, 3–5 GHz Fully-Integrated IR-UWB Transceiver with Multipath Equalization Using Pulse Injection-Locking for Receiver Phase Synchronization," *IEEE J. Solid-State Circuits*, vol. 46, pp. 1076-1088, May 2011.
[6] *ACUSON Freestyle Ultrasound System*, Siemens Medical Solutions: http://www.healthcare.siemens.com/ultrasound.

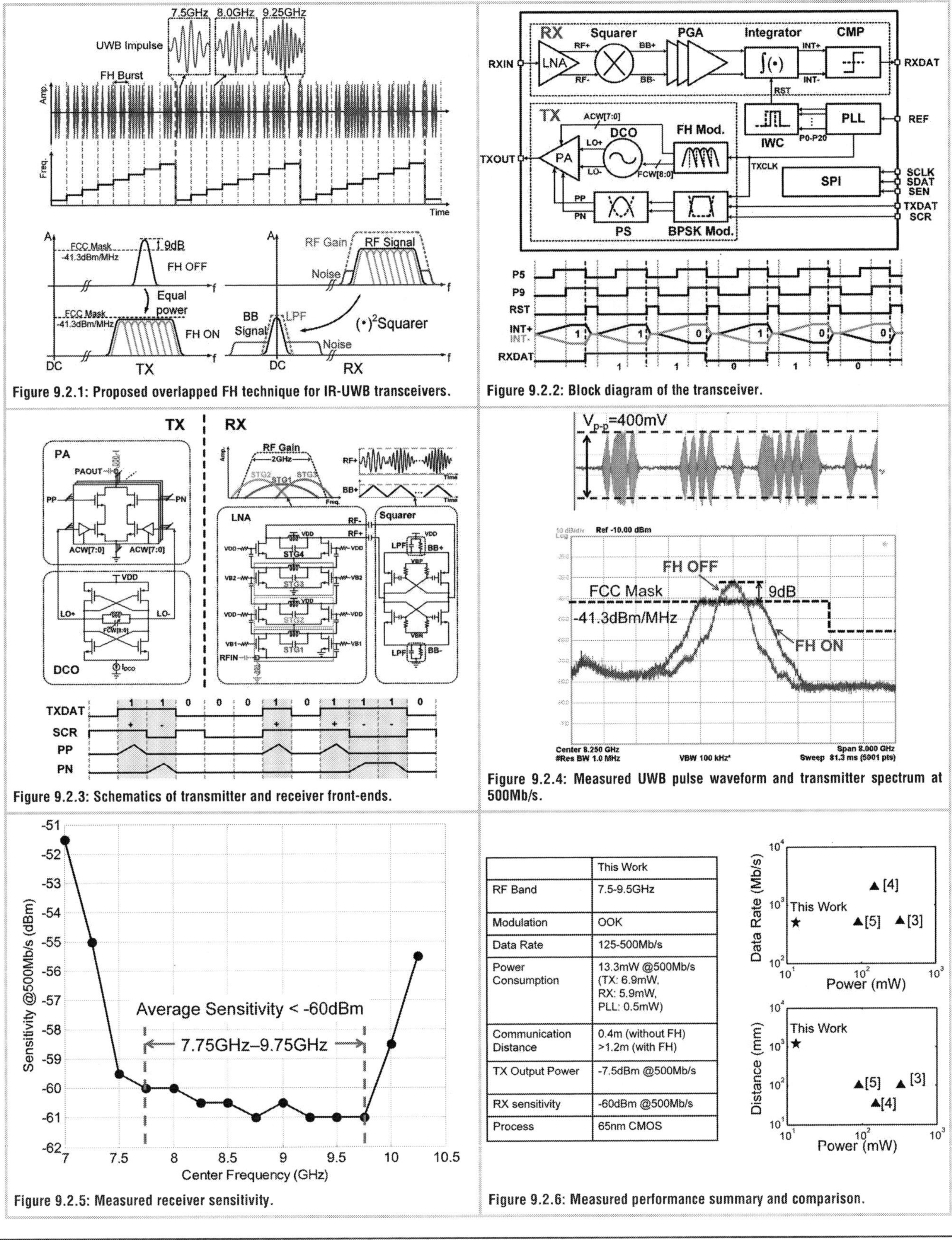

	This Work
RF Band	7.5-9.5GHz
Modulation	OOK
Data Rate	125-500Mb/s
Power Consumption	13.3mW @500Mb/s (TX: 6.9mW, RX: 5.9mW, PLL: 0.5mW)
Communication Distance	0.4m (without FH) >1.2m (with FH)
TX Output Power	-7.5dBm @500Mb/s
RX sensitivity	-60dBm @500Mb/s
Process	65nm CMOS

Figure 9.2.1: Proposed overlapped FH technique for IR-UWB transceivers.

Figure 9.2.2: Block diagram of the transceiver.

Figure 9.2.3: Schematics of transmitter and receiver front-ends.

Figure 9.2.4: Measured UWB pulse waveform and transmitter spectrum at 500Mb/s.

Figure 9.2.5: Measured receiver sensitivity.

Figure 9.2.6: Measured performance summary and comparison.

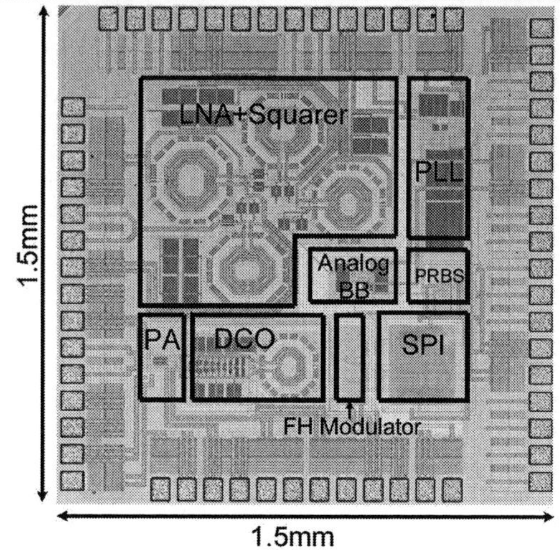

1.5mm
LNA+Squarer
PLL
Analog BB
PRBS
PA
DCO
SPI
FH Modulator
1.5mm

Figure 9.2.7: Chip micrograph.

9.3 A 1mW 1Mb/s 7.75-to-8.25GHz Chirp-UWB Transceiver with Low Peak-Power Transmission and Fast Synchronization Capability

Fei Chen[1], Yu Li[1], Dang Liu[1], Woogeun Rhee[1], Jongjin Kim[2], Dongwook Kim[2], Zhihua Wang[1]

[1]Tsinghua University, Beijing, China,
[2]Samsung Advanced Institute of Technology, Suwon, Korea

Future binaural hearing-aid devices face severe energy constraints on the wireless links where the ear-to-ear link enables signal processing of sound for both ears to enhance speech intelligence and the ear-to-device link provides an audio channel to commercial electronics such as smart TVs, MP3 players, and smart phones. Due to the limited size of the battery (35 to 90mAh) especially for in-the-ear (ITE) and in-the-canal (ITC) types, a sub-mW transceiver is required for long operational time. A low-data-rate impulse-radio ultra-wideband (IR-UWB) transceiver [1] achieves low power consumption with aggressive duty-cycled operation, namely 1%, but suffers from the bit-level synchronization problem by the baseband. Moreover, the low-data-rate IR-UWB exhibits a high peak transmission power to maintain a sufficient average transmission power for the given link margin. The constant-envelope frequency-modulated ultra-wide-band (FM-UWB) system in [2,3] features a low peak voltage and a steep roll-off spectrum, but the lack of duty-cycled operation makes it difficult to achieve low power. Narrowband WBAN transceivers such as the Bluetooth Low Energy (BLE) transceiver [4,5] offer good compliance with existing wireless SoCs, but the sub-mW power consumption is not feasible and they have a potential coexistence problem with existing Bluetooth devices. In this paper, a chirp-FSK-based UWB transceiver is proposed to significantly reduce the peak transmission power with relaxed duty-cycled operation for noninvasive, energy-efficient, and agile short-range communications.

Figure 9.3.1 illustrates the comparison of the chirp UWB (C-UWB) method with the IR-UWB and FM-UWB methods. Compared to the IR-UWB transceiver system, the C-UWB transceiver system can provide lower peak transmission power for the same link margin. Since an accurate timing window for the pulse-level synchronization is not needed for the 2-FSK chirp modulation and the duty cycle is higher than that of the IR-UWB by at least 10 times, a fast and reliable baseband synchronization time can be obtained. Compared to the FM-UWB transceiver system, the duty cycle of 10% can reduce the overall power of the transceiver by nearly ten times, thus achieving much higher energy efficiency.

Figure 9.3.2 shows the block diagram of the proposed C-UWB transceiver with 2-FSK modulation. Different from [6], the 2-FSK modulation is done with the starting frequency in the middle of the 500MHz band and the opposite frequency direction, which can achieve better spectrum efficiency and relax the VCO tuning range. In addition, a high-band UWB spectrum for the mandatory channel (7.75 to 8.25GHz) is utilized to accommodate a smaller antenna for miniaturized hearing-aid devices and to mitigate the interference problem by other wireless standards such as Bluetooth or WiFi. For high energy efficiency, duty cycles of 10% and 15% are set for the transmitter and the receiver respectively. In the transmitter, a digital-gradient generator (DGG) with a 100MHz system clock generates a 10-step gradient profile in 100ns with a time resolution of 10ns and an amplitude resolution of 4 bits. The DGG directly modulates a DCO which generates the chirp pulses. The upward gradient corresponds to chirp frequency from 8.025GHz to 8.25GHz with 25MHz frequency resolution and the downward gradient to frequency from 7.975GHz to 7.75GHz. Since the startup time of the DCO is important for the 10% duty-cycled transmitter operation, a turn-on time of 3ns is designed with a two-phase gating control.

The noncoherent receiver architecture is based on the dual-balanced-FM-demodulation topology [7]. Different from the conventional regenerative FM-UWB receiver having a narrowband LNA, the proposed receiver employs a wideband LNA and dual band-pass filters (BPFs) to improve FSK demodulation with a relaxed Q requirement of the BPF. The dual-BPF structure can also suppress the narrowband interference since the FM-to-AM conversion by the dual BPFs retains the gradient information of the UWB chirp pulses, while taking the narrowband interference as a dc component to be filtered out. Thanks to the relatively long pulse duration of 100ns for the data-rate of 1Mb/s, the receiver obtains fast bit-level synchronization by oversampling the FSK demodulation output with the 100MHz system clock. For instance, for the 100ns high-level FSK demodulation, the edge of the data clock is located at 85±5ns by considering the tolerance of the system clock skew between the transmitter and the receiver. Hence, the minimum synchronization delay is the bit period for one-bit synchronization beacon '1'. The oversampling method is simpler and faster than the conventional window-sliding method used in the IR-UWB system in which variable timing windows with different locations have to be tried even with the preamble code.

Figure 9.3.3 shows the schematic of the receiver front-ends. A wideband LNA converts the UWB signal to a differential signal with a balun at the bottom stage. The staggered structure of the two resonators provides a flat broadband gain of 18dB from 7.6GHz to 8.3GHz with a noise figure less than 5.2dB. The FSK demodulator consists of two BPFs, two envelope detectors and a comparator. The BPF is based on a 4th-order Butterworth filter. The upper-band BPF has a center frequency of 8.3GHz, and the lower-band BPF has a center frequency of 7.7GHz. The gain difference of 13dB between the two BPFs at the high end of 8.25GHz and at the low end of 7.75GHz is obtained. The envelope difference between the BPFs is extracted by the envelope detectors whose output is sampled by the data slicer with the synchronized data clock. A duty cycle of 15% is set for the receiver front-end circuits with extra bias setup time.

The prototype 1Mb/s C-UWB transceiver was implemented in 65nm CMOS. A chip micrograph is shown in Fig. 9.3.7. The active area of the transceiver is 0.7mm^2. Figure 9.3.4 shows the measured transmitted 1Mb/s chirp pulses with a duty cycle of 10% and a pulse duration of 100ns. The chirp UWB spectrum occupies the frequency band from 7.75GHz to 8.25GHz and complies with the UWB spectrum mask. In the measurement, the gain of the upper frequency band is lower than designed, showing slight distortion in the spectrum. However, the spectral efficiency is still higher than that of the IR-UWB system because of the inherent steep spectral roll-off of the wideband FM modulation. The free-running DCO has a phase noise of –107.2dBc/Hz at 1MHz offset. Figure 9.3.5 shows the measured receiver performance. The waveforms clearly show that the FSK demodulator with the synchronized clock successfully recovers the data. The LNA and the dual BPFs achieve a gain of 38dB at the high and low ends of chirp frequencies. The receiver sensitivity at 1Mb/s is –76 dBm for the bit error rate of 10^{-3}.

The transceiver system consumes an average power of 1mW from a 1V supply for a data-rate of 1Mb/s. The peak power consumption is 7.5mW where the DCO and the PA consume 2.8mW and the LNA and the dual BPFs consume 4.0mW. The performance comparison with other low-data-rate transceivers is summarized in Fig. 9.3.6.

References:

[1] X. Wang *et al.*, "A Meter-Range UWB Transceiver Chipset for Around-the-Head Audio Streaming," *ISSCC Dig. Tech. Papers*, pp. 450-451, Feb. 2012.

[2] N. Saputra and J. R. Long, "A Fully-Integrated, Short-Range, Low-Data-Rate FM-UWB Transmitter in 90nm CMOS," *IEEE J. Solid-State Circuits*, vol. 46, no. 7, pp. 1627-1635, July 2011.

[3] N. Saputra and J. R. Long, "A Short-Range Low-Data-Rate Regenerative FM-UWB Receiver," *IEEE Trans. Microwave Theory and Techniques*, vol. 59, no.4, pp. 1131-1140, Apr. 2011.

[4] Y. Liu *et al.*, "A 1.9nJ/b 2.4GHz Multistandard (Bluetooth Low Energy/Zigbee/IEEE802.15.6) Transceiver for Personal/Body-Area Networks," *ISSCC Dig. Tech. Papers*, pp. 446-447, Feb. 2013.

[5] A. Wong *et al.*, "A 1V 5mA Multimode IEEE 802.15.6/Bluetooth Low-Energy WBAN Transceiver for Biotelemetry Applications," *ISSCC Dig. Tech. Papers*, pp. 300-301, Feb. 2012.

[6] M. U. Nair *et al.*, "A Low SIR Impulse-UWB Transceiver Utilizing Chirp FSK in 0.18μm CMOS," *IEEE J. Solid-State Circuits*, vol. 45, no. 11, pp. 2388-2403, Nov. 2010.

[7] F. Chen *et al*, "A 3.8mW 3.5-4GHz Regenerative FM-UWB Receiver with Enhanced Linearity by Utilizing a Wideband LNA and Dual Band-Pass Filters," *IEEE Trans. Microwave Theory and Techniques*, vol. 61, no.9, pp. 3350-3359, Sept. 2013.

978-1-4799-0917-9/14 $31.00 © 2014 IEEE

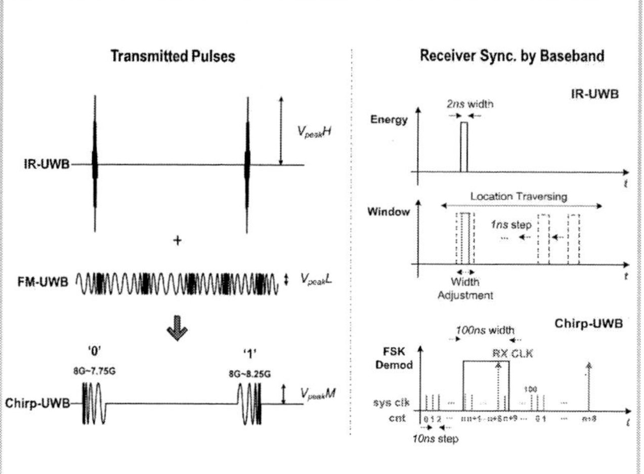

Figure 9.3.1: Chirp-UWB with reduced peak power and fast synchronization.

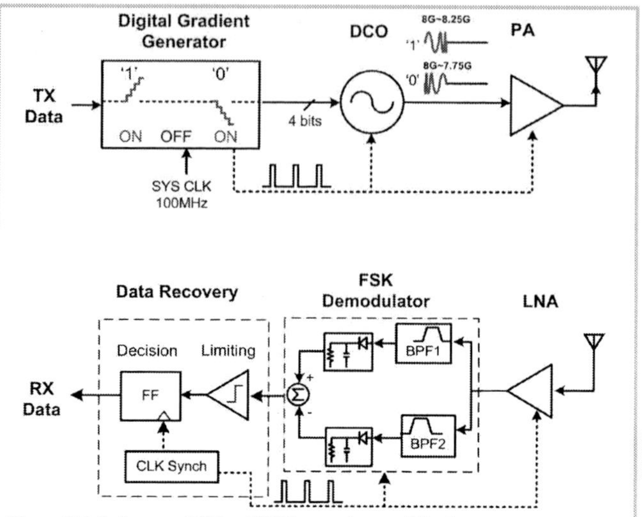

Digital Gradient Generator

DCO

PA

TX Data

'1'

'0'

ON OFF ON

4 bits

SYS CLK 100MHz

8G~8.25G

8G~7.75G

Data Recovery

FSK Demodulator

LNA

RX Data

Decision

Limiting

FF

CLK Synch

BPF1

BPF2

Figure 9.3.2: Proposed Chirp-UWB transceiver architecture.

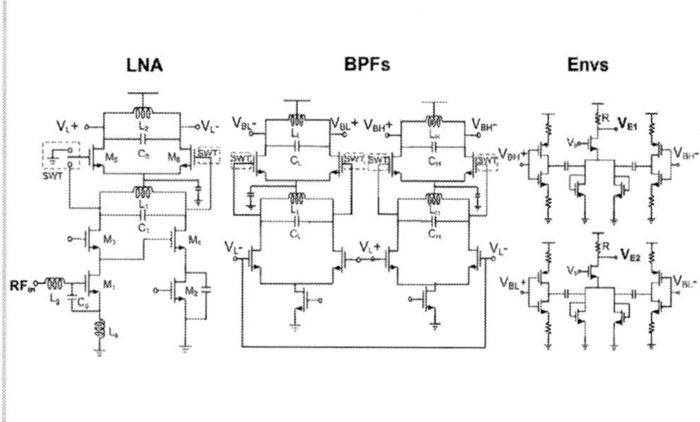

Figure 9.3.3: Schematic of receiver front-ends.

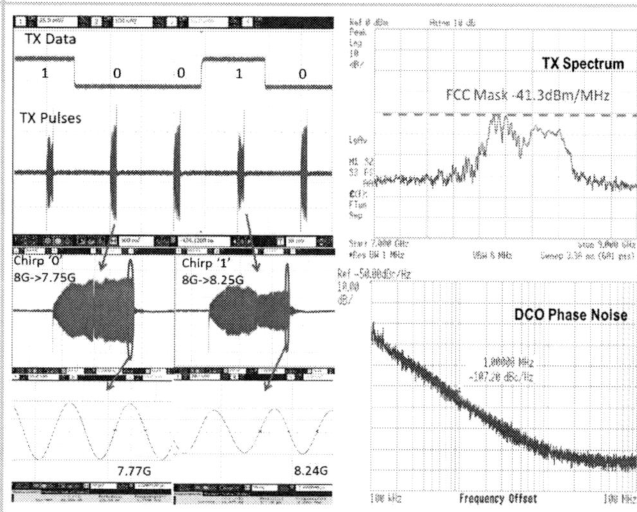

Figure 9.3.4: Measured transmitter performance.

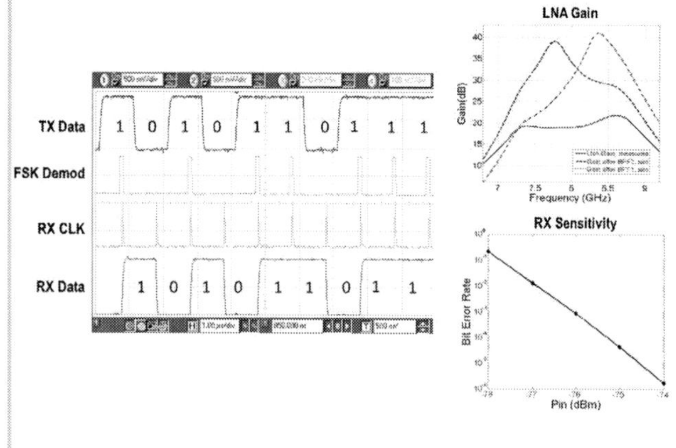

Figure 9.3.5: Measured receiver performance.

	[1]	[2] + [3]	[4]	[5]	[6]	This work
PHY	IR-UWB	FM-UWB	Bluetooth LE	Bluetooth LE	Chirp-UWB	Chirp-UWB
Frequency	6-9GHz	3.75-4.25GHz	2.4GHz	2.4GHz	3.15-3.9GHz	7.75-8.25GHz
Data rate	0.85Mb/s	0.1Mb/s	1Mb/s	1Mb/s	4Mb/s	1Mb/s
Power	6.5mW	3.1mW	9.2mW	9.4mW	56.8mW	1mW
Energy/bit	7.6nJ/bit	31nJ/bit	9.2nJ/bit	9.4nJ/bit	14.2nJ/bit	1nJ/bit
Sensitivity	-88dBm	-84dBm	-98dBm	-94dBm	-84dBm	-76dBm
Technology	90nm	90nm/65nm	90nm	130nm	180nm	65nm
Active area	2mm^2	0.77mm^2	2mm^2	5.9mm^2	6.7mm^2	0.7mm^2

Figure 9.3.6: Performance comparison with other WBAN transceivers.

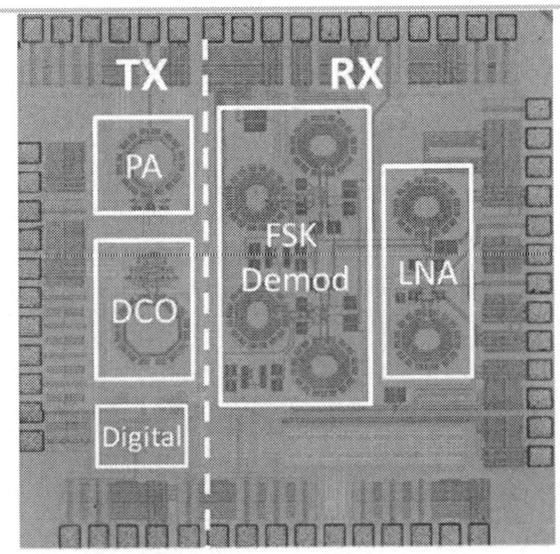

Figure 9.3.7: Chip micrograph.

9.4 A 0.5V 1.15mW 0.2mm² Sub-GHz ZigBee Receiver Supporting 433/860/915/960MHz ISM Bands with Zero External Components

Zhicheng Lin[1], Pui-In Mak[1,2], Rui Martins[1,2,3]

[1]University of Macau, Macao, China,
[2]UMTEC, Macao, China,
[3]Instituto Superior Tecnico, Lisbon, Portugal

The rapid proliferation of Internet of Things has urged the development of ultra-low-power (ULP) radios at the lowest possible cost, while being universal for worldwide markets. Both current-reuse [1,2] and ultra-low-voltage [3] receivers are promising solutions. [1] unifies most RF-to-BB functions in one cell for current-mode signal processing, resulting in a high IIP3 (–6dBm) at small power (2.7mW) and area (0.3mm²). However, outside the current-reuse cell, another supply is required for other circuits, complicating the power management [1,2]. [3] facilitates single-0.3V operation of the entire receiver at 1.6mW for energy harvesting, but the limited voltage headroom and transistor f_T call for bulky inductors/transformers to assist the biasing and to tune out the parasitics, penalizing the IIP3 (–21.5dBm) and area (2.5mm²). In both cases, a fixed LC network was adopted for input matching and pre-gain to lower the NF, which is costly and inflexible for multi-band designs.

Aiming for a single-0.5V ULP receiver for sub-GHz ZigBee (IEEE 802.15.4c/d) products (e.g., [4]), three circuit techniques are proposed: 1) An RF-to-BB-recycled front-end concurrently amplifies the RF (in common mode) and BB (in differential mode) signals under the same set of gain stages, squeezing the power by *frequency separation* and *signal orthogonality.* 2) An N-path (N=4) tunable LNA, embedded into the front-end, realizes low-noise input impedance matching while offering area-efficient blocker filtering to enhance the out-of-band linearity. 3) A VCO with extensively-distributed negative-gain cells for current-reuse with the BB complex low-IF filters is employed. With 1.15mW of power and 0.2mm² of area, the receiver shows 8.1dB NF and –20.5dBm IIP3 over the 433/860/915/960MHz ISM bands APT for China, Europe, North America and Japan, respectively, with zero external components.

The RF-to-BB-recycled front-end (Fig. 9.4.1) is described using the I channel. With C_i and C_o considered as short circuits at RF and ignoring their memory effects (detailed later), *Path A* amplifies the common-mode RF signal (and blockers) from V_i to V_o, where the two G_m stages are in parallel. *Path B* routes V_o to the two passive mixers for single-to-differential downconversion. *Path C* returns the differential BB-signals $V_{B1,I\pm}$ to the two G_m stages individually, recycling their gain orthogonally for BB amplification. Elegantly, BB filtering is inherent with C_i and C_o, as the differential BB signals and blockers see V_i and V_o as virtual grounds. Together with the Q channel, a functional view of the front-end (Fig. 9.4.2) is a single-ended $4G_m$ inverter-based LNA self-biased by $R_F/4$, followed by four I/Q passive mixers loaded by C_i, and finally by four *virtual* $1G_m$ BB amplifiers loaded by C_o. This topology not only nullifies the BB power, but also avoids the RF balun and balances the NF ($4G_m$ at RF) with linearity ($1G_m$ at BB).

When the memory effects of C_i and C_o are taken into account, the passive mixers become a 4-path switched-capacitor (SC) network, advancing the LNA into an *equivalent 4-path tunable LNA* (Fig. 9.4.3). For simplicity, we assume C_o is a short circuit at RF, but keep C_i since it dominates the frequency-translated filtering effect. After one LO cycle ($1/f_{LO}$), V_i is sampled and held by C_i building the 4-phase voltages (V_{ci}, $-V_{ci}$, jV_{ci}, $-jV_{ci}$). For the in-band RF signal, those voltages are in-phase-summed at V_o in the steady state. For the out-of-band RF blockers, those voltages are out of phase and cancelled when appearing at V_o. This bandpass effect can be modeled as an R_p-L_p-C_p resonator in series with the mixer's on-resistance (R_{sw}), and the center frequency is tunable by f_{LO} via L_p. It can be proven that such a resonator can be equivalently placed as the feedback network of the $4G_m$ stage (Fig. 9.4.3), rendering three benefits when comparing it with the passive N-path filter [5]: i) a closed-loop gain ($A_{v,LNA}$) much greater than 1 is feasible and bandpass filtering occurs twice at both V_i and V_o, enhancing the out-of-band linearity. ii) The $4G_m$ weakens the effect of R_{sw} to stopband rejection (i.e., β at V_i and $A_{v,LNA}/\gamma$ at V_o), given that R_{sw} is divided by $(1+(V_o/V_i))$ when reflecting back to V_i at the blocker frequencies, where L_p or C_p is considered as a short (Fig. 9.4.3). This feature saves the LO power for a given R_{sw}. The filtering effect at V_i is, to the first order, irrelevant to R_{sw}, and goes up with G_m that should be high for low NF. iii) Given an LNA's BW_{-3dB}, a smaller C_p is allowed due to the boosting factor $1+2A_{v,LNA}$, when referring to V_i. For instance, $A_{v,LNA}$=10 V/V can boost the effective C_p by ~20x.

The LNA's in-band input impedance (R_{in}) is $\sim[(R_F/4)//R_p]/4G_mR_L$ at L_pC_p resonance. Unlike the traditional R_F-feedback-only inverter-based LNA [6] that suffers from a tight tradeoff between S_{11} and NF, here R_p offers a freedom for input matching while contributing negligible noise (R_p is the equivalent resistance of the 4-path SC network).

A VCO filter is tailored for current reuse even at 0.5V (Fig. 9.4.4). The loss in the LC-tank of the VCO is compensated by a negative transconductor ($-G_{mT}$) pieced together from T number of M_v cells, i.e., $G_{mT}=T(4g_{mv})$, where g_{mv} is from M_v. The aim is to distribute the bias current of the VCO to all BB gain stages (A_1, A_2... A_{18}) that implement the filter. For the VCO, M_v operates at $2f_{LO}$ or $4f_{LO}$ for dividing out a 4-phase LO at f_{LO}. Thus, the VCO signal leaked to the source nodes of M_v ($V_{F1,I+}$, $V_{F1,I-}$) is pushed to very high frequencies ($4f_{LO}$ or $8f_{LO}$) and can be easily filtered by BB capacitors. For the filter's gain stages such as A_1, M_b (g_{mb}) is loaded by an impedance of $\sim 1/2g_{mv}$ when L_p is considered as a short at BB. Thus, A_1 has a ratio-based voltage gain of roughly g_{mb}/g_{mv}, or as given by $4Tg_{mb}/G_{mT}$. The latter shows how the distribution factor T can enlarge the BB gain, but is a tradeoff with its input-referred noise and can add more layout parasitics to $V_{VCOp,n}$ (i.e., narrower VCO's tuning range). The –R cell added at $V_{F1,I+}$ and $V_{F1,I-}$ boosts the BB gain without loss of voltage headroom. For the BB complex poles, $A_{2,5}$ and C_{f1} determine the real part while $A_{3,6}$ and C_{f1} yield the imaginary part. There are 3 similar stages cascaded for higher channel selectivity and image rejection ratio (IRR). R_{blk} and C_{blk} were added to avoid the large input capacitance of $A_{1,4}$ from degrading the gain of the front-end.

The receiver was fabricated in 65nm CMOS. Measurements (Fig. 9.4.5) showed that the gain (50±2dB), NF (8.1±0.6dB) and IRR (20.5±0.5dB) are stable over the four ISM bands. A two-tone test at [f_{LO}+12MHz, f_{LO}+22MHz] shows an $IIP3_{out\text{-}of\text{-}band}$ of –20.5±1.5dBm. All S_{11} are <–8dB and the VCO phase noise is –117.4±1.7dBc/Hz at 3.5MHz offset. Owing to the merged VCO filter, the BB signal should be <50mV_{pp} for not degrading the phase noise by 1dB. The 2MHz-IF gain response shows 18/38dB rejection at the adjacent/alternate channel. Other results (not shown) are the out-of-band P_{1dB} (–20dBm), and blocker-NF (13.7dB) for a single-tone blocker of –20dBm applied at 50MHz offset from the 860MHz RF. This blocker resilience is reasonably high for 1.15mW receiver power at 0.5V.

Benchmarking with the recent art [1,3,7] in Fig. 9.4.6, this work succeeds in covering multi-ISM bands with LO-defined input matching and RF filtering, while advancing the power and area efficiencies with zero external components. Figure 9.4.7 shows the die micrograph of the receiver.

Acknowledgements:
This work was funded by the Macao FDCT and UM - MYRG114-FST13-MPI.

References:
[1] Z. Lin, P.-I. Mak and R. P. Martins, "A 1.7mW 0.22mm² 2.4GHz ZigBee RX Exploiting a Current-Reuse Blixer + Hybrid Filter Topology in 65nm CMOS," *ISSCC Dig. Tech. Papers*, pp. 448-449, Feb. 2013.
[2] F. Lin, P.-I. Mak and R. P. Martins, "An RF-to-BB Current-Reuse Wideband Receiver with Parallel N-Path Active/Passive Mixers and a Single-MOS Pole-Zero LPF," *ISSCC Dig. Tech. Papers*, paper 3.9, Feb. 2014.
[3] F. Zhang, K. Wang, J. Koo, Y. Miyahara and B. Otis, "A 1.6mW 300mV Supply 2.4 GHz Receiver with –94 dBm Sensitivity for Energy-Harvesting Applications," *ISSCC Dig. Tech. Papers*, pp. 456-457, Feb. 2013.
[4] CC1200 SimpleLink Low Power, High Performance RF Transceiver: http://www.ti.com/lit/ds/symlink/cc1200.pdf
[5] A. Ghaffari, E. Klumperink, M. Soer and B. Nauta, "Tunable High-Q N-Path Band-Pass Filters: Modeling and Verification," *IEEE J. Solid-State Circuits*, vol. 46, pp. 998-1010, May 2011.
[6] J. Sinderen, G. Jong, F. Leong, *et al.*, "Wideband UHF ISM-Band Transceiver Supporting Multichannel Reception and DSSS Modulation," *ISSCC Dig. Tech. Papers*, pp. 454-455, Feb. 2013.
[7] A. Liscidini, M. Tedeschi and R. Castello, "Low-Power Quadrature Receivers for ZigBee (IEEE 802.15.4) Applications," *IEEE J.Solid-State Circuits*, vol. 45, pp. 1710-1719, Sep. 2010.

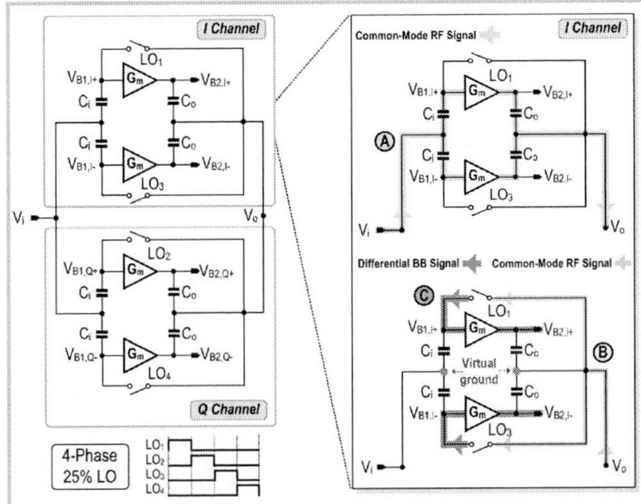

Figure 9.4.1: Proposed RF-to-BB-recycled front-end.

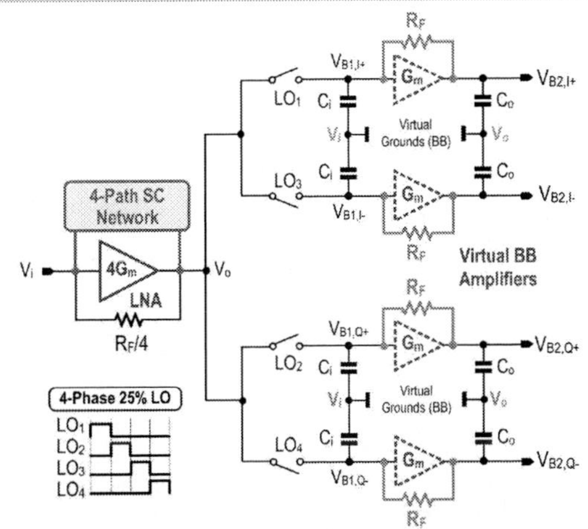

Figure 9.4.2: Functional view of the RF-to-BB-recycled front-end. The memory effects of C_i and C_o associated with the 4-path SC network are detailed in Fig. 9.4.3.

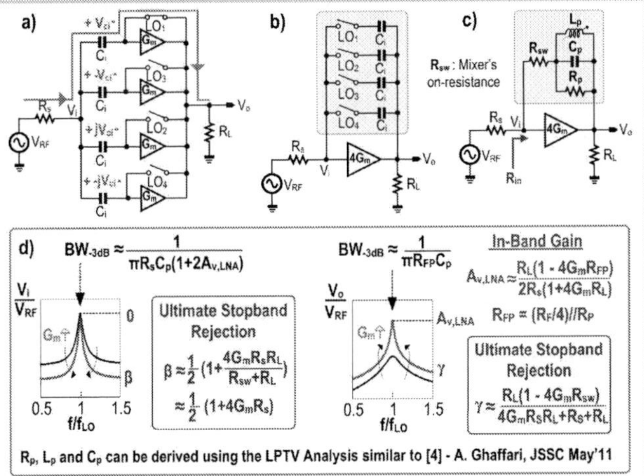

Figure 9.4.3: a) An equivalent 4-path tunable LNA embedded inside the front-end. a) and b) are mathematically equivalent and modeled as c). d) The filtering profiles of c).

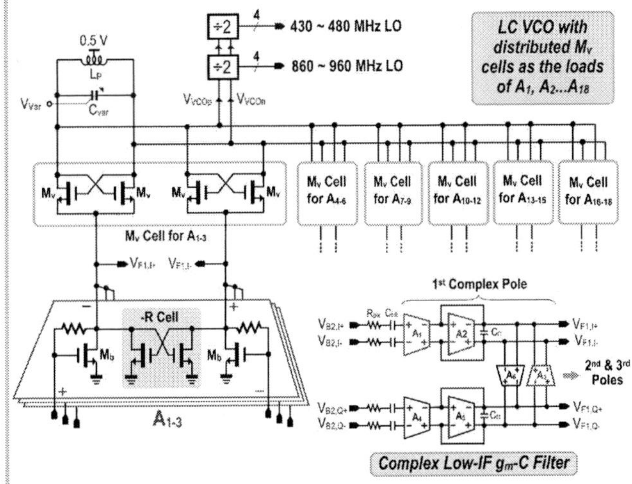

Figure 9.4.4: 0.5V current-reuse VCO filter and LO generation.

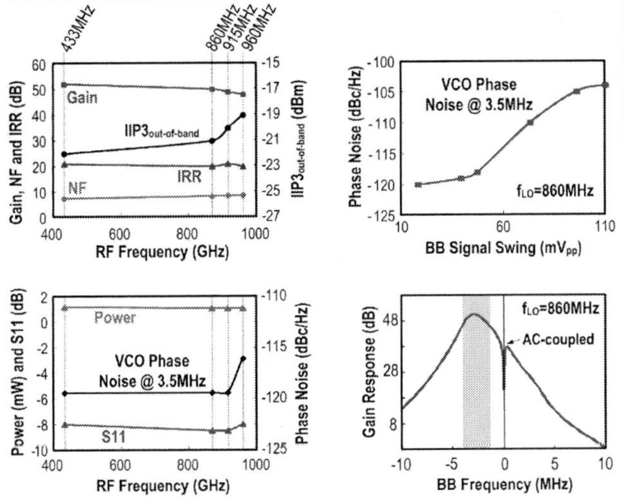

Figure 9.4.5: Measured key performance metrics.

	This Work	ISSCC'13 [1] (w/ VCO)	ISSCC'13 [3]	JSSC'10 [7]
Application	**433/860/915/960 MHz (ZigBee/IEEE802.15.4c/d)**	2.4 GHz (ZigBee/ IEEE 802.15.4)	2.4 GHz (Energy Harvesting)	2.4 GHz (ZigBee/ IEEE 802.15.4)
Architecture	RF-to-BB-Recycled Front-End + N-path Tunable LNA + Current-Reuse VCO-Filter	Blixer + Hybrid Filter + Passive RC-CR Filter + LC VCO	CG LNA + Passive Mixers + N-Path SC IF Filter + LC VCO	LNA-Mixer-VCO Merged Cell + Complex Filter
BB Filter	3 complex poles	1 Biquad, 4 complex poles	2 real poles	3 complex poles
Input Matching Technique	**On-chip N-path SC (tunable by LO, high Q)**	On-chip LC (fixed, low Q)	Off-chip LC (fixed, low Q)	Off-chip LC (fixed, high Q)
External Components	**zero**	zero	2 caps, 1 inductor	1 caps, 1 inductor
Input Matching BW and Tunability	**433 to 960 MHz (tunable by LO)**	2.25 to 3.55 GHz (fixed)	~ 2 to 2.6 GHz (fixed)	2.3 to 2.6 GHz (fixed)
Active Area (mm^2)	**0.2**	0.3	2.5	0.35
Power (mW) @V_{DD}	**1.15 ± 0.05 @ 0.5 V**	2.7 @ 0.6/1.2 V	1.6 @ 0.3 V	3.6 @ 1.2 V
Gain (dB)	50 ± 2	55	83	75
NF (dB)	8.1 ± 0.6	9 [spec.: 15.5]	6.1	9
$IIP3_{out\text{-}of\text{-}band}$ (dBm)	–20.5 ±1.5	–6 [spec.: –32]	–21.5	–12.5
IRR (dB)	20.5 ± 0.5	28 [spec.: 4]	N/A	35
VCO Phase Noise (dBc/Hz)	–117.4 ± 1.7 @ 3.5 MHz	–115 @ 3.5 MHz [spec.: –102]	–112 @ 1 MHz	–116 @ 3.5 MHz
Technology	65 nm CMOS	65 nm CMOS	65 nm CMOS	90 nm CMOS

Figure 9.4.6: Chip summary and benchmark with the state-of-the-art.

978-1-4799-0917-9/14 $31.00 © 2014 IEEE

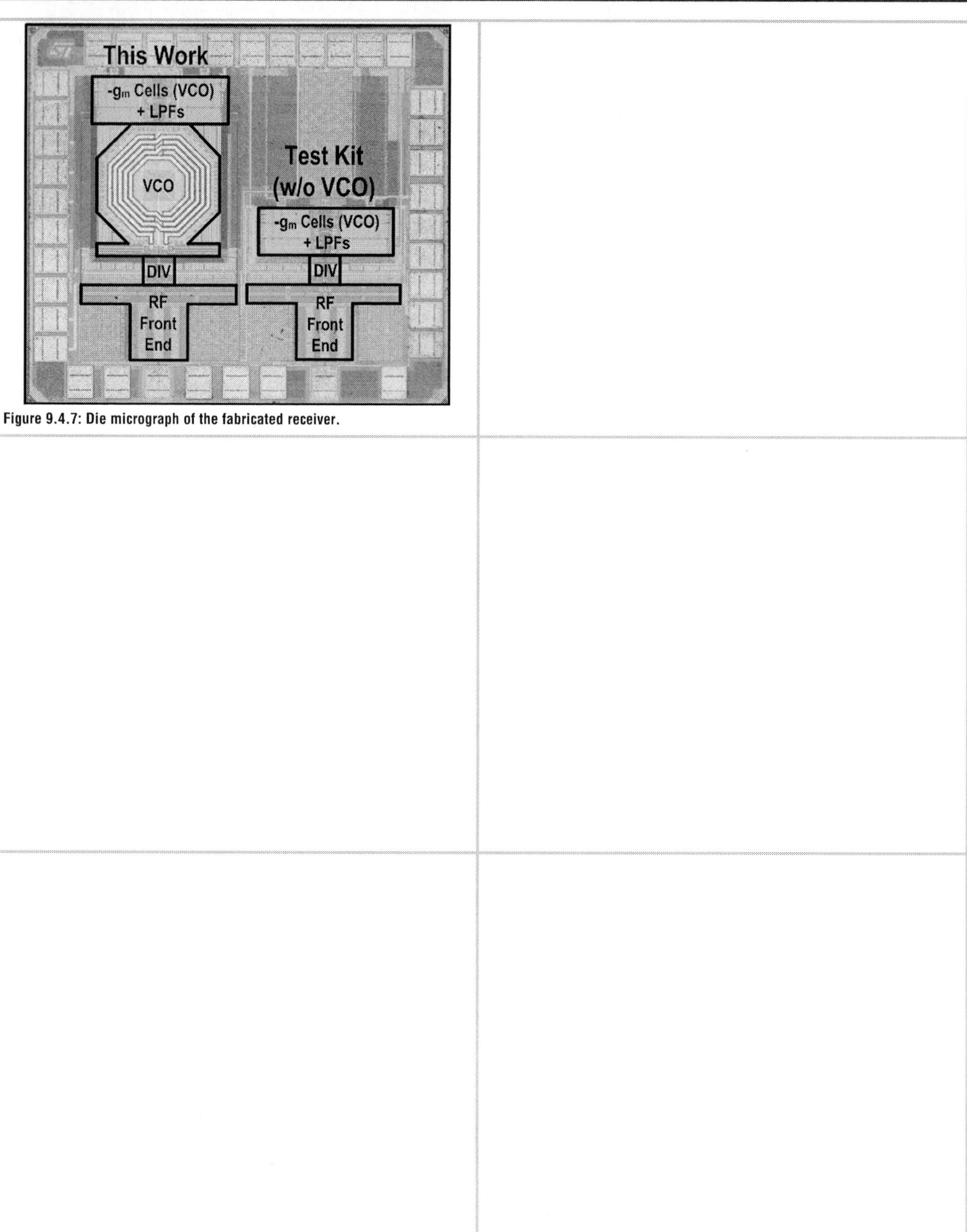

Figure 9.4.7: Die micrograph of the fabricated receiver.

9.5 A 1.2nJ/b 2.4GHz Receiver with a Sliding-IF Phase-to-Digital Converter for Wireless Personal/Body-Area Networks

Yao-Hong Liu, Ao Ba, J.H.C van den Heuvel, Kathleen Philips, Guido Dolmans, Harmke de Groot

Holst Centre/imec, Eindhoven, The Netherlands

This paper presents an ultra-low-power (ULP) 2.4GHz RX for short-range wireless personal and body-area networks. In such applications, the RF transceiver consumes up to 90% of the total battery energy in a remote sensor node. In order to extend the operation lifetime, it is a primary design goal for such transceivers to improve the energy efficiency, expressed as power consumption/data-rate, to below 1nJ/bit. Although energy-detection or super-regenerative ASK RXs [1] are very efficient, they are vulnerable to interference and this leads to a poor quality of the wireless link in a crowded 2.4GHz ISM band. On the other hand, FSK/PSK-type modulations are popular in the target applications because of their power-efficient hardware and higher immunity to interference. They are also widely adopted in many short-range wireless standards like IEEE802.15.4 and Bluetooth Smart. Thanks to the constant-envelope nature of FSK/PSK-type modulations, e.g., HS-OQPSK, they only modulate data on the carrier frequency or phase, so the TX hardware can be simplified, while the efficiency can be enhanced by driving the circuits into a saturated mode, e.g., PLL-based FSK TXs [2,5]. Similarly, in the RX counterpart, instead of processing the signal in the I/Q domain, it can be demodulated in the phase domain by using a phase-ADC [3]. However, such approach still requires a power-hungry high-frequency multi-phase LO generation and "2-dimensional" downconversion and filtering circuits. In the single-channel RX of [4], the VCO is part of the carrier-recovery and phase-demodulation loop. Therefore, the VCO frequency can be easily "pulled away" by an interferer, and the RX has a poor sensitivity because phase noise is deteriorated by the signal chain when input level is low.

In this work, a single-channel sliding-IF RX with a phase-to-digital converter (SIF-PDC) is proposed, as illustrated in Fig. 9.5.1. The sliding-IF architecture is adopted because it can effectively reduce the power consumption of multi-phase LO generation [2,5]. Unlike the frequency tracking/recovery loop in [4], a fractional-N PLL based on a 32MHz reference clock is employed to generate a stable carrier, which guarantees the carrier frequency will not be pulled by the interference. The VCO is locked at 8/9 of the input center frequency (e.g., f_C=2.475GHz) as the LO of the RF mixer (e.g., f_{LO1}=2.2GHz). Then it is further divided by 8 (e.g., f_{LO2}=275MHz)) to provide a 16-phase LO for the IF mixer which also performs phase detection. The detected phase difference ($\Delta\Phi$) is then filtered and digitized by a comparator. The 1b comparator output represents the binary frequency-demodulated data (f_{OUT}), and a 4b digital phase integrator output data (Φ_{OUT}) represents the demodulated phase. The phase selector selects one of 16 phases according to Φ_{OUT}. This RX demodulation loop forces the selected phase to be synchronized with the input modulated phase (Φ_{IN}), which is independent from the carrier generation loop. Hence, the LO phase noise will not be polluted by the signal path. Since the data demodulation is performed by digital phase switching, it equivalently transforms the voltage-domain processing into the digital-phase domain. Therefore, the dynamic range of the LPF and ADC (which now is implemented with a comparator) is not constricted by the supply voltage.

Similar to I/Q RXs, amplitude fluctuation due to fading and carrier frequency offset (CFO) degrade the sensitivity of the proposed RX. This is especially critical in certain standards e.g. Bluetooth Smart that demands the digital baseband to regulate the signal amplitude and detect a large CFO within a short preamble. The proposed RX is less sensitive to the amplitude fluctuation, so the requirements of the automatic-gain-control algorithm are relaxed. Furthermore, since the CFO translates to a unidirectional phase drift, it can be easily detected within preamble by monitoring the output digitized phase Φ_{OUT} of the SIF-PDC. Then it can be immediately compensated during payload by adding the compensation phase to the phase command word.

Figure 9.5.2 shows the phase-domain mathematical model of the SIF-PDC. Similar to a $\Delta\Sigma$ ADC, the SIF-PDC performs a continuous-time 1st-order $\Delta\Sigma$ phase-to-digital conversion. The phase detector implements a phase subtraction and has a signal-dependent phase-difference-to-voltage gain of "$A*K_{PD}$," where "A" is the signal amplitude at the phase detector input. The comparator is modeled as a signal-dependent gain of "K_{CMP}/A." The phase integration and selection equivalently implement a numerical-controlled oscillator (NCO) that has a transfer function of "K_{NCO}/s," and adds phase quantization noise "Φ_{QN}" due to the discrete phases. The low-pass signal transfer function ($\Phi_{OUT}(s)$ / $\Phi_{IN}(s)$) helps to suppress high-frequency unwanted components. In-band quantization noise Φ_{QN} is suppressed by a high-pass noise transfer function ($\Phi_{OUT}(s)$ / $\Phi_{QN}(s)$). Therefore, it is sufficient to digitize continuous-phase modulation with 16 discrete phases.

A single-ended LNA and a push-pull RF mixer [2] are employed for their low power consumption. The IF signal is further downconverted by a single-channel passive mixer. A dynamic comparator sampled at 32MHz provides a 1b demodulated frequency data. The implementation of the 4-b digital-to-phase converter (DPC) is shown in Fig. 9.5.3. The 16 LO phases are generated through a frequency division of 8 that is performed in two cascaded stages (/2 and /4). Compared to a single-stage implementation with 8 DFFs connected as a ring, this approach relaxes the timing requirements and reduces power consumption by half. The first divide-by-2 stage (/2) is realized with dynamic-load DFFs for their high-speed, low-power and low-voltage features. Its 4-phase output triggers parallel divide-by-4 stages (/4) to further produce 16 phases. A phase-sequence reset eliminates the ambiguous state by guaranteeing that the even-phase sequence {0,2... 14} is always triggered before the odd-phase sequence {1,3,... 15} during start up.

This RX was implemented in a 90nm CMOS technology, and occupies a core area of 0.9mm^2. The measured time-domain waveform is shown in Fig. 9.5.4. A 2-point PLL-based TX [2] is also integrated to provide a realistic modulated signal. IEEE802.15.4 2Mb/s HS-OQPSK and Bluetooth Smart 1Mb/s GFSK signals are generated with a fixed-pattern of "1111-0000-1010". The proposed SIF-PDC can track the frequency/phase modulations and directly provide demodulated digital outputs. Figure 9.5.5 shows the measured raw bit-error-rate of a 2Mb/s HS-OQPSK signal with a pseudo-random bit stream. The proposed RX achieves a sensitivity level of -92dBm. The adjacent-channel interference rejection (ACR) is measured with an un-modulated CW tone as interference and a modulated desired signal with a level 3dB higher than the sensitivity. The measured ACR is -3/12/17dB at the offset frequency of 2/4/6 MHz, respectively. The 2nd and 3rd ACR are around 8 to 10dB lower than conventional I/Q RXs [2] but 20 to 30dB better than super-regenerative RXs [1], which offer sufficient selectivity for the target applications.

Figure 9.5.6 summarizes and compares the performance with the state-of-the-art ULP 2.4GHz RXs. The proposed RX further reduces the power consumption by almost 40% compared to the previous work [2], thus leading to an excellent energy efficiency of 1.2nJ/bit, but without dramatically degrading the sensitivity as in [3] or selectivity as in [1]. Meanwhile, it directly provides phase/frequency demodulated bits, which significantly reduces the digital baseband complexity and power. Finally, the proposed SIF-PDC transforms the signal processing from the analog amplitude domain to the digital phase domain, making it favorable for low-voltage operation and future technology scaling.

A die micrograph of the chip is shown in Fig. 9.5.7.

References:

[1] M. Vidojkovic, et al., "A 2.4GHz ULP OOK Single-Chip Transceiver for Healthcare Applications," *ISSCC Dig. Tech Papers*, pp. 458-459, Feb. 2011.

[2] Y.-H. Liu, et al., "A 1.9nJ/bit 2.4GHz Multistandard (Bluetooth Low Energy/Zigbee/IEEE802.15.6) Transceiver for Personal/Body Area Networks," *ISSCC Dig. Tech Papers*, pp. 446-447, Feb. 2013.

[3] J. Masuch, et al., "A 1.1-mW-RX -81.4-dBm Sensitivity CMOS Transceiver for Bluetooth Low Energy," *IEEE Trans. Microwave Theory and Techniques*, vol. 61, pp. 1660-1673, Apr., 2013.

[4] W. Chen, et al., "A 2.4GHz Reference-less Receiver for 1 Mbps QPSK Demodulation," *IEEE Tran. Circuits and Systems-I*, pp. 505-514, Mar. 2012.

[5] A. Wang, et al., "A 1V 5mA Multimode IEEE 802.15.6/Bluetooth Low-Energy WBAN Transceiver for Biotelemetry Applications," *ISSCC Dig. Tech Papers*, pp. 300-301, Feb. 2012.

[6] S. Chakraborty, et al., "An Ultra Low Power, Reconfigurable, Multi-standard Transceiver Using Fully Digital PLL," *Proc. Symp. VLSI Circuits*, pp. 148-149, June 2013.

978-1-4799-0917-9/14 $31.00 © 2014 IEEE

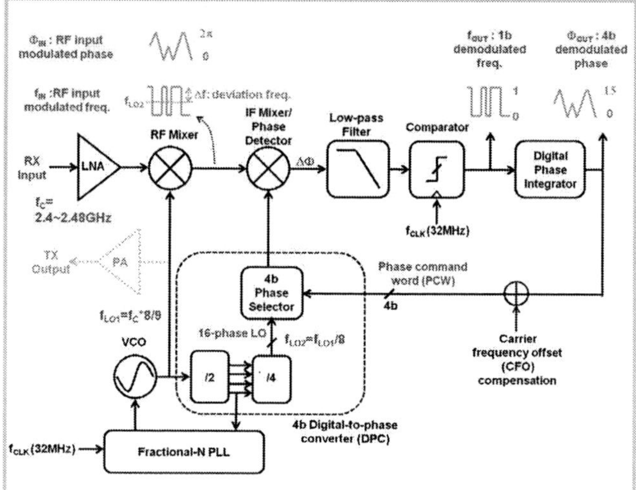

Figure 9.5.1: Simplified block diagram of the proposed RX with SIF-PDC.

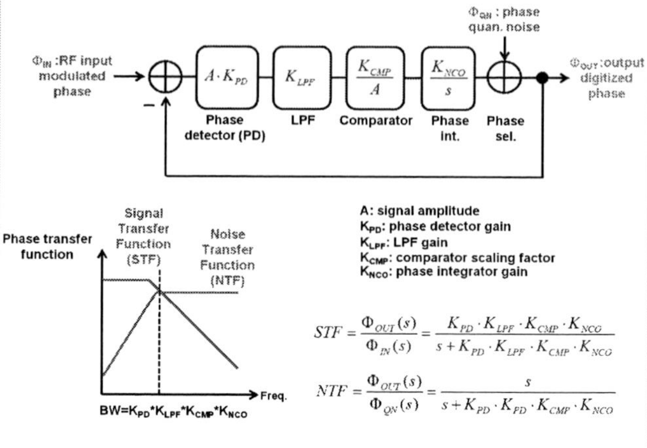

Φ_{QN} : phase quan. noise

Φ_{IN} :RF input modulated phase → Phase detector (PD) $A\cdot K_{PD}$ → LPF K_{LPF} → Comparator $\frac{K_{CMP}}{A}$ → Phase int. $\frac{K_{NCO}}{s}$ → Phase sel. → Φ_{OUT} :output digitized phase

Phase transfer function: Signal Transfer Function (STF), Noise Transfer Function (NTF); Freq.

BW=K_{PD}*K_{LPF}*K_{CMP}*K_{NCO}

A: signal amplitude
K_{PD}: phase detector gain
K_{LPF}: LPF gain
K_{CMP}: comparator scaling factor
K_{NCO}: phase integrator gain

$$STF=\frac{\Phi_{OUT}(s)}{\Phi_{IN}(s)}=\frac{K_{PD}\cdot K_{LPF}\cdot K_{CMP}\cdot K_{NCO}}{s+K_{PD}\cdot K_{LPF}\cdot K_{CMP}\cdot K_{NCO}}$$

$$NTF=\frac{\Phi_{OUT}(s)}{\Phi_{QN}(s)}=\frac{s}{s+K_{PD}\cdot K_{PD}\cdot K_{CMP}\cdot K_{NCO}}$$

Figure 9.5.2: A phase-domain mathematical model of the presented SIF-PDC.

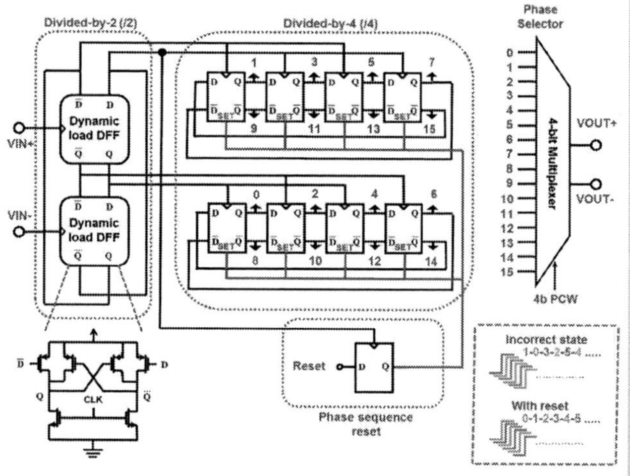

Figure 9.5.3: The 4-b digital-to-phase converter (DPC).

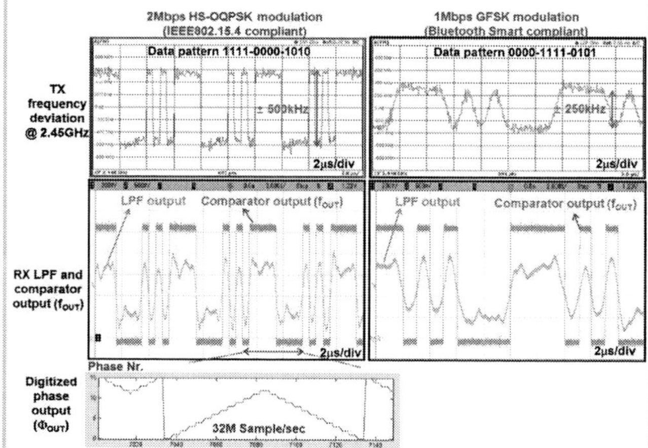

Figure 9.5.4: Measured time-domain demodulated waveforms of 2Mb/s HS-OQPSK and 1Mb/s GFSK signals.

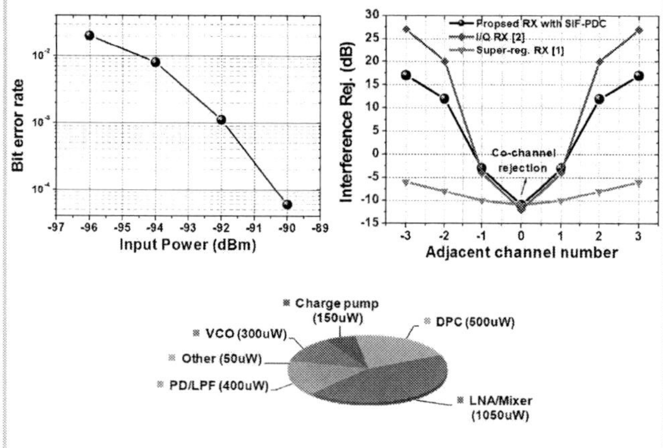

Bit error rate vs. Input Power (dBm); Interference Rej. (dB) vs. Adjacent channel number (Propsed RX with SIF-PDC, I/Q RX [2], Super-reg. RX [1]; Co-channel rejection)

Charge pump (150uW), DPC (600uW), VCO (300uW), Other (50uW), PD/LPF (400uW), LNA/Mixer (1050uW)

Figure 9.5.5: Measured bit error rate with 2Mb/s HS-OQPSK signal, adjacent-channel rejection and power breakdown of the presented RX.

	This work	[1]	[2]	[3]	[5]	[6]
Data rate & modulation	2-Mbps HS-OQPSK	5-Mbps OOK	2-Mbps HS-OQPSK	1-Mbps GFSK	1-Mbps GFSK	1-Mbps GFSK
Technology	90nm	90nm	90nm	130nm	130nm	65nm
Architecture	SIF-PDC	Super reg.	Sliding-IF	Phase ADC	Sliding-IF	Zero-IF
Noise figure	6dB	N.A.	6.1dB	16dB	6dB	N.A>
Max. input pwr.	5dBm	N.A.	5dBm	N.A.	N.A.	N.A.
Supply voltage	1 V	1/1.2V	1.2 V	1V	1/1.5 V	1.3V
Power cons.	2.4mW	0.53mW	3.8mW	1.1mW	6.5mW	8.2mW
Sensitivity*	-92dBm	-75dBm	-96dBm	-81dBm	-94dBm	-94dBm
ACR ($2^{nd}/3^{rd}$)**	12/17dB	-8/-6dB	20/27dB	N.A.	N.A.	N.A.
RX energy eff.	1.2nJ/b	0.1nJ/b	1.9nJ/b	1.1nJ/b	6.5nJ/b	8.2nJ/b
FOM***	181	175	183	171	175	174

* Based on BER of 10^{-3} without error corrections
** ETSI EN 300 440-1 V1.3.1 (2001-09) page 27
*** RX FOM = - Sensitivity -10*log(P_{DC} / Data rate)

Figure 9.5.6: Performance summary and comparison with the state-of-the-art low-power 2.4GHz RXs.

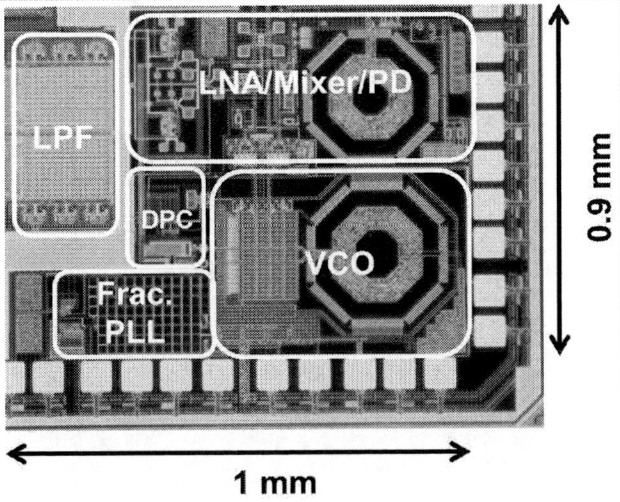

Figure 9.5.7: Chip Micrograph.

9.6 A 1.3mW 0.6V WBAN-Compatible Sub-Sampling PSK Receiver in 65nm CMOS

Jiao Cheng[1], Nan Qi[1], Patrick Yin Chiang[1,2], Arun Natarajan[1]

[1]Oregon State University, Corvallis, OR
[2]Fudan University, Shanghai, China

The release of the IEEE802.15.6 standard has led to increased interest in low-power technologies for wireless body-area-networks (WBAN). The power dissipation, supply voltage, and IC area are some of the most important criteria for successful WBAN implementations. Analog-intensive heterodyne receivers (RX) have been previously demonstrated, consuming 4 to 5mW of power from a 1-to-1.2V supply while occupying large silicon area, due to the presence of area-intensive analog building blocks such as low-pass filters at the IF [1,2]. Digital-intensive RX architectures can potentially result in sub-1V operation with significant reductions in power consumption and area, but require system and circuit-level innovations to achieve desired sensitivity and linearity. This paper presents a mostly-digital 2.4GHz RX architecture that uses a sub-sampling technique with digital IF/baseband signal processing to enable low-power (1.3mW) and low-voltage (0.6V) operation, resulting in ~3x reduction in power consumption. Early analog-to-digital conversion leads to the IC occupying only 0.35mm^2 of active silicon area. While the IC focuses on WBAN demodulation, the presented techniques are applicable to other low-power standards as well.

Figure 9.6.1 shows the block diagram of the presented sub-sampling PSK receiver. The input signal is amplified by a Q-enhanced LNA and fed to a sample-and-hold circuit (SHC) that sub-samples the 2.4GHz RF signal at a sampling frequency of f_S = 309.7MHz. Aggressive Q-enhancement in the LNA mitigates impact of noise-folding in the sub-sampling SHC. System-level studies show that an effective Q above 100 in the LNA output tank ensures that excess noise due to noise-folding is negligible, as the overall RX noise is dominated by LNA noise at the RF frequency. The SHC output is directly fed to dynamic comparators that perform 1-b A-to-D conversion, with the sampling frequency of f_S = 309.7MHz implying that the 1-b ADC oversamples the WBAN baseband signal with an oversampling ratio of 128. Decimation filtering of the oversampled baseband leads to an equivalent ADC resolution of 4.5b.

As shown in Fig. 9.6.1, the 2.4GHz RF signal is downconverted to a 77.42MHz IF when f_S = 309.7MHz. The sub-sampling architecture eliminates the need for a high-frequency LO and the SHC sampling clock is generated by an on-chip 100MHz-to-360MHz subharmonic injection-locked ring oscillator that requires a 33.3MHz-to-120MHz input. The IF frequency is chosen to be exactly 1/4th the sampling frequency, simplifying I/Q signal separation and IF-to-baseband down-conversion. I/Q signal separation is achieved by assigning odd quantized samples to the I-path and even quantized samples to the Q-path. For both I and Q paths, downconversion from IF to baseband is achieved by inverting the sign of every other sample. The resulting PSK symbol is oversampled by ~8X in the baseband (operating at 4.84MHz), providing adequate data-transition timing resolution for symbol timing recovery. For WBAN, system-level simulations indicate that a baseband SNR of 17dB is sufficient to demodulate the 971kb/s π/4-DQPSK modulation with 10^{-3} BER.

While sub-sampling architecture ensures early low-power analog-to-digital conversion, noise folding in the SHC must be addressed by the preceding LNA. As shown in Fig. 9.6.2, a pseudo-differential common-source LNA structure with symmetric center-tapped inductors is utilized to achieve high signal swing, low noise figure, and small area. Q-enhancement with the desired low-voltage operation is provided using positive feedback loops formed with transistors M3, M4, and the corresponding capacitors. PVT variations are calibrated by deliberately increasing the LNA bias current until positive feedback overcomes passive LC resonator losses, resulting in oscillation. The LNA output in oscillator mode is measured using integrated dividers and counters and desired LNA output tank center frequency is achieved using the switched-capacitor bank. During calibration, the inputs of the LNA are grounded through M5 and M6, eliminating interference from the antenna. Subsequently, the LNA DC biasing current is set to 80% of the oscillation value resulting in an effective Q ~ 150. Measured LNA gain from 2.2GHz to 2.6GHz demonstrates that an effective Q of 220 is both achievable and measurable on-die (Fig. 9.6.2). In this implementation, the switched-capacitor bank can tune the measured center frequency from 2395MHz to 2735MHz, while the LNA DC biasing current can be varied from 0.66mA to 3mA (44-b thermometer code).

A two-stage structure is adopted in the dynamic comparator to reduce kick-back to the preceding SHC (Fig. 9.6.3). This differential structure also rejects common-mode interference due to parasitic coupling between the sense amplifier latch and the sampling capacitors. Since the SHC is DC-coupled to the comparator, the 1-b over-sampled ADC approach requires the comparator DC offset to be smaller than the minimum input signal amplitude, in order to achieve higher resolution through decimation. On-chip offset calibration [3] is performed by analyzing the comparator output with the LNA inputs grounded through M5 and M6. Assuming that comparator noise satisfies a zero-mean Normal distribution, the probability of '1's at the output of the comparator depends upon the offset. Therefore, the internal DC offset calibration control is incremented until the accumulator output indicates equal probabilities of '1's and '0's at the comparator output.

The resolution of the on-chip offset calibration scheme and the comparator input-referred noise are measured by applying an external differential input directly to the comparator. The measured distribution of 1's at the comparator output across external inputs is used to calculate rms noise and DC offset, which are compared to results from the internal calibration procedure. As shown in Fig. 9.6.3, the measured DC offset and input-referred rms noise are 8.9mV and 0.88mV, respectively. Comparison with internal calibration settings shows that the 40-b thermometer-coded current sources can compensate for DC offsets with a measured resolution of 1.02mV and a total measured compensation range of 40mV, which is sufficient to account for the measured 8.9mV offset. The worst-case offset residue is ½ of the measured resolution or 0.51mV. Device scaling, larger transistors and lower bandwidth enable significantly smaller offset residue than the ~4mV achieved in [3] with similar calibration techniques.

Figure 9.6.4 shows the measured BER of the sub-sampling RX at 2.4GHz for π/4-DQPSK and π/2-DBPSK modulation based on 32767b of PRBS input. A Tektronix AWG7062B generates the I/Q modulation inputs for an Agilent E8267D signal source, resulting in a 2.4GHz I/Q modulated input signal. In measurement, tuning the LNA tank to 2.4GHz requires nearly all of the capacitors in the switched-capacitor bank to be enabled. The increased passive losses in the tank at 2.4GHz lead to 27% higher LNA bias current for Q-enhancement at 2.4GHz when compared to Q-enhancement at 2.7GHz, when the bank capacitors are switched off. At 2.4GHz, RX sensitivities of -91dBm and -96dBm (for 10^{-3} BER) are achieved at the nominal 309.7MHz sub-sampling frequency for π/4-DQPSK and π/2-DBPSK modulation. Decreasing the sub-sampling frequency to 157.4MHz results in 8% reduction in power with a 2dB degradation in measured sensitivity.

Receiver linearity is measured by determining the maximum input signal power that achieves better than 10^{-3} BER. Measured compression points for π/4-DQPSK and π/2-DBPSK are -35dBm and >10dBm, respectively. Q-enhancement in the LNA results in measured 33dB image rejection at 2.4GHz. Adjacent channel rejection (ACR) is measured by adding a single-tone interferer 1MHz away from the channel of interest. As shown in Fig. 9.6.5, 10dB and 14dB ACR are measured for π/4-DQPSK and π/2-DBPSK, respectively, when f_S = 309.7MHz.

The chip is fabricated in a 65nm CMOS technology and occupies 1mm x 1.2mm, with 0.35mm^2 of active silicon area. Figure 9.6.6 summarizes the measured performance of the WBAN RX for a 2.4GHz carrier and a 2.7GHz carrier, and compares this RX with prior art [1,2,4]. The presented Q-enhanced sub-sampling scheme with 1-b quantization achieves comparable sensitivity for WBAN data rates, while consuming 1.3mW at 2.4GHz (1.05mW at 2.7GHz), achieving a measured energy of 1.34nJ/b (1.08nJ/b) and RX FOM of 227dB (228dB). The die micrograph is shown in Fig. 9.6.7.

Acknowledgements:
This work was sponsored, in part, by the Center for the Design of Analog-Digital Integrated Circuits (CDADIC), the Catalyst Foundation, and the National Science Foundation (IIP-1127853, IIS-1118017).

References:

[1] Y. H. Liu et al., "A 1.9nJ/b 2.4GHz Multistandard (Bluetooth Low Energy/Zigbee/IEEE802.15.6) Transceiver for Personal/Body-Area Networks," *ISSCC Dig. Tech Papers*, pp. 446-447, 2013.
[2] A. Wong et al., "A 1V 5mA Multimode IEEE 802.15.6/Bluetooth Low-Energy WBAN Transceiver for Biotelemetry Applications," *ISSCC Dig. Tech Papers*, pp. 300-301, 2012.
[3] M. Lee et al., "A 90mW 4Gb/s Equalized I/O Circuit with Input Offset Cancellation," *ISSCC Dig. Tech Papers*, pp. 252-253, 2000.
[4] F. Zhang et al., "A 1.6mW 300mV-Supply 2.4GHz Receiver with -94dBm Sensitivity for Energy-Harvesting Applications," *ISSCC Dig. Tech Papers*, pp. 456-457, 2013.

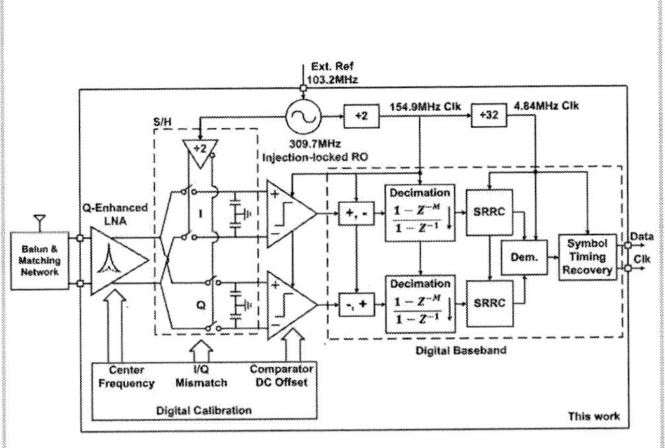

Figure 9.6.1: Block diagram of the 2.4GHz sub-sampling PSK receiver for WBAN.

Figure 9.6.2: Schematic and measured gain of the 2.4GHz LNA with Q-enhancement.

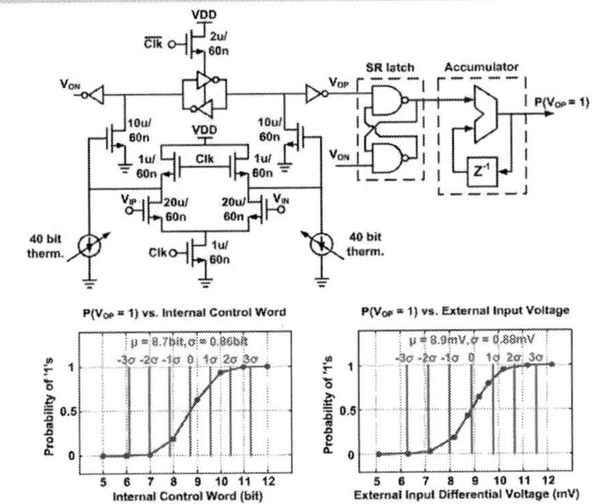

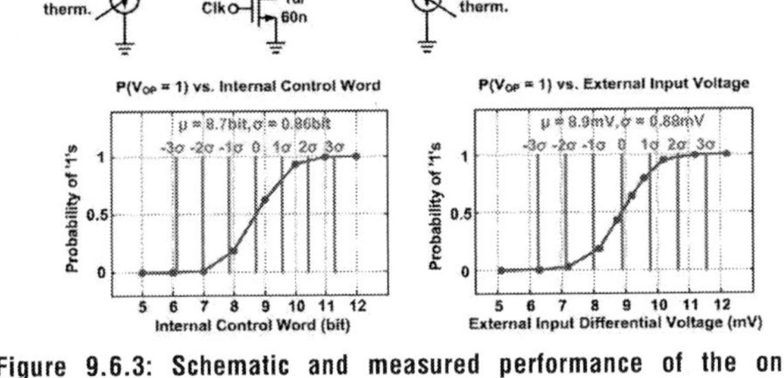

Figure 9.6.3: Schematic and measured performance of the on-chip comparator DC offset calibration.

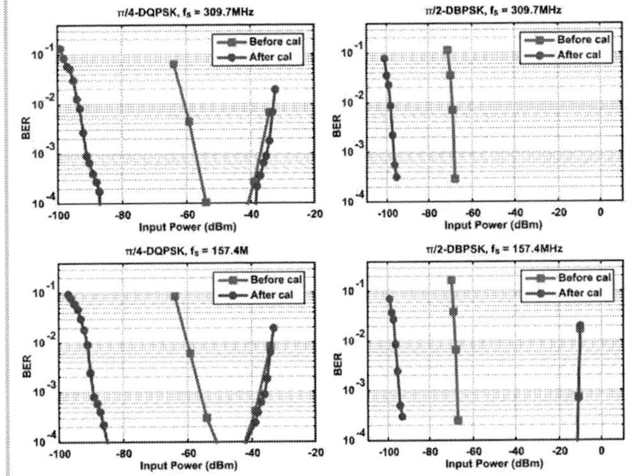

Figure 9.6.4: Measured BER and sensitivity of the 2.4GHz RX for π/4-DQPSK and π/2-DBPSK modulated inputs (before and after DC offset calibration).

Power Dissipation Breakdown		
Circuit block	2.4GHz	2.7GHz
LNA	1066uW	840uW
S/H & comparator	123uW	111uW
Ring oscillator	41uW	38uW
Digital baseband	71uW	65uW
Total	1301uW	1054uW

Figure 9.6.5: Measured adjacent-channel rejection at 2.4GHz (1MHz channel spacing) and RX power breakdown at 2.4GHz and 2.7GHz.

	This work		ISSCC 13 [1]	ISSCC 12 [2]	ISSCC 13 [4]
RX architecture	Sub-sampling		Sliding IF	Sliding IF	Low IF
Modulation	π/2-DBPSK / π/4-DQPSK		π/4-DQPSK	π/2-DBPSK / π/4-DQPSK	BFSK
Data rate	486 kbps / 971 kbps		971 kbps	121 kbps / 971 kbps	200 kbps
Active / die area	**0.35 / 1.2 mm² (65nm)**		**2 / 3.7 mm² (90nm)**	**- / 5.9 mm² (130nm)**	**- / 2.5 mm² (65nm)**
Supply	**0.6 V**		**1.2 V**	**1 V**	**0.3 V**
Frequency band	2.4 GHz	2.7 GHz	2.4 GHz	2.4 GHz	2.46 GHz
Noise figure	6 dB	5 dB	6 dB	6 dB	6.1 dB
Sensitivity	-96 / -91 dBm (0.1% BER)	-97 / -92 dBm (0.1% BER)	-96 dBm (10% PER)	-104 dBm / -96.5 dBm (10% PER)	-91.5 dBm (0.1% BER)
Compr. (10^{-3} BER)	>10 / -35 dBm	>10 / -38 dBm	-	-	-
ACR	14 / 10 dB	14 / 10 dB	-	-	-
Image rejection	33 dB	33 dB	35 dB	-	-
Power dissipation	**1301 uW**	**1054 uW**	**3.8 mW [1]**	**4.8 mW [2]**	**1.6 mW [1,2]**
RX energy effi.	**1.34 nJ/b**	**1.09 nJ/b**	**3.9 nJ/b**	**4.9 nJ/b**	**8 nJ/b**
RX FOM [3]	**227 dB**	**228 dB**	**222 dB**	**221 dB**	**216 dB**

1. Power dissipation excluding digital baseband
2. Power dissipation excluding ADC / off chip drivers
3. RX FOM = - 10*log(kTBF) - 10*log(P_{DC}/Data rate), based on [1]

Figure 9.6.6: Summary of measured performance and comparison with state-of-the-art.

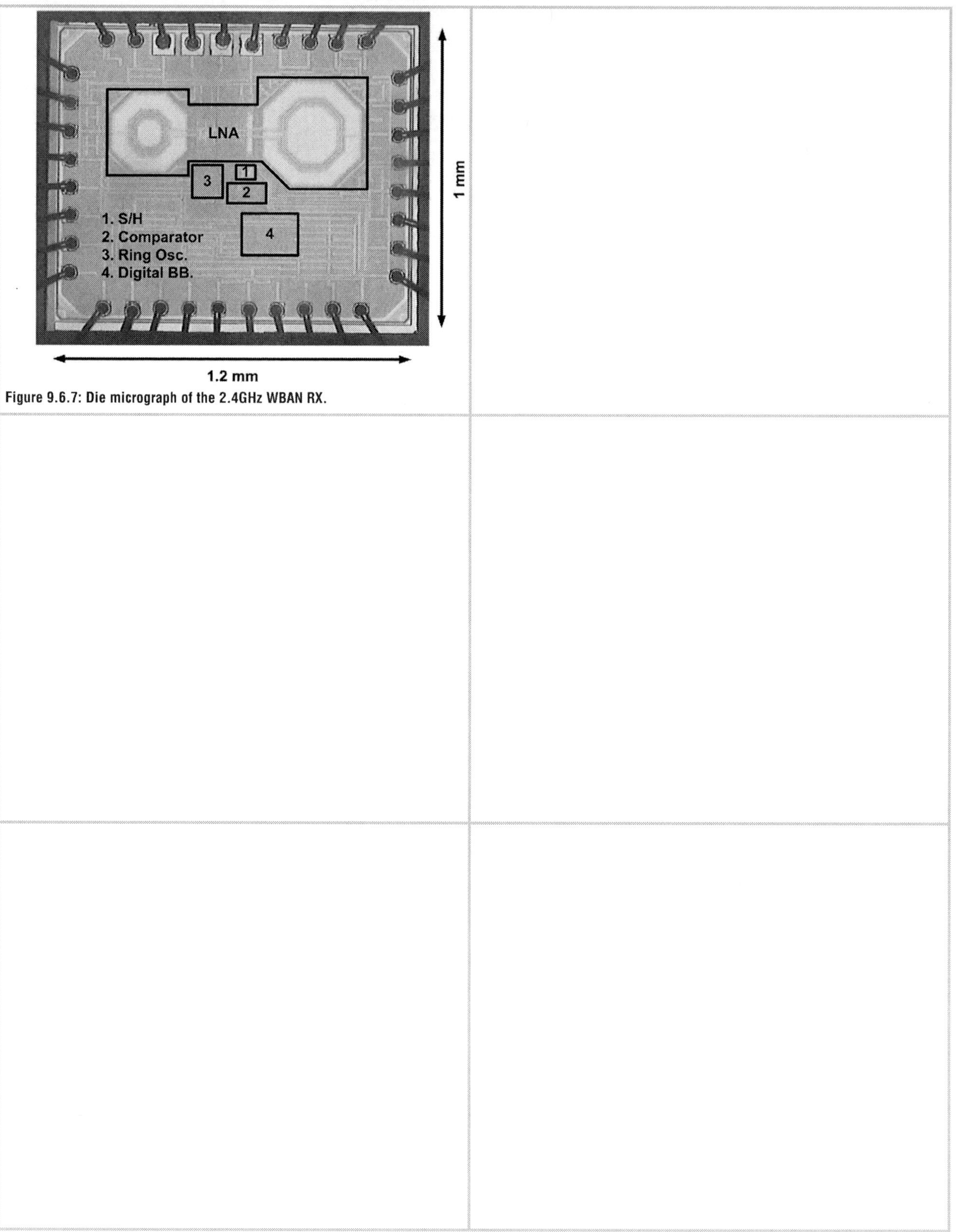

Figure 9.6.7: Die micrograph of the 2.4GHz WBAN RX.

9.7 A 0.33nJ/b IEEE802.15.6/Proprietary-MICS/ISM-Band Transceiver with Scalable Data-Rate from 11kb/s to 4.5Mb/s for Medical Applications

Maja Vidojkovic[1], Xiongchuan Huang[1], Xiaoyan Wang[1], Cui Zhou[1], Ao Ba[1], Maarten Lont[1], Yao-Hong Liu[1], Pieter Harpe[2], Ming Ding[1], Ben Busze[1], Nauman Kiyani[1], Kouichi Kanda[3], Shoichi Masui[3], Kathleen Philips[1], Harmke de Groot[1]

[1]Holst Centre/imec, Eindhoven, The Netherlands,
[2]Eindhoven University of Technology, Eindhoven, The Netherlands,
[3]Fujitsu Laboratories, Kawasaki, Japan

The introduction of the IEEE802.15.6 standard (15.6) for wireless-body-area networks signals the advent of new medical applications, where various wireless nodes in, on or around a human body monitor vital signs. Radio communication often dominates the power consumption in the nodes, thus low-power transceivers are desired. Most state-of-the-art low-power transceivers support only proprietary modes with OOK or FSK modulations, and have poor sensitivity or low data rate [1,2]. In this work, a 15.6-compliant transceiver with enhanced performance is proposed. First, the data-rate is extended to 4.5Mb/s to cover multi-channel EEG applications. Second, while a best-in-class energy efficiency of 0.33nJ/b is achieved in the high-speed mode, a dedicated low-power mode reduces the RX power further in low-data-rate operation. Third, a sensitivity 5 to 10dB better than the 15.6 specification is targeted to accommodate extra path loss due to shadowing effects from human bodies.

The MICS band (402 to 405MHz) is chosen to support medical implantable applications together with the 420-to-450MHz ISM band. The low-power mode has 11kb/s data rate, where the RX NF is relaxed by 12dB to save power by increasing the spreading factor from 1 (15.6) to 16. In the high-speed mode, eight 300kHz channels are combined to have 8x higher data-rate. A programmable low-pass filter (LPF) and multi-mode frequency/phase modulation (FM/PM) are introduced to cover the wide data-rate range. High sensitivity (-110dBm) and low power (1.19mW RX and 1.77mW TX) are enabled by system-level optimization such as architecture selection and LO frequency planning, and circuit-level techniques such as a low-power VCO, divider and LO distribution.

The transceiver IC illustrated in Fig. 9.7.1 consists of polar TX, zero-IF RX, a 24 MHz crystal oscillator and a digital baseband interface. The polar TX with two-point modulation PLL (for FM/PM) and direct-modulation PA (for AM) is better-suited for low power than mixer-based architectures. An LC-VCO is chosen over ring oscillators to achieve the same phase noise with 10-to-100 times lower power. Co-optimization of the VCO and the divider is essential to minimize their total power for two reasons : 1) the full-swing LO buffers required by a static CMOS divider consume a non-negligible amount of power, 2) the power of the LC-VCO and the divider have opposite dependence on frequency, which means frequency planning is important. Three different VCO frequencies (3.2/1.6/0.8GHz) were considered and 1.6GHz was selected together with a divide-by-4 circuit. The sense-amplifier divider directly connected to the VCO accepts a low input swing of 250mV_{DIFF}, which helps to further reduce VCO power. To guarantee this voltage swing over PVT variations, an LO swing-calibration loop is introduced to adjust the VCO bias automatically. The division ratio of the fractional divider is dynamically changed at 24MHz rate by a 5-b code supplied from a 3rd-order MASH 1-1-1 $\Delta\Sigma$ modulator. A 4th-order PLL loop with 135kHz bandwidth suppresses the high-frequency $\Delta\Sigma$ noise and reference spurs while achieving 17µs settling time. The PLL measurement showed integrated phase error of 1° and -123dBc/Hz phase noise at 1MHz offset from 450MHz with 520µW power consumption.

The required FM range and resolution (±50kHz and 7b for 187.5kb/s GMSK, and ±6MHz and 9b for 4.5Mb/s π/8-D8PSK) were derived from system-level simulations. To cover this wide range, output voltage swing of the DAC and K_{VCO} are programmable (Fig. 9.7.2). The DAC bias is set to 0.4µA and C_1 is used for the ±50kHz range, while the bias is set to 4µA and both C_1 and C_2 are used for the ±6MHz range. Placing the VCO far from the PA to reduce noise coupling results in highly-capacitive loads for LO buffers. Single-ended LO signal transmission and differential signal reconstruction using a two-stage edge aligner saves 40% of the LO buffer power compared with differential transmission. A 4-b AM code oversampled at 24MHz is given to the PA by a 2nd-order MASH 1-1 $\Delta\Sigma$ modulator, which keeps the quantization noise floor below the required ACPR level. Dynamic element matching is adopted to further reduce the ACPR. Unlike [3] where the AM code is applied to the gate of the cascode transistors, this PA is controlled by the AND gates placed at the input of each unit cell, which relaxes the gate-drain overstress of the cascode transistor (M_C).

Figure 9.7.3 shows the highly-reconfigurable receiver. In the single-ended LNA, g_m of the cascode transistor M_1 and the load resistor R_1 are variable so that the LNA current and the RX NF can be traded (400µA for 5dB and 100µA for 8.4dB). Inductive degeneration together with variable R and C between the gate and source of the input transistor M_2 keeps the input S_{11} below -10dB from 325MHz to 518MHz. The mixer is composed of a variable-g_m input stage, passive switches driven by 25% duty-cycled LO signals, and a variable-trans-impedance amplifier. To achieve 17dB adjacent-channel rejection, an active-RC 3rd-order Butterworth LPF is used. The capacitors C_1, C_2 and C_3 adjust the LPF bandwidth to 150kHz (MICS) and 160kHz (ISM) for the 15.6 mode and 1.2MHz for the high-speed mode (MICS/ISM). The transistors in the opamp are biased in the weak-inversion region where the best g_m/I_D is obtained, while gate length and width are carefully chosen to lower the 1/f noise. An 8-b differential DAC injects current to the input of the second LPF stage to cancel the DC offset voltage. The optimum code is chosen during the 90-b preamble period. Two 8x-oversampling SAR ADCs generate 9-b I/Q data at 1.5MS/s in the 15.6 mode and at 12MS/s in the high-speed mode.

The transceiver fabricated in 40nm CMOS (Fig. 9.7.7) occupies 1.8mm x 1.7mm on chip. The measured TX EVM is 3.5% and 7% for the 15.6 and high-speed mode (both in π/8 D8PSK), respectively, and ACPR in the 15.6 mode is -27dBc at -17.2dBm output (Fig. 9.7.4). The RX sensitivity at 0.1% BER is -110dBm in the 15.6 mode (GMSK, 187.5kb/s), while it is -100dBm (π/2 DBPSK, 1.5Mb/s) and -83dBm (π/8 D8PSK, 4.5Mb/s) in the high-speed mode. The RX sensitivity is 9dB better than the requirement of the 15.6 standard. The measurements were done without the BCH code of rate 51/63. With the BCH coding, the net data-rate decreases by 20% while the RX sensitivity improves by ~1dB. Figure 9.7.5 shows the measured power consumption breakdown. The TX consumes 1.77mW in GMSK mode with -10dBm output power, where 900µW is from the PA. In π/8 D8PSK mode, the PA power is reduced to 500µW as the output power is lowered to -17dBm, but the PLL power needs to be increased from 641µW to 988µW to mitigate the frequency-pulling effect from the PA to the VCO and to keep the ACPR low. RX power consumption is 1.49mW both in the 15.6 and high-speed modes, while it is 1.19mW in the low-power mode thanks to the 300µW LNA power-saving. The energy efficiency for the high-speed mode is 0.33nJ/b for RX, and 0.4nJ/b for TX. The simulated digital baseband power dissipation (0.5mW for TX and 0.7mW for RX) is added to compare the power and the energy efficiency of the radio with the existing products and published ICs [4-6] (Fig. 9.7.6). This work achieves the highest data-rate of 4.5Mb/s and the best energy efficiency of 0.51nJ/b and 0.49nJ/b for TX and RX, respectively. The high sensitivity, energy efficiency and scalability of this transceiver enable its use in various medical and healthcare applications.

Acknowledgement:
The authors would like to thank Christian Bachmann, Gert-Jan van Schaik and Hans Pflug from IMEC-Holst Centre, and Makoto Hamaminato and Hiroyuki Sato from Fujitsu Laboratories for support during the design and evaluation.

References:

[1] J. Pandey, et al., "A 90 µW MICS/ISM Band Transmitter with 22% Global Efficiency", *IEEE RFIC Symposium*, pp. 285-288, May 2010.
[2] J. L. Bohorquez, et al., "A 350µW CMOS MSK Transmitter and 400µW OOK Super-Regenerative Receiver for Medical Implant Communications", *IEEE J. Solid-State Circuits*, vol. 44, no. 4, April 2009.
[3] Y.H.Liu, et al., "A 1.9nJ/b 2.4GHz Multistandard (Bluetooth Low Energy/Zigbee/IEEE 802.15.6) Transceiver for Personal/Body-Area Networks", *ISSCC Dig. Tech Papers*, pp. 446-447, Feb. 2013.
[4] P.D. Bradley, et, al., "An Ultra Low Power, High Performance Medical Implant Communication System (MICS) Transceiver for Implantable Devices", *IEEE BioCAS*, pp.158-161, Dec. 2006.
[5] V. Peiris, et al., "A 1V 433/868MHz 25kb/s-FSK 2kb/s-OOK RF Transceiver SoC in Standard Digital 0.18µm CMOS", *ISSCC Dig. Tech Papers*, pp. 258-259, Feb. 2005.
[6] F. Carrara et al., "A 400-MHz CMOS Radio Front-End for Ultra Low-Power Medical Implantable Applications", *IEEE ESSCIRC*, pp. 232-235, Sept. 2009.

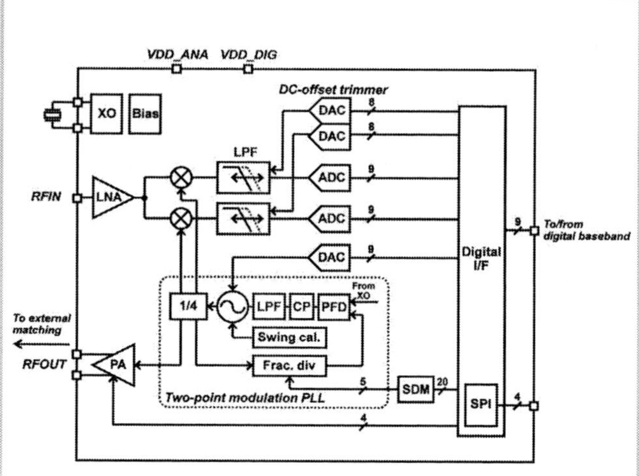

Figure 9.7.1: 400MHz transceiver block diagram.

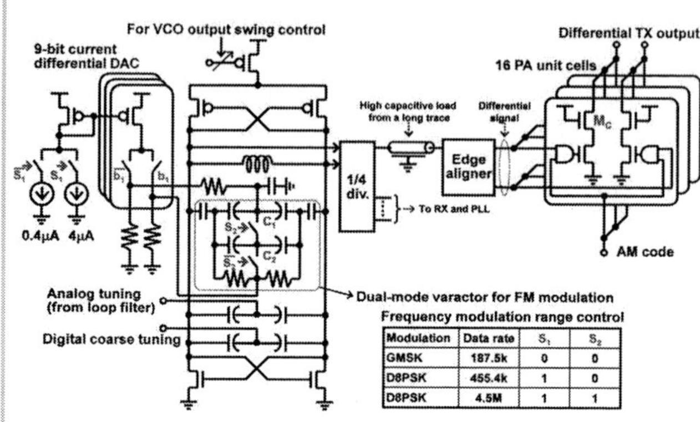

Modulation	Data rate	S_1	S_2
GMSK	187.5k	0	0
D8PSK	455.4k	1	0
D8PSK	4.5M	1	1

Figure 9.7.2: TX schematic from VCO to PA.

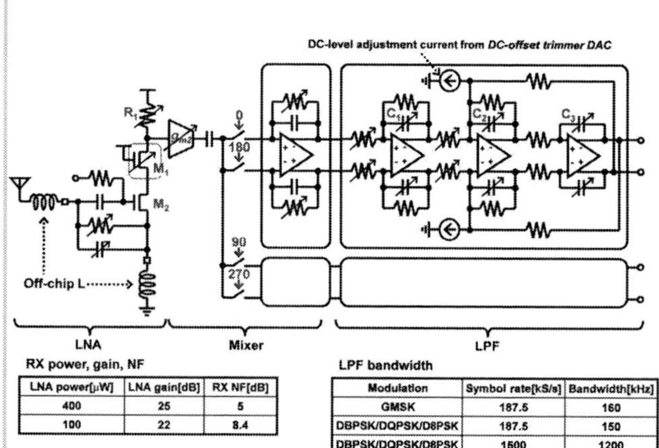

LNA power[μW]	LNA gain[dB]	RX NF[dB]
400	25	5
100	22	8.4

Modulation	Symbol rate[kS/s]	Bandwidth[kHz]
GMSK	187.5	160
DBPSK/DQPSK/D8PSK	187.5	150
DBPSK/DQPSK/D8PSK	1500	1200

Figure 9.7.3: RX schematic from LNA to low-pass filter.

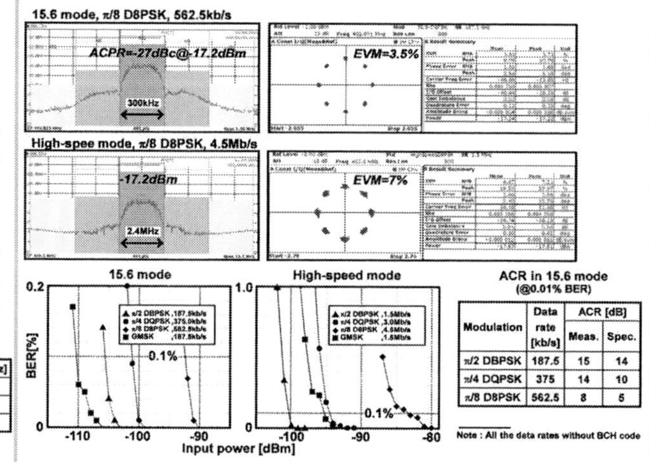

Modulation	Data rate [kb/s]	ACR [dB] Meas.	ACR [dB] Spec.
π/2 DBPSK	187.5	15	14
π/4 DQPSK	375	14	10
π/8 D8PSK	562.5	8	5

Note : All the data rates without BCH code

Figure 9.7.4: Measured TX ACPR and EVM, RX BER and ACR.

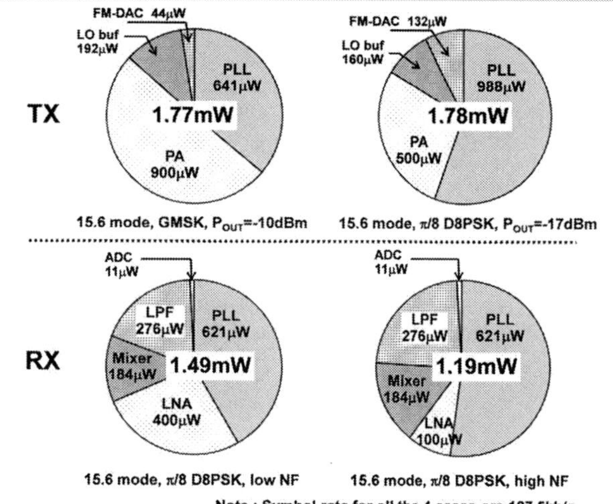

Figure 9.7.5: TX / RX power-consumption breakdown.

			ADF7021-V Analog Devices	nRF9E5 Nordic	CC1101 TI	[4]	[5]	[6] (*3)	This work				
Standard			-	-	-	-	-	-	IEEE 802.15.6	-	-	-	-
Modulation			2/3/4-FSK MSK	GFSK	2/4-FSK GFSK, MSK ASK, OOK	2/4/8-FSK	FSK OOK	GFSK π/4-DQPSK	GMSK	π/8 D8PSK	π/2 DBPSK	π/2 DBPSK	π/8 D8PSK
Frequency band		MHz	431-475	433	387-464	402-405 432-434	433	401-406	402-405 420-450				
Data rate, min/max		kb/s	0.05 / 24	100	0.6 / 600	200 / 800	2 / 100	- / 400	10 / 4500				
TX output power, min/max		dBm	-16 / +13	-10 / 10	-30 / +10	-17 / -4.5	- / 10.5	- / 0	-20 / -5				
Technology			-	-	-	0.18μm	0.18μm	0.13μm	40nm				
Supply voltage, min/max		V	2.3 / 3.6	1.9 / 3.6	1.8 / 3.6	2.1 / 3.5	1.0 / 1.6	1.2 / -	1.0 / -				
TX power consumption	Power	mW	31.74	20.9	23.58	11.55	27.6	5.4	2.27(*2)	2.28(*2)			
	@P_{OUT}	dBm	0	-10	-6	-4.5	10	0	-10	-17			
RX power consumption	Power	mW	42.09	23.75	30.78	11.55	2.1	3.6	2.19(*2)				
	@sensitivity	dBm	-114	-100	-95	-83(*1)	-111	-95	-110	-93	-112	-100	-83
	@data rate	kb/s	9.6	100	250	800	25	400	187.5	562.5	11.7	1500	4500

(*1) Sensitivity in dBm was calculated with 90uV and 1600ohm given in the paper.
(*2) Simulated DBB power of 0.5mW(TX) and 0.7mW(RX) were added to the measured analog-front end
(*3) This IC does not have digital base-band on the same die of the analog front-end.
(*4) 6dB improvement from spreading factor increase from 1 to 6 is taken into account

● = This work
▲ = Prior arts (above)

Energy efficiency [nJ/bit]
1nJ/bit
TX output power [dBm]
RX sensitivity [dBm]

Figure 9.7.6: Performance comparison.

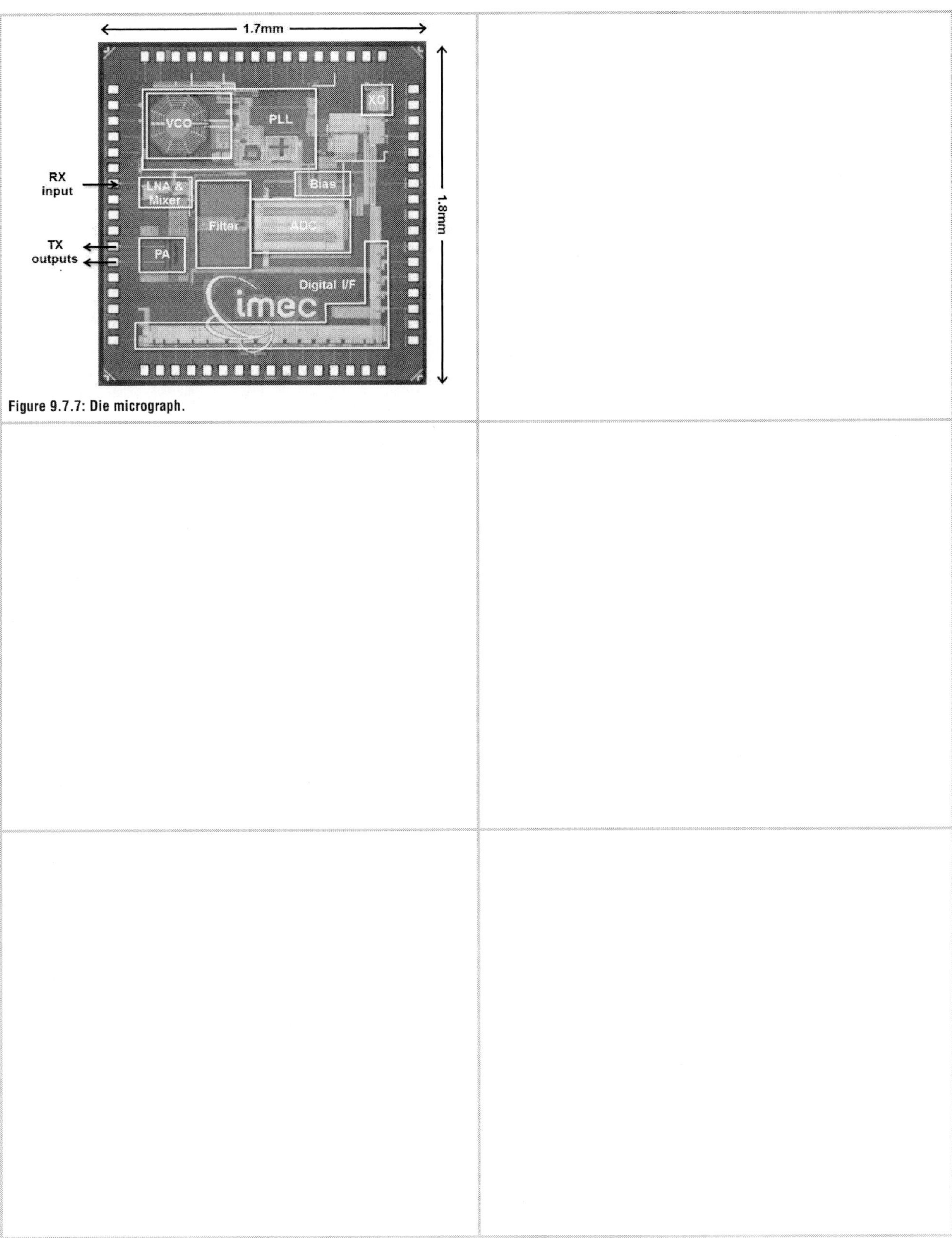

Figure 9.7.7: Die micrograph.

9.8 An 860μW 2.1-to-2.7GHz All-Digital PLL-Based Frequency Modulator with a DTC-Assisted Snapshot TDC for WPAN (Bluetooth Smart and ZigBee) Applications

Vamshi Krishna Chillara[1,2*], Yao-Hong Liu[1], Bindi Wang[1,2], Ao Ba[1], Maja Vidojkovic[1], Kathleen Philips[1], Harmke de Groot[1], Robert Bogdan Staszewski[2]

[1]Holst Centre/imec, Eindhoven, The Netherlands,
[2]Delft University of Technology, Delft, The Netherlands
*now at Analog Devices, Limerick, Ireland

Ultra-low-power (ULP) transceivers enable short-range networks of autonomous sensor nodes for wireless personal-area-network (WPAN) applications. RF PLLs for frequency synthesis and modulation consume a significant share of the total transceiver power, making sub-mW PLLs key to realize ULP WPAN radios. Compared to analog PLLs [1], all-digital PLLs (ADPLLs) are preferred in nanoscale CMOS as they offer benefits of smaller area, programmability, capability of extensive self-calibrations, and easy portability [2]. However, analog PLLs dominate the field of ULP WPAN radios [1], since the time-to-digital-converter (TDC) of an ADPLL has traditionally been power hungry. We present a 2.1-to-2.7GHz 860μW fractional-N ADPLL in 40nm CMOS for WPAN applications, which breaks the 1mW barrier and consumes at least 5× lower power compared to state-of-the-art ADPLLs.

Figure 9.8.1 shows the presented ADPLL architecture with 2-point FM capability. The TX modulation data is added to FCW in the low-frequency path (FM_{LF}) and the DCO control word in the high-frequency path (FM_{HF}), allowing the modulation bandwidth to exceed the PLL bandwidth. To meet stringent power constraints, three low-power techniques are employed. The biggest power saving is obtained by using a digital-to-time-converter (DTC)-assisted snapshot TDC for fractional phase detection. It reduces power by ~200× compared to the conventional approach [2]. Secondly, a power-efficient DCO buffer with a tunable voltage-transfer characteristic (VTC) is employed. It is DC-coupled to the low-swing DCO's output to avoid driving bulky resistor-biased de-coupling capacitors. Finally, a frequency divider (/2) reduces the operation speed of both integer and fractional phase detection to half the DCO rate, CKVD2. This saves power at the expense of doubling the required detection range.

A conventional TDC in a counter-based ADPLL [2] needs to cover one full DCO period (T_V) sensing the DCO clock at its full rate, thus consuming several mW. In this work, TDC snapshotting reduces the sampling rate from F_{CKVD2} to FREF, while the DTC reduces the TDC detection range to less than 1/10 of T_V, leading to a significant power reduction. The accumulated fractional part of the frequency command word, FCW_{frac}, controls the DTC to delay the reference signal FREF such that the delayed reference clock $FREF_{dly}$ is almost aligned with CKVD2, once the loop is locked [3]. $FREF_{dly}$ also triggers the snapshot to catch the first CKVD2 edge so that only one CKVD2 edge, $CKVD2_S$, per reference period is fed to the TDC. A reduced-range TDC operating at the reference frequency (32MHz) then compares the edge of $CKVD2_S$ with $FREF_{dly}$ to provide the fractional phase error, PHE_F. Moreover, since the $FREF_{dly}$ and $CKVD2_S$ are synchronized, retimed reference (CKR) is generated by directly sampling $FREF_{dly}$ with $CKVD2_S$ without concerns of metastability [2]. For correct delay prediction, the scaling factor, $1/K_{DTC}(= T_{CKVD2}/\Delta t_{DTC})$ is tracked over PVT variations by an LMS-based DTC gain-calibration circuitry that corrects the estimated DTC step by observing the phase error. To generate the integer value of variable phase, PHV, an asynchronous counter clocked by CKVD2 is used as a phase incrementer.

Figure 9.8.2 shows the implementation of the DCO, DCO buffer, divider, and phase incrementer/sampler. The DCO is realized by a complementary cross-coupled LC oscillator with digitally tunable tail resistor and capacitors. Using resistors instead of current biasing reduces the flicker noise upconversion. A large inductance (7.7nH) with Q factor of 14 is chosen to minimize the power consumption. The DCO is segmented into three banks: coarse, medium, and fine, to cover a 25% tuning range of 2.1 to 2.7GHz. Switched MOM capacitors, rather than MOS varactors, are used to implement all three banks as they are better modeled, and less sensitive to supply pushing and temperature variations. The resolution of the fine bank, $\Delta C = ½\, C_s^2 / (C_s + C_b) \approx ½\, C_s^2/C_b$, is determined by the ratio of capacitors, C_b and C_s ($C_b >> C_s$).

Using a DC-coupled buffer instead of an AC-coupled one, power consumption and noise feed-through into the DCO are reduced. The VTC of the DC-coupled DCO buffer can be varied by digitally controlling the device ratio of PMOS to NMOS (W_p/W_n) of the inverter to cover process variation, and is calibrated by monitoring the duty-cycle of the output signal. A transmission-gate-based dynamic divider is used for its high-speed, low-power, and low-voltage operation. Since the divider (/2) reduces the ADPLL operation rate, the phase incrementer can be realized by an asynchronous counter to reduce power consumption. The outputs of the counter are then synchronized by adding appropriate delays before being sampled by CKR to generate the integer part of the variable phase, PHV.

Figure 9.8.3 shows the implementation of the DTC/TDC combination. The DTC was adopted in [3] and [4] to replace the power-hungry TDC with a bang-bang phase detector. However, bang-bang ADPLLs either require a complicated frequency detector (e.g., a sampler-based counter in [3]) or warrant additional circuitry for frequency acquisition and loop bandwidth regulation [4], thereby increasing power consumption. In this paper, the DTC is used as a coarse 1st stage, which assists TDC—a fine 2nd stage—to reduce the detection range. The DTC with 64 stages covers one CKVD2 period with sufficient margin. A 16-stage TDC covering 1/5 of the CKVD2 period is implemented to avoid long settling time and to help with FM. The digitally controlled delay line in the DTC is similar to the one in [5] except that a cascade of two inverters, instead of one, is used as the delay element to eliminate even-odd mismatches. Moreover, the racing issues are avoided by turning off the preceding inverters when unused. The delay elements are carefully sized for low power consumption while ensuring the mismatch performance meets the fractional-spur requirements in the target applications, i.e. <-30dBc. As shown in Fig. 9.8.3, $FREF_{dly}$ triggers the snapshot circuit to catch the first CKVD2 edge [6]. Compared to the time-windowed architecture [7], this structure is power-efficient as it does not require the full range of the TDC. CKR is generated two CKVD2 periods after the rising edge of $FREF_{dly}$, to provide enough processing time for the TDC. The TDC is realized by pseudo-differential inverter-based delay lines and sense-amplifier-based flip-flops with identical rising and falling edge metastability windows [2]. The DTC/TDC combination consumes only 43μW.

Figure 9.8.7 shows the micrograph of the presented ADPLL, which is implemented in TSMC LP 40nm CMOS. It occupies a core active area of 0.2mm². Figure 9.8.4 shows the measured settling behavior, fractional-mode phase noise, and spur level. The PLL settles in 20μs, and has in-band and 1MHz-offset phase noise values of -90 and -109dBc/Hz, respectively. The reference spur is -70dBc, and the worst-case fractional spur for Bluetooth Smart channels is -38dBc. The ADPLL has an rms jitter of 1.71ps (integrated from 1k to 100MHz) and consumes 860μW at 1V supply, leading to a state-of-the-art FoM of -236dB. Figure 9.8.5 shows *ZigBee* 2Mc/s HS-OQPSK and Bluetooth Smart 1Mb/s GFSK modulation provided by the ADPLL with 2.3% EVM and 5.2% FSK error, while fulfilling all spectrum mask requirements. Figure 9.8.6 shows the comparison of this ADPLL with state-of-the-art low power PLLs. The presented low-power techniques enable ADPLLs to break the 1mW power consumption barrier and thus to be employed in the emerging ULP WPAN applications.

References:

[1] Y.-H. Liu, et al., "A 2.7nJ/b Multi-Standard 2.3/2.4GHz Polar Transmitter for Wireless Sensor Networks," *ISSCC Dig. Tech. Papers*, pp. 448-450, Feb. 2012.

[2] B. Staszewski, et al., "All-Digital Phase-Domain TX Frequency Synthesizer for Bluetooth Radios in 0.13μm CMOS," *ISSCC Dig. Tech. Papers*, pp. 272–273, Feb. 2004.

[3] N. Pavlovic, et al., "A 5.3GHz Digital-to-Time-Converter-Based Fractional-N All-Digital PLL," *ISSCC Dig. Tech. Papers*, pp. 54-56, Feb. 2011.

[4] D. Tasca, et al., "A 2.9-to-4.0GHz Fractional-N Digital PLL with Bang-Bang Phase Detector and 560fs rms Integrated Jitter at 4.5mW Power," *ISSCC Dig. Tech. Papers*, pp. 88-90, Feb. 2011.

[5] M. Park, et al., "An Amplitude Resolution Improvement of an RF-DAC Employing Pulse-Width Modulation," *IEEE Trans. Circuits and Systems–I*, pp. 2590–2603, Nov. 2011.

[6] J. Zhuang, et al., "A Low-Power All-Digital PLL Architecture Based on Phase Prediction," *ICECS'12*, pp. 797–800, Seville, Spain, Dec. 2012.

[7] T. Tokairin, et al., "A 2.1-to-2.8-GHz Low-Phase-Noise All-Digital Frequency Synthesizer with a Time-Windowed Time-to-Digital Converter," *IEEE J. Solid-State Circuits*, pp. 2582-2590, Dec. 2010.

978-1-4799-0917-9/14 $31.00 © 2014 IEEE

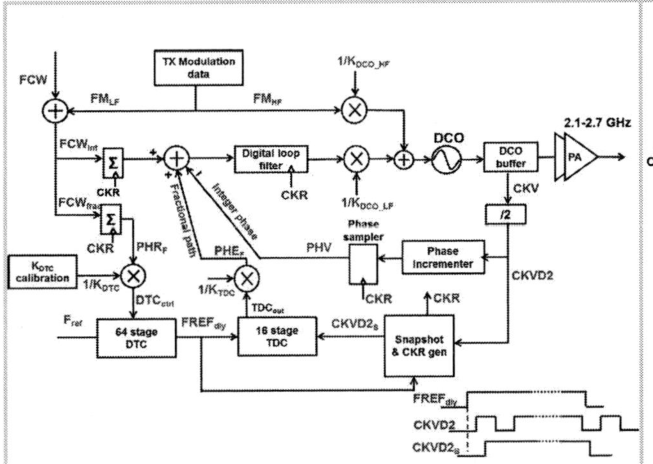

Figure 9.8.1: Block diagram of the presented ADPLL with 2-point FM.

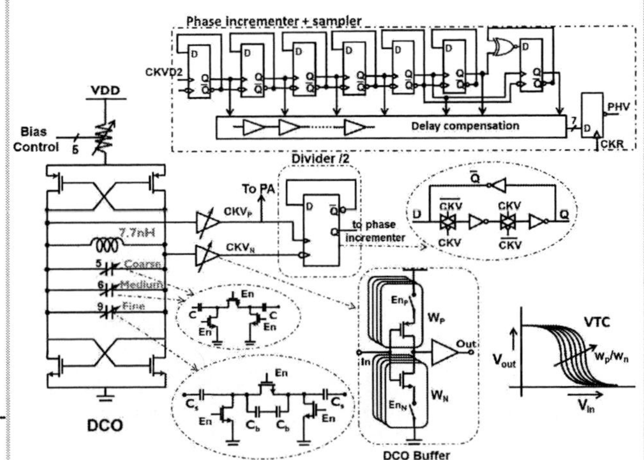

Figure 9.8.2: DCO, buffer, divider, and phase incrementer/sampler.

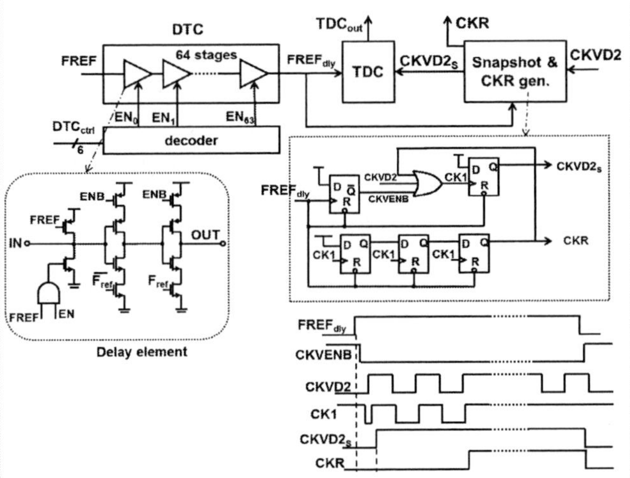

Figure 9.8.3: DTC-assisted snapshot TDC.

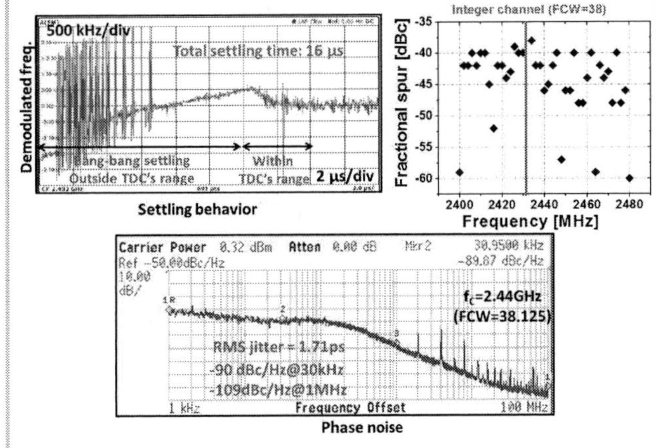

Figure 9.8.4: Measured settling behavior, phase noise, and fractional spur level.

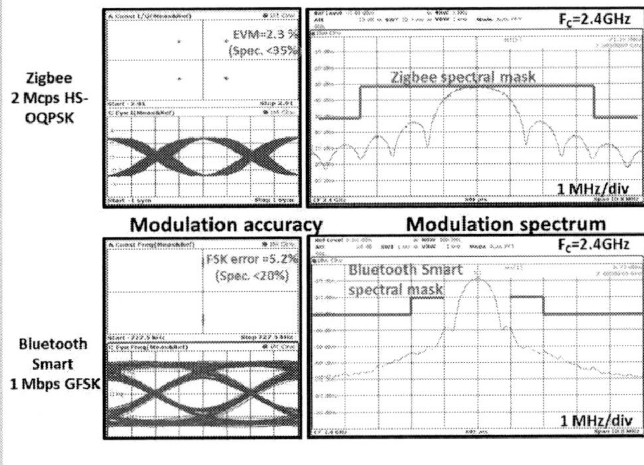

Figure 9.8.5: ZigBee and Bluetooth-Smart modulation accuracy and spectrum.

	This work	[1] ISSCC'12	[2] ISSCC'04	[3] ISSCC'11	[4] ISSCC'11	J-W. Lai 19.8 ISSCC'13
Architecture	**ADPLL DTC+TDC**	**CP-PLL Analog**	**ADPLL TDC-based**	**ADPLL Bang-Bang**	**ADPLL Bang-Bang**	**ADPLL TDC-based**
Applications	Bluetooth Smart (V4)/ Zigbee	Bluetooth Smart (V4)/ Zigbee	Bluetooth (V1)	N.A.	4G cellular	Bluetooth EDR(V2)
Technology (nm)	40	90	130	65	40	40
Reference (MHz)	32	24	13	48	40	26
Output (GHz)	2.1–2.7	1.7–2.48	2.4	4.9–6.9	2.9–4.0	2.4
RMS Jitter (ps)	**1.71**	2.66	1.02	0.61	0.56	0.98
Power (mW)	**0.86**	1.1	30	24	4.5	4.55
Supply	**1 V**	1.2 V	1.5 V	1.2 V	1.2 V	1.3 V
Reference spur	-70dBc	-62dBc	-80dBc	-67dBc	-72dBc	-80dBc
Core Area (mm^2)	0.2	0.75	0.8	0.91	0.22	0.075
PN @ 1 MHz offset	-109	-111	-86@10k	-116	-110	-113
FM capability	Two-point	Two-point	Two-point	No	Two-point	Two-point
FoM[1](dB)	-236	-231	-225	-230	-238.3	-233.6

[1]$FoM=10*\log[(\sigma^2_{jitter})*(P_{DC}/1mW)]$

Figure 9.8.6: Comparison with low-power PLLs.

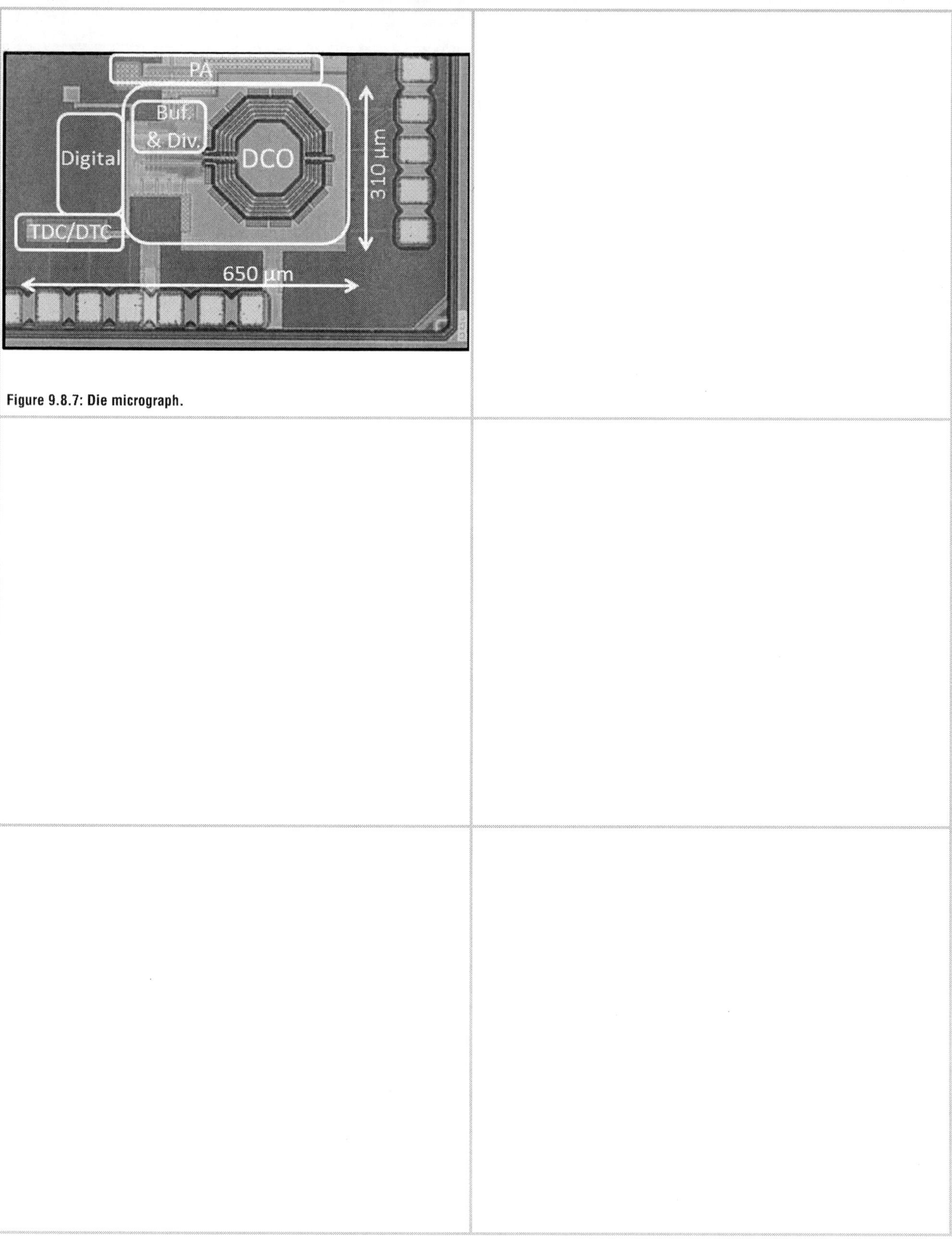

Figure 9.8.7: Die micrograph.

Session 10 Overview: *Mobile Systems-on-Chip (SoCs)*

ENERGY-EFFICIENT DIGITAL SUBCOMMITTEE

Session Chair: *Vasantha Erraguntla*
Intel, Bangalore, India

Session Co-Chair: *Takashi Hashimoto*
Panasonic, Fukuoka, Japan

Mobile SoCs are becoming extremely complex to meet the relentless demand for more compute power and a richer user experience in mobile applications. Qualcomm's Hexagon DSP in 28nm CMOS consumes 58mW/MHz at 0.6V, while Renesas and MediaTek present multicore heterogeneous processors with low power to achieve long battery life and boost performance by exploiting thermal headroom. A multicore processor from KAIST delivers a rich user experience with a 30fps 720p augmented reality chip for wearable electronics. To minimize standby-power, NEC introduces a 16b MTJ-based non-volatile microcontroller with low wakeup time among NV SoCs. Additionally, power-efficient digital baseband SoCs from imec, the Technical University of Dresden and Ericsson are presented, which support multi-standard communication at low power using fine-grain power management techniques.

10.1 A 28nm DSP Powered by an On-Chip LDO for High-Performance and Energy-Efficient Mobile Applications — **8:30 AM**

M. Saint-Laurent, Qualcomm, Austin, TX

In Paper 10.1, Qualcomm presents their 28nm Hexagon™ DSP optimized for mobile heterogeneous computing with an on-chip LDO. The DSP operates from 255MHz at 0.6V, up to 1.20GHz (5640 DMIPS) at 1.05V. With the LDO, the DSP consumes 58mW/MHz at 0.60V. The leakage is 4.08mW at 0.90V.

10.2 A 28nm HPM Heterogeneous Multi-Core Mobile Application Processor with 2GHz Cores and Low-Power 1GHz Cores — **9:00 AM**

M. Igarashi, Renesas Electronics, Tokyo, Japan

In Paper 10.2, Renesas describes a 28nm heterogeneous quad/octa-core mobile application processor. The high-performance cores and the low-power cores operate at 2GHz and 1GHz, respectively, to achieve a maximum performance of 35600 DMIPS. A 24% reduction in SRAM leakage is achieved using a long-length transistor in the word driver and the timing generator. 20% and 29% reductions in dynamic and leakage power, respectively, are accomplished by employing adaptive-voltage-scaling techniques.

10.3 Heterogeneous Multi-Processing Quad-Core CPU and Dual-GPU Design for Optimal Performance, Power, and Thermal Tradeoffs in a 28nm Mobile Application Processor — **9:30 AM**

S. Ouyang, MediaTek, San Jose, CA

In Paper 10.3, MediaTek presents a 28nm heterogeneous multi-processor in a mobile SoC, consisting of a dual-core ARM Cortex™-A15 and a dual-core Cortex-A7. Various power, thermal and performance optimizations techniques deliver 23% higher clock speed, or up to 41% power savings. A new adaptive thermal management algorithm is also introduced. The chip operates between 0.85 to 1.25V, with the A15 cores reaching 1.8GHz and the A7 cores reaching 1.4GHz.

978-1-4799-0917-9/14 $31.00 © 2014 IEEE

10.4 A 1.22TOPS and 1.52mW/MHz Augmented Reality Multi-Core Processor with Neural Network NoC for HMD Applications **10:15 AM**

G. Kim, KAIST, Daejeon, Korea

In Paper 10.4, KAIST describes a real-time markerless 1.22TOPS augmented reality multicore processor with a neural network NoC for wearable applications. This 4×8mm^2 65nm processor consumes 1.52mW/MHz for a 30fps 720p augmented-reality operation. The operating range with DVFS is 65MHz at 0.7V to 250MHz at 1.2V.

10.5 A 90nm 20MHz Fully Nonvolatile Microcontroller for Standby-Power-Critical Applications **10:45 AM**

N. Sakimura, NEC, Tsukuba, Japan and Tohoku University, Sendai, Japan

In Paper 10.5, NEC and Tohoku University presents a 4.79×4.79mm^2 fully non-volatile 16b microcontroller in 90nm MTJ technology that achieves zero standby power and full performance. The three-terminal magnetic spin SRAM MCU achieves 145mW/MHz with a 1V power supply, and 4.5mW when operating intermittently exhibiting 0.1% duty cycle with 120ns wakeup time.

10.6 A 0.74V 200μW Multi-Standard Transceiver Digital Baseband in 40nm LP-CMOS for 2.4GHz Bluetooth Smart / ZigBee / IEEE 802.15.6 Personal Area Networks **11:00 AM**

C. Bachmann, Holst Centre/imec, Eindhoven, The Netherlands

In Paper 10.6, Holst Centre/imec describes a multi-standard digital baseband transceiver supporting Bluetooth Smart / ZigBee / IEEE 802.15.6 for 2.5GHz personal networks, with a single RF front-end and consuming 200mW (RX) / 80mW (TX), with adaptive voltage scaling down to 0.74V. The complete digital baseband architecture, including the physical and lower data-link layer processing occupies a silicon footprint of 0.2mm^2 in 40nm.

10.7 A 105GOPS 36mm^2 Heterogeneous SDR MPSoC with Energy-Aware Dynamic Scheduling and Iterative Detection-Decoding for 4G in 65nm CMOS **11:15 AM**

B. Noethen, Technische Universität Dresden, Dresden, Germany

In Paper 10.7, the Technical University of Dresden presents a 105GOPS 36mm^2 heterogeneous software-defined radio (SDR) multi-processor SoC for 4G in 65nm CMOS. This SoC achieves a maximum frequency of 446MHz at 1.2V and 69.2mW, using fine-grained power management with ultra-fast DVFS combined with AVS.

10.8 A Multi-Standard 2G/3G/4G Cellular Modem Supporting Carrier Aggregation in 28nm CMOS **11:45 AM**

M. Breschel, Ericsson, Lund, Sweden

In Paper 10.8, Ericsson presents a 28nm multi-standard 2G/3G/4G cellular modem based on a unified baseband architecture supporting carrier aggregation. Operating at 1.03V, the modem features an LTE Cat4 downlink of 150Mb/s (10MHz+10MHz aggregated) together with an uplink of 25Mb/s.

10.1 A 28nm DSP Powered by an On-Chip LDO for High-Performance and Energy-Efficient Mobile Applications

Martin Saint-Laurent[1], Paul Bassett[1], Ken Lin[2], Yuhe Wang[2], Son Le[2], Xufeng Chen[2], Maen Alradaideh[1], Tom Wernimont[1], Kartik Ayyar[3], Dan Bui[1], Dwight Galbi[1], Allan Lester[1], Willie Anderson[1]

[1]Qualcomm, Austin, TX, [2]Qualcomm, San Diego, CA,
[3]Qualcomm, Bangalore, India

A very-long instruction word (VLIW) Hexagon™ DSP is fabricated using a 28nm high-κ metal-gate process technology optimized for mobile applications [1]. The DSP is designed for a heterogeneous computing environment. It targets high performance and low power across a wide variety of multimedia and modem applications, under aggressive area targets. Its architecture pursues high IPC as opposed to high frequency [2]. It includes a 32kB L1 data cache (D$), a 16kB L1 instruction cache (I$), and a 256kB L2 cache.

The DSP uses several power switches for leakage control and an on-chip low-dropout (LDO) voltage regulator to support dynamic voltage scaling. As shown in Fig. 10.1.1, a block head switch (BHS) is connected in parallel with the LDO to produce the DSP voltage (VDDQ6) from a chip power rail (VDDCX). VDDQ6 is routed over the DSP using a metal layer in the package and then connected to an on-chip power grid. This scheme avoids having to share the on-chip wiring resources between VDDQ6 and VDDCX, which helps improve power-supply noise and timing [3]. VDDCX is driven by a high-efficiency external switching regulator. Since VDDCX also supports voltage scaling, using the LDO to supply power to the DSP is not always optimal to maximize energy efficiency.

The BHS is used whenever VDDCX is at (or slightly above) the minimum voltage required by the DSP to support a given clock frequency. Using the BHS is advantageous because it avoids the energy loss associated with the IR drop across the LDO's pass transistor. In high-performance mode, the low resistance of the BHS and of the package power grid allows the DSP to run with minimal IR drop. In low-power mode, the LDO can reduce the voltage of the DSP while the rest of the system stays at a higher voltage. During power down, the BHS can cut the DSP leakage to practically zero.

Fig. 10.1.2 shows a block diagram of the LDO, where an analog and a digital loop operate in parallel. The controller receives information from the DSP about future pipeline events likely to cause large voltage droops. Switching between the LDO and the BHS is allowed while the DSP is running. When transitioning from the LDO to the BHS, the BHS is progressively turned on. The lowering of the impedance of the BHS gradually pinches the headroom of the LDO to nearly zero. Then, the LDO no longer regulates its output voltage and can be turned off. When transitioning from the BHS to the LDO, the LDO is forced to its minimum impedance state by the digital controller. Then, the BHS is turned off. Finally, the controller gradually increases the impedance of the LDO until the output voltage drops to its target value. The LDO does not require an external capacitor to be stable. The BHS and LDO occupy 0.015mm² and 0.027mm², respectively.

The clock distribution network is designed for low dynamic power consumption. The clock is routed to the middle of the core through a voltage level shifter and a duty-cycle correction circuit. It is distributed using four horizontal clock bays. When the DSP is idle, the entire clock distribution network can be gated off at its root. However, when the core is active, the global clock bays are free running. Regional clock gating cells (CGCs) are aligned under the clock bays to minimize the free-running capacitance associated with the global clock (gclk). The regional CGCs use the low-power topology shown in Fig. 10.1.3. This topology adds a second enable to the circuit presented in [4] to enhance testability. The CGCs are designed to minimize variations in electrical performance due to layout context effects. They drive 526 regional clocks (rclk). These clocks are not shielded, but are routed with extra spacing to reduce power. The last level of the clock distribution network has 1565 local clocks. The simulated global clock distribution power is 3.96μW/MHz at 0.90V. When running a typical workload, 79.8% of the regional clocks and 94.3% of the local clocks are gated off.

The pulsed latches, also shown in Fig. 10.1.3, are used to implement most sequential elements. Statistical optimization techniques ensure robust operation across a wide range of process, voltage, and temperature conditions. Compared to conventional master-slave flip-flops, these pulsed latches consume significantly less clock and data power [5]. They also offer more data transparency, which improves performance. To reduce the variability caused by layout context effects, the pulse generator and up to 32 latches are bundled together in a single cell. The scan chain is internally connected from one bit to the next (so[n] to si[n+1]) using high-V_t gates. This ensures that the delay between successive bits exceeds the pulse width, in particular at low voltages.

The core clock of the DSP is not synchronized to any other clock and, as discussed earlier, the core voltage can be different than the chip voltage. Asynchronous FIFOs with 4 or 8 entries bridge these clock and voltage boundaries and move data across the bus interfaces of the DSP. Fig. 10.1.4 shows the structure of the FIFOs. The memory array is implemented using a custom bitcell with a tristate output. Level shifters are placed between the input flip-flops and the array. A level shifter is also placed on each write wordline. This makes the write timing relatively insensitive to the difference between the core and chip voltages. An entry can be read by enabling the appropriate row of tristate drivers. The read and write pointers are synchronized using flip-flops optimized for low metastability. They are managed externally to avoid underflows and overflows.

In order to support operation at low voltages, the bitcells of the low-swing 6T arrays (L1 D$ and L2$) are connected to a dedicated memory voltage (VDDMX). As shown in Fig. 10.1.5, dynamic level shifters are embedded with the wordline drivers to reduce the area and delay penalty. Independent PMOS power switches are added to the bitcells and the peripheral logic to support different leakage saving modes. A source biasing circuit reduces the retention leakage while the memory is idle. Finally, a circuit tracking the level shifters automatically shifts the time at which the sense amps fire to maintain a sufficient sense margin. The 8T arrays (L1 I$) connect the bitcells to VDDQ6 and boost the write wordlines to VDDMX to ensure functionality at low voltages.

When the LDO is powering the DSP, the measurements displayed in Fig. 10.1.6 show that the DSP is operational from 255MHz at 0.60V up to 1.20GHz (5640 DMIPS) at 1.05V. At 1.05V, the BHS allows the DSP to reach 1.24GHz. Fig. 10.1.6 also shows the power measured on VDDCX for the LDO and the DSP, when VDDCX is fixed at 1.05V and the DSP is executing a typical workload. At 230MHz, the voltage scaling made possible by the LDO reduces the total power to 23.4mW. This is 2.25× smaller than what is measured when the DSP is operating with the BHS from the fixed supply. The best energy efficiency is achieved when VDDCX is reduced to 0.70V and the LDO regulates VDDQ6 down to 0.60V. There, the measured power for VDDCX is 13.4mW at 230MHz (58μW/MHz), including the internal power consumption of the LDO. At 25°C, the average of the VDDQ6 leakage measured on 5 typical parts is 4.08mW at 0.90V. A micrograph of the DSP and a comparison to other low-power processors are shown in Fig. 10.1.7. For the Hexagon DSP, the minimum energy per cycle is at least 2× lower than for the other processors, including the ones optimized for operation in or near the sub-threshold region.

References:

[1] S.Y. Wu, et al., "A Highly Manufacturable 28nm CMOS Low Power Platform Technology with Fully Functional 64Mb SRAM Using Dual/Tripe Gate Oxide Process," *IEEE Symp. VLSI Circuits*, pp. 210-211, 2009.

[2] L. Codrescu, et al., "Qualcomm Hexagon DSP: An Architecture Optimized for Mobile Multimedia and Communications," *Hot Chips*, 2013.

[3] M. Saint-Laurent and M. Swaminathan, "Impact of Power-Supply Noise on Timing in High-Frequency Microprocessors," *IEEE Trans. on Advanced Packaging*, vol. 27, no.1, pp. 135-144, 2004.

[4] M. Saint-Laurent and A. Datta, "A low-power clock gating cell optimized for low-voltage operation in a 45-nm technology," *ACM/IEEE International Symp. on Low-Power Electronics and Design*, pp. 159-163, 2010.

[5] P. Bassett and M. Saint-Laurent, "Energy Efficient Design Techniques for a Digital Signal Processor," *IEEE International Conf. on IC Design and Tech.*, pp. 41-44, 2012.

978-1-4799-0917-9/14 $31.00 © 2014 IEEE

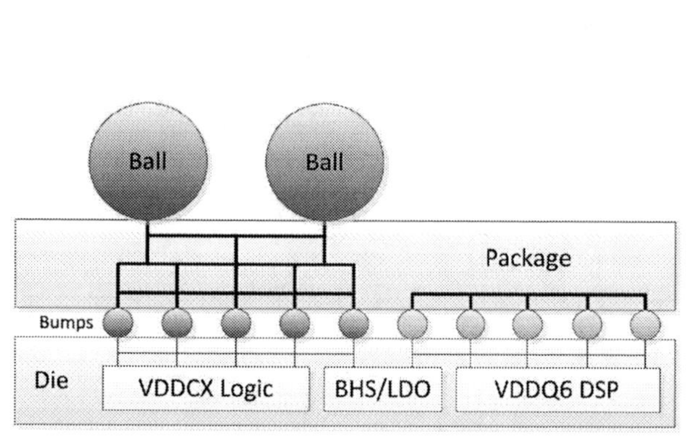

Figure 10.1.1: Package connectivity for the BHS and LDO.

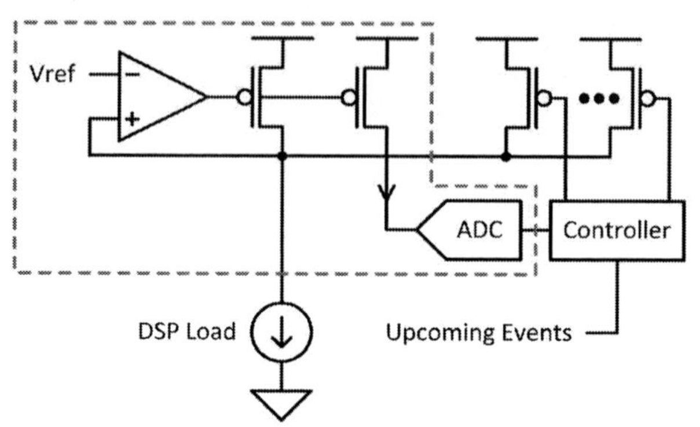

Figure 10.1.2: Block diagram of the LDO.

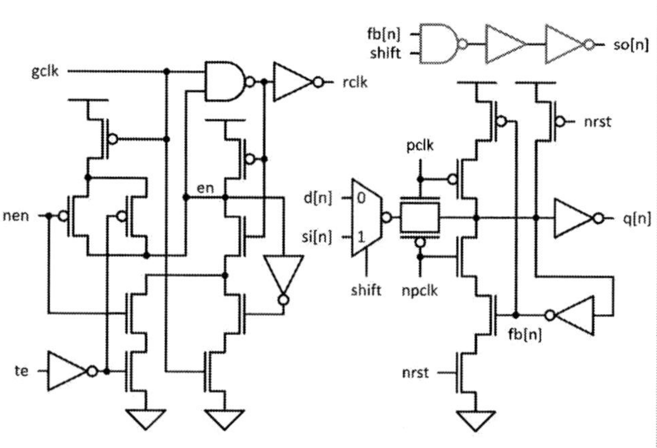

Figure 10.1.3: Low-power clock gating cell and pulsed latches. The scan out path uses high-V_t logic gates.

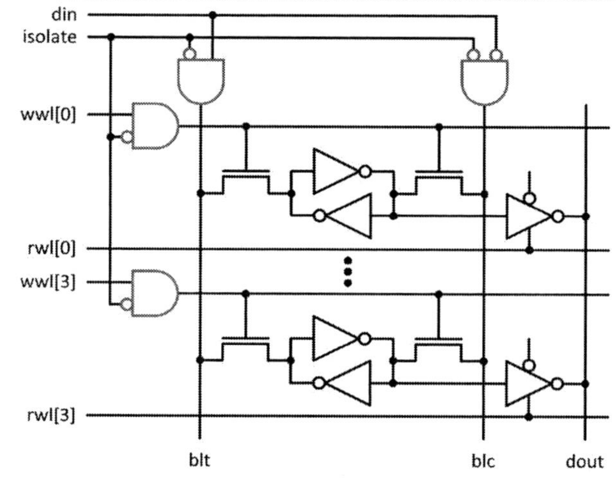

Figure 10.1.4: Asynchronous FIFO. Combined level shifter and isolation cells drive the bitlines and the write wordlines.

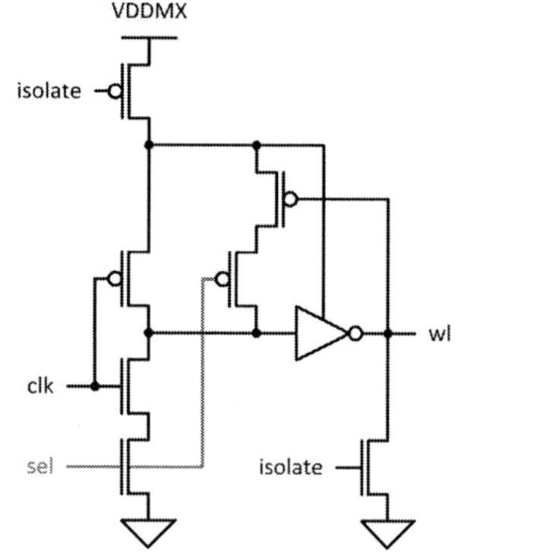

Figure 10.1.5: Dynamic level shifter used in arrays. The select is the only VDDQ6 signal.

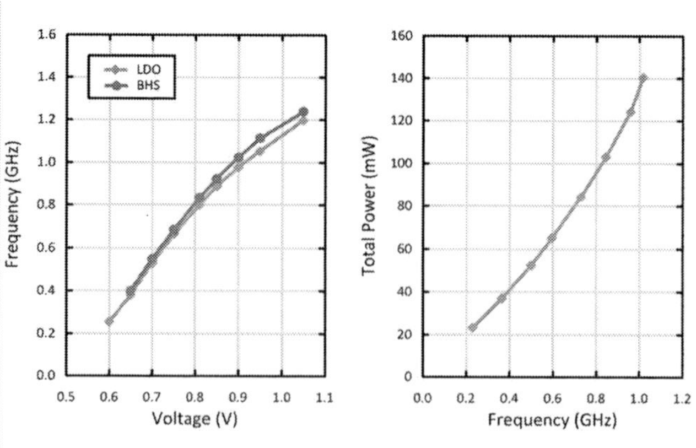

Figure 10.1.6: Measured performance and power at 25°C for typical silicon.

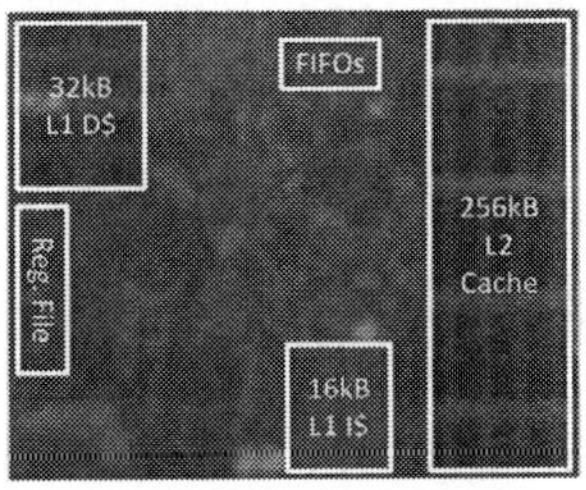

	This work ISSCC 2014	M. Fujigaya et al. ISSCC 2013	Y. Shin et al. ISSCC 2013	S. Jain et al. ISSCC 2012	G. Gammie et al. ISSCC 2011 (1)
Technology	28nm	28nm	28nm	32nm	28nm
I$/D$/L2 (kB)	16/32/256	N/A	32/32/512	8/8/0	32/32/128
On-Chip LDO?	Yes	No	No	No	No
Pulsed Latches?	Yes	N/A	Yes	No	No
Voltage Range	0.60V to 1.05V	1.10V	N/A	0.28V to 1.20V	0.60V to 1.00V
Fmax	1200MHz	1500MHz	1800MHz	915MHz	331MHz
Power at Fmax	197mW	800mW	N/A	737mW	145mW
Best Efficiency	58μW/MHz	533μW/MHz	N/A	170μW/MHz	114μW/MHz
Leakage at 0.90V	4.08mW	N/A	N/A	38mW	20mW

(1) Caches on.

Figure 10.1.7: Die photo and comparison to other low-power processors.

10.2 A 28nm HPM Heterogeneous Multi-Core Mobile Application Processor with 2GHz Cores and Low-Power 1GHz Cores

Mitsuhiko Igarashi, Toshifumi Uemura, Ryo Mori, Noriaki Maeda, Hiroshi Kishibe, Midori Nagayama, Masaaki Taniguchi, Kohei Wakahara, Toshiharu Saito, Masaki Fujigaya, Kazuki Fukuoka, Koji Nii, Takeshi Kataoka, Toshihiro Hattori

Renesas Electronics, Tokyo, Japan

The worldwide demand for high-performance mobile or car infotainment application processors (AP) is increasing. This demand coexists with the need for low power to achieve long battery life and avoid thermal runaway. A heterogeneous CPU configuration is an effective solution. The proposed heterogeneous quad/octa-core AP has a combination of high-performance 2GHz cores and energy-efficient 1GHz cores. The maximum performance in the octa-core configuration is 35600 DMIPS. The key design highlights are: 1) Using a dedicated PLL and H-tree clock in the high-performance CPU achieves both 2GHz operation and reduced dynamic power. 2) A low-leakage SRAM in a 28nm HPM process is used and the leakage current of the peripheral circuits of the SRAM macro is optimized via multiple threshold voltages (V_t) and gate lengths (L_g). 3) The effects of process and voltage variations are accurately corrected by an on-chip process sensor and direct sensing of the voltage in the power mesh of the chip. 4) An enhanced CPU clock control mechanism is employed, which uses an on-chip delay sensor to reduce AC IR drop. 5) The heterogeneous CPU architecture maintains high performance even during thermal throttling.

Figure 10.2.1 shows the chip's features and a block diagram of the octa-core configuration. Fig. 10.2.7 is a micrograph of the die. The AP has two different types of CPU core: high-performance Cortex-A15 cores (CA15) operating at up to 2GHz, and energy-efficient Cortex-A7 cores (CA7) operating at up to 1GHz. It uses both types of CPUs simultaneously, whereas the conventional approach [1] is to switch between the two types of CPU according to the workload. The CA7 and CA15 have six power domains each, and each domain can be powered off by a power switch to reduce leakage power when idle. The CA7 is active and the CA15 is powered off when the workload is light, whereas both the CA7 and CA15 become active when the workload is heavy. Measured power results for the CA15 and CA7 are 5.4W and 0.6W, respectively, at the maximum frequency when all four cores in each block are operating.

Figure 10.2.2 shows the clock distribution and synchronizing scheme for the CA15. In general, a grid/mesh [2] or hybrid mesh [1] clock structure is used to reduce clock skew in high-frequency operation. However, it has been observed that clock power due to grid capacitance can comprise 21% of the overall chip power [2]. An H-tree clock structure was therefore used in the CA15 to reduce dynamic power. The H-tree structure reduces the wiring capacitance of the clock tree and improves the drivability of the clock buffer cells, compared with a grid/mesh clock structure. Although an H-tree clock structure does not minimize timing variations to the maximum extent possible, using a dedicated PLL to minimize clock latency and jitter for the CA15 achieved a clock latency of 0.9ns, and clock jitter of 30ps, enabling 2GHz operation. Since the dedicated PLL results in the interface with the CA15 becoming asynchronous, a synchronizer using a gray code was developed to retain throughput almost equal to that of a synchronous interface. This also limits the increase in latency relative to the bus clock to one cycle. The area overhead of applying DVFS was reduced by using a pseudo level-shifter with a single rail.

Figure 10.2.3 shows the features and performance of the SRAM for the L1 cache of the CA15. We used a low-leakage SRAM (LL-SRAM) in a 28nm HPM process to reduce leakage within the SRAM bitcells. Transistors having multiple V_t and multiple L_g are used to optimize the leakage current of the peripheral circuits. Long-L_g transistors are used in the word drivers and timing generators. A 24% leakage current reduction is realized compared to normal SRAM (GL-SRAM) with single V_t transistors, while achieving 2GHz operation. The L1 cache has 512 78b words, and consumes 0.00974 mm^2.

We used a modified form of adaptive voltage scaling (AVS). Schematic diagrams of conventional AVS, our modified AVS, and measured results for the latter are shown in Fig. 10.2.4. In conventional AVS [3], on-chip sensors are used to reduce the effect of process, voltage and temperature variations by optimizing supply voltages (V_{DD}). However, voltage control by the traditional method requires a voltage margin because the sensors are not perfectly accurate. The voltage resolution of the power management IC (PMIC) is also restricted to a certain value. In addition, if communication with the PMIC is through an I2C interface, the response time is slow compared to the switching frequency of the PMIC. This is an obstacle to dynamic fine-grained voltage control. Our modified form of AVS improves the accuracy of voltage control and the speed of feedback. Process variations are assessed via on-chip process sensors at the time of testing, and voltage settings for the PMIC are written to fuses. Thus, the impact of process variations is reduced statistically by a coarse adjustment of the PMIC's voltage. The measured results for minimum V_{DD} of at-speed test and voltage setting is shown in Fig. 10.2.4. The PMIC voltage of the fast-corner chip is set to a low V_{DD}, reducing worst-case dynamic and leakage power by 29% and 20%, respectively. The PMIC also reduces voltage variation by directly sensing the pin driving the power mesh, detecting variations, and then controlling V_{DD} finely and dynamically. The improvement in minimum V_{DD} during program execution was measured at around 40 to 50mV.

Operation of a high-performance CPU at 2GHz leads to a large di/dt that produces excessive AC IR drop. The sampling rate of a conventional power saver [4] is relatively slow at ~1μs, and thus it cannot detect an AC IR drop at several tens of MHz. Accordingly, we developed a real-time power saver mechanism with a 20× faster sampling rate. Fig. 10.2.5 shows its block diagram and results. An on-chip delay sensor is used, which samples voltage at 50ns intervals. If the sensor detects an IR drop that exceeds a threshold (e.g. due to a sudden increase of activity), a request for the clock controller to step down the frequency is issued to suppress the IR drop. The clock controller changes its frequency after a delay of 100ns. The frequency is incrementally increased several microseconds after a drop that exceeds the threshold. Simulations indicate that this approach achieves a 20mV reduction in the AC IR drop.

Mobile devices typically have a small form factor, making cooling difficult in an environment where an expensive cooling system is not possible. Moreover, the power consumed under a heavy workload by a high-performance CPU operating at 2GHz is significant and may lead to thermal runaway. A mobile AP thus requires a thermal control technique [5]. Fig. 10.2.6 shows the model we used to analyze thermal control, and simulation results under heavy workloads. We assume worst-case leakage conditions and lack of a special cooling system. Junction temperatures (T_j) increase while the CPU is running at full throttle, indicated by mode-A in the figure. When the on-chip temperature sensor detects a temperature exceeding a threshold, operating frequency is decreased and/or some CPU cores are powered off to decrease T_j. In a homogeneous CPU architecture, high-performance CPUs must continue to be used during the cool-down period, despite their large leakage and dynamic power. This dramatically reduces the average performance to only 3600 DMIPS. In contrast, with the heterogeneous CPU architecture, the energy-efficient CPUs operate while the chip is cooling down. The heterogeneous octa-core architecture maintains an average performance of 11000 DMIPS in worst-case conditions.

References:

[1] Y. Shin, *et al.*, "28nm High-κ Metal-Gate Heterogeneous Quad-Core CPUs for High-Performance and Energy-Efficient Mobile Application Processor", *ISSCC Dig. Tech. Papers*, pp. 154-155, 2013

[2] T. Singh, *et al.*, "Jaguar: A next-generation low-power x86-64 core", *ISSCC Dig. Tech. Papers,* pp. 52-53, 2013

[3] Y. Ikenaga, *et al.*, "A 27% active-power-reduced 40-nm CMOS multimedia SoC with adaptive voltage scaling using distributed universal delay lines", *IEEE Symp. VLSI Circuits*, pp. 186-187, 2011.

[4] M. Fujigaya, *et al.*, "A 28nm High-κ Metal-Gate Single-Chip Communications Processor with 1.5GHz Dual-Core Application Processor and LTE/HSPA+ Capable Baseband Processor", *ISSCC Dig. Tech. Papers*, pp. 156-157, 2013.

[5] S. Yang, *et al.*, "A 32nm High-κ Metal Gate Application Processor with GHz Multi-Core CPU", *ISSCC Dig. Tech. Papers*, pp. 214-216, 2012.

978-1-4799-0917-9/14 $31.00 © 2014 IEEE

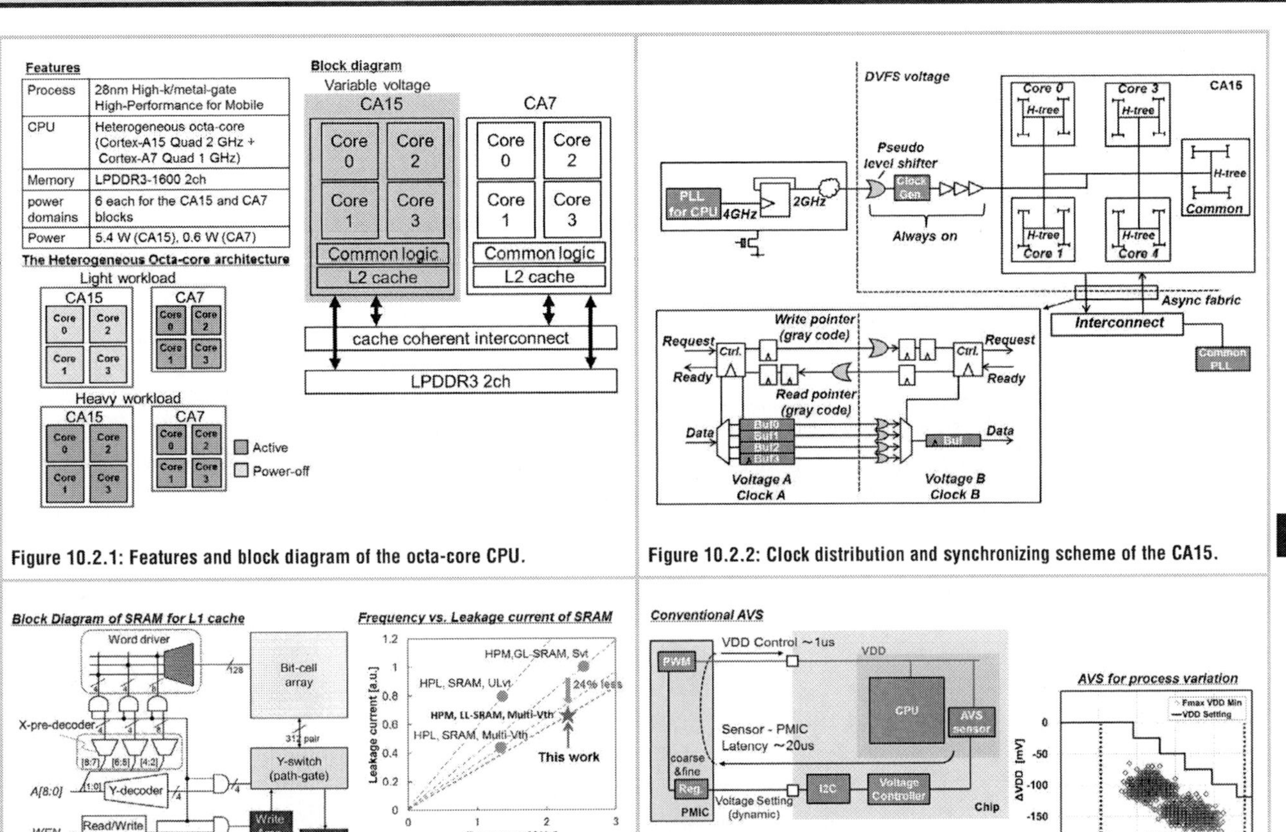

Figure 10.2.1: Features and block diagram of the octa-core CPU.

Figure 10.2.2: Clock distribution and synchronizing scheme of the CA15.

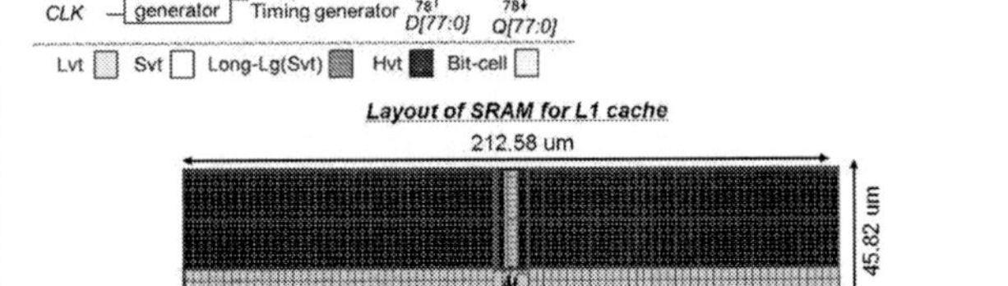

Figure 10.2.3: Feature and performance of SRAM for L1 cache of the CA15.

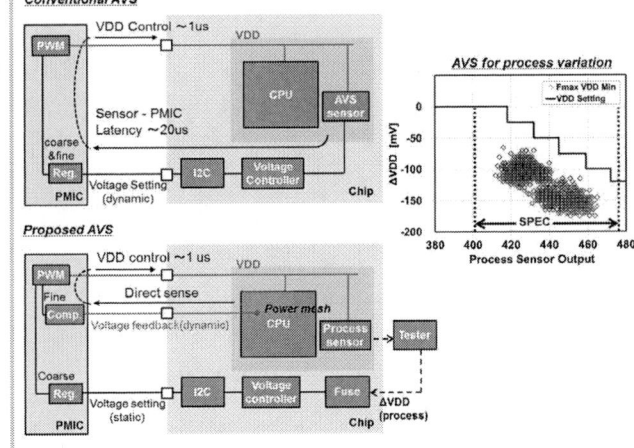

Figure 10.2.4: Schematic diagrams of AVS and measured results.

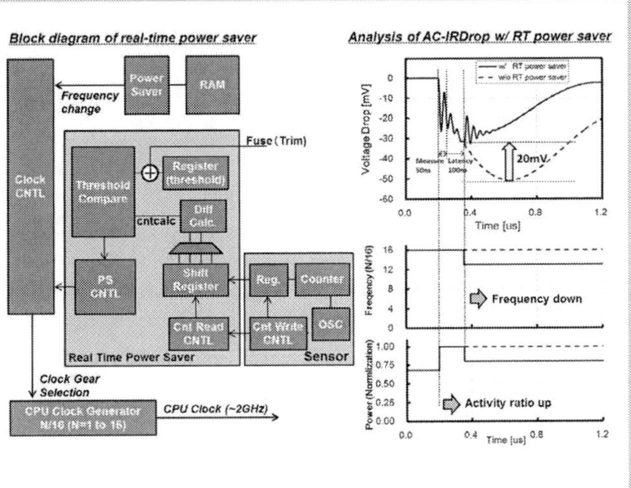

Figure 10.2.5: Real-time power saver.

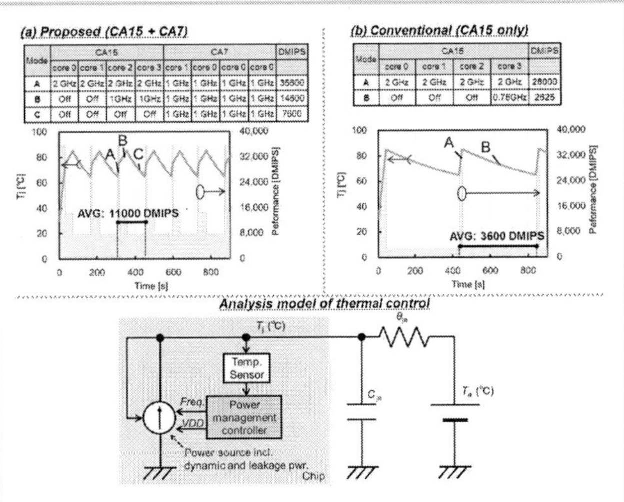

Mode	CA15 core 0	CA15 core 1	CA15 core 2	CA15 core 3	CA7 core 1	CA7 core 0	CA7 core 0	CA7 core 0	DMIPS
A	2 GHz	2 GHz	2 GHz	2 GHz	1 GHz	1 GHz	1 GHz	1 GHz	35800
B	Off	Off	1GHz	1GHz	1 GHz	1 GHz	1 GHz	1 GHz	14600
C	Off	Off	Off	Off	1 GHz	1 GHz	1 GHz	1 GHz	7600

Mode	CA15 core 0	CA15 core 1	CA15 core 2	CA15 core 3	DMIPS
A	2 GHz	2 GHz	2 GHz	2 GHz	28000
B	Off	Off	Off	0.78GHz	2525

Figure 10.2.6: Thermal control for the heterogeneous CPU architecture.

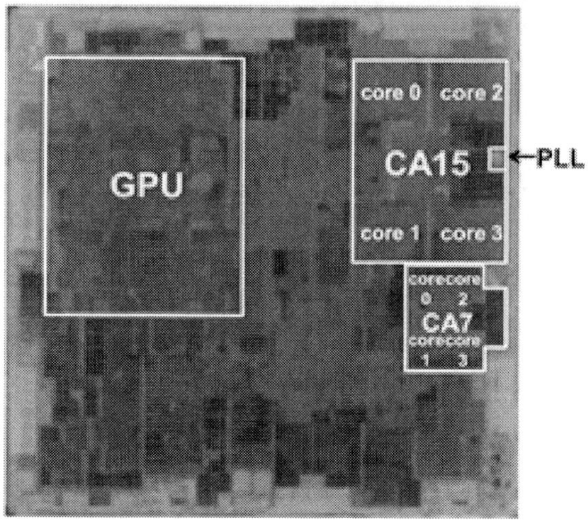

Figure 10.2.7: Die micrograph.

10.3 Heterogeneous Multi-Processing Quad-Core CPU and Dual-GPU Design for Optimal Performance, Power, and Thermal Tradeoffs in a 28nm Mobile Application Processor

Alice Wang[1], Tsung-Yao Lin[2], Shichin Ouyang[3], Wei-Hung Huang[2], Jidong Wang[1], Shu-Hsin Chang[2], Sheng-Ping Chen[2], Chun-Hsiung Hu[2], Jim C. Tai[2], Koan-Sin Tan[2], Meng-Nan Tsou[2], Ming-Hsien Lee[2], Gordon Gammie[1], Chi-Wei Yang[2], Chih-Chieh Yang[2], Yeh-Chi Chou[2], Shih-Hung Lin[2], Wuan Kuo[2], Chi-Jui Chung[2], Lee-Kee Yong[2], Chia-Wei Wang[2], Kin Hooi Dia[2], Cheng-Hsing Chien[2], You-Ming Tsao[2], Nitin Kumar Singh[1], Rolf Lagerquist[1], Chih-Cheng Chen[2], Uming Ko[1]

[1]MediaTek, Austin, TX,
[2]MediaTek, Hsinchu, Taiwan,
[3]MediaTek, San Jose, CA

Driven by consumer demand, mobile devices such as smartphones and tablets are offering more desktop-like capabilities. High-performance CPUs and GPUs, which handle compute-intensive tasks, are key to enhancing the user experience in applications such as 3D gaming, high-definition video and internet browsing. A CPU and GPU on a tablet device, however, can together consume up to 90% of the total SoC power. As the number of CPU and GPU cores on mobile devices continues to grow, it will require innovation to keep within fixed power and thermal budgets, while providing high performance.

Figure 10.3.1 shows a block diagram of a 28nm application processor SoC, which includes heterogeneous quad-core ARM-v7A processors and a dual-core GPU. The CPU cluster is made up of a dual-core CA15 cluster and a dual-core CA7 cluster, both with L1 I$/D$ 32kB/32kB. The CA15 has 1MB L2 and the CA7 has 256kB L2. The CA15 runs at 1.8GHz, while the CA7 runs at 1.4GHz. The dual-core GPU is based on the Imagination Technologies' PowerVR Series6 GPU running at 400MHz. The SoC supports a 4-in-1 connectivity package that includes Wi-Fi, Bluetooth 4.0, GPS and FM radio. Other multimedia support includes a 13M image signal processor (ISP) and a full HD 1080p video codec.

As shown in Fig. 10.3.2, Mediatek adopts a heterogeneous multi-processor (HMP) [1] software model, which eliminates the limitations of previous models, where only one CPU pair can be active at a time. This technology allows application software to access all of the processors in the asymmetric CPU subsystem simultaneously for a truly heterogeneous experience. While inherently superior to the previous models, the HMP performance remains highly dependent on the quality of the heterogeneous scheduler embedded in the SoC solution. An advanced scheduler algorithm, combined with the asymmetric CPU architecture, and an adaptive thermal and interactive power management system, maximize both performance and energy efficiency.

The SoC uses low-power hardware techniques to maximize performance, while staying within the thermal budget. Power, thermal and performance (PTP) detectors are distributed within the CPU to detect when operating conditions have changed. This technology allows the device to use the available voltage margin to increase performance or lower power consumption when possible. A controller monitors the detector data and dynamically adjusts the device for DC voltage bias, aging and temperature, as the device is exposed to different conditions. This PTP technology allows for a 23% increase in clock speed or up to 41% power savings, depending on the SoC operating conditions (Fig. 10.3.3 shows the PTP controller adapting to temperature and providing 20-to-25% power savings, on average). After CPU calibration, the GPU silicon "strength" (i.e. the process corner) can be assessed and faster silicon is operated at a lower voltage for additional power savings.

When the CPU and GPU are both configured to run at full speed simultaneously, the surface temperature will ramp up and exceed allowed limits for a mobile device. Dynamic power management must be used to keep the power consumption within the thermal limit. Most traditional approaches use a fixed threshold for thermal throttling control. However, an adaptive thermal management (ATM) algorithm is used in this work. The ATM technology monitors device temperature, as well as related context, to dynamically adjust the power budget, with minimal user-noticeable performance degradation and without causing the device to exceed the skin thermal limit. Compared with traditional thermal throttling techniques, the ATM provides an improved user experience with a 10% performance boost. For applications running continuously, like 3D gaming, the ATM provides speed control to maintain a stable user experience. For applications with short performance bursts, the ATM temporarily boosts the running speed. For applications utilizing both continuous and burst mode, the ATM monitors application behavior and decides the optimal thermal and power budget policy. Fig. 10.3.4 shows that the ATM allows a higher average CPU running frequency, under the same surface temperature limit.

At run-time, the CPU and GPU workload is continuously changing, causing voltage droop on the PDN and reducing the voltage margin available. To minimize voltage droops, a remote sensing technology is employed, as shown in Fig. 10.3.5. A feedback line using a PCB trace to a probe location close to the SoC pins permits measurement of the actual voltages delivered to the power management IC (PMIC). The voltage delivered from the PMIC continuously compensates for any losses, thus the power delivered to SoC is provided with a reduced error margin. Since the feedback mechanism is continuous, the power supplied is constantly tracked and adjusted during operation. Fig. 10.3.5 shows oscilloscope plots before and after remote sensing.

A high-density (HD) 6T bitcell is chosen for both the high-speed L1 and L2 CPU cache with small swing differential sensing to achieve low dynamic power and robust data sensing. Although the high current (HC) bitcell delivers larger cell current than the HD bitcell, the HC bitcell introduces larger word-line wire RC delay, which offsets the cell current gain. To cope with the lower cell current of HD bitcell, the gate bias (V_{gs}) on 6T pass-gate and pull-down devices are boosted to a higher voltage, without contributing to the total SRAM dynamic power significantly. A 64Kb SRAM macro with 64b/bitline can achieve a 315ps access time. SRAM leakage power is also a hefty contributor to the application processor's power consumption. Based on each subcircuit's power consumption, multiple built-in power gating switches are carefully sized and optimized to balance speed, IR drop and shut-down leakage. Fig. 10.3.6 shows that in sleep mode, leakage is cut by 13×, and in shut-down mode by 425×, compared to previously reported power gating results of 3× and 6× leakage reduction, respectively [2].

Clock gating is a commonly used methodology to reduce power. However, clock gating paths are frequently speed bottlenecks in high-speed CPU implementations. A high-speed clock-gating cell, shown in Fig. 10.3.6, is devised to replace the traditional clock-gating cells (4 inverter delays) in timing-critical paths. The new cells offer data-to-output timing of only two inverter delays. A clocking scheme to include test enable (TE) pin functionality allows for a two-stack pull-up network that further improves the speed.

Low standby-mode leakage is essential to prolong battery life in mobile devices. An external shutdown scheme is adopted to completely turn off the CPU and GPU cluster when in standby mode. This technique gates the leakage power consumed by power switches and always-on logic so the total SoC leakage can be further reduced by 30% in standby mode. Other power-reduction techniques include leakage optimization by employing a wide range of channel lengths and high-threshold voltage cells in paths that are non-critical and a quad-height grid power switch optimized to maximize the current delivered per switch resistance (41% improvement compared to the previous double-height implementation in terms of $\mu A/\mu m^2$).

The die photo of the 28nm heterogeneous quad-core CPU and dual-GPU mobile application processor is shown in Fig. 10.3.7.

References:

[1] Brian Jeff, "Advances in big.LITTLE Technology for Power and Energy Savings", *ARM White Paper,* Sept. 2012.

[2]. J. Chang, *et al.*, "A 20nm 112Mb SRAM in High-κ Metal-Gate with Assist Circuitry for Low-Leakage and Low-V_{min} Applications", *ISSCC Dig. Tech. Papers,* pp. 316-317, 2013.

978-1-4799-0917-9/14 $31.00 © 2014 IEEE

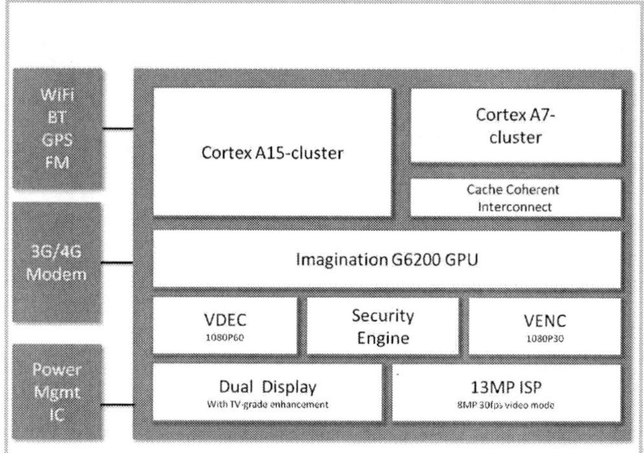

Figure 10.3.1: Block diagram of 28nm application processor SoC.

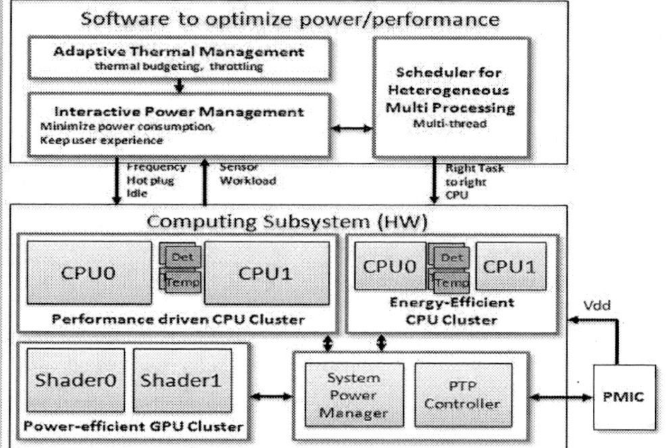

Figure 10.3.2: Low-power architecture that employs heterogeneous multi-processing (HMP), performance, thermal and power (PTP) control and adaptive thermal management (ATM).

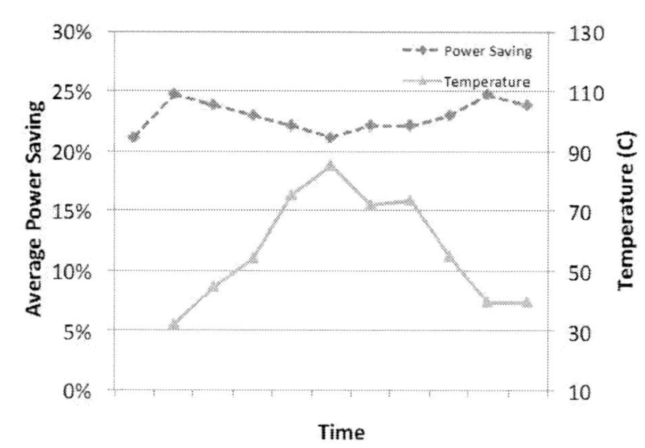

Figure 10.3.3: Power saving using PTP monitor showing adaptation to process and temperature.

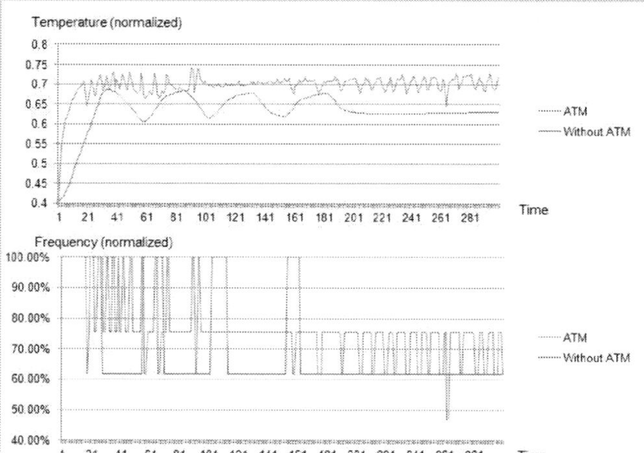

Figure 10.3.4: Adaptive thermal management (ATM) for dynamic thermal management responds quickly to maximize performance for a given temperature threshold.

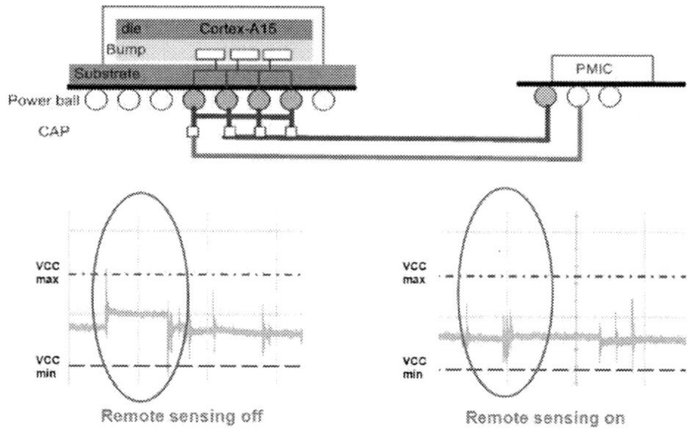

Figure 10.3.5: Oscilloscope plots of voltage measured at probe location on PCB closest to IC before and after remote sense (V_{CCmax} and V_{CCmin} are the high and low voltage specs, respectively).

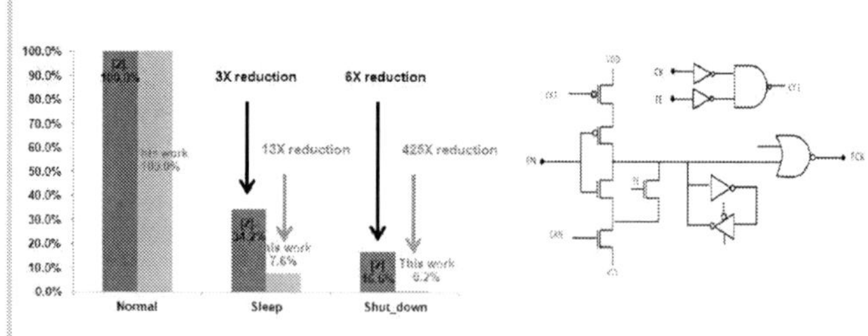

Figure 10.3.6: Optimized SRAM leakage reduction in sleep and shutdown mode and high-speed clock-gating cell with two-inverter delay and a two-stack at the first stage.

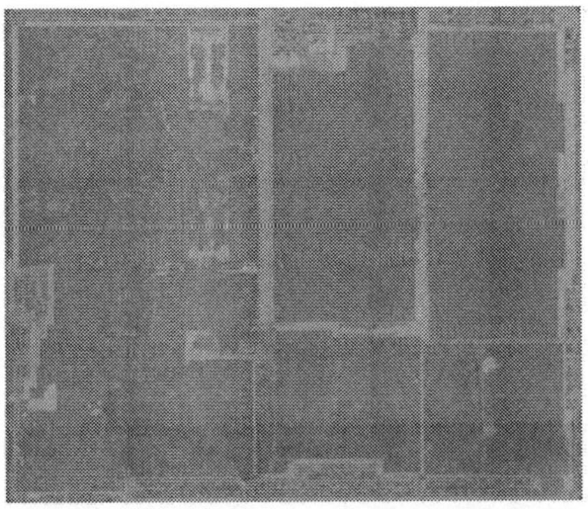

Process	28nm
Area	45mm2
Package	FCCSP
Voltage	0.85 V to 1.25V
CPU	Dual core CA15 L1 32K/32K L2 1MB 1.8Ghz Dual core CA7 L1 32K/32K L2 256KB 1.4Ghz
GPU	Dual core G6200 400Mhz

Figure 10.3.7: Die photo of SOC with quad-core CPU and dual-GPU.

10.4 A 1.22TOPS and 1.52mW/MHz Augmented Reality Multi-Core Processor with Neural Network NoC for HMD Applications

Gyeonghoon Kim, Youchang Kim, Kyuho Lee, Seongwook Park, Injoon Hong, Kyeongryeol Bong, Dongjoo Shin, Sungpill Choi, Jinwook Oh, Hoi-Jun Yoo

KAIST, Daejeon, Korea

Augmented reality (AR) is being investigated in advanced displays for the augmentation of images in a real-world environment. Wearable systems, such as head-mounted display (HMD) systems, have attempted to support real-time AR as a next generation UI/UX [1-2], but have failed, due to their limited computing power. In a prior work, a chip with limited AR functionality was reported that could perform AR with the help of markers placed in the environment (usually 1D or 2D bar codes) [3]. However, for a seamless visual experience, 3D objects should be rendered directly on the natural video image without any markers. Unlike marker-based AR, markerless AR requires natural feature extraction, general object recognition, 3D reconstruction, and camera-pose estimation to be performed in parallel. For instance, markerless AR for a VGA input-test video consumes ~1.3W power at 0.2fps throughput, with TI's OMAP4430, which exceeds power limits for wearable devices. Consequently, there is a need for a high-performance energy-efficient markerless AR processor to realize a real-time AR system, especially for HMD applications.

We propose a high-throughput low-energy AR processor mainly targeted for advanced 3D AR HMD applications. It has 4 key features. For high throughput, it includes: 1) task-level pipelined SIMD-PE clusters, and 2) a neural network network-on-chip (NoC). Both of these features exploit the high data-level parallelism (DLP) and task-level parallelism (TLP) of the pipelined multi-core architecture. For low energy consumption, the processor includes: 3) the vocabulary forest accelerator (VFA), and 4) mixed-mode support vector machine (SVM)-based dynamic resource management (DRM) to reduce unnecessary external memory accesses and core activations.

The proposed AR processor accelerates attention-based markerless AR, which adopts phase spectrum of quaternion Fourier transform (PQFT) as visual attention to choose the region-of-interest (ROI). This reduces the amount of computation by focusing on a small portion of the image, as shown in Fig. 10.4.1. In addition, since we reuse extracted keypoint descriptors utilized in recognition for camera tracking and mapping operations, the two main AR operations can be processed in parallel to increase system throughput by 1.45×. To realize this AR model in wearable applications, with real-time and power constraints, the processor employs a high-throughput task-level pipelined architecture and dedicated accelerators for the various stages of object recognition and camera pose estimation.

The block diagram of the processor is shown in Fig. 10.4.2. A total of 36 IP cores are connected by a 2D mesh NoC, and they are merged into 6 different SIMD PE-clusters, where each PE cluster is dedicated to a different vision or 3D graphics operation, and 2 dedicated accelerators assist with resource management or NoC bandwidth regulation. The scale-space generation engine (SSGE) performs Gaussian filtering for a 16×16 image tile to make multi-scale images, and then the visual attention engine (VAE) selects the ROIs from background clutter. The keypoint detection engine (KDE) and the keypoint descriptor generation engine (KDGE) perform feature extraction. The VFA recognizes target objects by matching keypoint descriptors from the database. The camera pose estimation engine (CPEE) calculates the relative position of the camera and the unified shaders in the rasterization engine (RE) augment 3D graphics into the output video.

By exploiting TLP of the proposed multi-stage AR algorithm, a task-level pipeline which consists of 5 sub-tasks is implemented, as shown in Fig. 10.4.3. Since each task processes only one image tile at a time instead of the whole image (tile-based processing), on-chip SRAM size is reduced. Also, the pipeline has deeper stages compared to the previous vision processors [4-5], from 2-stage to 4-stage, and SIMD PE clusters are more tightly pipelined with over 92% of utilization by running them simultaneously. It outperforms a previous ARM Cortex-A9 NEON and this processor (without the pipeline) by 4.2× and 2.7× in terms of throughput, respectively. However, since massive inter-stage communication between task-producing/consuming cores exists at the boundary of task-level pipeline stages, network congestion may negatively impact throughput and energy efficiency.

The neural network NoC is implemented to control the inter-stage communication to increase chip performance. It monitors the workload history of all processing cores, and manages producer-consumer data transactions between tasks in the case of task-level pipelines distributed over the multi-core processor. For example, as shown in Fig. 10.4.4, when PEs (P_0-P_7) of the KDE finish keypoint detection and transfer network packets containing the detection results to PEs (C_0-C_9) of the KDGE, the multi-layer perceptron (MLP) neural network operates as a workload predictor which can trace the workload history of ROI tiles that gradually change for frame sequences. The prediction of the next workloads of ROI tiles enables the dispersed mapping of producer-consumer pairs to avoid network congestion by prohibiting the unbalanced (locally concentrated) core assignment of conventional approaches before the actual workloads are known. With the proposed near-optimal scheduling based on the neural network, the data transfers on two dominant communication paths, the SSGE–KDE stage, and the KDE–KDGE stages, are reduced by 9.7% and 19.4%, respectively, compared to the first-come first-served scheduling method, and overall chip throughput is increased by 29.1%.

Figure 10.4.5 shows the proposed VFA architecture adopted to eliminate external memory accesses to realize a tile-based task-level pipeline for real-time AR implementation. In the VFA, database (DB) vectors are quantized during the learning process of vocabulary trees (VTs), so as to make VTs hold only seed DB vectors in the on-chip memory, thereby massive external memory accesses are entirely removed. Four VTs are implemented as binary tree structures with 7 levels to realize the best hardware efficiency. The VT hardware is composed of 32KB of node table memory, a memory controller, a histogram generator and the stage where the distance between the query vector and two centroid vectors are calculated. We utilize AdaBoost to combine 4 differently learned VTs to enhance matching accuracy of the final VT decision, and an inhibition scheme to realize multi-object recognition. The VFA achieves 94.3% matching accuracy under a low-cluttered test video with 2,066 kilo-vectors/s throughput and 35.7% less power consumption compared to previous art [6].

In order to control the dynamic power consumption of the multi-core processor, the DRM unit performs power management by using three analog SVM circuits – each implementing a 4D-Gaussian kernel computation, and a digital controller for input/output configuration of the analog SVM, as shown in Fig. 10.4.6. The DRM monitors the number of selected ROIs, the number of extracted keypoints, and the thermal/power headroom to perform per-frame power-mode control on the SIMD PE-clusters, which consume 87% of the overall power. The 4D-Gaussian kernel operation is implemented with a cascading of 4 analog 1D-Gaussian circuits to obtain 2.7× and 1.6× power and area efficiency, respectively, vs. a digital implementation. With the help of the high classification accuracy provided by the analog SVM circuits, the SVM chooses the operating mode of dynamic voltage and frequency scaling (DVFS) based on 4 monitoring parameters, and as a result, 51% of power consumption can be reduced compared to a multi-core processor without DRM.

The proposed AR processor is fabricated using 65nm CMOS technology, integrating 8.32M equivalent gates and 693KB of SRAM, for a battery-powered HMD platform with 30fps real-time performance. It consumes 381mW average power, and 778mW peak power at 250MHz, 1.2V. With 1.22TOPS peak performance, the processor achieves 1.57TOPS/W power efficiency, representing 76% improvement over a state-of-the-art augmented reality processor [3], and 1.52mW/MHz energy efficiency.

References:

[1] G. Klein, *et al.*, "Parallel Tracking and mapping for small AR workspaces," *In IEEE/ACM Int. Symp. On Mixed and Augmented Reality*, pp. 225-234, 2007.
[2] J.-H. Woo, *et al.*, "Mobile 3D Graphics SoC from Algorithm to Chip," *Wiley Press*, 2010.
[3] J. Yoon, *et al.*, "A Unified Graphics and Vision Processor With a 0.89μW/fps Pose Estimation Engine for Augmented Reality," *IEEE Trans. on Very-Large Scale Integration Systems*, pp. 206-216, 2013.
[4] J. Oh, *et al.*, "A 320mW 342GOPS Real-Time Moving Object Recognition Processor for HD 720p Video Streams," *ISSCC Dig. Tech. Papers*, pp. 220-221, 2012.
[5] J. Park, *et al.*, "A 646 GOPS/W Multi-classifier Many-core Processor with Cortex-like Architecture for Super-Resolution Recognition," *ISSCC Dig. Tech. Papers*, pp. 168-169, 2013.
[6] Y-C. Su, *et al.*, "A 52 mW Full HD 160-Degree Object Viewpoint Recognition SoC With Visual Vocabulary Processor for Wearable Vision Applications," *IEEE J. Solid-State Circuits*, vol. 47, no. 4, pp.797-809, 2012.

978-1-4799-0917-9/14 $31.00 © 2014 IEEE

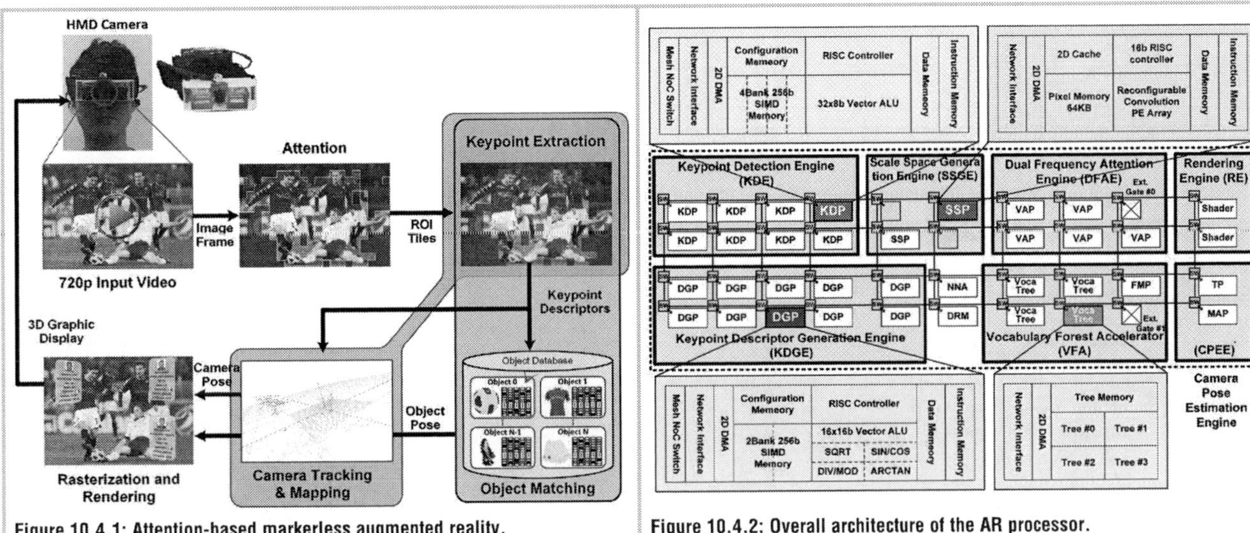

Figure 10.4.1: Attention-based markerless augmented reality.

Figure 10.4.2: Overall architecture of the AR processor.

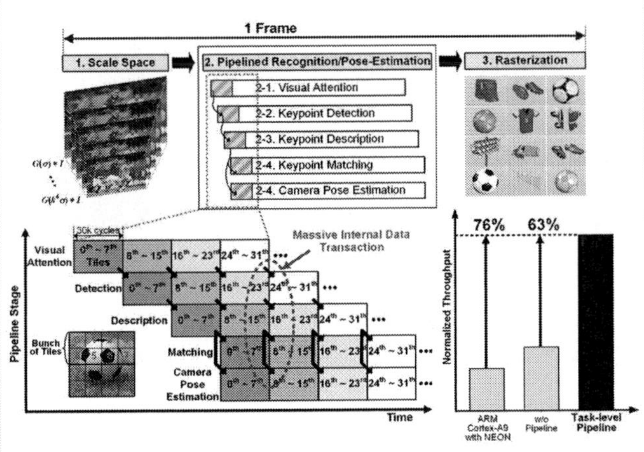

Figure 10.4.3: 4-stage task-level pipelined augmented reality.

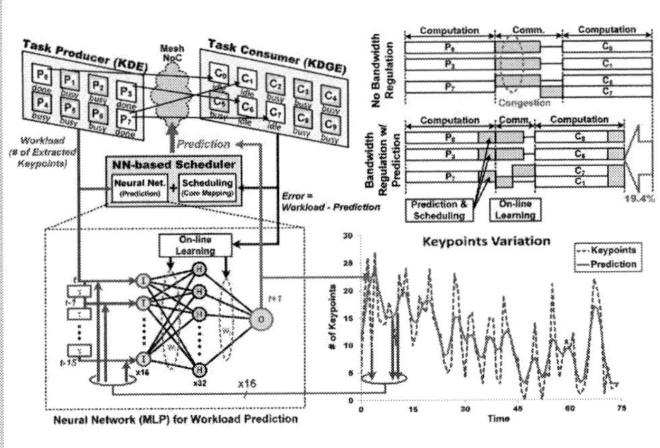

Figure 10.4.4: Neural network scheduler for 2D mesh NoC.

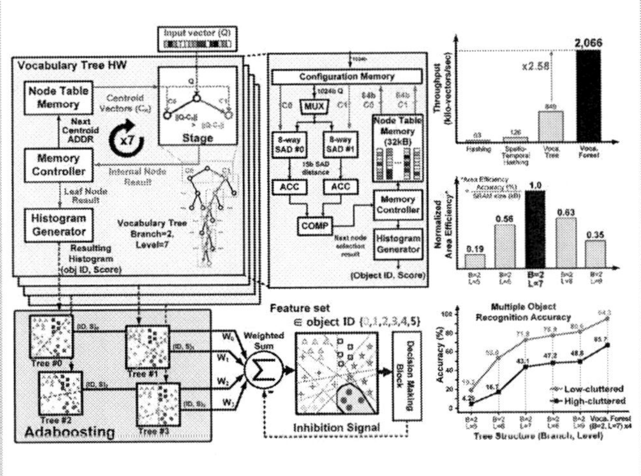

Figure 10.4.5: Vocabulary forest with four vocabulary trees.

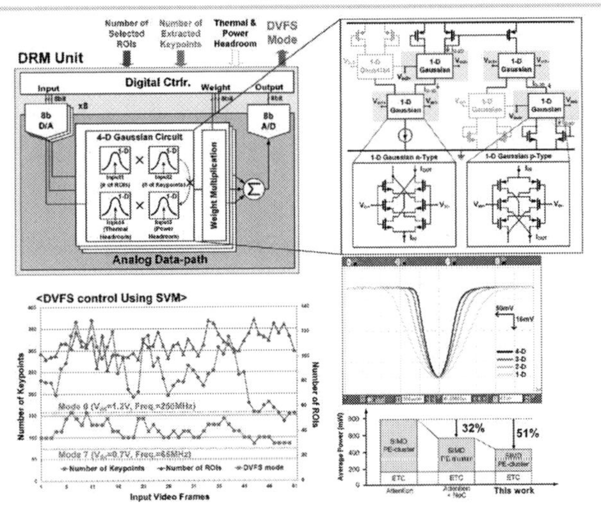

Figure 10.4.6: Measurement results of workload prediction model.

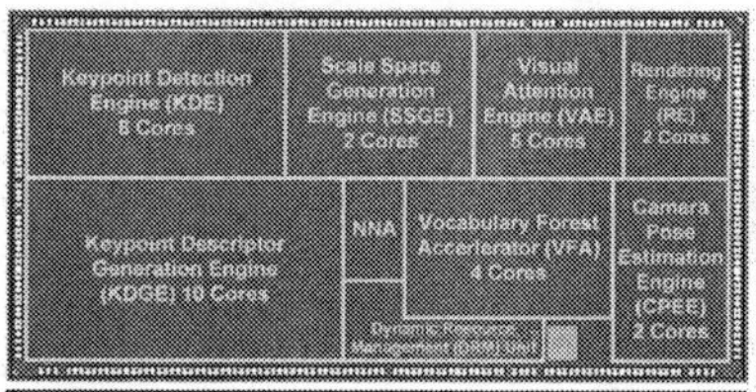

Keypoint Detection Engine (KDE) 8 Cores
Scale Space Generation Engine (SSGE) 2 Cores
Visual Attention Engine (VAE) 5 Cores
Rendering Engine (RE) 2 Cores
Keypoint Descriptor Generation Engine (KDGE) 10 Cores
NNA
Vocabulary Forest Accelerator (VFA) 4 Cores
Camera Pose Estimation Engine (CPEE) 2 Cores
Dynamic Resource Management (DRM) Unit

	ISSCC'12[4]	ISSCC'13[5]	TVLSI'13[3]	This Work
Functions	Object Recognition	Object Recognition	Marker-based Augmented Reality	Markerless Augmented Reality
Resolution	HD (720P)	HD (720P)	VGA (640x480)	HD (720P)
Technology	130nm	130nm	180nm	65nm
Area	$32mm^2$	$25mm^2$	$28.7mm^2$	$32mm^2$
Power	534mW	420mW	413mW	778mW
Performance	352GOPS	271.4GOPS	153.6GOPS	1.225TOPS
Power Efficiency	619GOPS/W	646GOPS/W	371GOPS/W	1.58TOPS/W

Technology		65nm 1P8M Logic CMOS	
Chip Size		4.0mm x 8.0mm	
Gate Count		8.32M	
SRAM		693 kByte	
Supply Voltage		Nominal	1.2V
		DVFS	0.7 ~ 1.2V
Clock Frequency		Nominal	250MHz (90FO4)
		DVFS	65 ~ 250MHz
Power Consumption		778 (Peak) 381 (Average)	
Peak Performance	Fixed-Point	VAE	108 GOPS
		SSGE	187 GOPS
		KDE	252 GOPS
		KDGE	365 GOPS
		RE	123 GOPS
		VFA	191 GOPS
	Floating-Point	CPEE	27.5 GFLOPS
	Total	1.22 TOPS/27.5 GFLOPS	
Power Efficiency		1.57 TOPS/W	

Figure 10.4.7: Chip micrograph and performance summary.

10.5 A 90nm 20MHz Fully Nonvolatile Microcontroller for Standby-Power-Critical Applications

Noboru Sakimura[1,2], Yukihide Tsuji[1], Ryusuke Nebashi[1], Hiroaki Honjo[1], Ayuka Morioka[1], Kunihiko Ishihara[1], Keizo Kinoshita[2], Shunsuke Fukami[2], Sadahiko Miura[1], Naoki Kasai[2], Tetsuo Endoh[2], Hideo Ohno[2], Takahiro Hanyu[2], Tadahiko Sugibayashi[1]

[1]NEC, Tsukuba, Japan, [2]Tohoku University, Sendai, Japan

Recently there has been increased demand for not only ultra-low power, but also high performance, even in standby-power-critical applications. Sensor nodes, for example, need a microcontroller unit (MCU) that has the ability to process signals and compress data immediately. A previously reported 130nm CMOS and FeRAM-based MCU features zero-standby power and fast wakeup operation by incorporating FeRAM devices into logic circuits [1]. The 8MHz speed, however, was not sufficiently high to meet application requirements, and the FeRAM process also has drawbacks: low compatibility with standard CMOS, and write endurance limitations. A spintronics-based nonvolatile integrated circuit is a promising option to achieve zero standby power and high-speed operation, along with compatibility with CMOS processes. In this work, we demonstrate a fully nonvolatile 16b MCU using 90nm standard CMOS and three-terminal SpinRAM technology. It achieves 20MHz, 145μW/MHz operation with a 1V supply in the active state, and 4.5μW intermittent operation with 120ns wakeup time and 0.1% active ratio, without forwarding of re-boot code from memory. The features provide sufficiently long battery life to achieve maintenance-free sensor nodes.

An overview of the nonvolatile microcontroller is shown in Fig. 10.5.1. The microcontroller comprises a 16b RISC architecture CPU core (compatible with the MSP430 instruction set [2]), a 64KB RAM/ROM-unified SpinRAM macro, a power-management module (PMM), a unified clock system (UCS), two timers (A and B), a 12b ADC, a 32b multiplier (MPY), two universal serial interfaces (USCI-A and B), a direct memory access (DMA) module, and eight I/O ports (P1-8). The SpinRAM macro consists of 2-transistor 1-magnetic-tunnel-junction (2T1MTJ) memory cells [3], and it has some redundant words and columns to replace defective cells, and an error check and correction (ECC) circuit for write failures. To eliminate backup/restore overhead through the memory bus, 4,072 nonvolatile magnetic flip-flops (MFFs [4]) are employed to capture the context of the CPU. Two instructions, "SAVE" and "LOAD", allow software to flexibly backup/restore to/from the MFFs. The supply voltage range (DV_{CC}) is 1.8 to 3.3V and the 1V internal power supply (V_{CORE}) is provided by a DC/DC converter in the PMM. The chip has 11 power domains (PDs), where the power supply (V_{PD}) in each PD is isolated from V_{CORE} by an internal power switch. Each logic core belongs to a PD so that cores can be individually turned off when they are not needed. The microcontroller supports three low-power modes: standby mode, power-gating (PG) mode, and sleep mode. In standby mode, all cores remain on, and the main clock is gated, and leakage power is 117μW. In PG mode, the power-switch status of the power domains is controlled by the PMM, and 1.6μW static power is consumed when all cores are turned off. In sleep mode, the DC/DC converter is turned off and the static power is zero.

The schematic of our three-terminal MTJ is shown in Fig. 10.5.2. The MTJ uses spin-torque switching, such as domain-wall-motion or spin-Hall-effect-induced switching [5, 6]. It comprises three perpendicular magnetic layers: a free layer, a reference layer and a sense layer. The magnetization of the free layer is switched by the write current (I_{WRITE}), whose direction is associated with the data written, between terminals T2 and T3. The magnetization of the sense layer is switched by stray field from the free layer. The MTJ resistance changes when the magnetization of the sense layer flips, and the stored data is read out by detecting the read current (I_{READ}) flowing from T1 to T2. The MTJ structure has the advantage of being read-disturbance free, which provides fast operation with a wide operating margin, since the I_{READ} path is different from the I_{WRITE} path. The switching time is 4ns with a switching energy of 6pJ/b, and the magnetoresistive (MR) ratio is 90%.

To minimize switching energy, the microcontroller supports two modes for backing up CPU state to the MFFs: a logging mode and a software-controlled mode. In the logging mode, the CPU activates the backup-enable signal (MWE) for every register access, even if no "SAVE" instructions are fetched. In the software-controlled mode, the CPU activates the MWE when it fetches the "SAVE" instruction, and this mode significantly reduces the backup energy for those registers whose data is frequently updated. To avoid unnecessary backup, each register has an active-high dirty bit (DT) to designate that the register's data differs from MTJ's one. When a "SAVE" instruction is fetched, the backup operation is skipped when the dirty bit is asserted, as shown in Fig. 10.5.2. The estimated backup energy is shown in Fig. 10.5.3, as a function of the average toggle frequency of the registers. In registers with high toggle frequency, the software control can reduce backup energy by an order of magnitude compared to the logging mode. In peripheral registers having low toggle frequency, the fewer number of toggled registers, the more energy reduction can be expected by software control with the dirty-bit functionality. More than 50% reduction in backup energy, for example, was obtained when 10% of peripheral registers toggled. In backing up the registers whose data hardly changes (e.g., such as I/O ports), the logging mode reduces energy relative to the software control.

A block diagram of the PMM is shown in Fig. 10.5.4. The PMM is responsible for the PG mode, and achieves both low standby leakage and a fast wake-up operation. It comprises a DC/DC converter, two level monitors (LMs) to monitor the D_{VCC} and V_{CORE} levels, PMM registers, PG control registers and a logic circuit for power management. When the CPU sets the PG-mode enable signal (PGME) in a PMM register to "1", the PG mode starts. In the PG mode, the PMM controls the on/off states of the power switches according to four different "power statuses" (PS#0-3) via the PG control registers. The power status information includes the current power status, two trigger signals (trigger A and B) to move the next state and the next-state IDs. In the example shown in Fig. 10.5.4, UCS (PD1) and Timer-A (PD6) are turned on, and all the other peripheral modules are turned off in power status PS#0. The microcontroller is in a "nap" state where the CPU's power supply is off. If Timer-A activates PGFLG6, the power state changes to PS#1, where the CPU, MRAM, MPY, and USCI-A modules are turned on. Fig. 10.5.5 shows the measured waveforms of a gated V_{PD}; a wake-up time of 120ns was observed. When V_{PD} rises, the backup data for each register is automatically restored with 10ns delay by the restore enable signal (PONLE) activated by the LM, which is placed in each power domain to monitor the V_{PD} level (see Fig. 10.5.2). It is not necessary to spend large numbers of cycles to re-boot via the memory bus. As a result, the CPU can immediately start a task. When the task is finished, the CPU informs the PMM of completion by activating PGFLG2, and the power state moves to PS#0 again. If I/O-port P1 activates PGFLG12, the power state changes to PS#3, where all power domains are turned on. In this way, the PMM can provide multi-peripheral-driven intermittent operation and scheduling with power-state flexibility, ultra-low-power consumption, and fast wake-up time.

Figure 10.5.6 shows the estimated power dissipation of the chip in the case of intermittent operations, assuming a sensor-node application. The measured dynamic and static power was used for these estimates. The horizontal axis represents the effective frequency defined as $f_{EFF} = f_{PG} \cdot N_{INST}$, where f_{PG} is frequency of the gated power-domain supply, and N_{INST} is the number of clock cycles executed during an active state. The total power dissipation in timer and I/O-driven power gating is 18μW and 4.5μW, respectively, where 88% and 99% of the standby power of 122μW in an always "on" state can be eliminated by geometric and temporal power gating, when f_{PG} = 4Hz and N_{INST} = 5K cycles (i.e., f_{EFF} = 20kHz, 0.1% CPU operation ratio). The specifications of the SpinRAM-embedded MCU and the die photo are given in Fig. 10.5.7.

Acknowledgement:
This research is supported by the JSPS through the FIRST program.

References:

[1] S. C. Bartling, *et al.*, ""An 8MHz 75μA/MHz Zero-Leakage Non-Volatile Logic-Based Cortex-M0 MCU SoC Exhibiting 100% Digital State Retention at VDD=0V with <400ns Wakeup and Sleep Transitions," *ISSCC Tech. Dig. Tech Papers*, pp. 432-433, 2013.

[2] http://www.ti.com/lsds/ti/microcontroller/16-bit_msp430/tech_docs.page

[3] R. Nebashi, *et al.*, "A 90nm 12ns 32Mb 2T1MTJ MRAM," *ISSCC Tech. Dig. Papers*, pp. 462-463, 2009.

[4] N. Sakimura, *et al.*, "Nonvolatile Magnetic Flip-Flop for Standby-Power-Free SoCs" *IEEE J. Solid-State Circuits*, vol. 44, no. 8, pp.2244-2250, Aug. 2009.

[5] S. Fukami, *et al.*, "High-speed and Reliable Domain Wall Motion Device: Material Design for Embedded Memory and Logic Application," *IEEE Symp. VLSI Tech.*, pp. 61-62, 2012.

[6] L. Liu, *et al.*, "Spin-Torque Switching with the Giant Spin Hall Effect of Tantalum," *Science*, vol. 336, p. 555, 2012.

978-1-4799-0917-9/14 $31.00 © 2014 IEEE

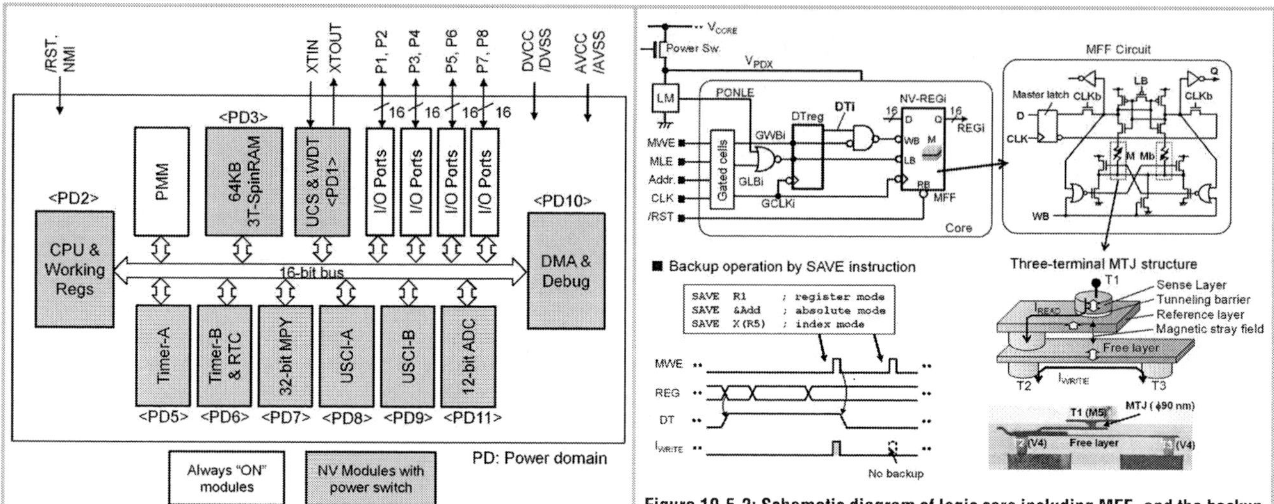

Figure 10.5.1: Block diagram of SpinRAM-embedded microcontroller.

Figure 10.5.2: Schematic diagram of logic core including MFF, and the backup operation by software control.

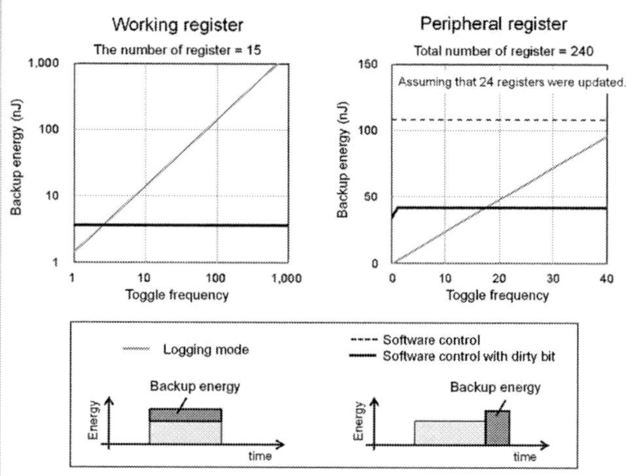

Figure 10.5.3: Backup energy of MFF-based nonvolatile register.

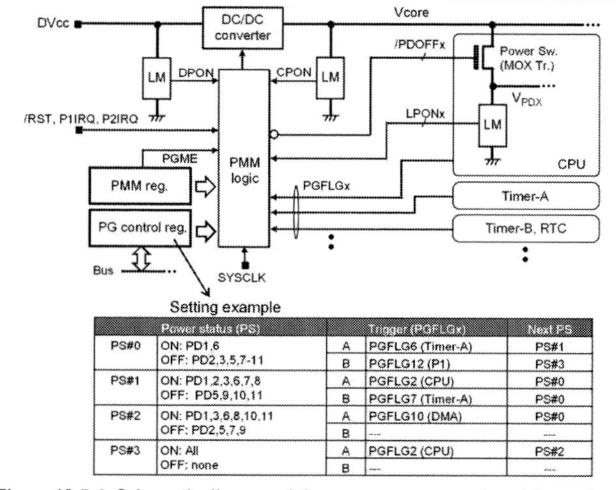

Power status (PS)		Trigger (PGFLGx)		Next PS
PS#0	ON: PD1,6 OFF: PD2,3,5,7-11	A	PGFLG6 (Timer-A)	PS#1
		B	PGFLG12 (P1)	PS#3
PS#1	ON: PD1,2,3,6,7,8 OFF: PD5,9,10,11	A	PGFLG2 (CPU)	PS#0
		B	PGFLG7 (Timer-A)	PS#0
PS#2	ON: PD1,3,6,8,10,11 OFF: PD2,5,7,9	A	PGFLG10 (DMA)	PS#0
		B	---	---
PS#3	ON: All OFF: none	A	PGFLG2 (CPU)	PS#2
		B	---	---

Figure 10.5.4: Schematic diagram of the power management module, and an example of power-domain scheduling.

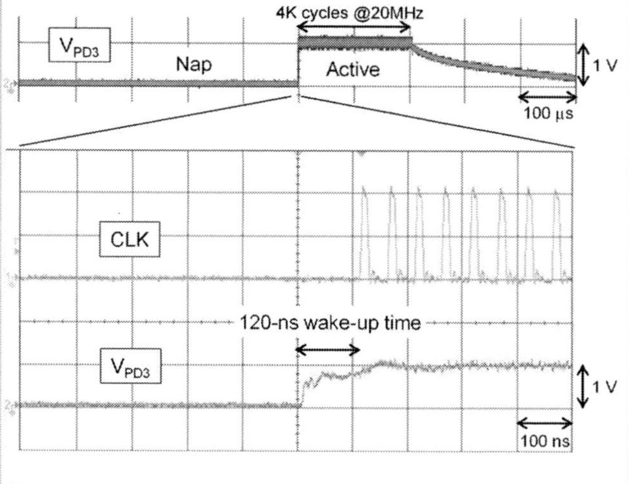

Figure 10.5.5: Measured waveforms of gated V_{PD3}.

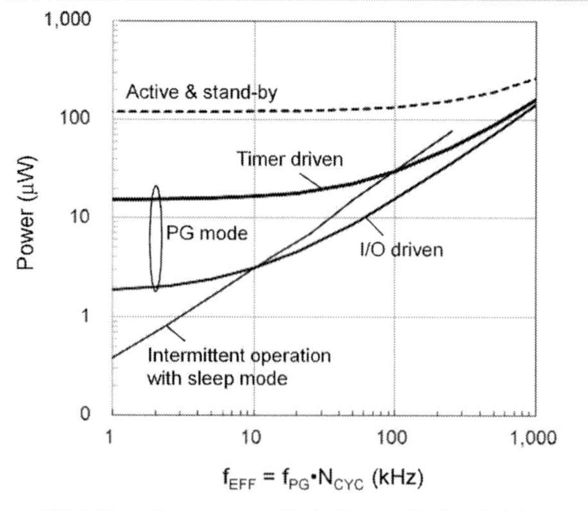

Figure 10.5.6: Dependence of power dissipation on effective clock frequency.

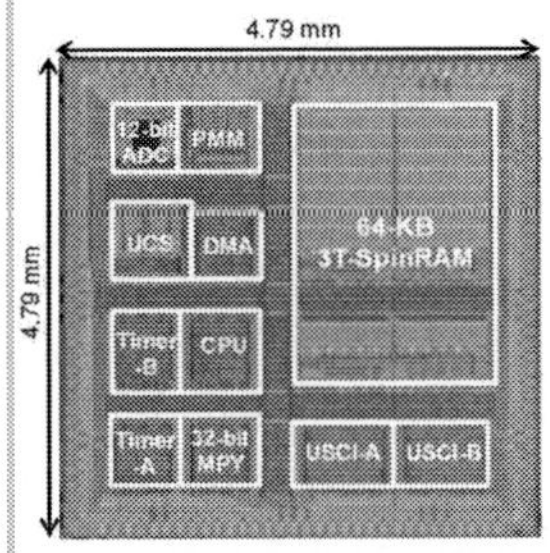

		This work	Bartling [1]
Technology node		90 nm, MVT	130 nm, HVT
Memory technology		3T-SpinRAM	FeRAM
Supply voltage		1.8~3.3V(D_{VCC}), 1.0V (V_{CORE})	1.5 V (Single supply)
Clock frequency		20 MHz	8 MHz
Dynamic power	SRAM mode	—	75 μW/MHz
	NV-RAM mode	145 μW/MHz	170 mW/MHz
PG mode leakage power		1.6-117 μW	0.28 μW
Sleep mode leakage power		< 0.1 μW	0
Wakeup time		120 ns	384 ns
Backup time / word		4 ns	320 ns
Backup energy / bit		6 pJ	2.2 pJ
Restore time / word		5 ns	384 ns
Restore energy / bit		0.3 pJ	0.66 pJ

Figure 10.5.7: Prototype chip die microphotograph and features.

10.6 A 0.74V 200µW Multi-Standard Transceiver Digital Baseband in 40nm LP-CMOS for 2.4GHz Bluetooth Smart / ZigBee / IEEE 802.15.6 Personal Area Networks

Christian Bachmann, Gert-Jan van Schaik, Benjamin Busze, Mario Konijnenburg, Yan Zhang, Jan Stuyt, Maryam Ashouei, Guido Dolmans, Tobias Gemmeke, Harmke de Groot

Holst Centre/imec, Eindhoven, The Netherlands

Ultra-low-power (ULP), short-range wireless connectivity is becoming increasingly relevant to a wide range of sensor and actuator node applications, ranging from consumer lifestyle to medical applications. In recent years, a multitude of wireless standards has been proposed to meet differing requirements of individual application domains such as data rates, range, QoS, peak and average power consumption. From a commercial perspective, a single radio component that is capable of supporting multiple wireless standards – targeting multiple application domains/markets – while reducing integration costs is highly preferable. At the same time, the multi-standard support may not compromise low-power operation or silicon area.

In this paper, we present a multi-standard digital baseband (DBB) ASIC for a 2.3/2.4GHz radio RF front-end (RFFE) transceiver [1], supporting the Bluetooth Smart (low energy, BTLE), IEEE 802.15.4 (ZigBee) and IEEE 802.15.6 standards. Ultra-low power consumption (200µW (RX) / 80µW (TX)) and low silicon area (0.2mm^2) are achieved by optimization of baseband algorithmic complexity, such as correlation-free symbol timing synchronization, and consequent resource sharing across standards (~85% of total area). Power dissipation is further decreased by utilizing data-rate-dependent clock scaling and fine-grained power mode control of sub-modules, based on the radio's operating state. An all-digital on-chip DC-DC converter [2] allows the DBB to operate at minimum supply voltage, independent of the rest of the radio system, to meet its target mode-dependent performance at reduced energy.

The DBB architecture as shown in Fig. 10.6.1 consists of transmitter (TX) and receiver (RX) digital baseband modules, as well as sub-modules required for physical (PHY) and lower data link (DL) processing. On transmitter side, the TX DL performs the payload data bitstream encoding before transmission. Depending on software configuration the bitstream is either processed by the symbol-to-chip-mapping/spreading (802.15.4), spreading-interleaving-scrambling (802.15.6), or whitening (BTLE) modules. Additionally, cyclic-redundancy-checks (CRC) (all standards) and BCH forward error correction (FEC, 802.15.6) data are calculated and appended to the transmission data. The encoded data are afterwards converted to amplitude- and frequency-path modulation control data to feed to the polar transmitter analog front-end by the TX PHY. For this purpose digital, FSK (802.15.4 and BTLE) and PSK (802.15.6) modulators are employed. On receiver side, the RX PHY contains multi-standard digital automatic gain control (AGC), synchronization (signal detection and timing synchronization) and demodulation modules. The received bitstream is afterwards passed on to the RX DL which implements the reverse operations of the TX DL to recover the received data bitstream. Furthermore, it implements CRC calculation and checking for all three standards, as well as a low-complexity BCH FEC decoding unit (802.15.6). Both the TX and RX DL also allow for SW-configurable non-standard combinations of processing blocks, such as enabling BCH for BTLE or 802.15.4 in proprietary modes targeting low SNR scenarios. The DBB also contains peripherals (PER) for host communication via SPI, interrupt control and software (SW) protocol stack/MAC layer hardware support. The latter ensures that standard-defined turnaround timing requirements are met by means of timer-based transmission/reception control of both DBB and RFFE.

Due to the complexity linked to synchronization and demodulation, the RX PHY contributes a dominant part to the total power consumption and chip area of the DBB architecture. Hence, a low-complexity and – in view of multi-standard support – resource sharing implementation is highly desirable for reducing silicon area, as well as leakage. Fig. 10.6.2 depicts functionality of the RX PHY module that is shared across all three standards. Instead of utilizing a higher-complexity correlation-based approach, the timing synchronization module uses a low-overhead timing error detector function. In contrast to traditional approaches operating on I/Q baseband data, the oversampled and pre-filtered I/Q data at the ADC rate are first converted to phase difference data by a differential detector. As a consequence, the same phase difference data can be used as an input to the non-data-aided, decision-directed Gardner timing error function to detect timing offsets. This error is evaluated by means of a low complexity, counter-based control loop that decides on whether to adjust the down-sampling process. The down-sampled phase difference data are afterwards used for frame synchronization (SFD detection), as well as data demodulation. By changing the configuration settings of the module, such as delay line lengths, threshold factors, the SFD correlation sequence and demodulation parameters, the same hardware module is reused for synchronization and demodulation in all three standards.

The effective hardware resource sharing across different standards, minimizing the standard-specific hardware overhead, is illustrated in Fig. 10.6.3. Of the total chip-area-dominant RX DBB functionality (~60% of total area), over 93% (55% of total area) is reused by all three standards. The remaining 5% belongs to standard-specific functionality, mainly in the RX DL (e.g. standard-dependent despreading, error detection/correction, etc.). Also for the TX DBB (~25% of total chip area), a large part of the HW resources are shared. Of the total TX DBB area, around 60% (~14% of total area) is reused by all three standards. Due to additional processing steps required for DPSK modulation in the 802.15.6 standard, around 40% of TX DBB area (~9.6% of total area) is standard specific.

To further reduce power dissipation, the DBB implements a clock management unit (CMU) and an on-chip DC-DC converter. The CMU allows the DBB to operate from the same 24MHz crystal oscillator clock as the RFFE, but also to derive the lowest-possible clock rates for the DL/PHY modules, considering the baseband symbol rates, as well as effective data bitrates. The on-chip DC-DC converter is used to generate a dedicated digital supply voltage enabling dynamic voltage scaling as a function of applied clock frequency, as well as adaptive voltage scaling. This allows operation at an application-dependent optimal supply voltage and enables compensation of process as well as temperature variations with the benefit of reducing the power consumed in the DBB. The implemented DC-DC converter (Fig. 10.6.4) is a fully digital design apart from the off-chip inductor. It comprises a tristate I/O cell with a pulse width modulation (PWM) scheme to generate the required core voltage. The converter has an efficiency above 70% at relevant output currents, and features a settling time of less than 40µs and a voltage ripple of less than 6mV.

The transceiver DBB has been implemented in TSMC 40nm LP CMOS technology and occupies a core area of 0.2mm^2. While minimizing power dissipation and chip area overhead introduced by the DBB and showing higher energy efficiency than the state-of-the-art, packet-error-rate measurements (Fig. 10.6.5) show that the radio system also exceeds the sensitivity requirements of the standards. Fig. 10.6.6 summarizes DBB, as well as radio system (i.e., DBB and RFFE [1]) power dissipation and performance results, and shows comparison results with other 2.4GHz short-range radios [3-6]. DBB power dissipation at a supply voltage of 0.74V in RX mode is measured to be 180µW (BTLE), 200µW (802.15.4) and 140µW (802.15.6), in TX mode, 80µW (BTLE), 80µW (802.15.4) and 60µW (802.15.6). Furthermore, interoperability with commercial devices has been successfully demonstrated by Bluetooth Smart wireless link functionality, enabled by an external MCU running the software stack in conjunction with the DBB-internal protocol support hardware.

Acknowledgements:
The authors would like to thank H. Giesen, J. Gloudemans, N. Kiyani, A. Sbai, G. Squillace, IMEC ULP DSP / RADIO teams for their contributions to this work.

References:
[1] Y.-H. Liu, *et al.*, "A 1.9nJ/b 2.4GHz multistandard (Bluetooth Low Energy/Zigbee/IEEE802.15.6) transceiver for personal/body-area networks," *ISSCC Dig. Tech. Papers*, pp. 446-447, 2013.
[2] M. Konijnenburg, *et al.*, "Reliable and energy-efficient 1MHz 0.4V dynamically reconfigurable SoC for ExG applications in 40nm LP CMOS," *ISSCC Dig. Tech. Papers*, pp. 430-431, 2013.
[3] W. Kluge, *et al.*, "A Fully Integrated 2.4-GHz IEEE 802.15.4-Compliant Transceiver for Zigbee Applications", *IEEE J. of Solid-State Circuits*, vol. 41, no. 12, pp. 2767-2775, 2006.
[4] G. Retz, *et al.*, "A Highly Integrated Low-Power 2.4GHz Transceiver Using a Direct-Conversion Diversity Receiver in 0.18µm CMOS for IEEE802.15.4 WPAN," *ISSCC Dig. Tech. Papers*, pp. 414-415, 2009.
[5] A. Wang, *et al.*, "A 1V 5mA Multimode IEEE 802.15.6/Bluetooth Low-Energy WBAN Transceiver for Biotelemetry Applications," *ISSCC Dig. Tech Papers*, pp. 300-301, 2012.
[6] Nordic Semi., "nRF8001 Product Spec.-Bluetooth 4.0," Jan. 2012.

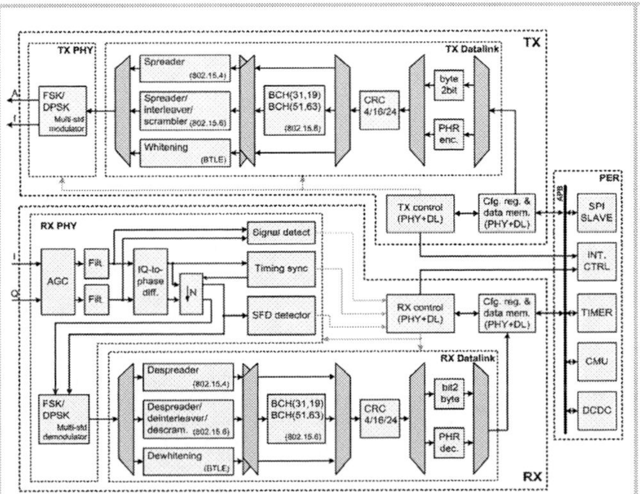

Figure 10.6.1: Multi-standard TRX DBB architecture.

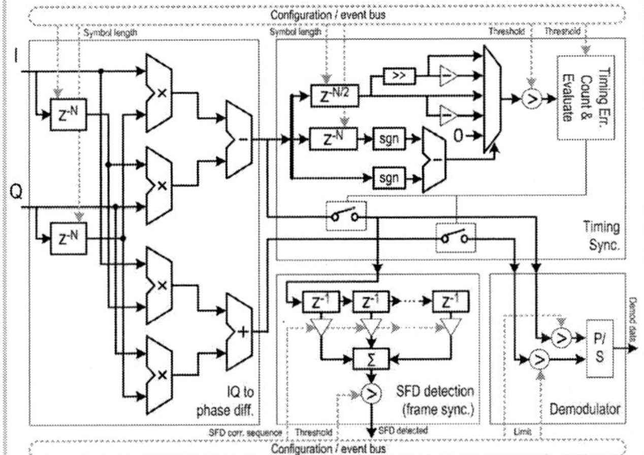

Figure 10.6.2: RX PHY: Synchronization and demodulation.

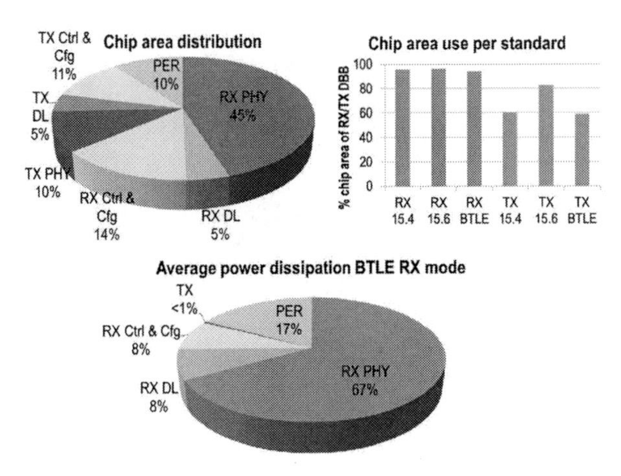

Figure 10.6.3: Chip area and RX PHY power distribution.

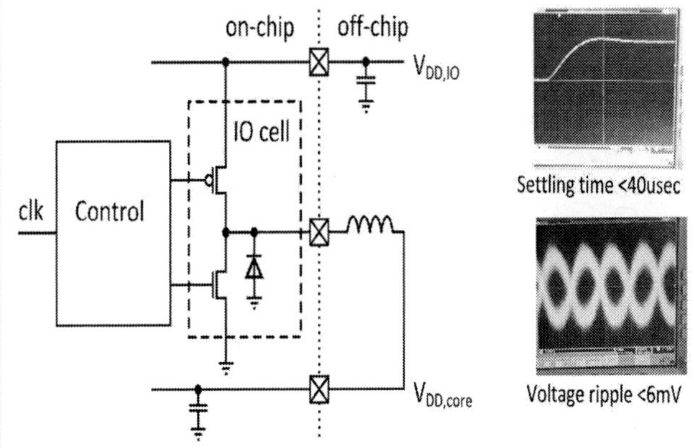

Figure 10.6.4: All-digital on-chip DC-DC converter.

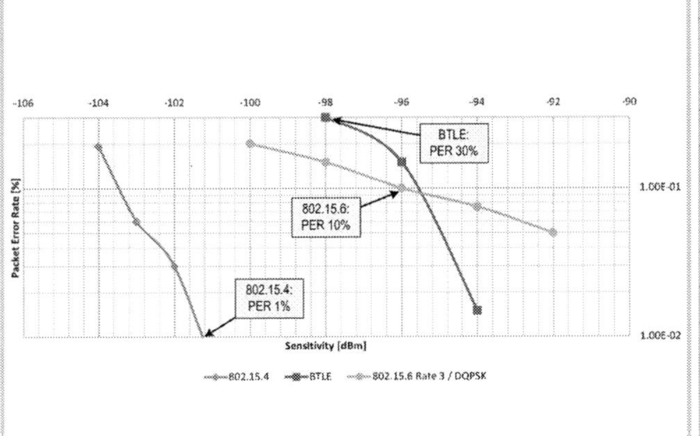

Figure 10.6.5: Measured packet error rates with RFFE [1].

	This work (DBB) and RFFE [1]			[3]	[4]	[5]		[6]
Technology	40nm (DBB) and 90nm (RFFE [1])			0.18μm	0.18μm	0.13μm		NA
Standard	BTLE	802.15.4	802.15.6	802.15.4	802.15.4	BTLE	802.15.6	BTLE
Modulation	GFSK	HS-OQPSK	pi/4-DQPSK	HS-OQPSK	HS-OQPSK	GFSK	pi/4-DQPSK	GFSK
Symbol rate [kS/s]	1000	1000	600	1000	1000	1000	600	1000
Data rate [kb/s]	1000	250	971.4	1000	250	1000	971.4	1000
RX sensitivity* [dBm]	-98	-101	-96	-101	-96	-94	-96.5	-86
RX DBB / digital power diss. [μW]	180	200	140	6660 (~ 250)**	5760 (~ 216)**	NA	NA	NA
TX DBB / digital power diss. [μW]	80	80	60	5040 (~ 189)**	4140 (~ 155)**	NA	NA	NA
RX RFFE power diss. [mW]	3.80	3.80	3.80	19.8	24.48	6.50	6.50	NA
TX RFFE power diss. [mW]	5.40	5.40	4.60	23.22	28.26	8.90	5.90	NA
RX total power diss. [mW]	3.98	4.00	3.94	26.46 (~ 20.05)**	30.24 (~ 24.70)**	NA	NA	21.00
TX total power diss. [mW]	5.48	5.48	4.66	28.26 (~ 23.41)**	32.40 (~ 28.42)**	NA	NA	17.80
RX energy eff. [nJ/b]	4.0	16.0	4.1	105.8 (~ 80.2)**	121.0 (~ 98.8)**	6.5***	6.7***	21.0

* Sensitivity definition: BER 10^{-3} (PER 30%) for BTLE, PER 1% for 802.15.4, and PER 10% for 802.15.6
** Considering geometric (0.18μm to 40nm) and supply voltage (1.8V to 0.74V) scaling of digital hardware blocks for comparison of power dissipation.
*** Without DBB.

Figure 10.6.6: Performance and power dissipation comparison.

978-1-4799-0917-9/14 $31.00 © 2014 IEEE

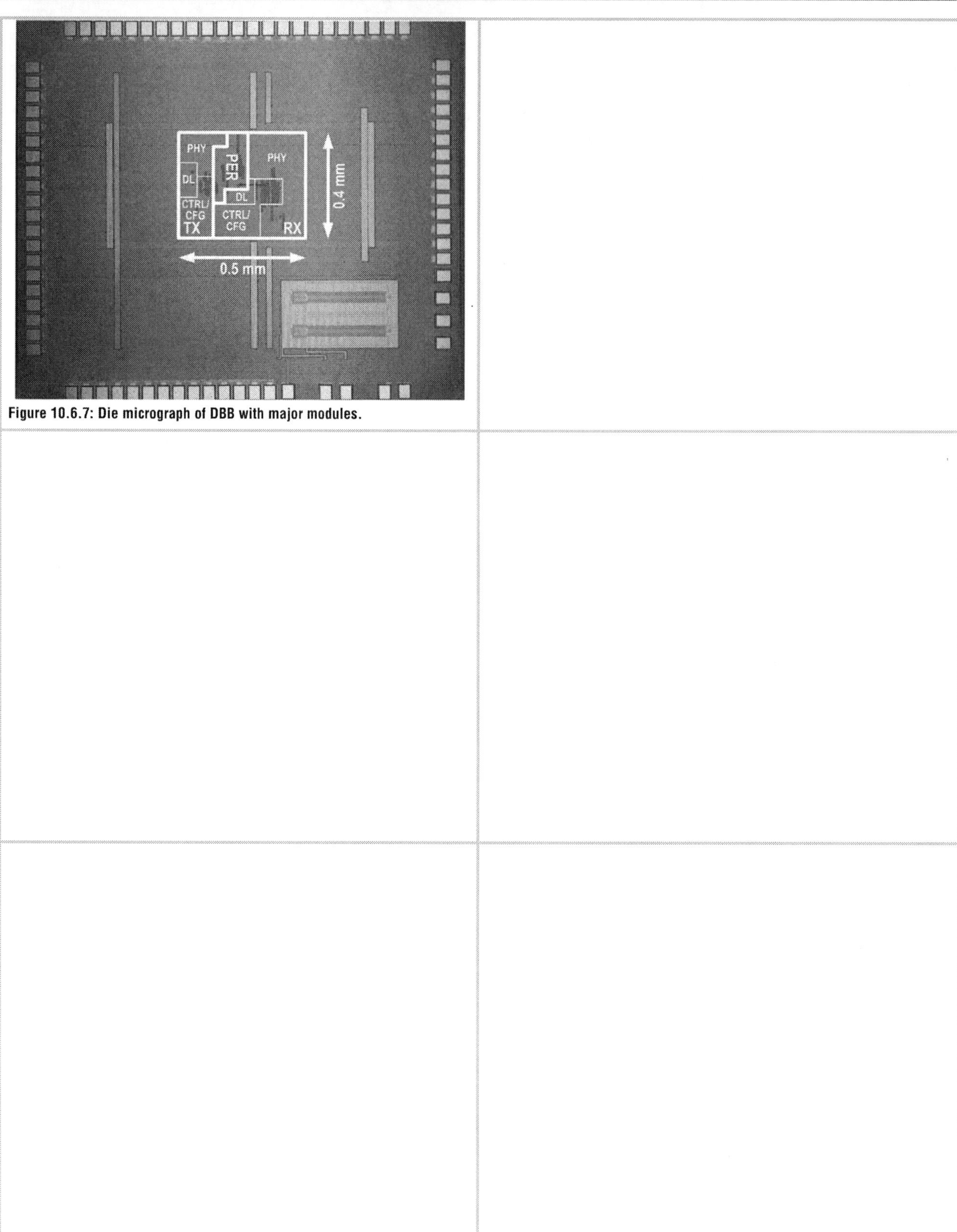

Figure 10.6.7: Die micrograph of DBB with major modules.

10.7 A 105GOPS 36mm² Heterogeneous SDR MPSoC with Energy-Aware Dynamic Scheduling and Iterative Detection-Decoding for 4G in 65nm CMOS

Benedikt Noethen, Oliver Arnold, Esther Pérez Adeva, Tobias Seifert, Erik Fischer, Steffen Kunze, Emil Matúš, Gerhard Fettweis, Holger Eisenreich, Georg Ellguth, Stephan Hartmann, Sebastian Höppner, Stefan Schiefer, Jens-Uwe Schlüßler, Stefan Scholze, Dennis Walter, René Schüffny

Technische Universität Dresden, Dresden, Germany

Modern mobile communication systems face conflicting design constraints. On the one hand, the expanding variety of transmission modes calls for highly flexible solutions supporting the ever-growing number and diversity of application requirements. On the other hand, stringent power restrictions (e.g., at femto base stations and terminals) must be considered, while satisfying the demanding performance requirements. In order to cope with these issues, existing SDR platforms, e.g. [1-2], propose an MPSoC with a heterogeneous array of processing elements (PEs). MPSoC solutions provide programmability and parallelism yielding flexibility, processing performance and power efficiency. To schedule the resources and to apply power gating, a static approach is employed. In contrast, we present a heterogeneous MPSoC platform (*Tomahawk2*) with runtime scheduling and fine-grained hierarchical power management. This solution can fully adapt to the dynamically varying workload and semi-deterministic behavior in modern concurrent wireless applications. The proposed dynamic scheduler (*CoreManager*, CM) can be implemented either in software on a general-purpose processor or on a dedicated application-specific hardware unit. It is evident that the software approach offers the highest degree of flexibility; however, it may become a performance-bottleneck for complex applications. A high-throughput ASIC was presented in [3], but this solution does not permit scheduling algorithms to be adjusted. In this work, these limitations are overcome by implementing the CM on an ASIP.

The *Tomahawk2* MPSoC is composed of 20 heterogeneous cores, connected by a hierarchical packet-switched star-mesh NoC, as depicted in Fig. 10.7.1. The NoC is clocked at 500MHz and provides a throughput of 80Gb/s per link, partly employing serial high-speed links [4]. This results in a compact top-level floorplan realization (Fig. 10.7.7). An ADPLL is attached to each unit, allowing individual adjustment of the clock frequency within the 83-666MHz range. The DDR2 interface connecting 2×128MB global memory at 400MHz provides a data rate of 12.8Gb/s. An FPGA I/O interface delivers 10Gb/s. The application processor (APP) is implemented as a Tensilica 570T RISC core with 16kB data and 16kB instruction caches. It executes application control code and sends task scheduling requests to the CM. The CM is based on a Tensilica LX4 core extended with a scheduling-specific instruction set [7], enabling efficient implementation of adaptive power management and dynamic task scheduling (including resource allocation, data-dependency checking and data management). For this purpose, the CM analyzes at runtime the scheduling requests and exploits the results to maximize data locality and to configure the dynamic voltage and frequency scaling (DVFS) performance levels of the PEs according to current system load, priorities and deadlines.

The Duo-PE is comprised of a vector DSP and a RISC core, connected to a shared local memory. This arrangement increases area efficiency and data locality. Each Duo-PE is equipped with a DMA unit, enabling concurrent data prefetching and task execution. The dual nature of these PEs allows high-performance 16b fixed-point signal processing on the 4-fold SIMD VDSP, as well as high-precision floating-point computing on the Tensilica LX4 RISC processor. To support fine-grained fast power management, each Duo-PE is equipped with DVFS (Fig. 10.7.3). The core domain is connected to one of 3 global V_{DD} rails by a set of PMOS switches. A LUT contains a set of fractional frequency multipliers (N_1, N_2 and N_3) for the ADPLL, associated with the 3 supply levels. The power rails are controlled by an AVS scheme, which tracks temperature variations and finds the minimum V_{DD} guaranteeing error-free operation for the selected performance level. Therefore, each PE contains 3 ring-oscillator hardware performance monitors (HPMs) replicating the critical timing of the design by its oscillation period T_{HPM}. The central AVS controller adjusts V_{DD} such that the oscillator period equals N times the reference period ($T_{HPM}=N*T_{REF}$). By configuring the HPMs with N_1, N_2 and N_3 multipliers, respectively, the DVFS target frequencies can be emulated. The voltages of all three V_{DD} rails can be hence slowly regulated by the AVS while the core is running at only one DVFS level, with the capability of rapid change to another level.

In order to accelerate computationally intensive SDR baseband algorithms, two programmable application-specific cores are included: specifically, a sphere detection (SD) core and a multi-mode forward error correction (FEC) core for convolutional, Turbo, and LDPC codes [6]. In the particular case of coded systems, the communications performance can be significantly enhanced by means of iterative detection-decoding. For this purpose, the SD architecture presented in [6] has been largely extended by implementing the algorithms proposed in [5], allowing the SD to process a-priori information generated by the FEC.

The Tomahawk2 SDR MPSoC was fabricated in TSMC 65nm LP-CMOS technology (Fig. 10.7.7). It integrates 10.2M NAND gate equivalents and occupies 6mm×6mm = 36mm² (Fig. 10.7.6). The MIMO iterative detection-decoding engine occupies 1.68mm², including 93kB of SRAM. Measurement results corresponding to SD and FEC modules can be found in Fig. 10.7.4. Each Duo-PE occupies 1.36mm², of which 0.8mm² is contributed by the two dual-port 32kB memories. The RISC core achieves a maximum frequency of 445MHz at 1.2V. For this configuration, 7.1GOPS are delivered by 8 cores. The VDSP reaches a maximum frequency of 500MHz at 1.2V, which yields a performance of 80GOPS for all 8 PEs. Executing a 2048-point complex FFT under these conditions, the VDSP and the RISC consume 98.1mW and 61mW, respectively. Due to its higher throughput for this application, the frequency and voltage of the VDSP can be downscaled, thus doubling the energy efficiency in comparison to the RISC.

The CM occupies 1.36mm², including 64KB memory for data and 32KB for instructions. It reaches a maximum frequency of 445MHz at 1.2V, resulting in scheduling-throughput of 1.1Mtasks/s with a power dissipation of 69.2mW. To show the advantage of the hardware-accelerated scheduler, different CM implementations are compared in Fig. 10.7.2.

The presented heterogeneous SDR MPSoC integrates 8 Duo-PEs and an iterative detection-decoding engine. It features a hierarchical power management (at system and PE levels) combined with flexible dynamic task scheduling. System-level power management is integrated in a programmable CM-ASIP which out-performs other scheduler implementations by a factor of 7 (ASIC) and 216 (SW) in terms of area-time-energy (ATE) product (Fig. 10.7.2). At the PE-level, the ultra-fast DVFS follows the dynamically adapting control of the CM, to further increase the energy efficiency. The flexible iterative multi-mode SD-FEC engine improves area-performance, as well as the energy efficiency by a factor of 3 vs. recent implementations (Fig. 10.7.4). This SDR MPSoC doubles the processing power compared to other SDR solutions (Fig. 10.7.5), enabling support of future communications standards. For a 4×4 MIMO 3GPP-LTE baseband application scenario, the throughput is increased by nearly a factor of 6 at the same power consumption (Fig. 10.7.5).

Acknowledgements:
This work has been supported by the German Ministry of Education and Research BMBF under grant number 13N10788 (CoolBaseStation), the state of Saxony under grant of ESF 100098198 and the German Research Foundation within cfaed. Furthermore, we would like to thank Synopsys and Tensilica for software and IP.

References:

[1] F. Clermidy, *et al.*, "A 477mW NoC-based digital baseband for MIMO 4G SDR," *ISSCC Dig. Tech. Papers*, pp. 278-279, 2010.

[2] D. Ilitzky, *et al.*, "Architecture of the Scalable Communications Core's Network on Chip," *IEEE Micro*, vol. 27, no. 5, pp. 62-74, 2007.

[3] T. Limberg, *et al.*, "A Fully Programmable 40 GOPS SDR Single Chip Baseband for LTE/WiMAX Terminals", *European Solid-State Circuits Conf.*, pp. 466-469, 2008.

[4] D. Walter, *et al.*, "A Source-Synchronous 90Gb/s Capacitively Driven Serial On-Chip Link Over 6mm in 65nm CMOS," *ISSCC Dig. Tech. Papers*, pp. 180-181, 2012.

[5] E. P. Adeva, *et al.*, "VLSI Architecture for MIMO Soft-Input Soft-Output Sphere Detection", *J. of Signal Processing Systems*, vol. 70, is. 2, pp. 125-143, 2013.

[6] M. Winter, *et al.*, "A 335Mb/s 3.9mm² 65nm CMOS Flexible MIMO Detection-Decoding Engine Achieving 4G Wireless Data Rates," *ISSCC Dig. Tech. Papers*, pp. 216-218, 2012.

[7] O. Arnold, *et al.*, "Instruction Set Architecture Extensions for a Dynamic Task Scheduling Unit", *IEEE Symp. VLSI Circuits*, pp. 249-254, 2012.

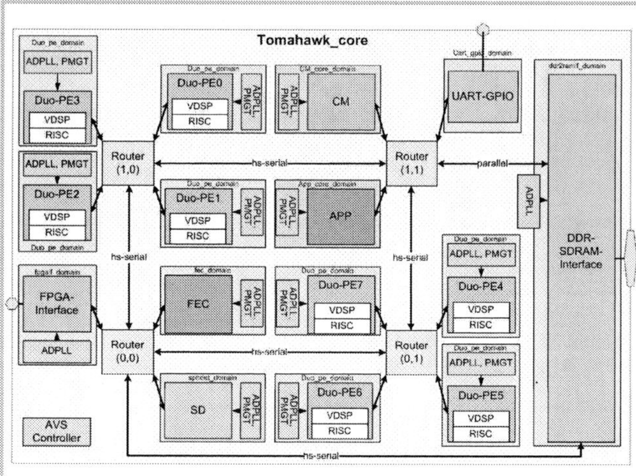

Figure 10.7.1: Block diagram of the Tomahawk2 MPSoC.

	ASIC [3]	RISC[a]	ASIP (this work)
Scheduling Algorithm	As-soon-as-possible list-based		
Scheduling Configurability	Fixed	Flexible	Flexible
Task Queue size	16	16-256	16-256
Supported I/O transfers	8	12	12
Frequency [MHz]	175	445	445
Task scheduling [cycle]	60	7533	402
Task scheduling [us]	0.4	16.9	0.9
Technology [nm]	130	65	65
Supply Voltage [V]	1.3	1.2	1.2
Power Consumption [mW] @fmax	282	68	74.6
Energy per Task [nJ] @ fmax	113	1149	67
Area (logic) [mm²]	4.51	0.34[b]	0.49
ATE product [mm²*us*nJ]	204	6602	29

[a] Measured on the Tomahawk2 (this work) without application-specific instruction set
[b] Based on synthesis with Synopsys Design Compiler for 65nm LP TSMC process

Figure 10.7.2: Comparison of CoreManager realizations.

Figure 10.7.3: Combined power management, DVFS and AVS.

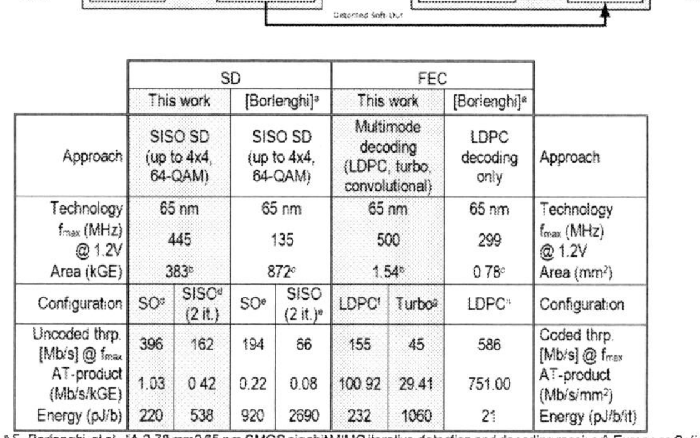

	SD				FEC			
	This work		[Borlenghi][a]		This work		[Borlenghi][a]	
Approach	SISO SD (up to 4x4, 64-QAM)		SISO SD (up to 4x4, 64-QAM)		Multimode decoding (LDPC, turbo, convolutional)		LDPC decoding only	Approach
Technology	65 nm		65 nm		65 nm		65 nm	Technology
f_{max} (MHz) @ 1.2V	445		135		500		299	f_{max} (MHz) @ 1.2V
Area (kGE)	383[b]		872[c]		1.54[b]		0.78[c]	Area (mm²)
Configuration	SO[d]	SISO[d] (2 it.)	SO[e]	SISO (2 it.)[e]	LDPC[f]	Turbo[g]	LDPC[h]	Configuration
Uncoded thrp. [Mb/s] @ f_{max}	396	162	194	66	155	45	586	Coded thrp. [Mb/s] @ f_{max}
AT-product (Mb/s/kGE)	1.03	0.42	0.22	0.08	100.92	29.41	751.00	AT-product (Mb/s/mm²)
Energy (pJ/b)	220	538	920	2690	232	1060	21	Energy (pJ/b/it)

[a] F. Borlenghi, et al., "A 2.78 mm2 65 nm CMOS gigabit MIMO iterative detection and decoding receiver", European Solid-State Circuits Conference, pp. 65- 68, 2012
[b] Memories for data exchange between SD and FEC included. [c] Memories for data exchange between SD and FEC excluded.
[d] 4x4 MIMO, 64-QAM, at 10^{-5} BER. [e] 4x4 MIMO, 64-QAM, at 1% BLER. [f] 768b code block, 3/4 rate, 10 iterations. [g] 1028b code block, 1/3 rate, 6 iterations. [h] 1944b code block, 5/6 rate, 10 iterations.

Figure 10.7.4: Comparison of detection-decoding engines.

	Tomahawk2	Magali [1]	Tomahawk [3]	Intel SCC [2]
Platform Scope	MIMO 3GPP-LTE, WiMAX, 802.11n, SDR	MIMO 3GPP-LTE, WiMAX, 802.11n, SDR, Cognitive Radio	MIMO 3GPP-LTE, WiMAX, H264, SDR	SISO WiMAX, 802.11-04, SDR
Clocking and Power Management	GALS, local DVFS and AVS, power-gating	GALS, local DFS	Global frequency	Global frequency
Scheduling	Dynamic (flexible, energy adaptable algorithm)	Static	Dynamic (fixed algorithm)	Static
Memory Organization	Distributed and shared	Distributed	Distributed and shared	Local
Peak Performance	105 GOPS (3.6 GFLOPS)	37 GOPS	40 GOPS	-
Application for power measurements	4x4 MIMO 3GPP-LTE Rx baseband, 60 Mbit/s	4x2 MIMO 3GPP-LTE Rx, 2x2 MIMO Tx, MAC, 10.8 Mbit/s	LTE/WiMAX	-
Power consumption	480 mW @1.15V[a]	477 mW @ 1.2V	1.2W @ 1.3V	-
NoC Throughput (per link)	80 Gbit/s	17 Gbit/s	5.47 Gbit/s	8Gbit/s
Die size	36mm²	29.6 mm²	100 mm²	25mm²
Technology	65nm	65nm	130nm	65nm

[a] See application scenario power measurement results in Fig. 6

Figure 10.7.5: Comparison of state-of-the-art SDR chips.

		Area [mm2] total	Area [mm2] mem	Mem size [bit]	f_{max} [MHz] @VDD =1.2 V	Throughput @f_{max}	P [mW] @f_{max}, VDD=1.2 V	$P_{Application-Scenario}$ [mW]
APP		0.582	0.245	274432	445	890 MOPS	49.7	off
CM		1.360	0.870	786432	445	1.1 MTasks/s	74.6	14.1 @200MHz, 0.9 V
Duo -PE[a]	Xtensa	1.357	0.800	532480	445	890 MOPS	61.5	off
	VDSP				500	10 GOPS	98.1	35.0 @282 MHz, 0.9 V
SD		0.522	0.260	292864	445	396 Mb/s	87.0	36.5 @200 MHz, 1.15 V
FEC		1.154	0.618	479232	500	155 Mb/s	360.0	132.2 @200 MHz, 1.15 V
FPGA-IF		0.602	-	-	500	10 Gb/s	-	-
DDR-IF		4.552	-	-	400	12.8 Gb/s	-	-
NoC		3.417	-	-	500	80 Gb/s[b]	32.0[c]	18.0 @286 MHz, 1.15 V

[a] Per instance. [b] Per link, max throughput. [c] Averaged over several test cases, NoC not fully utilized

Figure 10.7.6: Tomahawk2 components performance summary.

10

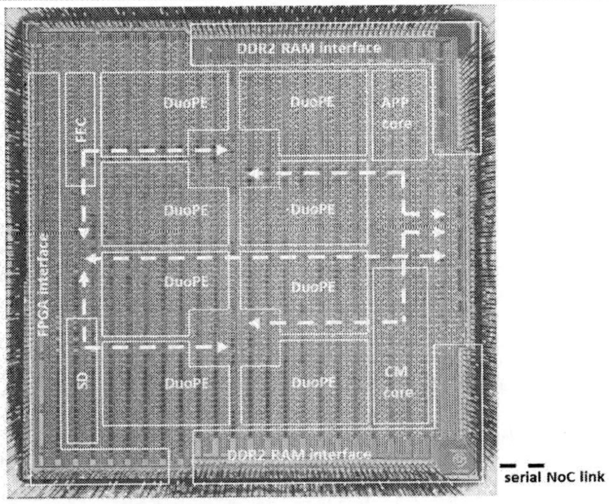

Figure 10.7.7: Tomahawk2 MPSoC die photo.

10.8 A Multi-Standard 2G/3G/4G Cellular Modem Supporting Carrier Aggregation in 28nm CMOS

Michael Breschel[1], Peter Almers[1], Fredrik Angsmark[1], Alberth Arvidsson[1], Harald Bauer[2], Kees van Berkel[3], Joaquin Canovas[1], Minh Do[1], Anders Ekelund[1], Torsten Larsson[1], Bo Lincoln[1], Magnus Malmberg[1], Masao Naruse[4], Masashi Onishi[4], Christer Östberg[1], Jean-Paul Smeets[3], Mario Vergara Escobar[1], Juergen Voelkl[2], Emma Wittenmark[1]

[1]Ericsson, Lund, Sweden, [2]Ericsson, Nuremberg, Germany,
[3]Ericsson, Eindhoven, The Netherlands, [4]Ericsson, Yokohama, Japan

Mobile networks today are divided into multiple radio access technologies (RATs) scattered over a variety of frequencies and functionality depending on the network region. The scattered networks require that the digital baseband for mobile user equipment handle multiple RATs, multiple bands, as well as seamlessly transition between these. In 3GPP release 10 [1] the problem with scattered frequency bands has been addressed by the possibility to aggregate spectrum from two separated carriers to create a wider aggregated total bandwidth. Which carriers to combine depends on the spectrum available to the specific operator.

The DB7450R system-in-package (SiP) includes three separate dies for the digital baseband (DBB), power management unit (PMU) and a LPDDR2 SDRAM. The DBB is implemented in 28nm CMOS technology. In this paper, we focus on the modem sub-system (MSS) (Fig. 10.8.1). The MSS supports multiple RATs within a unified HW and SW architecture, using a single radio IC RF7450. The supported RATs include:

- TDD/FDD LTE category 4 (up to 150 Mb/s), LTE carrier aggregation up to 20 MHz,
- FDD HSDPA category 28 (up to 84Mb/s), FDD HSUPA category 8,
- TDD HSDPA category 15, TDD HSUPA category 6,
- GSM, EGPRS2-A, and EGPRS multi-class 33.

The MSS architecture (Fig. 10.8.2) is built around an on-chip common memory (CM), allowing flexible memory allocation per RAT and per mode, as well as a flexible data flow. The CM comprises 2×16 SRAM banks that are 16b-word interleaved at 260MHz. This offers a predictable bandwidth and latency. The aggregate bandwidth per second to the connected processors and accelerators is 2×16×260MHz×16b. The added latency associated with bank interleaving can be afforded, because the CM is primarily used for large batch-oriented accesses. Two ARM™ Cortex-R4 control CPUs and an embedded vector processor (EVP, see below) can access the CM via a bridge. Some of the HW accelerators connected to the CM are RAT specific (e.g. a RAKE despreader for WCDMA), and others are RAT generic (e.g., a turbo decoder, a trace and debug unit, and an RF interface).

All processors and accelerators attached to the CM are also connected by a ring bus (RB) – a common multi-master interconnect to exchange messages between the hardware units. RB messages have a fixed length (128b), comprising a header (type and unit address) and a payload (typically specifying a task to be executed and a pointer to the CM). A local configurable-control interface contains the RB interface and one or more mailboxes for the messages. The control interface governs the execution of the task according to a list of messages stored in the CM. Besides setting control parameters and initiating the hardware function, the messages can be forwarded to other clients on the RB. This enables the chaining of multiple hardware processing tasks, as well as software processing tasks (e.g., on the EVP) without any control-software steps in between. All MSS infrastructure (CM, RB, etc.) is shared among the RATs, as are the programmable processors and many of the hardware accelerators. Compared to a multi-RAT modem architecture with dedicated sub-systems for GSM, WCDMA, TD-SCDMA, and LTE (e.g. [2]), this saves die area. Furthermore, the uniform and scalable architecture of Fig. 10.8.2 allows a flexible hardware-software partitioning: it is easy to move functional units from software to hardware (saving power) or vice versa (reducing cost). This enables hardware evolution from one generation to the next with minor impact on the software framework. For example, addition of a third carrier merely requires the introduction of an extra RF interface block in the architecture of Fig. 10.8.2, and the re-scheduling of the signal-processing tasks on the accelerators and processors.

The EVP is a programmable DSP supporting up to 100 operations per clock cycle at a rate of 416MHz. This high throughput is the result of executing multiple single-instruction-multiple-data (SIMD) operations (256b wide), as well as multiple scalar operations in a very-long instruction word (VLIW) fashion, as illustrated in Fig. 10.8.3. Compared to [3], the current EVP not only operates on fixed-point data, but also on floating-point data. The 32+4 floating-point multipliers support 8+1 complex multiply-accumulate operations every clock cycle, whereas other SIMD DSPs proposed for handsets such as [4] support fixed-point only. This EVP is power efficient (below 0.5mW/MHz, including program and data memory), and supports a large diversity of algorithms, as well as enabling algorithm improvement after tape-out.

The DB7450 supports a range of power management features [5, 6]. In sleep/flight mode only the *always-on* domain is powered, corresponding to less than 1% of the die area and running at only 1MHz and 32kHz. In active modes, the entire circuit is powered at a minimal duty cycle and transitions between the sleep and active modes are fully hardware controlled. A separate voltage is supplied to the RF interface, but only during data transfers to and from the RF IC. In active mode, dynamic voltage and frequency scaling (DVFS) is applied: when performance requirements pass pre-defined trigger levels, clocks and supply voltages are adjusted accordingly under hardware control. The switching can be controlled directly by a timer that is aligned with the RAT timing so that performance can be periodically increased during a small part of each sub frame. This allows critical deadlines to be met, while at the same time minimizing power consumption for scenarios with low average load. A few selected memories have retention capabilities to avoid save and restore operations in transitions to and from sleep-mode. During retention, the SRAMs are powered at 0.65V by on-chip bias generators. Throughout the circuit an advanced and extensive clock gating methodology is applied. The clock gating is done in four stages; 1) at the PLLs output, 2) in the central clock control unit, 3) at the block interface, 4) to specific functions within blocks. Clock enables are hardware controlled and handled locally by the blocks. The on-chip interconnect is segmented, so that only the active segments are clocked. Furthermore, adaptive voltage scaling (AVS) is applied on each individual device, thereby adjusting the voltage supply level based on the process characteristics of the specific device.

Figure 10.8.5 provides the measured power consumption levels in mW of the DBB (Fig. 10.8.1) for six critical use cases. The temperature is approximately 25°C, and the supply voltage is 1.03/0.90V. In Fig. 10.8.6, the M7450 is bench-marked vs. similar offerings. The table reflects the consensus in the industry on the 2014 requirements for smartphone and tablet modems.

With a HW/SW architecture that is RAT-unified, scalable, and programmable where it counts, the MSS is prepared for the future evolution of 3GPP standards, including carrier aggregation for more and wider carriers and heterogeneous networks (HetNet), and also for more advanced receiver algorithms for improved link performance.

References:

[1] 3GPP TR 36.808: Evolved Universal Terrestrial Radio Access (E-UTRA); Carrier Aggregation; Base Station (BS) Radio Transmission and Reception.

[2] G. Bublitz, *et al.*, "Power Management Challenges in Wireless WAN SoCs", *Hot Chips HC25*, 2013.

[3] K. van Berkel, *et al.*, "Vector Processing as an Enabler for Software-Defined Radio in Handheld Devices", *EURASIP J. on Applied Signal Processing*, issue 16, pp. 2613-2625, 2005.

[4] C. Rowen, *et al.*, "A DSP Architecture Optimized for Wireless Baseband", *IEEE System-on-Chip Conf.*, pp. 151-156, 2009.

[5] V. Venkatachalam and M. Franz, "Power Reduction Techniques for Microprocessor Systems", *ACM Computing Surveys*, vol. 37, no. 3. pp. 195-237, 2005.

[6] T. Sakurai, *et al.*, "Power Gating: Circuits, Design Methodologies, and Best Practice for Standard-Cell VLSI Designs", *ACM Trans. on Design Automation of Electronic Systems*, vol. 15, no. 4, article 28, 2010.

978-1-4799-0917-9/14 $31.00 © 2014 IEEE

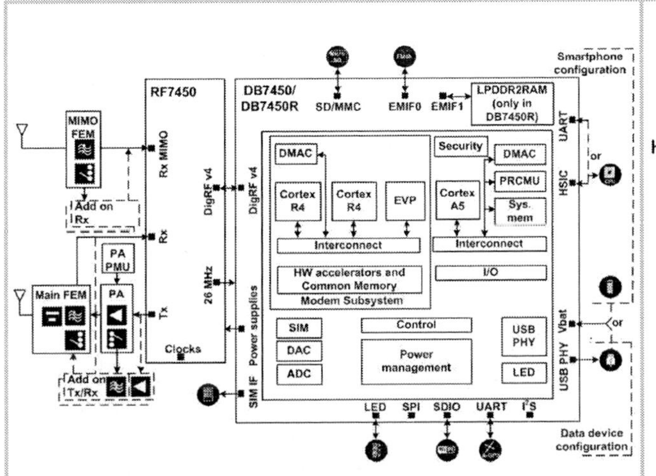

Figure 10.8.1: DB7450R baseband IC in context.

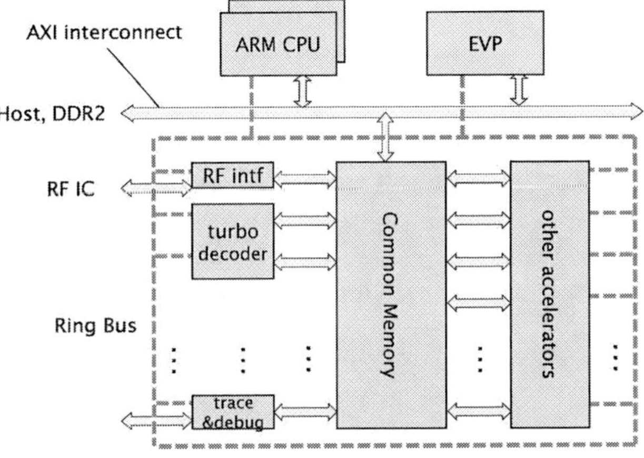

Figure 10.8.2: MSS hardware architecture.

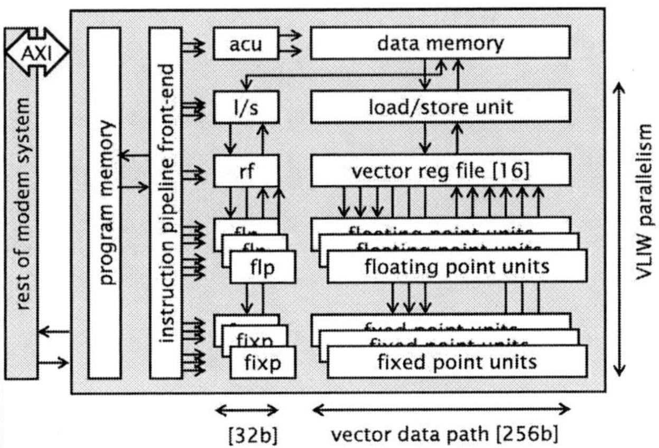

Figure 10.8.3: EVP architecture.

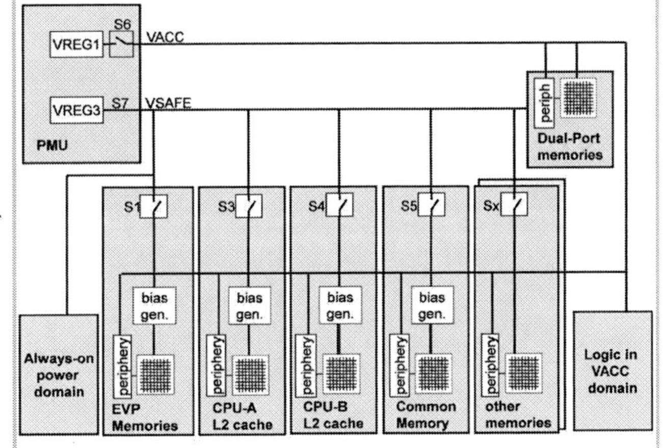

Figure 10.8.4: DB7450/MSS power architecture.

RAT	use case	power consumption [mW]
2G	GSM idle DRX9	1.0
3G	HSDPA cat 14; downlink: 21Mbps	183
LTE	Cat 3; 10MHz; downlink: 63Mbps; uplink: 25Mbps	321
	Cat 3; 20MHz; downlink: 100Mbps; uplink: 50Mbps	400
	Cat 4; 20MHz; downlink: 150Mbps; uplink: 50Mbps	430
	Cat 4; 10+10MHz; downlink: 150Mbps; uplink: 25Mbps	438

Figure 10.8.5: DBB power consumption (mW).

	Ericsson M7450	Competitor A	Competitor B	Competitor C
Availability	2012	2012	2013	2013
LTE FDD/TDD	Cat4	Cat 4	Cat 4	Cat4
LTE carrier aggregation	√	√	√	√
LTE VoLTE	√	√	√	√
FDD-HSPA+	√	√	√	√
TDD-HSPA+	√	√		
GSM/GPRS/EDGE	√	√	√	√
CMOS [nm]	28	28	28	28
#RF ICs	1	2	?	?

Figure 10.8.6: Benchmarking of M7450.

Figure 10.8.7: Micrograph of DB7450R.

Session 11 Overview: *Data Converter Techniques*

DATA CONVERTERS SUBCOMMITTEE

Session Chair: *Jan Mulder*
Broadcom, Bunnik, The Netherlands

Session Co-Chair: *Stéphane Le Tual*
STMicroelectronics, Crolles, France

Power efficiency and high dynamic range are of crucial importance for contemporary data converters, which are essential in a wide range of demanding applications, such as medical, sensor, and advanced wireless mobile systems. Circuits and architectures tailored to low-voltage deep-submicron CMOS technologies, data-driven conversion and optimized calibration algorithms are presented in this session. These innovative data converter techniques and algorithms are advancing the state of the art in dynamic range and power efficiency.

11.1 An Oversampled 12/14b SAR ADC with Noise Reduction and Linearity Enhancements Achieving up to 79.1dB SNDR **8:30 AM**

P. Harpe, Eindhoven University of Technology, Eindhoven, The Netherlands

In Paper 11.1, Eindhoven University of Technology uses chopping, dithering, oversampling and data-driven noise reduction to obtain a power-efficient SAR ADC that can be configured for 12b or 14b operation. Implemented in 65nm CMOS, the converter achieves an SNDR up to 79.1dB in a 4kHz bandwidth.

11.2 A 0.85fJ/conversion-step 10b 200kS/s Subranging SAR ADC in 40nm CMOS **9:00 AM**

H-Y. Tai, National Taiwan University, Taipei, Taiwan

In Paper 11.2, National Taiwan University reaches a record efficiency of 0.85fJ/conversion-step for a 200kS/s 10b ADC in 40nm CMOS. Combining subranging and SAR architectures, the DAC switching energy and comparator power consumption can be reduced significantly, while achieving 55.6dB SNDR.

978-1-4799-0917-9/14 $31.00 © 2014 IEEE

11.3 A 10b 0.6nW SAR ADC with Data-Dependent Energy Savings Using LSB-First Successive Approximation **9:30 AM**

F. M. Yaul, Massachusetts Institute of Technology, Cambridge, MA

In Paper 11.3, Massachusetts Institute of Technology presents a 10b 0.5V SAR ADC in 0.18µm CMOS exploiting the characteristics of ECG input signals to save energy. Using an LSB-first algorithm, a power efficiency down to 2.9fJ/conversion-step is achieved for a low-activity input signal.

11.4 A 1.5mW 68dB SNDR 80MS/s 2× Interleaved SAR-Assisted Pipelined ADC in 28nm CMOS **10:15 AM**

F. van der Goes, Broadcom, Bunnik, The Netherlands

In Paper 11.4, Broadcom describes a 28nm CMOS pipelined SAR ADC, which features a dynamic residue amplifier with noise-filtering properties. At 80MS/s, the converter achieves an SNDR of 68dB, while consuming only 1.5mW.

11.5 A 100MS/s 10.5b 2.46mW Comparator-less Pipeline ADC Using Self-Biased Ring Amplifiers **10:45 AM**

Y. Lim, University of Michigan, Ann Arbor, MI and Samsung Electronics, Yongin, Korea

In Paper 11.5, the University of Michigan introduces a comparator-less pipeline ADC using a self-biased ring amplifier in 65nm CMOS. It presents 56.3dB SNDR at 100MS/s for 2.46mW power consumption.

11.6 A 21mW 15b 48MS/s Zero-Crossing Pipeline ADC in 0.13µm CMOS with 74dB SNDR **11:15 AM**

D-Y. Chang, Maxim Integrated, San Jose, CA

In Paper 11.6, Maxim Integrated (with MIT) uses a dual-ramp zero-crossing-based circuit in a 0.13µm CMOS pipelined ADC to reach 74dB SNDR and 95dB linearity at 48MS/s.

11.7 A 240mW 16b 3.2GS/s DAC in 65nm CMOS with <-80dBc IM3 up to 600MHz **11:45 AM**

H. van de Vel, Integrated Device Technology, Eindhoven, The Netherlands

In Paper 11.7, Integrated Device Technology uses a 3D calibration algorithm to minimize amplitude, delay and duty-cycle mismatches in a 16b 3.2GS/s DAC fabricated in 65nm CMOS. Consuming 240mW, the converter achieves a linearity better than -80dB up to 600MHz.

11.1 An Oversampled 12/14b SAR ADC with Noise Reduction and Linearity Enhancements Achieving up to 79.1dB SNDR

Pieter Harpe, Eugenio Cantatore, Arthur van Roermund

Eindhoven University of Technology, Eindhoven, The Netherlands

Autonomous wireless sensor nodes for cloud networks require ultra-low-power electronics. In particular, sensor readout interfaces need low-speed high-precision ADCs for capturing, e.g., bio-potential signals, environmental information, or interactive multimedia. For these applications, state-of-the-art SAR ADCs can provide highly power-efficient solutions (<10fJ/conversion-step) but with limited accuracy (SNDR <63dB) [1,2]. Alternatively, ΔΣ ADCs offer higher precision at the cost of lower efficiency (e.g. 84dB SNDR with 54fJ/conversion-step [3]). This work bridges the existing performance gap by extending the accuracy of low-power SAR ADCs to SNDRs in the order of 70-to-80dB. Feedback-controlled data-driven noise reduction [1], oversampling, chopping [4] and dithering [5] techniques are combined to increase both SNR and linearity in a power-efficient way. Various ADC modes are supported by making these techniques individually programmable, thereby extending the application range.

The 12b/14b SAR ADC is shown in Fig. 11.1.1. The asynchronous architecture allows the use of a single external clock at the sample-rate frequency. While nominally 14b, the 2-LSB cycles can be disabled to provide a 12b mode at lower power. The unit element in the differential capacitive DAC equals 0.55fF, leading to a total sampling capacitance of 9pF per side. This results in 88dB kT/C-related SNR at 0.8V supply for a rail-to-rail input. The 4 MSBs are thermometer-encoded to save switching energy. As 14b matching cannot be achieved intrinsically with such small capacitors, chopping and dithering are applied in combination with oversampling to improve the linearity.

Apart from suppressing DC offset and 1/f noise, chopping also modulates distortion components. Thus, by chopping at half the sampling rate and using oversampling, also the dominant even-order distortions are moved out of the signal bandwidth. This helps in particular to counteract the even-order distortions due to mismatch in the thermometer-encoded MSBs. The implementation of this scheme can be done with little power overhead (Fig. 11.1.1): the sampling clock f_s is divided by a factor of two, and two boosted clocks φ1, φ2 are generated to drive the NMOS sampling switches that also implement the input chopping. While requiring two clock boosters, each of them operates at half the sample rate; thus it does not increase the overall power consumption. The output chopping is performed in the digital domain and is implemented with a MUX that selects the output data either from the non-inverted or from the inverted output of the SAR register.

Whereas chopping cancels out even-order distortions, it does not reduce the odd-order terms. Due to the small unit capacitors, undesirable layout parasitics especially influence the binary-scaled LSBs as these parasitics are usually not perfectly binary-scaled. The thermometer bits are less prone to this problem as they re-use identical layouts rather than scaled ones. To reduce the distortion related to the binary part of the DAC, dithering is applied to randomize these errors (Fig. 11.1.1). A deterministic dither sequence with 4 or 16 levels is injected at the input of the ADC after sampling but before the actual AD conversion. The dither logic is a simple counter to create the desired fixed sequence, and a 4-capacitor DAC adds the actual sequence to the sampling node. As the dither is a deterministic pattern, it results in spurious tones at multiples of $f_s/4$ or $f_s/16$, dependent on the selected length of the sequence. In combination with oversampling, these tones will be outside the critical baseband. Therefore, it is possible to use large-range dither without requiring dither subtraction after the ADC. The dither range is set to approximately 700 LSB to average over all 10b of the binary array rather than over the LSBs only. The power consumption of the dither circuit is dominated by its capacitors. Since these are a small fraction of the total DAC capacitance (about 4%), the power consumption is similarly low.

Next to linearity, power-efficient low-noise performance is a second requisite to enable efficient high-precision SAR ADCs. While oversampling helps to improve SNR, the required 4× speed per 6dB improvement is rather costly. As the comparator is the dominant contributor for overall noise, a power-efficient data-driven noise reduction (DDNR) technique is applied to the comparator as shown in Fig. 11.1.2 [1]. In brief, the noise reduction is selectively done on the noise-critical bit-cycles by voting on multiple repetitive comparator decisions. The selection is done by triggering a reference delay cell together with the comparator. For large input signals that are not noise-critical, the comparator generates an output before the reference delay, and the decision is immediately provided to the SAR logic. In case the comparator is slow, which indicates a small input level that could be corrupted by noise, the decision is taken several times and the voting logic produces the majority value as the final output. The amount of noise reduction depends on two parameters: first, the number of votes (Nv) used in the voting process, as a higher number results in better noise averaging; and second, the number of voting cycles (Nc) per conversion. For example, when voting is only applied in the most noise-critical case, Nc equals 1. When voting is also applied in the second-most noise-critical case, Nc equals 2 and additional noise reduction is achieved. In [1], the number of votes Nv is fixed to 5 while the number of voting cycles Nc is manually set by tuning the bias voltage V_{bias} that controls the reference delay. For that reason, the approach is not autonomous and sensitive to PVT variations.

This work introduces a feedback loop around the DDNR method to enable reliable autonomous operation (Fig. 11.1.2). Moreover, the noise reduction is digitally programmable by setting the two critical parameters Nv and Nc. Nv is simply used in the digital voting logic to count the number of repetitive decisions. Nc is used to control the reference delay by means of a feedback loop that drives V_{bias}. The actual number of voting cycles is determined by counting the number of times a slow, noise-critical decision is detected during a conversion. This number is compared against the desired value Nc. Dependent on the comparison result V_{bias} is either increased or decreased by a charge or discharge pulse on C_1 through M_1, M_2. To achieve a slow time-constant in the loop without needing an excessively large capacitor C_1, transistors M_3 and M_4 are added. These transistors are biased in sub-threshold and thus create a large RC constant for the loop.

The implemented ADC occupies 0.18mm² in a 65nm CMOS technology (Fig. 11.1.7) and operates at 0.8V supply. Out of 5 measured samples, the chip with the lowest SNDR was selected for the measurements presented in this paper. Figure 11.1.3 shows the measured impact of the various enhancement techniques. In 14b mode at 128kS/s with 16× oversampling and f_{in} = 169.22Hz, the application of chopping, 16-level dithering and DDNR improves the SFDR by about 8dB and the SNDR by about 6dB. The figure also shows the measured programmability of the DDNR at 10kS/s. The input-referred noise (IRN) of the ADC reduces while the power consumption increases as more votes are used or more voting cycles are allowed during the conversion. In 12b mode, the IRN improvement is limited by the quantization noise. Figure 11.1.4 shows the measured INL and DNL for 12b and 14b mode without chopping, dithering and oversampling. The measured SFDR and SNDR versus input frequency are given in Fig. 11.1.5 for Nyquist-rate 12b/14b operation as well as 4×/16× oversampled 14b operation. The detailed settings and measured results for these 4 modes are summarized in Fig. 11.1.6 together with a comparison to prior art. The power-efficient enhancement techniques enable reaching an SNDR between 67.8dB and 79.1dB, dependent on the selected mode. This is relatively high as compared to previous low-power SAR ADCs [1,2]. Moreover, all 4 modes of operation achieve state-of-the-art power-efficiencies in both Walden and Schreier-based FoMs as compared to ADCs with similar SNDRs [6].

References:

[1] P. Harpe, E. Cantatore, and A. van Roermund, "A 2.2/2.7fJ/conversion-step 10/12b 40kS/s SAR ADC with Data-Driven Noise Reduction," *ISSCC Dig. Tech. Papers*, pp. 270–271, Feb. 2013.

[2] C.-Y. Liou and C.-C. Hsieh, "A 2.4-to-5.2fJ/conversion-step 10b 0.5-to-4MS/s SAR ADC with Charge-Average Switching DAC in 90nm CMOS," *ISSCC Dig. Tech. Papers*, pp. 280–281, Feb. 2013.

[3] A.P. Perez, E. Bonizzoni, and F. Maloberti, "A 84dB SNDR 100kHz Bandwidth Low-Power Single Op-Amp Third-Order ΔΣ Modulator Consuming 140μW," *ISSCC Dig. Tech. Papers*, pp. 478 – 479, Feb. 2011.

[4] K.-C. Hsieh and P. Gray, "A Low-Noise Chopper-Stabilized Differential Switched-Capacitor Filtering Technique," *ISSCC Dig. Tech. Papers*, pp. 128–129, Feb. 1981.

[5] R.A. Wannamaker, *et al.*, "A Theory of Nonsubtractive Dither," *IEEE Trans. on Signal Processing*, pp. 499–516, Feb. 2000.

[6] B. Murmann, "ADC Performance Survey 1997-2013," [Online]. Available: http://www.stanford.edu/~murmann/adcsurvey.html, June 2013.

978-1-4799-0917-9/14 $31.00 © 2014 IEEE

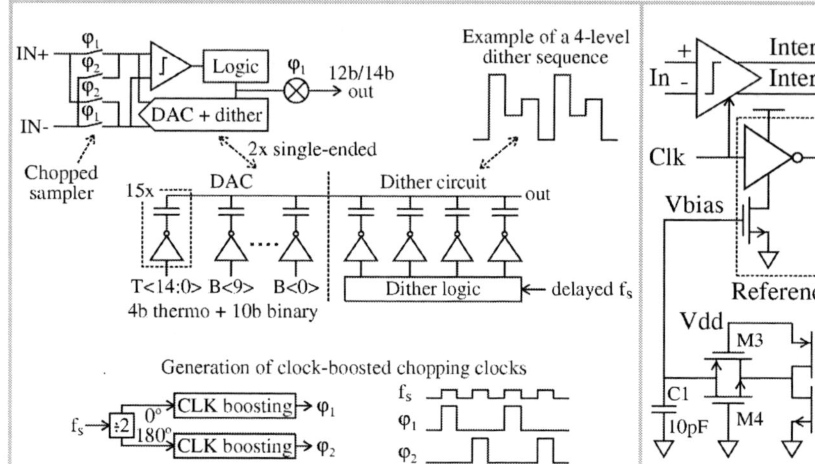

Figure 11.1.1: 12b/14b SAR ADC with chopping, dithering and segmented DAC.

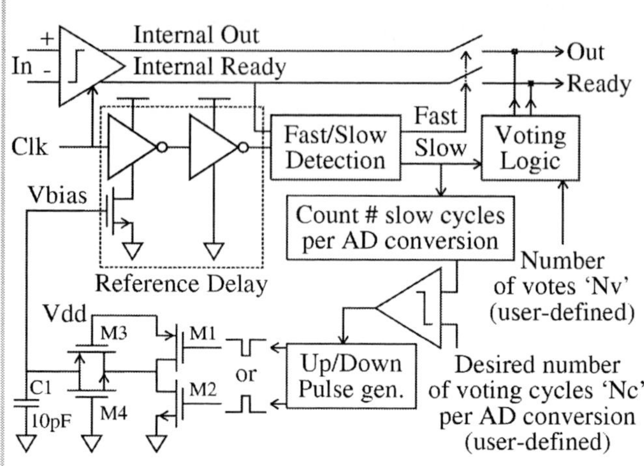

Figure 11.1.2: Closed-loop programmable Data-Driven Noise Reduction (DDNR) technique.

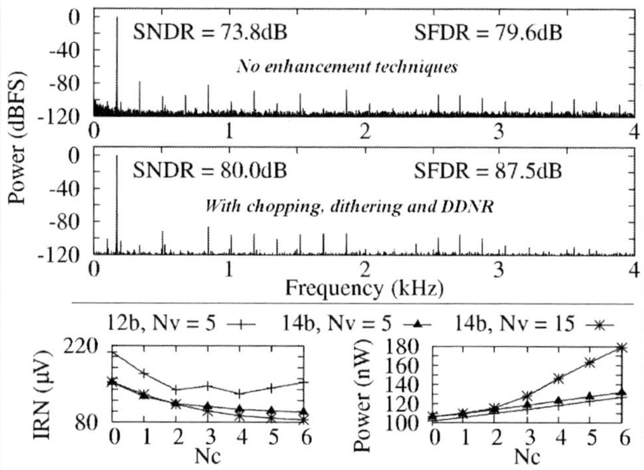

Figure 11.1.3: Spectrum without and with enhancement techniques, and influence of DDNR on IRN and power consumption.

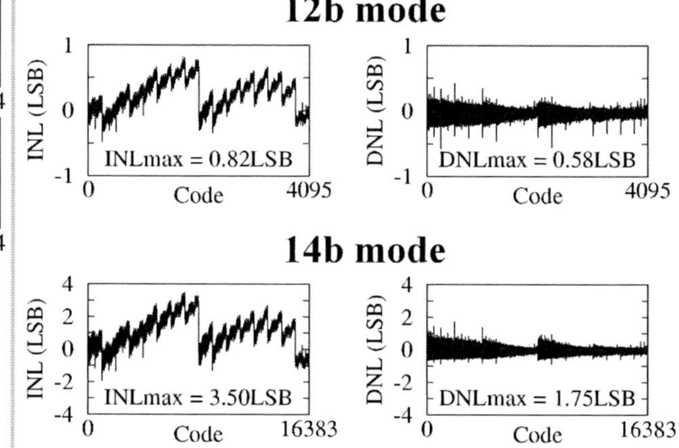

Figure 11.1.4: INL and DNL for 12b and 14b mode (without chopping, dithering and oversampling).

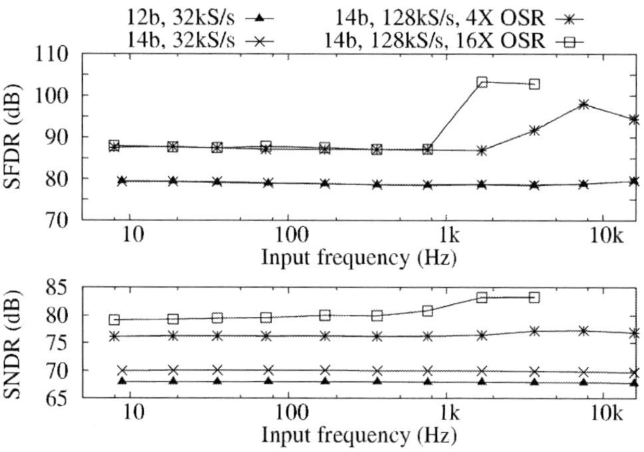

Figure 11.1.5: SFDR and SNDR versus input frequency for the different modes.

	[1]	[2]	[3]	This work			
Technology (nm)	65nm	90nm	180nm	65nm			
Area (mm²)	0.076	0.042	0.492	0.18			
Supply voltage (V)	0.6	0.4	1.5	0.8			
Resolution (bit)	12	10	-	12	14	14	14
Sample rate (kS/s)	40	500	3200	32	32	128	128
Oversampling ratio	-	-	16	-	-	4x	16x
Bandwidth (kHz)	20	250	100	16	16	16	4
#votes Nv, #cycles Nc	5, -	-	-	5, 2	5, 4	5, 4	5, 4
Chopping	-	-	-	Off	Off	On	On
Dithering	-	-	-	Off	Off	4-level	16-level
Power (μW)	0.097	0.5	140	0.310	0.352	1.367	1.370
INL (LSB)	1.90	0.5	-	0.82	3.50	-	-
DNL (LSB)	0.97	0.3	-	0.58	1.75	-	-
SFDR (dB)	68.8	81.3	96.0	78.4 *	78.5 *	86.9 *	87.1 *
SNDR (dB)	62.5	54.3	84.0	67.8 *	69.7 *	76.1 *	79.1 *
FOMW (fJ/conv.step)	2.2	2.4	54.0	4.8 *	4.4 *	8.2 *	23.2 *
FOMS (dB)	175.7	171.3	172.5	174.9 *	176.3 *	176.8 *	173.8 *

* Worst value across entire bandwidth

Figure 11.1.6: Performance summary and comparison.

978-1-4799-0917-9/14 $31.00 © 2014 IEEE

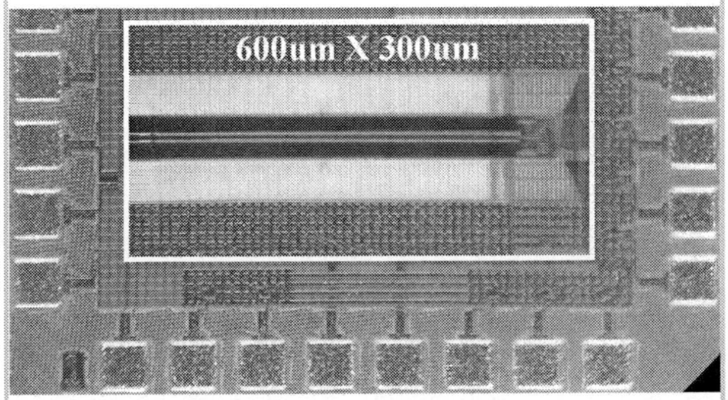

600um X 300um

Figure 11.1.7: Die photo of the ADC in 65nm CMOS.

11.2 A 0.85fJ/conversion-step 10b 200kS/s Subranging SAR ADC in 40nm CMOS

Hung-Yen Tai, Yao-Sheng Hu, Hung-Wei Chen, Hsin-Shu Chen

National Taiwan University, Taipei, Taiwan

Analog-to-digital converters (ADC) are extensively used in wireless sensor networks and healthcare electronic devices to monitor long-term signal conditions. It is essential to prolong battery life in these applications by using an energy-efficient ADC. A successive-approximation register (SAR) architecture, mostly composed of digital circuits, can achieve low power under low supply voltages [1,2]. Power consumption can be decreased by using either an energy-efficient capacitive-DAC switching method [1] or a low-power comparator with a majority voting technique [2]. In this work, a small coarse ADC resolves the MSB bits. Then, a detect-and-skip algorithm and an aligned switching technique are used to reduce the big fine DAC switching energy. The comparator power is also decreased by utilizing a low-power comparator during coarse conversion and a low-noise comparator during fine conversion. As a result, its FoM performance is as low as 0.85fJ/conversion-step, which is about 3 times better than that of the state-of-the-art work [2].

Figure 11.2.1 shows the architecture of our subranging SAR ADC. It comprises a 5b coarse SAR ADC, a 10b fine SAR ADC, and skipping control logic. The main idea is to relieve the requirements of the fine ADC by a coarse ADC whose accuracy constraint is greatly reduced by redundancy. During the sampling phase, the clock signal CLK turns on 4 bootstrap sampling switches. The input signals are sampled onto the coarse and fine capacitor arrays. During the conversion phase, the coarse ADC detects the sampled signal with the results of B1 to B5, and then the skipping control logic skips unnecessary capacitor switching. An asynchronous signal C_DONE passes the aligned switching signals to switch the MSB capacitors of the fine DAC. Next, the fine ADC continues to resolve the remaining bits of B6 to B11. The gains of coarse and fine DACs are matched by adding a C_D in the coarse DAC. An additional bit resolving [3] in the fine ADC with the redundancy range of ±16 LSB relieves the accuracy requirement on the coarse ADC. In addition, the DAC gain and comparator offset mismatches between the coarse and fine ADCs can also be compensated by the redundancy. The coarse DAC and the MSB capacitors of the fine DAC adopt a split-capacitor switching method [4] to keep the comparator common-mode voltage constant. Thus, the comparator dynamic offset due to common-mode variation does not affect the linearity. Moreover, the LSB capacitors in the fine DAC apply a monotonic switching method [3] to save switching energy. Because the DAC output common-mode variation with the LSB bits is relatively small, the dynamic offset error is negligible.

Figure 11.2.2 illustrates the detect-and-skip algorithm. The coarse ADC resolves B1 to B5. It detects the difference between the fine DAC output and the sampled signal. When the difference is small, unnecessary switching of MSB capacitors in the fine DAC can be skipped and then the DAC output would directly approach the sampled signal. Otherwise, energy is wasted by MSB capacitor switching with the conventional binary-search algorithm of SAR ADC. Since the MSB bits control large capacitors, most of energy is wasted during the resolving of the MSB bits. Our detect-and-skip algorithm saves much switching energy of the fine DAC with small switching energy of the coarse DAC. If the difference is small, B1 and B2 are 01 or 10. The MSB capacitor switching of the fine DAC can be skipped to save power. On the other hand, if the difference is large, B1 and B2 are 00 or 11. The MSB capacitor should be switched. Similarly, when B1 and B3 are 01 or 10, the MSB-1 capacitor switching can be skipped. The same principle can be applied to MSB-2 and MSB-3 capacitors as well. Figure 11.2.2 also depicts the aligned switching technique, which reduces switching energy when the difference is large. The plot of switching energy versus the ratio of C_{SW} and C_T shows that the largest switching energy occurs when half of capacitors are switched as in conventional successive switching algorithm. For example, when 3 unit capacitors (2C and 1C) have to be switched successively, the total switching energy is calculated to be $1.25CV^2$. However, if 2C and 1C are switched simultaneously by our aligned switching technique, the switching energy is as small as $0.75CV^2$. Hence, to switch the MSB capacitors of the fine DAC with the aligned switching technique is more energy-efficient than successive switching.

Figure 11.2.3 shows the switching energy versus output code. The average switching energy in the conversion phase is only $69.8CV^2$ (coarse $9.7CV^2$ and fine $60.1CV^2$), which is lower than the other switching methods. The total average switching energy is $229.7CV^2$ which includes a pre-charge energy of $159.9CV^2$ (coarse $15CV^2$ and fine $144.9CV^2$) in the sampling phase. It relaxes the settling requirement of the reference voltage by setting small portion of switching energy to the conversion phase. Also, the reference disturbance resulting from pre-charge is not an issue for the long settling time in the sampling phase. The total average switching energy of this work is larger than that of the MCS switching method [5]. But the MCS switching method uses $0.5V_{DD}$ in addition to V_{DD} and ground as reference voltages, which is hard to implement in low-voltage designs.

A 2-stage dynamic comparator [6] is used in both coarse and fine ADCs. Its noise level is the bottleneck of accuracy in low-voltage design. The accuracy requirement of the coarse comparator can be relieved from 10b to 5b with redundancy. Thus, a low-power comparator is implemented in the coarse ADC and a low-noise comparator is implemented in the fine ADC. The first-stage size of the coarse comparator is identical to that of the fine comparator to reduce the offset mismatch. The input-referred noise and comparator power are proportional to $1/\sqrt{C_L}$ and C_L, respectively, where C_L is the output capacitance loading of the first stage. Hence, C_L in the coarse comparator could be smaller than that of the fine comparator. The power-per-bit-conversion of the coarse comparator is designed to be 3 times less than that of the fine comparator. As a result, it saves 23% of power compared with the traditional SAR that uses only one 10b noise-level comparator.

Figure 11.2.7 shows the chip micrograph of the ADC fabricated in 40nm CMOS. The core circuit occupies an area of $0.0065mm^2$ (110μm×59μm). The capacitor array is constructed by metal–oxide–metal (MOM) type capacitors with unit capacitance of 1.5fF. The measured DNL and INL are shown in Fig. 11.2.4. At conversion rate F_S of 200kS/s, the DNL and INL are +0.29/-0.44 and +0.45/-0.29 LSB, respectively. The Nyquist rate FFT plot and dynamic performance versus input frequency are shown in Fig. 11.2.5. With Nyquist input frequency, the measured SNDR, SFDR, SNR, and THD are 55.63dB, 76.25dB, 55.75dB, and 71.3dB, respectively. The measured power dissipation is 84nW with a 0.45V supply. It can be broken down as follows: the DACs use 34% of the power; the comparators and the bootstraps use 30%; the digital circuits use 36%. Figure 11.2.6 shows the performance summary and comparison table of the energy-efficient ADCs in recent years.

Acknowledgements:
The authors would like to thank Taiwan National Science Council and TSMC University Shuttle Program for the support of this work. They would also like to thank HYCON technology for the help of measurement.

References:

[1] C.-Y. Liou and C.-C. Hsieh, "A 2.4-to-5.2fJ/conversion-step 10b 0.5-to-4MS/s SAR ADC with Charge-Average Switching DAC in 90nm CMOS," *ISSCC Dig. Tech. Papers*, pp. 280-281, Feb. 2013.
[2] P. Harpe, *et al.*, "A 2.2/2.7fJ/conversion-step 10/12b 40kS/s SAR ADC with Data-Driven Noise Reduction," *ISSCC Dig. Tech. Papers*, pp. 270-271, Feb. 2013.
[3] C.-C. Liu, *et al.*, "A 10b 100MS/s 1.13mW SAR ADC with Binary-Scaled Error Compensation," *ISSCC Dig. Tech. Papers*, pp. 386-387, Feb. 2010.
[4] M. Yip and A. P. Chandrakasan, "A Resolution-Reconfigurable 5-to-10b 0.4-to-1V Power Scalable SAR ADC," *ISSCC Dig. Tech. Papers*, pp. 190-191, Feb. 2011.
[5] V. Hariprasath, *et al.*, "Merged Capacitor Switching Based SAR ADC with Highest Switching Energy-Efficiency," *IET Electronics Letters*, Apr. 2010.
[6] M. van Elzakker, *et al.*, "A 1.9μW 4.4fJ/Conversion-step 10b 1MS/s Charge-Redistribution ADC," *ISSCC Dig. Tech. Papers*, pp. 244-245, Feb. 2008.

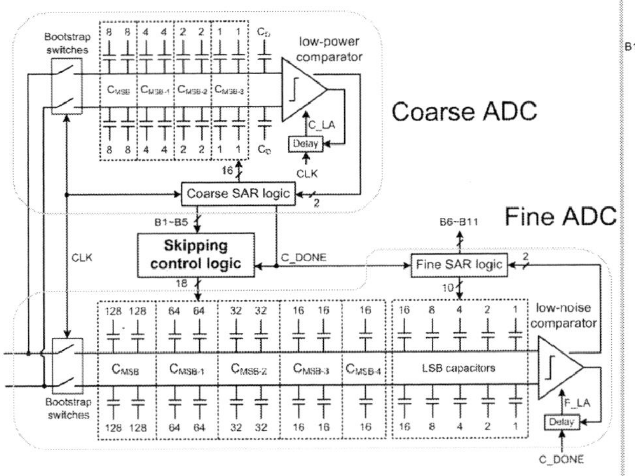

Figure 11.2.1: Subranging SAR ADC.

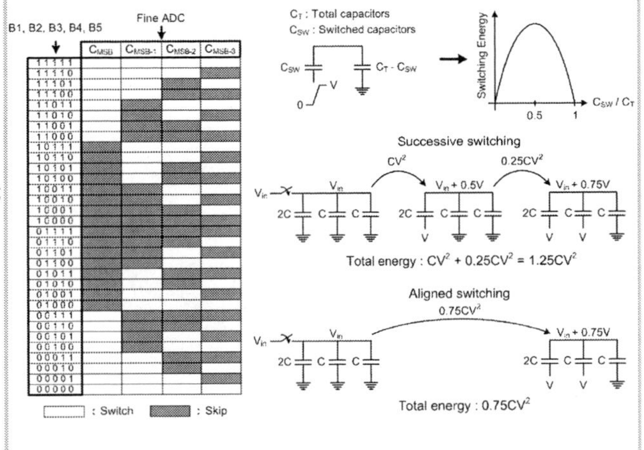

Figure 11.2.2: Detect-and-skip algorithm and aligned switching technique.

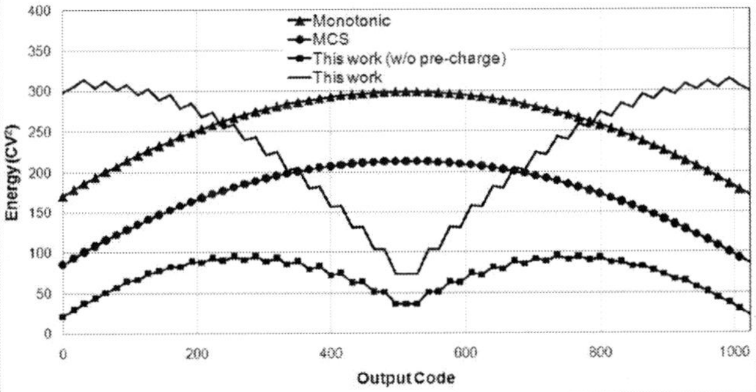

Switching method	Average energy (CV^2)
Monotonic	255.5
MCS	170.2
This work	229.7
This work (w/o pre-charge)	69.8

Figure 11.2.3: Switching energy versus output code.

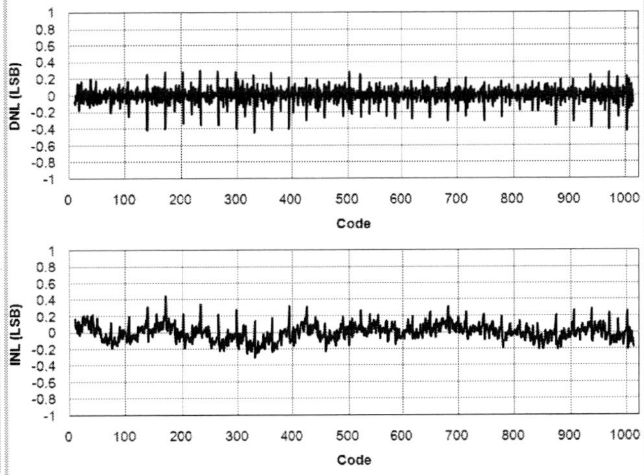

Figure 11.2.4: DNL and INL.

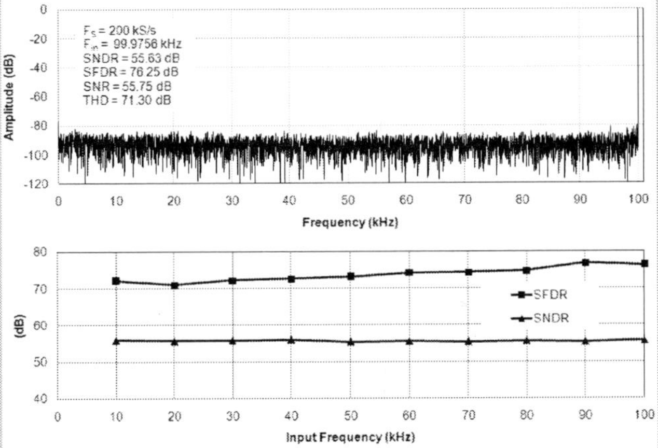

Figure 11.2.5: Nyquist rate FFT and dynamic performance versus input frequency at 200kS/s.

	[1]	[2]	[6]	This work
Technology	90nm	65nm	65nm	40nm
Supply Voltage (V)	0.4	0.6	1	0.45
Sample rate (kS/s)	500	40	1000	200
Resolution (bit)	10	12	10	10
DNL (LSB)	0.34	0.97	0.5	0.44
INL (LSB)	0.62	1.9	2.2	0.45
Power (nW)	500	97	1900	84
ENOB (bit)	8.72	10.1	8.75	8.95
FOM (fJ/c.-s.)	2.37	2.2	4.4	0.85
Active Area (mm^2)	0.042	0.076	0.026	0.0065

Figure 11.2.6: Performance summary and comparison.

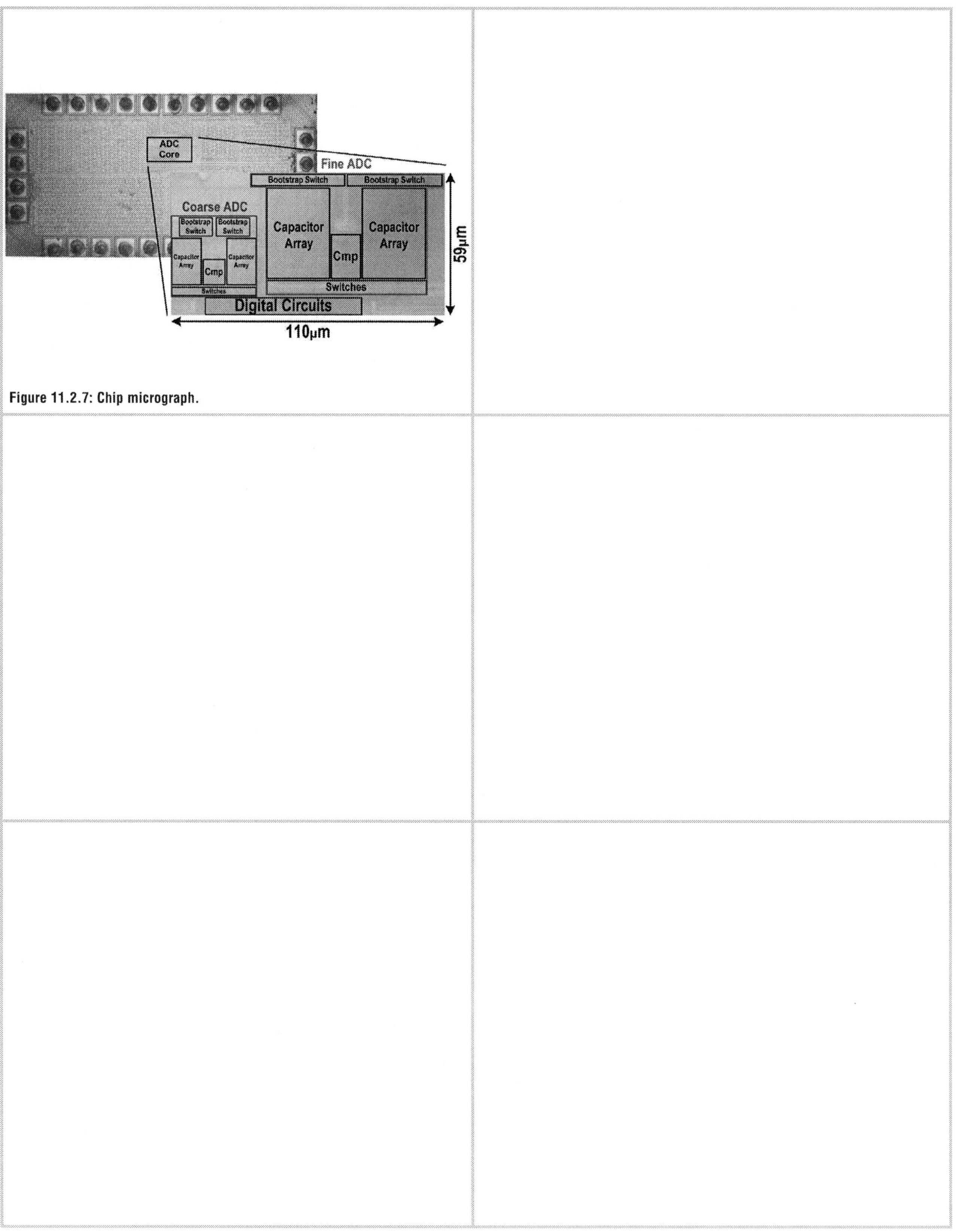

Figure 11.2.7: Chip micrograph.

11.3 A 10b 0.6nW SAR ADC with Data-Dependent Energy Savings Using LSB-First Successive Approximation

Frank M. Yaul, Anantha P. Chandrakasan

Massachusetts Institute of Technology, Cambridge, MA

ADCs used in medical and industrial monitoring often transduce signals with short bursts of high activity followed by long idle periods. Examples include biopotential, sound, and accelerometer waveforms. Current approaches to save energy during periods of low signal activity include variable resolution and sample rate systems [1], asynchronous level-crossing ADCs [2], and ADCs that bypass bitcycles when the signal is within a predefined small window [3]. This work presents a signal-activity-based power-saving algorithm called LSB-first successive approximation (SA) that maintains a constant sample rate and resolution; scales logarithmically with signal activity, and does not inherently suffer from slope overload.

LSB-first SA starts with an initial guess D of the current sample's value and bitcycles the LSBs of D first, instead of the MSBs as in conventional SA. Because this ADC is targeted at signals with low average activity, an initial guess equal to the value of the previous sample D_{PREV} is effective since the error will be Δcode, the change in output code from sample to sample. On average, the algorithm uses a number of bitcycles that scales logarithmically with the error between the guess code and the final output code. Since each bitcycle involves a DAC transition, an analog comparison, and logic transitions, LSB-first SA saves energy in all system blocks of the ADC by performing an N-bit conversion in fewer than N bitcycles when the signal activity lessens. A previous work [4] saves bitcycles when many of the MSBs are zero, as in a low-amplitude signal. It also proposes using the previous sample to predict which of the current sample's MSBs are zero and provides simulation results. In comparison, LSB-first SA saves bitcycles for signals with small mean Δcode, including signals that fill the fullscale range of the ADC.

Figure 11.3.1 shows the DAC waveform during three example 5b conversions using LSB-first SA. LSB-first SA has 3 phases. In the INIT Phase, the differential input voltage is sampled onto the bottom plates of the capacitive DAC, shown in Fig. 11.3.2. D, the 10b input code to the DAC, is set to the initial guess of D_{PREV}. If the output of the comparator is false, then the initial guess was too low so the direction of bitcycling DIR is set to 1, denoting the need to increase the guess. The inverse holds if the comparator returns true. Note that DIR sets the bottom plate voltage of a LSB capacitor, as shown in Fig. 11.3.2. When DIR is 0, the DAC outputs the code D's lower bound. When DIR is 1, the DAC is able to output D's upper bound without using a digital adder. Conversion uses only 2 bitcycles when the initial guess is exactly correct, as in the third example conversion.

Let D_i be the bit with index i, where D_0 is the LSB. Let Q be the index of the bit currently being tested. In the ToMSB Phase, Q is set to the index of the least significant bit that is not currently set to DIR. D_Q is inverted to move D in the desired direction. This is repeated until the comparator output flips, indicating that D has overshot the target value. The MSBs D[9:Q] are now finalized for this conversion. An example is given in the first conversion in Fig. 11.3.1, where the initial guess assigned to D is 11010 which is too high, so the ToMSB phase takes 3 bitcycles to invert the 1's in D at positions 1, 3, and 4. The DAC step sizes do not have a uniform radix since bits in D already set to DIR are skipped. Simulations over all possible code transitions show this bit-skipping saves bitcycles on average in comparison to a fixed radix-2 ToMSB phase, which additionally requires a digital adder and toggles multiple bits per bitcycle.

In the ToLSB Phase, the bitcycling proceeds as in conventional SA, except it must handle both polarities of DIR. If DIR is 0, then the bits D[Q-1:0] are currently all 0, so the conventional SAR algorithm involving setting D_{Q-1} to a test value of 1 can proceed. The inverse holds if DIR is 1. When bitcycling has gone back down to the LSB, conversion is finished, and the DACs are purged of charge. The best-case conversion uses only 2 bitcycles and the worst case uses 2N+1 bitcycles for N bits. Figure 11.3.2 shows a flowchart of the algorithm. Note that an error in the initial guess of over 512 LSB causes V_{DAC} to exceed the supply rails during the first bitcycle, causing DAC charge loss. This is not an issue for most signals, but when performing ENOB tests with fullscale sinusoids above $f_S/2\pi$, an initial guess of 511 rather than D_{PREV} is used to limit the maximum error.

Figure 11.3.3 shows the ADC architecture. The 10b capacitive DAC comprises a 5b main-DAC coupled to a 5b sub-DAC with the capacitor C_C as in [1], reducing the number of unit capacitors by 16× over a conventional capacitive DAC. The unit capacitor value C of 72 fF was conservatively chosen to keep DNL well below 1 LSB. A 2-stage dynamic comparator with a gain stage and a latch is used as in [5]. The sampling switches are driven with charge pumps as in [1] to lower on-resistance. Long transistor lengths are used to reduce leakage mean and variance at the cost of speed and energy, allowing linear power scaling down to the sub-kHz sample rates of typical sensor signals.

To demonstrate LSB-first SA's data-dependent power consumption, fullscale sinusoids of varying frequency were input to the test chip. The power and number of bitcycles/sample were averaged over many periods of each input, and the mean Δcode of each input was computed from the ADC output. Figure 11.3.4 shows the mean bitcycles/sample and mean energy/sample scaling logarithmically with respect to mean Δcode. Throughout the dataset, digital energy remains between 52% and 57% of the total energy.

The fullscale sinusoids exercise all codes of the ADC so that each datapoint approximately represents average behavior over the entire ADC input range. Thus, the ADC's performance for an arbitrary input signal may be estimated based on the signal's mean Δcode. The exact performance varies with the signal's probability density function since code transitions that flip higher bits require more energy, but the variation is small since higher bit transitions occur exponentially less frequently. To demonstrate this, Fig. 11.3.4 shows that datapoints taken with electrocardiogram (ECG) inputs of varying amplitude line up with the data from the fullscale sinusoids. Both the sinusoid and the ECG inputs were sampled at 10 kHz to maintain constant leakage energy per sample. Because 10 kHz is 20× a typical ECG sample rate of 500Hz, the ECG input was also sped up to 20× real time to compensate.

Figure 11.3.5 shows the LSB-first SAR ADC test chip's response to an ECG input at 1/5 fullscale amplitude, a typical signal level that safeguards against interferers which could overrange the ADC. At 1kS/s, the ADC consumes 3.9nW and is able to use an average of 3.7 bitcycles/sample, a 2.7× improvement over a conventional SAR. Also note the correlation between the power and bitcycles/sample waveforms.

The 10b ADC is implemented in 0.18µm CMOS and occupies 0.12mm². A die photo is shown in Fig. 11.3.7. Over 10 test chips, the DNL is bounded at +0.1/-0.1 LSB and the INL at +0.22/-0.14 LSB. At 0.6V, the leakage power mean is 0.58nW and upper bound is 0.63nW. The ADC power scales linearly with sample rate down to the leakage level. A sample output spectrum and ENOB versus input frequency for one test chip are shown in Fig. 11.3.6. In 0.6V mode, the sampling switch resistance degrades the ENOB near the Nyquist frequency.

Figure 11.3.6 summarizes the performance of one test chip. Compared to other recent low-leakage 10b SAR ADCs [1,3,5], this ADC achieves a high 9.73b ENOB and sub-nW leakage in 0.6V mode while maintaining a low Walden FoM ranging from 3.5 to 20fJ, depending on the input signal activity. The supply voltage can be scaled to 1V to reach 450kS/s or to 0.5V to achieve a 2.9-to-17fJ FoM range. Overall, the LSB-first SA logic presented in this work may be used as a drop-in replacement for conventional SA logic with minimal changes to the DAC, allowing ADC power to smoothly scale down with signal activity.

Acknowledgement:
This work was supported by the NDSEG fellowship, Shell, and TI. The authors thank M. Yip for technical assistance, and the TSMC University Shuttle Program for chip fabrication.

References:

[1] M. Yip and A. P. Chandrakasan, "A Resolution-Reconfigurable 5-to-10b 0.4-to-1V Power Scalable SAR ADC," *ISSCC Dig. Tech. Papers*, pp. 190-191, Feb. 2011.

[2] M. Trakimas and S. Sonkusale, "An Adaptive Resolution Asynchronous ADC Architecture for Data Compression in Energy Constrained Sensing Applications," *IEEE J. Solid-State Circuits*, vol. 58, no. 5, May 2011.

[3] G. Y. Huang, *et al.*, "A 1µW 10-bit 200-kS/s SAR ADC With a Bypass Window for Biomedical Applications," *IEEE J. Solid-State Circuits*, vol. 47, no. 11, Nov. 2012.

[4] H. Lampinen, *et al.*, "Novel Successive Approximation Algorithms," *IEEE Int. Symp. Circuits and Systems*, vol. 1, pp 188-191, May 2005.

[5] P. Harpe, *et al.*, "A 2.2/2.7fJ/conversion-step 10/12b 40kS/s SAR ADC with Data-Driven Noise Reduction," *ISSCC Dig. Tech. Papers*, pp. 270-271, Feb. 2013.

978-1-4799-0917-9/14 $31.00 © 2014 IEEE

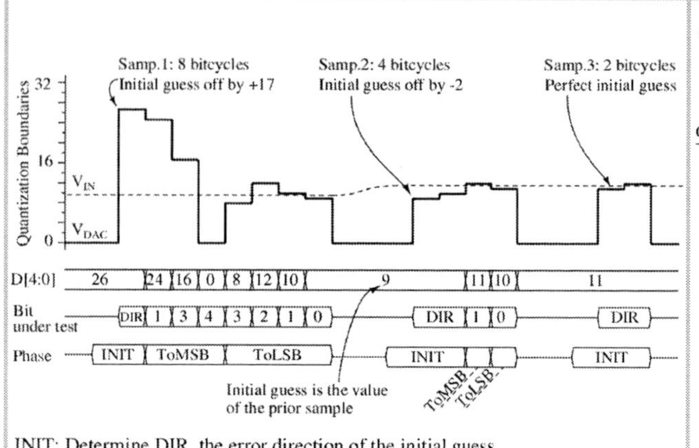

Figure 11.3.1: Example 5b conversions using LSB-first SA.

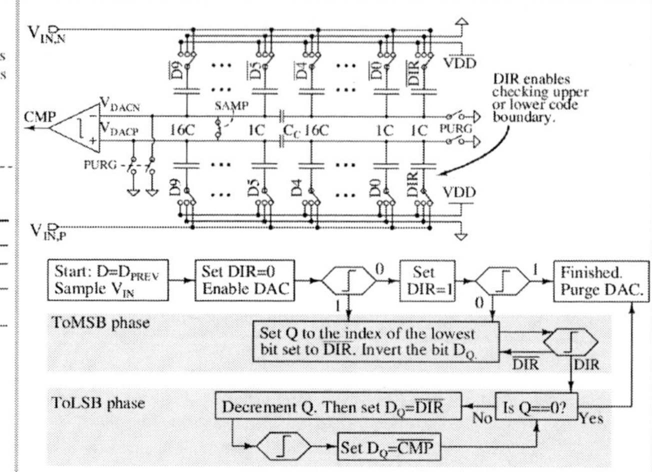

Figure 11.3.2: DAC schematic and flowchart illustrating LSB-first SA.

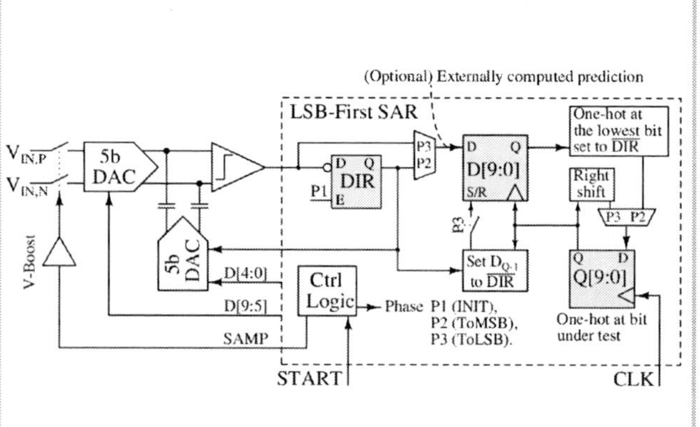

Figure 11.3.3: LSB-first SAR ADC architecture.

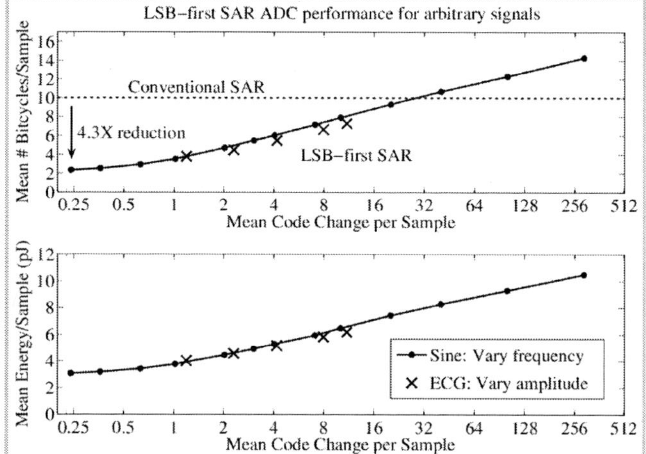

Figure 11.3.4: Measured mean bitcycles/sample and mean energy consumption at V_{DD}=0.6V and f_S=10kHz as a function of mean output code change per sample, for differently shaped inputs.

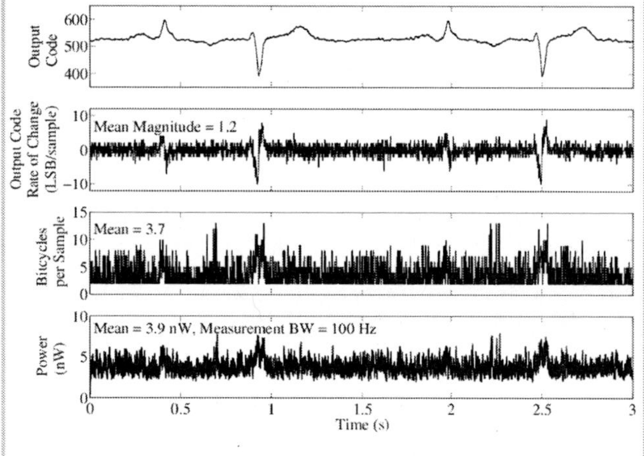

Figure 11.3.5: ADC response to ECG test input signal at V_{DD}=0.5V sampled at 1kHz.

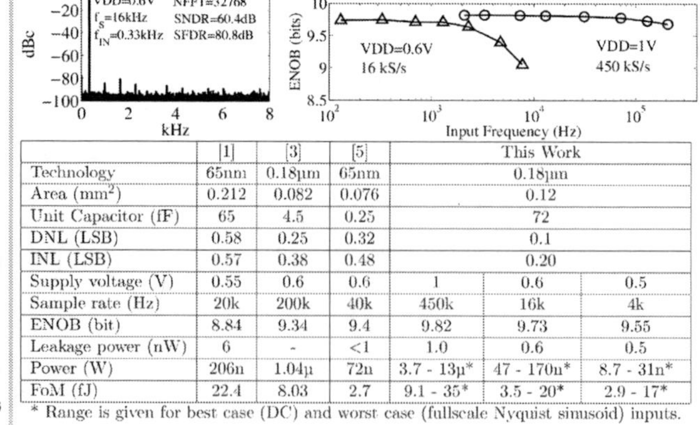

	[1]	[3]	[5]	This Work		
Technology	65nm	0.18μm	65nm	0.18μm		
Area (mm^2)	0.212	0.082	0.076	0.12		
Unit Capacitor (fF)	65	4.5	0.25	72		
DNL (LSB)	0.58	0.25	0.32	0.1		
INL (LSB)	0.57	0.38	0.48	0.20		
Supply voltage (V)	0.55	0.6	0.6	1	0.6	0.5
Sample rate (Hz)	20k	200k	40k	450k	16k	4k
ENOB (bit)	8.84	9.34	9.4	9.82	9.73	9.55
Leakage power (nW)	6	-	<1	1.0	0.6	0.5
Power (W)	206n	1.04μ	72n	3.7 - 13μ*	47 - 170n*	8.7 - 31n*
FoM (fJ)	22.4	8.03	2.7	9.1 - 35*	3.5 - 20*	2.9 - 17*

* Range is given for best case (DC) and worst case (fullscale Nyquist sinusoid) inputs.

Figure 11.3.6: FFT and ENOB versus input frequency, and summary/comparison table.

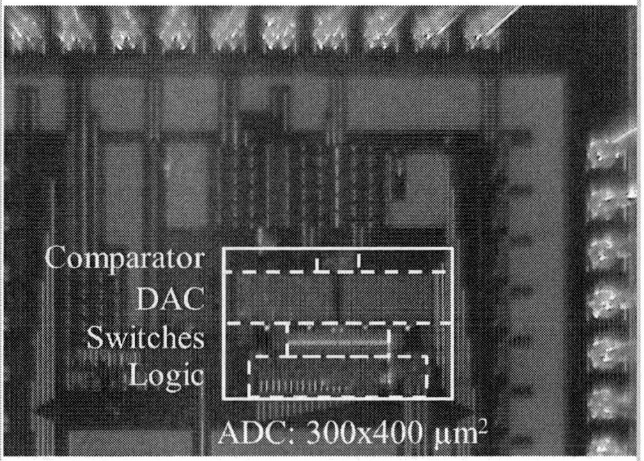

Figure 11.3.7: LSB-first SAR ADC chip micrograph.

11.4 A 1.5mW 68dB SNDR 80MS/s 2× Interleaved SAR-Assisted Pipelined ADC in 28nm CMOS

Frank van der Goes[1], Chris Ward[1], Santosh Astgimath[2], Han Yan[1], Jeff Riley[1], Jan Mulder[1], Sijia Wang[1], Klaas Bult[1]

[1]Broadcom, Bunnik, The Netherlands,
[2]Wolfson Microelectronics, Edinburgh, United Kingdom

The resolution and sampling speed of recently reported SAR ADCs have increased to 11+ ENOB at 50 to 100MS/s [1,2]; however, power efficiency has unfortunately suffered when compared to lower-resolution, lower-speed ADCs. This design targets the same high speed and resolution while simultaneously achieving power efficiency previously associated only with low-speed, low-resolution ADCs. Furthermore, the power reported includes the consumption from the active reference generator, clock generator and encoder (since this is an industrial SoC), differentiating it from the majority of reported SAR ADCs. A dynamic residue amplifier with excellent noise-filtering properties, embedded in a pipelined architecture, is a key power-saving technique. In addition, an energy-efficient switched-capacitor (SC) DAC is obtained by using a small fraction of the total DAC capacitance during the initial SAR steps. The realized Walden FOM is 9.1fJ/conv-step while the Schreier FoM is 172.3dB, currently the highest reported number to date for sampling speeds greater than 0.1Ms/s, based on the extensive list of recent data converters compiled in [3].

The ADC architecture is shown in Fig. 11.4.1. It uses 2 interleaved ADC lanes. Each lane is a 2-stage pipelined ADC that uses a 7b switched-capacitor DAC (SC-DAC) in the first stage, a fully dynamic 16×-gain residue amplifier between stages, and an 8b SC-DAC in the second stage. A 2× overrange between the stages results in 14b quantization. The sampling capacitor consists of unit capacitors to implement the DAC function. Pipelining provides an obvious speed advantage but is also necessary to achieve a good power-efficiency at higher resolutions. This is because the low-noise requirement now applies to the residue amplifier, and not to the comparator. Substantial power is saved because the (dynamic) residue amplifier is activated only once per full SAR conversion, while the comparator would be latching 14 times. On top of that, the residue amplifier filters more noise from the reference buffer (REF) and DAC switches than a comparator due to its inherent lower bandwidth. The stage-1 reference buffer is efficiently being used by both interleaved ADC lanes. Its reference voltages are designed at 1V and 0V, allowing the DAC switches to operate in their optimal region. Since the ADC is driven by on-chip circuits (SoC), the ADC full-scale range is limited to $1.4V_{pp\text{-}diff}$. This is done by assigning only 70% of the sampling capacitor to the DAC. The remaining 30% is not part of the DAC and switches to a fixed voltage. The on-chip encoder contains all functions that run at the clock speed.

A fully dynamic residue amplifier is applied, since these structures can achieve very energy-efficient amplification [4]. The amplifier in this work performs 2 cascaded integration steps. This combines excellent noise-filtering properties with high gain. The schematic is shown in Fig. 11.4.2. It consists of an input differential pair (M1, M2), cascodes (M3, M4), and integration capacitors C1 and C2. Capacitor C2 represents the SC-DAC of the SAR in the second pipelined stage. During the first integration step, the cascode devices are off and noise and residue voltage are integrated only on C1. After crossing the threshold voltage of a cascode transistor, step 2 starts by transferring the differential charge from C1 to C2 and integration continues on C2. The stop signal for the second step, required to set the gain, is generated using a common-mode (CM) detection circuit [5]. Since the overall gain equals that of a single integrator (note the dotted line through the origin in Fig. 11.4.2), the equivalent filter has the smallest possible noise bandwidth $1/2T_{int}$, where T_{int} is the total integration time. This filter reduces the effect of noise from the transistors of the residue amplifier itself, as well as from noise from the DAC switches and the reference buffer. The 2-step approach results in a gain equal to (V_{dCM}/V_{gt}) (1+C1/C2), V_{dCM} being the change in CM voltage and V_{gt} the transistor overdrive voltage. A large 16× gain, obtained by choosing C1=500fF and C2=200fF, has two advantages. First, the power consumption of the comparator in the second pipelined stage can go down significantly. The second advantage is that a large gain reduces the sensitivity to noise from the CM detection circuit because a small value of C2 increases the slope of the CM signal, allowing a power reduction of the CM detection circuit.

Many power-saving DAC switching techniques are known from literature. In this work, DAC power is saved by using only 25% of the DAC capacitance during the first 2 SAR steps while the remaining 75% is floating. Figure 11.4.3 shows the position of the switches directly after sampling. Only the capacitors in cell c9a, c8a and C_{fix_1} (denoted by the thick boxes) are used during the first 2 SAR steps. After that, all capacitors are used for noise reduction. This technique reduces the power consumption of the DAC by about 40%. Note that the weight of cells c9 and c8 remains unchanged. Further power saving is obtained by applying the split-capacitor technique [6] to the capacitors in cells c9..c3. Cells c2..c0 use the set-and-down approach [7] to avoid using a 2× smaller unit capacitor. The use of floating capacitors in the first SAR cycles adds a constraint to the off-impedance of the input switches because the moving input signal might interfere with SAR operation. Improved isolation is obtained by adding cross-coupled capacitors to the input switches. The DAC in the second pipelined stage uses the set-and-down approach.

The reference buffer normally is the most power-hungry block of the converter. Its power consumption in this design is substantially reduced by applying a reduced-radix DAC. The DAC in the first pipelined stage is designed with a radix of 1.72, resulting in nine SAR steps to reach 7b resolution. Figure 11.4.3 shows the actual values of the DAC cells. The reduced radix creates overrange inside the DAC, and this allows a very limited settling of the reference buffer, thus saving bias current. Class-AB techniques, applied to handle the current spikes from the SC-DAC efficiently, reduce the bias current to 60μA. These low current levels typically come with a noise penalty; however, reference-buffer noise is filtered by the residue amplifier. The reference buffer uses a 1.8V supply. Since the actual reference voltage is 1.0V, cascoding devices are applied to obtain a good PSRR, which is a stringent SoC requirement.

The capacitive DAC in stage 1 is calibrated during start-up. This allows the size of the sampling capacitor (1.4pF) to be set by noise and not by matching requirements. The offsets of all comparators and residue amplifiers are also calibrated during start-up. A background gain calibration for the residue amplifier uses a pseudorandom binary sequence (PRBS) and the reference capacitor in cell c0. The calibration values for the capacitive DAC and gain are obtained off-chip. Both pipeline stages use a self-timed SAR conversion. A DLL driving an analog delay cell in the SAR loop ensures that maximum time is allocated to DAC settling while guaranteeing a certain BER (metastability) at the comparator output. The input sampling network uses bootstrapped switches and bottom-plate sampling techniques.

The ADC is realized in a 28nm digital CMOS process. Figure 11.4.4 shows the FFT of 3×-decimated ADC output data for a 3MHz sinewave input. The SNDR, THD, SFDR and interleave tone vs. input frequency are shown in Fig. 11.4.5. A die photo is shown in Fig. 11.4.7. Figure. 11.4.6 shows a comparison between state-of-the-art designs [1,2,4] and this work. The realized Walden FoM is 9.1fJ/conv-step, which is better than other designs having a similar SNDR. The Schreier FoM of this work is 172.4dB and is the highest reported number to date for sampling speeds larger than 0.1MS/s [3].

References:

[1] R. Kapusta, *et al.*, "A 14b 80MS/s SAR ADC with 73.6dB SNDR in 65nm CMOS," *ISSCC Dig. Tech. Papers,* pp. 472-474, Feb. 2013.

[2] T. Morie, *et al.*, "A 71dB-SNDR 50MS/s 4.2mW CMOS SAR ADC by SNR Enhancement Technique Utilizing Noise," *ISSCC Dig. Tech. Papers,* pp.272-274, Feb. 2013.

[3] B. Murmann, "ADC Performance Survey 1997-2013," [Online]. Available: http://www.stanford.edu/~murmann/adcsurvey.html.

[4] B. Verbruggen, *et al.*, "A 1.7mW 11b 250MS/s 2× Interleaved Fully Dynamic Pipelined SAR ADC in 40nm Digital CMOS," *ISSCC Dig. Tech. Papers,* pp.466-468, Feb. 2012

[5] J. Lin, et al., "A 15.5dB Wide Signal Swing Dynamic Amplifier Using a Common-Mode Voltage Detection Technique," *IEEE Int. Symp. on Circuits and Systems,* pp. 21-24, 2011.

[6] B. Ginsberg, *et al.*, "An Energy-Efficient Charge Recycling Approach for a SAR Converter with Capacitive DAC," *IEEE Int. Symp. on Circuits and Systems,* pp. 184-187, 2005.

[7] C. Liu, *et al.*, "A 10-bit 50-MS/s SAR ADC with a Monotonic Capacitor Switching Procedure," *J. Solid State Circuits,* vol. 45, no. 4, pp. 731-740, 2010.

978-1-4799-0917-9/14 $31.00 © 2014 IEEE

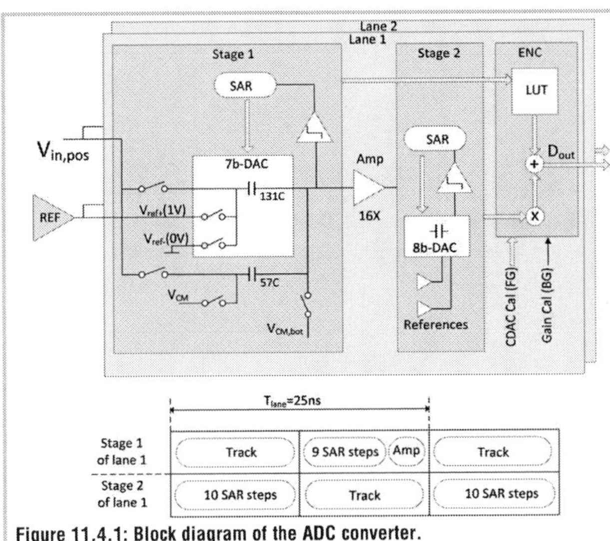

Figure 11.4.1: Block diagram of the ADC converter.

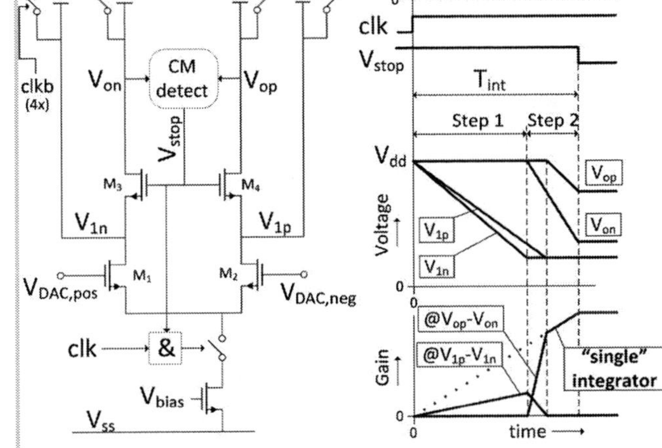

Figure 11.4.2: The dynamic residue amplifier.

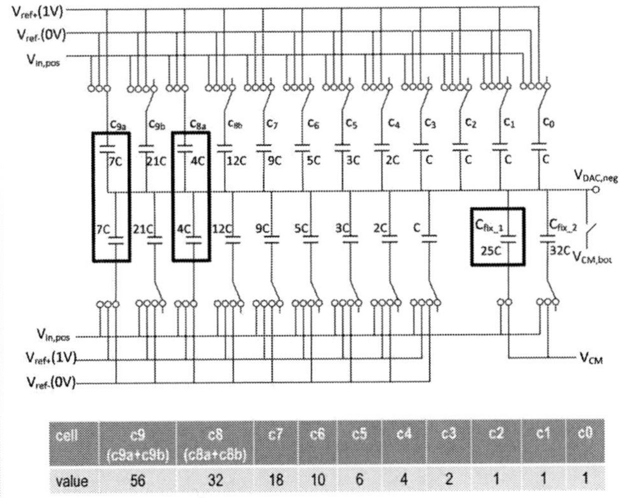

cell	c9 (c9a+c9b)	c8 (c8a+c8b)	c7	c6	c5	c4	c3	c2	c1	c0
value	56	32	18	10	6	4	2	1	1	1

Figure 11.4.3: The sampling capacitor / DAC of stage 1.

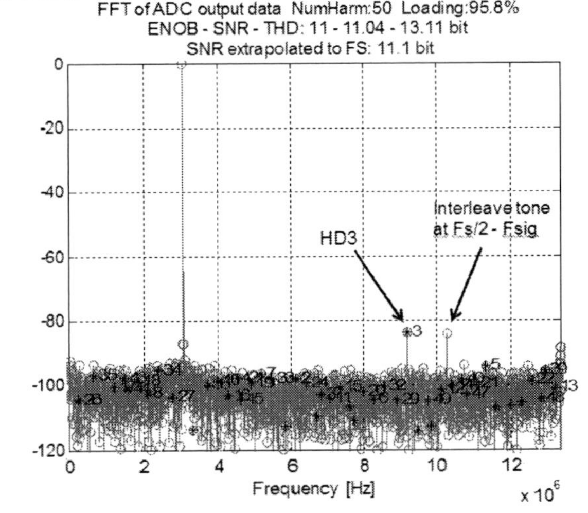

Figure 11.4.4: FFT of ADC output data.

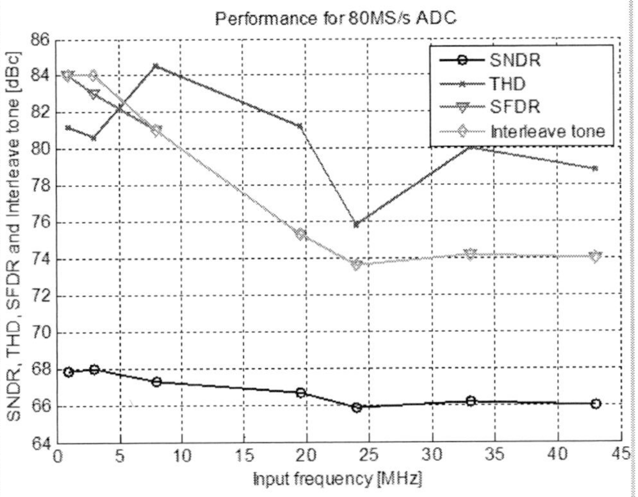

Performance for 80MS/s ADC

SNDR
THD
SFDR
Interleave tone

SNDR, THD, SFDR and Interleave tone [dBc]

Input frequency [MHz]

Figure 11.4.5: SNDR, THD, SFDR and interleave tone.

	[1] Kapusta (ISSCC 2013)	[2] Morie (ISSCC 2013)	[4] Verbruggen (ISSCC 2012)	This work
Technology	65nm CMOS	90nm CMOS	40nm CMOS	28nm CMOS
Architecture	SAR	SAR	SAR-assisted pipelined	SAR-assisted pipelined
Supply [V]	1.2	1.2	1.1	1.0 (ADC core) 1.8 (REF)
Max. input swing [$V_{pp\text{-}diff}$]	2.4	2.2	-	1.4
Fs [Ms/s]	80	50	250	80
Peak SNDR [dBc]	73.6	71	58.7	68
ENOB [bit] (SNDR-1.76)/6.02	11.9	11.5	9.5	11.0
Power [mW]	31.1	4.2	1.7	1.5
Walden FoM [fJ/conv-step] P/(2^ENOB*Fs)	99.4	29.7	9.7	9.1
Schreier FoM [dB] SNDR + 10*log10(Fs/(2*P))	164.7	168.7	167.4	172.3
Power / SNDR include Reference Buffer ?	No	No	No	Yes

Figure 11.4.6: Performance and comparison.

978-1-4799-0917-9/14 $31.00 © 2014 IEEE

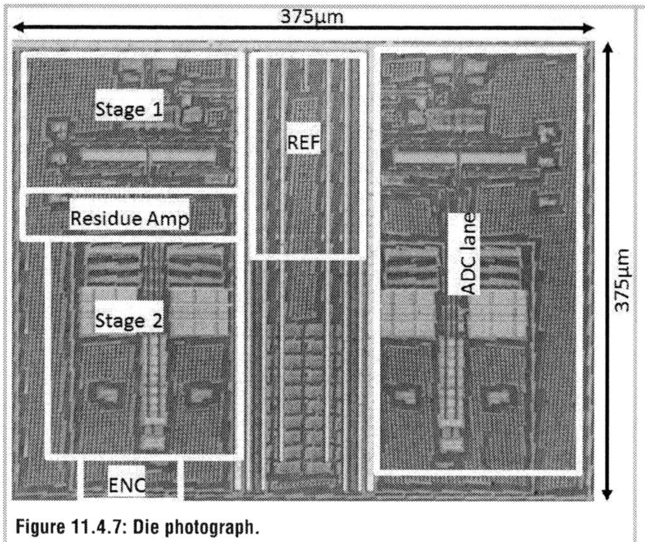

Figure 11.4.7: Die photograph.

11.5 A 100MS/s 10.5b 2.46mW Comparator-less Pipeline ADC Using Self-Biased Ring Amplifiers

Yong Lim[1,2], Michael P. Flynn[1]

[1]University of Michigan, Ann Arbor, MI,
[2]Samsung Electronics, Yongin, Korea

Pipelined ADCs require accurate amplification; however traditional OTAs limit power efficiency since they require high quiescent current for slewing. In addition, it is difficult to design low-voltage OTAs in modern, scaled CMOS. The ring amplifier [1-4] provides an intriguing alternative to traditional OTAs. This work improves the power efficiency and practicality of the ring amplifier by introducing a self-biasing scheme and by eliminating the comparators.

The ring amplifier is comprised of three inverter stages (Fig. 11.5.1(a)), stabilized in a feedback configuration. To prevent oscillation, [3] splits the second stage into two separately biased AC-coupled inverters. The bias voltages V_{RP} and V_{RN} are tuned to ensure that the 3rd-stage transistors M_{CP} and M_{CN} enter deep sub-threshold as V_{IN} approaches the desired virtual ground voltage, so that the output-stage resistance increases dramatically, and forms a dominant pole that stabilizes the amplifier. (A related consideration for stability [3] is that the peak overdrive voltage applied to the output transistors should decrease during each successive oscillation period.) The ring amplifier has high gain, thanks to its three stages. Furthermore, the ring amplifier can slew very efficiently because the 3rd-stage inverter acts as a pair of digital switches during slewing. A drawback, especially considering process and supply voltage variation, is that the bias voltages V_{RP} and V_{RN} must be set within a small voltage window. If the quiescent overdrive voltages are too high, then the ring amplifier can oscillate because the output resistance of the third inverter stage is never sufficiently large to create positive phase margin. On the other hand, if the quiescent overdrive is too low, then once the ring amplifier settles the 2nd-stage inverters operate in triode resulting in a low overall three-stage gain. This paper presents a self-biased ring amplifier that is robust to transistor variation. Furthermore, a comparator-less pipeline ADC structure uses the characteristics of the ring amplifier to replace the sub-ADC in each pipeline stage.

Our self-biased ring amplifier structure (Fig. 11.5.1(b)) replaces the two AC-coupled 2nd-stage inverters with a single self-biased DC-coupled inverter, and eliminates the two external biases and the switches. We implement the 3rd-stage inverter with high-V_{TH} devices, which present a much higher output resistance for a given gate-source voltage. An important change is that a resistor, R_B, placed between drains of the 2nd-inverter transistors, dynamically sets the last-stage gate voltages, V_{CP} and V_{CN}. This dynamic biasing maximally drives the last-stage transistors when the ring amplifier is slewing, and provides an offset voltage to set the last-stage transistors in deep sub-threshold when V_{IN} is close to the virtual ground. An important advantage of this resistor-based offset over the previous use of a reference voltage is that the offset tracks variation of the power supply voltage (V_{DD}).

The first inverter is optimized for low noise and power efficiency. A diode-connected transistor M_{NR} lowers the effective power supply voltage of the first inverter, composed of M_{P1} and M_{N1}. The reduced inverter power supply allows us to use a larger W/L ratio for M_{P1} and M_{N1}, so that a higher transconductance is achieved with a given current consumption, thereby resulting in a lower thermal noise without an increase in the static power consumption. This technique is only applicable to the first inverter because its output does not require rail-to-rail swing.

Since the self-biased ring amplifier is essentially a cascade of inverters, we use an auto-zero offset canceling technique to set a bias point without any external bias. The auto-zero self-biasing scheme in a 1.5b flip-around MDAC gain stage is shown in Fig. 11.5.2 (The actual implementation is pseudo-differential). During Φ_1, input signal is sampled onto the C_1 and C_2 capacitors, and the amplifier offset is sampled on C_C. An auxiliary loading capacitor, C_{LA}, connected to the output during auto-zeroing stabilizes the ring amplifier and also samples the input offset voltage. C_{LA} is required because the next-stage sampling capacitors are disconnected during the Φ_1 sampling phase. C_{LA} is made twice as large as the loading from the next stage because the feedback factor during auto-zero/sampling is larger than that during the gain phase. However, the use of C_{LA} does not increase the dynamic power consumption since the sampled voltage on C_{LA} stays almost constant for every cycle.

Significantly, we also use the sampled offset voltage stored on C_{LA} to mitigate the gain error caused by the parasitic capacitance across the feedback auto-zero switch S_{AZ1}. As shown in Fig. 11.5.2, we form this switch as the series combination of two switches, S_{AZ1} and S_{AZ2}, instead of single switch. During the sampling phase Φ_1, both S_{AZ1} and S_{AZ2} are on and S_{AZ3} is off to achieve auto-zeroing. Then in the gain phase Φ_2, S_{AZ1} and S_{AZ2} are disconnected and S_{AZ3} connects the intermediate node V_{AZ} to the sampled offset voltage on C_{LA}. This method isolates the S_{AZ1} source-drain parasitic from changes in the output voltage during slewing and settling. By holding V_{AZ} constant at the auto-zero voltage during amplification, the parasitic capacitance is effectively grounded.

We exploit the characteristics of the ring amplifier to eliminate the comparators in a 1.5b-per-stage pipeline (Fig. 11.5.3). Our scheme uses the direction of the amplification to replace the sub-ADC comparators in the subsequent stage. At the beginning of the amplification phase, the output, V_{CP}, of the second inverter hits ground when the amplification result is higher than V_{CM}, and hits V_{DD} when the amplification result is lower than V_{CM}, then goes near to the self-bias voltage. This information is captured by a clock-gated low-threshold NOR gate and a low-power set flip-flop shown in Fig. 11.5.3(a). The flip-flop is preset low at the end of the sampling phase and goes high when the amplification result is expected higher than V_{CM}. The opposite situation is captured by the same circuit of the other pseudo-differential path. When the amplification result is close to V_{CM}, neither of the flip-flops is set. The flip-flop information is decoded by a simple logic circuit (Fig. 11.5.3(b)), and used as the next-stage coarse ADC result. The auto-zeroing ensures the correct decision direction. The threshold is set by the NOR window and the output slew rate. The pulse V_{CP} (Fig. 11.5.3(c)) is low while the amplifier is slewing positively, but NOR is not immediately enabled thereby setting a threshold relative to V_{CM}. The accuracy of the threshold is well within the redundancy of the 1.5b-per-stage pipeline. The switches S_{R1} and S_{R2} in Fig. 11.5.2 empty the capacitor before the sampling phase to ensure the amplification of the previous stage starts from V_{CM} in every cycle.

The self-biased ring amplifier is used in all stages of a 100MS/s, 10.5b pipeline ADC, implemented in 1P9M 65nm 1.2V CMOS (Fig. 11.5.4). The ADC is composed of a pseudo-differential flip-around ring amplifier based SHA with bootstrapped input switches, nine pseudo-differential 1.5b MDAC stages, and dummy loading for the last stage. Pseudo-differential CMFB in [1] is implemented in every stage. The MDAC references are set at 0.1V and 1.1V thanks to the wide swing of the ring amplifier. The SHA stage also provides the first MDAC digitization. To further optimize the power efficiency, the seven last MDAC stages are scaled down by half compared to the first two MDAC stages.

The ADC has a measured (Fig. 11.5.5) SNDR, SNR and SFDR of 56.3dB (9.06b), 56.7dB and 67.6dB, respectively, for a Nyquist frequency input sampled at 100MS/s and consumes 2.46mW, which results in a Figure-of-Merit of 46.1fJ/conv-step. The power consumption is measured for a Nyquist input and includes the clock buffer, digital correction, and references power. The prototype is more than 3× faster and has a 2× better FoM than [2], which uses a similar pipeline ADC structure (Fig. 11.5.6). The measured INL and DNL from the 10b output are +0.30/-0.34 LSB and +0.75/-0.81 LSB respectively. The SNDR remains stable up to V_{DD} of 1.28V, proving that the self-bias effectively tracks the power supply voltage variation. Figure 11.5.7 shows a die micrograph. The ADC occupies 0.1mm^2.

Acknowledgements:
The authors thank Berkeley Design Automation for simulation software.

References:

[1] B. Hershberg, *et al.*, "Ring Amplifiers for Switched-Capacitor Circuits," *ISSCC Dig. Tech. Papers*, Feb. 2012.

[2] B. Hershberg, *et al.*, "A 61.5dB SNDR Pipelined ADC Using Simple Highly-Scalable Ring Amplifiers," *VLSI Circ. Symp. Dig. Tech. Papers*, June 2012.

[3] B. Hershberg, *et al.*, "Ring Amplifiers for Switched Capacitor Circuits," *IEEE J. Solid-State Circuits*, Dec. 2012.

[4] B. Hershberg and U.-K. Moon, "A 75.9dB-SNDR 2.96mW 29fJ/conv-step Ringamp-Only Pipelined ADC," *VLSI Circ. Symp. Dig. Tech. Papers*, June 2013.

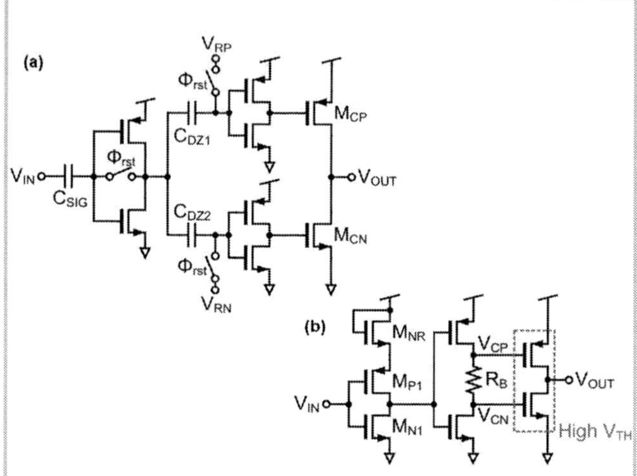

Figure 11.5.1: (a) Conventional ring amplifier, (b) self-biased ring amplifier.

Figure 11.5.2: Ring amplifier flip-around MDAC gain stage with auto-zero scheme.

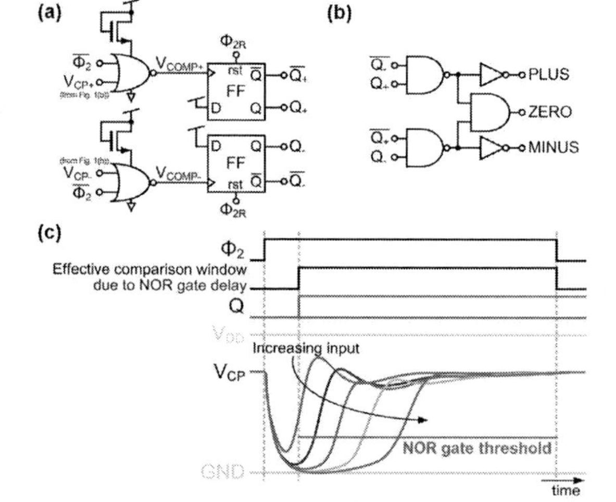

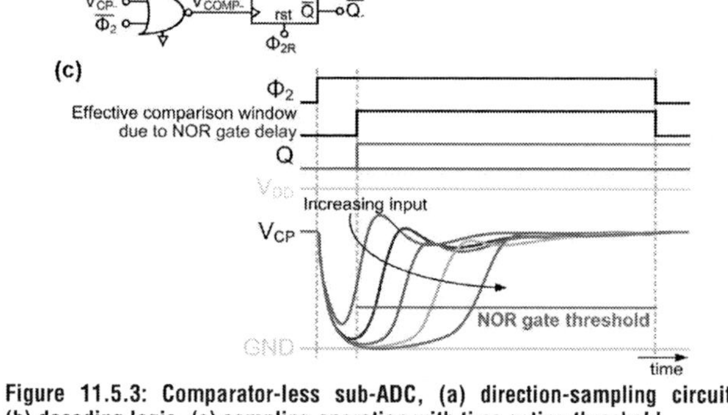

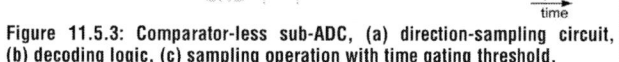

Figure 11.5.3: Comparator-less sub-ADC, (a) direction-sampling circuit, (b) decoding logic, (c) sampling operation with time gating threshold.

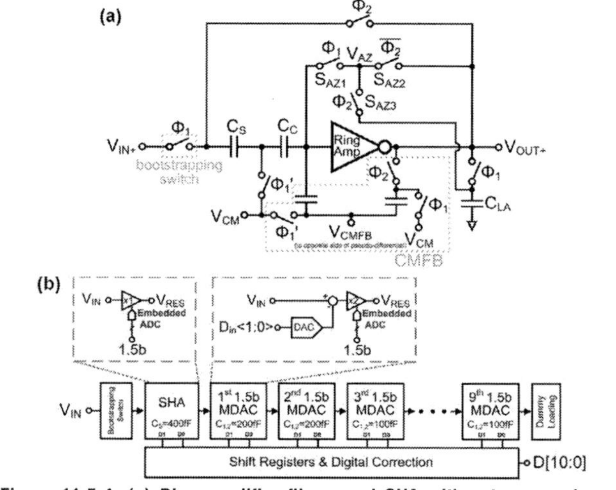

Figure 11.5.4: (a) Ring amplifier flip-around SHA with auto-zero scheme, (b) ADC structure.

Figure 11.5.5: Measurement results.

	This Work	ISSCC2012 [1, 3]	VLSI2012 [2, 3]	VLSI2013 [4]
Resolution	10.5bits	15bits	10.5bits	15bits
Analog Supply	1.2V	1.3V	1.3V	1.3V
Amp. Structure	Self-biased ring amplifier only	Ring amplifier + split-CLS	Ring amplifier only	Coarse + fine ring amplifier
Sampling Rate	100MSPS	20MSPS	30MSPS	20MSPS
Technology	65nm 1P9M CMOS	180nm 1P4M CMOS	180nm 1P4M CMOS	180nm 1P4M CMOS
Active Area	0.097mm^2	1.98mm^2	0.50mm^2	1.98mm^2
Input Range	2V$_{pk\text{-}pk}$ differential	2.5V$_{pk\text{-}pk}$ differential	2.2V$_{pk\text{-}pk}$ differential	2.5V$_{pk\text{-}pk}$ differential
SNDR at Nyquist	56.3dB	~73dB (from [3] graph)	~57dB (from [3] graph)	-
SFDR at Nyquist	67.6dB	~85dB (from [3] graph)	~72dB (from [3] graph)	-
ENOB at Nyquist	9.06bits	~11.83bits	~9.18bits	-
Total Power	2.46mW	5.1mW	2.6mW	2.96mW
FoM (with the best result)	46.1fJ/conv-step	45fJ/conv-step	90fJ/conv-step	29fJ/conv-step
FoM (with Nyquist freq.)	46.1fJ/conv-step	70fJ/conv-step	149.4fJ/conv-step	-

Figure 11.5.6: Performance comparison with conventional ring amplifier based pipeline ADCs.

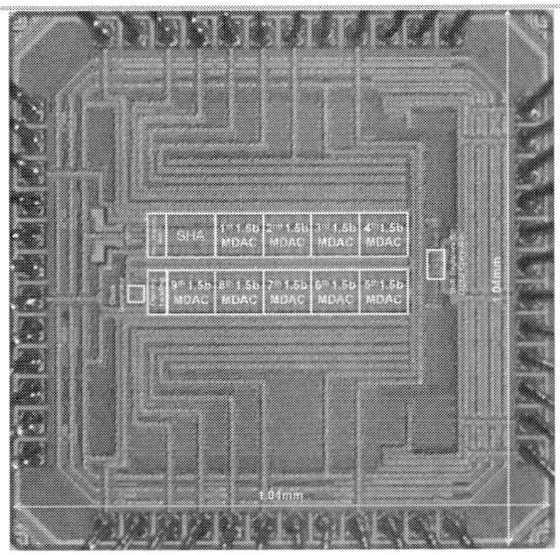

Figure 11.5.7: Die micrograph.

11.6 A 21mW 15b 48MS/s Zero-Crossing Pipeline ADC in 0.13μm CMOS with 74dB SNDR

Dong-Young Chang[1], Carlos Muñoz[2], Denis Daly[2], Soon-Kyun Shin[2], Kevin Guay[2], Thomas Thurston[2], Hae-Seung Lee[3], Kush Gulati[1], Matthew Straayer[2]

[1]Maxim Integrated, San Jose, CA,
[2]Maxim Integrated, North Chelmsford, MA,
[3]Massachusetts Institute of Technology, Cambridge, MA

Pipeline ADCs have traditionally served as a general-purpose architecture for high-speed and high-resolution applications such as medical and wireless receivers. Recently, achieving the highest levels of linearity with ultra-low power consumption has proven to be extremely challenging using modern CMOS technology with limited headroom. While zero-crossing-based circuits (ZCBC) have proven to be a power-efficient alternative to opamps in pipeline ADCs [1], performance using zero-crossing techniques have to-date only been demonstrated with ENOB ≤11. This paper presents a 15b 48MS/s zero-crossing-based pipeline ADC that achieves low power consumption of 99fJ/step and high linearity performance of 73.1dB SNDR and >80dB SFDR at Nyquist, demonstrating state-of-the-art FoM for thermal-noise-limited designs of 165.1dB.

Figure 11.6.1 shows a block diagram of the ADC, which is comprised of two 4.5b stages, each with a gain of 16 followed by a 7b FLASH/SAR quantizer, the clock generator circuits, and an on-chip reference buffer. It is well known that large gain is desirable in front-end stages of a pipeline ADC architecture to improve linearity and power efficiency. However, unlike a closed-loop operational amplifier that has a fixed gain-bandwidth product and must trade increased gain with reduced bandwidth, a zero-crossing decision is especially well-suited for amplification with large gain, and the ceiling for setting the gain then becomes the precision of the sub-ADC.

To maintain a reasonable amount of area and power consumption in the gain stage sub-ADC while achieving 33 levels and >6b precision, the work employs a coarse-fine sub-ADC architecture. First, a 4b coarse estimate of the input signal is made by a flash sub-ADC. The stage outputs are then reset and the reference voltage is applied to the MDAC capacitors in a traditional manner. Second, a 2b fine decision is made with calibrated comparators that monitor the residue voltage at the input of the zero-crossing detector (ZCD). 1b of overlap allows for relaxed offset requirements of the coarse decision, and only the fine sub-ADC thresholds require calibration to the full required precision. In this work, fine comparator calibration is performed first in the foreground and then in the background to ensure proper operation by observing the stage output residue range. Also note that an additional side benefit to the coarse-fine approach is that the fine decision is not sensitive to timing mismatch between the MDAC and bit-decision input sampling.

To address non-linearity in ZCBC, a dual ramp trajectory can be used that significantly reduces the ramp rate and the associated overshoot [1]. As depicted in Fig. 11.6.2, at the beginning of the charge-transfer phase, the fast coarse ramp current directly charges the output load (C_L) and feedback capacitor (C_{FB}). When V_X crosses a threshold less than zero (V_{TH1}), the coarse current is disabled and the ramp continues monotonically at a much slower rate until it crosses the second final threshold that is nominally zero. This approach requires both sufficient time for ZCD settling before the second slow ramp crosses zero, and also that the crossing occur before the end of the half-period charge transfer phase. Therefore, a challenge in ZCBC that has not been addressed in the literature is how to correctly set V_{TH1} over PVT to result in a desired second ramp time.

This paper proposes to adjust the first phase threshold (V_{TH1}) with a simple digital background control circuit that works by comparing the time of the actual zero-crossing with a desired time window. To explain, when V_{TH1} is too close to zero there will be a very short fine ramp period, and this is an event that can easily be detected by comparing to the output of a replica delay cell. Similarly, if V_{TH1} is too far away from zero, the zero-crossing detection will happen after the half-clock period has ended, an event that is also easily detected. These two error signals are digitally integrated and a control DAC adjusts V_{TH1} appropriately.

Compared to opamp-based circuits, a primary difference in ZCBC is that a small amount of current is flowing during the moment of stage 1 output sampling. As shown in Fig. 11.6.3, the constant charging current is switched into the reference buffer based on the bit-decisions, and therefore an IR drop (V_{ERR}) across the reference voltage source impedance (R_{REF}) modulates the effective reference voltage presented to the MDAC, causing primarily 2nd-order distortion. To eliminate this IR drop, we use a set of capacitors that are cross-coupled to the differential stage outputs. These additional capacitors compensate the reference current with an equal amount of current flowing in the opposite direction. Further, the compensation current is ratiometric, allowing for effective cancellation regardless of the shape or slope of the output waveform. The total penalty of increased stage 1 output load capacitance is <30%.

After the two zero-crossing-based gain stages have efficiently and precisely processed the input residue for thermal-noise limited stages, the task then becomes how to most efficiently quantize the stage-2 output. In this work, a 15-level FLASH is used for the first decisions, and then 4 conventional SAR decisions follow with 1b of overlap to result in 7b of quantization. This approach offers several advantages over a final FLASH quantizer, which is traditionally used in pipeline ADCs. First, a SAR ADC is very efficient when quantizing in the non-thermal noise region compared to active amplification of any kind. Second, there is a free half-clock cycle available for further processing before the gain stage charge transfer. Finally, the use of a single comparator threshold for final quantization eliminates comparator offset from contributing to DNL and INL.

The ADC is designed in 0.13μm CMOS technology, and occupies 1100×875μm^2 active die area for each of the I and Q channels. Figure 11.6.4 shows the measured power spectrum of a typical ADC output for a 2V peak-to-peak differential input sine wave using a high quality 50MHz crystal for the clock. A 1.8V supply is used for the reference buffer, and 1.2V is used for the ADC core. Current consumption is 3.9mA from the 1.8V supply and 12.2mA from the 1.2V supply for a total of 21.6mW at 50MS/s including on-chip reference buffer. Silicon die from 3 corners (TT, SS, FF) were obtained and better than 73dBFS SNDR was measured up to Nyquist input. The SNDR/SFDR performance curves for 15 parts (5 SS, 5 TT, 5 FF) over ±10% supply variation and temperature from -40 to 105°C are shown in Fig. 11.6.5 with 10MHz input frequency and 48MS/s sampling rate. For these measurements, power-on capacitor calibration for each part was conducted once at nominal conditions and then held across voltage and temperature. All the measured parts demonstrate very robust performance of >72.5dBFS SNDR.

In this work, we introduce multiple techniques to establish robust performance: a coarse-fine bit decision with efficient calibration to allow for only 2 pipeline gain stages, a control circuit and algorithm for establishing proper dual-ramp waveforms, a reference current compensation circuit that relaxes the reference buffer impedance requirements, and a backend FLASH/SAR quantizer. The performance of this work compared to other ADCs in the range of 40 to 80MS/s is summarized in Fig. 11.6.6. In relation to other zero-crossing ADCs, this work represents an improvement in SNDR of more than 6dB, and in relation to other ADCs we demonstrate state-of-the-art SNDR and FoM across PVT. The die photograph of the IC is shown in Fig. 11.6.7.

References:

[1] S. Shin, et al., "A 12b 200MS/s Frequency Scalable Zero-Crossing Based Pipelined ADC in 55nm CMOS," *Proc. IEEE CICC*, pp. 1-4, 2012.

[2] R Kapusta, et al., "A 12b 80MS/s SAR ADC with 73.6dB SNDR in 65nm CMOS," *ISSCC Dig. Tech. Papers*, pp. 472-473, Feb 2013.

[3] J. Brunsilius, et al., "A 16b 80MS/s 100mW 77.6dB SNR CMOS Pipeline ADC," *ISSCC Dig. Tech. Papers*, pp. 186-187, Feb 2011.

[4] S. Lee, et al., "A 12b 5-to-50MS/s 0.5V-to-1V Voltage Scalable Zero-Crossing Based Pipelined ADC," *Proc. IEEE ESSCIRC*, pp. 355-358, 2011.

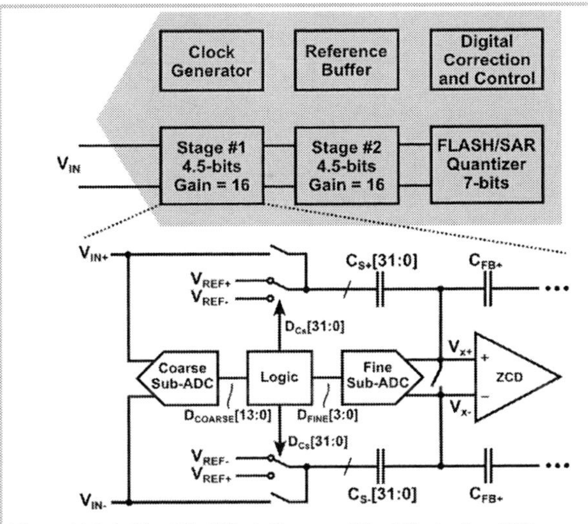

Figure 11.6.1: Simplified block diagram of the 15b pipeline ADC.

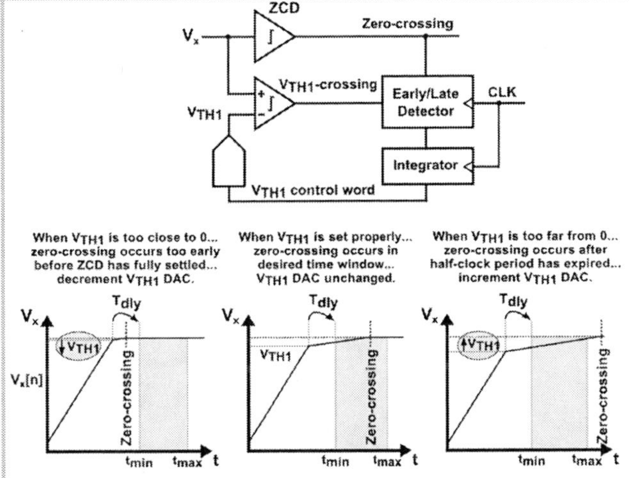

Figure 11.6.2: Coarse phase threshold (VTH1) control algorithm using a time window.

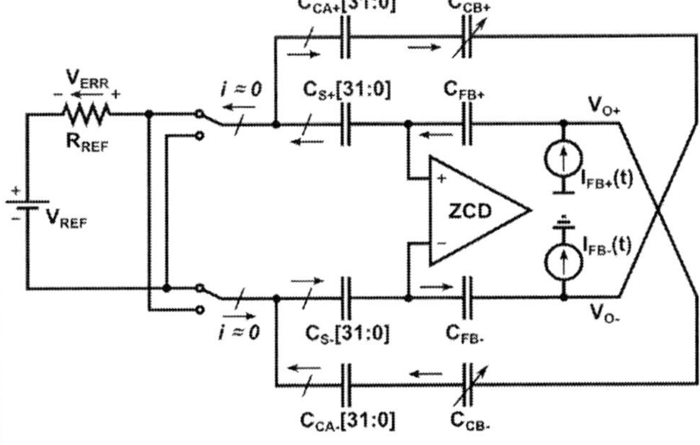

Figure 11.6.3: Schematic of reference-current compensation.

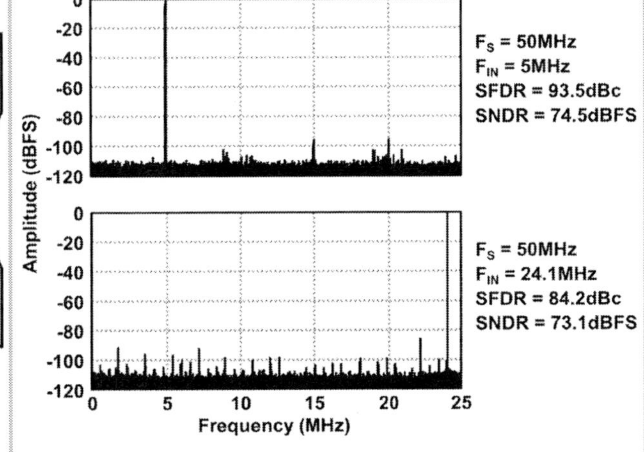

Figure 11.6.4: Measured output spectrum with FIN = 5MHz, 24MHz and FS = 48MHz.

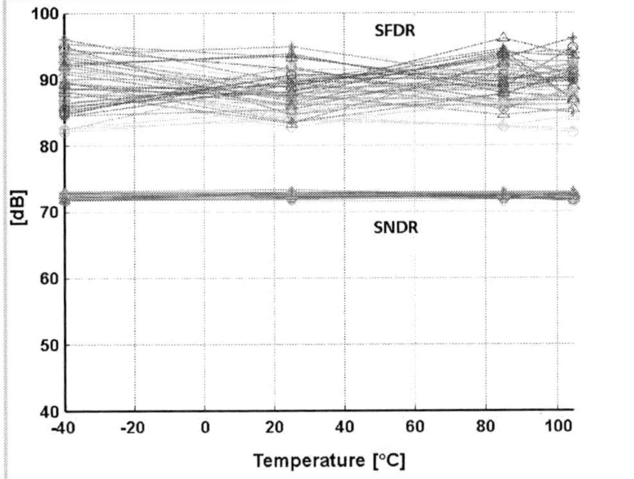

Figure 11.6.5: Performance of 15 parts (5 SS, 5 TT, 5 FF) over ±10% supply and temperature.

	This Work	ISSCC 2013 [2]	ISSCC 2011 [3]	ESSCIRC'11 [4]
Technology	130nm CMOS	65nm CMOS	180nm CMOS	65nm CMOS
Architecture	ZCBC Pipeline	Interleaved SAR	Opamp Pipeline	ZCBC Pipeline
Resolution	15b	14b	16b	12b
Conversion Rate	50MS/s	80MS/s	80MS/s	50MS/s
Input Signal Range	$2V_{p-p,diff}$	$2.4V_{p-p,diff}$	-	$1.67V_{p-p,diff}$
SNDR	74.5dBFS @ 5MHz 73.1dBFS @ 24MHz	73.6dB @ 5MHz 71.3dB @ 40MHz	77.6dB @ 9.7MHz	67.7dB @ 23.9MHz
SFDR	95dBc @ 5MHz 84.2dBc @ 24.1MHz	85.7dB @ 5MHz ≥80.3dB @ 40MHz	95dB @ 9.7MHz	85.8dB @ 23.9MHz
Power Consumption	21.6mW	31.1mW	100mW	4.1mW
FOM1 $[= SNDR + 10\log_{10}(f_s/2/P)]$	165.1dB	164.7dB	163.6dB	165.6dB
FOM2 $[= P/(2^{ENOB} \times f_s/2)]$	99fJ/step	99fJ/step	201fJ/step	41fJ/step

Figure 11.6.6: Performance summary and comparison.

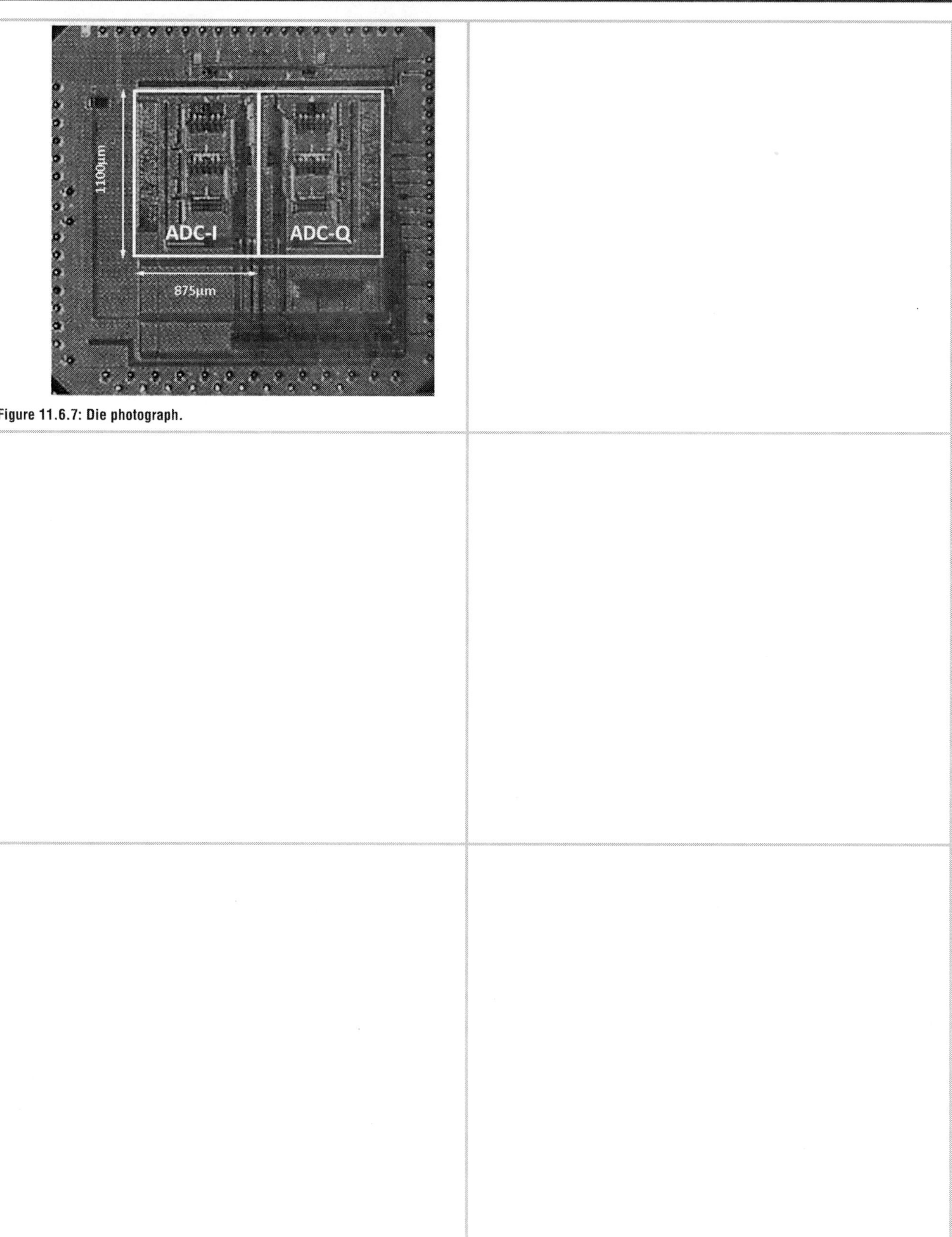

Figure 11.6.7: Die photograph.

11.7 A 240mW 16b 3.2GS/s DAC in 65nm CMOS with <-80dBc IM3 up to 600MHz

Hans Van de Vel, Joost Briaire, Corné Bastiaansen, Pieter van Beek, Govert Geelen, Harrie Gunnink, Yongjie Jin, Mustafa Kaba, Kerong Luo, Edward Paulus, Bang Pham, William Relyveld, Peter Zijlstra

Integrated Device Technology, Eindhoven, The Netherlands

Advanced wireless cellular infrastructure systems require DACs with high spectral purity over a wide bandwidth and which are fit for integration of multiple transmit channels with DSP. This calls for IM3 linearity better than -80dBc up to high frequencies and low power dissipation. In this paper, a high-speed current-steering DAC is reported that combines low power and high linearity, enabled by a 3-dimensional sort-and-combine (3D-SC) calibration technique, CML switch-driving circuitry, and switch cascoding. It achieves similar linearity at significantly reduced power compared to a state-of-the-art high-linearity DAC [1] while its IM3 performance is more than 10dB better than the multi-GS/s low-power DAC in [2]. The 16b 3.2GS/s DAC is implemented in a 65nm CMOS process and achieves -80dBc IM3 up to 600MHz while dissipating 240mW from 1.2V and 3.3V supplies.

Figure 11.7.1 shows a block diagram of the DAC. The 16b binary-coded digital input data is decoded in a programmable decoder towards 63 MSB units and 10 LSB units. In each unit the data is re-clocked in a CML flip-flop followed by a switch driver. The unit's output stage is a stack of a cascoded current source, a switch pair, and two sets of switch cascodes. The two sets of cascodes allow the DAC to switch between normal operation mode with V_{NOM} high and V_{EMM} low, and error measurement mode with V_{NOM} low and V_{EMM} high. In normal operation mode the output currents are summed and sent to off-chip resistors. In error-measurement mode each unit's matching error is measured on-chip relative to a reference unit. An AC error signal is generated by switching the unit to be measured and the reference unit as square waves with opposite phase and then summing their output currents. This combined output current is down-converted to DC with a passive mixer followed by a feedback amplifier and digitized in a $\Delta\Sigma$ ADC [3].

Achieving high linearity requires minimizing of unit matching errors. Amplitude matching errors limit low-frequency linearity while mismatch in timing dominates at high frequencies. Timing mismatch in a current-steering DAC is determined by mismatch in delay and duty-cycle of the switched-current waveforms. To minimize by design these 3 dimensions (3D) of unit matching errors, amplitude, delay and duty-cycle, requires the use of large current sources and fast switches, and consequently high power dissipation in switch drivers. This matching-limited power dissipation scales poorly with technology. In this work calibration reduces the linearity degradation due to 3D matching errors, enabling the use of smaller current sources and lower driving power. A 3D calibration technique through sorting only has been demonstrated earlier in the DAC design in [3], and a 1D sort-and-combine calibration technique to improve low-frequency linearity is shown in theory in [4]. The DAC design presented here employs a 3D sort and combine (3D-SC) calibration technique that sorts-and-combines DAC units with opposing 3D matching errors. The on-chip error measurement circuit is able to measure 3D matching errors by using minimally 3 different values for phase and frequency of the mixer's LO input as described in [3].

Figure 11.7.2 illustrates the 3D-SC technique where for clarity a 3b thermometer-coded DAC and 1D optimization are considered. 1D matching errors in current-source amplitude are represented as differences in height of the units, while 3D matching errors are shown as 3D error vectors. In a conventional 3b thermometer-coded DAC each thermometer-coded data-bit T_1 to T_7 is connected to a unit current source I_1 to I_7, where I_{AVG} is the average current source amplitude. The first step in sorting-and-combining is to select the unit with current source amplitude closest to I_{AVG} (I_7 in this case). Then all other units are sorted-and-combined such that their combined level becomes as close as possible to $2\times I_{AVG}$. Next, binary-coded LSB data is assigned to I_7 and new thermometer-coded data T_1 to T_3 is assigned to the remaining combined units. As a result, the thermometer data segmentation level decreases with 1b. A programmable decoder translates the 3b binary input data into 3 thermometer-coded MSBs and 1 binary-coded LSB, and assigns this data to the individual units in a MUX-array that is programmed according to the sorting-and-combining outcome. For 3D matching errors the principle remains, but now the units should be combined such that their 3D error vectors are minimized. This combining of units can be extended by repeating it. So, in a second round, the previously combined units can be combined again based on their remaining errors such that one arrives at sets of 4 units. The thermometer data segmentation level then decreases with an extra 1b and the programmable decoder translates into 1 thermometer-coded MSB and 2 binary coded LSBs. This sorting-and-combining into sets of 4 units is used in the 16b DAC described here. The 63 MSB units as shown in Fig. 11.7.1 are then programmed to be driven as 15 combinations of four, 1 combination of two and 1 single unit. The programmable decoder translates the 16b binary input data into 15 thermometer-coded MSBs and 12 binary-coded LSBs, reflecting the 2b decrease in thermometer data segmentation level.

Linearity degradation other than mismatch is reduced through the use of CML switch-driving circuitry and switch cascoding. The flip-flop and switch driver in Fig. 11.7.1, are implemented as low-swing, differential CML circuits. The static biasing with a DC tail current source, results in low data-dependent supply bounce. The high crossing point of the switch driver minimizes the glitch at the switch's source node and the associated inter-symbol interference. Furthermore using low-swing switch driver signals reduces gate-drain charge feed-through and channel charge injection.

The switch cascodes in Fig. 11.7.1 are implemented as 3.3V thick-oxide devices. The cascoding improves the screening of the switch source node from output signal feed-through. The cascodes enable a 3.3V supply for the output stage such that a maximal $2V_{pp\text{-}diff}$ output signal swing can be accommodated with sufficient voltage headroom. They are biased such that the drain node of the thin-oxide switches does not exceed 1.2V. In off-state the switch cascodes maintain a DC bias current thereby reducing the modulation of the DAC's output impedance [2].

The DAC is realized in a 1P7M 65nm CMOS process and dissipates 240mW from 1.2V and 3.3V supplies. It achieves -81dBc IM3 at 610MHz output frequency and 3.2GS/s sampling rate, as shown in the spectrum plot in Fig. 11.7.3. A plot of IM3 versus output frequency is shown in Fig. 11.7.4, with a comparison to state-of-the-art CMOS DACs with similar sampling rates. The uncalibrated DAC achieves similar IM3 as the low-power DAC in [2]. After 3D-SC calibration the IM3 improves by approximately 10dB over the complete Nyquist frequency range and remains below -80dBc up to 610MHz. Figure 11.7.5 shows a plot of SFDR versus output frequency. The DAC's performance is summarized in Fig. 11.7.6, where it can be seen that the DAC presented here dissipates significantly less power than [1] while it achieves a similar bandwidth for which IM3 is lower than -80dBc. Furthermore at 600MHz it achieves 13dB better IM3 than [2] for similar power. Figure 11.7.7 shows a die micrograph of the DAC which is part of an IC with 2 transmit channels and DSP, illustrating the DAC's integration capability.

Acknowledgements:
The authors would like to thank Peter Munter, François Beschon, Vincent Fresnaud, Patrice Géfart, Hugues Malezieux and Xavier Sergent for their support and contributions.

References:
[1] G. Engel, S. Kuo, and S. Rose, "A 14b 3/6GS/s Current-Steering RF DAC in 0.18um CMOS with 66dB ACLR at 2.9GHz," *ISSCC Dig. Tech. Papers*, pp. 458-459, Feb. 2012.
[2] C.-H. Lin, F.M.L. van der Goes, J.R. Westra, *et al.*, "A 12 bit 2.9 GS/s DAC With IM3 < -60 dBc Beyond 1 GHz in 65 nm CMOS," *IEEE J. Solid-State Circuits*, vol. 44, no. 12, pp. 3285-3293, Dec. 2009.
[3] Y. Tang, J. Briaire, K. Doris, *et al.*, "A 14 bit 200 MS/s DAC with SFDR >78dBc, IM3 <-83dBc and NSD <-163dBm/Hz Across the Whole Nyquist Band Enabled by Dynamic Mismatch Mapping", *IEEE J. Solid-State Circuits*, vol. 46, no. 6, pp. 1371-1381, June 2011.
[4] T. Zeng, D. Chen, "New Calibration Technique for Current-Steering DACs", *Proc. IEEE ISCAS*, pp. 573-576, May 2010.

978-1-4799-0917-9/14 $31.00 © 2014 IEEE

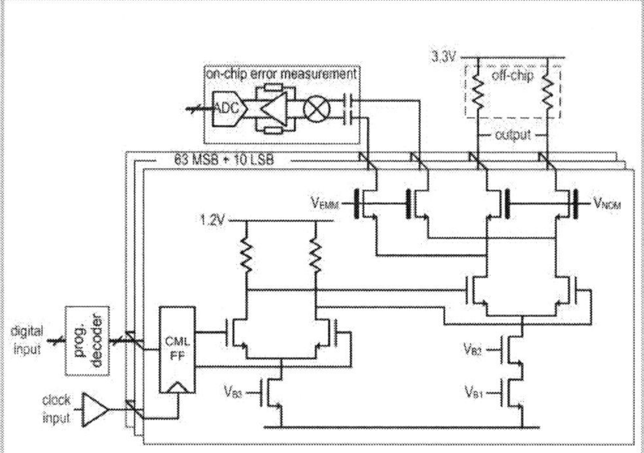

Figure 11.7.1: DAC block diagram and MSB unit cell circuit details. Biasing circuitry is not shown.

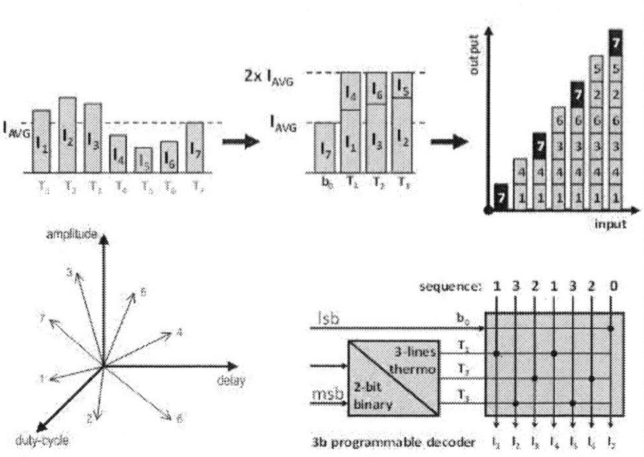

Figure 11.7.2: Illustration of the 3D-SC calibration technique.

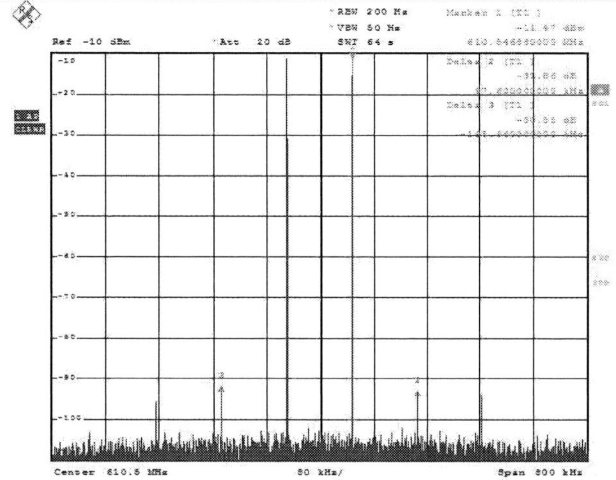

Figure 11.7.3: Two-tone spectrum plot at 610MHz output frequency. IM3 equals -80.85dBc.

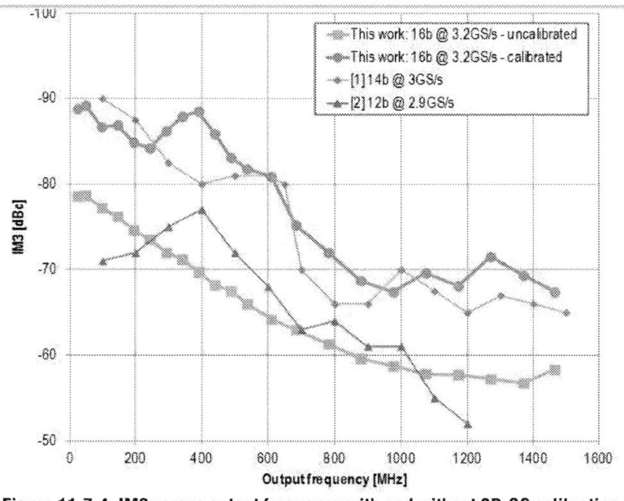

Figure 11.7.4: IM3 versus output frequency with and without 3D-SC calibration compared to state-of-the-art multi-GS/s CMOS DACs.

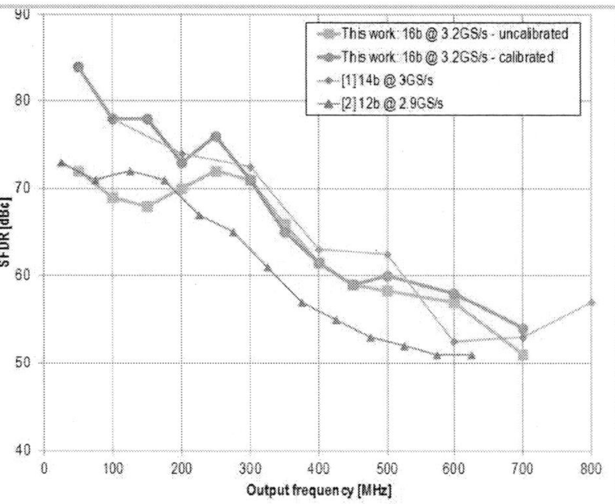

Figure 11.7.5: SFDR versus output frequency with and without 3D-SC calibration compared to state-of-the-art multi-GS/s CMOS DACs.

	This work	[1]	[2]
Technology	65nm CMOS	0.18µm CMOS	65nm CMOS
Supply (V)	1.2 / 3.3	1.8 / 3.3	1 / 2.5
Resolution	16	14	12
Sampling rate (GS/s)	3.2	3	2.9
Power (mW)	240	600	188
I_{load} (mA)	20 (max. 40)	20	50
$V_{pp\text{-}diff}$ (V)	1 (max. 2)	1	2.5
$BW_{IM3<-80dBc}$ (MHz)	610	650	n/a
IM3 at 600MHz (dBc)	-81	-81	-68
SFDR at 600MHz (dBc)	58	53	51
NSD (dBm/Hz)	-168	/	/

Figure 11.7.6: Performance summary and comparison to prior work.

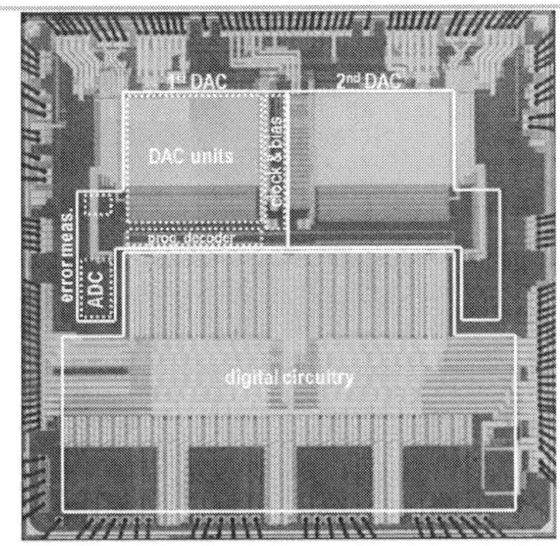

Figure 11.7.7: Die micrograph of a chip comprising 2 channels of the presented DAC, integrated with DSP.

Session 12 Overview: *Sensors, MEMS, and Displays*

IMMD SUBCOMMITTEE

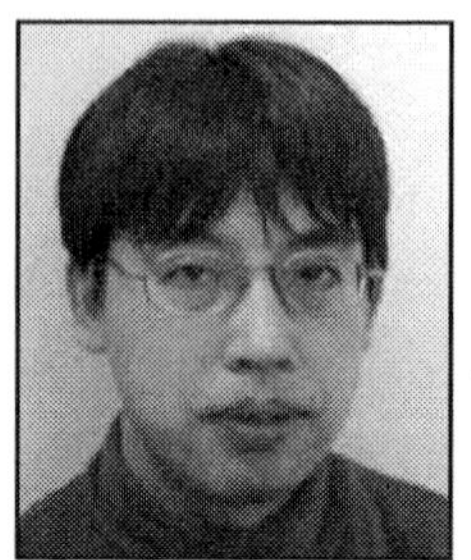

Session Chair: *Ralf Brederlow*
Texas Instruments, Freising, Germany

Session Co-Chair: *Yoshiharu Nakajima*
Japan Display, Ebina, Japan

This session presents recent achievements in the area of gesture sensing, capacitive touch displays, time, pressure and temperature sensors. Two advanced gesture-recognition systems are extending the touch experience into the 3rd dimension. These presentations are followed by three papers showing resolution- and size-related improvements in touch-screen display interfaces. An innovative, ultra-low-power capacitive interface circuit for pressure sensors is followed by two presentations on novel temperature-sensing principles pushing barriers of low-voltage operation and energy efficiency. Last but not least, an extremely accurate MEMS-based real-time clock is presented.

12.1 3D Ultrasonic Gesture Recognition **8:30 AM**

R. J. Przybyla, University of California, Berkeley, CA

In Paper 12.1, the University of California at Berkeley (with UC Davis) presents an ultrasound-time-of-flight-based 3D gesture-recognition system. The system enables object localization with up to 0.4mm resolution to a distance of up to 0.5m.

12.2 3D Gesture-Sensing System for Interactive Displays Based on Extended-Range Capacitive Sensing **9:00 AM**

Y. Hu, Princeton University, Princeton, NJ

In Paper 12.2, Princeton University demonstrates 30cm out-of-plane sensing for a 40x40cm^2 gesture-sensing system that relies on the minimization of the effects of stray capacitance.

12.3 A 240Hz-Reporting-Rate 143×81 Mutual-Capacitance Touch-Sensing Analog Front-End IC with 37dB SNR for 1mm-Diameter Stylus **9:15 AM**

M. Hamaguchi, Sharp, Fukuyama, Japan

In Paper 12.3, Sharp talks about a record-size 70-inch display capacitive touch-sensing system with 224 channels.

12.4 A 1mm-Pitch 80×80-Channel 322Hz-Frame-Rate Touch Sensor with Two-Step Dual-Mode Capacitance Scan 9:30 AM

N. Miura, Kobe University, Kobe, Japan

In Paper 12.4 Kobe University (with Panasonic) combines self and mutual capacitance sensing principles to achieve 1mm touch resolution at 322Hz frame rate, thus getting rid of ghosting artifacts.

12.5 2D Coded-Aperture-Based Ultra-Compact Capacitive Touch-Screen Controller with 40 Reconfigurable Channels 10:15 AM

H. Jang, KAIST, Daejeon, Korea

In Paper 12.5 KAIST (with Sentron) shows an extremely small, capacitive touch sensor chip consuming only 2.6mW of power with a total area of 0.46mm^2.

12.6 A 160nW 63.9fJ/conversion-step Capacitance-to-Digital Converter for Ultra-Low-Power Wireless Sensor Nodes 10:45 AM

H. Ha, Pohang University of Science and Technology, Pohang, Korea

In Paper 12.6 Pohang University (with the University of Michigan) presents a capacitive-to-digital converter using successive-approximation principles using a predictive baseline cancelation scheme to achieve a record figure-of-merit of 63.9fJ per conversion-step.

12.7 A 0.85V 600nW All-CMOS Temperature Sensor with an Inaccuracy of ±0.4°C (3σ) from -40 to 125°C 11:15 AM

K. Souri, Delft University of Technology, Delft, The Netherlands

In Paper 12.7 TU Delft (with NXP Semiconductors and Yonsei University) demonstrates a sub-1V temperature sensor based on dynamic threshold MOSFETs as reference with an accuracy of 0.4°C from a single-temperature trim.

12.8 A BJT-Based CMOS Temperature Sensor with a 3.6pJ·K^2-Resolution FoM 11:30 AM

A. Heidary, Smartec, Breda, The Netherlands;
Delft University of Technology, Delft, The Netherlands
and Guilan University, Rasht, Iran

In Paper 12.8 Smartec BV (with TU Delft, Guilan University, and SenseArt) presents a new temperature sensor principle resulting in 3× better energy efficiency compared to the state of the art.

12.9 A 1.55×0.85mm^2 3ppm 1.0μA 32.768kHz MEMS-Based Oscillator 11:45 AM

S. Zali Asl, SiTime, Sunnyvale, CA and Oregon State University, Corvallis, OR

In Paper 12.9 SiTime (with Oregon State University, UCLA, and TU Delft) describes a temperature-compensated MEMS oscillator system with a frequency stability of 3ppm for time-keeping applications.

12.1 3D Ultrasonic Gesture Recognition

Richard J. Przybyla[1], Hao-Yen Tang[1], Stefon E. Shelton[2], David A. Horsley[2], Bernhard E. Boser[1]

[1]University of California, Berkeley, CA, [2]University of California, Davis, CA

Optical 3D imagers for gesture recognition suffer from large size and high power consumption. Their performance depends on ambient illumination and they generally cannot operate in sunlight. These factors have prevented widespread adoption of gesture interfaces in energy- and volume-limited environments such as tablets and smartphones. Wearable mobile devices, too small to incorporate a touchscreen more than a few fingers wide, would benefit from a small, low-power gestural interface. Gesture recognition using sound is an attractive alternative to overcome these difficulties due to the potential for chip-scale size, low power consumption, and ambient light insensitivity. Using pulse-echo time-of-flight, MEMS ultrasonic rangers work over distances of up to a meter and achieve sub-mm ranging accuracy [1,2]. Using a 2-dimensional array of transducers, objects can be localized in 3 dimensions.

This paper presents an ultrasonic 3D gesture-recognition system that uses a custom transducer chip and an ASIC to sense the location of targets such as hands. The system block diagram is shown in Fig. 12.1.1. Targets are localized using pulse-echo time-of-flight methods. Each of the 10 transceiver channels interfaces with a MEMS transducer, and each includes a transmitter and a readout circuit. Echoes from off-axis targets arrive with different phase shifts for each element in the array. The off-chip digital beamformer realigns the signal phase to maximize the SNR and determine target location.

The 450μm diameter piezoelectric micromachined ultrasound transducers (pMUTs) used in this work are made up of a 2.2μm thick AlN/Mo/AlN/Al stack deposited on a Si wafer and released with a back-side through-wafer etch. The bottom electrode is continuous, while each pMUT has a top electrode lithographically defined to actuate the trampoline mode. Each pMUT can transmit and receive sound waves, and is operated at its resonance of 217kHz ± 2kHz with a bandwidth of 12kHz. The impedance of the transducers is dominated by the 10pF transducer capacitance, and the motional resistance at resonance is ~2.4MΩ. The resonant frequencies of the pMUTs vary due to fabrication, temperature, and packaging stress, so online frequency tracking is used to maintain maximum SNR during operation.

Two pMUTs are used for transmission and seven for reception as illustrated in Fig. 12.1.1. The receive array is 3.5 wavelengths wide in the x-angle axis, allowing targets separated by more than 15° to be distinguished. In the y-angle axis the array is only 0.16 wavelengths wide, sufficient to determine the y-angle to the target by measuring the average phase difference along the y axis of the array. The center element of the receive array and the element 900μm above it are used to launch a 138μs ≈ 24mm long pulse of sound into the environment. The transmit configuration illuminates a wide field of view, permitting the capture of an entire scene in a single measurement. Applications requiring better target resolution or greater maximum range can also use transmit beamforming at the expense of reduced measurement rate.

Each cycle begins with the launch of an acoustic pulse. Figure 12.1.2 shows the schematic of a single channel. High-voltage level shifters actuate the S_{TX} transmit switches, setting the transducer's bottom electrode to 16V to permit bi-polar actuation of the transducer. The transmitter then excites the transducer with a $32V_{pp}$ square wave for 30 cycles at the transmit frequency f_{TX} which is locked to $1/16^{th}$ of the sampling frequency f_s. At the end of the transmit phase, the mechanical energy stored in the inertia of the pMUT dissipates and the pMUT rings down at its natural frequency. The S_{RX} receiver isolation switches are turned on, and a resistor converts the ringdown current to a voltage that is subsequently amplified and digitized by the receiver normally. The ringdown signal is then I/Q demodulated with f_{TX}. The slope of the phase signal during the ringdown indicates the frequency offset and is used to update the f_s and f_{TX} used in the next measurement. Figure 12.1.3 shows the offset measured by the frequency autotuning loop as it is enabled. An initial 57kHz offset frequency is nulled to 1kHz within 30 measurement cycles.

After 86μs, the ringdown signal has decayed sufficiently for the S_{ring} switch to be opened, beginning the processing of received echoes. At this point, the signal from the transducer is integrated on the transducer's capacitance, and the front-end measures a voltage that is proportional to the displacement of the transducer's membrane.

The front-end amplifier consists of an open-loop current-reuse OTA with both NMOS and PMOS differential pairs biased near subthreshold for current efficiency. The front-end current is integrated onto the integrating capacitor of the second stage, which also makes up an integrator in the first of two switched capacitor resonators. Although the second stage is a switched capacitor integrator, the front-end current is processed in a continuous-time fashion before it is sampled at the output of the second integrator. As a result, the second integrator acts as an anti-aliasing filter for the wideband noise generated by the front-end and prevents this noise, the dominant noise source in the receiver architecture, from being aliased into the band of interest.

The signal then passes through a second switched-capacitor resonator and is quantized by a comparator. The high in-band gain provided by the 4th-order bandpass filter shapes the wideband quantization noise to be away from the signal at f_{TX}. The SC resonators are designed to resonate at 1/16 of the sampling frequency f_s, which is locked to the transducer's resonance by the ringdown autotuning circuit. This centers the bandpass ΔΣ's noise notch on the signal at f_{TX}.

The output of each ΔΣ ADC is I/Q demodulated, filtered, and downsampled off-chip. A digital beamformer [3] processes the received signals to maximize the receive SNR and determine the x-angle location of the target. This process can be repeated in the orthogonal angle axis to implement 3D beamforming; in this work we forgo 3D beamforming since the tiny y-axis aperture does not provide any y-axis resolution.

Thermal noise in the front-end amplifier and the thermal motion of air limit the minimum detectable echo. The input-referred noise of the amplifier is 11nV/√Hz, and the noise voltage of the transducer is 6nV/√Hz at resonance. Figure 12.1.4 shows the measured signal-to-noise ratio vs. range for a 127mm×181mm flat rectangular target. Figure 12.1.4 also shows the rms error in the range and direction measurement. Amplitude noise in the received signal limits the accuracy of the time-of-flight estimate. Figure 12.1.5 shows the output of the digital beamformer from a single measurement, which captures the echoes from a user's hands and head as he poses as shown. The system tracks objects between 45mm to 1m away and over an angular range of ±45°. Echoes from targets at a range of 1m return after 5.8ms, and this sets the maximum measurement rate of the system at 172 frames per second (fps).

Figure 12.1.7 shows a micrograph of the readout IC, which is fabricated in a 0.18μm CMOS process with 32V transistors. For a 1m maximum range, the system presented here uses 13.6μJ per measurement. At 30fps, the receive power consumption is 335μW and the transmit power consumption is 66μW. The energy consumption scales roughly linearly with maximum range. For a maximum range of 0.3m, the energy per frame is reduced to <0.5μJ per channel per frame. Single-element range measurements can be conducted at 10fps using only 5μW.

Figure 12.1.6 compares the performance of this system to an earlier MEMS ultrasonic 1D rangefinder [4] and two recent optical 3D rangers [5,6]. This ultrasonic 3D rangefinder offers dramatically reduced energy consumption compared to optical methods while permitting 3D target tracking. The energy consumption trades off with performance, permitting continuous operation in even tiny mobile devices. These characteristics enable energy-efficient gestural interfaces in applications such as smartphones and tablets, and permit gestural user interfaces in tiny mobile devices too small to accommodate a conventional touchscreen.

References:

[1] R. Przybyla, *et al.*, "In-air Ultrasonic Rangefinding and Angle Estimation using an Array of AlN Micromachined Transducers," in *Proc. Hilton Head Workshop*, pp. 50-53, 2012.

[2] R. Przybyla, *et al.*, "A Micromechanical Ultrasonic Distance Sensor With >1 Meter Range," in *Transducers Dig. Tech. Papers*, pp. 2070-2073, 2011.

[3] M. Skolnik, *Introduction to Radar Systems*. 3rd edition, McGraw-Hill, 2001.

[4] C. Kuratli and Q. Huang, "A CMOS Ultrasound Range-Finder Microsystem," *IEEE J. Solid-State Circuits*, vol.35, no.12, pp. 2005-2017, Dec. 2000.

[5] W. Kim, *et al.* "A 1.5Mpixel RGBZ CMOS Image Sensor for Simultaneous Color and Range Image Capture," *ISSCC Dig. Tech. Papers*, pp. 392-393, Feb. 2012.

[6] O. Shcherbakova, *et al.*, "3D Camera Based on Linear-Mode Gain-Modulated Avalanche Photodiodes," *ISSCC Dig. Tech. Papers*, pp. 490-491, Feb. 2013.

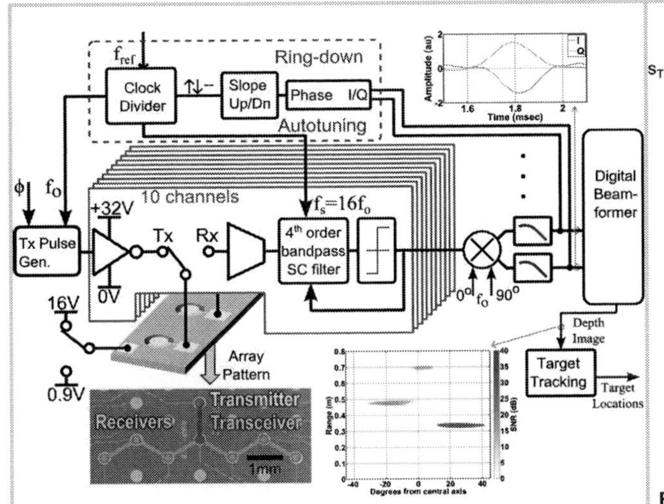

Figure 12.1.1: System block diagram.

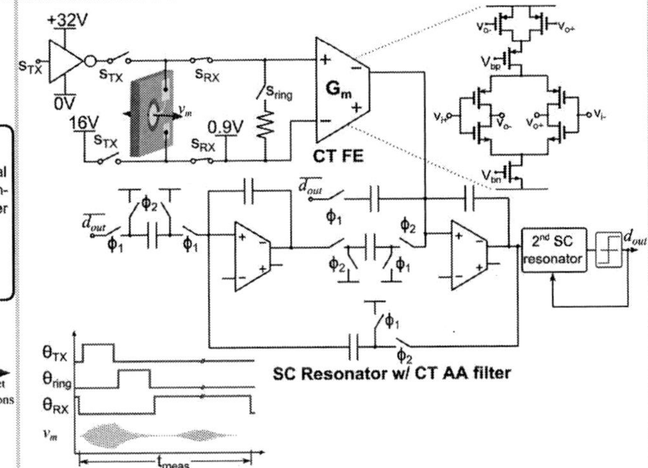

Figure 12.1.2: Readout circuit with mixed CT/SC architecture for inherent antialiasing. All structures are implemented differentially.

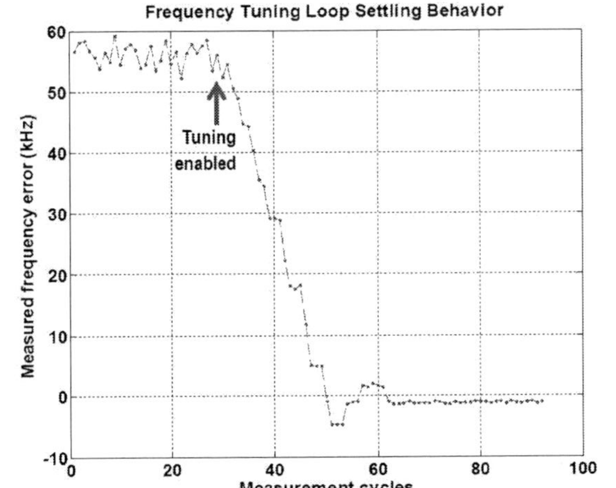

Figure 12.1.3: Ringdown frequency offset measurement and tuning loop settling behavior.

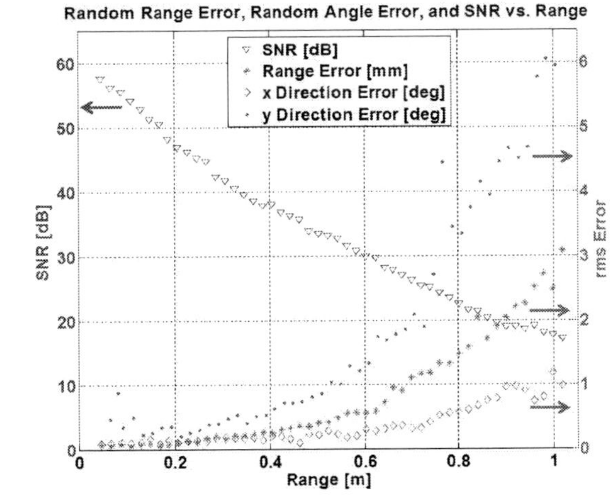

Figure 12.1.4: Signal-to-noise ratio and target localization accuracy vs. range for 127mm×181mm flat rectangular target.

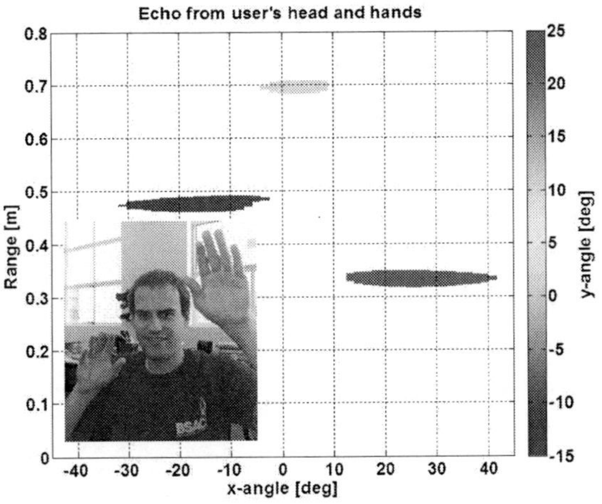

Figure 12.1.5: Echo from user's hands and head when posing as shown. Color axis shows y-angle position of the targets. Beamformed data is thresholded at 12dB SNR.

	This work	[4]	[5]	[6]
Method	Time-of-flight sound	Phase-shift sound	Phase-shift light	Phase-shift light
Transducers	AlN pMUT array	Si Thermal Ultrasound	Pinned Photodiode	Avalanche Photodiode
Carrier wavelength / Modulation wavelength	1.6mm / 48mm	3.6mm / 250mm	850nm / 15m	850nm / 12m
CMOS Process	0.18μm	0.8μm	0.13μm	0.35μm
Min / Max Range Demonstrated	45mm / 1m	18mm / 0.11m	200mm / 7m	500m / 5m
Range error	$0.41mm_{rms}$ @ 0.5m	2.5mm @ 0.1m	$7mm_{rms}$ @ 1m	$19mm_{rms}$ @ 2m
Field of View (x / y) (°)	90° / 90°	No angle measurement	Not specified	18° /18°
Multi-target angular resolution (x / y) (°)	14.8° / NA	Single target only	(480px x 360px)	0.28° / 0.28° (64px x 64px)
Max Rate of Measurement	170 fps	50 fps	100 fps	200 fps
Energy per Measurement	**13.64μJ/frame**	6.5mJ/frame	20mJ/frame	>2.9mJ/frame

Figure 12.1.6: Comparison table.

12

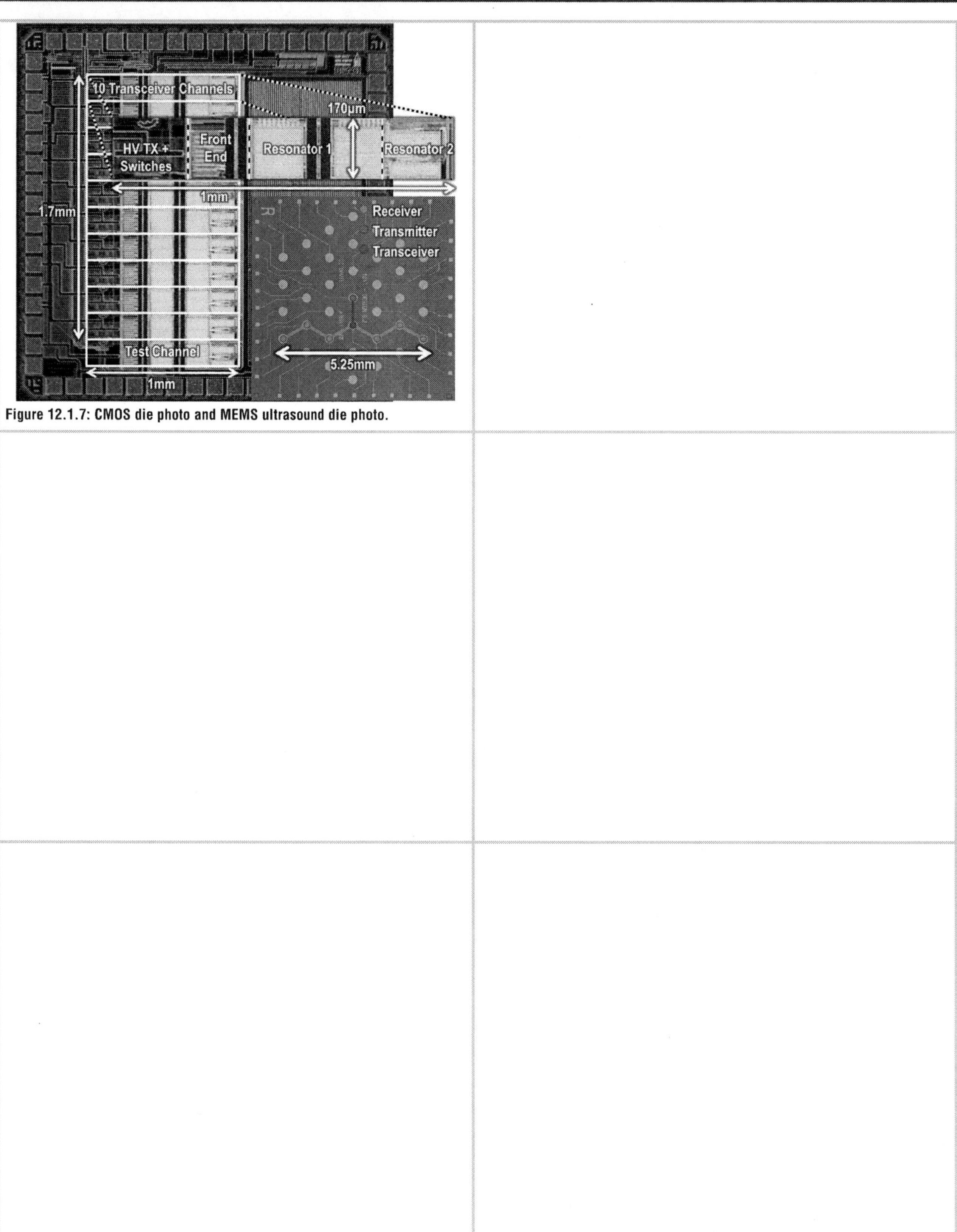

Figure 12.1.7: CMOS die photo and MEMS ultrasound die photo.

12.2 3D Gesture-Sensing System for Interactive Displays Based on Extended-Range Capacitive Sensing

Yingzhe Hu, Liechao Huang, Warren Rieutort-Louis, Josue Sanz-Robinson, Sigurd Wagner, James C. Sturm, Naveen Verma

Princeton University, Princeton, NJ

Capacitive touch screens have enabled compelling interfaces for displays [1]. Three-dimensional (3D) sensing, where user gestures can also be sensed in the out-of-plane dimension to distances of 20 to 30cm, represents new interfacing possibilities that could substantially enrich user experience. The challenge is achieving sensitivity at these distances when sensing the small capacitive perturbations caused by user interaction with sensing electrodes. Among capacitive-sensing approaches, self capacitance enables substantially greater distance than mutual capacitance (i.e., between electrodes), but can suffer from ghost effects during multi-touch. For gesture recognition, however, processing via classifiers can overcome such effects, enabling a rich dictionary of gestures [2]. Nonetheless, the sensing distance of such systems has been too limited for 3D sensing.

In this work, we present a 3D sensing system with 40×40cm² area and sensing distance to 30cm. This distance is achieved via two approaches. First, capacitance sensing is performed via frequency modulation, and the sensitivity of frequency readout is enhanced by high-Q oscillators capable of filtering noise sources in the readout system as well as stray noise sources from display coupling. Second, the capacitance signal is enhanced by eliminating electrostatic coupling between the sensing electrodes and surrounding ground planes. Figure 12.2.1 illustrates the concept. As shown on top, in order to minimize display thickness, ITO sensing electrodes are integrated with increasingly minimal separation to the display's common-electrode plane. This gives large electrostatic coupling from the sensing electrodes to the display, substantially degrading the coupling achievable to a user at a distance. In this system, however, the sensing electrodes are isolated from the display's common electrode by an ITO oscillating plane (OP). As shown, the sensing electrodes are connected one-by-one to a sensing LC oscillator (SO), such that their self-capacitance perturbs the tank capacitance, causing a frequency shift. Meanwhile, the OP is driven to the same voltage as the SO (and hence the connected electrode) by a unity-gain buffer (source follower). Consequently, electric field due to oscillatory charge redistribution on the electrode does not interact with the OP, resulting in much stronger coupling to a user even at great distances. In addition to sensing distance, this enables several benefits. First, since coupling between the electrodes and the OP is not a factor, their separation distance can be aggressively reduced (<1mm is used in this work). Second, separation between the OP and the display common electrode can also be reduced at the cost of increased OP capacitance and thus higher power in the unity-gain buffer; however, the OP driver consumes less than 19mW in this work with a separation of 1mm, making its overhead acceptable. A benefit of frequency-modulated readout is also that minimal noise is imposed on the display since the amplitude is not critical for increasing distance and is thus fixed at a value (0.75V). Third, extended sensing distance enables electrodes to provide later-displacement information (characterized below), allowing fewer electrode channels for covering large display areas, thus significantly mitigating power consumption and scan-rate constraints.

Figure 12.2.2 shows the readout channel architecture. Scanning of the sensing electrodes is controlled by a shift register. The SO's nominal center frequency of f_C=5MHz (tunable via MOS varactor) is perturbed by an amount Δf due to the sensed capacitance. The SO output is then fed to a differential Gilbert mixer and modulated down using a fixed local oscillator (LO). A low-frequency output f_{SENSE} is then derived from low-pass filtering (via a 2nd-order filter). The nominal SO and LO frequencies are offset by f_{OFFSET} (tunable by varactor) to give a minimum f_{SENSE}, which sets both the maximum output range of the time-to-digital converter (TDC) as well as the maximum scan rate. In this design f_{OFFSET} can be set from 5 to 20kHz. f_{SENSE} is amplified via a 2-stage preamp and a comparator before being provided to the TDC. The resulting digital signal controls an enable signal EN for a 16b counter through a period-control block. Since f_{SENSE} is a fairly non-linear function over sensing distance, the period-control block helps address TDC dynamic range by allowing multiples of the f_{SENSE} period to be selected for the counter EN signal; when f_{SENSE} is at high frequencies (due to short sensing distances), multiples N=2,4,8,16 can be selected. Such cases can be determined from the TDC code, and a digital controller can readily respond since high f_{SENSE} frequencies correspond to reduced readout delay. The sensed frequency shift, for a TDC count C, is thus $\Delta f = N{\times}f_C/C - f_{OFFSET}$. Readout noise is a key factor affecting sensitivity and is dominated by the SO/LO, mixer, and preamp, as discussed below.

Figure 12.2.3 shows the SO (LO has same structure) and mixer. Oscillator phase noise is a critical aspect and is set by device noise (1/f and white) as well as stray coupling from the display. Low phase noise is achieved thanks to substantial filtering of all these sources provided by the tank [3]. This requires high tank quality factor (Q), primarily limited by the inductor. We use an 0805 inductor of 33μH, giving Q=400 at 5MHz. In addition to tank Q, biasing-current noise is also a critical factor. We add a 100pF capacitor at the drain of the tail device, and also set the tail-current magnitude to ensure current-limiting, rather than voltage-limiting, conditions [3], giving a phase noise improvement of 21dB (@100Hz from f_C). Mixer linearity is also a critical factor for sensitivity. Since the SO and LO frequencies are offset, harmonics raise the possibility of in-band beat frequencies in the output at multiples of the ideal f_{SENSE}. To mitigate non-linearity, the SO is provided via a capacitor divider, as shown, to reduce its swing to ~100mV. The low-pass filter following the mixer has cut-off frequency of 50kHz to filter high frequencies and mixer clock feed through.

Figure 12.2.4 shows the preamp and comparator. With f_{SENSE} modulated to a low frequency, amplitude noise with respect to a zero-crossing reference can substantially degrade sensitivity, causing noise in the TDC output. To mitigate amplitude noise, the 2-stage preamp with diode-connected PMOS loads provides 6× gain per stage with noise filtering at a cutoff frequency of 200kHz per stage, set by 5pF output capacitors. The preamp feeds a hysteretic comparator. Hysteresis is adopted to ensure a digital output free of transient glitches, which is important for the operation of the TDC period-control block. The total input-referred noise of the mixer, preamp and comparator stages is $1.4\mu V_{RMS}$, corresponding to a frequency readout noise of σ_f=$16Hz_{RMS}$.

The system is prototyped, with the frequency-readout IC implemented in a 0.13μm CMOS process from IBM, and the sensing electrodes and OP patterned in-house using ITO-clad PET. The sensing electrodes are 1cm wide and spaced with 10cm pitch. For testing, we use 4 channels in each of the X and Y dimensions (8 channels total), giving a sensing area of 40×40cm². Figure 12.2.5 shows sensitivity measurements. On the left, the readout SNR and TDC code (with RMS bars) is plotted versus distance for a finger positioned above a sensing electrode; as shown substantial SNR is maintained out to 30cm (with 30dB SNR at 16cm). Though SNR is a widely used metric, in fact it is not representative of sensitivity in the presence of stray noise, such as from the display. On the right, we show the TDC code (with RMS bars) when display noise, varied from zero to various peak-peak values, is driven directly onto the OP (by a capacitively coupled amplifier whose input is fed from a display's common electrode); minimal impact on readout is observed even with large noise values. Figure 12.2.6 shows the measurement summary and a comparison with the state of the art. While other systems are touch-based, the presented system achieves the highest reported SNR for distances to 30cm. As an example, the worst-case resolution for lateral-displacement sensing is shown at 20cm above the electrode (resolution is defined as the displacement at which the difference in mean TDC code equals the code RMS). The digital circuits and OP driver are powered from 1.2V while the analog circuits are powered from 2.5V, giving total power consumption less than 20mW (475μW for frequency readout, 19mW for OP driver). The readout time is 500μs per channel, enabling a 240Hz scan rate.

Acknowledgements:

This work is funded by the Qualcomm Innovation Fellowship and NSF (grants ECCS-1202168 and CCF-1218206). We also thank MOSIS for IC fabrication.

References:

[1] H.-R. Kim, *et al.*, "A Mobile-Display-Driver IC Embedding a Capacitive Touch-Screen Controller System," *ISSCC Dig. Tech. Papers*, pp. 114-115, Feb. 2010.

[2] Zytronic. http://www.zytronic.co.uk/news/white-papers/. (online).

[3] A. Hajimiri and T H. Lee, "Design Issues in CMOS Differential LC Oscillators," *IEEE J. Solid-State Circuits*, vol. 34, no. 5, pp. 717-724, May 1999.

978-1-4799-0917-9/14 $31.00 © 2014 IEEE

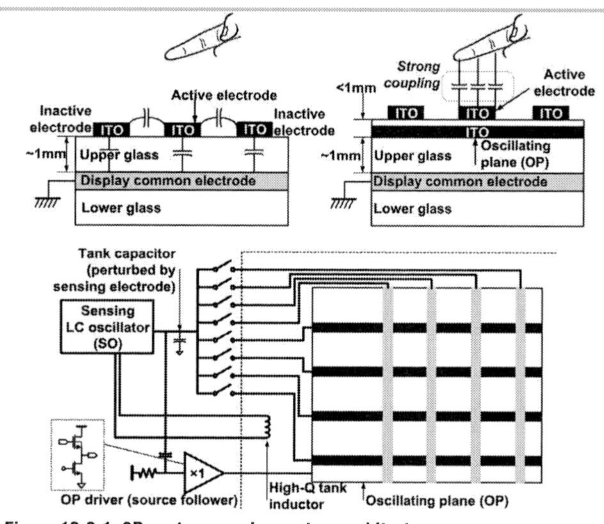

Figure 12.2.1: 3D gesture-sensing system architecture.

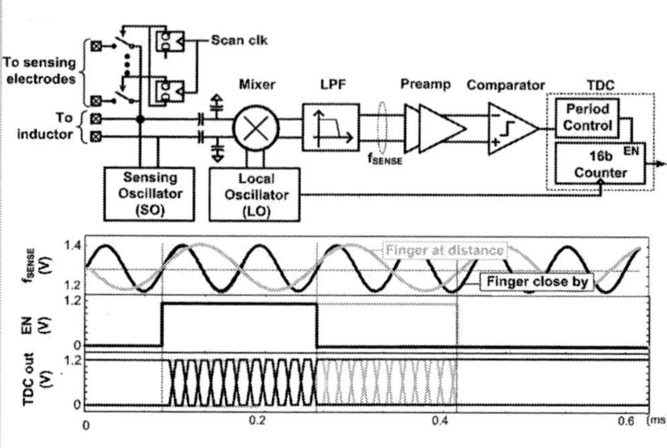

Figure 12.2.2: Frequency-modulation readout system based on low-noise oscillators and TDC; simulation waveforms illustrate frequency-modulation response by system for finger close by and at distance.

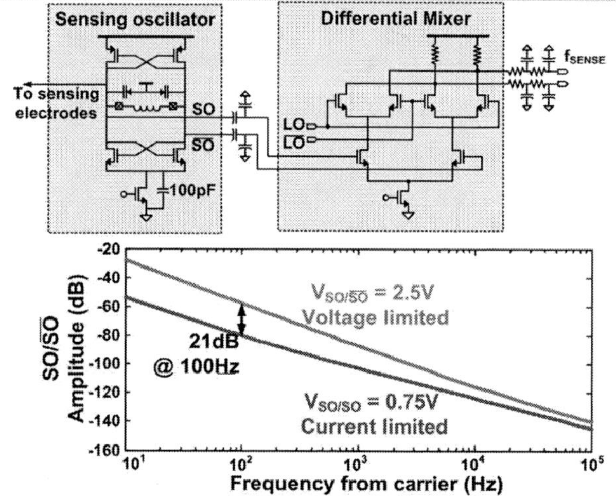

Figure 12.2.3: Sensing oscillator (SO) and mixer; oscillator phase noise is compared for current- versus voltage-limiting conditions.

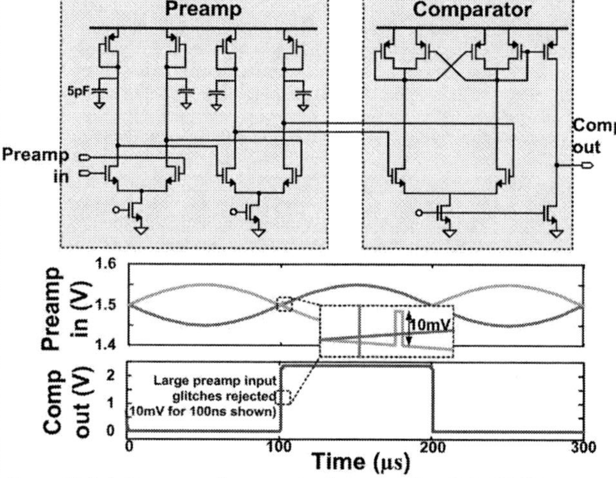

Figure 12.2.4: Preamp and comparator for generating digital TDC input from f_{SENSE}; glitch-free output is achieved with hysteretic comparator.

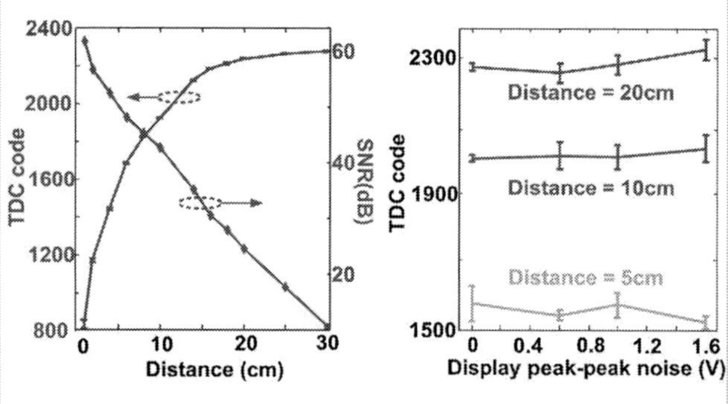

Figure 12.2.5: Prototype sensitivity measurements, showing SNR and TDC code (with RMS bars) versus distance (out to 30cm) as well as TDC code error due to stray coupling of display noise with varying peak-peak noise applied to oscillating plane.

	H.-R. Kim ISSCC10	K.-D. Kim ISSCC12	J.-H. Yang ISSCC13	This work
Process	1.5/5.5/30V 90nm CMOS	1.5/5.5/30V 90nm LDI	3.3V 350nm CMOS	1.2/2.5V 130nm CMOS
Channels	24 (X+Y)	30 (X+Y)	Tx 27; Rx 43	8 (X+Y)
Panel size	6.5cm x 4.9cm	6.5cm x 4.9cm	20.5cm x 15.4cm	40cm x 40cm
3D sensing	X	X	X	To 30cm
Capacitance type	Self capacitance	Self capacitance	Mutual capacitance	Self capacitance
Scan frequency	120Hz	120Hz	120Hz	240Hz
SNR	30dB	35dB	39dB	50dB@5cm 30dB@16cm 20dB@23cm
Resolution	--	0.9mm†	--	7.1mm(x) 7.1mm(y) 10mm(z)@20cm(hand)
Power consumption	12mW	10.6mW	18.7mW	210μW (readout) 19mW (OP)

†Precise definition of resolution not available

Figure 12.2.6: Performance summary and comparison with the state of the art.

12

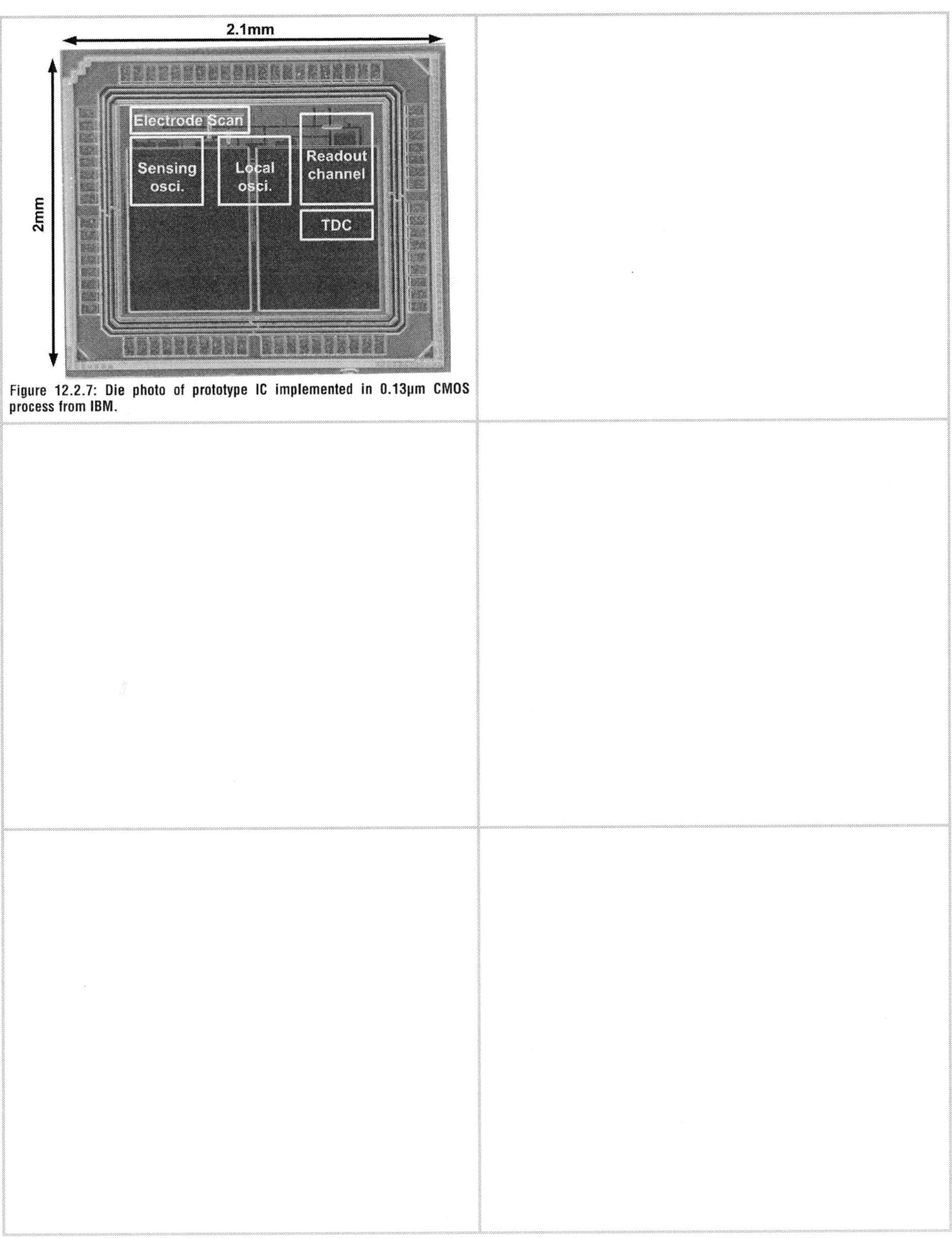

Figure 12.2.7: Die photo of prototype IC implemented in 0.13µm CMOS process from IBM.

12.3 A 240Hz-Reporting-Rate 143×81 Mutual-Capacitance Touch-Sensing Analog Front-End IC with 37dB SNR for 1mm-Diameter Stylus

Mutsumi Hamaguchi[1], Akira Nagao[2], Masayuki Miyamoto[2]

[1]Sharp, Fukuyama, Japan,
[2]Sharp, Tenri, Japan

Realization of a mutual-capacitance touch-sensing system spanning over 30 inches is not a straightforward task, because the SNRs of conventional sequential drive controllers degrade as the number of sensor channels increases. One common way to overcome this drawback is to increase the driving voltage, which however results in an increase in system complexity and cost because it requires high-voltage circuits and devices. This SNR issue is resolved by driving the sensor channels in parallel [1,4] as shown in Fig. 12.3.1. Although the parallel drive mixes up the signals from the multiple channels driven at the same time, the original signals can be reconstructed from the sequence of mixed signals if the drive sequences are linearly independent from each other. By appropriately designing the parallel drive sequences, the SNR is enhanced by $\sqrt{M}$ times compared to that of the sequential drive [1], where M is the number of drive channels. An analog front-end (AFE) IC capable of driving and sensing a 143×81 mutual-capacitance sensor is developed in 0.18μm 1P5M CMOS. A 32-inch and a 70-inch touch system are realized with the use of the AFE and an SNR over 37dB for 1mm diameter stylus is attained in either system.

Noise from a display paired with a touch sensor has to be reduced so as not to degrade the high SNR attained with the use of the parallel drive method. A differential sensing scheme between adjacent channels shown in Fig. 12.3.2 is adopted to cancel the strong LCD noise, which is commonly injected to adjacent channels. Original capacitance signals with the common-mode noise rejected are recovered in the digital domain after ADC by summing up the differential signals. A fully differential charge-to-voltage converter (CVC) is designed to have a tolerance against the input common-mode shift caused by the LCD noise and the parallel drive operations. 71 CVCs are laid out in parallel to receive signals from 143 sense channels with a switching operation shown in Fig. 12.3.2. In phase-1 (the first differential signal sensing), the (2i+1)-th and the 2i-th sense channel are connected to the i-th CVC's Inp node and Inn node respectively, while in phase-2 (the second differential signal sensing), the (2i+2)-th and the (2i+1)-th sense channel are connected to the i-th CVC's Inp node and Inn node respectively, where i=0, ..., 70. The two phase operations settle in 4μsec, if the time constant of an accompanied touch sensor is small enough.

The AFE has 224 sensor channel connections, 143 channel drivers with 3.3V driving voltage, and 71 CVCs. Using the channel switches shown in Fig. 12.3.2, the first 81 channels are connected to the drivers, each of the next 62 channels is connected to either a driver or a CVC via a selectable switch, and the last 81 channels are connected to CVCs. The output voltage signals are transferred from the CVCs to the dual 12b 20MHz pipeline ADCs through two multiplexers, where ADCs take 1.8μs (=36×1/20MHz) to convert all the signals from 71 CVCs. The capacitance distribution of the touch sensor is reconstructed in the decoder with a linear algebra algorithm and then transferred to the following digital back-end (DBE) circuitry through a 200MHz low-voltage differential signaling (LVDS) interface. The reconstruction calculation of 143×81 mutual capacitances takes L×4μs, where L is the length of the parallel drive sequences. DBE is an image signal processor that filters out unwanted noise, finds peaks in the capacitance distribution, calculates properties of touches such as positions, strength, and manages the consistency of the touch identifications. LVDS is selected to ensure reliable data transmission between AFE and DBE through long wiring up to 3m for large format applications. The supply voltages of the analog circuit block and digital circuit block are 3.3V and 1.8V.

Figure 12.3.3 shows the schematic of a CVC block and its operation. The first stage is a charge integrator with switchable feedback capacitances. The next stage is a correlated double sampler (CDS) to cancel the DC offset and low frequency noise. The third stage is a sample-and-hold circuit to feed the differential signal to the ADCs.

Two large application systems shown in Fig. 12.3.4 are built with the use of the AFE and a 7μm thickness copper mesh sensor technology. Copper mesh technology is suited for the realization of capacitive touch sensors with large formats and small time constants at the same time. The sheet resistance of the 7μm thickness copper is less than 0.003Ω/square, which is smaller than that of ordinary ITO by four orders. The 32-inch system is built up from one AFE, a companion DBE IC, 5.09mm channel pitch 138×78 metal mesh sensor laminated to a 1.8mm thickness cover glass, and mounted on a 4K2K LCD with 1.9mm air gap. The 70-inch system is built up from three AFEs, a DBE implemented in an FPGA, 6.25mm channel pitch 248×140 metal mesh sensor laminated to a 1.9mm thickness cover glass, and mounted on a FHD LCD with 3.24mm air gap. The air gap is designed so that the cover glass does not touch the LCD surface when deformed by self-weight and human touches. The width of the copper mesh is designed to be 7μm, which is narrow enough to reduce the unwanted visual effects caused by stacking the mesh sensors on the LCDs. The width of the lines and spaces in the edge region for addressing the sensor channels are 20μm and 60μm respectively, where the address wirings are placed in one edge of each drive and sense channels. The resistance of the longest channel of the 70-inch touch sensor is less than 300Ω, which is low enough for the AFE to operate at its highest frequency. In both systems a maximum length sequence (MLS) with a code length of 255 is used for the parallel drive sequences. Therefore the cycle time to construct each capacitance frame is 1.02ms (= 255×4μs).

Figure 12.3.5 shows the capacitance change induced by a touch of 1mm diameter stylus on the 32-inch system, where the peak signal is around 10fF with 37.4dB SNR. The definition of the SNR [2] is described in the figure.

Specifications and measurement results are compared with the state of the art [3,4] in Fig. 12.3.6, where all the SNRs are measured while the LCD is on. In the normal mode four capacitance frames are summed to obtain enough SNR and the reporting rate is 240Hz; however one capacitance frame without summation is reported at 120Hz in the slow mode. The SNR in the normal mode is higher than the one in the slow mode by around 6dB as expected. Based on the table showing the state of the art, the number of channels of this work is the largest reported and the SNR for 1mm diameter stylus is the highest. The SNR for a finger, measured with a 9mm diameter artificial finger, is also near the highest and the power consumption per node (= total power consumption divided by the total number of mutual capacitances) is near the lowest. The independence of the SNRs (especially for 1mm diameter stylus) from the number of the channels is the result of adopting the parallel drive method. Thanks to the differential sensing, the SNR's dependence on the LCD's on/off status is less than 2dB.

Figure 12.3.7 shows the micrograph of the AFE. The die size is 7.55×9.43mm² including 0.2mm scribe line. It is assembled in an 18×18mm² ball grid array (BGA) package with 364 balls.

Acknowledgements:
The authors would like to thank all the members of Sharp touch R&D team for their contributions in the development and evaluation of the chip and systems.

References:
[1] M. Miyamoto, "Linear Device Value Estimating Method, Capacitance Detection Method, Integrated Circuit, Touch Sensor System, and Electronic Device," JP Patent 4927216, Feb. 2012.
[2] S.-H. Ko, *et al.*, "Low Noise Capacitive Sensor for Multi-Touch Mobile Handset's Applications," *IEEE A-SSCC*, pp. 1-4, Nov. 2010.
[3] J.-H. Yang, *et al.*, "A Highly Noise-Immune Touch Controller Using Filtered-Delta-Integration and a Charge-Interpolation Technique for 10.1-inch Capacitive Touch-Screen Panels," *ISSCC Dig. Tech. Papers*, pp. 390-391, Feb. 2013.
[4] H. Shin, *et al.*, "A 55dB SNR with 240Hz Frame Scan Rate Mutual Capacitor 30×24 Touch-Screen Panel Read-Out IC Using Code-Division Multiple Sensing Technique," *ISSCC Dig. Tech. Papers*, pp. 388-389, Feb. 2013.

978-1-4799-0917-9/14 $31.00 © 2014 IEEE

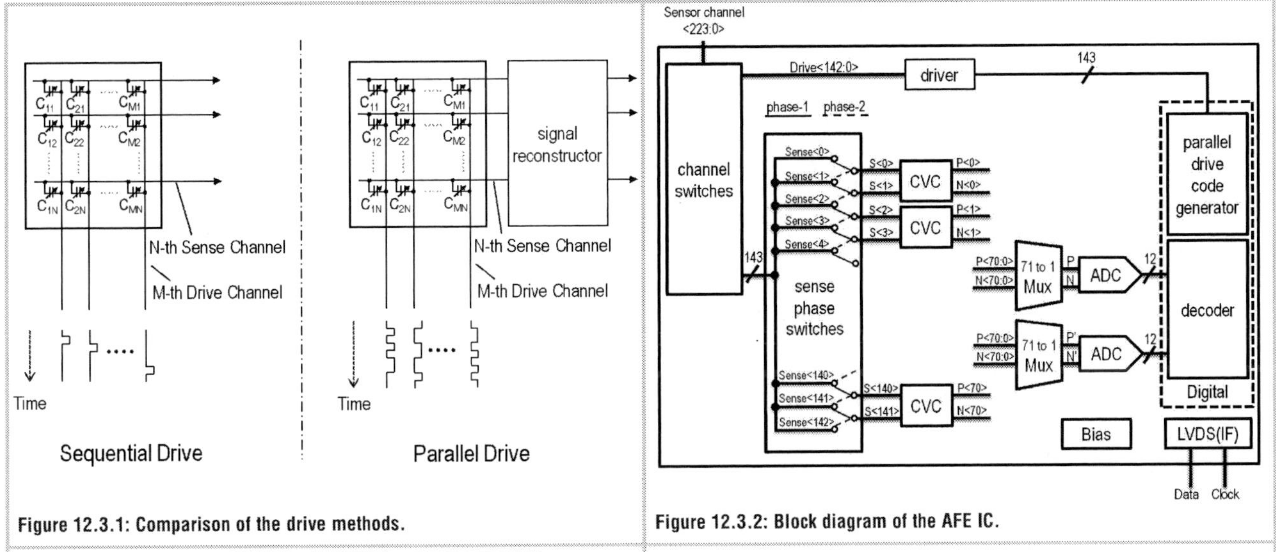

Figure 12.3.1: Comparison of the drive methods.

Figure 12.3.2: Block diagram of the AFE IC.

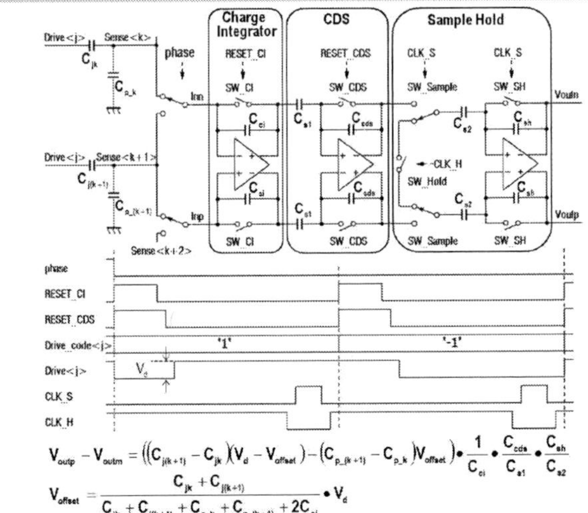

$$V_{outp} - V_{outm} = \left((C_{j(k+1)} - C_{jk})(V_d - V_{offset}) - (C_{p\_(k+1)} - C_{p\_k})V_{offset}\right) \bullet \frac{1}{C_{ci}} \bullet \frac{C_{cds}}{C_{s1}} \bullet \frac{C_{sh}}{C_{s2}}$$

$$V_{offset} = \frac{C_{jk} + C_{j(k+1)}}{C_{jk} + C_{j(k+1)} + C_{p\_k} + C_{p\_(k+1)} + 2C_{ci}} \bullet V_d$$

Figure 12.3.3: Schematic of the CVC.

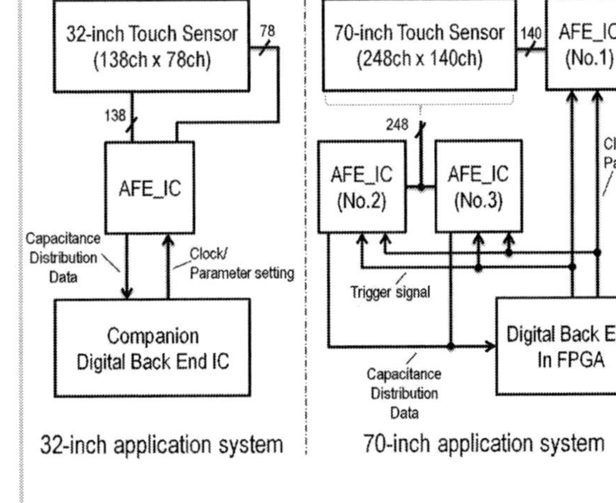

Figure 12.3.4: Application block diagrams.

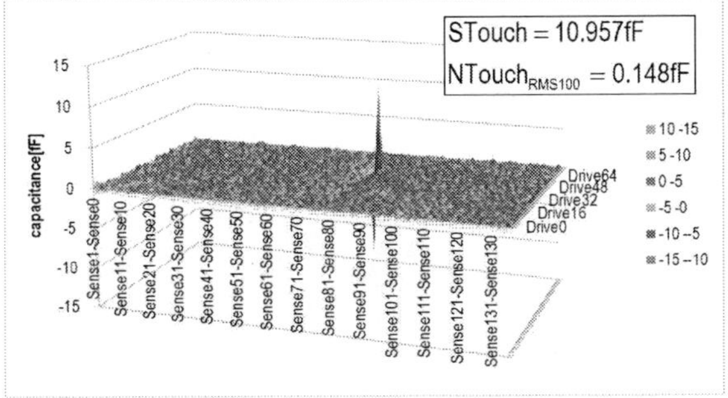

STouch = 10.957fF

$NTouch_{RMS100} = 0.148fF$

$$SNR(dB) = 20\log\frac{STouch}{NTouch_{RMS100}}$$

$$STouch = Signal_{TouchAVG100} - Signal_{UntouchAVG100}$$

$$NTouch_{RMS100} = \sqrt{\frac{\sum_{n=0}^{n=99}(Signal[n] - Signal_{Touch})^2}{100}}$$

Figure 12.3.5: Touch signal of a 1mm diameter stylus.

		[3]	[4]	This work			
Size[inch]		10.1	5	**32**		**70**	
Number of channels		27ch (Tx) x 43ch(Rx)	30ch(Tx) x 24ch(Rx)	**78ch(Tx) x 138ch(Rx)**		**140ch(Tx) x 248ch(Rx)**	
Mode		Undescribed	High SNR (CDMS DS-SC)	Slow	Normal	Slow	Normal
Reporting rate[Hz]		120	240	120	240	120	240
SNR[dB]	Finger	39	55	**50.8**	**56.6**	**47.5**	52.2
	Pen (1mmΦ)	Undescribed	35	**31.6**	**37.4**	**31.5**	**37.7**
Process		0.35μm CMOS	2P6M 0.18μm CMOS EEPROM	1P5M 0.18μm CMOS			
Power consumption	Total [mW]	18.7	52.8	214.7	559.9	562.8	1247.0
	Per node [uW/node]	16.1	73.3	19.9	**52.0**	**16.2**	**35.9**
Supply voltage		3.3V	3.3V	3.3V, 1.8V			
Die size		4mm x 4mm	4.06mm x 3.66mm	7.55mm x 9.43mm			

Figure 12.3.6: Specification and performance summary.

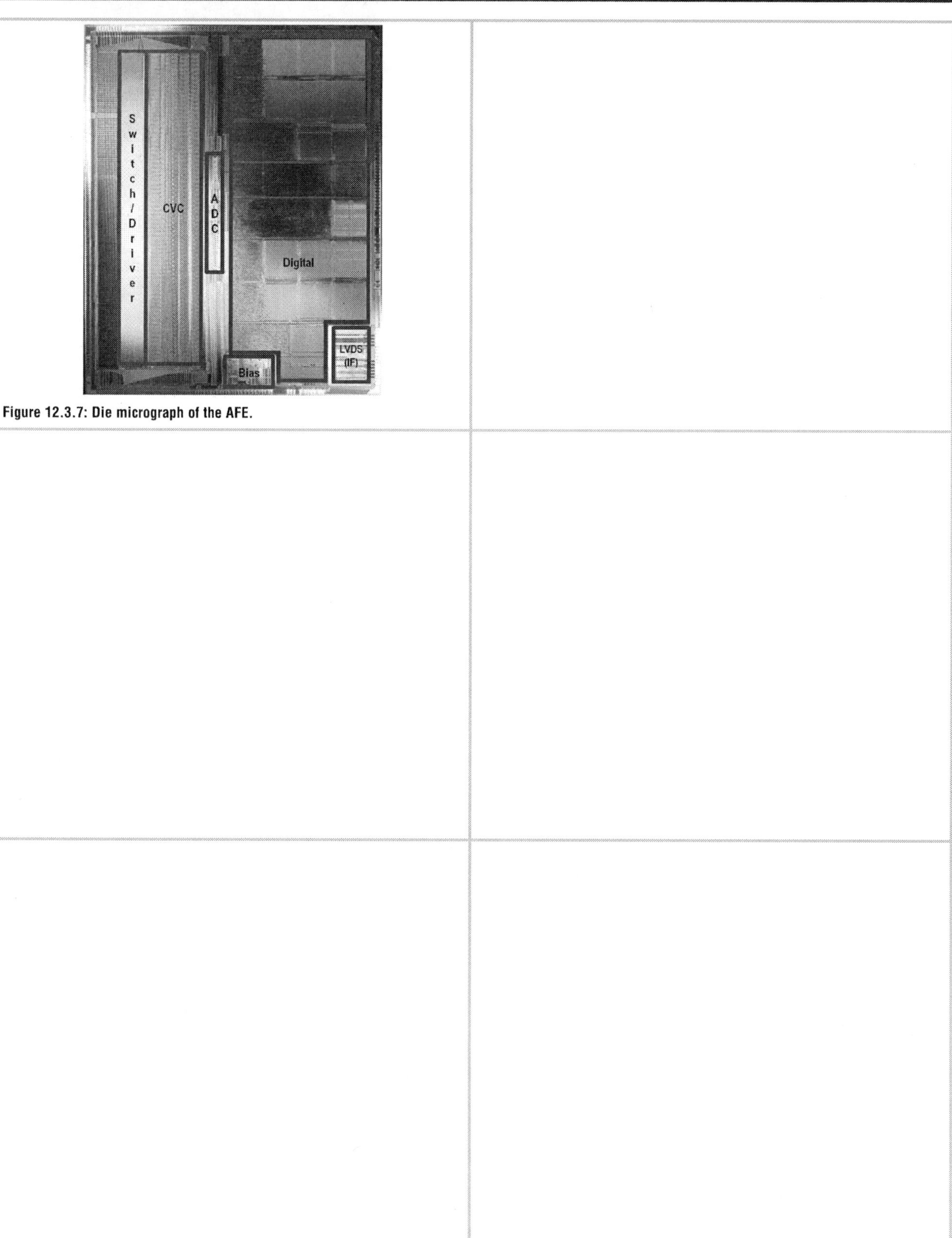

Figure 12.3.7: Die micrograph of the AFE.

12.4 A 1mm-Pitch 80×80-Channel 322Hz-Frame-Rate Touch Sensor with Two-Step Dual-Mode Capacitance Scan

Noriyuki Miura[1], Shiro Dosho[2], Satoshi Takaya[1], Daisuke Fujimoto[1], Takumi Kiriyama[1], Hiroyuki Tezuka[2], Takuji Miki[2], Hiroto Yanagawa[2], Makoto Nagata[1]

[1]Kobe University, Kobe, Japan, [2]Panasonic, Osaka, Japan

A 1mm-pitch 80×80-channel 322Hz-frame-rate touch sensor is reported. Multiple touch points are detected by a two-step dual-mode capacitance scan, where self- and mutual-capacitance measurements are hierarchically performed in two steps to reduce scan time that is otherwise increased due to high resolution. 160 dedicated row and column ADCs are used for the parallel read-out to further reduce scan time. A time-domain digital conversion that uses a counter-based slope ADC significantly reduces power and area for the parallel ADC approach. The signal attenuation due to the sensor capacitance reduction in the 1mm fine-pitch electrode is compensated by using thorough noise-reduction techniques in the sensor analog front-end (AFE). A 0.35μm CMOS prototype demonstrates 41dB SNR with >3× higher pitch resolution, >10× faster touch-point scan, 12× and 4× higher energy and area efficiency compared to state-of-the-art touch sensors [1,2].

The touch-sensor market is rapidly growing, especially for smart-phone and tablet-PC applications where high-resolution and large-area touch sensors are in strong demand for: more natural human interfaces, very accurate expressions required for medical applications, or to provide additional feature values (e.g. more artistic expressions such as brush art and painting for design and education purposes) (see Fig. 12.4.1). However, state-of-the-art touch sensors can only cover very limited resolution [1,2]. There are two main technical issues for high-resolution touch sensors: 1) a slow frame scan rate due to the increased number of sensing points, and 2) a limited sensing capacitance size and hence limited capacitance change due to the fine-pitch channel arrangement. This paper presents a 1mm fine-pitch 80×80-channel high-resolution touch sensor. A capacitance scan scheme is reported that solves the first issue and a low-noise AFE is presented that addresses the second issue.

Figure 12.4.1 depicts the two-step dual-mode capacitance scan scheme. In Step 1, possible touch regions including ghost touches are identified by a flash self-capacitance measurement (SCM) utilizing dedicated parallel row and column ADCs. In Step 2, actual touch points are precisely detected by local mutual-capacitance measurements (MCMs) in the limited regions identified by Step 1. Even though the pitch resolution and/or the touch-panel size are increased, the scan time can be significantly reduced as compared to the conventional exhaustive global MCM approaches [1,2], where the scan time is basically proportional to the square of the panel size. Figure 12.4.2 depicts the block diagram of the touch-panel system with the scan scheme. Each electrode (both top and bottom) is connected to a dedicated self-capacitance sensor (SCS) operating in parallel for the flash SCM. For the local MCM, selected mutual-capacitance drivers (MCDs) and mutual-capacitance sensors (MCSs) are activated by sensor control logic. Dedicated row and column ADCs digitize the capacitance values, which are read out through shift registers, and the multiple touch points are detected. A time-domain counter-based slope ADC is used for area and power saving in the parallel ADC approach.

Figure 12.4.3 depicts the block diagram of the sensor circuit. In SCS, a slope generator provides constant current I_B to an ITO electrode to charge up the self-capacitance and generates a ramp voltage in V_{SCB}. The total self-capacitance increases by the touch and hence the V_{SCB} slope is reduced. By comparing V_{SCB} with a certain reference voltage V_{REF}, the capacitance change is converted into a time difference. A counter is used to finally digitize the capacitance change. In MCS, a bottom electrode is driven by MCD. Pulse-shaped voltages are induced in V_{MC} through the mutual capacitance C_M. The pulses are integrated to generate a reference voltage for MCM in V_{Integ}. The reference level is increased by the touch, and the mutual capacitance change can be digitally measured by the same time-domain counter-based ADC.

For the 1mm fine pitch, the electrode size is limited to 0.5mm in width, resulting in a lower signal level due to the small sensor capacitance size and hence the small capacitance change due to touch. A low-noise sensor AFE is required to keep SNR constant for precise detection of touch points. A dominant noise source comes from a display stacked underneath of the touch panel, which appears as a shield ground (GND) bounce V_{NOISE} and generates noise in the sensor input through the parasitic capacitance of the electrodes C_{PT} and C_{PB} (Fig. 12.4.3). A matched RC network (details shown in Fig. 12.4.4) replicates this V_{NOISE} in the reference voltage V_{REF} to cancel out the noise using common-mode rejection in the frontend opamp. The PVT variations of the current bias I_B in the slope generator are critical to keep high SNR. A PTAT cascode current bias generator (Fig. 12.4.4) produces stable V_{PTAT} for low-noise constant I_B generation in the slope generator. This cascode topology enhances stability. The simulated I_B variation is less than 0.2% over a 2.6~3.6V supply voltage V_{DD} range. A delay-based spike-noise remover in SCS is used to suppress instantaneous noise effects in V_{SCB} appearing at near the V_{REF} level. The delay is clock-synchronous so that the delay variation has no impact on the sensitivity variation between the channels. In MCS, a chopper integrator is employed to remove the opamp offset and reduce low-frequency noise such as from the display, HUM, and SMPS. The integrator operates in 4 phases (P1~P4) for integration and chopper (Fig. 12.4.4) since the positive and negative bipolar pulse voltages are induced in V_{MC}. By combination with the input and the output chopper switches, the pulse signal is integrated in each phase and the chopper is performed at the same time for the low-frequency noise removal. The output chopper switches are hidden inside the opamp (Fig. 12.4.4). No chopper switch exists in the integration capacitor feedback paths, which further suppresses the noise and also reduces the settling time of the integrator. For the last stage of the slope AD conversion, a gray-code counter is used to suppress the timing-related instantaneous quantization noise within a single digit code [3].

A test chip is fabricated in 0.35μm CMOS. Forty sensor channels are integrated in the chip. Four chips are mounted on a evaluation board to build an 80×80-channel touch panel sensor system. An 8×8cm^2 touch panel is fabricated with 80×80 ITO electrodes arranged in a pitch of 1mm. The line and space are 0.5mm. The effective sensing capacitor size is therefore only 0.25mm^2, which is almost two orders of magnitude smaller than that of [1,2].

Figure 12.4.6 presents measured data. Successful operation is confirmed in both self- and mutual-capacitance measurements. The waveform snapshots of the integrator output show that successful finger-touch detection by 90mV voltage shift. Also it can be seen in the snapshot that the noise cancellation works effectively. The noise cancellation is effective up to 400Hz. Further high-frequency noise components can be removed by relying on the BPF characteristics of the chopper integrator. The measured SNR is 41dB for a finger touch and 32dB for a 1mm-ϕ metal pillar touch. The two-step dual-mode capacitance scan was performed on the 8×8cm^2 touch panel with two different object touches; one is a metal "A" plate and the other a finger-size pillar. Both objects were successfully touch-detected. A very smooth capacitance distribution (including about 200 capacitance sensing points) can be measured at 1mm fine pitch. Based on the measured capacitance profile with the 1mm pillar, the effective resolution was measured to be 0.1mm. The resolution is fine enough even for the direct brush arts on the touch panel. The frame scan rate can be potentially up to 500Hz. This prototype operates with a slow serial I/O at 1MHz and the actual frame scan rate is limited to 322Hz. The touch sensor operating at the frame scan rate of 322Hz consumes 21.8mW at 3.3V V_{DD}. The sensor performance is summarized in Fig. 12.4.7 and compared to state-of-the-art touch sensors.

References:

[1] J.-H. Yang, *et al.*, "A Highly Noise-Immune Touch Controller Using Filtered-Delta-Integration and a Charge-Interpolation Technique for 10.1-inch Capacitive Touch-Screen Panels," *ISSCC Dig. Tech. Papers*, pp. 390-391, Feb. 2013.

[2] H. Shin, *et al.*, "A 55dB SNR with 240Hz Frame Scan Rate Mutual Capacitor 30×24 Touch-Screen Panel Read-Out IC Using Code-Division Multiple Sensing Technique," *ISSCC Dig. Tech. Papers*, pp. 388-389, Feb. 2013.

[3] C. Shoushun, *et al.*, "Robust Intermediate Read-Out for Deep Submicron Technology CMOS Image Sensors," *IEEE Sensors Journal*, vol.8, no.3, pp. 286-294, Mar. 2008.

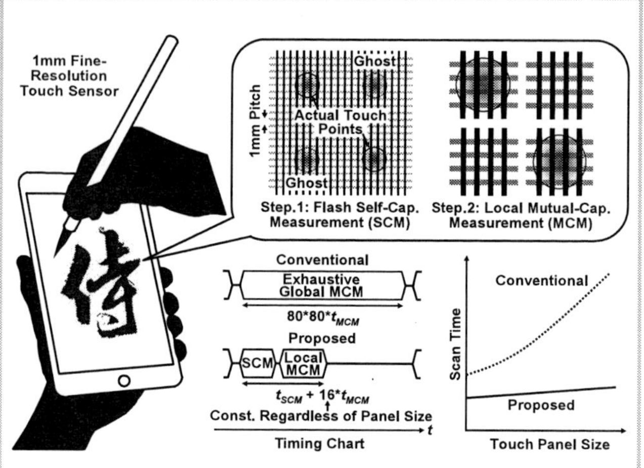

Figure 12.4.1: 1mm-pitch fine-resolution touch sensor with two-step dual-mode capacitance scan.

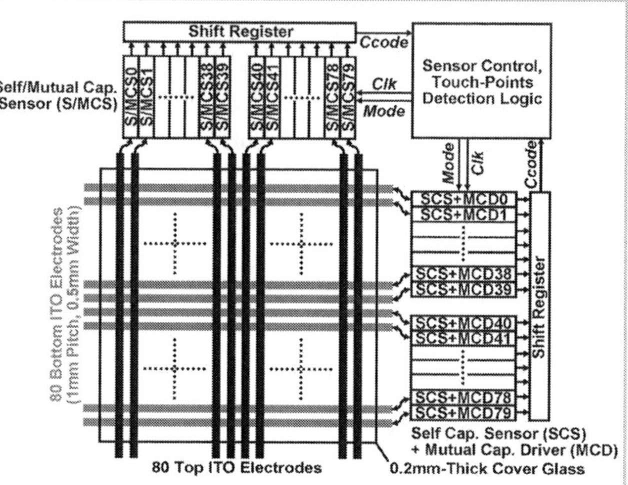

Figure 12.4.2: Block diagram of fine-resolution touch-panel system.

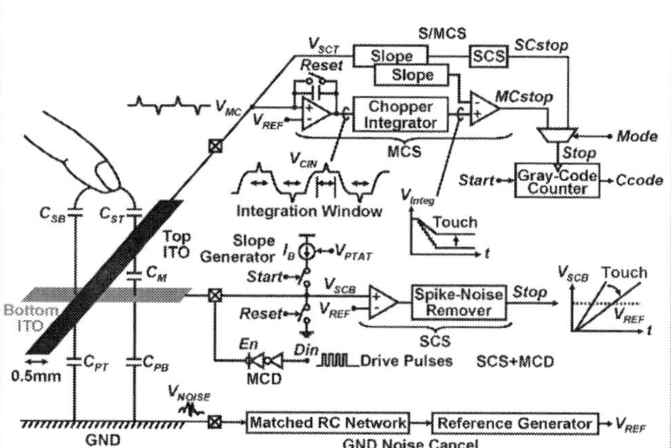

Figure 12.4.3: Block diagram of dual-mode capacitance sensor.

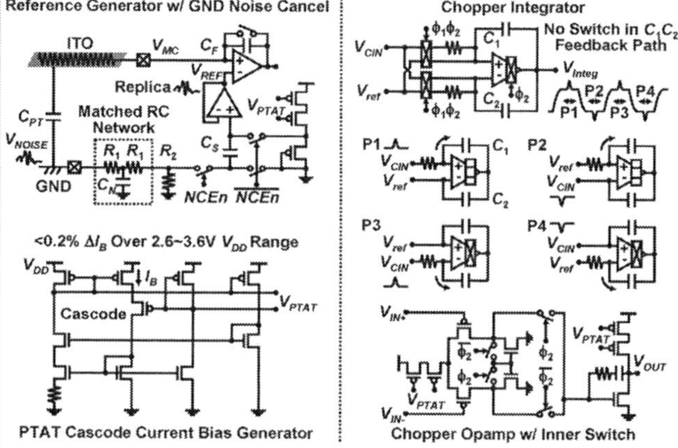

Figure 12.4.4: Detailed circuit schematics of key blocks.

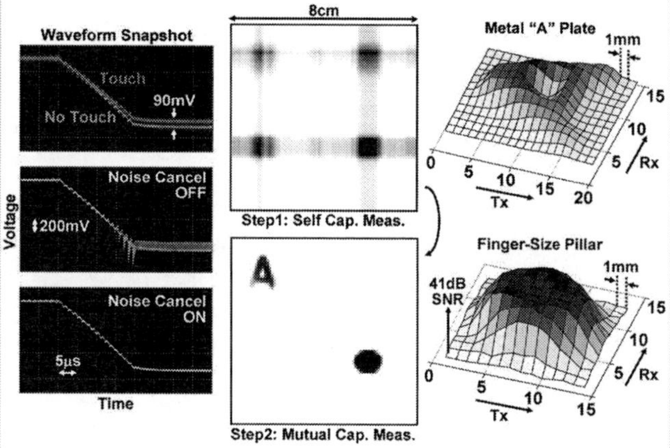

Figure 12.4.5: Snapshot of integrator output waveforms and measured touch capacitance map.

	[3]J.-H. Yang, (ISSCC'13)	[4]H. Shin, (ISSCC'13)	This Work
Capacitance Sensing Type	Mutual	Mutual	Self/Mutual Dual Mode
Channel Pitch	5mm	3mm	1mm (1/3)
Number of Channels	27x43	24x30	80x80 (5.5x)
Touch Panel Size	10.1"	5"	4.5" (8cm x 8cm)
Frame Rate (Scan Points / Sec.)	120Hz (0.14Mpoints/s)	240Hz (0.17Mpoints/s)	322Hz (2.06Mpoints/s)
Power Consumption	18.7mW	52.8mW	21.8mW
Energy / Channel	134nJ/ch	306nJ/ch	11nJ/ch (1/12)
Chip Layout Area	10.4mm²	14.9mm²	13.7mm²
Area / Channel	896μm²/ch	2069μm²/ch	214μm²/ch (1/4)
SNR	39dB (Finger)	55dB (Finger) 35dB (1mm-ϕ)	41dB (Finger) 32dB (1mm-ϕ)
Process	0.35μm CMOS	0.18μm CMOS	0.35μm CMOS

Figure 12.4.6: Performance summary and comparison.

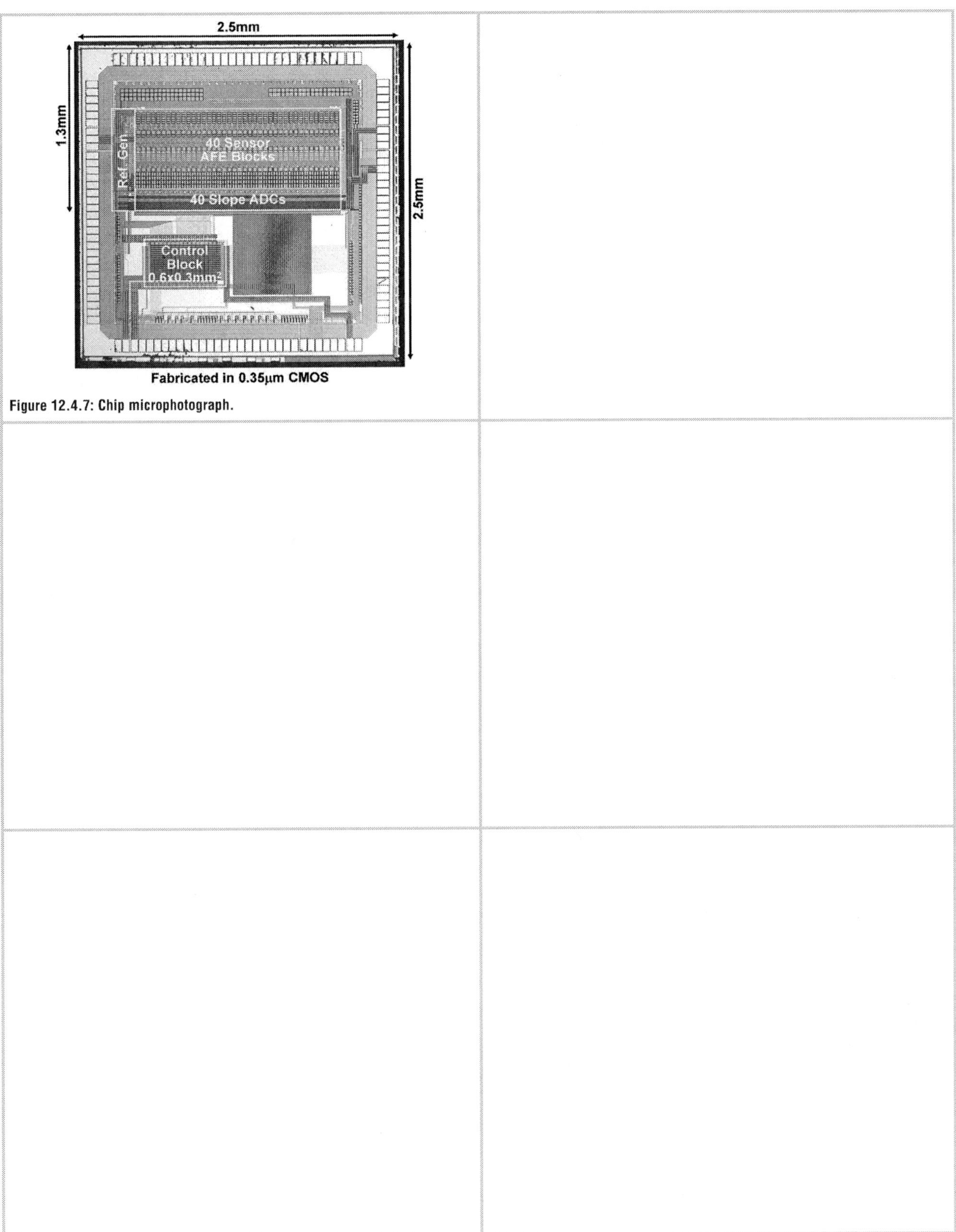

Figure 12.4.7: Chip microphotograph.

12.5 2D Coded-Aperture-Based Ultra-Compact Capacitive Touch-Screen Controller with 40 Reconfigurable Channels

Hongjae Jang[1], Hyungcheol Shin[2], Seunghoon Ko[1], Ilhyun Yun[2], Kwyro Lee[1]

[1]KAIST, Daejeon, Korea, [2]Zinitix, Daejeon, Korea

As mutual-capacitive touch-screens expand their application area to various information devices, better controllers are in demand for larger, thinner, lower-cost touch-screen panels (TSP), and in-cell/on-cell touch displays [1]. In order to gain higher sensitivity from such large and noisy TSPs, numerous parallel analog circuits are integrated on TSP controllers. Many solutions to cope with harsh noise environments, such as continuous-time implementation rather than DT [2] or adding aggressive filtering schemes [3], require sophisticated signal-conditioning circuits to be duplicated on every channel, easily consuming higher analog power and increasing silicon cost.

This paper presents an ultra-compact, low-power and noise-immune mutual-capacitance TSP controller based on a 2D coded-aperture read-out architecture. Figure 12.5.1 shows conceptual diagram of the read-out method, which obtains a TSP's response to a set of 2D basis patterns, based on the widely known notion that all kinds of 2D images can be transformed into other orthogonal basis domains, e.g., frequency or pseudo-random noise. Sensing whole TSP channels projected into a single-channel circuit prevents hardware redundancy due to a requirement for parallel circuits and enables designers to focus on maximizing noise-filtering performance with less concern about hardware cost. Note that even with a single-channel circuit, this method facilitates all mutual-capacitor-nodes to be driven simultaneously, therefore fully retaining high noise-immunity advantages of the previously reported CDMS method [4].

Figure 12.5.2 shows the operation principle of the controller architecture. The 2D basis pattern set is obtained by outer products of two 1D orthogonal code sets [5], one for TX channels and the other for RX's. Hence, the code sign of each node is defined by multiplication of the intersecting TX and RX codes. For each pattern, the TX drives an AC test voltage with 0 or π phase, according to the TX code, inducing a signal current on the sensor capacitor nodes. The signal currents on the same RX line are automatically superposed, and summed again as the RX code steers the currents into two directions: RX+ or RX-. The resulting single-valued signal difference between RX+ and RX- indicates the TSP response to the projected pattern. By repeating this operation for all basis patterns, the capacitance image can be fully specified and reconstructed.

A controller for the architecture includes reconfigurable 40 TRX I/Os connected to a single TX and a single RX core via shared rings, to provide wide TSP compatibility. The basis patterns are dynamically scheduled by an MCU through a serial programming link, to flexibly apply advanced coding schemes such as adaptive resolution scaling. In this mode, the scan event is divided into two steps where a full-spanned low-resolution basis approximates the rough area of a touch event, followed by local read-out of the area using a high-resolution basis to calculate fine touch coordinates. This scheme maximizes the scan rate by using a shorter basis length, with reduced TX driving power using a TX code reduction ratio.

Figure 12.5.3 shows a detailed circuit implementation and timing diagram. The TX core is differentially driven by inverters, and the RX circuit includes a pseudo-differential front-end TIA, PGA, and two 2nd-order incremental $\Delta\Sigma$ ADCs embedding a wave-shaping mixer. The TIA converts the incoming AC current through RX+ and RX- to a voltage. From the TX to the TIA output, a differential bandpass transfer function is formed where its passband gain at carrier frequency (f_C) is proportional to the total sum of the coded sensor capacitances, which approximates to $\Sigma g_{xy}C_0/C_F$ where g_{xy}, C_0, and C_F denote the code (+1 or -1) of each node, TSP base capacitance, and TIA feedback capacitance, respectively.

An offset current canceller block is added on TIA to minimize output voltage swing due to base capacitance (C_0) offset. Since the binary orthogonal code set cannot be perfectly balanced in principle, i.e., the total number of +1 and -1 cannot be equal, driving typical TSPs having a few pFs C_0 causes an AC offset current proportional to the TX code unbalance, flowing into each RX channel. When these currents are summed into the TIAs, the large output voltage swings of each TIA waste head-/leg-room and require large C_F. For example, the 4×4 basis shown in Fig. 12.5.2 has a TX code unbalance of +2 and an RX steering ratio of 3:1, requiring C_F larger than $6C_0$ even without room for noise margin. The offset canceller cancels out the common-mode input current using 0.5-to-4pF C_{OS} in series with a 1-to-4× programmable gain class-AB current amplifier, driven by a dummy TX channel coded with the opposite sign of the TX code unbalance, TX_D.

A PGA adds proper gain to the differential mode of the TIA output while eliminating the remaining common mode with a rail-to-rail input op-amp. The following SC $\Delta\Sigma$M, working at sampling rate (f_S) of 8 to 15MHz, embeds a wave-shaping mixer function by dynamically programming its 5b (s for sign and b[3:0] for binary weight) input capacitor to realize sinusoid-like sampling coefficients. Figure 12.5.4 shows that, using $16f_C$ rate coefficients, this down-conversion scheme performs enhanced frequency selectivity to f_C and rejection of harmonic components of f_C, resulting in better immunity to high-frequency external noise sources, e.g., charger switching noise, than square-wave mixing [3] or DT circuits [4]. Additionally, a spectral leakage effect is also minimized by windowing the coefficient at its head and tail for better low-frequency noise rejection. The solution eliminates the necessity of a linear analog multiplier or direct digital synthesizer [1], easily achieving superior linearity performance for robustness to large noise blockers. The unwanted conversion gain occurring at $15f_C$, $17f_C$ and aliasing at f_S are suppressed by the preceding low-pass poles at PGA and series resistor (R_S) with sampling capacitor (C_S). An incremental mode $\Delta\Sigma$ operation by reset (p_{RST}) in sync with $f_C=f_S/16$-$f_S/64$ makes the following sinc decimation filter simply remove redundant f_C harmonics after down-conversion [6]. An optional Q-path modulator is added to compensate different phase delays among the nodes due to the panel RC time-constant.

Figure 12.5.5 shows captured read-out images of full-scan mode and adaptive resolution-scaling mode, from commercial 4.8-inch TSP with 24 TX and 16 RX channels. For full-scan using 24×16-long basis at f_C=312kHz, four carrier pulses with $1/f_C$ reset time are allocated for each pattern measurement, yielding up to a 160Hz frame scan rate. For adaptive resolution-scaling mode, a 6×4 pixel basis is used for a low-resolution scan. In the case of a 3-finger touch event (shown in Fig. 12.5.5), the local high-resolution basis is 12×10 pixels. The frame scan time is reduced to 2.7ms from 6ms for that of full-scan mode with about half the TX driving power. The measured time-plot of all node capacitance values shows that the finger-touch signal strength ratio to untouched and touched noise rms, i.e., SNR, are 53dB and 38 dB, respectively.

The prototype controller is implemented and tested in 0.18μm 2P6M CMOS technology. The area measures 0.14mm² for the TX and RX core, plus 0.32mm² for 40 TRX I/O pads. The total power consumption of the analog part is less than 2.6mW, excluding TX driving power. As shown in Fig. 12.5.6, the controller features 15× smaller size and 6× less power than recently reported works, while maintaining comparable noise performance.

Acknowledgements:
This work was partially supported by Zinitix and the National Research Foundation of Korea (NRF) grant funded by the Korea government (No. 2010-0015119). The authors thank Chang-won Yoo for helping on digital implementation.

References:

[1] K.-D. Kim, *et al.*, "A Capacitive Touch Controller Robust to Display Noise for Ultrathin Touch Screen Displays," *ISSCC Dig. Tech. Papers*, pp. 116-117, Feb. 2012.
[2] S.P. Hotelling, *et al.*, "Multipoint Touch Surface Controller," U.S. Patent application US2007/0257890A1, Nov. 8, 2007.
[3] J.-H. Yang, *et al.*, "A Highly Noise-immune Touch Controller Using Filtered-Delta-Integration and a Charge-Interpolation Technique for 10.1-inch Capacitive Touch-Screen Panels," *ISSCC Dig. Tech. Papers*, pp. 390-391, Feb. 2013.
[4] H.-C. Shin, *et al.*, "A 55dB SNR with 240Hz Frame Scan Rate Mutual Capacitor 30×24 Touch-Screen Panel Read-Out IC Using Code-Division Multiple Sensing Technique," *ISSCC Dig. Tech. Papers*, pp. 388-389, Feb. 2013.
[5] R. Robucci, *et al.*, "Compressive Sensing on CMOS Separable Transform Image Sensor," in *Proc. of the IEEE*, vol. 98, no. 6, pp. 1089-1101, June 2010.
[6] J. Markus, J. Silva, and G. C. Temes, "Theory and applications of incremental $\Delta\Sigma$ converters," *IEEE Trans. Circuits and Systems-I*, vol. 51, no. 4, pp. 678-690, Apr. 2004.

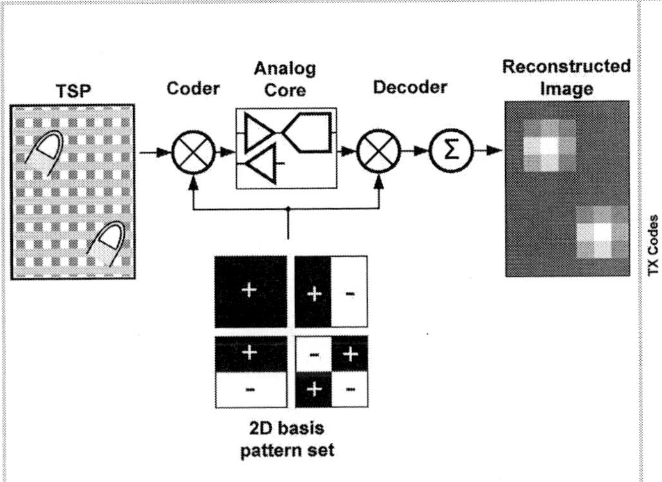

Figure 12.5.1: Conceptual diagram of the coded-aperture-based TSP read-out.

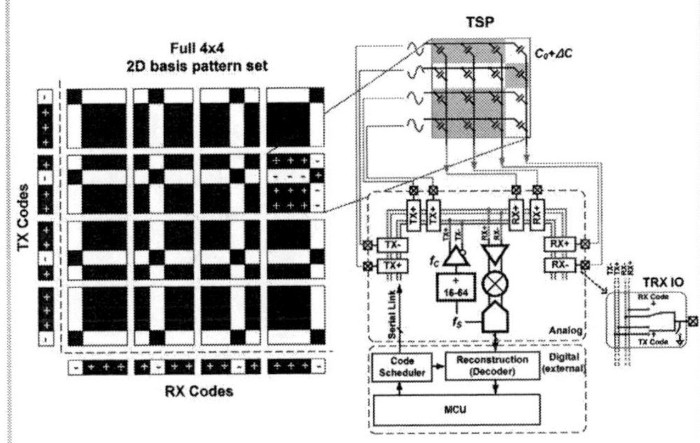

Figure 12.5.2: Code projection example and circuit architecture.

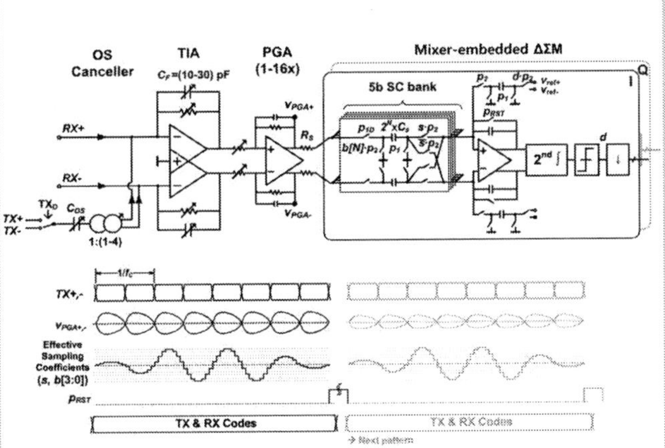

Figure 12.5.3: Circuit implementation details and operation timing diagram.

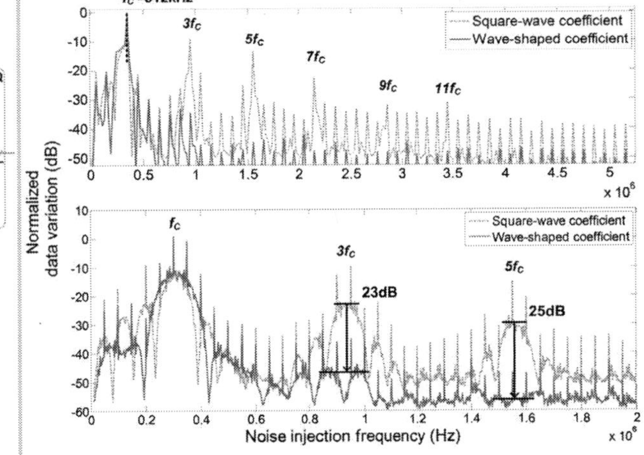

Figure 12.5.4: Normalized output data variation by intentional frequency-swept noise injection.

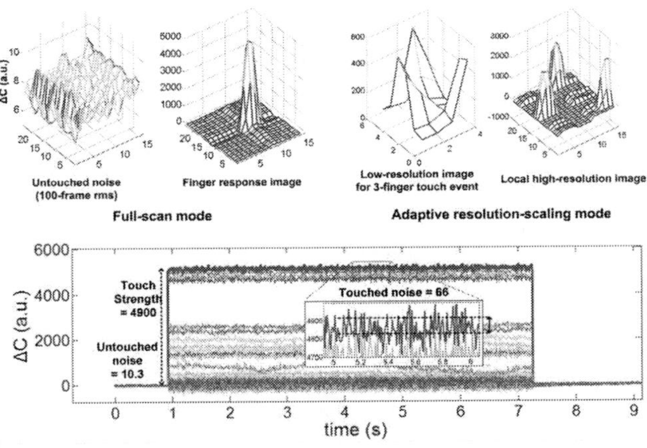

Figure 12.5.5: Captured read-out images (top) and time-plot of all-node finger response in full-scan mode (bottom).

	[3]	[4]	This Work
Process	0.35μm 3.3V CMOS	0.18μm 3.3/1.8V CMOS	0.18μm 3.3/1.8V CMOS
Channels (TX×RX)	27×43	30×24	24×16 (40 TRXs)
Scan rate	120Hz	240Hz	160Hz
SNR	39dB	55dB	53dB
Analog Power	18.7mW	20mW	2.6mW
Analog Area	10.4mm^2	7mm^2	0.46mm^2

Figure 12.5.6: Performance summary and comparison.

978-1-4799-0917-9/14 $31.00 © 2014 IEEE

TX & RX
Core

Figure 12.5.7: Chip microphoto.

12.6 A 160nW 63.9fJ/conversion-step Capacitance-to-Digital Converter for Ultra-Low-Power Wireless Sensor Nodes

Hyunsoo Ha[1], Dennis Sylvester[2], David Blaauw[2], Jae-Yoon Sim[1]

[1]Pohang University of Science and Technology, Pohang, Korea,
[2]University of Michigan, Ann Arbor, MI

Recent advances in nW-level wireless sensor nodes have created opportunities in emerging applications such as bio-implantable telemetry, smart healthcare, and environmental monitoring [1]. At the same time, there are many circuit and system design challenges to achieving high functionality in such ultra-low-power microsystems. One of the key sensing modalities in these systems is capacitive sensing. With zero static current during signal readout, capacitive sensing is well suited to ultra-low-power microsystems and has been widely adopted in the sensing of pressure [2,3], displacement [4], and humidity [5].

Previous research in capacitive-to-digital converter (CDC) design aimed at achieving high resolution with low power consumption. Switched-capacitor $\Delta\Sigma$ converters have been shown to achieve high resolution [4,5]. However, their oversampling procedure repeatedly charges and discharges the large sensor capacitor (5-to-70pF range), resulting in poor conversion energy. Furthermore, they are poorly suited to burst conversion with intermittent operation, which is a key feature in duty-cycled ultra-low-power sensor platforms, because they require numerous cycles to produce the first conversion result after initialization. Successive approximation is another candidate for CDCs [2], offering both low-power and good intermittent operation. However, the direct connection of the large sensor capacitor to the capacitor DAC (CDAC) in a SAR ADC places the large fixed capacitance parallel to the CDAC, which significantly reduces voltage swing at the input of the comparator, limiting the achievable effective number of bits. Because of these limitations, previous work achieves figures of merit (FoMs) of several pJ/conversion-step at best, while ADCs, which have a similar structure, can reach sub-10fJ/conversion-step [6].

To address this challenge, we present a general-purpose, wide-range CDC that combines a correlated double sampling (CDS) approach with a differential asynchronous SAR ADC. Since the sensor capacitor is sampled only twice per conversion, energy per conversion is low. Furthermore, since the CDS separates the sensor capacitor from the CDAC, a full differential input voltage range is preserved, resulting in a >13.3b ENOB. The CDC has a 2.5-to-75.3pF conversion range and an FoM of 63.9fJ/conversion-step while consuming 160nW. A demonstration with a 0.85×1.65 mm² pressure sensor [7] shows a resolution of 0.4mmHg and linearity of R^2=0.9997.

Figure 12.6.1 illustrates the concept of the readout scheme with correlated double sampling (CDS). The circuit consists of a sensor capacitor (C_{SENS}), a reference capacitor (C_{REF}), a sampling capacitor (C_{SAMPLE}), and an amplifier. C_{SENS} and C_{REF} are connected in series with both ends selectively switchable to VDD or VSS. The center node is set to virtual ground (V_{REF} or VDD/2) through the negative feedback configuration of the amplifier. The readout process then extracts charge proportional to difference between C_{SENS} and C_{REF} and stores it on C_{SAMPLE}, which is performed in four steps (*Pre-charge1*, *Sample1*, *Pre-charge2* and *Sample2*). The sequence of *Pre-charge1* and *Sample1* generates a voltage output proportional to C_{SENS} and C_{REF}. The same operation is repeated in the sequence of *Pre-charge2* and *Sample2* but with the role of C_{SENS} and C_{REF} switched. Therefore, the two output voltages after *Sample1* and *Sample2* are equal but with opposite polarity, providing a differential output with double the amount of signal. A key advantage of this approach is that the effect of variations on V_{REF} and offset voltage (V_{OS}) are all canceled in this differential output. Furthermore, parasitic capacitance to ground of off-chip C_{SENS}, which can be several tens of pF, does not affect the sampled result, because the center node voltage does not change during the CDS process.

Figure 12.6.2 shows the circuit diagram of the CDC, consisting of the readout frontend, a differential CDAC, the asynchronous SAR logic, a comparator, C_{REF} selection logic, and timing controller. Each conversion is performed with CDS followed by an A-D conversion. During CDS, the two capacitor banks of CDAC play the role of C_{SAMPLE} in *Sample1* phase and in *Sample2* phase, respectively. Therefore, the differential voltage output is sampled in CDAC. In A-D conversion phase, the amplifier is turned off to reduce power consumption and asynchronous SAR logic runs, generating a 13b output. A transition of the external clock (250Hz) initiates the conversion process. The timing controller generates all control signals with predefined timing. Though the SAR ADC has a differential input range of –VDD to VDD, the readout frontend cannot support this entire rail-to-rail output swing due to limited linear output range of the amplifier. In this work, the valid output range of the readout frontend is taken to be from –VDD/2 to VDD/2, corresponding to a C_{SENS} range from C_{REF} - C_{SAMPLE}/4 to C_{REF} + C_{SAMPLE}/4.

To increase the range of conversion, the C_{REF} select logic chooses one of eight cases of C_{REF} (from 7.5pF to 70.5 pF in 9pF steps) with a 3b code. Figure 12.6.3 illustrates how the optimal C_{REF} can be found. The C_{REF} select logic monitors the CDC output code and checks if C_{SENS} is in the valid range of the current C_{REF} value. If not, C_{REF} is updated by incrementing or decrementing. This procedure is repeated until the CDC output code falls in the valid range or C_{REF} reaches its maximum or minimum value. The input ranges of neighboring C_{REF}'s are overlapped by 1pF to avoid bang-bang updating at the boundaries. The overlapped range is also used in calibration of discontinuity caused by process variations on C_{REF} [6].

The CDC is implemented in 0.18µm standard CMOS. With dual supply voltages of 1.2V and 0.9V for analog and digital parts, respectively, the CDC consumes 119 to 160nW as C_{SENS} varies from minimum to maximum (Fig. 12.6.4). To characterize linearity error for the entire range of C_{SENS}, the input voltage was swept with fixed C_{SENS} to generate equivalent capacitance instead of varying capacitance. Error in C_{REF} was extracted by measuring the C_{SENS} in the overlapped range. There were −0.1pF and −0.2pF errors at the two largest cases of C_{REF}, 61.5pF and 70.5pF, respectively. After discontinuity calibration, the linearity errors for the eight cases of C_{REF} were obtained separately and combined to plot the error for the full conversion range. The CDC shows an effective number of bits (ENOB) of more than 13.3b with a resolution of 6.0fF. Figure 12.6.5 compares performance with previously published works. FoM is 63.9fJ/conversion-step. The CDC, when used with a 0.85×1.65 mm² pressure sensor [7], demonstrates a resolution of 0.4mmHg (Fig. 12.6.6). The active area is 0.49mm² (Fig. 12.6.7).

Acknowledgement:
This work was partly supported by MKE/KEIT [10039159] and NRF of Korea under grant No. 2008-0062617.

References:

[1] Y. Lee, G. Kim, S. Bang, *et al.*, "Modular 1mm³ Die-Stacked Sensing Platform with Optical Communication and Multi-Modal Energy Harvesting," *ISSCC Dig. Tech. Papers*, pp. 402-403, Feb. 2012.
[2] K. Tanaka, Y. Kuramochi, T. Kurashina, K. Okada, and A. Matsuzawa, "A 0.026mm² Capacitance-to-Digital Converter for Biotelemetry Applications Using a Charge Redistribution Technique," in *Proc. ASSCC*, pp. 244-247, 2007.
[3] H. Danneels, K. Coddens and G. Gielen, "A Fully-digital, 0.3V, 270nW Capacitive Sensor Interface Without External References," in *Proc. ESSCIRC*, pp. 287-290, 2011.
[4] S. Xia, K. Makinwa, and S. Nihtianov, "A Capacitance-to-Digital Converter for Displacement Sensing with 17b Resolution and 20µs Conversion Time," *ISSCC Dig. Tech. Papers*, pp. 198-199, Feb. 2012.
[5] Z. Tan, Y. Chae, R. Daamen, A. Humbert, Youri, V. Ponomarev, and M.A.P. Pertijs, "A 1.2V 8.3nJ Energy-Efficient CMOS Humidity Sensor for RFID Applications," *VLSI Circuits Dig. Tech. Papers*, pp. 24-25, 2012.
[6] M.A.P. Pertijs and Z. Tan, "Nyquist AD Converters, Sensor Interfaces, and Robustness: Chapter 8. Energy-Efficient Capacitive Sensor Interface," Springer/New York, pp. 129-147, 2013.
[7] R. M. Haque and K. D. Wise, "An Intraocular Pressure Sensor Based on a Glass Reflow Process," *Solid-State Sensors, Actuators, and Microsystems Workshop*, pp. 49-52, 2010.

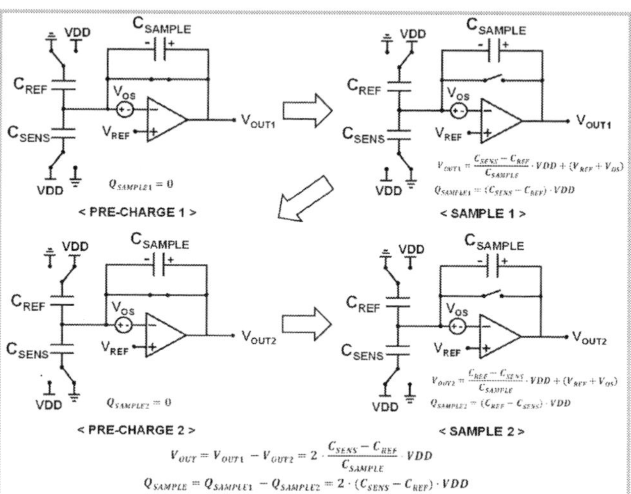

Figure 12.6.1: The concept of the readout frontend with correlated double sampling.

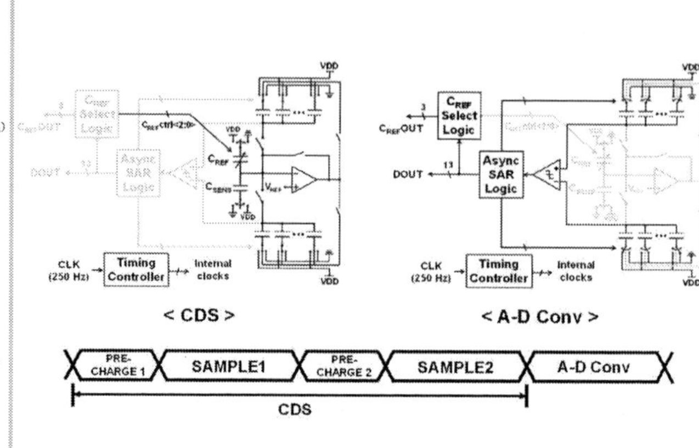

Figure 12.6.2: Circuit and timing diagram of the CDC.

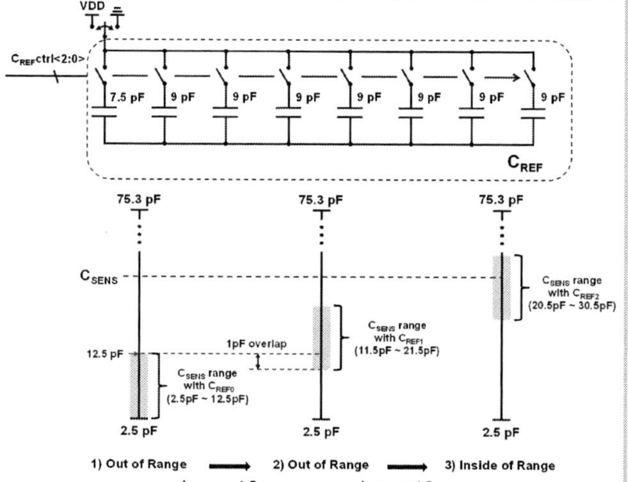

Figure 12.6.3: C_{REF} array and procedure of finding optimal C_{REF} for C_{SENS}.

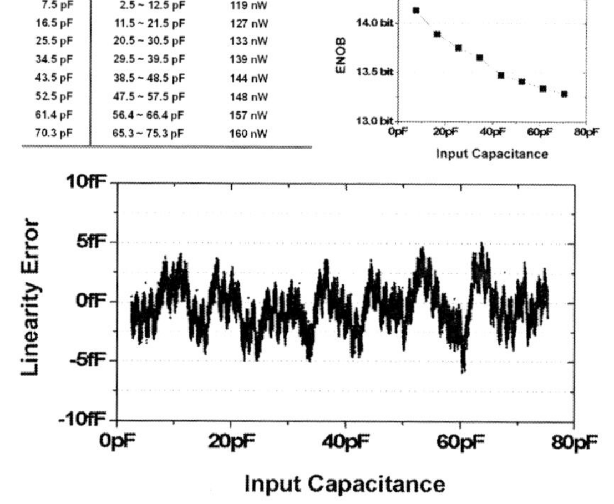

C_{REF}	Csensor range	Power
7.5 pF	2.5 ~ 12.5 pF	119 nW
16.5 pF	11.5 ~ 21.5 pF	127 nW
25.5 pF	20.5 ~ 30.5 pF	133 nW
34.5 pF	29.5 ~ 39.5 pF	139 nW
43.5 pF	38.5 ~ 48.5 pF	144 nW
52.5 pF	47.5 ~ 57.5 pF	148 nW
61.4 pF	56.4 ~ 66.4 pF	157 nW
70.3 pF	65.3 ~ 75.3 pF	160 nW

Figure 12.6.4: Power consumption and C_{SENS} range with given C_{REF}, and ENOB and linearity error according to C_{SENS}.

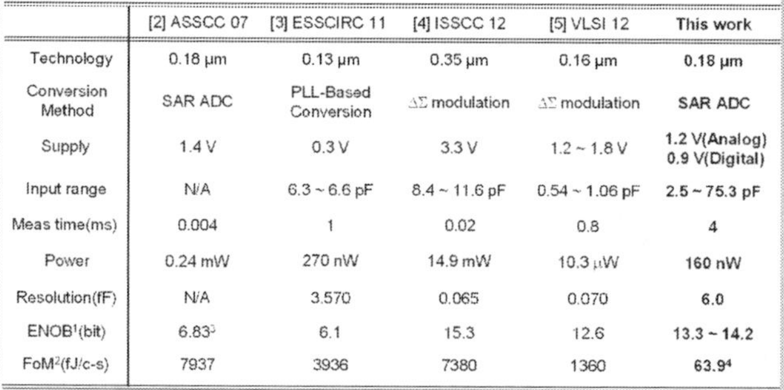

	[2] ASSCC 07	[3] ESSCIRC 11	[4] ISSCC 12	[5] VLSI 12	This work
Technology	0.18 µm	0.13 µm	0.35 µm	0.16 µm	0.18 µm
Conversion Method	SAR ADC	PLL-Based Conversion	ΔΣ modulation	ΔΣ modulation	SAR ADC
Supply	1.4 V	0.3 V	3.3 V	1.2 ~ 1.8 V	1.2 V(Analog) 0.9 V(Digital)
Input range	N/A	6.3 ~ 6.6 pF	8.4 ~ 11.6 pF	0.54 ~ 1.06 pF	2.5 ~ 75.3 pF
Meas time(ms)	0.004	1	0.02	0.8	4
Power	0.24 mW	270 nW	14.9 mW	10.3 µW	160 nW
Resolution(fF)	N/A	3.570	0.065	0.070	6.0
ENOB[1](bit)	6.83[3]	6.1	15.3	12.6	13.3 ~ 14.2
FoM[2](fJ/c-s)	7937	3936	7380	1360	63.9[4]

[1]ENOB = $(20 \times \log_{10}(Range/Resolution) - 1.76)/6.02$, [2]FoM = $(Power \times Tmeas)/2^{ENOB}$
[3]This ENOB is obtained by applying sinusoidal signal to fixed capacitor.
[4]This FoM is calculated by using lowest ENOB.

Figure 12.6.5: Comparison table with recently published works in major conferences.

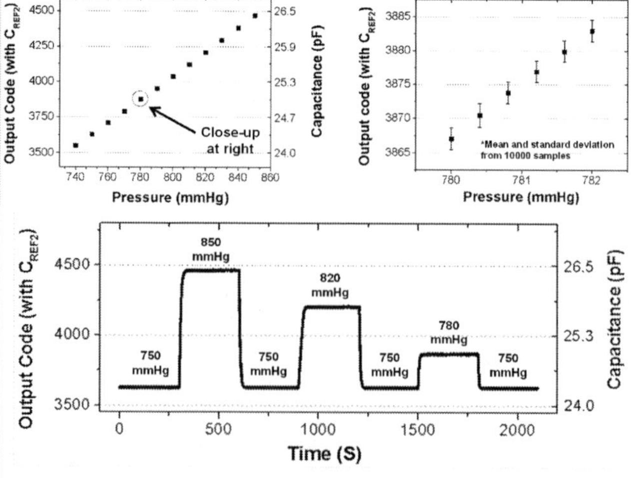

Figure 12.6.6: Conversion result of the CDC with pressure sensor.

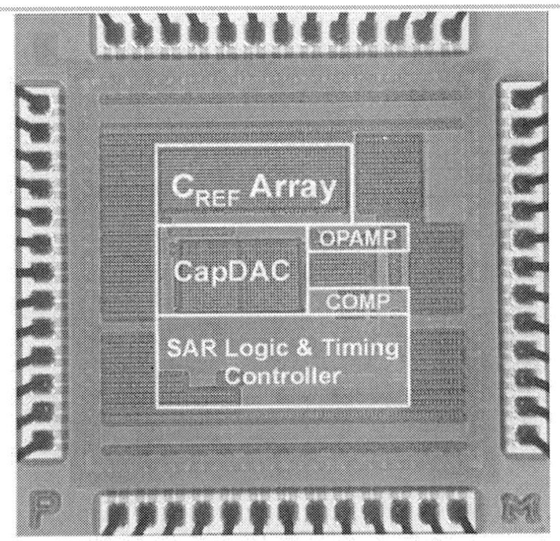

Figure 12.6.7: Microphotograph (0.7×0.7 mm²) .

12.7 A 0.85V 600nW All-CMOS Temperature Sensor with an Inaccuracy of ±0.4°C (3σ) from -40 to 125°C

Kamran Souri[1], Youngcheol Chae[2], Frank Thus[3], Kofi Makinwa[1]

[1]Delft University of Technology, Delft, The Netherlands,
[2]Yonsei University, Seoul, Korea,
[3]NXP Semiconductors, Eindhoven, The Netherlands

This paper describes an all-CMOS temperature sensor intended for RFID applications that achieves both sub-1V operation and high accuracy (±0.4°C) over a wide temperature range (-40 to 125°C). It is also an ultra-low-power design: drawing 700nA from a 0.85V supply. This is achieved by the use of dynamic threshold MOSTs (DTMOSTs) as temperature-sensing devices, which are then read out by an inverter-based 2nd-order zoom ADC. Circuit errors are mitigated by the use of dynamic error-correction techniques, while DTMOST spread is reduced by a single room temperature (RT) trim. The latter feature constitutes a significant advance over previous all-CMOS designs [5,6], which require two-point trimming to approach the same level of accuracy.

In most CMOS processes, a diode-connected DTMOST can be readily realized by connecting the gate, bulk and drain of a standard PMOST together (Fig. 12.7.1). The resulting device approximates an ideal diode, with an extrapolated gate-source voltage V_{GS}~ 0.6V at 0K and a linear temperature coefficient of about -1mV/°C [2]. Connecting the gate to the bulk reduces the influence of gate-oxide thickness on the resulting dynamic threshold voltage, and thus the V_{GS} spread of a DTMOST is significantly less than that of a normal PMOST [1,2]. Diode-connected DTMOSTs can thus be used to replace the BJTs of a conventional band-gap voltage reference [2] or temperature sensor [1]. However, since the magnitude of V_{GS} (~ 0.3V at RT) is only about half that of a BJT's base-emitter voltage V_{BE} (~ 0.6V at RT), the resulting circuit can be operated at supply voltages below 1V over a wide temperature range, e.g., from -40 to 125°C.

The sensor's front-end is shown in Fig. 12.7.1. A pair of DTMOSTs with a 1:2 area ratio that are biased by identical currents I=90nA (at RT). The same currents also power a so-called current-voltage mirror (CVM) [3], which forces a proportional-to-absolute-temperature (PTAT) voltage ΔV_{GS} across a resistor. As a result, the biasing currents will also have a well-defined PTAT dependency. To minimize the effect of DTMOST mismatch, which would otherwise impact the accuracy of ΔV_{GS}, the 1:2 area ratio is established by incorporating three unit DTMOSTs into a dynamic element matching (DEM) scheme. Since the associated DEM switches carry bias current, Kelvin connections are used to accurately read out V_{GS} and ΔV_{GS}. Another source of error is the CVM's offset and 1/f noise, which add directly to ΔV_{GS} and thus impact the accuracy of the bias currents, and hence of both V_{GS} and ΔV_{GS}. Such errors are mitigated by chopping the CVM (Fig. 12.7.1).

The sensor's block diagram is shown in Fig. 12.7.2. It consists of the DTMOST front-end, a 2nd-order incremental zoom ADC, a voltage doubler and some control logic. As in [4], the zoom ADC uses a power-efficient coarse/fine algorithm to convert the front-end's output voltages V_{GS} and ΔV_{GS} into a temperature-dependent ratio $X = V_{GS}/\Delta V_{GS}$. In this design, X varies from 5 to 28 over the temperature range –40 to 125°C. An off-chip digital backend then computes a PTAT function of temperature $\mu = \alpha/(\alpha+X)$, where α is a gain factor, which can be trimmed to compensate for V_{GS} spread. The sensor has two supply voltages: an analog supply AVDD, which powers the front-end and the ADC, and a digital supply DVDD, which powers the voltage doubler. The output of the doubler drives the logic that, in turn, drives the switches that sample V_{GS} and ΔV_{GS}, thus facilitating the use of sub-1V supply voltages. To minimize its residual offset, the entire ADC is chopped over two conversions.

As shown in Fig. 12.7.2, the zoom ADC digitizes the output of the DTMOST front-end in a two-step manner [4]. Each zoom ADC conversion begins with a coarse SAR conversion followed by a fine ΔΣ conversion to generate the ratio $X = V_{GS}/\Delta V_{GS}$. The coarse conversion determines the integer part of X, or n, by using a 5b SAR algorithm to compare V_{GS} with integer multiples of ΔV_{GS} (Fig. 12.7.2). The fractional part of X, or μ', is then determined by a 2nd-order incremental ΔΣ-ADC, whose reference voltages are arranged to straddle V_{GS} by setting them to $n \cdot \Delta V_{GS}$ and $(n+2) \cdot \Delta V_{GS}$. The resulting $2\Delta V_{GS}$ input range provides redundancy, thus relaxing the requirements on the coarse conversion, and ensuring that the modulator is not overloaded.

The heart of the zoom ADC is a feed-forward 2nd-order SC ΔΣ-ADC (Fig. 12.7.3). At its input is a capacitive-DAC (cap-DAC) with 30 unit elements (each 60fF), which can sample either V_{GS} or $k \cdot \Delta V_{GS}$, where k = 1..30. In contrast to [4], both integrators are formed around pseudo-differential inverter-based amplifiers, thus fully exploiting the reduced integrator swing conferred by zooming. The first integrator draws 135nA while the, less critical, 2nd integrator draws only 66nA. These current levels are defined with the help of a dynamic biasing technique that simultaneously auto-zeros each amplifier [4]. During the coarse conversion, the first integrator computes $V_{GS} - k \cdot \Delta V_{GS}$, while its output is connected directly to the comparator via the switch S_{bp}. Off-chip logic then implements the SAR algorithm by applying trial values k to the chip and monitoring the comparator's output.

During the fine conversion, the mismatch between the unit elements of the cap-DAC is mitigated by the use of DEM. In contrast to [4], the required DEM logic is implemented on-chip (Fig. 12.7.4). It consists of a 30b circular shift-register (SR), with an effective length (defined by a periodic reset signal) of n+3 bits, where n is the result of the coarse conversion. Resetting the SR loads it with a single logic "1," which then circulates on every succeeding clock pulse. This bit (via the m_i outputs) is used to select the capacitor that samples V_{GS}, while the other bits define the n+2 capacitors that *may* be used to sample ΔV_{GS}. Depending on the bitstream output (bs), either n or n+2 capacitors will be selected (via the t_i outputs). The SR is reset during the coarse conversion, so that the same capacitors are always used for the SAR conversion. The DTMOST's DEM logic is implemented by a separate 3b SR.

The prototype sensor is realized in a standard 0.16μm CMOS process (Fig. 12.7.7). It occupies 0.085mm², and draws 700nA from a 0.85V supply. The front-end and ADC draw 560nA, while the voltage doubler and the rest of the on-chip digital circuitry draw 140nA. For flexibility, the SAR logic and the sinc² decimation filter are implemented off-chip. However, simulations show that implementing them on-chip would only incur an extra 10nW per conversion. With DVDD fixed at 0.9V, AVDD was varied from 0.85V to 1.2V. The corresponding supply sensitivity of the front-end and ADC was 0.45°C/V.

A total of 16 devices in ceramic DIL packages were characterized over the temperature range from –40 to 125°C. As shown in Fig. 12.7.5 (top), their batch-calibrated inaccuracy was ±1°C (3σ, 16 devices), with a residual curvature of only 0.03°C. After an alpha trim at 30°C, the inaccuracy improves to ±0.4°C (3σ), as shown in Fig. 12.7.5 (bottom). Offset trimming, as in [1], is slightly worse, resulting in an inaccuracy of ±0.5°C (3σ). These results show that DTMOSTs, like BJTs, can be effectively trimmed at a single temperature.

While running at a clock frequency of 25kHz, the sensor requires only 3.6nJ to achieve a kT/C-limited resolution of 63mK (rms) in a conversion time of 6ms. This corresponds to a resolution FoM of 14.1pJK², which is in line with the state of the art [4]. The sensor's performance is summarized in Fig. 12.7.6 and compared to that of other state-of-the-art low-voltage designs. It can be seen that, except for the BJT-based design [4], this sensor is 2-to-3× more accurate than the rest, while also achieving the best energy efficiency.

References:

[1] K. Souri, Y. Chae, Y. Ponomarev and K.A.A. Makinwa, "A Precision DTMOST-Based Temperature Sensor," *Proc. ESSCIRC*, Sept. 2011.
[2] A.-J. Annema, "Low-Power Bandgap References Featuring DTMOSTs," *IEEE J. Solid-State Circuits*, vol. 34, no. 7, pp. 949-955, July 1999.
[3] Y.-H. Lam and W.-H. Ki, "CMOS Bandgap References With Self-Biased Symmetrically Matched Current-Voltage Mirror and Extension of Sub-1-V Design," *IEEE Trans. VLSI*, vol. 18, no. 6, pp. 857-865, June 2010.
[4] K. Souri, Y. Chae and K.A.A. Makinwa, "A CMOS Temperature Sensor With a Voltage-Calibrated Inaccuracy of ±0.15°C (3s) From -55 to 125°C," *IEEE J. Solid-State Circuits*, vol. 47, no. 12, pp. 292-301, Jan. 2013.
[5] M. Law, A. Bermak, and H. Luong,"A Sub-μW Embedded CMOS Temperature Sensor for RFID Food Monitoring Application," *IEEE J. Solid-State Circuits*, vol. 45, no. 6, pp 1246-1255, June 2010.
[6] Y.S. Lin, D. Sylvester and D. Blaauw, "An Ultra Low Power 1V, 220nW Temperature Sensor for Passive Wireless Applications," *Proc. CICC*, pp. 507-510, Sept. 2008.
[7] C-K. Wu, *et al.*, "A 80kS/s 36μW Resistor-based Temperature Sensor using BGR-free SAR ADC with a Unevenly-weighted Resistor String in 0.18μm CMOS," *VLSI Circ. Symp. Dig. Tech. Papers*, June 2011.

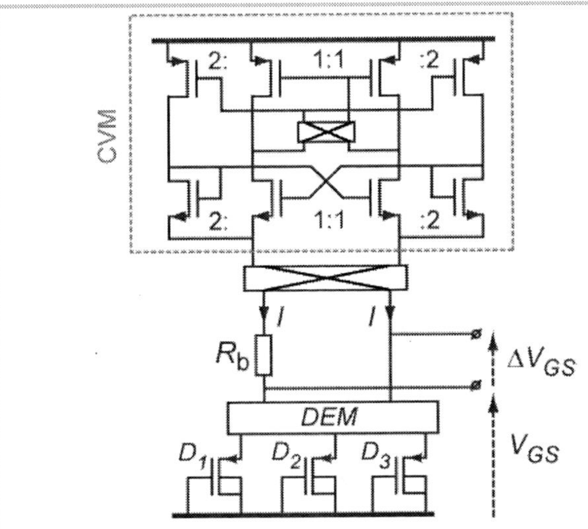

Figure 12.7.1: DTMOST-based sensor front-end.

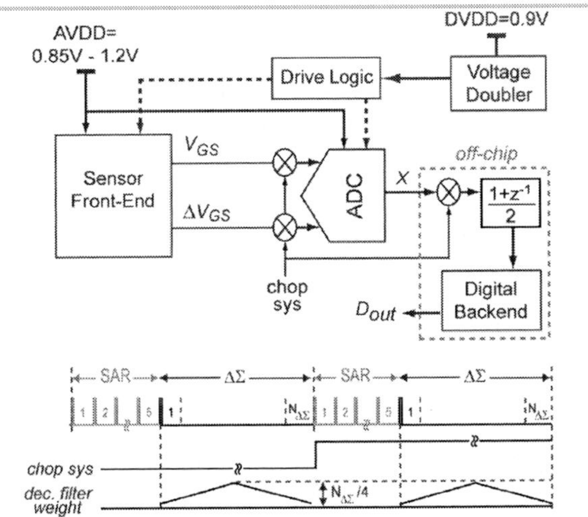

Figure 12.7.2: Top: the sensor's block diagram. Bottom: timing diagram of a temperature conversion.

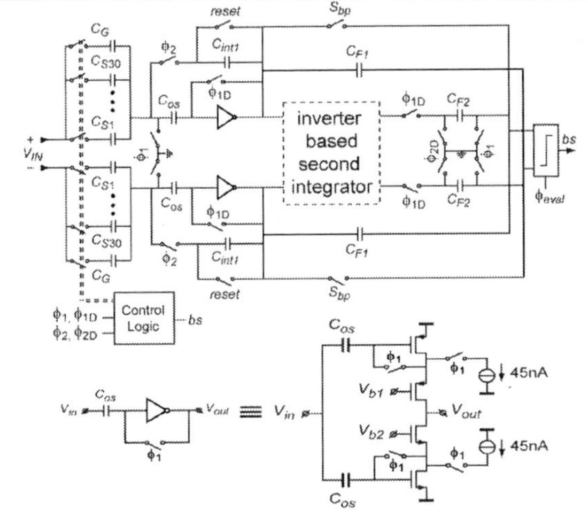

Figure 12.7.3: Top: Circuit diagram of the 2nd-order inverter-based zoom-ADC. Bottom: Circuit diagram of the inverter-based OTA.

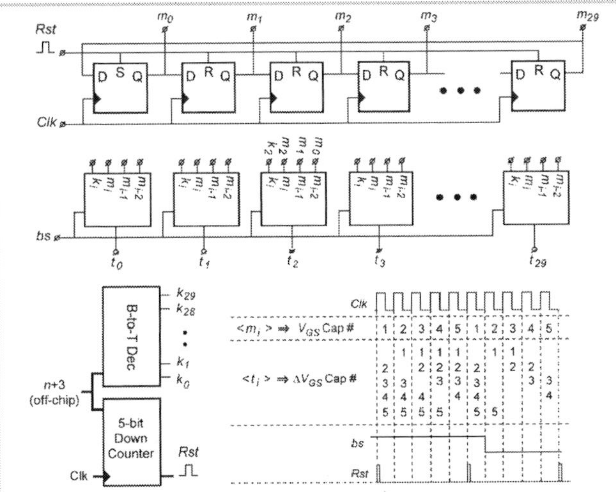

Figure 12.7.4: On-chip DEM logic for the zoom ADC's Cap-DAC (timing is shown for a 5-element Cap-DAC).

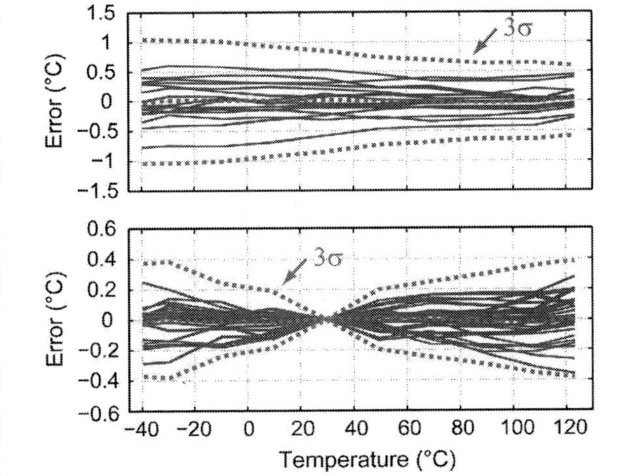

Figure 12.7.5: Measured temperature error of 16 samples after alpha-trimming at 30°C.

Parameter	This work	JSSC'13 [4]	JSSC'10 [5]	CICC'08 [6]	VLSI'11 [7]
Technology	**0.16μm**	0.16μm	0.18μm	0.18μm	0.18μm
Chip area	**0.085mm²**	0.08mm²	0.042mm²	0.05mm²	0.18mm²
Sensor type	**DTMOST**	BJT	MOST	MOST	Resistor
Supply current	**700nA**	3.4μA	190nA	220nA	20μA
Supply voltage	**0.85 - 1.2V**	1.5-2.0V	0.5V (sensor) 1.0V (digital)	1V	1.2-2V
Supply sensitivity	**0.45°C/V**	0.5°C/V	-	Supply referenced	0.625°C/V
Temperature range	**−40°C to 125°C**	−55°C to 125°C	−10°C to 30°C	0°C to 100°C	0°C to 100°C
Inaccuracy (Trim method)	**±0.4°C[1] (1-point)**	±0.15°C[1] (1-point)	−0.8°C/+1°C[2] (2-point)	−1.6°C/+3°C[2] (2-point)	±0.5°C[2] (1-point)
Relative inaccuracy	**0.48%**	0.2%	4.5%	4.6%	1%
Number of samples	**16**	18	9	5	5
Resolution (T_{conv})	**0.063°C (6msec)**	0.02°C (5.3msec)	0.2°C (30msec)	0.1°C (100msec)	0.25°C (12.5μsec)
Res.FOM	**14.1 pJK²**	11 pJK²	140 pJK²	220 pJK²	19pJK²

1: 3σ, 2: Maximum

Res. FOM=Energy/Conversion×(Resolution)²

Relative inaccuracy(%)=100×Max Error/Specified temperature range

Figure 12.7.6: Performance summary and comparison with previous work.

12

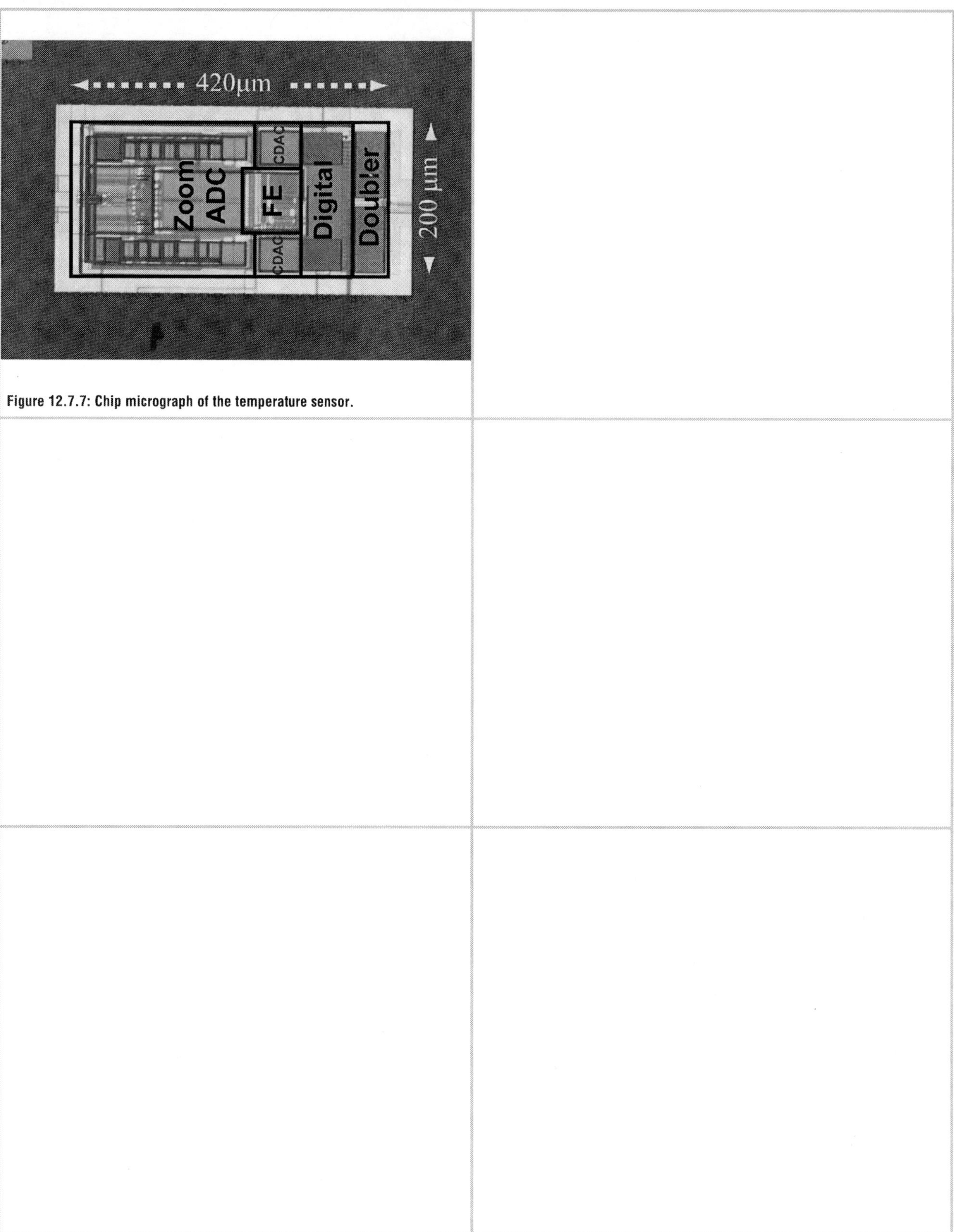

Figure 12.7.7: Chip micrograph of the temperature sensor.

12.8 A BJT-Based CMOS Temperature Sensor with a 3.6pJ·K²-Resolution FoM

Ali Heidary[1,2,3], Guijie Wang[1,2], Kofi Makinwa[2], Gerard Meijer[1,2,4]

[1]Smartec, Breda, The Netherlands,
[2]Delft University of Technology, Delft, The Netherlands,
[3]Guilan University, Rasht, Iran,
[4]SensArt, Delft, The Netherlands

This paper presents a precision BJT-based temperature sensor implemented in standard CMOS. Its interface electronics consists of a continuous-time duty-cycle modulator [1], whose output can be easily interfaced to a microcontroller, rather than the discrete-time $\Delta\Sigma$ modulators of most previous work [2-4]. This approach leads to high resolution (3mK in a 2.2ms measurement time) and high energy efficiency, as expressed by a resolution FoM of 3.6pJK², which is a 3× improvement on the state of the art [4,5]. By employing chopping, dynamic element matching and a single room temperature trim, the sensor also achieves a spread of less than ±0.15°C (3σ) from -45 to 130°C.

The sensor's basic operating principle is illustrated in Fig. 12.8.1. Under the control of a Schmitt trigger (ST), a capacitor C is alternately charged by a current I_1 that is proportional-to-absolute-temperature (PTAT) and discharged by a current I_2 that is complementary-to-absolute-temperature (CTAT). The duty-cycle D of the resulting oscillation is then given by $I_1/(I_1+I_2)$. To first order, this is independent of the exact value of the ST's threshold voltages. D is a linear function of temperature if $I_{ref}= I_1+I_2$ is temperature independent. In a CMOS process, I_2 can be derived from the base-emitter voltage V_{BE} of a substrate PNP, while I_1 can be derived from the difference between the base-emitter voltages ΔV_{BE} of two appropriately biased PNPs. As shown in Fig. 12.8.1, D then varies by about 30% over the desired temperature range: -45 to 130°C.

Figure 12.8.2 shows a simplified block diagram of the actual sensor. Substrate PNPs Q_1 and Q_2 are biased at a 1:9 current density ratio, and an opamp (OP_1) forces the resulting voltage $\Delta V_{BE} = V_T \ln(9)$ across a resistor R_{PTAT} to generate a PTAT current $I_{PTAT}=\Delta V_{BE}/R_{PTAT}$ (0.8μA at room temperature). Similarly, OP_2 and another resistor R_{BE} convert the base-emitter voltage V_{BE3} of Q_3 into a CTAT current $I_{CTAT}=V_{BE3}/R_{BE}$. These currents are then linearly combined such that the capacitor C is charged by a current $3I_{PTAT}$-$0.5I_{CTAT}$ and is discharged by a current I_{CTAT}-I_{PTAT}. To ensure accuracy over a wide supply range, all the associated current mirrors/sources are cascoded. As in [3], the sum of the charging and discharging currents, i.e., $2I_{PTAT}+0.5I_{CTAT}$, is designed to have a slightly positive temperature coefficient, which effectively compensates for the curvature in V_{BE3}. As shown in Fig. 12.8.1, this scheme ensures that D now varies from about 10% to 90% over the desired temperature range [1].

An important source of error is the PTAT spread of V_{BE3} [1]. This can be corrected by trimming the bias current and emitter area of Q_3. The total trimming range is about 10°C, with a worst-case step of about 50mK. Another source of error is device mismatch, which causes the ratios between the various charging and discharging currents to spread. The ratio R_{BE}/R_{PTAT} is set by using large devices (R_{BE} ~200kΩ) and careful layout, while the effect of opamp offset (and 1/f noise) is mitigated by chopping. Errors in the various current-mirror and resistor ratios, as well as in the emitter area ratio of the substrate PNPs are mitigated by DEM. Since the DEM switches around Q_1 and Q_2 carry their bias currents, and so drop some voltage, Kelvin connections are used to accurately sense ΔV_{BE} [4]. The DEM and chopping state machines are self-clocked (by the Schmitt trigger), and so no external clock is required. A full DEM and chopping cycle corresponds to 8 oscillator periods.

The finite gain of OP_1 and OP_2 also causes errors in I_{PTAT} and I_{CTAT}, respectively. In order to keep the resulting temperature-sensing errors below 50mK, the gains of OP_1 and OP_2, must be greater than 90dB and 70dB, respectively. Moreover, they must be able to handle input voltages (V_{BE}) down to about 0.3V at 130°C. Both requirements are met by implementing OP_1 and OP_2 as folded-cascode amplifiers with PMOS input pairs, which draw 20μA and 7μA, respectively.

As shown in Fig. 12.8.3, the ST is based on two inverters in series with a positive feedback path that controls the threshold voltages of the first one. When the ST's output is HIGH, M_1 is bypassed by M_6, and so its lower threshold (V_1 in Fig. 12.8.1) is set by the threshold voltage of M_2, i.e., V_{TH2}. When its output is LOW, M_4 is bypassed by M_5, and so its upper threshold voltage (V_2 in Fig. 12.8.1) is close to V_{DD}-$|V_{TH3}|$. The amount of positive feedback is determined by the relative W/L ratios of $M_{2,3}$ and $M_{1,4}$. The large swing (~ V_{DD} − 2V) at the ST's input ensures that its input-referred noise has negligible impact on the resulting duty-cycle. Making C large (150pF), makes the oscillation frequency low enough (less than 4kHz) to ensure that the error caused by the ST's own switching time (a few nanoseconds) is less than 50mK.

Only 8 cycles of the sensor's duty-cycle-modulated output are necessary for an accurate temperature measurement. As a result, the sensor is faster, and thus more energy-efficient, than sensors based on discrete-time $\Delta\Sigma$ modulators, which typically require hundreds of clock cycles to achieve the same thermal-noise-limited resolution [2-4]. Although accurately digitizing a duty-cycled signal requires a counter driven by a high frequency clock, this can be done in a micro-controller or FPGA realized in nanometer CMOS, thus making the associated power and area overhead negligible.

The sensor occupies 0.8mm² and was implemented in a 0.7μm CMOS process. It operates from supply voltages ranging from 2.9 to 5.5V, and draws 55μA at 3.3V. The sensor outputs a rail-to-rail square-wave, whose frequency varies from about 0.5 to 4kHz over temperature and supply voltage. These outputs were buffered and applied to an FPGA which digitized the high- and low-time intervals, t_H and t_L, with the help of an internally generated 100MHz sampling clock. The requirements on this clock are quite relaxed, since the sensor's jitter is in the order of a few tens of nanoseconds. To achieve an accurate result, 8 periods must be averaged, i.e. $D_{avg1}=\Sigma(t_H/(t_H+t_L))/8$. A simpler approach, analogous to low-pass filtering the sensor's output, involves summing the various periods over 8 cycles, i.e. $D_{avg2}=\Sigma t_H/\Sigma(t_H+t_L)$. However, this approach does not completely cancel the DEM and chopping residuals, and is thus less accurate.

A total of 15 devices in metal TO-5 packages were tested over the temperature range -45 to 130°C. A linear fit shows that the duty-cycle can be expressed as $D=AT+B$, where A=0.0046, B=0.30, and T is the temperature in degrees Celsius. However, the residual curvature causes a systematic non-linearity of about 0.2°C. After batch-calibration (using the default trim setting), the sensors' spread is less than ±0.2°C (3σ), as shown in Fig. 12.8.4 (top). The chips were then trimmed by post-processing, rather than by definitively storing trim data in their one-time programmable memory. A PTAT trim at 25°C only reduces the spread to ±0.15°C (3σ), but is required to correct for the expected batch-to-batch spread of V_{BE3} [3]. Using the simplified average, i.e. D_{avg2}, results in more spread, especially at low temperatures (Fig. 12.8.4).

The sensor's resolution was determined by logging the results of 40,000 measurements at a stable temperature (~25°C). As shown in Fig. 12.8.5 (top), the sensor achieves a resolution of 3mK (rms) in a measurement time t_m of 2.2ms (8 periods). This corresponds to a resolution FoM of 3.6pJK². Averaging over 100ms improves the resolution to 0.45mK (rms). In order not to degrade the sensor's intrinsic resolution, the sampling clock must be high enough to minimize the effects of quantization noise. As shown in Fig. 12.8.5 (bottom), reducing the sampling frequency from 100 to 20MHz increases the noise floor, which translates into lower resolution: 5.5mK (rms) in 2.2ms.

The sensor's performance is summarized in Fig. 12.8.6 and compared with that of other energy-efficient precision temperature sensors. It can be seen from the table that this design achieves the highest energy efficiency, as well as the highest reported resolution for a BJT-based temperature sensor.

References:

[1] G.C.M. Meijer, *et al.*, "A Three-Terminal Integrated Temperature Transducer with Microcomputer Interfacing," *Sensors and Actuators*, vol. 18, pp. 195-206, June 1989.

[2] A. L. Aita, *et al.*, "Low-Power CMOS Smart Temperature Sensor with a Batch-Calibrated Inaccuracy of ±0.25°C (±3σ) from -70 °C to 130 °C," *Sensors J.*, vol. 13, no. 5, pp. 1840-1848, May 2013.

[3] K. Souri, *et al.*, "A CMOS Temperature Sensor with a Voltage Calibrated Inaccuracy of ±0.15°C (±3σ) from -55 °C to 125 °C," *J. Solid-State Circuits*, vol. 48, no. 1, pp. 292-301, Jan. 2013.

[4] S. Shalmany, "A Micropower Battery Current Sensor with ±0.03% (3σ) Inaccuracy from -40 to +85°C." *ISSCC Dig. Tech. Papers*, pp. 1308-1313, Feb. 2013.

[5] M. Perrot, *et al.*, "A Temperature to Digital Converter for a MEMS Based Programmable Oscillator with Better than ±0.5ppm Frequency Stability," *J. Solid-State Circuits*, vol. 48, no. 1, pp. 206-207, Jan. 2013.

[6] Smartec BV, "Datasheet SMT 160-30 Digital Temperature Sensor," www.smartec-sensors.com, Nov. 2010. [online]

978-1-4799-0917-9/14 $31.00 © 2014 IEEE

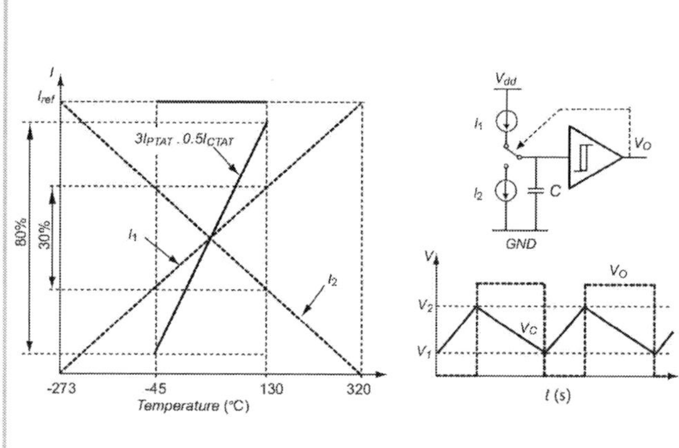

Figure 12.8.1: Operating principle.

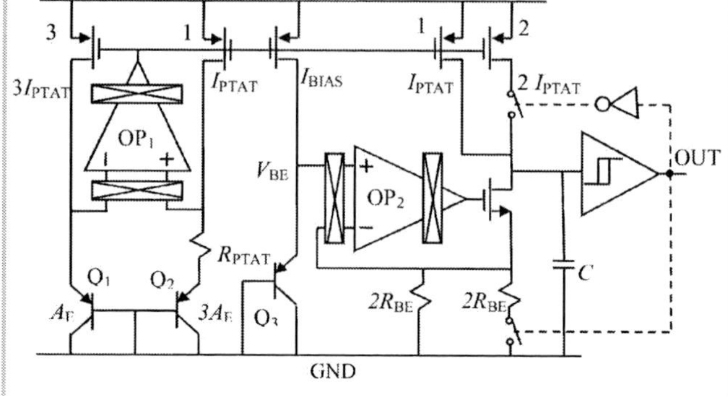

Figure 12.8.2: Simplified sensor block diagram.

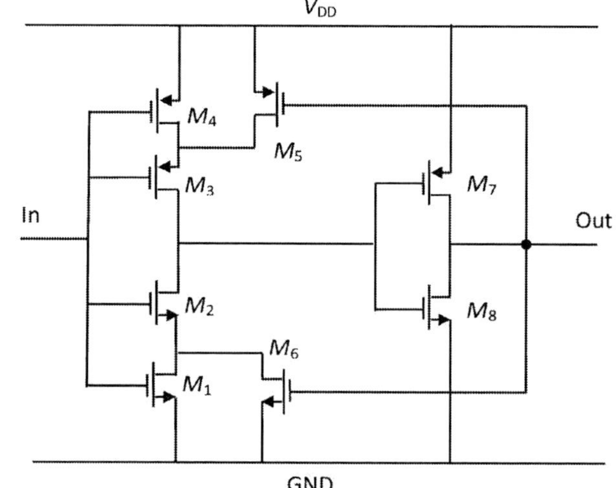

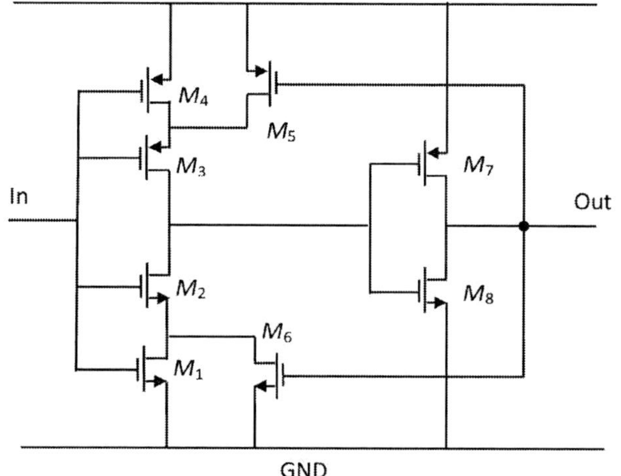

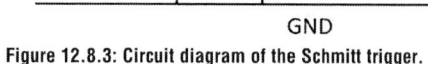

Figure 12.8.3: Circuit diagram of the Schmitt trigger.

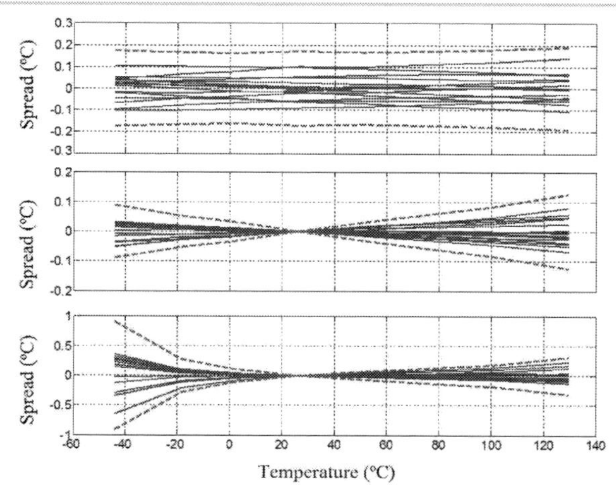

Figure 12.8.4: Measured spread of 15 samples after batch calibration (top); ideal PTAT trimming (middle) and ideal PTAT trimming and simplified averaging (bottom). Dashed lines indicate the ±3σ bounds.

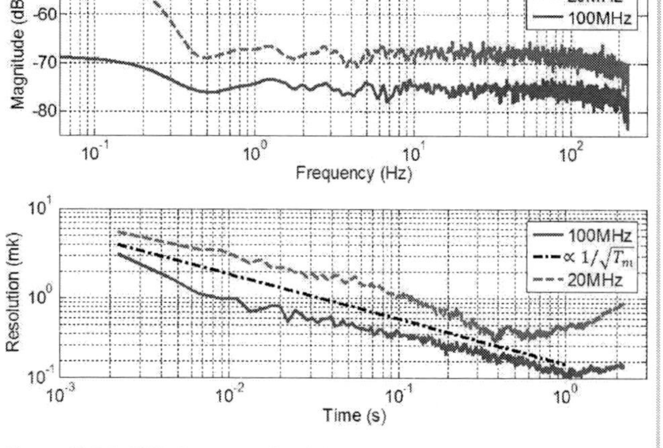

Figure 12.8.5: FFT of sensor noise (40,000 conversions) (top); resolution vs. measurement time (bottom).

	This work	JSSC13 [4]	ISSCC13 [3]	JSSC13 [2]	Sensors. J [1]	Smartec [6]
Technology	0.7μm	0.18μm	0.13μm	0.16μm	0.7μm	-
Chip area	0.8mm²	0.15mm²	1mm²	0.08mm²	4.5mm²	-
Supply current	55μA	4mA	55μA	3.4 μA	25 μA	180 μA
Supply voltage	2.9-5.5V	3.3V	1.4-1.6V	1.5-2V	2.5-5.5V	4.75-7.2V
Supply sensitivity	0.05°C/V	-	-	0.5°C/V	-	0.1°C/V
Temperature rage	-45°C to 130°C	-40°C to 85°C	-55°C to 85°C	-55°C to 125°C	-55°C to 125°C	-45°C to 130°C
Spread (trim method)	±0.15°C [1] (1-point)	±0.03°C [1] (6-point)	±0.15°C [1] (1-point)	±0.15°C [1] (1-point)	±0.1°C [1] (1-point)	±1°C [2] (1-point)
Resolution (measurement time t_m)	3.0mK [3] (2.2ms)	0.1mK (100ms)	5.0mK (20ms)	20mK (5.3ms)	25mK (100ms)	5.0mK (20ms)
Res. FoM	3.6 pJ°C²	13 pJ°C²	37 pJ°C²	11 pJ°C²	3.9nJ°C²	430 pJ°C²

Res. FoM=(resolution)²× Power ×t_m ; [1] 3σ, [2] Maximum, [3] V_{dd}=3.3V.

Figure 12.8.6: Performance summary and comparison with previous work.

12

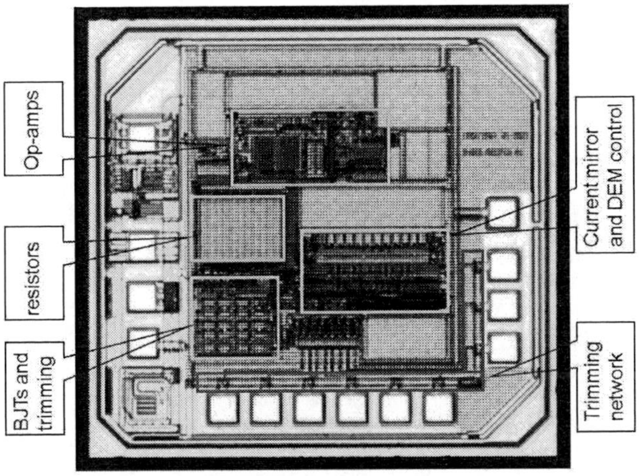

Op-amps

resistors

BJTs and trimming

Current mirror and DEM control

Trimming network

Core size: 0.85mmx0.95mm

Figure 12.8.7: Chip micrograph.

12.9 A 1.55×0.85mm² 3ppm 1.0µA 32.768kHz MEMS-Based Oscillator

Samira Zali Asl[1,2], Shouvik Mukherjee[1], Will Chen[1], Kimo Joo[1], Rajkumar Palwai[1], Niveditha Arumugam[1], Preston Galle[1], Meghan Phadke[1], Charles Grosjean[1], Jim Salvia[1], Haechang Lee[1], Sudhakar Pamarti[3], Terri Fiez[2], Kofi Makinwa[4], Aaron Partridge[1], Vinod Menon[1]

[1]SiTime, Sunnyvale, CA,
[2]Oregon State University, Corvallis, OR,
[3]University of California, Los Angeles, CA,
[4]Delft University of Technology, Delft, The Netherlands

Mobile time-keeping applications require small form-factor, tight frequency stability, and micro-power 32.768kHz clock references. Today's 32kHz quartz resonators and oscillators are facing challenges in size reduction [1,2]. Previously described MEMS-based oscillators can achieve tight accuracy but operate at high frequency with power unsuitable for mobile applications [3]. This paper introduces a 32kHz MEMS-based oscillator. Based on a comparison table of recent oscillators shown in Fig. 12.9.6, it offers the smallest size, 1.55×0.85mm², with the best frequency stability, 100ppm (XO) and 3ppm (TCXO) over the industrial temperature range of -40 to 85°C. Supply current is 0.9 and 1.0µA for XO and TCXO, respectively, at supply voltages from 1.5 to 3.6V.

A simplified block diagram of the MEMS-based oscillator is shown in Fig. 12.9.1. A 524kHz MEMS resonator and sustaining amplifier provide a frequency reference to a programmable fractional-N synthesizer, which in turn generates an accurate 32kHz output. The oscillator can be configured in either XO or TCXO mode. In XO mode, the fractional-N synthesizer compensates for the frequency inaccuracy due to process variations through a 2nd-order digital Delta-Sigma Modulator (DSM). In TCXO mode, a temperature-to-digital converter (TDC) and a 3rd-order polynomial additionally compensate frequency variation over temperature.

The capacitively transduced 524kHz MEMS resonator, shown in Fig. 12.9.2, has the following electrical characteristics: nominal quality factor (Q) of 52,000, nominal motional impedance (R_m) of 40kΩ, and resonant frequency variation of <100ppm over -40 to +85°C. The resonator is biased using a Charge Pump (CP) that triples a 1.2V regulated supply. A Pierce sustaining circuit maintains oscillation with a sub-threshold inverter. An Automatic Gain Control (AGC) [1] adjusts the g_m of this inverter at start up and over PVT variations. A series drive capacitor (C_{Drive}) is trimmed to compensate R_m variation over production. The total current consumption of the MEMS sustaining circuit and the CP is 240nA. Figure 12.9.3 shows the block diagram of the PLL and temperature-compensation path. The MEMS frequency is divided down to 32kHz by DSM controlled pre-divider. The PLL bandwidth is set to 1kHz to minimize the noise contribution from the pre-divider and the VCO. The VCO is a current-controlled ring oscillator with a nominal frequency of 262kHz. The current consumption of the PLL, including DSM, is 290nA.

In low-power mode, shown in Fig. 12.9.3, the PLL can be disabled and the output derived from the DSM controlled pre-divider. This introduces additional output jitter, but is not detrimental in applications that count pulses, e.g., 32,768 pulses to define one second. This low power XO mode (LPM) reduces chip current to 0.6µA for 1Hz rail-to-rail output clock with no external load.

The TDC shown in Fig. 12.9.3 employs a BJT-based temperature-sensing element [4]. This produces a PTAT voltage $\Delta V_{BE}=V_{BE2}-V_{BE1}$, using two equally sized BJTs biased with different currents I_2 and I_1 with a ratio of ρ. A 2nd-order ΔΣ modulator provides a digital bitstream whose average value is proportional to $(\alpha.\Delta V_{BE})/V_{BG}$, where $V_{BG}=V_{BE1}+\alpha.\Delta V_{BE}$. α and ρ are chosen to create a temperature-independent V_{BG}. To improve its accuracy and robustness, the TDC employs dynamic element matching in the BJT current sources, correlated double sampling in the first integrator, and chopping at the system level.

The TDC operates at a sampling frequency of 262kHz with an over-sampling ratio of 192. Each conversion takes 6ms with a current consumption of 4.5µA. This current consumption includes that of the decimation filter, digital filter, and clock generator. The resolution of the stand-alone TDC is 25mK/Conversion leading to a figure of merit (Energy/Conversion x Resolution²) of 24pJ°C² [5]. The TDC update rate is reduced by duty-cycling, but still guarantees <1ppm error during a 1°C/s temperature ramp. Clock gating in the digital circuitry is used to save power when the TDC is inactive. Figure 12.9.4 shows the current profile of the temperature-compensation engine. It takes 1ms for the BJT core and modulator to initialize, 6ms for two back-to-back conversions at 25°C, and 2ms to evaluate the polynomial. Two conversions are necessary for the system-level chopping. Duty cycling reduces TDC conversion rate to 3S/s and the average current to 100nA.

Replica biasing is employed in the digital regulator to guarantee timing closure over process and temperature. The replica structure in Fig. 12.9.3 consists of a series stack of NMOS and PMOS that matches the gates used in the standard cells. Driving a constant current into this series stack generates a voltage (VGS_{ref}) that results in a constant slew rate, and hence speed, in the digital gates. A resistor (R_{set}) adds output voltage margin to this replica structure. VDD_{dig} is generated using an open-loop unity gain buffer that ensures stability over all load conditions.

There are two options for the output driver: a rail-to-rail CMOS and a low-swing driver. With a low-swing driver, the CVF current can be reduced by V_{swing}/V_{DD}. As shown in Fig. 12.9.3, the driver has two regulators which together control its output swing. CMOS transmission gates are used to alternate between independently programmable V_{top} and V_{bottom}. Capacitive charge sharing generates fast output transitions followed by slow single pole settling from the regulators. To further reduce the output driver power consumption, the output frequency can be divided down to 1Hz in powers of 2.

Shown in Fig. 12.9.5, measured frequency stability over temperature (-40 to +85°C) is less than 100ppm for 85 XO devices and less than 3ppm for 45 TCXO devices. TCXOs are trimmed individually at several temperature points. Figure 12.9.5 also shows the hysteresis measured over 14 temperature cycles and tracking performance in the presence of a temperature transient as fast as 1.7°C/s on a TCXO device.

System performance is tabulated in Fig. 12.9.6 and compared to that of existing XO and TCXO devices. As shown, frequency stability over industrial temperature range is improved compared to other works. The total current consumption of this oscillator is 0.9µA and 1.0µA in XO and TCXO mode, respectively. The low-swing driver can be enabled to save power when driving external load capacitance. The low-power mode, with XO with PLL disabled, reduces the supply current further to 0.6µA. Figure 12.9.7 shows the 0.18µm CMOS die with area of 1.2mm² and MEMS die with area of 0.17mm². The MEMS resonator is flipped-chip bonded to the CMOS die in a 1.55×0.85mm² chip-scale package (CSP).

References:

[1] E. Vittoz, *et al.*, "High-Performance Crystal Oscillator Circuits: Theory and Application," *IEEE, J. Solid State Circuits*, vol. 23, no.3, pp. 774-783, June 1988.
[2] S. Dalla Piazza, "Quartz Tuning Forks: A High-Volume, Low-Cost, High-Tech MEMS Product," [Online].
Available: http://www.go4time.eu/publications/37-general.html.
[3] M. Perrott, *et al.*, "A Temperature-To-Digital Converter for a MEMS-based Programmable Oscillator with < ±0.5-ppm Frequency Stability and < 1-ps Integrated Jitter," *IEEE J. Solid-State Circuits*, vol. 48, no.1, pp. 276 - 291, Jan. 2013.
[4] M. A. P. Pertijs, *et al.*, "A CMOS Temperature Sensor with a 3σ Inaccuracy of 0.1°C from 55°C to 125°C," *IEEE J. Solid-State Circuits*, vol. 40, no. 12, pp. 2805–2815, Dec. 2005.
[5] K. Souri, *et al.*, "A CMOS Temperature Sensor with a Voltage-Calibrated Inaccuracy of ±0.15°C (3σ) from −55 to 125°C," *IEEE J. Solid-State Circuits*, vol. 48, no.1, pp. 292-301, Jan. 2013.

978-1-4799-0917-9/14 $31.00 © 2014 IEEE

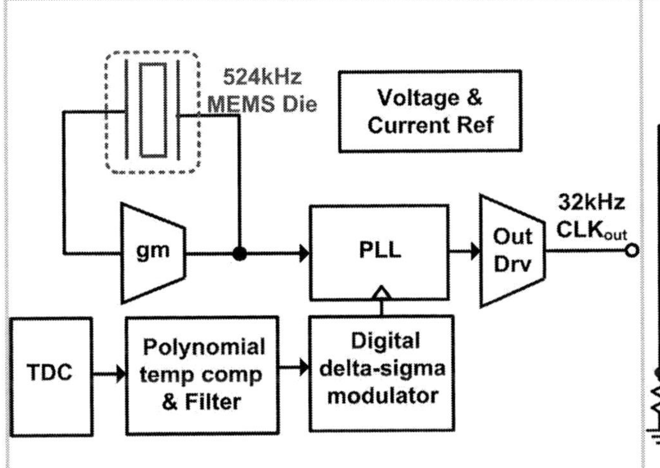

Figure 12.9.1: Simplified block diagram of MEMS-based TCXO. In XO mode, TDC and polynomial temperature compensation blocks are disabled.

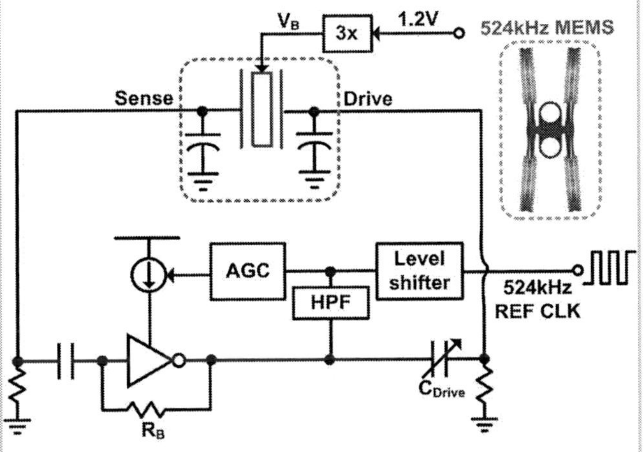

V_B
3x
1.2V
524kHz MEMS
Sense
Drive
AGC
Level
shifter
HPF
524kHz
REF CLK
C_{Drive}
R_B

Figure 12.9.2: MEMS resonator, its bias (VB) and 524kHz sustaining oscillator block diagram. RB is replica-biased transistor operating in sub-threshold.

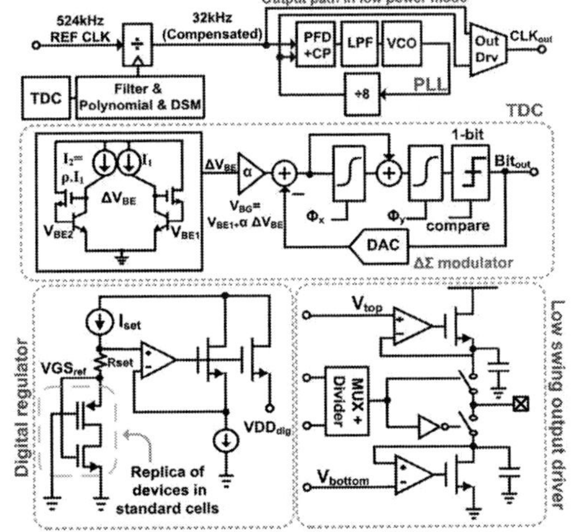

Figure 12.9.3: The Fractional-N synthesizer including temperature-compensation path, TDC, digital regulator, and low-swing driver.

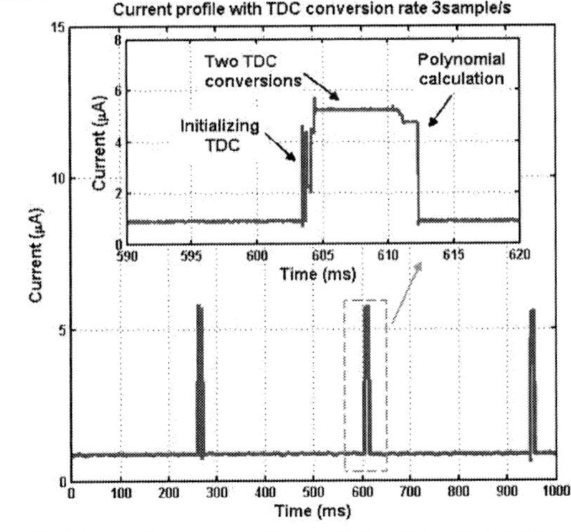

Figure 12.9.4: Supply current profile in TCXO mode. TDC conversion rate is 3S/s.

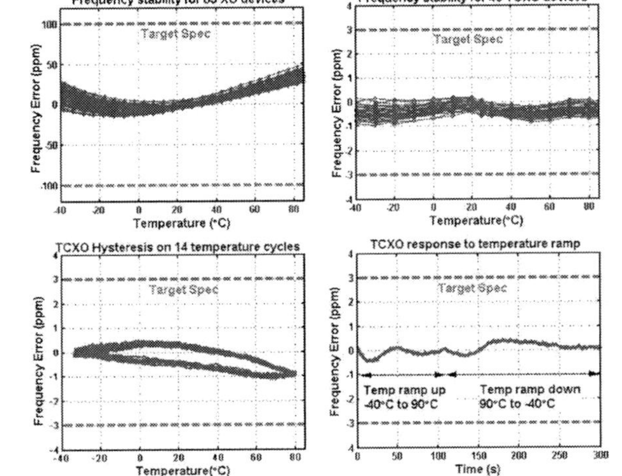

Figure 12.9.5: Frequency stability vs. temperature for XO and TCXO, TCXO Hysteresis, and its response to temperature ramp.

	XO			TCXO			
Parameter	This work	Epson SG-3050	Micro crystal OV-7604	This work	Maxim DS32KHz	Epson TG-3530	Kyocera KT3225T
Supply Voltage (V)	**1.2 to 3.63**	1.2 to 5.5	1.2 to 5.5	**1.5 to 3.63**	2.7 to 5.5	2.2 to 5.5	2 to 5.5
Temperature Range (°C)	**-40 to 85**	-20 to 70	-40 to 85	**-40 to 85**	-40 to 85	-20 to 70	-40 to 85
Frequency Stability vs. Temp (ppm)	**100 max**	120 max	160 max	**±3**	±7.5	±5	±5
Supply sensitivity (ppm/V)	**±0.25**	±3	±1.5	**±0.25**	2.5	±1	±1
OSC Start up (s)	**0.2**	1	0.5	**0.5**	1	3	3
Current (μA) Clock enabled, no load	**0.9 typ 1.4 max 0.6 LPM**	-Typ 2 max	0.5 typ 0.7 max	**1 typ 1.5 max**	1.85 typ 4 max	1.7 typ 4 max	1.5 typ 4 max
Package size (mm²)	**1.55x0.85**	2.2x1.4	3.2x1.5	**1.55x0.85**	18.5x6.35	5x10.1	3.2x2.5

Figure 12.9.6: Performance comparison with previous low-power 32kHz XOs and TCXOs.

Figure 12.9.7: Micrograph of MEMS and CMOS dies and final CSP package.

Session 13 Overview: *Advanced Embedded Memory*

MEMORY SUBCOMMITTEE

Session Chair: *Jonathan Chang*
TSMC, Hsinchu, Taiwan

Session Co-Chair: *Hugh Mair*
MediaTek, Austin, TX

Embedded memory continues to be a critical technology enabler for a wide range of applications from high-performance computing to mobile applications. This year's conference highlights significant increases in on-chip capacity and bandwidth along with a continued drive towards advanced technology nodes while maintaining a strong focus on low-power operation. A 1Gb embedded DRAM using 22nm tri-gate CMOS logic technology is presented to meet the demands of bandwidth-intense applications. Papers in 14nm FinFET, 16nm FinFET, and 20nm planar technologies demonstrate state-of-the-art read/write assist techniques in order to challenge V_{MIN} limitations. Various aspects of power and performance optimizations are highlighted in the session, including leakage power, dynamic power, latency, and throughput.

13.1 A 1Gb 2GHz Embedded DRAM in 22nm Tri-Gate CMOS Technology **1:30 PM**

F. Hamzaoglu, Intel, Hillsboro, OR

In Paper 13.1, Intel demonstrates a 1Gb 2GHz embedded DRAM using a 22nm tri-gate logic process. A 128Mb macro achieves a density of 17.5Mb/mm² using a cell size of 0.029μm². 2GHz operation at 1.05V and 1GHz operation at 0.7V are demonstrated.

13.2 A 14nm FinFET 128Mb 6T SRAM with V_{MIN}-Enhancement Techniques for Low-Power Applications **2:00 PM**

T. Song, Samsung Electronics, Yongin, Korea

In Paper 13.2, Samsung describes the first 14nm FinFET SRAM with the smallest bit-cell size reported to date (0.064μm²). By using innovative read/write assist schemes, up to a 200mV V_{MIN} improvement for the 0.064μm² cell is shown.

13.3 20nm High-Density Single-Port and Dual-Port SRAMs with Wordline-Voltage-Adjustment System for Read/Write Assists **2:30 PM**

M. Yabuuchi, Renesas Electronics, Tokyo, Japan

In Paper 13.3, Renesas describes a 20nm high-density single-port/dual-port SRAM with a wordline-voltage-adjustment system for read/write assist. This scheme achieves 100mV V_{MIN} improvement for 64Kb single-port and dual-port SRAMs. A bit density of 8.74Mb/mm² is achieved for the SP-SRAM while 3.08Mb/mm² is achieved for the DP-SRAM.

13.4 A 7ns-Access-Time 25μW/MHz 128kb SRAM for Low-Power Fast Wake-Up MCU in 65nm CMOS with 27fA/b Retention Current **2:45 PM**

T. Fukuda, Toshiba, Kawasaki, Japan

In Paper 13.4, Toshiba has developed a new SRAM bitcell in 65nm CMOS that achieves 27fA/b standby leakage for low power MCU products. Using SRAM over conventional nonvolatile approaches enables fast wake-up from deep sleep. Operating power is reduced though quarter-array activation and hierarchical bitlines.

13.5 A 16nm 128Mb SRAM in High-κ Metal-Gate FinFET Technology with Write-Assist Circuitry for Low-V_{MIN} Applications **3:15 PM**

Y-H. Chen, TSMC, Hsinchu, Taiwan

In Paper 13.5, TSMC describes the write-assist techniques used for their 0.07μm² 16nm FinFET SRAM bit cell. Silicon results demonstrating over 300mV improvement in V_{MIN} are shown.

13.6 A 28nm 400MHz 4-Parallel 1.6Gsearch/s 80Mb Ternary CAM **3:45 PM** **13**

K. Nii, Renesas Electronics, Kodaira, Japan

In Paper 13.6, Renesas describes a large, 80Mb/1Mentry, 1.6Gsearch/s TCAM implemented using four parallel 400MHz units in 28nm. The key design features include flexible search mode, row and column shift redundancy, and search omission with valid-bit.

13.7 A Reconfigurable Sense Amplifier with Auto-Zero Calibration and Pre-Amplification in 28nm CMOS **4:15 PM**

B. Giridhar, University of Michigan, Ann Arbor, MI

In Paper 13.7, University of Michigan describes a variation-tolerant differential sensing scheme to improve sense margin in advanced technologies, that can in turn be traded for read performance. Silicon results in 28nm are presented to show both margin and performance improvement.

13.8 A 32kb SRAM for Error-Free and Error-Tolerant Applications with Dynamic Energy-Quality Management in 28nm CMOS **4:45 PM**

M. Alioto, National University of Singapore, Singapore, Singapore

In Paper 13.8, National University of Singapore and University of Michigan present an SRAM that can trade-off power and effective precision. Compared to a conventional approach of reducing bit width, the design uses lower-order bits for ECC-based correction of high-order bits and bit-level write assist, enabling lower-voltage operation and lower total power.

13.1 A 1Gb 2GHz Embedded DRAM in 22nm Tri-Gate CMOS Technology

Fatih Hamzaoglu, Umut Arslan, Nabhendra Bisnik, Swaroop Ghosh, Manoj B. Lal, Nick Lindert, Mesut Meterelliyoz, Randy B. Osborne, Joodong Park, Shigeki Tomishima, Yih Wang, Kevin Zhang

Intel, Hillsboro, OR

CMOS technology scaling continues to drive higher levels of integration in VLSI design, which adds more compute engines on a die. To meet the overall performance-scaling needs, high-speed and high-bandwidth memory is becoming increasingly important. Conventional VLSI systems often rely on on-die SRAMs to address the performance gap between CPU and main memory, DRAM. However, with the rapid growth in capacity needs for high-performance memory, SRAM is not always sufficient to meet the demands of bandwidth-intense applications. Embedded DRAM (eDRAM) has been explored as an alternative to satisfy the high-performance and density needs in memory [1-3]. In this paper, a high-performance eDRAM based on a 22nm tri-gate CMOS technology is introduced. This eDRAM technology enables the integration of an eDRAM cell into the logic technology platform [4]. The design features a well-balanced configuration to achieve both optimal array efficiency and bandwidth. By leveraging the high-performance and low-voltage tri-gate transistor at 22nm generation, the eDRAM achieves a wide range in operating voltage, from 1.1V down to 0.7V, which is essential for low-power logic applications.

Figure 13.1.1 shows a 4th-generation Core™ processor, where the CPU is connected to a 1Gb eDRAM die through on-package-IO (OPIO) [5]. This multi-chip-package (MCP) product, the Iris Graphics Pro™, uses eDRAM as L4 cache and provides low-power high-bandwidth memory access to meet high-performance graphics segment needs. The eDRAM bitcell features a low-leakage access transistor in high-k tri-gate bulk technology and a MIM storage capacitor with capacitance of greater than 13fF [4,6]. The eDRAM cell area is 0.029μm², less than one-third of the high-density 6T-SRAM bitcell offered in the same 22nm technology [7], enabling design of high-density memory. The bitcell adopts the capacitor-over-bitline (COB) architecture to maximize the surface area of the capacitor. To support high-performance logic in the eDRAM design for GHz operation, the COB is embedded into the high-performance Cu-metallization interconnect layers. Negative word-line voltage (VSS_WL) with wide programmable range is employed to reduce access-transistor leakage (Fig. 13.1.2). To achieve high data-retention time, VSS_WL and threshold voltage of the access transistor are co-optimized to balance subthreshold leakage and gate-induced drain leakage at the storage node. The access transistor is turned on with the wordline overdriven to VCC_WL by an on-die charge pump (CP) to allow the storage capacitor to be fully charged to VCC and to achieve fast sensing and data write-back operations. The range of wordline swing during off and on states is largely determined by the reliability requirement of the access transistor. A NOR-based wordline driver is shown in Fig. 13.1.2, where the input signal voltage swings are designed to limit the gate bias to VCC_WL level and below to meet reliability needs. The final WL-driver voltage for the groups of drivers that are not accessed is kept at VCC in order to minimize the CP loading and leakage power. This design also avoids dual level-shifting circuitry in the same gate, which further reduces the design complexity.

The 256Kb-subarray architecture is shown in Fig. 13.1.3. The array has an open-bitline architecture with 128+ cells on each side, including redundant rows. Similarly, each wordline has a total of 1024+ columns, including redundant columns. The subarray achieves 65% area efficiency. The subarray reads or writes 128+ bits and each bit-slice contains its own set of half-VCC local bitline precharge circuitry, sense amplifier, and 8:1 column mux, as described in Fig. 13.1.3. Subarrays also contain local half-VCC generators, which are programmable for optimal sensing margin. Four bitcell operations are also shown in Fig. 13.1.3, including sense, write-back, wordline-turn-off and local bitline precharge.

Figure 13.1.4 shows the 1Gb array configuration and data-path for read, write and refresh operations. The chip contains 128 independent banks for read and write and 64 bank-groups for refresh, where bank random cycle time (RCT) is equal to six array clock cycles. By providing large number of banks and short RCT, we minimize bank conflict for high-bandwidth random accesses and maximize performance. Four vertical 256Mb quarters are activated simultaneously during each operation, where each bank reads out 64×2 bits in two consecutive cycles after column and row repairs to get 512b-wide word size. The OPIO is clocked at twice the array frequency and double data-rate to meet area and bandwidth requirements. The array has separate data buses for read and write operations but shares a common address bus, hence, it supports read and write operations in alternating array clock cycles to different banks. The refresh operation can occur during a read or write since it has a separate refresh bank-group address. There are two copies of CPs and regulators, each supporting the top or bottom 512Mb, which occupy less than 2% of the die area. The chip also contains fuses, programmable built-in self test (PBIST), test access port (TAP) and a digital thermal sensor (DTS).

Figure 13.1.5 describes the CP circuits that support wordline over- and under-drive voltages. A portion of the CPs always run to support leakage and the remaining portions are activated only when there is array access (read, write or refresh) to compensate the wordline-activation charge. Although positive and negative CPs can generate up to 2*VCC and –(VCC/2) output voltages, respectively, output voltages are regulated to a programmable value through on-die-generated reference voltages. The cumulative distribution of both VCC_WL and VSS_WL measurement data at hot (95°C) and cold (-10°C) are also shown in Fig. 13.1.5 for unregulated and two different regulator setting cases. Silicon data show that regulated voltages have less die-to-die variation and better control across temperatures. By introducing programming control over the CP output voltages, the design is able to provide a large window to compensate process and temperature variation while achieving a balance between performance and reliability.

Figure 13.1.6 shows the voltage-frequency shmoo of the 1Gb eDRAM array tested at 95°C and 100μs retention time. The design achieves 2GHz operation frequency at a supply voltage of 1.05V, hence 3ns of RCT. The array also supports a wide range in power supply, down to 0.7V at 1GHz frequency.

Figure 13.1.7 shows die micrograph of the 1Gb eDRAM and the feature summary table. The chip is fabricated in a 22nm high-performance tri-gate CMOS technology. The die size is 77mm² with 0.029μm² eDRAM bitcell area and an array density of 17.5Mb/mm² at the 128Mb macro level.

Acknowledgements:
The authors gratefully acknowledge many members of PTD and IDG technical staffs for their contributions to this work.

References:

[1] J. Barth et al., "A 45nm SOI Embedded DRAM Macro for POWER7™ 32MB On-Chip L3 Cache", *ISSCC Dig. Tech. Papers*, pp. 342-344, Feb. 2010.
[2] K. Hijioka et al., "A Novel Cylinder-Type MIM Capacitor in Porous Low-k Film (CAPL) for Embedded DRAM with Advanced CMOS Logics," *IEDM Technical Digest*, pp. 756-759, Dec. 2010.
[3] S. Romanovsky et al., "A 500MHz Random-Access Embedded 1Mb DRAM Macro in Bulk CMOS", *ISSCC Dig. Tech. Papers*, pp. 270-271, Feb. 2008.
[4] R. Brain et al., "A 22nm High Performance Embedded DRAM SoC Technology Featuring Tri-Gate Transistors and MIMCAP COB", *VLSI Tech. Symp.*, June 2013.
[5] N. Kurd et al., "Haswell: A Family of IA 22nm Processors," *ISSCC Dig. Tech. Papers*, Feb. 2014
[6] Y. Wang et al., "Retention Time Optimization for eDRAM in 22nm Tri-Gate CMOS Technology," *IEDM Technical Digest*, Dec. 2013.
[7] E. Karl et al., "A 4.6GHz, 162Mb SRAM Design in 22nm Tri-Gate CMOS Technology with Integrated Active Vmin-Enhancing Assist Circuitry," *ISSCC Dig. Tech. Papers*, pp. 230-232, Feb. 2012.

978-1-4799-0917-9/14 $31.00 © 2014 IEEE

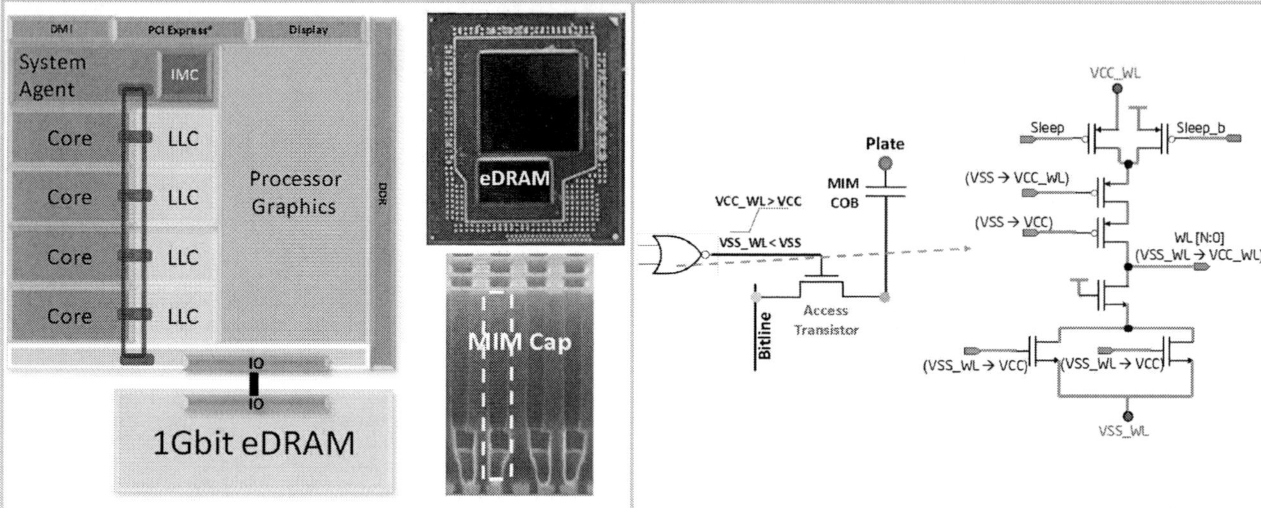

Figure 13.1.1: Intel Iris Pro™ with 1Gb eDRAM in 22nm tri-gate technology.

Figure 13.1.2: NOR WL driver with V_{max} protection and leakage reduction.

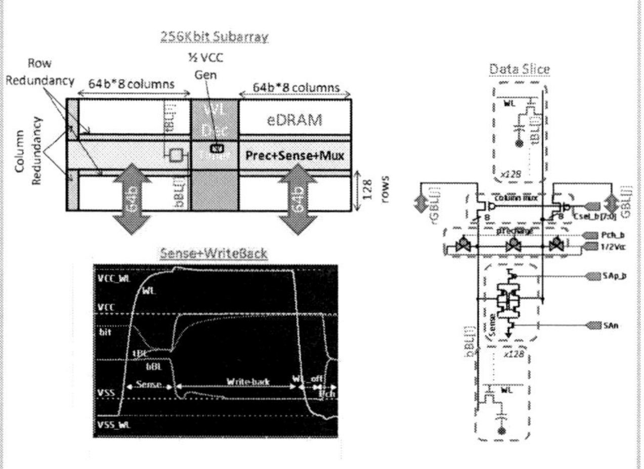

Figure 13.1.3: Floorplan of 256Kb eDRAM subarray and bit-slice architecture.

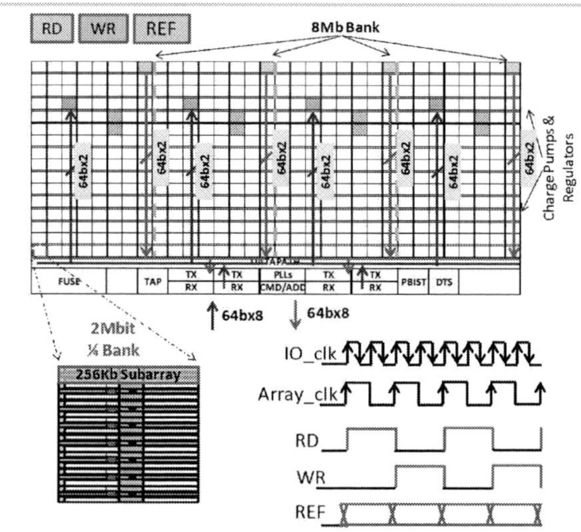

Figure 13.1.4: 1Gb eDRAM architecture and RD/WR/REF operations.

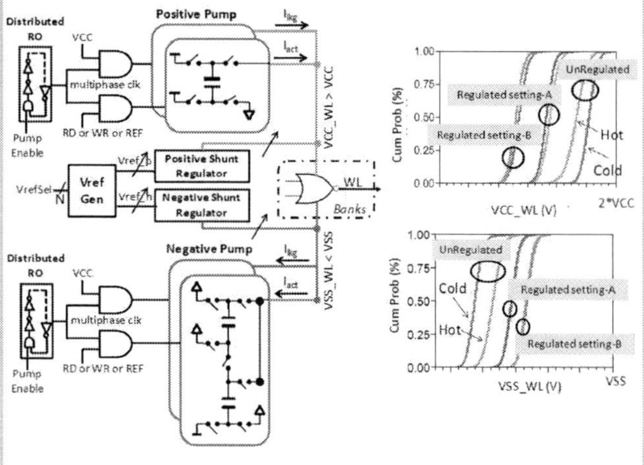

Figure 13.1.5: Charge pumps for WL voltages and measured silicon data.

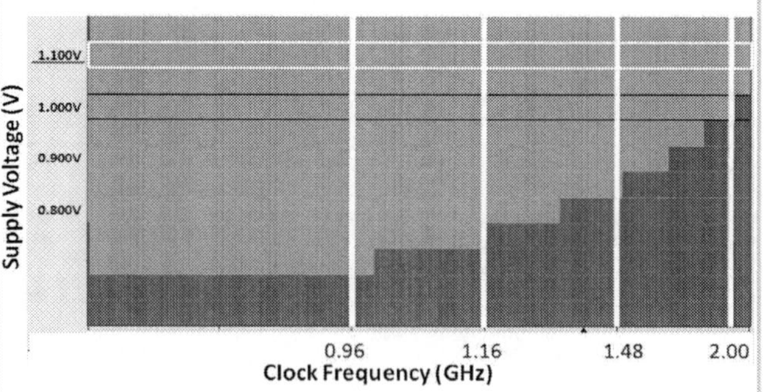

Figure 13.1.6: Frequency shmoo of 1Gb eDRAM array at 95°C and 100μs retention time.

13

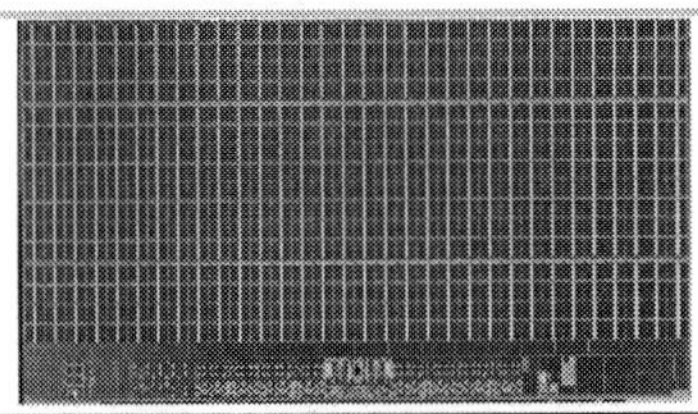

Technology	22nm Tri-gate CMOS
Cell Size	0.029μm²
Macro Area	17.5 Mb/mm² @ 128Mbit Macro
Chip Organization	¼ Bank: 8 Subarrays (2Mb) Bank: 4 Quarter Banks (8Mb) Chip: 128 Banks (1Gb)
Subarray Configuration & Array Efficiency	256 Word-line x 1024 bit-lines & 65%
Chip Size	77mm²
Supply	1.05V
Clock, Random Cycle Time	2GHz, 3ns
Retention Time	100μs @95C

Figure 13.1.7: Die micrograph and feature table of 1Gb eDRAM die.

13.2 A 14nm FinFET 128Mb 6T SRAM with V_{MIN}-Enhancement Techniques for Low-Power Applications

Taejoong Song, Woojin Rim, Jonghoon Jung, Giyong Yang, Jaeho Park, Sunghyun Park, Kang-Hyun Baek, Sanghoon Baek, Sang-Kyu Oh, Jinsuk Jung, Sungbong Kim, Gyuhong Kim, Jintae Kim, Youngkeun Lee, Kee Sup Kim, Sang-Pil Sim, Jong Shik Yoon, Kyu-Myung Choi

Samsung Electronics, Yongin, Korea

With the explosive growth of battery-operated portable devices, the demand for low power and small size has been increasing for system-on-a-chip (SoC). The FinFET is considered as one of the most promising technologies for future low-power mobile applications because of its good scaling ability, high on-current, better SCE and subthreshold slope, and small leakage current [1]. As a key approach for low-power, supply-voltage (V_{DD}) scaling has been widely used in SoC design. However, SRAM is the limiting factor of voltage-scaling, since all SRAM functions of read, write, and hold-stability are highly influenced by increased variations at low V_{DD}, resulting in lower yield. In addition, the width-quantization property of FinFET device reduces the design window for transistor sizing, and increases the failure probability due to the un-optimized bitcell sizing [1]. In order to overcome the bitcell challenges to high yield, peripheral-assist techniques are required. In this paper, we present 14nm FinFET-based 128Mb 6T SRAM chips featuring low-V_{MIN} with newly developed assist techniques.

Figure 13.2.1 shows a 6T SRAM bitcell in a 14nm FinFET technology. Figure 13.2.2 summarizes the assist techniques to improve SRAM functional success for higher bitcell yield. The write margin (WM) can be improved by increasing the PG to PU strength ratio; i.e., by increasing the strength of PG (WL overdrive or WLOD), driving a negative BL voltage (*NBL*), or decreasing the strength of the PU (cell supply-voltage lowering or VDCL). Meanwhile, access-disturbance margin (ADM) can be improved by reducing the amount of charge injection from the precharged BL to the '0' node of the active bitcell. WL underdrive (WLUD) is a commonly used method to reduce PG strength.

However, there is a trade-off between write-assist and stability-assist in some cases. For example, WLOD-based write-assist increases disturbance failures in half-selected bitcells, since it increases the amount of injected charge to half-selected bitcells. Therefore, additional assist techniques must be applied to overcome the disturbance instability with WLOD. Since a high-performance (HP) bitcell has larger NMOS (PG) than PMOS (PU), it typically exhibits a larger WM, but lower ADM. Therefore, in case of an HP bitcell, disturbance-stability assist should be explored. In contrast, a high-density (HD) bitcell, which is optimized for integration-density rather than speed, requires a write-assist technique due to weak write-ability in addition to disturbance-stability assist.

Figure 13.2.3 shows an access-disturbance failure and a write-failure with V_{DD} scaling in an HP and an HD bitcell, respectively. After fast turn-on of PG, the charge stored on BLs flows into the latch-nodes of the bitcells. If the data-'0'-node voltage reaches the trip voltage of the cell inverter via data-'1'-noise through PG, the bitcell flips destroying the stored data. Disturbance failures dramatically increase as V_{DD} is reduced, since the noise margin of the inverter reduces faster than the '0'-node-voltage disturbance. To mitigate disturbance instability, the induced-charge reduction technique has been explored in the literature. However, the conventional WLUD scheme has an intrinsic drawback in terms of timing penalties, since it also reduces both read and write current of the selected bitcell in addition to noise-injection of the half-selected bitcells [11]. An alternative solution is the *suppressed* BL [7], which precharges BL to V_{DD}-α instead of V_{DD}. It is an effective method to suppress disturbance noise from the BL. However, it incurs the design overheads of a voltage-generator circuit and voltage follower, which cause additional static-current dissipation and area overhead to guarantee the optimal BL.

Figure 13.2.4(a) illustrates the developed disturbance-noise reduction (DNR*)* scheme implemented with NBL and WLUD techniques. The DNR scheme does not require the additional analog circuit and has minimum area and static-current consumption overhead. The DNR circuit includes a simple clamping circuit, cross-coupled PMOSes, and a BL-discharge circuit. In an access operation, the BL starts to be lowered through DNR circuitry after precharge-disable. Since the BL is pulled-down with a strong NMOS device in DNR, the BL is lowered enough to decrease BL noise at WL-enable time. Meanwhile, clamping circuitry is necessary to prevent BL/BLb from being lower than the safe BL high voltage (V_{SAFE}), which is the minimum voltage level not to induce '0' noise through NMOS (PG). Meanwhile, we adopt NBL to improve WM for HD, since it is free of the half-selected problem of WLOD and retention problem of VDCL. Then, WLUD is selected for ADM, since HD does not require high speed like HP. Figures 13.2.4 (b) and (c) show the corresponding waveforms for DNR for HP, and NBL and WLUD for HD, respectively. The area overhead of DNR, WLUD, and NBL in a 1Mb SRAM macro is 0.87%, 0.323%, and 0.372%, respectively.

Two 128 Mb 6T SRAM chips are implemented in a 14nm FinFET technology, which configures tileable 128×1Mb SRAM macros with integrated read- and write-circuitry for HD and HP bitcell. Each 1Mb macro, as shown in Figure 13.2.5(a), contains 32-array banks of 128-row and 256-cloumn bitcell array. Figure 13.2.5(b) shows simulated ADM for HP bitcell and WM for HD bitcell. Simulation results show that the DNR can improve ADM in HP, and NBL improves WM for HD, respectively.

Figure 13.2.6 shows silicon test results with 2-D shmoo plot of 128Mb 6T SRAM chips. To explore the V_{MIN} improvement, assist circuits are turned on and off for comparison. Error-free functionality of 128Mb HP is achieved from 1.2V down to 0.51V without assist. With DNR assist, V_{MIN} can be reduced to 0.47V for a 40mV V_{MIN} reduction. Similarly, 128Mb 6T HD achieves 0.5V V_{MIN} with NBL and WLUD assists showing 200mV V_{MIN} reduction. Figure 13.2.7 summarizes the features and characteristics of the two test-chips with die micrograph. The 128Mb SRAM contains eFUSE, 1.8V GPIO, and 78nm gate-pitched 100K standard cells for glue logic, additionally. The die area of test-chip is 75.6mm^2.

References:

[1] E. J. Nowak *et al.*, "Turning silicon on its edge, Overcoming silicon scaling barriers with double-gate and FinFET technology", *IEEE Circuit and Devices Magazine*, vol. 20, no. 1, pp. 20-31, Jan. 2004.

[2] F. Hamzaoglu *et al.*, "A 153Mb-SRAM Design with Dynamic Stability Enhancement and Leakage Reduction in 45nm High-k Metal-Gate CMOS Technology", *ISSCC Dig. Tech. Papers*, pp. 376-621, Feb. 2008.

[3] H. Pilo *et al.*, "A 450ps Access-Time SRAM Macro in 45nm SOI Featuring a Two-Stage Sensing-Scheme and Dynamic Power Management", *ISSCC Dig. Tech. Papers*, pp. 378-621, Feb. 2008.

[4] O. Hirabayashi *et al.*, "A process-variation-tolerant dual-power-supply SRAM with 0.179μm^2 Cell in 40nm CMOS using level-programmable wordline driver", *ISSCC Dig. Tech. Papers*, pp. 458-459, Feb. 2009.

[5] Y. Wang *et al.*, "A 4.0 GHz 291Mb voltage-scalable SRAM design in 32nm high-κ metal-gate CMOS with integrated power management", *ISSCC Dig. Tech. Papers*, pp. 456-457, Feb. 2009.

[6] Y. Fujimura *et al.*, "A configurable SRAM with constant-negative-level write buffer for low-voltage operation with 0.149μm^2 cell in 32nm high-k metal-gate CMOS", *ISSCC Dig. Tech. Papers*, pp. 348-349, Feb. 2010.

[7] H. Pilo *et al.*, "A 64Mb SRAM in 32nm High-k metal-gate SOI technology with 0.7V operation enabled by stability, write-ability and read-ability enhancements", *ISSCC Dig. Tech. Papers*, pp. 254-256, Feb. 2011.

[8] E. Karl *et al.*, "A 4.6GHz 162Mb SRAM design in 22nm tri-gate CMOS technology with integrated active VMIN-enhancing assist circuitry", *ISSCC Dig. Tech. Papers*, pp. 230-232, Feb. 2012.

[9] H. Pilo *et al.*, "A 64Mb SRAM in 22nm SOI technology featuring fine-granularity power gating and low-energy power-supply-partition techniques for 37% leakage reduction", *ISSCC Dig. Tech. Papers*, pp. 322-323, Feb. 2013.

[10] J. Chang *et al.*, "A 20nm 112Mb SRAM in High- Metal-Gate with Assist Circuitry for Low-Leakage and Low-V_{MIN} Applications", *ISSCC Dig. Tech. Papers*, pp. 316-371, Feb. 2013.

[11] K. Nii *et al.*, "A 45-nm Bulk CMOS Embedded SRAM with Improved Immunity Against Process and Temperature Variations", *IEEE J. Solid-State Circuits*, vol. 43, no. 1, pp. 180-191, Jan. 2008.

978-1-4799-0917-9/14 $31.00 © 2014 IEEE

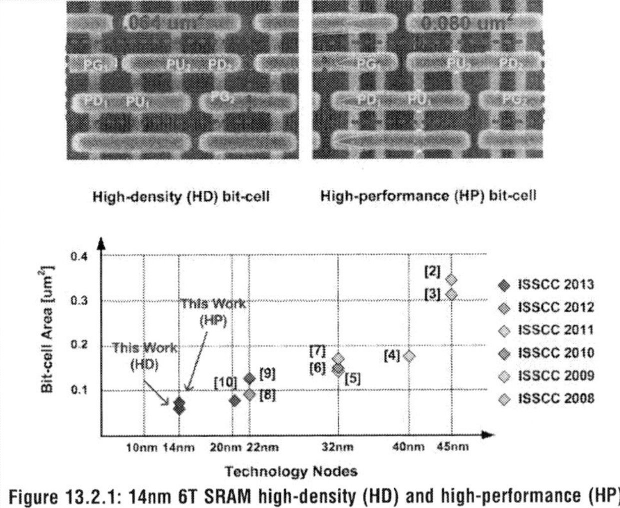

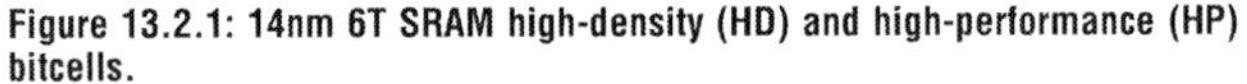

Figure 13.2.1: 14nm 6T SRAM high-density (HD) and high-performance (HP) bitcells.

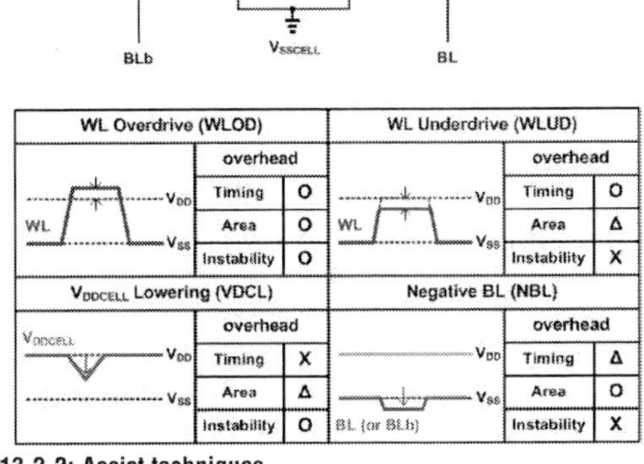

WL Overdrive (WLOD)			WL Underdrive (WLUD)		
	overhead			overhead	
WL, V_{DD}, V_{SS}	Timing	O	WL, V_{DD}, V_{SS}	Timing	O
	Area	O		Area	Δ
	Instability	O		Instability	X
V_{DDCELL} Lowering (VDCL)			Negative BL (NBL)		
	overhead			overhead	
V_{DDCELL}, V_{DD}, V_{SS}	Timing	X	V_{DD}, V_{SS}, BL (or BLb)	Timing	Δ
	Area	Δ		Area	O
	Instability	O		Instability	X

Figure 13.2.2: Assist techniques.

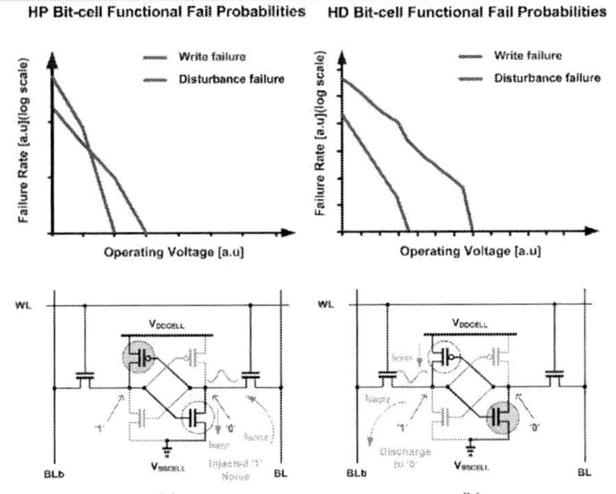

Figure 13.2.3: Failures probabilities with V_{DD} scaling for (a) HP bitcell and (b) HD bitcell.

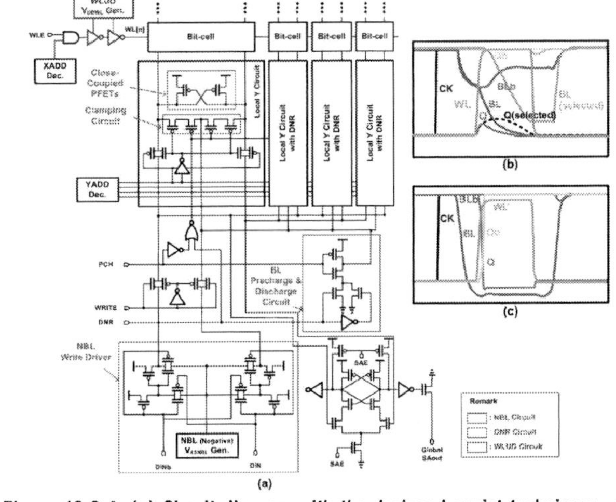

Figure 13.2.4: (a) Circuit diagram with the designed assist techniques, (b) waveform with the DNR assist, and (c) waveform with NBL and WLUD assists.

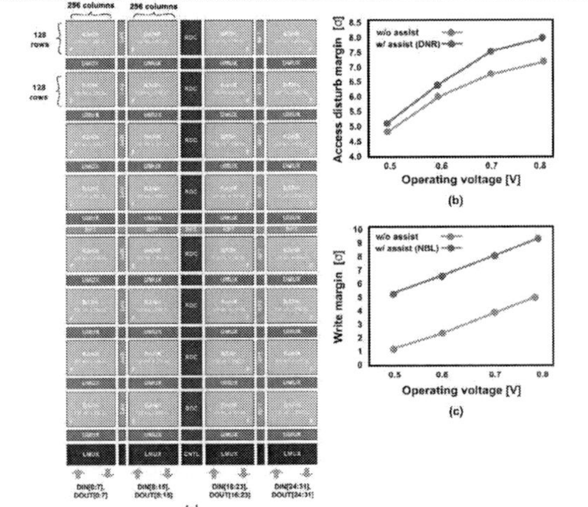

Figure 13.2.5: (a) 1Mb SRAM macro architecture, (b) simulated access-disturb margin, and (c) write margin.

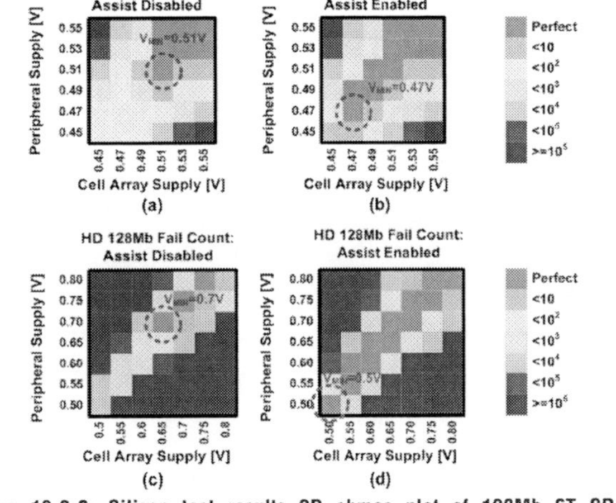

Figure 13.2.6: Silicon test results 2D shmoo plot of 128Mb 6T SRAM (a) without DNR assist in HP, (b) with DNR assist in HP, (c) without assists in HD, and (d) with NBL and WLUD in HD.

13

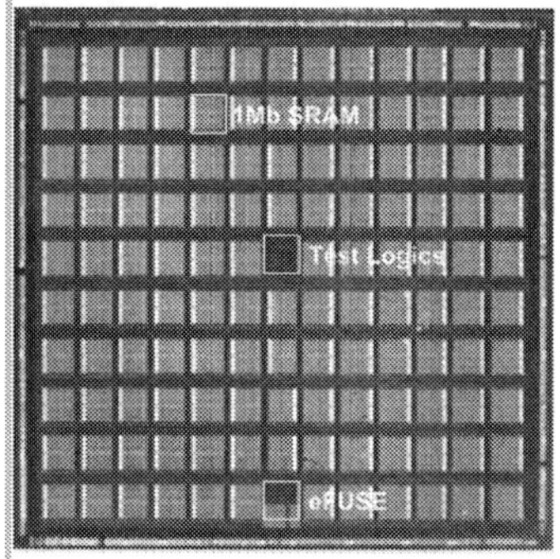

Technology	14nm Bulk FinFET
Chip Size	75.6mm^2
Density	128Mb (8 muxed IO in chip, 32 IO per macro)
Metal	9 metal for chip (5 metal probe-available
IPs	1Mb SRAM x 128, eFUSE, 1.8V general purpose IO, 78nm standard cells
Power Supply	0.8V (Nominal), 0.48V_{MIN} (HP), 0.52V_{MIN} (HD)

Figure 13.2.7: 128Mb 6T SRAM chip die micrograph (same floorplan for HD and HP) and its features.

13.3 20nm High-Density Single-Port and Dual-Port SRAMs with Wordline-Voltage-Adjustment System for Read/Write Assists

Makoto Yabuuchi, Yasumasa Tsukamoto,
Masao Morimoto, Miki Tanaka, Koji Nii

Renesas Electronics, Tokyo, Japan

Scaling of process technology is inevitably accompanied by the increase of local variation in transistor characteristics, which has been deteriorating the operation margin of SRAM. This trend necessitates assist circuits for SRAM to increase the immunity against variations, and many papers in this area [1-4] have been published. In this paper, we present an assist circuit suitable for the SRAMs in 20nm generation. Figure 13.3.1 compares local variations of SRAM cell transistors, pass-gate NMOS (PG), pull-down NMOS (PD) and pull-up PMOS (PU) for 28 and 20nm, showing degradation as the process advances. Noticeably, the NMOS transistors become worse than PMOS, which causes degradation in SRAM operating margin since SRAM characteristics such as static noise margin (SNM) are more sensitive to NMOS than PMOS. Figure 13.3.1 also shows the operational window enclosed by read and write immunity against local variations in 28 and 20nm. This indicates assist circuits must perform beyond the level established in previously published work to address SRAM variation in advanced technology nodes. Lowering wordline (WL) voltage level is one of the read-assist approaches. Lowering the supply voltage of PU in a cell (ARVDD) and negative bitline (BL) techniques are known to be effective for the write operation. These techniques, however, have side-effects: lowering the WL voltage degrades write margin and lowering ARVDD leads to higher power consumption and a long cycle-time. Furthermore, the negative BL technique can cause write errors in non-selected columns. Thus, it is necessary to select which assist technique should be applied depending on each process technology. In addition, the SRAM used in production generally include single-port SRAM (SP-SRAM) and dual-port SRAM (DP-SRAM), so the assist circuits to be applied should be effective for whole SRAM family.

Figure 13.3.2 depicts the simulated results of minimum operation voltage, V_{MIN}, against threshold voltage (V_{th}) of SRAM cell transistors. The horizontal axis is calculated by $2V_{thn}$-V_{thp}, indicating that the left side is the fast-slow (FS) process corner while the right side is the slow-fast (SF) process corner. Note that the V_{MIN} versus V_{th} plot depends on the fact that the failure lines in Fig. 13.3.1 can be approximately expressed by $V_{thp} = 2V_{thn} + \alpha$. The worst V_{MIN} cases are determined by SNM at high temperature, whereas by write operation at low temperature. We develop a simple method to improve V_{MIN} by only operating on the WL voltage, depending on temperature and process variation. We use a lower WL voltage at high temperature and use a higher WL voltage at low temperature.

Figure 13.3.3 shows an example of a collection of SRAM blocks on a chip and also summarizes the scheme for controlling the WL voltage across process and temperature conditions. Each SRAM macro is powered via V_{DD} and V_{SS} and each macro also has access to a selection of V_{DDW} voltages that operate as an independent power supply to control the WL voltage. The implemented on-chip voltage regulator generates the V_{DDW} voltage based on the fuse setting. As mentioned above, since assist circuits cause side effects such as increasing power consumption, assist is activated only on macros that need it. Each macro is controlled by the AST signal, turning assist on or off depending on the environment. The temperature-monitoring device and the assist controller connected to the fuse determine the AST signal. The code to be written in the fuse is decided before shipping. For instance, if the V_{th} condition of chip #1 is found to be close to the FS corner, the fuse is written with a code that turns the assist on to $V_{DDW} = V_{DD}$ - 0.1 (V) if the temperature is higher than 80°C. If the chip test detects failures for some macros in the system, this information is written to the fuse so that the AST can be asserted. The assist controller monitors whether the measured temperature is above or below the one written in the fuse. If the current temperature reaches the pre-determined temperature, the corresponding AST is enabled. The table in Fig. 13.3.3 shows some examples that would be written to the fuse, including information regarding V_{DDW} voltage, temperature threshold, and the assist-enable/disenable signal for each SRAM macro. Although the test procedure for fuse coding increases in complexity due to many temperature conditions, the procedures are not complicated because they are similar to a conventional redundancy-programming procedure. The voltage difference between V_{DD} and V_{DDW} is fixed; V_{DDW} should vary with V_{DD} as selected by a DVFS controller. One option is to provide the output of the temperature monitor with on-chip regulator and to control V_{DDW} dynamically depending on temperature. However, this is not effective since SRAM operation is not allowed during a relatively long time until the changing supply-voltage level stabilizes. Our scheme has several merits due to lower power consumption in V_{DDW} and large wire capacitance of V_{DDW}, which enables fast stabilization of the V_{DDW} voltage level with a small-area on-chip regulator.

Figure 13.3.4 shows the circuit diagram of our method. The source terminal of the PMOS in WL driver (VWL) includes a switch that selects V_{DD} or V_{DDW}. Since the voltage difference between V_{DD} and V_{DDW} is 0.1V, the switch transistor does not turn on even though the gate-overdrive voltage is not 0V. The AST signal is synchronized with clock signal. If the AST is not designed to be synchronous, V_{WL} might experience voltage transfer during its activation time, which would be a source of noise. Using a synchronous AST allows VWL to be stabilized at the expected voltage before the WL is activated. Figure 13.3.4 indicates the waveforms to confirm this operation. The logic delay for the WL activation through the address decoder is longer than the AST signal to be activated, indicating stable transfer of V_{WL} voltage with enough setup time.

Figure 13.3.5 shows how the V_{MIN} of the write operation (V_{MIN_W}) for DP-SRAM behaves in the case of skew between two separate port clocks, where one port is writing and the other is reading. In DP-SRAM, V_{MIN_W} degradation due to *write-disturb* should be examined [5]. In the case where the same row is accessed, the write of the bitcells via one WL is disturbed by the dummy read operation via the other WL, deteriorating V_{MIN_W}. The degradation is dependent on the timing skew between the two activated WLs. A higher WL voltage can improve dc write margin for the writing WL but may cause degradation due to the disturbing WL. Figure 13.3.5 indicates the measured V_{MIN_W} with assist is lower than without assist, confirming that increased WL voltage improves write margin for DP-SRAM.

Fig 13.3.6 plots the measured V_{MIN} of several test-chips for the temperature range of -40 to 125°C. The lower-right plots of Fig. 13.3.6 indicate drain current of PU (I_{dsp}) and PG (I_{dsn}), respectively. Group (b) has well balanced I_{ds} for PG and PU, in which we could achieve less than 0.7V of V_{MIN} without assist. Group (a) has relatively larger I_{dsn} than I_{dsp} compared to group (b), where read margin is small whereas write margin is sufficient, observing V_{MIN} improvement due to lower V_{DDW}. Group (c) is characterized with smaller I_{dsn} than I_{dsp} compared to group (b), which degrades the write operation. For this group, higher V_{DDW} effectively reduces V_{MIN_W}. Thus, by optimizing V_{DDW} depending on process variation and temperature, we achieve 0.1V V_{MIN} improvement for the whole system including both SP- and DP-SRAMs. Figure 13.3.6 shows the power consumption measured from our test circuits. For group (a), since a lower V_{DDW} for read assist suppresses the charging/discharging power of BL, total dynamic power is decreased. In contrast, group (c) incorporates higher V_{DDW} for write assist, which results in the increase of power consumption. Note that the designed assist circuit eliminates the variation in cell current by operating WL voltage, thus reducing the variation of dynamic power. Figure 13.3.7 summarizes the test-chip features based on 20nm high-k metal-gate bulk CMOS process. The bit density of SP-SRAM is 8.74Mb/mm^2 and that of DP-SRAM is 3.08Mb/mm^2.

References:

[1] E. Karl, et al., "A 4.6GHz 162Mb SRAM Design in 22nm Tri-Gate CMOS Technology with Integrated Active VMIN-Enhancing Assist Circuitry, " in *ISSCC Dig.*, pp. 230-231, Feb. 2012.

[2] J. Chang, et al., "A 20nm 112Mb SRAM in High-k Metal-Gate with Assist Circuitry for Low-Leakage and Low-VMIN Applications, " in *ISSCC Dig.*, pp. 316-317, Feb. 2013.

[3] H. Pilo, et al., "A 64Mb SRAM in 32nm High-k metal-gate SOI technology with 0.7V operation enabled by stability, write-ability and read-ability enhancements," in *ISSCC Dig.*, pp. 254-256, Feb. 2011.

[4] H. Fujiwara, et al., "A 20nm 0.6V 2.1µW/MHz 128kb SRAM with no half select issue by interleave wordline and hierarchical bitline scheme," in *Symp. VLSI Circuits*, pp. 118-119, June 2013.

[5] Y. Ishii, et al., "A 28nm Dual-Port SRAM Macro With Screening Circuitry Against Write-Read Disturb Failure Issues," *JSSC*, pp. 2535-2544, Nov. 2011.

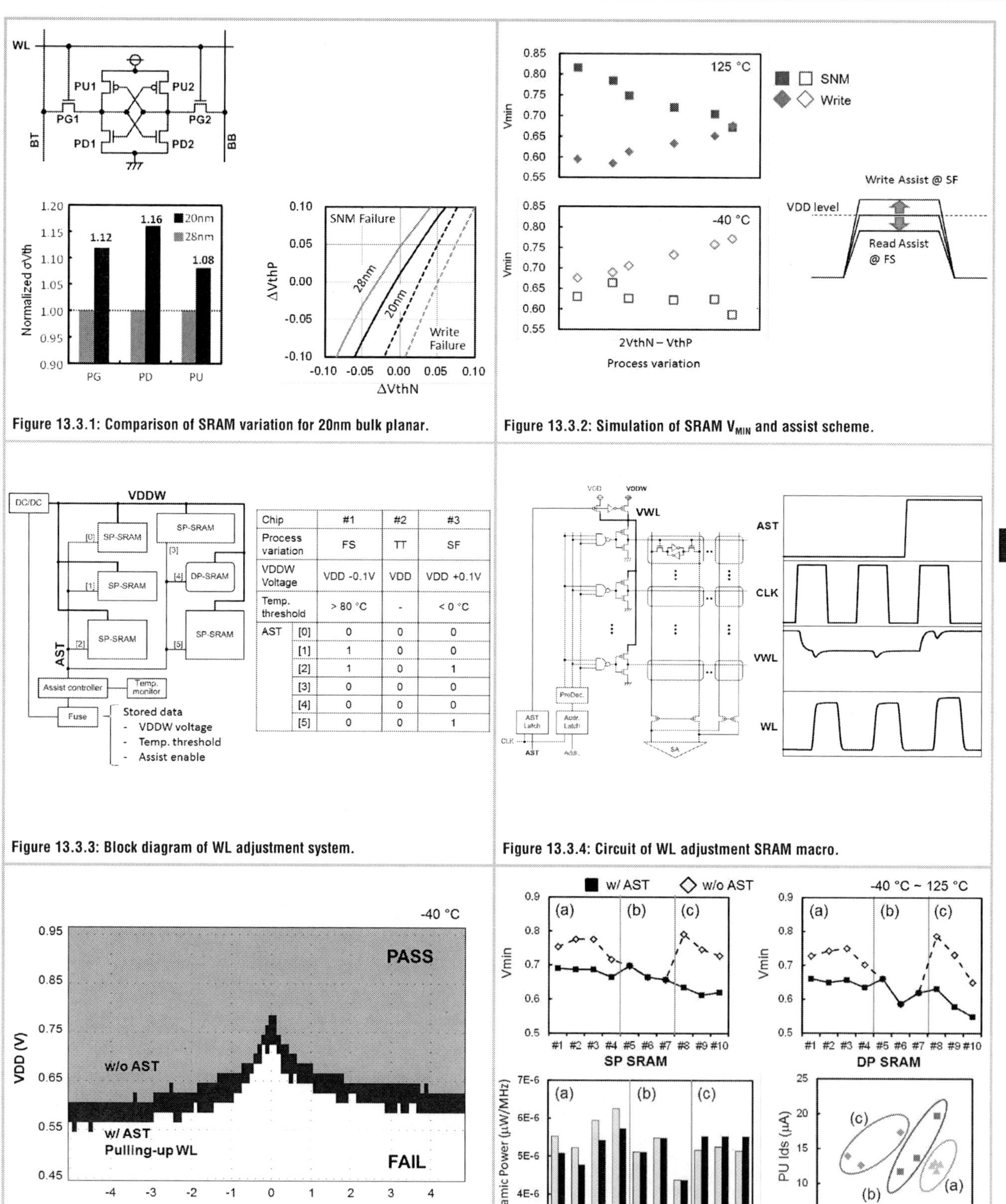

Figure 13.3.1: Comparison of SRAM variation for 20nm bulk planar.

Figure 13.3.2: Simulation of SRAM V_{MIN} and assist scheme.

Chip		#1	#2	#3
Process variation		FS	TT	SF
VDDW Voltage		VDD -0.1V	VDD	VDD +0.1V
Temp. threshold		> 80 °C	-	< 0 °C
AST	[0]	0	0	0
	[1]	1	0	0
	[2]	1	0	1
	[3]	0	0	0
	[4]	0	0	0
	[5]	0	0	1

Figure 13.3.3: Block diagram of WL adjustment system.

Figure 13.3.4: Circuit of WL adjustment SRAM macro.

Figure 13.3.5: Measurement of dual-port clock skew.

Figure 13.3.6: Measurement of SRAM V_{MIN} and dynamic power.

13

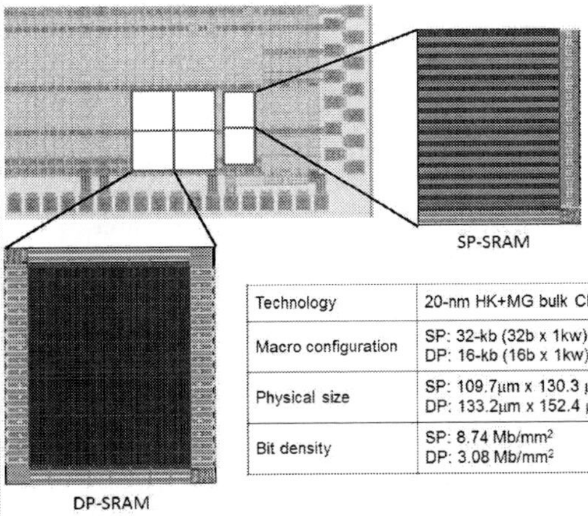

Technology	20-nm HK+MG bulk CMOS (planar)
Macro configuration	SP: 32-kb (32b x 1kw) x 2; Total 64-kb DP: 16-kb (16b x 1kw) x 4; Total 64-kb
Physical size	SP: 109.7μm x 130.3 μm @ 32-kb DP: 133.2μm x 152.4 μm @ 16-kb
Bit density	SP: 8.74 Mb/mm^2 DP: 3.08 Mb/mm^2

Figure 13.3.7: Test-chip micrograph and features.

13.4 A 7ns-Access-Time 25μW/MHz 128kb SRAM for Low-Power Fast Wake-Up MCU in 65nm CMOS with 27fA/b Retention Current

Toshikazu Fukuda[1], Koji Kohara[1], Toshiaki Dozaka[1], Yasuhisa Takeyama[1], Tsuyoshi Midorikawa[1], Kenji Hashimoto[2], Ichiro Wakiyama[2], Shinji Miyano[1], Takehiko Hojo[1]

[1]Toshiba, Kawasaki, Japan, [2]Toshiba Microelectronics, Kawasaki, Japan

Battery lifetime is the key feature in the growing markets of sensor networks and energy-management system (EMS). Low-power MCUs are widely used in these systems. For these applications, standby power, as well as active power, is important contributor to the total energy consumption because active sensing or computing phases are much shorter than the standby state. Figure 13.4.1 shows a typical power profile of low-power MCU applications. To achieve many years of battery lifetime, the power consumption of the chip must be kept below 1μA during deep sleep mode. Another key feature of a low-power MCU for such applications is fast wake-up from deep-sleep mode, which is important for low application latency and to keep wake-up energy minimal. For fast wake-up, the system must retain its state and logged information during sleep mode because several-hundred microseconds are needed for reloading such data to memories. Conventional SRAM consumes much higher retention current than the required deep-sleep-mode current as shown in Fig. 13.4.1. Embedded Flash memories have limited write endurance on the order of 10^5 cycles making them difficult to use in applications that frequently power down. Embedded FRAM [1,2] has been used for this purpose and it could be used as a random-access memory as well as a nonvolatile memory. However, as a random-access memory, its slow operation and high energy consumption [1,2] limits performance of the MCU and battery lifetime. Furthermore, additional process steps for fabricating FRAM memory cells increase the cost of MCU. SRAM can operate at higher speed with lower energy without additional process steps, but high retention current makes it difficult to sustain data in deep-sleep mode. To solve this problem, we develop low-leakage current SRAM (XLL SRAM) that reduce retention current by 1000× compared to conventional SRAM and operate with less than 10ns access time. The retention current of XLL SRAM is negligible in the deep-sleep mode because it is much smaller than the amount of the deep-sleep-mode current of MCU, which is dominated by active current of the real-time clock and control logic circuits. By using XLL SRAM, the store and reload process during mode transitions can be eliminated and wake-up time from deep-sleep mode of MCU is reduced to few microseconds. This paper describes a 128kb SRAM with 3.5nA (27fA/b) retention current, 7ns access time, and 25μW/MHz active energy consumption. Its low retention current, high-speed, and low-power operation enable to activate SRAM in the deep-sleep mode, and also provides fast wake-up, low active energy consumption and high performance to MCU.

Since complexity and required performance of MCU has been increasing, a more advance process has to be used. As the process geometry becomes smaller, the leakage current of transistors increases. In addition to channel leakage, other leakage mechanism become significant, such as gate-oxide leakage and GIDL. To realize XLL SRAM, these three types of leakage current must be suppressed. Low-leakage transistors with long gate length and thick gate oxide for SRAM memory cells have been developed. Their GIDL is also decreased by introducing lightly doped region in the active layer as shown in Fig. 13.4.2(c). Generally, adapting long-gate-length and thick-gate-oxide transistors for memory cells causes an increase in memory macro area and power dissipation. We use several techniques to shrink the memory cell size, avoiding the increase in active power dissipation. The memory cell layout is shown in Figs. 13.4.2 (a) and (b). Since the supply voltage of memory cells is 1.2V, the space between p-type and n-type well and between gate poly and adjacent diffusion area are shrunk compared to the original design rule for the transistors with the same gate-oxide thickness. As a result, the memory cell size is reduced about 20%. The cell height in the vertical direction is extended by adopting long gate length transistors. This allows four wordlines to be routed over the memory cell as shown in Fig. 13.4.2(b). A block diagram of the developed 128kb SRAM is shown in Fig. 13.4.3(a). Peripheral circuits consist of conventional transistors to achieve high performance and to reduce SRAM size. The supply voltage of the peripheral circuits is cut-off, and the NMOS source node of the memory cells (VSSB) is reverse biased via source-bias circuits in retention mode, as shown in Fig. 13.4.3(b). As a result, 27fA/b of leakage current is achieved with the fabricated XLL SRAM at room temperature, shown in Fig. 13.4.4(a). The leakage current of conventional SRAM is also shown for comparison. The XLL SRAM achieves 1000× lower leakage current at room temperature compared to the conventional SRAM due to lower gate leakage current and larger back-gate-bias effect. Low-power MCU in the deep-sleep mode consumes low energy, so that retention current at room temperature determines the battery lifetime. The leakage current of XLL SRAM is lower than the required deep-sleep-mode current of low power MCU, even though memory capacity increases up to several Mb. Therefore all of SRAMs in a low-power MCU can be active in deep-sleep mode, and the state and the logged information can be retained in them. Comparison of our SRAM leakage current to published leakage for SRAM in 65nm and smaller processes is shown in Fig. 13.4.4(b). It shows XLL SRAM reduces leakage current by more than 10× compared to FD-SOI SRAMs.

To compensate for the increase in active power due to relatively large SRAM area, several low-power techniques are adopted. Bitline-charging current is the dominant portion of active power consumption of SRAM. To reduce the bitline-charging current, we adopt quarter array activation scheme (QAAS) and charge-shared hierarchical bitline (CSHBL) [3,4]. Four wordlines are routed over a memory cell, and one of four wordlines connects to a memory cell in every 4 columns as shown in Fig. 13.4.5. An SRAM architecture where two word lines are routed over a memory cell has been reported [3]. By taking advantage of the extended memory cell height, the number of wordlines passed over a memory cell is doubled. Then 3/4 of bitlines remain inactive in active cycles, and bitline-charging current is reduced. The SRAM also employ CSHBL. It has been reported that CSHBL is effective for reducing the active power increase due to random variation of transistors [4]. We find that CSHBL is also effective for reducing the active power increase due to process and temperature variations. Waveforms of signals in CSHBL operation are shown on the right side of Fig. 13.4.5. The local bitlines are fully swung when the corresponding wordline is selected. Before the pass transistors are turned on, the selected wordline falls. Then stored charge on the local bitlines are transferred to the global bitlines by charge sharing. Since the amount of stored charge on the local bitlines is determined by the capacitance of the local bitlines and supply voltage, the bitline-charging current of the SRAM stays constant regardless temperature and process condition. In a conventional SRAM, bitline level varies substantially with changing temperature or process condition. Designing for the minimum bitline swing that can be sensed by the sense amplifiers in the slowest condition causes the bitline swing to become excessive in fast conditions. This causes excessive bitline-charging power dissipation. Figures 13.4.6 (a) and (b) show the bitline-charging current by measuring current to the memory cell ground in several process conditions. The charging current of the XLL SRAM decreases more than 40% compared to SRAM with conventional bitline architecture and timing-control circuits. Figure 13.4.6(c) shows measured active energy of the XLL SRAM. Active energy of the XLL SRAM is reduced about 40% by adapting QAAS and CSHBL, and 25μW/MHz of active energy at 1.2V is achieved. The achieved active energy is only 9% larger than the conventional SRAM.

Figure 13.4.7 shows the chip micrograph and key features of the test chip fabricated in a 65nm CMOS process. Memory cell size is 2.159μm^2 and macro area of the 128kb XLL SRAM is 0.443mm^2. Retention current and active energy at 1.2V is 3.5nA (27fA/b) and 25μW/MHz, respectively. The leakage current is negligible compared to deep-sleep-mode current of low power MCU. Because of this, all the SRAMs in MCU can be awake in deep-sleep mode, shortening wake-up time from deep-sleep mode and reducing the energy consumption during the mode transition.

References:

[1] A. Baumann et al., "A MCU platform with embedded FRAM achieving 350nA current consumption in real-time clock mode with full state retention and 6.5μs system wakeup time," *VLSI Cir. Symp.*, pp. 202-203, 2013.

[2] M. Zwerg et al. "An 82μA/MHz Microcontroller with Embedded FeRAM for Energy-Harvesting Applications," *ISSCC*, pp. 334-335, 2011.

[3] H. Fujiwara et al., "A 20nm 0.6V 2.1μW/MHz 128kb SRAM with No Half Select Issue by Interleave Wordline and Hierarchical Bitline Scheme", *VLSI Cir. Symp.*, pp 118-119, 2013.

[4] S. Miyano et al., "Highly Energy-Efficient SRAM With Hierarchical Bit Line Charge-Sharing Method Using Non-Selected Bit Line Charges," *JSSC vol. 48*, pp. 924-931, April 2013.

978-1-4799-0917-9/14 $31.00 © 2014 IEEE

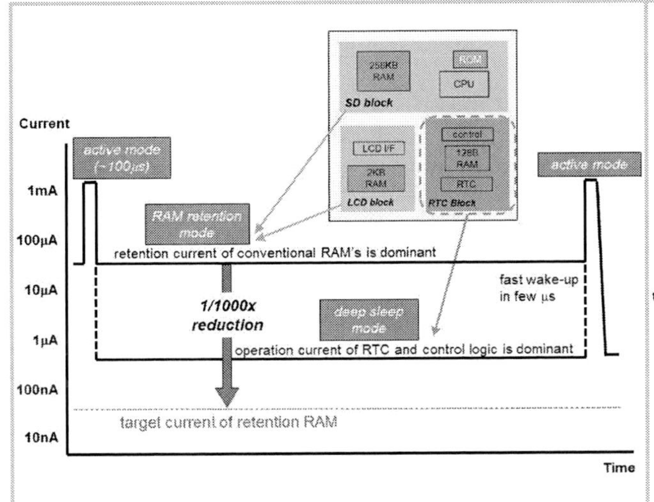

Figure 13.4.1: Typical power profile of low power MCU.

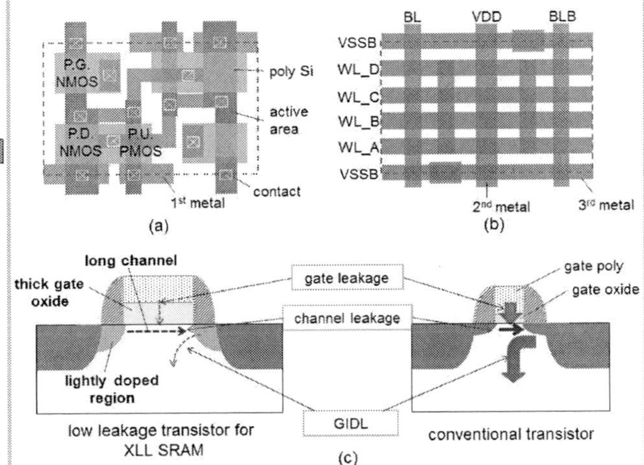

Figure 13.4.2: XLL SRAM bitcell and MOSFET leakage path.

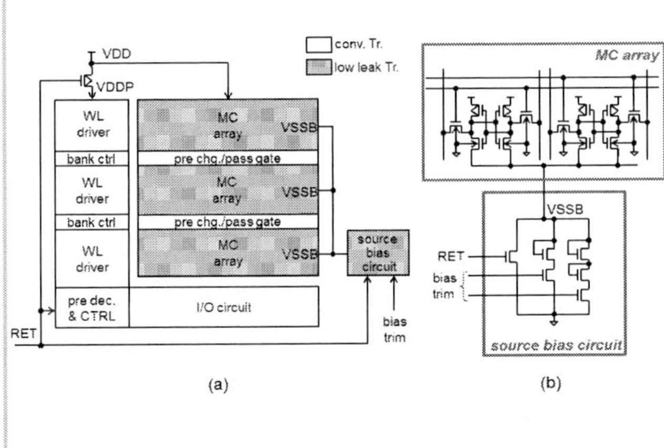

Figure 13.4.3: Block diagram and source bias circuit.

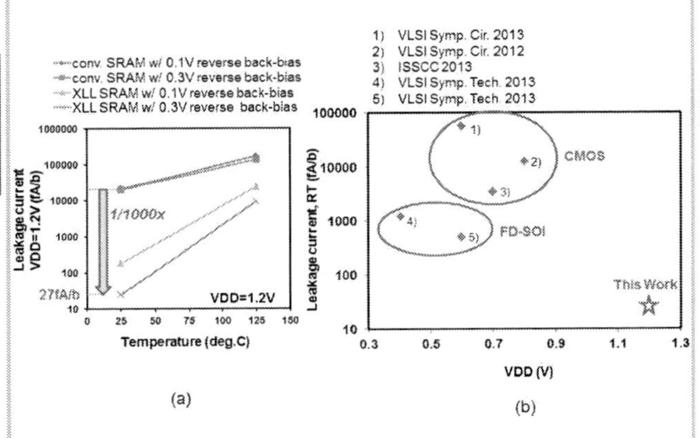

Figure 13.4.4: XLL SRAM leakage current measured results.

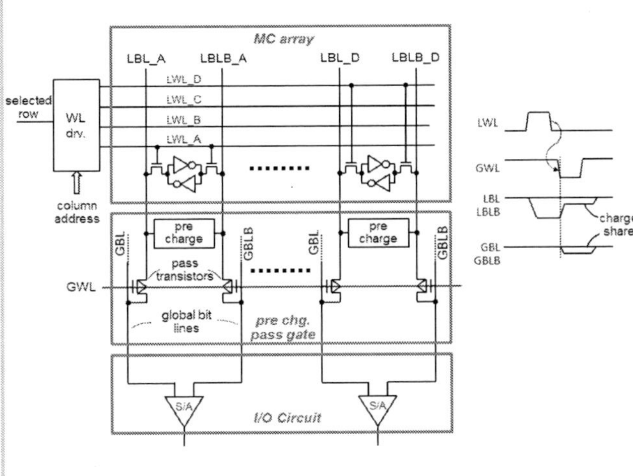

Figure 13.4.5: QAAS and CSHBL circuit and timing chart.

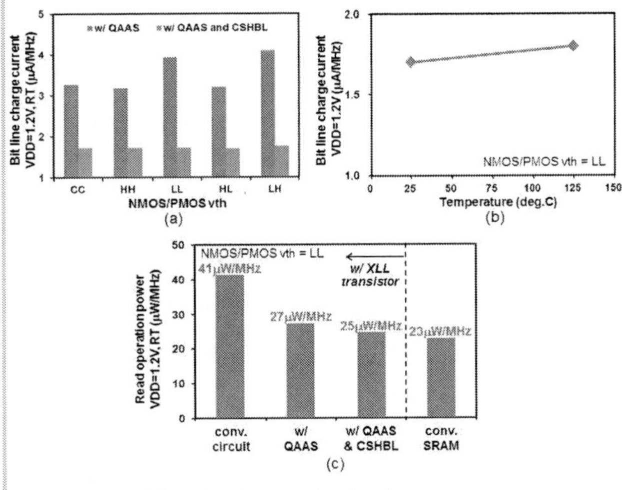

Figure 13.4.6: Bitline-charging current and active power.

Technology		65nm CMOS
Power supply		1.2V
Cell structure		6T
Cell size		$2.159\mu m^2$
Capacity		128kbit
Macro area		$0.443mm^2$
Read access time		7ns
Power consumption	Operation	25μW/MHz
	Standby	3.5nA (27fA/b)

Figure 13.4.7: XLL SRAM test-chip micrograph and key features.

13.5 A 16nm 128Mb SRAM in High-κ Metal-Gate FinFET Technology with Write-Assist Circuitry for Low-V_{MIN} Applications

Yen-Huei Chen, Wei-Min Chan, Wei-Cheng Wu, Hung-Jen Liao, Kuo-Hua Pan, Jhon-Jhy Liaw, Tang-Hsuan Chung, Quincy Li, George H. Chang, Chih-Yung Lin, Mu-Chi Chiang, Shien-Yang Wu, Sreedhar Natarajan, Jonathan Chang

TSMC, Hsinchu, Taiwan

FinFET technology has become a mainstream technology solution for post-20nm CMOS technology [1], since it has superior short-channel effects, better sub-threshold slope and reduced random dopant fluctuation. Therefore, it is expected to achieve better performance with lower SRAM V_{DDMIN}. However, the quantized sizing of the channel width and length has drawbacks for conventional 6T-SRAM bitcell scaling. To minimize the bitcell area of the high-density SRAM bitcell, the number of fins (setting the channel width, W) of the pull-up PMOS (PU), pass-gate NMOS (PG) and pull-down NMOS (PD) transistors must be selected as 1:1:1. Since PU, PG, and PD have the same channel length (L), the ratio in geometry between the PU transistor and the PG transistor is equal to one. With the process variations, the strength of PU transistor can be much stronger than the PG transistor. A stronger PU transistor increases read stability of the SRAM bitcell but it degrades the write margin significantly and results in worse write-V_{DDMIN} issue. Figure 13.5.1(a) shows a contention condition between PU and PG transistors of a 6T-SRAM bitcell for the write operation. During the write operation, the PU transistor impedes the ability of the PG transistor to pull the storage node (S) from V_{DD} to ground. The bitcell may suffer a write failure at the stronger PU with weaker PG condition caused by the device variations. Two techniques have been proposed to improve the high density SRAM bitcell write V_{DDMIN}: 1) negative bit-line voltage (NBL) to increase the strength of PG transistor and 2) lower cell V_{DD} (LCV) to weaken PU transistor strength [1-5]. Compared to the conventional techniques, this work develops a suppressed-coupling-signal negative bitline (SCS-NBL) scheme and a write-recovery-enhancement lower-cell-V_{DD} (WRE-LCV) scheme for write assist without the concern of reliability at higher V_{DD} operating region. A comparison of the effectiveness of the two design techniques is also performed. Figure 13.5.1(b) shows the layout view of the high-density 6T-SRAM bit-cell with 0.07μm² area in a 16nm high-k metal-gate FinFET technology. To minimize area, we set the geometric ratio of PU, PG, and PD transistors all equal to one. With the two developed write-assist circuits, the overall V_{DDMIN} improvement can be over 300mV in a 128Mb SRAM test-chip.

Figure 13.5.2 shows simulated results of the required negative bitline bias (blue curve) and coupling NBL voltage levels with and without negative-bias-suppression (clamping) circuit for the high-density SRAM bitcell. Since aggressive negative bias is needed to achieve the V_{DDMIN} target, the coupled negative-bias level has to be more negative to provide the required bitline write voltage. Due to the coupling technique, the negative-bias voltage level is proportional to the voltage level of the coupling signal. In the SRAM write operation, the negative bitline bias is applied to the source terminal of the PG transistor and wordline activation pulse signal is applied to the gate terminal of the PG transistor in the selected SRAM bitcell. The greater negative bias at higher V_{DD} region leads to more stress on the SRAM PG transistor, thus reliability is a concern for the SRAM operation at higher V_{DD}. Therefore, a suppressed coupling signal at high V_{DD} region for negative-bit-line bias (SCS-NBL) generation circuitry is applied to suppress the coupling signal voltage and reduce the generated negative bias level at higher V_{DD}. With the SCS-NLB scheme, the reliability concern is mitigated without degrading the negative-bias level generation at the lower V_{DD} region.

Figure 13.5.3 illustrates the SRAM design equipped with SCS-NBL write-assist scheme. The SCS-NBL scheme is directly implemented into read/write block with small area penalty. During the write operation, the write signal pulse triggers the replica write buffer to pull low the replica bitline (RBL) to generate a negative-bitline enable signal (ENB_NBL). Then the ENB_NBL signal propagates to become the coupling signal (NBL_FIRE). The coupling signal voltage suppression block is composed of the stacked NMOS transistors that can suppress the coupling signal (NBL_FIRE) voltage level to the designed suppressed voltage level that can be below the power rail (V_{DD}) at higher V_{DD} region. The falling edge of the suppressed NBL_FIRE signal couples to a capacitor (C1) to generate a negative coupling signal (NVSS). With the lower coupling NBL_FIRE signal voltage level, the coupled negative-bias signal (NVSS) is limited and kept constant below the higher V_{DD}, as shown in the waveforms of Fig. 13.5.3. Then the negative bias will be propagated to the selected bitline (BL[n]) and transferred into the selected SRAM bitcell through the write driver (WD1) and the write MUX (N1).

Figure 13.5.4 shows the write-recovery-enhancement lower-cell-V_{DD} (WRE-LCV) write-assist scheme. In order to save area, a shared WRE-LCV voltage generator is placed in each read/write block. During write operation, the generated LCV voltage signal (lower than the power rail V_{DD}) from the shared WRE-LCV voltage generator is propagated to the selected column power track (CVDD[n]) and to the selected bitcell through a transmission gate (T1) that is controlled by the LCV enable (LCV_EN) and the column selection (Y[n]) signals. For unselected columns, the CVDD[n-1] will remain at V_{DD} since the transmission gate (T2) is off and the pre-charging PMOS (P2) turns on to prevent the half-selection read-disturbance issue. Unlike the previous work [1,4] using voltage collapse as write assist, to prevent the transient data-retention issue, a slightly lower cell V_{DD} (LCV) with write-recovery-enhancement LCV pulse (WRE-LCV) technique is applied in this work. The WRE-LCV pulse has to be recovered before the wordline-pulse-signal turn off to prevent the write recovery failure issue as shown in the waveforms of Fig. 13.5.4. The WRE-LCV pulse write-recovery timing control is centralized at the WRE-LCV controller for the recovery timing adjustment. The WRE-LCV pulse voltage level options (75%, 50%, 25% of V_{DD}) are also offered for post-fabrication tuning to mitigate process variations.

Figure 13.5.5 shows the floorplan and the area of 128kb SRAM macro with 0.07μm² SRAM bitcell. The WRE-LCV scheme is placed at the boundary of SRAM array and read/write block (RWBLK). In order to reduce the area penalty, the WRE-LCV write-recovery timing-control block is placed in the main control (CTRL) block. The SCS-NBL scheme is located at the bottom of read/write block (RWBLK). The area overheads of the WRE-LCV and SCS-NBL schemes are 3% and 2%, respectively.

Figure 13.5.6 shows the cumulative distribution plot of the overall V_{DDMIN} improvement from a 128Mb test-chip at 25°C. The SCS-NBL and WRE-LCV improve the overall V_{DDMIN} by over 300mV at 95 percentile for the 0.07μm² SRAM bitcell in a 128Mb test-chip. With the reduced LCV voltage level, the V_{DDMIN} can be pushed even lower, but the improvement saturates at 50% LCV voltage level.

Figure 13.5.7 shows the micrograph of the 128Mb SRAM test-chip, which is equipped with electrically programmable fuses for post-silicon tuning on the write-assist options. The test-chip is built from 1024 128kb (4096×32) SRAM macros. The die area of the test-chip is 42.6mm².

Acknowledgements:
The authors thank the physical design team John Hung, R.S. Chen, Hanson Hsu, and L.J. Tyan for layout and chip implementation; the RD team for wafer manufacturing; the test department for chip measurements on this work.

References:

[1] Y. Wang et al., "Dynamic Behavior of SRAM Data Retention and a Novel Transient Voltage Collapse technique for 0.6V 32nm LP SRAM", *IEDM Dig. Tech. Papers*, Dec. 2011, pp 32.1.1-32.1.4.

[2] Y. Fujimura et at., "A Configurable SRAM with Constant-Negative-Level Write Buffer for Low-Voltage Operation with 0.149μm² Cell in 32nm High-K Metal-Gate CMOS", *ISSCC Digest of Technical Papers*, Feb. 2010, pp 348-349.

[3] H. Pilo et al., "A 64Mb SRAM in 32nm High-k Metal Gate SOI Technology with 0.7V Operation Enabled by Stability, Write-Ability and Read-Ability Enhancements," *ISSCC Digest of Technical Papers*, Feb. 2011, pp 254-256.

[4] Eric Karl et al., "A 4.6GHz 162Mb SRAM Design in 22nm Tri-Gate CMOS Technology with Integrated Active Vmin Enhanced Assist Circuitry," *ISSCC Digest of Technical Papers*, Feb. 2012, pp 230-231.

[5] Jonathan Chang et al., "A 20nm 112Mb SRAM in High-k Metal-Gate with Assist Circuitry for Low-Leakage and Low-Vmin Applications" *ISSCC Digest of Technical Papers*, Feb. 2013, pp 316-317.

978-1-4799-0917-9/14 $31.00 © 2014 IEEE

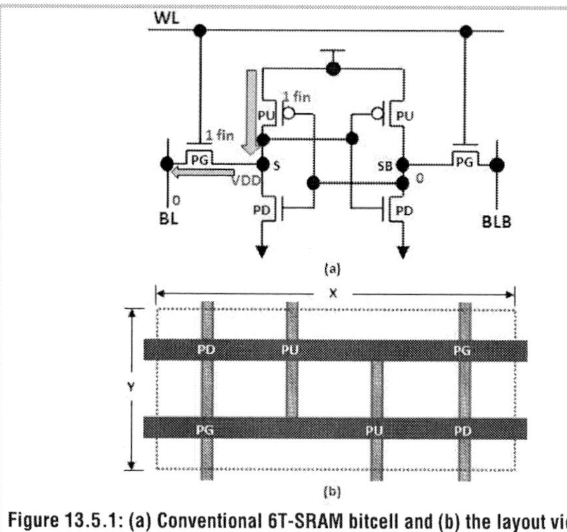

Figure 13.5.1: (a) Conventional 6T-SRAM bitcell and (b) the layout view of the high-density SRAM bit-cell with an area of 0.07µm².

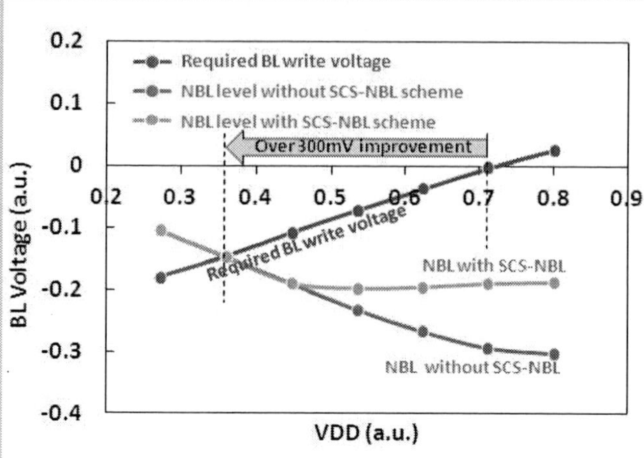

Figure 13.5.2: Negative bitline voltage versus required write bitline voltage.

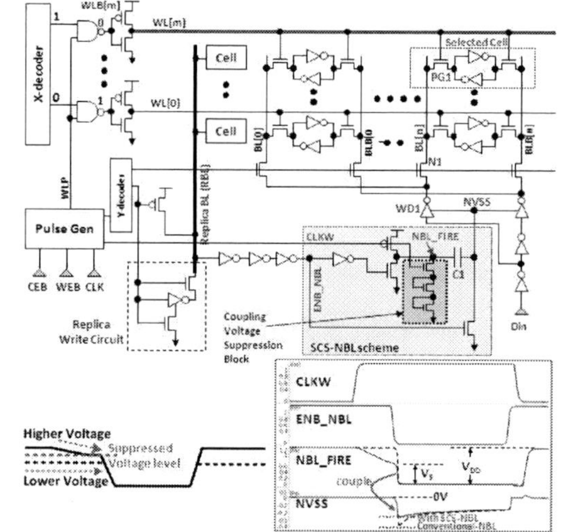

Figure 13.5.3: SRAM design equipped with SCS-NBL write assist scheme.

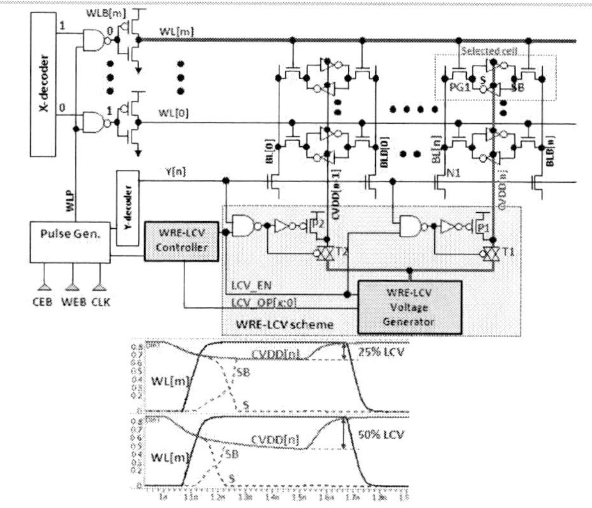

Figure 13.5.4: SRAM design equipped with WRE-LCV write assist scheme.

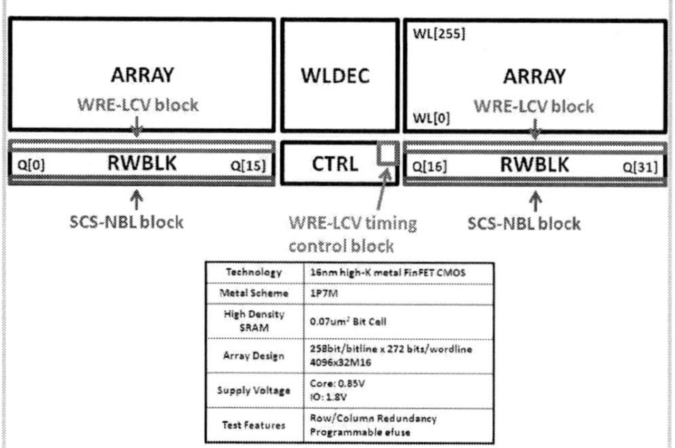

Technology	16nm high-K metal FinFET CMOS
Metal Scheme	1P7M
High Density SRAM	0.07um² Bit Cell
Array Design	258bit/bitline x 272 bits/wordline 4096x32M16
Supply Voltage	Core: 0.85V IO: 1.8V
Test Features	Row/Column Redundancy Programmable efuse

Figure 13.5.5: Floorplan of the WRE-LCV and SCS-NBL blocks for write-assist techniques.

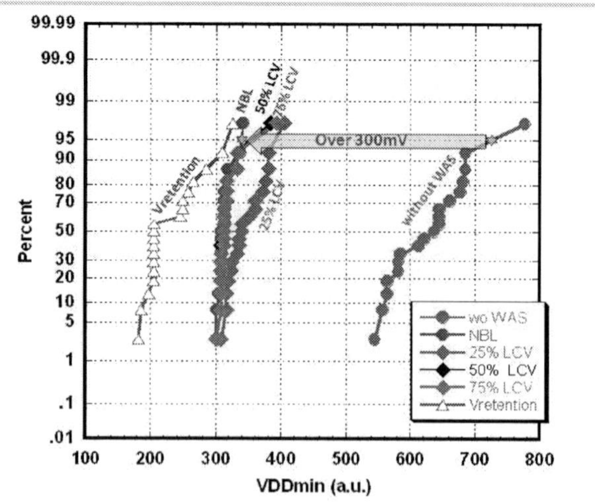

Figure 13.5.6: Cumulative distribution plot of overall V_{DDMIN} improvement for WRE-LCV and SCS-NBL write-assist techniques.

13

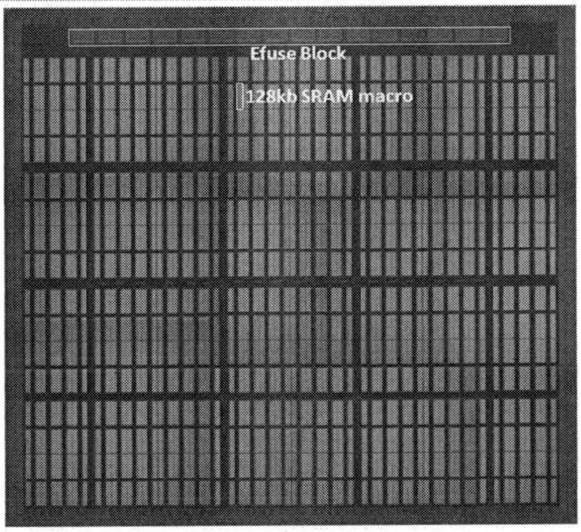

Figure 13.5.7: Die micrograph of 128Mb SRAM test-chip.

13.6 A 28nm 400MHz 4-Parallel 1.6Gsearch/s 80Mb Ternary CAM

Koji Nii[1], Teruhiko Amano[2], Naoya Watanabe[2], Minoru Yamawaki[3], Kenji Yoshinaga[3], Mihoko Wada[3], Isamu Hayashi[2]

[1]Renesas Electronics, Kodaira, Japan,
[2]Renesas Electronics, Itami, Japan,
[3]Renesas Design, Itami, Japan

The number of IPv4 routing table entries was around 460k in 2012 and is growing at a rate of 10% per year. The continuing increase of network-connected devices risks depletion of IPv4 addresses. As for IPv6, the number of routing-table entries is about to reach 16k with a rapid annual growth rate of 90%. This growth trend will continue in the Internet-of-Things era. In growing network applications with large amounts of traffic, there is a demand for routers and switches capable of packet processing for applications such as forwarding, quality of service, classification and access control. Fully parallel ternary content-addressable memories (TCAMs) are the key devices to handle the large number of routes in the lookup table with high-throughput low-latency header processing [1-6]. Realizing both high-density (high-capacity) and high-speed search operation is a challenge due to high power consumption and diminishing signal integrity for search operations. Process scaling is important to achieve a high-density high-performance TCAM, but it incurs the decrease of operating margin caused by threshold-voltage variation. This paper presents a 400MHz 4-paralleled search-operation 80Mb TCAM test-chip in a 28nm process that can accommodate 1M entries. This chip has three key features: flexible search mode, row and column shift redundancy, and search omission with valid-bit. With flexible search mode, the TCAM can be adapted to both IPv4 and IPv6.

Figure 13.6.1 shows a block diagram of the developed quad-search fully ternary CAM. The 80Mb CAM array consists of four 20Mb macros that are divided into 64 subarray blocks with 320kb (80b × 4k entries each). Based on application requirements, these blocks can be configured into a multiple lookup tables. In addition to the multiple-table option, each table can be 80b, 160b, 320b or 640b via a flexible priority encoder. The maximum throughput in single-search mode is 400Msearch/s with 400MHz clock frequency. In addition to the normal search mode, there are two optional operating modes: dual-search mode and quad-search mode up to 320b wide. Dual-search mode provides 2 parallel searches with a maximum search rate of 800Msearch/s whereas quad-search mode provides 4 parallel searches with a maximum search rate of 1.6Gsearch/s. Each 320kb subarray has row and column shift redundancy to improve yield. The redundant codes obtained by probe tests are written in the on-die electrical fuse blocks.

Figure 13.6.2 shows the circuit diagram of a multi-V_t differential match amplifier with a valid-bit cell. In this design, matchline (ML) lengths are 80b wide with one valid-bit cell. In order to reduce unnecessary discharge/charge power of the MLs in the invalid entries, the pre-charge pull-up PMOS connected to ML is gated by the stored data of the valid-bit cell. If the stored data in the valid-bit cell is set to the invalid status (stored "0") the pull-up PMOS is cut-off and does not pre-charge the corresponding ML to the V_{DD} level. In our previous work [6], the medium voltage level for ML pre-charging is used to reduce dynamic power, however, the lower pre-charge level induces significant speed penalty at the more advanced 28nm technology node, in which supply voltage has been scaled down to 0.85V. To enhance both search speed and leakage, low-V_t, middle-V_t and high-V_t MOSs are adaptively used for the match-amplifier circuit. The multi-V_t design ensures sense margin against process, voltage and temperature variations with a short sensing time. In test mode, the signal TST in the valid-bit cell is set to "1" level to pre-charge all MLs (ignored valid-bit status) as a worst-case condition. The simulated waveforms are shown in Fig. 13.6.3. The first cycle is 1b miss search (worst search vector) while the second cycle is match search with 2.5ns clock cycle at 0.85V, 25°C and worst process condition. After the ML pre-charge enable signal (MLPRE) is negated, the ML is discharged (search miss) or maintains the V_{DD} level (match). The ML reference voltage REFVD is generated by charge sharing between REFVD and REFVS nodes, lowering to 700mV and obtaining a 150mV deferential voltage for both miss and match search results. When the match-amplifier enable (MAE) is triggered, the match amplifier detects the differential voltage and presents the result at output terminal MAOUT.

Figure 13.6.4 shows the simulated and measured dynamic power consumption of a 20Mb TCAM macro. The power consumption increases linearly according to the number of the effective entries. The three measured samples and simulated data have a good correlation. At the 100% effective entry condition, the average of measured dynamic power consumption is 10.6W in the 250Msearch/s single-search mode at 0.85V and 25°C. If the number of effective entries occupied is half of the maximum capacity, the measured dynamic power consumption reduces to 8.4W, a 21% reduction from the full-entry condition.

Figure 13.6.5 indicates the distribution of measured minimum operating voltage (V_{MIN}) of fabricated 80Mb TCAM test-chips. Good V_{MIN} data in 250Mseach/s single-search mode at 25°C and 85°C are obtained. The worst V_{MIN} of measured samples (n = 9) are 0.73V at 25°C and 0.70V at 85°C, respectively. Figure 13.6.5 also shows a measured typical shmoo plot of 80Mb TCAM at 25°C. Functional pass regions in the search-operation mode with a worst-key vector are confirmed over the 400Msearch/s (below 2.5ns cycle time) over 780mV supply voltage. This means 1.6Gsearch/s is achieved with 2.5ns clock cycle in the quad-search mode where the four 20Mb macros operate simultaneously.

Figure 13.6.6 summarizes the designed and fabricated test-chips. We use TSMC 28nm high-k metal-gate high-performance (HP) bulk CMOS technology with 10 copper metals and aluminum top metal. The customized and optimized TCAM bitcell consists of 16 MOSFETs including two 6T-SRAM bitcells. The physical macro size of 320kb subarray is 793.1×665.3mm^2 (0.53mm^2) with density over 0.61Mb/mm^2. Figure 13.6.7 depicts the die micrograph of a fabricated 80Mb TCAM test-chip. Four 20Mb TCAM macros are tiled on the die. This work is over quadruple the size of our previous work [6].

Acknowledgements:
The authors would like to thank TCAM design teams.

References:

[1] H. Miyatabe, M. Tanaka, and Y. Mori, "A Design for High-Speed Low-Power CMOS Fully Parallel Content-Addressable Memory Macros," *IEEE J. Solid State Circuits*, vol. 36, no. 6, pp. 956-968, June 2001.

[2] G. Kasai, Y. Takarabe, K. Furumi, and Masato Yoneda, "200MHz/200MSPS 3.2W at 1.5V Vdd, 9.4Mbits Ternary CAM with New Charge Injection Match Detect Circuits and Bank Selection Scheme," in *Proc. IEEE Custom Integrated Circuits Conf. (CICC)*, pp. 387-390, 2003.

[3] H. Noda et al., "A cost-efficient high-performance dynamic TCAM with pipelined hierarchical searching and shift redundancy architecture," *IEEE J. Solid-State Circuits*, vol. 40, pp. 245-253, Jan. 2005.

[4] K. Pagiamtzis and A. Sheikholeslami, "Content-addressable memory (CAM) circuits and architectures: A tutorial and survey," *IEEE J. Solid-State Circuits*, vol. 41, no. 3, pp. 712-727, Mar. 2006.

[5] I. Arsovski, T. Hebig, D. Dobson, and R. Wisort, "A 32 nm 0.58-fJ/Bit/Search 1-GHz ternary content addressable memory compiler using silicon-aware early-predict late-correct sensing with embedded deeptrench capacitor noise mitigation," *IEEE J. Solid-State Circuits*, vol. 48, no. 4, Apr. 2013.

[6] Isamu Hayashi, Teruhiko Amano, Naoya Watanabe, Yuji Yano, Yasuto Kuroda, Masaya Shirata, Katsumi Dosaka, Koji Nii, Hideyuki Noda, and Hiroyuki Kawai, "A 250-MHz 18Mb Full Ternary CAM with Low-Voltage Matchline Sensing Scheme in 65-nm CMOS," IEEE J. Solid-State Circuits, vol. 48, no. 11, pp. 2671-2680, Nov. 2013.

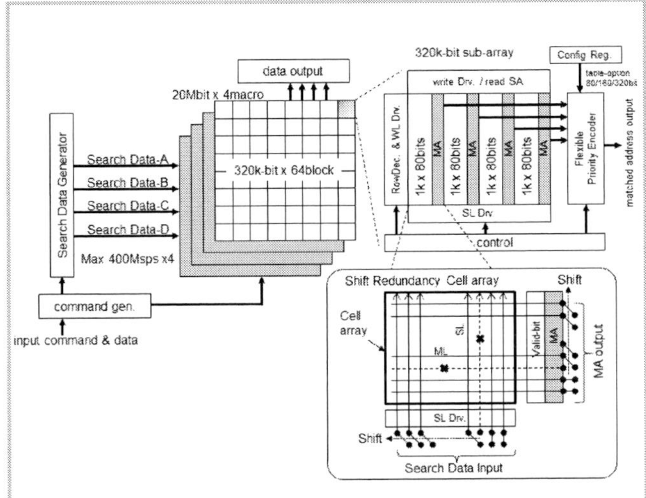

Figure 13.6.1: TCAM block diagram with single/dual/quad search modes.

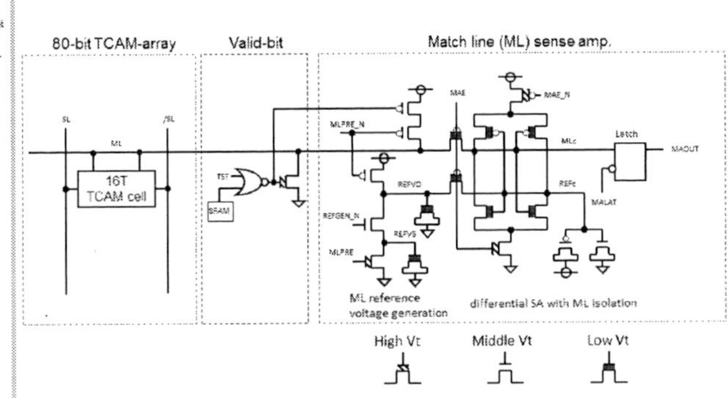

Figure 13.6.2: Circuit diagram of multi-V_t match-sense amplifier with valid-bit cell.

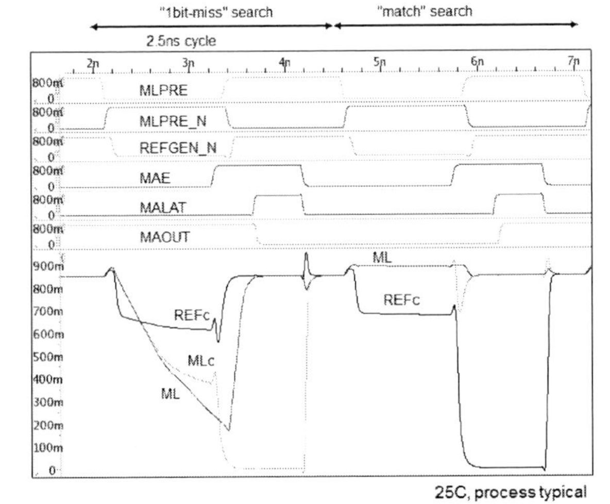

Figure 13.6.3: Simulated waveform of match-sense amplifier.

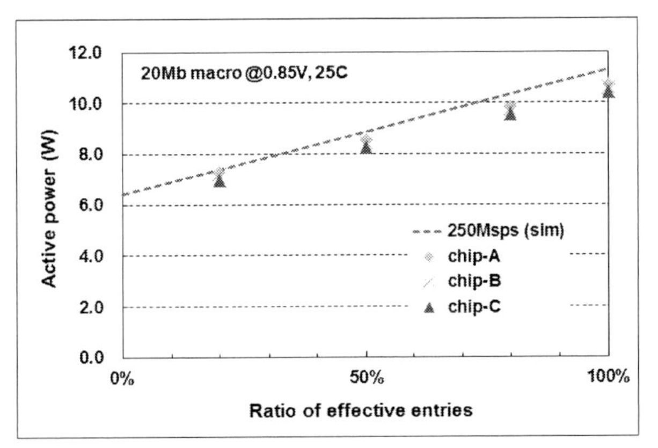

Figure 13.6.4: Simulated and measured search power consumption.

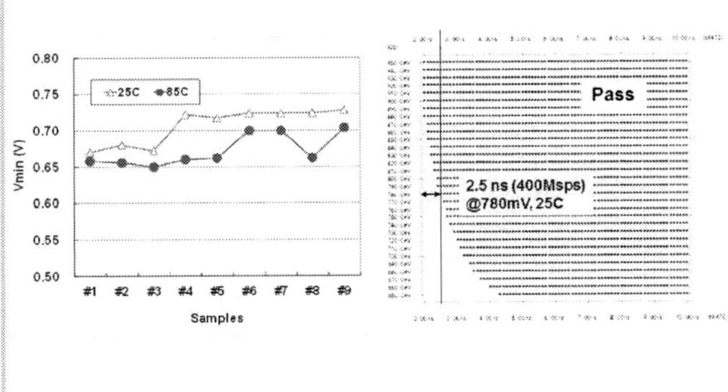

Figure 13.6.5: Measured V_{MIN} and shmoo plot.

Technology:	28nm HKMG high-performance bulk CMOS with 10-Cu metals and AL-top-metal
Supply voltage:	0.85 V
Clock frequency:	Max 400MHz
Total capacity:	80-Mbit
Configuration:	320k (4k x 80b) x 64 blocks x 4 macros
Physical macro size:	0.53 mm^2 @ 320-kbit sub-array
TCAM bitcell:	16T-bitcell type (including two 6T SRAMs) [6]
Max search speed:	400M-sps (single mode), 800M-sps (dual mode), 1.6G-sps (quad mode) @ 0.85 V, 25C
Power consumption:	8.4 W (Meas.) @ 20Mb, 250M-sps, 50% entries, 0.85 V, 25C

Figure 13.6.6: Feature of CAM test-chip.

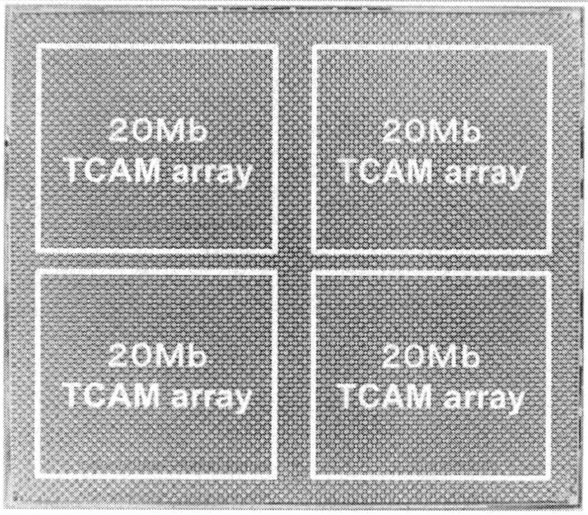

Figure 13.6.7: Micrograph of test-chip.

13.7 A Reconfigurable Sense Amplifier with Auto-Zero Calibration and Pre-Amplification in 28nm CMOS

Bharan Giridhar, Nathaniel Pinckney, Dennis Sylvester, David Blaauw

University of Michigan, Ann Arbor, MI

High-performance SRAMs are critical elements in microprocessors and SoCs. Fast and robust bitline sensing is a key requirement in such memories. With process scaling, increased mismatch in the sense amplifier (SA) circuit and increased I_{read} variation in the bitcell [1] have degraded sensing robustness. The fundamental tradeoff between sensing time and bitline read failures (Fig. 13.7.1) forces designers to heavily margin sensing time in order to guarantee sufficient bitline differential voltage prior to SA triggering. Previous research has improved SA robustness using pre-amplification circuits [2], capacitance-based offset cancellation [3-4], and redundancy [5]. However, most of these schemes target single-ended sensing (losing the benefit of common-mode rejection), incurring up to 60% area overhead or post-silicon tuning costs.

This work presents an area-efficient and variation-tolerant small-signal differential sensing (VTS) scheme that modifies the conventional SA circuit to include: 1) a structure for on-the-fly, auto-zeroing offset compensation, 2) pre-amplification of bitline differential by reconfiguring the SA inverter pair as amplifiers, and 3) latching of the amplified voltage differential by returning the SA to its conventional cross-coupled configuration. The approach is demonstrated to improve SA robustness over conventional sensing at iso-sensing time without area overhead (Fig. 13.7.1). Conversely, sensing time can be reduced at iso-robustness and area. Measurements of a 28nm CMOS test chip show that an iso-area VTS scheme improves offset noise tolerance by ~$1.2\sigma_{Vth}$ or sensing speed by up to 42% at iso-robustness (<0.3% failure rate).

The VTS scheme reconfigures the inverter pair of the SA, effectively putting it to use during all phases of operation to provide offset cancellation and additional amplification (Fig. 13.7.1): 1) During bitline precharge, the SA does not have to detect bitline droop allowing the inverters to be decoupled from each other and biased in their high-gain regions close to their ideal trip-points. AC-coupling capacitors C1 and C2 enable independent biasing of the bitlines and inverters. In addition, the capacitors also compensate for mismatch in the inverter trip-points via auto-zeroing. 2) During reads, the bitcell wordline is activated and the inverters function as offset-compensated pre-amplifiers for the bitline differential (in contrast to conventional SAs, where they remain idle). 3) Finally, the inverters are cross-coupled to further amplify and latch the data using regenerative feedback, as in a conventional SA.

Figure 13.7.2 shows the circuit schematic of the VTS-SA. The 2:1 bitline mux, precharge and output driver circuits are similar to those in the conventional SA. The 10-T reconfigurable inverter circuits are coupled to the multiplexed bitlines using capacitors C_{MOM1} and C_{MOM2}. Transistors M3-4 and M5-6 form the SA inverters and NMOS switches M7-10 are used to reconfigure inverter connections for auto-zeroing, pre-amplification, and latching modes. NMOS switches M11 and M12 isolate the MOM capacitors during regeneration, preventing full rail voltage swing at nodes BL_MX/BL_MX_B that could turn on bitline mux switches and severely degrade performance. Since this scheme incorporates automatic offset compensation, the SA is not highly sensitive to mismatch. Hence, all devices in the VTS-SA are near minimum-sized and can leverage density improvements from technology scaling. This is in contrast to conventional SAs, which require large devices to reduce mismatch and therefore have not tracked with feature size improvements [6].

Figure 13.7.2 also shows simulated waveforms for the selected bitlines and SA inverter outputs through various phases of operation. During biasing/offset storage, the input and output of the SA inverters are shorted together, which creates a 14μA (measured) short-circuit current that would increase power consumption in this scheme. However, biasing and offset storage require only ~60% of the precharge phase to complete and are therefore duty-cycled, resulting in 26% measured SA power savings (compared to no duty-cycling). Headers and footers for duty cycling are shared across 16 SAs. During the bitcell read phase, the capacitors connect the bitlines to the inverter inputs while their outputs are disconnected. This compensates for inverter trip-point offset and enables pre-amplification of bitline droop (~3.2× larger bitline swing at 60ps sensing time, simulated at TT corner, 1V, and 27°C). Finally, the inverters are cross-coupled when SA_EN is enabled for latching.

C_{MOM1}/C_{MOM2} size is a critical design parameter in the VTS scheme. Increasing these capacitances degrades sensing time (due to larger bitline capacitance) and requires upsizing of the inverter transistors (M3-M6) to charge the capacitors within a given precharge time. In contrast, smaller capacitors result in reduced coupling, attenuating the input bitline swing and negating the benefit of pre-amplification. Figure 13.7.3 shows the simulated design-space that was used to determine capacitor size. In the test-chip implementation, ~5fF capacitors are used to maximize gain-bandwidth product, striking a balance between coupling ratio and total bitline capacitance while minimizing area. The capacitors are implemented as 7.8×0.76μm² metal-oxide-metal (MOM) devices, rather than: 1) metal-insulator-metal (MIM) capacitors, that have larger minimum size constraints, or 2) metal-oxide-semiconductor (MOS) capacitors, that undergo weak inversion during auto-zeroing, increasing coupling loss.

The VTS-SA is implemented in an 8kb SRAM array composed of high-density 6T bitcells (Fig. 13.7.3). The bitlines are interleaved 2:1 with 128 bits on each column. The MOM capacitors are pitch-matched to the SA and placed on top of two bitcell columns in metals 5 and 6. Figure 13.7.3 also shows the timing diagram for read-control signals in the VTS scheme. To evaluate robustness and speed improvements, a conventional SA-based array is also implemented, where the SA is sized for 4.5σ yield and has an area of 4.62μm². Placing the MOM capacitors over the bitcells and using near-minimum sized devices enables an iso-area implementation of the VTS-SA, despite 2× higher transistor count. Because of the additional 2 routing layers used by the MOM capacitor placement strategy, it may not be feasible in some routing-resource-limited cases (e.g., generic memory compilers). However, custom memory design applications such as processors often have sufficient (≥ 9) metal layers where a similar implementation can be achieved with little impact on overall routing.

The test harness used to characterize the SAs is shown in Fig. 13.7.4. The arrays are programmed with pseudo-random data using a 32b LFSR. To measure sensing speed, the WL_EN to SA_EN delay is swept using a two-stage delay chain and any read failures are recorded over 2^{32} experiments operating at 1.8GHz. Similarly, SA robustness (offset noise tolerance) is characterized by skewing the supply voltages of the cross-coupled inverters (to induce mismatch) at a fixed nominal sensing time. Figure 13.7.5 shows measured SA sensing speed and robustness characterization for conventional and VTS implementations across 22 dice. For a typical die, VTS improves sensing speed by 34% over conventional sensing at a fixed read failure rate (<0.3%). Alternatively, this corresponds to ~$0.9\sigma_{Vth}$ higher offset noise tolerance. Across dice, sensing-speed improvements range from 25% to 42%, corresponding to robustness improvements of $0.6\sigma_{Vth}$ to $1.2\sigma_{Vth}$ (Fig. 13.7.6). Figure 13.7.6 also shows that VTS-based sensing-speed improvement is relatively stable across temperatures. The table in Fig. 13.7.6 compares the key characteristics of VTS and conventional sensing approaches.

Acknowledgements:

Funding support of NSF and DARPA (agreement HR0011-13-2-0006), and IC fabrication support of STMicroelectronics are gratefully acknowledged.

References:

[1] J. Wang, *et al.*, "Non-Gaussian Distribution of SRAM Read Current and Design Impact to Low Power Memory using Voltage Acceleration Method," *IEEE Symp. VLSI Technology*, pp. 220-221, 2011.

[2] D. Schinkel, *et al.*, "A Double-Tail Latch-Type Voltage Sense Amplifier with 18ps Setup+Hold Time," *ISSCC Dig. Tech. Papers*, pp. 314-315, 2007.

[3] N. Verma, *et al.*, "A High-Density 45nm SRAM Using Small-Signal Non-Strobed Regenerative Sensing," *ISSCC Dig. Tech. Papers*, pp. 380-381, 2008.

[4] M. Qazi, *et al.*, "A 512kb 8T SRAM Macro Operating Down to 0.57V with An AC-Coupled Sense Amplifier and Embedded Data-Retention-Voltage Sensor in 45nm SOI CMOS," *ISSCC Dig. Tech. Papers*, pp. 350-351, 2010.

[5] N. Verma, *et al.*, "A 65nm 8T Sub-Vt SRAM Employing Sense-Amplifier Redundancy," *ISSCC Dig. Tech. Papers*, pp. 328-329, 2007.

[6] K. Zhang, *et al.*, "The scaling of data sensing schemes for high speed cache design in sub-0.18μm technologies," *IEEE Symp. VLSI Circuits*, pp. 226-227, 2000.

978-1-4799-0917-9/14 $31.00 © 2014 IEEE

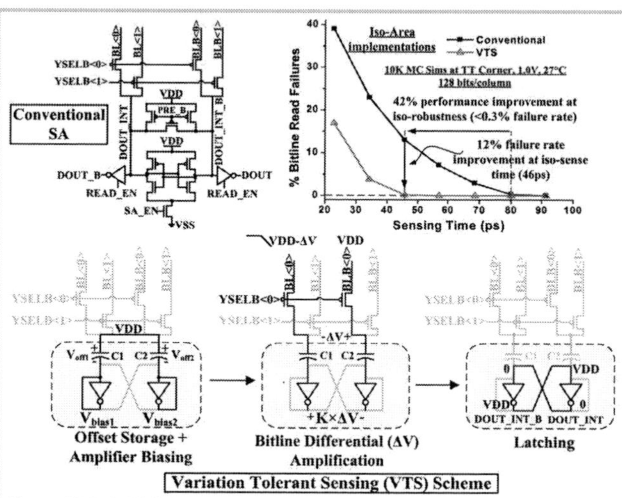

Figure 13.7.1: High-level operation of VTS and its sensing speed/robustness advantage over conventional sensing (simulated).

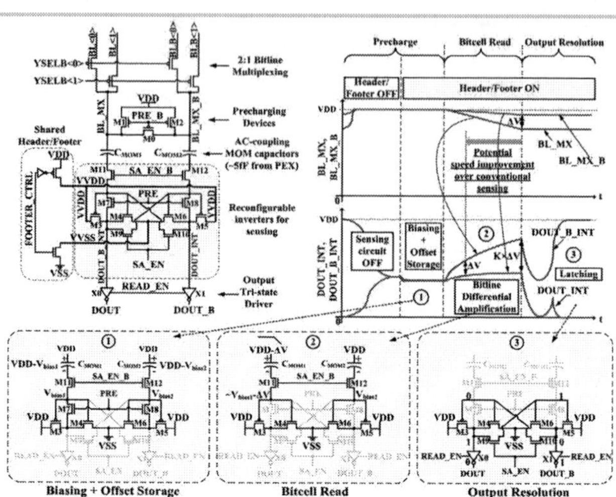

Figure 13.7.2: VTS-SA circuit schematic and operation phases showing relevant waveforms and circuit configuration in each phase.

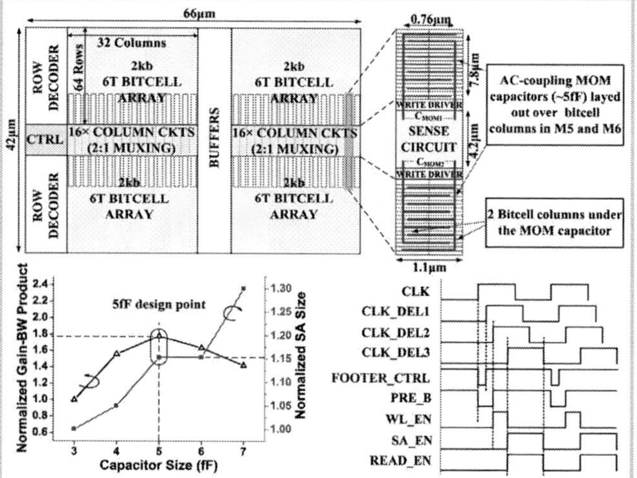

Figure 13.7.3: 8kb SRAM array with VTS showing capacitor placement strategy. Simulated capacitor sizing design space and VTS read timing.

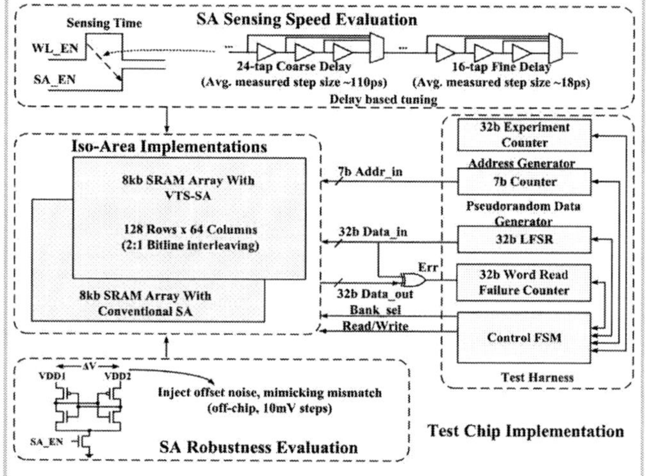

Figure 13.7.4: Test-chip implementation to characterize SA sensing speed and robustness.

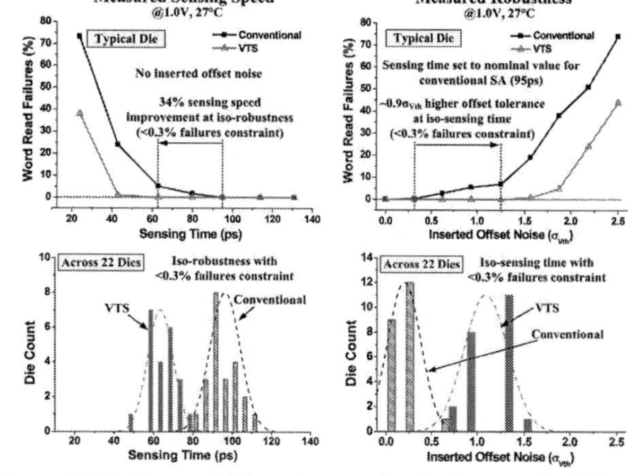

Figure 13.7.5: Measured SA sensing speed and robustness for VTS and conventional implementations. For a typical die, VTS improves sensing speed by 34% (corresponding to $0.9\sigma_{Vth}$ higher offset noise tolerance).

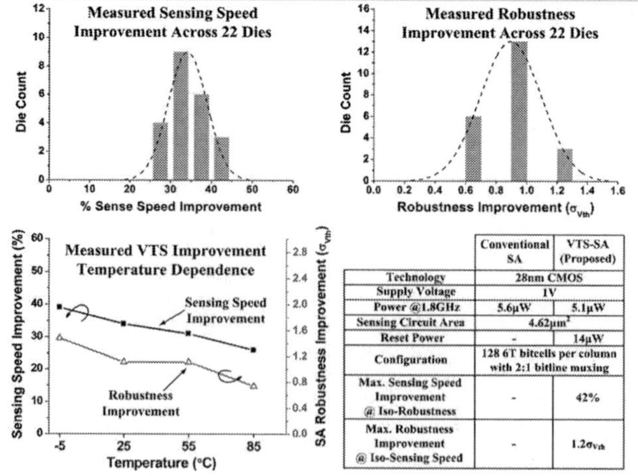

	Conventional SA	VTS-SA (Proposed)
Technology	28nm CMOS	
Supply Voltage	1V	
Power @1.8GHz	5.6μW	5.1μW
Sensing Circuit Area	4.62μm²	
Reset Power	-	14μW
Configuration	128 6T bitcells per column with 2:1 bitline muxing	
Max. Sensing Speed Improvement @ Iso-Robustness	-	42%
Max. Robustness Improvement @ Iso-Sensing Speed	-	$1.2\sigma_{Vth}$

Figure 13.7.6: Measured VTS-SA sensing speed and robustness improvements across 22 dice. Measured temperature dependence of VTS improvement and comparison summary.

13

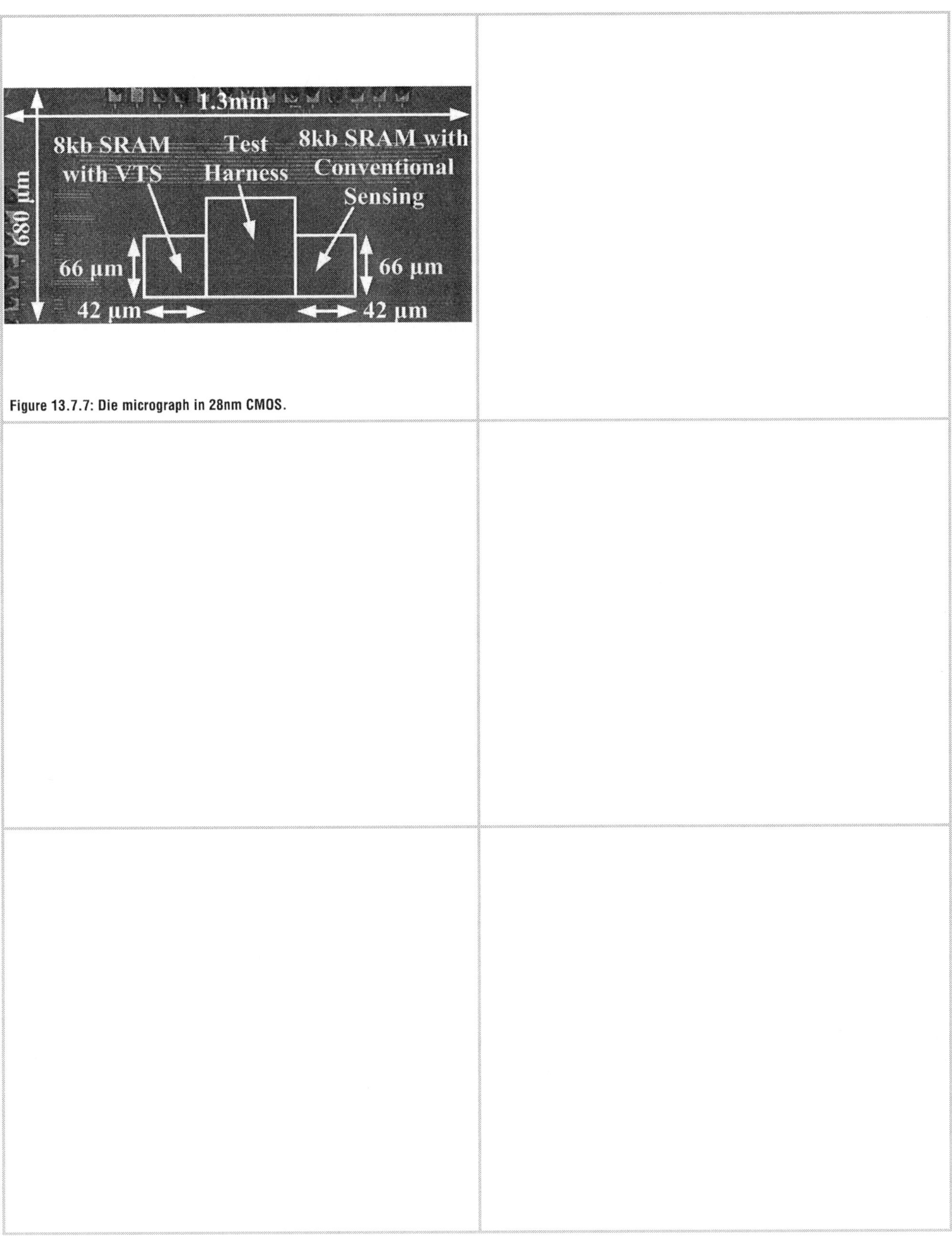

Figure 13.7.7: Die micrograph in 28nm CMOS.

13.8 A 32kb SRAM for Error-Free and Error-Tolerant Applications with Dynamic Energy-Quality Management in 28nm CMOS

Fabio Frustaci[1,2], Mahmood Khayatzadeh[2], David Blaauw[2], Dennis Sylvester[2], Massimo Alioto[3]

[1]University of Calabria, Rende, Italy,
[2]University of Michigan, Ann Arbor, MI,
[3]National University of Singapore, Singapore, Singapore

Voltage scaling is widely used to improve SRAM energy efficiency [1-2], particularly in mobile systems with tight power budgets. The resulting energy benefits are limited by the minimum voltage ensuring error-free operation, V_{min}, which has stagnated due to growing process variation in advanced technology nodes [3]. Error-tolerant applications and systems (e.g., multimedia) allow more aggressive voltage scaling by operating below V_{min}, which is acceptable if errors due to bitcell write/read failures do not perceptibly reduce application quality (e.g., image quality). Unfortunately, in traditional SRAMs bit error rate degrades rapidly for $V_{DD} < V_{min}$ [4], limiting energy gains. Under a given quality target, further energy reduction is possible through application-specific methods that exploit the features of data stored in a given application [4-5]. However, these approaches are not reusable across applications, and further the energy-quality trade-off is fixed at design time, which degrades energy savings in applications with lower quality targets and in chips near typical corner.

In this work, a new and highly flexible SRAM is developed for use in both error-free and error-tolerant applications, enabling a dynamic energy-quality trade-off. This work focuses on memories targeting video applications. The ideas can be generalized to different error-tolerant applications as well. The techniques are based on the observation that higher-order bits require more aggressive protection than lower-order bits. Two specific approaches are taken: 1) to address read stability, the array dynamically reconfigures lower-order bits to act as error-correcting code (ECC) bits to correct higher-order bits in the same word; 2) for write stability, the array selectively boosts bitlines of higher-order bits only. These techniques have low area overhead (2%) and are configurable in terms of aggressiveness to provide dynamic application-dependent adjustment of the energy/robustness trade-off. Energy is reduced by up to 35% based on measurements of a 28nm testchip.

Figure 13.8.1 shows the measured impact of supply-voltage reduction on bitcell error rate (BER) for a 28nm SRAM, and the resulting quality degradation of a 128×128-pixel grey-scale stored image, expressed as peak signal-to-noise-ratio (PSNR) [4]. Although all bit positions contribute equally to energy/access, quality is most strongly determined by the most-significant bits (MSBs), as shown by the PSNR increase in Fig. 13.8.1 when errors are disallowed in higher-order bit positions. In our SRAM, variation-resilient techniques are selectively introduced to improve the robustness of only those bitcells storing the most significant data. This reduces energy overhead and leads to more graceful quality degradation as bit errors mount at $V_{DD} < V_{min}$, enabling more aggressive voltage scaling. The selective robustness techniques can be independently applied to each bit position, hence our SRAM targets a wide range of energy/quality trade-offs, from low energy and low quality (protecting only MSBs) to error-free operation (protecting all bits for standard computation).

In the SRAM architecture (Fig. 13.8.2), negative bitline boosting (NBL) is used to reduce write failures in cells sharing the same column, while read failures are addressed through ECC, both of which are configurable. For example, in the error-free mode, NBL is enabled in all columns, and a traditional ECC equally protects all the bits within a 32b word D[31:0] (4 adjacent 8b pixels); in error-tolerant modes, NBL and ECC are enabled only in a desired subset of bit positions. As in Fig. 13.8.2, NBL is enabled (disabled) in columns having *boost*=1 (*boost*=0), with *boost* stored in a register, which entails an overhead of only one flip-flop every four columns (i.e., for 8b pixels in a 32b word) and two additional transistors per column for NBL enable. Column multiplexing (2:1) is used and controlled by C_{sel}. Similarly, *ECC_sel*=1 in Fig. 13.8.2 enables selective ECC, which corrects errors occurring in several MSBs in a pixel, as opposed to traditional ECC.

Previous work has adapted to lower quality targets by simply reducing bit width via dropping one or more lower-order bits [6]. This approach renders LSBs inactive, achieving a linear reduction in energy/access. In contrast, our approach re-uses LSBs as redundant bits that are then used by selective ECC to improve MSBs robustness. This enables further voltage scaling, yielding a quadratic reduction in energy/access. As described in Fig. 13.8.2, one LSB of each 8b pixel in the data word is used as a check bit. Selective ECC (a Hamming(15,11) code) protects three MSBs in each of pixels 0 to 2 and two MSBs of pixel 3, for a total of 11 bits protected in a 32b word. Only these MSBs (including check bits) are used as inputs to the ECC encoder. Fig. 13.8.3 shows an example of the selective ECC and traditional LSB dropping when a read error occurs on bit D[23] (i.e., the first MSB of pixel 2).

The above techniques are implemented in a 32kb SRAM in 28nm CMOS, comprising four 128×64 subarrays of traditional 6T cells. Negative bitline boosting voltage is set to –130mV to ensure writeability over 5σ, as appropriate for this 32kb array. The energy/quality trade-off of a test-chip near the SF corner (i.e., write critical, emulated by tuning WL voltage to skew the pull-up ratio) is shown in Fig. 13.8.4. The image testbench *peppers* (128×128 grayscale) is used. When scaling V_{DD}, selective NBL [7-4] on the first 4 MSBs reduces energy (voltage) by up to 35% (from 0.75 to 0.55V) compared to pure voltage scaling at the same quality. Other NBL schemes offer different energy/quality trade-offs: boosting [7-6] has the minimum advantage over pure voltage scaling (24%) due to its worse quality (PSNR = 25dB), while [7-2] has a 33% energy advantage due to the larger number of boosted bitlines and better quality (PSNR = 46dB). From Fig. 13.8.4, such advantage is consistently obtained within the range of practical PSNRs of ≥ 30 dB (see sample images in Fig. 13.8.5). Selective NBL also reduces energy by 18% compared to the error-free case. As expected, selective ECC does not provide significant benefit over pure voltage scaling, since it only corrects read failures (failures are mostly due to writes at SF corner). The same test-chip is used to emulate a read critical corner (FS) by tuning wordline voltage. In this case, using voltage scaling and selective ECC (Hamming(15,11) code), energy (V_{DD}) is reduced by 28% (from 0.7 to 0.6V) compared to pure voltage scaling at iso-quality. The core concept of using the dropped LSB to protect the MSB reduces energy by 19% through added voltage scaling compared to the simple case of keeping the dropped LSB inactive. As expected, selective NBL (omitted in Fig. 13.8.4) does not bring any energy advantage, as failures are mostly due to read.

Figure 13.8.5 shows the total energy advantage for the best combination of the schemes. In write-critical arrays, it is advantageous to progressively increase the number of boosted bitlines from [7-6] to [7-3] for higher PSNR targets, and the energy advantage over pure voltage scaling can be as high as 35% (28% on average) for practical PSNR ≥ 30dB. Similarly, in read-critical arrays, the energy saving enabled by ECC is as high as 28% (23% on average). The impact of process variation is evaluated across 19 dice for boosting [7-4] and PSNR = 30 dB. As in Fig. 13.8.6, average energy saving in write (read) critical case is 25% (27%), which is better (slightly worse) than the value of 20% (28%) obtained for the single chip measurements in Figs. 13.8.4 and 13.8.5. As in Fig. 13.8.6, our techniques significantly reduce V_{min} when operating in an error-tolerant mode compared to conventional voltage scaling. At a PSNR of 30dB, write-critical (read-critical) arrays can voltage scale by an additional 220mV (100 mV).

Acknowledgements:
The authors thank STMicroelectronics for chip fabrication. This work was supported in part by the NSF Variability Expedition and DARPA (agreement HR0011-13-2-0006).

References:

[1] J. Chang, et al., "A 20nm 112Mb SRAM in High-κ Metal-Gate with Assist Circuitry for Low-Leakage and Low-V_{MIN} Applications," *ISSCC Dig. Tech. Papers*, pp. 316-317, 2013.
[2] H. Pilo, et al., "A 64Mb SRAM in 22nm SOI Technology Featuring Fine-Granularity Power Gating and Low-Energy Power-Supply-Partition Techniques for 37% Leakage Reduction" *ISSCC Dig. Tech. Papers*, pp. 322-323, 2013.
[3] M. Yabuuchi, et al., "A 45nm Low-Standby-Power Embedded SRAM with Improved Immunity Against Process and Temperature Variations," *ISSCC Dig. Tech. Papers*, pp. 326-327, 2007.
[4] I. J. Chang, et al, "A Priority-Based 6T/8T Hybrid SRAM Architecture for Aggressive Voltage Scaling in Video Applications," *IEEE TCSVT*, vol. 21, no. 2, pp. 101-112, 2011.
[5] M.E. Sinangil, A. Chandrakasan, "An SRAM Using Output Prediction to Reduce BL-Switching Activity and Statistically-Gated SA for up to 1.9× Reduction in Energy/Access," *ISSCC Dig. Tech. Papers*, pp. 318-319, 2013.
[6] H. Kaul, et al., "A 1.45GHz 52-to-162GFLOPS/W Variable-Precision Floating-Point Fused Multiply-Add Unit with Certainty Tracking in 32nm CMOS," *ISSCC Dig. Tech. Papers*, pp. 182-183, 2013.

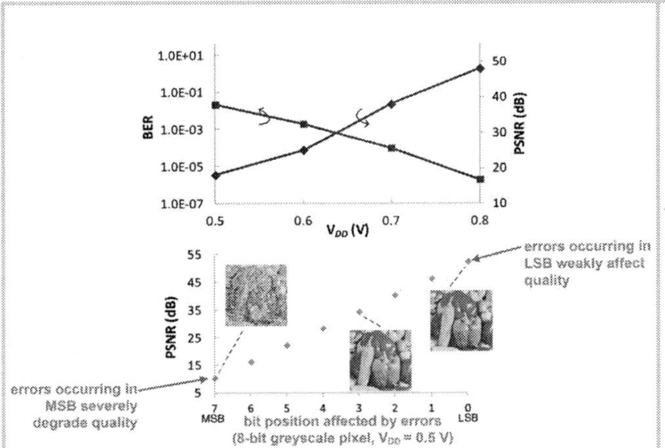

Figure 13.8.1: Aggressive SRAM voltage scaling rapidly degrades bit error rate (BER) and image quality (PSNR). Errors in MSBs impact PSNR more than LSBs (bottom).

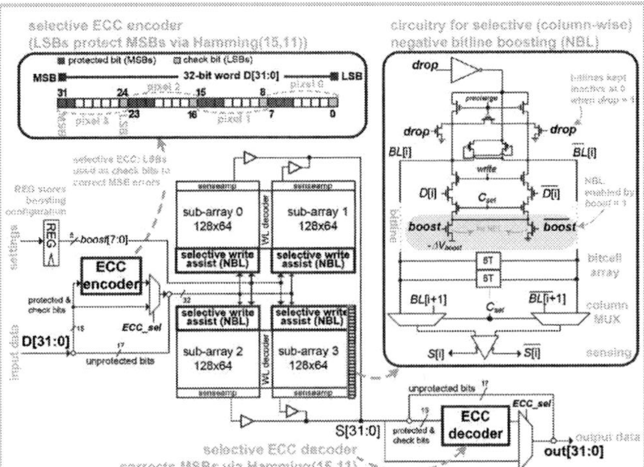

Figure 13.8.2: SRAM architecture (32b word, i.e., 4 8b pixels). MSBs of each pixel are protected via selective ECC (read) and selective NBL (write) for graceful quality degradation and lower energy.

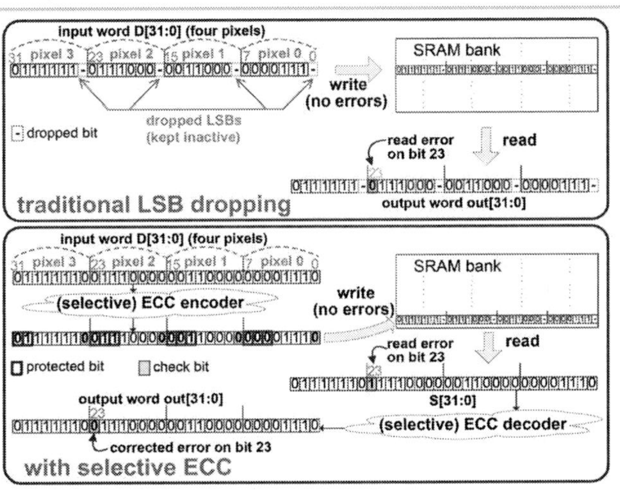

Figure 13.8.3: Operation of selective ECC (LSB employed as check bits of MSBs) as compared to traditional bit dropping.

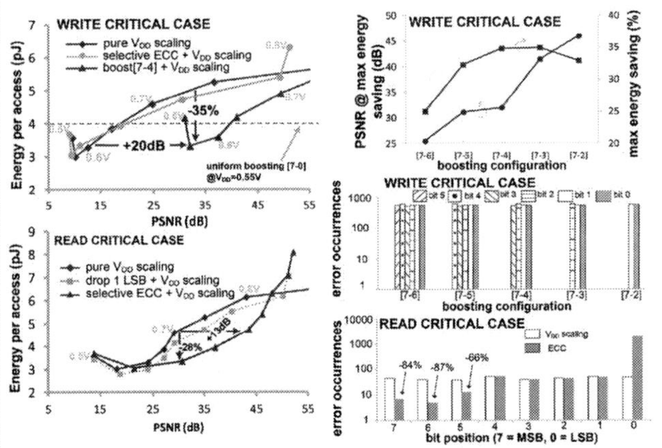

Figure 13.8.4: Measured energy/quality trade-off (left). Energy saving versus boosting configuration (top-right). Selective NBL and ECC suppress errors in MSBs (write: center-right, read: bottom-right).

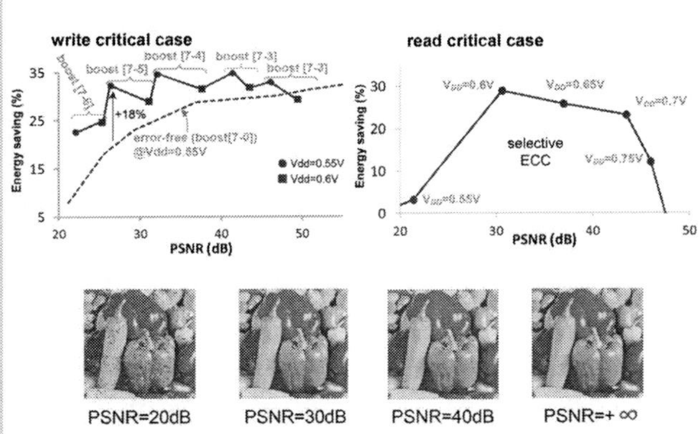

Figure 13.8.5: Measured energy saving in energy-optimal boosting configuration versus PSNR with respect to pure V_{DD} scaling for write-critical and read-critical cases.

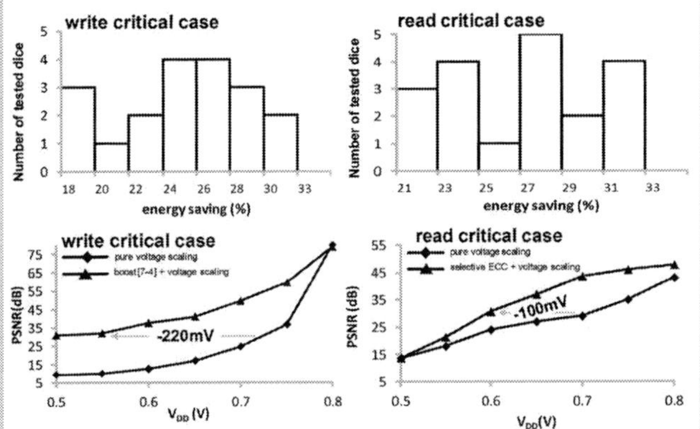

Figure 13.8.6: Energy savings on 19 dice for boost[7-4] (write critical, top-left) and selective ECC (read critical, top-right) with V_{DD} adjusted for PSNR=30dB. For write (read) critical case, average energy saving is 25% (27%). The techniques allow for more aggressive V_{DD} reduction at iso-quality (bottom): -220mV in the write-critical, -100mV in the read-critical case.

978-1-4799-0917-9/14 $31.00 © 2014 IEEE

13

Technology	28nm CMOS
Area	252×202 um^2
Operating voltage	0.5V – 1V
Data retention voltage (DRV) @ 22 °C	325mV
Leakage @ DRV, 22 °C	11uA

Figure 13.8.7: Die micrograph of the 28nm test-chip and data.

Session 14 Overview: *Millimeter-Wave and Terahertz Techniques*

RF SUBCOMMITTEE

Session Chair: *Ullrich Pfeiffer*
University of Wuppertal, Wuppertal, Germany

Session Co-Chair: *Brian Floyd*
North Carolina State University, Raleigh, NC

Efficient generation of mm-Wave and THz signals is a critical need for the enablement of radar, communications, and imaging applications. The first four papers in this session present state-of-the-art solutions for mm-Wave power amplifiers and transmitters at 60, 79, and 28GHz. The final four papers in the session present harmonically-generated 0.25-to-0.5THz sources and arrays in advanced silicon.

14.1 A 0.9V 20.9dBm 22.3%-PAE E-Band Power Amplifier with Broadband Parallel-Series Power Combiner in 40nm CMOS **1:30 PM**

D. Zhao, KU Leuven, Leuven, Belgium

In Paper 14.1, KU Leuven presents a 40nm CMOS power amplifier for licensed E-band that employs series-parallel power combining to achieve 17.5dBm output compression point, 20dBm saturated output power, and 20% peak PAE over 71 to 86GHz.

14.2 A 79GHz Phase-Modulated 4GHz-BW CW Radar TX in 28nm CMOS **2:00 PM**

V. Giannini, imec, Leuven, Belgium

In Paper 14.2, imec demonstrates a 79GHz phase-modulated CW radar transmitter in 28nm CMOS. The circuit architecture uses digital pseudo-random noise modulation at a 4GHz RF bandwidth for a low-power implementation with 10.4% power efficiency.

14.3 A Push-Pull mm-Wave Power Amplifier with <0.8° AM-PM Distortion in 40nm CMOS **2:30 PM**

S. Kulkarni, KU Leuven, Leuven, Belgium

In Paper 14.3, KU Leuven presents a 60GHz push-pull power amplifier in 40nm CMOS suitable for 64-QAM modulation. An inverter-based architecture is used to cancel AM-to-PM conversion and achieve 16.4dBm saturated output power with 23% peak PAE.

978-1-4799-0917-9/14 $31.00 © 2014 IEEE

14.4 A Class F-1/F 24-to-31GHz Power Amplifier with 40.7% Peak PAE, 15dBm OP_{1dB}, and 50mW P_{sat} in 0.13µm SiGe BiCMOS 2:45 PM

S. Y. Mortazavi, Virginia Polytechnic Institute, Blacksburg, VA

In Paper 14.4, Virginia Tech describes a 28GHz class-F/F^{-1} power amplifier in 0.13µm SiGe BiCMOS that achieves 15dBm output compression point, 17dBm saturated output power, and 40% peak PAE.

14.5 A 0.53THz Reconfigurable Source Array with up to 1mW Radiated Power for Terahertz Imaging Applications in 0.13µm SiGe BiCMOS 3:15 PM

U. R. Pfeiffer, University of Wuppertal, Wuppertal, Germany

In Paper 14.5, The University of Wuppertal and IHP present a programmable 0.53THz 16-pixel source array with up to 1mW radiated power in 0.13µm SiGe BiCMOS for active-THz imaging applications. Each asynchronously operated triple-push oscillator delivers up to -12dBm with a 25dBm EIRP.

14.6 A Scalable Terahertz 2D Phased Array with +17dBm of EIRP at 338GHz in 65nm Bulk CMOS 3:45 PM

Y. Tousi, Cornell University, Ithaca, NY

In Paper 14.6, Cornell University presents a lensless scalable 16-element phased array at 338GHz with 17.1dBm EIRP and 0.8mW radiated power in a 65nm CMOS process.

14.7 A 300GHz Frequency Synthesizer with 7.9% Locking Range in 90nm SiGe BiCMOS 4:15 PM

P-Y. Chiang, University of California, Irvine, CA

In Paper 14.7, The University of California, Irvine and The University of California, Davis describe a 300GHz phase-locked loop with 7.9% locking range using a triple-push VCO with an active varactor in a 90nm SiGe BiCMOS process.

14.8 A 247-to-263.5GHz VCO with 2.6mW Peak Output Power and 1.14% DC-to-RF Efficiency in 65nm Bulk CMOS 4:45 PM

M. Adnan, Cornell University, Ithaca, NY

In Paper 14.8, Cornell University shows a 247-to-263GHz VCO with 1.14% DC-to-RF efficiency and 2.6mW peak output power using coupled Colpitts oscillators in a 65nm CMOS process.

978-1-4799-0917-9/14 $31.00 © 2014 IEEE

14.1 A 0.9V 20.9dBm 22.3%-PAE E-Band Power Amplifier with Broadband Parallel-Series Power Combiner in 40nm CMOS

Dixian Zhao, Patrick Reynaert

KU Leuven, Leuven, Belgium

The 71-to-76GHz and 81-to-86GHz bands (known as E-band) exhibit low atmospheric attenuation and are allocated by FCC and CEPT for long-haul transmission. They enable multi-Gb/s fixed-link services such as fiber extension/replacement and cellular backhaul. It is beneficial to have a device operating in both 5GHz bands for high data throughput (due to full usage of the 5GHz band) and low interference (due to the 10GHz spacing between two bands). This requires a PA that delivers uniform gain and output power from 71 to 86GHz. An output power of more than 20dBm is also desired for sufficient link margin to accommodate rain attenuation. These requirements are not satisfied by prior reported 70/80GHz PAs in silicon-based technology [1-4].

This paper reports a fully integrated 40nm CMOS PA that utilizes a broadband parallel-series power combiner to achieve an output power (P_{OUT}) of 20.9dBm with more than 15GHz small-signal 3dB bandwidth (BW_{-3dB}) and 22% PAE at 0.9V supply. The in-band variation of P_{1dB} is only ±0.25dB. This silicon-based PA covers both 71-to-76GHz and 81-to-86GHz bands with uniform gain, output power and PAE.

Transformer-based series combiners are popular in mm-Wave PA design due to their high power transfer efficiency and compact layout. However, two crucial issues limit the efficiency and effectiveness of such structures to combine more than four signal paths (i.e., two differential signal paths) to achieve high output power at mm-Wave. 1) The parasitic interwinding capacitance between the primary and secondary coils will distort the amplitude and phase of the signals to be combined and reduce the combining efficiency. 2) The input impedance in each path of the series combiner decreases in proportion to the number of the combining paths. Considering that the performance of mm-Wave transistors is constrained by the loss of the peripheral intraconnects, the typical value of the transistor width above 60GHz is limited to 100 to 150µm, which leads to an optimum load impedance (Z_{OPT}) of a common-source (CS) differential amplifier in the range of 30 to 50Ω. Therefore, to accomplish beyond 2-way differential combining, either the transistor size needs to be considerably increased for a much lower Z_{OPT} (i.e., causing much higher intraconnect loss) or an additional matching circuit is required to perform the impedance transformation. The above two issues will limit the output power, combining efficiency and bandwidth of the whole network.

In this design, a combination of parallel and series combiners is used, which efficiently combines four differential amplifiers with low insertion loss and at the same time minimizes the impedance transformation ratio of the whole passive network to maximize the bandwidth. Figure 14.1.1 shows the E-band PA schematic that incorporates two unit PAs with a slow-wave T-line-based parallel combiner. Each unit PA consists of two neutralized CS amplifiers and a transformer-based series combiner. The slow-wave T-line in the parallel combiner has a differential-mode characteristic impedance of 79Ω. It transforms the 50Ω load (100Ω seen by each differential T-line) and pad parasitics to an equivalent resistance of 52Ω seen by the two unit PAs. The series combiner reduces the impedance by a factor of 2 and provides the Z_{OPT} (i.e., 28Ω, close to half of 52Ω) for the output stage. With this parallel-series power combining approach, the transistor size of each PA stays optimal for high power gain and efficiency, while four of these PAs are combined efficiently with low impedance transformation ratio as required for broadband operation. Figure 14.1.2 plots the power contours of the output stage at 74 and 83GHz and the input impedance of series combiner (i.e., Z_{OPT} seen by the output stage) from 70 to 95GHz. It is shown that optimum power matching is achieved in the entire band. The load resistance maintains around 28Ω and the load inductance increases at lower frequency which is desired for broadband matching. The insertion loss of the complete parallel-series combiner is less than 1dB from 65 to 86GHz.

The broadband power matching for the drivers will be constrained by the interstage matching network. Therefore, the driver stage is sized sufficiently large to provide enough linear power for the output stage in the 71-to-86GHz band. In this design, each driver drives two output stages and they have the same transistor size (176µm/40nm). The transistor layout is optimized to minimize the intraconnect loss. Compared to the PDK transistor model, the maximum power gain of the neutralized CS amplifier is reduced by only 0.7dB (C_N=43fF). In addition, the parallel T-line-based power divider at the input also lowers the overall input impedance by a factor of 2, which simplifies the input matching design and preserves the bandwidth.

A compact floor plan is crucial to mm-Wave circuits where the placement of all the interconnects has to be considered carefully. These interconnects should not only route the signals between stages but also be employed as part of the matching circuits to minimize the overall loss. Besides, the unwanted magnetic coupling that cannot be avoided in a compact design has to be characterized accurately. Figure 14.1.3 illustrates the layout of the output series combiner, the power divider and associated interconnects. The distance between the combiner and divider is only 34.2µm. There are two ways to connect the power divider and the output stages, referred as interconnects A and B in Fig. 14.1.3. Both differential interconnects act as part of matching circuits and are optimized to minimize the imbalance and associated loss. The major difference of using these two interconnects is the polarity of coupling between the power combiner and power divider. Simulations predict that the complete PA achieves 1.7dB higher gain with interconnect A and it even outperforms the case when no coupling exists.

The PA prototype is fabricated in a 40nm bulk CMOS process. The chip micrograph is shown in Fig. 14.1.7. Due to the compact floor plan, the chip occupies only 0.19mm² including the input and output RF pads. Thanks to the parallel combining/splitting structures at the PA output/input, the two unit PAs are perfectly symmetrical against the central dashed line. This facilitates the TX integration as the magnetic couplings of the two unit PAs to the preceding stages will be cancelled and the VCO pulling can be alleviated.

Figure 14.1.4 shows the measured small-signal S-parameters. The PA achieves a peak S_{21} of 18.1dB at 78.5GHz and small-signal BW_{-3dB} of 15.2GHz (70.3 to 85.5GHz). The S_{11}, S_{22} and S_{12} are lower than -8, -10 and -37dB respectively from 71 to 86GHz. The amplifier is unconditionally stable over the entire measured frequency range (0.1 to 110GHz). Consuming 375mW from a 0.9V supply, the PA has a measured P_{1dB} of 17.4dBm, P_{SAT} of 20.4dBm with 22% PAE_{MAX} at 80GHz. Figure 14.1.5 shows that the PA achieves a measured P_{1dB} of 17.55±0.25dBm, P_{SAT} of 20.55±0.35dBm and PAE_{MAX} of 20.5±1.8% from 71 to 86GHz. The variation of P_{1dB} is even smaller than that of P_{SAT} thanks to the design techniques discussed, which maximize the bandwidth of the output combiners and ensure sufficient power delivered by the drivers. The measured EVM in the 70GHz band (f_c=74GHz) is also shown in Fig. 14.1.5. The input data generated by the AWG and external SSB up-convertor limit the performance. The PA achieves 2Gb/s 16QAM and 5Gb/s QPSK at 12.5 and 13dBm average P_{OUT} respectively with a measured EVM slightly higher than the one directly measured from the setup. For long-term reliability, we operated the PA with 20dBm P_{OUT} at 0.9V for more than 10 hours. No obvious degradations of output power (<0.1dB) and drain current (<1%) were observed.

Figure 14.1.6 summarizes the comparison with the state-of-the-art 70/80GHz PAs in silicon. This work achieves comparable small-signal BW_{-3dB} to a distributed topology [3] that compromises in efficiency and silicon area. Among the PAs in the comparison table, the proposed 0.19mm² CMOS PA achieves highest and nearly uniform P_{1dB}, P_{SAT} and PAE_{MAX} in the 71-to-86GHz band at 0.9V supply.

Acknowledgments:
This work is supported by the ERC Advanced Grant (DARWIN) and Analog Devices Inc., Limerick. The authors would like to thank Mike Keaveney from Analog Devices.

References:
[1] K.-Y. Wang et al., "A 1V 19.3dBm 79GHz Power Amplifier in 65nm CMOS," *ISSCC Dig. Tech. Papers*, Feb. 2012.
[2] J. Oh, B. Ku, S. Hong, "A 77-GHz CMOS Power Amplifier with a Parallel Power Combiner Based on Transmission-Line Transformer," *IEEE Trans. on Microwave Theory and Techniques*, vol. 61, pp. 2662-2669, July 2013.
[3] E. Afshari et al., "Electrical Funnel: A Broadband Signal-Combining Method," *ISSCC Dig. Tech. Papers*, Feb. 2006.
[4] M. Thian, M. Tiebout et al., "A 76-to-84GHz SiGe Power Amplifier Array Employing Low-Loss Four-Way Differential Combining Transformer," *IEEE Trans. on Microwave Theory and Techniques,* vol. 61, no. 2, pp. 931-938, Feb. 2013.

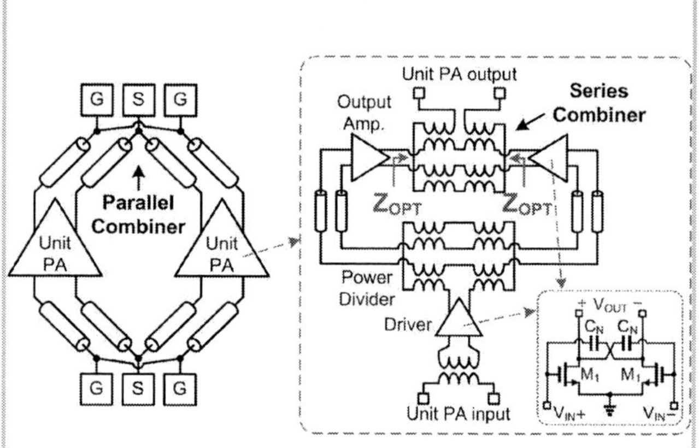

Figure 14.1.1: Schematic of the complete PA and the unit PA.

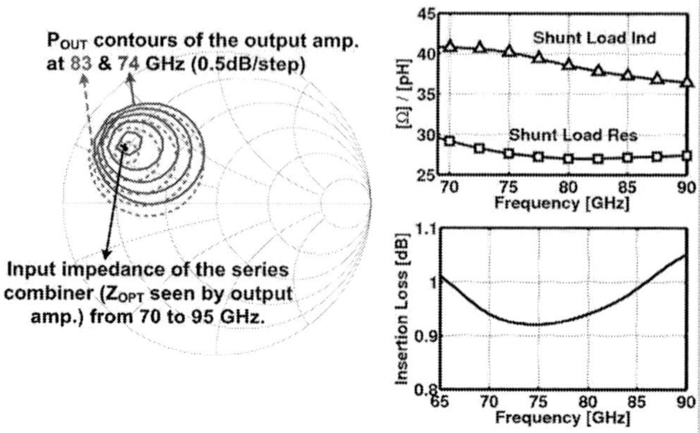

Figure 14.1.2: Input impedance of the series combiner and the simulated insertion loss of the parallel-series combiner (including output pads).

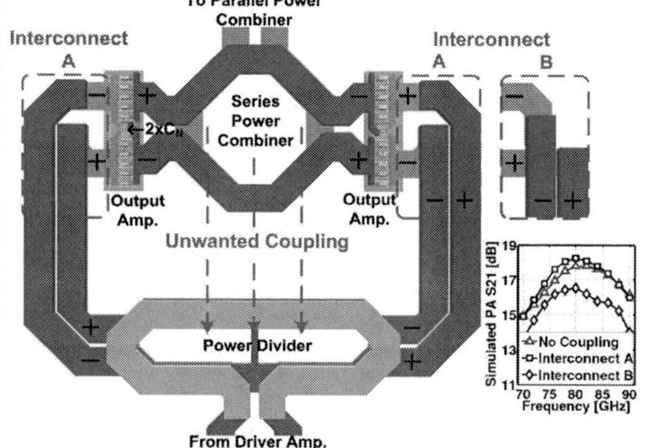

Figure 14.1.3: Unwanted magnetic coupling between series combiner and power divider (supply and ground lines not shown for simplicity but included in the simulation) and simulated S_{21} of the complete PA with different coupling effects.

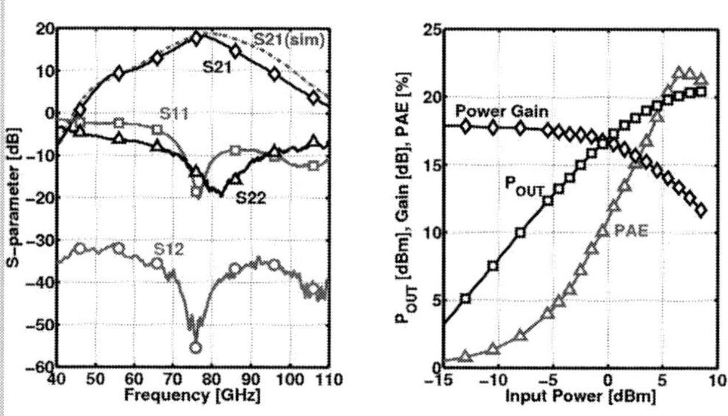

Figure 14.1.4: Measured S-parameters vs. frequency and measured power gain, output power, PAE vs. input power at 80GHz.

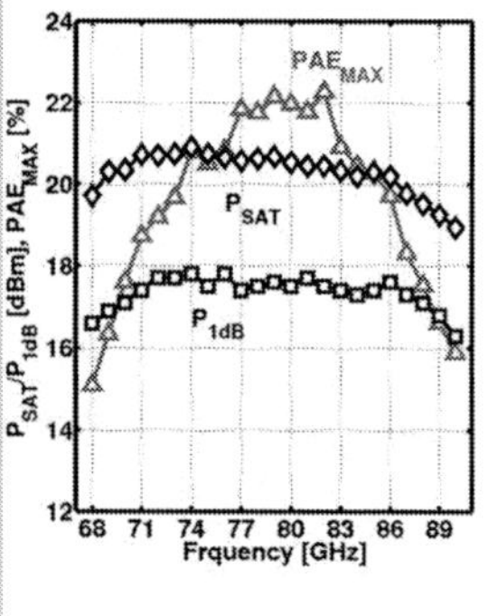

Modulation	QPSK	16QAM
Constellation [2048 points]	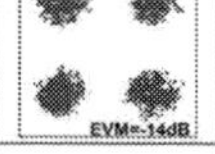 EVM=-14dB	 EVM=-18dB
Spectrum [dB]	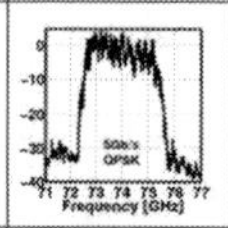	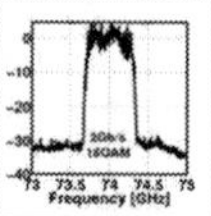
Data Rate	5Gb/s	2Gb/s
EVM	-14dB	-18dB
EVM of setup	-16dB	-18.5dB
Average P_{OUT}	13dBm	12.5dBm

Figure 14.1.5: Measured P_{SAT}, P_{1dB} and PAE_{MAX} vs. frequency, and measured EVM (w/ & w/o DUT) in 70GHz band (f_c=74GHz).

	This Work	K.-Y. Wang ISSCC12 [1]	J. Oh TMTT13 [2]	E. Afshari ISSCC06 [3]	Y. Zhao JSSC12	A.Y.-K.Chen TMTT13	M. Thian TMTT13 [4]
Technology	40nm CMOS	65nm CMOS	65nm CMOS	130nm SiGe	130nm SiGe	180nm SiGe	180nm SiGe
V_{DD} [V]	0.9	1.0	2	-2.5/0.8	2.5	4	3.2
Freq. [GHz]	70.3-85.5	79	77	85	84	83	78
Max. S_{21} [dB]	18.1	24.2	20.9	9	27	25	18.3
S_{21} BW_{-3dB} [GHz]	15.2	10	N/A	>18‡	8	9.6	8.9 ($BW_{-2.7}$)
P_{1dB} [dBm]	17.8	16.4 #	13	N/A	16	12.5	12.5
P_{1dB} variation [dB] Freq. range [GHz]	0.5 71-87	N/A	2 75-82	N/A	2 75-90	1 75-88	N/A
P_{SAT} [dBm]	20.9	19.3 #	15.8	21	18	14.7	14
P_{SAT} variation [dB] Freq. range [GHz]	0.7 69-86	N/A	2.3 75-82	3 74-90	1.5 75-90	1 75-88	4 76-84
PAE_{MAX} [%]	22.3	19.2 #	15.2	4 (Drain Eff)	9	8.1	2
Area [mm^2]	0.19	0.855	0.21	2.4†	0.68†	0.34	0.85
Topology* Path combined	2-stage CS 4-way diff.	4-stage CS 8-way	2-stage CA 2-way diff.	DA /CA 4-way	3-stage CB 2-way diff.	2-stage CA 2-way	2-stage CA 4-way diff.

The loss of the on-chip output balun (> 1.6dB) was de-embedded from the measured P_{SAT}, P_{1dB} and PAE_{MAX}.
‡ The gain is only shown from 72 to 90 GHz. † Include the area of DC pads.
* CS: common source, CB: common base, CA: cascode, DA: distributed amplifier.

Figure 14.1.6: Comparison of PAs in 70/80GHz bands.

978-1-4799-0917-9/14 $31.00 © 2014 IEEE

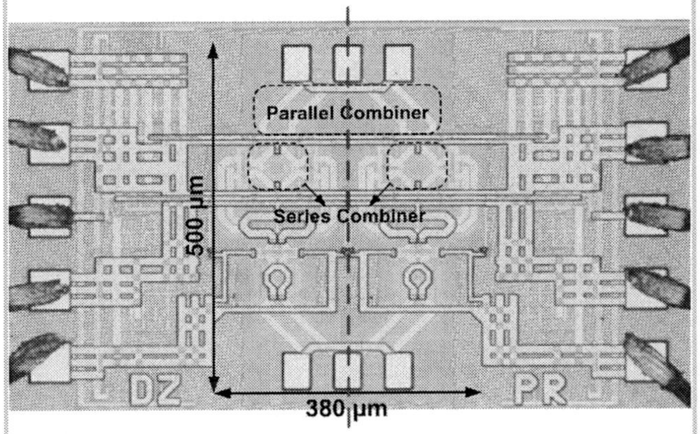

Figure 14.1.7: Die micrograph. The two unit PAs are symmetrical against the central dashed line and their magnetic coupling to the previous stages will be cancelled in an integrated TX.

14.2 A 79GHz Phase-Modulated 4GHz-BW CW Radar TX in 28nm CMOS

Vito Giannini[1], Davide Guermandi[1], Qixian Shi[1,2], Kristof Vaesen[1], Bertrand Parvais[1], Wim Van Thillo[1], André Bourdoux[1], Charlotte Soens[1], Jan Craninckx[1], Piet Wambacq[1,2]

[1]imec, Leuven, Belgium,
[2]Vrije Universiteit Brussel, Brussels, Belgium

Millimeter-Wave radar sensors perform accurate and robust remote motion detection with short latencies. Requirements are especially challenging for person detection [1]: a bandwidth higher than 1.5GHz is needed to achieve finer than 10cm depth resolution. Wide field of view combined with high angular resolution are also needed, which at mm-Waves translate into large and power-hungry antenna arrays. In classical FMCW radars [2-6], when bandwidth requirements exceed 1GHz, depth resolution is eventually limited by the linearity of the frequency slope of the FM PLL.

This paper describes a phase-modulated Continuous-Wave (CW) Radar TX that operates in the 79GHz band. With a sampling rate of 2GS/s and 4GHz RF bandwidth, we target a resolution of 7cm. We rely on a Pseudo-Random-Noise (PRN) maximum length sequence of 511 bits and BPSK modulation targeting three benefits. First, SNR is increased by 27dB thanks to compression gain. Second, a code-domain MIMO radar is now possible yielding both higher angular resolution for a given number of antenna elements and higher combined output power than a similar time-domain MIMO radar. Finally, interference rejection techniques can be applied in the digital domain. Such radar architecture is digital intensive so that a CMOS technology allows for a more power-efficient implementation compared to equivalent SiGe state-of-the-art [7-8]. This paper proposes the first 79GHz radar TX in 28nm CMOS.

Figure 14.2.1 shows the block diagram of the radar TX: a 15.8GHz input is multiplied by 5 by a Sub-Harmonically-Injection-Locked Oscillator (SH-ILO) VCO. After amplification, the 79GHz VCO output is fed into an RF modulator, which multiplies the signal with a PRN sequence. As the baseband signal is a digital stream, we use it to drive the switching gates: this allows significant power savings because we do not need a rail-to-rail 79GHz signal. The chosen PRN code is supplied at baseband with a scalable sampling frequency up to 2GS/s. However such architecture generates side-lobes, which do not comply with the spectral requirements [1] when transmitting at the maximum allowed EIRP of -3dBm/MHz. Differently from [7-8], we solve this problem by rejecting the closest side-lobe and by filtering the others. Overall, this lowers the out-of-band emissions and allows to maximize EIRP. The modulated signal goes through a 3-stage power amplifier. Figure 14.2.1 shows the expected output with and without side-lobe rejection: the spectrum is centered around 79GHz with spectral components every Fc/Lc, where Fc is the BPSK chip rate and Lc is the PRN code length.

Figure 14.2.2 shows the local oscillator section: it consists of a 15.8GHz differential-input buffer stage, a core 5th-order SH-ILO VCO and a programmable gain amplifier. Compared to a direct 79GHz LO generation, this choice requires a simpler 15.8GHz integer-N PLL [7]. Furthermore, compared to a 3rd-order SH-ILO, we can more efficiently distribute the LO in phased-array systems. On the other hand, achieving sufficient locking range for robustness becomes more complicated. Classical SH-ILO VCOs using inductive-peaking buffers are both area- and power-consuming and require a nonlinear core transistor biased in weak inversion, which is too large to be efficiently driven. In this work, we rely on simple CMOS inverters (Fig. 14.2.2) to generate a harmonic-rich square wave at 15.8GHz: compared to an inductive-peaking buffer. The fundamental input amplitude will be smaller but the odd-order harmonic components will be much bigger. Two mechanisms on the SH-ILO input transistors M_{I1} ensure a wide-locking range on the 5th harmonic: the linear behavior amplifies the 5th harmonic of the input square wave whereas the nonlinear behavior contributes to further increase the 5th-order component also through 3rd-order intermodulation of the fundamental and the 3rd harmonic. As the three components generated from the linear and non-linear mechanisms are approximately in-phase, the current-sum vector has a larger amplitude and the locking range increases accordingly. Transistors M_{I2} partially set the DC current of the cross-coupled pair, trading VCO core amplitude for locking range where needed.

The modulator core is sketched in Fig. 14.2.3 together with the baseband and the power amplifier simplified schematics. The input signal from the SH-ILO VCO is applied to the sources of the mixer switches M_{MX} via a transformer. A second transformer at the drains of M_{MX} collects the output current and transfers it to the PA. The current in the switching pairs is controlled by a current DAC. With a straightforward square-wave BPSK modulation, side-lobe peaks are expected to be 13.4dB and 17.9dB below the main-lobe peak at about ±3/2 F_c & ±5/2 F_c respectively. In this work, out-of-band emissions are reduced by using a side-lobe suppression technique in combination with pulse-shaping. The mixer is split into two parallel paths, half of which is fed with the digital sequence that is delayed (red in Fig. 14.2.3) by one-third of the bit period using a programmable delay line (Fig. 14.2.3). The remaining side-lobes are attenuated by means of analog filtering implemented as programmable degeneration of the final buffer driving the mixer. The modulator provides a simulated voltage gain up to 2dB that can be lowered by means of a programmable resistor at the high impedance output. The PA is based on a cascade of three transformer-coupled pseudo-differential common-source amplifiers with neutralization capacitances C_{nPA} that improve gain, bandwidth and stability. The PA is optimized by load-pull simulations to provide up to +9dBm output P_{1dB} and 17% PAE. The constant-envelope phase-modulation scheme does not contribute to spectral regrowth allowing us to transmit at saturated power.

The prototype has been fabricated in a 28nm HPM CMOS technology with a 0.9V supply voltage and consumes a total 121mW (25mW SH-ILO, 1mW baseband at 2GS/s, 3mW modulator, 92mW PA). The design was performed using custom RF models characterized at mm-Wave. The chip micrograph is shown in Fig. 14.2.7. The digital and DC IOs are bonded on PCB, whereas the 15.8GHz input and the 79GHz output RF GSG pads are probed with 150μm pitch Cascade Infinity probes. The SH-ILO has a locking range of 5GHz around 79GHz. By tuning its capacitor bank C_{ILO}, we can always lock the VCO when an input LO frequency between 14.2GHz and 16.8GHz is applied: Figure 14.2.4-(a) shows the TX output power versus the VCO locked frequency for three different samples. The upconverted RF bandwidth is always higher than 13GHz and centered around 78GHz. This allows robust operation in both the 79GHz and 77GHz bands. Within this band, the output power is higher than +11dBm. Figure 14.2.4-(b) shows that the SH-ILO phase noise follows closely the phase-noise of an R&S SMR40 input source with a 14dB offset. Figure 14.2.5 shows the modulated signal with and without out-of-band lobe rejection. The integrated output power over the 4GHz RF bandwidth is higher than +11.5dBm. The 3rd-order side-lobe drops from -13dBc to about -25dBc. Overall, the out-of-band emissions drop to about -20dBc with a negligible power consumption penalty. Such results allow to still comply with the spectral mask requirements [1] even with a 4-TX code-domain MIMO system. Figure 14.2.6 shows that this work achieves at least 2 times higher bandwidth compared to published FMCW TX and significantly improved power efficiency (Pout/Pdc). This work paves the way towards high-resolution power-efficient code-domain MIMO radars at 79GHz.

Acknowledgements:
The authors thank A. Medra, S. Brebels, K. Raczkowski, M. Libois, L. Pauwels, I. Ocket, the imec BODI team, INVOMEC and Integrand Software for EMX.

References:

[1] J. Hasch, et al., "Millimeter-Wave Technology for Automotive Radar Sensors in the 77GHz Frequency Band," *IEEE MTT*, vol. 60, no. 3, pp. 845-860, 2012.
[2] Y. Kawano, et al., "A 77GHz Transceiver in 90nm CMOS," *ISSCC Dig. Tech. Papers*, 2009.
[3] T. Mitomo, et al., "A 77GHz 90nm CMOS Transceiver for FMCW Radar Applications," *IEEE JSSC*, vol. 45, pp. 928-937, Apr. 2010.
[4] J. Lee, et al., "A Fully-Integrated 77-GHz FMCW Radar Transceiver in 65-nm CMOS Technology," *IEEE JSSC*, vol. 45, pp. 2746 -2756, Dec. 2010.
[5] W. Wu, et al., "A mm-Wave FMCW Radar Transmitter Based on a Multirate ADPLL," *IEEE RFIC*, 2013.
[6] K.-H. To, et al., "A 76-81GHz Transmitter with 10dBm Output Power at 125°C for Automotive Radar in 65nm Bulk CMOS," *IEEE CICC*, 2011.
[7] S. Trotta, et al., "A 79GHz SiGe-Bipolar Spread-Spectrum TX for Automotive Radar," *ISSCC Dig. Tech. Papers*, 2007.
[8] E. Ragonese, et al., "A Fully Integrated 24GHz UWB Radar Sensor for Automotive Applications," *ISSCC Dig. Tech. Papers*, 2009.

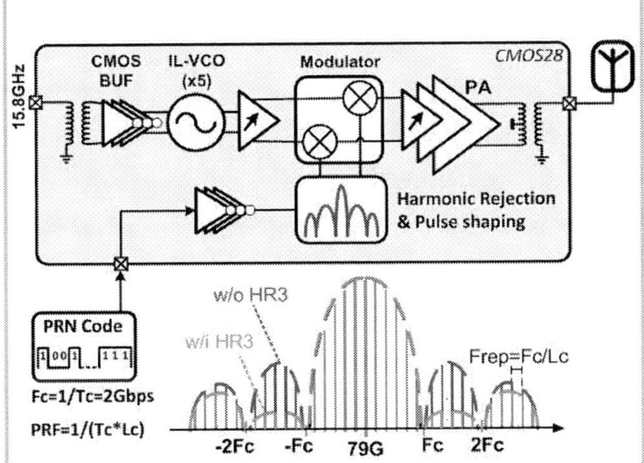

Figure 14.2.1: Block diagram of the proposed phase-modulated radar transmitter.

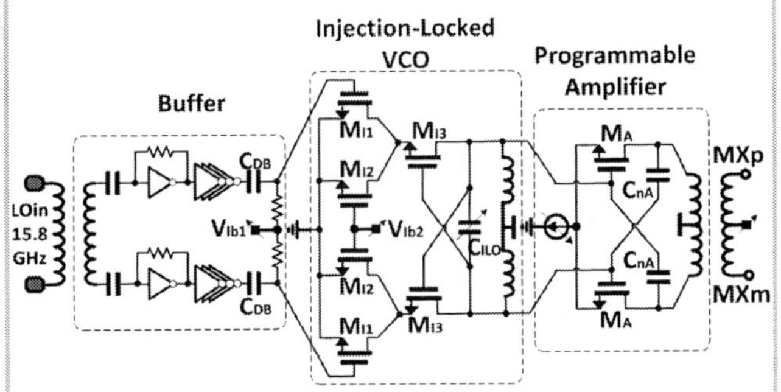

Figure 14.2.2: Schematic of the sub-harmonically-injection-locked VCO and programmable buffer.

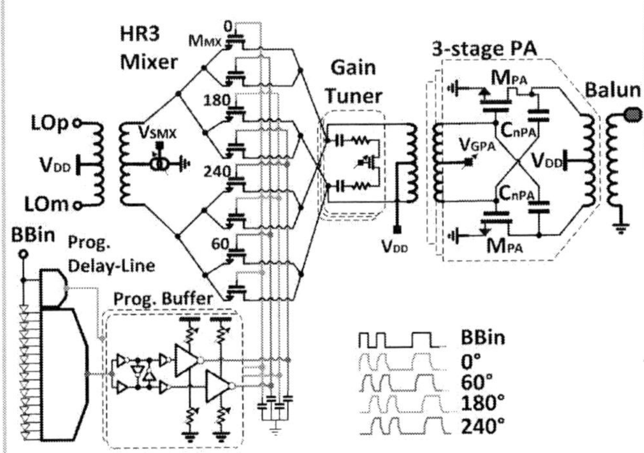

Figure 14.2.3: Schematic of the side-lobe-rejection baseband-mixer and the multi-stage PA.

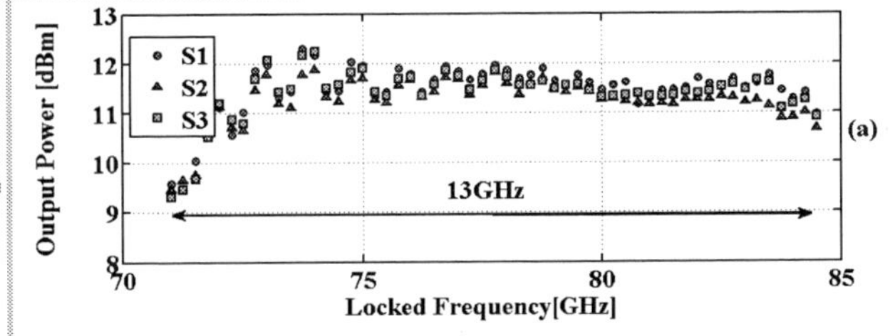

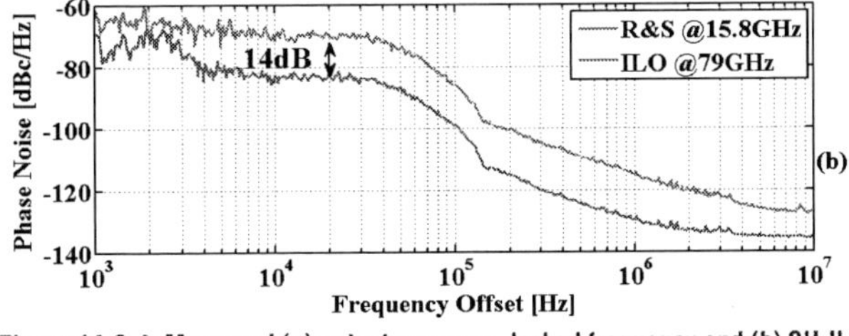

Figure 14.2.4: Measured (a) output power vs. locked frequency and (b) SH-ILO phase-noise.

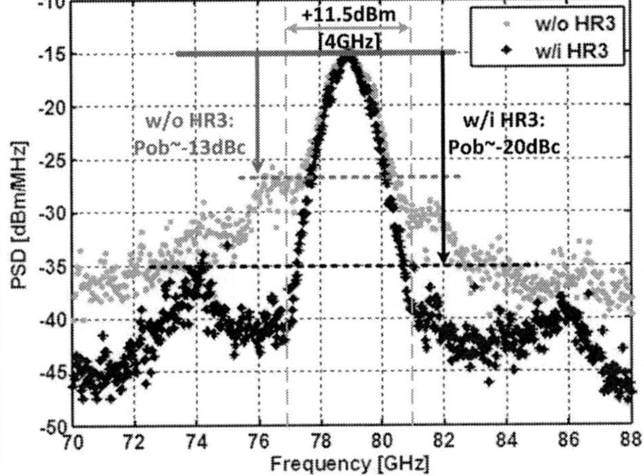

Figure 14.2.5: PRN-modulated output spectrum @ 79GHz with and without side-lobe rejection and filtering.

Ref	Tech	VDD [V]	BW [GHz]	Radar	Fc [GHz]	Pout [dBm]	Pdc [mW]	TX eff [%]
[2]	CMOS90	1.2	0.2	FMCW	77	6.3	660	0.6
[3]	CMOS90	1.2	0.6	FMCW	77	-2.8	406	0.13
[4]	CMOS65	1.2	0.7	FMCW	77	5.1	188	1.7
[5]	CMOS65	1.2	1.2	FMCW	62	5	89	3.5
[6]	CMOS65	1	n/a	n/a	77/79	13.5	420	5.3
[7]	SiGe 200G	5.5	1.235	PMCW	79	1.5	4100	0.03
[8]	SiGe 230G	2.5	2.1	PMCW	24	3	80	3.3
This work	CMOS28	0.9	2	PMCW	77/79	>11	121	>10.4

Figure 14.2.6: Performance summary and comparison with state-of-the-art 24/62/77/79GHz circuits.

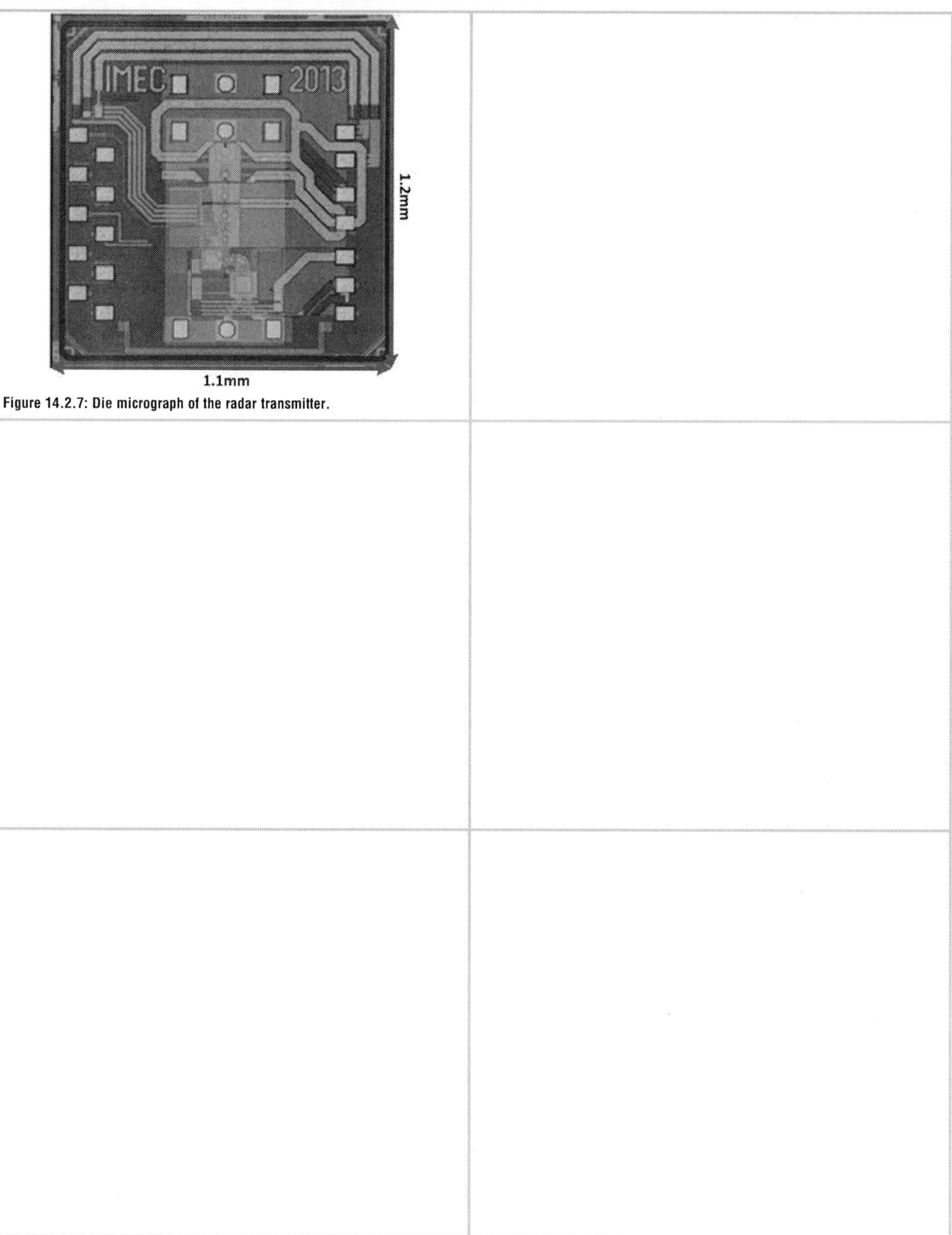

Figure 14.2.7: Die micrograph of the radar transmitter.

14.3 A Push-Pull mm-Wave Power Amplifier with <0.8° AM-PM Distortion in 40nm CMOS

Shailesh Kulkarni, Patrick Reynaert

KU Leuven, Leuven, Belgium

Millimeter-Wave standards like IEEE 802.15.3c and the new 802.11ad have classifications of their PHY to support single-carrier mode and more complex OFDM mode (high-speed interface) with high peak-to-average ratio (PAPR). To improve the efficiency of power amplifiers (PA), the trend is towards Class-AB and Class-B PAs that exhibit better energy efficiency compared to Class-A. However, Class-AB and -B biasing brings along large amplitude-to-phase-modulation (AM-PM) distortion which degrades EVM and ACPR. At the same time, PMOS transistors become attractive in nanometer CMOS as their f_{MAX} exceeds 140GHz. This makes it possible to use both NMOS and PMOS transistors at mm-Wave frequencies. This paper presents a 60GHz complementary Push-Pull PA, using both NMOS and PMOS transistors. An inverter-like architecture which uses both PMOS and NMOS results in the cancellation of AM-PM distortion which is particularly important in high-fidelity amplification of OFDM systems and high-order modulation schemes like 16- and 64-QAM, which are very sensitive to phase distortion. Furthermore, the complementary nature allows deep Class-AB operation, giving a high power efficiency at power back-off comparable to state-of-the-art 60GHz PA structures based on NMOS only.

Figure 14.3.1 shows the complete architecture of the PA. At the output, a transformer-based combiner is employed, which has two differential excitation ports on the primary side and a standard secondary winding connected to the output pads [1]. The combiner has an insertion loss of only 1dB at 62GHz. The PA consists of two stages of Push-Pull unit amplifiers, making use of both NMOS and PMOS, progressively sized with interstage transformers. A common—gate (CG) buffer stage is used at the input as it provides lower input impedance and can be matched to 50Ω easily [2]. The CG buffer only compensates for the loss in the input power divider.

The Push-Pull PA consists of a differential inverter-like structure shown in Fig 14.3.2. To enable deep Class-AB operation and cancellation of the AM-PM, the bias levels of the PMOS and NMOS are set independently. Therefore, an interstage power splitter that consists of a transformer with two secondary windings with separate DC center-tap access is designed as shown in Fig 14.3.2. As the PMOS is sized 30% larger than the NMOS, its input susceptance is larger than the NMOS. Hence it is tuned by the inner secondary winding. The input impedance of the PMOS and NMOS are different and they are matched to the previous stage by optimizing the width and overlap of the two secondary windings with the primary winding. In each Push-Pull unit PA, individual neutralization capacitors are adopted for the PMOS and NMOS (C_P,C_N) separately so as not to disturb their DC bias. This helps to improve the stability and reverse isolation at mm-Wave frequencies.

The two major benefits of using complementary Push-Pull PAs is their ability to minimize AM-PM distortion and their low quiescent current for a given RF output power. One of the primary causes for AM-PM distortion is the nonlinear input capacitance of the transistor [3]. As the input RF amplitude increases the average value of C_{GS} will change, thus causing AM-PM distortion. To counter this type of distortion, additional circuitry is often added to balance the phase change [3,4]. However, the changes of input capacitance for PMOS and NMOS are in opposite directions because of their complementary nature as shown in Fig. 14.3.3. As such, both the NMOS and PMOS show a fairly large amount of AM-PM distortion which is often a problem in mm-Wave NMOS-only deep Class-AB PAs [5]. But when the NMOS and PMOS drain currents are summed up, the distortion is cancelled and the effective AM-PM can be minimized up to and beyond the 1dB compression point (P_{1dB}). This is a major benefit of the proposed PA, which enables amplification of large PAPR signals like 64-QAM with low EVM. Figure 14.3.3 also shows the measured AM-AM and AM-PM at 63GHz which is below 0.25° up to P_{1dB}. This demonstrates the excellent phase linearity of the PA, which is required to support a large number of constellation points.

The last-stage transistors are biased very close to their threshold voltage for deep Class-AB operation. As a result the quiescent current consumption is small for low output power and goes up only when more RF output power is to be delivered. This change in current is more than three times for low-to-high RF output power levels as shown in Fig. 14.3.4. This second benefit is a direct manifestation of the complementary behavior of the Push-Pull PA. Because of this, the measured peak drain efficiency (DE) is larger than 40% as shown in Fig. 14.3.4.

The prototype PA is fabricated in a 0.9V 40nm GP 1P10M digital CMOS process. The 60GHz power measurement results are shown in Fig. 14.3.4. The quiescent voltage at the drains of the Push-Pull stage are at half of their supply voltage. Due to stacking of two transistors in the Push-Pull stage, their supply is maintained at 1.8V. In this configuration, the transistors face similar stress as conventional NMOS-only-based PAs, whereas the CG buffer stage uses 0.9V supply for reliable operation. Under these conditions, the measured gain is 22.4dB, P_{1dB} is 13.9dBm and the saturated output power (P_{SAT}) is 16.4dBm at 63GHz. The measured peak DE is 40.9%, peak PAE is 23% and PAE at P_{1dB} is 18.9%. The DC power profile clearly shows that the PA operates in deep Class-AB mode. Figure 14.3.4 also shows the individual current consumption of all the stages. The DC power consumption more than halves in the low-power region compared to saturated power. The PA maintains P_{SAT} above 15dBm, P_{1dB} above 10dBm and peak PAE above 16% in the band from 59 to 67GHz.

Figure 14.3.5 shows the measured AM-PM distortion at P_{1dB}, which is below 0.8° across the band of 59 to 67GHz. It also shows the measurement result when a high-order constellation modulated signal is applied to the PA. With a 64-QAM signal at 63GHz, an EVM of -25.2dB is measured for a data-rate of 3Gb/s with 7dBm average output power. This signal uses a raised-cosine shaped filter with a rolloff factor of 0.35 and has 8.1dB of PAPR. Achieving such wideband linearity is due to the complementary Push-Pull technique that minimizes AM-PM distortion.

Figure 14.3.6 summarizes the comparison with state-of-the-art mm-Wave PAs employed in transmitters. The proposed PA, which employs a complementary Push-Pull technique, uses AM-PM cancellation and is capable of amplifying 64-QAM signal and at the same time maintaining state-of-the-art PAE at 5dB backoff from P_{1dB}. Figure 14.3.7 shows the die photo of the PA with a total area of $0.4mm^2$ including the pads, while the active area is only $0.0812mm^2$.

Acknowledgments:
The authors thank Prof. D. Schreurs and F. Daenen from KU Leuven for their support during measurements.

References:

[1] J. Chen, A. Niknejad, "A Compact 1V 18.6dBm 60GHz Power Amplifier in 65nm CMOS," *ISSCC Dig. Tech. Papers*, pp. 432-433, Feb. 2011.
[2] D. Zhao, S. Kulkarni, P. Reynaert, "A 60-GHz Outphasing Transmitter in 40-nm CMOS," *IEEE J. Solid-State Circuits*, vol.47, pp. 3172-3183, Dec. 2012.
[3] Y. Palaskas et al., "A 5GHz 20dBm Power Amplifier with Digitally Assisted AM-PM Correction in a 90nm CMOS Process," *IEEE J. Solid-State Circuits*, vol.37, pp.1757-1763, Aug. 2006.
[4] J. Chen et al., "A Digitally Modulated mm-Wave Cartesian Beamforming Transmitter with Quadrature Spatial Combining," *ISSCC Dig. Tech. Papers*, pp. 232-233, Feb. 2013.
[5] D. Zhao, P. Reynaert, "A 60-GHz Dual-Mode Class AB Power Amplifier in 40-nm CMOS," *IEEE J. Solid State Circuits*, vol.48, pp. 2323-2337, Oct. 2013.
[6] V. Vidojkovic et al., "A Low-Power Radio Chipset in 40nm LP CMOS with Beamforming for 60GHz High-Data-Rate Wireless Communication," *ISSCC Dig. Tech. Papers*, pp. 236-237, Feb. 2013.
[7] K. Okada et al., "A 60-GHz 16QAM/8PSK/QPSK/BPSK Direct-Conversion Transceiver for IEEE802.15.3c," *IEEE J. Solid-State Circuits*, vol.46, pp. 2988-3004, Dec. 2011.
[8] A. Siligaris et al., "A 65nm CMOS Fully Integrated Transceiver Module for 60GHz Wireless HD Applications," *ISSCC Dig. Tech. Papers*, pp.162-164, Feb. 2011.

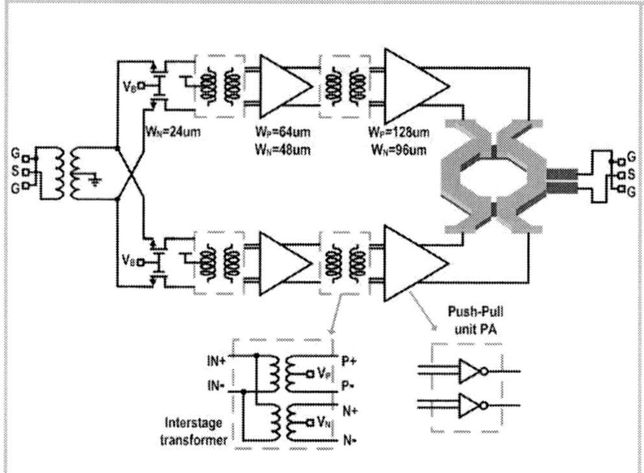

Figure 14.3.1: Architecture of the complementary Push-Pull power amplifier along with the output power combiner.

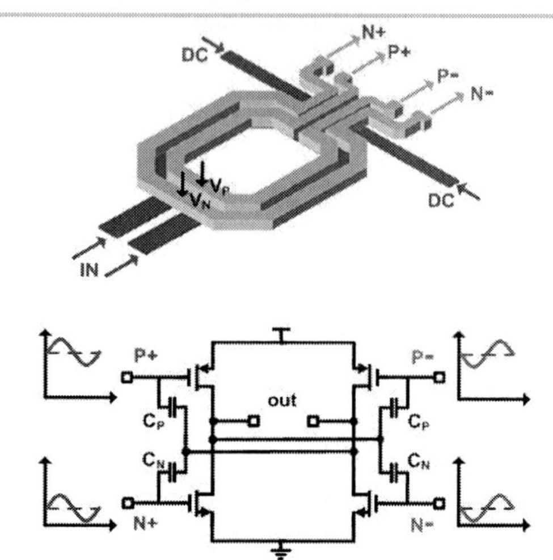

Figure 14.3.2: Layout of the interstage transformer; schematic of the complementary Push-Pull unit PA.

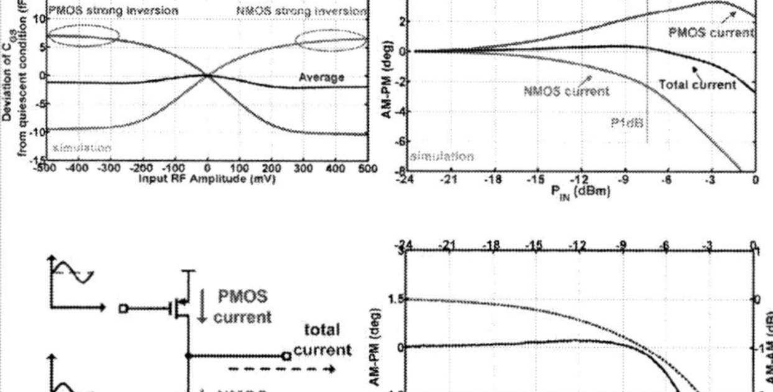

Figure 14.3.3: Variation of C_{GS} due to input RF amplitude; and its impact on AM-PM; measured AM-AM and AM-PM at 63GHz.

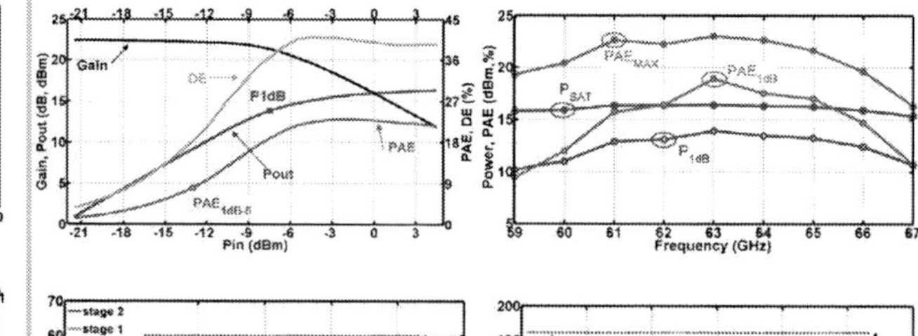

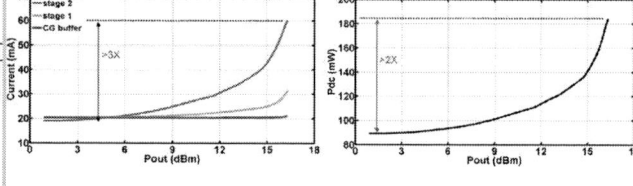

Figure 14.3.4: Measured gain, output power, DE and PAE over the band 59 to 67GHz; DC current and power consumption measured at 63GHz.

Figure 14.3.5: Measured AM-PM at P_{1dB} across the band 59 to 67GHz, 64-QAM constellation at a data-rate of 3Gb/s at 63GHz and 7dBm P_{AVG}; modulated signal performance.

	This work	[2]	[6]	[7]	[8] (w/o offchip pa)	[1]	[5]	
Tech (nm)	**40**	40	40	65	65	65	40	
Gain (dB)	**22.4**	26	22.5	18.3	16.4	20.3	17	
P_{SAT} (dBm)	**16.4**	15.6	10	10.9	13	18.6	17	
P_{1dB} (dBm)	**13.9**	15.6	8	9.5	8	15	13.8	
PAE_{max} (%)	**23**	25	22.5	8.8	8	15.1	30.3	
PAE_{1dB} (%)	**18.9**	25	16	< 8.8	< 8	6.8	21.6	
PAE_{1dB-6} (%)	**8**	10	7.4	2.5	NA	2	8.4	
AM-PM (deg)	**0.2°**	15°	-	-	-	-	0.2° AB	> 3° deep AB
modulated signal	**QAM-64**	QAM-16	QAM-16	QAM-16	OFDM QAM16	-	-	

Figure 14.3.6: Comparison with state-of-the-art 60GHz PAs with modulated signal measurements.

978-1-4799-0917-9/14 $31.00 © 2014 IEEE

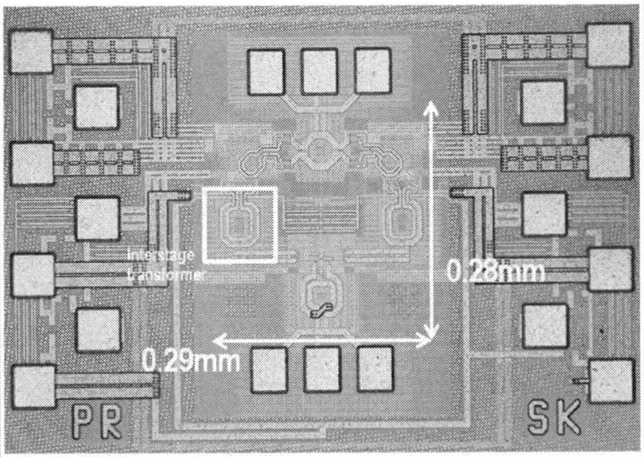

Figure 14.3.7: Die micrograph.

14.4 A Class F-1/F 24-to-31GHz Power Amplifier with 40.7% Peak PAE, 15dBm OP_{1dB}, and 50mW P_{sat} in 0.13µm SiGe BiCMOS

Seyed Yahya Mortazavi, Kwang-Jin Koh

Virginia Polytechnic Institute, Blacksburg, VA

The output power of power amplifiers (PAs) can be increased by using PA arrays and by combining individual PA output powers either on chip with linear combiners or in free space with antenna arrays [1,2]. Therefore, the output power of a PA element can be compromised by the array size. The overall array's power-added-efficiency (PAE) is predominantly determined by the unit PA, which is the most critical performance metric to contain thermal and reliability issues, especially in highly integrated large PA arrays, since thermal density will be proportional to the increase of power density. Class-F and Class-F^{-1} topologies are promising candidates for a high PAE at microwave and mm-Wave frequencies because on-chip high-Q passive components with a small form-factor will be readily available for a tuned load to terminate multiple harmonics appropriately [3]. Maintaining a high efficiency over a wide bandwidth, however, is challenging because of the narrowband nature of the high-Q networks. This paper presents a highly efficient Class-F^{-1}/F PA in 0.13µm SiGe BiCMOS achieving 39.3-40.7% PAE over 25 to 30GHz. In the PA, integrated on-chip passive load networks operate cooperatively to shape the load impedance for an in-band mode transition from Class-F^{-1} to Class-F, maintaining greater than 36.3% PAE over the entire operation band (24 to 31GHz, 25.5% of fractional bandwidth).

In the PA schematic shown in Fig. 14.4.1, the Q_1 is sized for a peak-f_T current density with respect to the *rms* current at target P_{sat} (50mW) and biased at a Class-AB point (V_{BE}=0.85V, I_{CE}=9mA). The bias path provides less than 300Ω seen from the base of Q_1, which increases effective collector-emitter breakdown voltage over 5V (BV_{CEO}=1.7V, BV_{CBO}=5.5V), allowing 2.2V supply voltage (V_{CC}) with a safe margin. The C_M, L_M and C_{Pi} constitute a π-matching network and provide S_{11} better than -10dB over 23 to 31GHz. The C_M absorbs about 18fF of input pad capacitance and the C_{Pi} includes C_π and Miller capacitance of C_{bc} of Q_1. Interconnects are characterized with an EM-field solver and absorbed in the passive network. A small resistor of 2.5Ω is in series with Q_1 to guarantee stability at all frequencies.

For Class-F/F^{-1} operations, the PA's load networks need to provide a low/high impedance for even harmonics and a high/low impedance for odd harmonics, while keeping an optimum impedance for fundamental inputs. The load design is mainly focused on shaping appropriate load impedance up to the 3rd harmonics of the input band because of insignificant harmonic power beyond the 3rd-order of input frequencies. In Fig. 14.4.1, the L_1, L_2, C_1 and C_{Po} form a multi-resonance parallel impedance load (Z_{PL}), and L_{H2}, C_{H2} and load-resistance R_L comprise a series-impedance load (Z_{SL}) resonating at the 2nd harmonics of the input frequencies. The C_{Po} is a nonlinear capacitance: namely, at fundamental input frequencies and their 3rd harmonics, the L_{H2}-C_{H2} tank provides a low impedance and a collector parasitic capacitance (32fF) plus output-pad capacitance (18fF) comprise the C_{Po}, while at the 2nd harmonics the C_{Po} is mainly set by the collector parasitic capacitance because of an impedance isolation from the load by the L_{H2}-C_{H2} tank. The collector-node capacitance is estimated through large-signal load-pull simulations including layout parasitic capacitance.

The impedance change of the L_2-C_1 tank over harmonic frequencies plays a key role in modulating the load impedance to transfer the PA's operational mode continuously from Class-F^{-1} to Class-F over the increase of input frequencies. Figure 14.4.2 shows magnitude response of the load impedances. At lower frequencies, the Z_{PL} can be approximated to an LC-tank composed of L_1+L_2 and C_{Po}, and its resonant frequency is set to around midpoint between the input band and the 2nd-harmonic band (48 to 62GHz) so that it can provide a high impedance at both bands (Fig. 14.4.2-(1)). At the center of the 3rd-harmonic band (72 to 93GHz), the dominant impedance of the L_2-C_1 tank is capacitive and can be approximated to C_{eq}≈25fF, which resonates with L_1 making an impedance null at around 82GHz (Fig. 14.4.2-(2)). As frequency increases further, after passing the series resonance, Z_{PL} becomes inductive, equivalently L_{eq}≈70pH, and makes a parallel resonance with C_{Po}, thereby creating an impedance peak at the edge of the 3rd–harmonic band (Fig. 14.4.2-(3)). Taking into account the Z_{SL} (Fig. 4.4.2-(4)), the magnitude of overall composite impedance $|Z_{TL}|=|Z_{SL}//Z_{PL}|$ is displayed as a solid line in Fig. 14.4.2. At the fundamental frequencies the $|Z_{TL}|$ maintains a constant impedance close to 50Ω. However, the $|Z_{TL}|$ trajectory exhibits impedance maxima at the 2nd-harmonic band and at the edge of the 3rd-harmonic band, while experiencing a steep impedance null at the center of the 3rd-harmonic band. Consequently, the PA operation will be dominated by Class-F^{-1} mode until the 3rd-harmonic impedance reaches its minimum point. After passing the null point, the 3rd-harmonic impedance starts to increase whereas the 2nd-harmonic impedance decreases, which gradually changes the operation mode from Class-F^{-1} to Class-F as the input frequency increases to 31GHz. The optimum load is set to be 42 to 45Ω close to the fundamental impedance of the composite load networks, which eliminates the necessity of an output matching network and provides a wideband output matching.

Figure 14.4.3 shows simulated time-domain waveforms of V_{CE} and I_{CE} at 27GHz (Class-F^{-1}) and 31GHz (Class-F) as input power increases from 0dBm to 8dBm with 2dB steps (V_{CC}=2.2V, V_{knee}=0.5V). At 27GHz, due to the Class-F^{-1} load impedance the voltage waveforms exhibit sinusoidal overshoot with a half-sinusoid waveform while the current waveforms show an odd-harmonic-rich response as input power increases. At 31GHz, because of the load impedance modulation toward Class-F mode the voltage waveforms get saturated to maximum 4V with square-wave-like shape and the 2nd harmonic is the dominant harmonic in the current waveforms. After final optimization of the load networks including post-layout parasitics, the PA performs peak PAE of 42.2% in Class-F^{-1} mode at 26 to 28GHz and minimum PAE of 36% in Class-F mode at 31GHz in simulations.

The PA is characterized with on-wafer testing using GSG probes for RF signal transitions after SOLT calibration. The small-signal S-parameter measurements with a Class-AB biasing (V_{CC}=2.2V, I_{CE}=9mA) show that S_{11}<-10dB over 22.5 to 32GHz, S_{22}<-10dB over 15 to 33GHz and S_{21}=9-10.8dB over 24 to 31GHz (Fig. 14.4.4). The measured k-factor is greater than 1 and the PA is stable over all measured frequencies. For large-signal measurements, a signal generator followed by an amplifier provides high-power input signals to the PA, and the input and output power of the DUT are sampled using directional couplers and measured by a dual-channel power meter. Figure 14.4.5 shows measured PAEs, collector efficiencies, output powers and power gains. At 27GHz with the Class-AB biasing, the PA achieves peak PAE of 40.7%, equivalent to 52.7% collector efficiency, with corresponding output power of 16.5dBm, OP_{1dB} of 15dBm, and P_{sat} of 17.1dBm. The peak PAE is greater than 36.3% over 24 to 31GHz and 39.3 to 40.7% PAE is maintained over 25 to 30GHz. The output power at the peak PAE ranges over 15.3 to 16.2dBm at 24 to 31GHz. The PAE is also measured for different supply voltages at 27GHz and is better than 40% over the V_{CC} range of 1.4 to 2.3V (Fig. 14.4.5). This is one of the highest PAEs reported so far in integrated silicon technologies at microwave and mm-Wave frequencies, on a par with III-V PAs, as shown in Fig. 14.4.6. It is notable that the measured linear-mode PAE at 6dB back-off from the P_{sat} is 26% (Fig. 14.4.5), still comparable to the peak PAE of the state-of-the-art silicon designs in Fig. 14.4.6. Without pads the PA occupies only 0.48x0.3mm^2 (Fig. 14.4.7), conducive for further integration into a large PA array for a higher P_{out}.

Acknowledgements:
This work was supported by the DARPA program BAA-11-50 (subcontract to UCSD). The authors would like to thank Dr. G.M. Rebeiz (UCSD) and Dr. M. Chang (Univ. of Michigan) for their helps in PAE measurements.

References:
[1] W. Tai, et al., "A 0.7W Fully Integrated 42GHz Power Amplifier with 10% PAE in 0.13µm SiGe BiCMOS", *ISSCC Dig. Tech. Papers*, pp. 142-443, Feb. 2012.
[2] K.-J. Koh, et al., "A Millimeter-Wave (40 to 45 GHz) 16-Element Phased-Array Transmitter in 0.18µm SiGe BiCMOS Technology", *IEEE JSSC*, vol. 44, no. 5, pp. 1498-1509, May 2009.
[3] V. Carrubba, et al., "The Continuous Inverse Class-F Mode With Resistive Second-Harmonic Impedance", *IEEE T-MTT*, vol. 60, no. 6, pp. 4107-4116, June 2012.

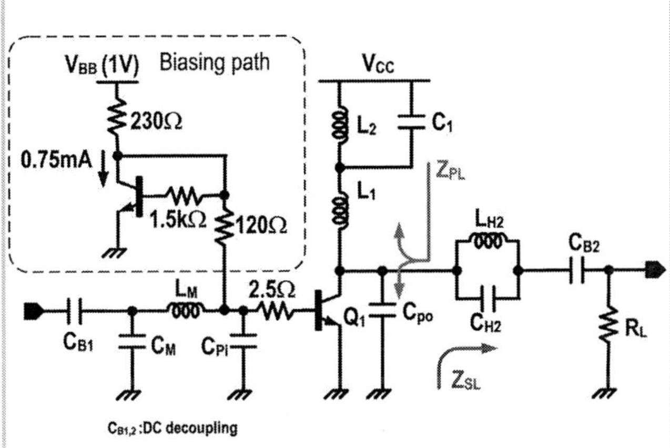

Figure 14.4.1: Class-F^{-1}/F power amplifier schematic in 0.13μm SiGe BiCMOS, employing a continuous mode transition from Class-F^{-1} to Class-F over 24-31GHz.

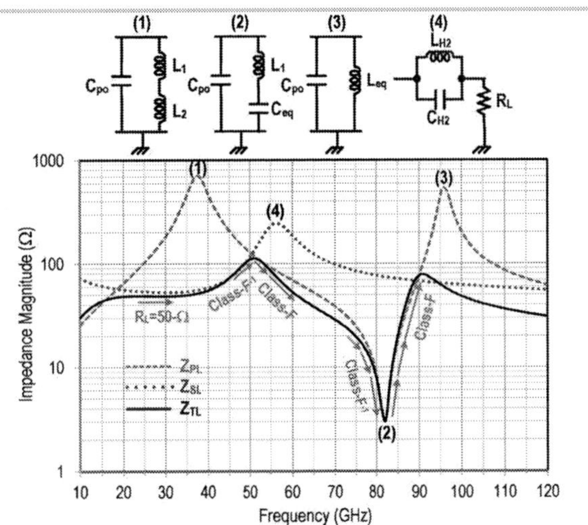

Figure 14.4.2: Simulated magnitude response of the load networks impedances.

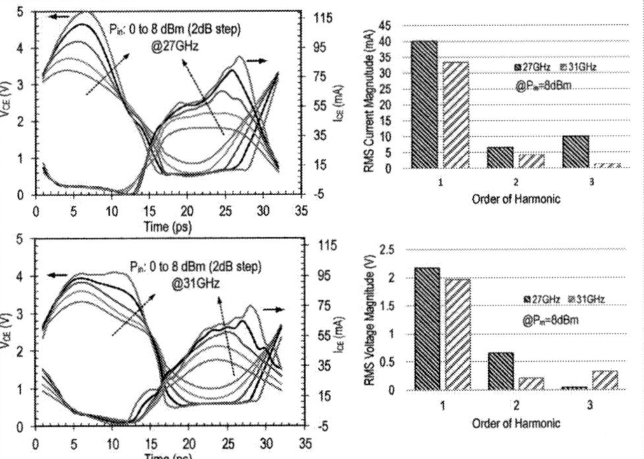

Figure 14.4.3: Simulated time-domain collector voltage and current waveforms @27GHz (Class-F^{-1}) and @31GHz (Class-F); and rms magnitude of the spectrum of the waveforms up to the 3rd harmonic.

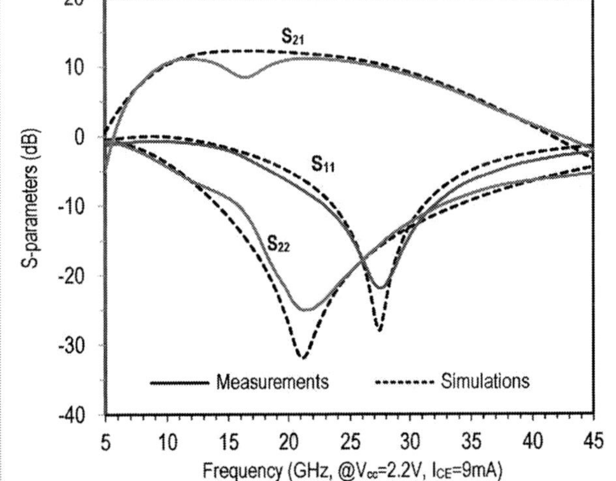

Figure 14.4.4: Simulated and measured S-parameters.

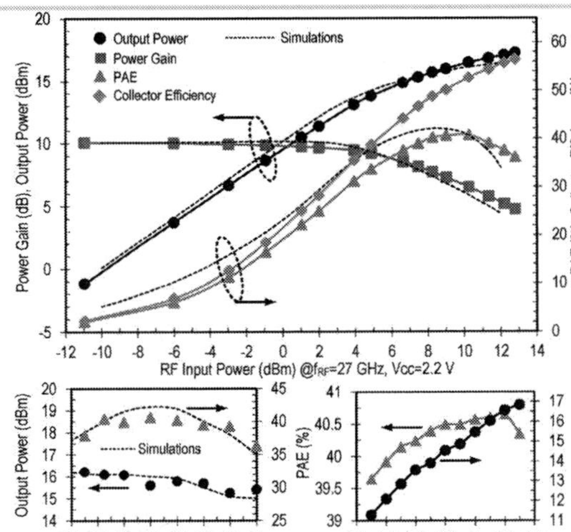

Figure 14.4.5: Simulated and measured PA performance.

Authors	Freq.(GHz)	PAE (%)	P_{sat} (dBm)	OP_{-1dB}(dBm)	Gain (dB)	Size (mm²)	Supply(V)	Process	Feature
This Work	25-30 24-31	39.3-40.7 36.3-40.7	17.1	15	10.3	0.27(0.14*) * w/o pad	2.2	0.13μm SiGe	1-stage Class-F^{-1}/F
JSSC 2005 A. Komijani, *et al.*	24	6.5	14.5	11	7	1.26	2.8	0.18μm CMOS	2-stage Class-AB
RFIC 2007 M. Chang, *et al.*	33	11.2	17	15.5	13	1.83	1.4	0.13μm SiGe	2-stage Class-AB
JSSC 2005 T.S.D. Cheung, *et al.*	22 24	19.7 13	20-23	NA	15-19	6	1.8	0.2μm SiGe	3-stage Class-AB
RFIC 2005 N. Kinayman, *et al.*	24	2.9	12	11	18	1.1	5	0.5μm SiGe	3-stage Class-AB
T-MTT 2012 N. Kalantari, *et al.*	38	20	23	NA	18.7	1.04	3	0.12μm SiGe	3-stage Class-AB
CICC 2012 A. Chakrabarti, *et al.*	47	34.6	17.6	NA	13	0.12	2.5	45nm SOI CMOS	2-stack Class-E
CICC 2012 K. Datta, *et al.*	45	31.5	20.2	NA	10.5	1.3	2.4	0.13μm SiGe	2-stage Class-E
ISSCC 2013 W. Tai, *et al.*	42	10	28.4 (16-array)	NA	18.5	5.56	4 / 2.4	0.13μm SiGe	3-stage, Class-AB
MTTs 2012 A. Agah, *et al.*	45	23	18	NA	7	0.64	2.5	45nm SOI CMOS	3-stack Doherty
T-MTT 2011 B.-H. Ku, *et al.*	9.5	20.3	21.5	20.2	25.3	0.63	3.6	0.18μm CMOS	2-stage Push-Pull
ISSCC 2010 H. Wang, *et al.*	8	21.6	25.2	22.6	17.5	0.7	NA	90nm CMOS	2-stage Class-AB
ISSCC 2013 M. Fathi, *et al.*	5.3	20.6	28.2 (2x4-array)	NA	23.8	5.88	4.8	65nm CMOS	3-stack Class-AB
T-MTT 2012 P.-C. Huang, *et al.*	24-26	40	23.5	22	9	1.5	4	GaAs HEMT	1-stage, Class-AB
RFIC 2013 N. Kinayman, *et al.*	26.4	38	25.3	NA	10.3	25	5	GaAs HEMT	2-stage Doherty
MTTs 2012 C.F. Campbell, *et al.*	29	30	37	NA	25	4.8	20	GaN HEMT	3-stage Class-AB

Figure 14.4.6: PA performance comparison with silicon and III-V PAs.

978-1-4799-0917-9/14 $31.00 © 2014 IEEE

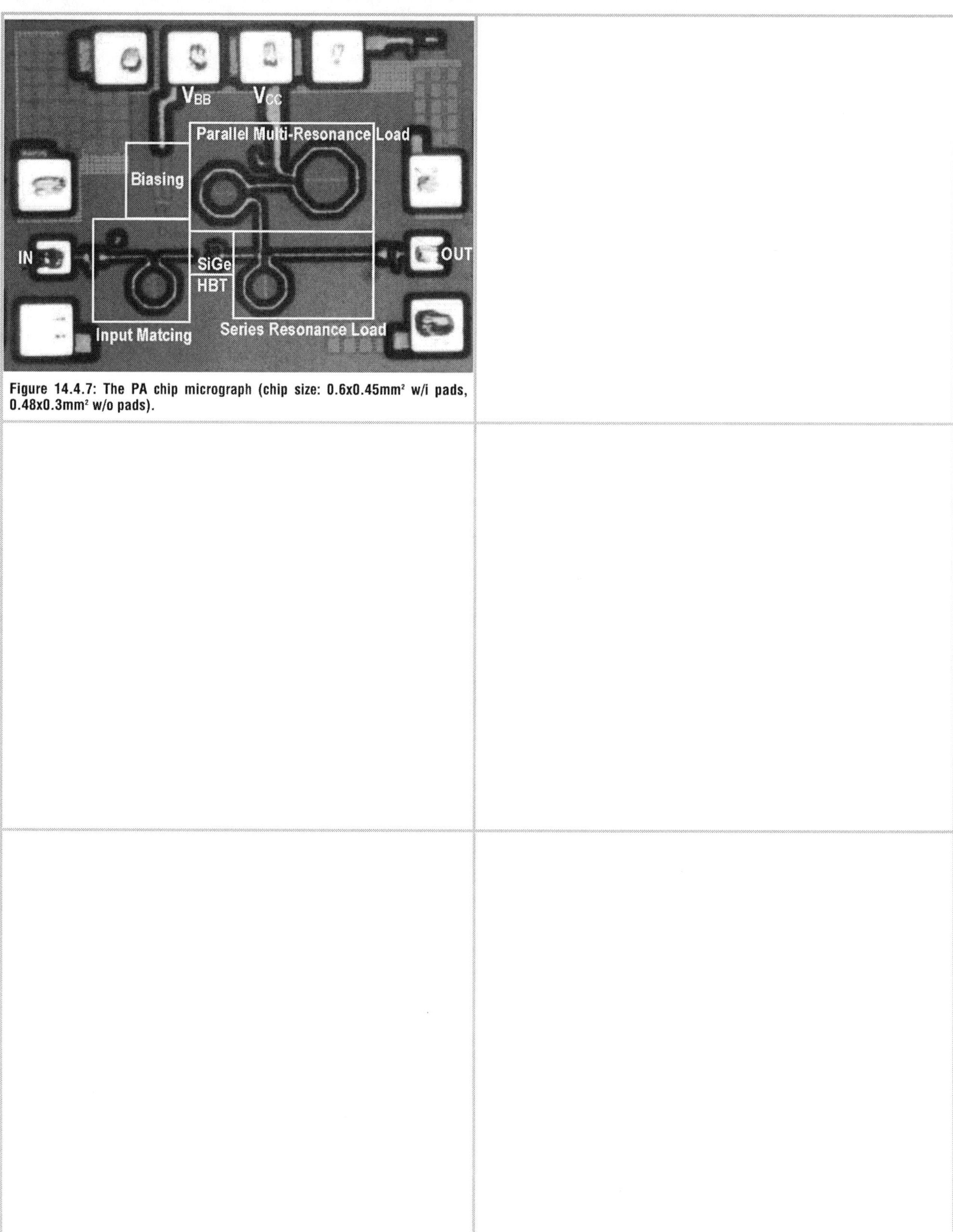

Figure 14.4.7: The PA chip micrograph (chip size: 0.6x0.45mm² w/i pads, 0.48x0.3mm² w/o pads).

14.5 A 0.53THz Reconfigurable Source Array with up to 1mW Radiated Power for Terahertz Imaging Applications in 0.13µm SiGe BiCMOS

Ullrich R. Pfeiffer[1], Yan Zhao[1], Janusz Grzyb[1], Richard Al Hadi[1], Neelanjan Sarmah[1], Wolfgang Förster[1], Holger Rücker[2], Bernd Heinemann[2]

[1]University of Wuppertal, Wuppertal, Germany,
[2]IHP, Frankfurt (Oder), Germany

Recently, silicon-based THz video cameras have been demonstrated for industrial, surveillance, scientific, and medical applications in the THz range (300GHz to 3THz) [1]. Such camera implementations favor pixels with antenna-coupled direct detectors for a low power dissipation and a high pixel count. Despite this progress, they lack the required sensitivity for passive imaging and imagers are in the need of artificial illumination to provide the required image quality. The choice has been to use expensive high-power focused illumination with a single direction for the incoming beam, which seriously limits the image quality due to its specular nature. Additionally, detectors in a focal-plane array configuration share the available source power and the image SNR further drops with the camera resolution. Like imaging at visible light, where shades or reflectors are commonly used, active THz imaging would greatly benefit from incoherent artificial light sources to adjust brightness, phase/frequency, and the direction of light to obtain the desired lighting conditions.

In order to address this, we present a high-power THz 16-pixel source array with programmable diversity for active THz imaging. The scalable circuit architecture includes control circuitry and in-pixel memory to configure the THz lighting conditions in real-time as indicated in Fig. 14.5.1. The core of a single source pixel consists of a ring-antenna and two triple-push oscillators (TPOs) locked 180° out-of-phase in order to efficiently generate radiation at 0.53THz. Conventional THz sources coherently lock all oscillators in phase to create a single THz beam. Unlike this, the circuit scheme locks oscillators only on the pixel-level to drive a differential on-chip antenna. A synchronous operation of all source pixels is not desired. Each pixel can be powered down independently such that arbitrary pattern configurations can be loaded. With all pixels activated, the array provides a total radiated power of up to 1mW (0dBm) with 62.5µW (-12dBm) per pixel on average.

A simplified circuit schematic of a differential oscillator is shown in Fig. 14.5.2. The ring antenna is differentially driven by two TPOs and the operation at the 3rd harmonic is realized by coupling two TPOs through the emitter capacitors C_e (13fF) at the fundamental mode. Each TPO consists of three Colpitts oscillators with common push output through the base inductors L_b (40pH). Their oscillation frequency is set by the base inductor L_b, the emitter capacitor C_e and the b-e junction capacitance of the transistor. The collector inductor L_c (10pH) is used to enhance the 3rd-harmonic output power. The emitter inductor L_e provides a DC current path and blocks the AC current. The decoupling capacitor C_{dc} at the VCC node serves as a bypass capacitor. In order to obtain sufficient model accuracy for passive devices, MOM capacitors were used and inductors were implemented as strip-lines.

The array control circuit is implemented in the CMOS portion of the technology. Array patterns can be loaded serially and updated at runtime. The digital circuitry of a 16-bit synchronous latched shift register is spread across the array in a meander-type structure making the circuit layout scalable in size and power. Upon receipt of a load signal the state of the shift register is copied into 16 output registers which drive a TPO power-down switch connected at the antenna center tap.

The on-chip antenna is designed to feed an extended hyper-hemispherical silicon lens (D=15mm) through the backside of the silicon die. The ring consists of 2 wire semi-rings coupled along the center feed point and it is driven differentially by all odd harmonics. The antenna was designed to provide a complex conjugate impedance match for a maximum oscillation swing. The common-mode harmonic rejection was achieved by proper termination on the antenna port. The common-to-differential mode conversion is below -40dB and the fundamental rejection is 20dB. The simulated radiation efficiency into a semi-infinite silicon substrate is around 86% at 520GHz. EM simulation results show a -35dB inter-element coupling between all antennas from 350 to 800GHz and practically negligible levels below 350GHz (-60dB at the fundamental).

The chip was fabricated in a 0.13µm SiGe BiCMOS technology SG13G2 [2]. The technology features SiGe HBTs with peak f_T/f_{max}=300GHz/500GHz, an open-base collector-emitter breakdown voltage of 1.6V, and a peak current gain of 650. The HBTs are integrated in a 0.13µm CMOS process with seven aluminum interconnect layers. The 4x4-pixel array exhibits a close-packing layout with a center-to-center pixel spacing of 520µm in a honeycomb-like tessellation. To demonstrate low-cost packaging solutions for handheld applications, the overall assembly was mounted and wire-bonded onto an FR4 PCB (see Fig. 14.5.3).

Frequency measurements were made with a spectrum analyzer and an 18th-harmonic mixer equipped with a diagonal horn antenna in close proximity of the lens (see Fig. 14.5.3). The mixer was pumped with a signal source followed by a doubler and a diplexer with an IF bandwidth of 3GHz. This setup was used to measure the tuning-range and to observe the spectral response under different source-pattern configurations. The key challenge with respect to accurate power measurements is to precisely measure the total radiated power emitted from multiple pixels under different angles in the presence of a broadband thermal emission background. The thermal emission background is caused by the heat generated on chip. For this reason, the total radiated power was measured with a photo-acoustic absolute power meter (TK) in the presence of electronic chopping. The on-chip digital circuitry thereby provides a convenient way for electronic chopping. The single pixel power was also verified with an Erikson calorimeter (PM4). The array was positioned in the pivotal point of a two-axis computer-controlled rotational joint to precisely measure array pattern configurations. The radiation pattern was scanned by a lens-coupled SiGe HBT direct detector positioned in line-of-sight at a 1-meter far-field zone and the array was chopped electronically at 80Hz in order to facilitate the readout of the SiGe detector. The measured directivity derived from the FWHM of a single beam is 37dBi and all 16 beams cover a ±15° field-of-view. The measured radiated output power and tuning-frequency versus VCC are shown in Fig. 14.5.4 (a) for a single source pixel. At VCC=2.4V a single pixel can deliver up to 85µW (-11.3dBm). The tuning range spans over 17GHz (3.2%) from 536 to 519GHz for a 1 to 2.4V supply. The output power and DC-to-RF conversion efficiency of the entire array is shown in Fig. 14.5.4 (b). Overall the 16-pixels radiate 1mW (0dBm) at a 0.4‰ efficiency with 62.5µW (-12dBm) per pixel on average (EIRP per pixel is 26dBm). The theoretical expected limit of 1.2mW is not reachable due to an asymmetrical distribution of wiring resistance in the power supply grid. A separate oscillator breakout circuit was used to measure the fundamental and 2nd-harmonic leakage on-wafer without antenna. The total leakage power on-wafer was 2µW/pixel after probe-loss de-embedding. This power is further rejected below the sensitivity-level due to the on-chip antenna selectivity.

Arbitrary source pattern configurations can be loaded as shown in Fig. 14.5.6 and the radiation can be recorded with a CMOS 1k-pixel camera as indicated in Fig. 14.5.7. The chip consumes up to 2.5W from a 2.4V supply and 3.2mW from a digital 1.2V supply respectively. A comparison table with state-of-the art silicon sources (radiated only) and the chip micrograph are shown in Fig. 14.5.6. Among the implementations shown in the table, this source array demonstrates the highest radiated power from a single silicon chip above 300GHz and the highest radiated output power for a single CW source above 500GHz.

Acknowledgments:
This work was supported in part by the European Science Foundation through a European Young Investigator Award.

References:

[1] R. Al Hadi, et al., "A 1 k-pixel Video Camera for 0.7-1.1 Terahertz Imaging Applications in 65nm CMOS," *IEEE J. Solid-State Circuits*, vol. 47, no. 12, pp. 2999-3012, Dec. 2012.
[2] H. Rücker, et al., "Half-Terahertz SiGe BiCMOS technology", *IEEE SiRF Symp. Digest*, pp. 129-132, 2012.
[3] R. Han, et al., "A CMOS High-Power Broadband 260GHz Radiator Array for Spectroscopy," *IEEE J. Solid-State Circuits*, vol. 48, no. 12, pp. TBD, Dec. 2013.
[4] K. Sengupta, et al., "A 0.28 THz Power-Generation and Beam-Steering Array in CMOS Based on Distributed Active Radiators," *IEEE J. Solid-State Circuits*, vol. 47, no. 12, pp. 3013-3031, Dec. 2012.
[5] J. Grzyb, et al., "A 288-GHz Lens-Integrated Balanced Triple-Push Source in a 65nm CMOS Technology," *IEEE J. Solid-State Circuits*, vol. 48, no. 7, pp. 1751-1761, July 2013.
[6] D. Shim, et al., "553-GHz signal generation in CMOS using quadruple-push oscillator," *Proc. VLSI Circuits Symp.*, pp. 154–155, 2011.
[7] E. Ojefors, et al., "An 820GHz SiGe Chipset for Terahertz Active Imaging Applications," *ISSCC Dig. Tech. Papers*, pp. 224-226, Feb. 2011.

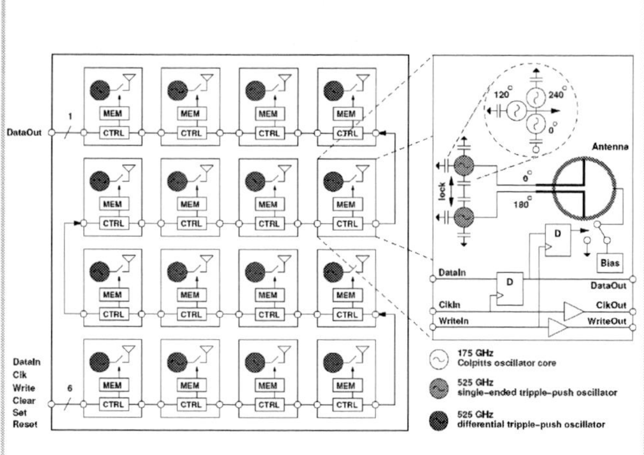

Figure 14.5.1: Reconfigurable THz source-array block diagram.

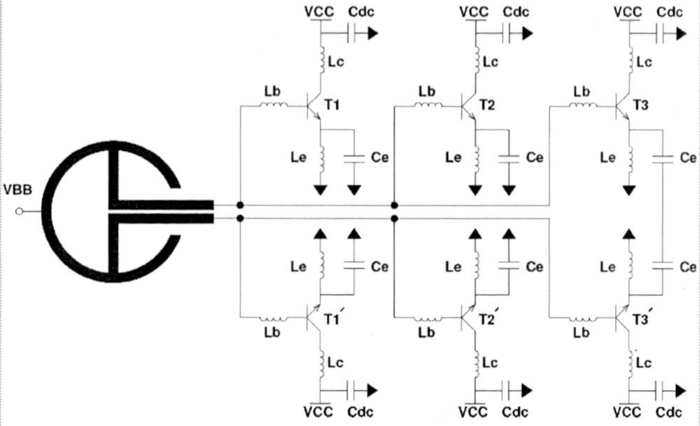

Figure 14.5.2: Simplified circuit schematic of the differential triple-push oscillator.

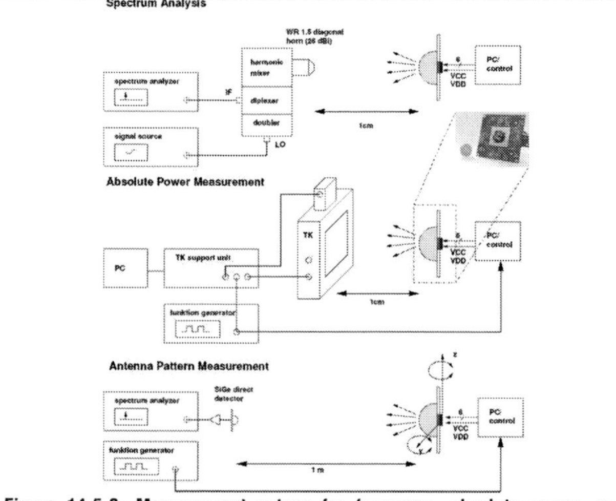

Figure 14.5.3: Measurement setups for frequency, absolute power, and radiation pattern measurements.

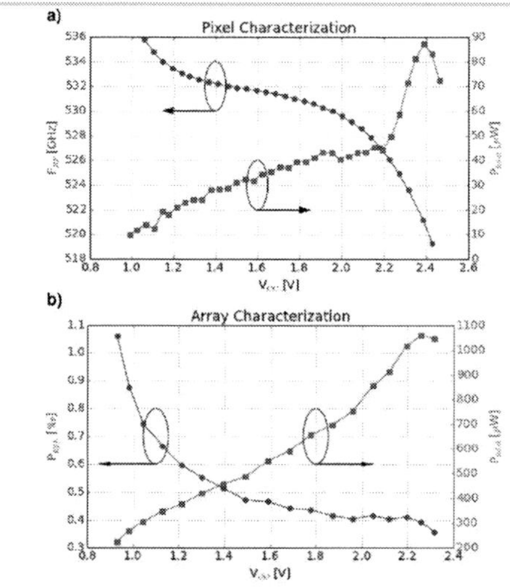

Figure 14.5.4: Measured results for single source pixel (a) and full array (b).

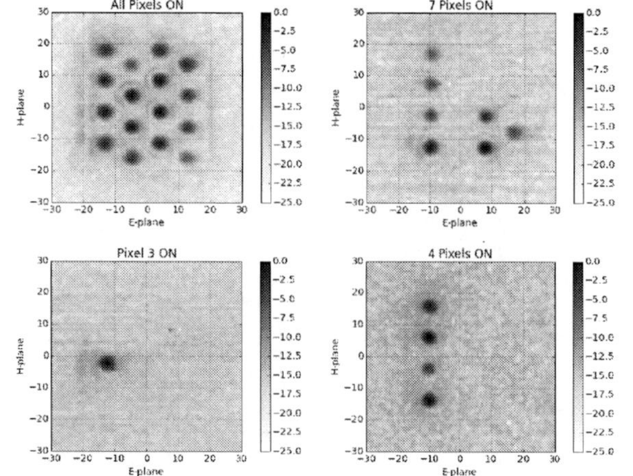

Figure 14.5.5: Arbitrary source patterns can be programmed at runtime. Measured patterns with 16, 7, 4, or 1 active pixel are shown for illustration purpose (in dB).

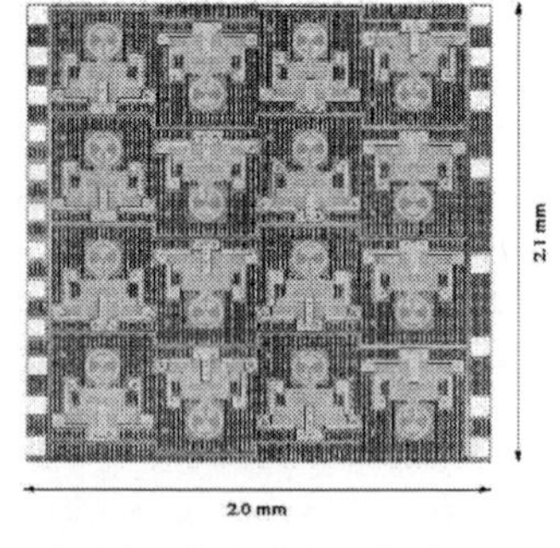

Technology	Frequency [GHz]	Radiated Power [dBm]	EIRP [dBm]	Radiators	Bandwidth [%]	DC-to-RF Eff. [‰]	Area [mm²]	Reference
65nm bulk	260	0.5	15.7	8	9.5	1.4	2.3	[3]
45nm SOI	280	-7.1	9.4	16	3.2	-	7.3	[4]
65nm bulk	288	-4.1	14.2	1	0.7	1.4	0.29	[5]
45nm bulk	553	-36.5	-	1	-	3.4E-4	0.29	[6]
250nm SiGe	825	-29	-17	4	1.8	2.7E-4	3.22	[7]
130nm SiGe	519-536	0, -11.3[1], -12[2]	25[1]	16	3.2[1]	0.4	4.2	This Work

[1] single source pixel
[2] average per single source pixel

Figure 14.5.6: Chip micrograph and comparison table of state-of-the-art THz radiation sources in silicon.

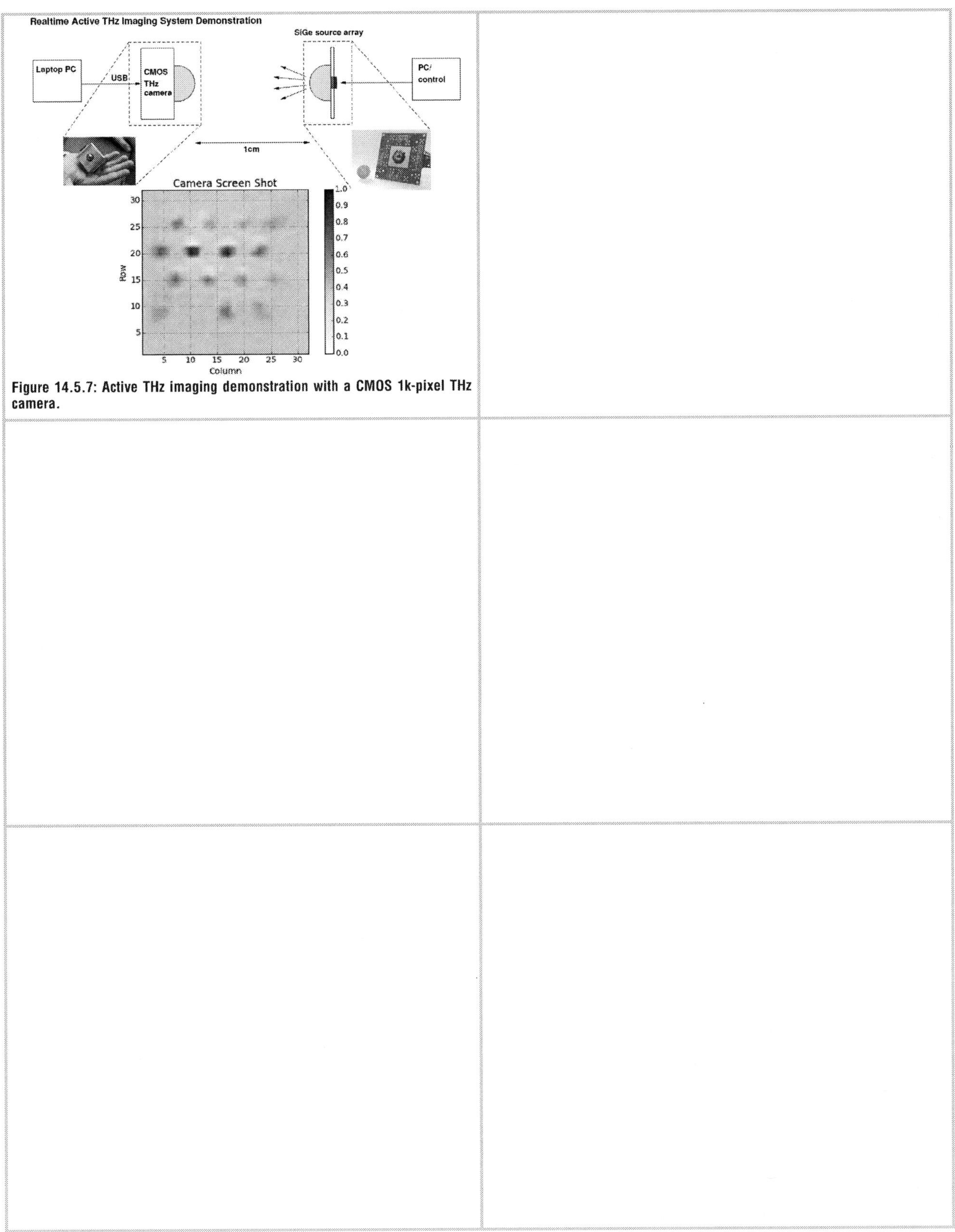

Figure 14.5.7: Active THz imaging demonstration with a CMOS 1k-pixel THz camera.

14.6 A Scalable THz 2D Phased Array with +17dBm of EIRP at 338GHz in 65nm Bulk CMOS

Yahya Tousi, Ehsan Afshari

Cornell University, Ithaca, NY

There is an untapped market for integrated high-resolution imaging and spectroscopy at mm-Wave and THz frequencies. Some novel approaches have been recently proposed to render on-chip signal generation and transmission at these frequencies [1-4]. However, as the frequency approaches the physical limitations of the device, self-sustained oscillation becomes increasingly difficult and the generated power fades away. Fortunately, the limited power from individual devices can be offset by coherent power-combining of multiple sources. A desirable mechanism to generate high power levels above the f_{max}, has to provide a solution that first, enables a scalable approach to generate and combine the maximum available power from individual sources and second, ensures controllable phase shift between multiple sources in order to provide beam steering at any desired direction.

In traditional phased arrays as the number of rows and columns increases, the complexity of array connections and phase shifters becomes a major obstacle. This challenge is even more detrimental at mm-Wave and THz frequencies where conductive loss, undesired couplings, phase/gain mismatch, and high power consumption are among many adverse effects of such lengthy connections.

To address this issue, this work presents a novel scalable system for THz signal generation and radiation. Figure 14.6.1 shows the architecture consisting of a 2-D array of coupled oscillating elements. Each oscillator with its antenna forms a small THz radiator. While independently radiating, each element is also unidirectionally connected to its neighboring elements in both horizontal and vertical directions through variable phase shifters, ψ_{row} and ψ_{col}, respectively. This network is inherently scalable because it only relies on couplings with the nearest neighbors and there is no high-frequency global routing to any oscillator. The purpose of this topology is twofold: first to synchronize all the oscillators to a single frequency and next, to set a desired phase shift between the adjacent elements ($\Delta\varphi_{row}$ and $\Delta\varphi_{col}$). We can show that by employing this particular coupling structure only a small subset of all the theoretical coupling modes are physically stable. By proper control of the couplings one can ensure the system settles into the desired coupling mode [1].

In this coupled network, the phase and frequency of each element of the array are related to their neighboring elements through a two-dimensional version of Adler's equation shown in Fig. 14.6.1. Based on this equation two scenarios are conceivable: If all coupling phases (ψ_{row} and ψ_{col}) are equally changed, the locking frequencies of oscillators change while their respective phases, $\varphi_{i,j}$ remain the same. This dynamic is similar to the delay-coupled oscillators we introduced in [1]. For example, if all coupling phases increase by the same amount, the frequency of all oscillators decreases to keep the phase difference between oscillators constant. On the other hand, as shown in Fig. 14.6.2, by changing the coupling phase shifts in a differential manner, the relative phase of adjacent oscillators changes while the frequency remains constant. A differential control of the horizontal and vertical phase shifters changes the phase shift of the columns and rows of the network, respectively. Since these differential phase shifts are independent, by combining them it is possible to simultaneously control the phase difference of adjacent rows and columns of the coupled system. This enables the radiated beam to steer in both polar directions. Hence, by changing the phase shifts together or differentially, we can independently control the frequency and the direction of the radiated beam.

To ensure a symmetric array, the injected energy into each oscillator and the output load should be the same among all oscillators. This is achieved by doubling the magnitude of the coupling blocks (by placing two of them in parallel) on the edges of the array as shown in Fig. 14.6.1. In this way while the oscillators are located in different parts of the lattice, they are all subject to the same frequency and phase shift dynamics.

Each oscillator is a cross-coupled pair that delivers its output power to a matched radiator designed at the fourth harmonic of the fundamental frequency. As shown in Fig. 14.6.3, the lines connecting the gates and sources of the core devices and the output load compose a distributed multi-port circuit. At the fundamental frequency, this network is optimized to maximize the voltage swing and the outgoing coupled energy. At the fourth harmonic the network is designed to deliver the maximum available harmonic power to the antenna.

The coupling phase shifters inject energy at the fundamental frequency from each given oscillator to two of its four neighbors. The other two adjacent nodes inject energy into that oscillator. In order to effectively absorb the half-wave-length separation between adjacent oscillators, we implement a distributed phase shifter shown in Fig. 14.6.3. The phase shifter consists of an artificial transmission line that is only impedance-matched at the input side, resulting in a distributed resonator. The phase shift of the resonator is tunable using the varactors along the transmission line. To maximize the tuning capability of the phase shifters, we design the center frequency of the resonator to match the fundamental frequency of the core oscillators.

Effective radiation is a crucial aspect of our proposed THz signal source. Integrated sources at mm-Wave and THz frequencies have widely used dipole-based antennas, which are double-sided radiators. As the frequency of operation approaches the THz band, substrate thickness becomes comparable to the wavelength, guiding most of the energy into lossy substrate modes. Either a matched silicon lens or extra wafer thinning is required to cancel these undesired modes [2–4]. In our proposed design, we eliminate the need for extra components and post-processing by using a patch antenna with broad-side radiation only from the top of the chip. However, at our target frequency of 340GHz, the largest distance between two metal layers is still considerably smaller than the wavelength. As a result, the radiation bandwidth of the antenna is limited and careful sizing is crucial. In order to maximize the radiation efficiency, the antenna central frequency is designed to properly match the center frequency of the oscillators.

The chip is fabricated in a TSMC 65nm bulk CMOS process. The prototype consists of a 4 x 4 phased array along with the integrated antennas. As shown in Fig. 14.6.4, we measure the radiated frequency spectrum using a WR2.2 horn antenna connected to a VDI WR-2.2EHM harmonic mixer. The 12th harmonic of the LO is mixed with the received signal and downconverted into the IF band. The measured central frequency of the oscillator is 338GHz and the measured phase noise is -93dBc/Hz at 1MHz offset frequency. The low phase noise is a direct result of the topology of the system that couples multiple sources. Adjusting the coupling phase shifters provides 2GHz of frequency tuning. The frequency tuning range can be increased to 7GHz (333GHz to 340GHz) by changing the supply voltage by ±10%. To measure the radiation patterns shown in Fig. 14.6.5, we use the VDI WR2.2-ZBD detector and rotate the chip in both angles with respect to the detector. Our measurement confirms that the differential control of the phase shifters keeps the frequency constant while steering the radiated beam across the 2-D angles. The measured beam-steering is 45 degrees along the φ axis and 50 degrees along the θ axis as shown in Fig. 14.6.2. By utilizing multiple active stages, one can increase the injected energy from the phase shifters and achieve a wider tuning range or beam steering. To measure the radiated power, we use the Ericson PM4 power detector. As shown in Fig. 14.6.4, we measure the radiated power both as a function of the control voltage in a fixed distance to the antenna and as a function of distance for a fixed control voltage. The measured peak EIRP of the chip is +17.1dBm at 338GHz. By reducing the antenna gain derived from the measured beam pattern, the peak total radiated power from the chip is 0.8mW. Figure 14.6.6 compares the performance of the circuit with prior art. This lensless source demonstrates the highest radiated power and EIRP as well as the lowest phase noise among the CMOS arrays above 200GHz shown in the Table. Moreover, the proposed technique provides an effective way to realize a scalable THz phased array with independent frequency control and beam steering.

Acknowledgment:
The authors would like to acknowledge the National Science Foundation and Dr. Paul Maki of ONR for supporting this project. They would also like to thank the TSMC university shuttle program for the chip fabrication.

References:

[1] Y. M. Tousi, O. Momeni, and E. Afshari, "A Novel CMOS High-power THz VCO Based on Coupled Oscillators: Theory and Implementation," *IEEE J. Solid-State Circuits*, vol. 47, pp. 3032–3042, Dec. 2012.

[2] R. Han and E. Afshari, "A 260GHz Broadband Source with 1.1mW Continuous-Wave Radiated Power and EIRP of 15.7dBm in 65nm CMOS," *ISSCC Dig. Tech. Papers*, pp. 138-139, Feb. 2013.

[3] K. Sengupta and A. Hajimiri, "A 0.28 THz 4x 4 Power-Generation and Beam-Steering Array", *ISSCC Dig. Tech. Papers*, pp. 256-258, Feb 2012.

[4] Y. Zhao, J. Grzyb and U. R. Pfeiffer, "A 288GHz Lens-Integrated Balanced Triple-Push Source in a 65nm CMOS Technology," *European Solid-State Circuits Conf.*, pp. 289-292, Sept. 2012.

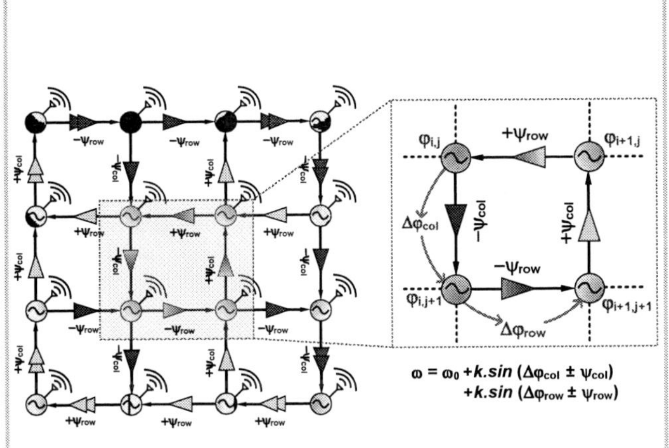

Figure 14.6.1: A 4 x 4 THz phased array based on locally coupled oscillators.

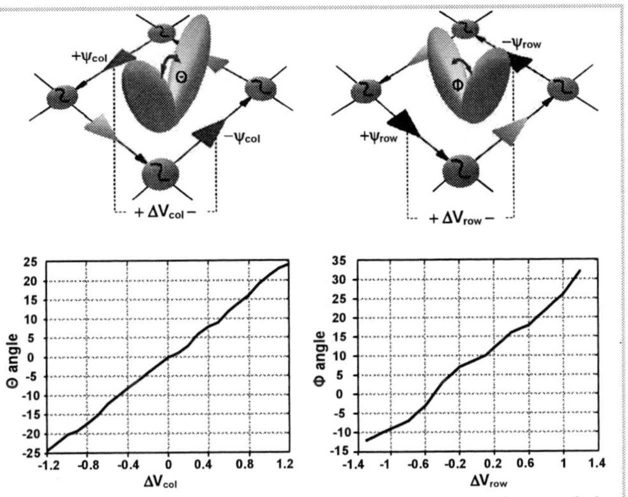

Figure 14.6.2: The concept of beam steering in a coupled loop, and the measured beam steering in the 4x4 structure.

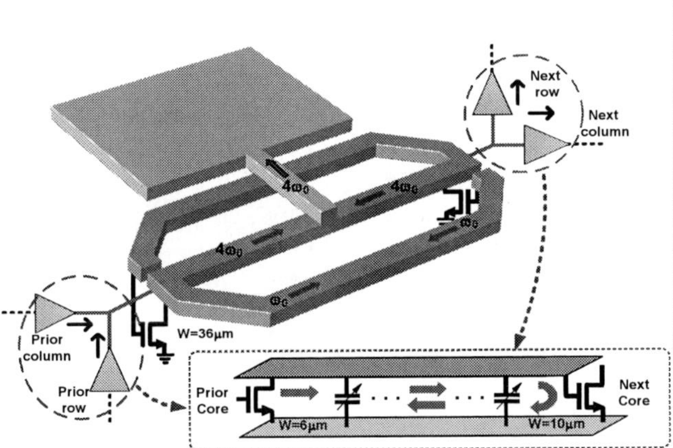

Figure 14.6.3: The core oscillator along with inter-connecting phase shifters.

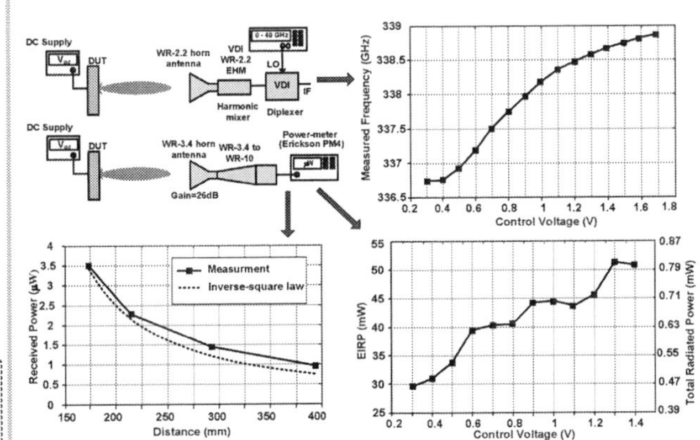

Figure 14.6.4: Chip measurement setup and the measured output power and frequency.

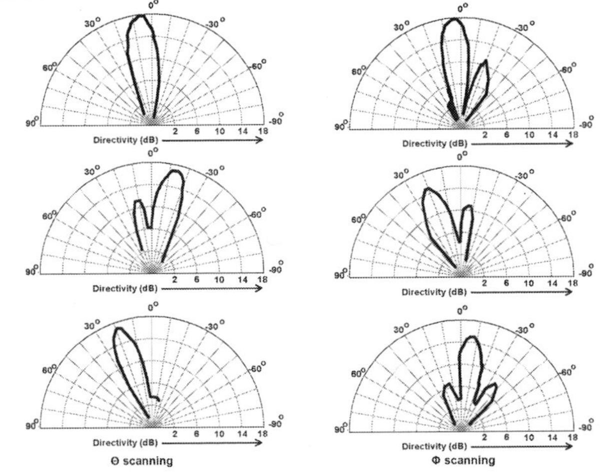

Figure 14.6.5: Measured beam pattern and directivity at different steering angles.

Ref.	This Work	[1]	[2]	[3]	[4]
Frequency (GHz)	338	290	260	280	288
Total Power (dBm)	-0.9	-1.2	0.5††	-7.2†††	-4.1††
Peak EIRP (dBm)	17.1	N/A†	15.7	9.4	N/A
Frequency Tuning (%)	2.1	4.5	1.4	3.2	Non-tuning
Phase Noise (dBc/Hz) (@ 1MHz offset)	-93	-78	-78.3	N/A	-87
Beam Steering (degree of each angle)	45/50	N/A	Fixed	80/80	Fixed
DC Power (W)	1.54††††	0.33	0.8	0.81	0.28
Technology	65nm bulk CMOS	65nm bulk CMOS	65nm bulk CMOS	45nm SOI CMOS	65nm bulk CMOS
Area (mm^2)	3.9	0.36	2.25	7.2	0.32

† Power measured by probing.
†† A Hemispheric lens is used for back-side radiation.
††† Substrate-thinning used for front-side radiation.
†††† 1.3W for radiators and 240mW for couplings.

Figure 14.6.6: Performance comparison with prior art.

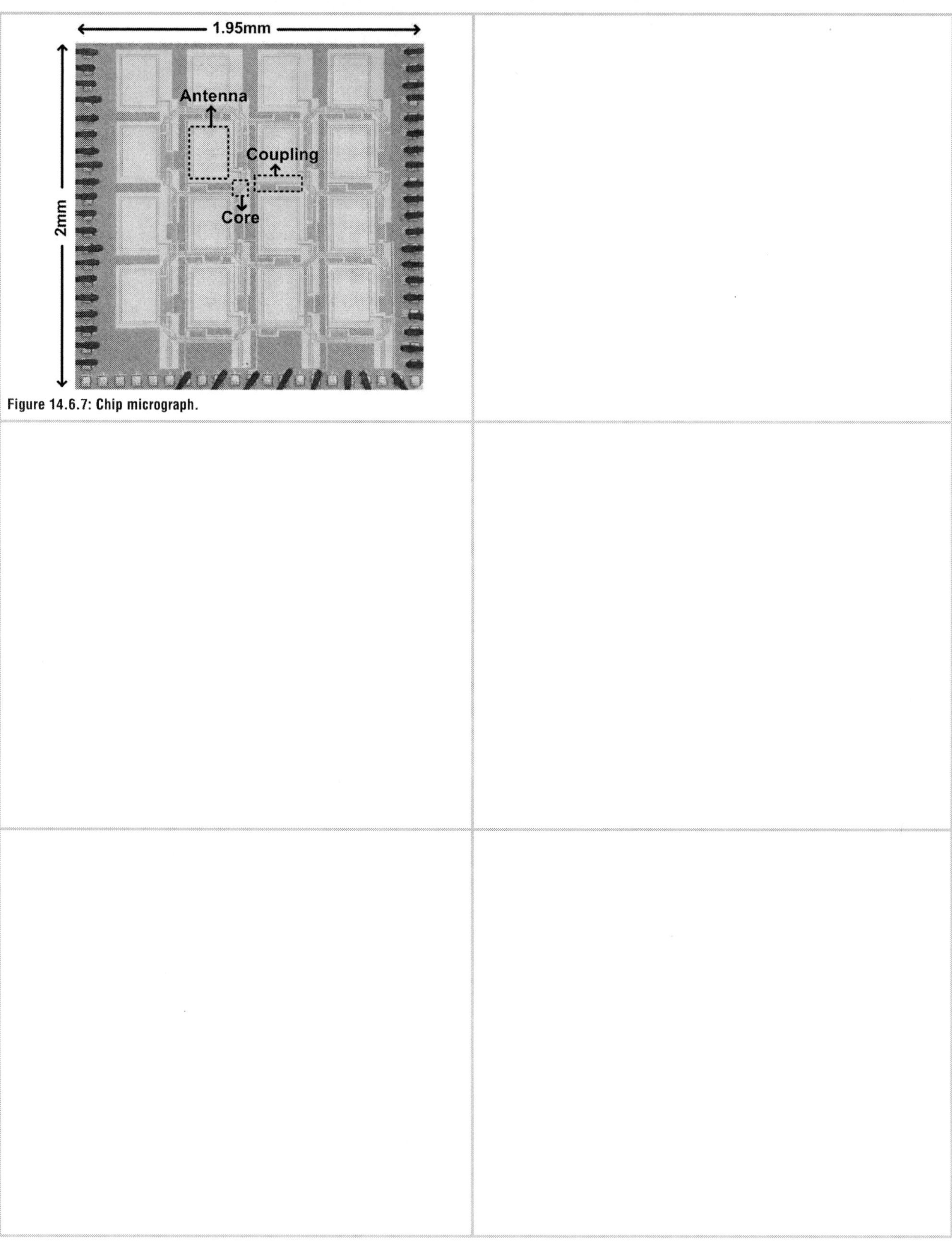

Figure 14.6.7: Chip micrograph.

14.7 A 300GHz Frequency Synthesizer with 7.9% Locking Range in 90nm SiGe BiCMOS

Pei-Yuan Chiang[1], Zheng Wang[1], Omeed Momeni[2], Payam Heydari[1]

[1]University of California, Irvine, CA,
[2]University of California, Davis, CA

The THz/sub-mm-Wave band is known to provide unique applications in spectroscopy, imaging and high-data-rate wireless communication. An accurate THz source is essential in coherent communications, radar systems, and frequency metrology. Recently, THz sources based on coupled VCOs with harmonic generation have been proposed [1]. However, open-loop signal sources exhibit severe frequency fluctuation, and are vulnerable to temperature/process/supply-induced frequency drift. The need for precise oscillation frequency with wide tuning range and low close-in phase noise calls for closed-loop topologies. Millimeter-Wave PLLs incorporating push-push VCOs have been demonstrated up to 164GHz [2] in silicon technology. [3] presented a 300GHz PLL with 0.12% locking range in III-IV technology.

In a high-frequency synthesizer, the VCO and the divider exert dominant effects on the locking range, phase noise, and output power. Starting with the VCO, achieving wide tuning range, low phase noise, and high output power simultaneously is challenging. This is because of the limited f_T/f_{MAX} of transistors and low Q factor of varactors at high frequencies. The output power degrades severely in harmonic VCOs, as the varactor loss is higher at harmonic frequencies. Furthermore, the first-stage divider suffers from low frequency range and high DC power dissipation. To alleviate these problems, a 300GHz phase-locked-based synthesizer incorporating a triple-push VCO with Colpitts-based Active Varactor (CAV) and a frequency divider with three-phase injection is introduced. The synthesizer, implemented in 90nm SiGe BiCMOS with f_T/f_{MAX} of 240/315GHz, achieves 7.9% of locking range (280.32 to 303.36GHz) and generates –14dBm of output power at 290GHz. Based on the measurement result, the frequency-scaled phase noise of a 294.9GHz signal is –77.8dBc/Hz (–82.5dBc/Hz) at 100kHz (1MHz) offset (not accounting for the noise contribution of the divider and the VCO harmonic generation [4]). Shown in Fig. 14.7.1 is the proposed 300GHz synthesizer with $1/4f_0$ and f_{FS}=$3f_0$ testing ports (f_0=100GHz). This synthesizer is comprised of a triple-push VCO with CAVs and a three-phase injection-locked divider (÷4), followed by a ÷256 divider chain, a 2-bit programmable (R_2-C_2) 3rd-order loop filter, a PFD and a tunable charge pump (I_{CP}: 150 to 300μA). The loop is locked at f_0 with a 96MHz reference generated from a signal generator or a crystal oscillator. The inner three gain stages and inductors L_{VB} form a three-stage LC ring oscillator with oscillation frequency of f_0. The f_0 signal travels along the ring, and its 3rd-harmonic signals generated by each stage are added in-phase at the RF common-mode pad, f_{FS} = $3f_0$. Frequency tuning is achieved using three CAVs connected to the inner ring. The CAV has three important roles: (1) It provides a wide-tuning-range variable capacitor with almost no loss to the inner ring for frequency tunability; (2) It isolates the varactor loss from the inner ring to increase the f_0 signal voltage swing and its harmonic power; and (3) It buffers and injects the f_0 signal to the divider without loading the oscillator. The outer ring in Fig. 14.7.1 represents the ÷4 divider composed of transistors T_{a2}~T_{c2} for signal amplification and T_{a1}~T_{c1} for signal injection and mixing. By injecting three phase signals (f_0, $f_0\angle120°$, $f_0\angle240°$), the outer ring is locked to the inner ring at $1/4f_0$. The output of the ÷4 divider is then fed to a ÷256 divider chain and divided down to the reference frequency range.

Figure 14.7.2 shows the detailed schematics of the VCO and the three-phase injection-locked divider with its conceptual loop. All the inductors are implemented using shielded CPWs. In the VCO, L_{VB} is used to adjust the signal amplitude and phase difference between the base and the collector of transistors T_1~T_3 so as to enhance f_0 voltage swing, and hence, maximize the 3rd-harmonic generation [5]. Furthermore, L_{VB} blocks the flow of 3rd-harmonic signal from transistors T_1-T_3 to the CAVs, resulting in higher harmonic power. The CAV provides a lossless voltage-controlled tunable capacitance to the ring. The loss-cancellation mechanism is achieved using a modified version of a Colpitts oscillator. In addition, the CAV buffers and injects the f_0 signal to the divider efficiently. In the CAV schematic, a large capacitor C_b and a 3kΩ resistor are added to level-shift V_{CTRL} within the voltage range required by the charge pump. The ÷4 divider is based on a ring oscillator with oscillation frequency close to $1/4f_0$. The 3rd harmonic, $3/4f_0$, generated by transistors T_{a2}~T_{c2}, flows to transistors T_{a1}~T_{c1}. The injected signals (f_0, $f_0\angle120°$, $f_0\angle240°$) are mixed with $3/4f_0$, and the frequency-difference component ($1/4f_0$) passes through the filter and reinforces the oscillation frequency. The mixing efficiency is a strong function of the harmonic power and is sufficiently large that, together with the high loop gain of the ring, results in locking at $1/4f_0$. Similar to the VCO, L_{DB} is added to maximize the voltage swing at $1/4f_0$, and hence, boost the 3rd-harmonic generation at $3/4f_0$. Moreover, L_{DB} blocks the $3/4f_0$ and along with L_{DM} ensures an efficient power transfer of this signal to transistors T_{a1}~T_{c1}. The enhancement of the 3rd-harmonic generation and stronger signal injection (three-phase injection) increase the mixing conversion gain and results in a wide locking range.

A standalone divider was fabricated and verified. The measured input sensitivity is shown in Fig. 14.7.3. With two bias settings, the operation range varies from 91.9 to 101.8GHz with DC power dissipation of 48.4mW. Another break-out circuit comprising of the VCO+divider (÷4) was fabricated to measure the VCO with its V_{CTRL} varying from 0 to 2.6V. The tuning range of the VCO $3f_0$ output spans from 279.96 to 303.36GHz. Figure 14.7.3 depicts the measured phase noise of the $1/4f_0$ (25.12GHz) signal using a signal source analyzer. The frequency-scaled phase noise at 301.4GHz is –80.28dBc/Hz at 1MHz offset. The VCO output power, directly measured from the $3f_0$ port using a PM4 *Ericson* power meter, is –14dBm at 290GHz. The synthesizer performance is verified by measuring both $1/4f_0$ and f_{FS} output ports. Figure 14.7.4 shows measured output spectra at 24.64GHz and 300.835GHz using an *OML* WR-3 harmonic mixer with conversion loss of 75 to 80dB. The measured reference spur at 23.45GHz is –64dBc. Also, shown in Fig. 14.7.4 is the synthesizer measured phase noise at $1/4f_0$ (24.32GHz) with the 95MHz reference generated from a signal generator. The phase noise profile shows that the synthesizer's output at 291.84GHz is locked to the 95MHz reference signal. The measured phase noise of the 24.32GHz signal at 1MHz (10MHz) offset is –96dBc/Hz (–109.8dBc/Hz). Due to high conversion loss of the harmonic mixer, it is not possible to directly measure the phase noise at f_{FS}. Therefore, the measured phase noise at $1/4f_0$ in Fig. 14.7.4 was used to obtain the synthesizer phase noise at f_{FS} [4]. The phase noise at 291.84GHz is –74.4dBc/Hz (–88.2dBc/Hz) and at 1MHz (10MHz) offset. The frequency-scaled measured phase noise across the synthesizer locking range is shown in Fig. 14.7.5. With the reference generated from a signal generator, the synthesizer achieves a locking range of 280.32 to 303.36GHz (7.9%) where the minimum phase noise is –75.4dBc/Hz at 1MHz offset. Moreover, using a 96MHz crystal oscillator as the reference, the phase noise is improved to –82.5dBc/Hz (–77.8dBc/Hz) at 1MHz (100kHz) offset. The measured phase noise profiles for the 96MHz signal generator (SG), the 96MHz crystal oscillator (XTAL), the 294.9GHz VCO and the synthesizer with both SG and XTAL inputs are shown in Fig. 14.7.5. Within the synthesizer loop BW, the phase noise at 1MHz offset is dominated by the input source, and is thus improved by a lower phase noise source (e.g., XTAL). Tables in Fig. 14.7.6 summarize the performance of the synthesizer, VCO, and divider; and compare the synthesizer with prior work.

Acknowledgements:
The authors thank Prof. E. Afshari and Agilent for test equipment support, TowerJazz for chip fabrication and for providing measured device data, Francis Caster and all NCIC Lab members for technical assistance.

References:

[1] Y. Tousi and E. Afshari, "A 283-to-296GHz VCO with 0.76mW Peak Output Power in 65nm CMOS," *ISSCC Dig. Tech. Papers*, pp. 258–260, Feb. 2012.
[2] S. Shahramian *et al.*, "Design of a Dual W- and D-band PLL," *IEEE J. Solid-State Circuits*, vol. 46, pp. 1011–1022, May 2011.
[3] M. Seo *et al.* "A 300 GHz PLL in an InP HBT Technology," *Int. Microwave Symp.*, Baltimore, June 2011.
[4] C.-C. Wang, Z. Chen, and P. Heydari, "W-Band Silicon-Based Frequency Synthesizers using Injection-Locked and Harmonic Triplers," *IEEE T-MTT*, vol. 60, pp. 1307–1320, May 2012.
[5] O. Momeni and E. Afshari, "High Power Terahertz and Millimeter-Wave Oscillator Design: A Systematic Approach," *IEEE J. Solid-State Circuits*, vol. 46, pp. 583–597, March 2011.

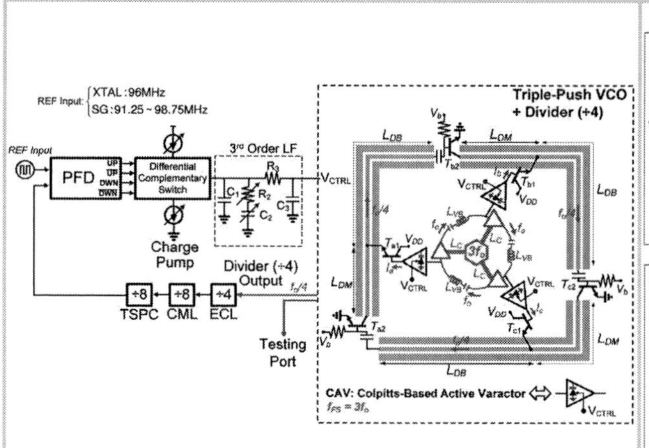

Figure 14.7.1: The 300GHz frequency synthesizer.

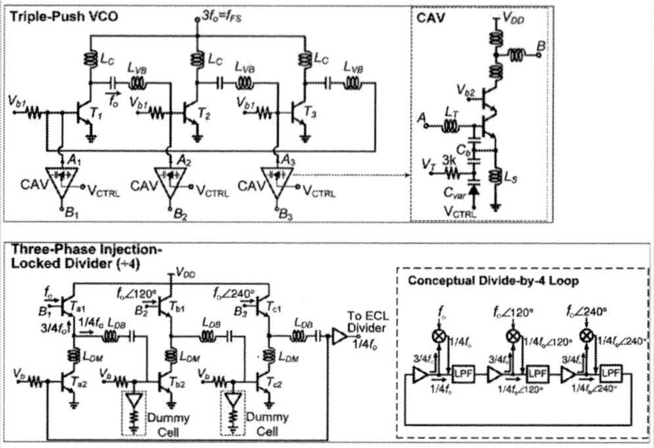

Figure 14.7.2: The proposed schematics of the triple-push VCO with CAVs, and three-phase injection-locked divider (÷4).

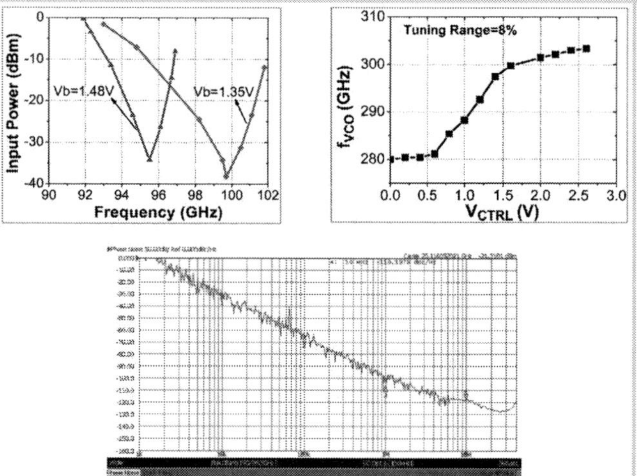

Figure 14.7.3: The measured divider (÷4) input sensitivity, triple-push VCO oscillation frequency and its phase noise at 25.12GHz ($1/4f_0$).

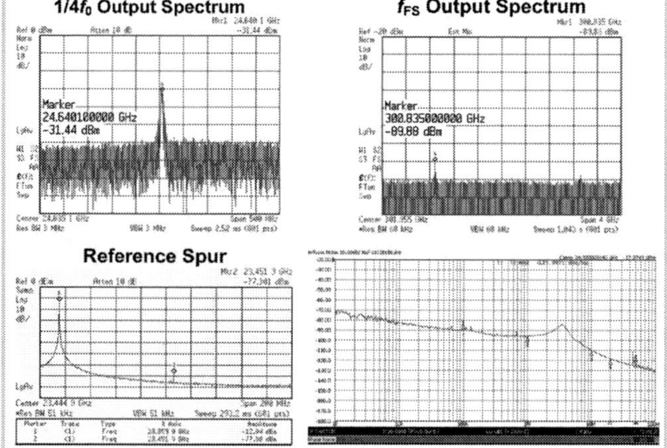

Figure 14.7.4: The synthesizer measurement results: output spectra of $1/4f_0$ and f_{FS}, input reference spur, and PN of $1/4f_0$.

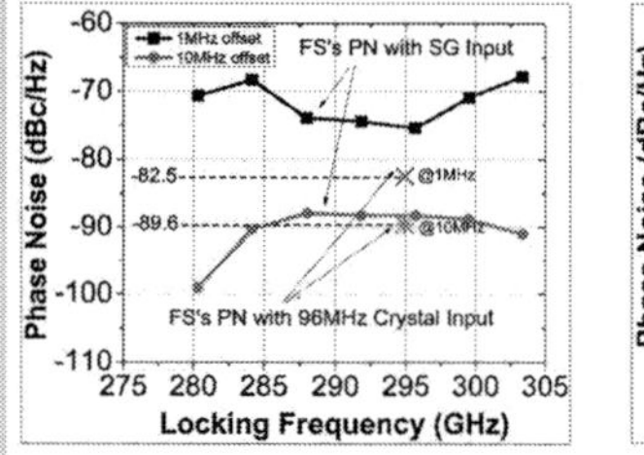

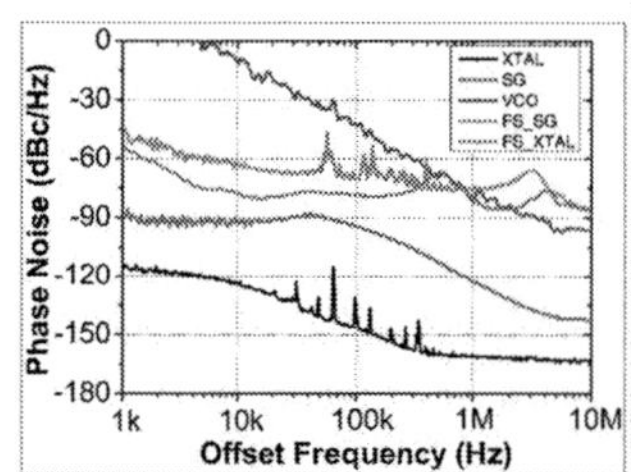

Figure 14.7.5: The synthesizer frequency scaled f_{FS} phase noise vs its locking frequencies, and the measured phase noise profile for the synthesizer, reference source, and VCO.

Performance summary of the proposed divider, VCO and synthesizer

Three-Phase Injection-Locked Divider		Triple-Push VCO		300GHz Frequency Synthesizer	
Frequency (GHz)	91.9 ~ 101.8	Frequency (GHz)	279.96 ~ 303.36	Frequency (GHz)	280.32 ~ 303.36
Divider Ratio	4	Tuning Range	8%	Divider ratio	1024
Locking Range	10.2%	Output Power	-14 dBm	Locking Range	7.9%
Input Power	< 0 dBm	P.N. @ 1MHz	-80.28 dBc/Hz	Ref. Spur	-64 dBc
Supply (V)	2	Supply (V)	1.8	PN @ 100kHz/1MHz offset (dBc/Hz)	-77.8 / -82.5 @ 294.9GHz
DC Power	48.4 mW	DC Power	105.6 mW	DC Power (P_D)	376 mW

Comparison table of 300GHz frequency synthesizer

	This work	*MTT-S* 2011 [3]**	*JSSC* 2011 [2]**
Frequency (GHz)	280.32 ~ 303.36 (3rd)	300.76 ~ 301.12 (fund.)	160 ~ 169 (2nd)
Divider ratio	1024	10	128
Locking Range	7.9%	0.12%	5.5%
Ref. Spur	-64 dBc	NA	NA
PN @ 100kHz/1MHz offset (dBc/Hz)	-77.8 / -82.5 @ 294.9GHz	-78 / -85 @ 300.96 GHz	-75 / -78 @ 163 GHz
DC Power (P_D)	376 mW	301.6 mW	1250 mW
FOM_T* @ 100kHz/1MHz offset	-179.4 / -163.9 dBc/Hz	-144.4 / -131.36 dBc/Hz	-163.1 / -146.1 dBc/Hz
Technology (f_{max})	90nm SiGe BiCMOS (315 GHz)	InP HBT (600 GHz)	130nm SiGe BiCMOS (280 GHz)

$$*\,FOM_T = PN - 20\log\left(\frac{f_0}{\Delta f}\cdot\frac{Locking\ Range}{10}\right) + 10\log\left(\frac{P_D}{1mW}\right)$$

** To the best of our knowledge, [2] [3] are the highest frequency synthesizers in silicon and InP HBT, respectively.

Figure 14.7.6: The performance summary and comparison tables for the VCO, divider and 300GHz frequency synthesizer.

978-1-4799-0917-9/14 $31.00 © 2014 IEEE

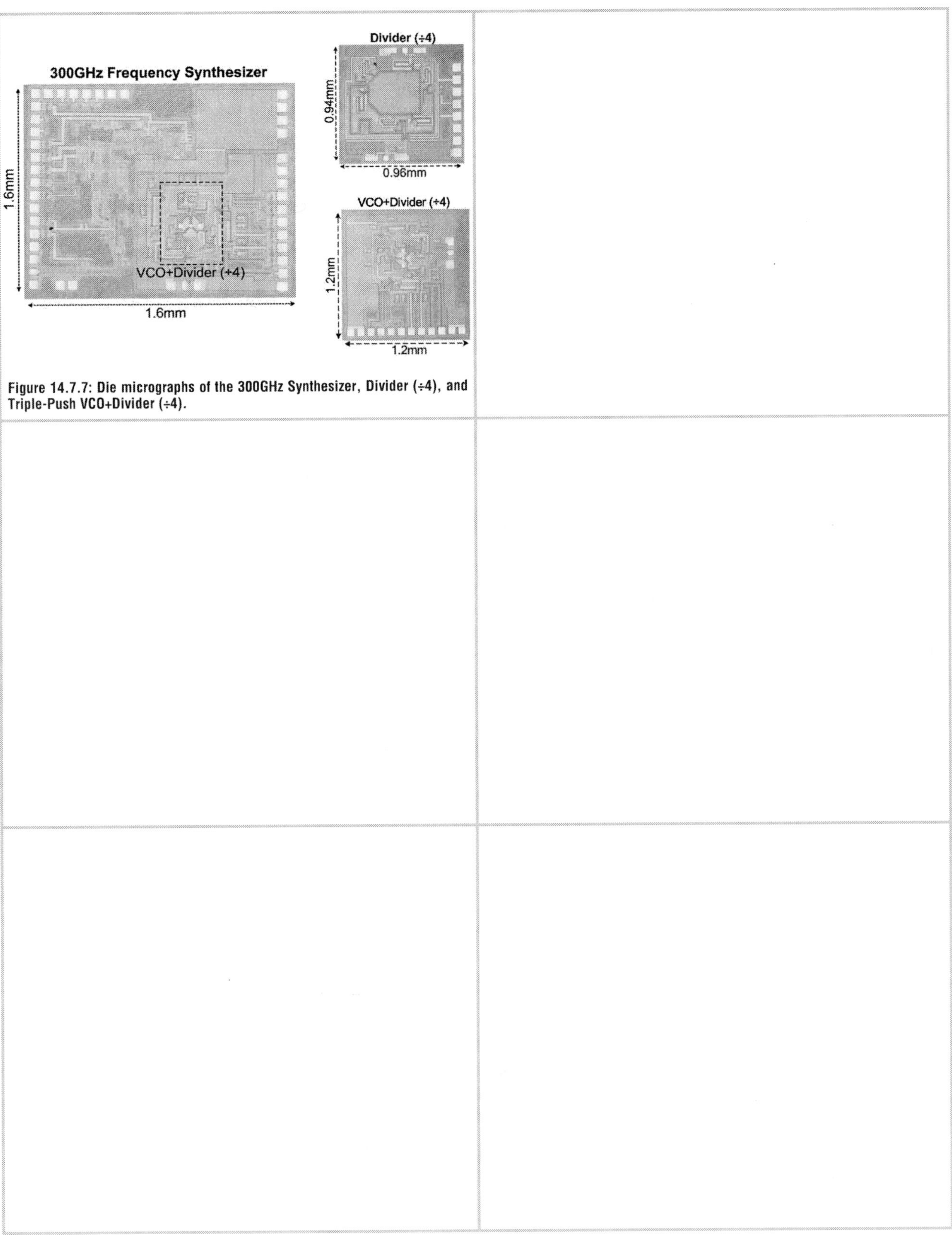

Figure 14.7.7: Die micrographs of the 300GHz Synthesizer, Divider (÷4), and Triple-Push VCO+Divider (÷4).

14.8 A 247-to-263.5GHz VCO with 2.6mW Peak Output Power and 1.14% DC-to-RF Efficiency in 65nm Bulk CMOS

Muhammad Adnan, Ehsan Afshari

Cornell University, Ithaca, NY

Signal generation at mm-Wave-to-THz frequencies is attractive because of its applications in bio-sensing, spectroscopy, detection of concealed weapons, as well as high-data-rate communication. CMOS is considered a potential platform to implement a low-cost and high-yield signal generation solution at this frequency range. Despite continuous scaling, effective f_{max} of active devices is not high enough and hence either harmonic oscillators or frequency multipliers are employed for high-frequency signal generation. In recent CMOS VCO designs, reasonable power levels (~1mW) and tuning range (~10GHz) have been reported at around 300GHz but with very low DC-to-RF efficiency (<0.4%), which is undesirable for portable applications [1,4]. Moreover, power-level fluctuation is more than 3dB across the frequency range [1]. In this work, we introduce a scalable VCO architecture that efficiently generates and extracts the 2nd harmonic at 256GHz and hence simultaneously achieves high tuning range (16GHz), high output power (2.6mW), high DC-RF efficiency (1.14%), and low phase noise (-94dBc/Hz to -85dBc/Hz at 1MHz offset frequency across the tuning range). When compared to published state-of-the-art, this work demonstrates the highest output power, tuning range, and DC-to-RF efficiency, and the lowest phase noise among all CMOS VCOs above 200GHz.

Figure 14.8.1(top) shows our proposed system that consists of a loop of oscillators, coupled via variable delay elements (i.e., phase shifters) to simultaneously enhance tuning range and output power. In steady state, each oscillator oscillates at the fundamental frequency, ω_0, and the total phase shift across the loop is a multiple of 2π at this frequency. At a given oscillation mode, if the phase of coupling elements is changed the frequency of each oscillator has to shift to conserve total phase shift across the loop as a multiple of 2π [1]. If the coupling delay is increased, the overall frequency of the loop decreases and vice versa. Besides this delay-based tuning mechanism, the varactor inside each core also changes the frequency (capacitive tuning). Hence our system has two fundamentally different methods of frequency tuning. The exact expression of total frequency tuning is calculated using Adler's equation and is shown in Fig. 14.8.1. There are three major factors resulting in high DC-to-RF efficiency. First, each core oscillator is carefully designed to generate maximum power. Secondly, the coupling between core oscillators is achieved through passive implicit phase shifters. Finally, the core oscillators and the coupling blocks are designed to keep the fundamental power in the loop while funneling the desired harmonic current to the output power combiner.

To generate the maximum fundamental power at each core oscillator, we propose a novel capacitive-degenerated self-feeding topology as shown in Fig. 14.8.1 (bottom left). It consists of a transmission line (TL_0) from gate to drain along with a varactor (C_s) at the source. By selecting the length and impedance of TL_0 as well as C_s, we can control the gain and phase shift between the gate and drain of M_1. To extract maximum power at the fundamental frequency (ω_0) out of a given transistor, [2] proved that an optimum gain (A_{opt}) and phase (ϕ_{opt}) between gate and drain is necessary. In our structure, the expression to achieve these optimum values is shown in Fig. 14.8.1 [6]. At the target ω_0 of 130GHz, for a 26μm/65nm transistor with a 20fF varactor (Q~5), TL_0 with a length of 69° and impedance of 58Ω satisfies the expression.

Figure 14.8.2 (top left) shows the passive coupling mechanism of multiple oscillators via TL_C along with harmonic extraction. The varactor at the source not only tunes the frequency of individual oscillators but also changes the phase of injected current into the next oscillator of the loop. This way, a single varactor performs both capacitive as well as delay-based tuning. To verify our approach, the simulated tuning range for each stand-alone core (only capacitive tuning) as well as the overall tuning range (both mechanisms) for a VCO with 2-stage loop is plotted in Fig. 14.8.2 (top right). It is evident that the two tuning mechanisms work constructively and hence the total tuning range is increased.

Figure 14.8.2 (top left) also shows extraction of $2\omega_0$ current using quarter-wave ($\lambda/4$) transformers (TL_d). The $\lambda/4$ transformer enforces differential oscillation mode of adjacent cores at ω_0. This is because if the two consecutive stages are in-phase, node "C" will have high impedance at ω_0, which after TL_d will be transformed to low impedance at the drains of transistors. Similarly, if the two stages oscillate in differential mode, node "C" will be a virtual ground at ω_0, which is transformed to an open circuit at the drains of the transistors. This differential operation is desirable for $2\omega_0$ extraction. To enhance the efficiency, proper matching at ω_0 and $2\omega_0$ is necessary. The $2\omega_0$ current generated at the drain of each transistor can travel in multiple paths. In a standard ring oscillator (e.g. double-push or triple-push), the gate of the next stage presents itself with a very small impedance at the drain of the transistor at the harmonic frequency. This becomes even worse in the presence of small interconnect and parasitic gate inductance (Fig. 14.8.2 bottom). On the other hand, in our proposed architecture, the TL_0 is long enough to block the drain from its own gate and the coupling line increases the small impedance of the gate of the next stage by a factor of 2.5. In simulation, this technique enhances harmonic extraction by more than 3dB compared to standard double-push architecture.

We implemented a loop of eight oscillators in a 65nm CMOS process using the proposed technique (Fig. 14.8.3). The chip area is 0.67mm x 0.65mm. All passives are carefully simulated in HFSS. The matching is performed at various levels of power combining. The last stages of the combiner are very important for two reasons. First, as can be seen in Fig. 14.8.3, the transmission lines in later stages are very long, so any standing wave in these lines results in additional loss. Second, these combining stages carry a large DC as well as AC current so any resonance behavior in these lines can cause reliability issues. We use an open stub for intermediate matching within the combiner. The 2nd-last combining stage consists of 46Ω lines. Hence the last stage matches 23Ω to 50Ω. The transmission lines in the last stage are implemented with an 8μm-wide coplanar waveguide that is gradually tapered into a 45μm×45μm output pad. The pad is implemented without any ground shielding for small capacitance. The last stage matching is shown in Fig. 14.8.3.

Figure 14.8.4 shows the measurement setup. The output is probed using a Cascade-I325 probe with built-in bias-T to power up the chip. Based on post-layout simulation, there is a DC voltage drop of 0.2 to 0.4V from output pads to drains of transistors. The tuning range and phase-noise measurements are performed at various supply voltages using an external WR-3.4 even-harmonic mixer and 16th harmonic of the LO source. By changing voltage across the varactor and supply voltage, continuous frequency tuning was achieved from 247 to 263.5GHz (Fig. 14.8.5). The phase noise varies between -85dBc/Hz to -94dBc/Hz across various control voltages at 1MHz offset frequency. The low phase noise level is the inherent result of our proposed topology that couples multiple oscillators. To accurately measure power a different setup using an Erikson PM4 calorimeter is used. Figure 14.8.5 plots output power at various supply voltages. The maximum achieved output power is 4.1dBm at 248.6GHz with a power supply of 1.6V and DC current of 142mA. Moreover, as we change the control voltage, the output power fluctuation remains less than 3dB for all supply voltages. From the output power and DC power consumption, we calculate DC-to-RF efficiency. The maximum DC-to-RF efficiency is 1.14%. The DC-to-RF efficiency is always better than 0.54% across all control and supply voltages. Figure 14.8.6 shows the comparison of our results with existing state of the art. It is evident that our proposed structure achieves record output power, DC-to-RF efficiency, and phase noise with comparable tuning range.

Acknowledgements:
The authors would like to acknowledge the National Science Foundation and Dr. Paul Maki of the Office of Naval Research for supporting this project. They also like to thank the TSMC university shuttle program for the chip fabrication.

References:
[1] Y. Tousi, O. Momeni, E. Afshari, "A 283-to-296GHz VCO with 0.76mW peak output power in 65nm CMOS," *ISSCC Dig. Tech. Papers*, Feb. 2012.
[2] O. Momeni, E. Afshari, "High Power Terahertz and Sub-Millimeter-Wave Oscillator Design: a Systematic Approach," *IEEE J. Solid-State Circuits*, Dec. 2011.
[3] J. Grzyb, Yan Zhao, U.R. Pfeiffer, "A 288GHz Lens-Integrated Balanced Triple-Push Source in a 65nm CMOS Technology," *IEEE J. Solid-State Circuits*, July 2013.
[4] P. Chiang, O. Momeni, P. Heydari, "A Highly Efficient 0.2THz Varactor-Less VCO with -7dBm Output Power in 130nm SiGe," *Compound Semiconductor Integ. Circuit Symp.*, Oct. 2012
[5] K. Sengupta, A. Hajimiri, "A 0.28THz 4×4 Power-Generation and Beam-Steering Array," *ISSCC Dig. Tech. Papers*, Feb. 2012.
[6] R. Han, E. Afshari, "A 260GHz Broadband Source with 1.1mW Continuous-Wave Radiated Power and EIRP of 15.7dBm in 65nm CMOS," *ISSCC Dig. Tech. Papers*, Feb. 2013.

978-1-4799-0917-9/14 $31.00 © 2014 IEEE

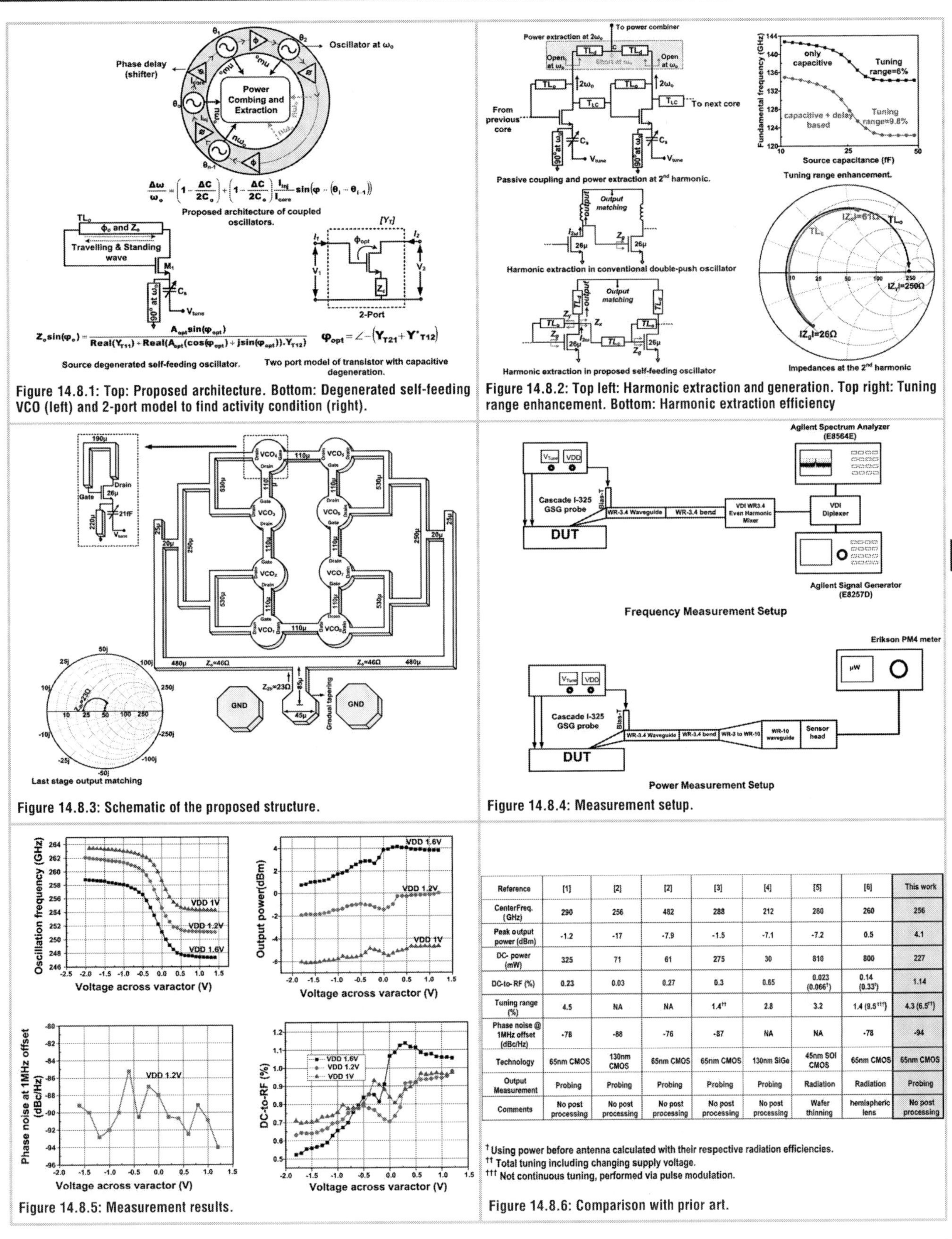

Figure 14.8.1: Top: Proposed architecture. Bottom: Degenerated self-feeding VCO (left) and 2-port model to find activity condition (right).

Figure 14.8.2: Top left: Harmonic extraction and generation. Top right: Tuning range enhancement. Bottom: Harmonic extraction efficiency

Figure 14.8.3: Schematic of the proposed structure.

Figure 14.8.4: Measurement setup.

Figure 14.8.5: Measurement results.

Reference	[1]	[2]	[2]	[3]	[4]	[5]	[6]	This work
CenterFreq. (GHz)	290	256	482	288	212	280	260	256
Peak output power (dBm)	-1.2	-17	-7.9	-1.5	-7.1	-7.2	0.5	4.1
DC- power (mW)	325	71	61	275	30	810	800	227
DC-to- RF (%)	0.23	0.03	0.27	0.3	0.65	0.023 (0.066†)	0.14 (0.33†)	1.14
Tuning range (%)	4.5	NA	NA	1.4††	2.8	3.2	1.4 (9.5†††)	4.3 (6.5††)
Phase noise @ 1MHz offset (dBc/Hz)	-78	-88	-76	-87	NA	NA	-78	-94
Technology	65nm CMOS	130nm CMOS	65nm CMOS	65nm CMOS	130nm SiGe	45nm SOI CMOS	65nm CMOS	65nm CMOS
Output Measurement	Probing	Probing	Probing	Probing	Probing	Radiation	Radiation	Probing
Comments	No post processing	No post processing	No post processing	No post processing	No post processing	Wafer thinning	hemispheric lens	No post processing

† Using power before antenna calculated with their respective radiation efficiencies.
†† Total tuning including changing supply voltage.
††† Not continuous tuning, performed via pulse modulation.

Figure 14.8.6: Comparison with prior art.

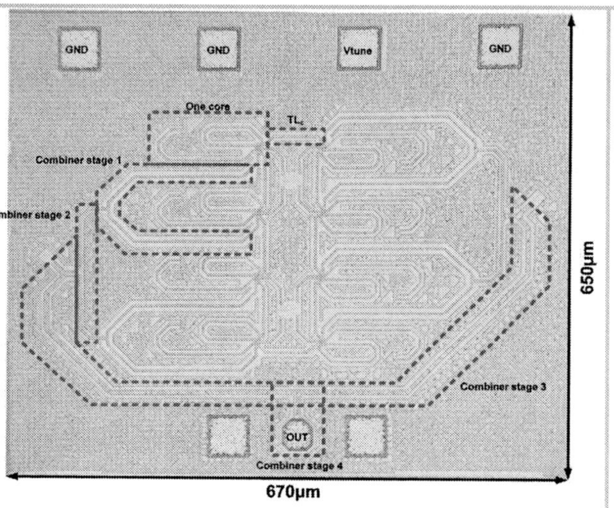

Figure 14.8.7: Die micrograph.

Session 15 Overview: *Digital PLLs*
HIGH-PERFORMANCE DIGITAL SUBCOMMITTEE

Session Chair: *Anthony Hill*
Texas Instruments, Dallas, TX

Session Co-Chair: *Hiroo Hayashi*
Toshiba, Kawasaki, Japan

The four papers presented in this session highlight developments in clock generation and distribution. These papers demonstrate the growing trend toward fully-synthesizable digital PLLs. Solutions presented relate to digital PLL integration, including power-supply noise rejection, temperature compensation, and fast frequency switching required in modern SoCs.

978-1-4799-0917-9/14 $31.00 © 2014 IEEE

15.1 A 0.0066mm² 780μW Fully Synthesizable PLL with a Current-Output DAC and an Interpolative Phase-Coupled Oscillator Using Edge-Injection Technique **1:30 PM**

W. Deng, Tokyo Institute of Technology, Tokyo, Japan

In Paper 15.1, the Tokyo Institute of Technology presents a 780mW fully synthesizable PLL with a current-output DAC and an interpolative phase-coupled oscillator using an edge-injection technique. This design introduces a current digital-to-analog converter composed of standard CMOS NAND gates, where the PLL is completely assembled using standard-cell components. The PLL consumes 0.0066mm² in 65nm CMOS operating at 380MHz to 1.41GHz.

15.2 A 0.012mm² 3.1mW Bang-Bang Digital Fractional-N PLL with a Power-Supply-Noise Cancellation Technique and a Walking-One-Phase-Selection Fractional Frequency Divider **2:00 PM**

J. Liu, Samsung Electronics, Yongin, Korea

In Paper 15.2, Samsung Electronics presents a 0.012mm² 3.1mW bang-bang digital fractional-N PLL with power-supply-noise cancellation and a walking-one-phase-selection fractional frequency divider in 20nm CMOS. This digital PLL offers a solution to power-supply noise and the high energy consumption of a time-to-digital converter (TDC). The noise-cancellation technique shown in this paper suppresses jitter degradation to <10% across a supply-noise frequency of 5kHz to 50MHz and supports a frequency range of 25MHz to 1.6GHz.

15.3 A 2.4GHz ADPLL with Digital-Regulated Supply-Noise-Insensitive and Temperature-Self-Compensated Ring DCO **2:15 PM**

Y-C. Huang, MediaTek, Hsinchu, Taiwan

In Paper 15.3, MediaTek presents a 2.4GHz all-digital PLL with a digitally regulated supply-noise-insensitive temperature-self-compensated ring DCO. This 40nm PLL consumes just 6.4mW of power, with an integrated jitter of 3.29ps at 2.418GHz. The supply sensitivity after calibration is 0.0087, with a frequency drift of 2% across temperatures from 20°C to 100°C. The PLL area is 0.013mm².

15.4 A 20-to-1000MHz ±14ps Peak-to-Peak Jitter Reconfigurable Multi-Output All-Digital Clock Generator Using Open-Loop Fractional Dividers in 65nm CMOS **2:30 PM**

A. Elkholy, University of Illinois, Urbana, IL

In Paper 15.4, the University of Illinois at Urbana-Champaign and Intel present a 20MHz-to-1GHz, 14ps peak-to-peak jitter reconfigurable multi-output all-digital clock generator using open-loop fractional dividers in 65nm CMOS. This clock-generation unit occupies an area of 0.12mm².

978-1-4799-0917-9/14 $31.00 © 2014 IEEE

15.1 A 0.0066mm² 780μW Fully Synthesizable PLL with a Current-Output DAC and an Interpolative Phase-Coupled Oscillator Using Edge-Injection Technique

Wei Deng, Dongsheng Yang, Tomohiro Ueno, Teerachot Siriburanon, Satoshi Kondo, Kenichi Okada, Akira Matsuzawa

Tokyo Institute of Technology, Tokyo, Japan

Phase-locked loops (PLLs) are widely used for clock generation in modern digital systems. All-digital PLLs have been proposed to address design issues in conventional analog PLLs. However, current all-digital PLLs require custom circuit design, and therefore cannot fully leverage advanced automated digital design flows. While fully synthesizable PLLs [1, 2, 3] have been reported, they suffer from high power consumption and large area. This arises because each stage of the ring needs to have a large number of parallel tristate buffers/inverters in order to achieve the necessary frequency resolution. Moreover, custom-designed cells are required in prior synthesizable PLLs [2, 3], introducing additional place-and-route (P&R) steps, leading to poor portability, integration, and scalability. To address these issues, this paper proposes a fully synthesizable PLL based solely on a standard digital library, with a current-output digital-to-analog converter (DAC) for maintaining frequency linearity and duty balance, an interpolative phase-coupled oscillator for minimizing the output phase imbalance from automatic P&R, as well as an edge injection technique for avoiding injection-pulse width issues.

A block diagram of the proposed PLL with a DAC and an interpolative phase-coupled oscillator is given in Fig. 15.1.1. All circuits that makeup the PLL, including the DAC and DCO, are implemented using standard cells and an automated design flow. A dual-loop PLL architecture [5] is employed and improved in this design to provide continuous tracking of voltage/temperature variations and to avoid the timing problems associated with conventional injection-locked PLLs. The frequencies of two oscillators are digitized by two counters. A signed adder/subtractor compares the digitized frequencies with a predefined frequency-control word (FCW) and provides a frequency difference to a digital loop filter consisting of a proportional path and an integral path. The filter output determines whether to increase, decrease, or hold the DCO oscillation frequency. In addition, an edge-injection technique is incorporated into this design.

Figure 15.1.2 illustrates the proposed oscillator architecture that is used in both the main and replica VCOs. In order to relieve the oscillator output phase imbalance caused by the automatic P&R, an interpolative phase-coupled oscillator built by three 3-stage oscillators is developed, based on the concept in [6]. Due to its internal feedback and feed-forward control using phase interpolators for the oscillator, the phase difference within the ring and between all adjacent rings will remain fixed in time, leading to balanced output phases. To maintain the control-code resolution and extend the operating frequency range, the oscillator is designed to operate with a coarse, medium, and fine tuning. A standard cell-based DAC controls the coarse tuning of the oscillator. As shown in Fig. 15.1.2, a digitally-controlled varactor using NAND gates is adopted in the medium-tuning circuitry. The fine-tuning circuitry is realized by another type of digitally controlled varactor using inverters and NAND gates. A digitally controlled NAND gate introduces a capacitance difference at the inverter output node, which alters the rising and falling slopes, thereby changing the Miller effect sensitivity. Thus, the effective capacitance seen from the ring oscillator is also changed. This capacitance difference is adopted for the fine-tuning stage.

Conventionally, a path-selection method is applied for coarse tuning in synthesized oscillators [4], helping to reduce power consumption and chip area. However, this method suffers from unbalanced output phases due to different loading at each stage, degrading injection efficiency. In our design, an analog-equivalent DAC built from standard CMOS NAND gates is used in the coarse-tuning circuitry, reducing power consumption and area, without sacrificing phase balance. Tristate buffers are not used, as they may not be present in some standard cell libraries. To make it easier to understand, Fig. 15.1.3 depicts a diagram of a 4b binary-weighted current-output DAC constructed from five two-input NAND gates. The DAC consists of a PMOS-current-source array and an NMOS current mirror. The PMOS-current-source array is built by connecting the outputs of 4 NAND gates together, and with an input of each NAND gate connected to a digital control bus and the other input connected to D4. The NMOS current mirror is built by connecting one input of a NAND gate to its output and the other input to logic-1. The proposed current-output DAC is combined with a current-starved ring oscillator using NAND gates as its delay cell to form a digitally controlled oscillator, which maintains high current-domain linearity. The simulated DAC power consumption is less than 60μW.

Injecting a stream of narrow pulses into the oscillator is a widely used approach to improve its jitter performance by resetting the oscillator at every reference cycle. However, the injection pulse width requires additional calibration to guarantee robust operation over environmental variations, since an excessively narrow pulse width causes a failure to lock or an over-designed pulse width causes a strong periodic disturbance to oscillator phase and amplitude-degrading deterministic jitter. As shown in Fig. 15.1.4, an edge-injection technique is applied to avoid the injection pulse width issue [7]. Assuming that the oscillator frequency is N times higher than the reference frequency, an inverted injection-window signal forces the oscillator to stop oscillating at the rising edge of the injection-window signal. Then, an injection signal with a clean edge is forwarded to the oscillator and replaces the noisy edge V_y to prevent jitter accumulation for several cycles. Finally, at the falling edge of the injection-window signal, the oscillator starts to oscillate again with its phase aligned to the phase of injection signal. Compared to conventional pulse injection, which requires a carefully timed injection-pulse width, the proposed edge injection style does not imply strict timing on the injection-window width. Phase replacement and alignment can be done in one reference cycle.

The fully synthesizable PLL is fabricated in a 65nm digital CMOS process. The PLL occupies 110×60μm². The phase noise is evaluated by using a signal-source analyzer (Agilent E5052B) and the spectrum is measured using a spectrum analyzer (Agilent E4407B). The measured frequency-tuning range of the PLL is 0.39GHz to 1.41GHz. At 0.9GHz output, the power consumption is 780μW excluding output buffers from a 0.8V power supply. Fig. 15.1.5 shows the measured phase noise and output spectrum at 0.9GHz output using a 150MHz reference clock. The phase noise corresponds to a 1.7ps jitter when integrated from 10kHz to 40MHz.

Figure 15.1.6 shows a performance comparison table between this work and previously published synthesizable PLLs. The PLL presented in this paper achieves the best performance in terms of power, jitter, and area. The figure of merit (FOM) is -236.5dB at 0.9GHz output frequency, where the FOM is defined as $10\log[(\sigma_t/1s)^2 \cdot (P_{DC}/1mW)]$, where σ_t is the integrated jitter, and P_{DC} is the DC power consumption. The chip micrograph is shown in Fig. 15.1.7.

Acknowledgments:
This work was partially supported by STARC, SCOPE, MIC, MEXT, Canon Foundation, and VDEC in collaboration with Cadence Design Systems, Inc., and Agilent Technologies Japan, Ltd.

References:
[1] Y. Park and D.D. Wentzloff, "An All-Digital PLL Synthesized from a Digital Standard Cell Library in 65nm CMOS," *IEEE Custom Integrated Circuits Conf.*, 2011.
[2] W. Kim, J. Park, J. Kim, T. Kim, H. Park, and D. Jeong, "A 0.032mm² 3.1mW Synthesized Pixel Clock Generator with 30psrms Integrated Jitter and 10-to-630MHz DCO Tuning Range," *ISSCC Dig. Tech. Papers*, pp. 250-251, 2013.
[3] M. Faisal and D.D. Wentzloff, "An Automatically Placed-and-Routed ADPLL for the MedRadio Band using PWM to Enhance DCO Resolution," *IEEE Radio Frequency Integrated Circuits Symposium*, pp. 115-118, 2013.
[4] D. Sheng, C.-C. Chung, and C.-Y. Lee, "An Ultra-Low-Power and Portable Digitally Controlled Oscillator for SoC Applications," *IEEE Trans. Circuits and Systems-II*, vol. 54, no. 11, pp. 954-958, 2007.
[5] W. Deng, A. Musa, T. Siriburanon, M. Miyahara, K.Okada, and A. Matsuzawa, "A 0.022 mm² 970μW Injection-Locked PLL with -243 dB FOM using Synthesizable All-Digital PVT Calibration Circuits, " *ISSCC Dig. Tech. Papers*, pp. 248-249, Feb. 2013.
[6] A. Matsumoto, S. Sakiyama, Y. Tokunaga, T. Morie, and S. Dosho, "A Design Method and Developments of a Low-Power and High-Resolution Multiphase Generation System," *IEEE J. Solid-State Circuits*, vol. 43, no. 4, pp. 831–843, 2008.
[7] D. Park and S. Cho, "A 14.2mW 2.55-to-3GHz Cascaded PLL with Reference Injection, 800MHz Delta-Sigma modulator and 255fs_{rms} Integrated Jitter in 0.13μm CMOS," *ISSCC Dig. Tech. Papers*, pp. 344-345, 2012.

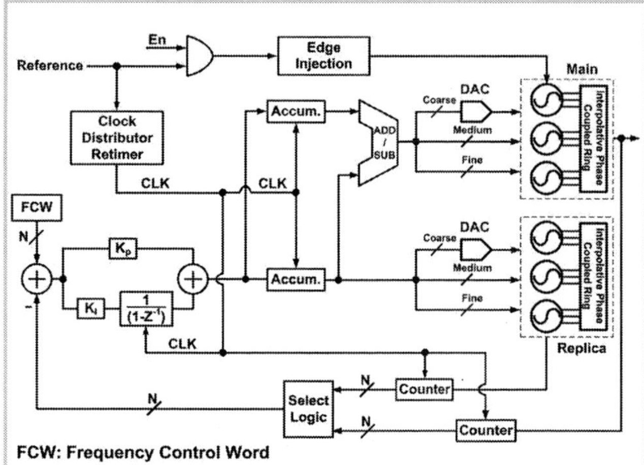

Figure 15.1.1: Block diagram of the proposed fully synthesizable PLL with a DAC and an interpolative phase-coupled oscillator.

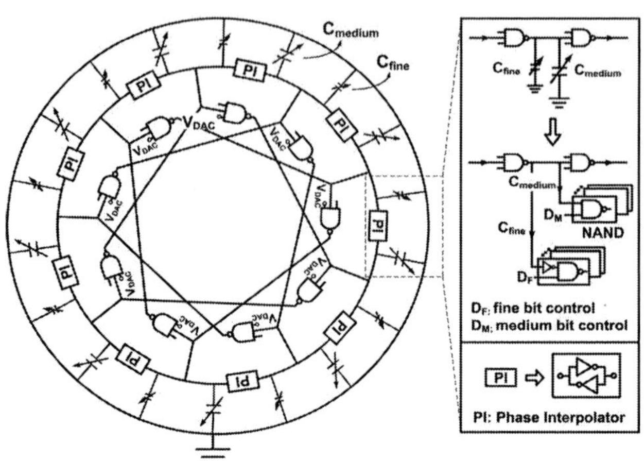

Figure 15.1.2: Block diagram of the interpolative phase-coupled oscillator.

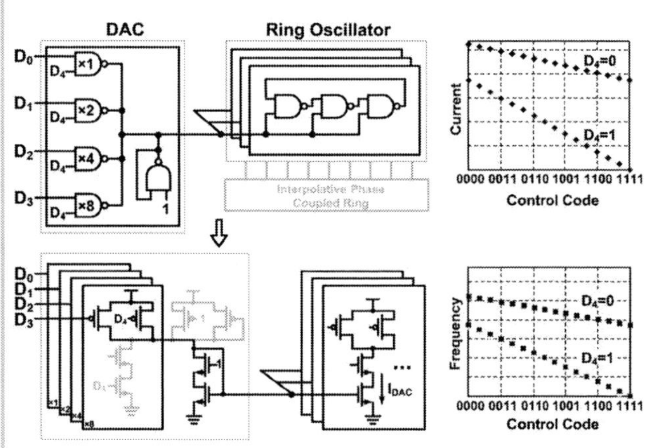

Figure 15.1.3: Conceptual diagram of the synthesizable DAC with a current-linear output.

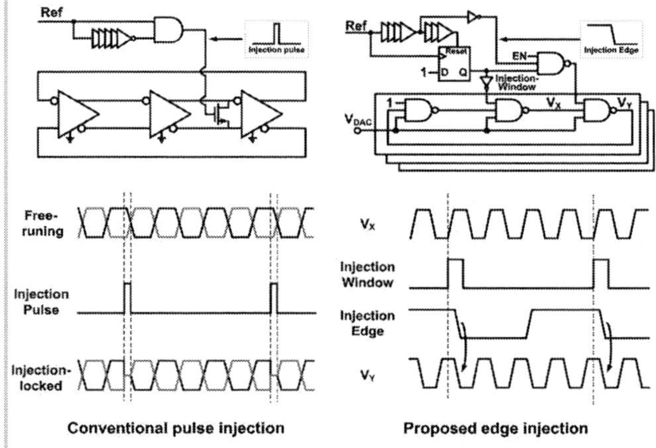

Figure 15.1.4: Block diagram and locking transient of the conventional pulse-injection locking and the proposed edge-injection method.

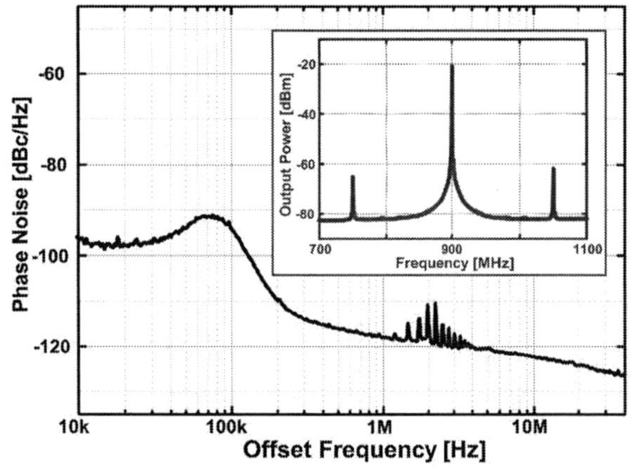

Figure 15.1.5: Measured phase noise and spectrum characteristics at a carrier of 0.9GHz.

	This work	[1]	[2]	[3]
Freq. [GHz]	0.39-1.41	1.5-2.7	0.25-1.65	0.4-0.46
Ref. [MHz]	40-350	10	25	40.3
Power [mW]	0.78 @900 MHz	13.7 @2.5 GHz	3.1 @250MHz	2.1 @403MHz
Area [mm^2]	**0.0066**	0.042	0.032	0.1
Normalized Area	1	6.36	4.84	15.15
Integ. Jitter [ps]	1.7	N.A.	30	N.A.
Jitter RMS [ps]	2.8	3.2	N.A.	13.3
FOM [dB]	**-236.5**	-218.6*	-205.5	-214*
CMOS Technology	65nm	65nm	28nm	65nm
W/ custom cells?	No	No	Yes	Yes
Coarse Frequency Control	DAC	None	None	None

*FOM is calculated based on RMS jitter.

Figure 15.1.6: Performance summary and comparison with state-of-the-art synthesized PLLs.

978-1-4799-0917-9/14 $31.00 © 2014 IEEE

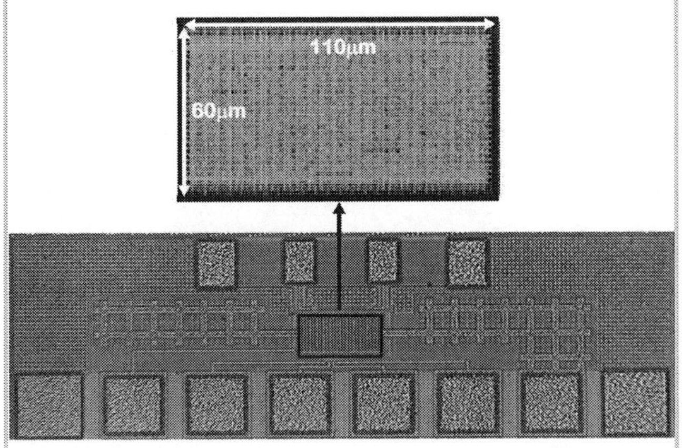

110μm

60μm

Figure 15.1.7: Die micrograph.

15.2 A 0.012mm² 3.1mW Bang-Bang Digital Fractional-N PLL with a Power-Supply-Noise Cancellation Technique and a Walking-One-Phase-Selection Fractional Frequency Divider

Jenlung Liu, Tae-Kwang Jang, Yonghee Lee, Jungeun Shin, Seunghoon Lee, Taeik Kim, Jaejin Park, Hojin Park

Samsung Electronics, Yongin, Korea

Digital phase-locked loops (DPLLs) [1-7] have received considerable attention recently due to their compatibility with advanced CMOS technology. However, there are two critical factors hindering their uptake in SoC products. One factor is that a digitally controlled oscillator (DCO) is highly sensitive to supply noise. A common solution is to apply voltage regulation or to adopt digital calibration [2] at the cost of larger area, higher power consumption or both. The other factor is a power-hungry time-to-digital converter (TDC), which typically requires complex auxiliary circuitry to overcome sensitivity to process, voltage and temperature [3]. A bang-bang phase/frequency detector (BBPFD) is a good alternative to the TDC for low-power small-size applications. A fractional-N implementation, however, still demands a fractional frequency divider with high design complexity [5].

This paper presents a bang-bang all-digital fractional-N PLL, which occupies a small area, consumes low power and addresses the aforementioned issues. A block diagram of the fractional-N DPLL is shown in Fig. 15.2.1. An automatic frequency controller (AFC) tunes a DCO frequency in the foreground for fast locking. The DCO employs a supply-noise canceling architecture to address the power supply noise issue with negligible power and area overhead. We adopt a phase-interpolator-based fractional divider controlled by a walking-one phase selector for low power and compactness.

Supply-noise immunity of the DCO is achieved by canceling variations in the frequency-controlling current that arise from power supply fluctuation. The DCO is composed of a current-controlled oscillator (CCO) whose frequency is controlled by a current-mode DAC. A constant-G_m bias circuit is widely used to generate a reference current for a DAC, as it is free from frequency compensation over process corners and it requires small supply headroom. However, its poor supply sensitivity ($\Delta I_{OUT}/\Delta V_{DD}$) is a bottleneck for the DCO operation and the situation gets worse in advanced process nodes. In our bias circuits (Fig. 15.2.2), we include an additional constant-G_m bias branch, BIAS2, whose current variation cancels out the variation in the original constant-G_m bias branch, BIAS1. Current variation, ΔI, in BIAS1 due to ΔV_{DD} depends on the finite output resistance of M1 and $1/G_m$ of a diode-connected transistor, M3, therefore, its supply sensitivity is inversely proportional to the output resistance of M1. BIAS2 generation uses transistors whose lengths are double that used to generate BIAS1, so that its output resistance becomes approximately twice as large, and hence, its current variation due to ΔV_{DD} is half of the one of BIAS1. As the current of BIAS2 is multiplied by two when it is mirrored to M_{10} (M_{12}), the current variation of M_9 and M_{10} (M_{11} and M_{12}) cancel each other out. As a result, I_{M16} (I_{M13}) is independent of supply voltage fluctuation and this reference current is mirrored to the current source of the DAC, M_{14} and M_{17}. To reduce mismatch among the transistors in the circuits for BIAS1 and BIAS2 all transistors are designed using a single unit cell and placed close to each other. We also adopted the CCO architecture in [4] to further minimize the power supply sensitivity of the DCO by providing a negative gain from the supply voltage to the output frequency.

The block diagram of our phase-interpolator-based fractional frequency divider is shown in Fig. 15.2.1. A division ratio is defined by 9b integer and 16b fractional-frequency words. 11 LSBs of the fractional frequency word, DIV_{F_LSBs}, are randomized by a delta-sigma modulator (DSM) and then added to the integer frequency word to compute a 14b divider-control signal. 9 MSBs of the control signal directly control the integer frequency divider, while the remaining 5b select a signal out of 32 clock phases generated by the 8:32 phase interpolator (PI). The phase interpolator provides a fractional-frequency-division ratio, and thus, the maximum phase dithering is decreased by a factor of 32 [5].

The phase selection is controlled by a walking-one phase selector, as shown in Fig. 15.2.3(a). It forms a closed chain of 32 unit cells, which mainly act as DFFs and propagate the selection signal of the 32:1 MUX, MX, in one direction of the chain depending on the one-hot-coded input signal, SEL. Only the output of one unit cell in the chain is 1'b1, and this 1'b1 will be propagated to the output of the next unit cell based on the controlling signals, S and S-1. The unit cell does not propagate 1'b1 from its input to its output when S is 1'b0 and S-1 is 1'b1. For instance, Fig. 15.2.3(b) describes the phase-selection process using an 8-phase case for simplicity. Initially, the output of the sixth unit cell is 1'b1, by setting SEL[7:0] to 8'b00100000, which implies that S and S-1 of the sixth cell are 1'b1 and 1'b0, respectively, and 1'b0 and 1'b1 for the seventh cell. When SEL[7:0] is changed to 8'b10000000, the 1'b1 at the output of the sixth cell will propagate to the output of the eighth cell and then stop. During these transitions, MXOUT is changed from P5 to P7 and the distance between the rising edges of FEED represents the fractional frequency multiplication ratio of 99.75, in this case, where 100 is realized by the number of divided pulses while -0.25 is realized by phase traveling. As MX propagates one cell at a time, the phase of MXOUT can move by one unit phase step, which is $2\pi/32$ in our design, and hence MXOUT is guaranteed to be glitch-free. To reduce the power associated with 32 DFFs on the clock signal, clock gating is applied to turn off a DFF if its input and output are both zero.

The DPLL is implemented in 20nm CMOS with an area of 0.012µm². Its output frequency range is from 25MHz to 1.6GHz, consuming 3.1mW at 1.6GHz from a 0.9V power supply. To validate the supply-canceling topology, square, triangular and white noise sources were added to a quiet power supply using an Agilent N6705B and Agilent 33250A. Fig. 15.2.4 shows the peak-to-peak jitter performance with/without supply noises. When $50mV_{PP}$ square and triangular supply fluctuations are applied with frequencies from 5kHz to 50MHz, Fig. 15.2.4(a) shows that the measured period jitter variation (peak-to-peak, 100k hits) is less than 10% relative to using a clean 0.9V supply at the output frequency of 0.6GHz. Fig. 15.2.4(b) illustrates that our noise-canceling scheme can suppress the period jitter degradation to less than 4% within the output frequency range from 0.4GHz to 0.8GHz with $20mV_{RMS}$ white noise. Fig. 15.2.5 is the phase-noise graph of both the integer-N and fractional-N operations. It demonstrates that the BBPFD fractional operation is identical to the integer operation and the performance degradation is negligible with the phase-interpolator-based fractional divider. A comparison with state-of-the-art digital fractional-N PLLs is given in Fig. 15.2.6. The current design achieves 1.3 to 6.3× normalized power efficiency with 2.17 to 16.67× less area.

References:

[1] A. Rylyakov, *et al.*, "Bang-Bang Digital PLLs at 11 and 20GHz with Sub-200fs Integrated Jitter for High-Speed Serial Communication Applications," *ISSCC Dig. Tech. Papers*, pp. 94-95, 2009.

[2] A. Elshazly, *et al.*, "A 0.4-to-3GHz Digital PLL with Supply-Noise Cancellation using Deterministic Background Calibration," *ISSCC Dig. Tech. Papers*, pp. 92-94, 2011.

[3] R.B. Staszewski, *et al.*, "All-Digital PLL and Transmitter for Mobile Phones," *IEEE J. Solid-State Circuits*, pp. 2469-2482, vol. 40, no. 12, 2005.

[4] T-K. Jang, *et al.*, "A 0.026mm² 5.3mW 32-to-2000MHz Digital Fractional-N Phase Locked-Loop using a Phase-Interpolating Phase-to-Digital Converter," *ISSCC Dig. Tech. Papers*, pp. 254-255, 2013.

[5] R. Nonis, *et al.*, "A $2.4ps_{rms}$-jitter Digital PLL with Multi-Output Bang-Bang Phase Detector and Phase-Interpolator-Based Fractional-N Divider," *ISSCC Dig. Tech. Papers*, pp. 356-357, 2013.

[6] M. S.W. Chen, *et al.*, "A Calibration-Free 800MHz Fractional-N Digital PLL with Embedded TDC," *ISSCC Dig. Tech. Papers*, pp. 472-473, 2010.

[7] D.S. Kim, *et al.*, "A 0.3–1.4 GHz All-Digital Fractional-N PLL With Adaptive Loop Gain Controller," *IEEE J. Solid-State Circuits*, vol. 45, no. 11, pp. 2300-2311, 2010.

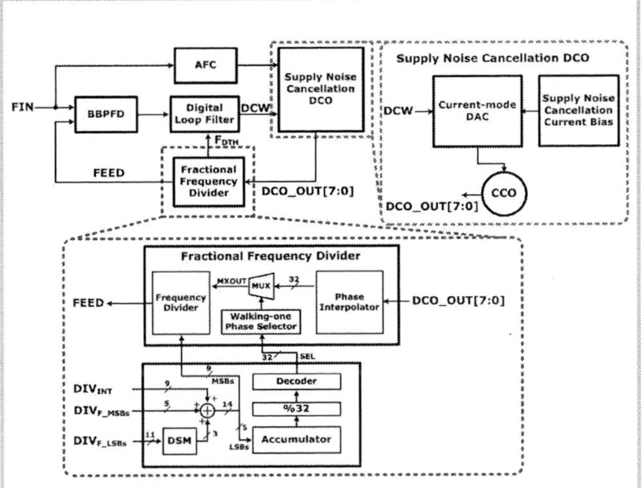

Figure 15.2.1: A block diagram of the digital phase-locked loop.

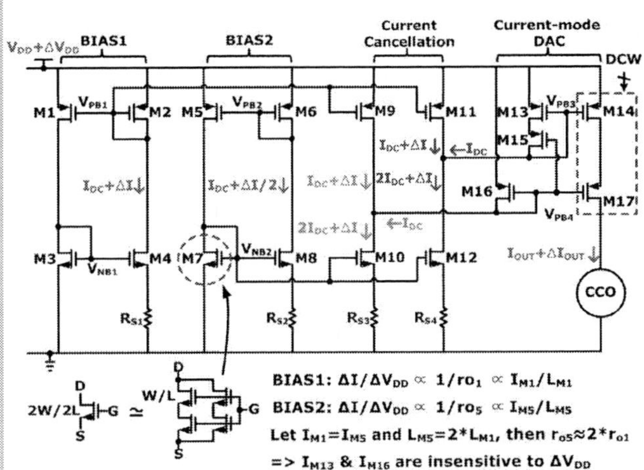

Figure 15.2.2: Schematics of the supply-noise cancellation DCO.

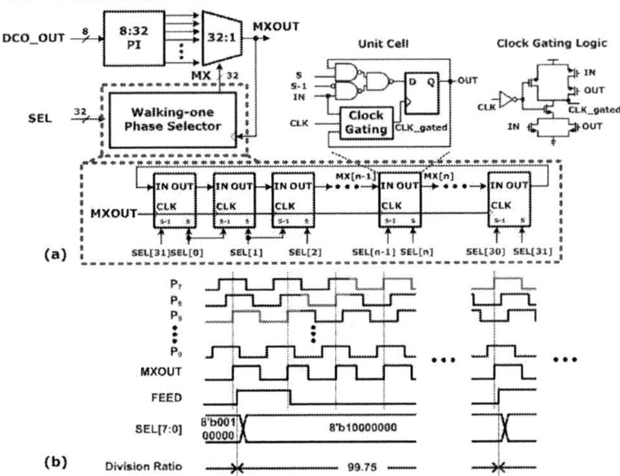

Figure 15.2.3: (a) Schematics of the walking-one phase selector, (b) functional operations with an 8-phase example for simplicity.

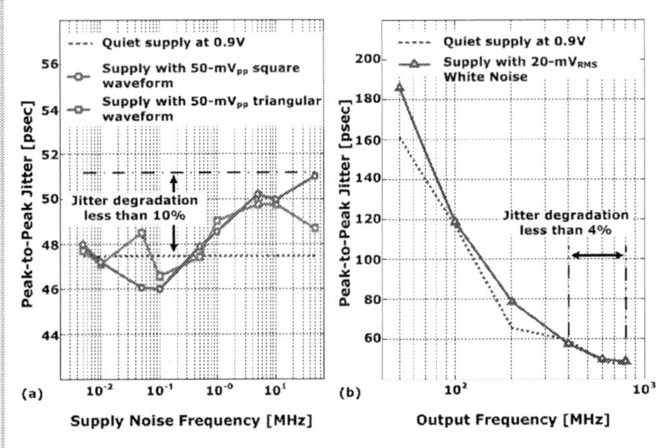

Figure 15.2.4: Measured jitter performances (a) with square/triangular noises on the supply, (b) with white noise on the supply.

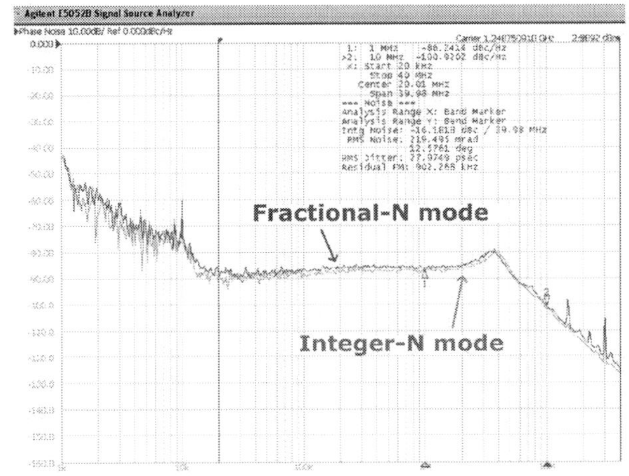

Fractional-N mode

Integer-N mode

Figure 15.2.5: Measured phase noise at integer and fractional modes with the output frequencies of 1.23GHz and 1.24875GHz, respectively.

Technology		20nm		
Freq. Range [GHz]		0.025 – 1.6		
Period Jitter RMS/Pk-Pk [ps]	Output/DCO Freq. [GHz]	0.4/0.8	0.6/1.2	0.8/1.6
	Quiet	6.78/59.48	5.63/48.32	5.83/47.39
	Supply with White Noise	7.08/57.57	6.20/49.67	5.89/49.87
Power [mW]	Analog	1.48	2.03	2.55
	Digital	0.28	0.43	0.55
	Total	1.76	2.46	3.10
Area [mm²]	Analog	0.005		
	Digital	0.007		
	Total	0.012		

	This Work	ISSCC 2013[4]	ISSCC 2010[6]	JSSC 2010[7]
Technology	20nm	28nm	65nm	130nm
Freq. Range [GHz]	0.025 - 1.6	0.032 - 2.0	0.6 - 0.8	0.3 - 1.4
Area [mm²]	0.012	0.026	0.05	0.2
Power [mW]	3.1@1.6GHz	5.3@2.0GHz	3.2@0.8GHz*	16.5@1.35GHz
Normalized Power [mW/GHz]	1.94	2.65	4	12.22
Jitter (RMS) [ps]	4-7 / 28**	19.3	20-30	3.7

* Assuming 1.2V supply

** Period jitter / integrated jitter

Figure 15.2.6: Performance summary.

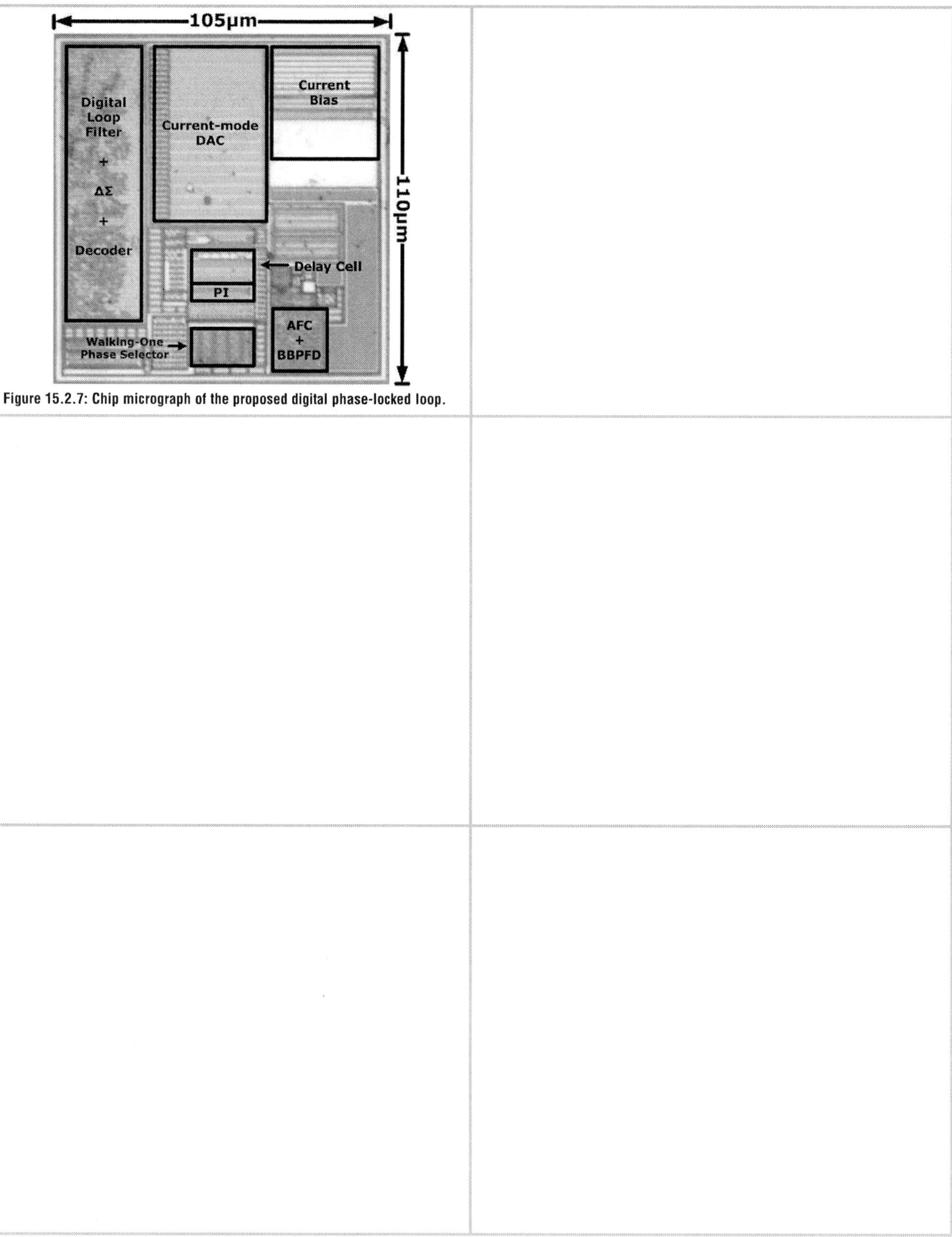

Figure 15.2.7: Chip micrograph of the proposed digital phase-locked loop.

15.3 A 2.4GHz ADPLL with Digital-Regulated Supply-Noise-Insensitive and Temperature-Self-Compensated Ring DCO

Yi-Chieh Huang, Che-Fu Liang, Hsien-Sheng Huang, Ping-Ying Wang

MediaTek, Hsinchu, Taiwan

Due to the high supply sensitivity of ring voltage-controlled oscillators (RVCOs) ([oscillation frequency change %] / [V_{DD} change %] typically lies in the range from 1 to 2 [1]), an LDO has to provide over 40dB power-supply-rejection ratio (PSRR) to maintain VCO phase noise. However, the voltage dropout of an LDO consumes extra power and voltage headroom, which is unacceptable in low-voltage design. Moreover, the device noise from the LDO degrades the phase-noise performance. Recently published works [1-5] employ analog compensation techniques to lower supply sensitivity, and [2] incorporates a hybrid background calibration scheme for robustness. However, the additional current sources and active devices embedded in the oscillator [1-5] increase power and noise. In this work, a DCO with passive devices and all-digital calibration mitigates supply sensitivity under PVT variation, while maintaining phase noise and power consumption. The digital background-calibration logic regulates the oscillator supply to an optimally insensitive point by monitoring a digital loop filter (DLF) code, leveraging an advantage of an ADPLL [6].

Figure 15.3.1 shows the block diagram of the ADPLL. It is composed of conventional ADPLL basic blocks and an additional digital background-calibration loop. The foreground calibration logic, composed of a frequency meter, band calibration and supply-sensitivity foreground calibration, ensures the PLL can operate at the correct sub-band and supply voltage. The high-speed digital block, composed of a 2nd-order $\Delta\Sigma$ modulator and a binary-to-thermometer decoder, is to enhance the resolution by shaping the quantization noise given the limited resolution of the integral-path $\Delta\Sigma$ modulator. The intrinsic noise in the loop provides enough dither for the 2nd-order $\Delta\Sigma$ modulator to avoid fractional spur. The supply-sensitivity background calibration logic reads the DLF code to regulate the VCO supply, i.e. VCO_VDD. The implementation is based on the measured curve in Fig. 15.3.3. It is observed that the search direction of the RES code is towards the point with the highest frequency in open loop, or with the lowest DLF code in closed loop. Based on this observation, the detailed implementation of supply-sensitivity background calibration is as follows: The FIR filter is used to remove the DLF noise. Then, the digital comparator compares current DLF code with the previous one. It operates like a differentiator with a binary output to guide the search direction of the RES code. If the current DLF code is smaller than previous one, the searching direction of the RES code is correct, and the integrator keeps increasing or decreasing in the same direction. Otherwise, the direction is reversed. Note that the update rate is configurable from 100Hz to 400kHz.

Figure 15.3.2 shows the schematic of the DCO. It is composed of a VCO, a DAC, and a digitally controlled resistor (DCR). The VCO has a 2-stage pseudo-differential ring structure. The delay cell is in series with a 380Ω resistor in order to lower the positive-supply sensitivity. On the other hand, the latch naturally has a negative supply sensitivity. Therefore, one can equalize the positive and negative supply sensitivity by choosing a specific value for the series resistor. The DCR schematic comprises 256 units of PMOS controlled by a thermometer code to guarantee monotonicity. The integral-path DAC is composed of 512 units of a 20Ω silicide resistor array and 256 units of PMOS and NMOS controlled by a thermometer code. The subsequent 2nd-order filter is to filter out quantization noise with a bandwidth of 100MHz. Note that the supply of DAC is connected to the digitally regulated supply VCO_VDD, rather than V_{DD} because the DCO frequency is also sensitive to the DAC supply.

Figure 15.3.3 shows measured VCO frequencies as function of VCO_VDD. Here, the pushing factors are measured at 4 different operating points, and their values are listed in the table. The supply-insensitive point A is found by background calibration, and the B, C and D points are manually set. Fig. 15.3.4 (top) shows the dynamic-supply-sensitivity test for these 4 operating points. In this test, $1mV_{pp}$ noise is injected onto VCO_VDD, with a frequency range from 1MHz to 80MHz. The spur levels are measured and compared with those predicted based on the spur equation in Fig. 15.3.4. As we can see in the measured results and the spur equation, the spur level is proportional to the pushing factor and inversely proportional to noise frequency. As background calibration settles to the optimal point A, the spur measured at 1MHz is -45dBc and its pushing factor is calculated as 22.5MHz/V, which is improved by 42.5dB in comparison to the D point. The output buffer chain coupled by injected noise causes around -50dBc floor. For the B, C and D cases, the spur improvement is almost the same over 50MHz, demonstrating that this structure has wideband immunity to supply noise. Note that the ADPLL bandwidth was set to 200KHz in order not to filter the injected noise for these measurements. The static-supply-sensitivity test is shown in Fig. 15.3.4 (bottom). The VCO_VDD is digitally regulated to lock to the supply-insensitive point (point A) as V_{DD} varies in the range from 0.95V to 1.45V.

The frequency-drift measurement in closed loop vs. temperature is shown in Fig. 15.3.5 (left). The supply-sensitivity background calibration is enabled to track the supply-insensitive point as temperature varies. It can be shown that the temperature sensitivity is minimized under the condition of zero-supply sensitivity. The VCTRL_I varies within 200mV, which is equivalent to 2% frequency drift under the measured integral path gain of 250MHz/V. The measured results for phase noise with and without supply-sensitivity calibration are shown in Fig. 15.3.5 (right). Without calibration, the measured pushing factor is 3GHz/V, roughly the same value as a conventional ring VCO. The supply noise degrades the phase noise by 20dB compared with the calibrated case. After background calibration, the RMS jitter integrated from 10kHz to 40MHz is 3.29ps.

Figure 15.3.6 shows a comparison with prior art and summarizes performance. In this work, the additional series resistor in the DCO improves the supply sensitivity to 0.0087. The background calibration guarantees the robustness under PVT variation. The measured analog and digital power consumptions are 4.2mW and 2.2mW, respectively. The reference spur is -75dBc. Fig. 15.3.7 shows the die photo. In summary, this work presents a supply-noise-insensitive and temperature-self-compensated ring DCO with digital background calibration, making low supply, low power and PVT-invariant design possible.

Acknowledgements:
The authors would like to thank Ching-San Wu and Jason Hsu for resource support. We also thank Archer Chang and Yen-Ting Tu for digital circuit synthesis, James Lin for PCB design. We acknowledge Yan-Bin Luo, Brian Liu and all ACD colleagues for valuable discussion and review.

References:
[1] E.J. Pankratz, and E. Sánchez-Sinencio, "Multiloop High Power Supply Rejection Quadrature Ring Oscillator," *IEEE J. Solid-State Circuits*, vol. 47, no. 9, pp. 2033-2048, 2012.
[2] A. Elshazly, R. Inti, W. Yin, B. Young, and P. Hanumolu, "A 0.4-to-3GHz Digital PLL with Supply-Noise Cancellation Using Deterministic Background Calibration," *IEEE ISSCC Dig. Tech. Papers*, pp.92-94, 2011.
[3] P.H. Hsieh, J. Maxey, and C.-K. K. Yang, "Minimizing the Supply Sensitivity of a CMOS Ring Oscillator Through Jointly Biasing the Supply and Control Voltages," *IEEE J. Solid-State Circuits*, vol. 44, no. 9, pp. 2488-2495, 2009.
[4] T. Wu, K. Mayaram, and U. Moon, "An On-chip Calibration Technique for Reducing Supply Voltage Sensitivity in Ring Oscillators," *IEEE J. Solid-State Circuits*, vol. 42, no. 4, pp. 775-783, 2007
[5] M. Mansuri and C.-K. K. Yang, "A Low-power Adaptive Bandwidth PLL and Clock Buffer with Supply-noise Compensation," *IEEE J. Solid-State Circuits*, vol. 38, no. 11, pp. 1804-1812, 2003.
[6] P.-Y. Wang, J.-H. Conan Zhan, H.-H. Chang, and H.-M. Sherman Chang, "A Digital Intensive Fractional-N PLL and All-Digital Self-Calibration Schemes," *IEEE J. Solid State Circuits*, vol. 44, no. 8, pp. 2182-2109, 2009.

978-1-4799-0917-9/14 $31.00 © 2014 IEEE

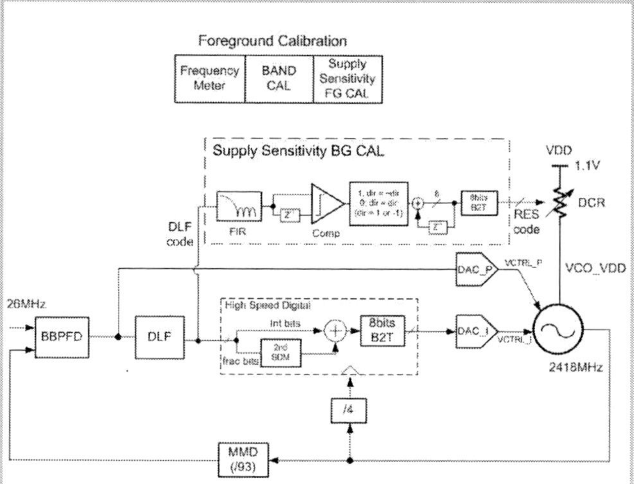

Figure 15.3.1: Block diagram.

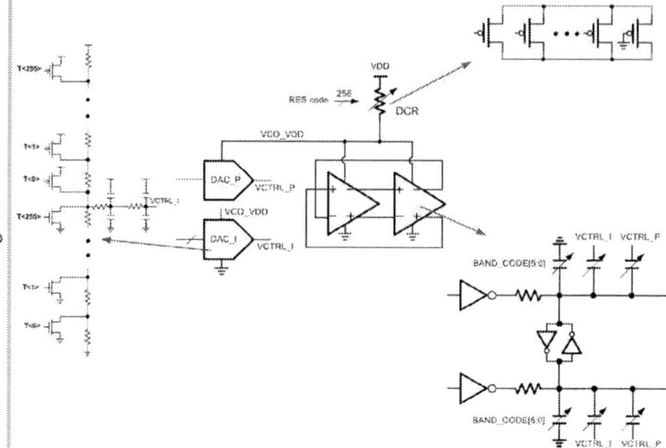

Figure 15.3.2: DCO schematic.

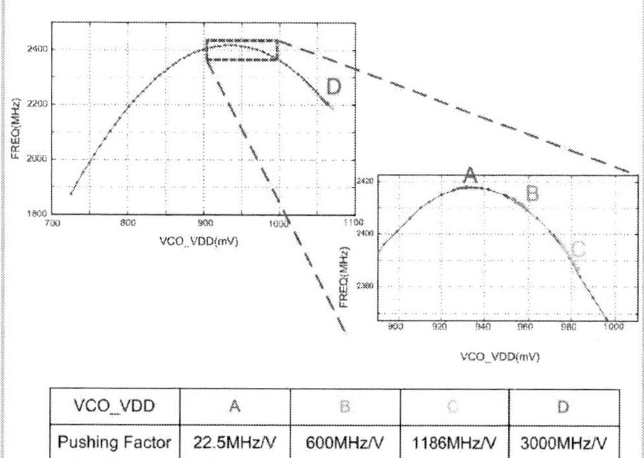

VCO_VDD	A	B	C	D
Pushing Factor	22.5MHz/V	600MHz/V	1186MHz/V	3000MHz/V

Figure 15.3.3: VCO frequency as a function of VCO_VDD measurement.

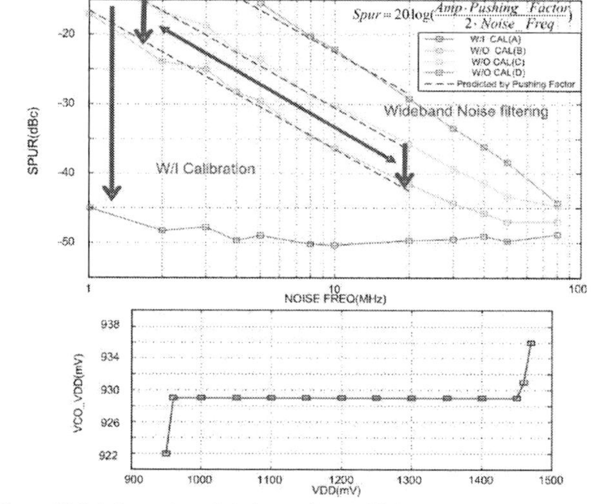

Figure 15.3.4: Dynamic and static supply-sensitivity test.

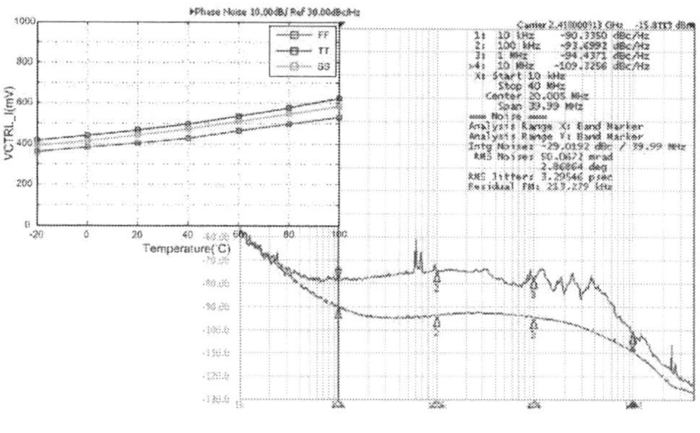

Figure 15.3.5: VCTRL_I drift vs. temperature and phase noise without and with supply-sensitivity calibration.

	JSSC03 [5]	JSSC07 [4]	JSSC09 [3]	ISSCC11 [2]	JSSC12 [1]	This Work
Technology	0.25um	130nm	65nm	130nm	90nm	40nm
Supply	2.5	1.0	1.1	1.1	1.0	1.1
Ref CK(MHz)	250	N/A	1280	N/A	N/A*	26
Fosc(MHz)	1000	1400	5120	2500	5650	2418
RMS Jitter(ps)	3.28	3.9	1.56	4.6	N/A*	3.29
Power(mW)	10	9.6	39.7	3.1	7~26**	6.4
Supply Sensitivity	0.03	N/A	0.0077	N/A	0.019	0.0087
Calibration	No	Foreground Calibration	No	Background Calibration	No	Background Calibration

*VCO only **operate at 0.63~8.1GHz

Supply	1.1
Power	Analog 4.2mW Digital 2.2mW Total 6.4mW
RMS Jitter(10kHz~40MHz)	3.29ps
Supply Sensitivity	0.0087
Pushing Factor	22.5MHz/V
Reference Spur	-75dBc
Frequency Drift(-20~100°C)	2%

Figure 15.3.6: Comparison with prior art and performance summary.

15

Figure 15.3.7: Die photo.

15.4 A 20-to-1000MHz ±14ps Peak-to-Peak Jitter Reconfigurable Multi-Output All-Digital Clock Generator Using Open-Loop Fractional Dividers in 65nm CMOS

Ahmed Elkholy[1], Amr Elshazly[2], Saurabh Saxena[1], Guanghua Shu[1], Pavan Kumar Hanumolu[1]

[1]University of Illinois, Urbana, IL, [2]Intel, Hillsboro, OR

Modern systems-on-chips (SoCs) perform many diverse analog, digital, and mixed-signal functions. They contain a wide variety of modules such as multicore processors, memories, I/O interfaces, power management, and wireless transceivers. Each module has its own unique clock requirements to maximize the overall system performance. For example, dynamic frequency scaling (DFS) saves processor power, spread spectrum clocking (SSC) reduces electromagnetic interference (EMI), and rapid power cycling between idle and active states allows energy-proportional operation. A conventional analog integer-N phase-locked loop (PLL)-based clock generation unit (CGU) occupies large area, has a long lock time, and its output frequency resolution is limited by the reference clock frequency. While the digital fractional-N PLL-based CGU in [1] overcomes some of these drawbacks, it suffers from an intrinsic tradeoff between the time-to-digital converter (TDC)/fractional-divider quantization error and oscillator phase noise. As a result, it requires either a high-resolution TDC or a low-noise oscillator both of which incur power penalty. Further, narrow PLL bandwidth limits SSC modulation frequency and increases lock time making it unsuitable for energy-proportional operation. Open-loop frequency generation using direct-digital synthesis (DDS) overcomes the drawbacks of closed-loop PLLs but it consumes a significant amount of power [2]. This paper presents an all-digital CGU using open-loop fractional dividers. Unlike [1], the proposed CGU, using only one integer-N PLL and a single reference clock, can provide multiple low-jitter outputs over a wide frequency range with fine frequency resolution. It also has SSC capability with programmable modulation depth and achieves instantaneous frequency switching.

Figure 15.4.1 shows the block diagram of the proposed all-digital CGU. It is composed of a digital integer-N PLL followed by multiple fractional dividers. The PLL, consisting of a TDC, a digital loop filter, a digitally controlled oscillator, and a feedback divider, generates a 5GHz output from a 100MHz reference clock. Each fractional divider is independently controlled by 7b integer and 14b fractional frequency control words denoted as FCW_I and FCW_F, respectively. FCW_I controls the integer division ratio (N) from 4 to 127, while FCW_F controls the fractional part (α), resulting in a division ratio of N+α. As depicted in Fig. 15.4.1, FCW_F is truncated to 1b using a first-order $\Delta\Sigma$ modulator and added to FCW_I. The resulting output controls an extended range multi-modulus divider (MMD), which switches seamlessly between two consecutive integer divider values to produce the desired output fractional frequency, $F_{IN}/(N+\alpha)$. However, due to open-loop behavior, the truncation error (e_q) introduced by the $\Delta\Sigma$ modulator is not filtered and directly appears as output phase noise. The resulting deterministic jitter can be as large as one input period (T_{IN} = 200ps). However, we note that an instantaneous quantization error $e_q[k]$ causes an output phase error of $e_q[k]\times T_{IN}$, which we propose to cancel by adding an equal and opposite amount of phase shift to the divider output. To this end, a digitally controlled delay line ($DCDL_M$), whose gain is calibrated to be T_{IN}, is used to cancel phase noise caused by e_q.

The detailed block diagram of the proposed fractional divider with the phase noise cancellation scheme is shown in Fig. 15.4.2. The least-mean-square (LMS)-based algorithm reported in [3] requires a reference clock at the output frequency ($F_{IN}/N+\alpha$), which prohibits its use in this work. Hence, we perform the desired calibration using a digital delay-locked loop (DLL), whose reference is generated within the CGU. Matched flip-flops, DFF_R and DFF_P, are used to generate two edges that are T_{IN} apart. A bang-bang phase detector measures the sign of the phase error between the DFF_P output and the $DCDL_C$ output and drives an 11b accumulator. The control words to tune $DCDL_M$ and $DCDL_C$ are generated by scaling $e_q[k]$ and its complement ($1-e_q[k]$) with the 7 MSBs of the calibration loop filter output (K_G). In steady state, the DLL establishes the sum of the $DCDL_C$ and the $DCDL_M$ delays to be equal to the T_{IN} time period (i.e. $e_q[k]\times K_G\times T_{DCDLM}+(1-e_q[k])\times K_G\times T_{DCDLC}=T_{IN}$). Thus, choosing $T_{DCDLC}=T_{DCDLM}$ leads to $K_G.T_{DCDLM}=T_{IN}$ and the exact amount of phase shift needed to perfectly cancel $e_q[k]$ at the output is $e_q[k]\times K_G\times T_{DCDLC}$. Because a non-zero minimum DCDL delay corresponds to a zero-input code, it appears as an offset and degrades calibration accuracy. Therefore, another DCDL ($DCDL_{OFF}$) is added to compensate for this offset delay, and the desired offset digital control word (DCW_{OFF}) is set at start-up using an initial foreground calibration step.

The DCDL is implemented using a cascade of 8 identical digitally controlled CMOS inverter-based delay cells (Fig. 15.4.2). Delay tuning is achieved by varying load capacitance. Using 8 stages, as opposed to one large delay cell, ensures fast rise/fall times and reduces the DCDL sensitivity to supply/thermal noise. To improve DCDL linearity, we use segmented control to distribute the desired delay equally among all the delays cells. Each delay cell consists of a CMOS inverter loaded with a tunable 16-unit capacitor bank. The DCW's 4MSBs control 15 capacitors in all the delay cells, while each of the 3LSBs control one unit capacitor in different delay cells, as illustrated in Fig. 15.4.2. This ensures a maximum load capacitance deviation of at most one unit capacitor over the entire range of the DCW, which improves the DCDL linearity and translates to less than 3ps of delay deviation. The MMD is composed of a cascade of a 6 divide-by-2/3 cells to provide an extended division range from 4 to 127. In this design, reset is added to the divide-by-2/3 cells to ensure seamless switching of the fractional division at extension boundaries.

The proposed CGU with multiple independent outputs is fabricated in 65nm CMOS process and occupies an active area of 0.12mm^2 (each fractional divider occupies only 0.017mm^2). It generates a wide range of frequencies from 20MHz to 1GHz with 6kHz resolution. At a 5GHz divider input and 1GHz output frequency the proposed fractional divider consumes 3.2mW from 0.9V supply. Fig. 15.4.3 shows the peak-to-peak absolute jitter at 975MHz for different calibration codes at two different supply voltages. Because of the supply sensitivity of the DCDL, the optimal calibration code shifts with supply voltage. The proposed calibration always converged to the optimal K_G (98 at V_{DD} = 0.9 and 106 at V_{DD} = 1.0), which reduces long-term jitter from $50ps_{pp}$ to better than $13ps_{pp}$ (for 10k hits). The jitter histograms with and without calibration are also shown in Fig. 15.4.3. Calibration reduces deterministic jitter (DJ) caused by the leakage of $\Delta\Sigma$ truncation error to within 2ps (DCDL resolution). Absolute output jitter is plotted as a function of output frequency (with fixed FCW_F = 0.25) and FCW_F (with fixed FCW_I = 5) in Fig. 15.4.4. When FCW_I is varied, jitter is less than $20ps_{pp}$ over the entire range and it increases to $27ps_{pp}$ when FCW_F is varied. The increased jitter is attributed to the integral non-linearity of $DCDL_M$. Fig. 15.4.5 shows the measured output spectrum when spread spectrum modulation is enabled. With a 33kHz modulation frequency and 2% modulation amplitude, the measured EMI reduction is 22dB. The performance of the proposed prototype is compared with other low-jitter state-of-the art fractional-N clock generators in Fig. 15.4.6. The proposed fractional-N clock generator consumes the lowest power and occupies smallest area compared to state-of-the-art designs. The die micrograph is shown in Fig. 15.4.7.

Acknowledgments:
Intel Labs University Research Office, SRC under task 1836.124, and NSF under CAREER Award EECS-0954969 supported this research. We thank Berkeley Design Automation for providing Analog Fast Spice (AFS) simulator.

References:

[1] Y. Li, *et al.*, "A Reconfigurable Distributed All-Digital Clock Generator Core with SSC and Skew Correction in 22nm High-κ Tri-Gate LP CMOS," *ISSCC Dig. Tech. Papers*, pp. 70–72, 2012.

[2] D. De Caro, *et al.*, "A 1.27GHz, All-Digital Spread Spectrum Clock Generator/Synthesizer in 65 nm CMOS," *IEEE J. Solid-State Circuits*, vol.45, no.5, pp. 1048-1060, 2010.

[3] D. Tasca, *et al.*, "A 2.9-4.0-GHz Fractional-N Digital PLL with Bang-Bang Phase Detector and 560-fs_{rms} Integrated Jitter at 4.5-mW Power," *ISSCC Dig. Tech. Papers*, pp. 88-90, 2011.

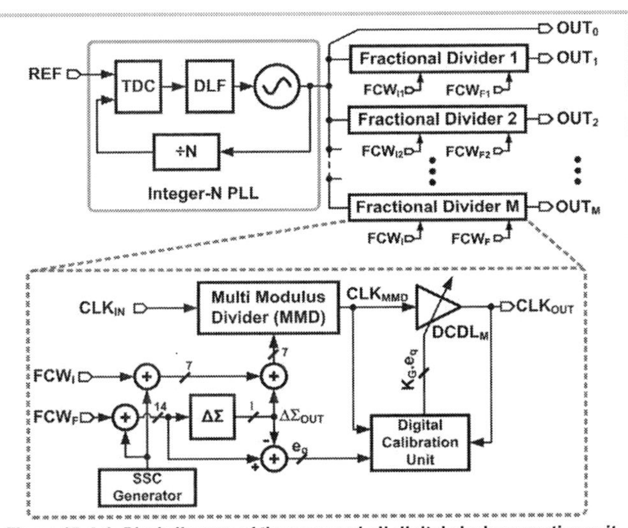

Figure 15.4.1: Block diagram of the proposed all-digital clock generation unit.

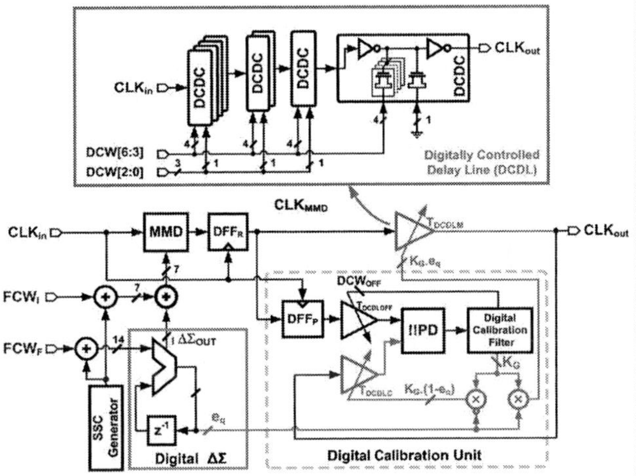

Figure 15.4.2: Fractional divider phase noise cancellation scheme with digital background calibration.

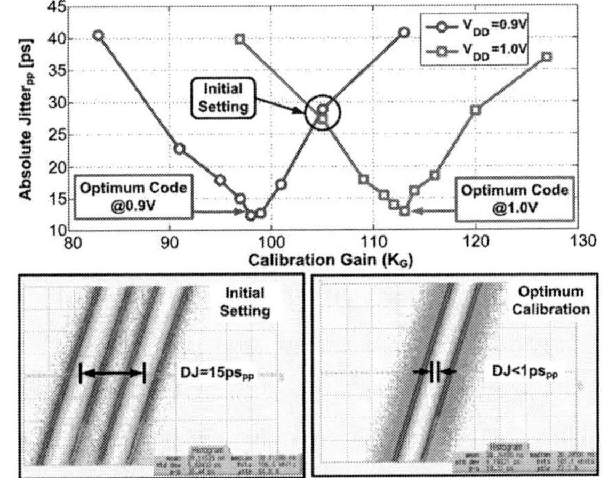

Figure 15.4.3: Jitter histograms at 975MHz output and measured absolute peak-to-peak jitter plotted as a function of calibration gain (K_G) for two different supply voltages.

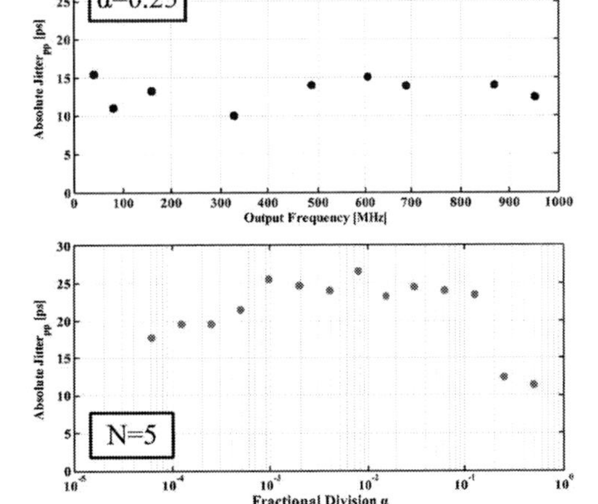

Figure 15.4.4: Measured peak-to-peak absolute jitter as a function of output frequency and fractional division ratio α.

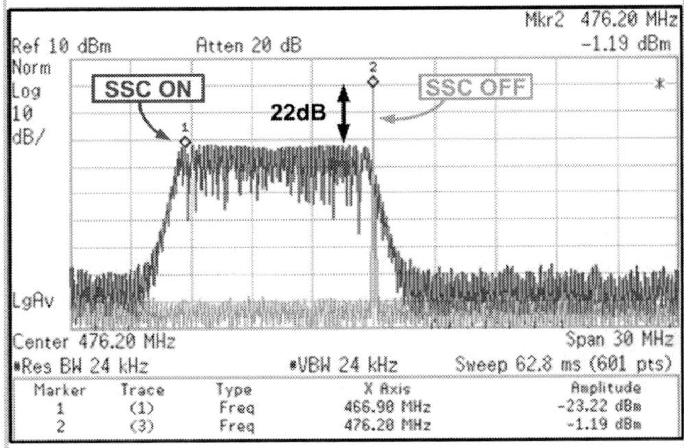

Figure 15.4.5: Measured output spectrum with and without SSC.

	This Work	Park ISSCC'12	Li [1] ISSCC'12	De Caro [2] JSSC'10	Grollitsch ISSCC'10	Si labs 5338
Technology	65nm	65nm	22nm	65nm	65nm	N/A
Supply [V]	0.9	1	1	1.2	1.1	1.2
Freq. Range [MHz]	20-1000	580	600-3600	180-1270	375-3025	10-350
RMS Jitter [ps]	3	8.05	10	12.8	5.51	0.7
PP Jitter [ps]	27	N/A	N/A	93	54	13
Instantaneous Switching	Yes	No	No	Yes	No	N/A
SSC Capability	Yes	No	Yes	Yes	Yes	Yes
Power Consumption	3.2mW @1GHz	10.5mW @580MHz	18.4mW @3.6GHz	19.8mW @1.27GHz	3.4mW @751MHz	~13.2mW @350MHz
Power Efficiency [mW/GHz]	3.2	18.1	5.1	15.6	4.1	~37.7
Architecture	Frac-N Divider	Frac-N ILO	Frac-N PLL	DDS	Frac-N PLL	Frac-N Divider
Area [mm^2]	0.017	0.03	0.03	0.044	0.038	N/A

Figure 15.4.6: Summary and comparison with state-of-the art.

15

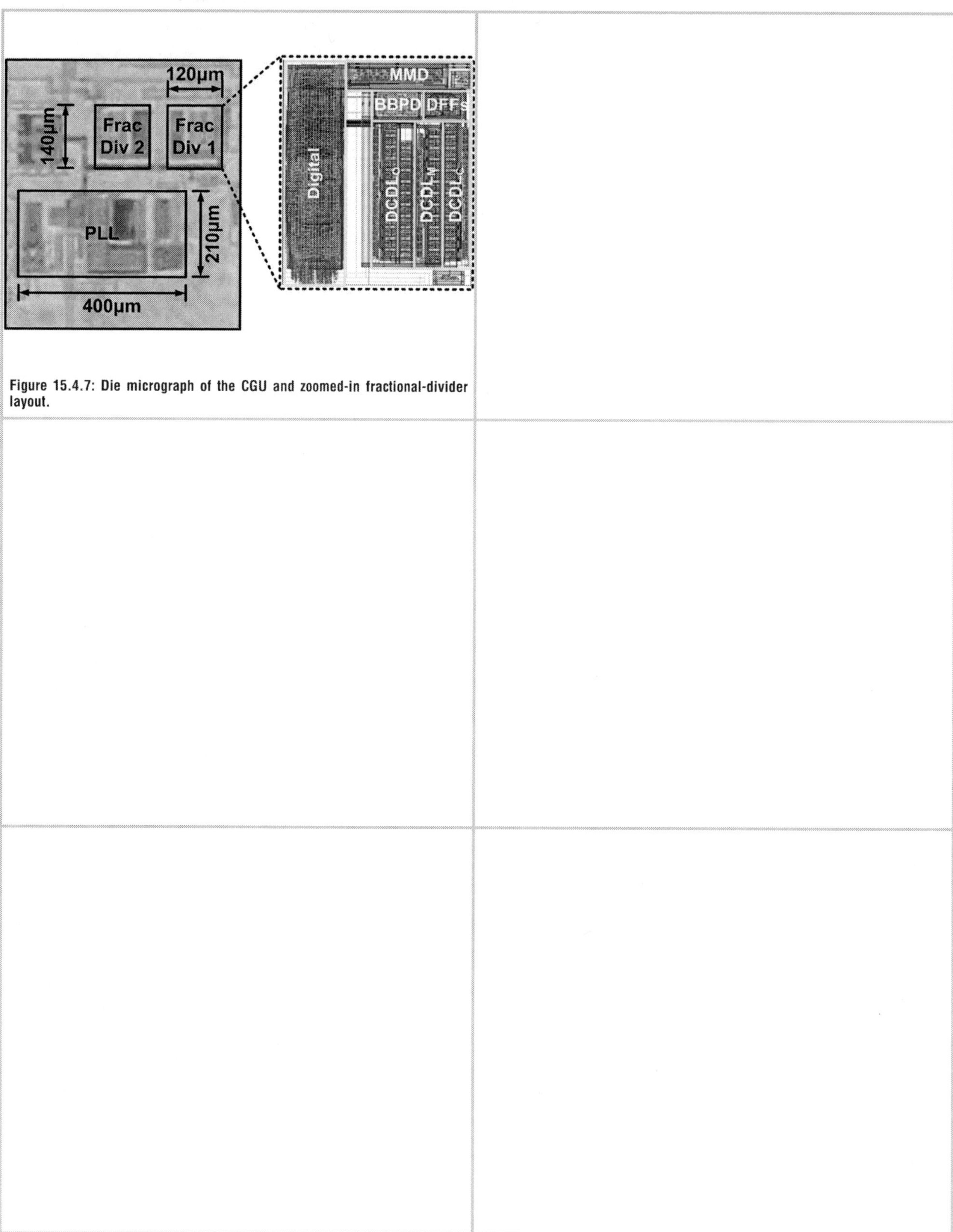

Figure 15.4.7: Die micrograph of the CGU and zoomed-in fractional-divider layout.

Session 16 Overview: *SoC Building Blocks*

HIGH-PERFORMANCE DIGITAL SUBCOMMITTEE

Session Chair: *Kathy Wilcox*
AMD, Boxborough, MA

Session Co-Chair: *Yasuhisa Shimazaki*
Renesas Electronics, Tokyo, Japan

The four papers in this session highlight developments in system-on-chip (SoC) blocks essential to continue scaling trends. These include fabrics required for high throughput, as well as blocks needed for secure systems and temperature sensors to achieve target power requirements. These designs are demonstrated in recent technologies with low area and low power overhead that enables ease of use in SoC design.

978-1-4799-0917-9/14 $31.00 © 2014 IEEE

16.1 A 340mV-to-0.9V 20.2Tb/s Source-Synchronous Hybrid Packet/Circuit-Switched 16×16 Network-on-Chip in 22nm Tri-Gate CMOS **3:15 PM**

G. Chen, Intel, Hillsboro, OR

In Paper 16.1, Intel presents a 256-node network-on-chip (NoC) with 20.2Tb/s throughput. Industry leading energy efficiency of 18.3Tb/s/W at 430mV is achieved through the use of source-synchronous hybrid packet/circuit-switched operation. The 23mm^2 design is implemented in a 22nm tri-gate CMOS process.

16.2 A 0.19pJ/b PVT-Variation-Tolerant Hybrid Physically Unclonable Function Circuit for 100% Stable Secure Key Generation in 22nm CMOS **3:45 PM**

S. K. Mathew, Intel, Hillsboro, OR

In Paper 16.2, Intel introduces a variation-tolerant physically unclonable function (PUF) circuit that occupies 4.66mm^2/b in 22nm CMOS. The hybrid cross-coupled/delay PUF has a low energy consumption of 0.19pJ/b at 0.9V, and a high operating frequency of 2GHz, while maintaining immunity to probing attacks.

16

16.3 A 23Mb/s 23pJ/b Fully Synthesized True-Random-Number Generator in 28nm and 65nm CMOS **4:15 PM**

K. Yang, University of Michigan, Ann Arbor, MI

In Paper 16.3, the University of Michigan presents true-random-number generators (TRNGs) implemented in 28nm and 65nm CMOS, which utilize a 3-edge ring oscillator synthesized entirely with standard cells. The proposed TRNG utilizes the thermal noise in a 3-edge ring oscillator as a source of randomness. It passes all NIST randomness tests, provides 23.16Mb/s throughput at 0.9V and occupies 375μm^2 in 28nm.

16.4 0.6-to-1.0V 279μm^2, 0.92μW Temperature Sensor with Less Than +3.2/-3.4°C Error for On-Chip Dense Thermal Monitoring **4:45 PM**

T. Yang, Columbia University, New York, NY

In Paper 16.4, Columbia University presents a temperature sensor, which utilizes 2-transistor voltage reference circuits and achieves an accuracy of +3.2/-3.4°C after 1-point calibration. The proposed differential reading achieves a DC-PSRR better than -67dB by eliminating sensitivity to V_{DD}. The temperature sensor occupies 279μm^2 in 65nm CMOS and consumes 0.92mW at V_{DD}=0.6V.

16.1 A 340mV-to-0.9V 20.2Tb/s Source-Synchronous Hybrid Packet/Circuit-Switched 16×16 Network-on-Chip in 22nm Tri-Gate CMOS

Gregory Chen, Mark A. Anders, Himanshu Kaul, Sudhir K. Satpathy, Sanu K. Mathew, Steven K. Hsu, Amit Agarwal, Ram K. Krishnamurthy, Shekhar Borkar, Vivek De

Intel, Hillsboro, OR

Energy-efficient networks-on-chip (NoCs) are key enablers for exa-scale computation by shifting power budget from communication toward computation. As core counts scale into the 100s, on-chip interconnect fabrics must support increasing heterogeneity and voltage/clock domains. Synchronous NoCs require either a single clock distributed globally or clock-crossing data FIFOs between clock domains [1]. A global clock requires costly full-chip margining and significant power and area for clock distribution, while synchronizing data FIFOs add power, performance, and area overhead per clock crossing. Source-synchronous NoCs mitigate these penalties by forwarding a local clock along with each packet, but still suffer from high data storage power due to packet switching. Circuit switching removes intra-route data storage, but suffers from low network utilization due to serialized channel setup and data transfer [2]. Hybrid packet/circuit switching parallelizes these operations for higher network utilization. A 16×16 mesh, 112b data, 256 voltage/clock domain NoC with source-synchronous operation, hybrid packet/circuit-switched flow control, and ultra-low-voltage optimizations is fabricated in 22nm tri-gate CMOS [3] to enable: i) 20.2Tb/s total throughput at 0.9V, 25°C, ii) a 2.7× increase in bisection bandwidth to 2.8Tb/s and 93% reduction in circuit-switched latency at 407ps/hop through source-synchronous operation, iii) a 62% latency improvement and 55% increase in energy efficiency to 7.0Tb/s/W through circuit switching, iv) a peak energy efficiency of 18.3Tb/s/W for near-threshold operation at 430mV, 25°C, and v) ultra-low-voltage operation down to 340mV with router power scaling to 363μW.

Energy-efficient circuit-switched data transfers eliminate intra-route storage, packet-switched channel reservation and sideband transfers increase network utilization, and source-synchronous operation adapts to delay imbalances while reducing clock power (Fig. 16.1.1). Each hybrid packet/circuit-switched transfer begins with a packet-switched request phase to reserve a circuit-switched channel, followed by data transfer and acknowledgement. Phase pipelining enables concurrent transfers. Source-synchronous signals initiate phase transitions and propagation through the network. Request and Grant signals control packet propagation. Sideband 32b packet-only transfers improve utilization by 72% for equal narrow and wide transfers by not leaving part of the 80b circuit-switched channel idle. Reservation packets allocate channels by queuing path directions in a FIFO, with FIFO outputs controlling the channel crossbar. Path direction queuing hides packet delay from the channel reservation phase, for 62% lower circuit transfer delay. Circuit-switched data propagates with Stream/Tail signals along the reserved channel. At the destination, a returned Ack signal either initiates another transfer from the source along the same channel or de-allocates the channel. Source-synchronous operation avoids global clock distribution power and adapts to different clock frequencies, path delays, and PVT variations.

Routers relay packets through the network following x-first, y-second ordering with source-synchronous credit-2 flow control (Fig. 16.1.2). Between routers, 855μm interconnect links with three segments include ultra-low-voltage split-output level shifters to enable multiple supply voltages [4]. In Ports forward packets to Out Ports based on the destination address. Request arbiters in Out Ports select a packet to send to the next router using a double-edge flip-flop (DEFF) for 25% lower clock power. Request transitions at an In Port indicate packet arrivals and Grant transitions sent to the previous router indicate completed packet processing. Credit-2 flow control hides Grant latency for 89% higher throughput by using In Port latches to hold packets during processing. Circuit Enable stores Direction in FIFOs for a future circuit-switched transfer. Out Ports block request packets when FIFOs are full, but sideband packet-only transfers from other In Ports continue propagating. Packets waiting for selection by the Out Port benefit from reduced delay (Path 2), resulting in higher throughput than synchronous operation with margins for the critical path that includes the link (Path 1). Credit tracking circuits bound these timing paths (Fig. 16.1.3). A C-element in the In Port tracks incoming Request and outgoing Grant transitions to control packet latches. A C-element in the Out Port blocks the toggle flip-flop slave stage until credit is available, while the arbiter select controls the master. Interruptible feedback removes contention in the C-element to improve V_{MIN} by 100mV. In/Out Port credit control state transitions based on Request, Grant, and C-element Q ensure consistent behavior despite unpredictable signal arrival times. Tree-based request arbitration implements optimized second-level mutex circuits for 33% reduced delay. In Ports de-assert RClock upon receiving Grant, causing the arbiter to select the next waiting request while preventing starvation.

The highest-priority FIFO entry configures the circuit-switched channel for a data transfer (Fig. 16.1.4). Stream/Tail transitions indicate data transfers and are converted to level-sensitive signals within the routers only, preventing glitches from direction changes. When Stream or Tail reach the destination, Ack is returned to the source with the Tail Ack switching FIFOs along the channel to the next circuit-switched transfer. Request packets allocate the next available FIFO entry, which may shift across FIFOs within routers only. Data transfer occurs when reservations reach the highest priority along an entire channel. Source-synchronous operation enables proximity-dependent transfer period and is not limited by synchronous corner-to-corner delay [5]. Increasing FIFO entries from 1 to 4 decreases packet blockage for 75% higher throughput. Rotating 3b Gray-counter read/write pointers prevent glitches from unpredictable read/write timing, while detecting a full FIFO.

A 16×16 mesh connects 138μm×109μm 5-port routers using 855μm 112b data links for an equivalent area of 13×13mm² with unfolded interconnect (Fig. 16.1.7). Core ports interface through dual-clock FIFOs to independently clocked traffic generation and measurement circuits. Network energy efficiency increases as the traffic insertion interval decreases, with packet-switched 4B sideband data transfers reaching 4.48Tb/s/W (Fig. 16.1.5). Circuit-switched transfers and streaming further improve energy efficiency by 55% to 7.0Tb/s/W at nominal 0.9V, 25°C. Increasing stream length amortizes arbitration overhead down to 8%. Comparing to the NoC operating in synchronous mode demonstrates source-synchronous advantages. Packet-switched latency improves by 14% to 1.04ns/hop, and a circuit-switched latency of 407ps/hop enables up to 93% reduction compared to a synchronous corner-to-corner clock period. These latency improvements result in 2.7× bisection bandwidth increase to 2.8Tb/s at 0.9V. Source-synchronous advantages further increase with timing asymmetry such as multi-supply operation, with two supply voltages assigned in a checkerboard pattern yielding up to 3.8× improved bisection bandwidth. The NoC achieves 20.2Tb/s total throughput for nearest-neighbor maximum-throughput traffic at 0.9V (Fig. 16.1.6). Intra-route storage removal for circuit-switched data improves energy efficiency by 55% and lowers one-hop latency by 62% compared to packet switching at 0.9V. Robust ultra-low-voltage optimized circuits [4] enable the NoC to operate across a wide-supply-voltage range of 340mV to 0.9V. As supply voltage is reduced, NoC energy efficiency increases to a peak 18.3Tb/s/W with 3.4Tb/s throughput at 430mV (near-threshold operation). Router power scales down to 363μW with 946Gb/s throughput at 340mV (ultra-low-voltage operation).

Acknowledgments:
The authors thank R. Forand, M. Haycock, W. H. Wang, G. Taylor, I. Young, C. Webb, R. Yavatkar, S. Rusu, J. Held, J. Schutz, and S. Pawlowski for encouragement and discussions. This research was, in part, funded by the U.S. Government under contract number HR0011-10-3-0007. The views and conclusions contained in this document are those of the authors and should not be interpreted as representing the official policies, either expressed or implied, of the U.S. Government.

References:
[1] S. Bell, *et al.*, "TILE64 – Processor: A 64-Core SoC with Mesh Interconnect," *ISSCC Dig. Tech. Papers*, pp. 88-89, 2008.
[2] P. Ou, *et al.*, "A 65nm 39GOPS/W 24-core Processor with 11Tb/s/W Packet-controlled Circuit-switched Double-layer Network-on-chip and Heterogeneous Execution Array," *ISSCC Dig. Tech. Papers*, pp. 56-57, 2013.
[3] C.-H. Jan, *et al.*, "A 22nm SoC Platform Technology Featuring 3-D Tri-Gate and High-k/Metal Gate, Optimized for Ultra Low Power, High Performance and High Density SoC Applications," *IEDM Dig. Tech. Papers*, pp. 4-7, 2012.
[4] S. Hsu, *et al.*, "A 280mV-to-1.1V 256b Reconfigurable SIMD Vector Permutation Engine with 2-Dimensional Shuffle in 22nm CMOS," *ISSCC Dig. Tech. Papers*, pp. 178-179, 2012.
[5] M. Anders, *et al.*, "A 4.1Tb/s Bisection-Bandwidth 560Gb/s/W Streaming Circuit-switched 8×8 Mesh Network-on-chip in 45nm CMOS," *ISSCC Dig. Tech. Papers*, pp. 110-111, 2010.

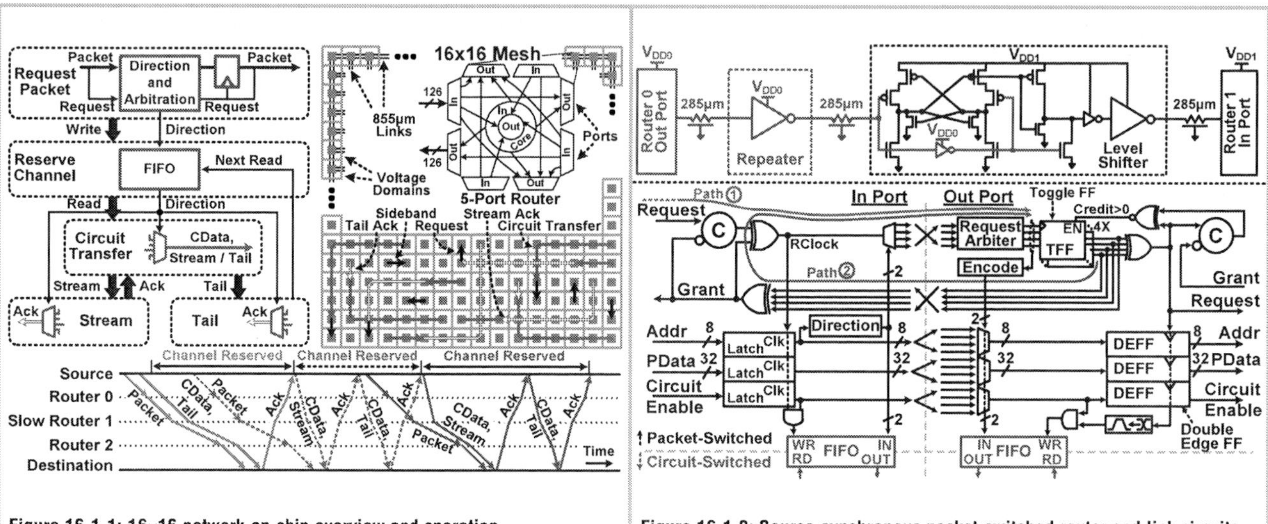

Figure 16.1.1: 16×16 network-on-chip overview and operation.

Figure 16.1.2: Source-synchronous packet-switched router and link circuits.

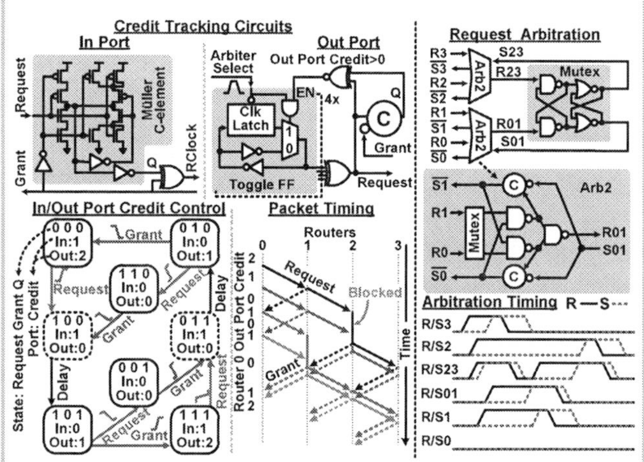

Figure 16.1.3: Credit tracking and request arbitration circuits.

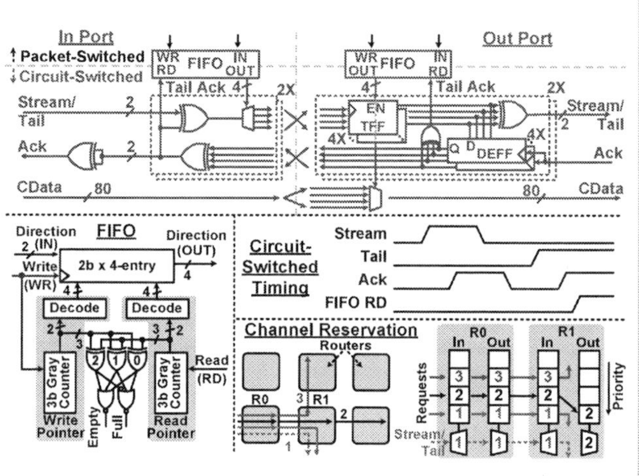

Figure 16.1.4: Circuit-switched operation and FIFO circuits.

16

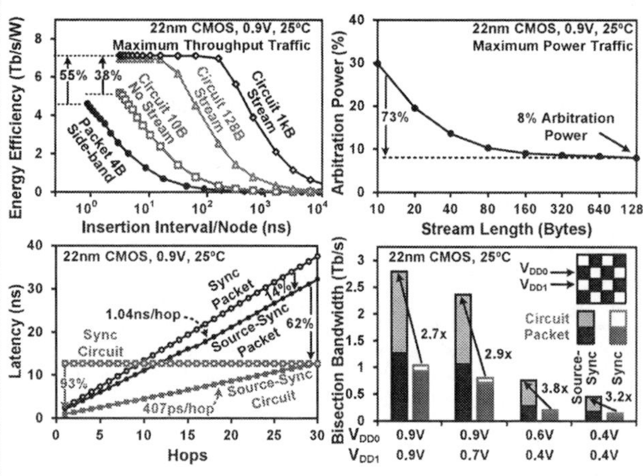

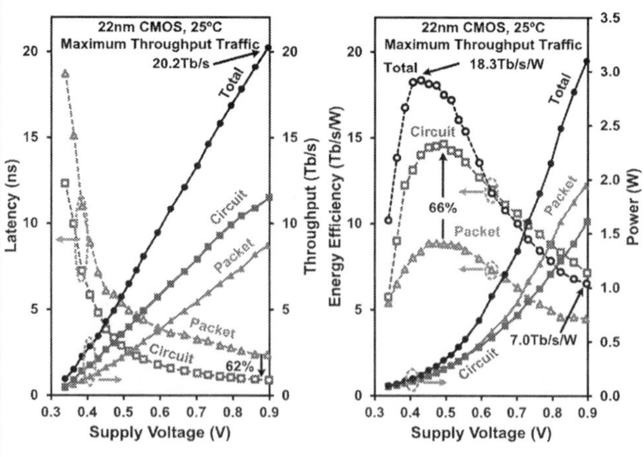

Figure 16.1.5: Traffic rate, streaming, hop distance, and multi-voltage measurement results.

Figure 16.1.6: Measured network latency, throughput, energy efficiency, and power vs. supply voltage.

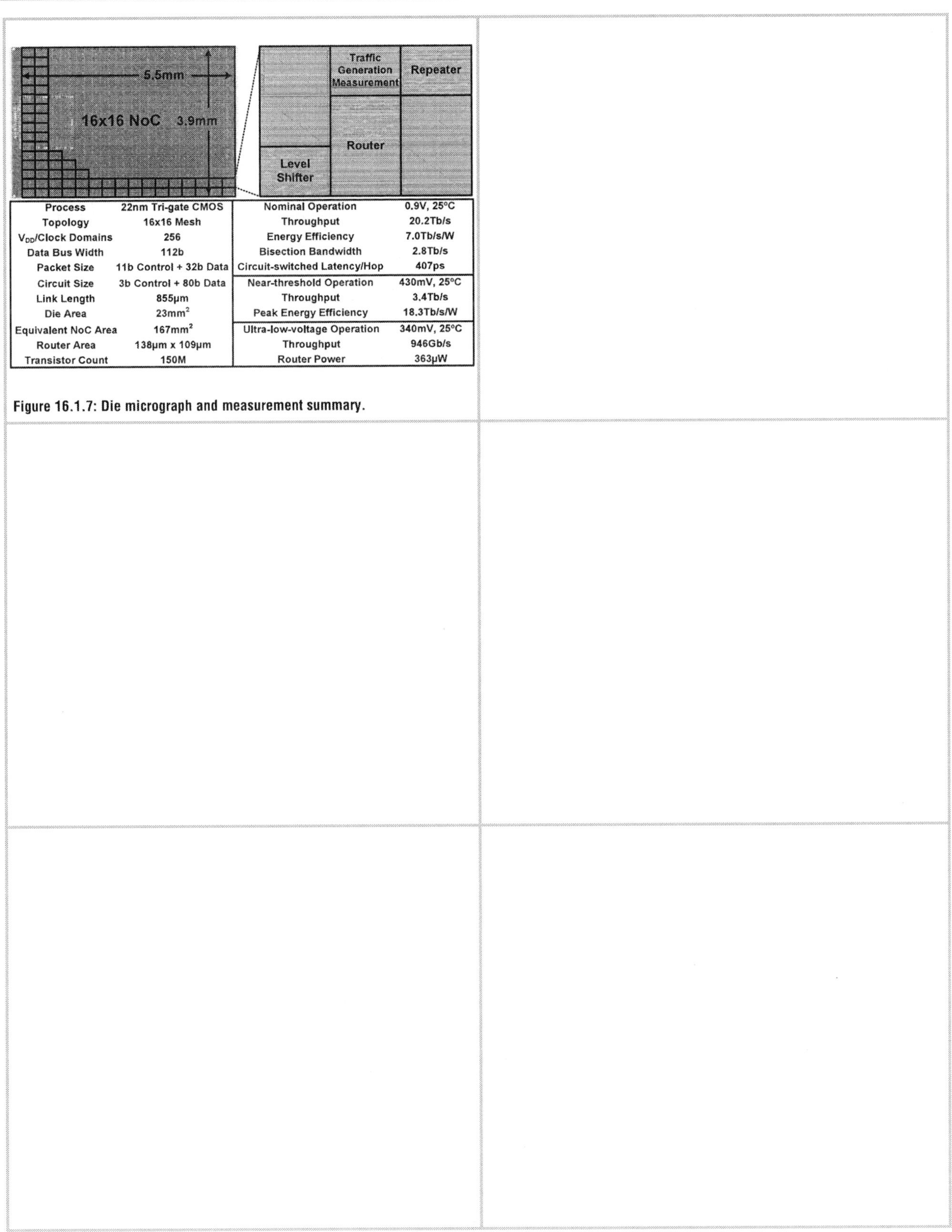

Process	22nm Tri-gate CMOS	Nominal Operation	0.9V, 25°C
Topology	16x16 Mesh	Throughput	20.2Tb/s
V_{DD}/Clock Domains	256	Energy Efficiency	7.0Tb/s/W
Data Bus Width	112b	Bisection Bandwidth	2.8Tb/s
Packet Size	11b Control + 32b Data	Circuit-switched Latency/Hop	407ps
Circuit Size	3b Control + 80b Data	Near-threshold Operation	430mV, 25°C
Link Length	855µm	Throughput	3.4Tb/s
Die Area	$23mm^2$	Peak Energy Efficiency	18.3Tb/s/W
Equivalent NoC Area	$167mm^2$	Ultra-low-voltage Operation	340mV, 25°C
Router Area	138µm x 109µm	Throughput	946Gb/s
Transistor Count	150M	Router Power	363µW

Figure 16.1.7: Die micrograph and measurement summary.

16.2 A 0.19pJ/b PVT-Variation-Tolerant Hybrid Physically Unclonable Function Circuit for 100% Stable Secure Key Generation in 22nm CMOS

Sanu K. Mathew, Sudhir K. Satpathy, Mark A. Anders, Himanshu Kaul, Steven K. Hsu, Amit Agarwal, Gregory K. Chen, Rachael J. Parker, Ram K. Krishnamurthy, Vivek De

Intel, Hillsboro, OR

Physically unclonable function (PUF) circuits are low-cost cryptographic primitives used for generation of unique, stable and secure keys or chip IDs for device authentication and data security in high-performance microprocessors [1][2][3][7]. The volatile nature of PUFs provides a high level of security and tamper resistance against invasive probing attacks compared to conventional fuse-based key storage technologies [4]. A process-voltage-temperature (PVT) variation-tolerant all-digital PUF array targeted for on-die generation of 100% stable, device-specific, high-entropy keys is fabricated in 22nm tri-gate high-κ metal-gate CMOS technology [5], featuring: i) a hybrid delay/cross-coupled PUF circuit where interaction of 16 minimum-sized, variation-impacted transistors determines resolution dynamics, ii) a temporal majority voting (TMV) circuit to stabilize occasionally unstable bits, resulting in 53% reduction in instability, iii) burn-in hardening to reinforce manufacturing-time PUF bias, resulting in 22% reduction in bit-errors, iv) soft dark bits for run-time identification and sequestration of highly unstable bits during field operation, resulting in 78% lower bit-errors, v) 19× separation between inter- and intra-PUF Hamming distance, enabling die-specific keys, vi) autocorrelation factor≈0 and entropy=0.9997, while passing NIST randomness tests, vii) high tolerance to voltage and temperature variation with 82% reduction in average Hamming-distance using a 100-cycle dark bit window, viii) in-situ PUF hardening by leveraging directed NBTI aging to improve stability during field operation, and ix) ultra-low energy consumption of 0.19pJ/b with compact bitcell layout of 4.66μm² (Fig. 16.2.7a).

Reliable secure key generation requires a 100% stable PUF value that is resilient to PVT variations. This is achieved by creating a golden key at first array power-up during tester-time operation (Fig. 16.2.1(a)) and thereafter recreating this golden value from the inherently noisy raw PUF value during regular field operation. The golden key is used to compute an ECC signature, which is stored on-die as fuses. During regular operation, error correction circuits mix raw PUF bits with ECC fuse values to regenerate the golden key with 100% accuracy [6]. The hybrid PUF circuit leverages meta-stability resolution of cross-coupled inverters along with delay variations in the clock path to generate static entropy in the PUF output (Fig. 16.2.1(b)). The bi-stable inverter circuit is pre-charged to an unstable state (bit=bit#=1) when clock is low. During evaluation phase, (clock=1), the circuit resolves to 1 of 2 stable states, bit=0 or bit=1, based on relative strengths of 16 minimum-sized transistors. In addition to inverter V_t mis-matches, differences in rise/arrival times of clocks clk0 and clk1, and gate capacitances C_0 and C_1 of load devices impact the direction of meta-stability resolution. Compared to conventional SRAM PUFs [1], this hybrid approach introduces runtime clock delay transients into PUF resolution dynamics, thereby providing tamper protection against chip decapsulation and invasive power-up probing attacks. PUF cells that have insufficient random variation generate unstable bits that resolve to 0 or 1 based on thermal noise or voltage/temperature conditions (Fig. 16.2.5(b)). These unstable bits (30% of total bits) undermine reliability of the key and are handled using 3 circuit techniques: TMV, burn-in hardening and soft dark bits.

Temporal majority voting stabilizes noisy bits that have a low, non-zero probability of going unstable by computing the quantized mean of responses within a voting window. A PUF cell with initial error probability of δ uses n-way TMV to reduce effective error rate to δ', where $\delta' = 1-\sum_{i=1}^{(n-1)/2} \delta^i \cdot (1-\delta)^{n-i} \cdot {}^nC_i$. TMV stabilizes occasionally unstable bits with error probabilities <8% using 15-way voting, reducing effective error rate to 10^{-6} (Fig. 16.2.2(a)). Cells with δ>0.1 require exponentially larger voting windows, reducing TMV effectiveness for highly unstable bits. The TMV circuit uses a 4b counter to keep track of the number of times the cell evaluates to 1 over a period of 15 consecutive cycles (Fig. 16.2.2(b)). The TMV threshold may be set to 2, 4 or 8 to implement TMV3, TMV7 or TMV15 respectively by selecting cnt[1], cnt[2] or cnt[3] as the PUF output (Fig. 16.2.2(c)). 22nm CMOS measurements at 0.7V to 0.9V, 25°C demonstrate that TMV15 reduces the percentage of unstable bits by 53% (Fig. 16.2.2(d)), down to 15% of total bits (Fig. 16.2.5(c)). However, not all unstable bits produce incorrect key values (Fig. 16.2.3(a)). At least 57% of the unstable bits evaluate to a similar logic state as they did during golden key generation, leaving only 6% bit errors that require ECC correction (Fig. 16.2.5(d)).

Burn-in hardening further reduces bit errors by directed accelerated aging during tester-operation. Subsequent to first power-up and TMV evaluation (*eval=1*), the PUF cell is subjected to high-temperature and high-voltage stress conditions (*eval=0*) while holding the complement of the golden key value. This state biases 10 of the cell transistors in a direction such that NBTI aging reinforces existing PUF bias. In addition to reinforcing inverter device mismatches, burn-in hardening also injects a bias into the delay path by aging pre-charge and clock NAND transistors in the direction that favors PUF stability. Burn-in hardening reduces unstable bit count by 13% (Fig. 16.2.3(b)), with a resultant bit-error rate of 4.6% (22% reduction), measured from 0.7V to 0.9V, 25°C (Fig. 16.2.5(e)). In-situ hardening is also employed to counter NBTI-induced long-term degradation of PUF stability by writing the opposite PUF value into the cell during regular field operation (Fig. 16.2.2(c)).

While TMV and burn-in hardening are effective at stabilizing 55% of the unstable bits, they are ineffective for highly unstable bits (δ>0.1). These bits manifest themselves as unstable bits within a few cycles of PUF operation and are identified as 'dark bits' during regular operation (Fig. 16.2.4a). In contrast to the use of fuses to identify dark bits, a soft dark bitmask is regenerated at the start of each power-up. Dark bit measurements with a window width of 100 cycles at 0.7V and 0.9V, 25°C identifies 11% of the PUF array as highly unstable (Fig. 16.2.4(b)). Forcing these bits to a pre-determined value ensures that they do not contribute to PUF instability, reducing overall bit-error rate to 0.97% (Fig. 16.2.5(f)). ECC circuits use BCH decoding to correct remnant bit errors, regenerating the golden key with 100% accuracy.

Intra- and inter-PUF Hamming distances are key metrics that quantify temporal stability and spatial correlation of bits across multiple dies. Inter-PUF Hamming distance measurements of 600K PUF bits across 6 dies at 0.7V to 0.9V, 25°C has a normalized mean value of 0.49, with a tight variation coefficient of 4.6%. The 19× separation of intra-PUF vs. inter-PUF mean distance confirms uniqueness of keys across PUF instances (Fig. 16.2.6(a)). Soft dark bits provide tolerance to voltage and temperature variations by masking cells that shift value with changes in operating conditions, reducing intra-PUF Hamming distance by 82% (Fig. 16.2.4(d)). The PUF bitstream passes NIST randomness tests, with an average p-value>0.0006 and measured static entropy/b of 0.9997 (Fig. 16.2.6(b)). A histogram of 1-count of 600K bits, grouped as 256b keys forms a normal distribution with μ=123 and σ=9, indicating random distribution of 1's across the key space (Fig. 16.2.6(c)). Autocorrelation of 50K bits with lags between ±25K demonstrate the absence of spatial correlations with measured ACF≈0 within 95% confidence bounds of a Gaussian distribution with μ=0 and σ^2=0.0088 (Fig. 16.2.6(d)). The all-digital hybrid PUF array operates at a maximum frequency of 2GHz consuming 25μW/b (measured at nominal 0.9V, 25°C), resulting in TMV15 energy consumption of 0.19pJ/b (Fig. 16.2.7(b)). An average Hamming distance of 0.51 between PUF power-up and evaluation values (Fig. 16.2.6(e)) indicates strong influence of clock delay transients on metastability resolution in 51% of cells, providing 2× higher immunity to invasive power-up probing attacks compared to conventional SRAM PUFs.

Acknowledgments:
The authors thank R. Forand, M. Haycock, W.H. Wang, S. Iyengar, G. Taylor, J. Li, G. Cox, A. Rajan, R. Maes for encouragement and discussions.

References:

[1] Y. Su, *et al.*, "A 1.6pJ/bit 96% Stable Chip-ID Generating Circuit Using Process Variations", *ISSCC Dig. Tech. Papers*, pp. 406-407, 2007.
[2] K. Lofstrom, *et al.*, "IC Identification Circuit Using Device Mismatch", *ISSCC Dig. Tech. Papers*, pp. 372-373, 2000.
[3] N. Liu, *et al.*, "OxID: On-chip One-Time Random ID Generation Using Oxide Breakdown", *IEEE Symp VLSI Circuits*, pp. 231-232, 2010.
[4] S. Rosenblatt, *et al.*, "Field Tolerant Dynamic Intrinsic Chip ID Using 32nm High-k/Metal Gate SOI Embedded DRAM", *IEEE J. Solid-State Circuits*, vol. 48, no. 4, pp. 940-947, 2013.
[5] C-H. Jan, *et al.*, "A 22nm SOC Platform Technology Featuring 3-D Tri-Gate and High-k/Metal Gate Optimized for Ultra-Low Power, High-Performance and High Density SOC Applications", *IEDM Dig. Tech. Papers*, pp. 4-7, 2012.
[6] J. Guajardo, *et al.* "FPGA Intrinsic PUFs and Their Use for IP protection", *Workshop on Cryptographic Hardware and Embedded Systems*, pp. 63-80, 2007.
[7] J.W. Lee, *et al.*, "A Technique to Build a Secret Key in ICs for ID and Authentication Applications", *IEEE Symp. VLSI Circuits*, pp. 176-179, 2004.

978-1-4799-0917-9/14 $31.00 © 2014 IEEE

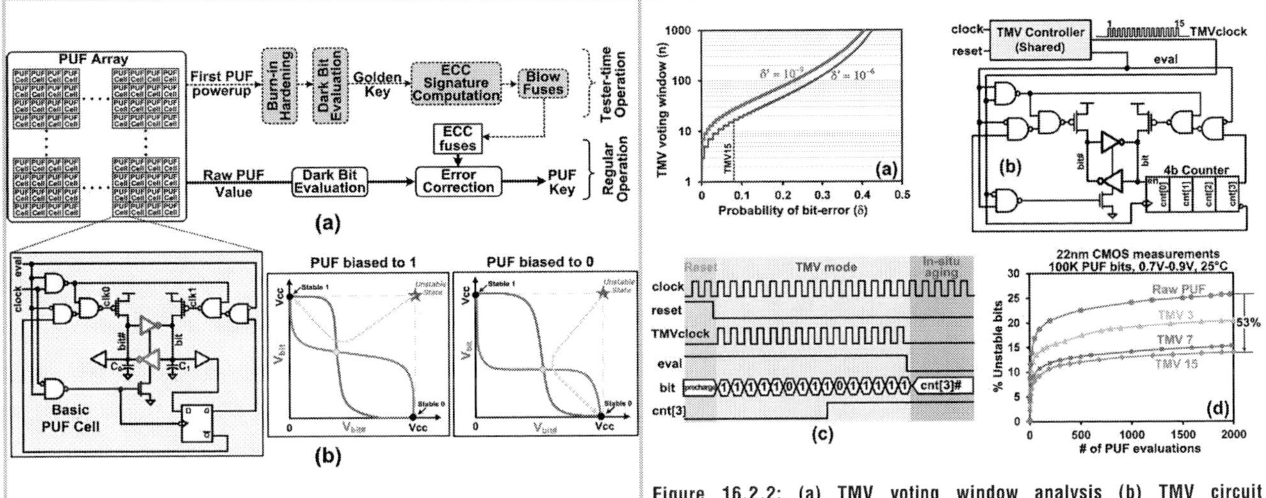

Figure 16.2.1: (a) Secure key generation system (b) Basic PUF circuit.

Figure 16.2.2: (a) TMV voting window analysis (b) TMV circuit (c) Timing diagram (d) TMV measurement results.

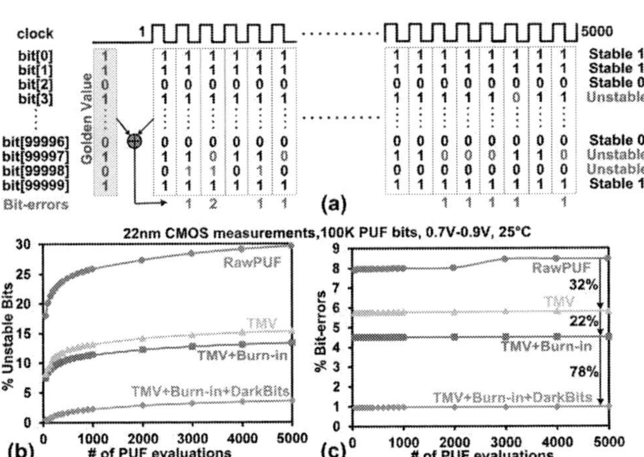

Figure 16.2.3: (a) Unstable bits and bit-errors; (b) Unstable bit measurements.

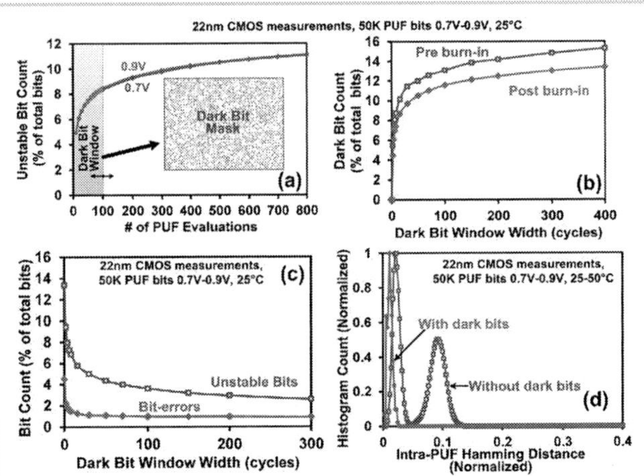

Figure 16.2.4: (a) Dark bit mask generation; (b) Dark bit count measurements; (c) Bit-error measurements; (d) Countering VT variations using dark bits.

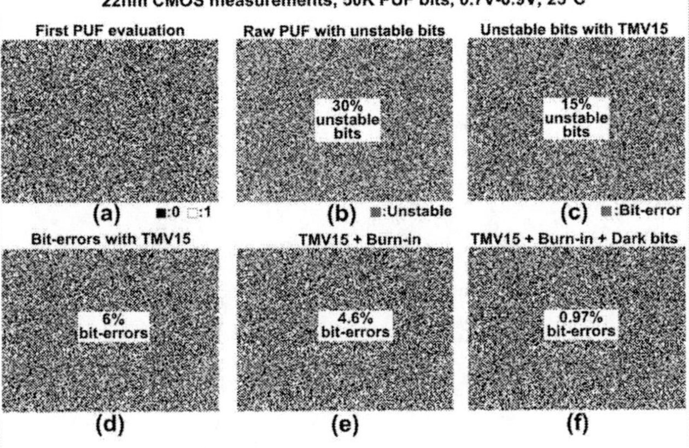

Figure 16.2.5: (a) Raw PUF bitstream; (b) Raw bitstream with unstable bits; (c) TMV bitstream with unstable bits; (d) TMV bitstream with bit-errors; (e) Bitstream after TMV and burn-in; (f) Bitstream after TMV, burn-in and \dark bits.

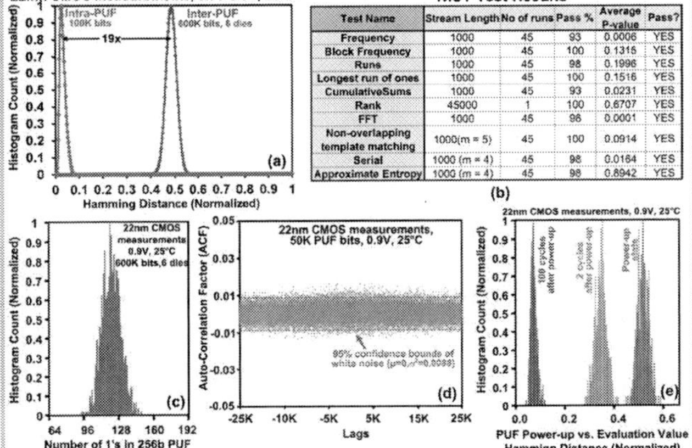

NIST Test Results

Test Name	Stream Length	No of runs	Pass %	Average P-value	Pass?
Frequency	1000	45	93	0.0006	YES
Block Frequency	1000	45	100	0.1315	YES
Runs	1000	45	98	0.1996	YES
Longest run of ones	1000	45	100	0.1516	YES
CumulativeSums	1000	45	93	0.0231	YES
Rank	45000	1	100	0.6707	YES
FFT	1000	45	96	0.0001	YES
Non-overlapping template matching	1000(m = 5)	45	100	0.0914	YES
Serial	1000 (m = 4)	45	98	0.0164	YES
Approximate Entropy	1000 (m = 4)	45	98	0.8942	YES

(b)

Figure 16.2.6: (a) Inter- & Intra-PUF hamming distance measurements; (b) NIST test results; (c) 1-count histogram measurements; (d) Autocorrelation measurements; (e) Power-up vs. evaluation state measurements.

16

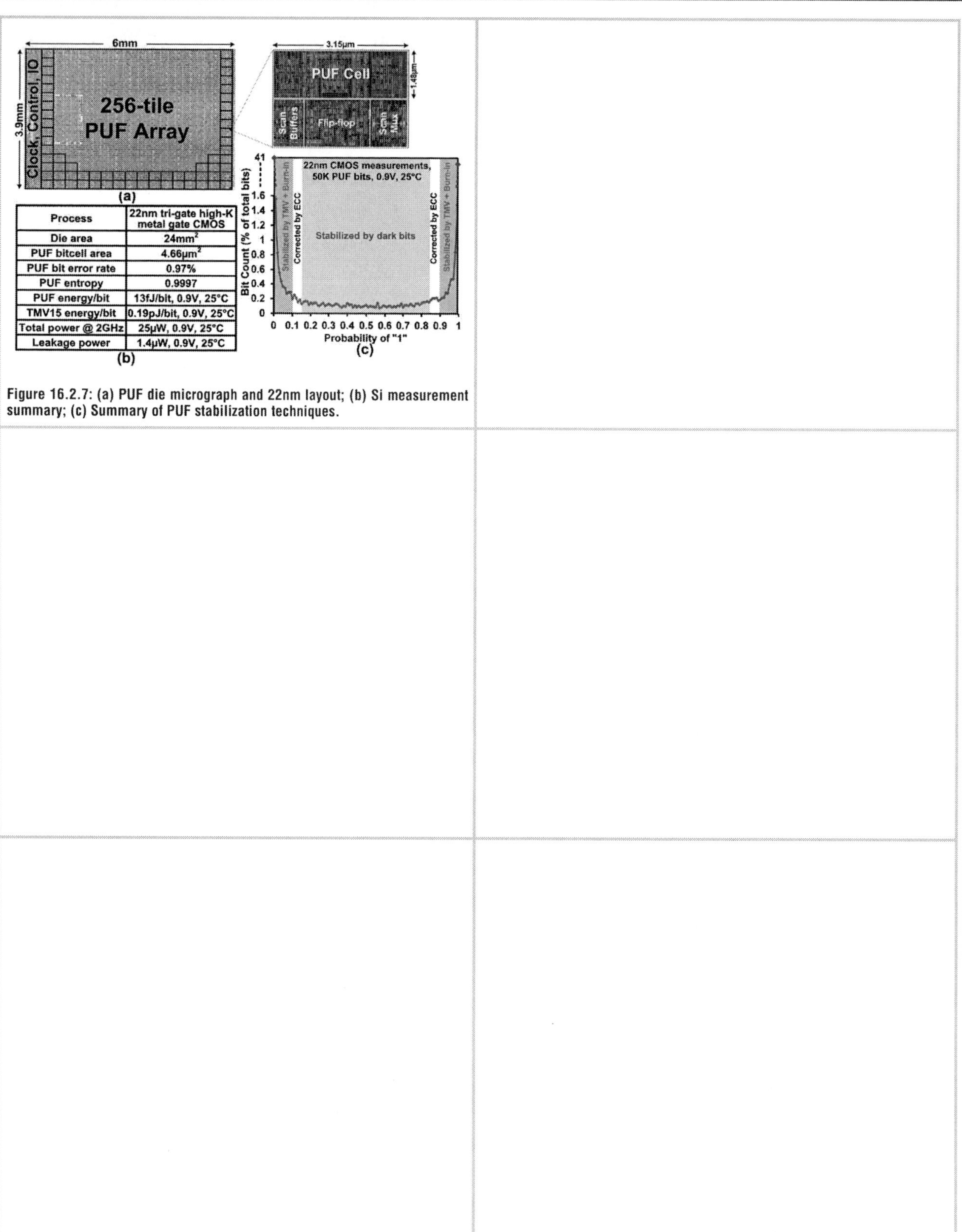

Process	22nm tri-gate high-K metal gate CMOS
Die area	$24mm^2$
PUF bitcell area	$4.66\mu m^2$
PUF bit error rate	0.97%
PUF entropy	0.9997
PUF energy/bit	13fJ/bit, 0.9V, 25°C
TMV15 energy/bit	0.19pJ/bit, 0.9V, 25°C
Total power @ 2GHz	25µW, 0.9V, 25°C
Leakage power	1.4µW, 0.9V, 25°C

(b)

Figure 16.2.7: (a) PUF die micrograph and 22nm layout; (b) Si measurement summary; (c) Summary of PUF stabilization techniques.

16.3 A 23Mb/s 23pJ/b Fully Synthesized True-Random-Number Generator in 28nm and 65nm CMOS

Kaiyuan Yang, David Fick, Michael B. Henry, Yoonmyung Lee, David Blaauw, Dennis Sylvester

University of Michigan, Ann Arbor, MI

True random number generators (TRNGs) use physical randomness as entropy sources and are heavily used in cryptography and security [1]. Although hardware TRNGs provide excellent randomness, power consumption and design complexity are often high. Previous work has demonstrated TRNGs based on a resistor-amplifier-ADC chain [2], oscillator jitter [1], metastability [3-5] and other device noise [6-7]. However, analog designs suffer from variation and noise, making them difficult to integrate with digital circuits. Recent metastability-based methods [3-5] provide excellent performance but often require careful calibration to remove bias. SiN MOSFETs [6] exploit larger thermal noise but require post-processing to achieve sufficient randomness. An oxide breakdown-based TRNG [7] shows high entropy but suffers from low performance and high energy/bit. Ring oscillator (RO)-based TRNGs offer the advantage of design simplicity, but previous methods using a slow jittery clock to sample a fast clock provide low randomness [1] and are vulnerable to power supply attacks [8]. In addition, the majority of previous methods cannot pass all NIST randomness tests.

To simultaneously achieve ease of design, high randomness, good throughput and energy/bit, we present a TRNG based on 3-edge multimode RO synthesized entirely with standard cells (Fig. 16.3.1). A conventional RO injects 1 edge that propagates through the ring to form pulses. The proposed 3-edge RO has 3 input nodes that inject 3 edges into the ring simultaneously. Each edge propagates as in a conventional RO; the period of each edge is the same, but the three edges are 120° phase shifted and overall frequency is boosted by 3×. However, the 3 edges independently accumulate jitter from thermal noise, causing an increasing variation of the pulse width between two neighboring edges with each completed cycle. Given time, two neighboring edges will eventually collapse, forcing the RO back to its nominal 1× frequency mode. The time to collapse reflects the accumulation of jitter and is used as the entropy source for random number generation. Process variation is inherently cancelled, since all 3 edges pass through the same RO stages. The design was fabricated in 28nm and 65nm and consistently passes all NIST randomness tests.

Figure 16.3.2 shows the TRNG consisting of 2 ROs, a counter, and control logic. A conventional RO (RO_REF) with ~2/3rd as many stages as the 3-edge RO (RO_RNG) acts as a reference for the phase frequency detector (PFD) to determine the edge collapse event. Since the frequency change is large (3×), a conventional digital implementation of the PFD is used, which enables a fully synthesizable design. To avoid setup and hold-time violations in the sampling registers, a glitch removal stage and 2b shift register is added. This ensures that a collapse event is flagged only after two consecutive pulses. A 14b cycle counter triggered by the 3-edge RO records the number of cycles until collapse. An intermediate counter bit, COUNT[3], is used to prevent false triggers in the first few cycles. Random number generation is initiated by a master clock, which is set sufficiently slow to ensure that the vast majority of collapse events (e.g. >90% in the tested design) complete within the active phase duration. The TRNG throughput is determined by the master clock frequency and the number of random bits harvested from each collapse event. The capture register reads the cycle counter when triggered by the PFD. As expected, the collapse cycle count displays a log-normal distribution (Fig. 16.3.5). To transform this into a uniform distribution, the collapse cycle counter is truncated, retaining the lower *p* bits, while the LSB is dropped to eliminate sensitivity to mismatch in the counter sampling flip-flop.

All hardware TRNGs must cope with interference from a potentially noisy environment as well as dedicated attacks. ROs are known to be sensitive to frequency injection, which can introduce errors in RO-based TRNGs [8]. The proposed TRNG uses accumulated jitter rather than jitter at a specific time point, making it more robust to noise injection. We tested the sensitivity to a deliberate attack with off-chip noise sources and also created on-chip test structures to inject and measure noise (Fig. 16.3.3). A programmable noise generator controlled by an on-chip VCO introduces substantial noise on the TRNGs supply, locking the oscillation and impacting collapse event time. To measure noise amplitude on-chip, an asynchronous clock samples the supply voltage, compares it with an external reference voltage, and increments a counter accordingly. With sufficient samples (2^{14} here) the noise amplitude can be determined from the counter value. In addition, an RC filter with a 210MHz corner frequency was designed to mitigate the impact of supply noise (Fig. 16.3.3).

The TRNG is evaluated using two test chips: one in 28nm CMOS with 8 different rings, the other in 65nm CMOS with 48 different TRNGs. The NIST Pub 800-22 RNG testing suite is used to evaluate the randomness of generated bits with 112Mb in total across 15 tests. Both 28nm and 65nm TRNGs pass all 15 NIST tests as shown in Fig. 16.3.4. Shorter rings with higher frequency collapse faster but have a narrower distribution, reducing the number of random bits obtained per cycle (i.e. they require higher truncation). Longer rings provide more random bits but overall throughput is limited by the slower master clock.

Using an RF signal generator, up to 600mV$_{pp}$ noise is injected on the power supplies (after removing PCB decoupling caps) to test TRNG robustness against off-chip attack. The 65nm TRNGs retain randomness up to 360mV$_{pp}$ noise without filter and up to the 600mV$_{pp}$ generator limit with filter. To compensate for filter IR drop, TRNGs with filters operate at 5% increased supply voltage, incurring a slight power penalty. Since ROs in 28nm TRNGs operate at a higher frequency they are less sensitive to external attack; even unfiltered versions did not suffer randomness degradation at the generator limit. EMI emitted by an antenna also did not cause failure in any randomness tests.

Figure 16.3.5 shows the impact of supply noise on TRNG performance using on-chip noise generation. Even though a deliberate attacker will not have access to such a noise source, this test can demonstrate how readily a 3-edge TRNG can be integrated with noisy circuits on an SoC. TRNGs showed sensitivity to supply noise at frequencies near 1× and 4× nominal RO frequencies, reducing collapse-time mean and variance. Randomness degrades at >125mV noise amplitude and 4× frequency without a filter, but is recovered using a filter. Denial of service occurs when a TRNG cannot generate outputs due to external an influence. This is observed only in unprotected TRNGs with on-chip noise at exactly 3× nominal frequency since the ring locks to its 3× frequency mode, preventing collapse. In this case, yield (the % of master cycles that generate outputs bits) drops to 7.37%. Generated bits remain random (passing all tests). Fig. 16.3.6 summarizes measurement results with comparisons to prior work. In 28nm, the TRNG generates random bits at 23.16Mb/s, while consuming 0.54mW and 375μm^2. Constructed entirely using a standard cell library and conventional place and route tools, this design presents a 'soft IP' TRNG that passes all NIST randomness tests without post processing.

Acknowledgments:

The authors acknowledge STMicroelectronics for IC fabrication support.

References:

[1] M. Bucci, *et al.*, "A High Speed Oscillator Based True Random Number Generator for Cryptographic Applications on a Smart Card IC," *IEEE Trans. Computers*, vol. 52, no. 4, pp. 403-409, 2003.

[2] C. Petrie, *et al.*, "A Noise-Based IC Random Number Generator for Applications in Cryptography," *IEEE Trans. Circuits and Systems-I*, vol. 47, no. 5, pp. 615-621, 2000.

[3] R. Brederlow, *et al.*, "A Low-Power True Random Number Generator using Random Telegraph Noise of Single Oxide Traps," *ISSCC Dig. Tech. Papers*, pp. 536-537, 2006.

[4] C. Tokunaga, *et al.*, "True Random Number Generator with a Metastability-Based Quality Control," *ISSCC Dig. Tech. Papers*, pp. 404-405, 2007.

[5] S. Mathew, *et al.*, "2.4Gbps, 7mW All-Digital PVT-variation Tolerant True Random Number Generator for 45nm CMOS High-Performance Microprocessors", *IEEE J. Solid-State Circuits*, vol. 47, no. 11, pp. 2807-2821, 2012.

[6] M. Matsumoto, *et al.*, "1200μm^2 Physical Random-Number Generators Based on SiN MOSFET for Secure Smart-Card Application," *ISSCC Dig. Tech. Papers*, pp. 414-415, 2008.

[7] N. Liu, *et al.*, "A true random number generator using time-dependent dielectric breakdown," *IEEE Symp. VLSI Circuits*, pp. 203-204, 2010.

[8] A. Markettos, *et al.*, "The Frequency Injection Attack on Ring-Oscillator-Based True Random Number Generators," *CHES*, pp. 317-331, 2009.

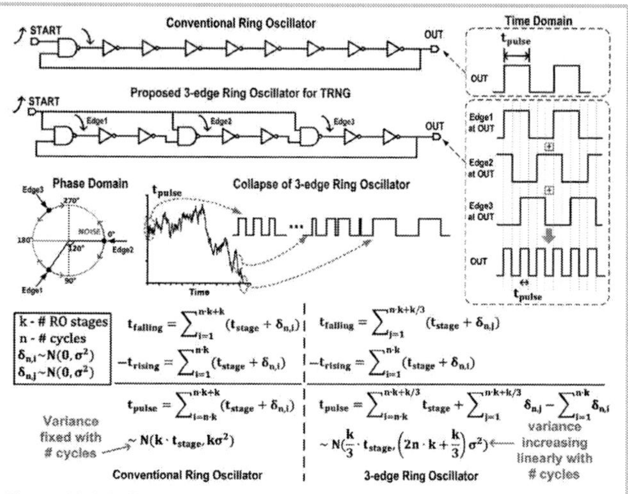

Figure 16.3.1: Frequency collapse in the 3-edge ring oscillator in time and phase domains.

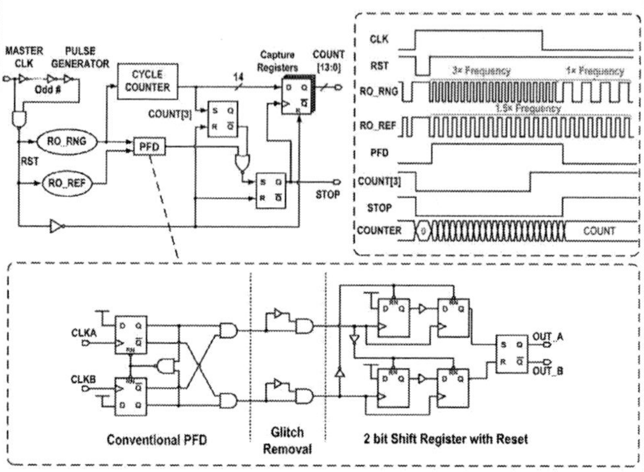

Figure 16.3.2: TRNG system block diagram and phase frequency detector (PFD) implementation.

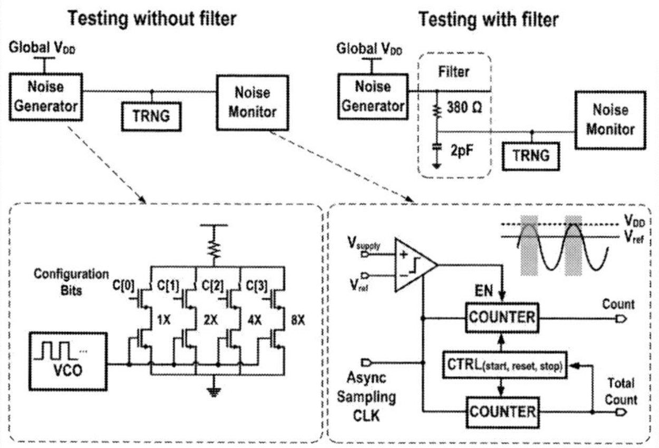

Figure 16.3.3: On-chip supply noise testing setups for protected and unprotected TRNGs.

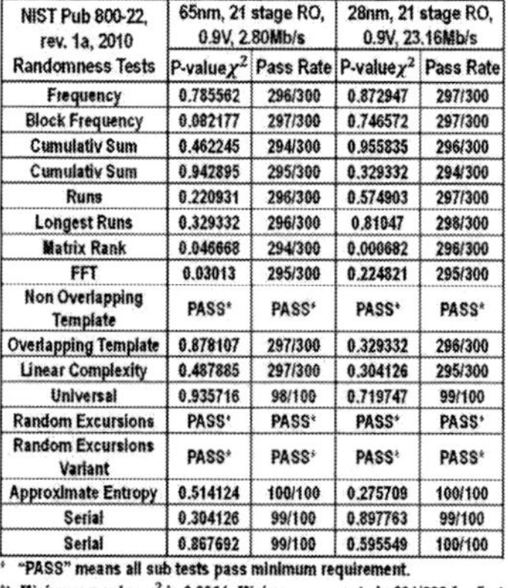

NIST Pub 800-22, rev. 1a, 2010 Randomness Tests	65nm, 21 stage RO, 0.9V, 2.80Mb/s		28nm, 21 stage RO, 0.9V, 23.16Mb/s	
	P-valueχ^2	Pass Rate	P-valueχ^2	Pass Rate
Frequency	0.785562	296/300	0.872947	297/300
Block Frequency	0.082177	297/300	0.746572	297/300
Cumulativ Sum	0.462245	294/300	0.955835	296/300
Cumulativ Sum	0.942895	295/300	0.329332	294/300
Runs	0.220931	296/300	0.574903	297/300
Longest Runs	0.329332	296/300	0.81047	298/300
Matrix Rank	0.046668	294/300	0.000682	296/300
FFT	0.03013	295/300	0.224821	295/300
Non Overlapping Template	PASS*	PASS*	PASS*	PASS*
Overlapping Template	0.878107	297/300	0.329332	296/300
Linear Complexity	0.487885	297/300	0.304126	295/300
Universal	0.935716	98/100	0.719747	99/100
Random Excursions	PASS*	PASS*	PASS*	PASS*
Random Excursions Variant	PASS*	PASS*	PASS*	PASS*
Approximate Entropy	0.514124	100/100	0.275709	100/100
Serial	0.304126	99/100	0.897763	99/100
Serial	0.867692	99/100	0.595549	100/100

* "PASS" means all sub tests pass minimum requirement.

** Minimum p-value χ^2 is 0.0001. Minimum pass rate is 291/300 for first 10 tests (using 300 × 40K bits) and 96/100 for the other 5 tests (using 100 × 1M bits).

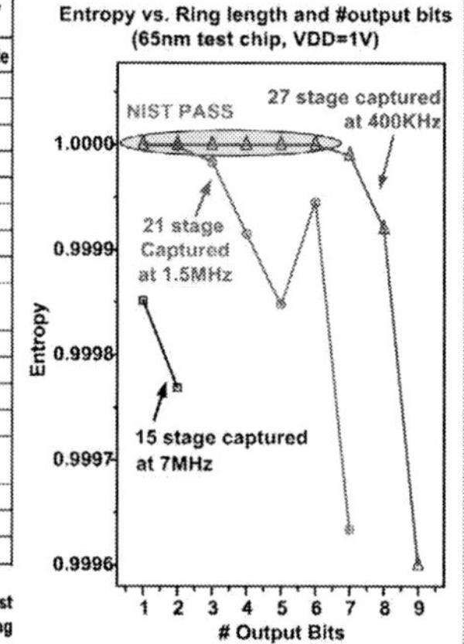

Figure 16.3.4: Measured NIST randomness test results and impacts of RO length and the number of harvested random bits on output data entropy.

Randomness under attack (noise amplitude = 200mV)
1x freq: 288MHz (w/ filter)
1x freq: 259MHz (w/o filter)
w/o filter
w/ filter
Entropy
\# Test Pass
Noise Frequency (MHz)

Distribution of cycles to collapse under attack (noise amplitude = 200mV)
Under 4x Freq Noise w/o filter
Under 4x Freq Noise w/ filter
No Attack
Occurrences($\times 10^4$)
Cycles to Collapse

Randomness under attack (noise freq = 4x nominal freq)
w/ filter
w/o filter
Noise Amplitude (mV)

Denial of service at 3x nominal frequency (noise amplitude = 200mV)
All cases pass NIST tests
Yield (%)
Noise Frequency (MHz)

Figure 16.3.5: Measured impacts of on-chip noise frequency and amplitude on randomness of protected and unprotected TRNGs (65nm, 21-stage RO TRNG).

	This work (25°C, 0.9V core supply)		JSSC' 12 [5]	VLSI' 11 [7]	ISSCC' 08 [6]	ISSCC' 07 [4]	ISSCC' 06 [3]	Trans. Computers' 03 [1]
Technology	28nm	65nm	45nm	65nm	0.25µm	0.13µm	0.12µm	0.18µm
Entropy Source	Jitter in 3-edge RO		Metas-tability	Oxide breakdown	SiN MOS-FET Noise	Metas-tability	Metas-tability	Oscillator jitter
Bit Rate (Mb/s)	23.16	2.8	2400	0.011	2	0.2	0.2	10
NIST Pass	All	All	All	All	not reported[b]	5	not reported	not reported[b]
TRNG Core Area (µm²)	375	960 (1080[a])	4004	1200	1200	36300	9000	16000
Power (mW)	0.54	0.159	7	2	1.9	1	0.05	2.3
Efficiency (nJ/bit)	0.023	0.057	0.0029	181.81	0.95	5	0.25	0.23
Post Processing	No	No	No	No	Yes	No	Yes	no
Resistance to Attack	Yes	Yes	Not reported	Not reported	Not reported	Not reported	Not reported	No[c]

[a] Including 1/8th of filter area (MIM cap and poly resistor); 1 filter is shared by 8 TRNGs and MIM cap is placed above TRNGs.

[b] NIST FIPS 140-2 test result is provided, which is older, less rigorous than NIST Pub 800-22 with 4 tests and only 20,000 bits required.

[c] Commercial TRNG based on similar RO approach is successfully attacked in [8].

Figure 16.3.6: Summary of measurement results and a comparison with state-of-the-art hardware TRNG designs.

16

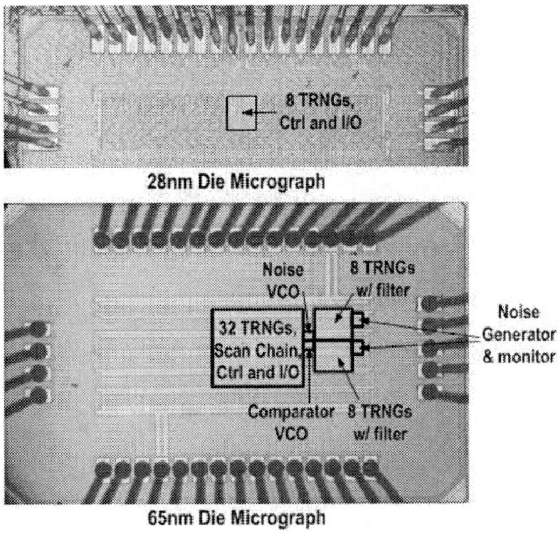

Figure 16.3.7: Die micrographs of 28nm and 65nm TRNG test chips.

16.4 0.6-to-1.0V 279μm², 0.92μW Temperature Sensor with Less Than +3.2/-3.4°C Error for On-Chip Dense Thermal Monitoring

Teng Yang, Seongjong Kim, Peter R. Kinget, Mingoo Seok

Columbia University, New York, NY

On-chip temperature sensors are key building blocks for the thermal management of multi-core microprocessors. The sensors are embedded at multiple locations in a microprocessor and monitor temperatures that are used to manage the operation of the microprocessor under local and global thermal constraints. While existing sensors achieve impressive area and accuracy, emerging technology trends such as multi-core architectures, 3D integration, tri-gate devices, and low-voltage operation demand even better sensors with difficult-to-meet requirements. Those requirements are three-fold: 1) Sensors need to be area efficient. By increasing the number of cores and hot spots, there are more locations that require thermal monitoring. To reduce the overhead, the sensor footprint needs to be minimized. A compact footprint is further critical for design flexibility, as the exact sensor locations (e.g. near hot spots) are often only identified in later stages of the design process [1]. 2) Sensors need to have low calibration cost, while achieving sufficient accuracy. The requirements of <8°C in absolute inaccuracy and <3°C in relative inaccuracy have been outlined in [2]. 3) Finally, the sensors need better supply voltage (V_{DD}) scalability. Sub-1V operation for digital systems is being explored to reduce power. The conventional sensors often cannot operate below 1V, necessitating additional power distribution or regulation. Sub-1V scalability eliminates such overhead.

This paper presents two temperature sensor designs that satisfy the above requirements. The sensors are designed utilizing the principles of 2T voltage-reference circuits (2T-VR) ([6], Fig. 16.4.1). The output of 2T-VR can be calculated with the equation in Fig. 16.4.1. The first term is mainly a difference of threshold voltages (V_t). The temperature coefficient of the second term can be configured to be proportional-to-absolute-temperature (PTAT) or complementary-to-absolute-temperature (CTAT) via sizing. The output, therefore, can be designated as either CTAT or PTAT. Two sensor designs have been developed in 65nm CMOS. The first design is optimized for a balanced area-accuracy performance, while the second design is optimized for small area (Fig. 16.4.1). Measured results show that the balanced design, which will be the main focus of this paper, achieves 1) a compact footprint of 279μm², 2) an absolute error less than +3.2/-3.4°C across 64 sensors in 8 chips after one-temperature-point calibration (OPC) at 50°C, 3) a relative error less than 5.2°C among 8 instances in each chip for a total 8 chips after the same OPC, 4) DC-PSRR better than -67dB, 5) V_{DD} scalability down to 0.6V, and 6) a power consumption of 0.92μW at V_{DD}=0.6V and 100°C.

Our designs require only 6 to 7 transistors per sensor, significantly reducing the area footprint. Each transistor is sized 1) to achieve a sufficient amount of temperature coefficient in Vp and Vc to ease design of the readout circuitry, 2) to mitigate local process variations, 3) to reduce short-channel effects, and 4) to achieve a sufficiently short settling time of the output voltages: simulations for the balanced sensor with 1.5pF load show a settling time of 20 and 3μs at 0 and 100°C, respectively. All transistors are laid out as multiples of a unit device (W=0.6μm, L=0.3μm) to share the same layout-dependent effects. The balanced and the area-optimized designs have footprints of only 279μm² and 115μm², respectively, which is 6.78-to-16.5× smaller than previous work [1].

The design achieves competitive accuracy across process variations. In order to minimize the impact of variations, a differential reading scheme is used, which digitizes Vp and Vc and takes the difference, eliminating the *shared* impact of process variations on both read-outs, particularly global V_t variation (top left of Fig. 16.4.2). The analytical solution for Vp-Vc confirms the reduced effect of V_t (bottom of Fig. 16.4.1). Measured results show that the differential reading can reduce the slope variations, improving the worst-case error by ~6.5°C, compared to reading either Vp or Vc (bottom left of Fig. 16.4.2).

64 sensors in 8 chips are measured and OPC-ed at 50°C. A linear model from simulation is used for calibration and error calculation. Measurements show a worst-case error less than +3.2/-3.4°C (bottom right of Fig. 16.4.2). The worst-case relative error is measured to be less than 5.2°C across 8 chips, each of which has 8 sensors (top of Fig. 16.4.3). Errors can be further reduced by modulating calibration temperature. For example, OPC can be performed at 80°C if higher accuracy around that temperature is desired. The worst-case error is reduced to less than +1.4/-1.9°C from 60 to 100°C (bottom left of Fig. 16.4.3). If a 2-point calibration is performed at 20°C and 80°C, the error can be further reduced to +1.5/-1.6°C across 64 sensors in 8 chips (bottom right of Fig. 16.4.3).

Achieving a good PSRR is critical to maintain the accuracy of sensors in noisy digital environments. In the 2T-VR, device length was increased to reduce the sensitivity to V_{DD}. This approach, however, incurs a large area overhead. The sensor achieves high PSRR by employing a cascode device and using a different reading scheme, which eliminated the shared sensitivity to V_{DD}. The PSRR is simulated to be less than −44dB across DC-to-1GHz. The PSRR measurements at DC are roughly matched to the simulation, with -67 to -71dB across 8 instances in a single chip (top of Fig. 16.4.4). It is worthwhile to note that the required accuracy of the bias voltage for the cascode device (V_b) is not high, easing the on-chip generation of the bias voltage. Measurements show that the PSRR can be maintained better than -50dB with the Vb of 400±40mV (bottom left of Fig. 16.4.4).

The conventional BJT-pair-based sensors cannot operate at V_{DD} < ~1V since the large diode turn-on voltage severely limits the voltage headroom [1,2]. Our sensor design relies on a different operating principle, exhibiting better V_{DD} scalability down to 0.6V (bottom right of Fig. 16.4.4). In addition, the scalability to low V_{DD}, coupled with the good PSRR, can enable the operation of the sensor across V_{DD} with a calibration setting done at a single V_{DD}/temperature point. A sensor, calibrated at 50°C and V_{DD}=0.6V, can operate at V_{DD}=1.0V with a worst-case error < 0.58°C.

Figure 16.4.5 shows the organization of the test chip in 65nm CMOS. The chip has 32 sensors, which include 8 balanced and 8 area-optimized sensors. The output voltages of sensors can be transferred off-chip through a 32-to-1 analog multiplexer, input switch network (ISN) and on-chip switched-capacitor amplifier (SC-AMP). In this paper, all the measurements are performed only through the analog multiplexer and ISN due to limited test time.

We compare the two sensors to several previous designs (Fig. 16.4.6). The balanced sensor achieves a 6.78× smaller footprint than [1], and exhibits error less than +3.2/-3.4°C after OPC across 64 sensor instances in 8 chips. The worst-case measured relative inaccuracy is 5.2°C across 8 chips. The sensor can operate at V_{DD} from 0.6 to 1V, while none of the compared designs can operate below 1V. In addition, the area-optimized core has a footprint of 115μm², 16.5× smaller than [1] with error less than +3.8/-4.7°C across 64 instances in 8 chips after OPC.

Acknowledgments:
This work was supported by the Catalyst Foundation.

References:

[1] J. Shor, *et al.*, "Ratiometric BJT-Based Thermal Sensor in 32nm and 22nm Technologies," *ISSCC Dig. Tech. Papers,* pp. 210-211, 2012.
[2] Y.W. Li, *et al.*, "A 1.05V 1.6mW 0.45°C 3σ-resolution ΔΣ-based Temperature Sensor with Parasitic-Resistance Compensation in 32nm CMOS," *ISSCC Dig. Tech. Papers,* pp. 340-341, 2009.
[3] G. R. Chowdhury, *et al.*, "A 0.001mm² 100μW on-chip Temperature Sensor with ±1.95 °C (3σ) Inaccuracy in 32nm SOI CMOS," *IEEE International Symp. On Circuits and Systems,* pp. 1999-2002, 2012.
[4] S. Paek, *et al.*, "All-Digital Hybrid Temperature Sensor Network for Dense Thermal Monitoring," *ISSCC Dig. Tech. Papers,* pp. 260-261, 2013.
[5] M. Sasaki, *et al.*, "A Temperature Sensor with an Inaccuracy of -1/+0.8°C Using 90-nm 1-V CMOS for Online Thermal Monitoring of VLSI Circuits," *IEEE Trans. on Semiconductor Manufacturing,* vol. 21, no. 2, pp. 201-208, 2008.
[6] M. Seok, *et al.*, "A Portable 2-Transistor Picowatt Temperature-Compensated Voltage Reference Operating at 0.5 V," *IEEE J. of Solid-State Circuits,* vol. 47, no. 10, pp. 2534-2545, 2012.

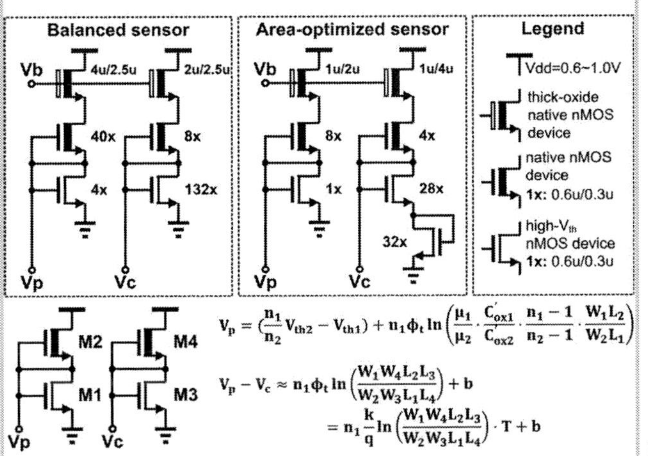

Figure 16.4.1: Two temperature sensor designs, exploiting the fundamental of the compact 2T voltage reference circuits, are shown.

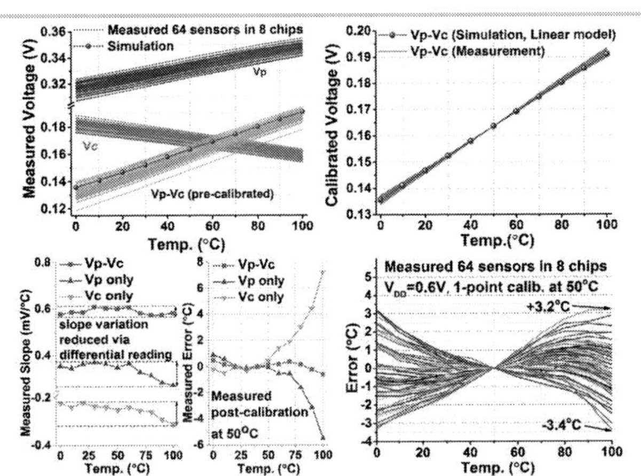

Figure 16.4.2: Differential reading can mitigate slope variations, enabling 6.5°C smaller error, as compared to single reading of either Vp or Vc. The worst-case error is less than +3.2/-3.4°C after OPC using the simulation-based linear model.

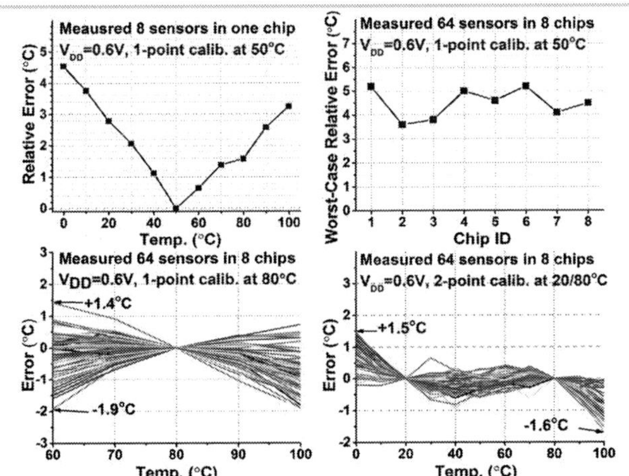

Figure 16.4.3: Per-chip relative error is ≤4.5°C. Relative error across 8 chips is ≤5.2°C. Absolute error after OPC-ed at 80°C is ≤3.3°C in 60 to 100°C. The error with 2 point calibration is ≤3.1°C.

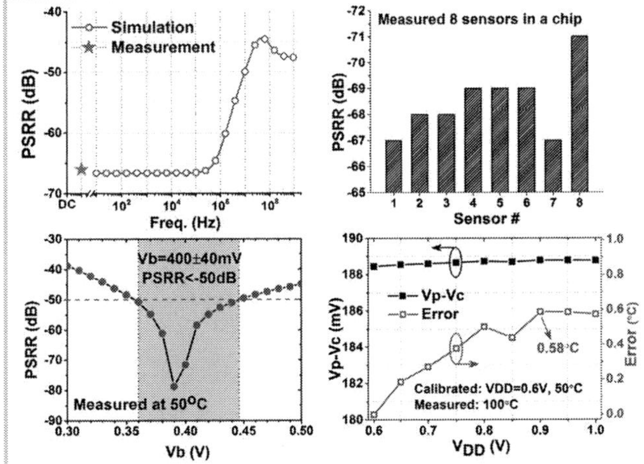

Figure 16.4.4: The balanced sensors exhibit good PSRR across chips. The design can maintain the DC-PSRR of <-50dB with the Vb of 400±40mV. OPC-ed at 0.6V and 50°C, the sensor can operate at 1V with the error of <0.58°C.

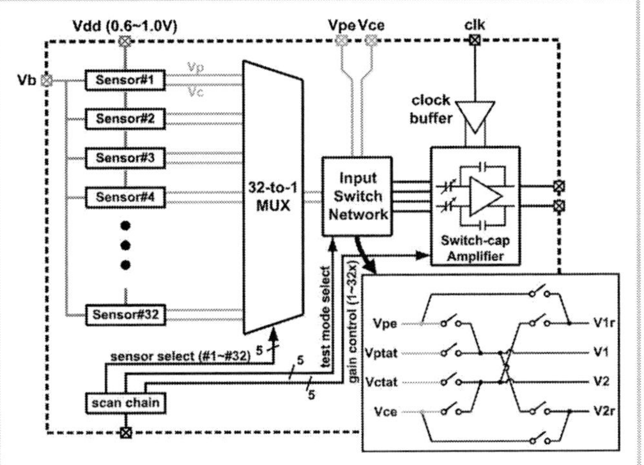

Figure 16.4.5: The chip-level block diagram. The outputs of the sensors can be transferred to an off-chip ADC via a 32-to-1 multiplexer, an ISN and a SC-AMP.

	[1]		[2]	[3]	[4]	[5]	**Balanced**	**Area-opt.**
Tech.	32nm	22nm	32nm	32nm	0.13μm	90nm	**65nm**	**65nm**
V_{DD} (V)	1.3-1.8	/	1.05	1.65	/	1	**0.6~1.0**	**0.6~1.0**
Power (total, mW)	3.78	1.4	1.6	0.1	/	/	/	/
Power (sensor)	/	/	/	/	/	25μW	**0.92μW**	**0.21μW**
Area[1] (μm^2)	20000	6100	20000	1053	165007	/	/	/
Area[2] (μm^2)	7700*	1900*	8000*	/	2164**	48	**279**	**115**
Temperature Coefficient	4.23 [c/°C]	3.82 [c/°C]	/	/	/	1.8 [mV/°C]	**0.57** [mV/°C]	**0.72** [mV/°C]
Range (°C)	-10~110		-10~110	5~100	/	50~125	**0~100**	**0~100**
Error[3] (°C)	/	/	<5	/	/	/	/	/
Error[4] (°C)	<4.5	/	/	/	/	/	**-3.4/+3.2**	**-3.8/+4.7**
Error[5] (°C)	<0.59	<1.5	/	±1.95	/	-1~0.8	**+1.5/-1.6**	**-1.5/+3**
DC-PSRR (dB)	50	/	/	/	/	/	**67**	**65**
DC-Line Sensitivity (°C/V)	0.7	/	/	/	/	2.2	**1.2**	**0.8**

[1]: area includes all read-out circuitry; [2]: area per sensor; [3]: error without calibration; [4]: error with 1-temperature point calibration; [5]: error with 2-temperature point calibration; *: estimated from die photo; **: average sensor area = (ATS+TDDL+RTS)/64.

Figure 16.4.6: The balanced sensor is 6.8× smaller than [1] at a moderate accuracy penalty. The proposed designs can operate at V_{DD}=1 to 0.6V while none of the previous works can operate at V_{DD}<1V.

16

978-1-4799-0917-9/14 $31.00 © 2014 IEEE

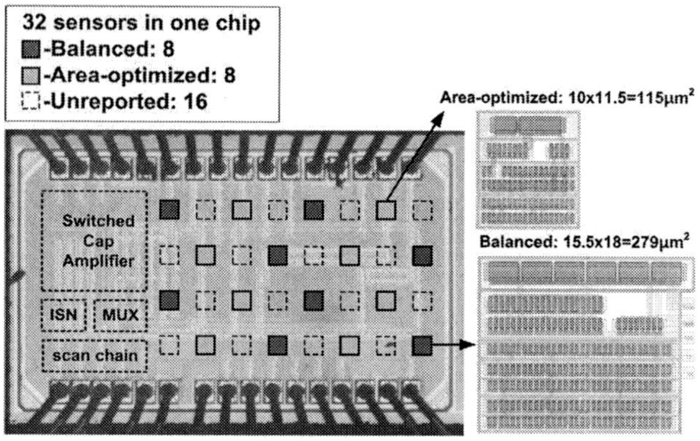

32 sensors in one chip
-Balanced: 8
-Area-optimized: 8
-Unreported: 16
Area-optimized: 10x11.5=115μm²
Switched Cap Amplifier
ISN
MUX
scan chain
Balanced: 15.5x18=279μm²

Figure 16.4.7: The die photo and the layout views for the two sensor designs.

Session 17 Overview: *Analog Techniques*

ANALOG SUBCOMMITTEE

Session Chair: *Marco Berkhout*
NXP Semiconductors,
Nijmegen, The Netherlands

Session Co-Chair: *Edgar Sanchez-Sinencio*
Texas A&M University,
College Station, TX

Analog technology continues to push the state of the art. This session illustrates the diversity and vitality of modern analog circuitry. Papers in this session span the range of filters, amplifiers, audio, voltage regulators, and oscillators. New frontiers of accuracy, power, and performance are established.

17.1 An Integrated 80V 45W Class-D Power Amplifier with Optimal-Efficiency-Tracking Switching Frequency Regulation **1:30 PM**
H. Ma, University of Twente, Enschede, The Netherlands
In Paper 17.1, by University of Twente presents an 80V class-D power amplifier implemented in a 0.14μm SOI BCD process. The amplifier adaptively regulates its switching frequency for optimal power efficiency and achieves best-in-class 93% efficiency at 45W output power, >80% power efficiency down to 4.5W output power and >49% efficiency down to 0.45W output power.

17.2 A 0.0013mm² 3.6μW Nested-Current-Mirror Single-Stage Amplifier Driving 0.15-to-15nF Capacitive Loads with >62° Phase Margin **2:00 PM**
Z. Yan, University of Macau, Macao, China
In Paper 17.2, by University of Macau, a single-stage amplifier with rail-to-rail output swing for LCD column driver ICs in 0.18μm CMOS is presented, that uses a Nested-Current-Mirror technique. The amplifier has 84dB DC gain, and 0.013-to-1.24MHz GBW over 0.15-to-15nF capacitive load with >62° phase margin.

17.3 A 0.9V 6.3μW Multistage Amplifier Driving 500pF Capacitive Load with 1.34MHz GBW **2:15 PM**
W. Qu, KAIST, Daejeon, Korea
In Paper 17.3, by KAIST, a 0.9V 6.3μW three-stage amplifier capable of driving 500pF capacitive load is presented. This amplifier, fabricated in a 0.18μm CMOS process, achieves 1.34MHz gain-bandwidth product, 52.7° phase margin and 0.62V/μs average slew rate.

17.4 CMOS Impedance Analyzer for Nanosamples Investigation Operating up to 150MHz with Sub-aF Resolution **2:30 PM**
D. Bianchi, Politecnico di Milano, Milan, Italy
In Paper 17.4, by Politecnico di Milano, an architecture in 0.35μm CMOS is presented that performs impedance spectroscopy in the frequency range from 1kHz to 150MHz with 180dB loop gain at all frequencies. Two DC outputs directly related to the real and imaginary parts of the input admittance achieve a resolution better than 1pS and 1aF, respectively.

17.5 A 0.07mm² 2-Channel Instrumentation Amplifier with 0.1% Gain Matching in 0.16µm CMOS — 3:15 PM

F. Butti, NXP Semiconductors, Eindhoven, The Netherlands
and Delft University of Technology, Delft, The Netherlands

In Paper 17.5, by NXP Semiconductors, a 2-channel instrumentation amplifier in 0.16µm CMOS is presented with 0.1% gain matching and a 13.3x area improvement with respect to state-of-the-art designs with similar gain accuracy. Dynamic element matching (DEM) and a high chopping frequency yields 18.7nV/ÖHz noise, 17µV offset, and 12.9 NEF.

17.6 Envelope Modulator for Multimode Transmitters with AC-Coupled Multilevel Regulators — 3:30 PM

P. Arnò, STMicroelectronics, Grenoble, France

In Paper 17.6, by ST Microelectronics, a buck-boost Envelope Modulator based on AC-coupled multilevel switching regulators is presented. Power added efficiency when loaded by an RF power amplifier reaches 39% for an output power of 26.3dBm and ACLR <-40dBc. The circuit is manufactured in a 0.13µm, 4.8V CMOS process.

17.7 A 1.89nW/0.15V Self-Charged XO for Real-Time Clock Generation — 3:45 PM

K-J. Hsiao, MediaTek, Hsinchu, Taiwan

In Paper 17.7, by MediaTek, an ultra-low power self-charged XO is proposed for 32.768kHz real-time clock generation. A direct-charging scheme with logic-intensive design is proposed to reduce the power and supply voltage. It consumes 1.89nW under 0.15V supply and occupies 0.03mm² active area in a 28nm CMOS process.

17.8 A 190nW 33kHz RC Oscillator with ±0.21% Temperature Stability and 4ppm Long-Term Stability — 4:00 PM

D. Griffith, Texas Instruments, Dallas, TX

In Paper 17.8, by Texas Instruments, a fully integrated 33kHz RC oscillator in 65nm CMOS for sleep timers in wireless sensor networks is presented. The oscillator uses an inverter as comparator and a supply that tracks the inverter parameters. It achieves ±0.21% frequency stability over –20°C to +90°C and better than 4ppm long term frequency stability over intervals above 2s while consuming only 190nW.

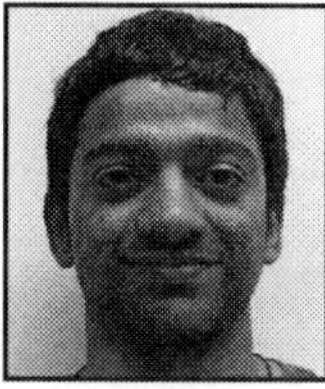

17.9 A 0.6V 70MHz 4th-Order Continuous-Time Butterworth Filter with 55.8dB SNR, 60dB THD at +2.8dBm Output Signal Power — 4:15 PM

J. N. Kuppambatti, Columbia University, New York, NY

In Paper 17.9, by Columbia University, a 0.6V 70MHz 4th Order Continuous-Time Butterworth filter in 65nm CMOS is presented where the output stages of OTAs are replaced with power-efficient rail-to-rail Class-D stages. The filter achieves 58dB dynamic range, 55.8dB SNR and 60dB THD at +2.8dBm output signal power, while dissipating 26.2mW from a 0.6V supply.

17.10 0.65V-Input-Voltage 0.6V-Output-Voltage 30ppm/°C Low-Dropout Regulator with Embedded Voltage Reference for Low-Power Biomedical Systems — 4:45 PM

W-C. Chen, National Chiao Tung University, Hsinchu, Taiwan

In Paper 17.10, by National Chiao Tung University, a 0.6V output voltage low-drop regulator with embedded voltage reference in a 21nm CMOS process is proposed for low-power biomedical systems. A 30ppm/°C temperature coefficient and 10mA driving capability is achieved at 0.65V input voltage.

17.11 A 0.65ns-Response-Time 3.01ps FOM Fully-Integrated Low-Dropout Regulator with Full-Spectrum Power-Supply-Rejection for Wideband Communication Systems — 5:00 PM

Y. Lu, Hong Kong University of Science and Technology, Hong Kong, China

In Paper 17.11, by HKUST, a fully-integrated tri-loop LDO architecture is proposed and verified in 65nm CMOS for wideband communication systems. It achieves a transient response time of 0.65ns with 50µA quiescent current and 140pF load capacitor. Measured PSR is -15.5dB at 1GHz.

17.1 An Integrated 80V 45W Class-D Power Amplifier with Optimal-Efficiency-Tracking Switching Frequency Regulation

Haifeng Ma, Ronan van der Zee, Bram Nauta

University of Twente, Enschede, The Netherlands

Piezoelectric actuators are widely used in smart materials for vibration and noise control, precision actuators, etc. [1]. These actuators are largely capacitive and the reactive power applied on them can go to several tens of Watts. High-voltage, high-power class-D amplifiers [2]-[5] are ideal drivers for such loads, because of their high power efficiency. Preferably, efficiency should be high both at maximum power and at average output power. Obtaining high power efficiency over the full output power range of a class-D amplifier is the main focus of this work.

Figure 17.1.1 shows a typical high-voltage class-D power stage, where two identical n-type DMOS FETs are used as both high-side (HS) and low-side (LS) power switches with their gate-driver supply voltage V_{DD} being much lower than V_{DDP} [2]-[5]. The three main dissipation sources in the power stage are then: 1) Conduction loss P_{con} caused by the output current I_{out} due to r_{on} switch resistance; 2) Ripple loss P_{Irip} caused by the inductor ripple current I_{rip} due to r_{on} and magnetic core loss of L_{out}; 3) Switching loss P_{sw} at the V_{pwm} node caused by M_{HS}/M_{LS} having to charge/discharge C_p in Fig. 17.1.1. This can be significant for high V_{DDP}, since the energy stored in C_p is proportional to V_{DDP}^2.

There are two scenarios for P_{sw}, depending on I_{rip} and I_{out}. In the first case, for low I_{out}, the inductor ripple current I_{rip} is large enough for the total inductor current $I_L=I_{rip}+I_{out}$ to be bidirectional. Then, when $I_{rip}>|I_{out}|+C_pV_{DDP}/t_d$, I_L can fully charge and discharge C_p during the dead time t_d without resorting to M_{HS}/M_{LS}. This is the soft switching case where P_{sw} is eliminated. P_{Irip} is now the main dissipation source, and the switching frequency, f_{sw}, should be high to reduce I_{rip} and thus P_{Irip}. In the second case, when $|I_{out}|>I_{rip}$, I_L is unidirectional and one of the V_{pwm} switching transitions has to be finished by M_{HS}/M_{LS}. This is the hard switching case where P_{con} and P_{sw} are dominant. In this case, the power MOSFET sizing for balanced P_{con} and P_{sw} plays a role, which benefits from choosing a low f_{sw} to reduce P_{sw}. The two cases above have contradicting demands on f_{sw}. Common practice is to set f_{sw} in between as a compromise [3], but this is not optimal.

Varying f_{sw} can achieve higher efficiency over a larger output power range as in [6] and [7], but both techniques choose f_{sw} based on output current only. This is suboptimal since the dissipation is highly dependent on both I_{rip} and I_{out}, and there are numerous factors causing I_{rip} variation, apart from external factors like V_{DDP} and L_{out} values. This is especially the case for class-D designs where I_{rip} changes by a factor >5 in the 0.05-0.95 duty cycle (D) range.

We propose to regulate the I_{rip} amplitude such that both P_{sw} and P_{Irip} are minimized by changing f_{sw} based on the V_{pwm} level at the turn-on transition of the power switches. This information is directly related to the dissipation sources and can be used to obtain the optimal f_{sw}, independent of circuit operating conditions affecting I_{rip}. The result is a class-D amplifier with its f_{sw} adapted to achieve minimal dissipation from idle to maximum output power.

Figure 17.1.2 shows the working principle. On the left are the soft switching waveforms, with I_{rip} larger than necessary for eliminating P_{sw}. Both V_{pwm} transitions finish within the dead time t_d and are already at the other supply rail when M_{HS}/M_{LS} turns on. This means I_{rip} (and consequently P_{Irip}) could be smaller by increasing f_{sw}. In the right part of Fig. 17.1.2, I_L is too small to charge C_P during t_d, and the remaining V_{pwm} rising transition is provided by M_{HS}. V_{pwm} is not yet at V_{DDP} when M_{HS} turns on, indicating the existence of P_{sw} and that therefore f_{sw} should decrease. By adapting f_{sw} such that either one of the V_{pwm} switching transitions is at the boundary of being lossless while the other is fully lossless, minimization of both P_{sw} and P_{Irip} is achieved. By setting an f_{sw} lower limit, the system naturally shifts to hard switching at high output power, with minimized P_{sw}.

The implementation of the amplifier is shown in Fig. 17.1.3. In this realization, the amplifier is based on a 1st-order hysteretic self-oscillating loop. Alternative implementations can also use carrier-based topologies [2] by changing f_{sw} of the triangle carrier, either continuously or through a frequency plan to control the spectral content. An f_{sw} regulation loop is added to the basic amplifier structure by tuning the hysteretic window voltage V_{tune}, which is generated by a charge pump/loop filter (CP/LF) that receives UP/DN 1-shots depending on the timing between the V_{pwm} level and the V_{HS}/V_{LS} rising edges.

The output stage works with a V_{DDP} of 80V, an on-chip regulated 3.3V driver supply and has a 2-step level shifter that can handle supply bounce higher than the internal supply [8]. Figure 17.1.4 (upper part) shows the V_{pwm} level detection circuit. At the beginning of a transition, when V_{pwm} is far (up to 80V) from the supply rail, M_{LSC}/M_{HSC} shield the clamps M_{LSD}/M_{HSD} from V_{pwm}. When V_{pwm} is close to the supply rail, M_{LSC}/M_{HSC} are in the linear region, such that M_1/M_4 can detect if V_{pwm} is less than a V_{TH} from the supply rail, which is close enough not to cause significant P_{sw}. Control signals $V_{LS\_detect}/V_{HS\_detect}$ are generated in the output stage with a time shift compared to V_{LS}/V_{HS} such that they only activate M_{LSC}/M_{HSC} for half the switching cycle to prevent cross current flow from the supply. M_4 level shifts to logic levels referred to V_{SSD}. M_1-M_3 level shift in 2 steps to deal with the large (>3.3V) on-chip PGND bounce. The lower part of Fig. 17.1.4 shows the UP/DN decision logic. The V_{pwm} status is sampled at the rising edge of V_{HS}/V_{LS}. The 1-shot for an f_{sw} increase is activated if both V_{pwm} transitions are finished in time while the 1-shot for an f_{sw} decrease is activated if either transition is not. Since V_{tune} is at 2× the signal frequency f_{sig} (when I_{out} increases in either direction), V_{tune} generation is fully differential for minimal 2nd-order distortion.

The amplifier is implemented in a 0.14μm SOI BCD process. For power efficiency measurements, a series-connected 23μF + 1.6Ω is used to model the piezo-actuator [1]. Because this load is mostly capacitive at f_{sig}, efficiency is defined here as $P_{out}/(P_{out}+P_d)$, where P_{out} is the apparent output power $V_{out,rms}*I_{out,rms}$ (VA) processed by the amplifier and P_d is the total amplifier dissipation. Figure 17.1.5 shows the measured efficiency of the amplifier for a 500Hz sine wave for three fixed V_{tune} settings and one with f_{sw}-regulation enabled. Figure 17.1.5 clearly shows that the amplifier can adjust its f_{sw} for best efficiency across the whole output power range. Idle power consumption is 360mW while for the two lower f_{sw} cases it is 440mW and 690mW, achieving a reduction of 18% and 48%. The peak efficiency of the amplifier is 93% while for the two higher f_{sw} cases it is 91% and 89%, achieving a power loss reduction of 19% and 31%. In idle, the adaptive f_{sw} is 500kHz while for 45VA output power, the adaptive f_{sw} is from 200kHz at D=0.5 to 100kHz at D=0.05 or 0.95.

A comparison with other high-voltage, high-power class-D designs is shown in Fig. 17.1.6. For better comparison, efficiency with a non-capacitive load (12Ω resistor) is also measured. The V_{pwm}-level-based f_{sw}-regulation technique enables this design to achieve peak efficiency better than those reported in Fig. 17.1.6, while significantly outperforming the other amplifiers at lower output powers. THD+N is 0.015% @ 100Hz, 9VA and 0.94% @ 500Hz, 45VA. For applications that require lower distortion, a higher-order feedback loop can be used. The chip photograph is shown in Fig. 17.1.7, with the die measuring 3.4×2.5mm². To conclude, this amplifier offers the high peak efficiency of existing class-D designs, keeping heat sinks small, while offering significant energy savings at lower, much more prevalent, output powers.

Acknowledgements:
We thank STW for project funding and NXP for silicon donation.

References:

[1] C. Wallenhauer, et al., "Efficiency-Improved High-Voltage Analog Power Amplifier for Driving Piezoelectric Actuators," *IEEE Trans. Circuits Syst. I*, vol. 57, no. 1, pp. 291–298, Jan. 2010.
[2] M. Berkhout, "An Integrated 200-W Class-D Audio Amplifier," *IEEE J. Solid-State Circuits*, vol. 38, no. 7, pp. 1198–1206, Jul. 2003.
[3] P. Morrow, E. Gaalaas, O. McCarthy, "A 20-W Stereo class-D Audio Output Power Stage in 0.6-μm BCDMOS Technology," *IEEE J. Solid-State Circuits*, vol. 39, no. 11, pp. 1948–1958, Nov. 2004.
[4] F. Nyboe, et al., "A 240W Monolithic Class-D Audio Amplifier Output Stage," in *ISSCC Dig. Tech. Papers*, pp.1346-1355, Feb. 2006.
[5] J. Liu, et al., "A 100 W 5.1-Channel Digital Class-D Audio Amplifier With Single-Chip Design," *IEEE J. Solid-State Circuits*, vol. 47, no. 6, pp. 1344–1354, June 2012.
[6] T. Y. Man, P. K. T. Mok, M. Chan, "An Auto-Selectable-Frequency Pulse-Width Modulator for Buck Converters with Improved Light-Load Efficiency," *ISSCC Dig. Tech. Papers*, pp. 440-441, Feb. 2008.
[7] S. Zhou, G.A. Rincón-Mora, "A High Efficiency, Soft Switching DC-DC Converter with Adaptive Current-Ripple Control for Portable Applications," *IEEE Trans. Circuits Syst. II*, vol. 53, no. 4, pp. 319–323, Apr. 2006.
[8] H. Ma, R. van der Zee, B. Nauta, "An Integrated 80-V Class-D Power Output Stage with 94% Efficiency in a 0.14μm SOI BCD Process," *Proc. ESSCIRC*, Sept. 2013.

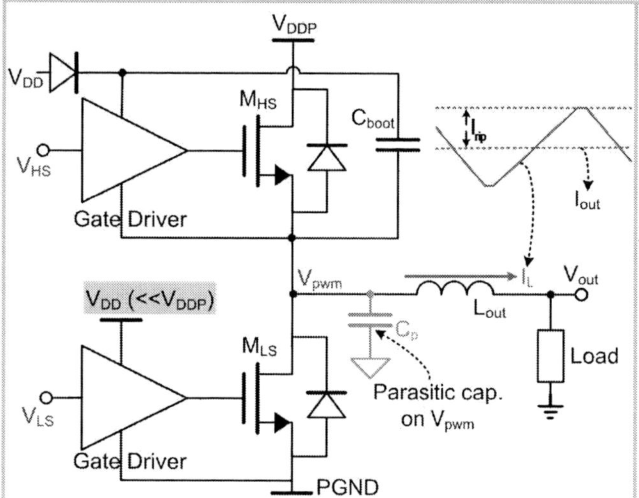

Figure 17.1.1: Basic class-D power output stage topology.

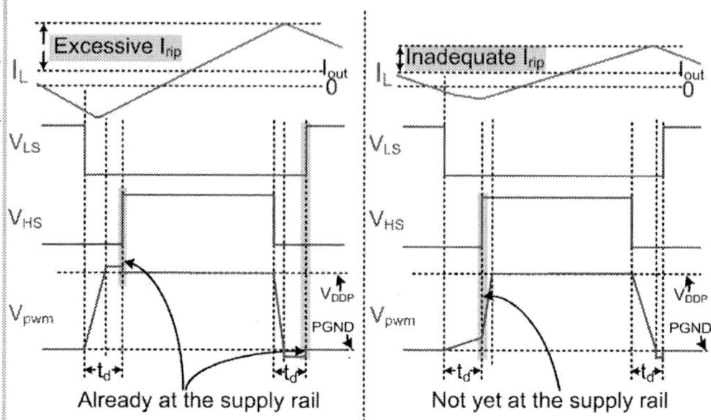

Figure 17.1.2: V_{pwm} level for excessive I_{rip} (left) and inadequate I_{rip} (right). V_{pwm} and t_d are not to scale.

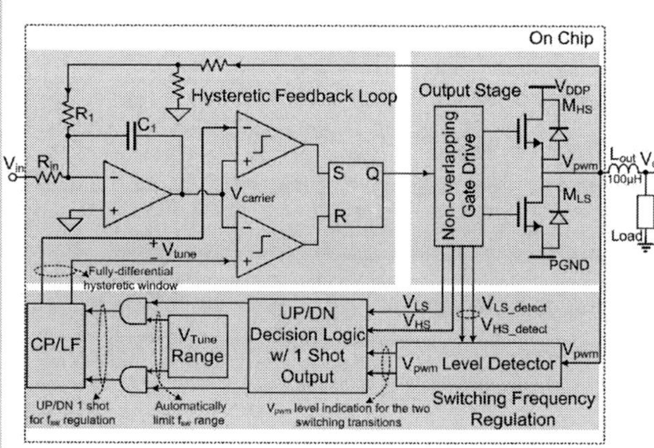

Figure 17.1.3: Implementation of the class-D power amplifier with f_{sw} regulation.

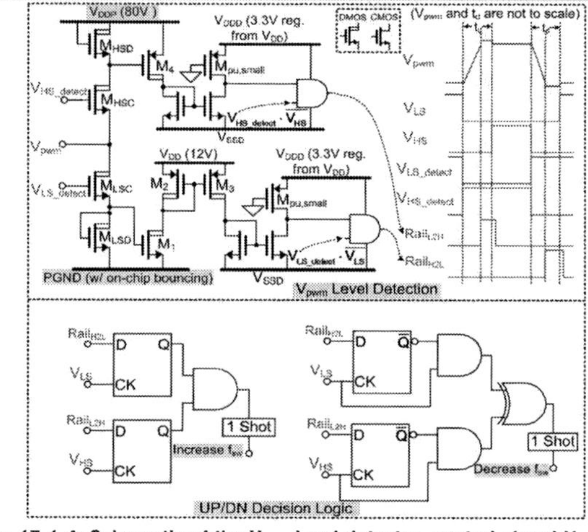

Figure 17.1.4: Schematic of the V_{pwm} level detector, control signal V_{HS_detect} is referred to V_{pwm} with level shifting (Upper); schematic of the UP/DN decision logic with 1-shot output (Lower).

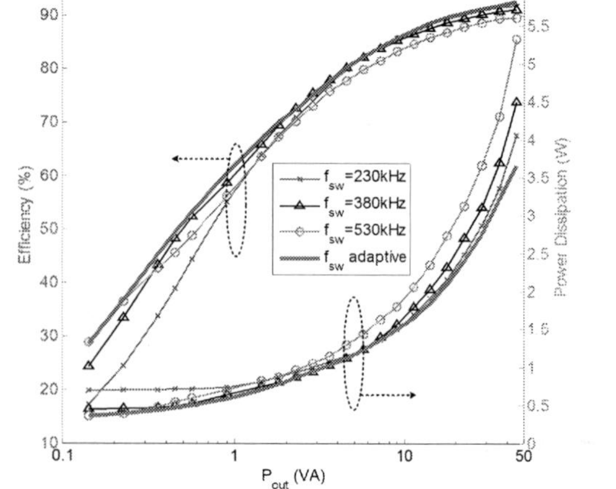

Figure 17.1.5: Efficiency and dissipation measurements for f_{sw} regulation enabled and for fixed V_{tune} settings. For the fixed V_{tune} cases, f_{sw} is measured in idle.

Parameters	This work		[2]	[3]	[4]	[5]
Type	Piezo Driver		Audio Amp.	Audio Amp.	Audio Amp.	Audio Amp.
V_{DDP}	80V		60V	20V	50V	18V
$P_{out,max}$/Channel	45VA[(1)]	45W[(2)]	100W	20W	240W	13W
Efficiency @ $P_{out,max}$	93%	91%	>90%	89%	N/A	88%
Efficiency @ 0.1* $P_{out,max}$	80%	84%	N/A	<75%	N/A	<70%
Efficiency @ 0.01* $P_{out,max}$	49%	51%	N/A	<30%	N/A	<30%
Idle Loss/Channel (w. output filter)	0.36W		1.6W	0.5W	2.1W	N/A
THD+N	0.015% (@9VA, f_{sig}=100Hz) 0.94% (@45VA, f_{sig}=500Hz)		0.017% (@1W, f_{sig}=1kHz)	0.01% (@10W, f_{sig}=1kHz)	<0.1%	0.7% (@13W, f_{sig}=1kHz)

(1) Load = 23µF+1.6Ω in series
(2) Load = 12Ω

Figure 17.1.6: Comparison with other high-voltage, high-power class-D power amplifiers.

17

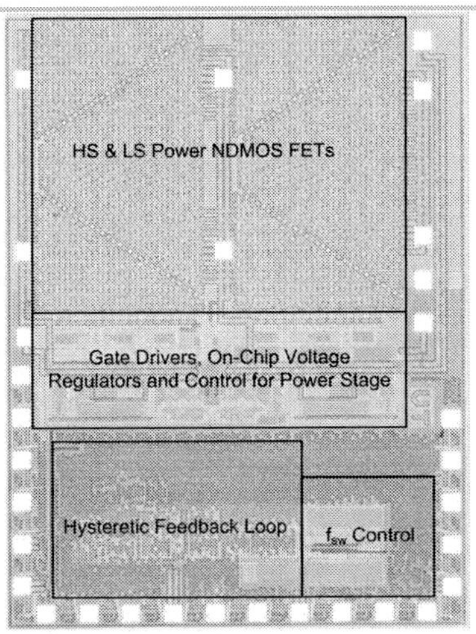

Figure 17.1.7: Chip photograph of the class-D amplifier, the die measures 3.4mm×2.5mm.

17.2 A 0.0013mm^2 3.6μW Nested-Current-Mirror Single-Stage Amplifier Driving 0.15-to-15nF Capacitive Loads with >62° Phase Margin

Zushu Yan[1], Pui-In Mak[1,2], Man-Kay Law[1,2], Rui Martins[1,2,3], Franco Maloberti[4]

[1]University of Macau, Macao, China,
[2]UMTEC, Macao, China,
[3]Instituto Superior Tecnico, Lisbon, Portugal,
[4]University of Pavia, Pavia, Italy

For active-matrix LCDs [1] that have thousands of buffer amplifiers integrated in its column-driver ICs, ultra-low power and area circuit solutions are continuously urged to meet the market pressure on cost, image quality and display size. Multi-stage amplifiers have dominated those buffers due to their reliable DC gain, output swing, gain-bandwidth product (GBW) and slew rate (SR). Yet, different kinds of frequency compensation are also useful for stability, bottlenecking the capacitive-load (C_L) drivability, power and area efficiencies.

Classical single-stage amplifiers were underused in those buffers due to their limited capability in most metrics despite being almost unconditionally stable at any C_L and tiny in size. In view of this, it is beneficial to revisit the fundamental limits of single-stage amplifiers and deal with them differently. This paper introduces a nested-current-mirror (NCM) single-stage amplifier to advance its GBW-to-power/area efficiency and C_L drivability beyond the multi-stage designs, while preserving a rail-to-rail output swing. The fabricated NCM amplifier demonstrates 33× higher GBW and 47dB higher DC gain than those of a typical differential-pair (DP) amplifier at equal power and area. By benchmarking with the recent three-stage amplifiers [2]-[4], this work improves FOM_1 [=GBW·C_L/(Power·Area)] by >6.6×, and upholds a comparable FOM_2 [=SR·C_L/(Power·Area)]. The C_L drivability is >10× wider than [2]-[4], while avoiding the stability limit at the heavy-C_L side. These results justify advanced single-stage amplifiers as a potential replacement for multi-stage designs in traditional (e.g. 100pF/m coaxial cable) and advanced (e.g. low temperature polysilicon LCD) buffer interfaces.

Most single-stage amplifiers suffer from a tight tradeoff between power and performance. Telescopic amplifiers feature a GBW-to-power efficiency as high as that of the DP amplifier (Fig. 17.2.1), but sacrifice output swing. Folded-cascode amplifiers partially surmount such a limit, but at the expense of power. For LCD column drivers, current-mirror amplifiers are favored for their rail-to-rail output swing, and extra design flexibility via adjusting the mirror ratio, K. A large K benefits most metrics (i.e., effective transconductance ($G_{m,eff}$), GBW and SR), but at the expense of noise and phase margin (PM). Yet, no matter how large K is, most metrics of the current-mirror amplifier still lag behind those of the DP amplifier.

The basic principle of the NCM amplifier (Fig. 17.2.1) is to subdivide a current mirror into a number of pieces with different ratios, and sequentially combine their outputs to concurrently advance $G_{m,eff}$ and output resistance (R_{out}) beyond those of the DP and current-mirror amplifiers. Specifically, by sharing the current I_{b2} (for the left-half side) with N divided differential-input transistors [(I_1, M_1),(I_2, M_2)...(I_N, M_N)], their outputs can be combined via N nested current mirrors with ratios [(1:K_1),(1:K_2)...(1:K_N)]. Since M_1-M_N are located in the signal path, all of their transconductances, which contribute to $G_{m,eff}$, are multiplied and customizable via choosing K_1 to K_N. To achieve high DC gain and GBW, more mirror stages and bigger ratios are preferred. To lower the noise, the largest amount of current can be allocated to the 1st mirror with a small K_1. To enhance SR, most of the current can be assigned to the 2nd-last mirror with enlarged K_{N-1} and K_N. Indeed, the mirror stages and ratios are only limited by the PM and transistor mismatches. If a large C_L is imposed, PM is no longer the stability constraint. Any mismatches generate a voltage offset. Upsizing W and L of the DP amplifier transistors and mirrors improves matching and intrinsic gain. Both are important to the expected values of $G_{m,eff}$ and R_{out}. Along such a NCM process, R_{out} is improved as well since less current goes to the output stage. Thus, the DC gain can be as high as that of a folded-cascode amplifier, but without the output swing penalty. Moreover, unlike the folded-cascode and current-mirror amplifiers, cutting the current of the output stage does not essentially degrade SR. In fact, as long as $K_N I_N > I_{b2}$, the SR of the NCM can still outperform that of the DP amplifier.

This work implements a 4-step NCM amplifier (Fig. 17.2.2) with the sub mirror ratios used for design flexibility. The total bias current (3μA) is divided into 60 unit currents (I_u=50nA). On the right-half side, the DP amplifier transistors are split into M_1-M_4. Their outputs are summed via the mirrors realized by M_5-M_{12}. M_{13} collects the output of the left-half, to form the single-ended output with M_{12}. The analytical equations in Fig. 17.2.2 show how the mirror ratios K_1 to K_7 contribute to each performance metric. The DC gain is mainly given by $2K_2K_4K_6/(K_3K_5)$. This, together with K_7, roughly defines the GBW. Large $K_{6,7}$ is set to enhance SR. The noise is a tradeoff under small K_1 (and large K_2/K_1), as the transistor's noise is also amplified by the mirror ratios. To leverage them, the 1st mirror M_5-M_6 uses a moderate ratio of 3 (K_1=2 and K_2=6). The 2nd mirror M_7-M_8 draws less current under a larger ratio of 4 (K_3=1 and K_4=4) to boost the DC gain and GBW, as they contribute negligible noise. The 3rd mirror M_9-M_{10} also uses a ratio of 4 (K_5=2 and K_6=8), but the given current is doubled to enhance SR. The 4th mirror M_{11}-M_{12} is given the largest ratio (K_7=6) to benefit the SR and $G_{m,eff}$. Overall, the DC gain, GBW and SR are theoretically improved by 47.6dB, 48× and 1.8×, respectively, when compared to the DP amplifier. Though the noise voltage of NCM amplifier is 1.8× higher than that of the DP amplifier, it is still 1.44× better than that of the current-mirror amplifier under K=6 (equivalent to K_7 in Fig. 17.2.2).

The multipath feedforward nature of NCM creates three left-half-plane (LHP) zeros, ω_{z1} to ω_{z3}, reducing the negative phase shift caused by the four non-dominant poles ω_{p2} to ω_{p5} (Fig. 17.2.3). Since ω_{p2}-ω_{p5} and ω_{z1}-ω_{z3} are all beyond the unity-gain frequency, ω_u, the stability is unconditional during startup, large transients or saturation recovery. This fact differentiates it from explicit multipath feedforward compensation [5] that consumes extra power and is conditionally stable. Here, the stability is bounded by the light-C_L condition.

The NCM and DP amplifiers are fabricated in 0.18μm CMOS for comparison at equal power and area. Their measured AC and small-step responses are plotted in Fig. 17.2.4. The NCM shows 0.013-to-1.24MHz GBW, linearly scalable with C_L from 15 to 0.15nF, which are >33× higher than those of the DP amplifier. The extrapolated DC gain (84dB) of the NCM amplifier also compares favorably with that (37dB) of the DP amplifier. The stability of the NCM amplifier is limited at the light-C_L side (62.4° PM at 0.15nF C_L). The large-step responses are plotted in Fig. 17.2.5. The chip summary and photos are given in Fig. 17.2.6. The NCM amplifier has better overall FOM_1 (>33×) and FOM_2 (>1.8×).

Comparing with the recent three-stage amplifiers of Fig. 17.2.7, this work shows the highest FOM_1 (>6.6×), widest C_L drivability (>10×) and better stability (PM>62°). Before adopting any dynamic-biasing SR-enhancement technique [6], FOM_2 is still 2.2× higher than [2], but 1.2 to 1.47× lower than [3], [4].

Acknowledgements:
This work was funded by University of Macau MYRG, and Macao FDCT (015/2012/A1) and SKL fund.

References:

[1] C.-W. Lu, et al., "A 10-bit Resistor-Floating-Resistor-String DAC (RFR-DAC) for High Color-Depth LCD Driver ICs," *IEEE J. Solid-State Circuits*, vol. 47, pp. 2454-2466, Oct. 2012.
[2] X. Peng, et al., "Impedance Adapting Compensation for Low-Power Multistage Amplifiers," *IEEE J. Solid-State Circuits*, vol. 46, pp. 445-451, Feb. 2011.
[3] S. Chong, et al., "Cross Feedforward Cascode Compensation for Low-Power Three-Stage Amplifier with Large Capacitive Load," *IEEE J. Solid-State Circuits*, vol. 47, pp. 2227-2234, Sep. 2012.
[4] Z. Yan, et al., "A 0.016mm^2 144μW Three-Stage Amplifier Capable of Driving 1-to-15nF Capacitive Load with >0.95MHz GBW," *ISSCC Dig. Tech. Papers*, pp. 368-369, Feb. 2012.
[5] A. Thomsen, et al., "A Five Stage Chopper Stabilized Instrumentation Amplifier Using Feedforward Compensation," *Symp. on VLSI Circuits, Dig. Tech. Papers*, pp. 220-223, Jun. 1998.
[6] A. Martin, et al., "Low-Voltage Super Class AB CMOS OTA Cells with Very High Slew Rate and Power Efficiency," *IEEE J. Solid-State Circuits*, vol. 40, pp. 1068-1077, May 2005.

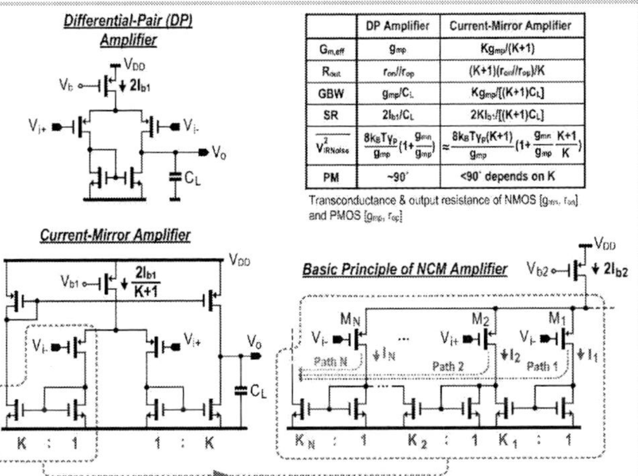

	DP Amplifier	Current-Mirror Amplifier
$G_{m,eff}$	g_{mp}	$Kg_{mp}/(K+1)$
R_{out}	$r_{on}//r_{op}$	$(K+1)(r_{on}//r_{op})/K$
GBW	g_{mp}/C_L	$Kg_{mp}/[(K+1)C_L]$
SR	$2I_{b1}/C_L$	$2KI_{b1}/[(K+1)C_L]$
$\overline{V^2_{iRNoise}}$	$\frac{8k_BT\gamma_P}{g_{mp}}(1+\frac{g_{mn}}{g_{mp}})$	$\approx\frac{8k_BT\gamma_P(K+1)}{g_{mp}}(1+\frac{g_{mn}}{g_{mp}}\frac{K+1}{K})$
PM	~90°	<90° depends on K

Transconductance & output resistance of NMOS [g_{mn}, r_{on}] and PMOS [g_{mp}, r_{op}]

Figure 17.2.1: DP amplifier outperforms the Current-Mirror amplifier in terms of $G_{m,eff}$, GBW, SR and noise, for any K. The NCM amplifier outperforms both by using the current to boost $G_{m,eff}$ and R_{out}.

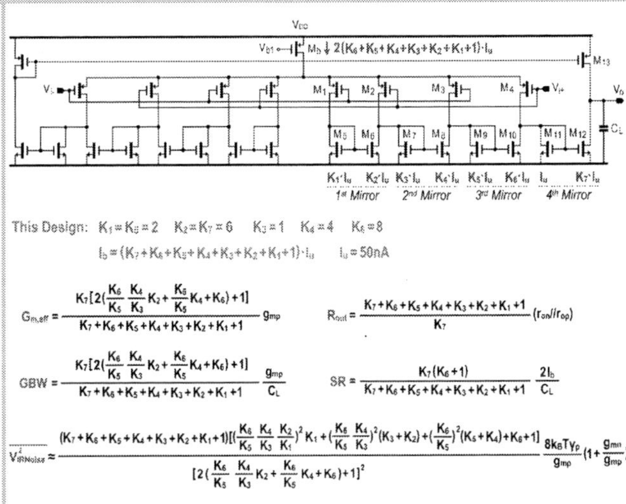

Figure 17.2.2: Implemented 4-step NCM single-stage amplifier.

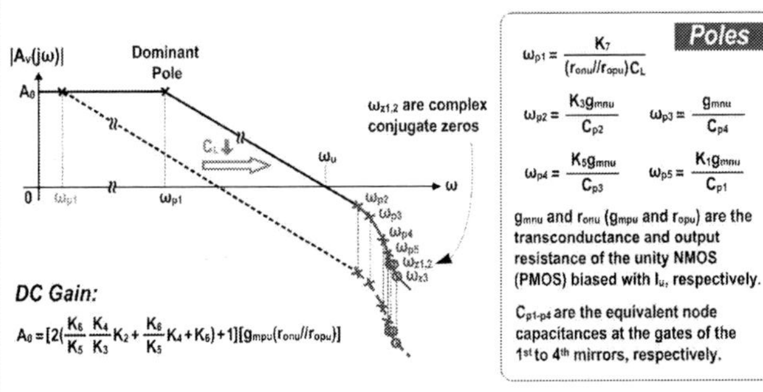

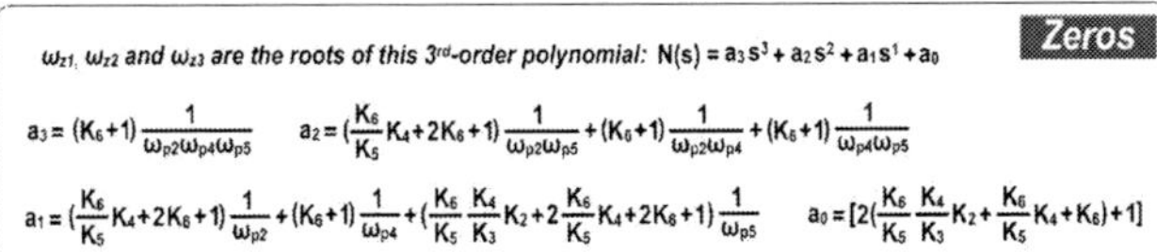

Figure 17.2.3: Bode plot of the 4-step NCM single-stage amplifier.

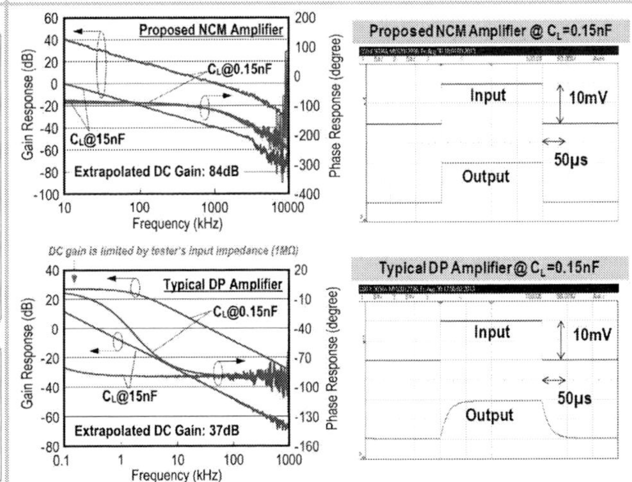

Figure 17.2.4: Measured AC (left) and small-step (right) responses of the NCM and DP amplifiers at equal power and area.

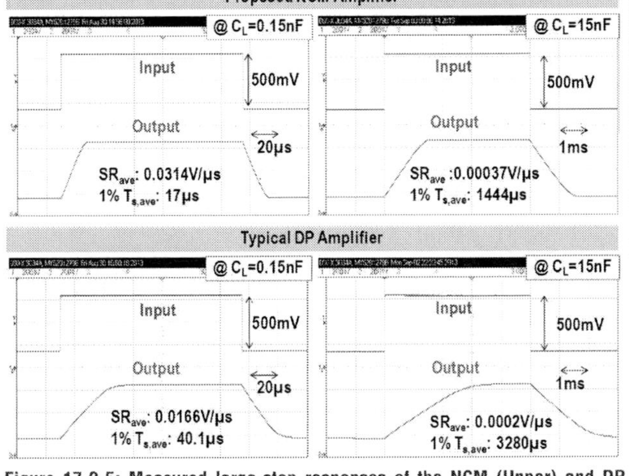

Figure 17.2.5: Measured large-step responses of the NCM (Upper) and DP (lower) amplifiers at equal power and area.

	Single-Stage Amplifiers					
Die Photo	Proposed NCM Amplifier			Typical DP Amplifier		
Load C_L (nF)	0.15	0.5	15	0.15	0.5	15
GBW (MHz)	1.24	0.396	0.013	0.0371	0.0116	0.00038
Phase Margin (degree)	62.4	81.4	90.2	91.7	92.5	92.9
Gain Margin (dB)	15.9	23.7	56.1	>60	>60	>60
SR_{ave} (V/µs)	0.0314	0.0115	0.00037	0.0166	0.0058	0.0002
1% $T_{s,ave}$ (µs)	17	47.1	1444	40.1	109	3280
DC Gain (dB) (extrapolated)	84			37		
Input-Referred Noise (nV/√Hz)	1470@ 0.1kHz	440@ 1kHz	140@ 10kHz	680@ 0.1kHz	270@ 1kHz	106@ 10kHz
Power (µW) @ V_{DD} (V)	3.6@1.2			3.6@1.2		
Chip Area (mm²)	0.0013			0.0013		
CMOS Technology	0.18µm			0.18µm		
FOM_1 [(MHz·pF)/µW/mm²]	39,744	42,308	41,667	1,189	1,239	1,218
FOM_2 [(V/µs·pF)/µW/mm²]	1,006	1,229	1,186	532	620	641

Figure 17.2.6: Comparison of NCM and DP amplifiers under the same power and area.

17

	Single-Stage Amplifier			*Three-Stage Amplifiers*		
	This Work (*NCM Amplifier*)			[2] JSSC Feb'11	[3] JSSC Sep'12	[4]* ISSCC'12
C_L Drivability ($C_{L,max}/C_{L,min}$)	100x			N/A	N/A	10x
C_L (nF)	0.15	0.5	15	0.15	0.5	15
GBW (MHz)	1.24	0.396	0.013	4.4	2	0.95
Phase Margin (degree)	62.4	81.4	90.2	57	52	52.3
Gain Margin (dB)	15.9	23.7	56.1	5#	8#	18.1
SR_{ave} (V/μs)	0.0314	0.0115	0.00037	1.8	0.65	0.22
1% $T_{s,ave}$ (μs)	17	47.1	1444	1.9	1.23	4.49
Input-Referred Noise (nV/√Hz)	1470 @ 0.1kHz	440 @ 1kHz	140 @ 10kHz	N/A	N/A	N/A
DC Gain (dB) (extrapolated)	84			110	>100	>100
Power (μW) @ V_{DD} (V)	3.6 @ 1.2			30 @ 1.5	20.4 @ 1.2	144 @ 2
Chip Area (mm^2)	0.0013			0.02	0.0088	0.016
CMOS Technology	0.18μm			0.35μm	65nm	0.35μm
FOM_1 [(MHz·pF)/μW/mm^2]	39,744	42,308	41,667	1,100	5,571	6,166
FOM_2 [(V/μs·pF)/μW/mm^2]	1,006	1,229	1,186	450	1,810	1,432

Figure 17.2.7: Performance comparison. * The data of [4] at 1.5nF is omitted here due to space limits. # means extracted values from figures.

17.3 A 0.9V 6.3μW Multistage Amplifier Driving 500pF Capacitive Load with 1.34MHz GBW

Wanyuan Qu[1], Jong-Pil Im[2], Hyun-Sik Kim[1], Gyu-Hyeong Cho[1]

[1]KAIST, Daejeon, Korea, [2]ETRI, Daejeon, Korea

As process scales down, low-voltage, low-power, multistage amplifiers capable of driving a large capacitive load with wide bandwidth are becoming more important for various applications. The conventional frequency compensation methods, however, are based on cumbersome transfer function derivations or complicated local loop analysis [1]-[3], inhibiting intuitive understanding. An approach is presented in this paper, which generates insight for the poles and zeros through distinctive compensation analysis, and is applicable to large-number-stage amplifiers. The approach applies feedback theory and simplifies high-frequency Miller amplifiers, thereby reducing orders of circuits and improving insight.

Since Miller compensation generates a large equivalent input capacitance by input-shunt-feedback, a simple Miller compensation amplifier equates a control system with plant H_2 and feedback element C_{m2}, as shown at the top of Fig. 17.3.1. Using feedback theory [4], the closed-loop gain of the control system is given by $V_o(s)/V_1(s)=A_{vf}=A_\infty T/(1+T)+A_0/(1+T)\approx A_\infty T/(1+T)$, where T is the loop gain of Miller feedback loop and A_∞ is the ideal closed-loop gain assuming $T=\infty$. The A_0 is usually negligible since it is the direct feed-through when T=0. Here, we can see that $A_\infty = g_{m2}/sC_{m2}$ and

$$T=\frac{H_2(R_{o2}\|1/sC_{p2})}{(R_L\|1/sC_L)+1/sC_{m2}+(R_{o2}\|1/sC_{p2})}$$

To get more insight, it is justifiable to focus only on the high-frequency loop gain T_{HF} because an amplifier's bandwidth and gain/phase margins only depend on the high-frequency behavior near unity gain-bandwidth (GBW), while the low-frequency behavior affects an amplifier's DC accuracy. By ignoring all low-frequency components, T_{HF} reduces to H_2. That is, $T_{HF}=\lim_{f\to GBW}T=H_2$ if $GBW>>1/R_{o2}C_{m2}$, $C_{m2}>>C_{p2}$ and $C_L>>C_{p2}$. Then, the amplifier's gain at high frequency can be rewritten as

$$\left.\frac{V_o(s)}{V_1(s)}\right|_{HF}=A_{vf\_HF}=A_{\infty 2}\frac{T_{HF}}{1+T_{HF}}=A_{\infty 2}\frac{H_2}{1+H_2}$$

This equation presents an insightful result. The original complex second-order equation is decomposed into the product of two first-order equations: $A_{\infty 2}$ and $H_2/(1+H_2)$. Here $A_{\infty 2}$ has a pole at the origin, while $H_2/(1+H_2)$ is the unity-feedback behavior of H_2. In the bode plot of Fig. 17.3.1 top right, $H_2/(1+H_2)$ has a band-limiting pole at g_{mL}/C_L. Therefore, setting $GBW=g_{m2}/C_{m2}=g_{mL}/2C_L$ is desirable and the resulting phase margin is around 60°.

By applying the above method recursively, simple design equations can be obtained for nested Miller compensation (NMC) amplifiers without delving into lengthy equations. For the NMC shown at the bottom of Fig. 17.3.1, the inner-loop can be designed in the same manner as above and $H_1=A_{\infty 2}H_2/(1+H_2)$ from the block diagram. Thus, $H_1/(1+H_1)$ introduces two limiting poles (strictly, a pair of complex poles with damping factor of $1/\sqrt{2}$) at the outer loop, as shown in Fig. 17.3.1 bottom right. Hence, setting $GBW=gm_1/C_{m1}=g_{m2}/2C_{m2}=g_{mL}/4C_L$ achieves 60° phase margin yielding a 3rd-order Butterworth frequency response. Similarly, applying the method more times gives a five-stage NMC yielding a 5th-order Butterworth frequency response:

$$GBW=\frac{g_{m1}}{C_{m1}}=\frac{g_{m2}}{2C_{m2}}=\frac{g_{m3}}{3.2C_{m3}}=\frac{g_{m4}}{5.2C_{m4}}=\frac{g_{mL}}{10.4C_L}$$

Clearly, the more Miller feedback applied to the multistage amplifier, the larger the expected GBW reduction. Although prior studies have proposed various solutions to extend the GBW [1]-[3], the principles behind those circuits can be difficult to grasp. Using the above method, it is easy to note that all prior studies extend GBW by reducing the number of Miller feedbacks and insert a zero into T_{HF} to enlarge the Miller loop bandwidth. For example, impedance-adapting-compensation (IAC) adds a passive zero [1], while active-zero-compensation (AZC) [2] and damping-factor-control [3] insert an active zero. That is, multistage amplifier design reduces to T_{HF} design for maximum unity-feedback bandwidth to reduce complexity.

Figures 17.3.2 and 17.3.3 compare the block diagrams and T_{HF} bode plots of prior studies and this work. For IAC of Fig. 17.3.2 top left, $T_{1\_HF}\approx H_1$ and T_{1_HF} is stabilized by adding a zero $1/R_aC_a$ and pushing the original second pole $1/R_{o2}C_{p2}$ down to $1/R_{o2}C_a$. Bandwidth ω_1 is enlarged by $g_{m2}R_a$ times, while a band-limiting pole occurs at $\omega_{L1}=1/R_aC_{p2}$. There is a trade-off in choosing R_a since a larger R_a enlarges ω_1 at the cost of a less stable Miller feedback loop.

In the AZC of Fig. 17.3.2 top right, $T_{2\_HF}\approx H_2C_m/C_{p1}$ because current-buffer Miller compensation forms a common-gate amplifier with high-frequency gain of C_m/C_{p1}. Here, an active zero is added to H_2 with the second pole untouched, so the bandwidth ω_2 is extended. However, ω_2 is limited by its third pole, $\omega_{L2}=g_{mb}/C_z$, which results from the loading effect of active zero circuit. Also, AZC is not applicable to low-voltage design due to the current-buffer transconductance-boosting scheme, which aims to push the current-buffer pole g_{mc}/C_m beyond ω_2.

In Fig. 17.3.2 bottom, we present a local-feedback-enhanced compensation (LEC) scheme for three-stage amplifiers with large capacitive loads. LEC also adds an active zero but pushes the band-limiting pole higher, making a larger bandwidth ω_3 achievable. Here, the active zero is generated by a local feedback loop. The effect of the zero can be evaluated from the impedance of node *p*, Zp, which is $1/g_{mb}$ at low frequencies and gradually increases as shunt-feedback weakens at high frequencies. Since a direct connection of R_z to node *p* severely limits the achievable Z_p, thereby limiting the effect of the zero, a no-loading local feedback has been implemented. This scheme isolates R_z from node *p* and pushes the band-limiting pole to $\omega_{L3}=P1=g_{mb}R_{op}/C_zR_z$. As shown in Fig. 17.3.3, ω_{L3} is extended by R_{op}/R_z times compared to that of [2]. Therefore, ω_3 can exceed ω_1 and ω_2 under the same C_L and power constraints.

Figure 17.3.4 shows an example of the circuit implementation of the proposed three-stage LEC amplifier. The first, second and third stages are implemented by M100-108, M201-202, and M301-302, respectively. The local feedback stage consists of M401-408, R_z and C_z. The slew helper contains M501-503. M403-406 is the no-loading sensing circuit, which is a unity-gain buffer using current mirror. Node *p'* equates the node *p*. R_z and C_z form the desired zero. M402 shields the large M401 from node *p* and reduces C_{pb}. Current bleeder M408 increases the g_{ma}/g_{mb} gain. M202 and M302 generate feed-forward paths and improve slew-operation for the 2nd and 3rd stages. The 3rd stage, however, needs further slew enhancement because M302 gate is a slow-transition node. By sensing the voltage of *p'* through M501, M503 turns on quickly during the output rising transition and turns off in steady state. The LEC amplifier can operate at low power supply where multistage amplifiers are mostly required.

Figure 17.3.5 shows the measured loop gain, step response, load capacitance and supply voltage variations. At C_L=500pF, the amplifier achieved a GBW of 1.34MHz and phase margin of 52.7° with extrapolated DC gain >100dB. In the input step response test, the LEC in a unity-feedback configuration without slew helper showed an average slew rate (SR) and average 1% settling time (T_S) of 0.48V/μs and 2.53μs, while the LEC with slew helper achieved 0.62V/μs and 1.12μs, respectively. The rising overshoot was also greatly suppressed with the slew helper. The stable load capacitance C_L range is 100pF to 500pF. When C_L is small gain peaking occurs at around 40MHz and degrades the amplifier's stability. For example, when C_L=33pF, a high frequency ringing (~40MHz with 70mV$_{pp}$) is superimposed on the step response. This result is consistent with our simplified analysis method. The gain peaking comes from the unity-feedback behavior of T_{3_HF}. Compared to prior studies with a single band-limiting pole, T_{3_HF} contains a pair of complex limiting poles (P1&P2) because the poles are maximally extended near the vicinity of optimal design. The complex poles can generate excessive gain peaking when C_L is small. Also as shown, the amplifier is functional from 0.7V to 1.8V supplies and the GBW is almost unchanged from 0.9V to 1.8V supplies. The results verify the applicability of LEC to low-voltage applications.

Figure 17.3.6 presents a performance summary. Compared with other works, LEC shows improved FOM_S (1.07×), FOM_L (2.15×), LC-FOM_S (2.87×) and LC-FOM_L (4.08×). The chip is fabricated using a 0.18μm CMOS process with an area of 0.007mm².

References:

[1] X. Peng, et al., "Impedance adapting compensation for low-power multistage amplifiers," *IEEE J. Solid-State Circuits*, vol. 46, no. 2, pp. 445-451, Feb. 2011.

[2] Z. Yan, et al., "A 0.016mm² 144μW three-stage amplifier capable of driving 1-to-15nF capacitive load with > 0.95MHz GBW," *ISSCC Dig. Tech. Papers*, pp. 368-370, Feb. 2012.

[3] K.N. Leung, et al., "Damping-factor-control frequency compensation technique for low-voltage low-power large capacitive load applications," *ISSCC Dig. Tech. Papers*, pp. 158-159, Feb. 1999.

[4] R.D. Middlebrook, "Design-Oriented Analysis of Feedback Amplifiers," *Proc. National Electronics Conference*, vol. 20, pp.234-238, Oct. 1964.

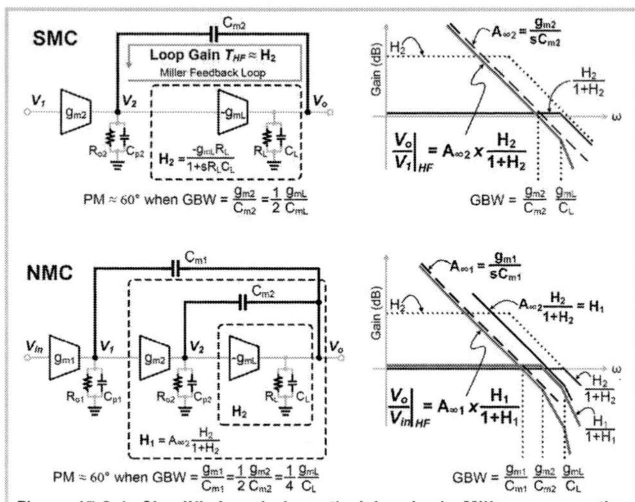

Figure 17.3.1: Simplified analysis method for simple Miller compensation (SMC) and nested Miller compensation (NMC).

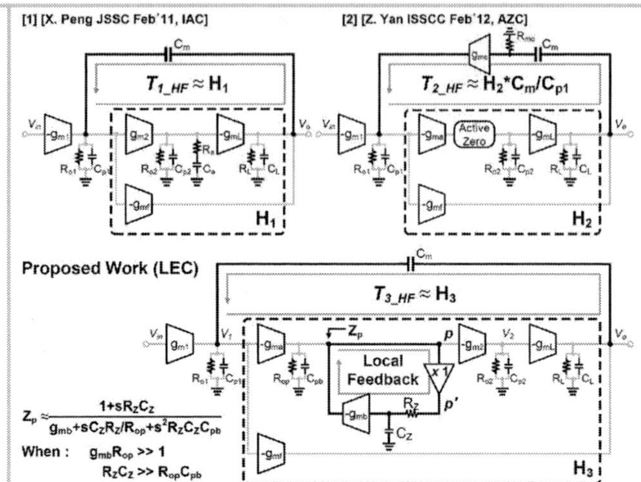

Figure 17.3.2: Block diagrams of the three-stage amplifiers of [1], [2] and this work.

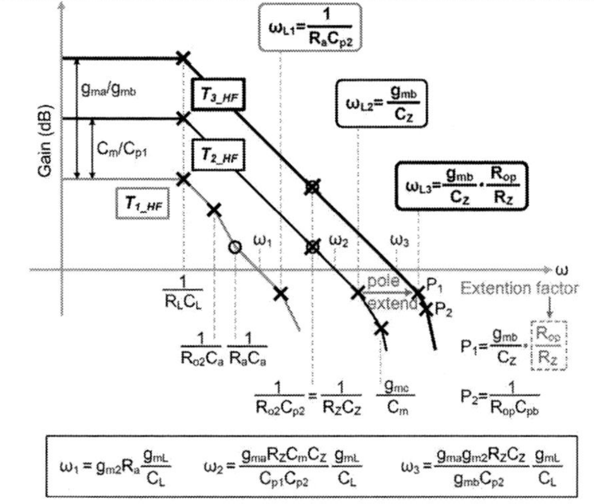

Figure 17.3.3: Bode plots of simplified high-frequency loop gain T_{HF} of the Miller feedback loop for [1], [2] and this work.

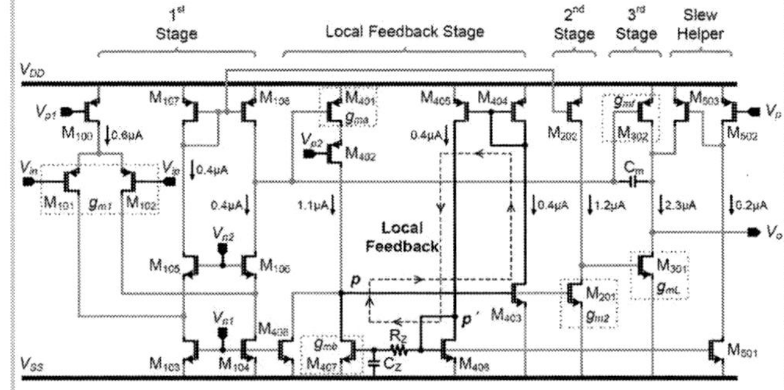

Device	Size (μm/μm)	Device	Size (μm/μm)	Device	Size (μm/μm)
M100	20/1	M403~M407	1/0.5	M502	10/1
M101/M102	8/1	M408	2/0.5	M503	12/0.2
M103/M104	2/0.5	M201	3/0.5	C_m	0.626 pF
M105/M106	1/0.2	M202	3/0.5	R_Z	252.9 kΩ
M107/M108	1/0.5	M301	6/0.2	C_Z	0.245 pF
M401	20/0.2	M302	35/0.2		
M402	1/0.2	M501	1/1		

Figure 17.3.4: Circuit schematic and device sizes of the LEC amplifier.

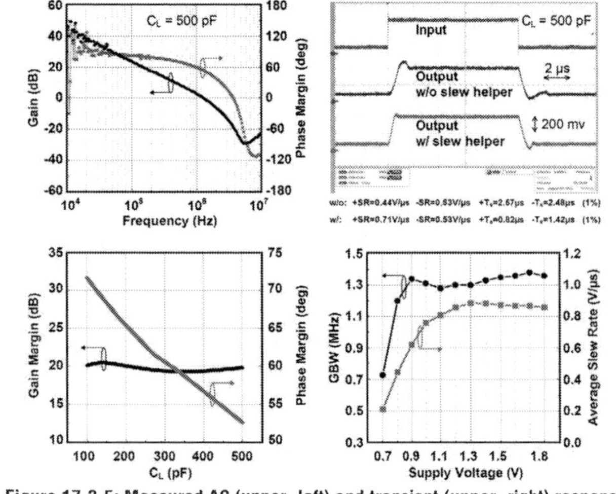

Figure 17.3.5: Measured AC (upper, left) and transient (upper, right) responses, load capacitance (lower, left) and supply voltage (lower, right) variations.

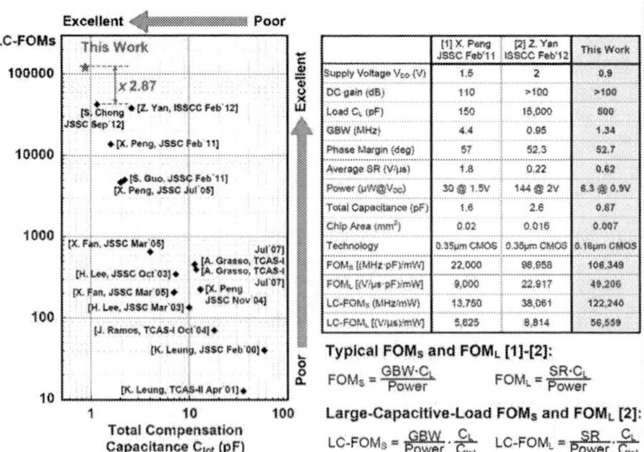

	[1] X. Peng JSSC Feb'11	[2] Z. Yan ISSCC Feb'12	This Work
Supply Voltage V_{DD} (V)	1.5	2	0.9
DC gain (dB)	110	>100	>100
Load C_L (pF)	150	15,000	500
GBW (MHz)	4.4	0.95	1.34
Phase Margin (deg)	57	52.3	52.7
Average SR (V/μs)	1.8	0.22	0.62
Power (μW@V_{DD})	30 @ 1.5V	144 @ 2V	6.3 @ 0.9V
Total Capacitance (pF)	1.6	2.6	0.87
Chip Area (mm²)	0.02	0.016	0.007
Technology	0.35μm CMOS	0.35μm CMOS	0.18μm CMOS
FOM_S [(MHz·pF)/mW]	22,000	98,958	106,349
FOM_L [(V/μs·pF)/mW]	9,000	22,917	49,206
$LC\text{-}FOM_S$ (MHz/mW)	13,750	38,061	122,240
$LC\text{-}FOM_L$ [(V/μs)/mW]	5,625	8,814	56,559

Typical FOM_S and FOM_L [1]-[2]:

$$FOM_S = \frac{GBW \cdot C_L}{Power} \qquad FOM_L = \frac{SR \cdot C_L}{Power}$$

Large-Capacitive-Load FOM_S and FOM_L [2]:

$$LC\text{-}FOM_S = \frac{GBW}{Power} \cdot \frac{C_L}{C_{tot}} \qquad LC\text{-}FOM_L = \frac{SR}{Power} \cdot \frac{C_L}{C_{tot}}$$

Figure 17.3.6: Performance summary and benchmark of the state-of-the-art multistage amplifiers.

978-1-4799-0917-9/14 $31.00 © 2014 IEEE

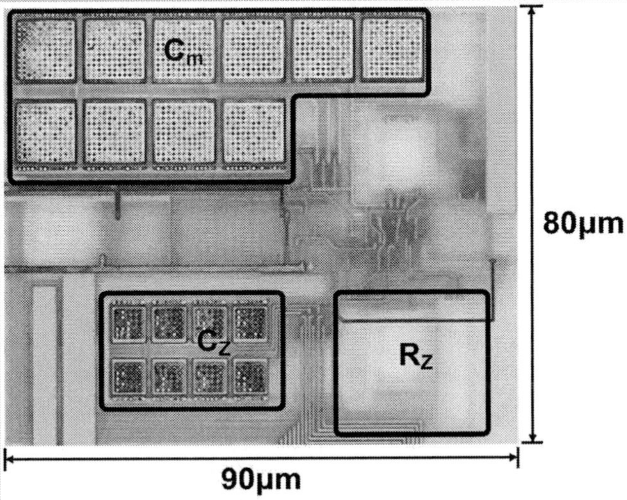

Figure 17.3.7: Chip micrograph of the LEC amplifier.

17.4 CMOS Impedance Analyzer for Nanosamples Investigation Operating up to 150MHz with Sub-aF Resolution

Giorgio Ferrari, Davide Bianchi, Angelo Rottigni, Marco Sampietro

Politecnico di Milano, Milan, Italy

Impedance analyzers find an important role in nanoscience and in biological research as a tool to access electrical and physical parameters of the matter as well as to enhance the read-out performance in sensor applications. Needs are emerging to perform impedance spectroscopy on a wide frequency range. Electrical assessment of the cell metabolism, for example, requires a frequency of investigation of about 100MHz for the signal to traverse the cell membrane and to access the cytoplasm [1]. Bench-top impedance analyzers exist that cover such a wide frequency range but they are bulky, expensive and have inadequate resolution for the high impedance shown by many nanosamples and semi-insulating biological molecules. Recent compact analyzers based on custom CMOS chips are mainly focused on low-power solutions with sub-MHz ranges [2]-[4] or highly multichannel applications [5].

Here we present a fully-integrated current-to-admittance converter operating from 1kHz to 150MHz with the low noise level of other solutions operating at much lower frequencies [6]. These results are obtained using a downconversion/amplification/upconversion architecture inside a feedback structure, as shown in Fig. 17.4.1. This is the same principle being used in bench-top impedance analyzers [7]. Here it is implemented in a single chip with the benefit of avoiding an additional lock-in system, as well as advantages in terms of sensitivity and compactness. The signal from the DUT at the exciting frequency f_s is buffered by single-stage amplifier B1, chopped at ϕ_{ch}=100kHz by passive mixer M1, and modulated at $f_{mod} = f_s$ by the Gilbert cell multiplier M2. This results in signal information at 100kHz at the input of amplifier G1, thus avoiding its 1/f noise. A further passive mixer, M3, operating at ϕ_{ch} brings the signal to DC for amplification and low-pass filtering by G2. The phase information of the input signal is preserved by using two mod/dem paths operating with in-phase and quadrature reference signals. Finally, the analog multiplier, M4, upconverts the signal to f_{mod} to close the feedback loop at the signal frequency. Capacitive feedback is preferred to resistive feedback for noise reasons. The two paths directly provide two DC outputs, V_R and V_I in Fig. 17.4.1, respectively proportional to the in-phase and quadrature components of the input current I_{DUT} (i.e. the DUT admittance). This profitably compensates the complexity of this realization when compared with an additional two-channel lock-in system operating on the high-frequency output V_{HF}, as in bench-top analyzers.

The main advantage of this architecture is a large loop gain ($\approx 10^9$ in our case, mainly given by G2), which is constant at all frequencies of operation, thereby overcoming the classical bandwidth limitation related to the gain-bandwidth product of the OpAmp. The high-frequency limit is mostly set by the technology used in the multipliers and by their phase shifts along the loop, potentially extending well over the 150MHz demonstrated in the prototype that uses a 2P4M-0.35μm CMOS process. The admittance of the DUT can be found using

$$\mathrm{Re}\{Y\}=\frac{-V_R\cdot 2\pi f_{mod} C_F\cdot G_{M4}}{A};\ \mathrm{Im}\{Y\}=\frac{V_I\cdot 2\pi f_{mod} C_F\cdot G_{M4}}{A}$$

where A is the amplitude of the forcing voltage and G_{M4} is the gain of multiplier M4. In order to achieve linearity over an extended range of V_R and V_I, multiplier M4 (shown in Fig. 17.4.2) uses a ±1V differential input dynamic range voltage-to-current converter (block A) coupled to a four-quadrant Gilbert cell. Transistors T1 and T2, which switch on only for large input signals, prevent feedback loop instability by avoiding non-monotonic behavior of the M4 multipliers.

The integrator G2 has a unity-gain frequency as low as 10Hz to effectively filter out the spurious harmonics produced by modulators M1, M2 and M3. It operates as if a high-value resistor of 10GΩ would be present, implemented by dividing the current from a physical resistor of R=1MΩ, as shown in Fig. 17.4.3. The current divider has been obtained through two pairs of properly matched capacitors. DC current, required only during transient response, is provided by the T3-T4 path. This structure has high linearity at the spurious harmonics, avoiding non-linear injection of spurious DC components into capacitor C_{int}, thus preserving the accuracy of the DC outputs V_R and V_I.

The overall capacitive feedback of the impedance analyzer requires a reset network to draw away the DC current at the input node that would otherwise saturate the amplifier. Figure 17.4.4 shows the details of the implemented DC feedback loop. It is connected to mixer M1 of the high-frequency path using a chopped amplifier, G3, for offset reduction at the input of the M2 modulators, thus limiting the spurious harmonic at f_{mod}. A differential-to-single-ended converter (G4) drives a filter that provides stability and a limited bandwidth (150Hz) in the DC feedback loop. A physical resistor R_{ATT}=2MΩ, followed by an active current divider [8], operates as an equivalent resistor of 20MΩ that closes the DC loop. The DC feedback loop effectively removes the input DC current up to ±60nA with a reduced injected noise equivalent to a resistor of 200MΩ.

Figure 17.4.7 shows the die micrograph of the overall chip, occupying 1.6 mm² and consuming 135mW at 3V supply.

Figure 17.4.5 (top) shows the measured transfer functions V_{HF}/V_X of the chip obtained for different modulating/demodulating frequencies, f_{mod}, and spanning the frequency f_S of the injected current by means of a test capacitance $C_{test}=C_F$=100fF. The expected gain $G=-C_{test}/C_F=-1$ at $f_S=f_{mod}$ is reached within ±1dB of the full frequency range from 1kHz to 150MHz and within ±0.1dB in the 3kHz-130MHz range. The small bandwidth of ≈50Hz at a fixed f_{mod} guarantees an efficient noise filtering. The bandwidth can be increased up to ≈50kHz for faster impedance measurements by digitally modifying the divide ratio in the equivalent resistor of G2 shown in Fig. 17.4.3.

Figure 17.4.5 (bottom) shows the impedance spectrum of known components connected to the input pad of the chip. Continuous line is the analytical expected curve and squares are the experimental values as extracted directly from the DC voltages V_R and V_I up to a working frequency of 150MHz, showing the effectiveness of the presented scheme even without calibration.

In Fig. 17.4.6 (top), we show the equivalent input current noise spectral density for the output V_I at f_{mod}=1MHz. The low level of the flicker noise is granted by the 2-step downconversion, which avoids the 1/f noise of multiplier M2 and reduces the 1/f term of integrator G2. By measuring the noise for different f_{mod} and assuming an applied voltage of 1V, the achievable capacitance resolution is obtained and plotted in Fig. 17.4.6 (bottom). This level of sensitivity over such an extended frequency range surpasses the performance of available impedance analyzers.

In conclusion, this paper presents an integrated system for impedance tracking or impedance spectroscopy. It is based on a forward amplifier organized as a two-channel modulator/demodulator circuit with two DC outputs proportional to real and imaginary components of the input admittance. The architecture features high loop gain at frequencies from 1kHz to 150MHz, ensuring stability and high linearity at all measuring frequencies. The noise figure of merit of the resulting impedance analyzer (about 0.3 aF in terms of capacitance resolution from 100kHz to 100MHz) is one order of magnitude better than available solutions operating over a similar frequency range.

References:

[1] T. Sun, H. Morgan, "Single-cell microfluidic impedance cytometry: a review," *Microfluidics and Nanofluidics*, Vol. 8, pp. 423–443, Mar. 2010.

[2] C. Yang, et al., "Compact Low-Power Impedance-to-Digital Converter for Sensor Array Microsystems," *IEEE Journal of Solid-State Circuits*, Vol. 44, pp. 2844–2855, Oct. 2009.

[3] H. Jafari, L. Soleymani, R. Genov, "16-Channel CMOS Impedance Spectroscopy DNA Analyzer With Dual-Slope Multiplying ADCs," *IEEE Trans. on Biomedical Circuits and Systems*, Vol. 6, pp. 468–78, Oct. 2012.

[4] J. Guo, W. Ng, J. Yuan, M. Chan, "A 51fA/Hz$^{0.5}$ low power heterodyne impedance analyzer for electrochemical impedance spectroscopy," *IEEE VLSIC*, pp. 56–57, June 2013.

[5] A. Manickam, et al., "A CMOS electrochemical impedance spectroscopy biosensor array for label-free biomolecular detection," *ISSCC Dig. Tech. Papers*, pp. 130–131, Feb. 2010.

[6] F. Gozzini, G. Ferrari, M. Sampietro, "An instrument-on-chip for impedance measurements on nanobiosensors with attoFarad resoution," *ISSCC Dig. Tech. Papers*, pp. 346–347, Feb. 2009.

[7] R. Sakiyama, "Narrow-band amplifier and impedance-measuring apparatus," *US Patent App. 20,040/257,093*, 2004.

[8] G. Ferrari, F. Gozzini, M. Sampietro, "A Current-Sensitive Front-End Amplifier for Nano-Biosensors with a 2MHz BW," *ISSCC Dig. Tech. Papers*, pp. 164–165, Feb. 2007.

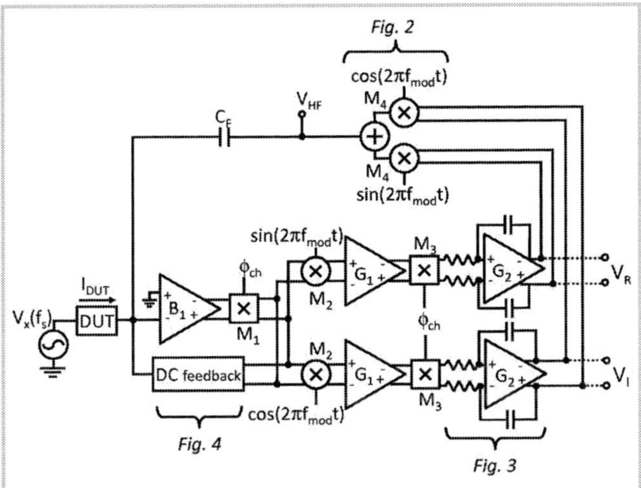

Figure 17.4.1: Schematic of the impedance analyzer based on modulating/demodulating architecture.

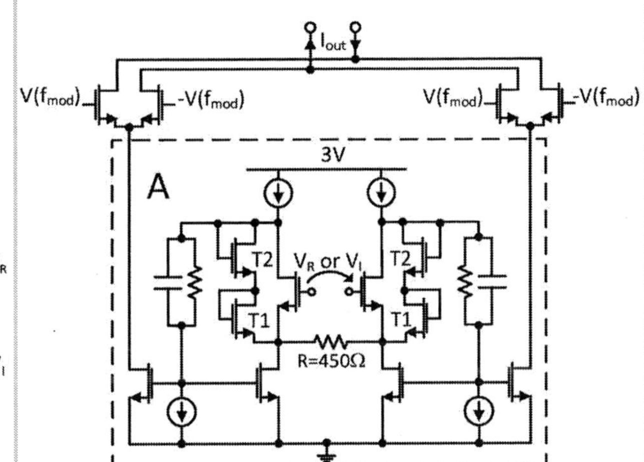

Figure 17.4.2: Schematic of the multiplier M4.

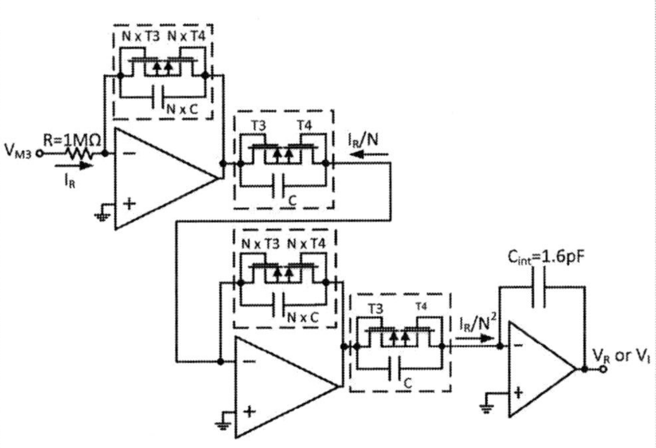

Figure 17.4.3: Detail of the equivalent resistor of 10GΩ implemented in the integrator G2 using a current divider of $N^2=10^4$.

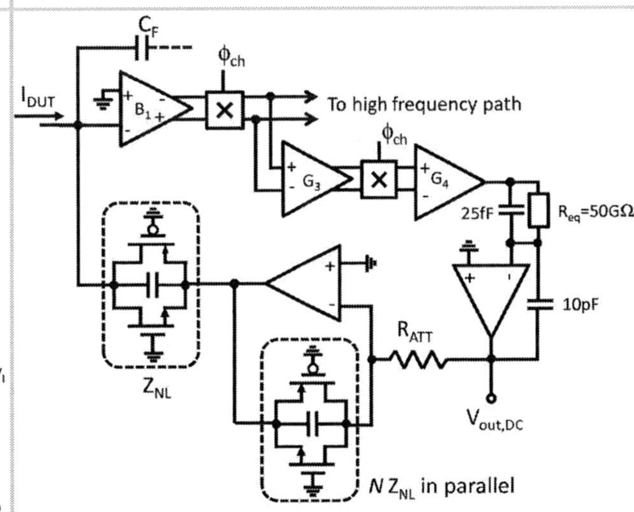

Figure 17.4.4: Schematics of the DC restoring network.

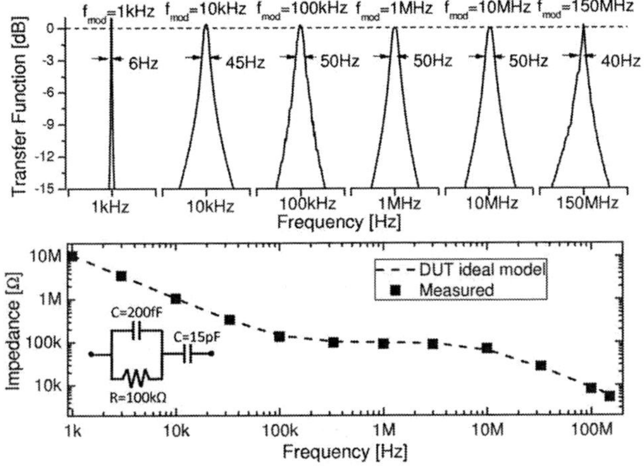

Figure 17.4.5: Transfer function measured for various f_{mod} (top) and impedance spectrum of a known DUT (inlet) connected at the input of the chip (bottom).

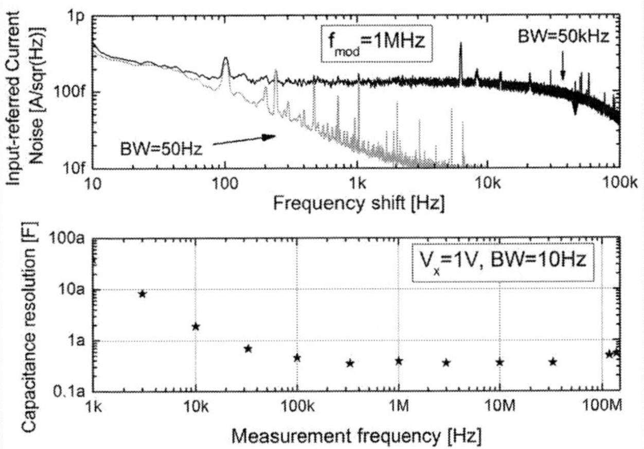

Figure 17.4.6: Equivalent input noise spectral density at f_{mod}=1MHz measured for two different bandwidths (top) and corresponding capacitance resolution (bottom).

17

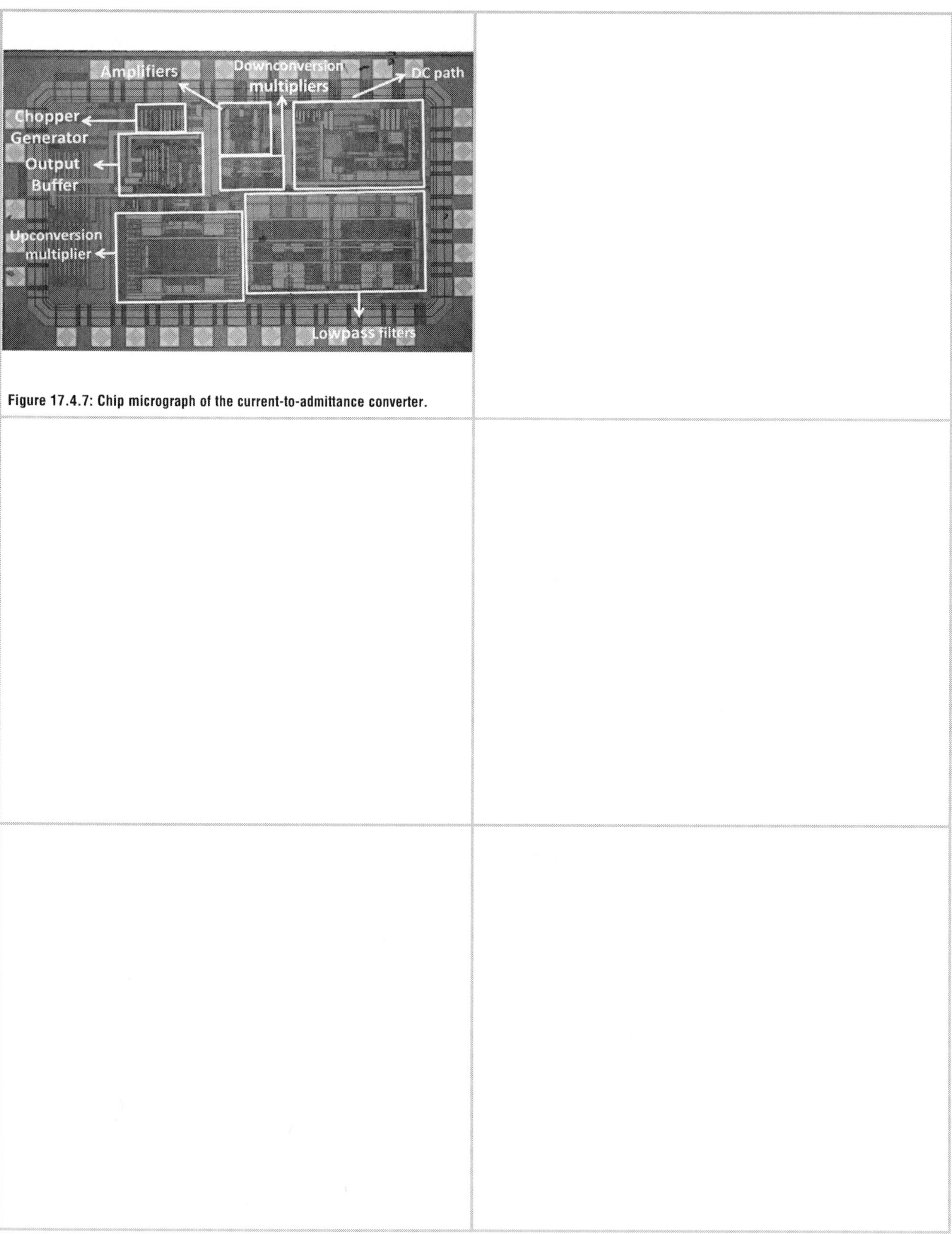

Figure 17.4.7: Chip micrograph of the current-to-admittance converter.

17.5 A 0.07mm² 2-Channel Instrumentation Amplifier with 0.1% Gain Matching in 0.16µm CMOS

Fabio Sebastiano[1,2], Federico Butti[3], Robert van Veldhoven[1], Paolo Bruschi[3]

[1]NXP Semiconductors, Eindhoven, The Netherlands,
[2]Delft University of Technology, Delft, The Netherlands,
[3]University of Pisa, Pisa, Italy

Extremely small-area sensor front-ends are required for cost-constrained automotive applications. Instrumentation amplifiers (IA) for such front-ends must process multi-channel sensor outputs and provide gain matching over the channels for proper sensor operation. Angular sensors are a typical example, in which the sine and cosine outputs of a resistive magnetic sensor must be processed with adequate gain matching to avoid unacceptable angular errors. This paper presents a 2-channel instrumentation amplifier in 0.16µm CMOS with 0.1% gain matching and occupying 0.035mm² per channel. This represents a 13.3× area improvement with respect to state-of-the-art designs with similar gain accuracy [1]-[4], while maintaining low noise (18.7nV/√Hz), low offset (17µV) and high power efficiency (NEF=12.9). The accurate gain matching in a limited area is enabled by the adoption of a dynamic element matching (DEM) scheme and by the use of a high chopping frequency.

Figure 17.5.1 shows the block diagram of the proposed IA. A current-feedback architecture is employed for both channels. For each channel, the input ($V_{in1,2}$) and feedback voltages ($V_{fbA,B}$) are applied to the input ($Gm_{inA,B}$) and feedback transconductors ($Gm_{fbA,B}$), whose output currents are summed at the input of the Miller-integrator built around $Gm_{2A,B}$. The overall feedback ensures the cancellation of the sum of the currents, so that the gain is

$$\frac{V_{out1,2}}{V_{in1,2}}=\frac{Gm_{inA,B}}{Gm_{fbA,B}}\frac{\left(R_{1A,B}+R_{2A,B}\right)}{R_{1A,B}}.$$

The gain accuracy of each channel can be optimized by setting $Gm_{inA,B}$=$Gm_{fbA,B}$ and by dynamically matching the input and feedback transconductors by periodically exchanging them. However, a residual gain error proportional to $(Gm_{inA,B}-Gm_{fbA,B})^2/Gm_{inA,B}^2$ would still be present due to the non-linear dependence of the gain from $Gm_{fbA,B}$ [2,3]. Moreover, precision resistors or DEM of the feedback resistive network would also be needed. In the latter case, the large number of unit resistors to be swapped increases the circuit complexity, and thus the area. Since the DEM frequency is inversely proportional to the number of resistive elements, it would also result in very low DEM frequencies, which in turn leads to in-band output ripple [4], or impractically high switching frequencies. These limitations can be overcome by choosing to directly maximize the gain matching, with no need of seeking high accuracy for the gain of the individual amplifiers.

In this paper, the IA's first stage, i.e. the input and feedback transconductors, and the resistive feedback networks are periodically exchanged between the channels with a frequency f_{DEM}, thus ensuring perfect channel-gain matching (neglecting switching non-idealities). DEM is implemented by switch matrices S1-S6 in Fig. 17.5.1. The Miller-integrator output stage ($Gm_{2A,B}$ and C) is not dynamically matched to avoid settling issues and the resulting channel crosstalk. Since mismatch in the common-mode feedback (CMFB) circuits causes differences between the output common-modes of each channel, the CMFB circuits are also swapped to avoid significant degradation of the gain matching. To limit offset and in-band flicker noise, chopping at frequency f_{chop}=2f_{DEM} is implemented by switch matrices S1-S4 and S7-S10. Chopping and DEM cause output ripples at f_{chop} and f_{DEM}. Those ripples are often attenuated by a ripple-reduction loop that increases the circuit complexity, and thus area and power [2], [3]. Instead, high-frequency chopping and DEM (f_{chop}=200kHz, f_{DEM}=100kHz) are used in this work. The chopping frequency is low enough to limit the residual offset due to charge injection mismatch and high enough to move the output ripple out of the band of interest (BW=20kHz) and to reduce the input stage area by allowing a higher input-stage flicker-noise corner frequency. The high-frequency output ripple can be removed in the digital domain after the analogue-to-digital conversion (ADC) in the sensor front-end chain [5]. For example, a digital sinc filter with first notch at f_{DEM} would remove the ripples while attenuating the edge of the signal band by only 0.6dB, thanks to the large ratio between f_{DEM} (and f_{chop}) and BW.

A folded-cascode amplifier is used for the first stage (M_{1-19} in Fig. 17.5.2). Input transistors are source degenerated (R=550Ω) to increase the linear input range, thus trading-off power consumption for better linearity at a constant noise level. Two folded-cascode gain-boosting amplifiers (A_p and A_n) ensure a typical DC gain of 138dB. The demodulation is implemented by the choppers in Fig. 17.5.2, which also modulate the offset and flicker-noise contributions of the current sources ($M_{5,6}$ and $M_{11,12}$) and of the boosting amplifiers. Contributions from cascode transistors (M_{7-10}) are negligible due to the gain-boosting topology [6]. Level shifters (M_{16-19}) at the output of the first stage bias the pseudo-differential class-AB output stage (M_{20-23}) that can be used to drive the typical switched-capacitor load of an ADC input sampler. Because of the wide output swing, bootstrapped switches are used in the feedback network switch matrix (S5 in Fig. 17.5.1) to preserve the IA's linearity. In addition, the low resistance of bootstrapped switches is needed for high gain matching, since their resistance is in series with $R_{2A,B}$ of the feedback network. Minimum-length NMOS switches are used in S1-S4 to limit the residual offset due to charge-injection.

The 0.07mm² IA (Fig. 17.5.7) is fabricated in an SSMC 0.16µm 1P5M CMOS process, and is packaged in both ceramic and plastic dual in-line packages. Approximately half of the area is occupied by the first (27%) and second (28%) stages while the remaining area is occupied by the current reference (14%), the bootstrapped switches (10%), the CMFB (6%) and the feedback network (6%). The gain was set to 13 ($R_{1A,B}$=2kΩ, $R_{2A,B}$=24kΩ) to fit the output voltage swing (1.6$V_{pp-diff}$) at the maximum input signal (120m$V_{pp-diff}$). Samples were characterized using signal frequencies up to 20kHz, which represents a typical requirement for automotive sensor front-ends. Measurements of 40 samples show that the offset is less than 17µV (Fig. 17.5.3) and that gain matching measured for sinusoidal signals at 20kHz is better than 0.1%. The offset of the two channels shows high correlation due to the connection of the inputs to the same switches S1-S4. No significant differences are observed between samples in ceramic and plastic packages. The crosstalk between the channels is -73dB at 20kHz. The measured gain matching and crosstalk are sufficiently low to achieve angular errors below 0.03°. CMRR is higher than 99dB up to 30kHz and PSRR is higher than 102dB up to 50kHz. A THD of -60dB was measured with a maximum input signal amplitude at 21kHz (to avoid the harmonics to be masked by the output ripple). The high loop gain enforcing equality of input ($V_{in1,2}$) and feedback ($V_{fbA,B}$) voltages, and the matching of the input and feedback transconductors, ensures compensation of the transconductors' non-linearity in the sum of their output currents. At large input frequencies, such compensation is limited by the low-pass frequency response of the loop gain. The input-referred noise density is 18.7nV/√Hz (measured by a HP3585A spectrum analyzer). The noise density is flat down to 2Hz (the lower frequency limit of the R&S UPD audio analyzer used for the measurement in Fig. 17.5.5). The IA draws 641µA from a 1.8V supply (including both channels and a reference for the bias current), resulting in a NEF of 12.9. A summary of performance and the comparison with the state-of-the-art are presented in Fig. 17.5.6. Achieving a 13.3× area improvement with respect to designs with similar gain accuracy, the proposed IA occupies the smallest area per channel, demonstrating that accurate gain matching can be achieved in an extremely small area, as required in cost-driven sensor applications.

References:

[1] S. Sakunia, et al., "A Ping-Pong-Pang Current-Feedback Instrumentation Amplifier with 0.04% Gain Error," *Symp. VLSI Circuits*, pp. 60-61, June 2011.
[2] R. Wu, J.H. Huijsing, K.A.A. Makinwa, "A Current-Feedback Instrumentation Amplifier with a Gain Error Reduction Loop and 0.06% Untrimmed Gain Error," *IEEE J. of Solid-State Circuits*, vol. 46, no. 12, pp. 2794-2806, Dec. 2011.
[3] Q. Fan, J.H. Huijsing, K.A.A. Makinwa, "A 21 nV/√Hz Chopper-Stabilized Multi-Path Current Feedback Instrumentation Amplifier with 2 µV Offset," *IEEE J. Solid-State Circuits*, vol. 47, no. 2, pp. 464-475, 2012.
[4] F. Michel, M. Steyaert, "On-Chip Gain Reconfigurable 1.2V 24µW Chopping Instrumentation Amplifier with Automatic Resistor Matching in 0.13µm CMOS," *ISSCC Dig. Tech. Papers*, pp. 372-373, Feb. 2012.
[5] F.Sebastiano, R. van Veldhoven, "A 0.1-mm² 3-Channel Area-Optimized Sigma-Delta ADC in 0.16-µm CMOS with 20-kHz BW and 86-dB DR," *Proc. ESSCIRC*, pp. 375-378, Sep. 2013.
[6] S.M. Kashmiri, S. Xia, K.A.A. Makinwa, "A Temperature-to-Digital Converter Based on an Optimized Electrothermal Filter," *IEEE J. Solid-State Circuits*, vol. 44, no. 7, pp. 2026-2035, 2009.
[7] I. Akita, M. Ishida, "A 0.06 mm² 14nV/√Hz Chopper Instrumentation Amplifier with Automatic Differential-Pair Matching," *ISSCC Dig. Tech. Papers*, pp. 178-179, Feb. 2013.

978-1-4799-0917-9/14 $31.00 © 2014 IEEE

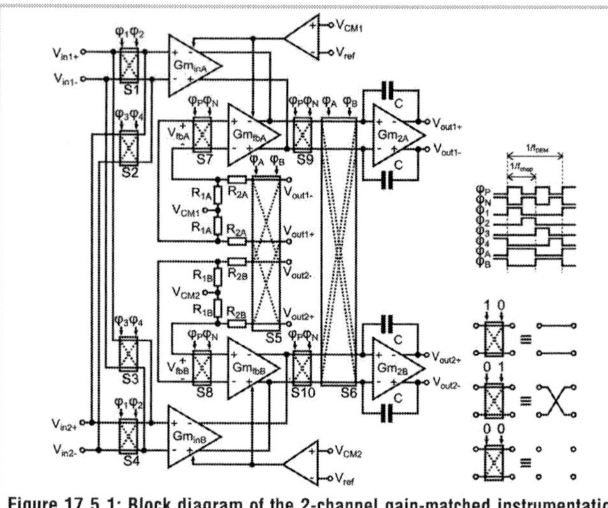

Figure 17.5.1: Block diagram of the 2-channel gain-matched instrumentation amplifier.

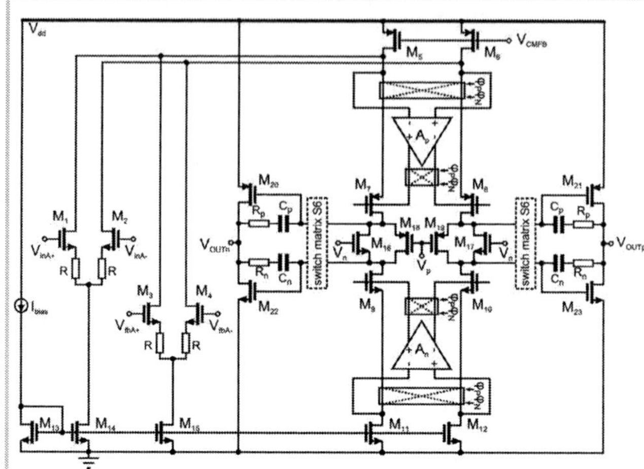

Figure 17.5.2: Schematic diagram of one channel of the instrumentation amplifier.

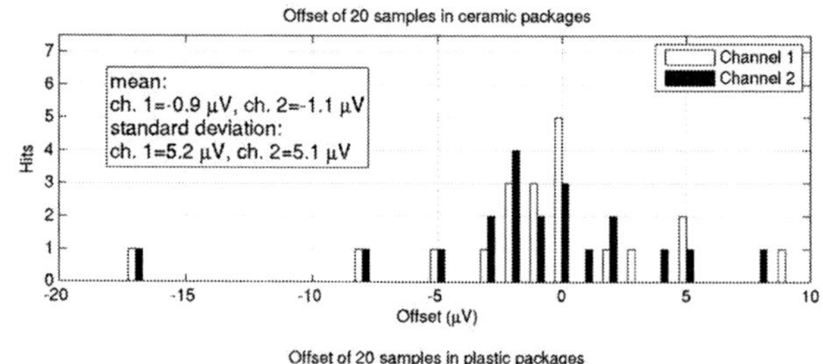

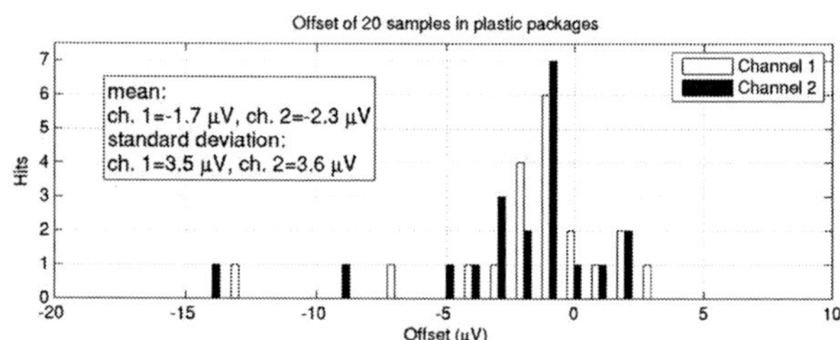

Figure 17.5.3: Measured input-referred offset voltage.

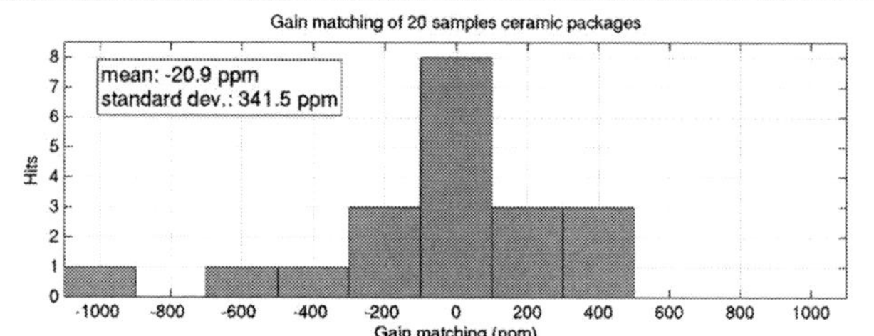

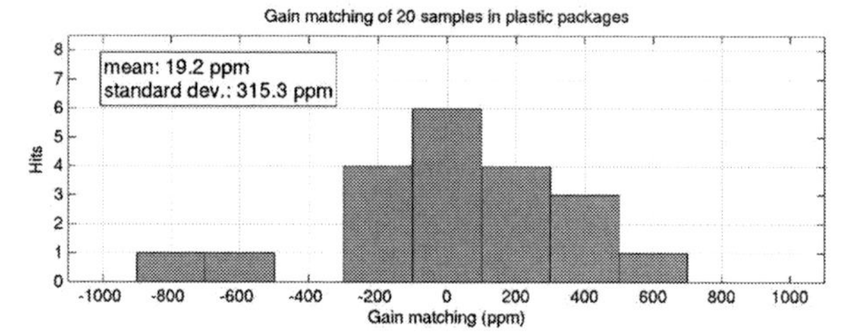

Figure 17.5.4: Gain matching measured for a 20kHz input signal.

Figure 17.5.5: Output noise spectrum (screenshot of R&S UPD audio analyzer); the RMS value (8.95μVrms) over the measurement BW (1.37kHz) results in 242nV/√Hz. Tones are supply-line interference coupling to the measurement setup.

	Unit	This work	Akita [7]	Michel [4]	Fan [3]	Wu [2]	Sakunia [1]
Year	-	2014	2013	2012	2010	2011	2011
Technology	μm	0.16	0.18	0.13	0.7	0.7	0.5
f_{chop}	kHz	200	500	25	30	32	-
f_{DEM}	kHz	100	-	0.025	-	8	-
Supply voltage	V	1.8	1.5	1.2	5	5	3.3-5.5
Supply current	μA	320	194	31	143	290	480
Input noise PSD	nV/√Hz	18.7	13.5	40	21	17	28
CMRR	dB	> 99	102	116	137	127	122
PSRR	dB	> 102	101	108	120	130	128
Linearity	-	THD=-60dB @ f_{in}=21kHz	-	-	-	INL= 4ppm	-
Crosstalk	dB	-73	-	-	-	-	-
Offset	μV	< 17	< 3.5	< 5	< 2	< 3	< 4
Gain accuracy	-	-	-	±0.5%	0.53%	0.06%	0.04%
Gain matching†	-	0.1%	-	±0.7%†	0.75%†	0.08%†	0.06%†
Number of samples	-	40	12	8	12	30	26
GBW	MHz	14	32	1	0.9	0.8	-
NEF	-	12.9	7.2	7.5	9.6	11.2	23.6
Die Area	mm^2	0.07 (2 channels)	0.06	0.465	1.8	5	1.48

† Estimated when not available as: gain matching = gain accuracy * √2

Figure 17.5.6: Performance summary and benchmark.

17

Figure 17.5.7: Chip micrograph.

17.6 Envelope Modulator for Multimode Transmitters with AC-Coupled Multilevel Regulators

Patrik Arnò[1], Matthieu Thomas[2,3], Vladimír Molata[2,3], Tomáš Jeřábek[2]

[1]STMicroelectronics, Grenoble, France,
[2]STMicroelectronics, Prague, Czech Republic,
[3]Czech Technical University, Prague, Czech Republic

Modern wireless communication systems, such as high-speed uplink packet access (HSUPA) or long term evolution (LTE), employ highly spectral-efficient modulations with inherent non-constant envelope signals having high peak-to-average power ratio (PAPR). These signals require the RF Power Amplifiers (PAs) to be backed off from saturation to satisfy the stringent linearity requirements. Unfortunately, linear operation leads to very low overall system efficiency. The envelope tracking (ET) technique has been identified as one possible solution to improve the global efficiency of the RF transmission chain. This technique has raised interest in the optimization of the fast supply envelope modulator (EM), the most critical component in the system. Several topologies for the EM exist such as fast buck converters, multilevel buck converters [1] and parallel hybrid structures with a class-AB amplifier AC-coupled to a buck converter [2].

In AC-coupled architectures, the high PAPR envelope signal is decomposed into low and high-frequency (LF, HF) spectral components. The LF part that carries a larger portion of the power is efficiently provided by a switching buck regulator, whereas the remaining HF power is supplied by the less efficient class-AB amplifier. AC coupling gives the additional benefit that instantaneous peak envelope voltages (higher than battery level) can be generated without requiring a dedicated boost converter. Nevertheless, this structure also has some drawbacks. Firstly, the class-AB amplifier becomes less efficient as the output power decreases. Secondly, as the DC output voltage cannot be boosted, the maximum average value for the PA supply stays lower than battery voltage.

This paper presents an EM that improves both high and mid-range power efficiency by replacing class-AB amplifier with a multilevel buck regulator (fast regulator). Furthermore, it extends the exploitable battery voltage range using a low-noise multilevel buck-boost converter (slow regulator) instead of a standard buck. The complete modulator has been designed to supply multi-mode transmitters. Besides ET operation, it is compatible with both 3G/4G average power tracking (APT) operation and GSM power ramping.

Figure 17.6.1 shows the architecture of the parallel AC-coupled EM. When the system is configured for ET operation, the fast regulator output is controlled by the differential envelope reference V_{ref}. On the other hand the slow regulator is configured to regulate the voltage across the AC-coupling capacitor (C_{fly}) according to V_{ref_slow} reference. The slow regulator feedback signal (FB_{slow}) is originated by sensing the voltage across C_{fly}. Thanks to the differential sensing, FB_{slow} signal has only a DC component and the loop does not require an additional low-pass filter to remove the AC part of the modulation. Both fast and slow loops use voltage-mode regulation and the pairs L_{slow}, C_{fly} and L_{fast}, C_{pa} are their respective LC output filters. Moreover, C_{fly}>>C_{pa} and L_{slow}>>L_{fast}. The multilevel fast regulator power stage can switch between ground, $V_{bat}/2$ and V_{bat}, whereas the slow regulator can switch between ground, V_{bat} and $2V_{bat}$. When needed, the system can be reconfigured for APT operation and, in this case, only the slow buck-boost regulator stays on while the APT switch shunts the bottom plate of C_{fly} to ground. Thus, the inductor L_{slow} and capacitor C_{fly} form its output LC low-pass filter. The PA supply voltage V_{pa} is sensed by the feedback and the voltage-mode pulse-width modulation (PWM) controller produces a constant output voltage at V_{pa} proportional to the reference voltage V_{ref_slow}.

There are several advantages of using a three-level power stage instead of a standard two-level architecture. Firstly, switching losses are lowered by a factor of 2 [3]. Furthermore, the current ripple in the output coil is divided by 2 and lastly, the output voltage ripple is attenuated by a factor of 4, leading to an improved frequency spectrum [1].

The controller of the fast regulator employs a voltage-mode PWM scheme extended with an end point prediction (EPP) path. There are several benefits of this fixed-frequency control scheme over variable-frequency ones such as lower near-channel noise, well-controlled spurious frequencies and a linear transfer function from PWM modulator input to output voltage [4]. In this architecture, the differential input reference V_{ref} is converted into a single-ended signal that becomes the reference for both parallel-connected EPP path and error amplifier. For a PWM controlled buck converter, the duty cycle can be estimated knowing V_{ref}, V_{bat} and k, the ratio V_{ref}/V_{pa}. In our system, this estimation is performed by the EPP through a feed-forward path from V_{ref} to V_{pa} as shown in [4]. In addition to this open-loop path, a standard proportional, integral, derivative (PID) closed-loop compensation inside the error amplifier ensures the stability of the fast regulator. Outputs of the EPP path and error amplifier are summed into a V_{err} signal. The comparison of V_{err} with two 180°-shifted triangular ramps results in 2 PWM signals having the same duty cycle but opposite phases. Each signal drives one pair of power switches M0, M3 and M1, M2 as in [1]. Furthermore, the 2 ramps used for modulation are proportional to V_{bat} variations, leading to an additional direct feed-forward to the PWM signals, which ensures independence of the regulator loop gain from V_{bat}.

Simulation results in Fig. 17.6.2 show that the addition of the EPP path to the PID loop achieves flatter gain and group delay. In the 0.2-10MHz band, the EPP helps reduce gain variation from 3.3dB to 1dB and group delay variation from 27ns to 3.8ns. This brings a benefit to whole ET architecture, as the group delay variation participates to global distortion of the system. Results are supported by measurement for gain, whereas extracted phase was too inaccurate to be used. The EM has been designed and manufactured leveraging 0.13μm, 4.8V, n-well CMOS technology. The package is a 400μm pitch wafer level chip scale package (WLCSP). A micrograph is shown on Fig. 17.6.7. For slow regulator an inductance L_{slow} of 4.7μH and a capacitance C_{fly} of 0.47μF are used while for the fast regulator L_{fast} of 22nH and C_{pa} of 6.8nF form the output filter. C_{mid} value is 1μF. The Slow buck-boost regulator operates from 1.9MHz to 7MHz (2MHz typ), whereas the switching frequency of the fast regulator can reach up to 140MHz (80MHz typ).

An LTE 10MHz (6.7dB of PAPR) test signal is used for all the measurements. Figure 17.6.3 shows measured waveforms of the envelope reference signal (delayed and scaled to V_{pa}) and modulator output voltage with resistive load. Fast modulator three-level output (V_{lx_fast}) is also presented. Figure 17.6.4 shows ET modulator efficiency with variable resistive load, while operating with 3.7V supply. The maximum efficiency is 86.2% and the maximum peak voltage is 4.7V.

To evaluate the performances with an RF power amplifier, a two-stage PA is used and both envelope pre-shaping and time alignment with RF signal are employed to minimize the distortion. The measured overall power added efficiency (PAE) reaches 39% for an output power of 26.3dBm and ACLR lower than -40dBc (Fig. 17.6.5, 17.6.6).

Acknowledgements:
The authors would like to thank Nicolas Marty, Jean-Pierre Covillers, Henri Ahde, Ondřej Tláskal, Jiří Hejda, Marek Kijovský, Vlastimil Kotě, Adam Kubačák, Milan Lžičař, Patrik Vacula and Radek Zelený.

References:
[1] V. Yousefzadeh, E. Alarcon, D. Maksimovic, "Three-level buck converter for envelope tracking in RF power amplifiers," in *Proc. IEEE Appl. Power Electron. Conf., vol 3. pp. 1588–1594*, Mar. 2005.
[2] P. Riehl, et al., "An AC-coupled hybrid envelope modulator for HSUPA transmitters with 80% modulator efficiency," in *ISSCC Dig. Tech. Papers*, Feb. 2013.
[3] M. Rodriguez, et al., "Fast dynamic response multilevel converter for voltage tracking applications," in *Power Electronics and Applications. EPE '09. 13th European Conf. on*, Barcelona, Sep. 2009.
[4] P. Y. Wu, P. K. T. Mok, "Comparative Studies of Common Fix-Frequency Controls for Reference Tracking and Enhancement by End-Point Prediction," *IEEE Trans. Circuits Syst. I, Reg. Papers,* vol. 57, no. 11, pp. 3023-3034, Nov. 2010.
[5] M. Hassan, et al., "A Combined Series-Parallel Hybrid Envelope Amplifier for Envelope Tracking Mobile Terminal RF Power Amplifier Applications," *Solid-State Circuits, IEEE J.*, vol. 47, no. 5, pp. 1185-1198, May 2012.
[6] D. Kim, et al., "Envelope-Tracking Two-Stage Power Amplifier With Dual-Mode Supply Modulator for LTE Applications," *Microw. Theory Techn., IEEE Trans.,* vol. 61, no. 1, pp. 543-552, Jan. 2013.

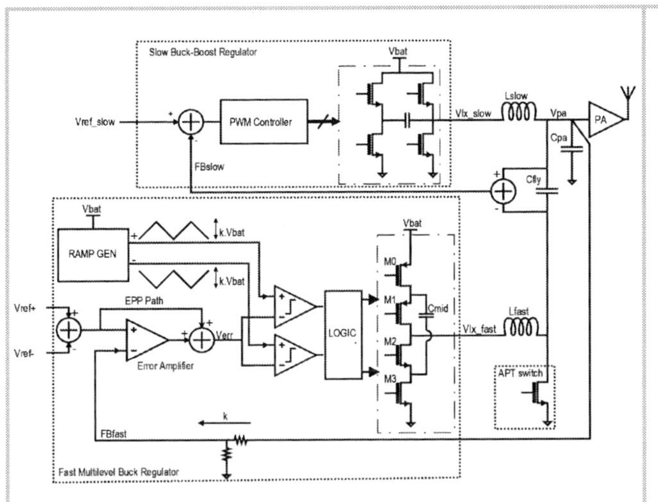

Figure 17.6.1: Envelope Modulator architecture.

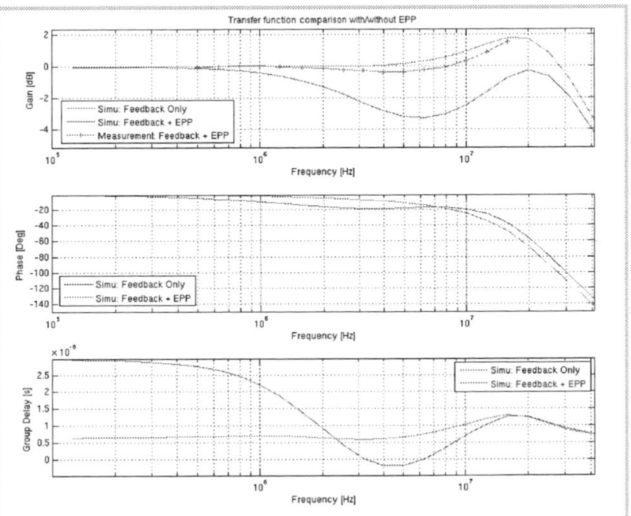

Transfer function comparison with/without EPP

Figure 17.6.2: Transfer function comparison of PID and PID+EPP.

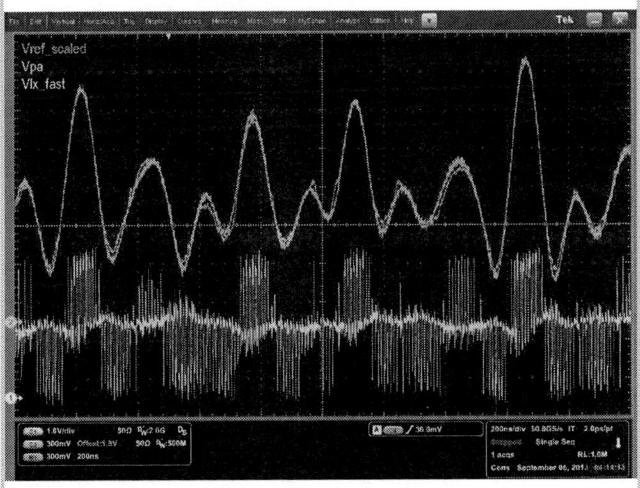

Figure 17.6.3: Tracking performance for LTE10 signal.

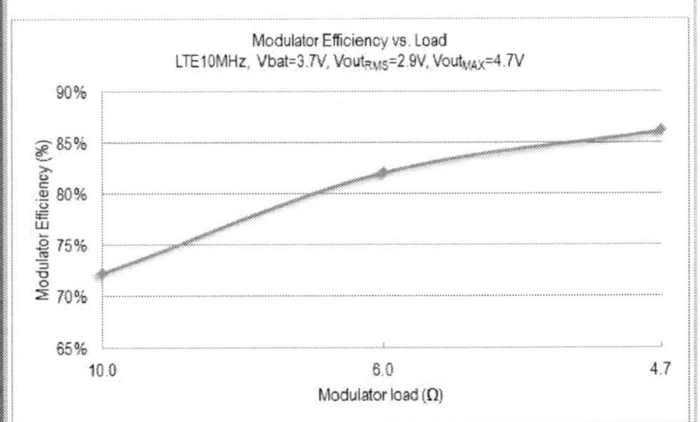

Figure 17.6.4: Modulator efficiency vs. load. LTE10 signal.

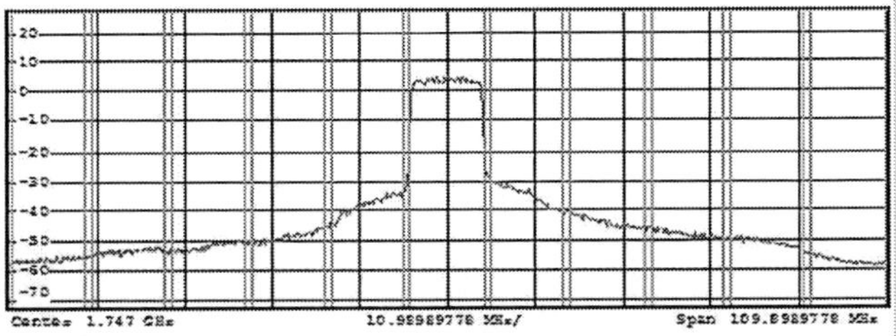

Center 1.747 GHz | 10.9989778 MHz/ | Span 109.989778 MHz

E-UTRA/LTE Square

Channel	Bandwidth	Spacing	Lower	Upper
Tx Channel	9.015 MHz		26.37 dBm	
Adjacent	9.015 MHz	10.000 MHz	-40.82 dB	-37.30 dB
Alternate	9.015 MHz	20.000 MHz	-52.17 dB	-47.28 dB
2nd Alt	9.015 MHz	30.000 MHz	-55.37 dB	-51.62 dB
3rd Alt	9.015 MHz	40.000 MHz	-57.20 dB	-54.45 dB
4th Alt	9.015 MHz	50.000 MHz	-60.04 dB	-60.74 dB

Figure 17.6.5: PA output spectrum @ 26.3dBm, Vbat=3.8V.

Reference	Protocol	RF Output Power [dBm]	PAPR 0.01% [dB]	Technology [μm]	Modulator Efficiency [%]	Supply Voltage [V]	Combined Efficiency [%]	ACLR [dBc]
This work	LTE 10MHz	26.3	6.7	0.13	86.2 ##	3.8	39 **	-40
[2]	HSUPA R6 5MHz	26	6.7	0.15	80	3.8	37	-40
[5]	LTE 10MHz	29.7	6	0.15	82	4.5	N/A	N/A
[6]	LTE 10MHz	27	6.44	0.18	76.3	N/A	39.8	-35.7

Maximum efficiency achieved on resistive load (4.7Ohm)

** 2 stages PA. First stage directly connected to 3.8V, second stage supplied by ET modulator

Figure 17.6.6: Performance comparison with the previous works.

978-1-4799-0917-9/14 $31.00 © 2014 IEEE

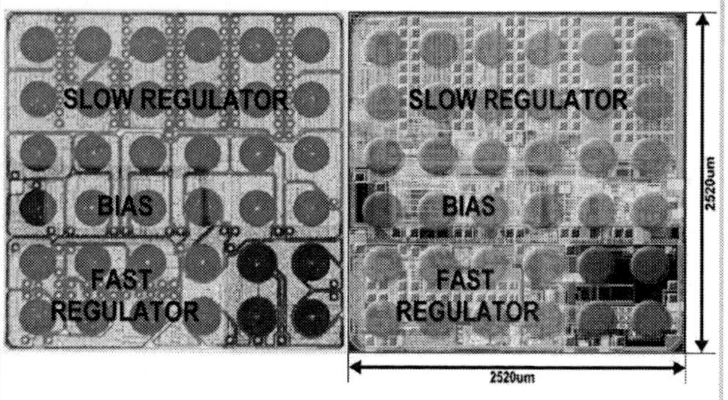
SLOW REGULATOR
BIAS
FAST REGULATOR
SLOW REGULATOR
BIAS
FAST REGULATOR
2520um
2520um

Figure 17.6.7: Micrograph & layout.

17.7 A 1.89nW/0.15V Self-Charged XO for Real-Time Clock Generation

Keng-Jan Hsiao

MediaTek, Hsinchu, Taiwan

A 32.768kHz crystal (XTAL) with its oscillation circuit is widely adopted for the generation of the real-time keeping and system-standby clock. Both functions are universally demanded by various systems such as cellular phones, smart wearable devices, GPS, etc. High frequency stability against environmental variations is necessary to meet system requirements. To increase the system stand-by time under limited battery capacity, an ultra-low power crystal oscillator (XO) is strongly demanded.

The Pierce oscillator, which adopts the XTAL as an inductive device in the loop and manages to make the loop unstable, is a common architecture for XO and sub-μW power consumption is achieved through amplitude control [1]. To reduce the extra power consumption induced by the large load capacitors required by the Pierce oscillator, a differential oscillator is proposed in [2]. Both works need amplifiers that operate in the linear region and consume static power, making it difficult to further reduce power. A DLL-assisted XO in [3] proved that power consumption could be greatly reduced by the separation of the gain and power stage of a single-stage inverting amplifier. A DLL is utilized to provide the necessary 180° phase shift typically introduced by an inverting amplifier. With the same oscillation principle as the Pierce oscillator in [3], load capacitors and a feedback resistor are inevitable. Although the amplifier is replaced, the complex design, which requires high supply voltage in [3], needs assistance from the separation of multiple power domains to reduce power to the nW scale. A higher supply voltage and a substantial amount of chip area are required to realize such designs. In this work, a self-charged XO (SCXO) is proposed to radically reduce the power consumption, supply voltage and chip area. In contrast to a Pierce oscillator, the SCXO excites the XTAL via charging the power into it directly. Without the extra power consumed by load capacitors and the feedback resistor, the power consumption is greatly reduced. The stable output clock usually provided by a conventional XO is still obtained.

Figure 17.7.1 shows the architecture of the SCXO and the oscillation waveform of an XTAL. The SCXO directly charges C_p and C_{par}, denoted with C_{pl} for short, periodically to replenish the energy loss in R_s and the XO's oscillation is sustained. The direct-charging scheme also alleviates the demands of the linear amplifier and makes the SCXO a logic-intensive design with power saving. The SCXO consists of a pair of dual-coupled clock slicers, pulse boosters and charging transistors, as shown in Fig. 17.7.1. The dual-coupled clock slicers extract positive clock edges and the proceeding pulse boosters up-convert control pulses to turn-on charging transistors. C_{pl} is charged to the supply voltage when V_{xi} or V_{xo} are in the 0° to 90° region, which is designed to conform to the inherent energy flow of the XTAL. The inductor current, I_{Ls}, is also diminishing while the capacitor voltage, V_{cp}, is increasing. Via this re-charging scheme, the energy loss due to R_s is compensated to maintain the XO's oscillation. The amount of the delivered energy is shown in eq. (1)

$$\frac{1}{2}\cdot(V_{DD}-V_{cp}(t))^2\cdot C_{pl},\quad(1)$$

where $V_{cp}(t)$ means the voltage across C_{pl}. The charged energy increases as $V_{cp}(t)$ decreases and vice versa. Consequently, the driving level of an XTAL is self-regulated without extra amplitude control circuits. The differential charging scheme also establishes a DC balance path between V_{xo} and V_{xi}, which is typically created by a large feedback resistor, R_{fb}. The extra power consumption induced by R_{fb} is hence eliminated. A conventional XO with R_{fb} and C_L is implemented for the start-up purpose only and could be completely turned off once the oscillation begins.

The wide charging window (0° ~ 90°) greatly increases the design flexibility and facilitates the adoption of logic-intensive designs. The latency from V_{xi}/V_{xo} to V_{jxi}/V_{jxo} must be less than a quarter period of the clock, which could be easily met under all PVT variations. Precise timing or phase control circuits are unnecessary and the design complexity is greatly simplified. The supply voltage and chip area are thus reduced. Figure 17.7.2 shows the schematic of the dual-coupled low-power clock slicer and the pulse booster. To reduce both the static and dynamic short-circuit current of M_p and M_n to zero, the input signal is dual-coupled to V_{gp} and V_{gn} respectively. With this technique, when one transistor is on, the other transistor's V_{gs} becomes negative. The sub-threshold conduction current is decreased, which effectively reduces both short and leakage current. The rail-to-rail clock V_{ck} is sliced-out with minimum current consumption. Subsequent CMOS logic extracts the positive clock edge, V_{eg}, for the XTAL's charging control. The pump-sense pulse booster raises the maximum level of V_j to 4-to-5 times the supply voltage and the completion of the level-boosting process is detected through the feedback of partially-boosted voltage, V_{rs}. The PMOS control signal V_{sp} remains low until V_{rs} toggles to high, which indicates that the boosting of V_j is complete. The pulse width of V_{rs} is determined jointly by the delay of the booster and logic gates. This closed-loop design guarantees a good pulse width of V_j, which is critical for the successful charging of the XTAL under all PVT conditions.

Figure 17.7.3 shows the frequency variation and the power consumption of the SCXO under the supply voltage variations from 0.15V to 0.5V, confirming that the SCXO works correctly over this range. Figure 17.7.4 shows the performance of the SCXO at different temperatures and 0.2V supply voltage. The frequency variation of the SCXO is compared with the XTAL's inherent temperature instability and a typical XO's frequency variation. Shown in Fig. 17.7.4, a frequency drift of less than 50ppm is achieved, which is smaller than the 74ppm frequency drift in a conventional XO. Figure 17.7.5 shows the measured wave-form of the SCXO. The signature of power-charging can be seen by V_{xo} and V_{xi} in the upper portion of this figure. The bottom part of Fig. 17.7.5 shows the measured Allan deviation plot of the SCXO and the conventional XO. The measurement result verifies that the SCXO provides a clock that is just as stable as a conventional XO. Figure 17.7.6 compares the SCXO with previous works. The SCXO consumes 1.89nW from a 0.15V supply, which is 34% of the power reported by previous work. Figure 17.7.7 is the micrograph of the chip and the active area is 0.03mm², including a start-up XO with R_{fb} and C_L. The SCXO occupies 0.0028mm², which is 9.3% of the total area. The SCXO works under a single supply without the demands of switched-cap networks for power partitioning thus saving chip area. This SCXO is fabricated in a 28nm CMOS process.

Acknowledgments:
The author would like to thank CM Chen, WL Lee for valuable technical discussions and CY Chien, Jeff Yen for integration assistance.

References:

[1] W. Thommen, "An Improved Low Power Crystal Oscillator," *Proc. Of the 25th ESSCIRC*, pp. 146-149, Sept. 1999.

[2] D. Ruffieux, "A High-Stability, Ultra-Low-Power Quartz Differential Oscillator Circuit for Demanding Radio Applications," *Proc. Of the 28th ESSCIRC*, pp. 85-88, Sept. 2002.

[3] Dongmin Yoon, Dennis Sylvester, David Blaauw, "A 5.58nW 32.768kHz DLL-Assisted XO for Real-Time Clocks in Wireless Sensing Applications," *IEEE Intl. Solid-State Circuit Conf. Dig. Tech. Papers*, Feb. 2012, pp. 366–367.

[4] D. Lanfranchi, E. Dijkstra, and D. Aebischer "A Microprocessor-Based Analog Wristwatch Chip with 3 Seconds/year Accuracy," *IEEE Intl. Solid-State Circuit Conf. Dig. Tech. Papers*, Feb. 1994, pp. 92–93.

[5] Micro Crystal - RV-2123-C2. Available:
http://www.microcrystal.com/Products/Oscillators/Real-Time-Clock-Modules.aspx

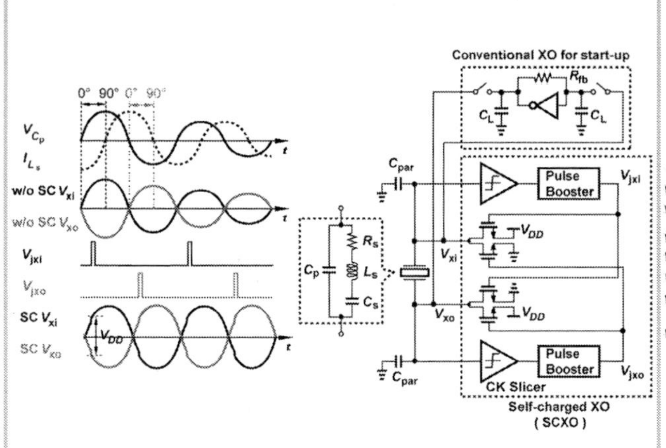

Figure 17.7.1: Operation waveform of the XTAL and the architecture of the SCXO.

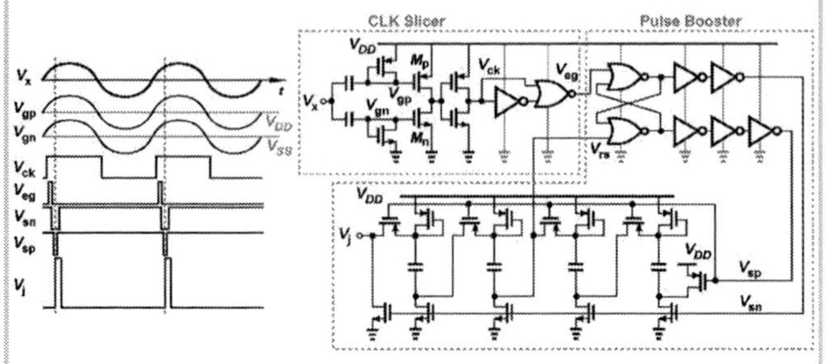

Figure 17.7.2: Schematic of the SCXO building blocks.

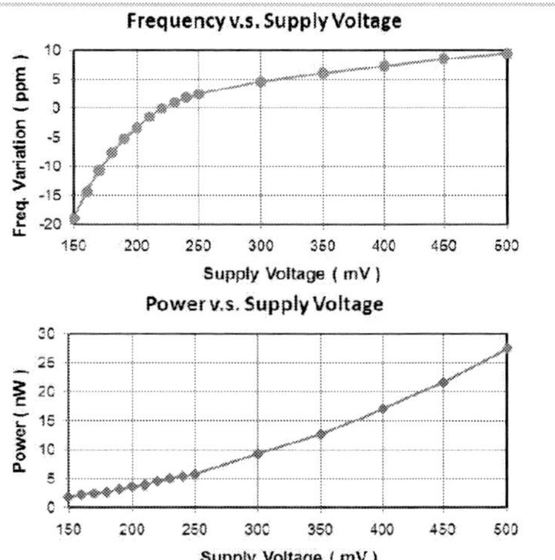

Figure 17.7.3: Measured frequency variation and power consumption from 0.15V to 0.5V supply.

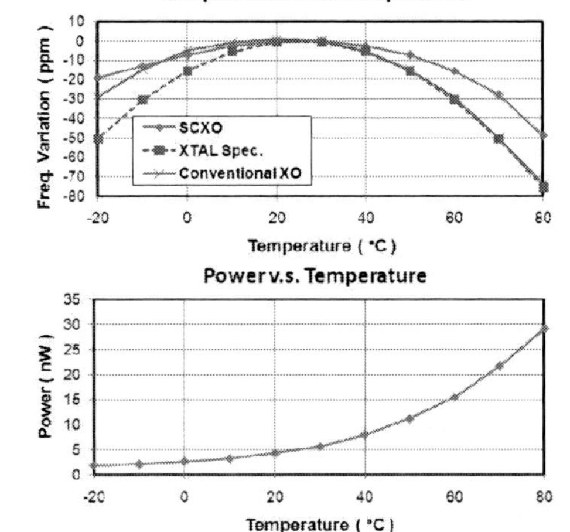

Figure 17.7.4: Measured frequency variation and power consumption from -20°C to 80°C at 0.2V supply.

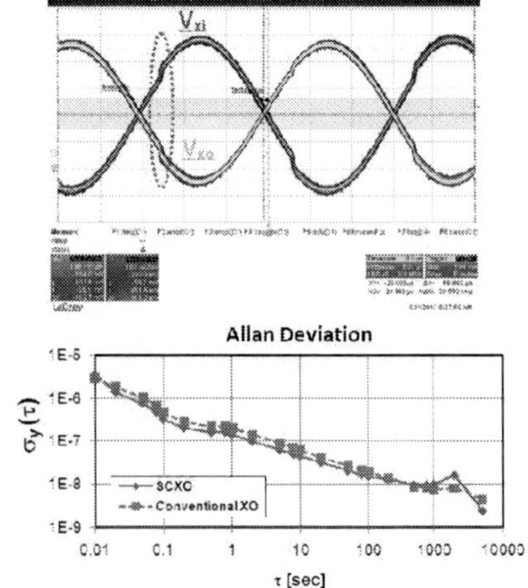

Figure 17.7.5: Measured waveform of the SCXO and Allan deviation plot of the SCXO and a conventional XO at 0.2V supply.

	Micro Crystal RV-2123-C2 [5]	ISSCC '94 [4]	ESSCIRC '99 [2]	ISSCC '12 [3]	This Work
Process	-	2 μm	2 μm	0.18 μm	**28 nm**
Area [mm²]	-	-	-	0.3	**0.03**
Frequency [Hz]	32.768k	32.768k	32.768k	32.768k	**32.768k**
Min. Power [nW]	110	220	27	5.58	**1.89**
Supply Voltage [V]	1.1 - 5.5	1.1 – 2.2	1.2	0.92 – 1.8	**0.15 – 0.5**
Temperature Range [°C]	-40 - 85	-10 - 70	-	-20 - 80	**-20 - 80**
Freq. variation with Temp. [ppm]	147 @ -40~85°C	24 @ 0~50°C	-	133.3 @ -20~80°C	**48.8 @ -20~80°C**

Figure 17.7.6: Performance comparison with previous works.

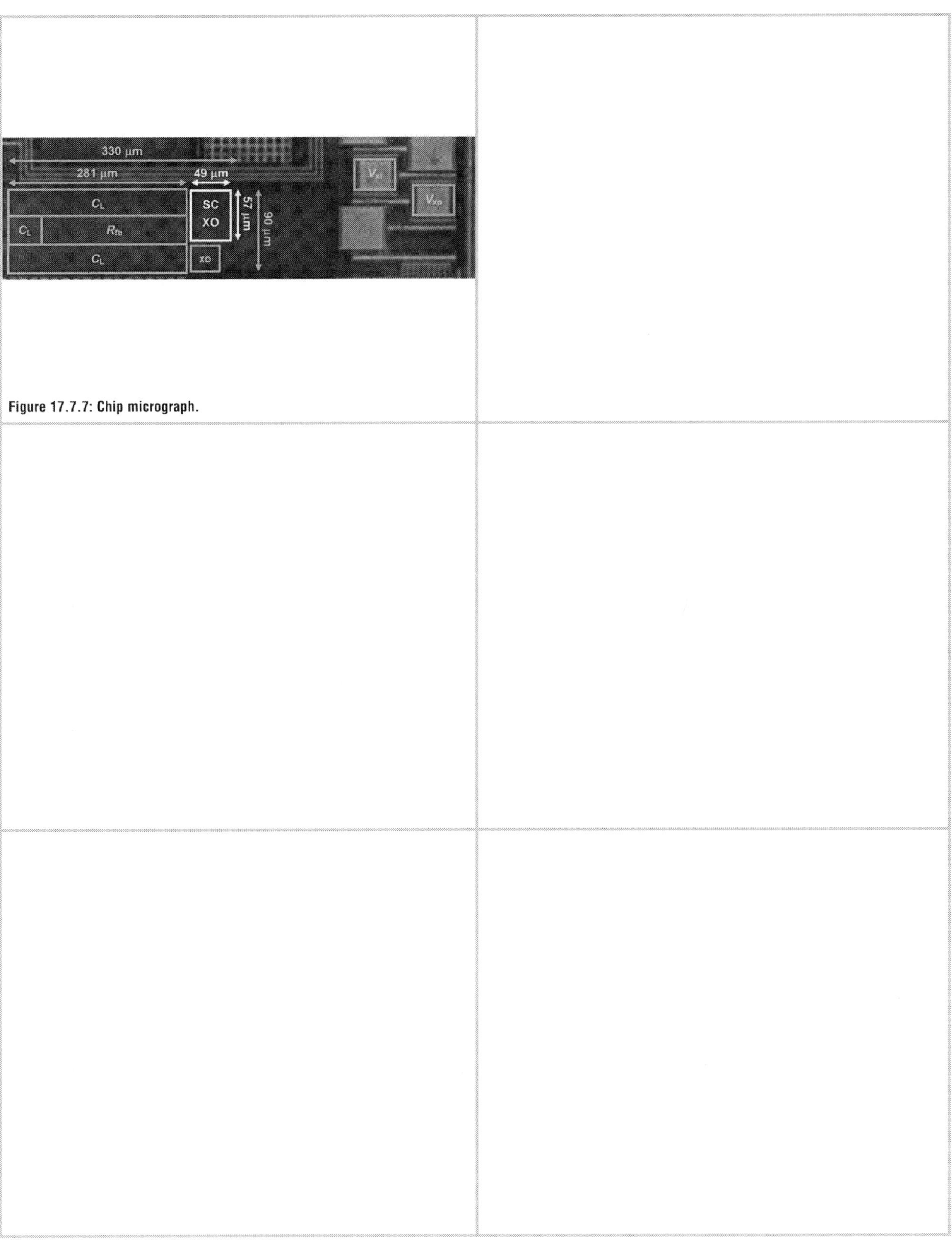

Figure 17.7.7: Chip micrograph.

17.8 A 190nW 33kHz RC Oscillator with ±0.21% Temperature Stability and 4ppm Long-Term Stability

Danielle Griffith[1], Per Torstein Røine[2], James Murdock[1], Ryan Smith[1]

[1]Texas Instruments, Dallas, TX, [2]Texas Instruments, Oslo, Norway

In wireless networks with a low duty cycle, the radio is operational for only a small percentage of the time. A sleep timer is used to synchronize the data transmission and reception. The total system power is then limited by the sleep power and the sleep timer frequency stability. Low-frequency crystal oscillators are a common choice for sleep timers due to their excellent long-term stability, frequency stability over temperature, and very low power consumption. However, the external crystal cost and board area are undesired. If an integrated oscillator is used as an alternative, the frequency variation must be minimized so the sleep time can be maximized.

Several integrated oscillators with good frequency stability have been proposed in recent publications. In [1], feed-forward correction is used to achieve a stable frequency over temperature, but the voltage sensitivity and die area are large. In [2], self-chopping is used and excellent stability results are achieved at the cost of relatively high power consumption. In [3], offset cancellation is used to allow lower power operation while achieving stable temperature performance. However, long-term stability is limited by flicker noise to 20ppm for intervals over 0.5s. Also, low power systems may have supplies that are allowed to droop while in sleep, so improved frequency stability over supply variation is desired. In this work, an RC oscillator with improved long-term stability and frequency stability over supply voltage is presented.

As shown in Fig. 17.8.1, the implemented RC oscillator consists of an RC network, and an inverting gain element from a resistor terminal to a capacitor terminal and another inverting gain element from the common resistor/capacitor terminal back to the resistor terminal. For high gain, the two inverting elements consist of three and five inverters, respectively. A simple regulator, consisting of an NMOS voltage follower and a replica inverter that is flipped and biased by a reference current, produces a local regulated supply for the inverters. This local supply is well below the standard core voltage for the technology. Therefore the inverters have very low average current because they are biased in weak inversion at their switching point.

To analyze the frequency variation due to noise and other non-idealities, we define the relative inverter switching point as $K_{SW}=V_{SW}/VDD_{LOCAL}$ or, in other words, the ratio between the input=output inverter switching voltage V_{SW} and the local supply voltage VDD_{LOCAL}. Ignoring inverter delays and output resistance, the voltage at the common RC node will swing from $(K_{SW}+1)VDD_{LOCAL}$ to $(K_{SW}-1)VDD_{LOCAL}$, as shown in Fig. 17.8.2. It can be shown that the ideal oscillation period equals

$$t_0=(RC)\ln\left[\frac{(1+K_{SW})(2-K_{SW})}{K_{SW}(1-K_{SW})}\right].$$

When K_{SW}=0.5, we get 50% duty cycle, oscillation period t_0 = RC·ln(9) ≈ 2.2RC and the lowest sensitivity to K_{SW} variation.

For a low frequency oscillator like this, the inverters driving the resistor and the capacitor will have much lower output resistance than the resistor R, and therefore contribute little to the oscillation period. Because the slew rate at the RC node is much slower than at the other circuit nodes, the transistor noise in the following inverter (INV1 in Fig. 17.8.1) contributes more to the total delay variation than transistor noise in the other inverters. However, noise in the INV1 transistors will mainly result in KSW variation, which affects the rising and falling delays with the opposite magnitude, and therefore only slightly changes the period. A positive offset will, for example, increase the positive charging time, but also decrease the negative charging time so that the oscillation period remains almost constant. For example, K_{SW}=0.45 or 0.55 gives 53%/47% duty cycle, but only 0.4% increased oscillation period compared to K_{SW}=0.5. In reality, the oscillation period equals $t=t_0+t_{INV}$, where t_{INV} represents the propagation delay through the inverting gain stages. Because the local supply voltage tracks the threshold of the NMOS and PMOS devices, the t_{INV} through the inverting elements is almost constant over temperature and supply voltage. This, in combination with low temperature coefficient for the resistor and capacitor, results in good frequency stability.

The voltage waveform out of the RC network is shown in Fig. 17.8.2. Note that VDD_{LOCAL} is less than the standard supply voltage so the maximum voltage seen, $1.5VDD_{LOCAL}$, does not cause any reliability issues. Due to the $2{\cdot}VDD_{LOCAL}$ swing at the output of the RC node, the required capacitance is cut in half for the same oscillation frequency compared to architectures that do not have a negative charging event. This is an advantage for low-frequency designs, where significant silicon area is needed for the resistor and the capacitor. Furthermore, the large voltage swing reduces the 1/f noise of the following inverter [4]. At stable temperature, 1/f noise in this inverter (INV1 in Fig. 17.8.1) is the main source of K_{SW} variation, and therefore limits long-term stability. The power consumption is reduced compared to [2] for similar frequency stability. Since inverters are used in place of comparators, the current consumption is dominated by the charging of the capacitor.

Frequency stability versus temperature and supply voltage is shown in Fig. 17.8.3 and Fig. 17.8.4, respectively. Five measured samples have frequency variation between ±0.11% and ±0.21% over the −20°C to +90°C temperature range. Due to the regulated local supply for the oscillator core, the frequency variation with supply voltage is less than 0.09%/V for five measured samples.

As well as frequency stability over temperature and voltage, long-term stability is required for a sleep timer to be able to wake the system up at a precise time. Circuit noise, both white and flicker, limits the long-term frequency stability. Allan deviation is often used to capture long-term stability. Figure 17.8.5 shows the measured Allan deviation over averaging periods up to 20 seconds for this RC oscillator. For intervals of up to 2 seconds, the Allan deviation is limited by white noise. Long-term Allan deviation is limited by 1/f noise, which in this architecture is minimized due to the absence of current sources, and by using an inverter with high input voltage swing as the voltage comparator. This allows a noise floor below 4ppm for intervals above 2 seconds, which is in contrast to [5], where a ≈1000ppm Allan deviation noise floor is reported, and [3] which reports Allan deviation below 20ppm beyond 0.5 seconds. The results show that the oscillator can meet the ±500ppm sleep timer requirement of low-power mode in Bluetooth 4.0.

A final advantage of this architecture is ease in scaling the frequency. A 48MHz version of the oscillator was designed by changing the resistor array and local supply regulator, but using the same capacitor array.

Figure 17.8.6 summarizes the RC oscillator performance and compares it to previous published works in the kHz range. The RC oscillator is fabricated in a 65nm CMOS process. A die micrograph is shown in Fig. 17.8.7. The RC oscillator consumes an area of 0.015mm², and is designed to operate at close to 32.768kHz after production trim. The RC oscillator frequency is determined to the first order by an unsalicided poly resistor using a dedicated temperature coefficient implant, and an array of fringe metal capacitors. Both the resistor and the capacitor can be trimmed. Before taking the measurements shown in Fig. 17.8.3 and Fig. 17.8.4, the capacitor array was trimmed as close as possible to the 32.768kHz target at 27°C.

In conclusion, an RC oscillator is implemented with an inverter as a comparator biased from a supply that tracks the inverter characteristics. The oscillator also uses the same circuit elements for positive and negative charging events to reduce frequency variation caused by offset and noise. These techniques help achieve excellent stability over temperature and voltage, with improved short-term and long-term Allan deviation when compared to other recently published low power RC oscillators.

Acknowledgements:
The authors would like to acknowledge Srividya Sundar for characterization support.

References:

[1] T. Tokairin, et al., "A 280nW, 100kHz, 1-cycle start-up time, on-chip CMOS relaxation oscillator employing a feedforward period control scheme," *Dig. Symp. VLSI Circuits*, pp. 16-17, June 2012.
[2] K.-J. Hsiao, "A 32.4ppm/°C 3.2-1.6V self-chopped relaxation oscillator with adaptive supply generation," *Dig. Symp. VLSI Circuits*, pp. 14-15, June 2012.
[3] A. Paidimarri, et al., "A 120nW 18.5kHz RC oscillator with comparator offset cancellation for ±0.25% temperature stability", *ISSCC Dig. Tech. Papers*, pp. 184-185, Feb. 2012.
[4] S. L. J. Gierkink, et al., "Intrinsic 1/f device noise reduction and its effect on phase noise in CMOS ring oscillators", *IEEE Journal of Solid-State Circuits*, Vol. 34, No. 7, pp. 1022-1025, July 1999.
[5] F. Sebastiano, et al., "A low-voltage mobility-based frequency reference for crystal-less ULP radios," *IEEE J. Solid State Circuits*, vol. 44, no. 7, pp. 2002-2009, July 2009.

978-1-4799-0917-9/14 $31.00 © 2014 IEEE

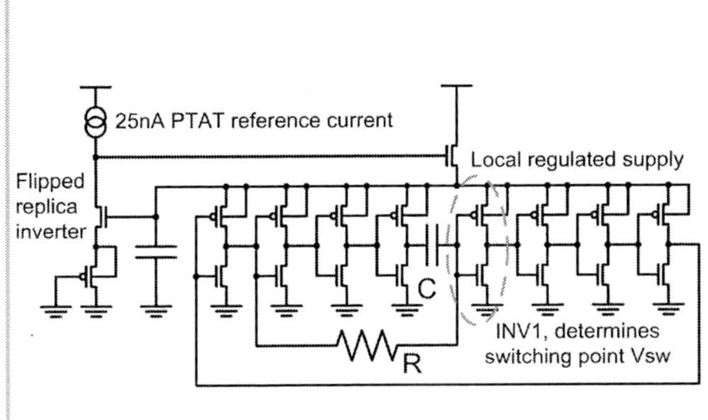

Figure 17.8.1: RC oscillator with local supply that tracks threshold voltage.

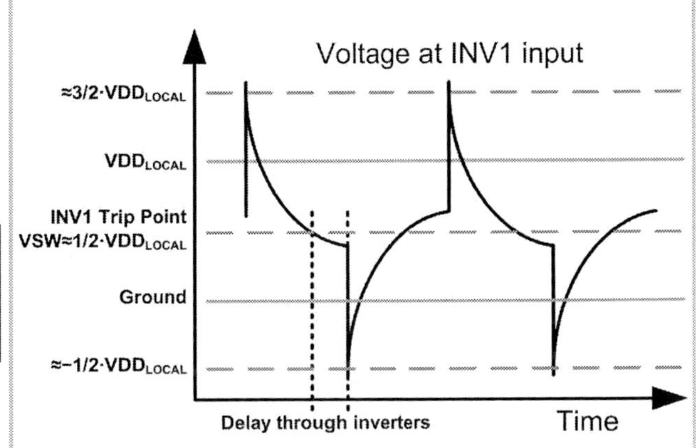

Figure 17.8.2: Voltage waveform showing operation of the RC oscillator.

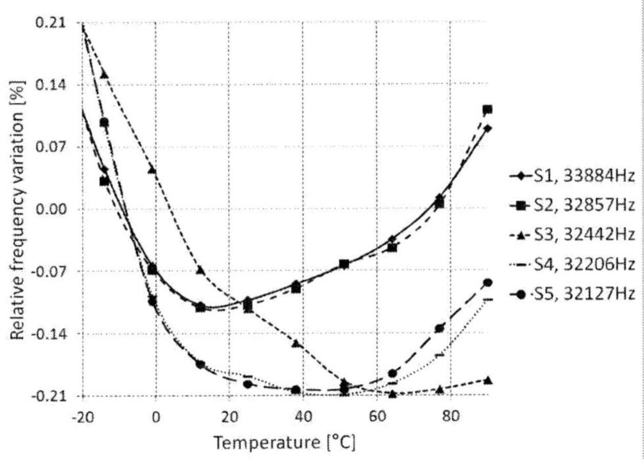

Figure 17.8.3: Measured RC oscillator frequency versus temperature for five samples.

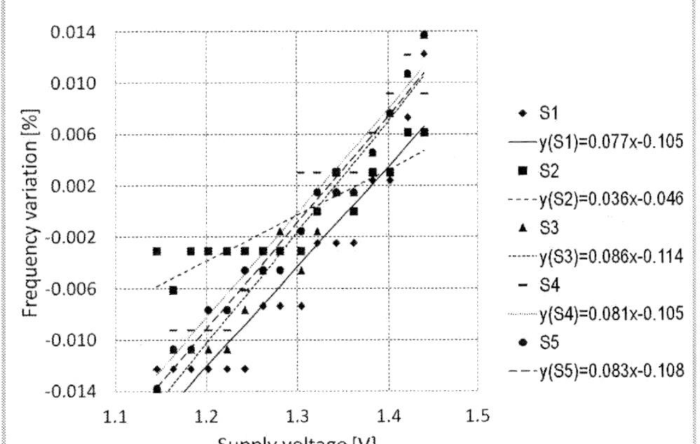

Figure 17.8.4: Measured RC oscillator frequency versus supply voltage for five samples.

17

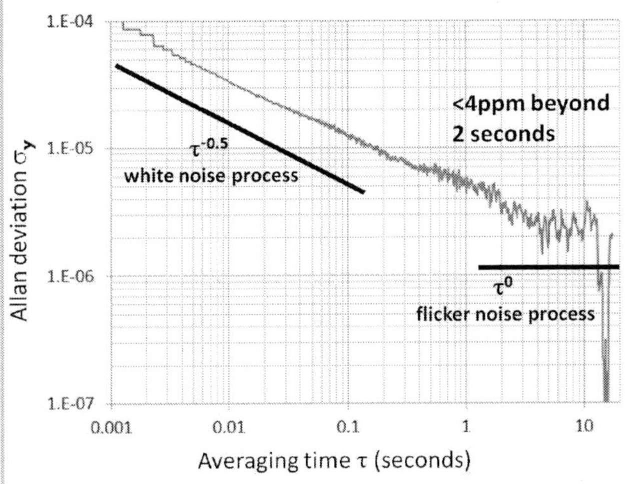

Figure 17.8.5: Measured Allan deviation for time intervals up to 20 seconds.

	[1] VLSI'12	[2] VLSI'12	[3] ISSCC'13		[5] JSSC'09	This work
Process	90nm	60nm	65nm		65nm	65nm
Area [mm²]	0.12	0.048	0.032		0.11	0.015
Freq (kHz)	100	32.768	18.5		100	33
Power (μW)	0.28	4.48	0.12		20.8	0.19
Temp Accuracy (%)	±0.68	±0.1	±0.25	±0.1	±1.1	±0.21
Temp Range (°C)	-40 to 90	-20 to 100	-40 to 90	0 to 90	-22 to 85	-20 to 90
Voltage Accuracy (%/V)	9.4	0.06	1		0.37	0.09
Allan Deviation Floor	N/A	N/A	< 20 ppm		≈1000ppm	< 4 ppm

Figure 17.8.6: Summary of measured results and comparison to previous work.

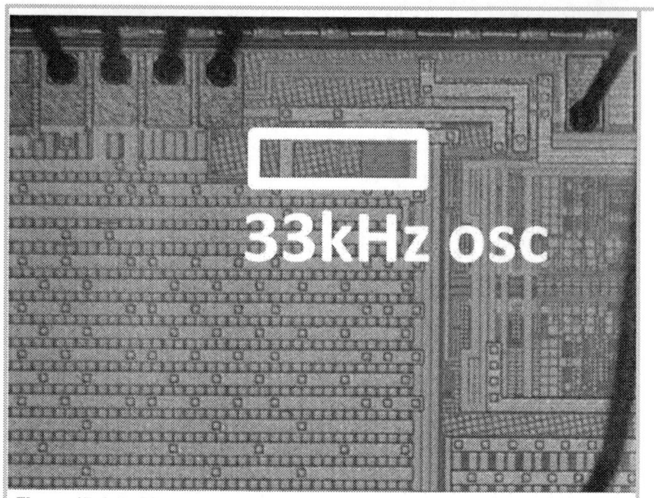

Figure 17.8.7: Die photo of the 33kHz RC oscillator on a 65nm CMOS radio SoC.

17.9 A 0.6V 70MHz 4th-Order Continuous-Time Butterworth Filter with 55.8dB SNR, 60dB THD at +2.8dBm Output Signal Power

Jayanth N. Kuppambatti, Baradwaj Vigraham, Peter R. Kinget

Columbia University, New York, NY

Technology scaling is leading to supply voltage reduction and shrinking voltage headroom, making it very challenging for analog circuits to achieve high signal-to-noise-and-distortion ratios (SNDRs). For example, at $V_{dd} \leq 0.6V$, the peak signal swing is 30% of V_{dd} in [1]-[3]. Since timing accuracy is improving in nanoscale CMOS, techniques have been proposed representing analog information in the time or phase domain. However, the linearity of the transformation between time or phase domain and the voltage or current domain remains a bottleneck at low supply voltages. In [2], ring-oscillator based integrators replace conventional OTA-based integrators, but the output signal swing is still limited to 30% of V_{dd} by the voltage headroom of the charge pump, which performs phase-to-voltage conversion. We propose switched-mode signal processing where analog information is represented in terms of pulse widths at the amplifier output stage, which can be replaced with power-efficient rail-to-rail class-D stages, thus producing switched-mode operational amplifiers (SMOAs). We present a 0.6V 70MHz 4th-order continuous-time Butterworth filter designed with SMOAs that achieves a dynamic range of 58dB, an SNR of 55.8dB and a THD of 60dB at +2.8dBm output signal power, while dissipating 26.2mW.

In a traditional two-stage CMOS OTA (Fig. 17.9.1), the output signal swing is limited by the output stage's (G_{m2}) headroom to $V_{pp,s/e}=V_{dd}-2V_{dsat}$ with $V_{dsat}\approx150mV$. To achieve a targeted SNR at low V_{dd}, the reduced output swing forces a large increase in the power consumption of G_{m1}. The first stage of the proposed SMOA (Fig. 17.9.1) is a transconductor G_{m1s} and the output stage is a pulse-width modulator (PWM) converting the signal information from the voltage domain to pulses with varying widths that drive class-D inverter output stages. The SMOA output swing is limited only by the minimum pulse width, t_{min}, in the switching stages; t_{min} is as low as 100ps in 65nm which enables large output swings, up to 73% at peak THD at V_{dd}=0.6V in the presented SMOAs. This higher signal swing leads to a lower power design for G_{m1s} for the same SNR and BW. The open loop output impedance of SMOA is very low, thus forming a true operational amplifier, and the output loading does not affect the unity-gain bandwidth (UGB) and DC gain. Representing the rail-to-rail signal information with a PWM stream is inherently more linear than class-A stages and hence the SMOA displays better open-loop linearity [4] than OTAs. This open-loop linearity of the PWM output stage is further enhanced by the gain of G_{m1s} when the SMOA is used in a closed-loop configuration, making highly linear amplifiers with rail-to-rail output swings possible. Given the switched nature of PWM signals, signal replication or (off-chip) buffering is easily implemented with inverters. A small-signal model for the PWM (Fig. 17.9.1) is a gain block K with a delay (of the modulator) T_d. K is set by the ratio of the peak-to-peak voltage of the PWM reference ramp signal V_{ref} to the supply V_{dd}; K=4 in this design. T_d sets the maximum SMOA UGB for a given stability constraint and hence T_d must be minimized by design.

The SMOA output spectrum (Fig. 17.9.1) contains the signal tone and also modulation spurs around the PWM clock frequency F_{pwm}. A higher F_{pwm} moves these spurs away from the signal band, but reduces the signal swing, since t_{min} is technology limited. Using a multi-phase PWM design with N modulators with phases staggered by $2\pi/N$ alleviates this problem by pushing the modulation spurs to $N*F_{pwm}$. When the N binary signals are summed, they form an (N+1)-level signal. This is very beneficial for SMOAs in feedback, as it avoids large voltage jumps at the input of G_{m1s}, thus preserving the linear behavior. An N-phase design further dramatically improves the SMOA power efficiency when driving capacitive loads, when compared to a single PWM running at $N*F_{pwm}$, since the capacitors are charged and discharged at an N-times lower frequency.

The 4th order continuous-time Butterworth filter prototype consists of two Tow-Thomas biquads (Fig. 17.9.2). $SMOA_1$ and $SMOA_2$ are 8-phase switched-mode operational amplifiers. The 8 clock phases are generated on-chip by dividing an externally fed 2.4GHz clock. Each of the SMOAs is made of 8 identical unit cells in parallel, each operating on a different clock phase ϕ<0:7> to implement the 8-phase PWM. Each unit cell (Fig. 17.9.3) consists of a trans-conductor, G_{m1s}, followed by the PWM. The G_{m1s}s are sized appropriately for noise requirements. The PWM reference ramp is generated by charging and discharging a capacitor connected to the output of G_{m1s}, using a current-steering differential pair. The output signal current of G_{m1s} is steered to the same capacitor. The PWM is a continuous-time comparator, implemented as a cascade of 3 differential pairs, followed by inverters to drive the output load. SMOAs use natural PWM, where the input signal is compared in continuous time with the ramp reference. In contrast, in sampled PWM the input signal is compared to the PWM reference only at specific times and the PWM delay is set by the comparison frequency. This would reduce the maximum UGB that can be achieved for stable operation compared to the presented SMOAs.

The closed loop UGB for the SMOA integrators is designed to be around 400MHz. The 8-phase PWM runs at 300MHz, pushing the first modulation spur to 2.4GHz. In the prototype, the 8 phases are tunable with programmable delay cells to enable calibration for phase mismatches between the PWMs. One-tap finite impulse response (FIR) filtering is further performed with delay cells at the output of each phase resulting in 16 signal streams (Fig. 17.9.2). The FIR delays are tuned to create a notch at 2.4GHz for the combined output, so that ideally the first significant modulation spur occurs at 4.8GHz. In a phase and gain matched 8-phase PWM system, the output of the SMOA can be followed by a simple RC low-pass filter to attenuate the 4.8GHz component, thus providing a very linear broadband spectrum.

To enable operation down to 0.6V, G_{m1s} is implemented (Fig. 17.9.3) as a pseudo-differential telescopic OTA, without a tail-current source, but the bias current is controlled by body-mirroring [1]. A digitally tunable series RC at the output of G_{m1s} forms the dominant pole and a stabilizing zero. The UGB-limiting capacitors at the output of G_{m1s} (Fig. 17.9.3) are connected to appear in common mode for the PWM current source, but in differential mode for the input signal, so that a smaller, lower-noise PWM current can be used. The PWM clock phases ϕ<0,3>, ϕ<1,4> etc. are 180° out of phase. By connecting the UGB-capacitors between the nodes V_{opi}<0>, V_{oni}<3> and V_{opi}<3>, V_{oni}<0>, they do not load the PWM current source, while the input signal path remains unchanged.

The active area on the 65nm CMOS die (Fig. 17.9.7) is 0.38mm². The output of the filter drives an on-chip 565Ω load (with no attenuation) and is then fed to an off-chip TIA (with 0.9dB gain) for broadband measurements with a spectrum analyzer. Figure 17.9.4 shows the measured filter transfer function, closely matching the ideal superimposed 70MHz Butterworth response. The small peaking in the frequency response well into the filter stop-band is due to a reduced gain margin in the SMOAs. Figure 17.9.5 shows the output spectrum from 10kHz to 4MHz and from 0 to 2GHz for a +2.8dBm 200kHz input. Due to incomplete spur cancellation between the PWM phases, modulation spurs can be observed at multiples of F_{pwm} (=300MHz). For a full-scale (+2.8dBm) output, the SFDR is 65dB and the THD is 60dB. The total integrated noise in the absence of an input signal is 365μV rms, thus giving a dynamic range of 58dB. Figure 17.9.6 shows the output SNR, THD and SFDR for varying input signal amplitude at 200kHz. At the peak output level of +2.8dBm, the SNR is 55.8dB, which is 2.2dB less than the dynamic range, due to noise folding back into the filter pass-band. Additional measurements have confirmed that PWM clock (2.4GHz) jitter up to 40ps rms does not degrade the filter SNR, enabling the use of low-power on-chip ring oscillators in future implementations. The THD for an in-band (-1dBm, 200kHz) signal is not affected by the presence of a blocker (up to +1dBm, 75MHz) in the transition band, demonstrating one of the benefits of the OTA-RC architecture of the SMOA filter. The filter consumes 26.2mW from a 0.6V supply, which also includes the power consumed to buffer the output signal.

Acknowledgments:
We thank Didier Belot and ST Microelectronics for silicon donation, NSF Grant EECS 1309721 for partial funding, Marianne Santangelo of Electrorent for test equipment and Berkeley Design Automation for their Analog FastSPICE Platform.

References:
[1] S. Chatterjee, et al., "A 0.5 V filter with PLL-based tuning in 0.18 um CMOS technology," *ISSCC*, 2005.
[2] B. Drost, et al., "A 0.55V 61dB-SNR 67dB-SFDR 7MHz 4th-Order Butterworth Filter Using Ring-Oscillator-Based Integrators in 90nm CMOS," *ISSCC*, 2012.
[3] M. De Matteis, et al., "A 0.55 V 60 dB-DR Fourth-Order Analog Baseband Filter," *JSSC*, Sept. 2009.
[4] W. Shu, et al., "IMD of Closed-Loop Filter-less Class D Amplifiers," TCAS-I, Feb. 2010.

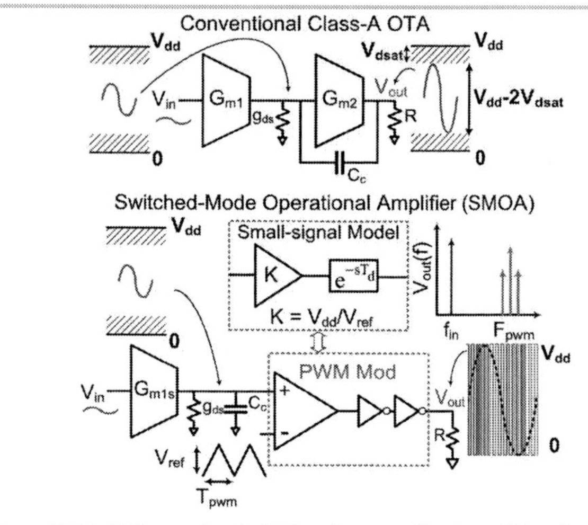

Figure 17.9.1: Voltage swing limitations in conventional amplifiers (top) and proposed rail-to-rail SMOA (bottom).

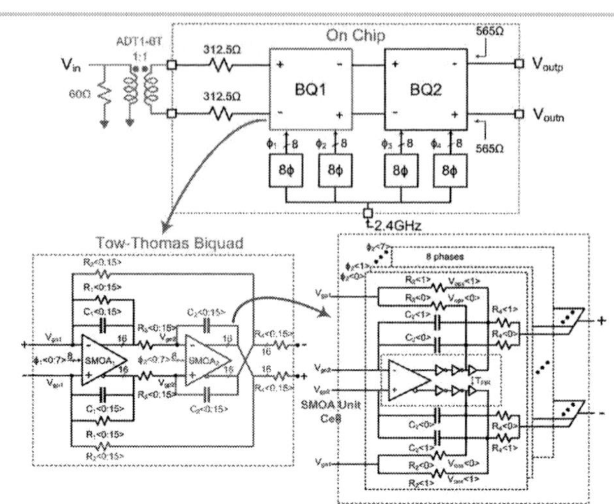

Figure 17.9.2: Architecture of the 4th-order Butterworth filter using proposed SMOAs.

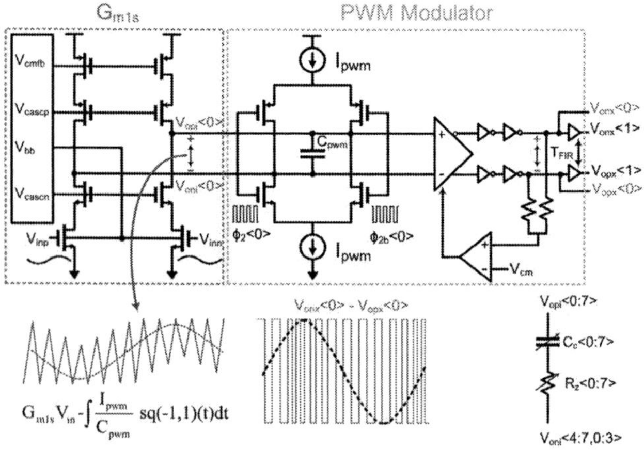

Figure 17.9.3: Single unit cell (phase 0) of the 8-phase Switched-Mode Operational Amplifier ($SMOA_2$) with UGB-limiting RC network.

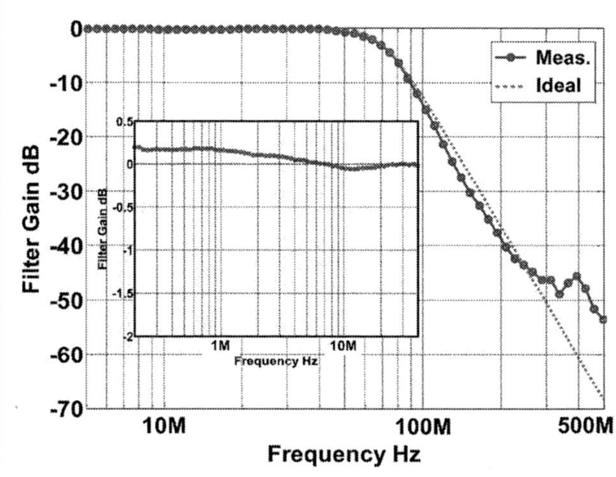

Figure 17.9.4: Measured 4th-order Butterworth filter transfer function.

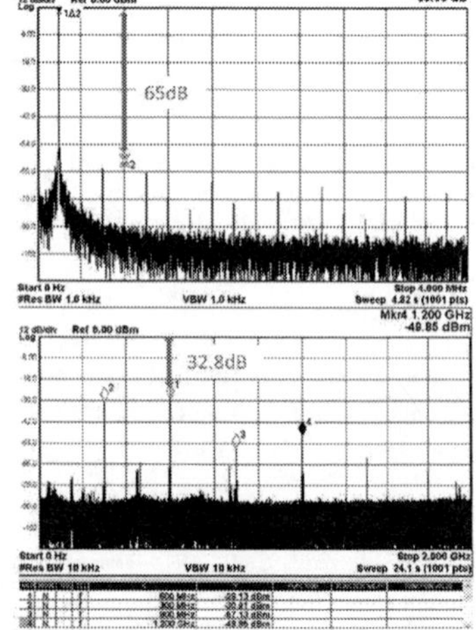

Figure 17.9.5: In-band output spectrum for a 200kHz +2.8dBm output and complete spectrum showing PWM spurs at NF_{pwm} (F_{pwm}=300MHz).

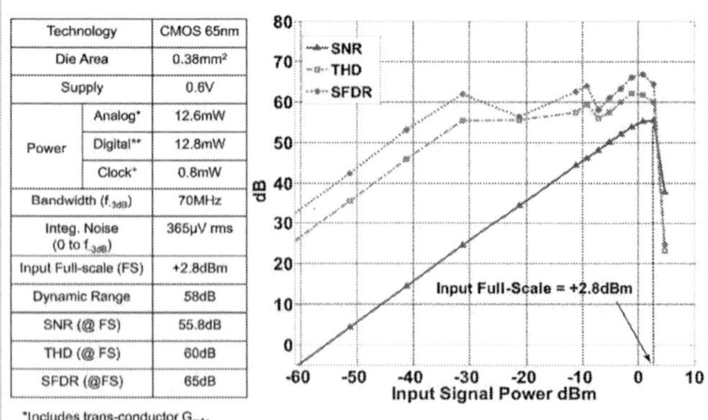

Technology		CMOS 65nm
Die Area		0.38mm²
Supply		0.6V
Power	Analog*	12.6mW
	Digital**	12.8mW
	Clock+	0.8mW
Bandwidth (f_{-3dB})		70MHz
Integ. Noise (0 to f_{-3dB})		365μV rms
Input Full-scale (FS)		+2.8dBm
Dynamic Range		58dB
SNR (@ FS)		55.8dB
THD (@ FS)		60dB
SFDR (@FS)		65dB

*Includes trans-conductor G_{m1s}
**Includes PWM-modulator, 8-phase clock gen and drivers
+Clock-chip interface

Figure 17.9.6: Performance summary and filter performance as a function of input signal amplitude at 200kHz.

17

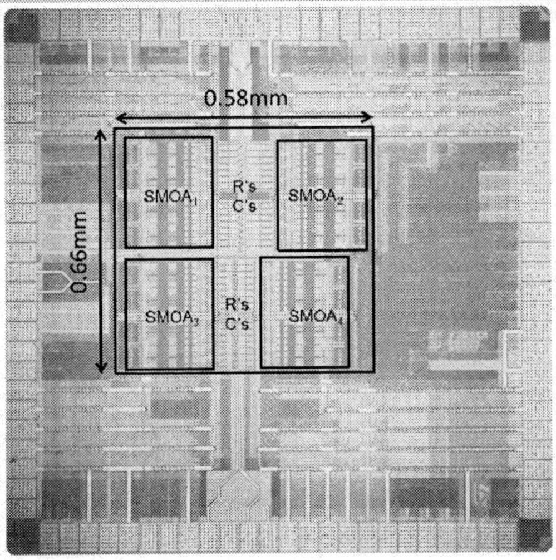

Figure 17.9.7: Die photo.

17.10 0.65V-Input-Voltage 0.6V-Output-Voltage 30ppm/°C Low-Dropout Regulator with Embedded Voltage Reference for Low-Power Biomedical Systems

Wei-Chung Chen, Yi-Ping Su, Yu-Huei Lee, Chin-Long Wey, Ke-Horng Chen

National Chiao Tung University, Hsinchu, Taiwan

Supplying a regulated 0.6V to biomedical systems requires a low dropout (LDO) regulator with a maximum driving current capability of 10mA. One sub-1V voltage reference circuit is commonly used in the conventional LDO design to generate the reference voltage V_{REF} with a low temperature coefficient (TC), as shown in Fig. 17.10.1. V_{REF} is sent to the inverting terminal of the error amplifier (EA) to regulate the output voltage V_{OUT}. The critical path of the voltage headroom exists between V_{REF} and V_{IN} through the inverting terminal and the tail current of the EA. That is, $V_{IN}>V_{SG}+V_{OV}+V_{REF}\approx|V_{tp}|+2V_{OV}+V_{REF}$, where V_{IN} is the input supply voltage, V_{tp} is the threshold voltage of the p-type MOSFET and V_{OV} is the overdrive voltage. If the minimum value of V_{IN} is reduced to 0.65V, the derived V_{REF} should be smaller than 50mV when $|V_{tp}|$ is 0.4V and $|V_{OV}|$ is 0.1V. Such a sub-1V voltage reference circuit is difficult to design [1]–[4]. Even if V_{REF} can be derived, the offset voltage in the EA will seriously affect the exact value of V_{REF} (≤50mV). In addition, the low noise immunity is another disadvantage that severely affects the performance of the biomedical system.

With the 21nm CMOS process, V_{IN} is reduced to 0.9V. The ratio of the threshold voltage V_t to V_{IN} is about 0.45. Thus, low-voltage logic in the sub-threshold region can be widely used for low power consumption. This paper presents an LDO regulator in a 21nm CMOS process, in which an EA with an embedded voltage reference (EVR) is employed. As illustrated in Fig. 17.10.1, the LDO regulator merges the functions of a sub-1V voltage reference and an EA to achieve a low TC and high driving current capability at the same time even if the V_{IN} is reduced to 0.65V.

Figure 17.10.2 shows the circuit implementation of the proposed LDO regulator, including the proposed EA with EVR, power MOSFET M_{PS}, and start-up circuit. All the MOSFETs, excluding M_{PS}, operate in the sub-threshold region for low power consumption. The MOSEFTs of M_{N2}, M_{N3}, M_{N4}, M_{P3}, and M_{P4} and the resistor R_2 can generate the reference current with a value of $V_T\cdot\ln(N)/R_2$, where V_T is the thermal voltage and N is equal to $(S_{MN4}\cdot S_{MP3})/(S_{MN3}\cdot S_{MP4})$. S is the aspect ratio of the MOSFET, and V_T is equal to kT/q, where k is Boltzmann's constant and T is the absolute temperature. Therefore, M_{N2} can work as the current source with a value dependent only on V_T and R_2 and independent of the characteristics of MOSFETs. Moreover, the composition of M_{N1}, M_{N2}, M_{P1}, M_{P2}, M_{NT}, M_{PS}, and R_1 forms two controlled paths, the current-mirror path and the feedback-loop path, to obtain V_{OUT} with a low TC. In the current-mirror path, the N-type current mirror pair (M_{NT} and M_{N1}) and the P-type current mirror pair (M_{P1} and M_{P2}) cause the current I_{P2} of M_{P2} to be K times the current I_{NT} of M_{NT}. On the other hand, the feedback-loop path provides negative feedback with the compensation capacitor and resistor, C_{MOM} and R_{CM}, to obtain $I_{N2}=I_{P2}\cdot[A_C/(1+A_C)]$ to confirm stability and ensure the regulation of V_{OUT}. A_C is the current loop gain from I_{N2} to I_{P2}. That is, the two paths can force I_{N2} and I_{P2} to have the same value if A_C is large enough. Consequently, V_{OUT} is determined by the value of $V_{GS\_MNT}+K\cdot[V_T\cdot\ln(N)/R_2]\cdot R_1$, where V_{GS_MNT} is the gate-source voltage of M_{NT}. Moreover, the start-up circuit provides instant current by M_{ST2} to pull up the node of V_{NT}. As a result, V_{OUT} can be settled to its desired value and not be trapped at zero. After the start-up operation, V_{ST} is pulled high to turn off M_{ST2} to shut down the start-up circuit.

V_{GS_MNT} has a complementary to absolute temperature (CTAT) coefficient because the threshold voltage V_t dominates the TC, whereas the voltage across the resistor R_1 has a proportional to absolute temperature (PTAT) coefficient. The measurement results of voltage versus temperature for V_{GS_MNT} and R_1 are depicted in Fig. 17.10.3. Thus, V_{OUT} with a low TC can be derived by setting $\partial V_{OUT}/\partial T$ equal to zero. That is, $G+K\cdot R_1/R_2$ is equal to 0. In this paper, $K\cdot R_1/R_2$ is equal to 175 when G=–175ppm/°C, which is the TC of V_{GS_MNT}. The right-side plots of Fig. 17.10.3 show the measured output voltage versus temperature from –40°C to 120°C when V_{IN} varies from 0.65V to 0.90V. In summary, the measured TC is 30ppm/°C when V_{IN} is 0.65V and I_{LOAD} is 10mA. In addition, TC is 28.5ppm/°C and 28ppm/°C when V_{IN} is 0.8V and 0.9V, respectively. The statistical histogram for the mean V_{OUT} of about 0.6V is determined by estimating 50 samples when temperature is 40°C and V_{IN} is 0.65V.

A small-signal model of the proposed LDO with the EVR can be viewed as a shunt–shunt feedback configuration, as illustrated in Fig. 17.10.4. gm_{PS} represents the transconductance of the power MOSFET. gm_f, equal to $1/(R_1+1/gm_{NT})$, is the transconductance from v_{out} to i_{NT}. $r_{o\_p2}//r_{o\_n2}$ is the equivalent resistance at V_g, where r_{o_p2} and r_{o_n2} are the output resistances of M_{P2} and M_{N2}, respectively. C_{p_o} and C_{p_g} are the equivalent capacitances at V_{OUT} and V_g, respectively. Thus, the loop gain, $T(s)$, under different load conditions can be derived to illustrate system stability. The dominated pole P_{dom} is located at the gate V_g of the power MOSFET because of the Miller compensation capacitor C_{MOM}. The first non-dominated pole P_O at V_{OUT} is generated by the output parasitic capacitance C_{p_o} and the equivalent resistance, which is composed of R_1+1/gm_{NT} and the output equivalent resistance R_{LOAD}. The DC gain and P_O depend on output loading current. To extend bandwidth (BW), the compensation zero Z_{RC} contributed by the Miller capacitor C_{MOM} and null resistor R_{CM} is utilized to cancel the pole P_O to increase stability. The zero is designed at the frequencies of P_O under medium load conditions to ensure a phase margin (PM) larger than 45 degrees and to avoid superfluous PM when the P_O moves toward low frequencies at light loads and high frequencies at heavy loads, respectively. The proposed LDO with the EVR is implemented in a 21nm CMOS process. The R_1, R_2, R_{CM}, and C_{MOM} are designed as 380kΩ, 400kΩ, 200kΩ, and 0.08pF, respectively. Because the proposed LDO regulator is capacitor-free, the allowable equivalent output capacitance ranges from 0pF to 100pF to ensure system stability. The table in Fig. 17.10.4 verifies the stability under different load conditions when V_{IN} is reduced to 0.65V and V_{OUT} is 0.60V.

Figure 17.10.5 shows the measured load transient response. The load changes from 0mA to 10mA and vice versa with a slew rate of 1µs rise/fall time. Adequate PM and BW optimize the transient response so that the overshoot and undershoot voltages are 7mV and 10mV, respectively. Figure 17.10.6 tabulates all the performance values of the proposed LDO with the EVR compared with LDO [5] and voltage reference [3], [4]. Although the design in [5] can operate with V_{IN} from 0.75V to 1.20V, it is biased because of an extra voltage reference circuit, which requires one buffer circuit with a level-shift structure and a large capacitor for stability. On the other hand, the voltage reference circuits in [3] and [4] provide good reference voltage and low TC but not driving capability. The die micrograph is shown in Fig. 17.10.7, and the die area is 0.015mm².

References:

[1] H.-W. Huang, et al., "A 1V 16.9ppm/°C 250nA Switched-Capacitor CMOS Voltage Reference," *ISSCC Dig. Tech. Papers*, pp. 438 - 626, Feb. 2008.

[2] L. Magnelli, et al., "A 2.6 nW, 0.45 V temperature-compensated subthreshold CMOS voltage reference," *IEEE J. Solid-State Circuits*, vol. 46, no. 2, pp. 465–474, Feb. 2011.

[3] A.-J. Annema, et al., "A sub-1V bandgap voltage reference in 32nm CMOS technology," *ISSCC Dig. Tech. Papers*, pp. 332-334, Feb., 2009.

[4] A.-J. Annema, G. Goksun, "A 0.0025mm² bandgap voltage reference for 1.1V supply in standard 0.16µm CMOS," *ISSCC Dig. Tech. Papers*, pp. 364-366, Feb. 2012.

[5] J. Guo, K.-L. Leung, "A 6-µW chip-area-efficient output-capacitorless LDO in 90-nm CMOS technology," *IEEE J. Solid-State Circuits*, vol. 45, no.9, pp. 1896-1905, Sep. 2010.

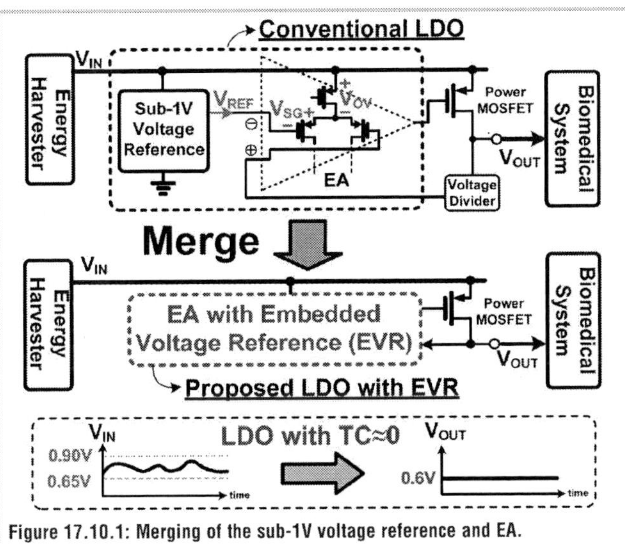

Figure 17.10.1: Merging of the sub-1V voltage reference and EA.

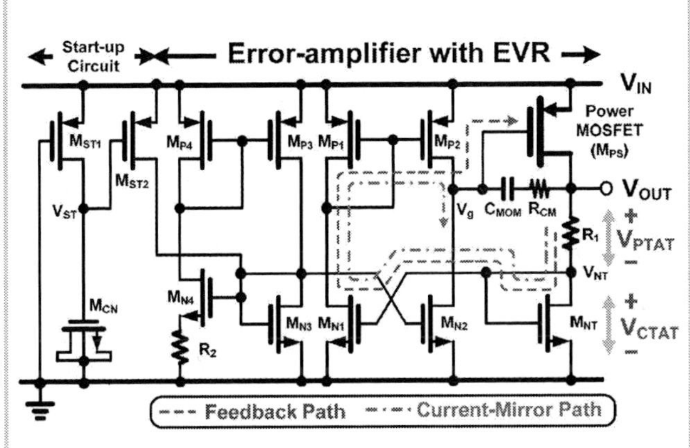

Figure 17.10.2: Schematic of proposed LDO with the EVR.

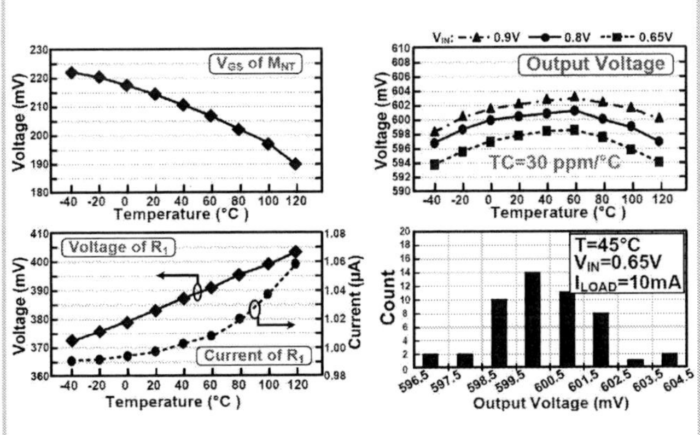

Figure 17.10.3: Output voltage versus temperature under different supply voltage and output voltage.

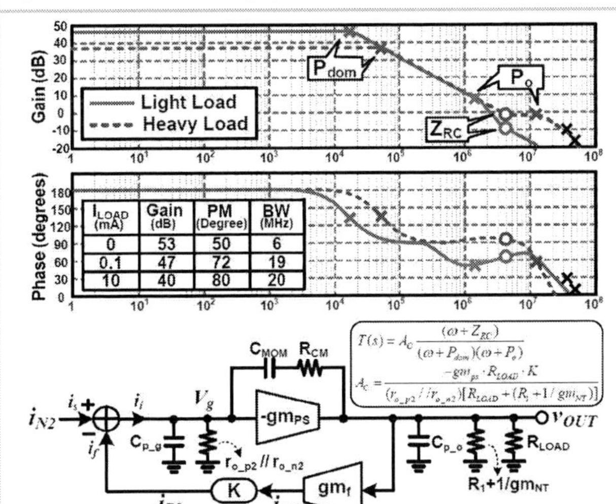

Figure 17.10.4: Small signal model and frequency response.

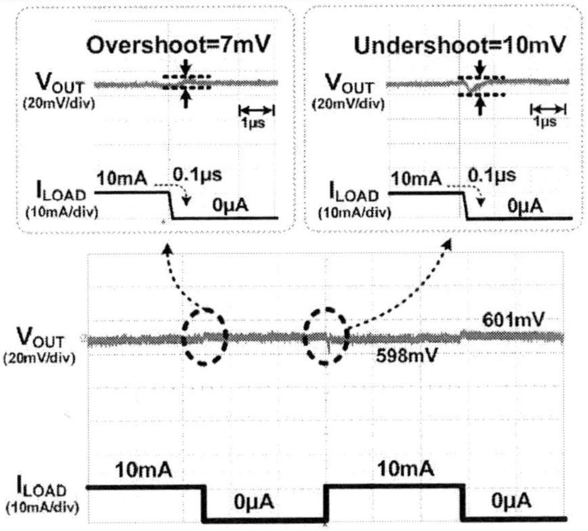

Overshoot=7mV
Undershoot=10mV
V_{OUT} (20mV/div)
1μs
I_{LOAD} (10mA/div)
10mA 0.1μs 0μA
601mV
598mV

Figure 17.10.5: Measured load transient response.

	Proposed	[5] LDO	[3] Voltage Reference	[4] Voltage Reference	unit
Technology	21nm CMOS	90nm CMOS	32nm FinFET	0.16μm CMOS	-
Voltage Reference	Embedded	#Not implement	-	-	-
Supply Voltage	0.65~0.9	0.75~1.2	0.9	1.1V	V
Supply Current	5	8	14	1.4	μA
Output Voltage	0.6	0.5~1	0.54	0.944	V
Output Capacitor	<100	<50	N/A	N/A	pF
Load Range	0~10	0~100	No	No	mA
Line Regulation	16	3.78	N/A	N/A	mV/V
Load Regulation	0.5	0.1	No	No	mV/mA
TC	30	N/A	560	30	ppm/°C
Area	0.015	0.019	0.016	0.0025	mm²
#Need extra buffer with capacitor to bias reference voltage					

Figure 17.10.6: Performance Summary.

17

978-1-4799-0917-9/14 $31.00 © 2014 IEEE

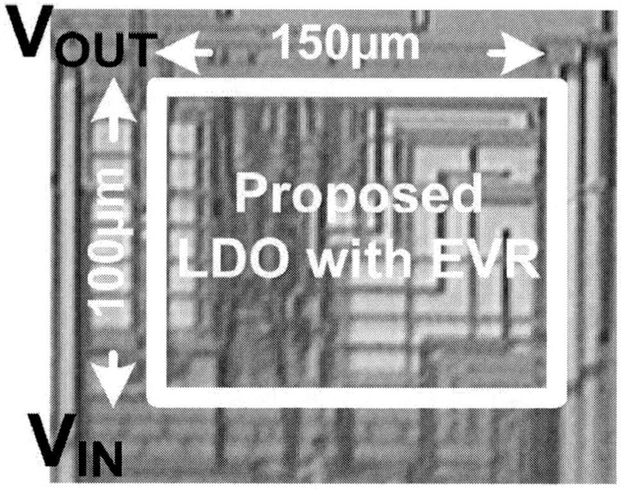

Figure 17.10.7: Die micrograph.

17.11 A 0.65ns-Response-Time 3.01ps FOM Fully-Integrated Low-Dropout Regulator with Full-Spectrum Power-Supply-Rejection for Wideband Communication Systems

Yan Lu, Wing-Hung Ki, C. Patrick Yue

Hong Kong University of Science and Technology, Hong Kong, China

High performance low-dropout regulators (LDOs) are indispensable in a system-on-a-chip (SoC) due to their low output noise, fast transient response and good power supply rejection (PSR) characteristics. In general, differential analog circuit loads need an LDO with high PSR, digital circuit loads need an LDO with fast load transient response, while single-ended analog/RF circuit loads need an LDO with both high PSR and fast transient response. Figure 17.11.1 shows an LDO embedded in an optical receiver that helps improve the sensitivity of the front-end system. On-chip LDOs with PSR in the GHz range are in high demand for wideband optical communication systems because there is only one photo detector in the optical receiver and supply voltage variations would degrade its sensitivity severely [1].

Off-chip capacitors are conventionally connected to supplies for filtering purposes. With a large output capacitor C_L, say 1µF, small ripples due to load transients and good PSR can be achieved [2]. However, for a fully-integrated LDO, large C_L is no longer available, so both transient responses and PSR performance will degrade significantly. Many fully-integrated LDOs with limited on-chip capacitance (a.k.a. capacitor-less LDOs) have been proposed in the past decade [3]–[6]. A figure-of-merit (FOM) of LDOs, shown in Fig. 17.11.1, is defined in [3], where I_Q is the quiescent current, and the response time, T_R, is a function of on-chip capacitance, C, load-transient of output voltage, ΔV_{OUT}, and maximum load, I_{MAX}. A considerably large current (6%) was used in [3] to move the non-dominant poles to high frequencies. A single-transistor-control LDO based on the flipped voltage follower (FVF) topology provided stable regulation at various C_L conditions including the capacitor-less case in [4], but it was sensitive to PVT variations, and was not fast enough with undershoots of 160mV observed. The FVF was also employed in [5] with a slew-rate enhancement circuit that responded to load-transient edges of 100ns. However, its PSR degraded to 0dB before reaching 1MHz. An ultra-fast-response comparator-based regulator in 45nm SOI process was proposed in [6] that consumed 12mA of I_Q and required a deep-trench capacitor of 1.46nF, and its intrinsic 10mV ripple was not suitable for RF/analog front-end circuits.

For an LDO, the largest capacitors are the output capacitor C_L and the power MOS gate capacitor C_G. Hence, there are at least two low-frequency (LF) poles: the output pole, p_{OUT}, and the pole at the gate of the power MOS, p_G. The pole p_{OUT} would shift to a lower frequency when R_L increases and vice versa. Basically, LDOs with an off-chip C_L are designed to be p_{OUT} dominant [2], while all previous fully-integrated analog LDOs have an internal dominant pole [3]–[5]. Therefore, LDOs can be classified by the need for an off-chip C_L, or by being output-pole dominant or internal-pole dominant. Thus, there are 4 combinations for which the pros and cons are summarized in the table of Fig. 17.11.1. An output-pole dominant LDO puts most of the available capacitors at the output, which could have intrinsically smaller ΔV_{OUT} and better PSR. The drawback is that a relatively high I_Q is needed to push the internal poles to high frequencies.

In this research, a fully-integrated tri-loop LDO designed in a 65nm GP CMOS is proposed that achieves 0.65ns T_R and full spectrum PSR. The transistor-level schematic is shown in Fig. 17.11.2. This circuit includes an error amplifier (EA) with V_{SET} generation and a buffered FVF. The signal paths of each loop are super-imposed on the schematic. Each loop has a different function: loop-1 is an ultra-fast low-gain loop with p_{OUT} being its dominant pole, while non-dominant poles p_A and p_G are pushed to the GHz range by the buffer impedance attenuation technique; loop-2 is composed of the EA and a diode-connected M_7 and is a slow loop that generates V_{MIR} and V_{SET}; and loop-3 has V_{OUT} fed back to the EA to improve the DC accuracy. The buffer, which consists of M_9 through M_{11}, presents low input capacitance to V_A and low output impedance to V_G, pushing p_A and p_G to very high frequencies. To save static current, the ratio of M_7 and M_8, and that of their bias currents, is set to be 1:4, as V_{SET} is in the low-speed path that does not need much current, while V_A is in the high-speed path and needs a larger current. In this design, C_L=130pF, I_2=20µA, and the buffer consumes another 20µA. All the above help pushing p_A and p_G to the GHz range. To increase the DC accuracy, a third loop is introduced through using a tri-input EA. The EA compares V_{REF} with both V_{MIR} and V_{OUT}, and the W/L ratios of the three input transistors M_1, M_2 and M_3 are 4:1:3. A bypass capacitor C_B=7pF is added at the V_{SET} node to improve the PSR by filtering out the ripple from V_{MIR} to V_{OUT}.

To simulate the loop response of each loop, two simulation setups are configured. In setup 1 loop-1 is broken between V_A and the buffer input. To isolate the influence from loop-2 and loop-3, the path from M_7 to M_8 is also broken. To maintain the DC bias, a DC V_{SET} is applied to the gate of M_8, and to account for the loading effect, a replica buffer is added to V_A to mimic the buffer. In setup 2 loop-2 and loop-3 are broken from V_{MIR} to M_2 and from V_{OUT} to M_3, respectively. The AC response of loop-2 can be obtained at V_{MIR}, and the response of loop-3 can be obtained at V_{OUT}, simultaneously. Simulation results of these two setups at I_O=10mA (the worst case) are combined in Fig. 17.11.3. When V_{OUT} is 1.0V, loop-1 has a DC gain of 21dB and a unity gain frequency (UGF_1) of 600MHz, with a phase margin (PM_1) of 60°. Loop-2 has one dominant pole at V_{SET} and a non-dominant pole at V_{EA}, and PM_2=80°. Loop-3 has two non-dominant poles at V_{EA} and V_{OUT}, respectively, and PM_3 is 20°. Nevertheless, the combined loop, with loop-2 very stable and loop-3 stable but with small phase margin, is stable. To gain more design margin for stability, the weighting of M_2 and M_3 could be set to 2:2 instead of 1:3 with a lower DC accuracy.

Figure 17.11.4 shows the measured transient response with on-chip load change from 0µA to 10mA within 200ps. With an I_Q of 50µA, the measured undershoot was 43mV, and V_{OUT} recovered to its steady-state value in 100ns with the help of loop-3 regulation. The measured overshoot was 82mV, and V_{OUT} was gradually discharged by the bias current of M_8, and then regulated by loop-3 to its steady-state value. The well-behaved waveforms of V_{OUT} confirmed the stability of the proposed tri-loop LDO. The FOM achieved is 3.01ps, and the T_R is 0.65ns. FOM is expected to be improved further with process scaling. Note that FOM improvement is not necessarily true for an internal-pole dominant LDO because it requires a minimum load current $I_{O,Min}$ for stability that results in a low loop bandwidth.

Figure 17.11.5 shows the measured PSR of the LDO up to 20GHz. For R_L=100Ω, PSR is better than -21dB at low frequencies; and the worst case occurs at 5MHz with -12dB rejection. Note that ripples generated by DC-DC converters could be higher than 100MHz [7], while noise generated by digital circuits is in the GHz range. PSR at 1GHz is -15.5dB. For frequencies higher than 1GHz, PSR would start to be dominated by the equivalent series resistance (ESR) of the filtering capacitor. As our design does not rely on the ESR to be zero for stability, ESR is minimized in the layout design for good PSR.

A performance comparison with state-of-the-art LDOs is listed in Fig. 17.11.6. Compared to previous ultra-fast response designs [3] and [6], sub-ns T_R is achieved with much smaller I_Q and C_L, hence resulting in a 3.01ps FOM. Furthermore, the full-spectrum PSR characteristic is presented, while other fully-integrated LDOs only measure PSR at specific frequencies. The chip area is 0.023mm² including the on-chip load, as shown in Fig. 17.11.7.

References:

[1] T. Takemoto, et al., "A 4x 25-to-28Gb/s 4.9mW/Gb/s -9.7dBm High-Sensitivity Optical Receiver Based on 65nm CMOS for Board-to-Board Interconnects," *ISSCC Dig. Tech. Papers*, pp. 118–119, Feb. 2013.

[2] M. Al-Shyoukh, H. Lee, R. Perez, "A Transient-Enhanced Low-Quiescent Current Low-Dropout Regulator with Buffer Impedance Attenuation," *IEEE J. Solid-State Circuits*, vol. 42, no. 8, pp. 1732–1742, Aug. 2007.

[3] P. Hazucha, et al., "Area-Efficient Linear Regulator with Ultra-Fast Load Regulation," *IEEE J. Solid-State Circuits*, vol. 40, no. 4, pp. 933–940, Apr. 2005.

[4] T. Y. Man, et al., "Development of Single-Transistor-Control LDO Based on Flipped Voltage Follower for SoC," *IEEE Trans. Circuits Syst. I: Regular Papers*, vol. 55, no. 5, pp. 1392–1401, May 2008.

[5] J. Guo, K. N. Leung, "A 6-µW Chip-Area-Efficient Output-Capacitorless LDO in 90-nm CMOS Technology," *IEEE J. Solid-State Circuits*, vol. 45, no. 9, pp 1896–1905, Sep. 2010.

[6] J. F. Bulzacchelli, et al., "Dual-Loop System of Distributed Microregulators with High DC Accuracy, Load Response Time Below 500 ps, and 85-mV Dropout Voltage," *IEEE J. Solid-State Circuits*, vol. 47, no. 4, pp. 863–874, Apr. 2012.

[7] C. Huang, P. K. T. Mok, "An 82.4% Efficiency Package-Bondwire-Based Four-Phase Fully Integrated Buck Converter with Flying Capacitor for Area Reduction," *ISSCC Dig. Tech. Papers*, pp. 362–363, Feb. 2013.

978-1-4799-0917-9/14 $31.00 © 2014 IEEE

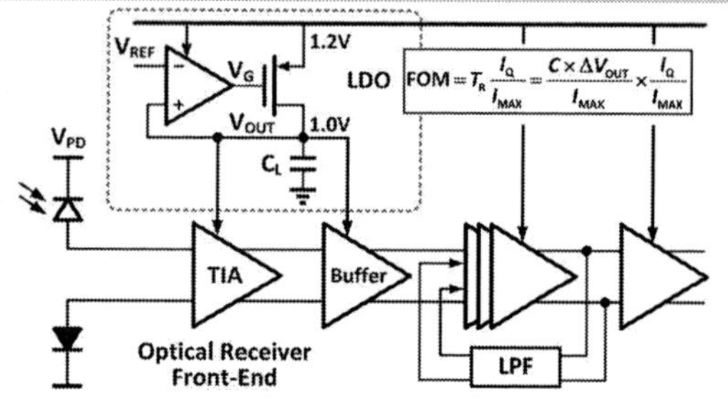

LDO CATEGORIZATION BY INTEGRATION LEVEL AND DOMINANT POLE LOCATION

Integration	w/ Off-Chip Cap.		Fully-On-Chip	
Dominant pole	Internal	Output	Internal	Output
Transient ΔV_{OUT}	Small	Small	large	Med.
PSR	Med.	High	Low	Med.
UGF	Low	Med.	Med.	High
Limit on $I_{O,Min}$	Yes	No	Yes	No
Process Scaling	No	Yes	No	Yes

Figure 17.11.1: Block diagram of an LDO providing a clean supply for the trans-impedance amplifier (TIA) in an optical receiver, and a table of LDO categorization.

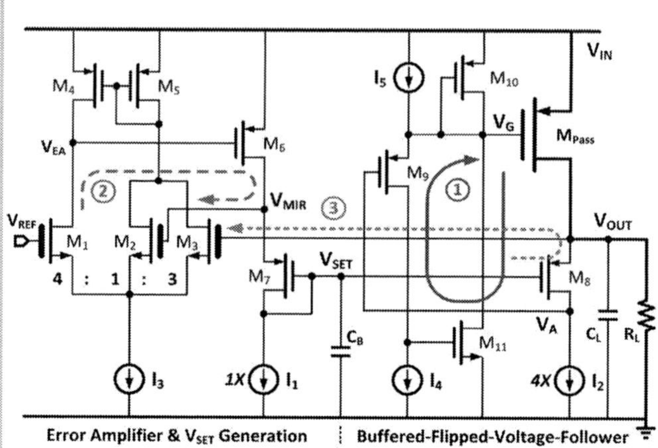

Figure 17.11.2: Schematic of the fully-integrated tri-loop LDO.

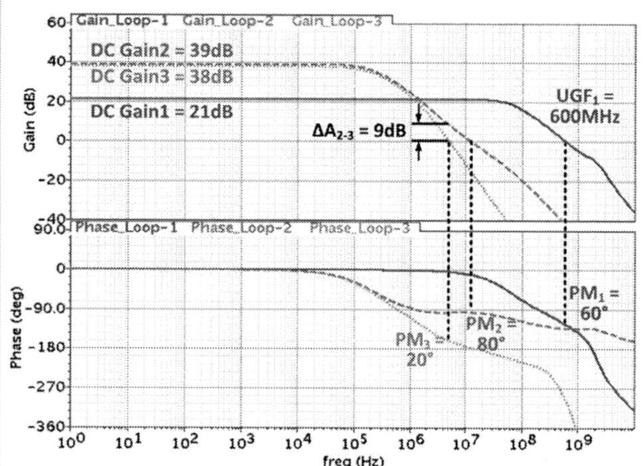

Figure 17.11.3: Simulated frequency response of the three loops with V_{IN}=1.2V, V_{OUT}=1.0V and R_L=100Ω.

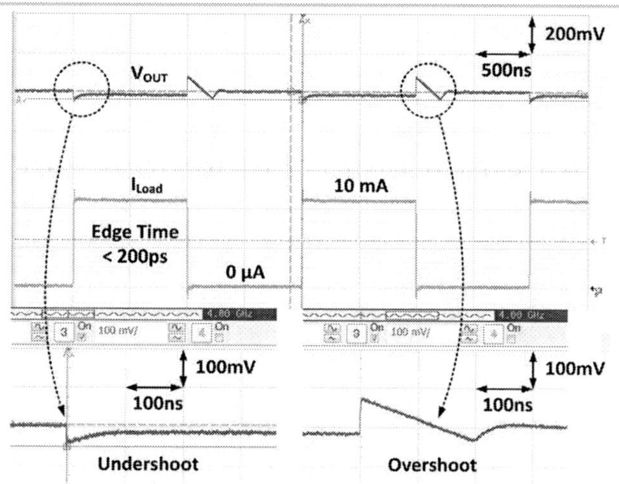

Figure 17.11.4: Measured transient response with an on-chip load current change from 0µA to 10mA within 200ps.

Figure 17.11.5: Measured PSR of the proposed LDO up to 20GHz.

Publication	[3] JSSC 05'	[5] JSSC 10'	[6] JSSC 12'	This Work
C_L	On-Chip			
Technology	90nm	90nm	45nm SOI	65nm
V_{OUT}	0.9V	0.5 to 1V	0.9 to 1.1V	1V
Dropout	300mV	200mV	85mV	150mV
I_Q	6mA	8µA	12mA	50µA
I_{MAX}	100mA	100mA	42mA	10mA
Total Cap.	600pF	50pF	1.46nF	140pF
PSR	N/A	0dB @1MHz	N/A	-15.5dB @1GHz
$\Delta V_{OUT}@T_{Edge}$	90mV @100ps	114mV @100ns	N/A	43mV @200ps
DC Load Reg.	90mV	10mV	3.5mV	11mV
T_R	0.54ns	N/A	0.288ns*	0.65ns
FOM	32ps	N/A	62.4ps*	3.01ps

*Simulated results.

Figure 17.11.6: Comparison of state-of-the-art LDOs.

17

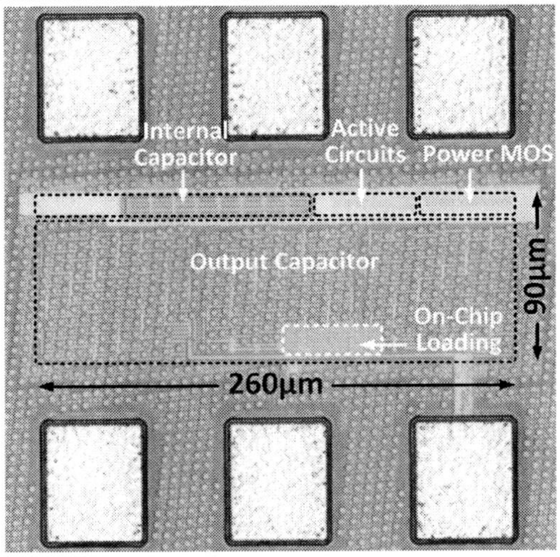

Figure 17.11.7: Chip micrograph of the proposed LDO.

Session 18 Overview: *Biomedical Systems for Improved Quality of Life*

TECHNOLOGY DIRECTIONS SUBCOMMITTEE

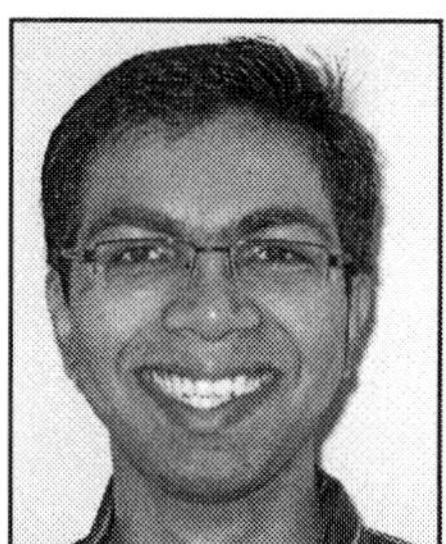

Session Chair: *Yogesh Ramadass*
Texas Instruments, Dallas, TX

Session Co-Chair: *David Ruffieux*
CSEM, Neufchatel, Switzerland

Advanced biomedical systems-on-chip combined with miniaturized sensors and actuators offer new developments in the field of mobile, low-cost healthcare applications. Implants and lightweight systems that bring a high level of comfort to the patient truly enable early diagnosis, personal point-of-care and therapy outside of a traditional clinical setting. Overall, this brings a higher quality-of-life to the patient while reducing costs.

This session presents recent advances in biomedical devices for point-of-care diagnosis and therapy. It includes SoCs for assisted hearing, SoCs for connected personal health applications and analysis of physiological signals such as mental health monitoring and early breast cancer detection.

18.1 A 1V 3mA 2.4GHz Wireless Digital Audio Communication SoC for Hearing-Aid Applications in 0.18µm CMOS **1:30 PM**

A. El-Hoiydi, Phonak Communications, Murten, Switzerland

In Paper 18.1 from Phonak Communications, a 1V 2.4GHz wireless digital audio communication SoC for hearing-aid applications is described. With the help of an embedded DC-DC converter using delta-sigma control, the SoC is able to reject noise in the audible frequency band. The SoC consumes 3mA while receiving a 7kHz audio signal in good link situations.

18.2 A Fully-Implantable Cochlear Implant SoC with Piezoelectric Middle-Ear Sensor and Energy-Efficient Stimulation in 0.18µm HVCMOS **2:00 PM**

M. Yip, Massachusetts Institute of Technology, Cambridge, MA

Paper 18.2 from the Massachusetts Institute of Technology presents a fully-implantable cochlear implant in a 0.18um HVCMOS process that uses a piezoelectric middle ear sensor interface. The SoC consumes 572uW including the stimulator, exploiting a 0.6V reconfigurable sound processor with adaptive channels and a neural stimulator providing energy-optimal waveforms.

978-1-4799-0917-9/14 $31.00 © 2014 IEEE

18.3 A Multi-Parameter Signal-Acquisition SoC for Connected Personal Health Applications **2:30 PM**

M. Konijnenburg, Holst Centre / imec, Eindhoven, The Netherlands

In Paper 18.3 from imec, a multi-parameter biomedical SoC with integrated AFE and DSP is presented. The AFE has 3 biopotential readout channels consuming 31uW each and 1 bioimpedance readout consuming 58uW. The DSP includes real-time motion artifact reduction with a general-purpose microcontroller and a dedicated hardware accelerator for improved energy efficiency.

18.4 A 4.9mΩ-Sensitivity Mobile Electrical Impedance Tomography IC for Early Breast-Cancer Detection System **3:15 PM**

S. Hong, KAIST, Daejeon, Korea

In Paper 18.4 from KAIST, a mobile electrical impedance tomography IC consuming 53.4mW for a personal early breast cancer detection system is presented. The system incorporates 92 electrodes for current stimulation and voltage sensing. The differential sinusoidal current stimulator injects up to 400μA of current, and voltage-sensing amplifier can measure breast tissue impedance with a maximum gain of 60dB thereby achieving a sensitivity of 4.9mΩ.

18.5 A 2.14mW EEG Neuro-Feedback Processor with Transcranial Electrical Stimulation for Mental-Health Management **3:45 PM**

T. Roh, KAIST, Daejeon, Korea

In paper 18.5 from KAIST, a mental health management system relying on EEG signal acquisition, data processing and classification is presented. The system implements closed-loop operation via neuro-feedback stimulation. The IC, integrated in a 0.13mm CMOS process, measures 5mm x 2.35mm and consumes 2mW.

18

18.6 2.5D Heterogeneously Integrated Bio-Sensing Microsystem for Multi-Channel Neural-Sensing Applications **4:15 PM**

P-T. Huang, National Chiao Tung University, Hsinchu, Taiwan

In paper 18.6 from National Chiao Tung University, a 670μW, 2.5D, 10mm x 5mm 16-channel hybrid microsystem is designed for neural sensing applications. It consists of a high-density micro-probe array that is connected via a silicon interposer with TSVs to a signal acquisition and processing/classification chain. The signal processing chain consists of several 180nm and 65nm dies and an MCU.

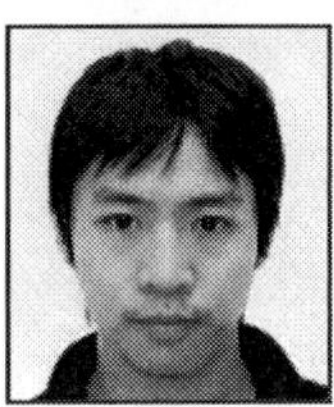

18.7 A Remotely Controlled Locomotive IC Driven by Electrolytic Bubbles and Wireless Powering **4:45 PM**

P-H. Kuo, National Taiwan University, Taipei, Taiwan

In Paper 18.7 from National Taiwan University, a batteryless remotely controlled locomotive IC that uses electrolytic bubbles as the propulsion mechanism is presented. The IC is wirelessly powered and controlled by a 10MHz ASK signal allowing movement in four orthogonal directions with two speed controls. The IC reaches a speed of 0.3mm/s. The circuit and electrolysis actuator consume 125.4uW and 82uW, respectively.

18.1 A 1V 3mA 2.4GHz Wireless Digital Audio Communication SoC for Hearing-Aid Applications in 0.18µm CMOS

Amre El-Hoiydi[1], François Callias[1], Yves Oesch[1], Christoph Kuratli[2], Robert Kvacek[3]

[1]Phonak Communications, Murten, Switzerland,
[2]EM Microelectronic, Marin, Switzerland,
[3]ASICentrum, Prague, Czech Republic

Hearing aids are complex integrated and highly miniaturised systems based on digital signal processor and wireless communication chips. There is a growing need for wireless connectivity between hearing aids and external devices such as cellular phone, PC, sound players (TV, radio, car navigation), remote microphones or even alarm systems. Assistive listening devices based on miniaturized FM receivers [1] operating around 200MHz have been in use for more than 15 years by hearing impaired people. A wireless microphone worn by the speaker (e.g. the teacher in a classroom) captures the speaker's voice clearly and transmits it directly into the hearing aids. Assistive listening devices are essential for increasing speech understanding in noisy situations such as a restaurant, a classroom or a professional meeting. The evolution of FM systems for hearing impaired people has reached its limits and there is a need for a better system, eliminating frequency management by the user, providing worldwide operation, privacy of communication and better audio quality. This paper describes a digital wireless radio SoC for hearing aid applications, operating in the worldwide available 2.4GHz ISM band.

Hearing aid systems are supplied with a Zinc-Air battery (operating voltage 1V - 1.4V) and consume between 0.5mA and a few milliamperes. Minimizing the average current consumption added by the wireless interface is hence crucial in order to keep battery lifetime within acceptable limits. Other key requirements for the wireless interface are a high signal-to-noise ratio, low distortion in a 7kHz audio bandwidth, and an overall wireless transmission delay of less than 20ms (from audio input at transmitter to audio output at receiver). The latter one in particular is important to guarantee lip-synchronization.

Wireless audio receivers are either integrated in a hearing aid or available as modules plugged into hearing aids of any brand via the 3-pin Europlug industry standard. This analog interface consists of one pin for the supply to the module, one pin for the audio low-level signal from the module and a shared ground. The fact that the ground pin is shared between supply and audio signal implies that any variation to the consumed current will translate into a disturbing noise voltage across the parasitic contact resistance of the ground pin, which will add to the audio signal voltage and will be rendered by the hearing aid (see Fig. 18.1.1).

As can be seen in this figure, such modules must be very small – and consequently also the die size. The chip must not only fit into the module, it should also be small enough to go into a hearing aid directly.

In order to satisfy the low power, low delay and broadcast requirements in the interference-prone 2.4GHz band, a packet radio audio transmission protocol has been designed based on an adaptive frequency hopping scheme, where each compressed audio frame is transmitted three times at different frequencies. A receiver needs only to listen for re-transmissions if the first transmission was not received successfully (see Fig. 18.1.2). This scheme minimizes the current consumption at the receiver.

The wireless audio SoC is based on a DSP running at 26, 13 or 6.5MHz executing both protocol and signal processing tasks. A hardware scheduler is responsible for executing periodic real-time tasks related to radio and audio path operations. Hardware accelerators are available for CRC, encryption, audio compression and decompression, for packet loss concealment and sample rate conversion. The system has a PWM (Pulse Width Modulation) audio amplifier for driving an earphone, and an analog audio output for driving the external audio input (AI) of a hearing aid; this output has programmable output level and impedance. A 26MHz clock reference is generated using a miniaturized standard quartz. Clock gating is used whenever possible for power saving. The chip block diagram is shown in Fig. 18.1.3.

In the chosen 0.18µm CMOS technology, the designed radio needs a higher voltage (VCC=2V) than provided by the hearing aid battery. A DC/DC converter using an external coil increases the supply voltage. A similar problematic is described in [2]. The current peaks of the radio and of the whole logic are delivered by an external 220µF capacitor. The radio operates at a frame rate of 4ms, and thus produces current ripple on its supply at a fundamental frequency of 250Hz with a wide range of harmonics, potentially producing the aforementioned audible noise. Figure 18.1.4 shows the current consumption bursts at the radio voltage VCC and at the digital voltage VDD during protocol operation. In order to avoid the current ripple of the radio and logic to modulate the battery current of the hearing aid, the DC/DC converter presents a constant current sink characteristic to the battery. This is achieved with a SIDO (Single Inductor, Dual Output) DC/DC architecture like shown in Fig. 18.1.5, using a digital delta-sigma modulator.

The DC/DC converter [3] is based on a bridge structure with two outputs (VCC and VDD). In a first step it pumps energy in non-continuous mode to a big tank capacitor on VCC, to supply the radio. VDD is then generated from VCC in a second step, to feed the overall logic part. While such scheme may be considered as a non-efficient power conversion, it has the advantage of complete isolation from the battery - only the pumping pattern for the VCC domain can be "seen" by the battery. The DC/DC control is performed by a slowly varying regulator, which sets the pulse density of the DC/DC through a digital delta-sigma modulator. Long regulation time constants are needed for rendering the changes inaudible, while keeping VCC between a minimum and a maximum voltage. To generate VCC, the DC/DC converter operates on 3 phases; in a first phase the coil is charged with current from the battery - the other connection of the coil being tied to ground; in a second phase the coil sends its current to VCC with its second end connected to the ground; the third phase is a reset phase: the remaining energy in the coil and in associated parasitic capacitances is damped with a parallel resistance. VDD is generated in an interleaved manner with VCC. At the end of each pump cycle from VBAT to VCC there is the possibility to pump (or not) from VCC to VDD, depending of the actual VDD voltage. If such a pumping pulse is required the coil is first connected between VCC and VDD to launch current into it, then it gets connected between ground and VDD to deliver the remaining current to VDD. The regulator controls the DC/DC pumping activity for VCC through the digital delta-sigma modulator, which has the consequence that the density of the pumping pulses is noise-shaped. The delta-sigma quantization noise is pushed above the audio frequency range, and the remaining noise associated with the regulation is low enough to be non-audible as well as non-visible in the spectrum of Fig. 18.1.6. The overall noise level produced at audio frequencies is then below the hearing aid microphone noise, even with a parasitic serial resistance in the ground of up to 2Ω. This dedicated DC/DC converter architecture is less efficient than usual solutions and has only 75% efficiency. The low efficiency results from operating in non-continuous mode, which is the only way to precisely control and shape the battery current, and also from the limited quality factor of the small SMD inductor.

In summary, a low voltage (1-to-1.4V) and low power 2.4GHz wireless digital audio communication SoC (Fig. 18.1.7) was developed for wireless audio receiver or transmitter applications. Mono, stereo or duplex audio exchange is supported with an audio bandwidth of either 7kHz or 14kHz. Although the current consumption of the radio and digital processing is bursty, the power management system rejects its noise outside the audible frequency band using a dual output DC/DC converter with digital delta-sigma control. A 7kHz audio signal is received with an average current consumption of 3mA in good link situations and a few hundred µA more when repetitions need to be received in bad link situations.

Although this chip has been mainly designed to satisfy the stringent requirements of audio reception in hearing aids, it can also be exploited for many other power and size constrained audio communication applications. The more energy demanding optional modes of this chip (audio transmission, 14kHz audio bandwidth, stereo) can be exploited in products having for example rechargeable batteries.

Acknowledgements:
We would like to acknowledge the contributions of all team members at Phonak and EM Microelectronic – Marin SA as well as from external contractors.

References:

[1] Nico Boom, et al., "A 0.9V 2.2mA Multi-Channel Programmable FM Receiver for Hearing-Aid Applications in 0.25µm CMOS", *ISSCC Dig. Tech. Papers,* pp. 436-537, Feb. 2004.
[2] P. Dal Fabbro, et al., "A 0.8V 2.4GHz 1Mb/s GFSK RF Transceiver with On-Chip DC-DC Converter in a Standard 0.18µm CMOS Technology", *Proc. ESSCIRC 2010,* pp. 458-461, Sept. 2010.
[3] François Callias, et al., "Hearing Assistance System and Method", *Patent Application WO2011131241.*

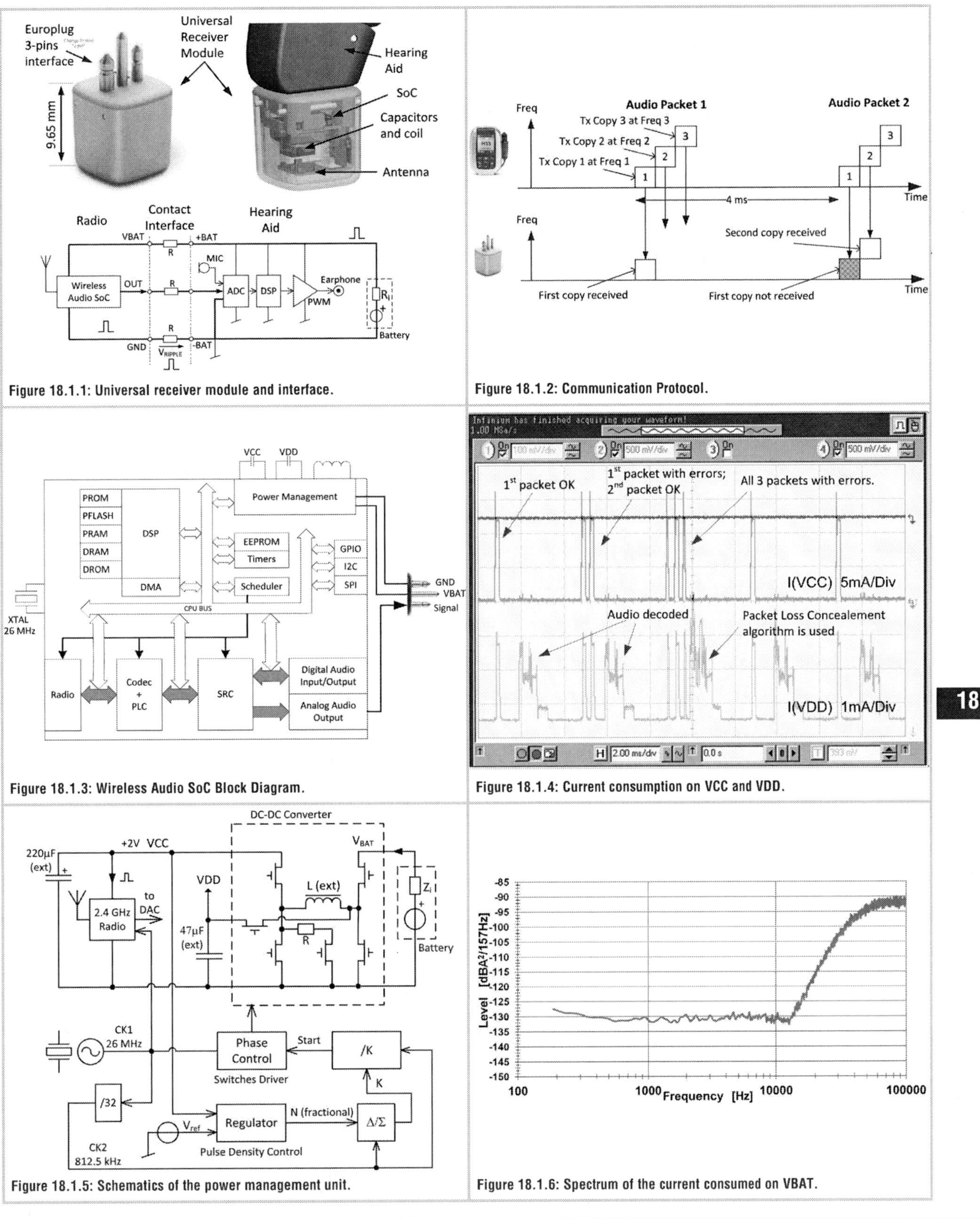

Figure 18.1.1: Universal receiver module and interface.

Figure 18.1.2: Communication Protocol.

Figure 18.1.3: Wireless Audio SoC Block Diagram.

Figure 18.1.4: Current consumption on VCC and VDD.

Figure 18.1.5: Schematics of the power management unit.

Figure 18.1.6: Spectrum of the current consumed on VBAT.

18

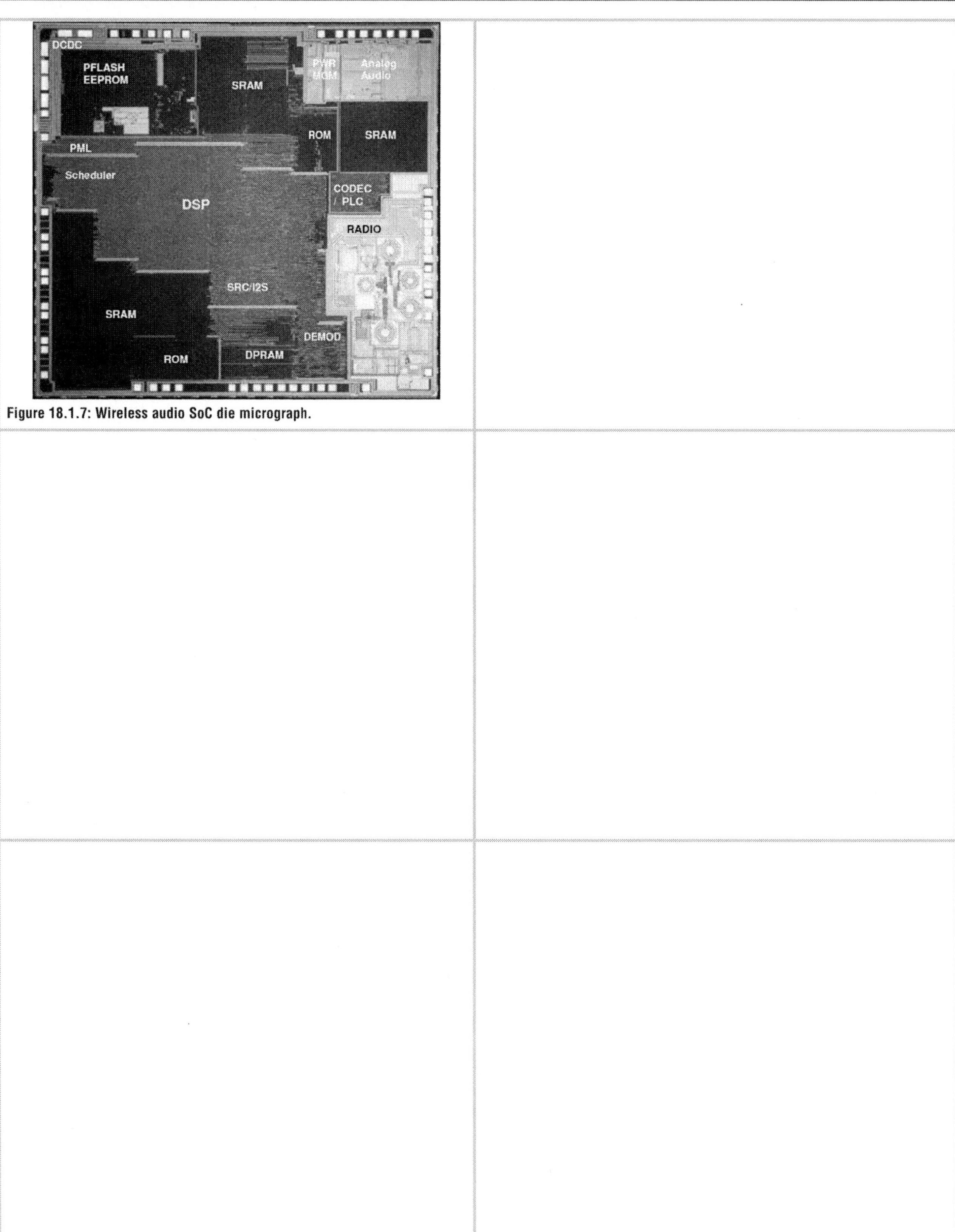

Figure 18.1.7: Wireless audio SoC die micrograph.

18.2 A Fully-Implantable Cochlear Implant SoC with Piezoelectric Middle-Ear Sensor and Energy-Efficient Stimulation in 0.18μm HVCMOS

Marcus Yip[1], Rui Jin[1], Hideko Heidi Nakajima[2,3], Konstantina M. Stankovic[2,3], Anantha P. Chandrakasan[1]

[1]Massachusetts Institute of Technology, Cambridge, MA,
[2]Harvard Medical School, Boston, MA,
[3]Massachusetts Eye and Ear Infirmary, Boston, MA

A cochlear implant (CI) is a device that electrically stimulates the auditory nerve to restore hearing in people with profound hearing loss. Conventional CIs rely on an external unit comprising a microphone and sound processor to pick up and encode sound. The external unit raises concerns with social stigma and limits usage in the shower or during water sports, motivating the need for a fully-implantable (i.e., *invisible*) cochlear implant (FICI). The limited energy storage capacity of the implanted system requires low-power (<1mW total power) sound processing and auditory nerve stimulation to enable operation from an implanted battery that is wirelessly recharged only once daily. Recent state-of-the-art ICs are typically designed for external microphone-based CIs and do not require the neural stimulator to be on the same chip [1]. Prior implantable acoustic sensors such as accelerometers sense the sound-induced vibration of the middle ear, but this approach has limited sensitivity and requires several mW of power for the sensor itself [2].

To address the above issues, this paper presents a system-on-chip (SoC) for a FICI that advances the state-of-the-art in several ways. First, low-power implantable acoustic sensing is achieved by interfacing the SoC to a piezoelectric sensor that is mounted at the umbo of the malleus within the middle ear, and this is demonstrated with measurements from human cadaveric temporal bones. Second, a highly-reconfigurable digital sound processor enables system power scalability by scaling the number of spectral channels. Third, simulations with an auditory nerve fiber model from [3] are used to determine energy-optimal biphasic stimulation pulses [4], which are delivered to the nerve through an arbitrary waveform neural stimulator. The resulting stimulation power savings (over the conventional rectangular pulse) transfer directly to overall system power savings because stimulation power typically dominates [4]. The SoC prototyped in 0.18μm HVCMOS integrates implantable acoustic sensing, sound processing, and neural stimulation on one chip to minimize the implant size and demonstrate proof-of-concept for a FICI.

Fig. 18.2.1 shows a block diagram of the FICI SoC. A piezoelectric sensor front-end (PZFE) conditions the signal from the sensor, which is a measure of the sound-induced motion of the umbo. The signal is digitized by a low-power 9b 16 kS/s SAR ADC, and a reconfigurable sound processor implements the well-known Continuous Interleaved Sampling (CIS) sound processing strategy. It has been shown that the speech recognition scores of CI users improve with the number of electrodes but plateaus after 7 or 8 [5]. Therefore, in this work, the number of channels can be reconfigured between 8, 6, or 4 to enable a power-performance tradeoff, and all processor parameters are programmable to enable a patient-specific fit. Finally, a neural stimulator is implemented with a single current source that is interleaved among all electrodes at 1000 pulses/sec per electrode, and a high-voltage switch matrix selects the active electrode. Arbitrary waveforms are generated with a digital arbitrary waveform controller.

The details of the 3-stage PZFE are shown in Fig. 18.2.1. The first stage is a charge amplifier with a gain of C_P/C_{1f} and bandwidth of $1/(C_PR_{1f})$, where C_P is the sensor capacitance, and C_{1f} and R_{1f} are a programmable capacitor and resistor to accommodate a range of sensor sizes. The second stage provides additional programmable gain and filtering, and the third stage level shifts and performs single-ended to differential conversion. The PZFE is tested by mounting the sensor on a human cadaveric temporal bone as shown in Fig. 18.2.2. Sound is generated in the ear canal with a speaker, and a probe microphone, laser Doppler vibrometer, and the SoC are used to measure the ear canal pressure, umbo velocity, and piezoelectric sensor output voltage respectively to characterize the channel. The PZFE consumes only 10.3μW from 1.5V and can detect sounds from 300Hz to 6kHz over a 50dB dynamic range from 40 to 90dB SPL (Fig. 18.2.2), which is adequate for speech.

Figure 18.2.3 shows a block diagram of the reconfigurable CIS sound processor. A digital approach is chosen to provide more programmability over analog approaches, and allow voltage scaling down to 0.6V for digital power savings. The processor spectrally decomposes the signal with a log-spaced filter bank to emulate natural hearing, and the output of each channel represents its log spectral energy and is used to modulate the pulse train on the corresponding electrode. In this work, the gain, rectification type, amount of logarithmic compression, and patient threshold and most-comfortable-level are all programmable. Furthermore, since the channel bandwidths are log-spaced, a 3-stage decimation filter and multi-rate filter bank are used to minimize the number of taps while achieving the required selectivity. Finally, the number of channels can be reconfigured by clock-gating a subset of channels, which is crucial for enabling linear power reduction in the stimulator. For example, only channels A/C/E/G are active in 4-channel mode. In order to adjust the filter bandwidths for each mode, the filter bank uses 3 types of reconfigurable FIR filters: FIR Type 1, 2, and 3 (e.g., Type 3 has 3 levels of reconfigurability).

To address the need for energy-efficient stimulation, energy optimization of the stimulation waveform using nerve fiber simulations based on [4] is performed, but with a focus on CI specific parameters and without constraint on either phase of the biphasic pulse. An arbitrary waveform stimulator shown in Fig. 18.2.4 is designed to realize the desired energy-optimal waveform. The core of the stimulator is a 6b current source with high output impedance and voltage compliance using a voltage-controlled resistor (VCR) operating from 3.3V [6]. A low-power 0.6V digital controller drives a high-speed current-steering DAC to generate near-arbitrary waveforms with 8 steps/phase. The shape of the pulse can be programmed with 16 4b values (w00[3:0] to w15[3:0]) representing the weight of each step, and the phase width is determined by the period of ϕ_{HI}. The stimulation current flows from a high voltage supply (V_{MID}=5-10 V) to accommodate high electrode impedances, and switches S_A/S_{iA}, S_C/S_{iC}, S_i, and S_{IPG} of the switch matrix are used to control the anodic and cathodic phases, shorting period, and inter-phase gap respectively for each electrode E_i. Fig. 18.2.4 (bottom) shows measured current pulses on each electrode, and the energy-optimal waveform at 31.25μs/phase.

Figure 18.2.5 (left) shows the measured spectrogram of the sound processor reconfigured between 4 and 8-channel modes with an exponential chirp input. Since the channels are log-spaced, the spectrograms are linear with time as expected. Hearing with a cadaver ear is also demonstrated by using the mounted sensor and SoC to process a speech clip generated in the ear canal. Figure 18.2.5 (right) shows the input speech spectrogram and the measured chip output spectrogram. A die micrograph of the SoC is shown in Fig. 18.2.7, and the measured scalability of the stimulation power with chip settings, and measured SoC power breakdown are shown in Fig. 18.2.6. The stimulation power increases with the pulse width and number of channels, and decreases by 20-to-30% when using the optimal waveform from nerve fiber simulations compared to the rectangular waveform. Under typical conditions, the PZFE and sound processor consume 12μW, while the 8-channel stimulator consumes 560μW with typical speech input, meeting the 1mW requirement [1]. Since the stimulation power represents 98% of the system power, the power savings from the energy-optimal waveform transfer directly to the system. Overall, hearing with a cadaver ear is demonstrated with the SoC, which integrates implantable sensing, processing, and efficient stimulation to show feasibility for a FICI.

References:

[1] R. Sarpeshkar, et al., "An Analog Bionic Ear Processor with Zero-Crossing Detection", *ISSCC Dig. Tech. Papers*, Feb. 2005, pp. 78-79.
[2] D.J. Young, et al., "MEMS Capacitive Accelerometer-Based Middle Ear Microphone", *IEEE TBME*, vol. 59, no. 12, pp. 3283-3292, Dec. 2012.
[3] D.M. Whiten, "Electro-anatomical models of the cochlear implant," Ph.D. thesis, Massachusetts Institute of Technology, Cambridge, MA, Feb. 2007.
[4] A. Wongsarnpigoon, et al., "Energy-efficient waveform shapes for neural stimulation revealed with genetic algorithm", *JNE*, vol. 7, no. 4, Aug. 2010.
[5] Q.J. Fu, et al., "Effects of noise and spectral resolution on vowel and consonant recognition: Acoustic and electric hearing," *J. Acoust. Soc. Am.*, vol. 104, no. 6, pp. 3586-3596, Dec. 1998.
[6] M. Ghovanloo, et al., "A Compact Large Voltage-Compliance High Output-Impedance Programmable Current Source for Implantable Microstimulators", *IEEE TBME*, vol. 52, no. 1, pp. 97-105, Jan. 2005.

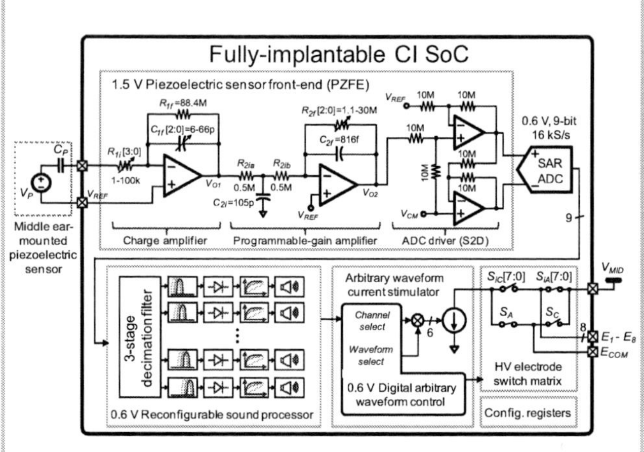

Figure 18.2.1: Block diagram of the fully-implantable cochlear implant SoC.

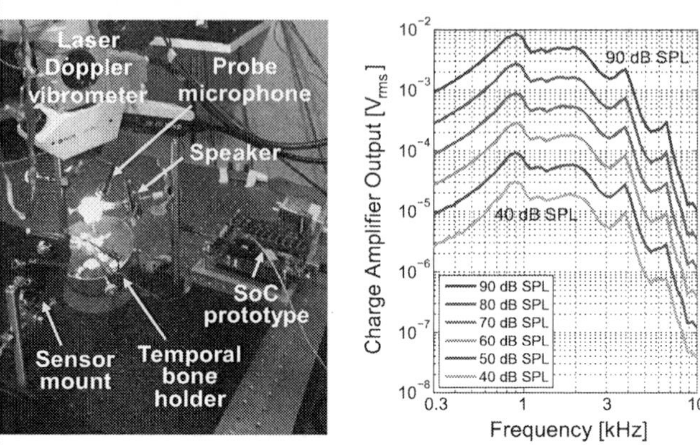

Figure 18.2.2: Test setup and characterization results from a piezoelectric sensor mounted on a human cadaveric temporal bone.

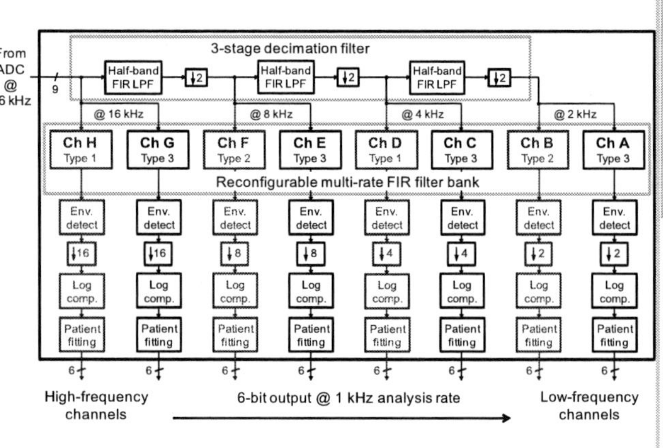

Figure 18.2.3: Block diagram of the 0.6V digital reconfigurable CIS sound processor.

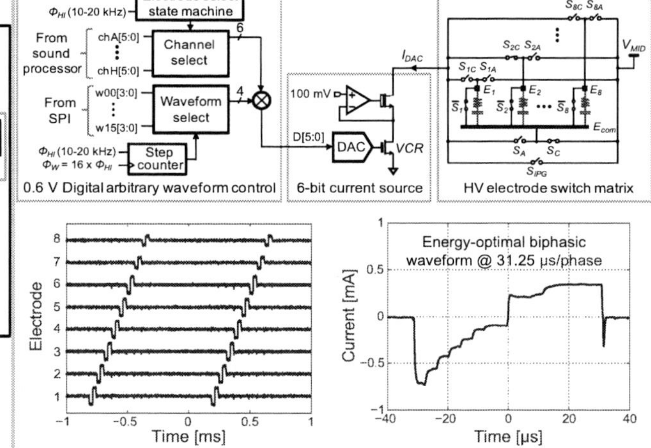

Figure 18.2.4: Arbitrary waveform stimulator and HV electrode switch matrix with measured current waveforms.

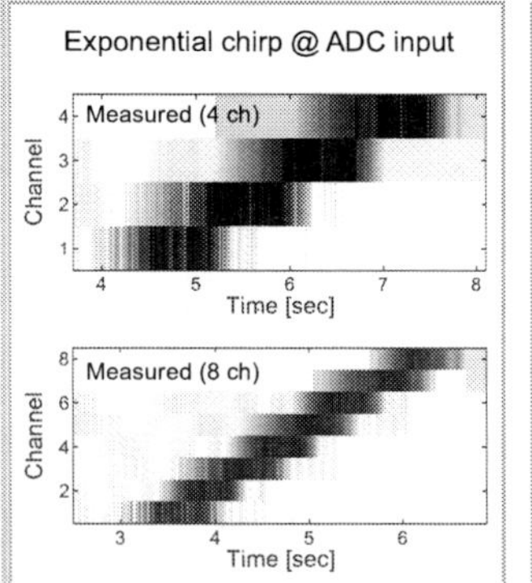

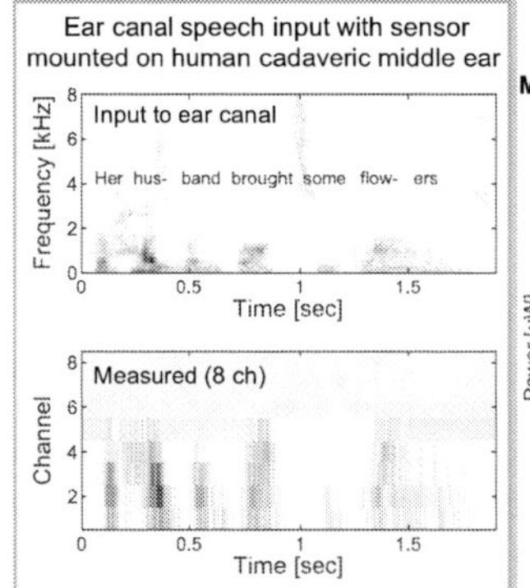

Figure 18.2.5: Measured spectrogram with a chirp input (left), and demonstration with a sensor mounted on a human cadaveric middle ear (right).

Measured neural stimulation power

Rectangular / Optimal

25 μs/phase — 50 μs/phase

Power [μW] — Channel Mode

Measured SoC power breakdown		
System component	Supply voltage [V]	Power [μW]
Piezoelectric sensor front-end	1.5	10.3
SAR ADC	0.6	0.3
Digital CIS sound processor	0.6	(8 / 6 / 4 chan) 1.6 / 1.3 / 1.1
Stimulator and switch matrix		(8 / 6 / 4 chan)
Digital waveform control	0.6	0.35 / 0.32 / 0.30
Level shifters	1.8	0.38 / 0.29 / 0.20
Current source circuits	3.3	75 / 57 / 40
Stimulator supply (V_{MID}=5-10V)	7	479 / 351 / 226
HV switch matrix control (7-12V)	9	5.4 / 4.1 / 2.7
Total (8-channel mode)		572
Total (6-channel mode)		425
Total (4-channel mode)		281

Figure 18.2.6: Measured stimulator power during typical speech, and overall SoC power breakdown.

18

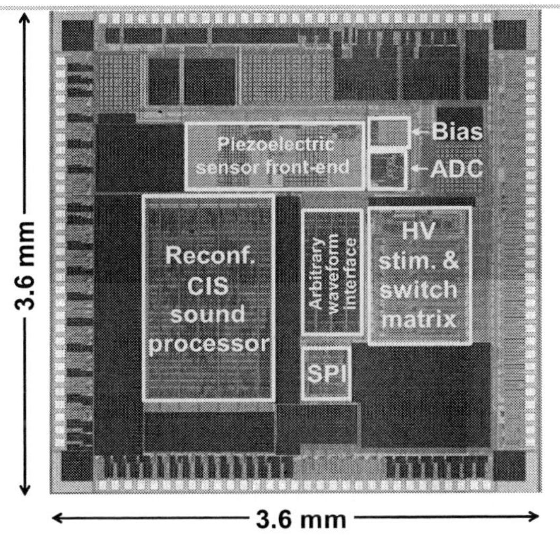

Figure 18.2.7: Die micrograph of the prototype SoC.

18.3 A Multi-Parameter Signal-Acquisition SoC for Connected Personal Health Applications

Nick Van Helleputte[1], Mario Konijnenburg[2], Hyejung Kim[1], Julia Pettine[2], Dong-Woo Jee[1], Arjan Breeschoten[2], Alonso Morgado[1], Tom Torfs[1], Harmke de Groot[2], Chris Van Hoof[1], Refet Firat Yazicioglu[1]

[1]imec, Leuven, Belgium, [2]Holst Centre / imec, Eindhoven, The Netherlands

Connected personal healthcare, or Telehealth, requires smart, miniature wearable devices that can collect and analyze physiological and environmental parameters during a user's daily routine. To truly support emerging applications (Fig. 18.3.1), a generic platform is needed that can acquire a multitude of sensor modalities and has generic energy-efficient signal processing capabilities. SoC technology gives significant advantages for miniaturization. But meeting low-power, medical grade signal quality, multi-sensor support and generic signal processing is still a challenge. For instance, [1] demonstrated a multi-sensor interface but it lacks support for efficient on-chip signal processing and doesn't have a high performance AFE. [2] showed a very low power signal processor but without support for multi-sensor interfacing. [3] presented a highly integrated SoC but lacking power efficiency. This paper demonstrates a highly integrated low-power SoC with enough flexibility to support many emerging applications. A wide range of sensor modalities are supported including 3-lead ECG and bio-impedance via high-performance and low-power AFE. The ARM Cortex™ M0 processor and matrix-multiply-accumulate accelerator can execute numerous biomedical signal processing algorithms (e.g. Independent Component Analysis (ICA), Principal Component Analysis (PCA,) CWT, feature extraction/classification, etc.) in an energy efficient way without sacrificing flexibility. The diversity in supported modalities and the generic processing capabilities, all provided in a single-chip low-power solution, make the proposed SoC a key enabler for emerging personal health applications (Fig. 18.3.1).

Figure 18.3.2 shows a block diagram of the AFE. It has 3 ECG readouts, each also extracting the electrode-tissue impedance (ETI). The latter is used for the integrated motion artifact reduction (MAR) with feedback to the analog domain to relax the dynamic range requirements of the ΣΔ-ADCs [4]. These operate on a 32kHz sampling clock (OSR=64) eliminating the need for power-consuming high-speed clock generation. A CIC filter followed by a decimator (DEC) provides the final 500S/s output. To reduce the latency in the MAR, the LMS filter operates on the CIC output. The LMS filter will reject the high frequency noise content present at the output of the CIC. The full ECG channel consumes only 31μW (50μW with ETI/MAR) while achieving high performance (605nVrms noise (0.5-to-150Hz), >110dB CMRR, >500MΩ $Z_{in@50Hz}$). The AFE also has a 77dB DR bio-impedance readout with pseudo-sine current generator (CG) for respiration monitoring [5]. A time-multiplexed 12b SAR-ADC provides the in-phase and quadrature output. The bio-impedance channel consumes 58μW with a resolution of 8.6mΩ√Hz. Finally a general-purpose analog input (8μW) supports external analog sensors.

The signal quality is mostly determined by the instrumentation amplifier (IA). Since the 3-opamp IA [6] is not suited for low power applications, Current-Feedback IAs (CFBIA) emerged as an interesting alternative, but suffer from limited capability to deal with DC-offsets [4]. This work proposes a 16μW chopper-modulated IA combining the best features of the above mentioned IA types. It consists of 2 CFBIAs, similar to [4] and a 300nW DC servo-loop, which removes up to ±400mV DC-offset (Fig. 18.3.3). The whole IA is chopper-modulated to suppress 1/f noise and mismatches between the two CFBIAs. The core CFBIAs consist of a gm-stage that converts the input to a current I_{sig} which in turn is copied to the TI output. The DC-servo consists of a similar TI stage but with high DC-gain (TI_{fb}) and capacitors C_{dm} and C_{cm}. For differential mode (DM) signals, the HPF is defined by C_{dm}, but for common mode (CM) signals, C_{dm} acts as an open (same CM voltage across both terminals) and C_{cm} defines the HPF cut-off. This results in an architecture where the CM HPF can be controlled independently from the DM one. By selecting C_{dm} large and C_{cm} as small as possible, the DM HPF can be set to 0.5Hz and CM HPF to >60Hz. Such arrangement rejects the differential polarization voltages below 0.5Hz as well as CM mains interference at the input stage before amplification. As such the IA can have a large gain since it will not saturate due to the DC-offset. This relaxes the DR and noise requirements of the ADC without requiring additional HPF and gain stages after the IA. Since no additional circuitry is added to the input lines in_p/in_n, the proposed architecture achieves very high input impedance. Finally since the DC-offset isn't cancelled after the input differential pair of the gm-stage the proposed method doesn't face the power vs. noise vs. DC-offset range trade-off from CFBIAs with DC-servos.

Various very low power signal processors have been demonstrated [2]. This is usually achieved through an optimized implementation dedicated to specific algorithms sacrificing flexibility, which limits the field of application. The proposed SoC has a Cortex™ M0 processor, HW accelerators, 128KB SRAM and a 16-channel DMA that supports the AFE and digital interfaces (Fig. 18.3.1). The ARM Cortex™ M series is popular for embedded applications (ease of programming and tool-availability). The Cortex™ M0 has low energy per clock cycle compared to other cores. However, the energy it needs to execute DSP tasks (e.g. filtering and MAR) is often worse than application specific processors because the instruction set architecture is optimized for control code. In the case of biomedical signal processing significant cycles are spent in functions as ICA and PCA (used in MAR), which are dominated by vector and matrix operations. In the proposed SoC, these are mapped to an accelerator (Fig. 18.3.4), while the main application is still mapped on the M0. The accelerator has a generic matrix and vector multiply-accumulate unit and an ICA specific calculation unit to support the most energy consuming calculations in biomedical filtering (i.e. ICA, PCA). Measurements showed that executing an ICA-based MAR on the accelerator yields 10x less energy compared to the running it on the M0 (Fig. 18.3.4).

Figure 18.3.5 shows a benchmarking of the proposed SoC as well as a detailed performance summary. Compared to other solutions the proposed SoC supports diverse sensor modalities combined with energy efficient digital processing. Thanks to the IA architecture, which provides high performance in terms of noise, input impedance and CMRR while still rejecting large DC-offsets before amplification, the dynamic range requirements of the ADC can be relaxed significantly. As a result the AFE provides high-quality biopotential readouts at up to 10x lower power for similar performing ICs [6].

Figure 18.3.6 shows the outputs of the presented SoC while ECG (170Hz BW), bio-impedance (80Hz BW) and accelerometer data are monitored which could be used for heart rate variability analysis and energy expenditure measurements. The top 2 plots show ECG and accelerometer data while the DBE was performing a CWT-based R-peak detection algorithm on the ECG data after motion artifact reduction by the LMS filter. The SoC maintains very good signal quality even during motion and R-peak detection remains reliable. The bottom 2 plots show a longer term ECG and bio-impedance recording where the respiration can be clearly identified.

Fig. 18.3.7 shows a die micrograph of the complete ASIC which measures 7mm x 7mm (TSMC180nm) and a power breakdown. The SoC consumes about 374μW when executing a multi-sensor application that acquires ECG, bio-impedance and accelerometer data while executing MAR and R-peak detection. This paper presented a SoC with high-performance AFE including 3-lead ECG and bio-impedance readouts combined with a generic low-power DBE consisting of an ARM Cortex™ M0, HW motion artifact reduction, multiply-accumulate accelerator and a variety of analog and digital interfaces supporting additional sensor modalities. By combining all these features into a single-chip low-power solution, the proposed SoC is a key enabler for numerous emerging personal health applications.

References:

[1] A. C.-W. Wong, et. al., "A 1V, micropower system-on-Chip for vital-sign monitoring in wireless body sensor networks", *ISSCC Dig. Tech. Papers*, pp. 138-139, Feb. 2008.

[2] K. H. Lee, et al., "A Low-power Processor with Configurable Embedded Machine-learning Accelerators for High-order and Adaptive Analysis of Medical-sensor Signals," *IEEE JSSC*, vol. 48, no. 7, pp. 1625-1637, July 2013.

[3] Infineon M8710, "Mobile measurement platform", Product overview, Feb. 2011.

[4] N. Van Helleputte, et. al., "A 160μA biopotential acquisition ASIC with fully integrated IA and motion-artifact suppression," *ISSCC Dig. Tech. Papers*, pp. 118-120, Feb. 2012.

[5] Sunyoung Kim, et al., "A 20μW intra-cardiac signal-processing IC with 82dB bio-impedance measurement dynamic range and analog feature extraction for ventricular fibrillation detection," *ISSCC Dig. Tech. Papers*, pp. 302-303, Feb. 2013.

[6] Texas Instruments ADS1292R, "Low-Power, 8-Channel, 24-Bit AFE for Biopotential Measurements", SBAS502B, Sept. 2012.

978-1-4799-0917-9/14 $31.00 © 2014 IEEE

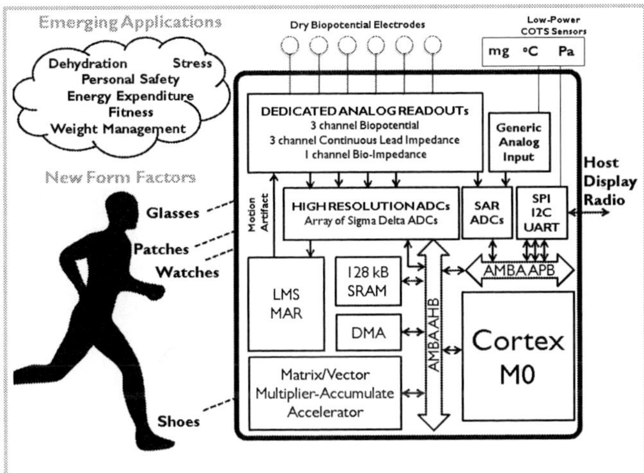

Figure 18.3.1: The SoC is a single-chip solution for emerging personal health applications. It provides signal acquisition, on-chip processing and supports COTS sensors.

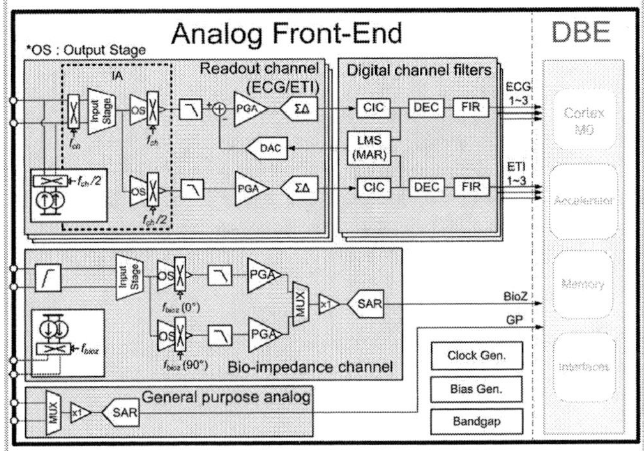

Figure 18.3.2: Analog Front-End block diagram.

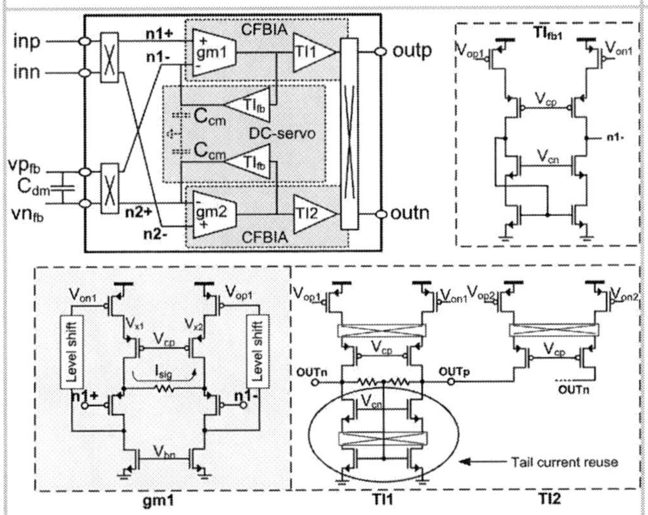

Figure 18.3.3: IA architecture and detailed schematic.

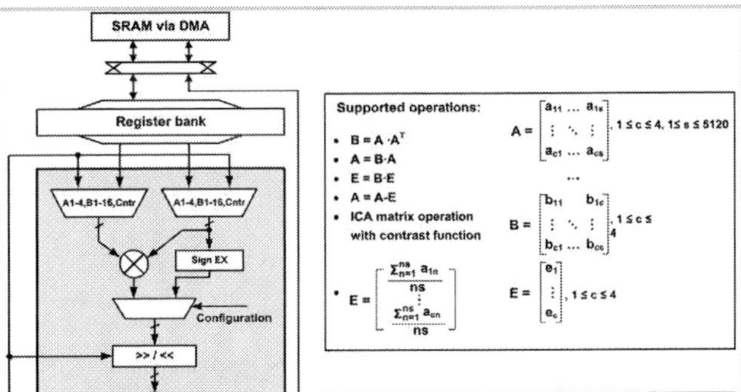

	Cortex M0	Accelerator	Saving Factor
Cycles (M cycles)	2.7	0.5	6.0
Power (uW/MHz)	362.4	188.1	1.9
Energy (uJ)	978.5	84.6	11.6

Figure 18.3.4: High level block diagram of the accelerator for vector and matrix operations, generically applicable, but also optimized for ICA and PCA operations. The accelerator saves more than a factor 10 of energy per ICA operation.

Features		This work	[1]	[3]	[5]	[6]
AFE	ECG	√	√	√	√	√
	Lead-off	√	-	-	-	√
	MAR	√	-	-	-	-
	Bio-impedance	√	-	-	√	√
	General purpose readout	√	√	√	-	-
DSP	Micro-processor	**Cortex M0**	8051	Cortex R4	-	-
	Accelerators	√	-	-	-	-
Bio-potential readout	Tech supply	**0.18 μm 1.2V**	0.13 μm 1V	n.a 3.3V	0.18 μm 1.8V	n.a. 3V
	Input Noise (0.5-150Hz)	**605nVrms**	n.a	n.a	2000nVrms	900nVrms²
	CMRR	**>110dB**	n.a	n.a	100	>105dB
	Diff. input range	**±400mV (DC) 30mVpp (AC)**	n.a	540mVpp¹	60mVpp	600mVpp²
	ADC	**13.5b ΣΔ-ADC**	10b ΣΔ-ADC	16b ΣΔ-ADC	9.3b SAR	17.14b ΣΔ-ADC²
	Power ECG/channel	**31 μW 56μW** (incl. bias)	10 μW	n.a	0.82μW	335μW
	Power ECG+ETI/ch	**50μW 75μW** (incl. bias)	n.a	n.a	n.a	n.a.
Bio-impedance Readout	Resolution	**8.6 mΩ/ √Hz @ 36 μApp, 20kHz pseudo-sine**	n.a	n.a	8.8 mΩ/ √Hz @ 40 μApp, 20kHz pseudo-sine	4.7 mΩ/ √Hz @ 30 μApp, 32kHz square
	Power	**58μW** (incl. CG, ADC)	n.a	n.a	56.2μW	335 μW (excl. CG)
Digital	Processor clock	**1-20 MHz**	32k or 16MHz	96MHz	n.a.	n.a.
	On-chip memory	**128 KB**	64KB	384 KB	n.a	n.a
	Power (sleep)	**10 μW**	5 μW	~30 μW	n.a	n.a
	Power (active)	**120 μW/MHz**	500 μW/MHz		n.a.	n.a.

¹ For gain 5, 3V supply, 2.5V reference high resolution mode and 1kS/s (for fair comparison)
² For gain 6, 3V supply, 2.4V reference high resolution mode and 500S/s (for fair comparison)

Figure 18.3.5: Performance overview of the SoC.

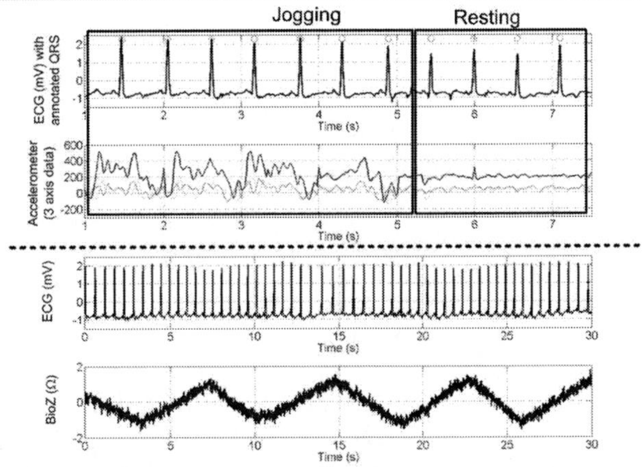

Figure 18.3.6: The top 2 curves show ECG and 3-axis accelerometer data when the subject was jogging and at rest. The dots show the detected R-peaks (CWT running on the processor). The bottom traces show a ECG and AC bio-impedance (respiration).

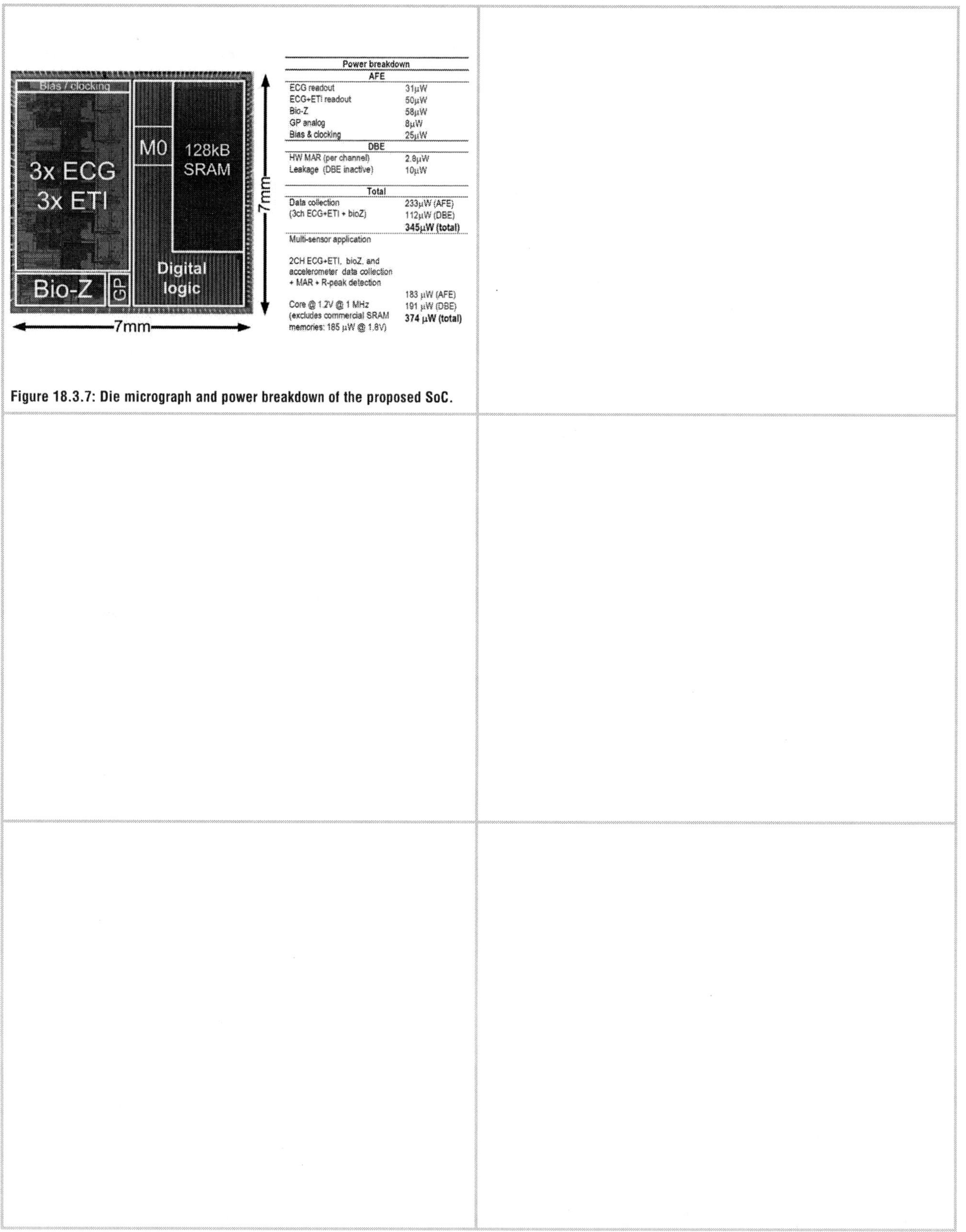

Power breakdown	
AFE	
ECG readout	31μW
ECG+ETI readout	50μW
Bio-Z	58μW
GP analog	8μW
Bias & clocking	25μW
DBE	
HW MAR (per channel)	2.8μW
Leakage (DBE inactive)	10μW
Total	
Data collection (3ch ECG+ETI + bioZ)	233μW (AFE) 112μW (DBE) **345μW (total)**
Multi-sensor application 2CH ECG+ETI, bioZ, and accelerometer data collection + MAR + R-peak detection Core @ 1.2V @ 1 MHz (excludes commercial SRAM memories: 185 μW @ 1.8V)	183 μW (AFE) 191 μW (DBE) **374 μW (total)**

Figure 18.3.7: Die micrograph and power breakdown of the proposed SoC.

18.4 A 4.9mΩ-Sensitivity Mobile Electrical Impedance Tomography IC for Early Breast-Cancer Detection System

Sunjoo Hong, Kwonjoon Lee, Unsoo Ha, Hyunki Kim, Yongsu Lee, Youchang Kim, Hoi-Jun Yoo

KAIST, Daejeon, Korea

Approximately 1 in 8 U.S. women will develop breast cancer over the course of her lifetime, and breast cancer death rates are higher than those for any other cancer, besides lung cancer. In 2013, an estimated 232,340 new cases of invasive breast cancer are expected to be diagnosed in women in the U.S. and about 39,620 women in the U.S. are expected to die from breast cancer [1]. According to the World Health Organization (WHO), if breast cancer can be detected and treated early, one-third of these cancer deaths could be prevented. For the early detection of breast cancer, X-ray mammography and ultrasonic screening are mainly used in hospitals. However, for personal cancer detection at home, currently, only unscientific palpation can be used, which is not particularly effective for early detection of tumors.

Since malignant breast tissue has about 3-times higher electrical conductance than normal fatty tissue and has different impedance characteristics as a function of frequency [2], Electrical Impedance Tomography (EIT) has been studied to measure the impedance distribution around the breast as an alternative cancer detection method. It requires more than 100 electrodes physically or electrically in contact with the breast. [3] used a large bed-like equipment setup with a breast-sized opening and a bucket of liquid electrolyte. The electrode array on the side of the bucket measures the impedance from the breast dipped into the bucket through liquid electrolyte. This equipment is too large to be portable and also the image resolution is not high due to the current diffusion in the liquid before entering into the breast. Another method is to use an array of metallic rods on the bottom of a probe plate system [4]. A skilled practitioner presses the breast with the probe plate to make the electrical contacts, but it provides only low-resolution 2D-slices of a breast image.

A compact and convenient breast cancer detection system is proposed, which can be personally used at home with scientific objectiveness and accuracy. The system illustrated in Fig. 18.4.1 can capture precise EIT images using a portable smart device while just wearing a brassiere-shaped electrode array fabricated by P-FCB [5]. The key features of this system are: 1) many electrodes are formed on the soft fabric that enables secure contact on the contoured surface of the breast, 2) all of the functional blocks of EIT system are integrated on a single IC to assemble the entire system as a simple brassiere shape, 3) the tomographic images can be reconstructed in portable smart devices so that the breast cancer can be detected personally at home. This system has a total 92 electrodes: 90 EIT electrodes, 1 voltage reference electrode and 1 current reference electrode with 8mm diameter and 2kΩ of typical contact impedance at 1kHz. All electrodes are printed on a fabric with silver paste and the circuits connecting the electrodes with the chip are printed on a separate fabric in the same way. The EIT chip is placed at the center and connected to 92 electrodes to stimulate current and measure the voltage across pairs of electrodes. It has an UART interface to transfer the measured voltage data with other related information to an external smart device through a USB port. Application software on the smart device can control the image capture and process the data to show the 3D tomographic images on its display screen.

Figure 18.4.2 shows the overall block diagram of the EIT chip architecture which is composed of: 1) switching network to select pairs of electrodes out of 92 with the help of a digital controller, 2) differential sinusoidal current stimulator (DSCS) to inject current through the selected pair of electrodes, 3) 6 channels of read-out front-end circuits each with a 14b ΔΣ ADC to measure the voltage across pairs of remaining electrodes, and 4) a digital controller with 12KB on-chip SRAM for switching control, pre-processing, data storing and data transferring through the UART interface.

The system has 3 operation modes; Gain Scanning (GS) mode, Contact Impedance Monitoring (CIM) mode and the main EIT mode as shown in Fig. 18.4.3. In GS mode, the digital controller detects the gain mismatch of each read-out channel by injecting 10μA through a 100Ω sensing resistor and storing its value to calibrate later measured data. In CIM mode, 10μA flows between 2 electrodes to measure contact impedance ignoring the tissue impedance so that digital controller can determine the stimulation current level according to the measured contact impedance values. In EIT mode, the current is injected between one EIT electrode and current reference electrode while 6 channels of voltage sensing amplifier are connected to pairs of remaining electrodes sequentially. The digital controller collects data and gain information of each channel, and it compensates for channel gain mismatch to reduce measurement variation between channels to less than 0.2mV.

Figure 18.4.4 shows the circuit schematic diagram and measured waveforms of DSCS. 2 differential OTA and R-C network are used to generate differential sinusoidal voltage, and the 2nd harmonic is suppressed by adopting the fully differential architecture [6]. To suppress 3rd harmonic further, an envelope detector is added to continuously adjust the loop gain by controlling the gate voltage of M5. By switching the positive and negative feedback loop, 4 kinds of frequencies, 0.1, 1, 10, 100kHz, are generated to make frequency-difference EIT images of the breast. The output voltage of the differential sinusoidal voltage generator (V_{out}) is converted into differential sinusoidal current (V_{out}/R_I) through V/I converter. The output current amplitude is controlled by the value of R_I according to the 2b digital code, which selects 10, 100, 200, 400μA_{pp} of current to inject into electrodes. As a result, 400μA_{pp} of current can be provided up to 2kΩ of load impedance, and the measured 4 different frequency spectra of DSCS show that all harmonics are less than -59dBc with the load impedance of 2kΩ when the current amplitude is 200μA_{pp}.

Figure 18.4.5 shows the circuit diagram of the voltage-sensing amplifier. The capacitive-coupled IA is adopted as an LNA with a gain of 18dB (C_1/C_2), and the input referred noise is 35.8nV/√Hz. The gain of the PGA is automatically controlled by the adaptive gain controller (AGC) because of the large variation of input voltage differences. The AGC generates a 3b gain control signal by comparing the differential input voltage of the PGA with 4 reference voltages so that the gain of the PGA can be set from 0 to 42dB with 6dB steps. The peak-to-peak detector (PPD) is inserted to lighten the burden imposed on the ADC and digital controller in case of fast monitoring such as contact impedance monitoring. In the PPD, the input S/H circuit creates a time delay on the input signal so that the comparator can detect the highest and lowest peak of the input. The clock frequency is chosen about 50 times faster than the input frequency, which is optimized to avoid error due to noises or spikes. Therefore, the amplitude of sine voltage is detected and converted into a DC voltage as an input to the ADC.

Figure 18.4.6 shows the measurement and imaging results of the breast model made with agar and carrot, which has impedance characteristics similar to normal breast tissue and a malignant tumor. The weighted back-projection algorithm is used for 3D image reconstruction, which assumes that the electric current is injected through one of the electrodes in a surface array and equipotential surfaces from the electrode are spherical. The reconstructed 3D image also can be visualized as several hemispheric slices. As a result, the system with the EIT IC can detect as small as 1cm of carrot in the 12cm diameter breast-shaped agar. For comparison, an x-ray mammography image of the same agar phantom is also shown.

The 2.5x5mm^2 chip is fabricated in a 0.18μm 1P6M CMOS process. Figure 18.4.7 shows the chip micrograph and its performance summary. The DSCS provides 4 kinds of frequency spectra and current amplitude, and a THD of less than 0.2% at 200μA_{pp}. The 6-channel read-out front-end has adaptively controlled gain of 18-to-60dB, and can measure impedance as small as 4.9mΩ. This compact and convenient system with fully integrated EIT IC enables early detection of breast cancer at home.

References:

[1] "http://www.breastcancer.org/symptoms/understand_bc/statistics".

[2] T. Morimoto, et al., "A Study of the Electrical Bio-impedance of Tumors," *Journal of Investigative Surgery*, vol. 6, no. 1, pp. 25-32, Jan. 1993.

[3] G. Ye, et al., "3D EIT for Breast Cancer Imaging: System, Measurements, and Reconstruction," *Microwave and Optical Technology Letters*, vol. 50, no. 12, pp. 3261-3271, Dec. 2008.

[4] D. D. Pak, et al., "Diagnosis of Breast Cancer Using Electrical Impedance Tomography," *Biomedical Engineering*, vol. 46, no. 4, pp. 154-157, Nov. 2012.

[5] J. Yoo, et al., "A 5.2mW Self-Configured Wearable Body Sensor Network Controller and a 12μW 54.9% Efficiency Wirelessly Powered Sensor for Continuous Health Monitoring System," *ISSCC Dig. Tech. Papers*, pp. 290-291, Feb. 2009.

[6] L. Yan, et al., "A 3.9mW 25-Electrode Reconfigured Thoracic Impedance/ECG SoC with Body-Channel Transponder," *ISSCC Dig. Tech. Papers*, pp. 490-491, Feb. 2010.

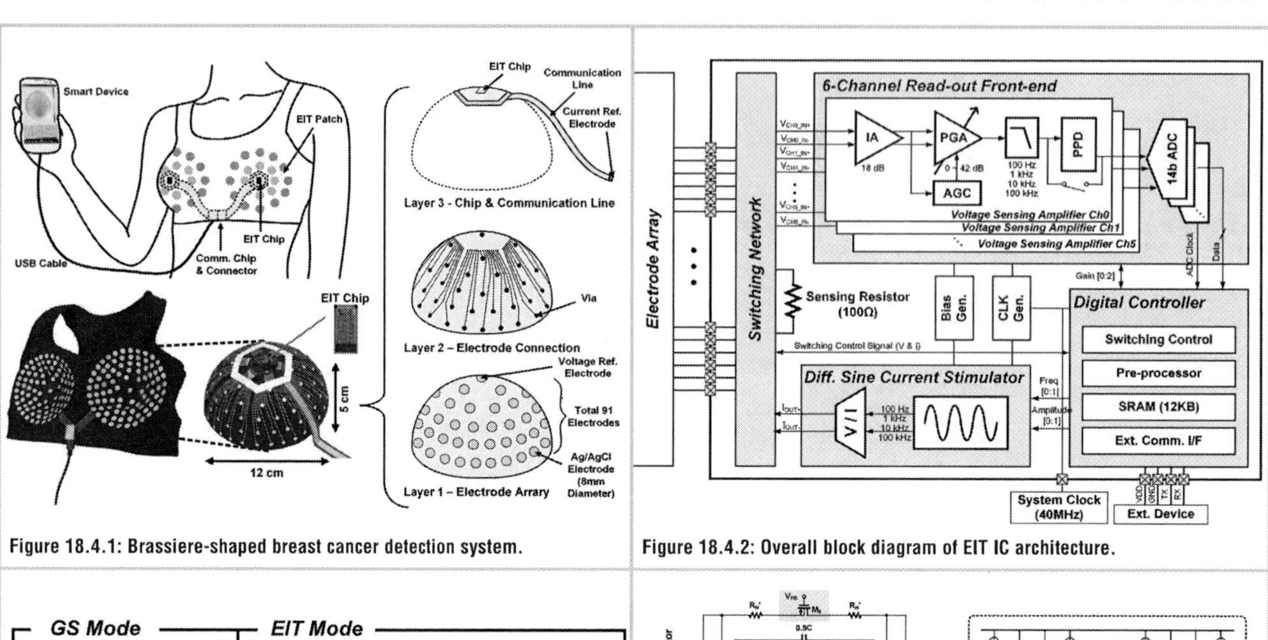

Figure 18.4.1: Brassiere-shaped breast cancer detection system.

Figure 18.4.2: Overall block diagram of EIT IC architecture.

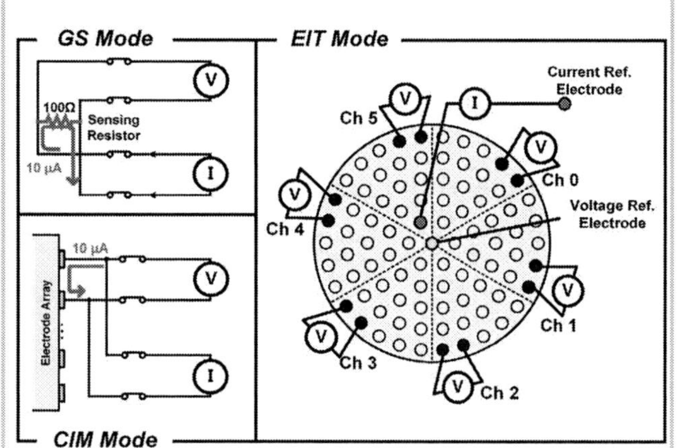

Figure 18.4.3: Reconfigurable switching for 3 operation modes.

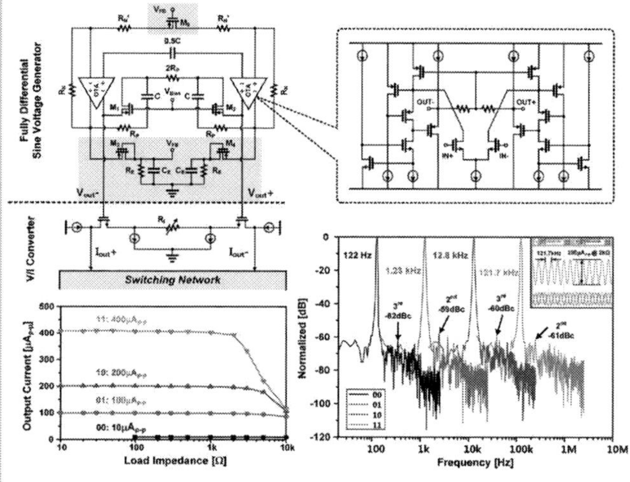

Figure 18.4.4: Differential sinusoidal current stimulator (DSCS).

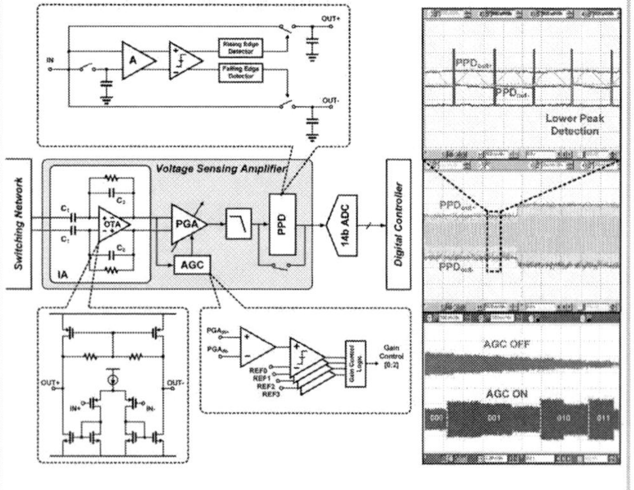

Figure 18.4.5: Voltage sensing amplifier with AGC and PPD.

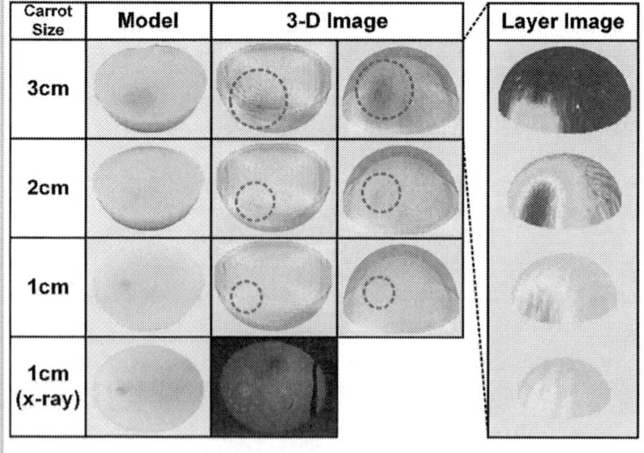

Figure 18.4.6: Breast model imaging result.

18

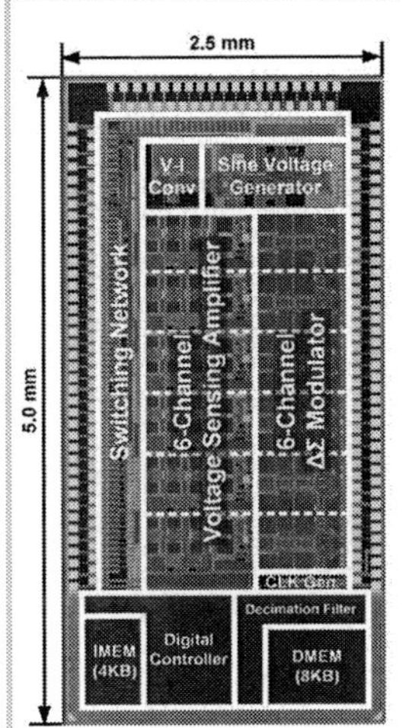

Process		0.18 μm 1P6M CMOS
Die Size		2.50 mm × 5.00 mm
Supply Voltage		1.8 V
Max. Power Consumption		53.4 mW (USB supply)
DSCS	Frequency	0.1 ~ 100 kHz (4 step)
	Amplitude	10 ~ 400 μA_{p-p}
	THD	< 0.2% @ 200 μA_{p-p}
Read-Out Front-End	# of Channels	6
	Gain	18 ~ 60 dB (6 dB step)
	Bandwidth	0.1 ~ 100 kHz
	Input Noise	36 nV/√Hz
	Sensitivity	4.9 mΩ
	ADC Clock Freq.	10 kHz ~ 40 MHz
Digital Controller	Operating Freq.	40 MHz
	On-chip SRAM	DMEM: 8 KB IMEM: 4 KB

Figure 18.4.7: Chip micrograph and performance summary.

18.5 A 2.14mW EEG Neuro-Feedback Processor with Transcranial Electrical Stimulation for Mental-Health Management

Taehwan Roh, Kiseok Song, Hyunwoo Cho, Dongjoo Shin, Unsoo Ha, Kwonjoon Lee, Hoi-Jun Yoo

KAIST, Daejeon, Korea

Recently, mental diseases have been successfully treated by neuro-feedback therapy based on Quantitative EEG (QEEG) and Event Related Potential (ERP) online data measurements. The U.S. Food and Drug Administration (FDA) approved the first EEG test for diagnosing attention deficit hyperactivity disorder (ADHD) in 2013 [1]. The EEG signals are measured by an EEG cap and analyzed by a high performance computer to extract not only the EEG power at a predetermined frequency and site combinations, but also the degree of coherence between all sites. Based on these results, brain stimulation is performed to modulate brain rhythms (EEG) toward the normal values for the therapy.

Portable and wearable monitoring and analysis devices for brain functions were reported for personal assessment of mental health and stress level at any time at any convenient place without visiting psychiatrists. An EEG headset with 1-channel electrode was well accepted in the toy and game devices but it was not suitable for medical applications due to its functional limitation [2]. More complicated wireless EEG headsets are widely used for brain research, but it requires a high-performance computer to operate the device [3]. Recently, a mental healthcare system with 4-channel EEG electrodes used a custom IC for signal processing to extract the features to detect user's mental health status [4]. However, it did not integrate the classifier used to make a decision whether the obtained data is signal or noise, which was instead done by a healthcare expert so that its practical mental healthcare applications were limited. In addition, all of previous approaches did not provide the brain stimulation tools for treatment of neuropsychiatric disorders.

In this paper, a wearable mental healthcare system is proposed for real-time regulation, which has not only mental health status monitoring function but also mental health therapeutic purposes such as neuro-feedback and non-invasive stimulations. Non-invasive stimulations such as transcranial magnetic stimulation (TMS) and transcranial electrical stimulation (tES) are used for mental health problem treatments or "brain-state alteration" because they are safer and easier than deep brain stimulation (DBS). Because of the huge size of the inductor coil, TMS is inappropriate for the mobile applications. Recent neuro-modulation research with tES shows respectable results compared to TMS as a medical therapy [5]. Headset-type tES stands out as one of the simplest in design and the most cost-effective amongst neuro-modulators, and has been increasingly investigated as a clinical tool for the treatment of neuropsychiatric disorders.

The neuro-therapeutic procedure is composed of 3 main functions: Recording, Analyzing and Stimulating. In the proposed neuro-feedback system of Fig. 18.5.1, there are a solid headgear module with a neuro-feedback stimulation (NFS) chip for Analyzing and Stimulating and a flexible active sensor module for Recording. The active sensor chip at each electrode senses and amplifies the EEG signals and then converts them into digital signals to reduce the noise and increase signal integrity. 16 electrodes are connected to the NFS chip through a serial peripheral interface (SPI) bus. The Analyzing subsystem extracts the EEG signals by independent component analysis (ICA), analyzes the features by fast Fourier transform (FFT) and diagnoses automatically by support vector machine (SVM) [6]. Finally, tES block stimulates the user based on the diagnosed results.

Figure 18.5.2 shows the NFS operating flow and chip architecture. The NFS chip consists of: 1) sensor interface for gathering EEG data from the active sensor SoCs which include programmable gain amplifiers and 10b SAR ADCs. 2) digital signal processor including RISC, ICA accelerator, FFT accelerator and SVM classifier, 3) stimulating circuits for non-intensive current stimulation for therapeutic procedure and 4) wireless communication module for data transfer between the proposed system and external devices such as active sensors, smart device and computer. In this study, only a wired interconnection is used between NFS and active sensors.

Figure 18.5.3 describes the proposed EEG analyzing datapath architecture. EEG analysis is divided into three parts: ICA accelerator for signal enhancement and feature extraction, FFT accelerator for spectrum analysis of EEG signals, and SVM accelerator for decision of QEEG characteristics. In addition to the previous forward ICA datapath [4], the feedback path from SVM is added to the step of the ICA iteration loop in the NFS SoC. From the results of the SVM, the ICA can be reconfigured in order to extract essential components to avoid unnecessary redundant analysis later. The sensor interface generates an interrupt signal when the 512-depth FIFO buffer is full, and then activates the ICA. The ICA outputs the independent component waveforms and the electrode locations with the independent component values. A 64-point FFT accelerator, which consists of a dual radix-8 architecture, transforms time-series data into the spectral domain to obtain QEEG.

Figure 18.5.4 shows the SVM architecture and its operation. There are two SVM operation modes with 64 support vectors of 16x8b stored in 4 different memory spaces. The first classification mode makes a decision whether the independent component output is a signal or noise as a binary classifier. In the other diagnosis mode, the SVM diagnoses the degree of the mental health status as a regression. The 16 dimensions of supporting vectors are composed of alpha-/beta-/theta-/delta-band 4 powers, 2 peak frequencies and electrode locations where the high independent component is observed. The SVM is initially trained with external data and it can achieve 91% classification accuracy. During analysis phase, the execution time of FFT and SVM is hidden by the ICA operation and it can reduce the overall time by 34% compared with serial operation.

Figure 18.5.5 shows the waveforms from the stimulation circuit. The amplitude, frequency and polarity of the stimulation current can be programmed with a 6b control signal by the RISC. For example, the reconfigurable anode and cathode electrodes, and various waveforms such as DC, AC and random noise signals can be defined. I_{out} can be varied from 0-to-2mA with a 32μA (I_{bias}) step. Current is multiplied from x1 to x32 by current mirror proportional to PMOS size.

Figure 18.5.6 shows the measurement results during the following experimental protocol: 10-minute analysis and 20-minute stimulation. During the NFS analysis period in an ADHD case, the delta-band power is measured larger than the alpha-band power in the frontal electrodes. Then, the SVM turns on the stimulator at the bi-frontal electrodes. After 10 sessions with 2mA of direct current tES, the peak delta-band power decreases dramatically to 1/5 and recovers to normal level, which is consistent with the other previous reports [5]. Each session is composed of 10-minute analysis and 20-minute stimulation and consumes 6.9J of energy. To reduce power consumption, DVFS is applied for 4 different modes. For the high-performance mode consuming 4.45mW, a 1V supply and a 20MHz operating frequency are used. For the low-power mode dissipating 120μW, only stimulation and sensor acquisition are activated with 700mV supply and 32kHz.

Figure 18.5.7 shows die micrograph and performance summary table. The NFS chip is fabricated in a 0.13μm CMOS process with 5×2.35mm^2 chip size. The processor dissipates 4.45mW peak power and 2.14mW average power for mental health diagnosis with a maximum stimulation current of 2mA. The sensor chip, which is fabricated in a 0.18μm CMOS process, converts EEG signals into 9.3b of digital data with 54dB-to-74dB of gain. The measured power consumption of the sensor chip is 75μW from a 1.5V supply.

References:

[1] U.S. FDA, "Evaluation of Automatic Class III Designation (De Novo) Summaries," 2013. [Online]. Available at: http://www.fda.gov/.

[2] Y. Yasui, "A Brainwave Signal Measurement and Data Processing Technique for Daily Life Applications," *Journal of Physiological Anthropology*, vol. 28, no. 3, pp. 145-150, 2009.

[3] H. Ekanayake, "P300 and Emotiv EPOC: Does Emotiv EPOC capture real EEG?," Oct. 7, 2011. [Online].
Available at: http://neurofeedback.visaduma.info/emotivresearch.htm.

[4] T. Roh, S. Hong, H. Cho and H.-J. Yoo, "A 259.6μW nonlinear HRV-EEG chaos processor with body channel communication interface for mental health monitoring," *ISSCC Dig. of Tech. Papers*, pp. 294-296, Feb., 2012.

[5] S. Zaghi, M. Acar, B. Hultgren, P. S. Boggio and F. Fregni, "Noninvasive Brain Stimulation with Low-Intensity Electrical Currents: Putative Mechanisms of Action for Direct and Alternating Current Stimulation," *The Neuroscientist*, vol. 16, no. 3, pp. 285-307, 2010.

[6] A. Mognon, J. Jovicich, L. Bruzzone and M. Buiatti, "ADJUST: An automatic EEG artifact detector based on the joint use of spatial and temporal features," *Psychophysiology*, vol. 48, no. 2, pp. 229-240, 2010.

978-1-4799-0917-9/14 $31.00 © 2014 IEEE

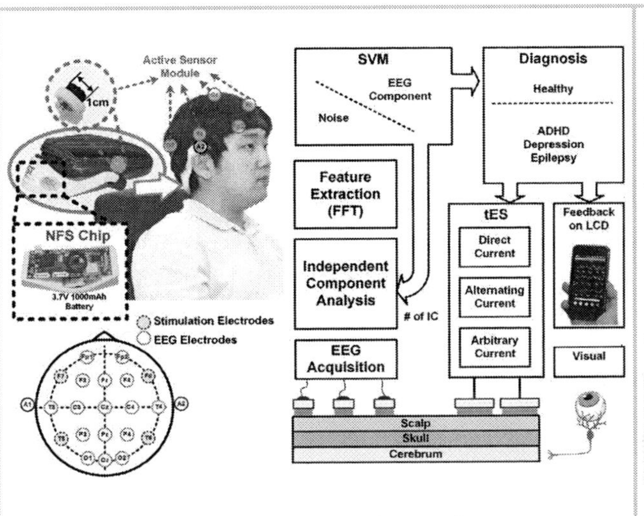

Figure 18.5.1: Mental health management system overview.

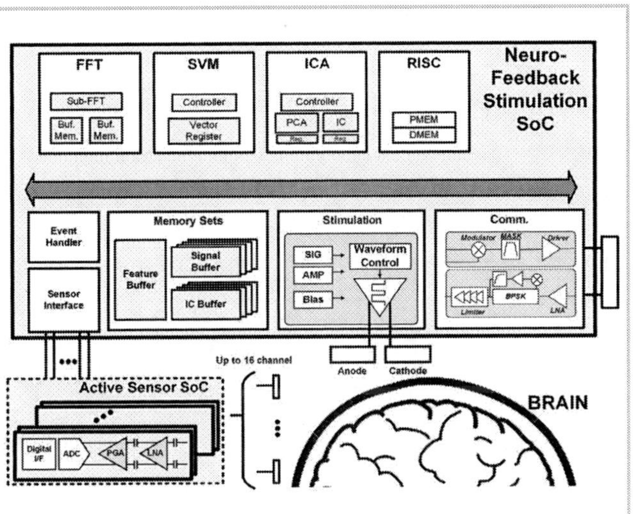

Figure 18.5.2: Overall block diagram of NFS SoC.

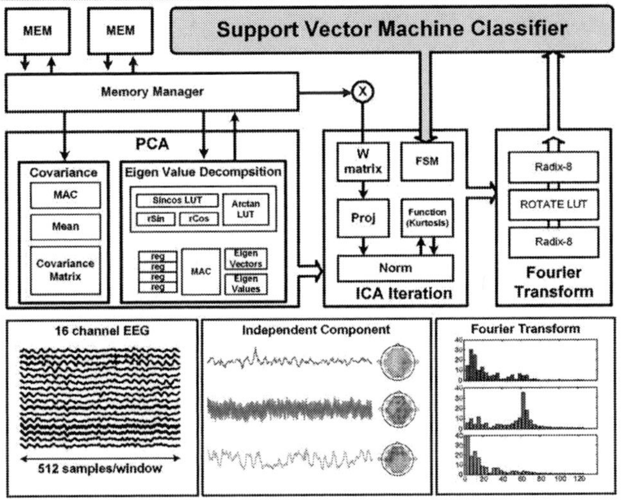

Figure 18.5.3: Operating flow of data analysis.

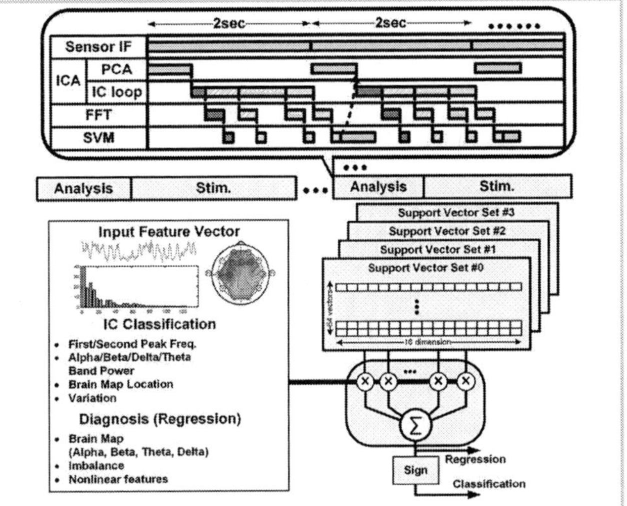

Figure 18.5.4: SVM architecture and its operation.

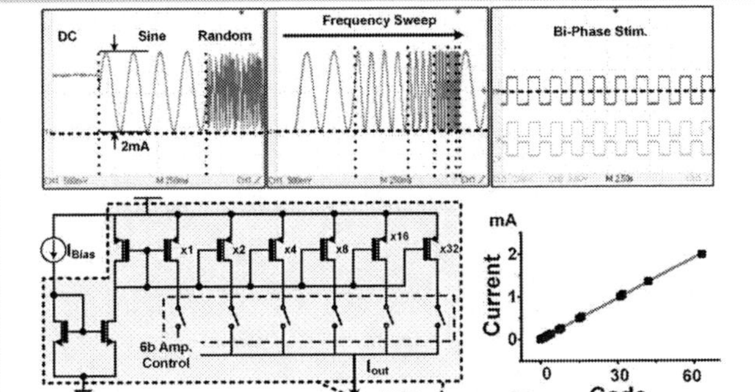

Figure 18.5.5: Stimulation circuit with DC, AC and random noise.

Mode	Supply	Freq.	Power
Mode3	1000mV	20MHz	4.45mW
Mode2	900mV	10MHz	2.14mW
Mode1	800mV	5MHz	1.28mW
Mode0	700mV	32kHz	0.12mW

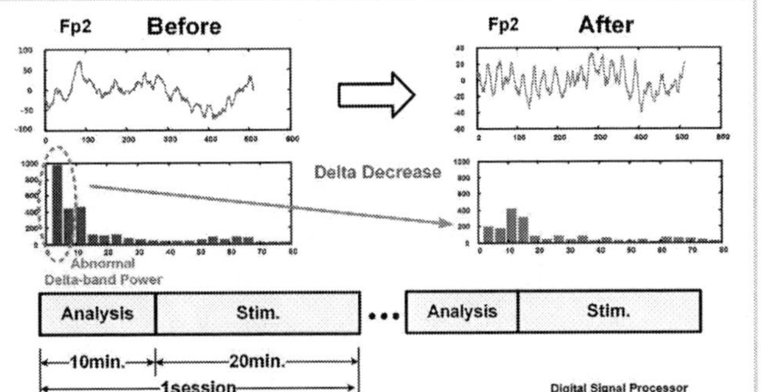

Figure 18.5.6: Measurement results during experiment protocol.

18

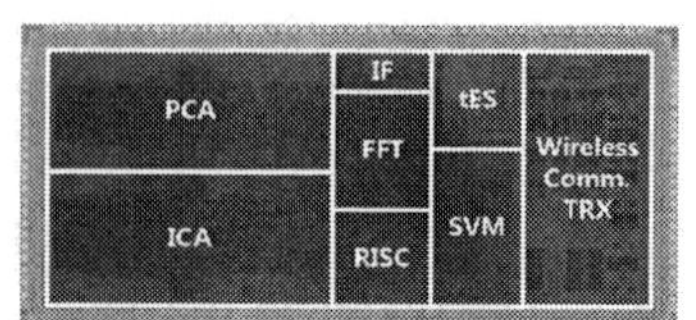

Process		0.13μm 1P8M CMOS	
Die Size [mm x mm]		2.35 x 5.0	
Supply Voltage		3.3V (IO) 0.7-1.0V (core)	
Operating Frequency		20MHz, 32kHz	
Power	Digital Proc.	Mode3	4.45mW
		Mode2	2.14mW
		Mode1	1.28mW
		Mode0	0.12mW
	Comm.	1.4mW(TX), 6mW(RX)	

tES	LSB Current	32μA
	Max. Current	2mA
ICA	Algorithm	FastICA
	# of Channel	Up to 16 channel
	Window	Max. 512 samples
FFT	Window	64points
	Structure	Radix-8
SVM	Input	16 dimension
	# of SV	64 Vectors

Process		0.18μm 1P6M CMOS
Die Size [mm x mm]		2.35 x 2.35
Supply		1.5V
Power		75μW
AFE	Gain	34dB(LNA) 20-40dB(PGA)
	Bandwidth	0.1Hz ~ 93kHz
ADC	Sampling Rate	1.82kS/s
	ENOB	9.3b

Figure 18.5.7: Chip micrograph and its performance summary.

18.6 2.5D Heterogeneously Integrated Bio-Sensing Microsystem for Multi-Channel Neural-Sensing Applications

Po-Tsang Huang[1], Lei-Chun Chou[1], Teng-Chieh Huang[1], Shang-Lin Wu[1], Tang-Shuan Wang[1], Yu-Rou Lin[1], Chuan-An Cheng[1], Wen-Wei Shen[1], Kuan-Neng Chen[1], Jin-Chern Chiou[1,2], Ching-Te Chuang[1], Wei Hwang[1], Kuo-Hua Chen[3], Chi-Tsung Chiu[3], Ming-Hsiang Cheng[3], Yueh-Lung Lin[3], Ho-Ming Tong[3]

[1]National Chiao Tung University, Hsinchu, Taiwan,
[2]China Medical University, Taichung, Taiwan,
[3]Advanced Semiconductor Engineering Group, Kaohsiung, Taiwan

Heterogeneously integrated and miniaturized neural sensing microsystems for accurately capturing and classifying signals are crucial for brain function investigation and neural prostheses realization [1]. Many neural sensing microsystems have been proposed to provide small form-factor and biocompatible properties, including stacked multichip [2, 3], microsystem with separated neural sensors [4], monolithic packaged microsystem [5] and through-silicon-via (TSV) based double-side integrated microsystem [6]. These heterogeneous biomedical devices are composed of sensors and CMOS circuits for biopotential acquisition, signal processing and transmission. However, the weak signals detected from sensors in [2-5] have to pass through a string of interconnections to the CMOS circuits by wire bonding. In view of this, TSV-based double-side integration [6] uses TSV arrays to transfer the weak signals from μ-probe arrays to CMOS circuits for reducing noises. Nevertheless, the double-side integration requires preserving large area for separate μ-probe arrays and TSV arrays, and the TSV fabrication process may induce damage on CMOS circuits.

This paper presents a 2.5D heterogeneously integrated bio-sensing microsystem with TSV-inside μ-probes for high-density multi-channel neural sensing applications. Figure 18.6.1 presents the structure of 2.5D integration with TSV-inside μ-probes and block diagrams of the bio-sensing microsystem. This microsystem is composed of of TSV-inside μ-probes, 4 dies and 1 interposer. These 4 dies are designed for biopotential acquisition, signal processing and feature extraction and classification via analog front-end (AFE) readout circuits, analog-to-digital converters (ADCs), configurable discrete wavelet transform (DWT) circuits, filters, and 1 microcontroller unit (MCU- Renesas RX210). Additionally, an on-interposer bus (μ-SPI, Serial Peripheral Interface) is designed for transferring data in the bio-sensing microsystem.

The MEMS TSV-inside μ-probes are fabricated in an 8" wafer and diced to 5x 5mm^2. At the beginning of post-process, two 8" wafers are Chemical Mechanical Polished (CMP) to 200μm thickness with fully-filled Cu plating process to fabricate Cu-TSVs with 200μm depth (height) and 30μm diameter. Next, a deep ion-coupled plasma (ICP) etching process is applied on the front side of 1 wafer to form TSV-inside μ-probes. Then the Pt is sputtered on the top of μ-probe, and a 5μm thick biocompatible parylene-C is deposited to isolate different μ-probes. The other wafer is diced by 16.5mm x 10.4mm as the interposer for the 2.5D integration. Then, the 4 dies and 5mm x 5mm TSV-inside μ-probe array are bonded on the top side and bottom side, respectively. On the top side, Redistribution Layers (RDLs) are fabricated for the connections between the 4 dies and TSVs. Moreover, the sensing channels can be defined by the routing of RDLs. Figure 18.6.2 shows the measured impedance and Scanning Electron Microphotograph (SEM) of the fabricated MEMS TSV-inside μ-probe array, including 1 cross-section view and 1 60-degree view. The measured impedance is 0.17Ω with a phase of -0.5° at 1kHz. The white vertical metal tube in the middle of the μ-probe denotes the Cu-TSV.

A 16-channel AFE with 4 ADCs chip is designed and implemented using 0.18μm CMOS process as shown in Fig. 18.6.3. The energy-efficient and low-noise 16-channel AFE circuitry comprises two differential difference amplifiers (DDAs) and DC offset rejection components. Additionally, the DDA is designed using a double input G_m-stage and a class-AB output for achieving high common-mode rejection ratio (CMRR), low-noise and energy efficiency. The measurement results show that the AFE can realize 60.3dB gain with only 20.7μW for each channel. The bandwidth of the AFE is from 2.32Hz to 6.61kHz. The noise efficiency factor (NEF) is 2.78. The area-power-efficient 11b ADCs are designed by 8b successive approximation register (SAR) ADCs with 3b delay-line enhanced tuning. To reduce the total amount of capacitance, the delay-lined enhanced tuning block is designed to detect the three most significant bits by a modified vernier structure. To relax the accuracy requirement of the delay-line enhanced tuning, the lifting-based searching algorithm and re-comparison procedure are proposed for the SAR ADC. For further achieving energy saving, split capacitor array and self-timed control are utilized in the SAR ADC. An ENOB of 10.4b at 8KS/s can be achieved with only 0.6μW power consumption and 0.032mm^2 area. The FoM of this ADC is 49.4fJ/conversion-step.

Multi-channel configurable lifting-based DWT is used to extract features of neural signals and to cluster the signals for medical diagnosis by filtering the neural signal into different frequency bands as shown in Fig. 18.6.4. For 16-channel neural sensing, four parallel 4-channel DWTs are implemented with two dies in a 65nm low-power CMOS process. Both the area and power consumption can be reduced by reducing the computation circuits using lifting-based DWT algorithm. For 1-level iteration of each channel, 10 cycles are required with two multipliers and two adders in the computation core. Additionally, both the time window and mother wavelets (Haar, Db2, Symlet4, Symlet6) can be adjusted for different neural sensing applications in this configurable lifting-based DWT. Moreover, power-gating and clock-gating techniques are utilized to further reduce the leakage currents and clock power for the energy-limited bio-systems, respectively. Clock gating is utilized during inactive iterations, and power gating is only applied during inactive sampling periods.

Figure 18.6.5 presents the on-interposer bus (μ-SPI) for providing low power data communication in the 2.5D integrated bio-sensing microsystem. The protocol of μ-SPI is designed by a hierarchical packetization technique based on the physical layer of SPI. The signal, SS (Slave Selection in SPI), is used to indicate the valid period of each packet and is controlled by the master. Additionally, the data width of μ-SPI is from 1b to 8b, and a 4b width is selected in this bio-sensing microsystem. To reduce the overhead of the header, the header of a packet is divided into two levels by the hierarchical packetization technique. The length of the 1st level header is fixed for indicating the functionality of this packet. Based on the information of the 1st level header, the 2nd level header is variable for providing wide range of the burst length, broadcasting selection and variable address. Moreover, a pseudo multi-master is proposed to replace the arbitration circuits via master passing. Only 1 master can exist by controlling MS_Flag in master/slave modules.

Figure 18.6.6 presents the specifications of the bio-sensing microsystem. The DWTs and filters are implemented with 2 ultra-low power FPGA dies fabricated in 65nm LP CMOS process for early evaluation with both clock/power gating. The overall power of this microsystem is only 676.3μW for 16-channel neural sensing. Figure 18.6.7 shows the micrographs of the microsystem in 2.5D integration.

Acknowledgements:
This work was supported in part by the National Science Council NPIE and I-RiCE Program, Taiwan, R.O.C. and "Aim for the Top University Plan" of the National Chiao Tung University and Ministry of Education, Taiwan, R.O.C. The authors would like to thank National Chip Implementation Center (CIC) for chip fabrication.

References:

[1] K. C. Smith, et al., "Through the Looking Glass: Trend Tracking for ISSCC 2012," *IEEE Solid-State Circuits Magazine*, vol. 4, no. 1, pp. 4-20, March 2012.
[2] B. Gosselin, et al., "A Mixed-Signal Multichip Neural Recording Interface With Bandwidth Reduction," *IEEE Trans. on Biomedical Circuits and Systems*, vol. 3, no. 3, pp. 129-141, June 2009.
[3] B.K. Thurgood, et al., "A Wireless Integrated Circuit for 100-Channel Charge-Balanced Neural Stimulation," *IEEE Trans. on Biomedical Circuits and Systems*, vol. 3, no. 6, pp. 405-414, Dec. 2009.
[4] A.M. Sodagar, et al., "A Wireless Implantable Microsystem for Multichannel Neural Recording," *IEEE Trans. on Microwave Theory and Techniques*, vol. 57, no. 10, pp. 2565-2573, Oct. 2009.
[5] C.-P. Cong, et al., "A wireless and batteryless 130mg 300μW 10b implantable blood-pressure-sensing microsystem for real-time genetically engineered mice monitoring," *ISSCC Dig. Tech. Papers*, pp. 428-429, Feb. 2009.
[6] C.-W. Chang, et al., "Through-Silicon-Via Based Double-Side Integrated Microsystem for Neural Sensing Applications," *ISSCC Dig. Tech. Papers*, pp.102-103, Feb., 2013.

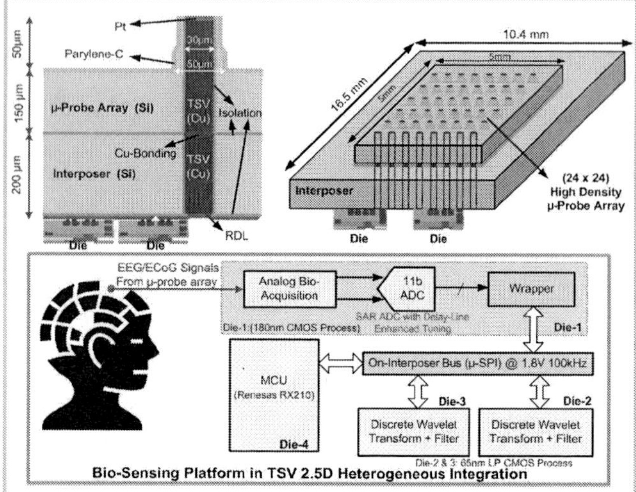

Figure 18.6.1: 2.5D heterogeneously integrated bio-sensing microsystem.

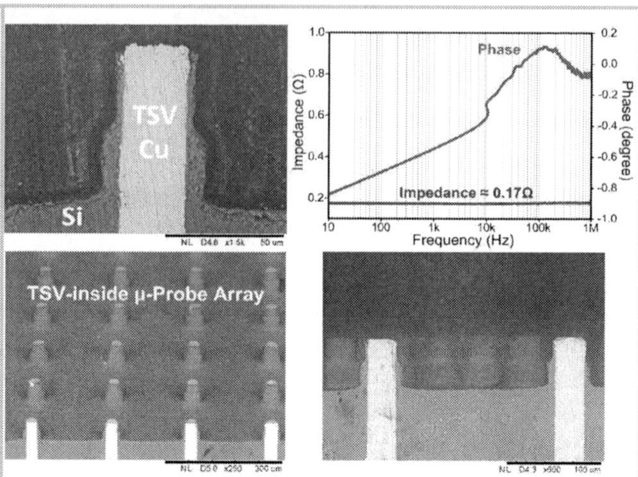

Figure 18.6.2: SEM and measured impedance of the fabricated MEMS TSV-inside μ-probes.

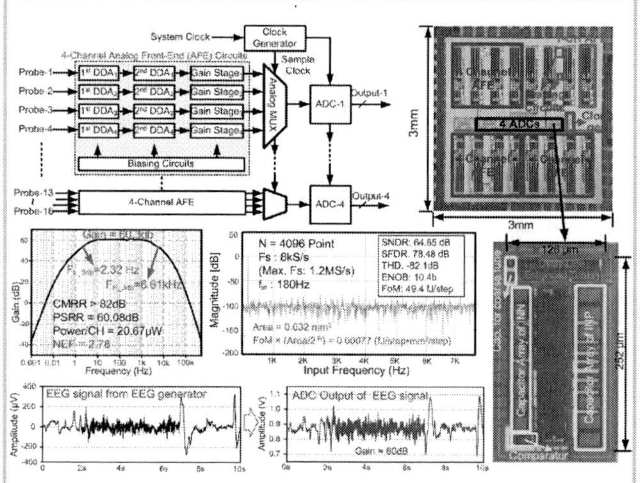

Figure 18.6.3: Low-noise 16-channel AFE + 4 area-power-efficient 11b ADCs.

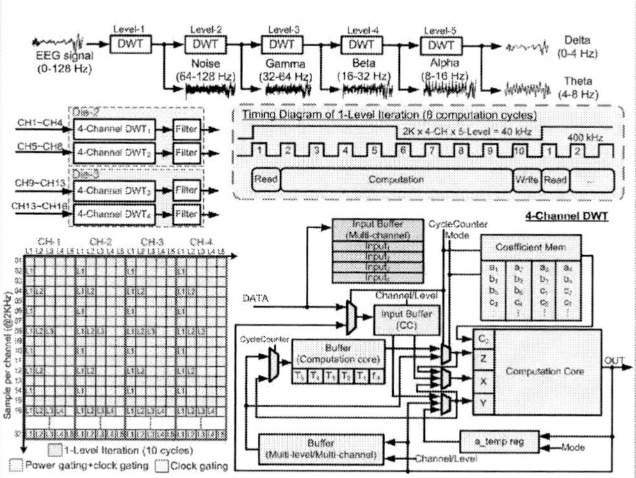

Figure 18.6.4: Multi-channel configurable lifting-based DWT.

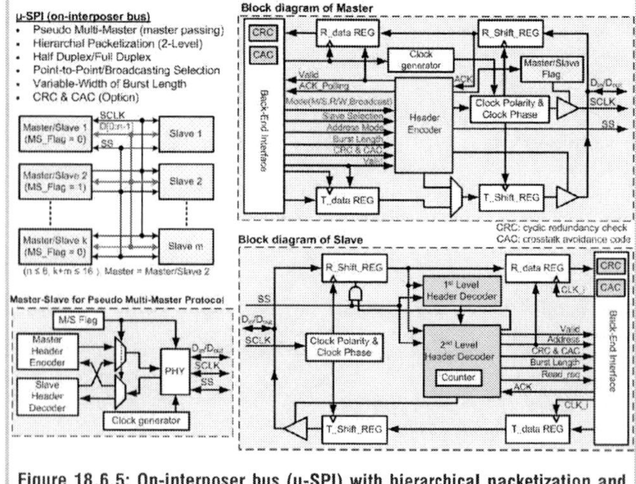

Figure 18.6.5: On-interposer bus (μ-SPI) with hierarchical packetization and pseudo multi-master.

		ISSCC'13 [6]	This Work
Integration Type		**TSV-Based Double-Side Integrated Microsystem**	**TSV-Based 2.5D Heterogeneously Integrated Microsystem**
Specification of Neural-signal Acquisition	μ-Probe Array	Diameter/Height: 25μm/150μm 5x6 Probes for 1-Channel	Diameter/Height: 30-50μm/50μm Configurable Probes for 1-Channel
	TSV	Via Last (Cu filling) Diameter/Height: 30μm /200μm 3x14 TSVs for 1-Channel	Post Process (Cu filling) Diameter/Height: 30μm/200μmx2 1 TSV *inside* 1 μ-Probe
Neural-Signal Read-out Circuits		***AFE*** (with off-chip capacitors) 0.18μm CMOS Process@1.8V Gain: 54.8dB, CMRR > 108dB Bandwidth: 0.41Hz – 7.3kHz Power/Channel: 22.0μW, NEF: 3.3	***AFE (w/o off-chip capacitors)*** 0.18μm CMOS Process@1.8V Gain: 60.3dB, CMRR > 82dB Bandwidth: 2.32Hz – 6.61kHz Power/Channel: 20.7μW, **NEF: 2.78**
		N.A.	***11b ADC*** SAR with delay-line enhanced tuning 0.18μm CMOS Process@1.8V ENOB: 10.4b, Area:0.032mm^2 FoM:49.4fJ/step, 0.6μW@2kS/s FoM×(Area/2^R):0.00077fJ·mm^2/step^2
Neural-Signal Processing Circuits		N.A.	***DWT + Filter (Die-2 & Die-3)*** 65nm CMOS Low Power Process 1.2V @ 400kHz, Power: Ultra-Low Power FPGA: 105.6x2 μW ASIC : 13.8x2 μW (simulated)
			MCU - Renesas RX210 (Die-4) Control + Feature Classification 1.62V @ 400kHz, Power: 108.3μW
On-Interposer Bus (μ-SPI)		N.A.	1.8V @ 100kHz, Power: 23.2μW

Figure 18.6.6: Specification of the 2.5D heterogeneously integrated bio-sensing microsystem.

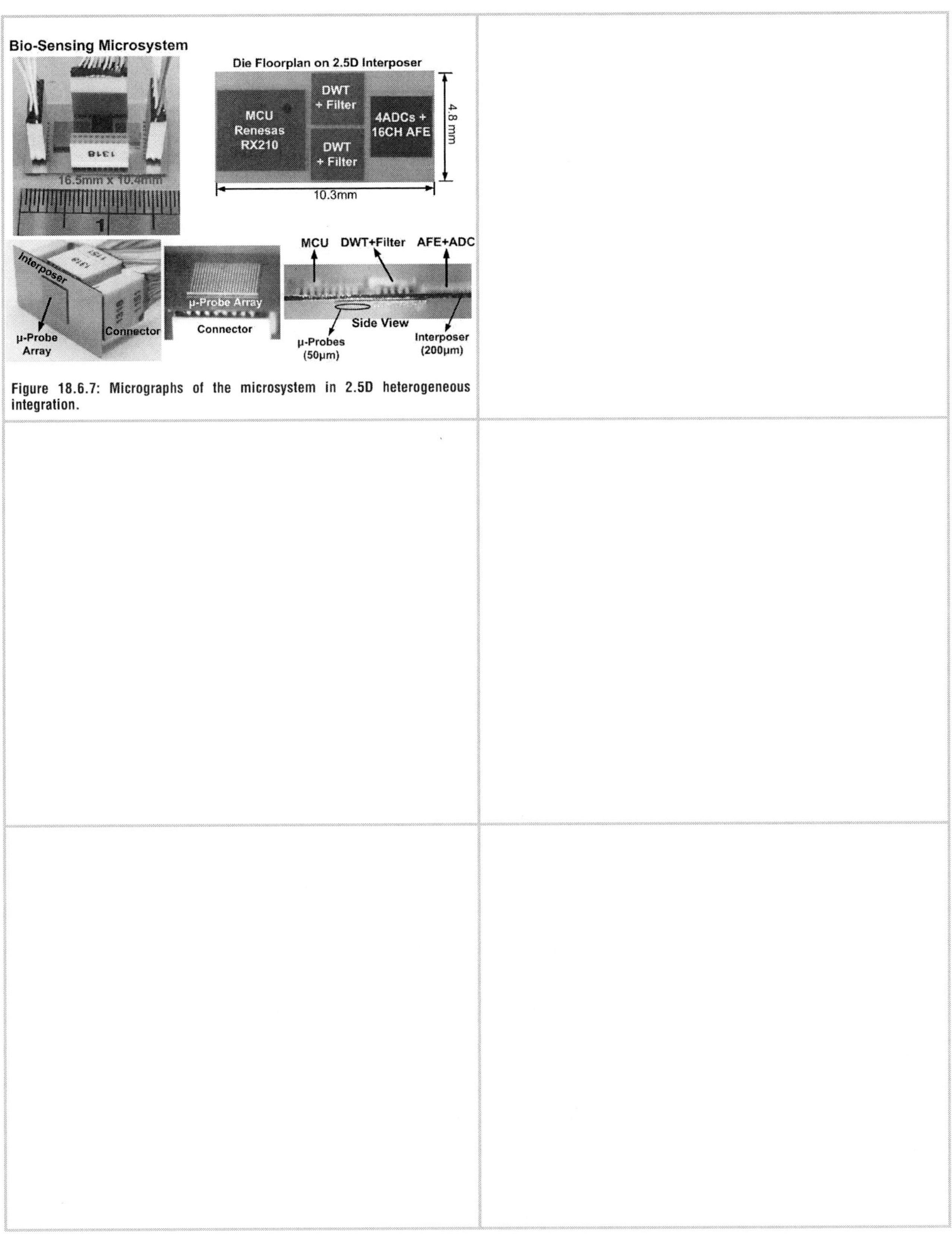

Figure 18.6.7: Micrographs of the microsystem in 2.5D heterogeneous integration.

18.7 A Remotely Controlled Locomotive IC Driven by Electrolytic Bubbles and Wireless Powering

Po-Hung Kuo[1], Jian-Yu Hsieh[1], Yi-Chun Huang[1], Yu-Jie Huang[1], Rong-Da Tsai[2], Tao Wang[3], Hung-Wei Chiu[2], Shey-Shi Lu[1]

[1]National Taiwan University, Taipei, Taiwan,
[2]National Taipei University of Technology, Taipei, Taiwan,
[3]Chang Gung University, Taoyuan, Taiwan

As implantable medical CMOS devices become a reality [1], motion control of such implantable devices has become the next challenge in the advanced integrated micro-system domain. With integrated sensors and a controllable propulsion mechanism, a micro-system will be able to perform tumor scan, drug delivery, neuron stimulation, bio-test, etc, in a revolutionary way and with minimum injury. Such devices are especially suitable for human hollow organs, such as urinary bladder and stomach. Motivated by the art reported in ISSCC 2012 [2], we demonstrate a remotely-controlled locomotive CMOS IC which is realized in TSMC 0.35μm technology. As illustrated in Fig. 18.7.1, a bare CMOS chip flipped on a liquid surface can be moved to the desired position without any wire connections. Instead of Lorentz forces [2], this chip utilizes the gas pressure resulting from electrolytic bubbles as the propulsive force. By appointing voltages to the on-chip electrolysis electrodes, one can decide the electrolysis location and thereby control the bubbles emissions as well as the direction of motion. With power management circuits, wireless receiver and micro-control unit (MCU), the received signal can be exploited as the movement control as well as wireless power. Experiments show a moving speed of 0.3mm/s of this chip. The total size is 21.2mm^2 and the power consumption of the integrated circuits and the electrolysis electrodes are 125.4μW and 82μW, respectively.

Electrolyzing water creates micro-bubbles of oxygen and hydrogen that can be adopted as a source of propulsion force. The micro-bubble can be absorbed by the human body, owing to its small volume [3]. Using electrolytic bubbles as a force in an integrated system was demonstrated in our previous work; however, to our knowledge, moving an IC by such force has not been reported in literature. The basic controls of motion include the direction and the speed. To make the direction of motion adjustable, four electrolysis electrodes are integrated on the four sides of the chip. In addition to the adjustable directions, a high bubble generation rate is important for making a strong enough propulsion force to move a chip. Theoretically, the bubble generation rate is proportional to the applied voltage and the reaction area [4]. Therefore, a high-density inter-digitated electrode pattern is adopted to enlarge the reaction area for higher bubble generation rate. Furthermore, in order to provide an adjustable speed of movement, two voltage levels (EV1 and EV2) are connected to the electrodes and selected by MCU. An experiment of bubble generation is performed and recorded in Fig. 18.7.2. As can be seen, under a fixed electrolytic voltage the volume of generated bubbles increases linearly with time, which means the bubble generation rate is nearly constant.

The circuit architecture of the IC is shown in Fig. 18.7.3. An external 10MHz signal with 1Mb/s amplitude-shift keying (ASK) modulated command is inductively coupled from a transmitter to the chip through on-chip coils. The ASK signal is received for system commands, wireless powering and clock regeneration. The received ASK signal is rectified by a bridge rectifier to direct current (DC). After rectification, the DC voltage is stabilized by a regulator creating the VDD (2V) that is supplied globally to all circuits. As a linear low drop-out topology, the regulator is composed of the start-up circuit, bandgap reference and error amplifier to stabilize the DC voltage converted from AC. Two DC voltages, EV1 (1.3V) and EV2 (2V), are provided from the regulator outputs and connected to the analog switches. As a result, two different bubble generation rates are provided to the chip as two shifts for the chip speed controls.

The wireless command is received and restored by the ASK demodulator [3]. After demodulation, a bit stream is sent to the MCU for further processing. The digital commands are in RS232 (8/N/1) format, including four bits for enabling four electrodes respectively, one bit for regulating electrolytic voltage level and three error-check bits. To improve command signal decoding correctness, two algorithms, dynamic sampling and voting sampling are adopted [3]. The digital clock for the MCU is provided by a clock regenerator, which generates a system clock for the MCU operation from the ASK signal. The analog switches are implemented by transmission gates, which are controlled by the MCU and switching currents to the desired electrodes.

To verify the function of the electrical circuits, commands that activate the four electrodes individually are transmitted wirelessly to the chip. The responsive waveforms are measured as illustrated in Fig. 18.7.4. The upper diagram shows that the ASK signal is successfully regulated to supply voltage VDD and demodulated to commands DO. The lower diagram indicates that the four electrolysis electrodes EP0, EP1, EP2 and EP3 are activated according to the RS232 commands.

In order to examine the locomotive function, a bare chip is put on water surface, powered wirelessly, and moved in one direction through a transmitting coil as illustrated in Fig. 18.7.5 (upper-left). A time-domain response of this experiment is recorded in Fig. 18.7.5 (lower views), where a track of the chip movement and distance of ~1.8cm can be observed in 60s. This indicates a velocity of 0.3mm/s. The substrate of this chip is thinned as shown Fig. 18.7.5 (upper-right) so that chip mass can be reduced for higher velocity. A performance comparison table is shown in Fig. 18.7.6 with the die micrograph shown in Fig. 18.7.7. This experiment has successfully demonstrated a proof-of-principle for the concept. Since the receiving coil is integrated, this chip is totally bond wire free. Furthermore, no special environmental conditions, such as a biasing magnetic field, are required.

Acknowledgments:
The authors would like to thank CIC and NDL-High Frequency Technology Center for chip fabrication and circuit measurement. This research is supported by National Science Council (NSC 100-2221-E-002-247-MY3), ROC and partially supported by National Taiwan University (NTU-CESRP-103R7624-3), ROC.

References:
[1] G. Ciuti, A. Menciassi., P. Dario, "Capsule Endoscopy: From Current Achievements to Open Challenges", *IEEE Reviews in Biomedical Engineering*, vol. 4, pp. 59-72, 2011.
[2] A. Yakovlev, D. Pivonka, T. Meng, A. Poon., "A mm-sized wirelessly powered and remotely controlled locomotive implantable device", *ISSCC Dig. Tech. Papers*, pp. 302-304, Feb., 2012.
[3] P.-L. Huang, P.-H. Kuo, Y.-J. Huang, H.-H. Liao, Y.-J. J. Yang, T. Wang, Y.-H. Wang, S.-S. Lu, "A Controlled-Release Drug Delivery System on a Chip Using Electrolysis", *IEEE Trans. Industrial Electronics*, vol. 59, no. 3, pp. 1578-1587, March 2012.
[4] S.-C. Chan, C.-R. Chen, C.-H. Liu, "A bubble-activated micropump with high-frequency flow reversal", *Sensors and Actuators A: Physical*, vol. 163, no. 2, pp. 501-509, October 2010.

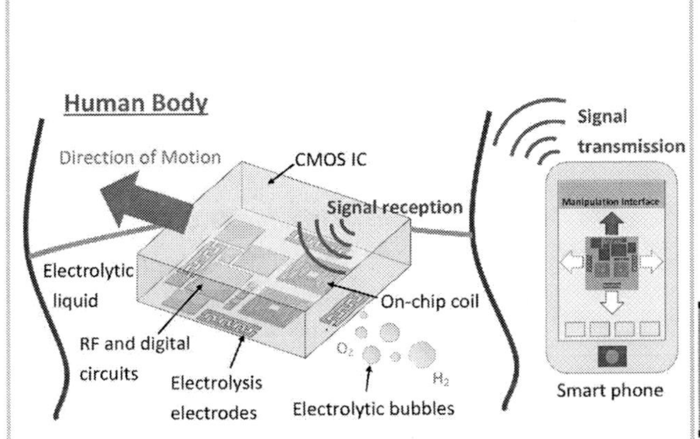

Figure 18.7.1: Schematic of the remotely-controlled locomotive CMOS IC.

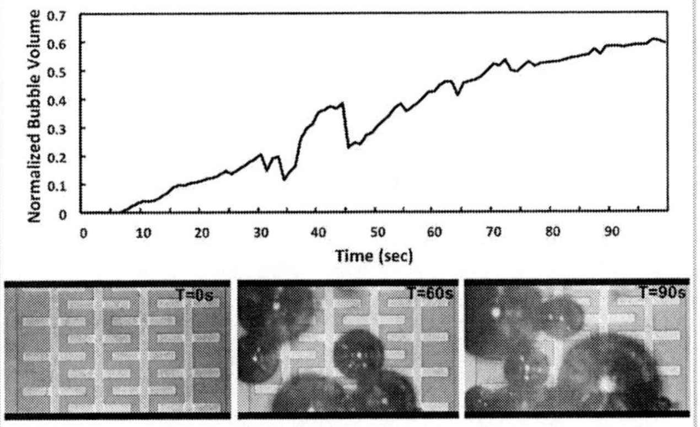

Figure 18.7.2: Measured transient response of generated bubble volume and its corresponding recorded images.

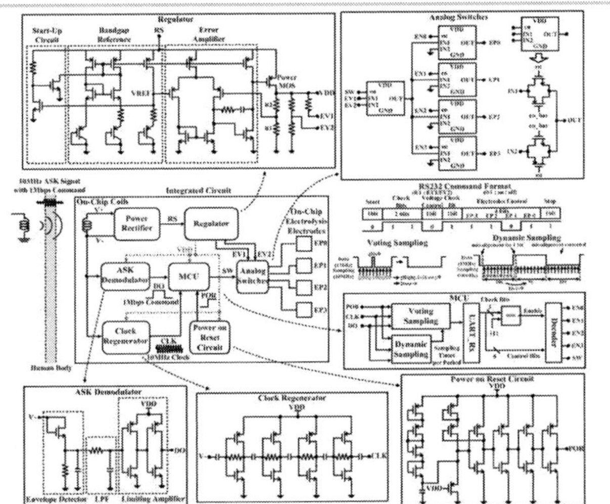

Figure 18.7.3: Circuit architecture.

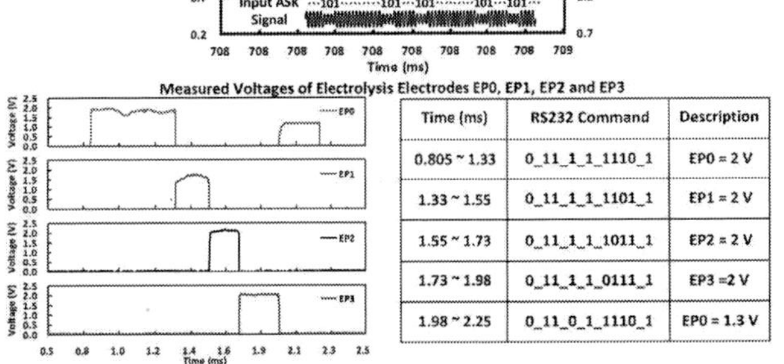

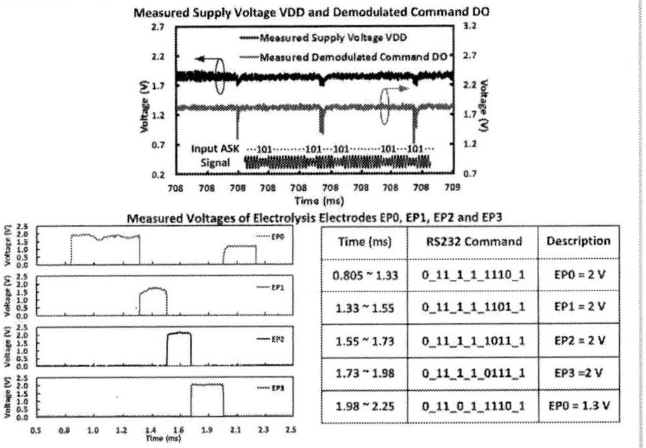

Time (ms)	RS232 Command	Description
0.805 ~ 1.33	0_11_1_1_1110_1	EP0 = 2 V
1.33 ~ 1.55	0_11_1_1_1101_1	EP1 = 2 V
1.55 ~ 1.73	0_11_1_1_1011_1	EP2 = 2 V
1.73 ~ 1.98	0_11_1_1_0111_1	EP3 =2 V
1.98 ~ 2.25	0_11_0_1_1110_1	EP0 = 1.3 V

Figure 18.7.4: Measured supply voltage VDD and demodulated command DO while the ASK modulated RF signal inputs. Measured voltages of the electrolysis electrodes EP0, EP1, EP2 and EP3 based on demodulated RS232 commands.

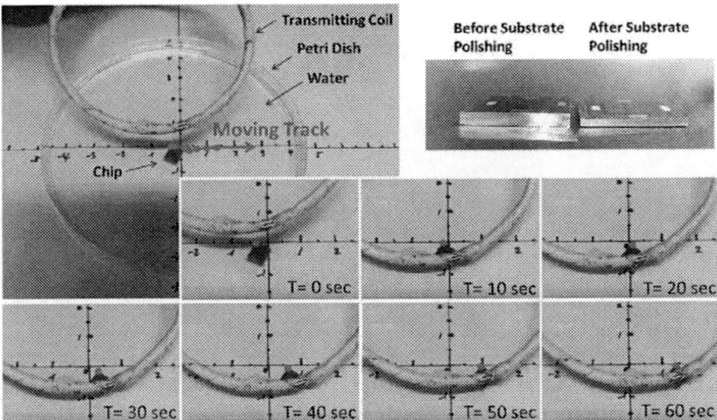

Figure 18.7.5: Pictures of the chip movement with red dots marking the cruising track. The moving distance is 1.8cm in 60 seconds, indicating a moving speed of 0.3mm/s. Substrate polishing is used to reduce mass of chip for higher velocity.

	This Work	ISSCC 2012 [2]
Technology	0.35μm CMOS	65nm CMOS
Battery	No	No
Propulsive Force Type	Electrolytic Bubbles	Lorentz Force*
Regulated Voltage	2V (VDD)	0.7V
Modulation Type	ASK	ASK+PWM
RF Carrier Frequency	10MHz	1.86GHz
Command Rate	1Mbps	2.5 - 25Mbps
Power Consumption	125.4μW (Chip Circuit) 82μW (Electrolysis)	17μW (Chip Circuit) 250μW (Fluid Propulsion System)
Chip Size	4.6mm x 4.6mm (including on-chip coils, electrolysis electrodes)	2mm x 2mm (Receive Antenna) 0.6mm x 1mm (Die)
External Component	No	Yes (PCB, Antenna, Magnet)
Electrolytic Voltage	EV2=2V EV1=1.3V	N/A
Moving Directions	4 directions	N/A
Velocity	~0.3mm/s	~0.53cm/s

* External magnet is required.

Figure 18.7.6: Comparison of the measured performance.

18

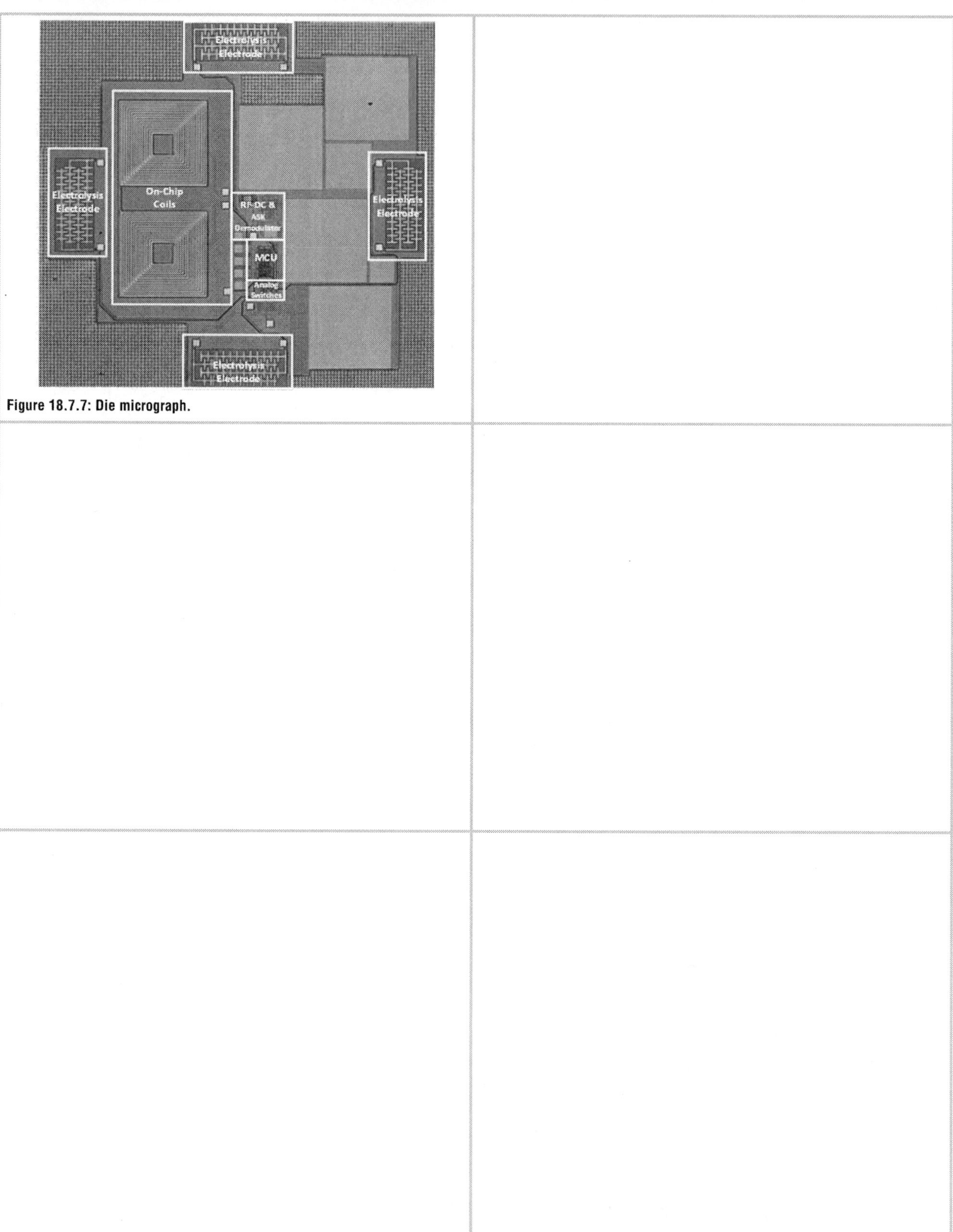

Figure 18.7.7: Die micrograph.

Session 19 Overview: *Nonvolatile Memory Solutions*

MEMORY SUBCOMMITTEE

Session Chair: *Jin-Man Han*
Samsung, Hwaseong, Korea

Session Co-Chair: *Tadaaki Yamauchi*
Renesas Electronics, Tokyo, Japan

Strong market demands of diverse non-volatile memory technologies show continuing increase in density, reliability, and performance. This year the leading edge process node for NAND Flash is scaled down to the minimum feature size of 16nm, and three-dimensional vertical NAND has been demonstrated. In addition, Flash controllers contribute to the higher reliability and performance on such advanced node. Emerging memories such as Resistive RAM (ReRAM) are continuing to show significant performance progress.

19.1 A 128Gb MLC NAND-Flash Device Using 16nm Planar Cell **8:30 AM**

M. Helm, Micron, San Jose, CA

In Paper 19.1, Micron reports 128Gb MLC NAND Flash memory using 16nm planar cell with high-k dielectric/metal gate cell architecture. 400MT/s/pin using a 1.8V DDR interface is enabled by an advanced data path design. New design features are employed to compensate memory cell degradation due to scaling.

19.2 A 93.4mm^2 64Gb MLC NAND-Flash Memory with 16nm CMOS Technology **9:00 AM**

S. Choi, SK Hynix, Icheon, Korea

In Paper 19.2, SK Hynix describes a 64Gb MLC NAND Flash memory implemented in 16nm process. It supports 400Mb/s NV-DDR2 interface for high-speed data transaction. The delayed P1 program pulse scheme makes the cell distribution tighter without additional overhead, and the cell current screen scheme both reduces the operation power and enhances the read accuracy.

19.3 66.3KIOPS-Random-Read 690MB/s-Sequential-Read Universal Flash Storage Device Controller with Unified Memory Extension **9:30 AM**

K. Watanabe, Toshiba, Yokohama, Japan

In Paper 19.3, Toshiba reports an embedded NAND-storage device controller with unified memory which contributes to 2× and 10× faster random read and write performance, respectively. Low-power dual-lane 5.8Gb/s M-PHY with random read command processor attains performance of 66.3KIOPS random read and 690MB/s.

978-1-4799-0917-9/14 $31.00 © 2014 IEEE

19.4 Embedded 1Mb ReRAM in 28nm CMOS with 0.27-to-1V Read Using Swing-Sample-and-Couple Sense Amplifier and Self-Boost-Write-Termination Scheme **10:15 AM**

M-F. Chang, National Tsing Hua University, Hsinchu, Taiwan

In Paper 19.4, National TsingHua U describes an embedded 1Mb ReRAM in 28nm process with 0.27-to-1V read. A swing-sample-and-couple sense amplifier improves read margins/speeds by over 1.8×. A self-boost-write-termination scheme contributes to a reduction of 99% SET energy with below 0.5% area penalty.

19.5 Three-Dimensional 128Gb MLC Vertical NAND Flash-Memory with 24-WL Stacked Layers and 50MB/s High-Speed Programming **10:45 AM**

K-T. Park, Samsung Semiconductor, Hwasung, Korea

In Paper 19.5, Samsung reports three dimensional 128Gb MLC vertical NAND Flash Memory with 24 WL stacked layers. The chip accomplishes 50MB/s write throughput with 3K endurance for typical embedded applications and extended endurance of 35K with 33MB/s write throughput for data center and enterprise SSD applications.

19.6 Hybrid Storage of ReRAM/TLC NAND Flash with RAID-5/6 for Cloud Data Centers **11:15 AM**

S. Tanakamaru, Chuo University, Tokyo, Japan and University of Tokyo, Tokyo, Japan

In Paper 19.6, Chuo U describes a hybrid storage of ReRAM/TLC NAND Flash with RAID-5/6 for cloud data centers. The bit-error rate of 50nm ReRAM and the failure rate of 2xnm TLC NAND are improved by 69% and 98%, respectively. It is achieved by applying 5 techniques involving adaptive asymmetric coding for ReRAM and bits/cell optimization for TLC NAND.

19.7 A 16Gb ReRAM with 200MB/s Write and 1GB/s Read in 27nm Technology **11:45 AM**

R. Fackenthal, Micron, Folsom, CA

In Paper 19.7, Micron reports 16Gb ReRAM with 200MB/s write and 1GB/s read in 27nm process. A 1GB/s DDR interface and an 8-bank concurrent core architecture are employed with high parallelism and pipelined data path architecture.

19.1 A 128Gb MLC NAND-Flash Device Using 16nm Planar Cell

Mark Helm[1], Jae-Kwan Park[1], Ali Ghalam[1], Jason Guo[1], Chang wan Ha[1], Cairong Hu[1], Heonwook Kim[1], Kalyan Kavalipurapu[1], Eric Lee[1], Ali Mohammadzadeh[1], Dan Nguyen[1], Vipul Patel[1], Ted Pekny[1], Bill Saiki[1], Daesik Song[1], Jeff Tsai[1], Vimon Viajedor[1], Luyen Vu[1], Tinwai Wong[1], Jung Hee Yun[1], Ramin Ghodsi[1], Andrea D'Alessandro[2], Domenico Di Cicco[2], Violante Moschiano[2]

[1]Micron, San Jose, CA, [2]Micron, Avezzano, Italy

The aggressive scaling of NAND Flash memory technology—one that is even outpacing Moore's Law—has enabled very rapid cost-per-bit reduction, resulting in an explosion of systems utilizing this versatile memory technology. From removable media and personal music players to smart phones, tablets, and now personal computers and data center applications employing client and enterprise solid state drives (SSDs), NAND technology is making solid-state memory-based storage affordable.

Maintaining this rapid pace in cost reduction faces many challenges, especially taking into consideration the increasing requirements in performance and reliability demanded by these sophisticated systems. To address the challenges and maintain the cost-reduction pace, we introduce a 128Gb MLC NAND Flash device using planar cell technology at a minimum feature size of 16nm and a die size of 173.3mm^2 which advances the state of NAND Flash technology beyond the work in [1,2].

The device is configured as a two-plane, 16kB page/plane utilizing shielded-bitline architecture to offer high bandwidth with low latency, low bit error rate, and low power consumption. A center-page buffer architecture cuts total bitline capacitance and resistance in half, achieving fast read and write performance optimized specifically for SSD applications with access times of 45µs and 1185µs, respectively. The data-path design enables 400MTransfer/s per pin using a 1.8V double data-rate (DDR) interface compliant with the ONFI 3 specification. Figure 19.1.7 shows the die micrograph of this device and Fig. 19.1.1 illustrates the key device metrics.

Advanced process technology is imperative in achieving the performance and reliability goals of this device on a 16nm technology node. Planar cell technology, first introduced with our 20nm technology node [3], has inherent advantages over the traditional floating gate cell structure such as dramatically reduced cell-to-cell interference. Further optimization of the high-k dielectric/metal gate cell architecture compared to the first generation of this technology enables scaling to 16nm feature size with minimal degradation in cell reliability.

Figure 19.1.2 shows a cross-section of the device wordline showing the high-k dielectric and metal-gate stack, and wordline air-gap construction used to minimize wordline resistance and capacitance as well as cell-to-cell interference. Through the use of immersion lithography and spacer-assisted pitch-multiplication techniques, a pattern printed above the resolution limit of the lithography tool can be reduced to the final 32nm pitch shown.

Bit-error-rate management is critical to achieving the reliability goals of this highly scaled memory technology. In addition, the use of error-correction techniques such as low-density parity-check (LDPC) code places new constraints not only on the total bit error rate but also on the characteristics of bits in error [4]. Figure 19.1.3 shows a typical programming algorithm in a MLC NAND Flash device where the lower page must be read from the NAND array and then combined with user-provided upper page data to determine the final state assignment. Any errors in reading the lower page data from the array will result in cells being misplaced into the wrong final V_t distribution, leading to a highly undesirable bit error characteristic. To minimize these misplacement bit errors, lower-page pre-read error mitigation (LPEM) is designed. LPEM compensates the read of the target lower-page data (victim) based on the adjacent cells' data state (aggressors). LPEM first reads the adjacent wordline to classify neighbor cells as aggressors or non-aggressors. By performing the victim read at two different read levels based on the neighbor cells' data state on a bit-by-bit basis as shown in Fig. 19.1.3, bit errors induced by adjacent cell interference can be eliminated with minimal programming time penalty.

As NAND cell dimensions shrink, it becomes more important that the V_t window of operation be matched to the natural V_t of the cell to minimize built-in electric fields that can degrade bit error rate through mechanisms such as program disturb or charge loss during retention stress. Application of source and bitline biases can be used to shift the cell's perceived V_t into the proper operating window of the sense amplifier [5]. However, the amount of V_t shift is limited by the internal V_{CC} voltage used to set the trip point of the sense latch. An on-chip negative-bias generator can expand the amount of negative V_t shift available, but this comes at the cost of increased die size and design complexity. Boosted bitline negative sense (BBNS) is designed to allow further extension of negative V_t window shift without requiring a negative-bias generator. The simplified circuit diagram of BBNS is shown in Fig. 19.1.4. The wmux control circuit chooses whether even or odd bitlines are selected and connected to the sense amplifier. Unselected bitlines are connected to the source terminal. Figure 19.1.5 shows the BBNS waveform where after initial bitline precharge, the selected bitline is floated and then boosted via capacitive coupling from the unselected bitline when the source terminal is raised. After the appropriate bitline develop time, the source is returned to ground to de-boost the selected bitlines back into the operating window of the sense amp. Also shown in Figure 19.1.6 are V_t distributions of cells for varying source voltages indicating a capability of up to 4V V_t window shift with no impact on the V_t distribution width.

Power management must be addressed and flexibility is key when addressing markets as varied as battery-operated mobile applications and high-bandwidth, high-parallelism data-center applications with the same NAND Flash device. This flexibility is provided by the user-selectable power-management (USPM) architecture that allows a user to set the maximum I_{CC} the NAND die is allowed to consume during a program operation. In a program operation, peak power consumption occurs when charging the bitlines for program or inhibit prior to the program pulse. USPM provides multiple options a user can choose for maximum I_{CC} to be applied during this charging operation, enabling the die to automatically manipulate the charging time to guarantee the same bitline bias independent of the peak current option chosen. Figure 19.1.6 shows the measured waveform of this feature where different maximum I_{CC} settings result in different bitline-charging time, thus performing the prescribed power-performance trade-off desired by the user.

The 128Gb MLC device built with the second generation of our planar cell technology utilizes a high-k dielectric metal-gate stack to greatly reduce cell-to-cell interference while maintaining required gate coupling ratio. A pre-read of lower-page data during the upper-page program operation compensates for cell-to-cell coupling, improving reliability with minimal performance impact. Boosted bitline negative sense is utilized as a low-cost way to provide greater flexibility for negative V_t window shift and minimize built-in electric fields that degrade bit error rate. A user-selectable I_{CC} power-management capability allows customization of the power-performance envelope, enabling the device to be used in a wide range of systems and markets.

References:

[1] N. Shibata, et. al, "A 19nm 112.8mm2 64Gb Multi-Level Flash Memory with 400Mb/s/pin 1.8V Toggle Mode Interface," *ISSCC Dig. Tech Papers*, pp. 422-423, Feb. 2012.

[2] D. Lee, et. al, "A 64Gb 533Mb/s DDR Interface MLC NAND Flash in Sub-20nm Technology," *ISSCC Dig. Tech Papers*, pp. 430-431, Feb. 2013.

[3] G. Naso, et. al, "A 128Gb 3b/cell NAND Flash Design Using 20nm Planar-Cell Technology," *ISSCC Dig. Tech Papers*, pp. 218-219, Feb. 2013.

[4] R. Motwani, et. al, "Robust decoder architecture for multi-level flash memory storage channels", *ICNC Dig. Tech Papers*, pp. 492-496, Jan. 2012.

[5] D.W. Lee, et. al, "The Operation Algorithm for improving the Reliability of TLC NAND Flash Characteristics", *IMW Dig. Tech Papers*, pp. 1-2, May 2011.

978-1-4799-0917-9/14 $31.00 © 2014 IEEE

Technology	16nm planar cell
Density	128Gb
Bits per cell	2
Architecture	2 plane 16kB page/plane Shielded bitline Center page buffer
Die Size	173.3mm^2
Average read time	45μs
Average program time	1185μs
I/O	400 MT/s/pin ONFI 3 @ 1.8V

Figure 19.1.1: 128Gb NAND Flash design attributes.

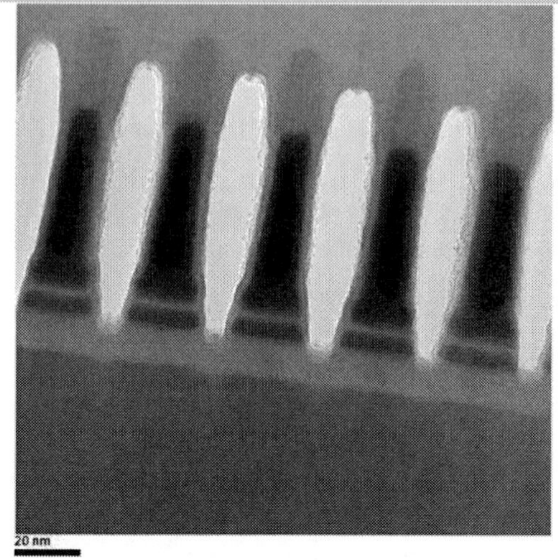

Figure 19.1.2: 16nm planar NAND cell cross-section.

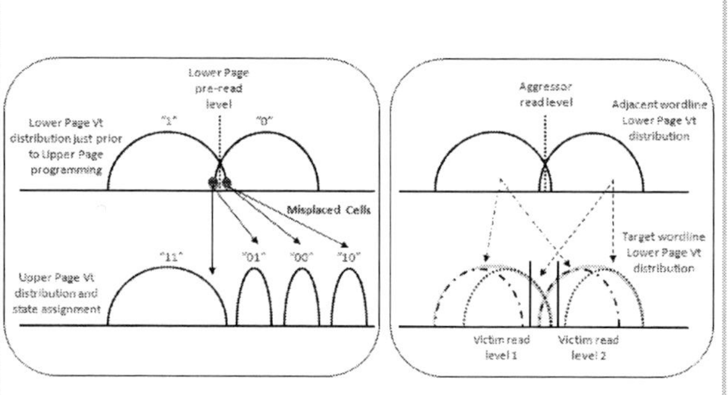

Figure 19.1.3: Lower page pre-read error mitigation.

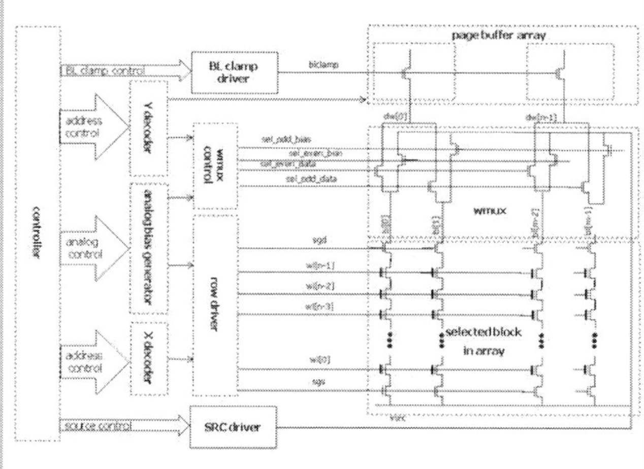

Figure 19.1.4: Boosted bitline negative sense circuit.

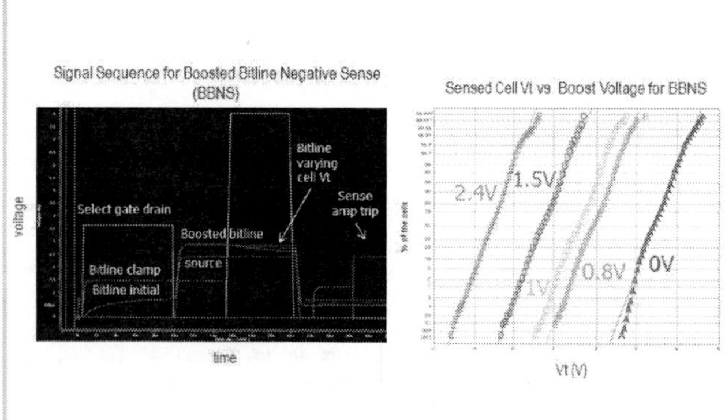

Figure 19.1.5: BBNS waveform and V_t distribution.

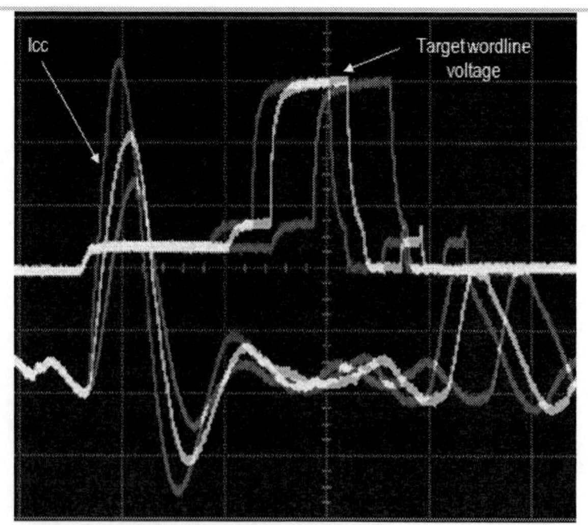

Figure 19.1.6: Peak power management waveform.

19

Figure 19.1.7: 16nm MLC 128Gb NAND Flash die micrograph.

19.2 A 93.4mm² 64Gb MLC NAND-Flash Memory with 16nm CMOS Technology

Sungdae Choi, Duckju Kim, Sungwook Choi, Byungryul Kim, Sunghyun Jung, Kichang Chun, Namkyeong Kim, Wanseob Lee, Taisik Shin, Hyunjong Jin, Hyunchul Cho, Sunghoon Ahn, Yonghwan Hong, Ingon Yang, Byoungyoung Kim, Pilseon Yoo, Youngdon Jung, Jinwoo Lee, Jaehyeon Shin, Taeyun Kim, Kunwoo Park, Jinwoong Kim

SK Hynix, Icheon, Korea

The demand for high-density low-cost NAND-Flash memory devices is growing due to the increase in the NAND-Flash application market such as SSD for tablet PCs and ultra-books as well as conventional mobile applications such as USB drives and digital still cameras. Various approaches to implement high-density NAND Flash with small area have been introduced to address the market. Moving from a single-level cell (SLC) to 2 bits per cell (MLC) or to 3 bits per cell is one of the approaches to increasing memory density with the same die area. Lithographic shrinking with conventional 2D technology is the most mature technology although 3D stacking technologies [1] are being developed.

This paper presents a 64Gb MLC NAND-Flash memory fabricated with 16nm CMOS process technology to achieve high density and as small as 93.4mm² die area. Figure 19.2.1 shows the implemented die micrograph and summary table. The chip consists of two planes of 1072 blocks each. A block consists of a string with 128 cells and a page size with 16KB and spare area for error-correction coding (ECC), totaling 4MB of capacity. The chip supports negative-level wordline drivability to increase cell V_{th} margin as shown in Figure 19.2.6.

One of the biggest issues of NAND Flash with technology shrink as small as 16nm is coupling issues between wordlines (WL) and bitlines (BL) [2]. Increase in WL-to-WL capacitive coupling raises interference between cells, and the corresponding cell V_{th} distribution becomes wider. Wide cell distribution is caused not only by WL-to-WL coupling but also by BL-to-BL coupling due to horizontal neighboring cells. The increased BL-to-BL coupling, which is caused by shorter physical inter-BL distance, affects the voltage level if the neighboring BL is in the floating state. This BL-coupling noise has been addressed in NAND-Flash memory using the BL shielding structure [3], which has the drawback of reducing the access bandwidth. Although current sensing with the all-bitline (ABL) [4] scheme fully accesses the whole page size at once, short-circuit current through the on-cells causes high power consumption. The lithographic shrink also affects the signal interconnection lines and makes the loading of the control signals higher, requiring larger signal drivers and corresponding area overhead.

To solve the problems addressed above, this paper presents the cell-current-controlled screen-out sensing scheme, delayed P1 PGM pulse scheme for cell distribution control. It also presents the peak-current control scheme and the load-balanced control signal block for efficient signal drive.

Current-sensing schemes consume high power through the strongly on cells to keep the level of every bitline constant. Corresponding source-line (SL) noise generated by the current sink prohibits the cell from drawing the expected amount of cell current due to the reduced string voltage. The conventional screen-out phase, shown in Fig. 19.2.2(a), filters BLs with large cell current, and minimizes the SL noise during the sensing phase. But the amount of the current sink during the screen-out phase does not decrease. To minimize the amount of short-circuit current during the screen-out phase, low BL voltage (V_{BL}) and correspondingly compensated trip-point are used as shown in Fig. 19.2.2(b). With the lowered V_{BL}, the screen-out accuracy is raised and more BLs with large cell current (I_{CELL}) are filtered, which affects lower SL level at sensing phase. This leads to smaller current consumption during the screen-out phase and higher sensing accuracy in the sensing phase.

Figure 19.2.3 shows the program pulse sequence with the proposed delayed P1 PGM pulse scheme. The parallel MSB program scheme simultaneously develops P1 and P2/P3 distribution from P0 and LP, respectively [5]. A typical PGM pulse sequence starting from P1 and P2/P3 completes the development of P1 earlier than P2/P3, and P1 distribution gets wider due to the rest of the P2/P3 pulse. In Fig. 19.2.3, the starting point of P1 pulse is delayed so that the PGM is finished at the same point as P3. With the proposed PGM sequence, measured cell distribution shows 5% reduction of P1 distribution width as shown in Figure 19.2.6, without additional timing overhead.

Charging large capacitance BLs is one of the essential operations at read and program the cell values, and also the main contributor to peak current, as shown in Fig. 19.2.4. The peak-current control scheme, also shown in Fig. 19.2.4 limits the maximum current (I_{LIMIT}) that flows through the PMOS MP_i in each sensing circuit. The current-mirror-based current-control block generates corresponding reference voltage (V_{REF}) and distributes it to all sensing circuits. To prevent additional timing overhead, I_{LIMIT} is set to about 1/3rd of the entire BL charging time by an *n*-bit digital signal I_{CTL}. With this scheme, a 25-to-40% reduction in peak current is measured, with the amount depending on the data pattern.

A technology shrink that reduces cell size also affects the critical dimension of vertical interconnection signal lines. It increases the resistance and capacitance, requiring a large signal driver. In the conventional approach shown in Fig. 19.2.5, all signals are generated from upper side of the sense-and-latch block, and every signal requires vertical interconnection. In our approach, signals are categorized by their signal level, digital level or analog level, and signal drivers with digital level ($CNTL_{DIGITAL}$) are placed on the side of the sense-and-latch block so that the signals are directly driven through small-RC-time-constant horizontal interconnection lines. Since the signal loading is divided by the number of the sense-and-latch stack, the $CTNL_{DIGITAL}$ block requires small inverters instead of a large driver. Although the analog signals are still driven from the upper side, more vertical lines are assigned to drive each analog signal because digital signals do not need vertical interconnection lines as shown in Figure 19.2.5. The block placement occupies 50% area and signal skew is 25% compared with conventional block.

References:

[1] Takashi Maeda, et al., "Multi-stacked 1G cell/layer Pipe-shaped BiCS Flash Memory," *Symposium on VLCI Circuits*, pp. 22-23, 2009.

[2] Koichi Fukuda, et al., "A 151mm2 64Gb MLC NAND Flash Memory in 24nm CMOS Technology," ISSCC Dig. Tech. Papers, pp.198-199, Feb., 2011

[3] Hiroshi Nakamura, et al., "A Novel Sense Amplifier for Flexible Voltage Operation NAND Flash Memories," *Symposium on VLSI Circuits*, pp. 71-72, 1995.

[4] Raul Cernea, et al., "A 34MB/s-Program-Throughput 16Gb MLC NAND with All-Bitline Architecture in 56nm," *ISSCC Dig. Tech. Papers*, pp.420-421, Feb. 2008.

[5] Hyunggon Kim, et al., "A 159mm2 32nm 32Gb MLC NAND-Flash Memory with 200MB/s Asynchronous DDR Interface," *ISSCC Dig. Tech. Papers*, pp.442-443, Feb. 2010.

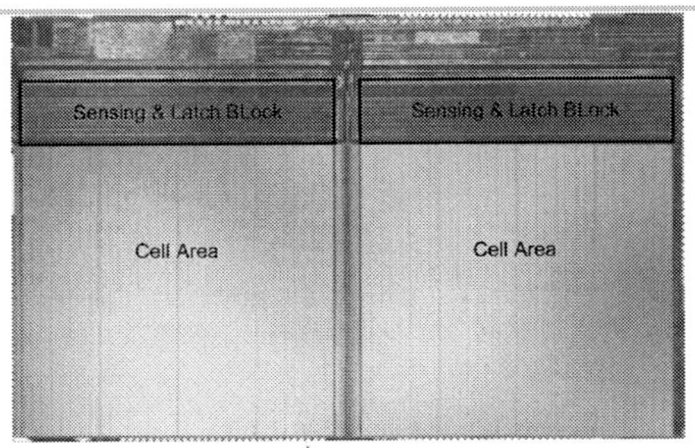

Technology	16nm 3-Metal CMOS
Density	64Gb (2bit/cell)
Die Size	93.43mm^2
Organization	16K bytes x 256 pages x 2K blocks
Block Size	4MB
Power Supply	2.7V ~ 3.6V
Data Transfer Rate	400Mb/s
Program Throughput	25MB/s (Typical)

Figure 19.2.1: Chip micrograph with key features.

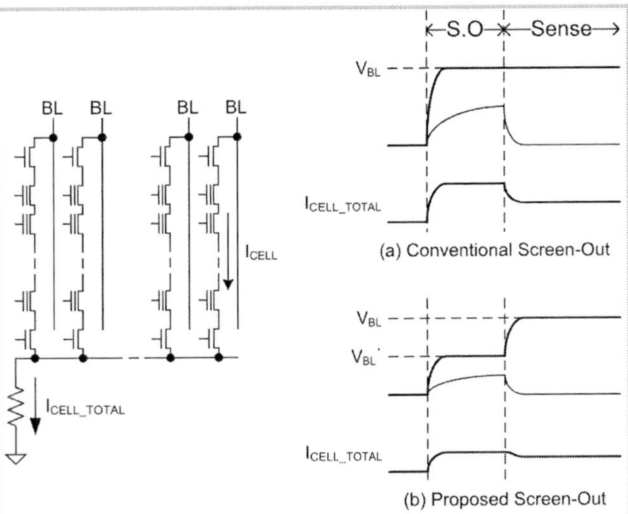

Figure 19.2.2: Cell-current-controlled screen-out sensing scheme.

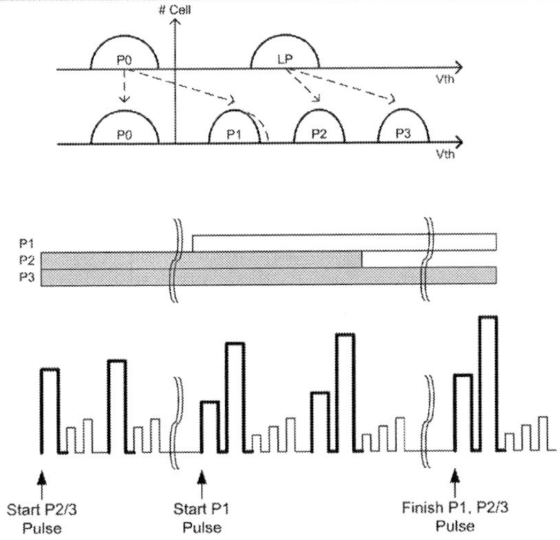

Figure 19.2.3: Delayed P1 PGM pulse.

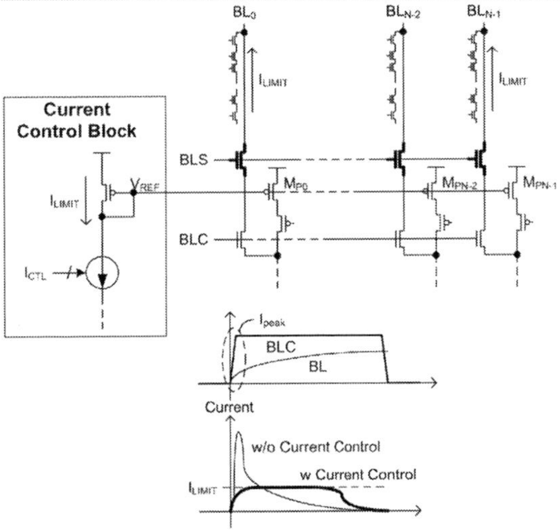

Figure 19.2.4: Peak-current control in BL controller.

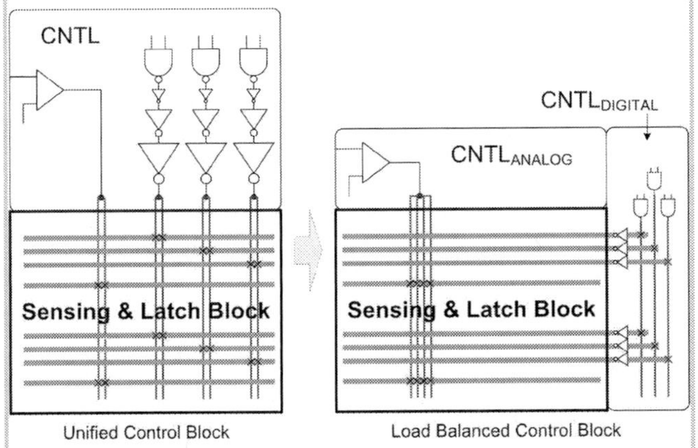

Figure 19.2.5: Load-balanced control signal block P&R.

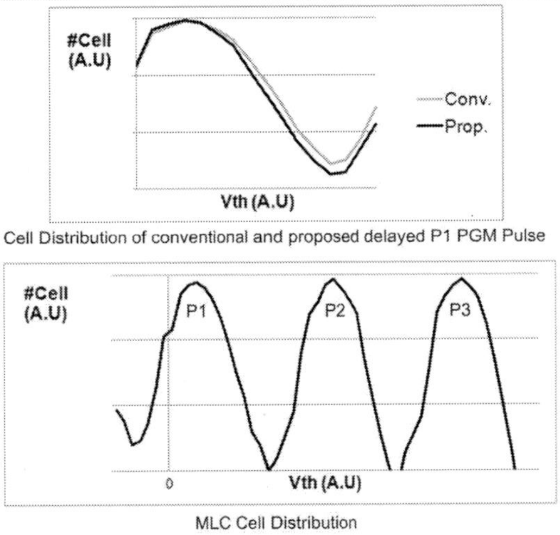

Figure 19.2.6: MLC cell distribution.

This Page is a Continuation of the Previous and is Intentionally Unpaginated
Page Intentionally Left Blank

Figure

19.3 66.3KIOPS-Random-Read 690MB/s-Sequential-Read Universal Flash Storage Device Controller with Unified Memory Extension

Konosuke Watanabe[1], Kenichiro Yoshii[1], Nobuhiro Kondo[1], Kenichi Maeda[1], Toshio Fujisawa[1], Junji Wadatsumi[2], Daisuke Miyashita[2], Shouhei Kousai[2], Yasuo Unekawa[2], Shinsuke Fujii[2], Takuma Aoyama[2], Takayuki Tamura[1], Atsushi Kunimatsu[1], Yukihito Oowaki[1]

[1]Toshiba, Yokohama, Japan, [2]Toshiba, Kawasaki, Japan

Mobile devices have made remarkable advances in recent years. They generally use embedded NAND storage devices, which are tiny (10s of millimeters square) and low-power (around 1W in the active state) single BGA packages that contain both a controller and NAND chips. Figure 19.3.1 shows read performance of recent embedded NAND storage device products and the maximum link speeds in their standards. The figure indicates that more powerful embedded NAND storage devices are desired by the market. In particular, universal Flash storage (UFS) 2.0, the latest standard, defines high link speed, which is 3× faster than the recent embedded multimedia card (eMMC). In this context, we develop a UFS 2.0 device that introduces new features to the conventional embedded NAND storage device controller architecture to improve read performance. Figure 19.3.2 shows a block diagram of our controller. We improve the read performance in the following ways: 1) suppress the number of NAND read accesses and reduce the read latency by introducing unified memory (UM) and caching data for address translations on it, 2) increase the number of NAND chips activated simultaneously with dedicated hardware and new command scheduling, and 3) maximize bandwidth by supporting 5.8Gb/s 2-lane M-PHY link with low-power analog circuits.

In general, a single random-read command is accomplished by 2 or more NAND read accesses because of the need to translate the logical address of the requested data (user data) into a NAND physical address in the Flash translation layer (FTL), which is stored in the NAND. The average number of NAND read accesses to obtain FTL data has increased over time. Every NAND read contains a fixed delay, called tR, which is several 10s of microseconds. Thus, a latency of a read always includes nearly 100μs latency caused by tRs, reducing random-read performance. Some SSDs solve this problem by replacing NAND accesses to FTL data with accesses to cached FTL data in a large amount of RAM[3,4]. This solution is simple and effective but increases power consumption and chip size. To obtain the benefit of FTL data caching without a large amount RAM, we make the controller access a part of the host memory as a dedicated external RAM, called UM, and use it to implement FTL data caching. To support UM access, we add to the UFS protocol new commands to read/write/manage UM and extend the host interface to support them. Figure 19.3.3 shows data flows and breakdown of read command processing sequences of a conventional embedded NAND controller and this work's controller. While the conventional controller makes n+1 NAND read accesses, our controller makes a single NAND read access for user data with n UM read accesses for cached FTL data. UM read-access latency is lower than NAND read latency. In our controller, this reduces total latency by an average of about 51%, making random-read performance 2× higher. Moreover UM provides a large buffer to prevent degradation of response time of the device during a rush of a large number of random write commands. This technique enables a random-write performance of 60.6KIOPS, which is about 10× faster than the random-write performance without UM.

Read performance can be improved by keeping multiple NAND chips in the active state [1]. For random-read commands, it is possible to process multiple read commands simultaneously. However, the commands sometimes cause contention for the target NAND chips and not all of the NAND access requests can be issued even though the number of commands is equal to or smaller than the number of NAND chips. To activate as many NAND chips as possible, more of the random read commands must be queued. For example, if both the number of queued commands and the number of NAND chips are 8, an expected number of NAND read requests that can be issued at the same time is 5.25 probabilistically. When the number of the queued commands increases to 16, the expected number issued improves to 7.0. The read performance of the NAND controller in [2] is greatly affected by the depth of its command queue. Improving random-read performance by increasing the number of active NAND chips requires a deep command queue that supports out-of-order command processing and adequate performance for processing the multiple random-read commands in parallel. Consequently, we design dedicated hardware called the random-read command processor (RRCP). Figure 19.3.4 shows a block diagram of the RRCP and 2 schedule charts that present commands processing timings of an in-order scheduling and the new NAND-activation-based out-of-order scheduling with 7 random read commands that have some contention with respect to the target NAND chips. The RRCP consists of a front-end and a back-end. The front-end provides out-of-order command processing with 3 cascaded pipelines that operate asynchronously and process up to 16 commands simultaneously. Reordering of the command may occur at the border of these pipelines. The back-end issues NAND read requests and arbitrates chip-level and channel-level contention of the NAND-access requests. The charts indicate that the out-of-order scheduling achieves better performance than the in-order scheduling. By using the RRCP, the random-read performance increases over n-fold, where n is the expected number of NAND requests that can be issued to NAND chips at the same time. As mentioned above, if the NAND storage device has 8 NAND chips, the expected number of NAND requests of out-of-order scheduling is ~7.0 while that of the in-order scheduling is ~3.24. Thus, the out-of-order scheduling improves performance by 116% over in-order scheduling.

Unlike random-read performance, sequential-read performance can be improved by increasing the bandwidth of data-paths in the controller. To achieve over 600MB/s sequential-read bandwidth, a 5.8Gb/s 2-lane M-PHY is adopted. In the receiver, two types of CDR are developed. A PLL-based CDR meets M-PHY specification and is a good candidate for production. In addition, a new injection CDR for faster locking capability is implemented. A previous injection-based CDR eliminates edge detector that is subject to PVT variation [5]. However the CDR has low jitter tolerance at high frequencies because a data-transition edge is injected simultaneously to 2 gated VCO cells, resulting in asynchronous injection. To overcome this issue we develop a synchronized-injection CDR, as shown in Fig. 19.3.5. Since the same cell is used for the delay line and the gated VCO, a transition edge can be injected 4 times synchronously, resulting in high jitter tolerance. The measured jitter tolerance with 5.8Gb/s continuous jitter pattern is also shown, which meets M-PHY specification. The CDR consumes 7.0mW per channel and a power efficiency of 1.2mW/Gb/s is achieved.

Figures 19.3.6 and 19.3.7 show the features and a micrograph of the controller, respectively. With FTL data caching using UM and the RRCP, a random-read performance of 66.3KIOPS is achieved. In addition, with the high bandwidth of the data-path, sequential-read performance of 690MB/s is achieved. Compared with existing embedded NAND storage devices [6-7], the device achieves higher read performance with typical package sizes and power consumption of an embedded NAND storage device.

References:

[1] F. Chen, et al., "Essential Role of Exploiting Internal Parallelism of Flash Memory based Solid State Drives in High-Speed Data Processing," *IEEE International Symp. on High Performance Computer Architecture*, pp.266-277, Feb. 2011.

[2] M. Jung, et al., "Physically Addressed Queueing (PAQ): Improving Parallelism in Solid State Disks," *International Symp. on Computer Architecture*, pp.405-415, Jun. 2012.

[3] Y. Hu, et al., "Achieving Page-Mapping FTL Performance at Block-Mapping FTL Cost by Hiding Address Translation," *IEEE Symp. on Mass Storage Systems and Technologies*, pp.1-12, May. 2010.

[4] A. Gupta, et al., "DFTL: a flash translation layer employing demand-based selective caching of page-level address mappings," *International Conf. on Architectural Support for Programming Languages and Operating Systems*, pp.229-240, Mar. 2009.

[5] C. F. Liang, et al., "A 20/10/5/2.5Gb/s Power-scaling Burst-Mode CDR Using GVCO/Div2/DFF Tri-mode Cells," *ISSCC Dig. Tech. Papers*, pp.224-225, Feb. 2008.

[6]http://www.samsung.com/global/business/semiconductor/news-events/press-releases/detail?newsId=12984

[7] http://www.sandisk.com/products/embedded/issd/i100

978-1-4799-0917-9/14 $31.00 © 2014 IEEE

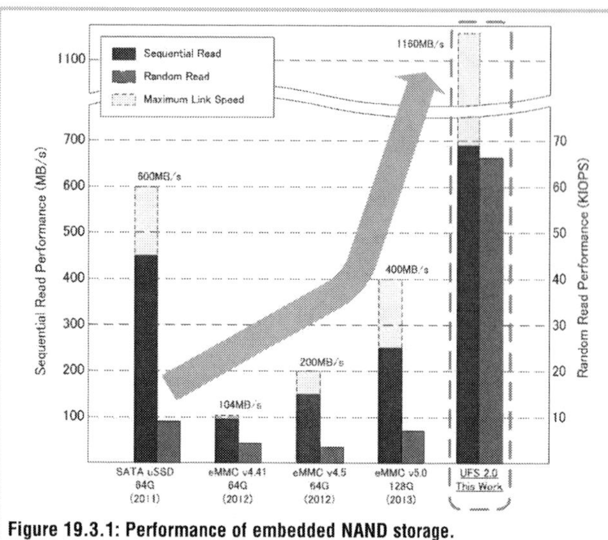

Figure 19.3.1: Performance of embedded NAND storage.

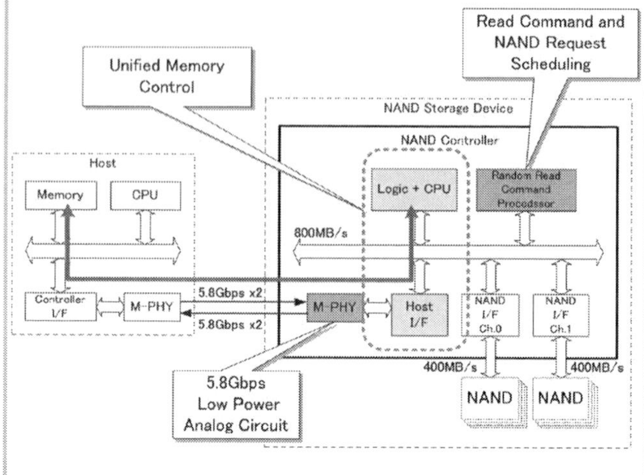

Figure 19.3.2: Outline of NAND controller.

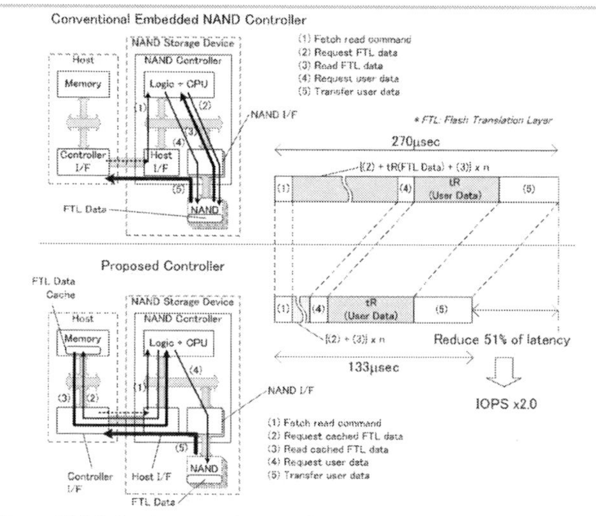

Figure 19.3.3: Read command processing sequences.

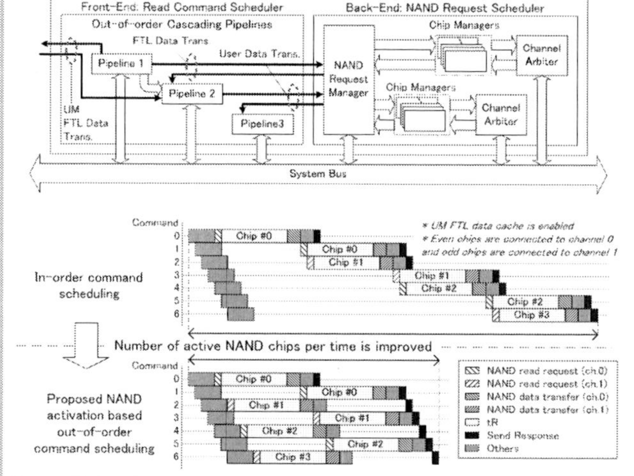

Figure 19.3.4: Random-read command processor.

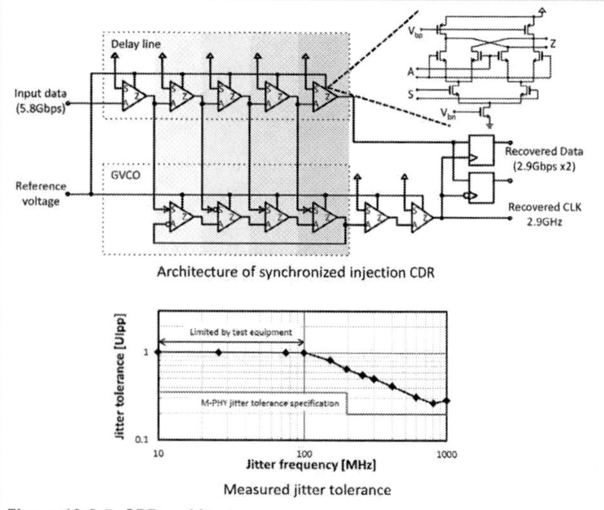

Figure 19.3.5: CDR architecture.

Performance

		This work (UFS v2.0, NAND x 8)	eMMC V5.0 [6]	SATA μSSD [7]
4KB Random Read	w/o UM	32.3 KIOPS	7 KIOPS	9 KIOPS
	w/ UM	66.3 KIOPS† (max. 94 KIOPS*)		
4KB Random Write	w/o UM	6.4 KIOPS	7 KIOPS	1 KIOPS
	w/ UM	60.6 KIOPS†		
Sequential Read		690 MB/s	250 MB/s	450 MB/s

† Depends on UM size, access range and host latency
* Optimized data alignment simulation

Device Features

Process		40nm CMOS
Controller Chip Size		2.688mm x 6.451mm = 17.34 mm²
Package Size (Controller + NAND)		11.5mm x 13.0mm x 1.2mm
Package Power (Controller + NAND)	Active (Seq. Write)	< 1.5 W‡
	Active (Seq. Read)	< 1.5 W‡
	Idle	< 1.45 mW‡
	Sleep	< 0.30 mW

‡ Depends on storage capacity

Figure 19.3.6: Performance and features.

19

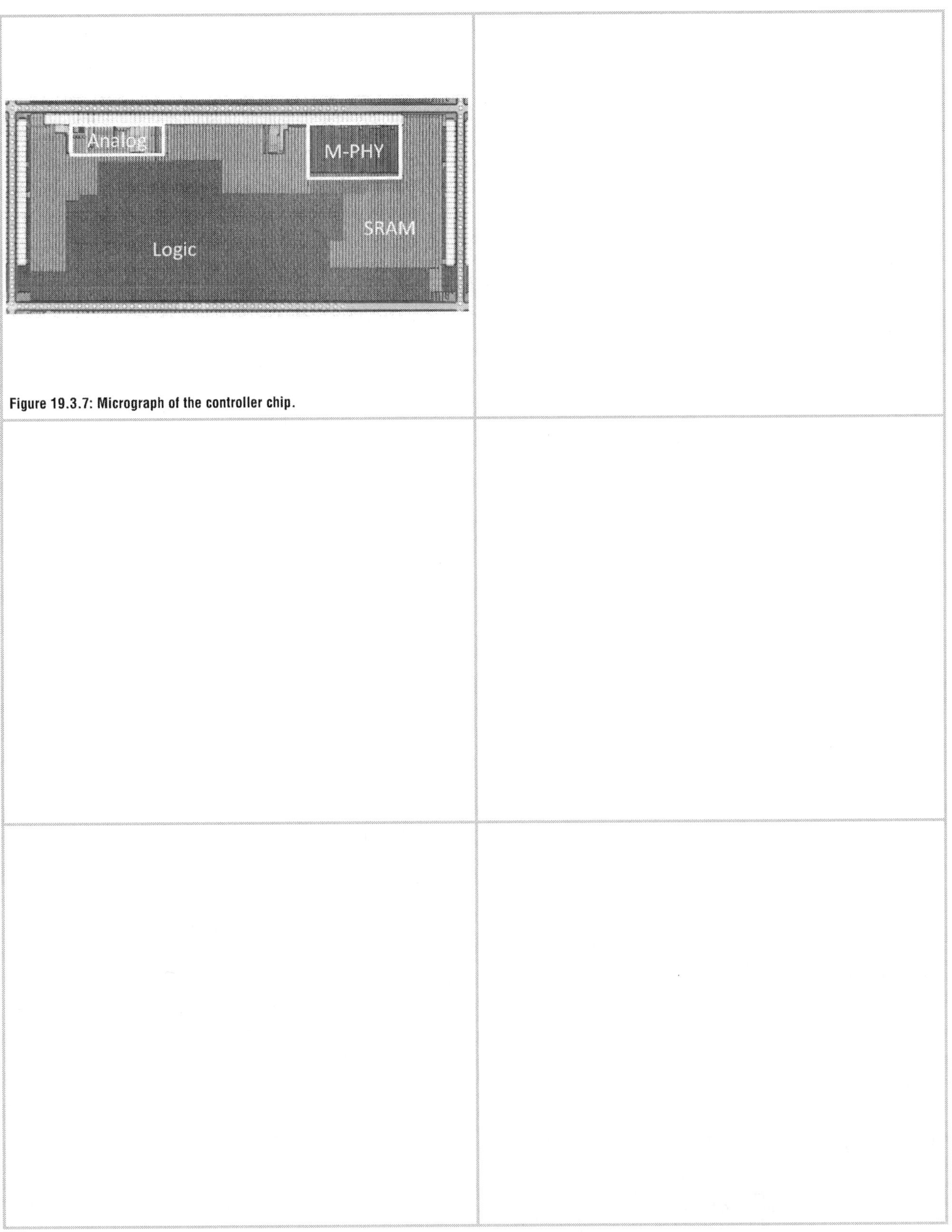

Figure 19.3.7: Micrograph of the controller chip.

19.4 Embedded 1Mb ReRAM in 28nm CMOS with 0.27-to-1V Read Using Swing-Sample-and-Couple Sense Amplifier and Self-Boost-Write-Termination Scheme

Meng-Fan Chang[1], Jui-Jen Wu[1], Tun-Fei Chien[1], Yen-Chen Liu[1], Ting-Chin Yang[1], Wen-Chao Shen[1], Ya-Chin King[1], Chorng-Jung Lin[1], Ku-Feng Lin[2], Yu-Der Chih[2], Sreedhar Natarajan[2], Jonathan Chang[2]

[1]National Tsing Hua University, Hsinchu, Taiwan, [2]TSMC, Hsinchu, Taiwan

Resistive RAM (ReRAM) is a promising nonvolatile memory with low write energy, logic-process compatibility, and compact cell area. The 1T1R ReRAM [1-3] fits embedded applications requiring fast read (RD) access time (T_{AC}) and low RD-V_{DDMIN}, particularly for devices powered by batteries or energy harvesters. The cross-point ReRAM [4-6] is meant for high capacities with high RD-V_{DDMIN} and slow T_{AC}. As devices shrink, ReRAMs have higher cell resistance (R) and greater variations in write time and R, which reduces the R-ratio (R_H/R_L) between the high-R state (HRS, R_H) and low-R state (LRS, R_L). ReRAM also have a high R_L, which enables a larger voltage drop across ReRAM to reduce write voltage and cell-switch (CS) size. Thus, ReRAM macro designs suffer: (1) small sensing margin (SM), limited RD-V_{DDMIN}, and slow T_{AC} due to high-R_L and small R-ratio; (2) increase in energy due to large set DC-current (I_{DC-SET}) resulting from wide set-time (T_{SET}) distribution. This study develops a swing-sample-and-couple (SSC) voltage-mode sense amplifier (VSA) to overcome (1), enabling 1.8× greater SM for lower RD-V_{DDMIN} and 1.7× faster T_{AC} across various V_{DD}, compared to conventional differential-input (CD) VSAs. To reduce >99% set energy, we use a 4T self-boost-write-termination (SBWT) scheme to cut off I_{DC-SET} of faster-T_{SET} devices, with an area penalty below 0.5%. A fabricated 28nm 1Mb ReRAM macro achieves T_{AC} = 404ns at V_{DD} = 0.27V and confirms the I_{DC-SET} cut-off by SBWT.

Figure 19.4.1 shows the ReRAM macro. This ReRAM cell [7] is logic-process compatible, capable of reaching 0.0308μm² using 28nm logic rules. The *set* and *reset* operations alter the ReRAM from HRS to LRS and LRS to HRS, respectively. Due to small R-ratio, the difference in read cell current (I_{CELL}) between RD-LRS and RD-HRS is small, particularly at a low V_{DD}. This prevents the use of current-mode sense amplifiers (CSAs) for low-V_{DD} applications. For VSA, BL voltages (V_{BL}) for RD-HRS (V_{BL-H}) and RD-LRS (V_{BL-L}) decrease with an increase in BL-developing time (T_{BL}). CD-VSAs establish the reference voltage (V_{REF}) at the mid-point between the V_{BL-H} (V_{BL-H_TAIL}) of tail-HRS (with min. R_H) and the V_{BL-L} (V_{BL-L_TAIL}) of tail-LRS (with max. R_L) cells, yielding SMs (V_{SM}) for tail-HRS and tail-LRS are (V_{SM-H_MIN} = V_{BL-H_TAIL} − V_{REF}) and (V_{SM-L_MIN} = V_{REF} − V_{BL-L_TAIL}). Thus, V_{SM} (= $0.5\Delta V_{BLS\_MIN}$) of CD-VSA is half the difference in minimum BL voltage swing (V_{BLS}), ΔV_{BLS_MIN} (= V_{BL-H_TAIL}−V_{BL-L_TAIL}). Unlike CSAs, VSAs can increase V_{SM} by using a longer T_{BL}, at the price of slower T_{AC}. However, V_{DDMIN} is limited by the need for a large ΔV_{BLS_MIN} to overcome SA offset (V_{OS}). Unfortunately, the effectiveness of V_{OS} suppression is degraded at low V_{DD}, necessitating a new VSA with smaller ΔV_{BLS_MIN}.

Figure 19.4.2 presents the circuitry of SSC-VSA, comprising a CD-VSA with one additional PMOS transistor (T1), two switches (SW1 and SW2), and one capacitor (C1). The V_{REF} for SSC-VSA is generated by the replica BLs (R-BLs) with tail-HRS and tail-LRS cells on each R-BL to track the sum of the V_{BLS} of tail-HRS and tail-LRS, resulting in V_{REF} = V_{DD} − (V_{BLS-H_TAIL} + V_{BLS-L_TAIL}). The replica tail-HRS/LRS cells are programmable using MLC program-verify operations. Unlike mid-point V_{REF} with V_{SM} = $0.5\Delta V_{BLS-MIN}$ as in CD-VSA, SSC-VSA places the V_{REF} below $V_{BL-L-TAIL}$ and increases the usage of $\Delta V_{BLS-MIN}$ (by up to 99%) as the SM for lower RD-V_{DDMIN} and faster T_{AC}.

Figure 19.4.3 presents the waveforms of SSC-VSA. In phase-1 (V_{BLS} sampling), for a given T_{BL} period, the BL develops V_{BLS} (= V_{DD} − V_{BL}) and V_{BL} is applied to IN1 of CD-VSA (V_{IN1} = V_{BL}) and node-A of C1, while node-B of C1 is biased at V_{REF}. In phase-2 (V_{BLS} coupling), SW1 is switched off to isolate node-A from BL. T1 is then turned on to pull node-A from V_{BL} to V_{DD}. This V_{BLS} shift at node-A boosts the voltage at IN2 (V_{IN2}) of the VSA from V_{REF} to (V_{REF} + V_{BLS}). In phase-3 (comparison), the SAEN turns on the VSA and detects the voltage difference between its two inputs (ΔV_{SA} = V_{IN2} − V_{IN1} = V_{SM}), resulting in ΔV_{SA} = (V_{REF} + V_{BLS}) − (V_{DD}-V_{BLS}) = $2V_{BLS}$ − (V_{BLS-H_TAIL} + V_{BLS-L_TAIL}). Reading an HRS cell ($V_{BLS-H} \leq V_{BLS-H\_TAIL}$) gives $\Delta V_{SA-H} \leq$ (V_{BLS-H_TAIL} − V_{BLS-L_TAIL}) = $-\Delta V_{BLS\_MIN}$. Reading an LRS cell ($V_{BLS-L} \geq V_{BLS-L\_TAIL}$) gives $\Delta V_{SA-L} \geq$ (V_{BLS-L_TAIL} − V_{BLS-H_TAIL}) = ΔV_{BLS_MIN}. Finally, the output is generated at DOUT.

Figure 19.4.4 presents the performance of SSC-VSA. Even with four additional devices (1T+1C+2SWs), the SSC-VSA imposes an area penalty below 1% for 1Mb ReRAM macros. Note that SSC-VSA can adopt a small-V_{OS} VSA to further shorten T_{AC}, and C1 should be chosen to assure faster T_{AC} despite increased BL load (T_{BL}). As T_{BL} increases, both ΔV_{SA-H} and ΔV_{SA-L} increase to achieve 1.8 to 2× larger SM than CD-VSA at the same T_{BL}. Using the ΔV_{BLS_MIN} as ΔV_{SA} enables SSC-VSA to achieve >100mV lower V_{DDMIN} and 1.7× faster T_{AC} than CD-VSA for reading tail bits with R-ratio = 5 at BL-length = 512.

Figure 19.4.5 presents the circuitry and performance of a SBWT write-driver (WD), comprising a 2T (P1 and N1) bias-voltage generator (VBG), set driver (NSD), and reset driver (NRSD). The gate of P1 is connected to BL, while the gate of N1 is controlled by a set-reference signal (SREF). During the reset period (T_{RESET}), reset current (I_{RESET}) is provided by NRSD (SEL = 1). At the beginning of a set operation, a short SEL pulse turns on NRSD and sets BL to 0V, while node SREF is biased at V_{SREF} to generate bias voltage (V_{FB}) at node FB. The V_{FB} enable NSD to provide current up to I_{NSD_MAX}, which is k times larger than I_{SET}. Before cell switching, the drain current of NSD (I_{NSD}) is equal to the SET current (I_{SET}) of the accessed HRS cell. When the accessed cell switches from HRS to LRS, the low R_L generates a large I_{DC-SET} (>>I_{NSD_MAX}), thereby increasing V_{BL} and lowering V_{FB} (reducing I_{NSD}). Positive feedback between NSD and VBG increases V_{BL} from 0V to V_{WL-SET}−V_{TH-CS}, and then turns off CS (I_{DC-SET} = 0), where V_{TH-CS} is the V_{TH} of CS of the accessed cell. I_{DC-SET} cut-off enables SBWT to reduce SET power by 99.9% for a 128b IO ReRAM macro with 100× difference in T_{SET} between cells. Using only 4T, SBWT-WD increases area overhead by less than 0.5% for a 1Mb macro, which is 15× smaller than a previous scheme [8].

Figure 19.4.6 presents the measurement results. A 1Mb ReRAM macro, including four 512-rows 256Kb subarrays and test modes, is fabricated in a 28nm CMOS process. To extract T_{AC}, this test-chip uses D flip-flops to latch the output (DOUT) of macro, mask out the wiring delay time due to on-chip buffers and test-board parasitic. The measured T_{AC} at V_{DD} = 0.85V and 0.27V are 6.8ns and 404.4ns, respectively. To monitor the I_{DC_SET} transition during the operation of SBWT, an external resistor (R1) is connected to the source terminal of NSD to translate I_{DC_SET} into the voltage drop across R1 (V_{R1}). As shown in the captured waveform, V_{R1} increases to 128mV with large I_{DC_SET} when the ReRAM switches from HRS to LRS and SBWT is off. When SBWT is on, V_{R1} drop to 0V within 5ns (excluding RC delay of testboard). Figure 19.4.7 shows the die micrograph.

Acknowledgement:

The authors would like to thank CIC, NSC-Taiwan, TSMC for various supports.

References:

[1] S.-S. Sheu, et al., "A 4Mb embedded SLC resistive-RAM macro with 7.2ns read-write random access time and 160ns MLC-access capability," *ISSCC*, pp. 200-201, 2011.

[2] W. Otsuka, et al., "A 4Mb conductive-bridge resistive memory with 2.3GB/s read-through and 216MB/s program throughput," *ISSCC*, pp. 210-211, 2011.

[3] M.-F. Chang, et al., "A 0.5V 4Mb Logic-Process Compatible Embedded Resistive RAM (ReRAM) in 65nm CMOS Using Low Voltage Current-Mode Sensing Scheme with 45ns Random Read Time," *ISSCC*, pp. 434-435, 2012.

[4] C. J. Chevallier et al., "A 0.13μm 64Mb Multi-Layered Conductive Metal-Oxide Memory," *ISSCC*, pp. 260–261, 2010.

[5] A. Kawahara, et al. "An 8Mb Multi-Layered Cross-Point ReRAM Macro with 443MB/s Write Throughput," *ISSCC*, pp. 432-433, 2012.

[6] T.-Y. Liu, et al., "A 130.7mm² 2-layer 32Gb ReRAM memory device in 24nm technology," *ISSCC*, pp. 210-211, 2013.

[7] W.-C. Shen, et al., "High-k metal gate contact RRAM (CRRAM) in pure 28nm CMOS logic process," *IEDM*, pp. 31.6.1-31.6.4, 2012.

[8] X.-Y. Xue, et al., "A 0.13μm 8Mb logic based CuSiO resistive memory with self-adaptive yield enhancement and operation for power reduction," *Symp. VLSI Tech.*, pp. 42-43, 2012.

978-1-4799-0917-9/14 $31.00 © 2014 IEEE

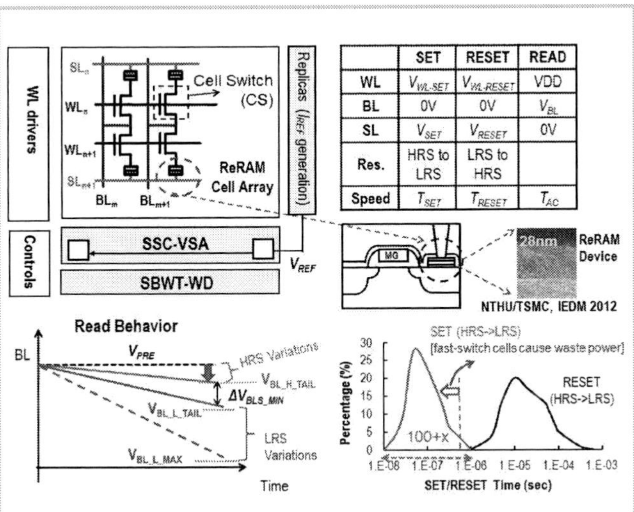

	SET	RESET	READ
WL	$V_{WL\text{-}SET}$	$V_{WL\text{-}RESET}$	VDD
BL	0V	0V	V_{BL}
SL	V_{SET}	V_{RESET}	0V
Res.	HRS to LRS	LRS to HRS	
Speed	T_{SET}	T_{RESET}	T_{AC}

Figure 19.4.1: Structure and cell behaviors of ReRAM macro.

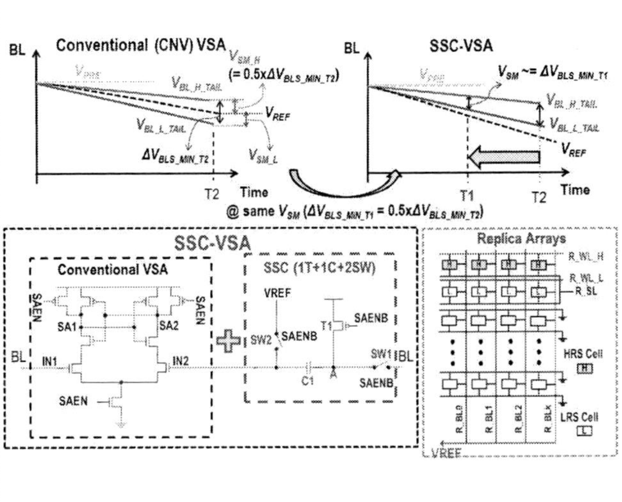

Figure 19.4.2: Concept and circuits of SSC-VSA.

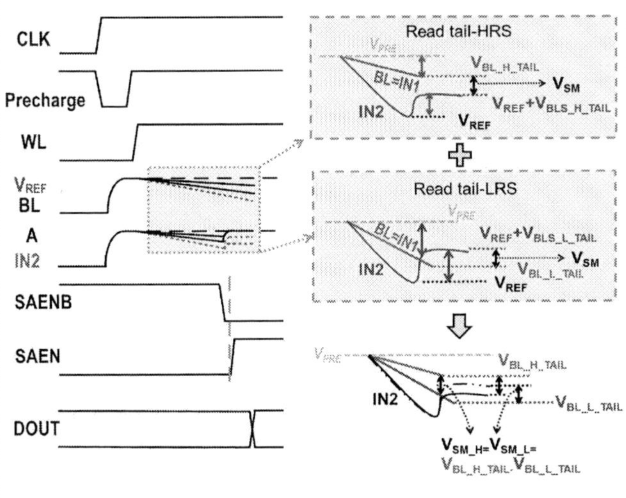

Figure 19.4.3: Waveforms of SSC-CSA.

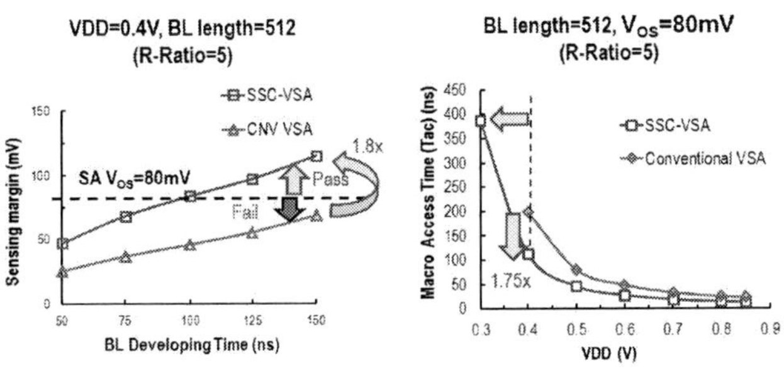

Figure 19.4.4: Performance of SSC-VSA.

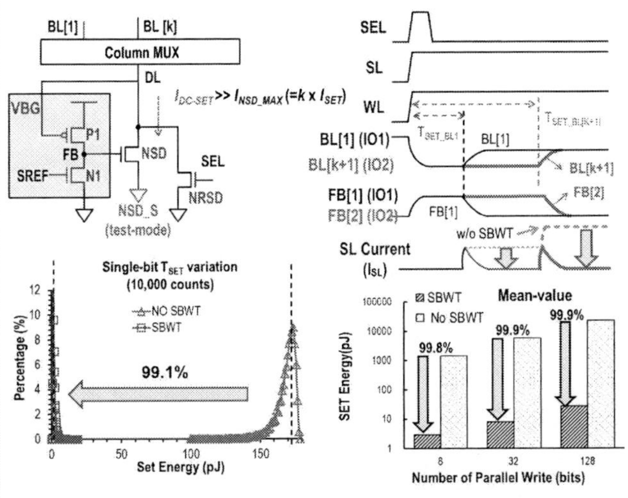

Figure 19.4.5: Circuitry and performance of SBWT.

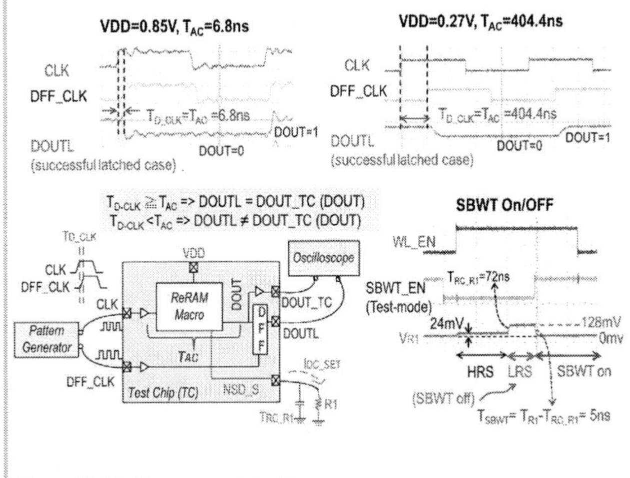

Figure 19.4.6: Measurement results.

978-1-4799-0917-9/14 $31.00 © 2014 IEEE

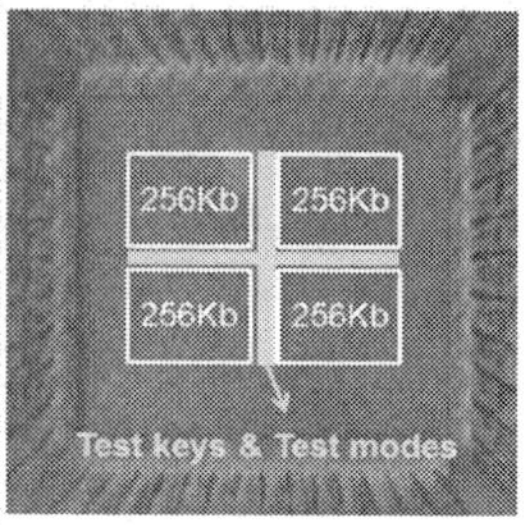

Technology	28nm Hi-K MG CMOS Process
Capacity	1Mb (4 x 256Kb)
Sub-Array (256Kb)	512 rows x 512 columns
ReRAM Cell Area	0.0308um^2
Sun-block Area (256Kb)	0.56mm^2 (include test-modes)
Read Speed	6.8ns at VDD=0.85V 404ns at VDD=0.27V (at LRS>75K ohm)
Write Speed	SET: ~500ns (at 25uA) RESEST: ~100us (at 50uA)
Read Power	VDD=0.85V: 77uA at 10Mhz VDD=0.27V: 3.2uA at 1Mhz

Figure 19.4.7: Die micrograph and summary table.

19.5 Three-Dimensional 128Gb MLC Vertical NAND Flash-Memory with 24-WL Stacked Layers and 50MB/s High-Speed Programming

Ki-Tae Park, Jin-man Han, Daehan Kim, Sangwan Nam, Kihwan Choi, Min-Su Kim, Pansuk Kwak, Doosub Lee, Yoon-He Choi, Kyung-Min Kang, Myung-Hoon Choi, Dong-Hun Kwak, Hyun-wook Park, Sang-won Shim, Hyun-Jun Yoon, Doohyun Kim, Sang-won Park, Kangbin Lee, Kuihan Ko, Dong-Kyo Shim, Yang-Lo Ahn, Jeunghwan Park, Jinho Ryu, Donghyun Kim, Kyungwa Yun, Joonsoo Kwon, Seunghoon Shin, Dongkyu Youn, Won-Tae Kim, Taehyun Kim, Sung-Jun Kim, Sungwhan Seo, Hyung-Gon Kim, Dae-Seok Byeon, Hyang-Ja Yang, Moosung Kim, Myong-Seok Kim, Jinseon Yeon, Jaehoon Jang, Han-Soo Kim, Woonkyung Lee, Duheon Song, Sungsoo Lee, Kye-Hyun Kyung, Jeong-Hyuk Choi

Samsung Semiconductor, Hwasung, Korea

In the past few years, various 3D NAND Flash memories have been demonstrated, from device feasibility to chip implementation, to overcome scaling challenges in conventional planar NAND Flash [1-3]. The difficulties include shrinking the NAND cell and increasing manufacturing costs due to quadruple patterning and extreme ultraviolet lithography, motivating the development of the next-generation node beyond 16nm-class NAND Flash [4]. In this paper, as a new 3D memory device with lower manufacturing cost and superior device scalability, we present a true 3D 128Gb 2b/cell vertical-NAND (V-NAND) Flash. The chip accomplishes 50MB/s write throughput with 3K endurance for typical embedded applications such as mobile and personal computer. Also, extended endurance of 35K is achieved with 33MB/s of write throughput for data center and enterprise SSD applications.

This work uses a vertical NAND cell array technology with damascened metal-gate SONOS type cell [3]. Figure 19.5.1 shows the die micrograph of the 3D V-NAND flash. The chip has two 64Gb memory planes, each consisting of 2732 blocks with 8KB page size and 3MB block size. We employ shared-WL-block scheme for row circuit [5] and one-side page buffer for column circuit [5-6] that are also used in planar 2D NAND to reduce area. The majority of data latches and control logic used for the 533Mb/s DDR interface of [4] are adopted in I/O area. This new placement is well suited for the wave-pipeline datapath architecture, which is optimized for NAND device, because no pipeline flip-flops are required for the datapath in the periphery. These techniques help obtain a 133mm^2 chip area with 80% array efficiency and 0.96Gb/mm^2 bit density.

Figure 19.5.2 shows the schematic diagram and cross-sectional SEM image of the fabricated 3D V-NAND. A 3D V-NAND string consists of 26 WL stacked layers including 2 dummy WLs and 2 select-gate layers. Eight V-NAND strings are shared by a bitline (BL) and all WLs of a layer in a block are connected and driven by a single row decoder. In order to selectively drive the V-NAND string in the shared BL structure, string select lines (SSL) are separately formed. One of the key features in the 3D V-NAND is a bulk-erase operation using FN tunneling mechanism. When compared to an erase operation by GIDL current [2], the bulk erase increases erase speed, lowers power and offers more reliable operation [3]. Thus, conventional operations used in the planar 2D NAND can also be applied to the 3D V-NAND with simple modifications.

The use of barrier-engineered materials and all-around gate structure in the 3D V-NAND cell exhibits advantages over 2D planar NAND, such as small V_{th} shift due to cell coupling and narrow natural V_{th} distribution of cells, improving device performance and reliability. A better cell characteristic of 84% reduced V_{th} shift due to coupling and 33% narrower natural V_{th} distribution are obtained in the 3D V-NAND compared to 1xnm planar NAND, as shown in Fig. 19.5.3(a). Due to fast detrapping, which usually occurs in the charge trap device [7], a larger V_{th} variation in programmed cells in the 3D V-NAND is observed compared to the planar 2D device. To overcome the problem, we develop a counter-pulse program using self-boosting shown in Fig. 19.5.3(b). By using a negative channel self-boosting operation during program, a large negative field is generated so that the fast detrapping process is accelerated and detrapped cells are reprogrammed in the next program pulse. As a result, the wider programmed V_{th} distribution due to fast detrapping is reduced. Additionally, by incorporating the cell characteristics mentioned above, a tightly programmed cell is realized with performance of 50MB/s write throughput.

Another important design consideration is signal integrity for driving 3D memory array, especially for WL crosstalk caused by large capacitive coupling between WLs. Unlike line-to-line capacitive coupling between WLs in 2D planar NAND arrays, area-to-area capacitive coupling between WL plates is dominant in the 3D V-NAND because of the vertical structure, as shown in Fig. 19.5.2. Consequently, a large glitch caused by the coupling between WLs during operation can cause unintended program and read disturbance reliability problems. Therefore, as one of the methods to reduce the coupling glitch, we adopt a degeneration scheme for WL driver that uses a controlled current sink, as shown in Fig. 19.5.4(a). The degeneration circuit selectively discharges the victim WL when the glitch voltage is greater than a predetermined target level. In addition to the degeneration scheme, coupling-predicted-offset scheme is also used. Depending on the target voltage of the aggressor WL, the target voltage of victim WL is less driven with an adjusted offset level. Figure 19.5.4(b) shows simulated waveforms of the WL signals for the coupling glitches during program operation. As shown in the figure, the overshoot of the victim WL voltage is drastically reduced.

Lowering power dissipation of the NAND device boosts overall system performance because it allows an increase in the number of interleaving NAND operations. In general, on-chip high-voltage-pump circuits generate many high voltages used in a memory array for switching circuits and heavily loaded signal lines in core and peripheral circuitry. Hence, the pump circuits are active in all NAND operations. In the SSD application, a high voltage of 12V is typically available via the system board such that NAND device utilizes the high voltage from the external supply source. In this work, we adopt an external high-voltage-supply scheme in order to further reduce power consumption, as shown in Fig. 19.5.5(a). It requires an extra pad for the external high-voltage supply and a simple circuit for switching between the external and internal supply sources. A level-detecting circuit is also implemented to avoid any high-voltage failure from unexpected power-down. The user can invoke the scheme by sending a specific command. By using the scheme, over 50% active power reduction for program operation is achieved with an overall system performance increase, as shown in Fig. 19.5.5(b).

Figure 19.5.6(a) shows a comparison of measured V_{th} distributions between planar 1xnm NAND and the 3D V-NAND. For higher endurance NAND applications such as enterprise and datacenter SSDs, the performance of the 3D V-NAND can be throttled down to 33MB/s that allows us to achieve over 10× endurance compared to that of normal operation. As shown in the figure, better V_{th} distribution of the 3D V-NAND than 1xnm planar NAND is achieved with an endurance of 35K. The device features of the 3D V-NAND are summarized in Fig. 19.5.6(b).

References:

[1] K.-T. Park et al., "A 45nm 4Gb 3-Dimensional Double-Stacked Multi-Level NAND Flash Memory with Shared Bitline Structure", *ISSCC Dig. Tech. Papers*, pp. 9-10, Feb., 2008.

[2] T. Maeda et al., "Multi-stacked 1G cell/layer Pipe-shaped BiCS Flash Memory", *Dig. Symp. VLSI Circuits*, pp. 22-23, June 2009

[3] J. Jang et al., "Vertical Cell Array using TCAT (Terabit Cell Array Transistor) Technology for Ultra High Density NAND Flash Memory", *Dig. Symp. VLSI Tech.*, pp. 192-193, June 2009

[4] D. Lee et al., "A 64Gb 533Mb/s DDR Interface MLC NAND Flash in Sub-20nm Technology", *ISSCC Dig. Tech. Papers*, pp. 430-431, Feb. 2012.

[5] K.-T. Park et al., "A 7MB/s 64Gb 3-Bit/Cell DDR NAND Flash Memory in 20nm-node Technology", ISSCC Dig. Tech. Papers, pp. 212-213, Feb. 2011.

[6] H. Kim et al., "A 159mm^2 32nm 32Gb MLC NAND-Flash Memory with 200MB/s Asynchronous DDR Interface", ISSCC Dig. Tech. Papers, pp. 442-443, Feb. 2010.

[7] C.-P. Chen et al., "Study of Fast Initial Charge Loss and It's Impact on the Programmed States V_{th} Distribution of Charge-Trapping NAND Flash", *IEDM Tech. Dig.*, pp. 118-121, Dec. 2010

978-1-4799-0917-9/14 $31.00 © 2014 IEEE

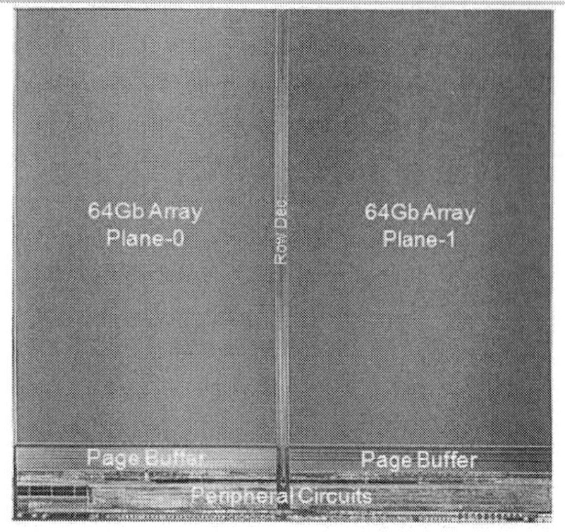

Figure 19.5.1: Die micrograph of 128Gb 2b/cell 3D V-NAND Flash.

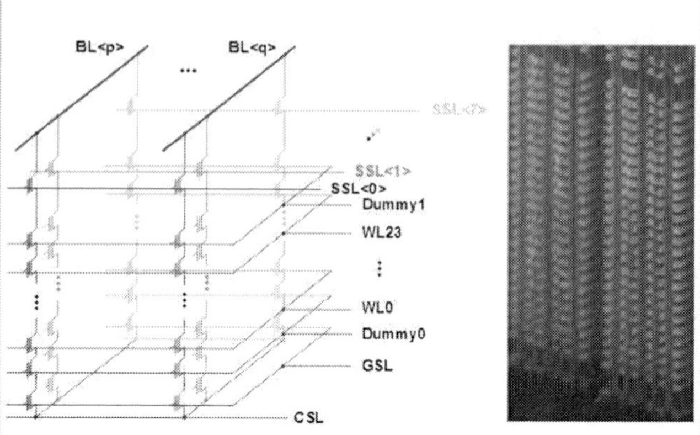

Figure 19.5.2: Schematic diagram and cross-sectional view of 3D V-NAND array.

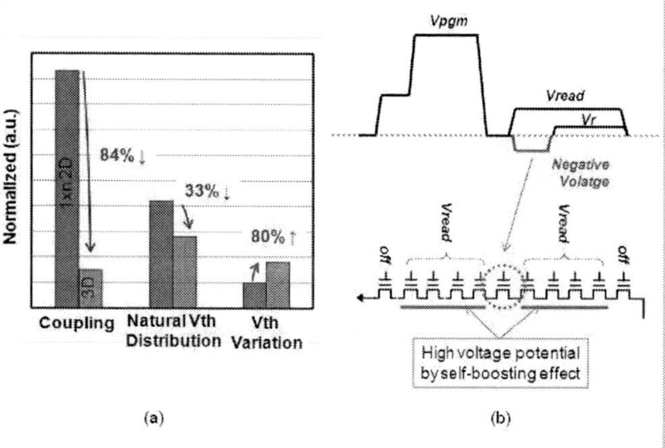

Figure 19.5.3: (a) Characteristic comparisons between planar 2D and 3D V-NAND, and (b) counter-pulse program scheme.

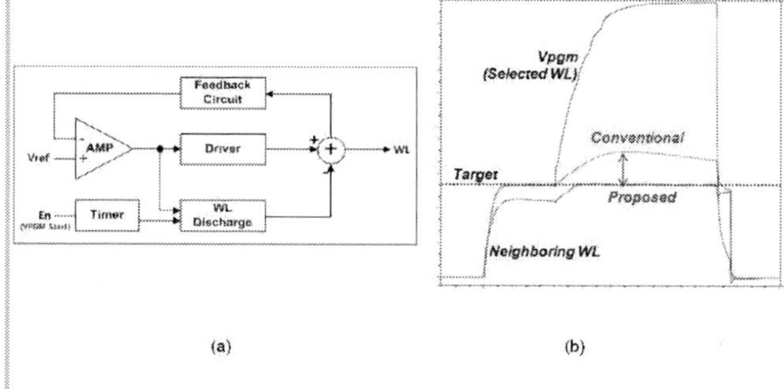

Figure 19.5.4: (a) Degeneration WL scheme, and (b) simulated waveform of WL signal in 3D V-NAND array.

19

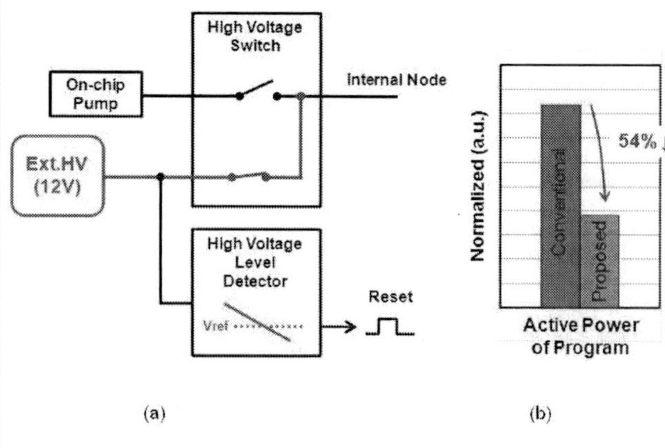

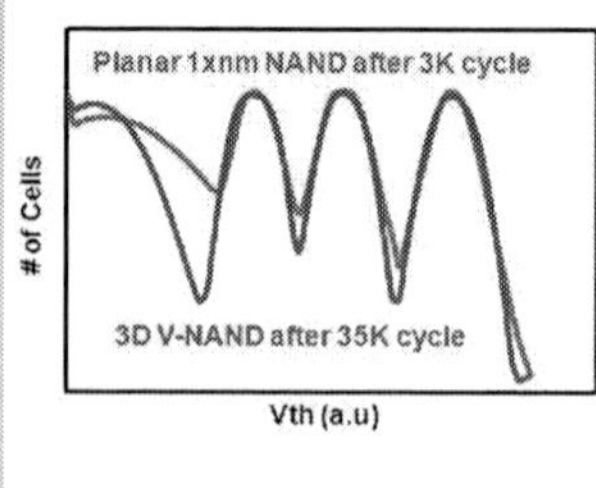

(a)

Bits per Cell	2
Density	128Gb
Chip Size	133mm² (0.96Gb/mm²)
Technology	Three Dimensional Vertical NAND, 3-metals
Organization	8kB × 384 pages × 5464 blocks × 8
Program Performance	50MB/s for Embedded App., 33MB/s for Enterprise SSD
Data Interface Speed	533Mbps (Async. DDR)
Power Supply	Vcc=3.3V / Vccq=1.8V

(b)

Figure 19.5.5: (a) External high-voltage-supply scheme, and (b) measured active power dissipation of program operation.

Figure 19.5.6: (a) Comparison of measured V_{th} distributions, and (b) device features of 128Gb 2b/cell 3D V-NAND.

This Page is a Continuation of the Previous and is Intentionally Unpaginated
Page Intentionally Left Blank

Figure

19.6 Hybrid Storage of ReRAM/TLC NAND Flash with RAID-5/6 for Cloud Data Centers

Shuhei Tanakamaru[1,2], Hiroki Yamazawa[1], Tsukasa Tokutomi[1], Sheyang Ning[1,2], Ken Takeuchi[1]

[1]Chuo University, Tokyo, Japan, [2]University of Tokyo, Tokyo, Japan

A hybrid storage architecture of ReRAM and TLC (3b/cell) NAND Flash with RAID-5/6 is developed to meet cloud data-center requirements of reliability, speed and capacity. The storage controller enhances reliability and performance through five techniques with minimal area overhead. The first three approaches, (i) flexible R_{Ref} (FR), (ii) adaptive asymmetric coding (AAC), and (iii) verify trials reduction (VTR), are applied to 50nm ReRAM to improve the bit-error rate (BER) by 69% and performance by 97%. Techniques (iv) balanced RAID-5/6 and (v) bits/cell optimization (BCO) are applied to 2Xnm TLC NAND to reduce the failure rate by 98% and extend the lifetime (write/erase (W/E) cycles) by >22×, respectively.

Conventionally, in hybrid ReRAM/MLC (2bits/cell) NAND, high-speed ReRAM is paired in small ratios with high capacity NAND, and data is allocated based on access frequency (i.e., frequently written hot data to ReRAM, cold data to NAND) and data size (i.e., fragmented random data to ReRAM, sequential data to NAND) to enhance the overall system performance, reliability and power [1]. Exchangeable TLC/MLC NAND storage arrays have been proposed [2], as well as application of duplicate data requiring RAID-1 to MLC NAND, for reliability improvement in enterprise servers [3].

This work presents hybrid storage of ReRAM and TLC NAND Flash with cost-effective RAID-5/6 (Fig. 19.6.1). RAID-5/6 is widely used in cloud storage and data warehouses due to its lower parity overhead (<10%) compared to RAID-1. Our architecture provides significant improvements in the reliability and performance of ReRAM and TLC NAND. Data to ReRAM is encoded by AAC and then written to ReRAM with VTR. During read, FR determines the optimum read reference resistance, R_{Ref}, to minimize the BER. In the NAND, balanced RAID-5/6 evenly allocates the data among the different page types in order to minimize the worst case RAID failure rate, and BCO decides the mode (TLC/MLC/SLC) to extend the NAND chip's lifetime.

Figure 19.6.2 describes FR in ReRAM, based on a 50nm, 64Mb Al_xO_y prototype, in which verify programming is applied on write units of 1Kb. After each set/reset pulse, verify read checks that the resistances satisfy the threshold levels ($<R_{Set}$, $>R_{Rst}$). The failed cells in the write unit are subjected to increasingly stronger set/reset pulses (voltage-increase for set, and pulse-width increase for reset) until the number of failed cells falls below the acceptable error threshold (E_{TH}), or the maximum number of verify trials is reached. The average and variance of measured set/reset resistances during cycling are also shown. At each set/reset cycle, the optimum R_{Ref} to minimize the BER is calculated. By adopting a more flexible 2-step R_{Ref} scheme to track with the optimum R_{Ref}, the measured BER is decreased by 65%.

Figure 19.6.3 describes AAC in ReRAM. Conventionally, in NAND, data retention causes the dominant error of threshold voltage (V_{th}) reduction; therefore, simple asymmetric coding is effective to reduce the population of high V_{TH} states [4]. However, we observe that ReRAM BER behavior is more complicated. Measurement of BER during set/reset cycling with FR shows switching of the dominant error type, as well as asymmetry between set and reset errors. This switching behavior is possibly due to combined effects from device wear-out and temporal imbalances between the set/reset algorithms. When the cycle count is below 5×10^6, increasing the set population reduces overall BER. Thus, if the write data has only a few set ("1") bits, the AAC encoder flips the data bits to increase the set population, raises the section flag, and then writes the flipped data to the ReRAM. After 5×10^6 cycles, because set error becomes higher, the reset population is increased instead. By combining FR and AAC with 60% asymmetry, 69% BER reduction is experimentally demonstrated with 6.3% overhead for the flag-cell area (see FR+AAC in Fig. 19.6.2).

VTR reduces set/reset times in ReRAM, and reset is shown in Fig. 19.6.4. To extend endurance, low reset voltages are maintained (~1.5V) and reset pulse widths are increased after every verify trial. As expected, the number of bits that fail verify decreases as the verify trials increase. Conventionally, reset completes when all of the write unit's selected cells reach the resistance threshold, R_{Rst}. Irregular cells with extra-long or branched filaments can cause unacceptably long (>15μs) reset times. One approach to reduce reset time or eliminate verify programming is to implement strong ECC to correct all of the irregular or unsuccessful bits [5]. In Fig. 19.6.4, measurements show that if 20 errors can be corrected (E_{TH} = 20), then only a single reset verify trial is needed. However, although write time is reduced, this level of ECC increases read power and read time. Since the access time in ReRAM is fast, ECC calculation time becomes a performance bottleneck. In the VTR, E_{TH} is dynamically adjusted, (see line in bottom Fig. 19.6.4), to minimize ECC calculation time and power, for only 1 reset verify trial, which reduces the reset time by 97%.

Figure 19.6.5 shows the balanced RAID-5/6 architecture for NAND. First, measured BER for TLC mode and MLC mode-1 are shown, based on page type (upper/middle/lower for TLC, upper/lower for MLC). In TLC mode, the middle-page BER is highest by 1.8×. The NAND Flash is organized in RAID-5/6, in which a single NAND chip represents one replacement unit. Generally, RAID-5 corrects one unit failure, and RAID-6 corrects up to two unit failures. Both approaches are more cost effective than mirroring (RAID-1) [3] to improve reliability, however, because stripes are conventionally assigned to the same type of page, the RAID failure rate is uneven depending on the page type (conv. RAID-5/6 in Fig. 19.6.5). Stripes allocated to the middle pages have the highest RAID failure rates. The developed balanced RAID distributes data within a stripe evenly among the upper/middle/lower pages (developed balanced RAID-5/6 in Fig. 19.6.5). As a result, each stripe has a similar failure rate, and the overall failure rate is decreased. The worst-case RAID failure rate improves by 98% with 2Xnm TLC NAND, and in MLC-mode, the improvement is 91%.

Figure 19.6.6 shows BCO for NAND, in which a worn TLC can be re-allocated as MLC or SLC to extend the useful lifetime of the chip and boost the storage-system performance. In cloud storage, TLC, MLC and SLC memories co-exist in the data system. Memory management software such as VMware and NetApp dynamically control allocation depending on the workload. Typically, SLC is used as cache, and MLC/TLC function as primary data storage. TLC NAND can operate in TLC mode for about 900 W/E cycles, due to the small V_{th} margins between memory states. In BCO, before reaching the TLC BER limit, the NAND is switched to MLC mode-1 {E P3 P5 P7}. At 2.5k cycles, MLC mode-2 {E P2 P4 P6} improves BER further, by shifting down the V_{th} of the highest state. Finally, after 7k W/E cycles, the NAND is operated as a SLC memory. At 10k cycles in SLC mode, changing the read reference from V_{Ref_A} to V_{Ref_B} is effective to track the minimum BER. By reusing the worn TLC NAND, the measured NAND chip's lifetime is extended by >22×.

Figure 19.6.7 shows the measurement system. FR and AAC decrease the BER of ReRAM by 69%, VTR decreases the ReRAM's program time by 97% and balanced RAID-5/6 improves the RAID failure rate by 98%. BCO extends the lifetime (W/E cycles) of NAND by >22×.

Acknowledgements:

This work is partially supported by CREST/JST. The authors appreciate T.O. Iwasaki, K. Johguchi, S. Hachiya, and C. Sun for their support.

References:

[1] H. Fujii, et al., "×11 Performance Increase, ×6.9 Endurance Enhancement, 93% Energy Reduction of 3D TSV-Integrated Hybrid ReRAM/MLC NAND SSDs by Data Fragmentation Suppression," *Dig. Symp. VLSI Circuits*, pp. 134-135, June 2012.

[2] S. Hachiya, et al., "TLC/MLC NAND Flash Mix-and-Match Design with Exchangeable Storage Array," *SSDM*, pp. 894-895, Sept. 2013.

[3] S. Tanakamaru, et al., "Unified Solid-State-Storage Architecture with NAND Flash Memory and ReRAM that Tolerates 32× Higher BER for Big-Data Applications," *ISSCC Dig. Tech. Papers*, pp. 226-227, Feb. 2013.

[4] S. Tanakamaru, et al., "95%-Lower-BER 43%-Lower-Power Intelligent Solid-State Drive (SSD) with Asymmetric Coding and Stripe Pattern Elimination Algorithm," *ISSCC Dig. Tech. Papers*, pp. 204-205, Feb. 2011.

[5] A. Kawahara, et al., "An 8Mb Multi-Layered Cross-Point ReRAM Macro with 443MB/s Write Throughput," *ISSCC Dig. Tech. Papers*, pp. 434-433, Feb. 2012.

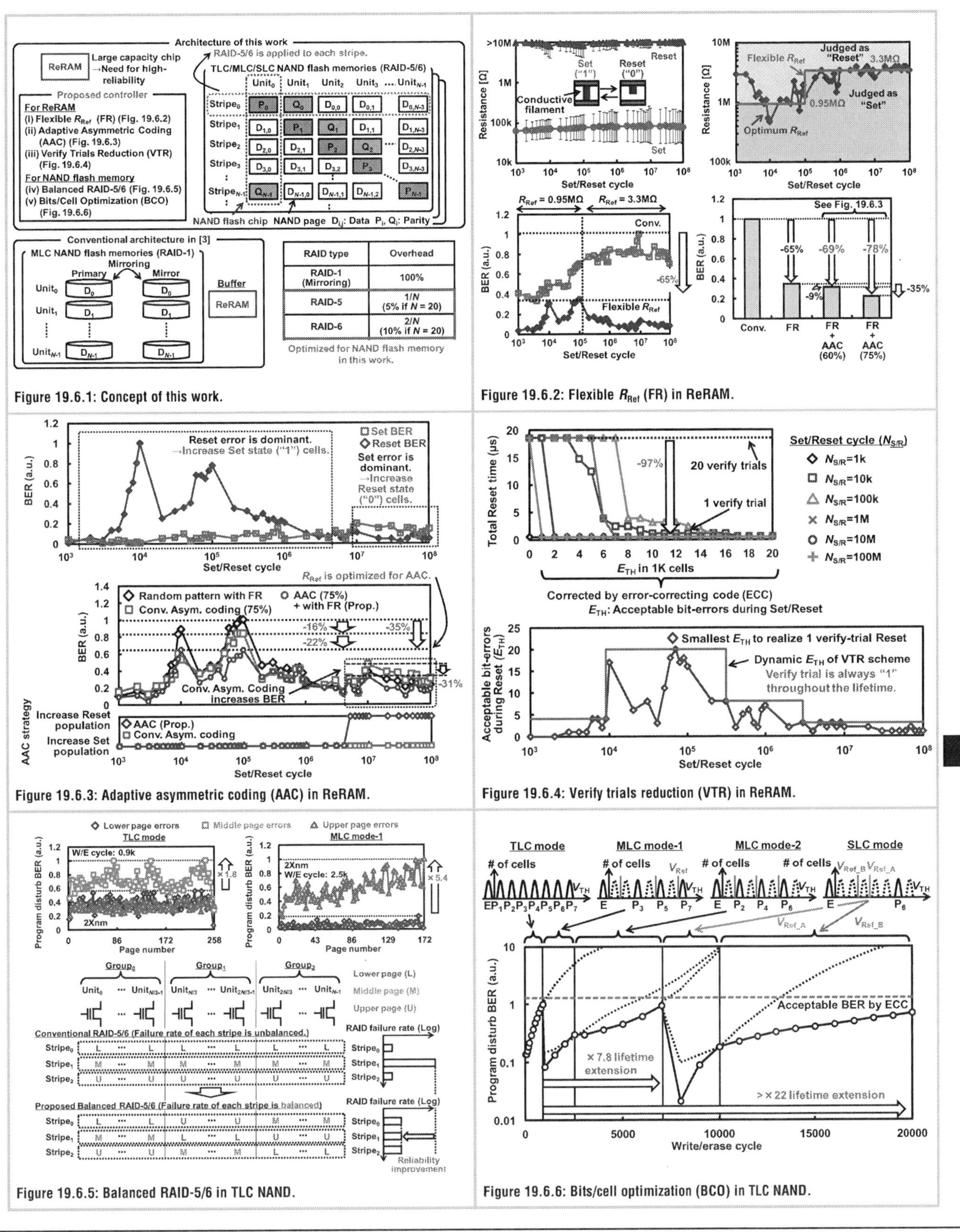

RAID type	Overhead
RAID-1 (Mirroring)	100%
RAID-5	1/N (5% if N = 20)
RAID-6	2/N (10% if N = 20)

Figure 19.6.1: Concept of this work.

Figure 19.6.2: Flexible R_{Ref} (FR) in ReRAM.

Figure 19.6.3: Adaptive asymmetric coding (AAC) in ReRAM.

Figure 19.6.4: Verify trials reduction (VTR) in ReRAM.

Figure 19.6.5: Balanced RAID-5/6 in TLC NAND.

Figure 19.6.6: Bits/cell optimization (BCO) in TLC NAND.

19

978-1-4799-0917-9/14 $31.00 © 2014 IEEE

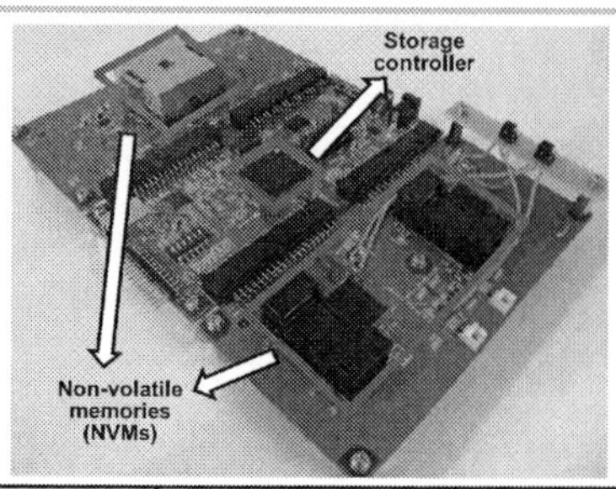

	Application	Improvement	
Flexible R_{Ref} (FR)	ReRAM	BER -65%	BER -69% (FR & AAC)
Adaptive Asymmetric Coding (AAC)	ReRAM	BER -9%	
Verify Trials Reduction (VTR)	ReRAM	Total Reset time -97%	
Balanced RAID-5/6	TLC NAND flash memory	RAID failure rate -98% (TLC RAID-6)	
Bits/Cell Optimization (BCO)	TLC NAND flash memory	Lifetime (W/E cycle) > × 22	

Figure 19.6.7: Micrograph of the measured storage system and summary of this work.

19.7 A 16Gb ReRAM with 200MB/s Write and 1GB/s Read in 27nm Technology

Richard Fackenthal[1], Makoto Kitagawa[2], Wataru Otsuka[2], Kirk Prall[3], Duane Mills[1], Keiichi Tsutsui[4], Jahanshir Javanifard[1], Kerry Tedrow[1], Tomohito Tsushima[2], Yoshiyuki Shibahara[4], Glen Hush[3]

[1]Micron, Folsom, CA,
[2]Sony, Boise, ID,
[3]Micron, Boise, ID,
[4]Sony, Kanagawa, Japan

Resistive RAMs (ReRAMs) have emerged as leading candidates to displace conventional Flash memories due to their high density, good scalability, low power and high performance. Previous ReRAM designs demonstrating high performance have done so on low density arrays (<1Gb) while those reporting high-density arrays (>8Gb) were accompanied by relatively low read and write performance [1-5]. This work describes a 16Gb ReRAM designed in a 27nm node, with a 1GB/s DDR interface and an 8-bank concurrent DRAM-like core architecture. High parallelism, a pipelined data-path architecture and innovations such as concurrent set/reset verify combine to achieve 200MB/s write and 1GB/s read throughputs in a high-density device.

The chip is fabricated in a 27nm node with 3-layer-Cu interconnect. The 6F2 memory cell consists of 1 selector transistor and 1 resistive memory element (Fig. 19.7.1). The resistive memory element is dual-layered with a CuTe conductive material and a thin insulator. The resistive element with top and bottom electrodes (TE, BE) operates through bipolar switching. The high-resistance state (HRS) changes to the low-resistance state (*set*, LRS) when the TE is positively biased, and the reverse (*reset*) when the BE is positively biased. The selector employs a buried-wordline architecture and is also used for current control during *set* (transition to LRS).

Supply voltages are V_{CC} = 1.2V and V_{PP} = 5V. The 5V V_{PP} is chosen as a standard system voltage and used internally to supply a high-efficiency charge pump that outputs 6.5V for programming operations. The chip has two interfaces, one high-speed LPDDR-like 1GB/s user interface, and a low-pin-count test interface used to enable massive parallelism for wafer-level testing. The chip is architected in a modular way such that the common core (CC), including periphery and arrays, can be easily interchanged with other interface link and physical layers. The CC is specified with its own command interface consisting of primitive operations such as loading, reading or writing a *sub-page* (unit of 66 bits) to the page buffer and issuing program or sense commands. The link layers assemble common core commands into higher-level functions such as burst write followed by program, or sense followed by burst read.

The 16Gb array is divided into 8 banks, each with 8 *Y-strips*, vertical groups of tiles with a common global bitline (GBL) (Fig. 19.7.2). A Y-strip is 16 tiles plus one redundant tile, and each tile is a matrix of 8192+256 local bitlines and 2048 wordlines (not including redundant elements) for a tile size of over 16Mb. During sense and program in a bank, 8 tiles (one per Y-strip) are activated simultaneously, each accessing a sub-page for a total sense/program concurrency of 512+16 cells. Since the page size is four times the number of sense/program circuits, there are 4 *nibbles*, defined as serial accesses to successive bitline addresses during the sense and program sequence. The complete sense/program unit is a *page* of 8 sub-pages (one per Y-strip) × 4 nibbles × (64+2) bits = 256+8 bytes per bank. The 8 byte per page side data is available to the system for purposes such as ECC. All 8 banks can be enabled concurrently to sense or program a total of 2048+64 bytes simultaneously in 10μs (program) and 2μs (sense), achieving bandwidths of over 200MB/s and 1GB/s, respectively.

The memory cell is a 3-terminal device with a common-source plate (CSP). The tile is divided into four *sub-tiles*, defined as the unit sharing a CSP. The sense and program circuits and X-decoder are shared across the 4 sub-tiles. This achieves a reduction in programming power due to plate charging, while minimizing the area penalty for tile and Y-strip-level circuits.

1GB/s read throughput is achieved by interleaving page access to the 8 banks and outputting the data through a 16b 500MT/s DDR data bus (DQ[15:0]) (Fig. 19.7.3). The *external read* command is converted into two internal CC commands, *sense* and *internal read* by the command decoder (DEC) and FSM in the interface. The sense operation transfers 66b sub-pages × 32 = 264B page data from the array to the page buffer in a bank (PBx[65:0]). Bank-level logic generates a 125MHz tri-state driver enable pulse (TPx) just after read is issued and the tri-state driver transfers sub-pages to the shared data-path (DP[65:0]). The enable pulses of the 8 banks are non-overlapping to avoid collision on the shared data path and to allow sequential reads of the banks. The sequential 66 bit × 32 cycle data is converted into sets of 64 bit × 33 cycle gapless data, then serialized to 16 bit × 500Mb/s DDR signaling and sent to the controller through HSUL-12 IOs.

Figure 19.7.4 shows the sense amplifier, which employs a differential current topology featuring an NMOS transistor to set a reference current (m0), cascode devices to control the bitline voltage (m1a,b), a current mirror to compare currents (m3a,b) and a comparator to restore logic levels. The cascode devices are thin-oxide transistors to reduce mismatch (thus reducing bitline voltage random offset) and speed up bitline charging. Thick-oxide protection transistors (m2a,b) shield the cascodes from the 5V supply. The protection transistors also reduce the drain capacitance presented to the current mirrors thus speeding up differential signal development. The sense amplifier is used in program operations to verify the cell state. Hence it can be configured into one of three modes: sense, set-verification, or reset-verification. Both the bitline voltage and reference current are adjusted for each mode. The bitline voltage is raised in verify modes to increase cell current for faster verification. Two muxes in each sense amplifier select the voltages according to the desired mode. Placing the mux in each sense amplifier allows concurrent set and reset verification to further speed up the verification step during programming.

Programming and sensing occurs in 8 tiles simultaneously, each with one active sub-tile. Each sub-tile can program or sense up to 66 cells simultaneously. This is limited by CSP resistance drop during programming and area due to sense/program circuits. Raising the CSP for the *set* operation consumes significant energy. Therefore we inhibit CSP charging on sub-tiles in which all bits have passed set-verify. This inhibit is likely to occur on a second or third pulse since most bits pass after the first pulse. CSP coupling to the wordlines during slew is managed by a trimmable driver strength to optimize the tradeoff of coupling magnitude and recovery time with program performance.

This device uses column redundancy to enhance yield. In traditional column redundancy architectures the solution matching and physical-to-logical steering occurs in the speed path, which for this chip is an 8ns cycle time from page buffer (PB) to main data-path (DP). Alternatively in this design, the microcontroller performs matching and steering during the programming and sensing algorithms, respectively (Fig. 19.7.5). This architecture re-uses the 66+2 latches per nibble in the page buffer to store logical (physical) data and overwrites them with physical (logical) data during programming (sensing). For optimal performance, the steering for nibble *N* is performed during the programming or sensing of nibble *N*+1. During programming, the bad columns are inhibited but their logical data is maintained in the original location as well as copied to the redundant one. This allows reuse of the same logical data to write at multiple addresses to optimize speed in the case of repeated patterns.

A summary table is shown in Figure 19.7.6 and the die micrograph is shown in Figure 19.7.7.

References:

[1] T. Liu, T. Yan, et al., "A 130.7mm² 2-Layer 32Gb ReRAM Memory Device in 24nm Technology", *ISSCC Dig. Tech. Papers*, pp. 210-211, Feb. 2013.
[2] A. Kawahara, R. Azuma, et al., "An 8Mb Multi-Layered Cross-Point ReRAM Macro with 443MB/s Write Throughput", *ISSCC Dig. Tech. Papers*, pp. 432-433, Feb. 2012.
[3] W. Otsuka, K. Miyata, et al., "A 4Mb Conductive-Bridge Resistive Memory with 2.3GB/s Read-Throughput and 216MB/s Program-Throughput", *ISSCC Dig. Tech. Papers*, pp. 210-211, Feb., 2011.
[4] J. Yi, H. Choi, et al., "Highly Reliable and Fast Nonvolatile Hybrid Switching ReRAM Memory Using Thin Al2O3 Demonstrated at 54nm memory Array", *Symposium on VLSI Technology (VLSIT)*, pp. 48-49, 2011
[5] X.Y. Xue, W.X. Jian, et al., "A 0.13μm 8Mb logic based Cu_xSi_yO resistive memory with self-adaptive yield enhancement and operation power reduction", *Symposium on VLSI Technology (VLSIT)*, pp. 42-43, 2012.

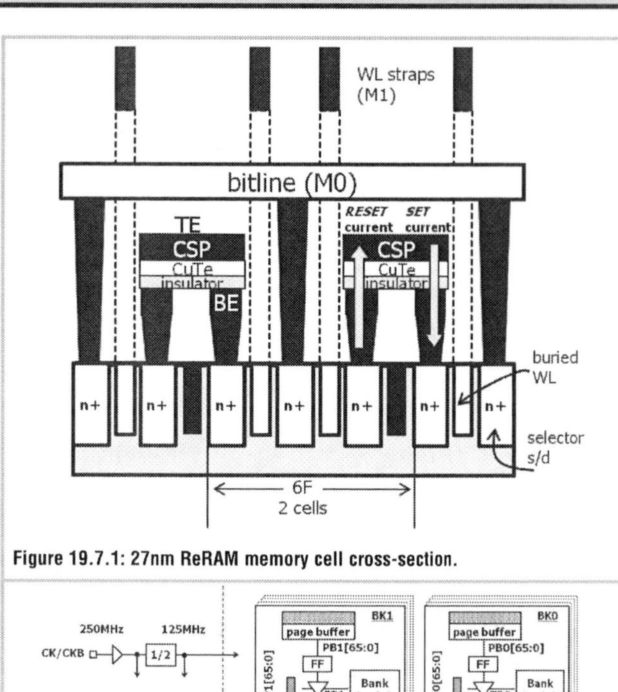

Figure 19.7.1: 27nm ReRAM memory cell cross-section.

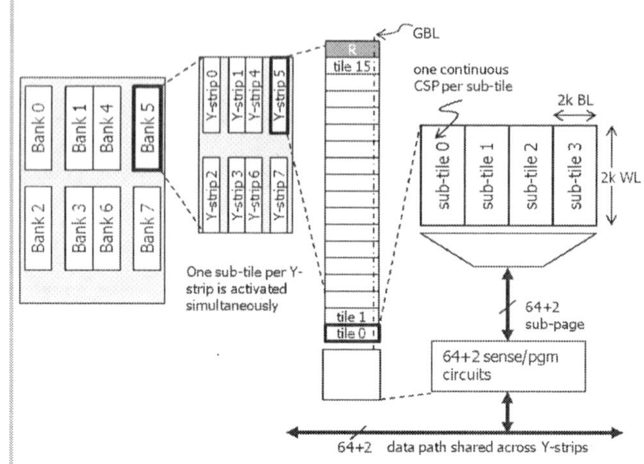

Figure 19.7.2: Array architecture block diagram.

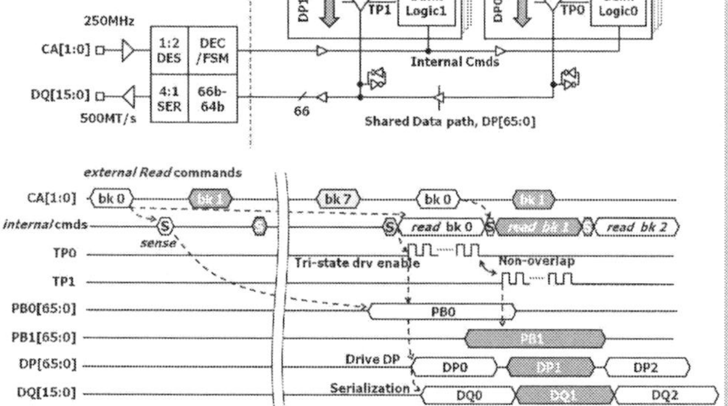
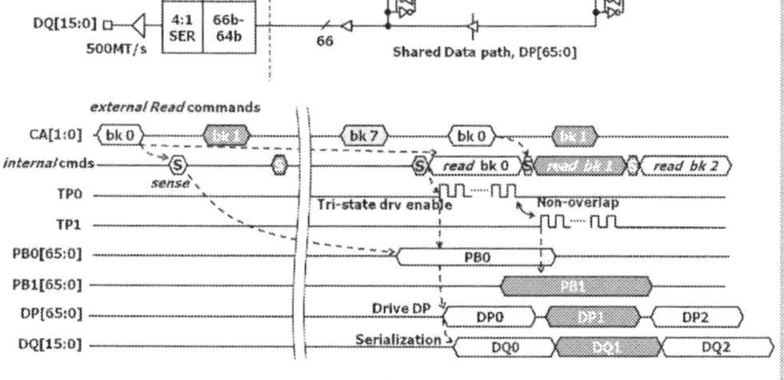

Figure 19.7.3: Data-path and high-speed interface.

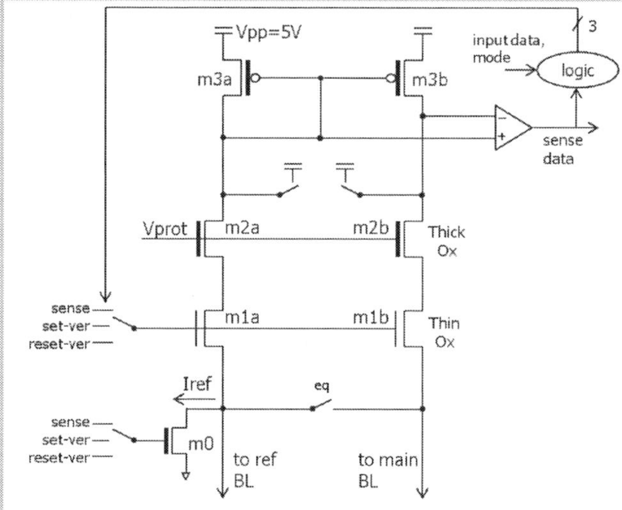

Figure 19.7.4: Sense amplifier schematic diagram,

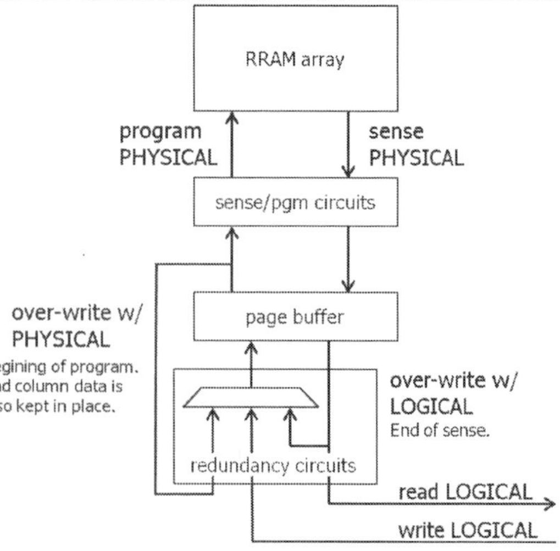

RRAM array

program PHYSICAL

sense PHYSICAL

sense/pgm circuits

over-write w/ PHYSICAL

Begining of program. Bad column data is also kept in place.

page buffer

over-write w/ LOGICAL

End of sense.

redundancy circuits

read LOGICAL

write LOGICAL

Figure 19.7.5: Column redundancy architecture.

Summary Table		
Density		16 Gb
Tech node (nm)		27
Cell Size (nm^2)		4374 ($6F^2$)
Die Size (mm^2)		168
Selector		Buried WL MOS selector
Read Performance	BW (MB/s)	1000
	Latency (uS)	2
Write Performance	BW (MB/s)	200
	Latency (uS)	10

Figure 19.7.6: Summary table.

19

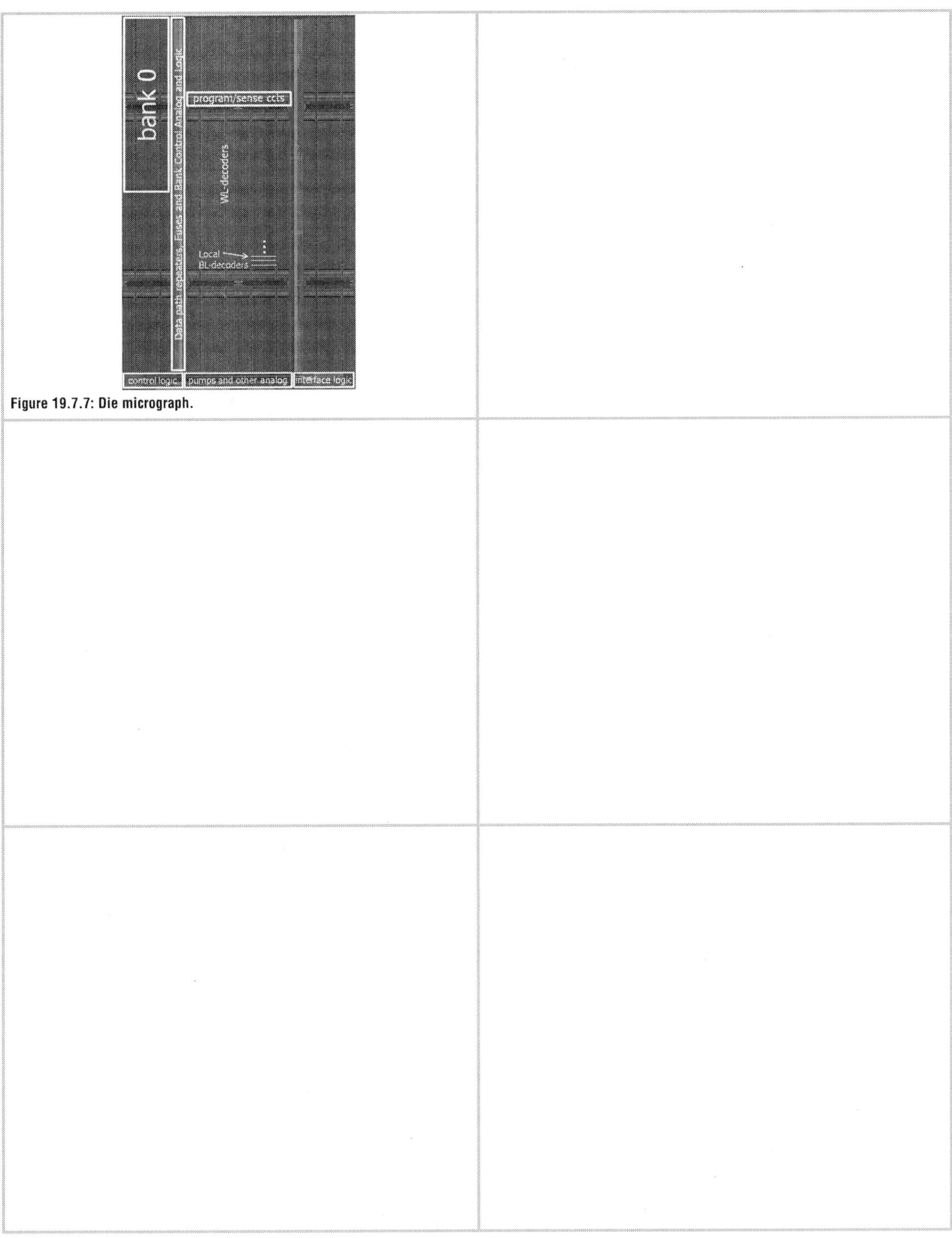

Figure 19.7.7: Die micrograph.

Session 20 Overview: *Wireless Systems*

WIRELESS SUBCOMMITTEE

Session Chair: *Iason Vassiliou*
Broadcom, Alimos, Greece

Session Co-Chair: *Myung-Woon Hwang*
FCI, Sungnam, Korea

State-of-the-art wireless systems implemented in low-cost, deep-sub-micron CMOS processes support a wide range of applications including mm-Wave ranging, Gb/s communications in 60GHz/5GHz bands and cost-sensitive cellular communications. This session includes one radar receiver paper, three state-of-the-art 60GHz transceivers supporting 2 to 28Gb/s, the first reported fully integrated 802.11a/b/g/n/ac SoC supporting over 1Gb/s and three cellular receivers implementing blocker-tolerant techniques intended to eliminate the need for external filters.

20.1 A 40nm CMOS Receiver for 60GHz Discrete-Carrier Indoor Localization Achieving mm-Precision at 4m Range **8:30 AM**

T. Redant, KU Leuven, Leuven, Belgium

In Paper 20.1, KU Leuven presents a 40nm CMOS receiver for 60GHz mm-precise localization that uses a combination of multi-tone signals and sub-sampling to reduce the required bandwidth in baseband, thus enabling fast and power-efficient ranging measurements. Multi-burst averaging leads to 4mm precision across 4-meter distances at 50kHz update rate.

20.2 A 16TX/16RX 60GHz 802.11ad Chipset with Single Coaxial Interface and Polarization Diversity **9:00 AM**

M. Boers, Broadcom, Irvine, CA

In Paper 20.2, Broadcom presents a highly integrated 60GHz 802.11ad, 16x16 phased-antenna-array transceiver chipset with polarization diversity in 40nm CMOS, including RF/PHY and MAC. It supports up to 64-QAM modulation and throughput of over 3.5Gb/s at 10m. Radio power consumption is 960/1190mW (RX/TX), while a single coaxial cable is used to interface between chips to ease integration to PC platforms.

20.3 A 64-QAM 60GHz CMOS Transceiver with 4-Channel Bonding **9:30 AM**

K. Okada, Tokyo Institute of Technology, Tokyo, Japan

In Paper 20.3, Tokyo Institute of Technology presents the first 64-QAM 60GHz CMOS RF transceiver that achieves a TX-to-RX EVM of -26.3dB, enabling 10.56Gb/s in regular 802.11ad mode and 28.16Gb/s 4-channel bonding mode using 16QAM. The transceiver is implemented in 65nm CMOS and consumes 251mW and 220mW from a 1.2V supply in transmit and receive mode respectively.

20.4 A Fully Integrated Single-Chip 60GHz CMOS Transceiver with Scalable Power Consumption for Proximity Wireless Communication **10:15 AM**

S. Saigusa, Toshiba, Kawasaki, Japan

In Paper 20.4, Toshiba presents the first fully-integrated single-chip 60GHz CMOS transceiver for proximity systems including RF, PHY and MAC, achieving MAC throughput of 2.0Gb/s at 4cm distance. Digital noise-suppression techniques and broadband noise-tolerant RF/analog blocks reduce noise from digital blocks to less than -35dBc. Total power consumption of the TX/RX chip, implemented in 65nm CMOS, is 1268mW, which is scaled to 36% when operating at 0.2Gb/s.

20.5 A 40nm Dual-Band 3-Stream 802.11a/b/g/n/ac MIMO WLAN SoC with 1.1Gb/s Over-the-Air Throughput **10:30 AM**

M. He, Marvell, Santa Clara, CA

In Paper 20.5, Marvell presents the first reported dual-band 3-stream 802.11a/b/g/n/ac MIMO WLAN SoC in 40nm CMOS. By employing an all-digital fractional-N PLL with an FoM of -244dB, and a wideband low-impedance bias circuit that minimizes pre-PA driver memory effect, it achieves an EVM floor of -37dB and an over-the-air TCP/IP throughput of 1Gb/s at 5775MHz in 802.11ac MCS9 80MHz mode.

20.6 A Blocker-Resilient Wideband Receiver with Low-Noise Active Two-Point Cancellation of >0dBm TX Leakage and TX Noise in RX Band for FDD/Co-Existence **10:45 AM**

J. Zhou, Columbia University, New York, NY

In Paper 20.6, Columbia University presents a blocker-resilient software-defined receiver in 65nm CMOS with a low-noise active TX leakage cancellation scheme that can cancel up to +2dBm peak TX leakage at the RX input, enabling an effective IIP3 of +33dBm (enhancement of 19dB) with an associated increase in NF of <0.8dB.

20.7 A Multi-Band Inductor-Less SAW-Less 2G/3G-TD-SCDMA Cellular Receiver in 40nm CMOS **11:15 AM**

M-D. Tsai, MediaTek, Hsinchu, Taiwan

In Paper 20.7, MediaTek presents the most compact inductor-less, SAW-less TDD cellular six-band receiver for GSM/EDGE/TDSCDMA. It is implemented in 40nm CMOS, occupies 0.57mm^2 and achieves a 1.7dB NF while consuming 26mA.

20.8 A 20mW GSM/WCDMA Receiver with RF Channel Selection **11:45 AM**

J. W. Park, University of California, Los Angeles, CA

In Paper 20.8, UCLA presents a wideband receiver by employing N-path filters around the LNA in 65nm CMOS. The receiver provides an attenuation of 15dB in the middle of the alternate adjacent channel and a noise figure of 5.4dB with a 0dBm blocker at 23MHz offset.

978-1-4799-0917-9/14 $31.00 © 2014 IEEE

20.1 A 40nm CMOS Receiver for 60GHz Discrete-Carrier Indoor Localization Achieving mm-Precision at 4m Range

Tom Redant, Tuba Ayhan, Nico De Clercq, Marian Verhelst, Patrick Reynaert, Wim Dehaene

KU Leuven, Leuven, Belgium

The globally available large unlicensed frequency spectrum around 60GHz has recently gained a lot of attention. Its broad bandwidth, combined with a high allowed transmitted power level, provides an excellent opportunity for numerous applications, among others high-precision ranging and localization. Despite being readily available at 60GHz, high bandwidths come with a significant power penalty in the baseband. The presented work brings a solution that delivers high ranging precision at heavily reduced processing bandwidths and sparse-bandwidth power allocations.

Classic ToA ranging and localization applications, such as FMCW, wideband equivalent-time sampling and successive-approximation radar (SAR), use single-carrier (SC) schemes [1-3] with power spread over a wide bandwidth to achieve precision. Moreover, they often require TX-RX carrier phase-locking (homodyne detection), which in its turn comes at a severe system complexity and thus energy cost. [4] shows that a less-dense power allocation by means of multiple discrete tones does not reduce the performance of a ranging system significantly with respect to a fully occupied bandwidth. This opens doors to wideband OFDM-like signal structures, easy to generate in TX and optimizable for a low crest factor to increase transmitter efficiency. Moreover, multi-tone frequency domain multiplexing instead of time-domain multiplexing as in FMCW [1] and SAR [2] enables higher update rates. Last but not least, the multi-tone approach enables smart signal sub-sampling for maximum efficiency with negligible precision loss, as exploited here.

The System's block diagram is presented in Fig. 20.1.1, top. The multi-tone 1.56GHz bandwidth, 60GHz burst is first processed by a wide bandwidth on-chip direct downconversion receiver front-end [5], resulting in a wideband quadrature baseband signal. Before digitization, all I/Q decomposed tones are folded to a small bandwidth of 60MHz by means of sub-sampling as shown in Fig. 20.1.1, right. The sub-sampling frequency is adjustable by division of the receiver's system clock of 3GHz by a sub-sampling factor, varying between 1 (for no subsampling) and 32. After sub-sampling, the frequency-folded I/Q signal is digitized at low rate [6]. The digitized data is then ready to be processed by the ToA estimation algorithm.

The time-based range-estimation algorithm [4] is based on detecting the value of the phase shifts of each of the discrete signal tones. As illustrated in Fig. 20.1.1 (left) the sub-sampling factor does not change the phases of the tones. Even sampling rates below Nyquist rate (2 times signal bandwidth, 2xB) can be used, provided a careful frequency planning of the multi-tone ranging measurement signal ensures the tones are folded to non-overlapping frequencies after sub-sampling. These frequency-folded tones hence preserve the information as provided by their original tones, enabling reconstruction of the original signal in the digital domain. However, this comes at the unavoidable penalty of aliased out-of-band noise, decreasing *SNR*. However, *SNR* loss is largely compensated by the advantage of the higher signal bandwidth that can be digitized using the technique: A typical noise-corrupted ranging system's precision, defined as the RMS value of distance estimation, is inversely proportional to $SNR^{1/2}$ and $B^{3/2}$ [4], shown theoretically and experimentally in Fig. 20.1.2 (top left). As confirmed by measurements (stars in figure), a 20x sub-sampling system can provide over a magnitude more precise estimations compared to a Nyquist rate system with identical bandwidth after sampling and similar complexity.

Figure 20.1.3 provides details on the IC. The 60GHz burst is first amplified by a wide bandwidth LNA followed by quadrature downconversion mixers, having a high bandwidth of 4GHz [5]. The LNA uses a fully differential 2-stage topology. Transformers are employed to perform the single-ended to differential conversion at the input, to couple both stages and to couple the LNA to the mixers. Matching is achieved by slow-wave transmission lines (SWTL), placed in series with the transformers. The combination of transformers and SWTL's allows for a good impedance match across a wide bandwidth. A double-balanced mixer topology is used to achieve a better port-to-port isolation and higher suppression of spurious mixing products. The quadrature LO signals, driving the downconversion mixers, are generated by means of an on-chip poly-phase filter (PPF). A cascade of 2 PPFs is used to obtain a good I/Q phase relationship, resilient to CMOS processing variations. Buffers, employing capacitive gate-drain neutralization, are used to compensate for the high insertion loss caused by the PPF. The downconverted differential signals are applied directly to the sub-sampling stage.

The sub-sampling is carried out by means of a wideband transmission gate and a MOM S&H capacitor. After amplification and AC-coupling, the folded signal is forwarded to the I/Q ADC. The dual ADC is an open-loop VCO-based topology, shaping quantization noise to higher frequencies, as is beneficial after the bandwidth reduction operation. The ADC uses internal AC coupling to set the VCO's input common mode close to the rail, improving linearity. The VCO's state is sampled at the receiver's system clock of 3GHz. This high clock frequency enables suppression of the ADC's quantization noise. The digital samples of a signal burst are stored in an on-chip 96kB SRAM memory, accessible by the ToA estimation algorithm.

The circuit was processed in a general purpose 0.9V 40nm CMOS technology (Fig. 20.1.7). The entire receiver has a total area of 2.93mm^2. The indoor lab-measurement setup is shown in Fig. 20.1.4. The transmitted tones, equal in power and phase, are generated at an IF by means of a Tektronix AWG. A Sivers IMA converter module takes care of upconversion to 60GHz and power amplification. Millitech SGH-15 Horn antennas are employed to transmit and receive the signal. An external 40dB LNA is added, compensating for channel attenuation effects and limited transmit power. LNA addition enables measuring the performance as limited by the IC and not by channel or transmitter.

Six discrete tones are transmitted at frequencies [59.22 59.6 59.8 60.2 60.4 60.78]GHz. The exact carrier allocation or phase may change in order to optimize the crest factor. The transmitted signal's crest factor is calculated to be 7.8dB. The total chip input power at 80cm separation equals -2.8dBm. Due to measurement setup non-linearities, the SFDR of the signal at the receiver's input is only 15.5dB. Sampled at 187.5MHz (subsampling factor 16), the discrete tones frequency-fold to the frequencies [-30.029 -20.142 -10.253 10.253 20.142 30.029] MHz. Wireless measurements are performed at different distances between transmitter and receiver, varying between 1m and 5m. At each distance, 60 measurements using a multi-tone 4µs burst are performed. After synchronization, the frequency domain representations of multiple bursts are added to further average out channel noise, drastically improving precision. The measured estimation precision and accuracy (mean value of the estimation error) using statistical averaging with 5 burst combinations (20µs packages) is plotted in Fig. 20.1.5, demonstrating overall cm-level precision, enabling a 50kHz update rate. Precisions in the order of a few millimeters are even achieved at 3.6 meter distance.

Figure 20.1.6 compares the design to state-of-the-art CMOS implementations of ranging receivers. Power consumption of the IC is measured to be 195mW during reception and 166mW in idle mode, excluding the SRAM modules which only serve for measurement purposes. This comparison clearly shows that sub-cm-precision, high-update-rate ranging can be achieved, based on wideband discrete tones around 60GHz. This performance is competitive to state-of-the-art, with the additional strong advantage of not requiring a fixed TX-RX LO phase relation nor baseband circuits with excessive bandwidth requirements.

Acknowledgements:
This work was funded and assisted by The Flemish Agency for Innovation by Science and Technology (IWT) and ERC Advanced Grant DARWIN. Thanks to FMTC Leuven, F. Daenen, N. Gaethofs and N. Deferm.

References:

[1] Lee et al., "A Fully-Integrated 77GHz FMCW Radar Transceiver in 65nm CMOS Technology," *IEEE J. Solid-State Circuits,* 2010.

[2] Tang et al., "A 144GHz 0.76cm-Resolution Sub-Carrier SAR Phase Radar for 3D Imaging in 65nm CMOS," *ISSCC Dig. Tech. Papers,* 2012.

[3] Lai et al., "A Scalable Direct-Sampling Broadband Radar Receiver Supporting Simultaneous Digital Multibeam Array in 65nm CMOS," *ISSCC Dig. Tech. Papers,* 2013.

[4] T. Ayhan et al., "Towards a Fast and Hardware-Efficient sub-mm Precision Ranging System", *IEEE Workshop on Signal Processing Systems (SiPS),* pp. 203-208, 2012.

[5] N. De Clercq et al., "A 60GHz Wideband Direct Downconversion Receiver in 40nm CMOS", *European Microwave Integrated Circuits Conference (EUMIC),* October 2013, Accepted.

[6] T. Redant and W. Dehaene, "A 40nm, High-Bandwidth, VCO-based Burst-Mode Receiver Backend for EHF Multi-Carrier Wireless", accepted for publication at *A-SSCC* 2013.

978-1-4799-0917-9/14 $31.00 © 2014 IEEE

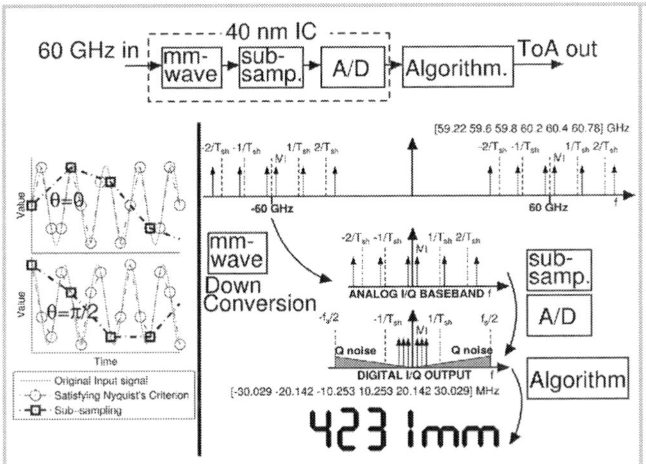

Figure 20.1.1: Core building blocks (top) and functionality of the IC + algorithmic processing. Explanation of the sub-sampling in both time (bottom left) and frequency domain (bottom right).

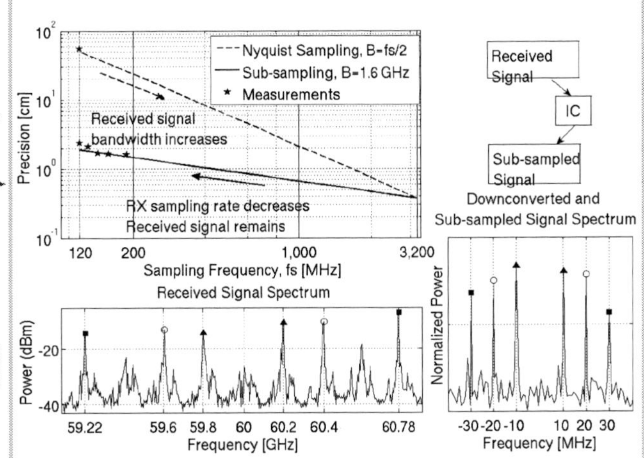

Figure 20.1.2: Impact of sampling rate and signal bandwidth on the range-estimation precision.

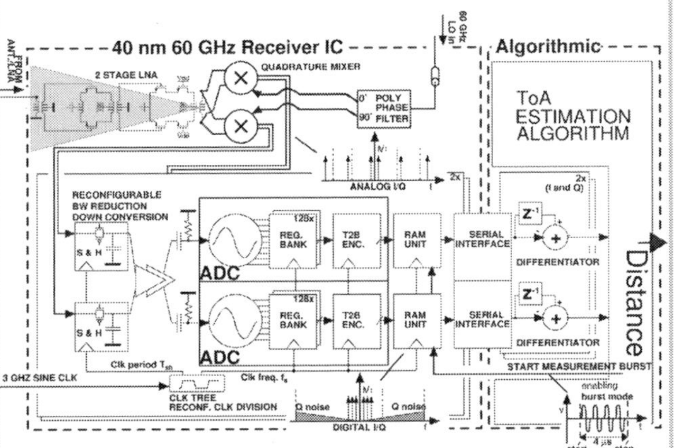

Figure 20.1.3: System block diagram overview. The IC is represented by the dashed box on the left.

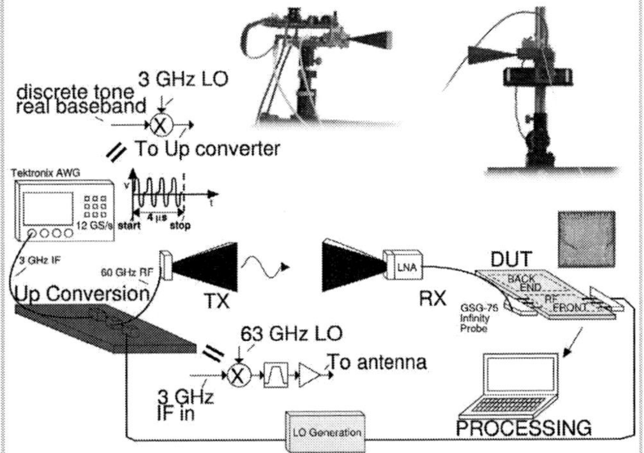

Figure 20.1.4: 60GHz discrete-tone ranging setup. Left: Transmitter + antenna, Right: DUT and receive antenna + LNA. Data is transferred by means of an Agilent ParBERT 81250 system to the algorithm.

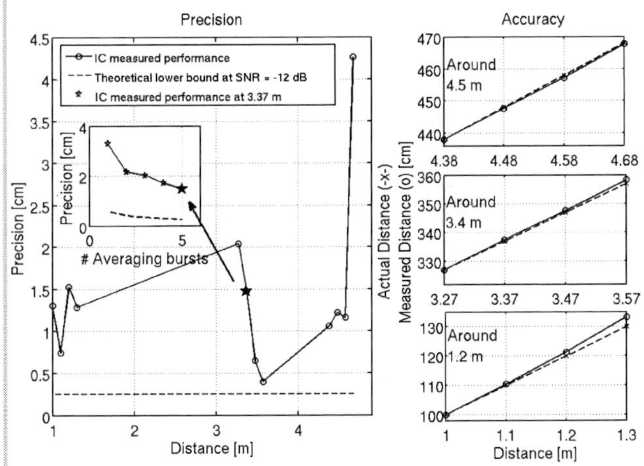

Figure 20.1.5: Measured performance of the ToA-based ranging IC setup: precision and absolute accuracy.

	This Work	[1] (JSSC2010)	[2] (ISSCC2012)	[3] (ISSCC2013)
Features	Multi-tone, bandwidth-reducing	FMCW	SAR frequency sweep	Equivalent time-sampling
Max measured range	5m	≈100m	1.2m	N/A
Precision (RMS value)	4mm @ 3.6m (measured)	< 100mm @ 10 m (measured)	7.6mm @ >1m (measured)	Depth resolution 1.5 mm (theoretical)
Receiver Power Consumption	195mW	> 100mW	214mW	76mW
Frequency Band	60GHz (V-band)	77GHz (W-band)	144GHz (D-band)	10GHz
Bandwidth	2GHz	700MHz	400MHz	5.4GHz
Package Duration (Integration time)	20µs	1.5ms	2µs	1.5µs
Signal Type	Multi-Tone	FMCW-sweep	Single Tone	Wideband
Tx-Rx LO phase relation theoretically needed?	NO	NO	YES	YES
Usage of external LO	YES	NO	NO	NO
CMOS Tech.	40nm	65nm	65nm	65nm

Figure 20.1.6: Comparison table. Sub-cm-precision high-update-rate ranging can be achieved, based on the wideband discrete tones around 60GHz that were employed, though the baseband bandwidth is small.

20

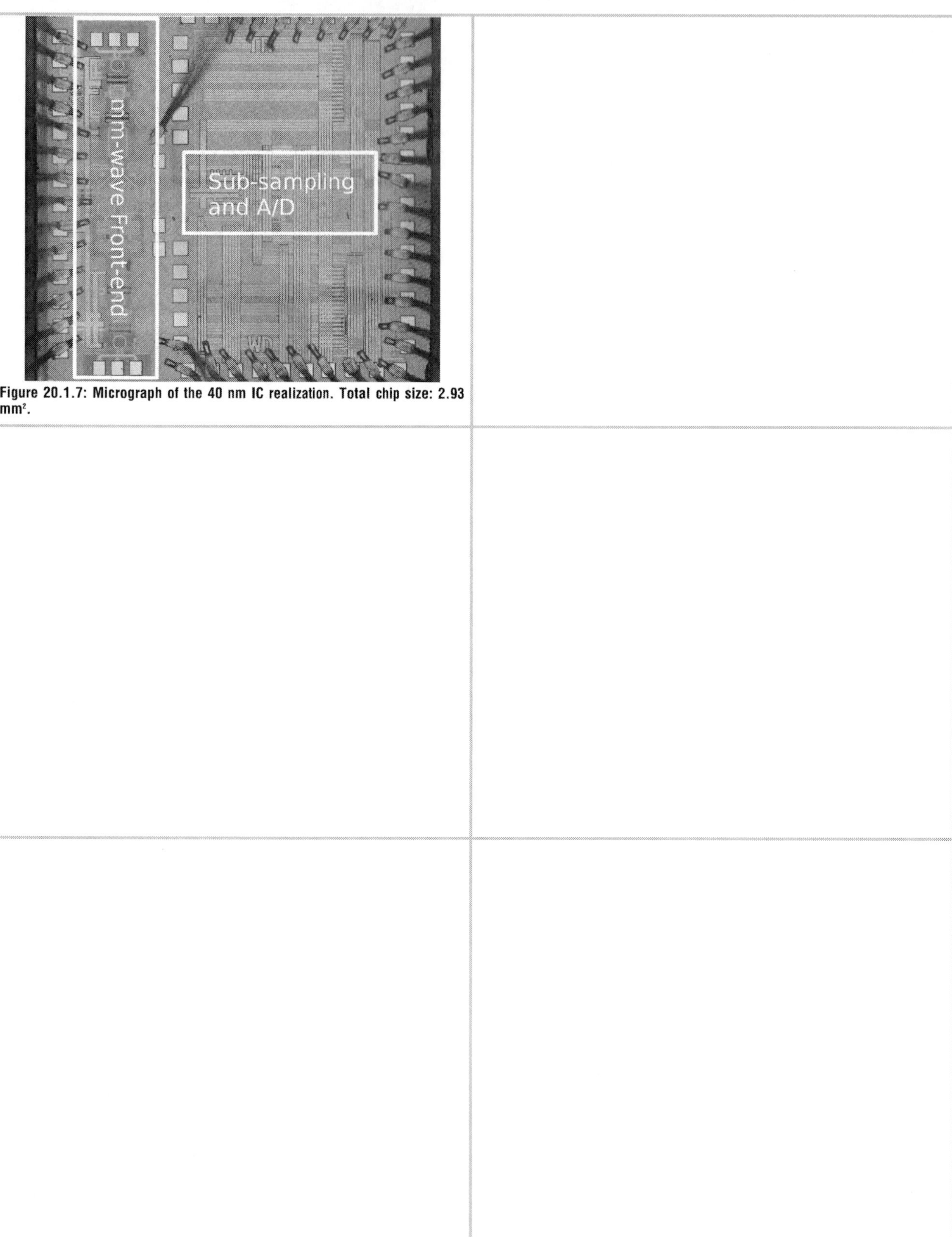

Figure 20.1.7: Micrograph of the 40 nm IC realization. Total chip size: 2.93 mm^2.

20.2 A 16TX/16RX 60GHz 802.11ad Chipset with Single Coaxial Interface and Polarization Diversity

Michael Boers[1], Iason Vassiliou[2], Saikat Sarkar[1], Sean Nicolson[1], Ehsan Adabi[1], Bagher Afshar[1], Bevin Perumana[1], Theodoros Chalvatzis[2], Spyros Kavadias[2], Padmanava Sen[1], Wei Liat Chan[1], Alvin Yu[1], Ali Parsa[3], Med Nariman[1], Seunghwan Yoon[1], Alfred Grau Besoli[1], Chryssoula Kyriazidou[2], Gerasimos Zochios[2], Namik Kocaman[1], Adesh Garg[1], Hans Eberhart[1], Phil Yang[1], Hongyu Xie[1], Hea Joung Kim[1], Alireza Tarighat[1], David Garrett[1], Andrew Blanksby[1], Mong Kuan Wong[4], Durai Pandian Thirupathi[1], Siukai Mak[3], Radha Srinivasan[1], Amir Ibrahim[1], Ersin Sengul[1], Vincent Roussel[1], Po-Chao Huang[1], Tsuifang Yeh[3], Murat Mese[1], Jesus Castaneda[1], Brima Ibrahim[1], Tirdad Sowlati[1], Maryam Rofougaran[1], Ahmadreza Rofougaran[1]

[1]Broadcom, Irvine, CA, [2]Broadcom, Alimos, Greece, [3]Broadcom, San Diego, CA, [4]Broadcom, Sunnyvale, CA

The IEEE 802.11ad standard supports PHY rates up to 6.7Gb/s on four 2GHz-wide channels from 57 to 64GHz. A 60GHz system offers higher throughput than existing 802.11ac solutions but has several challenges for high-volume production including: integration in the host platform, automated test, and high link loss due to blockage and polarization mismatch. This paper presents a full-featured 802.11ad chipset capable of SC and OFDM modulation using a 16TX-16RX beamforming RF front-end, complete with an antenna array that supports polarization diversity. To aid low-cost integration in PC platforms, a single coaxial cable interface is used between chips. The chipset includes MAC, PHY, and RF with a PCIe™ interface and is capable of maintaining a link of 4.6Gb/s (PHY rate) at 10m.

The radio employs a fixed-IF super-heterodyne architecture split across two chips. The baseband chip shown in Fig. 20.2.1 contains an IF radio with dual 6-b 2.64GS/s ADC's, dual 7-bit 2.64GS/s DAC's and quadrature up/down conversion stages. A 4.32GHz PLL and x2 multiplier provides an 8.64GHz quadrature LO to the mixers. A 2.64GHz PLL supplies a low-jitter (<2ps) clock to the ADC and DAC as well as clocks to the digital baseband.

The RF front-end chip shown in Fig. 20.2.2 handles IF-to-RF (57-to-64GHz) conversion and is designed to be co-located on a module with the antenna array. 8.64GHz is chosen for the IF as it is low enough that a low-cost coaxial cable can be used and high enough that the upconverted IF image at 41 to 48GHz meets out-of-band spurious emissions. The RF front-end integrates IF to RF up/down-conversion chains, a 24-to-28GHz PLL with x2 multiplier for mixer LO, PMU, bandgap, and IF amplifiers. The TX chain is optimized for high linearity. A 1:2 splitter after the mixer followed by dual pre-PA's boosts the signal before it is split into 16 paths. The divider network uses differential 100Ω CPW transmission lines and area-efficient lumped-element Wilkinson dividers. Each TX front-end path consists of a passive Cartesian phase shifter (PS) followed by PA and TSSI. The receive path has 16 LNAs followed by a PS and 16-to-1 combining network made up of similar components as the TX chain. A post-combiner attenuator provides up to 20dB attenuation, and a final post-combiner LNA provides additional gain when only a few paths are coherently combined or the input signal is low. A mixer provides downconversion to a fixed 8.64GHz IF followed by an 8.64GHz wideband amplifier.

Shared TX/RX antennas are preferred for multi-element (>4) implementations due to the loss associated with routing to separate arrays. In addition, given a certain module area, it is more efficient to use the space to achieve 3dB larger array gain using a single array rather than separate arrays with no switch (assuming switch loss is <3dB). Support for polarization diversity is desired due to the large link loss (up to 15dB) in the case of a cross-polarized link. Figure 20.2.3 shows co-polarized and cross-polarized measurements for antennas in V/H configuration. The block diagram for a single front-end TRX+P cell is shown in Fig, 20.2.2. Each TRX+P cell has two outputs feeding one antenna element along the x and y axes and thus delivering horizontal and vertical polarizations in either TX or RX mode. The DPDT switch has a loss of 2dB. High-efficiency low-k 4-stage transformer-coupled PAs with MOS neutralization achieve a wide bandwidth frequency response, 8dBm P_{sat}, and low variation across temperature. The PAs run directly from a switching regulator at 1.0V for increased reliability and consume 45mW each. The single-ended LNAs consume 35mW each from 1.2V and achieve a noise figure less than 5.5dB with an adjustable gain from 18 to 25dB. The PS's have 11dB loss and a resolution of 5 degrees. Figure 20.2.3 shows the dual linerarly polarized antenna configuration as well as measurements of corresponding radiation patterns including steering out to 60 degrees.

The single coaxial link shown in Fig. 20.2.4 enables simple integration of 60GHz in the host platform. The baseband chip can be placed on the system motherboard (typically HMC) with easy access to the PCIe interface while the front-end chip can be placed anywhere in the platform where it can radiate effectively. The coaxial interface multiplexes 3.3V DC, a 270MHz reference for the RFPLL, a 2.16GHz bidirectional control link, and IF signal at 8.64GHz (with 2GHz bandwidth). All the RF signals are multiplexed with an on-chip triplexer, DC is split off-chip using an inductor to improve DC efficiency. The combined port achieves a match of better than -10dB at 2.16GHz, -8dB from 7.5 to 9.5GHz and -5dB at 270MHz. A loss of up to 12dB at IF does not degrade system performance.

The control link is designed for error-free operation, has a fail-safe mode, and does not impact system performance. 2.16GHz is chosen to minimize harmonic energy in the IF band, (the 3rd harmonic falls at 6.48GHz - 5th at 10.8GHz). It is configurable in two modes: 1) Standard register read-write which allows the baseband chip to control and obtain the status of the front-end chip, 2) Streaming mode for low-latency operation of the front-end AGC and antenna weight vectors during communication. A digital controller on both sides of the link manages contention and includes clock data recovery and error correction. On the front-end chip, an additional state machine runs several calibrations at power-up, sets the transceiver into sleep mode, and configures the 2.16GHz control radio in a low-power RX IDLE mode, ready to receive control data. Figure 20.2.4 shows the block diagram of the control interface as well as a measurement of the spectral content on the coaxial cable.

An on-chip microprocessor calibrates both chips and includes support for ADC, PLL, BG, R/C, frequency flatness, TX/RX I/Q mismatch and LO leakage using an integrated 8GHz TSSI, filter response using integrated loopback paths, RF gain-step calibration using on-chip IF source derived from the PLL divided by 3, DC offset, gain, and output power calibrations. A unique feature of the RF front-end module is the lack of access to the 60GHz port for testing. To maximize link performance transmitting at the highest EIRP that meets EVM and mask requirements is preferred. To achieve this across PVT, the output power needs to be calibrated. This calibration is done with 60GHz TSSIs attached to each PA. Absolute accuracy of the TSSI is not required; rather it needs to accurately measure a 1dB drop in output power. The output power calibration sweeps the TX gain to find the P_{1dB} point for each path. P_{1dB} information is used to set TX gain based on back-off requirements and can be set per MCS.

This chipset implements hardware support for 802.11ad beamforming protocols including SLS and BRP. On-chip memories for phase and beam look-up tables and digital state machine enables high-speed switching during SLS and BRP timeslots. The 16 element TX/RX+V/H antenna array is implemented on a low-cost organic laminate substrate. The array consists of dual polarized patch antennas in a triangular lattice to enable better scanning.

Figure 20.2.5 shows the TX EVM performance of the chipset including PHY and DACs. The output signal meets +/-2dB gain flatness and spectral mask requirements as specified by the 11ad standard. The chipset demonstrates TX EVM of better than -23dB and EIRP of 24dBm when operating at MCS12. When power is reduced, the TX can achieve better than -25dB EVM. In receive mode, the sensitivity for MCS0 is better than -81dBm at chip input and better than -57dBm at MCS12 (single-port equivalent – does not include antenna gain). In addition, RX radio EVM is better than -23dB (-21dB with ADCs) and a state-of-the-art -22dB EVM is measured when TX and RX are cascaded.

Acknowledgements:
Authors thank Gary Ford, Xuefeng Fan and the layout team as well as Bo Pan, Shelley Xu, Badria Elnour and RF-DVT team.

References:

[1] T. Tsukizawa, et al., "A Fully Integrated 60GHz CMOS Transceiver Chipset Based on WiGig/IEEE802.11ad with Built-In Self Calibration for Mobile Applications", *ISSCC Dig. Tech. Papers*, pp. 230-231, Feb. 2013.
[2] V. Vidojkovic, et al., "A Low-Power Radio Chipset in 40nm LP CMOS with Beamforming for 60GHz High-Data-Rate Wireless Communication", *ISSCC Dig. Tech. Papers*, pp. 236-237, Feb. 2013.
[3] S. Emami, et al., "A 60GHz CMOS Phased-Array Transceiver Pair for Multi-Gb/s Wireless Communications", *ISSCC Dig. Tech. Papers*, pp. 164-165, Feb. 2011.
[4] E. Cohen, et al., "A CMOS Bidirectional 32-element Phased-Array Transceiver at 60GHz with LTCC Antenna", *IEEE Radio Frequency Integrated Circuits Symposium (RFIC)*, pp. 439, 442, June 2012.

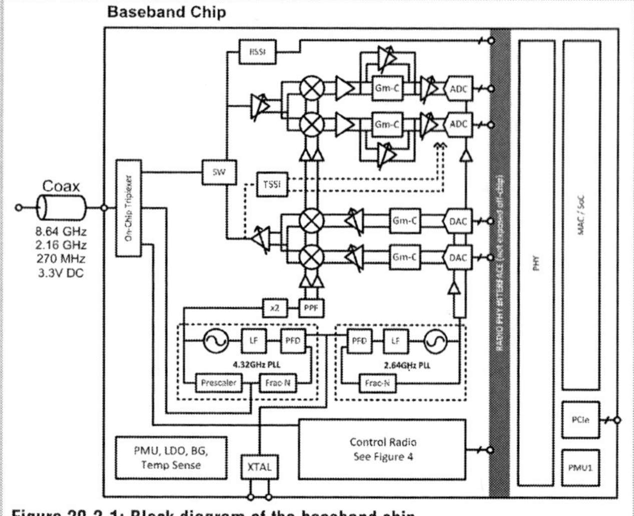

Figure 20.2.1: Block diagram of the baseband chip.

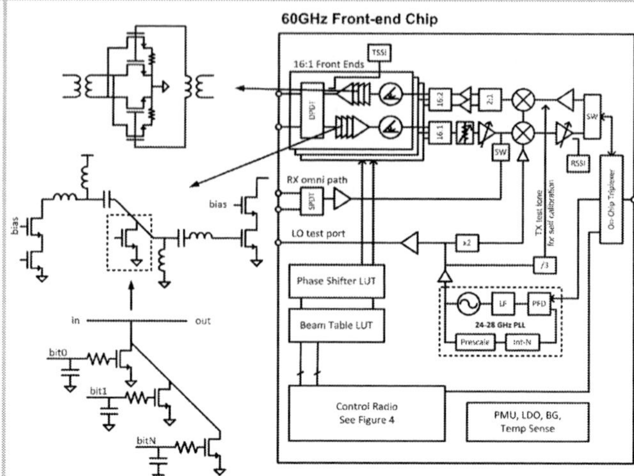

Figure 20.2.2: Block diagram of the 60GHz RF front-end chip and PA/LNA breakout.

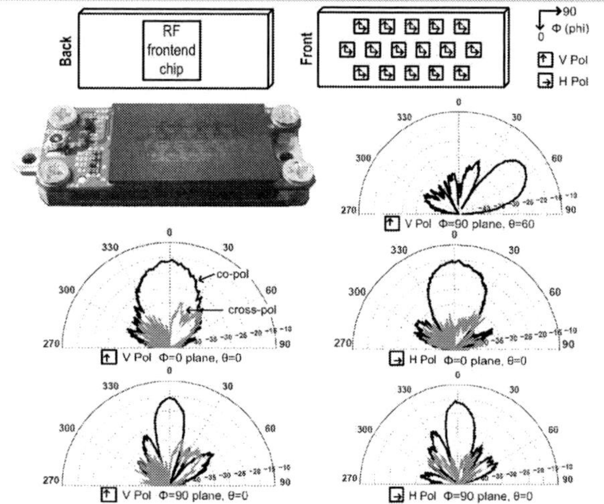

Figure 20.2.3: 60GHz RF front-end module and measured array patterns (co-pol and cross-pol).

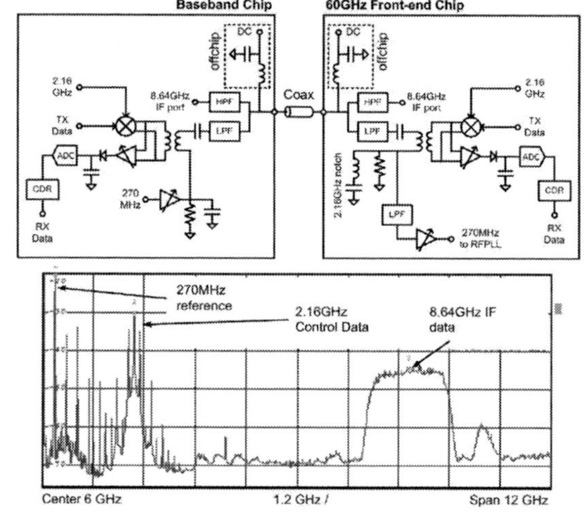

Figure 20.2.4: Block diagram of the coaxial interface and plot of spectral components on coaxial cable.

	MCS9 – CH2	MCS12 – CH2	MCS21 – CH2	MCS22 – CH2
Frequency response	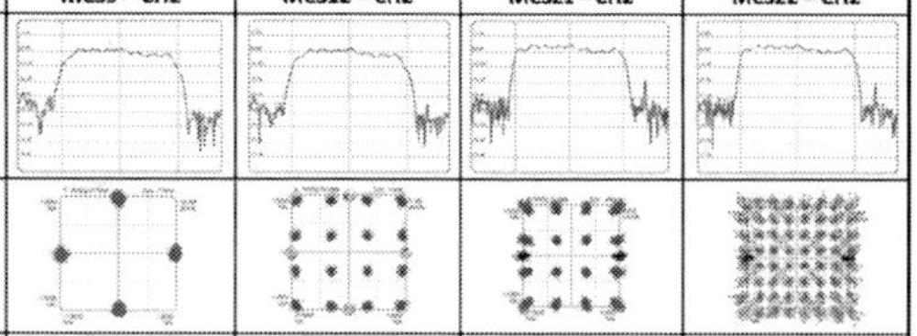			
Constellation				
Modulation	π/2-QPSK	π/2-16QAM	16QAM	64QAM
SC/OFDM	SC	SC	OFDM	OFDM
SNR	30dB	29.5dB	25.2dB	25.7dB
EVM spec	-15dB	-21dB	-20dB	-22dB
EVM meas[1]	-25.2dB	-25.5dB	-22dB	-22dB
EVM meas[2]	-30.3dB	-28.1dB	-27.4dB	-27dB

	MCS12 EVM per channel			
Constellation	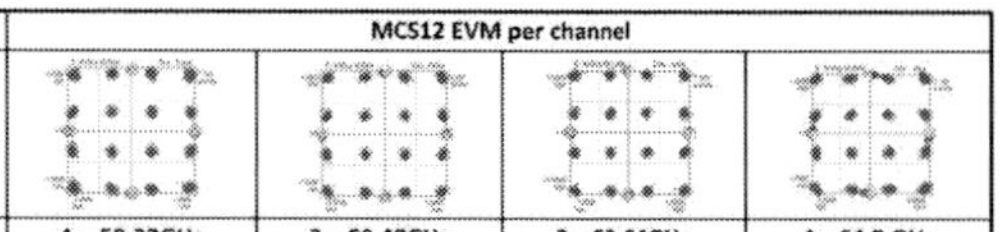			
Channel	1 – 58.32GHz	2 – 60.48GHz	3 – 62.64GHz	4 – 64.8 GHz
EVM	-24.8dB	-25.5dB	-26.0dB	-24.5dB

[1] EVM measurement with PHY generated data, DAC included

[2] EVM measurement of RF chain only (analog I/Q test-port to 60GHz RF out) after de-embedding of RX test EVM (-33dB)

Figure 20.2.5: Measurement results.

	This work	[1]	[2]	[3]	[4]
Technology	40nm LP	40nm / 90nm	40nm	65nm	90nm
Interface	Single Coax	Analog I/Q	Analog I/Q	Analog I/Q	IF
Supported Channels	4	4	3	2	2
Integration	PCIe, PHY, MAC, RF	USB3, PHY, MAC, RF	RF only		IF to RF only
Total Radio Area	33.2	13.5	12.5	72.6	29[1]
Architecture	16TX 16+1RX	1TX 1RX	4TX 4RX	32TX 8 RX	32TX 32+1RX
Beamforming	Yes	No	Yes	Yes	Yes
Single wire interface	Yes	No	No	No	No
OFDM support	Yes	No			
Polarization diversity	Yes	No	No	No	No
802.11ad MCS support	MCS 0-22	MCS 0-9			
EIRP	24 dBm @ -23dB EVM	8.5 dBm @ -22dB EVM	19.3 dBm @ -15.2dB EVM	28 dBm @ -19dB EVM	29 dBm @ -19dB EVM
TX EVM	-23 dB @ EIRP	-22 dB @ EIRP	-15.2 dB @ EIRP	-19 dB @ EIRP	-19 dB @ EIRP
Radio RX die power	960 mW	274 mW	496 mW		850 mW[1]
Radio TX die power	1190 mW	347 mW	584 mW	1800 mW	1200 mW[1]
Radio mm^2 / element	1.00	not phased array	1.56	1.82	0.45[1]
Link distance (PHY rate)	4.6Gbps @ 10m 3.0Gbps @ 20m	1.8Gbps @ 0.8m	4.6Gbps @ 0.7m 2.3Gbps @ 3.6m	3.8 Gbps @ 50m	

Notes [1] IF to RF chain only

Figure 20.2.6: Comparison of millimeter-Wave transceivers.

20

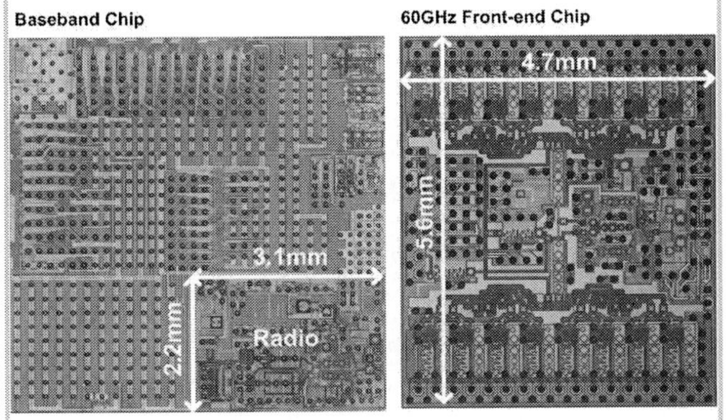

Figure 20.2.7: Die micrographs.

20.3 A 64-QAM 60GHz CMOS Transceiver with 4-Channel Bonding

Kenichi Okada, Ryo Minami, Yuuki Tsukui, Seitaro Kawai, Yuuki Seo, Shinji Sato, Satoshi Kondo, Tomohiro Ueno, Yasuaki Takeuchi, Tatsuya Yamaguchi, Ahmed Musa, Rui Wu, Masaya Miyahara, Akira Matsuzawa

Tokyo Institute of Technology, Tokyo, Japan

This paper presents a 64-QAM 60GHz CMOS transceiver, which achieves a TX-to-RX EVM of -26.3dB and can transmit 10.56Gb/s in all four channels defined in IEEE802.11ad/WiGig. By using a 4-bonded channel, 28.16Gb/s can be transmitted in 16QAM. The front-end consumes 251mW and 220mW from a 1.2-V supply in transmitting and receiving mode, respectively.

Figure 20.3.1 shows the 60GHz direct-conversion front-end design. The transmitter consists of a 6-stage PA, differential preamplifiers, I/Q passive mixers and a quadrature injection-locked oscillator (QILO). The receiver consists of a 4-stage LNA, differential amplifiers, I/Q double-balanced mixers, a QILO, and baseband amplifiers. A direct-conversion architecture is employed for both TX and RX because of wide-bandwidth capability [1]. The LO consists of the 60GHz QILO and a 20GHz PLL. The 60GHz QILO works as a frequency tripler with the integrated 20GHz PLL. It can generate 7 carrier frequencies with a 36/40MHz reference, 58.32GHz(ch.1), 60.48GHz(ch.2), 62.64GHz(ch.3), and 64.80GHz(ch.4) defined in IEEE802.11ad/WiGig, 59.40GHz(ch.1-2), 61.56GHz(ch.2-3), and 63.72GHz(ch.3-4) for the channel bonding.

Figure 20.3.2 shows the downconversion mixer and 3-stage baseband amplifiers on the receiver chain. For mm-Wave receiver design, the noise figure, linearity, gain flatness, and frequency-dependent I/Q mismatch should be considered. Especially for the 4-channel bonding using 7.04GS/s, the noise floor becomes at least 98dB higher from -174dBm/Hz, and peak SNDR of the RX also suffers from the non-linearity of the receiver chain. In addition, it is difficult to use a closed-loop baseband amplifier for improving linearity and gain flatness due to the wide bandwidth. Thus, an open-loop baseband amplifier based on the flipped voltage follower (FVF) is employed to maintain both gain flatness and linearity. The voltage gain is simply determined by the current-mirror ratio N and R_L/R_S, which contributes to improving the linearity. A mismatch in cut-off frequencies of baseband amplifiers causes frequency-dependent I/Q phase mismatch, which is mainly caused by the low-pass characteristics of baseband amplifiers. This FVF amplifier can also relax the mismatch due to its high cut-off frequency. A current-bleeding downconversion mixer with capacitive cross coupling is employed to reduce the LO power and the power consumption of LO buffers. The input matching block of the LNA has a shunt-grounded structure for electrostatic discharge (ESD) protection. The matching block is designed with low-impedance and 50Ω transmission lines and parallel-line transformers. The differential mismatch of the transformer as a balun is compensated by the common-mode rejection of differential amplifiers.

Figure 20.3.3 shows the upconversion mixer and RF differential amplifier. The wideband gain flatness and linearity are also important for the transmitter design. In this work, a mixer-first topology is employed for improving the gain flatness and linearity. An input buffer maintaining the 50Ω impedance matching usually has a narrow-band characteristic, which degrades the gain flatness. In this design, the baseband input impedance is directly maintained by the mixer impedance. This is a known technique for the mixer-first receiver but applied to a transmitter in this work. The input impedance of the RF differential amplifier is downconverted to the baseband, and the wideband impedance flatness around 60GHz is maintained by a shunt-feedback matching technique in the RF differential amplifier [6]. A 200Ω shunt resistor at the mixer input is also used for impedance compensation. The LO leakage can be minimized by a DC offset cancellation, which is adjusted by current sources. In addition, the mixer-first topology can relax the linearity requirement of the mixer by increasing the RF path gain. Thus, the LO power for mixers can be reduced, which contributes to reducing the power consumption of I/Q LO buffers considerably, *e.g.*, 84mW [4] becomes 37mW, even though increasing RF path gain.

Figure 20.3.4 shows the measured characteristics of the RF front-end. Both TX and RX cover 4 channels. The TX conversion gain is about 15dB, excluding the PCB loss. The saturated output power is 10.3dBm at the center frequency of 61.56GHz for the 4-channel bonding. The output power is measured for both a stand-alone PA and a transceiver chip implemented on a PCB, and the PCB loss is estimated from the difference between these saturated output powers depending on the frequency. The LO leakage is less than -45dBc as shown in Fig. 20.3.3. Due to the proposed upconversion architecture, a sideband rejection ratio of more than 40dB is achieved at the 0.5GHz offset after the I/Q calibration [5]. The TX EVM is less than -27.1dB for every channel and every mode, and the best TX EVM of -29.7dB is achieved in QPSK (ch.4). The PA consumes 115mW, and the two differential amplifiers and mixers consume 16mW.

The RX conversion gain is more than 20dB, excluding the PCB loss. SNDR at the center frequency of 61.56GHz is estimated from the measured IM3 of the RX PCB, and the measured noise figure of the stand-alone LNA. A peak SNDR is 35dB excluding the PCB loss. The power consumptions of LNA, two differential amplifiers, two mixers, and two BB amplifiers are 41mW, 19mW, 23mW, and 30mW, respectively.

The phase noise measured at the TX output is –96.5dBc/Hz@1MHz offset from the center frequency of 61.56GHz. The measured free-running frequency of QILO is from 58 to 66GHz. The 20GHz PLL consumes 64mW. QILOs for TX and RX consume 18mW and 15mW, and I/Q LO buffers consume 37mW and 28mW, respectively. Both QILO and LO buffers can be turned off in TDD operation.

Figure 20.3.5 shows the measured constellation and performance summary. Two PCBs are used. One is for TX mode and the other is for RX mode with on-board 36MHz TCXOs. Single-carrier modulated I/Q signals are generated by an arbitrary waveform generator (Tektronix AWG70002A) with symbol rates of 1.76GS/s (for 1 channel) and 7.04GS/s (for 4-bonded channel), and a roll-off factor of 25%. The spectrum is measured with a spectrum analyzer and a downconversion mixer, which satisfies the IEEE802.11ad spectrum mask for every channel. An oscilloscope (Tektronix DSA73304D) is used to evaluate the constellation and EVM. The measured TX-to-RX EVM (= - SNR) in 64QAM is less than -24dB at least for every channel with an RF data rate of 10.56Gb/s, and -26.3dB is achieved at channel 4. The phase noise has the largest influence on the EVM performance in this design. By using the 4-bonded channel, 14.08Gb/s in QPSK and 28.16Gb/s in 16QAM have been achieved within a BER of 10^{-3} (EVM is less than -9.8dB and -16.5dB at least). The maximum communication distances with 14-dBi horn antennas, within an EVM of -9.8dB, -16.5dB, and -22.5dB, are 2.4m, 2.0m, 2.6m, and 0.9m in QPSK, 0.7m, 0.6m, 0.6m, and 0.4m in 16QAM, 0.08m, 0.08m, 0.13m, and 0.06m in 64QAM for channels 1 to 4, respectively.

Figure 20.3.6 shows a comparison table for 60GHz CMOS transceivers [1-9]. This paper reports a 64-QAM 60GHz transceiver with a 28.16Gb/s wireless data rate, the highest among the transceivers listed.

Figure 20.3.7 shows the die micrograph. The transceiver is fabricated in a 65nm CMOS technology. The core areas of the transmitter, receiver, PLL and control logic are 1.035mm^2, 1.25mm^2, 0.90mm^2, and 0.67mm^2, respectively.

Acknowledgments:

This work was partially supported by MIC, SCOPE, MEXT, STARC, Canon Foundation, and VDEC in collaboration with Cadence Design Systems, Inc., and Agilent Technologies Japan, Ltd. The authors thank Dr. Hirose, Dr. Suzuki, Dr. Sato, and Dr. Kawano of Fujitsu Laboratories, Ltd., and Prof. Ando of Tokyo Institute of Technology for their valuable discussions and technical supports.

References:

[1] K. Okada, *et al.*, "A 60GHz 16QAM/8PSK/QPSK/BPSK Direct-Conversion Transceiver for IEEE802.15.3c," *ISSCC Dig. Tech. Papers*, pp. 160-161, Feb. 2011.

[2] A. Siligaris, *et al.*, "A 65nm CMOS Fully Integrated Transceiver Module for 60GHz Wireless HD Applications," *ISSCC Dig. Tech. Papers*, pp. 162-163, Feb. 2011.

[3] S. Emami, *et al.*, "A 60GHz CMOS Phased-Array Transceiver Pair for Multi-Gb/s Wireless Communications," *ISSCC Dig. Tech. Papers*, pp. 164-165, Feb. 2011.

[4] K. Okada, *et al.*, "A Full 4-Channel 6.3Gb/s 60GHz Direct-Conversion Transceiver with Low-Power Analog and Digital Baseband Circuitry," *ISSCC Dig. Tech. Papers*, pp. 218-219, Feb. 2012.

[5] S. Kawai, *et al.*, "A Digitally-Calibrated 20Gb/s 60GHz Direct-Conversion Transceiver in 65-nm CMOS," *RFIC Symp.*, pp.137-140, June 2013.

[6] V. Vidojkovic, *et al.*, "A Low-Power 57-to-66GHz Transceiver in 40nm LP CMOS with -17dB EVM at 7Gb/s," *ISSCC Dig. Tech. Papers*, pp. 268-269, Feb. 2012.

[7] T. Mitomo, *et al.*, "A 2Gb/s-Throughput CMOS Transceiver Chipset with In-Package Antenna for 60GHz Short-Range Wireless Communication," *ISSCC Dig. Tech. Papers*, pp. 266-267, Feb. 2012.

[8] V. Vidojkovic, *et al.*, " A Low-Power Radio Chipset in 40nm LP CMOS with Beamforming for 60GHz High-Data-Rate Wireless Communication," *ISSCC Dig. Tech. Papers*, pp. 236-237, Feb. 2013.

[9] T. Tsukizawa, *et al.*, "A Fully Integrated 60GHz CMOS Transceiver Chipset Based on WiGig/IEEE802.11ad with Built-in Self-Calibration for Mobile Applications," *ISSCC Dig. Tech. Papers*, pp. 230-231, Feb. 2013.

978-1-4799-0917-9/14 $31.00 © 2014 IEEE

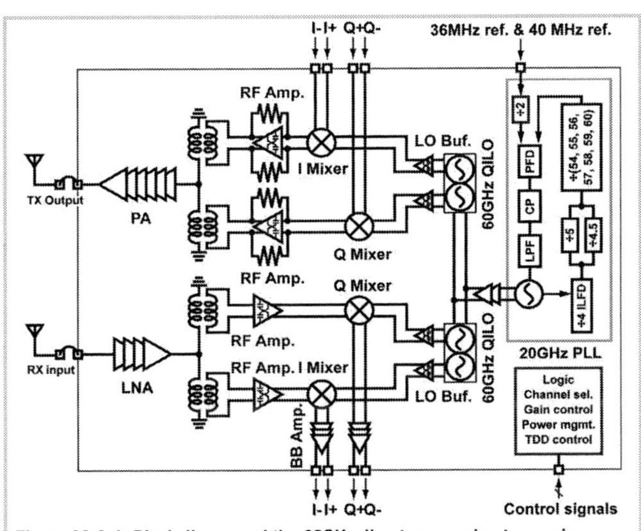

Figure 20.3.1: Block diagram of the 60GHz direct-conversion transceiver.

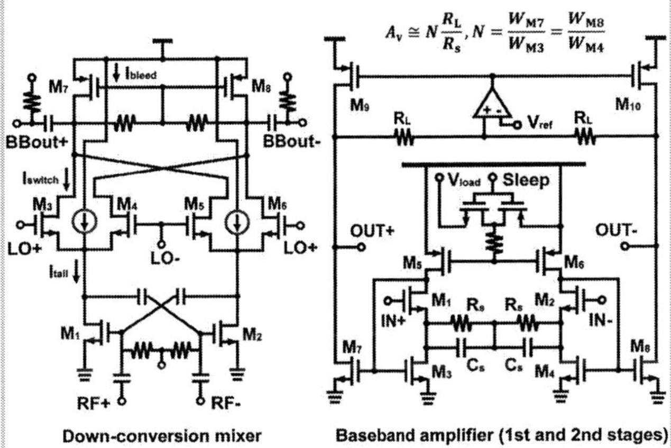

Figure 20.3.2: Receiver blocks.

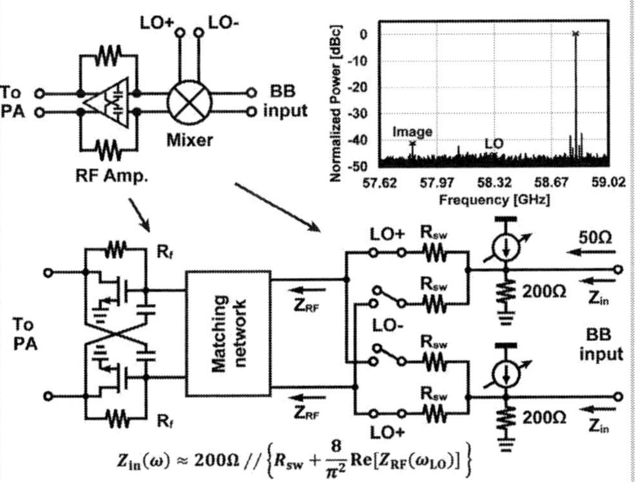

Figure 20.3.3: Transmitter blocks of the mixer-first topology.

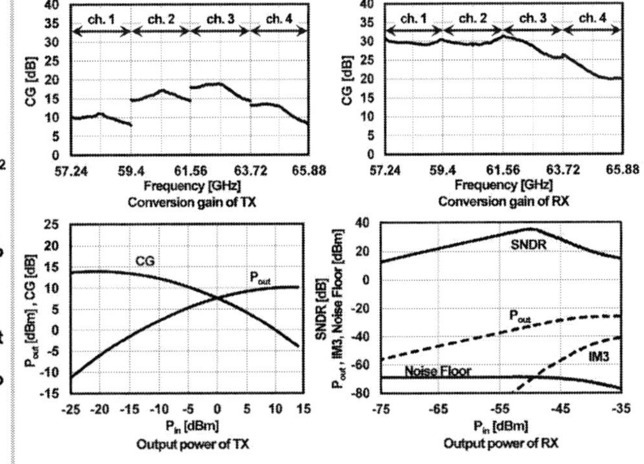

Figure 20.3.4: Measured characteristics of the RF front-end.

Channel/ Carrier freq.	ch.1 58.32GHz	ch.2 60.48GHz	ch.3 62.64GHz	ch.4 64.80GHz	ch.1-ch.4 Channel bond
Modula-tion	64QAM				16QAM
Data rate*	10.56Gb/s	10.56Gb/s	10.56Gb/s	10.56Gb/s	28.16Gb/s
Constella-tion**					
Spec-trum**					
TX EVM**	-27.1dB	-27.5dB	-28.0dB	-28.8dB	-20.0dB
TX-to-RX EVM***	-24.6dB	-23.9dB	-24.4dB	-26.3dB	-17.2dB

LO (20GHz PLL + 60GHz QILO)	
Frequency	58.32-64.80GHz (1.08GHz-step)
QILO range	58-66GHz
Phase noise @1MHz-offset	-95.3dBc/Hz (ch.1), -93.8dBc/Hz (ch.2), -92.1dBc/Hz (ch.3), -95.7dBc/Hz (ch.4) -96.5dBc/Hz (channel bond: 61.56GHz)

*The roll-off factor is 0.25. The bandwidth is 2.16GHz except for the channel bonding.
**Constellation, Spectrum, and TX EVM are measured with an external down-converter.
***TX-to-RX EVM is measured through TX and RX, which is equal to –SNR(MER).

Figure 20.3.5: Measured constellation and performance summary.

	Data rate / Modulation	TX-to-RX EVM	Integration	Power consumption
Tokyo Tech [1]	11Gb/s(16QAM)	-17dB	65nm, direct-conversion, TX, RX, LO, antenna	TX: 186mW RX: 106mW PLL: 66mW
CEA-LETI [2]	3.8Gb/s(16QAM)	-18dB	65nm, heterodyne, TX, RX, LO, antenna	TX: 1,357mW RX: 454mW
SiBeam [3]	7.14Gb/s(16QAM)	-19dB	65nm, 32x32-array heterodyne, TX, RX, LO	TX: 1,820mW RX: 1,250mW
Tokyo Tech [4, 5]	16Gb/s(16QAM) 20Gb/s(16QAM)[5]	-21dB	65nm, direct-conversion, TX, RX, LO, antenna, analog & digital BB	TX: 319mW RX: 223mW
IMEC [6]	7Gb/s(16QAM)	-18dB	40nm, direct-conversion, TX, RX, w/o PLL	TX: 167mW RX: 112mW
Toshiba [7]	2.62Gb/s(QPSK)	N/A	65nm, heterodyne, TX, RX, LO, antenna, analog & digital BB	TX: 160mW RX: 233mW
IMEC [8]	7Gb/s(16QAM)	-15dB	40nm, 4-array direct-conversion, TX, RX, LO, antenna	TX: 330mW RX: 284mW for 1 stream
Panasonic [9]	2.5Gb/s(QPSK)	-22dB	90nm, direct-conversion, TX, RX, LO, antenna, analog & digital BB	TX: 347mW RX: 274mW
This work	10.56Gb/s(64QAM) 28.16Gb/s(16QAM)	-26dB	65nm, direct-conversion, TX, RX, LO	TX: 251mW RX: 220mW

Figure 20.3.6: Performance comparison of 60GHz CMOS transceivers.

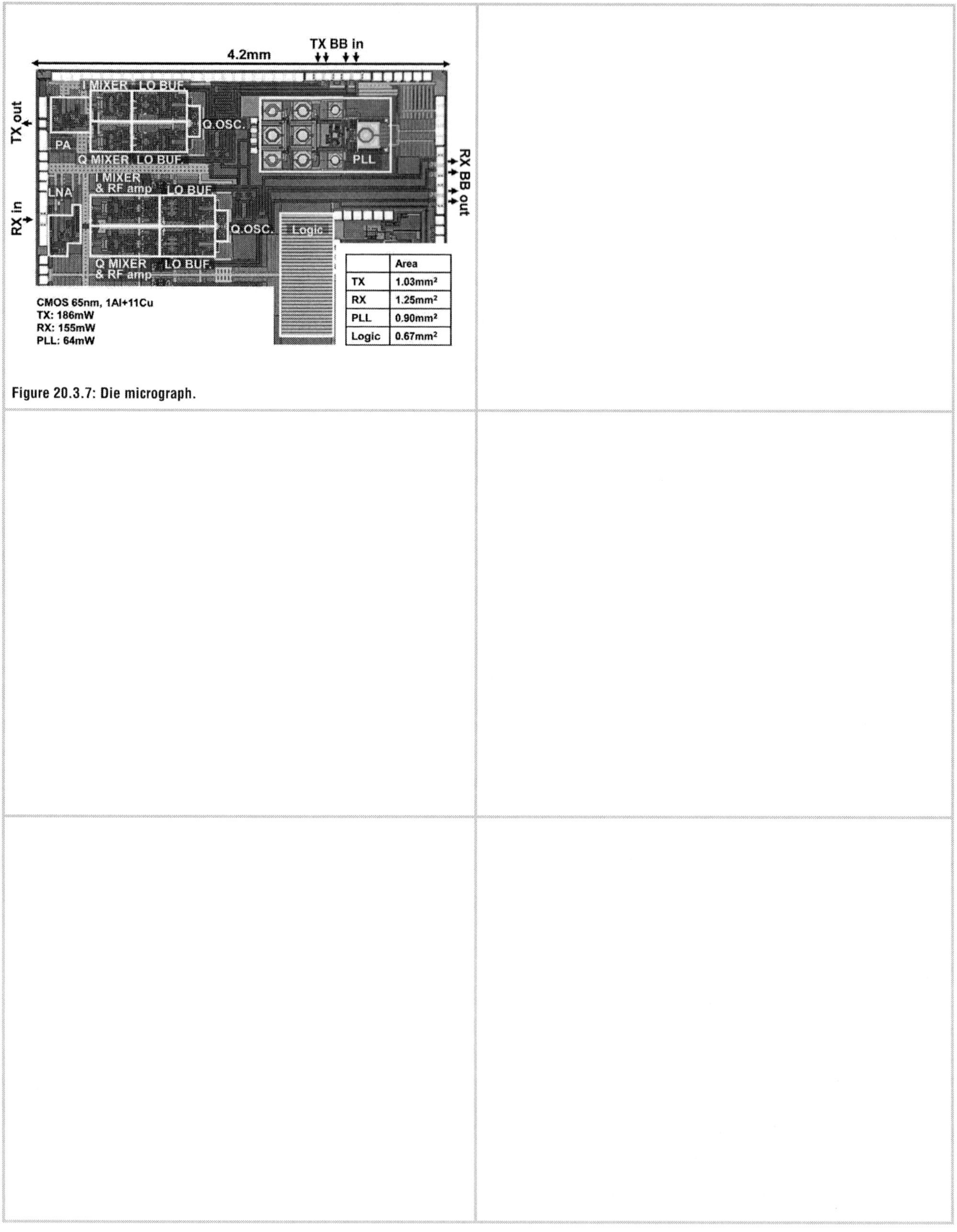

	Area
TX	1.03mm²
RX	1.25mm²
PLL	0.90mm²
Logic	0.67mm²

Figure 20.3.7: Die micrograph.

20.4 A Fully Integrated Single-Chip 60GHz CMOS Transceiver with Scalable Power Consumption for Proximity Wireless Communication

Shigehito Saigusa, Toshiya Mitomo, Hidenori Okuni, Masahiro Hosoya, Akihide Sai, Shusuke Kawai, Tong Wang, Masanori Furuta, Kei Shiraishi, Koichiro Ban, Seiichiro Horikawa, Tomoya Tandai, Ryoko Matsuo, Takeshi Tomizawa, Hiroaki Hoshino, Junya Matsuno, Yukako Tsutsumi, Ryoichi Tachibana, Osamu Watanabe, Tetsuro Itakura

Toshiba, Kawasaki, Japan

A fully-integrated single-chip CMOS transceiver with MAC and PHY for 60GHz proximity wireless communication is presented. A 60GHz wireless communication single-chip transceiver has not yet been reported due to large power consumption issues. However, by limiting the application to high-throughput proximity transmission, thermal issues arising in a single-chip have been overcome. A 2GHz broadband OFDM single-chip transceiver suffers from SNR degradation due to the reference clock (REFCLK) and baseband clock (BBCLK) spurs in RF/analog circuits. Low frequency spurs in the clock generator (CLKPLL) due to the mixing of the ADC/DAC sampling clock (SCLK) and other clocks such as REFCLK and BBCLK have been eliminated by careful frequency planning of those clocks. In addition to that, spur suppression in digital baseband and noise-tolerant RF/analog circuit designs are employed. The spurs have been successfully suppressed to less than -35dBc. The chip achieves a PHY data-rate of 2.35Gb/s and MAC throughput of 2.0Gb/s at a distance of 4cm. Power consumption is scalable to the throughput by the introduction of fast Sleep and Awake modes. The average power consumption at a throughput of 0.2Gb/s is reduced to 36% of that at 2.0Gb/s.

Figure 20.4.1 shows the block diagram of the chip. An in-package antenna is shared by TX and RX via a TX/RX switch (TRX-SW). Differential topology is utilized in all of the building blocks including the antenna. The sliding-IF architecture is used to realize high IQ accuracy and low carrier leakage. The IF frequency is chosen to be 12GHz. 48GHz and 12GHz LO signals are generated from a 24GHz LO PLL (LOPLL) via a frequency doubler and a balun and via a divider, respectively. Considering multipath fading channels, the PHY employs OFDM-QPSK modulation. The bandwidth of the baseband signal is 1.08GHz.

Figure 20.4.2 shows the frequency planning and digital baseband spur suppression. In an OFDM system, SNR degradation of sub-carriers lowers throughput. If the clock frequencies are not properly chosen as shown in the upper right figure, spurs generated by the mixing of REFCLK and BBCLK fall within the loop bandwidth of CLKPLL [1]. In this work, taking advantage of channel spacing of 2.16GHz, the frequencies of REFCLK (f_{REFCLK}), BBCLK (f_{BBCLK}) and SCLK (f_{SCLK}) are chosen to be 36MHz, 180MHz and 2.88GHz, respectively. Note that the latter two frequencies are integer multiples of 36MHz. Consequently, generated spurs move to DC and n x f_{REFCLK} (n=1,2,...), resulting in a spurious-free spectrum as shown in the upper left figure. Digital circuit implementation also affects the spurious level. The most power-consuming 64-point FFT/IFFT block in PHY is the largest spurious source. Thus, low power design of the FFT/IFFT block is required. 16-parallel 2-stage radix2 butterfly arithmetic circuits and a 16-point FFT/IFFT are employed in order to minimize the number of registers. Clock gating is applied to the entire digital baseband.

Figure 20.4.3 shows the broadband and noise-tolerant RF/analog circuits. The TRX-SW consists of $\lambda/4$ transmission lines (TLs) and shunt switches (Fig. 20.4.3(a)). The transistors at the output stage of the PA also work as ON-state shunt switches in RX mode and additional lossy shunt switches are eliminated. Measured insertion losses (ILs) of the TRX-SW at the center frequency of CH2 (60.48GHz) are 2.5dB in TX mode and 2.8dB in RX mode, respectively. A current-compensated bias circuit is utilized in LNA [4] and the PA to reduce temperature dependence. The LPFs/VGAs (LPFVGA) consists of a 3rd-order G_m-C LPF as an anti-alias filter and VGA (Fig. 20.4.3(b)). The transconductance amplifier (g_m cell) has a cascode configuration with resistor load. Bandwidth of the g_m cell is maximized by feeding common-mode feedback (CMFB) current and DC offset-cancelling (DCOC) current into low impedance nodes of the g_m cell. Capacitors in the LPF are connected between the signal terminal and AC GND in order to reduce common-mode noise. A 5th-order LC LPF (TX-LPF) suppresses aliases generated by the DACs (Fig. 20.4.3(c)). Making the most use of single-chip implementation, load impedance of the DAC can be chosen to be differential 50 ohms to maximize bandwidth of the TX-LPF. An RLC resonator-based equalizer is inserted at the output of the LPF to improve in-band gain flatness. For the DAC and LPFVGA, the values of decoupling capacitors (C_{DEC}) including parasitics related to pads, wire bonds, package redistribution layer and so on are optimized for lowering power supply impedance at f_{BBCLK}. The spurs and the intermodulation distortions caused by the ADC, the DAC and the PHY are mitigated.

A micrograph of the chip is shown in Fig. 20.4.7. The chip was fabricated in 65nm CMOS, and area including pads is 4.98mm x 3.40mm. The RF/analog blocks are carefully laid out such that differential signal lines are as isometric and straight as possible for high IQ accuracy and low carrier leakage.

Figure 20.4.4 shows measured RX and TX performances. At CH2, maximum RX gain is larger than 40dB and NF at the maximum gain is less than 8.9dB. The RX 3dB bandwidth from TRX-SW to the ADC is 2.2GHz. The maximum TX output power level is 4.9dBm. The TX 3dB bandwidth from DAC to TRX-SW is 1.9GHz. In-band digital noise is suppressed to less than -35dBc for both RX and TX modes. Minimum image-rejection ratio for both RX and TX modes is 30dB. Carrier leakage of TX mode is less than -30dBc.

Figure 20.4.5 shows measured MAC throughputs and data, ACK and current waveforms in RX mode. MAC throughput of 2.0Gb/s at a distance of 4cm is achieved. Standby [5] and Sleep functions in RX mode are integrated to reduce the average power consumption. A Standby function keeps RX blocks powered off until a connection request packet is received from the TX side. The RX blocks are intermittently activated for 150µsec to search for the packet. When designing the search interval, average power consumption while in Standby mode can be decreased by the ratio of its interval and the activated time. The Sleep function makes TX and RX synchronously power-off for the case of not requiring maximum throughput due to external interface limitations. Realizing this synchronization with information in data and ACK packets in 10µsec intervals, scalable power consumption as a function of throughput can be provided. Cases 2 and 3 are examples where the RX mode assumes the Sleep function. Power consumption with Sleep function in RX mode at MAC throughput of 0.2Gb/s and 0.4Gb/s can be reduced to 36% and 46% of that at 2.0Gb/s, respectively.

Figure 20.4.6 shows a performance summary of the chip and comparison with other works. Power consumptions of the PHY in RX and TX modes are 125mW and 82mW, which are 39% and 29% of that in [3], respectively. The chip achieves 2Gb/s MAC throughput at a distance of 4cm by utilizing the proposed frequency planning, digital baseband spur suppression and noise-tolerant RF/analog circuits. Total power consumption of the TX chip and the RX chip is 1268mW, which is less than other works including MAC [1,3]. The chip is a solution of a 60GHz single-chip transceiver achieving 2Gb/s MAC throughput and scalably, reducing the average power consumption by introducing fast Sleep and Awake modes.

Acknowledgments:
The research is partially supported by the Ministry of Internal Affairs and Communications of Japan.

References:
[1] T. Tsukizawa, et al., "A Fully Integrated 60GHz CMOS Transceiver Chipset Based on WiGig/IEEE802.11ad with Built-In Self Calibration for Mobile Application," *ISSCC Dig. Tech. Papers*, pp. 230-231, Feb. 2013.
[2] K. Okada et al., "A Full 4-Channel 6.3Gb/s Direct-Conversion Transceiver With Low-Power Analog and Digital Baseband Circuitry," *ISSCC Dig. Tech. Papers*, pp. 218-219, Feb. 2012.
[3] T. Mitomo, et al., "A 2Gb/s-Throughput CMOS Transceiver Chipset with In-Package Antenna for 60GHz Short-Range Wireless Communication," *ISSCC Dig. Tech. Papers*, pp. 266-267, Feb. 2012.
[4] S. Kawai, et al., "A Temperature Variation Tolerant 60GHz Low-Noise Amplifier with Current-Compensated Bias Circuit," *Proc. A-SSCC*, Nov. 2013.
[5] R. Matsuo, et al., "Energy-Efficient Standby-Mode Algorithm in Short-Range One-to-One mm-Wave Communications," *International Conference on Communications*, pp. 4033-4038, June 2012.

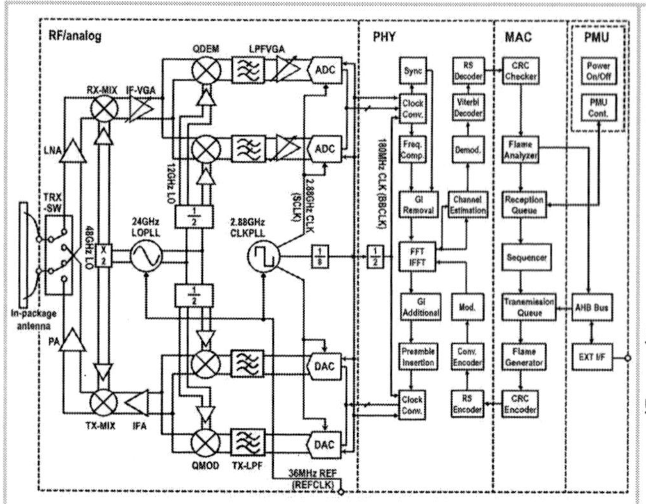

Figure 20.4.1: Block diagram of the single-chip TRX.

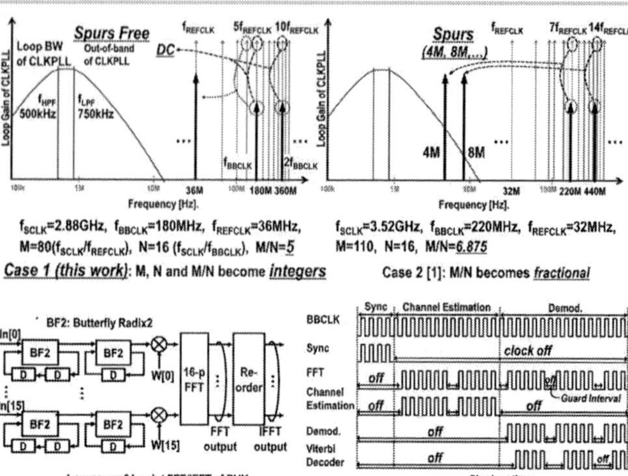

Figure 20.4.2: Frequency planning and digital spur suppression in CLKPLL and PHY.

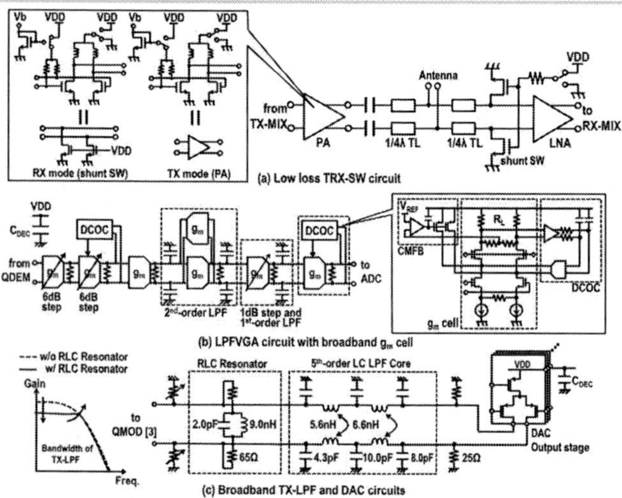

Figure 20.4.3: Broadband and noise-tolerant RF analog circuits.

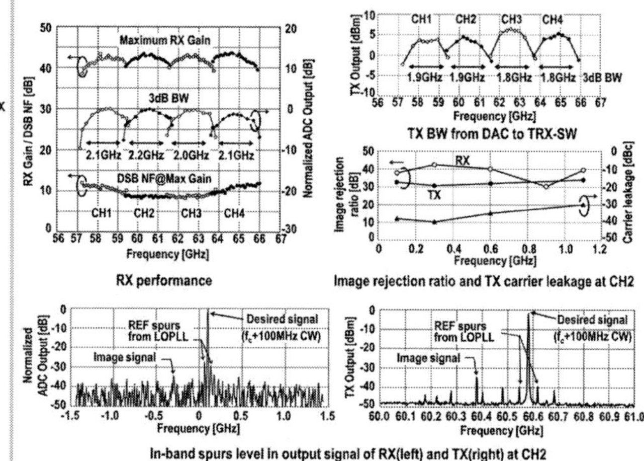

Figure 20.4.4: Measured RX/TX performance.

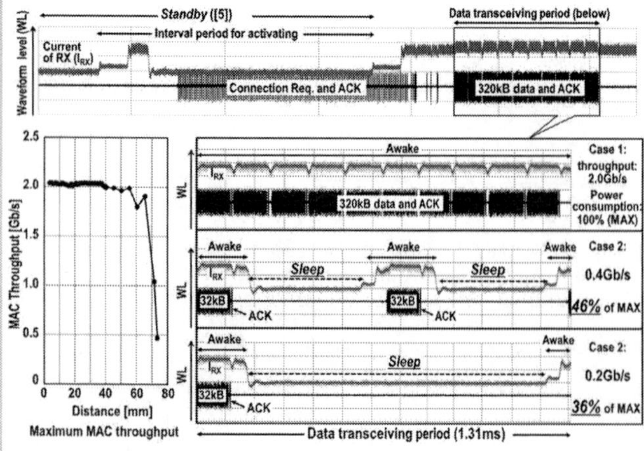

Figure 20.4.5: Measured MAC throughput, signal and current waveforms and power consumption.

Performance summary of the single-chip TRX	
Technology	65nm CMOS
Suppply Voltage	1.2 V / 1.8 V / 3.3V
Chip size	4.98 mm x 3.40 mm

RX Performances at CH2	
Voltage Gain	40 dB
DSB NF@Max Gain	8.9 dB
Image Rejection Ratio	30.3 dB
3dB bandwodth	2.2 GHz
Power Consumption	687 mW
(RF)	(468 mW)
(ADC&CLKPLL)	(94 mW)
(PHY & MAC & PMU)	(125 mW)

TX Performances at CH2	
Power Gain (estimated)	16.3 dB
P1dB output	2.8 dBm
Saturation output	4.9 dBm
Image Rejection Ratio	30.5 dB
Carrier Leakage	-29.8 dBc
3dB bandwidth	1.9 GHz
Power Consumption	581 mW
(RF)	(360 mW)
(DAC / CLKPLL)	(139 mW)
(PHY & MAC & PMU)	(82 mW)

1) PHY data rate, 2) w/o MAC

Comparison with other 60GHz chips							
	Num of chip	Chip integration	Package /Antenna(Ant.)	MAC/PMU function	Distance/MAC throughput(TH)	Total power consumption(PC) of TX and RX	Ratio of PC w/ sleep mode to that at max TH
This Work	1	RF/PHY/MAC /PMU	Standard BGA with bonding wire Ant.	Standby/ Sleep mode	4cm/2.0Gb/s	1268mW	46%@0.4Gb/s 36%@0.2Gb/s
[1]	2	RF/PHY/MAC/ USB 3.0/UHS-II	Ant. module	N/A	40cm/1.8Gb/s 1m/1.5Gb/s	1772mW	N/A
[2]	2	RF/PHY	BGA with waveguide Ant.	N/A	1.7m/3.1Gb/s 1) 10cm/6.3Gb/s 1)	1136mW 2)	N/A
[3]	2	RF/PHY/MAC	Standard BGA with bonding wire Ant.	N/A	4cm/2.1Gb/s	1348mW	N/A

Figure 20.4.6: Performance summary and comparison with other works.

978-1-4799-0917-9/14 $31.00 © 2014 IEEE

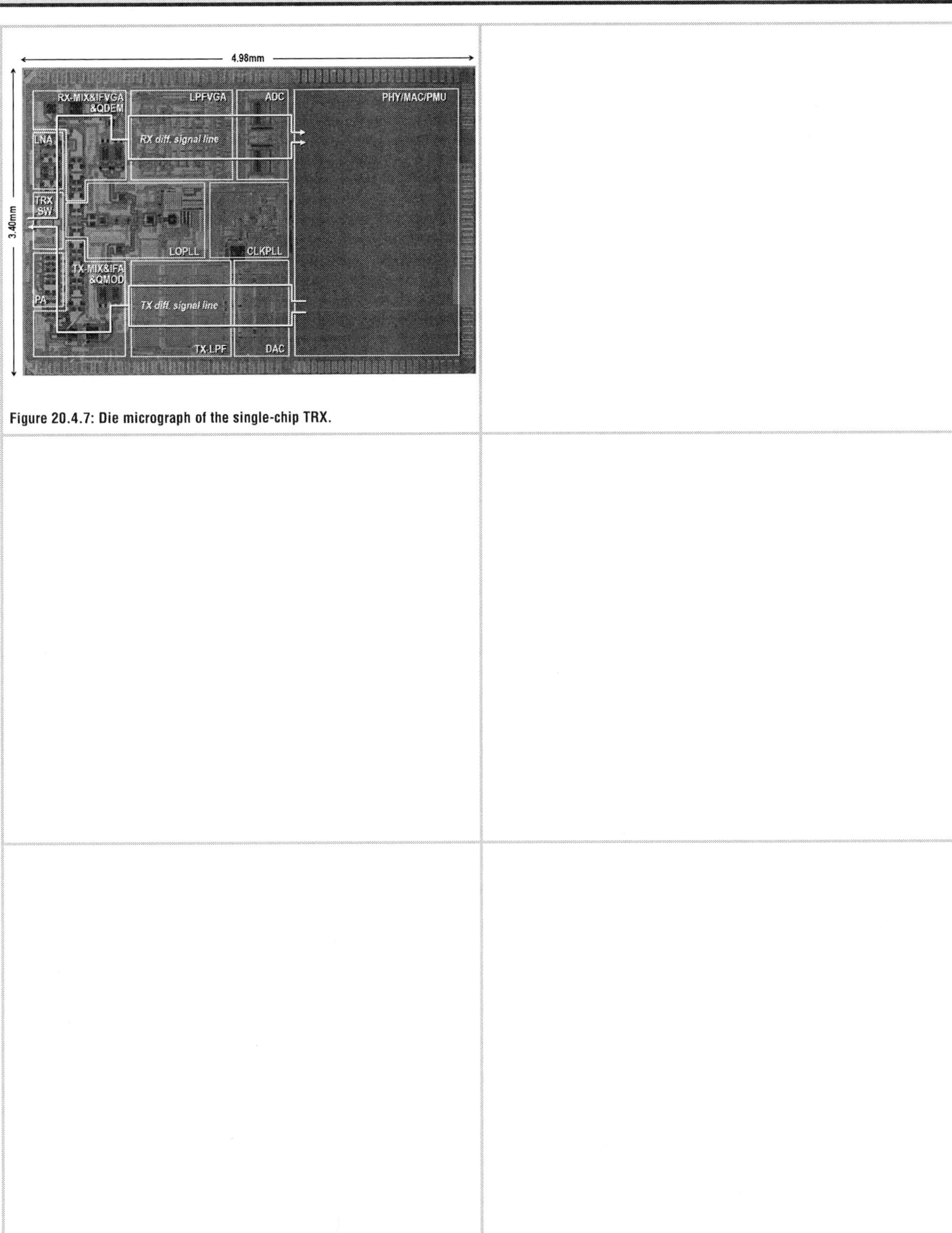

Figure 20.4.7: Die micrograph of the single-chip TRX.

20.5 A 40nm Dual-Band 3-Stream 802.11a/b/g/n/ac MIMO WLAN SoC with 1.1Gb/s Over-the-Air Throughput

Ming He, Renaldi Winoto, Xiang Gao, Wayne Loeb, David Signoff, Wai Lau, Yuan Lu, Donghong Cui, Kun-Seok Lee, Sai-Wang Tam, Philip Godoy, Yung Chen, Sanghoon Joo, Changhui Hu, Arvind Anumula Paramanandam, Xiaoyue Wang, Chi-Hung Lin, Li Lin

Marvell, Santa Clara, CA

The steep growth of digital-content consumption and increasing reliance on wireless networks has resulted in emerging standards such as IEEE 802.11ac. By employing spatial diversity, Multi-user MIMO and high-density modulation (up to 256-QAM), 802.11ac MIMO radios can provide significantly increased throughput, link robustness, and range while maintaining backward-compatibilities with existing 802.11a/n WLAN [1]. However, wide signal band-width and high-density modulation lead to significant challenges in all aspects of RF transceiver design, compared to previous WLAN standards. This paper introduces a fully integrated 3-stream MIMO WLAN SoC that integrates all of the functions of an 802.11a/b/g/n/ac WLAN with a record over-the-air TCP/IP throughput of 1.1Gb/s. The 40nm CMOS SoC integrates dual-band (2.4GHz and 5GHz) RF transceivers, data converters, digital physical layer, media access controller, and a PCI Express Gen-2 interface. The RF transceiver employs an all-digital fractional-N PLL with a record Figure-of-Merit (FoM) of -244dB, a wideband low-impedance bias circuit that minimizes pre-PA driver memory effect for 80MHz signal bandwidth, a dual-band receiver with 3dB/4.3dB NF, and a 5th-order Chebyshev low-pass filter with constant-G_m bias and pre-distorted filter coefficients to support up to 80MHz signal bandwidth.

Figure 20.5.1 shows the system block diagram of one dual-band transceiver in the 3-stream MIMO radio SoC. Both the transmitter and receiver use direct-conversion architecture. A single all-digital PLL (ADPLL) is used to generate the LO signals for all the 2.4GHz and 5GHz transceivers. A single analog baseband block is shared between the 2.4GHz and 5GHz paths. A single low-pass filter is used for both transmit and receive operations to reduce area.

The ADPLL architecture is shown in Fig. 20.5.1. A TDC resolution-compensation loop scales the output of a delay-line-based TDC to track process/voltage/temperature (PVT) variations. Background histogram calibration reduces TDC INL and quantization error. A simple, logic-gate-based circuit doubles the reference clock frequency to improve the in-band phase noise while a digital adaptive loop is used to dynamically adjust the feedback clock edges to compensate the doubler duty-cycle error. A wide range DCO with a 10-b coarse cap bank is designed to cover both 2.4GHz and 5GHz bands over PVT variations. The frequency calibration (FCAL) time is a concern for such a big cap bank as it would limit the channel switching time. A counter operating at f_{DCO} can be employed to speed up binary calibration [2]. However, this requires a digital counter running at GHz rate. In this work a fast cap-code search algorithm is proposed to drastically reduce calibration time. Figure 20.5.2(a) shows the data-path portion of the cap-code search circuit, in addition to the common Dctrl (loop filter output) bound check, TDC bound check is also performed. If the TDC input goes outside of its capture range, no information would be available to make a cap-code decision. Therefore, if the TDC output goes out-of-bound, the top priority is to bring the TDC input back to its capture range. This is done by temporarily changing division value for one cycle to pull in the feedback clock phase. In this way, the phase error can be kept within the TDC capture range while cap-code search is in progress. This helps to speed up phase locking after the cap-code search is finished. Shown in Fig. 20.5.2(b) is the flowchart of the cap-code search algorithm. Before each cap-code decision, the controller waits for M digital clock cycles. The value of M is adaptive. A small M value, such as 4, is used at the beginning of cap-code search when the initial frequency error is large. Both TDC and Dctrl outputs are expected to settle within a few cycles after cap-code update. When a change in cap-code search direction is detected, M is increased to allow longer settling time for the TDC and Dctrl to reduce the chance of making a wrong decision. During cap-code search, a wide loop bandwidth is used to reduce Dctrl settling time. Once the loop is locked, the bandwidth is tuned back to minimize integrated phase error.

Figure 20.5.2(c) shows the histogram of the measured cap-code search time over 100,000 random frequency tests. The minimum, average and maximum search times are 0.5μs, 5.1μs and 22μs, respectively. Figure 20.5.3 shows the phase noise measurement result and FoM [3] comparison with state-of-the-art fractional-N PLLs. At the highest 5GHz channel of 5825MHz with a 40MHz reference clock, the phase noise integrated from 10kHz to 10MHz is 0.37° or $175fs_{rms}$. The PLL consumes 12.9mW, which translates to an FoM of -244dB and is almost 4dB better than the previous record.

The TX output stage pre-PA driver (PPA) is biased in Class-AB for high linearity and efficiency, with an on-chip balun to convert the differential signal to single-ended. When wide-band signals are applied to the PPA, there can be memory effects which degrade EVM and cause asymmetry in the spectral mask [4]. One significant source of memory effects is bias voltage shift at the baseband modulation frequency. A new wideband bias scheme is proposed, which allows PPA bias to be low impedance over baseband frequencies and high impedance at carrier frequency. As shown in Fig. 20.5.4(a), the bias circuit consists of a replica circuit, master-slave source followers and a C-R-C filter. The PPA is driven through a transformer and the bias is connected to the center tap. In order to guarantee stability of the feedback network, a slave source follower is used to isolate the large non-linear capacitance of the PPA input device. The C-R-C filter is inserted to attenuate the second harmonic generated by the PPA. Figure 20.5.4(b) plots transmit spectral mask for a 256-QAM 5/6 80MHz bandwidth 802.11ac signal, centered at 5775MHz.

An inductive source-degenerated LNA is used for the 2.4GHz receiver while a resistive-feedback LNA is adopted in the 5GHz receiver to achieve wide impedance-matching bandwidth, high linearity, low noise figure and low power consumption. Both 2.4GHz and 5GHz LNAs are single-ended, which minimizes the number of RF input pins. It also reduces the overall size of the package, complexity of board design and bill of material (BOM) for MIMO applications. The measured receiver noise figures for 2.4GHz and 5GHz bands are 3dB and 4.3dB respectively.

A 5th-order Chebyshev low-pass filter (LPF) is implemented with programmable gain and bandwidth, supporting all the signal bandwidths for IEEE 802.11a/b/g/n/ac and public safety standards. The LPF is realized with active-RC bi-quads (Fig. 20.5.5(a)). To support 80MHz signal bandwidth, the high-Q bi-quad stage is designed with constant-G_m bias and pre-distorted filter coefficients, which relaxes the gain-bandwidth-product requirement for the operational transconductance amplifier (OTA). The wide filter bandwidth and gain range require the resistors to be programmable over a wide range. For example, the input resistor of the second LPF stage needs to vary by 7 octaves, which would require a large number of resistors. The number of resistors could be greatly reduced by synthesizing large resistance through a T-network, but the signal would be attenuated by T-network programmable resistors. Therefore, a hybrid T-network scheme is only applied to 2.5MHz and 5MHz filter bandwidth (Fig. 20.5.5(b)), where the resistance needed is the largest and the noise requirement is most relaxed.

A summary of the measured performance of the WLAN SoC is shown in Fig. 20.5.6 (a). It achieves an over-the-air TCP/IP throughput rate of 1.1Gb/s. Figure 20.5.6(b) shows 5GHz TX EVM over output power range. With 256-QAM 5/6 80MHz bandwidth 802.11ac signal centered at 5775MHz, TX EVM stays below -32dB over a 48dB range of output power and the EVM floor after I/Q calibration is -37dB. A chip micrograph is shown in Fig. 20.5.7. The die area is $46mm^2$, of which 47% is occupied by the analog and RF circuits.

Acknowledgements:
The authors gratefully acknowledge the contribution of the entire wireless team at Marvell, especially the DSP and system hardware groups.

References:

[1] S. Abdollahi-Alibeik, et al., "A 65nm Dual-Band 3-stream 802.11n MIMO WLAN SOC", *ISSCC Dig. Tech. Papers*, pp. 170-171, Feb. 2011.
[2] J. Shin and H. Shin, "A 1.9–3.8 GHz Fractional-N PLL Frequency Synthesizer With Fast Auto-Calibration of Loop Bandwidth and VCO Frequency", *IEEE J. Solid-State Circuits*, vol. 47, no.3, pp. 665-675, Mar. 2012.
[3] X. Gao, et al., "A Low Noise Sub-Sampling PLL in Which Divider Noise is Eliminated and PD/CP Noise is not Multiplied by N2", *IEEE J. Solid-State Circuits*, vol. 44, no.12, pp. 3253-3263, Dec. 2009.
[4] N. Borges de Carvalho and J.C. Pedro, "A Comprehensive Explanation of Distortion Sideband Asymmetries", *IEEE Trans. on Microwave Theory and Techniques*, vol. 50, no.9, pp. 2090-2101, Sept. 2002.

978-1-4799-0917-9/14 $31.00 © 2014 IEEE

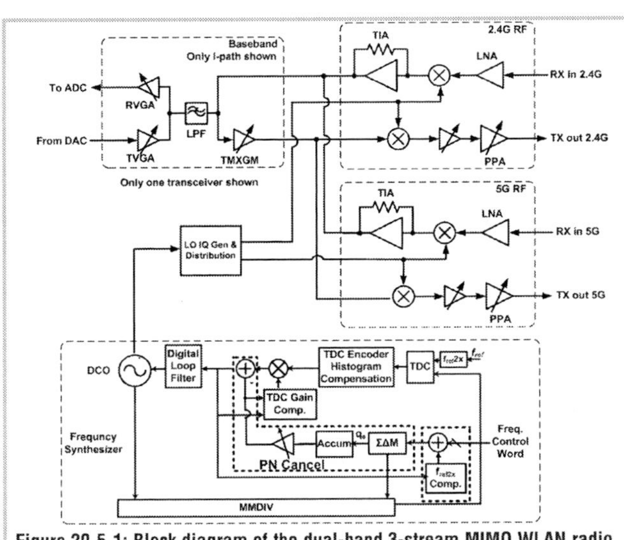

Figure 20.5.1: Block diagram of the dual-band 3-stream MIMO WLAN radio.

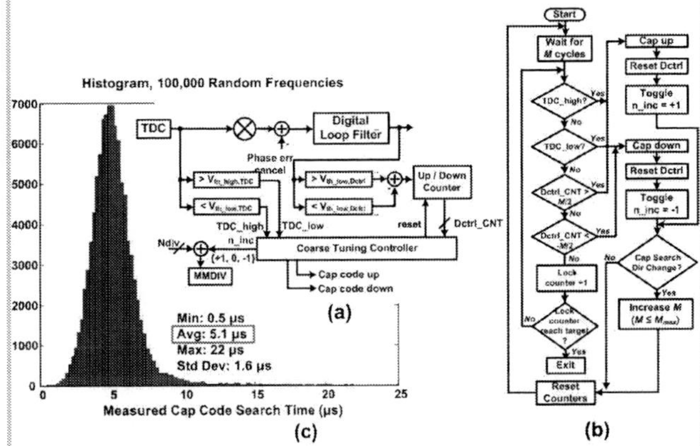

Figure 20.5.2: ADPLL DCO cap-code search a) data path, b) algorithm, c) measured cap-code search time histogram.

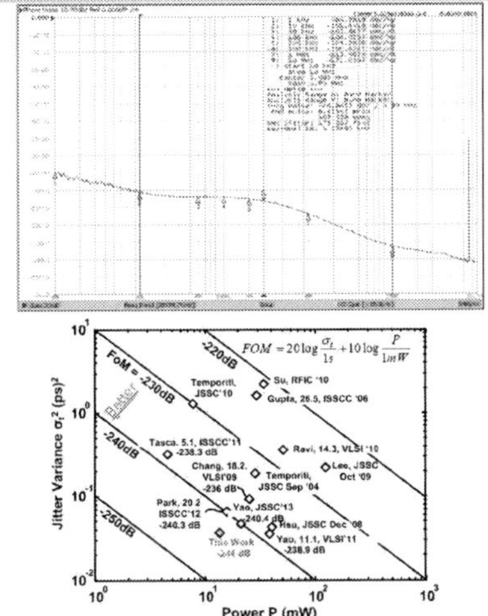

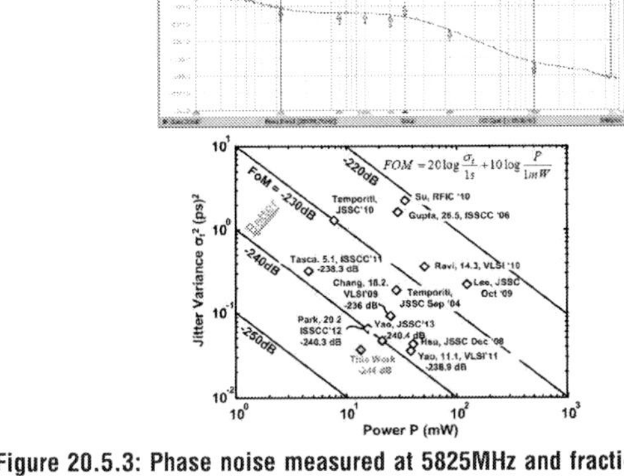

Figure 20.5.3: Phase noise measured at 5825MHz and fractional-N PLL FoM comparison.

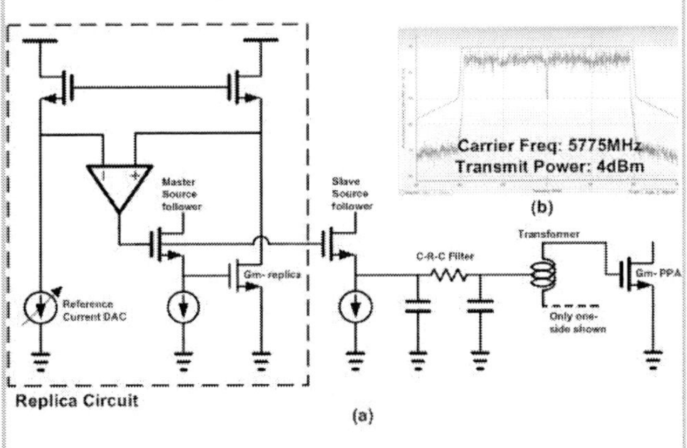

Figure 20.5.4: a) Low-impedance pre-PA driver bias schematic, b) 5GHz transmit spectral mask measurement.

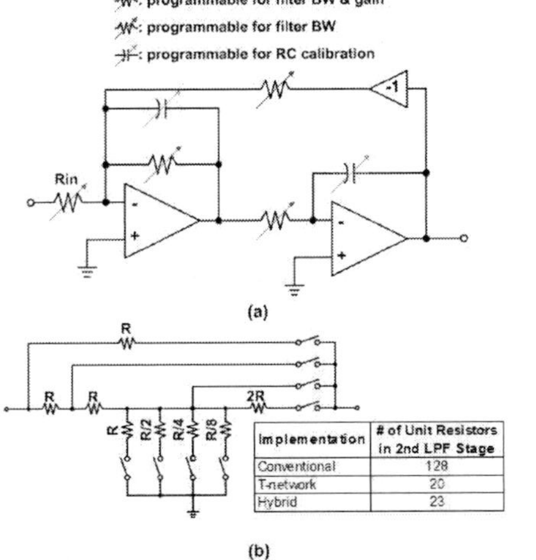

Implementation	# of Unit Resistors in 2nd LPF Stage
Conventional	128
T-network	20
Hybrid	23

Figure 20.5.5: a) Bi-quad used in 5th-order Chebyshev LPF, b) comparison of input resistor implementation schemes.

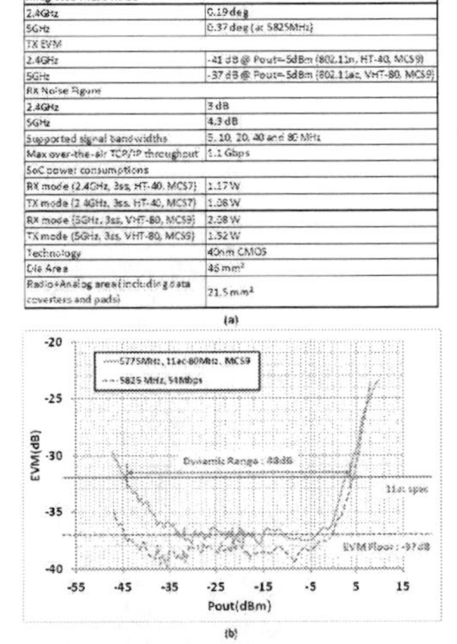

Integrated Phase Noise	
2.4GHz	0.19deg
5GHz	0.37deg (at 5825MHz)
TX EVM	
2.4GHz	-41 dB @ Pout=-5dBm (802.11n, HT-40, MCS9)
5GHz	-37 dB @ Pout=-5dBm (802.11ac, VHT-80, MCS9)
RX Noise Figure	
2.4GHz	3 dB
5GHz	4.3 dB
Supported signal bandwidths	5, 10, 20, 40 and 80 MHz
Max over-the-air TCP/IP throughput	1.1 Gbps
SoC power consumptions	
RX mode (2.4GHz, 3ss, HT-40, MCS7)	1.17 W
TX mode (2.4GHz, 3ss, HT-40, MCS7)	1.08 W
RX mode (5GHz, 3ss, VHT-80, MCS9)	2.08 W
TX mode (5GHz, 3ss, VHT-80, MCS9)	1.52 W
Technology	40nm CMOS
Die Area	46 mm²
Radio+Analog area (including data coverters and pads)	21.5 mm²

(a)

Figure 20.5.6: a) Transceiver performance summary, b) TX EVM measurements at the 5GHz transmitter output.

978-1-4799-0917-9/14 $31.00 © 2014 IEEE

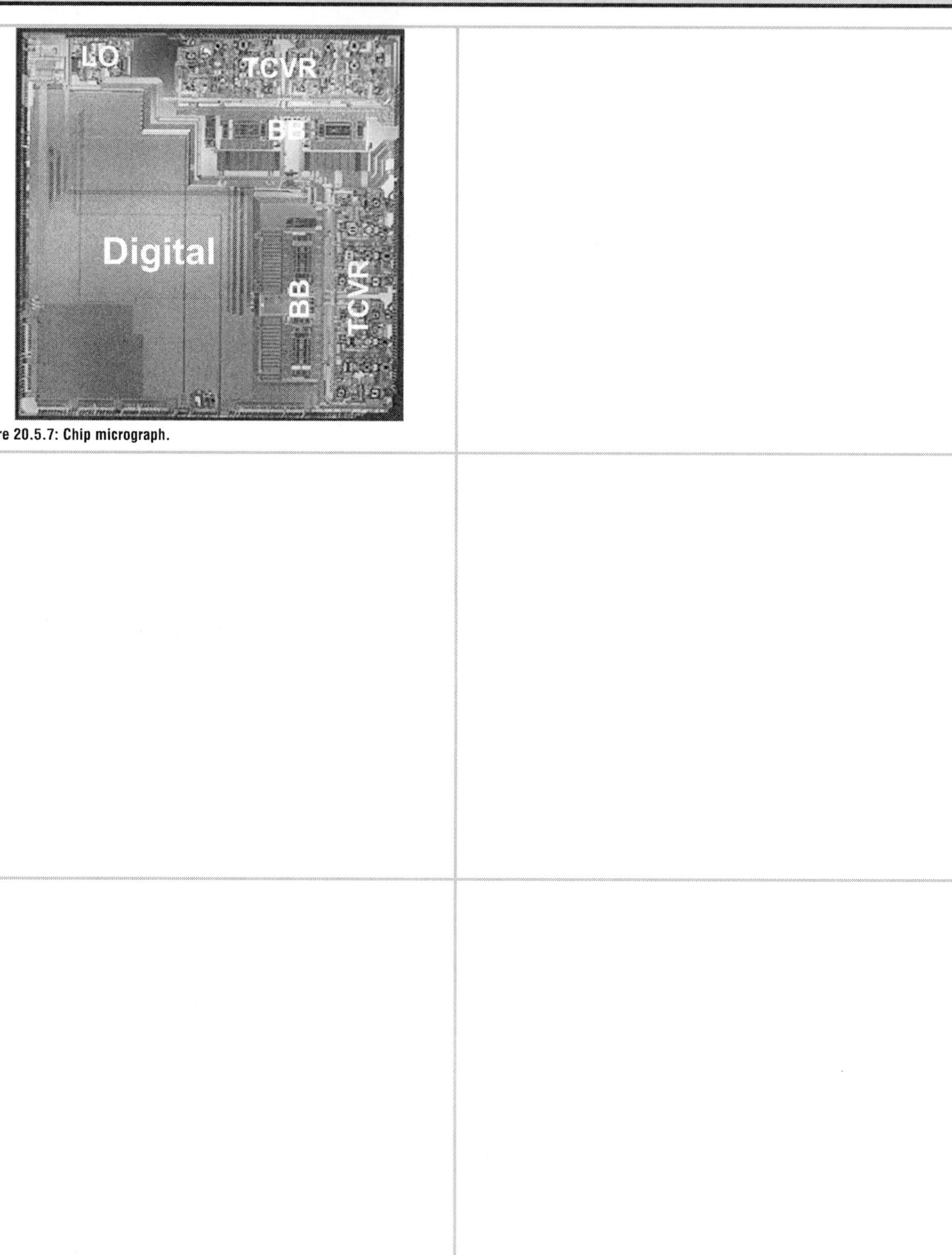

Figure 20.5.7: Chip micrograph.

20.6 A Blocker-Resilient Wideband Receiver with Low-Noise Active Two-Point Cancellation of >0dBm TX Leakage and TX Noise in RX Band for FDD/Co-Existence

Jin Zhou, Peter R. Kinget, Harish Krishnaswamy

Columbia University, New York, NY

The demand for lower cost and form factor and increased re-configurability in wireless systems has driven the investigation of blocker-tolerant software-defined radios [1-4]. While promising, a reduction in system form-factor will result in lower isolation among antennas due to the co-existence of multiple ratios or lower isolation within duplexers for FDD systems due to their reduced size and/or increased re-configurability. Out-of-band (OOB) powerful modulated TX-leakage due to limited antenna/duplexer isolation imposes challenges that are more severe than those posed by continuous-wave (CW) blockers by several orders of magnitude, including cross-modulation, second-order intermodulation and TX noise in the RX band (Fig. 20.6.1). Analysis shows that in the face of 0dBm peak TX leakage and a -30dBm CW in-band jammer, to achieve -90dBm sensitivity with 7dB SNR over 2MHz signal BW, a transceiver must exhibit +30dBm receiver OOB IIP3 and -160dBc/Hz TX noise in RX band.

In highly-linear current-mode receivers, the OOB linearity is dominated by the LNTA input. TX-leakage cancellation at the receiver input may be pursued using passive and active circuitry. Passive TX leakage cancellation requires bulky LC-based isolation structures that are not amenable to silicon integration and wideband/tunable operation. Active leakage cancellation at the receiver input suffers from NF degradation due to the noise of the cancellation circuitry. Here, we demonstrate a two-point low-noise active TX leakage-cancellation scheme that can (i) cancel up to +2dBm peak TX leakage at the receiver input (30dB higher than prior art [6-7]), enabling a triple beat at +2dBm peak TX leakage (TB_{2dBm}) of 68dB and an effective IIP3 of +33dBm (enhancements of 38dB and 19dB respectively) with an associated increase in receiver NF of <0.8dB, and (ii) effectively suppress TX noise in the RX band by up to 13dB, enabling transmitters with -148dBc/Hz RX-band noise to meet the -160dBc/Hz requirement. This is achieved by (i) embedding the leakage cancellation within a noise-cancelling LNTA so that the noise and distortion of the cancellation circuitry are cancelled, and (ii) performing a second-point cancellation of TX noise in the RX band after the LNTA so that the noise impact is reduced. This enables FDD/co-existence with as low as 25dB TX-RX isolation.

The proposed receiver (Fig. 20.6.1) uses a current-mode architecture with a common-gate (CG) common-source (CS) noise-cancelling LNTA (NC-LNTA) followed by current-driven passive mixers, baseband TIAs, and recombination circuitry. Leakage cancellation is pursued *right at the RX input by repurposing the CG device as a leakage canceller.* A TX signal replica coupled from the TX output is injected at the gate of the CG device of the LNTA with appropriate amplitude and phase scaling, creating a current that cancels the TX leakage at the RX input. This eliminates linearity (cross-modulation) issues in the CS device. *The advantage of this approach is that the noise of the cancellation path, namely the CG device, variable-gain amplifier and phase shifter, is completely cancelled through the noise-cancelling operation of the topology.* While the CS device is protected, the CG device still experiences a large TX-leakage (because of gate injection) and an incident CW jammer, and will generate cross-modulation products. *Interestingly, these products also get cancelled upon recombination through the distortion-cancelling property of the noise-cancelling architecture* [5] – the cross-modulation products are downconverted to baseband in the CG path, but also create a voltage at the RX input that is sensed and amplified by the CS path. Note that an equivalent TX leakage current does flow down the CG path to baseband as a result of the injection, but gets filtered out in the baseband TIA and will not create linearity issues due to the current-mode design. Thus, *cross-modulation challenges are mitigated with ideally no addition of noise.* The leakage current flowing down the CG path can degrade SNR due to TX noise in the RX band. So, a second injection of a TX replica signal is performed in the current domain in the CS path at the LNTA output. With appropriate scaling, TX noise in the RX band can be cancelled when the CS and CG paths are recombined. In general, two-point cancellation is needed as the transfer functions of the TX leakage and TX noise in the RX band through the duplexer/antenna pair will differ. Noise impact of the CS-path injection circuitry is reduced by the LNTA's CS path gain.

Figure 20.6.2 shows the detailed schematic of the receiver. A complementary structure is adopted for the NC-LNTA for high linearity. Thick-oxide transistors are used for the CG devices of the LNTA to enable cancellation of >0dBm peak TX leakage. The low-resistance current-driven passive mixers are ac-coupled to the LNTA and are driven by an integrated 4-phase 25%-duty-cycle LO generator. The baseband TIAs adopt a Rauch topology [3] for handling large TX leakage. Recombination circuits based on 5-b digitally-controlled binary-weighted differential g_m cells combine I/Q TIA outputs from the CG and CS paths, allowing adjustment of both amplitude and phase from each path. The cancellation path at the gate of the CG device consists of a 5-b digitally-controlled current-mode Cartesian variable-gain phase rotator (PR) and a variable-gain thick-oxide RF TIA buffer. The unit cell of the phase rotator uses a self-biased inverter-like complementary amplifier topology for high linearity. The RF TIA buffer provides low impedance at the output of the PR for wide bandwidth and at the input of the CG device of the LNTA. A scaled version of the 5-b variable-gain PR is used in current-mode in the second CS cancellation path without the RF TIA. The power consumption of the cancellation paths is scalable (13 to 72mW) based on the TX leakage level.

The prototype 65nm CMOS receiver (Fig. 20.6.7) operates over 0.3GHz to 1.7GHz, with peak gain setting of 34dB and widely programmable gain and baseband bandwidth. The measured NF ranges from 4.2dB to 5.6dB (with cancellation circuitry disabled), the out-of-band IIP3 is +12dBm and the input-referred out-of-band blocker power for 1dB compression is +2dBm (Fig. 20.6.3). The use of 8-phase mixing (not pursued here) would lower the NF even further and provide harmonic rejection. TX leakage cancellation experiments are performed using an attenuator to model the finite isolation as well as using a PCB-based planar antenna pair to model a co-existence environment (Fig. 20.6.4) with a transmitter featuring an off-the-shelf +30dBm power amplifier. Using the CG canceller, >30dB suppression at the input is measured across leakage levels and is not limited by isolation group delay. Figure 20.6.4 also shows the triple beat at the receiver output in the presence of an additional in-band -30dBm jammer for varying leakage levels and the equivalent IIP3 that results with and without cancellation. In the absence of cancellation (reducing our receiver to a generic highly-linear receiver with +12 to 14dBm out-of-band IIP3), the TB degrades to 30dB at +2dBm peak TX leakage. However, with the TX-leakage cancellation, 68dB TB_{2dBm} is measured (38dB improvement) enabling an effective IIP3 of +33dBm (19dB improvement). The receiver NF is measured with the CG canceller enabled using the same recombination settings that optimize TB, and the noise-figure degradation is <0.8dB thanks to noise cancellation. With noise cancellation disabled, the NF degrades to 12dB. Cancellation of TX noise in the RX band is also demonstrated (Fig. 20.6.5) for a constant -6dBm peak TX leakage level while varying the relative RX-band noise level (in dBc/Hz). Even though the CS canceller adds noise, the overall receiver noise still gets reduced significantly due to the TX noise cancellation. Up to 13dB effective suppression of TX noise in the RX band is measured. The receiver including the LO divider consumes 75mW-to-83mW power. When compared with the state of the art (Fig. 20.6.6), our work demonstrates the highest modulated leakage handling (TB_{2dBm} and equivalent IIP3), and is the only one that alleviates the TX's RX-band noise requirement.

Acknowledgment:
This work was sponsored by the DARPA RF-FPGA program.

References:
[1] D. Mahrof et al, "A Receiver with In-Band IIP3>20dBm, Exploiting Cancelling of OpAmp Finite-Gain-Induced Distortion via Negative Conductance," *IEEE RFIC Symp.*, pp. 85-88, June 2013.
[2] D. Murphy et al, "A Blocker-Tolerant Wideband Noise-Cancelling Receiver with a 2dB Noise Figure," *ISSCC Dig. Tech. Papers*, pp. 74-75, Feb. 2012.
[3] I. Fabiano et al, "SAW-Less Analog Front-End Receivers for TDD and FDD", *ISSCC Dig. Tech. Papers*, pp. 82-83, Feb. 2013.
[4] C. Andrews et al, "A Passive Mixer-First Receiver with Digitally Controlled and Widely Tunable RF Interface," *IEEE J. Solid-State Circuits,* vol. 45, no. 12, pp. 2696-2708, Dec. 2010.
[5] S. Blaakmeer et al, "Wideband Balun-LNA With Simultaneous Output Balancing, Noise-Canceling and Distortion-Canceling," *IEEE J. Solid-State Circuits,* vol. 43, no. 6, pp. 1341-1350, June 2011.
[6] H. Khatri et al, "An Active Transmitter Leakage Suppression Technique for CMOS SAW-Less CDMA Receivers," *IEEE J. Solid-State Circuits,* vol. 45, no. 8, pp. 1590-1601, Aug. 2010.
[7] V. Aparin et al, "An Integrated LMS Adaptive Filter of TX Leakage for CDMA Receiver Front-Ends," *IEEE J. Solid-State Circuits,* vol. 41, no. 5, pp. 1171-1182, May 2006.

978-1-4799-0917-9/14 $31.00 © 2014 IEEE

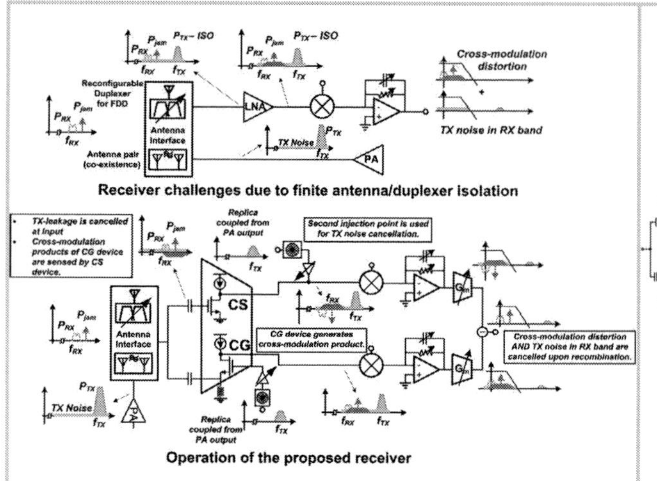

Figure 20.6.1: Challenges posed by modulated TX leakage and suppression of these mechanisms in our proposed receiver.

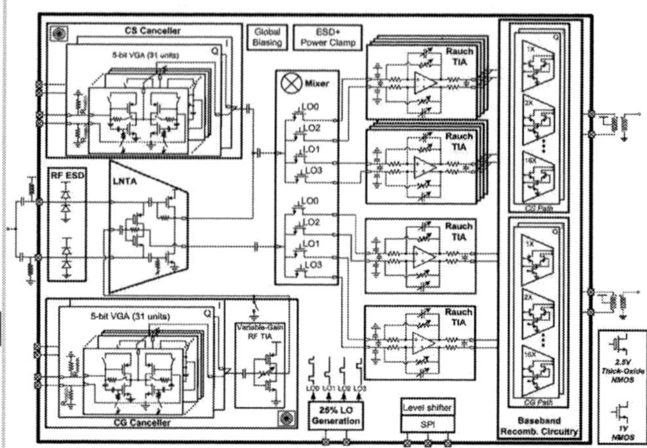

Figure 20.6.2: Proposed receiver with low-noise active two-point cancellation of TX leakage and RX-band noise.

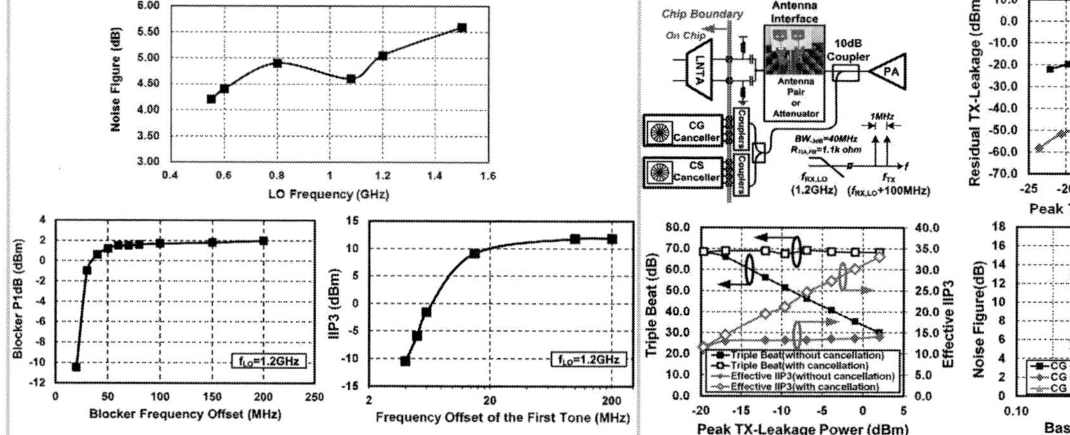

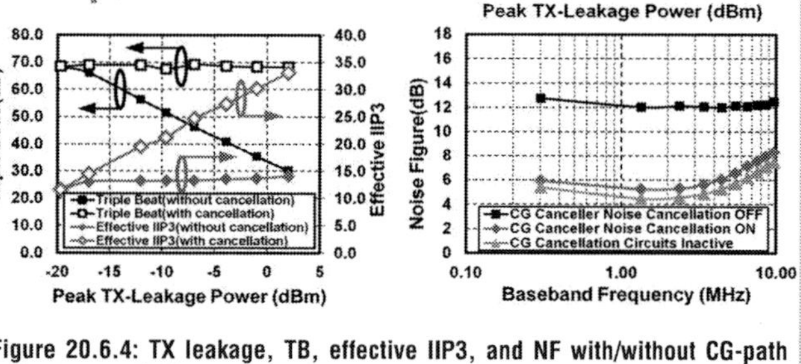

Figure 20.6.3: Measured NF (cancellation circuits inactive), input-referred blocker power for 1dB compression, and IIP3.

Figure 20.6.4: TX leakage, TB, effective IIP3, and NF with/without CG-path leakage cancellation.

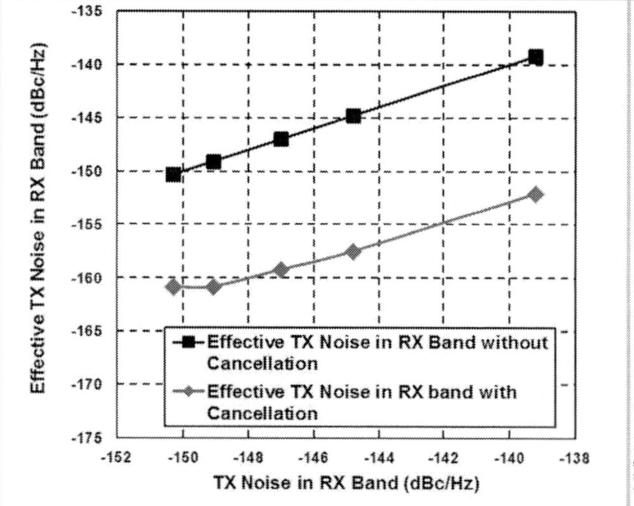

Figure 20.6.5: Measured TX noise in the RX band.

	[6]	[7]	[4]	[1]	[3]	This work
Architecture	TX Leakage Cancellation/Suppression		Highly-Linear Software-Defined Radio			Current-Mode Receiver with active two-point low-noise TX-leakage cancellation.
	Active TX Leakage Suppression After LNA	Active TX Leakage Cancellation After LNA	High Linearity Passive Mixer-First	Mixer First with Distortion Cancellation	Current-Mode RX with Highly-Linear LNTA	
CMOS Technology	180nm	250nm	65nm	65nm	40nm	65nm
RF Frequency	1.96GHz	800MHz	0.1-2.4GHz	0.2-2.6GHz	1.8-2.4(TDD)/ 1.8-2.1(FDD) GHz	0.3-1.7GHz
Gain	45dB	14.8dB (LNA)	40-70dB	26.5dB	45.5dB	19-34dB
Baseband BW	N/A	N/A	20MHz[1]	12MHz	1.4(TDD)/ 3.4(FDD) MHz	2-76MHz
DSB NF	3.1dB	1.4dB (LNA)	4dB	7.5dB	3.8(TDD)/ 3.1(FDD) dB	4.2dB
Blocker P1dB	N/A	N/A	+4dBm[2]	N/A	N/A	>+2dBm[3]
Out-of-band IIP3	N/A	N/A	+25dBm	+10/+18dBm[4]	+18(TDD)/ +16(FDD) dBm	+12dBm/+33dBm[5]
Maximum Handled Peak TX Leakage	-28dBm[6]	-28dBm[7]	N/A	N/A	N/A	+2dBm[6]
NF Degradation due to Leakage Cancellation	1.8dB	1.3dB (LNA)	N/A	N/A	N/A	<0.8dB
TB_{2dBm}[8]	9dB[9]	N/A	52dB[10]	22/38dB[10]	38(TDD)/ 34(FDD) dB[10]	30dB (without cancellation) 68dB (with cancellation)
TX Noise Cancellation	N/A	N/A	N/A	N/A	N/A	13dB
RX Power Consumption	114mW	N/A	37-70mW	17.3-36.7mW	30.7mW	74.6-83.0mW
Canceller Power Consumption	48mW	43mW	N/A	N/A	N/A	13-72mW
Active Area	N/A	N/A	2mm²	0.2 mm²	0.74 mm²	1.2 mm²

[1] Maximum BW [2] Blocker at 40MHz offset [3] Blocker at >60MHz offset [4] First tone at 135MHz/>450MHz offset
[5] Effective IIP3 for TX leakage under cancellation of +2dBm peak TX leakage [6] 3dB Peak-to-average ratio
[7] 6dB Peak-to-average ratio [8] Triple beat at +2dBm peak TX leakage power [9] Calculated from reported TB at -28dBm TX leakage level
[10] Calculated from reported IIP3 (IIP3=0.5×TB+$P_{TX,avg}$[6]) *Metrics related to the TX leakage cancellation are highlighted with*

Figure 20.6.6: Measurement summary and comparison table.

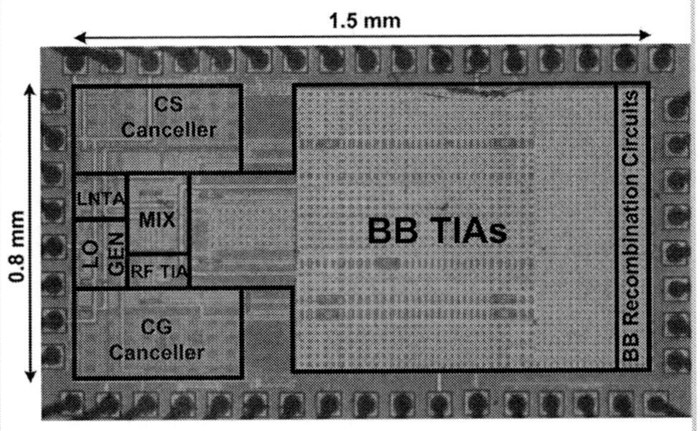

Figure 20.6.7: Die micrograph.

20.7 A Multi-Band Inductor-Less SAW-Less 2G/3G-TD-SCDMA Cellular Receiver in 40nm CMOS

Ming-Da Tsai, Chih-Fan Liao, Chi-Yun Wang, Yi-Bin Lee, Bosen Tzeng, Guang-Kaai Dehng

MediaTek, Hsinchu, Taiwan

The growing demand for high-speed wireless communication has driven the evolution of cellular phone networks. New-generation cellular standards use wider channel bandwidth and more sophisticated modulation to obtain higher data-rates. Due to various cellular standards, chip providers are required to offer highly integrated solutions that support 2G, 3G, and even 4G in one chip. This paper presents a receiver supporting 2G quad bands and 3G TD-SCDMA dual bands. Figure 20.7.1 shows the 2G/3G receiver, whose front-end current-mode outputs are combined at baseband CR filter and biquad PGA, which are shared between all bands. A dynamic gain-bandwidth-product-extension circuit technique is used to remove a transimpedance amplifier to save die area and current.

Multiple LNAs are necessary to support the large number of bands covering different areas and countries (up to 40 bands worldwide), which significantly increases die area and complexity. As a result, an inductor-less LNA topology is used in this design to save die area. Moreover, several LNAs are shared between different bands as follows: In contrast to WCDMA, TD-SCDMA is a time-division duplex (TDD) system and does not use any external duplexer in front of the receiver. Therefore, the LNAs between multiple bands of 2G and TD-SCDMA are shared. For example, a single LNA is used for DCS, PCS, B34, and B39 ranging from 1.8GHz to 2GHz instead of a dedicated LNA for each band. The SAW-less feature offers an option for phone manufacturers to upgrade their 2G platforms to 3G without increasing cost. In this chip, four high-band (HB) LNAs and one low-band (LB) LNA duplicated from SAW-less realizations are reserved for Dual-SIM Dual-Talk (DSDT) application and TD-SCDMA band 40.

The LNA in this work is pseudo-differential, as shown in Fig. 20.7.2. In contrast to [1], the auxiliary-path RF amplifier, M_1 and M_2, provides a negative voltage gain, -Av, which is needed to create the required real part for input matching. This allows the main receiver path to be purely current-mode, which is required for SAW-less operation. The noise from the auxiliary-path amplifier creates correlated noise voltage at nodes V_X and V_{IN}, which is cancelled by inserting a positive transconductor G_{mnc}, transistors M_7 and M_8, in parallel with the cascode transistors M_5 and M_6 respectively. The shared source nodes of M_5/M_7 and M_6/M_8 maintain a current-mode interface at the LNA output while the DC current of the main transconductor, M_3 and M_4, is reused. The signal current at the output of G_{mnc} is in-phase with the output signal while the noise current is out-of-phase and is cancelled at the output. The transconductance of G_{mnc} is well-defined because its DC current is determined by the size ratio of $M_{7,8}$ to $M_{5B,6B}$. The size ratio of $M_{7,8}$ to $M_{3,4}$ is around 1/20 for optimum noise cancellation. The G_{mnc} output common-mode voltage, 0.75V, is defined through R_B. Excessive voltage swing at node V_X degrades linearity at maximum LNA gain. Therefore C_1 and C_2 are properly sized to get a few dB of attenuation at the input of M_7 and M_8. This relaxes the linearity requirement of M_7 and M_8 and results in better IIP3. The LNA consumes a DC current of 8mA/10mA in LB/HB respectively.

Out-of-band blockers can reach 0dBm if no external SAW filter is used in front of the LNA. Such a large blocker creates a large voltage swing at the output of the auxiliary-path amplifier. To avoid saturating M_7 and M_8, a blocker detector is used at the LNA input. Consequently, M_7 and M_8 are turned off when blockers larger than –15dBm are detected, while the size of the cascode device is increased by $M_{5A/6A}$ to sustain strong blocker-induced current. Different from [2], the adaptive on/off control of M_7 and M_8 can still maintain current-mode operation of the LNA in the presence of strong blockers. The Class-AB operation of the LNA and the low-impedance nature along the receiver chain enhances the P_{1dB} significantly and hence dynamic range [3]. With the presence of a 2dBm out-of-band blocker, the resulting in-band SNR hardly exceeds 10dB due to reciprocal-mixing noise alone with a reasonable LO phase noise. Therefore, the RX NF plays a less important role in determining the overall SNR in this large blocker condition. As a result, when the blocker detector senses strong blockers, the LNA gain is decreased, resulting in lower blocker current flowing into the passive mixer, which allows smaller C_{LPF} in Fig. 20.7.3 and saves die area considerably compared to [3]. The LNA gain control is done by reducing the number of active LNA slices, which also results in reducing the DC current consumption when the LNA gain is dropped. This method of gain control is hardly available in an inductor-degenerated LNA [3] because the input matching changes considerably when the LNA slices change.

The receiver uses current-driven passive mixers for better NF and IIP2. The IF current is fed to a CR filter for unwanted blocker filtering followed by a Tow-Thomas biquad filter/PGA to perform channel selection as shown in Fig. 20.7.3. The CR filter, constructed by C_{LPF} and R_{LPF}, affects the feedback factor of the first opamp in the biquad filter, and the additional poles and zeros introduced by it should be carefully considered. Moreover, the CR pole is adjusted for different signal bandwidth. Specifically R_{LPF} is switched to a larger value in 2G, which allows larger out-of-band blocker current to be diverted into C_{LPF} to protect the current-mode biquad from strong blockers. On the other hand, R_{LPF} is switched back to a lower value in 3G to enlarge the bandwidth of the CR filter in order not to cause an in-band droop. The first opamp in the biquad filter utilizes a switchable compensation capacitor ($C_{c1/c2}$), as depicted in Fig. 20.7.3, because the differential-mode stability is affected by C_{LPF} and R_{LPF} while the common-mode stability is not. This approach maintains the stability in both differential and common mode while maximizing the gain-bandwidth product (GBW) of the opamp in 3G mode because the poles introduced by C_{LPF} and R_{LPF} are higher and the associated opamp compensation capacitor can be smaller. This technique is further illustrated in Fig. 20.7.3 by a pole-zero plot of feedback network β. The filter response, linearity, and the RX high-/low-side gain matching are improved with the aid of the GBW-maximizing technique in 3G mode. By avoiding a first-order TIA between the mixer and the channel-select filter, the blocker filtering is more efficient, which results in smaller voltage swings at the opamp inputs and outputs, and therefore better linearity and higher dynamic range are achieved.

The chip was fabricated in a 40nm 1P6M CMOS process. The receiver achieves a 1.9dB NF for operating frequencies below 1GHz, and 2.1dB for operating frequencies from 1.8GHz to 2.0GHz, as shown in Fig. 20.7.4. The receiver is tested with a –99dBm wanted signal and >0dBm 20M/80M-offset blockers for 2G (15dB more relaxed for TDS-CDMA) and the measured gain compression and SNR are shown in Fig. 20.7.5. With a 0dBm blocker, the worst-case gain compression is 1dB and the receiver NF is at most 11dB. The external balun approximately accounts for 1dB of loss, and the whole system achieves >7.7dB SNR for a blocker level up to +3dBm. The IF bandwidth is configured to 0.95MHz in TD-SCDMA mode and the measured in-band IIP3 is >+1.9dBm when the LNA is in high gain mode, and is > +7dBm when the LNA is in low gain mode. Given its high dynamic range and low power consumption, the proposed architecture is also suitable for WCDMA. The key indexes to quantify WCDMA RX include RX NF, IIP2 at duplex offset and IIP3 for 5MHz ACS, 10MHz/20MHz, and TX leakage/OOB inter-modulation tests. The IF bandwidth is increased to 2.3MHz in WCDMA mode. Figure 20.7.4 shows B1 NF is 2.4dB on average. Figure 20.7.5 shows that B1 IIP3 is +0.4dBm when the LNA is at high gain and +8.5dBm when the LNA is at middle gain. The average/minimum out-of-band IIP2 is measured to be +58dBm /+55dBm across 184 samples with 80MHz-offset and –30dBm two-tones. The performance is summarized in Fig. 20.7.6 and the die photo is shown in Fig. 20.7.7. The receiver is the most compact triple-mode and six-band TDD cellular receiver among the state-of-the-art receivers listed in the Table.

Acknowledgement:
The authors would like to thank many other colleagues who helped chip development, and Dr. Osama Shana'A for valuable paper reviews.

References:
[1] F. Beffa, et al., "A Receiver for WCDMA/EDGE Mobile phones with Inductorless Front-End in 65nm CMOS," *ISSCC Dig. Tech. Papers*, pp. 370–372, Feb. 2011.
[2] F. Bruccoleri, et al., "Noise Cancelling in Wideband CMOS LNAs," *ISSCC Dig. Tech. Papers*, pp. 406–407, Feb. 2002.
[3] I. S-C Lu, et al., "A SAW-less GSM/GPRS/EDGE Receiver Embedded in a 65nm CMOS SoC," *ISSCC Dig. Tech. Papers*, pp. 364–366, Feb. 2011.
[4] D. Murphy, et al. "A Blocker-Tolerant Wideband Noise-Cancelling Receiver with a 2dB Noise Figure," *ISSCC Dig. Tech. Papers*, pp. 74–76, Feb. 2012.
[5] I. Fabiano, et al., "SAW-less Analog Front-End Receivers for TDD and FDD," *ISSCC Dig. Tech. Papers*, pp. 82–83, Feb. 2013.

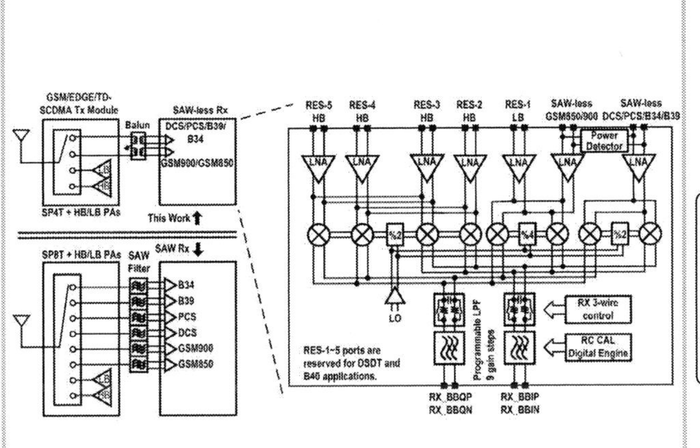

Figure 20.7.1: Block diagram of the 2G + 3G TDD cellular receivers.

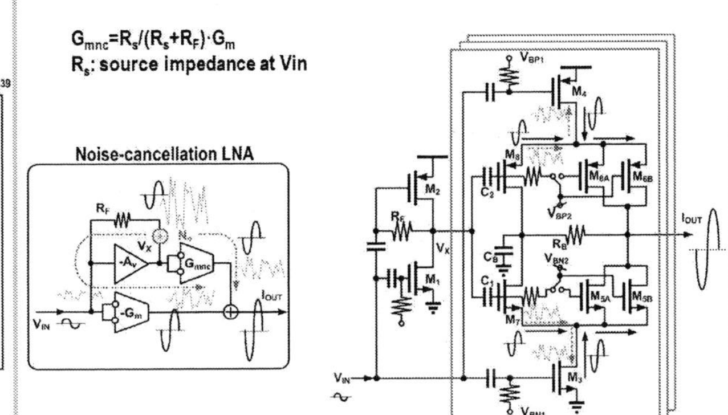

Figure 20.7.2: LNA half-circuit diagram.

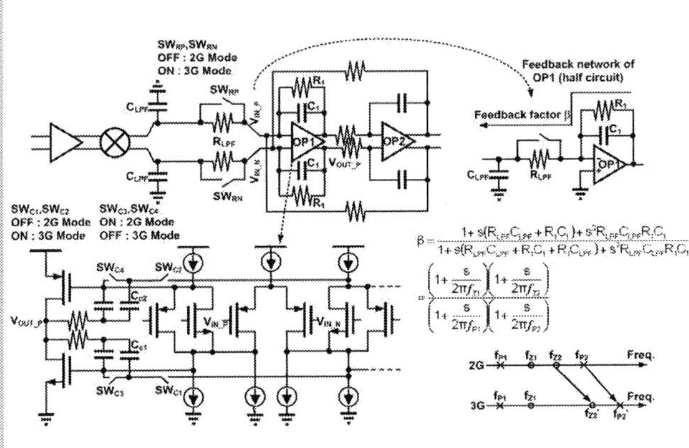

Figure 20.7.3: IF filter OP1 with switchable compensation and the analysis of OP1's feedback factor.

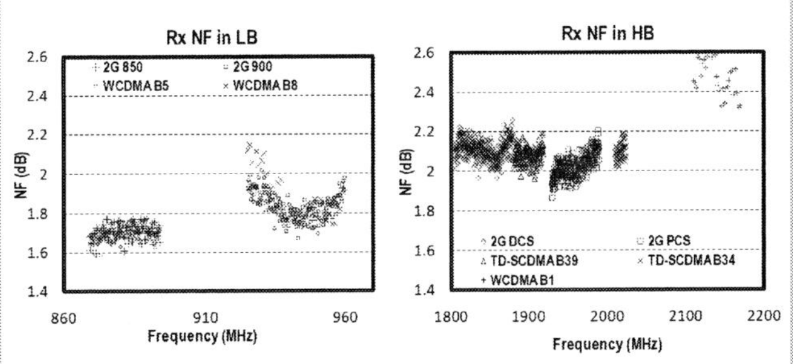

Figure 20.7.4: Measured NF of the receiver chain.

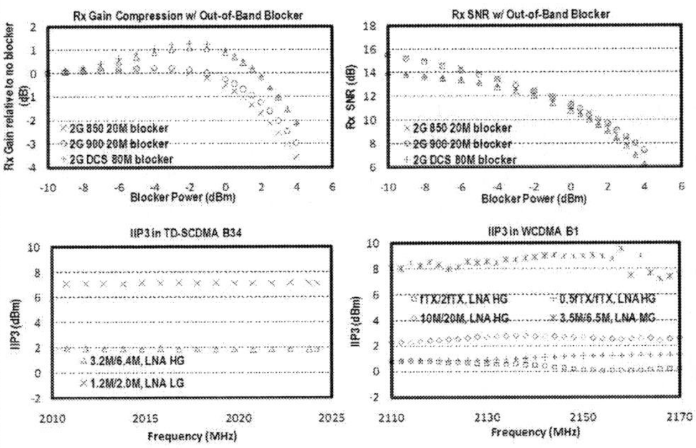

Figure 20.7.5: Measured RX linearity, including P_{1dB}, SNR with out-of-band blocker, and IIP3.

	ISSCC 2011[1]	ISSCC 2011 [3]	ISSCC 2012 [4]	ISSCC 2013 [5]	This work
Process	65nm CMOS	65nm CMOS	40nm CMOS	40nm CMOS	40nm CMOS
Need Rx SAW Filter	YES	NO	NO	NO	NO
On-Chip Inductor	NO	YES	NO	YES	NO
Number of LNAs	8	2	1	1	7
Applications	GSM/EDGE/WCDMA	GSM/EDGE	N/A	N/A	GSM/EDGE/TD-SCDMA/WCDMA
RF Frequency (MHz)	850/900/1800/1900/ 2100	850/900/1800/1900	80~2700	1800~2100	850/900/1800/ 1900/2000/2100/2400
Rx Gain (dB)	67	60	70	45.5	57
Rx NF (dB)	2.9/2.4	2.7/2.9	1.9	1.9	1.7/1.8/2.1/2.4
Rx Out-of-Band P_{1dB} (dBm)	N/A	1	>0	−1	>2
Rx NF with 0dBm Out-of-Band blocker	N/A	6.5/7.0	4.1	7.9	11
Out-of-Band IIP3 (dBm) (f_{tx}/$2f_{tx}$)	−3	>0	13.5	16	0.4
In-Band IIP3 (dBm) 10M/20M, LNA HG 3.2M/6.4M, LNA HG 3.5M/6.5M, LNA MG	 -3 N/A 4	 N/A N/A N/A	 10 -20 N/A	 N/A N/A N/A	 2.6 1.9 8.5
Out-of-Band IIP2 (dBm)	>50	>50	54	66	>55
Supply Voltage (V)	1.5	1.4/2.5	1.3	1.2/1.8	1.5
Rx Current (mA)	38.5*	37.1	27~60#	19.5	26/31
Area (mm^2)	0.88 (w/o IF filter)	1.05 1.4(w/ IF filter)	1.2 (w/o IF filter)	0.74 (w/o IF filter)	0.42 0.57(w/ IF filter)

*: using a DC-DC converter[1].

#: The LO path consumes 3 to 36 mA, the RF GM cell draws 8mA, and the baseband circuits consume 16mA[4].

Figure 20.7.6: Performance summary and comparison with prior art.

978-1-4799-0917-9/14 $31.00 © 2014 IEEE

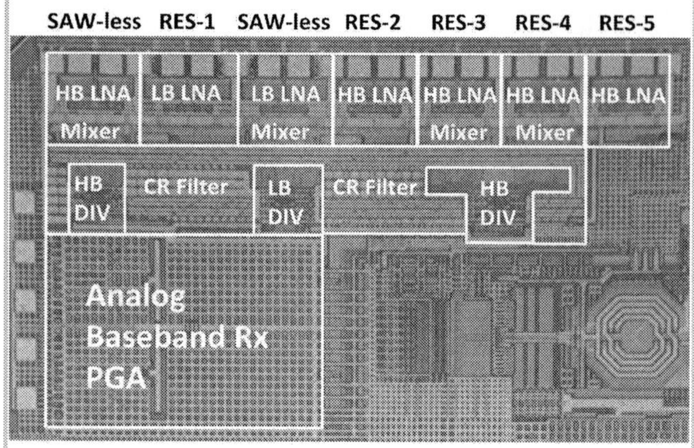

Figure 20.7.7: Die micrograph.

20.8 A 20mW GSM/WCDMA Receiver with RF Channel Selection

Joung Won Park, Behzad Razavi

University of California, Los Angeles, CA

Recent work on RF receivers has exploited N-path filters to address two critical issues, namely, blocker tolerance and high RF selectivity [1,2]. However, these designs face three drawbacks: (1) the low-noise amplifier (LNA) incorporates a G_m stage that, even with a virtual ground at its output nodes, must still withstand strong blockers at its input; (2) the low-order filter transfer function does not provide sufficient selectivity in narrow-band applications such as GSM or WCDMA; (3) they consume roughly 60mW around 2GHz.

This paper introduces a wideband receiver that realizes filtering at the LNA *input* (rather than output) so as to achieve selectivity even for GSM channels. The receiver can tolerate a 0dBm blocker at 23MHz offset and its RF channel selection devices can be readily configured to operate with WCDMA or IEEE802.11b/g as well.

In order to describe the receiver's operation, we begin with the simplified view in Fig. 20.8.1, where an LNA with resistive feedback ensures proper input impedance matching [3]. If an N-path notch filter [4,5] is placed around the amplifier, the resulting Miller multiplication of C_F manifests itself at the input *outside* the notch bandwidth, thereby providing selectivity. Unlike the passive topology in [5], we insert one switch on each side of C_F to allow its parasitic, C_P, to be upconverted and not directly load the LNA input or output. An important benefit of the Miller effect is that it reduces the on-resistance of the feedback switches proportionally, improving the out-of-notch rejection. As a result, the power consumption of the LO distribution network is significantly reduced. The notch bandwidth thus created still exceeds several MHz and hence is inadequate.

In the next step, we consider a three-stage implementation of the LNA [3] and recognize that the latter stages tend to saturate at high blocker levels, allowing less Miller multiplication of C_F. We therefore add a local notch filter, bank 2, around the first stage. Nonetheless, the Miller multiplication produced by the two notch filters still yields a gradual roll-off. In order to increase the selectivity, we seek a "super-Miller" effect, i.e., one creating a multiplication factor that *rises* with frequency. To this end, we add a third notch filter, bank 3, including a multi-stage amplifier, A_2, that contains two zeros near the origin. The overall arrangement produces a flat response within the channel bandwidth and a sharp drop beyond it. The bandwidth is programmed by the feedback capacitors in banks 1, 2, and 3, and the two zeros in A_2.

The various loops around the LNA raise stability concerns, particularly with respect to the bandwidth necessary for A_2. Fortunately, bank 1 and bank 2 help stabilize the loop even if each A_2 is biased at a current of 385μA. This is in contrast to the topology in [2], which requires 5mA of bias in each all-pass amplifier in the feedback path. Moreover, the greater stability permits a loop gain of 50dB for capacitor multiplication, another point of contrast to the 20dB loop gain in [2]. The S_{11} measurements of our prototype confirm the stability of the front end.

An interesting and unique issue that arises in the architecture of Fig. 20.8.1 is that the total delay around the loop - due to the LNA's stages and A_2's stages – slightly *shifts* the center frequency of the channel-selection bandpass response. This is because the delay alters the phase of the signal returning through the N-path filters, thus changing the Miller multiplication factor to a complex value and hence forcing the response to reach a peak slightly away from the LO frequency. Fortunately, this effect can be removed through the use of polyphase signaling: as depicted in Fig. 20.8.2, each A_2 amplifier is decomposed into two so that a fraction of the feedback signal can be injected from one branch to the adjacent branch and compensate for the phase shift.

The quadrature baseband signals are available within all three banks. Since the broadband LNA noise at the LO harmonics folds to the baseband, it is desirable to exploit harmonic-reject (HR) mixing [6]. As shown in Fig. 20.8.2, the downconvered signals in bank 1 are sensed by properly-ratioed G_m stages and combined. In contrast to conventional HR mixers that operate with overlapping phases, this design must perform harmonic rejection by means of nonoverlapping clocks and, consequently, employs weighting factors of 1 and $1+\sqrt{2}$ rather than 1 and $\sqrt{2}$. This arrangement suppresses the third and fifth harmonics, thus reducing the noise figure, according to simulations, by about 3dB. The Q-channel mixing can be performed similarly.

The generation and distribution of eight LO phases with 12.5% duty cycle for the N-path filters can potentially consume a high power, e.g., 36mA at 2.7GHz in 40nm technology [1]. We propose an approach that reduces the total power to 6mA at 2GHz in 65nm technology. Illustrated in Fig. 20.8.3, the idea is to combine the phases at the output of two cascaded ÷2 stages by means of simple combinational logic. For example, an AND gate, G_1, accepts $4f_{LO,0}$, $2f_{LO,90}$, and $f_{LO,135}$ signals to generate a clock with 12.5% duty cycle. Sharing some nodes for proper operation, eight such AND gates produce the eight phases that drive the switches. According to simulations, this arrangement exhibits a phase noise of -165.7dBc/Hz at 20MHz offset owing to its simplicity.

The receiver has been fabricated in TSMC's 65nm digital CMOS technology. Shown in Fig. 20.8.7 is a micrograph of the die, whose active area measures $0.82mm^2$. The capacitors in all banks are programmable through an on-chip serial bus. At 2GHz, the LNA draws 8.6mA, the Miller amplifiers (A_2's) a total of 1.5mA, the baseband combining amplifiers in Fig. 20.8.2 a total of 1mA and the LO generation circuit of Fig. 20.8.3, 6mA.

Figure 20.8.4 plots the measured RF-to-baseband gain as a function of the baseband frequency for three different settings. The 3dB bandwidth varies from 200kHz to 10MHz. We note that the GSM setting provides an attenuation of 15dB in the middle of the alternate adjacent channel. (A layout error does not allow switching in enough capacitors to further reduce the bandwidth.) For the WCDMA setting, the attenuation reaches 16dB in the middle of the alternate adjacent channel.

Figure 20.8.5 shows the measured noise figure (NF) vs. the input blocker level at 23MHz offset, indicating a degradation of 2.5dB as the latter reaches 0dBm. About 0.5dB of this degradation arises from the noise floor of the blocker. This figure also plots the NF vs. the baseband frequency for the exacting requirements of GSM and WCDMA. The measured LO leakage to the antenna falls below -67dBm at 2GHz.

Figure 20.8.6 summarizes the measured performance of our receiver and compares it with the state of the art.

Acknowledgments:
This research was supported by Lincoln Laboratory and Realtek Semiconductor. The authors are grateful to the TSMC University Shuttle Program for chip fabrication.

References:
[1] D. Murphy, A. Hafez, A. Mirzaei, M. Mikhemar, H. Darabi, M.F. Chang, A. Abidi, "A Blocker-Tolerant Wideband Noise-Cancelling Receiver with a 2dB Noise Figure," *ISSCC Dig. Tech. Papers*, pp. 74-76, Feb. 2012.
[2] S. Youssef, R. van der Zee, B. Nauta, "Active Feedback Receiver with Integrated Tunable RF Channel Selectivity, Distortion Cancelling, 48dB Stopband Rejection and >+12dBm Wideband IIP3, Occupying <0.06mm² in 65nm CMOS," *ISSCC Dig. Tech. Papers*, pp. 166-168, Feb. 2012.
[3] J.W. Park and B. Razavi, "A Harmonic-Rejecting CMOS LNA for Broadband Radios," *IEEE Journal of Solid-State Circuits*, vol. 48, pp. 1072-1084, Apr. 2013.
[4] M. A. Poole, "High Q Notch Filter," U.S. Patent 3,795,877, Mar. 5, 1974.
[5] A. Ghaffari, E. Klumperink, B. Nauta, "8-Path Tunable RF Notch Filters for Blocker Suppression," *ISSCC Dig. Tech. Papers*, pp. 76-78, Feb. 2012.
[6] J.A. Weldon, J.C. Rudell, L. Lin, R.S. Narayanaswami, M. Otsuka, S. Dedieu, L. Tee, K. Tsai, C. Lee, P.R. Gray, "A 1.75 GHz Highly-Integrated Narrow-Band CMOS Transmitter with Harmonic-Rejection Mixers," *ISSCC Dig. Tech. Papers*, pp. 160-161, Feb. 2001.
[7] I. Fabiano, M. Sosio, A. Liscidini, R. Castello, "SAW-Less Analog Front-End Receivers for TDD and FDD," *ISSCC Dig. Tech. Papers*, pp. 82-83, Feb. 2013.

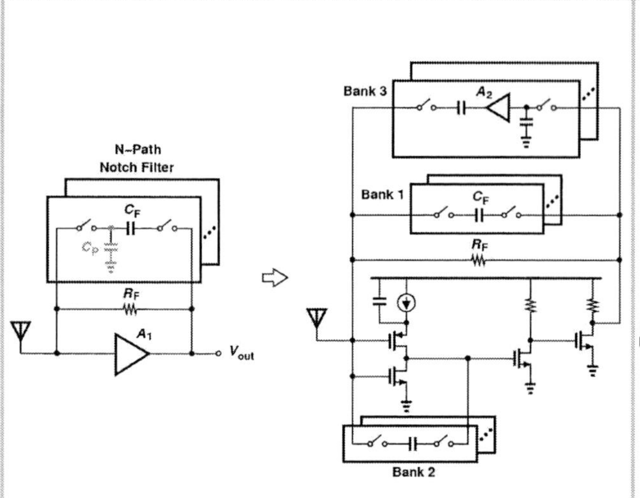

Figure 20.8.1: Channel-selection filtering in LNA by means of N-path filters.

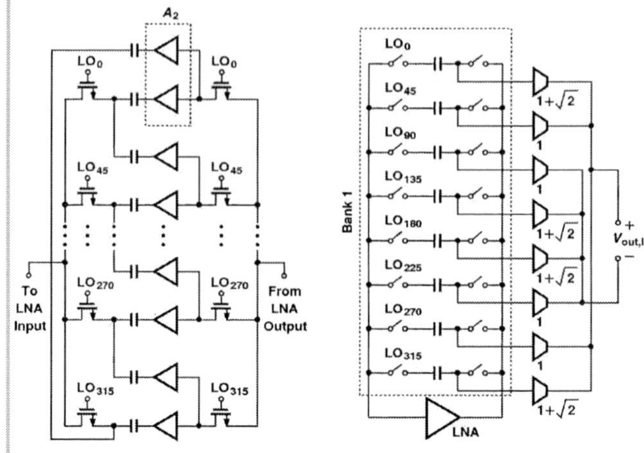

Figure 20.8.2: Bank 3 implementation and harmonic-reject mixing.

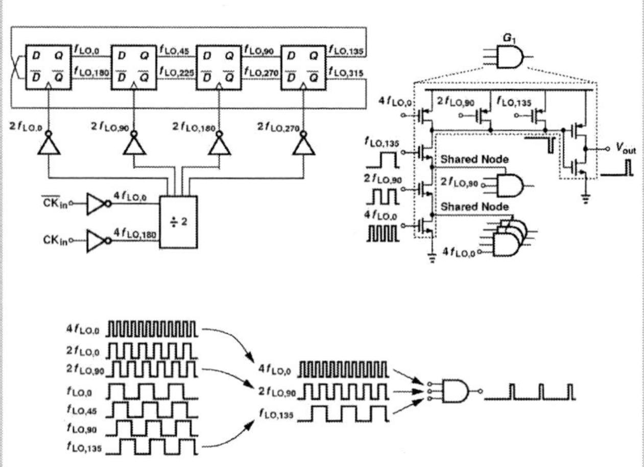

Figure 20.8.3: 8-phase clock-generation circuit.

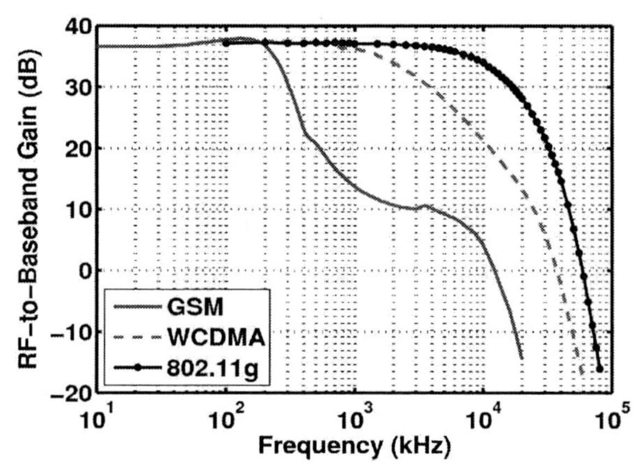

Figure 20.8.4: RF-to-baseband gain for various standards.

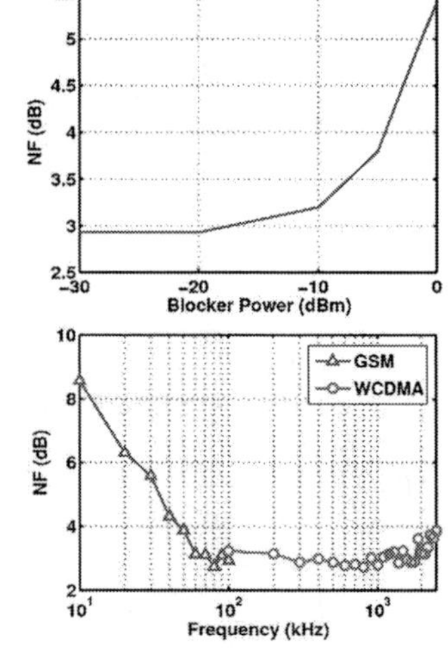

Figure 20.8.5: Measured NF vs. blocker level and vs. baseband frequency.

	[1]	[2]	[7]	This work
Input Frequency [MHz]	80 ~ 2700	1000 ~ 2500	1800 ~ 2400	50 ~ 2500
Channel Bandwidth [MHz]	N/A	5	N/A	0.4 ~ 20
Gain [dB]	72	30	45.5	38
NF [dB]	1.9	7.6	3.8	2.9
NF with 0-dBm Blocker [dB] (at Given Offset)	4.1 (80 MHz)	N/A	7.9 (20 MHz)	5.4 (23 MHz)
Out-of-Band-IIP3 [dBm]	13.5	12	18	10
Active Area [mm^2]	1.2	< 0.06	0.84	0.82
Supply Voltage [V]	1.3	1.2	1.2/1.8	1.2
Power Consumption [mW]	65 (2 GHz)	62[1]	35[2] (2 GHz)	20 (2 GHz)
CMOS Technology	40 nm	65 nm	40 nm	65 nm

[1]Excluding clock circuitry [2]With a 1.8 V supply for LO divider

Figure 20.8.6: Performance summary and comparison.

20

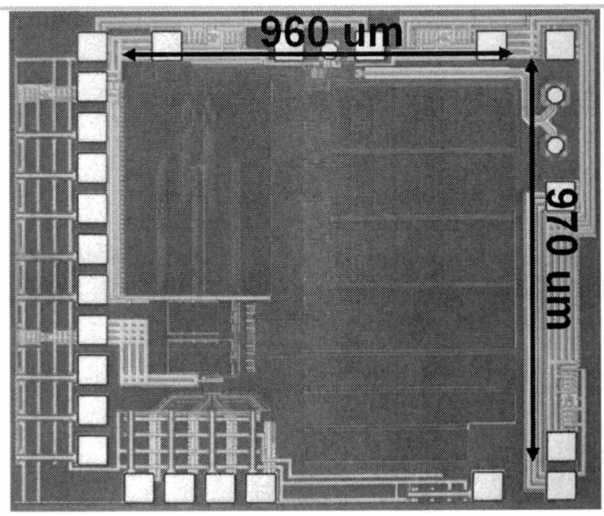

Figure 20.8.7: Die micrograph.

Session 21 Overview:
Frequency Generation Techniques
RF SUBCOMMITTEE

Session Chair: *Piet Wambacq*
imec, Heverlee, Belgium

Session Co-Chair: *Chih-Ming Hung,*
MStar Semiconductor, Taipei, Taiwan

VCOs and PLLs are at the heart of communication systems. They are critical for almost all RF systems such as receivers, transmitters, imagers, and radars. This session starts with three highly digitized frequency synthesizers, including multiplying DLL/PLLs with fractional-N operation and a direct-digital frequency synthesizer. The fourth paper demonstrates a mm-Wave PLL based on a subsampling technique. Interference-tolerant VCOs and PLLs are presented next, followed by a multichannel receiver whose LO frequencies are generated from a single stable FBAR oscillator without a PLL.

21.1 A 1.7GHz MDLL-Based Fractional-N Frequency Synthesizer with 1.4ps RMS Integrated Jitter and 3mW Power Using a 1b TDC — **8:30 AM**

G. Marucci, Politecnico di Milano, Milan, Italy

In Paper 21.1, Politecnico di Milano presents the first MDLL-based frequency synthesizer in 65nm CMOS with fine fractional-N resolution of 190Hz, low power consumption of 3mW and low integrated rms jitter of 1.4ps.

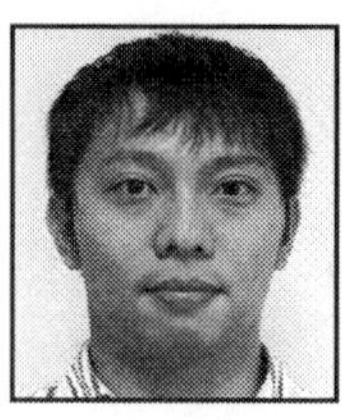

21.2 A 2.3GHz Fractional-N Dividerless Phase-Locked Loop with -112dBc/Hz In-Band Phase Noise — **9:00 AM**

P.-C. Huang, National Taiwan University, Taipei, Taiwan

In Paper 21.2, National Taiwan University presents an analog "divider-less" fractional-N PLL in 0.18μm CMOS, which modulates the edge of the reference clock. The design achieves a low in-band phase noise of -112dBc/Hz at 2.3GHz and 266fs rms jitter.

21.3 A 2GHz 130mW Direct-Digital Frequency Synthesizer with a Nonlinear DAC in 55nm CMOS — **9:30 AM**

T. Yoo, Chung-Ang University, Seoul, Korea

In Paper 21.3, Chung-Ang University, Analog Devices and KAIST present the first CMOS-based direct-digital frequency synthesizer that operates at 2GHz with a high spectral purity and up to 70.3dBc SFDR. Power consumption is only 130mW and the chip area is 0.1mm^2 in 55nm CMOS.

21.4 A 42mW 230fs-Jitter Subsampling 60GHz PLL in 40nm CMOS **10:15 AM**

V. Szortyka, imec, Heverlee, Belgium and Vrije Universiteit Brussel, Brussels, Belgium

In Paper 21.4, imec and Vrije Universiteit Brussel present the first mm-Wave PLL utilizing a subsampling phase detector instead of a classical divider chain. Implemented in 40nm CMOS, the PLL has a jitter as low as 230fs for a power consumption of 42mW.

21.5 A 3.24-to-8.45GHz Low-Phase-Noise Mode-Switching Oscillator **10:45 AM**

M. Taghivand, Stanford University, Stanford, CA and Qualcomm, San Jose, CA

In Paper 21.5, Stanford University presents a 3.24GHz-to-8.45GHz VCO in 40nm CMOS. The large tuning range is obtained via mode switching while preserving excellent phase noise, meeting the requirements for cellular applications.

21.6 A 2.4-to-5.3GHz Dual-Core CMOS VCO with Concentric 8-Shaped Coils **11:00 AM**

L. Fanori, Lund University, Lund, Sweden and Marvell, Pavia, Italy

In Paper 21.6, Lund University and Ericsson describe a dual-core VCO in 65nm CMOS. It uses two figure-8-shaped concentric coils to save area and simultaneously achieve an extremely large tuning range of 75% with a supply voltage of only 0.4V.

21.7 A 1.8mW PLL-Free Channelized 2.4GHz ZigBee Receiver Utilizing Fixed-LO Temperature-Compensated FBAR Resonator **11:15 AM**

K. Wang, University of Washington, Seattle, WA

In Paper 21.7, the University of Washington and Avago Technologies present a 2.4GHz ZigBee receiver that uses an RF oscillator based on a temperature-compensated thin-film bulk acoustic-wave resonator, thereby eliminating the need of a PLL. Using a 65nm CMOS technology, power consumption of the sliding-IF receiver architecture with careful LO frequency planning is only 1.8mW.

21.8 A Pulling Mitigation Technique for Direct-Conversion Transmitters **11:45 AM**

A. Mirzaei, Broadcom, Irvine, CA

In Paper 21.8, Broadcom presents a calibration technique that mitigates unwanted pulling effects in a direct-conversion transmitter. When the TX VCO is heavily pulled, the technique lowers the EVM from 11% to 2.4%. It also suppresses the unwanted sidebands caused by pulling from another nearby oscillator by more than 20dB.

21.1 A 1.7GHz MDLL-Based Fractional-N Frequency Synthesizer with 1.4ps RMS Integrated Jitter and 3mW Power Using a 1b TDC

Giovanni Marucci, Andrea Fenaroli, Giovanni Marzin, Salvatore Levantino, Carlo Samori, Andrea L. Lacaita

Politecnico di Milano, Milan, Italy

The introduction of inductorless frequency synthesizers into standardized wireless systems still requires a high level of innovation in order to achieve the stringent requirements of low noise and low power consumption. Synthesizers based on the so-called multiplying delay-locked loop (MDLL) represent one of the most promising architectures in this direction [1-3]. An MDLL resembles a ring oscillator, in which the signal edge traveling along the delay line is periodically refreshed by a clean edge of the reference clock. In this manner, the phase noise of the ring oscillator is filtered up to half the reference frequency and the total output jitter is reduced significantly. Unfortunately, the concept of MDLL, and in general of injection locking (IL), is inherently limited to integer-N synthesis, which makes it unacceptable in practical RF systems. A first extension of injection locking to coarse fractional-N resolution has been shown in [4], in which however the fractional resolution is bounded to the inverse of the number of ring-oscillator delay stages. This paper introduces a fractional-N MDLL-based frequency synthesizer with a 1b time/digital converter (TDC), which is able to outreach the performance of inductorless fractional-N synthesizers. The prototype synthesizes frequencies between 1.6 and 1.9GHz with 190Hz resolution and achieves RMS integrated jitter of 1.4ps at 3mW power consumption, even in the worst-case of near-integer channel.

Figure 21.1.1 shows the block schematic of the implemented frequency synthesizer. Five pseudo-differential inverter stages and a multiplexer driven by a selection logic realize an MDLL and allow to replace every N-th edge of the signal traveling through the MDLL with a clean edge of the reference ref1, N being the division factor of the integer divider. The delay of the inverters is finely tuned by a digital phase-locked loop (PLL) based on a 1b TDC. To avoid the noise contribution from the divider, the TDC directly compares the phases of the reference signal (ref2) and the output signal (out) [3]. Coarse frequency lock is achieved in background by an additional digital loop (not shown). Unfortunately, the mismatch between the two reference paths causes a phase offset between ref1 and ref2, which induces large jitter at the output [1]. To compensate this offset, two digital-to-time converters (DTCs) implemented as digitally-controlled delay stages are added to each path and one of them is regulated by the control loop introduced in [5], based on the time error detected by the 1b TDC. In contrast to PLLs, fractional-N operation cannot be achieved in MDLLs by simply dithering the modulus control (MC) of the integer-N divider via a $\Delta\Sigma$ modulator and subtracting the quantization noise from the loop. In fact, ref and out would not be aligned and no clean edge would be available to replace the edge of the signal traveling in the MDLL. This issue is solved here by subtracting the time error induced by $\Delta\Sigma$ quantization error from both ref1 and ref2 by means of the two DTC blocks. The amplitude of the digital input of the DTCs is regulated by a least-mean square (LMS) loop.

Figure 21.1.2 shows the signal diagram that clarifies the phase realignment in the case for fractional-N operation. In this example, the MC of the divider toggles between N and N+1. Proper MDLL operation is ensured when the MUX selection signal (sel), the output (out) and the injected reference signal (ref1) are synchronous. Signals sel and out are inherently synchronized, since sel is generated by the divider output. Unfortunately, the digital PLL nulls only the average value of the time error $t_{err}[k]$, while the waveform $t_{err}[k]$ is proportional to the quantization error q[k] introduced by the $\Delta\Sigma$ and has an amplitude equal to the output period T_{out} (or a multiple of its period, for higher-order $\Delta\Sigma$). Such a large phase error between ref and out would corrupt the refreshment of the MDLL signal edge, thus compromising its operation. The proposed solution solves this issue by realigning both ref1 and ref2 to out by means of two DTCs.

A practical implementation of the DTC is shown in Fig. 21.1.3. A second-order $\Delta\Sigma$ modulator is used to dither the MC of the divider. Thus, the DTC range has to cover the phase offset, as well as the time error induced by $\Delta\Sigma$ quantization, which is as large as $2T_{out}$ (i.e., 1.25ns at 1.6GHz). On the other hand, DTC resolution has to be as low as approximately 400fs to reduce quantization noise below thermal noise. Since the fractional quantization error is much larger than the typical ref1-ref2 offset to be cancelled out, the implemented DTC is divided into a coarse 6b DTC, common to both paths, which cancels $\Delta\Sigma$ quantization error, and a fine 8b DTC for each individual path, which cancels both the phase offset and the residual $\Delta\Sigma$ error. The additional bits are needed to accommodate the offset correction signal and to make the range of the fine DTC intentionally larger than the least significant bit of the coarse DTC (even in the presence of process, voltage and temperature variations). The coarse DTC is implemented as a digitally-controllable delay line shown in Fig. 21.1.3. Tri-state inverters in the delay line are employed to digitally tune the length of the signal path from in to out. The 8b DTCs are implemented as inverter stages loaded by an array of switched MOS varactors. The two digital correlators in Fig. 21.1.3 estimate in background the gains g_0 and g_1 of coarse and fine DTCs, respectively. Each cell of the MDLL is an inverter stage with tunable delay, whose schematic is shown in Fig. 21.1.1. Coarse tuning of each cell is obtained thanks to a 6b current-starving topology and fine tuning of delay is achieved via an 8b resistor-string DAC followed by an RC filter, which controls the gate voltage of an nMOS resistor in parallel to the coarse-tuning array.

The synthesizer has been fully integrated in 65nm CMOS and occupies a core area of about $0.4mm^2$ (a die micrograph is shown in Fig. 21.1.7). The worst case for the fractional spur levels occurs for near-integer channels. Figure 21.1.4 shows the measured phase noise of a channel 720kHz away from one of the integer channels at 1.65GHz. Thanks to the background cancellation of both $\Delta\Sigma$ error and phase offset, the level of the highest spur is -47dBc at 720kHz offset. Figure 21.1.5 shows the measured integrated jitter (from 30kHz to 30MHz) over the synthesized integer-N channels from 1.60 to 1.90GHz (main plot) and over the fractional-N channels from 1.65 to 1.70GHz (inset plot). The RMS jitter is under 0.55ps for the integer-N channels and under 1.4ps (including random noise and fractional spurs) for the fractional-N channels. In both cases, the level of the reference spur is -47dBc at 50MHz. The core power consumption (excluding pad driver and crystal oscillator) is 2.5 and 3.0mW, in integer- and fractional-N mode, respectively. For fractional-N channels, 60% of this power is used in the MDLL, 10% in the coarse DTC and 30% in the other blocks. The table in Fig. 21.1.6 summarizes the measured performance and compares this work with state-of-the-art inductorless fractional-N synthesizers. In the worst-case channel, the presented fractional-N synthesizer reaches -232dB figure of merit (FoM), defined in Fig. 21.1.6.

Acknowledgment:

The authors wish to thank M. Zuffada and E. Temporiti (STMicroelectronics) for supporting chip fabrication, C. Palattella and S. Vilardi for useful discussions.

References:

[1] B. Helal et al., "A Highly Digital MDLL-Based Clock Multiplier That Leverages a Self-Scrambling Time-to-Digital Converter to Achieve Subpicosecond Jitter Performance," *IEEE J. Solid-State Circuits*, vol. 43, no. 4, pp. 855-863, Apr. 2008.

[2] S. Gierkink, "Low-Spur, Low-Phase-Noise Clock Multiplier Based on a Combination of PLL and Recirculating DLL with Dual-Pulse Ring Oscillator and Self-Correcting Charge Pump," *IEEE J. Solid-State Circuits*, vol. 43, no. 12, pp. 2967–2976, Dec. 2008.

[3] A. Elshazly et al., "A 1.5GHz 890μW Digital MDLL with 400fsrms Integrated Jitter, -55.6dBc Reference Spur and 20fs/mV Supply-Noise Sensitivity Using 1b TDC," *ISSCC Dig. Tech. Papers*, pp. 242-243, 2013.

[4] P. Park et al., "An All-Digital Clock Generator Using a Fractionally Injection-Locked Oscillator in 65nm CMOS," *ISSCC Dig. of Tech. Papers*, pp.336,337, 2012.

[5] G. Marzin et al., "A Spur Cancellation Technique for MDLL-Based Frequency Synthesizers," *Proc. of ISCAS*, pp.165-168, 2013.

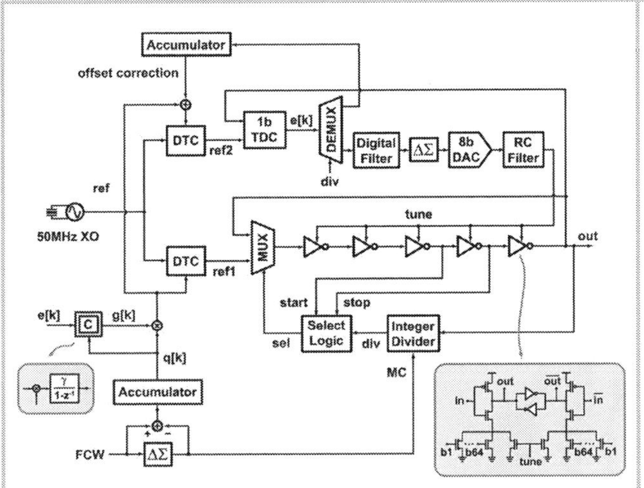

Figure 21.1.1: Implemented frequency synthesizer.

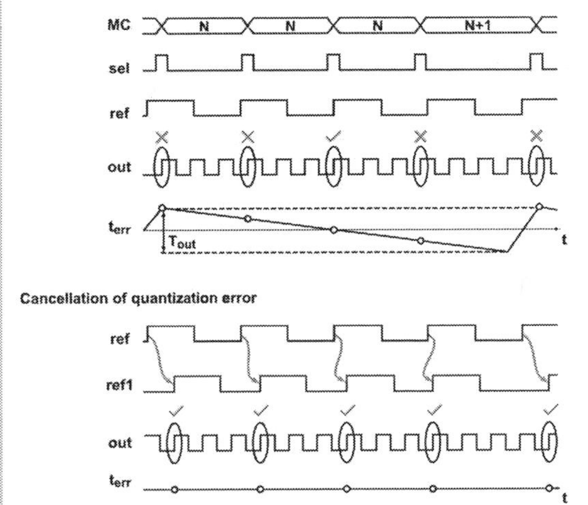

Figure 21.1.2: Signal diagrams in fractional-N mode.

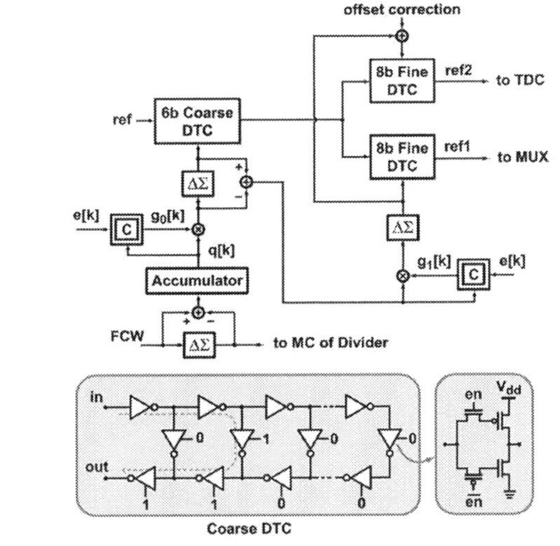

Figure 21.1.3: DTC schematic.

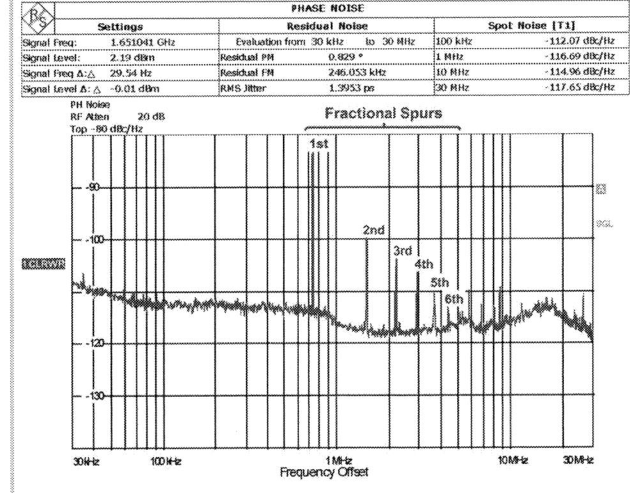

Figure 21.1.4: Measured phase noise.

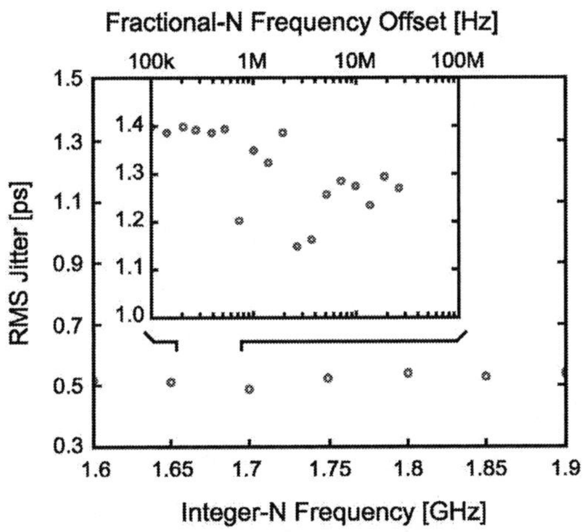

Fractional-N Frequency Offset [Hz]

RMS Jitter [ps]

Integer-N Frequency [GHz]

Figure 21.1.5: Measured jitter over the channels.

	This Work	T. Kao ISSCC13	[4] P. Park ISSCC12	S. Lee RFIC12
Architecture	**MDLL**	Ring-PLL	IL-PLL	IL-PLL
Frequency Resolution (kHz)	**0.19**	25.4	1000	40000
Frequency Multiplication (N)	**32**	72	18.125	22.5
Reference Frequency (MHz)	**50**	26	32	80
Output Frequency (GHz)	**1.6 - 1.9**	1.87 - 1.98	0.58	1.8
Reference Spur (dBc)	**-47**	N/A	N/A	-35
Near-Integer Frac. Spur (dBc)	**-47 (720kHz)**	-50	N/A	N/A
Far-Integer Frac. Spur (dBc)	**-51 (2MHz)**	N/A	N/A	N/A
Integ. RMS Jitter (ps)	**1.4**	3.4	8	N/A
Power Dissipation (mW)	**3**	10	10.1	22
Figure-of-Merit (FoM)* (dB)	**-232**	-219	-212	N/A
Area occupation (mm^2)	**0.4**	0.055	0.158	0.083
CMOS Process (nm)	**65**	40	65	90

$$^*\text{FoM} = 10\cdot\log_{10}\left[\left(\frac{\text{Jitter}}{1\text{s}}\right)^2\cdot\left(\frac{\text{Power}}{1\text{mW}}\right)\right]$$

Figure 21.1.6: State-of-the-art comparison table.

21

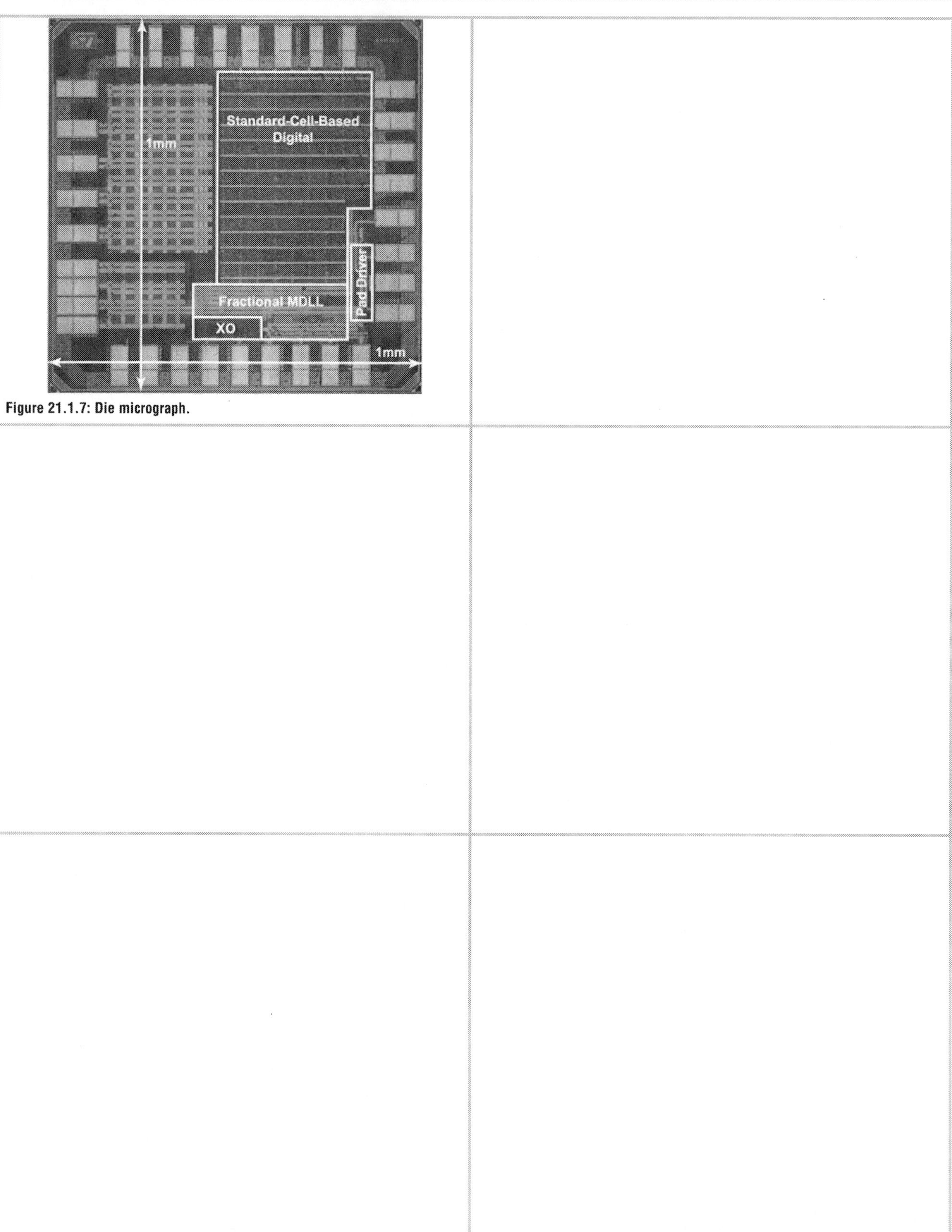

Figure 21.1.7: Die micrograph.

21.2 A 2.3GHz Fractional-N Dividerless Phase-Locked Loop with -112dBc/Hz In-Band Phase Noise

Po-Chun Huang, Wei-Sung Chang, Tai-Cheng Lee

National Taiwan University, Taipei, Taiwan

Recently, dividerless PLL architectures, including sub-sampling PLLs [1] and injection-locked PLLs [2], have been reported to achieve superior phase noise with respect to conventional PLL architectures. However, these dividerless architectures can only be operated in integer-N mode inherently. In order to operate in fractional-*N* mode, this work proposes a digital pulse-width modulator (DPWM) to modulate the pulse width of the input reference signal to synthesize the output frequency.

The block diagram of the proposed dividerless fractional-*N* PLL is shown in Fig. 21.2.1. The core circuit is a sub-sampling phase-locked loop (SSPLL) because it not only achieves a higher phase-to-current gain but also introduces no noise on the feedback divider. The sub-sampling-based PD (SSPD) directly samples the differential sinusoidal outputs of the VCO by the clock generated by the proposed DPWM. The sampler generates $V_{sam,up}$ and $V_{sam,dn}$ to control the charge pump (CP) output current. When the CP reaches zero average current, phase locking is achieved. Because the SSPD has a limited locking range, a frequency-locked loop (FLL) is added to ensure proper operation. In addition, an automatic frequency control (AFC) loop is used to select the VCO band to extend the output frequency range. The proposed DPWM modulates the rising edge of the input reference clock to tune the input pulse width and its corresponding instantaneous frequency. Because of the requirement of the linear operation, a digital modulator and a digital correlation loop (DCL) are employed to control the DPWM for better phase noise performance.

Figure 21.2.2 shows the detailed operation of the proposed DPWM. The DPWM takes the input reference clock, whose reference period is T_{ref}, and shifts its rising edge by $d[k]T_{vco}/2^p$, where d[k] is a 9-b digital-controlled word generated by a delta-sigma modulator, 2^p is the number of DPWM phases and T_{vco} is the period of the VCO output. When a negative fraction with respect to N is applied, the DPWM would rotate the phase clockwise by increasing d[k]. As long as d[k] has no overflow, the period of the DPWM output is equal to $T_{ref}+ d[k]T_{vco}/2^p$, which corresponds to NT_{vco}. However, due to the negative fraction, d[k] will eventually overflow such that the SSPD samples the $(N\text{-}1)^{th}$ VCO edge rather than N^{th} one. This inherent sub-sampling operation can perform the swallow-divider-like operation [3] without any feedback dividers. In the proposed architecture, the fractional-*N* divisor, T_{ref}/T_{vco}, is equal to $N\text{–}N_{ds}/2^r$, in which *N* is the integer part, r is the bit of the modulator and N_{ds} is the digital code of the modulator.

Because the SSPD/CP transfer characteristic is sinusoidally shaped due to the band-pass output characteristic of the LC-VCO, this nonlinearity problem between the SSPD and the DPWM will raise phase noise and deteriorate noise folding just as in conventional fractional-*N* PLLs. Based on the system simulations under the design target of -115dBc/Hz in-band phase noise level, the phase error between the DPWM output and the VCO output signal has to be maintained within 5°. In order to achieve such small phase error and to reduce the quantization error induced by the DPWM, an 8-b variable pulse width is realized to cover one VCO period. However, because of process variation and random mismatches, the pulse width of the DPWM still has two important sources of errors, namely mismatches and a gain error. These can severely degrade phase noise.

The gain error, defined as the deviation with respect to $T_{vco}/2^p$, would modulate V_{ctrl} periodically due to the erroneous CP current. To resolve this DPWM inaccuracy, a current DAC, which is controlled by a digital correlation loop (DCL) as shown in Fig. 21.2.3, is employed in the DPWM circuit to control the DPWM unit delay. Using a digital control word and phase error sign (P_{sign}) to perform an LMS algorithm, the DPWM unit delay can be tuned to its optimal value. In addition, a sign extractor circuit is proposed to generate the phase error sign. To obtain proper sign information, rather than sensing the tiny voltage variance in the control line [5], the proposed circuit can detect the error polarity based on the SSPD output voltage difference ($V_{sam,up}$ and $V_{sam,dn}$) because of the PM-to-AM operation in the SSPD. Nevertheless, the voltage difference at the SSPD output typically contains a significant dc offset voltage to compensate circuit non-idealities, such as CP current mismatch. The proposed sign extractor needs to remove such DC offset to avoid incorrect operation of the LMS algorithm by using a re-sampler to filter the VCO signal under the SSPD sampling mode, and then feeding it to a high-pass filter. Besides, the digital modulator [4], shown in Fig. 21.2.3, is used to generate the control signals. Note that the maximum tuning width of the DPWM is two VCO periods, because it needs a larger range to modulate the pulse width with MASH 1-1 on this modulator.

In addition to the gain error on the DPWM pulse width, the capacitor array, which is used to tune the delay, also introduces a random mismatch. Shown in Fig. 21.2.1, the pulse width of the DPWM is controlled by a 7-b MSB cap array and 2-b LSB caps, which can reduce area. A 2-D 7-b dynamic element matching (DEM) block is used to linearize the tuning caps. Shown in Fig. 21.2.4, the unit cap cell is controlled by a two-column word and one row word. The control method can average out the mismatch effect. In the controlled capacitor, in order to avoid memory effects of the preceding DPWM width unit, the control word is reset and then updated for each sampling of the SSPD, in which the timing diagram is shown in Fig. 21.2.4.

The proposed frequency synthesizer has been fabricated in a 0.18μm CMOS technology. The reference frequency is 48MHz. Figure 21.2.5 shows the plot of the measured phase noise of the 2.3GHz output. The in-band phase noise is -112dBc/Hz at a 50kHz offset frequency. The best-case rms jitter (integrated from 10kHz to 30MHz) is 266fs. By activating DEM, jitter can be reduced from 1310fs to 260fs. This is illustrated in Fig. 21.2.5, which shows the measured jitter over different fractional-N PLL (not PLLs) channels. The total power consumption is 9.6 mA, from which 6.3mA is consumed in the analog circuit and 3.3mA in the DPWM and the digital control. Shown in Fig. 21.2.6, the fractional spur is reduced from -23dBc to -53dBc with the aid of the DCL. Figure 21.2.6 summarizes the frequency synthesizer performance and the figure-of-merit (FoM) is compared with recent state-of-the-art fractional-*N* PLLs. The FoM of the proposed fractional-*N* PLL can be as good as -239.1dB. The die photo is shown in Fig. 21.2.7. It is believed that the performance can be greatly improved if a more downscaled technology can be used.

In the proposed architecture, the SSPLL suppresses the divider noise and the proposed DPWM is used to obtain fractional-*N* frequency locking. Through a DEM technique, the mismatch of the DPWM width units can be reduced. The gain error in the DPWM can be suppressed with the DCL. As a result, the fractional-*N* frequency synthesizer can yield a low in-band phase noise.

Acknowledgments:
The authors would like to acknowledge the NTU-MediaTek Lab. (MediaTek Inc.) and the National Science Council with contract number 101-2220-E-002-016(and -027) for supporting this research, and the Chip Implementation Center (CIC) for chip fabrication.

References:

[1] X. Gao, et al., "A 2.2GHz 7.6mW Sub-Sampling PLL with -126dBc/Hz In-Band Phase Noise and $0.15ps_{rms}$ Jitter in 0.18μm CMOS," *ISSCC Dig. Tech. Papers*, pp. 392-393, Feb. 2009.

[2] I.-T. Lee, et al., "A Dividerless Sub-Harmonically Injection-Locked PLL with Self-Adjusted Injection Timing," *ISSCC Dig. Tech. Papers,* pp. 414-415, Feb. 2013.

[3] I. A. Young, et al., "A PLL Clock Generator with 5-to-110MHz Lock Range for Microprocessors," *ISSCC Dig. Tech. Papers,* pp. 50-51, Feb. 1992.

[4] D. Tasca, et al., "A 2.9-to-4.0GHz Fractional-N Digital PLL with Bang-Bang Phase Detector and $560fs_{rms}$ Integrated Jitter at 4.5mW Power," *ISSCC Dig. Tech. Papers,* pp. 88-89, Feb. 2011.

[5] M. Gupta, et al., "A 1.8GHz Spur-Cancelled Fractional-N Frequency Synthesizer with LMS-Based DAC Gain Calibration," *ISSCC Dig. Tech. Papers,* pp. 1922-1931, Feb. 2006.

[6] P. Park, et al., "An All-Digital Clock Generator Using a Fractionally Injection-Locked Oscillator in 65nm CMOS," *ISSCC Dig. Tech. Papers,* pp. 336-337, Feb. 2012.

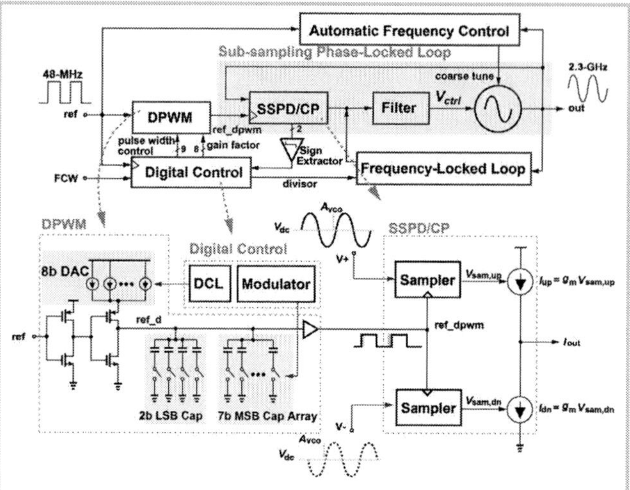

Figure 21.2.1: The proposed fractional-N dividerless PLL block diagram and principle of DPWM and SSPD/CP operation.

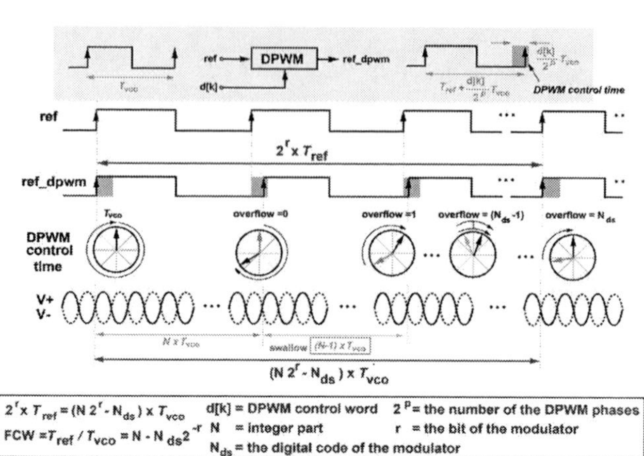

Figure 21.2.2: Principle of fractional-N operation.

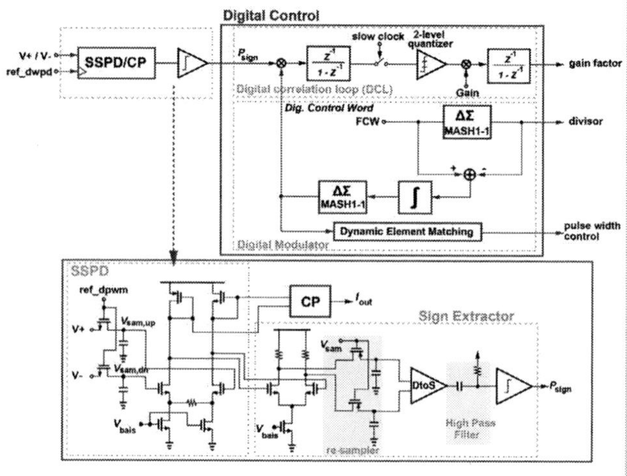

Figure 21.2.3: Schematic of digital control, SSPD and sign extractor.

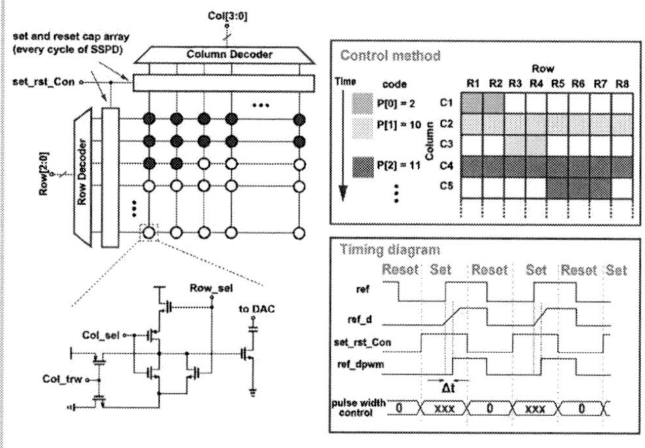

Figure 21.2.4: Implementation of dynamic element matching block.

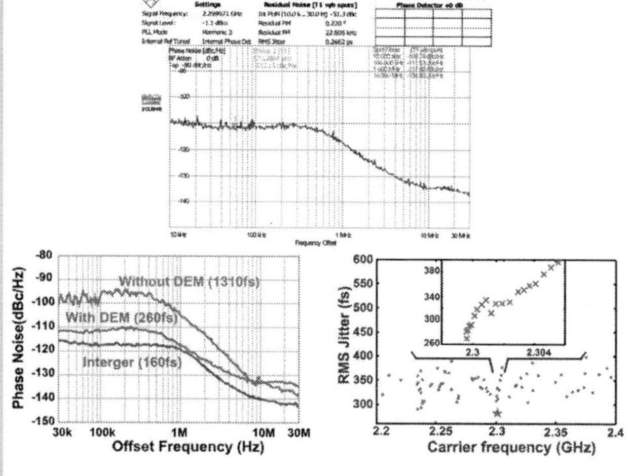

Figure 21.2.5: Measured phase noise and measured jitter over the fractional-N PLL channels.

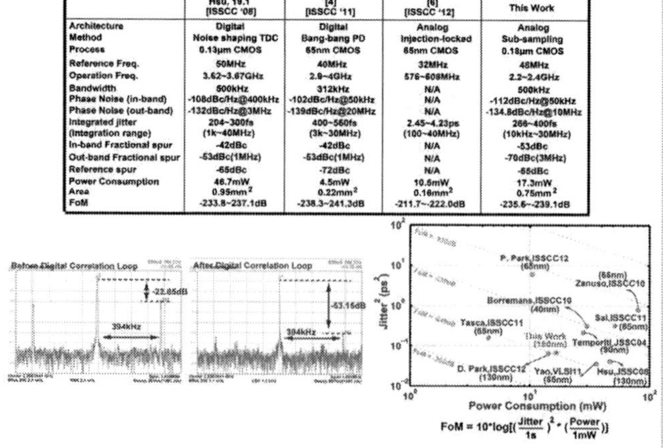

	Hsu, 19.1 [ISSCC '08]	[4] [ISSCC '11]	[6] [ISSCC '12]	This Work
Architecture	Digital	Digital	Analog	Analog
Method	Noise shaping TDC	Bang-bang PD	Injection-locked	Sub-sampling
Process	0.13µm CMOS	65nm CMOS	65nm CMOS	0.18µm CMOS
Reference Freq.	50MHz	40MHz	32MHz	48MHz
Operation Freq.	3.62~3.67GHz	2.9~4GHz	576~608MHz	2.2~2.4GHz
Bandwidth	500kHz	312kHz	N/A	500kHz
Phase Noise (in-band)	-108dBc/Hz@400kHz	-102dBc/Hz@50kHz	N/A	-112dBc/Hz@50kHz
Phase Noise (out-band)	-132dBc/Hz@3MHz	-139dBc/Hz@20MHz	N/A	-134.8dBc/Hz@10MHz
Integrated jitter (integration range)	204~300fs (1k~40MHz)	400~560fs (3k~30MHz)	2.45~4.23ps (100~40MHz)	266~400fs (10kHz~30MHz)
In-band Fractional spur	-42dBc	-42dBc	N/A	-53dBc
Out-band Fractional spur	-53dBc(1MHz)	-53dBc(1MHz)	N/A	-70dBc(3MHz)
Reference spur	-65dBc	-72dBc	N/A	-65dBc
Power Consumption	46.7mW	4.5mW	10.6mW	17.3mW
Area	0.95mm²	0.22mm²	0.16mm²	0.75mm²
FoM	-233.8~237.1dB	-238.3~241.3dB	-211.7~-222.0dB	-235.6~-239.1dB

Figure 21.2.6: Performance summary, comparison with other fractional-N PLLs and measured fractional spur.

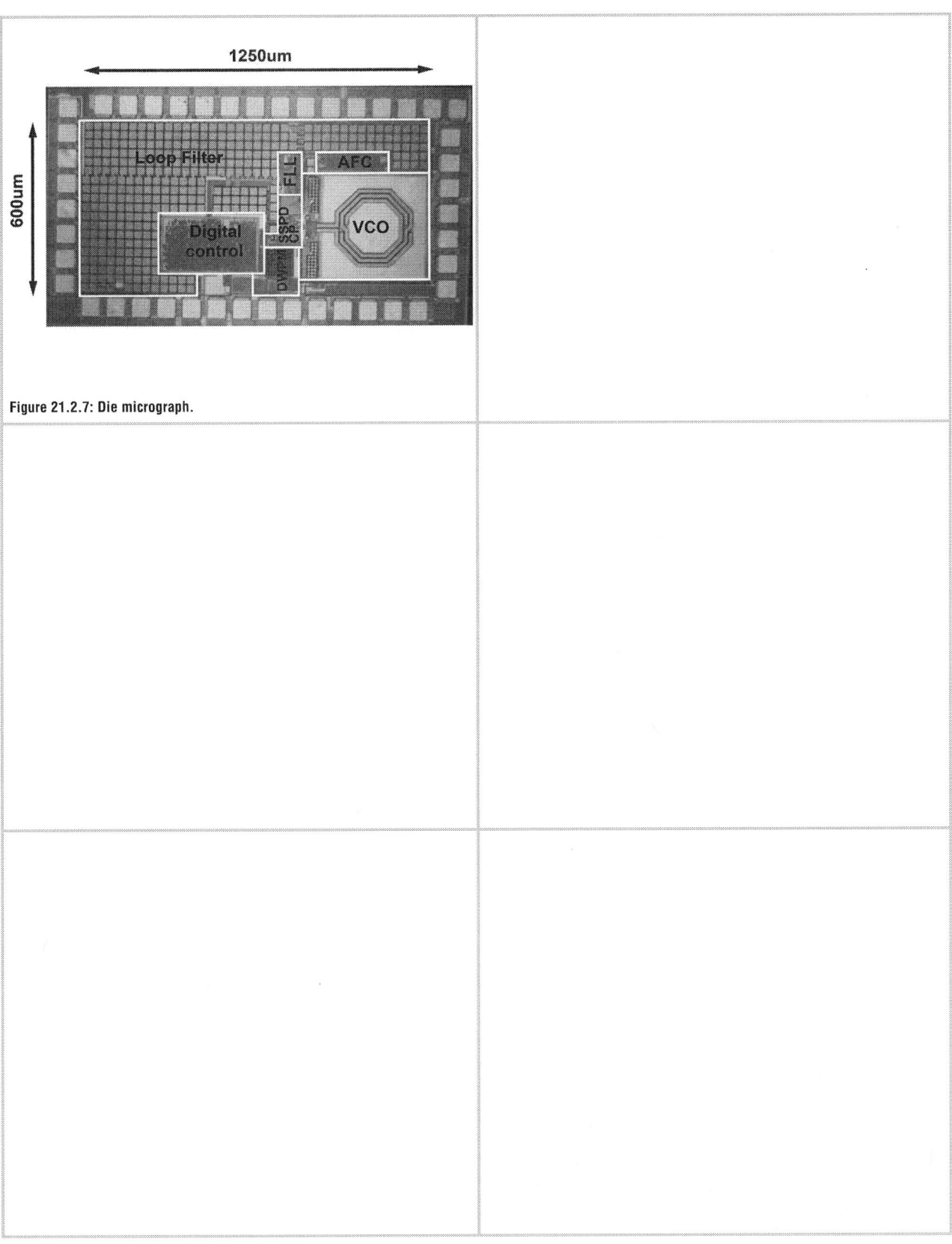

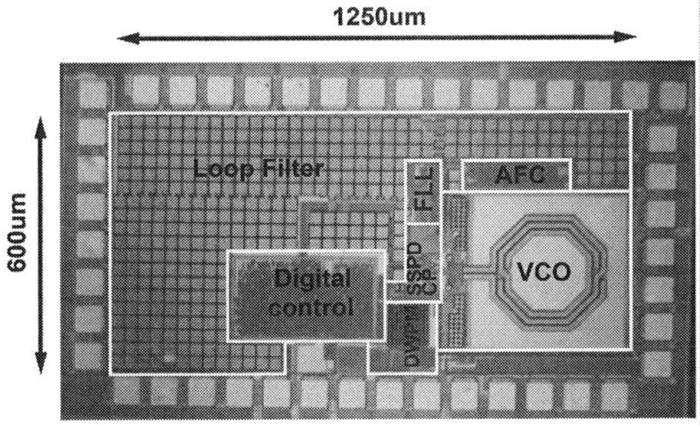

Figure 21.2.7: Die micrograph.

21.3 A 2GHz 130mW Direct-Digital Frequency Synthesizer with a Nonlinear DAC in 55nm CMOS

Taegeun Yoo[1], Yun-Hwan Jung[1], Hong Chang Yeoh[1,2], Yong Sin Kim[1], Sung-Mo Kang[3], Kwang-Hyun Baek[1]

[1]Chung-Ang University, Seoul, Korea, [2]Analog Devices, San Jose, CA, [3]KAIST, Daejeon, Korea

Direct-digital frequency synthesizers (DDFSs) have been employed in many frequency-agile communication systems because of their wide bandwidth, fine frequency resolution, and fast frequency-hopping characteristics. Recent developments in DDFSs are towards enhancing performances through reduction of both complexity and power consumption [1-3]. The segmented nonlinear DAC (NLDAC) structures in [1,2] require additional coarse phase information for fine amplitude decoding with a complex decoder. Moreover, the quarter-sine-wave technique incorporated into the segmented NLDACs in [1,2] degrades spectral purity due to the need of the MSB shift DAC that introduces additional offset. Another scheme in [3] reduces complexity and power consumption by replacing the digital-based phase-to-amplitude converter with an analog-based converter, resulting in limited spectral purity. Unlike previous schemes, this work presents comprehensive enhancements in all key areas of a DDFS, the pipelined phase accumulator (PACC), digital decoder, and NLDAC as shown in Fig. 21.3.1. First, the low-power PACC with multi-level momentarily activated bias (*M²AB*) is presented to reduce power dissipation. Second, the coarse phase-based consecutive fine-amplitude grouping (*C²FAG*) scheme reduces the hardware complexity and the power consumption in digital decoder circuits. Third, the mixed-wave conversion topology (*MCT*) in the NLDAC improves the output spectral purity.

Figure 21.3.2 illustrates the PACC with *M²AB*. The PACC is designed with a 32-b frequency control word (FCW) input, 10-b truncated output, and 8 pipeline stages. Pre-skewing F/Fs are used to synchronize the inputs for eight 4-bit accumulators. Since these F/Fs are based on current-mode logic for high-speed operations, static power is dissipated even when the FCW value does not change. Therefore, turning off tail current in the pre-skewing F/Fs once they have passed their input values to the next F/Fs can reduce overall current draw. Figure 21.3.2(a) shows that the bias voltages, V_{b1} ~ V_{b7}, for each column of pre-skewing F/Fs are sequentially activated and momentarily kept alive for only 2 clock cycles before turning off again unless the FCW is updated. As shown in Fig. 21.3.2(b), *M²AB* has four modes: sleep (ϕ_1), rapidly waking up (ϕ_2), normal operation (ϕ_3), and quickly turning off (ϕ_4). The *M²AB* includes two switches (SW_1 and SW_2) and two diode-connected transistors (M_A and M_B) to generate multi-level biasing, where W/L of M_A is 20 times larger than that of M_B and M_C. The switches are controlled by complementary signals (SC_1 and SC_2) with time delay through an inverter. As shown in Fig. 21.3.2(c), SW_1 is "on" and SW_2 is "off" during ϕ_1, thus settling the V_b of pre-skewing F/Fs at a fairly low voltage ('V_{L2}') and turning off M_C. Because V_b is not 0V in ϕ_1, M_C can be rapidly turned on when necessary. With both SW_1 and SW_2 turned off during ϕ_2, the full reference current I_{ref} abruptly raises V_b to 'V_{L4}' level, thereby rapidly turning on M_C by charging the total parasitic capacitance C_{par}. In ϕ_3, V_b becomes its nominal biasing voltage ('V_{L3}') to enable F/Fs. In ϕ_4, M_C is quickly turned off because V_b decreases to 'V_{L1}' level through SW_1 and SW_2.

The hardware complexity of the segmented NLDAC depends mostly on a phase-to-amplitude mapping technique. For NLDACs that use coarse and fine segmentations, the coarse amplitude can be easily mapped using a conventional NLDAC comprised of a thermometer decoder and current switches with sine-weighted current sources [1,2]. Because the fine decoding circuitry requires additional phase information from the coarse segments, the fine decoder is typically much more complex than the coarse decoder. Piecewise linear fine mapping with a linear DAC uses digital adders or subtractors to generate its fine amplitudes, which requires additional pipeline stages [2]. Figure 21.3.3(a) illustrates the current increment for sine-amplitude mapping and a table for fine phases (P_{F0} ~ P_{F15}) according to coarse phases (P_{C0} ~ P_{C15}) in the conventional 4:4 segmented approach. Note that the asterisk (*) indicates a position where the fine phase requires additional coarse phase information because of the change in current increment. As the number of asterisks increases, so does the complexity in fine decoding. For conventional 4:4 segmented designs, there exist 79 positions where additional coarse phase information is required to decode the fine amplitude. The recently reported segmented architecture in [1] adopts multiple fine NLDACs instead of using digital decoding circuitry, where the number of fine NLDACs equals the number of the coarse segmentations. However, the area overhead due to the use of multiple fine NLDACs and mismatches between NLDACs can limit the DDFS performance. The *C²FAG* allows for fine amplitude decoding with only one fine NLDAC. Because the *C²FAG* groups the current increment in descending order as shown in Fig. 21.3.3(b), the fine decoder can be simply implemented with twenty four 2-input AND gates and an additional 4-b thermometer decoder as shown in Fig. 21.3.3(b). For example C_6 (from the existing coarse thermometer decoder) and F_1 (from the fine thermometer decoder) can be simply connected to an AND gate for the position where the current increment in P_{F1} changes from 2 to 1 and coarse phase changes from P_{C5} to P_{C6}. As mentioned earlier, the conventional fine decoder requires complex circuitry due to 79 different decoding positions and the ungrouped current mapping.

The quarter-sine-wave technique (QST) uses one quadrant, then duplicates and transforms it, and finally completes the sine waveform. When the QST is in conjunction with a linear DAC, all the processing to make a full sine wave can be simply implemented by using digital circuitry. On the other hand, an MSB shift DAC that makes one half of the full sine wave is inevitable for the QST with an NLDAC. Even though this scheme is simple, the MSB shift DAC that generates one half of the sine amplitude can cause significant offset error due to mismatch as shown in Fig. 21.3.4(a). This offset causes a distorted waveform (sine combined with square), which introduces odd harmonics in the frequency domain. To overcome this problem, this work proposes *MCT* using hybrid-type conversion with both half sine-wave technique (HST) and QST. *MCT* generates the full sine amplitude (without using the MSB shift DAC) with HST in the coarse segment to avoid spectral performance degradation due to offset introduced by the MSB shift DAC. On the other hand, the QST is applied in the fine segment as shown in Fig. 21.3.4(b) in order to minimize hardware overhead. Note that the use of QST in the fine part does not degrade spectral purity because a shift DAC is not necessary for the fine conversion.

The proposed DDFS is fabricated using a 55nm standard CMOS process. All measurements have been performed with a 12mA output current into 50Ω load. The overall tuning latency of 5ns is obtained owing to the simple decoding scheme (*C²FAG*). Figure 21.3.5(a) shows the measured spectrum, and Fig. 21.3.5(b) summarizes the spectral performance from DC to Nyquist when operating at 2GHz. The best and the worst SFDRs of 70.3dBc and 55.1dBc are achieved at the output frequencies of 6.7MHz and 660MHz, respectively. With a power consumption of 130mW, active area of 0.1mm², and SFDR of 70.3dBc, the improved performance confirms the efficacy of the proposed *M²AB*, *C²FAG*, and *MCT* techniques. Comparisons with the prior state-of-the-art high-speed DDFSs are summarized in Fig. 21.3.6. FOM_1, FOM_2, and FOM_3 are cited from [1-3], [4] and [5], respectively. Figure 21.3.7 shows the micrograph.

Acknowledgement:

This research was supported by the MSIP (Ministry of Science, ICT & Future Planning), Korea, under IT/SW Creative research program supervised by the NIPA (National IT Industry Promotion Agency) (NIPA-2013-H0502-13-1110)

References:

[1] X. Geng, F. Dai, J. Irwin and R. Jaeger, "An 11-bit 8.6 GHz Direct-Digital Synthesizer MMIC with 10-bit Segmented Sine-Weighted DAC," *IEEE J. Solid-State Circuits*, pp. 300-313, Feb. 2010.

[2] H. Yeoh, J.-H. Jung, Y.-H. Jung, K.-H. Baek, "A 1.3GHz 350mW Hybrid Direct Digital Frequency Synthesizer in 90nm CMOS," *IEEE J. Solid-State Circuits*, pp. 1845-1855, Sept. 2010.

[3] C. Yang, J. Wang and H. Chang, "A 5GHz Direct-Digital Frequency Synthesizer using an Analog-Sine-Mapping Technique in 0.35um SiGe BiCMOS", *IEEE J. Solid-State Circuits*, pp. 2064-2072, Sept. 2011.

[4] T. Chen, et al., "A 14-bit 130MHz CMOS current-steering DAC with adjustable INL," *ESSCIRC*, pp. 167-170, Sept. 2004.

[5] W. Tseng, et al., "A 12-Bit 1.25GS/s DAC in 90nm CMOS With > 70dB SFDR up to 500MHz," *IEEE J. Solid-State Circuits*, pp. 2845-2856, Dec. 2011.

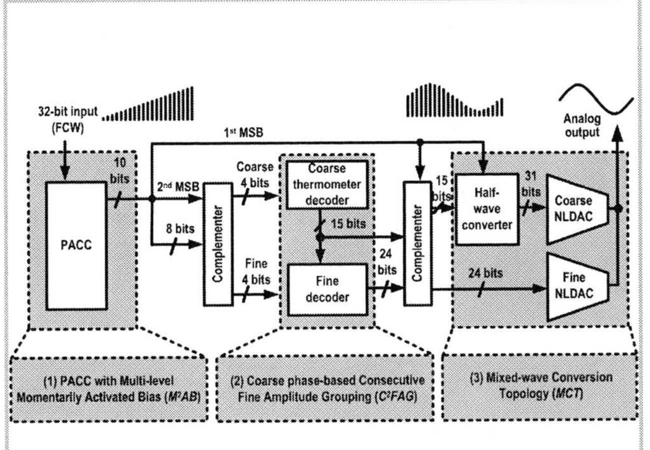

Figure 21.3.1: Block diagram of the proposed DDFS with three techniques.

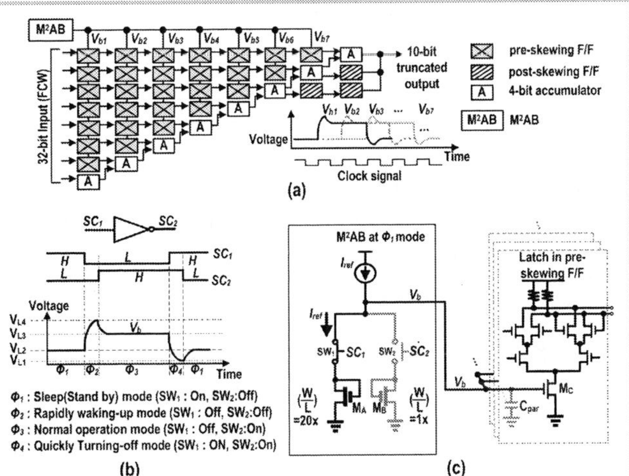

Figure 21.3.2: Pipelined phase accumulator with the *M²AB*. (a) Block diagram; (b) Timing diagram of the bias voltage; (c) Detailed schematic.

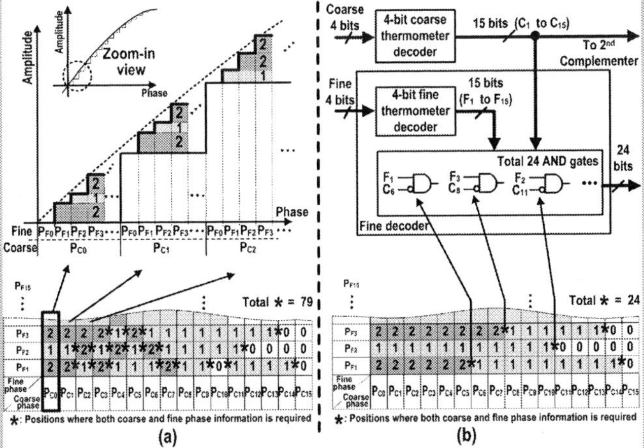

Figure 21.3.3: (a) Conventional fine phase-to-amplitude increment mapping table and waveform; (b) Fine phase-to-amplitude increment mapping table of the *C²FAG* and its implementation.

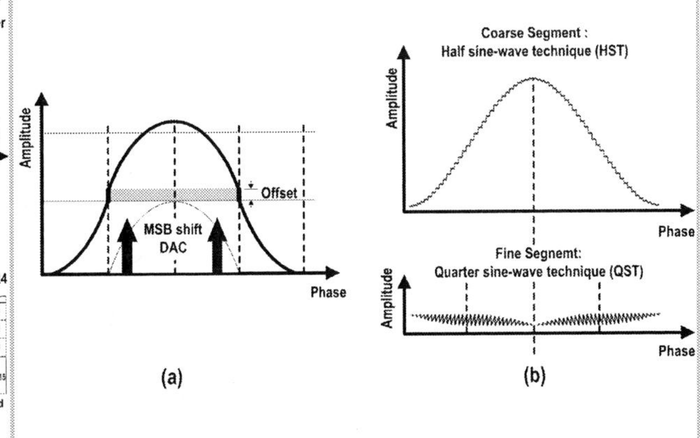

Figure 21.3.4: (a) Conventional quarter sine-wave technique with the MSB shift DAC; (b) Proposed *MCT*.

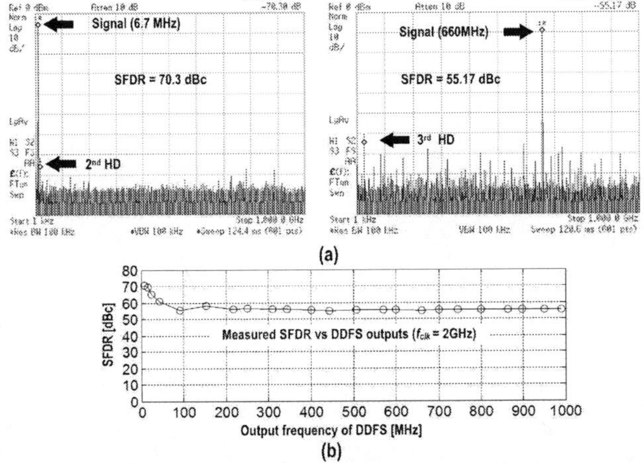

Figure 21.3.5: (a) Measured spectra for 6.7MHz and 660MHz signals at 2GHz operating frequency; (b) Measured SFDR according to various output frequencies.

	[1] JSSC 2010	[2] JSSC 2010	[3] JSSC 2011	This work
Technology	SiGe 200/250 GHz f_T/f_{MAX}	90nm CMOS	0.35um SiGe BiCMOS	55nm CMOS
Structure	Segmented NLDAC	Segmented NLDAC (Hybrid)	Linear DAC with Triangle-to-sine converter	Segmented NLDAC (M²AB + C²FAG+MCT)
Input (FCW) width [bits]	11	24	9	32
Output amplitude resolution [bits]	10	11	8	9
Operating clock frequency, f_{clk} [GHz]	8.6	1.3	5	2
Power consumption, P_{total} [W]	4.8	0.35	0.46	0.13
Power Efficiency, P_{total}/f_{clk} [W/GHz]	0.56	0.27	0.092	0.065
Max. output current, I_{load} [mA]	14.8	16	N/A	12
Load resistance, R_{load} [Ω]	15	50	N/A	50
$SFDR_{best}$ [dBc]	50	55	48.7	70.3
$SFDR_{worst}$ [dBc]	33	52	45.7	55.1
Active area [mm²]	7.5	0.9	2.1	0.1
FOM_1	81	1509	2133	8944
FOM_2	10.8	1677	1016	89442
FOM_3 (x10³)	16.9	565	N/A	19674

Term	FOM_1	FOM_2	FOM_3
Formula	$\frac{2^{SFDR_{worst}/6} \times f_{clk}}{P_{total}}$	$\frac{FOM_1}{\text{Active area}}$	$\frac{2^{(SFDR_{best}-1.76)/6.02} \times 2^{(SFDR_{worst}-1.76)/6.02} \times f_{clk}}{P_{total} - 0.5 \cdot I_{load}^2 \times R_{load}}$

Figure 21.3.6: Performance summary and comparison with previous work.

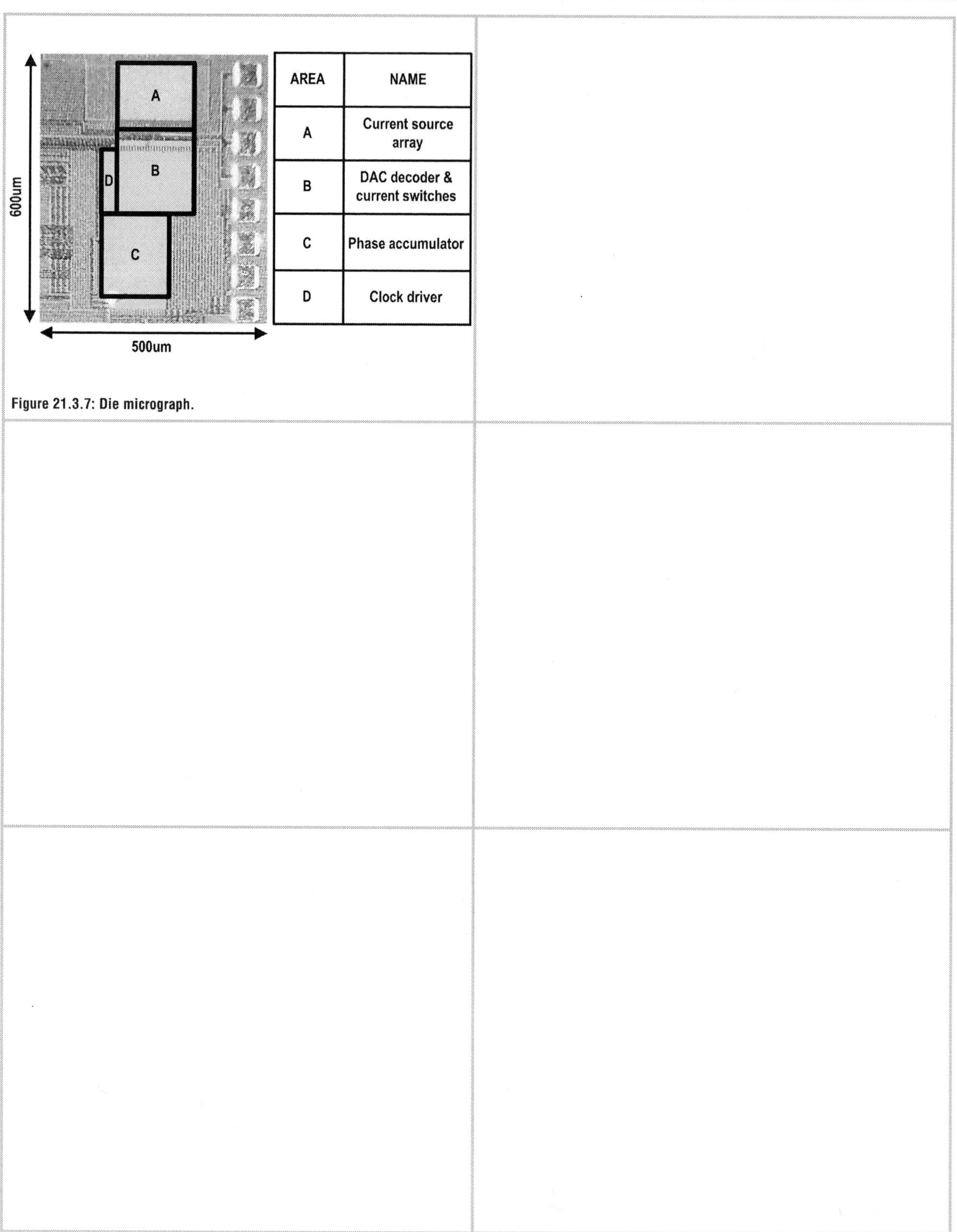

AREA	NAME
A	Current source array
B	DAC decoder & current switches
C	Phase accumulator
D	Clock driver

Figure 21.3.7: Die micrograph.

21.4 A 42mW 230fs-Jitter Sub-sampling 60GHz PLL in 40nm CMOS

Viki Szortyka[1,2], Qixian Shi[1,2], Kuba Raczkowski[1], Bertrand Parvais[1], Maarten Kuijk[2], Piet Wambacq[1,2]

[1]imec, Heverlee, Belgium, [2]Vrije Universiteit Brussel, Brussels, Belgium

For high data-rate communication at 60GHz using the IEEE 802.11ad standard, the LO synthesis needs both a low-noise VCO and low in-band phase noise. In the PLL shown in this paper, a QVCO with superharmonic passive coupling exhibits a large swing and low phase noise even with a 0.9V supply. In-band phase noise is reduced thanks to the use of a sub-sampling phase detector (SSPD), earlier introduced for low-GHz PLLs [1]. As most of the divider chain and the charge pump (CP) can be powered down in the sub-sampling mode, power consumption is also reduced.

The system architecture is shown in Fig. 21.4.1. The oscillator is a 60GHz QVCO for a direct-conversion transceiver. The synthesizer utilizes two loops, operated exclusively: classical PFD/CP integer-N and sub-sampling. The PFD/CP PLL forms a complete synthesizer with a static divide-by-2 and divide-by-3 circuit followed by a programmable divider. The division by 3 already in the second divider stage is essential for this integer-N PLL to synthesize the 802.11ad channels with a widely used 40MHz reference clock frequency rather than more exotic values, which can lead to expensive crystals. Previous solutions use injection-locked prescalers [3-5], but here static CML dividers are used for robustness. The CP is a current-steering topology. In the second mode, the conventional loop first locks the frequency and then the loop with a SSPD is enabled, while the PFD/CP loop is powered down. Separate divide-by-2 circuits are used for both loops. This approach to form a dual-path PLL fits naturally in a system with a QVCO, because the divider driven by the I output is a part of the PFD/CP loop, while the divider driven by the Q output is used in the sub-sampling loop. This second divider is used instead of a dummy load that is typically placed to avoid I/Q mismatch, so no extra capacitive loading is introduced, compared to a single-loop PLL. There is also no additional power consumption as only one prescaler is powered up for a given mode of operation. Together with the gain of the SSPD, the transconductor (g_m) and the pulser define the gain and bandwidth of this loop [1]. The loop filter that is common to both loops is an active opamp-based topology, to keep the CP output voltage constant with a 0.1V-to-0.8V range for V_{TUNE}, reducing reference spurs. The loop bandwidth is around 1MHz.

The QVCO (Fig. 21.4.2) uses superharmonic coupling to generate quadrature phases [2]. The coupled inductors (L_I/L_Q) enhance the impedance of the tail at second harmonic, forcing anti-phase 120GHz signals at the sources, leading to quadrature signals at 60GHz. Due to the symmetric topology, a small ring structure is added to eliminate mode ambiguity without adding much phase noise. Compared to a classical parallel QVCO, the superharmonic coupling provides lower phase noise, because no extra active devices are connected to the tank, except for small transistors of the ring structure. Moreover, both VCOs operate at resonance with the highest tank impedance and steepest phase change [2]. The alternative topology of a series-coupled QVCO is impractical with the 0.9V supply. The smaller coupling strength in the superharmonic coupling topology might introduce a phase error. To alleviate this, an automatic mixed-signal compensation loop is implemented on-chip. The buffered I and Q signals are mixed together, producing a differential DC voltage proportional to the quadrature error of the system that results from mismatches in the LC tanks. This voltage is sensed by a comparator, which provides a digital signal to on-chip logic. Compensation is applied by converting two 6-b digital signals into voltages that are applied to analog MOS varactors. A SAR loop converges to a small phase-error value. The PLL analog varactor is also a MOS capacitor. A small K_{VCO} of 1GHz/V is chosen to limit the loop-filter noise contribution. To cover the target tuning range of 9GHz, a 5-b capacitor bank is used. The bias voltage of the tail current source is RC filtered to eliminate noise from the reference bias current and the diode-connected transistor. Similar filtering is implemented in the VCO buffer.

The static divide-by-2 prescaler (Fig. 21.4.3) uses inductive peaking for increased speed. Pseudo-differential clocking transistors relax headroom requirements. Compact, low-Q, multi-turn peaking inductors use two top metal layers to minimize area. A similar inductor is used in the buffer at the prescaler output for a wideband response.

The SSPD (Fig. 21.4.3) operates at 30GHz and is driven by the buffered output of the divide-by-2. Placing the sampler after the divider output relaxes the design thanks to a lower frequency and it improves isolation between the QVCO and the sampler, reducing reference spurs. As the buffer's common-mode output voltage is V_{DD}, PMOS transistors are used to sample the differential voltage on 30fF capacitors (C_{SAMP}). This sampling occurs at the rising edge of the reference signal (see waveforms in Fig. 21.4.3) yielding a voltage proportional to the input phase error. In locked state the reference edge and the zero crossing of the differential signal coincide and the SSPD output is zero. The advantage of sub-sampling is that the in-band phase noise contributions of the divider and CP are eliminated [1]. The phase noise contribution of the divide-by-2 used here is marginal. The jitter from the rest of the divider chain is still eliminated. As the classical loop is designed for low in-band phase noise, the improvement in sub-sampling mode is limited. Power consumption, however, is decreased significantly as many power-hungry blocks (divider chain, CP) are powered down.

The chip (Fig. 21.4.7), fabricated in 40nm CMOS, consumes 75mW and 42mW in the PFD/CP and sub-sampling modes, respectively. Measured tuning range of the QVCO is from 53.8 to 63.3GHz. Due to a frequency shift, only three (58.32, 60.48 and 62.64GHz) of the four 802.11ad channels can be synthesized. The free-running QVCO phase noise is -92.5dBc/Hz at 1MHz offset. A crystal oscillator provides a 40MHz f_{REF}. The 60GHz sub-sampling PLL output spectrum and the worst-case phase noise measured at a 404MHz IF output of the spectrum analyzer are shown in Fig. 21.4.4. Integrated phase noise is -23.8dBc, corresponding to 230fs of RMS jitter. This meets the 802.11ad specs for 16-QAM modulation. The worst-case reference spur is below -40dBc, including clock harmonics.

I/Q imbalance is measured by downconverting an external 60GHz signal using the PLL I/Q signals as LO and measuring 100MHz baseband outputs using an oscilloscope. The uncompensated I/Q phase difference between 85.2° and 88.4° is corrected by the automatic compensation loop to a range between 88.6° and 92°, corresponding to an image rejection better than 35dB.

The integrated phase noise of the PFD/CP and sub-sampling PLLs is compared for 3 measured samples in Fig. 21.4.5. The sub-sampling PLL has only about 0.9dB better phase noise, as the PFD/CP PLL components (CP, divider) have originally been sized for a low-noise performance. The DC power reduction is more pronounced, with a decrease from 75mW to 42mW. Furthermore, if the QVCO consumption (core and buffer: 30mW) is subtracted, then the consumption of the loop components decreases from 45mW (dividers: 28mW, CP: 14mW, loop filter: <1mW, bias unit: 2mW) down to 12mW (divide-by-2: 10mW, loop filter: <1mW, bias unit: 2mW, sub-sampling part: <1mW). The consumption of the test buffers driving the chip output and the I/Q test mixers is excluded. Circuit performance is summarized and compared to state-of-the-art 60GHz LO generation systems in Fig. 21.4.6, with this sub-sampling PLL showing the lowest reported RMS jitter. This PLL is the only 60GHz CMOS PLL using a static divider among those shown in the Table. The power consumption is also competitive, as no VCO buffer to drive a transceiver is included in [3] and [4].

Acknowledgements:
The authors thank IWT, M. Libois, L. Pauwels, the imec BODI team, INVOMEC and Integrand Software for EMX.

References:

[1] X. Gao, et al., "A 2.2GHz 7.6mW Sub-Sampling PLL with -126dBc/Hz In-Band Phase Noise and $0.15ps_{rms}$ jitter in 0.18μm CMOS," *ISSCC Dig. Tech. Papers*, pp. 392-393, Feb. 2009.
[2] S. Gierkink, et al., "A Low-Phase-Noise 5GHz CMOS Quadrature VCO Using Superharmonic Coupling," *IEEE JSSC*, vol.38, pp. 1148-1154, July 2003.
[3] X. Yi, et al., "A 57.9-to-68.3GHz 24.6mW Frequency Synthesizer with In-Phase Injection-Coupled QVCO in 65nm CMOS," *ISSCC Dig. Tech. Papers*, pp. 354-355, Feb. 2013.
[4] W. Wu, et al., "A 56.4-to-63.4GHz Spurious-Free All-Digital Fractional-N PLL in 65nm CMOS," *ISSCC Dig. Tech. Papers*, pp. 352-353, Feb. 2013.
[5] D. Murphy, et al., "A Low Phase Noise, Wideband and Compact CMOS PLL for Use in a Heterodyne 802.15.3c Transceiver," *IEEE JSSC*, vol. 46, pp.1606-1617, July 2011.
[6] W. Deng, et al., "A Sub-Harmonic Injection-Locked Quadrature Frequency Synthesizer with Frequency Calibration Scheme for Millimeter-Wave TDD Transceivers," *IEEE JSSC*, vol. 48, pp.1710-1720, Jul. 2013.

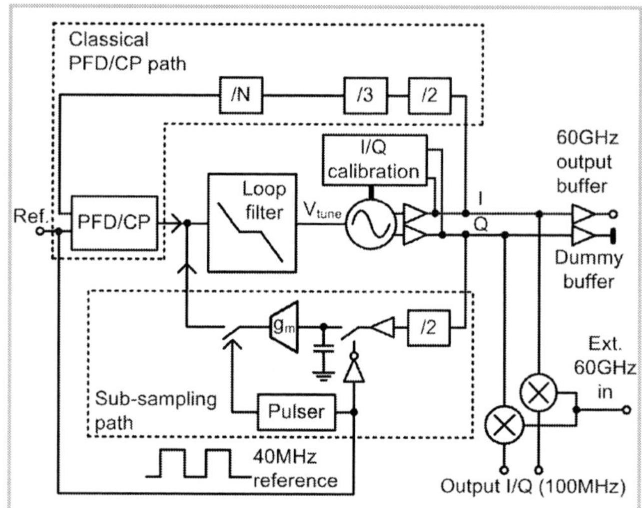

Figure 21.4.1: PLL architecture.

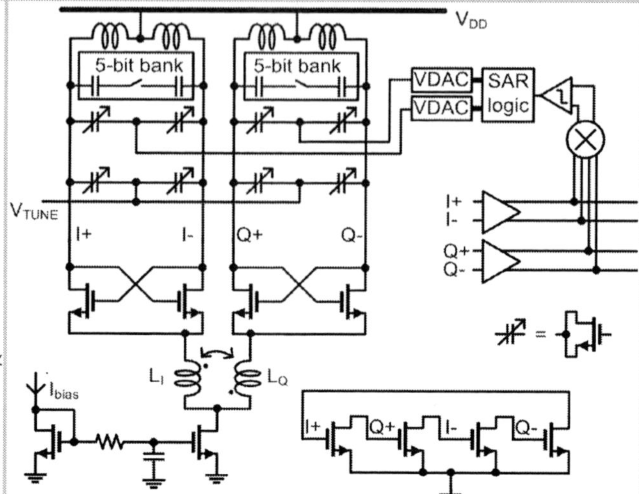

Figure 21.4.2: QVCO schematic.

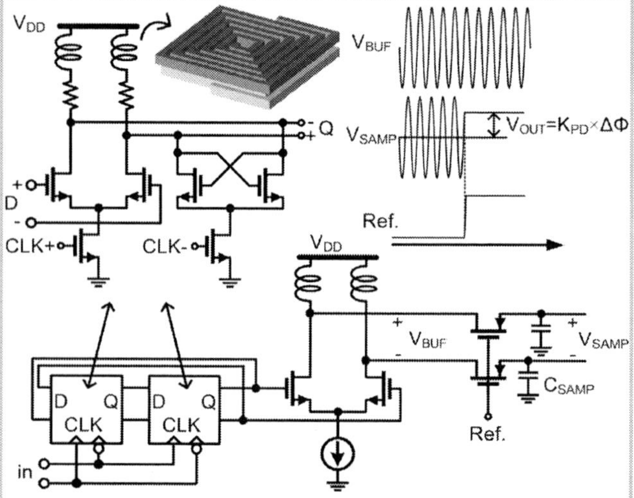

Figure 21.4.3: Divider and sub-sampling phase detector.

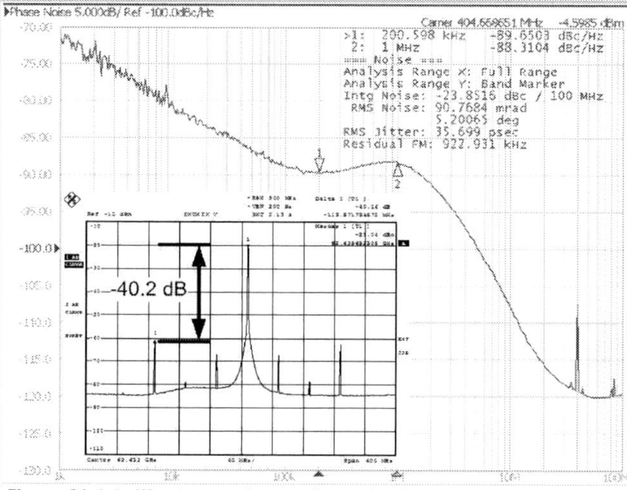

Figure 21.4.4: Worst-case phase noise and reference spurs for the sub-sampling PLL (@CH3, 62.64GHz).

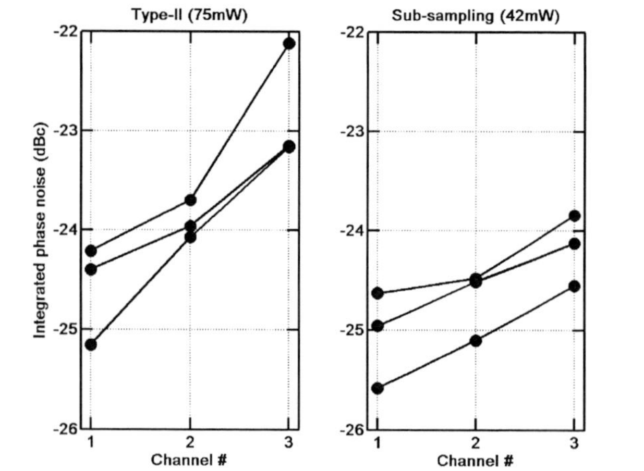

Type-II (75mW)

Sub-sampling (42mW)

Integrated phase noise (dBc)

Channel #

Figure 21.4.5: Performance comparison of the type-II and sub-sampling PLLs for three samples.

		This work		[3]	[4]	[5]	[6]
Topology		**60GHz Subsampling QPLL**	60GHz PFD/CP QPLL	60GHz PFD/CP QPLL	60GHz ADPLL	50GHz PFD/CP PLL	20GHz PLL + 60GHz QILO
Type		**Integer-N**	Integer-N	Integer-N	Fractional-N	Integer-N	Integer-N
CMOS technology		**40nm**	40nm	65nm	65nm	65nm	65nm
Tuning range (GHz)		**53.8-63.3 (16.2%)**	53.8-63.3 (16.2%)	57.9-68.3 (16.5%)	56.4-63.4 (11.7%)	42.1-53 (22.9%)	58.1-65 (11.2%)
f_{REF} (MHz)		**40**	40	135	10 .. 100	54	24
Normalized phase noise (dBc/Hz)	In-band	**-89.6**	-86.4	N/A	-72	-81	N/A
	@1MHz	**-88.3**	-89.7	-91	-90	-95.56	-96
RMS jitter (fs)		**203-230**	213-282	238(c)	522.9	N/A	N/A
Ref. spur (dBc)		**<-40**	<-42	<-45	<-74	N/A	<-52
V_{DD} (V)		**0.9/1(a)**	0.9/1(a)	1.2	1.2	1	1.2
Area (mm^2)		**0.16**	0.16	0.19	0.48	0.37	3.8
P_{DC} (mW)	VCO (quadrature?)	**30 (yes)**	30 (yes)	11.4 (yes)	13.2 (no)	N/A (no)	N/A (yes)
	Loop components	**12**	45	13.2	34.8	N/A	N/A
	Total	**42**	75	24.6(b)	48(b)	72	72

(a) Prescaler V_{DD}=1V.
(b) No VCO buffer included in the power consumption.
(c) Calculated from the integrated phase noise.

Figure 21.4.6: PLL performance summary and comparison.

21

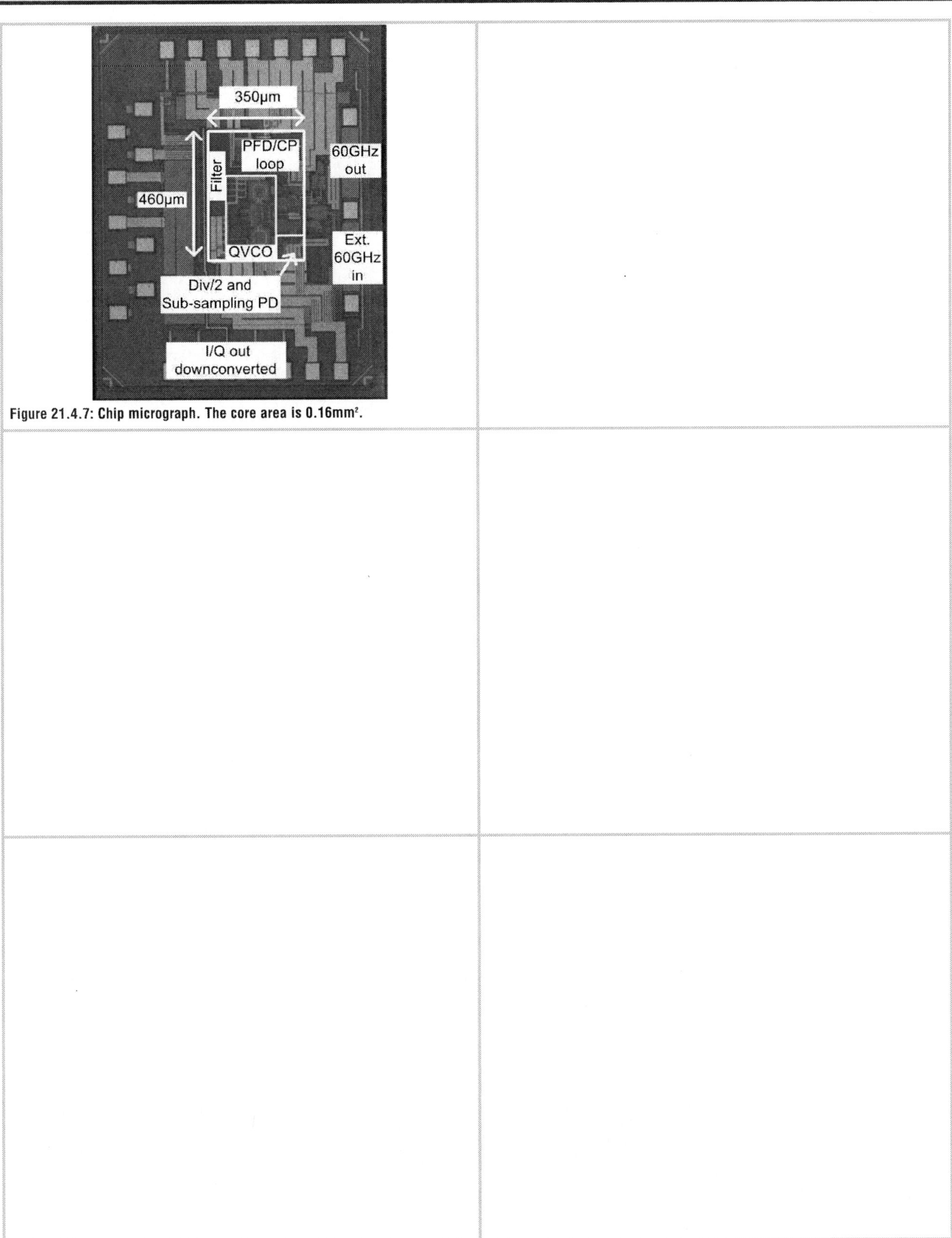

Figure 21.4.7: Chip micrograph. The core area is $0.16mm^2$.

21.5 A 3.24-to-8.45GHz Low-Phase-Noise Mode-Switching Oscillator

Mazhareddin Taghivand[1,2], Kamal Aggarwal[1], Ada S. Y. Poon[1]

[1]Stanford University, Stanford, CA,
[2]Qualcomm, San Jose, CA

VCO design for cellular applications to achieve universal coverage for a wide range of frequencies (400MHz to 3700MHz) in different standards and meeting stringent out-of-band and in-band phase-noise (PN) requirements is a challenging task. The simplest method to generate I and Q signals in the LO is to use a frequency divide-by-2 which requires the VCO frequency to be an even multiple of f_{LO}. This method is area efficient and superior for coexistence as it does not generate jammers in other bands. The technique in this work expands the VCO frequency range to ensure $2xf_{LO}$ for all cellular bands, notably 3700MHz, with sufficient margin, and meets the most stringent PN specification for a SAW-less GSM transceiver.

There is a trade-off between a VCO tuning range and its PN [1]. One way to obtain desired PN over a wide frequency range is to divide it into smaller sub-bands and assign a separate VCO to each sub-band. This significantly increases the die area and requires a multiplexer which increases the power consumption. Other techniques use transistors to switch in or out some of the tank inductance, which achieves good frequency range at the cost of poor PN performance due to Q degradation introduced by ohmic loss of inductor switches [2]. Recent works on mode-switching VCOs have significantly improved the trade-off between frequency range and PN [1,3]. A lossless mode-switching concept is demonstrated in [3], where a low/high oscillation frequency can be selected by switching in/out a coupling capacitor between odd/even modes of two coupled LC oscillators. In [1], a transformer in addition to a coupling capacitor switches modes between two LC tanks, thus providing good frequency tuning range and PN simultaneously. However, the Q of the secondary coil is at least 25% lower than that of the primary coil; and in the high-frequency mode, the effective inductance is reduced because of the magnetic flux cancellation between the weakly coupled coils. As a result, the Q of the effective inductance is degraded.

Figure 21.5.1 shows the architecture of the oscillator. It consists of a cross-shaped inductor with four identical quadrants, two pairs of cross-coupled NMOS transistors $M_{1,2,3,4}$ to provide negative resistance, a mode-switching network consisting of PMOS mode-switching transistors $M_{5,6,7,8}$, two coupling capacitors C_C, and a 4-b coarse capacitor bank. The oscillator has two modes, namely, low band (LB) and high band (HB). LB mode is selected when the voltage at the gates of M_5 and M_6 is pulled to ground, and the gates of M_7 and M_8 are pulled to high, while their source and drain are DC coupled to V_{DD}, shorting terminals P_1 and P_2 to P_3 and P_4, respectively. This arrangement eliminates any oscillation through the inductor in quadrants 1 and 3 as such oscillation requires a very large non-existing negative resistance to combat the loss introduced by transistors M_5 and M_6 operating in the deep-triode region. Figure 21.5.2 shows the direction of AC current through loops in quadrants 2 and 4. An axis of symmetry is shown for the LB in Fig. 21.5.2. This mirror symmetry explains the absence of AC current through quadrants 1 and 3. The oscillator frequency in this band is $1/2\pi/\sqrt{L_{LB}(C+C_C)}$ where C is the total capacitance of all the blocks in the oscillator, namely, two capacitor banks, two pairs of cross-coupled NMOS transistors, and all other parasitic capacitors. L_{LB} is the parallel combination of the loop inductors of quadrants 2 and 4. Joint terminals P_1 and P_3 can ideally be treated as a unified node, and similarly for P_2 and P_4. This mode has a frequency range of 3.24GHz to 4.9GHz.

The HB mode is selected when the voltage at the gates of M_7 and M_8 is pulled to ground, and the gates of M_5 and M_6 are pulled to high, while their source and drain are DC coupled to V_{DD}, shorting terminals P_1 and P_2 to P_4 and P_3, respectively. This switching triggers a mode in which all four loop inductors have a chance for oscillation. The direction of AC current is shown in Fig. 21.5.2. It is important to notice an axis of symmetry between quadrants 2, 3 and 1, 4. Just as in the LB mode, this is a mirror symmetry, which graphically explains why all the loop inductors are conducting AC current. In HB mode, each loop inductor is coupled to its two adjacent neighboring loops. Hence, the total inductance seen between terminals $P_{1,4}$ and $P_{2,3}$, namely L_{HB}, is larger than ¼ of a single loop inductor by a factor of (1+k), where k is the effective coupling coefficient of the system of 4 coupled loops. From EM simulation k is found to be 0.27. The positive sign of k is evident from the AC direction of currents in all 4 loops, where each loop contributes constructively to the magnetic flux of two of its adjacent neighbors. The oscillation frequency in HB mode is $1/2\pi/\sqrt{L_{HB}C}$. The frequency range of the HB mode is from 4.5GHz to 8.45GHz. The supply voltage is adjusted in Fig. 21.5.4 and Fig. 21.5.5 in order to obtain the best PN and FoM over the entire frequency range, respectively.

This technique is more immune to magnetic pulling, which is often caused by the PA's strong second harmonic that coincides with the VCO frequency. Due to mirror symmetry inherent in both HB and LB modes of operation, the oscillator ideally rejects any external magnetic interference. Substrate coupling and spurs performance are also improved as the guard ring (GR) around this oscillator can be closed, without the risk of forming an Eddy current in the GR which can subsequently reduce inductor Q. The mirror symmetry of the AC currents in both LB and HB modes ensures that no current is induced in the closed GR. As shown in the table of Fig. 21.5.2, small inductance values can be realized specially in the HB mode with a Q that is comparable with considerably larger inductance values. This improves the PN because at a given frequency when inductance is reduced by a factor of N and the capacitance increased by the same factor, and assuming the Q's remain unchanged, the PN improves by $10\log_{10}(N)$ dB.

The oscillator was taped-out in a 9-metal 40nm GP process with top UTM. The G_m core transistors utilized 1.8V IO devices to get better flicker noise performance at 100kHz offset frequency, which is extremely important for both cellular and WiFi MIMO applications. The VCO gave the best performance at a supply of 0.8V beyond which the PN improvement at far offset was minimal. For area reduction the entire active part was placed inside the inductor, which is completely filled with dummy metals to comply with strict metal density rules of advanced technology nodes. This mimics the scenario where the area inside the VCO can be utilized for placing other active and passive circuits. A 5052B Agilent Signal Source Analyzer (3MHz to 7GHz) was used for PN measurements. For measurements beyond 7GHz, an external mixer and LO at 7GHz were used to downconvert the signal. Figure 21.5.3 shows two snapshots taken at 3.72GHz and 7.759GHz, demonstrating the PN performance at all offset frequencies, especially at close-in 100kHz, and meeting cellular specification across the entire frequency band. PN beyond 10MHz offset becomes flat, hitting the noise floor of the output buffer transistors, which were matched to 50Ω and biased through an external Bias Tee. The external LO noise floor was also a limiting factor for the measurements beyond 10MHz offset.

In conclusion, a dual-band mode switching oscillator is shown to achieve high tuning range and low PN simultaneously. A prototype covering 3.24GHz to 8.45GHz with 400MHz of overlap between the HB and LB modes is measured. The measured PN values at an offset of 10MHz from 3.72GHz and 7.76GHz are −150.2dBc/Hz and −144.4dBc/Hz respectively, which correspond to FoMTs of 207.5dB/Hz and 208.2dB/Hz respectively.

Acknowledgements:
The authors thank the TSMC University Shuttle Program for chip fabrication, Lorentz Solution Inc. for providing EM tools to Stanford, and Denny Goetz from Qualcomm for helping in the measurement.

References:
[1] Guansheng Li; Li Liu; Yiwu Tang; Afshari, E., "A Low-Phase-Noise Wide-Tuning-Range Oscillator Based on Resonant Mode Switching," *IEEE J. Solid-State Circuits,* vol.47, no.6, pp.1295-1308, June 2012.
[2] Sadhu, B.; Jaehyup Kim; Harjani, R., "A CMOS 3.3-to-8.4GHz Wide Tuning Range, Low-Phase-Noise LC VCO," *IEEE Custom Integrated Circuits Conference,* pp. 559- 562, 13-16 Sept. 2009.
[3] Guansheng Li; Afshari, E., "A Distributed Dual-Band LC Oscillator Based on Mode Switching," *IEEE Trans. on Microwave Theory and Techniques,* vol.59, no.1, pp.99-107, Jan. 2011.
[4] Fanori, L.; Liscidini, A.; Andreani, P., "A 6.7-to-9.2GHz 55nm CMOS hybrid Class-B/Class-C cellular TX VCO," *ISSCC Dig. Tech. Papers*, pp. 354-356, 19-23 Feb. 2012.

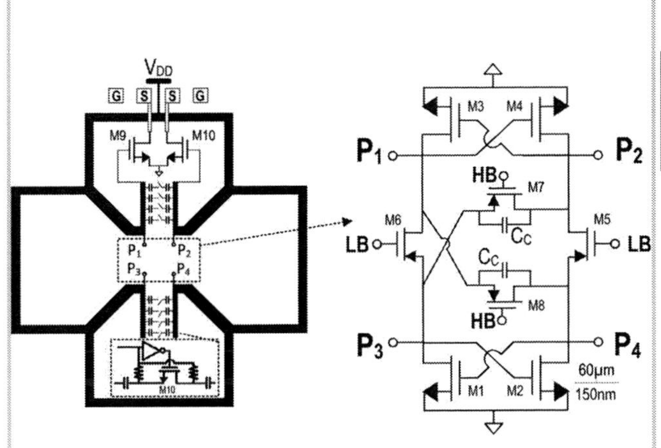

Figure 21.5.1: Circuit schematic of a mode-switching oscillator with a cross-shaped inductor.

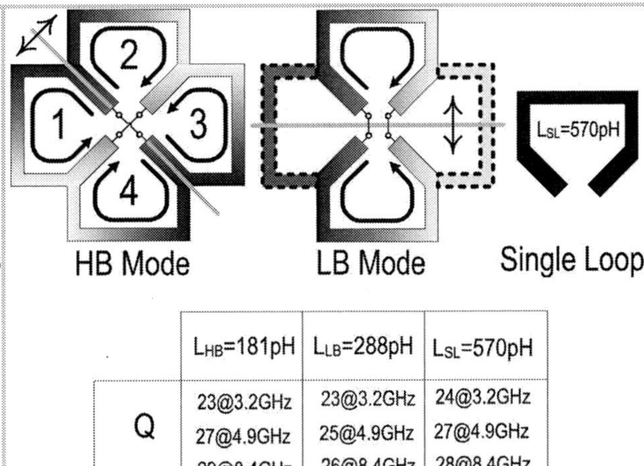

	L_{HB}=181pH	L_{LB}=288pH	L_{SL}=570pH
Q	23@3.2GHz	23@3.2GHz	24@3.2GHz
	27@4.9GHz	25@4.9GHz	27@4.9GHz
	29@8.4GHz	26@8.4GHz	28@8.4GHz

Figure 21.5.2: Two modes of the oscillator with its corresponding axis of symmetry. The table shows the effective inductance at LB and HB modes, compared to a single-loop inductor.

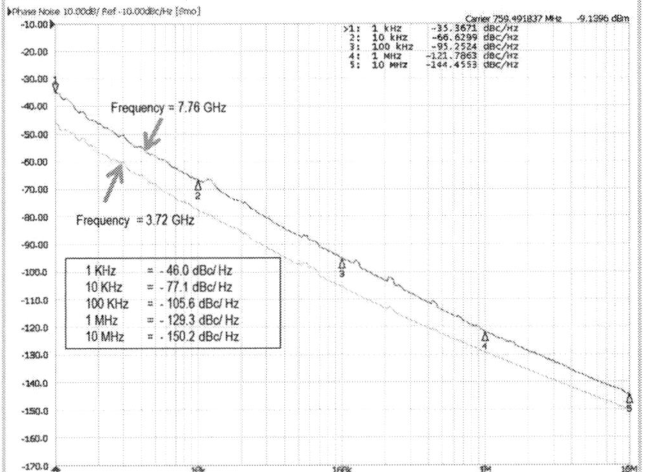

Figure 21.5.3: Phase-Noise measurement at different frequency offsets for 3.72 GHz and 7.76 GHz bands.

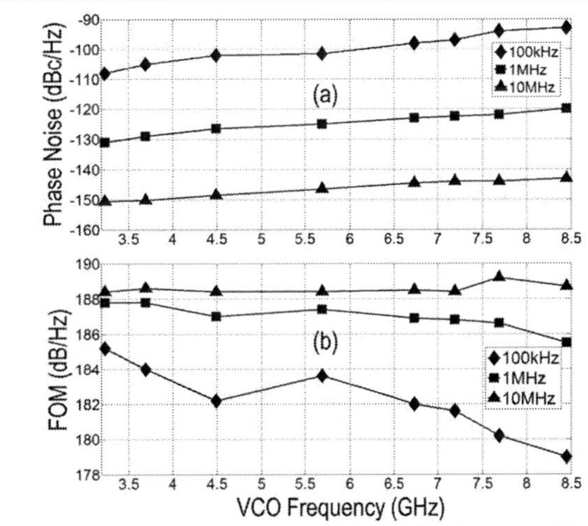

Figure 21.5.4: (a) Phase noise obtained by adjusting V_{DD} for best PN. (b) Corresponding FOM.

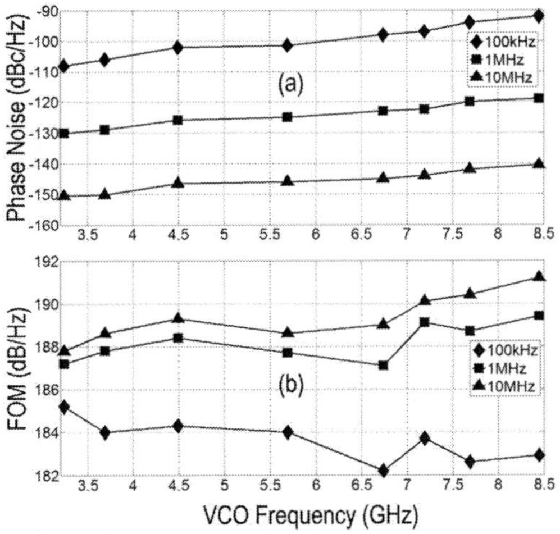

Figure 21.5.5: (a) Phase Noise obtained by adjusting V_{DD} for best PN. (b) Corresponding FOM.

	This Work		[4] ISSCC 2012	[1] JSSCC June 2012	[2] CICC 2009
Frequency (GHz)	**3.24 – 8.45 (88.7 %)**		6.72 – 9.2 (31%)	2.48 – 5.62 (77.5%)	3.28-8.35 (87.2%)
Core Voltage (V)	**0.8**		1.5	0.6	1.6
Core Power (mW)	**20**		27	9.8 - 14.2	6.5 - 15.4
Carrier (GHz)	**3.72**	**7.76**	3.72***	3.72***	3.72***
Estimated PN (dBc/Hz) @ 100KHz offset	**-105.56**	**-95.25**	-104**	-96**	-95**
FOM(dB/ Hz) @ 100KHz offset	**183.97**	**180.10**	180.66	177.16	178.24
FOMT(dB/Hz) @ 100KHz offset f	**202.92**	**199.05**	191	195	197
Estimated Phase Noise (dBc/Hz) @ 1MHz offset	**-129.36**	**-121.78**	-131	-128.27	-122
FOM(dB/Hz) @ 1MHz offset	**187.77**	**186.63**	187.66	189.43	184.14
FOMT(dB/Hz) @ 1MHz offset	**206.72**	**205.58**	198	207	203
Estimated Phase Noise (dBc/Hz) @ 10MHz	**-150.22**	**-144.45**	-151	-151.4	-141
FoM(dB/Hz) @ 10MHz offset	**188.63**	**189.30**	187.66	192.56	184.14
FoMT(dB/Hz) @ 10MHz offset	**207.58**	**208.25**	198	210	203
Area (mm²)	**0.432**		0.49	0.294	0.1
Technology	**40nm CMOS**		55nm CMOS	65nm CMOS	130nm CMOS
Description	**Dual Band Mode switching**		Hybrid Class-B/ Class-C	Dual Band Mode Switching	Dual Band Switch Inductor

* After divide by 2

**Estimated form PN plot

*** Normalized PN to the carrier frequency

Figure 21.5.6: Comparison with state-of-the-art.

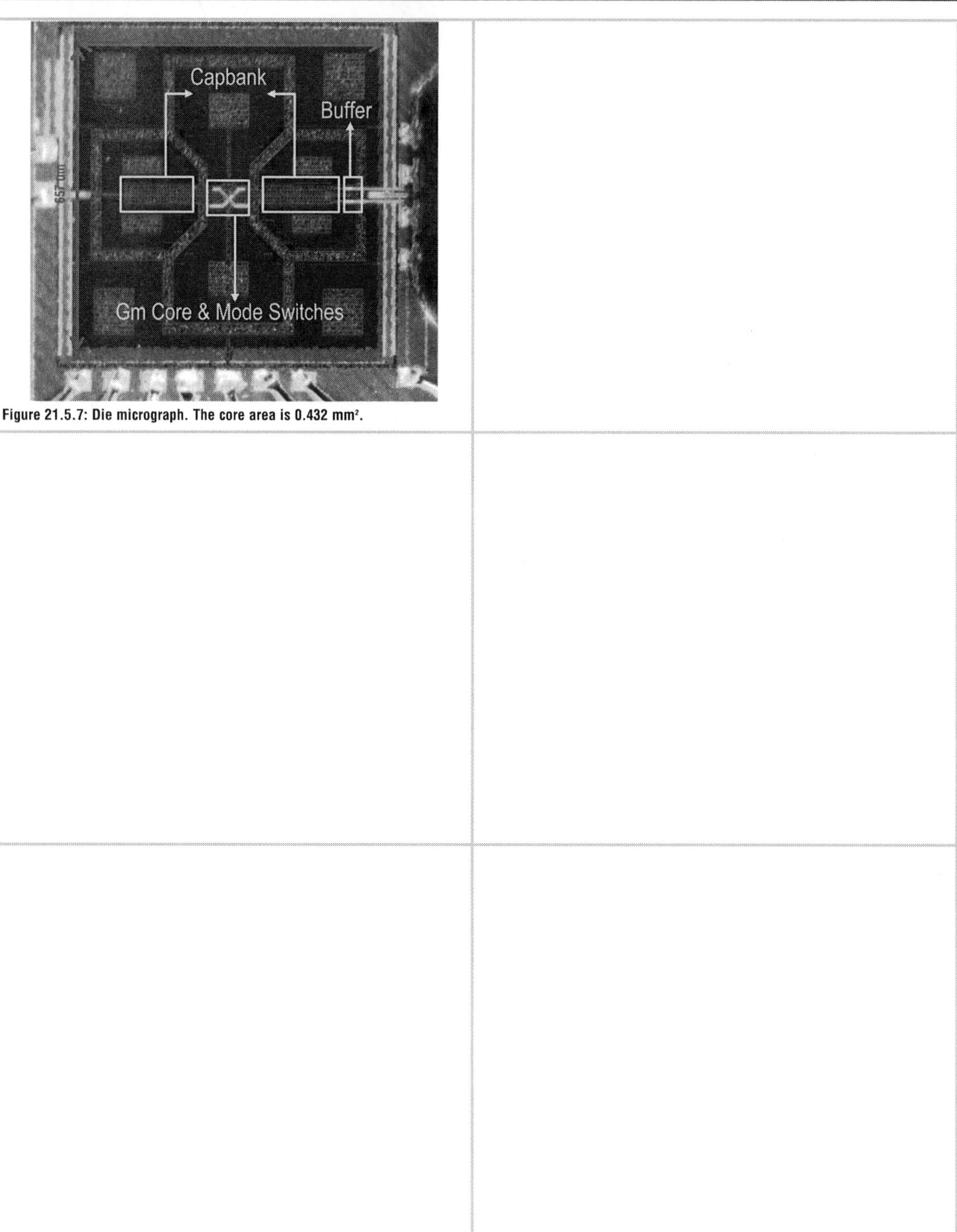

Figure 21.5.7: Die micrograph. The core area is 0.432 mm².

21.6 A 2.4-to-5.3GHz Dual-Core CMOS VCO with Concentric 8-Shaped Coils

Luca Fanori[1,2], Thomas Mattsson[3], Pietro Andreani[1,3]

[1]Lund University, Lund, Sweden,
[2]now at Marvell, Pavia, Italy,
[3]Ericsson Modems, Lund, Sweden

Despite recent attempts to relax the phase-noise demands on voltage-controlled oscillators (VCOs) for cellular communications [1], mainstream radios require harmonic VCOs capable of a very low phase noise with moderate power consumption, associated to a large tuning range (TR) and a high insensitivity to interfering signals. Ideally, the TR should be in excess of one octave, since this allows the easy synthesis of all frequencies below those directly generated by the VCOs via repeated frequency divisions by 2. At the same time, the oscillation spectrum should be affected as little as possible by spurious (common-mode) magnetic fields impinging on the inductor coil in the VCO tank. This is a crucial requirement in modern radios, where there are more PLLs active at the same time, and particularly when (non-contiguous) carrier aggregation is implemented, since in this case the signal bands may be very close to each other. If an individual PLL is used for each band, the VCOs may oscillate very close to each other, or at frequencies that are harmonically related to each other, posing a very serious issue of mutual pulling through the respective magnetic field. And even if a single VCO is used [2], or two (or more) VCOs that are not harmonically related [3], it is nevertheless a good practice to design the tank inductor as insensitive as possible to external magnetic fields, which abound in and close to the radio IC.

In this sense, a coil having the shape of an 8 is a much more favorable choice than the usual ring or octagonal layout, since it is immune to stray magnetic fields, as well as itself generating a vanishing (distant) magnetic field [4]. The penalty to be paid is a (minor) deterioration of the inductor quality factor (Q).

Turning to the TR issue, and targeting one octave at least (i.e., a TR in excess of 67%), we discard the straightforward adoption of a single inductor combined with a sufficient amount of switchable capacitors, since it leads to a waste of power in the lower frequency range [5]. A much more attractive architecture is based on a transformer [5,6], which, however, does not lend itself easily to a version with 8-shaped coils. The next obvious choice is simply employing two VCOs with two different inductors, covering together the desired TR with sufficient overlap between the two individual TRs. While this solution is perfectly reasonable and often used in practice, the total extra chip area required may become prohibitively large for an already-large IC featuring carrier aggregation. More specifically, the VCO area surplus is dominated by the two coils, which, moreover, cannot be placed too close to each other, lest the resulting mutual inductance deteriorate the Q of the LC tank of the active VCO (by reflecting onto it the losses induced in the LC tank of the inactive VCO).

The symmetry of the 8-shaped coil, however, comes to our rescue, as it induces only a vanishing magnetic field on another coil, either ring- or 8-shaped, if placed along the latter's axis of symmetry. In fact, depending on the actual dimensions, the smaller coil can even be placed inside the larger one, reducing the area occupation to basically the single VCO with the larger coil (of course, two cross-coupled transistor pairs and two capacitor banks are still needed, instead of single ones; however, the two capacitor banks together are not larger than the capacitor bank required in single-inductor or transformer-based designs covering the same TR).

Figure 21.6.1 shows a hybrid schematic/layout view of the proposed dual-core VCO, with the smaller 8-shaped coil rotated by 90° and placed inside the larger 8-shaped coil in a concentric, symmetric fashion. In reality, as is clear from the chip micrograph in Figure 21.6.7, the metal crossing in each coil has been offset from the geometrical center, thereby maximizing the number of parallel metal layers that can be used in each coil, while maintaining the layout symmetry.

Figure 21.6.2 shows the Momentum simulations of the Qs of the two coils, designed in a standard 65nm CMOS process without any thick metal layer, for three different configurations: 1) each coil in presence of the other one (capturing the real-life case); 2) each coil when the other one is removed (assessing the latter's impact on the former's Q); 3) each coil separately optimized for use in a single-inductor tank (assessing its Q if designed for a standalone VCO).

Clearly, the presence of the other coil has only a marginal impact on the Q of each coil, as clear from the small difference (contained within 10%) between the curves relative to alternatives 1) and 2) above. This shows that indeed only minor extra losses are due to a surviving coupling between the two concentric coils. The difference, however, is more significant between 1) and 3), indicating that a standalone coil, placed at a safe distance from the other, can be designed with a Q up to 30% higher than that obtained in the implemented tank. While this figure would in general vary with the number and thickness of the available metal layers, it is clear that there is a trade-off between Q (and, ultimately, power consumption) and chip area dedicated to frequency generation.

The dual-core VCO has been designed in the mentioned standard CMOS process. While a Class-D topology has been chosen for both high-band (HB) and low-band (LB) VCOs, it is clear that the same concentric 8-shaped inductors can be used with other topologies as well (e.g., Class-B or Class-C). Both VCOs use the same 5-b MIM capacitor array for discrete tuning, and an AMOS varactor continuously covering 2 discrete tuning steps.

As shown in Fig. 21.6.3, the LB VCO oscillates between 2.4GHz and 3.6GHz (TR = 40%), while the HB VCO covers the 3.4-to-5.3GHz band (TR = 44%), for an aggregate TR of 75%. The two TRs overlap by a safe 200MHz margin. Figure 21.6.4 displays the phase noise plots for both HB and LB VCOs at the respective extreme oscillation frequencies: the phase noise at 10MHz offset from the carrier varies from -139dBc/Hz to -149dBc/Hz, when 10 to 15mA are consumed from a power supply V_{dd} as low as 0.4V. As already in previous Class-D VCOs, the $1/f^3$ phase noise corner is relatively high, varying between 1MHz and 2MHz across the TR. Figure 21.6.5 reports a plot of current consumption, phase noise, and figure-of-merit (FoM) across the TR, still for V_{dd} = 0.4V. The FoM has a maximum of 189dBc/Hz at minimum frequency, and settles at 187dBc/Hz at medium and high frequencies. The table in Fig. 21.6.6 shows the state-of-the-art for VCOs covering a TR of at least 50%, either continuously or with two separate bands. The composite performance of the dual-core VCO here presented is topped only by the outstanding transformer-based VCO in [5], which, however, makes use of a thick metal layer, and, more importantly in the present context, does not use 8-shaped coils to prevent pulling hazards.

Acknowledgements:
We are very grateful to STMicroelectronics for silicon donation. This work has been partially funded by SSF under the DARE project, and by the European Marie Curie Project FP7-PEOPLE-2009-IAPP n° 251399.

References:
[1] M. Mikhemar et al., "A Phase-Noise and Spur-Filtering Technique Using Reciprocal-Mixing Cancellation", *ISSCC Dig. Tech. Papers*, pp. 86-87, Feb. 2013.
[2] L. Sundström et al., "A Receiver for LTE Rel-11 and Beyond Supporting Non-Contiguous Carrier Aggregation", *ISSCC Dig. Tech. Papers*, pp. 336-337, Feb. 2013.
[3] K. Chandrashekar et al., "A 32nm CMOS All-Digital Reconfigurable Fractional Frequency Divider for LO Generation in Multistandard SoC Radios with On-the-Fly Interference Management", *ISSCC Dig. Tech. Papers*, pp. 352-354, Feb. 2012.
[4] P. Andreani et al., "A TX VCO for WCDMA/EDGE in 90 nm RF CMOS", *IEEE J. Solid-State Circuits,* vol. 46, no. 7, pp. 1618-1626, July 2011.
[5] G. Li et al., "A Low-Phase-Noise Wide-Tuning-Range Oscillator Based on Resonant Mode Switching", *IEEE J. Solid-State Circuits,* vol. 47, no.6, pp. 1295-1308, June 2012.
[6] A. Bevilacqua et al., "Transformer-Based Dual-Mode Voltage-Controlled Oscillators", *IEEE Trans. Circuits Syst. II*, vol. 54, no. 4, pp. 293–297, Apr. 2007.
[7] B. Sadhu et al., "A CMOS 3.3–8.4 GHz Wide Tuning Range, Low Phase Noise LC VCO," *IEEE Custom Integrated Circuits Conf.*, pp. 559–562, Sept. 2009.
[8] P. Ruippo et al., "A UMTS and GSM Low-Phase-Noise Inductively-Tuned LC VCO," *IEEE Microwave Wireless and Component Letters*, vol. 20, no. 3, pp. 163–165, March 2010.
[9] J. Borremans et al., "A Compact Wideband Front-End Using a Single-Inductor Dual-Band VCO in 90nm Digital CMOS," *IEEE J. Solid-State Circuits*, vol. 43, no. 12, pp. 2693–2705, Dec. 2008.

978-1-4799-0917-9/14 $31.00 © 2014 IEEE

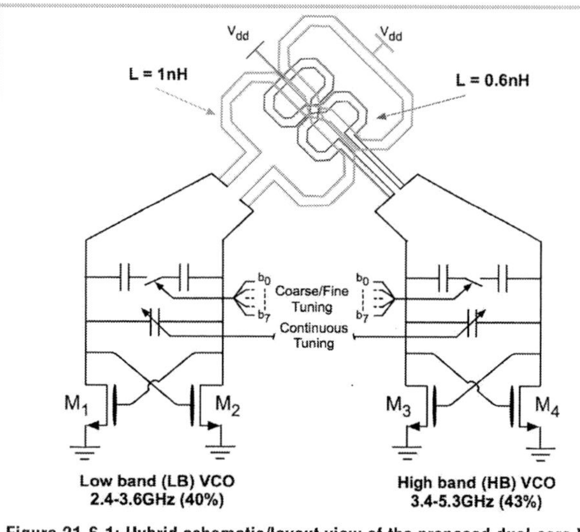

Figure 21.6.1: Hybrid schematic/layout view of the proposed dual-core VCO.

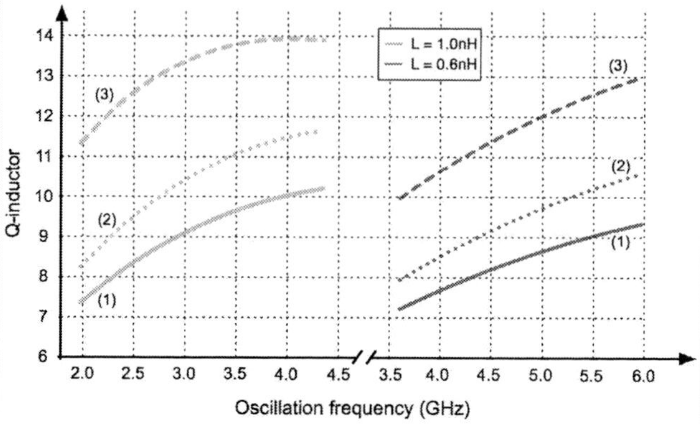

Figure 21.6.2: Plots of Q vs. frequency for the each 8-shaped inductor, when: 1) the other inductor is present (as in reality); 2) the other inductor is removed; 3) when the inductor is optimized for use in a standalone VCO.

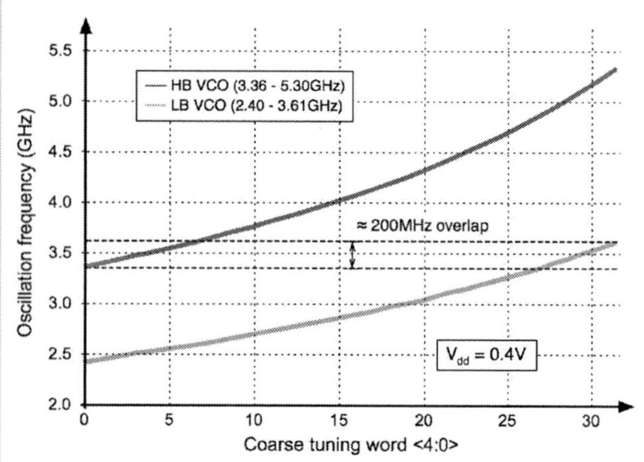

Figure 21.6.3: Measured tuning range of the dual-core VCO.

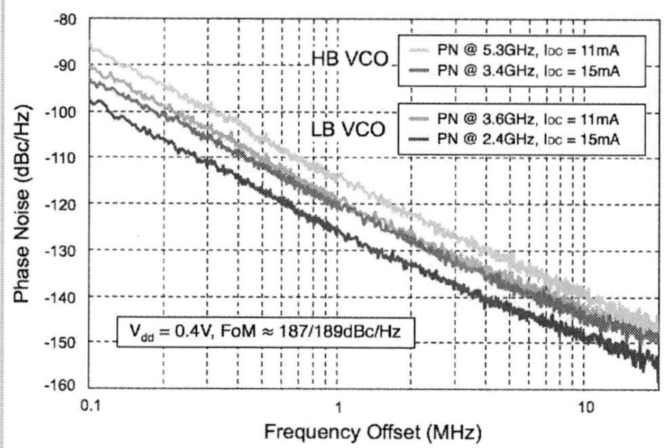

Figure 21.6.4: Phase noise plots at the TR limits of high-band and low-band VCOs.

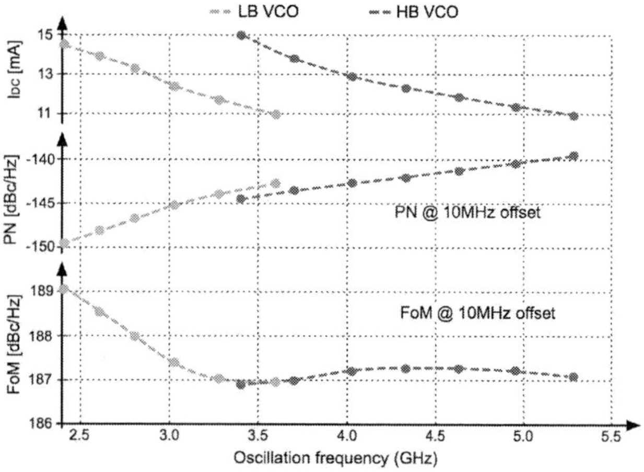

Figure 21.6.5: Current consumption, phase noise and figure-of-merit across the tuning range (at V_{dd} = 0.4V).

Parameter		This work	[5]	[7]	[8]	[4]	[9]
VCOs Frequency [GHz]		2.4 - 3.6 3.4 - 5.3 (75%)	2.5 - 3.9 3.3 - 5.6 (76%)	3.3 - 6.0 5.6 - 8.3 (86%)	2.9 - 4.8 4.7- 5.4 (60%)	2.5 - 4.0 4.9 - 5.8 (47%-17%)	3.1 - 4.0 8.8 - 11.2 (25%-24%)
Frequency Overlap [MHz]		200	600	400	100	-	-
Voltage Supply [V]		0.4	0.6	1.6	1.8	1.2	1.2
Current Consumption [mA]		15 - 11 15 - 11	16 - 23	9.5 - 4.0 9.0 - 4.0	7.5	19 - 21	3.5 - 1.8 8.3 - 5.6
Power [mW]		6.0 - 4.4	9.5 - 14	15.5 - 6.5	13.5 - 9.8	23 - 25.5	10 - 2.2
Phase Noise	PN [dBc/Hz]	-149 / -139	-157 / -152	-122 / -117	-128 / -122	-156 / -146	-122 / -117
	Offset [MHz]	10	20	1	1	20	2.5
	Carrier [GHz]	2.4 / 5.3	3.7 / 5.5	3.3 / 7.8	2.9 / 5.4	3.8 / 5.8	3.9 / 10.9
FoM PN [dBc/Hz]		187 - 189	188 - 192	181 - 187	185 - 190	188 - 182	181-182
Technology		65 nm CMOS	65 nm CMOS	130 nm CMOS	130 nm CMOS	90 nm RF CMOS	90 nm CMOS

Figure 21.6.6: Comparison with the state-of-the-art for very wide-band VCOs (TR ≥ 50%).

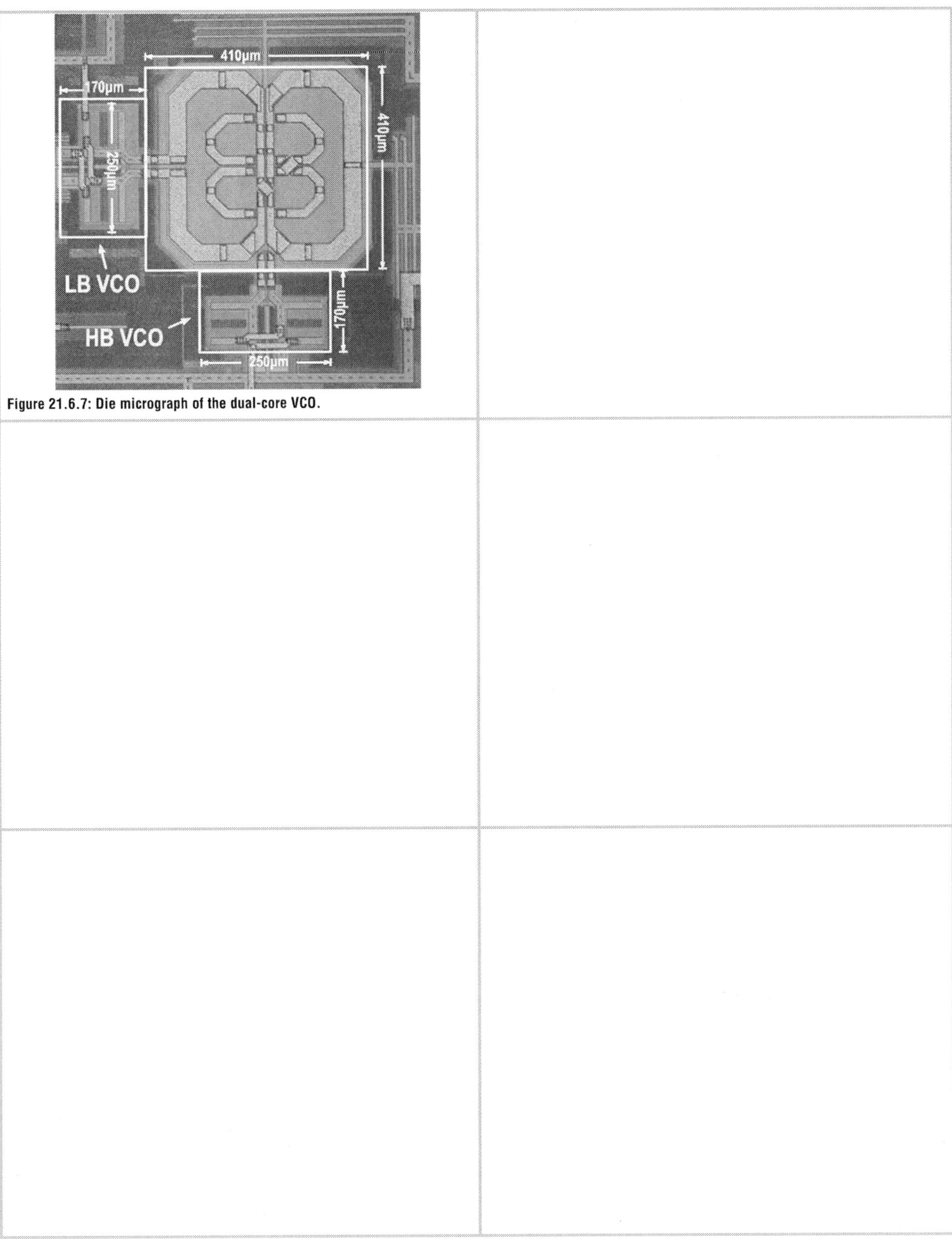

Figure 21.6.7: Die micrograph of the dual-core VCO.

21.7 A 1.8mW PLL-Free Channelized 2.4GHz ZigBee Receiver Utilizing Fixed-LO Temperature-Compensated FBAR Resonator

Keping Wang[1], Jabeom Koo[1], Richard Ruby[2], Brian Otis[1]

[1]University of Washington, Seattle, WA,
[2]Avago Technologies, San Jose, CA

This paper presents an ultra-low-power 2.4GHz receiver for the IEEE 802.15.4 (ZigBee) standard. Traditional short-range ISM-band radios require a PLL-based frequency synthesizer for channelization across the band of interest [1-3]. The lowest ZigBee power consumption found in the literature to date is 1.6mW (RX) and 1.8mW (PLL) by employing a sliding-IF architecture [1]. [4] proposes a BAW-based 2.4GHz ZigBee receiver that saves power by eliminating the off-chip quartz crystal with super-high IF architecture; however, 8.2mW DC power is consumed since a low frequency LC-PLL is necessary for channel select tuning. The Blixer in [5] reduced the RX DC power through current re-use; however, the LO generator suffers from high power consumption due to the high-frequency quadrature LO.

We propose a receiver architecture aimed at ZigBee that saves power using the following techniques: first, we propose a sliding-IF receiver architecture with a suitable LO frequency plan utilizing a temperature-compensated thin-film bulk-acoustic-wave resonator (FBAR). This strategy completely eliminates the need for a PLL by directly dividing down the fixed FBAR oscillator frequency. Second, we propose a current-reuse balun-LNA allowing a low-power wideband match without using inductors. Third, frequency translation is achieved by a hybrid mixer, which stacks a switching mixer on a switched-g_m mixer for current reuse. As a result, the performance of the proposed receiver exceeds the requirements of the ZigBee standard such as NF, linearity, phase noise and selectivity. It dissipates 0.86mW (RX) and 0.92mW (LO) from a single 1V supply.

Figure 21.7.1 shows the proposed receiver architecture. The RF signal is pre-filtered by a low-loss FBAR filter to ensure more than 40dB of image-rejection ratio (IRR). The balun-LNA amplifies the RF signal and provides a differential output to the hybrid mixer which downconverts the RF signal to a configurable IF. The proposed sliding-IF architecture avoids a high-frequency quadrature LO generator, thus reducing the DC power. Unlike previous designs [1], we use a fixed-frequency first LO signal produced by a temperature-compensated FBAR oscillator. The FBAR oscillator is temperature stable to within +/-50 ppm (200 kHz) over 0 to 100°C, which is sufficient for ZigBee applications [6]. A multi-mode divider (MMD) creates a 25% duty-cycle 4-phase second LO signal for channel selection. In order to convert all 16 ZigBee channels to baseband, we propose a configurable filter that can be programmed to a low-pass filter (LPF) and a complex band-pass filter (CBPF). The total RX gain can be controlled to accommodate a wide input range by using programmable-gain amplifiers (PGAs).

Figure 21.7.2 shows the proposed frequency plan. Generally, ZigBee receivers require PLL-based frequency synthesizers, which suffer from high power consumption as well as large area overhead due to on-chip inductors and off-chip quartz crystals. Instead, we propose a frequency plan using a PLL-free LO. FBAR oscillators exhibiting a power/phase-noise tradeoff 30dB better than LC oscillators can only tune across a limited (few MHz) bandwidth. Thus, we have designed a frequency plan allowing a fixed-frequency first LO. During the first frequency conversion, the entire RF band (2405 to 2480MHz) is converted to a wideband IF (5 to 80MHz) by a fixed 2.4GHz LO, while channelization is achieved in second frequency conversion. The second LO frequency is generated by the programmable MMD. We employ two different IF modes (zero IF and low IF) to ensure that the integer-N type MMD can cover all 16 ZigBee channels. Out-of-channel interferers are filtered by the FBAR RF filter and configurable IF filter simultaneously.

Figure 21.7.3 shows the proposed current-reuse balun LNA. Unlike active-gain-boost CG-CS LNA structures [5], we stacked the PFETs (M_5-M_8) above the NFETs (M_1-M_4) to further increase the current efficiency (g_m/I_d). We feed back the RF signal to M_3 and M_5 by using two stacked inverter-based amplifiers (A_x) to reduce R_{in} and to improve NF while keeping two branch currents small and equal. The RF input impedance is given by $R_{in}=R_P//R_N\approx 1/(2\times g_{m3}(1+A_P))$. We also feed forward the RF signal into M_4 and M_6 to reduce the differential signal imbalance. Simulated results show that the gain and phase imbalance of the proposed LNA are improved by 5.6dB and 2.2 degrees, respectively. Due to the severe headroom limitation, we bias the low V_t devices (M_1-M_8) in moderate inversion to optimize the trade-off between their g_m/I_d and f_T. We employ a self-biased common-mode feedback (CMFB) to stabilize the DC level [7]. Ideally, it generates finite common-mode resistance ($1/2\times g_{mc}$) and infinite differential resistance ($1/(g_{mc}-g_{mc})$) at the output nodes.

Figure 21.7.4 shows the proposed hybrid mixer and the PLL-free LO generator. We use a switched-g_m type mixer to achieve the first conversion. Compared to its Gilbert-type counterpart, it reduces the requirement of V_{dd} and LO amplitude. To further save power, we employed two techniques in the proposed mixer: 1) we stacked the second conversion mixer (switch stage) at the top of the switched-g_m mixer for current reuse, 2) I and Q signals in second conversion share the same switched-g_m cell. The 25% duty-cycle 4-phase second LO signals avoid the impedance loading between I and Q branches. The first LO signal induces only common-mode noise at the outputs that can be rejected differentially.

The first LO is a fixed-frequency oscillator utilizing a 0.1mm^2 temperature-compensated FBAR. We use a differential Colpitts oscillator, whose center frequency is determined by $\omega=\sqrt{((g_mL_X+R_L(C_1+C_2)))/(L_XR_LC_1C_2)}$, where L_X represents the motional inductance of the FBAR. Low V_t and triple-well NMOS transistors are used for M_5 and M_6 to maximize the transconductance and decrease the DC power. The total power consumption used in the oscillator is 660μW. The measured phase noise at 3.5MHz offset frequency is -144dBc/Hz with an amplitude of 250mV (peak-to-peak).

Figure 21.7.5 shows the measured receiver performance. The $|S_{11}|$ -10dB bandwidth is 0.7GHz looking directly into the chip. At the PGA output, we measure an overall gain of 57.8dB and a total NF of 15.7dB. The voltage gain is tunable for more than 50dB. We achieve an IIP_3 of -18.5dBm at a moderate gain of 25dB. The configurable filter response shows 41dB (44dB) rejection at the adjacent (alternate) channel. It also provides 40dB (37dB) image-rejection at the first (second) downconversion.

The chip is fabricated in a 65nm CMOS process. It occupies a core area of 0.45 mm^2. A performance summary and comparison with the state-of-the-art is presented in Fig. 21.7.6. Based on the SNR requirement of the ZigBee standard (7dB SNR at the demodulator for a 2MHz channel bandwidth), this receiver achieves -88dBm sensitivity. The CMOS die micrograph is shown in Fig. 21.7.7. This radio achieves the lowest power consumption compared to all other ZigBee radios in the table.

References:

[1] Y. Liu, et al., "A 1.9nJ/b 2.4GHz Multistandard (Bluetooth Low Energy/ZigBee/IEEE802.15.6) Transceiver for Personal/Body-Area Networks," *ISSCC Dig. Tech. Papers*, pp. 246-247, Feb. 2013.

[2] F. Zhang, et al., "A 1.6mW 300mV-Supply 2.4GHz Receiver with -94dBm Sensitivity for Energy-Harvesting Applications" *ISSCC Dig. Tech. Papers*, pp. 256-257, Feb. 2013.

[3] A. Liscidini, et al., "A 2.4GHz 3.6mW 0.35mm² Quadrature Front-End RX for ZigBee and WPAN Applications," *ISSCC Dig.Tech. Papers*, pp. 370-371, Feb., 2008.

[4] A. Heragu, et al., "A Low-Power BAW-Resonator-Based 2.4GHz Receiver with Bandwidth-Tunable Channel Selection Filter at RF," *IEEE J. Solid-State Circuits*, vol. 48, pp. 1343-1356, June 2013.

[5] Z. Lin, et al., "A 1.7mW 0.22mm² 2.4GHz ZigBee RX Exploiting a Current-Reuse Blixer + Hybrid Filter Topology in 65nm CMOS", *ISSCC Dig. Tech. Papers*, pp. 448-449, Feb. 2013.

[6] 802.15.4: Part 15.4: Wireless Medium Access Control (MAC) and Physical Layer (PHY) Specifications for Low-Rate Wireless Personal Area Networks (LR-WPANs), 2006. <http://standards.ieee.org/getieee802/download/802.15.4-2006.pdf>

[7] B. Nauta, "A CMOS Transconductance-C Filter Technique for Very High Frequencies", *IEEE J. Solid-State Circuits*, vol. 27, pp. 142-153, Feb. 1992.

[8] A. Balankutty, et al., "A 0.6V Zero-IF/Low-IF Receiver with Integrated Fractional-N Synthesizer for 2.4GHz ISM-Band Applications," *IEEE J. Solid-State Circuits*, vol. 45, pp. 538-553, Mar. 2010.

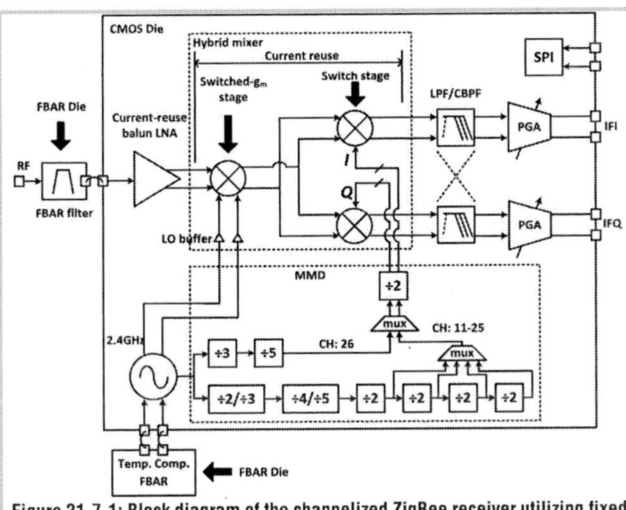

Figure 21.7.1: Block diagram of the channelized ZigBee receiver utilizing fixed LO temperature-compensated FBAR resonator.

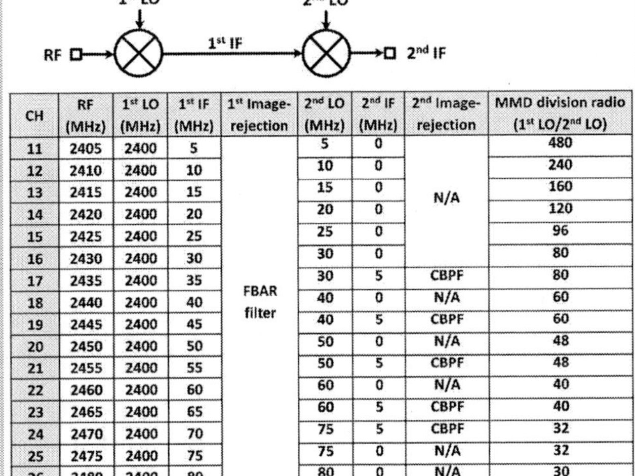

CH	RF (MHz)	1st LO (MHz)	1st IF (MHz)	1st Image-rejection	2nd LO (MHz)	2nd IF (MHz)	2nd Image-rejection	MMD division radio (1st LO/2nd LO)
11	2405	2400	5	FBAR filter	5	0	N/A	480
12	2410	2400	10		10	0		240
13	2415	2400	15		15	0		160
14	2420	2400	20		20	0		120
15	2425	2400	25		25	0		96
16	2430	2400	30		30	0		80
17	2435	2400	35		30	5	CBPF	80
18	2440	2400	40		40	0	N/A	60
19	2445	2400	45		40	5	CBPF	60
20	2450	2400	50		50	0	N/A	48
21	2455	2400	55		50	5	CBPF	48
22	2460	2400	60		60	0	N/A	40
23	2465	2400	65		60	5	CBPF	40
24	2470	2400	70		75	5	CBPF	32
25	2475	2400	75		75	0	N/A	32
26	2480	2400	80		80	0	N/A	30

Figure 21.7.2: Frequency plan and channelization.

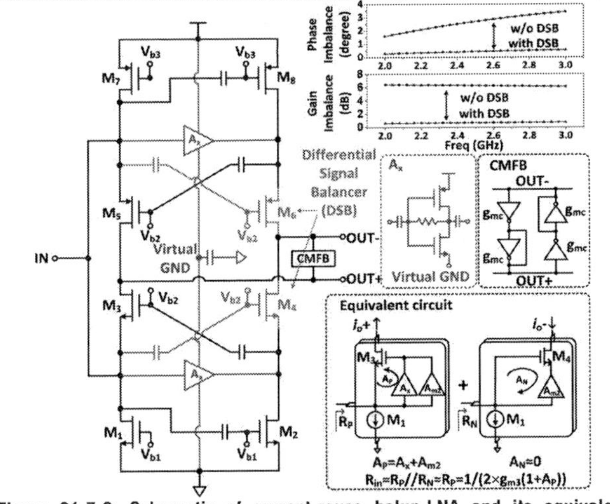

Figure 21.7.3: Schematic of current-reuse balun-LNA and its equivalent circuit.

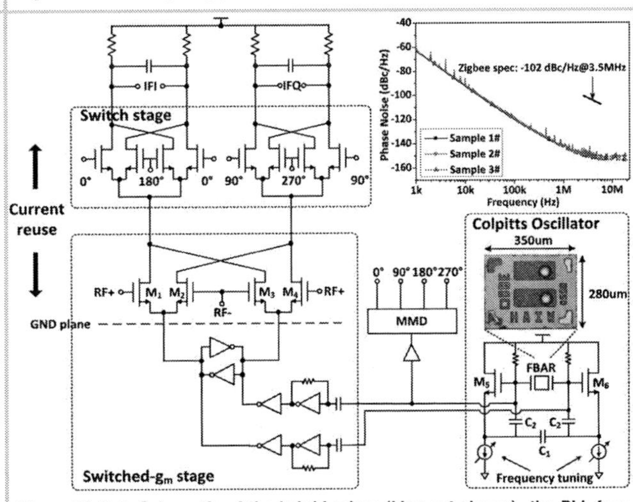

Figure 21.7.4: Schematic of the hybrid mixer (bias not shown), the PLL-free LO generator, and the measured phase noise.

Figure 21.7.5: Measured $|S_{11}|$, voltage gain and NF, IIP3, filter response.

Performance Comparison

	This work	[1] ISSCC '13	[3] ISSCC '08	[4] JSSC '13	[5] ISSCC '13	[8] JSSC '10
Standard	Zigbee	Zigbee/BT/ MBAN	Zigbee	NA	Zigbee	Zigbee/BT
Architecture	Sliding-IF RX+LO	Sliding-IF RX+TX+PLL	Low-IF RX	Sliding-IF RX+PLL	Low-IF RX	Low-IF/Zero-IF RX+PLL
XTAL	No	Yes	N/A	No	N/A	Yes
PLL-free	**Yes**	No	No	No	No	No
On-chip Inductors	**No**	Yes	Yes	Yes	Yes	Yes
Supply [V]	1	1.2	1.2	1.8	1.2	0.6
RX Power [mW]	**0.86**	>1.6	3.6	2.4	1.7	20
PLL/LO Power [mW]	**0.92**	1.8	N/A	15.4	N/A	12.5
Gain [dB]	57.8	NA	75	44.2	57	67
NF [dB]	15.7	6	12	14.8	8.5	16
IIP_3 [dBm]	-18.5	-19	-12.5	-28	-6	-10.5
Image-rejection [dB]	**1st = 40dB 2nd = 37dB**	35	35	N/A	36	32
PN @ 3.5MHz [dBc/Hz]	**-144**	N/A	-107	N/A	N/A	-127
Technology	65-nm	90-nm	90-nm	180-nm	65-nm	90-nm
Active Area [mm²]	0.45	2	0.35	N/A	0.22	1.45

Power Breakdown [mW]

LNA	Mixer	Filters	PGA	FBAR OSC	Dividers	Total
0.51	0.20	0.11	0.04	0.66	0.26	1.78

Figure 21.7.6: Performance summary and comparison.

21

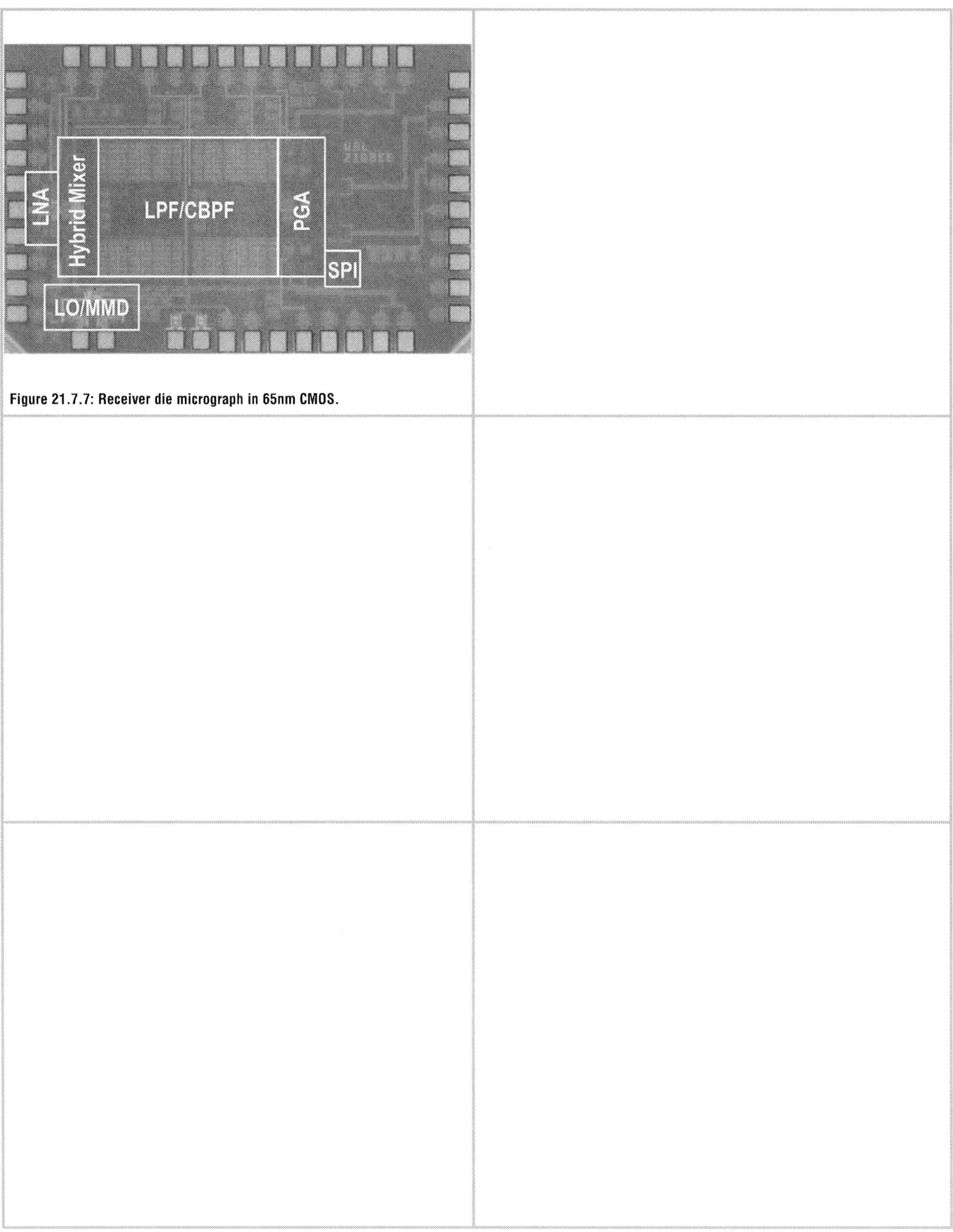

Figure 21.7.7: Receiver die micrograph in 65nm CMOS.

21.8 A Pulling Mitigation Technique for Direct-Conversion Transmitters

Ahmad Mirzaei, Mohyee Mikhemar, Hooman Darabi

Broadcom, Irvine, CA

Despite versatility and low power consumption, direct-conversion transmitters suffer from a fundamental drawback: the local oscillator disturbance by the power amplifier, through unwanted electromagnetic or capacitive coupling [1,2]. As shown in Fig. 21.8.1, the *pulled* oscillator spectrum is no longer a single-tone sinusoid, which can drastically degrade the transmitter EVM or spectrum mask. To alleviate this, *time-consuming* and often *unpredictable* optimization of the floor plan, package, and PCB is required to maximize the isolation between the PA and VCO. Ultimately, this issue may prohibit the use of this architecture for many applications, leading to higher power consumption. Moreover, in many modern radios it is common to have more than one VCO on-chip (Fig. 21.8.1) to support various features such as FDD, carrier aggregation, or coexistence, further exacerbating the problem through multiple-VCO cross-coupling. To address these concerns we propose a calibration scheme that corrects any pulling effect regardless of its source or magnitude. Our approach is *fully digital* and is *automatically calibrated*, leading to a reliable and robust solution, and has little impact on power consumption, size, or transmitter noise.

Figure 21.8.2 describes the proposed correction scheme in a linear transmitter with the VCO operating at double the frequency. Besides the fundamental, the mixer upconverts the baseband signal to all odd harmonics of the LO [3]. The mixer output passes through the PA and its driver, whose even-order nonlinearities create modulated frequency components around the VCO frequency. Due to limited isolation, these components leak into the VCO and pull it. The phasor of the final pulling element is generally given as:

$$(1)\quad \gamma_+ e^{j\psi_+}(x_{BB,I}(t)+jx_{BB,Q}(t))^2\times e^{j\theta}+\gamma_- e^{j\psi_-}(x_{BB,I}(t)-jx_{BB,Q}(t))^2\times e^{j\theta}$$

where θ is instantaneous phase of the VCO and $x_{BB,I}+jx_{BB,Q}$ is the complex baseband signal. The constants $\gamma_+exp(j\psi_+)$ and $\psi_-exp(j\psi_-)$ are two complex numbers that are functions of even-order nonlinearities, isolation from the VCO, and high-frequency phase shifts. Practically, only the first term is dominant and the second term can be safely ignored. Using Adler's generalized differential equation and the first term of (1) as the injection signal [4], it can be mathematically proven that there is always a correction voltage that can be added to the VCO control voltage to counterbalance the pulling effect from all aforementioned components, given as follows:

$$(2)\quad v_{correction}(t)=\frac{\omega_0}{2Q}\frac{1}{K_{VCO}}\frac{m_1(x_{BB,I}^2(t)-x_{BB,Q}^2(t))+m_2(2x_{BB,I}(t)x_{BB,Q}(t))}{1+m_2\left(x_{BB,I}^2(t)-x_{BB,Q}^2(t)\right)-m_1(2x_{BB,I}(t)x_{BB,Q}(t))}$$

where Q is the quality factor of the LC tank used in the TX VCO and K_{VCO} is the VCO gain. Fortuitously, the correction voltage in (2) is a function of the transmitted complex baseband signal $x_{BB,I}+jx_{BB,Q}$, which is known. The two constants m_1 and m_2 are proportional to the injection strength and are found through a calibration scheme described later. As most of synthesizers are fractional-N in practice, instead of modulating the control voltage directly, the calibration signal in (2) is applied by DSP through a $\Delta\Sigma$ modulator along with channel selection. No matter how strong the injection components are, and regardless of the PLL dynamics or type of modulation, the proposed scheme always corrects the VCO pulling.

Similarly, when two VCOs pull each other (Fig. 21.8.3), they produce sidebands separated from each other by a *beat frequency* equal to the difference between the frequencies of the two un-pulled VCOs. It can be shown that two correction voltages whose frequencies are equal to the beat frequency can be added through the $\Delta\Sigma$ modulator to counterbalance the mutual pulling between the two VCOs entirely. The proposed approach can be readily generalized to any number of locked VCOs concurrently operating on the same die.

As pulling strongly varies with process and temperature, as well as antenna mismatch, and is often a strong function of the package and platform design, it must be *automatically detected* and *periodically corrected*. The proposed *auto-calibration* scheme is illustrated in Fig. 21.8.4.

Suppose the complex baseband signal is chosen to be a tone at a frequency offset of ω_m, small enough to pass through the TX-LPF. The quadrature sequence of the complex baseband signal is assumed such that the upconverted RF signal resides at $\omega_{LO}+\omega_m$, where ω_{LO} is the TX-LO frequency. The upconversion mixer can also create higher-order frequency components namely at $3\omega_{LO}-\omega_m$, $5\omega_{LO}+\omega_m$, etc. The second-order nonlinearity of the subsequent power amplifier would create frequency components at $2\omega_{LO}+2\omega_m$ and $2\omega_{LO}-2\omega_m$, where both can pull the oscillator within the PLL bandwidth. The resulting injection pulling generates phase-modulated sidebands at frequency offsets equal to $\pm2\omega_m$ away from the VCO center frequency $2\omega_{LO}$. These phase-modulated sidebands would also appear at the output of the multi-modulus divide-by-N inside the PLL, only attenuated by $20logN$ with respect to the main carrier. Using a digital XOR, these sidebands are utilized for the pulling calibration, where the divider output is mixed with the *PLL reference* to frequency-shift the sidebands to $2\omega_m$. The XOR output is then filtered and digitized, and the calibration algorithm searches for a set of m_1 and m_2 to zero the detected signal at $2\omega_m$. In the general case of a modulated input at baseband, the XOR output is a modulated spectrum itself with approximately twice the bandwidth. In applications where the transmitter must operate continuously such as 3G, the *integrated power* of the XOR output then simply serves as a measure of pulling. Without interfering with the normal TX operation, this output is monitored and minimized to reduce pulling. Furthermore, this on-line calibration can be enabled with a duty cycle to cover temperature variations. The SNR of the detected signal sets the accuracy of the pulling calibration. The integrated noise floor at the XOR output must be below the corrected pulling sideband, so that the pulling mitigation is effective. The values found for m_1 and m_2 are used to compensate the pulling effect when the TX sends actual modulated data. It can be shown that only two *one-dimensional* (and not a two-dimensional) searches are needed to find m_1 and m_2, significantly lowering the calibration time. Shown in Fig. 21.8.4, a similar detection mechanism is applied to calibrate the mutual pulling between two PLLs, where two XORs, one for each PLL, detect the two tones at the beat frequency caused by the mutual pulling. The calibration algorithm searches for four constants, which are amplitudes and phases of the two sinusoidal signals at the beat frequency. For this case, it can be shown that only four simple one-dimensional searches are required.

For the proof of concept, a test chip was fabricated in 40nm CMOS (Fig. 21.8.7). It comprises two identical PLLs with the LC-VCOs laid out close to each other for maximum mutual pulling, as well as one direct conversion transmitter that obtains its LO from only one of the PLLs. Designed with a 25%-duty-cycle passive mixer followed by an on-chip PA driver, the transmitter can put out 4dBm modulated power. There is an on-chip block with variable gain and strong second-order nonlinearity connected to the very last output of the transmitter. The generated RF signal around the second-order harmonic is routed toward the TX VCO with a half turn around its differential inductor to be intentionally injected to the VCO through electromagnetic coupling. This leads to a strong injection pulling. Figure 21.8.5 shows the modulation spectrum of the transmitted 16-QAM signal before and after calibration. The EVM improves from 11% to 2.4%, while the modulation mask improves by over 15dB, almost identical to the case the pulling injection is turned off. To measure mutual pulling correction between two VCOs, DC signals are applied to the transmitter inputs, resulting in a single tone at the TX output, and the two PLL frequencies are set 3MHz apart. Figure 21.8.6 shows transmitter output before and after calibration. Prior to calibration, the strong sidebands around 20dBc at 3MHz offset indicate that the two PLLs heavily pull each other, whereas in the post-calibration output spectrum, the sidebands are suppressed by over 20dB, only limited by the sync accuracy of the external references to the two PLL's.

References:

[1] I. Bashir, et al., "A Novel Approach for Mitigation of RF Oscillator Pulling in a Polar Transmitter," *IEEE J. Solid-State Circuits*, pp. 403-415, 2011.

[2] C.H. Hsiao, et al., "Direct-Conversion Transmitter with Resistance to Local Oscillator Pulling in Non-Constant Envelope Modulation Systems," *IEEE International Microwave Symposium*, pp. 1-4, 2011.

[3] A. Mirzaei, et al., "Analysis of Direct-Conversion IQ Transmitters With 25% Duty-Cycle Passive Mixers," *IEEE Transactions on Circuits and Systems I: Regular Papers*, pp. 2318-2331, 2011.

[4] A. Mirzaei, et al., "Analysis of Oscillators Locked by Large Injection Signals: Generalized Adler's Equation and Geometrical Interpretation," *IEEE Custom Integrated Circuits Conference*, pp. 737-740, 2006.

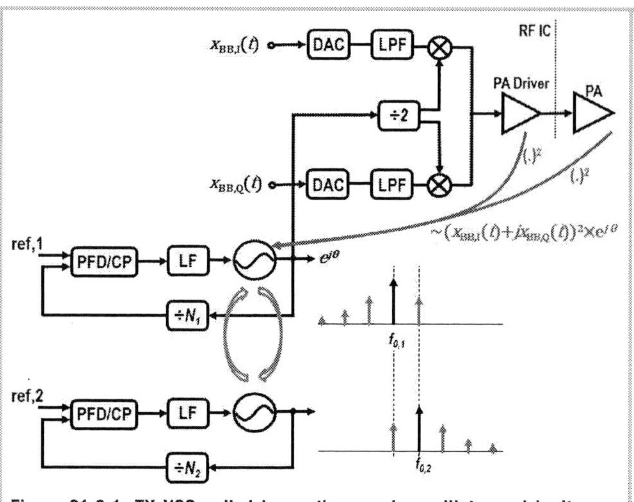

Figure 21.8.1: TX VCO pulled by another nearby oscillator and by its own transmitter.

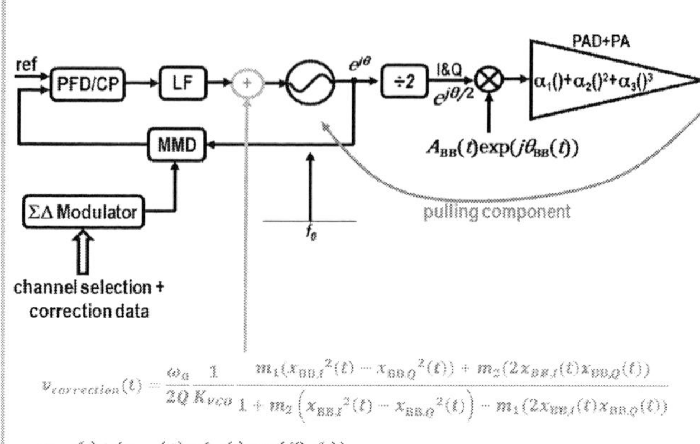

Figure 21.8.2: Pulling elimination by modifying VCO control voltage.

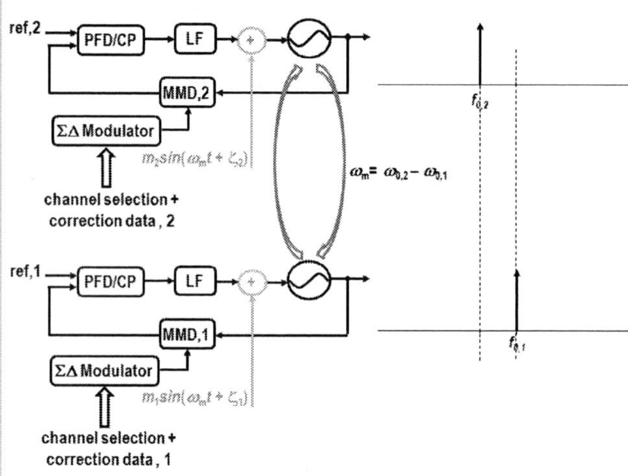

Figure 21.8.3: Elimination of mutual pulling by modifying VCO control voltages.

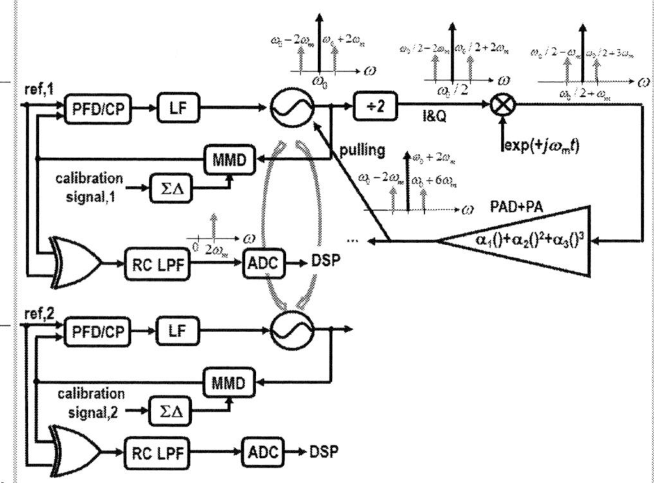

Figure 21.8.4: Detection mechanism in calibration of pulling.

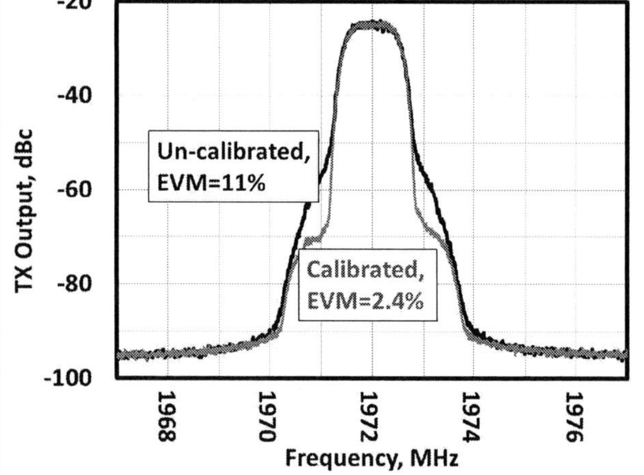

Figure 21.8.5: Modulation spectrum of the transmitted signal before and after calibration.

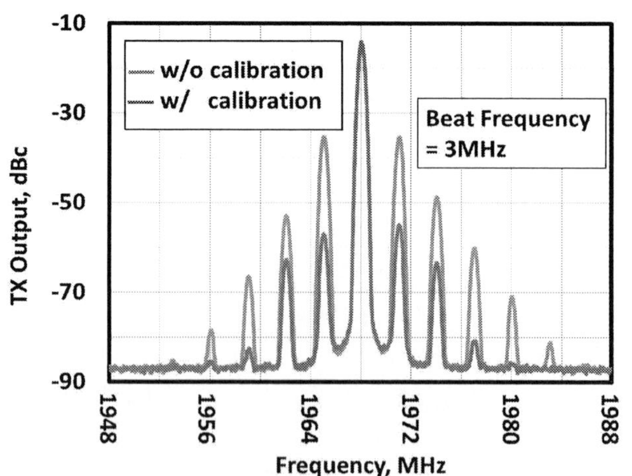

Figure 21.8.6: Transmitted signal for DC baseband inputs before and after calibration.

978-1-4799-0917-9/14 $31.00 © 2014 IEEE

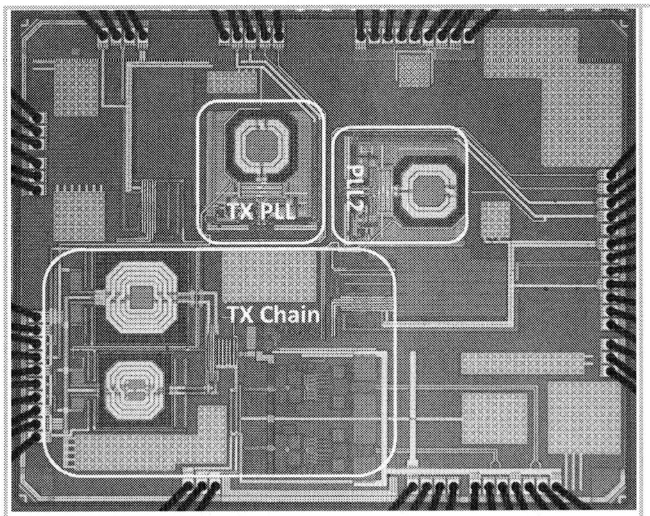

Figure 21.8.7: Die micrograph.

Session 22 Overview: *High-Speed Data Converters*

DATA CONVERTERS SUBCOMMITTEE

Session Chair: *Jieh-Tsorng Wu*
National Chiao-Tung University, Hsinchu, Taiwan

Session Co-Chair: *Seung-Tak Ryu*
KAIST, Daejeon, Korea

This session demonstrates design techniques to realize data converters with unprecedented combinations of speed, resolution, and power efficiency in advanced CMOS technologies. Papers in this session include a time-interleaved ADC with a sampling rate up to 90GS/s, a time-interleaved DAC at 4.6GS/s conversion rate, and a time-based 2.2GS/s ADC. These converters are essential for systems enhanced by digital signal processing, such as optical communications, wireline communications, broadband satellite receivers, and cable systems.

22.1 A 90GS/s 8b 667mW 64× Interleaved SAR ADC in 32nm Digital SOI CMOS **8:30 AM**

L. Kull, IBM Research, Rüschlikon, Switzerland and EPFL, Lausanne, Switzerland

In Paper 22.1, IBM Research (with EPFL) presents a 90GS/s 8b time-interleaved ADC in 32nm SOI CMOS. It contains 64 SAR ADC channels. It achieves above 33dB SNDR up to 19.9GHz input frequency and consumes 667mW from a 1.2V supply.

22.2 A 69.5mW 20GS/s 6b Time-Interleaved ADC with Embedded Time-to-Digital Calibration in 32nm CMOS SOI **9:00 AM**

V. H-C. Chen, Carnegie Mellon University, Pittsburgh, PA

In Paper 22.2, Carnegie Mellon University presents a 20GS/s 6b time-interleaved ADC in 32nm SOI CMOS. It contains 8 flash ADC channels and features on-chip calibration to reduce inter-channel mismatches in the background. It achieves 30.7dB SNDR up to Nyquist and consumes 69.5mW from a 0.9V supply.

22.3 A 20GHz-BW 6b 10GS/s 32mW Time-Interleaved SAR ADC with Master T&H in 28nm UTBB FDSOI Technology **9:30 AM**

S. Le Tual, STMicroelectronics, Crolles, France

In Paper 22.3, STMicroelectronics presents a 10GS/s 6b time-interleaved ADC in 28nm ultra thin body and BOX fully depleted SOI CMOS. It contains 8 SAR ADC channels. It uses a master T/H that enables 20GHz input sampling without the need for timing skew calibration. It achieves 4.6 ENOB at 20GHz input and consumes 32mW from a 1V supply.

22.4 A 1GS/s 10b 18.9mW Time-Interleaved SAR ADC with Background Timing-Skew Calibration **10:15 AM**

S. Lee, Massachusetts Institute of Technology, Cambridge, MA

In Paper 22.4, MIT presents a 1GS/s 10b time-interleaved ADC in 65 nm CMOS. It contains 8 SAR ADC channels. It features a full-rate 4b flash ADC for sub-ranging and timing skew calibration. It achieves 51.4dB SNDR at Nyquist and consumes 18.9mW from a 1V supply.

22.5 A 1.62GS/s Time-Interleaved SAR ADC with Digital Background Mismatch Calibration Achieving Interleaving Spurs Below 70dBFS **10:45 AM**

N. Le Dortz, STMicroelectronics, Crolles, France and Supélec, Gif-sur-Yvette, France

In Paper 22.5, STMicroelectronics (with Supélec) presents a 1.62GS/s 9b time-Interleaved ADC in 40nm CMOS. It contains 12 SAR ADC channels. It uses a digital calibration unit to estimate and correct inter-channel mismatches in the converter backend. It achieves 48dB SNDR and demonstrates interleaving spurs below -70dBFS up to 750MHz input frequency, while consuming a total power of 93mW.

22.6 A 2.2GS/s 7b 27.4mW Time-Based Folding-Flash ADC with Resistively Averaged Voltage-to-Time Amplifiers **11:15 AM**

M. Miyahara, Tokyo Institute of Technology, Tokyo, Japan

In Paper 22.6, the Tokyo Institute of Technology presents a 2.2GS/s 7b time-based folding ADC in 40nm CMOS. With no need for calibration, it achieves 37.4dB SNDR at Nyquist. It consumes 27.4mW from a 1.1V supply.

22.7 A 14b 4.6GS/s RF DAC in 0.18µm CMOS for Cable Head-End Systems **11:45 AM**

D. McMahill, Maxim Integrated, Woodstock, GA

In Paper 22.7, Maxim Integrated presents a 4.6GS/s 14b time-interleaved DAC in 0.18µm CMOS. It contains two current-steering DAC channels. It provides 80mA of output current and achieves -79dBc IMD with a 500MHz output without trimming or calibration. It consumes a total power of 2.3W.

22.1 A 90GS/s 8b 667mW 64× Interleaved SAR ADC in 32nm Digital SOI CMOS

Lukas Kull[1,2], Thomas Toifl[1], Martin Schmatz[1], Pier Andrea Francese[1], Christian Menolfi[1], Matthias Braendli[1], Marcel Kossel[1], Thomas Morf[1], Toke Meyer Andersen[1], Yusuf Leblebici[2]

[1]IBM Research, Rüschlikon, Switzerland,
[2]EPFL, Lausanne, Switzerland

Forthcoming optical communication standards such as ITU OTU-4 and 100/400Gb/s Ethernet require ADCs with more than 50GS/s and at least 5 ENOB to enable complex equalization in the digital domain. SAR ADCs and interleaved ADCs made impressive progress in recent years. First CMOS ADCs with at least 6b and conversion rates exceeding 20GS/s were presented [1-3], proving that interleaved SAR ADCs are an optimal choice for high-speed ADCs with moderate resolution. We present an interleaved CMOS ADC architecture based on an asynchronous redundant SAR ADC core element. It was measured up to a sampling rate of 100GS/s and can be operated from a single supply voltage. At 90GS/s, the measured SNDR stays above 36.0dB SNDR up to 6.1GHz and 33.0dB up to 19.9GHz input frequency while consuming 667mW. The ADC is implemented in 32nm digital SOI CMOS and occupies 0.45mm².

Figure 22.1.1 shows a top-level overview of the ADC. The differential input is terminated by 2×50Ω, protected with reduced ESD-diodes and directly connected to the 4 sampling and interleaving slices that feed buffered samples to 16 sub-ADCs. Therefore a total of 64 sub-ADCs convert the analog samples. The aggregated digital output is captured by a large high-speed memory block storing 8192 digitized samples. The architecture requires only 4 timing-critical clock phases, namely those connected to the first input sampling switches. These critical phases are derived from a half-rate differential clock *ck2* of up to 50GHz, which is divided by 2 in CML and converted to CMOS levels. An externally controlled digital signal *dskew* serves to adjust the skew between the 4 phases with a step size of less than 30fs. A separate sub-clock generation block receives the 4 clock phases and generates all non-timing-critical sub-clocks for the interleaver and sub-ADCs.

The core of the ADC consists of the sampling stage and interleaving architecture, as shown in Fig. 22.1.2. Sampling is implemented in voltage mode. Each of the 4 sampling switches M_1 is connected in series with a 1 by 4 demux stage, formed by transistors M_2, before connecting to the sampling capacitors C_S. Implementing 4 interleaved sampling switches and a 1:4 demux stage proved to be optimum for highest bandwidth while still providing sufficient hold time on the sampling capacitor to buffer the sampled voltage. Single NMOS sampling switches are fast and provide sufficient linearity for an 8b ADC. Size and operating point of the sampling transistors are optimized for high bandwidth and high linearity across corners, with operating temperatures up to 100°C. One of the 4 demux switches is enabled by *en16* before the rising edge of *ck4* and disabled after the falling edge of *ck4* to eliminate influences of *en16* on the sampling window. The sampling capacitor C_S is reset shortly before the sampling window by *res16*. The signal *en16* is enabled before *res16* is disabled to eliminate ISI by canceling remaining charges on transistor M_1 from the preceding sampling phase. This leaves about 120ps hold time of the sampled voltage on C_S. A source follower (M_3 and M_4) is chosen to buffer the sampled voltage because of its superior speed, noise figure and linearity. For high linearity, the source follower is operated with an output common mode of close to half the supply voltage. Its output common mode also defines the common mode of the sub-ADC and therefore the comparator. Control of the comparator common mode enables a good trade-off between conversion speed and comparator input-referred noise [4]. The buffered voltage is connected through a second demux stage (M_5) controlled by *en64* to the capacitive DAC of a SAR ADC. Signal *en64* is also used to trigger the conversion of the asynchronous SAR ADC. Both demux stages in the interleaver feature cross-coupled NMOS switches, with gates connected to GND to cancel signal feed-through.

The asynchronous SAR ADC shown in Fig. 22.1.3 features a redundant capacitive DAC with a constant common mode and alternate comparators for enhanced speed. The SAR ADC is described in detail in [4]. A falling edge on *en64* starts the asynchronous conversion and a rising edge of *en64* resets the ADC if it did not finish. The alternate comparators eliminate the reset phase of the comparator from the critical path, thus increasing speed by about 30%. While one comparator is active, the other comparator resides in reset state, and vice versa. The capacitive DAC is reset at the end of the conversion to eliminate ISI and zero the offset of the comparators. A low-power switch-capacitor reference buffer [5] is controlled with *Vgain* from an R-3R ladder to adjust the gain of each SAR ADC. The digital input *dgain* is set externally and enables per-channel gain calibration.

The ADC is manufactured in a 32nm SOI CMOS process with an area of 470x960μm². The interleaver/sampler and 64 SAR ADCs occupy 370x960μm² and the clock divider 90x100μm². The ADC supply on the interleaver, including sampler, CML clock divider and CML to CMOS stages (V_{DI}), is 1.2V for 90GS/s, and the SAR ADC supply (V_{DA}) is set to the same level for 90GS/s. To save power, V_{DA} can be lower than V_{DI} for lower conversion rates. At 1.2V and 90GS/s, the measured total power consumption of 667mW consists of 56mW for the CML clock divider and CML to CMOS stages, 112mW for the interleaver/sampler and sub-clock generation, and 499mW for the SAR ADCs including drivers for the memory block.

Gain calibration is performed off-chip with on-chip fine-grain adjustment on the R-3R ladders of each SAR ADC. The internal offset of the comparators is corrected in background in each SAR ADC, whereas the residual offset between the ADC channels is subtracted off-chip. For each supply voltage and conversion rate, gain and offset are calibrated only once at 2.1GHz input frequency. For better sensitivity, skew is calibrated once at a higher input frequency of 19.9GHz. Bandwidth mismatch is not calibrated.

Figure 22.1.4(a) shows SNDR vs. input frequency. More than 36.0dB is achieved up to 6.1GHz and 33.0dB up to 19.9GHz. The measurement series at 100GS/s is taken with skew calibration disabled, therefore the SNDR at higher input frequencies is limited by phase mismatch of the 4 input clock phases. With skew calibrated, lower SNDR at higher input frequencies mainly stem from the reduced amplitude due to the limited bandwidth, as shown in Fig. 22.1.4(b). Total jitter of individual measurements (8192 samples), including the external clock and input signal generators, is estimated to approx. 60fs based on modulo-time plot analysis and SNR comparison at different input frequencies at 90GS/s. This is in good agreement with jitter simulations that depending on corners, predict 30 to 50fs jitter. The ADC is measured with sampling frequencies from 56GHz to 100GHz (see Fig. 22.1.5(a)). Owing to the asynchronous design of the sub-ADC, the SNDR decreases gradually when some SAR cycles are no longer completed. Figure 22.1.5(b) shows the power consumption vs. sampling frequency. As can be seen at 70GS/s, power consumption highly depends on V_{DA}. 100GS/s was achieved at 1.27V on V_{DI} and V_{DA} with a reduced reset time of the capacitive DAC inside the SAR ADCs at the end of the conversion cycle. Best FoM is achieved at 70GS/s with 121fJ/conversion-step. FoM at 90GS/s is higher with 203fJ/conversion-step, mainly because of the increased voltage on the SAR ADCs.

SFDR of 41.4dB at 19.9GHz input frequency and full-scale input amplitude is limited by 3rd-order harmonic distortion (see Fig. 22.1.6(a)). Figure 22.1.6(b) compares the performance of high-speed CMOS ADCs with at least 6b resolution. As shown in the comparison table, this design exhibits the highest sampling frequency of previously reported 6b+ CMOS ADCs. The FoM is more than 50% lower and the technology-adjusted area is 4 times smaller than other 6b+, >20GS/s ADCs [1,7].

References:

[1] Fujitsu Semiconductor Europe, LUKE-ES 55–65 GSa/s 8 bit ADC, Mar. 2012, http://www.fujitsu.com/downloads/MICRO/fme/documentation/c63.pdf, Accessed Sept. 2013.

[2] Y. M. Greshishchev, *et al.*, "A 40GS/s 6b ADC in 65nm CMOS," *ISSCC Dig. Tech. Papers*, pp. 390–391, Feb. 2010.

[3] P. Schvan, *et al.*, "A 24GS/s 6b ADC in 90nm CMOS," *ISSCC Dig. Tech. Papers*, pp. 544–545, Feb. 2008.

[4] L. Kull, *et al.*, "A 3.1mW 8b 1.2GS/s Single-Channel Asynchronous SAR ADC with Alternate Comparators for Enhanced Speed in 32nm Digital SOI CMOS," *ISSCC Dig. Tech. Papers*, pp. 468–469, Feb. 2013.

[5] L. Kull, *et al.*, "A 35mW 8b 8.8GS/s SAR ADC with Low-Power Capacitive Reference Buffers in 32nm Digital SOI CMOS," *IEEE Symp. VLSI Circuits*, pp. 260–261, June 2013.

[6] E. Z. Tabasy, *et al.*, "A 6b 10GS/s TI-SAR ADC with Embedded 2-Tap FFE/1-Tap DFE in 65nm CMOS," *IEEE Symp. VLSI Circuits*, pp. 274-275, June 2013.

[7] B. Murmann, "ADC Performance Survey 1997-2013," [Online]. Available: http://www.stanford.edu/~murmann/adcsurvey.html.

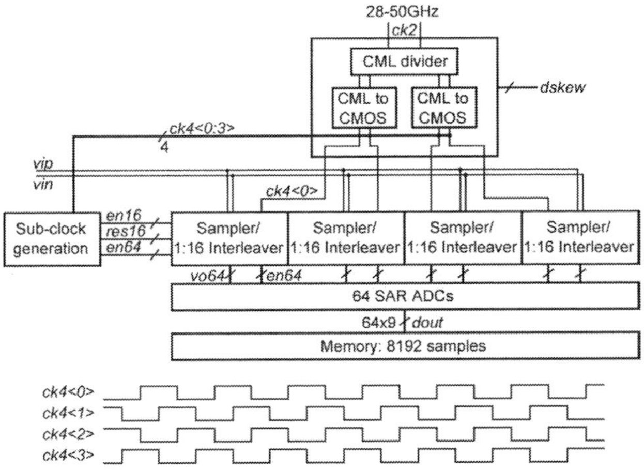

Figure 22.1.1: Architecture of the highly interleaved SAR ADC with a timing diagram of the clock signals defining the sampling time.

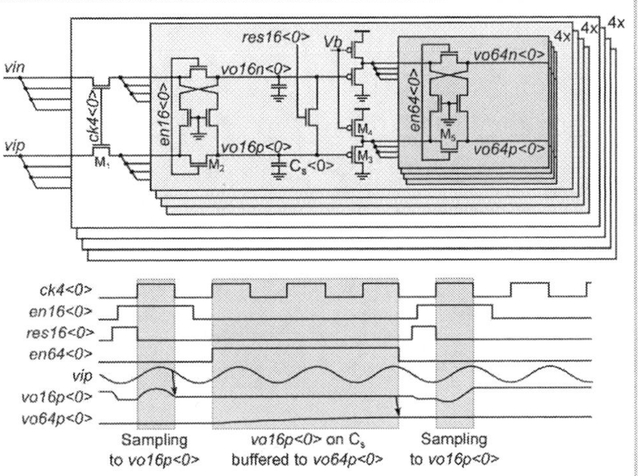

Figure 22.1.2: Schematic details of the differentially implemented 1:64 interleaver.

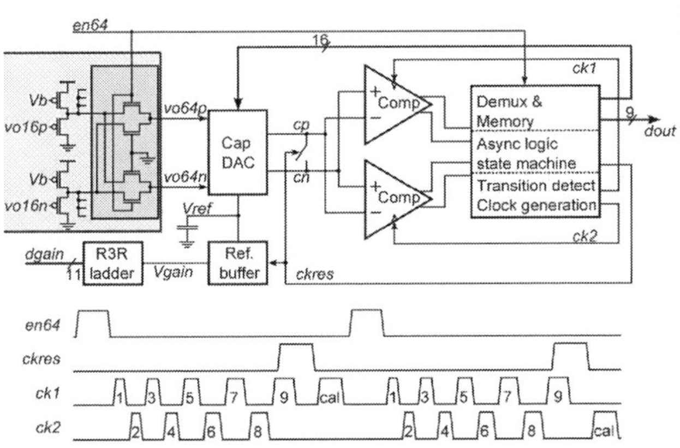

Figure 22.1.3: SAR ADC architecture [4] with last demux stage of the interleaver and corresponding timing diagram.

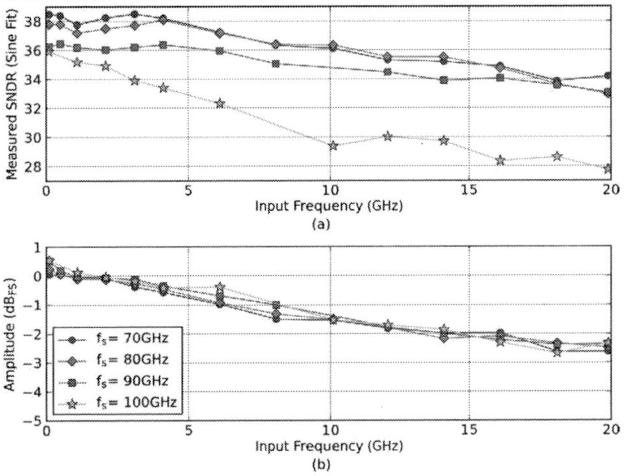

Figure 22.1.4: Measured SNDR and amplitude vs. input sine frequency for different sampling frequencies referred to 2.1GHz input frequency.

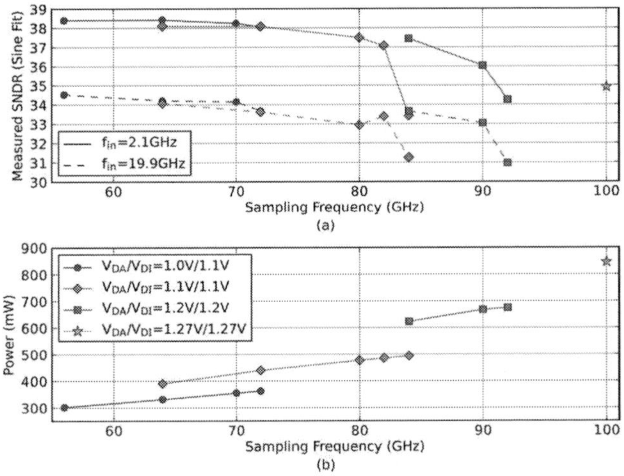

Figure 22.1.5: SNDR and power vs. sampling frequency for different supply voltages.

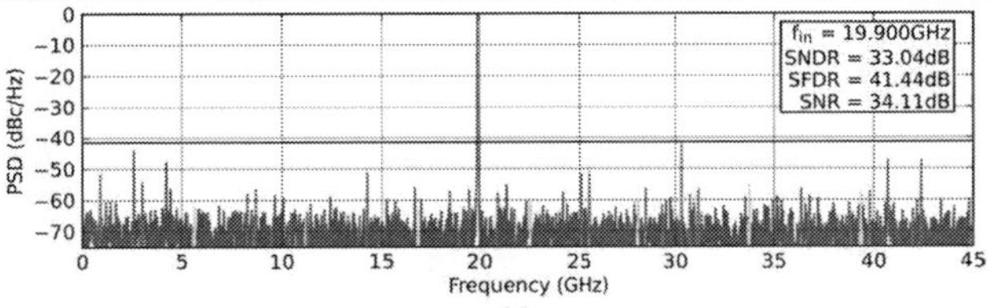

Specifications	[1]	[2]	[3]	[6]	[5]	This work			
Architecture	Ti-SAR	Ti-SAR	Ti-SAR	Ti-SAR	Ti-SAR	Ti-SAR			
CMOS Technology (nm)	40	65	90	65	32	32			
Resolution (bits)	8	6	6	6	8	8			
Sampling Speed (GHz)	65	40	24	10	8.8	70	80	90	100
Supply Voltage (V)	±0.9/1.8	1.0/1.2*	1.0/2.5*	0.9/1.1*	1.0	1.0/1.1*	1.1	1.2	1.27
Input Range ($V_{pp\text{-}diff}$)	0.7	1.2	1.2	0.5	0.5	0.7	0.7	0.8	0.85
SNDR low f_{in} (dB)	36.1**	34.9	34.9	29.2	39.1	37.7	37.2	36.0	34.9
SNDR high f_{in} (dB)		25.2	22.8	27	37	34.2	32.9	33.0	27.7
f_{in} for SNDR high f_{in} (GHz)		18	12	5	3.81	19.9	19.9	19.9	19.9
3dB Bandwidth (GHz)	20				4.2		22		
Power (mW)	1200	1500	1200	79	35	355	477	667	845
FoM low f_{in} (fJ/conv.-step)	355	829	1105	337	54	81	101	144	186
FoM high f_{in} (fJ/conv.-step)		2512	4419	434	69	121	165	203	426
Area (mm^2)	3***	16	16	0.33	0.025	0.45			

*SAR ADCs/Interleaver & Clocking ** SNDR(FS): 8GHz, -6dBFS ***Estimation based on 65nm

(b)

Figure 22.1.6: Spectrum of a 19.9GHz full-scale input signal and performance comparison table.

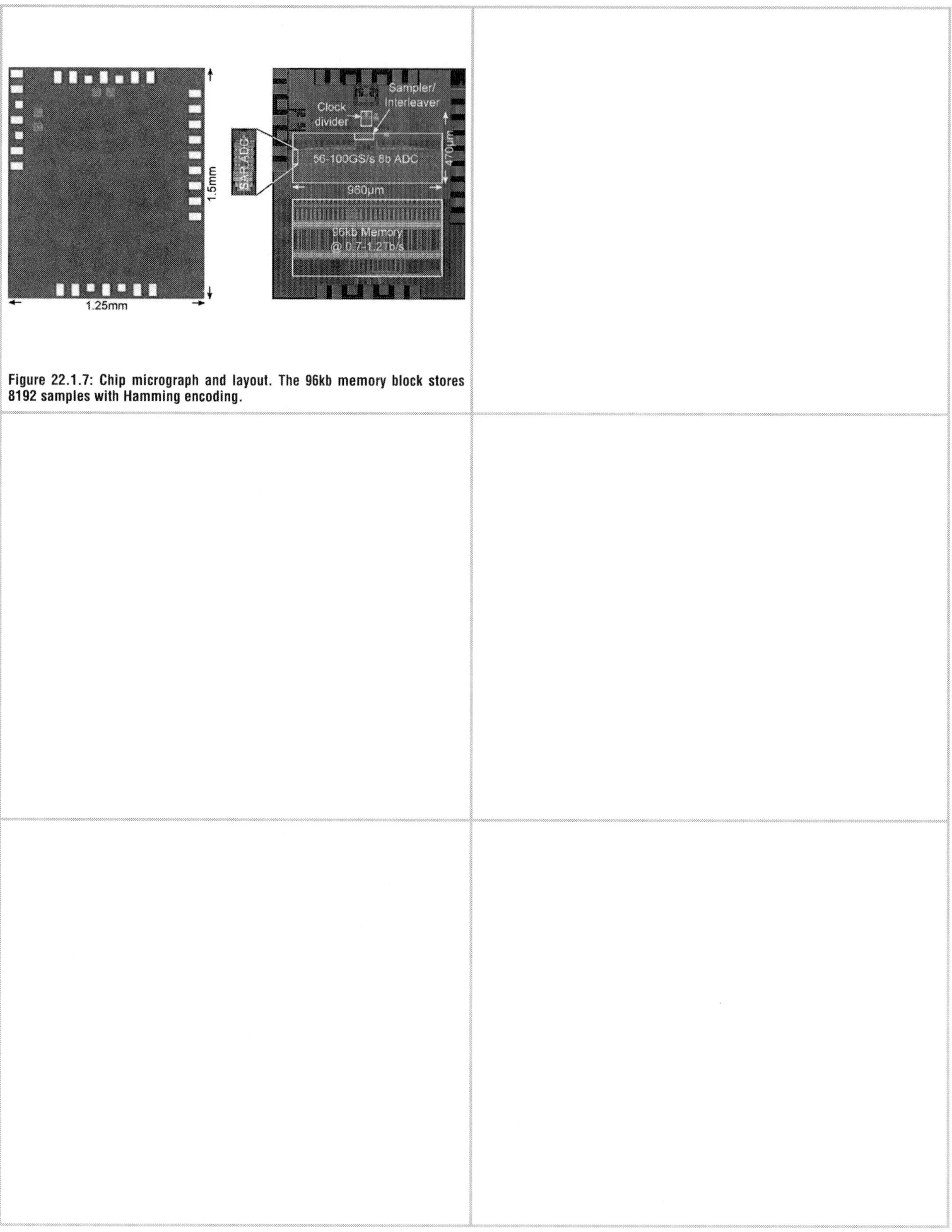

Figure 22.1.7: Chip micrograph and layout. The 96kb memory block stores 8192 samples with Hamming encoding.

22.2 A 69.5mW 20GS/s 6b Time-Interleaved ADC with Embedded Time-to-Digital Calibration in 32nm CMOS SOI

Vanessa Hung-Chu Chen, Lawrence Pileggi

Carnegie Mellon University, Pittsburgh, PA

Low-power time-interleaved ADCs with high sampling rates of over 10GS/s are in high demand for wireline communication systems. However, the time-interleaved channels suffer from process mismatch, particularly for timing skew. Although a power-consuming two-rank track-and-hold (T/H) can prevent such timing-skew problems, distributed T/Hs can be used for lower-power operation with timing-skew calibration to meet the skew specifications of $200fs_{rms}$ for 6b resolution and 10GHz input signals. Instead of using software calibration with Fourier analysis [1-3], requiring a special input reference signal [4], or relying on the statistics of the input signal [5], this work presents a low-complexity on-chip background calibration technique to reduce gain, offset, and delay mismatches between channels. This enables small-size transistors to be used in comparators and clock delivery circuits to avoid serious noise coupling and save considerable power for such an ultra-high-speed system. The presented 8-way time-interleaved 20GS/s 6b ADC achieves an SNDR of 30.7dB at Nyquist and consumes only 69.5mW.

Figure 22.2.1 illustrates the time-interleaved ADC architecture. After power optimization between the channel number and clock buffers, it consists of 8 flash-type sub-ADCs, one extra channel for background calibration, a delay-locked loop (DLL) as a phase generator, an 8b low-frequency current-DAC as a reference signal source, and an on-chip calibration processor. The sub-ADC operates at 2.5GS/s with 6b resolution and each sub-ADC is clocked 50ps after the previous channel. The DLL is driven by a 2.5GHz clock to generate 8 phases for each slice.

Figure 22.2.2 shows the sub-ADC architecture with timing diagram. The T/H utilizes bottom-plate sampling and the input switch is bootstrapped for better linearity and higher bandwidth. The fast-tracking feature that V_{GS} of M_1 always tries to go above V_{DD} speeds up signal tracking compared to the conventional bootstrapped switch [6]. The low input capacitance of the comparator array allows the use of a sampling capacitor of only 100fF. The comparator in the sub-ADC consists of a gain-enhanced pre-amplifier and a regenerative latching stage. The 63 pre-amplifiers are designed with unbalanced loading ratios to shift offset distributions to span the $300mV_{pp}$ full-scale range that eliminates the need for reference resistors. Each pre-amplifier input is constructed of 12 selectable minimum-sized differential pairs to enable dynamic offset calibration [7]. This effectively characterizes the comparator array with 126 checking steps to compensate for parasitic capacitor mismatches that occur during high-speed operation. The digital outputs from Wallace tree encoder are decimated for test.

Since all the built-in offsets of sub-ADCs including the extra channel, ADC_{REF}, are referred to the same transfer curve generated by an on-chip current DAC, the offset and gain mismatches among all the channels are calibrated by dynamic offset calibration along with the comparator offset variations. Then the timing skew is calibrated with ADC_{REF} to build reference timing windows. A moving signal, V_{CAL}, generated from the current DAC with a delay-variable clock is applied to ADC_{REF} as an input signal to generate a time-to-digital table, as shown in Fig. 22.2.3. ADC_{REF} is clocked externally at CLK_{REF} with a different clock period of $(8T_s+T_s)$, where T_s is the sampling period of the overall ADC, such that the edges of CLK_{REF} coincide with the ideal sampling points of each sub-ADC. The DAC input is a periodic signal with a period of $144T_s$ to generate a 2-level moving signal. DAC clock, CLK_{DAC}, is controlled by the programmable delay line with a step size ΔT of 100fs. While the sampling point falls within the range of t_k to t_{k+1}, ADC_{REF} outputs *HIGH* as CLK_{DAC} is programmed with delay t_k, and ADC_{REF} outputs *LOW* as CLK_{DAC} is programmed with delay t_{k+1}. 200 digital outputs are observed for each delay to reduce the jitter influence on the calibration. By sensing all the 8 transition points of CLK_{REF} with this embedded time-to-digital conversion, the corresponding delay-controlling information for calibrating the sampling point of each sub ADC is stored in the digital domain.

To calibrate sampling points of each sub-ADC sequentially, the corresponding delay-controlling codes are used to adjust CLK_{DAC} to generate moving signals for each sub-ADC, as shown in Fig. 22.2.4. CLK_{DAC} is programmed with 2 delay-controlling codes to generate a timing-checking window to detect the sampling point of the sub-ADC under calibration. As the sub-ADC outputs *HIGH* and *LOW* correspondingly, the sampling point of the sub-ADC is detected in the target range between t_k and t_{k+1}. The sub-ADC delay is feedback-controlled by choosing a combination set of 6 elements from 12 selectable units in the aligning buffer. Therefore, a large search space provides high-resolution tuning steps for delay alignment of sub-ADCs to CLK_{REF} edges. Without the need of knowing lead/lag of the clock, analysis of sub-ADC outputs only requires accumulators to calculate the *HGH/LOW* status. The calibration requires low-complexity hardware and ensures that the condition of the sub-ADC under calibration is similar to the situation during normal operation.

To allow background calibration, ADC_{REF} is also used to periodically substitute for the main channel that requires calibration next. ADC_{REF} is primarily used to collect the timing information of 8 sampling edges before starting timing skew calibration. Therefore, the low-complexity calibration only requires the overhead of one extra sub-ADC, one DAC, and calibration logic, which comprises only 25% of the overall area and 13.5% of power consumption when the calibration is on.

The 6b time-interleaved ADC is implemented in a 32nm CMOS SOI process. The calibrated DNL and INL of the ADC are +0.47/-0.42 and +0.42/-0.38 LSB, respectively. Figure 22.2.5 shows the output spectrum with 8.18GHz input before and after timing-skew calibration, and the measured SNDR versus input frequency at 20GS/s. The ADC achieves an SNDR of 34.8dB (5.49ENOB) at low input frequencies and 30.7dB (4.81ENOB) at Nyquist. The performance at high input frequencies degrades due to external clock jitter ($<0.1ps_{rms}$), the clock circuitry jitter, and residual timing skew. The estimated total jitter is $0.3ps_{rms}$ after removing harmonic tones, and the estimated timing skew after calibration is $<0.15ps_{rms}$. Without recalibration, over the 0-to-70°C temperature range the SNDR is degraded by 10dB. However, the ADC can be completely recalibrated at a rate of 2Hz or below without interrupting its normal operation in order to compensate for this temperature sensitivity. The total power consumption at 20GS/s while calibration is off is 69.5mW (excluding I/O and input clock buffers) from a 0.9V supply voltage, of which 57.6mW is dissipated in the 8 sub-ADCs, and 11.9mW in the 8-phase generator. The occasionally-on calibration consumes 7.2mW in ADC_{REF}, 3.2 mW in the DAC, and 0.5mW in the calibration logic. For comparison with other state-of-the-art works, the figure of merit of *power/*$(2^{ENOB}\times f_s)$ is used. The ADC achieves 123.9fJ/conv-step at Nyquist and 77.3fJ/conv-step at low input frequencies. Performance and comparisons are summarized in Fig. 22.2.6. The chip micrograph is shown in Fig. 22.2.7, and the ADC occupies an active area of <0.25 mm². For high-speed ADCs with over 10GS/s sampling rates, the presented design achieves the best reported power consumption and figure of merit among those in our table of recent state-of-the-art ADCs in this category (Fig. 22.2.6).

Acknowledgments:
The authors thank the support of FCRP C2S2 and R. Carley, J. Weldon, J. Paramesh, and E. Chen of CMU, J.-O. Plouchart of IBM, and C.-C. Lee of Realtek for discussion.

References:
[1] K. Poulton, *et al.*, "A 20GS/s 8b ADC with a 1MB Memory in 0.18μm CMOS," *ISSCC Dig. Tech. Papers*, pp. 318-319, Feb. 2003.
[2] P. Schvan, *et al.*, "A 24GS/s 6b ADC in 90nm CMOS," *ISSCC Dig. Tech. Papers*, pp. 544-545, Feb. 2008.
[3] Y. M. Greshishchev, *et al.*, "A 40GS/s 6b ADC in 65nm CMOS," *ISSCC Dig. Tech. Papers*, pp. 390-391, Feb. 2010.
[4] C.-C. Huang, C.-Y. Wang, and J.-T. Wu, "A CMOS 6-bit 16-GS/s Time-Interleaved ADC Using Digital Background Calibration Techniques," *IEEE J. Solid-State Circuits*, vol. 46, no. 4, pp. 848–858, Apr. 2011.
[5] M. El-Chammas and B. Murmann, "A 12-GS/s 81-mW 5-bit Time-Interleaved Flash ADC with Background Timing Skew Calibration," *IEEE J. Solid-State Circuits*, vol. 46, no. 4, pp. 838–847, Apr. 2011.
[6] E. Alpman, *et al.*, "A 1.1V 50mW 2.5GS/s 7b Time-Interleaved C-2C SAR ADC in 45nm LP Digital CMOS," *ISSCC Dig. Tech. Papers*, pp. 76-77, Feb. 2009.
[7] V. H.-C. Chen and L. Pileggi, "An 8.5mW 5GS/s 6b Flash ADC with Dynamic Offset Calibration in 32nm CMOS SOI," *Symp. VLSI Circuits*, pp. 264-265, June 2013.

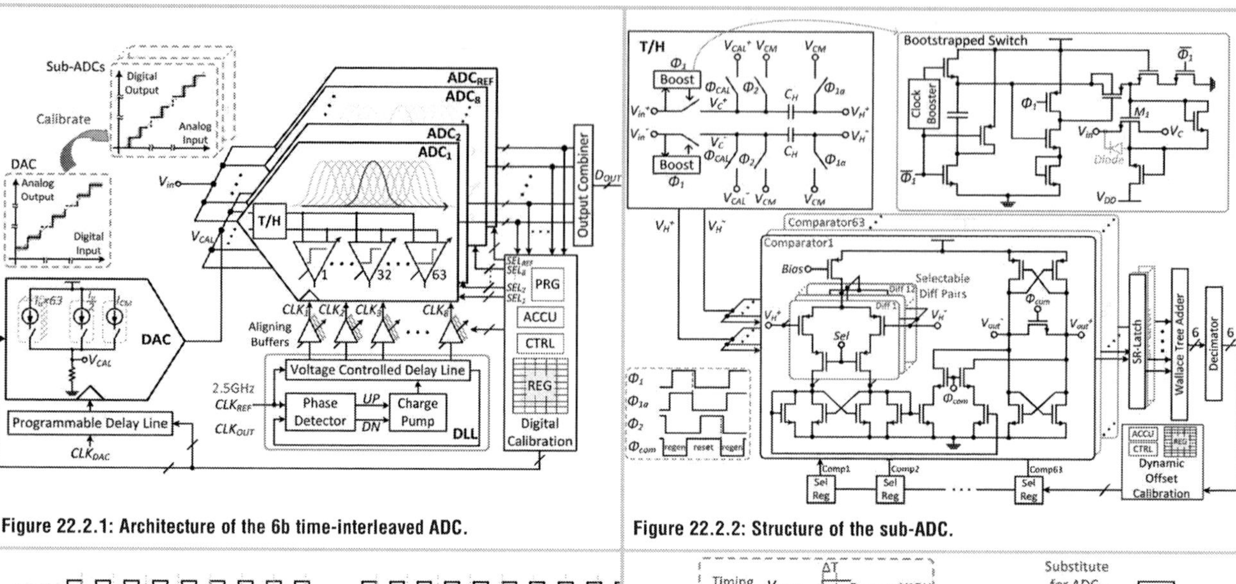

Figure 22.2.1: Architecture of the 6b time-interleaved ADC.

Figure 22.2.2: Structure of the sub-ADC.

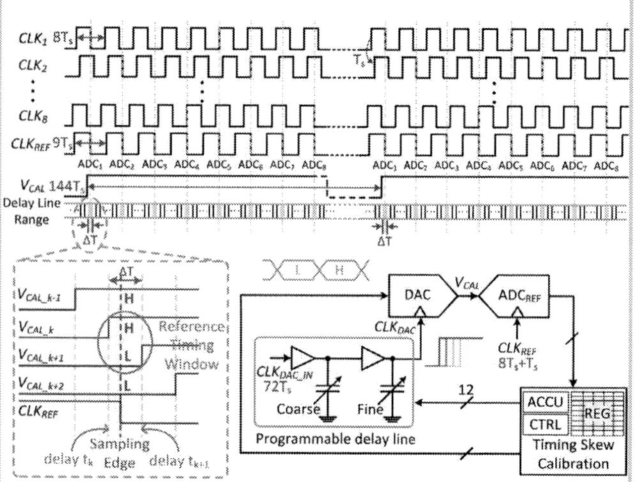

Figure 22.2.3: The moving signal to sense the sampling edges of the clocks.

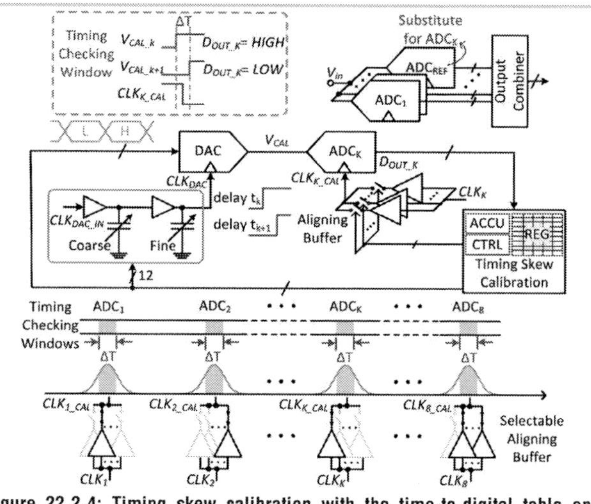

Figure 22.2.4: Timing skew calibration with the time-to-digital table and aligning buffers.

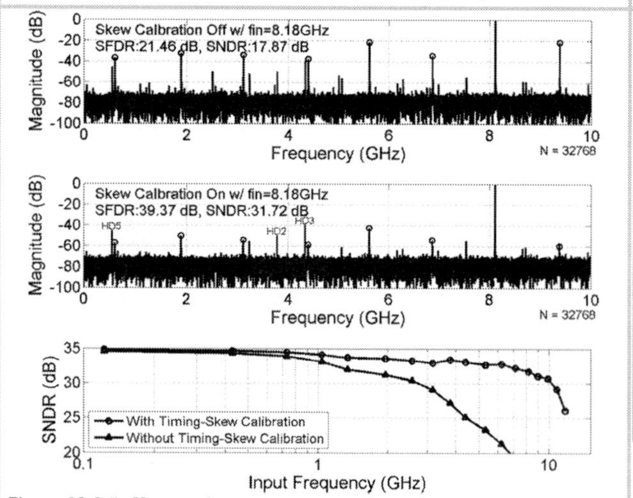

Figure 22.2.5: Measured output spectrum with F_{in}=8.18GHz and SNDR v.s. input frequency before and after timing-skew calibration at 20GS/s.

	Poulton ISSCC '03 [1]	Schvan ISSCC '08 [2]	Greshishchev ISSCC '10 [3]	Huang JSSC '11 [4]	El-Chammas JSSC '11 [5]	This Work
Structure	TI-Pipelined	TI-SAR	TI-SAR	TI-Flash	TI-Flash	TI-Flash
Technology	180nm	90nm	65nm	65nm	65nm	32nm SOI
Supply Voltage (V)	-	1/2.5	1/1.2/2.5	1.5	1.1	0.9
Sampling Rate (GS/s)	20	24	40	16	12	20
Resolution (bit)	8	6	6	6	5	6
Power (mW)	9000	1200	1500	435	81	69.5
DNL (LSB)	0.3	0.5	0.25	0.6	0.5	0.47
INL (LSB)	0.4	0.5	0.1	0.7	0.5	0.42
SNDR (dB) @ High Input Frequency	29.5	26.4	25.2	28	25.1	30.7
Active Area (mm²)	-	16	16	1.47	0.44	0.25
Input Capacitance (pF)	4.0	1.2	-	1.8	1.1	1.0
Figure of Merit (fJ/c-s)	18453	2916	2512	1325	459	124

Figure 22.2.6: Performance summary and comparison with the state of the art.

22

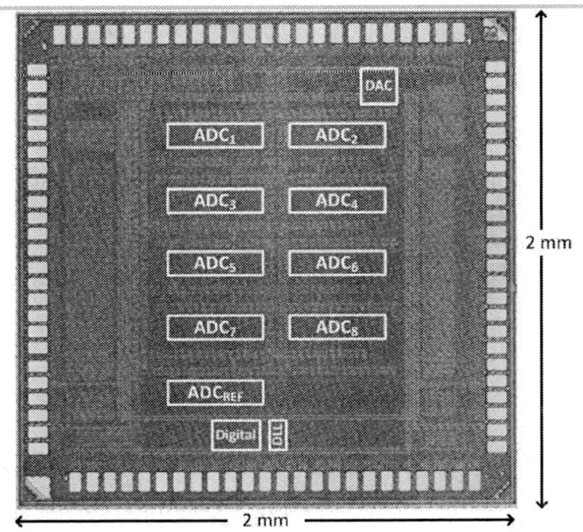

Figure 22.2.7: Chip micrograph.

22.3 A 20GHz-BW 6b 10GS/s 32mW Time-Interleaved SAR ADC with Master T&H in 28nm UTBB FDSOI Technology

Stéphane Le Tual[1], Pratap Narayan Singh[2], Christophe Curis[3], Pierre Dautriche[1]

[1]STMicroelectronics, Crolles, France,
[2]STMicroelectronics, Greater Noida, India,
[3]STMicroelectronics, Grenoble, France

To sustain ever-growing data traffic, modern wireline communication devices (over copper or fiber optic media) require a high-speed ADC in their receive path to do the digital equalization, or to recover the complex-modulated information. A 6b 10GS/s ADC able to acquire up to 20GHz input signal frequency and showing 5.3 ENOB in Nyquist condition is presented. It is based on a Master Track & Hold (T&H) followed by a time-interleaved synchronous SAR ADC, thus avoiding the need for any kind of skew or bandwidth calibration. Ultra Thin Body and BOX Fully Depleted SOI (UTBB FDSOI) 28nm CMOS technology is used for its fast switching and regenerating capability. The core ADC consumes 32mW from 1V power supply and occupies 0.009mm^2 area. The FoM is 81fJ/conversion step.

An input driver made of a PMOS GmR stage (Fig. 22.3.1) buffers the on-chip 50Ω matched differential input signal. Gm is 25mS and R is 40Ω to keep low impedance driving of the Master T&H. The current consumed in this stage is about 10mA. The high-frequency continuous-time signal is then sampled and held by a 20μm×30nm Low-V_{TH} (LVT) NMOS transistor using 1V Forward Body Bias (FBB) to further reduce its V_{TH} and thus lower its ON resistance (Fig. 22.3.3). A simple "top-plate sampling" scheme is used without the need for any switch bootstrapping or charge-injection cancellation. The sampling capacitor is made of 100fF routing and metal capacitor. During each hold phase, the sampled charges are successively shared with one of the 25fF sampling capacitors of the time-interleaved SARs (Fig. 22.3.2). This eliminates the need for intermediate buffering that would have been difficult to realize under 1V power supply with 50ps half-period for settling time. Signal attenuation penalty is 100fF/125fF, which is acceptable for 6b-resolution ADCs. The 10GHz 0° and 90° external clocks are used to create the 25ps Track pulses. When integrated onto an SoC, they may be derived from a 10GHz quadrature VCO.

A total of 8 clock cycles are needed to do the sampling, the 6b successive approximation, and the reset, leading to a SAR interleaving ratio of 8. Each SAR is made of a compact comb capacitor array, a comparator, and 6b registers. The capacitor array is made of binary-weighted fingers using the minimum pitch of the technology to limit propagation effects in the finger connections [1]. The resulting capacitor array width is less than 7μm. The comparator is made of a preamplifier followed by a clocked inverter [2] that drives a latch (Fig. 22.3.4). The speed performance of the comparator is mainly NMOS-based, and leverages the 1V FBB capability. The latch is not clocked and thus burns some current during its reset phase. As a result, it presents a fast regeneration time of <40ps for input smaller than 0.5mV, ensuring low BER at 10GS/s operation. The 6b registers are made of synchronous logic with a maximum of 4 gates delay in their critical path.

Two reference voltages Vtop and Vbot, together with an intermediate voltage Vc, are embedded on chip. Vc is applied during SAR charge-transfer and reset phases. As described in [3], depending on latch decision, the top plate of each capacitor is switched from Vc to either Vtop or Vbot on the positive side of the capacitor array, and from Vc to either Vbot or Vtop on the negative side of the array. By contrast to the classical 2-reference-voltage "set and correct" successive approximation algorithm, this 2-reference-voltage plus 1-intermediate-voltage "decide right" scheme makes the switching operation symmetrical, drawing the same current from the references whatever the decision is. It results in a very low dependency of the charge drawn from the references as a function of the input voltage, enabling a low-power design for the references (less than 2mA here). Additionally, only 5 capacitors instead of 6 in the classical way are needed to do the 6b quantization.

After quantization, each SAR stores its 6b word in 2 ping-pong 512-word RAMs running at 10GHz/8/2=625MHz. The total 8K words are finally read at low speed through a JTAG controller.

The chip is fabricated in 28nm CMOS UTBB FDSOI technology, using 10 metal layers and MIM capacitors for decoupling. Each SAR occupies 50×13μm^2 (0.0007mm^2) and the complete ADC core is 80×115μm^2 (0.009mm^2). It is mounted in a soft polymeric substrate (PTFE) with a cavity for the die gluing, such that bonding wires are kept as small as possible (Fig. 22.3.7).

Measurement results at 10GS/s are presented in Fig. 22.3.5. Offset mismatch between the 8 SARs is not calibrated on-chip and leads to 1.5dB penalty in dynamic range. In the following results, signal amplitude is thus set at -3dBFS and offset spurs are calibrated off-chip. At 4.8GHz input, ENOB is 5.3 and is thermal-noise limited. ENOB is still 4.6 at 20GHz input, limited by single-ended clock jitter (that can be calculated to 260fs_{rms}). THD and SFDR are maintained below 40dB over the whole bandwidth, and they are mainly due to input driver distortion. No gain or skew spurs are visible, demonstrating the efficiency of the Master T&H and the timing accuracy between the Master T&H and the 8 SAR sampling capacitors.

This work compares favorably with [4-7], as depicted in Fig. 22.3.6. It shows 81fJ/conversion step in Nyquist conditions with a very small 0.009mm^2 area and it has up to 20GHz sampling capability without any need for gain & skew calibration, which makes it ideal for further higher-speed interleaving (100Gb/s optical links for instance).

In conclusion, we demonstrate in this work the efficiency of the pure passive "sampling and redistribute" concept for signals up to 20GHz. Together with the low-power capability of the 28nm CMOS UTBB FDSOI technology, we can reach 10GS/s operation while keeping the power consumption at 32mW under 1V supply.

Acknowlegments:
The authors thank Alex Zabroda for initial concept discussions, Philipp Ritter and Martin Müller from Saarland University for module fabrication & chip assembly, and Dimitri Goguet for bench set-up and measurements.

References:
[1] J. Bach, "Capacitive Array", US Patent 7,873,191 B2, May 2005.
[2] D. Schinkel, *et al.*, "A Double-Tail Latch-Type Voltage Sense Amplifier with 18ps Setup+Hold Time," *ISSCC Dig. Tech. Papers*, pp. 314-315, Feb. 2007.
[3] S. Le Tual, *et al.*, "Differential Successive Approximation Analog to Digital Converter", US Patent 8,497,795 B2, June 2010.
[4] M. El-Chammas and B. Murmann, "A 12-GSps 81-mW 5-bit Time-Interleaved Flash ADC With Background Timing Skew Calibration," *IEEE J. Solid-State Circuits*, vol. 46, pp. 838-847, Apr. 2011.
[5] S. Verma, *et al.*, "A 10.3GS/s 6b Flash ADC for 10G Ethernet Applications," *ISSCC Dig. Tech. Papers*, pp. 462-463, Feb. 2013.
[6] E. Z. Tabasy, *et al.*, "A 6b 10GS/s TI-SAR ADC with Embedded 2-Tap FFE/1-Tap DFE in 65nm CMOS," *IEEE Symposium on VLSI Circuits*, pp. C274-C275, June 2013.
[7] L. Kull, *et al.*, "A 35 mW 8b 8.8 GS/s SAR ADC with Low-Power Capacitive Reference Buffers in 32nm Digital SOI CMOS," *IEEE Symposium on VLSI Circuits*, pp. C260-C264, June 2013.

978-1-4799-0917-9/14 $31.00 © 2014 IEEE

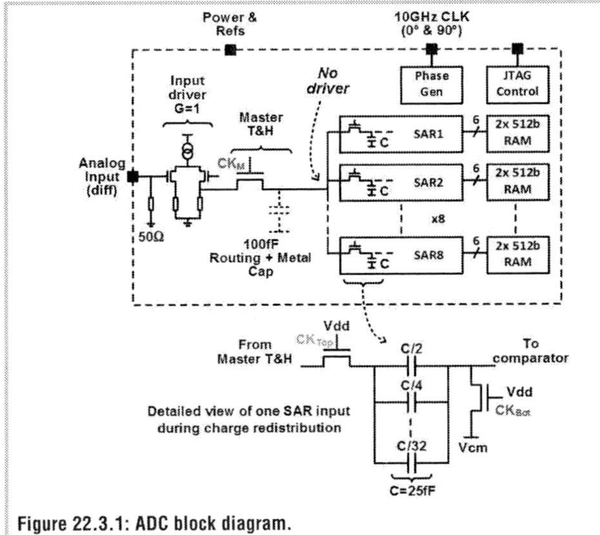

Figure 22.3.1: ADC block diagram.

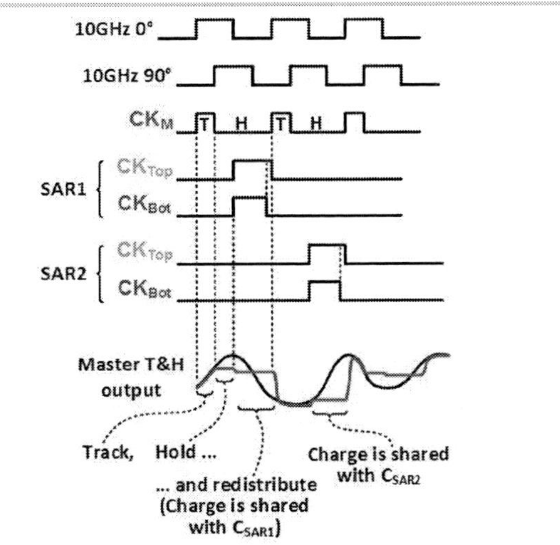

Figure 22.3.2: Master T&H and charge-redistribution timing diagram.

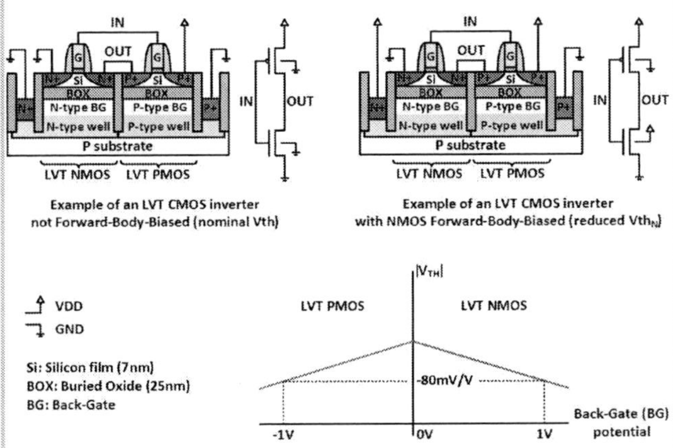

Figure 22.3.3: Fully Depleted Silicon-on-Insulator LVT transistors (flipped-well).

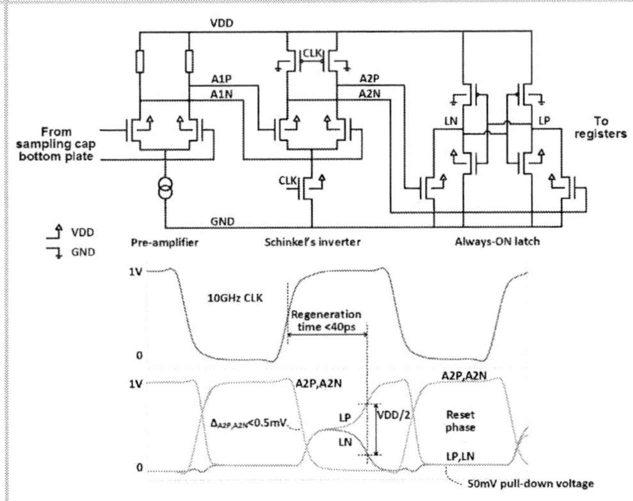

Figure 22.3.4: Comparator schematic.

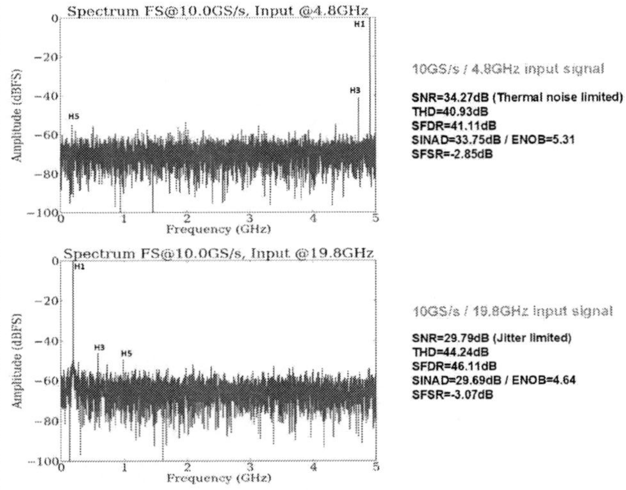

Spectrum FS@10.0GS/s, Input @4.8GHz

10GS/s / 4.8GHz input signal

SNR=34.27dB (Thermal noise limited)
THD=40.93dB
SFDR=41.11dB
SINAD=33.75dB / ENOB=5.31
SFSR=-2.85dB

Spectrum FS@10.0GS/s, Input @19.8GHz

10GS/s / 19.8GHz input signal

SNR=29.79dB (Jitter limited)
THD=44.24dB
SFDR=46.11dB
SINAD=29.69dB / ENOB=4.64
SFSR=-3.07dB

Figure 22.3.5: Output spectrum at 10GS/s.

	JSSC 2011 [4]	ISSCC 2013 [5]	VLSI 2013 [6]	VLSI 2013 [7]	This Work
Technology	65nm CMOS	40nm CMOS	65nm CMOS	32nm SOI	28nm UTBB FDSOI
Architecture	TI-FLASH	TI-FLASH	TI-SAR	TI-SAR	TI-SAR
Power Supply (V)	1.1	0.9	1.1 / 0.9	1	1
Sampling Rate (GS/s)	12	10.3	10	8.8	10
Resolution (bits)	5	6	6	8	6
Power Consumption (mW)	81	240	79.1	35	32
SNDR @ Nyquist (dB)	25.1	33	26	38.5	33.8
Active Area (mm^2)	0.44	0.27	0.33	0.025	0.009
FOM @ Nyquist (fJ/conv step)	460	700	480	58	81
Max Input Frequency (GHz)	8	6	4.5	4.2	20
Gain/Skew Calibration	Yes	Yes	Yes	Yes	No

Figure 22.3.6: Performance summary and state-of-the-art comparison.

Figure 22.3.7: Die photo in its chip-on-board cavity

22.4 A 1GS/s 10b 18.9mW Time-Interleaved SAR ADC with Background Timing-Skew Calibration

Sunghyuk Lee, Anantha P. Chandrakasan, Hae-Seung Lee

Massachusetts Institute of Technology, Cambridge, MA

SARs are one of the most energy-efficient ADC architectures for medium resolution and low-to-medium speed. To improve the limited bandwidth of SAR ADCs, the time-interleaved (TI) structure is often used [1,2]. However, TI ADCs have several issues caused by mismatches between channels, such as offset, gain, and timing-skew errors. Unlike the other errors, timing-skew causes errors that increase with input signal frequency. Considering that the TI structure is typically employed to increase bandwidth, timing-skew can be a dominant error source of TI ADCs. Recent works [1,3] have demonstrated a background timing-skew calibration using a dedicated additional channel as a timing reference. In this work, we present a TI SAR ADC that enables background timing-skew calibration without a separate timing reference channel and enhances the conversion speed of each channel.

Figure 22.4.1 shows the block diagram and timing waveform of the 8-way TI SAR ADC. The ADC is composed of a clock generator, a flash ADC, 8 SAR ADCs, and digital circuits for bit combining and multiplexing. The timing-skew estimator controls the programmable delay circuits in the clock generator to correct the timing-skew. The flash ADC, operating at the full sampling rate (Φ) of the TI ADC, is multiplexed to resolves 4 MSBs of each of the SAR channels [4]. Because the full-speed flash ADC does not suffer from timing-skew errors, the flash ADC output is also used as the timing reference to estimate timing-skew of SAR ADCs. SAR ADCs resolve 7 LSBs with 1b redundancy at 1/8 of the sampling rate (Φ_1~ Φ_8).

The implementation of the flash ADC is shown in Fig. 22.4.2. Each comparator samples input signals with a bottom plate switch on two capacitors with different sizes, which are then switched to the reference voltages. When the reference voltages settle, the comparators are enabled. The sampling comparators have several advantages over the conventional flash comparators that compare the input directly with a reference voltage [5]. These include rail-to-rail input range, zero static power, and true fully differential implementation. Although conventional flash comparators with two differential input pairs are topologically fully differential, they suffer from lower PSRR and CMRR when input or reference voltages are large. Also, because the sampling circuit of the flash comparator is a scaled version of a SAR ADC, the sampled signals between the SAR and the flash are closely matched. This is important in this work, because the sampling clocks of the SAR ADCs must be aligned as closely to the sampling clock of the flash ADC as possible to minimize the timing skew correction range.

Figure 22.4.3 shows the 10b SAR ADC composed of 1024 unit capacitors, a comparator, and SAR logic. MSB DACs have 14 unary-weighted capacitors (size of 64) that are controlled by the output of the flash ADC. LSB DACs are composed of binary-weighted capacitors (size of 1 to 64) that are controlled by the SAR logic. 1b redundancy is added between the flash ADC and the SAR conversion to cover the error from the flash ADC and to extract the timing-skew information. Since MSB DACs are set by the flash ADC output, only 7 successive approximations are required for a 10b SAR conversion. The final output of each channel ($D_{CHANNEL}$) is the weighted sum of the flash ADC output (D_{FLASH}) and the lower output bits of SAR conversion (D_{LSAR}), $D_{CHANNEL} = 64^{*}D_{FLASH} + D_{LSAR}$. To avoid using a high-frequency clock and to increase SAR conversion speed, an asynchronous SAR logic is implemented [6]. A dynamic latch with an offset control is used as the SAR comparator. An offset calibration block, which is a capacitor bank with switches, is added at the output of the comparator. To save area and power consumption, a 1fF unit capacitor, shown in Fig. 22.4.3, is custom designed. It is a combination of the MIM and MOM structures. The benefit of the MIM structure is that the top and bottom plates can be shielded from each other in the capacitor array and each capacitor can be accessed without adding parasitic capacitance. To increase the density of the capacitance, interdigitated MOM structures are added on M4 and connected to M3 and M5. The programmable delay circuit for the sampling clock is implemented with variable capacitors. 3b binary-weighted capacitors are added in the sampling clock path for coarse (~2ps/code) and fine (~0.8ps/code) delay control.

In this paper, we propose timing-skew calibration based on minimizing the variance of the difference between the flash ADC and channel outputs. The basic principle of the timing-skew calibration is to align the falling edge of the SAR ADC sampling clocks to the falling edge of the flash ADC sampling clock. Considering the threshold variation of the sampling switches, in actuality, the falling edge of the SAR sampling clock is adjusted so that the sampled signal of SAR ADCs is the same as the sampled signal of the flash ADC. When the sampled signal of the flash ADC is different from the sampled signal of the SAR ADC, the coarse estimation from flash ADC is inaccurate. The 1b redundancy in the SAR conversion corrects the flash ADC errors. In fact, the lower output bits of SAR conversion (D_{LSAR}) represent precisely the difference between the flash ADC output (D_{FLASH}) and the corresponding channel output ($D_{CHANNEL}$). Thus, the timing skew minima correspond to the minima of the D_{LSAR} variance. The D_{LSAR} is first sorted according the D_{FLASH}, then the delay of the SAR sampling clocks that minimizes the variance of the D_{LSAR}, VAR(D_{LSAR}), is computed. Because the computation processes only small signals represented in a short word (D_{LSAR}), we believe the computation is simpler than extending the correlation-based calibration to a multi-bit timing reference [3]. Since it does not have any constraint on the input signal and does not interrupt the normal ADC operation, this calibration can run in the background to track temperature and voltage variations. To save power, the calibration may be initiated only when temperature or voltage fluctuation is detected. Because each comparator in the flash ADC has a sampling circuit, the effective sampling instance of the flash ADC may vary depending on the input level. However, the sampling skew among flash comparators is statistically averaged out by the variance-based calibration.

Figure 22.4.4 shows measured data for the timing-skew calibration. VAR(D_{LSAR}) is plotted against coarse delay control codes of the SAR sampling clocks. 128K data are used to calculate each variance value. All channels show a smooth curve with one minimum. The same process is repeated for the fine delay code to complete the calibration. To demonstrate the possibility of background calibration and to avoid errors from a deterministic input signal, the calibration is performed over 4 different input frequencies and 2 different input amplitudes, and the differences were insignificant.

The ADC is implemented in 65nm CMOS using 0.78mm^2 (Fig. 22.4.7) area. The timing skew estimation algorithm is implemented off-chip. Offset errors of the SAR ADCs are calibrated first with a zero-differential DC input. Capacitive SAR ADCs have negligible gain errors, and the output spectrum with a low-frequency signal confirmed that gain errors are negligible. The measured output spectrum with a high-frequency input signal is shown in Fig. 22.4.5. Before the timing-skew calibration, the error tones from timing-skew limit the SNDR and SFDR. After background timing-skew calibration, typical results of 51.4dB SNDR (8.2 ENOB), 60.0dB SFDR, and ±1.0 LSB INL/DNL are achieved at 1GS/s with a 479MHz input signal. The power consumption is 18.9mW (clock 3.34mW, flash ADC 5.04mW, SAR ADCs 9.18mW, reference 1.35mW), which corresponds to 62.3fJ/step FoM. The SNDR versus input frequency, shown in Fig. 22.4.5, illustrates the effectiveness of the timing-skew calibration clearly. Figure 22.4.6 summarizes the performance with a comparison to the previously published works with f_s>0.8GS/s, SNDR >45dB, and FoM <180fJ/step.

Acknowledgements:
This work was supported by MIT Center for Integrated Circuits & Systems, Samsung Fellowship, and TSMC university shuttle program.

References:
[1] D. Stepanovic and B. Nikolic, "A 2.8 GS/s 44.6 mW Time-Interleaved ADC Achieving 50.9 dB SNDR and 3 dB Effective Resolution Bandwidth of 1.5 GHz in 65 nm CMOS," *IEEE J. Solid-State Circuits*, vol. 48, no. 4, pp. 971-982, Apr. 2013.
[2] H. Hong, *et al.* "An 8.6 ENOB 900MS/s Time-Interleaved 2b/cycle SAR ADC with a 1b/cycle Reconfiguration for Resolution Enhancement," *ISSCC Dig. Tech. Papers*, pp. 470-471, Feb. 2013.
[3] M. El-Chammas and B. Murmann, "A 12-GS/s 81-mW 5-bit Time-Interleaved Flash ADC With Background Timing Skew Calibration," *IEEE J. Solid-State Circuits*, vol. 46, no. 4, pp. 838-847, Apr. 2011.
[4] B. Sung, *et al.* "A Time-Interleaved Flash-SAR Architecture for High Speed A/D Conversion," *Proc. IEEE ISCAS*, pp. 984-987, May 2009.
[5] S. H. Lewis and P. R. Gray, "A Pipelined 5-Msample/s 9-bit Analog-to-Digital Converter," *IEEE J. Solid-State Circuits*, vol. 22, no. 6, pp. 954-961, Dec. 1987.
[6] S.-W. Chen and R. Brodersen, "A 6b 600MS/s 5.3mW Asynchronous ADC in 0.13μm CMOS," *ISSCC Dig. Tech. Papers*, pp. 574-575, Feb. 2006.

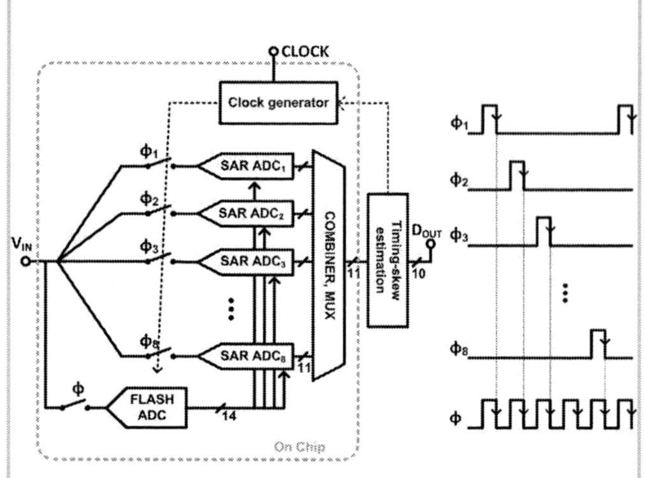

Figure 22.4.1: Block diagram of the TI SAR ADC.

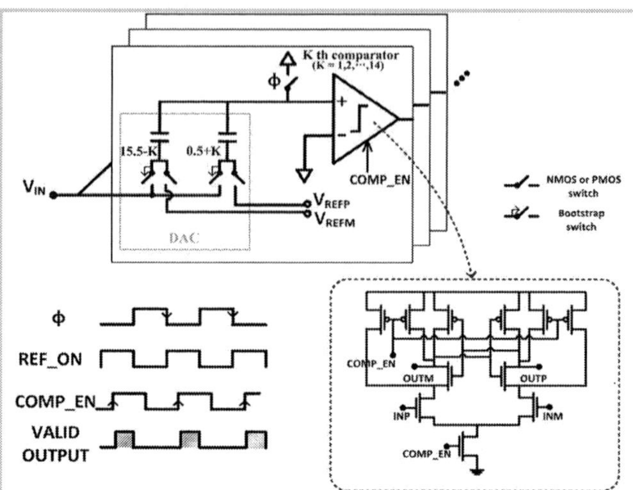

Figure 22.4.2: Implementation of the 4b flash ADC (A single-ended version is shown for simplicity).

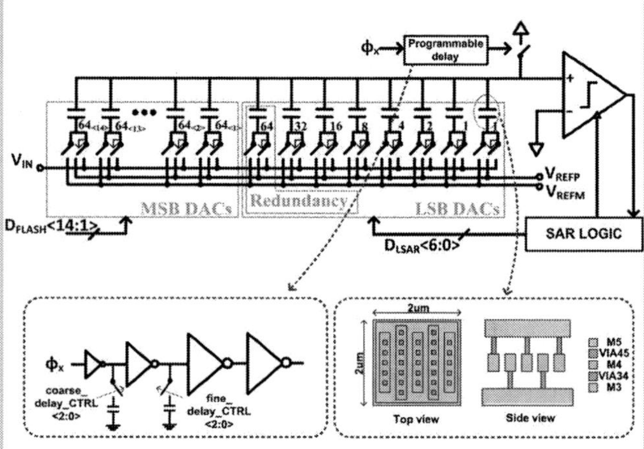

Figure 22.4.3: Implementation of the 10b SAR ADC (A single-ended version is shown for simplicity).

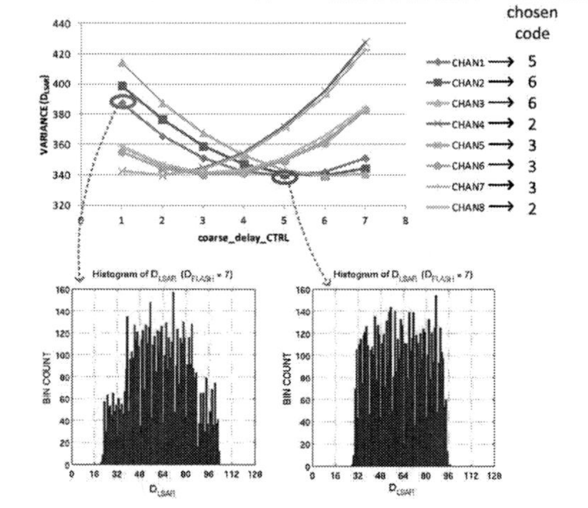

Figure 22.4.4: Measured variance of DLSAR against coarse delay of SAR ADC sampling clock (top) with the examples of the DLSAR histogram (bottom) for channel 1.

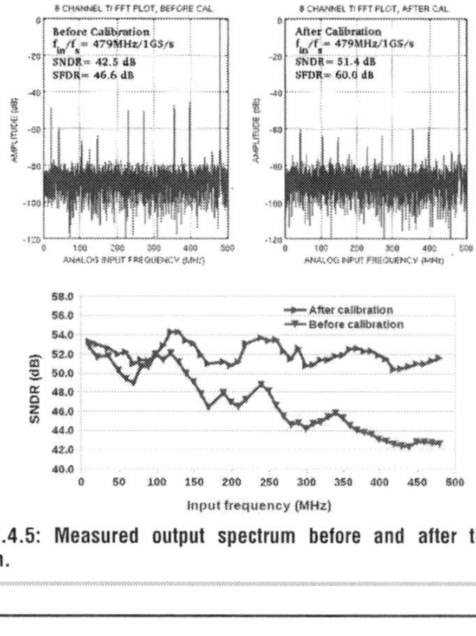

Figure 22.4.5: Measured output spectrum before and after timing-skew calibration.

	This work	ISSCC 2013 Hong [2]	VLSI 2013 Chiang	VLSI 2012 Stepanovic [1]	VLSI 2012 Sahoo	ISSCC 2011 Mulder
Architecture	TI SAR	TI SAR	PIPE	TI SAR	PIPE	TI PIPE
Technology	65nm	45nm	65nm	65nm	65nm	40nm
Supply Voltage (V)	1.0	1.2	1.0	1.2	1.2	1.0/2.5
Fs (GS/s)	1.0	0.9	0.8	2.8	1.0	0.8
Resolution (bit)	10	9	10	11	10	12
SNDR @ Nyquist (dB)	51.4	51.2	52.2	48.2	52.4	59.0
Power (mW)	19.8	10.8	19.0	44.6	32.9	105
FoM (fJ/step)	62.3	40.5	71.4	75.8	96.6	180.2

Figure 22.4.6: Performance summary and comparison table.

22

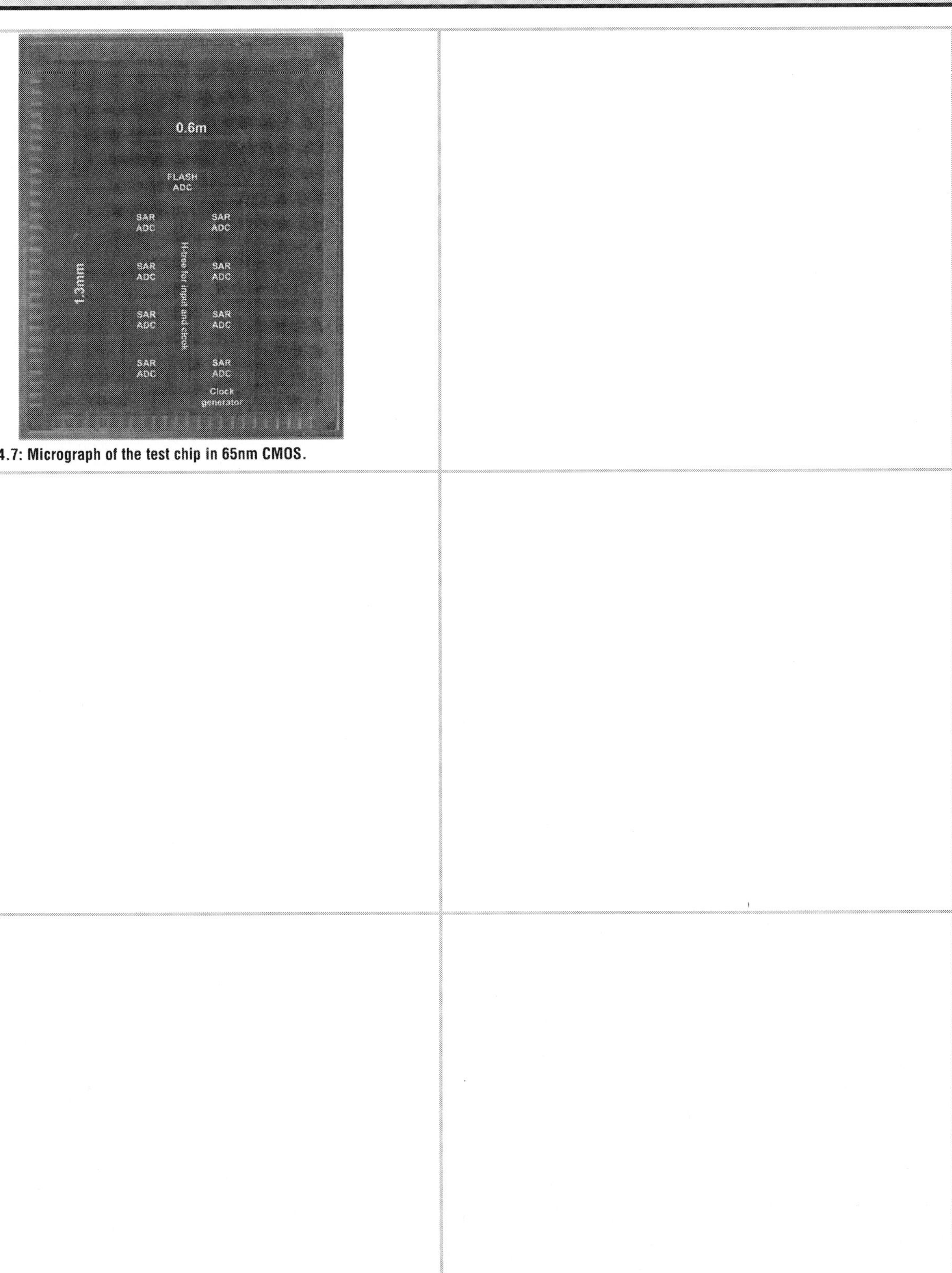

Figure 22.4.7: Micrograph of the test chip in 65nm CMOS.

22.5 A 1.62GS/s Time-Interleaved SAR ADC with Digital Background Mismatch Calibration Achieving Interleaving Spurs Below 70dBFS

Nicolas Le Dortz[1,2], Jean-Pierre Blanc[1], Thierry Simon[1], Sarah Verhaeren[1], Emmanuel Rouat[1], Pascal Urard[1], Stéphane Le Tual[1], Dimitri Goguet[1], Caroline Lelandais-Perrault[2], Philippe Benabes[2]

[1]STMicroelectronics, Crolles, France,
[2]Supélec, Gif-sur-Yvette, France

Today's applications such as broadband satellite receivers, cable TVs, and software-defined radios require highly efficient ADCs with high sampling rates and high resolutions. A time-interleaved ADC (TIADC) is a popular architecture used to achieve this goal. However, this structure suffers from mismatches between the sub-converters, which cause errors on the output signal, and more significantly, decrease the SFDR. These mismatches can be a severe limitation in applications such as satellite reception, where both narrowband and wideband signals are used. This paper introduces digital derivative-based estimation of timing mismatches. Gain, offset and skew mismatch calibrations are performed entirely in the digital domain through equalization.

Recently, multi-GS/s designs using mixed-signal techniques to reduce the effects of mismatches have been published [1-3]. Mixed calibration techniques exhibit good performance but they require additional development time. This prototype, done in 40nm CMOS technology, implements a 1.62GS/s 12-channel 9b TIADC with embedded digital background mismatch calibration. It demonstrates the efficiency of a generalizable and scalable co-design methodology where the analog core is designed without special care for the mismatches. The mismatches are suppressed by a separately designed digital background mismatch calibration unit. Over the range of 0 to 750MHz, the on-chip implementation of this architecture has interleaving spurs below 70dBFS and a SNDR higher than 48dB for a power consumption of 93mW.

The overall structure, shown in Fig. 22.5.1, consists of 12 interleaved 135MS/s 9b radix-2 SAR converters followed by a digital mismatch calibration unit. Each SAR resolves 9b in 12 clock cycles. The $1V_{pp\text{-}diff}$ input signal is delivered to the sub-ADCs via a 1.7V buffer in order to limit kickback noise. Each SAR embeds its own T/H circuit, and bottom-plate sampling is used to reduce charge injection. The top-plate sampling switch is a transmission gate that uses low-V_t high-performance analog (HPA) transistors to reduce R_{on} input signal dependency and thus, keep a good linearity (Fig. 22.5.2). In order to reduce the area while preserving good ratios between the capacitors, a compact, custom layout, lateral structure is used in the 9b radix-2 capacitive DAC as illustrated in Fig. 22.5.2 [4,5]. The common bottom plate consists of a metal comb connected to the input of a latch comparator. The capacitor top plates are made of several metal fingers inserted into the bottom-plate comb structure. The capacitor values are proportional to the number and the length of the fingers. The top plates can either be connected to the reference voltages V_n=250mV, V_p=750mV, and V_m=500mV, or the input signal V_{in}. The sequence of conversion steps shown in Fig. 22.5.1 is managed by the SAR logic that generates the control signals for the comparator and for the switches.

The 9b samples from the 12 interleaved ADCs are delivered in parallel to the digital mismatch calibration unit that runs in background in order to track mismatch variations. The calibration requires the input signal to be wide-sense stationary and band-limited to the Nyquist frequency. There is no adaptive feedback loop, meaning the signal is corrected after a single iteration, eliminating the potential stability issues encountered with adaptive techniques.

Offset mismatch is cancelled by equalizing the averages of each sub-ADC output to the average of sub-ADC_0 output. The relative offset of one sub-ADC is estimated as the difference between the modified moving averages of its output samples and the ones of sub-ADC_0. The estimated offsets are then subtracted from their respective sub-ADC outputs samples. A random sequence is added to the estimated offset before the subtraction to spread the residual offset mismatch spurs across the entire spectrum. The 12b offset-corrected samples are then transmitted to the gain mismatch calibration unit where gain mismatch is corrected. The relative gain each sub-ADC is calculated as the ratio between the modified moving average of its samples' absolute values and the modified moving average of sub-ADC_0 samples' absolute values. Gain mismatch is corrected by dividing each sub-ADC's output by its corresponding gain estimate.

Skew mismatch calibration, whose principle is illustrated in (Fig. 22.5.3), is performed in the last stage. Each sub-ADC output can be seen as a sum of an ideal signal and an error term proportional to the timing offset and the signal derivative. The ideal signal is orthogonal to its derivative, meaning that averaging the product between the sub-ADC output samples and their corresponding derivative samples eliminates the ideal signal component while leaving a skew-dependent term. This term, proportional to the derivative power and the sub-ADC timing offset, is used in the skew calibration unit to recover timing skew estimates. The estimated error signal, calculated as the product between the derivative and the timing skew estimate, is then subtracted from the sub-ADC output to recover the ideal samples [6]. The signal derivative is obtained by passing the TIADC gain-corrected samples through a differentiating FIR filter [6], whose coefficients are chosen such that its frequency response is accurate up to 750MHz. After skew mismatch calibration, the 12×12b digitally corrected signals are multiplexed and delivered at the output of the chip.

The test chip is fabricated using STMicroelectronics 40nm CMOS technology and is comprised of two of the previously described TIADCs (Fig. 22.5.7). The digital part has a 1.1V supply voltage similar to the analog core and the SAR logic. It is synthesized from a parametric C-code using Calypto Catapult C High-Level-Synthesis tool, making this flow quickly adaptable to any technology and any TIADC architecture.

Fig. 22.5.4 shows the measured performances versus input signal frequency. The SNDR progressively decreases from 51dB at DC to 48dB at a 750MHz input frequency, while the SFDR is higher than 62dBFS. The mismatch tones are kept below 70dBFS up to 750MHz. This is higher than previously published designs, which achieve, at best, 60dBFS [3] up to 90% of the Nyquist frequency. Above 750MHz, the performance deteriorates due to the limited accuracy of the derivative filter in that frequency range. For full-scale sine inputs, the SFDR is ultimately limited by the harmonic distortion caused by the input buffer non-linearity. The THD is maintained below -58dB up to the Nyquist frequency. This is as good as previously published designs [1-3] that use additional linearity calibration circuitry. Figure 22.5.5 shows the performances of the TIADC on a modulated signal, which is more likely found in practical applications. In this case, the buffer non-linearity is not a limitation, whereas mismatches create unwanted frequency components. At F_s=1.62GS/s the power consumption of the $0.83mm^2$ TIADC including the digital background calibration, the references, and the input buffer is 93mW. The digital unit occupies 40% of the total area and consumes 53% of the total power when running continuously (Fig. 22.5.4). This overhead is offset by the achieved high SFDR and the ability to easily resynthesize the digital unit for any TIADC architecture or technology. In addition, the digital calibration energy overhead will decrease with CMOS technology scaling due to improved digital efficiency.

Acknowledgement:
The authors would like to thank Andreia Cathelin and Borivoje Nikolić for their advice and support.

References:

[1] E. Janssen, *et al.*, "An 11b 3.6GS/s Time-Interleaved SAR ADC in 65nm CMOS," *ISSCC Dig. Tech. Papers*, pp. 464-465, Feb. 2013.
[2] K. Doris, *et al.*, "A 480mW 2.6GS/s 10b 65nm CMOS Time-Interleaved ADC with 48.5dB SNDR up to Nyquist," *ISSCC Dig. Tech. Papers*, pp. 180-181, Feb. 2011.
[3] D. Stepanovic and B. Nikolic, "A 2.8 GS/s 44.6 mW Time-Interleaved ADC Achieving 50.9 dB SNDR and 3 dB Effective Resolution Bandwidth of 1.5 GHz in 65 nm CMOS," *IEEE J. Solid-State Circuits*, vol. 48, no. 4, pp. 971-982, April 2013.
[4] J. Bach, "Capacitive Array," *US Patent 7,873,191 B2*, May 2005.
[5] S. Le Tual, *et al*, "Integrated Capacitive Device and Integrated Analog Digital Converter Comprising Such a Device," *US Patent Application 2013/0003255 A1*, June 2011.
[6] V. Divi and G. Wornell, "Blind Calibration of Timing Skew in Time-Interleaved Analog-to-Digital Converters," *IEEE J. Selected Topics in Signal Processing*, vol.3, no.3, pp.509-522, June 2009.

978-1-4799-0917-9/14 $31.00 © 2014 IEEE

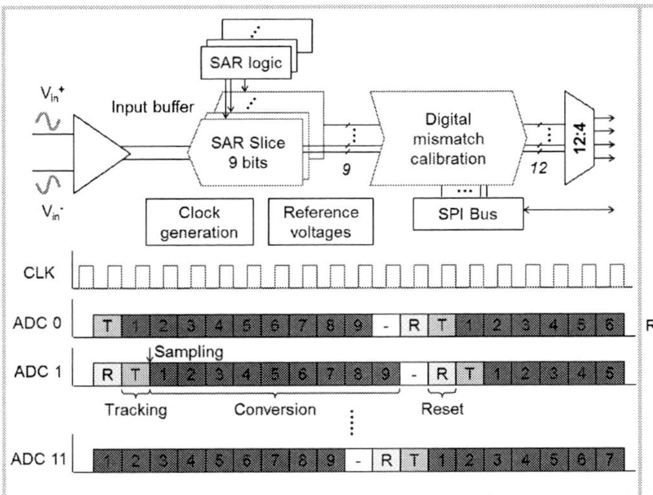

Figure 22.5.1: Overall structure of the Time-Interleaved ADC and sequence of conversion operations.

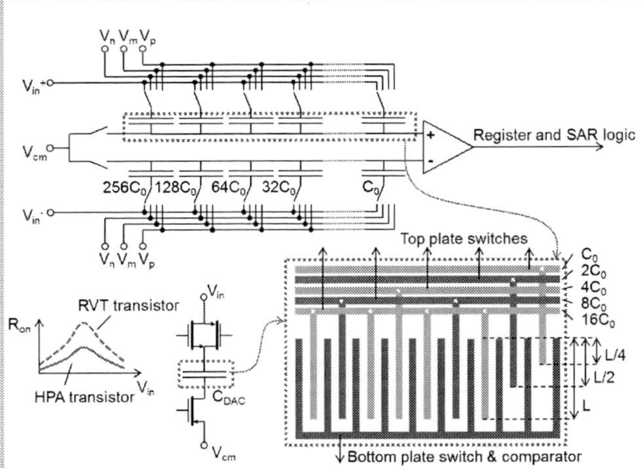

Figure 22.5.2: Bottom-plate sampling and capacitive DAC schematic with lateral capacitor array layout (not entirely represented).

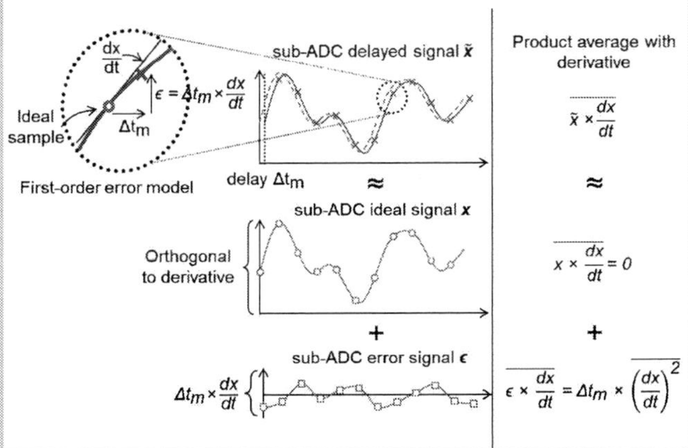

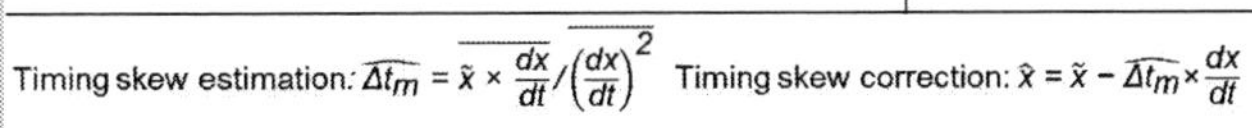

Figure 22.5.3: Skew mismatch estimation and correction principles.

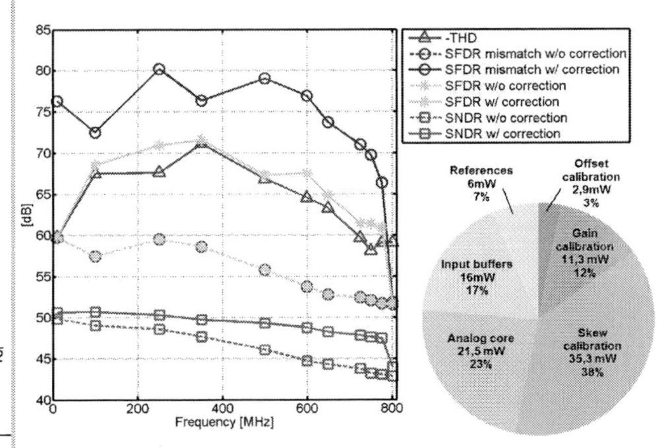

Figure 22.5.4: Performances of the TIADC over the Nyquist frequency range for a -1dBFS sine input and power consumption repartition.

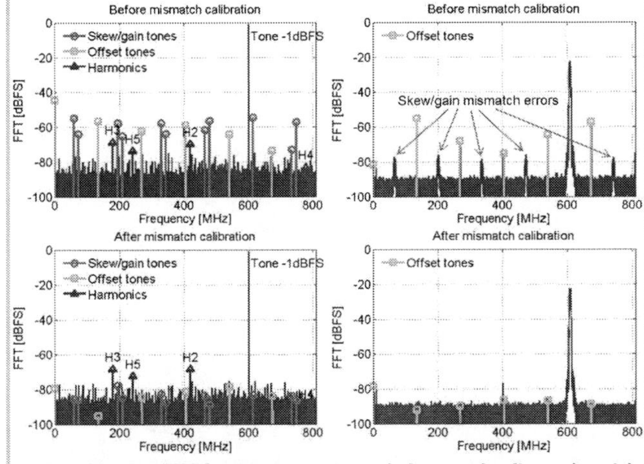

Figure 22.5.5: TIADC output spectrum before and after mismatch compensation with a sinusoid at 600MHz and a 10MHz bandwidth QAM16 modulated signal at $F_{carrier}$ = 607MHz.

	ISSCC 2013 [1]	ISSCC 2011 [2]	JSSC 2012 [3]	This work
Technology	65nm	65nm	65nm	**40nm**
Sampling rate [GS/s]	3.6	2.6	2.8	**1.6**
Mismatch tones [dBFS]	50	55	60	**70**
SFDR [dBFS]	50	55	55	**62**
THD [dB]	-55	-58	-55	**-58**
SNDR [dB]	47	49	48	**48**
Power [mW]	795	480	44.6 (1)	**93**
FOM [fJ/conv]	1207	801	76 (1)	**283**
Area [mm²]	7.4	5.1	0.63 (1)	**0.83**

(1) *without references and input buffer*

Figure 22.5.6: Performance comparison with state of the art for input frequencies up to 90% of the Nyquist frequency.

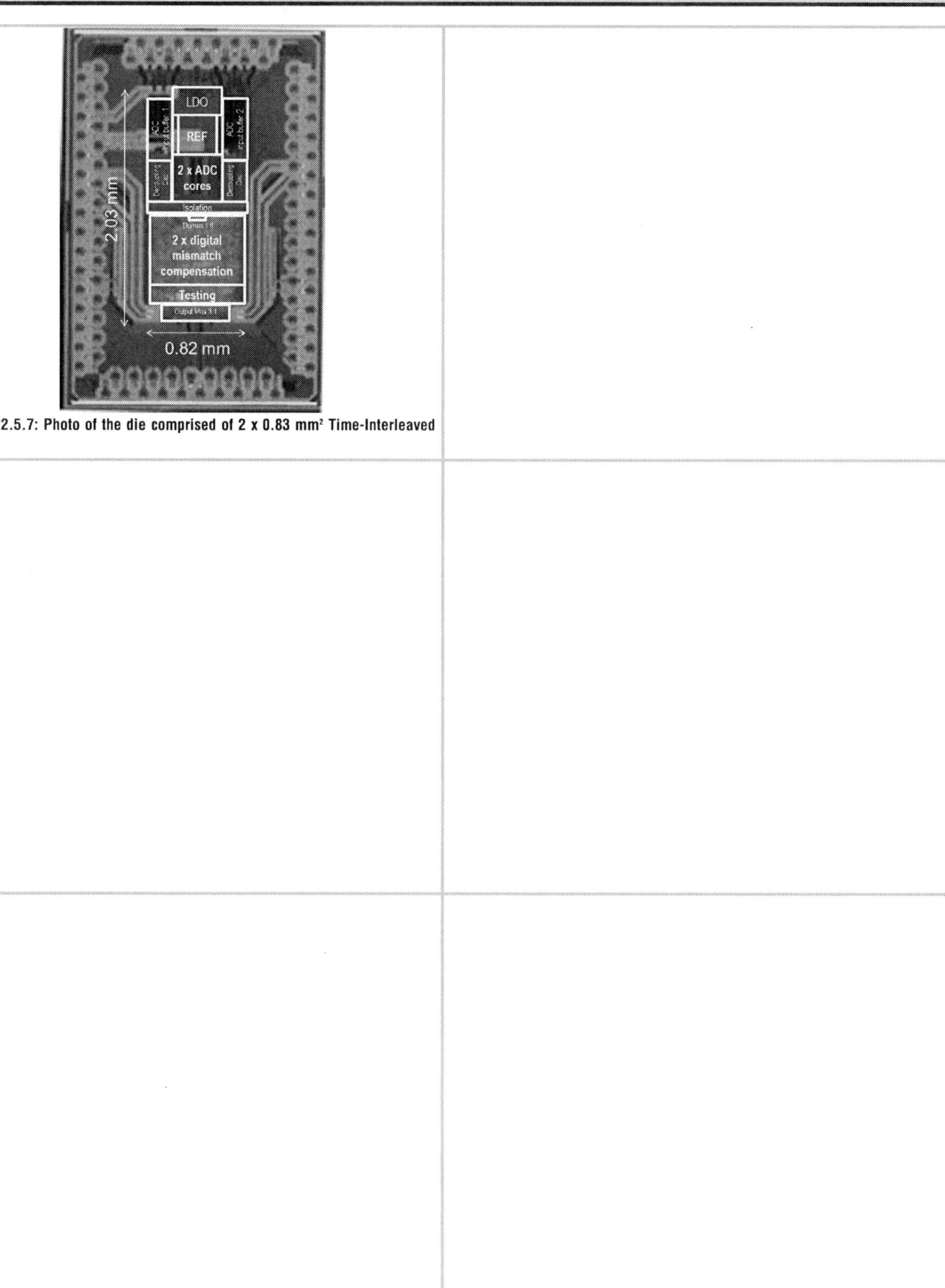

Figure 22.5.7: Photo of the die comprised of 2 x 0.83 mm² Time-Interleaved ADCs.

22.6 A 2.2GS/s 7b 27.4mW Time-Based Folding-Flash ADC with Resistively Averaged Voltage-to-Time Amplifiers

Masaya Miyahara, Ibuki Mano, Masaaki Nakayama, Kenichi Okada, Akira Matsuzawa

Tokyo Institute of Technology, Tokyo, Japan

High-speed low-resolution ADCs are widely used for various applications, such as 60GHz receivers, serial links, and high-density disk drive systems. Flash architectures have the highest conversion rate without employing time interleaving. Moreover, flash architectures have the lowest latency, which is often required in feedback-loop systems. However, the area and power consumption are exponentially increased by increasing the resolution since the number of comparators must be 2^N. A folding architecture is a well-known technique to reduce the number of comparators in an ADC while maintaining high sampling rate and low latency [1,2]. Folding architectures were previously realized by generating a number of zero crossings with folding amplifiers. However, the conventional folding amplifiers consume a large amount of power to realize a fast response. In contrast, a folding ADC with only dynamic power consumption and without using amplifiers is reported in [3]. However, only a folding factor of 2 is realized, and therefore the number of comparators is reduced by half.

This paper presents a 2.2GS/s 7b time-based folding ADC with resistively averaged voltage-to-time amplifiers (V-T amps). This time-based folding architecture consists of simple logic cells with a folding factor of 8 instead of the conventional static amplifiers. This reduces the number of comparators from 128 to 32 for a 7b resolution. Resistively averaged V-T amps have a low offset voltage and a high conversion gain to relax the offset requirement for the SR-latch. This ADC achieves an SNDR of 37.4dB at Nyquist frequency without any calibration technique.

Figure 22.6.1 shows the block diagram of the ADC. The input signal is sampled on passive S/H circuits with the total sampling capacitance of 300fF. An array of 25 V-T amps, 10 of which are dummies to generate extra folding points and overcome the edge effects in resistive averaging networks, converts the sampled voltages into pulse signals that have a delay time depending on input voltage levels. Next, pulse signals are inputted to the coarse SR latch and the time-based folder (TF) blocks. The coarse SR latches compare the delay time and determine the upper 4b digital code. Four TFs output folded delay signals to the fine interpolated SR latches. The fine interpolated SR latches convert the delay signal to the lower 4b digital code. The encoder corrects coarse and fine digital codes and converts 7b data with 1b redundancy. Finally, 7b data are retimed by D-FFs.

Figure 22.6.2 shows the dynamic V-T amps with resistive averaging. The V-T amp is based on the dynamic preamp of double-latch-type comparator [4] with a positive feedback circuit. During the reset phase (ϕ_L =0), M_7 and M_8 pre-charge the D_P and D_N nodes to the supply voltage. After the reset phase, ϕ_L turns to high, M_7 and M_8 are turned off and M_9 is turned on. At D_P and D_N nodes, the voltage drops with a rate determined by the input differential voltage, and an input-dependent Δt_d builds up in a short time of around 100ps. M_5 and M_6 form a positive feedback circuit to enhance the gain of the V-T conversion. M_3 and M_4 control the positive feedback to a suitable ratio because too much positive feedback degrades the linearity of the V-T conversion. In this design, the gains of the V-T conversion with and without positive feedback are 1.1ps/mV and 0.24ps/mV, respectively. The V-T conversion gain with positive feedback is enough to eliminate an offset calibration technique for coarse and fine SR latches. A resistive averaging technique is a well-known technique used to reduce the mismatch of conventional static amplifiers [5]. This architecture is also effective in a dynamic amplifier without increasing power consumption because it has no static current path. The offset voltage of the dynamic amplifier can be reduced to about 1/3.

Figure 22.6.3 shows the TF architecture and the circuit implementation. Delay times of V-T amps are slightly different depending on each reference voltage. The first mountain fold can be realized by using OR gates to select the faster delay signals; D_{1_1} is D_{N0} OR D_{P2} and D_{1_2} is D_{N4} OR D_{P6}. The second valley fold can be realized by using an AND gate to select the slower delay signals; D_{2_1} is D_{1_1} AND D_{1_2}. Finally, D_{F1} can be realized by D_{2_1} AND D_{2_2}. These logic cells have symmetrical input to obtain the same transition time from D_1 or D_2 to output. This architecture consists of simple logic cells with only dynamic power consumption. Moreover, this TF performs in very short time, about 100ps, because only 3 cascaded OR and AND cells are required to realize the folding factor of 8. TF block outputs four folding signals, D_{F1}, D_{F2}, D_{F3} and D_{F4}, to realize interpolated signals in the following fine SR latch block.

Figure 22.6.4 shows the fine SR latch circuit with a phase interpolator [6]. First, to realize the time-domain interpolation, the rising slope of the folding signal is moderated. A 3b phase interpolator consists of 2-stage cascaded inverters. The interpolation ratio of the 2b stage is realized by changing the number of the inverters connected to each input, with ratios of 1:3, 2:2, or 3:1. The next interpolator has a fixed ratio of 1:1. Finally, the phase-interpolated delay signal is converted into a digital code by the SR latch, which consists of NAND gates. The coarse SR latch has the same structure without interpolators. The offset voltage of the SR latch is negligible because the steep rising signals are less sensitive to Vt mismatch of the transistor. Moreover, the gain of V-T amps effectively reduces the input-referred offset voltage of the SR latch. Therefore, no offset calibration is required.

The chip is fabricated in a digital 40nm LP CMOS technology and the occupied area is 0.052mm^2, as shown in Fig. 22.6.7. The supply voltage is 1.1V with a power consumption of 27.4mW (3.3mW for reference ladder, 19.4mW for analog, and 4.7mW for digital) at a sampling rate of 2.2GS/s. The differential input signal range is 1.0Vpp. The output code is decimated by 8 in the measurement. The measured DNL and INL are +0.6/-0.6 and +1.0/-1.0 LSB, respectively. Figure 22.6.5 shows the measured SNR, SFDR, and SNDR vs. sampling rate and input frequency. An SNDR of 38.3dB is measured up to 2.2GS/s with a 100MHz input frequency. Also, an SNDR of 37.4dB is obtained at 2.2GS/s with Nyquist frequency of 1.1GHz. The FoMw of 210fJ/conv.-step and the FoMs of 143.3dB at Nyquist frequency are achieved.

Figure 22.6.6 shows the performance summary and comparison of flash and folding ADCs achieving a conversion rate of several GS/s [7-9].

Acknowledgments:

This work was partially supported by MIC, Berkeley Design Automation for the use of the Analog Fast SPICE(AFS) Platform, and VDEC in collaboration with Cadence Design Systems, Inc.

References:

[1] Y. Nakajima, *et al.*, "A Background Self-Calibrated 6b 2.7GS/s ADC With Cascade-Calibrated Folding-Interpolating Architecture," *IEEE J. Solid-State Circuits*, vol. 45, pp. 707-718, Apr. 2010.

[2] T. Yamase, *et al.*, "A 22-mW 7b 1.3-GS/s Pipeline ADC with 1-bit/stage Folding Converter Architecture," *Symp. VLSI Circuits*, pp. 124-125, June 2011.

[3] B. Verbruggen, *et al.*, "A 2.2mW 5b 1.75GS/s Folding Flash ADC in 90nm Digital CMOS," *ISSCC Dig. Tech. Papers*, pp. 252-253, Feb. 2008.

[4] M. Miyahara, *et al.*, "A Low-Noise Self-Calibrating Dynamic Comparator for High-Speed ADCs", *IEEE A-SSCC*, pp. 269-272, Nov. 2008.

[5] K. Makigawa, *et al.*, "A 7bit 800Msps 120mW Folding and Interpolation ADC Using a Mixed-Averaging Scheme," *Symp. VLSI Circuits*, pp. 124-125, June 2006.

[6] D. Miyashita, *et al.*, "A -104dBc/Hz In-Band Phase Noise 3GHz All Digital PLL with Phase Interpolation Based Hierarchical Time to Digital Convertor," *Symp. VLSI Circuits*, pp. 112-113, June 2011.

[7] B. Murmann, "ADC performance survey 1997-2013," [Online]. Available: http://www.stanford.edu/~murmann/adcsurvey.html.

[8] Y. -S Shu, "A 6b 3GS/s 11mW Fully Dynamic ADC in 40nm CMOS with Reduced Number of comparators," *Symp. VLSI Circuits*, pp. 26-27, June 2012.

[9] V. H. -C. Chen and L. Pileggi, "An 8.5mW 5GS/s 6b Flash ADC with Dynamic Offset Calibration in 32nm CMOS SOI," *Symp. VLSI Circuits*, pp. 264-265, June 2013.

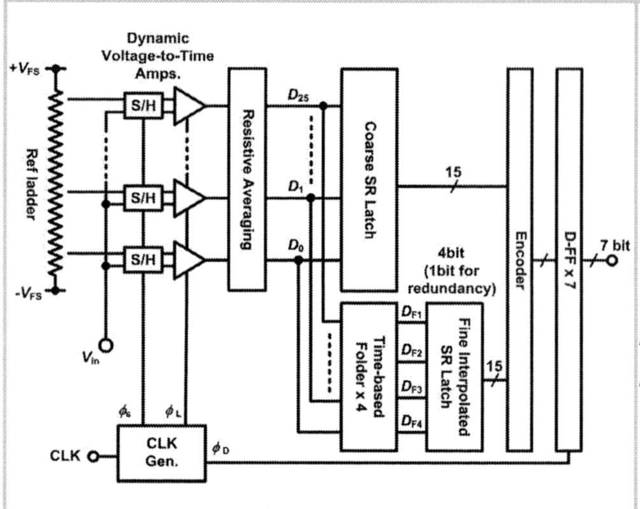

Figure 22.6.1: Block diagram of the time-based folding flash ADC.

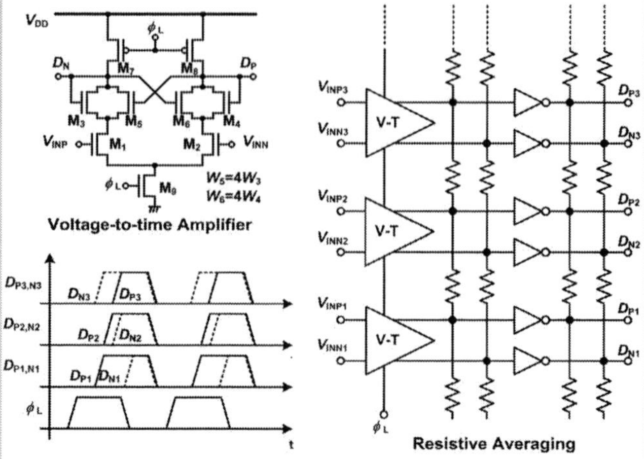

Figure 22.6.2: Voltage-to-time amplifier with positive feedback for gain boosting and resistive averaging.

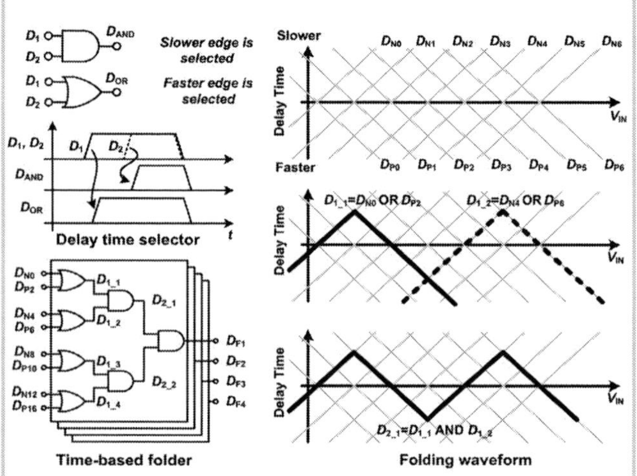

Figure 22.6.3: Time-based folder architecture and circuit implementation.

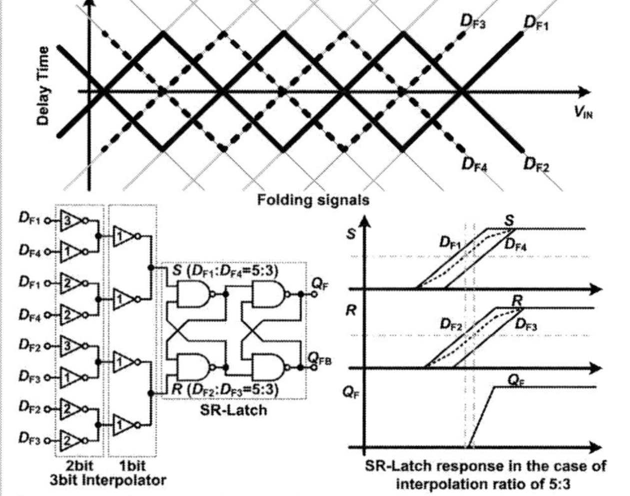

Figure 22.6.4: Interpolated fine SR latches.

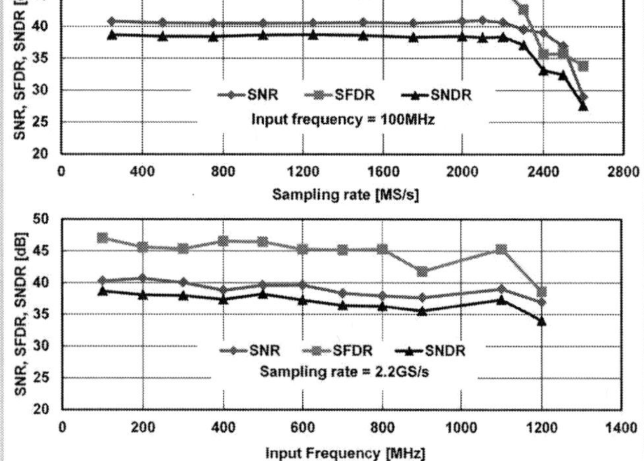

Figure 22.6.5: Measured SNDR vs. sampling frequency and input frequency.

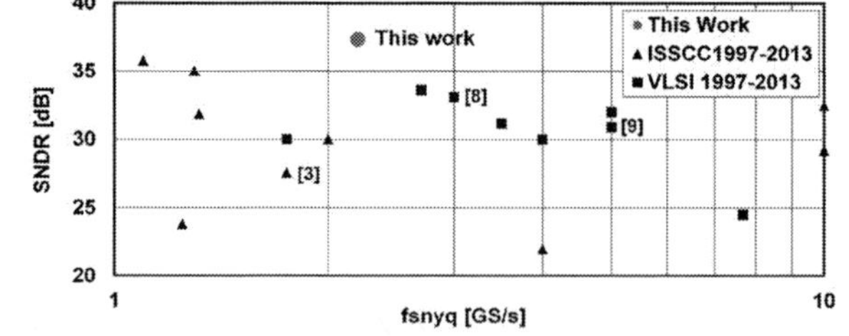

	ISSCC 2008 [3]	VLSI 2012 [8]	VLSI 2013 [9]	This work
Technology	90nm	40nm	32nm SOI	40nm LP
Resolution [bit]	5	6	6	7
Power Supply [V]	1	1.1	0.85	1.1
Sampling Frequency [GS/s]	1.75	3	5	2.2
Power Consumption [mW]	2.2	11	8.5	27.4
SNDR @Nyquist [dB]	27.6	33.1	30.9	37.4
FoMw [fJ/conv.-step]	64.5	99.3	59.4	210
FoMs [dB]	143.5	144.4	145.6	143.3
Core area [mm^2]	0.0165	0.021	0.02	0.052
Calibration	Off chip	Foreground	Off chip	No need

Figure 22.6.6: Performance comparison of flash and folding ADCs with sampling rate of several GS/s.

22

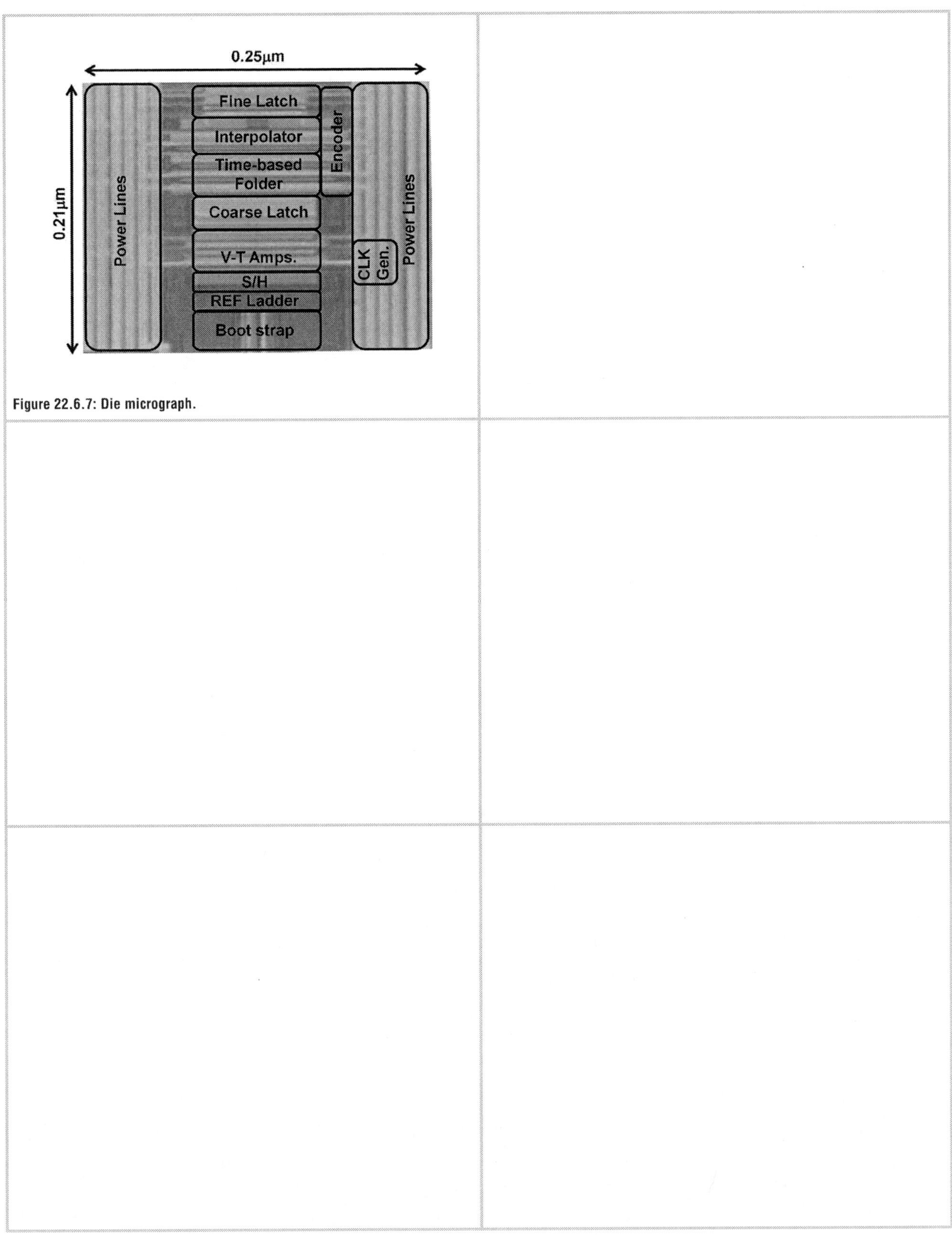

Figure 22.6.7: Die micrograph.

22.7 A 14b 4.6GS/s RF DAC in 0.18μm CMOS for Cable Head-End Systems

Brian Brandt[1], Dan McMahill[2], Miaochen Wu[1],
Paul Kalthoff[3], Ajay Kuckreja[4], Geir Ostrem[3]

[1]Maxim Integrated, North Chelmsford, MA, [2]Maxim Integrated, Woodstock, GA, [3]Maxim Integrated, Colorado Springs, CO, [4]Maxim Integrated, Boulder, CO

Cable head-end systems typically employ multiple QAM modulator signal chains to synthesize the entire TV spectrum for triple-play services, resulting in high power, cost, and size. The DAC reported here can directly synthesize 160 DOCSIS 3.0 compliant 6MHz QAM channels, spanning the entire cable band of 43 to 1003MHz. It thereby enables low-cost, low-power head-end systems for new cable standards such as the converged cable access platform (CCAP) for triple/quad play services.

Figure 22.7.1 shows the cable DAC system, comprising a 40nm digital up converter (DUC) and a 180nm DAC co-packaged in a 12×17mm^2 BGA multi-chip module. The DAC is the primary focus of this paper. The DUC performs QAM mapping, root raised cosine pulse shaping, resampling, and digital RF up-conversion of Forward Error Corrected (FEC) encoded data with full agility. A cascade of interpolation filters, complex modulators, and channel combiners allows modulation of the signal to any frequency from 43 to 1003MHz. The DUC output is passed to a simple digital pre-distortion (DPD) block to attenuate a particular residual frequency spur that is associated with the 2-channel interleaving employed in the DAC, as described later. The 4.6GS/s 14b output of the DPD is de-multiplexed by 4× and sent to the DAC at 1.15GS/s via 56 LVDS pairs. In the DAC, the LVDS data is latched using a synchronization clock produced by a DLL. The data is then multiplexed into two 2.3GS/s streams for input to a pair of time-interleaved 14b sub-DACs. Finally, the two sub-DAC outputs are re-timed by a butterfly switch operating at 4.6GS/s. A clock generator provides low-jitter timing signals to the butterfly and sub-DACs.

The interleaved DAC architecture allows 4.6GS/s operation in 180nm CMOS [1]. The butterfly circuit enables this interleaving while simultaneously achieving low distortion by retiming the sub-DAC outputs to both edges of a 2.3GHz clock. The butterfly serves a second important purpose of retiming the outputs of the segments within each sub-DAC core [2]. To reduce static and dynamic nonlinearity, each 14b sub-DAC is segmented into 4 unary current source arrays comprising the 5 MSBs, a 4b array, a 3b array, and a 2b LSB array. Because of timing mismatches, the outputs of these arrays can change at slightly different times. The resulting glitches, which normally degrade dynamic performance, are blocked from the main DAC output by the butterfly switch, which selects the other (settled) sub-DAC output during this time period.

A unique challenge resulting from the interleaved architecture is that the image of second harmonic distortion in each sub-DAC (at 2.3GHz – $2F_{SIG}$) will not generally be fully canceled in the combined DAC output spectrum. This spur has low amplitude, is very repeatable from part to part, and is constant over temperature because of small unavoidable systematic mismatches and couplings in the layout. As in interleaved ADCs, digital correction of interleave errors is advantageous. The simple DPD circuit shown in Fig. 22.7.2 reduces this spur by >8dB over the entire frequency band and temperature range. A fixed 6-tap FIR filter precedes an adjustable gain/phase block with 2 programmable coefficients. In practice, gain and phase calibration for individual parts is unnecessary. Next, the signal is squared and modulated by an alternating ±1 sequence before summation with the original signal. The DPD occupies 0.08mm^2 and does not reduce any other spurs, which are sufficiently controlled by careful design and layout.

Figure 22.7.3 depicts a single current source with switches and switch drivers. Two levels of latching are used to fully attenuate signal-dependent delays from the decoder. When conducting, each switch also serves as a cascode with its gate and drain biased at 1.8V. Segmentation, along with careful layout, non-minimum transistor sizes, and randomized current source placement, eliminate the need for trimming or calibration. Figure 22.7.4 shows a half-circuit of the differential butterfly switch in which 2 differential pairs operating 180° out of phase alternately steer currents from the 2 interleaved sub-DACs to either the output load or the dummy load. An identical half-circuit (not shown) produces the OUTM DAC output. A low-jitter differential clock drives the two inverters, whose outputs are ac coupled to M1/M2. One of the cross-coupled PMOS pull-up devices (M3/M4) pulls one differential pair input up to 1.8V+Vgs, while the other input is driven to about 0.25V+Vgs. As a result, the sources of M1/M2, which connect to the drain of the cascode switch in Fig. 22.7.3, are biased at 1.8V while their drains are set to 3.3V–Vgs by M7/M8. Accordingly, M1/M2 can be faster low-voltage 180nm transistors since they are protected from the 3.3V supply during normal operation. To protect them during startup, shutdown, or an inactive clock, M5/M6 are turned on by a simple detector (not shown) whenever the supplies or bias currents are not valid or the clock is inactive.

One of the distinguishing features of this DAC is its high output current capability of 80mA compared to the typical 16, 20, or 50mA of prior designs [3-6]. The QAM modulator in the cable head-end is followed by a chain of amplifiers. High DAC output power minimizes the noise contribution from these amplifiers and eliminates one stage of amplification, thereby reducing system power and cost. To achieve low distortion at the high output power level, the static 40mA current in Fig. 22.7.4 is added to the 0-to-80mA signal current from the sub-DACs. This improves the settling and large signal behavior of the butterfly switch by reducing current variation in critical signal path transistors. This additional dc current is absorbed in the external inductor typically connected to differential DAC outputs. The chip includes 25Ω resistors on each DAC output (Fig. 22.7.4) to minimize overshoot in the DAC output due to package inductance.

Figure 22.7.5 presents typical measured 2-tone intermodulation distortion (IMD), a better indicator of linearity performance than SFDR for communication systems. Low distortion is maintained across the entire cable band. Adjacent Channel Power Ratio (ACPR) is also of primary importance and this DAC achieves 55dBc for 160 DOCSIS channels. Notable in the measured results in Fig. 22.7.6 are the high update rate and resolution, the low distortion, and the high output current achieved in a 180nm technology without trimming or calibration. Several system advantages stem from the high update rate including simpler reconstruction filtering with smaller phase errors, reduced quantization noise density, and facilitation of DPD for correcting power amplifier distortion. Also the 4.6GS/s update rate provides gain flatness of 1.3dB across the 960MHz cable bandwidth, minimizing the need for analog equalization following the DAC. While power dissipation and die area are comparatively high, they are less important because its performance and output current capability enable a dramatic power and cost reduction at the system level of the cable head-end. This DAC signal chain replaces 20-to-40 octal or quad QAM modulator signal chains that dissipate 500 to 1000W in legacy systems. Each sub-DAC sinks 160mA at 3.3V (530mW) comprising an 80mA signal current and two 40mA dc currents, which improve distortion as described earlier. The low-jitter clock generator consumes 440mA at 1.8V (790mW) due in large part to driving the large switches needed to support the high output current and the comparatively large parasitic capacitances of the 180nm technology. Thus, most of the DAC's power dissipation is directly related to producing high output current at 4.6GS/s with low distortion in a 180nm technology. The remaining 450mW is consumed in digital logic, DLL and interface circuits, bias generation, and generally providing wide design margins to enable a robust product with high yield over process, supply, and temperature (–40 to 110°C) variations. The DPD dissipates about 60mW while the DUC dissipates 2.9W at 0.9V when processing 160 QAM channels. Figure 22.7.7 shows a die photo of the DAC.

Several techniques are key to achieving the DAC's performance level, including an interleaved architecture, a butterfly circuit for segment and interleave retiming, targeted and simple DPD, and design for high output current. The high speed, low distortion, and high output power achieved in a low-cost CMOS technology without trimming or calibration, enable a significant power, size, and cost reduction for cable head-end systems.

References:

[1] D. McMahill and A. Kuckreja, *High Speed Digital-to-Analog Converter with Low Voltage Device Protection*, US Patent 8,498,086, July 2013.

[2] G. Ostrem and A. Kuckreja, *Wide band digital to analog converters and methods, including converters with selectable impulse response*, US Patent 6,977,602, Dec. 2005.

[3] G. Engel, *et al.*, "A 14b 3/6GHz Current-Steering RF DAC in 0.18μm CMOS with 66dB ACLR at 2.9GHz," *ISSCC Dig. Tech. Papers*, pp. 458-459, 2012.

[4] W. Tseng, *et al.*, "A 12b 1.25GS/s DAC in 90nm CMOS with >70dB SFDR up to 500MHz," *ISSCC Dig. Tech. Papers*, pp. 192-193, Feb. 2011.

[5] C.-H. Lin, *et al.*, "A 12b 2.9GS/s DAC with IM3 <-60dBc Beyond 1GHz in 65nm CMOS," *ISSCC Dig. Tech. Papers*, pp. 74-75, Feb. 2009.

[6] W.T. Lin, *et al.*, "A 12b 1.6GS/s 40mW DAC in 40nm CMOS with >70dB SFDR over Entire Nyquist Bandwidth," *ISSCC Dig. Tech. Papers*, pp. 474-475, Feb. 2013.

978-1-4799-0917-9/14 $31.00 © 2014 IEEE

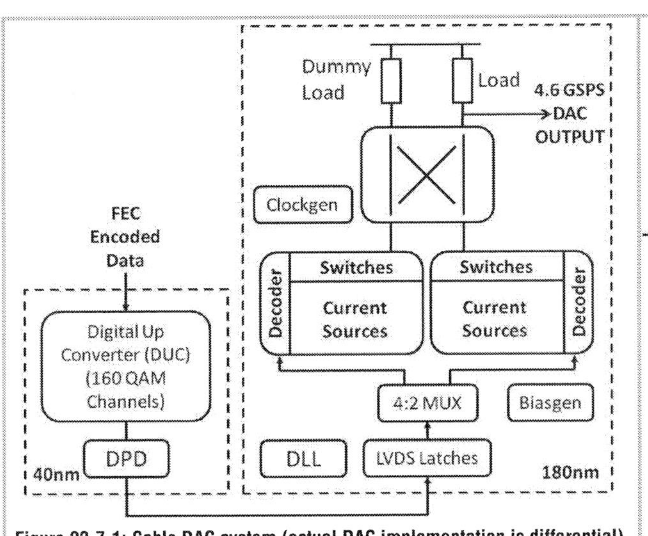

Figure 22.7.1: Cable DAC system (actual DAC implementation is differential).

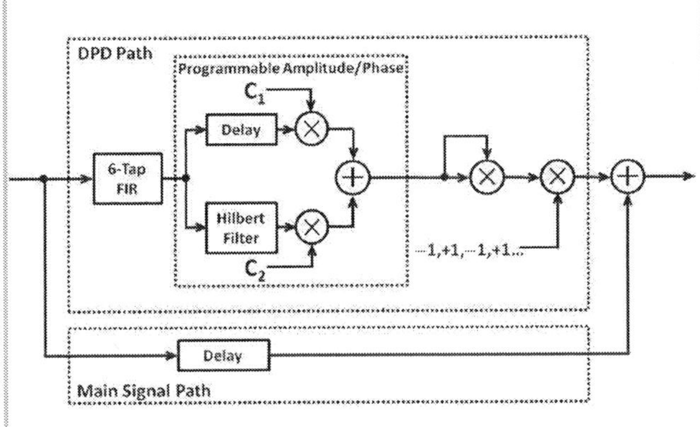

Figure 22.7.2: Digital pre-distortion block diagram.

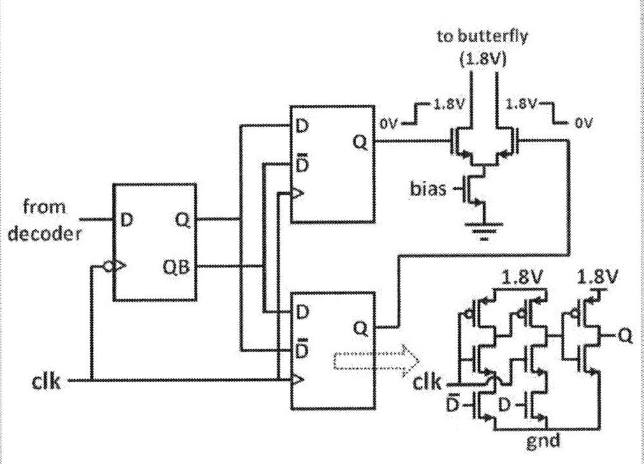

Figure 22.7.3: Current source, switches, and switch drivers.

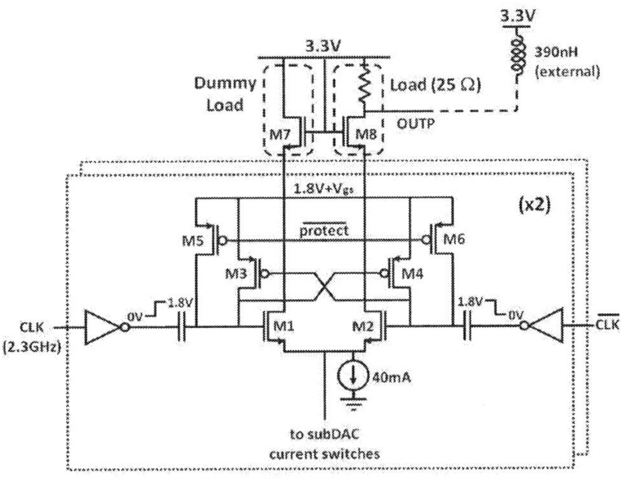

Figure 22.7.4: Butterfly switch half-circuit.

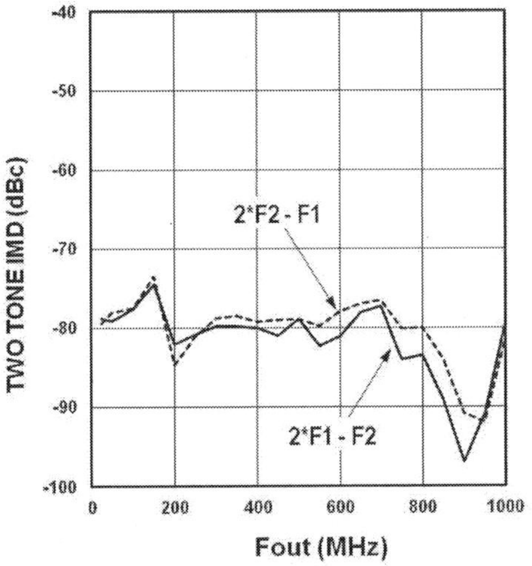

Figure 22.7.5: Typical measured IMD performance.

Parameter	This Work	[3]	[4]	[5]	[6]
Technology (nm)	180	180	90	65	40
Update Rate (GSPS)	4.6	3	1.25	2.9	1.6
Resolution (bits)	14	14	12	12	12
SFDR@500MHz (dBc)	69	60	70	52.5	73
IMD @500MHz (dBc)	-79	<-70		-72.5	<-70
DOCSIS ACPR (dBc)	55 (160 carriers)	52 (150 carriers)			
Output current (mA)	80	20	16	50	16
Power (mW)	2300	600	128	188	40
Vdd (V)	1.8, 3.3	1.8, −1.5	1.2, 2.5	1.0, 2.5	1.2
active area (mm^2)	5.2	4.0	0.83	0.31	0.016

Figure 22.7.6: Performance summary and comparison.

22

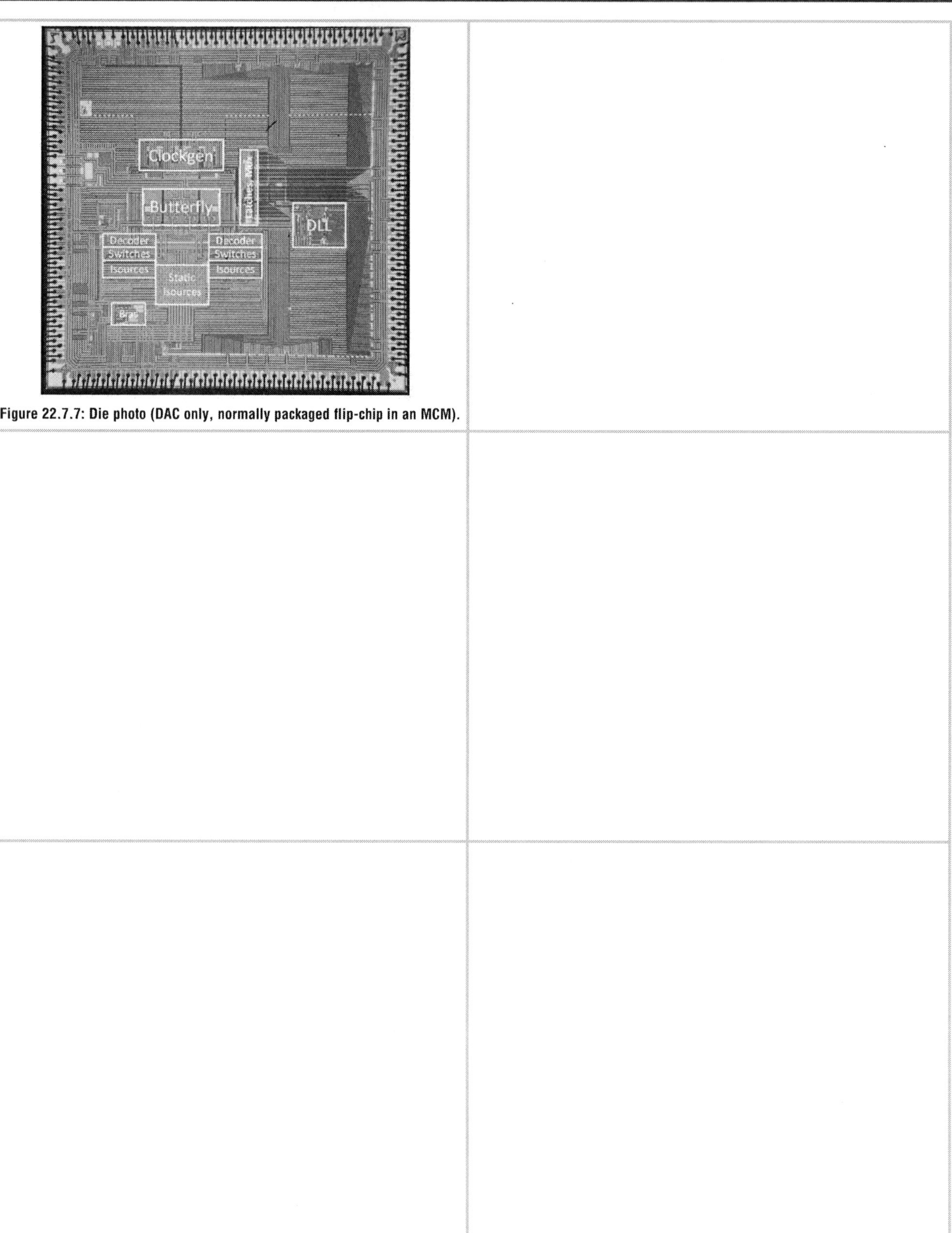

Figure 22.7.7: Die photo (DAC only, normally packaged flip-chip in an MCM).

Session 23 Overview: *Energy Harvesting*

ANALOG SUBCOMMITTEE

Session Chair: *Makoto Nagata*
Kobe University, Kobe, Japan

Session Co-Chair: *Tsung-Hsien Lin*
National Taiwan University, Taipei, Taiwan

For powering devices from ambient energy sources, innovative CMOS circuit techniques are devised to facilitate highly efficient energy harvesting. Depending on the applications, the loads can range widely from nW to mW and the required output voltages vary from sub-1V to tens of volts. The energy sources can be electrochemical gradient inside an ear, thermoelectric generators, piezoelectric transducers, fuel cells, solar cells, and others. This session starts with three papers presenting novel ultra low power circuits for charge pumping, DC-to-DC conversion and power management, followed by five papers on enabling techniques for dual source power management, energy pile-up resonance, maximum power point tracking and high-voltage charge pumps.

23.1 A 0.15V-Input Energy-Harvesting Charge Pump with Switching Body Biasing and Adaptive Dead-Time for Efficiency Improvement — **8:30 AM**

J. Kim, Hong Kong University of Science and Technology, Hong Kong, China and Korea University, Seoul, Korea

In Paper 23.1, Korea University and HKUST present an energy harvesting charge pump with switching body biasing and adaptive dead-time schemes for efficiency improvement. The charge pump achieves 240% improvement of maximum output current over conventional architecture and enables the operation at input voltage down to 0.15 V.

23.2 A 1.1nW Energy Harvesting System with 544pW Quiescent Power for Next-Generation Implants — **9:00 AM**

S. Bandyopadhyay, Massachusetts Institute of Technology, Cambridge, MA, now at Texas Instruments, Dallas, TX

In Paper 23.2, MIT, UCSD, Massachusetts Eye and Ear Infirmary and Massachusetts General Hospital, introduces a power management circuit that harvests energy from the endocochlear potential. The entire harvesting circuit consumes only 544pW with an efficiency of 56%.

23.3 A 3nW Fully Integrated Energy Harvester Based on Self-Oscillating Switched-Capacitor DC-DC Converter — **9:30 AM**

W. Jung, University of Michigan, Ann Arbor, MI

In Paper 23.3, University of Michigan presents a fully-integrated energy harvester, which consumes only 3nW and operates in an adaptive manner to loads ranging from 5nW to 5μW.

978-1-4799-0917-9/14 $31.00 © 2014 IEEE

23.4 Dual-Source Single-Inductor 0.18µm CMOS Charger-Supply with Nested Hysteretic and Adaptive On-Time PWM Control **10:15 AM**

S. Kim, Georgia Institute of Technology, Atlanta, GA

In Paper 23.4, Georgia Institute of Technology reports a dual-source single-inductor CMOS charger-supply. It achieves 62 to 83% power conversion efficiency over 0.1 to 8mA load current. This converter requires 65% less volume than the prior smallest design.

23.5 An Energy Pile-Up Resonance Circuit Extracting Maximum 422% Energy from Piezoelectric Material in a Dual-Source Energy-Harvesting Interface **10:45 AM**

Y-S. Yuk, KAIST, Daejeon, Korea

In Paper 23.5, KAIST presents an energy extracting scheme based on an energy pile-up technique which builds the internal voltage swing of the piezoelectric transducer up to 5.3 times of the original swing. The piezoelectric harvesting circuit achieves a peak efficiency of 74.9%.

23.6 A 43V 400mW-to-21W Global-Search-Based Photovoltaic Energy Harvester with 350µs Transient Time, 99.9% MPPT Efficiency, and 94% Power Efficiency **11:15 AM**

S. Uprety, University of Texas, Dallas, Richardson, TXIn Paper 23.6, University of Texas at Dallas presents a 43V 400mW-to-21W photovoltaic energy harvester with 94% power efficiency. With a global-search-based maximum-power-point tracking technique, the harvester tracks global maximum-power-point under partial shading conditions in 350µs transient time.

23.7 Self-Powered 30µW-to-10mW Piezoelectric Energy-Harvesting System with 9.09ms/V Maximum Power Point Tracking Time **11:45 AM**

M. Shim, Korea University, Seoul, Korea

In Paper 23.7, Korea University demonstrates a self-powered 30µW-to-10mW piezoelectric energy harvesting system with a maximum power conversion efficiency of 80%. This system realizes the fastest maximum-power-point tracking time of 9.09ms/V, over 9 times faster than the state-of-the-art.

23.8 A 34V Charge Pump in 65nm Bulk CMOS Technology **12:00 PM**

Y. Ismail, University of California, Los Angeles, CA

In Paper 23.8, UCLA and SiTime introduces a high-voltage charge pump producing 34V with 2.5V tolerant transistors in a 65nm bulk CMOS technology.

23.1 A 0.15V-Input Energy-Harvesting Charge Pump with Switching Body Biasing and Adaptive Dead-Time for Efficiency Improvement

Jungmoon Kim[1,2], Philip K. T. Mok[1], Chulwoo Kim[2]

[1]Hong Kong University of Science and Technology, Hong Kong, China
[2]Korea University, Seoul, Korea

Design of low-voltage and efficient energy-harvesting circuits is becoming increasingly important, particularly, for autonomous systems. Since the amount of energy that can be harvested from the surrounding environment is limited, the available output voltage of a harvester is low. Therefore, the design of a low-input-voltage (low-V_{IN}) up-converter is critical to self-powered systems [1-3]. Moreover, the form factor is very constrained in applications such as wearable electronic devices and sensor networks. Recently, low-V_{IN} charge pumps (CPs) for energy harvesting has been compared with DC-DC converters using a large inductor [1-3]. CPs introduced in [1] and [2] use the advanced process technology to push V_{IN} down to the subthreshold region. The CP in [1] introduces a forward-body-biasing (FBB) technique, which improves the voltage conversion efficiency (VCE) for low V_{IN} but shows poor power conversion efficiency (PCE). The CP in [2] achieves the lowest operation voltage. However, the design with a 10-stage CP provides low output power. This paper presents a CP with switching-body-biasing (SBB), adaptive-dead-time (AD), and switch-conductance (SW-*G*) enhancement techniques to improve the PCE for low V_{IN} as well as to extend the maximum load current.

The block diagram of a 3-stage CP is depicted in Fig. 23.1.1. Unit CP consists of a voltage doubler, a dual-series PMOS switch, and 4 switches for SBB. The 4 clock signals (CLK, CLK_B, N, and N_B) are generated from the input clock (V_{CLK}) by the AD circuit, which minimizes the dead-time for high PCE of the CP. The complementary signals, CLK and CLK_B, are overlapped while two pairs of complementary signals, N/N_B and E/E_B, are non-overlapped. The former 4 signals from buffers are swinging between 0 and V_{IN}. The latter signals from the SW-*G* enhancer are driven between $-V_{IN}$ and V_{OUT}. The SW-*G* enhancer improves the conductance of the dual series switch of each CP.

A detailed schematic of the CP is shown in Fig. 23.1.2. A voltage doubler, which includes a cross-coupled NMOS pair (M_{N1} and M_{N2}) in a deep n-well and two pumping capacitors (C_{P1} and C_{P2}), allows CT and CT_B to swin g between $V_{O,i-1}$ and $V_{O,i}$, where $V_{O,i-1}$ and $V_{O,i}$ are the input and output voltages of the unit CP, respectively. A FBB technique introduced in [1] decreases the threshold voltage of the NMOS pair to increase the on-current for low V_{IN}. However, when one of the NMOS pair is turned off, the "fixed" FBB voltage causes a reverse current to flow into the input of the unit CP. This means that the forward bias should be applied only when the MOSFET is turned on. Therefore, we use an SBB technique that avoids the reverse current. The body bias voltage (V_{B1}) for M_{N1} is set to zero when M_{N1} should be completely turned off. V_{B1} is connected to $V_{O,i}$ when the FBB is required. This dynamic biasing can be implemented by only four switches (M_{S1} to M_{S4}) and clock signals (E and E_B) used for the dual series switch after the voltage doubler. The conventional FBB technique for the low-voltage CP in [1] considerably improves the VCE but not the PCE, which limits the power throughput of the CP. Dual series switches M_{P1} and M_{P2} are controlled by E and E_B from the SW-*G* enhancer. In order to improve the conductance of the dual series switches, a level shifter (LS) is used for the low-voltage voltage doubler in [4]. We refer to this type of voltage doubler as CP-LS. The doubler with cross-coupled series switches is called CP-cross. In the first stage, the gates of the series switches are driven between V_{IN} and $2V_{IN}$ in the CP-cross, while the gates are driven between 0 and $2V_{IN}$ in the CP-LS. Therefore, the CP-LS has a better conductance than the CP-cross. However, the LS that fails during low-voltage operation causes a reverse current in the series switches. Moreover, when V_{IN} is around V_{TH} or even less than V_{TH} or the output voltage is decreased owing to a heavy load, there is no gain from the increased gate-source voltage (V_{GS}) of the series switch. The proposed CP has a dual series switch that is driven between $-V_{IN}$ and V_{OUT} without using an LS, instead using a negative CP [5]. Hence, a dual series PMOS switch has improved conductance even for low V_{IN}. It should be noted that the conductance improved by the SW-*G* enhancement technique is so large that SBB for the dual series switch is not used in this work.

Figure 23.1.3 shows the AD circuit. Control signals E and E_B for the dual series switches of the unit CP are non-overlapped because a low V_{IN} causes a severe reverse current through the series switches owing to the slow transition of the clock signals. The AD circuit generates 4 signals: CLK, CLK_B, N, and N_B. CLK and CLK_B, used for the unit CP, are overlapped. N and N_B are non-overlapping signals. The non-overlapping period, referred to as the dead-time, should be not only sufficiently long to avoid a large short-circuit current but also sufficiently short to maximize the current transfer for high PCE. In the conventional dead-time control technique using large fixed dead-times, the dead-times excessively increase with a decrease in V_{IN}. Hence, the delay of the delay cells in the dead-time circuit should be controlled. A possible method to accomplish this is to sharpen the slow rising and falling transitions of the delay cells by supplying more current as V_{IN} decreases. This method consumes a large amount of power, which is inappropriate for an efficient CP. Another possible solution is to simplify the method of controlling delay cells by a binary control: long or short dead-time. This method dissipates a small amount of power and can be very efficient in low-power applications. In this circuit, the position of the multiplexer (MUX) is important. If the MUX is placed within the delay cells in the dead-time circuit, the capacitance of delay cells will be increased and the dead-time cannot be shortened as intended. Thus, parallelism is exploited. This implies that 2 dead-time circuits, 1 with a short dead-time (τ_S) (with low SEL) and the other with a long dead-time (τ_L) (with high SEL), are supplied with power via sleep transistors. The MUX only forwards 1 output of the dead-time circuit into buffers. A low-V_{IN} detector is used to control the sleep transistors and the MUX. The PCE improvement of the proposed technique dominates the area and power overhead of added circuitries. Figure 23.1.3 shows the measured dead-time as a function of V_{IN} for 2 different fixed dead-times and the AD.

Figure 23.1.4 shows a detailed schematic of the SW-*G* enhancer. A two-phase negative CP used to increase V_{GS} of the series switches allows E and E_B to be driven down to $-V_{IN}$. The 2 small voltage doublers are supplied by the output of the second stage and generate control signals C1/C2 and L1/L2 to allow E and E_B to turn off the series PMOS switches completely, respectively.

Figure 23.1.5 shows the measured efficiency graphs. The PCE improvement by the AD circuit is 17% at the low V_{IN} of 0.2V and R_L of 10kΩ. The CP without the AD can work at a minimum V_{IN} of 0.2V. However, it can work at V_{IN} down to 0.15V when the AD is turned on. Moreover, the output power throughput can be extended. The PCE comparisons among the CP-cross, CP-LS, and suggested CP are also shown. Pumping capacitors (10nF), switches on the power path, and buffers driving pumping capacitors have same sizes for three CPs. The conventional CPs exhibit poor PCE at a low V_{IN} of 0.3V. The sunken region at the curve of the CP-LS is caused by the failure of LS. The CP-cross working at V_{IN} higher than 0.3V has slightly better PCE at a light load because the suggested CP has additional circuitry in comparison to the CP-cross. The circled A indicates the PCE improvement due to the AD under the same conditions of R_L of 10kΩ and V_{IN} of 0.3V. The suggested CP improves a maximum output current by 240% as compared to the CP design in [1].

A performance comparison table is presented in Fig. 23.1.6. The proposed CP has a lower V_{IN}, higher I_{OUT}, and better PCE than the CPs in [1] and [2]. The CP in [2] has the lowest V_{IN}, however, its output current is significantly smaller than the suggested CP. Moreover, techniques presented in this work can be flexibly applied to various types of CPs. Figure 23.1.7 shows the chip micrograph.

Acknowledgement:
This work is supported in part by the National Research Foundation of Korea through a grant funded by the Korean government (MEST) (No. 2011-0020128) and the Research Grant Council of Hong Kong SAR Government, China, under project No. 617010.

References:

[1] P.-H. Chen, K. Ishida, X. Zhang, *et al.*, "0.18-V Input Charge Pump with Forward Body Biasing in Startup Circuit Using 65nm CMOS," *IEEE CICC*, pp. 239-242, September 2010.
[2] P.-H. Chen, K. Ishida, X. Zhang, *et al.*, "A 120-mV Input, Fully Integrated Dual-Mode Charge Pump in 65-nm CMOS for Thermoelectric Energy Harvester," *Asia and South Pacific Design Automation Conference (ASP-DAC)*, pp. 469-470, January 2012.
[3] Y.-C. Shih and B. P. Otis, "An Inductorless DC–DC Converter for Energy Harvesting with a 1.2-μW Bandgap-Referenced Output Controller," *IEEE TCAS-II*, vol. 58, pp. 832-836, December 2011.
[4] P. Favrat, P. Deval, and M.J. Declercq, "A High-Efficiency CMOS Voltage Doubler," *IEEE JSSC*, vol. 33, no. 3, pp. 410-416, March 1998.
[5] J. Kim, P.K.T. Mok, C. Kim, and Y.K. Teh, "A Low-Voltage High-Efficiency Voltage Doubler for Thermoelectric Energy Harvesting," *IEEE Int'l Conf. of Electron Devices and Solid-State Circuits (EDSSC)*, June 2013.

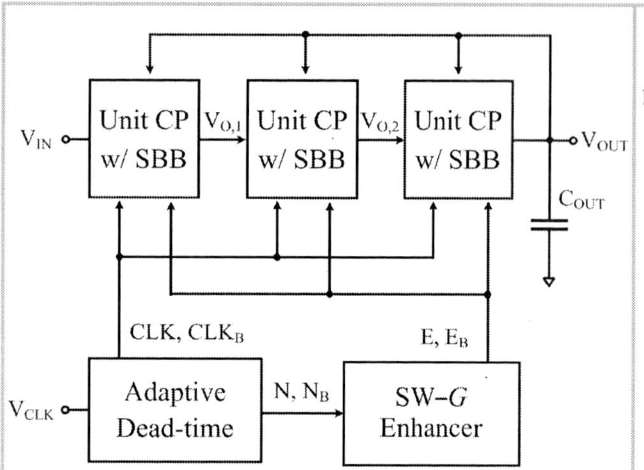

Figure 23.1.1: Block diagram of the proposed CP with switching-body-biasing, adaptive-dead-time, and SW-*G* enhancement techniques.

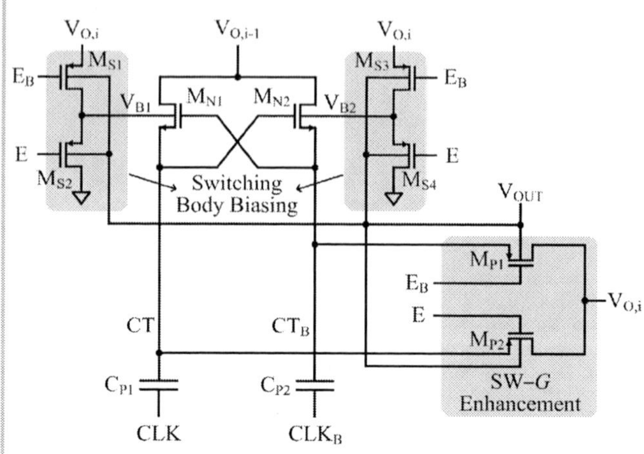

Figure 23.1.2: Unit CP with switching body biasing and conductance-enhanced dual-series switch.

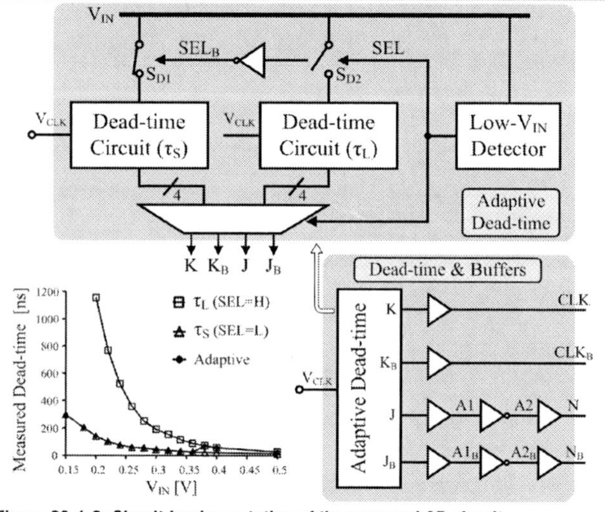

Figure 23.1.3: Circuit implementation of the proposed AD circuit.

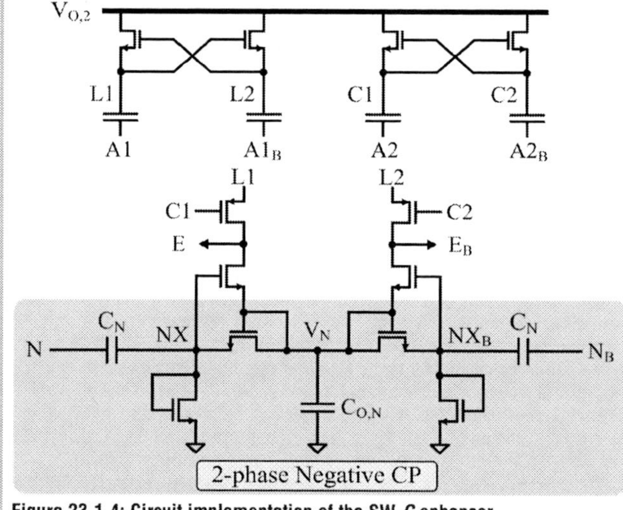

Figure 23.1.4: Circuit implementation of the SW-*G* enhancer.

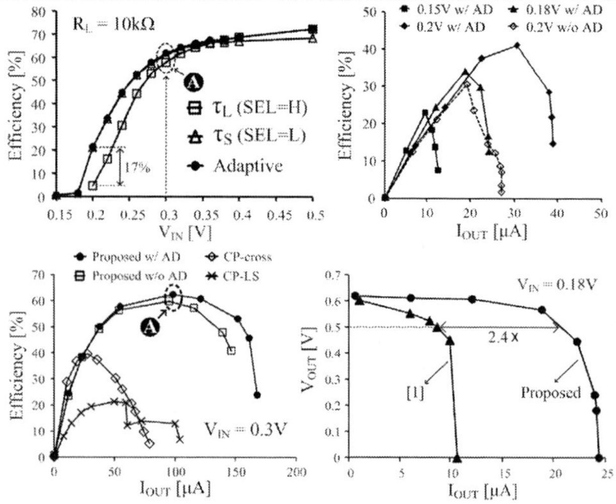

Figure 23.1.5: Measurement results.

Parameters	[1]	[2]	[3]	This work
Process	65nm CMOS	65nm CMOS	0.13μm CMOS	0.13μm CMOS
No. of Stages	3-stage doubler	10-stage CP	3-stage CP	3-stage doubler
Clock Frequency	10MHz	1MHz, 20MHz	800kHz	250kHz
Min. V_{IN}	0.18V	0.12V	0.27V	0.15V
V_{OUT} @ No load	0.6V @ V_{IN} = 0.18V	0.77V @ V_{IN} = 0.12V	1.3V @ V_{IN} = 0.35V	0.619V @ V_{IN} = 0.18V
Max. Efficiency	N/A	38.8% @ V_{IN} = 0.12V	56% @ V_{IN} = 0.45V	34% @ V_{IN} = 0.18V, 72.5% @ V_{IN} = 0.45V
I_{OUT} @ V_{OUT} = 0.5V	8.75μA @ V_{IN} = 0.18V	7μA @ V_{IN} = 0.12V (from extrapolation)	5μA @ V_{IN} = 0.45V	21μA @ V_{IN} = 0.18V
Pros	Low V_{IN}	Lowest V_{IN}, High efficiency	Low cost process	Low cost process, Low V_{IN}, Highest I_{OUT}, High efficiency
Cons	High cost process, Low efficiency, Low I_{OUT}	High cost process, Low I_{OUT}	Highest V_{IN}, Lowest I_{OUT}	

Figure 23.1.6: Performance comparison.

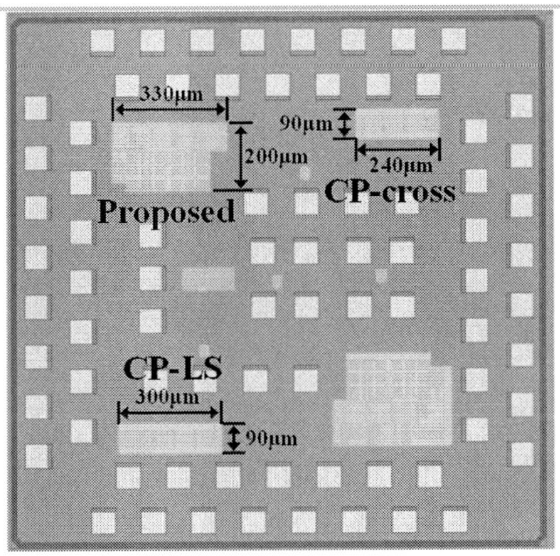

Figure 23.1.7: Chip micrograph.

23.2 A 1.1nW Energy Harvesting System with 544pW Quiescent Power for Next-Generation Implants

Saurav Bandyopadhyay[1,*], Patrick P. Mercier[1,2], Andrew C. Lysaght[3], Konstantina M. Stankovic[3,4], Anantha P. Chandrakasan[1]

[1]Massachusetts Institute of Technology, Cambridge, MA,
[2]University of California, San Diego, La Jolla, CA,
[3]Massachusetts Eye and Ear Infirmary, Boston, MA,
[4]Massachusetts General Hospital, Harvard Medical School, Boston, MA,
[*]Now at Texas Instruments, Dallas, TX,

A wireless sensor that is powered from the endocochlear potential (EP), a 70-to-100mV bio-potential inside the mammalian ear, has been demonstrated in [1]. Due to the anatomical size and physiological constraints inside the ear, a maximum of 1.1 to 6.25nW can be extracted from the EP. The nanowatt power budget of the sensor gives rise to unique challenges with power conversion efficiency and quiescent current reduction in the power management unit (PMU). While [1] presents the system aspects of the biomedical harvesting including the biologic interface and system measurements, this work presents the details of the nanowatt PMU required to power the electronics. More specifically, it focuses on the low-power circuit design techniques needed to realize a nW power converter that is applicable to a broad spectrum of emerging biomedical applications with ultra-low energy-harvesting sources.

Figure 23.2.1 shows the detailed schematic of the PMU that consists of a boost converter along with its associated control, drivers and timer circuits. Due to the ultra-low power budget, the boost converter operates in discontinuous-conduction mode (DCM). The EP and electrodes can be modeled as a voltage source (V_{EP} of 70 to 100mV) in series with a resistor (R_{elec}) of 400kΩ to 1.2MΩ [1]. The boost converter steps up the input voltage, V_{IN} (30 to 55mV, close to half the EP for maximum power extraction), to V_{DD} (0.8 to 1.1V in this implementation). In order to optimize the power conversion efficiency (PCE) within a nW power budget, the power FETs have been sized optimally to not only minimize switching and conduction losses that are normally considered in PMUs handling higher power levels, but also to minimize losses due to subthreshold leakage. The converter has the additional constraint arising from the input-impedance requirement for maximum power transfer. Since the converter input impedance is related to its switching frequency and power FETs on times [2], the switching frequency has been appropriately selected to minimize converter losses and meet the input impedance requirement. Figure 23.2.2 shows the optimization plots of the losses associated with the converter power train (conduction, switching and leakage loss) versus switching frequency and FET sizes. For a given input impedance, lower switching frequencies result in higher rms currents, hence higher conduction losses, and higher switching frequencies result in higher switching losses. In this implementation, the converter is made to operate at 12.8Hz, close to the optimal switching frequency shown in Fig. 23.2.2.

To ensure system sustainability, the control circuits, timer, reference and gate drivers in the PMU have been designed to have quiescent current in the 10 to 100s of pA range. A pW relaxation oscillator is used to generate the 12.8Hz clock required by the boost converter. A constant-g_m current reference in used to create the bias currents for the analog comparators and current sources in the relaxation oscillator. Additionally, the PMU employs a clock divider to create a *sub-Hz* clock to trigger the sensor RF TX. Duty cycling the sensor RF-TX enables the PMU to buffer the energy extracted from the EP and turn on the RF-TX periodically for short bursts.

Figure 23.2.1 also shows the circuits required to generate the gate signals for the boost converter power FETs *N0* and *P0*. A Φ_1-pulse-generation circuit creates the required pulse widths using delay elements for the converter Φ_1 phase. This circuit ensures the boost converter has close to optimal input impedance necessary for maximum power extraction from the EP. Since the impedance of the electrodes (R_{elec}) is known a priori, the input impedance of the converter does not need to be dynamically varied and is set to a fixed setting in the Φ_1-generation circuit. Additionally, a Φ_2-pulse-generation circuit is designed that uses similar delay elements. Since the converter operates in DCM, a zero current switching (ZCS) circuit [2] adjusts the delays in the Φ_2-pulse-generation circuit so that *P0* is turned off when the inductor current is close to zero. A dynamic comparator clocked with a delayed version of the Φ_2 pulse compares V_{DRAIN} with V_{DD}. By using an increment/decrement logic, a 3b code is adjusted that sets the Φ_2 pulse width. Digital implementations of these circuits help minimize the quiescent current. Logic transistors in the gate drivers and the pulse generation circuits have been sized to minimize leakage while meeting the desired speed requirements.

The PMU utilizes a voltage doubler to minimize losses arising from subthreshold leakage in the boost converter power train. Due to the low output current, the converter spends most of the time in the idle phase of DCM operation (for converter switching period T_{period} of 78ms, the durations of Φ_1 and Φ_2 are typically less than 5μs). Figure 23.2.3 shows the power-train leakage paths during the converter idle phase when both power FETs, *P0* and *N0*, are off. Assuming V_X (supply for the *P0* gate driver) is the system V_{DD} as is typically done in standard boost converters, the power loss associated with the output leakage path becomes *10×* higher than the loss associated with the input leakage path (20pW from input and 223pW from output). This is mainly due to the fact that V_{DD} is much higher than V_{IN} for the boost converter. Since the output leakage current is governed by the subthreshold leakage in *P0*, by using the voltage doubler, an elevated supply, V_{PUMP}, is generated to drive the gate of *P0*. When off, *P0* sees a negative source to gate voltage, putting it in super cut-off and reducing its subthreshold leakage [3]. Although the voltage doubler too has leakage paths of its own as shown in Fig. 23.2.3, the transistors are much smaller than the power FET *P0*. Even with the overhead of V_{PUMP} generation and increased switching losses in the *P0* gate driver, the overall power saving (simulated) due to the voltage doubler is 175 to 188pW (17% of the minimum power budget) in the typical corner and 950pW in the fast corner making the converter robust towards process variations.

Figure 23.2.4 shows the boost-converter output power and the corresponding efficiency with an electrode impedance of 1MΩ and EP of 80 to 100mV (emulated using a 1MΩ resistor and a power supply in this measurement) for a V_{DD} of 0.9V and boost converter inductance of 47μH. The converter is characterized for various input voltages by varying the boost converter input impedance. The maximum output power is achieved for an input voltage close to half of the EP which is when the converter input impedance is close to the electrode impedance. The converter achieves a peak PCE of 56%. For input power of close to 1.2nW (for EP of 80mV and electrode impedance of 1.28MΩ), the boost-converter efficiency is close to 53%. This translates to a total output power of 637pW from the boost converter. Figure 23.2.5 shows the power consumption of individual circuit blocks utilizing the power extracted by the boost converter. Overall, 544pW is consumed by the PMU circuits. As shown in Fig. 23.2.6, this work presents the lowest power PMU reported with the lowest quiescent current and highest efficiency at nW power levels as compared to the state-of-art ultra-low power PMUs [4,5,6]. The die micrograph of the PMU, fabricated in a 0.18μm CMOS process, is shown in Fig. 23.2.7.

Acknowledgements:
This work is funded by the C2S2 and the IFC, two of six research centers funded under the FCRP, a SRC entity, and by the US National Institutes of Health grants K08 DC010419 and T32 DC00038 and the Bertarelli Foundation.

References:
[1] P. P. Mercier, A. C. Lysaght, S. Bandyopadhyay, A. P. Chandrakasan, and K.M. Stankovic, "Energy extraction from the biologic battery in the inner ear," *Nature Biotechnology*, vol. 30, no. 12, pp. 1240-1243, Dec. 2012.
[2] S. Bandyopadhyay and A. P. Chandrakasan, "Platform Architecture for Solar, Thermal and Vibration Energy Combining with MPPT and Single Inductor," *IEEE J. Solid State Circuits*, vol. 47, no. 9, pp. 2199-2215, Sep. 2012.
[3] K. Roy, S. Mukhopadhyay, and H. Mahmoodi-Meimand, "Leakage Current Mechanisms and Leakage Reduction Techniques in Deep-Submicrometer CMOS Circuits," *Proceedings of the IEEE*, vol. 91, no. 2, pp. 305-327, Feb. 2003.
[4] K. Kadirvel, Y. Ramadass, U. Lyles, *et.al.*, "A 330nA Energy-Harvesting Charger with Battery Management for Solar and Thermoelectric Energy Harvesting," *IEEE ISSCC Dig. Tech. Papers*, pp. 106-108, Feb. 2012.
[5] Texas Instruments TPS62736, "Programmable Output Voltage Ultra-Low Power Buck Converter with up to 50mA/200mA Output Current," *Available Online http://www.ti.com/*.
[6] G. Chen, H. Ghaed. R. Haque, *et.al.*, "A Cubic-Millimeter Energy-Autonomous Wireless Intraocular Pressure Monitor," *IEEE ISSCC Dig. Tech. Papers*, pp. 310-311, Feb. 2011.

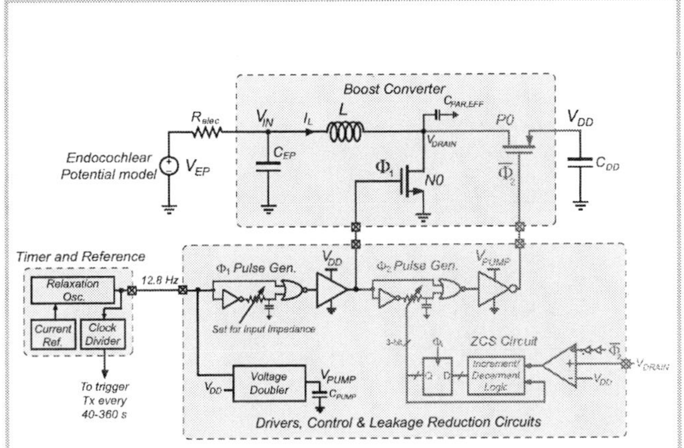

Figure 23.2.1: Power management unit (PMU) for harvesting energy from the endocochlear potential (EP).

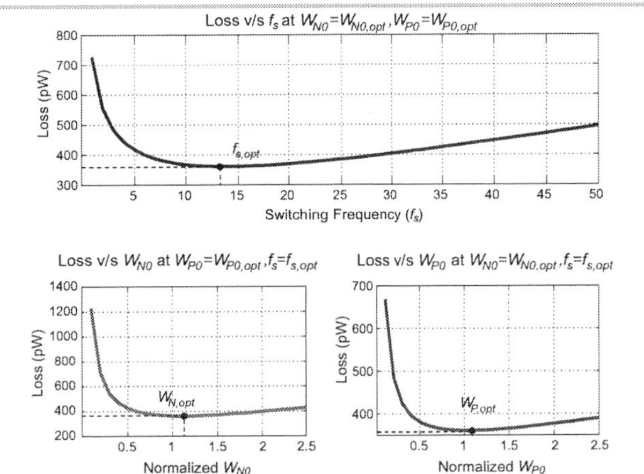

Figure 23.2.2: Loss (overall loss due to conduction, switching and leakage) optimization plots versus converter switching frequency and power FET size in DCM for a given input impedance (1.6MΩ) at input voltage of 40mV with boost converter inductance of 47µH.

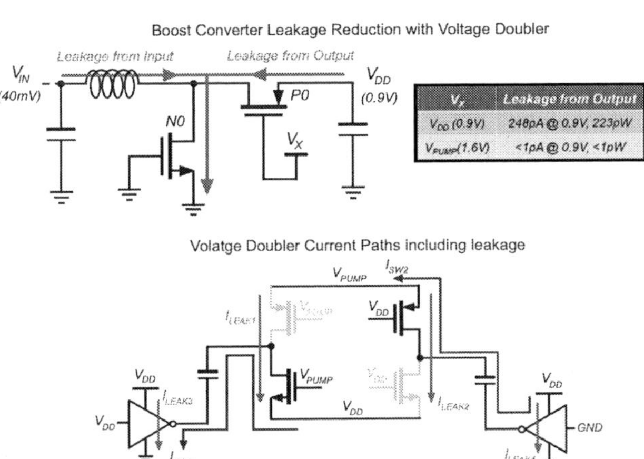

Figure 23.2.3: Boost converter leakage paths and voltage doubler current paths.

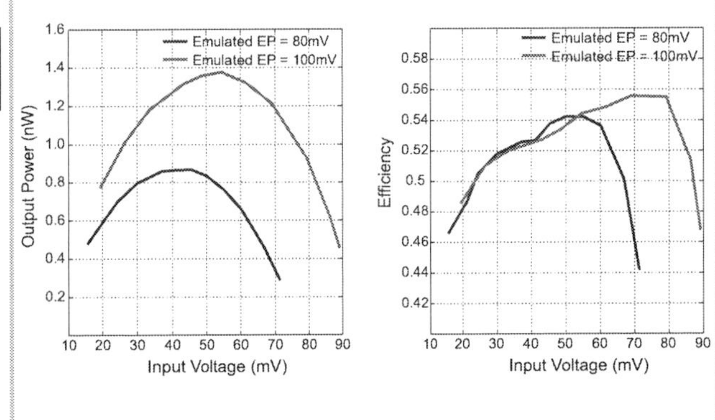

Figure 23.2.4: Boost converter measured results showing output power and corresponding efficiency.

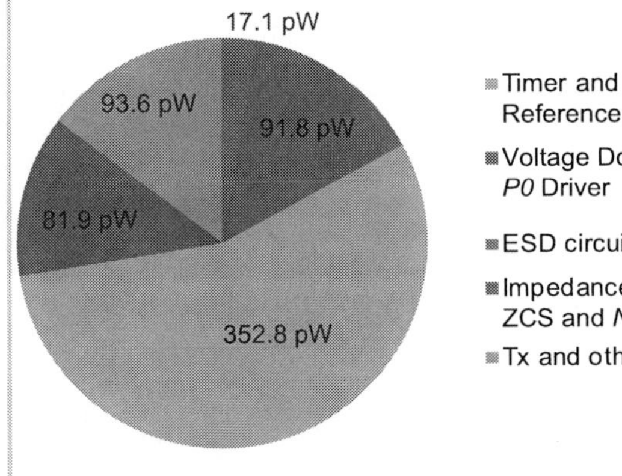

Figure 23.2.5: Measured power consumption of individual circuit blocks.

	ISSCC 2012 [4]	TPS62736 [5]	ISSCC 2011 [6]	This Work
Topology	Boost	Buck	Switched Capacitor Charge Pump (Boost)	Boost with Voltage Doubler
Voltage Conversion	80mV-2.5V boosted to 3-5V	2-5.5V step down to 1.3-5V	450mV boosted to 3.6V	Inductive Boost:- 20-70mV boosted to 0.8-1.1V Voltage Doubler:- 0.8-1.1V boosted to 1.5-1.9V
Output Power	1µW-30mW	2.5µW-125mW	10nW to 160nW	544pW-4nW
Efficiency at Ultra-Low Power	20% at 1µW	55% at 2.5µW	35% at 10nW	53% at 1.2nW
Quiescent Power	9.9nW	760nW	7.3nW	544pW

Figure 23.2.6: Comparison with previously published low-power PMUs in energy harvesting systems.

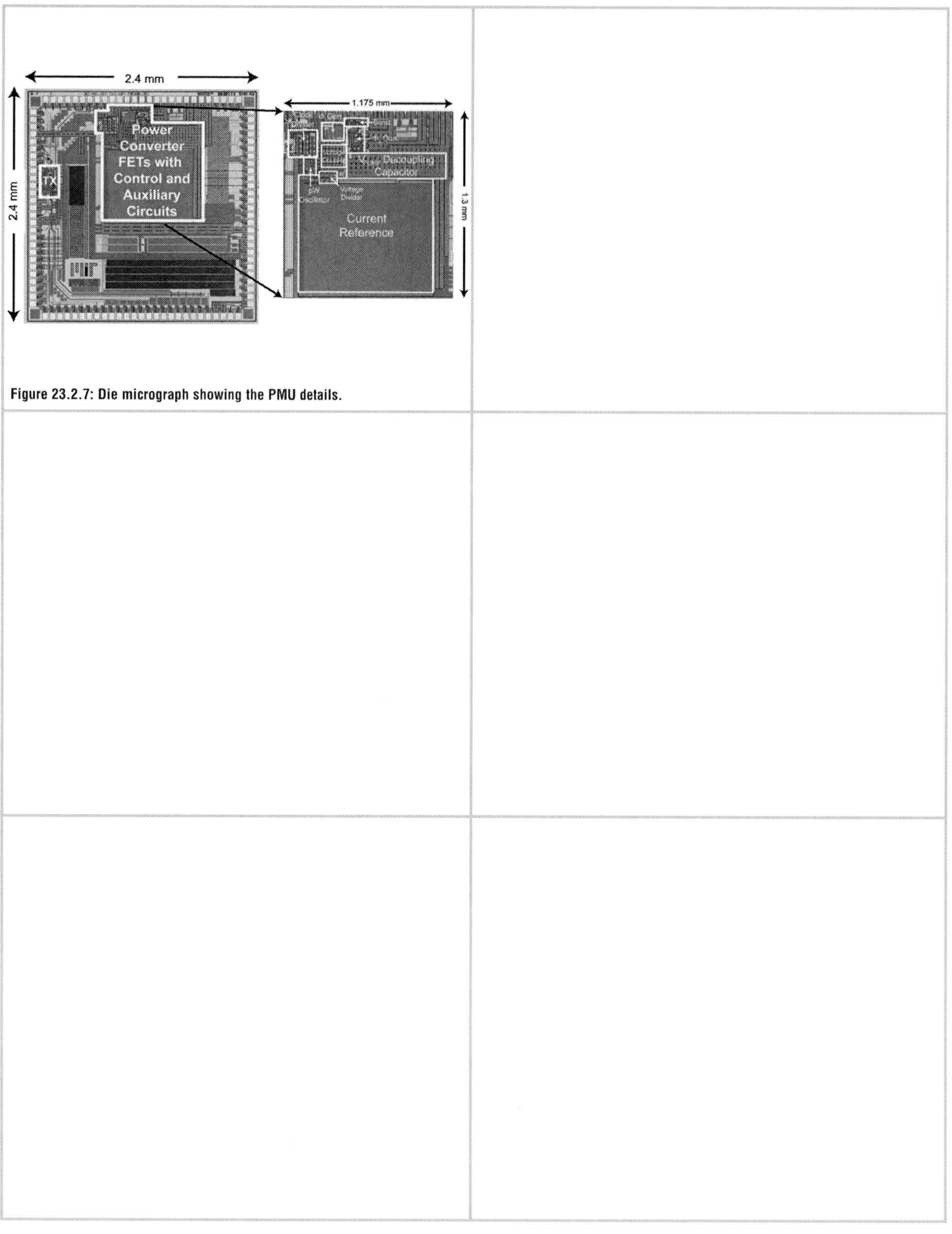

Figure 23.2.7: Die micrograph showing the PMU details.

23.3 A 3nW Fully Integrated Energy Harvester Based on Self-Oscillating Switched-Capacitor DC-DC Converter

Wanyeong Jung, Sechang Oh, Suyoung Bang, Yoonmyung Lee, Dennis Sylvester, David Blaauw

University of Michigan, Ann Arbor, MI

Recent advances in low-power circuits have enabled mm-scale wireless systems [1] for wireless sensor networks and implantable devices, among other applications. Energy harvesting is an attractive way to power such systems due to limited energy capacity of batteries at these form factors. However, the same size limitation restricts the amount of harvested power, which can be as low as 10s of nW for mm-scale photovoltaic cells in indoor conditions. Efficient DC-DC up-conversion at such low power levels (for battery charging) is extremely challenging and has not yet been demonstrated.

Boost DC-DC converters are widely used to harvest energy from DC sources and yield high conversion efficiency [2]. However, they require a large off-chip inductor at low harvested power levels, increasing system size. Alternatively, switched-capacitor (SC) DC-DC converters can be fully integrated on-chip and are favored for form-factor constrained applications [3-8]. However, at low power levels, SC converter efficiency has been constrained by overheads of clock generation and level-conversion to drive the switch capacitors. As a result, efficient SC converter operation has been limited to the μW range.

This paper presents a fully integrated energy harvester maintaining >35% end-to-end efficiency when harvesting from a 0.84mm^2 solar cell in low light condition of 260lux, converting 7nW input power from 250mV to 4V. The key contribution of this work lies in the proposed SC DC-DC voltage doubler structure, in which an oscillator is completely internalized within the SC network. This eliminates power overhead of clock generation and level shifting, and enables higher efficiency at lower power levels. Voltage doublers are cascaded to form a complete harvester with a wide load range from 5nW to 5μW and self-starting operation down to 140mV. Since each doubler is self-oscillating, the frequency of each stage can be independently modulated, thereby optimizing the overall conversion efficiency. The harvester conversion ratio is configurable from 9× to 23× and idle power consumption is less than 3nW.

Figure 23.3.1 shows a conventional SC DC-DC voltage doubler that includes a clock generator, level shifter, and clock buffers. Each of these blocks introduces power overhead, reducing efficiency. The proposed doubler consists of two stacked ring oscillators with the output nodes of corresponding stages connected through flying caps (C_{FLY}). In each stage, inverters from the top and bottom ring either charge or discharge the flying cap thereby transferring power to the upper ring. Simultaneously, the inverters drive the next stage in their ring, creating a multi-phase DC-DC converter with overlapping charge/discharge phases and self-sustaining operation. Because the two oscillators are synchronized at every stage, phase mismatch is <0.01×FO4, reducing contention loss to <1% of switching loss (simulation) and avoiding the need for non-overlapping clocks. R_{DLY} is a delay element that ensures matching of the flying cap charging/discharging time to the oscillation period. In the harvester, this delay is automatically tuned to balance switching and conduction losses and maintain optimum conversion efficiency across a range of load currents.

Figure 23.3.2 shows the detailed implementation of the voltage doubler. The delay element R_{DLY} consists of two coupled leakage-based delay elements [1] and a pass transistor T_P, followed by two buffers. When the inputs H_{NC} and L_{NC} of the stage switch from high to low, output nodes H_{1A} and L_{1A} (driven low) become isolated. T_P then provides a leakage path from L_1 to L_{1A} that slowly raises L_{1A} and, through C_{C1}, also H_{1A}. Back-to-back inverters of the delay element provide positive feedback and amplify the transition once it reaches V_{TH}, creating a sharp edge. This transition is then buffered and passed to the next stage. The opposite transition functions similarly. Due to the output isolation the structure can produce very long, synchronized delay while the coupled positive feedback creates a sharp edge that limits short-circuit current and contention loss, enabling ultra-low power operation.

For each doubler, the delay tuning voltage V_{CON} is dynamically adjusted to balance conduction and switching losses by examining the ratio of output to input voltage (R_{DIV}). A low R_{DIV} indicates a large voltage across the switches and dominant conduction loss. Conversely, high R_{DIV} indicates low conduction loss (zero as $R_{DIV}\rightarrow 2$) and dominant switching losses. This property is used to modulate the oscillation frequency of the doubler and maintain optimum efficiency across load currents. A clocked comparator, operating at a fraction of the internal oscillator frequency, takes in a divided form of the doubler output voltage ($V_{DIV} = V_{OUT}/R_{DIV\_DESIRED}$) and the input voltage (V_{MID}). The division ratio is configurable and is set to 1.73 in all measurements. A charge pump then takes in the corresponding pull-up/pull-down signals, and adjusts V_{CON} as needed to either speed or slow the oscillation. The voltage divider is implemented with a combination of a diode stack and capacitive divider to provide fast response and good low-frequency behavior.

Figure 23.3.3 shows the block diagram of the complete harvesting system, consisting of 4 stages of cascaded voltage doublers, a negative voltage generator, and circuits for conversion ratio control. A negative voltage is used to boost overall conversion ratio and to power control circuits, and is generated by connecting V_{HIGH} and V_{MED} of the doubler to V_{IN} and ground, respectively, resulting in $V_{NEG}\approx -V_{IN}$ at the V_{LOW} port of the doubler. To facilitate energy harvesting from a low voltage source (e.g., photovoltaic cell under low light), the first stage and negative voltage generator use low V_{TH} (~300mV) devices for their flying cap drivers. Bootstrapping is also used with these low V_{TH} switches to improve I_{ON}/I_{OFF} at low input voltages. I/O devices are used in the final stage to protect the circuit from high voltages used to charge energy storage devices (e.g., batteries).

The conversion ratio is adjusted by changing the number of cascaded stages. In addition, we propose a new adjustment scheme where the V_{LOW} of a doubler is switched among V_{IN}, GND and V_{NEG}. If V_{LOW} is set to $-V_{IN}$, the voltage across the flying cap increases, resulting in $V_{OUT} = (V_{MED}+V_{IN})\times 2-V_{IN} = 2\times V_{MED}+V_{IN}$. If V_{LOW} is set to GND for 4 cascaded stages, overall conversion ratio is 16×. However, if final stage V_{LOW} is set to V_{NEG}, overall conversion ratio increases by 1×, becoming 17×. Similarly, setting V_{LOW} to V_{IN} decreases conversion ratio. In a binary manner, the conversion ratio is controlled from 9× to 23×. Measured results in Fig. 23.3.4. demonstrate that a 0.35V input can be converted to a 2.2 to 5.2V voltage range with similar conversion efficiencies across settings.

To enable cold start of the complete system, the control logic operates between V_{NEG} and V_{IN} rails. Upon initial system startup, V_{NEG} and V_{2X} become available first and allow the control logic to turn on and configure the switches. As each stage is powered up, its internal frequency modulation is enabled and begins to control the frequency for optimum efficiency.

The voltage doubler and energy harvester are fabricated in 0.18μm CMOS (Fig. 23.3.7). Figure 23.3.4 shows a single doubler has >70% efficiency across 1nA to 0.35mA output current (>10^5 range), with low idle power consumption of 170pW. Figure 23.3.5 shows measured results of the complete harvester. When V_{IN}=0.45V, the harvester delivers 5nW to 5μW output power with >40% efficiency, and has an idle power consumption <3nW. For V_{IN}=0.25V, corresponding to a solar cell under very low light, the harvester can take in between 10nW and 120nW to charge a ~4V battery voltage with >35% efficiency. Measured results with a small silicon solar cell (0.84mm^2) show actual harvesting operation under dim light of <100lux. The harvester cold starts with <200lux of light and 6nW power source. Figure 23.3.6 summarizes the performance of the proposed design and compares it to prior work.

References:

[1] G. Chen, M. Fojtik, D. Kim, *et al.*, "Millimeter-Scale Nearly Perpetual Sensor System with Stacked Battery and Solar Cells," *ISSCC Dig. Tech. Papers*, pp. 288-289, Feb. 2010.

[2] J.-P. Im, S.-W. Wang, K.-H. Lee, *et al.*, "A 40mV Transformer-Reuse Self-Startup Boost Converter with MPPT Control for Thermoelectric Energy Harvesting," *ISSCC Dig. Tech. Papers*, pp. 104-105, Feb. 2012.

[3] H. Shao, C.-Y. Tsui, and W.-H. Ki, "An Inductor-Less MPPT Design for Light Energy Harvesting Systems," *Proc. ASP-DAC*, pp. 101-102, Jan. 2009.

[4] P.-H. Chen, K. Ishida, X. Zhang, *et al.*, "A 120-mV Input, Fully Integrated Dual-Mode Charge Pump in 65-nm CMOS for Thermoelectric Energy Harvester," *Proc. ASP-DAC*, pp. 469-470, Jan. 2012.

[5] D. Somasekhar, B. Srinivasan, G. Pandya, *et al.*, "Multi-Phase 1 GHz Voltage Doubler Charge Pump in 32 nm Logic Process," *IEEE J. Solid-State Circuits*, vol. 45, no. 4, pp. 751-758, Apr. 2010.

[6] L. Chang, R.K. Montoye, B.L. Ji, *et al.*, "A Fully-Integrated Switched-Capacitor 2:1 Voltage Converter with Regulation Capability and 90% Efficiency at 2.3A/mm^2," *IEEE Symp. VLSI Circuits*, pp. 55-56, June 2010.

[7] T.V. Breussegem and M. Steyaert, "A 82% Efficiency 0.5% Ripple 16-Phase Fully Integrated Capacitive Voltage Doubler," *IEEE Symp. VLSI Circuits*, pp. 198-199, June 2009.

[8] I. Doms, P. Merken, R. Mertens, C. Van Hoof, "Integrated Capacitive Power-Management Circuit for Thermal Harvesters with Output Power 10 to 1000μW," *ISSCC Dig. Tech. Papers*, pp. 300-301, Feb. 2009.

978-1-4799-0917-9/14 $31.00 © 2014 IEEE

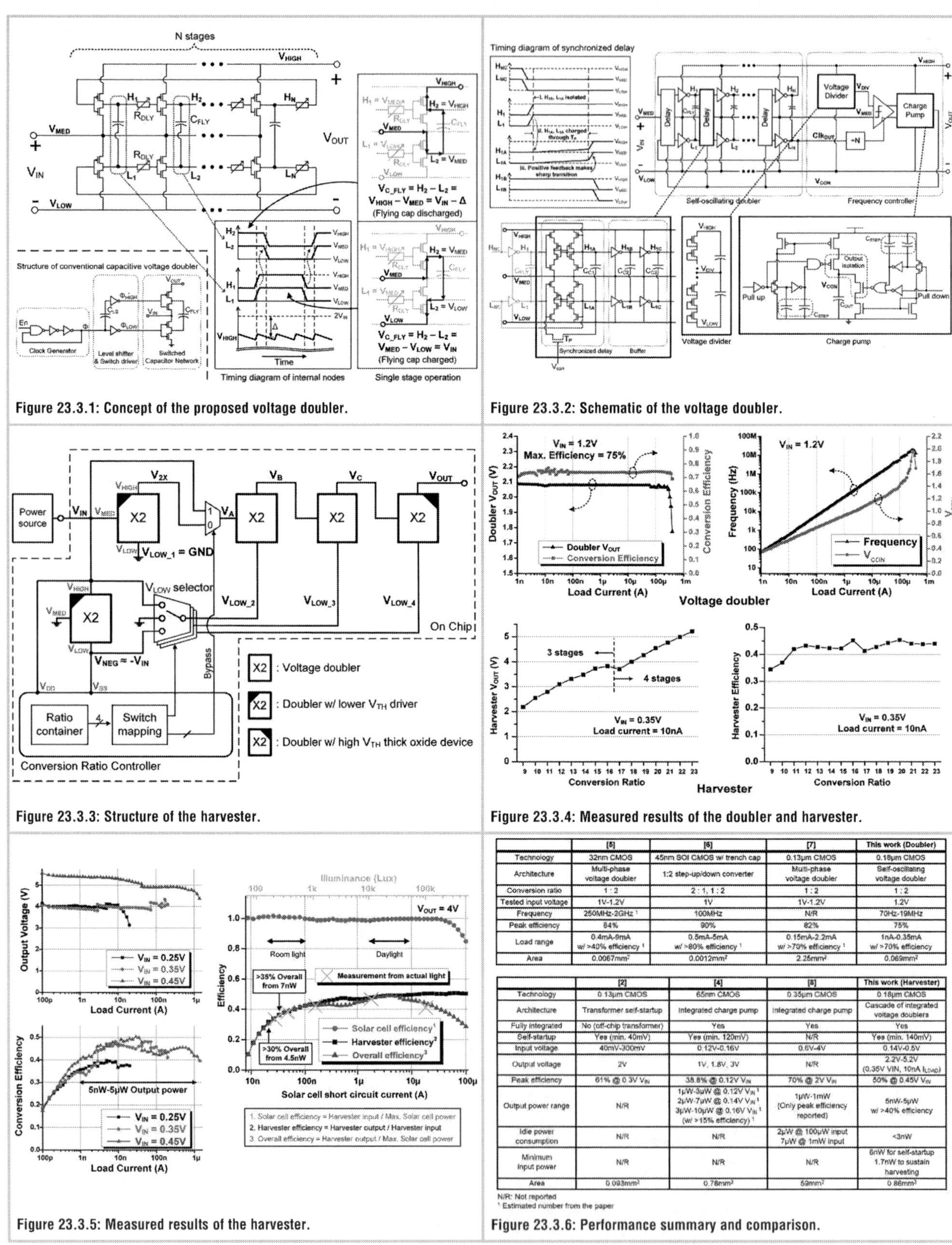

Figure 23.3.1: Concept of the proposed voltage doubler.

Figure 23.3.2: Schematic of the voltage doubler.

Figure 23.3.3: Structure of the harvester.

Figure 23.3.4: Measured results of the doubler and harvester.

Figure 23.3.5: Measured results of the harvester.

	[5]	[6]	[7]	This work (Doubler)
Technology	32nm CMOS	45nm SOI CMOS w/ trench cap	0.13µm CMOS	0.18µm CMOS
Architecture	Multi-phase voltage doubler	1:2 step-up/down converter	Multi-phase voltage doubler	Self-oscillating voltage doubler
Conversion ratio	1 : 2	2 : 1, 1 : 2	1 : 2	1 : 2
Tested input voltage	1V-1.2V	1V	1V-1.2V	1.2V
Frequency	250MHz-2GHz [1]	100MHz	N/R	70Hz-19MHz
Peak efficiency	64%	90%	82%	75%
Load range	0.4mA-9mA w/ >40% efficiency [1]	0.5mA-5mA w/ >80% efficiency [1]	0.15mA-2.2mA w/ >70% efficiency [1]	1nA-0.35mA w/ >70% efficiency
Area	0.0067mm^2	0.0012mm^2	2.25mm^2	0.069mm^2

	[2]	[4]	[8]	This work (Harvester)
Technology	0.13µm CMOS	65nm CMOS	0.35µm CMOS	0.18µm CMOS
Architecture	Transformer self-startup	Integrated charge pump	Integrated charge pump	Cascade of integrated voltage doublers
Fully integrated	No (off-chip transformer)	Yes	Yes	Yes
Self-startup	Yes (min. 40mV)	Yes (min. 120mV)	N/R	Yes (min. 140mV)
Input voltage	40mV-300mV	0.12V-0.16V	0.6V-4V	0.14V-0.5V
Output voltage	2V	1V, 1.8V, 3V	N/R	2.2V-5.2V (0.35V VIN, 10nA I_{LOAD})
Peak efficiency	61% @ 0.3V V_{IN}	38.8% @ 0.12V V_{IN}	70% @ 2V V_{IN}	50% @ 0.45V V_{IN}
Output power range	N/R	1µW-3µW @ 0.12V V_{IN} [1] 2µW-7µW @ 0.14V V_{IN} [1] 3µW-10µW @ 0.16V V_{IN} [1] (w/ >15% efficiency) [1]	1µW-1mW (Only peak efficiency reported)	5nW-5µW w/ >40% efficiency
Idle power consumption	N/R	N/R	2µW @ 100µW input 7µW @ 1mW input	<3nW
Minimum input power	N/R	N/R	N/R	6nW for self-startup 1.7nW to sustain harvesting
Area	0.093mm^2	0.78mm^2	59mm^2	0.86mm^2

N/R: Not reported
[1] Estimated number from the paper

Figure 23.3.6: Performance summary and comparison.

23

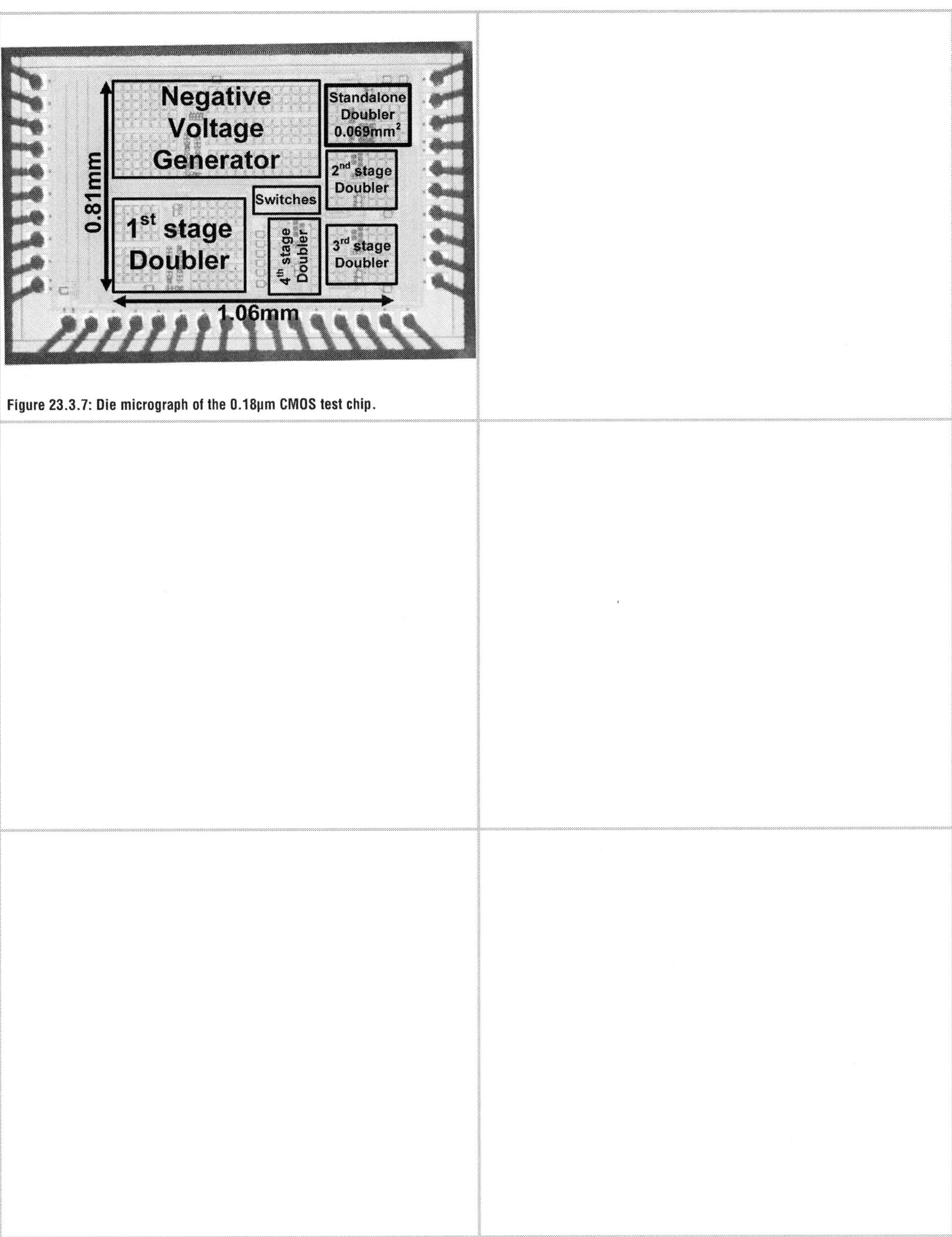

Figure 23.3.7: Die micrograph of the 0.18μm CMOS test chip.

23.4 Dual-Source Single-Inductor 0.18μm CMOS Charger-Supply with Nested Hysteretic and Adaptive On-Time PWM Control

Suhwan Kim, Gabriel A. Rincón-Mora

Georgia Institute of Technology, Atlanta, GA

A major challenge with emerging microsensors, biomedical implants, and other portable devices is operational life, because tiny batteries exhaust quickly. And even though 1g fuel cells store 5 to 10× more energy than 1g Li-ion batteries, fuel cells supply 10 to 20× less power [1]. This means fuel cells last longer with light loads and Li-ion batteries output more power across shorter periods. Therefore, when the peak power is far greater than the average power, which, for example, is typically the case in wireless sensors, a hybrid source can occupy less space than a single source [1]. Still, managing a fuel cell and a battery to supply a load and recharge the battery, which also acts as an output, with little space is difficult. Switched-inductor circuits are appealing in this context because they draw and supply more power with higher efficiency than their linear and switched-capacitor counterparts. Inductors, however, are bulky, and thus microsystems can only rely on one inductor [2-3]. Today, most single-inductor multiple-output systems derive power from one source [3, 4, 6], and therefore, the fuel cell and the battery require considerable space. Systems that use two sources either do not manage how much power each source should supply across loading conditions [5], or if they do [2] their efficiency is low. The advantage of the prototyped 0.18μm CMOS dual-source single-inductor system presented here is its less overall volume because it incorporates the functional intelligence of [2] with much higher efficiency.

Because the 1.1 to 1.3V energy source v_{ES} supplies more energy when delivering constant power, the prototyped system in Figs. 23.4.1 and 23.4.7 draws constant power P_{ES} from v_{ES} and supplementary power from the 1.8V power source v_{PS} to supply an 8mA 0.8V load. When P_{ES} exceeds the needs of the load in P_O, the system uses the excess to recharge v_{PS}. Since the hybrid supply uses and switches a 50μH 6×6×2mm³ inductor L_O to transfer power between v_{ES}, v_{PS}, and the output v_O, the purpose of capacitors C_{IN} and C_O is to suppress switching noise in v_{ES} and v_O. This way, when P_{ES} surpasses P_O, the voltage of C_O as well as v_O rise above the reference V_{REF} to such an extent that comparator CP_M trips to push the system into the light-load region. CP_M pulls the system back into the heavy-mode region when the opposite happens, when P_O exceeds P_{ES} to pull v_O below the lower hysteretic threshold of CP_M.

When lightly loaded, comparator CP_{LT} regulates v_O about V_{REF} and L_O conducts in discontinuous conduction mode (DCM) across the period of the 40kHz clock f_{CLK}. More specifically, CP_{LT} senses v_O to determine which output: v_O or v_{PS}, should receive the energy of L_O. For this, S_{ES} and S_E in Figs. 23.4.1 and 23.4.2 first energize L_O from v_{ES} to ground across τ_{EN}'s 1.2μs pulsewidth to raise L_O's current i_L from zero to 30mA. Afterwards, S_{ES} and S_E open and $v_{E.OFF}$ in Fig. 23.4.2 rises to close S_{DE} and either S_O or S_{PCHG}. If comparator CP_{LT} senses v_O is below V_{REF} by 10mV, S_O drains L_O into v_O; otherwise, S_{PCHG} depletes L_O into v_{PS}. Comparators CP_{IOZ} and CP_{IPZ} then disengage S_O and S_{PCHG} together with S_{DE} when S_O's and S_{PCHG}'s current nears zero, when L_O is close to empty, which happens at 2.7 and 27μs in Fig. 23.4.2. All switches remain open after that until f_{CLK} initiates another cycle.

CP_{LT}, CP_{IOZ}, and CP_{IPZ} need not operate across the entire 25μs period of f_{CLK}. CP_{LT}, for one, needs to sense v_O only at the end of τ_{EN}'s 1.2μs pulsewidth. This is why f_{CLK} in Fig. 23.4.2 engages CP_{LT} a short delay τ_D after τ_{EN} rises, to be ready by the end of τ_{EN}, and disengages CP_{LT} another short delay τ_D after τ_{EN} falls. Similarly, CP_{IOZ} and CP_{IPZ} must sense only when S_O and S_{PCHG} conduct i_L, so CP_{LT}'s output v_{LT} enables CP_{IOZ} and CP_{IPZ} and CP_{IOZ}'s and CP_{IPZ}'s flip flops disable them after they detect i_L nears zero. Duty-cycling CP_{LT}, CP_{IOZ}, and CP_{IPZ} this way reduces their power consumption by 90%.

When heavily loaded, L_O draws one energy packet from v_{ES} and one variable packet from v_{PS} that transconductor G_{HV} in Figs. 23.4.1 and 23.4.3 controls when regulating v_O about V_{REF}. As with light loads, L_O stops conducting after that until f_{CLK} starts another cycle. Comparator CP_{HV} compares G_{HV}'s slow-moving output v_G against a triangular saw-tooth voltage v_{SAW} to pulsewidth modulate (PWM) how long L_O energizes from v_{PS}. For all this, like before, S_{ES} and S_E first energize L_O from v_{ES} to ground across τ_{EN}'s 1.2μs pulsewidth to raise L_O's current i_L from zero to 30mA. Afterwards, S_{ES} and S_E open and S_{DE} and S_O close to drain L_O into v_O. S_{DE} then opens and, if G_{HV} senses that v_O still needs power, v_{SAW} starts ramping and S_{PE} closes to energize L_O from v_{PS} to v_O. When v_{SAW} falls below G_{HV}'s v_G, S_{PE} opens and S_{DE} closes to deplete L_O into v_O. S_{DE} and S_O then open when CP_{IOZ} in Fig. 23.4.3 senses that L_O's i_L is nearly zero, after which point all other switches remain open until the next f_{CLK} cycle.

As Fig. 23.4.4 illustrates, v_O ripples at ±2.5mV when lightly loaded with 0.1mA and ±40mV when heavily loaded with 8mA. The ripple is higher at 8mA because v_{ES} and v_{PS} deliver power early in the period and the load slews C_O afterwards, when disconnected from L_O. Since CP_M determines which mode to assert in hysteretic fashion, the system transitions through modes across rising and falling 0.1 to 8mA load dumps quickly and without ringing oscillations. When the load is light at 0.1 to 1 mA, the fraction of v_{ES} power that v_O and v_{PS} receive is 62 to 73%, as shown in Fig. 23.4.5. The fraction of power v_O receives from v_{ES} and v_{PS} when heavily loaded with 1 to 8 mA is 62 to 84%. Power-conversion efficiency, η_C, bottoms when the system transitions across 1mA and peaks to 73% under hysteretic control below 1mA and 84% under PWM control above 1 mA. η_C peaks at two points because switches are smaller in light mode than in heavy mode, thus, conduction and gate-drive losses balance at two load levels.

The key feature of the single-inductor 0.18μm CMOS charger-supply prototyped and validated here is managing two complementary sources with 62 to 84% power conversion efficiency. To achieve this, the system duty cycles circuit blocks, operates the inductor in discontinuous conduction mode, and employs hysteretic and PWM control schemes to regulate the output in and across light and heavy modes. The challenge with single-source systems when lightly loaded over extended periods and pulsed periodically with heavy loads, as in the case of wireless sensors, is that oversizing a fuel cell to output more power or a Li-ion battery to last longer demands more space than an efficient hybrid. To sustain a 0.1 to 10mW load for one month, for example, [4] and [6] in the table of Fig. 23.4.6 require a 1g fuel cell to supply 10mW or a 0.45g or 0.43g Li-ion battery to last one month. Since the circuit proposed in [5] cannot adjust how much power each source should supply according to the load, [5] needs a 1g fuel cell or a 0.43g Li-ion battery. [2] can manage a fuel cell and a Li-ion battery according to the load, but the cost of intelligence, robustness, and accuracy is unfortunately efficiency, so this hybrid system demands more space than [4-6]. The dual-source single-inductor charger–supply presented here, however, requires a 0.1g fuel cell and a 0.05g Li-ion battery, which at 0.15g combined results in 65% less weight than that of the smallest counterpart.

Acknowledgements:
The authors thank Pooya Forghani, Paul Emerson, and Texas Instruments for their support and for fabricating the prototyped IC.

References:

[1] M. Chen, J.P. Vogt, and G.A. Rincón-Mora, "Design Methodology of a Hybrid Micro-Scale Fuel Cell-Thin-Film Lithium Ion Source," *IEEE Int'l Midwest Symp. on Circuits and Systems*, pp. 674-677, Aug. 2007.
[2] S. Kim and G.A. Rincón-Mora, "Single-Inductor Dual-Input Dual-Output Buck-Boost Fuel Cell-Li Ion Charging DC-DC Converter," *ISSCC Dig. Tech. Papers*, pp. 444-445, Feb. 2009.
[3] D. Ma, W.H. Ki, C.Y. Tsui, and P.K.T. Mok, "Single-Inductor Multiple-Output Switching Converters with Time-Multiplexing Control in Discontinuous Conduction Mode," *IEEE J. Solid-State Circuits*, vol. 38, no. 1, pp. 89-100, Jan. 2003.
[4] M.H. Huang and K.H. Chen, "Single-Inductor Multi-Output (SIMO) DC-DC Converters with High Light-Load Efficiency and Minimized Cross-Regulation for Portable Devices," *IEEE J. Solid-State Circuits*, vol. 44, no. 4, pp. 1099-1111, Apr. 2009.
[5] K.W.R. Chew, Z. Sun, H. Tang, and L. Siek, "A 400nW Single-Inductor Dual-Input-Tri-Output DC-DC Buck-Boost Converter with Maximum Power Point Tracking for Indoor Photovoltaic Energy Harvesting," *ISSCC Dig. Tech. Papers*, pp.68-69, Feb. 2013.
[6] Y. Qiu, C.V. Liempd, B. Op het Veld, P.G. Blanken, and C.V. Hoof, "5μW-to-10mW Input Power Range Inductive Boost Converter for Indoor Photovoltaic Energy Harvesting with Integrated Maximum Power Point Tracking Algorithm," *ISSCC Dig. Tech. Papers*, pp. 118-119, Feb. 2011.

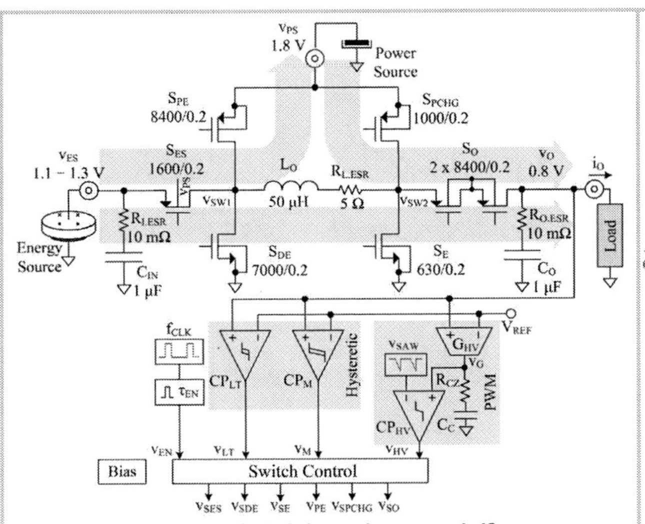

Figure 23.4.1: Dual-source single-inductor charger-supply IC.

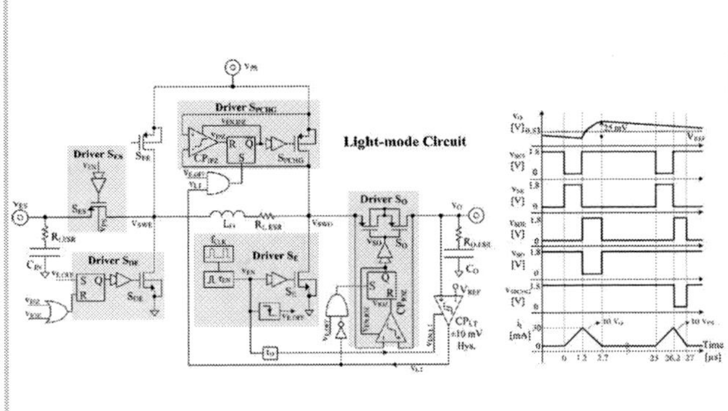

Figure 23.4.2: Light-load mode control and waveforms.

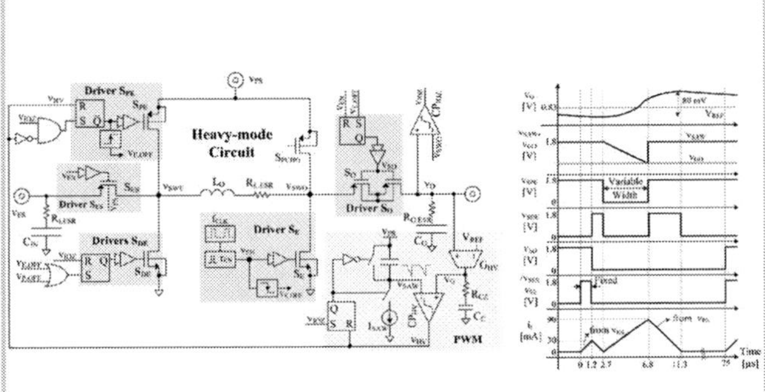

Figure 23.4.3: Heavy-load mode control and waveforms.

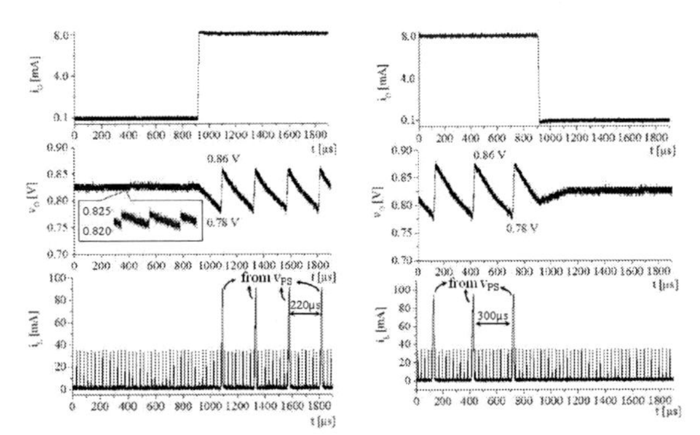

Figure 23.4.4: Mode transitions in rising and falling load dump responses.

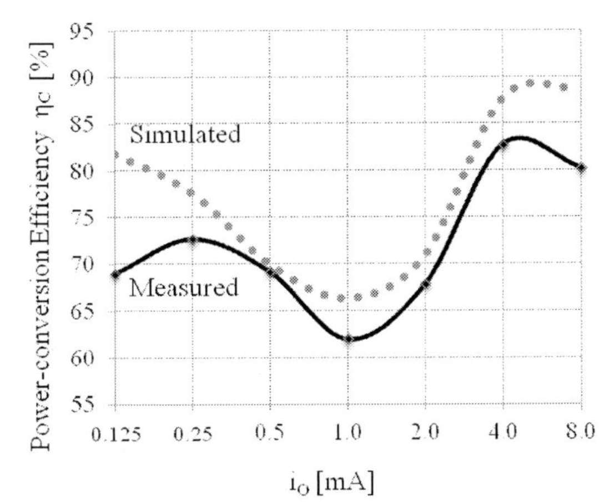

Figure 23.4.5: Simulated and measured efficiency of the prototype IC.

	[2] ISSCC 09	[4] JSSC 09	[5] ISSCC 13	[6] ISSCC 11	This work
Topology	SIDIDO Buck-Boost	SIMO Buck-Boost	SIDITO Buck-Boost	SISO Boost	SIDIDO Buck-Boost
Efficiency @ 0.1 mW / Peak	4% / 32%	80% / 93%	83% / 83%	83% / 87%	70% / 83%
Output Voltage	1.0	1.25	1, 1.8, 3	1.5, 3, 5	0.8
Load Dump/Filter cap	60 mV / 0.1 μF	25 mV / 33 μF	N/A	N/A	30 mV / 1 μF
Load Range	0 - 1 mW	0 - 125 mW	0 - 10 mW	0 - 10 mW	0 - 10 mW
Inductor	150 μH	10 μH	N/A	1000 μH	50 μH
Process	0.5 μm	0.25 μm	0.18 μm	0.25 μm	0.18 μm
Required sources' volume to supply 0.1 - 10 mW's loads for 1 month	1.8-g DMFC Or 9.1-g Li Ion	1-g DMFC Or 0.45-g Li Ion	1-g DMFC Or 0.43-g Li Ion	1-g DMFC Or 0.43-g Li Ion	0.1-g DMFC +0.05-g Li Ion Tot. 0.15 g

Figure 23.4.6: Performance summary and comparison.

23

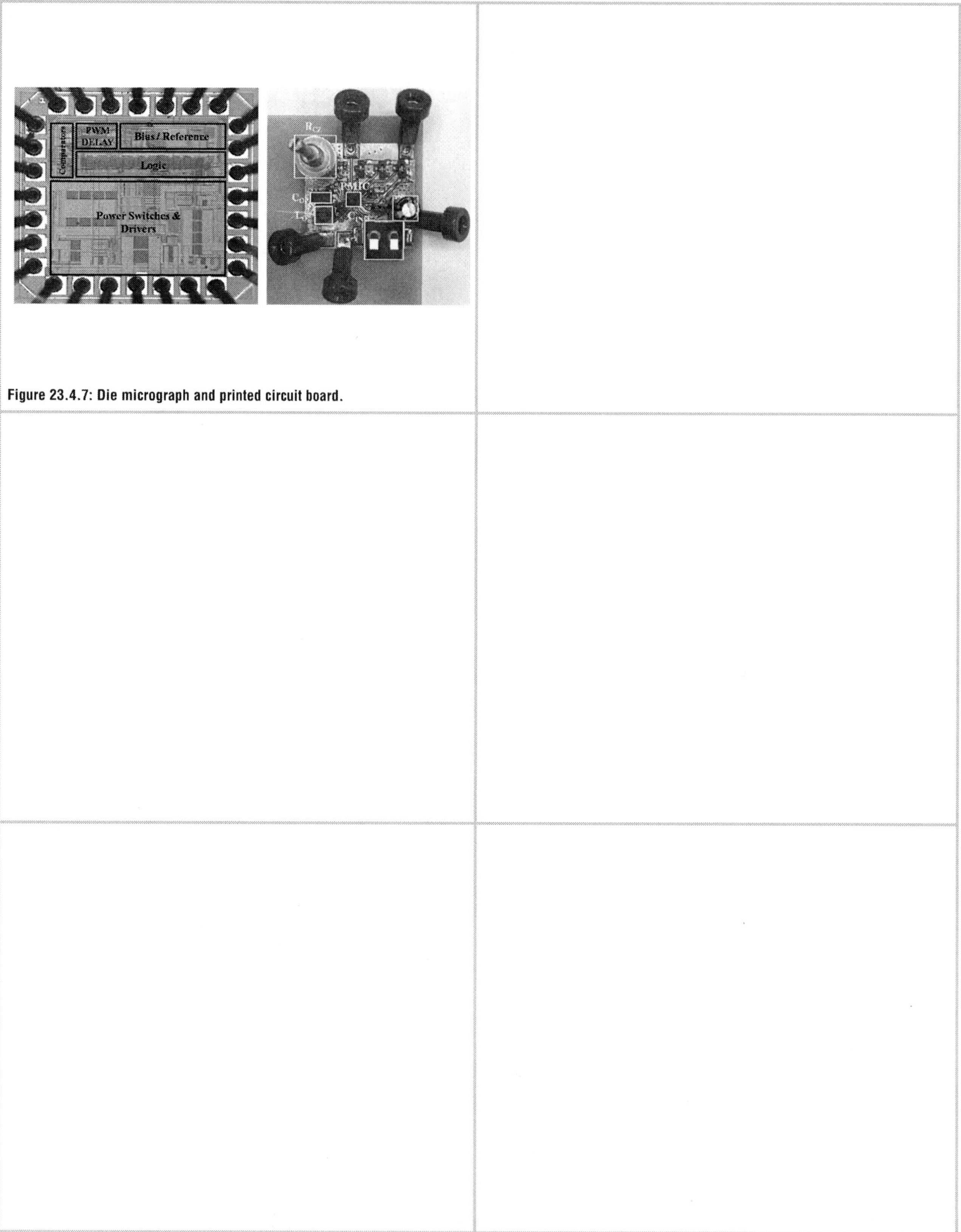

Figure 23.4.7: Die micrograph and printed circuit board.

23.5 An Energy Pile-Up Resonance Circuit Extracting Maximum 422% Energy from Piezoelectric Material in a Dual-Source Energy-Harvesting Interface

Young-Sub Yuk, Seungchul Jung, Hui-Dong Gwon, Sukhwan Choi, Si Duk Sung, Tae-Hwang Kong, Sung-Wan Hong, Jun-Han Choi, Min-Yong Jeong, Jong-Pil Im, Seung-Tak Ryu, Gyu-Hyeong Cho

KAIST, Daejeon, Korea

Energy harvesting is one of the key technologies used to realize self-sustaining systems such as wireless sensor networks and health-care devices. Much research on circuit design has been conducted to extract as much energy as possible from transducers, such as the *thermoelectric generator (TEG)* and the *piezoelectric transducer (PZT)*. Specifically, the energy in a PZT could be extracted more efficiently by utilizing resonance as [1] and [2] demonstrated. However, the maximum output voltage swing in those techniques are limited to twice of the original swing of the PZT, and thus, had a limited energy extraction capability in spite of more energy being available from the PZT. In [3], on the other hand, the large energy is obtained with higher voltage swing, but is limited up to 247% because the load energy is used to increase the output voltage swing of PZT. To obtain far more power from PZT, we propose an alternative resonance technique through which the PZT output swing can be boosted as high as CMOS devices can sustain. This technique is applied to a dual-energy- sourced (PZT and TEG) *energy-harvesting interface (EHI)* as a battery charger.

In Fig. 23.5.1, the equivalent electric model having dual energy sources, PZT and TEG, is shown with the detailed schematic of the proposed EHI. This circuit harvests energy from the TEG by a conventional boost converter operation through paths I and III in *discrete current mode (DCM)*, where the TEG generates 500mV for a temperature difference of about 10°K. The energy from the PZT is harvested through paths II and III simultaneously with the energy in the TEG by time multiplexing. The utilized PZT generates a peak AC power at 100Hz. Since the AC frequency of the PZT is low, the operational frequency of the boost converter is set at 2.5kHz.

Two waveforms are illustrated in the Fig. 23.5.2. Normally, the amount of energy extraction from the PZT is limited to just about 1% [4] due to the large parasitic capacitance C_z in the PZT. The half-cycle resonance technique [2], whose operation is illustrated in the left of Fig. 23.5.2, extracts energy with the help of resonance using an external inductor and increases the amount of energy extraction by about twice.

In contrast, we can obtain energy from PZT far more than twice in the proposed method using advanced resonance technique as shown in the right of Fig. 23.5.2. In this case, the waveform continuously grows until limited by external means as LC resonance appears at every edge of the waveform. Note that the resonant period is much shorter than the AC period of the PZT material. The resonance starts by closing M1 of Fig. 23.5.1 when the voltage of the capacitor C_z reaches the peak value of V_{zo} of the normal AC waveform of the PZT and the capacitor voltage is quickly inverted through the inductor. At the end of resonance, M1 is opened and the PZT current, i_{PZT}, charges C_z up again with the PZT vibration frequency until i_{PZT} reaches zero when its voltage has the next inverted peak. If this process is repeated in multiple cycles, the voltage V_z across C_z looks a square-like waveform with growing magnitude. The magnitude of V_z piles up cycle by cycle and reaches a very large value after multiple periods. This process is called the *energy pile-up mode (EPM)*, and thus we refer to the proposed scheme as *energy pile-up resonance (EPR)*. For a target magnitude of V_z in EPM, the peak value of V_z can be limited by extracting some amount of energy from the LC circuit when V_z reaches V_{Limit} and the extracted energy is transferred to the load in the form of current during the time t_{L2} by opening M1 and closing M3. This mode is called the *energy transfer mode (ETM)*. Note that the TEG operation must not interrupt the EPR operation. In order to prevent such a problem, as shown in the Fig. 23.5.1, RS, the resonance start signal, is delivered to the TEG controller to open M2 during the EPR. The "Peak Value Checker" block in Fig. 23.5.1 distinguishes whether the operation mode is EPM or ETM.

A proper control of the EPM and ETM is the crucial design point for large energy extraction. Figure 23.5.3 (bottom) shows the timing and circuit blocks for the M1 switch control in the EPR. The EPR operation starts by closing M1 when V_z hits its peak value. During the EPM, the "Full Inversion Detector" senses the voltage of C_z and opens M1 when V_{ci} reaches zero, where V_{ci} is the partial integration voltage of V_z by $(R_{I1}//R_{I2})C_I$ and represents the information of resonant current during the resonance interval. During the ETM, the "Resonance Interceptor" opens M1 when V_z hits the limit value of V_{Limit}. At this instant, V_{ci} is reset to zero by M_I to prepare for the next integration. Then, a part of the inductor energy is intercepted and transferred to the load when the mode is changed from the EPM to the ETM. The enlarged waveform illustrated at the bottom right of Fig. 23.5.3 shows that the energy transfer occurs during the period of t_{L2}. The amount of energy E_L flowing into the load during 't_{L2}' is given by $E_L = 1/2 \cdot C_z \cdot \{(|V_{max-}|+V_{IN})^2-(V_{Limit}-V_{IN})^2\}$,where V_{Limit} must be set up properly considering the breakdown voltage of the CMOS process because the maximum of V_z is determined by $R_Z \cdot i_{PZT}$ and it can be up to tens of volts.

To control the resonance that operates with much higher voltage V_z than V_{Load}, an attenuated signal of V_z is required. The "Sampling Attenuator and Peak Detector" shown in Fig. 23.5.3 (top) makes the attenuated signal 'V_{zs}' from V_z. For this purpose, it uses two series connected capacitors that reduces V_z by $C_1/(C_1+C_2)$. The DC level of V_{zs} is determined as $(C_3 \cdot VDD)/(C_3+C_4)$ by the capacitive voltage divider of C_3 and C_4 ($C_4=C_{41}+C_{42}$). The peak detector compares V_{zs} with its delayed signals, V_p and V_n, when the delayed signals cross over the V_{zs}, the comparator output transits and notifies the peak point. The delay circuit is designed using a capacitor and a switch-cap resistor with large equivalent resistance in small size even under a slow varying frequency of 100Hz. The small capacitors C_{42} and C_{43} in the peak detector exist to find the accurate peak point at each cycle by adding some amount of charge to C_n and removing some amount of charge from C_p , respectively. The comparator output transition is used for pulse generation in P.G whose outputs are 'ϕ_p' and 'ϕ_n' and are used for M1 control.

The top of Fig. 23.5.4 shows the detailed circuit of the "M1 Gate Driver". The supply of the "M1 Gate Driver", V_B, is made to be higher than the peak voltage of V_z by using V_z and V_{Load}. In order to control M1, two switches are used: M_{G1} and M_{G2}. The open state of the M1 switch is obtained by closing either M_{G1} or M_{G2} depending on the polarity of V_z. The bottom of Figure 23.5.4 shows the core circuit for the gate drivers of M_{G1} and M_{G2} to provide a negative gate diving voltage through the capacitor C_{MN} for each switch to be in the off state when turned off.

Figure 23.5.5 shows the measurement waveforms of V_z, i_L and V_{Load} when the proposed EPR operates, where the PZT is modeled as a capacitor, a resistor and a transformer. Figure 23.5.5 (top left) shows the ETM operation of the EPR when it generates 87µW output power from PZT. Considering that the conventional resonance technique [2] could achieve only 36µW output power for the same vibration amplitude, this shows an order of performance improvement. At the right of the top left of Fig. 23.5.5, the waveforms of V_z and i_L are magnified. The t_{R2} means the resonance duration between C_z and the inductor at the ETM. When the V_z meets the V_{Limit}, the resonance is intercepted and the current in the inductor flows into V_{Load} during the t_{L2}. The top right of Fig. 23.5.5 shows the transient waveform of V_z during the mode change from the EPM to the ETM. The bottom left of Fig. 23.5.5 shows that the proposed EHI charges the load from 2 to 4V with the proposed EPR. The bottom right of Fig. 23.5.5 shows the performance summary for this work. Figure 23.5.6 shows the output power increasing rate of the EPR specifically in comparison with previous designs. This shows that the available output power of EPR is controlled by V_z and reaches to 422% when the magnitude of V_z is boosted up to $7V_{pp}$ from the original PZT swing of $1.3V_{pp}$.

References:

[1] Y. K. Ramadass and A. P. Chandrakasan, "An Efficient Piezoelectric Energy-Harvesting Interface Circuit Using a Bias-Flip Rectifier and Shared Inductor," *ISSCC Dig. Tech. Papers*, pp. 296-297, Feb. 2009.

[2] D. W. Kwon and G. A. Rincon-Mora, "A Single-Inductor AC-DC Piezoelectric Energy-Harvesting/Battery-Charger IC Converting ±(0.35 to 1.2V) to (2.7 to 4.5V)," *ISSCC Dig. Tech. Papers*, pp. 494-495, Feb. 2010.

[3] D. W. Kwon and G. A. Rincon-Mora, "A Single-Inductor 0.35µm CMOS Energy-Investing Piezoelectric Harvester," *ISSCC Dig. Tech. Papers*, pp.78-79, Feb. 2013.

[4] H. A. Sodano, D. J. Inman, and G. Park, "Comparison of Piezoelectric Energy Harvesting Devices for Recharging Batteries," *Journal of Intelligent Material Systems and Structure*, vol.16, no. 10, pp. 799-807, Oct. 2005.

978-1-4799-0917-9/14 $31.00 © 2014 IEEE

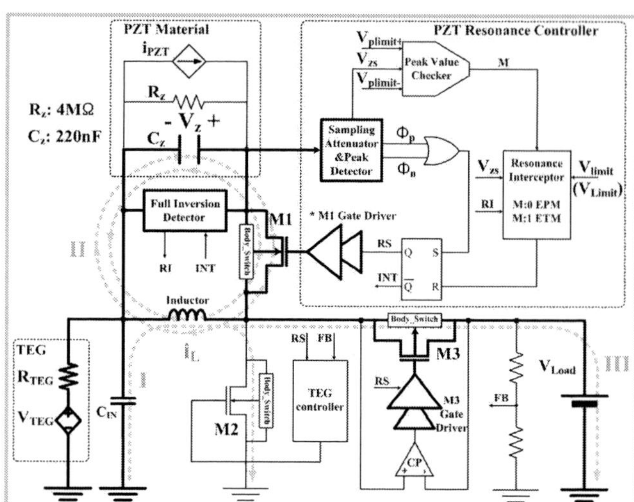

Figure 23.5.1: The proposed dual-source circuit for energy extraction from PZT material and TEG.

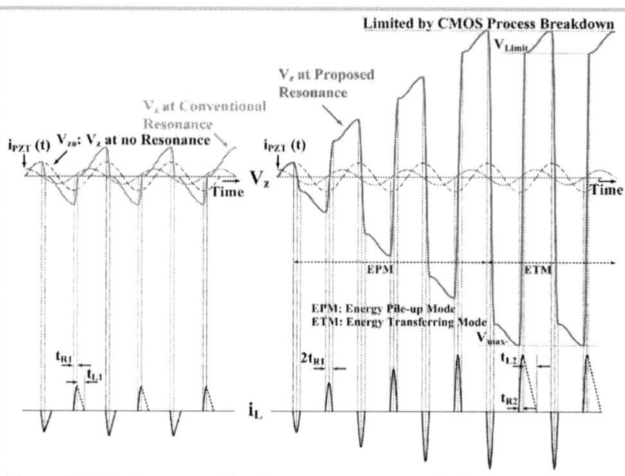

Figure 23.5.2: The conventional resonance concept (left) and the proposed Energy Pile-up Resonance concept (right) for extracting energy from the PZT material.

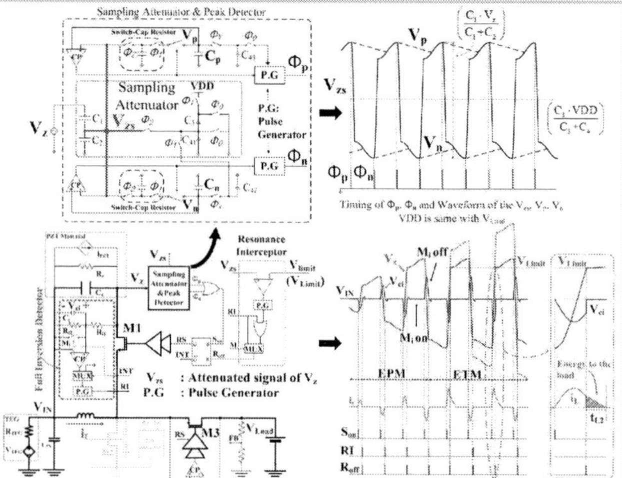

Figure 23.5.3: The full inversion detector and resonance Interceptor (bottom), and the sampling attenuator and peak detector (top).

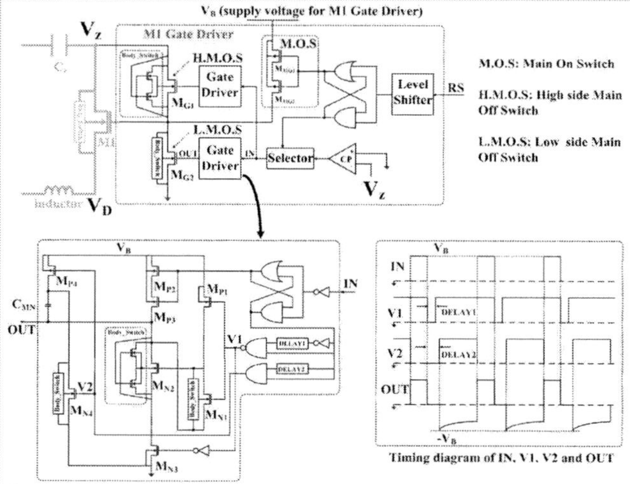

Figure 23.5.4: The proposed gate-driver scheme for M1 switch (top) and the core circuit of the gate driver (bottom).

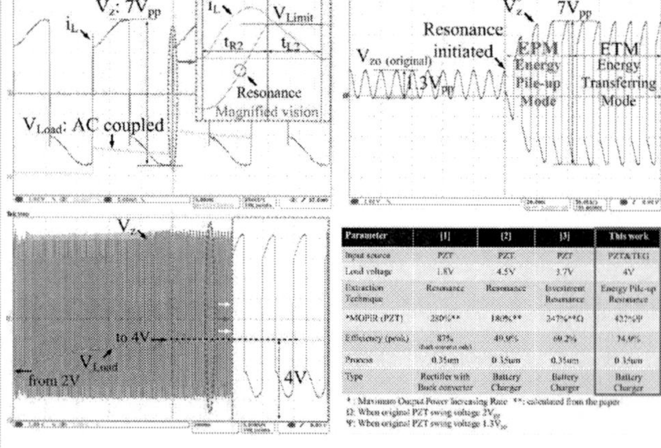

Parameter	[1]	[2]	[3]	This work
Input source	PZT	PZT	PZT	PZT&TEG
Load voltage	1.8V	4.5V	3.7V	4V
Extraction Technique	Resonance	Resonance	Investment Resonance	Energy Pile-up Resonance
*MOPIR (PZT)	280%**	180%**	247%**Ω	422%Ψ
Efficiency (peak)	87% (half-resonance only)	49.9%	69.2%	74.9%
Process	0.35um	0.35um	0.35um	0.35um
Type	Rectifier with Buck converter	Battery Charger	Battery Charger	Battery Charger

* : Maximum Output Power Increasing Rate **: calculated from the paper
Ω: When original PZT swing voltage $2V_{pp}$
Ψ: When original PZT swing voltage $1.3V_{pp}$

Figure 23.5.5: Measurement waveform of V_z, i_L, and V_{Load} at ETM with their magnified view (top left), mode change of V_z (top right), charging load with the proposed resonance (bottom left) and performance summary (bottom right).

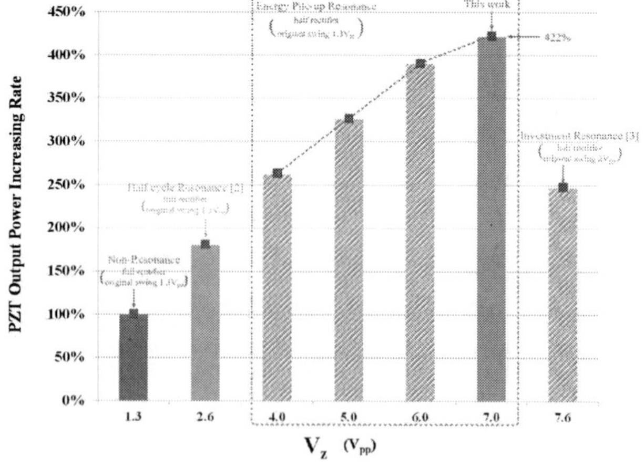

Figure 23.5.6: PZT output power increasing rate (OPIR) of energy pile-up resonance compared to previous works.

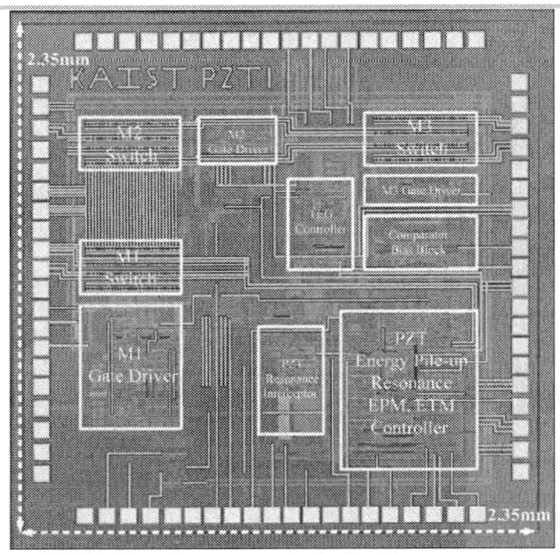

Figure 23.5.7: Chip micrograph of the proposed energy harvesting interface.

23.6 A 43V 400mW-to-21W Global-Search-Based Photovoltaic Energy Harvester with 350μs Transient Time, 99.9% MPPT Efficiency, and 94% Power Efficiency

Sandip Uprety, Hoi Lee

University of Texas, Dallas, Richardson, TX

Standalone systems such as outdoor lighting and remote monitoring stations can be self-powered by solar panels (SPs) having output powers of tens of watts. When the size of the SP is large, part of it can be shaded by birds, trees and other objects resulting in partial shading conditions (PSCs). Multiple local maxima in the power-versus-voltage (P-V) curve of the SP are generated during PSCs [1]. Based on a 10W SP measurement in Fig. 23.6.1, ~30% of the available power of the SP can be lost if an energy harvester only operates at the lower local maximum power point during PSC. This issue was previously addressed by using the approach of module-integrated converters (MIC), which are tied to smaller modules of the SP [2, 3]. Since none of the previously reported integrated maximum power point tracking (MPPT) techniques in MIC can track the global maximum power point (GMPP) of the P-V curve [2-5], the number of MICs would have to increase significantly in a large-size tens-of-watt SP. This increases the volume, weight, and cost of the photovoltaic system. In addition, previous harvester ICs have adopted the constant-frequency pulse-width-modulation control scheme [2-5], which has significant switching power loss under low solar irradiance.

This paper reports a global search algorithm (GSA) that enables an energy harvester (EH) to track GMPP from the SP P-V curve during PSC. Pulse-integration-based MPPT (PI-MPPT) is developed and incorporated with GSA for fast-tracking GMPP and maximizing GMPPT efficiency under partially shaded and unshaded conditions over a wide power range. An irradiance-aware adaptive frequency power controller (IAAFPC) is also designed to adaptively adjust power transistor off-time under low irradiance levels, allowing the EH to maintain high power efficiency for different power levels.

Figure 23.6.1 shows the system architecture. An input-regulated buck converter with 50V power switches (M_{N1}, M_{N2} and M_{P1}) serves as the power interface, extracting power from a SP and charging a battery. High-side M_{P1} is driven by a high-voltage gate driver. Low-side M_{N2} is used to prevent back-flow of current from the battery to the SP during SP under-voltage-lockout (UVLO) in very low irradiance. Sensing M_{N2} current also provides instantaneous SP current for global maximum power point tracking (GMPPT) without using any sensing resistor. The system is operated by GSA-based state-machine that controls PI-MPPT circuitry and IAAFPC. The state machine also incorporates battery management functions like cycle-to-cycle peak current limit and battery over-voltage protection (OVP) to support different battery chemistries.

Figure 23.6.2 illustrates GSA operation. GSA provides proper reference voltage V_{REF} for IAAFPC. The EH operates in two modes: GMPPT and steady state. When the EH is in standby ($Ø_{BCD}$ = 1), it allows capacitor C_{RAMP} to sample m_1V_{PV} and set $V_{REF} = V_{RAMP} = m_1V_{PV}$. When the converter enters GMPPT mode ($Ø_{MPPT}$ = 1 and $Ø_{BCD}$ = 0), C_{RAMP} is discharged by a constant current source and causes V_{RAMP} ramping down linearly from m_1V_{OC} to m_1V_{OCL}, where V_{OC} is the open cell voltage of the SP. The lower bound of V_{PV} (V_{OCL}) is programmable and is set to be larger than the maximum battery voltage to prevent UVLO at the end of GMPPT mode. During this period, instantaneous scaled voltage m_1V_{PV} and current information I_{PV} are sampled and processed by PI-MPPT block to identify voltage V_{MPP} that corresponds to the GMPP. After GMPPT phase, the converter settles to its steady state ($Ø_{SS}$ = 1) with its V_{REF} being set to V_{MPP}. In this design, the EH is programmed to stay in the steady state for 7s. The buck converter is then temporarily disabled to allow V_{PV} to go back to V_{OC}, and $Ø_{BCD}$ is set to 1. After that, GMPPT mode is enabled again and the same GSA operation cycle repeats unless V_{PV} UVLO or V_{BATT} OVP is triggered.

Figure 23.6.2 also explains IAAFPC operation with power transistor M_{P1} having constant on-time T_{ON}. When the irradiance on the SP is high, the EH operates in the continuous conduction mode (CCM). Since off-time t_{OFF} of M_{P1} depends on the regulated V_{PV}, the converter switching frequency f_{SW} is thus approximately the same as long as the converter stays in CCM. When the irradiance decreases to a low level, zero current detector of M_{N1} will enable the converter to enter the discontinuous conduction mode (DCM) automatically, eliminating the negative inductor current for power loss reduction. In DCM, t_{OFF} is inversely proportional to I_{PV}, so f_{SW} decreases with I_{PV}, reducing the converter switching power loss when irradiance level decreases. With relatively constant f_{SW} in CCM and linear fold-back of f_{SW} with decreasing irradiance in DCM, the EH can achieve high power efficiency across a wide power range.

Figure 23.6.3 shows the PI-MPPT circuitry which is operated at 500kS/s generated by an internal MPPT clock with non-overlapping $Ø_S$ and $Ø_I$ clock phases. Within an MPPT clock cycle, the first $Ø_S$ phase is used to sample m_1V_{PV} and m_2RI_{PV}, and discharge capacitors C_3 and C_5, whereas pulse-integration is performed in $Ø_{INT}$-$Ø_S$ phase. The PI-MPPT circuitry mainly consists of a pulse-width modulator (PWM), a pulse-amplitude modulator (PAM), a pulse integrator, and a power comparator. The M_7 current signal comprises of the pulse. The pulse amplitude is controlled by V_{PV} using PAM, while the pulse duration is modulated by I_{PV} using PWM. The pulse integrator is thus able to calculate the instantaneous power represented by voltage $V_{PRODUCT}$ across C_5. Voltage $V_{PRODUCT}(n)$ represents SP instantaneous power at n_{th} MPPT clock-cycle. In every clock cycle, $V_{PRODUCT}(n)$ is calculated and the power comparator compares its value with $V_{PRODUCT\text{-}MAX}(n)$, where $V_{PRODUCT\text{-}MAX}(n) = \max(V_{PRODUCT}(k))$ for $1 \le k < n$. In n^{th} cycle, if $V_{PRODUCT}(n) > V_{PRODUCT\text{-}MAX}(n)$, then $Ø_{Pmax}$ is asserted to turn on the corresponding S/H circuit, updating $V_{PRODUCT\text{-}MAX}(n)$ to $V_{PRODUCT}(n)$ and $V_{MPP}(n)$ to $V_{RAMP}(n)$. Since capacitor C_6 saves $V_{PRODUCT\text{-}MAX}$ during GMPPT period, V_{MPP}, which corresponds to the GMPP of the PV curve, can always be identified even if multiple local maxima exist and is provided as the reference voltage V_{REF} for the EH during steady-state operation. By sampling instantaneous voltage and current information of the SP in PI-MPPT, GMPP can be obtained with a fast speed. Since the clock period of PI-MPPT is much shorter than GMPPT operation phase, sufficiently large instantaneous power samples can be obtained to identify GMPP. High linearity and large input dynamic range of both PAM and PWM designs further guarantee the accuracy of each instantaneous power sample, thereby providing good tracking efficiency over a wide power range.

Figure 23.6.4 demonstrates different operation states of the EH IC during PSC. The IC test setup consists of L = 22μH, C_{IN} = 1μF, C_{OUT} = 10μF, V_{BATT} = 15V LiPo battery, and a SP with V_{OC} = 38 to 43V (depending on temperature). The PSC test is performed outdoors with 2 modules of 1 string shaded and rest of the modules and strings exposed to the sunlight. The output of PI-MPPT block V_{MPP}, shown in Fig. 23.6.4, illustrates the GMPP being tracked in presence of two local maxima. The GMPPT efficiency η_T of 99% is achieved. The measured transient time is 350μs for GMPPT and converter settling in normal operation. High GMPPT efficiency and short transient time can maximize energy capture from the SP. To further evaluate the effectiveness of IAAFPC and PI-MPPT in un-shaded conditions, Fig. 23.6.5 shows that the EH achieves 94.2% peak power conversion efficiency η_P at the switching frequency of 1MHz, and 99.9% peak tracking efficiency η_T. The quiescent current of the IC is 710μA. Figure 23.6.6 provides performance comparison with state-of-the-art designs. As can be seen, among these prior arts, this work features the widest output power range for concurrently achieving high power and tracking efficiencies. This EH IC also provides at least 6× improvement in power density compared to prior arts. The 0.35μm CMOS test-chip micrograph is shown in Fig. 23.6.7.

References:

[1] T. Esram and P.L. Chapman, "Comparison of Photovoltaic Array Maximum Power Point Tracking Techniques," *IEEE Trans. Energy Conversion*, vol. 22, no. 2, pp. 439-449, Jun. 2007.

[2] W. Liu, Y. Wang, and T. Kuo, "An Adaptive Load-Line Tuning IC for Photovoltaic Module Integrated Mobile Device with 470μs Transient Time, Over 99% Steady-State Accuracy and 94% Power Conversion Efficiency," *ISSCC Dig. Tech. Papers*, pp. 70-71, Feb. 2013.

[3] R. Enne, M. Nikolic, and H. Zimmermann, "A Maximum Power-Point Tracker without Digital Signal Processing in 0.35μm CMOS for Automotive Applications," *ISSCC Dig. Tech. Papers*, pp. 102-103, Feb. 2012.

[4] Linear Technology, LT3652 Data Sheet, "Power Tracking 2A Battery Charger for Solar Power," 2010, Accessed December 2013, <http://www.linear.com/product/LT3652>.

[5] T. Tsai and K. Chen, "A 3.4mW Photovoltaic Energy-Harvesting Charger with Integrated Maximum Power Point Tracking and Battery Management," *ISSCC Dig. Tech. Papers*, pp. 72-73, Feb. 2013.

978-1-4799-0917-9/14 $31.00 © 2014 IEEE

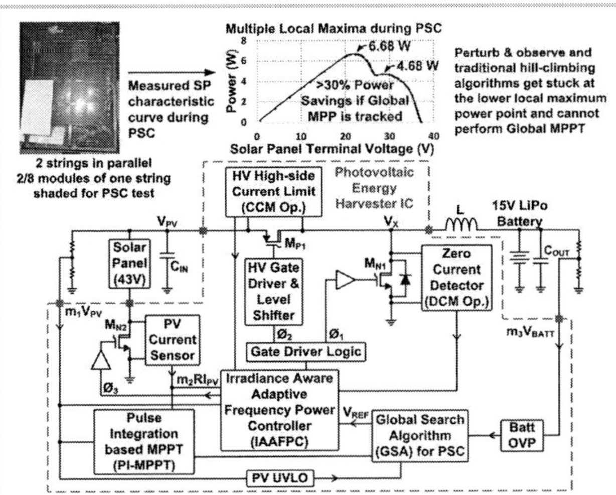

Figure 23.6.1: System architecture and SP P-V behavior under PSC.

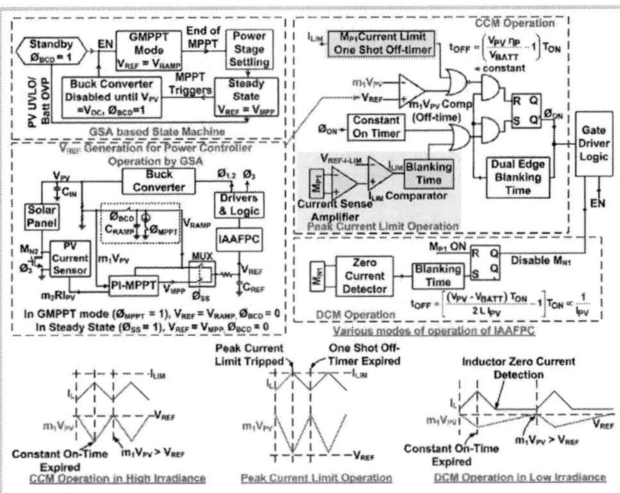

Figure 23.6.2: GSA-based state-machine operation, V_{REF} generation mechanism, and IAAFPC structure and operation.

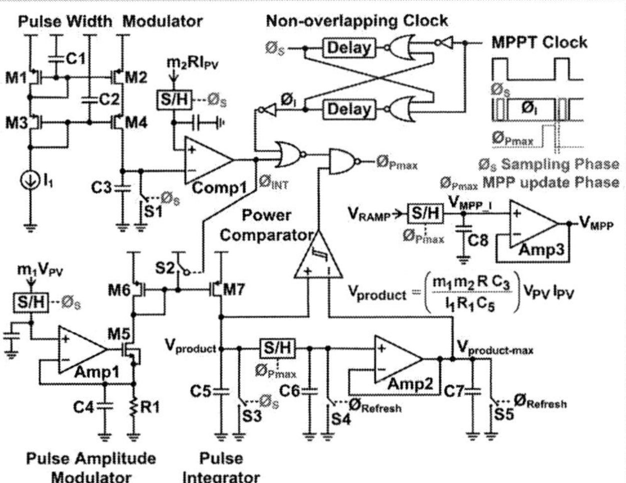

Figure 23.6.3: Schematic of PI-MPPT circuitry for GSA.

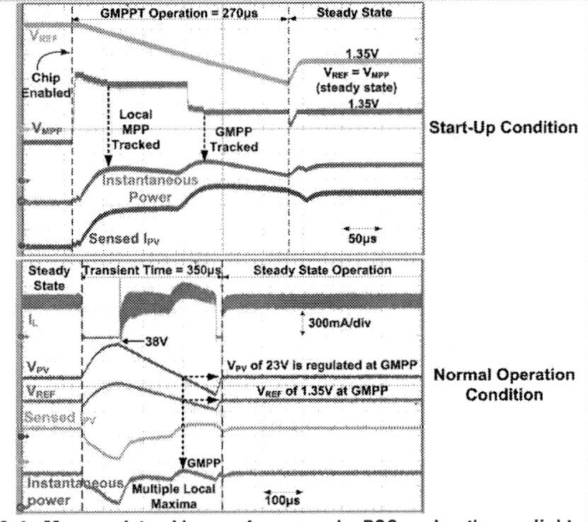

Figure 23.6.4: Measured tracking performance in PSC under the sunlight, achieving 270µs GMPPT time during start-up and total 350µs transient (GMPPT + settling) time in normal operation.

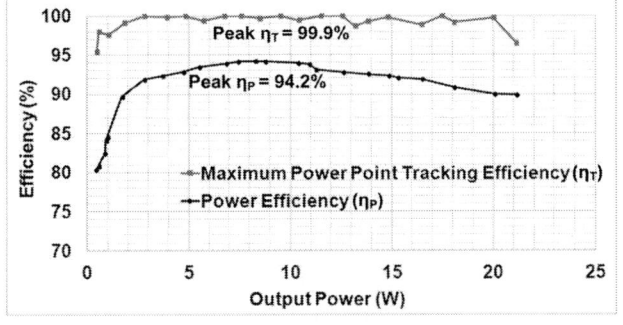

Peak η_T = 99.9%

Peak η_P = 94.2%

Efficiency (%)

Output Power (W)

Maximum Power Point Tracking Efficiency (η_T)

Power Efficiency (η_P)

V_{OUT} is a 15V Lithium polymer battery and V_{IN} is a 40V current-limited power source with source resistance ranging between 12Ω and 500Ω to emulate solar panel characteristics under different irradiance levels.

Figure 23.6.5: Measured MPPT efficiency and power conversion efficiency versus output power of the energy harvester.

	[5] ISSCC 13	[4] LT3652	[3] ISSCC 12	[2] ISSCC 13	This work
Process	TSMC 0.35µm	N.A.	0.35µm HV CMOS	0.5µm CMOS	0.35µm HV CMOS
Harvester Power Throughput (W)	~0.04 – 0.5(1)	7 – 28.8(1)	36 (off-chip power stage)	0.55 – 2.6(1)	0.4 – 21.1
V_{IN} (V)	0.5 – 2	5 – 32	3.4 – 5.5	4	7 – 43
Peak Power Efficiency	89% (f_{SW} = 500kHz)	88% (f_{SW} = 1MHz)	N.A.	94% (f_{SW} = 100kHz)	94.2% (f_{SW} = 1MHz)
Output Power Range for Power Efficiency (η_P) > 85%	~0.12 – 0.34(1)	N.A.	N.A.	0.55 – 2.2(1)	0.988 – 21.1
Power Density (2) @ Peak η_P (mW/mm²)	~158(3)	N.A.	N.A.	168	1022
Peak Tracking Efficiency (w/o Partial Shading)	N.A.	N.A. (pre-programmed V_{REF})	> 99%	99.9%	99.9%
Output Power Range (W) for Tracking Efficiency (η_T) > 95%	N.A.	N.A.	N.A. (~2 – 37.7 Input Power) (4)	0.55 – 2.6(4)	0.4 – 21.1
Capability of Global MPPT in Partial Shading (η_T)	No	No	No	No	Yes (99%)
Transient Time (Total Time for MPPT and Settling to reach P_{MAX}(6))	N.A.	N.A.	N.A.	~1.6ms(5) (470µs for settling within 50% P_{MAX} under single local maximum)	350µs (under multiple local maxima)

(1) Estimated from power efficiency plot
(2) Power density = (output power / chip area)
(3) Estimated from power efficiency plot and chip photo
(4) Estimated from tracking efficiency plots
(5) Estimated from measurement results
(6) P_{MAX} = Maximum power extracted from the solar panel during the steady state

Figure 23.6.6: Measurement summary and performance comparison with state-of-the-art photovoltaic energy harvesters.

23

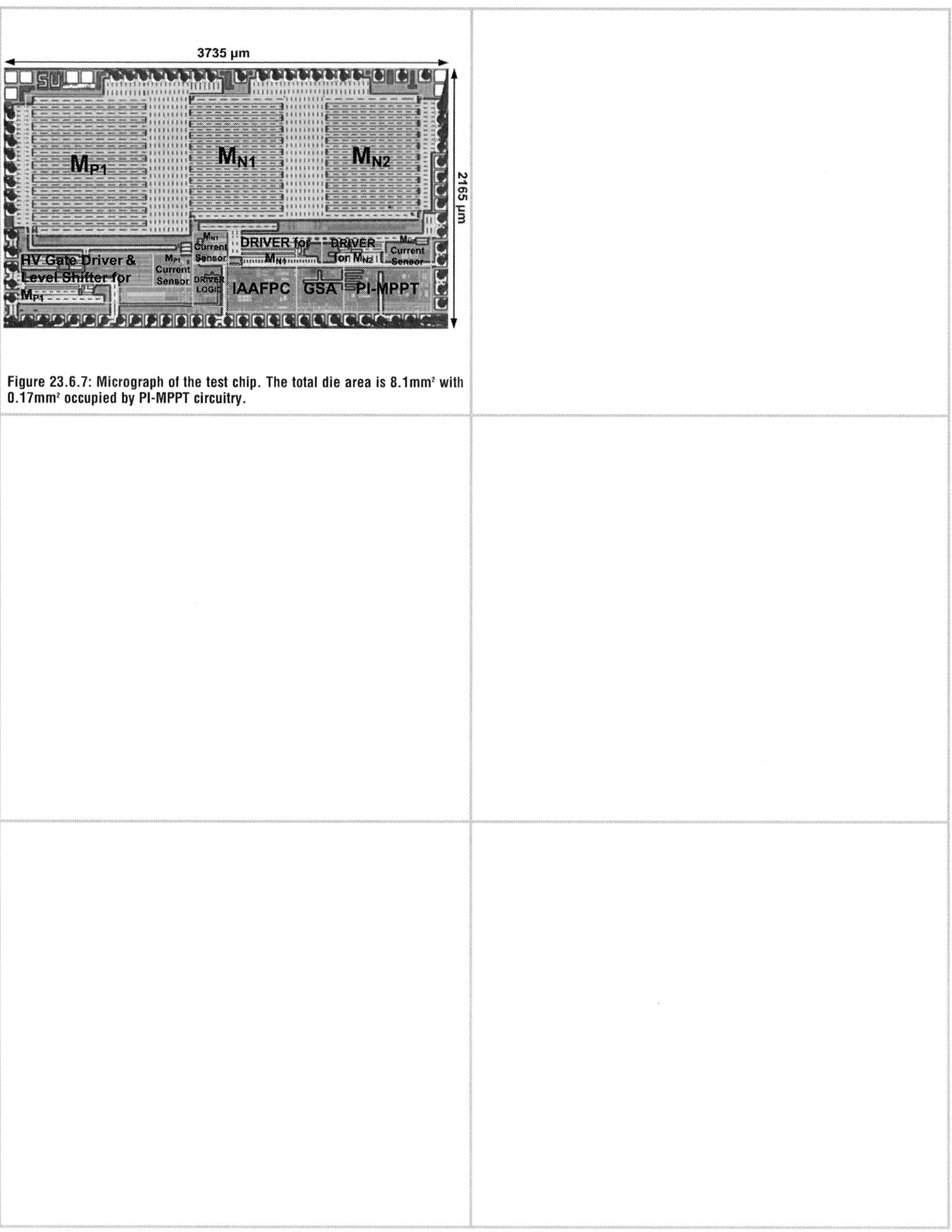

Figure 23.6.7: Micrograph of the test chip. The total die area is 8.1mm² with 0.17mm² occupied by PI-MPPT circuitry.

23.7 Self-Powered 30μW-to-10mW Piezoelectric Energy-Harvesting System with 9.09ms/V Maximum Power Point Tracking Time

Minseob Shim, Jungmoon Kim, Junwon Jung, Chulwoo Kim

Korea University, Seoul, Korea

Energy harvesting is a key technology in various small-size applications such as wireless sensor nodes, mobile devices, and implantable bio-devices to improve battery lifetime or to substitute for batteries. Piezoelectric (PE) transducers are popular energy harvesters that can be used to supply AC power at the μW to mW scale to electronic devices using ambient vibrational energy. To use PE energy effectively, the harvesting systems need a highly efficient AC-DC converter and a DC-DC converter with maximum power point tracking (MPPT). However, existing sub-mW converters do not include an integrated MPPT algorithm [1, 2] or use the perturb and observe MPPT method which has long tracking time [3, 4]. This long tracking time reduces the power extraction from the transducer because the maximum power point (MPP) of the input power can be changed frequently according to the environment. In this paper, a low-power harvesting system is presented that finds the MPP of the input power in one cycle of the vibration of the PE transducer.

Figure 23.7.1 shows the overall architecture of the proposed low-power PE harvesting system with fast MPPT. The PE transducer which generates an AC power is modeled as a current source I_P, in parallel with a parasitic capacitor, C_P, and a resistor, R_P. A comparator-based active rectifier can be used to achieve the high power conversion efficiency [5]. To harvest the maximum power from the rectifier, the harvesting system adopts a simple fractional open-circuit voltage (V_{OC}) method for MPPT because the MPP of the rectifier output is half of the V_{OC} as shown in Fig. 23.7.1. The fractional-V_{OC} MPPT method only evaluates the voltage; hence, the circuit complexity and power consumption are low. The V_{OC}-sensing block consists of a peak detector and a switching controller that periodically sample the V_{OC} value over one cycle. A ramp generator maintains the amplitude and frequency of the ramp signal with low power dissipation even if the supply voltage (V_{DDC}) is changed. Using the ramp signal, the SW controller generates the control signals for a buck-boost converter. Control blocks of the buck-boost converter also prevent reverse leakage current and the body-diode (BD) effect.

The proposed V_{OC}-sensing block uses a small sensing capacitor (C_2) in contrast to the conventional sensing circuits as shown in Fig. 23.7.2. When C_2 is used, the rectifier output voltage exhibits large ripples and reaches V_{OC} in one cycle. However, it is difficult to sense the peak voltage of the rectifier output because the peak voltage is the highest in the overall harvesting system. Therefore, the harvesting system uses a peak detector using a differentiator instead of a conventional peak detector. An open-circuit time controller generates a signal SW1 to periodically open the switch between the rectifier output (V_{RECT}) and the buck-boost input (V_{IN}) over one cycle. If the V_{RECT} reaches the V_{OC} value, the output (V_P) of the differentiator becomes smaller than the bias voltage (V_{BIAS}). At this point, the comparator and the rising edge detector generate the signal V_{PK}. Using the signal V_{PK}, the switching controller for sensing the half V_{OC} point generates signals SW2 and SW3. When the signal SW2 goes high, the V_{OC} value is copied from C_2 to C_{OC1} and discharges C_{OC2} to prevent charge accumulation. After storing the V_{OC} value, the signal SW3 closes the switch between C_{OC1} and C_{OC2}, which results in half of V_{OC} on the capacitors because of charge sharing. All processes for sensing half of the V_{OC} are completed over one cycle of the PE transducer. As a result, the harvesting system can find the MPP in a very short time and extract more power effectively.

The harvesting system employs pulse-frequency modulation (PFM) to harvest from a low input power source. Commonly used switch controllers for PFM DC-DC converters need a clock with a constant frequency and amplitude of V_{DDC}. However, a low-dropout regulator and a level shifter are needed to generate a suitable clock signal because the frequency and power dissipation of a voltage controlled oscillator can be changed significantly under V_{DDC} variation. These additional blocks also increase power and area requirements. Hence, a low-power ramp generator with V_{DDC} independence is proposed as shown in Fig. 23.7.3. In contrast with the conventional ramp generator [6], the proposed ramp generator consists of three ramp cells; each cell uses only 6 MOSFETs. The current sources in each ramp cell provide a constant current of 20nA, regardless of V_{DDC}. The constant current supplying the ramp cell makes the charging time for $C_{R1,2}$ constant and allows for low-power consumption. For this reason, the ramp generator produces a ramp signal that has a constant frequency and voltage amplitude, without requiring a comparator or other logic blocks. The SW controller compares the ramp signal and the reference voltage from the bias block and generates a clock signal ($V_{S1,2}$) with an amplitude of V_{DDC}, a constant frequency and a fixed duty cycle.

The control blocks of the system are self-powered circuits that use the input and output voltages of the DC-DC converter as their supply voltage. To supply the higher voltage to the self-powered control blocks, a voltage multiplexer (MUX) is employed, as shown in Fig. 23.7.4 [6]. The comparators in the MUX dissipate low power. However, reducing power consumption increases the delay for the comparison result. The delay of the comparator output causes a voltage spike when a conventional MUX is used. The proposed MUX uses an S-R latch and a rising edge detector not only to remove the spike but also to select the front edge of the comparator outputs to produce a smooth V_{DDC} curve.

Figure 23.7.4 shows a block that prevents reverse current and BD effect in the buck-boost converter. The buck-boost converter employs PFM and a discontinuous conduction mode (DCM) to control the V_{IN} value of the buck-boost converter. In the DCM, the inductor current I_L becomes zero during each period. When V_X is smaller than V_{OUT}, the control block opens all MOSFETs of the power stage to prevent reverse leakage current. On the other hand, the buck-boost converter can work in a continuous conduction mode (CCM) when the harvesting system tracks the MPP in the very low V_{OUT} condition. If V_{IN} reaches the MPP before I_L becomes zero, all power-MOSFETs are opened and a large conduction loss across the BD occurs. If V_{IN} is half of V_{OC} or less and V_X is higher than V_{OUT}, the control block closes M_2 and M_4 to remove the BD effect.

Figure 23.7.5 shows the measurement waveforms of the harvesting system. The SW controller compares V_{RAMP} to a reference voltage and generates a control signal with a suitable duty ratio. When the reference voltage is set to 0.4V, the output signal of the SW controller has a constant frequency of 90kHz and a duty ratio of 50%, regardless of V_{DDC} value (1.2 or 6.5V). These results indicate that the ramp signal also has a constant frequency and amplitude, regardless of V_{DDC}. The harvesting system senses a half V_{OC} value in one cycle and changes V_{RECT} and V_{IN} to produce the MPP. The MPP tracking time is 20ms when the MPP at the V_{IN} input is changed from 3.4 down to 1.2V (9.09ms/V). Figure 23.7.6 shows the measured power conversion efficiency of buck-boost converter and comparison table with other reported PE energy harvesting systems. The maximum power conversion efficiency of buck-boost converter is 80% including a power consumption of controller (V_{IN}=4.7V and 7kΩ load resistance). The system with maximum MPPT efficiency ($V_{MPP}/0.5V_{OC}$) of 99.9% dramatically reduces the MPPT time as compared to the state-of-the-art systems. All control blocks of the system are fabricated in a 0.35μm BCDMOS process. The system uses only two kinds of external components: two capacitors and an inductor. The micrograph of the die is shown in Fig. 23.7.7.

Acknowledgements:
This work is supported by the National Research Foundation of Korea grant funded by the Korea government (MEST) (No. 2011-0020128) and the MPW of IDEC.

References:
[1] S. Bandyopadhyay, Y.K. Ramadass, and A.P. Chandrakasan, "20μA to 100mA DC-DC Converter with 2.8 to 4.2V Battery Supply for Portable Applications in 45nm CMOS," *ISSCC Dig. Tech. Papers*, pp. 386-387, Feb. 2011.
[2] T.-C. Huang, C.-Y. Hseih, Y.-Y. Yang, *et al.*, "A Battery-Free 217 nW Static Control Power Buck Converter for Wireless RF Energy Harvesting with α-Calibrated Dynamic On/Off Time and Adaptive Phase Lead Control," *IEEE J. Solid-State Circuits*, vol. 47, no. 4, pp. 852-862, April 2012.
[3] S. Stanzione, C. van Liempd, R. van Schaijk, *et al.*, "A Self-Biased 5-to-60V Input Voltage and 25-to-1600μW Integrated DC-DC Buck Converter with Fully Analog MPPT Algorithm Reaching up to 88% End-to-End Efficiency," *ISSCC Dig.* Tech. Papers, pp. 74-75, Feb. 2013.
[4] N. Kong and D.S. Ha, "Low-Power Design of a Self-Powered Piezoelectric Energy Harvesting System with Maximum Power Point Tracking," *IEEE Trans. Power Electronics*, vol. 27, no. 5, pp. 2298-2308, May 2012.
[5] Y. Sun and D.S. Ha, "An Integrated High Performance Active Rectifier for Piezoelectric Vibration Energy Harvesting Systems," *IEEE Trans. Power Electronics*, vol. 27, no. 2, pp. 623-627, Feb. 2012.
[6] T.Y. Man, P. Mok, and M.J. Chan, "A 0.9-V Input Discontinuous-Conduction-Mode Boost Converter with CMOS-Control Rectifier," *IEEE J. Solid-State Circuits*, vol. 43, no. 9, pp. 2306-2346, Sep. 2008.

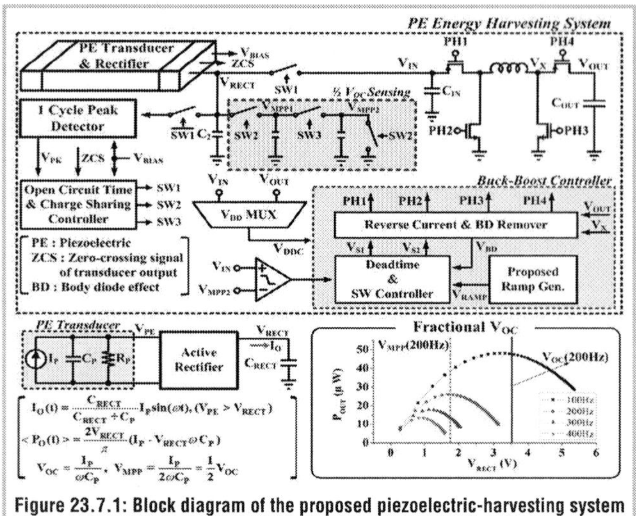

Figure 23.7.1: Block diagram of the proposed piezoelectric-harvesting system with one-cycle maximum power point tracking.

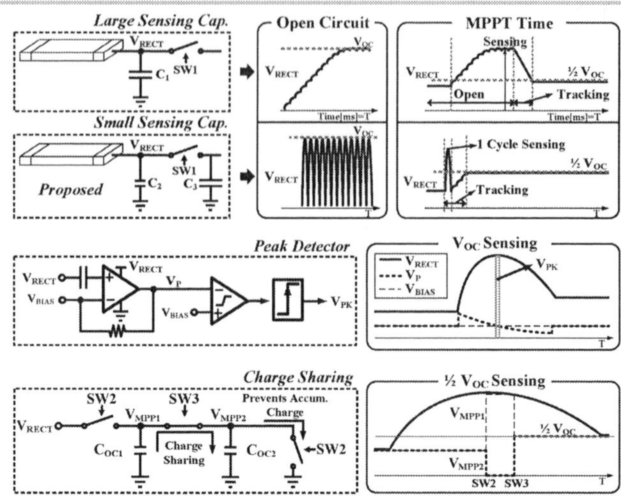

Figure 23.7.2: One-cycle half-V_{OC} (MPP) sensing block.

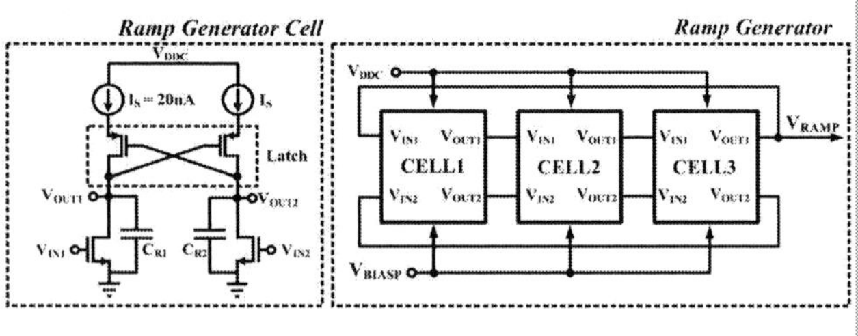

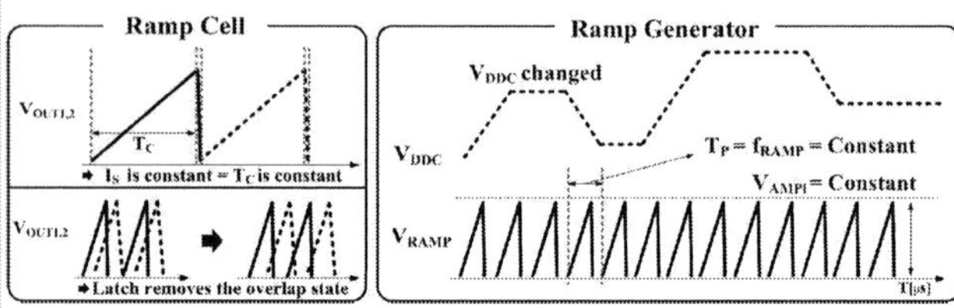

Figure 23.7.3: Low-power ramp generator with V_{DD} independence.

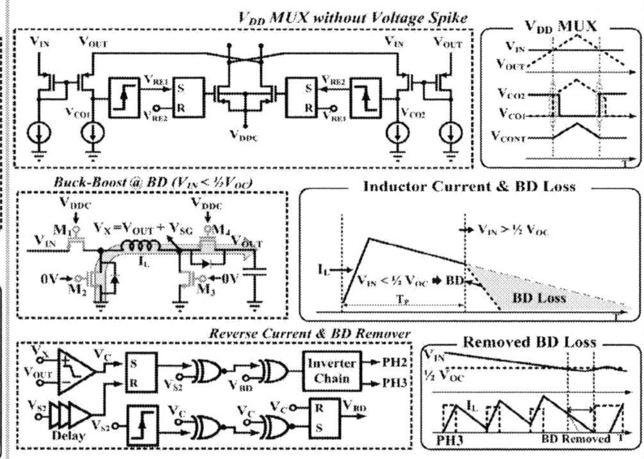

Figure 23.7.4: Schematics of voltage MUX without voltage spikes, and reverse current & BD effect remover.

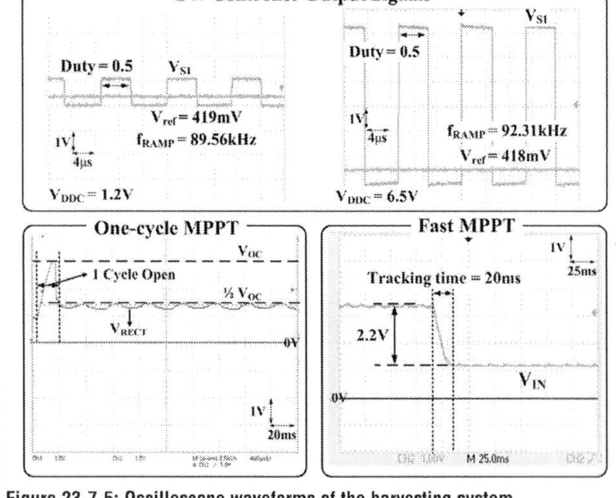

Figure 23.7.5: Oscilloscope waveforms of the harvesting system.

Power Conversion Efficiency of Buck-Boost

Efficiency (%) vs. P_{IN} (mW)

Parameters	ISSCC 2013 [3]	TPEL 2012 [4]	This work
Process	0.25μm BCD	Off-chip	0.35μm BCD
Input voltage	5~60V	3~25V	1~7V
Output voltage	2~5V	3V~	1~8V
Input power	25μW~1.6mW	N.A	33μW~10mW
Type of converter	Buck	Buck-Boost	Buck-Boost
Maximum converter efficiency	88.9%	76%	80%
MPPT algorithm	Variable Step-size P&O	P&O	Fractional V_{OC}
Maximum MPPT efficiency	99.9%	97%	99.9%
MPPT Time	800ms (21.5V to 11.5V)	47s (17V to 11V)	20ms (3.4V to 1.2V)

Figure 23.7.6: Measured conversion efficiency and comparison table.

Figure 23.7.7: Die micrograph of the piezoelectric energy harvesting system.

23.8 A 34V Charge Pump in 65nm Bulk CMOS Technology

Yousr Ismail[1], Haechang Lee[2,*], Sudhakar Pamarti[1], Chih-Kong Ken Yang[1]

[1]University of California, Los Angeles, CA, [2]SiTime, Sunnyvale, CA
[*]now with Altera, San Jose, CA

Recent advances in MEMS-based oscillators have resulted in their proliferation in timing applications that were once exclusive to quartz-based devices [1]. For applications requiring low phase noise — e.g., cellular, GPS and high-speed serial links — one possible approach is to bias the MEMS resonator at a higher DC voltage to reduce its motional impedance and increase signal energy [2]. Realizing high-voltage charge pumps in bulk CMOS technology is limited by the breakdown voltage of the well/substrate diodes shown in Fig. 23.8.1(a) and Fig. 23.8.1(b). This breakdown limit is even lower with technology scaling and is <10V in a 22nm CMOS node. Systems with high-voltage requirements often resort to older, high-voltage-tolerant nodes or exotic technologies that limit MEMS integration into SoCs. This work demonstrates a charge pump design in 65nm technology with a three-fold increase in the output voltage range. High-voltage tolerance is enabled by the proposed well-biasing arrangement and oxide isolation. The pump achieves 34V output by using three different charge pump cells that tradeoff achievable voltage range and power efficiency to achieve a peak efficiency of 38%. Additionally, finger capacitors are optimized to ensure reliability while maintaining efficiency.

This design employs two techniques in fabrication technology. First, for triple-well technologies, conventional charge pump designs are limited by the breakdown voltage of the deep-nwell/substrate diode (~12V in a 65nm technology). Instead, we propose to connect the NMOS bulk and deep-nwell terminals together and hold them at a fixed potential V_{mid}, ideally 12V, as shown in Fig. 23.8.1(c). This allows the NMOS source/drain terminals to be pumped up to voltages higher than V_{mid} by an additional n+/pwell diode breakdown voltage. Since the n+/pwell breakdown voltage in 65nm CMOS is 8V, this well-biasing scheme extends the output voltage range of bulk CMOS charge pumps from 12 to 20V. Second, to further extend achievable output voltages, instead of relying on reverse junction isolation, we leverage the field oxide isolation to build polysilicon diodes over shallow trench isolation (STI). Polysilicon PIN diodes are implemented with an intrinsic region length L_i and are surrounded by a deep-nwell ring to mitigate substrate coupling as shown in Fig. 23.8.1(d). The STI provides a much higher breakdown limit than reverse-biased junctions. While 43V output has been reported in 0.25μm CMOS [3], because of the knee voltage, diode based designs have poor efficiencies. By combining more efficient lower-voltage-tolerant cells with less efficient higher-voltage-tolerant cells, a 34V charge pump can be realized at improved efficiency.

The proposed charge pump is composed of a cascade of 3 distinct smaller sub-pumps as shown in Fig. 23.8.2. The first sub-pump consists of 4 cascaded stages. Each stage is implemented as a CMOS four-phase voltage doubler (FPVD), shown in Fig. 23.8.3. All devices in this FPVD have their bulk terminals connected to their respective source and deep-nwell terminals. Using separate capacitor pairs for the clock boosting and for the voltage pumping helps the design achieve smaller voltage droops at higher load currents [4]. The clock boosting capacitors can be very small. In this design, we introduce a voltage doubler cell that separates the gate drive for the NMOS switches from that for the PMOS switches. Such switch decoupling along with careful clock timing secures a break-before-make operation that eliminates reverse currents and improves efficiency. Clock phases φ_1 and φ_2 are overlapping while clock phases φ_3 and φ_4 are non-overlapping. Furthermore, clock phases φ_1 and φ_2 in all three sub-pumps are provided by tri-state buffers with a charge-equalization switch to reduce dynamic power losses caused by parasitic capacitances. The second sub-pump has an all-NMOS implementation and consists of five stages. Each stage is implemented as an NMOS FPVD, shown in Fig. 23.8.4. All devices in this FPVD have their deep-nwell terminal connected to V_{mid} which is also the input voltage provided by the first sub-pump output. This sub-pump has a lower efficiency due to the diode-connected devices and the increase in V_{th} caused by the back-gate bias. The third sub-pump is implemented as a Dickson charge pump with 8 stages. Each stage consists of 2 diode-capacitor pairs operating on opposite clock phases, φ_1 and φ_2.

An important concern in nanometer-scale high-voltage design is device reliability. All transistors in the design are thick-oxide devices with a 2.5V±10% voltage rating. In the FPVD, the gate-source and drain-source potentials of the transistors cannot exceed their clock amplitude (<2.75V). In the Dickson pump, the diodes are exposed to roughly twice the clock amplitude during their off state. Because a PIN diode breakdown voltage depends on L_i, in this design, L_i is chosen conservatively to be 1μm due to lack of prior characterization. The diode breakdown voltage as measured on the test chip is found to be ~30V, and a smaller L_i would have sufficed. Measurement results show that the current drive capability of the pump is not severely degraded because of this choice of L_i. All capacitors are metal-finger capacitors whose reliability is determined by the time-dependent dielectric breakdown (TDDB) failure rates. The TDDB is a function of the finger spacing and its lateral area. Wider finger spacing, for improved TDDB, results in higher parasitic capacitance for the same capacitor value. A unit 3.6pF pump capacitor designed with a minimum finger spacing of 100nm is estimated to sustain ~12V based on a TDDB cumulative failure rate of less than 100 ppm over 10 years at 85°C. The STI thickness is ~300nm and is estimated to sustain ~55V based on the same criteria. Note that only the STI areas underneath the polysilicon diodes are exposed to voltage stress. Our TDDB calculations are based on a Weibull statistical distribution, a field accelerated $\sqrt{E}$-model, an Arrhenius temperature relation, and a Poisson area-scaling model.

To optimize area and power efficiency, the finger spacing of the capacitors is gradually tapered depending on the voltage requirements of each sub-pump. Based on our model, the finger spacing does not need to increase linearly with the capacitor voltage requirement. The first sub-pump uses minimally spaced fingers (1× spaced) corresponding to a capacitance density of 1.4fF/μm². The second and third sub-pumps use 1.5× (0.93fF/μm²) and 2× (0.6fF/μm²) finger spacing, respectively. M1 is not used to cut down the bottom plate parasitic capacitance. In the case of the 2× capacitors, the vertical metal spacing becomes smaller than the lateral finger spacing, and capacitors with fringe-fields only are used. Also, careful layout observes sufficient spacing between low- and high-voltage interconnects for a reliable operation.

The measured output voltages of the 3 cascaded sub-pumps are plotted versus the pumping voltage as shown in Fig. 23.8.5. Measurements are performed at 8MHz pumping frequency and no load current. As expected, for higher pumping voltages, the first and second sub-pump outputs saturate at 12V and 20V, respectively. To characterize the breakdown limitation of the third sub-pump, instead of increasing the pumping voltage which risks permanently damaging the transistors, we fix its pumping voltage and slowly increase the input voltage using an external high-voltage supply. The third sub-pump is found to break down at 88V. This value is found to be consistent across different capacitor finger spacing. The charge pump I/V characteristics and efficiency are characterized at 8MHz pumping frequency and at different pumping voltages as shown in Fig. 23.8.6(a) and Fig. 23.8.6(b), respectively. The pump demonstrates a maximum output voltage of 34V at 2.75V pumping voltage and maintains an output voltage >30V for load current <10μA. Power efficiency is measured separately for each sub-pump and includes the power dissipated in the clock pre-drivers. Peak efficiencies of 68%, 42% and 29% are measured for the first, second, and third sub-pumps, respectively. The overall peak efficiency of the cascaded sub-pumps is 38% at 25μA load current and is still higher than the third sub-pump's efficiency alone. The pump output resistance is not limited by the diode series resistance and is plotted versus the pumping frequency as shown in Fig. 23.8.6(c). The die micrograph with all three sub-pumps highlighted is shown in Fig. 23.8.7. The areas of the first, second, and third sub-pumps are 0.028mm², 0.053mm², and 0.071mm², respectively.

Acknowledgments:
The authors would like to thank Professor M.-C. Frank Chang and TSMC for providing device fabrication.

References:
[1] Michael Perrott, J. Salvia, F. Lee, *et al.*, "A Temperature-to-Digital Converter for a MEMS-Based Programmable Oscillator with Better Than ±0.5ppm Frequency Stability," *ISSCC Dig. Tech. Papers*, pp. 206-207, Feb. 2012.
[2] H. Lee, A. Partridge, and F. Assaderaghi, "Low Jitter and Temperature Stable MEMS Oscillators," *IEEE Int'l Frequency Control Symposium*, May 2012.
[3] M.-D. Ker and S.-L. Chen, "Ultra-High-Voltage Charge Pump Circuit in Low-Voltage Bulk CMOS Processes with Polysilicon Diodes," *IEEE Trans. Circuits and Systems II*, vol. 54, no. 1, pp. 47-51, Jan. 2007.
[4] J.-Y. Kim, Y.-H. Jun, and B.-S. Kong, "CMOS Charge Pump with Transfer Blocking Technique for No Reversion Loss and Relaxed Clock Timing Restriction," *IEEE Trans. Circuits and Systems II*, vol. 56, no. 1, pp.11-15, Jan. 2009.

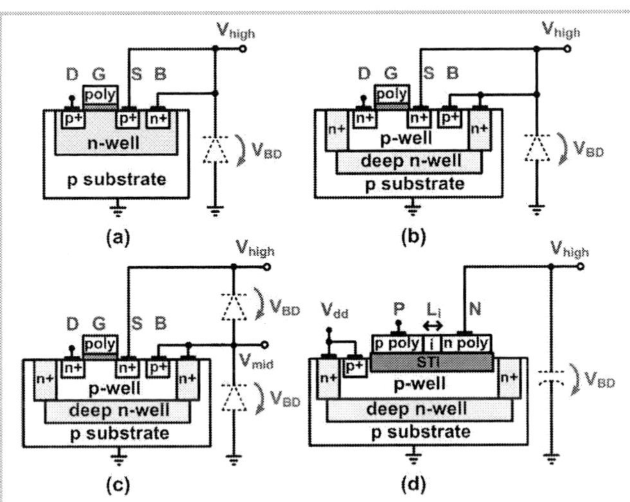

Figure 23.8.1: Illustration of device structure for: (a) PMOS device, (b) NMOS device, (c) the proposed NMOS device, and (d) polysilicon PIN diode.

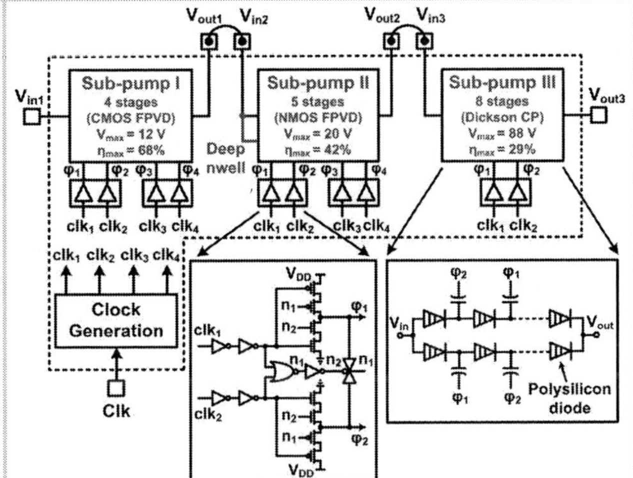

Figure 23.8.2: Block diagram of the 34V charge pump with circuit schematics for the clock drivers and the 3rd sub-pump.

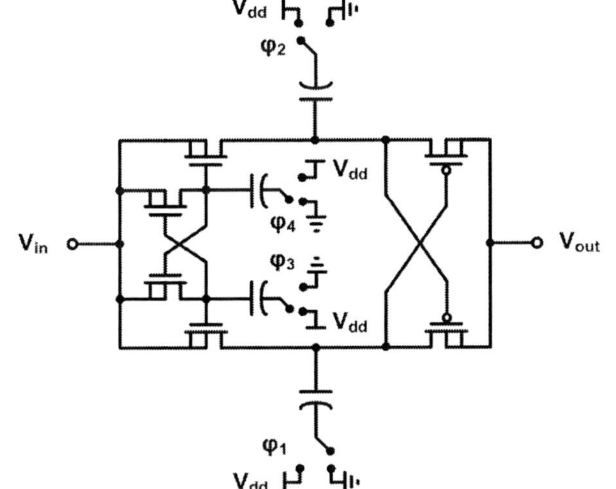

Figure 23.8.3: Circuit schematic of the CMOS FPVD in the 1st sub-pump.

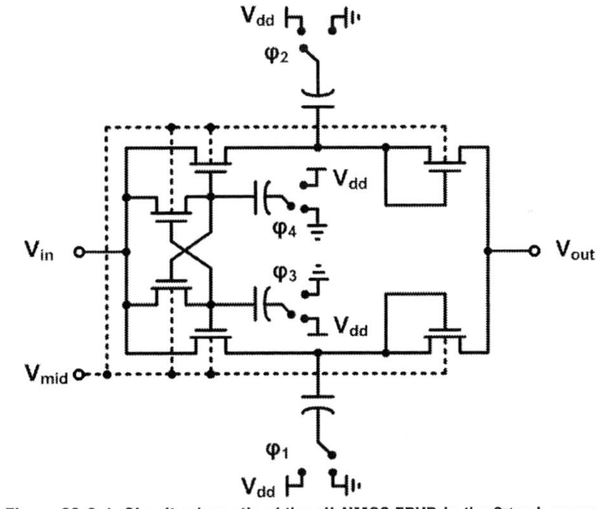

Figure 23.8.4: Circuit schematic of the all-NMOS FPVD in the 2nd sub-pump.

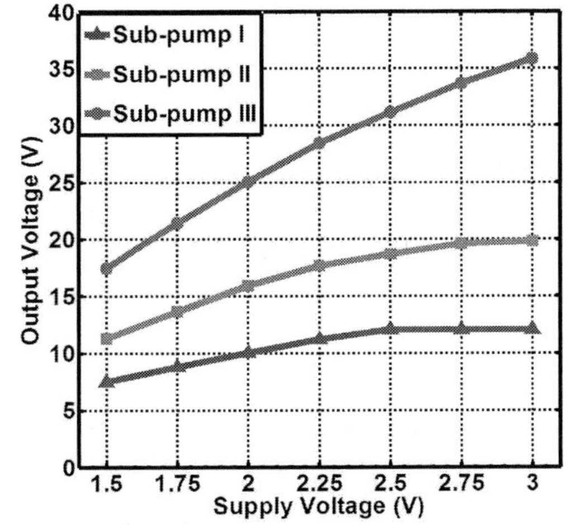

Figure 23.8.5: Measured output voltages of all 3 sub-pumps at pumping frequency f=8MHz and no load current.

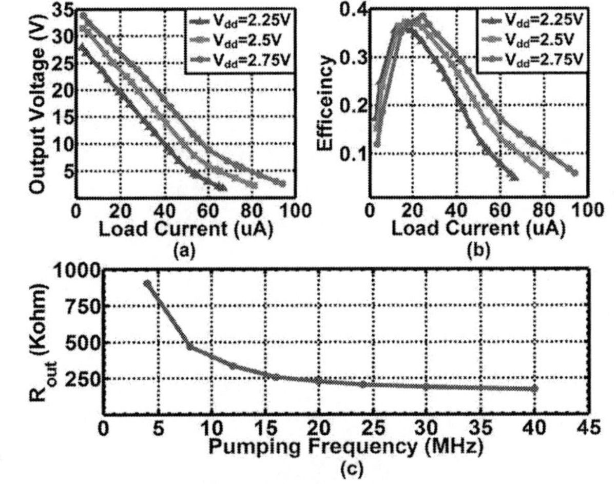

Figure 23.8.6: (a) I/V DC characteristics and (b) efficiency measured at pumping frequency f=8MHz, and (c) output resistance at V_{dd}=2.5V.

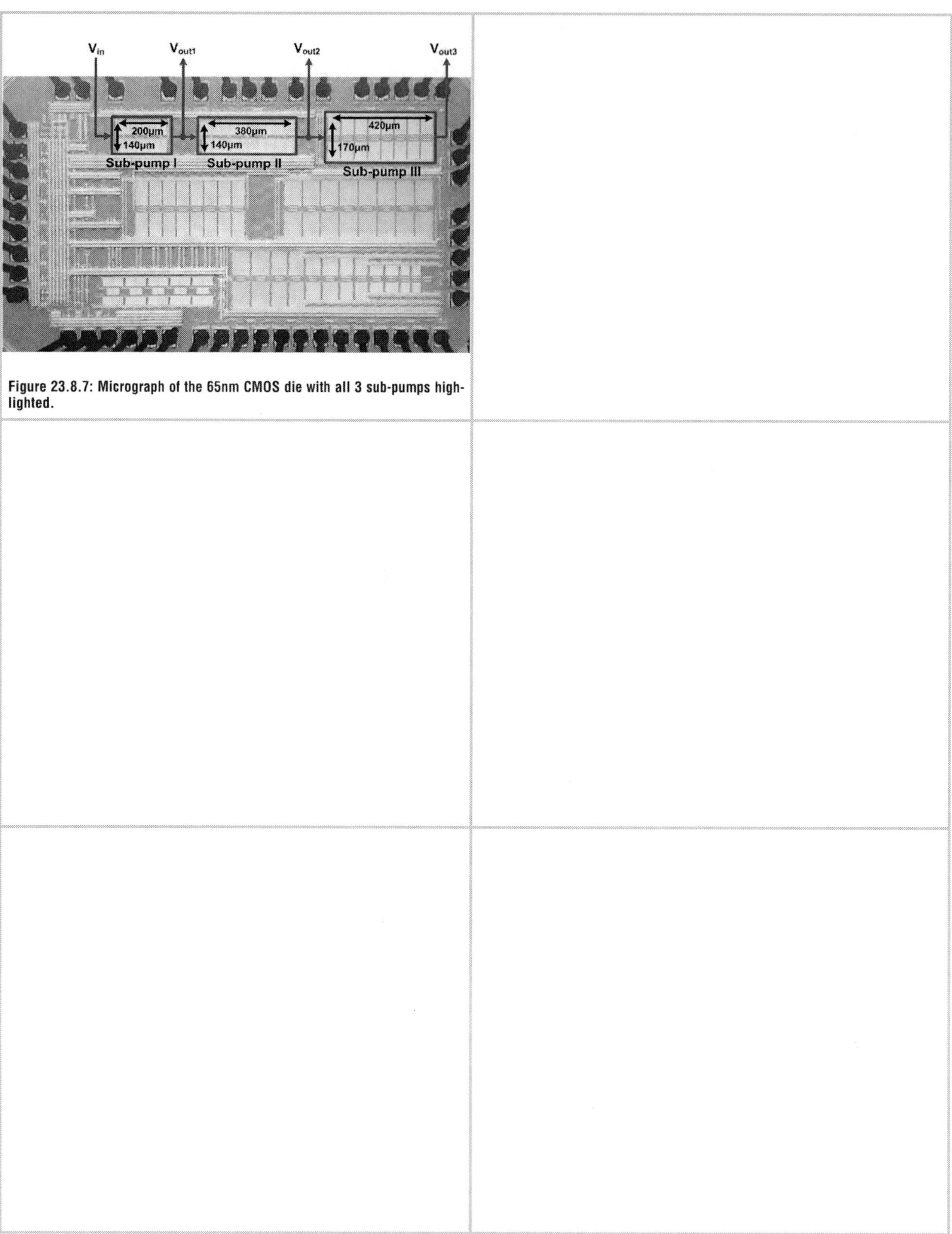

Figure 23.8.7: Micrograph of the 65nm CMOS die with all 3 sub-pumps highlighted.

Session 24 Overview: *Integrated Biomedical Systems*

IMMD SUBCOMMITTEE

Session Chair: *Maysam Ghovanloo*
Georgia Institute of Technology, Atlanta, GA

Session Co-Chair: *Wentai Liu*
University of California, Los Angeles, CA

This session presents state-of-the-art integrated biomedical systems for high-density neural recording, efficient neuromodulation, ultra-low-power cardiac monitoring, artificial noses, cell type detection, and 3D ultrasound imaging. The first paper focuses on a miniaturized, wirelessly powered, neural interface for acquisition of ECoG signals. The second paper presents a switched-capacitor-based wireless deep-brain stimulator that also supports optogenetic stimulation. The next two papers focus on ECG monitoring using ultra-low-power circuit-design techniques. The fifth paper describes an artificial nose-on-a-chip that can detect early-stage pneumonia. The next paper is about a flow cytometer-on-a-chip using magnetic bead labels. The seventh paper describes an interface chip for EEG recording with active dry electrodes. Finally, a 64-channel interface chip is presented for beamforming in 2D capacitive micromachined ultrasound transducers (CMUT).

24.1 A Miniaturized 64-Channel 225µW Wireless Electrocorticographic Neural Sensor — **1:30 PM**

R. Muller, University of California, Berkeley, CA
and University of Melbourne, Parkville, Australia

In Paper 24.1, the University of Melbourne (with UC Berkeley) presents a 225µW 64-channel implantable microsystem for recording of ECoG signals with wireless power delivery. It includes a microfabricated unit comprised of a planar electrode array, an antenna, and a 65nm CMOS IC.

24.2 A Power-Efficient Switched-Capacitor Stimulating System for Electrical/Optical Deep-Brain Stimulation — **2:00 PM**

H-M. Lee, Georgia Institute of Technology, Atlanta, GA

In Paper 24.2, Georgia Institute of Technology (with Michigan State University) presents a power-efficient wireless switched-capacitor stimulating (SCS) system for electrical/optical deep-brain stimulation. The SCS system improves the stimulator efficiency, reaching a peak of 80.4%, by generating charge-controlled decaying-exponential stimulus pulses.

24.3 An Implantable 64nW ECG-Monitoring Mixed-Signal SoC for Arrhythmia Diagnosis — **2:30 PM**

D. Jeon, University of Michigan, Ann Arbor, MI

In Paper 24.3, the University of Michigan presents an ECG-monitoring SoC suitable for arrhythmia diagnosis that can be injected using a syringe. The 65nm test chip consumes 64nW while continuously capturing abnormal cardiac events, achieving 100× lower power than prior work.

24.4 A 680nA Fully Integrated Implantable ECG-Acquisition IC with Analog Feature Extraction 3:15 PM

L. Yan, imec, Heverlee, Belgium

In Paper 24.4, imec (with TU Eindhoven, Olympus, and KU Leuven) presents a fully integrated implantable ECG-acquisition IC. It integrates a low-power wide-dynamic-range analog signal processor that extracts QRS features in the analog domain, while consuming 680nA.

24.5 A 0.5V 1.27mW Nose-on-a-Chip for Rapid Diagnosis of Ventilator-Associated Pneumonia 3:45 PM

K-T. Tang, National Tsing Hua University, Hsinchu, Taiwan

In Paper 24.5, National Tsing Hua University (with Taipei Medical University and National Chiao Tung University) presents a nose-on-a-chip for ventilators to monitor and detect pneumonia in early stage. The chip is fabricated in 90nm CMOS, consumes 1.27mW at 0.5V, and provides a promising solution for rapid diagnostics of ventilator-associated pneumonia.

24.6 A CMOS Micro-Flow Cytometer for Magnetic Label Detection and Classification 4:15 PM

P. Murali, University of California, Berkeley, CA

In Paper 24.6, UC Berkeley presents a flow cytometer on a chip embedded into a single-use microfluidic cartridge. The chip is implemented in 0.18μm CMOS, operates at 1.2GHz, and detects cell types using superparamagnetic micro-bead labels.

24.7 A 60nV/√Hz 15-Channel Digital Active Electrode System for Portable Biopotential Signal Acquisition 4:30 PM

J. Xu, Holst Centre/imec, Eindhoven, The Netherlands
and Delft University of Technology, Delft, The Netherlands

In Paper 24.7, Holst Center/imec (with TU Delft and KU Leuven) presents a digital active electrode for biopotential signal acquisition with dry electrodes. The 0.18μm CMOS ASIC achieves 60nV/√Hz input-referred noise, and ±350mV electrode offset rejection.

24.8 An Analog-Digital-Hybrid Single-Chip RX Beamformer with Non-Uniform Sampling for 2D-CMUT Ultrasound Imaging to Achieve Wide Dynamic Range of Delay and Small Chip Area 4:45 PM

J-Y. Um, Pohang University of Science and Technology, Pohang, Korea

In Paper 24.8, POSTECH (with Samsung Advanced Institute of Technology) presents an analog-digital-hybrid architecture for a single-chip 64-channel ultrasound beamformer to be used with a 2D CMUT array. The 0.13μm CMOS chip achieves a delay resolution of 6.25ns and maximum delay range of 8μs.

24.1 A Miniaturized 64-Channel 225μW Wireless Electrocorticographic Neural Sensor

Rikky Muller[1,2], Hanh-Phuc Le[1], Wen Li[1], Peter Ledochowitsch[1], Simone Gambini[2], Toni Bjorninen[3], Aaron Koralek[1], Jose M. Carmena[1], Michel M. Maharbiz[1], Elad Alon[1], Jan M. Rabaey[1]

[1]University of California, Berkeley, CA,
[2]University of Melbourne, Parkville, Australia,
[3]Tampere University of Technology, Tampere, Finland

Substantial improvements in neural-implant longevity are needed to transition brain-machine interface (BMI) systems from research labs to clinical practice. While action potential (AP) recording through penetrating electrode arrays offers the highest spatial resolution, it comes at the price of tissue scarring, resulting in signal degradation over the course of several months [1]. Electrocorticography (ECoG) is an electrophysiological technique where electrical potentials are recorded from the surface of the cerebral cortex, reducing cortical scarring. However, today's clinical ECoG implants are large, have low spatial resolution (0.4 to 1cm) and offer only wired operation.

To enable chronic and stable neural recording, we introduce a minimally invasive, wireless ECoG microsystem. Wireless powering and readout are combined with a microfabricated antenna and electrode grid that has >10× higher electrode density than clinical ECoG arrays, providing spatial resolution approaching today's penetrating electrodes. Area- and power-reduction techniques in the baseband and wireless subsystem result in over 10× IC area reduction with a simultaneous 3× improvement in power efficiency over the state of the art (see Fig. 24.1.4), enabling a minimally invasive platform for 64-channel recording. The low power consumption of the IC, together with the antenna integration strategy enables remote powering at 3× below established safety limits [2], while the small size and flexibility of the implant minimizes the foreign body response. The improved implant safety and longevity gives wireless ECoG excellent prospects to become the technology of choice for clinically relevant BMIs in the foreseeable future [1].

Figure 24.1.1 illustrates the concept of the implantable ECoG microsystem and a block diagram of the IC, which includes circuitry for signal acquisition, a matching network, clock recovery, communication and power management. To mitigate the implantation of a large rigid structure, the IC is bonded directly to thin-film platinum and gold electrodes that are patterned over Parylene C, a biocompatible polymer [3]. The highly flexible grid is 10μm thick and conforms to the cortical folds, further reducing neural damage. Platinum black is electroplated onto the electrode surface reducing the impedance to 10kΩ at 100Hz. A 6.5mm-diameter, single-loop antenna is monolithically integrated together with the electrodes and is used for both power and data telemetry. The antenna achieves -17.3dB link gain at 300MHz while transmitting across a human skull model [2]. A 1.5cm-diameter external antenna completes the link and powers the device, radiating 12mW of power, 3× lower than the IEEE and FCC recommendation.

The 64-channel front-end array (with an ADC per channel) dominates the IC power consumption, making a power-efficient design critical. The acquisition of useful ECoG signals necessitates an input-referred noise of ~1μV over 1 to 500Hz, which must be achieved in the presence of a large DC offset (up to ±50mV) at the electrode-chip interface. The key to power-efficient design lies in canceling the DC offset early in the signal chain while minimizing flicker noise. The architecture of Fig. 24.1.2 achieves this by using a chopper-stabilized amplifier to minimize 1/f noise, and an oversampled ΔΣ DAC with 15b resolution to cancel the upmodulated DC offset. To prevent instantaneously large amplifier inputs, 5 physical DAC bits are implemented as a 31-element, thermometer-coded capacitor array with unit capacitor C_{LSB}. $31C_{LSB}=0.1C_{in}$ is chosen to cancel the offset while keeping signal attenuation below 1dB. The large time constants necessary for filtering the offset are implemented digitally [4], enabling an area of 0.025mm² for each front-end. Using an open-loop amplifier improves input impedance, resulting in Z_{in}=28MΩ at 100Hz with f_{chop}=8kHz. After chopper demodulation and RC filtering (to suppress ΔΣ noise), the signal is digitized by a pseudo-differential, VCO-based ADC [4] operating at 1kS/s. The ADC has a raw resolution of 15b to suppress quantization noise while processing the ECoG signal, the chopper ripple, and the ΔΣ noise the DAC. By designing $f_{chop}=Nf_{ADC}$ (N is an integer) the chopper ripple falls in a notch of the ADC sinc transfer function [4] eliminating the need for a ripple-reduction loop.

The 1kS/s, 15b digital outputs are serialized into a 1Mb/s Miller-encoded data stream and back-scattered through a shunt-load modulation switch. We trade off modulation depth in order to support simultaneous data and power transfer and employ a dual-mode RF-to-DC rectifier to handle the input voltage variation. As shown in Fig. 24.1.3, a high-impedance passive rectifier (low-impedance active rectifier) is activated when the data modulated impedance is switched to high (low) impedance. The passive rectifier is implemented using diode-connected transistors, and the active rectifier utilizes a mixed-signal feedback loop to control the timing of the synchronous switches and prevent reverse conduction. This feedback loop replaces the asynchronous gate-driving comparators of conventional active rectifiers and uses clocked comparators operating at 8× lower frequency than the power carrier, reducing power. The dual-mode rectifier efficiency is modulated inversely to the data modulation, and therefore available input power, in order to maintain a constant output power. This technique reduces ripple by 10× at the rectifier output when compared to a single active rectifier and is exploited to reduce the supply decoupling capacitance to 4nF, eliminating the need for external capacitors.

The IC is fabricated in a 65nm 1P7M low-power CMOS process. A chip microphotograph is shown in Fig. 24.1.7. The total chip area is pad-limited to 2.4×2.4mm² and the active circuit area totals 1.72mm², with 1.6mm² occupied by the front-end array. The total power dissipation of the chip is 225μW, including the 60% power conversion efficiency (Fig. 24.1.5).

Figure 24.1.4 shows the measured closed-loop transfer functions of the ECoG front-end from the electrode input to the ADC output. The first-order high-pass pole frequencies are digitally configurable with four such configurations shown. The high-frequency roll-off is due to the sinc transfer function of the ADC. Input-referred noise spectral density is also shown in Fig. 24.1.4 with chopping disabled and for a range of digitally configurable chopper frequencies (and therefore also input impedance). Integrated over 500Hz, chopper stabilization decreases the noise floor by 400×. Comparing this design against state-of-the-art noise- and power-efficient ECoG and EEG front-ends [5-7], the reported techniques enabled a 16× area reduction and a 3× improvement in power efficiency factor (PEF) [4] while integrating an ADC per channel.

In Fig. 24.1.5, the performance of the wireless link is verified by wirelessly transmitting a PRBS-7 data pattern generated on-die. Zero errors were found in 5.9Mb of data resulting in a BER $<1.7\times10^{-7}$ with 1cm antenna separation in air and *in-vivo*. Figure 24.1.5 also shows the 10× reduction in rectifier output voltage (V_{RECT}) fluctuation when switched from single to dual mode operation.

The IC was assembled together with the microfabricated ECoG electrodes and antenna on a PCB and implanted in an anesthetized Long-Evans rat over the left cortical hemisphere. All experiments were performed in compliance with the regulations of the Animal Care and Use Committee at UC Berkeley. Electrical recordings were made on all channels prior to and 15 minutes after the administration of Pentobarbital, a sedative. It is known that anesthesia causes increased δ band (1 to 4Hz) oscillations and depressed high-γ (65 to 125Hz) activity [8]. A representative channel is plotted in Fig. 24.1.6 showing that results are consistent with deepened anesthetic state.

Acknowledgements:
The authors thank Filip Maksimovic, Lu Ye, Lingkai Kong, Nathan Narevsky, ST Microelectronics for IC fabrication, BDA for Analog FastSpice, MuSyC, and the sponsors of BWRC.

References:

[1] G. Schalk, "Can Electrocorticography (ECoG) Support Robust and Powerful Brain-Computer Interfaces?" *Frontiers in Neuroeng.*, vol. 3, Jan. 2010.
[2] T. Bjorninen, *et al.*, "Antenna Design for Wireless Electrocorticography." *IEEE AP-S/URSI*, July 2012.
[3] P. Ledochowitsch, *et al.*, "Fabrication and Testing of a Large Area, High Density, Parylene MEMS μECoG Array," *IEEE MEMS Conf.*, 2011.
[4] R. Muller, *et al.*, "A 0.013mm², 5μW, DC-Coupled Neural Signal Acquisition IC with 0.5V Supply." *J. Solid-State Circuits*, vol. 47, no. 1, Jan. 2012.
[5] T. Denison, *et al.*, "A 2μW 100nV/rtHz Chopper-Stabilized Instrumentation Amplifier for Chronic Measurement of Neural Field Potentials." *J. Solid-State Circuits*, vol. 42, no. 12, Dec. 2007.
[6] R.F. Yazicioglu, *et al.*, "A 200μW Eight-Channel EEG Acquisition ASIC for Ambulatory EEG Systems," *J. Solid-State Circuits*, vol. 43, no. 12, Dec. 2008.
[7] F. Zhang, *et al.*, "A Low-Power ECoG/EEG Processing IC With Integrated Multiband Energy Extractor," *IEEE Trans. CAS I*, vol. 58, no. 9, Sept. 2011.
[8] J. Borjigin, *et al.*, "Surge of Neurophysiological Coherence and Connectivity in the Dying Brain," *Proc. Nat. Acad. Sci.*, Aug 2013.

978-1-4799-0917-9/14 $31.00 © 2014 IEEE

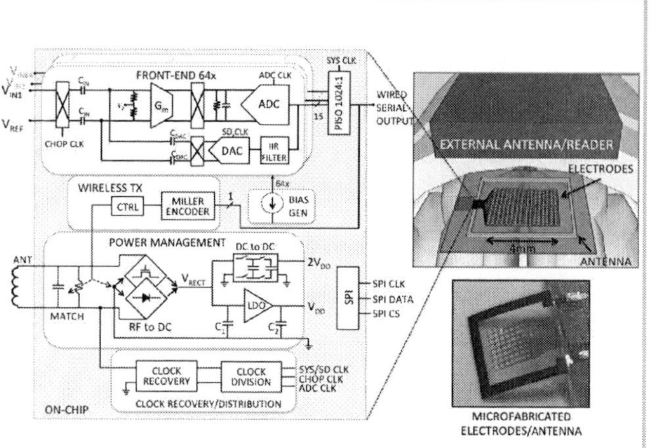

Figure 24.1.1: Device concept for Electrocorticographic (ECoG) neural sensor, IC block diagram, and photograph of the high-density microfabricated electrodes and antenna.

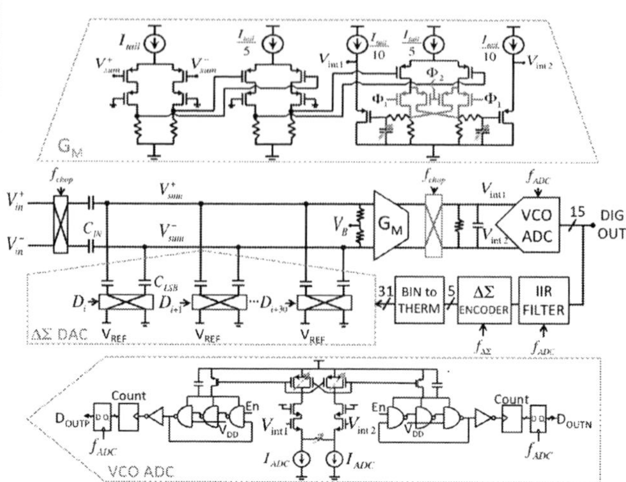

Figure 24.1.2: Detailed front-end circuit diagrams.

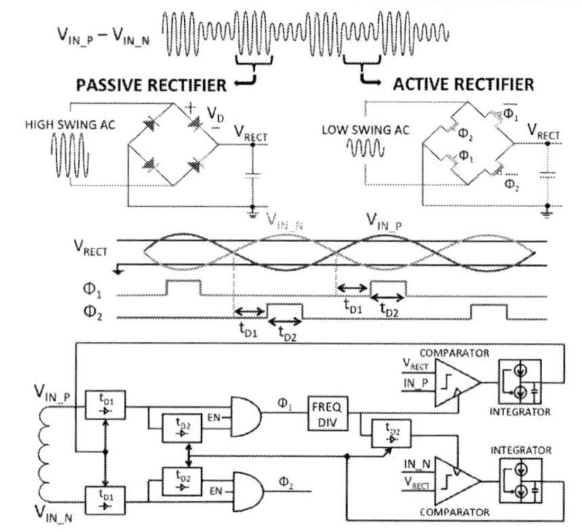

Figure 24.1.3: RF-to-DC rectifier circuit and timing diagrams.

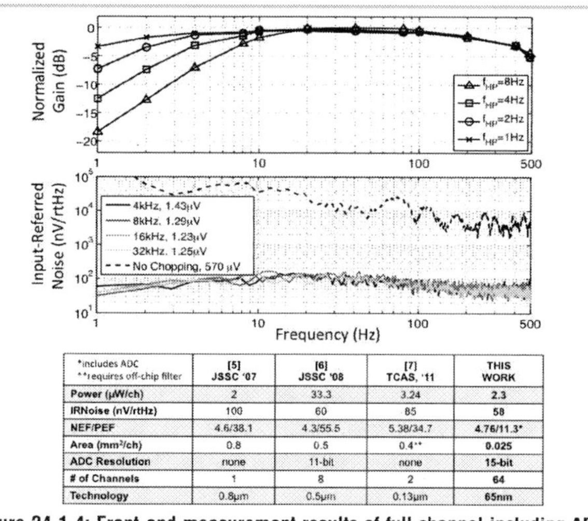

*includes ADC **requires off-chip filter	[5] JSSC '07	[6] JSSC '08	[7] TCAS, '11	THIS WORK
Power (μW/ch)	2	33.3	3.24	2.3
IRNoise (nV/rtHz)	100	60	85	58
NEF/PEF	4.6/38.1	4.3/55.5	5.38/34.7	4.76/11.3*
Area (mm²/ch)	0.8	0.5	0.4**	0.025
ADC Resolution	none	11-bit	none	15-bit
# of Channels	1	8	2	64
Technology	0.8μm	0.5μm	0.13μm	65nm

Figure 24.1.4: Front-end measurement results of full channel including ADC. State-of-the-art comparison is charted for f_{chop}=8kHz.

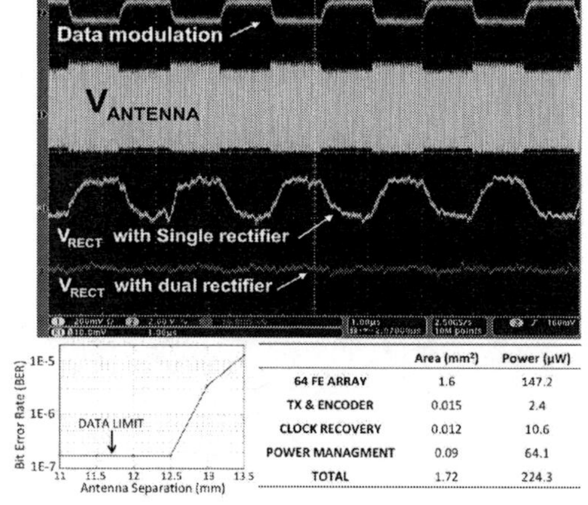

	Area (mm²)	Power (μW)
64 FE ARRAY	1.6	147.2
TX & ENCODER	0.015	2.4
CLOCK RECOVERY	0.012	10.6
POWER MANAGMENT	0.09	64.1
TOTAL	1.72	224.3

Figure 24.1.5: Wireless subsystem measurement results: RF-to-DC rectifier output ripple for single and dual-mode rectification during data modulation, BER vs. antenna separation, and power/area breakdowns.

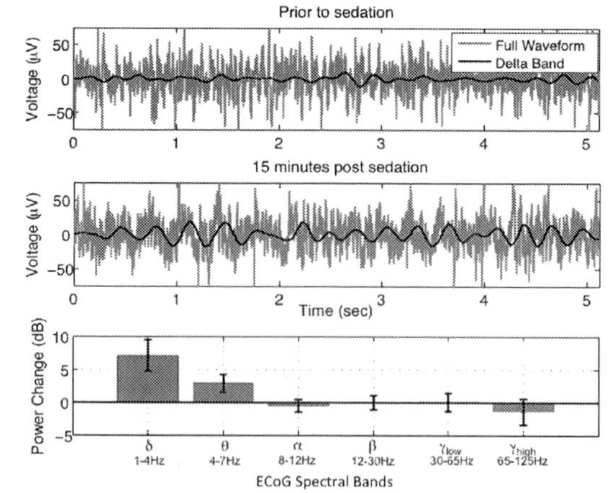

Figure 24.1.6: In-vivo system measurements: recordings from a single channel before and 15 min. after sedative administration. Spectral band power changes are plotted for all channels.

978-1-4799-0917-9/14 $31.00 © 2014 IEEE

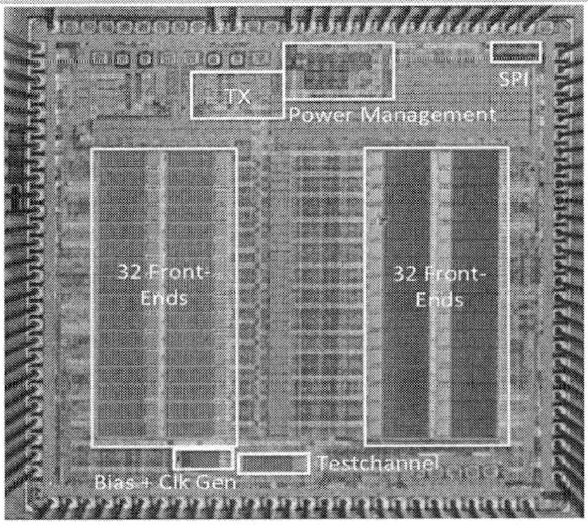

Figure 24.1.7: Chip microphotograph.

24.2 A Power-Efficient Switched-Capacitor Stimulating System for Electrical/Optical Deep-Brain Stimulation

Hyung-Min Lee[1], Ki-Yong Kwon[2], Wen Li[2], Maysam Ghovanloo[1]

[1]Georgia Institute of Technology, Atlanta, GA,
[2]Michigan State University, East Lansing, MI

Deep-brain stimulation (DBS) has been proven as an effective therapy to alleviate Parkinson's disease, tremor, and dystonia. Towards a less invasive head-mounted DBS, we utilize an inductive transcutaneous link to provide sufficient power without size, lifetime, and discomfort of chest-mounted battery-powered traditional DBS. The next step is to adopt aggressive power-management schemes to further improve the DBS efficiency. Current-controlled stimulation (CCS) enables precise charge control and safe operation, but it has low power efficiency due to the dropout voltage across current sources [1,2]. Switched-capacitor stimulation (SCS), proposed in [3], takes advantage of both high efficiency and safety using capacitor banks to transfer charge to the tissue, but it requires an efficient on-chip capacitor charging system, directly from the inductive link. We present an integrated wireless SCS system-on-a-chip with inductive capacitor charging and charge-based stimulation capabilities, which can improve both stimulator (before electrodes) and stimulus (after electrodes) efficiencies in DBS.

Figure 24.2.1 shows the overall architecture of the wireless SCS system. The inductive capacitor charger charges four pairs of positive/negative storage capacitors, C_{P1-4} and C_{N1-4}, sequentially. These capacitors are connected to the stimulation sites alternately through channel selectors for biphasic stimulation. A current limiter limits the stimulus amplitude to prevent large current flowing through the tissue. To ensure charge-balanced stimulation, a charge-monitoring (CM) circuit measures the amount of charge injected and withdrawn by observing storage capacitor voltages, and dynamically changes the stimulus pulse width to neutralize the residual charge in the tissue. An additional charge-balancing circuit further prevents residual charge accumulation. A power management block generates system supply and reference voltages, while a timing controller provides timing signals for capacitor charging and stimulation. In forward data telemetry, a pulse-position-modulated clock/data recovery (PPM-CDR) block extracts synchronized data and clock from an on-off-keying (OOK) modulated coil voltage, V_{COIL}, setting a 40b shift register, to store stimulation parameters. Load-shift-keying (LSK) back telemetry is adopted for handshaking and closed-loop power control by sensing V_{COIL} amplitude.

Figure 24.2.2 compares the conventional CCS with SCS while emphasizing the inductive capacitor charging and charge-based stimulation circuits. The CCS requires a rectifier, a regulator, and an array of current sources to generate rectangular stimuli. Power losses at each stage result in poor overall stimulator efficiency, which is defined as the stimulator output power over input power from the L_2C_2 tank. On the contrary, the inductively powered SCS efficiently charges the storage capacitors directly from the inductive link and delivers the stored charge to the tissue (series *RC*), creating a decaying-exponential stimulus. Since the capacitor-charging efficiency is the dominant factor in stimulator efficiency, we utilized the inductive capacitor charger reported in [4] plus additional safety measures for the SCS. The charger generates a fixed charging current through a series capacitor, C_S, reducing switching loss and improving charging efficiency. The improved charger benefits from dual-voltage control capability provided by comparators, CMP_P and CMP_N, and a dual-output DAC to guarantee that V_{CP1} and V_{CN1} are separately charged to target voltages, V_{TP} and V_{TN}, respectively. Even small residual voltage mismatch between C_{P1} and C_{N1} can otherwise be accumulated during long-term stimulation and saturate either V_{CP1} or V_{CN1}. There is also a reset function that can optionally discharge C_{P1} and C_{N1} before charging.

Figure 24.2.3 shows the schematics of the PPM-CDR and CM circuits. In the PPM-CDR, a PPM signal, S_{PPM}, extracted from V_{COIL}, is converted to the clock, *CLK*, through a frequency divider (DFF_1). *CLK* controls the timing and amplitude of V_{PPM} by alternately charging and discharging C_4. If positioning ratio among three pulses of S_{PPM} is 7:3, V_{PPM} exceeds V_{REF2} during *CLK*=1, and a demodulated signal, S_{PPD}, is sampled in DFF_2, leading to *DATA*=1. Otherwise, *DATA*=0 when the positioning ratio is 3:7. Since the stimulation parameters are set only once and the pulse width is narrow (3μs), the OOK-PPM offers a simple but robust programming method without costing in system efficiency.

The CM circuit in Fig. 24.2.3 integrates the discharged voltages of V_{CP} and V_{CN} during stimulation to detect the amount of charge transferred to tissue. A CM signal, S_{CM}, stays at 0 before stimulation, while amplifiers A_3 and A_4 operate as buffers, storing their offset voltages in C_7. When the negative stimulation starts first with S_{CM}=1 for a predefined period, A_3 becomes a capacitive-feedback amplifier, and A_4 operates as a comparator, while their offsets are cancelled through C_7. V_{SEN} decreases as V_{CN} increases in this period. When V_{CP} discharges for positive stimulation, V_{SEN} increases again. When the amounts of V_{SEN} decrement and increment are equal, S_{CM}=0, and the positive stimulation stops to ensure that the net injected and withdrawn charges are zero.

The ASIC is fabricated in the TSMC 0.35μm 4M2P standard CMOS process, occupying 12mm^2. Measured waveforms in Fig. 24.2.4 show the operation of CM circuit and overall SCS when a negative-first biphasic stimulus flows through a series *RC* model (500Ω-1μF). In the CM waveforms, the negative stimulation is applied for 512μs with one ±2V capacitor pair, discharging V_{CN1} by 850mV. Then, the positive stimulation period is dynamically adjusted at 228μs, discharging the same amount of ΔV_{CP1} to ensure charge balancing. In the overall SCS waveforms, the stimulus current, I_{STIM}, has a decaying exponential shape, supplied by four capacitor pairs, and its amplitude and time constant depend on $C_{P,N}$, tissue/electrode *RC*, and $V_{CP,N}$. After stimulation, four capacitor pairs are charged to target voltages (here ±2V) again, while the site is shorted to *GND* during a predefined period for additional charge balancing.

In addition to high stimulator efficiency, the SCS system's decaying exponential stimulus is proven to be more effective in activating the neural tissue compared to rectangular and ramp stimuli when consuming same amount of energy [5]. Figure 24.2.5 shows a tissue model with multi-compartment double-cable mammalian axon models in [5] to simulate the effects of stimulus waveforms. For extracellular stimulation, tissue potentials that vary depending on stimulus waveforms were calculated along the length of each axon and applied to the axon models, resulting in transmembrane potential increase and activation. It can be seen that at the same energy (=1nJ/Ω) and pulse width (=1ms), the decaying exponential can activate larger cross-sectional tissue area. Similarly, Fig. 24.2.5 shows that it requires smaller stimulus energy (40~70% less) and injected charge (30~78% less) to activate the same tissue area with 1.5ms pulse width compared to other stimulus waveforms, while all waveforms show similar stimulus efficiencies with small pulse width (<0.4ms).

The SCS system is also capable of providing high instantaneous current, which is a limiting factor in conventional inductively-powered devices. We have utilized the SCS system for power-efficient optogenetics by periodically discharging the capacitors into LEDs, which require high instantaneous power to emit sufficient light. C_P and C_N pairs were connected in series to provide higher LED voltage, V_{LED}, as shown in Fig. 24.2.6. *In vivo* local field potential (LFP) below 500Hz was recorded using an optrode array with waveguides in the brain of an anesthetized viral-transfected rat when the SCS system drove micro-LEDs with a 100ms pulse train at 1Hz, with V_{LED}=2.7V_{peak} and 3.2V_{peak}. The higher V_{LED} resulted in higher light intensity from micro-LEDs to deliver sufficient irradiance (≥1mW/mm^2) to the selective target tissue through micro-needle waveguides, leading to larger LFP variations, which verified the efficacy of optical stimulation via the SCS. The table in Fig. 24.2.6 benchmarks the wireless SCS against recently published inductively powered stimulating systems. Figure 24.2.7 shows the chip micrograph and specification summary.

References:

[1] K. Chen, Z. Yang, L. Hoang, J. Weiland, M. Humayun, and W. Liu, "An Integrated 256-Channel Epiretinal Prosthesis," *IEEE J. Solid-State Circuits*, vol. 45, no. 9, pp. 1946–1956, Sep. 2010.

[2] S. Arfin and R. Sarpeshkar, "An Energy-Efficient, Adiabatic Electrode Stimulator with Inductive Energy Recycling and Feedback Current Regulation," *IEEE Trans. Biomed. Circuits Syst.*, vol. 6, no. 1, pp. 1-14, Feb. 2012.

[3] M. Ghovanloo, "Switched-Capacitor Based Implantable Low-Power Wireless Microstimulating Systems," *IEEE ISCAS*, pp. 2197-2200, May 2006.

[4] H. Lee and M. Ghovanloo, "A Power-Efficient Wireless Capacitor Charging System Through an Inductive Link," *IEEE Trans. Circuits Syst. II*, vol. 60, no. 10, pp. 707-711, Oct. 2013.

[5] A. Wongsarnpigoon, J. P. Woock, and W. M. Grill, "Efficiency Analysis of Waveform Shape for Electrical Excitation of Nerve Fibers," *IEEE Trans. Neural Syst. Rehab. Eng.*, vol. 18, no. 3, pp. 319-328, June 2010.

[6] S. Kelly and J. Wyatt, "A Power-Efficient Neural Tissue Stimulator with Energy Recovery," *IEEE Trans. Biomed. Circuits Syst.*, vol. 5, no. 1, pp. 20-29, Feb. 2011.

978-1-4799-0917-9/14 $31.00 © 2014 IEEE

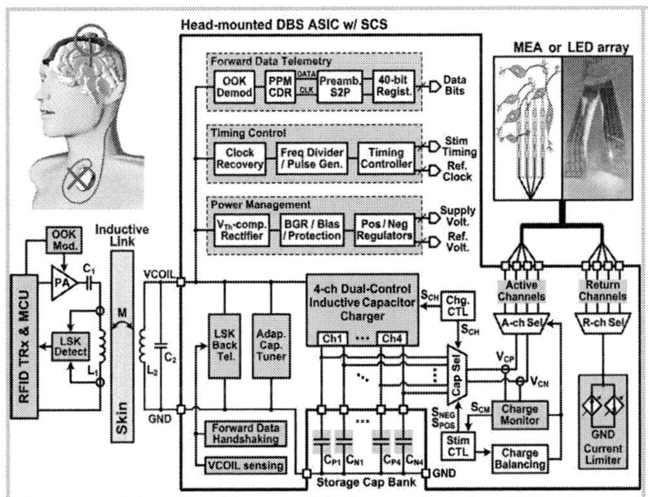

Figure 24.2.1: Overall architecture of the integrated wireless SCS system-on-a-chip for a head-mounted DBS.

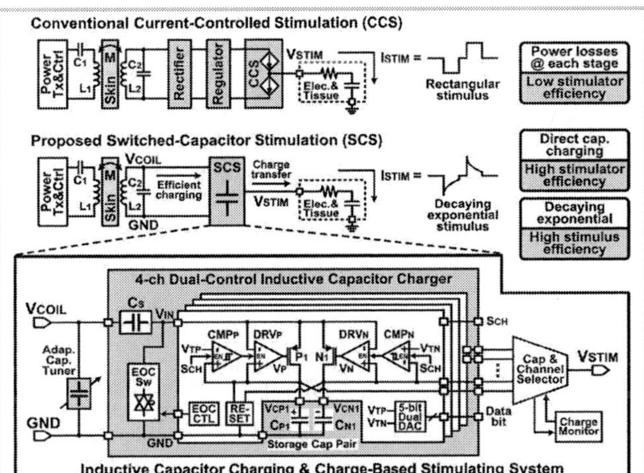

Figure 24.2.2: Conceptual diagrams of the conventional CCS and the SCS with emphasis on inductive capacitor charging function and charge-based stimulation.

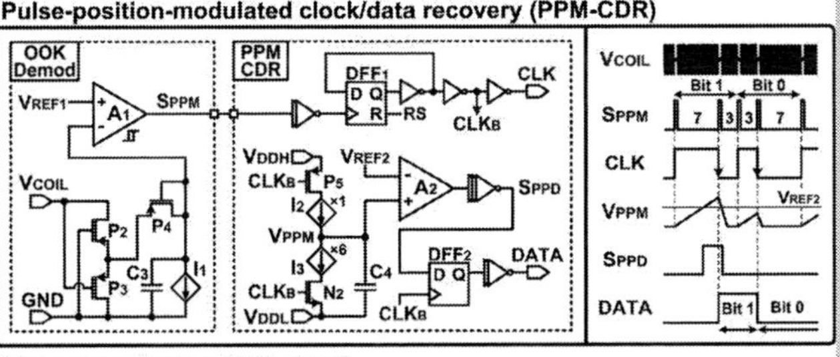

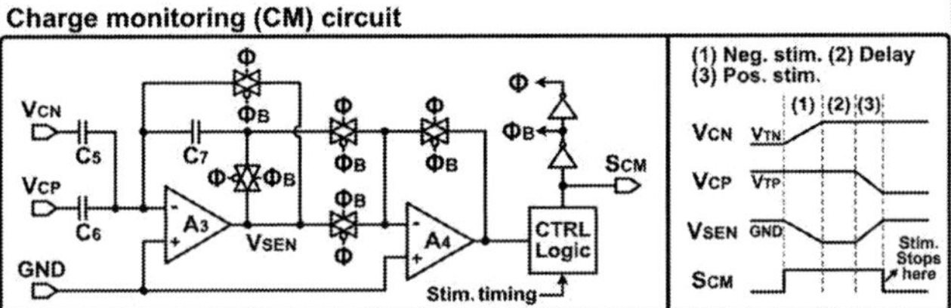

Figure 24.2.3: Schematic diagrams of the PPM-CDR (top) and the charge monitoring (CM) circuits (bottom).

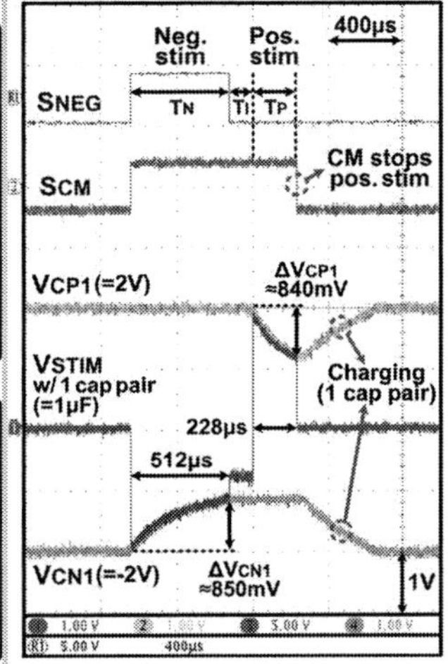

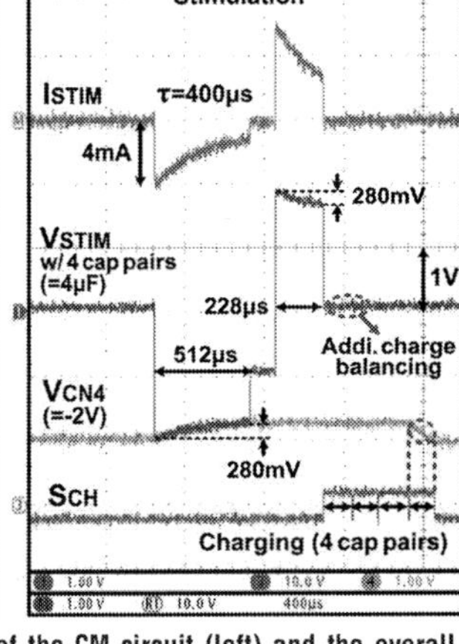

Figure 24.2.4: Measured waveforms of the CM circuit (left) and the overall SCS system (right).

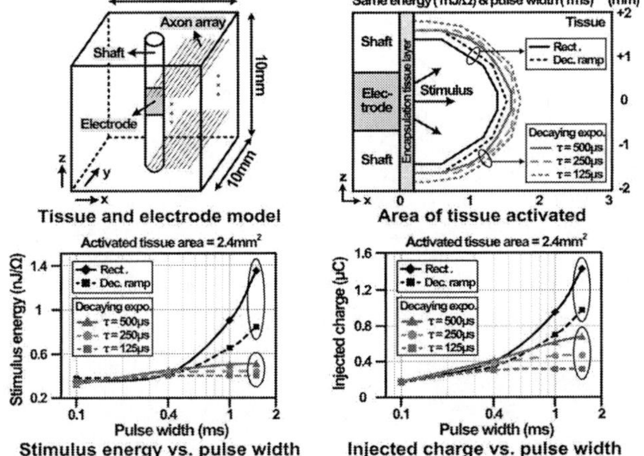

Figure 24.2.5: Tissue model and simulation of various stimulus waveforms to verify the effects of decaying exponential compared to other stimulus waveforms (rectangular and decaying ramp).

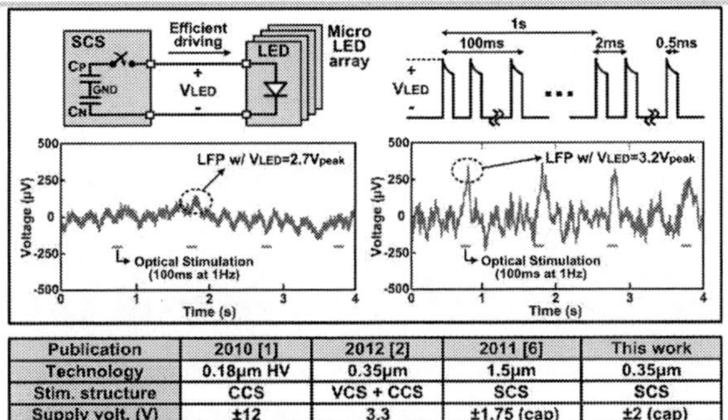

Publication		2010 [1]	2012 [2]	2011 [6]	This work
Technology		0.18µm HV	0.35µm	1.5µm	0.35µm
Stim. structure		CCS	VCS + CCS	SCS	SCS
Supply volt. (V)		±12	3.3	±1.75 (cap)	±2 (cap)
Stimulator power eff. (%)	Rec / Reg	85.6	80*	-	-
	DC-DC	-	55 ~ 94	-	-
	I-driver	41.6	-	-	-
	Charger+Sw	-	-	40**	80.4
	Total	35.6	44 ~ 75.2	40	80.4
Stimulus shape		Rectangular	Rectangular	Decaying expo.	Decaying expo.
Max. *ISTIM* (mA)		0.5	0.45	0.4 (peak)	4 (peak)
Series RC model		10kΩ + 100nF	1kΩ + 0.93µF	1.15kΩ + 0.98µF	0.5kΩ + 1µF

*Considered the rectifier only, **Including power consumption of other blocks.

Figure 24.2.6: SCS for power-efficient optogenetics with in vivo experimental results (top) and the benchmarking table (bottom).

24

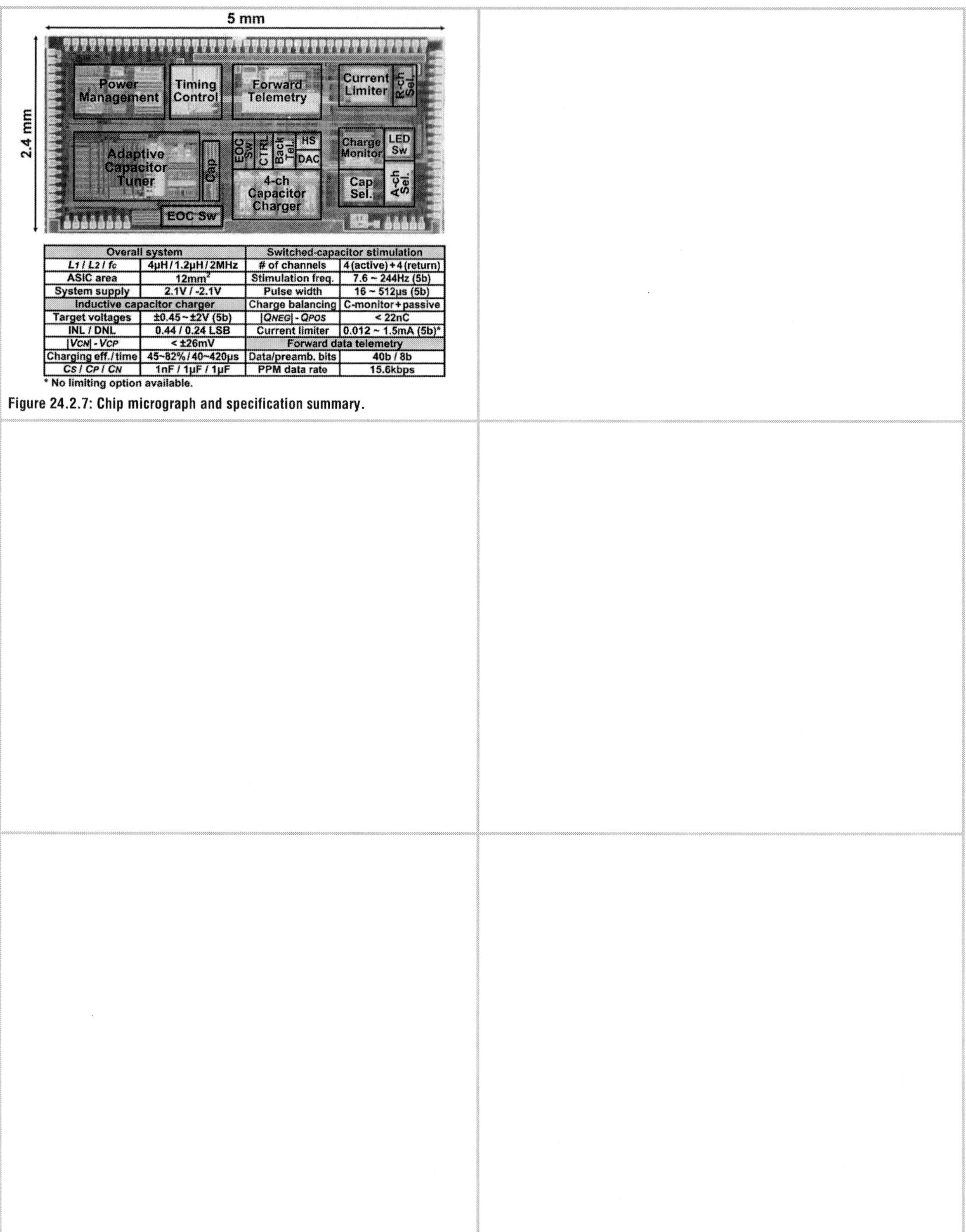

Overall system		Switched-capacitor stimulation	
L_1 / L_2 / f_c	4μH / 1.2μH / 2MHz	# of channels	4 (active) + 4 (return)
ASIC area	12mm^2	Stimulation freq.	7.6 ~ 244Hz (5b)
System supply	2.1V / -2.1V	Pulse width	16 ~ 512μs (5b)
Inductive capacitor charger		Charge balancing	C-monitor + passive
Target voltages	±0.45 ~ ±2V (5b)	$\lvert Q_{NEG}\rvert - Q_{POS}$	< 22nC
INL / DNL	0.44 / 0.24 LSB	Current limiter	0.012 ~ 1.5mA (5b)*
$\lvert V_{CN}\rvert - V_{CP}$	< ±26mV	Forward data telemetry	
Charging eff. / time	45~82% / 40~420μs	Data/preamb. bits	40b / 8b
C_S / C_P / C_N	1nF / 1μF / 1μF	PPM data rate	15.6kbps

* No limiting option available.

Figure 24.2.7: Chip micrograph and specification summary.

24.3 An Implantable 64nW ECG-Monitoring Mixed-Signal SoC for Arrhythmia Diagnosis

Dongsuk Jeon[1], Yen-Po Chen[1], Yoonmyung Lee[1], Yejoong Kim[1], Zhiyoong Foo[1], Grant Kruger[1], Hakan Oral[2], Omer Berenfeld[2], Zhengya Zhang[1], David Blaauw[1], Dennis Sylvester[1]

[1]University of Michigan, Ann Arbor, MI,
[2]University of Michigan Health System, Ann Arbor, MI

Electrocardiography (ECG) is a critical source of information for a number of heart disorders. In arrhythmia studies and treatment, long-term observation is critical to determine the nature of the abnormality and its severity. However, even small body-wearable systems can impact a patient's everyday life and signals captured using such systems are prone to noise from sources such as 60Hz power and body movement. In contrast, implanted devices are less susceptible to these noise sources and, while having closer-spaced electrodes, can obtain similar quality ECG signals due to their proximity to the heart [1]. In addition, implanted devices enable continuous monitoring without affecting patient quality of life. As in other implantable systems, low power consumption is a critical factor; in this case to provide a sufficiently long operating time between wireless recharge events.

This paper reports a syringe-implantable ECG recording and analysis device targeted primarily at arrhythmia monitoring (Fig. 24.3.1). In contrast to surgically implanted devices with large batteries such as pacemakers, the device is designed for daily wireless recharging, allowing for a much smaller battery. In order to pass through the needle canula during implantation, device width is limited to 1.5mm while overall system length is designed to be 2cm, providing sufficient distance between two electrodes to yield an acceptably large potential difference. The signal from electrodes is filtered, amplified, and converted to the digital domain by an analog front-end (AFE). A digital signal processing (DSP) module analyzes the waveform within a 10-second search window and detects abnormal cardiac events. When an abnormal event is detected, the device stores the current search window waveform into local memory; it can then be transferred to an external device through means such as a wireless transceiver for further analysis by clinicians. Assuming nightly wireless data readout and battery recharge, the design targets 5-day lifetime (providing a safety margin) when powered by an on-chip thin-film Li battery (5μA•hr, 4V). This translates to 167nW average system power consumption, presenting a challenging power constraint given that comparable systems in the literature typically consume 10 to 30μW [2, 5-7].

The AFE consists of a low-noise instrumentation amplifier, a variable-gain amplifier, and a successive-approximation register (SAR) analog-to-digital converter (ADC). To reduce power consumption, all building blocks except the ADCs clocked comparator are biased in the subthreshold regime. Due to the resulting high performance variability, the amplifier gain, bandwidth, and input-referred noise are all tunable by the subsequent digital blocks. Similar to other noise-limited amplifier designs [2], the first stage of the amplifier dominates total AFE power consumption. This design uses an inverter-based amplifier for high noise efficiency and its tail current can be tuned to match the desired noise level. Aided by the DSP algorithms, the system accurately detects arrhythmia with up to 15μV noise when tested under a database of ECG data collected from arrhythmia patients (Fig. 24.3.2). Since first-stage current consumption is largely dictated by input referred noise magnitude, we target a 9μV noise floor (excluding ADC distortion and margins), reducing AFE and system power by 6.7× and 2.15×, respectively, compared to typical ECG signal acquisition designs that require noise levels of ≤3μV [2]. Subsequent amplifiers are not noise-limited and therefore are designed to consume only 100s of pAs. A common problem of the inverter-based design is that the bias point is vulnerable to PVT variations. Therefore, a DC servo loop is used to stabilize the differential output to half V_{DD} (Fig. 24.3.2, left). Due to the large tissue-electrode impedance, the AFE input amplifier requires very high input impedance and, therefore, both AC coupling and an impedance-boosting loop are implemented.

Analog-to-digital conversion uses an 8b single-ended asynchronous SAR ADC with 500Hz sampling rate. Traditional asynchronous logic uses dynamic logic, which suffers from high leakage in a low voltage/frequency ECG application. Therefore, dynamic nodes are implemented with latches that are clocked by internal signals and delay lines (Fig. 24.3.3). To improve energy efficiency, a 10fF DAC unit capacitor and split capacitor array topology are used, enabling 1.97× ADC power reduction. Also, the comparator is a clocked 1-stage design chosen for low dynamic power consumption. However, the 1-stage clocked comparator and small capacitor array make the comparator input vulnerable to kickback noise. A split footer comparator [3] combined with cross-coupled compensation addresses this issue, reducing kickback noise by 84.9× (simulated). The measured amplifier current consumption is 31nA at 0.6V with input-referred noise at signal band of 6.52μV.

Figure 24.3.4 shows the digital processing back-end. The back-end detects the incoming signal amplitude and tunes AFE gain accordingly. Input samples from the ADC are first processed by a 600ms moving average filter (MAF) that removes the relatively slow baseline shift. To minimize power, we use frequency-domain processing with a lower sampling rate (yet comparable detection performance) than conventional QRS-peak detection algorithms. The frequency dispersion metric (FDM) block performs an FFT on the 10× downsampled ECG waveform and observes whether dominant clear peaks exist in a specific frequency range, which represents a stable heartbeat. A 512-point real-valued FFT accelerator takes in data from one of two ping-pong buffers and computes the FFT on a separate local buffer, thereby preserving the stored waveform. Once an arrhythmia is detected, the ping-pong buffer storing the last search window no longer accepts new samples until the waveform is fully read out through a data bus; meanwhile the other buffer acts as the primary input data channel. The actual arrhythmia detection algorithm is performed in an ARM Cortex-M0+ core and instruction memory can be programmed with different algorithms for flexibility.

Due to the low throughput requirement, the required operating voltage/frequency pair is located below the minimum energy point (Fig. 24.3.4). Therefore, the system operates at the minimum energy point (MEP) with the Cortex-M0+ core working in burst-mode (~6× faster than required). The core is then power-gated after processing of each 10s window is complete. This duty cycling ensures each operation consumes the minimum possible energy, reducing power of the duty-cycled block by 40%. The design can also perform standard QRS-peak detection (R-R block), which uses peak-to-peak distances to determine ECG signal regularity. A reconfigurable 80-tap FIR filter performs a band-pass filter on the input signal. Finally QRS peaks are detected using a threshold on the differentiated signal and the variance of peak-to-peak distance is used in detecting arrhythmia. A clinician can enable one of the two processing paths (FDM or R-R) with the other power gated.

The ECG monitoring SoC is fabricated in 65nm LP CMOS. It successfully communicates with other chips including a power management unit and external memory from [4] over a data bus; the complete system configuration is described in Fig. 24.3.5. The SoC is tested under different scenarios including an ECG simulator as well as an isolated sheep heart; measured waveforms are shown in Fig. 24.3.5. The SoC consumes 64nW and 110nW when running the FDM and RR algorithms, respectively, enabling >5 day lifetime with a 3.7mm^2 (5μA•hr) thin-film battery.

References:

[1] C. Zellerhoff, *et al.*, "How can we identify the best implantation site for an ECG event recorder?," *Pacing & Clinical Electrophys*, pp. 1545-1549, 2000.
[2] R. Yazicioglu, *et al.*, "A 30 μW Analog Signal Processor ASIC for Portable Biopotential Signal Monitoring," *J. Solid-State Circuits*, pp. 209-223, 2011.
[3] H. Zhang, *et al.*, "Design of an Ultra-Low Power SAR ADC for Biomedical Applications," *ICSICT*, pp. 460-462, 2010.
[4] Y. Lee, *et al.*, "A Modular 1mm^3 Die-Stacked Sensing Platform with Optical Communication and Multi-Modal Energy Harvesting," *ISSCC Dig. Tech. Papers*, pp. 402-403, 2012.
[5] F. Zhang, *et al.*, "A Batteryless 19μW MICS/ISM-Band Energy Harvesting Body Area Sensor Node SoC," *ISSCC Dig. Tech. Papers*, pp. 298-299, 2012.
[6] S.-Y. Hsu, *et al.*, "A Sub-100μW Multi-Functional Cardiac Signal Processor for Mobile Healthcare Applications," *VLSI Circ. Symp. Dig. Tech. Papers*, pp. 156-157, 2012.
[7] S. Kim, *et al.*, "A 20μW Intra-Cardiac Signal-Processing IC with 82dB Bio-Impedance Measurement Dynamic Range and Analog Feature Extraction for Ventricular Fibrillation Detection," *ISSCC Dig. Tech. Papers*, pp. 302-303, 2013.

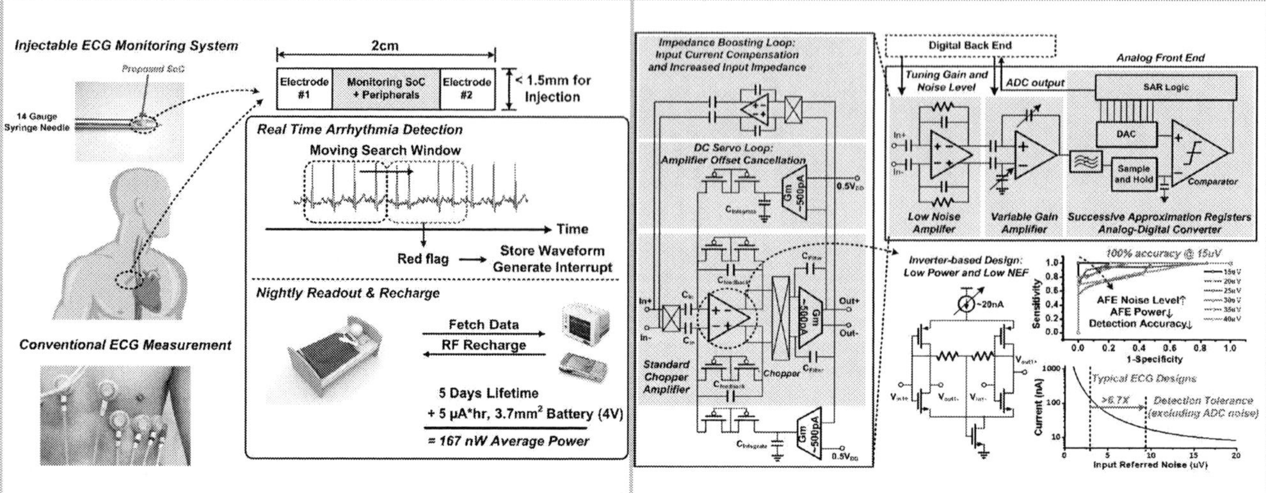

Figure 24.3.1: System overview of ECG monitoring SoC.

Figure 24.3.2: 31nA analog front-end.

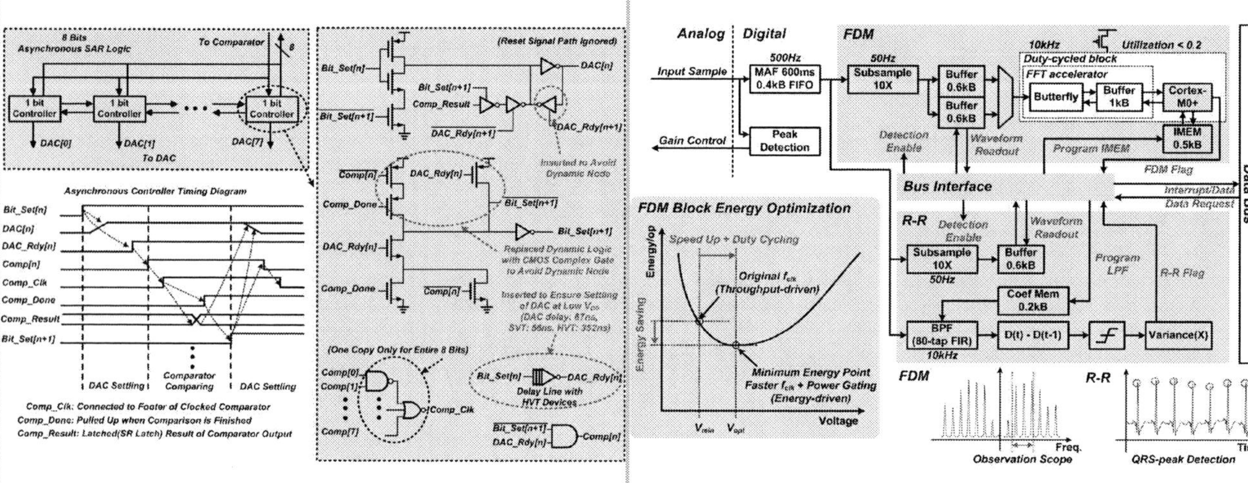

Figure 24.3.3: Robust asynchronous controller for ADC.

Figure 24.3.4: Digital back-end with two algorithms.

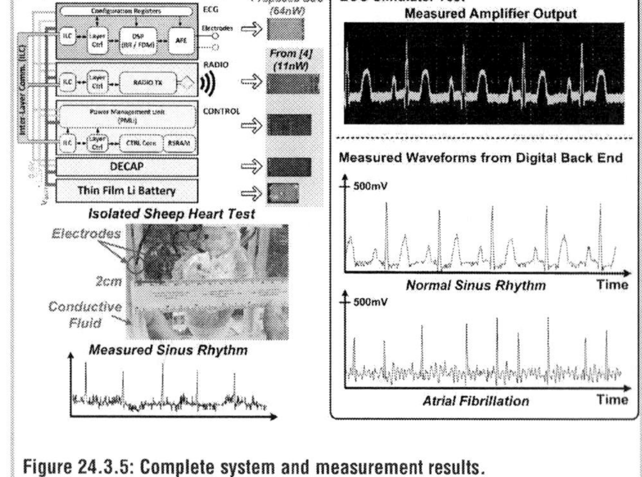

Figure 24.3.5: Complete system and measurement results.

		This Work	ISSCC '12 [5]	VLSI '12 [6]	ISSCC '13 [7]
Target Signals		**ECG**	ECG, EMG, EEG	ECG, VCG, PCG	ECG, Bio-Impedance
Technology		**65 nm**	130 nm	90 nm	180 nm
AFE	V_{CC}	**0.6 V**	1.2 V	0.5 V	1.8 V
	Current	**31 nA**	4 μA	20.44 μA (8-Bit, 2kHz Sampling)	-
	Gain	**51 ~ 96 dB**	40 ~ 78 dB	40 ~ 64 dB	40 ~ 64 dB
	Bandwidth	**250 Hz**	320 Hz	0.5 ~ 1 kHz	0.5 ~ 1 kHz
	Input Referred Noise	**253 nV/√Hz**	-	-	200 nV/√Hz
	ADC Bits	**8 Bits**	8 Bits	8/12 Bits	9.3 Bits (ENOB)
	ADC Sampling Frequency	**500 Hz**	-	250 Hz ~ 100 kHz	-
DSP	V_{CC}	**0.4 V**	0.3 ~ 1.2 V	0.5V (1.0V for SRAM)	- (Analog Signal Processing)
	Power Consumption	**45 nW**	2.1 μW	-	
	Clock Frequency	**10 kHz**	2 kHz ~ 1.7 MHz	25 MHz	
System Total Power Consumption		**64 nW**	6.9 μW	22.6 μW	11.3 μW
Power Calculation Configuration		**AFE + DSP Arrhythmia Detection (FDM)**	AFE + DSP R-R Extraction	AFE (BSI) + DSP + OSC Arrhythmia Detection	AFE (3ch ECG + RA) + ASP + OSC Arrhythmia Detection

Figure 24.3.6: Comparison with recent prior works.

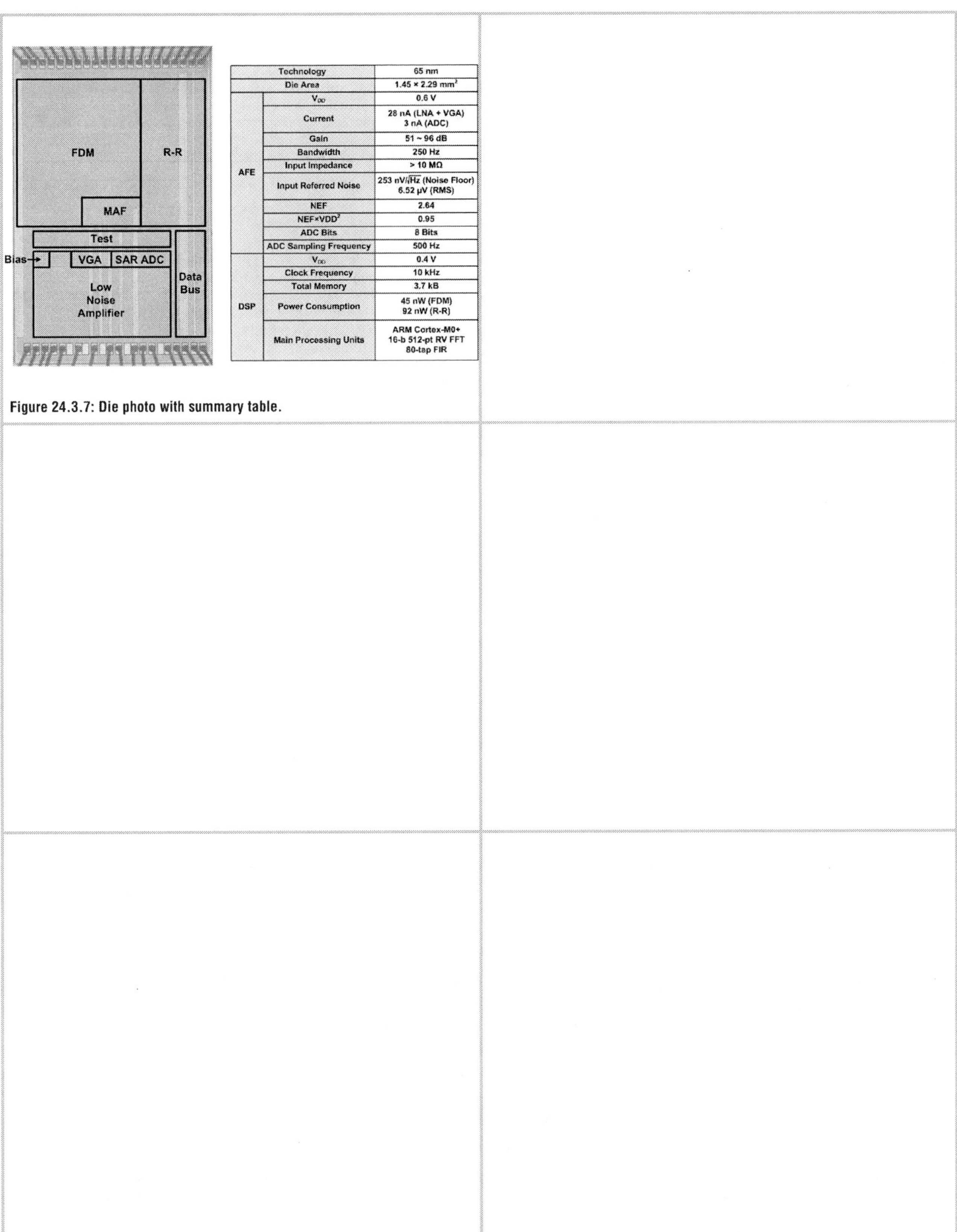

Technology		65 nm
Die Area		1.45×2.29 mm^2
AFE	V_{DD}	0.6 V
	Current	28 nA (LNA + VGA) 3 nA (ADC)
	Gain	51 ~ 96 dB
	Bandwidth	250 Hz
	Input Impedance	> 10 MΩ
	Input Referred Noise	253 nV/$\sqrt{Hz}$ (Noise Floor) 6.52 μV (RMS)
	NEF	2.64
	NEF×VDD^2	0.95
	ADC Bits	8 Bits
	ADC Sampling Frequency	500 Hz
DSP	V_{DD}	0.4 V
	Clock Frequency	10 kHz
	Total Memory	3.7 kB
	Power Consumption	45 nW (FDM) 92 nW (R-R)
	Main Processing Units	ARM Cortex-M0+ 16-b 512-pt RV FFT 80-tap FIR

Figure 24.3.7: Die photo with summary table.

24.4 A 680nA Fully Integrated Implantable ECG-Acquisition IC with Analog Feature Extraction

Long Yan[1], Pieter Harpe[2], Masato Osawa[3], Yasunari Harada[3], Kosei Tamiya[3], Chris Van Hoof[1,4], Refet Firat Yazicioglu[1]

[1]imec, Heverlee, Belgium,
[2]Eindhoven University of Technology, Eindhoven, The Netherlands,
[3]Olympus, Tokyo, Japan,
[4]KU Leuven, Leuven, Belgium

Ultra-low power consumption and miniature size are by far the most important design requirements for implantable pacemakers. In order to guarantee a long life span of the device, saving power in the sensing IC is a primary concern as cardiac rhythm disorders must be continuously monitored [1]. Shifting the functionality of QRS-band power parameter extraction to the analog domain can reduce system-level power consumption of heartbeat detection significantly through minimizing computational complexity of the DSP [2,3]. In addition, current biomedical ICs still require further improvement of power efficiency as their analog back ends consume significant power [2-4]. For low-power means, the presented analog signal processor (ASP) introduces a power-efficient analog feature extraction, a current-multiplexed ADC driver and a flexible ADC. This advances the state of the art by reducing the power consumption of the ASP below 1μW without compromising other specs, such as input SNR >70dB, CMRR >90dB, PSRR >80dB, and enables low-power heartbeat detection for implantable pacemakers.

Figure 24.4.1 shows the architecture of the fully integrated ASP that features a single-channel ECG readout and a flexible, low-power QRS feature extraction channel. The ASP uses a power-efficient fully differential capacitive-coupled instrumentation amplifier (IA) as a first stage. One branch of the IA goes to the analog feature extractor (FE), which consists of a programmable gain amplifier (PGA) and accurate filters, and is used to precisely monitor the signal activity in a selected frequency band. The ECG channel is similar, but omits the low-pass filter. To avoid the use of power-consuming channel buffers and ADC drivers, the 7-to-10b configurable ADC [7] digitizes ECG_{out} and FE_{out} via a current multiplexed (CMPX) buffer. A sub-1V bandgap reference is integrated to provide robust analog bias signals, and a clock generator delivers all necessary clock signals for ADC sampling and accurate filtering. The ASP can be configured through a serial-to-parallel interface and a parallel ADC output is provided for digital signal analysis.

Typical readout systems consume significant power in the back-end as they employ power-consuming channel buffers and analog switches (16% of channel power in [2]) to multiplex several signals and to drive the ADC [2][3][5]. In contrast, signal multiplexing before the filters and PGA can save power but limits the programmability between the channels [6]. A low-power buffer that multiplexes the input signals in the current domain is used (Fig. 24.4.2). The CMPX buffer employs a folded-cascode amplifier with complementary inputs to accommodate amplified large signals from the channels. Transistors P_1 and N_1 convert the ECG_{out} and FE_{out} signals into current and they are multiplexed by P_{SW} and N_{SW} at low-impedance nodes in the folded-cascode stage. Part of the bias current through P_1 and N_1 is recycled as the multiplexed branches are active alternatingly at a rate of f_{CHSEL} (half of the ADC sampling rate, f_{SMPL} = 1024Hz). The multiplexed current flows to the folded-cascode stage, creating an output voltage signal at ADC_{IN}. A single-ended 37fJ/conversion-step SAR ADC utilizing 2fF/unit-element capacitors in the DAC [7] is used to digitize the multiplexed signal. The fast conversion time (which allows more time for analog settling) and the low input capacitance of the ADC (2pF) reduce the required bias current of the CMPX buffer to only 100nA while the analog signal settles to within 0.1% accuracy.

The intra-cardiac signal is amplified by the IA (Fig. 24.4.3), which features a wide input dynamic range and on-chip rail-to-rail AC coupling. A fully differential amplifier with capacitive feedback provides a gain of 20dB. At the output, source followers N_{C1} together with 20MΩ resistors act as a common-mode detector. With 20nA bias current flowing through N_{C1}, the outputs of the IA support a large signal swing as high as $1.1V_{p-p}$ with a total harmonic distortion below 1%. The IA drives 2 switched-capacitor high pass filters (SC-HPF) in the ECG channel and FE channel, respectively. Unlike the ECG channel that provides a HPF corner frequency of 1Hz, the FE channel is normally configured to 10Hz to reject low-frequency T-waves as well as electrode motion artifacts. The SC-HPF utilizes a floating structure that has a unity gain at DC while providing an accurate and flexible bandwidth of 1Hz, 2Hz, 5Hz, 10Hz, or 20Hz by adjusting the value of C_R, and has an input impedance of more than 400MΩ. Thanks to the floating HPF structure, the IA not only can bias the HPFs and the following PGAs in ECG and FE channels, but can drive them directly without additional analog buffers and bias circuits. In addition, any common-mode noise from the IA is rejected by the floating HPF as it has very high input common-mode impedance.

Followed by the HPF, a PGA translates the differential signal to an amplified single-ended output (Fig. 24.4.4), thus saving power in the later stages. One differential pair in the differential-to-difference amplifier (DDA) receives the differential signal from the HPF. The other differential pair sets the DC output value to $VREF_{CH}$ and configures the gain $(1+C_g/C_f)$. The capacitor C_g is always connected to $VREF_{CH}$ instead of ground. According to the different gain settings, the feedback capacitors $C_{f1,2,3}$ are flipped from the side of C_{f0} to the side of C_g. The advantage to do so is that the noise from $VREF_{CH}$ is not anymore amplified by the high gain of the PGA. This is especially important in the case when PGA needs to provide high gain and reject relatively high noise presenting at power supply. The PGA provides 0dB to 24dB variable gain with 6dB/step while maintaining a constant bandwidth of 130Hz by adapting the capacitors at the output of the DDA. For unity gain, SW_{unity} is enabled. The switch consists of 2 PMOS transistors facing each other to ensure sufficiently high off-resistance not to deteriorate the low frequency cutoff of the PGA. In the FE channel, a 10-to-25Hz flexible (5Hz/step) switched-capacitor low-pass filter (SC-LPF, Fig. 24.4.1) is further integrated to remove high-frequency out-of-band interferences.

Figure 24.4.7 shows the ASP micrograph, which is implemented in 0.18μm CMOS. The input-referred noise of the entire ECG channel is $4.9\mu V_{rms}$ in a 130Hz bandwidth (Fig. 24.4.5a). In addition, the ECG channel has a CMRR more than 90dB with a channel gain of 31.2dB. With the help of the PGA improving supply noise rejection at a high gain, the entire ECG channel achieves a PSRR more than 80dB at the highest gain of 43.2dB, as shown in Fig. 24.4.5b. This is 14dB PSRR improvement by avoiding $VREF_{CH}$ noise amplification in the PGA without consuming additional power and area. A normalized ADC output power spectrum is shown in Fig. 24.4.5d. At 10b mode of the ADC, 9b ENOB is achieved, which is sufficient for accurate heartbeat detection. The on-chip analog feature extraction is verified with the help of an external DSP (Fig. 24.4.6). An intra-cardiac test signal from the Ann Arbor Electrogram Library [8] is acquired by the ASP with transfer functions shown in Fig. 24.4.5c. The ADC digitizes ECG_{out} and FE_{out} in 10b at a rate of 512Hz each. The power of FE_{out} (FE_{out}^2) is calculated and from this the threshold value (V_{th}) is extracted. The second half of the measurement is used to search the real peaks from ECG_{out}. As can be seen, the detected peaks correspond to real peaks correctly.

Compared to recently published state-of-the-art designs (Fig. 24.4.6), the presented ASP does not require any external passives, and moreover, it integrates an analog feature extractor. It consumes only 680nA without compromising other performance significantly.

References:

[1] L.S.Y. Wong, *et al.*, "A Very Low-Power CMOS Mixed Signal IC for Implantable Pacemaker Applications," *IEEE J. Solid-State Circuits*, vol. 39, no. 12, pp. 2246-2456, 2004.

[2] S. Kim, *et al.*, "A 20μW Intra-Cardiac Signal-Processing IC with 82dB Bio-Impedance Measurement Dynamic Range and Analog Feature Extraction for Ventricular Fibrillation Detection," *ISSCC Dig. Tech. Papers*, pp. 302-303, 2013.

[3] R.F. Yazicioglu, *et al.*, "A 30μW Analog Signal Processor ASIC for biomedical signal monitoring," *ISSCC Dig. Tech. Papers*, pp.124-125, 2010.

[4] M. Yip, *et al.*, "A 0.6V 2.9μW Mixed-Signal Front-End for ECG Monitoring," *IEEE Symp. on VLSI Circuits*, pp. 66-67, 2010.

[5] X. Zou, *et al.*, "A 1V 22μW 32-Channel Implantable EEG Recording IC," *ISSCC Dig. Tech. Papers*, pp.126-127, 2010.

[6] D. Han, *et al.*, "A 0.45V 100-Channel Neural-Recording IC with Sub-μW/Channel Consumption in 0.18μm CMOS," *ISSCC Dig. Tech. Papers*, pp. 290-291, 2013.

[7] P. Harpe, *et al.*, "A 7-to-10b 0-to-4MS/s Flexible SAR ADC with 6.5-to-16fJ/conversion-step," *ISSCC Dig. Tech. Papers*, pp. 472-473, 2012.

[8] Ann Arbor Electrogram Libraries, Chicago, IL, USA, [Online] http://www.electrogram/com.

978-1-4799-0917-9/14 $31.00 © 2014 IEEE

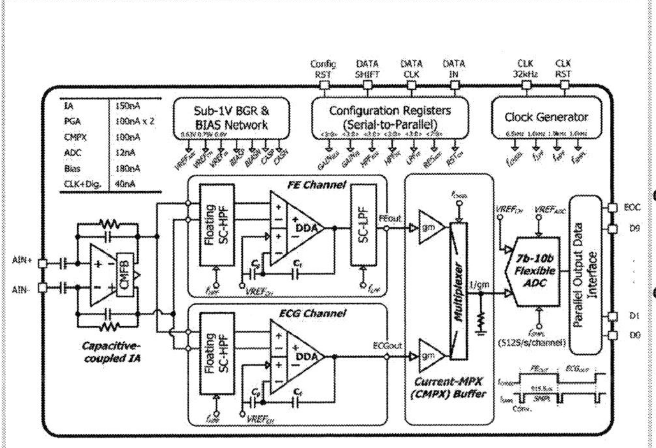

Figure 24.4.1: Architecture of fully integrated Analog Signal Processor (ASP).

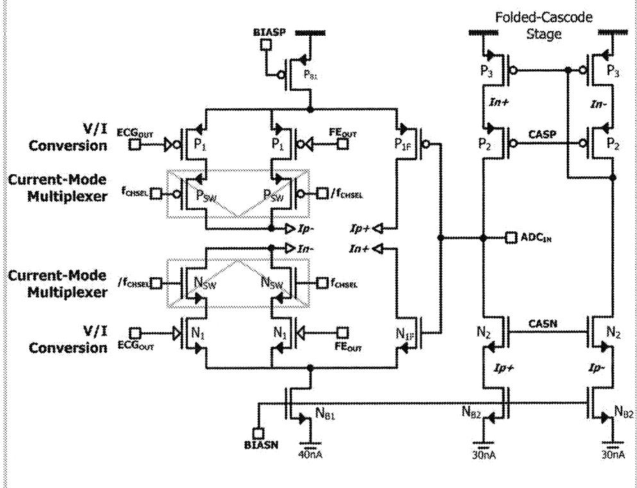

Figure 24.4.2: Current-Multiplexed (CMPX) ADC driver.

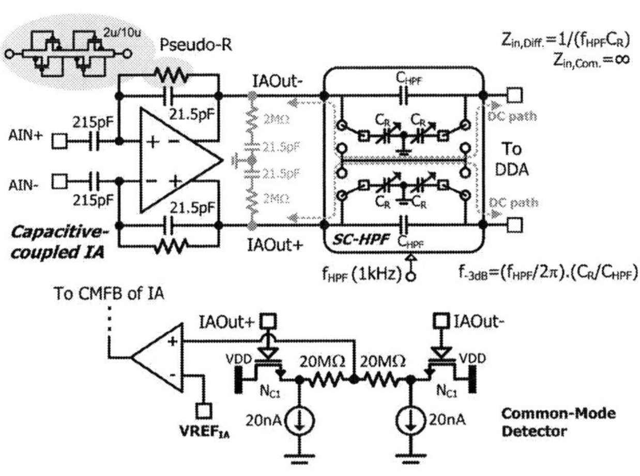

Figure 24.4.3: Capacitive-coupled Instrumentation Amplifier (IA) with floating switching-capacitor HPF.

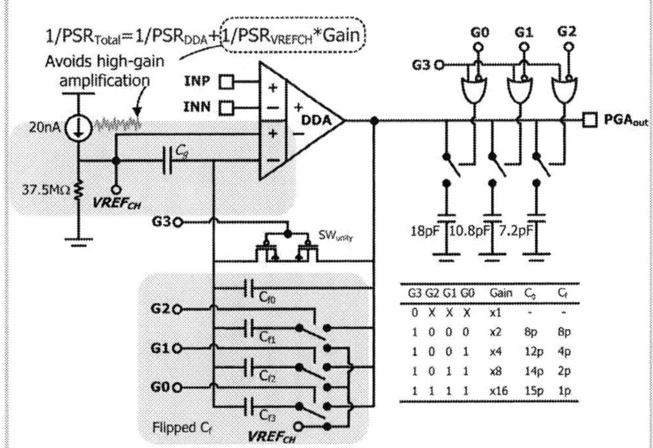

G3 G2 G1 G0	Gain	C_s	C_f
0 X X X	x1	-	-
1 0 0 0	x2	8p	8p
1 0 0 1	x4	12p	4p
1 0 1 1	x8	14p	2p
1 1 1 1	x16	15p	1p

Figure 24.4.4: Programmable Gain Amplifier (PGA) based on Differential-Difference Amplifier (DDA) with flipped capacitors.

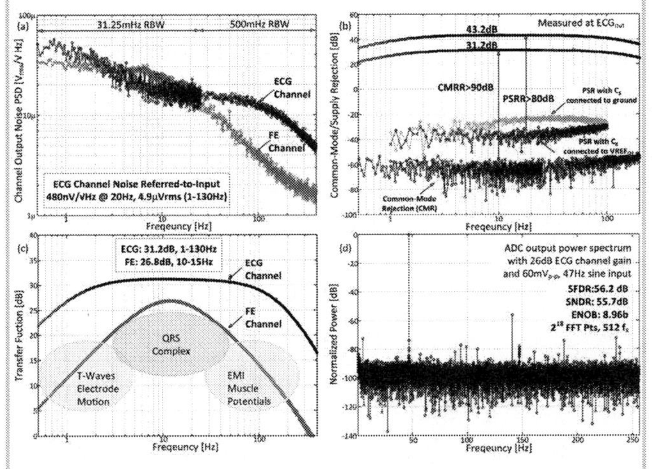

Figure 24.4.5: Measured performance of readout channels and ADC.

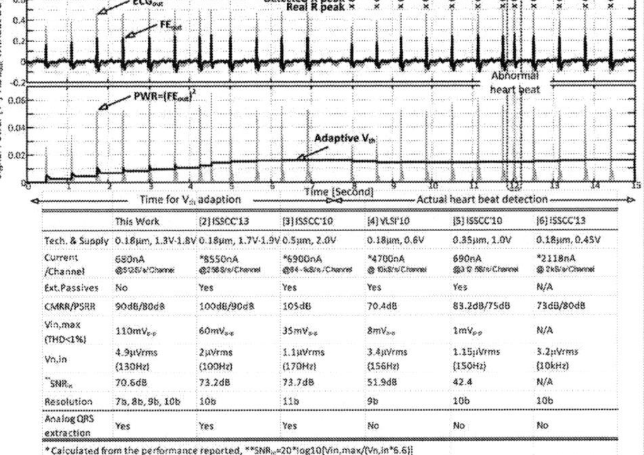

	This Work	[2] ISSCC'13	[3] ISSCC'10	[4] VLSI'10	[5] ISSCC'10	[6] ISSCC'13
Tech. & Supply	0.18μm, 1.3V-1.8V	0.18μm, 1.7V-1.9V	0.5μm, 2.0V	0.18μm, 0.6V	0.35μm, 1.0V	0.18μm, 0.45V
Current /Channel	680nA @512S/s/Channel	*8550nA @256S/s/Channel	*6900nA @64~1kS/s/Channel	*4700nA @10kS/s/Channel	690nA @[illegible]S/s/Channel	*2118nA @2kS/s/Channel
Ext.Passives	No	Yes	Yes	Yes	Yes	N/A
CMRR/PSRR	90dB/80dB	100dB/90dB	105dB	70.4dB	83.2dB/75dB	73dB/80dB
Vin,max (THD<1%)	$110mV_{p-p}$	$60mV_{p-p}$	$35mV_{p-p}$	$8mV_{p-p}$	$1mV_{p-p}$	N/A
Vn,in	4.9μVrms (130Hz)	2μVrms (100Hz)	1.1μVrms (170Hz)	3.4μVrms (156Hz)	1.15μVrms (150Hz)	3.2μVrms (10kHz)
$^{**}SNR_{in}$	70.6dB	73.2dB	73.7dB	51.9dB	42.4	N/A
Resolution	7b, 8b, 9b, 10b	10b	11b	9b	10b	10b
Analog QRS extraction	Yes	Yes	Yes	No	No	No

* Calculated from the performance reported, $^{**}SNR_{in}=20*log10(Vin,max/(Vn,in*6.6))$

Figure 24.4.6: Heartbeat detection measurements from the ASP and performance summary.

24

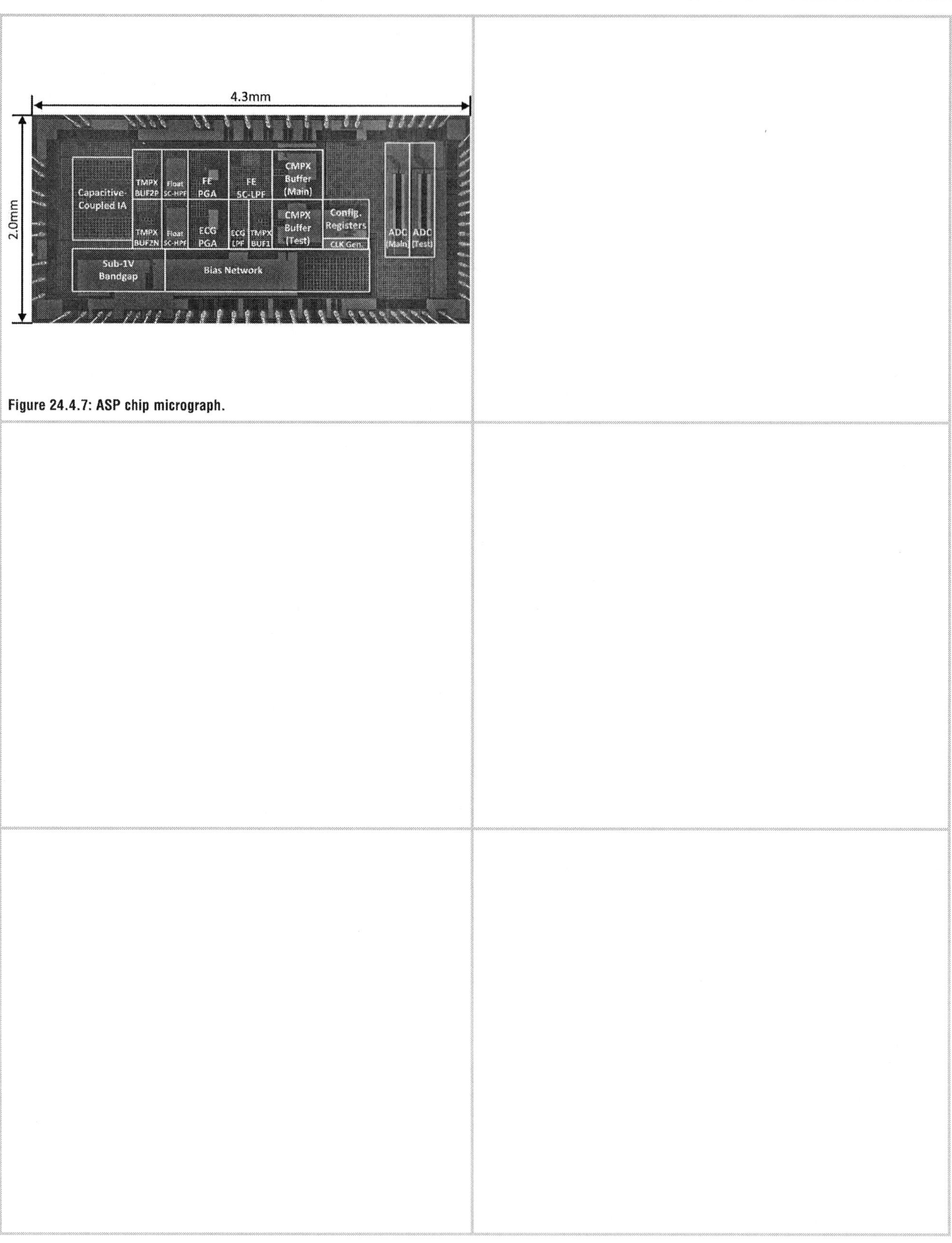

Figure 24.4.7: ASP chip micrograph.

24.5 A 0.5V 1.27mW Nose-on-a-Chip for Rapid Diagnosis of Ventilator-Associated Pneumonia

Kea-Tiong Tang[1], Shih-Wen Chiu[1], Chung-Hung Shih[2], Chia-Ling Chang[1], Chia-Min Yang[1], Da-Jeng Yao[1], Jen-Huo Wang[1], Chien-Ming Huang[1], Hsin Chen[1], Kwuang-Han Chang[1], Chih-Cheng Hsieh[1], Ting-Hau Chang[1], Meng-Fan Chang[1], Chia-Min Wang[1], Yi-Wen Liu[1], Tsan-Jieh Chen[3], Chia-Hsiang Yang[3], Herming Chiueh[3], Jyuo-Min Shyu[1]

[1]National Tsing Hua University, Hsinchu, Taiwan,
[2]Taipei Medical University, Taipei, Taiwan,
[3]National Chiao Tung University, Hsinchu, Taiwan

Ventilator-associated pneumonia (VAP) is the most frequently acquired infection among patients that receive mechanical ventilation in the intensive-care unit (ICU). The mortality rate for VAP lies in the 20-to-50% range and could be even higher in some ICUs. A standard operation procedure to VAP treatment includes a sequence of chest radiography, sputum gram stain, sputum culture, and empiric therapy, initially with antibiotics covering broad pathogens. However, collection of the gram stain and culture of lower respiratory tract specimen is usually not time-efficient (up to 5 days), delaying the initiation of therapy and unacceptable for critically ill patients. A rapid and accurate diagnosis for VAP is therefore crucial, but still unavailable. It is known that microorganisms generate complex metabolites during infection. Fast detection is feasible by examining metabolic wastes in proximal end of the expiratory device, demanding a miniaturized, battery-powered, gas-sensing device. In this work, a fully integrated low-power nose-on-a-chip with a robust learning kernel is developed for such a vital clinical need.

Figure 24.5.1 shows the target application scenario and a top-level system view of the nose-on-a-chip. With a 3D structure, the chip integrates 8 sensors on top and processing circuits at the bottom, completely in a standard CMOS process. The signal-processing circuits include an 8-channel adaptive sensor interface, a SAR analog-to-digital converter (ADC), a RISC processor core with an 8K×32b cache memory, and a dedicated continuous restricted Boltzmann machine (CRBM) kernel for data clustering. With the scalability to form a larger array of chips, massive sensor information can be processed efficiently in parallel to enhance sensing performance for a diversity of applications.

Figure 24.5.2 describes the principle of the nose-on-a-chip and the nanocomposite sensing materials. Distinct sensing materials deposited on the interdigitated electrodes (IDE) form an array of sensors. The collective response from the sensor array constructs unique gas fingerprints. Polymer-carbon composites are used for sensing materials. The carbon-based materials can be: carbon black, carbon nanotube, and mesoporous carbon, where mesoporous carbon has demonstrated superior sensitivity and reversibility. Mesoporous carbon is fabricated from platelet-shaped mesoporous SBA-15 silica and polymers are grown onto the carbon, as shown in the SEM images.

The sensing material is deposited on the IDE, as shown in Figure 24.5.3. The passivation layer is removed with the 400×400μm^2 opening windows. An 8-channel adaptive interface reads out the sensor signals. The interface circuit works as a negative-feedback loop to tune out long time constant signals such as temperature, humidity, and background odors. This sensor interface consumes 215μW. A 0.5V 10b SAR ADC with a charge-average switching (CAS) technique [1] is adopted. The CAS DAC generates top-plate voltage shift by charge averaging instead of conventional charging and discharging operation; it effectively reduces the switching energy to 88.6 CV_{ref}^2 (48% power saving) and minimizes the disturbance and noise of the reference supply voltage. Without the need of an extra voltage reference and common-mode shift issue during conversion, the CAS technique is robust and suitable for integration into a low-voltage SoC. This SAR ADC consumes 1.15μW.

Figure 24.5.4 shows the architecture, learning process flow, and functionality of the CRBM kernel [2]. The CRBM is a probabilistic neural network by injecting Gaussian noise to robustly generalize the variability of data of the same type. By adapting the connection weights $\{w_{ij}\}$ and the sigmoid gain $\{a_j\}$, the CRBM learns to regenerate training examples as the states of visible neurons. The CRBM kernel is trained to model three types of sensory data for 10,000 epochs. The similarity between the reconstruction and the training data indicates the data are properly modeled. Moreover, the learning trajectories of all $\{a_j\}$ and $\{w_{ij}\}$ reach equilibrium after 5,000 training epochs. The hidden neurons of the trained CRBM respond differently to sensory data. This helps to cluster the sensory data and to reduce the data dimensions for reliable classification.

Figure 24.5.5 shows the flow of data recognition and the processing unit for gas identification based on a scalar 32b reduced instruction set computing (RISC) core. Before the data enter the recognition model, the RISC core checks whether the response data are saturated, and then performs normalization against variation in gas concentration. Then, features are selected based on an objective function to train a K-Nearest Neighboring (KNN) classifier. A sequential backward selection (SBS) scheme is adopted here. The SBS has a computation complexity of $O(N^2)$, lower than $O(2^N)$ from exhaustive search. Because SBS extracts informative features, the accuracy on pneumonia detection increases from 74.3 to 85.6% and the accuracy on pneumonia bacteria identification increases from 76.4 to 91.8%.

L7T SRAM cells [3] are employed to achieve a low VDD_{min}. The L7T cells, however, suffer from limited read-bitline (RBL) voltage swing due to BL clamping current from unselected read-ports (RPs) of the accessed column. To increase the read margin, the gate-bias of 1T-RP is increased by boosting the cell-VDD (CVDD) in a read cycle through parasitic capacitors (PC) between metal-lines (BOOST-CVDD) on top of L7T cells. This parasitic boost-CVDD (PBV) scheme consumes only 1% area overhead. Measured read-waveform of PBV-L7T-SRAM probed through SRAM-Flash interface (SFIF) is shown. At SRAM-VDD=0.5V, this 256-rows 256Kb SRAM achieves 20.8ns read access time, including the path-delay due to SFIF, level-shifter, and IO-pads.

The functionality of the chip was verified in clinical trials. 74 samples infected with pneumonia (35 Klebsiella, 39 Pseudomonas aeruginosa) were categorized as experimental group and 43 samples as control group. Figure 24.5.6 shows the classification results performed in two steps: 1) recognizing whether the patient was infected, and 2) if infected, identifying which microorganism was the source. Although the raw data from these two groups overlap, the CRBM improves the accuracy from 89.74 to 95.73% by reducing Clustering Fisher Index (CFI) from 17.89 to 10.30. For the infected patients, the accuracy is improved from 91.89 to 100% by reducing CFI from 11.72 to 0.73. The chip occupies 10.49mm^2 in 90nm CMOS and dissipates 1.27mW. It achieves the highest level of integration and highest computation capability with the lowest power dissipation among state-of-the-art designs (see comparison table in Fig. 24.5.6). This chip fully integrates on-chip sensors, adaptive sensor interface, 10b SAR ADC, 32b RISC with a low-voltage L7T 8K×32b SRAM, along with a robust CRBM kernel, and scalability for higher-dimensional signal processing. This work provides a promising solution for a long-time unresolved issue — a rapid diagnostic strategy to VAP. Figure 24.5.7 shows the chip micrograph and summary.

Acknowledgements:
This work was supported by *NSC* under contract number NSC-102-2220-E-007-006. The authors thank *CIC* for chip fabrication.

References:
[1] C.-Y. Liou, *et al.*, "A 2.4-to-5.2fJ/conversion-step 10b 0.5-to-4MS/s SAR ADC with Charge-Average Switching DAC in 90nm CMOS," *ISSCC Dig. Tech. Papers*, pp. 280-281, Feb. 2013.
[2] H. Chen, *et al.*, "Continuous Restricted Boltzmann Machine with an Implementable Training Algorithm," *IEE P-VIS Image Sign*, pp. 153-158, Aug. 2003.
[3] M.-F. Chang, *et al.*, "A Sub-0.3 V Area-Efficient L-Shaped 7T SRAM With Read Bitline Swing Expansion Schemes Based on Boosted Read-Bitline, Asymmetric-VTH Read-Port, and Offset Cell VDD Biasing Techniques," *IEEE J. Solid-State Circuits*, pp. 2558-2569, Oct. 2013.
[4] C. Hagleitner, *et al.*, "A Gas Detection System on a Single CMOS Chip Comprising Capacitive, Calorimetric, and Mass-Sensitive Microsensors," *ISSCC Dig. Tech. Papers*, pp. 430-431, Feb. 2002.
[5] K.-T. Tang, *et al.*, "A Low-Power Electronic Nose Signal-Processing Chip for a Portable Artificial Olfaction System," *IEEE T. BioCAS*, pp. 380-390, Aug. 2011.
[6] V. Petrescu, *et al.*, "Power-efficient readout circuit for miniaturized electronic nose," *ISSCC Dig. Tech. Papers*, pp. 318-319, Feb. 2012.

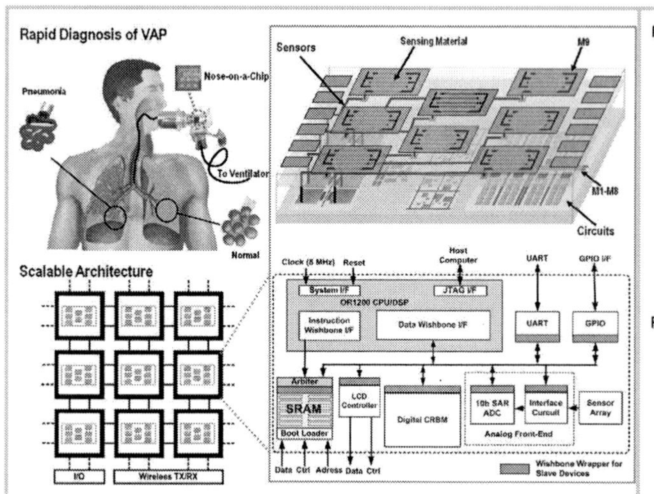

Figure 24.5.1: Target application scenario and top-level system view of the nose-on-a-chip.

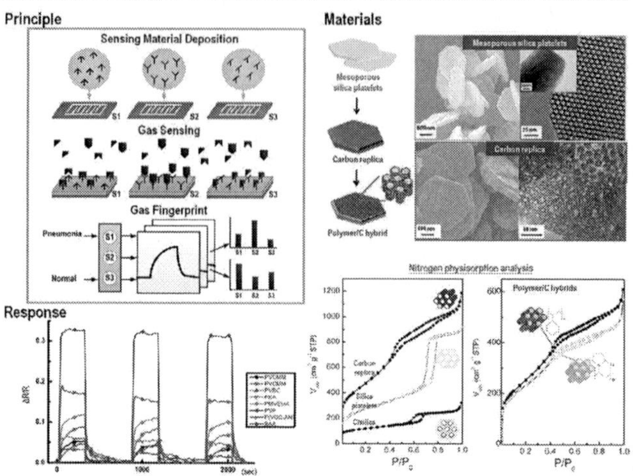

Figure 24.5.2: Principle of the nose-on-a-chip and nanocomposite sensing materials.

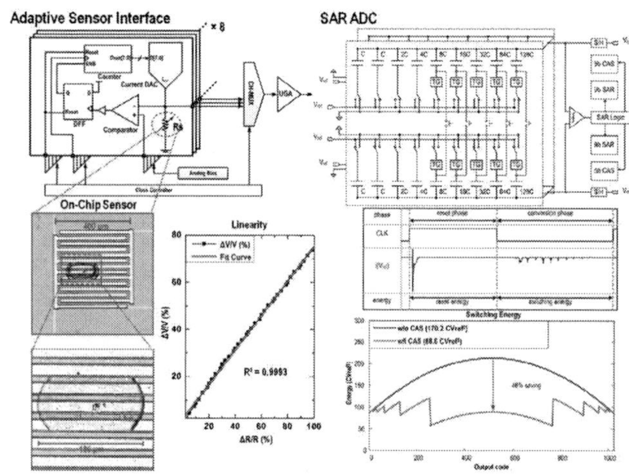

Figure 24.5.3: Adaptive sensor interface, on-chip sensor, and SAR ADC.

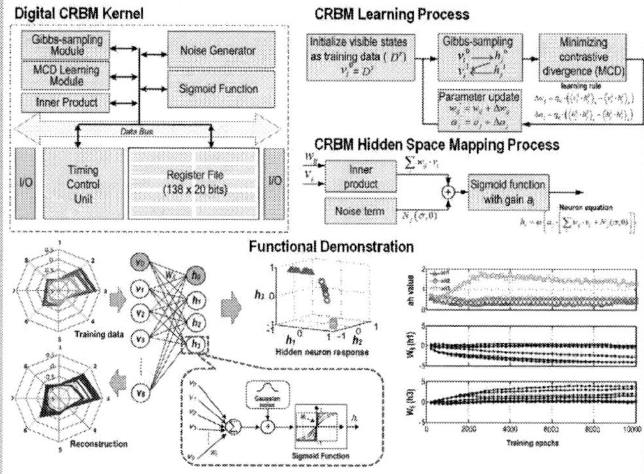

Figure 24.5.4: Architecture, learning process flow, and functionality of the CRBM kernel.

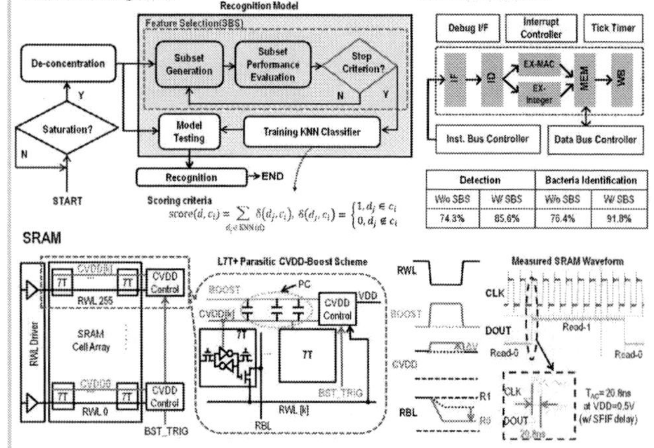

Detection		Bacteria Identification	
W/o SBS	W/ SBS	W/o SBS	W/ SBS
74.3%	85.6%	76.4%	91.8%

Figure 24.5.5: Flow of data recognition, RISC core architecture, and low-voltage SRAM.

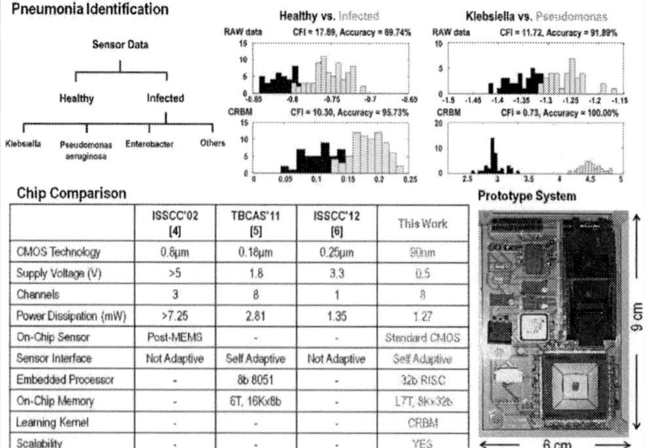

	ISSCC'02 [4]	TBCAS'11 [5]	ISSCC'12 [6]	This Work
CMOS Technology	0.8μm	0.18μm	0.25μm	90nm
Supply Voltage (V)	>5	1.8	3.3	0.5
Channels	3	8	1	8
Power Dissipation (mW)	>7.25	2.81	1.35	1.27
On-Chip Sensor	Post-MEMS	-	-	Standard CMOS
Sensor Interface	Not Adaptive	Self Adaptive	Not Adaptive	Self Adaptive
Embedded Processor	-	8b 8051	-	32b RISC
On-Chip Memory	-	6T, 16Kx8b	-	L7T, 8Kx32b
Learning Kernel	-	-	-	CRBM
Scalability	-	-	-	YES

Figure 24.5.6: Clinical results, chip comparison, and prototype system.

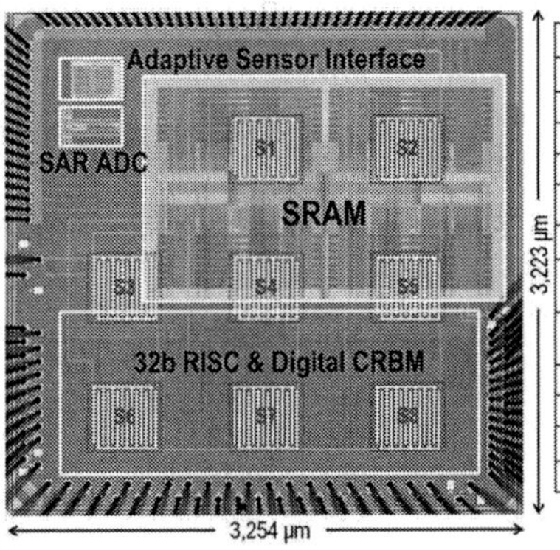

Chip Summary	
Technology	1P9M 90nm CMOS
Die Size	3,254 μmx3,223 μm
Supply Voltage	0.5V
Power Dissipation	1.27mW
Operation Frequency	8MHz
Sub-Blocks	
On-Chip Sensors	8 Polymer-Carbon Composites
Adaptive Interface	8 Channels, Range: 5k - 100kΩ
SAR-ADC	10b
Processor	32b RISC
SRAM	L7T, 8Kx32b
Learning Kernel	CRBM

Figure 24.5.7: Chip micrograph and summary.

24.6 A CMOS Micro-Flow Cytometer for Magnetic Label Detection and Classification

Pramod Murali, Igor Izyumin, Daniel Cohen, Jun-Chau Chien, Ali M. Niknejad, Bernhard Boser

University of California, Berkeley, CA

Flow cytometry is widely used in medicine for hematology, immunology, chemotherapy and pathology, as well as in food and water safety. While present instruments are predominantly used in laboratory environments, there is a growing and unmet need for devices that can be used in point-of-care (PoC) settings. The main impediments to PoC solutions are the fluorescent labels, which require sophisticated sample pre-processing and calibration to reduce background, and optics that are difficult to miniaturize. Label-free approaches such as reported in [1] eliminate this problem but have limited applications due to a lack of specificity. Substituting magnetic labels avoids this limitation and eliminates the need for preprocessing, but present solutions use μHall sensors [2] or giant-magneto-resistors (GMRs) [3], which are not available in standard CMOS technology. CMOS LC-tank spectrometry [4] is amenable to integration but does not achieve the bandwidth and sensitivity required for flow cytometry. In this paper, we present a magnetic flow cytometer integrated in standard CMOS technology assembled in a single-use microfluidic cartridge that meets all the above-mentioned requirements.

Many applications require differentiation between multiple biomarkers. Optical instruments employ fluorescent labels that fluoresce at different wavelengths to accomplish this. Magnetic labels with superparamagnetic nanoparticles (SPNPs) embedded in a polymer matrix exhibit frequency-dependent complex susceptibility. At low frequency, the susceptibility depends both on size and material composition, but at frequencies beyond 1GHz the material properties dominate [5]. For example, at 2GHz the phase of the susceptibility of iron oxide and cobalt differ by about 60 degrees. Therefore, variations of the complex susceptibility of SPNPs with different material compositions can be used to distinguish between label classes independent of label count or signal strength.

Figure 24.6.1 shows the magnetic sensor consisting of a primary excitation coil with embedded secondary pickup coils. The primary excitation coil generates about 0.35mT of magnetic flux density. The secondary coil consists of two pick-up coils each with area 30×30μm². Magnetic labels passing over either of them modulate the flux coupling between the primary and secondary coils, generating a pulse at the output as shown in the figure. A twisted pair interconnect carries the signal from the pick-up coils to the receiver. A center tap of the secondary coil provides the bias voltage to the input transistors of receiver circuit. Coupled quadrature oscillators (QOSC) excite the primary of the sensor and a dummy to generate both in-phase and quadrature signals for demodulation [6].

Figure 24.6.2 shows the architecture of the chip along with off-chip post-processing circuitry. Each arm of the QOSC is biased at 4mA to ensure adequate signal swing for the mixer. The negative transconductance (< -25mS) of the cross-coupled transistors is sufficient to sustain oscillation even in presence of the lossy saline solution used to carry live cells over the sensor. The nominal measured oscillation frequency is 1.2GHz. Each of the two V-I converters of the in-phase and quadrature channels is designed to have a transconductance of 20mS. The outputs of the V-I stages are self-mixed with the QOSC outputs by passive mixers implemented using NMOS transistors. The trans-impedance amplifiers (TIAs) are realized with two-stage fully differential Miller-compensated opamps and have a DC gain of 10kΩ and 3-dB bandwidth of 16MHz. A common-mode feedback circuit sets the opamp's output common mode to 1.1V. The input current to the TIA is chopped by an external square wave (2.85V amplitude at 100kHz) to suppress flicker noise. The in-phase and quadrature signal outputs from the chip are buffered with instrumentation amplifiers, digitized, and demodulated digitally. The chip is fabricated in a standard 2P6M 0.18μm CMOS process (Fig. 24.6.7). The active die area is 1.1×1.3mm², sufficiently small for use in a disposable cartridge. It dissipates 61.2 mW from a 1.8V supply.

Figure 24.6.3 shows the setup of the CMOS chip integrated with the microfluidics. The chip is epoxied into a hole of a carrier PCB. The chip is mounted planar on the PCB, easing integration with the micro-channel. The micro-channel is fabricated with polydimethylsiloxane (PDMS) using an SU-8 mold. It is 50μm high and 200μm wide to prevent clogging and compression sealed against the PCB with and an acrylic cover and metal screws to prevent leakage. All the pads of the CMOS chip are placed on one side and the wirebonds are covered with crystal bond to prevent any electrical shorts or electrolysis. Magnetic labels and cells are carried over the chip surface from a reservoir at the input port by suction pressure at the outlet port of the micro-channel.

The operation of the microfluidic device with embedded CMOS chip has been verified with mouse embryonic fibroblast cells (mEFs, 10-to-15μm diameter). Figure 24.6.4 illustrates a cell tagged with biotinlated anti-mouse/rat CD29 antibody (Biolegend) conjugated to streptavidin coated superparamagnetic Dynabeads (Invitrogen, 4.5μm diameter). Single-bead detection capability ensures high assay sensitivity. Labeled cells are carried over the chip in phosphate buffered saline as the carrier fluid. Throughput is 50 Dynabeads/s or up to 20 cells/s. The labeled cells and free Dynabeads can be distinguished by the difference in their time-of-flight over the pick-up coils. The time-of-flight of cells is longer than free Dynabeads due to frictional forces at the channel walls. Unlike GMR or μHall detector based flow cytometers [2,3], no external magnet is required for detection, eliminating the possibility of cell lysing due to magnetic shear forces. Cells remain viable throughout the detection process, enabling optional sorting and culturing required in some applications.

To demonstrate discrimination of multiple magnetic labels, we use Dynabeads with iron oxide nanoparticles and silica beads with cobalt nanoparticles (custom fabricated by OceanNanotech, 8-to-10μm diameter) in their magnetic cores. Figure 24.6.5 shows the scatter plot obtained for each of these two labels. The average SNR for Dynabeads and OceanNanotech (ONTech) beads is 16dB and 17dB, respectively. The measured phase angle is $< \emptyset_{Dyna} > = 91°$ ($\sigma = 9°$) and $< \emptyset_{ONTech} > = 56°$ ($\sigma = 6°$) and is independent of magnitude.

Figure 24.6.6 compares this work to prior art, which does not fully integrate a complete flow cytometer on-a-chip in CMOS technology. The use of magnetic labels combined with a single-use cartridge setup eliminates the need for sample preprocessing and sample cross-contamination, making the device ideally suited for point-of-care applications.

Acknowledgements:
The authors thank Mekhail Anwar, Paul Swirhun, Richard Przybyla of U.C. Berkeley, Oliver Hayden of Siemens for helpful discussions, Sandeep Prabhu for help with the measurements and OceanNanotech Inc. for providing samples of silica beads with cobalt nanoparticles. Chip fabrication by TSMC is gratefully acknowledged.

References:

[1] K.H. Lee, *et al.*, "A CMOS Impedance Cytometer for 3D Flowing Single- Cell Real-Time Analysis with ΔΣ Error Correction," *ISSCC Dig.Tech Papers,* pp. 304–305, Feb. 2012.
[2] D. Issadore, *et al.*, "Ultrasensitive Clinical Enumeration of Rare Cells Ex Vivo using a μHall Detector," *Science Translational Medicine*, vol. 4, no. 141, pp. 141ra92, 2012.
[3] M. Helou, *et al.*, "Time-of-Flight Magnetic Flow Cytometry in Whole Blood with Integrated Sample Preparation," *Lab on a Chip*, vol. 13, no. 6, pp. 1035–1038, 2013.
[4] C. Sideris, *et al.*, "An Integrated Magnetic Spectrometer for Multiplexed Biosensing," *ISSCC Dig. Tech. Papers*, pp. 300–301, Feb. 2013.
[5] P. C. Fannin, *et al.*, "Experimental and Theoretical Profiles of the Frequency-Dependent Complex Susceptibility of Systems Containing Nanometer-Sized Magnetic Particles," *Physical Review B*, vol. 55, no. 21, pp. 14423-14428, 1997.
[6] A. Rofougaran, *et al.*, "A 900 MHz CMOS LC-oscillator with Quadrature Outputs," *ISSCC Dig. Tech. Papers,* pp. 392-393, Feb. 1996.

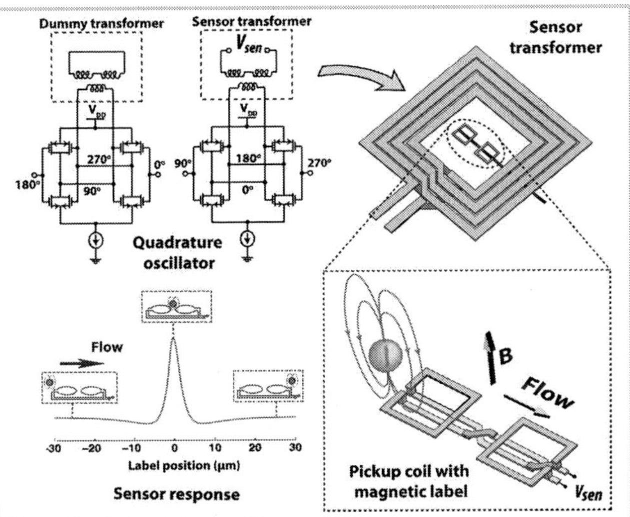

Figure 24.6.1: Quadrature oscillator and the sensor transformer.

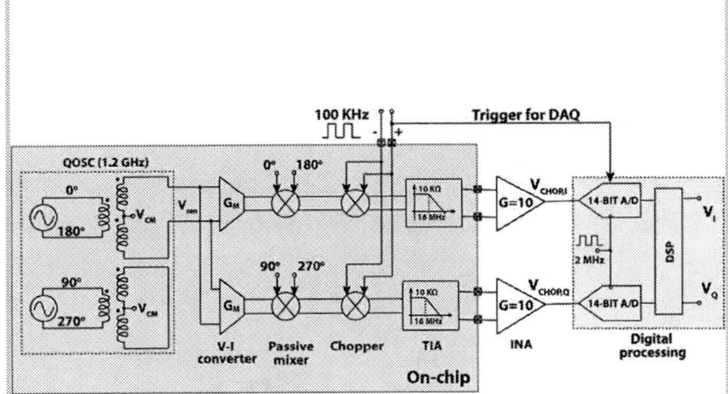

Figure 24.6.2: Chip block diagram showing quadrature excitation oscillators, pickup coils and demodulator.

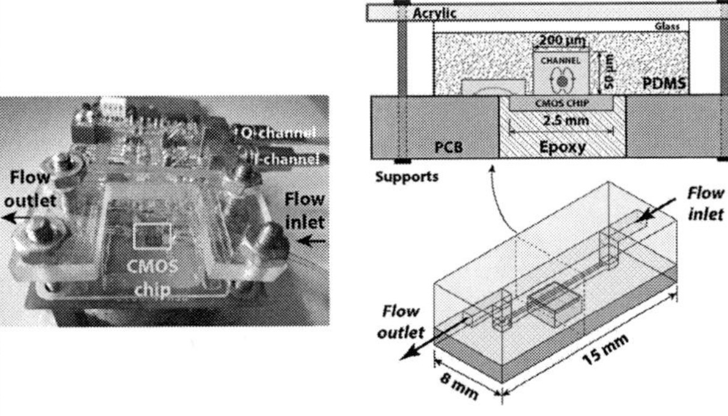

Figure 24.6.3: Microfluidics integration of the CMOS chip.

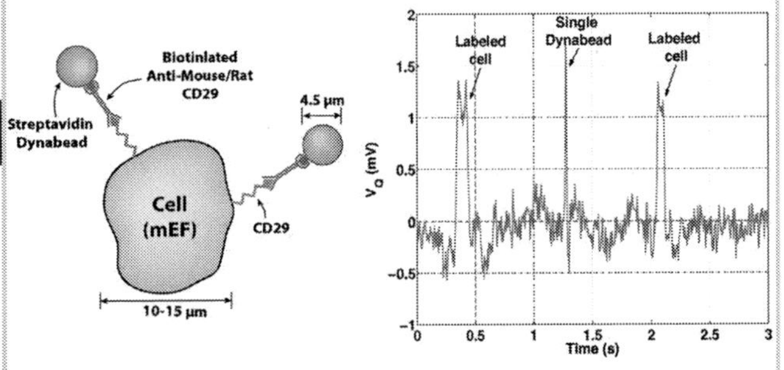

Figure 24.6.4: (Left) Cell labeled specifically with antibody coupled Dynabead. (Right) Electrical response. The duration of the response is used to differentiate between cells and free labels.

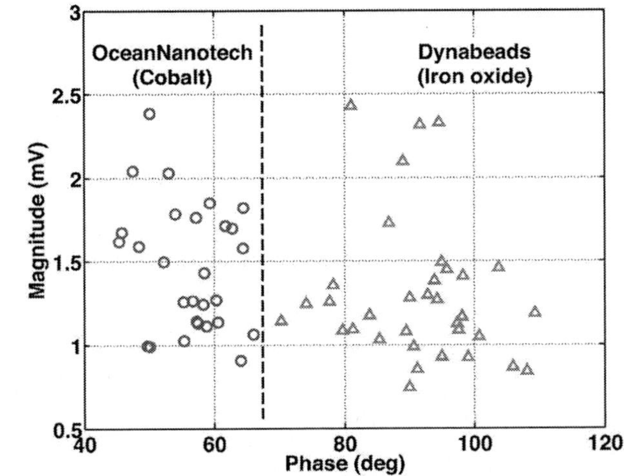

Figure 24.6.5: Phase permits unambiguous separation of label classes based on material.

	[1] 2012	[2] 2012	[3] 2013	[4] 2013	This work
Sensor	Electrode pair	μHall Detector	GMR	LC Tank	Spiral Transformer
Sensor integration	Off-chip	N/A	N/A	On-chip	On-chip
Application	Cell counter	Flow cytometry	Flow cytometry	Immunoassay	Flow cytometry
Other components	None	0.5 T Magnet	0.2 T Magnet	None	None
Labels	None	$MnFe_2O_4$	FeO_x	FeO_x	FeO_x and Co
Separate Label Classes	0	3	1	2	2
Detection Time	50 ms	0.02 ms	40 ms	120 s	20 ms

Figure 24.6.6: Comparison with recent prior art.

24

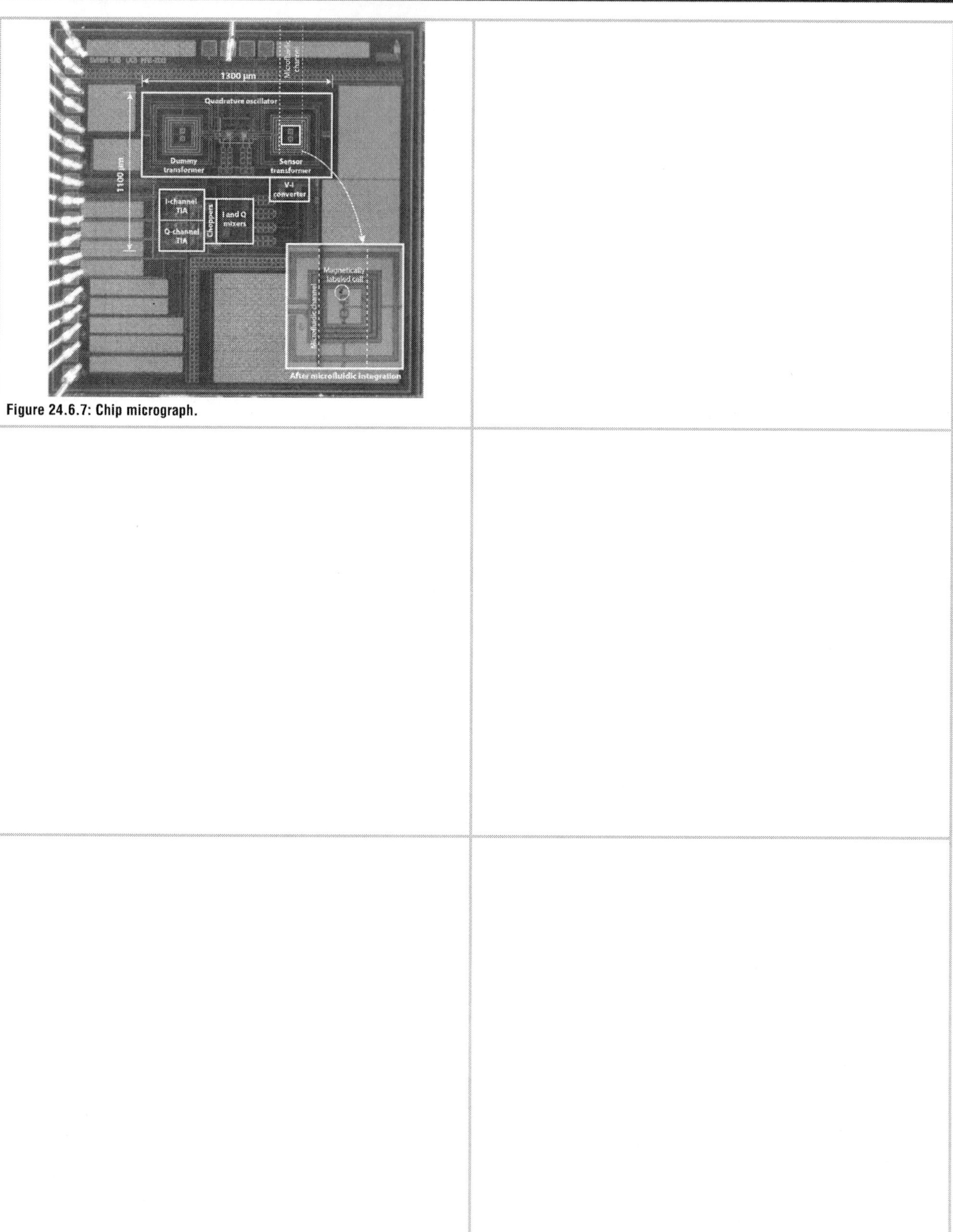

Figure 24.6.7: Chip micrograph.

24.7 A 60nV/√Hz 15-Channel Digital Active Electrode System for Portable Biopotential Signal Acquisition

Jiawei Xu[1,2], Benjamin Busze[1], Hyejung Kim[3], Kofi Makinwa[2], Chris Van Hoof[3], Refet Firat Yazicioglu[1,3]

[1]Holst Centre/imec, Eindhoven, The Netherlands,
[2]Delft University of Technology, Delft, The Netherlands,
[3]imec, Leuven, Belgium

Dry active electrodes (AE), i.e., the combination of dry electrodes with *in situ* amplification, are increasingly used for biopotential measurements in emerging healthcare and lifestyle applications [1]. Compared to gel-based wet electrodes, dry electrodes enable fast set-up time, greater user comfort, and long-term monitoring. AE amplifiers ensure local amplification providing improved robustness to noise interference and cable motion artifacts. However, current AEs have analog outputs requiring powerful analog buffers to drive biopotential signals over measurement cables. Furthermore, analog outputs must be digitized by the back-end (BE) system [2,3]. Besides, parameter mismatch between AEs limits the overall CMRR. CMFB [1] or CMFF [2] helps but comes at the expense of increased number cables between the BE and AEs. These problems significantly increase the overall system complexity and cost.

This paper presents a digital AE system (Fig. 24.7.1) that supports up to 16 AEs connecting to a generic MCU on the same I²C bus. Each AE consists of an analog front-end (AFE) with a built-in ADC, and an I²C interface. Digitally assisted CMFB and digital CMRR trimming mitigate AE mismatch and enable 78dB CMRR. In order to limit the number of cables, the CMFB signal is sent to AEs using supply voltage. Hence, the complete AE structure only needs 4 wires, which can be accommodated by commercially available medical-grade cables. For synchronization, a broadcast packet, containing a trigger is periodically sent to all AEs to realign their ADC sampling clocks.

Figure 24.7.2 shows the detailed architecture of one AE. Each AE can simultaneously monitor biopotential signals (EEG channel), complex electrode-tissue impedance (ETI) signals (IMP and IMQ channels), and infra-low frequency (ILF) signals (DC channel). The signal chain starts with a chopper instrumentation amplifier (IA) with a voltage gain of 70, which significantly reduces the noise/power requirements of the following programmable gain amplifier (PGA). The IA's output is split into two channels for ETI and EEG measurements. The ETI channel consists of two PGAs that demodulate the signals with in-phase f_I (0°) and quadrature-phase f_I (90°) clocks, with respect to a square-wave current injected into the electrode. The IMP and IMQ outputs then represent the real and imaginary components of ETI respectively [2]. Each EEG channel includes a chopper PGA with a notch filter that rejects the large ETI signal at f_I (0°) by 40dB. The CM channel is an auxiliary output of the EEG channel, and is only used for the CMFB loop. To avoid delay, which would destabilize the loop, it does not use the anti-aliasing low-pass filter (LPF) applied to all other signals. Aliasing of the CM signal is not critical, since such signals will appear to any pair of AEs as a common-mode signal. The DC channel acquires the DC and ILF signals present at the output of the DC-servo loop (DSL). The normalized gain and phase of the DC and EEG channels are complementary, which makes it possible to reconstruct the DC-coupled EEG signals in the MCU. This significantly relaxes the dynamic range requirements of the AE while implementing the same functionality as a DC-coupled IA [4]. A 12b SAR ADC digitizes the outputs of all channels at 1kS/s.

There are two different AE mismatch mechanisms limiting CMRR. The first is the ETI mismatch, which happens in a much larger scale in dry biopotential electrodes. The second is the amplifier parameter mismatch. Compensating for these mismatches can significantly improve CMRR and increase signal quality. Figure 24.7.3 shows the digitally assisted CMFB and CMRR trimming implementation. The former subtracts CM input signals before they are amplified by IA, and the latter compensates for gain mismatch between AEs, which is due to both amplifier parameter mismatch and electrode impedance mismatch.

The CMFB loop extracts the CM signal of all AEs by averaging their CM channel outputs in the MCU. A 12b CMFB DAC converts this digital CM signal back into analog domain, and then feeds it back to the inverting inputs of all IAs. This cancels input CM signals before they are amplified and so prevents AE saturation in the presence of large CM interference. To avoid adding an extra wire, the analog CMFB signal is superimposed on AE's power supply. V_{dd} is then extracted by a 1.8V low-dropout voltage regulator (LDO), while the CMFB signal is extracted via each IA's capacitively coupled DSL.

CMRR trimming operates by applying a CM input reference signal (15mV$_{pp}$ at 30Hz) to the AEs via the patient bias electrode (ground electrode). This signal appears at the input of each AE. After amplification and digitization, the MCU compares the output of each AE with that of a reference AE, and sends individual trimming values to PGAs for improved gain matching between AEs. The PGA employs a resistive feedback amplifier where its load resistance R_{DAC} can be trimmed with 50Ω resolution corresponding to 10ppm gain trimming capability or more than 100dB CMRR. The range of trimming can compensate for up to 4.1% gain mismatch between AEs, which is beyond the measured gain error (<0.5%) due to ETI mismatch and amplifier parameter mismatch.

The use of dry electrodes leads to increased electrode offset between AEs. DC-coupled amplifiers [5] are limited by the gain, thus causing large power dissipation due to the large dynamic range requirement. In contrast, AC-coupled amplifiers enable lower-power dissipation but reject ILF biopotential signals [1,6]. To overcome this, the AE amplifier implements a DSL (Fig. 24.7.4). The DSL realizes a high-pass filter characteristic to reject electrode offset. It consists of a gm-C integrator that monitors the output offset and then cancels it by driving the IA's inverting input. This architecture can reject up to ±350mV of electrode offset, and mainly defined by the amplifier's input range. In order to achieve a 0.5Hz low-pass corner and reduce the impedance at chopping node, an external capacitor (C_{ext}=1μF) is employed. However, tens of seconds are needed to charge C_{ext} at start-up. To overcome this, a parallel gm3 =100*gm2 is temporarily switched on to reduce the start-up time to less than 1s. A continuous-time ripple reduction loop (RRL) rejects output chopper ripple and so maximizes output headroom. Both the RRL and DSL run in the background during the operation of the circuit, and are chopped to reduce their 1/*f* noise contribution.

The 16mm² AE is implemented in a standard TSMC 0.18μm CMOS technology (Fig. 24.7.7). Each AE consumes 150μA from a 1.8V supply, including 92μA from the I²C interface. Figure 24.7.5 shows the ASIC characterization results. The AE system achieves 60nV/√Hz input-referred noise with functionally DC-coupled characteristics, and is able to measure ETI up to 400kΩ. The CMFB loop boosts the CMRR from 57 to 72dB, while CMRR trimming further improves the CMRR to 78dB at 30Hz. Figure 24.7.6 compares this ASIC to the state of the art and shows the results of a 2-channel EEG measurement using dry electrodes.

In conclusion, this paper presents a single-chip digital active electrode IC and the system enabling 15-channel (16-electrode) biopotential signal acquisition via only 4 wires, significantly reducing system complexity and enabling true modularity. The analog front-end achieves state-of-the-art performance, in terms of noise, input impedance, DC-coupled characteristics and electrode offset rejection, thus enabling the practical use of dry electrodes. The use of digital-assisted CMFB and digital CMRR trimming ensures a 21dB CMRR improvement.

References:

[1] J. Xu, *et al*, "A 160μW 8-channel Active Electrode System for EEG Monitoring," *ISSCC Dig. Tech. Papers*, pp. 300-301, Feb 2011.
[2] S. Mitra, *et al.*, "A 700μW 8-Channel EEG/Contact-impedance Acquisition System for Dry-electrodes," *VLSI Symposium*, pp. 68-69, June 2012.
[3] M. Guermandi, *et al.*, "Active Electrode IC Combining EEG, Electrical Impedance Tomography, Continuous Contact Impedance Measurement and Power Supply on a Single Wire," *Proc. ESSCIRC*, pp. 335-338, Sept. 2011.
[4] R. Muller, *et al.*, "A 0.013 mm² 2.5 μW, DC-Coupled Neural Signal Acquisition IC With 0.5 V Supply," *J. Solid-State Circ.*, pp. 232-243, Jan. 2012.
[5] TI-ADS1298 [Online]. <http://www.ti.com/lit/ds/symlink/ads1298.pdf>
[6] N. Verma, *et al.*, "A Micro-Power EEG Acquisition SoC With Integrated Feature Extraction Processor for a Chronic Seizure Detection System," *J. Solid-State Circ.*, pp. 804-816, Apr. 2010.
[7] R. F. Yazicioglu, *et al.*, "A 200 μW Eight-Channel EEG Acquisition ASIC for Ambulatory EEG Systems," *J. Solid-State Circ.*, pp. 3025-3038, Dec. 2008.

978-1-4799-0917-9/14 $31.00 © 2014 IEEE

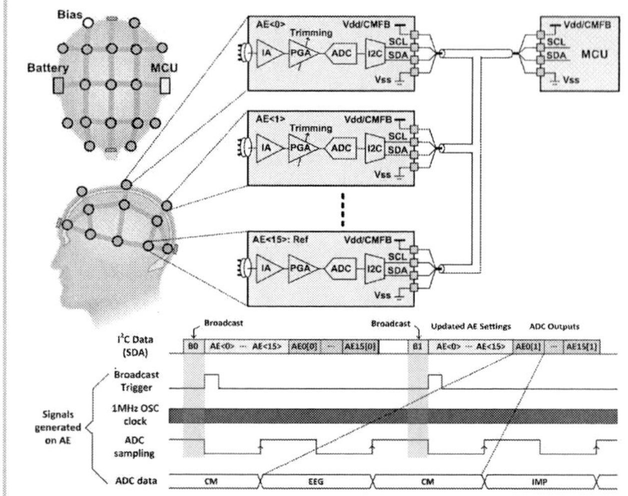

Figure 24.7.1: Block diagram of the digital AE system.

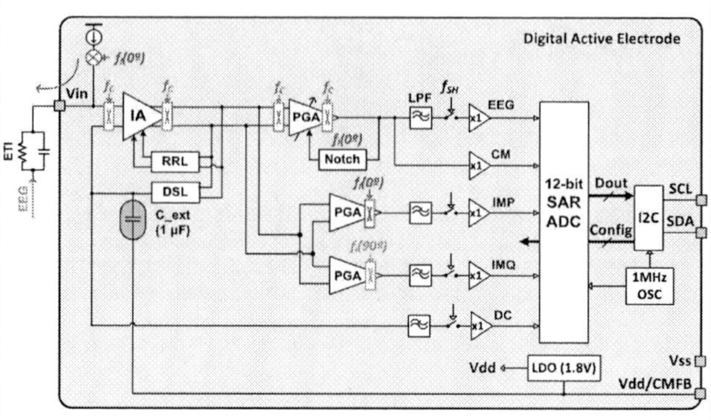

Figure 24.7.2: Architecture of the digital AE.

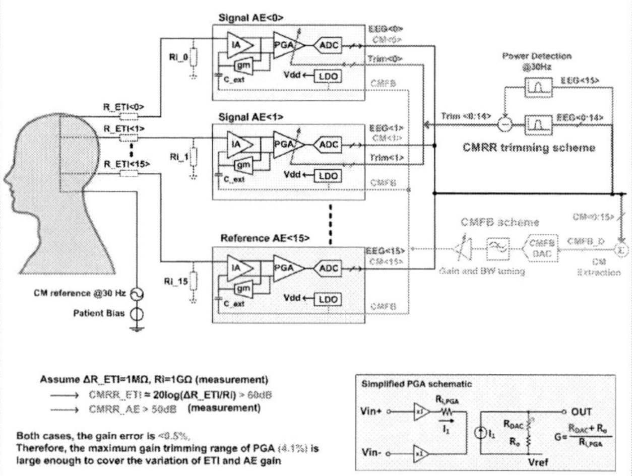

Figure 24.7.3: Digitally assisted CMFB and digital CMRR trimming.

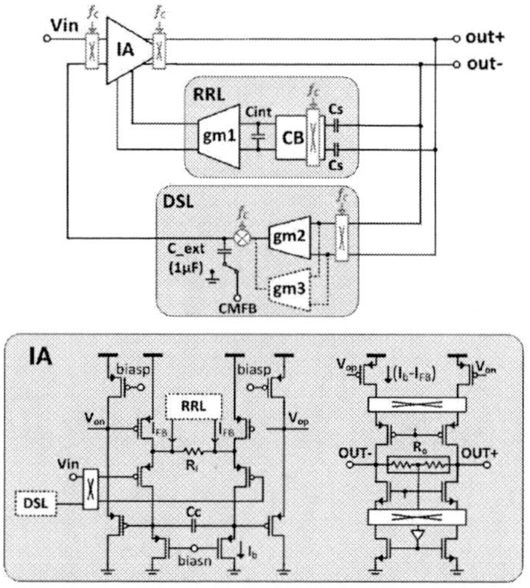

Figure 24.7.4: Schematics of IA.

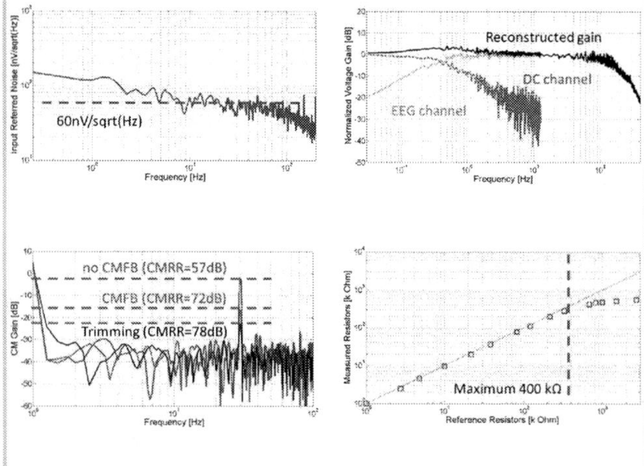

Figure 24.7.5: AE characterization results.

	Parameters	[1]	[2]	[3]	[4]	[5]	[6]	[7]	This work
Analog Front-End (AFE)	Active Electrode (AE)	yes	yes	yes	no	no	no	no	yes
	Input impedance (GΩ)	2	1.2	--	--	1	0.7	1	1
	Input referred noise (μVrms)/channel	1.2 (0.5-100Hz)	1.75 (0.5-100Hz)	0.9 (0.5-100Hz)	4.3 (DC-300Hz)	0.7 (DC-131Hz)	1.3 (0.5-100Hz)	0.6 (0.5-100Hz)	0.65 (0.5-100Hz)
	Offset rejection (mV)	Rail-to-Rail	±250	--	50	±250	Rail-to-Rail	45	±350
	CMRR (dB)	82	84	105	75	115	60	120	78
	Current (μA)/channel	12.5	48	330	10	250	>3.5	8.25	58
	DC-coupled ExG	no	no	no	yes	yes	no	no	yes
	ETI Measurement	no	yes	yes	no	yes	no	no	yes
System Integration	Digital interface	no	no	no	no	SPI	no	no	I²C
	Number of channels	7	8	--	2	8	18	8	15
User Comfort	Dry electrode	yes	yes	--	no	no	no	no	yes
	Number of cables (AE)	5	6	4	--	--	--	--	4

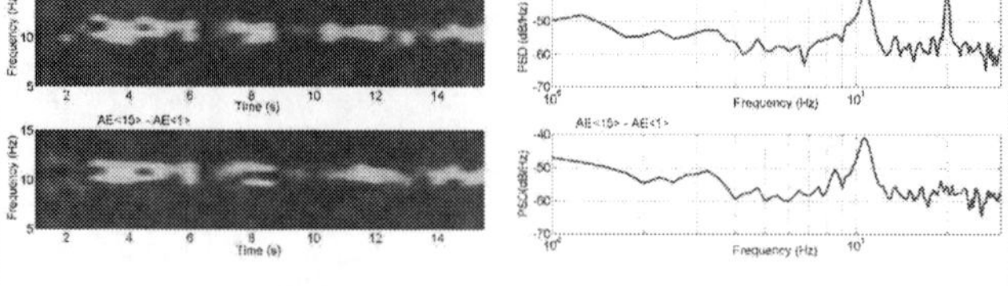

Figure 24.7.6: Comparison table and EEG measurement.

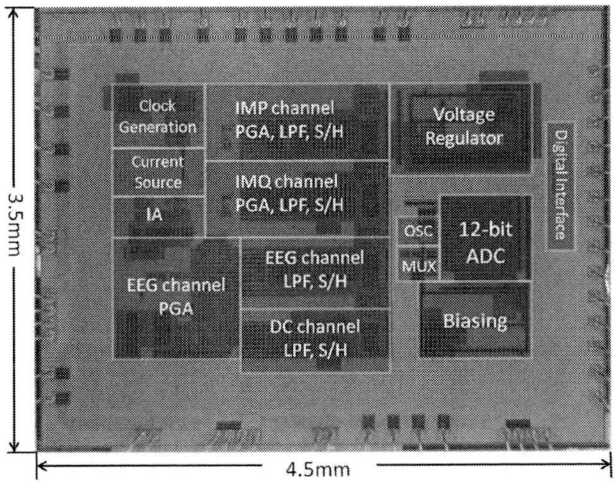

Figure 24.7.7: Photograph of ASIC.

24.8 An Analog-Digital-Hybrid Single-Chip RX Beamformer with Non-Uniform Sampling for 2D-CMUT Ultrasound Imaging to Achieve Wide Dynamic Range of Delay and Small Chip Area

Ji-Yong Um[1], Eun-Woo Song[1], Yoon-Jee Kim[1], Seong-Eun Cho[1], Min-Kyun Chae[1], Jongkeun Song[2], Baehyung Kim[2], Seunghun Lee[2], Jihoon Bang[2], Youngil Kim[2], Kyungil Cho[2], Byungsub Kim[1], Jae-Yoon Sim[1], Hong-June Park[1]

[1]Pohang University of Science and Technology, Pohang, Korea,
[2]Samsung Advanced Institute of Technology, Yongin, Korea

Ultrasound imaging is widely used for medical diagnosis, because it is harmless to the human body and has real-time processing capability. Usually the focusing (beamforming) operation is performed for both TX and RX. The RX focusing is performed by an RX beamformer [1-5], which consists of delay elements and adders. Nowadays, digital beamformers (DBF) are mostly used for conventional ultrasound imaging because of high SNR. Recently, 2D ultrasound transducers have been introduced for 3D imaging. Since the 2D transducer has a huge number of transducer elements (e.g., 9216 for a 72×128 array), it cannot use DBF because of the huge number of required ADCs and wires inside the probe cable. Therefore, analog beamforming must be performed, at least at the front stage of the 2D transducer.

In this work, where a 2D CMUT array is used, the maximum delay difference among transducer elements is 8μs with a maximum steering angle of 45° and a maximum focal depth of 15cm. The target sampling resolution is 6.25ns (λ_c / 53.3) with a carrier frequency of 3MHz. An analog-digital-hybrid architecture and a non-uniform sampling scheme are used for the RX beamformer of this work to achieve the wide dynamic range of delay time and small chip-area. The RX beamformer consists of 8 analog beamformers (ABF) followed by a single DBF, as shown in Fig. 24.8.1. An ABF performs the focusing operation for the input signals of the adjacent 8 channels to generate an analog output signal. The 8 analog output signals from the 8 ABFs are applied to the DBF. The DBF converts the 8 analog input signals into the 8 digital signals, and then performs the focusing operation on the 8 digital signals to generate a digital output signal for every focal point.

The hybrid beamformer (HBF) reduces the chip area by using a non-uniform sampling scheme in ABFs and an area-efficient digital memory (FIFO) in the DBF. Figure 24.8.2 shows a comparison of estimated chip area for 64-channel implementations of the HBF (this work) and conventional ABFs with uniform [4] and non-uniform [1,5] sampling schemes. It is assumed that the sample-and-hold type analog memory (capacitor) is used for all three cases. This work shows an area reduction of around 84% compared to the uniform-sampling ABF [4]. In the uniform sampling ABF [4], the sampling-time step determines the delay resolution. A non-uniform sampling scheme is used in this work, where the echo signal is sampled only at the time points when the echo signal arrives at the transducer [1,5]. To achieve the delay resolution of 6.25ns, uniform sampling requires a sampling frequency of 160MHz, while non-uniform sampling requires a sampling frequency in the range of 20 to 40MHz. The reduced sampling rate significantly lowers the number of capacitors required for the analog memory. In this work, an ABF and a DBF are designed to handle delay ranges larger than 1μs and 7μs, respectively, to cover the entire delay range of 8μs. The DBF is designed to cover the delay range of 11.2μs (>7μs), which requires a FIFO memory of 8.45kb. The FIFO memory is synthesized to reduce the chip area.

The circuit diagram of an ABF is shown in Fig. 24.8.3. It consists of an analog memory array, an analog adder, and a timing controller. Each analog memory cell can be represented by *AMC*[*i*, *j*], where *i* and *j* represent the channel number and the target focal point number, respectively. Each *AMC* consists of a sample-and-hold circuit, as shown in the insert of Fig. 24.8.3. Each echo signal (e.g., *Echo*[*i*]) from a transducer element is connected to the corresponding row (e.g., the *i*-th row) of the analog memory array. Each row of the array consists of 30 *AMC*s (e.g., *AMC*[*i*, *j*], *j* = 1, ..., 30) to cover the maximum delay difference up to 1.5μs in an ABF. Each *AMC*[*i*, *j*] receives a sampling signal *S*[*i*, *j*] and an addition signal *ADD*[*j*] from the timing controller. The timing controller receives the *Sample Code*[1:8] signal as input from an external FPGA. The echo signal from the *j*-th focal point (FP_j) arrives at the 8 adjacent transducers connected with an ABF at different times. The corresponding 8 electrical signals (*Echo*[*i*]) are sampled and stored at the *j*-th column memory cells (*AMC*[*i*, *j*], *i* = 1, ...,8) at the falling edges of *S*[*i*, *j*]. The *ADD*[*j*] signal is activated to get the average value of the *j*-th column memory.

The DBF of Fig. 24.8.4 consists of digital FIFOs, digital adders, and 8 branches of 10b SAR ADCs and *Retimers*. The sampling-time interval of the ADC is non-uniform in time and ranges from 25 to 50ns in 6.25ns steps. The *Retimer* re-samples the ADC output with a 40MHz clock signal and feeds the valid ADC output to the digital FIFO. The digital FIFO is divided into 3 stages. Each stage performs the addition of the adjacent two data points. The first stage consists of 8 FIFOs with the depth of 2^5. The second and the third stages consist of 4 FIFOs with the depth of 2^6, and 2 FIFOs with the depth of 2^7, respectively. Thus, the total length of the digital FIFO is set to 8.45kb to cover the delay difference up to 11.2μs ((2^5+2^6+2^7)×50ns). At each stage, as soon as two adjacent FIFOs both become non-empty, the topmost data of the two FIFOs are added and the sum data is sent to the next-stage FIFO. The third stage generates the 10b HBF output. The throughput of the DBF is either 40MHz or 20MHz, depending on the steering angle and the focal point.

The analog-digital-hybrid beamformer chip is fabricated in a 0.13μm CMOS process. The chip has an area of 19.4mm^2 and consumes 1.14W. A 2D CMUT probe [6] is used to evaluate the performance of the fabricated chip. The CMUT probe includes a 64×128 CMUT array and an ASIC chip, which includes TX pulsers and pre-amplifiers. One row of the 2D CMUT array is selected to apply 64 echo signals to the fabricated HBF chip. An ultrasound plane wave (un-focused wave) is transmitted from the CMUT probe to acquire the B-mode images of two phantom models. The left hand side of Fig. 24.8.5 shows the B-mode image of the phantom with the four uniformly spaced parallel plates, with the dynamic range set to 20dB. Four parallel plates can be clearly observed along with the bottom plane. From this image, the distance between two adjacent plates is measured to be 1.3cm, which is identical to the actual plate distance of the phantom. The right hand side of Fig. 24.8.5 shows the B-mode image of the phantom with two punched parallel plates, with the dynamic range set to 27dB. The image patterns of the upper punched plate are faint due to reverberation. The image patterns of the lower punched plate are clearly observed. The dynamic ranges used for the two images of Fig. 24.8.5 are relatively low, because a plane wave is used for the transmitted ultrasound wave without TX focusing.

Figure 24.8.6 shows a performance comparison of single-chip RX beamformers [2-5] with this work. It can handle a wide delay range of 8μs with a delay resolution of 6.25ns. Also, the supported number of channels is the maximum among the single-chip beamformers of Fig. 24.8.6, and it is the only one that incorporates HBF.

Acknowledgements:
This research was supported by the NRF funded by the MEST (2013029772), IT R&D program of MKE (10039159), and ITRC (NIPA-2013-H0301-13-1007)

References:

[1] T. K. Song, *et al.*, "Ultrasonic Dynamic Focusing Using an Analog FIFO and Asynchronous Sampling," *IEEE Trans. Ultraso., Ferroelec., Freq. Contr.*, vol. 41, no. 3, pp. 326-332, May 1994.

[2] G. Gurun, *et al.*, "An Analog Integrated Circuit Beamformer for High-Frequency Medical Ultrasound Imaging," *IEEE Trans. Biomed. Circuits Syst.*, vol. 6, pp. 454-467, Oct. 2012.

[3] J.R. Talman, *et al.*, "Integrated Circuit for High-Frequency Ultrasound Annular Array," *Proc. IEEE CICC*, pp. 477-480, Sept. 2003.

[4] B. Stefanelli, *et al.*, "An Analog Beam-Forming Circuit for Ultrasound Imaging Using Switched-Current Delay Lines," *IEEE J. Solid-State Circuits*, vol. 35, no. 2, pp. 202-211, Feb. 2000.

[5] J.Y. Um, *et al.*, "A Single-Chip Time-Interleaved 32-Channel Analog Beamformer for Ultrasound Medical Imaging," *IEEE Asian Solid-State Circuits Conf.*, pp. 193-196, Nov. 2012.

[6] J. Song, *et al.*, "Reconfigurable 2D cMUT-ASIC arrays for 3D ultrasound image," *Proc. SPIE Medical Imaging*, vol. 8320, Feb. 2012.

978-1-4799-0917-9/14 $31.00 © 2014 IEEE

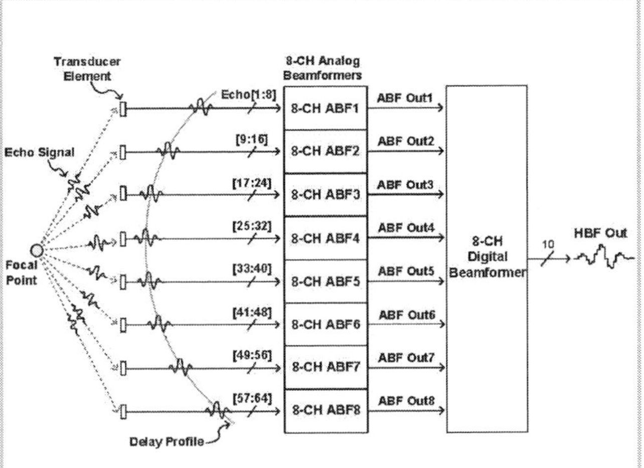

Figure 24.8.1: Single-chip HBF with 64 channel inputs.

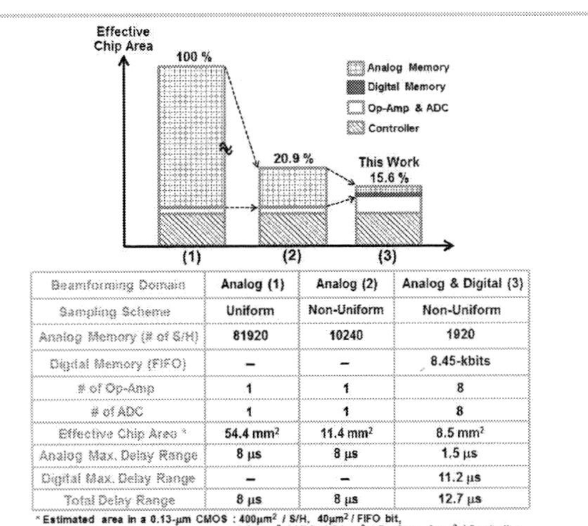

Beamforming Domain	Analog (1)	Analog (2)	Analog & Digital (3)
Sampling Scheme	Uniform	Non-Uniform	Non-Uniform
Analog Memory (# of S/H)	81920	10240	1920
Digital Memory (FIFO)	–	–	8.45-kbits
# of Op-Amp	1	1	8
# of ADC	1	1	8
Effective Chip Area *	54.4 mm^2	11.4 mm^2	8.5 mm^2
Analog Max. Delay Range	8 μs	8 μs	1.5 μs
Digital Max. Delay Range	–	–	11.2 μs
Total Delay Range	8 μs	8 μs	12.7 μs

* Estimated area in a 0.13-μm CMOS : 400μm^2 / S/H, 40μm^2 / FIFO bit, 0.09mm^2 / ADC, 0.24mm^2 / Op-Amp, 5mm^2 / Controller

Figure 24.8.2: Comparison of effective chip area between conventional ABFs and HBF (this work).

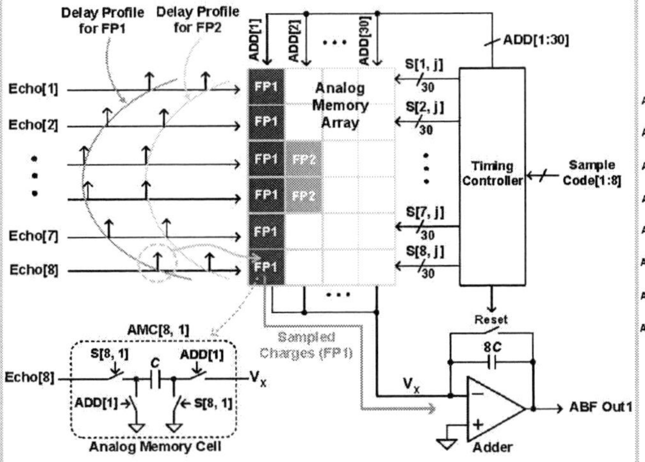

Figure 24.8.3: Circuit diagram of an ABF.

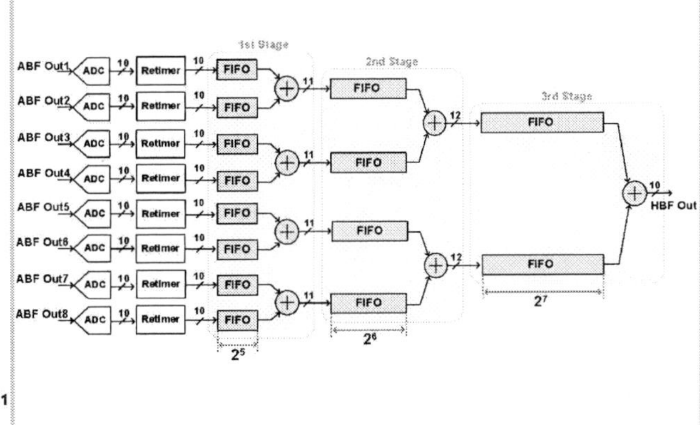

Figure 24.8.4: Block diagram of a DBF.

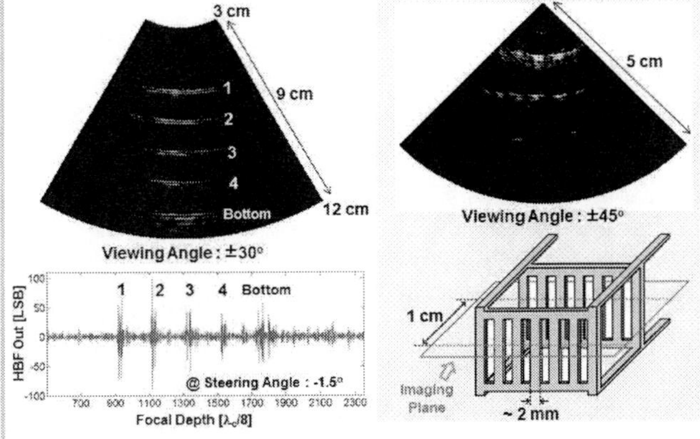

Figure 24.8.5: B-mode ultrasound images of four parallel plates (left) and two punched plates (right).

	This Work	TBCAS'12 [2]	CICC'03 [3]	JSSC'00 [4]	ASSCC'12 [5]
Beamforming Domain	Hybrid (Analog + Digital)	Analog	Analog	Analog	Analog
Delay Cell Type	S/H Circuit + FIFO	Analog Filter	Analog Filter	S/H Circuit	S/H Circuit
Number of Channels	64	8	5	16	32
Max. Delay	8 μs	35 ns	48 ns	1.094 μs	1.92 μs
Sampling Resolution	6.25 ns	1.75 ~ 2.5 ns	0.86 ns	31.25 ns	10 ns
Delay Dynamic Range (Max. Delay / Sampling Resolution)	1280	14	55.8	35	192
Power Consumption	1.14 W	67 mW	N.A.	1.12 W	150 mW
Active Area	19.4 mm^2	0.36 mm^2	0.5 mm^2	72 mm^2	16.4 mm^2
Process (CMOS)	0.13 μm	0.35 μm	0.35 μm	0.8 μm	0.13 μm
Target Transducer Array	2-D CMUT Array	Annular CMUT Array	Annular Array	2-D Array	Linear Array

Figure 24.8.6: Performance comparison of single-chip RX beamformers.

24

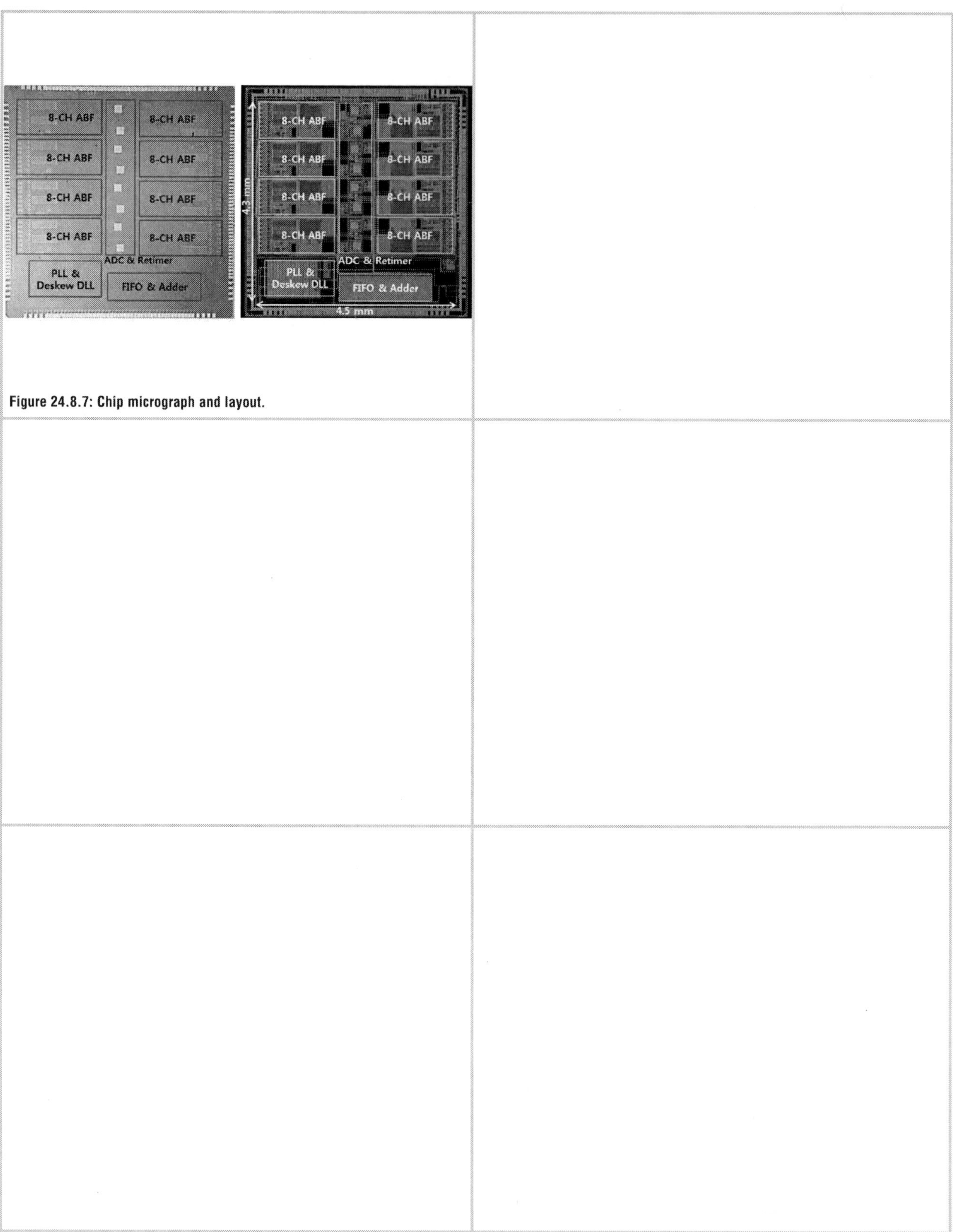

Figure 24.8.7: Chip micrograph and layout.

Session 25 Overview:
High-Bandwidth Low-Power DRAM and I/O
MEMORY SUBCOMMITTEE

Session Chair: *Uksong Kang*
Samsung Electronics,
Hwasung, Korea

Session Co-Chair: *James Sung*
Etron, Hsinchu, Taiwan

Requirements for high bandwidth and low power continue to increase in servers and consumer electronics. There are significant challenges in DRAMs to meet all such needs in various applications. In ISSCC 2014, the first LPDDR4 DRAM for mobile applications is demonstrated which has an integrated ECC engine for low-power operation. Next, the first High-Bandwidth Memory (HBM) with 4 TSV stacked layers achieving 128GB/s bandwidth is disclosed. Also, new circuits to reduce standby and I/O power in GDDR5M are shown. The papers in this session present the latest technologies and circuit techniques to improve the performance and power in DRAMs.

978-1-4799-0917-9/14 $31.00 © 2014 IEEE

25.1 A 3.2Gb/s/pin 8Gb 1.0V LPDDR4 SDRAM with Integrated ECC Engine for Sub-1V DRAM Core Operation 1:30 PM

T-Y. Oh, Samsung Electronics, Hwasung, Korea

In Paper 25.1, Samsung demonstrates LPDDR4 SDRAM for the first time. It has 8Gb density and is implemented in 25nm DRAM process technology. It achieves a data rate of 3.2Gb/s/pin at 1.0V supply voltage and has an integrated ECC engine for sub-1V operation.

25.2 A 1.2V 8Gb 8-Channel 128GB/s High-Bandwidth Memory (HBM) Stacked DRAM with Effective Microbump I/O Test Methods Using 29nm Process and TSV 2:00 PM

D. U. Lee, SK Hynix, Icheon, Korea

In Paper 25.2, SK Hynix discloses High-Bandwidth Memory (HBM) for the first time. It is implemented in 29nm DRAM process technology. It consists of 4 TSV stacked layers with total density of 8Gb. It shows high bandwidth of 128GB/s at 1.2V supply voltage with 8 channels and total number 1024 I/Os.

25.3 A 1.35V 5.0Gb/s/pin GDDR5M with 5.4mW Standby Power and an Error-Adaptive Duty-Cycle Corrector 2:30 PM

J. Kim, SK Hynix, Icheon, Korea

In Paper 25.3, SK Hynix presents a 1.35V 5Gb/s/pin GDDR5M using 29nm DRAM process technology. It achieves low standby power of only 5.4mW using WCK auto-sync mode. Additional circuit techniques including clock timing skewing and error-adaptive duty-cycle corrector are applied to improve signal integrity.

978-1-4799-0917-9/14 $31.00 © 2014 IEEE

25.1 A 3.2Gb/s/pin 8Gb 1.0V LPDDR4 SDRAM with Integrated ECC Engine for Sub-1V DRAM Core Operation

Tae-Young Oh, Hoeju Chung, Young-Chul Cho, Jang-Woo Ryu, Kiwon Lee, Changyoung Lee, Jin-Il Lee, Hyoung-Joo Kim, Min Soo Jang, Gong-Heum Han, Kihan Kim, Daesik Moon, Seungjun Bae, Joon-Young Park, Kyung-Soo Ha, Jaewoong Lee, Su-Yeon Doo, Jung-Bum Shin, Chang-Ho Shin, Kiseok Oh, Doohee Hwang, Taeseong Jang, Chulsung Park, Kwangil Park, Jung-Bae Lee, Joo Sun Choi

Samsung Electronics, Hwasung, Korea

The recent revolution in handheld computing with high-speed cellular network made mobile processors have multi-cores and powerful 3D graphic engines that support FHD (1920×1080) or even higher resolutions. Consequently, the memory bandwidth requirement has also been increasing, requiring a next-generation mobile DRAM standard. In this paper, we present a power-efficient LPDDR4 SDRAM operating at 3.2Gb/s/pin. Our LPDDR4 DRAM offers 2× bandwidth with improved power efficiency over LPDDR3 SDRAM's, due to the 2-channel architecture and low-voltage-swing terminated logic (LVSTL) [1]. Moreover, the supply voltage is further reduced to 1.0V in this work, 0.1V lower than the LPDDR4 standard, for extra power saving.

Another major improvement in LPDDR4 is that it is the first commodity DRAM standard defined with consideration of integrated error-correction coding (ECC) support. It is well known that ECC in DRAM improves yield [2,3] and data-retention time [3]. We use ECC because the DRAM core has to operate with 0.9V on-chip regulated power, generated from 1.0V external supply. At 0.9V, the mismatch in the bitline sense amplifiers and insufficient charge in the cell capacitors increase the number of fail bits, making a DRAM unrepairable by redundancy only. Pre-LPDDR4 commodity DRAMs are not able to integrate an ECC engine due to data-masked write (DM) operation. In a DM operation, ECC encoding is not possible without knowing the previously stored data at target address, because the host provides only partial write data with 8b granularity. Therefore, the DM write operation is implemented by the DRAM controller using a read-modify-write sequence in conventional systems with ECC, rather than by a DRAM-integrated ECC engine. LPDDR4 overcomes this limitation by introducing a command named MWR for data-masked write to distinguish it from normal write. When an MWR command is received, the LPDDR4 DRAM generates hidden internal DRAM read-during-write latency (WL) and combines the read data with non-masked write data to generate new ECC data.

Figure 25.1.1 shows the top-level block diagram and data-path structure of this work.The LPDDR4 SDRAM consists of two independent 16b DQ channels. The SDRAM has a 16*n* prefetch structure, doubled from LPDDR3, maintaining same DRAM core speed. The data-access granularity, 32-byte, is the same, because the DQ bus width of each channel is reduced to half (×16 DQ). Splitting a DRAM die into 2 channels also reduces DRAM die size and current by removing die end-to-end running data and clock paths. About 15% of current consumption saving is achieved by this 2-channel architecture. There are 16 physical banks in one channel. A0~H0 banks are for DQ byte 0, and A1~H1 banks are for DQ byte 1. The DQ I/O performs 8:1 de-serialization/serialization, resulting in a 64b datapath. This datapath is implemented as skewed logic [4] to minimize die size and current consumption.

The ECC encoding/decoding engine is placed between I/O sense amplifiers in each physical bank as shown in Fig. 25.1.2. It uses a (136,128) shortened hamming code with single-error-correction capability. It also performs 2:1 de-serialization/serialization to complete 16*n* prefetch core structure. Parity data and syndrome generation are performed by one unit controlled by PWR_BANK signal. The DRAM cells for parity data are located at the center of each bank to minimize the ECC engine layout size and delay. The required extra delay is 1.45ns for ECC parity generation in write operation, and 1.56ns for ECC decoding in read operation including syndrome decoding for error correction.

Figure 25.1.3 compares the MWR timing with normal write timing. In MWR operation, DRAM internal read command is generated 16 clocks before write, to prepare old data to combine with non-masked part of new write data. After the ECC engine completes error correction, it starts parity-bit generation 5 clocks before write, controlled by PWR_BANK signal. The write timings to the DRAM cell are exactly the same for normal write and MWR. The internal read operation is executed only inside of a target bank, allowing overlap of write or MWR operation in other banks. By this way, the additional timing restrictions caused by MWR command are minimized. In case of small write latency (WL ≤ 8) setting, the number of clocks used for MWR operation is reduced to 8. However, the absolute time for internal read-modify-write is still guaranteed because the input clock frequency is also low. Figure 25.1.3 shows the timing constraints between various commands to the same bank and different banks. There are two additional restrictions occurring from MWR: minimum delay from write or MWR to MWR in same bank is relaxed to $4t_{CCD}$. For different banks, the timing constraint is exactly same as that of conventional DRAM standard.

As power-supply voltage is reduced and input clock frequency is doubled from LPDDR3, the timing deviation created by PVT variation makes it difficult to design circuits that handle 1 clock-width pulses. To solve this problem, time interleaved latency control is introduced. Figure 25.1.4 shows the simplified block diagram of latency control. In each channel, there are two sets of command decoder (CMD_DEC), write/read latency and I/O control units, and each set is tied to even clock (CLK_E) or odd clock (CLK_O) only. As one control circuit set is driven by one divided clock, the timing margin is doubled. The die-size overhead caused by doubling these circuits is negligible because only command decoder and latency-related units are doubled, and they are shared between DQ I/Os. To reduce current consumption, each divided clock is enabled only when there is a command at the corresponding clock edge and disabled when command processing is completed. Considering that the t_{CCD} of LPDDR4 is $8t_{CK}$, even number, only one of the CLK_E and CLK_O is enabled and toggles during gapless write or read operation. The timing diagram in Fig. 25.1.4 illustrates internal divided clock control of this device.

There are three unique points in the LPDDR4 ZQ calibration. First, the pull-up driver strength is calibrated to set V_{OH} level, not the impedance. Unlike conventional DRAM I/Os, the LVSTL I/O pull-up is driven by an NMOS transistor, which cannot be in deep linear region. Due to the nonlinear pull-up I/O characteristic, the V_{OH} level is vulnerable to PVT variation, requiring V_{OH}-level calibration during ZQ calibration sequence. The V_{OH} level can be $V_{DDQ}/3$ or $V_{DDQ}/2.5$ depending on mode register setting. Second, the ZQ calibration is performed in the background, initiated by ZQ CAL start command, to allow frequent periodic calibration during normal DRAM operation. The calibration result is applied to DRAM I/Os only after ZQ CAL latch command is issued by the DRAM controller. In this way, the I/O downtime for ZQ calibration sequence is reduced from 1μs (LPDDR3) to 30ns. Third, only channel A has ZQ pin and calibration circuit to minimize cost, although the ZQ calibration command from channel B is also valid to initiate ZQ calibration sequence. The DRAM termination and pull-down driver strength can be programmed independently in channel B from channel A as well. Figure 25.1.5 shows our ZQ calibration circuit that satisfies above three requirements. Using two replica pull-down drivers and three 480Ω drivers for $V_{DDQ}/3$ and $V_{DDQ}/2.5$ V_{OH} level, respectively, in IO_ZQ_CAL circuit, the pull-down drivers are calibrated to set PD_EN[5:0]. Then PU_EN[5:0] are calibrated to make the replica output voltage same as reference V_{OH} level. A DQ output driver has 6 pull-up and pull-down driver sets, which are selectively enabled to support 6 different termination strengths.

Figure 25.1.6 is the chip summary and simulated I/O eye diagram in dual-die package (DDP) channel environment. The eye diagram shows 0.73UI eye opening at 3.2Gb/s. Figure 25.1.7 shows the micrograph of the DRAM die, which is designed in a 25nm 3-metal DRAM process.

References:

[1] Young-Chul Cho, Yong-Cheol Bae *et al.*, "A Sub-1.0V 20nm 5Gb/s/pin post-LPDDR3 I/O interface with Low Voltage-Swing Terminated Logic and adaptive calibration scheme for mobile application," *Symposium on VLSI Circuits, 2013*, pp. 240-241.

[2] C.H. Stapper and H.-S. Lee, "Synergistic fault-tolerance for memory chips," *IEEE Trans. Comput.*, vol.41, pp. 1078-1087, Sep. 1992.

[3] Saeng-Hwan Kim, Won-Oh Lee *et al.*, "A low power and highly reliable 400Mbps mobile DDR SDRAM with on-chip distributed ECC," *IEEE Asian Solid-State Circuits Conference, 2007*, pp 34-37.

[4] Tae-Young Oh, Young-Soo Sohn *et al.*, "A 7 Gb/s/pin 1 Gbit GDDR5 SDRAM With 2.5 ns Bank to Bank Active Time and No Bank Group Restriction," *IEEE J. Solid-State Circuits*, vol.46, pp.107-118, Jan. 2011.

978-1-4799-0917-9/14 $31.00 © 2014 IEEE

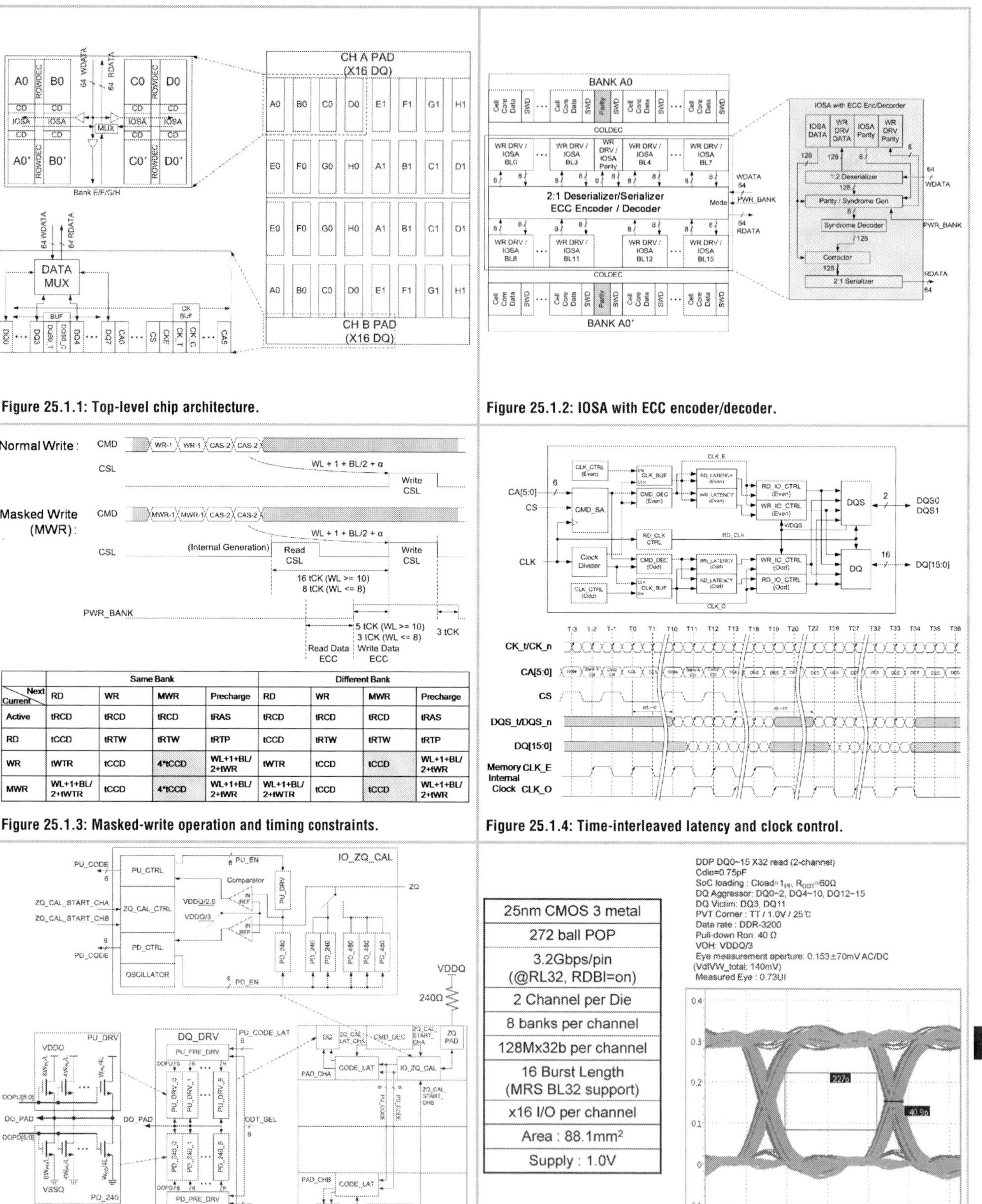

Figure 25.1.1: Top-level chip architecture.

Figure 25.1.2: IOSA with ECC encoder/decoder.

	Same Bank				Different Bank			
Current \ Next	RD	WR	MWR	Precharge	RD	WR	MWR	Precharge
Active	tRCD	tRCD	tRCD	tRAS	tRCD	tRCD	tRCD	tRAS
RD	tCCD	tRTW	tRTW	tRTP	tCCD	tRTW	tRTW	tRTP
WR	tWTR	tCCD	4*tCCD	WL+1+BL/2+tWR	tWTR	tCCD	tCCD	WL+1+BL/2+tWR
MWR	WL+1+BL/2+tWTR	tCCD	4*tCCD	WL+1+BL/2+tWR	WL+1+BL/2+tWTR	tCCD	tCCD	WL+1+BL/2+tWR

Figure 25.1.3: Masked-write operation and timing constraints.

Figure 25.1.4: Time-interleaved latency and clock control.

Figure 25.1.5: ZQ calibration.

Figure 25.1.6: Chip summary and eye diagram.

978-1-4799-0917-9/14 $31.00 © 2014 IEEE

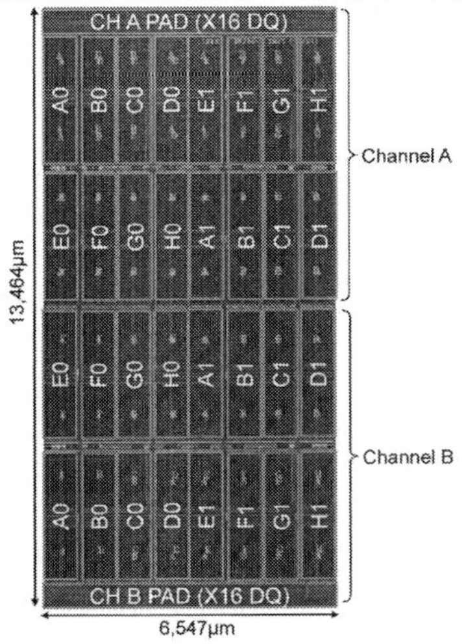

Figure 25.1.7: Die micrograph.

25.2 A 1.2V 8Gb 8-Channel 128GB/s High-Bandwidth Memory (HBM) Stacked DRAM with Effective Microbump I/O Test Methods Using 29nm Process and TSV

Dong Uk Lee, Kyung Whan Kim, Kwan Weon Kim, Hongjung Kim, Ju Young Kim, Young Jun Park, Jae Hwan Kim, Dae Suk Kim, Heat Bit Park, Jin Wook Shin, Jang Hwan Cho, Ki Hun Kwon, Min Jeong Kim, Jaejin Lee, Kun Woo Park, Byongtae Chung, Sungjoo Hong

SK Hynix, Icheon, Korea

Increasing demand for higher-bandwidth DRAM drive TSV technology development. With the capacity of fine-pitch wide I/O [1], DRAM can be directly integrated on the interposer or host chip and communicate with the memory controller. However, there are many limitations, such as reliability and testability, in developing the technology. It is advantageous to adopt a logic-interface chip between the interposer and stacked-DRAM with thousands of TSV. The logic interface chip in the base level of high-bandwidth memory (HBM) decreases the C_{IO}, repairs the chip-to-chip connection failure, and supports better testability and improves reliability.

The fundamental structure of HBM is composed of 4-Hi core DRAM and base logic die at the bottom, as depicted in Fig. 25.2.1(a). The core DRAM consists of 2 channels, where each channel has 1Gb density with 128 I/Os and 8 independent banks. Core DRAM has 2*n*-prefetch with minimum access granularity of 32 bytes [2]. Each channel of core DRAM die has independent address and data TSV with point-to-point (P2P) connection, isolating the operation of every channel. The power and ground of each channel are not isolated and has common plane. The layout in Figure 25.2.1(b) illustrates base logic die architecture. PHY located at the top of the die is for the interface between DRAM and memory controller. PHY has total of 8 channels where a channel consists of an AWORD (address buffer) and four channel-interleaved DWORD (data buffer). The center area is allocated for TSV that deliver signals and power to 4-Hi core DRAM. Similar to PHY, TSV has DDR address and data interface between base and core DRAM. The area between PHY and TSV is filled with signal lines for 1024 I/Os and decoupling capacitance. Some microbumps are depopulated for probe test at chip-on-wafer level. A direct-access (DA) port at the bottom is interface between the automatic test equipment (ATE) and stacked-DRAM. DA port is mainly for testing stacked-DRAM component circuits including cells.

HBM uses semi-independent row- and column-command interfaces, which allow RAS and CAS commands in parallel. A DDR command interface is adopted as well, for reducing pin count. The simplified row- and column-address decoder is shown in Fig. 25.2.2. R[0:5] pins for row addresses and C[0:7] for column addresses aligned by clock can be entered at the same time. In row active command case, there are 4b decoding signals, which are aligned to 4 phase pulse, and other row command (such as precharge, refresh) need 2b decoding signals aligned to 2 phase signal. The column-command operation uses a similar method to the 2b row command. Another function added in HBM is the single-bank refresh. In previous DRAM, the bandwidth efficiency is reduced by all-bank refresh period because DRAM cells cannot be accessed and the bus must idle. In the single-bank refresh function, the refresh cycle can be controlled per unit bank since every bank has a refresh counter and digital t_{RAS} (row active to precharge time) counter. Therefore, any bank that is not in refresh is accessible. HBM uses an unterminated interface, in which bus-switching power is dominant rather than dc power. To reduce this bus switching power, data-bus inversion (DBI-AC) is used.

The primary purpose of base die is the interface with controller and core DRAM. HBM communicates with memory controller through the interposer, which has 1024 I/Os. The interposer has large R and C, causing high power consumption and ISI. The usage of base die reduces C_{IO} significantly and the reduced distance of PHY between DRAM and memory controller decreases the total interface power. The PHY interface circuit is shown in Fig. 25.2.3. Differential clocks (CK_t, CK_c) and 15 DDR addresses are used in AWORD. Per-channel address-parity calculation result is returned to the controller via AERR pin for diagnostic purposes. Differential strobe pairs (WDQS_t, WDQS_c for write, RDQS_t, RDQS_c for read) are used for a DWORD. There is no skew-compensation circuit, such as DLL, between clock and data. Uncompensated clock strobe aligns data for read output, which saves standby power. There is also a per-DWORD data-parity function for both write and read. The calculated result is returned to the controller via DERR pin. The secondary purpose of base die is to support various test functions. External loopback function between memory and controller is designed for I/O link test and training after assembly. There are four states of loopback register function: resetting register, LFSR mode, register mode and MISR mode [3]. The figure shows a 4b polynomial composed of four flip-flops, XOR gates and feedback signal. In DWORD, there is 20b polynomial ($f(x)=X^{20}+X^{17}+1$) and there is a 30b polynomial ($f(x)=X^{30}+X^{6}+X^{4}+X+1$) in AWORD. HBM DRAM test feature also supports IEEE 1500 protocol interface, which is located in MIDSTACK in PHY region. Using IEEE 1500 protocol, DRAM can support basic connection test functions like boundary scan, chip reset, and internal VREF control, and also complex functions such as loopback, hard repair of DRAM cell after assembly, TSV scan, and memory built-in-self-test.

One of the issues in testing stacked chip is self-verification of the microbump interface. The microbump array has a fine pitch with wide I/O making it difficult to measure the full I/O characteristic directly. Impedance calibration is generally used to reduce PVT variation and this chip has internal resistor calibration, modified by register or fuse. However, it is difficult to measure the impedance of each microbump driver using the conventional method. An impedance-monitoring method is developed for measuring each microbump output drivability and integrity. Each microbump has a boundary scan cell to test and identify pin-to-pin connectivity after assembly. By monitoring enable signal and boundary scan function, one pullup/pulldown driver (of 1408) is enabled. By comparing the average difference of V_{DDQ} current between the case of all drivers off versus each driver on, every driver's drivability and integrity can be accurately measured. Test results of 1408 pins are shown in Fig 25.2.4. Using boundary scan cell and receiver, referenced to a programmable V_{REF}, the V_M level (mid level of pullup and pulldown) is also detected and stored in registers.

There are two ways to test DRAM write/read using DA port. The test circuit and brief flow chart are depicted in Fig. 25.2.5. The first way is pad-loopback, which is parallel write and read operation via microbump using DA port. In DA write case, DA addresses are transmitted through test driver. Simultaneously, DA write data are transmitted through TX and test register. Data are driven onto microbump, which is the same input condition as after assembly. The following write and read operations to DRAM cell are similar to conventional DRAM. In DA read case, core cell data is driven to read register, TX and microbump. To judge data pass/fail in DA mode, output data of read register is transmitted to comparator at the same time. The comparator detects and sends to DA the difference between the written data and read-out data. The second way to test DRAM is using both DA port and serial port. Using serial port WSI, various data pattern (different per DQ) are initialized. Writing these data to DRAM cell is through TX, RX and write register. Read data from DRAM cell goes through read register, TX and RX. The major usage of this method is to check the interface, especially the simultaneous operation of read and TX. Loopback MISR register value can be calculated after multiple read operations. Register value can be transmitted through serial output port WSO, and compared with the expected data. The test result of stacked-DRAM in Fig. 25.2.6 shows 8-channel read operation with 128GB/s at 1.2V and 111GB/s at 1.05V.

There are many design and test challenges in developing stacked-DRAM with microbump interface. System performance is higher using semi-independent row- and column-command interface and single-bank refresh. Parity, loopback and various IEEE 1500 functions support testing between memory and controller. With our methods such as pad-loopback test using DA and impedance-monitoring method, many problems of stacked-DRAM can be observed and distinguished. Four stacked 2Gb DRAM and base die are implemented using 29nm DRAM process. The chip micrograph of core die including TSV structure is shown in Fig. 25.2.7.

References:

[1] J. Kim, C. S. Oh, et al., "A 1.2V 12.8GB/s 2Gb Mobile Wide-I/O DRAM with 4×128 I/Os Using TSV-Based Stacking," *ISSCC Dig. Tech. Papers*, pp. 496-497, Feb. 2011.

[2] JEDEC Standard High Bandwidth Memory(HBM) DRAM Specification, 2013.

[3] B. Keller, T. Bartenstein, "Use of MISRs for Compression and Diagnostics," *Proc. International Test Conference*, pp. 735-743, 2005.

978-1-4799-0917-9/14 $31.00 © 2014 IEEE

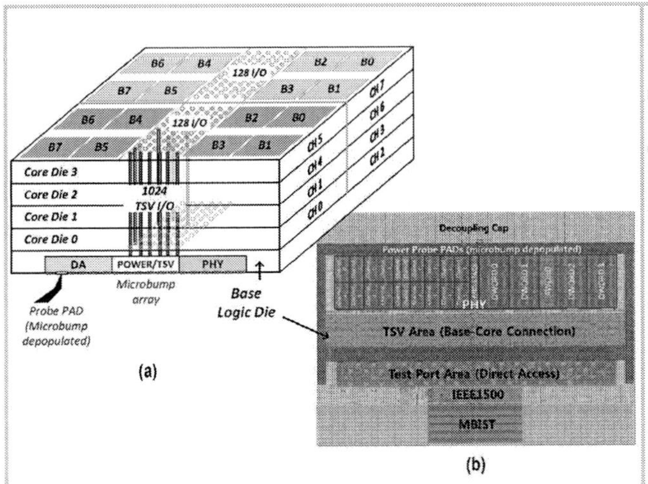

Figure 25.2.1: (a) HBM stacked-DRAM architecture, and (b) layout and functionality of base logic die.

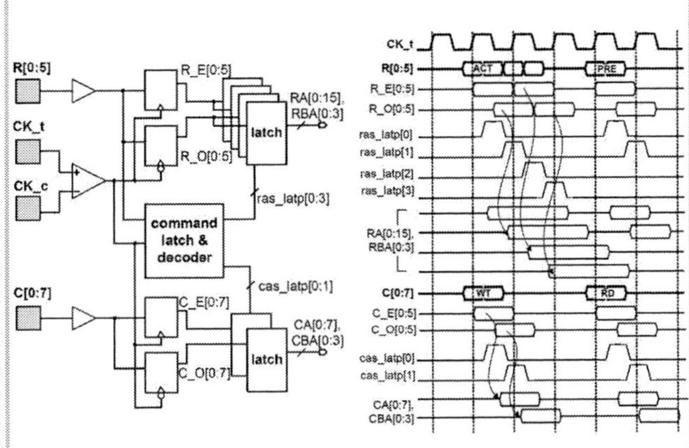

Figure 25.2.2: Semi-independent row/column command interface circuit and timing.

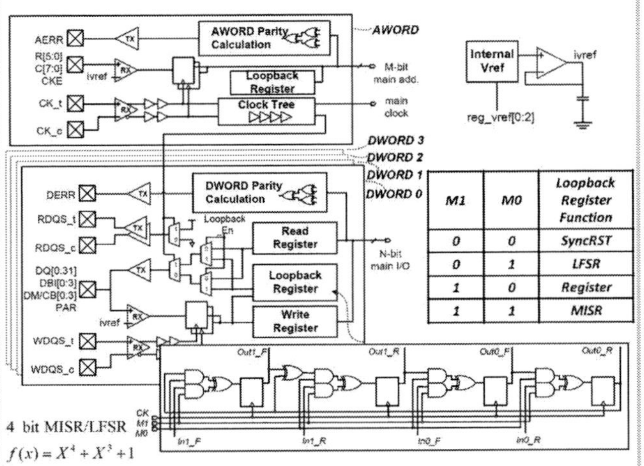

M1	M0	Loopback Register Function
0	0	SyncRST
0	1	LFSR
1	0	Register
1	1	MISR

Figure 25.2.3: Clocking scheme of HBM including parity and external loopback circuit.

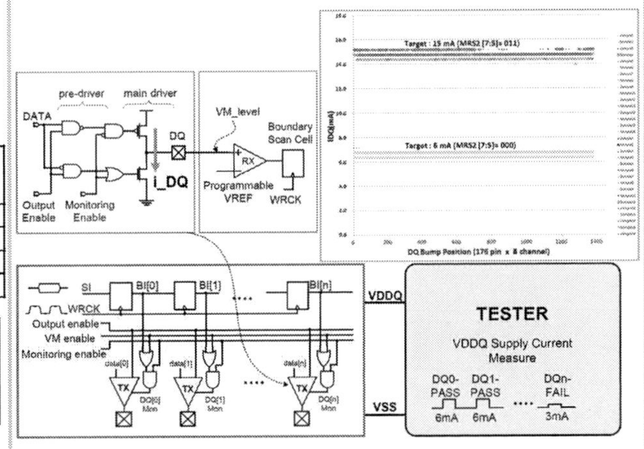

Figure 25.2.4: Microbump impedance-monitoring method and test results.

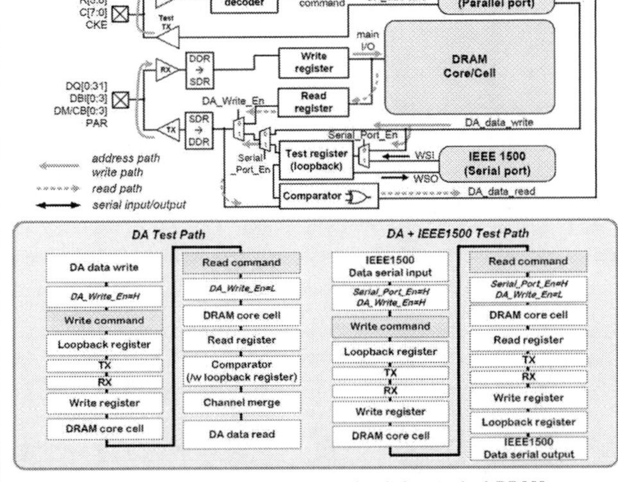

Figure 25.2.5: DA port and serial port test circuit for stacked-DRAM.

```
#          1.90NS____2.90NS____3.90NS____4.90NS_
#          v_________v_________v_________v_____
#          +----+----+----+----+----+----+----+-
1.500V     PPPPPPPPPPPPPPPPPPPPPPPPPPPPPPPPPPPPP
1.450V     PPPPPPPPPPPPPPPPPPPPPPPPPPPPPPPPPPPPP
1.400V     PPPPPPPPPPPPPPPPPPPPPPPPPPPPPPPPPPPPP
1.350V     PPPPPPPPPPPPPPPPPPPPPPPPPPPPPPPPPPPPP
1.300V     PPPPPPPPPPPPPPPPPPPPPPPPPPPPPPPPPPPPP
1.250V     PPPPPPPPP  1.2v 128GB/s  PPPPPPPPPP
1.200V     PPPPPPPPPPPPPPPPPPPPPPPPPPPPPPPPPPPPP
1.150V     ..PPPPPPPPPPPPPPPPPPPPPPPPPPPPPPPPPPP
1.100V     ..PPPPPPPPPP  1.05v 111GB/s  PPPP
1.050V     ....PPPPPPPPPPPPPPPPPPPPPPPPPPPPPPPPP
999.998M   .................PPPPPPPPPPPPPPPPPPPP
```

Process	29nm DRAM process
Chip Size	5.10mm x 6.91mm
Organization	8 bank x 8 channel x 128 I/O (total 1024 I/O)
Density	1Gb / channel
Microbump pitch (base die)	48 µm x 55 µm
Supply Voltage	VDD=1.2v, VPP=2.5v
Refresh	8k / 32ms
Page Size	2KB
Data Rate	1.0 Gbps (128GB/s)
C_{IO}	0.4pF

Figure 25.2.6: Stacked-DRAM test results (shmoo plot, 8-channel read operation, 25°C).

25

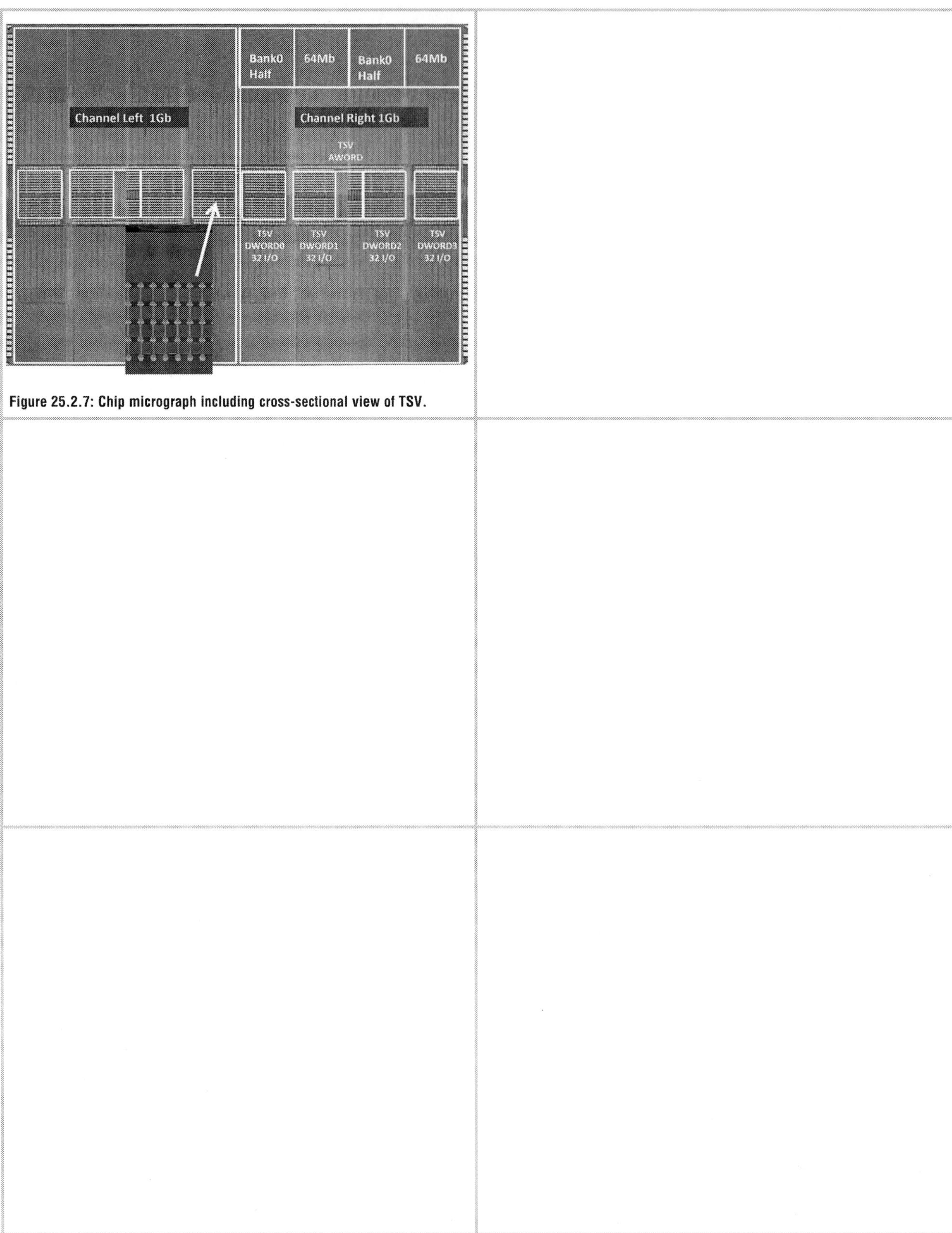

Figure 25.2.7: Chip micrograph including cross-sectional view of TSV.

25.3 A 1.35V 5.0Gb/s/pin GDDR5M with 5.4mW Standby Power and an Error-Adaptive Duty-Cycle Corrector

Hyun-Woo Lee[1,2], Junyoung Song[2], Sang-Ah Hyun[1], Seunggeun Baek[1], Yuri Lim[1], Jungwan Lee[1], Minsu Park[1], Haerang Choi[1], Changkyu Choi[1], Jinyoup Cha[1], Jaeil Kim[1], Hoon Choi[1], Seungwook Kwack[1], Yonggu Kang[1], Jongsam Kim[1], Junghoon Park[1], Jonghwan Kim[1], Jinhee Cho[1], Chulwoo Kim[2], Yunsaing Kim[1], Jaejin Lee[1], Byongtae Chung[1], Sungjoo Hong[1]

[1]SK Hynix, Icheon, Korea, [2]Korea University, Seoul, Korea

The demand for high-bandwidth memories is increasing with an increase in the need for high-performance systems. Wide-I/O memory and GDDR5 are two types of high-bandwidth memories. GDDR5 is more compatible than wide I/O for contemporary systems, such as graphics cards and game consoles. The data-rate of GDDR5 has reached 7Gb/s/pin. However, the power consumption and cost have increased owing to high-performance-oriented designs, die penalties, and additional test costs. DDR4 is an alternative low-cost memory with high performance in the range of 2.4 to 3.2Gb/s/pin [1]. However it is difficult for DDR4 DRAM to raise the 3.2Gb/s/pin bin portion to lower the cost. In DRAMs, the standby power and self-refresh power are more important than the operating power because DRAMs are mainly in the standby or self-refresh mode in systems. As the operating speed increases, the data window is narrowed, and the jitter increases. Therefore, a duty-cycle corrector (DCC) is employed to increase the data window when the external clock duty cycle is distorted in GDDR5 [2]. The bang-bang jitter caused by the DCC is inevitable even if the external clock duty ratio is exactly 50%. Sometimes the DCC may distort the data window because of an internal DCC offset. This paper presents a GDDR5M (mainstream) memory for graphics cards and a small-outline dual-inline memory module) (SO-DIMM). The standby power is managed by the auto-sync mode. Additionally, the architecture of GDDR5M is similar to that of DDR4, and not GDDR5. The error-adaptive DCC can remove the initial duty-cycle offset automatically and remove the bang-bang jitter when the duty cycle of the external clock is not distorted.

The simplified block diagram of a GDDR5M clocking system is depicted in Fig. 25.3.1. The command is sampled on the rising edge of CK, and the address is sampled on both edges of CK. WCK is used for the WRITE and READ operation. The WCK rate is twice the rate of CK. WCK is divided into 4-phase clocks and is connected to pipe registers for input and output. With a 4-phase IWCK<0,90,180,270>, DQ<0:15> (×16 mode) are sampled to receive data and aligned for transferring data. On the basis of the MF pin, the phase of WCK is determined according to the WCK and /WCK connectivity. The phases of a 4-phase clock (IWCK<0,90,180,270>) are determined by the alignment between WCK and CK during the WCK2CK training sequence. To maintain the alignment between WCK and CK, both must always be available, which makes dynamic power consumption unavoidable. GDDR5M does not have a DLL or a PLL. Therefore, it is easy to enter into and exit from the power down mode because it is not necessary for GDDR5M to maintain the locked condition of a DLL or PLL, unlike in the case of DDR3 or DDR4 [3]. GDDR5 having a PLL is affected by supply noise and requires additional circuit techniques [4].

Clock gating is a crucial technique to save dynamic power consumption [5]. As the operating frequency increases, the power dissipation of the clock distribution network (CDN) contributes 47% of the memory power under IDD2N (pre-charge standby mode) for the GDDR5M case. Moreover, GDDR5M has dual-clock networks, which consist of CK and WCK trees. WCK is divided at the divider inside of the DRAM and is split into 4-phase IWCK<0,90,180,270>. Because of the divide operation on WCK, the phase relationship between CK and WCK is important. From the training sequence that is processed at the initialization step, the phase relationship between CK and WCK should be aligned. Otherwise, to save the dynamic power of CDN, WCK should be turned off in the power-down mode. Because GDDR5M does not have a DLL, GDDR5M is free from the additional latency caused by a DLL. The block diagram of the clock tree is depicted in Fig. 25.3.2(a). GDDR5M has 16-DQs, 2-DBIs, and 2 EDCs. CK is divided by 2 at a divider block. By comparing the phase relationship between CK_DVD and IWCK_90, the auto-sync function is processed. To remove the simultaneous switching noise (SSN), a timing skew (Δt) is added to the lower DQs. Therefore, the data output timings of the upper DQs are slightly different from those of the lower DQs. Through the training sequence, the de-skew timing will be added and the output timing is calculated by the GPU. In this manner, the t_{DV} is slightly improved without complex SSO cancellation circuitry. To evaluate the effect of a skewed clock tree, we add additional switch logic to the WCK clock distribution network. As shown in Fig. 25.3.2(b), in the power-down mode when CKE transits to LOW, WCK will be stopped and may lose the phase relationship between CK and WCK. However, the phase relationship is recovered with the auto-sync mode. If the phase is aligned between them, the phase information that is sampled by both edges of CK_DVD divided from CK is always LOW. Then WCK_DVD_REVERSE stays LOW. When CKE transits to HIGH and clock uncertainty occurs, the phase relationship will be easily broken. If the phase information is always HIGH, then WCK_DVD_REVERSE goes to HIGH and the phase relationship is recovered by the auto-sync operation.

The block diagram of the error-adaptive duty-cycle corrector (EA-DCC) is illustrated in Fig. 25.3.3. EA-DCC has a successive-approximation register (SAR) control scheme before DCC lock and a linear correction scheme after DCC lock. When EA-DCC is turned on by the MRS command, it starts the duty-cycle correction. An asynchronous binary-search circuit finds each digital code in the interval from IWCK0 to IWCK90 and from IWCK90 to IWCK180, respectively. Latch1 and Latch2 store the digital codes at each step. By subtraction, the timing difference between codes is transferred to FA2. Four consecutive RESULTs are averaged. If the error code is greater than 2, then the error is reflected in the final DCC CODE<0:N>; otherwise, the error correction is rejected. Finally, if the error code is less than 2, it is inferred that duty-cylcle distortion has not occurred. In the case that the external duty-cycle is perfect, the duty-cyle distortion due to the offset of the asynchronous binary search circuit due to process variations is not generated. For a distorted external clock, when DCC is in the lock condition, the LSB is rejected, and only the sign bit is passed to FA3. The sign bit indicates that the duty-correcting direction should be wide or narrow, thus DCC operates in linear-correction mode after DCC locking.

Figure 25.3.4 shows the IDD measurement results of GDDR5 and GDDR5M fabricated with the same process technology. IDD2P of GDDR5 is 3.4× greater than that of GDDR5M because of dual WCK trees as part of the ×32 configuration and high-speed-oriented design schemes such as high-bandwidth buffers, a higher sink current of CML, and frequent usage of low-threshold-voltage transistors. Because of the WCK stop, IDD2P0 of GDDRM is only 15% of IDD2P of GDDR5M. GDDR5M has an IDD2P0 mode. The energy consumptions of WRITE and READ are 84pJ/b and 82pJ/b, respectively. The energy consumption of GDDR5M for write and read functions is <80% that of GDDR5. Figure 25.3.5(a) shows the performance measurement results. GDDR5M is tested with ATE according to the operating frequency at 1.35V. The operating speed is 5Gb/s/pin at 1.35V, and the t_{AC} window is 70ps. Our clocking scheme with additional timing skew has a wider t_{DV} than the previous scheme, as shown in Fig. 25.3.5(b). It is measured at 1.4V, and the shmoo plot is the sum of 16 DQs. The oscilloscope plots of the DQ6 and DCC outputs are shown in Fig. 25.3.6. With a distorted clock, EA-DCC corrects the duty-cycle error. It is measured under a WCK of 2.1GHz in a graphics-card application. The micropraph of the chip, fabricated with a 29nm CMOS DRAM process, is shown in Fig. 25.3.7. The active area of the chip is 51.2mm^2.

Acknowledgement:
The authors would like to thank Sugwan Jeong and Sungki Choi for their helpful supports and discussions.

References:
[1] K. Koo, *et al.*, "A 1.2V 38nm 2.4Gb/s/pin 2Gb DDR4 SDRAM with bank group and ×4 half-page architecture," in *ISSCC Dig. Tech. Papers*, pp. 40-41, Feb. 2012.
[2] D. Shin, *et al.*, "Wide-range fast-lock duty-cycle corrector with offset-tolerant duty-cycle detection scheme for 54nm 7Gb/s GDDR5 DRAM interface," in *Symposium on VLSI Circuits*, pp. 138-139, Feb. 2009.
[3] H.-W. Lee, *et al.*, "A 1.0-ns/1.0-V delay-locked loop with racing mode and countered CAS latency controller for DRAM Interfaces," *IEEE J. Solid-State Circuits*, vol. 47, no. 6, pp. 1436-1447, Jun. 2012.
[4] J. Song, *et al.*, "An adaptive-bandwidth PLL for avoiding noise interference and DFE-less fast precharge sampling for over 10Gb/s/pin graphics DRAM interface," in *ISSCC Dig. Tech. Papers*, pp. 312-313, Feb. 2013.
[5] H.-W. Lee, *et al.*, "A 1.6 V 1.4 Gb/s/pin consumer DRAM with self-dynamic voltage scaling technique in 44 nm CMOS technology," in *ISSCC Dig. Tech. Papers*, pp. 131-140, Feb. 2012.

978-1-4799-0917-9/14 $31.00 © 2014 IEEE

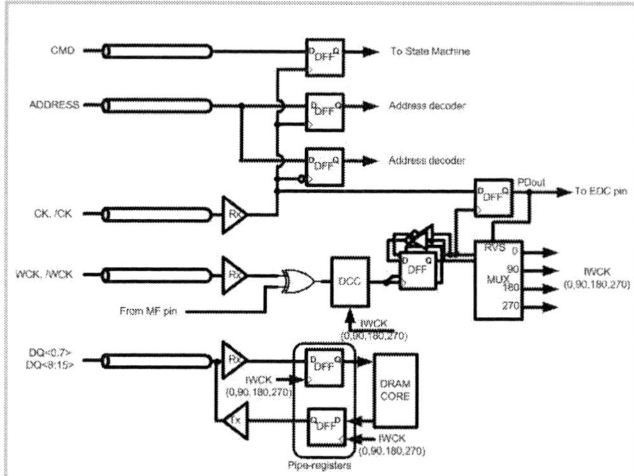

Figure 25.3.1: Simplified block diagram of GDDR5M clocking system with WCK MF function.

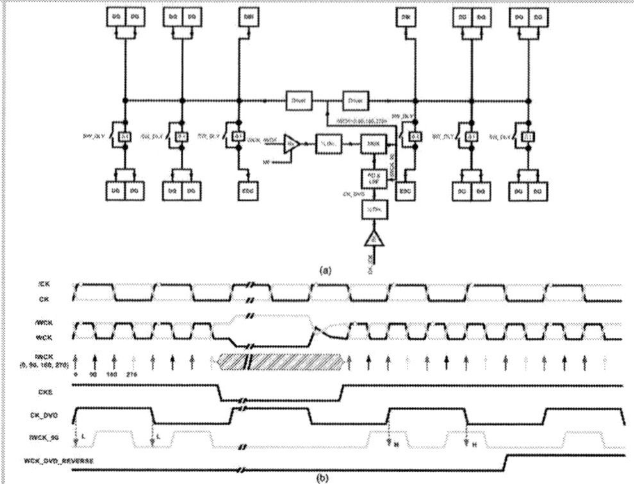

Figure 25.3.2: (a) Block diagram of clock tree, and (b) timing diagram of auto-sync with WCK stop mode.

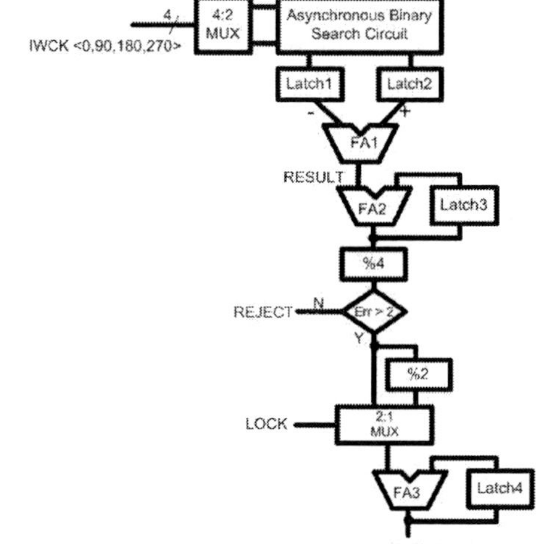

Figure 25.3.3: Flow chart diagram of error-adaptive duty cycle corrector.

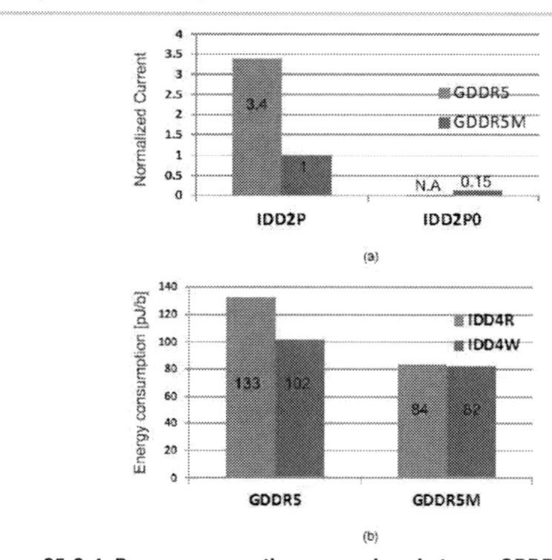

Figure 25.3.4: Power-consumption comparison between GDDR5 and GDDR5M of (a) precharged standby current and (b) read and write current with burst mode.

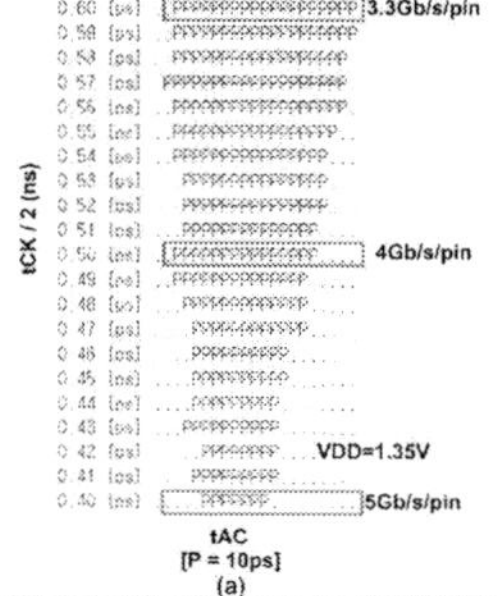

Shmoo plot of tCK vs. tAC		tDV [ps]	
Previous	Proposed	Previous	Proposed
tCK / 2 (ns) 0.55 [ps] ..PPPPPPPPPPPPPPPP...	tCK / 2 (ns) 0.55 [ps] ..PPPPPPPPPPPPPPPP..	150	160
0.54 [ps] ..PPPPPPPPPPPPPPPP...	0.54 [ps] ..PPPPPPPPPPPPPPPP..	150	160
0.53 [ps] ...PPPPPPPPPPPPP....	0.53 [ps] ..PPPPPPPPPPPPPPP...	130	150
0.52 [ps] ...PPPPPPPPPPPP.....	0.52 [ps] .PPPPPPPPPPPPPPP....	120	150
0.51 [ps] ...PPPPPPPPPPPPP....	0.51 [ps] .PPPPPPPPPPPPPP.....	130	140
0.50 [ps] ..PPPPPPPPPPPPP.....	0.50 [ps] ..PPPPPPPPPPPPP.....	130	130

(b)

Figure 25.3.5: Shmoo plot of (a) $t_{CK} - t_{AC}$ at 1.35V and (b) $t_{CK} - t_{AC}$ at 1.4V with developed clocking scheme.

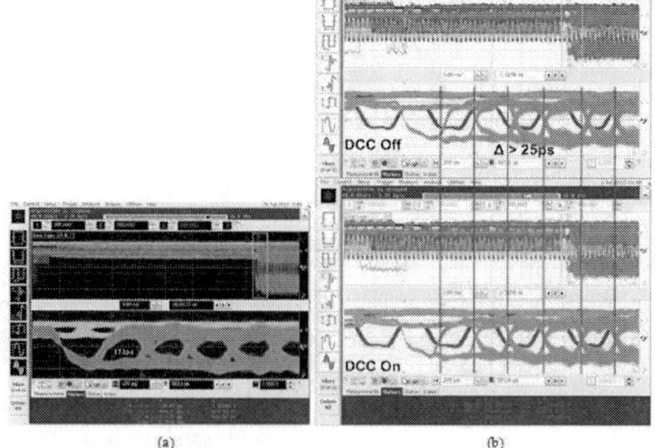

Figure 25.3.6: Measured oscilloscope plot of (a) DQ6 with worst pattern and (b) with 3% distorted WCK at 4.2Gb/s in graphics card.

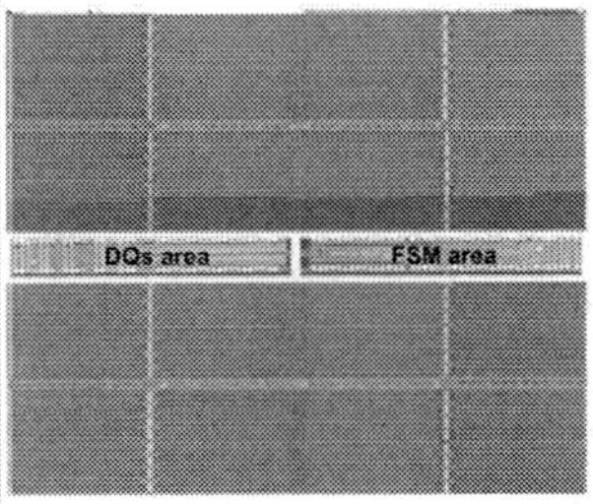

DQs area

FSM area

Process		29nm CMOS 3 metal
Capacity		4Gb
Chip configuration		512Mx8b or 256Mx16b
Supply voltage		1.35V VDD/VDDQ
Data rate		5Gb/p/pin @1.35V
Energy Consumption	Burst Read (IDD4R)	84pJ/b
	Burst Write (IDD4W)	82pJ/b
Powerdown Power	IDD2P	5.4mW
Chip size		51.2mm²

Figure 25.3.7: Chip micrograph of the GDDR5M DRAM.

Session 26 Overview:
Energy-Efficient Dense Interconnects
WIRELINE SUBCOMMITTEE

Session Chair: *SeongHwan Cho*
KAIST, Daejeon, Korea

Session Co-Chair: *Hisakatsu Yamaguchi*
Fujitsu Laboratories, Kawasaki, Japan

With the need for increased I/O bandwidth to support ever-increasing communication demands, the development of energy-efficient links that enable high-density interfaces is essential. This session presents 6 papers that introduce new high-speed aggregated serial-link techniques in advanced CMOS technologies. These designs address the demands of a range of key applications, from dense chip-to-chip communications to high-bandwidth memory access.

26.1 A 130mW 20Gb/s Half-Duplex Serial Link in 28nm CMOS **3:15 PM**

O. Oluwole, nVidia, Santa Clara, CA

In Paper 26.1, nVidia describes a 320Gb/s parallel transceiver capable of equalizing 20dB of channel loss using a single-tap reconfigurable FFE transmitter and a receiver with a single-stage continuous-time linear equalizer, 2-tap DFE and an IIR. The half-duplex transceiver power efficiency is 6.5pJ/b at 20Gb/s.

26.2 A 205mW 32Gb/s 3-Tap FFE/6-Tap DFE Bidirectional Serial Link in 22nm CMOS **3:45 PM**

J. Jaussi, Intel, Hillsboro, OR

In Paper 26.2, Intel introduces a scalable bidirectional serial link at 32Gb/s/lane, with 3-tap FFE transmitter, a continuous-time linear equalizer, and 6-tap DFE. The transceiver, implemented in a 22nm CMOS technology, has per-lane efficiency of 6.4pJ/b at 32Gb/s, while equalizing 16dB of channel loss at Nyquist.

26.3 A Pin- and Power-Efficient Low-Latency 8-to-12Gb/s/wire 8b8w-Coded SerDes Link for High-Loss Channels in 40nm Technology **4:15 PM**

A. Singh, Kandou Bus, Lausanne, Switzerland

In Paper 26.3, Kandou Bus presents a new coded signaling scheme that doubles pin efficiency, and is demonstrated in a design supporting 96Gb/s data transmission over an 8-wire bus at 4.3pJ/b/wire over 15dB loss channels. The circuit is implemented in 40nm CMOS.

26.4 A 25.6Gb/s Differential and DDR4/GDDR5 Dual-Mode Transmitter with Digital Clock Calibration in 22nm CMOS **4:45 PM**

T-C. Hsueh, Intel, Hillsboro, OR

In Paper 26.4, Intel presents a dual-mode transmitter that supports single-ended 1.2V-DDR4/1.5V-GDDR5 as well as 25.6Gb/s differential I/O in 22nm CMOS. The configurable pre-emphasis control, high-frequency current compensation for SST driver equalization, all-active DDR drivers and asynchronous digital sampling minimize power and area. Over a 24dB-loss channel, the bidirectional transceiver dissipates 4.8pJ/b at 25.6Gb/s.

26.5 An 8-to-16Gb/s 0.65-to-1.05pJ/b 2-Tap Impedance-Modulated Voltage-Mode Transmitter with Fast Power-State Transitioning in 65nm CMOS **5:00 PM**

Y-H. Song, Texas A&M University, College Station, TX

In Paper 26.5 from Texas A&M, Oregon State University, and Fudan University, a scalable quarter-rate transmitter is presented. This transmitter uses an analog-controlled impedance-modulated 2-tap equalizer and achieves 8-to-16Gb/s operation at 1pJ/b, supporting up to 12dB of equalization. The low-swing global clock distribution enables improved energy efficiency through aggressive voltage scaling, operating with V_{DD} from 0.75 to 1V. The circuit is implemented in a 65nm CMOS process.

26.6 A 2.667Gb/s DDR3 Memory Interface with Asymmetric ODT on Wirebond Package and Single-Side-Mounted PCB **5:15 PM**

S-P. Chen, MediaTek, Hsinchu, TaiwanIn Paper 26.6, MediaTek introduces a new DDR3 memory interface with asymmetric ODT to save I/O power dissipation on the controller side. It operates at data-rates to 2.667Gb/s/pin on a wire-bond BGA package and single-side mounted PCB.

26.1 A 130mW 20Gb/s Half-Duplex Serial Link in 28nm CMOS

Vishnu Balan, Olakanmi Oluwole, Gregory Kodani, Charlie Zhong, Sanjeev Maheswari, Ratnakar Dadi, Arif Amin, Gautam Bhatia, Peter Mills, Ahmed Ragab, Edward Lee

nVidia, Santa Clara, CA

As the processing power and clock rate of CPUs and GPUs increase, there is a need for increased I/O bandwidth to enable chip-to-chip communication [1]. I/O pin limitations demand faster links at low power to enable integration of high chip-to-chip bandwidth. However, the channel losses and impedance discontinuities increase at high data rates making it difficult to equalize the channel at low power. In this work, we target reliable, differential, bi-directional links at 20Gb/s over 6" FR4 PCB trace and flip-chip packages with a total loss budget of 20dB at Nyquist. In a half-duplex link, one TX and RX are connected on each side and the link direction can be turned around by the controller. A link-turnaround latency of <10ns is achieved by placing several key circuits on standby when not in use and by designing fast bias circuits. When fast turn-around is not required, the circuits not in use are powered down permanently and the link is reduced to the simplex case. Figure 26.1.1 shows the top-level transceiver architecture. An LC-VCO-based PLL oscillates at 20GHz and generates quadrature I/Q clocks at 10GHz. Both TX and RX use a half-rate architecture to optimize power. The clocks are distributed through an on-chip transmission line to 16 I/O lanes arranged in 2 rows. The links are capable of data rates as low as 14 Gb/s to save power when full bandwidth is not required.

While a voltage-mode SST TX line driver offers the potential for power savings compared to a CML design [2], the significant overhead in regulating the driver supply and data-dependent supply currents while using FFE, tilt the choice toward a traditional CML driver with CMOS pre-driver stages. In an application with hundreds of lanes, where every lane may not have the worst characteristics, it is beneficial to lower the swing (and save power) from a maximum value of $500mV_{ppdiff}$ (deemed sufficient for the worst case). Compared to a voltage-mode driver, the CML-based design scales power linearly as the output swing is reduced. The TX driver has a programmable termination impedance which is nominally 86Ω differential. The TX block diagram shown in Fig. 26.1.2, consists of a 16:2 multiplexor that creates two streams of data at 10Gb/s, and an FIR block that generates 1UI shifted streams to enable a 2-tap FIR. The FIR can be configured as either a pre-cursor or a post-cursor filter depending on which is more optimal for the given channel. The driver supports up to 9dB of programmable de-emphasis that is set by an adaptation loop closed through a backchannel from the receiver at link startup. A fast bias circuit sets the tail current to provide controlled TX amplitude and is capable of wake-up latency <4ns. The TX consumes 42mW of power from a 1.35V supply for a $500mV_{ppdiff}$ signal at 20Gb/s.

The receiver and equalizers are shown in Fig. 26.1.3. The receiver front-end has a CTLE that boosts high-frequency signals to compensate for channel loss. In order to meet the high bandwidth requirements without increasing power excessively, we adopt a TAS-TIA structure that uses a current-recycling technique. The first TAS stage with a source-degenerated PMOS input stage provides the boosting function. Unlike traditional resistive load at the output [3], an active shunt-feedback structure with a CMOS invertor based TIA structure provides a low-impedance node at the output of the TAS stage. The low impedance at the TAS output node also serves as a convenient place to combine the DFE h_2 tap and IIR currents (summing node). An inverter biased with a tail current as shown, provides a high g_m/I_d ratio compared to other NMOS-only or PMOS-only structures. A majority of the tail current used to bias the TAS structure is diverted into the folded load resistor to be re-used in the bias for TIA and to aid in setting the output common mode of the TIA stage. The low output impedance provided by the feedback in the TIA structure moves out the pole due to loading from the subsequent sampler stages. A high-bandwidth CMFB circuit sets a low output common mode that is appropriate for the pre-amplifier stages of the sampler. Each sampler consists of a pre-amplifier and a sense latch followed by a clocked SR latch. The first DFE tap, h_1, is directly added at the pre-amplifier output to ease the critical timing related to the first DFE tap. To avoid loop unrolling that increases the number of comparators and hence power [4], we deploy direct feedback of the first tap. In this design, the output of the SR latch is directly fed back to the input of the other sense latch well within one UI to meet other timing constraints. The meta-stability rate of the latch is reduced by careful layout to ensure that the time constant of the latch (τ) is less than 6ps over PVT variations. Since all equalization is completed before the latch, the smallest sampled voltage level estimated at an extrapolated BER of 10^{-18} is better than $50mV_{ppdiff}$. A time of 4.5τ is sufficient to amplify this signal to CMOS levels and leave time for h_1 tap settling and other clock delays. The recovered data after the SR latch is re-serialized back to full-rate. It is then scaled by h_2 tap before subtracting away from the boosted input signal at the summing node. In the targeted application for these links, a single-pole IIR shows better SNR compared to adding more DFE taps. A single-pole IIR [5] that cancels a slow tail is implemented to remove ISI at and beyond the third UI location. In addition to data latches, there are error latches to drive adaptation loops. There are also two edge latches clocked by quadrature phase to drive a bang-bang CDR loop. The error latch is time interleaved to spend equal time collecting errors around the target, $\pm h_0$. The adaptation loop bandwidths are set low to compensate for the longer time it takes to collect the errors due to the interleaving. Since the first DFE tap is not yet settled in half UI for the edge samplers, we have to account for residual ISI that may be present due to previous data. While one of the latches operates around an offset of zero as in a traditional case, the second latch operates around an offset value, $h_x{*}d_{k-1}$. The edge sample corresponding to zero offset is selected for CDR loop if the data pattern received is 101 (or 010), while the edge sample with an offset $+h_x$ $(-h_x)$ is selected for the case of 110 (001). The TX FIR settings, IIR pole location and CTLE LF gain settings are adapted initially at start-up and then frozen for the rest of the operation. The CDR, CTLE HF gain, IIR gain, DFE taps h_1 and h_2 are continuously adapted. Figure 26.1.4 shows the gradients used for various adaptation loops using the SS-LMS algorithm.

Figure 26.1.5 shows the TX eye diagram taken with 9dB post-cursor setting after passing through a 4" PCB trace with 13dB loss to SMP connectors. The phase-noise plot shown in the figure is for the same test repeated with a Nyquist bit pattern. The integrated jitter (100kHz to 100MHz) including spurious tones is about $0.4ps_{rms}$. Figure 26.1.6 shows an internal RX eye scan of a link, including packages and a 6" FR4 PCB trace that we use to report measurements. The RX bathtub plot shows an extrapolated eye opening at a BER of 10^{-18} to be $>0.2UI_{pp}$. Figure 26.1.7 shows the die micrograph where we integrate three 16× lane macros in a $45{\times}45mm^2$ flip-chip package representative of a large package typical in GPU applications. Each 16× macro occupies an area of $1.66{\times}1.6mm^2$. The power consumed by one simplex link (RX+TX+PLL/16) operating at 20Gb/s under nominal conditions of 1.35V supply and 0.9V logic supply is 130mW. The TX consumes 42 mW, the RX 78 mW and the shared PLL and clock distribution 10mW. For half-duplex mode with fast turnaround times where several key circuits are kept in standby mode the power increases to 180mW.

Acknowledgements:

The authors would like to thank S. Sudhakaran, S. Hwang and F. Lambrecht for SI simulation and correlation work. The layout team and package team also contributed extensively to this work.

References:

[1] M. Mansuri, et al, "A Scalable 0.128-to-1Tb/s 0.8-to-2.6pJ/b 64-Lane Parallel I/O in 32nm CMOS," *ISSCC Dig. Tech. Papers*, pp. 402, Feb 2013.

[2] C. Menolfi, et al., "A 16Gb/s source-series terminated transmitter in 65nm CMOS SOI," *ISSCC Dig. Tech. Papers*, pp. 446, Feb 2007.

[3] S. Parikh, et al., "A 32Gb/s Wireline Receiver with a Low-Frequency Equalizer, CTLE and 2-Tap DFE in 28nm CMOS," *ISSCC Dig. Tech. Papers*, pp. 28, Feb 2013.

[4] Y. Lu, et al., "A 66Gb/s 46mW 3-Tap Decision-Feedback Equalizer in 65nm CMOS," *ISSCC Dig. Tech. Papers*, pp. 30, Feb 2013.

[5] Y. Liu, et al., "A 10Gb/s Compact Low-Power Serial I/O with DFE-IIR Equalization in 65nm CMOS," *ISSCC Dig. Tech. Papers*, pp. 182, Feb 2009.

978-1-4799-0917-9/14 $31.00 © 2014 IEEE

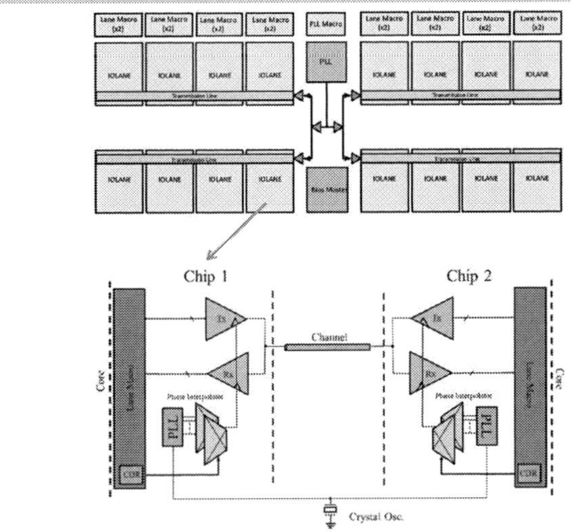

Figure 26.1.1: Top level overview of the link architecture.

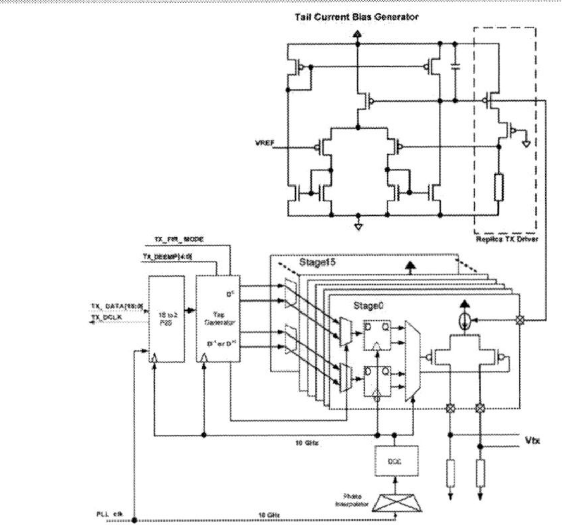

Figure 26.1.2: Transmitter block diagram.

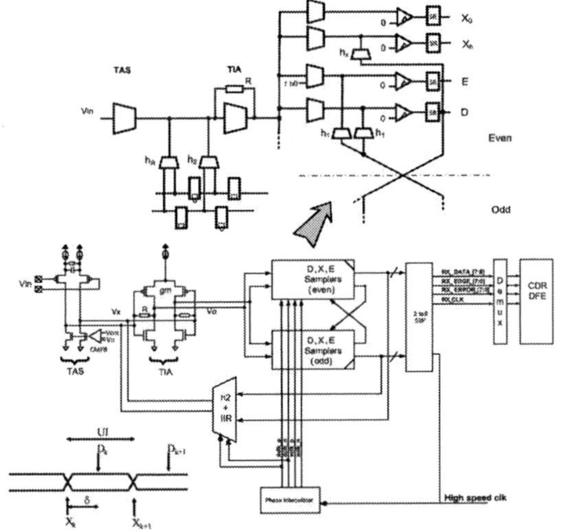

Figure 26.1.3: Receiver overview and CTLE schematic.

Loop	Gradient/Criterion	Comments
H0	$e_k.d_k$	Cursor tap
H1	$e_k.d_{k-1}$	First DFE tap
H2	$e_k.d_{k-2}$	Second DFE tap
Hx	$e_k.d_{k+1}$	Edge DFE.
IIR Gain	$e_k \cdot sgn(d_{k-3} + d_{k-4} + d_{k-5} + d_{k-6})$	IIR gain adjustment
IIR Pole	$e_k \cdot sgn(d_{k-3} - d_{k-4})$	Adjust tail of response. Fixed after initial training. Related to IIR Gain loop.
CTLE_C	$e_k \cdot sgn(d_{k-1} - d_{k-3})$	Adjust HF gain. Fixed after initial training.
CTLE_R	$e_k \cdot sgn(d_{k-1} + d_{k-2})$	Adjust LF gain.
Tx Amplitude	$h_{0,min} < h_0 < h_{0,max}$	Fixed after initial training.
Tx C_{-1}	$e_k.sgn(d_{k+1} + d_k)$	Fixed after initial training.
CDR	$x_k.d_{k-1}$	Bang-Bang CDR

Figure 26.1.4: Adaptation gradients for SS-LMS loops.

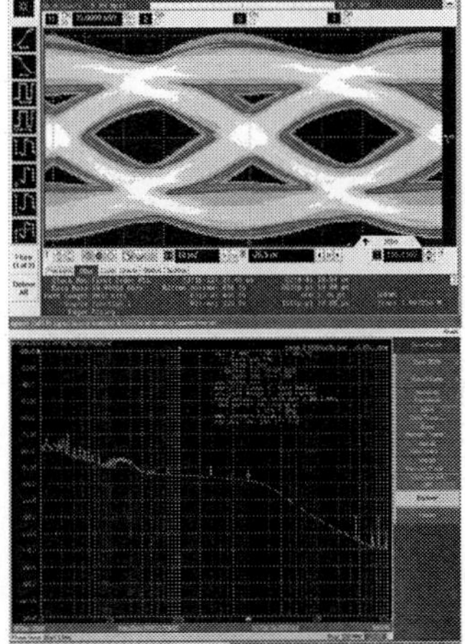

Figure 26.1.5: TX eye diagram with random pattern and phase noise with 1010 pattern.

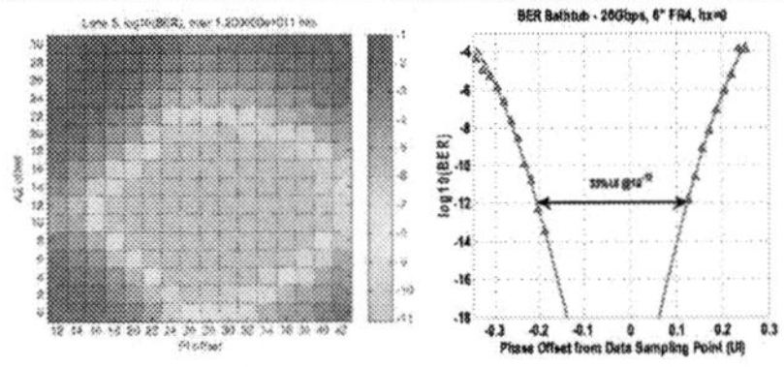

Technology	28nm CMOS
Power Supply	1.35V / 0.9V
Total bandwidth	320Gb/s (16 x 20 Gbps)
LC-PLL closed loop BW	1.2 MHz
PLL jitter (100 kHz - 100 MHz)	420 fs, rms
Power Efficiency at 20Gb/s	6.5pJ/bit
LF, HF Gain range of CTLE	LF: -4 to 6 dB, HF : 0 to 10 dB
Power efficiency of RX CTLE	0.3pJ/bit
Power efficiency of RX samplers	0.4pJ/bit
Total TX/RX pad cap	550fF
Worst case channel loss at Nyquist	20dB
RX input-referred noise	0.9mV, rms

Figure 26.1.6: RX eye scan and bathtub plot of 20Gb/s, PRBS31 across a 6" FR4 trace. Summary of the key results.

26

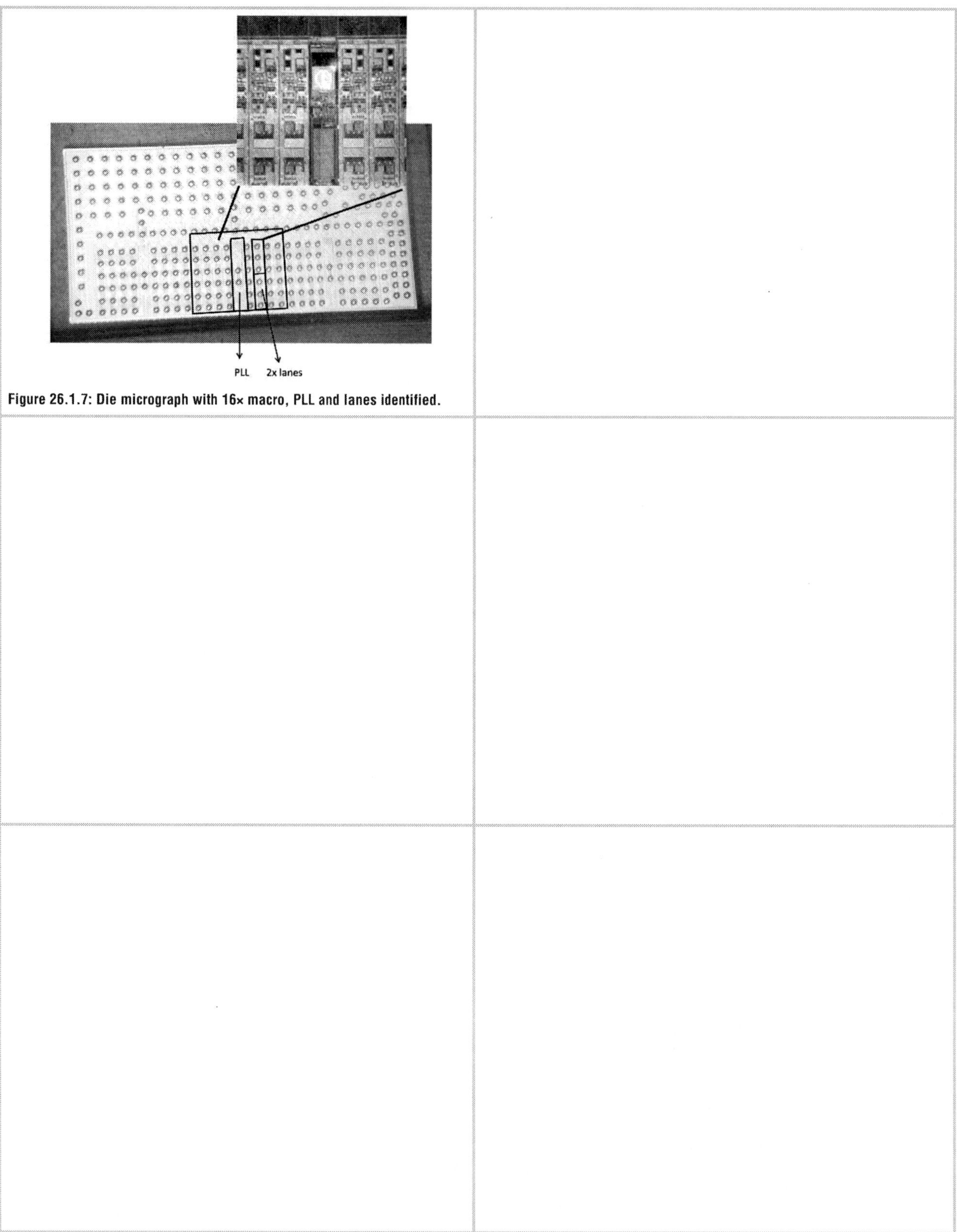

Figure 26.1.7: Die micrograph with 16× macro, PLL and lanes identified.

26.2 A 205mW 32Gb/s 3-Tap FFE/6-Tap DFE Bidirectional Serial Link in 22nm CMOS

James Jaussi[1], Ganesh Balamurugan[1], Sami Hyvonen[1], Tzu-Chien Hsueh[1], Tawfiq Musah[1], Gokce Keskin[1], Sudip Shekhar[1,*], Joseph Kennedy[1], Shreyas Sen[1], Rajesh Inti[1], Mozhgan Mansuri[1], Michael Leddige[1], Bryce Horine[1], Clark Roberts[1], Randy Mooney[2], Bryan Casper[1]

[1]Intel, Hillsboro, OR, [2]Intel, Mapleton, UT
*now with University of British Columbia, Vancouver, Canada

Peripheral I/O data-rates for PCs and mobile computing platforms continue to scale to meet high-bandwidth applications including high-resolution displays and large-capacity external storage. The bandwidth requirements will soon exceed the data-rates of current standards such as PCI Express and USB. A low-power low-cost serial link is needed for the next-generation peripheral interface that can scale to 32Gb/s per lane. Recent publications have demonstrated 28 to 32Gb/s rates [1-2]. However, the circuit power and channel characteristics are not suitable for mainstream PC and mobile markets. A low-profile connector and cable assembly prototype is developed for these markets, where the link architecture and design are optimized for the channel characteristics. This paper describes a data-rate-scalable 32Gb/s serial link that features a bidirectional transceiver, source-series terminated (SST) 3-tap FFE, a continuous-time linear equalizer (CTLE) with an active inductor, a 6-tap DFE, and clock calibration and adaptation circuitry.

Figure 26.2.1(a) illustrates the interconnect topology. The primary components are PCBs, packages, connectors and cable. The 8-lane cable assembly consists of consumer-grade 32AWG shielded twisted pair copper wires. The height and width of the connector are 3mm and 13mm, respectively. The bidirectional transceiver architecture is shown in Fig. 26.2.1(b). An LC-VCO-based PLL generates a quarter-rate clock that is distributed to each transceiver lane using regulated CMOS buffers similar to [3]. In order to save power, the transceivers are grouped into 4 lanes to form a *bundle*. The DLL and clock multiplier are common to the 4 lanes and are included in the bundle clock circuitry. The TX is based on a half-rate clock architecture, while the RX uses a quarter-rate clock architecture. After the DLL, the quarter-rate clock is multiplied to half-rate and distributed locally to 4 TXs. The output stage is a low-swing segmented SST driver using NMOS devices for channel termination. 3-tap TX pre-emphasis is implemented using switching devices to short differential outputs similar to [4]. The RX front-end consists of a CTLE and 6-tap DFE. The CTLE has an active inductor to provide up to 4dB of peaking while minimizing silicon area [5]. Other than the LC-VCO, no passive inductors are used for bandwidth extension or I/O-pad-capacitance reduction. A quadrature clock generator (Q-Gen) provides I/Q clocks for the 4-way interleaved DFE. Across <12dB loss channels and up to 12Gb/s, a separate 2-way interleaved low-power (LP) latch with 2× oversampled CDR is used to reduce power consumption by disabling the CTLE and DFE. There are two separate CDR logic blocks, one in the lane and one in the bundle, that independently control the quadrature phase interpolator (PI). The bundle CDR aggregates the phase-recovery information from all 4 lanes.

To minimize the area and pad capacitance of the bidirectional transceiver, the SST TX can be configured to be RX termination as shown in Fig. 26.2.2. When functioning as an output stage, the segmented legs are controlled by the 3-tap equalizer. Shorting devices between the differential pins enables low-power equalization while maintaining constant termination impedance [4]. When the RX is enabled, the driver segments are repurposed to generate an on-die common-mode voltage, which is required for AC-coupled channels. The equalizing devices are repurposed for channel termination.

The CTLE with variable gain and the DFE CML summer schematics are shown in Fig. 26.2.3. The CTLE uses active inductor PMOS devices, M1 and M2, with the gate terminals connected as a diode load through the variable resistance R_{IND}. The peaking zero frequency is set by R_{IND} and gate capacitance of M1 and M2. The current ratio between the input stage and M1/M2 determines the common-mode output voltage. Degeneration of the input pair is controlled by R_{DG}, which provides the VGA functionality for automatic gain control. Since the source degeneration and peaking frequency are controlled by independent devices, the CTLE and VGA can be combined into a single-stage amplifier without significant interaction. The CTLE drives 4 interleaved track-and-hold circuits followed by 4 CML summers, which sum the DFE currents from the non-speculative taps. The CML summer architecture enables a wide range of data rates from 4 to 32Gb/s. The voltage offsets of the CTLE and CML summer are cancelled with digitally controlled offset currents sourced into their respective outputs.

Following the CML summer are the speculative DFE tap latches and speculative DFE tap mux as shown in Fig. 26.2.4(a). The three parallel latches can be configured for baud-rate phase detection and to receive data with and without tap speculation. To support the various latch configurations, there are three input differential pairs with inputs from the CML summer, the voltage-reference generator (VRG) and the coarse speculative tap coefficient. The fine speculative tap coefficient is controlled by a 6b current DAC. The domino logic mux provides additional amplification and speculative tap selection. The total number of input latches is reduced by 50% through preconfiguring the phase-detection latch with a reference voltage and a speculative tap coefficient. The table and graph in Figs. 26.2.4(b) and (c), respectively, show the reference voltage (V_R) and speculative tap assignments based on the sampling phase of the 4 interleaved DFEs. These assignments, coupled with the phase-detector logic, reduce the average phase-detection rate to one quarter of the maximum available phase samples. On-die adaptation and calibration logic (ACL), shown in Fig. 26.2.1(b), controls foreground and background calibration including CTLE, CML summer, latch voltage-offset cancellation, and duty-cycle and quadrature error correction. The ACL also controls the VGA, V_R, DFE coefficient adaptation and continuous PI duty-cycle correction. The VGA, V_R and DFE tap adaptation are similar to [6].

Two 4-lane bidirectional transceivers (Fig. 26.2.7) are fabricated in a 22nm CMOS process. A 4-lane transceiver occupies 0.317mm^2. The measured TX output is shown in Fig. 26.2.5(a) at 32Gb/s operation over a short channel. Amortizing the PLL and clock distribution power over 4 lanes, the link power consumption is 205mW at 32Gb/s. Transceiver power efficiency is given in Fig. 26.2.5(b) for multiple data rates from 4 to 32Gb/s. Excluding the PLL power, the transceiver power efficiency ranges from 3.4 to 5.7pJ/b, while equalizing 24dB of channel loss at Nyquist up to 16Gb/s and 16dB loss at 32Gb/s. When the CTLE and DFE are disabled and the LP latch is enabled, the measured power efficiencies are 1.0pJ/b at 4Gb/s and 2.2pJ/b at 12Gb/s with 4.5dB and 10dB of channel loss, respectively. The cable assembly insertion loss measurements are given in Fig. 26.2.6(a) for 0.5m, 1m and 2m lengths. The total channel loss for the 0.5m cable is 16dB at 16GHz when including the PCB and package loss. Figure 26.2.6 (b) shows the on-die 10^{-10} BER eye at 32Gb/s over a 0.5m cable. With a TX voltage swing of 630mV$_{pp\text{-}diff}$ and lane CDR enabled, the 4-lane link demonstrates BER <10^{-12} transmission with PRBS10 data.

Acknowledgment:
The authors would like to thank V. Baca, K. Aygun, A. Zhang, P. Parmar, M. Balasubramanian, S. Thakare, A. Kornfeld, D. Duarte, P. Stolt, W.-H. Wang, T. Chan-Carusone, C. Thakkar, S. Spangler, Y. Fan, A. Mezhiba, M. Haycock, Y. Ling, H. Heck, D. Ackelson, M. Bell, and S. Spina.

References:
[1] J. Bulzacchelli, et al., "A 28Gb/s 4-Tap FFE/15-Tap DFE Serial Link Transceiver in 32nm SOI CMOS Technology," *ISSCC Dig. Tech. Papers*, pp. 324-325, Feb. 2012.
[2] S. Parikh, et al., "A 32Gb/s Wireline Receiver with a Low-Frequency Equalizer, CTLE and 2-tap DFE in 28nm CMOS," *ISSCC Dig. Tech. Papers*, pp. 28-29, Feb. 2013.
[3] M. Mansuri, et al., "A Scalable 0.128-to-1Tb/s 0.8-to-2.6pJ/b 64-Lane Parallel I/O in 32nm CMOS," *ISSCC Dig. Tech. Paper*, pp. 402-403, Feb. 2013.
[4] W. Dettloff, et al., "A 32mW 7.4Gb/s Protocol-Agile Source-Series-Terminated Transmitter in 45nm CMOS SOI," *ISSCC Dig. Tech. Papers*, pp. 370-371, Feb. 2010.
[5] M. Ramezani, et al., "An 8.4mW/Gb/s 4-lane 48Gb/s Multi-Standard-Compliant Transceiver in 40nm Digital CMOS Technology," *ISSCC Dig. Tech. Papers*, pp. 352-353, Feb. 2011.
[6] F. Spagna, et al., "A 78mW 11.8Gb/s Serial Link Transceiver with Adaptive RX Equalization and Baud-Rate CDR in 32nm CMOS," *ISSCC Dig. Tech. Papers*, pp. 366-367, Feb. 2010.

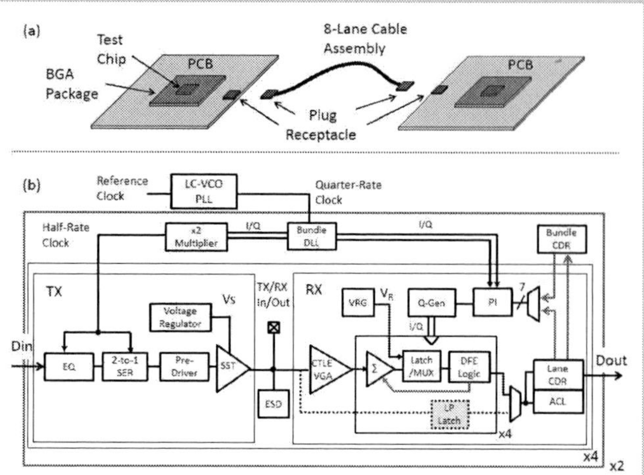

Figure 26.2.1: (a) Interconnect topology and (b) bidirectional transceiver architecture.

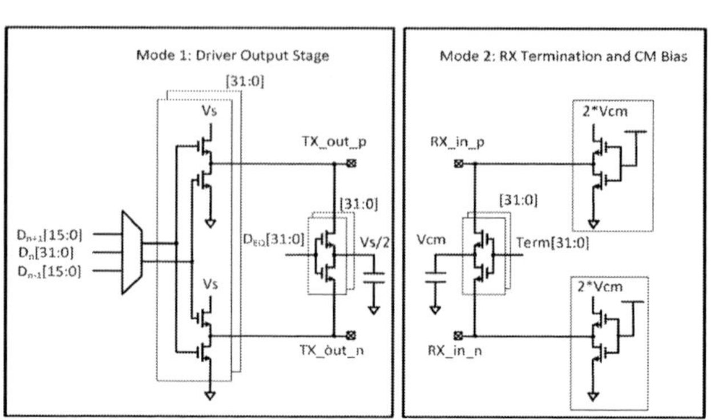

Figure 26.2.2: Configurable SST TX schematic.

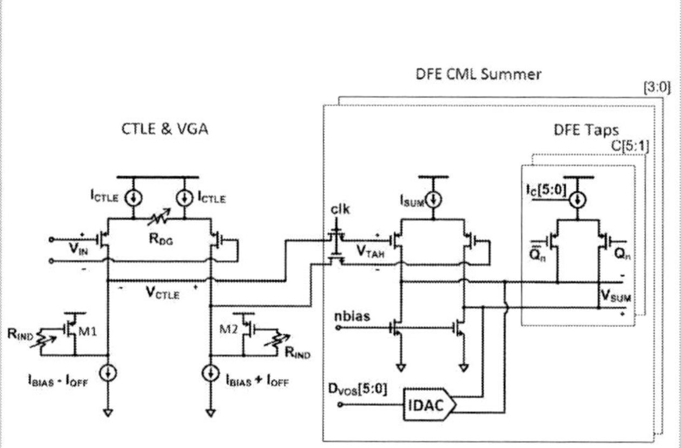

Figure 26.2.3: CTLE, VGA and DFE CML summer schematic.

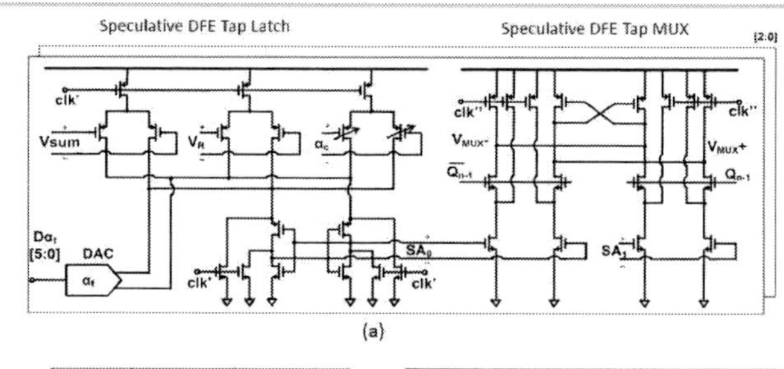

Phase Sample	Reference Voltage	Speculative Tap
0°	$-V_R$	$-\alpha$
90°	$+V_R$	$+\alpha$
180°	$+V_R$	$+\alpha$
270°	$-V_R$	$-\alpha$

(b)

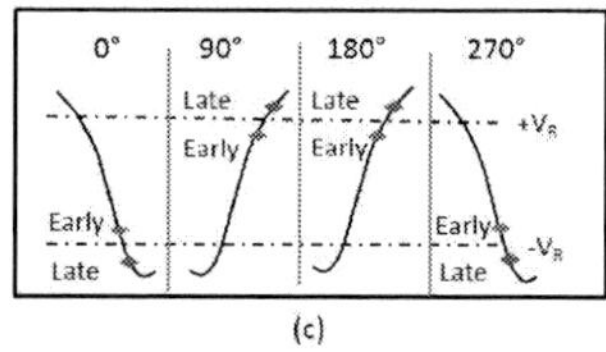

(c)

Figure 26.2.4: (a) Speculative DFE tap latch and MUX schematic, (b) phase-detection assignment for reference voltage and speculative tap and (c) CDR phase-detection graph.

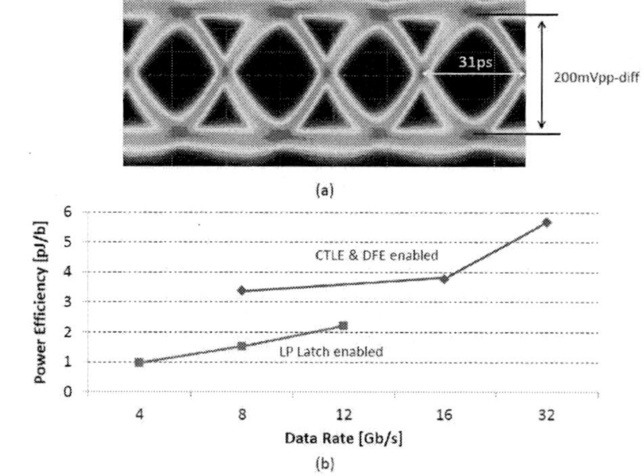

Figure 26.2.5: (a) Measured 32Gb/s TX output waveform and (b) measured transceiver power efficiency without PLL power.

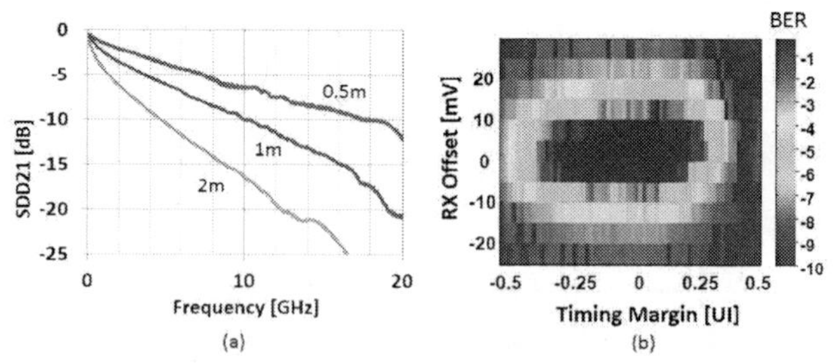

Maximum data rate	32Gb/s	Maximum TX swing	$630mV_{pp\text{-}diff}$
Link power efficiency	6.4pJ/b	TX jitter	$<540fs_{rms}$
I/O power supply	1.07V	RX input-referred noise	$2mV_{rms}$
TX/RX area	$0.079mm^2$	Total loss @ Nyquist	16dB

(c)

Figure 26.2.6: (a) Cable assembly insertion-loss measurements, (b) on-die RX BER eye at 32Gb/s and (c) summary table.

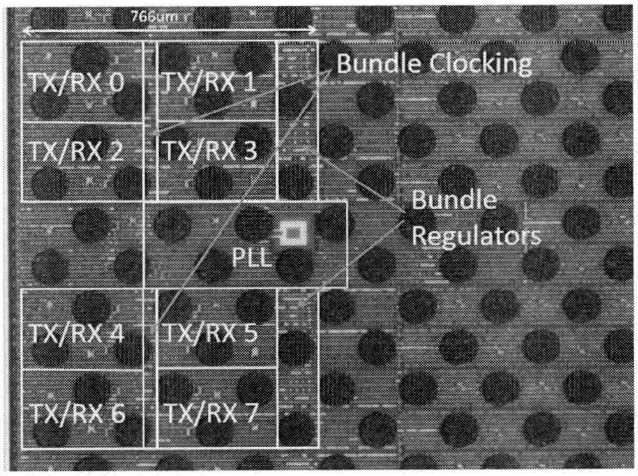

Figure 26.2.7: Die micrograph.

26.3 A Pin- and Power-Efficient Low-Latency 8-to-12Gb/s/wire 8b8w-Coded SerDes Link for High-Loss Channels in 40nm Technology

Anant Singh[1], Dario Carnelli[1], Altay Falay[1], Klaas Hofstra[1], Fabio Licciardello[1], Kia Salimi[1], Hugo Santos[1], Amin Shokrollahi[1], Roger Ulrich[1], Christoph Walter[1], John Fox[2], Peter Hunt[2], John Keay[2], Richard Simpson[2], Andy Stewart[2], Giuseppe Surace[2], Harm Cronie[3]

[1]Kandou Bus, Lausanne, Switzerland,
[2]Kandou Bus, Northampton, United Kingdom,
[3]Lausanne, Switzerland

The continuing demand for higher bandwidth in serial interconnects has pushed the symbol rate of differential lanes into the high-insertion-loss region of channels. Multi-level signaling such as differential PAM-4 [1] has been used to mitigate the loss of electrical channels by lowering the signal spectrum. Such an approach suffers from lower SNR tolerance as well as higher susceptibility to crosstalk and ISI as compared to differential signaling (DS). Coded differential approaches have been reported [2] to mitigate ISI. Our approach is a generalization of DS in which ternary values are transmitted on an 8-wire bus. The set of transmitted values belongs to a code consisting of 256 code-words called the 8b8w-code (8-bits-on-8-wires) [3]. The specific correlations in the code-words of the 8b8w-code eliminate transmit common-mode and simultaneous switching output (SSO) noise and allow for detection via self-referencing comparators (unlike PAM-4), which provides additional noise immunity. Compared to DS, the 8b8w-code offers twice the throughput at 50% of the line power. Compared to PAM-4, the code offers better SNR (3dB) at 38% of the line power with enhanced tolerance of ISI and lower crosstalk generation. The design and experimental verification of an 8b8w transceiver in 40nm CMOS is described. Transmission is achieved up to 12Gb/s per wire over 55cm of Rogers with up to 15dB loss.

The code-words are a permutation of the vector [+1, +1, −1, −1, 0, 0, 0, 0] and have the property that the sum of the ternary signals transmitted on the bus is 0 and thus have a balanced common mode voltage. Both the encoder and decoder have a maximum logic depth of 3 and are implemented with standard cells. The gate count for single instance of 8b encoder and decoder operating at 2GHz each is 81 and 63 respectively. The total power consumption and area for 8 such encoders is 3mW and 2000μm². The corresponding numbers for decoder are 4mW and 1330μm².

Figure 26.3.1 shows the architecture of the transmitter. The 8 wires of the output driver are terminated to a common point, which is connected to an external pin driven with a common node voltage (V_{cm}, nominally $V_{DD}/2$). A digital encoder block (8 instances of 8b encoder) transforms 64 bits of incoming data into two streams of coded signals that are serialized and drive the PMOS and NMOS devices in a switched current-mode driver. The '0'-level is driven passively with both devices turned-off. A replica-current circuit is used to bias the current sources, providing a mechanism to control the voltage swing at the output of the transmitter ($V_{cm} \pm \Delta V_{swing}$). A 1-tap programmable post-cursor FIR is included.

Any skew between the wires manifests itself as high-frequency common-mode noise at the receiver. Figure 26.3.2 shows the architecture of the 8b8w receiver. To allow realignment without signal distortion, the receiver front-end is designed to pass high-frequency common-mode signals. The CTLE (shown in Fig. 26.3.3) is a hybrid between a generalized differential pair and a common-source amplifier. The shared node *vcommon1* is stabilized at high frequencies by capacitors vca1, effectively turning the structure into a single-ended common-source amplifier with source degeneration. The receiver is architected as 16-phase time-interleaved system. Following the CTLE is a 4-phase sample-and-hold circuit operating at quarter rate. The sampling clocks can be adjusted per wire for de-skewing the incoming signals up to 1UI. A buffer drives the sampled signals into the next stage of samplers at the input of a signal-to-digital converter (SDC) circuit. An extra set of samplers/buffers on each wire are connected to an on-chip eye scope.

Figure 26.3.4 shows the SDC consisting of per wire, offset-corrected voltage-to-time converter circuit (VTC) that converts the sampled voltage to a time-proportional delayed edge by linearly discharging a pre-charged capacitor. It has a controlled current source with a common tail device across the 8 wires, which allows for different gain settings. The output is connected to a threshold buffer. A combination of 28 two-input latching arbiter circuits across the 8 wires compare the arrival times of the edges, thereby rank-ordering the wire-values. A digital logic block identifies the position of the wires with two maxima ('+1's) and two minima ('−1's) in order to decode the bits. The positions of the '0's are *don't care* for decoding. The overall ordering information is used to optimally adjust receiver sampling moment and per-wire deskew, to tune the equalizer response and to the set gain and offsets of various circuits in the receiver datapath in order to adapt the signal characteristics. An external clock generator (4 to 8GHz) is used as the clock source for both transmitter and receiver. A multi-phase clock generator in the receiver is used to produce 1UI-wide quarter-rate clocks with non-overlapping edges. The de-skew function is supported by per-wire phase interpolators (PI). A latch-based divider circuit generates the 1GHz clocks for the VTC and the decoders. A DLL aligns the clock edges of VTC and front-end samplers.

The 8b8w-coded SerDes link is fabricated in a 40nm 10LM process. The die micrograph is show in Figure 26.3.7. The design includes an option to operate in legacy quad-lane DS mode. The test-chip can be controlled with SPI and includes data generators (PRBS31, PRBS9 and custom patterns), error counters and other test features like eye-scope and analog test bus. The test setup is shown in Fig. 26.3.5 and uses multiple 2.4mm connectors, cables and board traces. The link achieves power efficiency of <4.3pJ/b at 12Gb/s/wire. Bathtub measurements are taken over multiple channel configurations and BER $<8\times10^{-15}$ is achieved at 12Gb/s/wire with PRBS31 data over channels with 15dB loss. Common-mode, power supply and crosstalk noise tests do not show any significant degradation in BER. Per-wire de-skew capability of up to 1UI is demonstrated by running the link on channels with mismatched cables. Measurements show 2× improvement in power efficiency when compared with DS mode at the same throughput. Silicon performance parameters are given in Fig. 26.3.6.

Acknowledgements:
We thank Ali Hormati, Terry Ward, Mehran Aliahmad, et al. from Faradesign Inc. for their contributions.

References:

[1] J. Lee, M. Chen, and H. Wang, "Design and Comparison of Three 20-Gb/s Backplane Transceivers for Duobinary, PAM4, and NRZ Data", *JSSC,* vol. 43, no .9, Sep. 2008.
[2] A. Amirkhany, et al, "4.1pJ/b 16Gb/s Coded Differential Bidirectional Parallel Electrical Link", *ISSCC Dig. Tech. Papers,* pp. 138-139, Feb. 2012.
[3] H. Cronie, A. Shokrollahi, and A. Tajalli, "Methods and Systems for Noise Resilient and Low Power Communications with Sparse Signaling Codes," *US Patent Application Number US2012/0213299 A1.*

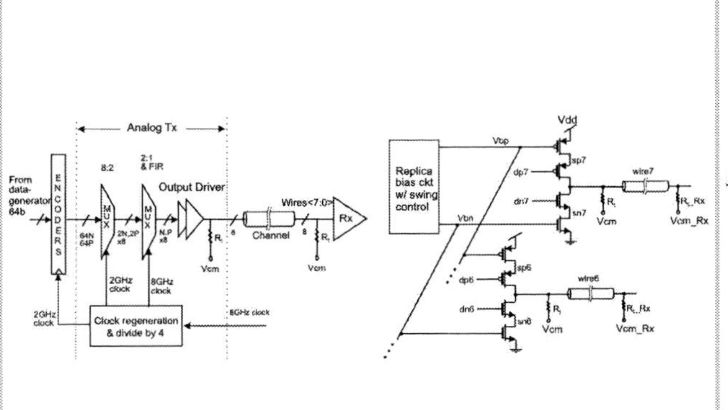

Figure 26.3.1: Transmitter architecture on the left showing the encoders and the serializer. The final stage is 2:1 mux that retimes the data-path with the edges of half-rate clock. The output driver structure for the 8 wires is shown on the right. It includes a 1-tap post-cursor FIR.

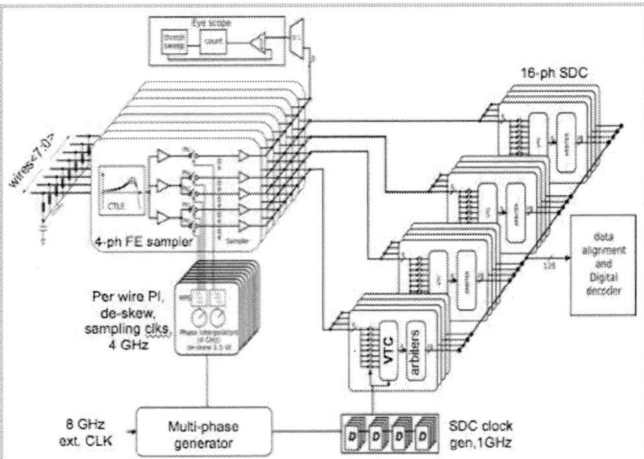

Figure 26.3.2: 16-phase receiver architecture. The FE samplers use non-overlapping quarter-rate clock, and the SDC block is clocked by 16 phases of 1GHz clock.

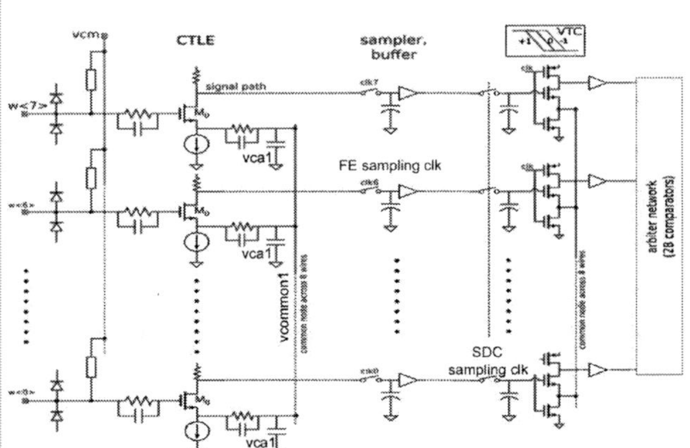

Figure 26.3.3: Receiver datapath showing the generalized differential CTLE across 8 wires. The FE sampling clock provides de-skew capability. The VTC current source is shared across 8 wires.

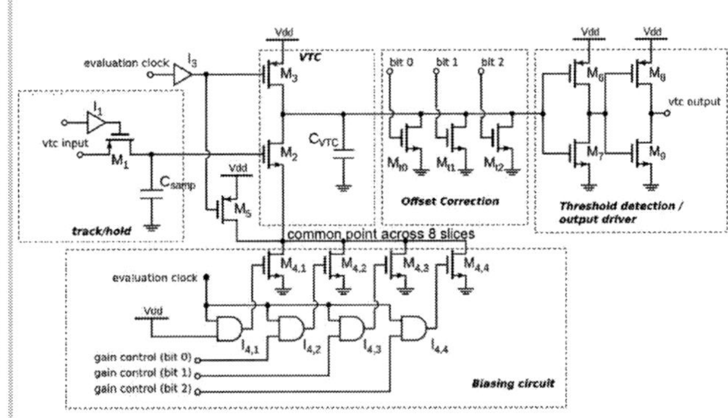

Figure 26.3.4: Describes the VTC circuit. Device M1 is the sampler switch. Capacitor CVTC is precharged to VDD through device M3 and discharged in the evaluation phase through the device M2. Three bit offset cancellation and gain-control schemes are used.

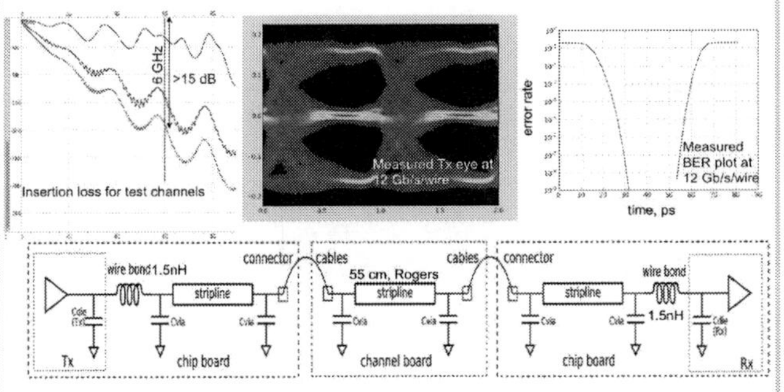

Figure 26.3.5: Channel configuration, insertion loss, TX eye and BER plot at 12Gb/s. The short, medium and long channels have a total loss in the range of 10 to 15dB. BER <8×10^{-15} is measured at 12 Gb/s/wire.

Technology	40nm CMOS GP, VDD=0.9V, 10M, DGO
Package	Wire bond (1.5-2.0 mm length), COB
Channels	78cm, 55cm & 36cm Rogers (RO4450F/RO4350B), four 2.4mm connectors, 12" cables
Loss at Nyquist	Up to 15dB
IO C_{die}	600fF, including ESD
Pads	Pitch 70μm, bond wire inductance = 1.5nH
Data Rate	8-12Gb/s/wire
Nominal Power	412mW
Energy Efficiency	4.29pJ/b at 12Gb/s/wire
BER	<8×10^{-15} at 12Gb/s/wire
64b-encoder latency, area, power	0.5ns, 2000μm^2, 3mW
64b-decoder latency, area, power	0.5ns, 1330μm^2, 4mW
Differential legacy mode	Yes
Testability	Pattern generators (prbs31, prbs9), on-chip Eye Scope, error counters, SPI, analog test bus, test software
Per wire RX de-skew	1UI

Figure 26.3.6: Performance summary of 8b8w test-chip.

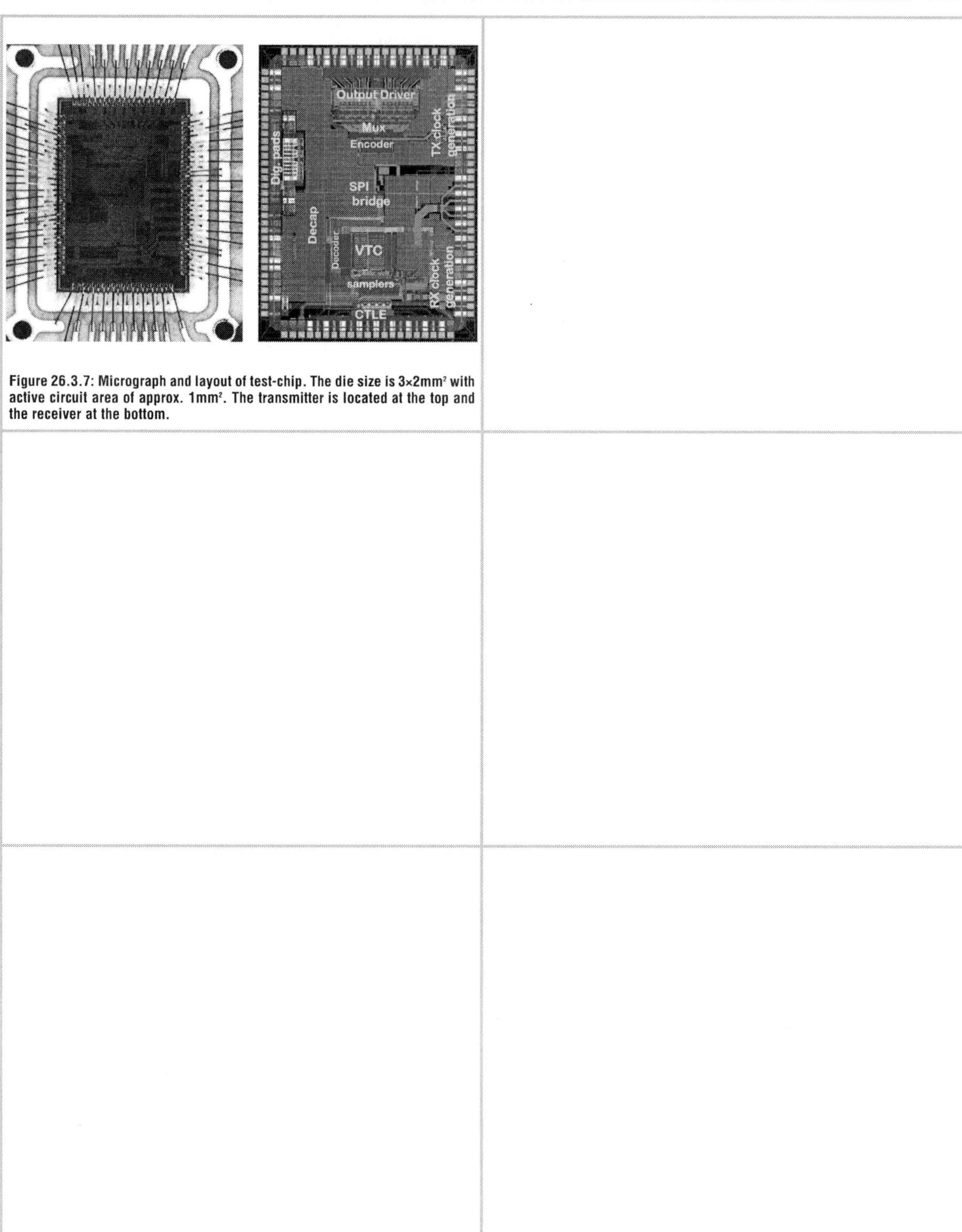

Figure 26.3.7: Micrograph and layout of test-chip. The die size is 3×2mm² with active circuit area of approx. 1mm². The transmitter is located at the top and the receiver at the bottom.

26.4 A 25.6Gb/s Differential and DDR4/GDDR5 Dual-Mode Transmitter with Digital Clock Calibration in 22nm CMOS

Tzu-Chien Hsueh, Ganesh Balamurugan, James Jaussi, Sami Hyvonen, Joseph Kennedy, Gokce Keskin, Tawfiq Musah, Sudip Shekhar*, Rajesh Inti, Shreyas Sen, Mozhgan Mansuri, Clark Roberts, Bryan Casper

Intel, Hillsboro, OR
*now with University of British Columbia, Vancouver, Canada

A wide range of memory configurations exist in today's high-speed digital systems to meet platform-specific bandwidth, power, capacity, and cost constraints. In the near term, DDR4 and GDDR5 are expected to meet the needs of server, client, graphics and mobile platforms [1]. Differential signaling with high-speed serial I/O enhancements will potentially continue I/O performance scaling for post-DDR4 and future buffered memory solutions. A unified memory interface that can meet the signaling requirements of all these memory standards offers several benefits: reduced cost and design time, greater platform design flexibility, and a smoother transition from DDR4/GDDR5 to a high-speed differential memory interface [2]. This paper presents a dual-mode TX that supports single-ended (SE) 1.2V-DDR4/1.5V-GDDR5 (hereafter referred to as DDR-mode) as well as high-speed differential signaling (hereafter referred to as HSD-mode), which is implemented using only thin-gate-oxide devices in 22nm CMOS. Other key design features include: (a) a DDR4/GDDR5 driver implemented using only active devices (no linearizing resistors), (b) enhanced voltage-mode driver supply regulation, (c) reconfigurable logic to support pre-emphasis in both TX modes, and (d) low-overhead digital clock-calibration techniques based on asynchronous digital sampling (ADS) to improve calibration coverage and accuracy.

A block diagram of the dual-mode bidirectional transceiver (XCVR) is shown in Fig. 26.4.1. Four TX lanes share clocking circuitry that includes an LC-VCO PLL, a DLL and an XOR frequency multiplier. While much of the digital circuitry (equalization logic, serializers and clock distribution) preceding the driver is shared between HSD and DDR modes, the final output driver is split to accommodate the unique signaling requirements of the two modes. In DDR-mode, the common TX datapath is configured to implement two individual 2-tap TX pre-emphasis equalizers (EQ) (single post-cursor) for the two SE DDR-mode drivers. In HSD-mode, the same datapath is configured to enable 3-tap TX pre-emphasis (one pre-cursor and one post-cursor) in the HSD driver with differential equalization (DEQ) [3]. While DEQ offers a lower power equalization solution for source-series-terminated (SST) drivers [3], it requires: (a) more complex pre-driver logic to control the equalizing switch, and (b) a way to handle high-frequency data-dependent driver current variation. Figure 26.4.2 shows the configurable circuit block used to implement the pre-driver logic functions required for an N-over-N SST HSD-mode driver with differential TX pre-emphasis. The parallel cascode CMOS circuits allow the realization of AND, BUFFER and XNOR functions with closely matched delays to meet 2:1 serializer timing constraints. To mitigate the problem of high-frequency driver current variation on the HSD-mode swing-control regulator, a data-dependent current compensation scheme is employed. Since the current variation (ΔI) occurs whenever the equalizing switch input, D_{SW}, toggles (and the variation magnitude is proportional to equalization coefficient, $\alpha = 0.5\times a/N$, in Fig. 26.4.2), D_{SW} is reused to enable a current compensation path in each driver leg. This compensation scheme enables a significant regulator output capacitor reduction by mitigating high-frequency current variations in the regulator load.

The dual-mode TX output drivers are shown in Fig. 26.4.3. The output stage uses only thin-gate-oxide devices: high-voltage tolerance (HVT) is enabled by cascoding and appropriate biasing. Termination in both modes is implemented using only active devices to minimize area and pad capacitance. The HSD-mode driver (redrawn in Fig. 26.4.3 with transistor-level details) is an N-over-N SST driver with NMOS equalizing switches controlled by D_{SW}. The equalizing switches can be configured as a differential termination when the bi-directional HSD XCVR is configured as an RX. The HSD-mode driver is segmented into 32 legs to enable 5b TX impedance and maximum 6b pre-emphasis [3] control. The all-active DDR-mode large-swing SST driver shown in Fig. 26.4.3 uses a parallel combination of complementary P/N, diode-/triode-region devices to achieve driver/on-die termination (ODT) linearity without the use of area-intensive passive resistors. Cascoding and level-shifting pre-drivers enable HVT using only thin-gate-oxide devices as well. Measured results shown in Fig. 26.4.3 indicate that this design can achieve less than ±10% resistance variation over (0.5 to 0.95)×V_{DDQ} DDR4 output swing range. The cascode triode-region devices also enable HVT without any supply-voltage reconfiguration between DDR- and HSD-modes. With independent 64-leg push-up (PU) and pull-down (PD) impedance controls, each SE DDR-mode driver can meet DDR4 specs for push/pull impedance matching, resolution and range requirements. By eliminating passive resistors, this design significantly reduces area overhead due to resistor variation and low resistance density. In the 22nm CMOS process, the DDR-mode driver area efficiency improvement is >2× with respect to a design using linearizing resistors. The HSD-mode swing-control regulator is configured as a V_{SSHI} (=V_{DDQ}-1V) generator for the level shifters (LS) required for HVT in DDR-mode. To enable this reuse, the regulator is designed to support 0.2 to 0.5V output voltage range in HSD- and DDR-modes.

To generate half-rate CMOS clocks for the reconfigurable TX, an XOR multiplier receives quadrature clock phases from the DLL and multiplies the frequency by two. Multiple calibration nodes exist in the clock path to ensure a robust clock distribution as shown in Fig. 26.4.4. Clock non-idealities, like duty-cycle distortion and quadrature error, are detected using an ADS technique to enable accurate calibration with low area overhead. For duty-cycle detection (DCD), the two polarities of a differential clock are individually sampled by an asynchronous clock using a digital MUX and a flip-flop, and the difference in 1's count is used as a measure of the duty cycle error (DCE). The all-digital DCD enables multiple sense nodes in the clock path with minimal loading of the clock distribution and area overhead. The differential nature of the measurement cancels the effect of sampler non-idealities. Similarly, the I-Q delay at DLL output is tuned by detecting the DCE at the multiplier output. Duty-cycle correction (DCC) with <0.25% resolution is accomplished by adding an offset from a current-DAC to the bias of a DLL replica delay buffer. A final stage of DCC, prior to the TX serializer, is performed by configuring the RX in the bi-directional XCVR as an asynchronous sampler. This extends TX clock calibration coverage to encompass all elements in the TX datapath up to and including the output stage with low circuit overhead.

The dual-mode TX test-chip is implemented in a 22nm CMOS technology. HSD and DDR4-mode operation is verified with BER $<10^{-12}$ (PRBS10) over representative channels using separate RXs in the bidirectional XCVR. DDR-mode TX output eyes after a package, HDI interposer, PGA socket, 2" FR4 and SODIMM connector are shown in Fig. 26.4.5. Comparing Figs. 26.4.5(a) and (b) indicates the eye-width degradation due to self-induced crosstalk in DDR-mode from the dual-mode driver design to be 0.077UI (24ps). Figures 26.4.5(c) and (d) show 6.4Gb/s 1.5V-GDDR5 TX eyes with and without TX pre-emphasis. To test the HSD-mode, a differential RX with a continuous-time linear equalizer, 3-tap decision-feedback equalizer and DFX circuitry is also included in the XCVR prototype. Over a 24dB loss differential channel composed of packages, interposers, sockets and 4" FR4, the bidirectional XCVR operates at a data-rate of 25.6Gb/s while dissipating 4.8pJ/b. Figure 26.4.6(a) shows the measured BER eye, and Figure 26.4.6(b) shows the 25.6Gb/s TX output probed after the package, interposer, socket and 2" FR4. These results (summarized in Fig. 26.4.6) demonstrate the feasibility of a dual-mode XCVR to enable aggressive memory I/O scaling post-DDR4 while maintaining backward compatibility. Figure 26.4.7 shows the die micrograph. The area of dual-mode TX (including regulator de-coupling capacitor) is less than 135×40μm^2.

Acknowledgements:
B. Horine, M. Leddige, P. Parmar, V. Baca, Y. Fan, A. Martin, S. Thakare, K. Aygun, A. Zhang, P. Stolt, D. Dunning, J. McCall, D. Conrow, V. Ragavassamy, M. Balasubramanian, R. Mooney, M. Haycock and W.-H. Wang.

References:

[1] A. Amirkhany, et al., "A 12.8-Gb/s/link Tri-Modal Single-Ended Memory Interface," *IEEE J. Solid-State Circuits*, vol. 47, no. 4, pp. 911-925, Apr. 2012.
[2] K. Kaviani, et al., "A Tri-Modal 20-Gbs/Link Differential/DDR3/GDDR5 Memory Interface," *IEEE J. Solid-State Circuits*, vol. 47, no. 4, pp. 926-937, Apr. 2012.
[3] W. Dettloff, et al., "A 32mW 7.4Gb/s Protocol-Agile Source-Series-Terminated Transmitter in 45nm CMOS SOI," *ISSCC Dig. Tech. Papers*, pp. 370-371, Feb. 2010.

978-1-4799-0917-9/14 $31.00 © 2014 IEEE

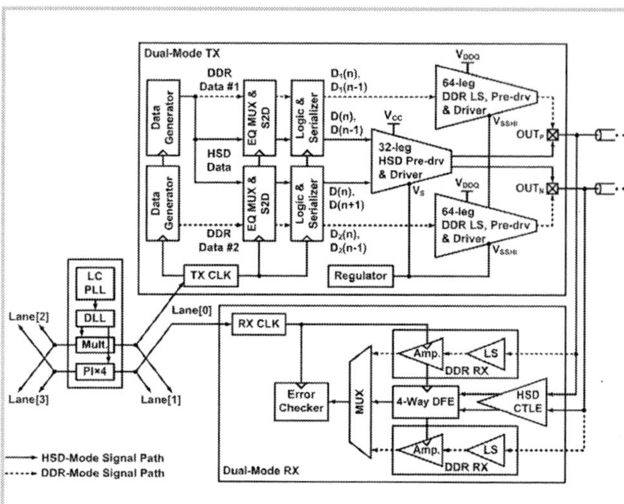

Figure 26.4.1: Dual-mode bidirectional transceiver block diagram.

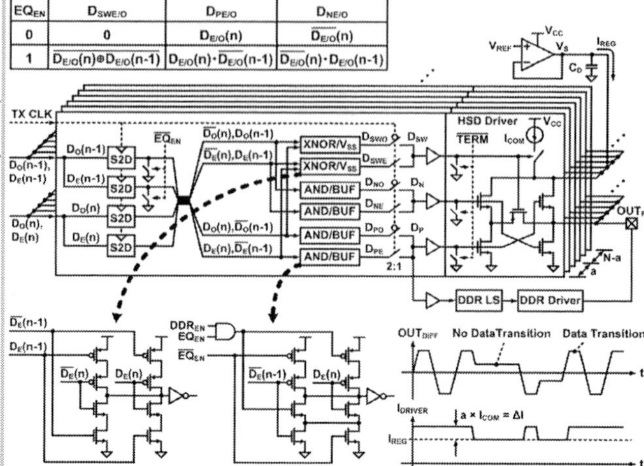

Figure 26.4.2: TX equalization control and current compensation scheme.

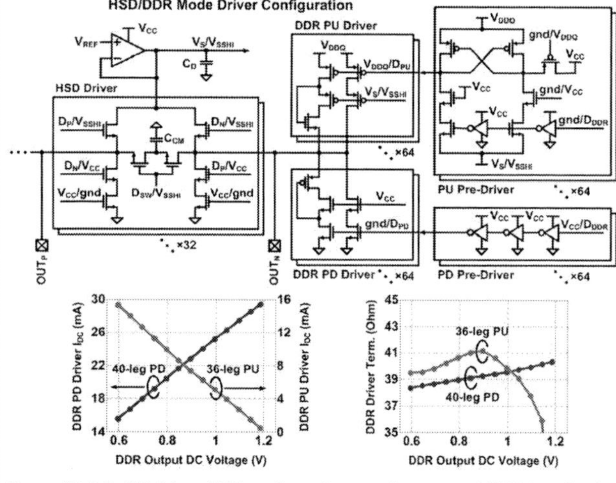

Figure 26.4.3: TX driver HVT configurations and measured DDR termination linearity.

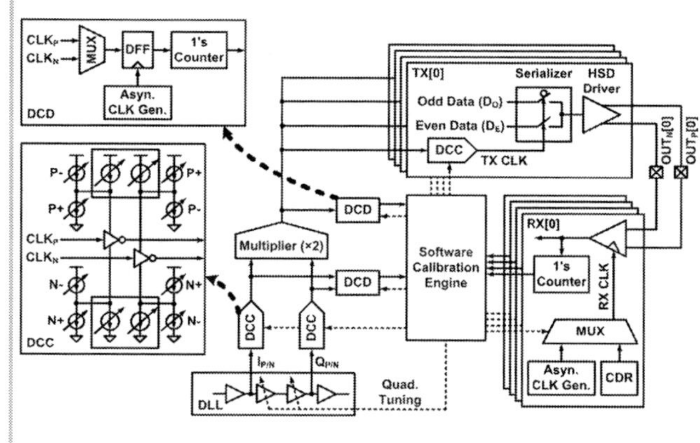

Figure 26.4.4: TX digital clock calibration block diagram.

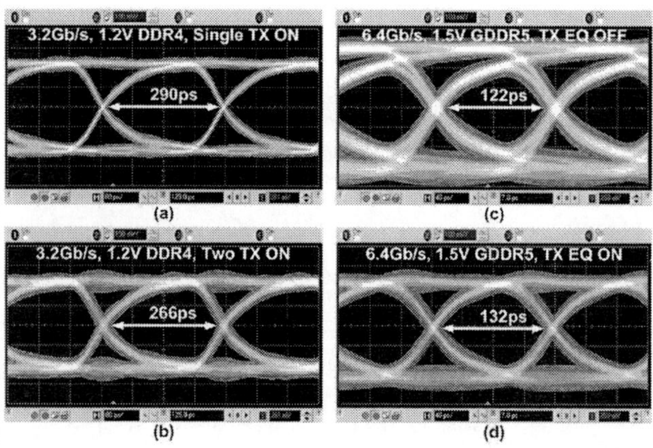

Figure 26.4.5: 3.2Gb/s DDR4 eye diagram (a) w/o and (b) w/ two TXs enabled simultaneously. 6.4Gb/s GDDR5 eye diagram (c) w/o and (d) w/ TX EQ enabled.

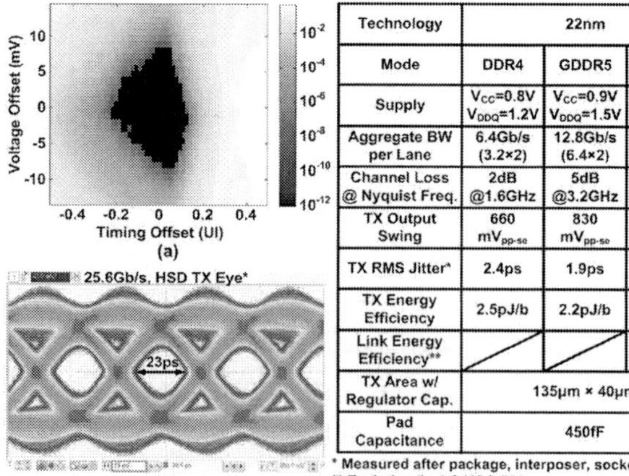

Technology	22nm		
Mode	DDR4	GDDR5	HSD
Supply	V_{CC}=0.8V V_{DDQ}=1.2V	V_{CC}=0.9V V_{DDQ}=1.5V	V_{CC}=1.05V
Aggregate BW per Lane	6.4Gb/s (3.2×2)	12.8Gb/s (6.4×2)	25.6Gb/s (25.6×1)
Channel Loss @ Nyquist Freq.	2dB @1.6GHz	5dB @3.2GHz	24dB @12.8GHz
TX Output Swing	660 mV_{pp-se}	830 mV_{pp-se}	500 $mV_{pp-diff}$
TX RMS Jitter*	2.4ps	1.9ps	820fs
TX Energy Efficiency	2.5pJ/b	2.2pJ/b	1.1pJ/b
Link Energy Efficiency**	/	/	4.8pJ/b
TX Area w/ Regulator Cap.	135μm × 40μm		
Pad Capacitance	450fF		

* Measured after package, interposer, socket & 2" FR4
** Excludes the LC-VCO PLL

Figure 26.4.6: Measured (a) HSD BER, (b) HSD TX eye, and performance summary.

26

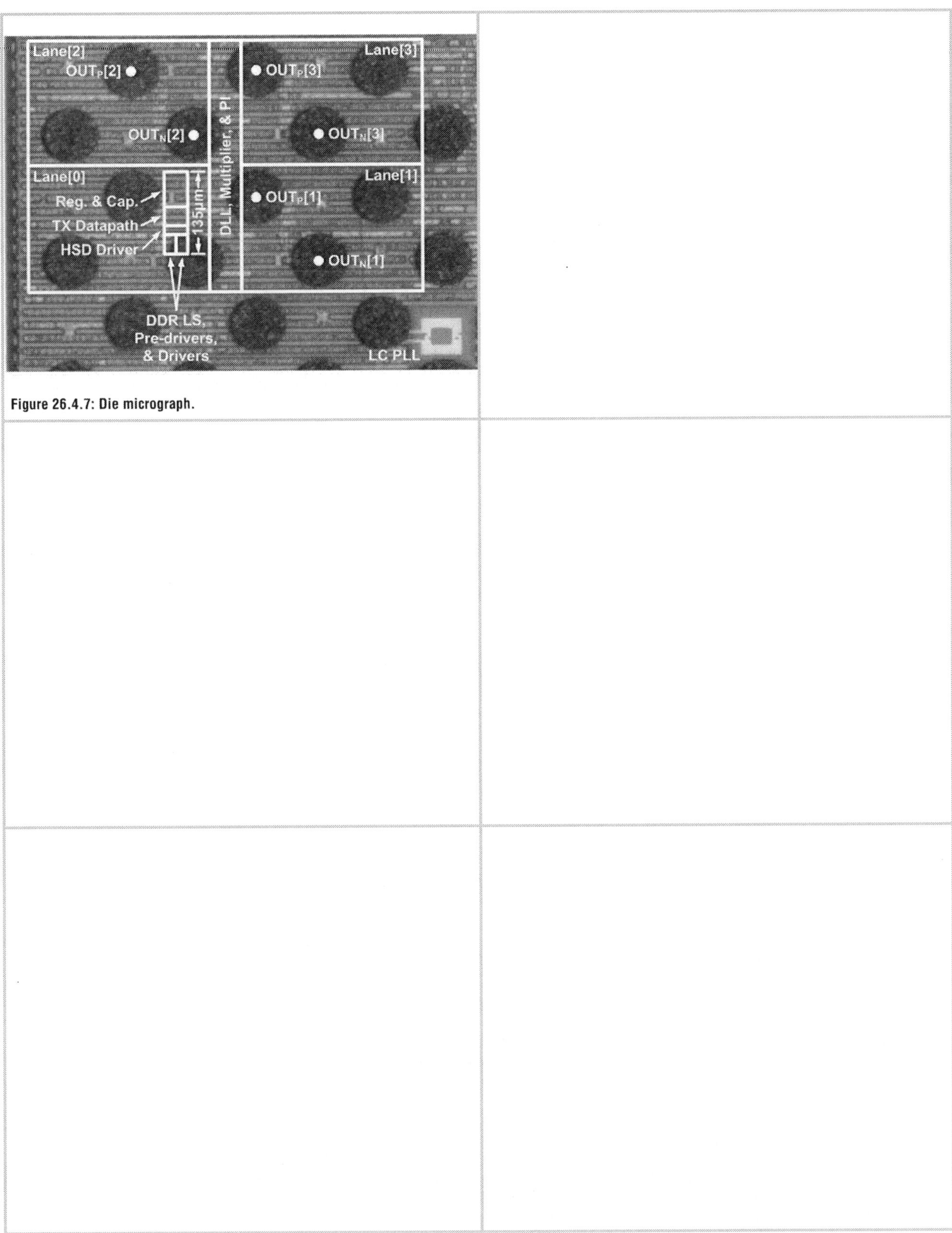

Figure 26.4.7: Die micrograph.

26.5 An 8-to-16Gb/s 0.65-to-1.05pJ/b 2-Tap Impedance-Modulated Voltage-Mode Transmitter with Fast Power-State Transitioning in 65nm CMOS

Young-Hoon Song[1], Hae-Woong Yang[1], Hao Li[2], Patrick Yin Chiang[2,3], Samuel Palermo[1]

[1]Texas A&M University, College Station, TX,
[2]Oregon State University, Corvallis, OR,
[3]Fudan University, Shanghai, China

Future processor I/Os must aggressively improve per-channel data-rates and energy efficiency to meet projected system bandwidth demands. These constraints necessitate the design of ultra-low-power serial-link transmitters that can efficiently incorporate equalization to compensate for channel losses, while enabling fast power-state transitioning to leverage dynamic power scaling. In this work, a scalable-data-rate voltage-mode transmitter is presented that introduces two main innovations. First, an impedance-modulated 2-tap equalizer is adopted that employs analog control of the equalizer taps, thereby obviating output driver segmentation. Second, fast power-state transitioning is achieved using a replica-biased voltage regulator to power the output stages of multiple channels and per-channel injection-locked oscillators (ILO) that can be rapidly disabled. Furthermore, capacitively driven low-swing global clock distribution and automatic phase calibration of the local ILO-generated quarter-rate clocks enables improved energy efficiency with aggressive supply scaling.

Figure 26.5.1 shows a conceptual diagram of our system, with 10 transmitter channels spanning across a 2mm distance. All transmitters share both a global regulator to set the nominal output swing, and two analog loops to set the driver output impedance during the maximum and de-emphasized levels of the 2-tap equalizer. In order to reduce dynamic clocking power, low-swing clocks are maintained throughout the global distribution and local generation of the four clocks used by the quarter-rate transmitters. A differential quarter-rate clock is distributed globally in a repeater-less manner via capacitively driven low-swing wires, while ac-coupled inverters with resistive feedback locally buffer these distributed clocks. Each ac-coupled inverter locally drives a two-stage ILO, producing four quadrature clocks that are shared by a two-channel bundle. ILO quadrature output phase spacing is improved by AC-coupling the injection clocks, adding dummy injection buffers, and optimizing the locking range via digital control of the injection-buffer drive strength. The ILO operating frequency is coarsely controlled via a dedicated power supply and finely set using the analog control voltage, EN_VCTL. This analog control voltage can also be rapidly switched between GND and its nominal value, enabling fast power-up/shut-down of the clock signals.

For each transmitter channel (Fig. 26.5.2), eight parallel input bits are first serialized by an initial 8:4 mux, followed by two sets of 4:1 muxes that drive the main- and post-cursor paths of the 2-tap impedance-modulated voltage-mode driver. The serialized data passes through a level-shifting pre-driver [1] that boosts the voltage swing by a fully scalable supply value, DVDD, above the nominal NMOS threshold voltage, enabling reduced transistor sizing for a given impedance value. Closed-loop phase calibration is implemented to correct for deterministic jitter that arises due to the quadrature clocks' static phase errors and duty-cycle distortion. In calibration mode, the transmitter output for two complementary fixed patterns is sampled with a comparator clocked by an asynchronous 100MHz signal. First, the duty cycle is corrected by comparing the count value obtained for a "1100" output pattern and its complement, followed by an FSM that adjusts the P/N strength of the local clock buffers. Second, quadrature correction is realized by using a "1010" pattern and its complement, with the FSM then adjusting the relative delay of the buffers through capacitive tuning. At 16Gb/s operation, enabling this entire phase calibration loop improves the eye width variation from an uncorrected 13.1% to 5.4%, limited by nonlinearities in the duty-cycle tuning range.

The transmitter exhibits two operating modes to provide transmitter equalization at higher data rates, while dramatically scaling energy efficiency at lower data rates when equalization is not required. In equalization mode, a new impedance-modulation technique [2] is introduced in the all-NMOS output stage (Fig. 26.5.3). During a transition bit, the maximum output swing is achieved with nearly a 50Ω output impedance, when both the higher-impedance single-transistor and 50Ω two-transistor paths are activated in parallel. Analog control of the impedance values allows for a non-segmented output stage, dramatically reducing pre-driver complexity and resulting in significant power savings. A global impedance loop sets the control voltages, VzceqUP and VzceqDN, of the top/bottom transistors of the two-transistor paths, realizing a 50Ω output impedance when combined in series with the two transistors that are switched by the main and post-cursor data bits. The area overhead of this effective three-transistor stack is minimized because the switch transistors see a large level-shifted overdrive voltage, VLS=DVDD+V_{thn}, when turned on. Only the single-transistor pull-up/pull-down path is activated for run-lengths greater than one, with the de-emphasis level set by the control voltages, VzmeqUP and VzmeqDN, provided by the global de-emphasis impedance modulation loop. In non-equalization mode, the output stage is placed in a standard configuration with a single series impedance-control transistor in the pull-up/pull-down paths, where the control voltages, VzcUP and VzcDN, are provided by the global impedance control loop. Furthermore, pre-drivers are disabled to save power.

A global voltage regulator with a replica output stage load (Figure 26.5.3) sets the output swing value and the transmitter output supply, VREG, enables amortization of the regulator power while providing a stable bias signal for fast power-state transitioning. A dual-supply topology is employed for the global regulator to improve accuracy and reduce power. The nominal 1V supply allows for a higher error amplifier gain, while a low 0.5V source-follower output stage provides a tunable output swing from 100 to 300mV_{ppd}. 2.9ns power-state transitioning is achieved on a per-channel basis through staggered switching of the output-stage decoupling capacitance. Delayed enabling of the output-stage decoupling capacitance by ~550ps allows for rapid charging of VREG and minimal charge sharing when the decoupling capacitance is reapplied.

Figure 26.5.7 shows a die micrograph of the transmitter, fabricated in a GP 65nm CMOS process. While chip area constraints prevent a full 10-channel prototype, the concept is accurately emulated by placing a two-transmitter bundle at the end of a snaked on-chip 2mm clock distribution. Each transmitter channel occupies 0.006mm^2, and the combined area of the injection-locked oscillator, global impedance control and modulation loop, bias circuitry, and voltage regulator is 0.014mm^2. A channel consisting of a 5.8" FR4 trace and a 0.6m SMA cable, with 15.5dB loss at 8GHz, is used to characterize the transmitter (Fig. 26.5.4). Low-frequency output patterns with a peak 300mV_{ppd} output swing verify the equalizer functionality up to the maximum 12dB setting. The transmitter transient performance at a maximum 16Gb/s data rate is verified for 2^7-1 PRBS eye diagrams, where a previously near-closed eye is opened to a 55mV height and 33.4ps width when the impedance-modulation equalization is enabled. As shown in Fig. 26.5.5, the transmitter achieves 8-to-16Gb/s operation at 0.65 to 1.05pJ/b energy efficiency by optimizing the transmitter's scalable supply and output swing for a minimum 50mV_{ppd} eye height and 0.5UI eye width at the channel output. By routing the internal power-state-enable signal off-chip using a delay-matched channel, rapid power-state transitioning is verified with 0.5ns disable and 2.9ns enable times. Figure 26.5.6 shows the measured transmitter power breakdown and compares this work with other voltage-mode transmitters that incorporate either 2-tap equalization [2-3] or fast power-state transitioning [4-6].

Acknowledgment:
This work was supported by the SRC-TxACE under grant 1836.060, Intel Labs Wireline Signaling Program, and a Department of Energy Early CAREER grant.

References:

[1] Y.-H. Song, *et al.*, " A 0.47-0.66 pJ/bit, 4.8-8 Gb/s I/O Transceiver in 65nm CMOS," *IEEE J. Solid-State Circuits*, vol. 48, no. 5, pp. 1276-1289, May 2013.
[2] R. Sredojevic, *et al.*, "Fully Digital Transmit Equalizer with Dynamic Impedance Modulation," *IEEE J. Solid-State Circuits*, vol. 46, no. 8, pp. 1857-1869, Aug. 2011.
[3] Yue Lu, *et al.*, "Design and Analysis of Energy-Efficient Reconfigurable Pre-Emphasis Voltage-Mode Transmitter," *IEEE J. Solid-State Circuits*, vol. 48, no. 8, pp. 1898-1909, Aug. 2013.
[4] B. Leibowitz, *et al.*, "A 4.3 GB/s Mobile Memory Interface With Power-Efficient Bandwidth Scaling," *IEEE J. Solid-State Circuits*, vol. 45, no. 4, pp. 889-898, April 2010.
[5] J. Zerbe, *et al.*, "A 5.6 Gb/s 2.4 mW/Gb/s Bidirectional Link With 8ns Power-On," *in Symp. VLSI Circuits Dig. Tech. Papers*, pp. 82-83, June 2011.
[6] F. O'Mahony, *et al.*, "A 47×10Gb/s 1.4 mW/Gb/s Parallel Interface in 45nm CMOS," in *ISSCC Dig. Tech. Papers*, pp. 156-157, Feb. 2010.

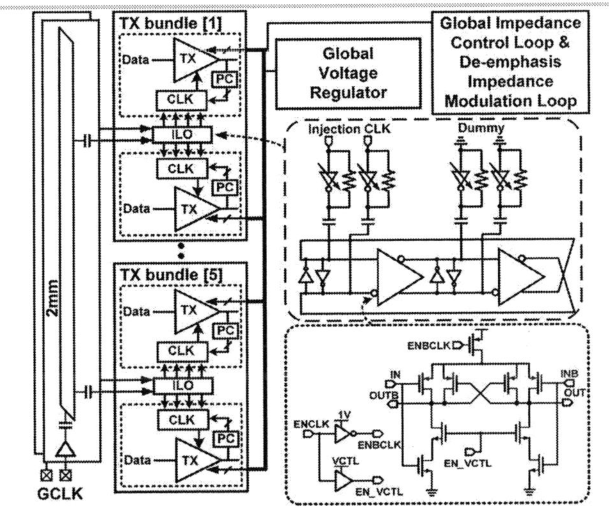

Figure 26.5.1: Multi-channel transmitter architecture.

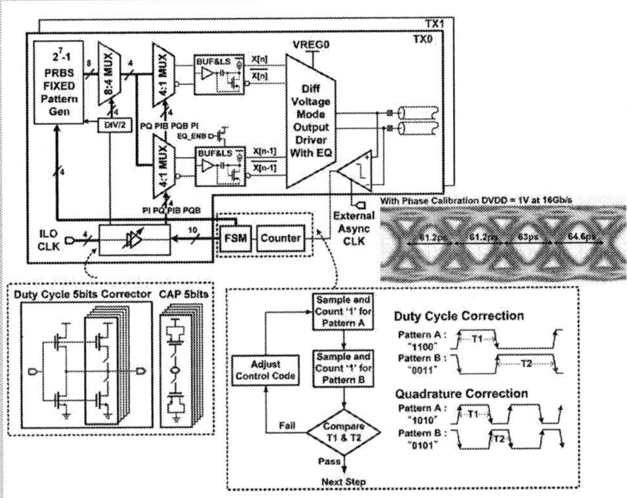

Figure 26.5.2: Transmitter block diagram with clock phase calibration details.

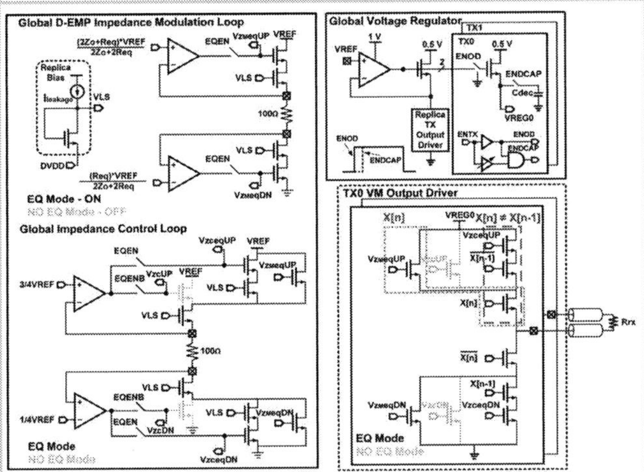

Figure 26.5.3: Transmitter output driver schematic with global impedance control loops and voltage regulator.

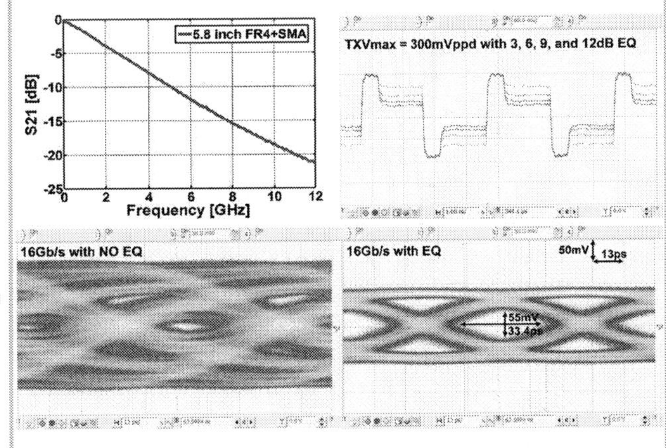

Figure 26.5.4: Measured channel S_{21} response, low-frequency transmitter output waveforms, and 16Gb/s eye diagrams without and with equalization.

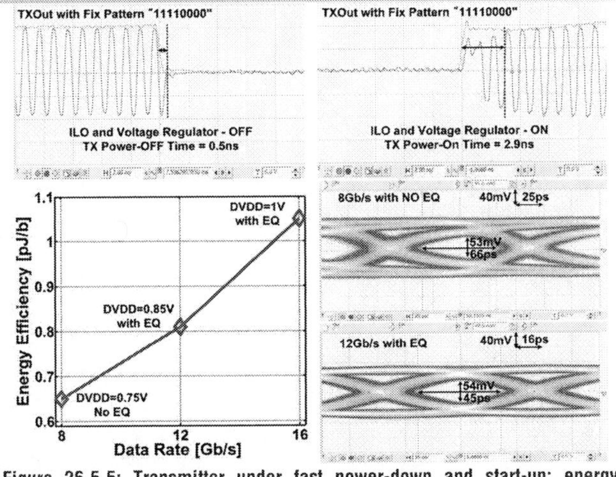

Figure 26.5.5: Transmitter under fast power-down and start-up; energy efficiency versus data-rate with eye diagrams after 5.8" FR4+ 0.6m SMA cable.

TRANSMITTER POWER BREAKDOWN (16Gb/s at 1V & 0.5V)	
LDO (amortized across 2 TX) & Output Driver (300mV$_{ppd}$ with EQ)	985uW
Serializer, Pre-drivers, Clocking	10.8mW
Global Impedance Control & Modulation Loop, Bias Circuit (amortized across 2 TX)	1.1mW
Global Clocking (amortized across 2 TX)	1.5mW
ILO (amortized across 2 TX)	2.4mW
Total Energy Efficiency	**1.05pJ/b**

TRANSMITTER PERFORMANCE COMPARISONS			
	[2]	[3]	This Work
Technology	90nm	65nm	65nm
Supply Voltage	1.15V	1.2V	1 & 0.5V
Data Rate	4Gb/s	10Gb/s	16Gb/s
TX Swing	0V-1Vppd	160mV~500mVppd	100mV~300mVppd
Channel Loss @ Nyqu. Freq	-8~-10dB	-13dB	-15.5dB
Equalization	2-Tap FIR	2-Tap FIR	2-Tap FIR
Energy Efficiency	1.25-4.25pJ/b	1pJ/b	1.05pJ/b

POWER STATE TRANSIENT TIME COMPARISONS				
	[4]	[5]	[6]	This Work
Technology	40nm	40nm	45nm	65nm
Data Rate	4.3Gb/s	5.6Gb/s	10Gb/s	16Gb/s
Power State Transient time	<5ns	8ns	<5ns	0.5ns(Off), 2.9ns(On)

Figure 26.5.6: Measured transmitter power breakdown at 16Gb/s and comparison to other voltage mode transmitters with 2-tap equalization and fast power-state transitioning.

26

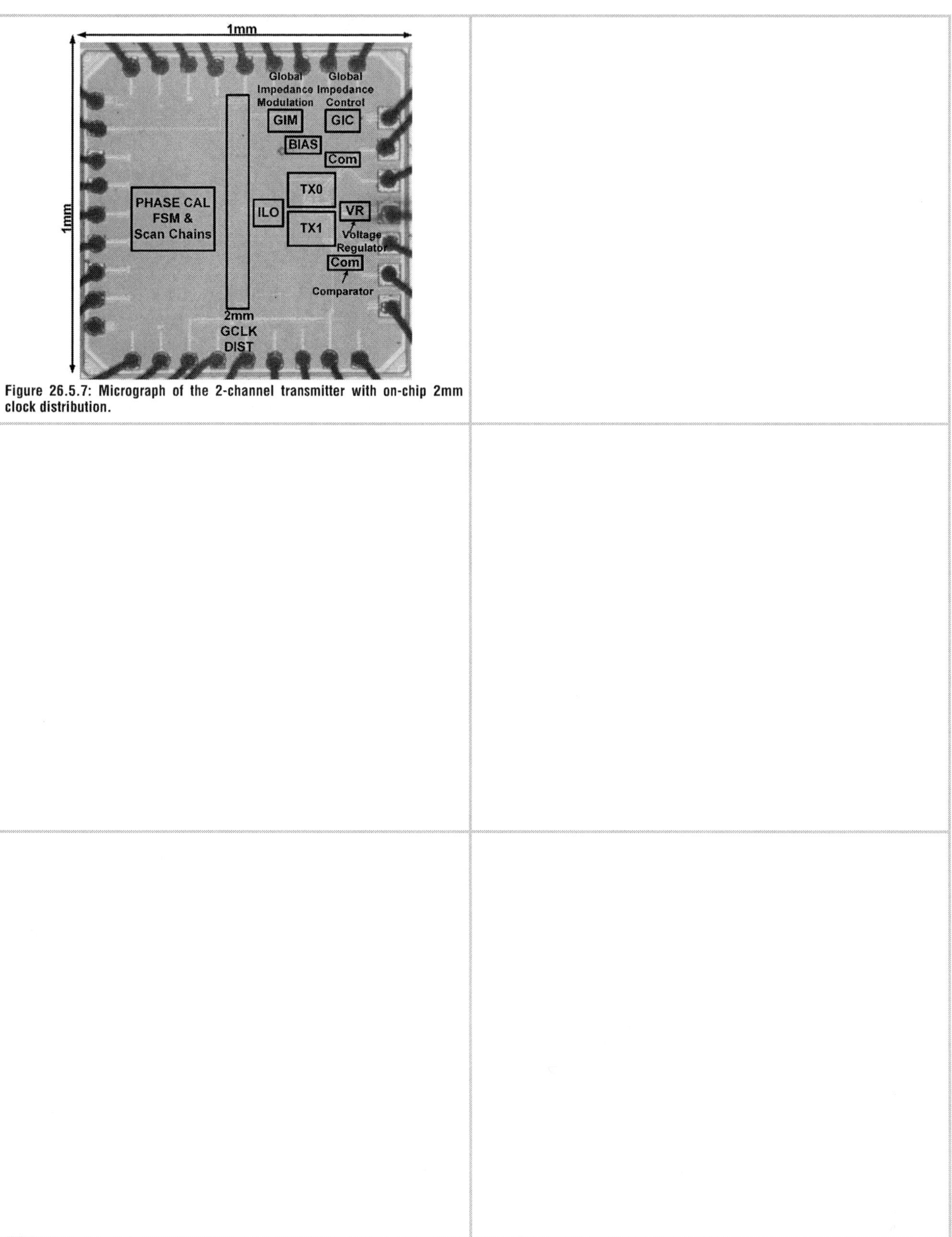

Figure 26.5.7: Micrograph of the 2-channel transmitter with on-chip 2mm clock distribution.

26.6 A 2.667Gb/s DDR3 Memory Interface with Asymmetric ODT on Wirebond Package and Single-Side-Mounted PCB

Shang-Pin Chen, Chih-Chien Hung, Qui-Ting Chen, Sheng-Ming Chang, Ming-Shi Liou, Bo-Wei Hsieh, Hsiang-I Huang, Brian Liu, Yan-Bin Luo

MediaTek, Hsinchu, Taiwan

DDR3 memory interface (I/F) with single-end signals is very sensitive to external environments, such as chip package type and system board design. In order to guarantee the system performance, IP providers often define the package and PCB design constraints to reduce product risks [1]. These design constraints may increase the package size and DDR3 PCB area to cost, increasing the whole system cost. Therefore, our DDR3 memory I/F design addresses this problem, relaxing the external environment requirements, especially relating to power integrity, and achieves 2.667Gb/s operation in a wirebond package and single-side mounted PCB. The difference between double-side and single-side mounted PCB is shown in Fig. 26.6.1. The capacitor on the PCB cannot be mounted directly near the SOC on the back side of PCB. This external environment increases the distance of the current return loop and also increases the inductance of power decouple capacitance equivalently.

General DDR3 memory interface is shown in Fig. 26.6.2(a). In the output stage, data signal traverses a programmable delay line, connects to level-shift-up circuit and then drives via the output driver. The supply bounce in DDR3 I/O 1.5V affects the output jitter so that the circuit speed and supply-noise generation of level-shift-up circuit and output driver are key factors. One method of output stage improvement is realized in the output driver by cascoded thin-gate-oxide device [2-3]. This method can reduce the latency of the output driver stage and reduce transient supply noise in its I/O power domain. However, it needs an additional overshoot/undershoot sense circuit [2] to decrease the risk of device reliability or lifetime issues and it also increases circuit complexity.

Our output driver is implemented using thick-gate-oxide devices to avoid the device reliability problem. The traditional level-shift-up circuit is sensitive to supply noise since the speed of thick-gate-oxide devices is not fast enough, especially in 2.5V I/O device process. Therefore, the clocking-based level-shift-up circuit with 2-1 serializer is designed to speed up the signal transition from the core power domain to the I/O power domain. In addition, the programmable-delay circuit is moved from the data path to clock path as shown in Figs. 26.6.2(b) and 26.6.2(c). The traditional programmable-delay circuit in the data path may suffer from supply noise and result in the data jitter increasing. An internal LDO can be implemented to suppress the supply noise but the capacitor area must be large enough to cover broadband data transition noise. In our method, the delay cell is moved to the clock path and its internal LDO design can be implemented easily without consideration for broadband data transition noise. Therefore, the output jitter of transmitter from supply noise is minimized.

Moreover, the power dissipation of SoC in home entertainment devices, such as DTV or DVD player, make it important to relax the thermal-resistance specification of chip package. Figure 26.6.3(a) shows the simplified block diagram of read path. The 60Ω-ODT structure of DDR3 I/F, i.e., center termination (CTT), dissipates 6.25mA (1.5V/240Ω) current per DQ pin. If the system needs a 16b DDR3 interface, the power dissipation supplied only for this CTT structure is around 125mA (16-pin data and 4-pin data strobes). In order to save this power dissipation, we implement a pseudo open-drain (POD)-termination method, which is similar to DDR4 or GDDRx, but it is applied only to the controller side. The termination of the DDR3 device side is always center termination following the JEDEC definition. Therefore, a custom DDR3 device is not necessary. The asymmetric ODT method, i.e., POD on the controller side and CTT for the DDR3 device, reduces around 125mA during read operation of the system with 16-bit DDR3 interface. The reference voltage of controller side also needs to be adjusted to fit the POD electrical requirement. The built-in programmable reference-voltage generator is implemented to support this feature without requirement for an additional PCB component.

The speed bottleneck of DDR3 memory I/F is not only in write path, but also in read path, especially with write and read operations that interact frequently. At the same time, the POD termination also degrades the eye opening since rising and falling edges are not symmetrical relative to center termination. The traditional input buffer with thick-oxide-gate device is realized in DDR3 I/O power domain. When the read and write commands interact, a larger supply bounce, generated from the output driver triggered by previous write operation, interferes with the start of read bursts, reducing the read margin. A circuit is designed to solve these problems and is shown in Figs. 26.6.3(b) and 26.6.3(c). In order to avoid the interference of the input buffer from supply noise, the circuit is operated only in the core power domain based on source-follower circuits. And the input device of this source follower is implemented using a native thick-gate-oxide NMOS device with low threshold voltage. A traditional input-buffer structure using thin-gate-oxide devices can be implemented directly after this source follower stage so that read operation performance can be improved.

In the ADDR/CMD/CK part, fly-by topology and V_{tt} termination guarantee signal integrity with multi-drop loads in DDR3 applications, as shown in Fig. 26.6.4(a). In non-DIMM applications, the number of multi-drop loads is predictable. A regulator topology without V_{tt} is illustrated in Fig. 26.6.4(b). When this feature is enabled, only CSN command supports 1T period time and the remaining ADDR/CMD commands support 2T period time to tolerate the effects of signal skew and reflection. The optional chip-select (CS) pin is also added to the DDR3 controller to compose the CS and CSN differential signals with 100Ω termination, but only the CSN signal is connected to DDR3 device. Therefore, this ADDR/CMD/CK termination topology without V_{tt} regulator saves bill-of-material (BOM) cost.

The read and write margins are listed in Fig. 26.6.5 and the DDR3 IP is tested from 1.6Gb/s to 2.667Gb/s. Figure 26.6.5(a) shows the timing margin of symmetric termination and read/write operation all have more than 0.5UI margin. Figure 26.6.5(b) shows the timing margin of asymmetric termination; its read margin is a little degraded relative to symmetric termination. However, it still has enough timing margin for the DDR3 2.2667Gb/s application. The read examples of maximum and minimum eye open for different termination methods at 2.667Gb/s are also shown in Fig. 26.6.6.

This DDR3 memory I/F can operate at 2.667Gb/s with a wirebond BGA package and single-side mounted PCB. Figure 26.6.7 shows the micrograph of CMD/ADDR/CK and 16b DQ macros, with their areas occupying 0.37mm^2 and 0.39mm^2, respectively, in a 40nm LP process with 2.5V I/O devices. It also reduces power dissipation to 176mW with asymmetric ODT in DQ signals and reduces BOM cost without V_{tt} regulator in CMD/ADDR signals.

Acknowledgements:
The authors would like to acknowledge Chun-Ping Chen for PCB design, Ying-Chih Chen for package design, Chao-Wei Tseng and Chia-Chuan Wu for measurement, and Chun-Yi Wu for layout.

References:
[1] J. Feng, R. Schmitt, H. Lan, et al., "Signal and Power Integrity for a 1600 Mbps DDR3 PHY in Wirebond Package," *DesignCon*, Feb. 2011.
[2] R., Kumar, G. Hinton, et al., "A Family of 45nm IA Processors," *ISSCC Dig. Tech. Papers*, pp. 58-59, Feb. 2009.
[3] N. Kurd, S. Bhamidipati, et al., "A Family of 32nm IA Processors," *JSSC*, pp. 119-130, Jan. 2011.

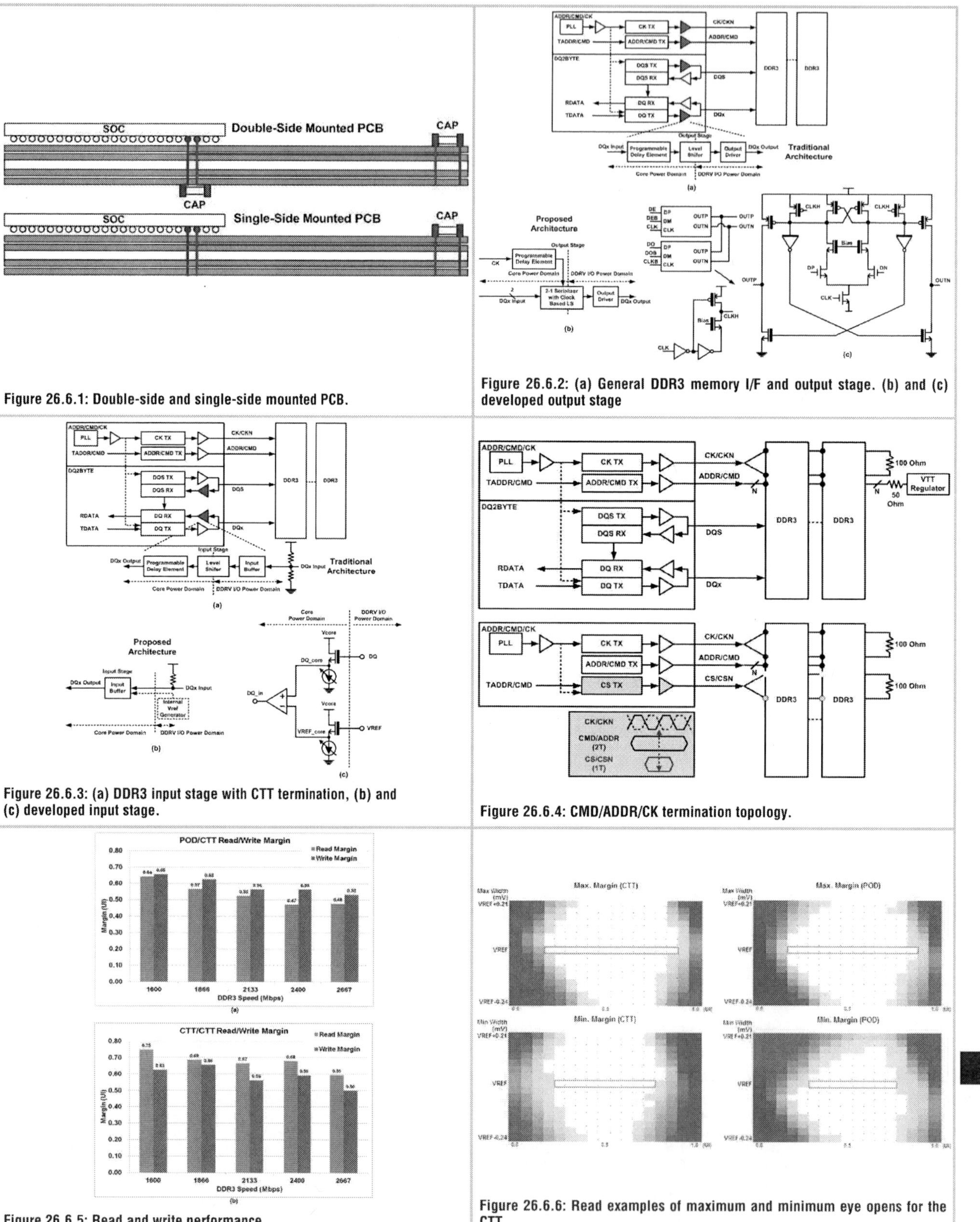

Figure 26.6.1: Double-side and single-side mounted PCB.

Figure 26.6.2: (a) General DDR3 memory I/F and output stage. (b) and (c) developed output stage

Figure 26.6.3: (a) DDR3 input stage with CTT termination, (b) and (c) developed input stage.

Figure 26.6.4: CMD/ADDR/CK termination topology.

Figure 26.6.5: Read and write performance.

Figure 26.6.6: Read examples of maximum and minimum eye opens for the CTT.

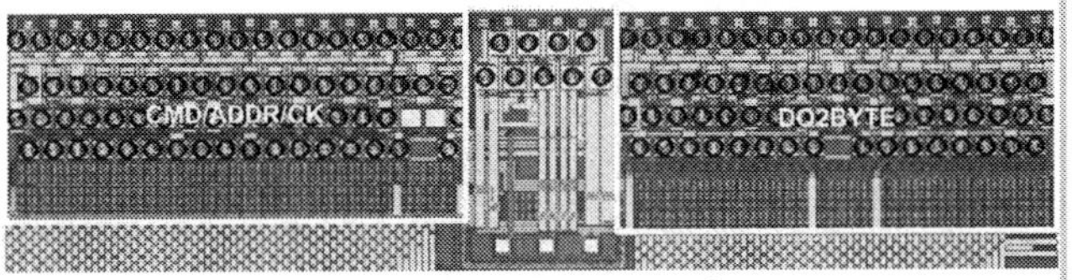

CMD/ADDR/CK

DQ2BYTE

Figure 26.6.7: Chip micrograph.

Session 27 Overview: *Energy-Efficient Digital Circuits*

ENERGY-EFFICIENT DIGITAL SUBCOMMITTEE

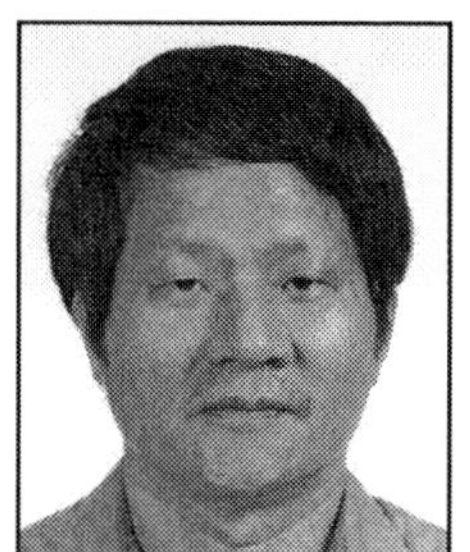

Session Chair: *Bing Sheu*
TSMC, Hsinchu, Taiwan

Session Co-Chair: *Marian Verhelst*
KU Leuven, Belgium

Next-generation computing devices for mobile applications require energy efficiency, while achieving increased system performance and programmability. This is enabled through ultra-low-voltage operation, application kernel hardware acceleration, and embedded reconfigurable logic. The papers in this session present several advanced implementation techniques to take these trends to the next level and bring them to the market. Examples include back-biasing techniques, variation-resilient near-threshold operation, charge-recovery schemes, and multi-granularity reconfigurable logic. These techniques are demonstrated in real-world applications such as a DSP processor, low-density parity check (LDPC) decoders, a speech-recognition processor, an FFT core, and a JPEG encoder.

27.1 A 460MHz at 397mV, 2.6GHz at 1.3V, 32b VLIW DSP, Embedding F_{MAX} Tracking **1:30 PM**

R. Wilson, STMicroelectronics, Crolles, France

In Paper 27.1, STMicroelectronics and CEA-LETI - MINATEC describe a 32b VLIW DSP fabricated in 28nm UTBB FDSOI technology capable of dynamically tracking the maximum frequency within 3.5% accuracy. The 1mm^2 chip achieves scalable performance and high energy efficiency in the 397mV-to-1.3V supply range, showing an operating frequency of 460MHz at the minimum 0.4V supply, and a minimum energy per operation of 62pJ at 0.46V.

27.2 A 6mW 5K-Word Real-Time Speech Recognizer Using WFST Models **2:00 PM**

M. Price, Massachusetts Institute of Technology, Cambridge, MA

In Paper 27.2, MIT presents a speech-recognition chip targeted at real-time decoding for medium vocabulary (5,000 word) continuous speech tasks. The 6.25mm^2 chip excels in memory usage efficiency, is implemented in 65nm CMOS, and consumes 6mW of core power during real-time decoding with a 13.0% word error rate.

27.3 A 210mV 5MHz Variation-Resilient Near-Threshold JPEG Encoder in 40nm CMOS **2:30 PM**

N. Reynders, KU Leuven, Leuven, Belgium

In Paper 27.3, KU Leuven describes the design of a near-threshold full JPEG encoder in a 40nm CMOS technology. The variation-resilient chip is functional down to a supply voltage of 210mV at a 5MHz clock frequency, and consumes 29pJ/pixel at the minimum-energy point, which occurs at 41MHz.

978-1-4799-0917-9/14 $31.00 © 2014 IEEE

27.4 A 0.75-Million-Point Fourier-Transform Chip for Frequency-Sparse Signals **3:15 PM**

O. Abari, Massachusetts Institute of Technology, Cambridge, MA

In Paper 27.4, MIT presents a 0.6mm² VLSI implementation of a 746,496-point FFT for up to 0.1% frequency-sparse signals in a 45nm SOI process, consuming 298 to 1206nJ of energy at a throughput of 36 to 109GS/s, utilizing scaled supply voltage of 0.66 to 1.18V, achieving an 88× reduction in run-time compared to a C++ implementation on an i7 CPU.

27.5 A Multi-Granularity FPGA with Hierarchical Interconnects for Efficient and Flexible Mobile Computing **3:45 PM**

C. C. Wang, University of California, Los Angeles, CA

In Paper 27.5, UCLA and Qualcomm describe a multi-granularity FPGA capable of efficiently mapping mobile-computing algorithms. The 20mm² core in 40nm CMOS, incorporates traditional configurable logic blocks, 2 coarse-grain kernels and an efficient, mixed-radix hierarchical interconnect, achieving a 4× interconnect area reduction and 10-50% lower active power over commercial FPGAs. Using the coarse-grain kernels, the chip's energy efficiency is within 4-5× of ASIC designs.

27.6 An 821MHz 7.9Gb/s 7.3pJ/b/iteration Charge-Recovery LDPC Decoder **4:15 PM**

T-C. Ou, University of Michigan, Ann Arbor, MI

In Paper 27.6, the University of Michigan presents a 65nm 576b LDPC decoder, using charge-recovery logic demonstrating a 1.7× gain in energy efficiency, drastically increasing the scale at which charge recovery logic has been applied, with a die area of 1.5mm². The lowest energy of 7.3pJ/b/iteration is achieved at a 7.9Gb/s throughput, when self-oscillating at 821MHz, recovering 51.4% of the supplied energy.

27.7 A Scalable 1.5-to-6Gb/s 6.2-to-38.1mW LDPC Decoder for 60GHz Wireless Networks in 28nm UTBB FDSOI **4:45 PM**

M. Weiner, University of California, Berkeley, CA

In Paper 27.7, UC Berkeley, STMicroelectronics and EPFL describe a 4mm² LDPC decoder chip that supports all codes and throughputs of the IEEE 802.11ad standard for Gb/s wireless LANs. The use of 28nm UTBB FDSOI with back bias allows for aggressive frequency and voltage scaling. Algorithmic-architectural co-optimization enables throughputs of 1.5, 3 and 6Gb/s, consuming 6, 14 and 38mW, respectively.

27.8 A Static Contention-Free Single-Phase-Clocked 24T Flip-Flop in 45nm for Low-Power Applications **5:00 PM**

Y. Kim, University of Michigan, Ann Arbor, MI

In Paper 27.8, the University of Michigan presents a static single-phase contention-free flip-flop (S2CFF) in 45nm SOI, targeting wide-range voltage scalable designs (0.4 to 1V). The static contention-free flip-flop offers single-phase clock operation with the same device count as a conventional transmission-gate flip-flop (TGFF). Measured active energy, clock power and leakage power are reduced by 32%, 41%, and 35%, respectively, compared to the TGFF.

27.1 A 460MHz at 397mV, 2.6GHz at 1.3V, 32b VLIW DSP, Embedding F_{MAX} Tracking

Robin Wilson[1], Edith Beigne[2], Philippe Flatresse[1], Alexandre Valentian[2], Fady Abouzeid[1], Thomas Benoist[2], Christian Bernard[2], Sebastien Bernard[2], Olivier Billoint[2], Sylvain Clerc[1], Bastien Giraud[2], Anuj Grover[1], Julien Le Coz[1], Ivan Miro Panades[2], Jean-Philippe Noel[1], Bertrand Pelloux-Prayer[1], Philippe Roche[1], Olivier Thomas[2], Y. Thonnart[2], David Turgis[1], Fabien Clermidy[2], Philippe Magarshack[1]

[1]STMicroelectronics, Crolles, France,
[2]CEA-LETI - MINATEC, Grenoble, France

Wide-voltage-range-operation DSPs bring more versatility to achieve high energy efficiency in mobile applications to increase signal processing complexity and handle a large range of performance specifications. This paper describes a 32b DSP fabricated in 28nm UTBB FDSOI technology [1]. Body-bias-voltage (V_{BB}) scaling from 0V up to ±2V (Pwell/Nwell) decreases the DSP core V_{DDMIN} to 397mV and increases clock frequency by +400% at 500mV and +114% at 1.3V. In addition to technology gains, dedicated design features are included to increase frequency over the full V_{DD} range, considering parameter variations. As depicted in Fig. 27.1.1, the 32b datapath VLIW DSP is organized around a MAC dedicated to complex arithmetic and two dedicated operators: a cordic/divider and a compare/select. Data enters the circuit through a serial interface and code is run from a 64×32b register file. It has been shown in [1] that a given operating frequency can be achieved at a lower V_{DD} in UTBB FDSOI compared to bulk by applying a forward-body bias. An additional design step is achieved in this work by (1) increasing the frequency at low V_{DD} thanks to a specific selection and design of standard cells with respect to power vs. performance and (2) dynamically tracking the maximum frequency to cope with variations.

Sixty combinational functions were selected through iterative synthesis optimizations. Multiple gate-length configurations (Fig. 27.1.2) were benchmarked through SPICE simulations and silicon measurements of quasi-FO4 ring oscillators [5], enabling us to discard redundant cell configurations and retain the power-delay optimal cells over the range (such as asymmetric poly biasing, APB4) offering 30% energy gain with less than 3% frequency cost. This resulted in a library of 320 low-V_t standard cells, characterized over the 0.275V-to-1.4V range, offering non-overlapping power-performance characteristics, and designed-for-variability tolerance using 6σ MC simulations. Fast pulse-triggered flip-flops (pulsed FFs) with small data-to-output (D-Q) delay were designed to complete the optimized library. Pulsed FFs contain a single latch, made transparent during a short period of time after the clock edge: the width of the transparent window is determined by a pulse generator. The difficulty in using pulsed FFs over a wide voltage range lies in the sensitivity of the pulse generator to process variations at low voltage, since large pulse widths lead to unacceptably large hold times, while short pulse widths can lead to functional failures. The smallest pulse width was determined with statistical simulations to ensure robustness to 3σ variations down to 275mV. Pulsed-FF silicon measurements were performed on a test scribe and showed correct functionality at 275mV and a measured D-to-Q delay of 1.3ns at 0.3V.

To maximize energy efficiency over the full voltage range under PVT variations, accurate and low-cost F_{MAX} estimation techniques are mandatory during circuit operation. We use two F_{MAX} tracking solutions based on (1) replica-path cloning (CODA) and (2) timing-slack monitoring (TMFLT), both of which offer good accuracy at 1.0V and the latter functional down to 0.6V. For both solutions, and over the full V_{DD} range, the objective is to prevent timing failures and to reduce margins required for PVT variations during DSP operation. A power-variability controller (CVP) is used to control, program and calibrate CODA and TMFLT and store the captured timing information. For each frequency, V_{DD} and V_{BB} are dynamically scaled to reach a minimum energy point.

CODA is a sensor-based circuit which accounts for local variations in critical paths impacting the DSP's F_{MAX}. As shown in Fig. 27.1.3, it contains 16 pairs of clones of representative critical paths (RCPs). RCP pairs are close to each other to capture local variation. One RCP is used as a canary path (delay monitor) and the other one as a loop (oscillation-frequency monitor). The propagation delay of the RCP (DRCP) is determined through the RCP-clone delay (DRCP-clone) – a warning appears when F_{CLK}=1/DRCP-clone (< 1/DRCP) – and is correlated in loop mode through direct measurement of oscillating frequency (FRCP-clone). Silicon measurements of the DSP were performed to determine F_{MAX} and correlation with frequency predictions of CODA was made. Multi-die measurements of 5 RCP pairs show that F_{MAX} can be predicted, using loop mode, with +4/-3% accuracy at 1V. At a fixed voltage and frequency target, this solution brings the benefit of being non-intrusive to the monitored circuit.

In the context of wide-voltage-range DSPs, a system (Fig. 27.1.4) composed of an on-line-programmable time-to-digital converter (TMFLT-R) correlated to off-line timing-slack sensors (TMFLT-S) is proposed. The system has a low area cost, and is functional and accurate, even at voltages down to 0.6V. 128 TMFLT-S instances, representing 0.9% of the total number of flip-flops, are directly instrumenting flip-flops inside the core. The TMFLT-S is based on a stability detector and delay lines. It warns of a slack-time violation 160ps (at 1V) before the register failure. Critical paths with a slack margin up to 160ps are instrumented to anticipate circuit failure, but there is no need to instrument the most critical paths of the circuit. The TFMLT-S responses are statistically correlated to the effective F_{MAX} through a post-silicon calibration step giving an accurate estimation. However, during regular processing, the TMFLT-S paths may not be activated. To cope with this, we embed an always-active (yet less accurate) TMFLT-R, based on a time-to-digital converter. F_{MAX} control relies on its digital response (SIG, see: Fig. 27.1.4) generated with a precision of 20ps. To improve the system accuracy, a calibration phase (e.g. at power-up) is used, in which (1) the TMFLT-S instances are activated by a core functional pattern at an increasing frequency and for multiple voltages, and sampled, (2) the response of the TMFLT-R is stored at the same points and for each of its programmable states, (3) an optimal piecewise TMFLT-R signature is computed by selecting the best segment (in a V_{DD}, V_{BB}, T° space) for each programmable state to fit the TMFLT-S F_{MAX} response. During computation, the TMFLT-R is programmed with respect to V_{DD}, V_{BB}, T° values to select the appropriate SIG value and raises a warning once this value is reached. The overall TMFLT monitoring results on silicon, and after calibration, in an F_{MAX} estimation error of +8/-6% at 0.6V, and +4.1/-2.9% at 1V.

Figure 27.1.5 shows frequency and energy/operation DSP silicon results obtained by performing a 1024 FFT using the register-array configuration. The frequency curve shows 2.6GHz at V_{DD}=1.3V and V_{BB}=2V; and also shows 460MHz at V_{DD}=397mV and V_{BB}=2V. The use of an efficient V_{BB} (boost) above 1.5V decreases V_{DDMIN} from 479mV down to 397mV. Energy-per-operation illustrates that, for a fixed energy budget at 100pJ/cycle, we can increase the frequency by 59% and reduce the energy by 17% to 20% for a fixed frequency target. Benchmark results are shown in Fig. 27.1.6. Our design methodology combined with the intrinsic advantages of the UTBB FDSOI technology, produces a high operating frequency at 0.4V, and a low-peak energy efficiency measured at 62pJ/op at 0.53V. The TMFLT tracking methodology, measured on silicon at 0.6V, shows a high energy gain of 40.6% with respect to a worst-case-corner approach.

References:

[1] D. Jacquet, *et al.*, "2.6 GHZ Ultra Wide Voltage Range Energy Efficient Dual A9 in 28 nm UTBB FD-SOI," *IEEE Symp. VLSI Circuits*, pp. C44-C45, 2013.

[2] G. Gammie, *et al.*, "A 28nm 0.6V Low-Power DSP for Mobile Applications", *ISSCC Dig. Tech. Papers*, pp. 131-134, 2011.

[3] S. Jain, *et al.*, "A 280mV-to-1.2V Wide-Operating-Range IA-32 Processor in 32nm CMOS," *ISSCC Dig. Tech. Papers*, pp. 66-68, 2012.

[4] S. Hsu, *et al.*, "A 280mV-to-1.1V 256b Reconfigurable SIMD Vector Permutation Engine with 2-Dimensional Shuffle in 22nm CMOS", *ISSCC Dig. Tech. Papers*, pp. 178-180, 2012.

[5] F. Abouzeid, *et al.*, "A 45nm CMOS 0.35v-Optimized Standard Cell library for Ultra-Low Power Applications", *IEEE International Symp. Low-Power Electronics and Design*, pp. 225-230, 2009.

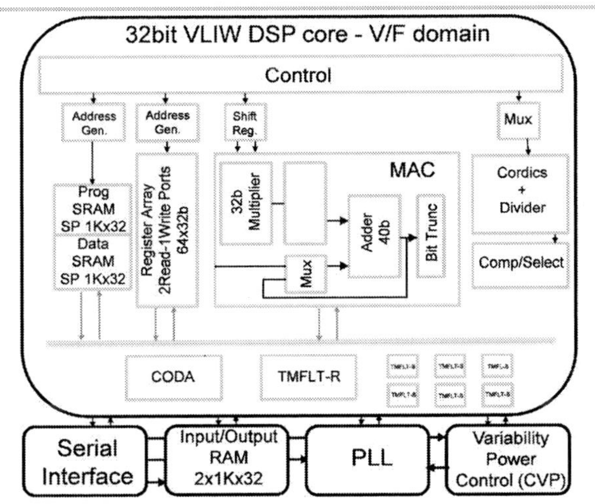

Figure 27.1.1: 32b datapath VLIW DSP architecture. CODA and TMFLT F_{MAX} tracking integration and its CVP control.

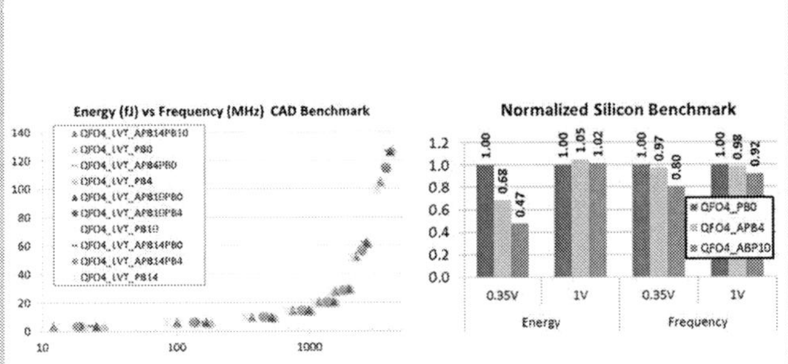

Figure 27.1.2: Energy vs. frequency CAD benchmark of qFO4 ring oscillators with various poly-biasing (PB) configurations and silicon benchmarking (25°C).

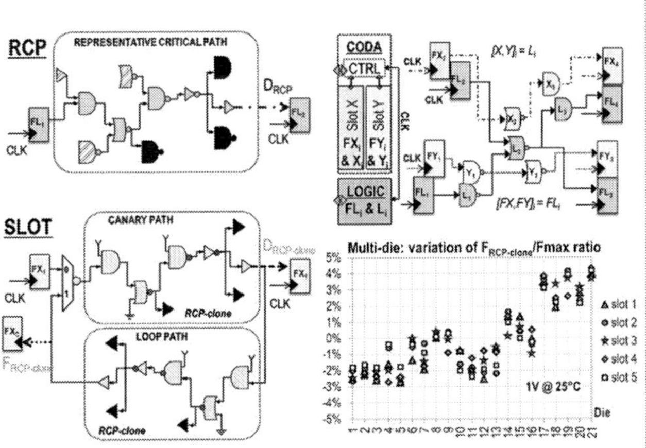

Figure 27.1.3: Replica-path cloning (CODA) F_{MAX} tracking architecture. Silicon variations measured across 21 dies at 1V, 25°C.

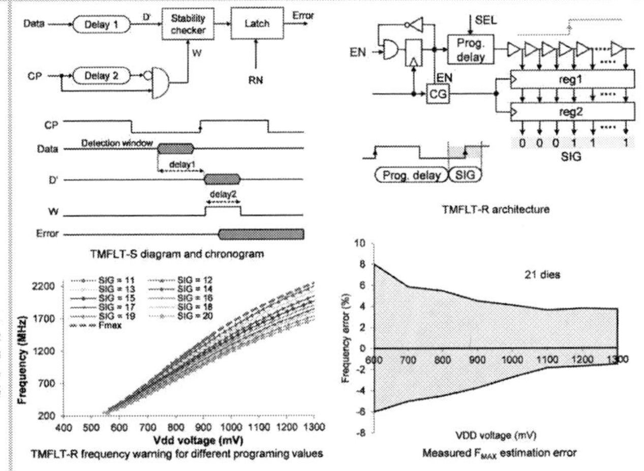

Figure 27.1.4: Timing-slack-monitoring (TMFLT) F_{MAX} tracking approach. Architecture of off-line timing-slack sensors (TMFLT-S) and on-line programmable time-to-digital converter (TMFLT-R).

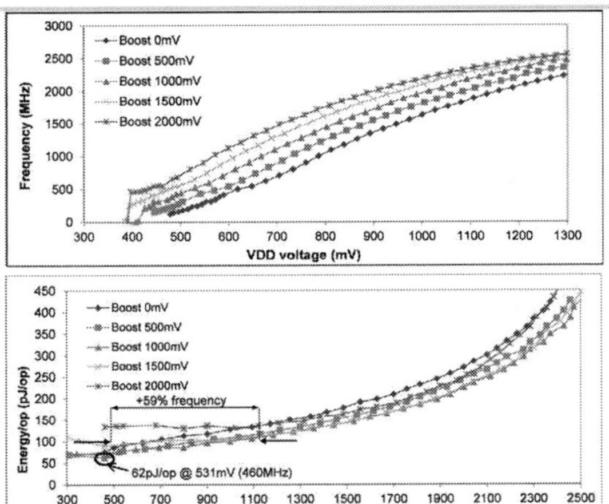

Figure 27.1.5: Frequency and energy/operation DSP silicon measurements obtained by performing a 1024 FFT using the register array configuration. Boost corresponding to V_{BB} values.

	[1]	[2]	[3]	[4]	This work
Technology	28nm UTBB FDSOI	28nm LP Bulk	32nm Bulk	22nm Trigate	**28nm UTBB FDSOI**
VDD operating range	0.6V-1.2V	0.34V-1V	0.28V-1.2V	0.28V-1.1V	**0.39V-1.3V**
Max measured Frequency	2.6GHz@1.3V	587MHz@1V	915MHz@1,2V	2.5GHz@1,1V	**2.6GHz@1.3V**
Frequency @Min voltage	1GHz@0.6V	3.6MHz@0,4V	3MHz@0.28V	16.8MHz@0.28V	**460MHz@0.397V**
Total power consumption	na	113mW@1V	400mW@1V	227mW@1V	**370mW@1V**
Peak energy efficiency	na	na	170pJ/cycle @0.45V	585GOPS/W @260mV	**62pJ/op @0.53V**

Figure 27.1.6: Comparison with recent wide-voltage-range DSPs or processors [1][2][3][4] in 32nm and below.

27

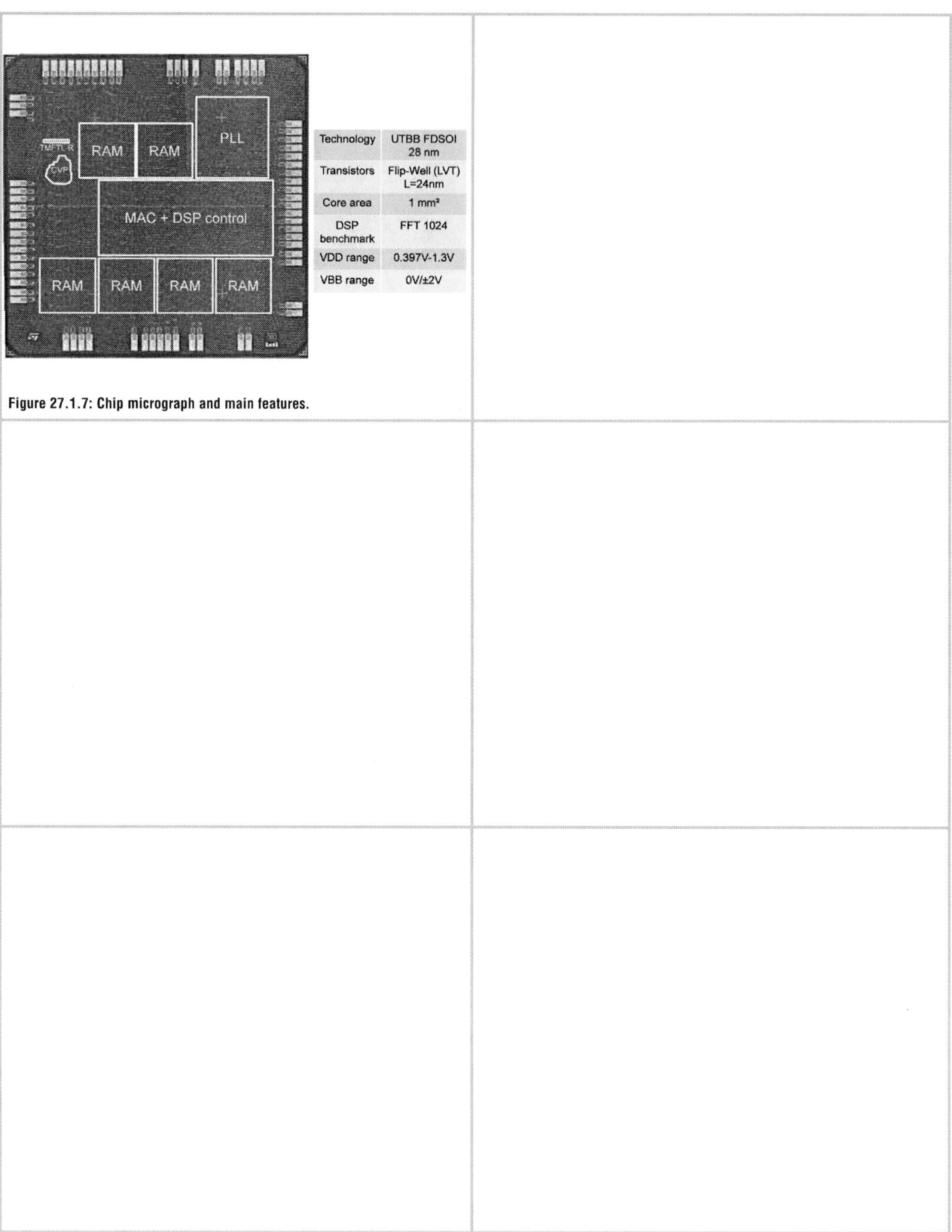

Technology	UTBB FDSOI 28 nm
Transistors	Flip-Well (LVT) L=24nm
Core area	1 mm²
DSP benchmark	FFT 1024
VDD range	0.397V-1.3V
VBB range	0V/±2V

Figure 27.1.7: Chip micrograph and main features.

27.2 A 6mW 5K-Word Real-Time Speech Recognizer Using WFST Models

Michael Price, James Glass, Anantha P. Chandrakasan

Massachusetts Institute of Technology, Cambridge, MA

Hardware-accelerated speech recognition is needed to supplement today's cloud-based systems in power- and bandwidth-constrained scenarios such as wearable electronics. With efficient hardware speech decoders, client devices can seamlessly transition between cloud-based and local tasks depending on the availability of power and networking. Most previous efforts in hardware speech decoding [1–2] focused primarily on faster decoding rather than low-power devices operating at real-time speed. More recently, [3] demonstrated real-time decoding using 54mW and 82MB/s memory bandwidth, though their architectural optimizations are not easily generalized to the weighted finite-state transducer (WFST) models used by state-of-the-art software decoders. This paper presents a 6mW speech recognition ASIC that uses WFST search networks and performs end-to-end decoding from audio input to text output.

Algorithms and data structures developed for software speech decoders are also applied in hardware. Fig. 27.2.1 outlines the operation of a speech recognition system, where annotated training data is used to generate statistical models of speech production. A hidden Markov model (HMM)-based Viterbi decoder searches for the most likely path through millions of hidden states. Different statistical models are needed for the transition $p(x_t \mid x_{t-1})$ and emission $p(y_t \mid x_t)$ probabilities. Most modern decoders store transition information in a WFST, which allows the system to combine and optimize several levels of knowledge about the training data from the sub-phonetic to the grammatical level in a single searchable network. The emission probabilities (acoustic model) are represented using a diagonal Gaussian mixture model (GMM), which is well-suited to modeling continuous distributions using relatively few parameters.

Our architecture includes the entire speech-to-text decoding chain and addresses the constraints of low-power embedded systems, including limited on-chip memory capacity and limited off-chip memory bandwidth. The block diagram shown in Fig. 27.2.2 illustrates the flow of information. Audio samples arriving at 16kHz pass through a chain of signal processing elements including an FFT, filter bank, and DCT to generate 39-dimensional mel-frequency cepstral coefficient (MFCC) feature vectors at 100Hz. The filter bank takes advantage of triangular response shapes to generate 26 outputs using only 2 multipliers, as shown in Fig. 27.2.3. We also decimate the real signal into a half-rate complex signal in order to complete the FFT in half as many clock cycles. These operations run at $^1/_{16}$ of the decoder clock frequency and consume an average of 110μW at 0.9V, 50MHz decoder clock.

Each new feature vector triggers a time-synchronous Viterbi search update propagating a set of hypotheses (WFST states with likelihood scores) forward by one time step. A list of hypotheses is read from the active state list (ASL) for the current frame, which is implemented as a hash table in on-chip SRAM. The WFST model is queried for reachable states, resulting in a large set of candidate hypotheses for the next frame. The GMM evaluates the likelihood of observing the feature vector under each of these hypotheses, and likelihood scores are used to select the most promising hypotheses for storage in the ASL for the next frame. Multiplexers swap read and write ports between the two ASLs and the cycle repeats until the end of the utterance. The control module traces backwards through ASL snapshots stored in memory to determine the most likely state sequence, which is converted to a text transcription using a table of WFST output labels.

On-chip memory capacity limits the number of hypotheses that can be stored in the ASL. A constant beam width (search pruning threshold) would have to be set fairly low in order to avoid overflowing the ASL, and desirable hypotheses would be rejected. As shown in [4], the overflow could be stored in off-chip DRAM, but this incurs additional latency and memory bandwidth. By adopting a feedback scheme (shown in Fig. 27.2.4), we are able to obtain a 13.0% word error rate (WER) with an ASL capacity of only 4096 states. We approximate the necessary beam width by regulating the fraction of candidate arcs accepted throughout a Viterbi update. The total number of candidates is predicted by accumulating outgoing arc counts from the previous frame. The histogram of ASL sizes under zero, moderate, and excessive feedback levels is shown on the right in Fig. 27.2.4. With an appropriate feedback gain and clamp range, beam width control will compress (but not eliminate) the natural variation in ambiguity that occurs throughout an utterance. This reduces the workload when there are few good hypotheses, and avoids discarding the best hypotheses when there are too many to be stored in the ASL.

Anticipating the use of slower non-volatile memories to reduce system power, we applied optimizations to reduce memory bandwidth and make memory accesses more sequential. To make related WFST arcs appear close to each other, we order the states in memory according to a breadth-first search. Our fully-associative WFST cache uses the pseudo-LRU eviction algorithm to maximize hit rate and prioritizes arcs at isolated memory locations to reduce page or bank activation penalties. In contrast, acoustic model reads of GMM parameters are highly sequential but can easily exceed 1GB/s; our architecture for reducing this bandwidth to a practical level is shown in Fig. 27.2.5. Repeated access of the same data is avoided by caching the likelihood of each mixture that has been evaluated against a given feature vector; this cache occupies just 98Kb and has an 86% hit rate. We also quantize the GMM means to 5b and variances to 3b, resulting in an 8:1 compression of parameters relative to 32b floating-point format [5]. The impact on decoding accuracy is minimized by selecting a nonlinear quantizer for each parameter according to its empirical distribution in the model. Separate quantization tables are stored for each dimension in order to accommodate nonwhite feature spaces. The combination of caching and parameter compression reduces the GMM memory bandwidth from 2.9GB/s to 54MB/s without requiring storage of GMM results for multiple frames.

This IC was fabricated on a 65nm low-power logic process, and all tests were performed in a real-world demonstration system using an FPGA for external memory access and including all communication latencies. Fig. 27.2.6 shows the measured WER vs. decoding time tradeoff for the Wall Street Journal (Nov. 1992) data set, using a WFST with 2.9M states and 9.1M arcs. The acoustic model contains 10.2M parameters, or 6 times the complexity of that used in [3]. Power consumption is closely correlated with the number of hypotheses evaluated; accuracy and decoding speed can be traded for the desired level of power consumption by adjusting the nominal beam width in conjunction with voltage/frequency scaling. There is no power gating, but we externally gate the clock between utterances for an idle (leakage) power of 42μW. SRAM accounts for 74% of core area and 77% of power consumption, highlighting the importance of memory in speech decoding architectures. Core power consumption averages 6.0mW during real-time decoding at 0.85V and 50MHz. The die photo and specifications are summarized in Fig. 27.2.7.

Acknowledgements:

This work was supported by Quanta Computer, Inc. as part of Project Qmulus, and by an Irwin and Joan Jacobs fellowship. The authors would like to thank the TSMC University Shuttle Program for providing chip fabrication and Xilinx for providing FPGA boards.

References:

[1] J. Choi, K. You, W Sung, "An FPGA Implementation of Speech Recognition with Weighted Finite State Transducers," *IEEE International Conf. on Acoustics Speech and Signal Processing,* pp. 1602–1605, 2010.

[2] J.R. Johnston, R.A. Rutenbar, "A High-rate, Low-power, ASIC Speech Decoder using Finite State Transducers," *IEEE International Conf. on Application-Specific Systems, Architectures and Processors,* pp. 77–85, 2012.

[3] G. He, Y. Miyamoto, K. Matsuda, S. Izumi, H. Kawaguchi, M. Yoshimoto, "A 40-nm 54-mW 3x-real-time VLSI Processor for 60-kword Continuous Speech Recognition," *IEEE Workshop on Signal Processing Systems,* pp. 147–152, 2013.

[4] Y. Choi, K. You, J. Choi, W. Sung, "A Real- Time FPGA-Based 20,000-Word Speech Recognizer With Optimized DRAM Access," *IEEE Trans. Circuits and Systems-I,* vol. 57, no. 8, pp. 2119–2131, 2010.

[5] I.L. Hetherington, "PocketSUMMIT: Small-Footprint Continuous Speech Recognition," *INTERSPEECH,* pp. 1465–1468, 2007.

978-1-4799-0917-9/14 $31.00 © 2014 IEEE

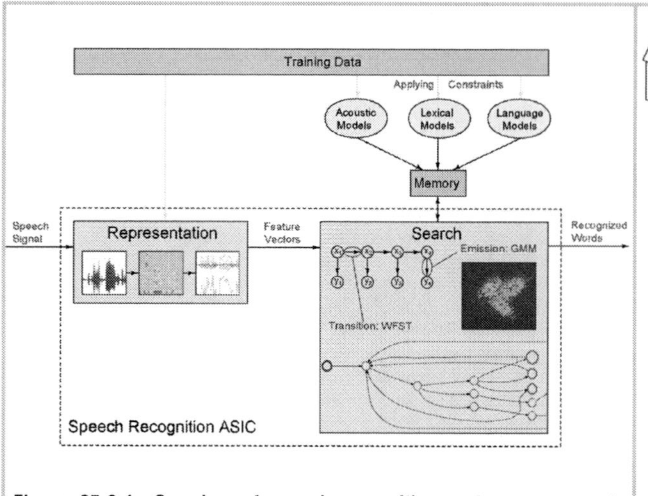

Figure 27.2.1: Overview of speech recognition system; components implemented on chip shown in dashed box.

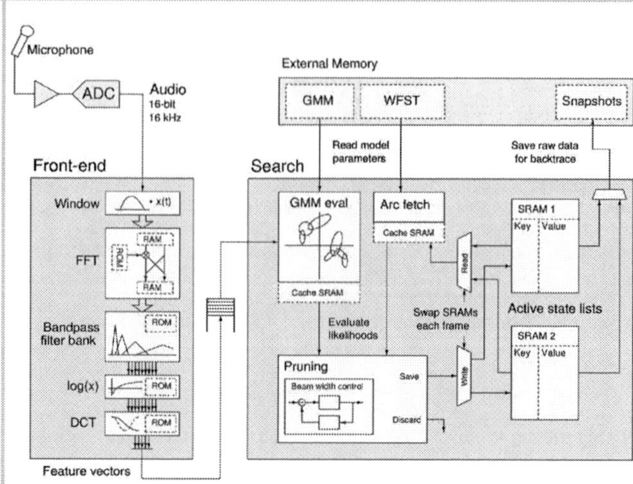

Figure 27.2.2: Block diagram of speech recognition chip.

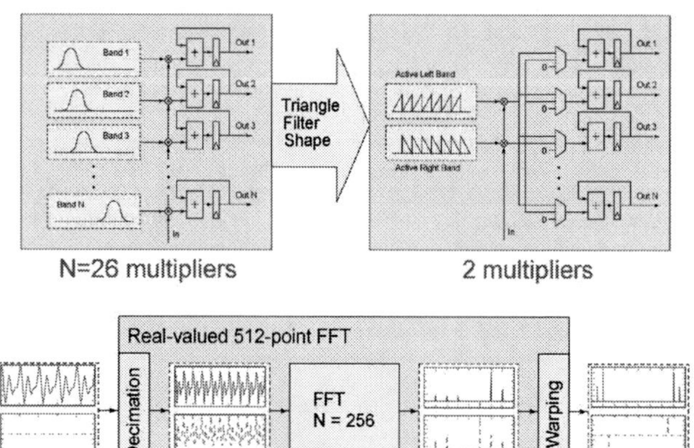

Figure 27.2.3: Bandpass filter bank and FFT optimizations applied to MFCC frontend.

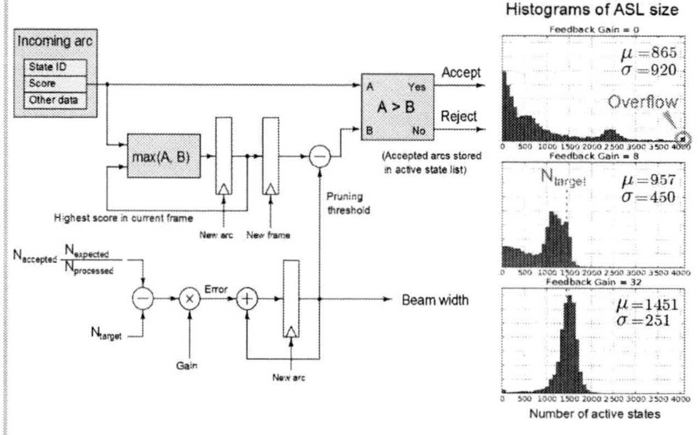

Figure 27.2.4: Architecture and behavior of beam width control via feedback to control variation in number of active states.

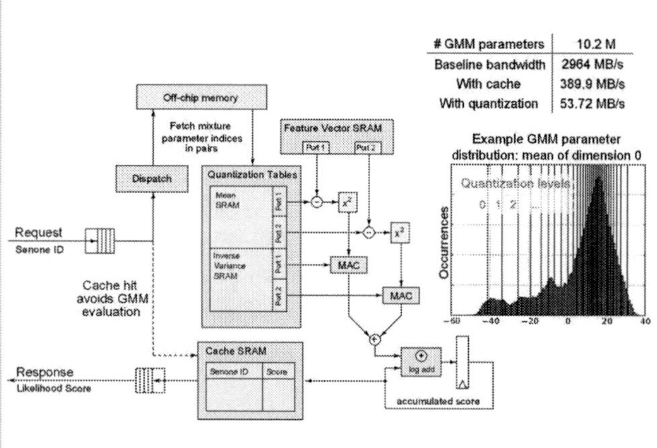

Figure 27.2.5: Quantization and caching of Gaussian mixture models to reduce memory bandwidth.

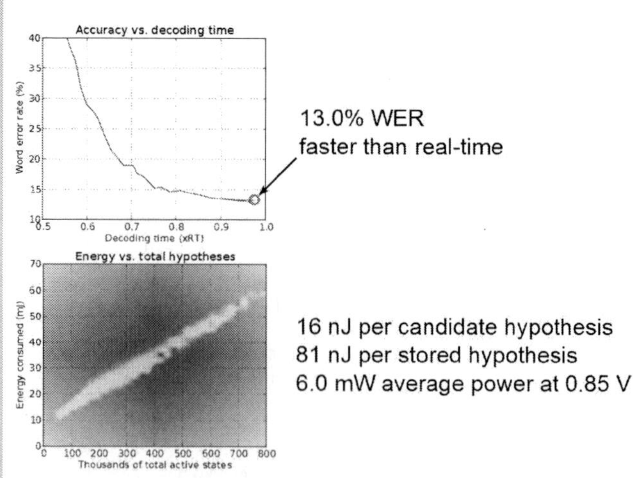

Figure 27.2.6: Word error rate and energy consumption trends at 50 MHz.

27

M1
M4
M6
M3
L1
L2
M2
M5
M7

Logic regions	Memories
L1: Decoder	M1: Active state list 1
L2: Frontend	M2: Active state list 2
	M3: GMM quantization tables and cache
	M4: WFST cache data table
	M5: Feature vector buffer
	M6: WFST cache hash table
	M7: Feature and audio log
	M8: Frontend and FFT scratch memories

Specification	Value
Process	TSMC 65 nm
Die size	2.5 x 2.5 mm
Package	128-pin LQFP
Logic gates	340k (NAND2 equiv.)
SRAM	2.4 Mb
Supply voltage	0.8 – 1.2 V
Power consumption	5 – 23 mW (core)
Clock frequency	40 – 110 MHz
# WFST states	2.9M
# GMMs	4k x 32 Gaussians

Figure 27.2.7: Die photo and summary of chip specifications.

27.3 A 210mV 5MHz Variation-Resilient Near-Threshold JPEG Encoder in 40nm CMOS

Nele Reynders, Wim Dehaene

KU Leuven, Leuven, Belgium

Operating circuits in the near-threshold region enables large energy savings. However, such circuits also pose many challenges, such as increased delay, unwanted leakage paths and high sensitivity to variations. Working in advanced nanometer CMOS technologies compromises the robustness of circuits even more due to the increased variability. Nonetheless, these technologies offer higher operating frequencies for ultra-low-voltage circuits. Transitioning to smaller technologies is attractive for digital near-threshold circuits, provided that the impact of the increased variability can be mitigated. Few prior works have considered the design of variation-resilient ultra-low-voltage circuits in CMOS technologies smaller than 65nm. The aim of this work is to design a large system that advances the state-of-the-art by not only reaching very low energy consumption, but also clock frequencies of tens of MHz, while providing high variation resilience. We present a full JPEG encoder fabricated in a 40nm CMOS technology, fully functional down to a supply voltage of 210mV. The JPEG encoder is able to operate at clock frequencies in a range from 5 to 275MHz for supplies from 210 to 550mV, and thus achieves very high ultra-low-voltage speed. At the minimum-energy point (MEP), the JPEG encoder consumes only 29.01pJ/pixel at an operating frequency of 41MHz (V_{DD}=330mV). The variation (σ/μ) of 26 dies at this point is only 9.4% for the frequency and 6.0% for the energy consumption.

The pipelined JPEG encoder, which is compliant with the baseline sequential mode of the JPEG image compression standard [1], consists of 3 main building blocks (Fig. 27.3.1). First, the discrete cosine transform (DCT) transforms blocks of 8×8 pixels to the frequency domain. Second, the quantization divides the transformed blocks by a specific quantization matrix, which determines the compression factor of the JPEG encoding. The quantization coefficients are stored in the quantization table. Third, the blocks are linearized in zigzag order to group similar frequencies. This data is then Huffman encoded according to the DC and AC Huffman tables.

The 2D-DCT (Fig. 27.3.2) is implemented as a sequence of two 1D-DCTs with a transpose matrix in between. The 1D-DCT is based on the algorithm proposed in [2] and consists of five 15b adder/subtractors and one 15b multiplier. The 2D-DCT output needs to be scaled by a scaling matrix afterwards, but this is performed without extra hardware, as it is incorporated in the quantization by scaling the quantization coefficients in advance. The quantization (Fig. 27.3.2) is computed by a multiplier that accesses the quantization table through a one-hot decoder. Fig. 27.3.1 also shows the different subblocks of the Huffman encoder. After the zigzagging, the DC component and the 63 AC components are encoded separately.

The JPEG encoder has a latch-based deeply pipelined architecture. The pipeline is controlled by non-overlapping clock signals, which are distributed to the entire chip by a clock tree. The timing block shown in Fig. 27.3.1 consists of the non-overlapping clock generator and the complete clock tree. The topology of the logic gates is critical for ultra-low-voltage designs, in terms of speed, energy consumption and variation resilience. Therefore, the topology of all logic gates used throughout this JPEG encoder is based on differential transmission gate (TG) logic extended with NMOS stacking [3]. A pipeline stage is at most 3 TG logic gates deep. This balances robustness and gate reliability on the one hand, and on the other hand, the averaging of timing variations obtained by cascading logic gates.

The quantization, DC and AC Huffman tables are implemented as register tables, where the desired content is serially shifted in at startup and the data can be accessed through a one-hot decoder in the case of quantization, and full address decoders in the case of the Huffman tables. The reason why the tables are not implemented as SRAM memories is because the energy consumption of SRAMs is not dominated by dynamic energy but rather, by standby leakage. Therefore, constructing the tables as sub-threshold SRAMs is not a good option. Furthermore, for the required numbers of bits in the design, the peripheral area and energy overhead would be too high, while the speed of the sub-threshold SRAM would be insufficient. Moreover, efficient SRAM requires ratioed logic, which is undesirable in the ultra-low-voltage domain due to its variation sensitivity. The transpose and zigzag matrices consist of two 8×8 blocks of registers. They are serially read in the original order in the 1st block, then copied in parallel to the 2nd block from which they are serially read out in a different, desired order. The dense layout of the JPEG encoder is carried out using a software tool (the Datapath Generator [4]), except for the layout of the 3 tables, which was done manually because their regularity allowed an optimized structure. The active area of the JPEG encoder is 0.557mm^2 (Fig. 27.3.7).

The JPEG encoder can function down to a supply of 210mV, at a clock frequency of 5MHz (Fig. 27.3.3). Frequencies of 25, 50 or 100MHz are achieved at supplies of 300, 350 and 410mV, respectively. Fig. 27.3.4 shows a comprehensive state-of-the-art frequency comparison. The present work exceeds the speed performance of previous ultra-low-voltage designs in advanced nanometer technologies. Only Seok *et al.* [5] achieve a similar frequency, but their FFT core consumes 15.8nJ/transform, which is 550× higher than this work's minimum energy consumption. This large difference in energy consumption cannot be explained by the difference in computational work between an FFT and a JPEG encoder. Thus, given the similar performance figures of both designs, we conclude that the design presented here is much more energy efficient.

The minimum-energy point occurs at 330mV (Fig. 27.3.5), with the chip consuming 29.01pJ/pixel at an operating frequency of 41MHz. Overall, the JPEG encoder achieves an energy consumption of less than 50pJ/pixel for clock frequencies below 275MHz. The percentage of energy consumed by each block (as shown in Fig. 27.3.1) is also given. Observe that the register tables contribute significantly to leakage, as the quantization and zigzag and Huffman blocks have a much higher percentage of leakage than the 2D-DCT block, which does not contain a table. A total of 26 dies were measured. Across a supply range of 210 to 550mV, the mean variation (σ/μ) in operating frequency is 8.6% and the mean variation in energy consumption/pixel is 5.4%.

Figure 27.3.6 provides a state-of-the-art comparison between this work and another ultra-low-voltage JPEG encoder, fabricated in 65nm CMOS [6]. This work is able to function at a minimum supply of 210mV, while [6] is only able to reach a minimum of 400mV. The architecture of [6] consists of 4 parallel engines and a Huffman encoder that runs in a different voltage and clock domain, i.e. with a clock that is 4× the engine clock. At 400mV, the engines are able to operate at a clock frequency of 2.5MHz and the Huffman encoder runs at 10MHz at 600mV. This work achieves a frequency of 5MHz at the minimum supply, and 41MHz at the MEP, significantly outperforming [6]. Unfortunately, a direct comparison between the energy consumptions cannot be made, because [6] only provides the energy consumption per cycle, i.e. per pipeline stage. Since the number of pipeline stages is not mentioned in [6], it is not possible to calculate the total energy consumption, nor the energy-delay product (EDP).

This paper presents the design of a near-threshold JPEG encoder in 40nm CMOS that is able to function at ultra-low supply voltages as low as 210mV. The chip achieves significantly better than state-of-the-art operating frequencies, well within the MHz range, combined with low energy consumption. The variation resiliency of this JPEG encoder is validated by the low variation results of the measurements. We expect that the design principles used for this chip can be applied to other processor designs and will result in similar ultra-low-voltage characteristics.

References:

[1] G.K. Wallace, "The JPEG Still Picture Compression Standard," *IEEE Trans. on Consumer Electronics*, vol. 38, no. 1, pp. xviii-xxxiv, 1992.

[2] M. Kovac and N. Ranganathan, "JAGUAR: A Fully Pipelined VLSI Architecture for JPEG Image Compression Standard," *Proceedings of the IEEE*, vol. 83, no. 2, pp. 247-258, 1995.

[3] N. Reynders and W. Dehaene, "Variation-Resilient Building Blocks for Ultra-Low-Energy Sub-Threshold Design," *IEEE Trans. Circuits and Systems-II*, vol. 59, no. 12, pp. 898-902, 2012.

[4] O. Weiss, M. Gansen and T.G. Noll, "A Flexible Datapath Generator for Physical Oriented Design," *European Solid-State Circuits Conf.*, pp. 393-396, 2001.

[5] M. Seok, *et al.*, "A 0.27V 30MHz 17.7nJ/transform 1024-pt Complex FFT Core with Super-Pipelining," *ISSCC Dig. Tech. Papers*, pp. 342-343, 2011.

[6] Y. Pu, *et al.*, "An Ultra-Low-Energy/Frame Multi-Standard JPEG Co-Processor in 65nm CMOS with Sub/Near-Threshold Power Supply," *ISSCC Dig. Tech. Papers*, pp. 146-147, 2009.

978-1-4799-0917-9/14 $31.00 © 2014 IEEE

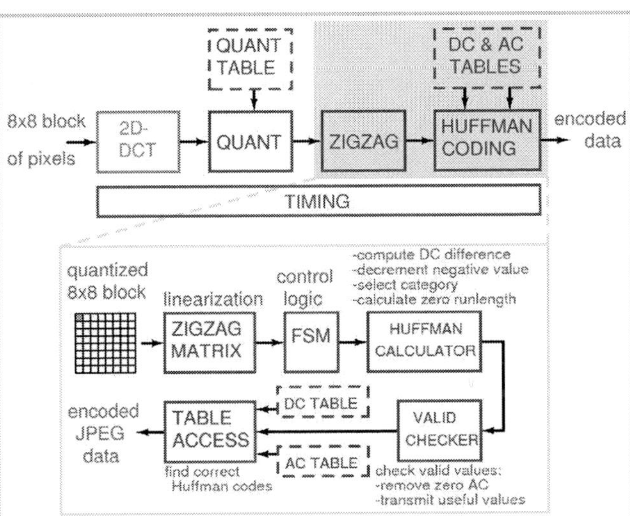

Figure 27.3.1: Block diagram of the JPEG encoder.

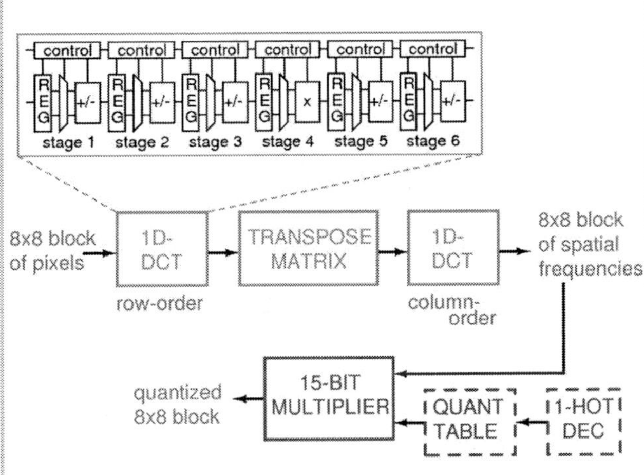

Figure 27.3.2: Implementation of the 2D-DCT and quantization.

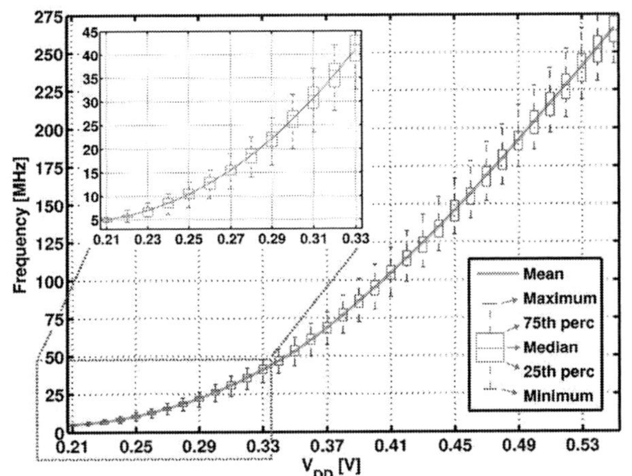

Figure 27.3.3: Boxplot of the measured maximum operating frequency as function of V_{DD}, obtained from 26 dies.

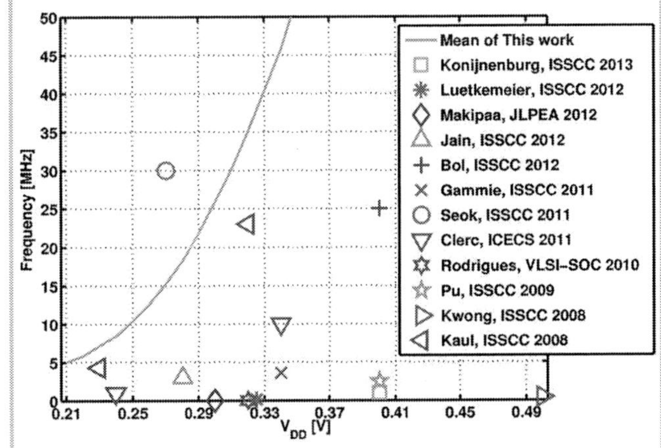

Figure 27.3.4: State-of-the-art frequency comparison between all other previously published ultra-low-voltage designs in advanced nanometer CMOS technologies.

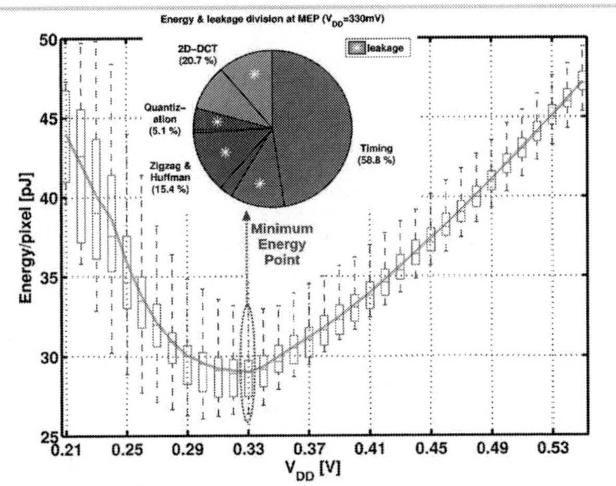

Figure 27.3.5: Boxplot of the measured energy consumption per operation as function of V_{DD}. The division of the energy consumption is also given at the MEP.

		This work	[6] [1]
CMOS Technology		40nm	65nm
Active area	[mm^2]	0.557	1.960
Minimal voltage	[V]	0.210	0.400 (engine) / 0.600 (Huff)
Voltage @ MEP	[V]	0.330	0.400 / 0.600
Frequency @ MEP	[MHz]	41.0	2.5 / 10
Energy/cycle @ MEP [2]	[pJ]	0.045	3 (for 4 engines) + 1.8
Energy/8x8 block @ MEP	[pJ]	1857	-
EDP @ MEP	[pJ.μs]	45.29	-

[1] In [6], 2D-DCT and quantization are joined in a so-called engine and such an engine is operating at 1 voltage domain, while the Huffman encoder is operating in a second higher voltage domain.

[2] This is design-dependent, but it is the only energy figure provided in [6].

Figure 27.3.6: State-of-the-art comparison between ultra-low-voltage JPEG encoders in advanced nanometer CMOS technologies.

27

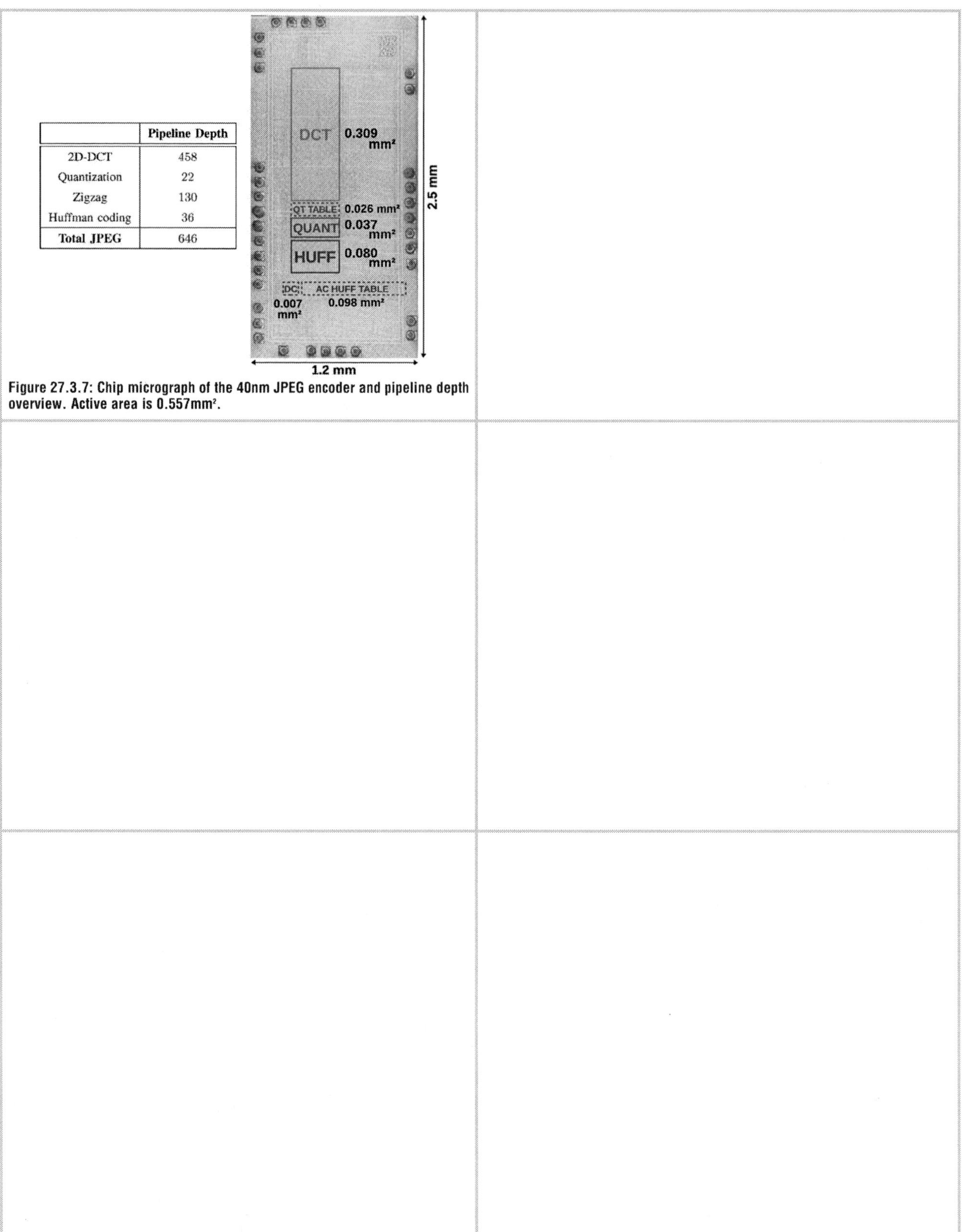

	Pipeline Depth
2D-DCT	458
Quantization	22
Zigzag	130
Huffman coding	36
Total JPEG	646

Figure 27.3.7: Chip micrograph of the 40nm JPEG encoder and pipeline depth overview. Active area is 0.557mm².

27.4 A 0.75-Million-Point Fourier-Transform Chip for Frequency-Sparse Signals

Omid Abari, Ezz Hamed, Haitham Hassanieh, Abhinav Agarwal, Dina Katabi, Anantha P. Chandrakasan, Vladimir Stojanovic

Massachusetts Institute of Technology, Cambridge, MA

Applications like spectrum sensing, radar signal processing, and pattern matching by convolving a signal with a long code, as in GPS, require large FFT sizes. ASIC implementations of such FFTs are challenging due to their large silicon area and high power consumption. However, the signals in these applications are sparse, i.e., the energy at the output of the FFT/IFFT is concentrated at a limited number of frequencies and with zero/negligible energy at most frequencies. Recent advances in signal processing have shown that, for such sparse signals, a new algorithm called the sparse FFT (sFFT) can compute the Fourier transform more efficiently than traditional FFTs [1].

This paper presents a VLSI implementation of the sFFT algorithm. The chip implements a 746,496-point sFFT, in 0.6mm² of silicon area. At 0.66V, it consumes 0.4pJ/sample and has an effective throughput of 36GS/s. The effective throughput is computed over all frequencies but frequencies with negligible magnitudes are not produced. The chip works for signals that occupy up to 0.1% of the transform frequency range (0.1% sparse). It can be used to detect a signal that is frequency hopping in a wideband, to perform pattern matching against a long code, or to detect a blocker location with very high frequency resolution. For example, it can detect and recover a signal that occupies 18MHz randomly scattered anywhere in an 18GHz band with a frequency resolution of ~24kHz.

The sFFT algorithm has three steps: bucketization, estimation, and collision resolution. *Bucketization*: The algorithm starts by mapping the spectrum into buckets as shown in Fig. 27.4.1. This is done by sub-sampling the signal and then performing an FFT. Sub-sampling in time causes aliasing in frequency. Since the spectrum is sparsely occupied, most buckets will be either empty or have a single active frequency, and only few buckets will have a collision of multiple active frequencies. Empty buckets are discarded and non-empty buckets are passed to the estimation step.

Estimation: This step estimates the value and frequency number (i.e. location in the spectrum) of each active frequency. In the absence of a collision, the value of an active frequency is the value of its bucket. To find the frequency number, the algorithm repeats the bucketization on the original signal after shifting it by 1 sample. A shift in time causes a phase change in the frequency domain of $2\pi f\tau/N$, where f is the frequency number, τ is the time shift, and N is the sFFT size. Thus, the phase change can be used to compute the frequency number.

Collision resolution: The algorithm detects collisions as follows: If a bucket contains a collision then repeating the bucketization with a time shift causes the bucket's magnitude to change since the colliding frequencies rotate by different phases. In contrast, the magnitude does not change if the bucket has a single active frequency. After detecting collisions, the algorithm resolves them by using bucketization multiple times with co-prime sampling rates (FFTs with co-prime sizes). Two numbers are co-prime if their greatest common divisor is one. The use of co-prime sampling rates guarantees that any two frequencies that collide in one bucketization do not collide in other bucketizations, as shown in Fig. 27.4.1.

The block diagram of the sFFT chip is shown in Fig. 27.4.2. A 12b 746,496-point ($2^{10}\times3^6$-point) sFFT is implemented. Two types of FFTs (2^{10} and 3^6-point) are used for bucketization. The input to the 2^{10}-point FFT is the signal sub-sampled by 3^6, while the input to the 3^6-point FFT is the signal sub-sampled by 2^{10}. FFTs of sizes 2^{10} and 3^6 were chosen since they are co-prime and can be implemented with simple low-radix FFTs. Three FFTs of each size are used with inputs shifted by 0, 1 or 32 time samples, as shown in Fig. 27.4.2. In principle, shifts of 0 and 1 are sufficient. However, the third shift is used to increase the estimation accuracy. One 1024-word and one 729-word SRAMs are used for three 2^{10}-point and three 3^6-point FFTs, respectively. SRAMs are triplicated to enable pipelined operation of the I/O interface, bucketization and reconstruction blocks. Thus, 3 sFFT frames exist in the pipeline.

The micro-architecture of the 2^{10}-point FFT is shown in Fig. 27.4.3. Each 2^{10}-point FFT uses one radix-4 butterfly to perform an in-place FFT, which is optimized to reduce area and power consumption as follows: First, the FFT block performs read and write operations at even and odd clock cycles, respectively, which enables the use of single port SRAMs. A single read operation provides three complex values, one for each radix-4 butterfly. The complex multiplication is computed over two clock cycles using two multipliers for each butterfly. Second, a twiddle factor (TWF) control unit is shared between the three butterflies. Third, the block floating point (BFP) technique is used to minimize the quantization error [2]. BFP is implemented using a single exponent shared between FFTs, and scaling is done by shifting in case of overflow. Round-half-away-from-zero is implemented by initializing the accumulator registers with 0.5LSB and truncating the results. The 3^6-point FFTs are similar, but use radix-3 butterflies.

The micro-architecture of estimation and collision detection is shown in Fig. 27.4.4. Phase shift and phase detector units use the CORDIC algorithm. The estimation block operates in two steps. First, time shifts of 1 and 32 samples are used to compute the MSBs and LSBs of the phase change, respectively. A 3b overlap is used to fix errors due to concatenation. Since the 5 MSBs of phase change are taken directly from the output of phase detectors, active frequencies have to be ~30dB above the quantization noise to be detected correctly. Frequencies below this level are considered negligible. The frequency number is estimated from the phase change. This frequency number may have errors in the LSBs due to quantization noise. The second step corrects any such errors by using the bucket number to recover the LSBs of the frequency number. This is possible because all frequencies in a bucket share the same remainder B (B=f mod M, where f is the frequency number and M is the FFT size), which is also the bucket number. Thus, in the frequency recovery block associated with the 2^{10}-point FFTs, the bucket number gives the 10 LSBs of the frequency number. However, in the frequency recovery for the 3^6-point FFTs, the LSBs cannot be directly replaced by the bucket number since M=3^6 is not a power of 2. Instead, the remainder of dividing the frequency number by 3^6 is calculated and subtracted from the frequency number. The bucket number is then added to the result of the subtraction. In our implementation, calculating and subtracting the remainder is done indirectly by truncating the LSBs of the phase change.

The collision detection block in Fig. 27.4.4 compares the values of the buckets with and without time-shifts. It uses the estimated frequency to remove the phase change in the time-shifted bucketizations and compares the three complex values to detect collisions. In the case of no collision, the three values are averaged to reduce noise. The result is used to update the output of the sFFT in SRAMs.

The testchip is fabricated in IBM's 45nm SOI technology. The sFFT core occupies 0.6mm² including SRAMs. At 1.18V supply, the chip operates at a maximum frequency of 1.5GHz, resulting in an effective throughput of 109GS/s. At this frequency, the measured energy efficiency is 1.2μJ per 746,496–point Fourier transform. Reducing the clock frequency to 500MHz enables an energy efficiency of 298nJ per Fourier transform at 0.66V supply. Energy and operating frequency for a range of supply voltages are shown in Fig. 27.4.5.

Since no prior ASIC implementations of sFFT exist, we compare with recent low-power implementations of the traditional FFT [3-5]. The measured energy is normalized by the Fourier transform size to obtain the energy per sample (the sFFT chip, however, outputs only active frequencies). Fig. 27.4.6 shows that the implementations in [3-5] work for sparse and non-sparse signals while the sFFT chip works for signal sparsity up to 0.1%. However, for such sparse signals, the sFFT chip delivers ~40× lower energy per sample for a $3^6\times$ larger FFT size. Fig. 27.4.6 also shows that the 746,496-point sFFT chip achieves an 88× reduction in run-time compared to a C++ implementation running on an i7 CPU [6].

References:

[1] H. Hassanieh, *et al.*, "Simple and Practical Algorithm for Sparse Fourier Transform," *ACM Symp. Discrete Algorithms*, pp. 1183-1184, 2012.

[2] G. Zhong, *et al.*, "A Power-Scalable Reconfigurable FFT/IFFT IC Based on a Multi-Processor Ring," *IEEE J. Solid-State Circuits*, vol. 41, no. 2, pp. 483-495, 2006.

[3] M. Seok, *et al.*, "A 0.27V 30MHz 17.7nJ/transform 1024-pt Complex FFT Core with Super-Pipelining," *ISSCC Dig. Tech Papers*, pp. 342-344, 2011.

[4] Y. Chen, *et al.*, "A 2.4-Gsample/s DVFS FFT Processor for MIMO OFDM Communication Systems," *IEEE J. Solid-State Circuits*, vol. 43, no. 5, pp. 1260-1273, 2008.

[5] C. Yang, *et al.*, "Power and Area Minimization of Reconfigurable FFT Processors: A 3GPP-LTE Example," *IEEE J. Solid-State Circuits*, vol. 47, no. 3, pp. 757-768, 2012.

[6] sFFT C++ code. http://groups.csail.mit.edu/netmit/sFFT/code.html

978-1-4799-0917-9/14 $31.00 © 2014 IEEE

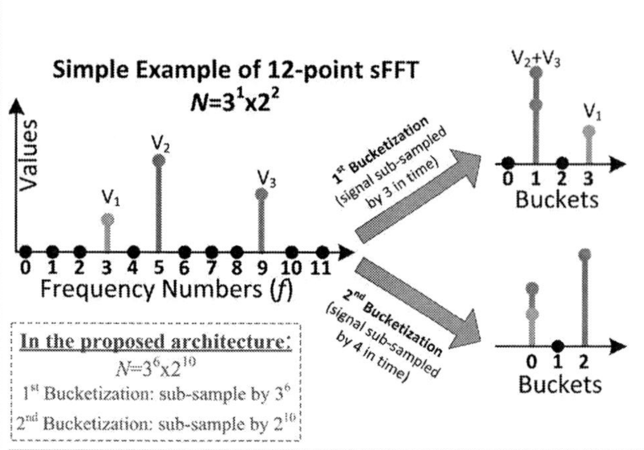

Figure 27.4.1: The sFFT algorithm performs bucketization by sub-sampling in the time domain then taking an FFT, which causes aliasing in the frequency domain.

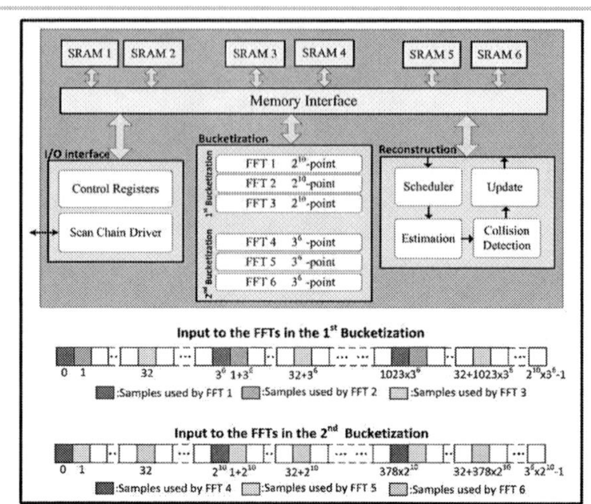

Figure 27.4.2: A block diagram of the 2^{10}x3^6-point sparse FFT and input samples to the 6 FFTs.

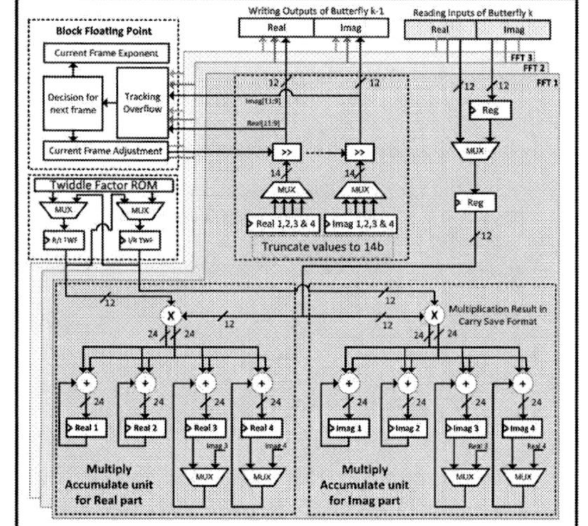

Figure 27.4.3: The micro-architecture of the 2^{10}-point FFTs.

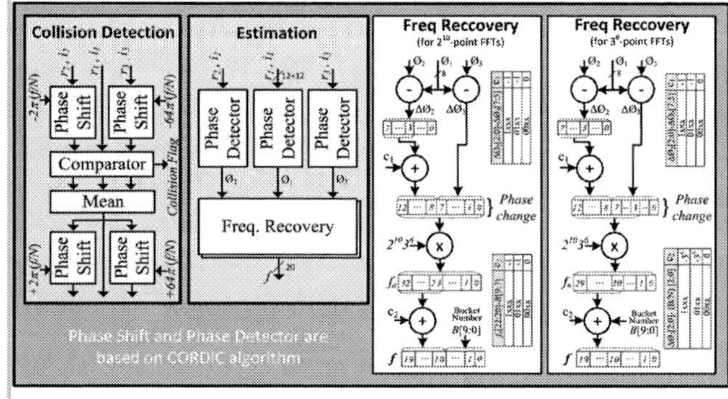

Figure 27.4.4: The micro-architecture of collision detection and estimation. The complex values (r_1, i_1), (r_2, i_2) and (r_3, i_3) are the output of bucketization for time-shifts 0, 1 and 32 samples.

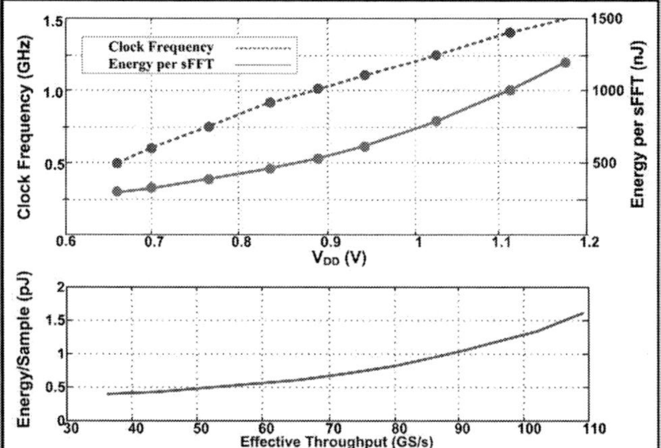

Figure 27.4.5: Measured energy and operating frequency for a range of voltage, and throughput versus energy per sample for computing a 746,496-point sparse Fourier transform.

	[3]	[4]	[5]	This work
Technology	65 nm	90 nm	65nm	45 nm
Signal Type	Any Signal	Any Signal	Any Signal	Freq.-Sparse signal
Size	2^{10}-point	2^8-point	2^7 to 2^{11}-point	3^6x2^{10}-point
Word width	16 bits	10 bits	12 bits	12 bits
Area	8.29mm²	5.1mm²	1.37mm²	0.6mm²
Effective Throughput	240MS/s	2.4GS/s	1.25-20MS/s	36.4-109.2GS/s
Energy/Sample	17.2pJ	50pJ	19.5-50.6pJ	0.4-1.6pJ

Processor	Intel Core i7 Quad Core (3.4GHz)	This work (1.5GHz)
Run-time	600μs	6.8μs

2^{10}x3^6 point sparse Fourier transform

Figure 27.4.6: Measured energy efficiency and performance of the sFFT chip compared to published FFTs.

27

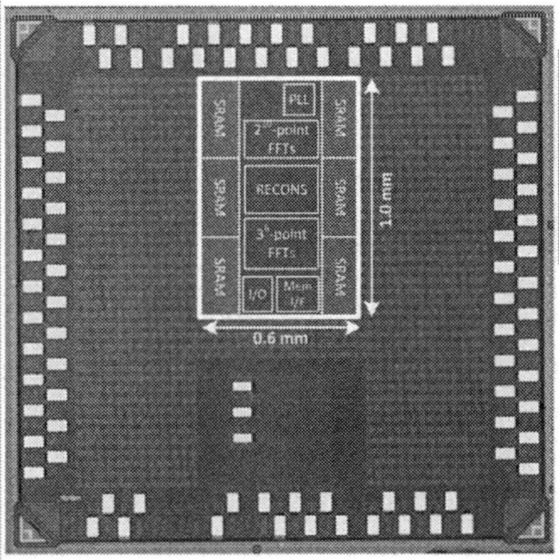

Chip Features

Technology	45nm SOI CMOS
Core Area	0.6mm x 1.0mm
SRAM	3x 75kbits 3x 54kbits
Core Supply Voltage	0.66-1.18 V
Clock Frequency	0.5-1.5 GHz
Core Power	14.6-174.8 mW (PLL in not included)

Figure 27.4.7: Die photo of the testchip.

27.5 A Multi-Granularity FPGA with Hierarchical Interconnects for Efficient and Flexible Mobile Computing

Cheng C. Wang[1], Fang-Li Yuan[1], Tsung-Han Yu[2], Dejan Markovic[1]

[1]University of California, Los Angeles, CA, [2]Qualcomm, Irvine, CA

Following the rapid expansion of mobile computing in the past decade, mobile system-on-a-chip (SoC) designs have off-loaded most compute-intensive tasks to dedicated accelerators to improve energy efficiency. An increasing number of accelerators in power-limited SoCs results in large regions of "dark silicon." Such accelerators lack flexibility, thus any design change requires a SoC re-spin, significantly impacting cost and timeline. To address the need for efficiency and flexibility, this work presents a multi-granularity FPGA suitable for mobile computing. Occupying 20.5mm^2 in 40nm CMOS, the chip incorporates 2,760 fine-grained configurable logic blocks (CLBs) with 11,040 6-input look-up-tables (LUTs) for random logic, basic arithmetic, shift registers, and distributed memories, 42 medium-grained 48b DSP processors for MAC and SIMD operations, 16 32K×1b to 512×72b reconfigurable block RAMs, and 2 coarse-grained kernels: a 64-8192-point fast Fourier transform (FFT) processor and a 16-core universal DSP (UDSP) for software-defined radio (SDR). Using a mix-radix hierarchical interconnect, the chip achieves a 4× interconnect area reduction over commercial FPGAs for comparable connectivity, reducing overall area and leakage by 2.5×, and delivering a 10-50% lower active power. With coarse-grained kernels, the chip's energy efficiency reaches within 4-5× of ASIC designs.

Although commercial FPGAs can come close to ASICs in performance, they are highly inefficient due to their high energy and a large area overhead. This is mainly due to the programmable interconnect. For over 20 years, a 2D-mesh network has been the backbone of FPGA interconnect, but full connectivity in a 2D mesh requires O(N^2) switches, requiring interconnects to grow faster than Moore's Law O(N). As a result, various heuristics are used to simplify switches at the cost of resource utilization, but the interconnect area is still ~4× the logic area in modern FPGAs. By effectively pruning a Beneš network, a hierarchical interconnect network is realized where the number of switches is less than O(N·logN), allowing us to maintain an interconnect-to-logic-area ratio of 1:1.

The O(N·logN) complexity of Beneš network is well-known in telecommunications, but such a network is seldom used in hardware primarily due to its implementation complexity. In a traditional Beneš, wirelength doubles for every stage. With an equal number of wires for all stages, this leads to long, congested wires in the upper hierarchies. An efficient implementation requires pruning the upper hierarchies, and we alternate the routing in the x- and y-directions so wirelength doubles every *two* stages [1]. Another drawback is the delay across radix boundaries. As shown in Fig. 27.5.1, communication between neighboring computing elements (CE) 4 and 5 requires 3 hierarchies. A boundary-less radix-3 network is created to restore spatial locality by shifting all local connections to the lower switch matrices (SMs). In the simplified illustration, radix-3 SMs are used in the lower stages to increase local bandwidth, allowing even fewer radix-2 SMs in the upper hierarchies. For improved timing and reduced power, fast-path routing allows hops directly to the required hierarchy level, routing only half the network on the return path. Our router automatically assigns fast-path interconnect based on congestion and timing.

Boundary-less radix-3 SMs are used in the lower 5 hierarchies (Fig. 27.5.2), and pruned radix-2 switches are used from stage 6 to 14, except stage 10 and 11. Stage 10 employs boundary-less radix-3 across the horizontal bisection to improve bisection bandwidth. The top-level connectivity (stage 14) is pruned to only 5%. This is a result of closed-loop optimization by mapping various FPGA benchmarks, then pruning or expanding each stage based on congestion and performance. To ease physical design, the chip is divided into 40 interconnect regions, each with 512 SM macros, with 9 to 14 stages per SM macro.

The fine-grained and medium-grained CLBs offer behavior identical to commercial FPGAs, allowing for a direct comparison of interconnects by executing identical netlists. To target common communications designs, two coarse-grained kernels were implemented. A 64-8192-point reconfigurable FFT is beneficial for digital baseband processing. It has a small dedicated memory, and interconnects to the FPGA memory to realize the long delay lines for 2048-8192-point FFTs. A 16-core UDSP targets a variety of SDR algorithms, where each core is reconfigurable for arbitrary 2×2 matrix operations using a flexible instruction-set architecture. Unlike the medium-grained DSP processor, the 2×2 butterfly core in the UDSP is very efficient for complex arithmetic, capable of many SDR functions, such as filtering, equalization, CORDIC, and sphere decoding by simply concatenating multiple butterfly stages. FFT and UDSP both connect to the interconnect network.

Power gating (PG) is desirable for large chips, but each interconnect signal often traverses many blocks, making block-level PG ineffective. A fine-grained PG is needed for individual switches. Traditional PG becomes very inefficient because the footer PG transistor is no longer shared by the entire block, so it cannot be made very large (Fig. 27.5.3), but a smaller footer can degrade performance by 30-50%. To power gate without a footer, a PG branch is added to the mux, and the pass-gate is separated into NMOS and PMOS segments, where enabling PG leaves the output floating, reducing the coupling capacitance on neighboring wires. When conducting, the NMOS segment is driven by PMOS pass-gates, thus it can rise much faster than the PMOS segment driven by NMOS pass-gates, which settles to V_{DD}-V_t (and vice versa). This results in larger transient leakage, but does not degrade performance significantly, because the output current is the *difference* of the pull-up and pull-down branches. A small high-V_t keeper pulls together the NMOS and PMOS voltages to overcome the V_t drop. This results in a 5-10% performance penalty, but reduces leakage by more than 50% (now gate-leakage dominated). The output floats during PG, so it cannot drive a CMOS gate, but can only enter a pass-gate that can be disabled during PG. Over 90% of the switches utilized this PG scheme, except those driving long wires that require buffer insertion.

With over 9 million configuration bits, an automated mapping tool is developed. The tool supports two modes (Fig. 27.5.4). Mode 1 maps an identical netlist as used by commercial FPGAs for a direct comparison of performance, power, and area utilization: the user design is first synthesized using commercial tools, then the output netlist is parsed into our custom tool, which performs timing analysis, floorplan, placement, routing, and bitstream generation for our FPGA. Mode 2 incorporates our coarse-grained kernels into the P&R flow. Although the configuration SRAM cells are distributed throughout our FPGA, their word-lines (WL) and bitlines (BL) are organized as one large memory for easy initialization. The FPGA core can only be powered on after configuration finishes.

Measurement results of our FPGA with CLBs, and with coarse-grained kernels are compared against processors, a commercial FPGA, and an ASIC (Fig. 27.5.5). Although the CLBs alone achieve over 1.5GOP/mW, an energy efficiency of 0.86GOPS/mW is achievable when mapping an FIR filter, which is 4× more efficient than commercial FPGA (both in 40nm). An 8× efficiency gain can be achieved by using UDSP kernels. FFT operations, which are dominated by memory and control, are 13× more energy efficient when mapped to the FFT kernel instead of CLBs. A 2-2.5× reduction in leakage is attained from smaller chip area and fine-grained PG, even with the disadvantage of dual-oxide transistors. Our chip is built with standard-cells, yet we are often within 20% of the performance of high-end FPGAs, though our software is still improving.

With efficient interconnect, our FPGA is within 20× of ASIC efficiency for most designs (Fig. 27.5.6). Coarse-grained kernels further improve the efficiency, bringing it within 4 to 5× of ASICs. The key to coarse-grained efficiency is to identify compact, reconfigurable kernels that improve efficiency, apply to a variety of applications, and leverage existing FPGA resources where possible. Our chip (Fig. 27.5.7) is able to attain the energy efficiency suitable for mobile applications while maintaining the full flexibility of an FPGA.

Acknowledgments:
The authors thank Dr. Sanjay Raman and DARPA for funding support.

References:

[1] C.C. Wang, *et al.*, "A 1.1 GOPS/mW FPGA Chip with Hierarchical Interconnect Fabric," *IEEE Symp. VLSI Circuits*, pp. 136-137, 2011.

[2] Z. Yu, *et al.*, "An 800 MHz 320 mW 16-Core Processor with Message-Passing and Shared Memory Inter-Core Communication Mechanisms," *ISSCC Dig. Tech. Papers*, pp. 64-65, 2012.

[3] "FFT Implementation on the TMS320C5535 DSP," *TI Technical Reference Manual*, pp. 111-134, 2012.

[4] T-H. Yu, *et al.*, "A 7.4 mW 200 MS/s Wideband Spectrum Sensing Digital Baseband Processor for Cognitive Radios," *IEEE J. Solid-State Circuits*, vol. 47, no. 9, pp. 2235-2245, 2012.

[5] F-L. Yuan, *et al.*, "A 256-Point Dataflow Scheduling 2x2 MIMO FFT/IFFT Processor for IEEE 802.16 WMAN," *Asian Solid-State Circuits Conf.*, pp. 309-312, 2008.

[6] J. Thompson, *et al.*, "An Integrated 802.11a Baseband and MAC Processor," *ISSCC Dig. Tech. Papers*, pp. 126-127, 2002.

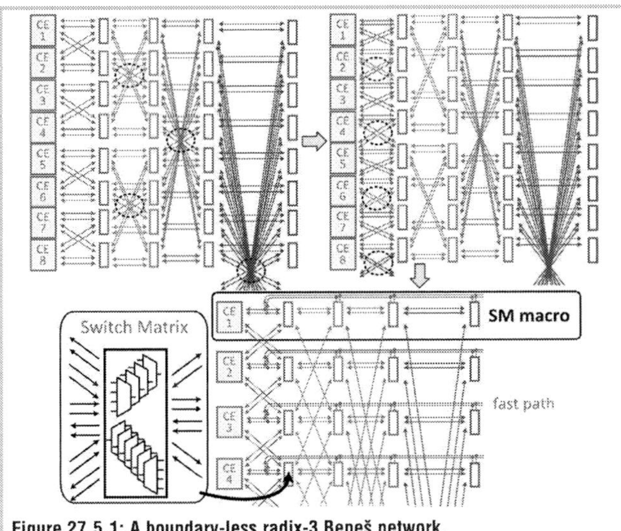

Figure 27.5.1: A boundary-less radix-3 Beneš network.

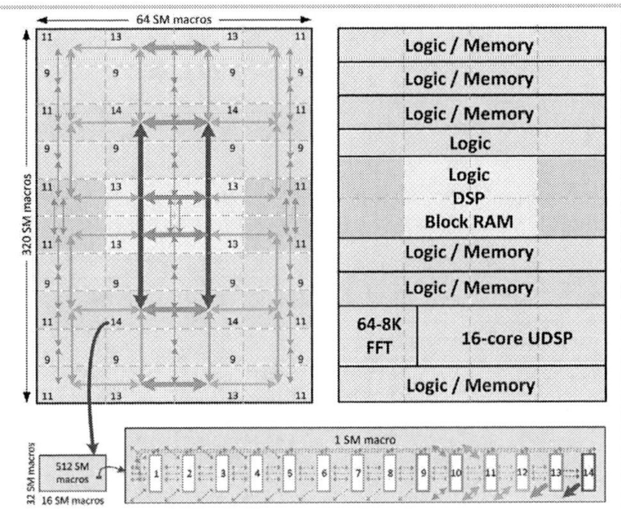

64 SM macros
320 SM macros
Logic / Memory
Logic / Memory
Logic / Memory
Logic
Logic
DSP
Block RAM
Logic / Memory
Logic / Memory
64-8K FFT
16-core UDSP
Logic / Memory
32 SM macros
512 SM macros
16 SM macros
1 SM macro

Figure 27.5.2: Interconnect and resource allocation.

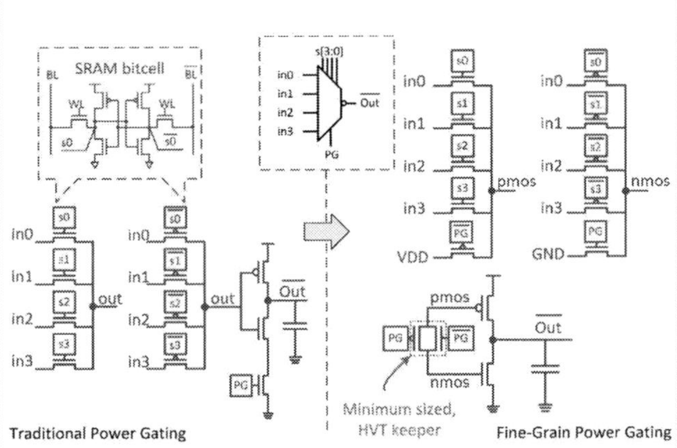

Figure 27.5.3: Multiplexer with traditional and fine-grain PG.

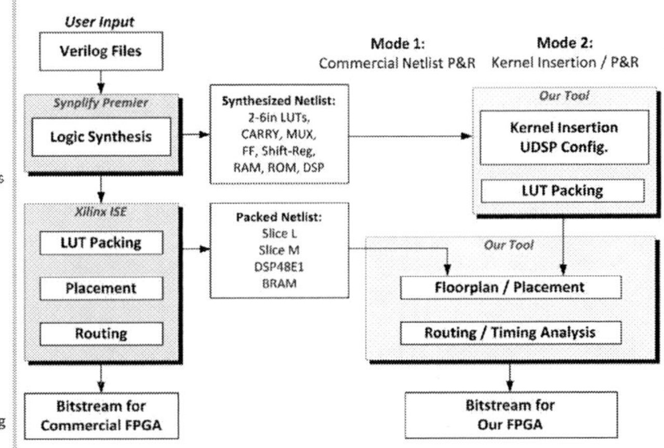

Figure 27.5.4: Automated place-and-route flow (2 modes).

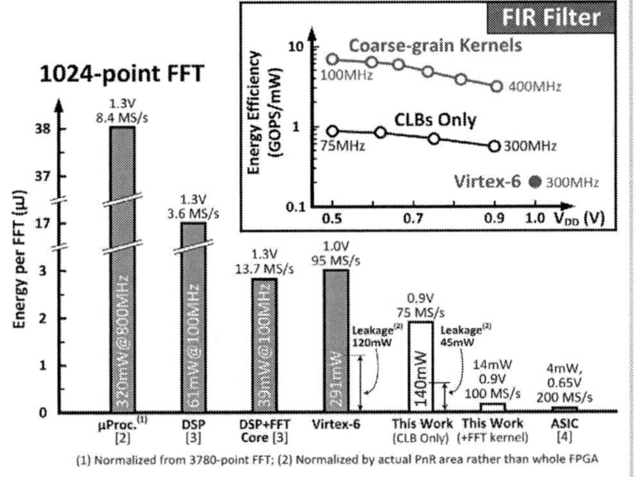

Figure 27.5.5: Comparisons of throughput and efficiency.

Fine-grain and Medium-grain CLBs Only

Resources	Utilization L/M 2760	Utilization DSP 42	Utilization BRAM 16	Virtex-6* Freq. (MHz)	Virtex-6* Power (mW)	This Work Freq. (MHz)	This Work Power (mW)	ASIC Freq. (MHz)	ASIC Power (mW)	ASIC Tech. (nm)	Energy Eff. gap vs. ASIC (norm.)
256-point MIMO FFT [5]	916	4	4	180	270	140 80	135 48	64	25	180	**12×**
K-best + depth first Sphere Decoder	1255	0	0	200	250	140 70	129 58	16	3	90	**15×**
Software-Defined Radio Tx Front End	1594	38	0	39	287	35 23	213 117	60	50	65	**13×**
802.11a Rx Baseband [6]	1824	0	6	66	271	55 20	219 60	20	203	250	**15×**

*The Virtex-6 (6VLX75T) FPGA has 11640 Slice L/Ms, 288 Slice DSPs, and 156 BRAMs.

CLBs and Coarse-grain Kernels

Resources	Utilization L/M 2760	Utilization DSP 42	Utilization BRAM 16	Utilization FFT 1	Utilization UDSP 16	This Work Freq. (MHz)	This Work Power (mW)	ASIC Freq (MHz)	ASIC Power (mW)	ASIC Tech. (nm)	Energy Eff. gap vs. ASIC (norm.)
Software-Defined Radio Tx Front End	664	26	0	0	12	80 30	166 50	60	50	65	**4×**
802.11a Rx Baseband [6]	701	0	6	1	5	60 20	88 21	20	203	250	**5×**

Figure 27.5.6: Example designs and ASIC efficiency gap.

27

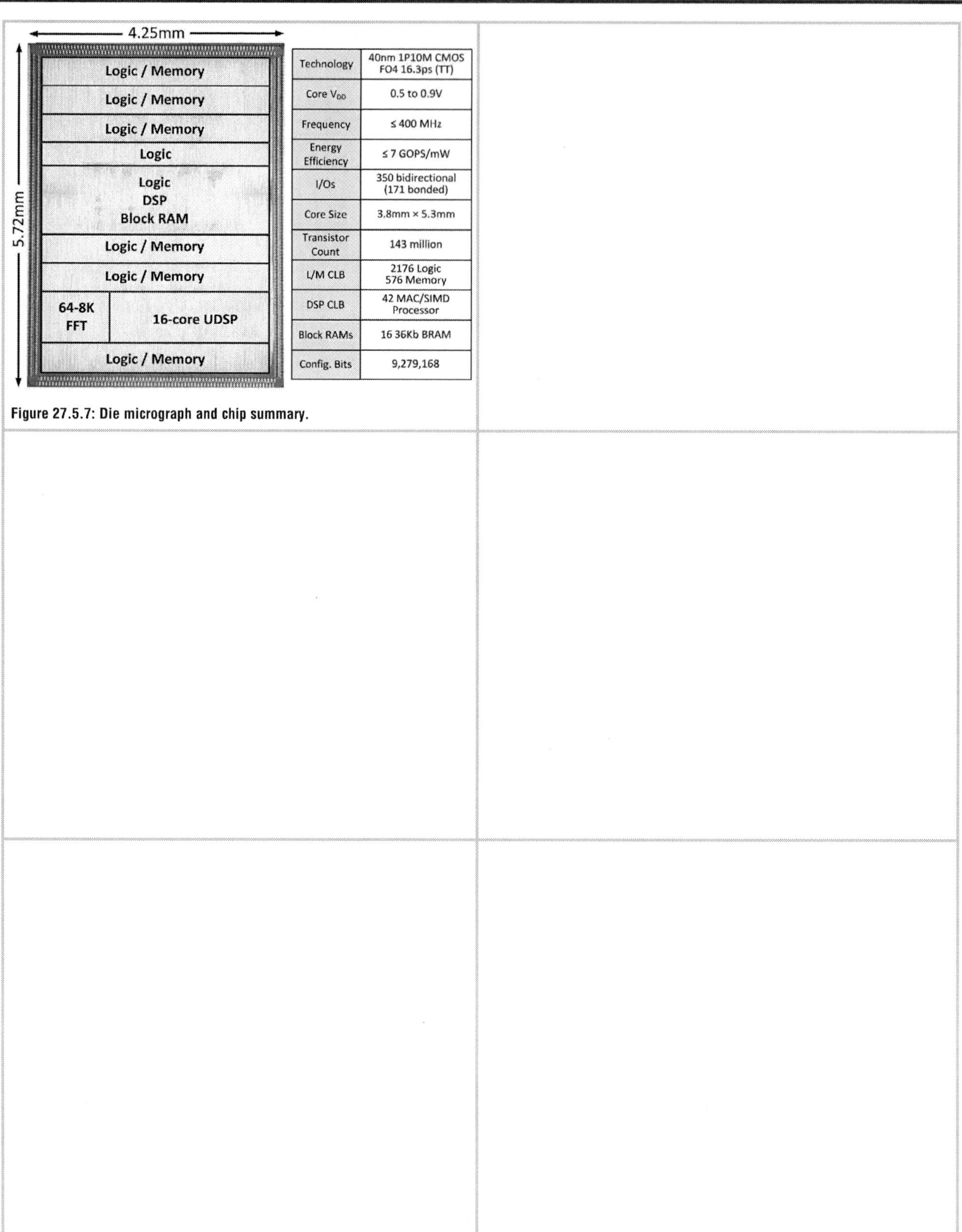

Technology	40nm 1P10M CMOS FO4 16.3ps (TT)
Core V_{DD}	0.5 to 0.9V
Frequency	≤ 400 MHz
Energy Efficiency	≤ 7 GOPS/mW
I/Os	350 bidirectional (171 bonded)
Core Size	3.8mm × 5.3mm
Transistor Count	143 million
L/M CLB	2176 Logic 576 Memory
DSP CLB	42 MAC/SIMD Processor
Block RAMs	16 36Kb BRAM
Config. Bits	9,279,168

Figure 27.5.7: Die micrograph and chip summary.

27.6 An 821MHz 7.9Gb/s 7.3pJ/b/iteration Charge-Recovery LDPC Decoder

Tai-Chuan Ou, Zhengya Zhang, Marios C. Papaefthymiou

University of Michigan, Ann Arbor, MI

This paper presents a 576b LDPC decoder test-chip designed using a charge-recovery logic family. The chip has been fabricated in a 65nm CMOS process and relies on 16 integrated inductors to achieve energy-efficient operation by recovering charge from gate fanouts. When self-oscillating at 821MHz, the chip recovers 51.4% of the energy supplied to it. In terms of device count, this chip is more than an order of magnitude larger than the largest previously-reported chips with charge-recovery logic [3-4]. When operating at 821MHz, it achieves a 7.9Gb/s throughput at 7.3pJ/b/iteration, improving on results in [1-2,5] by at least 1.7× in energy efficiency and 2.3× in area efficiency.

Figure 27.6.1 shows the schematic of a two-input four-bit charge-recovery logic comparator used in this chip. Compared to other similar charge-recovery logic families [3], this so-called subthreshold boost logic (SBL) [4] can achieve higher clock frequencies. The operation of a SBL gate is divided into two steps. The logic stage is similar to static CMOS logic operation, except that NMOS devices are used in both the pull-up and the pull-down network (PUN and PDN) to achieve gate-overdrive and perform functional evaluation with a subthreshold supply (V_{CC}) to develop an initial voltage difference between the dual-rail outputs. The boost stage, composed of a pair of cross-coupled inverters, then amplifies the voltage difference to full-rail during the rising transition of a power-clock waveform at pin PC. During the falling transition of the power-clock, charge is recovered from the output, and the output voltage returns back to about V_t. The dual-phase power-clock is generated using a so-called blip circuit that uses an inductive element and negative transconductance devices to resonate the parasitic capacitance of the network that distributes the power-clock to the charge-recovery gates. To achieve frequency scaling, an on-chip ring oscillator (RO), a pulse generator (PG), and frequency tuning circuits are included in the power-clock generator design. To operate the design off-resonance, a reference clock generated by the RO is supplied to the PG, and the PG then outputs a pair of 180-degree out-of-phase pulses with tunable duty cycle, enabling frequency tuning, and forcing the power-clock to run at the same frequency as the reference clock.

Figure 27.6.2 shows the charge-recovery LDPC decoder architecture for the 576b, rate-5/6 code specified by the IEEE 802.16e standard. Two columns in the code matrix are swapped for a regular structure to facilitate partitioning into four blocks. This partitioning results in regular relay interconnects between neighboring blocks, replacing complex and long global interconnects. The min-sum decoding begins with the check node operation on the first row of Block 1, the results of which are relayed in order to Blocks 2, 3, and 4. Variable node operations on the first row follow the check node operation, while Block 1 begins the check node operation on the second row in parallel. One decoding iteration takes 24 cycles (48 phases) for one complete check node and variable node operation in all four blocks. The check node operation in one block, shown as an example in Fig. 27.6.2, takes 2.5 cycles (5 phases). The deeply-pipelined relay architecture accommodates the processing of four streams in parallel without any pipeline stalls.

With over 57,000 SBL gates, the device count of this chip is more than an order of magnitude larger than previously-reported charge-recovery test-chips [3-4]. To accommodate this increased complexity, an automated standard-cell-like design flow has been developed that incorporates custom-designed dynamic cells and a two-phase power-clock. A SBL gate library has been created with 52 SBL gates of different drive strengths. Library gates have been characterized and used with commercial EDA tools for floorplanning, synthesis, placement, and routing.

Figure 27.6.3 shows the power-clock distribution network. To minimize clock skew, a clock mesh is employed for each of the two clock phases using top level metal (metal 9 for horizontal strips, and metal 8 for vertical strips). For each standard-cell row, two metal-3 horizontal strips are reserved for the power-clock waveform and its complement. These strips are tied to metal 8 of the clock mesh, so that the PC pin of each SBL gate can be connected to the mesh in a predictable manner using an automated place-and-route tool, while avoiding any possible large clock skew. 16 integrated 0.96nH (at 1GHz) inductors, along with 144 distributed negative transconductance devices are used to generate the two-phase power-clock in the chip by resonating the parasitic capacitance of the clock distribution network. A tree structure with supply and ground shielding distributes a pair of 180-degree out-of-phase pulses to frequency-scaling circuitry at each inductor that can be used to operate the power-clock off-resonance at a desired frequency.

The design of the power-clock network plays a key role in the efficiency of the charge-recovery LDPC decoder chip. Fig. 27.6.4 shows the energy consumption of the power-clock as obtained from one of the 16 inductors through simulations with the verified inductor models from the foundry and an extracted post-layout netlist that includes parasitic resistance, capacitance, and coupling capacitance. Simulation results show that the chip recovers 51.4% of the energy supplied from the specific inductor to the power-clock every cycle.

The chip has been fabricated in a 65nm CMOS process. The charge-recovery LDPC decoder logic occupies 1.54mm^2. Fig. 27.6.5 shows measured energy per cycle at each operating frequency. Minimum energy consumption is 702.9pJ per cycle when the power-clock is operating at its resonant frequency of 821MHz, with supply voltage V_{DC} = 0.64V and V_{CC} = 0.36V, yielding 576.8mW of power dissipation at room temperature. Correct functionality has been validated for clock frequencies ranging from 640MHz to 1.05GHz. Fig. 27.6.5 indicates that voltage supply V_{CC} consumes only 0.69-to-1.34% of total energy. In contrast, in the 5-to-187MHz FIR presented in [4], V_{CC} consumes about 3-to-10% of total energy. This difference can be explained by the distinctness in the operating frequency of the two chips. At lower operating frequencies, the boost stage recovers charge more efficiently, whereas the crowbar current in the logic stage increases, resulting in relatively higher energy consumption through V_{CC}. However, higher operating frequencies lead to higher energy consumption in the boost stage and lower crowbar current in the logic stage, thus decreasing the relative levels of energy consumption through V_{CC}.

Figure 27.6.6 gives the performance characteristics of the chip and compares it with the latest LDPC decoders. The charge-recovery LDPC decoder chip outperforms state-of-the-art designs with comparable code length and complexity [1-2,5], achieving at least 1.7× better energy efficiency and 2.3× better area efficiency. Compared to the decoder in [6] that uses an 18× smaller code length, this chip achieves higher area efficiency. It does not match its energy efficiency, however, as the 18× smaller code length avoids the significant overheads associated with scaling up to longer code lengths. Including the area overhead of inductors, the chip in this work is more area efficient compared with other LDPC decoder designs.

A die microphotograph is shown in Fig. 27.6.7. A built-in-self-test (BIST) circuit that is used to generate and process the input and output of the decoder, along with RO, PG, and frequency-tuning circuits are implemented with static CMOS logic and are distributed around the decoder core. To decrease eddy currents, the 16 214μm×237μm 0.96nH inductors are placed outside the staggered I/O pads, occupying 0.81mm^2. This charge-recovery LDPC decoder chip demonstrates the potential of charge-recovery logic for energy- and area-efficient high-performance design, as well as an accompanying design methodology that leverages automated EDA tools and is applicable to large-scale DSP applications.

Acknowledgments:
This work was supported in part by NSF under grant No. CCF-0916714. The authors thank Chia-Hsiang Chen, Jerry Kao, Jinwoo Kim, Suhwan Kim, and Wei-Hsiang Ma for their valuable contributions to this work.

References:

[1] B. Xiang, *et al.*, "An 847–955 Mb/s 342–397mW Dual-Path Fully-Overlapped QC-LDPC Decoder for WiMAX System in 0.13μm CMOS," *IEEE J. Solid-State Circuits*, vol. 46, no. 6, pp.1416-1432, 2011.
[2] X. Peng, *et al.*, "A 115mW 1Gbps QC-LDPC Decoder ASIC for WiMAX in 65nm CMOS," *Asian Solid-State Circuits Conf.*, pp.317-320, 2011.
[3] Y. Zhang, *et al.*, "A 1pJ/cycle Processing Engine in LDPC Application with Charge Recovery Logic," *Asian Solid-State Circuits Conf.*, pp.213-216, 2011.
[4] W.-H. Ma, *et al.*, "187 MHz Subthreshold-Supply Charge-Recovery FIR," *IEEE J. Solid-State Circuits*, vol. 45, no. 4, pp. 793-803, 2010.
[5] S.-W. Yen, *et al.*, "A 5.79-Gb/s Energy-Efficient Multirate LDPC Codec Chip for IEEE 802.15.3c Applications," *IEEE J. Solid-State Circuits*, vol. 47, no. 9, pp. 2246-2257, 2012.
[6] D. Miyashita, *et al.*, "A 10.4pJ/b (32, 8) LDPC Decoder with Time-Domain Analog and Digital Mixed-Signal Processing," *ISSCC Dig. Tech. Papers*, pp. 420-421, 2013.

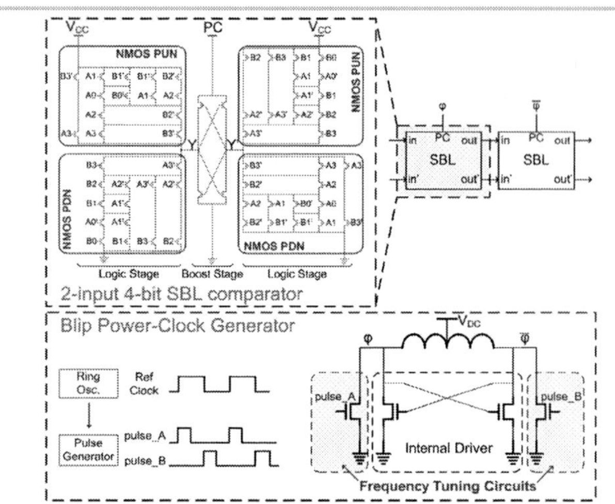

Figure 27.6.1: Schematic of a SBL two-input 4b comparator gate, cascade of SBL gates, and blip power-clock generator.

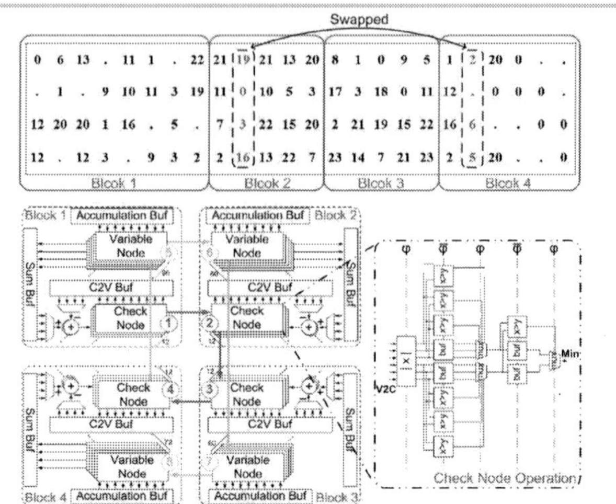

Figure 27.6.2: LDPC matrix, decoder architecture, and check node operation block diagram for 576b rate-5/6 IEEE 802.16e.

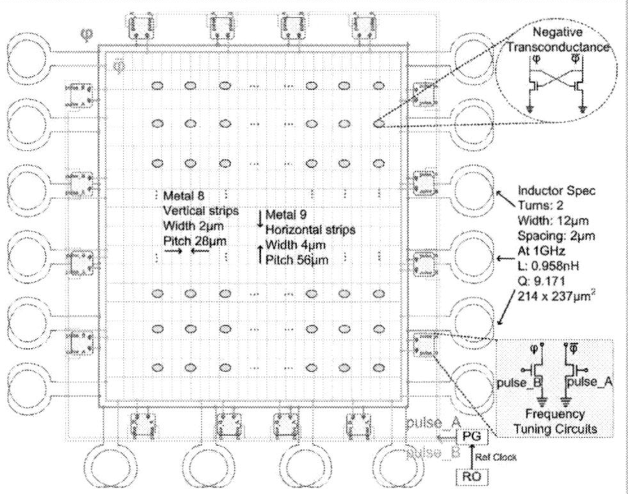

Figure 27.6.3: Power-clock generation and distribution.

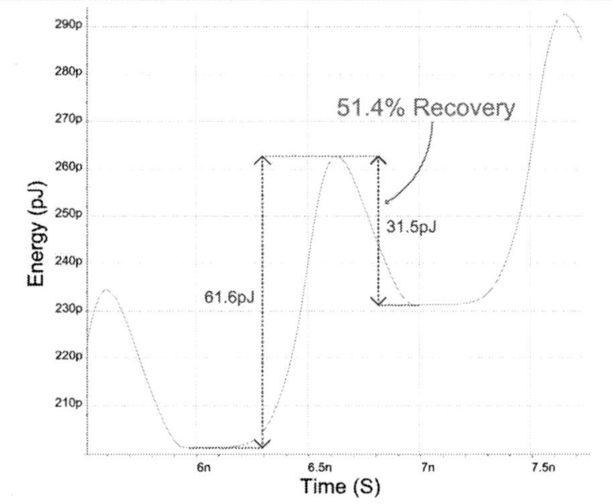

Figure 27.6.4: Simulated waveform of energy supplied to power-clock through one of the sixteen inductors.

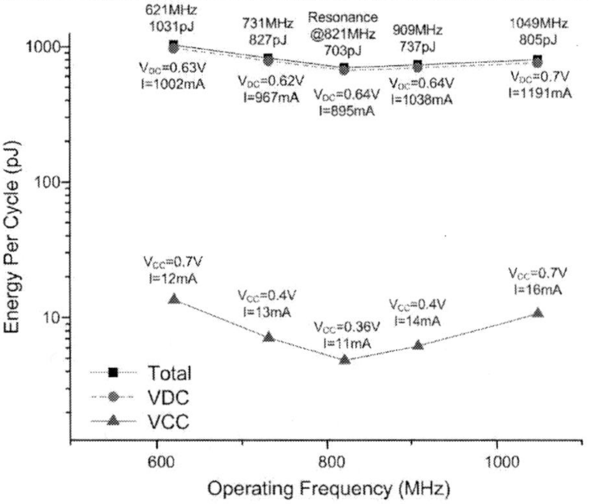

Figure 27.6.5: Measured energy per cycle versus operating frequency.

	This Work	JSSC'11 [1]	ASSCC'11 [2]	JSSC'12 [5]	ISSCC'13 [6]
Technology	65nm	0.13μm	65nm	65nm	65nm
Code Length	576	576~2304	576~2304	672	32
Core Area (mm^2)	1.54 (2.34 w/ inductors)	3.03	3.36	1.56	0.063
Frequency (MHz)	821	214	110	197	240
Iterations	10	10	10	5	5
Throughput (Gbps)	7.882	0.847	1.056	5.79	0.38
Power (mW)	576.8	342	115	361	4
Energy Efficiency (pJ/bit/iteration)	7.32	13.54 [1]	21.80	12.48 [2,3]	2.08 [3]
Area Efficiency (Gbps/mm^2)	5.12 (3.35 w/ inductors)	2.24 [1]	0.31	2.12 [2,3]	3.05 [3]

[1] Technology scaling from 0.13μm, V_{DD}=1.2V to 65nm, V'_{DD}=1.0V:

$$S = \frac{0.13\mu m}{65nm},\ U = \frac{V_{DD}}{V'_{DD}} = \frac{1.2V}{1.0V},\ \text{Delay} \sim \frac{1}{S},\ \text{Area} \sim \frac{1}{S^2},\ \text{Power} \sim \frac{1}{U^2}$$

[2] Information throughput is scaled to data throughput

[3] Normalized to 10 iterations

Figure 27.6.6: Chip summary and comparison with state-of-the-art.

27

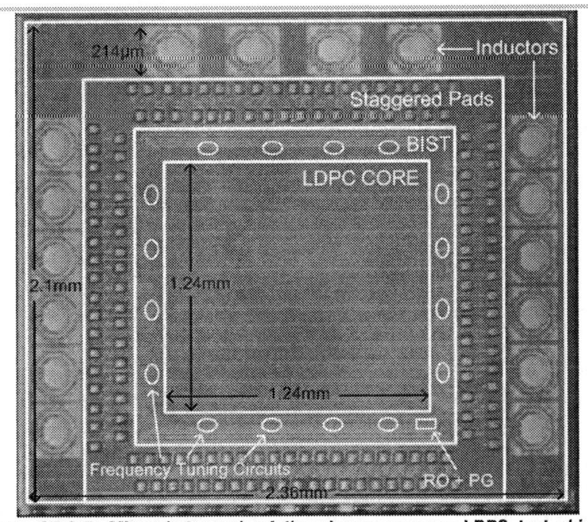

ISSCC 2014 PAPER CONTINUATIONS

Figure 27.6.7: Microphotograph of the charge-recovery LDPC test chip in 65nm CMOS.

27.7 A Scalable 1.5-to-6Gb/s 6.2-to-38.1mW LDPC Decoder for 60GHz Wireless Networks in 28nm UTBB FDSOI

Matthew Weiner[1], Milovan Blagojevic[1,2,3], Sergey Skotnikov[4], Andreas Burg[4], Philippe Flatresse[3], Borivoje Nikolic[1]

[1]University of California, Berkeley, CA,
[2]Institute Supérieur d'Electronique de Paris, Paris, France
[2]STMicroelectronics, Crolles, France,
[3]EPFL, Lausanne, Switzerland

Low-density parity-check (LDPC) codes in modern wireless communications are rate- and throughput-scalable, and despite their complexity, decoding them requires low power consumption. The IEEE 802.11ad standard for Gb/s wireless LANs in the 60GHz band requires an implementation of an LDPC encoder/decoder with throughputs of 1.5, 3, and 6Gb/s, with code rates of 1/2, 5/8, 3/4 and 13/16 [1]. Previous implementations of decoders for these throughputs and levels of reconfiguration have power consumptions on the order of the rest of the baseband processing [2,3]. This paper presents a fully compatible IEEE 802.11ad LDPC decoder in 28nm ultra-thin body and BOX fully-depleted SOI (UTBB FDSOI) technology with a power consumption that is a small fraction of the total baseband power. To achieve this, the decoder introduces an approximate marginalization technique and a simplified reconfiguration method. Forward body biasing of FDSOI technology allows for minimum energy consumption across all decoding modes.

Figure 27.7.1 shows the overall architecture of the LDPC decoder. The decoder fully parallelizes the variable nodes (VN), layer serializes the check nodes (CN), where a layer is a row of submatrices, and uses a five-bit quantization for all messages [4]. The flooding decoding schedule optimally utilizes the five-stage pipeline, and pipeline bubbles are eliminated by processing two subsequent frames simultaneously (Fig. 27.7.1). This deep pipelining shortens critical paths and permits aggressive voltage and frequency scaling, which allows the power to scale with throughput. To eliminate the need for an additional supply beyond the core and back-bias supplies, the decoder uses flip-flop-based memory. For the size of the memories required by the decoder, the flip-flops at a scaled voltage can have comparable efficiency to SRAM or eDRAM that must run on higher supply voltages. In addition, flip-flop-based designs transition well to new technologies in terms of reliability and time to market.

Memory dominates the power consumption of LDPC decoders, and the VNs (Fig. 27.7.2) and pipeline registers comprise a majority of this decoder's memory. The number of pipeline stages and the number of registers per stage has been optimized for power during the architecture exploration phase [4]. The largest number of flip-flops within the VN is contained in the shift registers that store check-to-variable (C2V) and variable-to-check (V2C) messages for marginalization of incoming or outgoing messages. Reducing the number of bits in these two blocks has a large effect on the total power since this architecture has 672 VNs. To ensure that this does not significantly affect the error-correcting performance of the decoder and to find the bits that should be removed, extensive simulations were performed where subsets of stored message bits were used for marginalization. The simulations showed that two bits of precision could be removed from both the V2C and C2V shift registers with less than 0.1dB loss in BER performance. From the C2V stored messages, the two least significant bits are removed, and from the V2C stored messages, the least significant and the most significant magnitude bits are removed. The loss in performance can be recovered by increasing the maximum number of iterations by 5, which increases the average number of iterations by less than 1% at E_b/N_0 values of interest. This decreases the decoder's power by 15% based on place-and-route results, and this technique can be applied to many LDPC decoder architectures.

Reconfigurable decoders can require more cycles to decode lower rate codes due to having more layers in the matrix, as well as longer critical paths from extra hardware for flexible routing. A low-overhead method for reconfiguration is a key requirement to minimize power. This is accomplished for each of the 4 codes defined in the standard by making the CNs switchable between acting as one 16-input CN to process a full-weight layer, or two 8-input CNs to process two half-weight layers. This switching is enabled by the non-overlapping layer structure of the lower rate IEEE 802.11ad codes (Fig. 27.7.3). It allows all code rates to be processed in the same number of cycles, but it requires an extra step of routing messages before and after the CNs. When non-overlapping layers are processed, the shuffler before the CN ensures that messages from the upper layer are routed to the top 8 inputs of the CN, and those from the lower layer are routed to the bottom 8 inputs. The multiplexers after the CN select the correct outputs to send to the VNs, which depends on whether the VN is in the upper or lower layer. The shuffler can be simplified significantly for the parity check matrices of interest (e.g. as compared to [4]) by observing, firstly, that a CN does not need a specific ordering of its inputs if it is acting as a 16-input CN, and secondly, that the interleaving pattern of the combined layers is nearly the same for each combined layer for all code rates (Fig. 27.7.3). Using these insights, much of the routing is accomplished by shuffling fixed wires, eliminating many of the long, complex global routes and multiplexers required in [4]. Since the path from the output of the VNs to the input of the CNs is the critical path in the architecture, this optimization allows additional voltage and frequency scaling to reduce overall power consumption and mitigate the overhead of reconfiguration.

Figure 27.7.7 shows a micrograph of the LDPC decoder, which was fabricated in a 28nm UTBB FDSOI process. The chip is 0.85×0.85mm^2, and the decoder core area is 0.63mm^2. It has four AWGN generators and an error collector in the periphery that can be used to measure the BER, FER, and average number of iterations. An early termination block detects when the decoding result has satisfied all parity constraints and then ends decoding.

Figure 27.7.4 shows the BER performance and the average number of decoding iterations as a function of the channel conditions. The decoder achieves a BER of less than 10^{-6} in the waterfall region for all code rates defined in the standard. For a given BER level, the rate-1/2 code takes the most iterations to converge, taking, for example, an average of 3.75 iterations to converge at an E_b/N_0 of 5.0dB. For the rate-1/2 code at an E_b/N_0 of 5.0dB and at the optimal core supply voltage, the decoder consumes 6.2mW, 14.4mW and 41mW for throughputs of 1.5Gb/s, 3Gb/s and 6Gb/s, respectively. The power can be further decreased by using the back biasing capability of the flip-well FDSOI devices (Fig. 27.7.5), which can implement LVT devices by biasing an n-well under NMOS devices and a p-well under PMOS devices [5]. This makes it possible to trade off leakage and dynamic power by adjusting the bias voltage on the n- and p-wells. Using optimal back bias and core supply voltages, the power can be decreased by 5.3% and 11% for the 3Gb/s and 6Gb/s throughputs, respectively (Fig. 27.7.5). The reason for the moderate decrease in power is that the thresholds are already close to optimal for this application; however, if the throughput required is higher, the savings would be larger. The decoder has efficiencies of 8.2pJ/b, 9.1pJ/b and 12.7pJ/b for the 1.5Gb/s, 3Gb/s and 6Gb/s throughputs. It consumes 3.0mW at a throughput of 0.7Gb/s at 15MHz with a core supply of 0.48V and no back bias, and it consumes 179.9mW at a throughput of 12Gb/s at 260MHz with a core supply of 1.07V and a back bias voltage of 0.8V. Fig. 27.7.6 compares this decoder to other state-of-the-art high-throughput decoder designs.

Acknowledgements:
The authors thank the BWRC sponsors, STMicroelectronics for chip fabrication, A. Cathelin, P. Urard, A. Vladimirescu, V. Heinrich, A. Cevrero, Y. Leblebici, N. Preyss, J. Kwak, B. Zimmer, E. Yeo and Z. Zhang.

References:
[1] T. Tsukizawa, *et al.*, "A Fully Integrated 60GHz CMOS Transceiver Chipset Based on WiGig/IEEE802.11ad with Built-In Self Calibration for Mobile Applications," *ISSCC Dig. Tech. Papers*, pp. 230-231, 2013.
[2] S.W. Yen, *et al.*, "A 5.79-Gb/s Energy-Efficient Multirate LDPC Codec Chip for IEEE 802.15.3c Applications," *IEEE J. Solid-State Circuits*, vol.47, no.9, pp. 2246-2257, 2012.
[3] X. Peng, *et al.*, "A 115mW 1Gbps QC-LDPC Decoder ASIC for WiMAX in 65nm CMOS," *Asian Solid-State Circuits Conf.*, pp. 317-320, 2011.
[4] M. Weiner, Z. Zhang, B. Nikolic, "LDPC Decoder Architecture for High-Data Rate Personal-Area Networks," *IEEE International Symp. Circuits and Systems*, pp. 1784–1787, 2011.
[5] P. Flatresse, et al., "Ultra-Wide Body-Bias Range LDPC Decoder in 28nm UTBB FDSOI Technology," *ISSCC Dig. Tech. Papers*, pp. 424-425, 2013.
[6] Y. Park, et al., "A 1.6-mm^2 38-mW 1.5-Gb/s LDPC Decoder Enabled by Refresh-Free Embedded DRAM," *IEEE Symp. VLSI Circuits*, pp. 114-115, 2012.

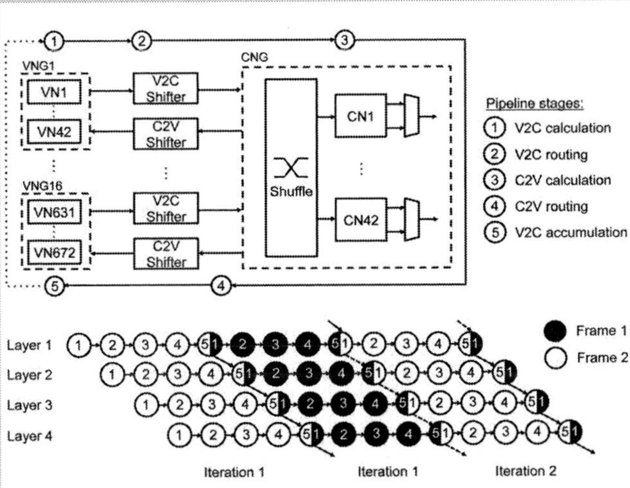

Figure 27.7.1: Decoder architecture and pipeline diagram.

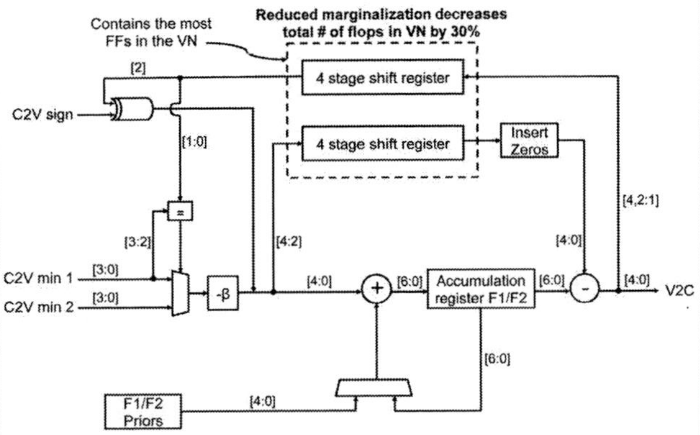

Figure 27.7.2: Variable node architecture with reduced marginalization.

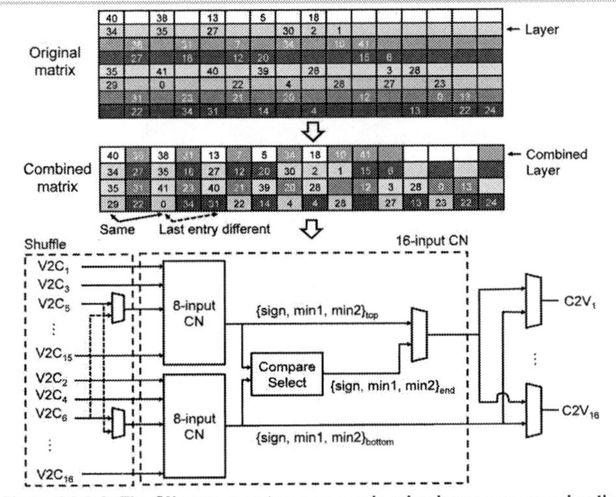

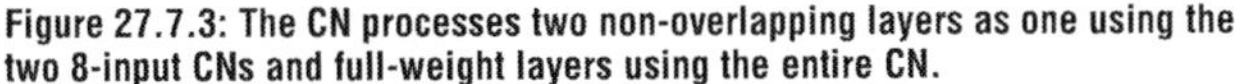
Figure 27.7.3: The CN processes two non-overlapping layers as one using the two 8-input CNs and full-weight layers using the entire CN.

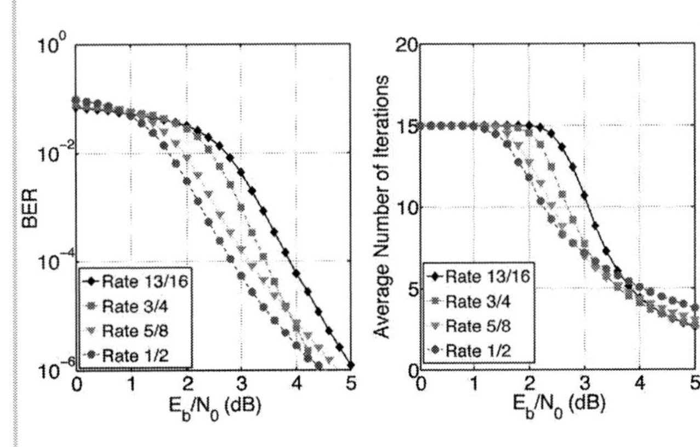

Figure 27.7.4: Measured BER and average number of iterations for all code rates with 5b quantization and reduced marginalization.

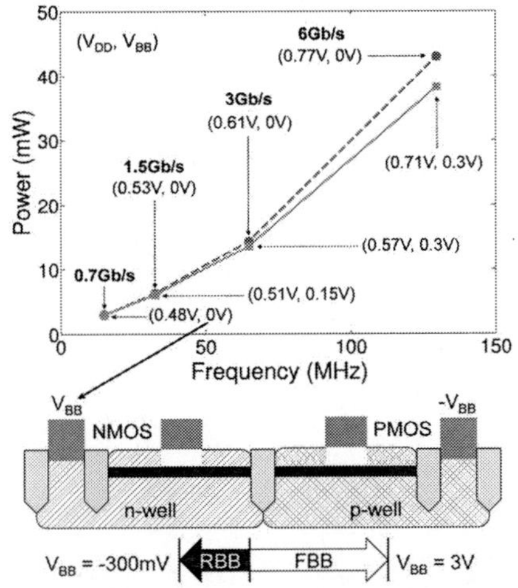

Figure 27.7.5: Measured power versus frequency at E_b/N_0 = 5.0dB with no and at optimal back biasing of the flip-well (LVT) FDSOI devices.

	This work				[6]				[2]		[3]
Technology	28nm FDSOI				65nm bulk				65nm bulk		65nm bulk
Standard	802.11ad				802.11ad				802.15.3c		802.16e
Blocklength	672				672				672		576-2304
Code rate	1/2, 5/8, 3/4, 13/16				1/2				1/2, 5/8, 3/4, 7/8		1/2, 2/3, 3/4, 5/6
Decoding Schedule	Flooding				Flooding				Layered		Layered
Core area (mm²)*	0.63				1.6				1.56		3.36
Throughput (Gb/s)	1.5	3	6	12	1.5	3	6	9	3.3	5.79	1.056
Core supply (V)	0.5	0.6	0.7	1.1	0.5	0.6	0.9	1.2	1.00	1.00	1.20
Memory supply (V)	-	-	-	-	0.8	0.9	1.1	1.3	-	-	-
Back bias supply (V)	0.2	0.3	0.3	0.8	-	-	-	-	-	-	-
Avg. Iterations	3.75				10				5		10
Frequency (MHz)	32	65	130	260	90	180	360	540	197	197	110
Power (mW)**	6.2	14	38	180	38	106	374	783	320	400	115
Energy Efficiency (pJ/decoded bit)	8.2	9.1	13	30	26	35	62	90	N/A	62.4***	109

* Area not normalized because decoders are routing limited
** Power consumption is for rate 1/2 codes at a BER of 1e-6 to 1e-7
*** For the rate 7/8 code, which will have higher energy efficiency than the rate 1/2 code

Figure 27.7.6: Comparison with state-of-the-art high-throughput LDPC decoders.

27

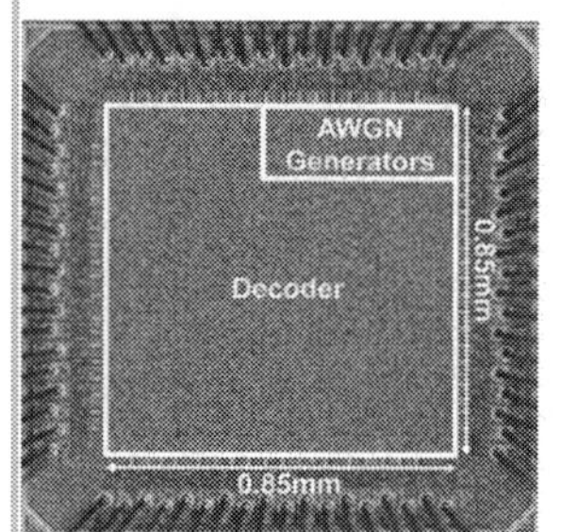

Technology	UTBB 28nm FDSOI
Transistors	Flip-well (LVT), L=24nm
Die area (mm^2)	2.56 (1.6mm x 1.6mm)
Pad Count	72
V_{DD} range (V)	0.41-1.07
V_{BB} range (V)	-0.3 to 0.8

Figure 27.7.7: Chip micrograph and technology summary.

27.8 A Static Contention-Free Single-Phase-Clocked 24T Flip-Flop in 45nm for Low-Power Applications

Yejoong Kim, Wanyeong Jung, Inhee Lee, Qing Dong, Michael Henry, Dennis Sylvester, David Blaauw

University of Michigan, Ann Arbor, MI

Near-threshold computing (NTC) is an attractive solution to stagnating energy efficiencies in digital integrated circuits, arising from slowed voltage scaling in nanometer CMOS [1-2]. The design of sequential elements for NTC, as well as in voltage-scaled systems operating at both near-threshold and super-threshold, has not been extensively studied. However, it is well known that sequential elements have a strong sensitivity to process variations in NTC [2], which can have a significant impact on system yield and power consumption. In order to achieve reliable energy-efficient operation across a wide operating voltage range, a flip-flop should have the following attributes: 1) static operation, since dynamic nodes are highly susceptible to PVT variations at low voltage; 2) contention-free transitions, since ratioed logic has poor robustness across the wide range of device I_{ON}/I_{OFF} ratios incurred with voltage scaling; 3) single-phase clocking, which avoids toggling of internal clock inverters and the corresponding power penalty; 4) minimum or no area penalty compared to conventional flip-flops.

While many flip-flops have been proposed, no prior design meets all these requirements for an energy-efficient, highly voltage-scalable sequential element. Fig. 27.8.1 highlights shortcomings in several common flip-flops. The widely-used conventional 24T TGFF exhibits high power consumption due to a large number of clocked nodes (i.e., it is not single-phase clocked). The ACFF [3] uses single-phase clocking operation and has fewer devices than the TGFF but experiences current contention in the slave latch. This contention can be suppressed at the expense of additional devices (area). The TGPL [4] is based on pulsed operation and achieves high performance at full V_{DD} but has poor robustness at low V_{DD} due to increased process-variation sensitivity. The TSPC [5] employs single-phase clock operation and uses only 11 devices. However, its dynamic operation degrades robustness, especially at low V_{DD}. In addition, Fig. 27.8.1 illustrates a non-negligible glitch at node *QN* in the TSPC whenever CK goes high while D remains 0. This arises since precharged *net2* begins to discharge *QN* before M5/M6 can pull *net2* low, resulting in unnecessary power consumption or even a system malfunction.

This work presents a new flip-flop, referred to as static single-phase contention-free flip-flop (S²CFF) that meets the requirements above: it is static, completely contention-free, and uses single-phase clocking. It has the same device count as a TGFF, with only a 7% layout size increase that corresponds to a one poly-pitch increase in 45nm technology, where fixed poly-pitch is enforced. Fig. 27.8.2 shows the S²CFF schematic and describes its operation. For CK=0, *net1* holds $\overline{D}$ value, *net2* precharges through M8, and the slave latch (M17~M22) stores the previous data. If D=0, the high *net1* starts discharging *net2* at the positive edge of CK. Then, discharged *net2* turns off M3, completely isolating the circuit from changes in D. Also, the low *net2* charges *QN* through M13, updating the data in the slave latch. Note that *net1* is held high by M5, while M9/M10 keep *net2* low during the high CK phase. If D=1, the positive edge of CK does not generate any dynamic transitions at *net1* and *net2*. During CK=1 phase, *net1* is kept low by M7/M10, and M6 holds *net2* high. If the previous Q value is the same as the current D input (i.e., *QN*=0), there is also no transition at *QN*. Otherwise, *QN* discharges through M14~M16. Signal *net1b* is also used to control M15; without this sub-circuit, *QN* will glitch when CK rises with D staying low in consecutive cycles, similar to the TSPC. M15 eliminates this glitch by cutting off the discharge path (M14~M16) depending on *net1*'s value. Note also that there is no contention throughout the operation, all internal nodes are fully static, and only one clock phase (CK) is used.

An additional benefit of the S²CFF topology is that it simplifies the "hold-time path" compared to a regular TGFF (Fig. 27.8.3). As described in [6], the worst-case hold time in a TGFF is when D changes from 1 to 0 just after the CK edge, and it is dictated by mismatch among the clock/data inverters (I1, I3, I4). Due to clock inversion in $PATH_{CK}$, the NMOS in I2 turns off earlier than its PMOS, while the PMOS in I5 turns on before its NMOS, weakening the pull-down strength at node *MN*. Hence, a TGFF shows severe hold-time degradation at low V_{DD} where mismatch is accentuated. On the contrary, the worst-case hold time in the S²CFF occurs when D changes from 0 to 1 just after the CK edge. The high *net1* starts discharging *net2*, and the discharged *net2* turns off M3, isolating the D input. If D becomes 1 before *net2* shuts off M3, and thus discharges *net1*, a hold failure may result. Only the discharging speed of *net2* through $PATH_{CK}$ dictates the hold time. As a result, the hold time of the S²CFF is much less prone to variability compared to the TGFF, which involves the time difference of several gate delays. Fig. 27.8.3 shows a substantial reduction (3.4×) in hold time at the 3σ value at 0.32V for S²CFF (from Monte Carlo simulations). This suggests a large potential benefit for NTC, since small hold-time variation reduces buffer-insertion overhead, reducing power and improving system yield.

The S²CFF is characterized in a 45nm SOI test chip that also includes TGFF, ACFF, and TGPL for comparison. 50 dies are measured. On-chip testing circuits are shown in Fig. 27.8.4, where the setup/hold-time measurement circuit is based on the structure in [7]. In the power measurement circuit, the activity ratio is controlled using the 20b initial pattern. To mimic a realistic scenario, the test chip has one clock buffer driving 10 DUTs. The current flowing into 'CLKBUF + 10 DUTs' is measured and then divided by 10. The C-Q delay measurement circuit incorporates a new flip-flop ring, where a short pulse at the EN input triggers the oscillation of DUT Ring with a period that is proportional to T_{CQ} with an offset value; this offset can be measured using Reference Ring. With a large N (= number of unit cells in a ring) value, local mismatch is effectively cancelled out making it possible to obtain accurate C-Q delays. This test chip uses 100 unit cells per ring (N=100). The DUT Ring also gives insight on DUT yield, since oscillation stops unless all 100 DUTs in the ring are functional.

Figure 27.8.5 shows measured total power and energy. The S²CFF does not require internal clock inverters, enabling a clock power (defined as total power at 0% activity ratio with D=0) reduction of 41% at 1V/1GHz compared to TGFF. Assuming that flip-flops in a typical system have 20% activity ratio, the S²CFF provides 39% and 38% improvement in total sequential power at 1V/1GHz and 0.4V/200MHz, respectively. Active energy consumption is also reduced by 32% and 34% (1.0V and 0.4V, respectively). The measured ACFF total power increases rapidly as activity rises due to contention in the slave latch. The TGPL has a delay element, which leads to higher total power consumption, even at 0% activity ratio. Fig. 27.8.6 shows measured C-Q delays and leakage power. The S²CFF shows modest improvement over the TGFF across V_{DD}. Missing points in the plot indicate that ACFF fails to have 100% yield at 0.4V due to contention. The TGPL fails at $V_{DD} \leq 0.6V$, mainly due to hold-time failures. This illustrates the importance of static and contention-free operation at low V_{DD}, since only the TGFF and S²CFF show 100% yield across the wide V_{DD} range. The S²CFF has 35% lower leakage power than the TGFF at 1.0V. The provided comparison table includes other recently proposed flip-flops, showing that the S²CFF is the only flip-flop with static, contention-free, single-phase clock operation with minimum area penalty (one poly-pitch) and same device count as TGFF. The S²CFF has 15.5% faster 'setup time + C-Q delay' at 1.0V vs. the TGFF, as shown in the table. Fig. 27.8.7 includes the die photo of the test chip.

Acknowledgment:

Funding support of DARPA (agreement HR0011-13-2-0006) and US Army Research Laboratory is gratefully acknowledged.

References:

[1] B. Zhai, *et al.*, "Energy Efficient Near-Threshold Chip Multi-Processing," *IEEE International Symp. Low-Power Electronics and Design*, pp. 32-37, 2007.

[2] S. Jain, *et al.*, "A 280mV-to-1.2V Wide-Operating-Range IA-32 Processor in 32nm CMOS," *ISSCC Dig. Tech. Papers*, pp. 66-67, 2012.

[3] C.-K. Teh, et *al.*, "A 77% Energy-Saving 22-Transistor Single-Phase Clocking D-Flip-Flop with Adaptive-Coupling Configuration in 40nm CMOS," *ISSCC Dig. Tech. Papers*, pp. 338-339, 2011.

[4] S. D. Naffziger, *et al.*, "The Implementation of the Itanium 2 Microprocessor," *IEEE J. Solid-State Circuits*, vol. 37, no. 11, pp. 1448-1460, 2002.

[5] J. Yuan, *et al.*, "High-Speed CMOS Circuit Technique," *IEEE J. Solid-State Circuits*, vol. 24, no. 1, pp. 62-70, 1989.

[6] C.-H. Chen, *et al.*, "Minimum Supply Voltage for Sequential Logic Circuits in a 22nm Technology," *IEEE International Symp. Low-Power Electronics and Design*, pp. 181-186, 2013.

[7] B. Giridhar, *et al.*, "Pulse Amplification Based Dynamic Synchronizers with Metastability Measurement using Capacitance De-rating," *IEEE Custom Integrated Circuits Conf.*, 2013.

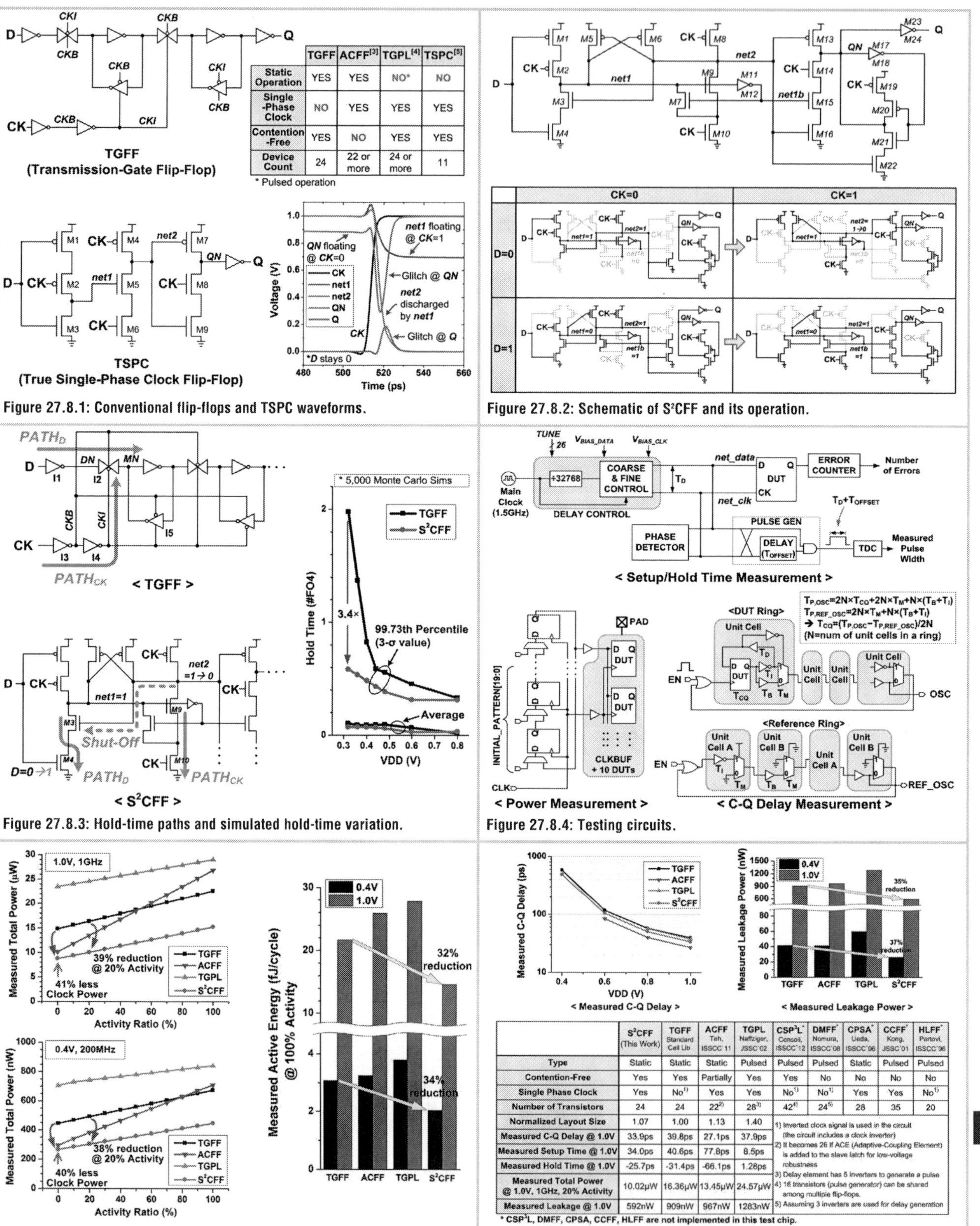

	TGFF	ACFF[3]	TGPL[4]	TSPC[5]
Static Operation	YES	YES	NO*	NO
Single-Phase Clock	NO	YES	YES	YES
Contention-Free	YES	NO	YES	YES
Device Count	24	22 or more	24 or more	11

* Pulsed operation

Figure 27.8.1: Conventional flip-flops and TSPC waveforms.

Figure 27.8.2: Schematic of S²CFF and its operation.

Figure 27.8.3: Hold-time paths and simulated hold-time variation.

Figure 27.8.4: Testing circuits.

Figure 27.8.5: Measured power and energy comparisons.

	S²CFF (This Work)	TGFF Standard Cell Lib	ACFF Teh, ISSCC'11	TGPL Naffziger, JSSC'02	CSP³L* Consoli, ISSCC'12	DMFF* Nomura, ISSCC'08	CPSA* Ueda, ISSCC'06	CCFF* Kong, JSSC'01	HLFF* Partovi, ISSCC'96
Type	Static	Static	Static	Pulsed	Pulsed	Pulsed	Static	Pulsed	Pulsed
Contention-Free	Yes	Yes	Partially	Yes	Yes	No	No	No	No
Single Phase Clock	Yes	No1)	Yes	Yes	No1)	No1)	Yes	Yes	No1)
Number of Transistors	24	24	22 2)	28 3)	42 4)	24 5)	28	35	20
Normalized Layout Size	1.07	1.00	1.13	1.40					
Measured C-Q Delay @ 1.0V	33.9ps	39.8ps	27.1ps	37.9ps					
Measured Setup Time @ 1.0V	34.0ps	40.6ps	77.8ps	8.5ps					
Measured Hold Time @ 1.0V	-25.7ps	-31.4ps	-66.1ps	1.28ps					
Measured Total Power @ 1.0V, 1GHz, 20% Activity	10.02µW	16.36µW	13.45µW	24.57µW					
Measured Leakage @ 1.0V	592nW	909nW	967nW	1283nW					

1) Inverted clock signal is used in the circuit (the circuit includes a clock inverter)
2) It becomes 26 if ACE (Adaptive-Coupling Element) is added to the slave latch for low-voltage robustness
3) Delay element has 5 inverters to generate a pulse
4) 16 transistors (pulse generator) can be shared among multiple flip-flops.
5) Assuming 3 inverters are used for delay generation

* CSP³L, DMFF, CPSA, CCFF, HLFF are not implemented in this test chip.

Figure 27.8.6: Measured C-Q delay, leakage, and comparison table.

27

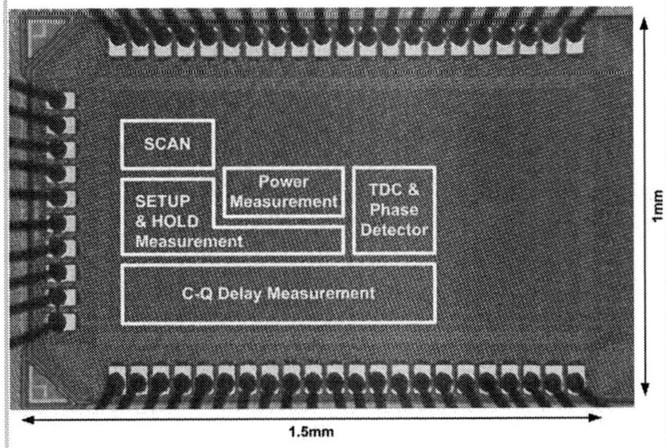

Figure 27.8.7: Die photograph of the 45nm SOI test chip.

Session 28 Overview: *Mixed-Signal Techniques for Wireless*

WIRELESS SUBCOMMITTEE

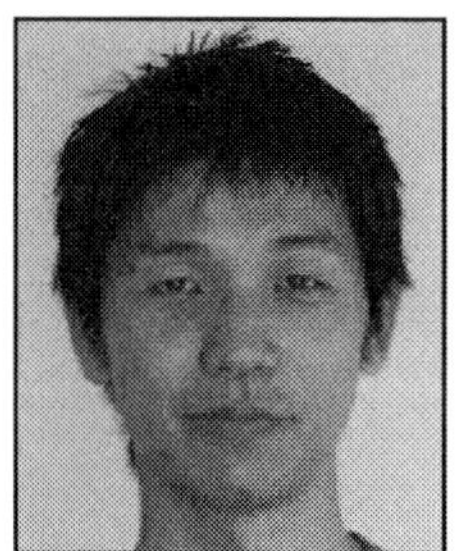

Session Chair: *Shouhei Kousai*
Toshiba, Kawasaki, Japan

Session Co-Chair: *Jan van Sinderen*
NXP Semiconductors,
Eindhoven, The Netherlands

Mixed-signal circuit techniques are essential for high-performance and energy-efficient wireless systems. This session covers circuit techniques of $\Delta\Sigma$-modulators and digital-to-time converters, as well as their application to software-defined radios, frequency synthesizers, and radar systems.

28.1 A Programmable 0.7-to-2.7GHz Direct ΔΣ Receiver in 40nm CMOS **1:30 PM**

K. Östman, Aalto University, Espoo, Finland

In Paper 28.1, Aalto University and Ericsson present a software-defined radio receiver that investigates the possibility of using ΔΣ-modulators for direct RF-to-digital conversion. The wideband ΔΣ-modulator-based receiver using 40nm CMOS is capable of handling up to 20MHz signal bandwidth with a narrowband NF of 4.2dB.

28.2 A 0.29mm² Frequency Synthesizer in 40nm CMOS with 0.19ps$_{rms}$ Jitter and <-100dBc Reference Spur for 802.11ac **2:00 PM**

Y-L. Hsueh, MediaTek, Hsinchu, Taiwan

In Paper 28.2, MediaTek describes a fractional-N frequency synthesizer for 802.11ac applications using 40nm CMOS. The presented circuit has an area of only 0.29mm² while fulfilling the stringent noise requirement of 802.11ac at a power consumption not exceeding 17.5mW. The fractional spurs are suppressed by minimizing charge-pump switching noise.

28.3 A Frequency-Defined Vernier Digital-to-Time Converter for Impulse Radar Systems in 65nm CMOS **2:30 PM**

Y-H. Kao, National Tsing Hua University, Hsinchu, Taiwan

In Paper 28.3, National Tsing Hua University presents a digital-to-time converter for an impulse radar system using 65nm CMOS. The proposed converter is robust to process, temperature and voltage variations, and can measure time-of-arrival ranges up to 10nsec with a 6ps resolution, which corresponds to 15m distance with 0.9mm accuracy.

978-1-4799-0917-9/14 $31.00 © 2014 IEEE

28.1 A Programmable 0.7-to-2.7GHz Direct ΔΣ Receiver in 40nm CMOS

Mikko Englund[1], Kim B. Östman[1], Olli Viitala[1], Mikko Kaltiokallio[1], Kari Stadius[1], Kimmo Koli[2], Jussi Ryynänen[1]

[1]Aalto University, Espoo, Finland, [2]Ericsson, Turku, Finland

The software-defined radio paradigm calls for increasingly digital-intensive programmable receivers, ideally placing the analog-to-digital converter (ADC) right at the antenna. Such an RF ADC should be tunable over several GHz, have programmable gain, low noise, be blocker-tolerant, and consume minimal power. As an attempt to satisfy these requirements, delta-sigma (ΔΣ) modulation close to the antenna interface has been proposed in both bandpass [1], [2] and downconverting [3], [4] configurations. The latter technique enables simpler GHz-range wideband (WB) operation with low power consumption, but such receivers navigate a tradeoff between sensitivity and blocker toleration. The narrowband (NB) direct ΔΣ structure introduced in [3] combined RF N-path filtering, upconverted ΔΣ RF feedback, and a second RF gain stage to obtain acceptable noise and linearity simultaneously. In this paper we present a WB direct ΔΣ receiver, designed for programmable, inductorless operation in the long-term evolution (LTE) frequency division duplexing bands from 0.7 to 2.7GHz. The 40nm CMOS circuit uses a supply of 1.1V and provides RF channel bandwidths up to 20MHz, 37dB maximum gain, NF of 5.9 to 8.8dB, and -2dBm IIP3. A design strategy that emphasizes ΔΣ coefficient programmability ensures good performance throughout the frequency range.

The wideband direct ΔΣ receiver structure embeds a direct-conversion RF front-end into a ΔΣ ADC. Channel filtering, signal downconversion, and quantization noise shaping are thus performed simultaneously, with the filtering characteristic being emphasized in wideband employment. The continuous-time 4-stage feedback modulator architecture depicted in Fig. 28.1.1 was selected as a good tradeoff between complexity, stability, and sufficient blocker filtering. The loop filter baseband bandwidth f_{BW} can be programmed to 1 or 10 MHz. In contrast to conventional ΔΣ modulators, the receiver emphasizes blocker filtering in the signal transfer function (STF) at the cost of SNDR when the channel bandwidth approaches f_{BW}. To minimize SNDR reduction, the internal feedback coefficient g_2 is used to create an NTF notch, which enhances noise shaping close to the band edge. The coefficients b_1 and a_2 can be adjusted to lower the gain of the receiver by a maximum of 15dB, while preserving the shape of the STF. A half clock cycle delayed feedback to the input of the quantizer (b_{ELDC}) is used to compensate for the delay in the feedback path.

In the RF section, the first ΔΣ integrator stage consists of a common-source LNA with adjustable active common-drain RC feedback, loaded by PMOS devices and an N-path filter, as shown in Fig. 28.1.2. This combination sets the ΔΣ coefficients a_1 and g_1, and the center frequency of the N-path filter controls the S_{11} notch of the receiver. Moreover, the combination simultaneously creates an inductorless blocker-filtering RF resonator and an RF integrator response at the LNA output node, with the center frequency of both being controlled by f_{LO}. At the highest operating frequencies, the digitally controlled transconductors G_{mx} may be used to cancel the bandpass resonator center frequency offset from f_{LO} that appears due to parasitic capacitance at the LNA output. This output node acts as a summing point for the forward and the upconverted ΔΣ feedback signals, and the closed-loop RF integrator gain around f_{LO} is reduced as a result. The second ΔΣ stage is a g_m-boosted source follower that drives the post-downconversion Miller integrator capacitor C_{int}. The programmable transconductance directly adjusts the ΔΣ receiver's coefficient a_2 and reduces LO leakage in the receiver. The low output impedance of the follower enables WB performance that trades off with sub-optimal functionality as a transconductor. The N-path and downconversion mixers are operated with an LO signal that has a nominal 25% duty cycle. As shown in Fig. 28.1.2, the post-divider LO signal is processed by a programmable buffer, where the duty cycle is adjusted by setting $V_{LO,ref}$ [5]. Moreover, the compensation offered by the feedback loop mitigates the loss of LO signal integrity at high frequencies.

In the baseband (BB) section, the design philosophy focuses on achieving low quantization noise and good signal selectivity simultaneously, which is in contrast to conventional ΔΣ modulators. The quantizer is realized with 1.5 bits to lower the total amount of quantization noise. Tunable reference levels in the quantizer allow additional flexibility in the choice of modulator coefficients. The BB integrators are realized as conventional opamp-RC stages with tunable capacitor matrices. The feedback to each integration stage utilizes a 1.5-bit tunable current-mode DAC (IDAC), shown in Fig. 28.1.2. Each corresponding feedback coefficient b_i can be programmed by one decade on both sides of the nominal value. This tunability allows bandwidth and gain adjustment, which can be applied to optimize blocker tolerance and noise shaping performance. The ratio f_{LO}/f_{CLK} affects the amount of quantization noise and spurious content that is upconverted to RF from the outermost loop (b_1) [3]. In this receiver, the noise is filtered by a tunable 2nd order FIR filter consisting of three parallel IDACs. The tunability of the FIR filter extends the range of usable f_{LO}/f_{CLK} ratios by allowing optimal placing of the notches in the low-pass sinc response of the filter. Another benefit of using FIR filtered feedback is an increased resilience against clock jitter [6]. The outermost feedback must be co-designed with the other ΔΣ coefficients so that the closed-loop LNA voltage gain reduction is limited to a few dB.

A test chip containing the fully integrated 0.7-to-2.7GHz receiver was fabricated in 40nm CMOS and packaged with flip-chip technology for access through a 61-pin BGA. A narrowband receiver version with an inductively degenerated 2.5GHz LNA with on-chip inductors for input matching and LNA loading was included on the same die for reference. The measured input matching for both is better than –9dB in the entire operating range. The WB receiver consumes 90mW from 1.1V, consisting of 15mW for the RF front-end, 33mW for the LO buffering, and 42mW for the BB section. The measured output spectra in Fig. 28.1.3 confirm consistent operation in both ends of the RF band, with an in-channel signal of P_{in} = –68dBm at f_{LO} + 100kHz being amplified and downconverted in both cases. An out-of-band signal with P_{in} = –43dBm at f_{LO} + 87.5MHz is filtered by the ΔΣ loop and disappears into the noise floor. Moreover, the two spectra exhibit the intended 6.5MHz NTF notch for SNDR improvement. Fig. 28.1.4 shows a maximum SNDR of 43dB for a 15MHz LTE RF channel. The low output impedance of the WB LNA weakens the noise shaping functionality of the first stage, and explains the difference to the higher 50dB SNDR of the narrowband reference setup. Fig. 28.1.5 plots the IIP3 of the receiver around f_{LO} = 2.5GHz. The gain exhibits minor peaking close to the ΔΣ loop filter edges, after which the receiver filters blockers. IIP3 is limited by the baseband chain at low offsets from f_{LO}, whereas the RF front-end limits the out-of-band performance to –2dBm. The measured 4.2dB NF of the narrowband receiver compares favorably to other delta-sigma-based receivers and shows the benefit of an on-chip load resonator in the LNA. The higher 5.9-to-8.8dB NF of the WB receiver is the natural consequence of using a non-resonator LNA with a lower load impedance level.

Figure 28.1.6 compares the reported circuit with other digital-intensive designs. The presented work demonstrates the WB direct ΔΣ receiver concept and exhibits lower noise than the wideband down-converting ΔΣ receiver [4], while exhibiting moderate SNDR and linearity. By exploiting direct-conversion functionality combined with baseband ΔΣ modulation, the receiver requires no inductors and uses less power than the LC bandpass solutions. Finally, Fig. 28.1.7 shows a micrograph of the test chip, in which the two front-ends use a combined BB signal path.

Acknowledgements:
This work was supported by the ENIAC JU under the ARTEMOS project.

References:
[1] H. Shibata, et al., "A DC-to-1GHz Tunable RF ΔΣ ADC Achieving DR = 74dB and BW =150 MHz at f_0 = 450MHz Using 550mW," *ISSCC Dig. Tech. Papers*, pp. 150–151, Feb. 2012.
[2] E. Martens, et al., "RF-to-Baseband Digitization in 40 nm CMOS With RF Bandpass ΔΣ Modulator and Polyphase Decimation Filter," *IEEE J. Solid-State Circuits*, vol. 47, no. 4, pp. 990–1002, Apr. 2012.
[3] K. Koli, et al., "A 900MHz Direct ΔΣ Receiver in 65nm CMOS," *ISSCC Dig. Tech. Papers*, pp. 64–65, Feb. 2010.
[4] C. Wu, et al., "A 0.4 GHz – 4 GHz Direct RF-to-Digital ΔΣ Multi-Mode Receiver," *Eur. Solid-State Circuit Conf.*, pp. 275–278, Sep. 2013.
[5] M. Kaltiokallio, et al., "A 0.7–2.7-GHz Blocker-Tolerant Compact-Size Single-Antenna Receiver for Wideband Mobile Applications," *IEEE Trans. Microw. Theory and Techn.*, vol. 61, no. 9, pp. 3339–49, Sept. 2013.
[6] V. Srinivasan, et al., "A 20mW 61dB SNDR (60MHz BW) 1b 3rd-order Continuous-Time Delta-Sigma Modulator Clocked at 6GHz in 45nm CMOS," *ISSCC Dig. Tech. Papers*, pp. 158–159, Feb. 2012.

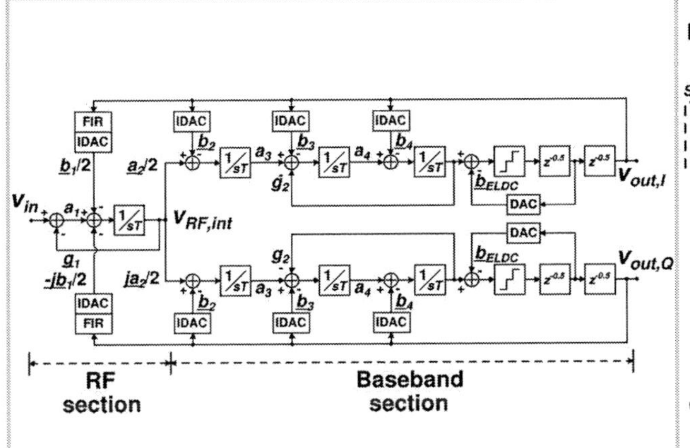

Figure 28.1.1: Block diagram of the fourth-order continuous-time wideband direct $\Delta\Sigma$ receiver. The underlined coefficients are programmable.

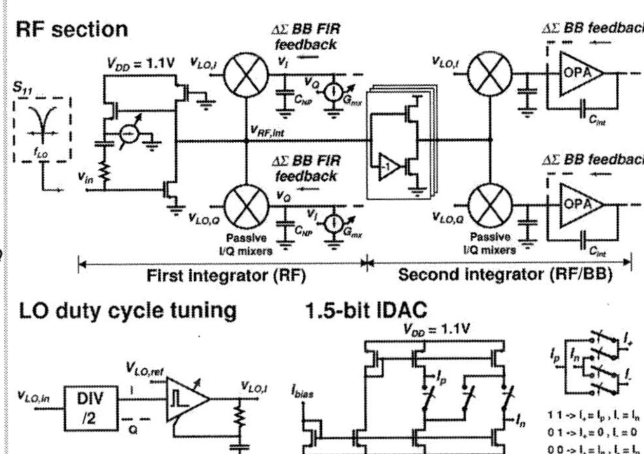

Figure 28.1.2: The WB integrator stages, the LO buffering block, and the 1.5-bit IDAC used in the BB section.

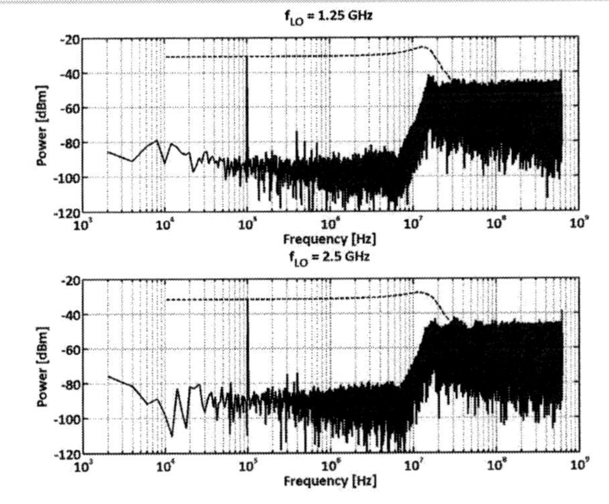

Figure 28.1.3: Receiver output spectrum at f_{LO} = 1.25GHz and 2.5GHz, f_{in} = f_{LO} + 100kHz, with overlaid normalized STF (dashed).

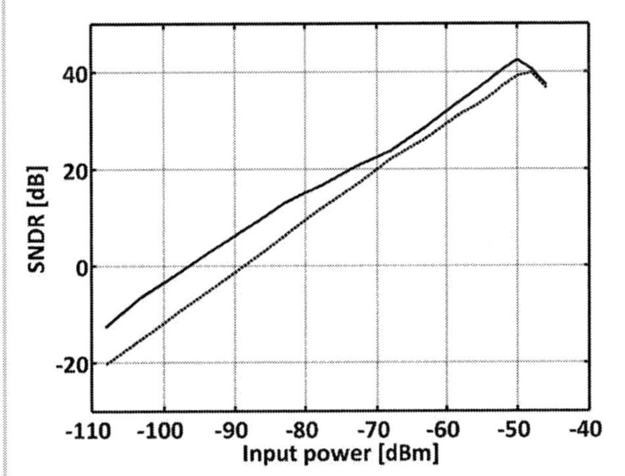

Figure 28.1.4: SNDR vs. P_{in} at f_{LO} = 1.25GHz (solid) and f_{LO} = 2.5GHz (dashed).

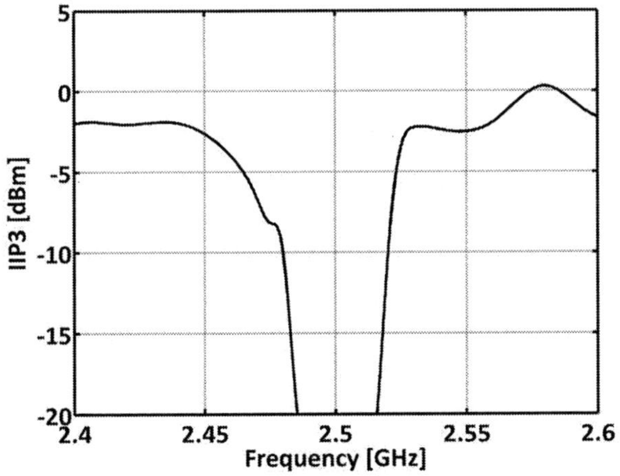

Figure 28.1.5: IIP3 around f_{LO} = 2.5GHz.

Param. / Ref.	[1]	[2]	[3]	[4]	This work		
System type	BP ΔΣ RX	BP ΔΣ RX	NB Direct ΔΣ RX	Downc. ΔΣ RX	WB Direct ΔΣ RX		NB Direct ΔΣ RX
Freq. [GHz]	0–1	2.22	0.9	0.4–4	0.7–2.7	0.7–2.7	2.5
f_{clk} [GHz]	4	8.88	1	-	1.25[1)]	1.25[1)]	1[1)]
Signal BW [MHz]	75	80	9	4	15	1.4	15
Gain [dB]	-	-	40	-	37	37	41
NF [dB]	-	-	6.2	16	5.9–8.8	5.9–8.2	4.2
IIP3[2)] [dBm]	-	+1[3)]	+4	+13.5[4)]	-2	-2	0
BCP-1dB[5)] [dBm]	-	-	-18[6)]	-	-12/ -15/-26	-14/ -14/-16	-15/ -16/-32
SNDR [dB]	76 (SNR)	42	56	>60	43	40	50
Power [mW]	550	164	80	40.3	90	90	90
Supply [V]	1.0/±2.5	1.1/1.5	1.2	1.1/1.5	1.1		
Act. area [mm²]	5.5	0.4	1.2	0.56	1.0		1.1
Tech. [nm]	65	40	65	65	40		

[1)] Max. used f_{CLK}, [2)] Closest interferer at f_{LO} + 95 MHz, [3)] Interferers in ADC in-band, [4)] Closest interferer at f_{LO} + 40 MHz, [5)] Blocker at f_{LO} + BW/2 + 85/60/15 MHz, [6)] Blocker at f_{LO} + 80 MHz

Figure 28.1.6: Performance summary and comparison with other digital-intensive receivers.

Figure 28.1.7: Test chip micrograph.
The die size including pads is 1.53x1.53mm^2.

28.2 A 0.29mm² Frequency Synthesizer in 40nm CMOS with 0.19ps$_{rms}$ Jitter and <-100dBc Reference Spur for 802.11ac

Yu-Li Hsueh, Lan-Chou Cho, Chih-Hsien Shen, Yi-Chien Tsai, Tzu-Chan Chueh, Tao-Yao Chang, Jui-Lin Hsu, Jing-Hong Conan Zhan

MediaTek, Hsinchu, Taiwan

Conventional analog PLLs do not scale well with process when compared to all-digital PLLs due to several substantial building blocks such as the loop filter and charge pump (CP). To achieve the required phase noise, the in-band noise is typically suppressed by increasing CP current and loop filter size, while the out-of-band noise is reduced by improving VCO tank Q; both lead to increased die area. This paper presents a fractional-N synthesizer targeting the relatively stringent phase noise requirement to support 256-QAM and MIMO in 802.11ac. It deploys the following techniques to simultaneously address requirements of compact area, low noise and fast calibration: reuse of VCO inductor area for the loop filter; a PFD and CP design that relaxes CP design constraints without sacrificing noise; inductor-less LO generation for 802.11bgn mode; and an area-efficient reference clock doubler and associated calibration scheme. The synthesizer block diagram is shown in Fig. 28.2.1. In 802.11ac/a mode a frequency tripler followed by I/Q dividers realizes the 3/2 frequency multiplication and I/Q generation. In 802.11bgn mode, an LO generation circuit performs the 2/3 frequency multiplication and I/Q generation. This frequency plan features overlapping VCO tuning ranges between 802.11ac/a (F_{LO}=4915~5825MHz) and 802.11bgn (F_{LO}=2412~2484MHz) modes, such that the VCO designed for 802.11ac/a can support 802.11bgn without additional tuning range.

The loop filter occupies considerable area in typical analog synthesizers. In this work, a significant portion of the loop filter is placed under the VCO inductor for space reuse [1]. Unlike [1], capacitors are covered by metal in a patterned ground shield fashion and placed both under inductor traces and at the center to fully exploit the area. Since the shielding does not block eddy currents, conductive layers including metal and polysilicon would degrade the inductor quality factor if any composite conductive loop were formed. A layout pattern, similar to a fish-bone structure, has been developed to place the large loop filter capacitor under the VCO inductor with the intension of avoiding such loops. EM simulations show negligible degradation of the inductor quality factor with the loop filter capacitor underneath. A 3rd order LPF with properly designed pole frequencies is used in the loop filter to help reduce high-order harmonics of the reference clock before entering the VCO inductor region. Figure 28.2.1 also shows the detailed schematic of the low-power VCO, which uses a CMOS topology; it can also be configured as a PMOS-only topology for high-performance mode [2]. An issue arises in the PMOS-only topology where M_{n1} and M_{n2} are disabled by shorting the gate to the source. They can momentarily conduct and cause un-intentional tank losses when the VCO swing is high. In this work the gates are biased with a voltage outside the normal supply range, specifically, a negative voltage realized by a switched-capacitor charge pump to ensure they are completely off when desired. In this particular VCO implementation, this technique improved the overall VCO tank quality factor by more than 10%.

To fulfill the low-noise requirement, the CP would occupy sizable area due to a large I_{CP}. This paper proposes a PFD/CP that significantly shrinks area by relaxing design constraints for the CP current I_{UP}, typically implemented with PMOS current mirrors, and avoiding the OpAmp and dummy capacitors. Figure 28.2.2 shows the CP employing a small offset current to shift the operating point, to avoid delta-sigma modulator ($\Delta\Sigma$M) noise folding due to nonlinearities. In conventional implementations, both I_{UP} and I_{DN} occupy sizable areas due to noise and matching requirements. On the timing diagram, I_{UP} always exhibits a fixed duration due to the PFD feedback delay for dead-zone avoidance and cancels the same amount of charge from I_{DN} in steady state. However, I_{UP} and the cancelled I_{DN} do nothing but generate noise. In the proposed CP, I_{DN} and I_{OFFSET} reach balance in normal operation, while I_{UP} is conditionally turned on only when unlocked. Since I_{UP} is off in normal operation, there is no noise requirement and it can be realized within a much smaller area. Additionally I_{OFFSET} is gated by the reference clock. The switch SW1 remains open when I_{DN} and I_{OFFSET} are active and is closed when both are silent, thus functioning in a sample-and-hold manner [3] to reduce reference spurs. The proposed PFD is capable of detecting the relative phase between CK_{REF} and CK_{DIV} to enable/disable I_{UP} for negative/positive delay values respectively. The transfer curve of charge vs. delay in Fig. 28.2.2 features a slope of $-I_{CP}$ and a charge gap of $\Delta Q_{CP}=I_{CP}\times t_{PFD}$ at delay=0, where t_{PFD} is the PFD feedback delay. When phase locked, it operates at positive delays and the offset current is chosen to avoid $\Delta\Sigma$-modulated clock phases crossing delay=0. Another area-consuming building block of the traditional current-steering CP is the OpAmp and its associated dummy capacitors, which are used to minimize charge sharing. This work employs the cascode topology at the output and dummy nodes to minimize charge sharing while avoiding the OpAmp and dummy capacitors.

Figure 28.2.3 shows the proposed inductor-less LO generation circuit for the 802.11bgn mode. Compared to conventional LC-tank-based frequency-mixing approaches, the area is much smaller, and the concern of magnetic coupling with nearby LC tanks is relaxed. It converts the incoming 3.6GHz VCO clocks to quadrature output clocks at 2.4GHz. A calibrated VCDL provides 8 clock signals CK<0:7> with evenly spaced phases. The LO generation circuit extracts specific clock edges to synthesize output LO signals by the multiplexers controlled with (S_1, S_2, S_3, S_4), (Z_I, Z_Q) signals. Complementary output signals (e.g. I and IB) operate in a cyclic manner by transferring the token signal (e.g. Z_I) between each other; only the one holding the token is allowed to output specified edges, and releases it afterwards. The synthesized clocks exhibit a 33% duty cycle and subsequent duty-cycle reshaping circuits provide 25% and 50% duty-cycle LOs required by RX and TX, respectively.

A reference clock doubler improves in-band phase noise, which necessitates duty-cycle correction to minimize the reference spur. This synthesizer adopts a duty-cycle correction scheme, which does not require integrators or an LPF [4]. It not only reduces area by avoiding integration capacitors but also significantly improves the calibration speed achieving, for example, 50ps resolution within 1.1us. As shown in Fig. 28.2.4, it consists of tunable delays DL0 to DL4, and all edge comparisons are implemented with DFFs. First, identical delays DL1 to DL4 are calibrated by aligning the rising edge of signal E with the rising edge of A. Afterwards, the multiplexer is configured such that its output has ≥50% duty cycle, and DL0 is calibrated by aligning the falling edge of A with the rising edge of C. The calibration procedure not only corrects the duty cycle of the input clock to minimize the spur at F_{REF} offset frequency, but also calibrates the duty cycle of the output clock at $2\times F_{REF}$ for the sample-and-hold CP to function at a well-defined operating point.

In summary, a low-noise and area-efficient frequency synthesizer is realized by space reuse and the presented sub-block designs. Fabricated in 40nm CMOS, the synthesizer core occupies 0.29mm². Current consumptions are 5.5mA and 13.5mA in 802.11bgn and 802.11ac/a modes, respectively, from 1.3V supplies. On the LO paths, the tripler occupies 39000um² (<0.04mm²) and consumes 10mA in the 802.11ac/a mode, while the LO generation circuit occupies 9800um² (<0.01mm²) and consumes 5.5mA in the 802.11bgn mode. Measured RMS jitter (from 1k to 10MHz) of 0.19ps and 0.30ps are achieved in 802.11ac/a and 802.11bgn modes, respectively. Figure 28.2.5 shows a measured phase noise profile. For F_{REF}=26MHz, measured reference spur levels at 26/52MHz offsets are <-100dBc. The measured worst-case fractional spur level is -65dBc. Figure 28.2.6 shows the performance summary table for comparison with recently published analog and all-digital PLLs. Compared to the prior art with equal jitter performance for 802.11ac [5], this work has a smaller area (43%) and lower power consumption (49%). Figure 28.2.7 shows the chip micrograph.

References:

[1] F. Zhang, et al., "Design of Components and Circuits Underneath Integrated Inductors," *IEEE J. Solid-State Circuits*, Vol. 41, pp. 2265-2271, Oct. 2006.

[2] A. Liscidini, "A 36mW/9mW Power-Scalable DCO in 55nm CMOS for GSM/WCDMA Frequency Synthesizers," *ISSCC Dig. Tech. Papers*, pp. 348-349, Feb. 2012.

[3] K. J. Wang, et al., "Spurious Tone Suppression Techniques Applied to a Wide-Bandwidth 2.4GHz Fractional-N PLL," *IEEE J. Solid-State Circuits*, Vol. 43, pp. 2787-2797, Dec. 2008.

[4] S. Abdollahi-Alibeik, et al., "A 65nm Dual-Band 3-Stream 802.11n MIMO WLAN SoC," *ISSCC Dig. Tech. Papers*, pp. 170-171, Feb. 2011.

[5] C.-W. Yao et al., "A Low Spur Fractional-N Digital PLL for 802.11 a/b/g/n/ac with 0.19 ps$_{rms}$ Jitter," *IEEE Symp. on VLSI Circuits*, pp. 110-111, June 2011.

[6] J. Borremans, et al., "A 86MHz-to-12GHz Digital-Intensive Phase-Modulated Fractional-N PLL Using a 15pJ/Shot 5ps TDC in 40nm digital CMOS," *ISSCC Dig. Tech. Papers*, pp. 480-481, Feb. 2010.

[7] N. Pavlovic, J. Bergervoet, "A 5.3GHz Digital-to-Time-Converter-Based Fractional-N All-Digital PLL," *ISSCC Dig. Tech. Papers*, pp. 54-55, Feb. 2011.

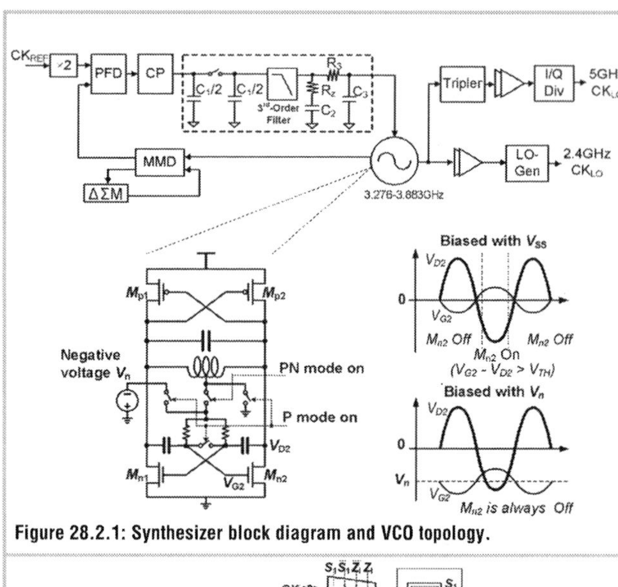

Figure 28.2.1: Synthesizer block diagram and VCO topology.

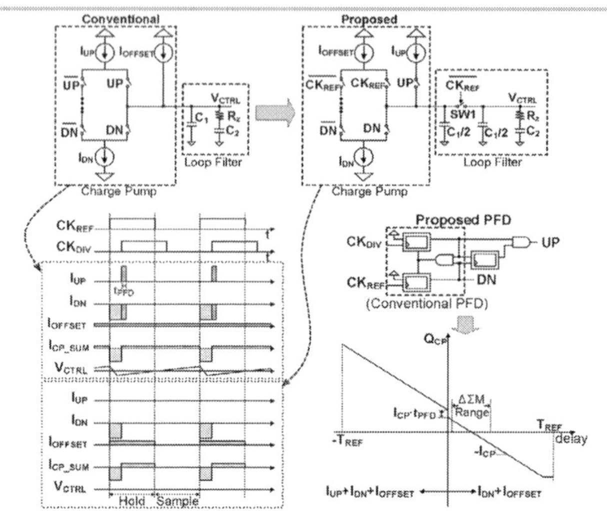

Figure 28.2.2: Proposed PFD/CP and operating principle.

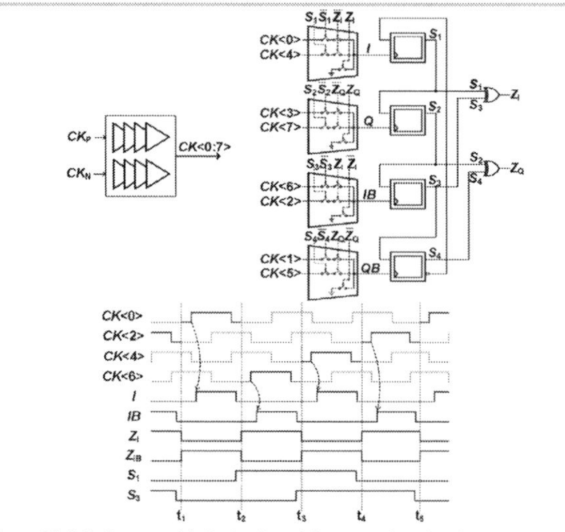

Figure 28.2.3: Proposed inductor-less LO generation circuit.

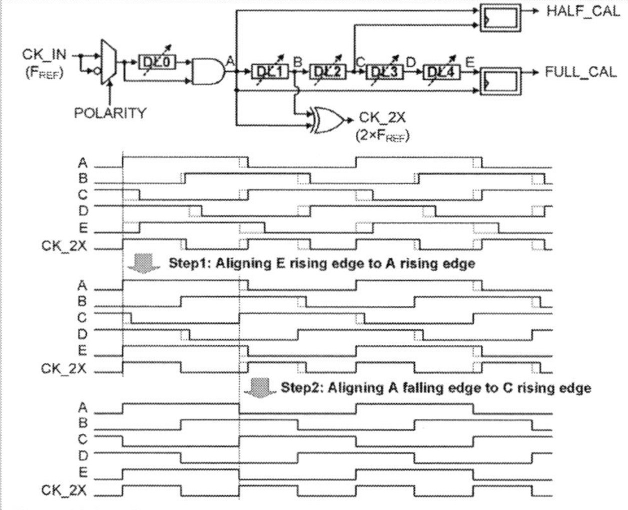

Figure 28.2.4: Reference doubler and duty-cycle calibration circuit.

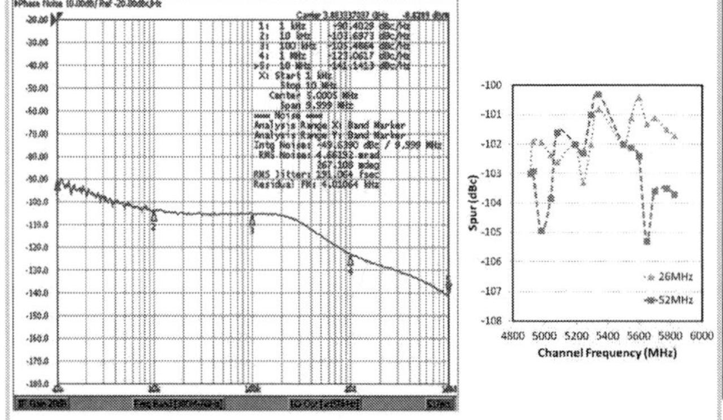

Figure 28.2.5: Measured phase noise profile and reference spur levels.

	Borremans. ISSCC'2010 pp.480-481 [6]	Pavlovic. ISSCC'2011 pp.54-55 [7]	Tasca. ISSCC'2011 pp.88-89	Yao. VLSI'2011 pp.110-111 [5]	Levantino. RFIC'2012 pp.177-180	Zhang. RFIC'2013 pp.119-122	Ahmadi. VLSI'2013 pp.C160-C161	This work
Process	40nm	65nm	65nm	55nm	65nm	130nm	40nm	**40nm LP w/o UTM**
Architecture	Digital	Digital	Digital	Digital	Analog	Analog	Analog	**Analog**
Reference Frequency	40MHz	48MHz	40MHz	40MHz	40MHz	32MHz	156.2MHz	**26MHz**
Reference Spur	-56dBc @ 7GHz	-67dBc @ 5.3GHz	-72dBc @ 4GHz	-94dBc @ 5.83GHz	-71dBc @ 4GHz	-75dBc @ 2.4GHz	N/A	**-100dBc @ 5.83GHz**
Normalized Reference Spur*	-50.1dBc	-55.5dBc	-61.2dBc	-86.5dBc	-60.2dBc	-63.7dBc	N/A	**-100dBc**
RMS Jitter	0.56ps	0.30ps	0.56ps	0.19ps	0.79ps	0.45ps	0.288ps	**0.19ps (ac/a) 0.30ps (bgn)**
Power	30mW	20mA from regulated supply	4.5mW	36mW	5mW	5.8mW	16.9mW	**17.5mW(ac/a) 7.2mW (bgn)**
FoM	-230.3dB	N/A	-238.5dB	-238.9dB	-235.1dB	-239.3dB	-238.5dB	**-242.0dB**
Area	0.28mm²	0.91mm²	0.22mm²	0.68mm²	0.22mm²	0.3mm²	0.39mm²	**0.29mm²**

* The reference spur levels are normalized to F_{LO}=5.83GHz and F_{REF}=26MHz for comparison, assuming 20dB/dec and 40dB/dec scalings respectively.

Figure 28.2.6: Performance summary table.

Figure 28.2.7: Chip micrograph.

28.3 A Frequency-Defined Vernier Digital-to-Time Converter for Impulse Radar Systems in 65nm CMOS

Yu-Hsien Kao, Chang-Ming Lai, Jen-Ming Wu, Po-Chiun Huang, Ping-Hsuan Hsieh, Ta-Shun Chu

National Tsing Hua University, Hsinchu, Taiwan

Impulse radar is a promising method for achieving high-range resolution and multi-path immunity for ranging and localization applications [1], [2]. Impulse radar sends signals with short duration and spreads signal power over a large bandwidth. Favorable features of impulse radar are minor interference with other wireless systems as well as high immunity to nearby radio signals. Thus, impulse radars are compatible with the harsh environments where intensive radio services are operating. In conjunction with the advantages of high integration and low cost in mass production from CMOS technology, more ubiquitous utilizations are feasible for CMOS impulse radars.

In impulse radar, the distance between the radar system and an object can be extracted from the flight time of pulses. A straightforward way to measure the timing difference between these pulses with fine resolution is by using time-to-digital converters (TDCs). However, the scattering signal has poor SNR, which fails to trigger the TDCs properly. Prior to time-of-arrival (TOA) estimation, the receiver can be gated with varying relative delays after transmission. TOA is acquired based on the active gate when the reflected signal is sensed. The distance of the object is derived from the speed of the electromagnetic wave times TOA/2. The timing generator controls the relative timing of one radar system, which governs the performance of minimum resolution, maximum scanning range, and scanning speed. Therefore, a timing generator with fine resolution over a wide full-scale range with fast transient response is important for impulse radar systems. In this work, a frequency-defined Vernier digital-to-time converter (DTC) for a high-performance radar system is proposed and implemented. The DTC supports a timing resolution of 1.5ps over the range of 100ns with instantaneous timing selection. When accompanying the implemented direct sampling radar system in [1], the ranging resolution is 0.24mm over a total distance of 15m.

In conventional radar systems, either the receiver or transmitter is operated under variable timing, while the other one uses a sequence of constant time intervals. The timing resolution is equal to the unit interval of the variable timing signal. In the proposed Vernier DTC, both transmitter and receiver are operated with variable timings (Fig. 28.3.1). The range of the two variable timings are equal to pulse repetition time (T_{PR}). The unit time interval of the transmitter (ΔT_{TX}) is designed to be $T_{PR}/(n-1)$, while the unit time interval of receiver (ΔT_{RX}) is T_{PR}/n. The effective timing resolution is shrunk aggressively and given by $\Delta T_E = T_{PR}/(n(n-1))$. The effective timing resolution is much smaller than that of the standalone transmitter or receiver. The operation of the proposed DTC is as follows. Since the difference in unit the time interval is accumulated along the timing train, the relative timings less than $T_{PR}/(n-1)$ can be generated from choosing the same order of the transmitter and receiver timing trains. Considering a relative timing of $T_{PR}/(n-1) \sim 2T_{PR}/n$, an additional relative timing of T_{PR}/n is required in this range. Therefore, the transmitter timing is selected with one time interval prior to receiver timing. Applying this concept, the relative timing difference between transmitter timing and receiver timing in the range of T_{PR} with a resolution of ΔT_E can be synthesized from the n and n-1 timing taps of the transceiver.

Conventional Vernier delay lines rely on the timing difference between two inverter delay lines. The inverter delay entails the high sensitivity towards PVT variation, which limits the effective timing resolution [3], [4]. The proposed frequency-defined Vernier DTC aligns the timing signal with the PLL transition edge through retiming which minimizes the delay mismatch from previous circuits. In the reported frequency-defined Vernier DTC, the 2 unit time intervals for the variable timings are defined by the frequencies of 2.55GHz and 2.56GHz. The designed effective timing resolution is 1.5ps.

Figure 28.3.2 demonstrates the block diagram of the proposed timing circuitry, which consists of three PLLs and two digital time shifters (DTSs). Wide PLL loop bandwidth is advantageous for synchronization of radar transmitter and receiver, since the relative timings between transmitter and receiver are highly correlated. The reference PLL is adopted to have higher input frequencies for transmitter PLL (TX PLL) and receiver PLL (RX PLL). The reference signal for TX PLL and RX PLL are synthesized from the reference PLL, which generates a frequency of 4.8GHz from an off-chip 80MHz source. The output of the reference PLL is divided by 16 and 15 to feed the TX PLL and RX PLL, respectively. The output frequency is 2.56GHz for the TX PLL. To prevent possible pulling or mutual injection of the two closely oscillating VCOs, the RX PLL generates 2.55GHz from a 5.1GHz oscillator and a divide-by-two circuit. Transmitter and receiver DTSs cover the range of 100ns with a time interval of 390.6ps and 392.2ps, such that a fine resolution (1.5ps) over the entire pulse repetition range (100ns) is accomplished. In the transmitter DTS, the fine range covers 6.25 ns with 15 intervals and the coarse range covers 100ns with 16 intervals. Multiplexers select the desired timing from a fine shift register and a coarse shift register. To combine the coarse and fine ranges, the output of the fine shift register is used to sample the output of the coarse shift register. The fine and coarse codes control the open-loop selection, which has instantaneous response time. This alleviates radar range scanning speed. The path delay mismatches from different codes are minimized by the retiming from a D flip-flop. The same structure is implemented in the receiver DTS with 17 codes in the fine range and 16 codes in the coarse range.

Figure 28.3.3 shows the implemented impulse radar system with the proposed timing circuitry. The frequency-defined Vernier DTC produces two signals with dedicated delay to trigger the transceiver. The programmable pulse generator (PPG) triggered by the TX DTS produces short pulses and the off-chip antenna driven by the driver transduces the pulses into an electromagnetic wave. The timing controller commanded by RX DTS triggers the sample-and-hold to sample the reflected scattering signals, which are received by the front-end LNA. A range of input powers is allowed thanks to the RFVGA. Integration of multiple reflected pulses rejects interference and uncorrelated noise, which improves SNR. The programmable gain amplifier (PGA) provides flexibility between the integration time and quantization error. Moreover, the chopper topology is adopted to cancel the DC offset and flicker noise, which comes from the analog block of the receiver chain. To meet the link power budget, an 8bit SAR ADC with low power emphasis is designed for the receiver chain.

The reported radar system with the proposed frequency-defined Vernier DTC is implemented in 65nm CMOS technology. Figure 28.3.4 shows the measured transfer curve and linearity (e.g., DNL and INL) of the DTC. The timing measurement is tested with 1 code=4LSB≈6ps which is limited by the use of a real-time scope. Figure 28.3.5 illustrates the measured results of the receiver frequency response and signal-to-noise and distortion ratio (SNDR), which is tested with 10ps timing resolution. Figure 28.3.6 summarizes the performance of the proposed timing circuitry and the implemented direct sampling radar. The chip microphotograph is shown in Fig. 28.3.7.

Acknowledgements:
This work was partially sponsored by MediaTek Inc. and National Science Council.

References:

[1] C.-M. Lai, et al., "A Scalable Direct-Sampling Broadband Radar Receiver Supporting Simultaneous Digital Multibeam Array in 65nm CMOS," *IEEE ISSCC Dig. Tech. Papers*, pp. 242-243, Feb. 2013.
[2] C.-M. Lai, et al., "A UWB IR Timed-Array Radar Using Time-Shifted Direct-Sampling Architecture," *IEEE Symp. VLSI Circuits*, pp. 54-55, June 2012.
[3] P. Dudek, S. Szczepanski, J. V. Hatfield, "A high-resolution CMOS time-to-digital converter utilizing a Vernier delay line," *IEEE J. Solid-State Circuits*, vol. 35, no. 2, pp. 240–247, Feb. 2000.
[4] T. Hashimoto, et al., "Time-to-Digital Converter with Vernier Delay Mismatch Compensation for High Resolution On-Die Clock Jitter Measurement" in *VLSI Circuits Symp. Dig. Tech. Papers*, pp. 166-167, Jan. 2008.

978-1-4799-0917-9/14 $31.00 © 2014 IEEE

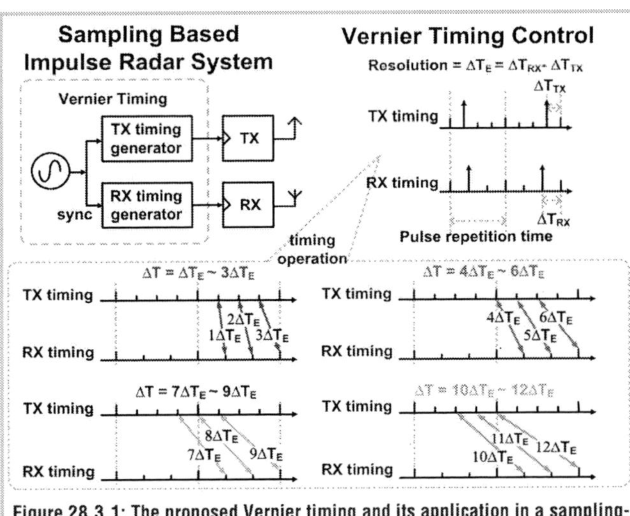

Figure 28.3.1: The proposed Vernier timing and its application in a sampling-based impulse radar system.

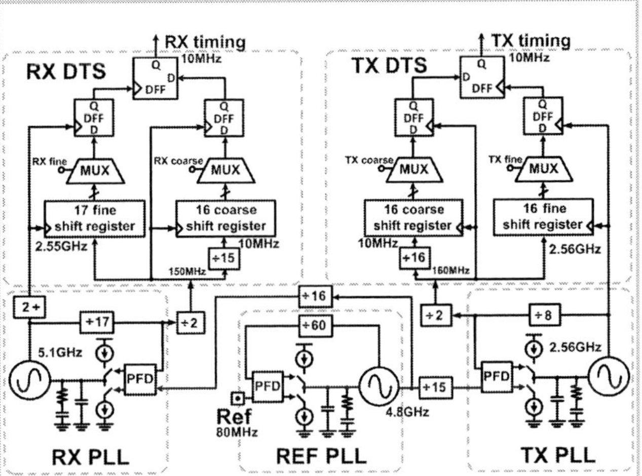

Figure 28.3.2: Block diagram of the proposed frequency-defined Vernier DTC.

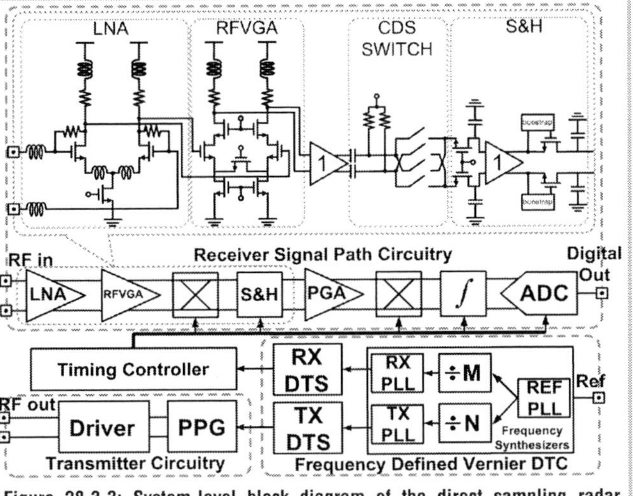

Figure 28.3.3: System-level block diagram of the direct sampling radar system.

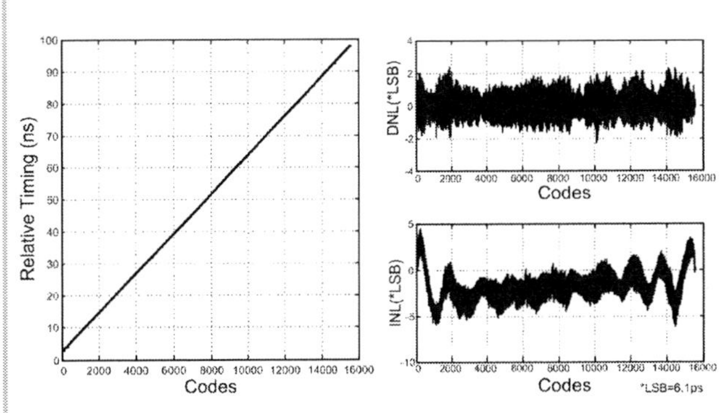

Figure 28.3.4: Measured transfer curve, DNL and INL of the digital-to-time converter.

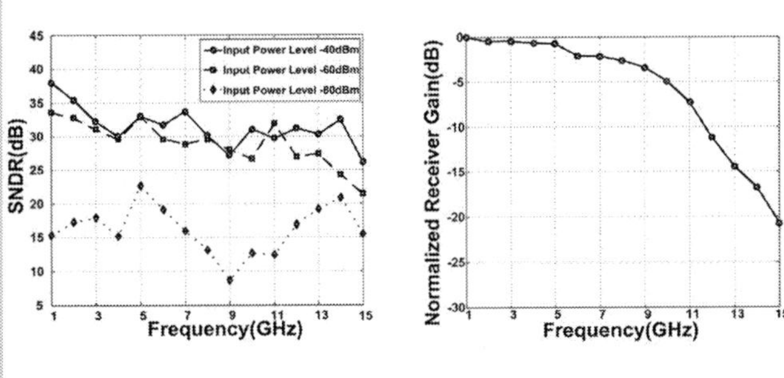

Figure 28.3.5: Performance results of the proposed direct-sampling receiver.

Implementation	
Technology	65nm CMOS
Die Area(Core)	2 mm²
Supply Voltage	1V
Digital to Time Converter	
Reference Clock Frequency	80MHz
Pulse Repetition Rate	10MHz
Timing resolution (design) (1psec = 0.15mm)	1.5ps (0.225mm)
Timing resolution (measurement) (1psec = 0.15mm)	6ps (0.9mm)
Maximum Scanning Range	100ns (15m)
DNL (LSB)	+2.2/-2.2
INL (LSB)	+4.2/-6
RMS jitter	6ps

Receiver And Transmitter Signal Path Circuitry	
Receiver gain bandwidth (-3dB)	8GHz
Receiver gain bandwidth (-10dB)	12GHz
MAX integration Time	6.5536ms
MAX Receiver SNDR	>40dB
MAX SNR improvement (Theory)	48dB
ADC resolution/ENOB	8bit/7.3bit
Current Consumption@1V	
RF front-end (LNA to S&H)	33.5mA
Analog blocks (include PGA, Integrator)	7.5mA
ADC	0.2mA
Frequency-defined Vernier DTC	44.5mA
Transmitter circuitry	0.5mA
Digital Controller and IO	2mA
Total	88.2mA

Figure 28.3.6: Performance summary.

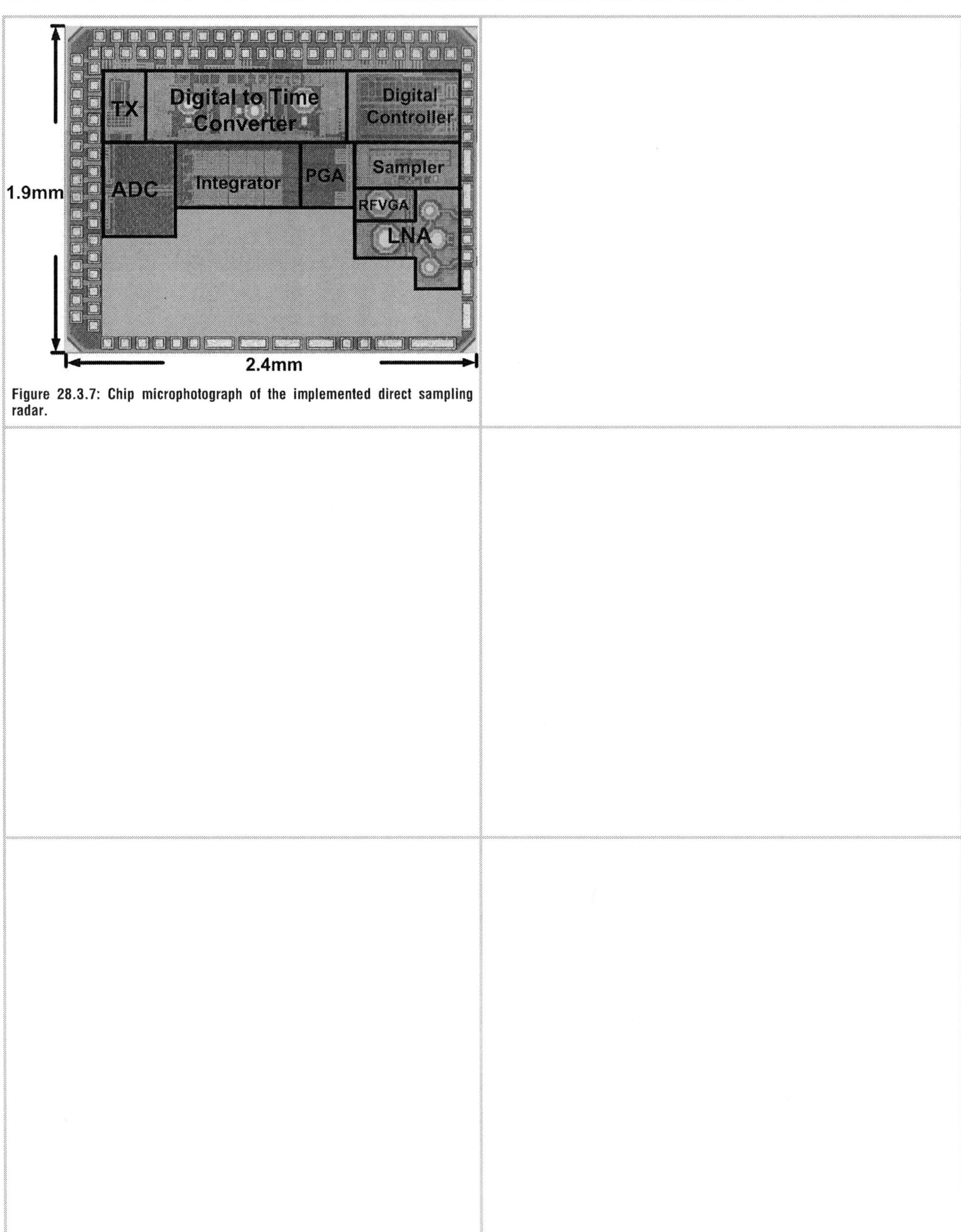

Figure 28.3.7: Chip microphotograph of the implemented direct sampling radar.

Session 29 Overview:
Data Converters for Wireless Systems
DATA CONVERTERS SUBCOMMITTEE

Session Chair: *Gerhard Mitteregger*
Intel Mobile Communications,
Munich, Germany

Session Co-Chair: *Brian Brandt*
Maxim Integrated, MA

Analog-to-digital converters for wireless systems continue to improve in noise and distortion performance, sample rate, and integration. This steady advance allows them to move closer to the antenna and digitize an increasing number of communication channels. Two of the ADCs in this session employ continuous-time delta-sigma modulators while one uses a pipelined architecture. The first paper addresses the challenges posed by large out-of-band blocker signals. The second paper tackles the high dynamic range and low noise floor requirements inherent to modern communication channels. The final paper employs new background calibration techniques, including one that compensates for nonlinear kickback resulting from the input sampler. The impressive performance advancements in these works confirm the effectiveness of the design techniques.

978-1-4799-0917-9/14 $31.00 © 2014 IEEE

29.1 A 5mW CT ΔΣ ADC with Embedded 2nd-Order Active Filter and VGA Achieving 82dB DR in 2MHz BW **3:15 PM**

R. Rajan, IIT Madras, Chennai, India

In paper 29.1, IIT Madras presents a 5mW CT ΔΣ ADC with an embedded 2nd-order active filter and VGA achieving 82dB DR over a 2MHz bandwidth. With the embedded Butterworth filter, the out-of-band IIP3 and IIP2 improve by 10dB and 15dB, respectively, compared to a system with the filter placed in front of the ADC. The combined system, VGA, Filter and CT ΔΣ ADC, achieves a DR of 92dB in 130nm CMOS.

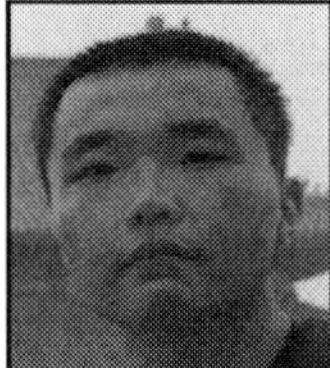

29.2 A 235mW CT 0-3 MASH ADC Achieving -167dBFS/Hz NSD with 53MHz BW **3:45 PM**

Y. Dong, Analog Devices, Toronto, ON, Canada

In paper 29.2, Analog Devices (with the University of Toronto) describes a ΔΣ 0-3 MASH ADC achieving -167dBFS/Hz noise spectral density with 53MHz bandwidth. The converter samples at 3.2GHz and achieves a DR of 88-90dB in a bandwidth of 53.3 to 45.7MHz. A novel 0-3 MASH ADC is used, which contains a 16-step flash ADC front-end, and a 3rd-order feed-forward continuous-time ΔΣ modulator back-end. The ADC is designed in 28nm CMOS and draws a total power of 235mW from 0.9/1.8/-1V supplies.

29.3 A 14b 1GS/s RF Sampling Pipelined ADC with Background Calibration **4:15 PM**

A. M. Ali, Analog Devices, Greensboro, NC

In paper 29.3, Analog Devices shows a 14b 1GS/s RF sampling pipelined ADC that relies on background calibration to correct for inter-stage gain, settling and memory errors. To improve the sampling linearity, it employs input distortion cancellation and digital calibration to compensate for the nonlinear charge injection from the sampling capacitors. The ADC is fabricated in 65nm CMOS and has an integrated input buffer. With a 140MHz and $2V_{pp}$ input signal, the converter maintains SNDR and SFDR of 69dB and 86dB, respectively while dissipating 1.2W.

29.1 A 5mW CT ΔΣ ADC with Embedded 2nd-Order Active Filter and VGA Achieving 82dB DR in 2MHz BW

Radha Rajan, Shanthi Pavan

IIT Madras, Chennai, India

Conventional continuous-time delta-sigma modulator (CTDSM) architectures do not allow independent control of the shape and bandwidth of the signal transfer function (STF), since the STF is simply a by-product of NTF synthesis. This is particularly troublesome when the input to the CTDSM consists of large out-of-band interferers; handling them without saturating the quantizer needs larger in-band DR, leading to increased power dissipation. A solution to this problem is to use a filter upfront to attenuate interferers. Alternatively [1], the filter can be moved into the CTDSM loop. In [1], a 1st-order RC filter was used to "tame" the STF peak of a cascade of integrators with feedforward summation (CIFF) DSM. Apart from the limited selectivity offered by a 1st-order filter, an active feedback path was necessary to stabilize the loop. The CTDSM in our work obtains an STF with a sharper transition band and a lower cutoff frequency (normalized to the desired signal bandwidth) compared to [1], with the aim of more effectively attenuating close-in interferers. This is realized by embedding a 2nd-order active filter into the CTDSM. We show that this has the same functionality as the filter upfront, but achieves better linearity (for the same noise and power dissipation) when compared to the filter-CTDSM cascade. Further, no extra active circuitry is necessary to stabilize the loop. Measurements of a CTDSM (signal BW=2MHz), with a built-in VGA (0 to 18dB) and a 2nd-order Butterworth filter (4MHz cutoff), show that the out-of-band IIP3 improves by about 10dB when compared to the CTDSM with the filter placed upfront. The filtering CTDSM+VGA, which uses a single-bit quantizer and a 4-tap FIR DAC, achieves a DR of 92dB in a 2MHz BW while consuming 5mW in a 0.13μm CMOS process. The peak instantaneous DR/SNR/SNDR are 82/80.5/74.5dB. With the VGA, the DR is 92dB.

A 1-bit quantizer necessitates a 4th-order NTF and OSR=64 (f_s=256MHz) to achieve the desired signal-to-quantization-noise ratio (SQNR). The CIFF-B architecture (Fig. 29.1.1(a)) is chosen as a prototype, as it combines the best features of the CIFF and cascade of integrators with feedback (CIFB) loops [2]. The STF rolls off as $1/s^3$ at high frequencies but peaks outside the signal band, which is problematic when there are interferers, due to the following. The maximum stable amplitude of the single-bit CTDSM at frequency f_{in} is $\approx 0.7V_{FS}/STF(f_{in})$, where V_{FS} is the full scale. A 6dB STF peak restricts the CTDSM input to $(1/2)0.7V_{FS}$, necessitating a 6dB smaller in-band noise, thereby increasing power dissipation.

A way of addressing the STF peak in the 4th-order CIFF-B CTDSM is illustrated with the help of the normalized modulator (1Hz sampling rate) of Fig. 29.1.1(a). The upfront 2nd-order low-pass filter, H_1, attenuates interferers and, as seen in Fig. 29.1.1(c), the effective STF of the H_1-CTDSM cascade is flat and has 5th-order roll off at high frequencies. Since the BW of H_1 is low, close-in interferers can be effectively attenuated, relaxing in-band SNR requirements. However, cascading H_1 with the CTDSM is not power efficient as *both* the filter and the CTDSM contribute to noise and distortion of the system. The technique proposed in this work moves the filter beyond the first integrator of the CTDSM as shown in Fig. 29.1.1(b). Comparing Fig. 29.1.1(a) and (b), we see that the transfer function from u to y is the same. Embedding H_1 into the CTDSM alters the NTF, necessitating a compensation path with gain $a_4(1-H_1)/s$ to restore it. Assuming the DC gain of H_1=1, this path has a zero at DC, which cancels the integrator pole. Fortunately, this can be realized without extra hardware by injecting v into the second integrator of the filter. This way, the gain from v to y is the same in Fig. 29.1.1(a) and (b), rendering the systems equivalent. However, the filter's noise and distortion are reduced by the gain of the first integrator when referred to the CTDSM input, thus the filter can be impedance-scaled, saving power. When compared to the H_1-CTDSM cascade, the linearity of the filtering-CTDSM of Fig. 29.1.1(b) is enhanced thanks to overall feedback. While this compensation technique is illustrated with a 2nd-order filter, the idea can be extended to LPFs of arbitrary order (which can be realized as a cascade of biquads).

Figure 29.1.2 shows the single-ended diagram of the CIFF-B CTDSM with the 2nd-order embedded filter drawn in blue. Active-RC integrators are used for better linearity and wide-swing. The high-speed path around the quantizer is through I_1. The RZ DAC_1 compensates for excess loop delay (~600ps). DAC_4 and DAC_3 are resistive 4-tap FIR DACs, implemented using semi-digital techniques. R_{31} implements the feedforward path across I_3. Optimized tap-weights for DAC_4 reduce jitter noise by 15dB when compared to a 1-bit CTDSM without an FIR DAC. Integrator I_4 is cascaded with a 2nd-order Butterworth filter H_1 (f_{3db}=4MHz). The FIR DAC_3 feeding into the 2nd integrator (I_{f2}) of H_1 compensates the loop for the effect of H_1, restoring the NTF. The delay introduced by F(z) in DAC_4 is compensated by FIR DAC_2 with filter $F_c(z)$. The OTAs are 2-stage designs with ac-coupled feedforward compensation [4]. Capacitors are implemented as digitally programmable banks to tune out time constant variations. Varying R_{in} implements the VGA; when R_{in} is reduced to increase gain, the bias current in the 2nd stage of I_4 is increased to maintain loop gain.

Intuition for power reduction resulting from embedding the filter into the CTDSM is given using Fig. 29.1.3. Part (a) shows the filter-CTDSM cascade. Assuming only resistor noise, the inband thermal noise density is ≈6R4kT. When the filter is embedded (Fig. 29.1.3(b)), the CTDSM input and DAC impedances are doubled, and the filter impedance is increased by ≈3×. In spite of this, the input noise density is smaller at ≈5R4kT. Thus, the power in OTAs $\hat{A}_1$, $\hat{A}_2$ and $\hat{A}_3$ can be reduced while achieving similar linearity, since 3rd-order distortion depends on $G_{OTA}R$ (G_{OTA} is the OTA transconductance). In this work, $\hat{A}_1$, $\hat{A}_2$ and $\hat{A}_3$ are scaled so that they consume 0.65, 0.33 and 0.75 times the currents of A_1, A_2 and A_3, respectively. This way, the $G_{OTA}R$ products for $\hat{A}_1$, $\hat{A}_2$ and $\hat{A}_3$ are about 2, 1 and 1.5 times the corresponding products for A_1, A_2 and A_3, respectively. Thus, not only is power dissipation lower when the filter is embedded into the CTDSM, linearity is also enhanced with respect to a filter-CTDSM cascade, while achieving similar noise.

Two CTDSMs, one with an embedded filter (called Δ-H_1-Σ) and another with a filter upfront (called H_1-ΔΣ) are fabricated in a 0.13μm CMOS technology. Figure 29.1.3(c) shows the measured STF and maximum stable amplitude (MSA), which remain almost the same for both designs. The MSA was measured with the VGA gain set to 8. The area and power dissipation of Δ-H_1-Σ are 0.33mm² and 5mW (including references and for the highest VGA gain setting), respectively. Thanks to impedance scaling, these are about 25% smaller compared to those of H_1-ΔΣ. Figure 29.1.4 shows the SN(D)R plots of Δ-H_1-Σ. The peak instantaneous DR is 82dB, which increases to 92dB with the VGA. Figure 29.1.5(a) shows the PSDs at the outputs of Δ-H_1-Σ and H_1-ΔΣ (gain set to 1) when the input consists of two -11dBFS tones at 4.75 and 10.5MHz, and a small desired signal at 2MHz. The filtering effect of the CTDSM is apparent. Further, the IM_3 product at 1MHz is 20dB smaller with the embedded filter, representing an IIP3 improvement of ≈10dB. Figure 29.1.5(b) shows the PSDs when the interferers are 30 and 61 MHz tones (gain set to 4). IM3 is better in the embedded case by about 13dB (IIP3 better by ≈6.5dB). Inband IIP3/IIP2 improve by about 3dB/10dB. Figure 29.1.6 compares the performance of this design with other filtering CTDSMs [1]-[3]. Higher power efficiency, linearity and DR are achieved with better close-in filtering. References [5] and [6] are stand-alone CTDSMs (which do not have a desirable STF) implemented in more advanced technologies. The design presented in this work achieves a comparable FoM, especially considering that it is a filter+CTDSM.

Acknowledgments:
This work was supported by the 2nd author's Swarnajayanthi Fellowship from the Department of Science and Technology, Government of India. The authors thank Europractice for chip fabrication, and N. Krishnapura for feedback on the manuscript.

References:
[1] K. Phillips, et. al, "A 2mW 89dB DR Continuous-time ΔΣ ADC with increased immunity to interferers " *Proc. of the ISSCC*, Feb. 2004.
[2] H. Munoz, et. al, "A 4.7mW 89.5dB DR CT Complex ΔΣ ADC with built-in LPF " *Proc. of the ISSCC*, Feb. 2005.
[3] M. Sosio, et. al, "A complete DVB-T/ATSC tuner analog base-band implemented with a single filtering ADC " *Proc. of the ESSCIRC*, Sept. 2011.
[4] P. Shettigar, S. Pavan, "A 15mW 3.6 GS/s CTΔΣ ADC with 36MHz bandwidth and 83dB DR in 90nm CMOS" *Proc. of the ISSCC*, Feb. 2012.
[5] J. Gealow, et. al, "A 2.8 mW ΣΔ ADC with 83 dB DR and 1.92 MHz BW using FIR outer feedback and TIA-Based integrator," *Proc. of the Symp. On VLSI Circuits*, June. 2011.
[6] S. Huang, et. al, "A 1.2 V 2MHz BW 0.084mm² CT ΔΣ ADC with -97.7dBc THD and 80dB DR using low-latency DEM," *Proc. of the ISSCC*, Feb. 2009.

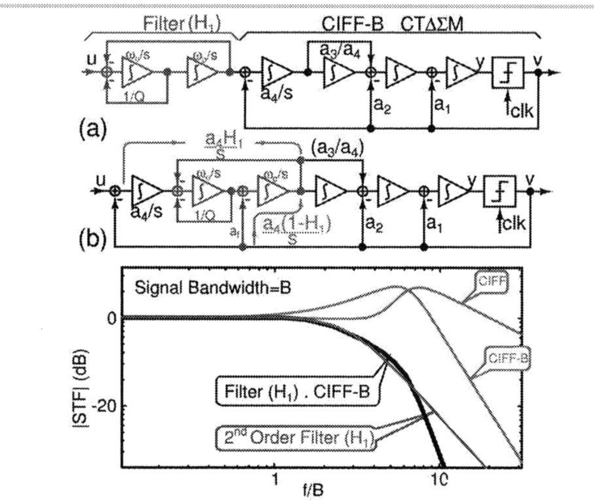

Figure 29.1.1: CIFF-B CTDSMs with (a) upfront filter (b) embedded filter (c) Resulting STF.

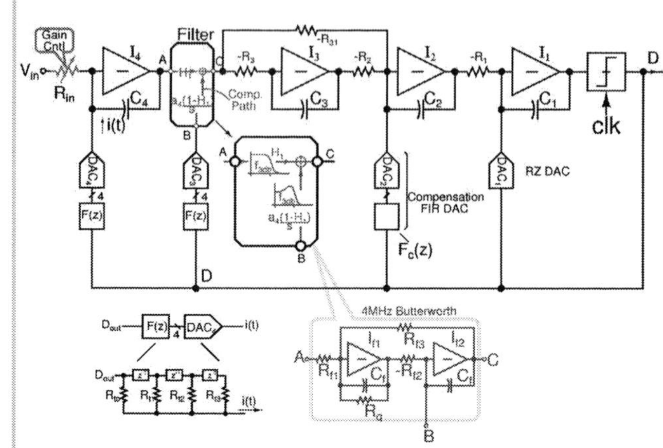

Figure 29.1.2: CTDSM with VGA and embedded second order Butterworth filter.

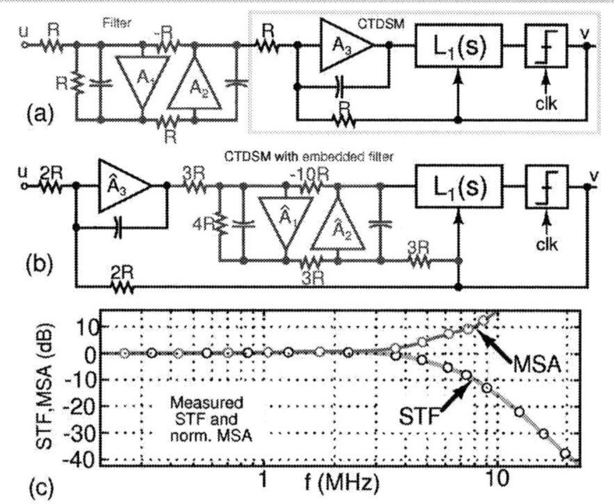

Figure 29.1.3: Impedance levels in a CIFF-B CTDSM with (a) upfront filter (b) embedded filter. (c) Measured STF and maximum stable amplitude (MSA) (normalized to the MSA @ DC).

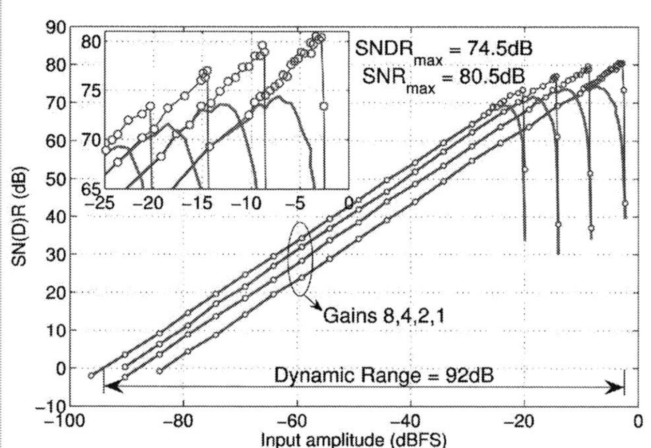

Figure 29.1.4: Measured SN(D)R for various gain settings, for the CTDSM with embedded filter.

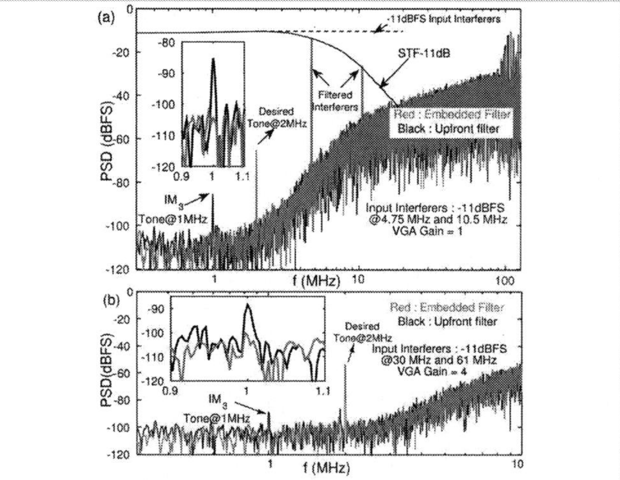

Figure 29.1.5: Response to out-of-band interferers at (a) 4.75 and 10.5 MHz (b) 30 and 61 MHz.

	This work Embedded Filter Δ-H_1-Σ	This work Upfront Filter H_1-$\Delta\Sigma$	[1]	[2]	[3]	[5]	[6]
BW (MHz)	2	2	1	1	6	1.92	2
fs (MHz)	256	256	64	64	405	245.76	128
DR with VGA (dB)	92	90.8	89	89.5	-	-	-
DR (dB)	82	81	65	71	75.6	83	80
SNRmax (dB)	80.5	79.5	59	68.5	74.6	78	79.1
SNDRmax (dB)	74.4	73.7	59	68.5	74.6	78	79.1
P_d (mW)	5	6.8	2	2.35	54	2.8	4.5
FoM_{SNDR} (fJ/lvl)	291	430	1373	540	1025	112	153
DR + $10\log_{10}(BW/P_d)$	168.0	165.7	152.0	157.3	156.1	171.4	166.5
Inband IIP3 / IIP2 (dBFS) 2 MHz input tones	25.5 / 83	22.3 / 73.5	-	-	-	-	-
Out-of-band IIP3 (dBFS) 4.75 & 10.5 MHz tones	35.7	26.2	5.7	-	-	-	-
Out-of-band IIP3 (dBFS) 30 & 61 MHz tones	40.8	33.7	-	-	-	-	-
(Filter 3dB BW)/Signal BW	2	2	3	-	1.43	-	-
Hi freq. roll off (dB/dec)	100	100	40	60	80	40	20
Active area (mm^2)	0.33	0.42	0.14	0.10	0.21	0.09	0.08
Technology (nm)	130	130	180	180	90	40	65
Supply (V)	1.4	1.4	1.8	1.8	1.8	2.4/1.4	1.2

Figure 29.1.6: Summary and Comparison: [1-3] are filtering ADCs, [5-6] are stand alone CTDSMs.

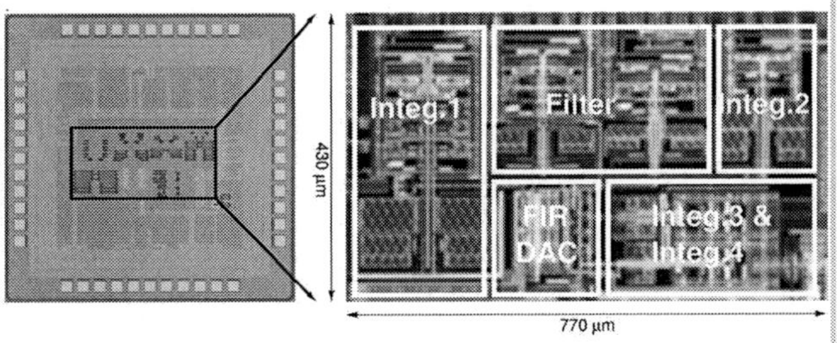

Figure 29.1.7: Die photo of the CTDSM with embedded filter (0.13μm CMOS). The active area is 27% smaller than the CTDSM with the filter placed upfront.

29.2 A 235mW CT 0-3 MASH ADC Achieving -167dBFS/Hz NSD with 53MHz BW

Yunzhi Dong[1], Richard Schreier[1], Wenhua Yang[2], Sudhir Korrapati[2], Ali Sheikholeslami[1,3]

[1]Analog Devices, Toronto, ON, Canada,
[2]Analog Devices, Wilmington, MA,
[3]University of Toronto, Toronto, ON, Canada

The trend for ADCs in wireless communication infrastructure is increased bandwidth with little or no relaxation in noise density or power consumption. The historical expectation of system designers is a noise spectral density (NSD) of -157dBFS/Hz with a power consumption of 0.5W. This expectation is a difficult one to meet with existing ADC architectures when the system bandwidth is 100MHz as demanded by standards such as LTE-A. The 0-3 continuous-time (CT) MASH [1-2] ADC described in this paper allows a direct-conversion receiver with the requisite bandwidth to be constructed, with 10dB lower noise than established benchmarks.

As shown in Fig. 29.2.1, the ADC consists of two stages: a residue-producing front end and a third-order feed-forward CT $\Delta\Sigma$ back end. The front end consists of a 16-step flash quantizer whose output V0 drives a 16-element current-mode DAC, IDAC0. The difference between the input current and the IDAC0 current is a residue signal that is digitized by the $\Delta\Sigma$ back end. The back end contains a 6-step flash quantizer whose output V1 drives the remaining 6-element current-mode DAC, IDAC1. A feed-forward structure is employed in the back end so that its sole feedback DAC is merged with IDAC0 into a single IDAC. As a consequence of this arrangement, the virtual ground established at the input of A1 by the $\Delta\Sigma$ loop ensures that in the high-gain region the output of IDAC is a faithful representation of the input signal. For a discrete-time implementation, V0 + V1 can be written as:

$$V0 + V1 = U + E0 + (-E0) * STF + E1 * NTF = U + (E0 + E1) * NTF$$

In the above equation, E0 and E1 are the quantization noise of FLASH0 and FLASH1, respectively, while STF and NTF are the signal and noise transfer functions of the $\Delta\Sigma$ back end. By virtue of the feed-forward topology of the loop filter, STF = 1 – NTF, which results in the final expression in the above equation. As this expression indicates, the signal transfer function of the system is unity and the NTF shapes E0 + E1. The digital signal V0 + V1 is therefore as faithful a representation of the input signal as NTF is small and IDAC is accurate. Furthermore, the system is free of out-of-band STF peaks even though the loop filter uses a feed-forward topology. Since the full scale (FS) of the back end is large enough to accommodate ~1LSB of front end over-load, the system can process input signals above the FS of the front-end flash quantizer. Also note that when the loop filter is implemented with CT circuitry, the system possesses inherent anti-aliasing despite the fact that sampling occurs in FLASH0 provided the additional aliasing component does not overload the $\Delta\Sigma$ back end.

Since the effective FS of the $\Delta\Sigma$ back end is a fraction (6/22) of the full ADC's FS, the band-edge gain of the first integrator is high (20dB) and thus the power efficiency of the ADC is maximized. Within the $\Delta\Sigma$ modulator, an R-C all-pass summer is utilized to provide wide-band summation in front of FLASH1. The LSBs of FLASH1 and the output swings of A1 to A3 are adjusted to allow such a passive summation.

A feed-forward amplifier is adopted in this design to achieve high bandwidth for given power and load requirements. Figure 29.2.2 shows the schematic of a single amplifier slice. A third-order topology is picked to achieve sufficient gain (~30dB) in the $\Delta\Sigma$ loop passband with large output swing (~$0.9V_{diff,p-p}$). The input stage is realized as a telescopic amplifier to achieve high voltage gain and low input-referred thermal noise. The signal swings at O1P and O1N are kept under tens of mV and their common-mode voltage is set to be the bias voltage for PMOS input pair M7-M8 of the second stage. M3 and M4 act as the direct feed-forward input pair for the second stage. Negative Miller compensation is applied to M3-M4 to reduce the capacitive loading on the input nodes. The outputs of the second stage, O2P and O2N, are used to drive the NMOS input pair M5-M6 of the third stage. A direct feed-forward input is applied to PMOS input pair M9-M10 through AC coupling. Compared to the second-order loop through M3-M4 and M5-M6, the first-order loop via M9-M10 has much less common-mode degeneration in order to improve common-mode stability. In addition, the source node of M1-M2, VS1, contains the common-mode voltage of the differential input and is used to drive M13 at high frequencies to further enhance common-mode stability. A1 contains 16 slices of the amplifier as shown in Fig. 29.2.2 while A2 and A3 contain 2 slices each. In post-layout simulations, each amplifier (A1-A3) achieves an in-band gain of 31 dB with a unity-gain frequency of 8.6 GHz.

A fully complementary current-steering DAC is used to implement IDAC. Figure 29.2.3 shows the schematic of a single IDAC element. Thick-oxide devices M1 and M8 together with cascode core devices M2 and M7 provide accurate P and N currents that are steered by the switching quad consisting of core devices M3 to M6. M1 and M8 are sized for static matching while M2 to M7 are sized for speed and desired output impedance across process corners. 1.8V and –1.0V supplies are utilized to provide large voltage headroom for both current sources to minimize thermal noise. Shown in the DAC timing diagram, the flash output data, T, is generated after the rising edge of the clock signal CK and gets latched at the end of the clock period by a pulse signal, CKD. This timing arrangement gives almost one clock period for the flash regeneration and logic delays in the feedback path. The one clock period delay in the flash-DAC path is compensated by the direct feedback to the input of FLASH1 provided through the VDAC element as shown in Fig. 29.2.1.

The ADC is fabricated in a digital 28nm bulk CMOS process with a 0.9V core supply voltage. The active area of the ADC is about 0.9mm by 1.0mm. The ADC is clocked at 3.2GHz and draws a combined DC power of 235mW from the 0.9/1.8/–1.0V power supplies. The effective signal bandwidth is 53.3MHz with an OSR of 30. The top and middle plots in Fig. 29.2.4 show the measured spectra of the combined output, V0 + V1, with a –2dBFS input tone at 30MHz. The bottom plot in Fig. 29.2.4 gives the measured spectra with two –8dBFS input tones at 30MHz and 32MHz. The IMD3 and IMD2 are –67.5dB and –66.3dB, respectively. Figure 29.2.5 shows the measured SNR as a function of the input signal amplitude. A peak SNR of 83.2dB is achieved at –2dBFS input. The measured passband noise floor with small input signal is at –167.2dBFS/Hz or –90dBFS integrated, resulting in a dynamic range (DR) of 88dB. The corresponding thermal-noise figure-of-merit (FOM) is thus 171.6dB. With a single-tone signal at 15MHz, the achieved peak SNDR is 71.4dB at –15dBFS input. The SNDR performance is believed to be limited by the timing skew and transition error among the 22 DAC elements in IDAC. Figure 29.2.6 compares this work with CT $\Delta\Sigma$ ADCs recently published at ISSCC. Figure 29.2.7 shows the microphotograph of the test chip. This work achieves a higher DR and thermal-noise FOM than others with a bandwidth of more than 50MHz.

Acknowledgements:
The authors would like to thank Chuanwei Li, Kevin Lam, Bill Harrington, Ziwei Zheng, Anthony Del Muro, and Abrar Ahmed Pathan for their hard work in layout, digital support and evaluation.

References:

[1] A. Gharbiya, D.A. Johns, "A 12-bit 3.125 MHz Bandwidth 0–3 MASH Delta-Sigma Modulator," *IEEE J. Solid-State Circuits*, vol. 44, no. 7, pp. 2010-2018, July 2009.
[2] N. Maghari, S. Kwon, U.K, Moon, "74 dB SNDR Multi-Loop Sturdy-MASH Delta-Sigma Modulator Using 35 dB Open-Loop Opamp Gain," *IEEE J. Solid-State Circuits*, vol. 44, no. 8, pp. 2212-2221, Aug. 2009.
[3] P. Shettigar, S. Pavan, "A 15mW 3.6GS/s CT-ΔΣ ADC with 36MHz bandwidth and 83dB DR in 90nm CMOS," *ISSCC Dig. Tech. Papers*, pp.156-158, Feb. 2012.
[4] H. Shibata, et al., "A DC-to-1GHz tunable RF ΔΣ ADC achieving DR = 74dB and BW = 150MHz at f0 = 450MHz using 550mW," *ISSCC Dig. Tech. Papers*, pp. 150-152, Feb. 2012.
[5] V. Srinivasan, et al., "A 20mW 61dB SNDR (60MHz BW) 1b 3rd-order continuous-time delta-sigma modulator clocked at 6GHz in 45nm CMOS," *ISSCC Dig. Tech. Papers*, pp. 158-160, Feb. 2012.
[6] Y-S. Shu, et al., "A 28fJ/conv-step CT ΔΣ modulator with 78dB DR and 18MHz BW in 28nm CMOS using a highly digital multibit quantizer," *ISSCC Dig. Tech. Papers*, pp. 268-269, Feb. 2013.

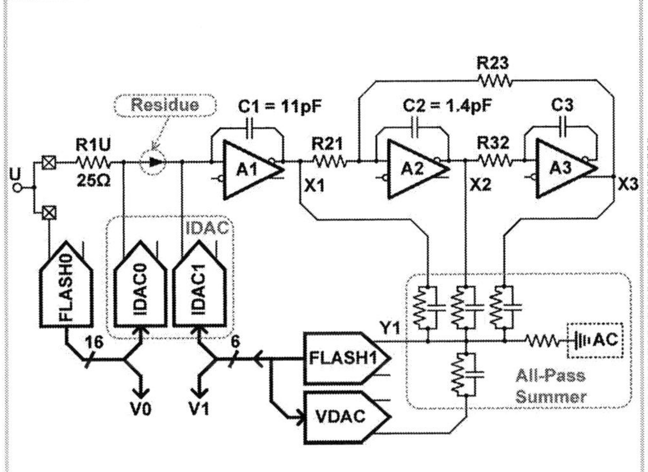

Figure 29.2.1: Single-ended representation of the 0-3 MASH CT ΔΣ ADC.

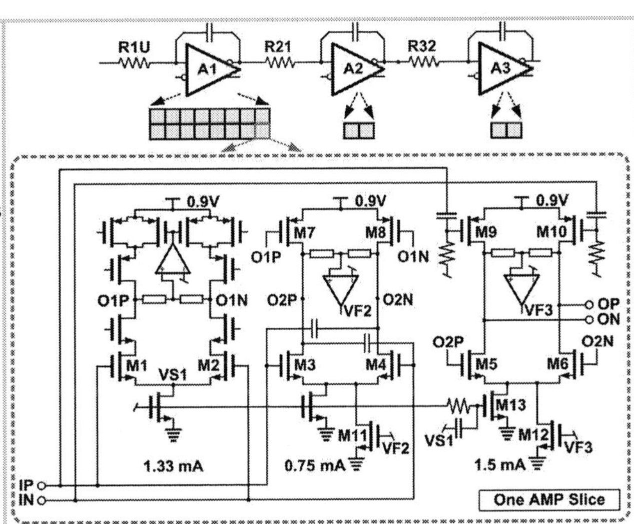

Figure 29.2.2: Amplifier slice allocation and slice schematic.

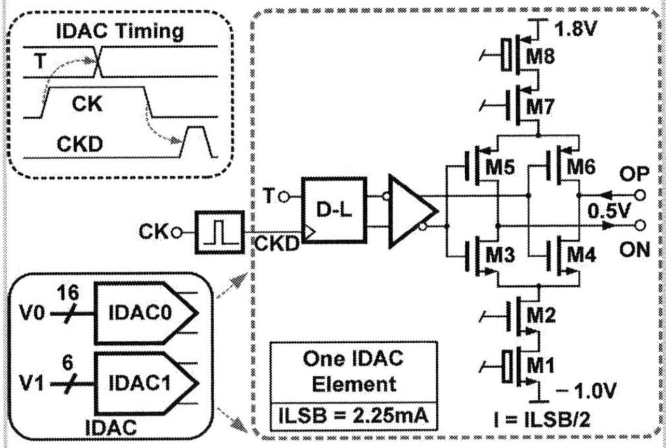

Figure 29.2.3: Schematic and timing diagram of the fully-complementary IDAC.

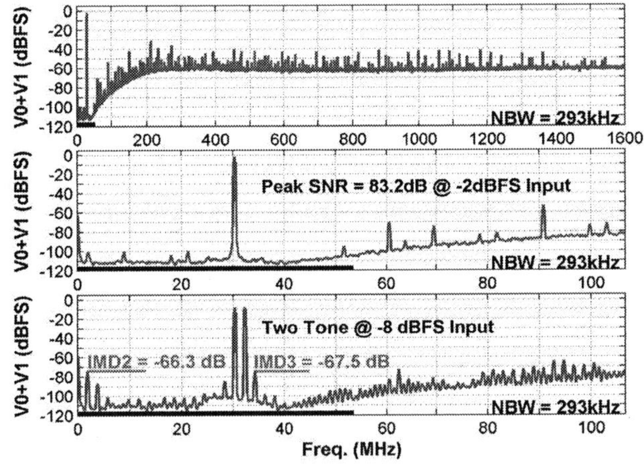

Figure 29.2.4: Measured spectra (top) -2dBFS tone at 30MHz, (middle) -2dBFS tone at 30MHz zoomed in, (bottom) -8dBFS tones at 30MHz and 32MHz.

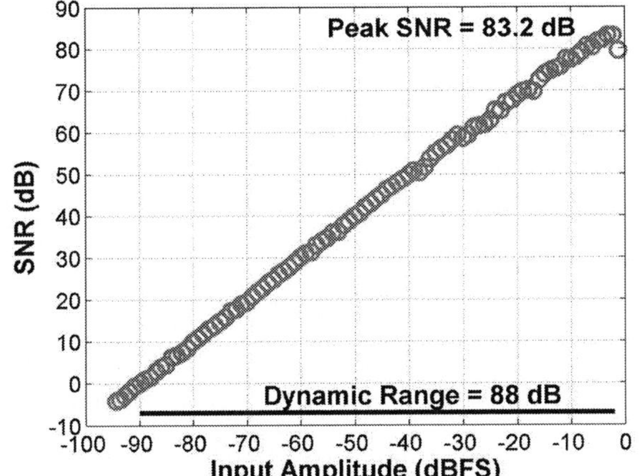

Figure 29.2.5: Measured SNR as a function of input signal amplitude at 30MHz.

Publication	This Work		[3] ISSCC 12	[4] ISSCC 12	[5] ISSCC 12	[6] ISSCC 13
CMOS L_{MIN}	**28nm**		90nm	65nm	45nm	28nm
Active Area	**0.9mm^2**		0.12mm^2	5.5mm^2	0.49mm^2	0.08mm^2
Supply Voltages	**0.9/1.8/–1.0V**		1.2V	1.2/±2.5V	1.4/1.8V	N.A.
Sampling Rate	**3.2GHz**		3.6GHz	4.0GHz	6.0GHz	0.64GHz
Power (*P*)	**235mW**		15mW	750mW	20mW	3.9mW
OSR	**35**	**30**	50	26.6	50	17.7
Bandwidth (*BW*)	**45.7MHz**	**53.3MHz**	36MHz	75MHz	60MHz	18MHz
*DR**	**90dB**	**88dB**	80dB*	79dB	62dB	78.1dB
SNRmax	**84.6dB**	**83.1dB**	76.4dB	N.A.	61.5dB	N.A.
SNDRmax	**72.6dB**	**71.4dB**	70.9dB	N.A.	60.6dB	73.6dB
$(DR-1.76)/6.02$	**14.7**	**14.3**	13	12.5	10	12.7
$DR+10\log_{10}(BW/P)$	**172.9dB**	**171.6dB**	173.8dB	158.2dB	156.8dB	174.7dB
$P/(2BW\cdot 2^{(DR-1.76)/6.02})$	**96.7fJ**	**107fJ**	25.5fJ	686fJ	162.8fJ	16.5fJ

* *DR* is measured as the input signal range corresponding to 0 dB SNR and peak SNR

Figure 29.2.6: Comparison of this work with state-of-the-art CT ΔΣ ADCs.

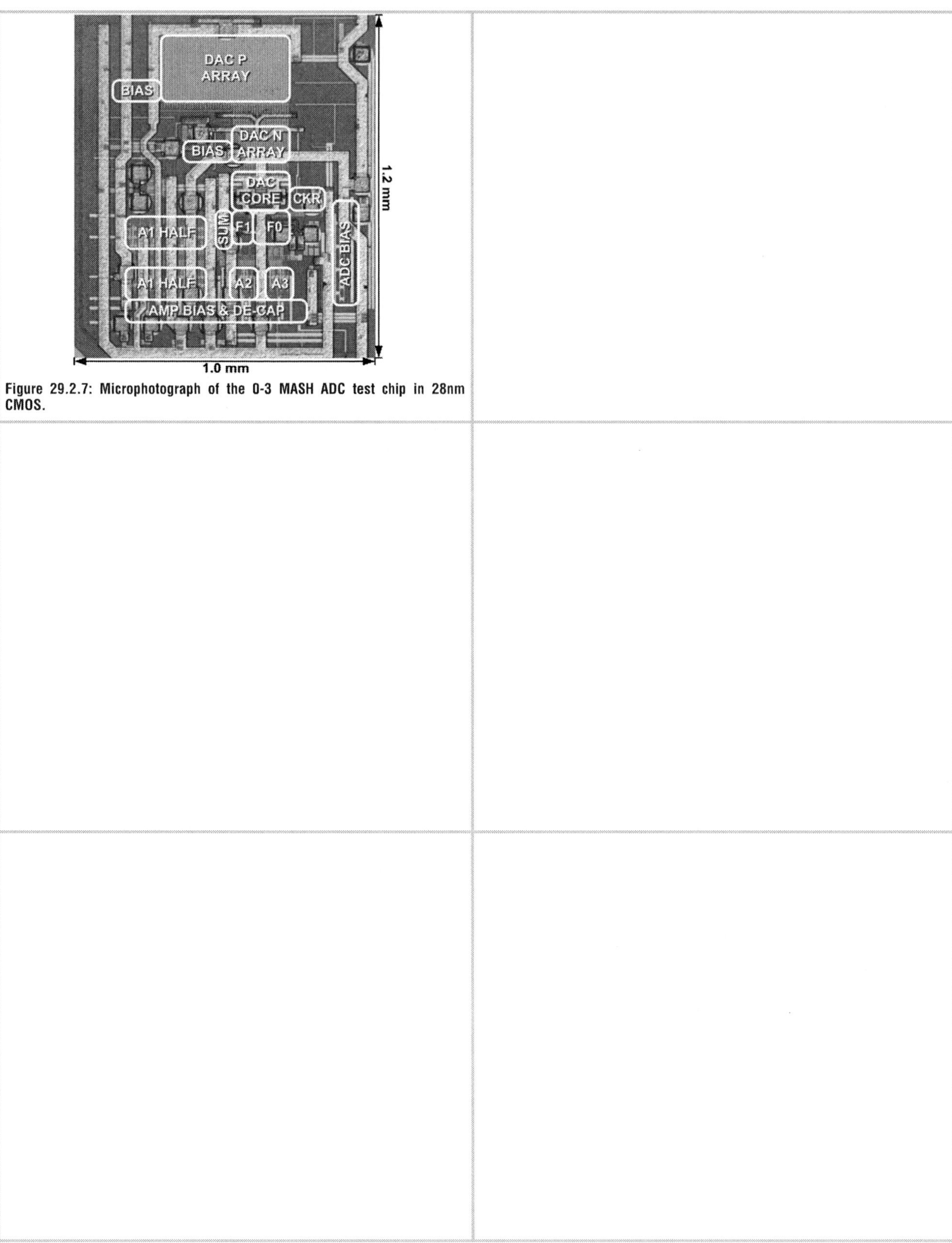

Figure 29.2.7: Microphotograph of the 0-3 MASH ADC test chip in 28nm CMOS.

29.3 A 14b 1GS/s RF Sampling Pipelined ADC with Background Calibration

Ahmed M. A. Ali[1], Huseyin Dinc[1], Paritosh Bhoraskar[1], Chris Dillon[1], Scott Puckett[1], Bryce Gray[1], Carroll Speir[1], Jonathan Lanford[1], David Jarman[1], Janet Brunsilius[2], Peter Derounian[1], Brad Jeffries[1], Ushma Mehta[1], Matt McShea[1], Ho-Young Lee[3]

[1]Analog Devices, Greensboro, NC, [2]Analog Devices, San Diego, CA, [3]Analog Devices, Wilmington, MA

We describe a 14-bit 1GS/s pipelined ADC that relies on correlation-based background calibration to correct the inter-stage gain, settling (dynamic) and memory errors. An effective dithering technique is embedded in the calibration signal to break the dependence of the calibration on the input signal amplitude. In addition, to improve the sampling linearity, the ADC employs input distortion cancellation and another digital calibration to compensate for the non-linear charge injection (kickback) from the sampling capacitors on the input driver. The ADC is fabricated in a 65nm CMOS process and has an integrated input buffer. With a 140MHz and 2Vpp input signal, the SNR is 69dB, the SFDR is 86dB, and the power is 1.2W.

A block diagram of the ADC is shown in Fig. 29.3.1, which shows a SHA-less pipelined architecture with five 3-bit stages and a 4-bit backend flash. To accommodate the 2Vpp input, the input buffer uses a 3.3V power supply, while the reference and MDAC amplifiers use a 2.5V supply. The flash comparators, clocks and digital circuitry use a 1.2V supply.

The first stage of the pipeline is shown in Fig. 29.3.2. At 1GS/s, the gain phase of the MDAC is about 450ps. To avoid non-linear settling errors in the amplifier, which are harder to calibrate, the comparators need to make their decisions in less than 100ps of propagation delay, leaving about 350ps for the MDAC amplifier.

To improve the comparator's speed in comparison with previous SHA-less implementations [1], a single set of capacitors is used to sample the reference value in the gain phase and connect to the input in the sample phase, thus eliminating the charge redistribution time. Additionally, the latch is composed of cross-coupled inverters with level-shifting capacitors that decouple the V_{gs} of the PMOS and NMOS devices, allowing them to be maximized independently beyond what is allowed by the low supply (1.2V), and hence improve the regeneration and propagation time. In addition, during the sample phase, these capacitors provide positive feedback on the latch nodes, thereby greatly improving the sampling bandwidth of the flash. In this work, acceptable bandwidth matching between the comparator and the MDAC sampling network is achieved up to 3GHz input bandwidth without exceeding the stage's correction range, which enables RF input sampling.

A high-bandwidth pre-amplifier is used to reset the latch during the sample phase and reduce the kickback from the latch to the input driver. In addition, the pre-amplifier is designed to achieve some distortion cancellation at the input by providing an opposite capacitive non-linearity to that generated by the parasitic capacitance of the MDAC's input switch. This reduces the non-linear current flowing in the input driver impedance and hence improves the input sampling distortion. The improvement in the third order harmonic level is about 6dB.

The MDAC residue amplifier (RA), shown in Fig. 29.3.3, is a two-stage Miller-compensated amplifier with active cascodes and positive feedback in the first stage, and a differential pair second stage. The back-gates of the PMOS devices of the second stage are AC floated to reduce the effective parasitic capacitance on the drain of the PMOS devices. In addition, an on-chip inductor is inserted in series with the PMOS load of the second stage. This inductor is optimized to "tune out" the capacitive load by forming an RLC circuit that has a pair of complex poles, which improves the phase margin and settling across process, supply and temperature, compared to a traditional RC circuit. Using this inductance lowered the power consumption of the amplifier by about 40%.

The RA's open loop gain is about 75dB and closed loop bandwidth is about 4GHz, which are insufficient to achieve the targeted performance. Therefore, background calibration is implemented for the first three pipeline stages to correct for inter-stage gain, settling (dynamic) and memory errors. The technique used in this work is a correlation-based calibration, where a pseudo-random calibration signal (dither) is injected into the MDAC, as shown in Fig. 29.3.4, and the LMS algorithm is used to estimate the correction coefficients using the recursive formula:

$$Ge_{n+1,k} = Ge_{n,k} - \mu \times Vd[n-k] \times (Vd[n-k] \times Ge_{n,k} - V_R[n])$$

Where Vd is the dither value, V_R is the residue of the stage being calibrated and μ is the algorithm step size. $Ge_{n,k}$ is the estimate of the gain coefficient of the kth past sample on the present n^{th} sample. If k=0, then $Ge_{n,0}$ is the coefficient for the inter-stage gain and settling error. If k>0, then $Ge_{n,k}$ is a memory coefficient.

The correction and equalization are done using FIR filters whose M taps are ($Ge_{n,k}$):

$$V_{R\_cal}[n] = \sum_{k=0}^{M} V_R[n-k] \times Ge_{n,k}$$

Another issue at high sample rates is the non-linear kickback from the sampling capacitors to the input driver. The MDAC input sampling network, shown in Fig. 29.3.2, uses the same capacitors for input sampling and DAC operation to improve the RA's feedback factor, and hence reduce the noise and power consumption. However, this creates non-linear charge injection on the input due to the DAC charge stored on the capacitors in the previous gain phase. To eliminate this kickback, a brief reset pulse (ϕ_rst) can be used to briefly discharge the sampling capacitors before the input sampling starts [1]. However, this pulse consumes a portion of the acquisition time and consumes considerable power. Alternatively, we employed a kickback calibration to correct for this in the digital domain. Since this non-linear kickback is proportional to the previous DAC values, the previous codes of the first flash are used to correct for the resulting distortion as follows:

$$V_{out\_kbcal}[n] = V_o[n] - \sum_{i=1}^{M_{kb}} D_1[n-i] \times G_{kb,i}$$

Where D_1 is the digital code of the first stage flash and V_o is the ADC output code. The M_{kb} kick-back coefficients ($G_{kb,i}$) can be obtained using the LMS algorithm by employing dither injection that "kicks" the input during the sample phase in parallel with the input capacitors as shown in Fig. 29.3.4.

A major problem with correlation-based calibration techniques has been the dependence of the estimated gain value on the input amplitude [2, 3]. This is due to the non-idealities of the back-end stages that create "jumps" in the transfer function, and hence cause the gain estimates to change with the input amplitude. Since dithering is known to improve the linearity proportionally to the number of dither levels, we use the calibration signals' dither to simultaneously improve the back-end accuracy. In this work, the calibration signals are composed of uniformly distributed multi-level dithers that are injected in the front-end stages and properly propagated to the back-end pipeline. However, unlike previous work, the number of dither levels used here is an odd number (N) that is greater than FS/DA, where FS is the input full-scale amplitude and DA is the dither amplitude. Also, the peak-to-peak amplitude of the dither injected in a certain stage is set to be equal to (N-1)/N of the sub-range size of the following stage. As shown in Fig. 29.3.5, this approach ensures dither propagation with the same number of uniformly distributed levels down the pipeline stages without losing any levels due to folding. In addition, the dither levels injected in the different stages are designed to complement each other by filling in the gaps between the levels of the other stages. We use 9 dither levels in stage one, 3 levels in stage two and 3 levels in stage three. Therefore, stages 4, 5 and 6 are dithered by 81 uniformly distributed dither levels, which translate into about 6.3 bits of additional accuracy. This practically eliminates the dependence of the convergence on the input amplitude. Dynamic element matching of the MDAC and dither capacitors is also employed.

The ADC is implemented, along with the digital processing needed for the calibration, on chip. The MDAC and dither capacitors are factory calibrated and the background calibration is verified to preserve the performance with temperature and supply variations. The measured results are summarized in Fig. 29.3.6.

References:

[1] A.M.A. Ali, et al., "A 16-bit 250-MS/s IF Sampling Pipelined ADC With Background Calibration" *J. Solid-State Circuits*, 45, no. 12, pp. 2602-2612, Dec. 2010.

[2] A. Panigada, I. Galton, "A 130mW 100MS/s pipelined ADC with 69dB SNDR enabled by digital harmonic distortion correction", *ISSCC Dig. Tech. Papers*, pp. 162-163, Feb. 2009.

[3] N. Rakuljic, I. Galton, "Suppression of Quantization-Induced Convergence Error in Pipelined ADCs with Harmonic Distortion Correction", *IEEE Trans. Circuits and Systems I*, 60(3), pp. 593-602, March 2013.

978-1-4799-0917-9/14 $31.00 © 2014 IEEE

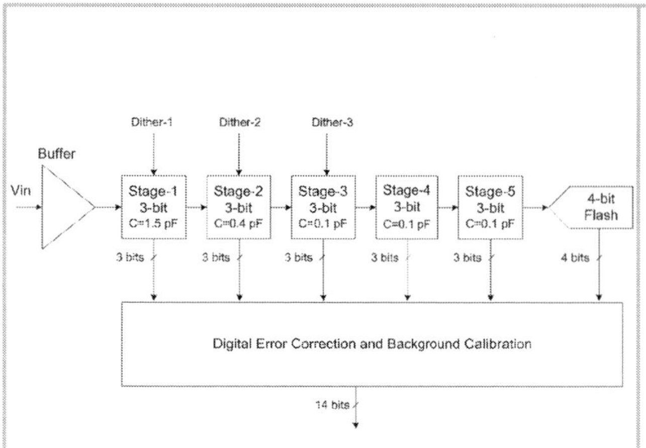

Figure 29.3.1: Block diagram of the pipelined ADC.

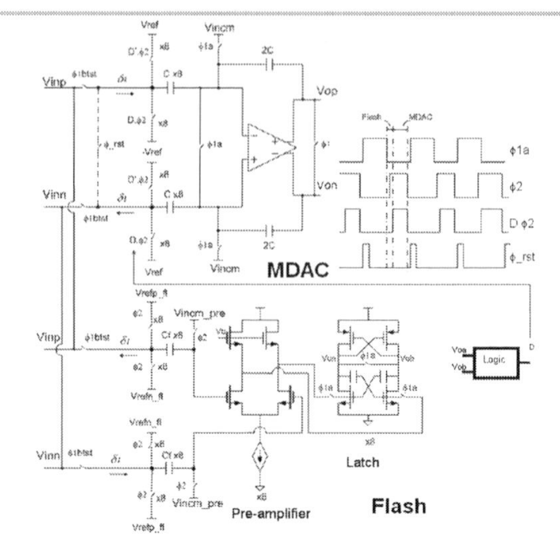

Figure 29.3.2: A schematic of the MDAC and flash stage.

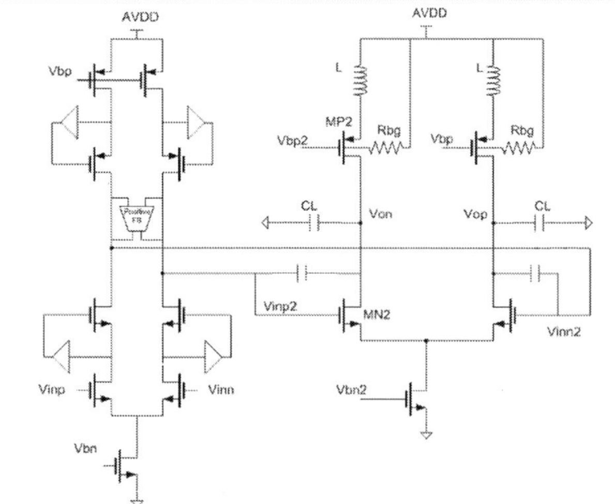

Figure 29.3.3: A schematic of the MDAC amplifier.

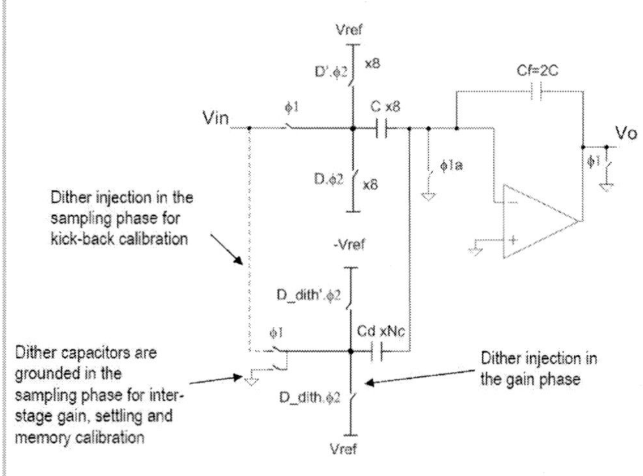

Figure 29.3.4: A schematic of the dither injection.

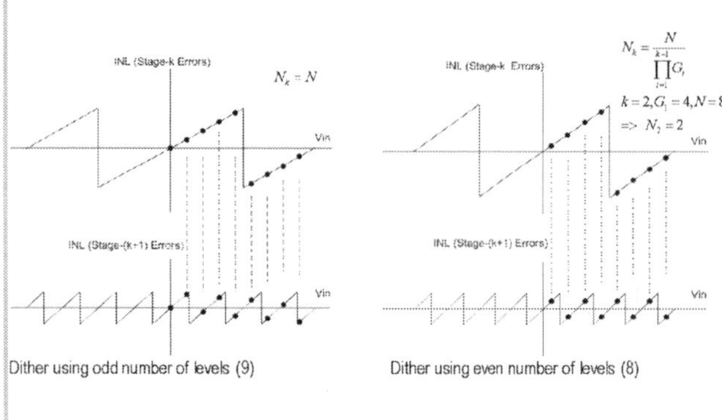

Figure 29.3.5: Dither propagation.

	Without Calibration	With Calibration
SNDR (Fin=140MHz)	<62 dB	69 dB
SFDR (Fin=140MHz)	<70 dB	86 dB
INL	+/-15 LSB	+/-1 LSB
DNL	-1 LSB	+/-0.3 LSB
Noise Spectral Density (NSD)	-149 dBFS/Hz	-156 dBFS/Hz
Power	1.2 W	
Sample Rate	1 GS/s	
FOM=SNDR+10log(BW/Power)	155.2 dB	
Input Span	1.2Vpp to 2Vpp	
Jitter	50 fs	
Dimensions	6mm x 3mm	
Process	65nm CMOS	
Package for 2 ADCs	64-pin LFCSP (QFN)	

Figure 29.3.6: Summary of measured results.

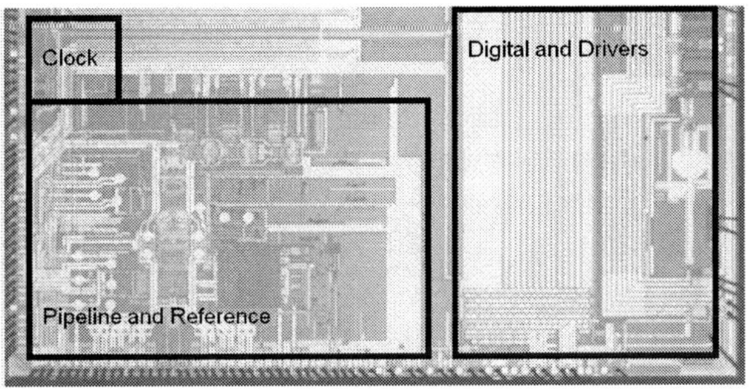

Clock
Digital and Drivers
Pipeline and Reference

Figure 29.3.7: Die micrograph.

Session 30 Overview:
Technologies for Next-Generation Systems
TECHNOLOGY DIRECTIONS SUBCOMMITTEE

Session Chair: *Fu Lung Hsueh*
TSMC, Hsinchu, Taiwan

Session Co-Chair: *Jan Genoe*
imec, Leuven, Belgium

New materials and technologies are enabling next generation systems and applications. The first four papers in this session combine different semiconductor materials (organic and/or oxide) on flexible foils to realize a variety of applications such as a microprocessor, a display, a sensor and an RFID tag. Papers 5, 8 and 9 combine in a hybrid manner crystalline high-bandgap semiconductors on top of the back-end-of-line (BEOL) of silicon. In papers 6 and 7 dedicated communication systems are implemented in automobiles and over the human body. Finally, paper 10 implements a learning machine mimicking the activity in the human brain.

30.1 8-bit Thin-Film Microprocessor Using a Hybrid Oxide-Organic Complementary Technology with Inkjet-Printed P²ROM Memory **1:30 PM**
K. Myny, imec, Leuven, Belgium
In Paper 30.1, imec, KULeuven, TNO, Panasonic and Evonik show a flexible 8b microprocessor on foil comprising solution-processed n-type oxide transistors and organic p-type transistors combined in a CMOS technology. The microprocessor instruction sequence is implemented by inkjet printing silver paste drops on the foil. The processor operates from 6.5V supply and has a maximum clock speed of 2.1kHz.

30.2 Digital PWM-Driven AMOLED Display on Flex Reducing Static Power Consumption **2:00 PM**
J. Genoe, imec, Leuven, Belgium and KU Leuven, Leuven, Belgium
In Paper 30.2, imec, TNO and Panasonic show new AMOLED scan drivers integrated with an a-IGZO backplane on flexible foil. The system enables PWM driving of the OLED pixels with a duty cycle of almost 100%. This PWM driving method enables up to 40% static power reduction of AMOLED displays.

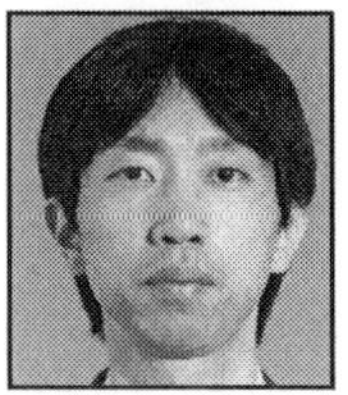

30.3 Organic-Transistor-Based 2kV ESD-Tolerant Flexible Wet Sensor Sheet for Biomedical Applications with Wireless Power and Data Transmission Using 13.56MHz Magnetic Resonance **2:30 PM**
H. Fuketa, University of Tokyo, Tokyo, Japan and JST/ERATO, Tokyo, Japan
In Paper 30.3, University of Tokyo discloses a wireless and flexible wet sensor sheet with organic transistors for urination detection in diapers. An ESD protection circuit with organic Schottky diodes using CuPc achieves 2kV ESD tolerance, which is indispensable for biomedical applications.

978-1-4799-0917-9/14 $31.00 © 2014 IEEE

30.4 A 13.56MHz RFID Tag with Active Envelope Detection in an Organic Complementary TFT Technology — 2:45 PM

V. Fiore, University of Catania, Catania, Italy

In Paper 30.4, the University of Catania, STMicroelectronics, University of Eindhoven and CEA-Liten show a 13.56 MHz RFID tag on foil using a complementary organic CMOS process. The tag can recognize ASK PWM signals up to 75bps with 25% modulation depth.

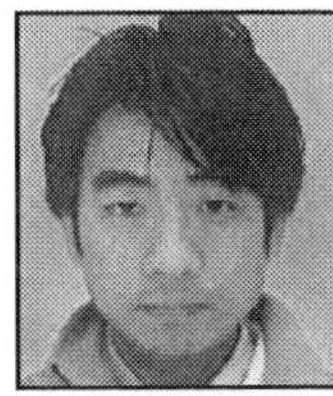

30.5 A GaN 3×3 Matrix Converter Chipset with Drive-by-Microwave Technologies — 3:15 PM

S. Nagai, Panasonic, Osaka, Japan

In Paper 30.5, Panasonic presents a three-phase AC-AC matrix converter realized by GaN/Si integrated chips. The extremely compact solution (19mm by 14 mm) can switch up to 10A under 600V

30.6 An Electromagnetic Clip Connector for In-Vehicle LAN to Reduce Wire Harness Weight by 30% — 3:45 PM

A. Kosuge, Keio University, Yokohama, Japan

In Paper 30.6, Keio University shows an electromagnetic clip connector that enables an in-vehicle LAN to reduce wire harness weight by 30%. Signaling schemes are implemented to improve noise immunity and satisfy EMC standards at 1.4GHz.

30.7 A 60Mb/s Wideband BCC Transceiver with 150pJ/b RX and 31pJ/b TX for Emerging Wearable Applications — 4:00 PM

J. Lee, Institute of Microelectronics, Singapore, Singapore

In Paper 30.7, the Institute of Microelectronics Singapore discloses a wideband Body Channel Communication (BCC) transceiver achieving 60 Mb/s data rate. The transceiver operates with high energy efficiency (31pJ/b Tx and 150 pJ/b Rx).

30.8 A 30GS/s Double-Switching Track-and-Hold Amplifier with 19dBm IIP3 in an InP BiCMOS Technology — 4:15 PM

T. D. Gathman, Qualcomm, San Diego, CA and University of California, San Diego, La Jolla, CA

In Paper 30.8, University of California, Qualcomm and HRL Laboratories demonstrate a compensated track-and-hold (THA) in InP BiCMOS technology providing low non-linear distortion at 30 GS/s sampling rate.

30.9 Normally-Off Computing with Crystalline InGaZnO-based FPGA — 4:30 PM

T. Aoki, Semiconductor Energy Laboratory, Kanagawa, Japan

In Paper 30.9, the Semiconductor Energy Laboratory elaborates a normally-off computing FPGA by implementing a 1um c-axis aligned crystal (CAAC) IGZO FET technology on top of a 0.5um CMOS wafer. Load and store times in the nonvolatile registers in the BEOL are 8ns and 40ns, respectively.

30.10 A 1TOPS/W Analog Deep Machine-Learning Engine with Floating-Gate Storage in 0.13μm CMOS — 4:45 PM

J. Lu, University of Tennessee, Knoxville, TN

In Paper 30.10, the University of Tennessee presents an analog feature extraction engine with a deep machine learning algorithm. It is implemented in a 0.13um CMOS technology and achieves 1 TOPS/s peak efficiency at 3V supply.

30.1 8b Thin-Film Microprocessor Using a Hybrid Oxide-Organic Complementary Technology with Inkjet-Printed P²ROM Memory

Kris Myny[1], Steve Smout[1], Maarten Rockelé[1,2], Ajay Bhoolokam[1,2], Tung Huei Ke[1], Soeren Steudel[1], Koji Obata[3], Marko Marinkovic[4], Duy-Vu Pham[4], Arne Hoppe[4], Aashini Gulati[5], Francisco Gonzalez Rodriguez[5], Brian Cobb[5], Gerwin H. Gelinck[5], Jan Genoe[1,2], Wim Dehaene[1,2], Paul Heremans[1,2]

[1]imec, Leuven, Belgium, [2]KU Leuven, Leuven, Belgium,
[3]Panasonic, Osaka, Japan, [4]Evonik Industries, Marl, Germany,
[5]Holst Centre/TNO, Eindhoven, The Netherlands

We present an 8b general-purpose microprocessor realized in a hybrid oxide-organic complementary thin-film technology. The n-type transistors are based on a solution-processed n-type metal-oxide semiconductor, and the p-type transistors use an organic semiconductor. As compared to previous work utilizing unipolar logic gates [1], the higher mobility n-type semiconductor and the use of complementary logic allow for a >50x speed improvement. It also adds robustness to the design, which allowed for a more complex and complete standard cell library. The microprocessor consists of two parts, a processor core chip and an instruction generator. The instructions are stored in a Write-Once-Read-Many (WORM) memory formatted by a post-fabrication inkjet printing step, called Print-Programmable Read-Only Memory (P²ROM). The entire processing was performed at temperatures compatible with plastic foil substrates, i.e., at or below 250°C [2].

Typical output characteristics of the 250°C hybrid organic/oxide complementary transistors are shown in Fig. 30.1.1. The use of this technology for complex designs has been proven already for a bi-directional RFID tag [3] and has been proven on flexible substrates [2]. The p:n transistor ratio for logic gates has been chosen to be 3:1, whereby the minimal device size for an oxide n-TFT equals 50/5 µm/µm and for an organic p-TFT 150/5 µm/µm. Typical inverter characteristics are also shown in Fig. 30.1.1. The circuit realizations in this work are based on bottom-gate top S/D contact oxide n-TFTs and bottom S/D contact organic p-TFTs, fabricated on a Si/SiO_2 substrate. More process details can be found in [4].

The architecture of the processor core chip is similar to our previous work [1,5], but has now been implemented in the complementary TFT technology rather than utilizing unipolar p-type dual-gate zero-V_{GS}-load logic [1,5]. The processor-core chip can perform logic (AND, OR, NOT), arithmetic (ADD, SUB, INC, DEC) and bit shift (LSR, LSL) functions, or execute a NOOP command. It can also store signals into the accumulator, into one of the 3 C-registers and into the output register. Due to the use of a more robust complementary technology in this work, the standard cell library has been expanded with more complex cells. Besides a basic inverter and 2-input NAND, our standard cell library consists of inverting buffers (x3, x4 and x9), a 2-input NAND x2 buffer cell, and a mirror adder [6]. For the P²ROM instruction generator chip, we also included a 2-input NOR cell. The mirror adder helps to minimize the critical path in the 8-bit ripple carry adder. The buffering of the signals has also been optimized with the introduction of extra buffer cells. The processor-core chip can be controlled by 6 opcode bits and 2 register select bits for all C-registers allowing execution all different possible operations. The fourth C-register is in fact a hard-wired digital 1, to ease the implementation of the INC and DEC functions.

Figure 30.1.2 shows the results of a general testbench that evaluates all different functions. The correct output signals are observed, directly compared to the generated output signals by our measurement setup that mainly consists of a PIC-microcontroller test board. Figure 30.1.2 also plots the obtained operational frequency versus the supply voltage. It starts operating correctly at a minimal supply voltage of 6.5V. The maximum obtained clock frequency in this measurement range is 2.1kHz, which is more than 52x better than previous state-of-art microprocessor fabricated directly on flexible foil [5]. This improvement stems from the switch from unipolar, dual-gate, zero-V_{GS}-load logic towards a hybrid oxide/organic complementary technology with improved charge carrier mobilities. Other key factors for this improvement are the optimized buffering in the core of the microprocessor and the implementation of a mirror adder [6] in the critical path.

The full microprocessor is consists of two separate chips, one being the processor core chip, as described in previous section, and the other a general-purpose instruction generator or P²ROM. The P²ROM chip is a one-time programmable ROM memory that is configured by means of inkjet printing using a conductive silver ink. This is a key improvement over our previously published microprocessor, which utilized a hardcoded instruction generator [1,5]. The block diagram of the general-purpose instruction generator is depicted in Fig. 30.1.3. It consists of a 4b program counter (PC), a 4-16 decoder to select each instruction line at once, a printable WORM memory and a 9b register that is updated each clock cycle with the next opcode to drive the microprocessor. Each printed connection will result in a logical 1, while unprinted connections result in a logical 0. The printable WORM memory is designed as a unipolar n-TFT NOR, with a 1:10 ratio between drive and load transistor, as depicted in Fig. 30.1.4. The drive transistor has a size of 140/5 µm/µm, while the load transistor equals 1400/5 µm/µm. In order to guarantee good NOR characteristics for the case that multiple select transistors are connected and required, up to 5 more load transistors can be added also by inkjet printing. This is also illustrated in Fig. 30.1.3.

Figure 30.1.4 depicts the layout of this P²ROM instruction generator chip, divided into a hybrid complementary part and a unipolar n-TFT part. In order to evaluate the P²ROM chip, we have chosen to print the instructions to execute a running averager algorithm (out_{new} = 0.5 round (in + out_{old})). The first twelve lines have been printed for the running averager algorithm, the other 4 lines in the instruction generator are not printed and therefore result in the NOOP command. This is also shown in Fig. 30.1.4. The instructions execute the algorithm twice before storing the value into the output register. Because we execute the LSR instruction only after the storage into the output register, the output code is a 7b code, which is one bit more accurate than the 6b input. Figure 30.1.5 depicts the correct behaviour of the P²ROM chip at a supply voltage of 10V and a maximum clock frequency of 650Hz. It generates the register select bits and the operational codes to drive the processor core chip in order to execute the running averager algorithm. The order of instructions are also detailed in Fig. 30.1.5.

Finally, we have connected both the processor core and P²ROM chips. Figure 30.1.6 shows the measured results when both chips are connected at a clock frequency of 500Hz. When the input switches from 0 to 7 (hexadecimal), the output averages between 7, C and E and remains constant at E (hexadecimal).

Figure 30.1.7 depicts micrographs of the 8b processor core and P²ROM chips and a comparison table to previous state-of-the-art. The 8b processor core chip comprises 1752 p-TFTs and 1752 n-TFTs with a die size of 1.20x1.88 cm². The P²ROM chip comprises 403 p-TFTs and 412 n-TFTs. The latter chip contains more n-TFTs because of the unipolar n-TFTs in the WORM memory. By inkjet printing, one can add up to 189 n-TFTs in order to execute different programs. For the running averager program, 37 n-TFTs are added by inkjet printing, employing in total to 852 TFTs.

Acknowledgements:
This work has been a collaboration between imec and TNO within the framework of the HOLST centre.

References:
[1] K. Myny, et al., "An 8b Organic Microprocessor on Plastic Foil", *ISSCC Dig. Tech. Papers*, pp. 322-323, Feb. 2011.
[2] M. Rockelé, et al., "Solution-processed and low-temperature metal oxide n-channel thin-film transistors and low-voltage complementary circuitry on large-area flexible polyimide foil", *J. of the Society for Information Display*, vol. 20, no. 9, pp. 499-507, 2012.
[3] K. Myny, et al., "Bidirectional Communication in an HF Hybrid Organic/Solution-Processed Metal-Oxide RFID Tag", *ISSCC Dig. Tech. Papers*, pp. 312-313, Feb. 2012.
[4] M. Rockelé, et al., "Low-temperature and scalable complementary thin-film technology based on solution-processed metal oxide n-TFTs and pentacene p-TFTs," *Organic Electronics*, vol. 12, no. 11, pp. 1909-1913, Nov. 2011.
[5] K. Myny, et al., "An 8-Bit, 40-Instructions-Per-Second Organic Microprocessor on Plastic Foil", *IEEE JSSC*, vol. 47, no. 1, pp. 284-291, Jan. 2012.
[6] J. Rabaey, et al., "Digital Integrated Circuits, A Design Perspective", Prentice Hall, 2003.

978-1-4799-0917-9/14 $31.00 © 2014 IEEE

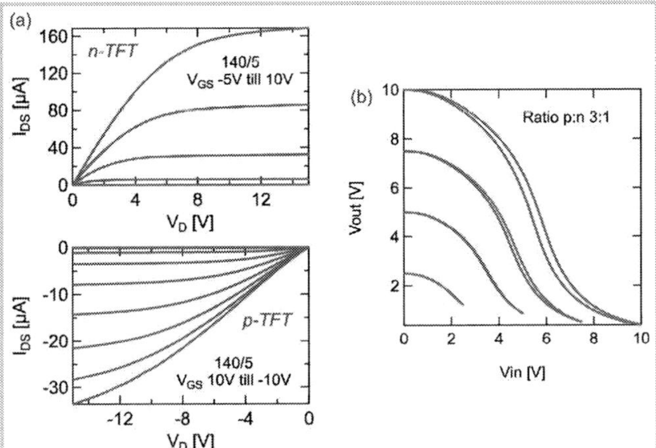

Figure 30.1.1: (a) Output characteristics of typical solution-processed oxide n-type and evaporated pentacene p-type transistors, and (b) inverter characteristics of the hybrid complementary technology at different power supply voltages.

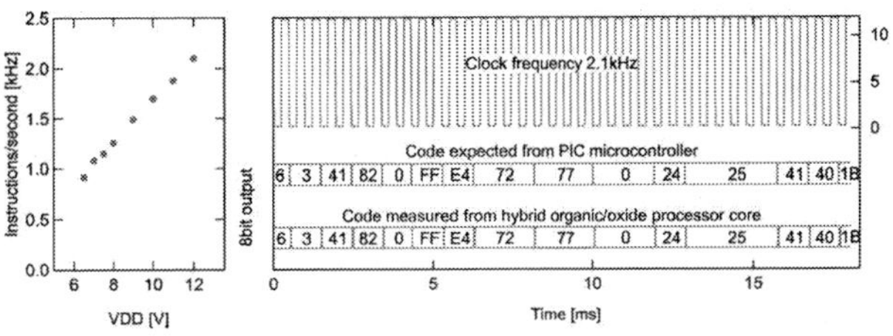

Figure 30.1.2: (left) The measured instructions per second of the processor core chip for different supply voltages. (right) A zoom of the general test bench examining the processor core at a supply voltage of 12V.

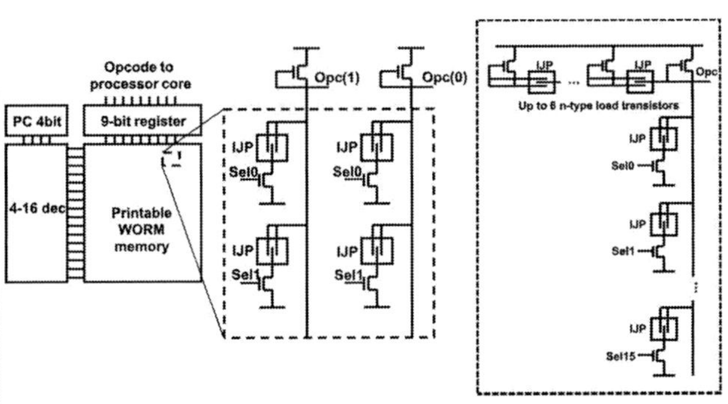

Figure 30.1.3: (left) Block diagram of the P²ROM instruction generator chip and details of the unipolar n-type printable WORM memory, and (right) details of a full column of 16 select transistors and the possibility to add 5 load transistors for the NOR. (IJP stands for inkjet print.)

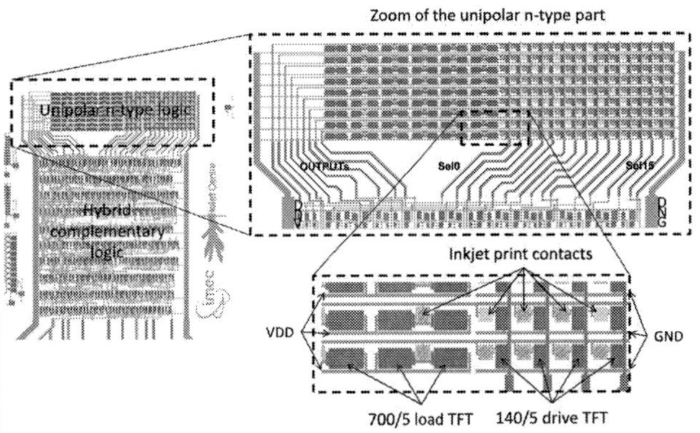

Figure 30.1.4: Detailed layout of the P²ROM instruction generator. The printed connections shown here result in the running averager algorithm.

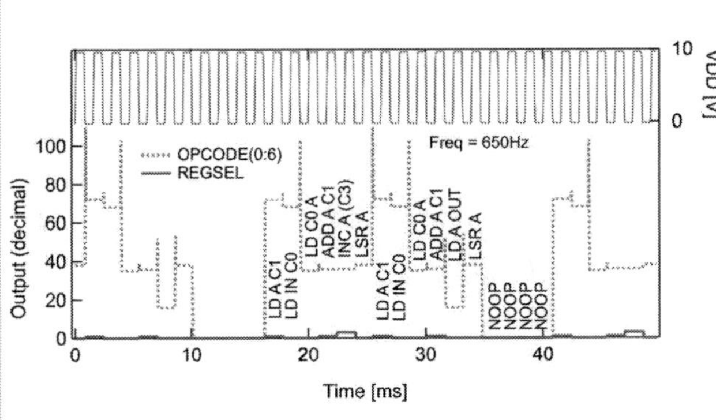

Figure 30.1.5: Measured signals of the P²ROM instruction generator when configured (printed) to execute the running averager algorithm. It consists of 12 instructions and 4 NOOP commands.

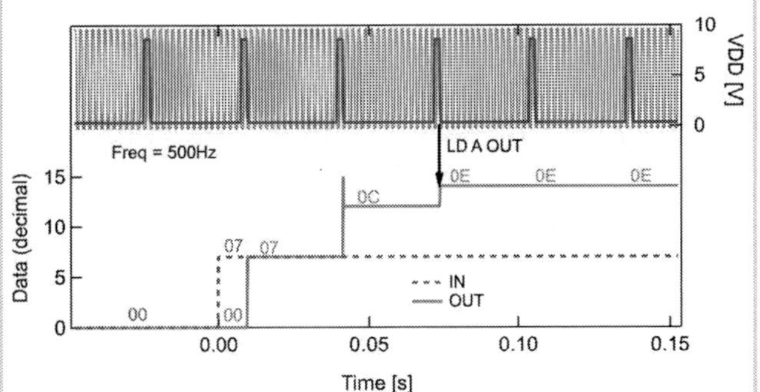

Figure 30.1.6: Measured signals of both the P²ROM and processor core chips while executing a running averager algorithm. The pulses in the top part of the figure correspond to the command "store in output register".

978-1-4799-0917-9/14 $31.00 © 2014 IEEE

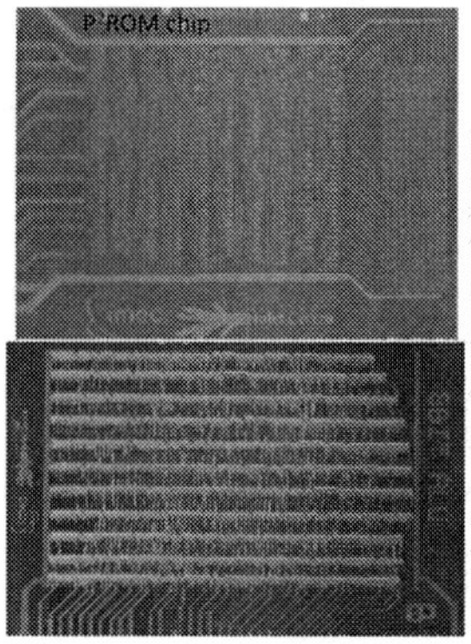

P²ROM chip

	Organic processor [5]	Hybrid organic/oxide complementary processor [this work]
Transistor-count	3381(processor core) + 612 (averager-foil)	3504 (processor core) + 852 (P²ROM averager)
Area	1.96 x 1.72 cm^2	1.20 x 1.88 cm^2
Pin-count	30	30
Min. supply voltage	10V	6.5V
Max. operational speed	40Hz @ 10V VDD	2.1kHz @ 12V VDD
Semiconductor	Pentacene	Pentacene (p-type) and metal-oxide (n-type)
Mobility	~0.15 cm^2/Vs	~0.15 cm^2/Vs (p-type) ~ 2.00 cm^2/Vs (n-type)
Logic family	Unipolar p-type	Hybrid organic/oxide complementary
Operation	accumulation	accumulation
Technology	5 μm	5 μm
Bus width	8 bit	8 bit

Figure 30.1.7: Micrographs of the processor core chip and the P²ROM instruction generator chip, and (right) comparison to the previously published organic microprocessor.

30.2 Digital PWM-Driven AMOLED Display on Flex Reducing Static Power Consumption

Jan Genoe[1,2], Koji Obata[3], Marc Ameys[1], Kris Myny[1], Tung Huei Ke[1], Manoj Nag[1], Soeren Steudel[1], Sarah Schols[1], Joris Maas[4], Ashutosh Tripathi[4], Jan-Laurens van der Steen[4], Tim Ellis[4], Gerwin H. Gelinck[4], Paul Heremans[1,2,4]

[1]imec, Leuven, Belgium, [2]KU Leuven, Leuven, Belgium,
[3]Panasonic, Osaka, Japan, [4]Holst Centre/TNO, Eindhoven, The Netherlands

The efficiency of small-molecule OLED devices increased substantially in recent years, creating opportunities for power-efficient displays, as only light is generated proportional to the subpixel intensity. However, current active matrix OLED (AMOLED) displays on foil do not validate this power-efficient advantage, as too much power is lost in the AM backplane. AMOLED displays use the analog voltage on the gate of a drive transistor (e.g. M1 in Fig. 30.2.1) to control the pixel current and hence the pixel brightness. Accurate and uniform pixel currents can only be obtained when transistor M1 is driven is saturation. In high-resolution technologies on foil, transistor parameters W, L and the mobility μ are limited by technology, imposing a minimal $V_{GS}-V_T$ to obtain sufficient current, i.e. $V_{GS}-V_T > 4V$ for a-IGZO on foil [1]. Subsequently, to obtain saturation, $V_{DS} >$ 4V, which translates in a static backplane power loss surpassing the OLED power consumption (see red stars in Fig 30.2.1). However, when the OLED pixel impedance around a specific reference current can be matched along a display column line, the accurate pixel current control can be imposed by current DACs implemented in external silicon display column drivers. In this work, we operate M1 as a switch and pixel intensity variations are obtained using Pulse Width Modulation (PWM) of a predefined pixel current, i.e. 2μA/pixel [80*80μm²] (which corresponds in our OLED technology to a light output of 2000Cd/m²). When, in a future implementation the external DACs are calibrated at 0.2μA/pixel, the full brightness would correspond to the typical display brightness of a portable PC, i.e. 200Cd/m². This concept enables us to reduce the display power voltage at full brightness from 8.2V in a classical AMOLED display on foil configuration to 5V (measured) and for future implementations even down to 4V (see Fig. 30.2.1). As the OLED current load remains equal, a corresponding static power reduction of the display (and increased battery lifetime) is obtained. Digital driving methods of AMOLED displays have been shown before. However, ΔΣ techniques [2] still integrate charge packets on the gate of M1 and hence do not solve the power issue on foil. Other PWM techniques [3] activate only a single active line in the linedriver yielding difficulties to obtain color depths above 6 bits. When multiple independent linedrivers are implemented and their output is multiplexed to alternately drive a single select line, a higher color depth can be obtained [4]. This leads however to a bulky linedriver, which is hard to get within an e.g. 80μm pitch. The design and implementation of a compact integrated linedriver on foil enabling multiple alternating active signals through a single shift register is demonstrated here.

To demonstrate the functionality of the select linedriver, it has been integrated in a small high-resolution red half-QQVGA display on PEN foil (64*160 pixels@80*80μm², 0.54 inch diagonal]. It comprises 22352 a-IGZO transistors, from which 1872 in the linedriver (see Fig. 30.2.2). Externally, the display is driven as a 64*320 pixel display. An etch-stop-layer stack is used to fabricate the a-IGZO backplane and a red small-molecule top-emitting OLED stack (at 2.6Cd/A) is evaporated on top. Devices are realized on top of PEN foil including top and bottom moisture barrier. The dedicated linedriver can be decompiled in 16 blocks, each comprising 10 linedrivers (see Fig. 30.2.2, bottom).

In order to obtain an 8b PWM at a duty cycle of almost 100%, we have split each frame into 8 subframes and during the first 7 subframes, the select lines are driven twice, with a fixed delay between drive cycles. This delay is alternated each subframe (see table Fig. 30.2.3). Figure 30.2.3 shows the pulses on a set of select lines during the first subframe. In the first subframe, the n-th select line is driven a second time after the n+10-th select line is driven for the first time. We implemented a dedicated single shift register with a 3-clocking scheme. Each 10 lines, the clocking scheme is rotated and the select line is driven on a different clock (see green dashed line in Fig. 30.2.3) to prevent that two select lines are driven at the same time.

Each time a select line is driven, M1 of Fig. 30.2.1 is switched on or off. This is done according to the table in Fig. 30.2.4, for an 8b pixel intensity equal to $b_{7..0}$ [4]. Figure 30.2.4 also shows the on-off switching of a pixel during the driving of 1 full frame (8 subframes) when the pixel intensity is 10011001. This implementation enables almost a 100% duty cycle. The table in Fig. 30.2.4 can be extended to 16b color depth when applied to larger size display. The table in Fig. 30.2.4 shows that subframes are scheduled with increasing delay between both moments the select lines are driven. This, combined with the fact that the select lines during the last subframe are only driven once, avoids overlap between subframes and enables driving at almost 100% duty cycle. As the backplane only comprises n-type a-IGZO transistors, we also designed the integrated linedrivers in unipolar n-type logic. In n-type logic, it is typically challenging to obtain short rise times for charging large capacitances, such as the select lines of a display. Recently reported techniques, such as bootstrapping enable fast rise times in unipolar technologies [5]. The top of Fig. 30.2.5 shows the schematic of the 3-phase clocked linedriver, comprising 2 clocked inverting stages and 1 bootstrapped stage. The overlap capacitance between the gate and source of M9 (510fF) acts as bootstrap capacitance, so no dedicated bootstrap capacitance has been added in the layout. To enable that a single shift register drives alternating two select lines, spatially separated by a multitude of 10 lines, the clocking scheme is rotated each 10 lines amongst the 3 clocks. As an interface between blocks operated with a specific clock scheme, we designed a 2-clock phase delay linedriver, which is inserted each 10 select lines to cycle the clock phase. It is comprised of two bootstrapped stages (see Fig. 30.2.5, bottom)

A copy of the linedriver has also been instantiated on the same foil to enable electrical characterization (see bottom of Fig. 30.2.7). Although the linedriver in the display is driven at 15V (and consuming less than 97μW) and at 200kHz clock speed, we explored the lowest voltage and highest clock frequency the linedriver continued to operate (see Shmoo plot of select line 157 in Fig. 30.2.6). Figure 30.2.6 shows also the select line 4, 10 and 157 voltage pulses when driven at 9V and 240kHz and loaded with the 0.1pF input capacitance and 1MΩ input resistance of a Picoprobe (model 12C from GGB industries). We notice correct timing, but the select line amplitude gets only to within 4V of the rail voltage. This is partially due to the Picoprobe load and partially due to the slew rate.

Finally, Fig. 30.2.7 shows at the top some images from the digital PWM-driven 320 ppi AMOLED display on flex operated at 200 subframes per second. Each display column is driven by a current DAC (DAC902 from TI), driving a multiple of 2μA. Five Xilinx FPGAs send the image data to these 64 current DACs and the data lines. The all-pixels fully-on static power consumption of this digital driven backplane is: 160*64*2μA*5V = 102.4mW, whereas the corresponding classic driving method would have consumed 168mW. As both data lines and select lines are driven at a higher rate, the dynamic power increases. Data lines are driven with 10V swing and have a capacitive load of 9.3pF each, select lines are driven with 15V swing and have a capacitive load of 5.6pF. The dynamic driving power at 30 frames/s increases as a consequence to a maximum of 2.5mW. The overall energy consumption for this digital driving implementation is 62% of the energy consumption of its analog counterpart. Finally, digital driving of AMOLED displays does not need to be restricted to only displays on foil, as digital driving also opens a whole set of other advantages, similar to the introduction of DLP for projection displays.

Acknowledgements:
This work has been done in collaboration between imec and TNO in the frame of the HOLST centre.

References:

[1] N. Saito, T. Ueda, K. Miura, S. Nakano, T. Sakano, Y. Maeda, H. Yamaguchi, I. Amemiya, "10.2-inch WUXGA Flexible AMOLED Display Driven by Amorphous Oxide TFTs on Plastic Substrate," *Proc. SID*, vol. 44, no. 1, pp. 443-446, June 2013.

[2] J.H. Jang, M. Kwon, E. Tjandranegara, K. Lee, B. Jung, "A PDM-Based Digital Driving Technique Using ΔΣ Modulation for QVGA Full-Color AMOLED Display Applications," *J. Display Technol.*, vol. 6, no. 7, pp, 269-278, July 2010.

[3] M. Mizukami, K. Inukai, H. Yamagata, T. Konuma, T. Nishi, J. Koyama, S. Yamazaki,T. Tsutsui, "6-Bit Digital VGA OLED," *Proc. SID*, vol. 31, no. 1, pp. 912-915, May 2000.

[4] J. Genoe, "Digital driving of AMOLED displays"' PCT and TW patent application, November 1, 2012.

[5] D. Geng, B.S. Kim, M. Mativenga, M.J. Seok, D.H. Kang, J. Jang, "40 um-pitch IGZO TFT Gate Driver for High-resolution Rollable AMOLED," *Proc. SID*, vol. 44, no. 1, pp. 927-930, June 2013.

978-1-4799-0917-9/14 $31.00 © 2014 IEEE

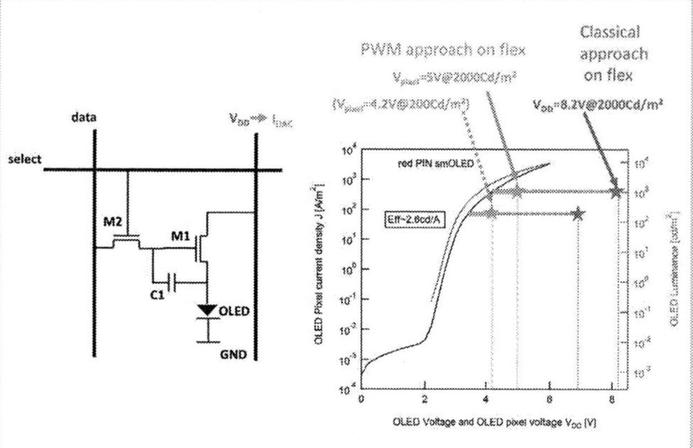

Figure 30.2.1: (left) 2T1C AMOLED pixel schematic, in which M1 can act as current control element or as a switch. (right) Current density-voltage and luminance-voltage characteristics of our red OLEDs. The overall pixel voltage drop is also indicated.

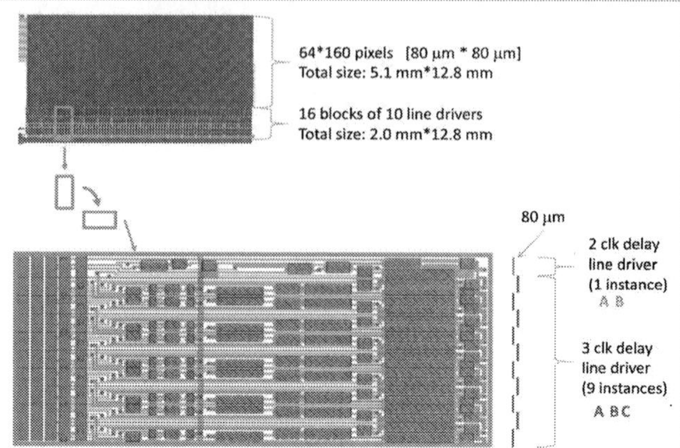

Figure 30.2.2: (top) Lay-out of the 320dpi a-IGZO AMOLED for digital PWM driving comprising the integrated line driver. (bottom) Block of 10 line drivers, each with 80μm pitch. A 5μm design rule is used.

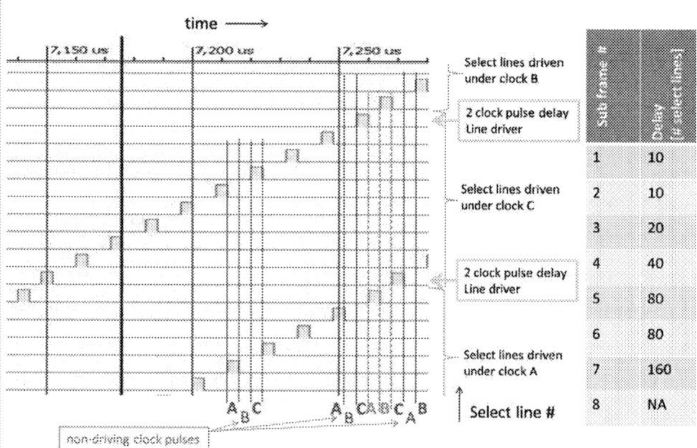

Sub frame #	Delay (# select lines)
1	10
2	10
3	20
4	40
5	80
6	80
7	160
8	NA

Figure 30.2.3: Subsequent select lines driven by the clocks A, B and C. Every 10 select lines, a non-driving clock pulse is skipped. This enables the shift of two active signals through the line driver, without simultaneous driving of a select line. The table indicates the delay of the select lines.

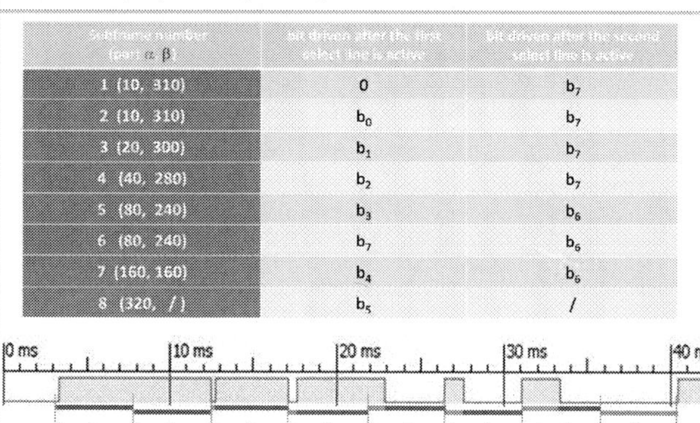

Subframe number (part α β)	bit driven after the first select line is active	bit driven after the second select line is active
1 (10, 310)	0	b_7
2 (10, 310)	b_0	b_7
3 (20, 300)	b_1	b_7
4 (40, 280)	b_2	b_7
5 (80, 240)	b_3	b_6
6 (80, 240)	b_7	b_6
7 (160, 160)	b_4	b_6
8 (320, /)	b_5	/

Figure 30.2.4: (top) Encoding table for the eight bit number b7..0, which ensures accurate pixel intensity control. (bottom) Encoding applied for 1 frame, i.e. 8 subframes for the binary number 10011001. The frame rate is 25 frames/s.

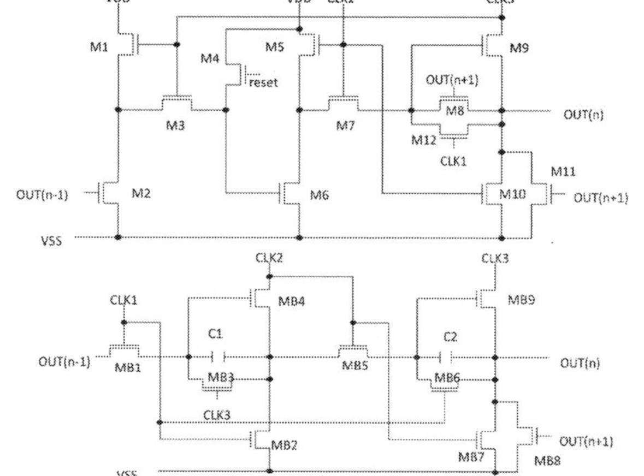

Figure 30.2.5: (top) Line driver schematic with a delay of 3 clock pulses. (bottom) Line driver schematic with a delay of 2 clock pulses to allow a skip of one cycle when going to the next block of 10 line drivers.

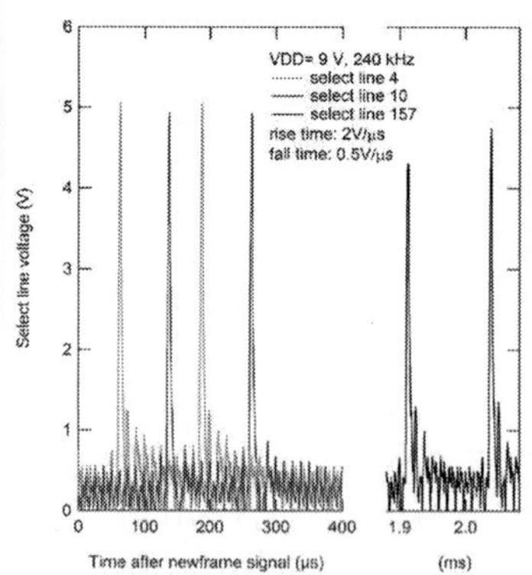

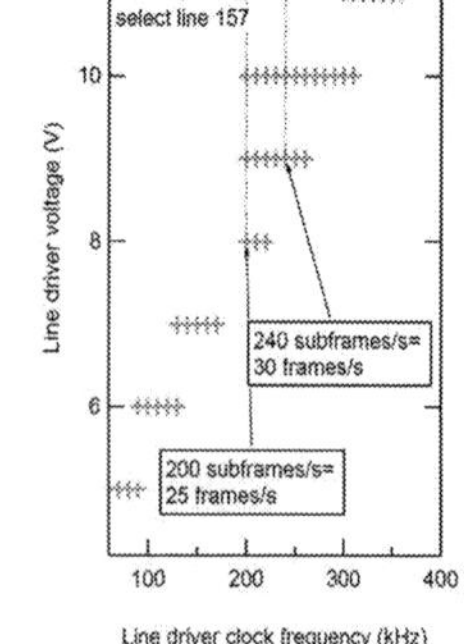

Figure 30.2.6: (left) Picoprobe measurements of the select line signals in the first subframe. (right) Shmoo plot of the select line 157 signal illustrating the line driver voltage versus the line driver clock frequency.

978-1-4799-0917-9/14 $31.00 © 2014 IEEE

Figure 30.2.7: (top) Photographs of the PWM 64*160 AMOLED display on foil with embedded line driver in the a-IGZO backplane. (bottom) Micrograph of the separated line driver on the same foil (for testing). Only select lines numbered 3n+1 have a bonding pad.

30.3 Organic-Transistor-Based 2kV ESD-Tolerant Flexible Wet Sensor Sheet for Biomedical Applications with Wireless Power and Data Transmission Using 13.56MHz Magnetic Resonance

Hiroshi Fuketa[1,2], Kazuaki Yoshioka[1,2], Tomoyuki Yokota[1,2], Wakako Yukita[1,2], Mari Koizumi[1,2], Masaki Sekino[1,2], Tsuyoshi Sekitani[1,2], Makoto Takamiya[1,2], Takao Someya[1,2], Takayasu Sakurai[1,2]

[1]University of Tokyo, Tokyo, Japan, [2]JST/ERATO, Tokyo, Japan

A wet sensor, which detects the presence or absence of liquid, is an important tool for biomedical, nursing-care, and elderly-care applications such as the detection of blood in bandages, sweat in underwear, and urination in diapers. A wet sensor should be a thin, mechanically flexible, large-area, and low-cost device with wireless power and data transmission, because constant monitoring with a rigid and wired wet sensor placed on human skin is annoying. Moreover, the wet sensor should be disposable from a hygiene perspective. In order to meet these requirements, an organic transistor based flexible wet sensor sheet (FWSS) with wireless power and data transmission using 13.56MHz magnetic resonance is developed to detect urination in diapers.

Figure 30.3.1 shows a photograph of the developed 78mm x 53mm FWSS. In its actual implementation, the FWSS is embedded in the cotton of a diaper, although FWSS is placed on the surface of the diaper in Fig. 30.3.1 for clarity. The FWSS is a passive transponder that is wirelessly powered by a reader and sends sensor data to the reader. In the FWSS, organic circuits fabricated on a 12.5μm-thick flexible polyimide film are stacked on 40mm square coil on 12.5μm-thick flexible PCB.

Figure 30.3.2 (a) shows a circuit schematic of the FWSS and reader. The reader is expected to be attached to the pants near the diaper. The reader wirelessly transmits power between the coils (L_1 and L_2) via magnetic resonance at 13.56MHz. In this work, instead of conventional electromagnetic induction [1,2], magnetic resonance is used to increase the distance between the reader and FWSS. The typical power supply voltage (V_{DD}) of the FWSS is 2V. In the FWSS, except for L_2 shown in Fig. 30.3.1, all the circuits including capacitors are integrated in the organic transistor film. In the FWSS, the resistance between two electrodes is measured to detect the presence or absence of liquid. Actually, the resistance is converted to the frequency of TXdata by an RC oscillator (Fig. 30.3.4) in the wet sensor. Then, TXdata performs the load modulation in the transponder and the demodulated signal is RXdata. Fig. 30.3.2 (b) shows the measured waveforms of V_2 and TXdata in the FWSS and RXdata in the reader. The 10Hz frequency of TXdata is successfully demodulated in the reader.

The design challenges of the FWSS are as follows. 1) The power consumption of the battery-operated reader is large because the reader must transmit a maximum power to deal with the loss of transmission efficiency due to the bending of the flexible coil (L_2) as well as change in the distance between L_1 and L_2. 2) ESD protection is essential in the FWSS because the electrodes can directly come in contact with wet human skin. ESD protection in organic transistors, however, is difficult because they are fabricated on an insulating film. To solve these problems, we propose the following two solutions. 1) Adaptive amplitude control (AAC) decreases the power consumption of the reader by reducing the amplitude by up to 92%. 2) ESD protection circuits with organic Schottky diodes using copper phthalocyanine (CuPc) are used to achieve 2kV ESD tolerance. Both AAC and ESD protection are essential technologies in wireless and flexible sensors for biomedical and elderly-care applications.

Figure 30.3.3(a) shows a circuit schematic of the wet sensor, including a four-input multiplexer and three RC oscillators. The four inputs are toggled with a preamble for the data receiver in the reader. The reference RC oscillator is used in the proposed AAC and the other two RC oscillators are used as the wet sensors. In AAC, as shown in Fig. 30.3.2 (a), the amplitude of the reader (V_1) is adaptively controlled by the adaptive amplitude controller to maintain the frequency of the reference oscillator at a target value, which regulates V_{DD} of the FWSS. Figures 30.3.3(b) and (c) show a circuit schematic and a photograph of the four-input multiplexer based on a pseudo-CMOS [3], respectively. The pseudo-CMOS is used to obtain a high gain in pMOS-only organic transistors. To demonstrate the multiplexer, Fig. 30.3.3(d) shows measured waveforms of TXdata, RC oscillator 3, and the reference RC oscillator 1. In this measurement, the reference RC oscillator 1 is 10Hz, RC oscillator 2 (= Sensor 1) is not oscillating, which corresponds to "Dry" (= no urination), and RC oscillator 3 (= Sensor 2) is 5Hz, which corresponds to "Wet" (= urination). Figure 30.3.3(d) shows a successful toggle operation of the multiplexer.

Figure 30.3.4(a) shows a circuit schematic of the RC oscillator with the pseudo-CMOS inverters [3]. The RC oscillator converts the resistance between electrodes (R_{MEA}) to the frequency of OSC. Fig. 30.3.4 (b) shows the measured waveform of OSC. Figure 30.3.4(c) shows the measured dependence of the oscillation period on R_{MEA}. In this work, the target range of R_{MEA} is from 2MΩ to 10MΩ because R_{MEA} in cotton immersed in normal saline is several megaohms. The oscillation period linearly depends on R_{MEA} with an offset. The sensitivity of the resistance sensor is 3.5%/MΩ, though such sensitivity is not required in the FWSS which detects the presence or absence of liquid.

In Fig. 30.3.5, the change in the transmission efficiency due to the bending of the flexible coil (L_2) as well as the changes in distance between L_1 and L_2 is discussed. Figure 30.3.5(a) shows the measurement setup used to investigate the relationship between the input voltage (V_{IN}) in the reader and the output voltage (V_{OUT}) in the FWSS. The bending (B) of the flexible coil (L_2) and the distance (D) between L_1 and L_2 are varied. Figure 30.3.5(b) shows a photograph of the coil L_1 and the bent coil L_2. Figure 30.3.5(c) shows the measured D dependence of V_{IN}/V_{OUT} at B of 0mm and 17mm. D of 40mm is equal to the diameter of L_1 and L_2. The required V_{OUT} for the FWSS is constant. Therefore, V_{IN} should be increased as D and B increase. In the conventional design, a constant V_{IN} (= 9.7 x V_{OUT}) is used to guarantee the required V_{OUT} in the worst case of D=40mm and B=17mm. A constant V_{IN}, however, is usually a waste of the energy, because the required V_{IN} is smaller than 9.7 x V_{OUT} in most cases, as shown in Figure 30.3.5(c). Therefore, AAC is proposed to track the minimum V_{IN}, which reduces the power consumption of the battery-operated reader. AAC reduces V_{IN} by up to 92%. To clarify the effect of B on V_{IN}, Figure 30.3.5(d) shows the measured D dependence of V_{IN} at B=17mm / V_{IN} at B=0mm extracted from Fig. 30.3.5(c). By changing B from 0mm to 17mm, V_{IN} should be increased by a factor of more than 1.5 when D is larger than 10mm. Therefore, AAC is more effective in applications where the coil is bent.

Figure 30.3.6(a) shows a schematic of the circuit developed for ESD protection. Instead of an organic pMOSFET, a vertical organic Schottky diode [4] is used as the diode for ESD as well as a rectifier, because the Schottky diode has large current drivability and a superior frequency characteristic compared with a MOSFET. Figures 30.3.6(b) and (c) show a cross section and a photograph of the organic Schottky diode using CuPc, respectively. The ESD tolerance is evaluated in accordance with the ESD standard IEC 61000-4-2 with a MiniZap ESD simulator [5]. Figure 30.3.6(d) shows the ESD measurement steps and the measured results. Without the ESD protection, the ESD tolerance is below 0.5kV. In contrast, with the ESD protection, the ESD tolerance is above 2kV, which successfully satisfies level 1 of IEC 61000-4-2. One of the issues of adding ESD protection to the signal pad is bandwidth degradation. Figure 30.3.6(e) shows the measured frequency dependence of the gain of the source follower with and without the ESD protection. By adding the ESD protection, the -3dB bandwidth is reduced from 500Hz to 300Hz, which is not a problem in the FWSS because the oscillation frequency is less than 4Hz, as shown in Fig.30.3.4(c). Figure 30.3.7 shows a micrograph of the organic full-wave rectifier and RC oscillator, and a summary of key features.

References:

[1] L. Yan, et al., "A 3.9mW 25-Electrode Reconfigured Thoracic Impedance/ECG SoC with Body-Channel Transponder," *ISSCC Dig. of Tech. Papers*, pp. 490-491, Feb. 2010.

[2] J. Yoo, et al., "A 5.2mW Self-Configured Wearable Body Sensor Network Controller and a 12μW 54.9% Efficiency Wirelessly Powered Sensor for Continuous Health Monitoring System," *ISSCC Dig. of Tech. Papers*, pp. 290-291, Feb. 2009.

[3] K. Ishida, et al.,"Insole Pedometer With Piezoelectric Energy Harvester and 2V Organic Digital and Analog Circuits," *ISSCC Dig. of Tech. Papers*, pp. 308-309, Feb. 2012.

[4] Y. Ai, et al., "14 MHz Organic Diodes Fabricated Using Photolithographic Processes," Appl. Phys. Lett. 90, 262105 (2007).

[5] Thermo Scientific, MiniZap ESD tester, Accessed Sep. 2013, <https://static.thermoscientific.com/images/D20078~.pdf>.

978-1-4799-0917-9/14 $31.00 © 2014 IEEE

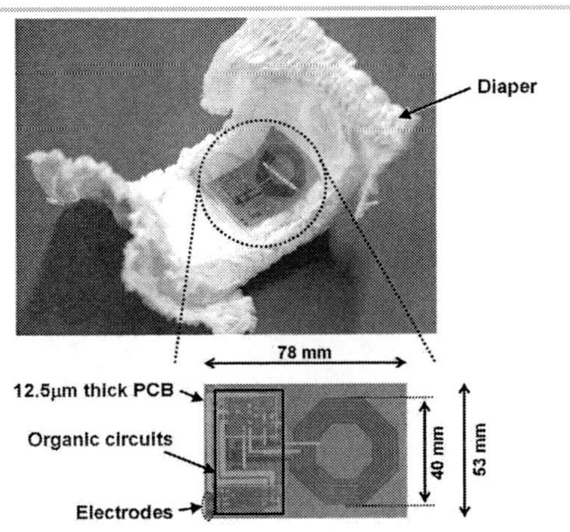

Figure 30.3.1: Developed flexible wet sensor sheet (FWSS).

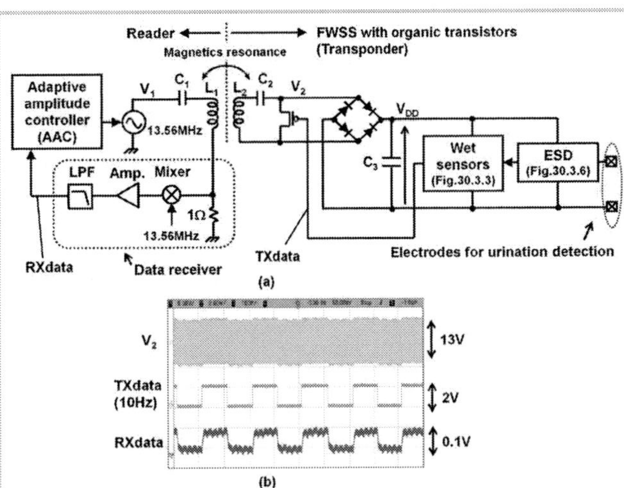

Figure 30.3.2: (a) Circuit schematic of FWSS and reader. (b) Measured waveforms of load modulation.

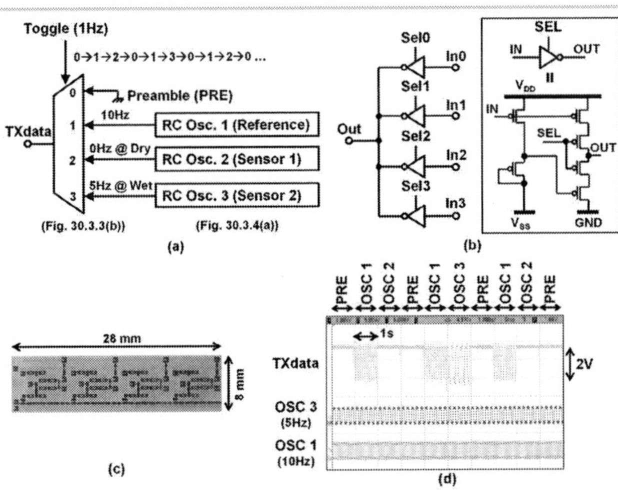

Figure 30.3.3: (a) Circuit schematic of wet sensor. (b) Circuit schematic, (c) photograph, and (d) measured waveforms of four-input multiplexer.

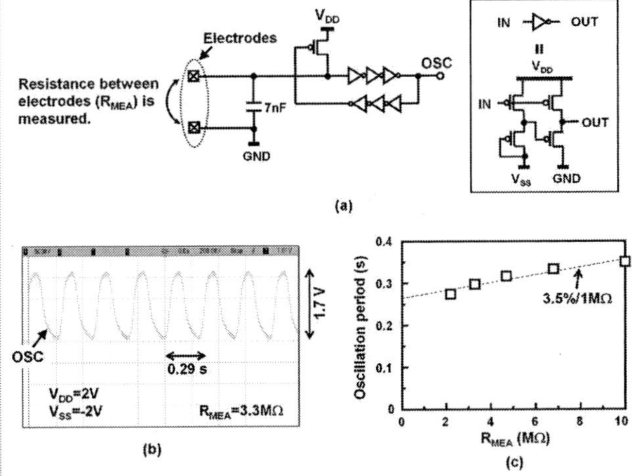

Figure 30.3.4: RC oscillator. (a) Circuit schematic. (b) Measured waveform. (c) Measured dependence of oscillation period on R_{MEA}.

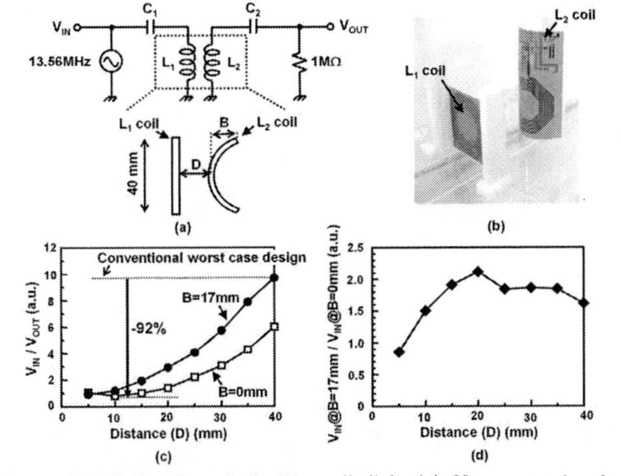

Figure 30.3.5: Bending of flexible coil (L_2). (a) Measurement setup. (b) Photograph. (c) Measured D dependence of V_{IN}/V_{OUT}. (d) Measured D dependence of V_{IN} at B=17mm / V_{IN} at B=0mm.

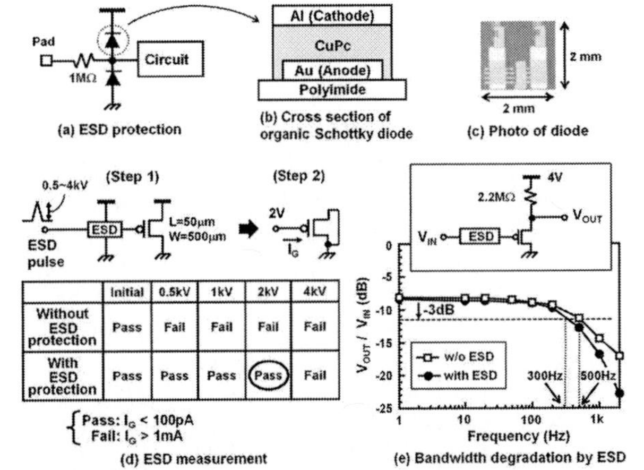

	Initial	0.5kV	1kV	2kV	4kV
Without ESD protection	Pass	Fail	Fail	Fail	Fail
With ESD protection	Pass	Pass	Pass	Pass	Fail

Figure 30.3.6: ESD protection. (a) Circuit schematic. (b) Cross section and (c) photograph of organic diode. (d) ESD measurement. (e) Bandwidth degradation caused by ESD protection.

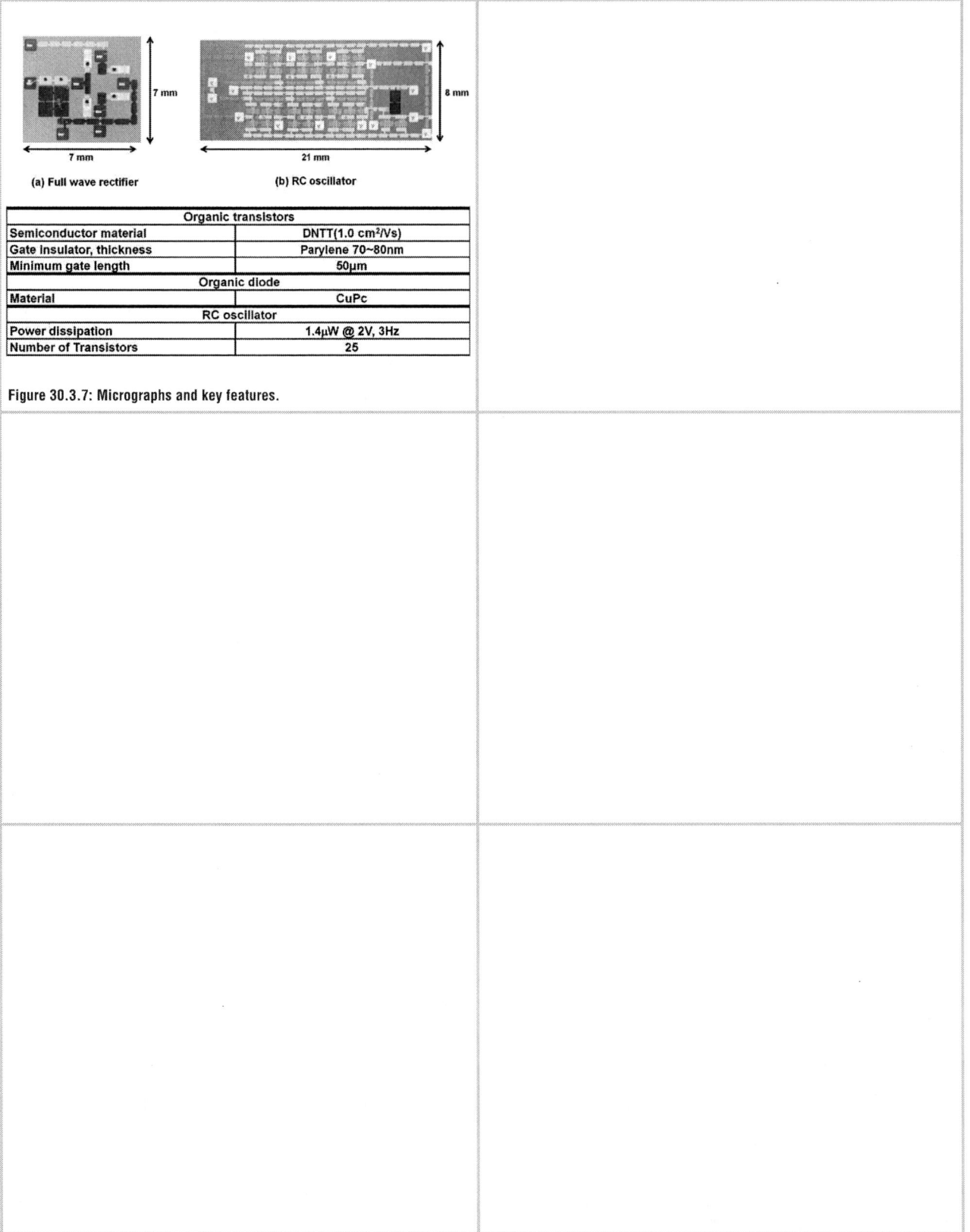

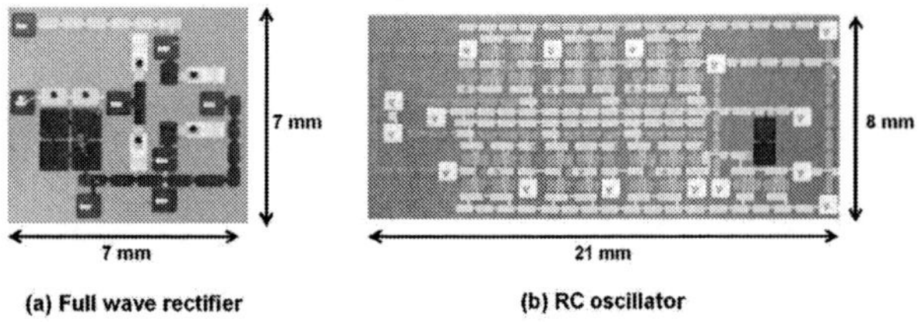

7 mm
7 mm
(a) Full wave rectifier

8 mm
21 mm
(b) RC oscillator

Organic transistors	
Semiconductor material	DNTT(1.0 cm^2/Vs)
Gate insulator, thickness	Parylene 70~80nm
Minimum gate length	50μm
Organic diode	
Material	CuPc
RC oscillator	
Power dissipation	1.4μW @ 2V, 3Hz
Number of Transistors	25

Figure 30.3.7: Micrographs and key features.

30.4 A 13.56MHz RFID Tag with Active Envelope Detection in an Organic Complementary TFT Technology

Vincenzo Fiore[1], Egidio Ragonese[2], Sahel Abdinia[3], Stephanie Jacob[4], Isabelle Chartier[4], Romain Coppard[4], Arthur van Roermund[3], Eugenio Cantatore[3], Giuseppe Palmisano[1]

[1]University of Catania, Catania, Italy, [2]STMicroelectronics, Catania, Italy,
[3]Eindhoven University of Technology, Eindhoven, The Netherlands,
[4]CEA-LITEN, Grenoble, France

In the last several years, organic electronics have gained increasing consideration as a cost-effective alternative to silicon, especially in RFID applications. An inductive-coupled organic RFID operating at 13.56MHz was demonstrated on foil using p-type organic technologies [1]. A complementary organic technology was used for a 13.56MHz transponder in [2]. Recently, a complementary hybrid organic/metal-oxide process was exploited to demonstrate bidirectional communication in an HF RFID [3]. It adopts passive envelope detection using traditional diode-based schemes with OOK modulation. However, OOK modulation usually reduces sensitivity and reading range. In this work, a complementary organic TFT (C-OTFT) technology [4] is used for the first time to implement a 13.56MHz RX front-end, which exploits an active detection scheme and is able to demodulate ASK PWM-coded signals with modulation depth (h) as low as 25%.

The adopted process is a printed C-OTFT technology manufactured on an 11×11cm^2 flexible foil. The TFTs are implemented in a top-gate bottom-contact multi-finger structure with a 20μm channel length on 125μm thick polyethylene-naphtalate (PEN) substrate. The C-OTFTs, based on small-molecule organic semiconductors, exhibit a typical carrier mobility of 1.5cm^2/V·s and 0.55cm^2/V·s for p-type and n-type, respectively. The screen-printed dielectric is a fluoropolymer with a thickness of 750nm. A 30nm Au source/drain layer and a 5μm silver-ink gate layer are used as first and second connection metal, respectively. The technology also provides metal-insulator-metal capacitors and carbon ink resistors.

The architecture of the proposed RFID tag is shown in Fig. 30.4.1. The tag includes an RX front-end, consisting of an active envelope detector (ED) and clock/data recovery circuitry, a rectifier, and a code recognition unit. Both ED and the rectifier are inductively coupled to the reader by means of a two-coil foil antenna on PEN. A PWM coding is adopted with bit "1" and "0" corresponding to a duty cycle of 70% and 30% high, respectively. The rectifier provides the supply voltage (VDD) to the other blocks. The ED extracts the 13.56MHz ASK envelope and drives the recovery circuitry. The clock generator detects the ENV signal rising edge, thus providing a synchronizing clock (CLOCK) to the code recognition unit. CLOCK is properly delayed and then used by the D-FF to extract DATA by sampling the ENV signal. The delay is implemented by a cascade of inverters charged with a capacitance. The code recognition unit is designed to receive a sequence of "reset" and "identity" codes and consists of two modules. The reset module (RM) synchronizes all the tags in the reader range with the start of the following identity transmission code. The identity verification module (IVM) compares the received code with the tag identity and in case of code matching enables the response back to the reader by using a load modulator.

Figure 30.4.2 shows transistor-level schematics of the main blocks of the RFID tag. ED is the key circuit of the RX front-end and determines its performance. The core of ED is made up of the p-type couple, M1-M2, the self-biased load including M3, R_L, C_L and the comparator, M4-M7. The input pair injects current pulses at 13.56MHz into the load. Thanks to the RC filter, M3 draws the average current, I_{AV}, produced by the input envelope and provides at its gate the output average value, ED_M, needed for the logic detection of the comparator. To this purpose, the RC time constant has to be set higher than the bit period (T_{BIT}). The variable component of the current pulses is mainly related to the input signal envelope. It flows into R_L producing the output voltage, ED_OUT, that is proportional to the envelope amplitude.

In comparison with traditional diode-based ED schemes, the proposed topology guarantees a large swing of ED_OUT, avoiding the need for a further amplification stage before the comparator. A simple pseudo-differential comparator, M4-M7, provides a quasi rail-to-rail signal at its output, $\overline{ENV}$, that is suitable for the digital circuits. Both the clock generator and the D-FF are implemented with dynamic circuits to reduce transistor count and current consumption [4]. The clock generator exploits a customized topology, which minimizes the transistor count with respect to the traditional D-FF with reset. This simplification is made possible since both the R and the IN input are low before the beginning of a new bit, according to the coding scheme.

Figure 30.4.3 depicts the measured waveforms of ED at different bit rates (B) and differential input voltages (VRF) with a modulation depth (h) of 50%. The detector is able to demodulate up to 75b/s with an RF input of 80V_{pp}. For this input level, the signal swing at ED_OUT is around 20V_{pp}, which in turn produces a 35V_{pp} excursion at the comparator output, $\overline{ENV}$. Finally, for an RF input signal of 40V_{pp} the swing of $\overline{ENV}$ is about 20V that still allows the ED to properly drive the subsequent clock/recovery circuitry. The envelope detector draws 8μA from a VDD of 40V.

The measurement of the RX front-end outputs is reported in Fig. 30.4.4 for a modulated input word of 8 bits (i.e., "01101001"). The overall current consumption is 20μA. Figure 30.4.4 also includes a benchmarking with the most relevant work previously reported [3]. The RX front-end discussed here is able to work with h as low as 25% at bit rate of 50b/s. Thanks to the active detection scheme, the proposed RFID demonstrates for the first time an ASK PWM receiver in a printed fully-organic complementary technology.

Figure 30.4.5 shows the measurement of the code recognition unit. The reset signal code is "0000" and the tag identity code is "0011". In the RM, a 2b counter adds up the number of zeros in the input code and sends a reset signal (RESET OK) to the IVM after receiving four consecutive zeros. This will announce the arrival of a new identity code to the IVM by clearing all its FFs. The identity comparison is performed through an XOR gate. At each clock, a 2b counter and a logic array generate the correct tag identity code corresponding to the bit being received. In case of un-match (e.g., code "0001" in Fig. 30.4.5), the comparator resets the counter. Otherwise, the operation will continue until all bits of the identification code are received and successfully compared. When this happens, the counter "Carry" bit (IDENTITY OK) will trigger the reply back. This is shown in Fig. 30.4.5 where the measured IDENTITY OK signal goes high only after "0000-0011" is received. The tag will answer with a message (using the load modulator) only after the correct code is received, according to an identification scheme called "silent tag" [5]. The advantage of the silent tag is that a search is possible only if a list of the tags that can be offered to the reader (e.g., the tags present in the shop) is known in advance. In this way tag security is ensured with no need for encryption.

Figure 30.4.6 shows the measured output voltage (V_{DC}) of the rectifier under different load and input frequency conditions. At 13.56MHz and with the full-tag-load of 1MΩ (R_L), the rectifier generates 40V and 24V V_{DC} for an input voltage (VRF) of 75V_{pp} and 60V_{pp}, respectively. Comparing an implementation on the same foil of the proposed 4-stage rectifier (Fig. 30.4.2) and of the cross-coupled full-wave topology in [6] confirms that the adopted solution in our technology has better performance. Figure 30.4.7 depicts the micrographs of the RFID circuits on foil along with the area occupied by the main macro-blocks.

Acknowledgments:
This work was funded in the framework of the European FP7 project COSMIC (grant agreement n°247681). The authors would like to thank P. Battiato for support in circuit simulations, S. Cantella and E. Cintolo for the layout assistance, and A. Castorina for the measurement set-up. A special thanks to G. Maiellaro for fruitful discussions.

References:
[1] K. Myny, et al., "An inductively-coupled 64b organic RFID tag operating at 13.56MHz with a data rate of 787b/s", *ISSCC Dig. Tech. Papers*, pp. 290-292, Feb. 2008.
[2] R. Blache, J. Krumm, and W. Fix, "Organic CMOS circuits for RFID applications," *ISSCC Dig. Tech. Papers*, pp. 208-210, Feb. 2009.
[3] K. Myny et al., "Bidirectional communication in an HF hybrid organic/solution-processed metal-oxide RFID tag," *ISSCC Dig. Tech. Papers*, pp. 312-314, Feb. 2012.
[4] S. Jacob, et al., "High performance printed N and P-type OTFTs enabling digital and analog complementary circuits on flexible plastic substrate," *Elsevier Solid-State Electronics*, pp. 167-178, June 2013.
[5] H. Moran, et al., "Selective addressing transponders" *US patent app.* 13/0154799A1, filed 9 June 2011.
[6] K. Ishida, et al., "100V AC power meter system-on-a-film (SoF) integrating 20V organic CMOS digital and analog circuits with floating gate for process-variation compensation and 100V organic PMOS rectifier," *ISSCC Dig. Tech. Papers*, pp. 218-220, Feb. 2011.

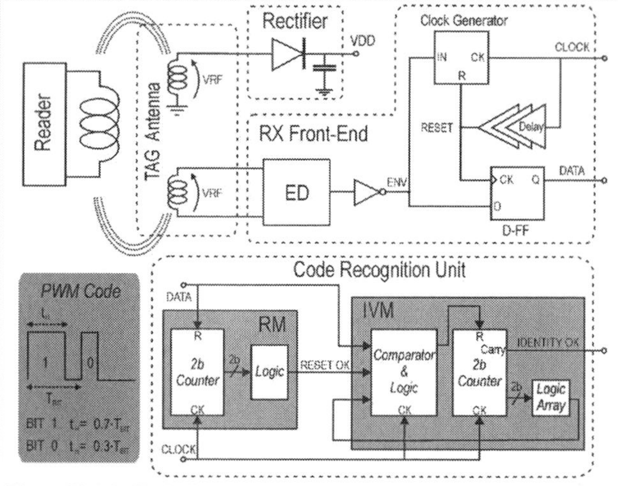

Figure 30.4.1: Block diagram of the 13.56MHz RFID tag and PWM coding scheme.

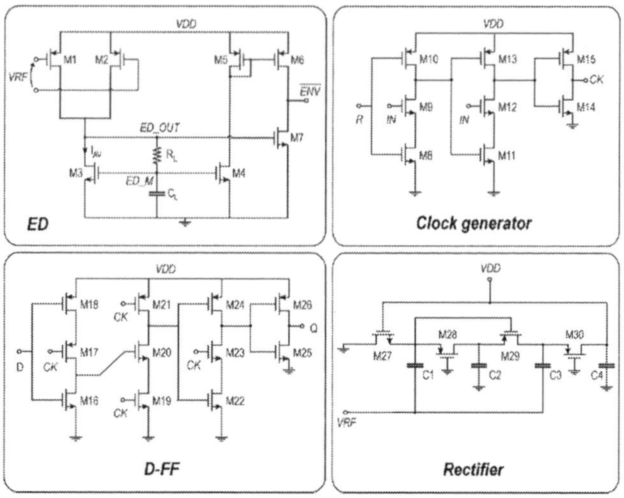

Figure 30.4.2: Schematics of the main blocks of the 13.56MHz RFID tag.

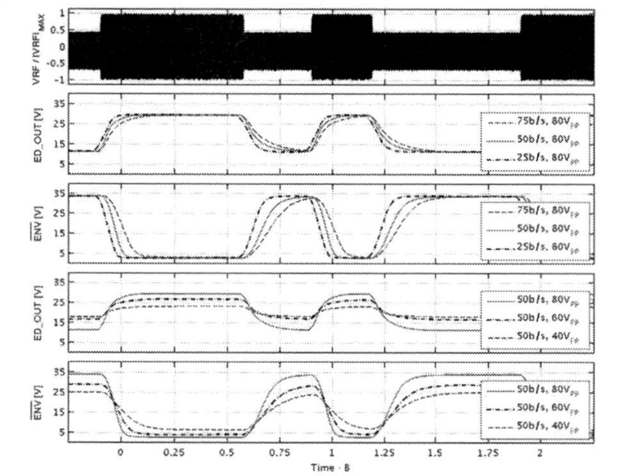

Figure 30.4.3: ED measurements at different B and peak-to-peak VRF (h=50%).

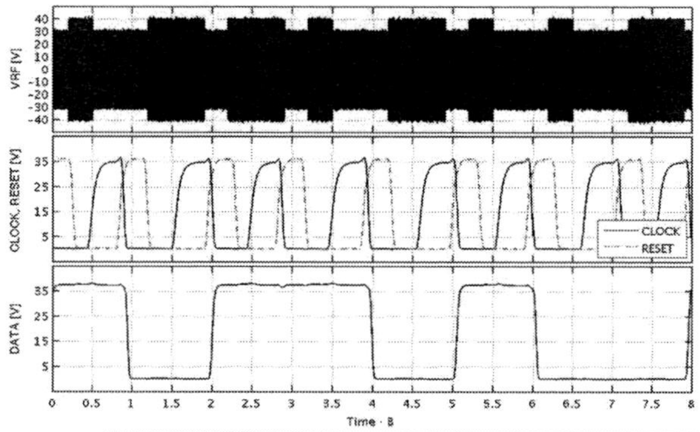

	Technology	VDD [V]	RF [MHz]	B_{MAX} [b/s]	h_{MIN} [%]
Myny [3]	Hybrid organic/metal-oxide	5	13.56	1200	100
This work	Printed C-OTFT	40	13.56	75	25

Figure 30.4.4: RX front-end measurements (B=50b/s, h=25%) and benchmarking.

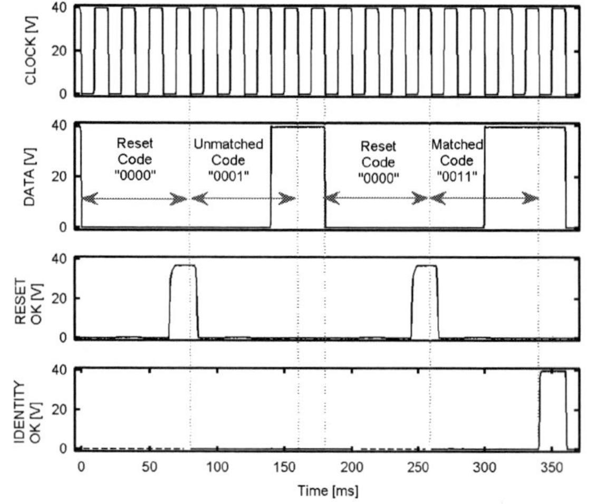

Figure 30.4.5: Measurements of the code recognition unit.

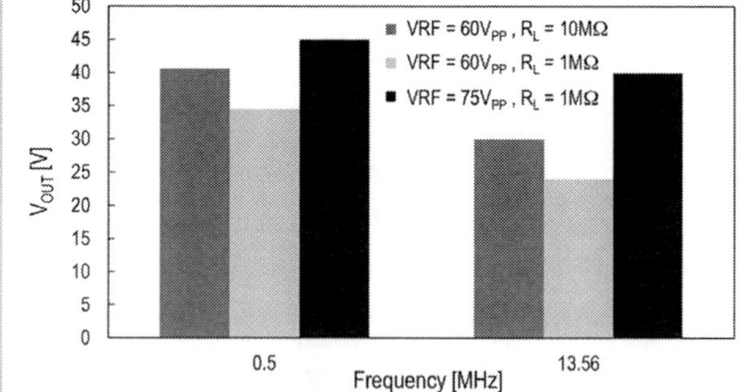

RF [MHz]	VRF [V_{pp}]	Load [MΩ]	V_{DC} [V] Full-wave rectifier	V_{DC} [V] 4-stage rectifier
0.5	60	10	21	40.6
13.56	60	10	16.5	30

Figure 30.4.6: Measurements of the 4-stage rectifier and comparison with measured full-wave rectifier in the same technology.

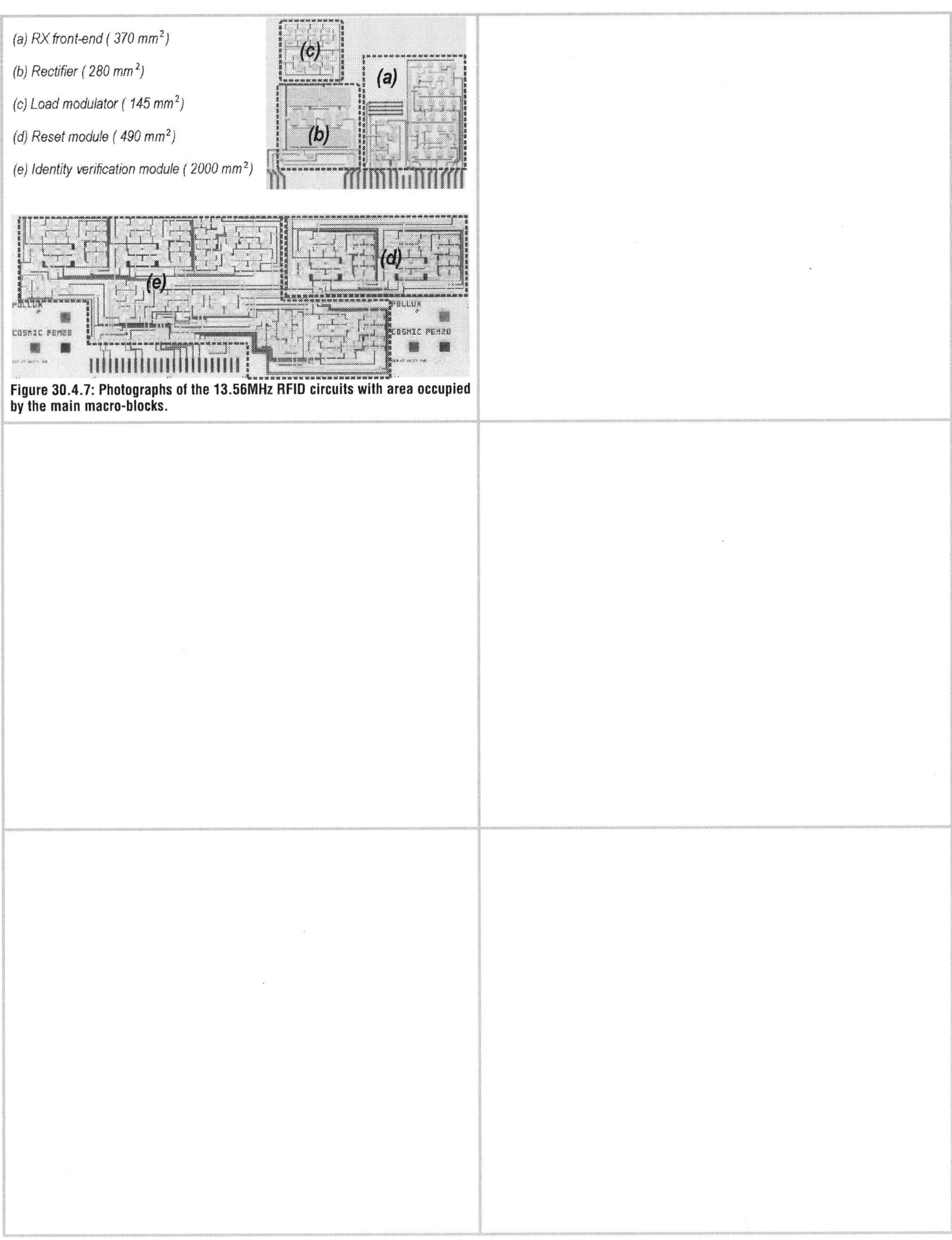

Figure 30.4.7: Photographs of the 13.56MHz RFID circuits with area occupied by the main macro-blocks.

30.5 A GaN 3×3 Matrix Converter Chipset with Drive-by-Microwave Technologies

Shuichi Nagai[1], Yasuhiro Yamada[1], Noboru Negoro[2], Hiroyuki Handa[2], Yuji Kudoh[1], Hiroaki Ueno[1], Masahiro Ishida[2], Nobuyuki Otuska[1], Daisuke Ueda[1]

[1]Panasonic, Osaka, Japan, [2]Panasonic, Kyoto, Japan

A matrix converter [1] that directly transduces power and frequency by bidirectional switches has been expected to be an ultimate AC-to-AC converter because it eliminates limited-lifetime capacitors and achieves high efficiency power conversion even without PFC (Power Factor Control) circuits. However, it has not been practically realized due to the following issues resulting from the abundant components it includes. There has been no existing monolithic bidirectional switch that offers a high blocking voltage with current-handling capability and no compact isolated gate driver that provides a gate signal against a positive/negative voltage reference at every moment. Consequently, the matrix converter with discrete components makes the system large and complicated because of numerous power switches and isolated drivers. Relevant to this discussion, GaN devices [2] are very attractive in power applications since the GaN bidirectional power switches with a high blocking voltage can be monolithically implemented by using its lateral device structure [3]. Additionally, a GaN Drive-by-Microwave (DBM) isolated gate driver [4] is the best candidate for a matrix converter, because it is very compact being both photo-coupler and transformer free and co-integratable with GaN power devices. These GaN integration technologies are very valuable in terms of a high-speed switching with less inductance among lines as well as offering a small system size.

In this paper, we describe a GaN 3x3 matrix converter chipset, which are composed of a GaN integrated bidirectional switching chip and a GaN integrated gate drive transmitter chip using 5.0GHz Drive-by-Microwave technology. The extremely compact three phase AC-AC matrix converter such as a 25x18mm^2 is realized by these GaN/Si integrated chips and novel isolated dividing couplers, which duplicate the gate signal with different references for dual-gate bidirectional switches and reduce gate lines and gate drive components by half. The proposed GaN 3x3 matrix converter is significantly more compact than the conventional one that requires numerous power switches, flywheel diodes, photo-couplers, isolated power supplies and gate drivers.

Figure 30.5.1 shows the system block of a proposed GaN 3x3 matrix converter for 3-phase AC-AC converter with the Drive-by-Microwave technology that supplies a power switching device an isolated signal and power by mixture using a microwave wireless power transmission with an electromagnetic resonant coupling. The proposed 3x3 matrix converter consists of the 3x3 integrated bidirectional switching chip, the DBM gate drive transmitter chip, and isolated dividing couplers in contrast to the conventional one with many discrete components. The GaN integration technologies contribute downsizing of the power electronic system by footprint reduction for lots of PWM gate lines and power lines.

Figure 30.5.2 shows the GaN integrated bidirectional switching chip, which is integrated with 9 GaN dual-gate bidirectional switches with gate injection transistors (GaN-GIT) [2] that has a normally-off operation. The GaN integration device with a high junction temperature can cope with the issues related to the high heat density that come from the beneficial compactness of the power device integration. Since each GaN bidirectional switch has co-integrated RF rectifier circuits with a single shunt diode for gate driving, it performs a switching operation when the RF modulated signals from the DBM transmitter are input through isolated couplers. As shown in Fig. 30.5.2, the fabricated GaN bidirectional switch with the RF rectifier circuits on a Si substrate successfully demonstrated bi-directional switching operation by 5.0GHz RF signal inputs and exhibited the blocking voltage over 600V. Although the GaN bidirectional switches enable a matrix converter to be compact by eliminating fast recovery diodes for flywheel currents, the gate drive system is still complicated with a lot of gate signal lines because the GaN bidirectional switch needs two identical isolated gate signals against each source ports due to its dual-gate structure.

Therefore, the isolated dividing coupler that duplicates an isolated gate signal with separated references for a dual-gate switch is newly developed. The structure and cross-section of the proposed isolated dividing coupler in a printed circuit board (PCB) are shown in Fig. 30.5.3, where the three ¼-wavelength resonators that are shorted with outer ground lines are stacked vertically. Because the middle resonator is coupled with the upper and the lower resonator by electromagnetic resonance coupling [5], the input signal is divided to two output signals with a low insertion loss and a very high DC-isolation voltage. In addition, the coupler size is very compact as 1.2x 1.2mm^2 for a 5.0GHz signal. The transmission and return loss of the fabricated isolated dividing coupler in a PCB are respectively around 4.1dB and 12.0dB at 5.0GHz as shown in Fig. 30.5.4. The relative permittivity and dielectric loss tangent of the PCB at 1.0GHz are around 10 and 0.003, respectively. The insertion loss as a 3dB power divider is very low as 1.1dB even though it includes the loss of input and output lines. And the thickness of the each layer is chosen to be 0.28mm to achieve a DC-isolation voltage over 5.0kV. The fabrication of the isolated coupler in the PCB results in a low-cost matrix converter.

As shown in Fig. 30.5.5, in order to transmit 9 sets of 5.0GHz signal that is envelop-modulated by the PWM (Pulse Width Modulation) gate signal, the DBM gate drive transmitter chip with 9 signals output is designed using GaN HFETs (HFET: Hetero junction Field-Effect Transistors) with the gate length of 700nm on a Si substrate. The DBM gate drive transmitter chip is implemented with 3 sets of a 5.0GHz oscillator and a 3-way switching mixer. By taking advantage of the algorism of a 3x3 matrix converter that two of three GaN switches are off, this gate driver with the 3-way switching mixer outputs 3 signals from the shared 5.0GHz oscillator by switching among three output ports. Because this driver shares the source of 5.0GHz signal, the power consumption is very low as a 1.95W (13V, 150mA) in comparison with the conventional system with individual 18 drivers that constantly operate. On top of this, it is a great feature of this new DBM gate driver that has the power saving function by means of reducing the internal power dissipation current by the current control transistor (Tmod) by 0.6W (13V, 105mA) to minimize the output power for only maintaining on-state of a GaN power switch. Figure 30.5.6 shows the output spectrum from the fabricated GaN DBM gate drive transmitter chip. The 17dBm output at 4.85GHz is enough power to generate a gate voltage of 2.0V at the RF rectifier circuit in the GaN integrated bidirectional switching chip and make it ON. As seen from Fig. 30.5.6, the fabricated DBM gate drive transmitter chip successfully outputs the PWM modulated 5.0GHz signal from 3 output ports in sequence.

Figure 30.5.7 shows the photograph of the fabricated GaN integrated bidirectional switching chip (1.74x3.5mm^2) and the fabricated 9 signals output DBM gate drive transmitter chip (4.5x2.0mm^2) mounded on the PCB that includes 9 sets of the isolated dividing coupler. The fabricated GaN 3x3 matrix converter has the capability to drive 5.0kW motor, since the GaN switches can switch up to 10A under 600V. The total size of only 25mm x 18mm is equivalent to about one thousandth of the conventional 3x3 matrix converter (30.0cm x 50.0cm). This is the first realization of an integrated matrix converter by GaN power and GaN RF devices beyond the scope of a discrete component system.

Acknowledgements:
The authors thank Dr. K.Mizutani, Mr. M.Nishijima, Mr. Y.Kawai, Mr. H.Fujiwara, and Mr. O.Tabata of Panasonic for their significant technical support.

References:

[1] P. Wheeler, et al.," Matrix converters: A technology review," *IEEE Trans. Ind. Electron.,* vol. 49, no. 2, pp. 276–288, Apr. 2002.

[2] M. Yanagihara, et al., "Recent advances in GaN transistors for future emerging applications," *Physica Status Solidi A*, vol. 206, no. 6, pp. 1221-1227, Jan. 2009.

[3] Y. Uemoto, et al., "GaN monolithic inverter IC using normally-off gate injection transistors with planar isolation on Si substrate" *IEDM Dig. Tech. Papers*, pp. 1-4, Dec. 2009.

[4] S. Nagai et al., "A DC-Isolated gate drive IC with Drive-by-Microwave technology for power switching devices," *ISSCC Dig. Tech. Papers*, pp. 404-405, Feb. 2012.

[5] I. Awai and A. K. Saha, "Open ring resonators applicable to wide-band BPF," *Asia-Pacific Microwave Conference,* pp. 167-172, Dec. 2006.

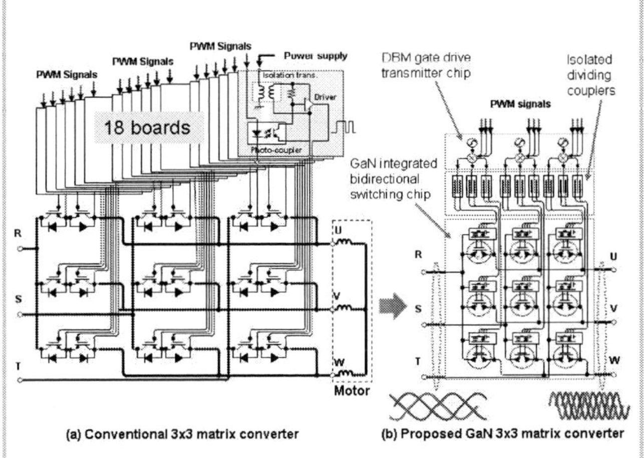

Figure 30.5.1: System blocks of (a) conventional matrix converter and (b) proposed GaN 3x3 matrix converter with Drive-by-Microwave technology.

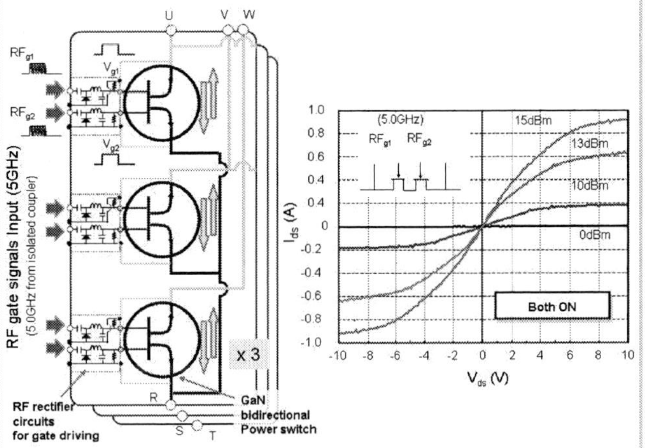

Figure 30.5.2: Proposed 3x3 integrated bidirectional switching chip and its characteristics.

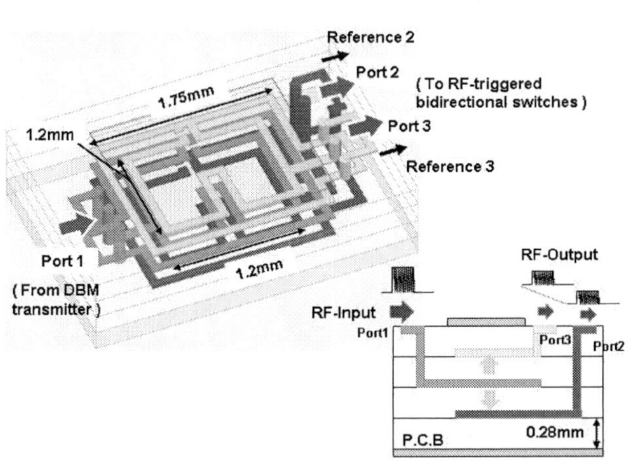

Figure 30.5.3: Proposed 5.0GHz isolated dividing coupler based on electromagnetic resonant coupling.

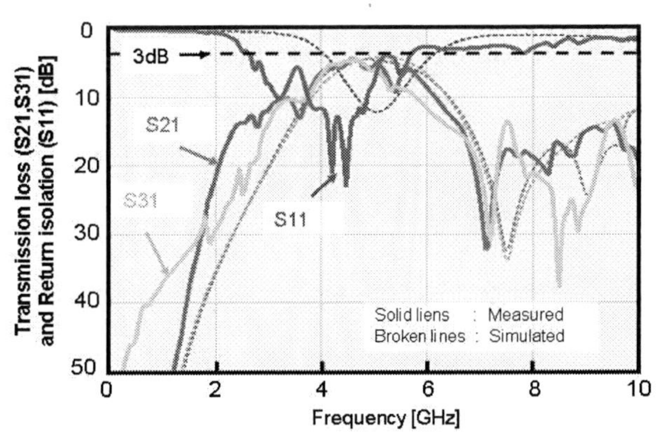

Figure 30.5.4: Measured and simulated small signal S-parameter of the fabricated isolated dividing coupler.

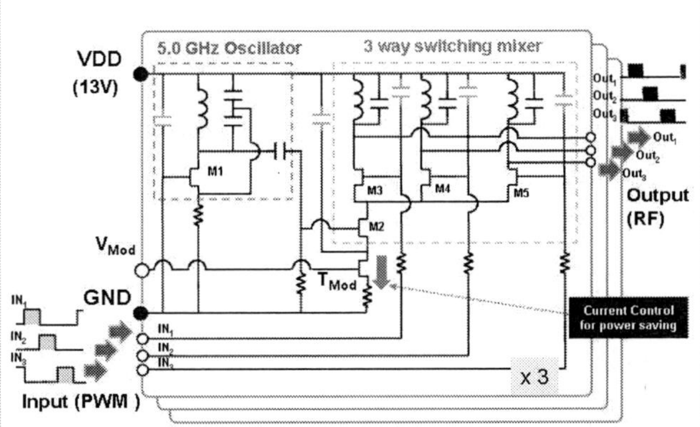

Figure 30.5.5: Circuit of the DBM gate drive transmitter chip with 9 output signals.

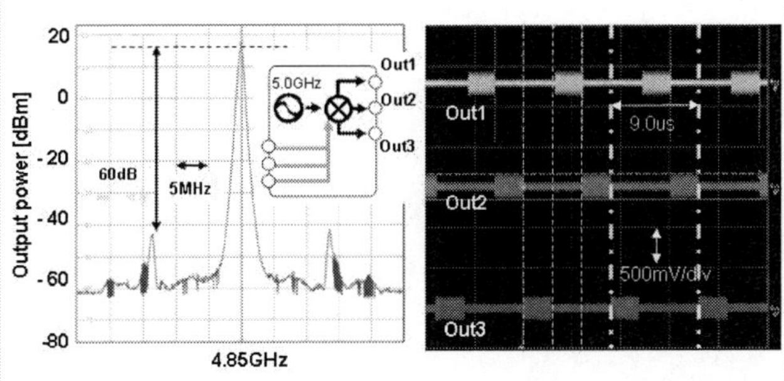

Figure 30.5.6: Output spectrum and waveform of the fabricated DBM gate drive transmitter chip with 9 output signals.

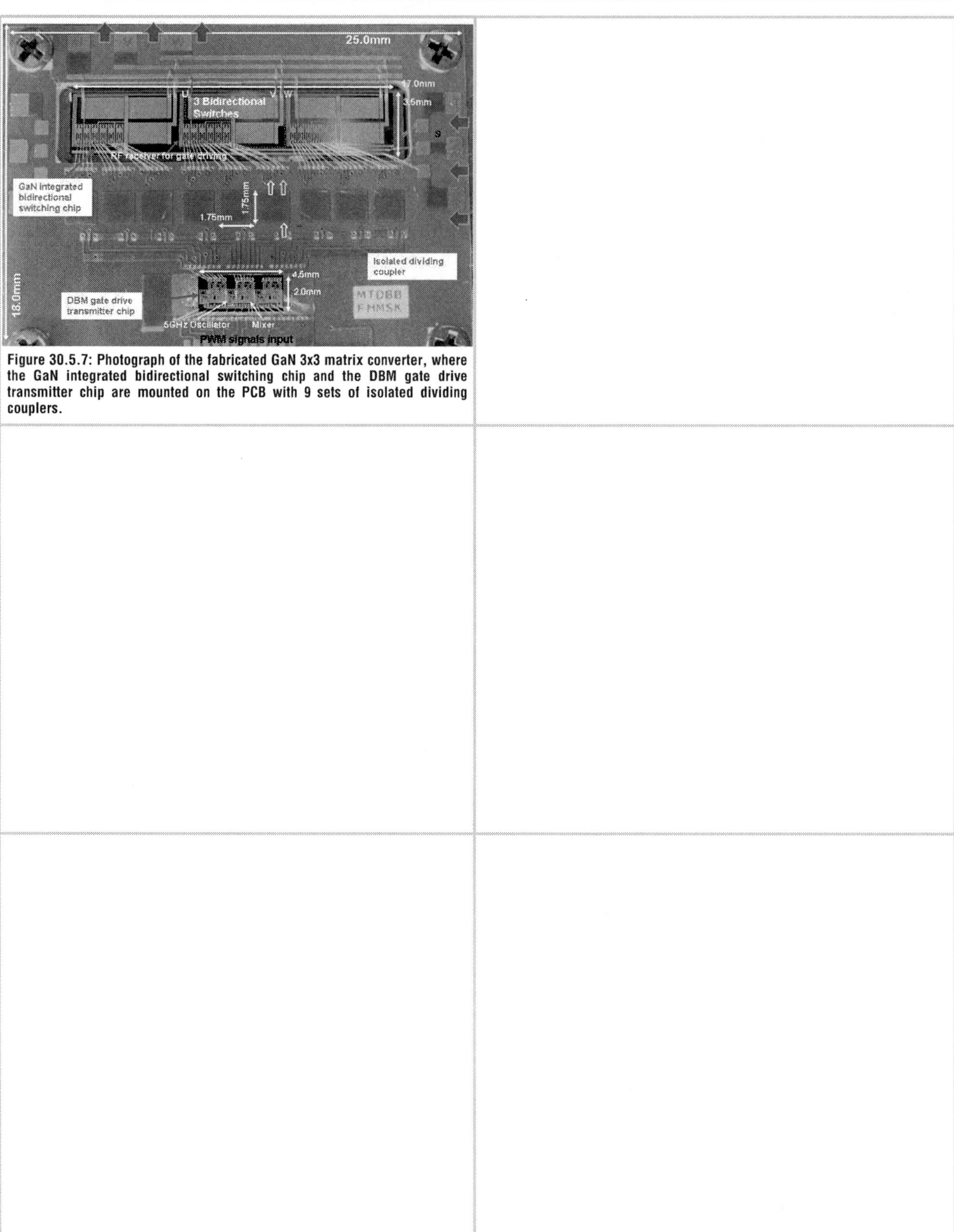

Figure 30.5.7: Photograph of the fabricated GaN 3x3 matrix converter, where the GaN integrated bidirectional switching chip and the DBM gate drive transmitter chip are mounted on the PCB with 9 sets of isolated dividing couplers.

30.6 An Electromagnetic Clip Connector for In-Vehicle LAN to Reduce Wire Harness Weight by 30%

Atsutake Kosuge, Shu Ishizuka, Lechang Liu, Akira Okada, Masao Taguchi, Hiroki Ishikuro, Tadahiro Kuroda

Keio University, Yokohama, Japan

Heavier wire harnesses decrease the fuel efficiency of vehicles. The number of wires has increased sharply with presently over one hundred electronic control units (ECUs). Connectors also add to the weight (Fig. 30.6.1(a)). The connectors used in vehicles occupy a considerable space because heavy protection is required against transmission interruption caused by vibration. In addition, a significant number of wires are used together in a junction box to increase total wire length (e.g. 55*L* when 10 ECUs are connected) and thus weight increases accordingly.

To reduce the weight of the wire harness, an electromagnetic clip connector (EM-Clip) is proposed, which uses electromagnetic coupling to make clip-like connections (Fig. 30.6.1(b)). Compared to wireless networking systems, there could be no security concerns in the non-contacting part. The EM-Clip has a cylindrical electrode to couple with the near electromagnetic field produced by the conductor of the twisted-pair cable. Even if the distance between the electrode and the cable varies due to the vibration, communication is unaffected and there is no issue with interruption. Accordingly, the protection can be simplified by reducing the weight and the volume by 50% and 65% respectively, compared to the D-sub connectors widely used in control area network (CAN) applications. The EM-Clip also makes it possible to create a signal branch anywhere without stripping the wire covering. By connecting ECUs to the LAN by the shortest distance (e.g. 9*L*) without a junction box, the weight of the signal wires can be reduced by 83% when 10 ECUs are connected. For a typical wire harness weight of which 35% is the signal lines and 3% is the signal connectors, use of the EM-Clip can reduce the overall harness weight by 30%. Another advantage of the EM-Clip is that it filters out the DC component. Therefore, using it prevents short circuit accidents affecting the LAN cable and also enables connection of ECUs that operate with different supply voltages without using isolators to reduce cost. To prevent collision on the bus, TDMA techniques can be used in the same way as Flexray which has been standardized for in-vehicle LAN.

If the conventional transmission line coupler (TLC) [1-2] is used for the EM-Clip, a signal propagates in only one direction at the connection point. When using the conventional TLC shown in Fig. 30.6.2(a), there is strong coupling between port3 and port1, but there is almost none between port3 and port2. Therefore it could not be used for an in-vehicle LAN where multiple ECUs need to transmit signals toward any directions on the LAN. To provide a TLC which can be used in an in-vehicle LAN, a bi-directional TLC (BD-TLC) which makes signal branches in both directions at the connection point is proposed (Fig. 30.6.2(b)). The signals are coupled at two places, positive polarity (A or B) and negative polarity (A' or B'), and the signal propagates in both directions at the connection point. As a result, the input signal from port3 is propagated as a -10dB signal in the direction of port1 and also as a -12dB signal in the direction of port2, as shown in Fig. 30.6.2(b). Although the polarity of the signal propagating in the direction of port2 is reversed, the polarity can be corrected by using the pilot signals before sending the data. Also, reflection is slight because the impedance can be adjusted, and communication is not impeded due to multiple reflections even when a multi-drop bus is used.

There are two issues to solve with the EM-Clip. One is that the signal is attenuated as a result of passing through multiple couplers and noise immunity is thus decreased; the other is that errors propagate in multiple bits, making it difficult to use forward error correction (FEC). If multiple EM-Clips are used, the signal attenuates exponentially and becomes susceptible to noise. Furthermore, if a pulse is generated only when there is a transition in digital data, received errors can propagate in multiple consecutive bits. Because vehicle applications require real-time operation, a resend request cannot be issued.

To overcome such concerns, two countermeasures are proposed. One is FEC by Manchester encoding with *N*-x oversampling on the transmitter side with majority voting on the receiver side (Fig. 30.6.3). Because alternative positive and negative pulses are received for each bit due to Manchester encoding, error propagation can be limited within two bits. Therefore, FEC techniques can be applied. If the Manchester encoding is done with oversampling at a frequency of *N*-times the data rate, one bit of the data is represented as *N* pulses. By passing the signal through an *N*-bit majority voting filter at the receiver, errors less than $(N-1)/2$ bits are corrected. Accordingly, if $N \geq 5$, the remaining two-bit errors are corrected by the majority voting filter. In this work, *N* is 5.

The other countermeasure is to improve noise immunity by extracting the clock from the received Manchester encoded data for synchronous receiving. For that purpose, a digital clock recovery circuit that is robust even where there are noise and variations in manufacturing is proposed (Fig. 30.6.4). The procedure for the clock extracting is as follows. First, the transmitter sends the preamble data used for clock recovery. The preamble data is a signal that consists of consecutive identical bits (e.g.'11...1' or '00...0'). When Manchester encoding is applied to that signal, it becomes a clock signal. The received preamble signal and the inverted clock signal (*Template*) generated at the receiver are input to an OR gate correlator. The correlation factor (ΣQ_i) is calculated by oversampling and summing the correlator output for each half of the period while the *Template* is low. The peak value of the correlation factor is detected by shifting the *Template* and stored. This cycle is repeated several times to reduce the influence of the noise. The highest value is selected from the stored peak values and the clock timing is locked. Using an OR gate in the correlator ensures correct calculation of the correlation factor, even if there is variation in the delay cells. It is then possible to simplify the circuit with delay cells instead of using complex phase control circuits.

Manchester encoding with *N*-x oversampling improves the EMI characteristics as well as the electromagnetic susceptibility (EMS) characteristics. In the standard specifications for in-vehicle systems, noise radiation is to be strictly regulated at frequencies below 1GHz, the region used for television and radio. By encoding with the clock signal at *N*-times the data rate, the radiation spectrum shifts to *N*-times higher frequency region, so unwanted emissions below 1GHz can be suppressed. In this work, the value of five times the clock signal frequency is set to 1.4GHz to avoid the 1.23 and 1.58GHz bands used by the GPS. The communication speed is then 280Mb/s. The coupler length is set to 2cm, which corresponds to the center frequency of 1.4GHz.

The test chip was fabricated in 65nm CMOS technology and 280Mb/s communication tests were conducted in which 10 ECUs were connected by a 10m AWG24 unshielded twisted-pair cable. The 2^7-1 PRBS signal was used for the evaluation. Photographs of the transceiver chip and the prototype EM-Clip are shown in Fig. 30.6.7. The experimental set-up and results are shown in Fig. 30.6.5. According to the bulk current injection (BCI) method specified by the ISO Test Standard [3], 30dBm noise was injected from a BCI probe in the range from 1 to 400MHz. A snapshot of the received waveform when 100MHz noise was injected is shown in the figure. The graph shows BER characteristics for various numbers of EM-Clip connections. Error-free communication ($BER<10^{-11}$) with 10 EM-Clips connected was confirmed. Both the measured radiation spectrum of the NRZ signal and the Manchester encoded signal with N=5 according to the CISPR Test Standard [4] are presented in Fig. 30.6.6. The radiation spectrum is shifted from 280MHz to a center around 1.4GHz. Emissions are sufficiently suppressed in the GPS signal band.

Acknowledgements:
This work was supported by CREST/JST.

References:

[1] W. Mizuhara, *et al.*, "A 0.15mm-Thick Non-Contact Connector for MIPI Using Vertical Directional Coupler," *ISSCC Dig. Tech. Papers*, pp. 200-201, Feb. 2013.
[2] W. –J. Yun, *et al.*, "A 7Gb/s/Link Non-Contact Memory Module for Multi-Drop Bus System Using Energy-Equipartitioned Coupled Transmission Line," *ISSCC Dig. Tech. Papers*, pp. 52-53, Feb. 2012.
[3] ISO 11452-4, "Road vehicles - Component test methods for electrical disturbances from narrowband radiated electromagnetic energy - Part 4: Harness excitation methods," Nov. 2011.
[4] CISPR 25 ed3.0, "Vehicles, boats and internal combustion engines - Radio disturbance characteristics - Limits and methods of measurement for the protection of on-board receivers," Jan. 2009.

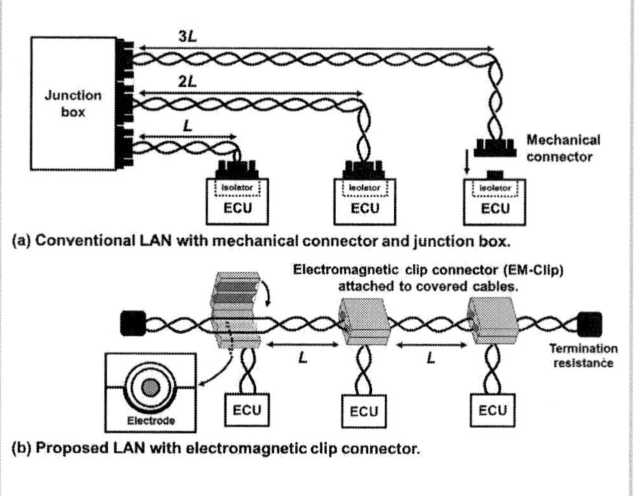

Figure 30.6.1: Electromagnetic clip connector for in-vehicle LAN.

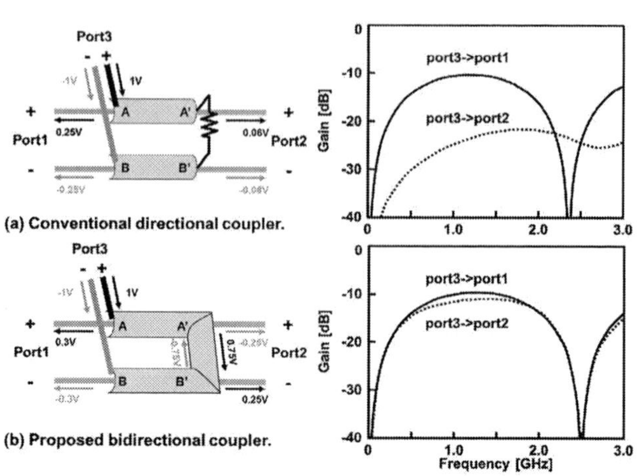

Figure 30.6.2: Bi-directional transmission line coupler (BD-TLC).

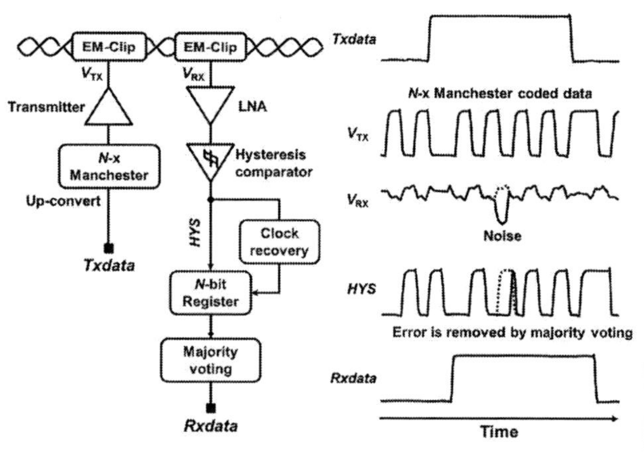

Figure 30.6.3: Forward error correction with N-x Manchester encoding and N-bit majority voting.

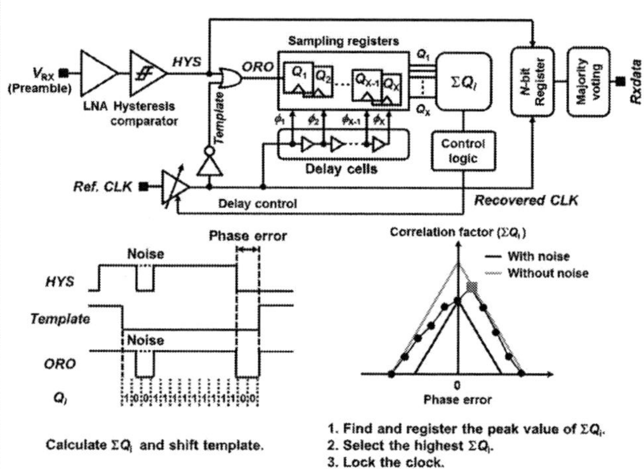

Figure 30.6.4: Clock recovery using correlator with OR gate.

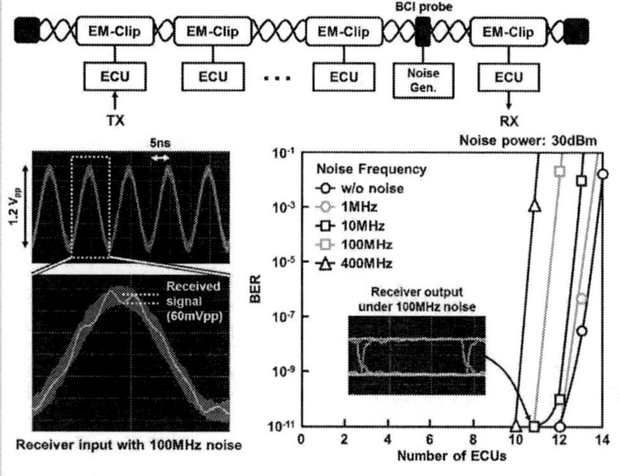

Figure 30.6.5: Measured BER dependence on noise injection.

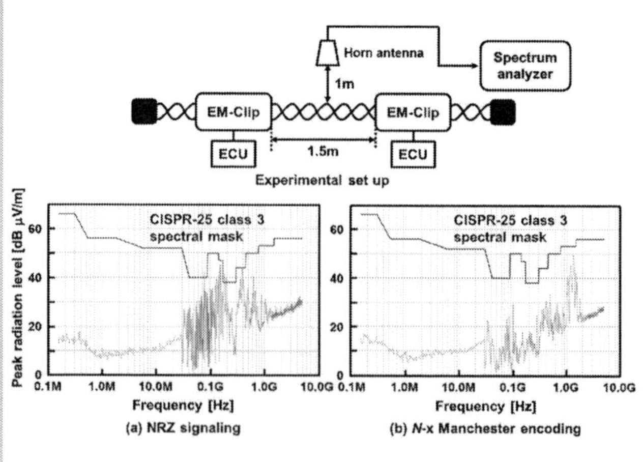

Figure 30.6.6: Measured spectrum and regulation.

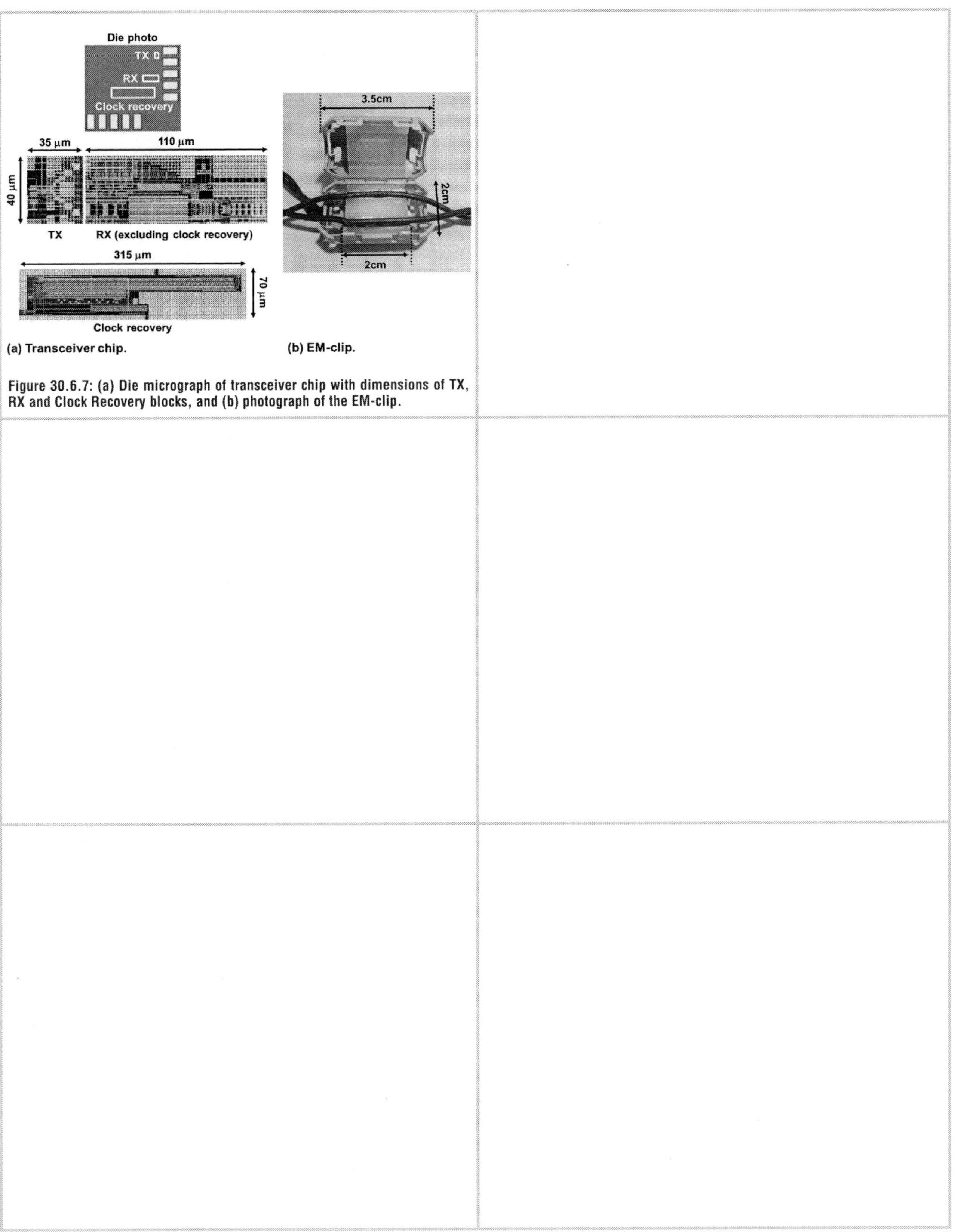

(a) Transceiver chip.

(b) EM-clip.

Figure 30.6.7: (a) Die micrograph of transceiver chip with dimensions of TX, RX and Clock Recovery blocks, and (b) photograph of the EM-clip.

30.7 A 60Mb/s Wideband BCC Transceiver with 150pJ/b RX and 31pJ/b TX for Emerging Wearable Applications

Junghyup Lee[1], Vishal Vinayak Kulkarni[1], Chee Keong Ho[1], Jia Hao Cheong[1], Peng Li[1], Jun Zhou[1], Wei Da Toh[1], Xin Zhang[1], Yuan Gao[1], Kuang Wei Cheng[2], Xin Liu[1], Minkyu Je[1]

[1]Institute of Microelectronics, Singapore, Singapore,
[2]National Cheng Kung University, Tainan, Taiwan

Wearable technology is opening the door to future wellness and mobile experience. Following the first generation wearable devices in the form of headsets, shoes and fitness monitors, second generation devices such as smart glasses and watches are making an entrance to the market with a great potential to eventually replace the current mobile device platform eventually (Fig. 30.7.1). Wearable devices can be carried by users in a most natural way and provide all-round connectivity 24-7 without the hassle of stopping all other activities, which enables a totally different mobile experience. For wearable devices, body channel communication (BCC) is an excellent alternative of conventional wireless communication through the air, to obviate the need of high-power transceivers and bulky antennas. However, present BCC transceivers [1]-[5] that mainly target biomedical and sensing applications offer rather limited data rates up to 10Mb/s, which is insufficient in transferring multimedia data for emerging wearable smart devices and content-rich information for high-end medical devices (e.g. multi-channel neural recording microsystems). In this paper, a highly energy-efficient and robust wideband BCC transceiver is presented, which achieves a maximum data rate of 60Mb/s by employing 1) a high input impedance and an equalizer at the RX front-end, 2) transient-detection RX architecture using differentiator-integrator combination coupled with injection-locking-based clock recovery, and 3) 3-level direct digital Walsh-coded signaling at the TX.

In order to achieve high-data-rate BCC with low power consumption, it is important to secure wideband body channel characteristics with low path loss. The channel characteristics can be optimized by selecting the right value for the RX input impedance R_{RX} as illustrated in Fig. 30.7.1. For two different paths P_T and P_A that respectively present the paths passing through the torso and formed outside the torso, the path losses are measured with varying R_{RX}. At high frequencies, both P_A and P_T show clear band-pass characteristics. Furthermore, it is observed that the path loss in the frequency band below 60MHz is significantly influenced by R_{RX}, creating differences up to 15dB. Since the larger R_{RX} results in lower path loss and wider useful bandwidth, a high input impedance of 10kΩ is used in our RX design, unlike the conventional designs having a lower input impedance [1], [2], [4], [5].

The proposed wideband BCC TX and RX are shown in Fig. 30.7.2 and Fig. 30.7.3, respectively. The transceiver architecture is based on direct digital data transmission to achieve high data rate and high energy efficiency at the same time, and employs Walsh-coded signaling that adds robustness and channel selectivity. Further, to compensate for the reduction in data rate caused by spread coding, the transceiver exploits 3-level signaling. To overcome strong interferers in noisy environment, the transceiver can also operate in a robust mode using 2-level Walsh coding with a band-stop filter (BSF) added to the front-end of the RX for blocking unwanted FM radio signals.

In the digital TX shown in Fig. 30.7.2, the data is spread with 16b Walsh codes prior to transmission through an inverter-based driver. Due to the limited bandwidth of the body channel, the TX can only use Walsh code 7 to 14 out of a set of 16 Walsh codes, to effectively confine a dominant portion of the signal energy within the optimum frequency band. The limited set of available Walsh-code sequences limits the maximum data rate to 30Mb/s. To enhance the data rate, the set of available Walsh codes is doubled by using the inverse of code 7 to 14. Due to the orthogonality of Walsh codes and the synchronization achieved with the aid of the preamble, the inverted Walsh-code sequences can be distinguished in the RX. The 128b preamble used for synchronization is generated by spreading the 32b M-sequence code with a 4b Walsh code. By using the inverted Walsh codes, the data rate is increased to 40Mb/s with 2-level Walsh-code signaling. To further enhance the data rate, the set of 16 Walsh codes is divided into 2 groups to generate 3-level Walsh-code signals, which can be transmitted by a 3-level driver [1]. With 3-level signaling, the data rate reaches 60Mb/s without increasing the clock frequency or the number of Walsh codes.

The RX shown in Fig. 30.7.3 consists of an analog front-end (AFE), a level detector, a 3-level Walsh-code demodulator, and a clock recovery circuit (CRC). In the AFE, there are a pre-amplifier, an equalizer, a differentiator and an integrator with reset. The 2nd-order equalizer is exploited to compensate for high-frequency path loss so that a wider channel band can be obtained. The differentiator-integrator structure enables the detection of signal transients. Since the body channel has band-pass characteristics, the signal received after passing through the body (C, D) has an undefined DC level, although the transmitted signal (B) has a well-defined DC level. This loss in DC reference level makes the received signal undetectable by the conventional amplitude-detection approach. The differentiator solves this problem by extracting data transitions in the received signal such that pulses with the amplitude proportional to the transition step size are generated (E). The differentiator output is then integrated, which nullifies the effect of interference and noise that degrade the input signal (F). The integrator uses a 25%-duty clock signal CLK_{INT} from the CRC after reset. The level detector generates 2-bit digital words LD by comparing the integrator output voltage with four threshold voltages V_{TH1} to V_{TH4} at every rising edge of CLK_{INT}. Depending on in which of five regions defined by 4 thresholds the integrator output falls, LD_n is set as -2, LD_{n-1}-1, LD_{n-1}, LD_{n-1}+1, or 2, where n presents the nth clock period. In the Walsh-code demodulator, the raw digital data D_{RX} is finally restored from LD by dispreading with two Walsh codes. The CRC provides the timing signal to the integrator and level detector by recovering the clock from the incoming data. By making use of pulses generated from the differentiator for injection locking, the clock can be recovered with consuming low power and low area, eliminating the need of a power-consuming PLL and a bulky crystal oscillator.

The proposed BCC transceiver has been fabricated in a 65nm CMOS process and occupies a core area of 0.24x0.3 mm^2 for TX and 1.5x0.7 mm^2 for RX as shown in Fig. 30.7.7. When operating at the maximum data rate of 60Mb/s, the TX and RX consume 1.85mW and 9.02mW, respectively. The measured RX sensitivity is -58dBm at BER lower than 10^{-5} and SIR is -20dB at 10^{-3} BER when 100kHz BW FM interference exists at 90MHz. Time-domain waveforms measured at critical nodes in the TX and RX are shown in Fig. 30.7.4. The waveforms are clearly improved by using the equalizer. Figure 30.7.5(a) shows the measured spectrum and phase noise characteristics of the injection-locked oscillator in the CRC. The recovered 160MHz clock has 170ps RMS jitter. Figure 30.7.5(b) shows the test setup for image data communication where the TX and RX are placed respectively on the left arm and right arm, together with the Lena image successfully transferred in the testing. The measured transceiver performance is summarized and compared with other state-of-the-art designs in Fig. 30.7.6. The proposed wideband BCC TX and RX respectively achieve 31pJ/b and 150pJ/b energy efficiencies, while offering unprecedented 60Mb/s data rate.

Acknowledgements:
This work is funded by SERC (Science and Engineering Research Council), A*STAR (Agency for Science, Technology and Research), Singapore, under the grant number 102 171 0163.

References:
[1] S.-J. Song, et al., "A 0.9 V 2.6mW body-coupled scalable PHY transceiver for body sensor applications," *ISSCC Dig. Tech. Papers*, pp. 366–367, Feb. 2007.
[2] N. Cho, et al., "A 60 kb/s–10 Mb/s adaptive frequency hopping transceiver for interference-resilient body channel communication," *IEEE J. Solid-State Circuits*, vol. 44, no. 3, pp. 708–717, Mar. 2009.
[3] A. Fazzi, et al., "A 2.75 mW wideband correlation-based transceiver for body-coupled communication," *ISSCC Dig. Tech. Papers*, pp. 204–205, Feb. 2009.
[4] J. Bae, et al., "A 0.24-nJ/b wireless body-area-network transceiver with scalable double-FSK modulation," IEEE J. Solid-State Circuits, vol. 47, no. 1, pp. 310–322, Jan. 2012.
[5] H. Lee, et al., "A 5.5mW IEEE-802.15.6 wireless body-area-network standard transceiver for multichannel electro-acupuncture application," *ISSCC Dig. Tech. Papers*, pp. 452–453, Feb. 2013.

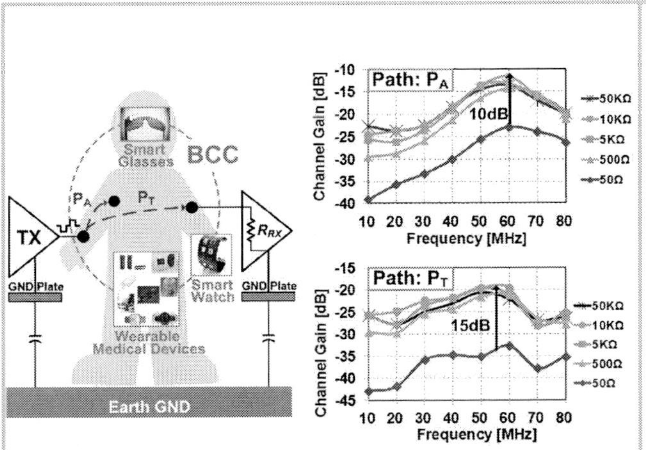

Figure 30.7.1: Emerging BCC applications and measured body channel characteristics.

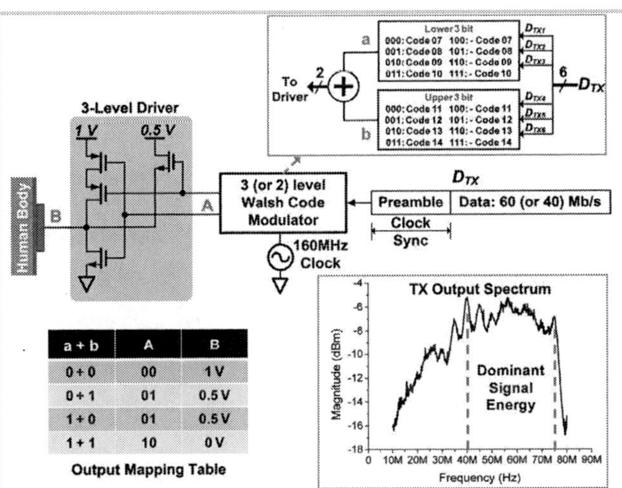

a + b	A	B
0 + 0	00	1 V
0 + 1	01	0.5 V
1 + 0	01	0.5 V
1 + 1	10	0 V

Figure 30.7.2: TX structure including the Walsh-code modulator and 3-level driver, and measured TX output spectrum.

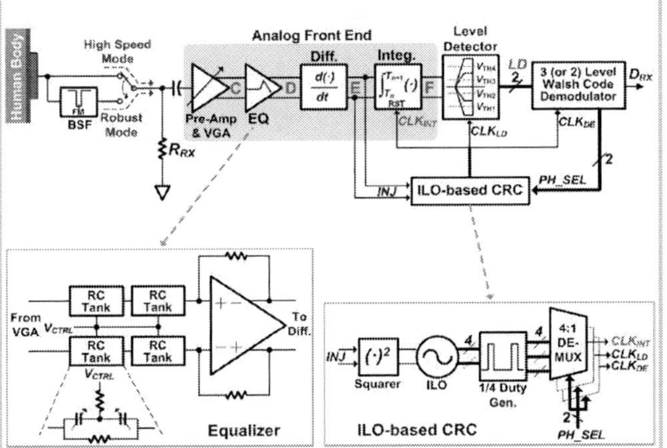

Figure 30.7.3: RX structure and schematics of the equalizer and CRC circuits.

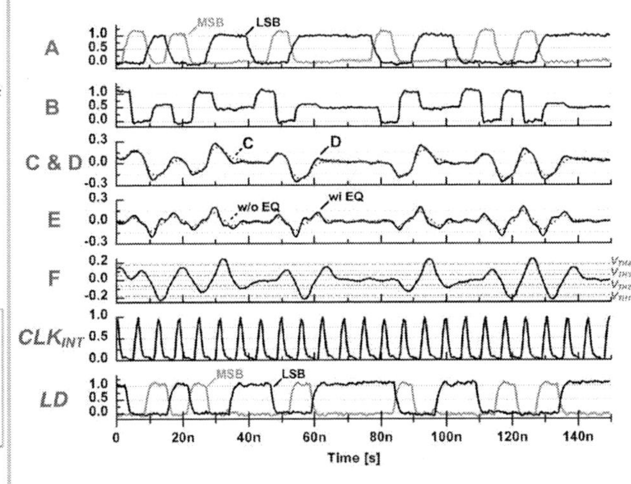

Figure 30.7.4: Measured time-domain waveforms.

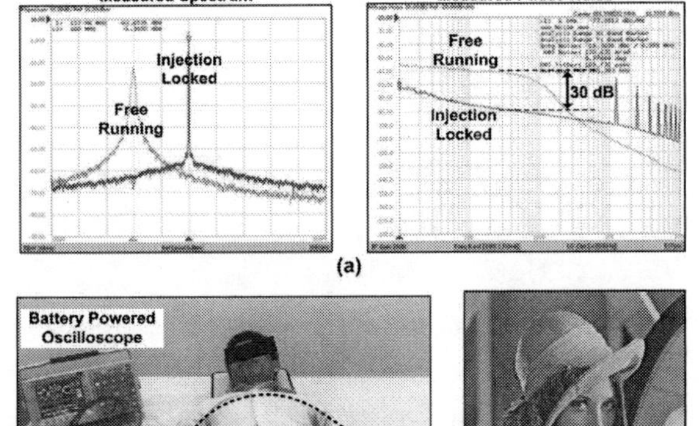

Figure 30.7.5: (a) Measured spectrum and phase noise characteristics of the injection-locked oscillator output. (b) Setup for the image data communication test, and transferred Lena image.

Parameters	ISSCC '07 [1]	JSSC '09 [2]	ISSCC '09 [3]	JSSC '12 [4]	This Work
Technology [nm CMOS]	180	180	130	180	65
Supply Voltage	0.9V	1V	1.2V	1V	1.1V
Modulation	3-Level PPM	AFH FSK	Correlation Direct Digital	Double FSK	3-Level Walsh Coding
Input Impedance	50Ω	< 100Ω	Capacitive load	100~600Ω	10KΩ
Maximum Data Rate	10Mb/s	10Mb/s	8.5Mb/s	10Mb/s	60Mb/s
Sensitivity	-30dBm @ < 10^{-4} BER	-65dBm @ 10^{-5} BER	-60dBm @ 10^{-3} BER	-62dBm @ 10^{-5} BER	-58dBm @ < 10^{-5} BER
SIR	N.A.	-28dB @ 10^{-3} BER	N.A.	-20dB @ 10^{-3} BER	-20dB[1] @ 10^{-3} BER
Power Consumption	TX: 0.71mW RX: 1.89mW	TX: 2.4mW RX: 3.7mW	TX: 0.6mW RX: 2.15mW	TX: 2mW RX: 2.4mW	TX: 1.85mW RX: 9.02mW
Energy/bit	TX: 71pJ/b RX: 189pJ/b	TX: 240pJ/b RX: 370pJ/b	TX: 70pJ/b RX: 250pJ/b	TX: 200pJ/b RX: 240pJ/b	TX: 31pJ/b RX: 150pJ/b

[1] Robust mode operation.

Figure 30.7.6: Performance summary and comparison.

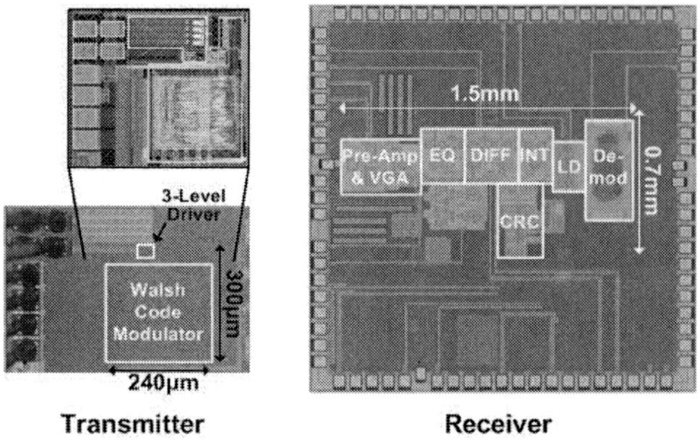

Figure 30.7.7: Die micrographs of the BCC TX and RX.

30.8 A 30GS/s Double-Switching Track-and-Hold Amplifier with 19dBm IIP3 in an InP BiCMOS Technology

Timothy D. Gathman[1,2], Kristian N. Madsen[2,3], James C. Li[4], Thomas C. Oh[4], James F. Buckwalter[2]

[1]Qualcomm, San Diego, CA,
[2]University of California, San Diego, La Jolla, CA,
[3]Peregrine Semiconductor, San Diego, CA,
[4]HRL Laboratories, Malibu, CA

High-speed track-and-hold amplifier (THA) circuits are critical for high-speed, high-resolution data converters, particularly in emerging 100Gb/s optical communication systems. High-speed CMOS analog-to-digital converters (ADCs) have been demonstrated to 56GS/s, but the linearity tends to rapidly degrade at high frequency [1]. The use of a high-speed THA can broaden the frequency response, improve the distortion performance, and mitigate some of the timing requirements for high-speed time-interleaved ADCs. Recent THA work has leveraged high-f_{max} Indium Phosphide (InP) HBTs to implement 50GS/s track-and-hold amplifiers [2]. Nonetheless, a two-chip solution consisting of an InP THA and time-interleaved CMOS ADCs is extremely difficult to package while retaining high-speed performance.

This work demonstrates a high-speed THA realized in an InP Bi-CMOS technology with wafer-scale device-level heterogeneous integration of 250nm, 300GHz f_T/f_{max} InP DHBTs fabricated by HRL Laboratories with IBM 90nm CMOS. SiGe BiCMOS technologies have become a successful platform for mixed-signal and RF/millimeter-wave circuits by allowing the integration of high-f_T HBTs alongside standard CMOS devices. However, InP HBTs operate at higher frequencies than SiGe HBTs because greater peak carrier saturation velocities are attainable in InP devices. The proposed concept for an InP BiCMOS chip is illustrated in Fig. 30.8.1 where an InP epitaxial layer is bonded to a standard 90nm CMOS wafer [3]. The resulting structure includes an InP HBT that sits above the CMOS BEOL and does not require any adjustments to the CMOS fabrication process. Heterogeneous interconnects are fabricated as an extension of the CMOS top metal to route connections between the CMOS FETs and InP HBTs. This allows an extremely high-speed InP HBT to be integrated with a bulk CMOS process, but requires some attention to the thermal environment and hence biasing conditions of the InP HBT. The peak f_T/f_{max} of the InP HBT occurs for a current density of 10mA/μm^2 [3].

Advancements in HBT f_T/f_{max} improve the bandwidth of switched emitter follower (SEF) based THAs, but don't necessarily improve the distortion performance of these switches. Previous work demonstrates architectures with sampling rates up to 40GS/s, with a focus primarily on open loop compensation techniques for resolution improvement [4,5]. A conventional high-resolution THA is depicted in Fig. 30.8.2 using a feedthrough attenuation buffer to improve the isolation between the input signal and the output voltage on the hold capacitor. This feedthrough attenuation buffer is typically a replica of the input buffer with cross-coupled outputs. This buffer provides additional cancellation of the input signal during the hold mode and improves the isolation, reducing feedthrough distortion on the sampled voltage. Unfortunately, as the differential input signal is cancelled while the SEF switches into hold mode, there is a differential voltage step seen by the SEF. The differential voltage step creates unequal voltage swings seen by the SEF during the track to hold transition. Differing voltage steps affect both the switching instant, creating an instantaneous timing skew between the voltages sampled by the SEF, and also the linearity of the sampling process, as unequal steps create signal-dependent pedestal error and ultimately nonlinear distortion [6].

A more desirable circuit approach improves isolation without creating additional pedestal distortion during the switching event. To meet both of these requirements, the double-switching architecture in Fig. 30.8.2 is introduced. During the track-to-hold transition, the input buffer is switched away from the input of the SEF to a dummy resistive load, while at the same time the sampled voltage is replicated across the input terminals of the SEF by an auxiliary feedback buffer. This double-switching action results in an equal voltage transition on the two differential input nodes of the SEF, which avoids the pedestal errors and nonlinear distortion of the conventional architecture [6]. An added benefit of double-switching is that the input signal is hard-switched away from the SEF during the hold phase. This provides higher isolation than the feedthrough attenuation technique, which is limited by the matching between the main and auxiliary buffers. The high-speed devices available in InP BiCMOS technology can be leveraged in the design of distortion-compensation THA architectures, improving the sampling resolution even at state-of-the-art sampling rates.

Fig. 30.8.3 represents the circuit level schematic of the double-switching THA. The input buffer consists of a switched transconductor to pass the load current through the load resistors (R_L) during the track phase and dummy resistors (R_{Dummy}) during the hold phase. The feedback buffer is a replica of the input buffer, but is switched with opposite phase as the input buffer. The SEF consists of a cascoded switching stage to allow for faster switching. A feedthrough cancellation capacitor is added from the output of each SEF to its complementary input. This capacitance (C_{ff}) is created by a quartet of series parallel-connected HBTs, which replicate the base-emitter capacitance of the SEF.

The InP BiCMOS THA is characterized via on-wafer probing and operates to 30GS/s. Harmonic distortion is measured for a 1GHz, 600mVpp-differential input tone sampled at 30GS/s. Fig. 30.8.4 shows the differential measurement of the 3rd harmonic distortion at -60.5dB relative to the fundamental, and results in a THD of less than -59dB. The result of a beat frequency test is also shown in Fig. 30.8.4 for an input tone of 31GHz. The fundamental tone, second, and third harmonics are shown aliasing down to 1, 2, and 3GHz, respectively. A two-tone measurement is shown in Fig. 30.8.5 with tones at 1 and 1.1GHz was performed on 4 die to capture the robustness of the process. A worst case OIP3 and IIP3 of 15.6dBm and 19dBm are measured, respectively. The linearity of the proposed THA exceeds other compensated THAs clocked at rates of 20GS/s or greater at comparable power consumption. This work also demonstrates an increase in sampling rate of more than 18X over the first implementation in CMOS with little degradation in linearity.

The power consumption of the system excluding the output buffer is 420mW and is broken down into its constituent contributions in Fig. 30.8.6 alongside a comparison of high-speed compensated and uncompensated THAs in CMOS, SiGe, and InP. The micrograph of the 1.1x0.7mm^2 IC is shown in Fig. 30.8.7.

Acknowledgments:
This work was supported through a subcontract from HRL Laboratories under the DARPA / AFRL COSMOS program. We thank Daniel Zehnder and Thomas Oh for help with wafer screening the test results.

References:

[1] I. Dedic, "56Gs/s ADC: Enabling 100GbE", *2010 Conference on Optical Fiber Communication (OFC)*, pp. 1-3, Feb. 2010.
[2] S. Deneshgar, *et al.*, "A High IIP3, 50 GSamples/s Track and Hold Amplifier in 0.25 μm InP HBT Technology", *Compound Semiconductor Integrated Circuits Sympsium*, pp. 1-4, Oct. 2012.
[3] Y. Royter, *et al.*, "Dense Heterogeneous Integration for InP Bi-CMOS Technology," *IEEE Conference on Indium Phosphide & Related Materials*, pp. 105-110, Aug. 2009.
[4] S. Yamanaka, *et al.*, "A 20-Gs/s Track-and-Hold Amplifier in InP HBT Technology," *IEEE Trans. on Microwave Theory and Techniques*, vol. 58, no. 9, pp. 2334-2339, Sept. 2010.
[5] X. Li, *et al.*, "A 40 GS/s SiGe track-and-hold amplifier," *IEEE Bipolar/BiCMOS Circuits Technology Meeting (BCTM)*, pp. 1–4, Oct. 2008.
[6] H. Dinc and P. E. Allen, "A 1.2 GSample/s Double-Switching CMOS THA With 62 dB THD," *IEEE Journal of Solid-State Circuits*, vol. 44, no. 3, pp. 848-861, March 2009.

978-1-4799-0917-9/14 $31.00 © 2014 IEEE

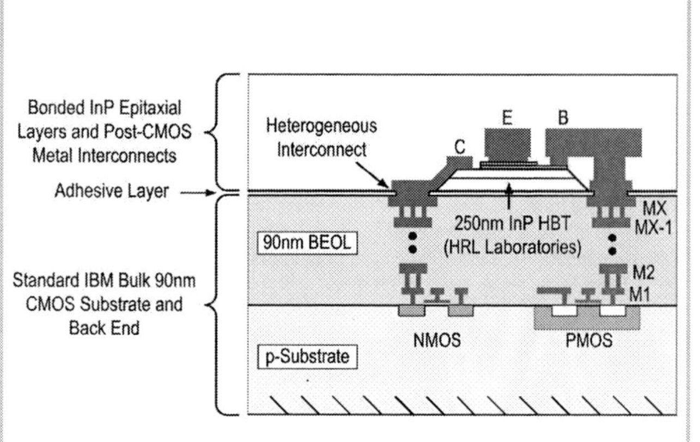

Figure 30.8.1: Cross section of the heterogeneously integrated BiCMOS process after [3].

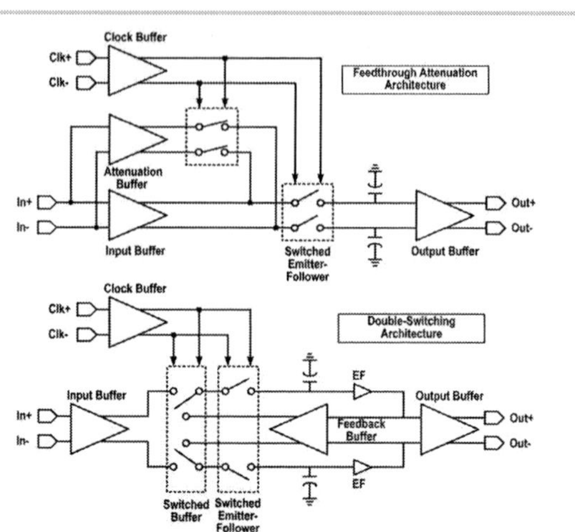

Figure 30.8.2: Conventional feedthrough attenuation and proposed double-switching THA architectures.

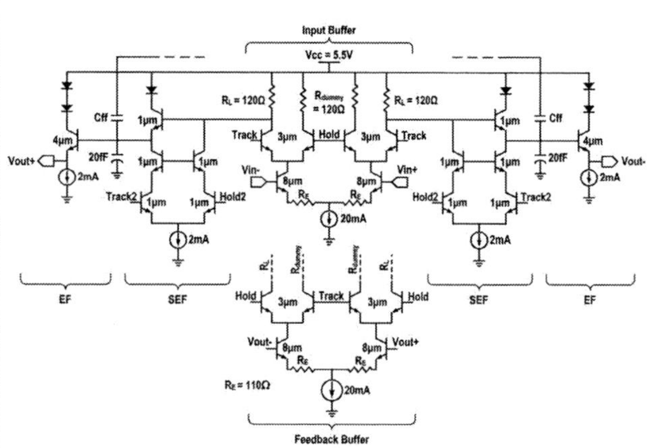

Figure 30.8.3: Schematic of the double-switching track-and-hold amplifier.

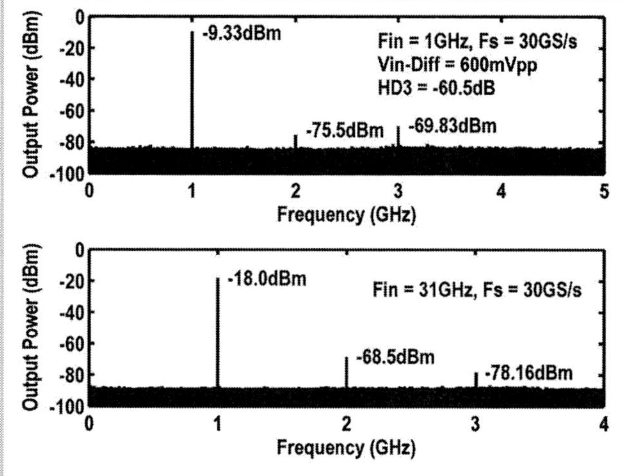

Figure 30.8.4: Differential output spectra with 1GHz and 31GHz inputs at 30GS/s.

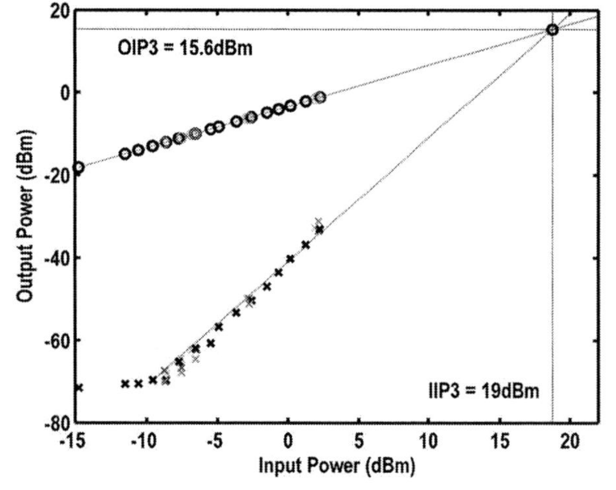

Figure 30.8.5: IIP3 curves for 4 die with 1 and 1.1GHz inputs at 30GS/s.

Stage	Power Consumption (mW)
Input Buffer	110
Differential SEF	22
Emitter Followers	22
Feedback Buffer	110
Clock Driver	154
Total	420

Clock Buffer
Input Buffer
SEF
EF
Feedback Buffer

Reference	Li [5] (BCTM '08)	Dinc [6] (JSSC'09)	Yamanaka [4] (TMTT '10)	Deneshgar [2] (CSICS '12)	This Work
Process / fT	SiGe / 200 GHz	CMOS 180nm	InP / 175 GHz	InP / 400 GHz	InP / 300 GHz
Architecture	Feedthrough Attenuation	Double Switching	Feedthrough Attenuation	Uncompensated	Double Switching
Supply	5.5V	1.8V, 3.3V	-5.2V	-2.5V, -5.0V	5.5V
Fsample	40GS/s	1.6GS/s	20GS/s	50GS/s	30GS/s
Input Amplitude	1.0Vpp	400mVpp	500mVpp	-	600mVpp
THD Fin	-32.4dB 10GHz	-60dB ≤ 0.8GHz	-45dB 0.9GHz	-	< -59dB 1GHz
IIP3 / OIP3 Fin	-	-	-	17.2dBm/6.2dBm 18GHz	19dBm/15.6dBm 1GHz
Power Consumption	560mW	258mW*	735mW	1025mW*	420mW*
Area	1.8x1.0mm²	1.47x1.36mm²	2.0x2.0mm²	0.675x1.075mm²	1.1x0.7mm²

* Excludes the power consumption of the output buffer.

Figure 30.8.6: THA power consumption and comparison table.

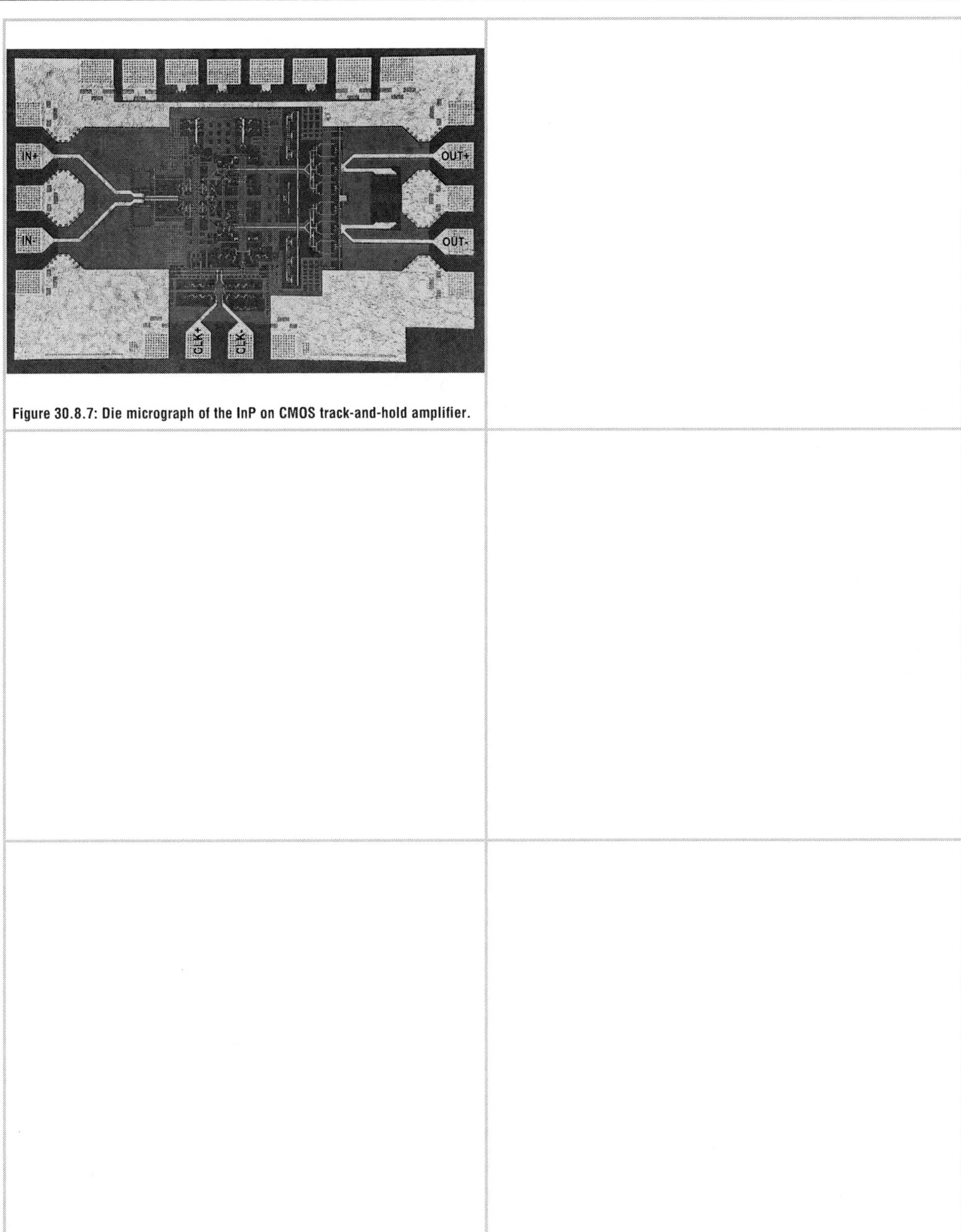

Figure 30.8.7: Die micrograph of the InP on CMOS track-and-hold amplifier.

30.9 Normally-Off Computing with Crystalline InGaZnO-based FPGA

Takeshi Aoki[1], Yuki Okamoto[1], Takashi Nakagawa[1], Masataka Ikeda[1], Munehiro Kozuma[1], Takeshi Osada[1], Yoshiyuki Kurokawa[1], Takayuki Ikeda[1], Naoto Yamade[1], Yutaka Okazaki[1], Hidekazu Miyairi[1], Masahiro Fujita[2], Jun Koyama[1], Shunpei Yamazaki[1]

[1]Semiconductor Energy Laboratory, Kanagawa, Japan,
[2]University of Tokyo, Tokyo, Japan

An FPGA employing c-axis aligned crystal In-Ga-Zn oxide (CAAC-IGZO) FET [1] based configuration memories (CMs) is known to need no reconfiguration thanks to nonvolatile CMs, shows high operation speed due to boosting effect of pass gates used in routing switches (RS) [2], and easily realizes fine-grained multi-context (FG-MC) architecture [2] because CMs which need very low power to keep the contents can be constructed with a small number of transistors. It would be very difficult to realize all of these features in FPGAs using MRAM [3] or RRAM [4]. These features are very unique to the CAAC-IGZO FPGA.

An FG-MC CAAC-IGZO FPGA has recently been reported and its effectiveness of power gating (PG) in each programmable logic element (PLE) associated with context switching (fine-grained PG) [5] has been shown. This paper presents an extension of such an FG-MC CAAC-IGZO FPGA by further employing nonvolatile registers (NV-Regs) in the PLEs in order to achieve fine-grained normally-off computing (FG-Noff computing). That is, the CAAC-IGZO FPGA presented in this paper can stop or resume tasks based on context switching and the power is supplied to the PLEs only where actual computation is performed in each task. The time to store/load with NV-Regs can be as short as the time for context switching of the FG-MC CAAC-IGZO FPGA. The 2-phase non-overlap clock scheme is effective in forming such CAAC-IGZO FPGAs without decreasing the advantages of fine-grained PG. A CAAC-IGZO FPGA has been fabricated by a hybrid process of a 1.0µm CAAC-IGZO FET/a 0.5µm CMOS FET. When data are stored to and loaded from the NV-Regs in the CAAC-IGZO FPGA driven at 2.5V, measured store/load time is 40ns/8ns and measured power consumed for a store/load is 1.6pJ/17.4pJ, respectively.

Figure 30.9.1 shows a typical sequence of FG-Noff computing presented in this paper. In the FG-Noff computing, tasks allocated to contexts are configured in accordance with switching of the contexts, and in each task, no power is supplied to PLEs unless actual computation is performed. To keep continuity of correct calculation between suspension and resumption of a task, register data in the idling PLEs keep their values even without supply of power. When reconfiguring the tasks, it is extremely important for the registers in the PLEs of the CAAC-IGZO FPGA to store/load their data to nonvolatile memory at high speed and low power consumption. Another important point is to supply stable clock signals to flip flops even when the distribution of power consumption in the die may greatly change depending on tasks.

The fabricated CAAC-IGZO FPGA employs a NV-Reg shown in Fig. 30.9.2. The NV-Reg has a volatile block of CMOS FETs and a nonvolatile block of CMOS FETs and CAAC-IGZO FETs. In normal operation, the volatile block can be used as a flip-flop utilizing 2-phase non-overlap clocks which are described later. To store/load data in registers to the nonvolatile block, control signals ϕ_S/ϕ_L are set "High" asynchronous with clock signals. Since data stored in nodes N1 and N1B of the nonvolatile block have always opposite logic values, this NV-Reg works as a pull-down circuit which supplies the GND potential to an inverter latch in the volatile block through M1 or M2 when loading data. Therefore, high-speed load operation can be expected.

Figure 30.9.3 shows the measurement results of timing for storing/loading data in the fabricated NV-Reg. In the NV-Reg, storage capacitance is 32.7fF, which realizes a one-year retention time. From Fig. 30.9.3, a store/load takes 128ns/24ns and the power consumed for data storing and loading is 1.6pJ/17.4pJ at a VDD of 2.5V, respectively. Storing data means charging capacitors C1 and C2. Here, overdriving the ϕ_S signal realizes an improvement of speed in writing "High" data to the capacitors. Figure 30.9.3 also shows the results on overdriving. Store/load time is at most 40ns/8ns with 0.5V overdrive, which is a significant reduction from 128ns/24ns above.

Figure 30.9.4 shows block diagrams of the entire FPGA and its PLE. There are 20 PLEs in the FPGA. By using the NV-Reg shown in Fig. 30.9.2, data in the register can be stored before the power supply stops, and it can be loaded after the power supply restarts. CM can be also used as a power switch (PSW) to control the power supply to the PLEs. The fabricated FPGA is based on a multi-context architecture (the number of contexts is 2). The CM can have two configurations of Context<1> and Context<2> corresponding to individual tasks. This enables the circuit configuration to be changed by switching a selection signal ϕ_{CTX} in accordance with a task schedule. Since the power supply to each PLE is specified by configuration data stored in the CM, PG in each PLE can be achieved without providing any complicated control circuit [5].

A fine-grained FPGA has wide flexibility in terms of circuit configuration. In such a FPGA, connections with and among the registers tend to be complex in order to realize flexible routing. As a result, it becomes highly difficult to realize such routing with fewer occurrences of data races. Since the CAAC-IGZO FPGA performs the fine-grained PG, load on a driven clock buffer is possibly changed depending on configurations. It is difficult to keep clock skews uniform through clock buffering with a clock tree structure. From these reasons, a 2-phase non-overlap clock structure is employed. Figure 30.9.5 shows the circuit configuration of a clock generator. 2-phase non-overlap clocks (ph1 and ph2) which can stop operation of flip-flops when storing and loading operations with NV-Regs are generated from control signals ϕ_S and ϕ_L, and storing/loading operations become asynchronous with clocks. Note that the clock generator can generate a clock signal of higher frequency when transistors have characteristics suitable for high-speed operation due to variation caused in manufacturing process by making the "High" period of a 2-phase clock shorter.

Figure 30.9.7 shows a die micrograph of a CAAC-IGZO FPGA fabricated by the hybrid process. The area overhead caused by using a NV-Reg increases the size of the register of the PLE by 40%, which corresponds to 0.6% increase in the PLE as a whole and 0.3% increase in the entire die. With this slight increase in area of the die, an FG-Noff function can be added to the FPGA and power consumption of the FPGA can be much lower.

Figure 30.9.6 shows the results of operation verification of the fabricated FPGA die. Operation verification has been performed under various conditions. The following is an example. The configuration in Context<1>/<2> corresponding to the 1st/2nd task is a shift circuit/a counter circuit. A table in Fig. 30.9.6 shows an event schedule. The shift circuit in Context<1> has 20 stages realized with 20 PLEs. Outputs from the 1st to 7th stages correspond to OUT<0> to OUT<6>. The counter circuit in Context<2> consists of 8 PLEs. In the counter circuit, the LSB is OUT<1>, and the MSB is OUT<6>. Note that the output of PLE<0> is OUT<0>, and PLE<0> performs PG in Context<2> as it is not used in the counter circuits in Context<2>. Here, by confirming continuity of the value OUT<0> when switching the contexts, correct operation has been verified. The timing diagram in Fig. 30.9.6 shows the observed output waveforms corresponding to Context<1> and Context<2>. As can be seen from the diagram, the values of OUT<0> to OUT<6> when storing data before the 1st context switching (including OUT<0> corresponding to the output of PLE<0> implementing PG in Context<2>) are respectively equal to the values of OUT<0> to OUT<6> when loading data after the 2nd context switching. The results prove that the proposed FG-Noff operation is working as intended.

References:

[1] S. Yamazaki, et al., "Research, Development, and Application of Crystalline Oxide Semiconductor," *Proc. SID*, pp. 183-186, Jun. 2012.

[2] Y. Okamoto, et al., "Novel Application of Crystalline Indium-Gallium-Zinc-Oxide Technology to LSI: Dynamically Reconfigurable Programmable Logic Device Based on Multi-Context Architecture," *ECS Transactions*, vol. 54, no. 1, pp. 141-149, Jun. 2013.

[3] D. Suzuki, et al., "Design of a Process-Variation-Aware Nonvolatile MTJ-Based Lookup-Table Circuit," *SSDM*, pp. 1146-1147, Sep. 2010.

[4] Y. Y. Liauw, et al., "Nonvolatile 3D-FPGA With Monolithically Stacked RRAM-Based Configuration Memory," *ISSCC Dig. Tech. Papers*, pp. 406-407, Feb. 2012.

[5] M. Kozuma, et al., "Crystalline In-Ga-Zn-O FET-based Configuration Memory for Multi-Context Field-Programmable Gate Array Realizing Fine-Grained Power Gating," *SSDM*, pp. 1096-1097, Sep. 2013.

978-1-4799-0917-9/14 $31.00 © 2014 IEEE

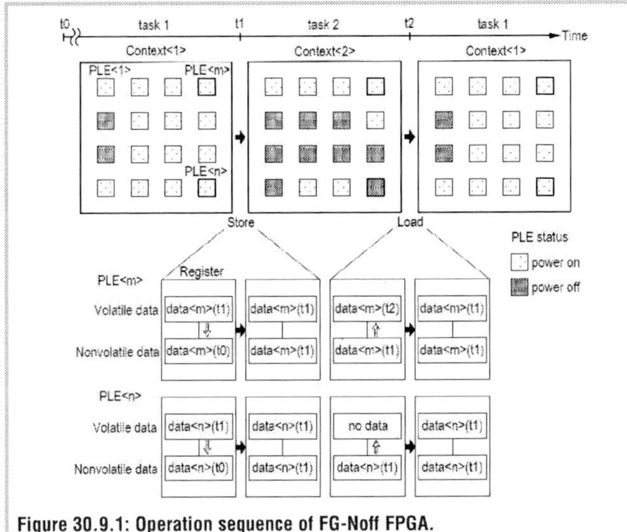
Figure 30.9.1: Operation sequence of FG-Noff FPGA.

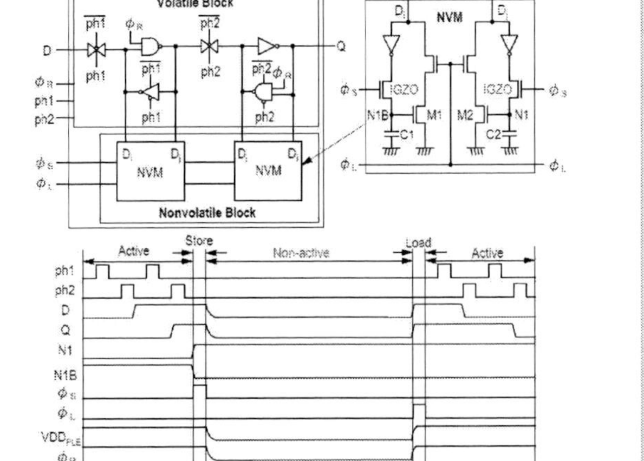
Figure 30.9.2: Nonvolatile register and its driving timing chart.

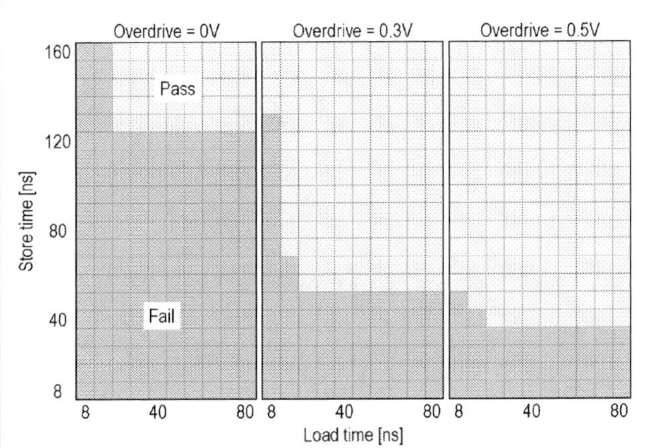

Figure 30.9.3: Store/load time of register data in nonvolatile register.

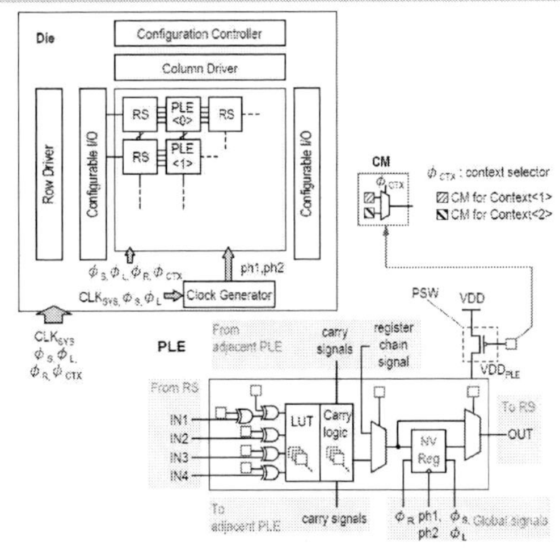
Figure 30.9.4: Block diagram of the FPGA employing PLE with a nonvolatile register.

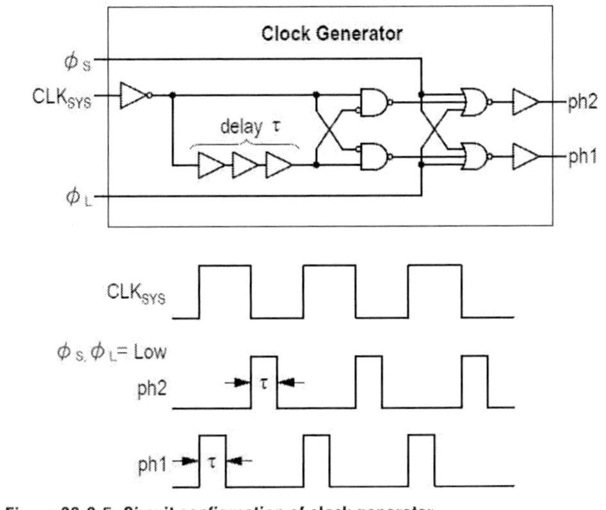

Figure 30.9.5: Circuit configuration of clock generator.

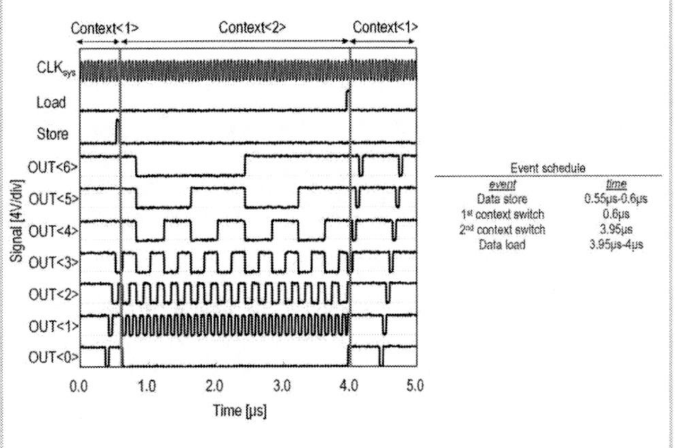
Figure 30.9.6: Operation waveforms and event schedule of an example verification of the fabricated FPGA.

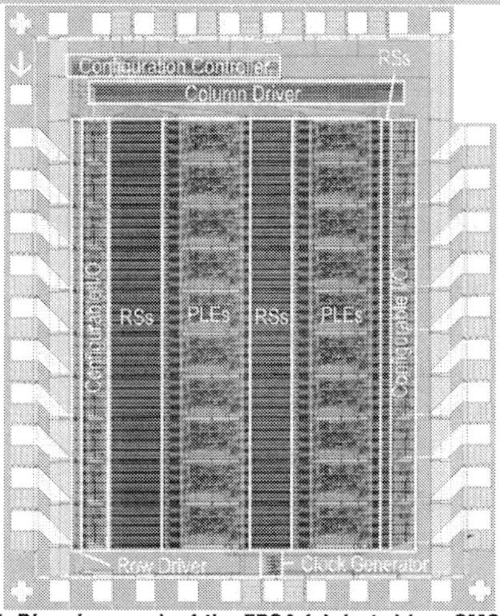

Figure 30.9.7: Die micrograph of the FPGA fabricated in a CMOS/CAAC-IGZO hybrid process (24.75mm²).

30.10 A 1TOPS/W Analog Deep Machine-Learning Engine with Floating-Gate Storage in 0.13µm CMOS

Junjie Lu, Steven Young, Itamar Arel, Jeremy Holleman

University of Tennessee, Knoxville, TN

Direct processing of raw high-dimensional data such as images and video by machine learning systems is impractical both due to prohibitive power consumption and the "curse of dimensionality," which makes learning tasks exponentially more difficult as dimension increases. Deep machine learning (DML) mimics the hierarchical presentation of information in the human brain to achieve robust automated feature extraction, reducing the dimension of such data. However, the computational complexity of DML systems limits large-scale implementations in standard digital computers. Custom analog or mixed-mode signal processors have been reported to yield much higher energy efficiency than DSP [1-4], presenting the means of overcoming these limitations. However, the use of volatile digital memory in [1-3] precludes their use in intermittently-powered devices, and the required interfacing and internal A/D/A conversions add power and area overhead. Nonvolatile storage is employed in [4], but the lack of learning capability requires task-specific programming before operation, and precludes online adaptation.

The feasibility of analog clustering, a key component of DML, has been demonstrated in [5]. In this paper, we present an analog DML engine (ADE) implementing DeSTIN [6], a state-of-art DML framework, and featuring online unsupervised trainability. Floating-gate nonvolatile memory facilitates operation with intermittent harvested energy. An energy efficiency of 1TOPS/W is achieved through massive parallelism, deep weak-inversion biasing, current-mode analog arithmetic, distributed memory, and power gating applied to per-operation partitions. Additionally, algorithm-level feedback desensitizes the system to errors such as offset and noise, allowing reduced device sizes and bias currents.

Figure 30.10.1 shows the architecture of the ADE, in which seven identical cortical circuits (nodes) form a 4-2-1 hierarchy. Each node captures regularities in its inputs through an unsupervised learning process. The lowest layer receives raw data (e.g. the pixels of an image), and continuously constructs belief states that characterize the sequence observed. The inputs of nodes on 2nd and 3rd layers are the belief states of nodes at their respective lower layers. The beliefs of the top layer are then used as rich features for a classifier.

The node (Fig. 30.10.2) incorporates an 8×4 array of reconfigurable analog computation cells (RAC), grouped into 4 centroids, each with 8-dimensional input. The centroids are characterized by their mean μ and variance σ^2 in each dimension, stored in their respective floating gate memories (FGM). In a training cycle, the analog arithmetic elements (AAE) calculate a simplified Mahalanobis distance (assuming a diagonal covariance matrix) D_{MAH} between the input observation $\boldsymbol{o}$ and each centroid. The 8-D distances are built by joining the output currents. A distance processing unit (DPU) performs inverse-normalization (IN) operation to the 4 distances to construct the belief states, which are the likelihood that the input belongs to each centroid. Then the centroid parameters μ and σ^2 are adapted using the online clustering algorithm. The centroid with the smallest Euclidean distance D_{EUC} to the input is selected (classification). The errors between the selected centroids and input are loaded to the training control (TC) and their memories are then updated proportionally. In recognition mode, only the belief states are constructed and the memories are not adapted. Intra-cycle power gating is applied to reduce the power consumption by up to 37%.

Figure 30.10.3 shows the schematic of the RAC, which performs three different operations through reconfigurable current routing. Two embedded FGMs provide nonvolatile storage for centroid parameters. Capacitive feedback stabilizes the floating gate voltage (V_{FG}) to yield pulse-width controlled update. Tunneling is enabled by lowering its supply to bring down the V_{FG}, increasing the voltage across the tunneling junction. Injection is enabled by raising the source of the injection transistor. This allows random-accessible bidirectional updates without the need for on-chip high-voltage switches or charge pump. A 2-T V-I converter then provides a current output and sigmoid update rule. The FGM consumes 0.5nA of bias current, and shows an 8b programming accuracy and a 46dB SNR at full scale. The absolute value circuit (ABS) in the AAE rectifies the bidirectional difference current between o and μ. Class-B operation and the virtual ground provided by amplifier A allow high-speed resolution of small current differences. The rectified currents are then fed into a translinear X^2/Y circuit, which simulations indicate operates with more than an order of magnitude higher energy efficiency than its digital equivalence.

In the belief construction phase, the DPU (Fig. 30.10.4) inverts the distance outputs from the 4 centroids to calculate similarities, and normalizes them to yield a valid probability distribution. The output belief states are sampled then held for the rest of the cycle to allow parallel operation of all layers. The sampling switch reduces current sampling error due to charge injection: a diode-connected PMOS provides a reduced V_{GS} to the switch NMOS to turn it on with minimal channel charge. The S/H achieved less than 0.7mV of charge injection error (2% current error), and less than 14µV of droop with parasitic capacitors as holding capacitor. In classification phase, the IN circuits are reused together with the winner-take-all network (WTA) to classify the observation to the nearest centroid. A starvation trace (ST) circuit is implemented to address unfavorable initial conditions wherein some centroids are starved of nearby inputs and never updated. The ST provides starved centroids with a small but increasing additional current to force their occasional selection and pull them into more populated areas of the input space. The lower right of Fig. 30.10.4 shows the TC circuit, which performs current-to-pulse-width conversion using a V_{DD}-referenced comparison. Proportional updates cause the mean and variance memories to converge to the sample statistics, respectively.

The ADE is evaluated on a custom test board with data acquisition hardware connecting to a host PC. The waveforms in Fig. 30.10.5 show the measured beliefs, one from each layer. The sampling of beliefs proceeds from the top layer to the bottom to avoid delays due to output settling. The performance of the node is demonstrated by solving a clustering problem. The input data consists of 4 underlying clusters, each drawn from a Gaussian distribution with different mean and variance. The node achieves accurate extraction of the cluster parameters (μ and σ^2), and the ST ensures a robust operation against unfavorable initial conditions.

We demonstrate feature extraction for pattern recognition with the setup shown in Fig. 30.10.6. The input patterns are 16×16 symbol bitmaps corrupted by random pixel errors. An 8×4 moving window defines the pixels applied to the ADE's 32-D input. First the ADE is trained unsupervised with examples of patterns. After the training converges, the 4 belief states from the top layer are used as rich features and classified with a neural network implemented in software, achieving a dimension reduction from 32 to 4. Recognition accuracies of 100% with corruption lower than 10%, and 95.4% with 20% corruption are obtained, comparable to a software baseline, demonstrating robustness to the non-idealities of analog computation.

The ADE was fabricated in a 0.13µm CMOS process with thick-oxide IO FETs. The die micrograph is shown in Fig. 30.10.7, together with a performance summary and a comparison with state-of-art bio-inspired parallel processors utilizing analog computation. We achieve 1TOPS/W peak energy efficiency in recognition mode. Compared to state-of-art, this work achieves very high energy efficiency in both modes. This combined with the advantages of nonvolatile memory and unsupervised online trainability makes it a general-purpose feature extraction engine ideal for autonomous sensory applications or as a building block for large-scale learning systems.

References:

[1] J. Park, et al., "A 646GOPS/W Multi-Classifier Many-Core Processor with Cortex-Like Architecture for Super-Resolution Recognition," *ISSCC Dig. Tech. Papers,* pp. 168-169, Feb. 2013.

[2] J. Oh, et al., "A 57mW Embedded Mixed-Mode Neuro-Fuzzy Accelerator for Intelligent Multi-Core Processor," *ISSCC Dig. Tech. Papers,* pp. 130-132, Feb. 2011.

[3] J.-Y. Kim, et al., "A 201.4GOPS 496mW Real-Time Multi-Object Recognition Processor With Bio-Inspired Neural Perception Engine," *ISSCC Dig. Tech. Papers,* pp. 150-151, Feb. 2009.

[4] S. Chakrabartty and G. Cauwenberghs, "Sub-Microwatt Analog VLSI Trainable Pattern Classifier," *IEEE J. Solid-State Circuits,* vol. 42, no. 5, pp. 1169-1179, May 2007.

[5] J. Lu, et al., "An Analog Online Clustering Circuit in 130nm CMOS," *IEEE Asian Solid-State Circuits Conference,* Nov. 2013.

[6] S. Young, et al., "Hierarchical Spatiotemporal Feature Extraction using Recurrent Online Clustering," Pattern Recognition Letters, Oct. 2013.

978-1-4799-0917-9/14 $31.00 © 2014 IEEE

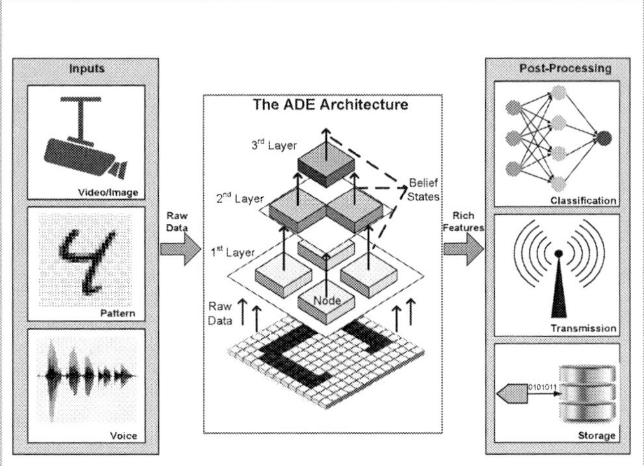

Figure 30.10.1: The analog deep learning engine (ADE) architecture.

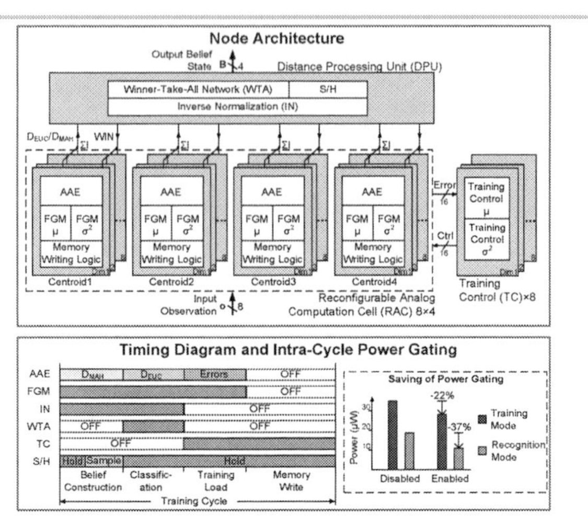

Figure 30.10.2: The node architecture and its timing diagram showing power gating.

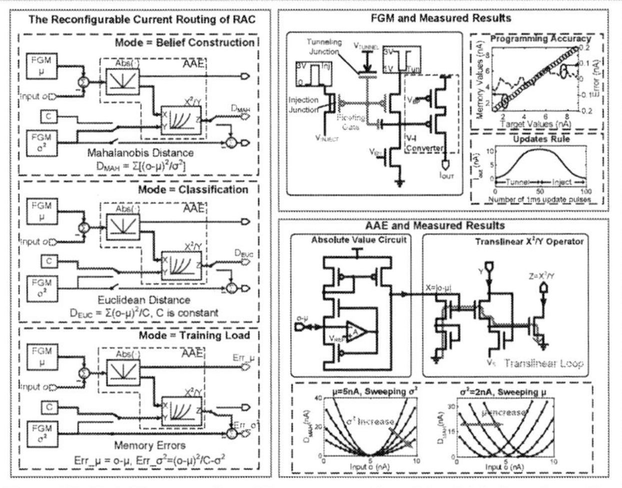

Figure 30.10.3: The reconfigurable current routing of the RAC, the schematics of the FGM and AAE and measurement results.

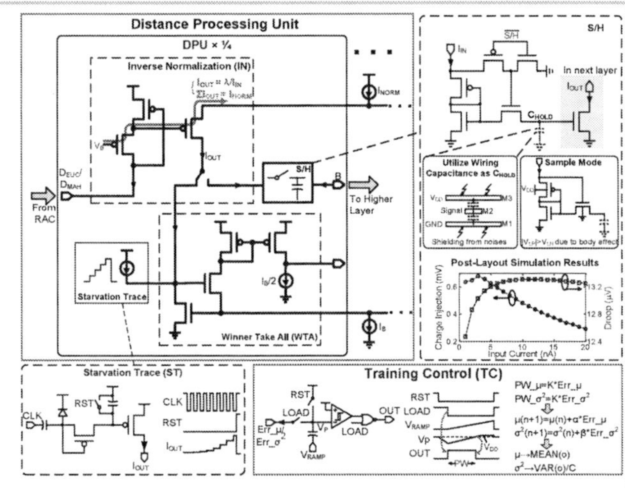

Figure 30.10.4: The schematic of the DPU and its sub-blocks. The training control is shown on the lower right.

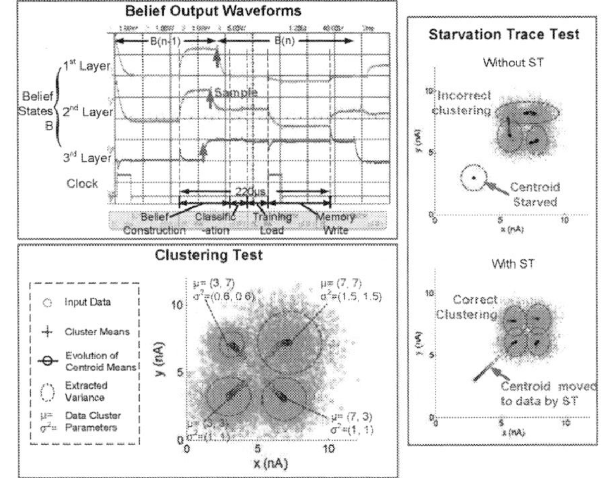

Figure 30.10.5: Measured output waveforms, clustering and ST test results. For clustering, 2-D results are shown for better visualization.

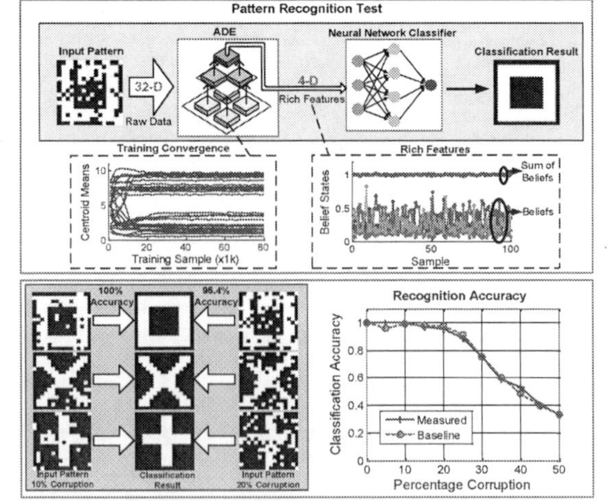

Figure 30.10.6: Pattern recognition test setup and results, demonstrating accuracy comparable to baseline software simulation.

Technology	1P8M 0.13μm CMOS	
Power Supply	3V	
Active Area	0.9mm×0.4mm	
Memory	Non-Volatile Floating Gate	
Memory SNR	46dB	
Training Algorithm	Unsupervised Online Clustering	
Output Feature	Inverse-Nomalized Mahanalobis Distances	
Input Referred Noise	$56.23pA_{rms}$	
System SNR	45dB	
I/O Type	Analog Current	
Operating Frequency	Training Mode	4.5kHz
	Recognition Mode	8.3kHz
Power Consumption	Training Mode	27μW
	Recognition Mode	11.4μW
Energy Efficiency	Training Mode	480GOPS/W
	Recognition Mode	1.04TOPS/W

	This work	ISSCC'13 [1]	ISSCC'11 [2]	ISSCC'09 [3]
Process	0.13μm	0.13μm	0.13μm	0.13μm
Purpose	DML Feature Extraction	Object Recognition	Neural-Fuzzy Accelarator	Object Recognition
Non-volatile Memory	Floating Gate	NA	NA	NA
Power (W)	11.4μW	260mW	57mW	496mW
Peak Energy Efficiency	1.04TOPS/W	646GOPS/W	655GOPS/W	290GOPS/W

Figure 30.10.7: Die micrograph, performance summary and comparison table.

Gap in pagination due to formatting issues.

Pages 506-507

T1: Filtering in RF Transceivers

Filtering is a fundamental function performed by RF transceivers. This tutorial starts with a review of filtering requirements for modern RF receivers and presents a review of traditional and recently rediscovered filtering techniques. Design of passive switched-capacitor filters is covered in detail along with recently rediscovered N-path filters. Examples of filter designs are provided.

Instructor: Borivoje Nikolic

Borivoje Nikolic is a Professor of Electrical Engineering and Computer Science at the University of California, Berkeley and a Scientific Co-Director of the Berkeley Wireless Research Center. He received the Ph.D. degree from the University of California at Davis in 1999. Dr. Nikolic has received best paper awards at ISSCC, the Symposium on VLSI Circuits, the IEEE International SOI Conference, the European Solid-State Device Research Conference, and ACM/IEEE ISLPE.

T2: V_{min} Constraints and Optimization in VLSI Circuit Design

With growing demand for energy efficiency, the ability to achieve lower minimum operating voltages (V_{min}) will be a key enabler in future VLSI systems. V_{min} for a given design depends on factors such as performance and energy targets, variability margins, and package limitations, the relative importance of which can vary widely between applications. This tutorial reviews V_{min} constraints that are relevant to VLSI circuit designers and summarizes methods to overcome these issues to improve V_{min}. Topics to be covered include V_{min} constraints and optimization techniques spanning frequency, functionality, package, energy/op, and reliability considerations.

Instructor: Leland Chang

Leland Chang received the Ph.D. degree in Electrical Engineering and Computer Sciences from the University of California, Berkeley and joined the IBM T. J. Watson Research Center in 2003, where he is now Manager of Design and Technology Solutions. His work focuses on power efficiency in high-performance systems, spanning the development of new technology elements, the design of high-performance memory and power management circuits, and the exploration of novel accelerator and memory system architectures. Key contributions include early demonstration of the FinFET structure for CMOS scaling, 8T-SRAM for voltage scaling in high-performance caches, high-speed register files with double-pumped access ports, and high-efficiency voltage conversion circuits using new passive device technologies. He is the author of 69 technical articles and 56 patents and is currently a member of the ISSCC technical program committee.

T3: 3D Integration, Power Delivery, and Contactless Interconnect by Near-Field Coupling

Data communication and power transfer by using near-field coupling have gained increasing attention as an attractive technology direction. With the recent introduction of standards for next-generation memory interfaces (Wide IO) and wireless power transfer (Qi), the commercial adoption of these technologies is progressing. Research on non-contact connecters using electromagnetic coupling has also started. In this tutorial, we introduce the basics of near-field coupling, together with circuit implementations, and applications to the following areas: (1) Wide IO for die stacking by inductive coupling (compared with TSV), (2) Power transfer by inductive coupling and magnetic resonant coupling, and (3) Connectors for reliable and compact assembly by electromagnetic coupling.

Instructor: Tadahiro Kuroda

Tadahiro Kuroda received the Ph.D. degree in Electrical Engineering from the University of Tokyo. In 1982, he joined Toshiba Corporation. In 2000, he moved to Keio University, where he has been a Professor since 2002. He was a Visiting MacKay Professor at the University of California, Berkeley in 2007. His research interests include low-power CMOS design, proximity communications and image recognition. He has published more than 200 papers, including 34 ISSCC papers, 21 VLSI Symposia papers, 19 CICC papers, and 16 A-SSCC papers. He wrote 22 books/chapters and filed >100 patents. He is an IEEE Fellow and an IEICE Fellow.

T4: Power-Optimized Processor Design

An increasing focus on design productivity is causing a gradual shift in the microprocessor design flow from custom design to synthesis-based methodologies. This reduces the power-optimization opportunities offered by previously developed custom-circuit-tuning techniques. This tutorial gives an overview of power-reduction techniques used in microprocessors, with a particular focus on digital design. It covers newly developed methodologies that emphasize broad design-space exploration to optimize designs for power at the structural level allowed by recent advances in the synthesis tools, and discusses the shift in the role and responsibilities of logic and circuit designers in the power-optimization efforts. In the summary, the tutorial shows how the power-optimization efforts across design disciplines come together in delivering a competitive product meeting power, performance and power-frequency limited yield goals.

Instructor: Victor Zyuban

Dr. Zyuban received a Ph.D. degree in Computer Science and Engineering from the University of Notre Dame. He joined the IBM T.J. Watson Research Center in 2000, where is he a Research Staff Member and manager. Within IBM, he is currently the pSeries server processor power lead. In this role, which he has held over the past 7 years, his responsibilities include leading the processor power-optimization and power-modeling flow, modeling the chip frequency and the Power-Frequency-Limited Yield (PFLY), as well as the chip power/frequency hardware-to-model correlation. Victor also leads several research projects in the area of low-power design at the IBM T.J. Watson Research Center. He has over 30 publications and more than 30 patents issued or filed.

T5: Peripheral Circuits for Analog-to-Digital Converters

The performance of modern analog-to-digital converters is often limited by the circuitry surrounding them. This tutorial explains how we measure analog-to-digital converters to ensure that we are not measuring the test equipment, and goes into detail on various circuit techniques often applied to ensure we get the best performance. I discuss topics ranging from DC precision to low-distortion RF. The digital side of an ADC often produces measurement artefacts and I describe common approaches to mitigate these issues. The firehose of data coming out of a GS/s ADC is a significant challenge when dealing with such devices. The tutorial helps both board-level engineers and chip designers to overcome issues that may limit circuit performance.

Instructor: Marco Corsi

Marco Corsi joined Texas Instruments in 1990. Shortly after, he designed breakthrough linear innovations including PCMCIA distribution switches, RS485 drivers, and custom PMICs for the then-new mobile phone industry. Marco led the high-speed ADC design group at Texas Instruments. He is responsible for many industry firsts including a 16b 200MS/s ADC as well as the first monolithic 12b 1GS/s ADC. He was elected TI Senior Fellow in 2013 and IEEE Fellow in 2012. Corsi earned a BA and MA from the University of Cambridge in England and holds 76 US patents.

T6: Analog Front-End Design for Gb/s Wireline Receivers

This tutorial introduces the design, analysis, and optimization of analog front-ends for state-of-the-art Gb/s wireline receivers. Specifically, the tutorial highlights key circuit/system requirements, tradeoffs, and gotchas inherent in the design of such receivers. In addition, the tutorial develops simple analytical formulas that accurately predict the power and performance characteristics of key circuit blocks such as continuous-time linear equalizers and decision-feedback equalizers.

Instructor: Elad Alon

Elad Alon received the Ph.D. degree in Electrical Engineering from Stanford University, Stanford, CA, in 2006. In 2007, he joined the University of California, Berkeley, where he is now an Associate Professor of Electrical Engineering and Computer Sciences, as well as a Co-Director of the Berkeley Wireless Research Center (BWRC). He has held consulting or visiting positions at Cadence, Xilinx, Wilocity, Oracle, Intel, AMD,

Rambus, Hewlett Packard, and IBM Research. His research focuses on energy-efficient integrated systems, including the circuit, device, communication, and optimization techniques used to design them. He received the IBM Faculty Award in 2008, the 2009 Hellman Family Faculty Fund Award, and the 2010 UC Berkeley Electrical Engineering Outstanding Teaching Award, and has co-authored papers that received the 2010 ISSCC Jack Raper Award for Outstanding Technology Directions Paper, the 2011 Symposium on VLSI Circuits Best Student Paper Award, and the 2012 CICC Best Student Paper Award.

T7: Self-Adapting Design Techniques for Power-Constrained Processors

In today's power-constrained processor designs, self-adaptive design techniques are essential to maximize the performance of a processor under growing variations in process, supply noise, clock skew, and activity. In addition, adaptive designs can reduce design-margin requirements and improve performance by avoiding the over-design that results from metric-based worst-case design methodologies. This tutorial overviews recent techniques for self-adapting design, from industry and academia. Emphasis is placed on various dynamic voltage- and frequency-scaling techniques that adapt to changes in supply noise and processor activities. It then discusses novel methods to improve yield by mitigating design guardband. The presentation reviews state-of-the-art circuit components necessary to implement these adaptive techniques.

Instructor: Jinuk Luke Shin

Jinuk Luke Shin has been with Oracle (formerly Sun Microsystems) since 2000. He is currently a Director of Hardware Development, responsible for circuit and physical design of next-generation SPARC processors. He has led design activities in technology, power distribution/management, clocking, memories, analog components, Si validation and chip integration for 8 SPARC processors. He received his M.S. degree from the University of Texas at Austin in 1995. Prior to joining Oracle, he was with Motorola, Austin from 1995 to 1997, where he was involved in the design of embedded flash memories for digital signal processors. From 1997 to 2000, he was with Hitachi Semiconductor America, San Jose, CA. His current research interest is in energy-efficient circuit and physical-design techniques and methodologies for high-end microprocessors. He is an author of 35 technical papers and holds 12 issued and pending U.S. patents. He serves as a member of the High-Performance Digital subcommittee of ISSCC.

T8: Interference-Robust CMOS Radio-Receiver Techniques

The radio spectrum is becoming more and more crowded, and radio receivers become interference-limited. As there is a demand for multimode flexible radio devices, traditional dedicated narrowband filtering no longer works. During the last decade, several new radio receiver architectures have been proposed that offer more flexibility than traditional receivers with dedicated fixed filtering, while maintaining good sensitivity and robustness to interference. Different names have been used to refer to these receivers: reconfigurable receiver, multiband receiver, wideband receiver, SAW-less receiver, software-defined radio receiver, or cognitive-radio receiver. These receivers all aim for a high dynamic range while relying less on fixed filters. This tutorial reviews several concepts, including: linearization techniques, noise and distortion cancelling, LNTAs followed by current-mode mixing, mixer-first receivers, frequency-translated filtering, harmonic rejection, and spatial interference rejection.

Instructor: Eric Klumperink

Eric Klumperink received his Ph.D. from Twente University in Enschede in 1997, where he is currently an Associate Professor, teaching Analog and RF CMOS IC Design. His research focus is on Cognitive Radio, Software-Defined Radio and Beamforming. Eric serves as Associate Editor for the Journal of Solid-State Circuits and is a TPC member of ISSCC and RFIC. He holds several patents, has authored or co-authored more than 180 international refereed journal and conference papers, and is a co-recipient of the ISSCC 2002 and the ISSCC 2009 Jan Van Vessem Outstanding Paper Award.

T9: Charge Pump and Capacitive DC-DC Converter Design

Capacitive charge pumps are gaining popularity due to their ease of integration and low EM radiation. Recent research and advances in process technology enable highly efficient, fully integrated DC-DC converters. After an overview of the application field for charge pumps, the tutorial overviews basic structures: the Dickson QP, the basic voltage doubler and the ladder converter, discussing their functionality and their components. Step-up and step-down converters are presented together with techniques to generate different voltage ratios. Advanced topics include ripple reduction by multiphase charge pumps, load and line regulation, control-loop design and design methodologies for optimal efficiency including reducing the influence of device parasitics.

Author: Tim Piessens

Tim Piessens received a Ph.D. degree in 2003 from the University of Leuven. From 1998 to 2003, he was a research assistant at the ESAT-MICAS laboratories under the supervision of Prof. Steyaert. His main research focus was on nonlinear system design and more specific line drivers for xDSL applications. In 2004, Tim cofounded ICsense as a spin-off of the University of Leuven. He works as CTO in the company, with a strong focus on analog and high-voltage IC design. He has authored several ISSCC and JSSC papers and holds several patents.

Presenter: Anton Bakker

Anton Bakker received M.Sc and Ph.D degrees in EE from Delft University in 1991 and 2000 resp. He is the (co)author of over 25 scientific publications and holds 15 patents. From 1991-2000 he was an Assistant Professor at Delft University and since then he has worked for many companies in the Power Management Field including NXP, ADI and MPS. In 2011 he returned to his Alma Mater to teach a course on Power Conversion. He is currently with IDT where he is leading a product development team in the field of Wireless Charging. He has been an ISSCC Analog Subcommittee team member since 2013.

T10: Design of Physical-to-Digital Converters

Modern electronic systems employ increasing numbers of sensors to gather information about the physical world around us. This information, which is inherently non-electrical and analog in nature, needs to be digitized, often with increasingly demanding requirements on accuracy and power efficiency. This tutorial presents an integral approach to designing suitable physical-to-digital converters that goes beyond the conventional approach of combining a sensor, front-end circuit and ADC in terms of accuracy and efficiency. Approaches for embedding sensors into data-converter architectures are presented, including ratiometric charge-balancing architectures and architectures that exploit feedback around the sensor. The need for suitable references and approaches for calibration and correction for cross-sensitivity are addressed. These concepts are illustrated using case studies of state-of-the-art temperature-to-digital, humidity-to-digital, flow-to-digital and light-intensity-to-digital converters.

Instructor: Michiel Pertijs

Michiel Pertijs is an Associate Professor at Delft University of Technology, where he heads a research group working on low-power integrated circuits for medical ultrasound and energy-efficient smart sensors. Prior to this, he was with imec / Holst Centre, and with National Semiconductor, where he designed precision operational amplifiers and instrumentation amplifiers. He has authored or co-authored one book, three book chapters, 11 patents, and over 50 technical papers. He is a program committee member of ISSCC, ESSCIRC, and the IEEE Sensors Conference. He received the ISSCC 2005 Jack Kilby Award and the JSSC 2005 Best Paper Award.

F1: *Digitally Assisted Analog and Analog-Assisted Digital in High-Performance Scaled CMOS Process*

Organizer: Xicheng Jiang, *Broadcom, Irvine, CA*

Committee: Xicheng Jiang, *Broadcom, Irvine, CA*
Piero Malcovati, *University of Pavia, Pavia, Italy*
Vladimir Stojanovic, *MIT, Cambridge, MA*
Iizuka, Tetsuya, *University of Tokyo, Tokyo, Japan*

Digitally assisted analog and analog-assisted digital techniques are increasingly needed in future circuit and system designs, as FinFET and FD-SOI replace planar CMOS technology at the advanced process nodes of 20nm and beyond. The intrinsic features of these new devices are lowering the barrier between the analog and the digital worlds, allowing unprecedented performance to be achieved by assisting digital circuits with analog techniques (e.g. body bias) or by assisting analog circuits with digital techniques (e.g. calibration and run-time control). The objective of the forum is to discuss practical design considerations in high-performance scaled CMOS processes, established circuit techniques that take advantage of scaled CMOS process technology in analog, digital, RF and SoC designs, and an outlook for the future in the context of challenges and solutions.

The forum covers, on one hand, advances in technology, by introducing features and challenges of FinFET and FD-SOI devices and comparing their performance, while, on the other hand, it discusses advances in both analog and digital design enabled by scaled process technologies, in different application fields, ranging from RF to data converters, microprocessors, and mobile SoC.

FD-SOI Technology for High-Performance and Low-Power Circuit Design

Nicolas Planes, *STMicroelectronics, Crolles, France*

UTBB FD-SOI technology can be employed to meet customer requirements for low-leakage, high-speed and low-voltage applications. At the 28nm node, this planar technology offers ease-of-portability for products implemented in bulk substrates. Thanks to excellent electrostatic control and low DIBL leakage, FD-SOI boosts speed by more than 30% compared to standard bulk technology at nominal voltage. Best-in-class threshold voltage variability ($AV_{th} = 1.1mV/\mu m$) permits a decrease of V_{min} for SRAM arrays to 100mV below bulk counterparts. Utilizing thin BOX, this technology offers an excellent body factor, allowing body-bias in an extended voltage range (up to $2{\times}V_{DD}$). The 28nm UTBB process is a gate-first high-κ metal-gate technology based on a 28nm LP bulk technology. Fast yield ramp-up allowed the manufacturing of a 2.6GHz dual-core ARM Cortex A9 in 2012. DVFS and AVS techniques can be used in a wide voltage range thanks to the low V_{min} of SRAM cells and the high V_{max} (thick gate oxide). This presentation describes all process features that enable scaling down to the 10nm node, fulfilling digital, analog and RF requirements. The focus is on high-performance applications.

Nicolas Planes is device manager in R&D Technology and Process Integration at STMicroelectronics in Crolles, France. He received his Ph.D. from the University of Montpellier in France in 1999. Prior to joining STMicroelectronics, he worked for SOFRADIR as R&D engineer on infrared detectors. He joined STMicroelectronics in 2002 in the extensive electrical characterization group working on CMOS logic, embedded DRAM, and SRAM devices. Since 2006, he has been working on the R&D of device process integration, focusing on dense SRAM cells. He is currently device manager for 28nm FD-SOI technology. He holds several patents in FD-SOI and has authored or co-authored more than 30 technical papers across a range of topics. In 2008, he received an ST Corporate Award for his work on the V_{ccmin} of SRAM cells.

FinFET Technology for High-Performance Circuit Design

Tatsuya Ohguro, *Toshiba, Yokohama, Japan*

FinFETs are attractive devices for logic, but also for mixed signal and RF circuits, because those application areas can benefit from higher f_t, better V_{th} matching, and lower noise. This forum focuses on the optimum design of FinFETs and their fabrication process. Additionally, mixed signal characteristics of InGaZnO TFTs, and the combination these with FinFETs are discussed. TFT devices have been attractive for high-voltage analog applications due to low leakage current, high-breakdown voltage and larger mobility.

Tatsuya Ohguro received the B.S. and M.S. degrees in physics from Hokkaido University, Sapporo, Japan, in 1987 and 1989, respectively. He joined Toshiba Corporation in 1989, where he has been engaged in the research and development of advanced CMOS devices and process for high-performance logic, mixed signal and RF circuits. He was a guest professor at Hiroshima University from 2003 to 2006. He is the recipient of the Nikkei BP Grand Prize (1994), the Yamazaki-Teiichi Prize (2007) and the Commendation for Science and Technology by the Minister of Education, Culture, Sports, Science and Technology (2009).

Digitally Assisted Analog Design Enables Mobile SoC Evolution

Todd Brooks, *Broadcom, Irvine, CA*

Mobile SoCs have evolved together with our ability to integrate features and reduce cost. We rely on an expanding list of complex multimedia and interface capabilities in today's 4G LTE era. Steady improvements in functionality and performance will enable continuing increases in mobile product complexity to make these devices ever more useful in our daily lives. The future of mobile products will depend on circuit and architecture innovations that take full advantage of advanced process nodes. Many mixed-signal analog challenges are addressed today using digital techniques to take advantage of high-performance process nodes. This talk will focus on established techniques for digitally assisted analog to address practical analog design considerations in Mobile SoCs. Practical case studies of peripheral functions and transceiver datapath circuits will be given, including circuits that utilize high-speed analog and digital processing, complex mixed-signal feedback loops, calibration and run-time control. The case studies will highlight performance, die area, and power consumption improvements and provide an outlook of future challenges and solutions.

Todd L. Brooks received the B.S. degree in engineering science from Trinity University, San Antonio, in 1987, and the M.S. degree in electrical engineering from Texas A&M University, College Station, in 1992. He is a Director at Broadcom in Irvine, CA in the Analog and RF Microelectronics group. At Broadcom, his research interests include audio and precision data converters, speaker drivers, and microphone interfaces used in VoIP, cellular, Bluetooth, DSL, set-top boxes and cable modem products. Prior to joining Broadcom in 1997, he worked at Analog Devices in Wilmington, MA, on high-speed pipelined and oversampled ADCs from 1993 to 1997, and at Motorola on mobile radio and telecommunication circuits from 1990 to 1993. Mr. Brooks was elected a Broadcom Fellow in 2009. He currently holds 38 patents, has authored or coauthored 13 conference and journal papers, and has served on the analog subcommittee of ISSCC. From 1997 to 2004, he lectured on high-speed ADCs for Mead Microelectronics. He is a co-recipient of Best Paper Awards for the 1997 IEEE Journal of Solid-State Circuits and the 2009 IEEE Custom Integrated Circuit Conference.

Analog-Assisted Digital Design in Mobile SoCs

Martin Saint-Laurent, *Qualcomm, Austin, TX*

This presentation gives an overview of some of the main digital design challenges for mobile SoCs in advanced process nodes, including: low-power operation, process variability, power-supply noise, and temperature management. It then describes a number of techniques with analog aspects to assist the operation of the digital processors in these SoCs.

Martin Saint-Laurent received the B.Eng. degree (with honors) in electrical engineering from McGill University, Montréal, Canada, and the M.S. and Ph.D. degrees, also in electrical engineering, from the Georgia Institute of Technology, Atlanta, GA. From 1998 to 2005, he was with Intel Corporation, where he worked on two generations of high-frequency IA-32 processors as a custom circuit designer. His responsibilities included clock distribution and sequential element design. In 2005, he joined Qualcomm, Austin, TX, where he currently works on developing and implementing power reduction techniques for wireless applications as a principal engineer. His technical interests include: power analysis, low-power design techniques, and power delivery.

978-1-4799-0917-9/14 $31.00 © 2014 IEEE

Charge-Steering Techniques for Gigahertz Analog and Digital Circuits

Behzad Razavi, *UCLA, Los Angeles, CA*

Charge-steering techniques provide an alternative design paradigm to conventional continuous-time digital and analog circuits. The use of charge steering in latches and flipflops can considerably reduce the power consumption at high speed, affording, as an example, clock and data recovery at 25Gb/s with a power drain of 5mW. Charge-steering operational amplifiers outperform their continuous-time counterparts in terms of noise, speed, and power consumption. With the aid of digital calibration techniques, such op amps allow the design of 10b gigahertz ADCs with competitive figures of merit.

Behzad Razavi is Professor of Electrical Engineering at the University of California, Los Angeles. His research deals with wireless and wireline transceivers, high-speed communication circuits, and data converters. An author of 150 papers and seven books, Prof. Razavi has received awards for his research at ISSCC, CICC, ESSCIRC, and the VLSI Circuits Symposium. He received the IEEE Donald Pederson Award in 2012 for his pioneering contributions to the design of high-speed CMOS communication circuits.

Digitally Assisted Data Converter Techniques

Ian Galton, *University of California, San Diego, CA*

The relentlessly improving performance and decreasing costs of electronic consumer communication and entertainment devices have been enabled by sophisticated digital signal processing techniques made practical by the continued scaling of CMOS technology. However, high-performance analog circuitry, such as data converters, are also required in these systems, and intense market pressure usually dictates that as much of it as possible be integrated along with the digital circuitry in the same technology. Unfortunately, the design of such circuitry is increasingly challenging as CMOS technology is scaled below 65nm. Low supply voltages, high device non-linearity, poor signal isolation, and device leakage limit the effectiveness of traditional analog circuit topologies. Increasingly, digital signal processing techniques and new digital-like analog circuits that take advantage of the strengths of highly scaled CMOS technology with respect to digital circuitry are used to enable high-performance analog functionality. This talk will describe several such digital enhancement techniques applied to data converters.

Ian Galton received the Sc.B. degree from Brown University in 1984, and the M.S. and Ph.D. degrees from the California Institute of Technology in 1989 and 1992, respectively, all in electrical engineering. Since 1996 he has been a professor of electrical engineering at the University of California, San Diego, where he teaches and conducts research in the field of mixed-signal integrated circuits and systems for communications. Prior to 1996, he was with UC Irvine, and prior to 1989, he was with Acuson and Mead Data Central. His research involves the invention, analysis, and integrated circuit implementation of critical communication system blocks, such as data converters and phase-locked loops. In addition to his academic research, he regularly consults at several semiconductor companies and teaches industry-oriented short courses on the design of mixed-signal integrated circuits. He is a Fellow of the IEEE, and has served on a corporate Board of Directors, on several corporate Technical Advisory Boards, as the Editor-in-Chief of the IEEE Transactions on Circuits and Systems II: Analog and Digital Signal Processing, as a member of the IEEE Solid-State Circuits Society Administrative Committee, as a member of the IEEE Circuits and Systems Society Board of Governors, as a member of the ISSCC Technical Program Committee, and as a member of the IEEE Solid-State Circuits Society Distinguished Lecturer Program.

Digitally Intensive CMOS Transmitters

Mark Ingels, *imec, Leuven, Belgium*

CMOS scaling has improved the processing power of integrated digital circuits tremendously in the last decades. It also brought high-speed wireless communication to the consumer market. Unfortunately, while digital performance improved, many important analog transistor parameters have tended to worsen, while transmitter design complexity increased. Fortunately, we have arrived at a point where a new family of transmitters that make full use of the strong points of nanoscale CMOS has become viable: the direct digital transmitters. This presentation will give an overview of several digital TX architectures. Practical realizations will be shown to demonstrate how this family of transmitters has a bright future ahead.

Mark Ingels received the M.Sc. and Ph.D. degrees in Microelectronics from the ESAT-MICAS Laboratories of the K. U. Leuven, Belgium, in 1990 and 2000, respectively. During his Ph.D., he designed high-speed optical transceivers in CMOS. In 1999, he joined Alcatel Microelectronics, later acquired by STMicroelectronics, as a Senior Researcher working on the integration of ADSL analog front-ends and Bluetooth RF transceivers. He joined the wireless research group in imec, Leuven, Belgium, in 2005 to work on RF transceivers for software-defined radio systems. Dr. Ingels was a member of the technical program committee for ISSCC from 2005 to 2010.

Multiband Radios in Deep-Scaled CMOS by Using On-Chip Self-Healing Algorithm and Digitally Controlled Artificial Dielectric

Frank Chang, *University of California, Los Angeles, CA*

Increased process variation in deep-scaled CMOS technology nodes has rendered modern radio-on-chip (RoC) designs more challenging to achieve high performance yield. Although variations due to matching of devices can be reduced by increasing their sizes, such design measures would inevitably increase circuit power consumption and die area. For mm-Wave circuits operating close to device cut-off frequencies of f_t and f_{max}, an increase in device sizes becomes even more impractical. Traditionally, calibration has been used extensively as means to counter performance deviations caused by process variations and device aging. These calibration algorithms, however, typically focus on localized tuning of circuit parameters and do not optimize multiple transceiver parameters concurrently. In this talk, we present a new and holistic self-healing transceiver architecture based on information collected from on-chip sensors to intelligently adjust tuning knobs for healing the complete transceiver and DiCAD-embedded frequency synthesizers, rather than only the RF building blocks, thereby boosting the RoC's post-healing performance yield. The effectiveness of such architectures in performance yield enhancement has been validated by a fully integrated 60GHz RoC in 65nm CMOS.

Mau-Chung Frank Chang is the Chairman and the Wintek Chair Professor in the Electrical Engineering Department at UCLA. Before joining UCLA, he was the Assistant Director and Department Manager of the High-Speed Electronics Laboratory at Rockwell Science Center, Thousand Oaks, California. Throughout his career, he has led many research programs with an excellent track record, including the successful development and commercialization of GaAs HBT power amplifiers under DARPA contracts for mobile system applications. He also led the initial multi-Gb/s ADC/DAC developments under both DARPA and Air Force Wright Patterson supports. He is also known for his CMOS circuit development for mm-Wave and sub-mm-Wave radio, radar and imagers. He was elected to the US National Academy of Engineering in 2008 for development and commercialization of GaAs power amplifiers and integrated circuits. He received the IEEE David Sarnoff Award in 2006, and became a Fellow of IEEE in 1996.

F2: *3D Stacking Technologies for Image Sensors and Memories*

Organizer: Yusuke Oike, *Sony, Kanagawa, Japan*

Committee: Yusuke Oike, *Sony, Kanagawa, Japan*
Makoto Ikeda, *University Tokyo, Tokyo, Japan*
Albert Theuwissen, *Harvest Imaging and TU Delft, Delft, Belgium*
Johannes Solhusvik, *Omnivision, Oslo, Norway*
Jonathan Chang, *TSMC, Hsinchu, Taiwan*
Tadahiro Kuroda, *Keio University, Kanagawa, Japan*

Three-dimensional (3D) stacking integration is offering many product benefits to SoC and memory: performance enhancements, product miniaturization and cost reduction. Besides, image sensors featuring 3D stacking of a specialized image sensor layer on the top of a deep submicron digital CMOS have just come to the market. The objective of this forum is to present applications and details of process integration, device techniques, circuits and system featuring 3D stacking integration. This will start with an overview of 3D stacking ICs, followed by a system perspective with scaling the memory wall. The next two talks will discuss challanges for power reduction with wide memory bandwidth, and performance gains through advanced packaging and chip stacking. This is followed by two talks covering challenges for foundry-specific issues and impact on device performance. The last two talks highlight how to integrate an imaging device on an SoC layer: technical issues and phenomenon of 3D stacked image sensor products, and evolution of 3D integration for imaging systems.

3D System Module with Stacked Image Sensors, Stacked Memories and Stacked Processors on a Si Interposer

Mitsumasa Koyanagi, *Tohoku University, Sendai, Japan*

We have developed a 3D system module with stacked image sensors, stacked memories and stacked processors. Circuit design and process technologies for such 3D stacked module are described. We introduce a concept of block parallel in designing stacked image sensors, stacked memories and stacked processors. In addition, to fabricate a 3D system module with 3D stacked ICs, we have developed a new heterogeneous 3D technology called a super chip technology where many conventional 2D ICs LSIs with different functions are simultaneously stacked by a new self-assembly and electrostatic (SAE) bonding technology after TSVs and microbumps are simultaneously formed on them in die level.

Mitsumasa Koyanagi received the B.S. degree in electrical engineering from Muroran Institute of Technology, Japan in 1969 and the M.S. and Ph.D. degrees in electronic engineering from Tohoku University, Sendai, Japan, in 1971 and 1974, respectively. He joined the Central Research Laboratory, Hitachi Ltd. in 1974. In 1985, he joined the Xerox Palo Alto Research Center, California. In 1988 he joined the Research Center for Integrated Systems, Hiroshima University, as a professor. Since 1994, he has been a professor in Tohoku University, Japan. He was a director of the venture business laboratory, Tohoku University from 1998 to 2000 and a Distinguished Professor in Tohoku University from 2008 to 2010. He has published more than 300 technical papers and given more than 100 invited talks.

He was awarded the 2006 IEEE Jun-ichi Nishizawa Medal, the 1996 IEEE Cledo Brunetti Award, the National Medal with Purple Ribbon in Japan in 2011, and the Award of Ministry of Education, Culture, Sports, Science and Technology in 2001, the SSDM (Solid-State Devices and Materials) Award in 1994 and the Okouchi Prize in 1990 owing to the invention of Stacked capacitor DRAM memory device. In addition, he was awarded the Optoelectronic Technology Achievement Award (Japan Society of Applied Physics) in 2004 owing to his outstanding contributions to optoelectronic technology, the JIEP (Japan Institute of Electronics Packaging) Achievement Award in 2008 owing to his contribution to three-dimensional integration technology, and the IEEE ECTC (Electronics Components and Technology Conference) outstanding paper award in 2011. He is an IEEE fellow and a Japanese Applied Physics Society fellow.

Scaling the Memory Wall with 3D-IC: A System Perspective

Shih-Lien Lu, *Intel, Hillsboro, Oregon*

DRAM has been the technology for computer main memory since Intel released the first commercial DRAM chip (i1103) in 1970. As technology scales and demand for memory performance, it seems DRAM is facing several challenges. Many other memory technologies are anticipated to replace it but none has emerged as a clear winner thus far. In this paper we pose the question. Is it possible to re-examine the design of DRAM to continue its life for another decade at least?

Shih-Lien Lu received his BS in EECS from UC Berkeley, and MS and PhD in CSE from UCLA. He is a Principal Researcher and leads the memory architecture team at Intel Labs. From 1984 to 1991 he was on the MOSIS project at USC/ISI which provides the research and education community VLSI fabrication services. He was on the faculty of the ECE Department at the Oregon State University from 1991 to 2001. His current research interests include computer microarchitecture, memory circuits and VLSI systems design.

3D-TSV Integration of Memory and SoC for Low Power Applications

Ho-Kyu Kang, *Samsung Electronics, Hwasung, Korea*

The idea of wide memory bandwidth is to increase the data traffic between the microprocessor and memory chip. To increase the bandwidth while minimizing the power consumption related with the data traffic, the number of interconnections needs to be increased up to several thousand while the length remains short. 3-D TSV (Through Silicon Via) integration of Memory chips and SoCs has been proposed as a potential solution for the wide memory bandwidth with low power consumption. In this study, 3D-TSV integration between a Wide I/O stacked DRAM and Application Processor for mobile devices has been investigated. Figure of merit and design, manufacturing, and cost challenges of 3-D TSV integration will be discussed.

Ho-Kyu Kang received the B.S. degree from Hanyang University, Seoul, Korea, the M.S. degree in Materials Science and Engineering from Korea Advanced Institute of Science and Technology (KAIST), Korea, and the Ph.D. in said field from Stanford University. He has joined Samsung Electronics, Semiconductor Division since 1985. Currently, as a Senior Vice President of Technology, he leads the Process Development Team of R&D Centre, focusing on advanced memory and logic process technology development. He has authored or co-authored over 80 international papers and holds 8 U.S. patents.

Orthogonal Scaling - the Role of Packaging and 3D Integration

Subramanian Iyer, *IBM, Hopewell Junction, New York*

The absence of cost effective lithography and patterning schemes is predicted to make the historical expectations of the cost–performance benefits of scaling (popularly known as Moore's Law) difficult to sustain. And while some companies are charging like the "Light Brigade" with eUV and 450 mm diameter wafers, a more judicious and sustainable approach is warranted. We examine critically the introduction of hi-K dielectrics, FINs, and the challenges of interconnect scaling and why they may be only part of the solution. In this talk we introduce the concept of orthogonal scaling. Orthogonal scaling refers to features that can be added to the technology which significantly enhance the technology and which are sustainable over several generations of technology. We will examine some such orthogonal features that have either been implemented or being actively worked on. We will focus on three dimensional integration and "packaging" which depending on its implementation can address die size, performance, process simplicity and cost well beyond the expectation of semiconductor scaling. We summarize this talk with where the fundamental limits are and what our long-term options are for the evolution of a systems-based scaling methodology.

978-1-4799-0917-9/14 $31.00 © 2014 IEEE

Subramanian S. Iyer is an IBM Fellow at the Systems & Technology Group, and is responsible for technology strategy and competitiveness, embedded memory and three-dimensional Integration. He obtained his B.Tech. at IIT-Bombay, and Ph.D. at UCLA. His key technical contributions have been the development of the world's first SiGe base HBT, Salicide, electrical Fuses, eDRAM and 45nm technology used at IBM and IBM's development partners. His current technical interests and work lie in the area of 3-dimensional integration for memory sub-systems and the long-term semiconductor and packaging roadmap. He is a Distinguished Alumnus of IIT Bombay and received the IEEE Daniel Noble Medal for emerging technologies in 2012. He also studies Sanskrit in his spare time.

Challenges and Opportunities of 3D Chips Stacking - A Foundry's Perspective

Douglas Yu, *TSMC, Hsinchu, Taiwan*

3D Stacking of sub-systems such as DRAM on Logic has the advantages of small form factor, high bandwidth and high performance with low system power. General challenges involve cost and yields, known-good-stack testing, and thermal solutions of the stacking. However, there are many foundry-specific challenges in addition to the general ones. For example, we need various sub-system architectures (a separate Logic controller chip under the top unit, or integrate the controllers in SoC chips, passives included or not), various structures (eg. DRAM bigger or Logic bigger), and supply chain management (possible needs for multiple chips, passives and DRAM, etc.). The opportunities to address these issues include a new "Grand Alliance" foundry model plus innovations in technologies and business processes.

Douglas Yu is a Senior Director of at TSMC, currently in charge of Backend R&D. On interconnect; he has been responsible for on-chip metallization technology development for many nodes from 0.5mm down to 28nm. The interconnect includes Al/SiO2, Al/FSG, Cu/FSG, Cu/Low-K, Cu/ELK and Low-R/ELK.

On packaging, Douglas' role is to deliver wide range solutions from advanced bumpings, fine-pitch and low cost flip-chip, wafer-level-packagings and 3D-stackings, such as TSV stacking with CoWoSTM flow.

Doug served as general co-chairs of IEEE IITC, IEEE EDAPS, and ITRS Interconnect conferences. He also serves as an advisory board member of IEEE IMPACT and Industrial Advisory Board member of Microsystems Industrial Group at Microsystems Technology Laboratories/MIT. He received his Ph.D. degree on Material Science and Technology from Georgia Institute of Technology. He holds 355 US issued patents with numerous publications. He is an IEEE Fellow.

3D System Integration – Mitigating the Impact on CMOS Device Performance

Eric Beyne, *imec, Leuven, Belgium*

3D System integration allows for effective integration of heterogeneous device technologies, allowing for high bandwidth, short interconnect between the devices. This results in smaller form factors but also better signal/noise ratio, lower power consumption and/or faster systems.

The technology steps required in 3D system integration - such as Through-Silicon-Vias (TSV), wafer thinning and µbump die stacking – may however induce thermo-mechanical stresses on the semiconductor devices, resulting in device performance degradation. It is therefore important to understand the mechanisms generating these additional stresses and improve the technology to mitigate these undesired effects. In this presentation, the impact of TSVs and µbump assembly technologies on advanced planar and finfet CMOS devices will be discussed. A methodology was developed to assess the impact on advanced node devices, even before actually integrating TSV's or performing 3D stacking.

Eric Beyne obtained a degree in electrical engineering in 1983 and the Ph.D. in Applied Sciences in 1990, both from the Katholieke Universiteit Leuven, Belgium. Since 1986 he has been with imec in Leuven, Belgium where he has worked on advanced packaging and interconnect technologies. Currently, he is program director of imec's 3D System Integration program.

He is an active member of the IEEE-CPMT society, president of the IMAPS-Benelux committee and member of the IMAPS-Europe Liaison committee. He served as general chair of the IEEE-CPMT European System Technology Conference, ESTC2012, 17-20 September in Amsterdam.

3D Stacked CMOS Image Sensor Exmor RS™

Taku Umebayashi, *Sony, Kanagawa, Japan*

Sony has launched 3D Stacked CMOS Image Sensor mass production branded as Exmor RS™ in 2012 and reported at ISSCC 2013. The presentation will show some discussions such as Exmor RS™ overview, development motivation, stacked structure merit and efficiency. We also would like to mention some technical issues and phenomenon that we have solved during the development.

Taku Umebayashi received the B.S., and M.S. degrees from Waseda University, Tokyo, Japan, in 1987 and 1989 respectively. He is currently involved in Research and Development Plat Form of Sony Corporation. He has been working on the CMOS and image sensor device and process development.

Evolution of 3D Integration for CMOS Image Sensor Cameras

Lindsay Grant, *STMicroelectronics, Edinburgh, United Kingdom*

3D integration of a sensor IC bonded and directly connected to readout/processing IC can enable improved sensor and camera performance through higher resolution, added capability and reduced camera size. This presentation will look at the motivation for 3D stacked architecture in CIS and illustrate the evolution from FSI through BSI and into 3D. Key sensor/camera performance parameters and metrics will be presented and how they are linked to technology/architecture choices. A review of some different approaches for 3D integration of CIS cameras will be discussed including Above IC, wafer level camera and as well as 3D stacked CIS. Enabling integration steps will be identified and put in context with BSI and TSV. The theme of the presentation will be to show the evolution of CIS technology over past decade and the potential future opportunities in 3D.

Lindsay Grant is a Pixel Technology Expert in STMicroelectronics with more than 25 years in semiconductor R&D. Since 1999 he has worked on multiple generations of CMOS Image Sensor pixel & process technology from 5.6um to 1.1um pitch including TSV and BSI developments. Since 2006 his R&D interest has also covered embedded SPAD devices in CMOS. He has been a member of the steering committee for two European funded programs in SPAD development. He has authored or co-authored more than 40 technical papers and conference presentations, been an invited speaker on imaging technology at the ISSCC Forum on Image Sensors, at the Fraunhofer IMS Workshop, and at ESSDERC. He sits on the Technical Program Committee for IISW since 2009 and has co-chaired the Image Sensors Europe conference in London since 2008. He has a degree in Physics from St. Andrews University, Scotland. Currently he manages a team working on pixel characterization, runs a program of sponsored PhD studentships and is a Director of STMicroelectronics R&D, UK Ltd.

F3: *Adaptive Design Techniques for Energy Efficiency*

Organizer: Eric Fluhr, *IBM, Austin, TX*

Committee: Eric Fluhr, *IBM, Austin, TX*
Michael Polley, *Samsung Mobile, Dallas, TX*
Se-Hyun Yang, *Samsung, Yongin-Si, Korea*
Vasantha Erraguntla, *Intel, Bangalore, India*
Tobias Noll, *RWTH Aachen University, Aachen, Germany*
Kees van Berkel, *Ericsson, Eindhoven, The Netherlands*

Silicon technology continues to shrink, allowing a greater density of devices per unit area. At the same time, process and chip variations increase, making it more difficult to build power-efficient, functional chips. We first review the physical issues that are driving such increasing variation. Then, we look at state-of-the-art approaches to adapt to this variation. Voltage management is a key element, both in innovative management circuits, as well as in integration across the hardware/software design stack for adaptive control of the system. Leveraging error tolerance, where available, is as critical as building new memories that can avoid or mitigate sensitivity to variation. Finally, innovative models to predict the effects of variations are critical to guiding choices in the design process.

Process Technology Variation Characteristics and Trends
Martin Giles, *Intel, Hillsboro, OR*

As advanced technologies continue the pursuit of Moore's Law, one key challenge is the management of process technology variation. Continued improvements in process technology and transistor architecture have enabled the development of advanced technologies with both scaled transistor area and improved device variation to support lower voltage operation. This presentation will consider the characteristics and trends of device variation sources across technology generations, test structures to quantify variation effects, some circuit implications of different kinds of variation, and the challenges and outlook for the future.

Martin D. Giles is a Senior Principal Engineer in the Process Technology Modeling Department at Intel Corporation, Hillsboro, OR. He conducts research and development on advanced logic and memory technologies, with a focus on variation issues, and leads strategic planning for process technology modeling. From 1998 to 2007, he was Program Manager for process/device modeling at Intel, and prior to that, conducted research and development in advanced process modeling for logic technologies. From 1990 to 1994, he was an assistant professor in the EECS department at the University of Michigan, working on silicon process modeling and related topics. From 1984 to 1990, he was a member of the VLSI device analysis and simulation group at AT&T Bell Laboratories. Dr. Giles received his B.A. and M.A. degrees in natural sciences from Cambridge University, England in 1981 and 1985, respectively, and the M.S.E.E. and Ph.D. degrees from Stanford University, CA, in 1983 and 1984, respectively. He is a Fellow of the IEEE.

Robust and Resilient Systems from the Bottom-Up: Circuits, Architecture and Software Integration
Vijay Reddi, *University of Texas, Austin, TX*

How do we design variation-tolerant processors that meet historically established high-reliability standards without exceeding a fixed power budget and cost constraint? This is the challenge facing present-day and future system architects. Recent work in academia and industry suggests that a tight coupling and interaction between the hardware and software layers can overcome this challenge. In this talk, I discuss the broad range of work in circuits, architecture and software to overcome one particularly important form of variation, voltage noise, which can lead to excessive guardbands that compromise performance and decrease energy efficiency. We begin with a quick primer on voltage noise, followed by an overview of a generic integrated hardware-software emergency management system, where the hardware is responsible for guaranteeing reliable operation in the event of a voltage emergency, and software is responsible for eliminating recurring emergencies. Hardware solutions are reactive and lack the global knowledge about execution that is necessary to enable root-cause analysis to eliminate recurring emergencies. Compilers and operating systems possess such global knowledge, and they can, therefore, morph execution and application code on-the-fly to eliminate recurring hardware penalties, but they require circuit and microarchitecture support. A holistic system can enable us to sustain operation at peak points without catastrophic failures, and severe fail-stop overheads.

Vijay Janapa Reddi is an Assistant Professor in the Department of Electrical and Computer Engineering at the University of Texas, Austin. He received a Ph.D. in Computer Science from Harvard University. His interests are in the area of computer systems, focusing on the interactions between hardware and software. He explores new opportunities and synergies for cross-layer solutions that improve processor- and system-level power, performance and reliability. He has authored several papers in these areas and has received IEEE Micro Top Picks and Best Paper Awards in Computer Architecture. In the past, he has also worked on architecture and compiler aspects at Intel, VMware, AMD Research and Microsoft Research.

Ultra-Low Power Computing Systems with Graceful Performance Degradation on Unreliable Silicon
Andreas Burg, *EPFL, Lausanne, Switzerland*

This talk describes the recent issues in the design of ultra-low power (ULP) computing systems due to the increasing process variation in nanometer technologies, and introduces the concept of inherent fault-tolerant applications. This new concept enables trade-offs between power, performance and area overhead at the system level for various types of inaccuracies, thanks to a tight link between the software and hardware layers, which jointly can suitably model output distortions due to failures. Thus, according to the primary points of impact and frequency of the hardware failures, it is possible to design ULP software-programmable computing architectures that support a graceful quality degradation at the system level. This new approach to designing ULP computing architectures is assessed with different applications from the areas of multimedia and bio-signals processing.

Andreas Burg was born in Munich, Germany, in 1975. He received his Dipl.-Ing. degree from the Swiss Federal Institute of Technology (ETH) Zurich, Zurich, Switzerland, in 2000, and the Dr. sc. techn. degree from the Integrated Systems Laboratory of ETH Zurich, in 2006.

In 1998, he worked at Siemens Semiconductors, San Jose, CA. During his doctoral studies, he was an intern with Bell Labs Wireless Research for a total of one year. From 2006 to 2007, he held positions as postdoctoral researcher at the Integrated Systems Laboratory and at the Communication Theory Group of the ETH Zurich. In 2007 he co-founded Celestrius, an ETH-spinoff in the field of MIMO wireless communication, where he was responsible for the ASIC development as Director for VLSI. In January 2009, he joined ETH Zurich as SNF Assistant Professor and as Head of the Signal Processing Circuits and Systems group at the Integrated Systems Laboratory. Since January 2011, he is a Tenure Track Assistant Professor at the École Polytechnique Fédérale de Lausanne (EPFL), Lausanne, Switzerland, where he is leading the Telecommunications Circuits Laboratory in the School of Engineering. In his professional career, he was involved in the development of more than 35 ASICs, he has been on the TPC of several conferences and served as the technical program chair for the IEEE VLSI-SoC 2012 conference. In 2013 he served as AE of the IEEE Transactions on Circuits and Systems and he is on the editorial team of the MDPI Journal on Energy Efficient Electronics and Applications.

In 2000, Mr. Burg received the Willi Studer Award and the ETH Medal for his diploma and his diploma thesis, respectively. He was also awarded an ETH Medal for his Ph.D. dissertation in 2006. In 2008, he received a four-year grant from the Swiss National Science Foundation (SNF) for an SNF Assistant Professorship. In 2011 he was the co-recipient of the Best Paper Award of the EURASIP Journal on Image and Video Processing and in 2013 he was a co-recipient of an ISCAS Best Demonstration Award.

978-1-4799-0917-9/14 $31.00 © 2014 IEEE

Enabling Adaptive Techniques in High-Volume Design

***Sam Naffziger**, AMD, Fort Colins, CO*

Variations significantly impact the performance levels that can be delivered in a high-volume design given a certain average silicon capability. These variations come in the form of process variations, DC voltage uncertainty, AC voltage variations, device aging, environmental conditions and workload changes. This talk will discuss how these variations manifest at the product level with traditional high-volume test flows, and then move into ways that adaptive techniques can be used to mitigate these effects, while maintaining the robustness and efficiency required for today's flexible manufacturing flows. The goal is to deliver as much of the true, average capability of the silicon as possible, while maintaining yield, minimum performance levels and robust operation. Some of the most promising circuits and methodologies for achieving this goal will be discussed along with supporting data.

Sam Naffziger received the B.S.E.E. degree from the California Institute of Technology, Pasadena, CA, in 1988, and the M.S.E.E. degree from Stanford University, Stanford, CA, in 1993. He joined Hewlett Packard in 1988, and spent eight years working on PA-RISC processor development, including floating point, out-of-order execution and circuit methodologies employed in the PA8000. He then became part of the Itanium2 Joint Development Team with Intel Corporation, Fort Collins, CO, and has led the design of both the first Itanium2 processor (McKinley) and also Montecito. In 2006, he helped start the Mile High Design Center of Advanced Micro Devices in Fort Collins, CO, to work on next generation processor designs. Sam is widely known for his contributions to innovative power management and reduction approaches on Intel and AMD processors. He holds over 100 US patents on processor circuits and architecture and has over 30 IEEE publications and presentations. Mr. Naffziger chaired the digital subcommittee of ISSCC for 5 years, and is a Corporate Fellow at AMD.

Design of Adaptive and Resilient Circuits for Power-Delivery Solutions

***Ramnarayanan Muthukaruppan**, Intel, Bengaluru, India*

This talk will outline the challenges for power delivery in deep sub-micron processes, including the following topics. 1) A look at the design of power gates for optimal power delivery and the future of power gating for optimal power at targeted performance. 2) Strategies for improving the performance of circuits with optimal power gating. 3) Partitioning of power domains into integrated VRMs and power gates. 4) Design of adaptive VRM architectures for reduced power and reduced noise. 5) An introduction to different VRM architectures and their applications, including linear regulators and switching regulators.

Ram Muthukarappan completed his master's degree in EE in 1998 from the Birla Institute of Technology, Pilani, India. He worked with National Semiconductor (now Texas Instruments) from 1998 to 2001, then with Analog Devices from 2001 to 2004. He has been with Intel since 2004 as a system architect and his field of interest is circuit design with a focus on high-speed serial IOs, power delivery circuits, and data converter circuits, with application emphasis on tablets and smartphone SoCs. He has worked on the development of system power-delivery solutions, including adaptively biased linear and switching regulators, including switched capacitor regulators. He has five patents filed in the field of power-delivery circuits and configurable memory circuits.

Variability and Design of SRAM in Nanoscale CMOS and Emerging Device Technologies

***Ching-Te Chuang**, National Chiao Tung University, Hsinchu, Taiwan*

This presentation reviews the design challenges and techniques for SRAMs in the "End of Scaling" nanoscale CMOS and emerging device technologies. The first part discusses design directions and leakage/variation/degradation tolerant SRAM circuit techniques to mitigate various stability, performance and reliability constraints in conventional planar CMOS technology. Examples are given to illustrate the evolution and relative merits of various read/write-assist circuit techniques. Alternative cell structures to improve static/dynamic margins and V_{min}, and subthreshold SRAMs for ultra-low-power applications are discussed. The second part addresses device variability and SRAM designs in UTB-SOI (Ultra-Thin-Body SOI), FinFET/trigate devices, hetero-channel devices, nanowire FET and tunneling FET technologies. The intrinsic device characteristics and impacts of variations on these emerging devices will be addressed with emphasis on the low voltage/power SRAM applications.

Ching-Te Chuang received the B.S.E.E. from National Taiwan University, Taipei, Taiwan in 1975 and Ph.D. degree in Electrical Engineering from University of California, Berkeley, CA in 1982. From 1977 to 1982, he was a research assistant in the Electronics Research Laboratory, University of California, Berkeley, working on bulk and surface acoustic wave devices. He joined the IBM T. J. Watson Research Center, Yorktown Heights, NY in 1982. From 1982 to 1986, he worked on scaled bipolar devices, technology, and circuits. He studied the scaling properties of epitaxial Schottky barrier diodes, did pioneering work on the perimeter effects of advanced double-poly self-aligned bipolar transistors, and designed the first sub-nanosecond 5Kb bipolar ECL SRAM. From 1986 to 1988, he was Manager of the Bipolar VLSI Design Group, working on low-power bipolar circuits, high-speed high-density bipolar SRAMs, multi-Gb/s fiber-optic data-link circuits, and scaling issues for bipolar/BiCMOS devices and circuits. Since 1988, he has managed the High-Performance Circuit Group, investigating high-performance logic and memory circuits. Since 1993, his group has been primarily responsible for the circuit design of IBM's high-performance CMOS microprocessors for enterprise servers, PowerPC workstations, and game/media processors. Since 1996, he has been leading the efforts in evaluating and exploring scaled/emerging technologies, such as PD-SOI, UTB-SOI, strained-Si devices, hybrid orientation technology, and multi-gate/FinFET devices, for high-performance logic and SRAM applications. Since 1998, he has been responsible for the Research VLSI Technology Circuit Co-design strategy and execution. His group has also been very active and visible in leakage/variation/degradation tolerant circuit and SRAM design techniques. He has received an Outstanding Technical Achievement Award, a Research Division Outstanding Contribution Award, 5 Research Division Awards, and 12 Invention Achievement Awards from IBM. He took early retirement from IBM to join National Chiao-Tung University, Hsinchu, Taiwan, as a Chair Professor in the Department of Electronics Engineering in February 2008. He is the founding Director of ASE/NCTU 3D-IC Joint Research Center at National Chiao-Tung University. He has received the Outstanding Scholar Award from Taiwan's Foundation for the Advancement of Outstanding Scholarship for 2008 to 2012. His current research interests are in the areas of nanoscale CMOS SRAM design and characterization, emerging devices (multigate/gate-all-around devices, hetero-channel devices, nanowire MOSFET and tunneling FET) for logic, SRAM and analog applications, 3D-IC and integrated MEMS/TSV/CMOS microsystems for bio-signal recording.

Dr. Chuang served on the Device Technology Program Committee for IEDM in 1986 and 1987, and the Program Committee for Symposium on VLSI Circuits from 1992 to 2006. He was the Publication/Publicity Chairman for the Symposium on VLSI Technology and Symposium on VLSI Circuits in 1993 and 1994, and the Best Student Paper Award Subcommittee Chairman for the Symposium on VLSI Circuits from 2004 to 2006. He was elected an IEEE Fellow in 1994 "For contributions to high-performance bipolar devices, circuits, and technology". He has authored many invited papers in international journals such as the International Journal of High Speed Electronics, Proceedings of IEEE, IEEE Circuits and Devices Magazine, and Microelectronics Journal. He has presented numerous plenary, invited or tutorial papers/talks at international conferences such as the International SOI Conference, DAC, VLSI-TSA, the ISSCC Microprocessor Design Workshop, VLSI Circuit Symposium Short Course, ISQED, ICCAD, APMC, VLSI-DAT, ISCAS, MTDT, WSEAS, VLSI Design/CAD Symposium, and International Variability Characterization Workshop. He was the co-recipient of the Best Paper Award at the 2000 IEEE International SOI Conference. He holds 50 U.S. patents with another 20 pending. He has authored or coauthored over 360 papers.

Predictive Simulation of Performance and Variability at and Beyond 14nm and Impact on Design

***Asen Asenov**, University of Glasgow, Glasgow, United Kingdom*

The statistical variability introduced to bulk transistors particularly by random discrete dopants (RDD) in the channel region is reaching an intolerable level in 20nm CMOS, impeding low-power SRAM design and area scaling and compromising circuit performance and yield. Better performing, variability-resilient devices like FinFETs and ultra-thin body (UTB) SOI transistors have been introduced to remedy bulk CMOS deficiencies. This talk forecasts the performance and statistical variability at 14nm CMOS and beyond via comprehensive predictive TCAD simulations. Alternative solutions to expand the life of bulk transistors, e.g. epitaxial bulk channel-first and channel-last technologies, are discussed. Compact models extracted based on TCAD simulations allow evaluation of the impact of the corresponding device performance and variability on design. Results for bulk, SOI FinFETs, and FD-SOI transistors in 22-28nm are shown along with 14nm performance and variability comparisons. Projecting to 10nm, can FD-SOI deliver, or is a switch to SOI FinFETs needed? What will new channel materials such as SiGe, GE and III-V bring?

Asen Asenov is a founder and CEO of Gold Standard Simulations Ltd., a leader in the physical simulation of statistical variability, statistical compact model extraction and generation technology, and statistical circuit simulation. He is also a James Watt Professor in Electrical Engineering at Glasgow University, leading the 30-member-strong Glasgow Device Modeling Group, and directing the development of advanced CMOS models and tool. He is an IEEE Fellow.

F4: *mm-Wave Advances for Active Safety and Communication Systems*

Organizer: Marc Tiebout, *Infineon Technologies, Villach, Austria*

Committee: Brian Floyd, *North Carolina State University, Raleigh, NC*
Mike Keaveney, *Analog Devices, Limerick, Ireland*
Marc Tiebout, *Infineon Technologies, Villach, Austria*
Pierre Busson, *STMicroelectronics, Crolles, France*
Kenichi Okada, *Tokyo Institute of Technology, Tokyo, Japan*

Recent advances in microwave and mm-Wave applications targeting existing and upcoming safety, radar and communication systems will be presented. Speakers from university and major industry companies will highlight both system aspects as well as implementation aspects, including packaging and high-volume production testing. Topics include car-to-car / car-to-x communications, FMCW and pulse radar, MIMO and novel CMOS-based architectures. Silicon implementations for frequencies from 5GHz to 240GHz in SiGe and CMOS will be presented. Emphasis is placed on automotive 77-to-79GHz radar, which is the highest-volume existing mm-Wave application.

Wireless Transceivers for Car2Car & Car2X Systems and Applications

Marc Klaassen, *BU Automotive - Car Entertainment Solutions, NXP Semiconductors, Nijmegen, The Netherlands*

Intelligent Transport Systems (ITS) is an emerging application area, opening opportunities for more safe and green driving. One of the enabling factors is the car-to-car (C2C) and car-to-infrastructure (C2I) wireless communication. Broadcasting the car position and velocity information allow cars to build a dynamic model of their surroundings. Worldwide, a significant effort is ongoing in this field, as is visible in industrial consortia, in projects and in standardization bodies. One of the important questions to be answered is: how is ITS going to be introduced into the cars? In order to have an effective network at least 10 percent of all cars need to be equipped with an ITS module in a scalable manner for low-, mid- and high-end cars. There are several aspects that determine scalability of the system. The location of the antenna(s) is an important aspect. Antenna diversity can improve reception performance. The radio front-end module, containing the IEEE802.11p/1609.x transceiver, can be connected close to the antenna if room permits, or is centralized in a box. Cable costs and performance can be impacted by this choice. Another aspect of system partitioning is the number of services received concurrently. As with other connected nodes, the Car will be equipped with multiple sensors (cameras, radar, ITS). These triggers will be used concurrently to enhance the safety of the driver.

Marc Klaassen was born in Nijmegen, The Netherlands, in 1968. He received the MSEE in 1992 at the Eindhoven University of Technology in the field of Wireless Communications. After completing the degree, he did a post graduate study until 1995 in the department of Wireless Engineering of the Eindhoven University of Technology in cooperation with Dr. Neher Laboratories (KPN) and ESA – ESTEC. From 1995 to 2011, he worked for NXP (formerly Philips Semiconductors) in the Netherlands as well as in the USA. The main work focus was on Digital Signal Processing and Communication Techniques. This has resulted in various applications in the field of Terrestrial Radio and TV (DVB-T, DVB-C, DVB-H, DAB), Satellite Reception (DVB-S, DVB-SH), Cable Modem (EuroDocsis and Docsis), Wireless communication (UWB, 802.11abg), Intelligent Transportation systems (IEEE 1609.x, 802.11p), and others.

Current and Future Application Requirements of mm-Wave Radar Sensors

Nils Pohl, *Fraunhofer FHR, Wachtberg, Germany*

In the last decade, modern SiGe technology began to enable the use of the mm-Wave frequency domain. In particular, their use in FMCW Radar systems, e.g. in 77GHz automotive radar and imaging systems, enabled high-performance systems even for mass market applications. This contribution will first summarize, at the system level, the challenging requirements, e.g. regarding output power, phase noise, tuning characteristic, and power dissipation. Especially, the signal synthesis is a key component for radar performance. The commonly used VCO circuits and frequency multiplier concepts and their limitations will be discussed. Finally ultra-wideband radar circuits in SiGe around 80 GHz and 240 GHz will be shown together with application examples.

Nils Pohl received his Dipl.-Ing. and Dr.-Ing. degrees from Ruhr-University Bochum in Electrical Engineering in 2005 and 2010, respectively. From 2006 to 2011, he was a Research Assistant with the Institute of Integrated Systems working on integrated circuits for mm-Wave radar applications. In 2011, he became assistant professor for Integrated Systems at Ruhr-University Bochum. Since 2013, he has been the Head of Millimetre-Wave Radar and High-Frequency Sensors Department of Fraunhofer FHR in Wachtberg. His main fields of research are concerned with the design and optimization of mm-Wave integrated SiGe circuits and system concepts with frequencies up to 100GHz and above (especially for wideband radar applications), as well as frequency synthesis and antennas. He is the author or coauthor of more than 50 scientific papers and has been awarded several patents.

Packaging Technologies and Production Test for Automotive Radar Front-End Products

Sergio Pacheco, *Freescale Semiconductors, Phoenix, AZ*

Ever since the advent of the seat belt, safety has become a key differentiator in the automotive industry. This trend continued with airbags, anti-lock braking systems, and now with stability control. Although these systems have been pervasive for the past 20 years, the number of accidents and fatalities in the US has remained steady. The next step on the road towards greater safety in automobiles is the use of active sensing for collision avoidance. This talk will cover the current state-of-the art automotive radar systems and recent advances in the development of packaging technologies to enable such systems. The thermal, mechanical, and electrical impact on the mm-Wave performance of radar front-end for various package types will be discussed. In addition, the challenges and achievements in the development and characterization of a fully automated mm-Wave electrical test system for high-volume production will be presented.

Sergio Pacheco is a Senior Member of the Technical Staff and Program Manager in the Technology Solutions Organization of Freescale Semiconductors. His current interest lies in the development of mm-Wave packaging technology for automotive radar products. Prior to that, he was the Automotive Radar Program Manager in the Sensor and Actuator Solutions Division. He first joined Motorola as a Principal Research Engineer in the Emerging Technologies Group where his research focused on integrated passives and microelectromechanical structures (MEMS) and their application to RF and wireless communication systems. He is a member of the IEEE Microwave Theory and Techniques Society (MTT-S) and the IEEE Electron Device Society (EDS). He has over 45 publications in internationally refereed journals and conferences as well as a chapter on "Microelectromechanical Switches for RF Applications" in the book RF Technologies for Low-Power Wireless Communications. He has currently six patents issued and/or applications. He received B.S.E.E. and M.S.E.E degrees from Auburn University, and a Ph.D. in Electrical Engineering from the University of Michigan.

CMOS Realization of 24/26/77/79 GHz FMCW and Pulse Radars for Automotive Applications

Jri Lee, *National Taiwan University, Taipei, Taiwan*

This talk covers the design principles and physical realization of modern FMCW and pulse radar systems in CMOS technology. Targeting mass production of automotive safety equipment such as automatic cruise control (ACC), blind-spot detection (BSD), and parking-assistant system (PAS), the commercial radars must be fabricated in a high-yield and low-cost way. CMOS technology provides unbeatable advantages in these aspects, making it suitable for fully-integrated solutions. It is especially true as the operational speed of the CMOS circuits can compete with its compound counterparts. In this talk, we discuss the design considerations as well as circuit details of advanced FMCW and Pulse radars. Two engineering prototypes will be demonstrated with complete measurement results.

Jri Lee received the B.Sc. degree in electrical engineering from National Taiwan University (NTU), Taipei, Taiwan, in 1995, and the M.S. and Ph.D. degrees in electrical engineering from the University of California at Los Angeles (UCLA), both in 2003. He joined National Taiwan University in 2004, where he is currently a Professor of electrical engineering. His current research interests include high-speed wireless and wireline transceivers, phase-locked loops and applications, and mm-Wave circuits. Prof. Lee received the Beatrice Winner Award for Editorial Excellence at the 2007 ISSCC, and the Takuo Sugano Award for Outstanding Far-East Paper at the 2008 ISSCC. He has served on the Technical Program Committees of ISSCC from 2007 to 2010, and the Symposium on VLSI Circuits since 2008. He was a guest editor of the IEEE Journal of Solid State Circuits in 2008, and served as a Distinguished Lecturer of the IEEE Solid-State-Circuits Society (SSCS) from 2011-2013.

Applications and Implementations of Integrated Direct-Sampling Impulse Radar Systems

Ta-Shun Chu, *National Tsing Hua University, Hsinchu, Taiwan*

Wireless non-contact sensors can enable detection, localization, and monitoring of humans along with specific features such as gait and cardiopulmonary activities. Wireless sensors can be embedded in the environment and networked with the existing wireless infrastructure to create an intelligent and responsive ambient where the health of children, patients, and the elderly can be monitored without intrusion. Impulse radars have more advantages than continuous-wave radars in terms of depth resolution, multi-path effect, material penetration, and power emission. A target can be characterized by a radar system through the knowledge of two parameters; one is the time of arrival (TOA) and the other is the direction of arrival (DOA).In impulse radar systems, the time of flight of the scattering pulse has to be measured for distance estimation. A digital-to-time converter (DTC) can generate a time interval between the trigger signals of samplers in the TX and RX with a fine resolution over a wide range. Therefore, a target can be tracked with a precise TOA. An antenna array that enables beamforming can achieve spatial selectivity in a radar system. DTCs in different channels can produce the trigger signals of samplers with different progression delays. The different time shifts among the different channels of the TX and RX in the antenna array can create a beam pattern for determining the DOA. The architectures and building blocks of direct-sampling radar transceivers will be discussed.

Ta-Shun Chu (S'06-M'10) received the B.S. degree in Civil Engineering and the M.S.degree in Applied Mechanics from the National Taiwan University, Taipei, Taiwan, R.O.C., in 2000 and 2002, respectively, and the Ph.D. in Electrical Engineering from the University of Southern California, Los Angeles, in 2010. In 2010, he joined the Department of Electrical Engineering, National Tsing Hua University as an Assistant Professor.

Millimeter-Wave MIMO Radar in CMOS for Vehicular Applications

Harish Krishnaswamy, *Columbia University, New York, NY*

Vehicular radar is perhaps the most compelling application of silicon-based mm-Wave circuits and systems. While multiple-antenna systems, such as phased arrays, have been explored for mm-Wave vehicular radar and enable operation under weak-SNR conditions, the potential of communications-inspired MIMO techniques as applied to radar (or the so-called MIMO radar concept) has yet to be significantly explored. This presentation will initially discuss MIMO radar principles at the system level, including space-time array processing, multi-beam MIMO radar, waveform trade-offs etc., and will then move on to CMOS implementations in the 22-to-29GHz frequency range.

Harish Krishnaswamy received the B.Tech. degree in Electrical Engineering from the Indian Institute of Technology-Madras, India, in 2001, and the M.S. and Ph.D. degrees in Electrical Engineering from the University of Southern California (USC) in 2003 and 2009, respectively. He joined the EE department of Columbia University as an Assistant Professor in 2009. His research group at Columbia, funded by various federal agencies, including NSF and DARPA, and industry, focuses on mm-Wave CMOS power amplifiers and transmitters, sub-mm-Wave devices, circuits and systems in CMOS, broadband RF receivers for cognitive radio, field-programmable, waveform-adaptive RF CMOS transmitters and communications-inspired radar and imaging systems. He received the IEEE International Solid State Circuits Conference (ISSCC) Lewis Winner Award for Outstanding Paper in 2007. He also received the Best Thesis in Experimental Research Award from the USC Viterbi School of Engineering in 2009, and the DARPA Young Faculty Award in 2011.

24GHz Versus 79GHz Automotive Radar Sensors, Applications and Implementations

Patrice Garcia, *STMicroelectronics, Crolles, France*

For years the automotive industry is developing sensors in order to improve security on road. Radar sensors and even more coupled with infrared imaging and communication systems offer the best solution to prevent from traffic causalities.

More generally the radar sensors are more and more involved to improve the vehicle occupant security. A lot of driving improvement can then be considered like the most well-known automotive cruise control, cross line detection, crash detection that helps to bring the safety through for example a faster air bag deployment and seat belts tensioner activation.

Today the radar sensors systems are either based on wide band frequency range or on Doppler effect. Based on these two different architectures the radar sensor design techniques for on Silicon integrated sensor will be presented. A specific design tool that allows to take into account for the frequency and proximity effects directly from the schematic allows a compact design. This also brings the opportunity to perform a complete Silicon integration that is a key point for these applications..

Patrice Garcia received the M.S.E.E. and Ph.D. degrees in Low-IF 900MHz BiCMOS Receiver for cellular communications from the National Polytechnic Institute of Grenoble, France, and STMicroelectronics, Grenoble, France in 1999. He joined STMicroelectronics, Crolles, France, in 1999, where he contributed to the development of GSM/WCDMA RFIC Front-end. From 2005 to 2012, he was leading designs in RF/mmW and he is now involved in 77GHz BiCMOS radar design for automotive.

978-1-4799-0917-9/14 $31.00 © 2014 IEEE

F5: *Low-Power Radios for Sensor Networks*

Organizer: Woogeun Rhee, *Tsinghua University, Beijing, China*

Chair: Gangadhar Burra, *Qualcomm-Atheros, San Jose, CA*

Committee: Kazutami Arimoto, *Okayama Prefectural University, Okayama, Japan*
Pieter Harpe, *Eindhoven University of Technology, Eindhoven, The Netherlands*
Brian Otis, *University of Washington, Seattle, CA*
David Ruffieux, *CSEM, Neuchatel, Switzerland*

Sensor node systems are expected to be a major growth field for semiconductor markets, and will connect to the cloud in a future cyber physical world. Wireless sensor networks (WSN) face multiple challenges from system design to low-power electronics and energy sources. Ultra-low-power radios are key elements in such systems, putting high demand on energy efficiency in different modes of operation (active, wake-up and sleep). This forum presents system perspectives and practical design aspects in various energy-efficient and short-range radio circuits and systems, including an introduction to various applications and their requirements for RF and digital signal processing in WSN systems. The technologies range from RF to digital signal processing and algorithms to give comprehensive understanding of recent advances. The forum begins with two high-level design talks on low power radio SoCs. The following four talks present ultra-low-power transceivers for body-area networks, sensor nodes, and health-monitoring applications. The last two talks cover efficient power management and emerging compression methodologies.

Wireless Medical Device Communication: Performance Considerations, System Design, and Recent Innovations

Peter Bradley, *Microsemi, Sydney, Australia*

Communicating within the body presents many unique challenges associated with the small size, very low power and attenuation of human tissue. The key requirements, regulations, standards, and system issues are presented. The capabilities of new technology for communicating to medical implants are presented. Facilitating developments include recent advances in integrated circuit technology and the acceptance of a worldwide band for medical implant communication in the 401-to-406 MHz (Medical Implant Communication Service) range. Such technology will enable new applications and improved healthcare with examples presented from a range of areas including endoscopic camera capsules, remote monitoring of implantable defibrillators, neuro-stimulators, bionic eyes, and dosimetric monitoring. The use of other ISM bands for medical communications are also discussed.

Peter D. Bradley received the B.E. (Elec. Eng. 1st class honors) in 1987 and the M.E. degree in Biomedical Engineering in 1996, both from the University of NSW, Sydney, Australia, and the Ph.D. degree in Medical Physics from the University of Wollongong, Australia in 2000. Peter has over 25 years experience in medical device design. He was the system architect for several of Zarlink-Microsemi's medical transceivers, which dominate the implanted medical transceiver market including the EEtimes Product of the Year in 2008. Peter also published the first paper detailing the effects of cosmic radiation on medical implants. He has over 20 published papers and 9 patents on low-power transceiver techniques. Peter is a medical device and low-power communications consultant actively involved in the development of systems for next-generation medical transceivers.

Low-Power Radio SoCs and Microsystems for Tiny-Battery-Operated Healthcare and Lifestyle Applications

Vincent Peiris, *CSEM, Neuchatel, Switzerland*

Ultra-low-power integrated radios for wireless sensors in body-area networks are driven by many stringent and sometimes contradictory requirements. A starting constraint is that many healthcare applications are, and will remain, operated from tiny batteries because they are reliable and well-known, but on the other hand come with a variety of pros and cons in terms of energy density, voltage curves and cost. Furthermore, a huge variety of applications can be foreseen, from fairly simple monitors to sophisticated systems embedding multi-sensing, signal conditioning, data processing and communication, thereby leading to specific compromises in terms of system breakdown, power consumption, and miniaturization. In addition, the miniaturization trade-space ranges from integration along the CMOS SoC path (More Moore) or else SiP implementation using MEMS-based approaches (more than Moore). Last but not least, a variety of high-level connectivity requirements can be considered, from simple standard-based networking topologies up to highly optimized proprietary approaches for more sophisticated multi-sensor networking. Within this broad landscape of requirements, this forum talk describes a selection of low-power integrated wireless developments – covering 1V RF CMOS and MEMS-based miniature radio implementations – that address specific healthcare and lifestyle scenarios.

Vincent Peiris holds a Ph.D. from the Swiss Federal Institute of Technology in Lausanne, and has been the Head of the RF and Analog IC Program at CSEM since 1999. His areas of interest are RF CMOS microelectronics and low-power transceiver design for wireless sensor networks and wireless body area networks. He has been project leader for several large size RF transceiver developments involving Swiss and European industrial partners and is currently the coordinator of the EU FP7-WiserBAN project.

Multistandard Transceiver and SoC Design for WBAN Applications

Alan Wong, *Toumaz Microsystems, Abingdon, United Kingdom*

In recent years, there has been significant interest and growth in low-power wireless technologies beyond traditional consumer use cases into medical applications and Wireless Body Area Networks (WBAN). To date, application-specific wireless proprietary solutions and protocols for specific biomedical devices or products have been the norm, limiting the extent of any one BAN. Hence the WBAN community has worked together to develop a new wireless communication standard, IEEE802.15.6, to promote interoperability between all devices in and around the body. In parallel, the consumer electronics industry has migrated existing standardized wireless protocols for personal-area networks such as Bluetooth to meet the demanding low-energy yet robust needs of WBAN. This talk discusses some of the key chip and wireless architecture considerations, circuit building blocks, and presents a WBAN SoC case study in 1V 65nm CMOS technology to allow multi-standard interoperability between ultra-low power body-worn WBAN sensor nodes.

Alan Wong received the M.Eng. (Hons) degree in Engineering Science from the University of Oxford, UK in 1997. He is currently with Toumaz Ltd. UK, where he leads the IC design team actively engaged in SoC development for medical WBAN and consumer wireless applications. He has been actively involved in the IEEE 802.15.6 task group working to develop a wireless standard for body-area networks, with a particular focus on medical monitoring applications. Prior to Toumaz, he held positions with Sony Semiconductor, Tokyo Electron Ltd, and the University of Oxford, and has been working in wireless IC and SoC design since 1998.

Wireless Communication for Cubic-mm Sensor Nodes

David D. Wentzloff, *University of Michigan, Ann Arbor, MI*

Over the last two decades, wireless computers have evolved from the laptop to the smartphone to cm-scale sensors. At each step, the volume shrinks by 2-to-3 orders of magnitude, as does the battery size, while the functionality and computing power remains constant or increases. With thin-film battery technology and CMOS scaling, we can now envision complete sensor nodes at the 1 cubic-mm scale. In each of these computing platforms, wireless communication has been essential, and at the cubic-mm scale, wireless may be the only means of communicating with a contact-less device. As volume has reduced, radios have consumed a larger fraction of the energy budget, a trend that cannot continue into the cubic-mm regime. This talk will give an overview of mm-scale sensor nodes, and describe the primary challenges facing the design of integrated radios at this scale. Recent results are presented from CMOS and SiGe ICs developed in the Wireless Integrated Circuits and Systems group at the University of Michigan that address these challenges.

David Wentzloff received the B.S. degree in Electrical Engineering from the University of Michigan in 1999, and the S.M. and Ph.D. degrees from MIT in 2002 and 2007, respectively. Since 2007 he has been at the University of Michigan, where he is currently an Associate Professor of Electrical Engineering and Computer Science. He is the recipient of the 2009 DARPA Young Faculty Award, the 2012 NSF CAREER Award, and several other teaching and best paper awards. His research group focuses on ultra-low power radios, body area networks, and mm-scale sensor nodes.

CMOS Radios for Health Monitoring: Closing the Gap Between Power and Performance

Jagdish Pandey, *Qualcomm, San Diego, CA*

As low-power radio circuits are enabling new technology avenues such as health monitoring, both industry and academia have taken up the design challenge of making extremely reliable yet lost-cost, low-power CMOS radios. Innovative system architectures and circuits are explored to achieve this goal. Over the last few years, we have seen sub-mW systems from academia that sacrifice performance and reliability to achieve low power. At the same time, commercial products that guarantee robust operation have power consumption in 10's of mWs. As far as implantable systems go, full energy autonomy is absolutely necessary and a drastic reduction in power is mandatory while providing robust operation over a highly variable environment. The twain need to meet to enable a technology that will perhaps have the most direct impact on human lives, yet! In this talk, we discuss some key ideas that have narrowed this gap between desired performance and power.

Jagdish Pandey received a Bachelor's degree in Electrical Engineering from the Indian Institute of Technology Madras in 2003, a Master's degree from the Indian Institute of Science Bangalore in 2007, and a doctorate from the University of Washington Seattle in 2011. He currently works at Qualcomm San Diego where he focuses on cellular RFIC development. His research interests are primarily in energy-efficient communication circuits and systems.

Challenges in ULP Event-Driven Transceiver Design

Guido Dolmans, *imec - Holst Centre, Eindhoven, The Netherlands*

Wireless sensor nodes are often battery powered and have to operate for a long period of time. A special class of ultra-low-power radio design in sensor nodes is event-triggered radio, i.e. , radios that are triggered by an external event, or by an internal (sensor) signal. An external event can be the reception of a wake-up call together with sensor payload data. An internal event can be a processor timer signal, a sensor that is being activated, or other internal trigger signals.

Two approaches to realize the low-power budget of an event-driven transceiver are being investigated: the first one is based on RF envelope detection and on/off keying (OOK) modulation, and the second one is a wideband FSK modulation and a mixer-first design. These two transceiver architectures can reduce the power budget with at least 1-to-2 orders of magnitude lower than state-of-the art radio transceivers. The challenges in the first RF envelope based design are interference rejection specs and 1/f noise sensitivity degradation. The second wideband FSK design needs careful system planning, to relax the phase noise requirements and lower the power consumption.

This forum talk focuses on the design and implementation of the two event-driven transceivers, both on system level and circuit level. Special attention is paid on power-reduction techniques. A comparison of figure-of-merit and power consumption between these two transceivers and state-of-the-art literature is given.

Guido Dolmans received the M.Sc. degree in electrical engineering in 1992 and the Ph.D. degree in 1997, from the Eindhoven University of Technology, The Netherlands. He worked at Philips Research Laboratories in Eindhoven from 1997 to 2006. Currently, he is a principal researcher/program manager for Holst Centre/imec in the ULP Wireless group. His primary research interest is system architecture/IC design of ultra-low-power radio transceivers. Other research interests are wireless communications PHY and MAC layer design, radio wave propagation, smart antenna design, and RF and microwave IC design. He has published over 40 papers in scientific and technical journals and conference proceedings and holds 12 US patents.

Normally-Off Computing for Sensor-Net Applications

Hiroshi Nakamura, *University of Tokyo, Tokyo, Japan*

The sensor network is an infrastructure that enables real-time monitoring of the status of the social and the natural environment. It also enables global optimization of the power supply to our electric systems. However, the sensor network itself consumes a lot more electric power when the number of the nodes increases. The power management of the sensor node is important as well as the power management of the sensor network. "Normally-off Computing" is a power-management technology that aggressively cuts off a power supply to components of computer systems when they do not need to operate, even under computation, while maintaining system requirements. The application of Normally-off Computing to sensor networks is discussed in this talk. The concept and target of normally-off sensor networks are described, and the development status is reviewed.

Hiroshi Nakamura is a Professor at the Department of Information Physics and Computing in the Graduate School of Information Science and Technology at the University of Tokyo. He received the Ph.D. degree in Electrical Engineering from the University of Tokyo in 1990. His research interests include power-efficient computer architecture and VLSI design for high-performance and embedded systems. He led a project on "Innovative Power Control for Ultra Low-Power and High-Performance System LSIs" supported by JST from 2007 to 2012, and is now leading the "Normally-Off Computing Project" supported by NEDO/METI. He served IEEE ISLPED 2011 as a general chair. He is a Senior Member of IEEE and ACM.

Mixed-Signal Processing for Low-Power WSN/BAN

Hyejung Kim, *imec, Heverlee, Belgium*

BAN/health/medical applications require continuous signal collection and low-power consumption, but also smartness with robust operation under daily circumstances. The use of multiple sensors and higher sampling rates leads to power-consumption issues. Since the wireless transmission and data storage are most power consuming, local data processing on the sensor node is required. This can be addressed by combining digital signal processing with analog-assisted processing in the mixed-signal domain. To reduce the raw data-rate from tens/hundreds of sensors, an adaptive sampling ADC is designed that adapts itself to the input signal features. Also, an activity-dependent wireless system for neural recording and compressive sampling is presented. The architecture of emerging compression methodologies and their benefits for the system are discussed. Another approach is to transmit only the essential information after feature extraction and local data analysis. In ECG monitoring, the accuracy of the R-peak detection is crucial for reliable analysis. Several algorithms are investigated such as derivative-based methods, digital and analog CWT-based methods, and band-power extraction methods. Filtering or motion-artifact reduction algorithms are applied to maximize signal integrity and reliability.

Hyejung Kim received the B.S., M.S., and Ph.D. degrees in electrical engineering from KAIST, Daejeon, Korea, in 2004 and 2006, and 2009, respectively. During the period in KAIST, she developed low-energy biomedical signal processors. She joined imec, Belgium, as a researcher in 2009, and is working on ultra-low-power integrated circuit design for biomedical applications. Her research focuses on VLSI implementation of low-energy digital signal processors, arithmetic units, microprocessors and integration of SoC for wearable biomedical applications.

F6: *Energy-Efficient I/O Design for Next-Generation Systems*

Organizer: Frank O'Mahony, *Intel, Hillsboro, OR*

Committee: Nicola da Dalt, *Infineon, Villach, Austria*
Ken Chang, *Xilinx, San Jose, CA*
Hisakatsu Yamaguchi, *Fujitsu, Kawasaki, Japan*
Chulwoo Kim, *Korea University, Seoul, Korea*
Elad Alon, *UC Berkeley, Berkeley, CA*

System power consumption will drive the architecture of future computing systems. From cloud-connected smart phones to the first exaFLOP supercomputers, systems that are the best at managing and minimizing power consumption will hold a key competitive advantage. At the same time, wireline communication bandwidth requirements within these systems will continue to grow exponentially, driving per-lane data rates beyond 25Gb/s and aggregate bandwidth past 1Tb/s while demanding dramatically improved energy efficiency. The objective of this Forum is to provide an overview of ultra-efficient parallel and serial interfaces, advanced memory applications, dense and high-speed optical communication, and platform-driven wired I/O for mobile. The Forum begins with two talks describing how innovative packaging and form factors along with co-design of I/O circuits and interconnects can improve the power/performance tradeoff by more than an order of magnitude. The next two talks address how memory I/O is adapting to meet the aggressive bandwidth and power requirements for systems ranging from cell phones to supercomputers. The next talk explores how to design serial I/O specifically for mobile products, including low-power equalization and clocking and low-latency standby states. The following talk also focuses on energy-efficient clocking and equalization, but explores analog and digital design options for very high-speed link standards. The final two talks highlight recent advances in both discrete and integrated optical transceivers and the power, performance, density and cost benefits for optical in high-performance computing systems.

Advanced Packaging for Low-Power I/O

Liam Madden, *Xilinx, San Jose, CA*

In recent years, chip-to-chip communication has become the critical determinant of overall system performance. Despite SerDes rates doubling every four years, the resulting bandwidth increase falls far short of the demands of wireless communications, where in-air bandwidth is tripling every eighteen months. Increasing SerDes rates also come at the expense of I/O power and complexity, to a degree where I/O power is once again becoming a dominant concern in thermally constrained systems. By taking advantage of recent developments in packaging technology, it is now possible to increase inter-chip connectivity by more than one order of magnitude, while reducing the energy per bit by a similar amount. This presentation looks at figures of merit for a range of chip-to-chip interconnect strategies in terms of energy efficiency, bandwidth and complexity. Technologies ranging from simple inverter-based designs all the way to advanced opto-electronic interfaces are considered.

BIO:

Liam Madden is corporate vice president of FPGA Development and Silicon Technology at Xilinx. He has responsibility for FPGA design, Advanced Packaging (including 3-D Chip Stacking) and Foundry Technology. Madden joined Xilinx in 2008, bringing more than 25 years of experience in design and technology leadership positions. He has contributed to a range of industry-leading products, including high-performance and low-power microprocessors (Alpha and StrongArm at DEC), embedded processors and IP (MIPS) and consumer devices (Xbox 360 at Microsoft). He holds a BE degree from University College Dublin, an M.Eng. degree from Cornell University and is an Adjunct Professor of Engineering at UCD.

Co-Designing Channel, Signal, and Circuits for High-Bandwidth Low-Power I/O

John Poulton, *NVIDIA, Durham, NC*

High-speed signaling over high-density interconnect on organic packages or silicon interposers offer attractive solutions to the off-chip bandwidth problems faced in modern digital systems. This talk describes a signaling system co-designed with the high-performance interconnect of an organic package to enable a high-speed low-area and low-power die-to-die link; the link is single-ended, so that it also uses the minimum number of pins and package traces. The system is fabricated in a standard 28nm process and exhibits 20Gb/s operation at 0.54pJ/b over 5 to 10mm of interconnect at a nominal 0.9V power supply. I conclude by outlining recent progress in low-energy signaling, investigating where and how signaling energy might be further reduced, and exploring the lower limit of energy per bit for short-haul die-to-die links.

John Poulton (M'85-SM'90) received the B.S. degree from Virginia Polytechnic Institute and State University, Blacksburg, VA in 1967, the M.S. degree from the State University of New York, Stony Brook, NY in 1969, and the Ph.D. degree from the University of North Carolina, Chapel Hill, NC (UNCCH) in 1980, all in physics. From 1981 to 1999, he was a researcher with the Department of Computer Science, UNCCH, where from 1995 he held the rank of Research Professor. He performed research on VLSI-based architectures for graphics and imaging and was a principal contributor to the design and construction of the Pixel-Planes and PixelFlow computer graphics systems, and designed custom beam-forming chips for the first commercial 3D medical ultrasound scanner. From 1999 to 2003 he was Chief Engineer with Velio Communications, where he developed multi-gigabit chip-to-chip signaling systems. From 2003 to 2009 he was a Technical Director with Rambus, Chapel Hill, NC where he led an effort to build power-efficient multi-gigabit I/O systems, demonstrating a system in 2006 with the lowest energy per bit published up to that time. Presently he is Senior Scientist at NVIDIA, Durham, NC, where he is working on low-energy on- and off-chip signaling. He is an IEEE Fellow.

Low-Power Memory for Mobile

Hyun-Woo Lee, *SK Hynix, Ichon, Korea*

DDR3L computing memory supports 1600Mb/s/pin at 1.35V and 11-11-11 (CL-tRCD-tRP). LPDDR2 supports 1066Mb/s/pin at 1.2V and LPDDR4 supports 2133Mb/s/pin at 1.2V. New GDDR5M products support 3200Mb/s/pin at 1.35V. High-performance memory moves to achieve low power consumption by TCSR, FGSR, and low standby consumption. Otherwise, LPDDR cannot avoid needs for high performance by the cost of additional power consumption. For example, ODT is considered to increase the channel speed. Techniques of lo- power control for LPDDR are employed in computing and graphics memories. And techniques of high performance for GDDR are also employed to computing and LPDDR memories. In order to make DRAM access more efficient, it is necessary to understand DRAM requirements of computing, graphics, and mobile applications. This presentation covers key features, method for efficient DRAM access, low power circuit techniques, and future trends for mobile memory including computing and graphics memory.

Hyun-Woo Lee was born in Cheong-Ju, Korea, in 1971. He received M.S. degrees in nano-semiconductor engineering from the Korea University, Korea, in 2012. In 1997, he joined the device development team at Hyundai Electronics (now SK Hynix semiconductor), Ichon, Korea, where he was involved in process integration, design of ESD protection devices and process failure analysis for DRAM (RAMBUS and DDR). In 2002, he joined the DRAM design team where he designed and developed DDR2, DDR3, GDDR3 and GDDR5M products. He has authored more than 20 papers in the field of electronic circuits and device, seven of them in ISSCC, and holds more than 70 US patents. His current focus includes low power memory design, wireline transceivers, high-speed I/O design for DRAM, equalizers, DLL, PLL and CDR. Mr. Lee received the 12th Korea Semiconductor Design Contest/ Minister of Ministry of Education, Science and Technology Award in 2011.

978-1-4799-0917-9/14 $31.00 © 2014 IEEE

Enabling the Next 10× Leap in Memory Bandwidth
Feng Lin, *Micron, Boise, ID*

Memory bandwidth is a limiting factor for high-performance computing (HPC). Current double-data rate (DDR) DRAMs follow an evolutionary path and face big challenges to address bandwidth, power and scalability issues. To enable the next 10× leap in memory bandwidth, a three-dimensional memory architecture, called hybrid memory cube (HMC) has been demonstrated to change the landscape. Equipped with through-silicon vias (TSVs), the heterogeneous integration between a memory stack and a logic layer greatly increases number of I/Os and simultaneously reduces the distance signals travel. Compared to other emerging memory interfaces, i.e., WideIO2 or LPDDR4, the HMC memory interface deploys small-swing and ultra-wide I/Os to achieve sub 1 pJ/b energy efficiency as well as more than 1Tb/s bandwidth. The third-generation HMC is in development to achieve an aggregated bandwidth of 320GB/s and single-digit pJ/b, well-suited for next generation X-scale HPC systems.

Feng Lin received BS and MS degrees in electrical engineering from the University of Electronic Science and Technology of China, Chengdu, China in 1992 and 1995, respectively. He received a Ph.D. degree in electrical engineering from the University of Idaho, Boise, ID in 2000. He joined DRAM Design R&D at Micron Technology, Boise in 2000 and currently is a Senior Member of Technical Staff (SMTS) for the development of high-speed leading-edge DRAM for high-performance computing, and most recently, for hybrid memory cube. Dr. Lin is a co-author of the textbook *DRAM Circuit Design, Fundamental and High-Speed Topics* (Wiley-IEEE Press, 2007). He also contributed a book chapter of the textbook *CMOS Processors and Memories,* published by Springer B.V. in 2010. Dr. Lin holds over 100 US and foreign patents. He also authored and co-authored one ISSCC paper, two IEEE Journal papers and several conference papers in the areas of clock synchronization and high-speed memory interfaces. His research interests include high-speed low-power I/O circuits, PLL/DLL, clock distribution, and mixed-signal circuit design.

Ultra-Efficient Mobile I/O
James Jaussi, *Intel, Hillsboro, OR*

As bandwidth demand in mobile systems continues to increase, optimizing I/O power efficiency is essential to maximize battery life while delivering a high-quality user experience. Examples of high-bandwidth mobile interfaces include high-resolution displays, cameras and storage. Ultra-efficient I/O architectures and circuits are required to simultaneously satisfy the need for performance and low power data transfer. Power consumption is minimized in two key power states: active and sleep. Additionally, low-latency transition between these states is crucial. This presentation highlights multi-gigabit I/O architectures and design techniques to improve mobile power efficiency. Clocking circuits, channel equalizers and low-voltage swing transmitters are discussed. Fast-lock CDR circuits to support low latency transition between power states are presented.

James E. Jaussi received the B.S. and M.S. degrees in Electrical Engineering from Brigham Young University, Provo, UT, in 2000. In January 2001, he joined a circuit research group at Intel, Hillsboro, OR. He is presently a member of the Electro-Photonics Lab. His research interests include high-speed CMOS transceivers and low-loss interconnect. He has worked with I/O industry standards efforts on link jitter budging, system clocking architectures and channel equalization. He currently co-chairs the MIPI Alliance M-PHY Electrical Sub-Group. In 2010 he received the Transactions on Circuits and Systems Darlington Best Paper Award and the Outstanding Speaker Award at Intel's Developer Forum.

Low-Power Equalization and CDR for 10-to-28Gb/s SerDes
Thomas Toifl, *IBM, Ruschlikon, Switzerland*

Power efficiency is one of the most important parameters in any high-speed I/O design. This talk explores low-power circuit implementations for high-speed I/O receivers, where I discuss both analog and digital I/O implementations. The talk starts with a short introduction to important wireline I/O standards, and describe the associated equalization requirements. I then turn to the implementation of a power-optimized data-path using a continuous-time linear equalizer (CTLE). An important part of the power budget of a high-speed RX goes in the decision-feedback equalizer (DFE), which is the focus of the next part of the talk. I then describe the design of a latency-optimized CDR logic, which is required for power minimization without penalizing jitter tolerance. Finally, we will turn to digital I/O implementations: Here, we will first display the design and recent results of high-speed low-power ADCs, which will then be followed by a discussion of low-power solutions for digital equalizer implementations.

Thomas Toifl (S'97-M'99-SM'09) received the Dipl.-Ing. (M.S.) degree and the Ph.D. degree (with highest honors) from Vienna University of Technology, Austria, in 1995 and 1999, respectively. In 1996, he joined the Microelectronics Group of the European Research Center for Particle Physics (CERN), Geneva, Switzerland, where he developed radiation-hard circuits for detector synchronization and data transmission, which were integrated into the four-particle detector systems of the new Large Hadron Collider (LHC). In 2001, he joined the IBM Zurich Research Laboratory in Ruschlikon, Switzerland, where since then he has been working on multi-gigabit per second, low-power communication circuits in advanced CMOS technologies. In that area he authored or co-authored nineteen patents and more than fifty technical publications. Since July 2008 he manages the I/O Link technology group at the IBM Zurich Research Laboratory. Dr. Toifl received the Beatrice Winner Award for Editorial Excellence at the 2005 IEEE International Solid-State Circuits Conference (ISSCC).

Energy-Efficient 25Gb/s Optical Transceivers
Takashi Takemoto, *Hitachi, Tokyo, Japan*

Recently, in accordance with the growth in software as a service, huge amounts of information are being collected in data centers through networks. Such a large amount of information requires a low-power 25Gb/s-class optical transceiver for short-reach-communication inside ICT systems. A CMOS-based optical interconnect has great potential for providing multi-functionality by integrating a logic circuit and for enhancing throughput by using equalization. Aiming to fulfill this potential, a power-efficient 25Gb/s CMOS optical transceiver has been developed. The optical transceiver has two key features: first, data-driven power-supply-variation-tolerant analog FE consisting of fully differential LD driver and high-sensitivity TIA; and second, ultra-low-power SerDes consisting of dynamic CMOS circuits for on-board electrical transmission. Moreover, the optical transceiver has a redundant data-format conversion, which improves reliability of optics without relying on redundant network topology at system level.

Takashi Takemoto received a B.S. degree in physics from Rikkyo University, Japan, in 2001, an M.S. degree in physics from University of Tokyo, Japan, in 2003, and a Ph.D. degree in science from University of Tokyo, Japan, in 2006. In 2006, he joined the Central Research Laboratory, Hitachi, Tokyo, Japan. From that time, he has been engaged in the development of analog integrated circuits, especially high-speed I/O interface circuits for wireline and optical communications. He currently serves on the Technical Program Committees of the SPIE photonic west. He is member of IEEE and the Institute of Electronics, Information, and Communication Engineers (IEICE).

Low-Power Si Photonics for Ultra-Dense Optical I/O
Brian Welch, *Luxtera, Carlsbad, CA*

With the advent of silicon photonics, the ability to design optical interconnects using CMOS design practices has become a reality, enabling many of the advancements that will be required for exascale computing. At both the circuit and system level, greater electro-optical integration and optimization is possible, enabling dramatic improvements in power, density, and cost. This presentation looks at the evolution of electro-optical circuit and system design in silicon photonics, and projects the types of solutions that will be deployed for exascale supercomputers.

Brian Welch is Director of Product Marketing at Luxtera, where he overseas new product development and planning. Prior to that role he held a variety of Marketing and Design positions within Luxtera, focusing on high-speed optical interconnect solutions. Brian has a PhD in Electrical Engineering from Cornell University.

ES1: Student Research Preview

Chair: Jan Van der Spiegel,
University of Pennsylvania, PA

Session I: Data Converters and RF/MMICs

Session Co-Chairs: Denis Daly, Patrick Reynaert

Yuan Zhou	*University of Texas at Dallas, United States*
Yongfu Li	*National University of Singapore, Singapore*
Kentaro Yoshioka	*Keio University, Japan*
Xinwang Zhang	*Tsinghua University, China*
Yang Shang	*Nanyang Technological University, Singapore*
Zhe Zhang	*National University of Singapore, Singapore*
Kaizhe Guo	*University of Electronic Science and Technology of China, China*
Suman Sah	*Washington State University, United States*

Session II: Circuits and Systems for Biomedical Applications and Analog Techniques

Session Co-Chairs: Andrea Baschirotto, Shahriar Mirabbasi

Walker Turner	*University of Florida, United States*
Jingjing Dong	*Tsinghua University, China*
Hideki Naganuma	*Tohoku University, Japan*
Mehran Bakhshiani	*Case Western Reserve University, United States*
Sabino Pietrangelo	*Massachusetts Institute of Technology, United States*
Xilin Liu	*University of Pennsylvania, United States*
Myeong-Jae Park	*Seoul National University, Korea*
Yaohua Zhao	*University of Macau, China*
Congyin Shi	*Texas A&M University, United States*

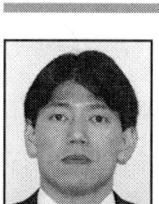

Session III: Digital, Imagers, and Si Photonics

Session Co-Chairs: Makoto Ikeda, Dejan Markovic

Hwa-Suk Cho	*POSTECH, Korea*
Yildiz Sinangil	*Massachusetts Institute of Technology, United States*
Wei-Chang Liu	*National Chiao Tung University, Taiwan*
Shuo-Hong Hung	*National Taiwan University, Taiwan*
Wai Chiu Ng	*The Hong Kong University of Science & Technology, Hong Kong*
Neale Dutton	*The University of Edinburgh, United Kingdom*
Xiwei Huang	*Nanyang Technological University, Singapore*
Chen Sun	*Massachusetts Institute of Technology, United States*

Poster Session

Session Co-Chairs:

Jeff Weldon, Carnegie Mellon University, United States

Marian Verhelst, K.U. Leuven, Belgium

ES2: Data Centers to Support Tomorrow's Cloud

Organizers: **Leland Chang,** *IBM, Yorktown Heights, NY*
Ajith Amerasekera, *TI, Dallas, TX*
Takashi Hashimoto, *Panasonic, Fukuoka City, Fukuoka, Japan*

Chair: **Leland Chang,** *IBM, Yorktown Heights, NY*

With the rise of cloud computing and Big Data, data centers are an important counterpoint to rapid growth in the mobile market. Building cost-effective, efficient computing infrastructures is a challenge that starts with technologies that ISSCC knows so well (processors, I/O, memory, etc.), but also encompasses system and customer-centric issues such as cooling, power delivery, and total cost of ownership. An outlook on the future of data centers, including recent trends such as open source models, energy-proportional computing, disaggregation, and software-defined data centers, will be discussed as it pertains to the ISSCC community.

Panelist's Statements

Today's Big (data) Is Small
Steve Pawlowski, *Intel, Hillsboro, OR*
Mankind has never seen so much data as it is seeing today. The knowledge extracted from this (big) data via analytics is essential for both fundamental discoveries as well as businesses transformation, small and big. However, the challenges in extracting this knowledge are immense, both in terms of computational requirements and data intensity and they are breaking the standard model by which data has historically been managed and analyzed. One of the most important real-world grand challenges is the understanding of complex biological systems. New technologies have allowed the sequencing of genomes at unprecedented rates and volumes. The roughly 2000 sequencing instruments in labs and hospitals around the world can generate about 15 petabytes of compressed genetic data. Storing, analyzing, and sharing such vast quantities locally is an immense challenge. This talk will use genomics as an example to illustrate how big data stresses today's paradigms and technologies along with the key actions needed to address big data challenges and how the 'cloud' paradigm could be crucial to ensuring generalized solutions for most big data opportunities to achieve huge economies of scale.

Healthcare in the Big Data Era
Yasunori Kimura, *Fujitsu, Sunnyvale, CA*
With the recent advances of computing power, network speeds, and sensing technologies, we are now in the era where anything can potentially be connected; home appliances, cars, houses, etc. and most importantly the human. If we take an example of Healthcare area, mHealth (mobile health) has become one of the promising infrastructures to improve medical systems and reduce its cost, which will benefit people. In mHealth, human vital signs such as blood pressure, pulse, EKG, etc. are continuously monitored and measured, and are sent to the cloud via mobile devices i.e. smartphones. Then the gathered data is analyzed to exploit meaningful information which could contribute to making people's life much more comfortable. But the size of the data becomes huge, it will be more than a few hundred MB of data if you measure the three of vital data above a day, for example, and where to store, how to represent, what to analyze, etc. are important issues to be solved. In the talk, a stress monitoring system is presented as an example of mHealth. Some of the issues we've found out while doing the preliminary experiments and our implementations are discussed. We argue the issues in terms of technology point of view, but also address issues such as security, privacy, and others.

Evolution of ARM Technology Towards the Data Center
John Goodacre, *ARM, Cambridge, UK*
Tomorrow's data center needs to scale-out their performance significantly to satisfy the rapid growth from both the ever growing mobile market and the expected explosion in the connected internet of things. Together their big-data requirements for storage and processing challenge significantly the need to lower the total cost of ownership of each data center, especially in terms of power efficiency and compute density. Technology from ARM is being deployed within the consumer device and embedded terminal at a rate expect to exceed 10 Billion processors in the next 12 months, driven significantly by its power efficiency and small size. This talk will look at some of the recent developments around the ARM technology, including ARM multiprocessing, virtualization and 64bit addressing capabilities that together make the ARM processor also technically suitable for the data center. In addition, this talk will look into a potential future where using the current ARM technology, a system on chip design will be able to offer the data center servers all the key open-source frameworks on a fully scalable and energy-proportional compute platform, founded around the same power efficient and low power processors used at client.

Landheld Computing
Luiz André Barroso, *Google, Mountain View, CA*
A few interesting phenomena arise in distributed systems when they go from very large to massively large. I will discuss two examples of such phenomena involving energy use and service level responsiveness in Google datacenters.

978-1-4799-0917-9/14 $31.00 © 2014 IEEE

EP1: Next-Generation Networked Systems-Challenges for Silicon

Organizer: **Hoi-Jun Yoo,** *KAIST, Daejeon, Korea*

Moderators: **Anantha Chandrakasan,** *MIT, Cambridge, MA*
Siva Narendra, *Tyfone, Portland, OR*

Semiconductor systems connected through wireless and wired networks are the mainstream approach in the semiconductor market to achieve technical innovations and business advantages. Systems are defined differently according to the application area, implementation technology and also the individual engineer's background. This presents challenges to circuit designers and circuit-design educators.

Does the design primarily belong to system architects or to circuit designers? Is domain-specific knowledge, for example biomedical or automotive, essential for system development? Can the domain expert alone, rather than the system architect, design and develop the system as the principal engineer? Or is it possible for the circuit designer to provide a "system platform" to the domain experts for their applications? Are there any good examples where the semiconductor technology and circuits contribute mainly to the innovation of the total system performance? Are MEMS integration, or integration of a temperature-sensing circuit on an SoC, good examples?

Experts on the system side and the traditional silicon side will share their experiences and vision about the development of innovative systems based on semiconductor technology and circuits.

Panelist's Statements

Alex Jinsung Choi, *SK Telecom, Seoul, Korea*
The massive growth of network services is increasing the demand on the device and infrastructure. Thus the semiconductor market today is open to significant opportunities for growth and technical innovations. As more mobile devices connect to the Internet, data usage for multimedia service, big data, and cloud service is increasing explosively. The growing volume of data causes heavy traffic in existing systems as it is transmitted, aggregated, and analyzed, thus creating new challenges for next-generation systems: systems must be more agile and service-driven and easier to manage and operate. In particular, the need for system re-architecting (including storage, server, networks- SDN/NFV, HetNet, high-throughput network appliance, IoT platform) is emerging and at the same time the importance for data trust and security is drawing attention. This system evolution requires support from low-power high-performance semiconductor technology and circuits. Thus, next generation systems must be vertically integrated from semiconductors, devices, and platforms, to networks and service applications in order to resolve the challenges it faces in the world of ever-increasing data.

William Dally, *NVIDIA, Santa Clara, CA*
Contemporary chips are power limited and supply-voltage scaling has nearly stopped. In this post-Dennard era, the main challenge for all systems—networked or not—is to improve efficiency through circuits and architecture. Process matters far less than in the past. Today a new process node gives about a 1.2× improvement in efficiency—compared to 2.8× under Dennard scaling. To make up for this 2.3× gap we look to improvements elsewhere. Better signaling circuits, for example, operate at 20pJ/b·mm—an order of magnitude less than full-swing signaling. Low-voltage-tolerant circuits enable more efficient operating points. Better architecture reduces the overhead associated with instruction and data supply. The 100× efficiency advantage of specialized hardware demonstrates the magnitude of typical overhead. The challenge is to eliminate this overhead without sacrificing flexibility and programmability. Networked systems offer the additional option of moving some processing to the cloud, dividing applications strategically between client and server.

Robert Gilmore, *Qualcomm, San Diego, CA*
An early Qualcomm employee once said that the technical lead of a project is a person who, given enough time, could complete the project by him or herself. Depending upon the deliverable, this requires that the technical lead be an individual with incredible breadth. This may work in start-ups or small organizations but is not a scalable model. Our current philosophy is to have multi-disciplinary teams with hardware, system and software expertise. The teams need to be collaborative with good leadership. A circuit designer who is not highly skilled in the application will not know the correct problems to target. The subject matter expert alone is not going to be able to optimally realize his or her solution: achieving performance, size, power and cost constraints. There is no doubt that the advance of software and platforms lowers the barrier for domain experts to create solutions with phone applications or integration of development boards into a new system.

Mike Muller, *ARM, Cambridge, United Kingdom*
Highly optimized networked systems, whether high-performance wired compute clusters, leading-edge wireless smart phones or ultra-low power sensors can only be designed by teams. These teams need a good understanding of the overall system requirements and use cases while also specializing in their own area of expertise from architecture to implementation. Even within a single domain, for example a low-power wireless sensor, the optimal circuit design and clocking architecture is very different for what is apparently the same system whose only change is whether the power source is a battery or any energy-scavenging system. System optimization must be done both globally and locally and this can only be achieved if all members of the team have a proper understanding of the systems use cases

Udo-Martin Gomez, *Bosch Sensortec, Stuttgart, Germany*
MEMS technology has been used in commercial production for more than 20 years. Product and application scope is expanding, enabled by a wide range of highly specialized technologies ("one product, one technology") with strong focus on miniaturization. With the second wave of MEMS commercialization towards consumer electronics applications, the focus on low cost, small size, low power consumption and overall system performance has driven major technological achievements such as combined sensing structures, integrated power management schemes and sensor data fusion on raw data level. The evolution is still ongoing, now with multiple sensors integrated within one single package or on single die. The technological capabilities of the sensor manufacturers are getting challenged further, leaving only a few players in the field that can cope with the required depth of manufacturing technologies and range of sensing elements. With the modularization and sensing scope extension, the foundation for the next wave of sensor applications is just being laid within the vision of "connected sensors everywhere".

Atsushi Takahara, *NTT, Yokosuka, Japan*
Smart phones are a specific example of a networked device that has an impact on the social realm. Smart phones make peoples' lives easier and more fruitful, by enabling connection to a rich variety of applications and information. Even in disaster situations, such as the Greater East Japan Earthquake in 2011, we learned that smart phones help people in dire circumstances. On the other hand, smart phones can easily bring down the system with a flood of traffic beyond the expectations of the system. This type of system trouble could occur more easily in the future because of the high computational and communication ability of each device and the large number of devices in the system. In networking technology, software-defined networks (SDN) and network functions virtualization (NFV) are proposed to provide more flexibility to the demands of the social realm. It is still unclear how these technologies can help the above problem. So, I address the benefit and danger of the evolution of networked devices. Then, I discuss how devices and the social realm can be harmonized to achieve safe and reliable networking technologies.

978-1-4799-0917-9/14 $31.00 © 2014 IEEE

ES3: Wearable Wellness Devices: Fashion, Health, and Informatics

Organizers: **Firat Yazicioglu**, *imec, Leuven, Belgium*
Sam Kavusi, *Bosch Research, Palo Alto, CA*

Chair: **Chris Van Hoof**, *imec, Leuven, Belgium*

Imagine using the same device for fashion/style and for monitoring your wellness? What about a tattoo of your child's name that also tracks your fitness and activity level. Can we make contact lenses that can change the color of your eyes but also see the calorie content of your lunch box?

Early prototypes of such devices are already emerging. Smart contact lenses can display words or images in your line of sight, and stretchable/tattoo electronics can monitor vital signs. However, there are significant design and technology challenges ahead to actually make these devices fully functional, autonomous, and reliable.

Let's hear from the experts what could be the future of wearable wellness devices and what technologies are being developed.

Panelist's Statements

Stretchable Electronics: Biointegrated Systems with Unusual Materials and Designs
Roozbeh Ghaffari, *MC10, Cambridge, MA*
Advances in soft biomaterials and microelectronics technologies have driven important advances in healthcare. However, there are significant mechanical and geometrical constraints inherent in all standard forms of rigid electronics. These constraints impose unique integration and therapeutic delivery challenges for noninvasive, minimally invasive and implantable medical devices. Here, we describe novel materials and design constructs for skin-based systems that incorporate physiological sensors (e.g. active electrodes, temperature sensors, and accelerometers) and therapeutic actuators configured in stretchable formats. Quantitative analyses of light diffusion, electronic/sensor performance and data transmission under mechanical stress, underpin the clinical utility of these systems in health monitoring, wellness and photomedicine. As demonstrations of this technology, we present representative examples of biointegrated systems that highlight previously unrealized functionality and performance coupled with extreme mechanical flexibility.

Smart Textiles for Wearable Electronics
Jerald Yoo, *Masdar Institute of Science and Tech, Abu Dhabi, UAE*
Smart textile that prints circuit board directly on wearable materials (e.g., natural or synthetic fabrics and textiles) will provide an ideal platform for wearable electronics such as sensing, computing, and a communication platform that can address several individual and societal needs in the areas of healthcare, lifestyle, and social networking. However, there are several design challenges including non-ideality and variability of the platform. This talk presents recent trends in smart textile and their efforts in circuit perspective to overcome these challenges. The electrical characterization of wearable electronic passives such as printed capacitors and inductors will be discussed, in which the supporting substrate is undergoing constant deformations, including stretching, wrinkling, compression and aging. Novel circuit techniques to compensate such non-idealities will be presented. We will also cover system examples.

Extremely Low-Power Circuit Design for Wearable Systems
Makoto Takamiya, *University of Tokyo, Tokyo, Japan*
In order to expand the use of wearable wellness devices, maintenance-free and wearing-unconscious devices are required. Extremely low power circuits and autonomous energy delivery systems are required for maintenance-free devices. Ultra-flexible or ultra-small-size devices are required so the user is ultimately not conscious of wearing the devices. This talk will cover (1) a sub-100mW 0.5-V RF transceiver for body area networks and power management circuits, (2) an 80mV input boost converter for thermoelectric energy harvesting, and (3) a "not conscious of wearing" surface electromyogram measurement sheet with 2V organic transistors on an ultra flexible 1mm thickness PEN film for a prosthetic hand control. Finally, future technical challenges of wearable wellness devices will be discussed.

Smart Contact Lenses and Integrated Circuits
Jelle De Smet, *Ghent University, Ghent, Belgium*
Contact lenses are widely used as a passive tool for ophthalmic corrections. Smart contact lenses go beyond their classic counterparts, having applications in augmented reality, biomedical sensing and active vision correction. This increased functionality encompasses the integration of electronic components such as RF antennas, IC chips, solar cells, sensors and electro-optic elements. A brief overview is given of this new research field and the challenges concerning IC design are presented. Amongst others, these include maximum die size, mechanical properties, power consumption and cost.

Energy Harvesting for Wearable Systems
Yiannos Manoli, *University of Freiburg and HSG-IMIT, Freiburg, Germany*
Energy Harvesting has established itself as a method of extracting energy from the environment such as light, vibration, thermal gradients or electrochemical reactions. Extending these concepts from the technical domain of machines, cars or appliances to the human body confronts the designer with a number of challenges: human actions provide only very small motions of low and non-resonant frequencies and photovoltaic or thermoelectric methods can be applied only under very demanding conditions.

These power sources deliver low levels of usable energy, thus any system will have to seek ways to maximize the effectiveness. Each transduction principle that converts the available energy to electrical power requires a different circuit technique for achieving an optimal conversion. Such an adaptive control has to be intelligent enough to always find the optimum without requiring too much energy. Here lies a major challenge for low-power and low-voltage electronics.

978-1-4799-0917-9/14 $31.00 © 2014 IEEE

EP2: Anatomy of Innovation: Bug or Feature?

Organizer: **Harry Lee,** *MIT, Cambridge, MA*
Co-organizer: **Ken Nishimura,** *Agilent Technologies, Santa Clara, CA*

Moderator: **Harry Lee,** *MIT, Cambridge, MA*

As process scaling slows down, circuit innovation is becoming one of the most important differentiators. We can point to great inventions of the past that were accidental, or failed attempts to solve other problems (bugs), as well as those from logical thinking (features). Which is more effective? In this panel, top analog circuit innovators describe the process by which their best innovations were conceived. They give interesting examples, such as turning a bug in the circuit into a feature. Then they argue whether innovation is more effective as a result of accidental discovery or logical thinking.

Panelist's Statements

Ali Hajimiri, *CalTech, Pasadena, CA*

Although imagination is often linked to creativity, in the absence of proper mental and educational discipline it only occasionally leads to useful innovations. The vast majority of "novel" ideas are not useful in solving a given problem. The key is to generate as many new potential solutions as possible and not to close the door on them too early, so they can be critically evaluated and pruned, hopefully leaving one with a non-empty set of solutions. On the other hand, the idea-generation process itself is truly affected by various factors such as the breadth of experiences and knowledge one has been exposed to and the amount of truly free reflection time one has had to connect those seemingly unrelated matters in one's mind. Orthogonal experiences and observations, such as bugs, enhance this process from time to time. Perhaps as Oscar Wilde said: "The imagination imitates. It is the critical spirit that creates."

Qiuting Huang, *Swiss Federal Institute of Technology (ETH), Zürich, Switzerland*

An integrated circuit, be it the pièce de résistance for your PhD or the crucial target of a multi-million-dollar development of your company, usually takes years to develop. The moment of truth when you finally power up the chip is, at least for me, one of the most suspenseful moments. A bug (or more) is among the most dreaded outcomes that can set you back a year, hundreds of thousands of dollars, and a lost window of opportunity that might never come back again. At such moments of despair, wouldn't we kill to find the alchemy that turns those insects into features! Alas, in my own experience fatal bugs only inflict so much pain on oneself that you try never to let it happen again. Nowadays I preach to my students to get it correct by construction rather than by exhaustive simulation. While we can try to minimize the occurrence of mistakes due to oversight by design reviews etc, the hardest is to prevent problems you didn't know existed. The solution to a bug, be it by a clever digital post-processing or a re-design based on the enlightenment through failure, sometimes becomes a feature that you may even be proud of.

Lawrence Loh, *MediaTek, Hsinchu, Taiwan*

Throughout the history of mankind, many great inventions were made accidentally. Fire-making techniques, for example, could be considered to be the very first kind of great accidental inventions made to contribute to the survival of human beings. Another example is in ancient China, where a Taoist priest performing alchemy accidentally discovered gunpowder but almost got himself killed in the explosion. In the world of integrated circuit design, especially in creations of modern SoCs, however, logical thinking plays a more important role than accidental findings. Billions of digital transistors are made with specific data/signal processing functions varying from demanding and sophisticated applications to dealing with realities and imperfections of the analog world. Bugs are never desired and must be avoided. Successes of such large-scale R&D activities can never depend on serendipity, rather they are made by engineering teams featuring strong analytical thinking and planning practices plus years of development experiences.

Kofi Makinwa, *Delft University of Technology, Delft, The Netherlands*

Circuit innovation is driven by the need to solve problems, meet challenges and improve performance. After decades of innovation, however, most of the low-hanging fruits have been plucked, which means that true circuit-level innovations are now few and far between. Instead, the most exciting opportunities for innovation are now at the system level and involve new algorithms and architectures. How does innovation come about? In my experience, usually not by accident. Instead, innovative ideas arise when the strengths and weaknesses of existing solutions can be clearly and succinctly formulated. Alternative ways of improving things then seem to present themselves. Or in other words, once a problem can be clearly formulated, other solutions become (almost) obvious.

Akira Matsuzawa, *Tokyo Institute of Technology, Tokyo, Japan*

Bugs, such as an unexpected bad result, a failure of development, coupled with a high pressure development schedule, and tough performance targets are always the triggers for innovative ideas. However they are created from the process of logical thinking. Seeking the essence of technology leads us to the novel view. For example: What is an analog-to-digital conversion? Why is a reference voltage needed? What is interpolation? Why is an amplifier needed here? What is the essential difference between a resistor and a capacitor? What sets an ultimate performance limitation or a bottleneck? Also, a high level abstraction of the system, such as a mathematical model sometimes reveals the essence. The other important factor is a sense of beauty. I believe an essential feature is always beautiful.

Dennis Monticelli, *Texas Instruments, Santa Clara, CA*

Corporate management would like to think that innovation could be scheduled like any other business process. While innovators do get results when they're focusing on a given problem, they should be allowed the freedom to explore interesting discoveries along the way, while keeping them out of the critical path of the current project. Sometimes a bug can be cleverly turned into a feature, but more often the bug is a seed that leads to something else interesting and useful. The serendipity that often occurs when smart and curious innovators get together is an even more fertile seed. Managers can do much to encourage and also to harm the innovation process, which is different (and messier) than any other process.

EP3: Perspectives on the Future of Semiconductor Innovation

Organizers / Moderators: Chris Nicol, *Wave Semiconductor, Sunnyvale, CA*
Ali Keshavarzi, *Cypress Semiconductor, San Jose, CA*

In a "call for leadership" panel at ISSCC, we will be seeking leaders' perspectives on the future of discontinuous innovation in the semiconductor industry. An ensemble of visionaries, experts and CEOs will discuss the opportunities and challenges for innovation in our industry. Of primary concern is the reduction in funding available for new semiconductor ventures. Are the escalating NRE costs of ASICs providing a barrier to new entrants? What advice would our distinguished panellists give to entrepreneurs thinking of starting a new semiconductor company?

We believe that significant innovation is needed across a broad spectrum of technologies to fuel the insatiable demand for enhanced user experiences in portable systems. For mobile computing, any slowdown in MIPS/W improvement will require a faster rate of improvement in the energy density (Wh/L) of battery technology. For cloud services, the cost of power delivery and cooling is a significant % of operating expenditures and needs to be reduced to sustain growth. In these applications, lack of semiconductor innovation will result in anaemic improvement in MIPS/W/$ and will therefore impact the growth rate of mobile and cloud computing. This panel contains representatives from service and system providers and we will ask them if their future needs are likely to be met by current silicon roadmaps or if increased investment in semiconductor innovation is required.

Many of the semiconductor platforms for future systems will be delivered by Multi-National Corporations (MNCs). The panel contains CEOs from semiconductor MNCs and we will ask them if they are increasing their investment in innovation or if a strategy of acquisition is a sustainable innovation strategy going forward. It is possible that the VCs will have something to say on this matter.

Are funding sources (and incentives) in offshore markets shifting the focus of semiconductor innovation to nations like China, Brazil and Russia or regions like The Middle East? Are there pitfalls with this trend? Should Government do more to support small companies? If public-funded institutes and universities provide the research, must they fund the commercialization of this research into semiconductor products? Is the fabless semi model being displaced by the IP model? And many more unlisted questions that will be discussed in front of an ISSCC audience consisting of experts in the field, inventors, startups, researchers, young professionals and the students that will fuel the future of the industry. We believe that this issue being debated in our panel is serious, and affects everybody working in, or considering a career in the semiconductor industry. We therefore have extended our call for leadership to discuss these issues that will shape the future of the innovation landscape in our industry.

Panelist's Statements

Nicky Lu, *Etron, Hsinchu, Taiwan*
Nicky will share his perspectives on the future of semiconductor innovation based on his current position as CEO and President of Etron Technology, Inc., a public Taiwanese fabless IC design and product company specializing in memory, DRAM and SoCs that he has founded in 1991. He will debate the panel controversy according to his broad range of experiences as a researcher, design architect, entrepreneur and chief executive. He has co-funded several other high-tech companies including Global Unichip Corporation, an IC design foundry and TM Tecnology Inc. in 3D-IC. He has worked at IBM Research. He serves on the board of the Taiwan Semiconductor Industry Association (TSIA) and as Chairman of the Global Semiconductor Alliance (GSA). He holds 24 patents and has published more than 50 papers. He holds a Ph.D. in EE from Stanford University. He is a Fellow of the IEEE and a member of National Academy of Engineering. He has collected many prestigious awards.

Scott McGregor, *Broadcom, Irvine, CA*
Scott is president and CEO of Broadcom Corporation. Scott will discuss his perspectives regarding the future of semiconductor innovation in line with his role guiding the vision and direction of Broadcom's growth strategy. Since he joined Broadcom in 2005, the company has expanded from $2.40 billion in revenue and 3,250 employees to $8.01 billion in 2012 and 11,750 employees. In addition, Broadcom's geographic footprint has grown from 13 countries in 2005 to 24 and its patent portfolio has expanded from 4,800 patents to more than 19,350. Scott joined Broadcom from Philips Semiconductors where he served as President and CEO from 2001 to 2004. He had joined Philips in 1998. Prior, Scott worked at Santa Cruz Operation Inc. and Digital Equipment Corporation. At Microsoft, he was Director of the Interactive Systems Group and architect for the original version of Microsoft Windows®. Prior to Microsoft, Scott worked at Xerox Corporation's Palo Alto Research Center in software for the first personal computers graphical user interfaces. Scott has degrees from Stanford and serves on the board of the Engineering Advisory Council for Stanford and is President of the Broadcom Foundation

T.J. Rodgers, *Cypress Semiconductor, San Jose, CA*
T.J. is founder, president, and CEO of Cypress Semiconductor since 1982. He is a former chairman of the Semiconductor Industry Association (SIA) and SunPower Corp., and currently sits on the board of directors of Agiga Tech and Bloom Energy. T.J. will discuss his perspectives regarding the future of semiconductor innovation based on his past experiences and his vision. Inside Cypress, T.J. has perpetuated a spirit of entrepreneurship by launching a series of autonomous businesses that have relied on the parent company for funding in much the same way as startup companies rely on venture capital. This "federation of autonomous subsidiaries" has delivered multiple successes including Cypress Microsystems, which developed Cypress's flagship PSoC products and SunPower which was spun out of Cypress in 2008 for $2.6 billion. T.J. first described the tenets of his "internal startup" concept in his 1992 book "No Excuses Management". Cypress has been called "a quintessential entrepreneurial company" by The Wall Street Journal. T.J. has a Ph.D. from Stanford and holds 14 patents. T.J. has received numerous awards and recognitions. He is a member of the board of trustees at Dartmouth College, his alma mater. T.J. has testified before Congress five times.

Simon Segars, *ARM, Cambridge, UK*
Simon is the CEO of ARM and a member of Board of Directors of ARM. Simon will describe his requirements for the future of semiconductor innovation based on his wide range of experiences at ARM. Simon was appointed CEO in 2013. Simon joined the Board of ARM in 2005. He was previously President of ARM. Earlier in his career Simon served various senior roles at ARM including: EVP of Engineering; EVP of Worldwide Sales; EVP of Business Development, and EVP and General Manager of the Processor and Physical IP Divisions. He joined ARM in 1991 and worked on many of the early ARM CPU products. He led the development of the ARM7 and ARM9 Thumb® families. He holds a number of patents in the field of embedded CPU architectures. Simon's experiences will shed light on what innovation is required and also whether the fabless semi model being displaced by the IP model and more interesting insights.

Luc Van den hove, *imec, Heverlee, Belgium*
Luc will elaborate his perspectives regarding the future of semiconductor innovation based on his current position as CEO of imec, a world leading nanoelectronics research organization who leverages global partnership to deliver industry relevant solutions, and his past experiences. Luc joined imec in 1984 and has been involved in variety of roles until becoming CEO in 2009. He started his research career in the field of silicide and interconnect technologies. In 1988, he became manager of imec's micro-patterning group; in 1996 department director of Unit Processes R&D; and in 1998 Vice President of Silicon Process and Device Technology. In January 2007, he was appointed as EVP and COO. Luc received his Ph.D. in EE from the University of Leuven, Belgium. He has authored or co-authored more than 100 publications and conference contributions.

Dado Banatao, *Tallwood, Menlo Park, CA*
Dado is the managing partner of Tallwood Venture Capital. With his past experiences as an entrepreneur, Dado provides Tallwood along with our panel with a unique perspective in technology investments. Tallwood invests in unique and hard-to-do semiconductor technology solutions for computing, communication, and consumer platforms. Dado has served as a venture partner at the Mayfield Fund. He co-founded three technology startups: S3, Chips & Technologies, and Mostron. He also held positions in engineering and general management at National Semiconductor, Seeq Technologies, Intersil and Commodore International. Dado pioneered the PC chip set and graphics acceleration architecture critical in every PC today. Dado is regarded as a Silicon Valley visionary. Dado serves as Chairman of Ikanos, InPhi Corporation and Quintic Corporation and is on the board of directors of Alphion Corporation, Wave Semiconductor and Wilocity. He also served as Chairman and led investments in Marvell, SiRF, Marvell, Acclaim Communications (to Level One and Intel), Newport Communications (acquired by Broadcom), and more. Dado holds an M.S. degree from Stanford University.

John Doerr, *KPCB, Menlo Park, CA*
John is a general partner at Kleiner Perkins Caufield & Byers (KPCB). John will elaborate his thoughts on the future of semiconductor innovation based on his track record and passion in helping entrepreneurs create the "Next Big Thing" in mobile and social networks, greentech innovation, education and economic development. Since joining KPCB in 1980, John and his partners have backed some of the world's most successful entrepreneurs at Google, Amazon.com, Intuit, Zynga and Twitter. Ventures sponsored by John have created more than 200,000 new jobs. John is a member of U.S. President Barack Obama's Council on Jobs and Competitiveness. John also serves on boards of several companies including: Google, Zynga, and Bloom Energy. John's technology career began in 1974 at Intel, just as the chipmaker was inventing the groundbreaking 8080 microprocessor. John holds degrees from Rice University and Harvard Business School. He also holds several patents. John is a member of the American Academy of Arts and Sciences.

Andy Rappaport, *August Capital, Menlo Park, CA*
Andy is a general partner in the August Capital funds. Andy will describe his thoughts on the future of semiconductor innovation based on his past experiences. Andy is currently on board of Alta Devices, Conexant, SuVolta, Luxtera, Magnum Semicondutor and Scintera. He raised August Capital VI in 2012 and decided to retire from the venture capital business. Andy remains active on management of prior funds and his portfolio companies while advises August new funds. Prior to joining August Capital in 1996, Andy was involved in the formation and success of several start-ups, including Actel, MMC Networks, Sequence Design Automation, Silicon Architects,Transmeta, and Viewlogic. In 1984, Andy founded The Technology Research Group that provided business-strategy counsel to executives at various electronics companies, including Alcatel, AT&T, EDS, IBM, and Intel. Andy is an often-cited authority on changing technologies and markets and has written and lectured extensively on the evolving structures of semiconductor, computer, and telecommunications industries. He co-wrote "The Computerless Computer Company," which won the 1991 McKinsey award for Article of the Year in the Harvard Business Review.

978-1-4799-0917-9/14 $31.00 © 2014 IEEE

SC1: Biomedical and Sensor Interface Circuits

Chair:
Willy Sansen,
K.U. Leuven, Leuven, Belgium

An increasing number of circuits interface with the human body and other bio-sensors. They have to interface with living tissues, which change in impedance, with temperature, and with time. Very high input circuit impedances are required to avoid the extraction of current, causing drift and infection. High sensitivity and low noise are required as well.

This short course aims to describe the difficulties and technical solutions for such bio-interface circuits. Successful realizations are presented for different kinds of medical applications.

OUTLINE

The Biomedical Electrode-Tissue Interface: A Simple Explanation of a Complex Subject

Eric McAdams, *INAS, Lyon, France*

Biomedical electrodes are used in various forms in a wide range of biomedical applications. Good electrode design is not as simple as is often assumed, and all electrode designs are not equal in performance. One must not simply choose an electrode with as conductive a sensing element as possible, which unfortunately, was and still is the case in many designs, especially in the emerging area of "wearable sensing".

Monitoring bioelectrodes, if they are not chosen/designed correctly, give rise to significant problems that make biosignal analysis difficult, if not impossible. Both the electrode–electrolyte interface and the skin under the electrode (collectively known as the contact) give rise to potentials and impedances that can distort the measured biosignal. The potentials and impedances of the electrode–electrolyte interface and of the skin will therefore be studied in some depth using a simple yet effective model to give a sound working knowledge of biomedical electrode technology and design.

About the presenter:
Prof. Eric McAdams is Head of the Medical Sensors Group within INL and works and teaches at INSA Lyon. Before that he was Head of the Medical Sensors Group at NIBEC, University of Ulster, and he co-founded and was a director of a successful spin-off company, Intelesens. Intelesens researches, designs, develops, and manufactures innovative, devices in pioneering areas such as wireless Vital Signs Monitoring. He is widely recognized as a leading specialist in the study and modeling of the electrical properties of materials and interfaces. He has successfully designed, developed, patented and commercialized a wide range of Biomedical Sensors and Electrodes.

Low-Power Interface Circuits for Bio-Potential and Physiological Signal Acquisition

Firat Yazicioglu, *IMEC, Leuven, Belgium*

With an increasing number of applications, smarter and smaller wearable/implantable biomedical devices have been one of the main drivers behind low-power analog IC design for sensor interfaces. In recent years, the adoption of wearable devices for wellness applications has boosted the need for such sensor interfaces making them even more interesting for the research community and industry.

This presentation focuses on the instrumentation amplifiers and readout circuits for the readout of different biomedical signal modalities. The course starts with minimum safety and performance requirements for different applications. Later, signal integrity problems are discussed and circuit and environmental noise sources are introduced. A large part of course is on the design of sensor interface circuits and instrumentation amplifiers for different signal modalities that are relevant for both wearable and implantable applications such as bio-potential, bio-impedance, and optical measurements. Finally, commercially available ICs in the field are introduced and future directions are discussed.

About the presenter:
Firat Yazicioglu received the Ph.D. degree in Electrical Eng. from Katholieke Universiteit Leuven in 2008. He is currently at IMEC, Belgium where he leads the "Biomedical Integrated Circuits" team focusing on Mixed Signal Integrated Circuit design for wearable and implantable biomedical applications. He has (co)authored over 70 publications and several patents in the field of IC design for biomedical applications. He has contributed/lead the development of several generations of integrated circuits for wearable and implantable healthcare applications, some of which are absorbed by the industry for commercialization. Dr. Yazicioglu serves on the technical program committees of ESSCIRC and ISSCC. He is the co-chair of BioCAS 2013.

Design Strategies for Wearable Sensor Interface Circuits - from Electrodes to Signal Processing

Jerald Yoo, *Masdar Institute of Science and Technology, Abu Dhabi, United Arab Emirates*

Wearable healthcare sensors provide attractive opportunities for the semiconductor sector. The target is to mitigate the impact of chronic diseases by providing continuous yet adequate low-noise *monitoring and analysis* of physiological signals. The wearable environment is challenging for circuit designers due to its unstable skin-electrode interface. Wet and dry electrodes have very different electrical characteristics that need to be addressed. Also, in a wearable environment, a trade-off between available resources and component performance (analog front-end and digital back-end) is crucial.

This short course covers the design strategies for bio-interface circuits for such wearable sensors. We first explore the difficulties, limitations and potential pitfalls in wearable interfaces, and strategies to overcome them. After that, system-level considerations for better key metrics such as energy efficiency are introduced. Several state-of-the-art instrumentation amplifiers that emphasize different parameters are also discussed. We then see how the signal analysis part impacts the analog interface circuit design. The talk concludes with interesting aspects and opportunities that lie ahead.

About the presenter:
Jerald Yoo received the Ph.D. degree in Electrical Engineering from the Korea Advanced Institute of Science and Technology (KAIST), Daejeon, Korea, in 2010. In May 2010, he joined the Faculty of Microsystems Engineering, Masdar Institute, Abu Dhabi, United Arab Emirates, where he is an Assistant Professor. Dr. Yoo developed low-energy Body Area Network (BAN) transceivers and wearable body sensor networks using Planar-Fashionable Circuit Board (P-FCB) for continuous health monitoring. His research focuses on low-energy circuit technology for wearable bio signal sensors, wireless power transmission, SoC design to system realization for wearable healthcare applications, and energy-efficient biomedical circuit techniques. He is an author of a book chapter in Biomedical CMOS ICs (Springer, 2010). Dr. Yoo is a co-recipient of the Asian Solid-State Circuits Conference (A-SSCC) Outstanding Design Awards in 2005.

System Architectures and Strategies for Bi-directional Interfacing of Circuits with Physiological Systems

Tim Denison, *Medtronic, Minneapolis, MN*

This talk presents state-of-the-art circuit techniques for *bi-directionally* interfacing to physiological systems with the goal of restoring function. To provide context for the design discussion, a general framework of sensors, classifiers, control policies and actuators is first covered. The circuit design constraints and implementations for each of these sub-blocks is then discussed, along with the relevant physiological principles that guide the overall design. Examples of how the sub-systems are integrated together with physiology are then drawn from cardiac, neural, and prosthetic system applications. To close the discussion, a commentary on potential future trends is given.

About the presenter:
Tim Denison is a Technical Fellow at Medtronic and Director of Core Technology in the Neuromodulation division, where he helps oversee the design of next generation neural interface and algorithm technologies for the treatment of chronic neurologic disease. In 2012, he was awarded membership to the Bakken Society, Medtronic's highest technical and scientific honor. Tens of thousands of patients receive implantable circuits each year from systems Tim helped design. He received a Ph.D. in Electrical Engineering from MIT. He is also a graduate of Harvard Business School's Program for Leadership Development. Tim's extracurricular pursuits include serving as an Adjunct Assistant Professor at Brown University, teaching "smart" medical sensor design short courses at TU Delft, and chairing the IEEE EMBS society Twin Cities chapter.

978-1-4799-0917-9/14 $31.00 © 2014 IEEE

INDEX TO AUTHORS

978-1-4799-0917-9/14 $31.00 © 2014 IEEE

F

G

H

I

J

K

L

M

978-1-4799-0917-9/14 $31.00 © 2014 IEEE

EXECUTIVE COMMITTEE

CONFERENCE CHAIR
Anantha Chandrakasan,
Massachusetts Institute of Technology,
Cambridge, MA

SECRETARY, PRESS COORDINATOR AND DATA TEAM CHAIR
Siva Narendra,
Tyfone,
Portland, OR

DIRECTOR OF FINANCE AND BOOK DISPLAY COORDINATOR
Bryant Griffin,
Penfield, NY

PROGRAM CHAIR
Trudy Stetzler,
Halliburton,
Houston, TX

PROGRAM VICE CHAIR
Hoi-Jun Yoo,
KAIST,
Daejeon, Korea

STUDENT RESEARCH CHAIR AND UNIVERSITY RECEPTIONS
Jan Van der Spiegel,
University of Pennsylvania,
Philadelphia, PA

DEMONSTRATION SESSION CO-CHAIR
Bill Bowhill,
Intel,
Hudson, MA

DEMONSTRATION SESSION CO-CHAIR
Uming Ko,
MediaTek,
Austin, TX

ITPC FAR EAST REGIONAL CHAIR
Kazutami Arimoto,
Okayama Prefectural University,
Okayama, Japan

ITPC FAR EAST REGIONAL VICE CHAIR
Jae-Youl Lee,
Samsung,
Yongin, Korea

ITPC FAR EAST REGIONAL SECRETARY
Tsung-Hsien Lin,
National Taiwan University,
Taipei, Taiwan

ITPC EUROPEAN REGIONAL CHAIR
Alison Burdett,
Toumaz Microsystems,
Abingdon, United Kingdom

ITPC EUROPEAN REGIONAL VICE CHAIR AND SECRETARY
Maurits Ortmanns,
University of Ulm,
Ulm, Germany

ADCOM REPRESENTATIVE
Bryan Ackland,
Stevens Institute of Technology,
Old Bridge, NJ

DIRECTOR OF PUBLICATIONS AND PRESENTATIONS
Laura Fujino,
University of Toronto,
Toronto, Canada

DIRECTOR OF AUDIOVISUAL SERVICES
John Trnka,
Rochester, MN

PRESS LIAISON AND AWARDS & RECOGNITION COMMITTEE (ARC) CHAIR
Kenneth C. Smith,
University of Toronto,
Toronto, Canada

FORUMS CHAIR
Bram Nauta, *University of Twente,*
Enschede, The Netherlands

EDUCATIONAL EVENTS CHAIR
Ali Sheikholeslami,
University of Toronto,
Toronto, Canada

DIRECTOR OF OPERATIONS
Melissa Widerkehr,
Widerkehr and Associates,
Montgomery Village, MD

TECHNICAL EDITORS

Jason H. Anderson, *University of Toronto, Toronto, Canada*
Dustin Dunwell *(Editor-at-Large)*, *Kapik Integration, Toronto, Canada*
Vincent Gaudet, *University of Waterloo, Waterloo, Canada*
Glenn Gulak *(Editor-at-Large)*, *University of Toronto, Toronto, Canada*
James W. Haslett, *University of Calgary, Calgary, Canada*
Shahriar Mirabbasi, *(Editor-at-Large)* *University of British Columbia, Vancouver, Canada*
Kostas Pagiamtzis, *Semtech, Gennum Products Group, Burlington, Canada*
Kenneth C. Smith *(Editor-at-Large)*, *University of Toronto, Toronto, Canada*

MULTI-MEDIA COORDINATOR

Dave Halupka, *Kapik Integration, Toronto, Canada*

INTERNATIONAL TECHNICAL PROGRAM COMMITTEE

PROGRAM CHAIR: Trudy Stetzler, *Halliburton, Houston, TX*
PROGRAM VICE CHAIR: Hoi-Jun Yoo, *KAIST, Daejeon, Korea*

ANALOG SUBCOMMITTEE

Chair: Axel Thomsen, Silicon Laboratories, Austin, TX
Anton Bakker, IDT, Morgan Hill, CA
Marco Berkhout, NXP Semiconductors, Nijmegen, The Netherlands
Minkyu Je, Institute of Microelectronics, A*STAR, Singapore
Xicheng Jiang, Broadcom, Irvine, CA
Wing Hung Ki, HKUST, Hong Kong, China
Kimmo Koli, Ericsson, Turku, Finland
Jae-Youl Lee, Samsung, Yongin, Korea
Tsung-Hsien Lin, National Taiwan University, Taipei, Taiwan
Saska Lindfors, Texas Instruments, Helsinki, Finland
Piero Malcovati, University of Pavia, Pavia, Italy
Makoto Nagata, Kobe University, Kobe, Japan
Tim Piessens, ICSense, Leuven, Belgium
Edgar Sanchez-Sinencio, Texas A&M University, College Station, TX
Christoph Sandner, Infineon Technologies Austria, Villach, Austria
Jafar Savoj, Xilinx, San Jose, CA
Ed (Adrianus JM) van Tuijl, University of Twente, Enschede, The Netherlands
Young-Jin Woo, Silicon Works, Daejon, Korea

DATA CONVERTERS SUBCOMMITTEE

Chair: Boris Murmann, Stanford University, Stanford, CA
Brian Brandt, Maxim Integrated Products, North Chelmsford, MA
Lucien Breems, NXP Semiconductors, Eindhoven, The Netherlands
Marco Corsi, Texas Instruments, Dallas, TX
Pieter Harpe, Eindhoven University of Technology, Eindhoven, The Netherlands
Tetsuya Iizuka, University of Tokyo, Tokyo, Japan
John Khoury, Silicon Laboratories, Austin, TX
Stéphane Le Tual, STMicroelectronics, Crolles, France
Hae-Seung Lee, Massachusetts Institute of Technology, Cambridge, MA
Takahiro Miki, Renesas Electronics, Hyogo, Japan
Gerhard Mitterregger, Intel Mobile Communications Austria, St. Magdalen, Austria
Jan Mulder, Broadcom, Bunnik, The Netherlands
Ken Nishimura, Agilent, Santa Clara, CA
Maurits Ortmanns, University of Ulm, Ulm, Germany
Shanthi Pavan, Indian Institute Of Technology, Chennai, India
Seung-Tak Ryu, KAIST, Daejon, Korea
Richard Schreier, Analog Devices, Toronto, Canada
Jieh-Tsorng Wu, National Chiao-Tung University, Hsin-Chu, Taiwan

ENERGY-EFFICIENT DIGITAL SUBCOMMITTEE

Chair: Stephen Kosonocky, Advanced Micro Devices, Fort Collins, CO
Kazutami Arimoto, Okayama Prefectural University, Okayama, Japan
Wim Dehaene, KU Leuven, Leuven, Belgium
Vasantha Erraguntla, Intel Technology India, Bangalore, India
Takashi Hashimoto, Panasonic, Fukuoka, Japan
Byeong-Gyu Nam, Chungnam National University, Daejeon, Korea
Yongha Park, Samsung, Yongin, Korea
Michael Polley, Samsung Mobile, Dallas, TX
Bing Sheu, TSMC, Hsinchu, Taiwan
Kees van Berkel, Ericsson, Eindhoven, The Netherlands
Marian Verhelst, KU Leuven, Leuven, Belgium
Victor Zyuban, IBM T.J. Watson Research Center, Yorktown Heights, NY

HIGH-PERFORMANCE DIGITAL SUBCOMMITTEE

Chair: Stefan Rusu, Intel, Santa Clara, CA
Vivek De, Intel, Hillsboro, OR
Christopher Gonzalez, IBM, Yorktown Heights, NY
Hiroo Hayashi, Toshiba Corporation, Kawasaki, Japan
Anthony Hill, Texas Instruments, Dallas, TX
Atsuki Inoue, Fujitsu, Kawasaki, Japan
Tobias Noll, RWTH Aachen University, Aachen, Germany
Yasuhisa Shimazaki, Renesas, Tokyo, Japan
Luke (Jinuk) Shin, Oracle, Santa Clara, CA
Vladimir Stojanovic, Massachusetts Institute of Technology, Cambridge, MA
Kathy Wilcox, AMD, Boxborough, MA
Se-Hyun Yang, Samsung, Yongin, Korea

IMAGERS, MEMS, MEDICAL AND DISPLAYS SUBCOMMITTEE

Chair: Roland Thewes, TU Berlin, Berlin, Germany
Ralf Brederlow, Texas Instruments, Freising, Germany
Roman Genov, University of Toronto, Toronto, Canada
Maysam Ghovanloo, Georgia Institute of Technology, Atlanta, GA
Makoto Ikeda, University of Tokyo, Tokyo, Japan
Robert Johansson, OmniVision, Oslo, Norway
Sam Kavusi, Bosch Research and Technology Center, Palo Alto, CA
Tae-Chan Kim, Samsung, Yongin, Kprea
Wentai Liu, UCLA, Los Angeles, CA
Young-Sun Na, LG Electronics, Seoul, Korea
Yoshiharu Nakajima, Japan Display, Kanagawa, Japan
Yusuke Oike, Sony, Kanagawa, Japan
Aaron Partridge, SiTime, Sunnyvale, CA
Michiel Pertijs, TU Delft, Delft, The Netherlands
Joseph Shor, Intel, Yakum, Israel
David Stoppa, Fondazione Bruno Kessler, Trento, Italy
Yong Ping Xu, National University of Singapore, Singapore
Refet Firat Yazicioglu, IMEC, Leuven, Belgium

MEMORY SUBCOMMITTEE

Chair: Kevin Zhang, Intel, Hillsboro, OR
Jonathan Chang, TSMC, Hsinchu, Taiwan
Leland Chang, IBM T. J. Watson Research Center, Yorktown Heights, NY
Sungdae Choi, SK hynix, Icheon-si, Korea
Michael Clinton, TSMC Technology, Austin, TX
Jin-Man Han, Samsung Electronics, Hwasung, Korea
Satoru Hanzawa, Hitachi, Tokyo, Japan
Uksong Kang, Samsung Electronics, Hwasung-si, Korea
Atsushi Kawasumi, Toshiba, Kawasaki, Japan
Chulwoo Kim, Korea University, Seoul, Korea
Hugh Mair, MediaTek, Fairview, Texas
James Sung, Etron Technology, Hsinchu, Taiwan
Daniele Vimercati, Micron Technology, Agrate, Italy
Tadaaki Yamauchi, Renesas Electronics, Itami, Japan
Takefumi Yoshikawa, Panasonic, Kyoto, Japan

RF SUBCOMMITTEE

Chair: Andreia Cathelin, STMicroelectronics, Crolles, France
Ehsan Afshari, Cornell University, Ithaca, NY
Brian Floyd, North Carolina State University, Raleigh, NC
Chih-Ming Hung, MStar Semiconductor, Taipei, Taiwan
Mike Keaveney, Analog Devices, Limerick, Ireland
Tae Wook Kim, Yonsei University, Seoul, Korea
Antonio Liscidini, University of Toronto, Toronto, Canada
Ahmad Mirzaei, Pennsylvania State University, Unversity Park, PA
Borivoje Nikolic, University of California, Berkeley, CA
Brian Otis, University of Washington, Seattle, WA
Ullrich Pfeiffer, University of Wuppertal, Wuppertal, Germany
Jussi Ryynanen, Aalto University, Espoo, Finland
Carlo Samori, Politecnico di Milano, Milano, Italy
Marc Tiebout, Infineon Technologies Austria, Villach, Austria
Piet Wambacq, imec, Heverlee, Belgium
Taizo Yamawaki, Hitachi, Tokyo, Japan
Masoud Zargari, Qualcomm-Atheros, Irvine, CA
Jing-Hong Zhan, MediaTek, HsinChu, Taiwan

TECHNICAL DIRECTIONS SUBCOMMITTEE

Chair: Eugenio Cantatore, Eindhoven University of Technology, Eindhoven, The Netherlands
Alison Burdett, Toumaz Microsystems, Abingdon, United Kingdom
Eric Colinet, CEA-LETI, Grenoble, France
Jan Genoe, imec, Leuven, Belgium
Fu-Lung Hsueh, TSMC, Hsinchu, Taiwan
Ali Keshavarzi, Cypress,Los Altos, CA
Donghyun Kim, Samsung Techwin, Gyeong, Korea
Tadahiro Kuroda, Keio University, Yokohama, Japan
Shinichiro Mutoh, NTT, Atsugi, Japan
Masaitsu Nakajima, Panasonic, Kyoto, Japan
Chris Nicol, Wave Semiconductor, Sunnyvale, CA
Koichi Nose, Renesas Electronics, Kawasaki, Japan
Pirooz Pavarandeh, Maxim Integrated Products, Sunnyvale, CA
Yogesh K. Ramadass, Texas Instruments, Dallas, TX
David Ruffieux, CSEM, Neuchated, Switzerland
Alice Wang, MediaTek, Austin, TX

WIRELESS SUBCOMMITTEE

Chair: Aarno Pärssinen, Broadcom, Helsinki, Finland
Gangadhar Burra, Texas Instruments, Dallas, TX
Pierre Busson, STMicroelectronics, Crolles, France
Jan Crols, AnSem, Heverlee, Belgium
Hossein Hashemi, University of Southern California, Los Angeles, CA
Myung-Woon Hwang, FCI, Sungnam, Korea
Albert Jerng, Apple, Cupertino, CA
Eric Klumperink, University of Twente, Enschede, The Netherlands
Shouhei Kousai, Toshiba, Kawasaki, Japan
Li Lin, Marvell Semiconductor, Santa Clara, CA
Sven Mattisson, Ericsson, Lund, Sweden
Alyosha Molnar, Cornell University, Ithaca, NY
Kenichi Okada, Tokyo Institute of Technology, Tokyo, Japan
Stefano Pellerano, Intel, Hillsboro, OR
Woogeun Rhee, Tsinghua University, Beijing, China
Koji Takinami, Panasonic, Yokohama, Japan
Jan Van Sinderen, NXP Semiconductors, Eindhoven, The Netherlands
Iason Vassiliou, Broadcom, Alimos, Greece

WIRELINE SUBCOMMITTEE

Chair: Daniel Friedman, IBM Thomas J. Watson Research Center, Yorktown Heights, NY
Elad Alon University of California, Berkeley, Berkeley,CA
Ajith Amerasekera, Texas Instruments, Dallas, TX
Gerrit den Besten, NXP Semiconductors, Eindhoven, The Netherlands
Ken Chang, Xilinx, San Jose, CA
SeongHwan Cho, KAIST, Daejon, Korea
Nicola Da Dalt, Infineon, Villach, Austria
Azita Emami, California Institute of Technology, Pasadena, CA
Ichiro Fujimori, Broadcom, Irvine, CA
Pavan Kumar Hanumolu, University of Illinois, Urbana, IL
Chewnpu Jou, TSMC, Hsinchu, Taiwan
Jack Kenney, Analog Devices, Somerset, NJ
Hideyuki Nosaka, NTT Photonics Laboratories, Atsugi, Japan
Frank O'Mahony, Intel, Hillsboro, OR
Hisakatsu Yamaguchi, Fujitsu Laboratories, Kawasaki, Japan
Koichi Yamaguchi, Renesas Electronics, Kawasaki, Japan

PROGRAM COMMITTEE

EUROPEAN REGIONAL COMMITTEE

ITPC EUROPEAN REGIONAL CHAIR
Alison Burdett, *Toumaz Microsystems, Abingdon, United Kingdom*

ITPC EUROPEAN REGIONAL VICE CHAIR AND SECRETARY
Maurits Ortmanns, *University of Ulm, Ulm Germany*

Gerrit den Besten, NXP Semiconductors, Eindhoven, The Netherlands
Marco Berkhout, NXP Semiconductors, Nijmegen, The Netherlands
Ralf Brederlow, Texas Instruments, Freising, Germany
Lucien Breems, NXP Semiconductors, Eindhoven, The Netherlands
Pierre Busson, STMicroelectronics, Crolles, France
Eugenio Cantatore, Eindhoven University of Technology, Eindhoven, The Netherlands
Andreia Cathelin, STMicroelectronics, Crolles, France
Eric Colinet, CEA-LETI, Grenoble, France
Jan Crols, AnSem, Heverlee, Belgium
Nicola Da Dalt, Infineon, Villach, Austria
Wim Dehaene, KU Leuven, Leuven, Belgium
Jan Genoe, imec, Leuven, Belgium
Pieter Harpe, Eindhoven University of Technology, Eindhoven, The Netherlands
Robert Johansson, OmniVision, Oslo, Norway
Mike Keaveney, Analog Devices, Limerick, Ireland
Eric Klumperink, University of Twente, Enschede, The Netherlands
Kimmo Koli, ST-Ericsson, Turku, Finland
Stéphane Le Tual, STMicroelectronics, Crolles, France
Saska Lindfors, Texas Instruments, Helsinki, Finland
Piero Malcovati, University of Pavia, Pavia, Italy
Sven Mattisson, Ericsson, Lund, Sweden
Gerhard Mitterregger, Intel Mobile Communications Austria, St. Magdalen, Austria
Jan Mulder, Broadcom, Bunnik, The Netherlands
Tobias Noll, RWTH Aachen University, Aachen, Germany
Aarno Pärssinen, Broadcom, Helsinki, Finland
Michiel Pertijs, TU Delft, Delft, The Netherlands
Ullrich Pfeiffer, University of Wuppertal, Wuppertal, Germany
Tim Piessens, ICSense, Leuven, Belgium
David Ruffieux, CSEM, Neuchated, Switzerland
Jussi Ryynanen, Aalto University, Espoo, Finland
Carlo Samori, Politecnico di Milano, Milano, Italy
Christoph Sandner, Infineon Technologies Austria, Villach, Austria
Joseph Shor, Intel, Yakum, Israel
David Stoppa, Fondazione Bruno Kessler, Trento, Italy
Roland Thewes, TU Berlin, Berlin, Germany
Marc Tiebout, Infineon Technologies, Villach, Austria
Kees van Berkel, Ericsson, Eindhoven, The Netherlands
Jan Van Sinderen, NXP Semiconductors, Eindhoven, The Netherlands
Ed van Tuijl, University of Twente, Enschede, The Netherlands
Iason Vassiliou, Broadcom, Alimos, Greece
Marian Verhelst, KU Leuven, Leuven, Belgium
Daniele Vimercati, Micron Technology, Agrate, Italy
Piet Wambacq, imec, Heverlee, Belgium
Refet Firat Yazicioglu, imec, Leuven, Belgium

FAR EAST REGIONAL COMMITTEE

ITPC FAR EAST REGIONAL CHAIR
Kazutami Arimoto, Okayama Prefectural University, Okayama, Japan

ITPC FAR EAST REGIONAL VICE-CHAIR
Jae-Youl Lee, Samsung, Yongin, Korea

ITPC FAR EAST REGIONAL SECRETARY
Tsung-Hsien Lin, National Taiwan University, Taipei, Taiwan

Jonathan Chang, TSMC, Hsinchu, Taiwan
SeongHwan Cho, KAIST, Daejon, Korea
Sungdae Choi, SK hynix, Icheon-si, Korea
Vasantha Erraguntla, Intel Technology India, Bangalore, India
Jin-Man Han, Samsung Electronics, Hwasung, Korea
Satoru Hanzawa, Hitachi, Tokyo, Japan
Takashi Hashimoto, Panasonic, Fukuoka, Japan
Hiroo Hayashi, Toshiba, Kawasaki, Japan
Fu-Lung Hsueh, TSMC, Hsinchu, Taiwan
Chih-Ming Hung, MStar Semiconductor, Taipei, Taiwan
Myung-Woon Hwang, FCI, Sungnam, Korea
Tetsuya Iizuka, University of Tokyo, Tokyo, Japan
Makoto Ikeda, University of Tokyo, Tokyo, Japan
Atsuki Inoue, Fujitsu, Kawasaki, Japan
Minkyu Je, Institute of Microelectronics, A*STAR, Singapore
Albert Jerng, Apple, Cupertino, CA
Chewnpu Jou, TSMC, Hsinchu, Taiwan
Uksong Kang, Samsung Electronics, Hwasung, Korea
Atsushi Kawasumi, Toshiba, Kawasaki, Japan
Wing Hung Ki, HKUST, Hong Kong, China
Chulwoo Kim, Korea University, Seoul, Korea
Donghyun Kim, Samsung Techwin, Gyeong, Korea
Tae-Chan Kim, Samsung, Yongin, Korea
Tae Wook Kim, Yonsei University, Seoul, Korea
Shouhei Kousai, Toshiba, Kawasaki, Japan
Tadahiro Kuroda, Keio University, Yokohama, Japan
Takahiro Miki, Renesas Electronics, Itami, Japan
Shinichiro Mutoh, NTT, Atsugi, Japan
Young-Sun Na, LG Electronics, Seoul, Korea
Makoto Nagata, Kobe University, Kobe, Japan
Masaitsu Nakajima, Panasonic, Kyoto, Japan
Yoshiharu Nakajima, Japan Display, Kanagawa, Japan
Byeong-Gyu Nam, Chungnam National University, Daejeon, Korea
Hideyuki Nosaka, NTT Photonics Laboratories, Atsugi, Japan
Koichi Nose, Renesas Electronics, Kawasaki, Japan
Yusuke Oike, Sony, Atsugi, Japan
Kenichi Okada, Tokyo Institute of Technology, Tokyo, Japan
Yongha Park, Samsung, Yongin, Korea
Shanthi Pavan, Indian Institute of Technology Madras, Chennai, India
Woogeun Rhee, Tsinghua University, Beijing, China
Seung-Tak Ryu, KAIST, Daejon, Korea
Bing Sheu, TSMC, Hsinchu, Taiwan
Yasuhisa Shimazaki, Renesas, Tokyo, Japan
James Sung, Etron Technology, Hsinchu, Taiwan
Koji Takinami, Panasonic, Yokohama, Japan
Young-Jin Woo, Silicon Works, Daejon, Korea
Jieh-Tsorng Wu, National Chiao-Tung University, Hsinchu, Taiwan
Yong Ping Xu, National University of Singapore, Singapore
Hisakatsu Yamaguchi, Fujitsu Laboratories, Kawasaki, Japan
Koichi Yamaguchi, Renesas Electronics, Kawasaki, Japan
Tadaaki Yamauchi, Renesas Electronics, Itami, Japan
Taizo Yamawaki, Hitachi, Tokyo, Japan
Se-Hyun Yang, Samsung, Yongin-Si, Korea
Takefumi Yoshikawa, Panasonic, Kyoto, Japan
Jing-Hong Zhan, MediaTek, Hsinchu, Taiwan

ISSCC 2015 Call for Papers

IEEE International Solid-State Circuits Conference

Sunday–Thursday, February 22-26, 2015 • San Francisco Marriott Marquis Hotel, San Francisco, CA

ISSCC 2015 Conference Theme:
"Silicon Systems – Small Chips for Big Data"

Big Data generated by IoT, healthcare, and WWW, are changing our lifestyle and our society. Small silicon chips are enabling this change through data sensing, gathering, processing, storing, and networking through wireless and wireline connection. Recent silicon-system technologies including ultra-low power systems, high performance circuits and systems, wireless power and data transmission, and 3D IC structures, will open the door to Big Data applications. Moreover, Big-Data applications such as healthcare, machine learning, and sensor systems, will challenge current circuit and system designers to consider new system architectures requiring advances in circuits and technology. ISSCC 2015 is looking for novel system and circuit solutions which open new vistas for society, with opportunities for new lifestyles, all driven by Big-Data technology.

Innovative and original papers are solicited in subject areas including (but not limited to) the following:

ANALOG — Op-amps and instrumentation amplifiers, comparators, power control and management circuits, voltage references, regulators, DC-DC converters, energy-scavenging circuits; filters, continuous-time and discrete-time; nonlinear analog circuits, switched-capacitor circuits; oscillators, synthesizers, PLLs; very-low-power and low-voltage analog circuits, digitally-assisted analog circuits.

DATA CONVERTERS — Nyquist-rate and oversampling A/D and D/A converters; time-to-digital converters.

ENERGY-EFFICIENT DIGITAL — Ultra-low-power heterogeneous embedded many-core systems; energy-optimized embedded function acceleration; sub-threshold and near-threshold systems; power-and clock gating circuits; low-power failure-resistant systems; reliable resilient and robust circuits and systems; integrated systems such as smart-phone ICs and application processors, digital baseband, innovative multimedia ICs, personal e-health ICs, processors for energy-efficient sensors.

HIGH-PERFORMANCE DIGITAL — Microprocessors; graphics processors; systems on chips integrating processor cores, graphics and peripheral controllers; many-core and thread-rich processors; network processors; high-speed digital circuits; intra-chip communication circuits; soft error, variation, and fault-tolerant circuits; reconfigurable logic arrays; security and encryption circuits; high-speed CAMs and register files; clock-generation and distribution circuits and architectures; power-and-leakage-management techniques for high-performance processors and graphics; adaptive digital circuits; thermal and wear-out sensors; 3D stacking techniques for processors; all-digital PLLs and DLLs; integrated voltage regulators and DC/DC converters for processors and graphics; system-level papers describing power, thermal, clock and interconnect challenges and their circuit solutions.

IMAGERS, MEMS, MEDICAL & DISPLAYS — Image sensors and companion chips; image-sensor SoCs; MEMS for analog, RF, and sensor applications; MEMS- and sensor interface circuits; smart sensors; integrated sensors and transducers; organic sensors; biosensors, microarrays and lab-on-a-chip devices; neural interfaces; environmental and wearable biomedical electronics; display drivers, controllers, and companion chips; organic LED and liquid-crystal-display interface circuits; flat-panel and projection displays.

MEMORY — Static, dynamic, and non-volatile memories with single and multi-ports for both stand-alone and embedded applications; memory subsystem and array architectures along with core circuits; memory I/O interface architecture and circuit techniques, including 3D memory and logic integration; memory design based on new technologies such as phase-change, magnetic, spin-transfer-torque, ferroelectric, and resistive; advanced architectures and circuits to improve low-voltage operation, power reduction, reliability and fault tolerance in memories; memory controllers and solid-state-disk controllers.

RF — From building blocks to subsystems for RF, mmW and THz design with focus on novel design techniques; design techniques for 1 to 10GHz bands (GPS, GSM, LTE, WiFi, WiMax, ...), mmW bands up to 300GHz (60GHz, 77GHz radar, 79GHz point to point, 120GHz high-data-rate wireless, mmW imaging), THz bands (above 300GHz, THz imaging, high-data-rate wireless communications).

TECHNOLOGY DIRECTIONS — Non-silicon-, carbon-, organic-, and nano-electronics; flexible substrates and printable electronics; heterogeneous 3D integration; compound-semiconductor, superconductive and micro-photonic technologies and circuits; energy sources and energy harvesting; biomedical SoCs, ambient-intelligence; artificial intelligence ICs, analog and optical processors, non-transistor-based analog and digital circuits and their system architectures; advanced memory technologies; spintronics; quantum storage; emerging sensor-network concepts such as body-area and body-sensor networks.

WIRELESS — Wireless systems (receivers, transmitters, transceivers, SoCs, SiPs) with focus on standards-based applications up to 100GHz; applications in cellular, WLAN, BT, GPS, BAN, GPS, TV tuner, UWB, ISM, WiMax, WiGig, broadcast, radar, imaging, low-power, multi-standard and multi-band; innovative system architectures and solutions for advanced wireless applications.

WIRELINE — Receivers/transmitters/transceivers for wireline systems, including backplane transceivers, optical links, chip-to-chip communications and 3D structures; applications in designs for (but not limited to) Ethernet, Fibre Channel, optical/electrical data transfer, PON, advanced serial memory, consumer product wired communication, SONET, SDH, FDDI, and xDSL; wireline transceiver building blocks such as AGC, equalization circuits, oscillators, PLLs, line drivers, and hybrids.

Submission Deadline is Monday, September 8, 2014 • 3:00PM Eastern Daylight Time (19:00 GMT)

STUDENT INITIATIVES

Graduate students are invited to participate in opportunities to showcase ongoing work and exchange experiences with other students and researchers from academia and industry. These include the Student Research Preview and the Silkroad Award (to a first-time student presenting author of a regular paper from an emerging region in the Far East).

Further information including submission procedures, formats, student initiatives and deadlines can be found at http://www.isscc.org

978-1-4799-0917-9/14 $31.00 © 2014 IEEE

Sunday, February 9th	ISSCC 2014 TUTORIALS		
8:30AM	**T1:** Filtering in RF Transceivers	**T2:** V(min) Constraints and Optimization in VLSI Circuit Design	**T3:** 3D Integration, Power Delivery, and Contactless Interconnect by Near Field Coupling
10:30AM	**T4:** Power Optimized Processor Design	**T5:** Peripheral Circuits for Analog-to-Digital Converters	**T6**: Analog Front-End Design for Gb/s Wireline Receivers
1:30PM	**T7:** Self-Adapting Design Techniques for Power Constrained Processors	**T8:** Interference Robust CMOS Radio Receiver Techniques	
3:30PM	**T9:** Charge Pump and Capacitive DCDC Converter Design	**T10:** Design of Physical-to-Digital Converters	

	ISSCC 2014 FORUMS	
8:00AM	**F1:** Digitally-Assisted Analog and Analog-Assisted Digital in High-Performance Scaled CMOS Process	**F2:** 3D Stacking Technologies for Image Sensors & Memories

Events below in Bold Box included in Conference registration

	ISSCC 2014 EVENING SESSIONS	
	7:30 PM ES1: Student Research Peview: Short Presentations with Poster Session	**8:00 PM ES2:** Data Centers to Support Tomorrow's Cloud

Monday, February 10th	ISSCC 2014 PAPER SESSIONS				
8:30AM	**Session 1:** Plenary Session				
1:30PM	**Session 2:** Ultra High-Speed Wireline Transceiver & Techniques	**Session 3:** RF Techniques	**Session 4:** DC-DC Converters	**Session 5:** Processors	**Session 6:** Technologies for High-Speed Data Networks **Session 7:** Image Sensors
5:15PM	Demonstration Session (4:00-7:00 PM), Author Interviews, Social Hour				

	ISSCC 2014 SESSIONS
8:00PM	**EP1: Plenary Roundtable : Next-Generation Networked Systems - Challenges for Silicon**

Tuesday, February 11th	ISSCC 2014 PAPER SESSIONS				
8:30AM	**Session 8:** Optical Links and Copper PHYs	**Session 9:** Low-Power Wireless	**Session 10:** Mobile System-on-Chip (SoCs)	**Session 11:** Data Converter Techniques	**Session 12:** Sensors, MEMs, and Displays
1:30PM	**Session 13:** Advanced Embedded Memory	**Session 14:** Millimeter-Wave / Terahertz Techniques	**Session 15:** Digital PLLs **Session 16:** SoC Building Blocks	**Session 17:** Analog Techniques	**Session 18:** Biomedical Systems for Improved Quality of Life
5:15PM	Author Interviews, Social Hour				

	ISSCC 2014 EVENING SESSIONS		
8:00PM	**ES3:** Wearable Wellness Devices: Fashion, Health, & Informatics	**EP2:** Anatomy of Innovation: Bug or Feature?	**EP3:** Perspectives on the Future of Semiconductor Innovation

Wednesday, February 12th	ISSCC 2014 PAPER SESSIONS					
8:30AM	**Session 19:** Nonvolatile Memory		**Session 20:** Wireless Systems	**Session 21:** Frequency Generation Techniques	**Session 22:** High-Speed Data Converters	**Session 23:** Energy Harvesting
1:30PM	**Session 24:** Integrated Biomedical Systems	**Session 25:** High-Bandwidth Low-Power DRAM and I/O **Session 26:** Energy-Efficient Dense Interconnect		**Session 27:** Energy-Efficient Digital Circuits	**Session 28:** Mixed-Signal Techniques for Wireless **Session 29:** Data Converters for Wireless Systems	**Session 30:** Technologies for Next-Generation Systems
5:15 PM	Author Interviews					

Thursday, February 13th	ISSCC 2014 SHORT COURSE
8:00 AM	**SC1:** Biomedical and Sensor Interface Circuits

	ISSCC 2014 FORUMS			
8:00AM	**F3:** Adaptive Design Techniques for Energy Efficiency	**F4:** mm-Wave Advances for Active Safety and Communication Systems	**F5:** Low Power Radios for Sensor Networks	**F6:** Energy Efficient I/O Design for Next-Generation Systems

978-1-4799-0917-9/14 $31.00 © 2014 IEEE